PHYSICAL CONSTANTS

Standard acceleration of terrestrial gravity	$g = 9.80665$ m s^{-2} (exactly)
Avogadro's number	$N_0 = 6.0221420 \times 10^{23}$ mol^{-1}
Bohr radius	$a_0 = 0.529177208$ Å $= 5.29177208 \times 10^{-11}$ m
Electron charge	$e = 1.60217646 \times 10^{-19}$ C
Faraday constant	$\mathscr{F} = 96{,}485.342$ C mol^{-1}
Universal gas constant	$R = 8.31447$ J mol^{-1} K^{-1}
	$= 0.0820574$ L atm mol^{-1} K^{-1}
Masses of fundamental particles:	
Electron	$m_e = 9.1093819 \times 10^{-31}$ kg
Proton	$m_p = 1.6726216 \times 10^{-27}$ kg
Neutron	$m_n = 1.6749272 \times 10^{-27}$ kg
Planck's constant	$h = 6.6260688 \times 10^{-34}$ J s
Speed of light in a vacuum	$c_o = 2.99792458 \times 10^{8}$ m s^{-1} (exactly)

Values are taken from National Institute of Standards and Technology Special Publication 961, Jan. 2001.

CONVERSION FACTORS

Atomic mass unit	1 u $= 1.6605387 \times 10^{-27}$ kg
	1 u $= 1.4924117 \times 10^{-10}$ J $= 931.49401$ MeV
	(energy equivalent from $E = mc_0^2$)
Pound	1 lb $= 16$ oz $= 0.4539237$ kg (exactly)
Metric ton	1 t $= 1000$ kg (exactly)
Standard atmosphere	1 atm $= 1.01325 \times 10^{5}$ Pa $= 1.01325 \times 10^{5}$ kg m^{-1} s^{-2} (exactly)
Torr	1 torr $= 133.3224$ Pa
Calorie	1 cal $= 4.184$ J (exactly)
Electron volt	1 eV $= 1.60217646 \times 10^{-19}$ J $= 96.485342$ kJ mol^{-1}
Rydberg	1 Ry $= 2.17987190 \times 10^{-18}$ J
	$= 1312.7498$ kJ mol^{-1}
	$= 13.6056917$ eV
Liter-atmosphere	1 L atm $= 101.325$ J (exactly)
Ångström	1 Å $= 10^{-10}$ m
Foot	1 ft $= 12$ in $= 0.3048$ m (exactly)
Liter	1 L $= 10^{-3}$ m^3 $= 10^{3}$ cm^3 (exactly)
Gallon (U.S.)	1 gallon $= 4$ quarts $= 3.78541$ L (exactly)

David W. Oxtoby · Wade A. Freeman · Toby F. Block

The University of Chicago   *The University of Illinois at Chicago*   *Georgia Institute of Technology*

CHEMISTRY:

Science of Change

fourth edition

THOMSON
™
BROOKS/COLE

Australia · Canada · Mexico · Singapore · Spain · United Kingdom · United States

THOMSON

BROOKS/COLE

Executive Editor: Angus McDonald
Acquisitions Editor: John Holdcroft
Developmental Editor: Peter McGahey
Technology Project Manager: Ericka Yeoman-Saler
Assistant Editor: Emily Levitan
Editorial Assistant: Elisabeth Hamilton
Marketing Manager: Jeff Ward
Marketing Assistant: Mona Weltmer
Print/Media Buyer: Karen Hunt

Permissions Editor: Jane Sanders
Production Manager: Charlene Catlett Squibb
Production Service: Ruth Cottrell
Text Designer: Lisa Adamitis
Cover Designer: Jacqui LeFranc
Cover Printer: Lehigh Press
Compositor: TechBooks
Printer: RR Donnelley & Sons Company

For more information about our products,
contact us at:
Thomson Learning Academic Resource Center
1-800-423-0563

For permission to use material from this text,
contact us by:
Phone: 1-800-730-2214
Fax: 1-800-730-2215
Web: http://www.thomsonrights.com

Library of Congress Control Number 2002106436

ISBN 0-03-033188-9

Brooks/Cole-Thomson Learning
511 Forest Lodge Road
Pacific Grove, CA 93950
USA

For more information about our products, contact us:
Thomson Learning Academic Resource Center
1-800-423-0563
http://www.brookscole.com

International Headquarters
Thomson Learning
International Division
290 Harbor Drive, 2nd floor
Stamford, CT 06902-7477
USA

UK/Europe/Middle East/South Africa
Thomson Learning
Berkshire House
168-173 High Holborn
London WC1V 7AA
United Kingdom

Asia
Thomson Learning
5 Shenton Way #01-01
UIC Building
Singapore 068808

Canada
Nelson Thomson Learning
1120 Birchmount Road
Toronto, Ontario M1K 5G4
Canada

In Memoriam

We dedicate this fourth edition of Chemistry: Science of Change *to the memories of Norman H. Nachtrieb and John Vondeling. The first, our original co-author, launched this book; the second, our publisher, made it a reality.*

David W. Oxtoby
Wade A. Freeman
Toby F. Block

Preface

Chemistry: Science of Change, Fourth Edition, is intended for use as the text in introductory chemistry courses taken by students of biology, chemistry, geology, physics, engineering, and related subjects. The text builds on students' knowledge of algebra but does not require calculus. Although some background in high-school science is assumed, no specific knowledge of topics in chemistry is presupposed; this book is a self-contained presentation of the fundamentals of chemistry. Its aim is to convey to students the dynamic and changing aspects of chemistry in the modern world.

The fundamental organization of the fourth edition of *Chemistry: Science of Change* is the same as that of the first three editions. Chemistry, as an experimental science, is introduced in terms of macroscopic concepts and principles that have their origins in the laboratory and everyday observations. The more abstract, theoretical, interpretative aspects of the subject then follow. The overall goals are to persuade students that both theory and experiment are indispensable to the progress of science and to help them appreciate how each aspect informs and guides the other.

Changes in This Edition

A number of general changes and additions have been made to improve the text within the "macro–micro" framework just described:

- The in-text worked examples now include "Strategy" statements where appropriate. These short sections assist students in thinking critically about how to solve the problem. Post-solution "Check" statements present alternate solutions or give the experimentally observed facts with the aim of encouraging students to ponder the reasonableness of their answers.
- Numerous data tables (for example, standard reduction potentials, average bond enthalpies) have been critically updated.
- References are made at various points to reliable Web sites at which additional chemical and physical data can be obtained.
- Additional numbered or bulleted lists recapitulate preceding discussion.
- New end-of-chapter problems appear in most chapters; many problems appearing in previous editions have been clarified, extended, or otherwise modified.
- A new appendix (Appendix C–6) lists the symbols and units recommended by the International Union of Pure and Applied Chemistry for the physical quantities encountered in general chemistry as well as common alternative symbols and units. The aim is to confront the problem of inconsistent use of symbols that so frequently confuses and vexes students.

Most chapters have been extensively rewritten. For example:

- Chapter 1 now emphasizes more than ever the historical problem of connecting mass ratios of different elements in a compound to ratios of numbers of atoms of the different kinds. Accordingly, the discussion of chemical formulas has been moved from Chapter 2 into Chapter 1. The chapter also now includes a section discussing modern analytical methods of arriving at chemical formulas that balances the largely historical account that precedes it.

- A theoretically correct, two-part method for balancing all chemical equations now appears in Chapter 2, which has been retitled "Stoichiometry." The treatment, thoroughly established for years in the chemical literature, is novel in general-chemistry textbooks. The discussion emphasizes equation *completion* as a matter of major chemical interest and properly positions equation *balancing* as a mainly mechanical skill. Students are taught to identify *invalid* chemical equations (i.e., equations that can be balanced but have no value in stoichiometric calculations). Mathematical background is developed separately in an expanded Appendix C.

- Chapter 2 now introduces formal "Tables of Changes" (used extensively in previous editions in treating chemical equilibria) as a means to organize data in solving stoichiometry problems. The chapter also now introduces solution stoichiometry, which sets the stage for an early move into chemically significant laboratory work.

- Chapter 6 includes new material on osmometry, osmolality, and osmolarity—topics of biochemical and clinical significance.

- Chapter 7 includes a new presentation of the mathematical relationships among the equilibrium constants of related equations and new tabular summaries to guide the use of LeChâtelier's principle.

- Chapter 8, which covers acid–base equilibria, has a substantially revised discussion of acid–base titration curves.

- The discussion of calorimetry in Chapter 10 now centers on experiments that many students perform in the laboratory. The logical motivation for the definition and use of thermodynamic states (Section 10–4), a topic that many students find difficult, has been clarified.

- Chapter 12 has been rearranged to put the material on completion and balancing of redox equations first (reflecting classroom experience). This material is also rewritten to reflect the approach to equation balancing that was established in Chapter 2. Positioning the discussion about metals and metallurgy at the end of the chapter permits teachers who do not want to treat the topic in detail to de-emphasize it more easily. Qualitative questions have been added to the problems in Chapter 12 (and Chapter 13) to accommodate teachers who wish to assign metallurgy, batteries, corrosion, and electrorefining as topics for self-study by their classes.

- The discussion of the size and shape of atomic orbitals in Chapter 16 has been simplified.

- Chapter 17 includes new graphics to show the build-up of electron configurations. Later sections clarify the definitions of the metallic, covalent, and ionic radius and use intrinsic atomic radii for the discussion of periodic trends.

- New or heavily revised special essays appear in Chapters 15 ("Superheavy Nuclides and the Island of Stability"), 20 ("Odor Visualization"), 21 ("Superconducting Ceramics"), and 22 ("New High-Energy Materials").

The art program has been enhanced, with the consistent integration of new, more vivid representations of molecular structure. Throughout the book, numerous small changes intended to clarify and expand explanations appear in both the figures and the text.

Organization

The underlying premise of this book is that chemistry is a science based on observations and measurements, from which deductions and principles can be derived. Our view is that the study of chemistry should start with the visible, tangible world to organize and rationalize daily experience. As students become comfortable with chemists' way of viewing natural phenomena, the more fundamental (but also less evident) theoretical basis of the subject can be introduced gradually. It is our hope that students will come to look on theory and experiment (or observation) as mutually reinforcing aspects of science. Each informs the other, and together they point the way to a deeper understanding of the workings of the real world. The organization of this text also allows students to begin meaningful laboratory work at an early stage.

Chapters 1 through 6 therefore begin by emphasizing macroscopic aspects of chemistry, moving from mass relationships in chemical reactions and the atomic theory of matter through the periodic table of the elements toward a description of the states of matter. The Lewis electron-dot model is introduced early as a useful organizing principle directly connected to the structure of the periodic table. Chapters 7 through 14 then treat chemical equilibrium and its thermodynamic and kinetic bases. In the next seven chapters, the emphasis shifts to a systematic development of the modern microscopic picture of the structure of matter: from electrons and nuclei through atoms and molecules to coordination complexes and extended solid structures. Chapters 22 through 25 then present an overview of chemical processes: industrial, atmospheric, geochemical, and biological. These chapters introduce no new principles; instead, they provide a useful review of principles from earlier chapters, such as the tradeoffs between thermodynamics and kinetics in process design. Our presentation of industrial chemistry has a strong historical component, illustrating the ways in which chemical processes are developed and improved. There is real value in concluding an introductory course with a broader look at chemistry in its "real-world" context, and that has been our goal in the final chapters.

Alternative Teaching Options

We recognize that some instructors may prefer a different order of presentation from that just described. We have therefore made the book flexible enough to accommodate many alternative ways of organization. For example, the material on atomic structure and chemical bonding in Chapters 17 through 19 could be introduced after Chapter 3 or after Chapter 10. Chapters 15 and 19 through 21 could be postponed until later in the course or omitted if time constraints are severe. Some instructors choose to split the coverage of thermodynamics into two parts. Those wishing to do so could cover Chapter 10 anytime after Chapter 5, while postponing Chapter 11 until later in the course. Other instructors may choose to postpone the discussion of intermolecular forces in Section 6–1 until after the quantum theory has been introduced in Chapters 16 and 17.

A number of individual sections of chapters can be omitted without loss of continuity. For example, the last three sections of Chapter 9 ("Dissolution and Precipitation Equilibria"), while important for any course covering qualitative analysis of metal ions, are not essential to the material that follows them in the book. Chapter 15 ("Nuclear Chemistry") could also be omitted or postponed until later in the course. Chapters 22 and 23 can be treated as "interchapters" and covered earlier. For example, Chapter 22 could be covered after Chapter 11 and Chapter 23 after Chapter 13. Chapters 24 and 25 provide a coherent introduction to organic chemistry (including polymers) at a level of detail appropriate to general chemistry and could be included after Chapter 18 or at the end of the course.

Appendices

The book contains six appendices. Appendix A treats scientific notation, experimental uncertainty (including the distinction between precision and accuracy), and significant figures. Appendix B gives an overview of the SI system of units and shows how unit-factors are created and used to assist in the conversion from one unit to another. It also presents an extensive table of recommended symbols and units for quantities encountered in general chemistry. This table is cross-referenced to the locations in the text at which the unit and symbol are introduced. It provides a valuable basis for study and review. Appendix C treats some mathematical operations used in general chemistry, including the drawing and reading of graphs, the solution of systems of linear equations, the solution of quadratic equations, and the use of logarithms. The material in these three appendices can be used by the instructor at any appropriate point in a course. All three contain problems for students to work to test their understanding of the material.

Appendices D and E consist of tables of thermodynamic and electrochemical data. Together with the tables inside the covers of the book, they provide all numerical data (beyond what is specifically quoted in the problems) required to solve the problems in the text. Appendix F contains the answers to selected odd-numbered end-of-chapter problems and to the cumulative problems that conclude chapters.

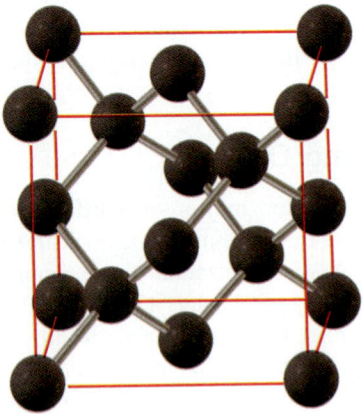

Acknowledgments

This revision has benefited from the help of the following individuals who helped to shape the revision and ensure accuracy:

Sharmistha Basu-Dutt, State University of West Georgia
Jon A. Booze, AccuText Publishing Support
E. David Cater, University of Iowa
Julie Ealy, Pennsylvania State University, Berks–Lehigh Valley
John P. Graham, Arkansas Tech University
John M. Halpin, New York University
Curtis Hare, University of Miami
David O. Harris, University of California, Santa Barbara
James F. Harrison, Michigan State University
Marie Nguyen, Indiana University–Purdue University Indianapolis
Shawn T. Phillip, Vanderbilt University
Andrew Pounds, Mercer University
William P. Reinhardt, University of Washington
B. Ken Robertson, University of Missouri–Rolla
Allan Smith, Drexel University
Joel B. Tellinghuisen, Vanderbilt University
Petra Van Koppen, University of California, Santa Barbara
Jess C. Vickery, Siena College
Marcy Whitney, The University of Alabama, Tuscaloosa
Donald J. Wink, University of Illinois, Chicago

David W. Oxtoby
Wade A. Freeman
Toby F. Block
March 2002

Supporting Materials

Instructor Resources

Instructor's Manual: **Jess C. Vickery, Siena College (ISBN: 0-03-033196-X).** Contains worked solutions to even-numbered problems using the problem-solving strategies described in the text.

Transparencies **(ISBN: 0-03-033221-4).** Contains 120 transparencies with important figures from the text.

Test Bank: **Joel B. Tellinghuisen, Vanderbilt University (ISBN: 0-03-033246-X).** Contains over 1000 multiple-choice questions keyed to the text contents.

ExamView® *ExamView®[1]* *Computerized Testing* **(Cross-platform CD-ROM ISBN: 0-03-033336-9).** Computerized testing! Create, deliver, and customize tests and study guides (both print and online) in minutes with this easy-to-use assessment and tutorial system. ExamView offers both a *Quick Test Wizard* and an *Online Test Wizard,* which guide the user step by step through the process of creating tests.

Student Resources

(For a preview of these titles or to purchase on-line, please visit the *Chemistry: Science of Change* **Web page at http://www.brookscole.com/chemistry)**

 *General Chemistry Interactive CD-ROM,* **Version 3.0,**[2] by William J. Vining, John C. Kotz, and Patrick Harman (Cross-platform ISBN: 0-03-035319-X). *General Chemistry Interactive CD-ROM* is one of the most successful pieces of science education software ever commercially published. It uses technology to enhance learning, by leading students to think and explore through interactive, computer-based exercises. *General Chemistry Interactive's* Intelligent Tutors mimic the interaction between instructor and student in a one-on-one tutoring session. The program's Intelligent Tutors do not simply give answers—they walk students through the steps required to arrive at a correct solution. The Intelligent Tutor works with students just as a live tutor would. Guided Simulations give students a deeper understanding of fundamental concepts by offering a detailed approach to designing and performing simulated laboratory and thought experiments. These guided simulations help with conceptual knowledge and scientific reasoning.

Student Solutions Manual: **Donald W. Wink, University of Illinois at Chicago (ISBN: 0-03-033238-9).** Contains worked solutions to all odd-numbered end-of-chapter problems using the problem-solving strategies described in the text.

Study Guide: **E. David Cater, The University of Iowa (ISBN: 0-03-033231-1).** This resource helps students organize their study and guides them through the topics in a systematic way. Each chapter of the text is covered by an introduction, a list of review topics, section-by-section study suggestions and questions, a list of key terms, and a practice exam with worked-out answers.

[1]ExamView® is a registered trademark of FSCreations, Inc. Used herein under license.

[2]*General Chemistry Interactive CD-ROM,* Version 3.0, can be packaged with *Chemistry: Science of Change,* Fourth Edition, for significant savings. Contact your local Brooks/Cole•Thomson representative for details or call 1-800-423-0563.

Internet Resources

OWL: Online Web-based Learning System.[3] (owl.harcourtcollege.com/demo) Developed at the University of Massachusetts over more than six years, OWL is a fully customizable, flexible learning system. The OWL General Chemistry System is a database of parameterized questions, discovery modules, and integrated tutors, correlated to the text. This powerful on-line homework, quizzing, and testing tool is conveniently organized by chapter and topic, in units that can easily be assigned or excluded, in any sequence an instructor desires. This flexibility, along with customized feedback, creates an ideal learning environment.

The Brooks/Cole Chemistry Resource Center (http://www.brookscole.com/chemistry). Users of *Chemistry: Science of Change* have access to a rich array of teaching and learning resources. This site is the ideal way to make teaching and learning an interactive and intriguing experience. It includes Suggested Course and Lecture Outlines; Syllabus Builder; Downloadable Microsoft® PowerPoint® presentation slides; text correlation to laboratory experiments; topic-by-topic Internet links; and so much more!

WebTUTOR Advantage PLUS

WebTutor™ Advantage on WebCT (WebCT ISBN: 0-03-033281-8). For students, WebTutor™ Advantage offers real-time access to a full array of study tools, including flashcards (with audio), practice quizzes, online tutorials, and Web links. Professors can use WebTutor™ Advantage to provide virtual office hours, post syllabi, set up threaded discussions, track student progress with quizzing material, and more.

Presentation Tools

MultiMedia Manager CD-ROM (ISBN: 0-03-033261-3). This digital library and presentation tool includes already-created text-specific presentations and a library of resources valuable to instructors such as art, photos, and tables. Electronic files are easily exported into other software packages so you can create your own materials.

Laboratory Resources

Chemical Education Resources (CER).

Customized Experiments for Introductory Chemistry, Health Sciences Chemistry, General Chemistry, and Organic Chemistry (Catalog ISBN: 0-534-97709-X). Users of the text can create their own laboratory manual program by selecting from more than 300 high-quality experiments available from CER and Brooks/Cole. They can work with Brooks/Cole or Thomson Custom Publishing representatives or work online at www.textchoice.com to view and select experiments, collate them in any order, and combine them with materials of their own course notes, lecture outlines, and articles. Custom laboratory manuals can be packaged with the text for even greater value and convenience.

Chemical Principles in the Laboratory, Seventh Edition (ISBN: 0-03-031167-5), by Emil J. Slowinski, Wayne C. Wolsey, and William L. Masterton.

Experiments in General Chemistry (ISBN: 0-534-36102-1), by Steven L. Murov.

Laboratory Experiments for General Chemistry, Fourth Edition (ISBN: 0-03-032906-X), by Harold R. Hunt, Toby F. Block, and George M. McKelvy.

Laboratory Inquiry in Chemistry (ISBN: 0-534-37694-0), by Rich Bauer, Jim Birk, and Doug Sawyer.

Standard and Microscale Experiments in General Chemistry, Third Edition (ISBN: 0-03-021017-8), by Carl B. Bishop, Muriel B. Bishop, and Kenneth W. Whitten.

---

[3]To order an OWL access code with *Chemistry: Science of Change,* Fourth Edition, contact your local Brooks/Cole•Thomson representative for details at 1-800-423-0563.

To the Student

Chemistry is an immensely practical subject that, properly understood, furnishes answers to many important problems. Like all of the sciences, chemistry is an experimental discipline. Doing coursework such as homework, quizzes, and examinations is very much like doing experiments. The difference is that someone else has asked the question (done the experiment) and collected the results in the form of data. It remains for you to devise a strategy for converting the data into an answer. How well you answer the question depends on how well you have "sized up" the problem.

Understanding exactly what is wanted is the first step. The second step is to call to mind the principle(s) that underlies the desired result. Find out what factors are involved. Decide whether the information provided is relevant and discard anything that is not. Plan a method of attack and carry it out. The way to do this is formally illustrated in "Strategy" statements in many of the worked-out examples in the book. Once you have an answer, consider whether it makes sense. Is it about the right amount? Are its units correct? Has it been determined experimentally? Is there another approach you can take to work the problem? (Often there are several ways to solve a given problem.) If so, does it give the same answer? This is illustrated in "Check" statements in worked-out examples in the book.

As you gain experience in solving problems, you may find that your way of looking at new situations has changed. You may find yourself "looking for the facts," wondering what lies behind an occurrence, questioning the validity of a piece of information. Solving problems is a *transferable skill* that applies not only to working chemistry problems (and getting good grades on examinations) but also to almost every activity in life.

As you are using this book, read thoughtfully for a few pages and then set the book aside. Reflect on what you've read. Is there a concept that might prove useful, a generalization that ties several ideas together, an analogy that is especially apt? Summarize it succinctly by making a note of it in your own words. The act of putting pen to paper is in itself of remarkable assistance to the memory, and expressing the concept in your own words is a powerful study aid.

Another suggestion is to study with a classmate, taking turns instructing each other. There is no surer way to learn a concept than to explain it to another person. Make up problems to test your study portion and, to overcome the examination anxiety that students sometimes experience, limit the time allotted to solve them. If you prefer, choose some of the odd-numbered problems at the end of the chapter you're studying. The answers are in Appendix F.

Above all, in your study of chemistry, develop a questioning, aggressive approach to what you read and hear. Ask for the evidence, and don't despair if you make mistakes. The wonderful edifice of science was built on trial and error, and persistence always wins!

About This Book

We have tried to make this a user-friendly book, and our principal concern has been to explain the subject in as clear and unambiguous a manner as possible. We have included a number of learning tools to help you increase your understanding of chemistry. We hope these features will stimulate your interest in the subject.

Figures

The book makes extensive use of **figures** to aid in explaining concepts and to illustrate chemical behavior. These include color photographs, diagrams, and drawings of molecular structures. Especially in the first 14 chapters, we use molecular structures with their figure captions to present brief "capsule summaries" of the properties and uses of important simple compounds. In these structural diagrams, carbon is black, hydrogen is light gray, oxygen is red, nitrogen is blue, sulfur is yellow-orange, phosphorus is orange, fluorine is yellow, chlorine is green, bromine is reddish brown, iodine is reddish purple, boron is brown, and xenon is blue-green.

Examples and Exercises

Each chapter contains numerous **worked examples** to illustrate the principles that have been discussed. Where appropriate, the worked examples include "Strategy" and "Check" statements to show how to attack a problem and how to think about the reasonableness of its solution. Each example is followed by a related **exercise** together with its answer.

Highlighted Text

Throughout each chapter, **key terms** are printed in **boldface** where they first appear. Important statements and equations are screened in sage and answers to examples are screened in yellow. Acids are color-coded in red and bases in blue; formal charges are shown in blue circles, while oxidation numbers appear in red.

Marginal Notes

The occasional **notes in the margin** serve several purposes: to emphasize or elaborate on a point, to provide a different perspective on a statement made in the text, and to provide "signposts" directing you to other parts of the book where subjects have been or will be covered.

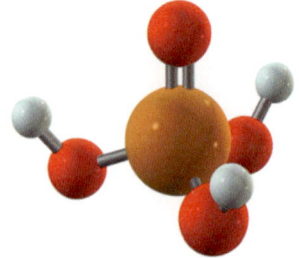

"Chemistry in Your Life" and "Chemistry in Progress"

Throughout the book we have included a number of **special boxed topics.** Those entitled "Chemistry in Your Life" focus on the chemistry of biological processes or else on the impact of chemistry in daily life. The boxes titled "Chemistry in Progress" show the ways in which chemistry contributes to and benefits from developments in science and technology that change the modern world.

Chapter Summaries

Each chapter ends with a **summary** of the material presented in that chapter. The purpose of these summaries is to place the new material in context and to review it. These summaries should be especially useful when you are studying for an examination.

End-of-Chapter Problems

End-of-chapter problems are arranged in two groups. The first group consists of paired problems, placed in order of presentation of the topics in the chapter. The answers to most odd-numbered problems in this group are given in Appendix F, allowing you to check the answer before undertaking the even-numbered problem, which parallels the first. These paired problems are followed by a group of **additional** problems, which may involve concepts from different parts of that chapter and which may draw on material from earlier chapters as well. More challenging problems are indicated with an asterisk.

Cumulative Problems

At the end of each of Chapters 1 through 20 there is a multipart **cumulative problem** that is built around a problem of chemical interest and draws on material from the entire chapter for its solution. Working through these exercises provides a useful review of material from the chapter, helps you to put principles into practice, and prepares you to solve the problems that follow. The answers to these problems appear in Appendix F.

Index/Glossary

The **index/glossary** at the back of the book gives a brief definition of each term and a reference to the pages on which that term appears. You should get into the habit of consulting it when you encounter a term that you do not fully understand or one you encountered earlier in the book and need to review.

Contents Overview

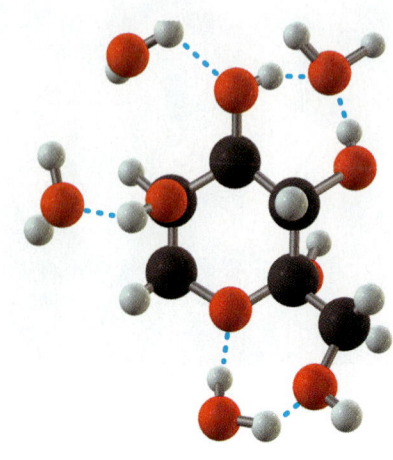

Contents

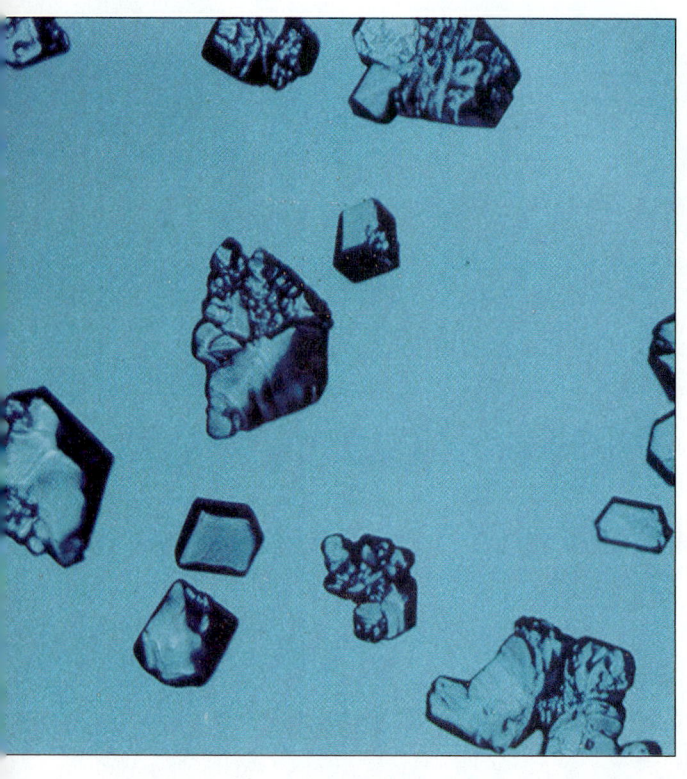

1

The Atomic Nature of Matter

CHAPTER OUTLINE

The minerals orpiment (yellow, As_2S_3) and realgar (red, As_4S_4) in heterogenous mixture in a rock. (*Charles D. Winters*)

Chemical change occurs everywhere: in the processes of respiration that permit us to live, in the slow geological transformations that shape the Earth's surface, in the formation of compounds deep in interstellar space. Some of these changes are slow, some fast; some are barely detectable, some, strikingly dramatic. Chemistry is the study of these changes.

1–1 Chemistry: Science of Change

A chemist carefully draws off a small amount of a yellow-green gas from a cylinder into a test-tube. The gas is dense. It settles to the bottom of the test-tube, displacing the air up and out. The chemist drops a small piece of a soft metal into the test-tube, using a pair of tongs. Nothing happens. Next comes a tiny drop of water that strikes the metal. Immediately, flames and sparks shoot out as the metal reacts with the gas (Fig. 1–1); the test-tube becomes noticeably hotter. After the reaction dies down, a white solid is found to have formed on the walls of the test-tube.

What has happened? The yellow-green gas is chlorine, a noxious irritant that can kill if inhaled. The metal is sodium, a substance so soft that it can be cut with a knife, like butter. Like chlorine, it must be handled with care because it can burst into flame on contact with water. These two highly reactive materials combine vigorously to give the white solid sodium chloride, which is the major constituent of ordinary table salt. Such transformations of substances into others with different properties are the subject of chemistry.

The first step in chemistry is observation—taking a close look at the properties of substances (color, shape, hardness, and so forth) and what happens to them when they are mixed under different conditions. Chemistry begins with the description of matter and change.

If the first question is "what?," then the second is "how much?" Chemists seek to make their knowledge quantitative—to understand how much sodium has reacted, how much sodium chloride was produced, and what volume of gaseous chlorine was consumed. The flask has become hotter. How can that heat be measured, and how does it compare with the heat evolved when, for example, iron instead of sodium reacts with chlorine?

The third question is "why?" Why does this particular mass of sodium react with this particular volume of chlorine? More fundamentally, why do sodium and chlorine react at all? Why is sodium so soft and sodium chloride so hard? Why does the drop of water start the reaction? Each question leads to further questions.

Although a single chemical reaction can raise many questions, each such reaction is also a part of a far larger set of issues, such as why the sea contains so much salt and why a dietary excess of salt leads to hypertension and other severe health problems in humans. Here the overlap of chemistry with other fields, such as geology and biology, provides a fruitful source for new discoveries.

Chemistry is an extremely practical subject. The reaction we have just discussed is not a very useful one because salt is readily available from salt deposits or from evaporated seawater. But if we were able to reverse the reaction and produce sodium and chlorine from sodium chloride, then we could use those products

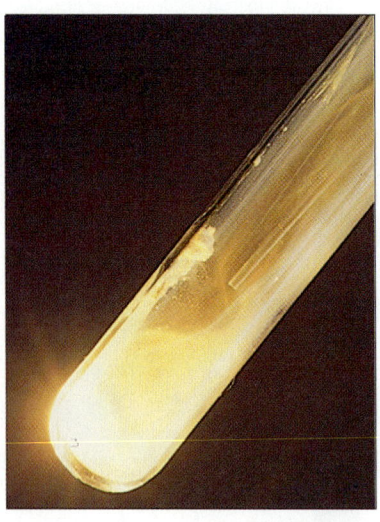

Figure 1–1 • The reaction between sodium and chlorine proceeds very slowly until the addition of a small drop of water sets it off. Sparks then fly, the reaction vessel becomes hot, and a white powder of sodium chloride forms on the walls of the flask. *(Larry Cameron)*

Figure 1–2 • The bright yellow-orange glow from these street lights is given off by sodium atoms raised to high-energy states. *(Science VU/GERL/Visuals Unlimited)*

as starting materials for chemical reactions to make new substances. In fact, chemists have learned how to do this. Sodium and chlorine can be obtained by passing an electric current through molten sodium chloride. Sodium has applications ranging from filling sodium vapor lamps (Fig. 1–2) to producing titanium for aircraft hulls. Chlorine finds an equal range of uses, from treating wastewater to producing plastic piping (Fig. 1–3). As chemists seek to understand chemical reactions better, they also undertake to make new materials with useful properties.

Figure 1–3 • More than 56% by mass of this plastic pipe is chlorine. *(Delve Communications)*

1–2   The Composition of Matter

Two traditions intertwine throughout the history of chemistry: **analysis** (taking things apart) and **synthesis** (putting things together). The two trace from the earliest periods of Greek natural philosophy and remain driving forces to the present day. When a useful product is discovered in a natural source (such as the antimalarial drug quinine in the bark of the cinchona tree), the first step is to analyze it to find its composition and structure. Then a strategy is designed to synthesize it from easily available starting materials. The procedure can be difficult. Penicillin was discovered in 1929 through the accidental contamination of a bacterial culture by mold; its laboratory synthesis was not achieved until 1957.

• Penicillin manufacture still uses a fermentation process, carried out in large vats. Direct chemical synthesis of penicillin is not economically competitive with this large-scale and efficient biological method.

CHEMISTRY IN YOUR LIFE

Semi-Synthesis, Total Synthesis, and Biosynthesis

The discovery in the early 1980s that paclitaxel, a compound extracted from the bark of the Pacific yew tree and trade-named Taxol, works astonishingly well in the treatment of even difficult cases of ovarian and breast cancer brought new hope to cancer sufferers. But there were problems. Only 0.1 to 0.2 g of Taxol could be extracted from a kilogram of bark, and the bark from three trees was required to treat a single patient for a year. The Pacific yew grows extremely slowly and is not abundant; stripping its bark kills it.

So, do human lives come first, even if it means depletion or extinction of an entire species? Should all the Pacific yew trees be sacrificed to help current cancer victims, with nothing being left for the future? Might the trees hold cures for other ailments that could be lost in a rush to extract Taxol? If authorities reserved some Pacific yews, would it be right to buy black-market Taxol extracted from bark taken by "tree-rustlers"?

Fortunately, synthetic research by chemists soon eliminated these questions as matters of practical concern. A semi-synthetic preparation of Taxol starting from the needles of the English yew, a related species, was announced. In this procedure, a compound called 10-deacetyl-baccatin III is extracted from the needles and converted to Taxol by chemical reactions. This substance is much more concentrated in these needles than Taxol is in Pacific yew bark, and harvesting needles (within reason) does not harm the trees. This process is called semi-synthetic because the tree does most of the work, and the chemist just finishes up. Semi-synthesis has proved useful in preparing many drugs, including other anticancer agents.

In 1994, two groups of chemists managed to achieve a **total synthesis** of Taxol from stock starting materials—in effect, from the constituent chemical elements themselves. However, their multistep procedures, required by the extreme complexity of the structure of Taxol, give yields too poor for practical use in making Taxol.

Figure 1–A • The bark of the Pacific yew tree, *Taxus brevifolia. (Tom & Pat Leeson/Photo Researchers, Inc.)*

One alternative to collecting yew needles as a cash crop is a **biosynthesis** of Taxol. Scientists are attempting to isolate the genes that code for the synthesis of Taxol in the bark cells of the Pacific yew. They hope to insert them into the cells of plants that are easily cultured in the laboratory and activate them. The cells would then grow, divide repeatedly, and secrete Taxol for collection. Genetically engineered cells of baker's yeast already produce high-quality supplies of human insulin to control diabetes; biosynthesis undoubtedly will be adapted in the future to make more and more lifesaving drugs.

Substances and Mixtures

Let us begin our study of chemistry with the process of analysis, which is the determination of what makes up matter (Fig. 1–4). Matter occupies space and has mass; it is all the stuff of the world around us. In daily life, we perceive matter either as objects or as materials, depending on our purposes. Materials endure, while objects come and go. Thus, chopping a wooden chair (an object) into splinters destroys the chair, but the wood (a material) remains and persists even if the splinters are reduced to sawdust. Chemical analysis focuses on materials, not objects.

Suppose then that we have before us a sample of a material for study. We examine its various specific properties, or distinguishing characteristics. We start with **physical properties,** such as color, odor, taste, hardness, melting point, boiling point, electrical conductivity, magnetism, and density. Physical properties are properties that can be defined and discussed without reference to any other material and thus differ from **chemical properties,** which involve behavior with respect to other materials. If the sample is a piece of wood, we notice immediately that its properties are not uniform throughout the sample; some portions are darker than others, whereas others (the knots, for example) are harder and resist indentation better. If the properties of a material, like those of wood, change from one point to another within a sample, then the material is **heterogeneous.** Heterogeneous samples usually contain

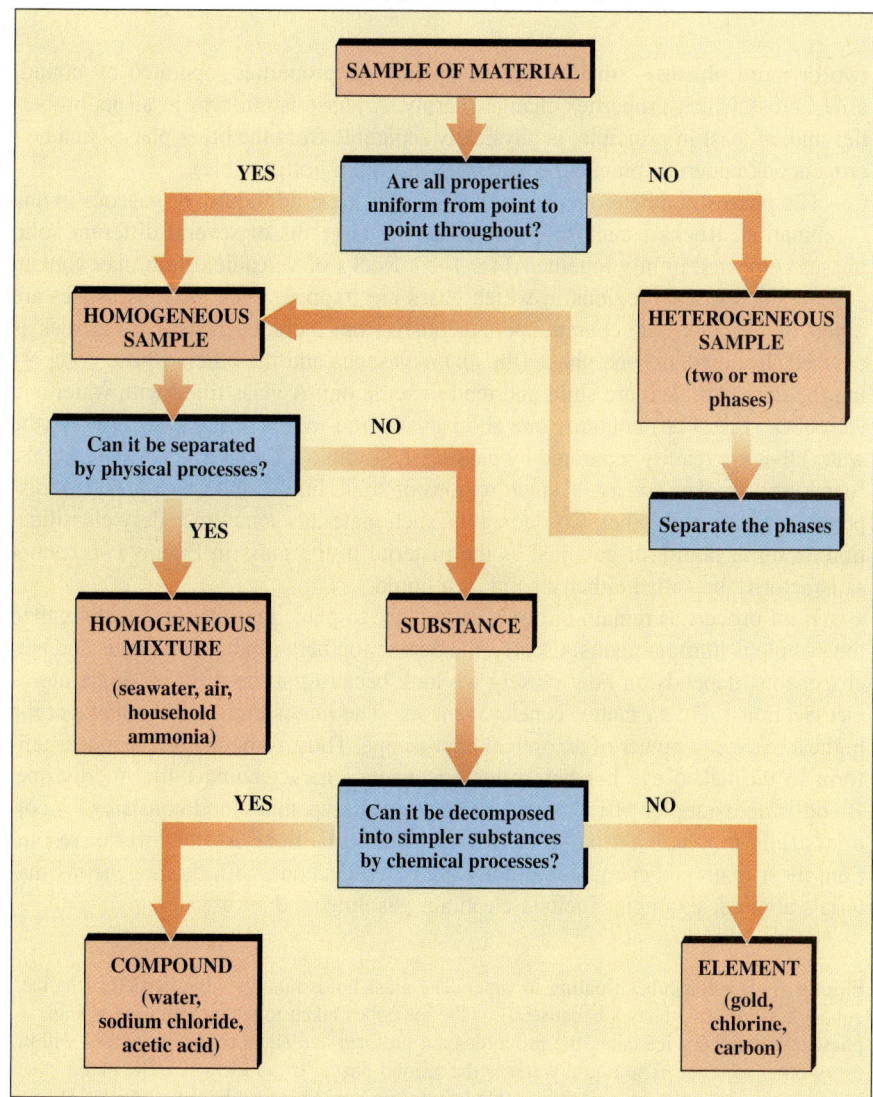

Figure 1–4 • An outline of the steps in the analysis of a material.

Figure 1–5 • Close examination of a piece of granite reveals that it is heterogeneous. At least four different kinds of grain can be distinguished in this close-up photograph. *(Albert J. Copley/Visuals Unlimited)*

two or more **phases**—smaller regions of uniform properties separated by boundaries across which properties change sharply. A phase is uniform in all its properties and, at least in principle, is physically separable from the other phases in a heterogeneous material (practical separations are often not possible).

The phases in a heterogeneous material may be solid, liquid, or gaseous in any combination. Rocks usually consist of numerous grains of several different solid phases cemented tightly together (Fig. 1–5). Rocks of volcanic origin may contain gaseous inclusions—regions in which gases are trapped. Such gaseous phases are a part of the rock but separate spontaneously from the solid phases if the rock is crushed. In dusty air, one phase (the air) is gaseous and the others (those composing the dust particles) are solid and tend to settle out. A glass filled with water and ice cubes (Fig. 1–6) contains one solid phase (the ice) and one liquid phase (the water) that are readily separated by passing the sample through a sieve. Pastes, gels, foams, and smokes consist of small regions of solid, liquid, and gaseous phases dispersed through each other. Consequently, such materials sometimes defy classification as solid, liquid, or gas, just as the material in the glass in Figure 1–6 cannot satisfactorily be called either a solid or a liquid.

If all properties remain uniform from place to place throughout a sample, then the sample is **homogeneous.** Clearly, the distinction between homogeneous and heterogeneous depends on how closely we look because, at the level of single atoms (see Section 1–3), *all* matter is heterogeneous. The line is traditionally drawn at the highest resolving power of an optical microscope. Thus, some materials appear uniform to the naked eye but betray heterogeneous character under the microscope. Blood is an example. Microscopic examination shows that blood consists of a colorless fluid in which a mixture of red and white cells float. At this level of resolution, most materials encountered daily are heterogeneous. Still, homogeneous materials abound. Examples include clean air, gasoline, and pure water.

• The categories solid, liquid, and gas apply best to phases, not to materials in general. See Problem 79 at the end of this chapter.

Figure 1–6 • Ice cubes floating in water. The glass holds many visible particles (the ice cubes) but only two phases because all of the ice cubes taken together comprise a single phase. Within every ice cube, the properties are uniform and equal to the properties within every other ice cube. The liquid water is the second phase. If the ice melts, the phase boundaries cease to exist, and the sample becomes a homogeneous liquid. *(Charles D. Winters)*

(a)

(b)

(c)

Figure 1–7 • (a) A solid mixture of blue $Cu(NO_3)_2 \cdot 6H_2O$ and yellow CdS is added to water. (b) While the $Cu(NO_3)_2 \cdot 6H_2O$ dissolves readily and passes through the filter, the CdS remains largely undissolved and is held on the filter. (c) Evaporation of the solution leaves crystals of pure $Cu(NO_3)_2 \cdot 6H_2O$. *(Leon Lewandowski)*

Even if a material consists of a single phase, it may be possible to separate it into two or more differing components by ordinary physical means. If so, it is a **homogeneous mixture,** or **solution.** Air, for example, is a homogeneous mixture of several gases (including nitrogen, oxygen, argon, and carbon dioxide). If we liquefy air and then warm it slowly, the gas having the lowest boiling point evaporates first from the mixture, leaving behind in the liquid the components with higher boiling points. Such a separation is not perfect, but we can repeat the process of liquefaction and evaporation on separated portions and improve the resolution of air into its component gases to any degree of purity we may require. We take advantage of the different physical properties (in this case, the boiling points) of the components of the mixture to separate them from one another.

Other physical operations can be used to separate both homogeneous mixtures and heterogeneous mixtures into components. Treating a mixture of solids with water or another solvent frequently allows the components to be separated on the basis of their different solubilities (Fig. 1–7). The salt and water in seawater can be separated to a considerable extent by slowly freezing the mixture; the first crystals to form are nearly pure water. Centrifuges are often used to separate a suspended solid from a liquid. "Molecular sieves" contain channels that permit small molecules to pass while larger molecules are blocked. Other materials trap and hold (adsorb) certain kinds of molecules on their surfaces. They are used to remove pollutants from gases or water supplies.

If all these physical procedures (and many more) fail to separate a material into portions that have different properties, we call the material a **chemical substance** (or simply a **substance**). Different samples of a chemical substance, when examined under the same conditions, have the same properties if they are pure.

What about the common material sodium chloride, which we call "table salt"? Is it a substance? The answer is "yes" if we use the term "sodium chloride," but "no" if we use the term "table salt." Table salt is a mixture that consists principally of sodium chloride, to which a small amount of sodium iodide usually has been added to provide for the requirements of our thyroid glands, along with a small quantity of magnesium carbonate that coats each grain of sodium chloride and prevents the salt from caking. Even if these two components were not deliberately added, table salt would contain other impurities that were not removed

• To a chemist, "pure" means "containing a single substance." Pure orange juice is unadulterated, but, chemically speaking, it is a mixture. It contains water, citric acid, ascorbic acid, and fructose, along with many other substances.

Figure 1-8 • Nearly pure elemental silicon is produced by pulling out solid cylinders called "boules" from molten silicon. Most of the impurities are left behind. The boules are then sliced into thin wafers for use in integrated circuits. *(Charles D. Winters)*

in its preparation; to that extent, table salt is a mixture. When we refer to "sodium chloride," however, we imply that all other materials are absent and that it is pure.

In practice, nothing is absolutely pure; the word "substance" is therefore an idealization. Among the most nearly pure materials ever prepared are silicon (Fig. 1–8) and germanium. For applications in electronic devices and solar cells, the electronic properties of these materials require very high purity (or precisely controlled concentrations of deliberately added impurities). By meticulous chemical and physical methods, germanium and silicon have been prepared with concentrations of impurities on the order of only 1 part in 10 million.

Elements and Compounds

• More exactly: by the end of 2000, over 17.2 million chemical substances had been registered by the Chemical Abstracts Service of the American Chemical Society (as reported at http://www.cas.org).

At present, literally millions of chemical substances have been discovered (or synthesized) and characterized (described in terms of their properties). Are these the fundamental building blocks of matter? If they were, then classifying them, much less understanding them, would prove an insurmountable task. In fact, however, all these substances are found to be composed of a much smaller number of building blocks, called elements. **Elements** are substances that cannot be decomposed into two or more simpler substances by ordinary physical or chemical means. Here, the word "ordinary" is important because we exclude the processes of radioactive decay (either natural or artificial) and of high-energy nuclear reactions that do transform one elementary substance into another. **Compounds** are chemical substances that contain two or more chemical elements combined in fixed proportions. Hydrogen and oxygen are elements because no further chemical separation has been found possible, whereas water is a compound because it can be separated into hydrogen and oxygen by passing an electric current through it (Fig. 1–9).

Binary compounds are substances (such as water) that contain two elements. Ternary compounds contain three elements; quaternary compounds contain four, and so on.

At present, 113 chemical elements are known. A few have been known since before recorded history because they occur in nature in elemental form rather than in combination with one another in compounds. Gold, silver, lead, copper, and sulfur are chief among this group. Gold was found in streams in the form of little granules (placer gold) or nuggets in loosely consolidated rock. Sulfur was associated with volcanoes; copper and, more rarely, lead could be found in their native states in shallow mines.

Each element is designated both by a name and by a one- or two-letter symbol. The origin of these names and symbols is a fascinating subject. Many have a Latin root, such as gold (aurum, symbol Au); copper (cuprum, Cu), a word that

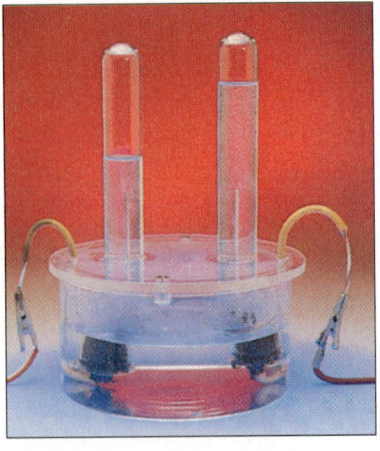

Figure 1-9 • As electric current passes through water containing dissolved sulfuric acid, gaseous hydrogen and oxygen form as bubbles at the electrodes. This method of electrolysis can be used to prepare very pure gases. Note that the volume of hydrogen produced (on the left) is twice that of the oxygen (on the right). This is a consequence of the law of combining volumes, which is discussed in Section 1–4. *(Charles D. Winters)*

echoes the Greek name for the island of Cyprus in the eastern Mediterranean, where the metal was found in large lumps; iron (ferrum, Fe); and mercury (hydrargyrum, Hg). Hydrogen (H) means "water-former." Potassium (kalium, K) takes its common name from potash (potassium carbonate), a useful chemical obtained in early times in impure form by leaching the ashes of wood fires with water. Many elements take their names from Greek and Roman mythology: cerium (Ce) from Ceres, goddess of plenty; tantalum (Ta) from Tantalus, a king condemned by the gods; and niobium (Nb) from Niobe, daughter of Tantalus. Some elements, such as europium (Eu) and americium (Am), are named for continents. Others are named after countries: germanium (Germany, Ge); francium (France, Fr); polonium (Poland, Po); and, less obviously, ruthenium (Russia, Ru). Cities provide the names of other elements: holmium (Stockholm, Ho) and berkelium (Berkeley, Bk). Some are named for the planets: uranium (U), neptunium (Np), and plutonium (Pu). Others take their names from colors: praseodymium (green, Pr), rubidium (red, Rb), and cesium (sky blue, Cs). The names of other elements honor great scientists: curium (Marie Curie, Cm), mendelevium (Dmitri Mendeleev, Md), bohrium (Niels Bohr, Bh), fermium (Enrico Fermi, Fm), and einsteinium (Albert Einstein, Es).

• Ytterby, a village in Sweden, gave its name to four elements: erbium, terbium, ytterbium, and yttrium.

1-3 The Atomic Theory of Matter

The modern understanding of matter has its roots in antiquity. In ancient Greek philosophy, all substances were thought to contain, in proportions to explain their properties, the four elements air, earth, fire, and water. The yellow luster of gold, in this view, derived from fire, its great density from earth, and so forth. What would happen, the ancient Greek philosophers wondered, if a quantity of a substance, such as gold, were divided in half, then divided again and again, without limit. Many believed that the substance would ultimately yield up its constituent earth, air, fire, and water. The philosopher Democritus (ca. 460–ca. 370 B.C.) opposed this view and postulated that repeated subdivision of a quantity of gold or other material would eventually lead to a smallest particle, an unchangeable and indestructible **atom** (named from the Greek *a* ("not") + *tomos* ("cut"), meaning "not divisible"). The question remained in the realm of speculative philosophy (with Democritus's idea largely discounted) until the advent of the modern scientific view that theories must be tested and refined by experiment.

When scientists use the formal **scientific method,** they first gather the facts of a situation, putting reliance on reproducible observations and experiments. Sets of related facts are summarized, if possible, in more or less general principles or **laws,** such as the law of gravity ("objects fall to earth if unsupported"). Next, they formulate a **hypothesis,** a provisional statement that fits the known facts of a situation and also suggests new experiments or observations. The hypothesis is an attempt to model reality. Carrying out new experiments then tests the hypothesis. Most frequently, the results of the experiments force changes in the hypothesis. Rarely, they confirm the hypothesis entirely or else oblige its complete abandonment. The cycle continues until no practicable new observations or experimental tests are available. A hypothesis or set of related hypotheses that explains diverse experiments and observations and does not contradict any solid factual findings is a **theory.**

• The cycle of hypothesizing and testing is familiar to anyone troubleshooting a stalled car.

Scientific progress toward understanding the microscopic nature of matter began in the 18th century. It arose out of an interest in the nature of heat and the way things burn, or undergo **combustion.** Fire of course had attracted speculation (and sometimes veneration) since its discovery, and the combustion of fuels was then, as now, essential for heat and light. It had been observed that wood, when burned, left behind a residue of ash, whereas metal, when heated in air, was transformed into a "calx," which we now call an oxide. The explanation for this phenomenon that was popular in the first part of the 18th century was that a substance called "phlogiston" was driven out of wood or metal by the heat of a fire. From the modern perspective this seems absurd because the ash weighs less than the original wood, whereas the calx weighs more than the metal. At the time, however, the principle of conservation of mass was not yet established, and people saw no reason why the mass of a material should not change upon heating.

A better understanding of combustion was gained through a careful study of the gases consumed or produced in such reactions. The great French chemist Antoine Lavoisier carried out an important experiment in 1775. He gently heated the liquid metal mercury in a closed vessel that contained air. After several days, a red substance (mercury(II) oxide) was produced that was identical to the calx that formed when mercury was burned in the open. Lavoisier found that the gas remaining in the apparatus was reduced to five sixths of its original volume during the experiment. Moreover, he discovered that the residual gas did not support combustion or life; it extinguished candles and suffocated animals. He next carefully measured the mass[1] of a quantity of the red oxide of mercury and heated it strongly. This treatment transformed the red oxide back into mercury (Fig. 1–10), but the mass of mercury was distinctly less than the mass of the red oxide. The strong heating also generated a previously unknown gas. The new gas caused a smoldering wooden splint to burst into bright flames and was breathable. When it was added to the residual gas from the first part of the experiment, the mixture was indistinguishable from ordinary air. Lavoisier concluded that the mercury first combined with, and then lost, a definite substance during the sequence of gentle and strong heating. He subsequently managed to measure the mass of the gas produced by the strong heating. This mass equaled the mass lost by the solid red oxide when it was converted to the metal. The appearance and properties of the materials changed greatly during the cycle of change, but their total mass did not. Lavoisier stated the **law of conservation of mass:**

> Mass is neither created nor destroyed in a chemical change.

Under this law, the small mass of ash left after the open burning of large piles of wood is explained by the loss of quantities of gases (principally carbon dioxide and water) that mix into the atmosphere. When these gases are captured and weighed, mass balance in the combustion reaction is restored.

Lavoisier was the first to observe that a chemical reaction is analogous to an algebraic equation. An algebraic equation involves equal quantities that are expressed in different mathematical terms; a chemical reaction involves equal masses of matter that are present in different chemical forms. As discussed in Section 2–1, we now represent Lavoisier's two reactions, which were each other's reverses, as

$$2\,\mathrm{Hg} + \mathrm{O_2} \longrightarrow 2\,\mathrm{HgO} \qquad \text{and} \qquad 2\,\mathrm{HgO} \longrightarrow 2\,\mathrm{Hg} + \mathrm{O_2}$$

• The system of naming compounds that leads to the designation mercury(II) oxide is discussed in Section 3–8.

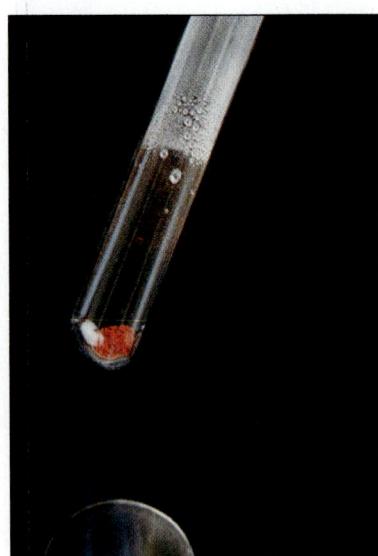

Figure 1–10 • When the red solid mercury(II) oxide is heated, it decomposes to mercury and oxygen. Note the drops of liquid mercury condensing on the side of the test tube. (*Leon Lewandowski*)

---

[1]Chemists sometimes use the term "weight" in place of "mass." Strictly speaking, weight and mass are not the same. The mass of a body is an invariant quantity, but its weight is the force exerted upon it by gravitational attraction (usually by the Earth). Newton's second law relates the two ($w = m \times g$, where g is the acceleration due to gravity). As g varies from place to place on the Earth's surface, so also does the weight of a body. In chemistry, we deal mostly with ratios, which are the same for masses and weights. We shall use the term "mass" exclusively, but "weight" is still in common chemical use.

In Lavoisier's experiment, oxygen was removed from the air by chemical reaction with mercury, leaving nitrogen, which did not react with the mercury, as the residual gas.

Lavoisier's extensive experiments confirmed not only that mass was, in every case, conserved (within the limits of experimental error) but also that combustion was the chemical combination of substances with oxygen from the air. Phlogiston did not exist.

The Law of Definite Proportions

Rapid progress ensued as chemists made increasingly accurate determinations of the masses of reactants and products in chemical reactions (confirming the law of conservation of mass) and also analyzed compounds as to the identity and mass of the elements comprising them. Naturally, experimental results obtained on the same compound by different workers differed somewhat. A controversy arose between two schools of thought, led by the French chemists Claude Berthollet and Joseph Proust. Berthollet believed that the proportions (by mass) of the elements in a particular compound were not fixed but could vary over a certain range; thus water, for example, rather than containing 11.1% by mass of hydrogen, might have somewhat less, or somewhat more, than this mass percentage. Proust disagreed, and he showed that the apparent variation was due to the presence of impurities and experimental errors. He also stressed the difference between homogeneous mixtures and chemical compounds. Through his careful work, Proust was able to demonstrate the important **law of definite proportions:**

> In a given chemical compound, the proportions by mass of the elements that compose it are fixed, independent of the origin of the compound or its mode of preparation.

Pure sodium chloride contains 39.34% sodium and 60.66% chlorine by mass whether we obtain it from salt mines; crystallize it from waters of the oceans; or synthesize it from its elements, sodium and chlorine.[2] Of course, the key word in this sentence is "pure." We must be certain that no elements other than sodium and chlorine are present.

The discovery of the law of definite proportions (also called the **law of constant composition**) was a crucial step in the development of modern chemistry. By 1808, Proust's conclusions had become widely accepted. In fact, we now recognize that this law is not strictly true in all cases. Certain solids exist over a range of compositions and are called **non-stoichiometric** compounds. An example is wüstite, which has the nominal chemical formula FeO (with 77.73% iron by mass) but which in fact ranges continuously in composition from $Fe_{0.95}O$ (with 76.8% iron) down to $Fe_{0.85}O$ (74.8% iron), depending on the method of preparation. This illustrates a common pattern of scientific progress: Observation and experiment lead to the establishment of a law or principle. Later, more accurate studies then reveal exceptions, the explanation of which leads to deeper understanding.

• Such compounds, also called bertholides in honor of Berthollet, are discussed further in Section 20–4.

---

[2]This statement needs some qualification. As we see later in this chapter, many elements have several isotopes, which are species having atoms of essentially identical chemical properties but different masses. Natural variation in the abundances of isotopes leads to small variations in the mass proportions of elements in a compound. Larger variations can be induced by artificial isotopic enrichment.

Dalton's Atomic Theory

John Dalton, an English scientist, proposed the first modern atomic theory in the early years of the 19th century. The idea of atoms was nothing new, but Dalton marshaled the evidence for their existence compellingly. He believed that the mass relationships found by Lavoisier and Proust were best interpreted by postulating the existence of different kinds of atoms for the different elements. Because approximately 50 different chemical elements were known at the time, Dalton was proposing a "particle zoo" of approximately 50 different fundamental particles. This moved sharply away from previous thinking, which had emphasized the fundamental identity of all substances.

In 1808, Dalton published *A New System of Chemical Philosophy,* in which he advanced the following five formal hypotheses about the nature of matter. These are the postulates of **Dalton's atomic theory of matter:**

1. All matter consists of solid and indivisible atoms.

2. All of the atoms of a given chemical element are identical in mass and in all other properties.

3. Different elements have different kinds of atoms; these atoms differ in mass from element to element.

4. Atoms are indestructible and retain their identity in all chemical reactions.

5. The formation of a compound from its elements occurs through the combination of atoms of unlike elements in small whole-number ratios.

The fourth postulate is clearly related to the law of conservation of mass. The fifth, which can be called "Dalton's rule for chemical synthesis," is an attempt to explain the law of definite proportions.

Suppose that one rejects the atomic theory and adopts a view in which compounds can be divided without limit. What then ensures the constancy of composition of a substance such as sodium chloride? It becomes in fact logically unexplainable. On the other hand, if sodium atoms are matched by a fixed number of chlorine atoms in sodium chloride, then constancy of composition is inescapable. Note that it is not necessary to know the sizes or numbers of the atoms of sodium and chlorine to understand why there should be a law of definite proportions. It is important merely that some lower bound to the subdivisibility of matter exist. The moment that such a lower bound arises, arithmetic steps in. Matter then becomes countable, and the units of counting are simply atoms. Believing in the law of definite proportions as an established experimental fact, Dalton postulated the atom.

New facts since 1808 have caused the modification of nearly all of the postulates of Dalton's original theory (see the following sections of this chapter and Problem 87 at the end of this chapter). The core of the theory, however, which is the essential graininess of matter, is stronger than ever. There is no doubt today of the existence of atoms.

Twentieth-century science developed a number of sophisticated techniques for measuring the properties of single atoms. A relatively recent development, the scanning tunneling microscope (STM), has allowed their direct observation. This device uses an incredibly fine-pointed electrically conducting probe that is passed over the surface of the sample being examined (Fig. 1–11). When it comes nearly in contact with the atoms of the sample, a small electric current (called the "tunneling current") can pass from the sample to the probe. The strength of this current is extraordinarily sensitive to the distance of the probe from the surface. It typically falls

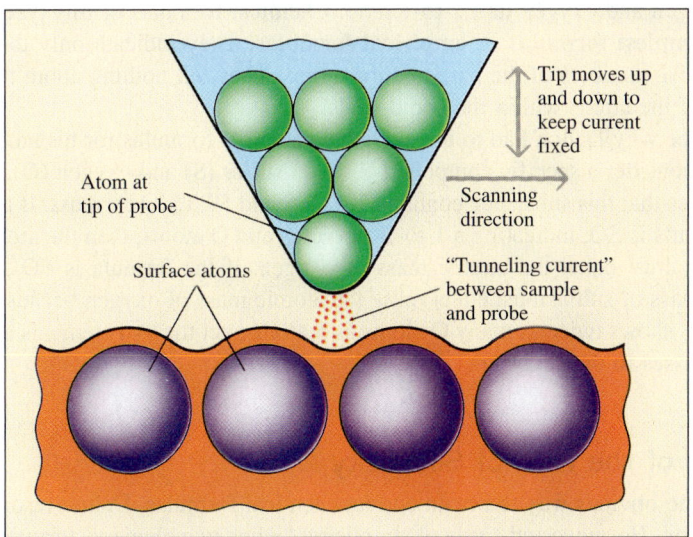

Figure 1–11 • In a scanning tunneling microscope, an electric current passes through a single atom or small group of atoms in the probe tip and then into the surface of the sample being examined. As the probe moves over the surface, its distance is adjusted to keep the current constant, allowing a tracing out of the shapes of the atoms or molecules on the surface.

off by a factor of 1000 as the probe moves out from the surface by a mere 10^{-8} cm. By means of a feedback circuit, the current is held constant while the probe is swept over the surface, moving in and out. The position of the probe is monitored, and the information is stored in a computer. Sweeping the probe tip along each of a series of closely spaced parallel tracks permits the construction of a three-dimensional image of the surface.

The images provided by the STM (Fig. 1–12) confirm visually many features, such as the sizes of atoms and the distances between them, known previously by indirect techniques. Much new information comes out as well. Scanning tunneling microscopy has revealed the positions and shapes of molecules undergoing chemical reactions on surfaces, helping to guide the search for new ways of carrying out such reactions. It has also been used to deduce the shape of the surface of molecules of deoxyribonucleic acid, DNA, which plays a central role in genetics.

1–4   Chemical Formulas and Relative Atomic Masses

John Dalton hypothesized that different elements had different kinds of atoms and that the masses of atoms (atomic masses) differ from one element to the next. Measurements of some atomic masses would strongly support this hypothesis. Although the absolute masses of all known types of stable atoms were measured in the 20th century to high precision, the necessary experiments were theoretically and technically inaccessible in the early 19th century. Dalton sought instead to determine the *ratios* of the masses of the atoms of the different elements. He set out on (in his words) "an inquiry into the relative weights of the ultimate particles of bodies … a subject as far as I know entirely new."

The main theoretical problem was to arrive at correct chemical formulas for at least some of the numerous compounds already known at that time. A **chemical formula** displays the symbols for the elements making up a compound with numerical subscripts that state the relative number of atoms of each kind. These subscripts equal the small whole numbers of Dalton's fifth postulate. The chemical formula P_4S_7 indicates a compound containing four atoms of P (phosphorus) for every seven atoms of S (sulfur). The formula H_2O means that the compound water contains atoms

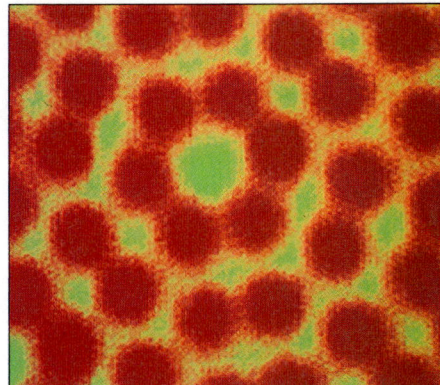

Figure 1–12 • Atoms of silicon imaged with a scanning tunneling microscope. *(Science VU/©IBMRL/ Visuals Unlimited)*

of hydrogen and oxygen in a 2-to-1 ratio. Chemical formulas of this type are now called **simplest formulas** or **empirical formulas.** They indicate only the *identity* and *relative* numbers of the participating atoms. They tell nothing about the organization of the atoms within the compounds.

To see why Dalton had to have reliable chemical formulas for his atomic mass project, consider a specific compound between sulfur (S) and oxygen (O). Analysis establishes that this substance contains 50.0% S and 50.0% O by mass. If the chemical formula is SO, indicating a 1-to-1 ratio of S and O atoms, then the atomic mass of sulfur must equal the atomic mass of oxygen. If the formula is SO_2, then the atomic mass of sulfur must equal *twice* the atomic mass of oxygen because the formula SO_2 shows twice as many O atoms as S atoms, yet the two elements contribute equal masses to the compound.

- Subscripts equal to 1 are customarily omitted from chemical formulas.

Failure of the Rule of Simplicity

Finding no obvious way to obtain correct chemical formulas, Dalton resorted to an assumption. He suggested a rule of simplicity: "when two elements form only a single compound, then their atomic ratio in the compound is 1 to 1." By this rule, Dalton assigned the formula HO to water, which was at the time the only known binary compound between hydrogen and oxygen. Analysis of 100.00 g of water gives 88.81 g of O and 11.19 g of H. The ratio of these masses is

- The gram (g) is a unit of mass in the "Système International d'Unités" (International System of Units, abbreviated SI). For discussion of these units and unit conversions, see Appendix B–1.

$$\frac{88.81 \text{ g O}}{11.19 \text{ g H}}$$

which can be expressed as

$$\frac{7.937 \text{ g O}}{1 \text{ g H}}$$

by dividing both the top and bottom of the fraction by 11.19. This mass ratio is the same for any amount of water, right down to 1 atom of O and 1 of H. Therefore,

$$\frac{\text{mass of 1 atom of oxygen}}{\text{mass of 1 atom of hydrogen}} = \frac{7.937}{1}$$

- Dalton actually obtained a different answer. The best analysis of water available to him was off a bit: 87.4% O and 12.6% H. This gave a ratio of 6.94 to 1, which he rounded off to 7 to 1.

If the formula of water is HO, then the mass of an O atom equals 7.94 times the mass of an H atom. This ratio of atomic masses was incorrect because the rule of simplicity was wrong. The rule led to some correct results but to errors as well (Fig. 1–13). The true solution to the problem of assigning chemical formulas lay elsewhere, with experiments on the combining volumes of gases.

The Law of Combining Volumes and Avogadro's Hypothesis

In 1808, a French chemist, Joseph Gay-Lussac, drew attention to recent experiments on the volumes of gases that react with one another to form new gases. He found that "gases combine amongst themselves in very simple proportions" and announced **the law of combining volumes:**

When two gases react, the volumes that combine (at the same temperature and pressure) stand in a ratio of small whole numbers. The ratio of the volume of each product gas to the volume of either reacting gas is also a ratio of small whole numbers.

We consider three examples. The experiments showed that:

2 volumes of hydrogen + 1 volume of oxygen $\longrightarrow$ 2 volumes of water vapor
1 volume of nitrogen + 1 volume of oxygen $\longrightarrow$ 2 volumes of nitrogen monoxide
1 volume of nitrogen + 3 volumes of hydrogen $\longrightarrow$ 2 volumes of ammonia

These results find ready interpretation in terms of combinations of atoms in ratios of small whole numbers and, in that respect, support Dalton's atomic theory. Gay-Lussac did not theorize on his findings, but in 1811, an Italian chemist, Amedeo Avogadro, used them to formulate an important idea known first as Avogadro's hypothesis and now as **Avogadro's law:**

Equal volumes of different gases (at the same temperature and pressure) contain equal numbers of particles.

The obvious question immediately arose: Are the particles of the gaseous elements the same as Dalton's atoms? Avogadro took the point of view that they were not. He proposed instead that the gaseous elements (such as hydrogen, oxygen, or nitrogen) exist as diatomic molecules.

Molecules are groupings of two or more atoms bound closely together by strong attractive forces that maintain them in a persistent combination.

The additional assumption that the gaseous elements occur in molecules containing two atoms worked neatly with Avogadro's main hypothesis to explain Gay-Lussac's law of combining volumes (Fig. 1–14). Avogadro would have represented the three gas-phase reactions just described as

$$2\,H_2 + O_2 \longrightarrow 2\,H_2O$$
$$N_2 + O_2 \longrightarrow 2\,NO$$
$$N_2 + 3\,H_2 \longrightarrow 2\,NH_3$$

in which the coefficients, which represent both the relative number of molecules and the relative volume, agree with experiment and in which the chemical formulas agree with modern results. Dalton, on the other hand, would have written

$$H + O \longrightarrow HO$$
$$N + O \longrightarrow NO$$
$$N + H \longrightarrow NH$$

in which the coefficients disagree with the observations.

Dalton did not welcome Avogadro's contributions and indeed questioned the validity of Gay-Lussac's experiments. He and others remained firmly convinced that the gaseous elements contained only single, free atoms. One reason for this was the belief that atoms joined together in compounds by the operation of an affinity between opposites, similar to the attraction between positive and negative electric charges. If this were true, why should two identical atoms bond together in a molecule? Moreover, if identical atoms somehow did hold together in pairs,

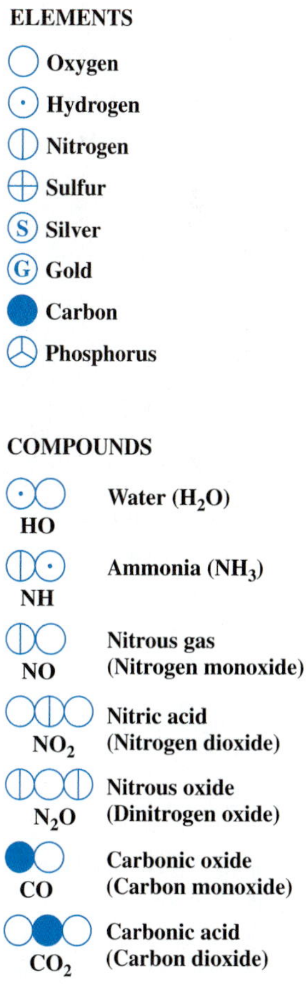

ELEMENTS

Oxygen
Hydrogen
Nitrogen
Sulfur
Silver
Gold
Carbon
Phosphorus

COMPOUNDS

Water (H_2O)
HO

Ammonia (NH_3)
NH

Nitrous gas
NO (Nitrogen monoxide)

Nitric acid
NO_2 (Nitrogen dioxide)

Nitrous oxide
N_2O (Dinitrogen oxide)

Carbonic oxide
CO (Carbon monoxide)

Carbonic acid
CO_2 (Carbon dioxide)

Figure 1–13 • The formulas of some simple compounds as written by Dalton. Dalton used a different system of symbolism for the elements and named many compounds differently as well. Some of the formulas are incorrect, such as HO instead of H_2O. Modern designations for all of the substances are given in parentheses.

Figure 1-14 • Each circle stands for a container of the same volume under the same conditions. The containers hold different gases but contain equal numbers of molecules (as provided by Avogadro's hypothesis). If, as shown, hydrogen, oxygen, and nitrogen exist as diatomic molecules, then the combining volumes observed by Gay-Lussac in the three reactions are fully explained. The number of atoms does not change in a chemical reaction, but the number of molecules does.

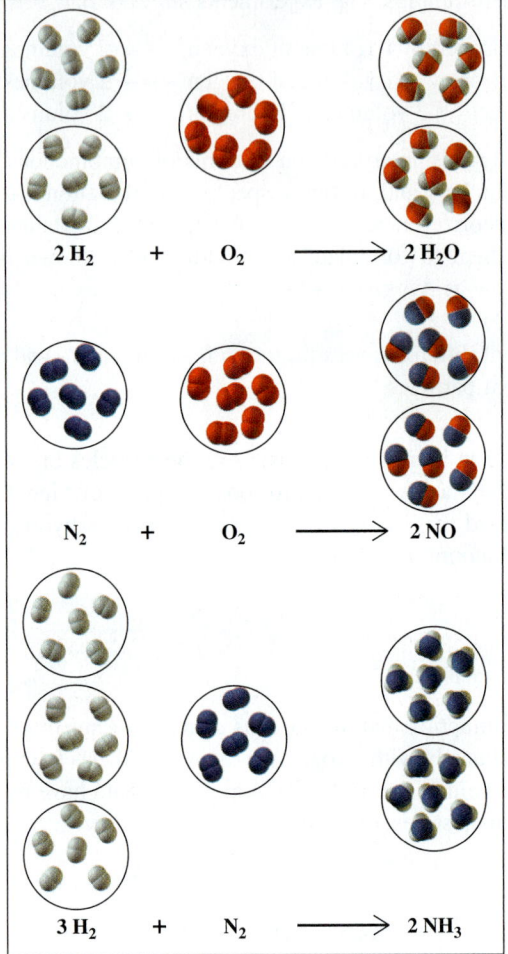

why should they not aggregate further? Hydrogen might then form H_3, H_4, and H_5 molecules, and so forth. As a result of Dalton's opposition, Avogadro's reasoning did not attract the attention it deserved. Rival tables of relative atomic masses based on conflicting assumptions about the chemical formulas of key substances proliferated, creating considerable confusion. It was not until some 50 years had passed that Avogadro's thinking was vindicated through the efforts of Stanislao Cannizzaro. Cannizzaro studied the elemental analyses of a large number of gaseous compounds and showed that the whole set of their chemical formulas could be established in a consistent manner that used Avogadro's hypothesis but avoided any extra assumptions about the formulas of the gaseous elements. Gaseous hydrogen, oxygen, and nitrogen (and fluorine, chlorine, bromine, and iodine as well) indeed consist of diatomic molecules under ordinary conditions. Identical atoms do form molecules. Widespread acceptance of Avogadro's hypothesis came in the 1860s, after Cannizzaro published his treatise. Since then, several other elements have been shown to exist as polyatomic (not just diatomic) molecules when prepared as gases: phosphorus (P_4), arsenic (As_4), carbon (C_{60}, C_{70}, and others), and sulfur (S_8 and others).

• Polyatomic molecules contain three or more atoms per molecule.

Changing the formula for water from HO to H_2O changed Dalton's computation of the relative atomic masses of H and O by the insertion of a factor of 2, as follows:

$$\frac{\text{mass of 1 atom of oxygen}}{2 \times \text{mass of 1 atom of hydrogen}} = \frac{7.937}{1}$$

$$\frac{\text{mass of 1 atom of oxygen}}{\text{mass of 1 atom of hydrogen}} = 2 \times \frac{7.937}{1} = \frac{15.87}{1} \approx \frac{16}{1}$$

The use of other formulas recommended by Cannizarro on the basis of Avogadro's hypothesis (such as NH_3 for ammonia instead of NH) finally allowed the construction of a consistent table of relative atomic masses for the elements.

EXAMPLE 1-1

A careful chemical analysis establishes that every 100.00 g of ammonia contains 82.24 g of nitrogen (N) and 17.76 g of hydrogen (H). Compute the ratio of the atomic mass of nitrogen to the atomic mass of hydrogen if the chemical formula of ammonia is NH_3.

Strategy

Taking a ratio requires dividing one number by another. Divide the mass of N in the 100.00 g of ammonia by the mass of H. Account for the $1:3$ ratio of N to H in the chemical formula by setting the ratio of masses equal to the mass of 1 atom of N divided by the mass of 3 atoms of H.

Solution

$$\frac{82.24 \text{ g}}{17.76 \text{ g}} = \frac{\text{mass of 1 atom of N}}{3 \times \text{mass of 1 atom of H}}$$

Multiplying both sides of the equation by 3 gives

$$3 \times \left(\frac{82.24 \text{ g}}{17.76 \text{ g}}\right) = \frac{\text{mass of 1 atom of N}}{\text{mass of 1 atom of H}}$$

$$13.89 = \frac{\text{mass of 1 atom of N}}{\text{mass of 1 atom of H}}$$

Check

A current table of relative atomic masses gives 14.007 and 1.00794 for the relative masses of N and H. Dividing 14.007 by 1.0079 gives 13.90.

EXERCISE

Every 100.00 g of the compound SiH_4 contains 87.45 g of Si and 12.55 g of H. Find the ratio of the atomic mass of S to the atomic mass of H.

Answer: 27.87

Shortcomings of Early Atomic Theory

Dalton's contribution was enormous, but questions remained. Most significantly for chemists, early atomic theory was silent as to *why* atoms combine in the ratios that they do. Why does oxygen combine with carbon to give compounds with formulas

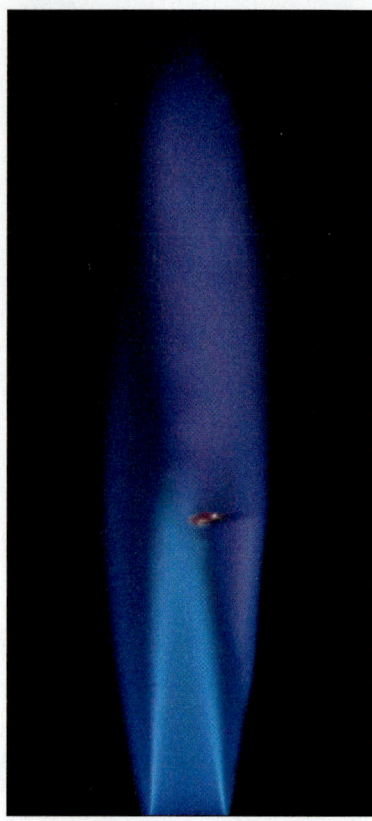

Figure 1–15 • Cesium gives a purplish blue color to a flame. *(Leon Lewandowski)*

CO and CO_2 but with sulfur to give SO_2 and SO_3? Why do some pairs of elements form numerous compounds, while other pairs form just one or fail entirely to combine? Why do some elements exist in molecules containing two (or more) identical atoms? Also, early atomic theory had no role for electricity. A flow of electricity from the newly discovered voltaic pile (the first battery) had been shown to decompose water into its constituent hydrogen and oxygen (Fig. 1–9). How and why?

The answers to these questions required insights about the *interior* structure of the atom. Dalton had not sought such insights because he had postulated that atoms were the smallest particles of matter. He could not have obtained them in any case from the scant experimental data that were available in his time. This situation was to change.

1–5 The Building Blocks of the Atom

Persuasive evidence accumulated during the 19th century that atoms are not the indivisible, tiny balls envisioned by Dalton. This evidence eventually led to the modern theory of chemical bonding discussed in Chapters 3 and 17. Here, we discuss the observations and experiments that revealed the interior structure of atoms.

In 1860, Robert Bunsen and Gustav Kirchhoff discovered the element cesium in certain mineral waters, and they named it for the sky-blue color of the light that it emitted in a flame (Fig. 1–15). Other elements also were found to emit light of different characteristic color when heated to high temperatures. These observations strongly implied, although they did not prove, that atoms possess an internal structure of some kind.

In other laboratories, scientists found that gases confined in a glass tube at low pressure were able to conduct electricity when a high voltage was applied across electrodes that had been sealed into the tube. It was evident that the electrical discharge was transported by charged particles of some kind. G. B. Stoney believed that the charge carriers in the gas-discharge tubes were derived from the molecules of the gas, and in 1874 he gave the name **electron** to the species that carried a negative charge. Because the originating atoms or molecules were electrically neutral, the very existence of negatively charged species, wrenched from atoms or molecules by the action of an electric field, meant that positively charged species also must exist in atoms.

Electrons

The passage of an electric current through a gas-discharge tube was observed to induce luminous regions in a confined gas, alternating with dark bands that lengthened toward the anode (positive electrode) as the pressure of the gas was reduced. At a sufficiently low pressure, the luminous discharge disappeared altogether, but an electric current continued to flow (Fig. 1–16)!

The vehicle that carried the electric current through a vacuum was something of a mystery. Gas molecules could hardly provide the charge carriers when essentially no gas was present. The charge carriers appeared to come directly from the negative electrode and came to be known as **cathode rays.** Cathode rays could transfer sufficient energy to raise a piece of metal foil in the tube to white heat. They also transferred a negative charge and could be deflected from their straight-line trajectories by both electric and magnetic fields. Some scientists believed that cathode rays were some kind of electromagnetic wave, a form of invisible light. Others believed that they were streams of charged particles—electrons, in fact. In 1897,

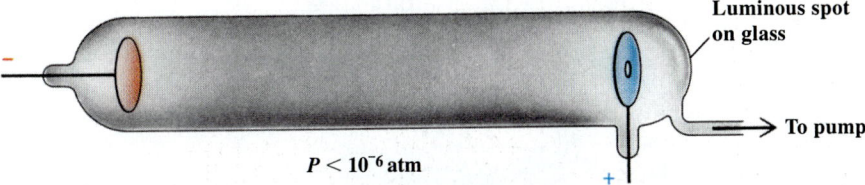

Figure 1–16 • An electric current passes through this high-voltage gas-discharge tube despite a nearly total absence of matter between the two electrodes. Charge carriers apparently emanate from the negative electrode (the cathode). The hole in the positive electrode (the anode) allows some of them to pass through to excite a visible glow where they strike the wall of the tube.

J. J. Thomson, a British physicist at the Cavendish Laboratory of Cambridge University, settled the issue by proving that cathode rays are negatively charged particles. Thomson built an apparatus (Fig. 1–17) in which the deflection of a beam of cathode rays caused by a magnetic field was countered by an opposing deflection caused by an electric field. In careful experiments, he varied the fields so as to balance exactly the opposing deflections. From the conditions of balance, he demonstrated that cathode "rays" have mass and are truly particles. Indeed, he went further and measured the ratio of the charge of these particles (electrons) to their mass. The currently accepted value of the ratio, e/m_e, is 1.7588202×10^{11} C kg^{-1}.

• The coulomb (C) is a unit for the measurement of quantity of electric charge. See Appendix B.

Thomson's experiments were a milestone in the search to understand the composition of the atom, but they raised two obvious questions: How much electric charge is carried by the electron, and what is the electron mass? The answers were not long in coming. In 1909, the American physicist Robert Millikan and his student H. A. Fletcher developed an elegant experiment that yielded e, the charge of the electron. Using a microscope, they observed the fall of tiny oil droplets that had

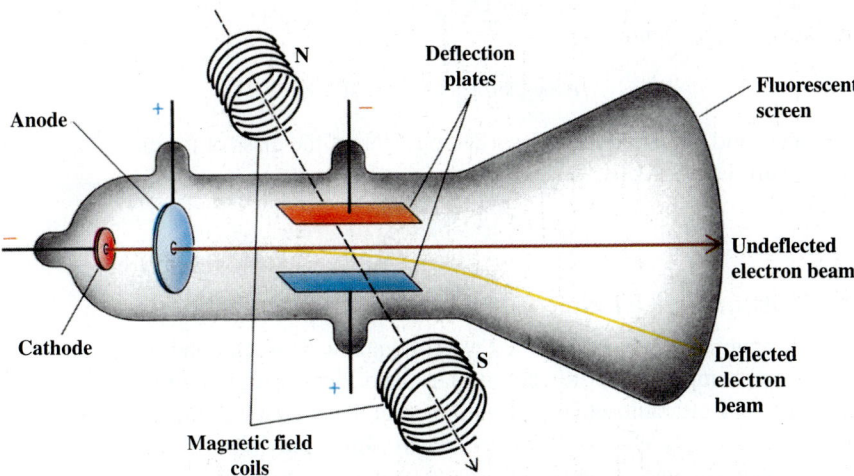

Figure 1–17 • Thomson's apparatus to measure the electron charge-to-mass ratio, e/m_e. Electrons (cathode rays) leave the cathode and accelerate toward the anode. Many pass through the hole in the anode and stream across the tube from left to right. The electric field alone deflects this beam down, and the magnetic field alone deflects it up. By adjusting the two field strengths, Thomson achieved zero net deflection. From the strengths of the two fields at this point, he could calculate the velocity of the beam. From that velocity and the deflection caused by the electric field alone, he arrived at the charge-to-mass ratio.

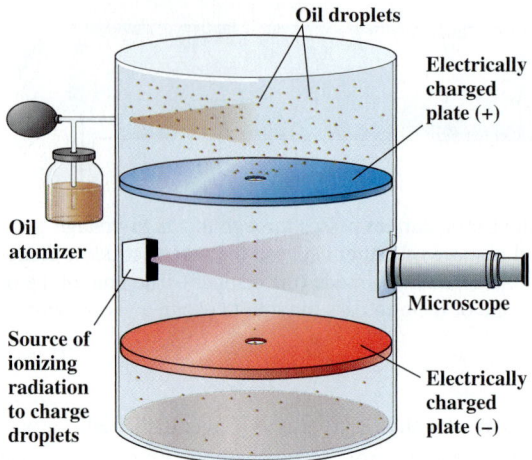

Figure 1–18 • Millikan's apparatus to measure the charge (e) on an electron. By adjusting the strength of the electric field between the charged plates, Millikan could slow down, halt, or even reverse the fall of negatively charged oil drops. From the way the rate of fall depended on the strength of the field, he determined the net charges on the drops, which were always whole-number multiples of the charge on the electron.

Oil droplets

Electrically charged plate (+)

Oil atomizer

Microscope

Source of ionizing radiation to charge droplets

Electrically charged plate (–)

acquired an electric charge (Fig. 1–18). If left to themselves, the droplets fell at a uniform rate, pulled down by gravity but slowed by the viscous drag of the air. When Millikan applied a vertical electric field in the region in which the oil droplets were observed, the field acted on their charge. Droplets could be slowed, held suspended, or even accelerated upward, just as a lock of dry hair can be suspended or moved upward by the "static electricity" (the electric field) surrounding a comb. Millikan could single out specific oil droplets and record their rates of descent in both the absence and the presence of the electric field. From these two rates and other pertinent facts (namely, the acceleration of gravity, the viscosity of the air, and the density of the oil), Millikan calculated the amount of charge that a single droplet carried. Repeated observations revealed that different oil drops bore different charges but that these charges were always integral multiples of some basic charge, which Millikan and Fletcher took to be the charge of a single electron. Their experimental value for the magnitude of this charge was 1.59×10^{-19} C, quite close to the currently accepted value of

$$e = 1.60217646 \times 10^{-19} \text{ C}$$

Knowing e and the ratio e/m_e allows computation of the mass of the electron, which has the currently accepted value

$$m_e = 9.1093819 \times 10^{-31} \text{ kg}$$

• These experimental values are subject to (slight) revision from time to time. The U.S. National Institute of Standards and Technology maintains an up-to-date listing of their current accepted value and the values of numerous other physical constants at http://physics.nist.gov/cuu/Constants/index.html on the World Wide Web.

The Rutherford Model of the Atom

In 1911, Ernest Rutherford reported experiments that revealed the arrangement of the charged components of the atom. Marie and Pierre Curie had recently isolated compounds of the radioactive elements radium and polonium and discovered that they emit positively charged particles, called **alpha particles.** Rutherford and his students characterized alpha particles as to charge and mass and then used them to probe atomic structure. They caused a beam of alpha particles from a radium source to bombard a thin gold foil and observed the deflections of the alpha particles. They detected the alpha particles by the scintillations that they caused as they struck a fluorescent screen (Fig. 1–19a). Rutherford expected the alpha particles to experience only small deflections, and small deflections were in fact observed most of the time. Occasionally, however, a particle was deflected through a large angle. Even

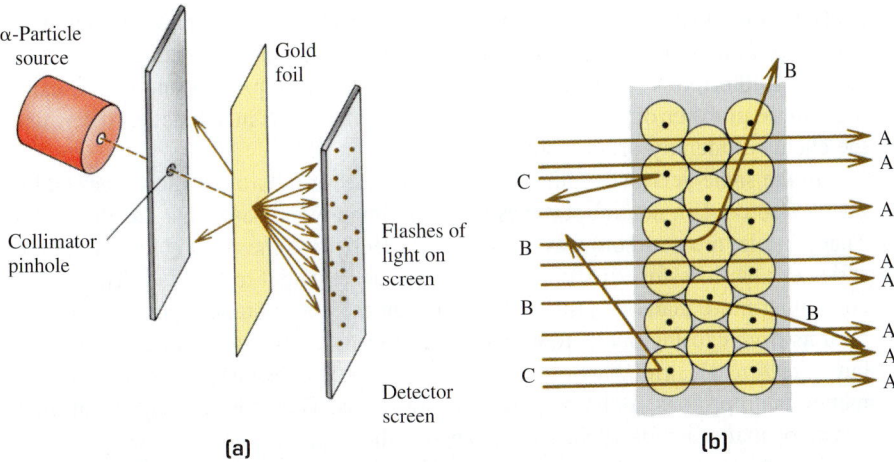

(a)   **(b)**

Figure 1–19 • (a) Flashes of light mark the impact of alpha particles on a fluorescent screen after their encounters with the target foil. In the Rutherford experiment, the rate of hits on the screen varied from approximately 20 per minute at high angles to nearly 132,000 per minute at low angles. (b) The interpretation of Rutherford's experiment. Most of the alpha particles pass through the space between nuclei and undergo only small deflections (A). A few pass close to a nucleus and are more strongly deflected (B). Some are even scattered backward (C).

less frequently, an alpha particle was observed to rebound nearly directly backward! Rutherford was astounded by these rare events because the alpha particles were both relatively massive and fast moving. He remarked, "It was almost as incredible as if you fired a 15-inch shell [heavy artillery] at a piece of tissue paper and it came back and hit you." Some compact, heavy kernel must lie at the heart of the atom. A mathematical analysis of the frequency and angular distribution of the deflections confirmed that most of the mass of the gold atoms in the foil was concentrated in very dense, extremely small, positively charged particles, which were called **nuclei.** Rutherford's group estimated the radius of the nucleus in gold to be less than 10^{-14} m and the positive charge on the gold nucleus to be 100 times larger than the negative charge possessed by the electron (actual value, $+79e$). Bombardment studies using other elements as targets gave similar results but with different values for the positive charge on the nucleus.

Rutherford proposed a **nuclear model** of the atom to explain the results. He theorized that the massive nucleus of an atom possesses a net charge of $+Ze$, with Z negatively charged electrons (held by electrical attractions) surrounding the nucleus out to a distance of approximately 10^{-10} m. Thus, in the nuclear model, a gold atom has 79 extranuclear electrons (each with a charge of $-1e$) arranged around a nucleus of charge $+79e$. Nearly all of the volume of an atom is occupied by its electrons; nearly all of its mass is concentrated in the nucleus.

Protons and Neutrons

The Rutherford nuclear model quickly gained acceptance as a valid picture of atomic structure, and much subsequent research has focused on gaining an understanding of the structure of the central nucleus itself. Hydrogen has the smallest and simplest nucleus. Rutherford himself showed that the hydrogen nucleus is the fundamental unit of positive charge within the nuclei of heavier atoms. He proposed the name

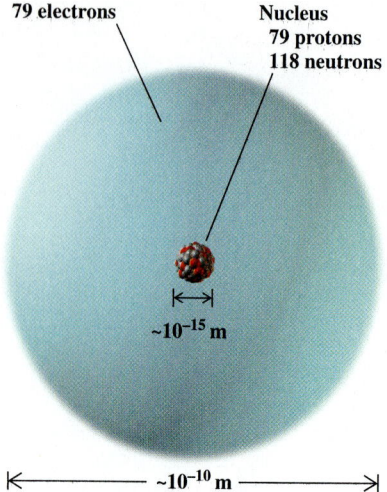

79 electrons   Nucleus
                      79 protons
                      118 neutrons

$\sim 10^{-15}$ m

$\longleftarrow \sim 10^{-10}$ m $\longrightarrow$

This atom has $Z = 79$ and is therefore an atom of gold. Its mass number A equals 197, the sum of 79 and 118. In the real atom, the relative size of the nucleus is far smaller than shown.

proton for the hydrogen nucleus in 1920. Different elements contain different numbers of protons in the nuclei of their atoms. The value of Z associated with a nucleus in Rutherford's original work equals this number of protons and is now called the **atomic number** of the element. The number of protons in its nucleus determines the chemical identity of an atom.

In the same year that Rutherford proposed the term "proton," he suggested that nuclei also contain neutral particles of approximately the same mass as the proton. That particle, the **neutron,** was discovered in 1932 by James Chadwick after a long series of unsuccessful experiments. Neutrons are essential to the stability of nuclei containing more than one proton. They prevent the electrostatic repulsions between protons from causing nuclei to dissociate. Neutrons contribute to the mass of atoms but have zero electric charge and, for that reason, do not interact as strongly with matter as do protons, alpha particles, and electrons. They pass through great thicknesses of matter unabsorbed, which explains the delay in their discovery.

The number of neutrons in a nucleus is called the **neutron number** (N) of an atom, and the combined number of neutrons and protons is called the **mass number** (A) because the protons and neutrons packed into the tiny nucleus account for essentially all of the mass of atoms. Clearly, $A = Z + N$.

1-6   Finding Atomic Masses the Modern Way

For most chemical purposes, it is sufficient to know the relative masses of atoms on a scale in which the mass of an atom of a particular element is assigned some arbitrary value. Throughout the 19th century, chemists established increasingly precise tables of relative atomic masses based strictly on chemical synthesis and analysis. Behind every value lay a calculation just like the one shown in Example 1–1. In the 20th century, knowledge of atomic structure enabled a new approach to the problem of relative atomic masses. In this section, we take up this physical method, which is capable of exceedingly high precision.

Mass Spectrometry and Isotopes

The precise determination of relative atomic (and molecular) masses by physical methods began with the development of **mass spectrometry** by J. J. Thomson, F. W. Aston, and others in the early 20th century. Mass spectrometry relies on the manipulation of beams of ions in a vacuum chamber. An **ion** results when an atom either loses or gains one or more electrons. Ionization occurs readily when suitable electrical arrangements are provided. The loss of electrons, which carry a negative charge, converts atoms into positively charged ions; the gain of electrons converts atoms into negatively charged ions. Ions are represented by the chemical symbol for the atom together with the net charge written as a right superscript. If chlorine (Cl) loses one electron, Cl^+ results; if Cl gains one electron, Cl^- results, as shown by the equations

• The symbol for the electron, a particle, is e^-. The electrical charge carried by the electron is symbolized e, and the mass carried by the particle is symbolized m_e.

$$Cl \longrightarrow Cl^+ + e^- \text{ (loss of an electron)}   Cl + e^- \longrightarrow Cl^- \text{ (gain of an electron)}$$

Whole molecules can, without breaking apart, also lose or gain electrons to create **molecular ions,** the charges of which are indicated similarly.

Moving ions interact strongly with electric and magnetic fields. In a mass spectrometer (Fig. 1–20), the same number of electrons is removed from each atom or

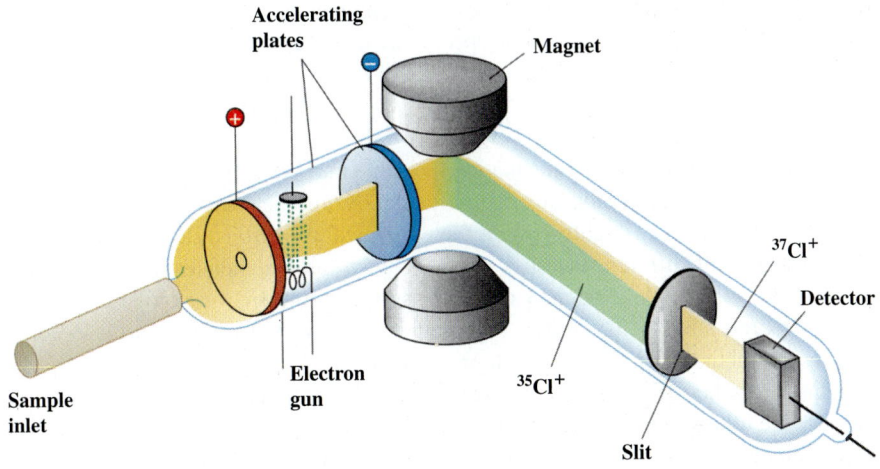

Accelerating plates

Magnet

$^{37}Cl^+$

Detector

Electron gun

$^{35}Cl^+$

Sample inlet

Slit

Figure 1-20 • A simp sentation of a modern m ter. Atoms or molecules are ionized and then accelerated by the electric field between the plates. The ion beam passes into a magnetic field, where it is separated into components, each containing particles having a characteristic charge-to-mass ratio. Here, the element chlorine, which has two naturally occurring isotopes, is under study. The spectrometer is adjusted to detect the less strongly deflected $^{37}Cl^+$ ions. By changing the magnitude of the electric or magnetic field, one can move the beam of $^{35}Cl^+$ ions to the exit slit so that it can be detected.

molecule in a sample. The resulting ions are accelerated by an electric field and then shot into a magnetic field, which causes their paths to curve. The curvature of the trajectories of the ions depends only on the ratios of their charges to their masses (once the strengths of the magnetic and electric fields are set). When the ions all have the same charge, the curvature depends only on the mass of the ions. The paths of more massive ions curve less than do the paths of less massive ions. This difference allows species of different mass to be separated and detected. Early experiments in mass spectrometry demonstrated, for example, a mass ratio of just under $16:1$ for oxygen relative to hydrogen, confirming by physical techniques a relationship deduced originally on chemical grounds.

Further investigations began to produce some real surprises. Because the relative atomic mass of chlorine (atomic number $Z = 17$) determined by chemical means equaled 35.45, atoms of chlorine were expected to give a signal at the relative mass of 35.45. Instead, pure Cl gave *two* signals (peaks): the first at a relative mass near 35, and the second, about one third as intense, near 37 (Fig. 1–21). This meant that approximately three quarters of all chlorine atoms had a relative atomic mass of 35, and one quarter of all chlorine atoms a relative atomic mass of 37. Ordinary chlorine

What does this mean?

^{35}Cl is 3× abundant as ^{37}Cl

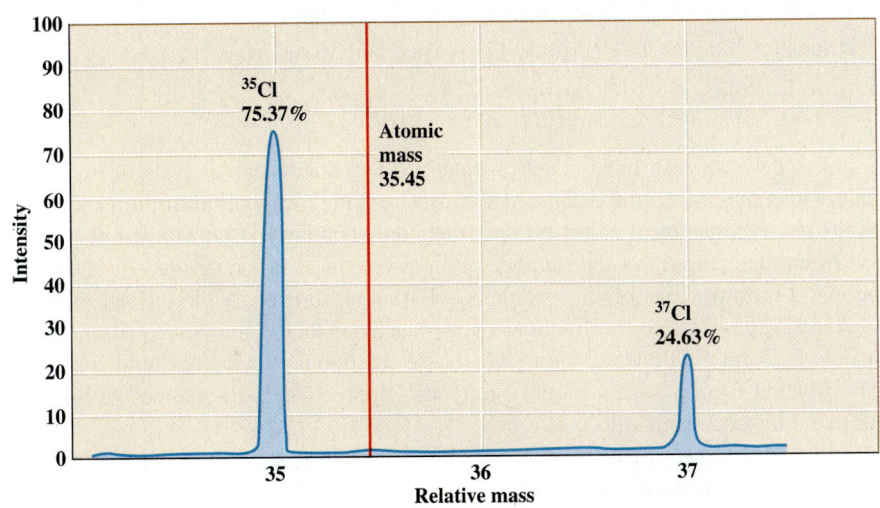

Atomic mass 35.45

^{35}Cl 75.37%

^{37}Cl 24.63%

Intensity

100
90
80
70
60
50
40
30
20
10
0

35   36   37
Relative mass

Figure 1-21 • Naturally occurring chlorine has a relative atomic mass of 35.45. If all chlorine atoms had the same mass, then a single peak would appear in the mass spectrum of chlorine. Instead, two peaks appear, one at a relative mass near 35 and the other near 37, showing that there are two naturally occurring isotopes of chlorine. The peak near 35 is approximately three times higher than the one near 37, meaning that the number of atoms of the lighter isotope is approximately three times greater than the number of atoms of the heavier isotope in chlorine.

- Twenty elements have only one naturally occurring isotope. Some examples are fluorine (F), sodium (Na), aluminum (Al), phosphorus (P), and iodine (I). Tin (Sn) has ten naturally occurring isotopes, the most of any element.

was revealed as a mixture of two types of atoms having nearly identical chemical properties but different masses! Such species are called **isotopes.** Further mass-spectrometric experiments showed that a majority of the elements occur in nature in two or more isotopes. Dalton's postulate that all atoms of a given element are identical in mass was wrong.

Isotopes exist because atoms having the same number (Z) of protons may have different numbers (N) of neutrons in their nuclei. Neutrons add to the mass of atoms but do not (except very slightly) affect their chemical properties. Because the chemical properties of different isotopes are extremely similar, their existence was not discovered until the development of the mass spectrometer. An isotope of an element is represented by its chemical symbol prefixed with a superscript to indicate its mass number. Thus, the two isotopes of chlorine are designated ^{35}Cl and ^{37}Cl. The mass number of an isotope is always close to its actual relative mass simply because the relative masses of both the proton and neutron on our current scale equal approximately 1. For example, the relative atomic mass of ^{35}Cl equals 34.96885272 on a scale in which the relative mass of ^{12}C is defined as exactly 12 (see subsequent sections in this chapter). The ten significant figures in this result illustrate the extremely high precision attainable in mass spectrometry, which is the most accurate method yet developed for determining relative atomic masses.

- In spoken references to isotopes, the mass number is given after the name of the element: "chlorine-35" and "chlorine-37." Isotopes may also be designated this way in writing.

EXAMPLE 1–2

An atom detected in a mass spectrometer has atomic number 47 and a relative mass of 106.9051. Write the symbol for this atom, and list the subatomic particles composing it.

Solution

The table on the inside back cover of this book lists silver (Ag) as element 47. The integer closest to 106.9051 is 107, so the mass number of this atom is 107. This isotope of silver therefore has the symbol ^{107}Ag. The atom contains 47 electrons, 47 protons, and $107 - 47 = 60$ neutrons.

EXERCISE

Write the symbol for the isotope of element 42 that has a relative mass of 97.9055. List the subatomic particles that compose this isotope.

Answer: ^{98}Mo has 42 electrons, 42 protons, and 56 neutrons.

The detector in a mass spectrometer in effect counts the arrivals of the individual ions making up the beam and so furnishes the **fractional abundance** as well as the relative masses of isotopes. The fractional abundance of an isotope in a sample of an element equals the number of atoms of that isotope divided by the total number of atoms. Ions of ^{35}Cl are detected at approximately triple the rate of ions of ^{37}Cl (Fig. 1–21), so the fractional abundances of these isotopes equal approximately 0.75 for ^{35}Cl and 0.25 for ^{37}Cl. These fractional abundances hold for chlorine isolated from a wide variety of terrestrial sources and are assumed to hold in all naturally occurring chlorine.

For some time, chemists used a scale of atomic masses in which the relative atomic mass of naturally occurring oxygen (a mixture of ^{16}O, ^{17}O, and ^{18}O) was set at exactly 16. In 1961, by international agreement, the atomic mass scale was revised, with the adoption of exactly 12 as the relative atomic mass of ^{12}C. This scale continues to be used today. There are two stable isotopes of carbon: ^{12}C and ^{13}C (^{14}C and other isotopes of carbon are radioactive and have very low terrestrial abundance). In natural carbon, 98.892% of the atoms are ^{12}C and 1.108% are ^{13}C; this corresponds to fractional abundances of $98.892/100 = 0.98892$ for ^{12}C and $1.108/100 = 0.01108$ for ^{13}C.

The relative atomic masses of the elements as found in nature can be obtained as averages over the relative masses of the isotopes of each element, weighted by their observed fractional abundances. If an element consists of n isotopes, of relative masses $A_1, A_2, \ldots, A_n$ and fractional abundances $p_1, p_2, \ldots, p_n$, then the average relative atomic mass of the element is

$$\text{average relative atomic mass} = A_1 p_1 + A_2 p_2 + \cdots + A_n p_n$$

Applying this formula to carbon gives

Isotope	Isotope Mass × Fractional Abundance
^{12}C	$12.000000 \times 0.98892 = 11.867$
^{13}C	$13.003354 \times 0.01108 = \underline{00.144}$
average relative atomic mass	$= 12.011$

Chemists nearly always use elements with the isotopic compositions that occur naturally. For this reason, relative atomic masses computed by this formula are called "chemical relative atomic masses" and are tabulated as essential data.

EXAMPLE 1-3

During 1982, the United States mint produced two kinds of one-cent coin, one with a mass of 2.52 g (light pennies) and the other with a mass of 3.09 g (heavy pennies). These coins, which are made of different metals, have the same appearance and value. They are monetary isotopes. In a group of one thousand 1982 pennies, 220 pennies were light. Compute the average mass of a penny in this group.

Strategy

We need a kind of average that accounts both for the different masses of the two types of pennies and their relative abundances. If we were interested in the average mass of a 2.52 g coin and a 3.09 g coin, then adding up the masses and dividing by the number of coins (2) would give the answer

$$\text{average mass} = \frac{2.52 \text{ g} + 3.09 \text{ g}}{2} = \tfrac{1}{2}(2.52 \text{ g}) + \tfrac{1}{2}(3.09 \text{ g}) = 2.80 \text{ g}$$

The problem, however, concerns the average mass of *unequal* numbers of 2.52 g coins and 3.09 g coins. There are 220 light coins and $1000 - 220 = 780$ heavy coins. We therefore weight (tilt) the average to give extra emphasis to the more numerous coin.

Solution

$$\text{weighted average mass} = \tfrac{220}{1000}(2.52 \text{ g}) + \tfrac{780}{1000}(3.09 \text{ g}) = 2.96 \text{ g}$$

Check

Compare the forms of the two computations. In the first, the equal fractions (1/2 and 1/2) give equal importance to the two masses. In the second, a larger fraction (780/1000) gives more emphasis to the more numerous heavy pennies, and a smaller fraction (220/1000) gives less emphasis to the less numerous light pennies.

EXERCISE

The element europium has two naturally occurring isotopes. The first has a relative mass of 150.9196 and a fractional abundance of 0.47820, and the second has a relative mass of 152.9209 and a fractional abundance of 0.52180. Compute the chemical relative atomic mass of europium.

Answer: 151.96.

The number of significant figures in a chemist's table of relative atomic masses (see the inside back cover of this book) is limited not only by the accuracy of the mass spectrometric data but also by the variability in the natural abundances of the isotopes. The isotopic composition of lead (Pb), for example, varies somewhat from location to location in the Earth's crust. If lead from one mine has a relative atomic mass of 207.18 and lead from another has a relative atomic mass of 207.23, then 207.2 must be tabulated as the overall relative atomic mass of lead. In fact, geochemists now use small variations in the $^{16}O : ^{18}O$ isotopic abundance ratio as a "thermometer" to deduce the temperatures at which different rocks were formed in the Earth's crust over geological time scales. They also find anomalies in the oxygen isotopic composition of certain meteorites, implying that their origin lies outside of our solar system.

• For a discussion of significant figures, accuracy, and precision in chemistry, see Appendices A–2 and A–3.

1–7 The Mole Concept: Counting and Weighing Atoms and Molecules

The very precise atomic masses discussed in Section 1–6 are relative atomic masses reported on a scale on which the atomic mass of ^{12}C is set at exactly 12. Relative atomic masses have no units because they are *ratios* of two masses measured in whatever units we choose (grams, kilograms, pounds, and so forth). When we say that the relative atomic mass of ^{19}F is 18.9984, we mean that the ratio of the mass of one atom of ^{19}F to that of one atom of ^{12}C is $18.9984 : 12.0000 = 1.58320$. This is the same as

$$\frac{m_{1 \text{ atom } ^{19}F}}{m_{1 \text{ atom } ^{12}C}} = \frac{18.9884}{12.0000} = 1.58320$$

What are the *actual* masses of individual atoms? In other words, what is the connection between the macroscopic scale of masses used in the laboratory and the microscopic scale of individual atoms? The link between the two is provided by **Avogadro's number:**

By definition, **Avogadro's number (N_0)** equals the number of atoms in exactly 0.012 kg (12 g) of ^{12}C.

Increasingly accurate techniques for determining Avogadro's number have been developed, and the currently accepted value is

$$N_0 = 6.0221420 \times 10^{23}$$

The mass of a single ^{12}C atom is then found by dividing exactly 12 g by N_0:

$$m_{\text{atom} \, ^{12}C} = 1 \text{ atom } ^{12}C \times \frac{12 \text{ g}}{6.0221420 \times 10^{23} \text{ atom } ^{12}C} = 1.9926465 \times 10^{-23} \text{ g}$$

This is truly a very small mass, reflecting the very large number of atoms in a 12 g sample of carbon.

Avogadro's number is defined using ^{12}C because that atom is the reference for the modern scale of relative atomic masses, but it can be used with any element. For example, sodium has a relative atomic mass of 22.98977, so a sodium atom is 22.98977/12 times as heavy as a ^{12}C atom. If N_0 atoms of ^{12}C have a mass of 12 g, then the mass of N_0 atoms of sodium must be

$$(12 \text{ g}) \times \frac{22.98977}{12} = 22.98977 \text{ g}$$

More generally:

> The mass, in grams, of Avogadro's number of atoms of an element is numerically equal to the relative atomic mass of that element.

EXAMPLE 1-4

One of the heaviest atoms found in nature is ^{238}U. It has a relative atomic mass of 238.0508 on a scale on which 12 is the atomic mass of ^{12}C. Calculate the mass (in grams) of one atom of ^{238}U.

Solution

Avogadro's number of ^{238}U atoms have a mass of 238.0508 g. Avogadro's number is 6.0221420×10^{23}; one atom of ^{238}U has a mass

$$m_{\text{atom} \, ^{238}U} = 1 \text{ atom } ^{238}U \times \left(\frac{238.0508 \text{ g}}{6.0221420 \times 10^{23} \text{ atom } ^{238}U} \right)$$
$$= 3.952926 \times 10^{-22} \text{ g}$$

Check

Divide the answer by $1.9926465 \times 10^{-23}$ g, which is the mass of one atom of C computed in the text. The result, 19.83756, should equal the ratio of 238.0508, the relative mass of ^{238}U, to 12.0000, the relative mass of ^{12}C. It does. In more general terms, the answer is a very small mass (much less than a billionth of a billionth of a gram), which is consistent with the mass of a single atom.

EXERCISE

A single atom of a certain element has a mass of 2.10730×10^{-22} g. Compute the relative atomic mass of this atom on a scale on which 12 is the relative atomic mass of ^{12}C, and identify the element, assuming that it has only one isotope.

Answer: 126.904, iodine.

An analogous development applies to molecules (groups of atoms bound together in persistent combinations). The **relative molecular mass** of a molecule equals the sum of the relative atomic masses of all of the atoms making up the molecule. The relative molecular mass of H_2O is

$$\text{relative molecular mass of water} = 2(\text{relative atomic mass of H}) +$$
$$\text{relative atomic mass of O}$$
$$= 2(1.00794) + 15.9994 = 18.0153$$

Thus, Avogadro's number of molecules of water has a mass of 18.0153 g, and the mass of a single molecule of water is

$$m_{\text{molecule } H_2O} = 1 \text{ molecule } H_2O \times \frac{18.0153 \text{ g } H_2O}{6.0221420 \times 10^{23} \text{ molecules } H_2O}$$
$$= 2.99151 \times 10^{-23} \text{ g}$$

More Complex Molecular Formulas

Chemists often use molecular formulas to show how the atoms are organized within the molecule. Instead of writing $C_4H_8O_2$ as the formula for ethyl acetate (an ingredient in perfumes), a chemist might write $CH_3COOC_2H_5$ or $CH_3CO_2CH_2CH_3$ to allow for quick recognition of common groupings of atoms (CH_3 is one) and their linkages. This practice requires the reader to scan alertly for recurrences of symbols when computing a molecular mass. Also, formulas may include parentheses around groupings of atoms and even nests of parentheses. For example, the formula of the explosive TNT very frequently is written $(NO_2)_3C_6H_2CH_3$, which is equivalent to $C_7H_5N_3O_6$. The subscript just to the right of the closing parenthesis operates on everything within. Working out the nested parentheses in the formula $((CH_3)_3CO)_3Al$ gives $C_{12}H_{27}O_3Al$.

EXAMPLE 1–5

Calculate the mass of one molecule of dimethyl sebacate $((CH_2)_4COOCH_3)_2$, a compound used in making plastics.

Strategy

Rewrite the molecular formula with all parentheses and repetitions removed. Add up the contributions of the different atoms. Then proceed as in Example 1–4.

Solution

The contents of the outermost set of parentheses equal $C_6H_{11}O_2$. The last subscript doubles this to $C_{12}H_{22}O_4$. The sum of the relative atomic masses of 12 C's, 22 H's, and 4 O's equals the relative molecular mass

$$\text{relative molecular mass} = 12 \times (12.011) + 22 \times (1.0079) + 4 \times (15.999)$$
$$= 230.302$$

N_0 molecules of $((CH_2)_4COOCH_3)_2$ have a mass of 230.302 g, and N_0 is 6.0221420×10^{23}. The mass of one molecule of $((CH_2)_4COOCH_3)_2$ equals

$$m_{C_{12}H_{22}O_4} = 1 \text{ molecule} \times \frac{230.302 \text{ g } C_{12}H_{22}O_4}{N_0 \text{ molecules } C_{12}H_{22}O_4}$$

$$= \left(\frac{230.302}{6.0221420 \times 10^{23}}\right) \text{g} = 3.82425 \times 10^{-22} \text{ g}$$

> **EXERCISE**
>
> A single molecule of a certain compound has a mass of 1.9624×10^{-22} g. Compute the relative molecular mass of this molecule on a scale on which 12 is the relative atomic mass of ^{12}C.
>
> **Answer:** 118.18.

The Mole

Atoms and molecules have very small masses. Chemical experiments with measurable amounts of substances involve enormous numbers of these "elementary entities." For convenience, chemists group atoms or molecules into counting units that contain Avogadro's number $(6.0221420 \times 10^{23})$ of entities. This counting unit, called the **mole,** measures the **chemical amount** of a substance.

> One mole of a substance equals the amount that contains Avogadro's number of atoms, molecules, or other entities.

In other words, one mole (abbreviation: mol) of ^{12}C contains N_0 ^{12}C atoms, one mole of water contains N_0 H_2O molecules, and so forth. We must be careful in some cases because a phrase such as "one mole of oxygen" is ambiguous. We should instead refer to "one mole of O_2" or "one mole of dioxygen" if we have N_0 ordinary oxygen molecules and "one mole of O" or "one mole of atomic oxygen" if we have N_0 oxygen atoms.

The mass of one mole of atoms of an element is called the **molar mass** (symbolized M). The molar mass of an element is numerically equal to the unitless relative atomic mass of the element, but it has units of grams per mole. For example,

$$M_C = 12.00 \text{ g mol}^{-1} \text{ and } M_{Au} = 196.97 \text{ g mol}^{-1}$$

The same relationship holds between the molar mass of a molecular compound and its relative molecular mass. Water has a relative molecular mass of 18.0153, so

$$M_{H_2O} = 18.0153 \text{ g mol}^{-1}$$

How do we determine the chemical amount of a given substance in a sample, a supply of sucrose (table sugar), say, in a jar? If we could literally count out atoms or molecules, we might choose to do so, arriving at a final count N. Then

$$\text{chemical amount} = n = \frac{\text{number of molecules}}{\text{number of molecules per mole}} = \frac{N}{N_0} \text{ mol}$$

But counting molecules in a 1 g sample of sucrose would take millions of years, even with the help of the entire population of the world, each person counting at the rate of one molecule per second! Instead, chemists use a most powerful tool, the laboratory balance. Suppose a sample of iron has a mass of 8.232 g (a measurement that can be completed in moments with a modern balance). This mass can be converted to a chemical amount by multiplying by a chemical conversion factor, using the unit-factor method (see Appendix B–2). One mole of iron is equivalent to 55.85 g of iron, so that when we multiply as follows

$$8.232 \text{ g Fe} \times \left(\frac{1 \text{ mol Fe}}{55.85 \text{ g Fe}} \right) = 0.1474 \text{ mol Fe}$$

Figure 1–22 • One-mole quantities of several substances. Clockwise, from the lower right: oxalic acid ($H_2C_2O_4$), oxalic acid dihydrate ($H_2C_2O_4 \cdot 2\,H_2O$), and copper sulfate pentahydrate ($CuSO_4 \cdot 5\,H_2O$), and mercury(II) oxide (HgO). At the center is water (H_2O). *(Charles D. Winters)*

we are, in effect, multiplying the term "8.232 g Fe" by 1 (unity). The calculation can be turned around as well. Suppose we need a certain chemical amount, 0.2000 mol, of water to use in a chemical reaction. We have

(chemical amount of water) $\times$ (molar mass of water) $=$ mass of water

which becomes

$$0.2000\ \text{mol}\,H_2O \times \left(\frac{18.0153\ \text{g}\ H_2O}{1\ \text{mol}\,H_2O} \right) = 3.603\ \text{g}\ H_2O$$

We simply measure out 3.603 g of water. In both cases, the molar mass is the conversion factor between mass of substance and chemical amount of substance. When mass (in grams) is changed to chemical amount (in moles), the molar mass (in units of g mol^{-1}) appears in the denominator of the conversion factor. When a chemical amount is converted to a mass, the molar mass appears in the numerator.

Although chemical amounts are frequently measured by obtaining the mass of samples, it is preferable to think of a mole as a fixed number of particles (Avogadro's number) rather than as a fixed mass. The term "mole" is thus analogous to terms such as "dozen" (12) or "score" (20). One dozen United States pennies weighs 30 g, substantially less than 60 g, which is the mass of one dozen nickels, but both contain 12 coins. Figure 1–22 shows mole quantities of several substances.

• Such "grouping terms" are common: a score is 20, a gross is 144, a great gross is 1728, and so forth.

EXAMPLE 1–6

Nitrogen dioxide (NO_2) is a major component of urban air pollution. A sealed tube contains 1.000 g of NO_2. Calculate (a) the chemical amount (the number of moles) of NO_2 and (b) the number of molecules of NO_2.

Strategy

Add up the molar mass of NO_2. Use the result to construct a unit-factor converting mass to chemical amount. Then use Avogadro's number to go from chemical amount to number of molecules.

Solution

(a) The tabulated molar masses of nitrogen and oxygen are 14.01 g mol^{-1} and 16.00 g mol^{-1}, respectively. Therefore

$$M_{NO_2} = (14.01) + 2 \times (16.00) = 46.01\ \text{g mol}^{-1}$$

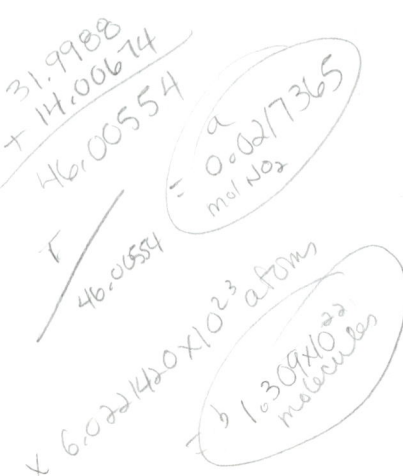

This means that

$$46.01 \text{ g NO}_2 = 1 \text{ mol NO}_2$$

Dividing both sides by 46.01 g NO_2 gives

$$\frac{1 \text{ mol NO}_2}{46.01 \text{ g NO}_2} = 1$$

This factor allows conversion from grams to moles:

$$n_{\text{NO}_2} = 1.000 \text{ g NO}_2 \times \left(\frac{1 \text{ mol NO}_2}{46.01 \text{ g NO}_2}\right) = 0.02173 \text{ mol NO}_2$$

(b) To convert from chemical amount to number of molecules, we multiply by Avogadro's number:

$$N_{\text{NO}_2} = 0.02173 \text{ mol NO}_2 \times \left(\frac{6.0221 \times 10^{23} \text{ molecules NO}_2}{1 \text{ mol NO}_2}\right)$$
$$= 1.309 \times 10^{22} \text{ molecules NO}_2$$

EXERCISE

Molecules of isoamyl acetate, a compound that contributes to the flavor of apples, have the formula $C_7H_{14}O_2$. Calculate (a) how many moles and (b) how many molecules are present in 0.250 g of isoamyl acetate.

Answer: (a) 0.00192 mol. (b) 1.16×10^{21} molecules.

The mole provides the essential conversion between the number of molecules or atoms in a sample of a substance and the mass of the sample. Chemists use the mole constantly because they measure *masses* in the laboratory but explain chemical interactions in terms of relative *numbers* of particles.

The Atomic Mass Unit

The absolute masses of single atoms and single molecules are far too small to be expressed in kilograms or grams without the use of 10 raised to some large negative exponent. We need a unit of mass that is better sized for measuring on the atomic scale.

> The **unified atomic mass unit** (abbreviation: u) equals, by definition, exactly one twelfth of the mass of a single atom of ^{12}C.

• The abbreviation "amu" for atomic mass unit is encountered frequently but is not recommended. The unit is also sometimes called the "dalton" (abbreviated Da), particularly in biochemistry.

We have seen that the mass of Avogadro's number (one mole) of ^{12}C atoms equals exactly 12 g. Now we find from the definition that a single atom of ^{12}C has a mass of exactly 12 u. The definition of the atomic mass unit is contrived purposely to give this numerical equivalence. The equivalence extends to all elements and compounds: The molar mass of a substance in grams per mole is numerically equal to its atomic mass or molecular mass in atomic mass units per atom or atomic mass units per molecule. Thus, a mole of F weighs 18.9984 g, and a single atom of F weighs 18.9984 u; a mole of PF_3 weighs 87.9728 g, and a single molecule of PF_3 weighs 87.9728 u, and so forth. The essence of the relationship is that a gram contains Avogadro's number of atomic mass units.

$$1 \text{ g} = 6.0221420 \times 10^{23} \text{ u}$$

1-8   Finding Empirical and Molecular Formulas the Modern Way

The existence of an accurate set of atomic masses and the availability of high-quality analytical methods make it routine for modern chemists to assign chemical formulas to new substances. We now examine the computations used to make these assignments.

Empirical Formulas and Molecular Formulas

Empirical formulas give the ratios of the numbers of atoms of different kinds that made up a substance. They are exactly the kind of formula that Dalton sought. Modern analytical practice also determines the molecular formulas, if they exist, of new substances. In a **molecular formula,** the subscripts specify the exact number of atoms that are present in one molecule of the substance. The molecular formula H_2O means that each molecule of water contains two atoms of hydrogen linked to one atom of oxygen. Similarly, the molecular formula O_2 (for oxygen) means that each oxygen molecule contains two atoms of oxygen bonded in a persistent combination, and the molecular formula $C_6H_{12}O_6$ (for glucose) means that each glucose molecule contains 6 atoms of carbon, 6 atoms of oxygen, and 12 atoms of hydrogen.

Meaningful molecular formulas exist only for **molecular substances,** such as H_2O, O_2, and $C_6H_{12}O_6$. Molecular substances consist of collections of identical, well-defined, and persisting molecules. Strong attractions called **chemical bonds** hold the atoms in such molecules together. Other forces operate between neighboring molecules, but these intermolecular forces (discussed in Section 6–1) are weaker than are full-fledged chemical bonds. The strength of the attractions within molecules always exceeds the strength of the attractions between molecules. Without this difference, molecular boundaries cannot be identified, which means that distinct molecules do not exist.

Indeed, many solid and liquid substances do *not* contain distinguishable molecules and have no molecular formulas. For example, sodium chloride, the main ingredient in table salt, consists as a solid of a collection of positively charged sodium ions (Na^+ ions) and negatively charged chloride ions (Cl^- ions) locked into a lattice. The lattice is constructed in such a way that interactions between positive and negative ions are maximized while interactions between ions of like charge are minimized. Each ion is surrounded at the same distance by six nearest neighbors of opposite charge. No special bond exists between any single sodium ion and any chloride ion. Solid sodium chloride contains no definable molecules, and only an empirical formula "NaCl" can be written. Other examples are KBr, Li_2O, and $CoCl_2$.

Chemists use the cautious term **formula unit** to refer to the grouping of atoms indicated by the empirical formula of a non-molecular substance or substance that has an unknown microscopic organization. Saying "one formula unit of NaCl" avoids the incorrect implication of "one molecule of NaCl."

The molecular formulas of molecular substances often differ from their empirical formulas. The empirical formula of glucose, for example, is CH_2O, indicating that the numbers of atoms of carbon, hydrogen, and oxygen stand in a ratio of $1:2:1$; and the empirical formula of O_2 is simply O, indicating that molecular oxygen contains only the element oxygen. Clearly, if the subscripts in a formula have a factor in common (all divisible by 2, for example), then the formula must be a molecular formula: The only possible reason not to use smallest whole-number subscripts in

Figure 1–23 • When cobalt(II) chloride crystallizes from solution, it brings with it six water molecules per formula unit, giving a red solid with the empirical formula $CoCl_2 \cdot 6H_2O$. This solid melts at 86°C; above approximately 110°C, it loses some water and forms a light purple solid. *(Leon Lewandowski)*

a formula is to state the contents of a molecule. If the subscripts in a chemical formula do *not* have a factor in common, then additional chemical knowledge is required to decide whether it is an empirical or molecular formula.

Some substances are partly molecular and partly non-molecular. This occurs when a definite proportion of molecules (usually small ones) is incorporated intact into non-molecular solids. The chemical formula is written to show this fact explicitly. Thus, cobalt and chlorine form not only the non-molecular compound $CoCl_2$, but also, when enough water is available under the right conditions, $CoCl_2 \cdot 6H_2O$, in which exactly six molecules of water are incorporated per $CoCl_2$ formula unit. $CoCl_2 \cdot 6H_2O$ is a distinct substance, differing in all its properties from $CoCl_2$ (Fig. 1–23). The dot in the formula $CoCl_2 \cdot 6H_2O$ sets off the water as a well-defined molecular component of the solid.

molec
d nonmolec

EXAMPLE 1–7

Ethylene is a gas under room conditions and is the chemical starting material for many plastics. Its molecular formula is C_2H_4.
(a) What is its empirical formula?
(b) Write other molecular formulas that correspond to the same empirical formula.

Solution

(a) The ratio of the number of carbon atoms to the number of hydrogen atoms in ethylene is 2 : 4. The smallest integers that have this ratio are 1 and 2, so the empirical formula is CH_2.
(b) Every integral multiple of CH_2 is a different molecular formula corresponding to this same empirical formula. Other molecular formulas are C_3H_6, C_4H_8, and so forth. Note that CH_2 itself is also a possible molecular formula.

EXERCISE

In the following list of six compounds of carbon, hydrogen, and oxygen, identify the pairs that have the same empirical formula and give that formula: acetic acid ($C_2H_4O_2$), sorbic acid ($C_6H_8O_2$), caproic acid ($C_6H_{12}O_2$), acrolein (C_3H_4O), propionaldehyde (C_3H_6O), and glucose ($C_6H_{12}O_6$).

Answer: Acetic acid and glucose (CH_2O), sorbic acid and acrolein (C_3H_4O), and caproic acid and propionaldehyde (C_3H_6O).

Percentage Composition from Empirical or Molecular Formula

If we know either the empirical or molecular formula of a substance, we can compute its percentage composition by mass. To do this, we calculate the mass of each element in one mole of the compound and then add those masses to find the total mass of one mole of the compound. The percentages by mass are then found by dividing the mass of each element by the total mass and multiplying by 100%. This computation is illustrated in the following example.

EXAMPLE 1-8

Iron forms a yellow crystalline compound with the empirical formula $Fe_2(SO_4)_3$, which is used in wastewater treatment to aid in the removal of suspended impurities. Calculate the mass percentages of iron, sulfur, and oxygen in this compound.

Strategy

Use the formula to obtain the chemical amount (number of moles) of Fe, S, and O in 1 mol of $Fe_2(SO_4)_3$. Convert, for each element, from chemical amount to mass. Divide the mass contributed by each element by the total mass of the elements and multiply by 100%.

Solution

$$m_{Fe} = 2 \text{ mol Fe} \times \left(\frac{55.847 \text{ g Fe}}{1 \text{ mol Fe}} \right) = 111.69 \text{ g Fe}$$

$$m_{S} = 3 \text{ mol S} \times \left(\frac{32.066 \text{ g S}}{1 \text{ mol S}} \right) = 96.198 \text{ g S}$$

$$m_{O} = 12 \text{ mol O} \times \left(\frac{15.999 \text{ g O}}{1 \text{ mol O}} \right) = 191.99 \text{ g O}$$

The total of these masses is 399.88 g. The percentage composition is

$$\% \text{ Fe} = \left(\frac{111.69 \text{ g Fe}}{399.88 \text{ g compound}} \right) \times 100\% = 27.931\% \text{ Fe}$$

$$\% \text{ S} = \left(\frac{96.198 \text{ g S}}{399.88 \text{ g compound}} \right) \times 100\% = 24.057\% \text{ S}$$

$$\% \text{ O} = \left(\frac{191.99 \text{ g O}}{399.88 \text{ g compound}} \right) \times 100\% = 48.012\% \text{ O}$$

Check

The three percentages add up to 100.000%, as they should.

EXERCISE

Tetrodotoxin, a potent poison found in the ovaries and liver of the globefish, has the empirical formula $C_{11}H_{17}N_3O_8$. Calculate the mass percentages of the four elements in this compound.

Answer: C, 41.38%; H, 5.37%; N, 13.16%; O, 40.09%.

- Why not write $Fe_2S_3O_{12}$ instead? Grouping symbols in parentheses implies that the atoms are grouped together in the compound. In this case, each sulfur atom is surrounded by four oxygen atoms to make a group called the sulfate ion (see Section 3–6).

Empirical Formula from Analytical Experiment

In principle, any compound can be broken down into its constituent elements and the elements then separated and weighed. Such an experiment is called an **elemental analysis.** Starting from such data, a reversal of procedure of Example 1–7 gives the empirical formula of the compound. This is illustrated by the following example.

EXAMPLE 1–9

Elemental analysis of a 60.00 g sample of a molecular compound used as a dry-cleaning fluid finds 10.80 g of carbon, 1.36 g of hydrogen, and 47.84 g of chlorine. Determine the empirical formula of the compound.

Strategy

Calculate the chemical amount (in moles) of each of the elements in the sample, using a table of atomic masses. Such quantities are represented with an n followed by the symbol of the element as a subscript. Find the ratios of the n's by dividing each in turn by the one that is smallest. Multiply the ratios by the smallest factor that clears any fractions that they contain.

Solution

$$n_C = 10.80 \text{ g C} \times \left(\frac{1 \text{ mol C}}{12.011 \text{ g C}} \right) = 0.8992 \text{ mol C}$$

✳ Compare to smallest mole amount

$$n_H = 1.36 \text{ g H} \times \left(\frac{1 \text{ mol H}}{1.008 \text{ g H}} \right) = 1.35 \text{ mol H}$$

$$n_{Cl} = 47.84 \text{ g Cl} \times \left(\frac{1 \text{ mol Cl}}{35.453 \text{ g Cl}} \right) = 1.349 \text{ mol Cl}$$

The ratios of each n to the smallest n are

$$\frac{n_C}{n_C} = \frac{0.8992 \text{ mol C}}{0.8992 \text{ mol C}} = \frac{1 \text{ mol C}}{1 \text{ mol C}}$$

$$\frac{n_H}{n_C} = \frac{1.35 \text{ mol H}}{0.8992 \text{ mol C}} = \frac{1.50 \text{ mol H}}{1 \text{ mol C}}$$

$$\frac{n_{Cl}}{n_C} = \frac{1.349 \text{ mol Cl}}{0.8992 \text{ mol C}} = \frac{1.500 \text{ mol Cl}}{1 \text{ mol C}}$$

Multiplying all three of these ratios by 2 clears the fractions. The three n's stand in the whole-number ratio of $2:3:3$, so the empirical formula is $C_2H_3Cl_3$. Further information is necessary to find the correct *molecular* formula from among such possibilities as $C_2H_3Cl_3$, $C_4H_6Cl_6$, and higher multiples $(C_2H_3Cl_3)_n$.

EXERCISE

Heating a 150.0 mg dose of a compound used to treat rheumatism decomposes it to its constituent elements, which are separated. There are 60.29 mg of gold, 21.10 mg of sodium, 29.37 mg of oxygen, and 39.24 mg of sulfur. Determine the empirical formula of this compound.

Answer: $AuNa_3O_6S_4$.

Figure 1–24 • Chromium(III) oxide (Cr_2O_3) is a green solid that can be reduced to shiny, metallic, elemental chromium. The major use of chromium is in the plating of steel and other metals to give a brighter and more resistant finish, as in the chrome trim on a stapler. *(Leon Lewandowski)*

An empirical formula also can be determined from the percentages by mass of the elements in a compound (rather than the actual masses found in the analyzed sample). The first step in this very common case is to assume some total mass (exactly 100 g is a convenient choice) and then figure the mass of the different elements that this total would supply. The procedure then follows the pattern just shown. The percentage composition is independent of the actual number of grams in the sample, so the answer is the same regardless of the choice of total mass.

Sometimes, separating a particular element from a compound and weighing it is difficult to do accurately. By measuring the masses of all the other elements and using the law of conservation of mass, we can still obtain the empirical formula of the compound. This tactic is particularly useful in cases in which a binary compound decomposes to give one of its elements in the pure state, as in the following example.

EXAMPLE 1–10

A certain green oxide of chromium gives elemental chromium when heated in a stream of gaseous hydrogen (Fig. 1–24). Complete reaction of 6.245 g of the oxide yields 4.273 g of pure chromium. Determine the empirical formula of the oxide.

Strategy

An "oxide of chromium" contains only Cr and O. Use the law of conservation of mass to find the mass of O in the original sample. Then proceed as in Example 1–9.

Solution

$$m_O = m_{compound} - m_{Cr} = 6.245 \text{ g} - 4.273 \text{ g} = 1.972 \text{ g}$$

The chemical amounts of chromium and oxygen atoms are

chemical amounts = moles

$$n_{Cr} = 4.273 \text{ g Cr} \times \left(\frac{1 \text{ mol Cr}}{51.996 \text{ g Cr}} \right) = 0.08218 \text{ mol Cr}$$

$$n_O = 1.972 \text{ g O} \times \left(\frac{1 \text{ mol O}}{15.999 \text{ g O}} \right) = 0.1233 \text{ mol O}$$

The ratio of these chemical amounts is

$$\frac{n_O}{n_{Cr}} = \frac{0.1233 \text{ mol O}}{0.08218 \text{ mol Cr}} = \frac{1.500 \text{ mol O}}{1 \text{ mol Cr}} = \frac{3 \text{ mol O}}{2 \text{ mol Cr}}$$

The empirical formula is Cr_2O_3 because the compound contains three moles of oxygen atoms for every two moles of chromium atoms.

EXERCISE

Moderate heating of 97.44 mg of a compound containing nickel, carbon, and oxygen and no other elements drives off all of the carbon and oxygen in the form of carbon monoxide (CO) and leaves 33.50 mg of metallic nickel behind. Determine the empirical formula of the compound.

Answer: NiC_4O_4.

Combustion Analysis

Organic compounds, a prominent and numerous category, contain carbon and hydrogen, very often in combination with additional elements such as nitrogen, oxygen, sulfur, chlorine, and phosphorus. Compounds containing C and H only are called **hydrocarbons.** The simplest hydrocarbon is methane (molecular formula CH_4), the major component of natural gas. Two other hydrocarbons are benzene (molecular formula C_6H_6), an important industrial solvent, and styrene (molecular formula C_8H_8), which is used to produce polystyrene for packaging and insulation. The empirical formulas of hydrocarbons can be determined by analysis in the combustion train shown in Figure 1–25. In this apparatus, a weighed amount of the

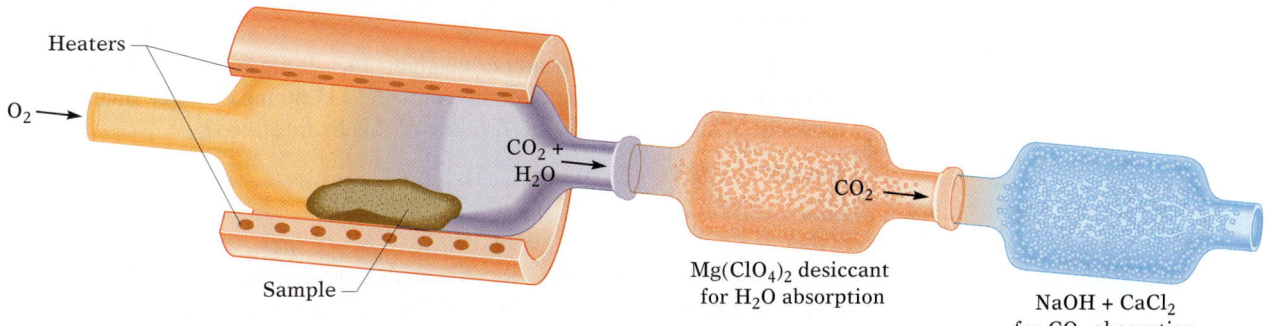

Figure 1–25 • A combustion train for measuring the amounts of carbon and hydrogen in a compound. The sample is burned in a flow of oxygen to form water and carbon dioxide. These products pass over a desiccant, such as magnesium perchlorate ($Mg(ClO_4)_2$), which absorbs the water. The carbon dioxide passes on to a second chamber, where it is absorbed on finely divided particles of sodium hydroxide (NaOH) mixed with calcium chloride ($CaCl_2$). The increases in mass of the two traps equal the amounts of water and carbon dioxide produced.

hydrocarbon is burned completely in oxygen, yielding carbon dioxide and water. The masses of water and carbon dioxide are then determined, and from these data the empirical formula is found, as in the following example.

EXAMPLE 1–11

A certain compound, used as a welding fuel, contains carbon and hydrogen only. Burning a 1.000 g sample of it completely in oxygen gives 3.38 g of carbon dioxide, 0.690 g of water, and no other products. Determine the empirical formula of the compound.

Strategy

Compute the chemical amounts of the carbon dioxide and water. Recognize that all of the C in the compound appears in the carbon dioxide, and all of the H, in the water. Write unit factors based on these facts to compute the chemical amounts of C and H that were in the unburned gas. Find the smallest whole-number ratio of these chemical amounts.

Solution

$$n_C = 3.38 \text{ g } CO_2 \times \left(\frac{1 \text{ mol } CO_2}{44.01 \text{ g } CO_2} \right) \times \left(\frac{1 \text{ mol C}}{1 \text{ mol } CO_2} \right) = 0.0768 \text{ mol C}$$

$$n_H = 0.690 \text{ g } H_2O \times \left(\frac{1 \text{ mol } H_2O}{18.015 \text{ g } H_2O} \right) \times \left(\frac{2 \text{ mol H}}{1 \text{ mol } H_2O} \right) = 0.0766 \text{ mol H}$$

The ratio of the n's is 0.0768 mol C divided by 0.0767 mol H, which is a 1 : 1 ratio within the precision of the data. The compound contains equal chemical amounts of carbon and hydrogen. Its empirical formula is CH . Its molecular formula may be CH, or C_2H_2, or C_3H_3, and so on.

Check

If the original sample contained 0.0768 mol of C, it contained 0.9224 g of C (computed by multiplying 0.0768 mol by 12.011 g mol^{-1}, the molar mass of C). Similarly, if the original sample contained 0.0766 mol H, it contained 0.0772 g H. The masses of C and H add up to 0.9996 g, which rounds off to 1.000 g, the correct mass of the original sample.

EXERCISE

A sample of a liquid hydrocarbon weighing 142.70 mg is burned in a combustion train to give 477.00 mg of carbon dioxide, 111.60 mg of water, and no other products. What is the empirical formula of this hydrocarbon?

Answer: C_7H_8.

Combustion analysis works for compounds containing elements other than carbon and hydrogen. A suitable experimental setup, for example, allows for the conversion of all the nitrogen in a sample to gaseous N_2 and all of the sulfur to gaseous SO_2, both of which can be "quantified" (their amounts measured) fairly easily. Modern practice typically uses flash combustion of small (2–10 mg) samples in pure oxygen at approximately 1000°C to ensure complete reaction. Clearly, incomplete reaction would lead to errors. Modern automatic analyzers (Fig. 1–26) accu-

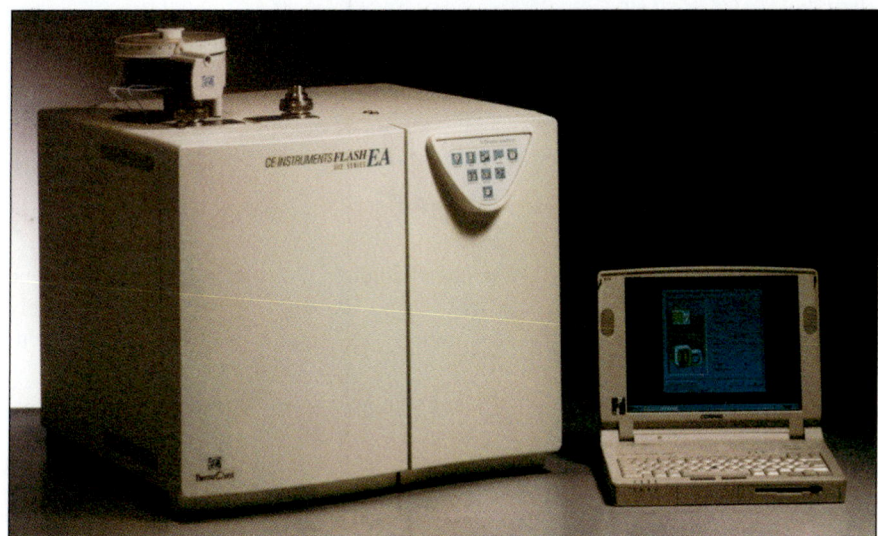

rately determine the percentages of C, H, N, and S in a single small sample in a matter of 10 minutes. These devices can be programmed to work unattended on a magazine of hundreds of samples, storing the results in a computer. The equipment is specialized, so synthetic chemists rarely perform their own elemental analyses nowadays. Instead, they send samples out to specialists.

Determining Molecular Formulas the Modern Way

Even the most rapid and accurate elemental analysis of a molecular substance does not furnish enough information to determine its molecular formula. The required missing information is the approximate molar mass of the substance. Fortunately, molar masses are available from a variety of experiments. Numerous molecular substances can be vaporized and the intact molecules ionized in mass spectrometers. The molar masses are then determined by accelerating the ions and observing the degree of curvature of their paths through a magnetic field, as described earlier in Section 1–6. Other experiments (described in Sections 5–4 and 6–6) give the approximate molar masses of other molecular substances. Once the molar mass and the empirical formula are known, the molecular formula follows quickly, as shown in the following example.

EXAMPLE 1–12

A molecular compound isolated from lemons is thought to confer important health benefits and is therefore studied. Its empirical formula is $C_5H_4O_2$. Mass spectrometry establishes that its molar mass is 288 g mol^{-1}. Determine the molecular formula of the compound.

Strategy

See how many times the molar mass that corresponds to the empirical formula fits into the known molar mass.

TABLE 1–1	
Densities of Some Substances[a]	
Substance	**Density (g cm⁻³)**
Hydrogen	0.000082
Oxygen	0.00130
Water	1.00
Magnesium	1.74
Sodium chloride	2.16
Quartz	2.65
Aluminum	2.70
Iron	7.86
Copper	8.96
Silver	10.5
Lead	11.4
Mercury	13.6
Gold	19.3
Platinum	21.4

[a]Measured under room-temperature conditions and at average atmospheric pressure near sea level.

Solution

$64 + 32 \approx 96$

If the molecular formula were $C_5H_4O_2$, then the molar mass would be 96.08 g mol⁻¹. The actual molar mass of 288 g mol⁻¹ is 288/96.08 = 3.00 times larger than this. Hence, the molecular formula of the compound is $C_{15}H_{12}O_6$, in which each subscript is three times the subscript in the empirical formula.

EXERCISE

A molecular compound called "anemone camphor" is isolated from the anemone plant. Its empirical formula is $C_5H_4O_2$, and its molar mass is approximately 192 g mol⁻¹ according to mass spectrometry measurements. What is its molecular formula?

$\dfrac{192 \text{ g/mol}}{96 \text{ g/mol}} = 2$

Answer: $C_{10}H_8O_4$.

1–9 Volume and Density

The density of a sample of a material is the ratio of its mass to its volume:

$$\text{density} = \frac{\text{mass}}{\text{volume}}$$

In the International System of Units (SI, discussed in Appendix B), the base unit of mass is the kilogram (kg), but this is inconveniently large for practical purposes in chemistry. Chemists very often use grams instead. Several units for volume are in common use. The natural SI unit, the cubic meter (m³), is quite unwieldy for laboratory purposes (1 m³ of water has a mass of 1000 kg, or 1 metric ton). Therefore, chemists prefer to use the liter (L), which is the same as a cubic decimeter (1 L = 1 dm³ = 10⁻³ m³), and the cubic centimeter, which is identical to the milliliter (1 cm³ = 1 mL = 10⁻³ L = 10⁻⁶ m³). Figure 1–27 shows some laboratory equipment used to measure volumes of liquids, and Table 1–1 lists the densities of a number of substances in units of g cm⁻³. It is important to recognize that the density of a substance depends on the conditions of pressure and temperature that exist at the time of the measurement. For some substances, especially gases and liquids, the volume may be more convenient to measure than the mass, and when the density is known, it provides the conversion factor to pass from volume to mass. This is illustrated by the following example.

EXAMPLE 1–13

Near room temperature, liquid benzene (C_6H_6) has a density of 0.8765 g cm⁻³. Suppose that 0.2124 L of benzene is measured into a container. Calculate (a) the mass and (b) the chemical amount of benzene in the container.

Strategy

Use the density as a unit-factor "bridge" to pass from volume to mass.

Solution

(a) Because 1 L = 10³ cm³, the volume of the benzene sample is

$$0.2124 \text{ L benzene} \times \left(\frac{10^3 \text{ cm}^3}{1 \text{ L}}\right) = 212.4 \text{ cm}^3 \text{ benzene}$$

Because 0.8765 g of benzene is equivalent to 1 cm^3 of benzene, we can multiply the volume by the ratio of the two

$$212.4 \ \text{cm}^3 \ \text{benzene} \times \left(\frac{0.8765 \ \text{g benzene}}{1 \ \text{cm}^3 \ \text{benzene}}\right) = 186.2 \ \text{g benzene}$$

(b) The relative molecular mass of benzene is found from the relative atomic masses of carbon (12.011) and hydrogen (1.00794):

molecular mass of benzene = (6 × 12.011) + (6 × 1.00794) = 78.114

so that the molar mass is 78.114 g mol^{-1}. The chemical amount is then found by the unit-factor method:

$$n_{\text{benzene}} = 186.2 \ \text{g benzene} \times \left(\frac{1 \ \text{mol benzene}}{78.114 \ \text{g benzene}}\right) = 2.384 \ \text{mol benzene}$$

EXERCISE

The density of liquid mercury at 20°C is 13.594 g cm^{-3}. A chemical reaction requires 0.560 mol of mercury. What volume (in cubic centimeters) of mercury should be measured out at 20°C?

Answer: 8.26 cm^3.

Figure 1-27 • Some volumetric equipment: 150 mL beaker (green liquid), 25 mL buret (red), 1000 mL volumetric flask (yellow), 100 mL graduated cylinder (blue), and 10 mL volumetric pipet (green). (*Charles D. Winters*)

The density of a substance, together with Avogadro's number, gives the volume occupied by a molecule. In Example 1–13, 2.384 mol of benzene occupied a volume of 212.4 cm^3. The number of benzene molecules in this volume is

$$N_{\text{benzene}} = 2.384 \ \text{mol benzene} \times \left(\frac{6.022 \times 10^{23} \ \text{molecule}}{1 \ \text{mol benzene}}\right)$$

$$= 1.436 \times 10^{24} \ \text{molecules}$$

$$\text{volume per molecule} = \left(\frac{212.4 \ \text{cm}^3}{1.436 \times 10^{24} \ \text{molecules}}\right) = 1.479 \times 10^{-22} \ \frac{\text{cm}^3}{\text{molecule}}$$

This volume is not necessarily the same as the volume of the benzene molecule itself. It is the room available per molecule. The volume available per molecule increases greatly when a substance becomes a gas. A benzene molecule in gaseous benzene at its boiling point has approximately 4.8 × 10^{-20} cm^3 available to it, approximately 300 times more than the volume that we just computed.

Depends on phase?

How can we interpret this fact on a microscopic scale? Note first that the volumes of liquids and solids do not change very much with changes in temperature or pressure, whereas gas volumes are quite sensitive to these changes. A hypothesis to explain this observation is that molecules in liquids and solids are close enough to touch one another, whereas in a gas they are separated by large expanses of empty space (Fig. 1–28). If this hypothesis is correct (and it has been borne out by further study), then the sizes of the molecules themselves can be estimated from the volume per molecule in the liquid or solid state. For benzene, this leads to a volume of approximately 1.5 × 10^{-22} cm^3, corresponding to a cube with edges about 5 × 10^{-8} cm long. This and other density measurements show that the characteristic size of atoms and small molecules is on the order of 10^{-8} cm. Avogadro's number provides the link between laboratory-scale measurements and the masses and volumes of single atoms and molecules.

• This is because the cube root of 1.5 × 10^{-22} cm^3 is approximately 5 × 10^{-8} cm.

Figure 1–28 • In a liquid or solid, molecules (shown here as spheres) are in close contact with one another, so the volume per molecule fairly closely approximates the volume of the molecule itself. In a gas, the distances between molecules are greater, and considerable empty space is present.

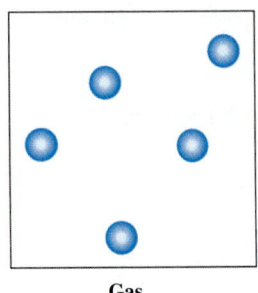
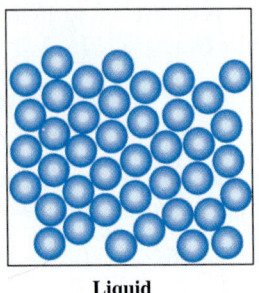
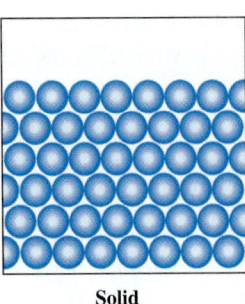

Gas   Liquid   Solid

SUMMARY

1–1   Chemistry begins with the description of matter and change. Chemists ask: What reacts? How much of one substance is needed to react with another? How much product is produced? and, Why do substances have the characteristics that they possess? Chemists seek to understand reactions and to synthesize useful new products.

1–2   Chemists classify materials as elements, compounds, or mixtures. Substances are defined as pure **elements** or pure **compounds. Mixtures** contain more than one substance. Samples may be **homogeneous** or **heterogeneous.** Even a pure substance can exist in more than one **phase.**

1–3   Scientific studies of combustion led to the formulation of the **law of conservation of mass,** and accurate analysis of compounds led to the **law of definite proportions. Dalton's atomic theory of matter** sought to explain these observed laws. Modern experimentation has confirmed the existence of atoms.

1–4   Dalton sought to find relative atomic masses and formulas of compounds. In order to do this, he had to guess or postulate the chemical formulas of some compounds. Some of his guesses were wrong. Gay-Lussac's **law of combining volumes** and **Avogadro's hypothesis** provided the tools to arrive at correct chemical formulas and a consistent set of relative atomic masses.

1–5   Experiments by Thomson, Millikan, and Rutherford in the late 19th and early 20th centuries led to the discovery of subatomic particles (**electrons, protons,** and **neutrons**) and a model of atomic structure: Negatively charged electrons surround a small, massive, positively charged **nucleus** composed of protons and neutrons.

1–6   Careful chemical analysis of a compound of known formula provides the **relative atomic masses** of the elements that make it up. **Mass spectrometry,** a physical method in which beams of ions are manipulated by electric and magnetic fields, provides more precise relative atomic masses. It also reveals the existence of **isotopes** (atoms that contain the same number of protons but different numbers of neutrons) in the majority of naturally occurring elements.

1–7   The **mole concept** and knowledge of Avogadro's number (6.0221420×10^{23} mol^{-1}) permit one to "count" the number of atoms in a sample of a substance of known formula by weighing the sample. The **molar mass (M)** of a substance equals its relative atomic or molecular mass expressed in grams and contains Avogadro's number of formula units.

1–8   The availability of reliable atomic masses makes it straightforward to determine **empirical formulas** from percent composition data obtained by chemical analysis. When molar masses are known, **molecular formulas** can be obtained from empirical formulas.

1–9   Consideration of the **density** of substances in the liquid or solid phase leads to an estimate of the sizes of atoms.

PROBLEMS

Note: Answers to blue-numbered problems are given in Appendix F. Problems that are more challenging are indicated with asterisks.

The Composition of Matter

1. Classify the following materials as homogeneous or heterogeneous: table salt, wood, mercury, air, water, lemonade made with fresh-squeezed lemons, sodium chloride, and ketchup. Are they substances or mixtures? If they are substances, are they compounds or elements?

2. Classify the following materials as homogeneous or heterogeneous: absolute (pure) alcohol, milk (as purchased in a store), copper wire, rust, calcium carbonate, concrete, baking soda, baking powder, and chalk. Are they substances or mixtures? If they are substances, are they compounds or elements?

3. Give an example of a heterogeneous sample that would gradually become homogeneous if left to itself.

4. Give an example of a homogeneous sample that would gradually become heterogeneous if left to itself.

5. A 10.00 g sample of impure silicon is 99.89% silicon by mass, and the rest is carbon. Compute the mass of carbon present in the sample.

6. A sample of 97.62% pure calcium phosphate by mass contains 136.7 g of the compound. Determine the total mass of the sample.

The Atomic Theory of Matter

7. A 10.0 lb bundle of newspapers is burned in an incinerator. The ashes weigh 0.14 lb. Explain what has become of the rest of the mass.

8. A copper penny has a mass of 3.1041 g. Its mass increases to 3.1063 after it is heated strongly in the air and then cooled. Explain the gain in the penny's mass.

9. A 1.75 g sample of solid potassium sulfite (K_2SO_3) is placed in a flask containing 19.75 g of hydrochloric acid. The two react, generating gaseous sulfur dioxide, which all escapes from the flask, and no other gases. After the reaction, the mass of the contents of the flask is 20.79 g. Determine the mass of the SO_2 produced.

10. Potassium chlorate $(KClO_3)$ decomposes to potassium chloride and oxygen (and no other products) when heated. In one experiment, 100.0 g $KClO_3$ generates 36.9 g of O_2 and 57.3 g of KCl. What mass of $KClO_3$ remains unreacted?

11. A sample of ascorbic acid (vitamin C) is synthesized in the laboratory. It contains 30.0 g of carbon and 40.0 g of oxygen. Another sample of ascorbic acid, isolated from lemons (an excellent source of the vitamin), contains 21.3 g of carbon. Compute the mass of oxygen in the second sample.

12. A sample of a compound synthesized and purified in the laboratory contains 25.0 g of hafnium and 31.5 g of tellurium. The identical compound is discovered in a rock formation. A sample from the rock formation contains 0.125 g of hafnium. Determine how much tellurium is in the sample from the rock formation.

Chemical Formulas and Relative Atomic Masses

13. Determine the ratio of the number of oxygen atoms to the number of calcium atoms in each of the following compounds:
    (a) CaO
    (b) $Ca_3(PO_4)_2$
    (c) $Ca(OH)_2$
    (d) $CaCr_2O_7$
    (e) $Ca_5(PO_4)_3OH$

14. Write formulas for the following:
    (a) A ternary compound containing four atoms of sodium and two atoms of phosphorus for every seven atoms of oxygen
    (b) A binary compound of nitrogen and oxygen in which the oxygen atoms are $2\frac{1}{2}$ times more numerous than the nitrogen atoms
    (c) A ternary compound in which the numbers of carbon and hydrogen atoms equal each other and also equal ten times the number of iron atoms

15. (See Example 1–1.) Analysis shows that 100.00 g of an oxide of iron (Fe) contains 72.36 g of Fe and 27.64 g of O.
    (a) Compute the ratio of the atomic mass of Fe to the atomic mass of O if the formula of the compound is FeO.
    (b) Compute the ratio of the atomic mass of Fe to the atomic mass of O if the formula of the compound is Fe_3O_4.

16. (See Example 1–1.) Analysis shows that 100.00 g of an oxide of zinc (Zn) contains 80.34 g of Zn and 19.66 g of O.
    (a) Compute the ratio of the atomic mass of Zn to the atomic mass of O if the formula of the compound is ZnO.
    (b) Compute the ratio of atomic mass of Zn to the atomic mass of O if the formula of the compound is Zn_3O_4.

17. Use the analytical data in Problems 15 and 16 to compute the ratio of the atomic mass of Zn to the atomic mass of Fe. Assume that the oxide of iron is Fe_3O_4 and the oxide of zinc is ZnO.

18. Use the analytical data in Problems 15 and 16 to compute the ratio of the atomic mass of Zn to the atomic mass of Fe. Assume that the oxide of iron is FeO and the oxide of zinc is Zn_3O_4.

19. In 1804, John Dalton suggested the *law of multiple proportions*: When two elements form more than one compound, the masses of the second that combine with a fixed mass of the first have the ratio of small whole numbers.
    (a) A binary compound between silicon (Si) and nitrogen (N) contains 2.005 g of Si per 1.000 g of N. A second binary compound between Si and N contains 1.504 g of Si per 1.000 g of N. Show that these data are consistent with the law of multiple proportions.

(b) If the second compound in part (a) has the formula Si_3N_4, then what is the formula of the first compound?

20. Iodine (I) and fluorine (F) form a series of binary compounds with the following compositions:

Compound	Mass % I	Mass % F
1	86.979	13.021
2	69.007	30.993
3	57.191	42.089
4	48.829	51.171

(a) Compute in each case the mass of fluorine that combines with 1.0000 g of iodine.
(b) Show that these compounds satisfy the law of multiple proportions (see Problem 19) by figuring out small whole-number ratios among the four answers in part (a).

21. Under certain conditions, pure nitrogen dioxide (NO_2) forms when gaseous dinitrogen monoxide (N_2O) and gaseous oxygen (O_2) are mixed. What volumes of N_2O and O_2 are needed to produce 4.0 L of NO_2 if all gases are held at the same conditions of temperature and pressure?

22. Gaseous methanol (CH_3OH) reacts with oxygen (O_2) to produce water vapor and carbon dioxide. What volume of water vapor and carbon dioxide is produced from 2.0 L of methanol if all gases are held at the same conditions of temperature and pressure?

The Building Blocks of the Atom

23. Complete the following table:

Symbol	Z	N	A	No. of Electrons
$^{12}C^-$	—	—	—	—
—	15	—	31	13
—	—	6	11	5
Bi^{3+}	—	126	—	—

24. Complete the following table:

Symbol	Z	N	A	No. of Electrons
$^{228}Ra^{2+}$	—	—	—	—
—	53	—	127	54
—	—	20	40	20
N^{3-}	—	8	—	—

25. Suppose that an atomic nucleus were scaled up to the diameter of a baseball (approximately 10 cm). The size of the entire atom would then roughly equal the size of __. Support your choice with a short calculation.
(a) a basketball
(b) a city bus
(c) a football stadium
(d) Manhattan Island
(e) the planet Earth

26. The mass of the proton is 1.672623×10^{-27} kg. How many electrons does it take to equal the mass of a single proton?

27. (See Example 1–3.) Plutonium-239 is used for nuclear fission. Determine (a) the ratio of the number of neutrons in a ^{239}Pu nucleus to the number of protons and (b) the number of electrons in a single Pu^- ion.

28. (See Example 1–3.) The element promethium was discovered in 1947. Determine (a) the ratio of the number of neutrons in a ^{145}Pm nucleus to the number of protons and (b) the number of electrons in a single Pm^{3+} ion.

Finding Atomic Masses the Modern Way

29. Before 1961, an atomic mass scale was used that was slightly different from the present one. It assigned exactly 16 as the atomic mass of the naturally occurring mixture of three oxygen isotopes. The relative atomic mass of this mixture on the present-day scale is 15.9994. Compute the atomic mass of ^{12}C on the pre-1961 scale.

30. Suppose that the atomic mass of naturally occurring phosphorus is assigned a value of exactly 10 "paulings." Determine the atomic mass of uranium in paulings.

31. (See Example 1–3.) In a certain collection of marbles, each red marble weighs 10.67 g, and each green marble weighs 13.53 g. A big sack contains 19,490 red marbles and 26,278 green marbles. Determine the mass of the average marble in this sack. Explain why the answer is not equal to $(10.67 + 13.53)/2 = 12.10$ g.

32. (See Example 1–3.) Naturally occurring iron consists of four isotopes of the following relative atomic masses and fractional abundances:

Isotope	Fractional Abundance	Relative Mass
^{54}Fe	0.05845	53.939613
^{56}Fe	0.91754	55.934939
^{57}Fe	0.02119	56.935396
^{58}Fe	0.00282	57.933277

Calculate the relative atomic mass of naturally occurring iron.

33. Only two isotopes of copper occur in nature, with the relative atomic masses and abundances given in the following table. Complete the table by computing the relative atomic mass of ^{65}Cu to four significant figures, taking the tabulated relative atomic mass of naturally occurring copper as 63.546.

Isotope	Fractional Abundance	Relative Mass
^{63}Cu	0.6917	62.929599
^{65}Cu	0.3083	?

34. Boron has only two naturally occurring isotopes: ^{10}B, with a relative mass of 10.013, and ^{11}B, with a relative mass of 11.009. Compute the fractional abundance of ^{10}B, given that the relative atomic mass of boron equals 10.811.

The Mole Concept: Counting and Weighing Atoms and Molecules

35. Compute the relative molecular masses or formula masses of the following compounds on the ^{12}C scale:
   (a) N_2O_5
   (b) IF
   (c) Na_2SO_3
   (d) $KMnO_4$
   (e) $(NH_4)_2SO_4$

36. Compute the relative molecular masses or formula masses of the following compounds on the ^{12}C scale:
   (a) $[Cu(NH_3)_4]Cl_2$
   (b) $Ca_3[Co(CO_3)_3]_2$
   (c) RuO_4
   (d) $HClO_4$
   (e) $Ca_3Al_2(SiO_4)_3$

37. (a) Determine the number of SF_2 molecules in 12.000 g of SF_2.
   (b) Compute the mass of sulfur present per 1.000 g of fluorine in SF_2.
   (c) Repeat parts (a) and (b) for the related compound S_2F_4.
   (d) Explain the similarities and differences between the answers for SF_2 and the answers for S_2F_4.

38. Arrange the following in order of increasing mass: 1.17 mol of SF_4; 152 g of CH_4; 9.3×10^{23} molecules of Cl_2O_7; 5.34×10^{23} atoms of argon (Ar).

39. (See Example 1–4.) Compute the mass, in grams, of a single iodine atom if the relative atomic mass of iodine is 126.90447 on the accepted scale of atomic masses (based on 12 as the relative atomic mass of ^{12}C). Determine the mass, in atomic mass units, of a single iodine atom.

40. (See Example 1–4.) Determine the mass, in grams, of exactly 100 million atoms of fluorine if the relative atomic mass of fluorine is 18.998403 on a scale on which exactly 12 is the relative atomic mass of ^{12}C. Express the mass of the same 100 million atoms of fluorine in atomic mass units.

41. (See Example 1–6.) Ferrocene has the formula $Fe(C_5H_5)_2$.
   (a) Calculate the chemical amount of ferrocene in 100.0 g of ferrocene.
   (b) Calculate the number of molecules of ferrocene in 100.0 g of ferrocene.

42. (See Example 1–6.) The flavoring agent vanillin ($C_8H_8O_3$) occurs naturally in the vanilla bean.
   (a) Calculate the chemical amount of vanillin in 10.00 g of vanillin.
   (b) Calculate the number of molecules of vanillin in 10.00 g of vanillin.
   (c) Compute the mass (in atomic mass units) of one molecule of vanillin.

   (d) Compute the mass (in grams) of one molecule of vanillin.

43. (a) Compute the mass (in grams) of lithium in 65.4 g of Li_2CO_3.
   (b) Compute the mass (in pounds) of lithium in 65.4 lb of Li_2CO_3.

44. (a) Compute the mass (in grams) of titanium in 91.2 g of $TiCl_4$.
   (b) Compute the mass (in tons) of titanium in 91.2 tons of $TiCl_4$.

45. You need 450,000 finishing nails for a big construction project you are managing. A supplier furnishes these nails, which weigh an average of 10 g each, in 100 lb kegs. How many kegs of nails should you order?

46. You get a job at a screw factory. As a joke, or test, one of the older workers tells you to get a count of the number of screws in a 250 lb lot of identical small machine screws. Explain how to get the answer without sitting down and counting all the screws.

47. A patient is found to have 6.2 nanograms (ng) of mercury per deciliter (dL) of blood (*Note*: 1 ng = 10^{-9} g; see Appendix B). Determine the mass of mercury in 500 mL (roughly a pint) of this patient's blood.

48. The United States Environmental Protection Agency (EPA) sets the maximum safe level for lead in the blood at 24 mg dL^{-1}. A 1.00 mL sample of a patient's blood contains 1.50×10^{-8} mol of lead. Express the patient's lead level in the unit used by the EPA to see if the patient is in danger of lead poisoning.

Finding Empirical and Molecular Formulas the Modern Way

49. Arrange the following compounds from left to right in order of increasing percentage by mass of hydrogen: H_2O, $C_{12}H_{26}$, N_4H_6, LiH.

50. Arrange the following compounds from left to right in order of increasing percentage by mass of fluorine: HF, C_6HF_5, BrF, UF_6.

51. The male sex hormone testosterone has the molecular formula $C_{19}H_{28}O_2$. Compute, to four significant figures, the mass percentage of carbon in this hormone.

52. The female sex hormone estradiol has the molecular formula $C_{18}H_{24}O_2$.
   (a) Compute, to four significant figures, the mass percentage of carbon in this hormone.
   (b) The male sex hormone (Problem 51) contains 19 atoms of carbon per molecule, and the female sex hormone contains only 18 atoms of carbon per molecule. Despite this, the female sex hormone contains the larger mass percentage of carbon. Explain how this is possible.

53. (See Example 1–8.) A newly synthesized compound has the molecular formula $ClF_2O_2PtF_6$. Compute, to four significant figures, the mass percentage of each of the four elements in this compound.

54. (See Example 1–8.) Acetaminophen, the pain reliever in Tylenol and some other headache remedies, has the molecular formula $C_8H_9NO_2$. Compute, to four significant figures, the mass percentage of each of the four elements in acetaminophen.

55. Chrome yellow ($PbCrO_4$) is a yellow pigment used in oil paints. Chrome red is the related compound $PbCrO_4 \cdot PbO$. Compute the percentage by mass of lead in each of these pigments.

56. Compute the percentage by mass of magnesium in a mineral if the formula is given as $MgF_2 \cdot (MgSiO_3)_2$.

57. A binary compound of silicon and chlorine is 28.37% silicon by mass. Determine the empirical formula of the compound.

58. A binary compound of silver and chlorine is 7.52% silver by mass. Determine the empirical formula of the compound.

59. Fulgurites are the products of the melting that occurs when lightning strikes the earth. Microscopic examination of a sand fulgurite reveals that it is a globule with variable composition containing some grains of the definite chemical composition Fe, 46.01%; Si, 53.99%. Determine the empirical formula of these grains.

60. A sample of a "suboxide" of cesium gives up 1.6907% of its mass as gaseous oxygen when gently heated, leaving pure cesium behind. Determine the empirical formula of this binary compound.

61. Barium and nitrogen form two binary compounds containing 90.745% and 93.634% barium, respectively. Determine the empirical formulas of these two compounds.

62. Carbon and oxygen form no fewer than five different binary compounds. The mass percentages of carbon in the five compounds are as follows: A, 27.29; B, 42.88; C, 50.02; D, 52.97; and E, 65.24. Determine the empirical formulas of the five compounds.

63. (See Example 1–10.) A pure compound that weighs 65.32 g and contains only sodium, sulfur, and oxygen is found on analysis to contain 21.14 g of sodium and 14.75 g of sulfur. Calculate the empirical formula of the compound.

64. (See Example 1–10.) A white crystalline compound that contains only iodine, potassium, and oxygen is purified. A mass of 87.41 g of pure compound is found on analysis to contain 15.97 g of potassium and 51.83 g of iodine. What is the empirical formula of the compound?

65. (See Example 1–11.) Burning a sample of a hydrocarbon (a compound containing hydrogen and carbon only) in oxygen produced 3.701 g of CO_2 and 1.082 g of H_2O. Determine the empirical formula of the hydrocarbon.

66. (See Example 1–11.) Burning a compound of calcium, carbon, and nitrogen in oxygen in a combustion train generates calcium oxide (CaO), carbon dioxide (CO_2), nitrogen dioxide (NO_2), and no other substances. A small sample gives 2.389 mg of CaO, 1.876 mg of CO_2, and 3.921 mg of NO_2. Determine the empirical formula of the compound.

67. A solid compound contains carbon, hydrogen, and sulfur and no other elements. A 9.051 mg sample of the compound is burned in a combustion analyzer to give 13.252 mg of carbon dioxide, 5.427 mg of water, and 9.646 mg of sulfur dioxide. Determine the empirical formula of the compound.

68. A 317.4 mg sample of a compound containing only carbon, hydrogen, and chlorine is burned completely in a combustion train. The carbon dioxide has a mass of 369.78 mg, and the water has a mass of 32.45 mg. Determine the empirical formula of the compound.

69. (See Example 1–12.) The empirical formula of squalene, which is isolated from shark-liver oil, is C_3H_5. Its molar mass is 410.7 g mol^{-1}. Determine the molecular formula of squalene.

70. (See Example 1–12.) A newly synthesized compound has the empirical formula $C_{10}H_{12}N_2Cu_3I_5$. Write the molecular formula of this compound if its molar mass is 11,824 g mol^{-1}.

Volume and Density

71. (See Example 1–13.) Arsenic trifluoride (AsF_3) is an oily liquid with a density at room temperature of 2.163 g cm^{-3}. Compute the mass of a 5.12 cm^3 sample of arsenic trifluoride. Compute the chemical amount of AsF_3 in this sample.

72. (See Example 1–13.) At 20°C, the density of liquid methanol (CH_3OH) is 0.7914 g cm^{-3}. Compute the mass of a 9.33 cm^3 sample of methanol. Compute the chemical amount of CH_3OH in this sample.

73. A Nerf ball has a mass of 49.4 g and a volume of 537 cm^3. Compute its density.

74. Rhenium is the fourth densest element at room conditions; its density is 21.02 g cm^{-3}. Compute the volume of rhenium that has a mass of 10.00 g.

75. Mercury is traded by the "flask," a unit that has a mass of 34.5 kg. Determine the volume of a flask of mercury if the density of mercury is 13.6 g cm^{-3}.

76. Gold costs $400 per troy ounce, and one troy ounce equals 31.1035 g. Determine the cost of 10.0 cm^3 of gold if the density of gold is 19.32 g cm^{-3} at room conditions.

77. Lutetium has a density of 9.84 g cm^{-3} under room conditions. Calculate the volume per atom in lutetium.

78. Lithium has a density of 0.534 g cm^{-3} under room conditions. Calculate the volume per atom in lithium.

ADDITIONAL PROBLEMS

79. A schoolbook asks pupils to classify steel wool and sawdust (among other items) as solid, liquid, or gas. The official answer is that both are solids. Several pupils reject the official answer. They say, "Solids have no holes," "Solids are not hollow," "You can squeeze down a wad of steel wool, just

like air in a balloon," "You can pour sawdust, just like water." Identify the source of the misunderstanding and justify the official answer in words a schoolchild would understand.

80. In the 1640s, the natural philosopher van Helmont filled a large pot with 200 lb of soil that had been dried in a furnace. He moistened the soil with rainwater and planted a small willow tree weighing 5 lb. He tended the tree for five years, adding only rainwater or distilled water and taking care to keep out dust and other solids. He then uprooted the tree, separated it from the soil, and found that it weighed 169 lb and approximately 3 oz. He recovered and dried all of the soil used in the experiment and found it to weigh the same 200 lb he had started with, less approximately 2 oz. He then reported that "164 lb of wood, bark, and roots arose out of water only."
(a) Is this conclusion valid? Why or why not?
(b) Was van Helmont using the law of conservation of matter? Explain.

81. A naturopath swears by the merit of rose-hip tea (which contains much vitamin C) in warding off colds. This person contends that laboratory-synthesized vitamin C is ineffective in this role. Explain how the naturopath's contention could be correct.

82. Raisin Bran cereal contains bran flakes and raisins. Crispix cereal contains double-layered biscuits, each of which is rice on one side and corn on the other. Which is more analogous to a mixture and which to a compound? Explain.

83. Soft-wood chips having a mass of 17.2 kg are placed in an iron vessel and mixed with 150.1 kg of water and 22.43 kg of sodium hydroxide. A steel lid seals the vessel, which is then placed in an oven at 250°C for 6 hours. Much of the wood fiber decomposes under these conditions; the vessel and lid do not react.
(a) Classify each of the materials mentioned as a substance or mixture. Subclassify all substances as elements or compounds.
(b) Determine the mass of the contents of the iron vessel after the reaction.

84. A binary compound of nickel and oxygen contains 78.06% nickel by mass. Is this a stoichiometric or a non-stoichiometric compound? Explain.

85. Krulls weigh 4.80 times more than grebes, and grebes possess 0.825 the mass of stoats. The lightest of these objects weighs 1.000 g. How much does a collection of 17 krulls, 19 grebes, and 11 stoats weigh?

86. Dalton stated a *law of multiple proportions:* "When two elements form more than one compound, the masses of the second that combine with a fixed mass of the first have the ratio of small whole numbers." Explain how this law is a logical consequence of the postulates of Dalton's atomic theory.

87. Dalton's 1808 version of the atomic theory of matter was based on five postulates (Section 1–3). According to mod-

ern understanding, four of these require amendment or extension. List the modifications that have been made to four of the five original postulates.

88. Suppose that the relative atomic mass of oxygen were 8 times larger than the relative atomic mass of hydrogen, instead of approximately 16 times larger.
(a) Determine the empirical formula of water (assume that the elemental analysis of water is unchanged).
(b) Hydrogen peroxide is 5.93% H and 94.07% O by mass. What would its formula be?

89. A chemistry student suggests that the formula of water is H_4O_2, that the formula of hydrogen is H_4, and that the formula of oxygen is O_4. Is this suggestion consistent with Gay-Lussac's observations on the combining volumes of hydrogen and oxygen? Explain.

90. Sir Humphrey Davy discovered several chemical elements but always insisted that any identification of a material as an element was tentative.
(a) Explain how, in logic, the negative nature of the definition of a chemical element requires this approach.
(b) In what sense did the discovery of isotopic forms of the chemical elements justify Davy's caution?

91. Synthetic zircon, a gemstone, consists of a single compound. One such gem contains 5.474 g of zirconium (Zr), 1.685 g of silicon (Si), and 3.840 g of oxygen (O). Determine the mass of zirconium in a second zircon that has a mass of 43.67 g.

92. You take a breath and inhale 1.16×10^{22} molecules of air, a mixture of several substances. Assume that the effective molar mass of air is 29.0 g mol^{-1}. Compute the mass of the air that you inhaled.

93. (a) Compute to five significant figures the average mass in grams of a molecule of UF_6.
(b) Paradoxically, actual molecules of UF_6 never have exactly the mass that is the correct answer to part (a). Explain why.

94. Identify what is wrong or ambiguous about each of the following statements.
(a) The molar mass of H_2O is 18.01.
(b) 1.0 mol of bromine reacted.
(c) The proportion of oxygen to hydrogen in H_2O is $2:1$.
(d) The mass of one atom of ^{12}C is 12.0000 g exactly.
(e) A mole of water contains 6.022×10^{23} atoms.

95. Oxygen has three naturally occurring isotopes. In a sample of oxygen taken from the sea, the ratios of the numbers of atoms are

$$N(^{16}O)/N(^{18}O) = 498.7/1$$
$$N(^{17}O)/N(^{18}O) = 0.1832/1$$

On the basis of these data, how many out of every 10 million oxygen atoms are ^{16}O, ^{17}O, and ^{18}O atoms?

96. Suppose that 1.00 mol of "doped" silicon is prepared for use in a specialized electronic device by adding germanium to hyperpure silicon until the final product contains $0.62 \times 10^{-7}\%$ germanium on the basis of mass. Estimate (to two significant digits) the number of atoms of germanium present in this sample.

*97. The inhabitants of a planet in a distant galaxy measure mass in units of "margs," where 1 marg = 4.8648 g. Their scale of atomic masses is based on the isotope ^{32}S (atomic mass on Earth = 31.972), so they define one "elom" of ^{32}S as the amount of sulfur in exactly 32 margs of ^{32}S. Furthermore, they define N_{or}, or "Ordagova's number" (after the well-known scientist Oedema Ordagova), as the number of atoms of sulfur in exactly 32 margs of ^{32}S.
  (a) Calculate the numerical value of N_{or} to four significant digits.
  (b) Calculate the atomic mass of phosphorus in margs per elom to five significant digits, assuming that the abundances of phosphorus isotopes are the same as those found on Earth.

98. Under room conditions, gaseous argon has a density of 1.64 g L^{-1}.
  (a) Compute the volume (in liters) occupied by 1.00 mol of gaseous argon under these conditions.
  (b) The density of liquid argon at its normal boiling point is 1.40 g cm^{-3}. Compute the volume (in liters) occupied by 1.00 mol of liquid argon.
  (c) Assume that no vacant space exists between the atoms in liquid argon. Estimate the percentage of the volume of gaseous argon that is empty space at room conditions.

99. Iridium has a density of 22.56 g cm^{-3}, and hydrogen has a density of $8.24 \times 10^{-5} \text{ g cm}^{-3}$, both at room conditions. In the following graph, identify which line corresponds to hydrogen and which to iridium. Explain your choice.

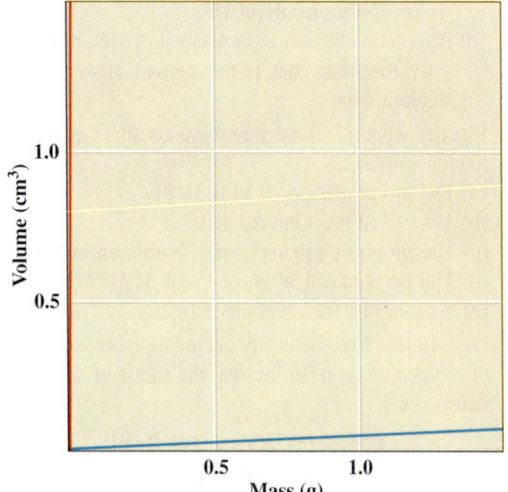

100. Sodium chloride (NaCl) has a density of 2.165 g cm^{-3} at 25°C. Calculate the number of atoms in 1.000 cm^3 of sodium chloride at 25°C.

101. Osmium(VIII) oxide (OsO_4) is a volatile crystalline solid. It is hazardous to the eyes because it vaporizes readily (which means it can get into the eye) and reacts with organic matter (the fabric of the eye itself) to leave an opaque, black, permanent stain of osmium. Compute the mass of osmium in 17.6 mg of OsO_4.

102. A binary compound between erbium and boron is difficult to prepare and analyze. It is reported, however, that it has the chemical formula $ErB_{65 \pm 4}$. Compute the range of the mass percentage of erbium in this compound.

103. At one time, it was thought that indium formed a chloride of formula $InCl_2$. More recent work shows that the compound in question is actually $In_3[In_2Cl_9]$. Determine the percentage by mass of indium according to each formula and explain how the error could be made.

104. Djenkolic acid is a sulfur-containing amino acid that is isolated from the djenkol and other beans and is possibly implicated in flatulence. It is 33.06% carbon, 5.55% hydrogen, 11.01% nitrogen, 25.16% oxygen, and 25.22% sulfur by mass. Each molecule contains two atoms of sulfur. Determine the molar mass of djenkolic acid and its molecular formula.

105. A binary compound between potassium and cesium is stable at temperatures below −90°C but otherwise decomposes to the elements. A 0.03151 g sample of the compound yields 0.02346 g of cesium. Determine the empirical formula of this compound. (*Hint*: Take particular care with significant figures.)

*106. A dark brown binary compound contains oxygen and a metal. It is 13.38% oxygen by mass. Heating it moderately drives off some of the oxygen and gives a red binary compound that is 9.334% oxygen by mass. Strong heating drives off more oxygen and gives still another binary compound that is only 7.168% oxygen by mass.
  (a) Compute the mass of oxygen that is combined with 1.000 g of the metal in each of these three oxides.
  (b) Assume that the empirical formula of the first compound is MO_2 (where M stands for the metal). Give the empirical formulas of the second and third compounds.
  (c) Name the metal.

107. A chemist finds that 0.7200 g of barium reacts when heated with a large excess of oxygen to form 0.8878 g of a solid compound.
  (a) What is the empirical formula of the compound?
  (b) Write a balanced equation for the reaction that forms the compound.

CUMULATIVE PROBLEM

Gallium Sulfides

Analysis of several samples of materials that are known to contain no elements other than gallium (Ga) and sulfur (S) gave these results:

Sample	Description	Mass of Ga (g)	Mass of S (g)	Total Mass (g)	Volume of Sample (cm³)
1	Yellow crystals	1.3509	0.9319	2.2828	0.625
2	Yellow crystals	0.6289	0.2892	0.9181	0.238
3	Gray powder	0.6441	0.1481	0.7922	0.189
4	Yellow crystals	0.2228	0.1537	0.3765	0.103
5	Yellow crystals	0.4949	0.2276	0.7225	0.187

(a) Each sample is a chemical compound, not a mixture. Show that the mass data are consistent with this fact.

(b) Determine the minimum number of different compounds present.

(c) Assume that the formula of the compound in sample 1 is GaS. Give possible formulas of the compounds in samples 2 through 5.

(d) Assume that the formula of the compound in sample 1 is Ga_2S_3. Give possible formulas of the compounds in samples 2 through 5.

(e) If the relative atomic mass of sulfur is taken to be 32.07 and the formula of the compound in sample 1 is taken to be GaS, then what is the relative atomic mass of gallium?

(f) The correct relative atomic masses of sulfur and gallium are 32.07 and 69.72, respectively. Determine the formulas of the compounds in samples 1 through 5.

(g) Sulfur occurs naturally in four isotopes: ^{32}S, ^{33}S, ^{34}S, and ^{36}S; gallium occurs naturally in two isotopes: ^{69}Ga and ^{71}Ga. Which isotope of sulfur is the most abundant; which isotope of gallium is more abundant?

(h) Use Avogadro's number and the correct relative atomic masses of sulfur and gallium to determine the number of atoms of gallium in sample 1, the number of atoms of sulfur in sample 1, and the ratio of these two numbers.

(i) Calculate the densities of the five samples.

(j) Calculate the average volume (in cubic centimeters) occupied by an atom in each of the samples.

2 Stoichiometry

CHAPTER OUTLINE

The smelting of copper involves a series of chemical reactions to convert the ore into elemental copper. Here, molten copper streams from an oven. (*Day Williams/Photo Researchers, Inc.*)

Chemistry became a true science only when it became quantitative. **Stoichiometry** is the study of quantitative mass relationships in chemical change; it is the study of *how much* reacts and *how much* is produced in chemical reactions. This immensely practical subject underlies chemical processes at all scales, from the largest industrial operations to the subtle transformations by which biological cells live and reproduce.

• The word "stoichiometry" comes from the Greek *stoicheion*, meaning "element," + *metron*, meaning "measure."

2-1 Writing Balanced Chemical Equations

Chemical reactions are events in which elements combine into chemical compounds, compounds decompose back into elements, and existing compounds transform into new compounds. One way to represent a chemical event is with a **word equation.** For example, the burning (combustion) of the fuel butane (Fig. 2–1) to give carbon dioxide and water can be described in these words:

"Gaseous butane and gaseous oxygen react to yield gaseous carbon dioxide and liquid water."

Figure 2–1 • Butane burns in a camp stove. It also fuels disposable cigarette lighters. *(Charles D. Winters)*

TABLE 2-1	
Some Symbols and Abbreviations Used in Chemical Equations	
$\longrightarrow$	Yields (from left to right)
$\longleftarrow$	Yields (from right to left)
$\rightleftharpoons$	Reaction may proceed in reverse as well as in forward direction.
$+$	"Reacts with" or "is formed with"
(g)	The substance is in the gaseous state or vapor state.
(ℓ)	The substance is in the liquid state.
(fl)	The substance is in a fluid state (either gas or liquid).
(s)	The substance is in the solid state.
(cr)	The substance is crystalline.
(aq)	The substance is dissolved in water (in aqueous solution).
(sln)	The substance is in solution.

This sentence becomes more compact if some of the symbols and abbreviations collected in Table 2–1 are employed:

$$\text{butane}(g) + \text{oxygen}(g) \longrightarrow \text{carbon dioxide}(g) + \text{water}(\ell)$$

With these changes, the word equation begins to look like an algebraic equation. An important difference is the directionality imposed by using an arrow ($\longrightarrow$) instead of an equals sign ($=$). The arrow points from the **reactants,** the starting substances, to the **products,** the ending substances. Exchanging the two sides of the preceding word equation gives its **reverse** by making the products into the reactants and the reactants into the products

$$\text{carbon dioxide}(g) + \text{water}(\ell) \longrightarrow \text{butane}(g) + \text{oxygen}(g)$$

whereas exchanging the sides of an algebraic equation does nothing. Reversing the arrow to point from right to left also signifies a reversal

$$\text{butane}(g) + \text{oxygen}(g) \longleftarrow \text{carbon dioxide}(g) + \text{water}(\ell)$$

Reversal of reactions is common.

Word equations do not provide quantitative information. The preceding statements offer no help in finding the mass of oxygen that reacts with 30 g of butane or the mass of carbon dioxide produced as the reaction generates 100 g of water. Obtaining such information requires more development. First, we put chemical formulas into the equation in place of the names of the substances. These formulas must represent the substances as they exist and take part in the reaction. For example, we must use the formula O_2 to represent diatomic oxygen, not O or O_3, which are chemical formulas of different substances with different chemical and physical properties. Taking this step with the word equation for the burning of butane gives the following **chemical equation:**

$$C_4H_{10}(g) + O_2(g) \longrightarrow CO_2(g) + H_2O(\ell)$$

This chemical equation has a serious defect: It represents *unequal* amounts of all three chemical elements on the left and right sides, specifically:

- 4 mol of C on the left, but 1 mol of C on the right;
- 10 mol of H on the left, but 2 mol of H on the right;
- 2 mol of O on the left, but 3 mol of O on the right.

The equation is an *unbalanced* equation. This is a concern because atomic theory and the law of conservation of mass require a **material balance** for each element. The two sides of a *balanced* equation represent equal numbers of moles of each element.

    We must not try to fix the problem by changing the subscripts in one or more of the chemical formulas. Doing this would change the chemical identity of a reactant or product, which is clearly a mistake because the identities of the reactants and products are the starting facts of the whole exercise. However, nothing prevents us from inserting multiplying coefficients in front of some or all of the formulas. In fact:

> Balancing a chemical equation consists of inserting coefficients in such a way that the same number of each type of atom is shown on each side of the arrow.

- It is admittedly a contradiction in terms to refer to an "unbalanced equation," but the practice is well entrenched in chemistry and causes no difficulties.

Balancing Equations by Inspection

How do we find a good set of coefficients? We begin by setting 1 as the coefficient of one of the formulas (usually the one that contains the most different elements). In the equation for the combustion of butane, that is the formula C_4H_{10}. Next, we scan (inspect) both sides of the equation for elements that appear in only one formula for which the coefficient is unassigned. As we find them, we insert coefficients that achieve the desired material balance for that element.

    In the combustion of butane, both C and H appear in only one reactant and one product. We obtain the required balance for C if one mole of butane produces four moles of CO_2; we obtain balance for H if one mole of butane produces five moles of H_2O. Inserting these coefficients gives

$$1\ C_4H_{10}(g) + \underline{\ \ } O_2(g) \longrightarrow 4\ CO_2(g) + 5\ H_2O(\ell)$$

The underscore in front of the O_2 emphasizes that the coefficient that goes there remains to be determined. The 1, 4, and 5 are locked in.

    To determine the final coefficient, we observe that 4 moles of CO_2 contains 8 moles of O and that 5 moles of H_2O contains 5 moles of O, giving a total of 13 moles of O on the product side. Thirteen moles of O is provided by $\frac{13}{2}$ moles of O_2 (6.5 mol O_2). This is confirmed by noting that the coefficient $\frac{13}{2}$ multiplied by the subscript 2 equals 13. Hence, the required coefficient of O_2 is $\frac{13}{2}$.

$$1\ C_4H_{10}(g) + \tfrac{13}{2}\ O_2(g) \longrightarrow 4\ CO_2(g) + 5\ H_2O(\ell)$$

The equation is now balanced. Fractions such as $\frac{13}{2}$ are perfectly meaningful as coefficients in chemical equations. They represent fractions of moles. However, fractions must be eliminated if the equation is to be interpreted on the microscopic (atomic or molecular) level, as explained in Section 2-2. Here, multiplying all coefficients by 2 gives

$$2\ C_4H_{10}(g) + 13\ O_2(g) \longrightarrow 8\ CO_2(g) + 10\ H_2O(\ell)$$

The equation remains balanced when all of the coefficients are doubled. It represents the same molar *ratio* among its reactants and products. Doubling the coefficients simply means the consumption of double quantities of reactants and the formation of double quantities of products. Balance is maintained if the coefficients are doubled, tripled, or even multiplied by 1000, as long as all of the coefficients are treated the same way. In summary, to balance an equation:

1. Assign 1 as the coefficient of one reactant or product. The best choice is usually the most complicated formula, with the largest number of different elements.
2. Identify an element that appears in only one chemical formula for which the coefficient is not yet determined. Assign that coefficient so as to balance the number of moles of atoms of that element.
3. Repeat for other elements until all coefficients are assigned.
4. It is often desirable to eliminate fractional coefficients. To do so, multiply all of the coefficients by the smallest integer that eliminates the fractions.

This method succeeds in a majority of practical cases, but becomes unworkable when the chemical equation involves several complex species. Techniques (soon to be explained) exist to balance even very complex chemical equations, but it is always wise to try to obtain a balance by inspection.

EXAMPLE 2-1

The Hargreaves process is an industrial procedure for producing sodium sulfate (Na_2SO_4) for use in papermaking. The starting materials are sodium chloride (NaCl), sulfur dioxide (SO_2), water (H_2O), and oxygen (O_2); hydrogen chloride (HCl) is generated as a by-product. Write a balanced chemical equation to represent the process.

Strategy

Write an unbalanced equation and use the method just outlined. Omit mention of the physical states of the reactants and products because they are not stated and are not needed in the procedure of balancing.

Solution

The unbalanced equation is

$$\_\_ NaCl + \_\_ SO_2 + \_\_ H_2O + \_\_ O_2 \longrightarrow \_\_ Na_2SO_4 + \_\_ HCl$$

where the underscores in front of the formulas hold places for the coefficients. Begin by putting a coefficient of 1 in front of any one of the formulas. The product Na_2SO_4 is a reasonable choice because it is the most complex species. Assigning this 1 means that there are two moles of sodium atoms on the right. Balancing this requires the same number on the left, giving 2 as the coefficient of the NaCl. By a similar argument, the coefficient of the SO_2 must be 1 to balance the one mole of sulfur on the right. This gives

$$2\,NaCl + 1\,SO_2 + \_\_ H_2O + \_\_ O_2 \longrightarrow 1\,Na_2SO_4 + \_\_ HCl$$

This equation shows two moles of chlorine atoms on the left-hand (reactant) side, so the coefficient of the HCl on the right must be 2. Hydrogen is the next element to balance. With two moles of hydrogen atoms on the right-hand side, the coefficient for the H_2O must be 1:

$$2\,NaCl + 1\,SO_2 + 1\,H_2O + \_\_ O_2 \longrightarrow Na_2SO_4 + 2\,HCl$$

Finally, the oxygen atoms must be balanced. Four moles of oxygen atoms appear on the right; the left shows two moles from the SO_2 and one mole from the H_2O. Therefore, one mole of oxygen atoms must come from the O_2. The coefficient of the O_2 equals $\frac{1}{2}$, because $\frac{1}{2}$ mole of O_2 contains one mole of O:

$$2\, NaCl + 1\, SO_2 + 1\, H_2O + \tfrac{1}{2}\, O_2 \longrightarrow 1\, Na_2SO_4 + 2\, HCl$$

Multiplying all coefficients in the equation by 2 eliminates the fraction:

$$4\, NaCl + 2\, SO_2 + 2\, H_2O + O_2 \longrightarrow 2\, Na_2SO_4 + 4\, HCl$$

Note that in balancing the equation, we considered oxygen last because it appears in several places on the left-hand side of the equation.

Check

Construct a table to confirm material balance for each element.

Element	Reactant Side	Product Side
Na	$4 \times 1 = 4$ mol	$2 \times 2 = 4$ mol
Cl	$4 \times 1 = 4$ mol	$4 \times 1 = 4$ mol
S	$2 \times 1 = 2$ mol	$2 \times 1 = 2$ mol
O	$2 \times 2 + 2 \times 1 + 1 \times 2 = 8$ mol	$2 \times 4 = 8$ mol
H	$2 \times 2 = 4$ mol	$4 \times 1 = 4$ mol

EXERCISE

Gaseous isocyanic acid (HNCO) reacts with gaseous nitrogen monoxide (NO), an air pollutant, to form nitrogen, carbon dioxide, and water (all as gases). The reaction may offer a feasible means to eliminate nitrogen monoxide from automobile exhaust. Write a balanced chemical equation to represent the reaction.

Answer: $4\, HNCO(g) + 6\, NO(g) \longrightarrow 5\, N_2(g) + 2\, H_2O(g) + 4\, CO_2(g)$.

Balancing Equations Algebraically

Some chemical equations cannot be balanced by inspection. In such cases, we resort to an algebraic method that gives good sets of coefficients for all valid chemical equations. In this method, coefficients that cannot be assigned by inspection are represented by symbols. The requirement of material balance is then used to generate equations relating these symbolic coefficients—one equation for each chemical element appearing in the equation. These equations compose a **system of linear equations** (see Appendix C–3). The coefficients are determined by solving the system of equations.

Consider the problem of balancing the equation

$$\_\, Cu(s) + \_\, HNO_3(aq) \longrightarrow \_\, Cu(NO_3)_2(aq) + \_\, NO_2(g) + \_\, H_2O(\ell)$$

We assign a 1 as the coefficient of $Cu(NO_3)_2(aq)$ (the most complicated formula). The coefficient of $Cu(s)$ also equals 1, based on material balance for Cu. Progress by inspection now stalls because neither H nor N nor O appears in just one chemical formula with an undetermined coefficient. We insert symbols (x, y, and z) as the coefficients of the other three formulas

$$1\, Cu(s) + x\, HNO_3(aq) \longrightarrow 1\, Cu(NO_3)_2(aq) + y\, NO_2(g) + z\, H_2O(\ell)$$

Next, we figure out, in terms of the symbols that represent the coefficients, the amounts of each element appearing on the two sides of the equation. The results are

Element	Reactant Side	Product Side
Cu	$1 \times 1 = 1$ mol	$1 \times 1 = 1$ mol
H	$1 \times x = x$ mol	$2 \times z = 2z$ mol
N	$1 \times x = x$ mol	$(2 \times 1) + (1 \times y) = (2 + y)$ mol
O	$3 \times x = 3x$ mol	$(6 \times 1) + (2 \times y) + (1 \times z) = (6 + 2y + z)$ mol

Material balance demands equal amounts of each element on reactant side and product side, so

$$
\begin{aligned}
1 &= 1 && \text{(material balance for Cu)} \\
x &= 2z && \text{(material balance for H)} \\
x &= (2 + y) && \text{(material balance for N)} \\
3x &= (6 + 2y + z) && \text{(material balance for O)}
\end{aligned}
$$

The material-balance equations for H, N, and O comprise a system of three linear equations in three unknowns. The system has a unique solution:

$$x = 4 \qquad y = 2 \qquad z = 2$$

Appendix C–3 reviews how to obtain this solution. Substitution of the answers for x, y, and z gives the balanced equation

$$1\,\text{Cu}(s) + 4\,\text{HNO}_3(aq) \longrightarrow 1\,\text{Cu(NO}_3)_2(aq) + 2\,\text{NO}_2(g) + 2\,\text{H}_2\text{O}(\ell)$$

Here is a step-by-step summary of the algebraic method for balancing equations:

1. Assign 1 as the coefficient of one reactant or product. As before, the best choice is usually the formula with the largest number of different elements.
2. Assign as many coefficients as possible by inspection.
3. Insert *symbolic coefficients* (x, y, z, ...) to represent the coefficients that cannot be determined by inspection.
4. Use the condition of material balance on each element to write algebraic equations relating the symbolic coefficients.
5. Solve this system of linear equations (use a calculator if necessary).
6. The resulting answers are often fractions. Clear fractions, if desired, by multiplying all coefficients by the smallest integer that eliminates fractions.

Paper-and-pencil balancing of chemical equations by the algebraic method often becomes laborious when more than five unknown coefficients must be determined. Fortunately, many models of hand-held calculators are preprogrammed to solve large systems of linear equations (for example, up to 30 equations in 30 unknowns). This is plenty of capacity because chemical equations rarely have more than seven or eight unknown coefficients. With a suitable calculator, the coefficients can be determined as fast as the material-balance equations can be written down and punched in (see Appendix C–3).

EXAMPLE 2-2

Balance the chemical equation

$$\_\_\,Al(s) + \_\_\,NaNO_3(aq) + \_\_\,NaOH(aq) + \_\_\,H_2O(\ell) \longrightarrow$$
$$\_\_\,NaAl(OH)_4(aq) + \_\_\,NH_3(g)$$

Solution

Set the coefficient of $NaAl(OH)_4$ equal to 1. Inspection shows that the coefficient of Al is then also 1. Further progress by inspection is not possible, so insert symbols for the other four coefficients

$$1\,Al(s) + x\,NaNO_3(aq) + y\,NaOH(aq) + z\,H_2O(\ell) \longrightarrow$$
$$1\,NaAl(OH)_4(aq) + w\,NH_3(g)$$

Construct a table giving the chemical amounts of each element in the reactants and in the products:

Element	Reactant Side	Product Side
Al	1 mol	1 mol
Na	$x + y$ mol	1 mol
N	x mol	w mol
O	$3x + y + z$ mol	4 mol
H	$y + 2z$ mol	$4 + 3w$ mol

Reading from the table, the material-balance equations (one for each element) are

$$1 = 1 \qquad \text{(material balance for Al)}$$
$$1x + 1y + 0z - 0w = 1 \qquad \text{(material balance for Na)}$$
$$1x + 0y + 0z - 1w = 0 \qquad \text{(material balance for N)}$$
$$3x + 1y + 1z - 0w = 4 \qquad \text{(material balance for O)}$$
$$0x + 1y + 2z - 3w = 4 \qquad \text{(material balance for H)}$$

The first equation contains no unknowns and can be disregarded. The remaining four equations relate the four unknown coefficients. The organized layout (with the x's in the first column, the y's in the second, and so on) makes it much easier to enter the equations into a calculator (see Appendix C–3). Details of entry of course vary with the make and model of calculator. The system of equations is also easily solved by hand (see Appendix C–3) to give

$$x = 0.375 \qquad y = 0.625 \qquad z = 2.25 \qquad w = 0.375$$

Replacing the symbolic coefficients with these answers gives

$$1\,Al(s) + 0.375\,NaNO_3(aq) + 0.625\,NaOH(aq) + 2.25\,H_2O(\ell) \longrightarrow$$
$$1\,Al(OH)_4(aq) + 0.375\,NH_3(g)$$

The equation is balanced. Multiplying through by 8 clears the decimal fractions:

$$8\,Al(s) + 3\,NaNO_3(aq) + 5\,NaOH(aq) + 18\,H_2O(\ell) \longrightarrow$$
$$8\,NaAl(OH)_4(aq) + 3\,NH_3(g)$$

Check

The final equation represents 8 mol Al, 8 mol Na, 3 mol N, 32 mol O, and 41 mol H on each side.

EXERCISE

Balance the equation

$$\_\,Cu(s) + \_\,HNO_3(aq) \longrightarrow \_\,Cu(NO_3)_2(aq) + \_\,NO(g) + \_\,H_2O(\ell)$$

Answer: $3\,Cu(s) + 8\,HNO_3(aq) \longrightarrow 3\,Cu(NO_3)_2(aq) + 2\,NO(g) + 4\,H_2O(\ell)$

Chemical Equations That Cannot Be Uniquely Balanced

The algebraic method quickly identifies chemical equations that are *invalid* for use in stoichiometry. In Step 4, the number of unknown coefficients may *exceed* the number of material-balance equations that can be written to relate them (one equation per chemical element). If so, the equation cannot be uniquely balanced. Chemical equations of this type result when a single equation represents two or more separable reactions taking place concurrently among the same reactants. For example, the combustion of carbon in oxygen under most conditions simultaneously produces both carbon dioxide (CO_2) and carbon monoxide (CO) (Fig. 2–2). It seems reasonable to represent this fact with the statement

$$C(s) + O_2(g) \longrightarrow CO_2(g) + CO(g)$$

After a 1 is assigned as the coefficient for one of the four formulas, three coefficients remain undetermined. But only two material-balance conditions (one for C and one for O) exist. The equation is easy to balance … *too* easy! An infinite number of nonequivalent sets of coefficients work (see Appendix C–3). Some examples are

$$4\,C(s) + 3\,O_2(g) \longrightarrow 2\,CO(g) + 2\,CO_2(g)$$
$$3\,C(s) + 2\,O_2(g) \longrightarrow 2\,CO(g) + CO_2(g)$$
$$8\,C(s) + 7\,O_2(g) \longrightarrow 2\,CO(g) + 6\,CO_2(g)$$

Figure 2–2 • The combustion of the carbon in the charcoal briquettes in a barbeque grill produces CO(g) and CO_2(g) concurrently. Representing the process requires two chemical equations. *(Charles D. Winters)*

Balancing an invalid equation is a waste of time because the outcome is useless for stoichiometric calculations. The chemical situation in which carbon burns to give both $CO(g)$ and $CO_2(g)$ is best represented by two balanced equations:

$$1 \; C(s) + 1 \; O_2(g) \longrightarrow 1 \; CO_2(g) \quad \text{and} \quad 2 \; C(s) + O_2(g) \longrightarrow 2 \; CO(g)$$

because two different reactions are taking place at the same time.

• Inadvertent invalid equations pop up fairly often in worksheets, examinations, and textbooks (see Problem 2–69).

The challenge for beginners in balancing equations is correctly interpreting chemical formulas. Once formulas are thoroughly understood, writing material-balance equations becomes routine, and the balancing of equations turns into a standard mathematical task (solving a set of simultaneous linear equations). This is so precisely because the balancing of chemical equations is a routine process of accounting once the reactants and products are known. The really interesting part (the part where chemistry comes in) is to know what substances react with each other and to determine what products are formed. This is a question to which we return many times throughout this book.

2-2 Using Balanced Chemical Equations

The balanced chemical equation for the combustion of butane

$$2 \; C_4H_{10}(g) + 13 \; O_2(g) \longrightarrow 8 \; CO_2(g) + 10 \; H_2O(\ell)$$

gives the ratios of the chemical amounts of the reactants as they are consumed and of the two products as they are produced:

$$2 \; \text{mol} \; C_4H_{10} + 13 \; \text{mol} \; O_2 \longrightarrow 8 \; \text{mol} \; CO_2 + 10 \; \text{mol} \; H_2O$$

This balanced equation, like all others, remains balanced if the coefficients are all multiplied by (or divided by) the same number. Neither operation alters the molar ratios. Dividing each coefficient by Avogadro's number, N_0, gives

$$\frac{2}{N_0} \; \text{mol} \; C_4H_{10} + \frac{13}{N_0} \; \text{mol} \; O_2 \longrightarrow \frac{8}{N_0} \; \text{mol} \; CO_2 + \frac{10}{N_0} \; \text{mol} \; H_2O$$

which is still balanced. Now, recall that a mole of C_4H_{10} or O_2 or CO_2 or H_2O (or any other molecular substance) contains Avogadro's number of molecules. We therefore may replace "mol" with "N_0 molecules" everywhere in the equation

$$\frac{2}{\cancel{N_0}} \; \cancel{N_0} \; \text{molecules} \; C_4H_{10} + \frac{13}{\cancel{N_0}} \; \cancel{N_0} \; \text{molecules} \; O_2 \longrightarrow$$

$$\frac{8}{\cancel{N_0}} \; \cancel{N_0} \; \text{molecules} \; CO_2 + \frac{10}{\cancel{N_0}} \; \cancel{N_0} \; \text{molecules} \; H_2O$$

The N_0's in each coefficient cancel out as shown to give

$$2 \; \text{molecules} \; C_4H_{10} + 13 \; \text{molecules} \; O_2 \longrightarrow 8 \; \text{molecules} \; CO_2 + 10 \; \text{molecules} \; H_2O$$

Division by N_0 has scaled down the original statement about relative numbers of moles to a statement about relative numbers of molecules. The point is completely general. Chemical equations may be interpreted on two levels:

- A *macroscopic* (laboratory) level, in which the coefficients are taken to refer to relative numbers of *moles* of substances; and
- A *microscopic* (atomic or molecular) level, in which the coefficients are taken to refer to relative numbers of *molecules*.

• For non-molecular substances, the coefficients refer on the microscopic level to relative numbers of formula units. See Section 1–8.

Chemists switch back and forth freely between the two levels, as circumstances dictate.

Mass Relationships in Chemical Reactions

A balanced, valid chemical equation provides a platform for a variety of quantitative statements about the relative amounts of the substances reacting and forming. If we compute the molar masses of the different substances in the reaction and then multiply each by the number of moles represented in the balanced equation, we obtain the mass of each substance involved in the reaction. For the combustion of butane:

$$m_{C_4H_{10}} = 2 \text{ mol } C_4H_{10} \times \left(\frac{58.12 \text{ g } C_4H_{10}}{1 \text{ mol } C_4H_{10}} \right) = 116.2 \text{ g } C_4H_{10}$$

$$m_{O_2} = 13 \text{ mol } O_2 \times \left(\frac{32.00 \text{ g } O_2}{1 \text{ mol } O_2} \right) = 416.0 \text{ g } O_2$$

$$m_{CO_2} = 8 \text{ mol } CO_2 \times \left(\frac{44.01 \text{ g } CO_2}{1 \text{ mol } CO_2} \right) = 352.1 \text{ g } CO_2$$

$$m_{H_2O} = 10 \text{ mol } H_2O \times \left(\frac{18.015 \text{ g } H_2O}{1 \text{ mol } H_2O} \right) = 180.1 \text{ g } H_2O$$

The equation then becomes, in terms of masses,

$$116.2 \text{ g of } C_4H_{10} + 416.0 \text{ g of } O_2 \longrightarrow 352.1 \text{ g of } CO_2 + 180.1 \text{ g of } H_2O$$

Note that the total mass of the products, 532.2 g, equals the total mass of the reactants, as it must under the law of conservation of mass.

A balanced chemical equation thus gives quantitative relationships among the masses of reactants and products. These "mass–mass" relationships are at the heart of stoichiometry. The preceding analysis explicitly established the masses of CO_2 and C_4H_{10} that are equivalent in the reaction: 352.1 g of CO_2 is produced for every 116.2 g of C_4H_{10} consumed. We can express this equivalence by means of these two unit-factors:

- Recall that a unit-factor is a fraction in which the numerator (top) is equivalent to the denominator (bottom). See Appendix B–2 for a discussion of the unit-factor method.

$$\left(\frac{352.1 \text{ g } CO_2}{116.2 \text{ g } C_4H_{10}} \right) = 1 \quad \text{and} \quad \left(\frac{116.2 \text{ g } C_4H_{10}}{352.1 \text{ g } CO_2} \right) = 1$$

just as we express the equivalence of 1 km and 1000 m by the unit-factors (1 km/1000 m) = 1 and (1000 m/1 km) = 1. Similar unit-factors can be constructed for every pair of substances taking part in this reaction (for a total of 12 unit-factors). Suppose that we need to know the mass of CO_2 that results when, say, 453.6 g (a pound) of C_4H_{10} burns according to this equation. (An environmental engineer might require this information to compare emissions of CO_2 from burning different fuels.) We multiply the 453.6 g of C_4H_{10} by one of the preceding unit-factors. The choice of which one is guided by observing the way that the units cancel out (shown in magenta):

$$m_{CO_2} = 453.6 \text{ g } C_4H_{10} \times \left(\frac{352.1 \text{ g } CO_2}{116.2 \text{ g } C_4H_{10}} \right) = 1374 \text{ g } CO_2$$

Usually it is not necessary to work out an explicit mass-to-mass unit-factor like the one just used. Instead, computations can be set up by stringing two or three other unit-factors together. The preceding computation could be conducted as a string of three unit-factors:

$$m_{CO_2} = 453.6 \text{ g } C_4H_{10} \times \left(\frac{1 \text{ mol } C_4H_{10}}{58.12 \text{ g } C_4H_{10}} \right) \times \left(\frac{8 \text{ mol } CO_2}{2 \text{ mol } C_4H_{10}} \right) \times \left(\frac{44.01 \text{ g } CO_2}{1 \text{ mol } CO_2} \right)$$

$$= 1374 \text{ g } CO_2$$

Figure 2–3 • The steps in a stoichiometric calculation. In a typical calculation, the mass of one reactant or product is known, and the masses of one or more other reactants or products are to be calculated, using the balanced chemical equation and a table of relative atomic masses.

The first unit-factor converts from grams of C_4H_{10} to moles of C_4H_{10}. The second converts from moles of C_4H_{10} to moles of CO_2, using the coefficients assigned to the two substances in the balanced equation. The third converts from moles of CO_2 to grams of CO_2. The cancellation of units, indicated in magenta, confirms the correctness of every step. Figure 2–3 summarizes this pattern, which applies in all mass–mass stoichiometric calculations.

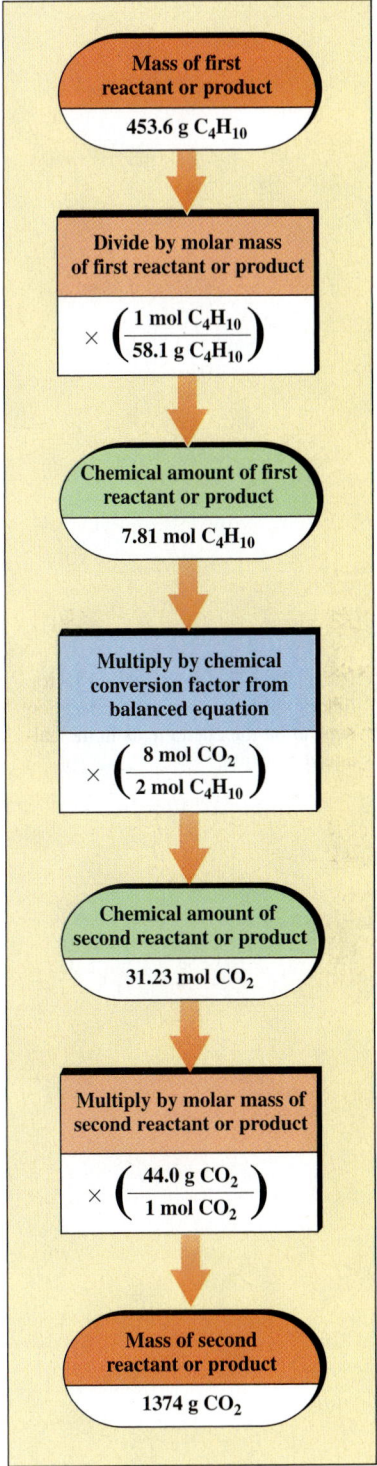

EXAMPLE 2–3

Boron carbide (B_4C) is a very hard, low-density solid with a high melting point. These properties make it useful in armor plates and other surfaces subject to wear and abrasion. Boron carbide is made by the reaction of boron oxide with carbon:

$$2\,B_2O_3 + 4\,C \longrightarrow B_4C + 3\,CO_2$$

(a) Calculate the chemical amount of C that reacts with 3.32 mol of B_2O_3, according to this equation, and the chemical amount of B_4C that is produced.
(b) Calculate the mass (in grams) of C needed to react with 232.0 g of B_2O_3.

Strategy

Use the balanced chemical equation to generate unit-factors that work the required conversions. Use molar masses to convert from grams to moles or from moles to grams.

Solution

(a) Set up the ratio

$$\frac{4\ \text{mol C}}{2\ \text{mol B}_2O_3}$$

as a chemical unit-factor to convert between moles of B_2O_3 and moles of C. Then

$$n_C = 3.32\ \text{mol B}_2O_3 \times \left(\frac{4\ \text{mol C}}{2\ \text{mol B}_2O_3}\right) = 6.64\ \text{mol C}$$

Set up a different unit-factor to calculate the chemical amount of B_4C produced

$$n_{B_4C} = 3.32\ \text{mol B}_2O_3 \times \left(\frac{1\ \text{mol B}_4C}{2\ \text{mol B}_2O_3}\right) = 1.66\ \text{mol B}_4C$$

(b) Use a string of three unit-factors, as shown in Figure 2–3

$$m_C = 232.0\ \text{g B}_2O_3 \times \left(\frac{1\ \text{mol B}_2O_3}{69.62\ \text{g B}_2O_3}\right) \times \left(\frac{4\ \text{mol C}}{2\ \text{mol B}_2O_3}\right) \times \left(\frac{12.01\ \text{g C}}{1\ \text{mol C}}\right)$$

$$= 80.04\ \text{g C}$$

Check

According to the balanced chemical equation, 4 mol of C ($\approx$ 48 g C) reacts with 2 mol of B_2O_3 ($\approx$140 g B_2O_3). To consume approximately 230 g of B_2O_3 requires approximately

$$230 \text{ g } B_2O_3 \times \left(\frac{48 \text{ g C}}{140 \text{ g } B_2O_3} \right) = 80 \text{ g C}$$

EXERCISE

The very hard compound TiB_2 is deposited as a thin protective film on cutting tools by running the following reaction at approximately 1000°C:

$$TiCl_4(g) + 2 \, BCl_3(g) + 5 \, H_2(g) \longrightarrow TiB_2(s) + 10 \, HCl(g)$$

(a) Compute the chemical amount (in moles) of H_2 required to react with 1.64 mol of $TiCl_4$. (b) Determine the chemical amount (in moles) of HCl that is produced. (c) Determine the mass (in grams) of TiB_2 that is produced.

Answer: (a) 8.20 mol $H_2(g)$. (b) 16.4 mol $HCl(g)$. (c) 114 g $TiB_2(s)$.

• Remember: The chemical unit-factors linking different species in a reaction depend on the coefficients in the balanced equation.

The solution to the last part of Example 2–3 uses a string of unit-factors to set up a continued multiplication, but each step could have been completed separately. We might have written

$$232.0 \text{ g } B_2O_3 \times \left(\frac{1 \text{ mol } B_2O_3}{69.62 \text{ g } B_2O_3} \right) = 3.332 \text{ mol } B_2O_3$$

$$3.332 \text{ mol } B_2O_3 \times \left(\frac{4 \text{ mol C}}{2 \text{ mol } B_2O_3} \right) = 6.664 \text{ mol C}$$

$$6.664 \text{ mol C} \times \left(\frac{12.01 \text{ g C}}{1 \text{ mol C}} \right) = 80.03 \text{ g C}$$

The intermediate answers are interesting, but the labor saved and errors avoided in not transcribing intermediate results usually make it preferable to set up a string of unit-factors and do the arithmetic all at once. Doing the calculation stepwise also may cause annoying round-off errors. (See Appendix A–3.)

EXAMPLE 2–4

Calcium hypochlorite $Ca(OCl)_2$ is used as a bleaching agent (Fig. 2–4). It is produced by treating a mixed aqueous solution of sodium hydroxide and calcium hydroxide with gaseous chlorine. The subsequent reaction is described by the balanced equation

$$2 \, NaOH(aq) + Ca(OH)_2(aq) + 2 \, Cl_2(g) \longrightarrow$$
$$Ca(OCl)_2(aq) + 2 \, NaCl(aq) + 2 \, H_2O(\ell)$$

What masses (in grams) of chlorine and sodium hydroxide react with 1067 g of $Ca(OH)_2$, and what mass (in grams) of $Ca(OCl)_2$ is produced according to this equation? What masses of the by-products NaCl and H_2O are produced?

Strategy

Determine the molar mass of calcium hydroxide and find the chemical amount of $Ca(OH)_2$ contained in the specified mass. Use the balanced chemical equation and the molar masses of the different reactants and products to create unit-factors leading to the desired answers.

Solution

The molar mass of calcium hydroxide is

$$M_{Ca(OH)_2} = 40.08 + 2(16.00) + 2(1.008) = 74.09 \text{ g mol}^{-1}$$

The chemical amount of $Ca(OH)_2$ consumed is

$$n_{Ca(OH)_2} = 1067 \text{ g } Ca(OH)_2 \times \left(\frac{1 \text{ mol } Ca(OH)_2}{74.09 \text{ g } Ca(OH)_2} \right) = 14.40 \text{ mol } Ca(OH)_2$$

According to the balanced equation, one mole of $Ca(OH)_2$ reacts with two moles of NaOH and two moles of Cl_2 to produce one mole of $Ca(OCl)_2$. If 14.40 mol of $Ca(OH)_2$ reacts, then

$$m_{NaOH} = 14.40 \text{ mol } Ca(OH)_2 \times \left(\frac{2 \text{ mol } NaOH}{1 \text{ mol } Ca(OH)_2} \right) \times \left(\frac{40.00 \text{ g } NaOH}{1 \text{ mol } NaOH} \right)$$
$$= 1152 \text{ g NaOH reacting}$$

$$m_{Cl_2} = 14.40 \text{ mol } Ca(OH)_2 \times \left(\frac{2 \text{ mol } Cl_2}{1 \text{ mol } Ca(OH)_2} \right) \times \left(\frac{70.91 \text{ g } Cl_2}{1 \text{ mol } Cl_2} \right)$$
$$= 2042 \text{ g } Cl_2 \text{ reacting}$$

$$m_{Ca(OCl)_2} = 14.40 \text{ mol } Ca(OH)_2 \times \left(\frac{1 \text{ mol } Ca(OCl)_2}{1 \text{ mol } Ca(OH)_2} \right) \times \left(\frac{142.98 \text{ g } Ca(OCl)_2}{1 \text{ mol } Ca(OCl)_2} \right)$$
$$= 2059 \text{ g } Ca(OCl)_2 \text{ produced}$$

Similar setups allow computation of the masses of the two by-products:

$$m_{NaCl} = 14.40 \text{ mol } Ca(OH)_2 \times \left(\frac{2 \text{ mol } NaCl}{1 \text{ mol } Ca(OH)_2} \right) \times \left(\frac{58.44 \text{ g } NaCl}{1 \text{ mol } NaCl} \right)$$
$$= 1683 \text{ g NaCl produced}$$

$$m_{H_2O} = 14.40 \text{ mol } Ca(OH)_2 \times \left(\frac{2 \text{ mol } H_2O}{1 \text{ mol } Ca(OH)_2} \right) \times \left(\frac{18.02 \text{ g } H_2O}{1 \text{ mol } H_2O} \right)$$
$$= 519.0 \text{ g } H_2O \text{ produced}$$

Check

The total mass of the reactants equals $1067 + 1152 + 2042 = 4261$ g. The total mass of the products equals $2059 + 1683 + 519.0 = 4261$ g.

Figure 2-4 • A solution of calcium hypochlorite ($Ca(OCl)_2$) is a powerful bleaching agent. A few drops of a solution quickly destroy the color of the blue dye on the cloth (*right*). (*Leon Lewandowski*)

EXERCISE

When heated in dry air, sodium reacts with oxygen to form sodium peroxide:

$$2 \, Na(s) + O_2(g) \longrightarrow Na_2O_2(s)$$

What masses (in kilograms) of $Na(s)$ and of $O_2(g)$ are required to prepare 457 kg of $Na_2O_2(s)$ by this reaction?

Answer: 269 kg of $Na(s)$; 188 kg of $O_2(g)$.

Volume Relationships of Gases in Chemical Reactions

The law of combining volumes (see Section 1–4) states that the volumes of gaseous reactants and products stand in ratios of simple integers, as long as those volumes are measured at the same temperature and pressure. These simple integers turn out to be none other than the coefficients in the balanced chemical equation representing the reaction. A balanced chemical equation therefore provides relationships among volumes of gases reacting, as illustrated in the following.

EXAMPLE 2–5

Calcium cyanamide, $CaCN_2$, is a solid. It reacts with gaseous O_2 under strong heating to give CaO, which is a solid, and two gases:

$$2\, CaCN_2(s) + 7\, O_2(g) \longrightarrow 2\, CaO(s) + 2\, CO_2(g) + 4\, NO_2(g)$$

A run of this reaction consumes 1.40 L of $O_2(g)$ (measured at room conditions of temperature and pressure). Predict the volume of $CO_2(g)$ and the volume of $NO_2(g)$ present when the products are examined at the same conditions.

Strategy

Under the conditions of the problem the ratios of the coefficients in the balanced equation equal the ratios of volumes of gases produced and reacting. Use this fact to construct unit factors.

Solution

For every 7 volumes of $O_2(g)$ consumed, 2 volumes of $CO_2(g)$ and 4 volumes of $NO_2(g)$ form. Therefore,

$$V_{CO_2} = 1.40\ \text{L O}_2 \times \left(\frac{2\ \text{L CO}_2}{7\ \text{L O}_2}\right) = 0.40\ \text{L CO}_2$$

$$V_{NO_2} = 1.40\ \text{L O}_2 \times \left(\frac{4\ \text{L NO}_2}{7\ \text{L O}_2}\right) = 0.80\ \text{L NO}_2$$

Setting up ratios of volumes in this way works *only* for gases and *only* when the volumes are measured at the same temperature and pressure.

Check

Two moles of $NO_2(g)$ are produced for each mole of $CO_2(g)$. Therefore, the volume of $NO_2(g)$ should be twice the volume of $CO_2(g)$. It is.

EXERCISE

Some $H_2(g)$ and $N_2(g)$ react to form 5.00 L of $NH_3(g)$, according to the equation

$$3\, H_2(g) + N_2(g) \longrightarrow 2\, NH_3(g)$$

What volume of $H_2(g)$ reacted, assuming that the pressure and temperature are the same after the reaction as before?

Answer: 7.50 L of $H_2(g)$.

2-3 Limiting Reactant and Percentage Yield

In practical chemistry, reactants are almost never mixed in the exact amounts necessary for them all to be completely consumed in forming products. If we mix several reactants in arbitrary amounts, their chemical reaction must stop when the supply of any one of them gives out. The reactant that is used up first is the **limiting reactant.** The other reactants remain after the reaction stops; these reactants are present **in excess** (Figs. 2–5 and 2–6). An additional supply of the limiting reactant restarts the reaction and forms more products. This is not true of the other reactants.

In an industrial process, the limiting reactant is commonly the most expensive reactant to ensure that none of it is wasted. In the preparation of silver chloride for photographic film

$$AgNO_3(aq) + NaCl(aq) \longrightarrow AgCl(s) + NaNO_3(aq)$$

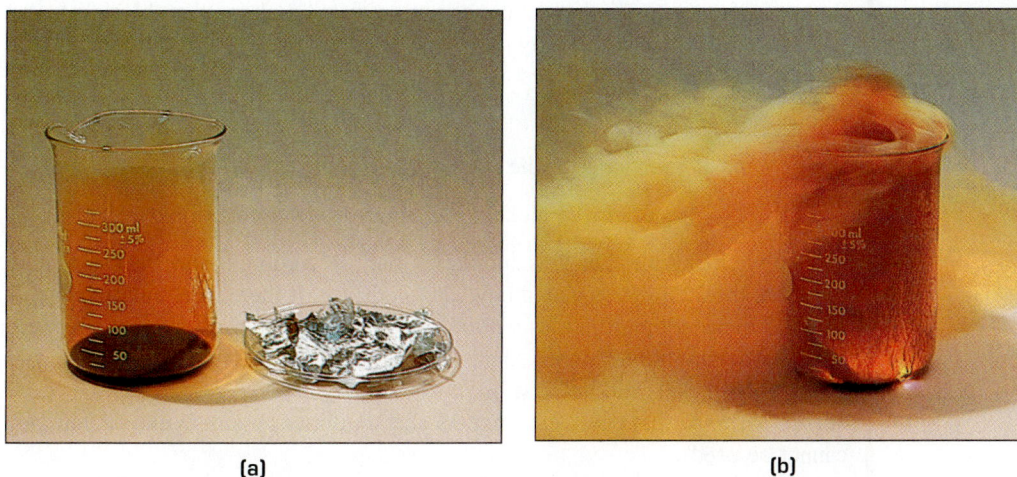

| (a) | (b) |

Figure 2–5 • (a) Aluminum, a shiny metal, and bromine, a reddish brown liquid. (b) The two react to give aluminum bromide according to the equation $2\,Al + 3\,Br_2 \longrightarrow Al_2Br_6$. Some of the product is ejected, along with excess bromine, by the vigor of the reaction. *(Charles D. Winters)*

REACTANTS   PRODUCTS

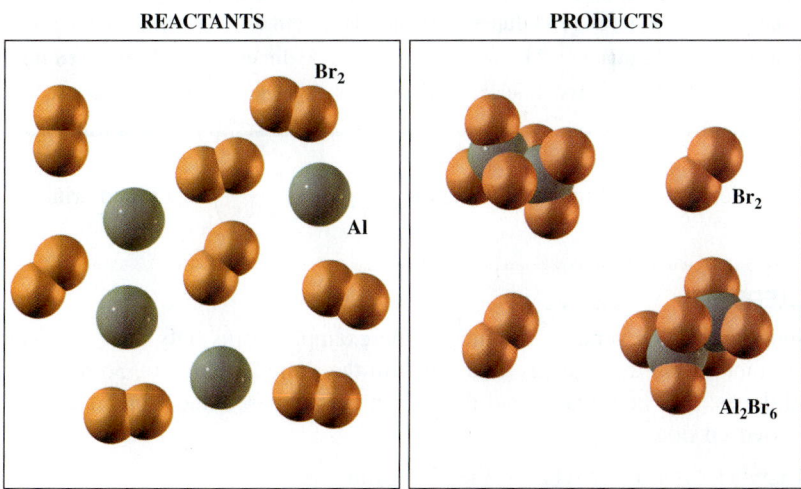

Figure 2–6 • Every two atoms of aluminum react with three molecules of bromine (Br_2) to form one molecule of Al_2Br_6. In the case diagrammed here, bromine is present in excess, and some remains after the reaction ends. Aluminum is the limiting reactant; the addition of more aluminum would enable the further production of Al_2Br_6.

the silver nitrate is far more expensive than the sodium chloride. It makes sense to carry out the reaction with an excess of sodium chloride to ensure that as much silver nitrate as possible reacts to form the product.

EXAMPLE 2-6

Cold drinks cost 60 cents from a (rather peculiar) vending machine that works only when one quarter, two dimes, and three nickels are deposited. Which coin runs out first if as many cold drinks as possible are purchased starting with 22 quarters, 36 nickels, and 30 dimes? How many purchases are possible?

Strategy

Find the number of drinks that could be purchased assuming that the quarters run out first (implying that the nickels and dimes are in excess). Repeat, assuming that the nickels run out first (implying that the quarters and dimes are in excess); Repeat again, assuming that the dimes run out first (implying that the nickels and quarters are in excess). Clearly, only one of these answers can be correct. It is the one that provides the fewest drinks.

Solution

The "chemical equation" for the transaction is

$$1 \text{ quarter} + 3 \text{ nickels} + 2 \text{ dimes} \longrightarrow 1 \text{ cold drink}$$

If there were unlimited nickels and dimes, then, according to the 1:1 ratio in the equation, the 22 quarters would buy 22 cold drinks. If there were unlimited quarters and nickels, the 30 dimes would buy 30/2 = 15 cold drinks. If there were unlimited quarters and dimes, then the 36 nickels would buy 36/3 = 12 cold drinks. The nickels run out first because the first 12 purchases use up the entire supply of 36. At that point, ten quarters and six dimes remain in excess, but they cannot be used.

Check

Make a chart to confirm how many of each *other* coin would be needed to go with the given supplies of quarters, nickels, and dimes:

To Consume:	You Need:	You Have:	Therefore:
22 quarters	66 nickels, 44 dimes	36 nickels, 30 dimes	Quarters are not used up.
36 nickels	12 quarters, 24 dimes	22 quarters, 30 dimes	**Nickels are used up first.**
30 dimes	15 quarters, 45 nickels	22 quarters, 36 nickels	Dimes are not used up.

$$\text{yield of cold drinks} = 36 \text{ nickels} \times \left(\frac{1 \text{ cold drink}}{3 \text{ nickels}} \right) = 12 \text{ cold drinks}$$

EXERCISE

Suppose that the vending machine in the example works only with one quarter, one nickel, and three dimes. Starting with the same 22 quarters, 36 nickels, and 30 dimes, find how many cold drinks can be purchased and which type of coin is used up first.

Answer: After ten purchases, the dimes run out.

The nickels, dimes, and quarters of Example 2–6 are analogous to chemical reactants as they combine in definite ratios to give products. The coefficients in balanced chemical equations give these ratios in terms of chemical amount (moles), never in terms of mass. Therefore, masses must be converted to chemical amounts, as the following example shows.

EXAMPLE 2-7

Sulfuric acid (H_2SO_4) forms in the chemical reaction

$$2\ SO_2(g) + O_2(g) + 2\ H_2O(\ell) \longrightarrow 2\ H_2SO_4(\ell)$$

Suppose that 400 g of SO_2, 175 g of O_2, and 125 g of H_2O are mixed and that the reaction proceeds until one of the reactants is used up. Which is the limiting reactant? How many grams of H_2SO_4 are produced, and how many grams of the other reactants remain?

Strategy

Convert the quantities of the SO_2, O_2, and H_2O from masses (grams) to chemical amounts (moles). Determine which of the three is the limiting reactant by figuring which would give the *least* sulfuric acid. The chemical amount of the limiting reactant determines the maximum possible yield of the H_2SO_4. It also determines the amounts of the two "in-excess" reactants used up to make that product.

Solution

$$n_{SO_2} = 400\ \text{g SO}_2 \times \left(\frac{1\ \text{mol SO}_2}{64.06\ \text{g SO}_2}\right) = 6.24\ \text{mol SO}_2$$

$$n_{O_2} = 175\ \text{g O}_2 \times \left(\frac{1\ \text{mol O}_2}{32.00\ \text{g O}_2}\right) = 5.47\ \text{mol O}_2$$

$$n_{H_2O} = 125\ \text{g H}_2\text{O} \times \left(\frac{1\ \text{mol H}_2\text{O}}{18.02\ \text{g H}_2\text{O}}\right) = 6.94\ \text{mol H}_2\text{O}$$

The following table shows the chemical amount of H_2SO_4 produced if each of the three reactants were completely used up.

Reactant	Unit-Factor	Amount of H_2SO_4 Produced If Reactant Is Consumed Completely
6.24 mol SO_2	$\times \dfrac{2\ \text{mol H}_2\text{SO}_4}{2\ \text{mol SO}_2} =$	6.24 mol H_2SO_4
5.47 mol O_2	$\times \dfrac{2\ \text{mol H}_2\text{SO}_4}{1\ \text{mol O}_2} =$	10.9 mol H_2SO_4
6.94 mol H_2O	$\times \dfrac{2\ \text{mol H}_2\text{SO}_4}{2\ \text{mol H}_2\text{O}} =$	6.94 mol H_2SO_4

The limiting reactant is SO_2 because it produces the *smallest* amount of H_2SO_4. Computing the mass of H_2SO_4 in grams proceeds this way:

$$m_{H_2SO_4} = 6.24\ \text{mol SO}_2 \times \left(\frac{2\ \text{mol H}_2\text{SO}_4}{2\ \text{mol SO}_2}\right) \times \left(\frac{98.07\ \text{g H}_2\text{SO}_4}{1\ \text{mol H}_2\text{SO}_4}\right)$$

$$= 612\ \text{g H}_2\text{SO}_4\ \text{produced}$$

• This is true even though the mass of the SO_2 exceeds the mass of the other two reactants put together.

The whole supply of SO_2 is consumed in the production of 612 g of sulfuric acid. The amounts of the other reactants that are left over when production of sulfuric acid has ceased are determined by subtracting the amounts consumed from the amounts initially present.

$$m_{O_2} = 6.24 \text{ mol } H_2SO_4 \times \left(\frac{1 \text{ mol } O_2}{2 \text{ mol } H_2SO_4} \right) \times \left(\frac{32.00 \text{ g } O_2}{1 \text{ mol } O_2} \right)$$

$$= 99.8 \text{ g } O_2 \text{ consumed}$$

$$m_{H_2O} = 6.24 \text{ mol } H_2SO_4 \times \left(\frac{2 \text{ mol } H_2O}{2 \text{ mol } H_2SO_4} \right) \times \left(\frac{18.02 \text{ g } H_2O}{1 \text{ mol } H_2O} \right)$$

$$= 112 \text{ g } H_2O \text{ consumed}$$

$$m_{O_2} = 175 \text{ g } O_2 - 100 \text{ g } O_2 = 75 \text{ g } O_2 \text{ remaining}$$

$$m_{H_2O} = 125 \text{ g } H_2O - 112 \text{ g } H_2O = 13 \text{ g } H_2O \text{ remaining}$$

Check

The total mass at the end is $612 + 13 + 75 = 700$ g, which equals the total mass originally present, $400 + 175 + 125 = 700$ g, as required by the law of conservation of mass.

EXERCISE

Suppose that 1.00 g of sodium and 1.00 g of chlorine react to form sodium chloride (NaCl). Which of the reactants is in excess, and what mass (in grams) of it remains when all of the limiting reactant is consumed?

Answer: Sodium is in excess, and 0.352 g of it remains.

In summary, identify the limiting reactant when specific amounts of substances react in a known way by these steps:

1. Select any one product.
2. Use the balanced equation to calculate the amount of the selected product that would form if the entire supply of the first reactant were used up.
3. Repeat with respect to every other reactant.
4. Identify the limiting reactant as the reactant that gives the smallest amount of the selected product.

Problem 2–77 gives a useful short-cut for identifying the limiting reactant.

The amounts of products calculated so far have been theoretical yields. The **theoretical yield** of a product equals the amount that forms when all of the limiting reactant reacts. The assumption is that the reaction goes cleanly and completely according to the equation that has been written to represent it. The only way to determine a theoretical yield is to do a stoichiometric calculation based on a balanced equation.

The **actual yield** of a product (the amount present after separating it from other products and reactants and purifying it) is usually less than the theoretical yield. There are several reasons for this. The reaction may stop short of completion, so that some of the limiting reactant remains unconsumed. There may be competing reactions (side-reactions) that give other products and so reduce the yield of the desired one. Finally, some of the product is inevitably lost in the process of separation and purification, although losses can be reduced by careful experimental techniques. The ratio of the actual yield to the theoretical yield (multiplied by 100%) gives the **percentage yield** for that product in the reaction.

EXAMPLE 2–8

Zinc is obtained from ores that contain zinc sulfide by roasting (heating in air) to give ZnO and then heating the ZnO with carbon monoxide. The equations for the two-step process are

$$ZnS(s) + \tfrac{3}{2}O_2(g) \longrightarrow ZnO(s) + SO_2(g)$$
$$ZnO(s) + CO(g) \longrightarrow Zn(\ell) + CO_2(g)$$

Suppose that 5.32 kg of ZnS is treated in this way and that 3.30 kg of pure zinc is obtained. Calculate the theoretical yield of zinc and the percentage yield of zinc.

Crystals of sphalerite or zinc sulfide (ZnS), a widely distributed mineral that is the principal source of zinc. *(Charles D. Winters)*

Strategy

Recognize that the limited supply of ZnS determines the amount of Zn that forms (O_2 and CO are in excess). Do a mass–mass stoichiometry calculation to find this amount, which is the theoretical yield of Zn. Obtain the percentage yield by dividing the actual yield by the theoretical yield and multiplying by 100%.

Solution

The theoretical yield of Zn is

$$m_{ZnS} = 5.32 \text{ kg ZnS} \times \left(\frac{1000 \text{ g}}{1 \text{ kg}}\right) \times \left(\frac{1 \text{ mol ZnS}}{97.46 \text{ g ZnS}}\right) \times \left(\frac{1 \text{ mol ZnO}}{1 \text{ mol ZnS}}\right)$$

$$\times \left(\frac{1 \text{ mol Zn}}{1 \text{ mol ZnO}}\right) \times \left(\frac{65.39 \text{ g Zn}}{1 \text{ mol Zn}}\right)$$

$$= 3570 \text{ g Zn}$$

This equals 3.57 kg of zinc. The actual yield is less, 3.30 kg of zinc. Therefore,

Ingots of newly refined elemental zinc. *(Nickelsberg/Gamma Liaison/Getty Images)*

$$\text{percentage yield of Zn} = \frac{3.30 \text{ kg Zn}}{3.57 \text{ kg Zn}} \times 100\% = 92.4\%$$

EXERCISE

One step in the production of vanadium uses the reaction of V_2O_5 with an excess of aluminum at high temperature to give vanadium and Al_2O_3. In a test run, 38.4 kg of vanadium is obtained from 72.0 kg of V_2O_5. Calculate the percentage yield of vanadium.

Answer: 95.2%.

It is clearly desirable in chemical reactions to achieve as high a percentage yield of product as possible in order to reduce consumption of raw materials and prevent the generation of possibly noxious by-products. In some synthetic processes (especially in organic chemistry), the final product is the result of many successive reactions. In such processes, the yields in the individual steps must be quite high if the synthetic

• Organic chemistry, the study of the chemistry of carbon compounds, is introduced in Chapter 24.

method is to be a practical success. Suppose, for example, that ten consecutive reactions must be carried out to reach the product and that each of these has a percentage yield of 50% (a fractional yield of 0.5). Then the overall yield equals the product of the fractional yields of the steps:

$$\underbrace{(0.5) \times (0.5) \times \cdots \times (0.5)}_{\text{10 terms}} = (0.5)^{10} = 0.001$$

This overall percentage yield of 0.1% makes the process useless for synthetic purposes. If all the individual percentage yields could be increased to 90%, however, the overall fractional yield would be $(0.9)^{10} = 0.35$, or a 35% overall yield. This is much more reasonable and might make the process feasible.

Following the Progress of Chemical Reactions

A formal device called a **table of changes** helps in tracking the progress of chemical reactions. Tables of change contain a column for every reactant and product in a chemical equation. Each column has three lines. The first lists the *initial* chemical amounts (in moles) of the substance. The second gives the *change* in the chemical amount of the substance as the reaction proceeds. The third gives the *final* amount. For any process, "final" equals "initial" plus "change." This means that the entry on the third line of a column in a table of changes must equal the sum of the two entries above it. Consider a mixture of 17.0 g of nitrogen (N_2) and 10.0 g of hydrogen (H_2). The two react to give ammonia (NH_3), according to the balanced equation

$$N_2(g) + 3\,H_2(g) \longrightarrow 2\,NH_3(g)$$

• Say the process is a reducing diet and the weight loss is 25 lb. The change is −25 lb. If the original weight were 200 lb, the final weight would be 200 + (−25) = 175 lb.

Ammonia is a valuable commodity that is in fact synthesized by this reaction. How do we construct a table of changes to track the progress of this synthetic reaction? The molar mass of N_2 is 28.02 g mol^{-1}. Therefore, the initial amount of N_2 is

$$17.0 \text{ g } N_2 \times \frac{1 \text{ mol } N_2}{28.02 \text{ g } N_2}$$

or 0.607 mol N_2. Similarly, the initial 10.0 g of H_2 (molar mass: 2.016 g mol^{-1}) equals 4.96 mol H_2. And the initial amount of the product ammonia equals zero. These facts occupy the first line of the table of changes:

	$N_2(g)$ +	$3\,H_2(g)$ ⟶	$2\,NH_3(g)$
	n_{N_2} (mol)	n_{H_2} (mol)	n_{NH_3} (mol)
Initial	0.607	4.96	0
Change	—	—	—
Final	—	—	—

Values on the "change" line depend on the degree of advancement of the reaction from reactants to products. Before the reaction starts, all changes equal zero. As the reaction runs, N_2 and H_2 are consumed, so each has a negative change. Ammonia is produced and so has a positive change. *The coefficients in the balanced chemical equation relate the magnitudes of the changes.* Suppose that x mol of N_2 is consumed (a change of $-x$). Then $3x$ mol of H_2 is consumed (a change of $-3x$) while $2x$ mol of NH_3 is produced (a change of $+2x$). Adding up the entries in each column fills in the bottom line. The complete table is

$N_2(g)$	$+$	$3\,H_2(g)$	$\longrightarrow$	$2\,NH_3(g)$
n_{N_2} (mol)		n_{H_2} (mol)		n_{NH_3} (mol)

	n_{N_2} (mol)	n_{H_2} (mol)	n_{NH_3} (mol)
Initial	0.607	4.96	0
Change	$-x$	$-3x$	$+2x$
Final	$0.607 - x$	$4.96 - 3x$	$2x$

Such tables have many uses. For example, to determine which reactant is the limiting reactant, simply set the expression for the final amount of each reactant equal to zero. A final zero clearly corresponds to the consumption of all of that reactant. Then solve separately for x. The reactant for which x is *smallest* is the limiting reactant because *less* advancement of the reaction, as measured by the size of x, suffices to use it up. For the preceding N_2–H_2 reaction mixture, the results are

for N_2: $\quad 0.607 - x = 0; \quad x = 0.607$ mol
for H_2: $\quad 4.96 - 3x = 0; \quad x = 1.65$ mol

Therefore, N_2 is the limiting reactant in this N_2–H_2 reaction mixture.

A reaction **reaches completion** when a reactant has been used up. Our reaction between N_2 and H_2 reaches completion at x equal to 0.607 mol because all of the N_2 has reacted. The amount of a product at completion equals its theoretical yield. In the N_2–H_2 reaction mixture, the theoretical yield of ammonia, the single product, is

$$n_{NH_3} = 2x_{completion} = 2(0.607 \text{ mol}) = 1.21 \text{ mol } NH_3 = \text{theoretical yield of } NH_3$$

Of course, yields may be expressed as masses as well as chemical amounts:

$$m_{NH_3} = 1.214 \text{ mol } NH_3 \times \frac{17.03 \text{ g } NH_3}{1 \text{ mol } NH_3} = 20.7 \text{ g } NH_3 = \text{theoretical yield of } NH_3$$

It is interesting to compute the amounts of the reactants that remain at completion:

$$n_{N_2} = 0.607 - 0.607 = 0 \text{ mol } N_2$$
$$n_{H_2} = 4.96 - 3(0.607) = 3.14 \text{ mol } H_2 \text{ in excess}$$

The excess amount of H_2 equals 6.33 g H_2 (using the fact that 1 mol of H_2 equals 2.016 g of H_2). This reaction started with 27.0 g of mixed N_2 and H_2. At completion, 27.0 g of matter was still present, but in the form of 20.7 g of ammonia and 6.3 g of unreacted H_2.

Chapters 7, 8, and 9 give additional applications of tables of changes.

2-4 The Stoichiometry of Reactions in Solution

The use of liquid solutions dominates practical laboratory work in chemistry for several reasons. Dissolved chemicals mix rapidly when stirred together and react evenly because intimate contact among reacting particles is ensured. Handling and storage are often easier, and dilution in a solvent often moderates or nullifies unpleasant or dangerous properties. Finally, dissolved chemicals are easily dispensed in known quantities because measuring liquid volumes is easy. The actual amount of a **solute** (the substance dissolved in the solvent) in any given volume of solution depends on how concentrated or dilute the solution happens to

• The other way to dispense known quantities of substances is to weigh them out. Weighing is easy too but usually less rapid and convenient.

• The term "concentration" by international agreement refers exclusively to ratios having the volume of the solution in the denominator; the word is often used more loosely, however, to refer to any measure of the composition of a solution.

be. This is expressed by a ratio. The **concentration** (c) of a solute in a solution equals the chemical amount of the solute (n) divided by the volume (V) of the entire solution:

$$c_{solute} = \frac{n_{solute}}{V_{solution}}$$

When the chemical amount of the solute is measured in moles and the volume of solution is measured in liters, the resulting unit of concentration is called the **molarity:**

$$molarity = \frac{moles_{solute}}{liters_{solution}} = mol\ L^{-1}$$

This is a very convenient unit of concentration for routine laboratory work. The symbol M stands for "moles of solute per liter of solution." A 1.0 M solution of fructose contains 1.0 mol of fructose per liter of solution and is said to be "1.0 molar in fructose." This fact is also expressed by writing

$$[fructose] = 1.0\ mol\ L^{-1}$$

The brackets are understood to mean the molarity of the species whose formula or name they enclose. A solution might contain several solutes. If so, it might be 1.0 M in solute A, 3.0 M in solute B, and so forth. Although the molarity is the most used unit of solution composition in chemistry, it has the disadvantage for accurate measurements that it depends slightly on temperature. If a solution is heated or cooled, its volume changes, so the number of moles of solute per liter of solution also changes.

EXAMPLE 2-9

Solutions of magnesium chloride ($MgCl_2$) are sometimes administered as purgatives. A solution is prepared by mixing 10.0 g of magnesium chloride with enough water to produce 100.00 mL of solution. Find the molarity of the solution.

Strategy

Express the amount of magnesium chloride as a chemical amount (change from grams to moles). Express the volume of the solution in liters. Divide chemical amount in moles by volume in liters to obtain molarity.

Solution

$$n_{MgCl_2} = 10.0\ g\ MgCl_2 \times \left(\frac{1\ mol\ MgCl_2}{95.21\ g\ MgCl_2}\right) = 0.105\ mol\ MgCl_2$$

$$V_{solution} = 100.00\ mL \times \left(\frac{1\ L}{1000\ mL}\right) = 0.10000\ L$$

$$c_{MgCl_2} = \frac{0.105\ mol\ MgCl_2}{0.10000\ L} = \frac{1.05\ mol\ MgCl_2}{1\ L} = 1.05\ mol\ L^{-1}$$

Check

9.521 g of $MgCl_2$ equals 0.1000 mol of $MgCl_2$. Dissolving 0.1000 mol of $MgCl_2$ in 0.1000 L (100.0 mL) of solution gives a 1.000 M solution. Here, 10.0 g of $MgCl_2$, a bit more than 0.1 mol, is dissolved, so the molarity should be a bit higher than 1 M. It is.

> **EXERCISE**
> Calculate the molarity of a solution prepared by dissolving 10.0 g of $Al(NO_3)_3$ in enough water to make 250.0 mL of solution.
>
> **Answer:** $c_{Al(NO_3)_3} = 0.188$ mol L^{-1}.

The Preparation of Solutions of Accurately Known Concentration

If asked to prepare a 1.000 M solution of ethanol (C_2H_5OH) in water, a beginner might dissolve 1.000 mol (46.07 g) of pure ethanol in 1.000 L (1000 mL) of water. This naïve approach gives a solution with the wrong molarity. It fails because it ignores the volume that the solute contributes. The total volume of the solution after mixing exceeds 1000 mL, and the molarity is accordingly less than 1.000 M. The correct method of mixing a solution of known molarity is not even to try to premeasure the volume of solvent, but instead to dissolve the required chemical amount of solute in somewhat less solvent than required for the target molarity, then to add just enough further solvent with continuous stirring to bring the volume "up to the mark," that is, to reach the required total volume.

Stirring (or swirling) in preparing solutions ensures thorough mixing and speeds dissolution. The latter is important because many dissolution reactions are discouragingly slow otherwise. Experienced laboratory workers also usually grind up solids before attempting to dissolve them; fine powders dissolve faster than large chunks. Gentle heating sometimes works to hasten dissolution but can cause problems. Undesired reactions may occur, and both the solution and container definitely change their volumes. Because an accurate final volume is crucial in knowing the molarity, hot solutions must be cooled fully before adding the final portion of solvent that brings the volume to the required total. For accurate work in preparing solutions, a **volumetric flask** is used (Fig. 2–7). These flasks are calibrated to contain precisely known volumes of solution at a specified temperature when filled to the mark etched on their necks.

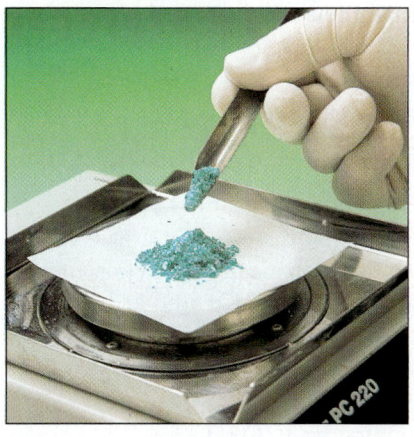  

(a) (b) (c)

Figure 2–7 • To prepare a solution of accurately known concentration: (a) weigh out an amount of the solid (here, nickel(II) chloride, $NiCl_2$), (b) dissolve it in less water than the final desired volume of solution, and (c) dilute carefully to the total volume marked on the neck of the volumetric flask. The molarity equals the number of moles of solute divided by the total volume (in liters). (*Charles D. Winters*)

Dilutions

On occasion, it is necessary to prepare a dilute solution of specified concentration from a more concentrated solution by adding pure solvent to it. Suppose that the initial concentration is c_i and the initial solution volume is V_i. The chemical amount of solute is

$$n_{solute} = c_i V_i$$

A similar equation applies after dilution to a final solution volume V_f

$$n_{solute} = c_f V_f$$

The amount of solute does not change upon dilution to V_f because only solvent, and not solute, is being added. Thus, $c_i V_i = c_f V_f$, and the final concentration c_f is

$$c_f = \frac{c_i V_i}{V_f}$$

This dilution equation can be used to calculate the final concentration after dilution to a known final volume. Alternatively, we can solve for V_f to determine what final volume must be aimed for to achieve a desired final concentration.

EXAMPLE 2-10

(a) Describe how to prepare 0.500 L of a 0.0250 M aqueous solution of potassium dichromate ($K_2Cr_2O_7$).
(b) Describe how to dilute this solution to obtain a solution that is 0.0125 M in $K_2Cr_2O_7$.

Strategy

(a) Find the mass of $K_2Cr_2O_7$ that the solution must contain. Consider how to dissolve the $K_2Cr_2O_7$ and dilute to the required volume. (b) Use the dilution equation to find the volume to which the original solution must be brought.

Solution

(a) $m_{K_2Cr_2O_7} = 0.500 \text{ L solution} \times \left(\dfrac{0.0250 \text{ mol } K_2Cr_2O_7}{\text{L solution}}\right) \times \left(\dfrac{294.2 \text{ g } K_2Cr_2O_7}{\text{mol } K_2Cr_2O_7}\right)$

$= 3.68 \text{ g } K_2Cr_2O_7$

To prepare the solution, weigh out 3.68 g of $K_2Cr_2O_7$, dissolve it in a small amount of water, and add enough water to make the final volume of the solution equal to 0.500 L. This is *not* the same as weighing out 3.68 g of $K_2Cr_2O_7$ and mixing it with 0.500 L of water.
(b) Rearranging the dilution equation gives

$$V_f = \frac{c_i V_i}{c_f} = \frac{(0.0250 \text{ M})(0.500 \text{ L})}{0.0125 \text{ M}} = 1.00 \text{ L}$$

Dilute the 0.500 L of solution from part (a) to a total volume of 1.00 L by adding water. Once again, this is not the same as adding 0.500 L of water!

Using the Molarity in Calculations

The procedure for stoichiometric calculations given in Figure 2–3 is readily modified to deal with concentrations of solutions. Instead of converting between masses and chemical amounts, using the molar mass as a conversion factor, the approach is to convert between *solution volumes* and chemical amounts, using the molarity as the conversion factor.

Consider as an example a reaction that is used commercially to prepare elemental bromine (Fig. 2–8):

$$2\,NaBr(aq) + Cl_2(aq) \longrightarrow 2\,NaCl(aq) + Br_2(\ell)$$

Suppose we have 50.0 mL of 0.0600 M $NaBr(aq)$. What volume of a 0.0500 M $Cl_2(aq)$ is needed to react completely with this solution? First, we must find the chemical amount of NaBr available:

$$n_{NaBr} = 0.0500 \text{ L solution} \times \left(\frac{0.0600 \text{ mol NaBr}}{1 \text{ L solution}} \right) = 3.00 \times 10^{-3} \text{ mol NaBr}$$

Figure 2–8 • Glass piping transfers reddish brown elemental bromine in a chemical plant in Arkansas where brominated flame retardants are synthesized. *(Ethyl Corporation)*

Next, we use the stoichiometric unit-factor "1 mol of Cl_2 per two moles of NaBr" to find the chemical amount of Cl_2 reacting:

$$n_{Cl_2} = 3.00 \times 10^{-3} \text{ mol NaBr} \times \left(\frac{1 \text{ mol } Cl_2}{2 \text{ mol NaBr}} \right) = 1.50 \times 10^{-3} \text{ mol } Cl_2$$

Finally, we find the volume of solution needed:

$$1.50 \times 10^{-3} \text{ mol } Cl_2 \times \left(\frac{1 \text{ L } Cl_2 \text{ solution}}{0.0500 \text{ mol } Cl_2} \right) = 3.00 \times 10^{-2} \text{ L } Cl_2 \text{ solution}$$

In this last step, the molarity of the Cl_2 solution appears "upside-down" in the conversion factor because the unit "mol Cl_2" must cancel out. The reaction requires 3.00×10^{-2} L, or 30.0 mL, of the aqueous Cl_2 solution. In practice, an excess of Cl_2 solution would be used to ensure more nearly complete conversion of the NaBr to Br_2.

• Turning the ratio upside-down is legitimate: A solution with 0.0500 mol Cl_2 per liter also has 1 L of solution per 0.0500 mol Cl_2.

We might also want to know the NaCl concentration after the reaction has been completed. Because each mole of NaBr that reacts gives 1 mol of NaCl in the products, the chemical amount of NaCl that is produced is 3.00×10^{-3} mol. In calculating the concentration of NaCl, however, we must use care. The final volume of the solution is the total volume, approximately equal to the sum of the 50.0 mL originally present and the 30.0 mL added, giving 80.0 mL, which equals 0.0800 L. The final concentration of NaCl is then

$$[NaCl] = \frac{3.00 \times 10^{-3} \text{ mol NaCl}}{0.0800 \text{ L solution}} = 0.0375 \text{ M}$$

where we have used the notation of square brackets around a chemical formula to refer to molarity.

In the preceding discussion, volumes were given in milliliters, not liters. Sometimes it is easier when working with milliliters to change concentration units from mol L^{-1} to mmol mL^{-1}. Because a millimole equals 0.001 mol and a milliliter equals 0.001 L, the numerical value stays the same. The first step in the preceding computation then becomes

$$50.0 \text{ mL solution} \times \left(\frac{0.0600 \text{ mmol NaBr}}{1 \text{ mL solution}} \right) = 3.00 \text{ mmol NaBr}$$

At the heart of any stoichiometric calculation is a relation between the chemical amounts (the numbers of moles) of two substances taking part in the reaction. This central relation comes from the coefficients assigned to the two substances in the balanced chemical equation. The first and last steps then involve conversions between chemical amount and other measures of amounts of substance, such as mass (see Section 2–2), gas volume (see Sections 2–2 and 5–5), or solution volume (this section).

EXAMPLE 2–11

When the orange solid potassium dichromate is added to a solution of concentrated hydrochloric acid, it reacts according to the balanced equation

$$K_2Cr_2O_7(s) + 14 \text{ HCl}(aq) \longrightarrow 2 \text{ KCl}(aq) + 2 \text{ CrCl}_3(aq) + 7 \text{ H}_2O(\ell) + 3 \text{ Cl}_2(g)$$

producing a mixed solution of chromium(III) chloride ($CrCl_3$) and potassium chloride (KCl), and evolving gaseous chlorine (Cl_2). Suppose that 6.20 g of $K_2Cr_2O_7$ reacts completely in this way in a solution with a total volume of 100.0 mL. Calculate the final concentration of $CrCl_3(aq)$ that results and the chemical amount (in moles) of chlorine produced.

Strategy

Convert the amount of potassium dichromate present before the reaction from grams to moles, using its molar mass as a conversion factor. Use unit factors from the balanced chemical equation to obtain the number of moles of $CrCl_3$ and the number of moles of Cl_2 formed by the reaction. Use the definition of molarity to obtain the concentration of the $CrCl_3$.

Solution

$$n_{K_2Cr_2O_7} = 6.20 \text{ g } K_2Cr_2O_7 \times \left(\frac{1 \text{ mol } K_2Cr_2O_7}{294.19 \text{ g } K_2Cr_2O_7} \right) = 0.0211 \text{ mol } K_2Cr_2O_7$$

$$n_{CrCl_3} = 0.0211 \text{ mol } K_2Cr_2O_7 \times \left(\frac{2 \text{ mol } CrCl_3}{1 \text{ mol } K_2Cr_2O_7} \right) = 0.0422 \text{ mol } CrCl_3$$

$$n_{Cl_2} = 0.0211 \text{ mol } K_2Cr_2O_7 \times \left(\frac{3 \text{ mol } Cl_2}{1 \text{ mol } K_2Cr_2O_7} \right) = 0.0633 \text{ mol } Cl_2$$

$$[CrCl_3] = \frac{0.0422 \text{ mol } CrCl_3}{0.100 \text{ L}} = 0.422 \text{ mol } L^{-1}$$

Check

$$\frac{0.0633 \text{ mol } Cl_2}{0.0422 \text{ mol } CrCl_3} = \frac{3 \text{ mol } Cl_2}{2 \text{ mol } CrCl_3}$$

EXERCISE

Suppose the chlorine remained dissolved in solution. Calculate its concentration. (In fact, the resulting concentration exceeds the solubility of chlorine in water; most of the chlorine must leave the solution.)

Answer: $[Cl_2] = 0.633$ M.

2–5 The Scale of Chemical Processes

Most experiments and lecture demonstrations in general chemistry are carried out on a scale of grams. Masses of substance ranging from 0.1 g to 100 g are easy to measure out and are appropriate for chemical operations performed in beakers and test tubes, on a laboratory table, or in a fume hood. With proper care, the heat that is liberated is controllable, and the waste products can be disposed of in a straightforward and safe manner. Not all chemistry is carried out at this scale, however. In this section, we take a brief look at chemical processes on two different scales: microscale and industrial scale.

Microscale processes arise in a number of areas of chemistry but are particularly prevalent in biochemistry, the study of the chemistry of life processes. The reason for this is that many biomolecules cause major effects in very small quantities. Hormones, for example, are chemical messengers that help to initiate, control, and coordinate chemical reactions in different organs in the body. Even at the level of parts per trillion, they can have significant effects on growth and development. Isolating a hormone and determining its structure require the separation of a very small amount of the substance from a large amount of starting material, followed by accurate analytical measurements and chemical transformations. One of the early triumphs of such microscale chemistry was the famous isolation and characterization of the hormone insulin (used in the treatment of diabetes) by Frederick Sanger, for which he received a Nobel Prize in chemistry in 1957. Another hormone, cortisol, is synthesized in the adrenal glands. A synthetic derivative, cortisone, and other related chemicals are used in the treatment of arthritis and allergies and have annual world sales on the order of $1 billion.

• If a hormone is present in a sample to the extent of one part per trillion by mass, then every 10^{12} g (or kg or lb) of the sample contains 1 g (or kg or lb) of that hormone. See Appendix C–4.

EXAMPLE 2-12

The characteristic "green herbaceous vegetative" aroma of Cabernet Sauvignon wine probably arises from the chemical 2-methoxy-3-(2-methylpropyl)-pyrazine, with chemical formula $C_9H_{14}ON_2$ and molar mass 166.2 g mol^{-1}. Two molecules of this compound are present in the wine for every 10 trillion (10^{13}) molecules of water. The predominant component of wine is water, with 55.5 mol of water per liter of wine. Estimate (a) the mass of $C_9H_{14}ON_2$ in a bottle (750 mL) of wine and (b) the number of molecules of $C_9H_{14}ON_2$ in a bottle of wine.

• This quantity of 55.5 mol comes from the mass of 1 L of water (1.00×10^3 g) divided by the molar mass of water (18.02 g mol^{-1}).

Strategy

Use suitable strings of unit-factors. The ratio "2 molecules of $C_9H_{14}ON_2$ per every 10 trillion molecules of water" gives "2 mol $C_9H_{14}ON_2$ per 10 trillion mol water," by scaling by a factor of Avogadro's number, as explained in Section 2–2.

Dozens of distinct chemical substances contribute to the natural flavor and bouquet of wines. *(George Semple)*

Solution

(a)

$$m_{C_9H_{14}ON_2} = 750 \text{ mL wine} \times \left(\frac{1 \text{ L}}{1000 \text{ mL}}\right) \times \left(\frac{55.5 \text{ mol water}}{1 \text{ L wine}}\right)$$

$$\times \left(\frac{2 \text{ mol } C_9H_{14}ON_2}{10 \times 10^{12} \text{ mol water}}\right) \times \left(\frac{166.2 \text{ g } C_9H_{14}ON_2}{1 \text{ mol } C_9H_{14}ON_2}\right)$$

$$= 1.4 \times 10^{-9} \text{ g } C_9H_{14}ON_2$$

(b)

$$N_{C_9H_{14}ON_2} = 750 \text{ mL wine} \times \left(\frac{1 \text{ L}}{1000 \text{ mL}}\right) \times \left(\frac{55.5 \text{ mol water}}{1 \text{ L wine}}\right)$$

$$\times \left(\frac{2 \text{ mol } C_9H_{14}ON_2}{10 \times 10^{12} \text{ mol water}}\right) \times \left(\frac{6.022 \times 10^{23} \text{ molecules } C_9H_{14}ON_2}{1 \text{ mol } C_9H_{14}ON_2}\right)$$

$$= 5.0 \times 10^{12} \text{ molecules } C_9H_{14}ON_2$$

Even this tiny mass of compound contains 5 trillion molecules.

Check

The mass of $C_9H_{14}ON_2$ in the bottle of wine divided by its chemical amount should give its molar mass. The mass is 1.4×10^{-9} g. The chemical amount is 8.3×10^{-12} mol (obtained by dividing 5.0×10^{12} molecules by Avogadro's number). Division gives 170 g mol^{-1}, which equals the given molar mass (166.2 g mol^{-1}) within the precision of the calculation. The number of moles of $C_9H_{14}ON_2$ is also obtained by simply ending the string of conversions in part (b) without doing the last one.

EXERCISE

The powerful antitumor medicine cisplatin, $PtCl_2(NH_3)_2$, is prepared for injection by suspending small particles of solid in iodinated poppy-seed oil as a vehicle. The average particle of $PtCl_2(NH_3)_2$ is a sphere that is 6.6×10^{-4} cm in diameter. Compute the number of atoms of platinum in such a particle if the density of cisplatin equals 3.86 g cm^{-3}.

Answer: 1.2×10^{12} atoms of platinum.

Figure 2–9 • A single chemical plant can produce more than 1 billion lb of a product per year. This Texas plant produces ethylene (C_2H_4), a starting material for many plastics. *(Courtesy of Occidental Petroleum)*

Let us move to the opposite extreme and consider the industrial-scale processing of chemicals (Fig. 2–9). Chemical processes are developed in the laboratory and then must be scaled up to make them into economical sources of large quantities of bulk chemicals. The scaling-up process is not a simple one because nothing guarantees that a reaction proceeds the same way when the size of the equipment is expanded. One aspect of this is the different way in which disposal of heat must be carried out. In most reactions in ordinary beakers and flasks, the heat produced dissipates rapidly to the surrounding air. In an industrial process using hundreds of kilograms of reactants, the heat can build up and lead to an explosion unless properly designed cooling equipment is used. A more extensive discussion of the nature of industrial-scale processes is presented in Chapter 22.

CHEMISTRY IN YOUR LIFE

Green Chemistry

Practical chemical reactions rarely proceed according to the tidy theoretical plan that balanced equations offer. Reactants mix incompletely and may never come in contact with each other despite the use of solvents and vigorous mixing. Unwanted side-reactions compete with the planned "official" reaction, reducing yields and creating impurities. Finally, many reactions naturally come to a stop well short of completion, requiring an often laborious separation of the product from the unconsumed reactants.

Unwanted by-products and used solvents cause huge problems in the chemical industry. Society benefits immensely from the desired products of chemistry (such as antibiotics, cheap durable cloth, and strong, corrosion-resistant materials), but questions arise as to whether "miracle" products are worth the price of polluted air, water, and land. In the United States, the Pollution Prevention Act of 1990 focused attention on pollution *avoidance* rather than pollution disposal and remediation. Green chemistry has since grown up "to develop innovative chemical technologies that reduce or eliminate the use or generation of hazardous substances in the design, manufacture, and use of chemical products." The U.S. National Science Foundation has funded research initiatives, such as "Environmentally Conscious Manufacturing" and "Environmentally Benign Chemical Synthesis and Processing." In 1997, the Green Chemistry Institute (GCI), a worldwide consortium of institutions representing academia, industry, government, and nongovernmental organizations, was established with the mission to promote and further the practice of environmentally friendly chemistry.

The production of ethylene glycol, widely used as an antifreeze, at a plant in Tennessee exemplifies the type of chemical production that is being encouraged by GCI. The process starts with coal and water. It generates the desired product along with carbon dioxide and a mixture of solid wastes. Carefully designed treatments make the plant's effluent cleaner than the river water it takes in. Solid wastes are not carted off to dump sites for burial but are incorporated into construction materials. The carbon dioxide is not immediately released into the atmosphere but is cleaned and used to produce carbonated beverages.

Green chemistry emphasizes "atom economy" in which the aim of the synthetic process is to see that all of the reactants' atoms are incorporated into the desired products. This concept favors addition reactions over displacement or decomposition reactions. (These and other types of chemical reactions are discussed in Chapter 4.) Green chemistry also favors the use of catalysts rather than heating to speed up chemical reactions. Catalysts increase the rate of reactions without themselves being consumed; they can be reused (a discussion of catalysis appears in Chapter 14). Heating accelerates reactions but often leads to unwanted by-products. Further principles of green chemistry include use of renewable raw materials, design of safer chemicals for specific purposes, use of low-hazard methods of synthesis (to avoid explosions and fires), avoidance of methods of synthesis that might generate toxic by-products, avoidance of toxic solvents, and emphasis on energy efficiency.

- More information on green chemistry appears on the World Wide Web at www.acs.org/education/greenchem/. The site www.chemsoc.org/networks/gcn/index.htm/ gives a European perspective.

EXAMPLE 2-13

In the contact process for making sulfuric acid (described in Chapter 22), the overall reaction is

$$2\,S(s) + 3\,O_2(g) + 2\,H_2O(\ell) \longrightarrow 2\,H_2SO_4(\ell)$$

A typical sulfuric acid plant makes approximately 1000 metric tons of sulfuric acid per day (1 metric ton is 10^3 kg). Estimate the volumes of sulfur and oxygen consumed by a plant in a *year*, assuming that the density of sulfur is 2.1 g cm^{-3} and the density of the gaseous oxygen is 1.3 g L^{-1}.

A small amount of sulfur burning in air. This reaction is the first step in the manufacture of sulfuric acid (H_2SO_4), the world's most important industrial chemical. (*Charles D. Winters*)

Strategy

Construct unit-factors to convert from metric tons of sulfuric acid to grams and from grams to moles. Obtain unit-factors from the balanced equation to convert from moles of sulfuric acid to moles of sulfur and oxygen. Use molar masses and densities to convert from moles of sulfur and oxygen to their volumes.

Solution

A metric ton (t) is 1000 kilograms. A kilogram is 1000 grams. Therefore, a metric ton is 1000 × 1000 grams, which equals 1 million (10^6) grams.

$$n_{H_2SO_4} = 1\text{ yr} \times \frac{365\text{ d}}{1\text{ yr}} \times \left(\frac{1.0 \times 10^3\text{ t }H_2SO_4}{1\text{ d}}\right) \times \left(\frac{10^6\text{ g}}{1\text{ t}}\right) \times \left(\frac{1\text{ mol }H_2SO_4}{98.1\text{ g }H_2SO_4}\right)$$

$$= 3.7 \times 10^9\text{ mol }H_2SO_4$$

$$V_S = 3.7 \times 10^9\text{ mol }H_2SO_4 \times \left(\frac{2\text{ mol S}}{2\text{ mol }H_2SO_4}\right) \times \left(\frac{32.07\text{ g S}}{1\text{ mol S}}\right) \times \left(\frac{1\text{ cm}^3\text{ S}}{2.1\text{ g S}}\right)$$

$$= 5.7 \times 10^{10}\text{ cm}^3\text{ S}$$

$$V_{O_2} = 3.7 \times 10^9\text{ mol }H_2SO_4 \times \left(\frac{3\text{ mol }O_2}{2\text{ mol }H_2SO_4}\right) \times \left(\frac{32.00\text{ g }O_2}{1\text{ mol }O_2}\right) \times \left(\frac{1\text{ L }O_2}{1.3\text{ g }O_2}\right)$$

$$= 1.4 \times 10^{11}\text{ L }O_2$$

Check

Compare the volumes. First get them into the same units. From Appendix B–1, $1\text{ cm}^3 = (10^{-2}\text{ m})^3 = 10^{-6}\text{ m}^3$. The S($s$) therefore occupies $5.7 \times 10^4\text{ m}^3$, which corresponds to a cube 38 m (or more than ten stories high) on a side. From Appendix B–1, 1 L equals $1\text{ dm}^3 = (0.1\text{ m})^3 = 10^{-3}\text{ m}^3$. The $O_2(g)$ therefore occupies $1.4 \times 10^8\text{ m}^3$, which fills a cube more than 500 m on a side. The numbers make sense. The plant consumes a much larger volume of gaseous oxygen because the density of the gas is much less than the density of the solid sulfur.

EXERCISE

The production of ammonium perchlorate, which is used in solid fuels for rockets, can be represented by the equation

$$NH_3(g) + HCl(g) + 4\,H_2O(g) \longrightarrow NH_4ClO_4(s) + 4\,H_2(g)$$

Calculate the volume of gaseous ammonia and the volume of gaseous hydrogen chloride (in liters) required to produce 8000 lb of NH_4ClO_4. Take the densities of the two gases as 0.759 and 1.63 g L^{-1}, respectively.

Answer: $V_{NH_3} = V_{HCl} = 6.9 \times 10^5$ L.

• Megagrams and megamoles could have been used advantageously in Example 2–13, which deals with metric tons of sulfuric acid.

Units of grams and moles are used so much in the preceding discussion that one might think them essential to the topic. This is not so. Industrial chemists might choose to use megagrams (Mg) and megamoles (Mmol), or kilograms (kg) and kilomoles (kmol). If 12 g of carbon equals 1 mol of carbon, then clearly 12 kg of carbon equals 1 kmol of carbon. And if 44 g of CO_2 equals 1 mol of CO_2, then 44 kg of CO_2 equals 1 kmol of CO_2, and so on for every other substance. Inserting the prefix "kilo-" scales up all quantities by a factor of 1000, just as inserting "mega-" scales up by a factor of 1 million. Neither affects the real essential, which is the ratio of the masses. A biochemist might prefer to work with milligrams and millimoles or with micrograms and micromoles, and a radiochemist might prefer nanograms and nanomoles. These prefixes scale quantities down by a factor of 1 thousand, 1 million, and 1 billion, respectively. In every case, the essential ratios are preserved:

$$12 \text{ milligram (mg) C} = 1 \text{ millimole (mmol) C}$$
$$12 \text{ microgram } (\mu g) \text{ C} = 1 \text{ micromole } (\mu mol) \text{ C}$$
$$12 \text{ nanogram (ng) C} = 1 \text{ nanomol (nmol) C}$$

SUMMARY

2-1 **Stoichiometry** is the study of mass relationships in chemical reactions. If the reactants and products in a chemical reaction are known, a balanced **chemical equation** can be written for that reaction. Chemical equations are balanced by a logical step-wise process in which coefficients are chosen to ensure that the same number of atoms of each element (or moles of atoms) is represented on each side of the equation.

2-2 The coefficients in the balanced equation are used to construct chemical **unit-factors** (conversion factors) to relate the amount of any substance shown in the equation to the amounts of all the others. The coefficients applying to gaseous reactants and products in balanced equations are the same simple integers mentioned in the law of combining volumes.

2-3 The reactant that is used up first in a chemical reaction is called the **limiting reactant.** Reactants other than the limiting reactant are said to be **in excess.** The **actual yield** of a product of a reaction is determined by separating the product and determining its mass. It usually is less than the **theoretical yield,** which is calculated on the assumption that the reaction goes cleanly and completely. The ratio of the actual yield of a product to its theoretical yield is multiplied by 100% to give the **percentage yield** for that product in a reaction.

2-4 The **molarity** of a solution equals the chemical amount in moles of solute dissolved per liter of solution. Solutions of known molarity are prepared by weighing out the desired chemical amount of solute and adding enough solvent to bring the solution to the required final volume.

2-5 In general chemistry laboratories, work is usually done with gram quantities, but the mole concept applies irrespective of scale. Thus, if 1 mol ^{12}C weighs exactly 12 g, 1 kmol ^{12}C weighs 12 kg, and 1 mmol ^{12}C weighs 12 mg.

PROBLEMS

Note: Answers to blue-numbered problems are given in Appendix F. Problems that are more challenging are indicated with asterisks.

Writing Balanced Chemical Equations

1. (See Example 2–1.) Balance the following chemical equations by inspection. In each case, confirm the balance by preparing a table showing the number of moles of each element on the left side and on the right side of the arrow.
    (a) $N_2 + O_2 \longrightarrow NO$
    (b) $N_2 + O_2 \longrightarrow N_2O$
    (c) $K_2SO_3 + HCl \longrightarrow KCl + H_2O + SO_2$
    (d) $NH_3 + O_2 + CH_4 \longrightarrow HCN + H_2O$
    (e) $CaC_2 + CO \longrightarrow C + CaCO_3$

2. (See Example 2–1.) Balance the following chemical equations by inspection:
    (a) $Al + FeO \longrightarrow Al_2O_3 + Fe$
    (b) $Br_2 + I_2 \longrightarrow IBr_3$
    (c) $PbO + NH_3 \longrightarrow Pb + N_2 + H_2O$
    (d) $Cl_2O_7 + H_2O \longrightarrow HClO_4$
    (e) $CaC_2 + CO_2 \longrightarrow C + CaCO_3$

3. (See Example 2–1.) Balance the following chemical equations by inspection:
    (a) $H_2 + P_4 \longrightarrow PH_3$
    (b) $K + O_2 \longrightarrow K_2O_2$
    (c) $PbO_2 + Pb + H_2SO_4 \longrightarrow PbSO_4 + H_2O$
    (d) $BF_3 + H_2O \longrightarrow B_2O_3 + HF$

(e) $KClO_3 \longrightarrow KCl + O_2$

(f) $K_2O_2 + H_2O \longrightarrow KOH + O_2$

(g) $PCl_5 + AsF_3 \longrightarrow PF_5 + AsCl_3$

(h) $KOH + K_2Cr_2O_7 \longrightarrow K_2CrO_4 + H_2O$

(i) $P_4 + NaOH + H_2O \longrightarrow PH_3 + Na_2HPO_3$

4. (See Example 2–1.) Balance the following chemical equations by inspection:

(a) $Al + HCl \longrightarrow AlCl_3 + H_2$

(b) $NH_3 + O_2 \longrightarrow NO + H_2O$

(c) $Fe + O_2 + H_2O \longrightarrow Fe(OH)_2$

(d) $HSbCl_4 + H_2S \longrightarrow Sb_2S_3 + HCl$

(e) $Al + Cr_2O_3 \longrightarrow Al_2O_3 + Cr$

(f) $XeF_4 + H_2O \longrightarrow Xe + O_2 + HF$

(g) $(NH_4)_2Cr_2O_7 \longrightarrow N_2 + Cr_2O_3 + H_2O$

(h) $NaBH_4 + H_2O \longrightarrow NaBO_2 + H_2$

(i) $I_2O_5 + CO \longrightarrow I_2 + CO_2$

5. (See Example 2–1.) Write balanced chemical equations to represent the following reactions. Avoid fractional coefficients in your answers.

(a) Liquid benzene (C_6H_6) burns in gaseous oxygen to give gaseous carbon dioxide and gaseous water.

(b) Gaseous fluorine (F_2) reacts with liquid water to give gaseous oxygen difluoride (OF_2) and gaseous hydrogen fluoride (HF).

(c) Solid calcium carbide (CaC_2) reacts with liquid water to give aqueous calcium hydroxide ($Ca(OH)_2$) and gaseous acetylene (C_2H_2).

(d) Gaseous oxygen oxidizes gaseous ammonia (NH_3) to give gaseous nitrogen dioxide (NO_2) and liquid water.

(e) Solid aluminum dissolves in aqueous sodium hydroxide (NaOH) with the release of gaseous hydrogen and the formation of aqueous sodium aluminate (Na_3AlO_3).

6. (See Example 2–1.) Write balanced chemical equations to represent the following reactions. Avoid fractional coefficients in your answers.

(a) The combustion of the liquid hydrocarbon styrene (C_8H_8) in gaseous oxygen gives gaseous carbon dioxide and liquid water.

(b) The reaction of liquid dimanganese heptaoxide (Mn_2O_7) with solid calcium oxide (CaO) gives solid calcium permanganate ($Ca(MnO_4)_2$).

(c) The burning of solid magnesium in gaseous nitrogen gives solid magnesium nitride (Mg_2N_3).

(d) The oxidation of solid magnetite (Fe_3O_4) by gaseous oxygen gives solid iron(III) oxide (Fe_2O_3).

(e) The reduction of solid magnesium oxide (MgO) with solid carbon gives solid magnesium and gaseous carbon dioxide.

7. If solid Fe_2O_3 is heated in a stream of gaseous hydrogen, the products are solid iron and water vapor.

(a) Express this fact in the form of a balanced chemical equation.

(b) The reaction is performed, but the temperature control is imperfect, and some solid Fe_3O_4 forms in addition to the iron and water vapor. Write a balanced chemical equation for a side-reaction that gives Fe_3O_4.

8. If solid zinc sulfide (ZnS) is heated in the air, gaseous oxygen replaces sulfur in combination with the zinc to form solid zinc oxide (ZnO), and the sulfur goes off in the form of gaseous sulfur dioxide (SO_2).

(a) Express these facts in the form of a balanced chemical equation.

(b) A student speculates that gaseous SO_3 forms instead of SO_2 under certain conditions. Write a balanced chemical equation to represent this reaction.

*__9.__ (See Example 2–2.) Balance the following chemical equations by any means:

(a) $NO_2(g) + H_2O(\ell) \longrightarrow HNO_3(aq) + NO(g)$

(b) $KOH(aq) + V(s) + KClO_3(aq) \longrightarrow$
$K_3HV_2O_7(aq) + KCl(aq) + H_2O(\ell)$

(c) $Ag_2S(s) + KCN(aq) + O_2(g) + H_2O(\ell) \longrightarrow$
$KAg(CN)_2(aq) + S(s) + KOH(aq)$

(d) $Cl_2(g) + NaI(aq) + NaOH(aq) \longrightarrow$
$NaIO_4(aq) + NaCl(aq) + H_2O(\ell)$

(e) $Cr(CO)_6(s) + KOH(aq) \longrightarrow KHCr_2(CO)_{10}(aq)$
$+ K_2CO_3(aq) + KHCOO(aq) + H_2O(\ell)$

*__10.__ (See Example 2–2.) Balance the following chemical equations by any means:

(a) $CrCl_2(s) + KOH(aq) + Cl_2(g) \longrightarrow$
$K_2CrO_4(aq) + KCl(aq) + H_2O(\ell)$

(b) $H_2SO_4(aq) + NaBr(aq) \longrightarrow$
$Br_2(\ell) + SO_2(g) + Na_2SO_4(aq) + H_2O(\ell)$

(c) $H_2SO_4(aq) + K_2MnO_4(aq) \longrightarrow$
$KMnO_4(aq) + MnO_2(s) + K_2SO_4(aq) + H_2O(\ell)$

(d) $K_2O(s) + KO_2(s) + Co_3O_4(s) \longrightarrow KCoO_2(s)$

Using Balanced Chemical Equations

11. The coefficients in the following balanced equations give the number of moles of each substance. Determine the masses in grams of all products and reactants. Confirm that the total mass of reactants equals the total mass of products in each reaction.

(a) $CO_2 + C \longrightarrow 2\,CO$

(b) $2\,C_6H_{14} + 19\,O_2 \longrightarrow 12\,CO_2 + 14\,H_2O$

(c) $Mn_2O_7 + H_2O \longrightarrow 2\,HMnO_4$

12. The coefficients in the following equations give the number of formula units of each substance. Determine the mass in atomic mass units of all products and reactants. Confirm that the total mass of reactants equals the total mass of products in each reaction. (*Note:* Compare to the previous problem.)

(a) $CO_2 + C \longrightarrow 2\,CO$

(b) $2\,C_6H_{14} + 19\,O_2 \longrightarrow 12\,CO_2 + 14\,H_2O$

(c) $Mn_2O_7 + H_2O \longrightarrow 2\,HMnO_4$

13. (See Example 2–3.) The reaction of lead sulfide with hydrogen peroxide to give lead sulfate is described by the equation

$$PbS(s) + 4\,H_2O_2(aq) \longrightarrow PbSO_4(s) + 4\,H_2O(\ell)$$

How many moles of hydrogen peroxide are required to give 2.47 mol $PbSO_4(s)$?

14. (See Example 2–3.) The preparation of hydrogen cyanide from ammonia, oxygen, and methane is described by the equation

$$2 NH_3(g) + 3 O_2(g) + 2 CH_4(g) \longrightarrow$$
$$2 HCN(aq) + 6 H_2O(\ell)$$

How many moles of HCN can be prepared from 6.48 mol NH_3? What is the minimum number of moles of O_2 that would be consumed?

15. (See Example 2–3.) What mass (in grams) of the first reactant in each of the following would be required to react completely with 1.000 g of the second reactant?
(a) $CH_4(g) + 2 O_2(g) \longrightarrow CO_2(g) + 2 H_2O(\ell)$
(b) $TiCl_2(s) + TiCl_4(\ell) \longrightarrow 2 TiCl_3(s)$
(c) $2 Na_3VO_4(aq) + H_2O(\ell) \longrightarrow$
$$Na_4V_2O_7(aq) + 2 NaOH(aq)$$
(d) $2 K_2O_2(s) + 2 H_2O(\ell) \longrightarrow 4 KOH(aq) + O_2(g)$

16. (See Example 2–3.) What mass (in grams) of the first reactant in each of the following would be required to react completely with 1.000 g of the second reactant?
(a) $2 C_5H_{12}(g) + 16 O_2(g) \longrightarrow 10 CO_2(g) + 12 H_2O(g)$
(b) $XeF_6(g) + 2 H_2O(g) \longrightarrow Xe(g) + 4 HF(g) + O_2(g)$
(c) $K_2Cr_2O_7(aq) + 2 KOH(aq) \longrightarrow$
$$2 K_2CrO_4(aq) + H_2O(\ell)$$
(d) $N_2H_4(aq) + 2 H_2O_2(aq) \longrightarrow N_2(g) + 4 H_2O(\ell)$

17. (See Example 2–3.) Potassium nitrate (KNO_3) is used as a fertilizer for certain crops. It is produced through the reaction

$$4 KCl(aq) + 4 HNO_3(aq) + O_2(g) \longrightarrow$$
$$4 KNO_3(aq) + 2 Cl_2(g) + 2 H_2O(\ell)$$

Calculate the minimum mass of KCl required to produce 567 g of KNO_3. What mass of Cl_2 is generated as well?

18. (See Example 2–3.) Elemental phosphorus can be prepared from calcium phosphate via the overall reaction

$$2 Ca_3(PO_4)_2(s) + 6 SiO_2(s) + 10 C(s) \longrightarrow$$
$$6 CaSiO_3(s) + P_4(s) + 10 CO(g)$$

Calculate the minimum mass of $Ca_3(PO_4)_2(s)$ required to produce 69.8 g of $P_4(s)$. What mass of $CaSiO_3(s)$ is generated as a by-product?

19. (See Example 2–5.) Gaseous acetylene reacts with gaseous hydrogen to produce gaseous ethane:

$$C_2H_2(g) + 2 H_2(g) \longrightarrow C_2H_6(g)$$

Assuming that the volumes of reactants and products are measured at the same temperature and pressure, calculate the volume of ethane generated by the reaction of 14.2 L of hydrogen.

20. (See Example 2–5.) One step in the production of nitric acid is the gas-phase reaction

$$4 NH_3(g) + 5 O_2(g) \longrightarrow 4 NO(g) + 6 H_2O(g)$$

Assuming that the volumes of reactants and products are measured at the same temperature and pressure, calculate the volume of $O_2(g)$ required to generate 16.4 L of NO(g).

21. An 18.6 g sample of K_2CO_3 was treated in such a way that all of its carbon was captured in the compound $K_2Zn_3[Fe(CN)_6]_2$. Compute the mass (in grams) of this product.

22. A chemist dissolves 1.406 g of pure platinum in an excess of a mixture of hydrochloric and nitric acids and then, after a series of subsequent steps involving several other chemicals, isolates a compound of molecular formula $Pt_2C_{10}H_{18}N_2S_2O_6$. Determine the maximum possible yield of this compound.

23. A hydrated compound of rhodium has the formula $Na_3RhCl_6 \cdot xH_2O(s)$, where x is an integer. When heated to 120°C, 2.1670 g of the compound loses 0.7798 g of water, leaving 1.3872 g of $Na_3RhCl_6(s)$. Determine the value of x.

24. A crystalline compound of formula $Na_2SiO_3 \cdot 9 H_2O(s)$ loses a portion of its water when it is heated to 100°C but undergoes no other chemical changes. When a 10.00 g sample is heated in this way, the residue weighs 6.198 g. What is the formula of the residue?

25. Disilane (Si_2H_6) is a gas that reacts with oxygen to give silica (SiO_2) and water. Calculate the maximum mass of silica that could be obtained if 25.0 cm^3 of disilane (with a density of 2.78×10^{-3} g cm^{-3}) reacted with excess oxygen.

26. Tetrasilane (Si_4H_{10}) is a liquid with a density of 0.825 g cm^{-3}. It reacts with oxygen to give solid silica (SiO_2) and water. Calculate the maximum mass of silica that could be obtained if 25.0 cm^3 of tetrasilane reacted completely with excess gaseous oxygen.

27. Sodium nitrate ($NaNO_3$) decomposes, upon moderate heating, to yield sodium nitrite ($NaNO_2$) and gaseous oxygen. A 3.4671 g sample of *impure* sodium nitrate was heated until the evolution of oxygen ceased. The sample residue was subsequently found to have a mass of 2.9073 g. What percentage of the original impure sample was actually sodium nitrate? Assume that none of the impurities evolved gases.

28. Sodium chlorate ($NaClO_3$) decomposes, upon moderate heating, to yield sodium chloride (NaCl) and gaseous oxygen. A 3.4671 g sample of *impure* sodium chlorate was heated until the evolution of oxygen ceased. The sample residue was subsequently found to have a mass of 2.9073 g. What percentage of the original impure sample was actually sodium chlorate? Assume that none of the impurities evolved gases.

Limiting Reactant and Percentage Yield

29. (See Example 2–6.). One jumbo cheeseburger contains two hamburger patties, one slice of cheese, four slices of tomato, and one hamburger bun. A restaurant has 600 hamburger patties, 400 slices of cheese, 1000 slices of tomato, and 350 hamburger buns. How many jumbo cheeseburgers can the restaurant make? How many hamburger patties will be left over?

30. (See Example 2–6.) Every airplane model kit must contain two wings, one fuselage, four engines, and six wheels. The manufacturer has 17,254 wings, 8713 fuselages, 31,008 engines, and 48,143 wheels. Determine how many complete kits can be assembled. Determine the number of leftover parts of each kind.

31. (See Example 2–6.) One kind of dining table consists of four legs, two leaves, one frame, and one top. At assembly there are 25 dozen legs, 20 dozen leaves, 11 dozen frames, and 7 dozen tops. How many tables can be assembled?

32. A ream is a bundle of paper consisting of 500 sheets. A booklet is to consist of 16 sheets of contents printed on "copy stock" weighing 1.628 kg per ream and a single-folded binding sheet printed on "cover stock" weighing 3.071 kg per ream. There are 5.00 kg of the cover stock and 50.00 kg of the copy stock on hand. Determine how many booklets can be printed.

33. (See Example 2–7.) A mixture of 34.0 g of ammonia and 50.0 g of oxygen reacts according to the equation

$$4\,NH_3(g) + 3\,O_2(g) \longrightarrow 2\,N_2(g) + 6\,H_2O(g)$$

Which is the limiting reactant? How many grams of water can form?

34. (See Example 2–7.) Given the equation

$$Si(s) + C(s) \longrightarrow SiC(s)$$

determine the limiting reactant and the theoretical yield (in grams) of silicon carbide when 2.00 g of silicon reacts with 2.00 g of carbon.

35. Barium hydroxide octahydrate and ammonium chloride react in an unusual solid state reaction to produce liquid water, solid barium chloride, and gaseous ammonia, as shown in the equation

$$Ba(OH)_2 \cdot 8H_2O(s) + 2\,NH_4Cl(s) \longrightarrow$$
$$BaCl_2(s) + 10\,H_2O(\ell) + 2\,NH_3(g)$$

Assume that 35.12 g of barium hydroxide octahydrate and 13.75 g of ammonium chloride are mixed.
(a) Identify the limiting reactant.
(b) Determine the maximum mass of barium chloride that can be produced.
(c) Determine the mass of the excess reactant that remains after the production of barium chloride ceases.

36. Many reactions are performed in solvents other than water. For example, boron oxide (B_2O_3) reacts with ammonium

tetrafluoroborate (NH_4BF_4) in sulfuric acid

$$B_2O_3(sln) + 6\,NH_4BF_4(sln) + 3\,H_2SO_4(\ell) \longrightarrow$$
$$8\,BF_3(g) + 3\,(NH_4)_2SO_4(sln) + 3\,H_2O(sln)$$

Suppose that 69.62 g of B_2O_3 and 104.84 g of NH_4BF_4 are mixed in excess H_2SO_4 and react according to this equation.
(a) Identify the limiting reactant.
(b) Determine the maximum mass of BF_3 that can be produced.
(c) Determine the mass of the excess reactant that remains unreacted after the production of BF_3 ceases.

37. Exactly 2.000 g of hydrogen and 1.000 g of oxygen are mixed, and the two gases react to form water. Compute the theoretical yield (in grams) of water.

38. Compute the maximum mass of AgCl that can form when 35.453 g of NaCl and 107.87 g of $AgNO_3$ are mixed in water solution.

39. Gaseous C_2H_4 reacts with O_2 according to the equation

$$2\,C_2H_4(g) + 6\,O_2(g) \longrightarrow 4\,CO_2(g) + 4\,H_2O(g)$$

Calculate the theoretical yield of CO_2, in liters, from a mixture of 40.0 L of C_2H_4 and 20.0 L of O_2. Assume that the final temperature and pressure equal the initial temperature and pressure.

40. A new process to produce N_2O is being tested. In it, 2.33 L of N_2 reacts with 2.70 L of O_2. Determine the theoretical yield of N_2O, in liters. Assume that the final temperature and pressure equal the initial temperature and pressure.

41. (See Example 2–8.) The iron oxide Fe_2O_3 reacts with carbon monoxide, CO, to give iron and carbon dioxide:

$$Fe_2O_3(s) + 3\,CO(g) \longrightarrow 2\,Fe(s) + 3\,CO_2(g)$$

The reaction of 433.2 g of Fe_2O_3 with excess CO yields 254.3 g of iron. Calculate the theoretical yield of iron (assuming complete reaction) and its percentage yield.

42. (See Example 2–8.) Titanium dioxide (TiO_2) reacts with carbon and chlorine to give gaseous $TiCl_4$:

$$TiO_2(s) + 2\,C(s) + 2\,Cl_2(g) \longrightarrow TiCl_4(g) + 2\,CO(g)$$

The reaction of 7.39 kg of titanium dioxide with excess C and Cl_2 gives 14.24 kg of titanium tetrachloride. Calculate the theoretical yield of $TiCl_4$ and its percentage yield.

43. You promise to bake 200 dozen cookies and deliver them to a bake sale. Experience shows that you break (and then eat) 8.0% of your cookies during the process of making them.
(a) How many cookies should you buy ingredients for?
(b) How many cookies will you be eating?

44. Silicon nitride (Si_3N_4), a valuable ceramic, is made by the direct combination of silicon and nitrogen at high temperature. How much silicon must react with excess nitrogen to prepare 125 g of silicon nitride if the yield of the reaction is 95.0%?

The Stoichiometry of Reactions in Solution

45. Determine the chemical amount, in moles or millimoles, of sucrose contained in 319 mL of a 0.375 M solution of sucrose.

46. What is the final concentration in a solution prepared by mixing 500.0 mL of 0.600 M fructose with 300.0 mL of 0.400 M fructose? State any assumptions you must make in obtaining your answer.

47. (See Example 2–10.) Suppliers sell concentrated hydrochloric acid that is nominally 12 M. How much of this concentrated acid should be added to water to produce 10.0 L of a solution that is approximately 0.1 M?

48. (See Example 2–10.) Suppliers sell concentrated nitric acid that is nominally 16 M. How much of this concentrated acid should be added to water to produce 10.0 L of a solution that is approximately 0.1 M HNO_3?

49. A 100.0 mL sample of 0.2516 M barium chloride $BaCl_2(aq)$ was treated with an excess of sulfuric acid $H_2SO_4(aq)$. The resulting precipitate of barium sulfate $BaSO_4(s)$ was filtered, dried, and found to have a mass of 4.9852 g. Compute the percentage yield of barium sulfate in this process.

50. The reaction of iodine (I_2) with potassium hydroxide (KOH) produces potassium iodate (KIO_3), potassium iodide (KI), and water (H_2O).
(a) Write a balanced chemical equation for this reaction.
(b) Determine the limiting reagent when 2.178 g of iodine is treated with 15.12 mL of 3.1 M potassium hydroxide.

51. (See Example 2–11.) Hydrazine, N_2H_4, is made by the reaction of ammonia (NH_3) with sodium hypochlorite (NaOCl) in aqueous solution:

$$2\ NH_3(aq) + NaOCl(aq) \longrightarrow$$
$$N_2H_4(aq) + NaCl(aq) + H_2O(\ell)$$

What mass of hydrazine is produced from 51.6 L of a 0.650 M solution of sodium hypochlorite that reacts completely according to this equation with an excess of ammonia?

52. (See Example 2–11.) When the blue liquid dinitrogen trioxide (N_2O_3) is added to a solution of sodium hydroxide (NaOH), it reacts to give sodium nitrite ($NaNO_2$):

$$N_2O_3(\ell) + 2\ NaOH(aq) \longrightarrow 2\ NaNO_2(aq) + H_2O(\ell)$$

What is the concentration of sodium nitrite if 2.13 g of N_2O_3 is added to an excess of aqueous sodium hydroxide and the resulting solution is diluted to a total volume of 500.00 mL?

53. (See Example 2–11.) When treated with nitric acid, lead(IV) oxide reacts as follows:

$$2\ PbO_2(s) + 4\ HNO_3(aq) \longrightarrow$$
$$2\ Pb(NO_3)_2(aq) + 2\ H_2O(\ell) + O_2(g)$$

What volume of a 7.91 M solution of nitric acid is just sufficient to react with 15.9 g of lead(IV) oxide according to this equation?

54. Phosphoric acid is made industrially by the reaction of fluorapatite ($Ca_5(PO_4)_3F$) in phosphate rock with sulfuric acid

$$Ca_5(PO_4)_3F(s) + 5\ H_2SO_4(aq) + 10\ H_2O(\ell) \longrightarrow$$
$$3\ H_3PO_4(aq) + 5\ (CaSO_4 \cdot 2H_2O)(s) + HF(aq)$$

What volume of 6.3 M phosphoric acid is generated by the reaction of 2.2 metric tons (2200 kg) of fluorapatite?

55. What mass, in grams or milligrams, of silver chloride (AgCl) should precipitate when 21.5 mL of 0.150 M silver nitrate ($AgNO_3$) is mixed with 15.0 mL of 0.100 M calcium chloride ($CaCl_2$)?

56. Nitrogen monoxide (NO) can be generated on a laboratory scale by the reaction of dilute sulfuric acid with aqueous sodium nitrite ($NaNO_2$):

$$6\ NaNO_2(aq) + 3\ H_2SO_4(aq) \longrightarrow$$
$$4\ NO(g) + 2\ HNO_3(aq) + 2\ H_2O(aq) + 3\ Na_2SO_4(aq)$$

What volume of 0.646 M aqueous $NaNO_2$ is just sufficient in this reaction to generate 6.07 g of nitrogen monoxide?

The Scale of Chemical Processes

57. (See Example 2–12.) A container holds 55.5 g of $MgSO_4 \cdot 7H_2O$. Determine the chemical amount (in moles) of water in the container. Determine the number of molecules of water in the container. Determine the chemical amount (in millimoles) of water in the container.

58. (See Example 2–12.) A container holds 10.90 g of $ZnCl_2 \cdot 4NH_3$. Determine the chemical amount (in moles) of ammonia in the container. Determine the number of molecules of ammonia in the container. Determine the chemical amount (in millimoles) of ammonia in the container.

59. The gold-containing drug disodium gold(I) thiomalate relieves suffering from rheumatoid arthritis. The molecular formula of the drug is $AuC_4H_4O_4Na_2S$. An injection of 50.0 mg of the drug is given each week for 20 weeks. Compute the total mass of gold administered. Assume that the molecular formula of the substance is not known. Show how to solve the problem knowing only the empirical formula of the drug.

60. The compound 3′-azido-3′-thymidine (AZT) helps in the treatment of AIDS. The molecular formula of AZT is $C_{10}H_{13}N_5O_5$. A patient weighs 60.5 kg. The daily experimental dose in a test is 3.00 mg per kilogram of body mass.
(a) Compute the chemical amount (in moles) of AZT that this patient receives each day.
(b) Explain why it is not possible to solve part (a) given only the empirical formula of AZT.

61. (See Example 2–13.) Cryolite (Na_3AlF_6) is used in the production of aluminum from its ores. It is made by the reaction

$$6\ NaOH(aq) + Al_2O_3(s) + 12\ HF(aq) \longrightarrow$$
$$2\ Na_3AlF_6(s) + 9\ H_2O(\ell)$$

Calculate the mass of cryolite that can be prepared by the complete reaction of 287 metric tons of Al_2O_3 (one metric ton equals 1000 kg).

62. (See Example 2–13.) Carbon disulfide (CS_2) is a liquid that is used in the production of rayon and cellophane. It is manufactured from methane and elemental sulfur by the reaction

$$CH_4(g) + 4\ S(s) \longrightarrow CS_2(\ell) + 2\ H_2S(g)$$

Calculate the mass of $CS_2(\ell)$ that can be prepared by the complete reaction of 67.2 lb of sulfur.

63. The chemical known as 2,3,7,8-tetrachlorodibenzo-*p*-dioxin (TCDD, commonly referred to as "dioxin," with chemical formula $C_2H_4O_2Cl_4$), is a contaminant of some agricultural herbicides. It is extremely lethal to certain laboratory animals. An oral dose of just 0.6 μg per kilogram of body mass kills half the test population of guinea pigs. Assuming that the body of a guinea pig consists entirely of water molecules, what percentage of the molecules in the guinea pig are TCDD molecules after such a dose is administered?

64. Dioxin, or TCDD (see previous problem), appears to be much less lethal to human beings than to guinea pigs. An experiment with human volunteers involved treatment with 107,000 ng per kilogram of body mass, externally applied to the skin. No long-term symptoms were observed. If 20% of the TCDD entered a volunteer's body in such an experiment, what fraction of molecules in the body were TCDD molecules? Assume that essentially all of the molecules in the human body are water molecules.

65. A new chlor-alkali plant produces 660 metric tons of chlorine daily. The process uses the decomposition of an aqueous solution of sodium chloride (NaCl) to give the chlorine (Cl_2) and sodium hydroxide (NaOH) as a by-product. Estimate (in metric tons) how much sodium hydroxide is produced daily.

66. Portland cement is approximately 46% calcium and 9% silicon by mass, with most of the rest being oxygen. It is made from lime (CaO) derived from limestone. Estimate the mass of lime used in one year's world production of portland cement (800 million metric tons).

67. How many kilomoles (kmol) of CO_2 molecules are contained in 88 kg of carbon dioxide? How many kilomoles of oxygen atoms are contained in the sample? How many atoms of oxygen are contained in the sample?

68. How many pounds of hydrogen (H_2) are needed to react completely with 8.0 lb of oxygen (O_2) to produce water? How many pounds of water are produced as a result of this reaction?

ADDITIONAL PROBLEMS

69. The following chemical equations all appear in chemistry publications as examples to be balanced. All are invalid (cannot be balanced uniquely). Prove this by balancing each equation in two ways that have different ratios among the coefficients.

(a) $KClO_3(s) + HCl(aq) \longrightarrow$
$$KCl(aq) + H_2O(\ell) + Cl_2(g) + ClO_2(g)$$

(b) $Ba(OH)_2(aq) + H_2O_2(aq) + ClO_2(aq) \longrightarrow$
$$Ba(ClO_2)_2(s) + H_2O(\ell) + O_2(g)$$

(c) $CrCl_2(s) + KOH(aq) + Cl_2(g) + KI(aq) \longrightarrow$
$$K_2CrO_4(aq) + KIO_4(aq) + KCl(aq) + H_2O(\ell)$$

70. A new compound with excellent potential for making temperature-resistant plastics has the empirical formula $C_6N_2O_3$. When treated with water, it reacts to form a compound of empirical formula C_3HNO_2. It requires three moles of water to perform this conversion on one mole of the original compound, and no other products are formed. Determine the molecular formula and molar mass of the original compound.

71. Aspartame (molecular formula $C_{14}H_{18}N_2O_5$) is a sugar substitute. Under certain conditions, one mole of aspartame reacts with two moles of water to give one mole of aspartic acid (molecular formula $C_4H_7NO_4$), one mole of methanol (molecular formula CH_3OH), and one mole of phenylalanine. Determine the molecular formula of phenylalanine.

72. Alkenes are hydrocarbons that have the general formula C_nH_{2n}, where *n* is a whole number. Suppose that 1.000 g of any alkene is burned in excess oxygen to give carbon dioxide and water. Compute the mass of carbon dioxide and the mass of water formed.

*73. 3'-Methylphthalanilic acid is used commercially as a "fruit set" to prevent premature drop of apples, pears, cherries, and peaches from the tree. It is 70.58% carbon, 5.13% hydrogen, 5.49% nitrogen, and 18.80% oxygen. If eaten, the fruit set reacts with water in the body to split into an innocuous product that contains carbon, hydrogen, and oxygen only and *m*-toluidine ($NH_2C_6H_4CH_3$), which causes anemia and kidney damage. Compute the mass of the fruit set that would produce 5.23 g of *m*-toluidine.

*74. If 1.000 g of a hydrated lanthanum phosphate is heated in a vacuum, it gives 0.2922 g of P_4O_{10}, 0.6707 g of La_2O_3, and 0.0371 g of H_2O.
(a) Determine the empirical formula of the hydrated lanthanum phosphate.
(b) Write a balanced equation to represent the decomposition described in part (a).

75. When 48.044 g of carbon reacts with 12.09 g of hydrogen to form a hydrocarbon, no part of either reactant is left in excess. Determine the empirical formula of the hydrocarbon.

76. A yield of 3.00 g of $KClO_4$ is obtained from the reaction

$$KClO_3 \longrightarrow KClO_4 + KCl$$

when 4.00 g of the reactant is used. What is the percentage yield of the reaction? (*Note:* The equation is not balanced.)

77. A student tells you this short-cut way to identify the limiting reactant when specific amounts of substances react in a known way (according to a single balanced chemical equation): "Figure out the number of moles of every reactant. Then divide each answer by the coefficient that the reactant has in the balanced equation. The reactant for which the answer is smallest is the limiting reactant." This method always works! Prove why.

78. Potassium superoxide (KO_2) absorbs gaseous carbon dioxide from the air according to the (unbalanced) equation

$$KO_2(s) + CO_2(g) \longrightarrow K_2CO_3(s) + O_2(g)$$

In one experiment, run at constant temperature and pressure, there was an excess of KO_2, the yield of O_2 was 98.5% of the theoretical yield, and 117 L of O_2 was produced. Find the volume (in liters) of CO_2 that was used.

79. If metallic niobium is covered with liquid niobium pentachloride in a sealed tube and heated to 300°C for a week, a chemical reaction occurs in which $NbCl_4$ is the only product.
(a) Write a balanced chemical equation for this reaction, taking $NbCl_5$ as the chemical formula of niobium pentachloride.
(b) The reaction is performed using 5.0 g of liquid niobium pentachloride and 0.50 g of niobium. Which reactant is the limiting reactant in this experiment?
(c) During the week of heating, it is learned that the molecular formula of niobium pentachloride is really Nb_2Cl_{10}, not $NbCl_5$ as suggested by the compound's name. How does this affect the conclusion in part (b)?

*80. The senior partner in your law firm calls you in on a case in which a drug dealer was arrested in possession of 5.0 U.S. gallons of piperidine (molecular formula $C_5H_{11}N$) and some other chemicals. Piperidine is a starting material in the synthesis of phencyclidine (molecular formula $C_{17}H_{25}N$), the notorious "angel dust." The senior partner asks roughly how many kilograms of angel dust could be synthesized from 5.0 gal of piperidine.
(a) List the facts you would need to look up to find out first the mass (in kilograms) of piperidine and then the chemical amount (in moles) of piperidine that was seized.
(b) There is one N in the molecular formula of each substance, so it seems reasonable to assume that one mole of phencyclidine is produced for every mole of piperidine used. However, you are aware that this assumption could be wrong. Explain exactly how this assumption could be wrong.
(c) Use the assumption in part (b) and the facts in part (a) to answer your senior partner's question.

*81. A newspaper article about the danger of global warming from the accumulation of greenhouse gases, such as carbon dioxide, states that "reducing driving your car by 20 miles a week would prevent release of over 1000 lb of CO_2 per year into the atmosphere." Check whether this is a reasonable statement. Assume that gasoline is octane (molecular formula C_8H_{18}) and that it is burned completely to CO_2 and H_2O in the engine of your car. Facts (or reasonable guesses) about your car's gas mileage, the density of octane, and other factors are also needed.

82. The addition of oxygen-containing compounds to gasoline reduces the amount of pollutants (hydrocarbons and carbon monoxide) emitted in automobile exhaust. Compute the mass of methyl *t*-butyl ether (MTBE) that must be added to 1.0 kg of gasoline to increase the mass percentage of oxygen in the gasoline from 0.0% to 2.7%. The molecular formula of MTBE is $C_5H_{12}O$.

83. Existing stockpiles of the refrigerant Freon-12 (molecular formula CF_2Cl_2) have to be destroyed because it (and other Freons) leads to the depletion of the ozone layer when released into the atmosphere. One method is to pass gaseous Freon-12 through a bed of powdered sodium oxalate at 270°C. The following reaction occurs:

$$CF_2Cl_2(g) + Na_2C_2O_4(s) \longrightarrow$$
$$NaF(s) + NaCl(s) + C(s) + CO_2(g)$$

(a) Balance the equation.
(b) Determine the mass (in kilograms) of sodium oxalate needed to destroy 1.47×10^3 kg of Freon-12.

84. The carbon dioxide produced (together with hydrogen) from the industrial-scale oxygenation of methane in the presence of nickel is removed from the gas mixture in a scrubber containing an aqueous solution of potassium carbonate:

$$CO_2(g) + H_2O(\ell) + K_2CO_3(aq) \longrightarrow 2\,KHCO_3(aq)$$

Calculate the mass (in kilograms) of carbon dioxide that reacts with 187 L of a 1.36 M potassium carbonate solution.

85. Phosphorus trifluoride is a highly toxic gas that reacts slowly with water to give a mixture of phosphorous acid H_3PO_3 and hydrofluoric acid (HF). Determine the concentrations (in moles per liter) of each of the acids that result from the reaction of 0.077 g of phosphorus trifluoride with water to give a solution volume of 872 mL.

86. Phosphorus pentachloride (PCl_5) reacts violently with water to give a mixture of phosphoric acid (H_3PO_4) and hydrochloric acid (HCl).
(a) Write a balanced chemical equation for this reaction.
(b) Determine the concentration (in moles per liter) of each of the acids that result from mixing 0.0293 g of phosphorus pentachloride with enough water to give a solution with a final volume of 697 mL.

87. Potassium oxide (K_2O) is prepared by carefully melting potassium nitrate (KNO_3) with elemental potassium:

$$2\,KNO_3(\ell) + 10\,K(\ell) \longrightarrow 6\,K_2O(s) + N_2(g)$$

Determine the minimum masses of KNO_3 and K necessary to produce 8.34 g of K_2O according to this equation.

88. The mixing of 25.35 mL of 0.1430 M silver nitrate ($AgNO_3$) with 18.29 mL of 0.09875 M barium iodide (BaI_2) resulted in the precipitation of silver iodide (AgI). What was the theoretical yield of silver iodide?

89. A 2.513 g sample of *impure* ammonium sulfate (($NH_4)_2SO_4$) was dissolved in water. Addition of 32.71 mL of 0.5148 M barium chloride ($BaCl_2$) resulted in the removal of all the sulfate ions in the sample in the form of a barium sulfate ($BaSO_4$) precipitate. What percentage of the original impure sample was actually ammonium sulfate?

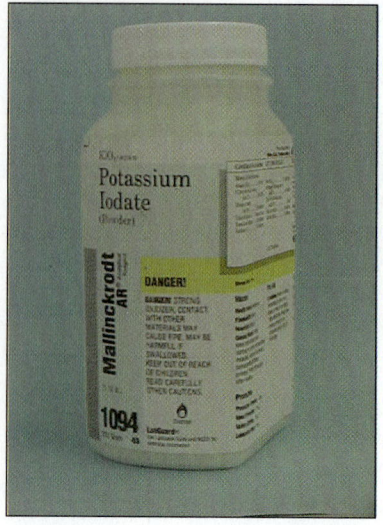

A bottle of analytical grade potassium iodate.

CUMULATIVE PROBLEM

Potassium Iodate

Potassium iodate (KIO_3) is a useful chemical. Although it is fairly reactive, it can be prepared and kept in very pure form. It does not decompose or gather moisture with age; it is easily dried in an oven if it gets wet; it gives solutions that do not change their concentration (if capped against evaporation) during reasonable periods of storage.

(a) Determine the concentration (in mol L^{-1}) of a solution prepared by dissolving 4.280 g of potassium iodate in enough water to make 2.00 L of solution.

(b) 75.0 mL of the solution in part (a) is diluted with water to a final volume of 500.0 mL. Determine the concentration (in mol L^{-1}) of this new solution.

(c) Treating aqueous potassium iodate with a mixture of aqueous potassium iodide (KI) and aqueous hydrochloric acid (HCl) produces aqueous iodine (I_2), liquid water (H_2O), aqueous potassium chloride (KCl), and no other products. Write and balance a chemical equation to represent this reaction.

(d) 40.00 mL of 0.0100 M potassium iodate, 1.00 g of potassium iodide, and 10.00 mL of 3 M hydrochloric acid react according to the equation in the previous part. Determine which reactant is the limiting reactant. Compute the mass (in grams) of the two reactants that remain in excess when the reaction stops.

(e) Silver iodate ($AgIO_3$) is not very soluble in water. In an experiment, 37.65 mL of 0.1003 M silver nitrate ($AgNO_3$) is mixed with 25.00 mL of 0.07964 M potassium iodate. The two react to give aqueous potassium nitrate (KNO_3) and solid silver iodate. The solid is collected, dried, and found to have mass of 508.73 mg. Compute the percentage yield of silver iodate.

(f) A quantity of solid potassium iodate reacts with 10.00 mL of gaseous fluorine (F_2). All of the fluorine is consumed to give gaseous iodine heptafluoride (IF_7), gaseous oxygen difluoride (OF_2), solid potassium fluoride (KF), and no other products. Determine the volume of IF_7 and the volume of OF_2 produced if the temperature and pressure are the same after the reaction as before.

(g) Is it possible in part (f) to determine the volume of $KF(s)$ that is produced? Explain why or why not.

3

Chemical Periodicity and the Formation of Simple Compounds

CHAPTER OUTLINE

Samples of bromine, a non-metal (*left*); silicon, a semi-metal (*top right*); and copper, a transition metal (*bottom*) arrayed on a sheet of aluminum, a main-group metal. (*Charles D. Winters*)

In Chapter 2, we showed how the law of conservation of mass, through the mole concept, allows us to establish and use quantitative mass relationships in chemical reactions. In that discussion, we assumed as prior knowledge the chemical formulas of the reactants and products in each equation. We also assumed that the indicated reaction would actually occur. We turn now to the far more open-ended question of what determines chemical reactivity. Why are some elements and compounds fiercely reactive and others inert? Why are there compounds with the chemical formulas H_2O and $NaCl$, but never an H_3O or an $NaCl_2$? Why do hydrogen atoms cling together in closely bound H_2 molecules under most conditions, whereas helium atoms remain resolutely unassociated? These questions can be answered on various levels, and they are addressed throughout this book.

As discussed in Chapter 1, every atom contains one or more electrons, which are fundamental particles possessing both electrical charge and mass. The behavior of electrons in atoms is complex, and we defer a full discussion to Chapters 16 and 17. Here, we focus on counting electrons and classifying them into two broad groups: outer electrons and core electrons. Many similarities in the physical properties of the elements echo similarities in the number of outer electrons, those farthest from the nucleus, in their atoms. Chemical bonding, which consists of the transfer or sharing of electrons among atoms, also depends strongly on the number of outer electrons, simply because outer electrons are the ones most affected when other atoms approach.

A pictorial device called a Lewis structure illustrates the transfer and sharing of outer electrons in a particularly simple way. In this chapter, we show how to set down, interpret, and evaluate these useful diagrams. We then employ Lewis structures in the prediction of the shapes of molecules. Finally, we present a system of naming compounds that uses the bonding concepts just developed and provides a language for describing the chemical reactions that we take up in Chapter 4.

3-1 Groups of Elements

By the late 1860s, more than 60 chemical elements had been identified, and much descriptive information on the physical and chemical properties of these elements had been accumulated. Similarities among their properties suggested various natural groupings. Elements were classified as **metals** or **non-metals,** for example, depending on the presence (or absence) of a characteristic metallic luster, their good (or poor) ability to conduct electricity and heat, and their malleability (or brittleness). The classification of an element as a metal or non-metal was generally straightforward, but certain elements (antimony, arsenic, boron, germanium, silicon, and tellurium) resembled metals in some respects and non-metals in others and were called **semi-metals** or **metalloids.**

Close scrutiny of other properties of the elements, in particular the empirical formulas of their binary compounds with chlorine (their *chlorides*), with oxygen (their *oxides*), and with hydrogen (their *hydrides*), allowed smaller groups of three to six chemically similar elements to be identified. With some changes, these groups have significance to this day. We briefly describe the characteristics of seven of the groups and then show in Section 3–2 how they were brought together in a single pattern with the discovery of the **periodic law.**

• With the exception of mercury, the metallic elements are all solids at room temperature; the non-metallic elements are solids, liquids, and gases.

Seven Groups of Elements

The **alkali metals** (lithium, sodium, potassium, rubidium, and cesium) are relatively soft metals (Fig. 3–1) with low melting points; they form 1 : 1 binary compounds with chlorine with chemical formulas such as NaCl, RbCl, and so forth. The alkali metals react with water to liberate hydrogen. Potassium, rubidium, and cesium do this particularly vigorously, but all alkali metals must be handled with care because heat from their reaction with water can cause the hydrogen generated by them to burst into flame (Fig. 3–2). All of these metals form oxides in 2 : 1 proportions (with formulas such as Na_2O and K_2O), but all also combine with oxygen in other ratios to form reactive compounds, such as KO_2 and Na_2O_2. The elements sodium and potassium are quite common in the form of dissolved NaCl and KCl in the oceans. Both are important in biological systems; a proper balance of dissolved sodium and potassium is essential in regulating the transport of molecules across cell walls. Dissolved lithium also can play a significant role in the cell; it is used in the treatment of manic-depressive behavior.

A second group of elements, the **alkaline earth metals** (beryllium, magnesium, calcium, strontium, barium, and radium), combine in a 1 : 2 mole ratio with chlorine to give compounds such as $MgCl_2$ and $CaCl_2$ and with oxygen in 1 : 1 proportions to give compounds such as MgO and CaO. Calcium is an abundant element. Its carbonate, $CaCO_3$, is widely distributed geologically, appearing as chalk, limestone, marble, or crystalline calcite and aragonite (Fig. 3–3) under different circumstances. When heated, it gives up carbon dioxide and leaves lime (CaO), a compound used for making glass, mortar, and cement. The carbonate of magnesium ($MgCO_3$) forms a plentiful mineral (dolomite) with calcium carbonate. Beryllium is a much less abundant alkaline earth metal, although masses of the mineral beryl ($Be_3Al_2Si_6O_{18}$) that weigh as much as a ton have been found. Pure beryl is colorless, but impurities

• The name "alkali" comes from the fact that some of these metals are found in compounds leached from wood ashes (*alqali* is Arabic for "ashes").

Figure 3–1 • Elemental sodium is a soft metal that is easily cut with a knife. It melts at a temperature of only 97.8°C (just below the normal boiling point of water). Sodium is stored under oil because it corrodes rapidly by reacting with water vapor in the air. (*Charles D. Winters*)

(a)

(b)

Figure 3–2 • (a) A fragment of sodium added to water starts to react instantly. The reaction produces hydrogen and sodium hydroxide and enough heat to melt the unreacted sodium. The molten globule skitters over the surface of the water, propelled by evolving bubbles of gaseous hydrogen. Sometimes the hydrogen catches fire. Here, the red indicator shows where sodium hydroxide has formed. (*Marna G. Clarke*) (b) The reaction of potassium with water is similar but more vigorous. Its heat generally ignites both the hydrogen that forms and the unreacted potassium, causing a burst of flames. (*Charles Steele*)

Figure 3–3 • Calcium carbonate ($CaCO_3$) occurs in two crystalline forms: calcite (a) and aragonite (b). Limestone is an aggregate of small crystals of the two plus a minor percentage of impurities, such as $MgCO_3$ and SiO_2. *(a and b, Leon Lewandowski)*

(a)

(b)

Figure 3–4 • An emerald is an impure crystal of beryl. Its deep green color depends on the presence of chromium atoms at approximately 20% of the beryllium atom sites in the crystal. Crystals of pure beryl are colorless. *(M. Clayel Jacana/Photo Researchers, Inc.)*

give it color. Chromium as an impurity makes beryl into emerald (Fig. 3–4), and iron makes it aquamarine (Fig. 3–5).

The **chalcogens** oxygen, sulfur, selenium, and tellurium form 1 : 1 compounds with the alkaline earth metals, but 2 : 1 compounds with the alkali metals; that is, the alkali metal chalcogenides have formulas such as Li_2O and Li_2S, whereas the alkaline earth chalcogenides have formulas such as CaO and CaS. The lightest chalcogen, oxygen, is the most abundant element at the Earth's surface. It is found free (as O_2) in the atmosphere, in combination with hydrogen (in H_2O) in the oceans, and combined with other elements in the Earth's crust. (The primary constituents of most rocks are oxygen, silicon, and aluminum.) Sulfur is also abundant. It is found in elemental form as a powdery or crystalline yellow solid, and in combination with metals in minerals such as iron pyrites (FeS_2, commonly known as "fool's gold"; Fig. 3–6) and gypsum ($CaSO_4 \cdot 2H_2O$). Selenium and tellurium are quite rare, although it has been established that traces of selenium are essential in the human diet.

Figure 3–5 • Aquamarine gemstones can be green, yellow, or blue. Aquamarines are all crystals of beryl with atoms of iron lodged as impurities in different types of sites. Upon heating, a green aquamarine turns blue as chemical changes occur at some of the iron sites. *(Charles D. Winters)*

Figure 3–6 • Iron pyrites, or fool's gold, has the chemical formula FeS_2. The mineral is an abundant and important source of elemental sulfur. Iron pyrites has a metallic luster, but whereas gold (a true metal) is malleable and is dented by pressure with a sharp tool, fool's gold is brittle and shatters. *(Charles D. Winters)*

(a)   (b)   (c)

Figure 3–7 • Chlorine (a), bromine (b), and iodine (c) differ in their colors and physical states. They are respectively gaseous, liquid, and solid at room temperature. As members of the halogen family (Group VII), the three have closely related chemical properties. (*Charles Steele*)

Fluorine, chlorine, bromine, and iodine are members of a family of elements called **halogens.** These four elements differ significantly in their physical properties (the first two are gases at room temperature, whereas bromine is a liquid and iodine a solid; Fig. 3–7), but on the whole their chemical behaviors are similar. Any alkali metal element combines with any halogen in 1 : 1 proportions to form a compound, such as LiF or RbI. As a group, these compounds are the "alkali-metal halides." They are relatively hard solids at room temperature and have high melting points; when dissolved in water, they give solutions that conduct electric current. Chlorine is quite abundant in seawater (as dissolved sodium chloride and potassium chloride); bromine is less abundant, and iodine still less so. Iodine, however, is concentrated in seaweed and other forms of aquatic life and can be recovered economically from such sources. Iodine deficiency in humans causes goiter and is avoided through the use of iodized salt. The chemistry of fluorine often differs from that of the other halogens and is discussed in Chapter 23. Fluorine is found primarily in minerals such as fluorite, CaF_2 (Fig. 3–8).

• Goiter is an enlargement of the thyroid gland. It is associated with low production of the iodine-containing hormone thyroxine.

Figure 3–8 • The mineral fluorite (calcium fluoride CaF_2) occurs in various shades of purple. The color arises from a small number of defects (see Section 20–4) in which fluoride ions are missing from their sites in the crystal and electrons take their places. (*Charles D. Winters*)

Figure 3–9 • A small mountain of borax awaits shipment from a mine in the Mojave Desert. Borax ($Na_2B_4O_5(OH)_4 \cdot 8H_2O$) is mined on a scale of millions of tons per year and is the major source of boron. Although boron is not abundant in the Earth's crust, it occurs in high concentration in deposits associated with former volcanic activity and can be mined economically. *(John Cunningham/Visuals Unlimited)*

Other elements fall into three additional groups that are less clearly defined in terms of their chemical and physical properties than those mentioned so far.

The first of these contains a semi-metal (boron) and four metals (aluminum, gallium, indium, and thallium). All form 1 : 3 chlorides (such as $GaCl_3$) and 2 : 3 oxides (such as Al_2O_3). This chemical similarity is the reason for the inclusion of boron in this group. Boron is found in large deposits of borax ($Na_2B_4O_5(OH)_4 \cdot 8H_2O$), especially in the California desert (Fig. 3–9). Of the metals of this group, aluminum is the most abundant, being found in many minerals in association with silicon, oxygen, and other elements.

Another group includes a non-metal as well as semi-metals and metals. It consists of the elements carbon, silicon, germanium, tin, and lead. All of these elements form 1 : 4 chlorides (such as $SiCl_4$), 1 : 4 hydrides (such as GeH_4), and 1 : 2 oxides (such as SnO_2). In this group, tin and lead are metals with low melting points, and silicon and germanium are semi-metals. Carbon, the non-metal, exists in several forms with dramatically different properties (Fig. 3–10). *Diamond* consists of very hard, transparent crystals that are non-conductors of electricity; graphite is a soft

• The hydride of lead (PbH_4) is unstable and has not been characterized definitively.

(a)

(b)

Figure 3–10 • Diamond and graphite (a), as well as buckminsterfullerene (b), are allotropic forms of carbon. The largest use of graphite is in electrodes for electrochemical cells, where advantage is taken of its ability to conduct electricity. Diamond does not conduct electricity. Smaller, non-gem-quality diamonds are used industrially in drill bits and in other applications in which the exceptional hardness of diamond is needed. Buckminsterfullerene is a recently discovered molecular form of carbon, with the formula C_{60}. It and related fullerenes have many potential uses. *(a, Gerard Vandystadt/Photo Researchers, Inc.; b, Donald Huffman. Photo by W. Kratschmer)*

Figure 3–11 • Fluorapatite ($Ca_5(PO_4)_3F$) is one of several apatite minerals that provide the commercial sources for phosphorus. Collectively, the apatites are mined on a scale of over a million tons per year and are massively used in the production of phosphate fertilizers. Major apatite deposits are found in Florida and in Morocco. Pure $Ca_5(PO_4)_3F$ is white. The purple in this sample results from manganese-containing impurities. *(Julius Weber)*

grayish black, flaky solid that conducts electricity; and the *fullerenes,* first synthesized in quantity in 1990, are dark-colored, soft, crystalline solids. These forms are **allotropes**—modifications of an element that differ because the atoms are organized in different ways. Silicon, the most abundant element of this group, is found in many minerals, such as quartz (SiO_2). Carbon is less abundant overall, but it plays a crucial role in the biosphere. Life as we know it on Earth would not be possible without carbon.

A seventh and final group includes nitrogen, phosphorus, arsenic, antimony, and bismuth. These elements form $1:3$ binary compounds with hydrogen (for example, PH_3) and $2:5$ binary compounds with oxygen, (such as N_2O_5). The hydrides become increasingly unstable as their molar masses increase, and BiH_3 can be kept only below $-45°C$. A similar trend exists for the oxides, and Bi_2O_5 has never been obtained in pure form. Nitrogen and phosphorus are quite abundant, the former as the major component of the atmosphere and the latter in minerals such as the apatites (Fig. 3–11), with formula $Ca_5(PO_4)_3X$, where X stands for F, Cl, or OH. Arsenic and antimony, which are poisonous, are often troublesome impurities in the smelting of ores, but bismuth is less bothersome because it is less abundant. All three are used to make alloys, which are mixtures created by melting metals together. They tend to confer hardness, the property of expanding upon solidification (useful in making castings), and low melting points upon their alloys. The lighter members of this group are non-metals (nitrogen and phosphorus); bismuth is a metal; and arsenic and antimony are classed as semi-metals.

• The "lead" in a pencil is graphite.

• Several alloys of bismuth melt at temperatures below the boiling point of water. The melting of plugs of such alloys releases the water in fire-extinguishing sprinkler systems.

3–2   The Periodic Table

The seven groups we have just outlined reflect modern understanding of the relationships among the elements. To a 19th-century chemist, some of the associations would have been evident (such as those in the alkali metal group), but others would have been much less obvious (such as placing nitrogen and bismuth in the same group). Moreover, many elements do not fall into these seven groups; several were well known in the 19th century. Silver, for example, resembles the alkali metals in forming a $1:1$ chloride (AgCl) and a $2:1$ oxide (Ag_2O), but its melting point is much higher (962°C, compared with 39°C for rubidium, for example), and it does not react with water. It therefore required considerable chemical judgment to determine which elements really belonged together in groups.

The Discovery of the Periodic Law

The next step came with the recognition of a connection between the group properties and atomic masses. When the lightest 17 elements known in the 1860s were arranged in order of increasing relative atomic mass,

Element: (H) Li Be B C N O F Na Mg Al Si P S Cl K Ca

Atomic Mass: (1) 7 9 11 12 14 16 19 23 24 27 28 31 32 35.5 39 40

repetitive sequences of the groups appeared. An element of the halogen group, for example, was repeatedly followed by one of the alkali metal group and preceded by one of the chalcogens. The similarities among these elements occurred with a periodicity of seven; that is, a resemblance was found between the second and ninth elements (alkali metals), another between the third and tenth (alkaline earth metals), and so forth. When the preceding series was rewritten to begin a new row with every seventh element, elements of the same groups were automatically assembled in columns in a **periodic table:**

• Obviously, the lightest element, hydrogen (H), does not fit the pattern.

(H)						
(1)						
Li	Be	B	C	N	O	F
7	9	11	12	14	16	19
Na	Mg	Al	Si	P	S	Cl
23	24	27	28	31	32	35.5
K	Ca	etc.				
39	40					

The theoretical basis of this table was summarized in the original **periodic law:**

> The chemical and physical properties of the elements are periodic functions of their atomic masses.

More complete periodic tables were constructed and introduced (independently) by the German chemist Lothar Meyer and the Russian Dmitri Mendeleev in 1869 and 1870. Figure 3–12 gives Mendeleev's 1872 version of the table. At that time, one third of the naturally occurring chemical elements had not yet been discovered, and both chemists were farsighted enough to leave gaps where their analysis of periodic physical and chemical properties indicated that unknown elements should be located. Mendeleev was bolder than Meyer was because he assumed that if a measured atomic mass put an element in the wrong place in the table, then the atomic mass was wrong. In some cases, this was true. Indium, for example, previously had been assigned a relative atomic mass near 76; however, there was no place for it in the periodic table between arsenic (with an atomic mass of 75) and selenium (with an atomic mass of 79). Mendeleev suggested that its relative atomic mass should be changed to $\frac{3}{2} \times 76 = 114$, near the currently accepted value of 114.82. In this position, its chemical properties fit the pattern defined by the known properties of aluminum and thallium. Subsequent work has shown that the elements are *not* strictly ordered in the periodic system by atomic mass and that four reversals exist. The chemical properties of tellurium, for example, demand that it come before iodine in the periodic table, even though its atomic mass is slightly greater. The other reversals are of argon and potassium, nickel and cobalt, and protactinium and thorium.

• These four anomalies arise from differing distributions in the natural abundances of the isotopes of the two neighboring elements. For example, the most abundant isotope of potassium is ^{39}K, which weighs less than ^{40}Ar, the most abundant isotope of argon.

TABELLE II

REIHEN	GRUPPE I. — R²O	GRUPPE II. — RO	GRUPPE III. — R²O³	GRUPPE IV. RH⁴ RO²	GRUPPE V. RH³ R²O⁵	GRUPPE VI. RH² RO³	GRUPPE VII. RH R²O⁷	GRUPPE VIII. — RO⁴
1	H=1							
2	Li = 7	Be = 9,4	B = 11	C = 12	N = 14	O = 16	F = 19	
3	Na = 23	Mg = 24	Al = 27,3	Si = 28	P = 31	S = 32	Cl = 35,5	
4	K = 39	Ca = 40	— = 44	Ti = 48	V = 51	Cr = 52	Mn = 55	Fe = 56, Co = 59, Ni = 59, Cu = 63.
5	(Cu = 63)	Zn = 65	— = 68	— = 72	As = 75	Se = 78	Br = 80	
6	Rb = 85	Sr = 87	?Yt = 88	Zr = 90	Nb = 94	Mo = 96	— = 100	Ru = 104, Rh = 104, Pd = 106, Ag = 108.
7	(Ag = 108)	Cd = 112	In = 113	Sn = 118	Sb = 122	Te = 125	J = 127	
8	Cs = 133	Ba = 137	?Di = 138	?Ce = 140	—	—	—	— — — —
9	(—)		—	—	—	—	—	
10	—	—	?Er = 178	?La = 180	Ta = 182	W = 184	—	Os = 195, Ir = 197, Pt = 198, Au = 199.
11	(Au = 199)	Hg = 200	Tl = 204	Pb = 207	Bi = 208	—		
12	—	—	—	Th = 231	—	U = 240	—	— — — —

Figure 3–12 • Dmitri Mendeleev published this early version of the periodic table in 1872. The German headings "GRUPPE" and "REIHEN" mean "group" and "rows," and the letter "R" is a generic symbol for all the elements of a group. Note how Mendeleev coped with the problem of the elements such as manganese (Mn) that do not fit well in the pattern of the seven main groups. He displayed them together with the main-group elements but pushed their symbols to the opposite side of the available space, seeking to establish a left–right alternation going down each group. The gap below boron was filled by the discovery of scandium; the gap below aluminum, by gallium; and the gap below silicon, by germanium. Mendeleev's superscripts mean the same as modern subscripts.

Mendeleev went further than Meyer in another respect. He predicted the properties of six elements yet to be discovered. A gap in his table just below aluminum, for example, implied the existence of an undiscovered element. Because he expected its properties to resemble those of aluminum, Mendeleev designated this element "eka-aluminum" (Sanskrit: *eka,* meaning "first") and predicted for it the set of properties given in Table 3–1. Just five years later, an element with the proper atomic

TABLE 3–1

Comparison of Eka-Aluminum and Gallium	
Prediction by Mendeleev for Eka-Aluminum	**Observed Properties of Gallium**
Relative atomic mass 68	Relative atomic mass 69.72
Low-melting metal	Metal, melting point = 29.8°C
Attacked slowly by acids and bases, does not oxidize readily in air	Observed as predicted
Density 6.0 g cm^{-3}	Density 5.9 g cm^{-3}
Formula of oxide E$_2$O$_3$	Formula of oxide Ga$_2$O$_3$
Density of oxide 5.5 g cm^{-3}	Density of oxide 5.88 g cm^{-3}
Oxide insoluble in water, soluble in bases or strong acids	Observed as predicted
Formula of chloride ECl$_3$, chloride volatile	Formula of chloride GaCl$_3$ (GaCl$_2$ is also found); melting point of chloride 78°C; boiling point 201°C

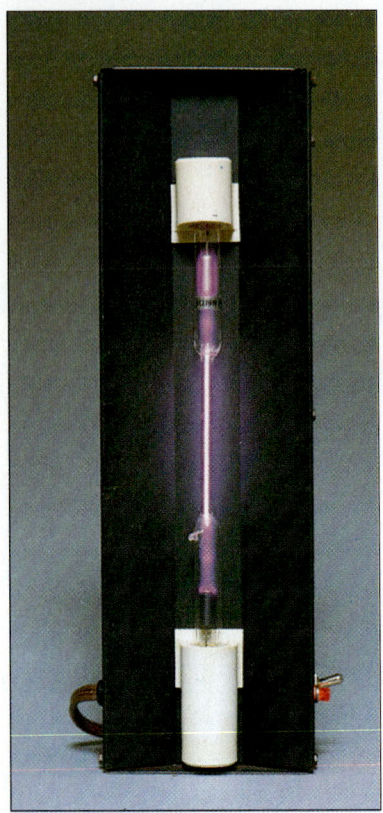

Figure 3-13 • Helium atoms emit light of a characteristic color after gaining energy from the passage of an electric current. *(Leon Lewandowski)*

• The magnitude of the electron's charge *e* equals $1.60217646 \times 10^{-19}$ coulombs (C). See Section 1–5 and Appendix B–1.

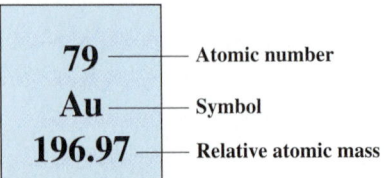

In the periodic table in the front of this book, the atomic number is placed directly above the element symbol, and the relative atomic mass is placed directly below.

mass was isolated and named "gallium" by its discoverer. The close correspondence (see Table 3–1) between the observed properties of gallium and Mendeleev's predictions for eka-aluminum lent strong support to the periodic law. Additional support came in 1885, when eka-silicon, which also had been described in advance by Mendeleev, was discovered and named "germanium."

The Noble Gases: An Eighth Group of Elements

The structure of the periodic table appeared to limit the number of possible elements. It was therefore quite surprising when John William Strutt, Lord Rayleigh, discovered a gaseous element in 1895 that did not fit into the previous classification scheme. A century earlier, Cavendish had noted the existence of a residual gas when oxygen and nitrogen are removed from air, but its importance had not been realized. Together with William Ramsay, Rayleigh isolated the gas and named it "argon." Ramsay then studied a gas isolated from a uranium-containing mineral and found that it was helium. Helium, which also occurs in certain natural-gas deposits, had been discovered (and named) previously, but only on the basis of an otherwise unexplainable yellow emission in the light from the corona of the sun (Fig. 3–13). Helium was not previously known on Earth. Rayleigh and Ramsay observed that it resembled argon in its properties and postulated the existence of a new group in the periodic table to accommodate these new elements. In 1898, other members of the series (neon, krypton, and xenon) were isolated. These elements are referred to as **noble gases,** or sometimes as inert gases, because of their relative inertness toward chemical combination.

The Modern Periodic Table

To understand the ordering of elements in the periodic table, we recall that, contrary to Dalton's original idea, atoms are *not* hard, indestructible spheres but have internal structure (see Section 1–5). Every atom consists of a positively charged nucleus that is surrounded by a swarm of negatively charged electrons. Opposite charges attract, and like charges repel; the electrons in an atom repel each other and would fly apart except for the attraction of the nucleus. An atom as a whole is neutral because the total negative charge of its electrons exactly balances the positive charge of its nucleus. The charge of the nucleus comes from the protons it contains. Each proton carries a positive charge exactly equal in magnitude to the negative charge of the electron. This magnitude of charge, symbolized *e*, provides a convenient natural unit for measurements. An electron has charge equal to -1 in such units and a proton $+1$. The number of protons in the nucleus of an atom equals its atomic number Z. Hence, the charge on the nucleus of an atom equals $+Z$, and the atom must contain Z electrons if it is to be electrically neutral.

All atoms of a given element have the same atomic number and the same number of electrons, even though their masses vary from one isotope to another. The chemistry of an atom depends negligibly on its mass and almost entirely on its number of electrons, which is, as just explained, dictated by its atomic number. Thus, the atomic number is the fundamental determinant of chemical behavior for an atom. For this reason, the modern periodic table is arranged strictly according to atomic number. Hydrogen, the first element in the table, has atomic number 1 and one electron per atom. The electrically neutral atoms of the following elements have atomic numbers that are successively larger by 1 and have successively more electrons. The synthesis of atoms with an atomic number as large as 114 has been reported. This corresponds to a nuclear charge of $+114$ and the presence of 114 electrons.

The periodic table, as we represent it today, encloses each element's symbol in a box and displays the boxes in **groups** (arranged vertically) and **periods** (arranged horizontally). The complete table is shown in Figure 3–14 and on the inside front cover of the book. There are eight groups of **representative elements** (or main-group elements), the properties of which we have already sketched. There are ten groups of **transition elements** (or "transition metals"). We shall discuss certain unique aspects of these important elements in Chapter 19. Among the transition elements is a set of 15 consecutive elements (with atomic numbers 57 through 71), called the **lanthanide elements,** after the first of the series, lanthanum, or the "rare-earth metals." Although it is now known that the rare-earth metals are not particularly rare (lanthanum and cerium, for example, are more abundant than tin), their remarkably similar chemical and physical properties still set them apart. They are

• The synthesis of three atoms of Element 118 was claimed in 1999. The claim was withdrawn in 2001.

• The lanthanide and actinide elements usually are placed below the rest of the table to conserve space.

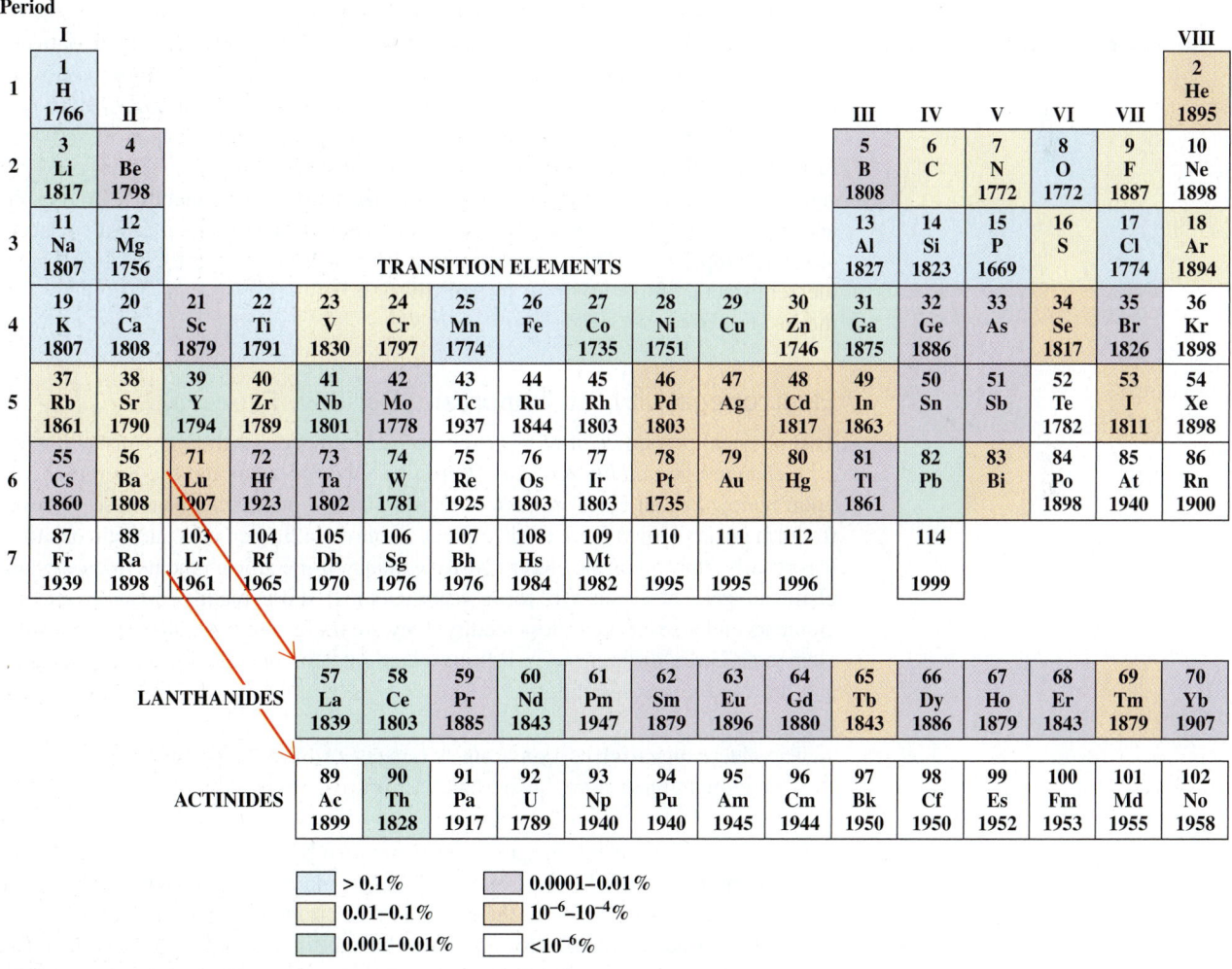

Figure 3–14 • The modern periodic table of the elements. Below each symbol is the element's year of discovery; elements with no dates have been known since ancient times. Above each symbol is the atomic number. The color coding indicates the relative abundance by mass of the elements in the world around us (the atmosphere, oceans, freshwater lakes and rivers, and the Earth's crust to a depth of 40 km). Oxygen alone makes up almost 50% of this mass, and silicon more than 25%.

usually found in association with one another and are difficult to separate. In the next period of the table, another set of 15 chemically similar elements (with atomic numbers 89 through 103) occurs. These **actinides** are all radioactive; that is, they have unstable nuclei and decay at varying rates to form less massive atoms. Only three of the actinides are found in nature; the rest must be produced artificially. The study of their chemistry poses special difficulties because they constantly contaminate themselves with decay products. Finally, elements beyond number 103 form a fourth period of transition elements. These radioactive elements are produced artificially and are so short-lived that few details of their chemistry are known.

The groups of representative elements are numbered (with Roman numerals) from I to VIII, with the letter "A" sometimes added to differentiate them from the transition-metal groups, which are labeled from IB to VIIIB. In this book, we use group numbers for the representative elements exclusively (and drop the "A"), and we refer to transition elements by the first element in the corresponding group. For example, we designate the elements in the carbon group (C, Si, Ge, Sn, Pb) as Group IV and the elements Cr, Mo, and W as the chromium group.

Physical as well as chemical properties of the elements vary systematically across the periodic table. Important physical properties include melting points and boiling points, thermal and electrical conductivity, density, and atomic size. In general, the elements on the left side of the table (especially in the later periods) are lustrous metallic solids and good conductors of both heat and electricity. On the right side (especially in earlier periods), the elements are non-metallic. In the solid and liquid states, they lack the characteristic luster of metals, and many are gaseous at room temperature. They are poor conductors of heat and electricity. In between, the semi-metals form a zigzag line of division between metal and non-metal (see the inside front cover of the book).

- The International Union of Pure and Applied Chemistry (IUPAC) and the American Chemical Society recommend designation of the main-group elements as Groups 1, 2, and 13 to 18 (with Arabic numerals) and of the transition elements as Groups 3 to 12.

Electronegativity, an Important Periodic Property

Two physical properties of atoms exert crucial influence on their chemistry: the energy change upon *addition* of an electron to a neutral atom and the energy change upon *removal* of an electron from a neutral atom. Metallic elements lose electrons (to form positive ions) more readily than do non-metallic elements and gain electrons less readily; they are more **electropositive** than non-metallic elements. Non-metallic elements gain electrons (to form negative ions) more readily than do metallic elements and lose electrons less readily; they are more **electronegative** than metallic elements. These differences in affinity carry into the chemical behavior of atoms and have led to this definition:

> The **electronegativity** of an atom is a measure of its power when in chemical combination to attract electrons to itself.

Large differences in electronegativity between two bonded atoms favor the transfer of electrons from the less electronegative to the more electronegative atom. Small differences are associated with the sharing of electrons between the atoms.

Although numerical electronegativities have been assigned to the elements (see Fig. 17–13), good use can be made of the concept on a wholly qualitative basis. With some exceptions, electronegativity increases from left to right across the periodic table and decreases from the top down the table. Elements of high electronegativity cluster toward the upper right-hand corner of the table; elements of low electronegativity (the most electropositive elements) occur toward the lower left-hand corner.

3-3 Ions and Ionic Compounds

The periodic table provides a useful way to describe systematically the properties of the elements, and we shall return to it frequently. In Chapter 17, we show how the structure of the table arises in the application of quantum mechanics to the electronic structures of atoms. Here we confine our rationalization of the chemical properties of the elements to a simple model, the **Lewis electron-dot model.** The American chemist G. N. Lewis proposed this model in 1916, well before modern quantum mechanics was fully developed.

Lewis Dot Symbols

The Lewis model begins by recognizing that not all the electrons in an atom participate directly in chemical bonding. Electrons can be divided into two groups: **core electrons,** which are held close to the nucleus and are not significantly involved in the formation of bonds, and **valence electrons,** which occupy the outer regions of the atom, the so-called **valence shell,** and take part in chemical bonds. Moreover, with the exception of helium, the number of valence electrons in a neutral atom of a representative element (those in Groups I through VIII) is equal to the element's group number in the periodic table. Atoms of bromine, for example, have seven valence electrons (like chlorine), and atoms of strontium have two (like beryllium). Similar chemical behavior within a group arises from the equal numbers of valence electrons in atoms of elements in that group. The maximum number of valence electrons is eight, and atoms with this full set of valence electrons (the noble gases from neon through xenon) are particularly stable and unreactive chemically. Helium is an unreactive element with only two valence electrons. The Lewis model represents valence electrons by dots; core electrons are not shown. The first four dots are displayed singly around the four sides of the symbol of the element. If there are more than four valence electrons, then dots are paired with those already present. The result is a **Lewis dot symbol** for that atom. The Lewis dot symbols for the elements of the first three periods are

• "Valence" means "having to do with the capacity for chemical combination."

• As we shall see, the number of unpaired dots tells the element's typical combining capacity in compounds.

$$· H \qquad\qquad\qquad\qquad :He$$

$$·Li \quad ·Be· \quad ·\overset{.}{B}· \quad ·\overset{.}{\underset{.}{C}}· \quad :\overset{.}{\underset{.}{N}}· \quad :\overset{.}{\underset{.}{O}}: \quad :\overset{.}{\underset{.}{F}}: \quad :\overset{..}{\underset{..}{Ne}}:$$

$$·Na \quad ·Mg· \quad ·\overset{.}{Al}· \quad ·\overset{.}{\underset{.}{Si}}· \quad :\overset{.}{\underset{.}{P}}· \quad :\overset{.}{\underset{.}{S}}: \quad :\overset{.}{\underset{.}{Cl}}: \quad :\overset{..}{\underset{..}{Ar}}:$$

EXAMPLE 3-1

How many electrons does an atom of tellurium have? Of these, how many are valence electrons and how many are core electrons? Draw the Lewis dot symbol for tellurium.

Solution

The atomic number of tellurium is given in the periodic table as 52, so each tellurium atom has 52 electrons. Because tellurium is in Group VI in the table, it has six valence electrons. This leaves $52 - 6 = 46$ core electrons. The Lewis dot symbol for tellurium resembles that of oxygen or sulfur:

$$:\overset{.}{\underset{.}{Te}}:$$

> **EXERCISE**
>
> How many electrons does an atom of germanium have? Of these, how many are valence electrons and how many are core electrons? Draw the Lewis dot symbol for germanium.
>
> **Answer:** 32, 4, 28, $\cdot \overset{\displaystyle .}{Ge} \cdot$.

The Formation of Binary Ionic Compounds

A positively charged ion (called a cation) forms when an atom loses one or more electrons. A negatively charged ion (called an anion) forms when an atom adds electrons. Atoms lose and gain electrons fairly easily both in physical processes (such as mass spectrometry, see Section 1–6) and in chemical processes. The creation of ions is indicated by adding or removing the proper number of dots from the Lewis dot symbol and also by writing the net electric charge of the ion as a right superscript. This charge is understood to be expressed in units of the charge on the electron; thus, "2−" means that the ion has a net charge equal to the charge of two electrons. For example,

Na·	Na⁺	·Ca·	Ca²⁺
sodium atom	sodium ion	calcium atom	calcium ion
$:\overset{\displaystyle .}{F}:$	$:\overset{\displaystyle ..}{\underset{\displaystyle ..}{F}}:^-$	$:\overset{\displaystyle .}{\underset{\displaystyle .}{S}}:$	$:\overset{\displaystyle ..}{\underset{\displaystyle ..}{S}}:^{2-}$
fluorine atom	fluoride ion	sulfur atom	sulfide ion

Special stability results when an atom, by either losing or gaining electrons, forms an ion with the same number of valence electrons as a noble-gas atom. Except for hydrogen and helium, which can have at most two valence electrons, atoms of the main-group elements of the periodic table have a maximum of eight valence electrons. Thus, we speak of a chloride ion ($:\overset{\displaystyle ..}{\underset{\displaystyle ..}{Cl}}:^-$) or an argon atom ($:\overset{\displaystyle ..}{Ar}:$) as having a completed **octet** of valence electrons. Ions such as Na⁺ and Ca²⁺ have the same number of electrons as the next lower noble-gas atom Ne and so have completed octets. Dots are however not used in the Lewis symbols for such ions because these ions are core electrons that were exposed only by the loss of valence electrons.

The tendency of atoms to attain valence octets explains much chemical reactivity. Atoms of the electropositive elements in Groups I and II achieve an octet by losing electrons to form cations, and atoms of the electronegative elements in Groups VI and VII achieve an octet by gaining electrons to form anions. The cations and anions then attract each other in **ionic bonds.** Ionic bonding is favored by large differences in electronegativity between the bonded atoms. If ionic bonds are the only type or the predominant type of bonds in a compound, it is an ionic compound. Reactions of the metallic elements on the left side of the periodic table with the non-metallic elements on the right side always transfer just enough electrons to form ions with completed octets. The following equations, in which e^- stands for an electron, use Lewis symbols to show the formation first of a cation and anion and then of a compound.

$$Na\cdot \longrightarrow Na^+ + e^- \qquad \text{loss of a valence electron}$$

$$e^- + :\overset{\displaystyle ..}{\underset{\displaystyle ..}{Cl}}: \longrightarrow :\overset{\displaystyle ..}{\underset{\displaystyle ..}{Cl}}:^- \qquad \text{gain of a valence electron}$$

$$Na^+ + :\overset{\displaystyle ..}{\underset{\displaystyle ..}{Cl}}:^- \longrightarrow Na^+:\overset{\displaystyle ..}{\underset{\displaystyle ..}{Cl}}:^- \qquad \text{combination to form the compound NaCl}$$

• Shuffling your feet across a carpet under the right conditions gives you a static electric charge; some of the atoms in your body are then ionized.

Another example is the formation of $CaBr_2$:

$$\cdot Ca \cdot + : \overset{\cdot}{\underset{\cdot\cdot}{Br}} : + : \overset{\cdot}{\underset{\cdot\cdot}{Br}} : \longrightarrow Ca^{2+} + : \overset{\cdot\cdot}{\underset{\cdot\cdot}{Br}} : ^- + : \overset{\cdot\cdot}{\underset{\cdot\cdot}{Br}} : ^- \longrightarrow Ca^{2+}(: \overset{\cdot\cdot}{\underset{\cdot\cdot}{Br}} : ^-)_2$$

| neutral atoms not having octets | positive ion with octet | negative ions with octets | compound |

The model predicts a 1:1 compound between Na and Cl and a 1:2 compound between Ca and Br. Both predictions are correct.

EXAMPLE 3–2

Predict the formula of the compound between rubidium and sulfur. Give Lewis symbols for the elements both before and after chemical combination.

Solution

Rubidium, from Group I of the periodic table, has one valence electron. Sulfur, from Group VI, has six valence electrons. Their Lewis symbols are Rb· and $: \overset{\cdot}{\underset{\cdot\cdot}{S}} :$. The transfer of one electron from each of two rubidium atoms to a sulfur atom gives two Rb^+ ions and one $: \overset{\cdot\cdot}{\underset{\cdot\cdot}{S}} : ^{2-}$ ion, all having octets. The compound is Rb_2S or, in Lewis symbols, $(Rb^+)_2 \, (: \overset{\cdot\cdot}{\underset{\cdot\cdot}{S}} : ^{2-})$.

EXERCISE

Predict the formula of a compound between magnesium and nitrogen, which attains an octet by gaining three electrons. Use Lewis symbols to keep track of the valence electrons.

Answer: Mg_3N_2.

Ionic compounds are often called **salts**, by analogy with NaCl, the main component of common salt, which is everyone's typical ionic compound. Ionic compounds are solids at room conditions and generally have high melting and boiling points (for example, NaCl melts at 801°C and boils at 1413°C). Solid ionic compounds usually conduct electricity poorly, but their melts (the molten liquids) conduct well. Whether in the solid or liquid state, binary ionic compounds contain no molecules. Their chemical compositions are given by empirical, not molecular, formulas.

Names and Formulas of Binary Ionic Compounds

Ionic compounds result from the combination of cations with anions. Their names consist of the name of the cation followed by a space and the name of the anion. How are cations and anions named? The full answer is complex, but the following points provide a start:

• The names of many *monatomic cations* consist simply of the unmodified name of the element from which they derive. Positive ions derived from the elements in Groups I and II, and the first three elements in Group III are named in this way. For example, Na^+ is sodium ion, Ca^{2+} is calcium ion, and H^+ is hydrogen ion. Other electropositive elements, particularly the transition elements, tend to form more than one monatomic cation, so that

TABLE 3–2

Formulas and Names of Some Monatomic Anions and Cations

F^-	fluoride ion	H^+	hydrogen ion
Cl^-	chloride ion	Li^+	lithium ion
Br^-	bromide ion	Na^+	sodium ion
I^-	iodide ion	K^+	potassium ion
O^{2-}	oxide ion	Rb^+	rubidium ion
S^{2-}	sulfide ion	Cs^+	cesium ion
Se^{2-}	selenide ion	Be^{2+}	beryllium ion
Te^{2-}	telluride ion	Ca^{2+}	calcium ion
N^{3-}	nitride ion	Sr^{2+}	strontium ion
P^{3-}	phosphide ion	Ba^{2+}	barium ion
As^{3-}	arsenide ion	B^{3+}	boron ion
Sb^{3-}	antimonide ion	Al^{3+}	aluminum ion
H^-	hydride ion	Ga^{3+}	gallium ion

a simple name of this type can be ambiguous. This complication is dealt with as explained in Section 3–8.

- *Monatomic anions* are named by adding the anion-designating suffix -*ide* to the first portion of the name of the element. By this rule, Cl^- is the chlo*ride* ion (derived from chlor*ine*), O^{2-} is the ox*ide* ion (derived from ox*ygen*), S^{2-} is the sulf*ide* ion (derived from sulf*ur*), and N^{3-} is the nitr*ide* ion (derived from nitr*ogen*). Monatomic anions of the other Group V, VI, and VII elements are named similarly.

Table 3–2 gives the names and formulas of many monatomic anions and cations for reference and study. Note that the Group I element hydrogen gives a monatomic anion H^-, the hydr*ide* ion, as well as a monatomic cation.

Generating the name of a binary ionic compound from its formula is straightforward (assuming familiarity with the symbols for the elements). The task is made easier by the conventional practice of writing the symbol of the more electropositive element first. The reverse operation, obtaining a chemical formula from a name, can be carried out as shown in Example 3–2. It can also be carried out using the formal **principle of charge neutrality,** which recognizes that chemical compounds are electrically neutral and that the positive charge contributed by cations is therefore always exactly balanced by the negative charge contributed by anions. Charge neutrality controls the relative number of the two kinds of ions in binary ionic compounds:

Name	Charge Neutrality Requires...	Therefore the Formula Is...
Sodium bromide	One +1 cation to balance one −1 anion	NaBr
Lithium sulfide	Two +1 cations to balance one −2 anion	Li_2S
Calcium fluoride	One +2 cation to balance two −1 anions	CaF_2
Aluminum sulfide	Two +3 cations to balance three −2 anions	Al_2S_3
Magnesium nitride	Three +2 cations to balance two −3 anions	Mg_3N_2
Gallium arsenide	One +3 cation to balance one −3 anion	GaAs

The charges on many monatomic cations and anions can be inferred from the position of the parent element in the periodic table. Group VII atoms become -1 anions, Group VI atoms become -2 anions, Group V atoms become -3 anions, Group I atoms become $+1$ cations, and Group II atoms become $+2$ cations. This point is illustrated by the order of the entries in Table 3–2. Knowledge of the periodic table allows one to go from name to formula for a large number of ionic compounds.

EXAMPLE 3-3

Give the chemical formulas of these ionic compounds: (a) barium oxide and (b) cesium nitride.

Strategy

Write the symbols of the elements appearing in the compound, and locate them in the periodic table. From the position of the element in the table, determine the charge of its monatomic ion. Apply the principle of charge neutrality to arrive at the correct subscripts to use in the chemical formula.

Solution

(a) Barium oxide contains Ba^{2+} ions (barium is in Group II) and O^{2-} ions (oxygen is in Group VI). Charge neutrality requires one O^{2-} ion for each Ba^{2+} ion. The chemical formula is BaO.

(b) Cesium nitride contains Cs^+ ions (cesium is in Group I) and N^{3-} ions (nitrogen is in Group V). To attain charge neutrality, three Cs^+ ions (total charge, $+3$) must be present for every one N^{3-} ion (total charge, -3). The chemical formula is Cs_3N.

EXERCISE

Give the chemical formulas of aluminum oxide, potassium sulfide, strontium chloride, and gallium arsenide.

Answer: Al_2O_3, K_2S, $SrCl_2$, and $GaAs$.

3-4 Covalent Bonding and Lewis Structures

Elements of intermediate electronegativity form ionic compounds far less readily than do highly electronegative or highly electropositive elements. Consider methane (CH_4), the simplest compound between carbon and hydrogen, which have comparable, intermediate electronegativities. Unlike any ionic compound, methane is a gas, not a solid, at room temperature. Cooling methane to low temperatures condenses it first to a liquid and then to a solid in which distinct molecules retain their identities. Unlike melted ionic compounds, liquid methane does not conduct electricity. It is thus not useful to think of methane as an ionic substance made up of C^{4-} and H^+ ions (or C^{4+} and H^- ions). It is instead a **molecular substance.** The bonding within the methane molecules arises from (approximately) equal sharing of electrons between the carbon and the hydrogen atoms, not from the transfer of electrons. Such bonds are **covalent bonds.**

Lewis Structures

The Lewis electron-dot model is quite effective in describing the covalent bonding in molecules formed by non-metallic main-group elements. Compounds that contain only covalent bonds are called **covalent compounds.** Hydrogen and chlorine combine, for example, to form the covalent compound hydrogen chloride. This can be indicated through a **Lewis structure** for the molecule of the product, in which the valence electrons from each atom are redistributed so that one electron from the hydrogen and one from the chlorine are shown as shared by the two atoms. The two dots representing this electron pair are placed between the symbols for the two elements:

$$\text{H} \cdot + \cdot \overset{\cdot\cdot}{\underset{\cdot\cdot}{\text{Cl}}} : \longrightarrow \text{H} : \overset{\cdot\cdot}{\underset{\cdot\cdot}{\text{Cl}}} :$$

The basic rule that governs the writing of Lewis structures is the **octet rule:**

> Whenever possible, the valence electrons are distributed in such a way that eight electrons (an octet of electrons) surround each main-group element (except hydrogen, which should have two electrons).

When the octet rule is satisfied, the atom attains the special stability of a noble-gas atom. In the structure for HCl shown earlier, the H nucleus, through sharing, is close to two valence electrons (like the noble gas helium), and the Cl has eight valence electrons near it (like the noble gas argon). Electrons that are shared between two atoms are counted as contributing to each atom.

Shared electrons hold the atoms together in covalent molecules. A covalent bond accordingly can be represented by a pair of dots positioned between the symbols of the atoms. A shared pair of electrons is also (and more frequently) represented by a short line (—):

$$\text{H} - \overset{\cdot\cdot}{\underset{\cdot\cdot}{\text{Cl}}} :$$

The unshared electron pairs around the chlorine atom in the Lewis structure are called **lone pairs,** and they make no contribution to the bond between the atoms. Lewis structures of some simple, but important, covalent compounds are

ammonia (NH₃)	water (H₂O)	methane (CH₄)

$$\text{H} : \overset{\cdot\cdot}{\underset{}{\text{N}}} : \text{H} \qquad \text{H} : \overset{\cdot\cdot}{\underset{\cdot\cdot}{\text{O}}} : \text{H} \qquad \begin{matrix} \text{H} \\ \text{H} : \overset{}{\underset{}{\text{C}}} : \text{H} \\ \text{H} \end{matrix}$$

$$\text{H} - \overset{\cdot\cdot}{\underset{|}{\text{N}}} - \text{H} \qquad \text{H} - \overset{\cdot\cdot}{\underset{\cdot\cdot}{\text{O}}} - \text{H} \qquad \begin{matrix} \text{H} \\ | \\ \text{H} - \overset{|}{\underset{|}{\text{C}}} - \text{H} \\ | \\ \text{H} \end{matrix}$$

Lewis structures indicate the way bonds connect the atoms in a molecule, but they do not show the three-dimensional molecular geometry. The ammonia molecule, for example, is not planar but pyramidal, with the nitrogen atom at the apex. The water molecule is bent rather than straight. Ball-and-stick models, such as those in Figure 3–15, can indicate three-dimensional geometry.

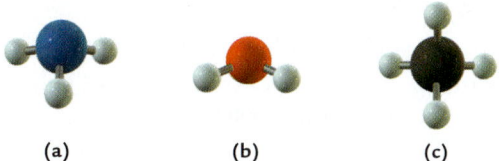

(a)   **(b)**   **(c)**

Figure 3–15 • The molecules of three familiar substances are shown in ball-and-stick drawings. (a) Ammonia (NH_3) is produced in amounts approaching 100 million tons per year to meet the demand for nitrogen-based fertilizers. (b) Water (H_2O) fills the oceans, lakes, and rivers and accounts for most of the mass of the human body. (c) Methane (CH_4) is the major constituent of natural gas and is an important starting material for the synthesis of substances ranging from ammonia to plastics.

Multiple Bonding

In some molecules, two atoms share more than one pair of electrons. The oxygen atoms in O_2, for example, each contribute six valence electrons to the molecule. Drawing a single bond between the atoms does not lead to valence octets. Octets can be attained on both atoms only by drawing the Lewis structure

$$\overset{..}{O}::\overset{..}{\underset{..}{O}} \quad \text{or} \quad \overset{..}{O}=\overset{..}{\underset{..}{O}}$$

in which two pairs of electrons are shared between the oxygen atoms. The bond is again a covalent bond but is a **double bond.** Similarly, the N_2 molecule has a **triple bond,** involving three shared electron pairs:

$$:N:::N: \quad \text{or} \quad :N{\equiv}N:$$

As the examples show, double bonds are written as double lines, and triple bonds, as triple lines. Carbon–carbon bonds can involve the sharing of one, two, or three electron pairs. A progression from single to triple bonding is found in the three hydrocarbons ethane (C_2H_6), ethylene (C_2H_4), and acetylene (C_2H_2):

$$H-\overset{\overset{\displaystyle H}{|}}{\underset{\underset{\displaystyle H}{|}}{C}}-\overset{\overset{\displaystyle H}{|}}{\underset{\underset{\displaystyle H}{|}}{C}}-H \qquad \overset{H}{\underset{H}{\Large{>}}}C=C\overset{H}{\underset{H}{\Large{<}}} \qquad H-C{\equiv}C-H$$

Multiple bonding to attain an octet most often involves the elements carbon, nitrogen, oxygen, and, to a lesser extent, sulfur. Double and triple bonds are shorter than single bonds between the same pair of atoms (Table 3–3).

Ethane (C_2H_6) occurs with methane in natural gas. It can be burned in oxygen, and, if strongly heated, it reacts to form hydrogen and ethylene.

Ethylene (C_2H_4) is the largest-volume organic (carbon-containing) chemical produced. Its end uses are predominantly in polymers, such as polyethylene plastic containers and polyvinyl chloride pipes and roofing materials.

TABLE 3–3			
Average Bond Lengths			
C—C 1.54	N—N 1.45	C—H 1.10	O—O 1.48
C=C 1.34	N=N 1.25	N—H 1.01	O=O 1.21
C≡C 1.20	N≡N 1.10	O—H 0.96	
C—O 1.43	N—O 1.43	C—N 1.47	
C=O 1.20	N=O 1.18	C≡N 1.16	

All values are in units of 10^{-10} m.

Acetylene (C_2H_2) has a triple bond that makes it highly reactive. The large amount of heat given off as it burns in oxygen in an oxy-acetylene torch makes the torch ideal for cutting and welding metals.

Carbon monoxide (CO) is a colorless, odorless, and toxic gas. It is produced by the incomplete burning of carbon-containing fuels in air. It is used in producing elemental metals from their oxide ores.

Formal Charges

Let us consider a molecule of carbon monoxide (CO). This molecule has ten valence electrons (four from the C and six from the O). The only Lewis structure that gives octets to both atoms uses a triple bond:

$$:C:::O:$$

As far as the positioning of electrons is concerned, this Lewis structure is just like the one given for N_2. If we assume, however, that the six bonding electrons are shared equally between the carbon and oxygen atoms, then the carbon atom owns five valence electrons (one more than its group number) and the oxygen atom owns five valence electrons (one less than its group number). If the sharing of the electrons in the bond is equal, then formally the carbon atom has gained an electron, attaining a **formal charge** of -1, and the oxygen atom has lost an electron, attaining a formal charge of $+1$.

$$:C:::O:$$

- We indicate the formal charge of an atom in a Lewis structure by a circled number in blue next to the atom.

Carbon monoxide is a covalent compound, and the assignment of these formal charges does not make it ionic. These charges would exist only if the carbon and oxygen atoms shared the six bonding electrons equally. In fact, the bonding electrons in carbon monoxide are not equally shared because the more electronegative oxygen atom attracts the shared electrons to itself more strongly than the carbon atom does. This makes the true charges on the oxygen and carbon atoms nowhere near as large as the formal values of $+1$ and -1.

Although they lack physical meaning, formal charges are useful in evaluating Lewis structures. A Lewis structure that gives large positive formal charges to some atoms and large negative formal charges to others is a poor model of covalent bonding. Often, two or more Lewis structures are possible for the same molecule. If so, the structure that assigns the lowest formal charges to the atoms generally provides the best description of the bonding. When non-zero formal charges appear, Lewis structures that assign negative formal charges to the more electronegative atoms and positive formal charges to the more electropositive atoms are preferable to the ones that do the reverse.

- Remember: A formal charge is determined only after a Lewis structure has been drawn.

The formal charge on an atom in a Lewis structure is simple to calculate. Determine the positive charge the atom would exhibit if it lost all of its valence electrons. For elements in the representative groups, this equals the group number in the periodic table: Elements in Group VI have six valence electrons and a charge of $+6$ if these electrons are removed, and so forth. Then, subtract the number of lone-pair valence electrons that the atom possesses in the Lewis structure. Finally, subtract one half of the number of bonding electrons associated with the atom in the Lewis structure:

formal charge = number of valence electrons −
number of electrons in lone pairs − $\frac{1}{2}$(number of electrons in bonding pairs)

EXAMPLE 3-4

Compute the formal charges on the atoms in the following Lewis structure for the azide ion N_3^-.

$$[:\overset{..}{N}=N=\overset{..}{N}:]^-$$

Solution

Nitrogen is in Group V; hence, each nitrogen atom contributes five valence electrons to the bonding, and the negative charge on the ion contributes one more electron. The Lewis structure correctly represents 16 electrons. We now use the formula

formal charge = number of valence electrons −
number of electrons in lone pairs − $\frac{1}{2}$(number of electrons in bonding pairs)

The group number of nitrogen is 5. The nitrogen on the left end of the Lewis structure has four electrons in lone pairs and four bonding electrons (which compose a double bond). Therefore,

$$\text{formal charge}_{(\text{left N})} = 5 - 4 - \tfrac{1}{2}(4) = -1$$

The nitrogen on the right end of the Lewis structure also has four electrons in lone pairs and four bonding electrons:

$$\text{formal charge}_{(\text{right N})} = 5 - 4 - \tfrac{1}{2}(4) = -1$$

The nitrogen in the center of the structure has no electrons in lone pairs. Its entire octet lies in the eight bonding electrons:

$$\text{formal charge}_{(\text{center N})} = 5 - 0 - \tfrac{1}{2}(8) = +1$$

Check

The sum of the three formal charges is −1, which is the true overall charge on this molecular ion. If this check fails, an error exists in either the Lewis structure or the arithmetic.

EXERCISE

Determine the formal charges on the atoms in an alternative Lewis structure for the same ion:

$$[\,\ddot{\text{N}} - \text{N} \equiv \text{N}\,\colon]^-$$

Answer: −2, +1, and 0, reading from left to right.

3–5 Drawing Lewis Structures

The drawing of Lewis structures strikes some as a game with no relation to chemical reality. In fact, however, Lewis structures work: They produce helpful predictions on matters of chemical importance. For example, double bonds are shorter and stronger than are single bonds, so their placement in molecules affects geometry and energy. Also, chemical reactivity differs strongly from single to double and triple bonds, so Lewis structures help to explain or predict trends in reactivity. Throughout this section, think of Lewis structures as a useful, even if imperfect, description of reality.

Some molecules can exist as two or more **isomers**—structures containing the same set of atoms, connected to each other in different orders. Formal charges can sometimes be used to explore possible isomers and determine which are likely to be less stable because they involve considerable separation of formal charge. For example, of the two isomers of the molecule HCN,

$$\overset{0\quad\;0\quad\;\;0}{H:C:::N:} \quad \text{and} \quad \overset{0\;\;+1\quad-1}{H:N:::C:}$$

the one on the left (with zero formal charges on all atoms) is favored. In evaluating isomers, it also helps to know that hydrogen and fluorine are always terminal atoms in Lewis structures, bonded to only one other atom.

Several equally good isomeric arrangements of atoms may exist. The correct **molecular skeleton** (the plan by which atoms are connected) for a given substance must then be determined by other means. Given a skeleton, the following systematic procedure furnishes a valid Lewis structure, based on the octet rule.

- A short version of this method for drawing Lewis structures is given in the Summary for Section 3–5 at the end of the chapter.

1. Count the total number of valence electrons *available* (symbolized by A) by first adding up the valence electrons contributed by all of the atoms present. If the species is a negative ion, *add* the absolute value of the total charge; if it is a positive ion, *subtract* it.
2. Count the total number of electrons *needed* (N) for each atom to have its *own* noble-gas set of electrons (two for hydrogen, eight for the elements from carbon on in the periodic table).
3. Subtract the number in step 1 from the number in step 2. This is the number of *shared* (or bonding) electrons present (S).

$$S = N - A$$

4. Assign two bonding electrons (as one shared pair) to each connection between two atoms in the molecule or ion.
5. If any of the electrons earmarked for sharing remain, assign them in pairs by making some of the bonds double or triple bonds. In some cases, there may be more than one way to do this. In general, double bonds form only between atoms of these elements: carbon, nitrogen, oxygen, and sulfur. Triple bonds usually are restricted to either carbon or nitrogen.
6. Assign the remaining electrons as lone pairs to the atoms, giving octets to all atoms except hydrogen.
7. Determine the formal charge on each atom, and write it next to that atom. Check that the sum of the formal charges equals the correct total charge on the molecule or molecular ion.

The last step identifies structures that are undesirable because they imply large separations of negative from positive charge; it also catches mechanical errors (using the wrong number of dots, and so forth).

The use of the rules is illustrated by the following examples.

EXAMPLE 3–5

Write a Lewis electron-dot structure for phosphoryl chloride ($POCl_3$), which consists of a central phosphorus atom bonded to three chlorine atoms and one oxygen atom. Assign formal charges to all of the atoms.

Solution

Calculate the total number of valence electrons available in the molecule

$A = 5$ (from P) $+ 6$ (from O) $+ 3 \times 7$ (from the Cl's) $= 32$ valence electrons

Next, calculate how many electrons are needed to give each atom its own noble-gas set of electrons. All five of these atoms require eight electrons (none of them is hydrogen). Therefore, the total is

$$N = 5 \times 8 = 40 \text{ valence electrons}$$

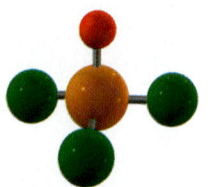

Phosphoryl chloride ($POCl_3$) is a reactive compound used to introduce phosphorus into organic molecules.

The difference between the number of valence electrons available and the number needed equals the number that must be shared:

$$S = N - A = 40 - 32 = 8 \text{ valence electrons}$$

The five atoms achieve valence octets only if eight electrons are shared in bonds. Eight electrons correspond to four electron pairs. It follows that each of the four linkages in $POCl_3$ must be a single bond. (If the number of shared electron pairs exceeded the number of linkages, double or triple bonds would be present.)

Draw the molecular skeleton (central P surrounded by four Cl's and one O) and put in lines to represent the four single bonds. Allocate the other 24 valence electrons as lone pairs to the atoms in such a way that each atom has an octet. The resulting Lewis structure is

Formal charges are already indicated in this diagram. The formal charge of $+1$ on the central phosphorus atom was computed by

5	−	0	−	$\frac{1}{2}(8)$	=	$+1$
number of valence electrons in P		lone-pair electrons		bonding electrons		formal charge

Each chlorine atom has a formal charge of 0, computed by

7	−	6	−	$\frac{1}{2}(2)$	=	0
number of valence electrons in Cl		lone-pair electrons		bonding electrons		formal charge

The formal charge on the oxygen atom is -1, computed by

6	−	6	−	$\frac{1}{2}(2)$	=	-1
number of valence electrons in O		lone-pair electrons		bonding electrons		formal charge

Check

The formal charges add up to zero, as they must for this electrically neutral species.

EXERCISE

Draw a Lewis dot structure for thionyl chloride ($SOCl_2$). Sulfur is the central atom. Indicate formal charges on all atoms.

Answer:

EXAMPLE 3-6

Draw a Lewis structure for the molecular ion NO_2^+, which has a central nitrogen atom. Estimate the bond lengths in this ion, using Table 3–3.

Solution

One N and two O's contribute 5 (from N, Group V) + 2 × 6 (from O, Group VI) or 17 valence electrons. Subtract 1 because of the +1 charge on the ion (taking one electron away from NO_2 gives NO_2^+). This leaves $A = 16$ valence electrons available.

Giving each atom its *own* noble-gas shell would require $N = 3 \times 8 = 24$ valence electrons. It follows that $24 - 16$, or 8, valence electrons must be shared. After two are assigned to each of the two N-to-O links, four remain. Use these valence electrons to make both bonds into double bonds:

$$O::N::O$$

Position the other eight electrons on the two oxygen atoms to give them octet configurations (the nitrogen atom already has an octet). The result is

$$[:\overset{..}{O}::N::\overset{..}{O}:]^+ \quad \text{or} \quad [:\overset{..}{O}=N=\overset{..}{O}:]^+$$

Finally, assign the formal charges:

$$\text{formal charge on O atoms} = 6 - 4 - \tfrac{1}{2}(4) = 0$$

$$\text{formal charge on N atom} = 5 - 0 - \tfrac{1}{2}(8) = +1$$

The Lewis structure has two N-to-O double bonds. According to Table 3–3, the length of N=O bonds is approximately 1.18×10^{-10} m.

Check

The formal charges add up to the total charge on the ion, +1, as they must. What about the following Lewis structure, which has one single bond and one triple bond and fulfills all the rules?

$$\left[:\overset{..}{\underset{..}{O}}:N:::O:\right]^+$$

$$\quad \text{(−1)(+1)} \quad \text{(+1)}$$

This structure is undesirable because it displays a larger separation of formal charge than does the other and forces a positive formal charge onto an oxygen, which is more electronegative than nitrogen.

EXERCISE

Draw Lewis structures for both ions in ammonium sulfate. Show all formal charges.

Answer:

$$\left[\begin{array}{c} H \\ H:N:H \\ H \end{array}\right]^+ \qquad \left[\begin{array}{c} :\overset{..}{O}: \\ :\overset{..}{O}:S:\overset{..}{O}: \\ :\overset{..}{O}: \end{array}\right]^{2-}$$

EXAMPLE 3-7

Draw a Lewis structure for the cyanide ion CN$^-$. Estimate the bond length.

Solution

The carbon and nitrogen atoms contribute $4 + 5 = 9$ valence electrons. The net negative charge of -1 on the ion means to *add* 1 to this, giving $A = 10$ valence electrons. Two separate octets would require $N = 16$. Therefore, the structure must have $S = 16 - 10 = 6$ shared electrons. The two atoms must have a triple bond between them:

$$C::\!:N$$

The remaining four electrons form lone pairs on the two atoms:

$$[:C::\!:N:]^- \quad \text{or} \quad [:C\!\equiv\!N:]^-$$

The bond length (Table 3-3) should equal approximately $1.16 \times 10^{-10}\,\mathrm{m}$.

Check

Compute the formal charges:

$$\text{formal charge on C} = 4 - 2 - \tfrac{1}{2}(6) = -1$$

$$\text{formal charge on N} = 5 - 2 - \tfrac{1}{2}(6) = 0$$

Their sum is -1, as it must be. The required negative formal charge is on the carbon rather than on the more electronegative nitrogen. This is undesirable, but it is not possible to do otherwise without breaking the octet rule.

EXERCISE

Draw two possible Lewis structures for the thiocyanate ion, SCN$^-$. Indicate formal charges on all atoms.

Answer:

$$\left[:\!\overset{..}{\underset{..}{S}}\!:C::\!:N:\right]^{-} \quad \text{and} \quad \left[:\!\overset{..}{\underset{..}{S}}\!::C::\overset{..}{\underset{..}{N}}\!:\right]^{-}$$

with formal charges $(-1)\ (0)\ (0)$ and $(0)\ (0)\ (-1)$ respectively.

Resonance Structures

For certain molecules or molecular ions, two or more equivalent Lewis structures can be drawn on the same skeleton. An example is sulfur dioxide (SO$_2$), for which two Lewis structures satisfy the octet rule:

$$\overset{(0)}{O}::\overset{(-1)}{\underset{(+1)}{S}}:\overset{}{O}: \quad \text{and} \quad :\overset{(-1)}{O}:\overset{(0)}{\underset{(+1)}{S}}::\overset{}{O}:$$

Each of these structures suggests that one S—O bond is a single bond and the other a double bond, so that the two bond lengths would differ. In fact, the lengths of the S—O bonds are found experimentally to be equal. The Lewis approach has apparently failed. The difficulty is remedied by saying that the true structure is a **resonance hybrid** of the two Lewis structures, in which each of the bonds is

Sulfur dioxide (SO$_2$) is produced by burning sulfur as the first step in the production of sulfuric acid, the most widely used industrial acid. Sulfur dioxide also is emitted in the burning of coal, oil, and natural gas, all of which contain small amounts of sulfur. In the atmosphere, it contributes significantly to the formation of acid rain.

intermediate between a single and a double bond. This is diagrammed by drawing both structures and connecting them with a double-headed arrow:

$$\left\{ \overset{0}{\underset{+1}{\overset{..}{O}}}::\overset{..}{S}:\overset{-1}{\overset{..}{\underset{..}{O}}}: \longleftrightarrow :\overset{-1}{\overset{..}{\underset{..}{O}}}:\overset{..}{\underset{+1}{S}}::\overset{0}{\overset{..}{O}} \right\}$$

The term "resonance" does not mean that the molecule resonates, or jumps, back and forth from one structure to the other. Rather, the true structure is a hybrid that simultaneously includes all the bonding features of the contributing Lewis structures. For some molecules, numerous resonance contributors are possible and must be diagrammed. This awkwardness can be avoided by treating bonds using molecular orbitals, as we show in Chapter 18.

EXAMPLE 3–8

Draw three resonance forms for sulfur trioxide (SO_3), in which the central sulfur atom is bonded to the three oxygen atoms and the octet rule is satisfied for all atoms.

Solution

The SO_3 molecule has $A = 24$ valence electrons. For each atom to have its own octet would require $N = 4 \times 8 = 32$ electrons. Therefore, $S = 32 - 24 = 8$ electrons must be shared between atoms, implying a total of four bonding pairs. These can be distributed in one double and two single bonds, leading to the equivalent resonance structures

$$\left\{ \begin{array}{ccc} \overset{0}{\underset{-1}{\overset{..}{O}}} & \overset{-1}{\overset{..}{O}} & \overset{-1}{\overset{..}{O}} \\ :\overset{..}{O}:\overset{..}{\underset{+2}{S}}:\overset{..}{O}: & \longleftrightarrow & \overset{..}{O}::\overset{..}{\underset{+2}{S}}:\overset{..}{O}: & \longleftrightarrow & :\overset{..}{O}:\overset{..}{\underset{+2}{S}}::\overset{..}{O} \end{array} \right\}$$

Check

A visual count of the dots surrounding each atom symbol confirms that the octet rule is satisfied. The sum of the formal charges in each resonance form equals zero, which is the correct overall charge on the molecule. Experimentally, the three bonds in SO_3 are equivalent.

EXERCISE

The ozone molecule contains three oxygen atoms. Draw two resonance structures that jointly represent the bonding in this molecule.

Answer:

$$\left\{ \overset{0}{\overset{..}{O}}::\overset{..}{\underset{+1}{O}}:\overset{-1}{\overset{..}{\underset{..}{O}}}: \longleftrightarrow :\overset{-1}{\overset{..}{\underset{..}{O}}}:\overset{..}{\underset{+1}{O}}::\overset{0}{\overset{..}{O}} \right\}$$

Sulfur trioxide (SO_3) is a reactive and corrosive gas that is produced by the oxidation of sulfur dioxide. It reacts with water to give sulfuric acid.

Ozone (O_3) is a pale blue gas with a pungent odor that condenses to a deep blue liquid below $-112°C$. In the upper atmosphere, it plays a crucial role in shielding the Earth's surface from the full effect of the sun's radiation.

Breakdown of the Octet Rule

Lewis structures constructed using the octet rule are useful in predicting whether a proposed molecule will be stable under ordinary conditions of temperature and pressure; for example, we can write a simple Lewis structure for water (H_2O):

$$H : \overset{\displaystyle ..}{\underset{\displaystyle ..}{O}} : H$$

in which each atom has a noble-gas configuration. The impossibility of doing this for OH or for H_3O suggests that these species are either unstable or highly reactive. Several situations, however, do come up in which the octet rule is not satisfied but the molecule or molecular ion is still stable.

Case 1: Odd-Electron Molecules.   The electrons in a Lewis structure that satisfies the octet rule must occur in pairs: bonding pairs or lone pairs. A molecule with an odd number of electrons cannot satisfy the octet rule on all of its atoms. Most stable molecules have an even number of electrons, but a few have an odd number. In this case, the best one can do is give up the octet rule for one of the atoms by leaving an unpaired lone electron on it and trying to reduce the separation of formal charge as much as possible. All bonding electrons should be kept paired in such Lewis structures.

• Just leaving off the troublesome odd electron is not legal.

   An example is nitrogen monoxide (NO), a stable (although reactive) molecule that is an important factor in air pollution. It has 11 valence electrons. In the two Lewis structures that can be drawn for it

$$\overset{\textcircled{\scriptsize 0}}{\underset{}{:}} N = \overset{\textcircled{\scriptsize 0}}{O} : \qquad \text{and} \qquad \overset{\textcircled{\scriptsize -1}}{:} N = \overset{\textcircled{\scriptsize +1}}{O} :$$

only one of the atoms has an octet of electrons. Of these two structures, the second, in which the odd electron resides on the oxygen atom, should be less favored because it leads to a separation of formal charge with positive formal charge assigned to the more electronegative atom. Some other stable odd-electron molecules (and molecular ions) are chlorine monoxide (ClO), nitrogen dioxide (NO_2), and superoxide ion (O_2^-).

Case 2: Octet-Deficient Molecules.   Covalent compounds of beryllium and boron are frequently octet deficient; that is, sufficient electrons are not available for each atom to achieve an octet, consistent with other constraints on the bonding. For example, applying the standard rules to BF_3 leads to the Lewis structure

• This "deficiency" is from the point of view of simple bonding theory.

$$\overset{\displaystyle :\overset{..}{F}:}{\underset{\textstyle \underset{\textcircled{\scriptsize -1}}{:F - B = F:}}{\vert}}\ \textcircled{\scriptsize +1}$$

Boron trifluoride (BF_3) is a highly reactive gas that condenses to a liquid at $-100°C$. Its major use is in speeding up a large category of reactions involving carbon compounds.

• The B in BF_3 still "wants" an octet. It reacts vigorously with NH_3 to form $F_3B—NH_3$, in which the new bond, which uses electrons from a lone pair on the N in NH_3, makes its octet complete.

The experimental evidence, however, strongly indicates that no double bonds exist in BF_3 (fluorine never forms double bonds). Moreover, the placement of a positive formal charge on the very electronegative fluorine atom is quite undesirable. Placing only six electrons around the central (boron) atom is more nearly correct:

$$\ddot{:}\overset{\displaystyle ..}{\underset{\displaystyle}{F}}\ddot{:}$$
$$:\!\ddot{F}\!-\!B\!-\!\ddot{F}\!:$$

Although this Lewis structure denies an octet to the boron atom, it does at least assign zero formal charges to all atoms.

Case 3: Valence-Shell Expansion.   Lewis structures become more complex in the compounds of elements from the third and subsequent periods of the periodic table. Sulfur, for example, forms some compounds that are easily described by Lewis structures that satisfy the octet rule. An example is hydrogen sulfide (H_2S), which resembles water in its Lewis representation. Other sulfur compounds cannot be described in this way. In sulfur hexafluoride (SF_6), the central sulfur atom is bonded to six fluorine atoms. A Lewis structure can be drawn only if more than eight electrons are allowed around the sulfur atom, a process called **valence-shell expansion.** The resulting Lewis structure is written as

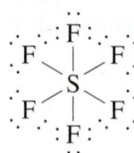

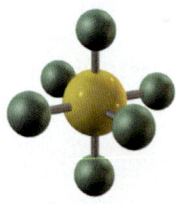

Sulfur hexafluoride (SF_6) is an extremely stable, dense, and unreactive gas. It is used as an insulator in high-voltage generators and switches.

in which the outer atoms have octets, and the central sulfur atom shares a total of 12 electrons.

The need for valence-shell expansion is signaled in the standard procedure for writing Lewis structures when the value of S calculated for the number of shared electrons is not large enough to place a bonding pair between each pair of atoms that are supposed to be bonded. In sulfur hexafluoride (SF_6), for example, $A = 48$ electrons available and $N = 56$ needed. This means that $S = N - A = 8$ electrons. Four electron pairs are not sufficient to make even single bonds between the central sulfur atom and the six outer fluorine atoms. In this case, rule 4 (assign one bonding pair to each bond in the molecule or ion) is still followed, even though doing so uses more than S electrons. Rule 5 becomes irrelevant because there are no extra shared electrons, and rule 6 is replaced with a new rule:

6'. Assign lone pairs to the outer atoms to give them octets. If any electrons still remain, assign them to the central atoms as lone pairs.

The effect of rule 6' is to abandon the octet rule for the central atoms but preserve it for the outer atoms.

EXAMPLE 3-9

Write a Lewis structure for the linear I_3^- (triiodide) ion.

Solution

The ion has $A = 22$ (seven valence electrons from each iodine atom, plus one from the overall charge of the ion). The three atoms need $3 \times 8 = 24$ electrons for each to have its own octet. Hence $S = 24 - 22 = 2$. Because two electrons are not sufficient to make two different bonds, valence expansion is necessary.

Therefore, use two pairs of electrons to make the two bonds. Then follow rule 6′ and complete the octets of the two outer iodine atoms. This gives

$$: \ddot{I} - I - \ddot{I} :$$

At this stage, a total of 16 valence electrons has been used. The remaining six are placed as lone pairs on the central iodine atom. A formal charge of -1 then resides on this atom:

$$: \ddot{I} - \overset{\overset{\text{(−1)}}{..}}{I} - \ddot{I} :$$

Check

Valence expansion has occurred only on the central atom, which is surrounded by a total of ten electrons, rather than the eight required by the octet rule. The octet rule is satisfied for the outer atoms.

EXERCISE

Write a Lewis structure for the XeF_4 molecule, with a central xenon atom.

Answer:

$$\ddot{F} \qquad \ddot{F} \\ \searrow \quad \swarrow \\ \ddot{X}e \\ \nearrow \quad \nwarrow \\ \ddot{F} \qquad \ddot{F}$$

Xenon tetrafluoride (XeF_4) is a white crystalline solid that melts at 117°C. It is one of several compounds of xenon with fluorine and other elements, whose preparation since 1962 has demonstrated that the noble gases are not chemically inert, despite their having complete octets of electrons.

Valence-shell expansion also can be used to avoid non-zero formal charges in certain Lewis structures. The bonding in $POCl_3$ can be described as in Example 3–5, or it can be written as

$$\begin{array}{c} : \ddot{O} : \\ \parallel \\ : \ddot{C}l - P - \ddot{C}l : \\ \mid \\ : \ddot{C}l : \end{array}$$

in which all formal charges are zero, but the central phosphorus atom is surrounded by ten, rather than eight, valence electrons. In a similar fashion, we can write the Lewis structure of the sulfate ion, SO_4^{2-}, as

$$\begin{array}{c} \overset{\text{(−1)}}{: \ddot{O} :} \\ \overset{\text{(−1)}}{:} \mid \overset{\text{(+2)}}{} \\ : O - S - O : \\ \mid \\ : \ddot{O} : \\ \overset{\text{(−1)}}{} \end{array} \quad \text{or} \quad \begin{array}{c} \overset{\text{(0)}}{: \ddot{O} :} \\ \overset{\text{(−1)}}{:} \parallel \overset{\text{(0)}}{} \\ : O - S - O : \\ \parallel \\ : \ddot{O} : \\ \overset{\text{(0)}}{} \end{array}$$

These two structures offer a choice between building up significant formal charges on some atoms and valence-shell expansion on the central sulfur atom. Which is preferable: formal charges or valence-shell expansion? It is hard to say. Valence expansion certainly should not be used for second-period elements, but it may play a

• Bond lengths do not help much in determining whether the S-to-O bond is single or double because comparison requires clear-cut examples of S—O and S=O that do not exist.

role in the bonding of elements in later periods. It is probably best to regard the true structure of the sulfate ion as a resonance hybrid to which *both* of the structures shown contribute. In fact, this type of question cannot be answered very well within the simple Lewis model. Its resolution requires the more powerful bonding theory introduced in Chapters 16 to 18.

3-6   Naming Compounds in Which Covalent Bonding Occurs

Covalent bonding frequently gives scope for several different compounds between the same two elements. Also, covalently bound compounds frequently contain three or more different elements. Avoiding ambiguous names in these situations requires new rules. We look first at the nomenclature of binary molecular compounds that contain only non-metals. Then we consider ionic compounds in which the cation or anion (or both) are polyatomic ions with covalent bonding within the ion.

Naming Binary Molecular Compounds

If two non-metals form only one compound, the compound is named according to this pattern:

(name of more electropositive atom)

+ (start of name of more electronegative atom + *ide*)

This rule is the same as the rule for naming binary ionic compounds. Just as we call the ionic compound NaBr "sodium bromide," so we give the following names to typical molecular compounds:

HBr hydrogen bromide	H$_2$S hydrogen sulfide
BeCl$_2$ beryllium chloride	BN boron nitride

If a pair of elements forms more than one molecular compound, which is very common, then the numerical prefixes listed in Table 3–4 are used to specify the number of atoms of each element. Use of the prefixes removes ambiguity. "Chlorine oxide" could refer to any of the several oxides of chlorine, but "chlorine dioxide" (ClO_2) and "dichlorine hexaoxide" (Cl_2O_6) are specific.

Coming up with names for binary molecular compounds requires simply reading the molecular formula of the compound, translating the subscripts (*di-* for two, *tri-* for three, *tetra-* for four, and so forth), and putting the *-ide* ending where it belongs. Only one additional point arises with these names: The prefix for one (*mono-*) generally is omitted when it would modify the name of the more electropositive element. Thus, CO is carbon monoxide (not monocarbon monoxide) and SO$_2$ is sulfur dioxide (not monosulfur dioxide). A few examples of binary molecular compounds and their names include

NO$_2$ nitrogen dioxide	IF$_5$ iodine pentafluoride
N$_2$O dinitrogen monoxide	As$_4$O$_6$ tetraarsenic hexaoxide
XeF$_4$ xenon tetrafluoride	P$_4$O$_{10}$ tetraphosphorus decaoxide

TABLE 3-4

Prefixes Used for Naming Binary Molecular Compounds	
Number	**Prefix**
1	mono-
2	di-
3	tri-
4	tetra-
5	penta-
6	hexa-
7	hepta-
8	octa-
9	nona-
10	deca-
11	undeca-
12	dodeca-

EXAMPLE 3-10

Give systematic names to both members in each of following pairs of binary molecular compounds:
(a) Cl_2O_7 and Cl_2O_3 (b) SiO_2 and SiO (c) NF_3 and N_2F_4 (d) P_4S_7 and P_4S_3
(e) H_2O and H_2O_2

Solution

Name the more electropositive element first with a suitable prefix, then the more electronegative element with a suitable prefix and an *-ide* ending:
(a) Dichlorine heptaoxide and dichlorine trioxide ;
(b) Silicon dioxide and silicon monoxide ;
(c) Nitrogen trifluoride and dinitrogen tetrafluoride ;
(d) Tetraphosphorus heptasulfide and tetraphosphorus trisulfide ;
(e) Dihydrogen monoxide and dihydrogen dioxide .

• Alternate names for dihydrogen monoxide (DHMO) and additional information appear at the web site www.dhmo.org/

EXERCISE

Write chemical formulas for the following binary compounds: (a) chlorine monoxide, (b) oxygen difluoride, and (c) tetraiodine dodecaoxide.

Answer: (a) ClO; (b) OF_2; (c) I_4O_{12}.

The oxides of nitrogen shown in Figure 3–16 illustrate the need for a systematic way of naming compounds. Despite this, a few nonsystematic names for

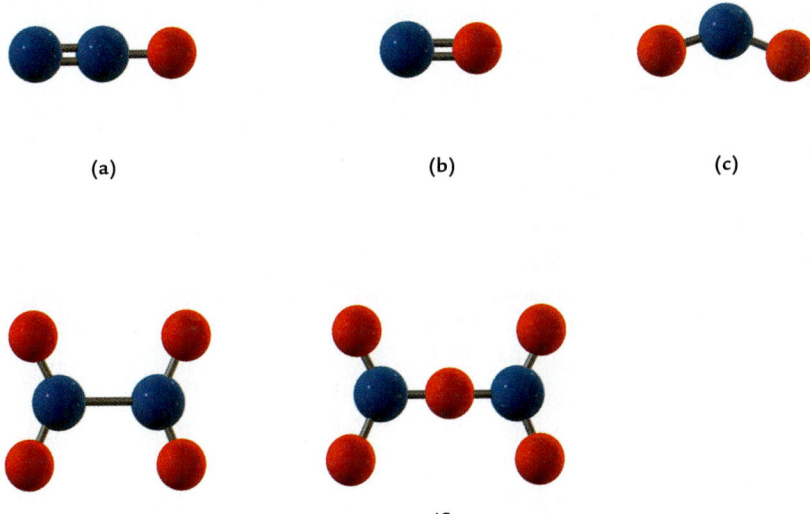

(a) (b) (c) (d)

(e) (f)

Figure 3–16 • (a) Dinitrogen monoxide (N_2O, also called "nitrous oxide") is a colorless and rather unreactive gas. It is used as an anesthetic and is referred to as "laughing gas." (b) Nitrogen monoxide (NO, also called "nitric oxide") is another colorless gas. It has an odd number of electrons. When warmed at high pressure, it reacts to form N_2O and NO_2. (c) Nitrogen dioxide (NO_2) is a brown gas formed when NO reacts with oxygen. Together with NO, NO_2 is a significant factor in urban air pollution. (d) Dinitrogen trioxide (N_2O_3) exists in pure form only in the solid state (in which it is a very pale blue) at temperatures below approximately $-100°C$. At higher temperatures, it breaks down extensively to NO and NO_2. (e) Dinitrogen tetraoxide (N_2O_4) is a colorless compound formed when pairs of NO_2 molecules link up (a process called "dimerization") at low temperature. (f) Dinitrogen pentaoxide (N_2O_5) is a white solid made up of nitronium (NO_2^+) and nitrate (NO_3^-) ions. The symmetric N_2O_5 molecule shown here exists in the vapor, although it dissociates fairly rapidly to form nitrogen dioxide and oxygen.

TABLE 3–5

Formulas and Names of Some Common Molecular Ions	
Formula	**Name**
Cations	
H_3O^+	hydronium
NH_4^+	ammonium
Anions	
OH^-	hydroxide
O_2^{2-}	peroxide
O_2^-	superoxide
BO_3^{3-}	borate
CO_3^{2-}	carbonate
HCO_3^-	hydrogen carbonate *or* bicarbonate
SiO_4^{4-}	silicate
NO_3^-	nitrate
NO_2^-	nitrite
PO_4^{3-}	phosphate
HPO_4^{2-}	hydrogen phosphate
$H_2PO_4^-$	dihydrogen phosphate
PO_3^{3-}	phosphite
SO_4^{2-}	sulfate
SO_3^{2-}	sulfite
HSO_4^-	hydrogen sulfate *or* bisulfate
HSO_3^-	hydrogen sulfite *or* bisulfite
$S_2O_3^{2-}$	thiosulfate
$S_2O_6^{2-}$	dithionate
$S_2O_4^{2-}$	dithionite
ClO_4^-	perchlorate
ClO_3^-	chlorate
ClO_2^-	chlorite
ClO^-	hypochlorite
IO_4^-	periodate
IO_3^-	iodate
MnO_4^-	permanganate
MnO_4^{2-}	manganate
CrO_4^{2-}	chromate
$Cr_2O_7^{2-}$	dichromate
AsO_4^{3-}	arsenate
AsO_3^{3-}	arsenite
CH_3COO^-	acetate
$C_2O_4^{2-}$	oxalate
CN^-	cyanide
CNO^-	cyanate
SCN^-	thiocyanate

important compounds are well established and must be memorized. These include

H_2O water	N_2H_4 hydrazine
NH_3 ammonia	PH_3 phosphine
AsH_3 arsine	SbH_3 stibine
NO nitric oxide	N_2O nitrous oxide

Naming Compounds that Contain Polyatomic Ions

Ions can be polyatomic as well as monatomic. In polyatomic ions, or **molecular ions,** covalent bonds maintain atoms in a tight group that carries a net charge. Because of the charge, the group can participate as a single entity in ionic bonding. Polyatomic cations are most frequently named as derivatives of molecular compounds. For example, among the relatively few polyatomic cations of importance in inorganic chemistry are the ammonium ion NH_4^+ (obtained by adding H^+ to ammonia) and the hydronium ion H_3O^+ (obtained by adding H^+ to water). The names of polyatomic cations often end in -*onium* or -*ium*.

A large number of polyatomic anions exist. Many have names, established by long practice, that are not fully systematic. These names must be learned. The names of the **oxoanions,** which contain oxygen in combination with another element and are very prevalent, are derived by adding the ending -*ate* to the stem of the name of the non-oxygen element. Several elements form two oxoanions. The -*ate* ending is then used for the oxoanion with the larger number of oxygen atoms (for example, NO_3^- nit*rate*), and the ending -*ite* is used to name the oxoanion with the smaller number (for example, NO_2^- nit*rite*). For elements such as chlorine, which form more than two oxoanions, the additional prefixes *per-* (largest number of oxygen atoms) and *hypo-* (smallest number of oxygen atoms) are used. Oxoanions containing hydrogen as a third element include that word in their name; for example, HCO_3^- ion is the hydrogen carbonate ion, and the HSO_4^- ion is the hydrogen sulfate ion. These two ions also have older names, "bicarbonate ion" and "bisulfate ion," that are still encountered. Many ions and compounds have more than one name, particularly ones of long-standing technical or industrial importance. The names and formulas of some of the most important cations and anions are listed in Table 3–5 on the facing page. In this table, the charges indicated by the right superscript are intrinsic parts of the formulas. They are as important as the subscripts showing the relative numbers of atoms.

As with binary ionic compounds, the compositions of ionic compounds containing polyatomic ions are determined by the principle of charge neutrality: The total positive charge on the cations must be exactly balanced by the total negative charge on the anions. The formulas of a few ionic compounds that contain polyatomic ions include

Name	Charge Neutrality Requires...	Therefore the Formula Is...
Sodium bromate	One +1 cation to balance one −1 anion	$NaBrO_3$
Potassium permanganate	One +1 cation, one −1 anion	$KMnO_4$
Ammonium sulfate	Two +1 cations, one −2 anion	$(NH_4)_2SO_4$
Calcium dihydrogen phosphate	One +2 cation, two −1 anions	$Ca(H_2PO_4)_2$

EXAMPLE 3–11

Name the following ionic compounds that contain polyatomic ions:
(a) NH_4ClO_3 (b) $NaNO_2$ (c) Li_2CO_3 (d) $Mg(HSO_3)_2$

Solution

Consulting Table 3–4 provides the names of the different polyatomic ions, which leads to the names
(a) Ammonium chlorate
(b) Sodium nitrite
(c) Lithium carbonate
(d) Magnesium hydrogen sulfite

EXERCISE

Write chemical formulas for the following ionic compounds: (a) ammonium chlorite, (b) sodium nitrate, (c) magnesium chromate, and (d) aluminum hydrogen carbonate.

Answer: (a) NH_4ClO_2.   (b) $NaNO_3$.   (c) $MgCrO_4$.   (d) $Al(HCO_3)_3$.

3–7 The Shapes of Molecules

Molecules are three-dimensional objects. Although Lewis structures show which atoms are connected to which and even provide a basis to estimate the distances between bonded atoms (as in the use of Table 3–2), they reveal nothing about the way the atoms are situated in space. An extension of Lewis theory is required to predict the angles defined at an atom by two or more covalently bonded neighbors.

The VSEPR Theory

The **valence shell electron-pair repulsion (VSEPR) theory** starts with the fundamental idea that electron pairs in the valence shell of an atom repel each other on a spherical surface formed by the underlying core of the atom. The repulsions involve both lone pairs, which are localized on the atom and are not involved in bonding, and bonding pairs, which are covalently shared with other atoms. The electron pairs occupy sites on the valence shell of each atom in such a way as to minimize their repulsions. They position themselves as far apart from each other as possible. The molecular geometry, which is defined by the relative positions of the atoms, is then traced from the relative locations of the electron pairs.

• Picture electron pairs as bugs that hate each other but are forced to dwell on the surface of the same billiard ball.

The arrangements that best minimize repulsions naturally depend on the number of electron pairs. Figure 3–17 shows the configuration of minimum energy for two to six electron pairs around a central atom. Two electron pairs locate themselves on opposite sides of the atom in a linear arrangement, three pairs form a trigonal planar structure, four arrange themselves at the corners of a tetrahedron, five define a trigonal bipyramid, and six form an octahedron. To find which geometry applies, one determines the **steric number** (*SN*) of the central atom.

• A central atom is one that has two or more other atoms bonded to it. The steric numbers of non-central atoms are not worth calculating because they do not influence the actual positions of atoms in molecules but only the (unobservable) positions of lone pairs.

"Steric" means "having to do with space." The steric number of an atom in a molecule can be determined by drawing the Lewis structure of the molecule and adding the number of atoms that are bonded to it and the number of lone pairs that it has. The Lewis structure need not be complete because it is not necessary to consider resonance structures and formal charges. Indeed, the Lewis structure often need not even be drawn out. All that is really required is to find out how many atoms are bonded to the central atom (the atom in question) and how many of the valence electrons on the central atom take part in covalent bonds. The remaining electrons are then parceled off into lone pairs, and the steric number is calculated:

$$SN = \text{(number of atoms bonded to central atom)} +$$
$$\text{(number of lone pairs on central atom)}$$

EXAMPLE 3-12

Calculate steric numbers for iodine in IF_4^- and for bromine in BrO_4^-. These molecular ions have central I or Br atoms bonded to the other four atoms.

Solution

The central I^- has eight valence electrons. Each F atom has seven valence electrons of its own and needs to share one of the electrons from the I^- to achieve a noble-gas configuration. Four of the I^- valence electrons thus take part in covalent bonds, leaving the remaining four to form two lone pairs. The steric number is given by

$$SN = 4 \text{ (bonded atoms)} + 2 \text{ (lone pairs)} = \boxed{6}$$

In BrO_4^-, each O atom needs to share two of the electrons from the Br^- in order to achieve a noble-gas configuration. Because this accounts for all eight of the Br^- valence electrons, there are no lone pairs on the central atom, and

$$SN = 4 \text{ (bonded atoms)} + 0 \text{ (lone pairs)} = \boxed{4}$$

EXERCISE

Calculate the steric numbers for the central Se in $SeOF_4$ and the central N in NO_2^-.

Answer: $SN = 5$ and $SN = 3$.

In determining the steric number, double-bonded or triple-bonded atoms count the same as single-bonded atoms. In the case of CO_2, for example, two double-bonded oxygen atoms are attached to the central carbon and no lone pairs are on that atom, so $SN = 2$.

The steric number is used to predict molecular geometries. For the simplest case, molecules of generic formula XY_n in which there are no lone pairs on the central atom X,

$$SN = \text{number of bonded atoms} = n \quad \text{(no lone pairs)}$$

The n bonding electron pairs (and, therefore, the outer atoms) locate themselves as shown in Figure 3–16 to minimize electron-pair repulsion. Thus, CO_2 is predicted (and found experimentally) to be linear, BF_3 trigonal planar, CH_4 tetrahedral, PCl_5 trigonal bipyramidal, and SF_6 octahedral (Fig. 3–18).

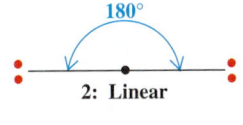

2: Linear

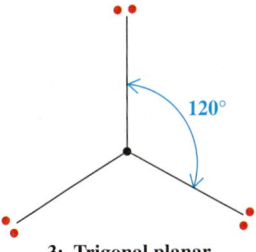

3: Trigonal planar

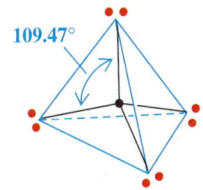

4: Tetrahedral

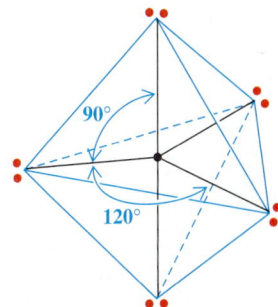

5: Trigonal bipyramidal

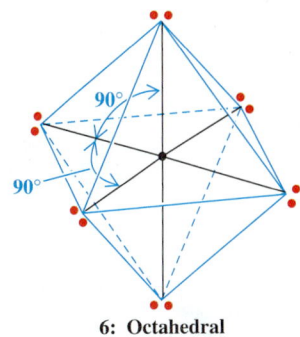

6: Octahedral

Figure 3-17 • Diagrams and names of the geometries favored for various numbers of mutually repelling electron pairs around a central atom. The angles between the electron pairs are indicated.

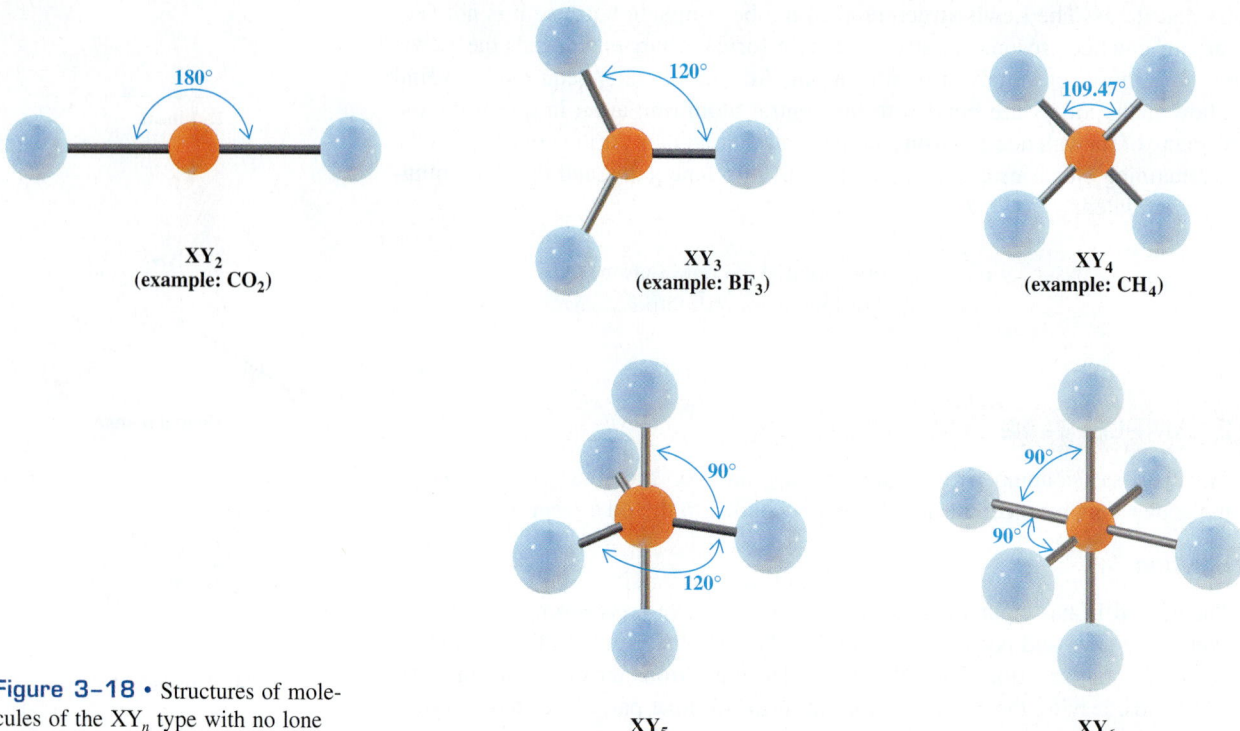

Figure 3–18 • Structures of molecules of the XY_n type with no lone pairs on the central atom X.

When lone pairs are present, the situation is more complicated. Three different types of repulsions are now possible: bonding pair against bonding pair, bonding pair against lone pair, and lone pair against lone pair. Consider the molecule of ammonia (NH_3), for example. It has three bonding electron pairs and one lone pair (Fig. 3–19). The steric number is 4, and the four electron pairs arrange themselves into an approximately tetrahedral structure. It is found that lone pairs tend to occupy more space than do bonding pairs (because they are held closer to the central atom). As a consequence, the angles between bonds opposite to lone pairs are reduced. The geometry of the *molecule,* as distinct from that of the electron pairs, is named for the sites occupied by actual *atoms.* In describing the molecular geometry, no consideration is given to lone pairs that may be present on the central atom, even though their presence affects the geometry. The structure of the ammonia molecule is thus predicted to be a trigonal pyramid in which the H—N—H bond angle is smaller than the tetrahedral angle of 109.5°. The observed structure is a trigonal pyramid with an H—N—H bond angle of 107.3°.

The greater steric demands of lone pairs imply that repulsive forces among valence pairs diminish in the following order:

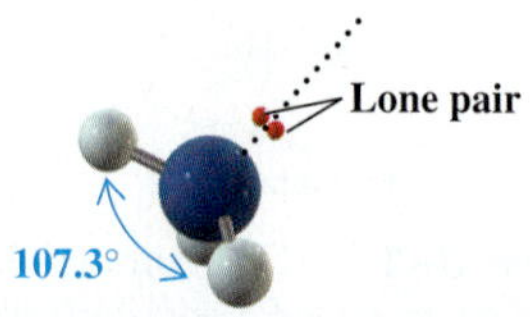

Figure 3–19 • Ammonia (NH_3) has a pyramidal structure in which the bond angles are smaller than 109.5°.

lone pair vs. lone pair > lone pair vs. bonding pair >

bonding pair vs. bonding pair

EXAMPLE 3-13

Predict the structure of water using the VSEPR theory.

Solution

The steric number on the central oxygen atom is 4, with two bonds and two lone pairs. The four electron pairs around the oxygen atom define a distorted tetrahedron. The molecule itself (not considering lone pairs) is bent . The bond angle should be somewhat *less* than the tetrahedral angle of 109.5° because the two lone pairs require more space.

Check

Experiment establishes that the H—O—H bond angle is 104.5° (Fig. 3–20).

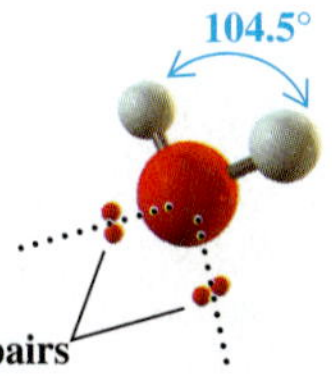

Figure 3-20 • Water (H_2O) has a bent structure with a bond angle smaller than 109.5°.

EXERCISE

Determine the steric number of the sulfur atom in SO_2, and predict the molecular structure of SO_2.

Answer: The steric number of the sulfur is 3 (two bonded atoms and one lone pair). The SO_2 molecule is bent, with a bond angle slightly less than 120°.

Molecules of the fluorides PF_5, SF_4, ClF_3, and XeF_2 all have steric number 5 but have different numbers of lone pairs (0, 1, 2, and 3, respectively). What shapes do these molecules have? We have already mentioned that PF_5 is trigonal bipyramidal. Two of the fluorine atoms in PF_5 occupy **axial** sites (Fig. 3–21a), and the

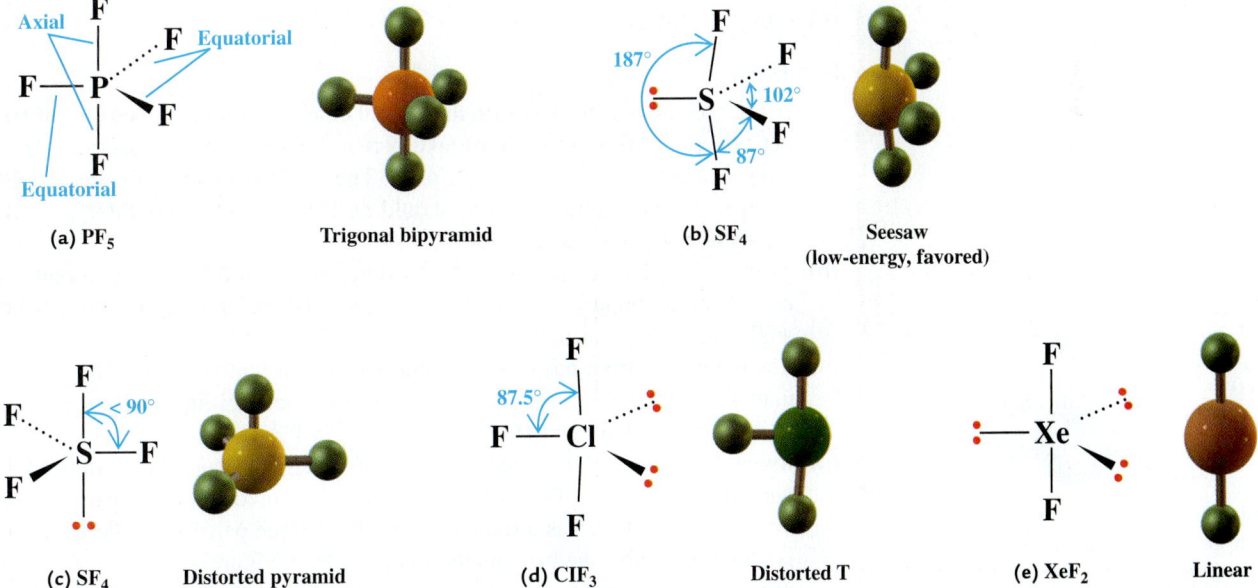

Figure 3-21 • Molecules with steric number 5. All five of the molecular geometries shown here derive from the trigonal bipyramid, but only one of them includes those words in its name. Molecular geometries are named for the sites occupied by the atoms, not for the underlying distribution of the electron pairs.

other three occupy **equatorial** sites. Because the two kinds of sites are not equivalent, there is no reason for all five P—F bond lengths to be equal. Experiment shows that the equatorial P—F bond length is 1.534×10^{-10} m, shorter than the axial P—F lengths, which are 1.577×10^{-10} m.

Sulfur tetrafluoride has four bonded atoms and one lone pair. Does the lone pair occupy an axial or an equatorial site? In the VSEPR theory, electron pairs repel each other more strongly when they form a 90° angle with respect to the central atom than when the angle is larger. A single lone pair therefore finds a position that minimizes the number of 90° repulsions it has with bonding electron pairs. It occupies an equatorial position with two 90° repulsions (Fig. 3–21b) rather than an axial position with three 90° repulsions (Fig. 3–21c). The direction of axial S—F bonds is skewed slightly away from the lone pair, so the molecular structure of SF_4 is a distorted seesaw. A second lone pair (for example, in ClF_3) also takes an equatorial position, leading to a distorted T-shaped molecular structure (Fig. 3–21d). A third lone pair (for example, in XeF_2 or I_3^-) occupies the third equatorial position, and the molecular geometry is linear (Fig. 3–21e). To summarize this discussion:

> Lone pairs occupy equatorial positions in preference to axial positions in structures of steric number 5.

EXAMPLE 3–14

Predict the geometry of the following molecules and ions:
(a) ClO_3^+ (c) SiH_4
(b) ClO_2^+ (d) IF_5

Solution

(a) All of the valence electrons on the central chlorine atom (six, because of the net positive charge on the ion) take part in bonds to the surrounding three oxygen atoms; there are no lone pairs. The steric number of the central Cl atom is 3. The central Cl atom should be surrounded by the three oxygen atoms in a trigonal planar structure.

(b) The central Cl atom in this ion has a steric number of 3; it has two bonded atoms and a single lone pair. The prediction is a bent species with a bond angle somewhat less than 120°.

(c) The central Si atom has a steric number of 4 and no lone pairs. The molecular geometry should consist of the silicon atom surrounded by a regular tetrahedron of hydrogen atoms.

(d) Iodine has seven valence electrons, five of which are shared in bonding pairs with fluorine atoms. This leaves two electrons to form a lone pair, so the steric number of the I is 5 (bonded atoms) + 1 (lone pair) = 6. The structure is based on the octahedron of electron pairs from Figure 3–17, with five fluorine atoms and one lone pair. The lone pair can be placed on any one of the six equivalent sites. It causes the four more closely neighboring fluorine atoms to bend away from it toward the more distant fifth fluorine atom, giving the distorted structure shown in Figure 3–22.

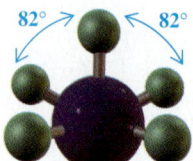

82°   82°

Figure 3–22 • The molecular structure of iodine pentafluoride (IF_5). The deviations of the F—I—F bond angles from the nominal value of 90° are caused by the spatial requirements of the lone pair of electrons (not shown) at the bottom.

EXERCISE

The ionic compound $(S_2N^+)(AsF_6^-)$ was recently isolated and studied. The geometry of the molecular anion and of the molecular cation both conform to the predictions of the VSEPR theory. What are those geometries?

Answer: The cation is linear and the anion is octahedral.

The VSEPR theory is a remarkably powerful model for predicting the geometries and approximate bond angles of molecules that have a central atom. In fact, in many cases it is more successful than much more elaborate theories that require extensive calculation.

Dipole Moments

Two bonded atoms share electrons unequally whenever they differ in electronegativity. In hydrogen chloride (HCl), the highly electronegative Cl atom attracts the bonding electrons more strongly than does the H atom. The Cl end of the HCl molecule consequently carries a slight negative electric charge, and the H end carries a positive charge of equal magnitude. The HCl molecule is an electric **dipole.** Dipolar (or **polar**) molecules are said to possess a **dipole moment** (symbolized μ) because they experience a torque in electric fields. They tend to rotate to align their negative ends as close as possible to the positive side of the field and their positive ends as close as possible to the negative side of the field. Hydrogen chloride is diatomic, but polyatomic molecules also often possess dipole moments. This molecular property greatly influences the bulk properties of substances because dipolar molecules interact strongly and directionally with each other.

• Torques are called "moments" or "turning-moments" in physics.

We can predict the presence or absence of a dipole moment in a molecule. First we draw the Lewis structure and figure out the molecular geometry using the VSEPR theory. Next, we assign a *bond dipole* (shown as an arrow) to each bond in the molecule, with the arrow pointing from the more electronegative to the more electropositive atom and the length corresponding to the magnitude of the bond dipole. Only identical bonds (those having the same atoms at each end) can have bond dipole arrows of identical length. Finally, the total dipole moment is obtained by adding the arrows of the bond dipoles vectorially, that is, by placing the arrows head to tail (without reorienting them) and examining the net arrow that connects the tail of the first arrow to the head of the last.

• In some books, dipole arrows are drawn in the reverse direction (from positive to negative). The convention used here follows long-standing international recommendation (see Appendix C-6).

In CO_2, for example, each C=O has a bond dipole pointing from O to C (Fig. 3–23a). Because the bonds are identical, the bond dipoles are equal in magnitude. They point in opposite directions, so the total molecular dipole moment vanishes. In OCS, which is also linear with a central carbon atom, the C=O and C=S bond dipole moments also point in opposite directions but have different magnitudes, leaving a net non-zero dipole moment for the O=C=S molecule (Fig. 3–23b). The water molecule has a non-zero dipole moment (Fig. 3–23c) because the two bond dipole moments, although equal in magnitude, do not directly oppose each other. They add vectorially to give a net molecular dipole moment that splits the angle between the bond dipoles. The more symmetric molecule CCl_4, on the other hand, has no net dipole moment (Fig. 3–23d). The four C—Cl bonds have dipole moments (pointing from the four corners of a tetrahedron), but when the four

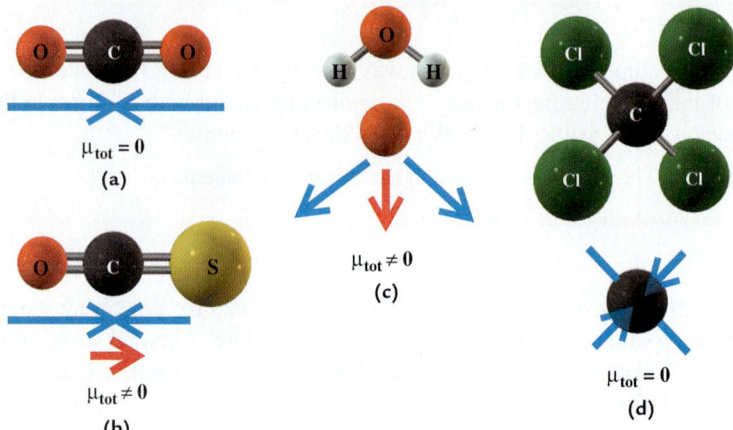

Figure 3-23 • The total dipole moment of a molecule is obtained by vector addition of its bond dipoles; this is performed diagrammatically by placing them head to tail. (a) CO_2, (b) OCS, (c) H_2O, (d) CCl_4.

CHEMISTRY IN YOUR LIFE

Dry Ice Versus Wet Ice

Perhaps you have amused yourself at picnics by throwing the blocks of solid carbon dioxide (dry ice) that kept the ice cream cold into a pond or lake and watching the water "boil" up and spew clouds of white vapor (Fig. 3–A). Dipole moments help to explain what was going on.

As the VSEPR theory predicts, molecules of carbon dioxide are linear. This means (Fig. 3–23a) that the bond dipoles in O=C=O cancel out. The molecule is non-polar. Non-polar molecules attract each other weakly. Consequently, substances composed of them are hard to condense to liquids or solids. Carbon dioxide is gaseous under room conditions and requires considerable cooling to solidify. Only at −78°C do the weak attractions between its non-polar molecules take hold sufficiently for dry ice to form. In contrast, the bend in water molecules means that the two O—H bond dipoles partially reinforce each other (Fig. 3–23c). The dipolar molecules in water attract their neighbors rather strongly. Water freezes at 0°C, a much higher temperature than −78°C. Dry ice is far colder than is the water in a pond. When it is thrown in, the surrounding water heats it rapidly. The solid CO_2 converts to gaseous CO_2, which churns to the surface, where it is still cold enough to cause the moisture in the air to condense as trails of white fog. Soon the activity

Figure 3–A • Bubbles of gaseous carbon dioxide (CO_2) erupt when dry ice is put into water. *(Charles D. Winters)*

abates, not because the dry ice is gone, but because it has frozen an insulating layer of wet ice (water ice) all around it.

If H—O were somehow linear, it would be non-polar. Its melting point would be far lower than 0°C and probably lower than −78°C. Linear water could not support life as we know it, but some other kind of life might then enjoy watching hot dry ice melt cold H—O—H away.

arrows are placed head to tail, their sum is zero. Molecules such as H_2O and OCS with non-zero dipole moments are polar, whereas those such as CO_2 and CCl_4 with no net dipole moment are nonpolar, even though they contain polar bonds.

EXAMPLE 3–15

Predict whether the molecules of NH_3 and SF_6 have dipole moments.

Strategy

Use the VSEPR theory to determine the geometry of the molecules. Then consider the way the bond dipoles add up in three dimensions. Make drawings of the type in Figures 3–17 or 3–18 as required.

Solution

The NH_3 molecule has a dipole moment because the three N—H bond dipoles add vectorially to give a net dipole pointing downward from the nitrogen atom through the center of the triangle defined by the three H atoms (see Fig. 3–19). The octahedral SF_6 molecule does not have a dipole moment because each S—F bond dipole is canceled by an equal bond dipole pointing in the opposite direction on the other side of the molecule (see Fig. 3–18).

EXERCISE

Predict whether the molecules of BF_3 and SO_2 have dipole moments.

Answer: BF_3 does not have a dipole moment; SO_2 does have one.

3–8   Elements Forming More than One Ion

The transition elements and the metals found late in Groups III, IV, and V differ from the Groups I and II metals in that they often form more than one stable ion in compounds and in solution. Although calcium compounds never contain Ca^+ ions or Ca^{3+} ions (always Ca^{2+}), the element iron forms both Fe^{2+} and Fe^{3+} ions, and the element thallium forms both Tl^+ and Tl^{3+} ions. Furthermore, the transition elements differ from the main-group elements (see Section 3–4) in that no simple octet rule allows one to write Lewis structures for their compounds. The additional ions of these metals offer rich possibilities for chemical combination, but their existence is not predicted by simple ideas concerning the periodic table.

Naming Multiple Ions of One Element

Cations of the same element that have different charges are differentiated by adding a Roman numeral in parentheses immediately after the name of the metal (with no space) to indicate the ionic charge. This removes all possible ambiguity as to the charge of the ion. Examples include

Cu^+, copper(I) ion	Fe^{2+}, iron(II) ion	Sn^{2+}, tin(II) ion
Cu^{2+}, copper(II) ion	Fe^{3+}, iron(III) ion	Sn^{4+}, tin(IV) ion
Au^+, gold(I) ion	Cr^{2+}, chromium(II) ion	Ce^{3+}, cerium(III) ion
Au^{3+}, gold(III) ion	Cr^{3+}, chromium(III) ion	Ce^{4+}, cerium(IV) ion

These names are pronounced "copper-one ion," "iron-two ion," and so forth. Such cation names are then combined with the names of anions, either monatomic or polyatomic, according to the standard pattern (name of cation, a space, name of anion) to give the names of ionic compounds.

An earlier system for distinguishing between pairs of ions such as Fe^{2+} and Fe^{3+} used the suffixes *-ous* and *-ic,* added to the first part of the name (often the Latin name) of the metal, to indicate the ions of lower and higher charge, respectively. In this system, the ions in the preceding table have these names:

Cu^+, cuprous ion	Fe^{2+}, ferrous ion	Sn^{2+}, stannous ion
Cu^{2+}, cupric ion	Fe^{3+}, ferric ion	Sn^{4+}, stannic ion
Au^+, aurous ion	Cr^{2+}, chromous ion	Ce^{3+}, cerous ion
Au^{3+}, auric ion	Cr^{3+}, chromic ion	Ce^{4+}, ceric ion

This method, although still sometimes used, is too limited and does not appear further in this book.

One interesting molecular ion is Hg_2^{2+}, the mercury(I) ion. This species must be carefully distinguished from Hg^{2+}, the mercury(II) ion. The Roman numeral "I" in parentheses means in this case that the average charge on each of the two mercury atoms is $+1$. A monatomic ion of the form Hg^+ is not found.

As always with ionic compounds, the requirement for charge neutrality determines the chemical composition. Thus, iron forms two different chlorides. Iron(III) chloride has the formula $FeCl_3$, in which three chloride ions (total charge $3 \times (-1) = -3$) balance the $+3$ charge on one iron(III) ion, and iron(II) chloride has the formula $FeCl_2$, in which the $+2$ charge on the iron is balanced by a -2 charge of two chloride ions. Similarly, the formulas of thallium(I) sulfate and thallium(III) sulfate are Tl_2SO_4 and $Tl_2(SO_4)_3$, respectively.

• Avoid the mistake of taking the two in a name like iron(II) chloride to mean that the subscript on Fe equals 2.

Oxidation State (Oxidation Number)

Some compounds of the transition elements have bonds that are more covalent than ionic. For example, both titanium(IV) chloride ($TiCl_4$) and vanadium(IV) chloride (VCl_4) are molecular substances. At room temperature, they are liquids that decompose in water. Despite this, they are named as if they contained stable Ti^{4+} or V^{4+} ions, that is, by the same system that gives the names iron(II) chloride, iron(III) chloride, and copper(II) chloride, all of which are solids with more or less typical ionic properties. Covalent character occurs in compounds of the heavier elements of Groups III, IV, and V, as well as the transition metals. Lead(II) chloride, for example, is a white ionic solid that dissolves (sparingly) in water to give Pb^{2+} and Cl^- ions. Lead(IV) chloride, on the other hand, is an oily yellow liquid that decomposes in water. Its bonds are largely covalent, and it somewhat resembles the molecular compound carbon tetrachloride (CCl_4) in its physical properties (Fig. 3–24).

Roman numerals used in naming covalent compounds do not state the true charge on the atom in question but rather refer to what that charge would be if the compound were ionic. They report what is called the **oxidation state** or **oxidation number** of the element in the compound. The oxidation state coincides with the actual ionic charge on a metal atom only when the charge is low. Thus, the compound Mn_2O_7 has the name manganese(VII) oxide, and manganese is in the $+7$ oxidation state, but is not present as Mn^{7+} ions. Rather, there is substantial covalent character to the bonds in this compound. Much important chemistry involves changes in the oxidation states of elements. We discuss this more extensively in Chapters 4 and 13.

Figure 3-24 • The tetrachlorides of titanium ($TiCl_4$), lead ($PbCl_4$), and carbon (CCl_4) have tetrahedral structures, although the central atom comes from three quite different places in the periodic table. All are liquids at room temperature, with very similar freezing points: $TiCl_4$, $-24°C$; $PbCl_4$, $-15°C$; and CCl_4, $-23°C$. Lead tetrachloride is the least stable of the three, decomposing to $PbCl_2$ and Cl_2 above 50°C. Carbon tetrachloride boils at 76.5°C, and $TiCl_4$ boils at 136.5°C.

EXAMPLE 3-16

Write chemical formulas for each of the following compounds:
(a) Osmium(VI) fluoride
(b) Vanadium(V) oxide
(c) Copper(I) sulfide

Solution

(a) Each fluoride ion carries a -1 charge, so there must be six such ions to balance the hypothetical $+6$ charge on the osmium. The formula is thus OsF_6.

(b) Each oxide ion has a -2 charge, and the vanadium ion has a hypothetical $+5$ charge. If each vanadium atom were actually to give up five electrons, they could be transferred to $\frac{5}{2} = 2\frac{1}{2}$ oxygen atoms to form oxide ions. To avoid fractions, we can multiply by 2 and say that two vanadium ions give up ten electrons, transferring them to five oxygen atoms. The formula is therefore V_2O_5.

(c) The sulfide ion has a -2 charge. This must be balanced by two copper(I) ions (Cu^+), so the chemical formula is Cu_2S.

EXERCISE

Write systematic names for the following: (a) CrO_3, (b) $TlCl_3$, (c) Mn_3N_2.

Answer: (a) Chromium(VI) oxide.   (b) Thallium(III) chloride.
(c) Manganese(II) nitride.

Coordination Complexes

Many ions of the transition elements form **coordination complexes** in solution or in the solid state. These species consist of a metal cation (the transition elements are all metals) surrounded by a group of anions or neutral molecules called **ligands.** The interaction involves a sharing of lone pairs of electrons from the ligand molecules onto the metal ion. The sharing creates partially covalent bonds between the ligands and the metal ion. Coordination complexes commonly have intense colors. When exposed to gaseous ammonia, greenish white crystals of copper sulfate ($CuSO_4$) react to give a deep blue crystalline solid with the chemical formula $Cu(NH_3)_4SO_4$. The anions in the solid are still sulfate ions (SO_4^{2-} ions), but the cations are now coordination complexes of the central Cu^{2+} ion with four ammonia molecules. The formula of these cations is $[Cu(NH_3)_4]^{2+}$. The ammonia molecules *coordinate* to the copper ion through their lone-pair electrons, as is shown in the Lewis structure

• Coordination complexes are discussed more extensively in Chapter 19. Systematic names for these complexes are presented there.

$$H_3N : \overset{\displaystyle NH_3}{\underset{\displaystyle NH_3}{Cu^{2+}}} : NH_3$$

The octet rule does not apply to the central metal ion in coordination complexes, although it does to the ligands. Also, one encounters difficulty in using the VSEPR theory to predict the geometry around the central metal ions in coordination complexes, although it works for the ligands. Additional ideas are needed to explain the bonding and molecular geometry in coordination complexes (see Chapter 19).

When solid $[Cu(NH_3)_4]SO_4$ is dissolved in water, the deep blue color remains. This is evidence that the complex persists in water, because when ordinary $CuSO_4$ (without ammonia ligands) is dissolved in water, a much paler blue color results. Similar color changes are seen in forming other complex ions. The soluble ionic compound $Fe(NO_3)_3$ has a pale yellow color in solution. Upon addition of colorless potassium cyanide, the complex ion $[Fe(CN)_6]^{3-}$ forms and makes the solution reddish brown (Fig. 3–25). The complex responsible for this color consists of an Fe^{3+} ion surrounded by six CN^- ligands, coordinated through the lone-pair electrons on the carbon atoms.

Charge, Formal Charge, and Oxidation State

Before closing this chapter, let us look back at the meanings and roles of the different measures of charge on ions and on atoms within molecules. An isolated ion in free space has a well-defined and measurable electric charge. This charge can be positive or negative, but it is always a whole-number multiple of the charge carried by an electron. This simple behavior carries over to ionic compounds formed between elements in the early groups of the periodic table and those in late groups: NaCl is very well described as an ionic compound composed of Na^+ and Cl^- ions.

Other compounds are not purely ionic but involve electron sharing between atoms. In compounds with covalent bonds, it is not possible to associate a measurable charge with a particular atom. In a purely artificial way, we define the formal charge to be the charge that an atom would have if electrons were divided exactly equally between the two atoms that share them. In other words, when the bonding is covalent, we simply assume that it is *perfectly* covalent (equal sharing). If the result is a Lewis structure with large positive or negative formal charges on atoms, then something is wrong. The artificial assignment of formal charge is a check for consistency because, in a true covalent compound, the atoms should not be highly charged. If they turn out to have high charges in our representation, we either look for a new Lewis structure or predict that the compound will not be stable. Formal charge also allows for conclusions about the nature of bonding. A C—N single bond, for example, has a different length, bond strength, and reactivity than a C=N double bond, so the choice of the best Lewis structure (based on formal charge) is often a very useful guide to molecular properties.

Oxidation state is just as arbitrary as formal charge, but it is defined in a completely different way and has a different purpose. To determine oxidation state, we suppose that the compound is purely *ionic* and ask what charge results on the different atoms. If the compound *is* ionic, then oxidation state coincides with the true charge, but if there is covalent character, this no longer is true. The concept of oxidation state is useful in understanding the chemistry of transition elements that form a variety of compounds of mixed ionic and covalent character. It leads to valid predictions of chemical and physical properties. For example, compounds with elements in high oxidation states tend to have lower boiling points than do compounds of the same elements in low oxidation states. The concept of oxidation state is also useful in naming compounds.

If every compound could be classed as purely ionic or purely covalent, then neither formal charge nor oxidation state would be necessary. In purely covalent molecules, formal charges would always be zero. In purely ionic compounds, oxidation state would always correspond to ionic charge, and there would be no point in defining it separately. The beauty of chemistry, however, is that borderline situations abound. Most bonds have partial ionic and partial covalent character. Formal charge and oxidation state, although they are arbitrary concepts, help to describe and predict real chemical behavior.

SUMMARY

3-1 Elements are classified broadly as metals, non-metals, or semi-metals and also fall naturally into smaller **groups** based on similarities in chemical and physical properties. The groups include (among others) the **alkali metals,** the **alkaline earth metals,** the **chalcogens,** and the **halogens.**

3-2 The modern periodic law states that the chemical and physical properties of the elements are periodic functions of their atomic numbers. The periodic table displays elements in groups (columns) and **periods** (rows). The eight **main groups** (designated I–VIII) are the **representative elements.** An additional ten groups of **transition** elements (all metals) include the **actinides** and **lanthanides.** The **electronegativity** (the power of an atom when in chemical combination to attract electrons to itself) is a periodic property; the relative electronegativity of two bonded atoms determines the type of bonding (ionic versus covalent) between them.

3-3 In ionic compounds **(salts),** metals (electropositive elements) lose electrons to form **cations,** and non-metals (electronegative elements) gain electrons to form **anions;** the two kinds of ions then interact. Lewis dot symbols show the outer (or **valence)** electrons of a species as dots. Atoms tend to lose or gain electrons to achieve an **octet** of valence electrons (doublet for hydrogen). Because transfers of electrons preserve **charge neutrality,** the formulas of ionic compounds can be predicted. The naming of simple ionic compounds includes pairing the name of the cation and anion in that order; the names of the ions derive from the names of the elements they contain.

3-4 Non-metals combine by sharing valence electrons in **covalent** bonds. Compounds that contain only covalent bonds are **covalent compounds.** A single bond is a pair of shared electrons; bonds may be single or multiple. Valence electrons not involved in bonding appear in **lone pairs. Formal charges** associated with atoms in **Lewis structures** can help one to decide whether a proposed structure is meaningful.

3-5 Drawing a Lewis structure requires first a plan of the connections of the atoms. Then count the valence electrons; join all covalently bonded atoms by single bonds; subtract the number of electrons used in this way from the total count; and distribute the remaining electrons in lone pairs and multiple bonds to give each atom a valence octet (a doublet for hydrogen). For some compounds and molecular ions, covalent bonding can be rationalized by assuming that two or more Lewis structures contribute as **resonance hybrids.** Exceptions to the octet rule include odd-electron molecules (an odd number of valence electrons), octet-deficient molecules (fewer than eight electrons around a non-H atom), and molecules showing valence-shell expansion (more than eight electrons around an atom).

3–6   The names of binary molecular compounds derive from the names of the elements, with prefixes to indicate the number of atoms of each element and the suffix *-ide* added to the first portion of the name of the more electronegative element. The names of ionic compounds that contain polyatomic ions (**molecular ions**) use the names of the molecular ions. Many of these names, particularly the names of the oxoanions, are only partially systematic. Many polyatomic cations are named as *-onium* derivatives of molecules; oxoanions are named using *-ate* and *-ite* suffixes (and, if necessary, *per-* and *hypo-* prefixes) to indicate the number of oxygens attached to a central atom. The necessity of charge neutrality determines the formula of all ionic compounds, whether or not they contain polyatomic ions.

3–7   Molecular geometries can be predicted by the **valence shell electron-pair repulsion (VSEPR) theory:** Bonded atoms and lone pairs around a central atom position themselves as far from one another as possible; the steric number of an atom equals the total number of such atoms and pairs; a different three-dimensional geometry is associated with each steric number. The geometries for different steric numbers are 2, linear; 3, trigonal planar; 4, tetrahedral; 5, trigonal bipyramidal; and 6, octahedral, each giving different sets of angles between the electron pairs.

3–8   The transition elements tend to exhibit mixed ionic and covalent behavior. Most of them occur in more than one **oxidation state,** and compounds of the higher oxidation states have more covalent character than those of the lower. Many transition-metal ions form **coordination complexes** with molecules or ions that serve as **ligands** by sharing lone pairs with the metal ion. The resulting ligand–metal bonds are partially covalent.

PROBLEMS

Note: Answers to blue-numbered problems are given in Appendix F. Problems that are more challenging are indicated with asterisks.

The Periodic Table

1. Hydrogen forms compounds with C, N, O, and F that have the formulas CH_4, NH_3, H_2O, and HF. Predict the formulas of the compounds hydrogen forms with P, Cl, Si, and S.

2. Carbon monoxide reacts with many metals to give metal carbonyls. Examples are chromium carbonyl ($Cr(CO)_6$), manganese carbonyl ($Mn_2(CO)_{10}$), iron carbonyl ($Fe(CO)_5$), and cobalt carbonyl ($Co_2(CO)_8$). Use the periodic table to predict the molecular formulas of tungsten carbonyl, molybdenum carbonyl, ruthenium carbonyl, and rhenium carbonyl.

3. Based on the periodic table, write the formula expected for binary compounds between the following elements:

(a) Rubidium and arsenic
(b) Strontium and fluorine
(c) Calcium and sulfur

4. Use the periodic table to predict the formula expected for binary compounds between the following elements:
(a) Magnesium and oxygen
(b) Cesium and bromine
(c) Aluminum and iodine

5. Plot the melting points (on the vertical axis) of the elements Li through Ca against their atomic numbers (on the horizontal axis). The melting points in °C: Li(181), Be(1283), B(2300), C(3550), N(−210), O(−218), F(−220), Ne(−249), Na(98), Mg(649), Al(660), Si(1410), P(44), S(113), Cl(−101), Ar(−189), K(64), Ca(839). Summarize the periodic pattern in a brief statement.

6. Plot the melting points (on the vertical axis) of the elements Li through Ca against their atomic *masses* (on the horizontal axis). Refer to Problem 5 for data. Discuss any violations of the periodic pattern in this plot.

7. A certain element M is a main-group metal that reacts with chlorine to give a compound with chemical formula MCl_2 and with oxygen to give the compound MO.
(a) To which group in the periodic table does element M belong?
(b) The chloride contains 74.5% chlorine by mass. Name the element M.

8. A certain non-metallic element Q reacts with potassium to give a compound with chemical formula K_2Q and with calcium to give the compound CaQ.
(a) To which group in the periodic table does element Q belong?
(b) The compound K_2Q contains 49.8% potassium by mass. Name the element Q.

9. (a) Name six elements that are gases at room temperature.
(b) Do these elements cluster in some part of the periodic table? Explain.

10. Metals are found on the left side or in the lower part of the periodic table. Most are solids at room temperature. Name one metallic element that is a liquid at room temperature.

11. Before the element scandium was discovered in 1879, it was known as "eka-boron." Predict the properties of scandium from averages of the corresponding properties of its neighboring elements in the periodic table, as listed below.

Element	Melting Point (°C)	Boiling Point (°C)	Density (g cm^{-3})
Calcium (Ca)	839	1484	1.55
Titanium (Ti)	1660	3287	4.50
Scandium (Sc)	?	?	?

12. The element technetium (Tc) is not found in nature but has been produced artificially through nuclear reactions. Use the data for several neighboring elements (given in the table below) to estimate the melting point, boiling point, and density of technetium.

Element	Melting Point (°C)	Boiling Point (°C)	Density (g cm^{-3})
Manganese (Mn)	1244	1962	7.2
Molybdenum (Mo)	2610	5560	10.2
Rhenium (Re)	3180	5627	20.5
Ruthenium (Ru)	2310	3900	12.3

13. Use the group structure of the periodic table to predict the empirical formulas for the binary compounds that hydrogen forms with each of the following elements: arsenic, iodine, lead, and tellurium.

14. Use the group structure of the periodic table to predict the empirical formulas for the binary compounds that hydrogen forms with each of the following elements: silicon, astatine, sulfur, and antimony.

Ions and Ionic Compounds

15. (See Example 3–1.) State the number of core electrons and valence electrons in an atom of each of the following elements. Draw a Lewis dot symbol for each of these atoms: (a) Sb (b) Br (c) B (d) Ra

16. (See Example 3–1.) State the number of core electrons and valence electrons in an atom of each of the following elements. Draw a Lewis dot symbol for each of these atoms: (a) Rn (b) Pb (c) In (d) As

17. Give the Lewis dot symbols for Xe, Se^{2-}, Se^+, and Be^+.

18. Give the Lewis dot symbols for S^-, Ba^+, Na, and Rb^-.

19. For each of the atoms or ions in Problem 17, give the total number of electrons, the number of valence electrons, and the number of core electrons.

20. For each of the atoms or ions in Problem 18, give the total number of electrons, the number of valence electrons, and the number of core electrons.

21. (See Examples 3–2 and 3–3.) Give the name and formula of the binary ionic compounds involving the elements in each pair listed below. Write Lewis symbols for the elements both before and after chemical combination.
(a) Calcium and sulfur   (d) Nitrogen and strontium
(b) Iodine and cesium   (e) Lithium and oxygen
(c) Francium and sulfur

22. (See Examples 3–2 and 3–3.) Give the name and formula of the binary ionic compounds involving the elements in each pair listed below. Write Lewis symbols for the elements both before and after chemical combination.
(a) Lithium and selenium
(b) Chlorine and barium
(c) Strontium and astatine
(d) Boron and chlorine
(e) Cesium and oxygen

23. Name the following binary ionic compounds:
(a) Al_2S_3 (d) Ca_3N_2
(b) Cs_2Se (e) Cs_2O
(c) Li_2S (f) KBr

24. Name the following binary ionic compounds:
(a) NaF   (d) Na_3P
(b) SrI_2 (e) $RaCl_2$
(c) MgO   (f) RbCl

Covalent Bonding and Lewis Structures

25. By counting electrons, determine what is wrong with each of the following Lewis structures:

(a) IF_2 (b) OCN   (c) CH_2O

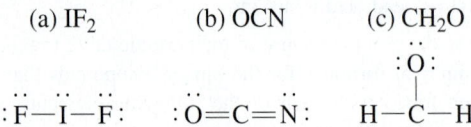

26. By counting electrons, determine what is wrong with each of the following Lewis structures:

(a) CO_2^- (b) $HBCl_3$ (c) C_2H_2

$$\left[\ddot{O}=C-\ddot{\ddot{O}}:\right]^-   :\ddot{\underset{\underset{\overset{|}{H}}{}}{\overset{:\ddot{Cl}:}{Cl}}-\overset{:\ddot{Cl}:}{\underset{}{B}}-\ddot{Cl}:   :\ddot{H}-C\equiv C-\ddot{H}:$$

27. (See Example 3–4.) Assign formal charges to all atoms in the following Lewis structures:

(a) SO_4^{2-} (b) $S_2O_3^{2-}$

$$\left[\underset{\underset{:\ddot{O}:}{|}}{\overset{:\ddot{O}:}{\underset{|}{:\ddot{O}-S-\ddot{O}:}}}\right]^{2-}   \left[\underset{\underset{:\ddot{S}:}{||}}{\overset{:\ddot{O}:}{\underset{|}{:\ddot{O}-S-\ddot{O}:}}}\right]^{2-}$$

(c) SbF_3 (d) SCN^-

$$:\ddot{F}-\underset{\underset{:\ddot{F}:}{|}}{Sb}-\ddot{F}:   \left[:\ddot{S}-C\equiv N:\right]^-$$

28. (See Example 3–4.) Assign formal charges to all atoms in the following Lewis structures:

(a) ClO_4^- (b) SO_2

$$\left[\underset{\underset{:\ddot{O}:}{|}}{\overset{:\ddot{O}:}{\underset{|}{:\ddot{O}-Cl-\ddot{O}:}}}\right]^-   :\ddot{O}-\ddot{S}=O$$

(c) BrO_2^- (d) NO_3^-

$$\left[:\ddot{O}-\ddot{Br}-\ddot{O}:\right]^-   \left[\underset{\underset{:\ddot{O}:}{|}}{O=N-\ddot{O}:}\right]^-$$

29. (See Example 3–4.) Determine the formal charge on all the atoms in the following Lewis structures:

$$H-N=\ddot{O}   \text{and}   H-\ddot{O}=N$$

Which one would probably be favored as the structure of the molecule HNO?

30. (See Example 3–4.) Determine the formal charge on all the atoms in the following Lewis structures:

$$:\ddot{Cl}-\ddot{Cl}-\ddot{O}:   \text{and}   :\ddot{Cl}-\ddot{O}-\ddot{Cl}:$$

Which one would probably be favored as the structure of the molecule Cl_2O?

31. In each of the following Lewis structures, Z represents a main-group element. Name the group to which Z belongs in each case, and give an example of such a compound or ion that actually exists.

(a)   (b)

$$\ddot{O}=Z=\ddot{O}   \underset{\underset{:\ddot{O}:}{|}}{:\ddot{O}-\overset{:\ddot{O}:}{\underset{|}{Z}}-\ddot{O}-\overset{:\ddot{O}:}{\underset{|}{Z}}-\ddot{O}:}$$

(c)   (d)

$$\left[\underset{}{\overset{:\ddot{O}:}{\underset{|}{O=Z-\ddot{O}:}}}\right]^-   \left[\underset{\underset{:\ddot{O}:}{|}}{H-\ddot{O}-\overset{:\ddot{O}:}{\underset{|}{Z}}-\ddot{O}:}\right]^-$$

32. In each of the following Lewis structures, Z represents a main-group element. Name the group to which Z belongs in each case, and give an example of such a compound or ion that actually exists.

(a)   (b)

$$\left[:C\equiv Z:\right]^-   \left[\underset{\underset{:\ddot{O}:}{|}}{\overset{:\ddot{O}:}{\underset{}{:\ddot{O}-Z-\ddot{O}:}}}\right]^-$$

(c)   (d)

$$\left[\underset{}{\overset{:\ddot{O}:}{\underset{|}{:\ddot{O}-Z-\ddot{O}:}}}\right]^{2-}   H-\underset{\underset{H}{|}}{Z}-\underset{\underset{H}{|}}{Z}-H$$

Drawing Lewis Structures

33. Draw Lewis electron-dot structures for the following species:
(a) H_2Se
(b) SbH_3
(c) HOI
(d) $(CNCN)^{2-}$ (atoms arranged in a row, as written)

34. Draw Lewis electron-dot structures for the following species:
(a) Methane, CH_4
(b) Carbon dioxide, CO_2
(c) Phosphorus trichloride, PCl_3
(d) Hypochlorite ion, ClO^-

35. Draw a Lewis structure for OF_2, with O as the central atom.

36. Draw a Lewis structure for ClI_2^-, with Cl as the central atom.

37. Urea (H_2NCONH_2) is an important chemical fertilizer, and it is also excreted in urine. The carbon atom is bonded to both nitrogen atoms and to the oxygen atom. Draw its Lewis structure, and use Table 3–3 to estimate its bond lengths.

38. Acetic acid (CH_3COOH) is the active ingredient of vinegar. The second carbon atom is bonded to the first carbon atom and to both oxygen atoms. Draw its Lewis structure, and use Table 3–3 to estimate its bond lengths.

39. (See Examples 3–5 through 3–7.) Draw a Lewis structure for each of the following ions, and assign formal charges to all atoms:
   (a) ClO_3^- (b) $AlCl_4^-$ (c) XeF^+

40. (See Examples 3–5 through 3–7.) Draw a Lewis structure for each of the following ions, and assign formal charges to all atoms:
   (a) OCN^- (b) PH_4^+ (c) $PO_2Cl_2^-$ (central P atom)

41. (See Example 3–8.) Draw Lewis structures for the two resonance forms of the nitrite ion, NO_2^-. Use Table 3–3 to estimate the approximate lengths of the nitrogen–oxygen bonds.

42. (See Example 3–8.) Draw Lewis structures for the three resonance forms of the carbonate ion, CO_3^{2-}. Use Table 3–3 to estimate the approximate lengths of the carbon–oxygen bonds.

43. The compound with molecular formula ONF may exist in two isomers: one with N as the central atom, and the other with O as the central atom in a three-atom chain. Draw Lewis structures to represent the bonding in both isomers, and indicate formal charges.

44. Some evidence suggests that the molecule S_3 exists, possibly in a bent geometry analogous to that of ozone (see Exercise 3–8) and also possibly in a cyclic form (an equilateral triangle of S atoms). Draw Lewis structures (with formal charge where necessary) to represent the bonding in both these isomers of S_3.

45. The stable form of gaseous sulfur under certain conditions consists of S_8 molecules whose skeleton is a ring of eight sulfur atoms. Draw the Lewis structure for such a ring.

46. White phosphorus (P_4) consists of four phosphorus atoms arranged at the corners of a tetrahedron.

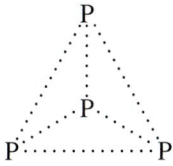

Draw in the valence electrons on this structure to give a Lewis structure that satisfies the octet rule.

47. (See Example 3–9.) Draw Lewis structures for the following compounds. In the formulas, the symbol of the central atom is given first. (*Hint*: The valence octet may be expanded for the central atom.)
   (a) PF_5 (b) SF_4 (c) XeO_2F_2

48. (See Example 3–9.) Draw Lewis structures for the following ions. In the formulas, the symbol of the central atom is given first. (*Hint*: The valence octet may be expanded for the central atom.)
   (a) BrO_4^- (b) PCl_6^- (c) XeF_3^+

Naming Compounds in Which Covalent Bonding Occurs

49. (See Example 3–10.) Name the following binary molecular compounds.
   (a) SeF_4
   (b) IBr_3
   (c) P_4O_6

50. (See Example 3–10.) Name the following binary molecular compounds.
   (a) P_5S_{10}
   (b) O_2F_2
   (c) P_2Cl_4

51. (See Example 3–11.) Complete the following table:

Name	Formula
Silver nitrate	—
—	$Ca(OCl)_2$
Potassium hydrogen sulfate	—
—	$Ga_2(SO_3)_3$
Potassium chlorate	—
—	$NaHCO_3$

52. (See Example 3–11.) Complete the following table:

Name	Formula
Lithium hydrogen sulfite	—
—	$Sr(ClO_4)_2$
Rubidium hydrogen sulfite	—
—	$Al(NO_3)_3$
Ammonium chromate	—
—	$(H_3O)NO_3$

53. The tartrate anion has the formula $C_4H_4O_6^{2-}$, and $Na_2C_4H_4O_6$ is sodium tartrate. Name the compound $NaHC_4H_4O_6$.

54. Monoammonium phosphate is a compound made up of NH_4^+ and $H_2PO_4^-$ ions that is employed as a flame retardant (its use for this purpose was first suggested by Gay-Lussac in 1821). Write its chemical formula. What is the systematic chemical name of this compound?

55. The last step in the industrial preparation of phosphoryl chloride is the reaction of tetraphosphorus decaoxide with phosphorus pentachloride to give $POCl_3$ as the sole product. Write a balanced chemical equation for this reaction.

56. Dichlorine heptaoxide is a colorless, oily liquid that boils at 81°C. Further heating causes it to decompose explosively to chlorine trioxide and chlorine tetraoxide. Write a balanced chemical equation for this reaction.

The Shapes of Molecules

57. (See Example 3–12.) Give the steric number for the central atom in each of the following molecules. In each case, the central atom is listed first, and the other atoms are all bonded directly to it.
(a) CI_4 (d) $SOCl_2$
(b) SO_2 (e) IBr_3
(c) SF_6

58. (See Example 3–12.) For each of the following molecules or molecular ions, give the steric number of the central atom. In each case, the central atom is listed first and the other atoms are all bonded directly to it.
(a) PCl_3 (d) ClO_2^-
(b) SO_2Cl_2 (e) $GeBr_4$
(c) TeF_6

59. (See Example 3–14.) For each of the molecules in Problem 57, sketch and name the approximate molecular geometry.

60. (See Example 3–14.) For each of the molecules or molecular ions in Problem 58, sketch and name the approximate molecular geometry.

61. For each of the following molecules or molecular ions, give the steric number of the central atom, sketch and name the approximate molecular geometry, and describe the direction of any *distortions* from the approximate geometry due to lone pairs. In each case, the central atom is listed first, and the other atoms are all bonded directly to it.
(a) ICl_4^-
(b) OF_2
(c) BrO_3^-
(d) CS_2

62. For each of the following molecules or molecular ions, give the steric number of the central atom, sketch and name the approximate molecular geometry, and describe the direction of any *distortions* from the approximate geometry due to lone pairs. In each case, the central atom is listed first, and the other atoms are all bonded directly to it.
(a) TeH_2 (c) PCl_4^+
(b) AsF_3 (d) XeF_5^+

63. Give an example of a molecule or ion having the following formula type and structure:
(a) AB_3 (planar)   (c) AB_2^- (bent)
(b) AB_3 (pyramidal)   (d) AB_3^{2-} (planar)

64. Give an example of a molecule or ion having the following formula type and structure:
(a) AB_4^- (tetrahedral)   (c) AB_6^- (octahedral)
(b) AB_2 (linear)   (d) AB_3^- (pyramidal)

65. (See Example 3–15.) For each of the answers in Problem 57, state whether the species is *polar* or *non-polar*.

66. (See Example 3–15.) For each of the answers in Problem 58, state whether the species is *polar* or *non-polar*.

67. The molecules of a certain compound contain one atom each of nitrogen, fluorine, and oxygen. Two structures, NOF (O as central atom) and ONF (N as central atom), are under consideration as the correct structure. Does the information that the molecule is bent allow for a decision between these two? Explain.

68. Mixing $SbCl_3$ and $GaCl_3$ in a one-to-one molar ratio (using liquid sulfur dioxide as a solvent) gives a solid ionic compound of empirical formula $GaSbCl_6$. A controversy erupts over whether this compound is $(SbCl_2^+)(GaCl_4^-)$ or $(GaCl_2^+)(SbCl_4^-)$.
(a) Predict the molecular structures of the two anions.
(b) It is learned that the cation in the compound has a bent structure. Based on this fact, which formulation is more likely to be correct?

69. (a) Use the VSEPR theory to predict the structure of the NNO molecule.
(b) The substance NNO has a small dipole moment. Which end of the molecule is more likely to be the positive end, given that O is more electronegative than N?

70. Ozone (O_3) has a non-zero dipole moment. In molecules of O_3, one of the oxygen atoms is directly bonded to the other two, which are not bonded to each other.
(a) Based on this information, which of the following structures are possible for the ozone molecule: symmetric linear, unsymmetric linear (for example, different O—O bond lengths), and bent? (*Note*: Even an O—O bond can have a bond dipole if the two oxygen atoms are bonded to different atoms, or if only one of the oxygen atoms is bonded to a third atom.)
(b) Use VSEPR theory to predict which of the structures of part (a) should be observed.

Elements Forming More than One Ion

71. (See Example 3–16.) Give names for the following compounds of transition-metal or late main-group elements:
(a) Fe_2O_3 (d) $PbCl_4$
(b) $TiBr_4$ (e) MnF_3
(c) WO_3 (f) $AgClO_3$

72. (See Example 3–16.) Give names for the following compounds of transition or late main-group elements:
(a) Mo_2S_5 (d) $NbCl_4$
(b) NiF_3 (e) TaS_2
(c) Hg_2Cl_2 (f) Er_5Sb_3

73. (See Example 3–16.) Write chemical formulas for the following compounds of transition-metal or late main-group elements:
(a) tin(II) fluoride   (d) tungsten(V) chloride
(b) rhenium(VII) oxide   (e) copper(II) nitrate
(c) cobalt(III) fluoride   (f) nickel(II) perbromate hexahydrate

74. (See Example 3–16.) Write chemical formulas for the following compounds of transition or late main-group elements:
    (a) molybdenum(III) iodide
    (b) thallium(I) chloride
    (c) manganese(IV) oxide
    (d) niobium(IV) oxide
    (e) copper(I) oxide
    (f) gadolinium(IV) silicide

75. Rhenium(VII) sulfide reacts with hydrogen to give rhenium(VI) sulfide and hydrogen sulfide. Write a balanced chemical equation for this process.

76. Vanadium(V) oxide reacts with carbon monoxide to give vanadium(III) oxide and carbon dioxide. Write a balanced chemical equation for this process.

ADDITIONAL PROBLEMS

77. Trends in physical and chemical properties are not always smooth across the periodic table.
    (a) Estimate the boiling point of zinc, based on the boiling points of its neighbors copper (2567°C) and gallium (2403°C). Compare your estimate with the measured value of 907°C.
    (b) Gold and thallium are solids at room temperature, whereas mercury is a liquid. Can you draw a general conclusion about the boiling temperatures of elements in the zinc group?
    (c) Where would you predict the boiling point of cadmium to lie relative to that of silver and indium?

78. Indium forms a pale yellow oxide that is 82.712% indium and 17.288% oxygen by mass.
    (a) At the time of Mendeleev, it was believed that the relative atomic mass of oxygen was 16 and the relative atomic mass of indium was 76. At that time, what formula was given to the pale yellow indium oxide?
    (b) Mendeleev suggested that the correct atomic mass of indium was 114 ($\frac{3}{2}$ of 76). What formula was he simultaneously suggesting for the pale yellow oxide?

79. In a recent science-fiction novel, the author has one character say, "You don't realize it, but oxygen is a corrosive gas. It's in the same chemical family as chlorine and fluorine, and hydrofluoric acid is the most corrosive acid known." Criticize this statement. Give the name and the formula of the compound of oxygen that is most analogous to hydrofluoric acid (HF).

80. Only some of the elements in the seventh period have been discovered.
    (a) Suppose scientists prepare element 113 and study its properties. What elements would you predict it to resemble in chemical properties? Do you expect it to be a metal?
    (b) What other elements should resemble element 117? Predict the chemical formula of the compound of element 117 with potassium.

81. Which of the following do you expect to exist as stable ionic compounds? In each unstable case, write the empirical formula of a stable compound containing the same ions. Name the five compounds.
    (a) Na_3CO_3
    (b) $MgCl$
    (c) $(NH_4)_2SO_4$
    (d) $BaCN$
    (e) $Sr(OH)_2$

82. Predict the chemical formula of sodium selenate, by comparison with the chemical formula of sodium sulfate.

83. Many important fertilizers are ionic compounds containing the elements nitrogen, phosphorus, and potassium, because these are frequently the limiting nutrients in soil for plant growth.
    (a) Write the chemical formulas for the following chemical fertilizers: ammonium phosphate, potassium nitrate, ammonium sulfate.
    (b) Calculate the mass percentage of nitrogen, phosphorus, and potassium for each of the compounds in part (a).

84. Two possible Lewis structures for sulfine (H_2CSO) are

$$\text{H} \diagdown \atop \text{H} \diagup \text{C}=\overset{..}{\underset{..}{\text{S}}}-\overset{..}{\underset{..}{\text{O}}}: \quad \text{and} \quad \text{H} \diagdown \atop \text{H} \diagup \text{C}-\overset{..}{\underset{..}{\text{S}}}=\overset{..}{\underset{..}{\text{O}}}$$

    (a) Compute the formal charges on all atoms.
    (b) Draw a Lewis structure for which all the atoms in sulfine have formal charges of zero.

85. A hydrogen ion (H^+) consists of a bare nucleus with no electrons. To satisfy the Lewis criterion, it needs an electron pair. Predict what happens when an H^+ ion encounters an H_2O molecule or an NH_3 molecule. Write the Lewis structures of the products. Does your analysis explain the common existence of NH_4^+, the ammonium ion?

86. Consider the three molecules N_2, N_2H_2, and N_2H_4. Each contains a central nitrogen–nitrogen bond. Write Lewis structures for all three molecules, and use Table 3–3 to estimate bond lengths for all the bonds.

87. There is persuasive evidence for the brief existence of the unstable molecule O—P—Cl.
    (a) Draw a Lewis structure for this molecule in which the octet rule is satisfied on all atoms and the formal charges on all atoms are zero.
    (b) The compound OPCl reacts with oxygen to give O_2PCl. Draw a Lewis structure of O_2PCl for which all formal charges are equal to zero. Draw a Lewis structure in which the octet rule is satisfied on all atoms.

88. A common error is to mix up the subscripts in a binary compound (for example, to write HO_2 when H_2O is intended). Sometimes both the compounds represented in such mix-ups actually exist.
    (a) Draw Lewis structures for ammonia (NH_3) and hydrazoic acid (HN_3). In HN_3, the three nitrogen atoms are attached in a row, with the hydrogen atom on one end.
    (b) Draw Lewis structures for sulfur dioxide (SO_2) and disulfur monoxide (S_2O). Both have central sulfur atoms. Do not overlook the possibility of resonance.

*89. The compound SF_3N has been synthesized.
   (a) Draw the Lewis structure of this molecule, supposing that the three F's and the N surround the S. Indicate the formal charges. Repeat, but assume that the three F's and the S surround the N.
   (b) On the basis of the results in part (a), speculate about which arrangement is more likely to correspond to the actual molecular structure.

90. In nitryl chloride (NO_2Cl), the chlorine atom and the two oxygen atoms are bonded to a central nitrogen atom, and all the atoms lie in a plane. Draw the two resonance structures that satisfy the octet rule and that together are consistent with the fact that the two nitrogen–oxygen bonds are equivalent.

91. The peroxonitrite ion is an isomer of the nitrate ion. It has the same charge and the same number and kind of atoms but uses a different molecular skeleton. In the peroxonitrite ion, the atoms are connected as follows.

$$O\cdots N\cdots O\cdots O$$

Draw a Lewis structure for this ion, and assign formal charges to the atoms.

92. The element xenon (Xe) is by no means chemically inert; it forms a number of chemical compounds with highly electronegative elements, such as fluorine and oxygen. The reaction of xenon with varying amounts of fluorine produces XeF_2 and XeF_4. Subsequent reaction of one or the other of these compounds with water produces (depending on conditions) XeO_3, XeO_4, and H_4XeO_6, as well as mixed compounds such as $XeOF_4$. Predict the structures of these six xenon compounds, using VSEPR theory.

93. The ion $SbCl_6^{3-}$ is prepared from $SbCl_5^{2-}$ by treating it with a source of the Cl^- ion.
   (a) Predict the geometry of the $SbCl_5^{2-}$ ion, using the VSEPR method.
   (b) Determine the steric number of the central antimony atom in $SbCl_6^{3-}$, and discuss the extension of the VSEPR theory to make a prediction of its molecular geometry.

94. Predict the arrangement of the atoms around the sulfur atom in $F_4S{=}O$, assuming that double-bonded atoms require more space than single-bonded atoms.

95. The "dicyanomethanide ion" has the formula $(NC-C-CN)^{2-}$; that is, it has a central carbon atom with two $-CN$ groups attached. A recent publication reports that this ion is bent at the central C atom. Determine the most likely steric number for the central C atom, and predict the bond angle at that atom.

96. Draw Lewis structures and predict the geometries of the following molecules. State which are polar and which nonpolar:
   (a) ONCl   (d) SCl_4
   (b) O_2NCl (e) CHF_3
   (c) XeF_2

*97. The reaction between H_2S and SO_2 under certain conditions gives a compound with the molecular formula $H_2S_2O_2$. One researcher suggests that the skeleton of the compound is

$$H\cdots O\cdots S\cdots S\cdots O\cdots H$$

but another thinks that it is

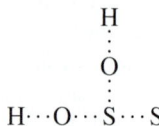

Draw Lewis structures, including resonance forms if they exist, for both of these isomers of $H_2S_2O_2$. What guidance do the Lewis structures offer in deciding which isomer is more likely to have been isolated?

*98. The dimerization of nitrogen dioxide (NO_2) to give dinitrogen tetraoxide is mentioned in the caption of Figure 3–16.
   (a) NO_2 is an odd-electron compound. Draw the best Lewis structures you can for it, recognizing that one atom cannot achieve an octet configuration. Use formal charges to decide whether that should be the nitrogen atom or one of the oxygen atoms.
   (b) From the structure given in Figure 3–16 for N_2O_4, draw resonance structures for this molecule that obey the octet rule.

99. Although magnesium and the alkaline earth metals located below it in the periodic table form ionic chlorides, beryllium chloride ($BeCl_2$) is a covalent compound.
   (a) Follow the usual rules to write a Lewis structure for $BeCl_2$, in which each atom attains an octet configuration. Indicate formal charges.
   (b) The Lewis structure that results from part (a) is an extremely unlikely structure because of the double bonds and formal charges it shows. By relaxing the requirement of placing an octet on the beryllium atom, show how a Lewis structure with all formal charges equal to zero can be written. (*Note*: The compound $BeCl_2$, like BF_3, in fact lacks an octet on the central atom.)

100. The first compound of a noble gas, prepared in 1962, was an orange-yellow ionic solid consisting of XeF^+ and $Pt_2F_{11}^-$ ions.
   (a) Draw a Lewis structure for the XeF^+ ion.
   (b) Shortly after the preparation of the ionic compound discussed in part (a), it was found that the irradiation of mixtures of xenon and fluorine with sunlight produced white crystalline XeF_2. Draw a Lewis structure for this molecule, allowing valence expansion on the central xenon atom.

101. Molecules of iodine trioxide actually have the molecular formula I_4O_{12}. The following structural diagram appears in a research report.

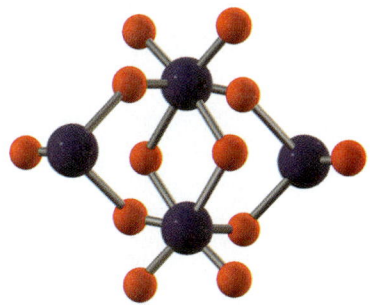

(a) Compute the formal charge on each atom in this structure, assuming that each line that connects two atoms represents a single shared pair of electrons and that the octet rule is satisfied on all oxygen atoms.

(b) Draw the Lewis structure in which the electrons are redistributed so that all of the oxygen atoms have exactly two bonds. Calculate the formal charges on all the atoms in this structure.

(c) Give this compound a better name than "iodine trioxide."

102. A theoretical chemist does a calculation that suggests that molecules of N_6 might exist.

    (a) Draw a Lewis structure for N_6 in which the six atoms are connected in a straight line and the highest and lowest formal charges on any atom are $+1$ and -1.

    (b) Suggest a name for this substance.

103. When indium and sulfur are dissolved in molten tin at a temperature between 650°C and 1100°C, they react to form In_5S_4.

(a) Write a balanced chemical equation for this reaction.

(b) Name the product of the reaction, treating both elements as non-metals.

104. At elevated temperatures, nitrogen trifluoride reacts with copper to give the colorless, reactive gas dinitrogen tetrafluoride and copper(I) fluoride. Write a balanced chemical equation for this reaction.

105. In the following coordination complexes, identify the transition metals and the ligands: $Na_2[(CuCl_4)]$, $[Co(H_2O)_6]Br_2$, $K_2[Pt(CN)_4]$, $[Cr(CO)_6]$.

106. You encounter the old-style name "ceric ammonium nitrate" in a chemistry paper from the early 20th century. Later on in the paper, the same compound is called "ammonium hexanitratocerate," which is a kind of name you have never seen before. Despite this, you are able to figure out the correct chemical formula of the compound. What is it?

107. At one time, the bonding in coordination complexes was not understood, and they were often formulated as "double salts"; for example, $K_2[Pt(CN)_4]$ was written as $Pt(CN)_2 \cdot 2KCN$. Translate the formulas of the following complexes from the outmoded double-salt notation to current notation: $CuCl_2 \cdot 2KCl$, $CoF_3 \cdot 3NaF$.

*108. The nitrate ion and the peroxonitrite ion (see Problem 91) are isomers. They are the only species of molecular formula NO_3^- so far known to exist. Amazingly, no fewer than eight different skeletons are conceivable by which a single N atom and three O atoms might bond together. Draw Lewis structures for all eight NO_3^- isomers. (*Hint:* The eight include four skeletons in which atoms are linked in a cycle.)

CUMULATIVE PROBLEM

Compounds of Oxygen

Consider the four compounds KO_2, BaO_2, TiO_2, and H_2O_3. All contain oxygen but in quite different bonding situations.

(a) The oxygen in TiO_2 occurs as O^{2-} ions. Give the Lewis dot symbol for this ion. How many total electrons does this ion have? How many of these are valence electrons and how many are core electrons? What is the chemical name for TiO_2?

(b) Recall that Group II elements form stable $+2$ ions. By referring to Table 3–5, identify the oxygen-containing ion in BaO_2, and give the name of the compound. Draw a Lewis structure for the oxygen-containing ion, and show formal charges for all atoms. Predict the length of the O — O bond in this ion.

(c) Recall that Group I elements form stable $+1$ ions. By referring to Table 3–5, identify the oxygen-containing ion in KO_2 and give the name of the compound. Show that the oxygen-containing ion is an odd-electron species. Draw the best Lewis structure you can for it.

(d) H_2O_3 is a molecular compound. Name this compound. Draw a Lewis structure for H_2O_3 in which every atom has a formal charge of zero. (*Hint:* Use a skeleton that links the three oxygen atoms in a chain with hydrogen atoms at each end.)

(e) By referring to Table 3–3, predict the bond distances in H_2O_3.

(f) Use the VSEPR theory to predict the three bond angles in H_2O_3.

On the right is a sample of rutile, TiO_2, which is the primary ore of titanium. Behind it is a large quartz crystal through which slender rutile hairs penetrate. Crystals of this type are called "Venus hairstone." (*Michael Dalton/Fundamental Photographs*)

4 Types of Chemical Reactions

CHAPTER OUTLINE

Hot steel wool is oxidized by pure oxygen to form iron(III) oxide. *(Charles D. Winters)*

At the heart of chemistry are chemical reactions, the dynamic processes that lead to the formation of new substances. No one can predict the course of chemical events when substances interact in all possible combinations under all conditions. However, patterns of reactivity do exist. In this chapter, we classify chemical reactions, particularly those taking place in solution, with an eye to predicting their outcomes.

4-1 Dissolution Reactions

In **dissolution,** two (or more) substances spread out, or disperse, into each other at the level of individual atoms, molecules, or ions. As this happens, formerly distinct phases merge into a new single phase. The result of dissolution is a homogeneous mixture called a **solution.** Thus, solid sucrose (sugar) disperses into liquid water to give a new liquid (sugar-water); liquid octane and liquid methanol dissolve each other to give a new liquid fuel; and gaseous hydrogen infiltrates into solid palladium to give a new solid.

Usually, one component in a solution is present in greater amount than any other. The major component is the **solvent,** and the minor components are **solutes.** Dilute solutions have smaller proportions of solute; concentrated solutions have larger proportions of solute. The physical and chemical properties of a solution differ from those of the pure solvent or pure solute and change if the relative amounts of the components are changed.

The phenomenon of dissolution is intermediate between a physical change and a chemical change. During dissolution, the attractions among the particles in the original phases (solvent-to-solvent and solute-to-solute attractions) are broken up and replaced, at least in part, by new solvent-to-solute attractions. In this respect, dissolution is a distinctly chemical event. On the other hand, a solution, unlike a compound, has its components in *indefinite* proportions and cannot be represented by the usual kind of chemical formula. On this basis, dissolution is a physical change, like the mixing of salt and sand. Equations for dissolution reactions simply omit the solvent as a reactant. They show the change by indicating the state of the solute in parentheses on the reactant side of the arrow and the name of the solvent in parentheses on the product side. Thus, solid (s) sucrose dissolves in water to give an aqueous (aq) solution of sucrose

$$C_{12}H_{22}O_{11}(s) \longrightarrow C_{12}H_{22}O_{11}(aq)$$

In principle, the solute and solvent can be any combination of solid (s), liquid (ℓ), and gaseous (g) phases. In practice, solid–solid dissolutions are mostly too slow to worry about, mixtures of gases are treated in a different way (see Section 5–6), and the best solvents are liquids. Liquid water is indisputably the world's best-known, most important, most nearly universal solvent; it readily dissolves many disparate solids, liquids, and gases. We therefore emphasize aqueous solutions with the understanding that dissolution also occurs in a host of other solvents, both inorganic ones, such as liquid ammonia (NH_3), and organic ones, such as ethanol (C_2H_5OH), diethyl ether ($C_2H_5OC_2H_5$), acetone (CH_3COCH_3), benzene (C_6H_6), and chloroform ($CHCl_3$).

• In a true solution, the particles are smaller than approximately 10^{-9} m in diameter. Dispersions with bigger particles are colloids (see Section 6–8). The dividing line between solution and colloid is necessarily vague.

• Dissolution also takes place in supercritical fluids, which are phases that are neither gaseous nor liquid. See Section 6–5.

(a) In ethanol (C_2H_5OH), one of the H atoms in water is replaced by a C_2H_5 group (the ethyl group). (b) In diethyl ether ($H_5C_2OC_2H_5$), both of the H atoms in water are replaced by C_2H_5 groups.

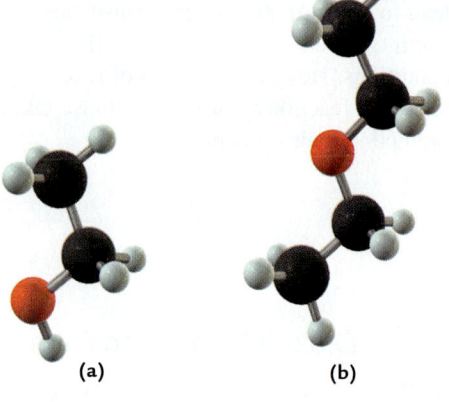

(a)   (b)

Ionic Compounds in Water

Ionic compounds are solids in the range of temperature in which water is a liquid. They have rigid lattices in which strong forces (ionic bonds) pin the constituent ions in place. The high melting points of ionic compounds indicate that a good deal of energy must be supplied to destroy the lattice and produce a liquid (molten) form in which the ions move more freely. How is it then that many ionic compounds dissolve freely in water? How does water overcome the forces of attraction (the solute–solute forces) among the ions in soluble ionic solids, such as NaCl or K_2SO_4? The answer is that it replaces those forces with strong and numerous new solvent–solute attractions, a process called **aquation.** Molecules of water are electrically polar (having sides bearing opposite but equal electric charges; see Fig. 3–20). When water touches an ionic compound such as K_2SO_4, the negative poles of some of its molecules attract the positive ions while the positive poles of others attract the negative ions. The combined strength of these electrostatic attractions pulls the ions out of the ionic lattice. Layers of water molecules cling to the freed ions, which then wander off into the aqueous phase (Fig. 4–1). The solute is said to **dissociate** into ions or to **ionize** upon dissolution. Equations for the dissolution of ionic substances recognize dissociation by showing aquated ions as products:

- If the solvent is not water, then the more general term "solvation" is used.

$$K_2SO_4(s) \longrightarrow 2\,K^+(aq) + SO_4^{2-}(aq)$$

Molecular Substances in Water

Molecular substances contain no ions to which water molecules can adhere, but their molecules are often polar—the differing electronegativities of the atoms lead to build-ups of negative and positive charge in different molecular regions. Sugars are notable examples of molecular substances having polar molecules. They are members of a broader class of compounds, the carbohydrates (see Section 25–2) and have the general formula, $C_m(H_2O)_n$. Typical sugars include sucrose, $C_{12}H_{22}O_{11}$ (table sugar); fructose, $C_6H_{12}O_6$ (fruit sugar); and ribose, $C_5H_{10}O_5$ (a sub-unit in the genetically important compounds known as ribonucleic acids).

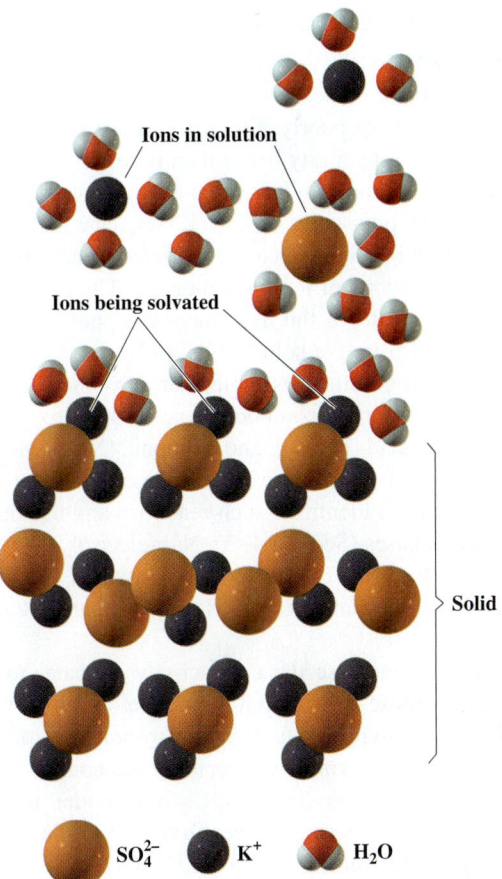

Ions in solution

Ions being solvated

> Solid

SO_4^{2-} K^+ H_2O

Figure 4–1 • When an ionic solid (in this case, K_2SO_4) dissolves in water, the ions move away from their sites in the solid, where they are attracted strongly by ions of opposite electric charge. New strong attractions replace those lost as each ion becomes surrounded by a group of water molecules. In a precipitation reaction, the process is reversed.

Despite the appearance of the general formula, sugars do not contain water molecules. They do contain O—H (hydroxyl) groups bonded to carbon atoms. Hydroxyl groups are bent at the oxygen atom and are therefore electrically polar. For this reason, hydroxyl groups provide sites for interactions with the polar water molecule. These interactions favor dissolution (Fig. 4–2). The molecules remain intact, so the dissolution of fructose in water is represented

$$C_6H_{12}O_6(s) \longrightarrow C_6H_{12}O_6(aq)$$

The dissolution of many other molecular substances in water follows the same pattern. No ions form, but the solute molecules are sufficiently polar to support strong attractions to the water molecules. These attractions replace existing solute–solute attractions, and aquated molecules move off into solution.

• The O—H group (uncharged) is not the same as the hydroxide ion (OH⁻). The latter has one more electron.

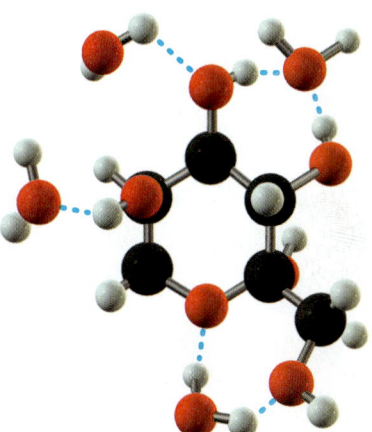

Figure 4–2 • A molecule of fructose in aqueous solution. Note the attractions between the hydroxyl (O—H) groups of the fructose and molecules of water. The fructose molecule is aquated; the exact number and arrangement of the aquating molecules fluctuate.

Electrolytes and Non-Electrolytes

• The terms "electrolyte" and "non-electrolyte" are often used to refer to the solution as well as to the solute.

The study of the **conductivity** of aqueous solutions (their ability to conduct an electrical current) provides major evidence for our ideas about dissolution in water. Pure water conducts electricity very poorly. **Electrolytes** are substances that dissolve to give solutions that conduct electricity better than pure water does; **non-electrolytes** do not increase the conductivity of water when they dissolve. Many electrolytes are ionic compounds, such as potassium sulfate (K_2SO_4) and sodium chloride (NaCl). The picture of dissolution into ions (charged particles) neatly explains the generally high conductivities of solutions of ionic compounds. The electric current is carried through the solution by means of the movement in opposite directions of positive and negative ions derived from the solute (Fig. 4–3).

Somewhat surprisingly, numerous molecular substances are also electrolytes. Examples include some important compounds, such as hydrochloric acid (HCl), nitric acid (HNO_3), sulfuric acid (H_2SO_4), and ammonia (NH_3). Molecular substances, which contain no ions, act as electrolytes by reacting with water to generate ions as they dissolve. Chemists have identified such reactions and use them to characterize compounds as acids and bases (Section 4–3). Non-electrolytes, which are generally molecular substances, furnish no ions as they dissolve. Examples are acetone (CH_3COCH_3), the alcohols ethanol (C_2H_5OH) and methanol (CH_3OH), and sugars such as fructose and sucrose.

Different electrolytes increase the conductivity of water to different degrees. **Strong electrolytes** increase conductivity much more than **weak electrolytes.** For example, the conductivity of a 0.2 molar solution of calcium chloride (0.2 M $CaCl_2(aq)$) is about 200 times greater than the conductivity of a 0.2 molar solution of mercury(II) chloride (0.2 M $HgCl_2(aq)$) under the same conditions. Calcium chloride is a strong electrolyte; mercury(II) chloride is a weak electrolyte. Other weak electrolytes are lead acetate ($Pb(CH_3COO)_2$), cadmium iodide (CdI_2), mercury(II) cyanide ($Hg(CN)_2$), acetic acid (CH_3COOH), and ammonia (NH_3).

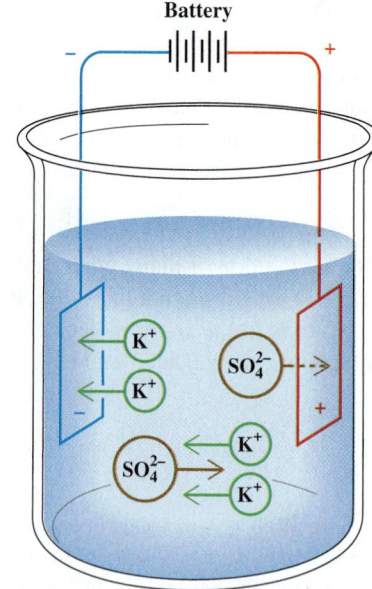

Figure 4–3 • An aqueous solution of potassium sulfate conducts electricity better than does pure water. When metallic plates (electrodes) charged by a battery are inserted into the solution, positive ions (K^+) migrate toward the negative plate and negative ions (SO_4^{2-}) migrate toward the positive plate. Thus, the solution conducts electricity. Potassium sulfate is a strong electrolyte.

Solutions of weak electrolytes conduct poorly because the production of ions during dissolution is not complete. A typical aqueous solution of the weak electrolyte ammonia contains less than 1% of the ions that it would contain if the reaction between ammonia and water that produces ions went to completion. Fewer current carriers mean less current. Despite this, solutions of weak electrolytes conduct electricity thousands of time better than does pure water.

Current theory takes the view that strong electrolytes dissociate completely into ions but that the ions do not act in complete independence in solution. Oppositely charged, aquated ions attract each other and even form loosely bound, temporary **ion-pairs,** especially in more concentrated solutions. The attractions exert a "drag" on the mobility of the ions when they are called upon to conduct electricity.

Even small amounts of dissolved electrolytes increase the conductivity of water substantially. Accidental electrocutions in which the fatal shock is conducted by rainwater or bathwater occur because even clean drinking water contains sufficient dissolved electrolytic impurities to make it a rather good conductor of electricity. Preparing and storing water that is free of such dissolved impurities requires specialized equipment and considerable effort.

Solubilities

Some substances dissolve in each other in all proportions. For example, ethanol (C_2H_5OH) and water form solutions of all possible compositions, ranging from slightly wet alcohol to an equal mixture (at which point the two trade places as solute and solvent) to slightly alcoholic water. Such pairs are **miscible** in all proportions. More commonly, however, substances are only partially miscible. Mixing diethyl ether (($C_2H_5)_2O$) with water gives two phases, one clear liquid floating on top of another. The bottom layer consists mainly of water with some dissolved diethyl ether, and the top layer consists mainly of diethyl ether with some dissolved water. To be more exact, at 25°C the water phase holds 6.41 g of ether per 100 g of water, and the ether phase holds 1.36 g of water per 100 g of ether. The water is **saturated** with respect to ether, and the ether is saturated with respect to water. Neither liquid can hold any more of the other.

> The **solubility** of a substance is the largest amount that can dissolve in a given amount of a solvent at a particular temperature.

• It is common to report solubilities in units of grams of dissolved substance (solute) per 100 g of solvent or in moles of solute per liter of solution. Other units are also used.

By this definition, the solubility of ethanol in water is infinite (saturation is impossible for fully miscible substances), but the solubility of diethyl ether in water is 6.41 g/100 g, and the solubility of water in diethyl ether is 1.36 g/100 g at 25°C. If additional solute is added to a saturated solution, it does not dissolve but forms a new phase that can be separated by decantation (pouring off), filtration, or other physical means.

Predicting Dissolution Reactions

Dissolution is a complex phenomenon, and absolute rules to predict solubilities do not exist. Some imperfect generalizations about solubility can be made based on experimental solubility data.

(a)   (b)

"Like dissolves like." (a) Ethylene glycol, the main ingredient in antifreeze, has polar molecules and dissolves readily in water. (b) Oil and water are nearly completely immiscible. The molecules in water are quite polar; those in oil are not. Here, the less dense oil rises to float on top of the water. *(Kip Peticolas/Fundamental Photographs, NYC)*

Methyl amine (CH_3NH_2) is ammonia (NH_3) with one of its H atoms replaced by a methyl (CH_3) group. It is soluble in water.

The most general statement is *"like dissolves like."* Substances having molecules with similar polarity tend to be mutually soluble. Water, which has markedly polar molecules, dissolves ionic and polar-molecular substances such as salt, ethanol, and the simple sugars but does not dissolve benzene (C_6H_6) and common oils, which have non-polar molecules. Carbon tetrachloride (CCl_4), which has non-polar molecules, dissolves benzene, oils, and other non-polar or slightly polar substances, but not water and ionic compounds. Some additional generalities follow:

- No gaseous or solid substances are infinitely soluble in (miscible in all proportions with) water; many liquids are.
- Common inorganic acids are soluble in water; organic acids (see Section 4–3) of low molar mass are soluble in water. Some are completely miscible.
- Organic compounds of low molar mass containing the OH group (these include sugars and alcohols such as methanol and ethanol) or the NH_2 group (the amine group) are usually soluble in water; several are completely miscible.
- The solubilities of gases in water and other solvents usually decrease as the temperature increases; the solubilities of solids and liquids usually increase with increasing temperature.
- The solubilities of ionic compounds in water vary. Nearly all ionic compounds containing lithium, sodium, potassium, rubidium, cesium, or ammonium ions dissolve well in water and less well in other solvents. Table 4–1 lists aqueous solubilities of other ionic compounds.

TABLE 4–1

Solubilities of Ionic Compounds in Water			
Anion	**Soluble**[a]	**Slightly Soluble**	**Insoluble**
NO_3^- (nitrate)	All	—	—
CH_3COO^- (acetate)	Most	—	$Be(CH_3COO)_2$
ClO_3^- (chlorate)	All	—	—
ClO_4^- (perchlorate)	Most	$KClO_4$	—
F^- (fluoride)	Group I[b], AgF, BeF$_2$	SrF_2, BaF_2, PbF_2	MgF_2, CaF_2
Cl^- (chloride)	Most	$PbCl_2$	$AgCl, Hg_2Cl_2$
Br^- (bromide)	Most	$PbBr_2, HgBr_2$	$AgBr, Hg_2Br_2$
I^- (iodide)	Most	—	$AgI, Hg_2I_2,$ PbI_2, HgI_2
SO_4^{2-} (sulfate)	Most	$CaSO_4, Ag_2SO_4,$ Hg_2SO_4	$SrSO_4, BaSO_4,$ $PbSO_4$
S^{2-} (sulfide)	Groups I and II[c] $(NH_4)_2S$	—	Most
CO_3^{2-} (carbonate)	Group I, $(NH_4)_2CO_3$	—	Most
SO_3^{2-} (sulfite)	Group I, $(NH_4)_2SO_3$	—	Most
PO_4^{3-} (phosphate)	Group I, $(NH_4)_3PO_4$	Li_3PO_4	Most
OH^- (hydroxide)	Group I, $Ba(OH)_2$	$Sr(OH)_2, Ca(OH)_2$	Most

[a]Soluble compounds have solubilities exceeding 1 g/100 g water. Slightly soluble compounds have solubilities between 0.01 and 1 g/100 g; insoluble compounds have solubilities less than 0.01 g/100 g at room temperature.
[b]Compounds of elements from the first column in the periodic table: Li, Na, K, Rb, Cs.
[c]Compounds of elements from the second column in the periodic table: Be, Mg, Ca, Sr, Ba.

EXAMPLE 4–1

Predict whether the following substances are soluble in water:
(a) Cobalt(II) chloride.   (b) Lanthanum(III) perchlorate.   (c) Maltose $(C_{12}H_{22}O_{11})$.

Strategy

Use the solubility rules of Table 4–1 and the preceding generalizations about solubility.

Solution

(a) According to Table 4–1, most chlorides are soluble in water; $CoCl_2$ is not among the listed exceptions. The prediction is that $CoCl_2$ is soluble.
(b) Most perchlorates are soluble, and $La(ClO_4)_3$ is not listed as an exception. The prediction is that $La(ClO_4)_3$ dissolves.
(c) From its name and formula, we expect maltose to be a sugar (like glucose, fructose, and sucrose). Maltose should be soluble, like other low-molar-mass sugars.

EXERCISE

Predict whether the following substances are soluble in water: (a) calcium carbonate, (b) mercury(II) sulfide, (c) isopropanol (C_3H_7OH).

Answer: (a) Insoluble. (b) Insoluble. (c) Soluble.

4-2 Precipitation Reactions

Whenever the concentration of a substance in solution exceeds its solubility, a new phase starts to separate. A new solid phase is usually dense enough to sink to the bottom of a liquid solution. That is, solids usually **precipitate** ("fall down") from solutions when their solubility is exceeded. Precipitation of solids can be effected in several ways:

1. By removing some solvent. Removal is easy when the solvent evaporates readily but the solute does not, a very common case. Evaporation of sea-water causes sodium chloride (and other salts) to deposit on the sides and bottom of the container. A saturated solution that deposits a precipitate in this way is called the "mother-liquor" of the precipitate. The equation for the precipitation of NaCl(s) from water is

$$Na^+(aq) + Cl^-(aq) \longrightarrow NaCl(s)$$

   Note that the equation for precipitation of NaCl is exactly the reverse of the equation for its dissolution.

2. By changing the solvent. At 25°C, the solubility of sodium chloride is 36 g/100 g of water, but only 0.12 g/100 g of ethanol. Ethanol and water are miscible in all proportions. Stirring ethanol into an aqueous solution of sodium chloride creates a "mixed solvent" that has less power to dissolve NaCl than water has. Unless the original aqueous solution of NaCl was very dilute, the added ethanol causes NaCl to precipitate:

$$Na^+(ethanol, aq) + Cl^-(ethanol, aq) \longrightarrow NaCl(s)$$

3. By changing the temperature. The solubility of silver nitrate ($AgNO_3$) in water is 654 g/100 g at 90°C, but only 235 g/100 g at 25°C. Cooling a solution that is saturated in $AgNO_3$ at 90°C to 25°C causes $AgNO_3$ to precipitate.

All three of these techniques are much used in chemistry.

Other precipitation reactions approach the limit of solubility in a different way. As Table 4–1 shows, most ionic compounds dissolve in water to some extent. However, actual solubilities vary (Fig. 4–4). For example, the solubilities at 25°C of $K_2SO_4(s)$ and $BaCl_2(s)$ differ by a factor of three: 12 g/100 g and 38 g/100 g, respectively. These compounds are both water-soluble. Meanwhile, the solubility of $BaSO_4$ is a mere 0.12 mg/100 g of water at the same temperature, approximately 100,000 times less. $BaSO_4$ is "insoluble" in water. Suppose that a solution of K_2SO_4 is mixed with a solution of $BaCl_2$. Both solutes are ionized in solution. The solubility of barium sulfate is quickly exceeded as barium ions ($Ba^{2+}(aq)$) encounter sulfate ions ($SO_4^{2-}(aq)$). Solid barium sulfate precipitates (Fig. 4–5):

$$BaCl_2(aq) + K_2SO_4(aq) \longrightarrow BaSO_4(s) + 2\,KCl(aq)$$

• Barium sulfate is routinely swallowed in suspended form in a "barium cocktail" to improve contrast in X-ray photographs of the alimentary tract. Soluble barium compounds are quite toxic, but $BaSO_4$ is harmless because nearly all of the Ba^{2+} ion stays tightly bound up with sulfate ion and is not absorbed by the body.

Figure 4–4 • The amounts of different substances that dissolve in 1 L of water at 20°C. Clockwise from the front are borax ($Na_2B_4O_5(OH)_4 \cdot 8H_2O$), potassium permanganate ($KMnO_4$), lead(II) chloride ($PbCl_2$), sodium phosphate hydrate ($Na_3PO_4 \cdot 12H_2O$), calcium oxide (CaO), and potassium dichromate ($K_2Cr_2O_7$). *(Leon Lewandowski)*

The reaction can be viewed as a switching of ionic partners driven by the precipitation of the insoluble barium sulfate: the (s) on the $BaSO_4$ tells why the reaction goes. It is a **metathesis** reaction (a reaction in which atoms or groups of atoms are interchanged).

• Metathesis reactions can be driven by processes other than precipitation, as discussed subsequently in this chapter.

Ionic Equations and Net Ionic Equations

Often, chemical equations are written to recognize the presence of aquated ions among the reactants or products. Rewriting the previous equation as an **ionic equation**

$$Ba^{2+}(aq) + 2\,Cl^-(aq) + 2\,K^+(aq) + SO_4^{2-}(aq) \longrightarrow$$
$$BaSO_4(s) + 2\,K^+(aq) + 2\,Cl^-(aq) \qquad \text{(ionic equation)}$$

reveals which ions take part in the reaction (the aquated barium and sulfate ions start out moving at large in solution but end up tied together in the precipitate) and which are **spectator ions** (the aquated potassium ions and chloride ions are unaffected by the reaction).

   Cancellation of species that appear on both the left and right sides of an ionic equation results in a **net ionic equation.** For example

$$Ba^{2+}(aq) + SO_4^{2-}(aq) \longrightarrow BaSO_4(s) \qquad \text{(net ionic equation)}$$

Net ionic equations include only the ions (and molecules) that actually react. They provide excellent models for the prediction of reactions because removal of the spectator ions focuses attention on what really drives the chemical change. In this case, the driving force is the insolubility of $BaSO_4$ in water. The net ionic equation emphasizes that a solution of *any* soluble barium salt mixed with a solution of *any* soluble sulfate salt gives the same result: a precipitate of barium sulfate.

Figure 4–5 • A solution of potassium sulfate is added to one of barium chloride. A cloud of white particles of solid barium sulfate forms and settles out (precipitates); the potassium chloride remains in solution. *(Leon Lewandowski)*

EXAMPLE 4-2

An aqueous solution of sodium carbonate is mixed with an aqueous solution of calcium chloride, and a white precipitate immediately forms. Write a net ionic equation to account for this precipitate.

Strategy

Identify the ions into which the two salts dissociate when dissolved. Look in Table 4–1 for combinations of these ions that are insoluble.

Solution

Mixing the two solutions puts $Na^+(aq)$ and $Cl^-(aq)$ ions and also $Ca^{2+}(aq)$ and $CO_3^{2-}(aq)$ ions in contact for the first time. The precipitate forms by the reaction

$$Ca^{2+}(aq) + CO_3^{2-}(aq) \longrightarrow CaCO_3(s)$$

Check

The $Na^+(aq)$ and $Cl^-(aq)$ ions do not appear in this equation because they are non-reacting spectator ions; sodium chloride is known to be soluble in water.

EXERCISE

Write a net ionic equation to represent the formation of the precipitate observed when aqueous solutions of $CaCl_2$ and NaF are mixed. Identify the spectator ions in this process.

Answer: $Ca^{2+}(aq) + 2 F^-(aq) \longrightarrow CaF_2(s)$. The spectator ions are $Na^+(aq)$ and $Cl^-(aq)$.

Weak electrolytes also take part in precipitation reactions. When a solution of the weak electrolyte mercury(II) chloride and a solution of potassium iodide are mixed, insoluble mercury(II) iodide forms immediately:

$$HgCl_2(aq) + 2 KI(aq) \longrightarrow 2 KCl(aq) + HgI_2(s) \quad \text{(overall equation)}$$
$$HgCl_2(aq) + 2 K^+(aq) + 2 I^-(aq) \longrightarrow$$
$$2 K^+(aq) + 2 Cl^-(aq) + HgI_2(s) \quad \text{(ionic equation)}$$
$$HgCl_2(aq) + 2 I^-(aq) \longrightarrow HgI_2(s) + 2 Cl^-(aq) \quad \text{(net ionic equation)}$$

The precipitation of insoluble $HgI_2(s)$ pulls the Hg^{2+} ion out of its association with Cl^- ions.

Predicting Precipitation Reactions

Precipitation is the reverse of dissolution. The various solubility rules (see Table 4–1) therefore serve to predict precipitations. To know the result when two ionic compounds are mixed in aqueous solution, simply check the solubilities of the compounds formed by switching partners:

• If either of the new compounds is insoluble (or slightly soluble), then precipitation occurs.

- If *both* new compounds are insoluble, then two precipitation reactions occur concurrently.
- If both new compounds are soluble, no precipitation occurs.

As an example of the last, consider mixing aqueous sodium chloride with aqueous calcium nitrate. "Exchanging partners" suggests the formation of sodium nitrate and calcium chloride:

$$2\,NaCl(aq) + Ca(NO_3)_2(aq) \longrightarrow 2\,NaNO_3(aq) + CaCl_2(aq) \quad \text{(overall equation)}$$

The solubility rules state, however, that sodium nitrate and calcium chloride are both soluble. Thus,

$$2\,Na^+(aq) + 2\,Cl^-(aq) + Ca^{2+}(aq) + 2\,NO_3^-(aq) \longrightarrow$$
$$2\,Na^+(aq) + 2\,NO_3^-(aq) + Ca^{2+}(aq) + 2\,Cl^-(aq) \quad \text{(ionic equation)}$$

in which the left and right sides of the equation are identical. Nothing visible happens. This is often indicated by writing "NR" (for "no reaction") after the arrow.

$$NaCl(aq) + Ca(NO_3)_2(aq) \longrightarrow NR$$

Writing a net ionic equation is not possible because no net change occurs.

Figure 4-6 • A yellow precipitate of lead iodide, PbI_2, is formed when KI solution is added to a solution of $Pb(NO_3)_2$. *(Charles D. Winters)*

EXAMPLE 4-3

Predict the result when an aqueous solution of KI is added to one of $Pb(NO_3)_2$. Write net ionic equations to represent any reactions that take place. Name the product or products.

Strategy

Find the chemical formulas of possible precipitates by requiring charge neutrality, as in Example 3–3. Check Table 4–1 for the solubility of these compounds. Expect insoluble compounds to precipitate. Use the rules in Section 3–3 to generate names.

Solution

The lead (Pb^{2+}) and iodide (I^-) ions combine to give PbI_2 (two -1 iodide ions balancing the $+2$ lead ion). According to Table 4–1, PbI_2 is insoluble. The other possible candidate for precipitation, KNO_3, is soluble. Thus, a precipitate of PbI_2, which is lead(II) iodide , appears (Fig. 4–6), according to the net ionic equation

$$Pb^{2+}(aq) + 2\,I^-(aq) \longrightarrow PbI_2(s)$$

EXERCISE

An aqueous solution of sodium sulfate is added to one of potassium nitrate. Write net ionic equations to represent any reactions that take place. Give the name or names of any product or products.

Answer: The solutions mix completely, but no precipitate forms.

In dissolution and precipitation, molecules or ions retain their identities and are simply shuffled in their associations. They leave one phase to take up a position in another. These reactions tend to be less vigorous (generating less heat, for example) than the acid–base and oxidation–reduction (redox) reactions that we examine in the next two sections.

4-3 Acids and Bases and Their Reactions

Acids and bases have been known and characterized since early times. Many fruits contain acids that contribute to their characteristic flavors, such as tartaric acid ($C_4H_6O_6$) in grapes, citric acid ($C_6H_8O_7$) in lemons, and malic acid ($C_4H_6O_5$) in apples. These acids are responsible for the sharp taste of the fruit, just as acetic acid (CH_3COOH) is responsible for the tang of vinegar. The ashes of plants yield substances, called **bases,** that counteract, or neutralize, the properties of acids. Potassium hydroxide (KOH, called caustic potash) is an important base. It was first prepared from potassium carbonate (K_2CO_3), obtained by leaching plant ashes with water and evaporating the solution in large pots (whence the name). Both potassium hydroxide and the closely related base sodium hydroxide (NaOH, caustic soda) are quite corrosive and destructive of organic tissue; they convert animal fats to soaps, a property exploited in commercial soap-making (see Section 25–1). Although most bases are less caustic than these two, bases generally have a soapy feel, because they act on oils in the skin, and a bitter taste. Bases are also called **alkalis.**

Many natural pigments signal the presence of an acid or base by changing color. They are acid–base **indicators.** Usually, vegetable blues turn red in acid and vegetable reds turn blue in base. This happens with grape juice (Fig. 4–7). Similarly, blueberry juice is dark red in its normal (acidic) state but turns blue in a neutral or basic environment. This explains why blueberry pie stains the tongue blue (saliva is nearly neutral) but stains a napkin red. Strips of paper impregnated with litmus, a blue pigment obtained from lichens, have long been used to test solutions; litmus turns red in acid and blue in base.

Acids and bases are easily recognized by taste, or feel, or the color changes they cause in a natural indicator. Also, their reactivity is simple: acid + base $\longrightarrow$ neutral. These facts led to an early acceptance of acid and base as fundamental chemical types. When atomic theory appeared, chemists sought a structural explanation for acid and base behavior. A great many acids, including all of the organic acids just listed and several important inorganic acids, such as nitric acid (HNO_3) and sulfuric acid (H_2SO_4), include oxygen in their chemical formulas. For a time in the late 18th century it was thought that oxygen was an essential constituent of any acid.

- The word "acid" comes from the Latin *acidus,* meaning "sharp."

- The term "litmus test" now means a decisive test on any issue.

- Oxygen got its name, which means "acid-giver" in Greek, during this period.

The lye in this can is sodium hydroxide.

Figure 4–7 • Grape juice is a natural acid–base indicator, changing from red in acidic solution to greenish blue in basic solution. *(Leon Lewandowski)*

Then, in 1810, the British chemist Sir Humphrey Davy analyzed hydrochloric acid (HCl dissolved in water) and showed that this notion was wrong. From then on, the role of *hydrogen* in the properties of acids was increasingly stressed, and a fresh understanding of acid–base reactivity developed. Note that hydrogen is the only element common to HNO_3, H_2SO_4, HCl, and acetic acid, CH_3COOH.

Arrhenius Acids and Bases

Over the years, chemists have used a variety of definitions of the terms "acid" and "base." The most recent (see Sections 8–1 and 8–8) aim for generality and succeed in encompassing earlier ideas as special cases. The earlier concepts, however, provide major insights. One still useful pair of definitions derives from work performed by the Swedish chemist Svante Arrhenius in the late 19th century. These definitions apply only for "wet chemistry"—that is, work carried out in aqueous solutions. Although acids and bases react in solvents other than water, and, indeed, in the absence of solvent, aqueous solutions are so important in everyday life that we restrict our attention to them at this point.

Small concentrations of aquated hydrogen ions ($H^+(aq)$) and aquated hydroxide ions ($OH^-(aq)$) exist in even the most carefully purified water. These ions come from the dissociation of a small proportion of the H_2O molecules. The ions become aquated, just as do ions from a soluble ionic compound. The net ionic equation is

$$H_2O(\ell) \longrightarrow H^+(aq) + OH^-(aq)$$

One can view the process as water dissolving in itself. This **autoionization** is an intrinsic property of water and cannot be suppressed. Water *always* contains $H^+(aq)$ and $OH^-(aq)$ ions. If ions are present in pure water, then pure water should conduct electricity. Experiment confirms that water is a weak electrolyte (Section 4–1), although a *very* weak one.

Arrhenius took $H^+(aq)$ ion and $OH^-(aq)$ ion as essential for acid and base behavior in aqueous solution:

> An **Arrhenius acid** is a substance that, when dissolved in water, delivers hydrogen ions ($H^+(aq)$ ions) to the solution.

> An **Arrhenius base** is any substance that, when dissolved in water, delivers hydroxide ions ($OH^-(aq)$ ions) to the solution.

• Use of the term "proton" in place of "hydrogen ion" is very common but slightly inexact. Ordinary hydrogen contains two isotopes; one gives a proton when it loses its electron, but the other does not.

By these definitions Arrhenius elevated water, which donates hydrogen ion and hydroxide ion equally, to a unique status: Water is simultaneously and equally both an Arrhenius acid and an Arrhenius base. A substance having both acidic and basic properties is said to be **amphoteric;** water is perfectly amphoteric in the Arrhenius scheme. However, insoluble substances and those containing no hydrogen atoms never satisfy either Arrhenius definition.

Arrhenius acids and Arrhenius bases release ions into solution. They thereby increase the electrical conductivity of water and must be electrolytes. **Strong acids,** such as hydrochloric acid (HCl), nitric acid (HNO_3), and sulfuric acid (H_2SO_4), and **strong bases,** such as sodium hydroxide (NaOH) and potassium hydroxide (KOH), are strong electrolytes. Strong acids and strong bases dissociate essentially completely in water:

$$HCl(g) \longrightarrow H^+(aq) + Cl^-(aq)$$
$$NaOH(s) \longrightarrow Na^+(aq) + OH^-(aq)$$

The Arrhenius model views the reaction between an acid and a base (called a **neutralization reaction**) as a metathesis. The H^+ from the acid and the OH^- from the base leave their previous partners to combine into neutral H_2O. The previous partners form a salt, which stays in solution or precipitates, depending on its solubility, or gets involved in other reactions. The following three chemical equations all represent the neutralization reaction between hydrochloric acid and sodium hydroxide.

$$HCl(aq) + NaOH(aq) \longrightarrow H_2O(\ell) + NaCl(aq) \qquad \text{(overall equation)}$$
$$H^+(aq) + Cl^-(aq) + Na^+(aq) + OH^-(aq) \longrightarrow$$
$$H_2O(\ell) + Na^+(aq) + Cl^-(aq) \qquad \text{(ionic equation)}$$
$$H^+(aq) + OH^-(aq) \longrightarrow H_2O(\ell) \qquad \text{(net ionic equation)}$$

The driving force in Arrhenius neutralization is the formation of the molecular substance water by the combination of H^+ and OH^- ions.

Weak Acids

Only a handful of acids are strong electrolytes. They include the seven common strong acids: nitric acid (HNO_3), sulfuric acid (H_2SO_4), chloric acid ($HClO_3$), perchloric acid ($HClO_4$), hydrochloric acid (HCl), hydrobromic acid (HBr), and hydroiodic acid (HI). Other common acids are all weak. Important weak acids include hydrofluoric acid (HF), acetic acid (CH_3COOH), and formic acid ($HCOOH$), all of which can donate one hydrogen ion per molecule; oxalic acid ($H_2C_2O_4$), which can donate two hydrogen ions per molecule; and phosphoric acid (H_3PO_4), which can donate three hydrogen ions per molecule. Acetic acid has four hydrogens but only

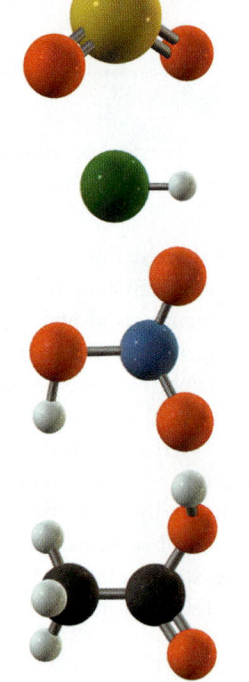

Sulfuric acid (H_2SO_4) is the industrial chemical produced on the largest scale in the world, in amounts exceeding 100 million tons per year. It is used for fertilizer production, as an intermediate in the manufacture of other chemicals, and in the metals and petroleum-refining industries.

Hydrogen chloride (HCl) is produced directly from the elements or as a by-product in making chlorinated polymers from petroleum and chlorine. The largest use of the aqueous acid is in the pickling of steel and other metals to remove oxide layers on the surface.

Nitric acid (HNO_3) is made largely from ammonia; once produced, most of it is reacted further with ammonia to make ammonium nitrate for use as a fertilizer.

Acetic acid (CH_3COOH) is a weak organic acid. It is the active ingredient in vinegar, which is a 3% to 5% solution of acetic acid in water. The single acidic hydrogen atom in acetic acid is the one bonded to the oxygen. Pure acetic acid is called "glacial acetic acid" because it freezes readily into crystals that resemble ice.

one "acidic hydrogen." Acidic hydrogens, the ones that can be donated, are generally bonded to oxygen or another electronegative element. Hydrogens bonded to carbon are rarely acidic.

Weak acids produce hydrogen ions in aqueous solution but to a limited extent that depends on their concentration (greater dilution *increases* dissociation). For example, the dissociation of acetic acid

$$CH_3COOH(aq) \longrightarrow CH_3COO^-(aq) + H^+(aq)$$

stops with less than 10% of the acetic acid molecules dissociated unless the solution is quite dilute. Naturally, weak acids are not all equally weak but differ in their degree of dissociation at the same concentration (see Section 8–3). Despite their incomplete dissociation, weak acids, like strong acids, neutralize strong bases to produce water and salts. The equations for such reactions represent the true nature of the reactants better if they show the acidic hydrogen as undissociated. The neutralization of acetic acid by sodium hydroxide is thus represented

$$CH_3COOH(aq) + NaOH(aq) \longrightarrow$$
$$H_2O(\ell) + NaCH_3COO(aq) \qquad \text{(overall equation)}$$

$$CH_3COOH(aq) + Na^+(aq) + OH^-(aq) \longrightarrow$$
$$H_2O(\ell) + Na^+(aq) + CH_3COO^-(aq) \qquad \text{(ionic equation)}$$

$$CH_3COOH(aq) + OH^-(aq) \longrightarrow$$
$$H_2O(\ell) + CH_3COO^-(aq) \qquad \text{(net ionic equation)}$$

The coefficients in these balanced equations show that the neutralization of a mole of base does not require more weak acid than strong acid. Weakness and strength in Arrhenius acids relate to degree of dissociation, not to the ultimate chemical amount of $H^+(aq)$ ion that can be furnished.

• Many organic acids contain the carboxylic acid group $-C\overset{\displaystyle O}{\underset{\displaystyle O-H}{\big|\big|}}$ in which the hydrogen atom is always acidic, and their formulas are written to show it.

Naming Acids

Acids are conveniently grouped into three categories: **binary acids,** which are formed by hydrogen with a second element; **oxoacids,** which are formed by the combination of hydrogen with oxoanions, and **organic acids,** which are compounds of carbon, hydrogen, and some electronegative element (usually oxygen). Some rules for naming each type of acid are summarized here.

1. *For binary acids.* Reduce the word "hydrogen" to the prefix *hydro-* and attach it to the stem of the name of the other element. Add *-ic* at the end of the stem. Complete the name with the separate word "acid." For example:

Compound	Named as a Covalent Compound	Named as a Binary Acid
HCl	hydrogen chloride	hydrochloric acid
HBr	hydrogen bromide	hydrobromic acid
HI	hydrogen iodide	hydroiodic acid
H_2S	dihydrogen sulfide	hydrosulfuric acid

2. *For oxoacids.* Chemically, oxoacids come from the combination of oxoanions with sufficient hydrogen ions for charge neutrality. Start with the name of the oxoanion. If it ends in *-ate,* replace the *-ate* with *-ic* and add the word "acid." If the name of the oxoanion ends in *-ite,* replace the *-ite* with *-ous* and add the separate word "acid." The prefixes *per-* (largest

number of oxygen atoms in the oxoacid) and *hypo-* (smallest number) are used, just as for oxoanions, in cases in which an element forms more than two oxoacids. Here are some examples:

Oxoanion	Corresponding Acid
ClO^- (hypochlorite ion)	$HClO$ (hypochlorous acid)
ClO_2^- (chlorite ion)	$HClO_2$ (chlorous acid)
ClO_3^- (chlorate ion)	$HClO_3$ (chloric acid)
ClO_4^- (perchlorate ion)	$HClO_4$ (perchloric acid)
SO_4^{2-} (sulfate ion)	H_2SO_4 sulfuric acid
SO_3^{2-} (sulfite ion)	H_2SO_3 sulfurous acid

3. *For organic acids.* The stems for the common organic acids are nonsystematic and must be memorized. The *-ate* to *-ic* relationship holds between the names of the organic anion and the acid. For example:

Organic Anion	Corresponding Acid
$HCOO^-$ (formate ion)	$HCOOH$ (formic acid)
CH_3COO^- (acetate ion)	CH_3COOH (acetic acid)
$(COO)_2^{2-}$ (oxalate ion)	$(COOH)_2$ (oxalic acid)
$CH_2(COO)_2^{2-}$ (malonate ion)	$CH_2(COOH)_2$ (malonic acid)

EXAMPLE 4–4

Name the following compounds as acids: H_2Te, H_2CrO_4. Write the overall, ionic, and net ionic equations for the neutralization of H_2Te by potassium hydroxide (KOH), a strong base.

Solution

H_2Te is a binary acid based on the element tellurium (stem, *tellur-*). Its name is thus hydrotelluric acid . In Table 3–5 the oxoanion corresponding to H_2CrO_4 is chromate ion (CrO_4^{2-}). The oxoacid is therefore chromic acid .

Like other inorganic acids (see Section 4–1), H_2Te dissolves in water. It is a weak acid (not being one of the seven strong acids). It should therefore appear among the reactants as undissociated and aquated. Neutralization gives water and a soluble salt (nearly all potassium salts are soluble [see Table 4–1]). The equations are

$$H_2Te(aq) + 2\,KOH(aq) \longrightarrow$$
$$2\,H_2O(\ell) + K_2Te(aq) \quad \text{(overall equation)}$$

$$H_2Te(aq) + 2\,K^+(aq) + 2\,OH^-(aq) \longrightarrow$$
$$2\,H_2O(\ell) + 2\,K^+(aq) + Te^{2-}(aq) \quad \text{(ionic equation)}$$

$$H_2Te(aq) + 2\,OH^-(aq) \longrightarrow$$
$$2\,H_2O(\ell) + Te^{2-}(aq) \quad \text{(net ionic equation)}$$

EXERCISE

Give names for the acids H_3PO_3 and H_2SeO_3. Write the overall, ionic, and net ionic equations for the complete neutralization of H_2SeO_3 by sodium hydroxide.

Answer: The acids are phosphorous acid and selenous acid. The equations are:

$$H_2SeO_3(aq) + 2\,NaOH(aq) \longrightarrow$$
$$2\,H_2O(\ell) + Na_2SeO_3(aq) \quad \text{(overall equation)}$$

$$H_2SeO_3(aq) + 2\,Na^+(aq) + 2\,OH^-(aq) \longrightarrow$$
$$2\,H_2O(\ell) + 2\,Na^+(aq) + SeO_3^{2-}(aq) \quad \text{(ionic equation)}$$

$$H_2SeO_3(aq) + 2\,OH^-(aq) \longrightarrow$$
$$2\,H_2O(\ell) + SeO_3^{2-}(aq) \quad \text{(net ionic equation)}$$

Weak Bases

Any ionic compound containing hydroxide ions serves as an Arrhenius base simply by dissolving in water. However, the only soluble ionic hydroxides are $Ba(OH)_2$ and the hydroxides of the alkali metals (most notably NaOH and KOH). All of these compounds dissociate in water when they dissolve. They are strong bases. Other ionic hydroxides are either slightly soluble or insoluble. They donate OH^- ion poorly because they keep it mostly tied up in the solid. They are weak bases. Experimentally, poorly soluble weak bases are recognized by their ability to neutralize acids. Magnesium hydroxide ($Mg(OH)_2$), for example, dissolves only very slightly in water (about 0.001 g/100 g at 25°C). A fine powder of magnesium hydroxide suspended in water ("milk of magnesia") is an effective antacid, however, and is drunk to counteract excess stomach acidity. The small proportion of $Mg(OH)_2$ that dissolves indeed dissociates to Mg^{2+} and OH^- ions, but these ions are too few to raise the conductance of the water significantly. Hydroxides that are poorly soluble in water dissolve in acidic solutions by undergoing neutralization reactions. Equations for the reaction of magnesium hydroxide with hydrochloric acid are

$$2\,HCl(aq) + Mg(OH)_2(s) \longrightarrow$$
$$2\,H_2O(\ell) + MgCl_2(aq) \quad \text{(overall equation)}$$

$$2\,H^+(aq) + 2\,Cl^-(aq) + Mg(OH)_2(s) \longrightarrow$$
$$2\,H_2O(\ell) + Mg^{2+}(aq) + 2\,Cl^-(aq) \quad \text{(ionic equation)}$$

$$2\,H^+(aq) + Mg(OH)_2(s) \longrightarrow$$
$$2\,H_2O(\ell) + Mg^{2+}(aq) \quad \text{(net ionic equation)}$$

White solid $Mg(OH)_2$ disappears into the clear aqueous solution of HCl as this reaction proceeds.

Modifying the Arrhenius Model

Many compounds, such as ammonia (NH_3), methylamine (CH_3NH_2), and pyridine (C_5H_5N), have a bitter taste and a soapy feel and neutralize acids but *do not contain hydroxide ions*. To include these substances as bases, the original Arrhenius definition was soon expanded as follows:

An Arrhenius base is a substance that, when dissolved in water, increases the concentration of hydroxide ion over what is present in the pure solvent.

Ammonia, like the other compounds just named, increases the number of hydroxide ions by reacting with water:

$$NH_3(aq) + H_2O(\ell) \longrightarrow NH_4^+(aq) + OH^-(aq)$$

The reaction occurs to only a slight extent. Although ammonia is quite soluble, it is a weak base; conductance measurements confirm that it is a weak electrolyte.

The definition of an Arrhenius acid is modified similarly:

> An Arrhenius acid is a substance that, when dissolved in water, increases the concentration of hydrogen ion over what is present in the pure solvent.

Acid and Base Anhydrides

The modification of the Arrhenius definitions qualifies numerous additional substances as acids and bases. Of particular interest and importance are the binary oxides. For example, the soluble gas sulfur trioxide (SO_3) gives solutions in which the concentration of hydrogen ion far exceeds the concentration present in pure water; sulfur trioxide fits the broadened definition of an acid well. In fact, SO_3 reacts with water as it dissolves, in a **hydration** reaction, to give sulfuric acid:

$$SO_3(g) + H_2O(\ell) \longrightarrow H_2SO_4(aq)$$

The sulfuric acid, which is a strong oxoacid, then donates hydrogen ions. As for bases, sodium oxide (Na_2O) becomes an Arrhenius base under the modified definition because it too undergoes a hydration reaction:

$$Na_2O(s) + H_2O(\ell) \longrightarrow 2\,NaOH(aq)$$

and the product, sodium hydroxide, is an excellent hydroxide-ion donor.

Many other oxides behave in a similar way. They are termed **acid anhydrides** and **base anhydrides** ("anhydrous" means "without water"). Acid anhydrides react with water to give acidic solutions; base anhydrides react with water to give basic solutions. In general, ionic oxides are base anhydrides, and covalent or partially covalent oxides are acid anhydrides. Therefore, with few exceptions, the oxides of the non-metals are acid anhydrides, and the oxides of alkali metals and alkaline-earth metals are base anhydrides (Fig. 4–8).

	I	II	III	IV	V	VI	VII
	Li_2O	BeO	B_2O_3	CO_2	N_2O_5	(O_2)	OF_2
	Na_2O	MgO	Al_2O_3	SiO_2	P_4O_{10}	SO_3	Cl_2O_7
	K_2O	CaO	Ga_2O_3	GeO_2	As_2O_5	SeO_3	Br_2O_7
	Rb_2O	SrO	In_2O_3	SnO_2	Sb_2O_5	TeO_5	I_2O_7
	Cs_2O	BaO	Tl_2O_3	PbO_2	Bi_2O_5	PoO_3	At_2O_7

Increasing acidity ⟶ (top); Increasing basicity ↓ (left); Increasing acidity ↑ (right); Increasing basicity ⟵ (bottom)

Figure 4–8 • Among the oxides of the main-group elements, behavior as an acid anhydride tends to strengthen from left to right and from bottom to top in the periodic table; behavior as a base anhydride does the reverse. Acidity also increases with increasing oxidation number of the element. Oxygen difluoride, however, has only weakly acidic properties. The oxides listed here have the element in the maximum oxidation state.

Oxides of semi-metals, such as boron, and oxides in which a non-metal is in a lower oxidation state serve as acid anhydrides, but the hydration reaction generates weakly acidic solutions:

$$Cl_2O(g) + H_2O(\ell) \longrightarrow 2\ HOCl(aq)$$

oxide of a non-metal in   weak acid
a low oxidation state

- Non-metals have electronegativities close to that of the non-metal oxygen; therefore, their oxides have predominantly covalent bonding character.

Oxides that have bonding on the borderline between ionic and covalent are commonly amphoteric (intermediate between acid and base) in their behavior as anhydrides and indeed dissolve to only a limited extent in water. Transition-metal oxides show a range of acidic and basic properties, with acidity usually increasing with the oxidation state of the element.

The formulas of acid and base anhydrides can be generated in a purely formal operation from the chemical formulas of oxoacids and hydroxides by removing molecules of water until only an oxide remains. This corresponds to the dehydration of the compound. Thus, *subtracting* a molecule of water from the formula H_2CO_3 (carbonic acid) identifies CO_2 (carbon dioxide) as an acid anhydride:

$$H_2CO_3 - H_2O \longrightarrow CO_2$$

which is equivalent to

$$H_2CO_3 \longrightarrow CO_2 + H_2O$$

When an oxoacid or hydroxide contains an odd number of hydrogen atoms in its chemical formula, the formula is doubled and *then* water molecules are removed until no hydrogen atoms remain.

EXAMPLE 4–5

What is the acid anhydride of nitrous acid (HNO_2)?

Solution

The dehydration of nitrous acid can be represented by

$$2\ HNO_2 \longrightarrow N_2O_3 + H_2O$$

Consequently, dinitrogen trioxide (N_2O_3) is the acid anhydride of nitrous acid.

EXERCISE

What is the acid anhydride of orthoperiodic acid (H_5IO_6)?

Answer: I_2O_7 (diiodine heptaoxide).

$$2H_5IO_6$$
$$-5H_2O$$
$$\overline{I_2O_7}$$

$$\begin{array}{r} 12 \\ -5 \\ \hline 7 \end{array}$$

The subtraction of water molecules from an oxoacid or hydroxide can sometimes be carried out experimentally, but not always. Although carbonic acid dehydrates readily to its acid anhydride,

$$H_2CO_3(aq) \longrightarrow CO_2(g) + H_2O(\ell)$$

any attempt to dehydrate liquid perchloric acid by heating it to induce the reaction

$$2\ HClO_4(\ell) \longrightarrow Cl_2O_7(\ell) + H_2O(\ell)$$

would not only fail but most likely cause an explosion. Other, less direct ways of dehydrating perchloric acid must be used.

$$2HClO_4$$
$$H_2O$$
$$Cl_2O_7$$

Limits of the Arrhenius Model

The neutralization of aqueous ammonia by hydrochloric acid

$$NH_3(aq) + HCl(aq) \longrightarrow NH_4Cl(aq) \qquad \text{(overall equation)}$$
$$NH_3(aq) + H^+(aq) + Cl^-(aq) \longrightarrow NH_4^+(aq) + Cl^-(aq) \qquad \text{(ionic equation)}$$
$$NH_3(aq) + H^+(aq) \longrightarrow NH_4^+(aq) \qquad \text{(net ionic equation)}$$

produces a salt, but no water. This fact does not fit neatly into even the expanded Arrhenius model. An early attempt to repair the difficulty was the suggestion that dissolved NH_3 reacts completely with water to form soluble "ammonium hydroxide" (NH_4OH) that is then neutralized according to

$$NH_4OH(aq) + HCl(aq) \longrightarrow H_2O(\ell) + NH_4Cl(aq) \qquad \text{(overall equation)}$$

Although concentrated ammonia solutions are still marketed under the name "ammonium hydroxide," few NH_4^+ and OH^- ions are present in them. Moreover, the reaction between NH_3 and HCl occurs even in the complete absence of water (between the two gases) to produce the same salt (ammonium chloride) as a solid and no water. No amount of stretching can make this gas-phase reaction and others like it into Arrhenius neutralizations. More recent acid–base concepts, some of which are explored in Chapter 8, accommodate gas-phase and other kinds of non-aqueous reactions easily.

Acid–Base Titrations

Analytical chemists determine the amounts of substances in solution by means of **titrations.** A titration consists of a controlled addition of measured volumes of a solution of known concentration to a second solution of unknown concentration under conditions in which the solutes react cleanly (without side reactions), completely, and rapidly. A titration is complete when the second solute is used up. Completion is signaled by a change in some physical property, such as the color of the reacting mixture or the color of an **indicator** that has been added to it.

As an illustration, consider the **standardization** of a solution of the strong base sodium hydroxide. Standardization is the accurate determination of the concentration of a "stock solution" (a solution for use in later experiments). It consumes a portion of the stock but gives essential information about the remainder. This standardization employs a neutralization reaction between the aqueous NaOH and an accurately known amount of an acid. Many acids would work. We choose the easily handled and purified weak acid potassium hydrogen phthalate, $KHC_8H_4O_4$, a solid that dissolves to give $K^+(aq)$ and $HC_8H_4O_4^-(aq)$ ions. It neutralizes aqueous NaOH according to

$$NaOH(aq) + KHC_8H_4O_4(aq) \longrightarrow H_2O(\ell) + NaKC_8H_4O_4(aq) \qquad \text{(overall equation)}$$
$$OH^-(aq) + HC_8H_4O_4^-(aq) \longrightarrow H_2O(\ell) + C_8H_4O_4^{2-}(aq) \qquad \text{(net ionic equation)}$$

A known mass of potassium hydrogen phthalate is dissolved, and a few drops of a solution of the indicator phenolphthalein is added. The solution is colorless. Some of the NaOH solution is poured into a **buret,** a tube with a delivery valve (a stopcock) and markings for measuring solution volumes (Fig. 4–9). The initial level of the solution in the buret is noted, and the titration is begun (Fig. 4–10). Small quantities of NaOH solution are dispensed from the buret into a flask containing the acid. The flask is swirled after each addition of base, and its interior walls are washed occasionally with water to flush any splashed droplets of solution into the main body

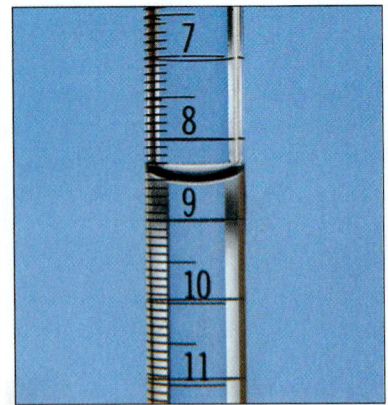

Figure 4–9 • The tube of this buret is marked in milliliters with ten subdivisions between each main marking. Readings are taken at the bottom of the lens-like meniscus. Careful workers attempt to estimate where the level lies between the finest markings. The reading here is 8.53 or 8.54 mL. (*Leon Lewandowski*)

(a)   (b)   (c)

Figure 4–10 • The standardization of a solution of NaOH. (a) A buret is filled with some of the NaOH solution. (b) The solution is dispersed slowly into a flask containing a known chemical amount of an acid; the two neutralize each other. (c) A change in the color of the indicator signals the end-point. In a carefully performed titration, the color change becomes complete with the addition of a single drop (or less) from the buret. *(Charles D. Winters)*

of solution. As the NaOH joins the solution in the titration flask, regions of pink appear and then fade. This is the indicator responding to locally high concentrations of NaOH. As long as the original chemical amount of $KHC_8H_4O_4$ exceeds the total chemical amount of NaOH added, the color fades with swirling. As more and more NaOH solution is added, the pink color fades more and more slowly. Finally, when all the acid has just been used up, the whole of the reaction mixture stays pink. This is the **end-point.** The final level of the solution in the buret is noted, and the volume of NaOH solution added during the titration is calculated by subtraction.

In a good titration, the observed end-point approximates the theoretical **equivalence point** very closely. The equivalence point in this titration is the point at which the chemical amount of NaOH dispensed exactly equals the chemical amount of $KHC_8H_4O_4$ originally put into the reaction flask. The concentration of the NaOH solution equals this number of moles divided by the volume of solution delivered from the buret to reach the equivalence point.

EXAMPLE 4–6

It requires 37.65 mL of a solution of NaOH to titrate 0.6135 g of dissolved potassium hydrogen phthalate to the phenolphthalein end-point. Compute the concentration of the NaOH solution, and state the assumptions behind the calculation.

Strategy

Find the chemical amount of potassium hydrogen phthalate contained in 0.6135 g of potassium hydrogen phthalate. Recognize from the balanced equation that this equals the number of moles of NaOH required to reach the equivalence point. The concentration of the NaOH solution equals the chemical amount of NaOH divided by the volume of the solution.

Solution

$$n_{\text{KHC}_8\text{H}_4\text{O}_4} = 0.6135 \text{ g KHC}_8\text{H}_4\text{O}_4 \times \left(\frac{1 \text{ mol KHC}_8\text{H}_4\text{O}_4}{204.22 \text{ g KHC}_8\text{H}_4\text{O}_4} \right)$$

$$= 3.004 \times 10^{-3} \text{ mol KHC}_8\text{H}_4\text{O}_4$$

$$n_{\text{NaOH}} = n_{\text{KHC}_8\text{H}_4\text{O}_4} \qquad \text{(at the equivalence point)}$$

$$= 3.004 \times 10^{-3} \text{ mol NaOH}$$

$$[\text{NaOH}] = \frac{3.004 \times 10^{-3} \text{ mol NaOH}}{37.65 \times 10^{-3} \text{ L}} = 0.07979 \text{ mol L}^{-1}$$

The crucial assumptions are (1) that the end-point is the equivalence point, (2) that all the $KHC_8H_4O_4$ was transferred into the titration flask without spilling, (3) that every drop of NaOH solution from the buret reacted with $KHC_8H_4O_4$ (and was not splashed out of the flask or reacted with an impurity), and (4) that the addition of base was stopped at the instant that the pink color became permanent.

EXERCISE

Compute the molarity of a solution of sodium hydroxide if 25.64 mL of solution must be added to a solution containing 0.5333 g of $KHC_8H_4O_4$ to reach the phenolphthalein end-point.

Answer: $0.1018 \text{ mol L}^{-1}$.

Once a sodium hydroxide solution has been standardized, it can be used in the titration of acid solutions. For instance, to determine the concentration of a solution of acetic acid, we add a few drops of phenolphthalein solution to the acid solution, fill a clean buret with standardized sodium hydroxide solution, and titrate to the phenolphthalein end-point. The base added from the buret reacts with the acetic acid according to the net ionic equation

$$CH_3COOH(aq) + OH^-(aq) \longrightarrow CH_3COO^-(aq) + H_2O(\ell)$$

EXAMPLE 4–7

A production lot of vinegar is being tested for its acetic acid content. A 50.00 mL sample is measured out and titrated with aqueous NaOH. It takes 31.66 mL of 1.3057 M NaOH to reach the phenolphthalein end-point. Calculate the concentration (in mol L^{-1}) of acetic acid in the vinegar. Assume that the vinegar contains no other acids.

Strategy

Find the chemical amount of NaOH that was used by multiplying the volume of the solution by its concentration. At the equivalence point, an equal chemical amount of acetic acid has reacted. Divide the chemical amount of acetic acid by the volume of the acetic acid solution to obtain the concentration of the acetic acid.

Solution

$$n_{NaOH} = 0.03166 \text{ L solution} \times \left(\frac{1.3057 \text{ mol NaOH}}{\text{L solution}}\right)$$

$$= 4.134 \times 10^{-2} \text{ mol NaOH}$$

$$n_{CH_3COOH} = n_{NaOH} \quad \text{(at the equivalence point)}$$

$$[CH_3COOH] = \frac{4.134 \times 10^{-2} \text{ mol CH_3COOH}}{50.00 \times 10^{-3} \text{ L solution}} = 0.8268 \text{ mol L}^{-1}$$

Check

The volume of NaOH solution that reacted was less than the volume of acetic acid solution. Therefore, the acetic acid solution should be less concentrated than the NaOH solution. It is.

EXERCISE

The indicator methyl red turns from yellow to red when the medium in which it is dissolved changes from basic to acidic. A 25.00 mL volume of a sodium hydroxide solution is titrated with 0.8367 M HCl. It takes 22.48 mL of this acid to reach the methyl red end-point. Find the concentration of the sodium hydroxide solution.

Answer: $[NaOH] = 0.7524 \text{ mol L}^{-1}$.

Many (but by no means all) acid–base reactions have $1:1$ stoichiometry; that is, the coefficients of the acid and base in the balanced neutralization equation are equal. At the equivalence point in a $1:1$ acid–base titration, the number of moles of base reacted equals the number of moles of acid reacted

$$n_{acid} = n_{base} \quad \text{(at equivalence in a 1:1 neutralization)}$$

These chemical amounts equal the concentrations of the solutions multiplied by the volumes used. Therefore

$$c_{acid}V_{acid} = c_{base}V_{base} \quad \text{(at equivalence in a 1:1 neutralization)}$$

This equation is very useful. It applies in Example 4–7 as follows:

• This relation resembles the equation for dilutions ($c_iV_i = c_fV_f$). However, it applies only at the equivalence point of a titration in which substances react in a $1:1$ ratio.

$$c_{CH_3COOH}V_{CH_3COOH} = c_{NaOH}V_{NaOH} \quad \text{(at equivalence)}$$

$$(c_{CH_3COOH}) \times (50.00 \text{ mL}) = (1.3057 \text{ mol L}^{-1}) \times (31.66 \text{ mL})$$

$$c_{CH_3COOH} = 0.8268 \text{ mol L}^{-1} = 0.8268 \text{ M}$$

Further Reactions of Acids and Bases

So far we know that solutions of Arrhenius acids and bases neutralize each other to give water plus a salt and cause color changes in indicators. What else do they do? The following typical reactions of acids can be used to predict the course and products of large numbers of reactions:

Figure 4–11 • When acetic acid is added to sodium carbonate (washing soda), bubbles of carbon dioxide form: $2\,CH_3COOH(aq) + Na_2CO_3(s) \longrightarrow 2\,Na^+(aq) + 2\,CH_3COO^-(aq) + CO_2(g) + H_2O(\ell)$. *(Charles D. Winters)*

1. Acids act on carbonates and hydrogen carbonates (such as Na_2CO_3 or $NaHCO_3$) to liberate gaseous CO_2 and produce a salt and water (Fig. 4–11). A typical reaction is

$$NaHCO_3(s) + HCl(aq) \longrightarrow NaCl(aq) + H_2O(\ell) + CO_2(g) \quad \text{(overall equation)}$$
$$NaHCO_3(s) + H^+(aq) \longrightarrow Na^+(aq) + H_2O(\ell) + CO_2(g) \quad \text{(net ionic equation)}$$

2. Acids react with the oxides of metals to form salts and water. A typical reaction is

$$CuO(s) + H_2SO_4(aq) \longrightarrow CuSO_4(aq) + H_2O(\ell) \quad \text{(overall equation)}$$
$$CuO(s) + 2\,H^+(aq) \longrightarrow Cu^{2+}(aq) + H_2O(\ell) \quad \text{(net ionic equation)}$$

3. Acids react with zinc, iron, and many other metallic elements to liberate gaseous H_2 and form a salt (Fig. 4–12).

$$Zn(s) + 2\,HCl(aq) \longrightarrow ZnCl_2(aq) + H_2(g) \quad \text{(overall equation)}$$
$$Zn(s) + 2\,H^+(aq) \longrightarrow Zn^{2+}(aq) + H_2(g) \quad \text{(net ionic equation)}$$

The first reaction is easy to summarize: Acids make carbonates fizz. Visualize a changing of ionic partners. Hydrogen ions from the acid join to the carbonate (or hydrogen carbonate) to produce carbonic acid (H_2CO_3). This unstable compound quickly decomposes to give water and carbon dioxide (CO_2), which is not very soluble in water. The fizzing occurs as gaseous CO_2 bubbles out of solution.

In the second reaction, hydrogen ions from the acid are attracted by the negative charge on the oxide ions. They link to the oxides, forming the stable molecular compound H_2O. This frees the metal ion to go into solution or to take part in further reactions on its own.

The third reaction is the displacement of hydrogen ions from solution by "active metals." This reaction is discussed along with other oxidation–reduction reactions in Section 4–4.

Figure 4–12 • As zinc dissolves in dilute hydrochloric acid, bubbles of hydrogen appear, and zinc chloride forms in solution. *(Charles D. Winters)*

EXAMPLE 4–8

Pickling is the removal of a surface oxide layer on a metal by its reaction with an acid or base to give a soluble salt. Write a balanced chemical equation representing the reaction between hydrochloric acid and nickel(II) oxide, $NiO(s)$.

Strategy

Realize that hydrochloric acid is an aqueous solution of hydrogen chloride ($HCl(aq)$). Predict its reaction with nickel oxide according to the pattern set by other metal oxides and acids.

Solution

$$NiO(s) + 2\,HCl(aq) \longrightarrow NiCl_2(aq) + H_2O(\ell) \quad \text{(overall equation)}$$
$$NiO(s) + 2\,H^+(aq) \longrightarrow Ni^{2+}(aq) + H_2O(\ell) \quad \text{(net ionic equation)}$$

The salt formed in such a reaction might precipitate. That precipitation would be represented by a separate net ionic equation. The nickel(II) chloride ($NiCl_2$) in this example is water-soluble. What is observed is the dissolution of the green-black $NiO(s)$ to give a green solution.

EXERCISE

One of the constituents of acid rain is nitric acid. Marble (used in monuments and statues) has the chemical composition of calcium carbonate. Write balanced chemical equations representing how rain containing nitric acid dissolves marble statues.

Answer:

$$CaCO_3(s) + 2\ HNO_3(aq) \longrightarrow$$
$$Ca(NO_3)_2(aq) + H_2O(\ell) + CO_2(g) \qquad \text{(overall equation)}$$

$$CaCO_3(s) + 2\ H^+(aq) \longrightarrow$$
$$Ca^{2+}(aq) + H_2O(\ell) + CO_2(g) \qquad \text{(net ionic equation)}$$

CHEMISTRY IN YOUR LIFE

Acids and Bases in and around You

Acids and bases are everywhere. Insoluble carbonates (in limestone) form the foundation stones of houses; dilute nitric acid in the rain (from the reaction of atmospheric nitrogen with oxygen during lightning flashes) weathers the stones away. Amino acids contain a basic group (the amino group NH_2) and an acidic group (the carboxylic acid group COOH) in every molecule; they are the building blocks of the proteins that regulate the operation of living cells. Nucleic acids make up the genetic material in cells (DNA stands for "deoxyribonucleic acid," and RNA stands for "ribonucleic acid"; both are derivatives of phosphoric acid). Most food is acidic; soaps, detergents, and cleansers for washing dishes and for cleaning the stove are basic. Citrus fruits, which all contain citric acid, furnish vitamin C, which is also called "ascorbic acid." Cola drinks contain phosphoric acid or lactic acid (read the labels), which give them their tartness; salad dressings contain vinegar (dilute acetic acid). Window-washing solutions contain the base ammonia, and the active ingredient in spray-on oven cleaners is caustic sodium hydroxide. Both promote the conversion of sticky, greasy materials into soapy, soluble residues that can be easily wiped away. Cooking tomato sauce in an aluminum saucepan can pit the metal—tomatoes are loaded with acids, and aluminum is an active metal.

The body depends on a complex web of acid–base chemistry. Normal blood remains nearly neutral thanks to several interacting acid–base reactions. Some conditions, such as severe diarrhea, diabetes, and prolonged starva-

Figure 4–A • Many common foods are acidic; most cleaning agents are basic. *(Charles D. Winters)*

tion, can lead to acidosis, a state in which the blood is too acidic. One treatment for acidosis is intravenous administration of solutions containing hydrogen carbonate (HCO_3^-) ions, which react with the acid. Prolonged vomiting or the ingestion of excessive amounts of hydrogen carbonate ion can lead to alkalosis, a state in which the blood is too basic. This sometimes occurs in self-treatment of stomach pain with sodium hydrogen carbonate ($NaHCO_3$, "bicarbonate of soda"). Carefully monitored intravenous administration of aqueous ammonium chloride ($NH_4Cl(aq)$) is sometimes used to treat alkalosis. Both acidosis and alkalosis are life-threatening medical emergencies. The treatments for both are straightforward uses of the typical reactions of acids and bases.

Like acids, bases have typical reactions:

1. With the exception of ammonia itself, bases act on ammonium salts, such as NH_4Cl and $(NH_4)_2SO_4$, to generate gaseous ammonia, a salt, and water. For instance:

$$NH_4Cl(s) + NaOH(aq) \longrightarrow NaCl(aq) + NH_3(g) + H_2O(\ell) \qquad \text{(overall equation)}$$
$$NH_4Cl(s) + OH^-(aq) \longrightarrow Cl^-(aq) + NH_3(g) + H_2O(\ell) \qquad \text{(net ionic equation)}$$

2. Bases react with the oxides of non-metals to produce salts and water. For example:

$$SO_3(g) + 2\,NaOH(aq) \longrightarrow Na_2SO_4(aq) + H_2O(\ell) \qquad \text{(overall equation)}$$
$$SO_3(g) + 2\,OH^-(aq) \longrightarrow SO_4^{2-}(aq) + H_2O(\ell) \qquad \text{(net ionic equation)}$$

The first reaction starts with a metathesis to give NH_4OH, which, if it ever actually exists, instantly breaks down to $NH_3(aq)$ and H_2O. Although ammonia is quite soluble in water, some (but not enough to fizz) still wafts out of the solution. Even small amounts of ammonia are easily detected because of ammonia's penetrating odor.

The second reaction completes a symmetry: Bases react with the oxides of *non-metals* to give a salt and water just as acids react with the oxides of *metals* to give a salt and water.

EXAMPLE 4–9

Predict the products of a reaction between solid ammonium sulfate and aqueous potassium hydroxide.

Strategy

Write the chemical formulas of ammonium sulfate and potassium hydroxide, and then the formulas if the cations and anions switch partners. Recall that most potassium compounds are soluble in water and that "ammonium hydroxide" decomposes to give ammonia and water, neither of which dissociates extensively.

Solution

$$(NH_4)_2SO_4(s) + 2\,KOH(aq) \longrightarrow$$
$$K_2SO_4(aq) + 2\,NH_3(g) + 2\,H_2O(\ell) \qquad \text{(overall equation)}$$
$$(NH_4)_2SO_4(s) + 2\,OH^-(aq) \longrightarrow$$
$$SO_4^{2-}(aq) + 2\,NH_3(g) + 2\,H_2O(\ell) \qquad \text{(net ionic equation)}$$

The products are aqueous potassium sulfate, ammonia, and water.

Check

One of the characteristic reactions of bases is that they react with ammonium salts to generate ammonia.

EXERCISE

In a certain combustion train used for the determination of carbon and hydrogen (see Fig. 1–25), the gaseous carbon dioxide produced by combustion is absorbed in a reaction with barium hydroxide. Predict the products of this reaction.

Answer: $BaCO_3$ and H_2O.

4–4 Oxidation–Reduction Reactions

In Chapter 3, we learned to predict the formulas of ionic compounds that result from the direct combination of two elements. The reactions that give these compounds are all **oxidation–reduction (redox) reactions,** an extensive and important class that is characterized by the transfer of electrons.

When sodium burns in chlorine to give sodium chloride (see Fig. 1–1),

$$2\,Na(s) + Cl_2(g) \longrightarrow 2\,NaCl(s)$$

the sodium is **oxidized:** it *gives up* electrons as the charge on its atoms increases from zero (in elemental Na) to $+1$ (in NaCl). Some species must always be present to accept these electrons. This electron-accepting species is said to be **reduced.** In this case, chlorine is reduced, as the charge on its atoms decreases from zero to -1 (i.e., becomes more negative). We can indicate the transfer of electrons by means of vertical arrows:

$$\overset{0}{2\,Na(s)} + \overset{0}{Cl_2(g)} \longrightarrow \overset{+1\ -1}{2\,NaCl(s)}$$
$$\downarrow \qquad\qquad \uparrow$$
$$2 \times 1e^- = 1 \times 2 \times 1e^-$$

The vertical arrows point *away* from the species being oxidized (giving up electrons) and *toward* the species being reduced (accepting electrons). The charges are indicated above the species. The electron bookkeeping beneath the equation ensures that the same number of electrons is taken up by chlorine as is given up by sodium:

no. of electrons lost by sodium = no. of electrons gained by chlorine

$$2\,\text{Na atoms} \times \frac{1e^- \text{ lost}}{1\,\text{Na atom}} = 1\,\text{Cl}_2 \text{ molecule} \times \frac{2\,\text{Cl atoms}}{1\,\text{Cl}_2 \text{ molecule}} \times \frac{1e^- \text{ gained}}{1\,\text{Cl atom}}$$

2 electrons lost = 2 electrons gained

Originally, the term "oxidation" referred only to reactions of a substance with oxygen, such as the burning of magnesium in air (Fig. 4–13).

$$\overset{0}{2\,Mg(s)} + \overset{0}{O_2(g)} \longrightarrow \overset{+2\ -2}{2\,MgO(s)}$$
$$\downarrow \qquad\quad \uparrow$$
$$2 \times 2e^- = 1 \times 2 \times 2e^-$$

At that time, the term "reduction" had a narrower meaning as well—the winning of a metallic element from chemical combination with non-metals such as oxygen or sulfur in an ore. Iron(III) oxide, for example, is reduced by carbon monoxide to elemental iron in the reaction

$$\overset{+3}{Fe_2O_3(s)} + 3\,\overset{+2}{CO(g)} \longrightarrow 2\,\overset{0}{Fe} + 3\,\overset{+4}{CO_2}$$
$$\downarrow \qquad\quad \uparrow$$
$$2 \times 3e^- = 3 \times 2e^-$$

The terms "oxidation" and "reduction" are now used to describe reactions of all species, as long as electron transfer takes place.

Oxidation Number

Sodium chloride and other compounds that are combinations between Group I or II elements and Group VI or VII elements are ionic. They are legitimately described as mixtures of ions with integral charges, and, as just shown, there is no difficulty

Figure 4–13 • The burning of magnesium in air gives off an extremely bright light. This led to the incorporation of magnesium into the flash powder used in early photography. Magnesium powder is still used in fireworks and flares. *(Charles D. Winters)*

in tracing the transfer of electrons in the formation of these compounds. In contrast, compounds formed from elements in the middle of the periodic table have mixed covalent and ionic character and are not composed of well-defined ionic units. Thus, tin(IV) chloride ($SnCl_4$) is a molecular compound that is very poorly described as a grouping of Sn^{4+} and Cl^- ions. When tin and chlorine combine to form $SnCl_4$, many valence electrons end up being shared in covalent bonds. What proportion (if any) of these electrons should be counted as transferred? This question is answered by the concept of the **oxidation number.** Oxidation numbers (also called **oxidation states**) are determined for the atoms in covalently bonded compounds by applying the following set of simple rules:

• Oxidation states were briefly introduced in Section 3–8.

---

1. The oxidation numbers of the atoms in a neutral molecule must add up to zero; those in an ion must add up to the charge on the ion.

2. Alkali metal (Group I) atoms have oxidation number $+1$, and alkaline earth (Group II) atoms have oxidation number $+2$ in their compounds; atoms of Group III elements usually have oxidation number $+3$ in their compounds.

3. Fluorine always has an oxidation number of -1 in its compounds. The other halogens have oxidation number -1 in their compounds, except in compounds with oxygen and with other halogens, in which they can have positive oxidation numbers.

4. Hydrogen is assigned an oxidation number of $+1$ in its compounds, except in metal hydrides such as LiH, in which rule 2 takes precedence and hydrogen has an oxidation number of -1.

5. Oxygen is assigned an oxidation number of -2 in compounds. There are two exceptions: in compounds with fluorine, rule 3 takes precedence, and in compounds that contain O—O bonds, rules 2 and 4 take precedence. Thus, the oxidation number of oxygen in OF_2 is $+2$; in peroxides (e.g., H_2O_2 and Na_2O_2), its oxidation number is -1, and in superoxides (e.g., KO_2), its oxidation number is $-\frac{1}{2}$.

• Fractional oxidation numbers, although not common, are allowed and in fact are necessary to be consistent with this set of rules.

Rule 1 is essential if the total number of electrons is to remain constant in chemical reactions. This rule also requires that the oxidation numbers of atoms in the uncombined (free) forms of the elements be zero. Rules 2 through 5 are conventions based on the principle that, in ionic compounds, the sum of the oxidation numbers must equal the charge on the ion.

Oxidation numbers must not be confused with the formal charges of Chapter 3. Their resemblance to formal charges arises because both are assigned by arbitrary rules to symbols in formulas for specific purposes. The purposes are different, however. Formal charges are used solely in connection with Lewis structures. Oxidation numbers are used in nomenclature, as in the names iron(II) chloride and iron(III) chloride (see Section 3–8) and in identifying electron transfer in oxidation–reduction reactions. In Chapter 17 and later chapters, we see how oxidation numbers help in exploring trends in chemical reactivity across the periodic table.

With the preceding rules in hand, chemists can assign oxidation numbers to the atoms in the vast majority of compounds. Apply rules 2 through 5 as listed, noting the exceptions given. Then assign oxidation numbers to the other elements in such a way that rule 1 is obeyed. Note that rule 1 applies not only to free ions but also to the ions making up an ionic solid. Chlorine has oxidation number -1 not only as a free Cl^- ion but also in the ionic solid AgCl. Tables 3–2 and 3–5 list the names and formulas of many common anions. Inspection of the table reveals that several elements exhibit different oxidation numbers in different anions. In Ag_2S, for example, sulfur in the sulfide anion has oxidation number -2, but in Ag_2SO_4 it appears as part of a sulfate (SO_4^{2-}) anion and must have oxidation number $+6$ in order to satisfy rule 1:

$$(\text{oxidation no. of S}) + 4\,(\text{oxidation no. of O}) = \text{total charge on ion}$$
$$6 + 4(-2) = -2$$

A convenient way to indicate the oxidation number of an atom is to write it directly above the corresponding symbol in the formula of the compound:

$$\overset{+1\ -2}{N_2O} \qquad \overset{+1-1}{LiH} \qquad \overset{0}{O_2} \qquad \overset{+6\ -2}{SO_4^{2-}}$$

EXAMPLE 4–10

Assign oxidation numbers to all the elements in the following chemical compounds and ions:
(a) NaCl, (b) ClO^-, (c) $Fe_2(SO_4)_3$, (d) SO_2, (e) I_2, (f) $KMnO_4$, (g) CaH_2.

Solution

(a) $\overset{+1\ -1}{NaCl}$, from rules 2 and 3.

(b) $\overset{+1\ -2}{ClO^-}$, from rules 1 and 5.

(c) $\overset{+3\ +6\ -2}{Fe_2(SO_4)_3}$, from rules 1 and 5. This can also be answered by recognizing the presence of sulfate (SO_4^{2-}) groups.

(d) $\overset{+4-2}{SO_2}$, from rules 1 and 5.

(e) $\overset{0}{I_2}$, from rule 1. I_2 is an element.

(f) $\overset{+1\ +7\ \ -2}{KMnO_4}$, from rules 1, 2, and 5. This also can be answered by recognizing the presence of the permanganate (MnO_4^-) group.

(g) $\overset{+2\ -1}{CaH_2}$, from rules 1 and 4 (metal hydride case).

EXERCISE

Determine the oxidation number of carbon in each of the following compounds: CH_4, CH_2O, CO_2, and C_6H_6.

Answer: $-4, 0, +4,$ and -1.

We can now state the following definitions:

An atom is oxidized (loses electrons) if its oxidation number increases in a chemical reaction; an atom is reduced (gains electrons) if its oxidation number decreases.

The amount of the change tells how many electrons are lost or gained. If any atom in a chemical reaction changes its oxidation state, then the reaction is a **redox** reaction.

Predicting Redox Reactions

In discussing redox reactions, chemists speak of the relative strengths of oxidizing and reducing agents. An **oxidizing agent** (or oxidant) causes the oxidation of another species by accepting electrons from it. It is itself reduced in the process. A **reducing agent** (or reductant) gives electrons to another species and is itself oxidized. A strong oxidizing agent can take electrons away from a poor reducing agent; a strong reducing agent can force electrons onto a poor oxidizing agent. Predicting the occurrence and outcome of redox reactions requires knowledge of the relative strengths of different atoms, molecules, and ions as oxidizing or reducing agents. Experiments have provided much information of this kind along with two realizations:

- Both the strength of an oxidizing or reducing agent and the identity of its products can change drastically with conditions. In particular, the course of a redox reaction in aqueous solution depends on whether the solution is acidic or basic.
- Redox strength is not the same as *rapidity* of action. An oxidizing agent (or reducing agent) that is weaker may sometimes react more rapidly than one that is stronger.

The first point is illustrated by some reactions of a well-known, potent oxidizing agent: the aqueous permanganate ion ($MnO_4^-(aq)$). If iron(II) sulfate and an acidic solution of potassium permanganate are mixed, they react according to the net ionic equation

$$5\,Fe^{2+}(aq) + MnO_4^-(aq) + 8\,H^+(aq) \longrightarrow$$
$$5\,Fe^{3+}(aq) + Mn^{2+}(aq) + 4\,H_2O(\ell) \qquad \text{(acidic solution)}$$

The oxidation number of manganese starts at $+7$ (in the permanganate ion) and drops to $+2$ (in the manganese(II) ion), showing that the permanganate ion is reduced, while the oxidation number of iron rises from $+2$ to $+3$, showing that it is oxidized. The same two reactants give *different* products under neutral and basic conditions:

$$3\,Fe^{2+}(aq) + MnO_4^-(aq) + 4\,H^+(aq) \longrightarrow$$
$$3\,Fe^{3+}(aq) + MnO_2(s) + 2\,H_2O(\ell) \qquad \text{(neutral solution)}$$

$$Fe^{2+}(aq) + MnO_4^-(aq) + 3\,OH^-(aq) \longrightarrow$$
$$Fe(OH)_3(s) + MnO_4^{2-}(aq) \qquad \text{(basic solution)}$$

A similar diversity of products is found with other oxidizing agents and with reducing agents as well. Nitrogen in ammonia (NH_3) has an oxidation number of -3. When ammonia serves as a reductant (reducing agent), N ends up in oxidation states ranging from -2 to $+5$, depending on the conditions. Such factors clearly complicate the prediction of the products even when a redox reaction is certain to take place.

The quite different rates at which nitric acid attacks aluminum and copper illustrate the second point. Aluminum ($Al(s)$) is fundamentally a far stronger reducing agent than copper ($Cu(s)$). Yet copper dipped into nitric acid immediately begins a vigorous redox reaction, whereas aluminum remains unaffected at first (Fig. 4–14). Eventually the aluminum does react as well.

Figure 4–14 • Redox strength is not the same as rapidity of action. Copper (*left*) reduces nitric acid more rapidly than does aluminum, although aluminum is a stronger reducing agent. *(Charles D. Winters)*

We examine the relative strengths of oxidizing and reducing agents in more detail in Chapters 12 and 13. In the meantime, we describe some important types of redox reactions and offer guidelines to decide whether they occur.

Types of Redox Reactions

Redox Combination and Decomposition Reactions

Most metallic elements react with most non-metallic elements to give ionic compounds. The metals are oxidized; the non-metals are reduced. Simple Lewis theory (see Section 3–3) predicts the products of some of these combination reactions. If the metal forms two or more ions (see Section 3–8), multiple products are usually possible. For example, the reaction of iron and oxygen gives iron(II) oxide (FeO) or iron(III) oxide (Fe_2O_3), depending on conditions. Usually, a higher oxidation state in the metal is favored by an excess of the non-metal.

Non-metallic elements also can combine directly with other non-metallic elements. Covalent compounds result. In these reactions, the oxidation number of the less electronegative element in the final compound depends on the conditions. For example, phosphorus reacts vigorously (Fig. 4–15) with chlorine. If the relative amount of chlorine is limited, the product is phosphorus trichloride (PCl_3)

$$\overset{0}{P_4}(s) + 6\,\overset{0}{Cl_2}(g) \longrightarrow 4\,\overset{+3\,-1}{PCl_3}(\ell)$$

but with an excess of chlorine, phosphorus is oxidized all the way to the +5 state, and the product is phosphorus pentachloride

$$\overset{0}{P_4}(s) + 10\,\overset{0}{Cl_2}(g) \longrightarrow 4\,\overset{+5\,-1}{PCl_5}(\ell)$$

• Many of these reactions are vigorous. Figure 2–6 shows the metal aluminum and the non-metal bromine as they combine.

• The concept of electronegativity is introduced in Section 3–2. Figure 17–13 on page 749 shows numerical electronegativities.

Figure 4–15 • White phosphorus (P_4) reacts vigorously with chlorine to give phosphorus pentachloride (PCl_5). If the relative amount of chlorine is restricted, the reaction gives the trichloride PCl_3. *(Charles D. Winters)*

• Burning charcoal (mostly carbon) in a brazier for heat in a closed room has led to many deaths by inhalation of the poisonous minor product carbon monoxide.

Similarly, carbon burns in oxygen to give both carbon dioxide (CO_2) and carbon monoxide (CO). Although an excess of oxygen favors CO_2, some CO almost always forms.

Combination reactions can be reversed by a proper choice of conditions. Such reversed reactions are **decompositions** because a more complex substance breaks down to two or more simpler substances. In Lavoisier's preparation of oxygen (see Section 1–3), strong heating decomposed mercury(II) oxide to its elements

$$\overset{+2\ -2}{2\ HgO(s)} \longrightarrow \overset{0}{2\ Hg(\ell)} + \overset{0}{O_2(g)}$$

although mild heating of the two elements generated the compound in the first place. Similar decomposition reactions occur when mercury(I) oxide and silver oxide are heated moderately

$$\overset{+1\ -2}{2\ Hg_2O(s)} \longrightarrow \overset{0}{4\ Hg(\ell)} + \overset{0}{O_2(g)}$$

$$\overset{+1\ -2}{2\ Ag_2O(s)} \longrightarrow \overset{0}{4\ Ag(s)} + \overset{0}{O_2(g)}$$

A sufficiently high temperature causes the redox breakdown of any compound to its constituent elements, although the temperatures required to decompose some strongly bonded compounds are truly extreme.

Often, decomposition to an element and a compound occurs upon heating. Hydrogen peroxide decomposes to give oxygen and water:

$$\overset{+1\ -1}{2\ H_2O_2(\ell)} \longrightarrow \overset{+1\ -2}{2\ H_2O(\ell)} + \overset{0}{O_2(g)}$$

at low temperature, but the H_2O resists further decomposition. Moderate heating of sulfuryl chloride ($SO_2Cl_2(g)$) decomposes it to sulfur dioxide ($SO_2(g)$) and $Cl_2(g)$; no O_2 or S is produced unless the temperature is much higher. Decomposition products can sometimes be predicted by looking for the formulas of simple stable compounds (e.g., H_2O, HCl, CO_2, SO_2, or NaCl) embedded in the formula of the substance being heated. On this basis, one predicts that heating $KClO_3$ would break it down to KCl (a simple stable compound) and O_2:

$$\overset{+1\ +5\ -2}{2\ KClO_3(\ell)} \longrightarrow \overset{+1\ -1}{2\ KCl(s)} + \overset{0}{3\ O_2(g)}$$

which is exactly what happens. Similarly, heating the salt ammonium nitrate melts it and then drives out the stable compound water:

$$\overset{-3+1+5-2}{NH_4NO_3}(\ell) \longrightarrow \overset{+1\ -2}{N_2O}(g) + 2\ \overset{+1\ -2}{H_2O}(g)$$

Some decompositions in which one compound breaks down to two or more new compounds are not redox reactions. This includes, for example, all dehydration reactions (see Section 4–3). The point is easily checked by determining all oxidation numbers on both sides of the equation.

Oxygenation

Oxygen occurs in two forms: ordinary atmospheric oxygen ("dioxygen," consisting of O_2 molecules) and the less stable ozone ("trioxygen," consisting of O_3 molecules). Both are powerful oxidants. In both forms, oxygen combines directly with most other elements and attacks many compounds to give binary oxides—compounds in which oxygen is in a negative oxidation state and everything else is in some positive oxidation state and combined with oxygen. These are **oxygenation** reactions (and also, of course, combination reactions). Examples are

$$4\ \overset{0}{Li}(s) + \overset{0}{O_2}(g) \longrightarrow 2\ \overset{+1\ -2}{Li_2O}(s)$$

and

$$2\ \overset{+2-2}{ZnS}(s) + 3\ \overset{0}{O_2}(g) \longrightarrow 2\ \overset{+2-2}{ZnO}(s) + 2\ \overset{+4-2}{SO_2}(g)$$

The electropositive elements sodium and barium can be more thoroughly oxygenated, if oxygen is plentiful, to form the **peroxides** Na_2O_2 and BaO_2 in reactions such as

$$\overset{0}{Ba}(s) + \overset{0}{O_2}(g) \longrightarrow \overset{+2\ -1}{BaO_2}(s) \qquad \text{(O}_2 \text{ in excess)}$$

Potassium, rubidium, and cesium are even more avid for oxygen and form the **superoxides** KO_2, RbO_2, and CsO_2 through reactions such as

$$\overset{0}{K}(s) + \overset{0}{O_2}(g) \longrightarrow \overset{+1\ -\frac{1}{2}}{KO_2}(s) \qquad \text{(O}_2 \text{ in excess)}$$

Oxides are more stable than peroxides or superoxides and usually are obtained either by allowing the peroxides and superoxides to react with additional metal or oxygenating the metal with a restricted supply of oxygen.

Practically every binary compound involving hydrogen can be oxidized with oxygen to water and an oxide. For example

$$4\ \overset{-3+1}{PH_3}(g) + 8\ \overset{0}{O_2}(g) \longrightarrow \overset{+5\ -2}{P_4O_{10}}(s) + 6\ \overset{+1\ -2}{H_2O}(g)$$

Most reactions of O_2 are slow at room temperature but become rapid at high temperature. When O_2 and another substance combine rapidly enough to generate light and heat, the process is a combustion. Heating oxygen-containing mixtures often initiates combustion. Complete combustion of hydrocarbons (organic compounds containing C and H only) yields carbon dioxide and water. For example, the fuel octane burns according to the equation

$$\overset{-\frac{9}{4}+1}{C_8H_{18}}(\ell) + \frac{25}{2}\ \overset{0}{O_2}(g) \longrightarrow 8\ \overset{+4-2}{CO_2}(g) + 9\ \overset{+1\ -2}{H_2O}(g)$$

Molten sulfur burns in oxygen to give sulfur dioxide: $S(\ell) + O_2(g) \longrightarrow SO_2(g)$. Sulfur trioxide forms under similar conditions in the presence of platinum. *(Charles D. Winters)*

• This reaction appears in Example 2–8.

• The combustion of hydrocarbon fuels refined from petroleum is essential to modern civilization.

The hydrogenation of bromine. Bromine, the brown gas in the flask, reacts vigorously as gaseous hydrogen is passed in through a tube. *(Charles D. Winters)*

Similar combustions take place with carbohydrates, with other compounds that contain C, H, and O only, and with a large number of compounds containing predominantly C and H. In all of these cases, complete combustion produces CO_2 and H_2O. Incomplete combustion occurs when the supply of oxygen is limited. It gives rise to new compounds that contain either more oxygen than the original compound or less hydrogen (if the oxygen takes out some hydrogen as water) or both.

Hydrogenation

Elemental hydrogen (H_2) is a good reducing agent. It **hydrogenates** other substances. Complete hydrogenation of a compound or element usually leads to new binary compounds having hydrogen in the $+1$ oxidation state and the other elements in negative oxidation states. For example, the hydrogenation of phosphorus gives PH_3, the hydrogenation of sulfur gives H_2S, and the hydrogenation of bromine gives HBr. These hydrogenations are also combination reactions, and the formulas of the products are readily predicted using Lewis theory.

In some hydrogenation reactions, H_2 is reduced, not oxidized. These involve very strong reducing agents, such as the alkali and heavier alkaline-earth metals, and give compounds in which hydrogen is in the -1 oxidation state. Hydrogen reacts at high pressures and elevated temperatures with liquid sodium, for example, to give sodium hydride

$$\overset{0}{2\,Na(\ell)} + \overset{0}{H_2(g)} \longrightarrow \overset{+1\,-1}{2\,NaH(s)}$$

The product is a *saline* (salt-like) hydride, an ionic solid composed of Na^+ cations and *negatively* charged hydride ions (H^-). Alkaline earth hydrides (such as CaH_2) are $1:2$ ionic compounds in which two H^- ions balance the charge of each $+2$ ion. Saline hydrides are oxidized by oxygen in air and by water. The reaction of NaH with water is quite violent, but that of CaH_2 is more controllable and can be used as a convenient laboratory source for hydrogen (Fig. 4–16):

$$\overset{+2\,-1}{CaH_2(s)} + \overset{+1\,-2}{2\,H_2O(\ell)} \longrightarrow \overset{0}{2\,H_2(g)} + \overset{+2}{Ca^{2+}(aq)} + \overset{-2\,+1}{2\,OH^-(aq)}$$

Note that CaH_2 is a base because it increases the concentration of hydroxide ion when put in water.

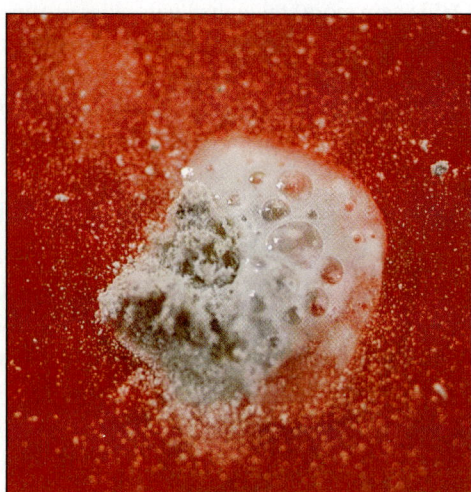

Figure 4–16 • Calcium hydride reacts vigorously with water to give hydrogen. *(Charles D. Winters)*

The hydrogenation of metal oxides usually leads to the removal of the oxygen in the form of water and the appearance of the uncombined metal:

$$\overset{+3\ -2}{Fe_2O_3}(g) + 3\ \overset{0}{H_2}(g) \longrightarrow 2\ \overset{0}{Fe}(s) + 3\ \overset{+1\ -2}{H_2O}(g)$$

or

$$\overset{+1\ -2}{Ag_2O}(s) + \overset{0}{H_2}(g) \longrightarrow 2\ \overset{0}{Ag}(s) + \overset{+1\ -2}{H_2O}(g)$$

while the hydrogenation of non-metal oxides can give water and the non-metal combined with hydrogen

$$\overset{+4\ -2}{SO_2}(g) + 3\ \overset{0}{H_2}(g) \longrightarrow \overset{+1\ -2}{H_2S}(g) + 2\ \overset{+1\ -2}{H_2O}(g)$$

$$\overset{+2-2}{CO}(g) + 3\ \overset{0}{H_2}(g) \longrightarrow \overset{-4+1}{CH_4}(g) + \overset{+1\ -2}{H_2O}(g)$$

or else water and the free element

$$\overset{+4\ -2}{SO_2}(g) + 2\ \overset{0}{H_2}(g) \longrightarrow \overset{0}{S}(s) + 2\ \overset{+1\ -2}{H_2O}(g)$$

depending on conditions.

Many organic compounds containing double and triple bonds take up hydrogen to give new compounds containing single bonds only. Hydrogen adds "across a multiple bond." This means that H atoms from H_2 appear on either side of the multiple bond as it becomes a single bond:

$$\overset{+1-2}{H_2C}{=}\overset{-2+1}{CH_2}(g) + \overset{0}{H_2}(g) \longrightarrow \overset{+1-3}{H_3C}{-}\overset{-3+1}{CH_3}(g)$$

$$\overset{+1-1}{HC}{\equiv}\overset{-1+1}{CH}(g) + 2\ \overset{0}{H_2}(g) \longrightarrow \overset{+1-3}{H_3C}{-}\overset{-3+1}{CH_3}(g)$$

$$\overset{+1\ 0}{H_2C}{=}\overset{-2}{O}(g) + \overset{0}{H_2}(g) \longrightarrow \overset{+1-2}{H_3C}{-}\overset{-2+1}{OH}(g)$$

Although hydrogen is a strong reducing agent, hydrogenation reactions are usually slow at room conditions. Rapid and complete hydrogenation is favored by the addition of a catalyst, by the use of high-pressure hydrogen, by raising the temperature, or by a combination of all three.

• Catalysts accelerate chemical reactions without themselves being consumed. See Section 14–7.

The production of water from its elements,

$$2\ \overset{0}{H_2}(g) + \overset{0}{O_2}(g) \longrightarrow 2\ \overset{+1\ -2}{H_2O}(g)$$

which can be a violent event once ignited, is simultaneously both the hydrogenation of O_2 and the oxygenation of H_2. Hydrides (from hydrogenation) and oxides (from oxygenation) are common compounds. Water ("dihydrogen monoxide") is the most common compound of all.

Displacement Reactions

Often one element displaces another from a compound. Thus, copper forces silver out of chemical combination and assumes its place, as shown in Figure 4–17, according to the equation

$$2\ \overset{+1+5-2}{AgNO_3}(aq) + \overset{0}{Cu}(s) \longrightarrow \overset{+2\ +5\ -2}{Cu(NO_3)_2}(aq) + 2\ \overset{0}{Ag}(s) \qquad \text{(overall equation)}$$

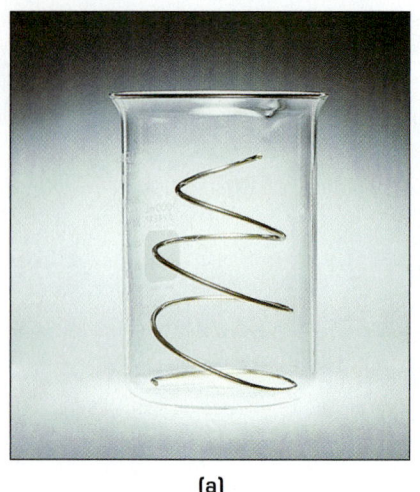

(a)   (b)   (c)

Figure 4–17 • Copper displaces silver ion from solution. (a) A length of clean copper wire. (b) After addition of a solution of silver nitrate, the Cu(*s*) reduces $Ag^+(aq)$ ions to metallic silver, which appears as granular crystals clinging to the wire. (c) The other product, $Cu(NO_3)_2(aq)$, makes the solution blue. (*Charles D. Winters*)

TABLE 4–2		
Activity Series for Some Electropositive Elements		
Element	**Product after Loss of Electrons**	
Li	Li^+	
K	K^+	
Ca	Ca^{2+}	
Na	Na^+	
Mg	Mg^{2+}	
Al	Al^{3+}	
Mn	Mn^{2+}	
Zn	Zn^{2+}	
Cr	Cr^{3+}	
Fe	Fe^{2+}	
Cd	Cd^{2+}	
Co	Co^{2+}	
Ni	Ni^{2+}	
Sn	Sn^{2+}	
Pb	Pb^{2+}	
H_2	**H^+**	
Cu	Cu^{2+}	
Ag	Ag^+	
Hg	Hg^{2+}	
Pt	Pt^{2+}	
Au	Au^{3+}	

(left margin: Increasing Ability to Displace, arrow pointing up)

The nitrate ion (NO_3^-) plays no part in this reaction. Omitting it from the equation ("canceling it out" on the two sides) gives

$$\overset{+1}{2\,Ag^+}(aq) + \overset{0}{Cu}(s) \longrightarrow \overset{+2}{Cu^{2+}}(aq) + \overset{0}{2\,Ag}(s) \qquad \text{(net ionic equation)}$$

Displacement reactions involve a change in oxidation numbers. Electropositive elements displace each other from compounds in aqueous solution according to the **activity series** in Table 4–2. Activity (ability to displace) increases from bottom to top, and the strongest reducing agent is at the very top. The second column gives the product formed by the displacing element. The position of hydrogen is important. Elements more active than hydrogen (above H_2 in the series) are "active metals." They can displace hydrogen from acidic solutions. The most active metals (Na, Ca, K, and Li) can displace hydrogen from water:

$$\overset{0}{2\,Na}(s) + \overset{+1\,-2}{2\,H_2O}(\ell) \longrightarrow \overset{+1\,-2\,+1}{2\,NaOH}(aq) + \overset{0}{H_2}(g)$$

The activity series in Table 4–2 was constructed for aqueous solutions at 25°C with all solutes having a concentration of $1\ mol\ L^{-1}$. It provides a good guide for predicting displacement reactions under any conditions, however. The activities of electronegative elements can be compared in a similar way. Among the halogens, for example, the elements higher up in the periodic table displace elements lower down from compounds:

		Electronegative Element	**Product after Gain of Electrons**
Increasing		F_2	F^-
Ability		Cl_2	Cl^-
to		Br_2	Br^-
Displace		I_2	I^-

(arrow pointing up indicating increasing ability to displace)

In this table, the strongest *oxidizing* agent is at the top of the left column. The table indicates that chlorine displaces iodide ion from, for example, potassium iodide by oxidizing it to iodine.

$$\overset{0}{Cl_2}(g) + 2\,\overset{+1-1}{KI}(aq) \longrightarrow \overset{0}{I_2}(s) + 2\,\overset{+1-1}{KCl}(aq) \qquad \text{(overall equation)}$$

$$\overset{0}{Cl_2}(g) + 2\,\overset{-1}{I}{}^{-}(aq) \longrightarrow \overset{0}{I_2}(s) + 2\,\overset{-1}{Cl}{}^{-}(aq) \qquad \text{(net ionic equation)}$$

The net ionic equation helps focus attention on the species that actually gain and lose electrons.

EXAMPLE 4–11

Predict the products of the following redox reactions. If no reaction occurs, write "NR" in place of products. Attach a descriptive name to each reaction.
(a) $Rb(s) + Cl_2(g) \longrightarrow$
(b) $Br_2(\ell) + NaCl(s) \longrightarrow$
(c) $Br_2(\ell) + NaI(s) \xrightarrow{\text{heating}}$
(d) $Fe_2O_3(s) + H_2(g) \longrightarrow$

Solution

(a) $2\,\overset{0}{Rb}(s) + \overset{0}{Cl_2}(g) \longrightarrow 2\,\overset{+1\,-1}{RbCl}(s)$. We expect rubidium to be strongly electropositive based on its position in the periodic table (in the same column with the other alkali metals such as sodium). Sodium is known to be capable of reducing chlorine to chloride. This is a combination reaction .

(b) NR ; bromine cannot displace chlorine from compounds.

(c) $\overset{0}{Br_2}(\ell) + 2\,\overset{+1-1}{NaI}(s) \longrightarrow \overset{0}{I_2}(s) + 2\,\overset{+1\,-1}{NaBr}(s)$. Bromine displaces iodine from the compound.

(d) $\overset{+3\,-2}{Fe_2O_3}(s) + 3\,\overset{0}{H_2}(g) \longrightarrow 2\,\overset{0}{Fe}(s) + 3\,\overset{+1\,-2}{H_2O}(g)$. A metal oxide is reduced to the metal by hydrogen, which is oxidized to water. This is both a displacement reaction (hydrogen displaces iron from a compound) and a hydrogenation reaction .

EXERCISE

Predict the products of the following redox reactions, and attach a descriptive name to each.
(a) $P_4(s) + O_2(g) \longrightarrow$
(b) $Te(s) + H_2(g) \longrightarrow$
(c) $Fe_2O_3(s) + Al(s) \longrightarrow$

Answer: (a) P_2O_5 in a combination reaction that is also an oxygenation. (b) H_2Te in a combination reaction that is also a hydrogenation.   (c) Al_2O_3 and Fe in a displacement reaction.

CHEMISTRY IN PROGRESS

Separation and Storage of Hydrogen

The hydrogenation of transition metals often gives non-stoichiometric compounds (see Section 1–3), with formulas such as $TiH_{1.7}$ and $ZrH_{1.9}$, in which the bonding is complex. Lanthanum reacts directly with hydrogen under room conditions to form LaH_2, and additional hydrogen can be put in until a composition of LaH_3 is reached. These hydrides retain the metallic character of the pure elements, although they are more brittle. Their electrical conductivity decreases with increasing hydrogen content.

The entry of hydrogen into transition metals and their alloys is often reversible. Gaseous hydrogen is taken up under pressure and flows out again if the pressure is reduced. This fact has led to proposals to store "dissolved" hydrogen in these metals. Hydrogen is attractive as a fuel because it burns cleanly in air to give an environmentally harmless product, water. Storage as a gas is impractical, however, because very large tanks would be required. Storage as a liquid is expensive because low temperatures

are needed (the normal boiling point of hydrogen is −253°C). Therefore, the high hydrogen densities that are possible under room conditions in transition metals appear attractive. Researchers have focused particular attention on the alloys FeTi and $LaNi_5$. Although the use of transition metals for hydrogen storage is largely speculative right now, one transition metal (palladium) is used for the separation of hydrogen from other gases in multiton quantities today. This method dates from the observation by Thomas Graham in 1866 that palladium can absorb up to 935 times its own volume of hydrogen (giving a hydrogen density approaching that of pure liquid hydrogen). Moreover, the hydrogen diffuses rapidly through the metal and the metal remains ductile. No other gases (not even helium) enter palladium in this way; thus, hydrogen can be separated from a gas mixture by passing it through palladium. Industrial plants produce volumes in excess of 250 million L (20,000 kg) of pure hydrogen per day.

Disproportionation

A single molecule or ion can, under the proper conditions, undergo a reaction in which a portion of one of the elements that it contains is oxidized and another portion of that same element is reduced. The species takes part in a redox reaction with itself. This phenomenon is called **disproportionation.** The decomposition of hydrogen peroxide (Fig. 4–18) to water and oxygen

$$2\, \overset{+1\ -1}{H_2O_2}(\ell) \longrightarrow 2\, \overset{+1\ -2}{H_2O}(\ell) + \overset{0}{O_2}(g)$$

is a disproportionation. Oxygen starts out on the left in the −1 oxidation state (see rule 5 concerning oxidation states on page 170). The reaction reduces half of it to the −2 state (in H_2O) and oxidizes the other half to the zero state (in O_2). Disproportionation is a possibility whenever a substance can act as both an oxidizing and a reducing agent; that is, whenever both lower and higher oxidation states are available to any element in the substance. Disproportionation is fairly common. Metal ions in intermediate oxidation states often disproportionate. For example:

$$2\, \overset{+1}{Cu^+}(aq) \longrightarrow \overset{0}{Cu}(s) + \overset{+2}{Cu^{2+}}(aq) \qquad \text{(net ionic equation)}$$

$$2\, \overset{+3}{Mn^{3+}}(aq) + 2\, \overset{+1\ -2}{H_2O}(\ell) \longrightarrow \overset{+2}{Mn^{2+}}(aq) + \overset{+4\ -2}{MnO_2}(s) + 4\, \overset{+1}{H^+}(aq) \qquad \text{(net ionic equation)}$$

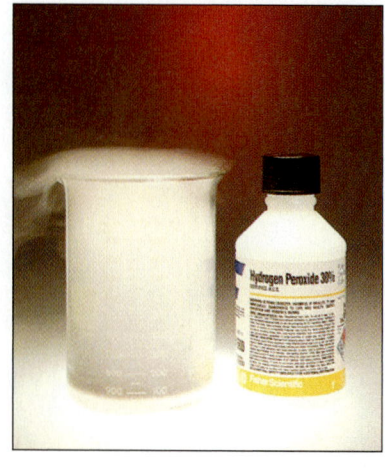

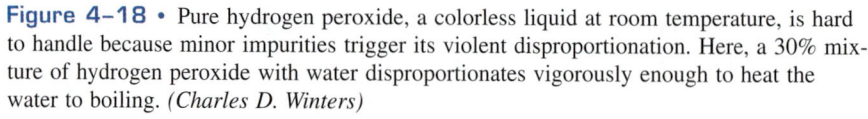

Figure 4–18 • Pure hydrogen peroxide, a colorless liquid at room temperature, is hard to handle because minor impurities trigger its violent disproportionation. Here, a 30% mixture of hydrogen peroxide with water disproportionates vigorously enough to heat the water to boiling. *(Charles D. Winters)*

Some non-metallic elements disproportionate in basic aqueous solutions:

$$\overset{0}{P_4}(s) + 3\overset{-2+1}{OH^-}(aq) + 3\overset{+1-2}{H_2O}(\ell) \longrightarrow \overset{-3+1}{PH_3}(g) + 3\overset{+1+1-2}{H_2PO_2^-}(aq) \quad \text{(net ionic equation)}$$

$$\overset{0}{I_2}(s) + 2\overset{-2+1}{OH^-}(aq) \longrightarrow \overset{-1}{I^-}(aq) + \overset{+1-2}{IO^-}(aq) + \overset{+1-2}{H_2O}(\ell) \quad \text{(net ionic equation)}$$

The first reaction is slow until the mixture is heated nearly to boiling. A single substance may disproportionate in more than one way, depending on conditions. Thus, iodine may disproportionate in basic aqueous solutions according to

$$3\overset{0}{I_2}(s) + 6\overset{-2+1}{OH^-}(aq) \longrightarrow 5\overset{-1}{I^-}(aq) + \overset{+5-2}{IO_3^-}(aq) + 3\overset{+1-2}{H_2O}(\ell) \quad \text{(net ionic equation)}$$

as well as the way shown in the preceding equation.

The structure of hydrogen peroxide (H_2O_2).

EXAMPLE 4-12

Complete and balance the following chemical equations, and state whether each represents a redox, acid–base, dissolution, or precipitation reaction.
(a) $H_2(g) + Cl_2(g) \longrightarrow$
(b) $HI(aq) + NH_3(aq) \longrightarrow$
(c) $RbH(s) + H_2O(\ell) \longrightarrow$

Solution

(a) This is analogous to the reaction of H_2 with I_2, and the balanced equation is

$$\overset{0}{H_2}(g) + \overset{0}{Cl_2}(g) \longrightarrow 2\overset{+1-1}{HCl}(g)$$

It is a redox reaction because the oxidation states of H and Cl change from 0 to $+1$ and -1, respectively.
(b) This is an acid–base reaction between the strong Arrhenius acid HI (hydroiodic acid) and the Arrhenius base NH_3 (ammonia).

$$\overset{+1-1}{HI}(aq) + \overset{-3+1}{NH_3}(aq) \longrightarrow \overset{-3+1-1}{NH_4I}(aq) \quad \text{(overall equation)}$$

$$\overset{+1}{H^+}(aq) + \overset{-3+1}{NH_3}(aq) \longrightarrow \overset{-3+1}{NH_4^+}(aq) \quad \text{(net ionic equation)}$$

Note that no oxidation numbers change.
(c) This is analogous to the reaction of sodium hydride with water and is written

$$\overset{+1-1}{RbH}(s) + \overset{+1-2}{H_2O}(\ell) \longrightarrow \overset{+1}{Rb^+}(aq) + \overset{-2+1}{OH^-}(aq) + \overset{0}{H_2}(g) \quad \text{(net ionic equation)}$$

It is a redox reaction because one hydrogen atom changes oxidation state from -1 to 0 and one changes oxidation state from $+1$ to 0. The reaction also increases the concentration of hydroxide ions in the water.

EXERCISE

Complete and balance the following chemical equations, and state whether each represents a redox, acid–base, dissolution, or precipitation reaction.
(a) $NH_3(g) + H_2O(\ell) \longrightarrow$
(b) $Li(\ell) + H_2(g) \longrightarrow$
(c) $C_4H_{10}(g) + O_2(g) \longrightarrow$

Answer: (a) $NH_3(g) \longrightarrow NH_3(aq)$ is a dissolution reaction.

(b) $2\,Li(\ell) + H_2(g) \longrightarrow 2\,LiH(s)$ is a redox reaction.

(c) $2\,C_4H_{10}(g) + 13\,O_2(g) \longrightarrow 8\,CO_2(g) + 10\,H_2O(g)$ is a redox reaction.

Titrations Using Redox Reactions

Redox reactions are often the basis of titrations. Consider the determination of the amount (or concentration) of $Fe^{2+}(aq)$ ions in a solution. Under acidic conditions, the $Fe^{2+}(aq)$ ions react quantitatively with the permanganate ion according to the net ionic equation

$$MnO_4^-(aq) + 5\,Fe^{2+}(aq) + 8\,H^+(aq) \longrightarrow Mn^{2+}(aq) + 5\,Fe^{3+}(aq) + 4\,H_2O(\ell)$$

The $MnO_4^-(aq)$ ion is intensely purple, whereas $Mn^{2+}(aq)$ and $Fe^{3+}(aq)$ have pale colors. If a little $MnO_4^-(aq)$ (in the form of dissolved potassium permanganate) is dispensed from a buret into $Fe^{2+}(aq)$, a dash of purple color appears in the flask (Fig. 4–19a) but then quickly disappears as the MnO_4^- and Fe^{2+} ions react to give the nearly colorless Mn^{2+} and Fe^{3+} products. Further additions of $MnO_4^-(aq)$ consume more $Fe^{2+}(aq)$ ion. Eventually, all of the $Fe^{2+}(aq)$ ions are consumed. The addition of just one drop of potassium permanganate solution beyond this point imparts a pale purple color to the solution from the excess $MnO_4^-(aq)$, and this color does not disappear (Fig. 4–19b). The titration is at its end-point.

Suppose that the solution of potassium permanganate has a concentration of $0.09625\ mol\ L^{-1}$ and that 26.34 mL (0.02634 L) of solution brings the titration to the end-point. Then

$$n_{MnO_4^-} = 0.02634\ \text{L solution} \times \left(\frac{0.09625\ \text{mol KMnO}_4}{\text{L solution}}\right) \times \left(\frac{1\ \text{mol MnO}_4^-}{1\ \text{mol KMnO}_4}\right)$$

$$= 2.535 \times 10^{-3}\ mol\ MnO_4^-$$

From the balanced chemical equation, reduction of one mole of MnO_4^- causes oxidation of five moles of Fe^{2+}:

$$n_{Fe^{2+}} = 2.535 \times 10^{-3}\ \text{mol MnO}_4^- \times \left(\frac{5\ \text{mol Fe}^{2+}}{1\ \text{mol MnO}_4^-}\right) = 1.268 \times 10^{-2}\ mol\ Fe^{2+}$$

What might be the purpose of this titration? Suppose that an analyst must determine the total iron content in a sample of rock. The iron may be in various forms, such as hematite (Fe_2O_3), limonite ($Fe_2O_3 \cdot H_2O$), or magnetite (Fe_3O_4). The analyst crushes the sample and determines its mass. Addition of concentrated hydrochloric acid brings the iron into solution as a mixture of $Fe^{2+}(aq)$ and $Fe^{3+}(aq)$. Reaction with tin(II) chloride then reduces all of the Fe^{3+} to Fe^{2+}

$$Sn^{2+}(aq) + 2\,Fe^{3+}(aq) \longrightarrow Sn^{4+}(aq) + 2\,Fe^{2+}(aq) \qquad \text{(net ionic equation)}$$

and excess tin(II) is oxidized to tin(IV), which does not interfere with the subsequent titration, by the addition of mercury(II) chloride

$$Sn^{2+}(aq) + 2\,HgCl_2(aq) \longrightarrow$$

$$Sn^{4+}(aq) + Hg_2Cl_2(s) + 2\,Cl^-(aq) \qquad \text{(net ionic equation)}$$

Finally, the analyst titrates as just described to find the amount of Fe^{2+} present. This procedure gives the amount of iron in the original sample, provided that the preparative reactions proceeded in close to 100% yield.

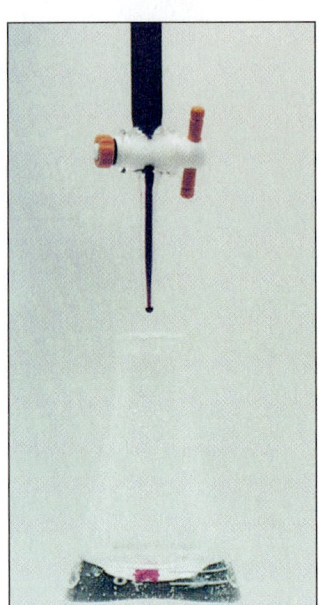

(a)

(b)

Figure 4–19 • (a) Addition of a small amount of potassium permanganate from the buret gives a dash of purple color to the solution containing Fe^{2+}. The color soon disappears as the permanganate ion is used up in oxidizing Fe^{2+} to Fe^{3+}. (b) At the end-point, all the Fe^{2+} ions have just been consumed. Additional drops of permanganate solution give a lasting pale purple color to the solution. *(George M. McKelvy/Georgia Institute of Technology)*

Many analytical determinations are indirect and rely on preparative reactions of the sample before the titration can be started; for example, a solution containing an unknown amount of a soluble calcium salt does not react with potassium permanganate. Instead, we add a solution of ammonium oxalate ($(NH_4)_2C_2O_4$), which causes the quantitative precipitation of the calcium as insoluble calcium oxalate:

$$Ca^{2+}(aq) + C_2O_4^{2-}(aq) \longrightarrow CaC_2O_4(s) \qquad \text{(net ionic equation)}$$

After filtering and washing the solid, we dissolve it in sulfuric acid to form oxalic acid:

$$CaC_2O_4(s) + 2\,H^+(aq) \longrightarrow Ca^{2+}(aq) + H_2C_2O_4(aq) \qquad \text{(net ionic equation)}$$

Finally, we titrate the oxalic acid with permanganate solution, taking advantage of the redox reaction

$$2\,MnO_4^-(aq) + 5\,H_2C_2O_4(aq) + 6\,H^+(aq) \longrightarrow$$
$$2\,Mn^{2+}(aq) + 10\,CO_2(g) + 8\,H_2O(\ell) \qquad \text{(net ionic equation)}$$

In other words, the unknown (calcium) reacts in the preparation stage, and the titration involves a second species (here, oxalic acid). The development and use of analytical techniques require a broad knowledge of reaction chemistry—the three reactions in this example are a precipitation reaction, an acid–base reaction, and a redox reaction, respectively.

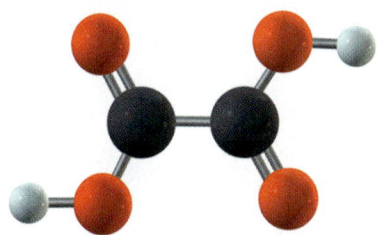

Oxalic acid ($H_2C_2O_4$) is an acid that can give up two hydrogen ions in aqueous solution. Spinach and rhubarb leaves contain large amounts of potassium oxalate ($K_2C_2O_4$) and calcium oxalate (CaC_2O_4).

SUMMARY

4-1   In **dissolution,** ions or molecules disperse from a pure phase into a homogeneous mixture. The polarity of the water molecule makes water an excellent solvent for ionic substances and **polar** molecular substances. Some substances, strong **electrolytes** or weak electrolytes, dissolve in water by complete or partial **ionization** (**dissociation** into ions); other substances (non-electrolytes) dissolve without forming ions. Electrolyte solutions conduct electricity better than pure water.

4-2   **Precipitation** is the reverse of dissolution. It occurs when the solubility of a solid solute is exceeded. A precipitation resulting from the mixing of solutions of ionic compounds is a **metathesis.** It can be represented by an ordinary balanced equation, by an **ionic equation,** or by a **net ionic equation.**

4-3   **Acids** are characterized by sour taste, ability to change the colors of indicators, and **neutralization** by **bases.** Bases are characterized by bitter taste, soapy feel, ability to change the colors of indicators, and neutralization by acids. **Arrhenius acids** donate hydrogen ions in aqueous solution; **Arrhenius bases** donate hydroxide ions. Neutralization of an Arrhenius acid by an Arrhenius base gives a salt plus water. Strong Arrhenius acids and bases are strong electrolytes; weak acids and bases are weak electrolytes. Metal oxides are **base anhydrides,** whereas non-metal oxides are **acid anhydrides. Hydration** of an anhydride gives an acid or base; **dehydration** of an oxoacid or a hydroxide gives an anhydride. **Titration** is an analytical technique used to determine the amount of a substance in solution by measuring the volume of a second solution (of known concentration) required to react quantitatively with the first substance in a well-defined reaction.

4-4   **Oxidation–reduction (redox)** reactions involve the transfer of electrons from one species to another. An atom is oxidized if its **oxidation number** becomes more positive; it is reduced if its **oxidation number** becomes more negative. In a redox reaction, an **oxidizing agent** takes electrons from another species and is itself reduced; a **reducing agent** furnishes electrons to another species and is itself oxidized. An **activity series** gives the relative strengths of oxidizing agents or reducing agents. Redox reactions include the **combination** of elements into compounds; the **decomposition** of compounds to elements; the **oxygenation** (reaction with oxygen) and the **hydrogenation** (reaction with hydrogen) of both compounds and elements; **displacement** reactions, in which more active elements displace less active elements from compounds; and **disproportionation** reactions, in which a species undergoes a redox reaction with itself. Redox titrations, like acid–base titrations, permit the determination of the amount of a substance in solution by measuring the volume of a second solution (of known concentration) required to react with the substance in a well-defined reaction.

PROBLEMS

Note: Answers to blue-numbered problems are given in Appendix F. Problems that are more challenging are indicated with asterisks.

Dissolution Reactions

1. Define and distinguish among "strong electrolyte," "weak electrolyte," and "non-electrolyte."

2. Ethanol (C_2H_5OH) is completely miscible with water, but the solutions do not conduct electricity. Write an equation for the dissolution of liquid ethanol in water.

3. Magnesium perchlorate has a strong affinity for water and is used as a drying agent to remove water vapor from gases. Write a chemical equation for the dissolution of $Mg(ClO_4)_2$ in water.

4. Soluble stannous fluoride (systematic name: tin(II) fluoride) has been added to toothpastes as an aid in combating tooth decay. Write two chemical equations for the dissolution of this compound, one assuming that it is a strong electrolyte and the other assuming that it is a weak electrolyte.

5. (See Example 4–1.) Predict whether the following compounds dissolve in water:
   (a) deoxyribose ($C_5H_{10}O_4$) (This compound is related to ribose by the removal of one oxygen atom.)
   (b) $KClO_4$
   (c) ethylene glycol ($HOCH_2CH_2OH$)

6. (See Example 4–1.) Predict whether the following compounds dissolve in water:
   (a) formic acid ($HCOOH$)
   (b) heptane (C_7H_{16})
   (c) potassium rubidium sulfate ($KRbSO_4$)

Precipitation Reactions

7. A solution is prepared by dissolving 1 mol of $NaCl(s)$ and 1 mol of $KBr(s)$ in 10 L of water. The solution is then left in an open container until all of the water evaporates. A solid residue remains. Name the compounds present in this mixture.

8. A solution is prepared by dissolving 0.1 mol of $KClO_4(s)$ and 0.1 mol of $NaCl(s)$ in sufficient water to get both the solids into solution. The solution is then allowed to evaporate until a solid starts to precipitate. Name the solid.

9. Lithium chloride is soluble in ethanol, but rubidium chloride is not. Suggest a method based on this fact to separate LiCl from RbCl if the two are mixed together in aqueous solution.

10. An experiment requires water that contains little or no dissolved oxygen. Suggest a way to remove dissolved oxygen (taken up from the air) from a supply of water.

11. Rewrite the following balanced equations as net ionic equations:
    (a) $NaCl(aq) + AgNO_3(aq) \longrightarrow AgCl(s) + NaNO_3(aq)$
    (b) $K_2CO_3(s) + 2\,HCl(aq) \longrightarrow$
    $$2\,KCl(aq) + CO_2(g) + H_2O(\ell)$$
    (c) $2\,Cs(s) + 2\,H_2O(\ell) \longrightarrow 2\,CsOH(aq) + H_2(g)$
    (d) $2\,KMnO_4(aq) + 16\,HCl(aq) \longrightarrow$
    $$5\,Cl_2(g) + 2\,MnCl_2(aq) + 2\,KCl(aq) + 8\,H_2O(\ell)$$

12. Rewrite the following balanced equations as net ionic equations:
    (a) $Na_2SO_4(aq) + BaCl_2(aq) \longrightarrow BaSO_4(s) + 2\,NaCl(aq)$
    (b) $6\,NaOH(aq) + 3\,Cl_2(g) \longrightarrow$
    $$NaClO_3(aq) + 5\,NaCl(aq) + 3\,H_2O(\ell)$$
    (c) $Hg_2(NO_3)_2(aq) + 2\,KI(aq) \longrightarrow$
    $$Hg_2I_2(s) + 2\,KNO_3(aq)$$
    (d) $3\,NaOCl(aq) + KI(aq) \longrightarrow$
    $$NaIO_3(aq) + 2\,NaCl(aq) + KCl(aq)$$

13. (See Example 4–2.) The silver bromide that is used in photographic film is made by mixing sodium bromide and silver nitrate in aqueous solution. Write a balanced overall equation and a net ionic equation for this reaction.

14. (See Example 4–2.) Nickel carbonate ($NiCO_3$) is a light green solid used to color glass. It is made by mixing aqueous sodium carbonate with nickel(II) nitrate (under controlled acidic conditions). Write a balanced overall equation and a net ionic equation for this reaction.

15. Explain how a saturated solution prepared by dissolving solid barium sulfate in water differs from a saturated solution of barium sulfate prepared by mixing aqueous barium chloride with aqueous sodium sulfate.

16. Solid sodium chloride dissolves in water but then precipitates when $HCl(g)$, which is quite soluble in water, is bubbled into the solution. Explain why.

17. (See Example 4–3.) Predict the products of the following reactions. Complete and balance the equations, and write a net ionic equation for each reaction. If no reaction occurs, write "NR" on the right side of the equation.
    (a) $Zn(NO_3)_2(aq) + K_2S(aq) \longrightarrow$
    (b) $AgClO_4(aq) + CaCl_2(aq) \longrightarrow$
    (c) $NaOH(aq) + Fe(NO_3)_3(aq) \longrightarrow$
    (d) $Ba(CH_3COO)_2(aq) + Na_3PO_4(aq) \longrightarrow$

18. (See Example 4–3.) Predict the products of the following reactions. Complete and balance the equations, and write a net ionic equation for each reaction. If no reaction occurs, write "NR" on the right side of the equation.
    (a) $CaI_2(aq) + NH_4F(aq) \longrightarrow$
    (b) $Hg_2(NO_3)_2(aq) + ZnSO_4(aq) \longrightarrow$
    (c) $NaBr(aq) + Mg(CH_3COO)_2(aq) \longrightarrow$
    (d) $NH_4I(aq) + Pb(ClO_3)_2(s) \longrightarrow$

19. (See Example 4–3.) Use the information in Table 4–1 to write balanced net ionic equations showing the result of mixing aqueous solutions of the following. If no reaction occurs, write "NR."
    (a) beryllium nitrate and sodium acetate
    (b) barium nitrate and silver sulfate
    (c) sodium hydroxide and calcium chloride
    (d) sodium chloride and potassium phosphate

20. (See Example 4–3.) Use the information in Table 4–1 to write balanced net ionic equations showing the result of mixing aqueous solutions of the following. If no reaction occurs, write "NR."
    (a) ammonium phosphate and calcium nitrate
    (b) copper(II) chloride and potassium sulfate
    (c) cesium hydroxide and barium bromide
    (d) lead chloride and potassium iodide

21. An aqueous solution of $NaNO_3$ is mixed with a solution of KCl. The reaction

    $$NaNO_3 + KCl \longrightarrow NaCl + KNO_3$$

    is anticipated, but nothing happens. Explain why. Use net ionic equations and the solubility data in Table 4–1.

22. Silver fluoride (AgF) and calcium chloride ($CaCl_2$) are both soluble in water. An aqueous solution of the first is mixed with an aqueous solution of the second. Write two balanced net ionic equations to represent the two precipitation reactions that occur. Use the solubility data in Table 4–1.

23. A mixture of magnesium nitrate and barium nitrate is dissolved in water (both are quite soluble). A chemist wants to add a second solution in order to obtain a solid barium salt without simultaneous precipitation of any magne-

sium. What should the second solution contain? Refer to Table 4–1.

24. A student accidentally contaminates a solution of sodium acetate by adding silver acetate. What can be done to remove the silver ions from the solution? Refer to Table 4–1.

Acids and Bases and Their Reactions

25. (See Example 4–4.) Name the following as acids:
    (a) H_2S, (b) HIO_4, (c) H_2CO_3, (d) HBr.

26. (See Example 4–4.) Name the following as acids:
    (a) HIO, (b) CH_3COOH, (c) HNO_2, (d) H_2Se.

27. (See Example 4–5.) Write the chemical formula and give the name of the anhydride corresponding to each of the following acids or bases. Identify each as an acid or base anhydride:
    (a) HIO_3, (b) $Ba(OH)_2$, (c) H_2CrO_4, (d) $H_2N_2O_2$.

28. (See Example 4–5.) Write the chemical formula and give the name of the anhydride corresponding to each of the following acids or bases. Identify each as an acid or base anhydride.
    (a) H_3AsO_4, (b) H_2MoO_4, (c) RbOH, (d) H_2SO_3.

29. Identify each of the following oxides as an acid or base anhydride. Write the chemical formula and give the name of the acid or base formed upon reaction with water.
    (a) MgO, (b) SO_3, (c) Cl_2O, (d) Cs_2O.

30. Identify each of the following oxides as an acid or base anhydride. Write the chemical formula and give the name of the acid or base formed upon reaction with water.
    (a) N_2O_3, (b) Li_2O, (c) P_4O_{10}, (d) BaO.

31. (See Example 4–6.) Compute the concentration of a solution of sodium hydroxide if 23.72 mL of it titrates a solution containing 0.5836 g of $KHC_8H_4O_4$ to the phenolphthalein endpoint.

32. (See Example 4–6.) Compute the concentration of a solution of sodium hydroxide if 24.48 mL of it titrates a solution containing 0.5621 g of $KHC_4H_4O_6$ to the phenolphthalein endpoint.

33. (See Example 4–7.) The indicator bromothymol blue turns from yellow to blue when the medium in which it is dissolved changes from acidic to basic. A 25.00 mL volume of a hydrochloric acid solution is titrated with 0.1025 M NaOH. It takes 26.94 mL of this base to reach a bromothymol blue end-point. Find the concentration of the HCl solution.

34. (See Example 4–7.) The indicator cresol red turns from yellow to red when the medium in which it is dissolved changes from acidic to basic. A 20.00 mL volume of a nitric acid solution is titrated with 0.1225 M KOH. It takes 16.49 mL of this base to reach a cresol red end-point. Find the concentration of the HNO_3 solution.

35. (See Example 4–9.) Predict the products of the following reactions. If no reaction occurs, write "NR."
    (a) $HNO_3(aq) + K_2CO_3(s) \longrightarrow$
    (b) $HBr(aq) + Zn(s) \longrightarrow$
    (c) $H_2SO_4(aq) + Zn(OH)_2(s) \longrightarrow$

36. (See Example 4–9.) Predict the products of the following reactions. If no reaction occurs, write "NR."
    (a) $HNO_3(aq) + LiOH(aq) \longrightarrow$
    (b) $Ni(s) + HCl(aq) \longrightarrow$
    (c) $(NH_4)_2CO_3(aq) + HI(aq) \longrightarrow$

37. (See Example 4–9.) Predict the products of the following reactions. If no reaction occurs, write "NR."
    (a) $HClO_3(aq) + KOH(aq) \longrightarrow$
    (b) $N_2O_5(g) + NaOH(aq) \longrightarrow$
    (c) $(NH_4)_2SO_4(aq) + Ba(OH)_2(aq) \longrightarrow$

38. (See Example 4–9.) Predict the products of the following reactions. If no reaction occurs, write "NR."
    (a) $NH_4Cl(aq) + RbOH(aq) \longrightarrow$
    (b) $SeO_3(s) + KOH(aq) \longrightarrow$
    (c) $HF(aq) + CsOH(aq) \longrightarrow$

39. Write a balanced equation to represent the reaction between each of the following pairs of substances in the presence of water:
    (a) hydrogen bromide and calcium hydroxide
    (b) ammonia and sulfuric acid
    (c) lithium hydroxide and nitric acid

40. Write a balanced chemical equation to represent the reaction between each of the following pairs of substances in the presence of water:
    (a) benzoic acid and ammonia
    (b) perchloric acid and magnesium hydroxide
    (c) nitric acid and cesium hydroxide

41. Write balanced equations for acid–base reactions that produce each of the following salts. Name the acid, base, and salt.
    (a) CaF_2 (b) Rb_2SO_4
    (c) $Zn(NO_3)_2$ (d) KCH_3COO

42. Write balanced equations for acid–base reactions that produce of each of the following salts. Name the acid, base, and salt.
    (a) Na_2SO_3
    (b) $Ca(C_6H_5COO)_2$
    (c) $PbSO_4$
    (d) $CuCl_2$

43. The gaseous compound PF_3 reacts slowly with water to give H_3PO_3 and HF.
    (a) Name the three compounds referred to by their formulas.
    (b) Write and balance an equation representing this reaction.

44. The acid $HClO_4$ reacts with P_4O_{10} to give Cl_2O_7 and H_3PO_4.
    (a) Name all four of these compounds.
    (b) Write and balance an equation representing the reaction.

45. Hydrogen sulfide can be removed from natural gas by reaction with excess sodium hydroxide. Name the salt that is produced in this reaction. (*Note:* Hydrogen sulfide donates two hydrogen ions in the course of this reaction.)

46. During the preparation of viscose rayon, cellulose is dissolved in a bath containing sodium hydroxide and later reprecipitated as rayon by the addition of a solution of sulfuric acid. Name the salt that is a by-product of this process.

Rayon production is in fact a significant commercial source for this salt.

47. Your tomato sauce eats a hole in an aluminum saucepan. Explain why, using a net ionic equation.

48. Manganese(II) oxide (MnO) dissolves when treated with concentrated $HCl(aq)$. Name the products, and write a balanced chemical equation for this reaction.

Oxidation–Reduction Reactions

49. (See Example 4–10.) Assign oxidation numbers to each element in the following compounds:
    (a) formaldehyde (CH_2O), (b) carbonic acid (H_2CO_3), (c) rubidium hydride (RbH), (d) dinitrogen pentaoxide (N_2O_5).

50. (See Example 4–10.) Assign oxidation numbers to each element in the following compounds:
    (a) ethane (C_2H_6), (b) acetic acid ($C_2H_4O_2$), (c) ammonium nitrate (NH_4NO_3), (d) potassium superoxide (KO_2).

51. A useful analytical technique for determining the amount of arsenic in a solution relies on its oxidation by iodine

$$As_2O_3(aq) + 2\,I_2(s) + 2\,H_2O(\ell) \longrightarrow$$
$$As_2O_5(aq) + 4\,HI(aq)$$

Write the oxidation number above the symbol for each element that changes oxidation state in the course of this reaction, identify which is reduced and which is oxidized, and verify that the total decrease in oxidation number is equal to the total increase in oxidation number.

52. When calcium sulfate is heated with coke (carbon) above 1200° C, it reacts according to

$$2\,CaSO_4(s) + C(s) \longrightarrow 2\,CaO(s) + 2\,SO_2(g) + CO_2(g)$$

This reaction is used industrially on a small scale to make SO_2, which is then converted into sulfuric acid. Write the oxidation number above the symbol for each element that changes oxidation state in the course of this reaction, identify which is reduced and which is oxidized, and verify that the total decrease in oxidation number is equal to the total increase in oxidation number.

53. For each of the following balanced equations, write the oxidation number above the symbol of each atom that changes oxidation state in the course of the reaction:
    (a) $2\,PF_2I(\ell) + 2\,Hg(\ell) \longrightarrow P_2F_4(g) + Hg_2I_2(s)$
    (b) $2\,KClO_3(s) \longrightarrow 2\,KCl(s) + 3\,O_2(g)$
    (c) $4\,NH_3(g) + 5\,O_2(g) \longrightarrow 4\,NO(g) + 6\,H_2O(g)$
    (d) $2\,As(s) + 6\,NaOH(\ell) \longrightarrow 2\,Na_3AsO_3(s) + 3\,H_2(g)$

54. For each of the following balanced equations, write the oxidation number above the symbol of each atom that changes oxidation state in the course of the reaction:
    (a) $N_2O_4(g) + KCl(s) \longrightarrow NOCl(g) + KNO_3(s)$
    (b) $H_2S(g) + 4\,O_2F_2(s) \longrightarrow SF_6(g) + 2\,HF(g) + 4\,O_2(g)$
    (c) $2\,POBr_3(s) + 3\,Mg(s) \longrightarrow 2\,PO(s) + 3\,MgBr_2(s)$
    (d) $4\,BCl_3(g) + 3\,SF_4(g) \longrightarrow$
$$4\,BF_3(g) + 3\,SCl_2(\ell) + 3\,Cl_2(g)$$

55. For each of the reactions in problem 53, identify the species that is oxidized and the one that is reduced. Verify that the total decrease in oxidation number is equal to the total increase in oxidation number.

56. For each of the reactions in problem 54, identify the species that is oxidized and the one that is reduced. Verify that the total decrease in oxidation number is equal to the total increase in oxidation number.

57. Selenic acid (H_2SeO_4) is a powerful oxidizing acid that dissolves not only silver (as does the related acid H_2SO_4) but also gold, through the reaction

$$2\,Au(s) + 6\,H_2SeO_4(aq) \longrightarrow$$
$$Au_2(SeO_4)_3(aq) + 3\,H_2SeO_3(aq) + 3\,H_2O(\ell)$$

Determine the oxidation numbers of the atoms in this equation. Which species is oxidized and which is reduced?

58. Diiodine pentaoxide reacts with carbon monoxide under room conditions:

$$I_2O_5(s) + 5\,CO(g) \longrightarrow I_2(s) + 5\,CO_2(g)$$

The reaction is used in an analytical method to measure the amount of carbon monoxide in a sample of air. Determine the oxidation numbers of the atoms in this equation. Which species is oxidized and which is reduced?

59. Aluminum(III) oxide is caused to react with carbon to give elemental aluminum and carbon dioxide. Write a balanced equation for this process, and indicate which species is oxidized and which is reduced.

60. During the recharging of a lead-acid storage battery (used in automobiles), lead(II) sulfate reacts with water to give metallic lead, lead(IV) oxide, and aqueous sulfuric acid. Write a balanced equation for the reaction that occurs, and identify which species is oxidized and which is reduced.

61. (See Example 4–12.) Give likely products of the following reactions. Balance the chemical equations in each case, and give a descriptive name for the reaction.
    (a) $HCl(g) + O_2(g) \longrightarrow$
    (b) $H_2C{=}O(g) + H_2(g) \longrightarrow$
    (c) $Mg(s) + HCl(aq) \longrightarrow$
    (d) $N_2(g) + H_2(g) \longrightarrow$

62. (See Example 4–12.) Give likely products of the following reactions. Balance the chemical equations in each case, and give a descriptive name for the reaction.
    (a) $Br_2(g) + H_2(g) \longrightarrow$
    (b) $NH_2NH_2(g) + H_2(g) \longrightarrow$
    (c) $CH_4(g) + O_2(g) \longrightarrow$
    (d) $Ca(s) + Br_2(g) \longrightarrow$

63. Chloric acid ($HClO_3$) cannot be prepared outside of solution. Even in solution, it is not stable and can decompose in several ways, depending on conditions. The modes of decomposition include

$$HClO_3(aq) \longrightarrow ClO_2(g) + O_2(g) + H_2O(\ell)$$
$$HClO_3(aq) \longrightarrow HCl(aq) + O_2(g)$$
$$HClO_3(aq) \longrightarrow HClO_4(aq) + HCl(aq)$$

(a) Balance all three of the equations, and assign oxidation numbers to the elements in every substance.
(b) Which of these reactions is a disproportionation?

64. Aqueous nitrous acid reacts to give nitric acid, gaseous nitrogen monoxide, and water. Write a balanced chemical equation for this reaction, and show that it is a disproportionation.

65. Heating ammonium nitrite (NH_4NO_2) gives water and nitrogen (N_2).
    (a) Write and balance a chemical equation to represent this change.
    (b) Determine the oxidation number of nitrogen in the ammonium ion (NH_4^+) and in the nitrite ion (NO_2^-); determine the "overall" oxidation number of nitrogen in this compound. Discuss whether this is a redox reaction or a dehydration reaction, or both.

66. Heating lead(II) nitrate gives lead(II) oxide, O_2, and one other product, the gas X, which contains no lead. The volume of X is four times the volume of the O_2 when the two are held at the same temperature and pressure. Name X. Is this a redox reaction?

67. Write balanced equations for the formation of barium hydride from the elements and the reactions of barium hydride with water, with hydrochloric acid, and with aqueous zinc sulfate.

68. Write balanced equations for the formation of hydrogen bromide from the elements and the reactions of hydrogen bromide with water, with potassium carbonate, and with magnesium.

69. (See Example 4–12.) Complete and balance each of the following chemical equations. State whether each represents a redox, acid–base, dissolution, or precipitation reaction or does not fit into any of these categories:
    (a) $H_2Te(g) + H_2O(\ell) \longrightarrow$
    (b) $SrO(s) + CO_2(g) \longrightarrow$
    (c) $HI(aq) + CaCO_3(s) \longrightarrow$
    (d) $Na_2O(s) + NH_4Br(aq) \longrightarrow$

*70. (See Example 4–12.) Complete and balance each of the following chemical equations. State whether each represents a redox, acid–base, dissolution, or precipitation reaction or does not fit into any of these categories:
    (a) $HBr(aq) + ZnO(s) \longrightarrow$
    (b) $Cs(s) + O_2(g) \longrightarrow$
    (c) $MgO(s) + SO_3(g) \longrightarrow$
    (d) $NaH(s) + Cu(NO_3)_2(aq) + H_2O(\ell) \longrightarrow$

71. An aqueous solution contains iron in several different chemical forms. A 10.00 mL portion of the solution is treated with oxidizing agents to convert all the iron to the Fe(III) oxidation state. Addition of a reducing agent then converts all of the Fe(III) to Fe(II). Finally, the solution is titrated according to the balanced equation

$$Cr_2O_7^{2-}(aq) + 6\,Fe^{2+}(aq) + 14\,H^+(aq) \longrightarrow$$
$$2\,Cr^{3+}(aq) + 6\,Fe^{3+}(aq) + 7\,H_2O(\ell)$$

A total of 27.15 mL of 0.01667 M $K_2Cr_2O_7$ solution brings the titration to its end-point. Compute the mass of iron that was in the 10.00 mL sample.

72. You convert all of the iron in a 0.6010 g ore sample to iron(II) in aqueous solution and titrate the solution with potassium permanganate. Compute the percentage of iron (by mass) in the sample if 21.98 mL of 0.01210 M $KMnO_4$ is needed for reaction according to the equation

$$MnO_4^-(aq) + 5\,Fe^{2+}(aq) + 8\,H^+(aq) \longrightarrow$$
$$Mn^{2+}(aq) + 5\,Fe^{3+}(aq) + 4\,H_2O(\ell)$$

ADDITIONAL PROBLEMS

73. All of the following substances are soluble in water. State whether the solutions conduct electricity well, poorly, or not at all:
(a) acetic acid, (b) acetone, (c) $HgCl_2$, (d) $NaClO_3$.

74. Lithium chlorate is among the most soluble of all salts. A saturated aqueous solution contains 315 g/100 g of water at room temperature.
(a) Compute the ratio of the number of water molecules to the number of ions in this solution, assuming complete ionization.
(b) Is lithium chlorate completely dissociated in a saturated solution? Why or why not?

75. When ammonia and carbon dioxide are added to water, an aqueous solution of ammonium carbonate forms. This solution reacts with calcium sulfate to give a solid and an electrolyte solution. Identify the solid and the solution, and write a balanced chemical equation for the reaction. This reaction is carried out industrially on a small scale; the salt in the electrolyte solution is a useful fertilizer.

76. Dissolution of approximately 0.6 g of mercury(I) sulfate saturates 1.0 L of water at room temperature. Suppose that a saturated solution of mercury(I) sulfate is mixed with aqueous barium chloride. Tell what will be seen, and write net ionic equations to represent the reaction or reactions taking place.

77. The autoionization of water

$$H_2O(\ell) \longrightarrow H^+(aq) + OH^-(aq)$$

cannot be prevented. This means that even the purest water contains quantities of $OH^-(aq)$ and $H^+(aq)$ ions. Does water then fulfill the definition of a substance given in Chapter 1? Explain.

78. Aluminum sulfate hydrate ($Al_2(SO_4)_3 \cdot 18H_2O$) is used in two major ways: by the paper industry to coat paper pulp and give it a hard surface (its aqueous solutions are referred to as "papermaker's alum") and in water treatment to remove bacteria and impurities.
(a) Propose a method for making this hydrate from the aluminum ore bauxite ($Al(OH)_3$). Write a balanced chemical equation for the reaction.

(b) In aqueous solution, aluminum sulfate reacts with calcium hydroxide to form a loose solid hydroxide precipitate (floc) that removes impurities from the drinking water as it settles. Write a balanced equation for this reaction.

79. A researcher adds a solution of potassium hydroxide to one of perchloric acid in order to neutralize the strong acid. The reaction generates a large amount of heat, which was expected, and a white precipitate appears, which surprises the researcher. Write chemical equations that explain both observations.

80. Workers in paint factories making the pigment white lead ($Pb(OH)_2 \cdot 2PbCO_3$) at one time customarily added dilute sulfuric acid to their drinking water for protection against lead poisoning. Explain the chemistry behind this practice.

81. A compound is said to be *hygroscopic* if it has a strong affinity for water. The very hygroscopic compound P_4O_{10} dehydrates many oxoacids to give the corresponding acid anhydrides (while it in turn forms a phosphorus oxoacid). Write a balanced chemical equation for the reaction between P_4O_{10} and nitric acid. In this case, the oxoacid of phosphorus produced is HPO_3 rather than the fully hydrated H_3PO_4.

82. Oxygen difluoride reacts with water to give an acid, but not an oxoacid. Rather, its reaction is described by the equation

$$OF_2(g) + H_2O(\ell) \longrightarrow O_2(g) + 2\,HF(aq)$$

(a) Is this a redox reaction? If so, name the species being oxidized and the species being reduced.
(b) The oxoacid that might be predicted to form from the addition of water to OF_2 was first prepared in 1968 through the reaction of F_2 with water at very low temperatures and under special conditions. It is a white solid that melts to a pale yellow liquid at $-117°C$. Give its chemical formula and its name, based on the formula and name of the parallel oxoacid of chlorine.

83. A total of 222 mL of 0.0450 M hydrobromic acid (HBr) is mixed with 125 mL of a 0.0540 M solution of the base barium hydroxide ($Ba(OH)_2$). They react.
(a) Determine which reactant is the limiting reactant.
(b) Determine the additional volume of acid or base that must be added to exactly neutralize the excess base or acid.

84. Bismuth forms an ion with the formula Bi_5^{3+}. Arsenic and fluorine form a complex ion $[AsF_6]^-$, with fluorine atoms arranged around an arsenic atom. Assign oxidation numbers to each of the atoms in the bright yellow crystalline solid with the formula $Bi_5(AsF_6)_3 \cdot 2SO_2$.

*85. The best method to prepare very pure oxygen on a small scale is the decomposition of $KMnO_4$ in a vacuum above 215°C:

$$KMnO_4(s) \longrightarrow K_2MnO_4(s) + MnO_2(s) + O_2(g)$$

Balance this equation. Assign oxidation numbers to each atom, and verify that the total decrease in oxidation number equals the total increase in oxidation number.

*86. (a) Determine the oxidation number of lead in each of the following oxides: PbO, PbO_2, Pb_2O_3, Pb_3O_4.
   (b) The only known lead ions are Pb^{2+} and Pb^{4+}. How can you reconcile this statement with your answer to part (a)?

87. The hydrogen xenate ion reacts with base according to the equation

$$2\,HXeO_4^-(aq) + 2\,OH^-(aq) \longrightarrow$$
$$XeO_6^{4-}(aq) + Xe(g) + O_2(g) + 2\,H_2O(\ell)$$

   Give the oxidation numbers for all the atoms in this equation. Which species are being reduced? Which oxidized?

88. Balance the following equations, and classify them as representing dissolution, precipitation, acid–base, or redox reactions or as reactions that do not fit any of the categories listed:
   (a) $KClO_3(s) \longrightarrow KCl(s) + KClO_4(s)$
   (b) $H_2S(g) \longrightarrow H^+(aq) + HS^-(aq)$
   (c) $H_2O_2(g) + HI(g) \longrightarrow I_2(g) + H_2O(g)$
   (d) $P_4O_{10}(s) + C(s) \longrightarrow P_4(g) + CO(g)$

89. The following descriptive names of reactions have all been used in the chemical literature. Use a balanced equation to give a likely example of each type of reaction. Compare and contrast each with those discussed in the chapter:
   (a) chlorination, (b) dehydrogenation, (c) hydrochlorination.

*90. The following descriptive names of processes have all been used in the chemical literature. Give an example of each. Compare and contrast each with those discussed in the chapter.
   (a) deoxygenation, (b) comproportionation, (c) fluoridation.

91. Barium perxenate (Ba_2XeO_6) is a colorless salt of the perxenate ion.
   (a) Americium(II) perxenate has also been prepared. Write its chemical formula.

   (b) The corresponding oxoacid, perxenic acid, has not been prepared because it is spontaneously reduced in aqueous solution. Give its chemical formula.
   (c) Give the chemical formula and name for the anhydride of perxenic acid.

92. Arsenic forms two important oxoacids, with chemical formulas H_3AsO_3 and H_3AsO_4.
   (a) Give the chemical formulas and names of the acid anhydrides of these two acids.
   (b) Follow the usual naming procedure for an element that forms two oxoacids to determine the names of the oxoacids of arsenic.
   (c) Arsenic is in the same group as nitrogen. How are the chemical formulas of the oxoacids of arsenic and nitrogen different?
   (d) Can you give an explanation for the result in part (c) based on Lewis structures of the arsenic oxoacids? (*Hint*: Recall that second-period elements readily form multiple bonds, but multiple bonding is unusual in the fourth period.)

93. When coal that contains sulfur is burned to generate electricity, sulfur dioxide and sulfur trioxide are produced. These air pollutants contribute significantly to acid rain. One possible way to prevent their emission is to line the smokestacks of power plants with lime (CaO). What would be the products of its reaction with the sulfur oxides in the stacks?

94. Balance the following equations, and classify them as representing dissolution, precipitation, acid–base, or redox reactions or as reactions that do not fit any of the categories listed:
   (a) $H_3PO_4(aq) + NaOH(aq) \longrightarrow NaH_2PO_4(aq) + H_2O(\ell)$
   (b) $Fe_3O_4(s) + C(s) \longrightarrow Fe(s) + CO(g)$
   (c) $Ca(NO_3)_2(aq) + K_3PO_4(aq) \longrightarrow$
   $$Ca_3(PO_4)_2(s) + KNO_3(aq)$$
   (d) $K_2Cr_2O_7(aq) + HI(aq) \longrightarrow$
   $$CrI_3(s) + I_2(s) + KI(aq) + H_2O(\ell)$$

CUMULATIVE PROBLEM

The Solvay Process

The Solvay process is used industrially to make sodium carbonate from the inexpensive starting materials sodium chloride and calcium carbonate. It can be represented in the following six steps, which are discussed in greater detail in Chapter 23:

1. $2\,NH_3(aq) + 2\,CO_2(g) + 2\,H_2O(\ell) \longrightarrow 2\,NH_4HCO_3(aq)$
2. $2\,NH_4HCO_3(aq) + 2\,NaCl(aq) \longrightarrow 2\,NaHCO_3(s) + 2\,NH_4Cl(aq)$
3. $2\,NaHCO_3(s) \longrightarrow Na_2CO_3(s) + H_2O(g) + CO_2(g)$
4. $CaCO_3(s) \longrightarrow CaO(s) + CO_2(g)$
5. $CaO(s) + H_2O(\ell) \longrightarrow Ca(OH)_2(aq)$
6. $2\,NH_4Cl(aq) + Ca(OH)_2(aq) \longrightarrow 2\,NH_3(aq) + CaCl_2(aq) + 2\,H_2O(\ell)$

Most of the chemicals involved in the Solvay process are found in the home, including Na_2CO_3 (sodium carbonate, washing soda) and $NaHCO_3$ (sodium hydrogen carbonate, baking soda).

Answer the following questions about these reaction steps:

(a) Which oxoacid has CO_2 as its anhydride?

(b) What type of reaction is the second step? Write a net ionic equation for this step.

(c) Is step 3 an acid–base reaction? If so, what is the acid reacting and what is the base?

(d) Name the base that has CaO as its anhydride. In which step is this base prepared?

(e) Write a net ionic equation for step 6.

(f) Are any of the steps redox reactions? If so, which ones?

(g) Add the six reactions, canceling chemical species that appear on both sides of the equation. The result is the overall reaction. Suppose the two products were mixed in aqueous solution. Predict the results.

5 The Gaseous State

CHAPTER OUTLINE

Hot air is less dense than cool air. The difference gives these hot-air balloons their buoyancy. *(J.S. Sroka/ Dembinsky Photo Associates)*

There are two major ways of classifying matter. The first is to study the *chemical constitution* of a material and categorize it as an element, compound, or mixture of substances. Chapter 1 uses this approach. The second is to focus on the *physical state* of a material, calling it a gas, a liquid, or a solid (see Fig. 3–7). Gases, liquids, and solids are recognized quite easily because an important example of each is all around every day: air, water, and earth. Gases and liquids are fluid, but solids are rigid. A gas expands to fill any container it occupies; a liquid has a fixed volume but flows to conform to the shape of its container; a solid has a fixed volume *and* a fixed shape that resists deformation. These simple properties allow the classification of nearly all substances as to physical state, at least under ordinary terrestrial conditions. Changes of state are also matters of everyday experience—that is, ice melts and water freezes; water boils and steam condenses. Similar transitions occur with thousands of less familiar substances and are crucial in chemistry.

The goal of this chapter and the next is to describe the states of matter and the transitions between them and to develop some theoretical understanding of why they exist. Gases come first because they have played a central role throughout the history of chemistry and because their physical properties are easier to understand than those of liquids and solids.

5–1 The Chemistry of Gases

The ancient Greeks considered air to be one of the four fundamental elements in nature. As early as the 17th century, its physical properties (such as resistance to compression) were being studied. The chemical composition of air (Table 5–1) was not appreciated until late in the 18th century, when Priestley, Lavoisier, and others showed that it consists primarily of two different substances: oxygen and nitrogen. Oxygen is characterized by its ability to support life. Once the oxygen in air is used

TABLE 5–1	The Composition of Air[a]	
Constituent	**Formula**	**Volume Percentage[b]**
Nitrogen	N_2	78.110
Oxygen	O_2	20.953
Argon	Ar	0.934
Neon	Ne	0.001818
Helium	He	0.000524
Krypton	Kr	0.000114
Xenon	Xe	0.0000087
Hydrogen	H_2	0.00005
Methane	CH_4	0.0002
Dinitrogen monoxide	N_2O	0.00005

[a]Air also contains other constituents, the abundance of which varies. Examples are water (H_2O), 0–7%; carbon dioxide (CO_2), 0.01–0.1%; ozone (O_3), 0–0.000007%; carbon monoxide (CO), 0–0.000002%; nitrogen dioxide (NO_2), 0–0.000002%; and sulfur dioxide (SO_2), 0–0.0001%.
[b]Volume percentage is defined in Appendix C–4. The volume percentage of a component in a mixture of gases equals its mole percentage; see Section 5–6.

Figure 5–1 • Methane wells in a landfill in Chicago.

up (by the burning of a candle in a closed container, for example), the nitrogen that remains no longer keeps animals alive. More than 100 years elapsed before a careful re-analysis of air showed that oxygen and nitrogen account for only about 99% of the total volume, with most of the remaining 1% consisting of a new gas, which was given the name "argon." The other noble gases (helium, neon, krypton, and xenon) are present in air to a lesser extent.

Gases are found at and under the Earth's surface as well. Bacterial processes, especially in swampy areas, produce methane (CH_4). It is a major constituent of natural gas deposits formed over many millennia by the decay of plant matter under the Earth's surface. The recovery of methane from municipal landfills for use as a fuel gas is now a commercially feasible process (Fig. 5–1). Helium is found in natural gas sources as well, and the radioactive gas radon, which originates from the decay of trace amounts of actinide elements in certain rocks, can cause hazardous concentrations of radioactivity in the air in unventilated basements. The evaporation of liquids also produces gases. All liquids are **volatile** to some extent and therefore evaporate; the most familiar example is water. Water vapor rising from lakes and oceans causes the humidity of the air.

• "Volatile" means liable to evaporate at normal temperatures. It does not mean explosive, toxic, noxious, or dangerous.

Gases are often formed by chemical reactions. It is quite common, for example, for solids to decompose to give one or more gases when heated. We have already mentioned (see Fig. 1–10) the decomposition of mercury(II) oxide to mercury and oxygen:

$$2\,HgO(s) \xrightarrow{\text{heat}} 2\,Hg(\ell) + O_2(g)$$

Antoine Lavoisier used this reaction in establishing the law of conservation of mass. Even earlier (1756), Joseph Black had shown that marble, which is composed primarily of calcium carbonate ($CaCO_3$), would decompose on heating to give quicklime (CaO) and carbon dioxide:

Carbon dioxide (CO_2) dissolved in aqueous solution gives soft drinks their fizz. It also serves as an inexpensive acid through its reaction with water to form $H_2CO_3(aq)$. Solid carbon dioxide (dry ice) is used for refrigeration.

$$CaCO_3(s) \xrightarrow{\text{heat}} CaO(s) + CO_2(g)$$

CHEMISTRY IN YOUR LIFE

How an Air Bag Works

This is a true story. One rainy day, a motorist hit a bump at an intersection hard enough to cause an air bag to deploy. No one was hurt, and damage to the car was minor, but a police officer noticed some strange white powder on the air bag. He immediately cleared the area and called in the bomb squad! Traffic was snarled until the bomb squad determined that no danger existed.

Why did the police officer fear an explosion? He had learned at a training session that the "air" that inflates air bags is actually nitrogen from the decomposition of sodium azide:

$$2\,NaN_3(s) \longrightarrow 2\,Na(s) + 3\,N_2(g)$$

He had also learned that elemental sodium reacts violently with water:

$$2\,Na(s) + 2\,H_2O(\ell) \longrightarrow 2\,NaOH(s) + H_2(g)$$

If the white powder were elemental sodium from the first reaction that had escaped the air bag, then the rain might set off the second reaction, with disastrous results. Therefore, he took drastic action. Good chemistry as far as it went . . . but incomplete.

The material that generates the gas in an air bag is not pure sodium azide, but a mixture of sodium azide with (usually) potassium nitrate (KNO_3) and silicon dioxide (SiO_2). When sensors detect a collision and send an electric current through this mixture, the NaN_3 experiences a *deflagration,* or slow detonation, to give gaseous nitrogen and elemental sodium, just as the police officer had learned. The sodium, however, reacts nearly instantly with the potassium nitrate to generate more nitrogen:

$$10\,Na(s) + 2\,KNO_3(s) \longrightarrow$$
$$K_2O(s) + 5\,Na_2O(s) + N_2(g)$$

Thus, no elemental sodium accumulates when an air bag deploys. The potassium oxide and sodium oxide from the

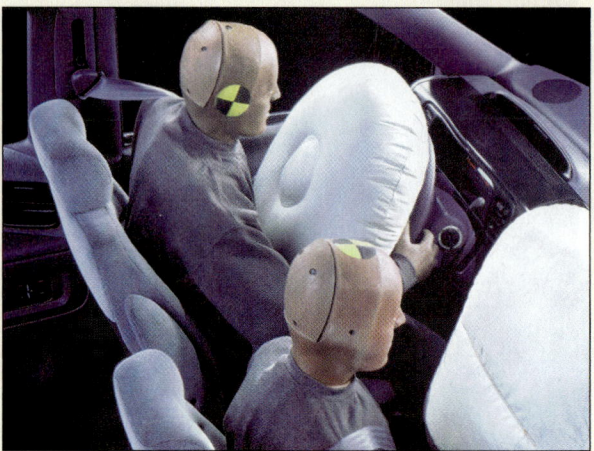

Figure 5–A • Deployment of air bags in a test. *(Don Johnson/Stone/Getty Images)*

second reaction are anhydrides of strong bases and might prove injurious if released. The acid anhydride SiO_2 neutralizes them to form silicates. Consequently, the only chemical products in an air bag deployment are the nitrogen that inflates the bag and a harmless glassy mixture of the salts Na_2SiO_3 and K_2SiO_3.

Although the products of an air bag deployment are innocuous, the gas-generating mixture is not; sodium azide is quite toxic. Because drivers try hard never to cause an air bag to deploy, the release of sodium azide during the junking of cars with unused air bags poses real problems. One possible solution is to develop inflator modules that are recyclable from old cars into new ones.

What about the suspicious white powder that fired the police officer's concern? It was cornstarch that had been sprinkled between the folds of the air bag to prevent sticking and ensure smooth inflation.

The salt ammonium chloride (NH_4Cl) decomposes when heated to produce both gaseous ammonia and gaseous hydrogen chloride:

$$NH_4Cl(s) \xrightarrow{\text{heat}} NH_3(g) + HCl(g)$$

Many ammonium salts give similar reactions. Some gas-forming reactions proceed explosively; the decomposition of nitroglycerin is a detonation in which all of the products are gaseous:

$$4\,C_3H_5(NO_3)_3(\ell) \longrightarrow 6\,N_2(g) + 12\,CO_2(g) + O_2(g) + 10\,H_2O(g)$$

Recall from Section 4–4 that several of the elements react with oxygen to form gaseous oxides. Carbon dioxide is produced during animal respiration but also, to a much greater extent, by combustion of materials that contain compounds of carbon, such as coal and oil. Oxides of sulfur are produced by burning elemental sulfur, and oxides of nitrogen are produced in the oxidation of elemental nitrogen that occurs inside the engines of motor vehicles:

$$S(s) + O_2(g) \longrightarrow SO_2(g)$$
$$2\,SO_2(g) + O_2(g) \longrightarrow 2\,SO_3(g)$$
$$N_2(g) + O_2(g) \longrightarrow 2\,NO(g)$$
$$2\,NO(g) + O_2(g) \longrightarrow 2\,NO_2(g)$$

• The role of these compounds in air pollution is considered further in Section 18–5.

Another important class of reactions that form gases is the action of acids on certain ionic solids. In Section 4–3, we mentioned the production of carbon dioxide from carbonates (Fig. 5–2)

$$CaCO_3(s) + 2\,HCl(aq) \longrightarrow CaCl_2(aq) + CO_2(g) + H_2O(\ell)$$

as a characteristic reaction for acids. Other examples of this type of reaction are

$$Na_2S(s) + 2\,HCl(aq) \longrightarrow 2\,NaCl(aq) + H_2S(g)$$
$$K_2SO_3(s) + 2\,HCl(aq) \longrightarrow 2\,KCl(aq) + SO_2(g) + H_2O(\ell)$$
$$NaCl(s) + H_2SO_4(aq) \longrightarrow NaHSO_4(aq) + HCl(g)$$

In these reactions, the identity of the metal (calcium, sodium, and potassium) is immaterial; similar reactions of the carbonates, sulfides, sulfites, and chlorides of other metals also occur.

Gases have extremely varied chemistries. Some, such as HCl and SO_3, are highly reactive; others, such as N_2O, N_2, and Kr, are weakly reactive or almost nonreactive. Some, such as O_2 and Cl_2, are oxidizing agents; others, such as H_2 and CO, are reducing agents. Their physical properties are much simpler to understand: At sufficiently low densities, all gases behave in the same way. Their properties approach the physical properties of an "ideal" gas, which we consider in the next sections.

• An interesting generalization: All colored gases are poisonous. It is *not* true that all colorless gases are nonpoisonous!

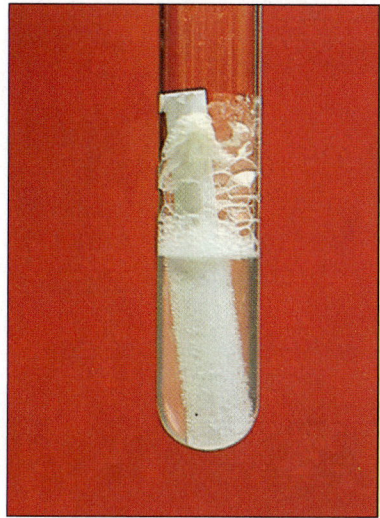

Figure 5–2 • The familiar material chalk serves as a source of carbon dioxide. Here, the calcium carbonate in a stick of chalk reacts with an aqueous solution of hydrochloric acid, producing bubbles of carbon dioxide. (*Charles D. Winters*)

5-2 Pressure and Boyle's Law

In physics, a **force** (F) is a simple push exerted by one object on another. Applying a force to a stationary object sets it in motion, unless the object pushes back with an equal force. Applying a force to an object that pushes back creates a **pressure** (P) on the object. The pressure equals the force divided by the area (A) over which the force is applied:

$$P = \frac{F}{A} \qquad \text{(by definition)}$$

A confined gas often exerts a large force on the walls of its container, as can be testified by anyone who has ever seen an overinflated automobile tire explode. The source of this force is the impact of molecules of the gas as they bombard the walls (see Section 5–7). The total force, divided by the area of the walls, equals the pressure of the gas. Unless the container is actively expanding or contracting, its walls exert an exactly equal and balancing pressure on the gas; the pressure *on* the gas equals the pressure *of* the gas.

The surface of the Earth lies at the bottom of an ocean of air that is many kilometers deep. The Earth attracts its atmosphere gravitationally, causing it to exert a pressure on every square centimeter of our skin and every other surface that it touches, even when no wind blows. Our skin and other surfaces exert an exactly counterbalancing pressure. This situation can be hard to accept because we grow up under atmospheric pressure and are accustomed to it. Clear evidence of the facts was first provided in an ingenious experiment devised in the 17th century by Evangelista Torricelli, an Italian scientist who had been an assistant to Galileo. Torricelli sealed a long glass tube at one end, filled it with mercury, closed the open end with a finger, turned it upside-down, and immersed the open end in a dish of mercury (Fig. 5–3a), taking care that no air leaked in. When he removed his finger,

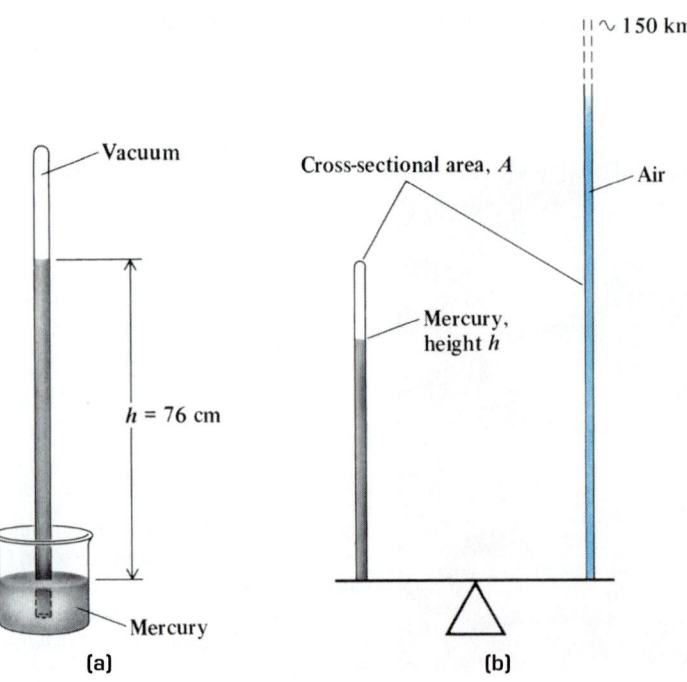

Figure 5-3 • (a) In Torricelli's barometer, the height of the top of the mercury in the tube is approximately 76 cm above that in the open dish. (b) The mass of mercury in height h exactly balances the mass of a column of air of the same diameter extending to the top of the atmosphere.

the level of mercury in the tube fell, leaving a nearly perfect vacuum at the closed end. Only a portion of the mercury flowed out of the tube. The flow stopped when the top surface of the mercury inside the tube was about 76 cm above the surface of the mercury in the dish. Torricelli showed that the exact height varied somewhat from day to day and from place to place.

This simple device, called a **barometer,** works like a balance. One arm is loaded with the mass of mercury in the tube, and the other, with a column of air of the same cross-sectional area that extends to the top of the atmosphere, approximately 150 km up (Fig. 5–3b). The height of the mercury column adjusts itself so that the mass of the mercury and the mass of the air (and thus the two forces on the surface of the mercury in the dish) become equal. Changes in the height of the column occur as the force exerted by the atmosphere varies with the weather. Barometers at high elevations have a portion of the atmosphere below them and always give lower readings than those at sea level, if the effect of the weather is excluded.

The relationship between the height h of the column of liquid in a barometer and the pressure P exerted by the atmosphere is

$$P = gdh$$

In this equation, g stands for the acceleration of gravity at the surface of the Earth, 9.80665 m s^{-2}, and d stands for the density of the liquid. According to this equation, a pressure that raises mercury in a barometer to a height of, say, 50 cm would support a shorter column if the strength of the Earth's gravitational attraction were somehow increased (g bigger, h smaller), and a longer one if the effect of gravity were somehow decreased (g smaller, h bigger). Furthermore, this same pressure would raise a column of a less dense liquid higher than 50 cm (d smaller, h bigger). In principle, a barometer can be constructed using almost any liquid if a long enough tube is available. Mercury is the best liquid for barometers because it does not evaporate very rapidly, it is fairly unreactive, and its high density keeps the instrument small. The last point is illustrated by the following example.

EXAMPLE 5–1

A barometer is constructed using water instead of mercury. Estimate the height of the column of water in this barometer on a day when the atmosphere supports a 76.0 cm column of mercury. The density of mercury is approximately 13.6 times greater than the density of water.

Strategy

Imagine the water and mercury barometers side by side. The equation $P = gdh$ applies to both. The pressure P and the acceleration of gravity g are the same for both, but the density d and height h differ. Write the governing equation for each barometer and compare the two. Use subscripts to keep the different d's and h's straight.

Solution

The two equations are

$$P = gd_{\text{water}}h_{\text{water}} \qquad \text{and} \qquad P = gd_{\text{mercury}}h_{\text{mercury}}$$

The right sides of the two equations equal the same thing, so

$$gd_{\text{water}}h_{\text{water}} = gd_{\text{mercury}}h_{\text{mercury}}$$

Dividing both sides of this equation by gd_{water} gives

$$h_{water} = \left(\frac{gd_{mercury}}{gd_{water}}\right)h_{mercury} = \left(\frac{d_{mercury}}{d_{water}}\right)h_{mercury}$$

But $d_{mercury} = 13.6d_{water}$. Also, $h_{mercury} = 76.0$ cm. Insert these facts into the equation

$$h_{water} = \left(\frac{13.6d_{water}}{d_{water}}\right)76.0 \text{ cm} = 1030 \text{ cm} = 10.3 \text{ m}$$

Check

The barometer containing water, the less dense liquid, is taller, as expected. It is 33.8 feet tall, the size of a three-story building, which makes it much less practical than a barometer containing mercury.

EXERCISE

The pressurized habitat of a Moon colony contains air at a pressure sufficient to raise a 76.0 cm column of mercury in a barometer on Earth. Estimate the height of the column of mercury in a barometer that operates inside the habitat. The acceleration of the Moon's gravity is only 0.165 times that of the Earth.

Answer: 461 cm.

The equation $P = gdh$ allows a direct calculation of the pressure exerted by the atmosphere, but care must be taken with units. The best plan is to convert the units of the quantities on the right side of the equation to combinations of SI base units (see Appendix B–1). This approach relies on the consistency of the SI units. The density of mercury at 0°C is 13.5951 g cm^{-3}. In base SI units, this density equals

- One *million* (10^6) cm^3 in 1 m^3 is correct. One meter equals 100 cm. Hence, a *cubic* meter equals $100 \times 100 \times 100$ cubic centimeters.

$$d_{mercury} = \frac{13.5951 \text{ g}}{cm^3} \times \left(\frac{10^6 \text{ cm}^3}{1 \text{ m}^3}\right) \times \left(\frac{1 \text{ kg}}{10^3 \text{ g}}\right) = 1.35951 \times 10^4 \text{ kg m}^{-3}$$

As Torricelli observed, the height h of the mercury column in a barometer stays close to 76 cm under ordinary weather conditions at sea level. Suppose that h equals exactly 76 cm. Convert this to 0.76 m because the meter is the SI base unit of length. The acceleration of gravity g equals 9.80665 m s^{-2}. The unit m s^{-2} is already a combination of SI base units, so the calculation can proceed:

$$P = gdh = 9.80665 \frac{m}{s^2} \times 1.35951 \times 10^4 \frac{kg}{m^3} \times 0.76 \text{ m}$$

$$= 1.01325 \times 10^5 \frac{kg}{m \, s^2} = 1.01325 \times 10^5 \text{ kg m}^{-1}\text{s}^{-2}$$

The combination of units kg m^{-1} s^{-2} is defined as the **pascal** (Pa), a unit of pressure (see Appendix B–1). The atmosphere exerts a pressure of 101,325 Pa in raising the column of mercury in a barometer to a height of 76 cm. Pressure is defined as force divided by area, and the pascal indeed does equal the SI unit of force divided by the SI unit of area, the **newton** (N) divided by the square meter (m^2)

- One newton is approximately the gravitational force exerted by the Earth on an apple; a penny resting flat on a table exerts a pressure (averaged over its surface) of about 100 Pa.

$$1 \text{ N m}^{-2} = 1 \text{ pascal} = 1 \text{ kg m}^{-1}\text{s}^{-2}$$

as proved in Appendix B–1.

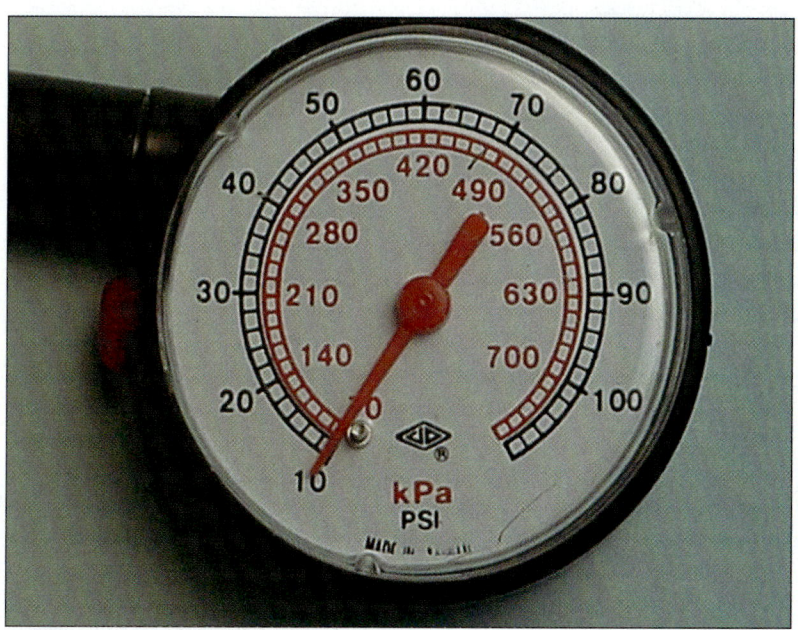

Figure 5–4 • This tire pressure gauge is calibrated in kilopascals (kPa, the red scale) and in pounds per square inch (psi, the black scale). On the dial, 100 psi equals just less than 700 kPa, which is consistent with the facts in Table 5–2. The pressure on this gauge was 1 atm when it was photographed, but it reads zero. The gauge measures the pressure in excess of atmospheric pressure, called *gauge pressure*. A gauge pressure is some-times indicated by adding a "g" to the unit: psig, or kPag. Gauge pressure equals the absolute pressure minus the atmospheric pressure.

Numerous other units are used to measure pressure. The pascal is small, so the larger kilopascal (1 kPa = 10^3 Pa) and megapascal (1 MPa = 10^6 Pa) find frequent use. Atmospheric pressure varies but stays fairly close to 100,000 Pa. One **standard atmosphere** (1 atm) is defined as exactly 1.01325×10^5 Pa. This non-SI unit of pressure arose historically because of the importance of atmospheric pressure in daily life. A pressure of 1 atm supports a column of mercury in a barometer to a height of 0.76 m or 760 mm or 29.92 inches at 0°C. Weather observers generally report pressures in millimeters of mercury (**mm Hg**) or in inches of mercury. Because the density of mercury depends slightly on temperature, it is necessary to specify the temperature when reporting accurate pressures in mm Hg. This complication is avoided with the **torr,** which is defined as exactly 1/760 atm at *any* temperature. Only at 0°C do the torr and the mm Hg equal each other exactly. In the customary U.S. system of units, the most common unit of pressure (used in measuring tire pressure, for example) is the pound of force per square inch (psi). One atmosphere equals approximately 14.7 psi. The tire pressure gauge in Figure 5–4 shows the relationship between the psi and the kPa. Finally, the **bar** is defined as exactly 10^5 (100,000) Pa, an exact power of 10 that is close to 101,325 Pa, the standard atmosphere. Pressures in the kilobar (1 kbar = 10^3 bar) and even megabar (1 Mbar = 10^6 bar) range can be achieved experimentally. Table 5–2 summarizes some facts about different units of pressure.

TABLE 5-2	Units Used to Measure Pressure
Unit Name and Abbreviation	**Definition or Equivalency**
Pascal (Pa)	$1 \text{ kg m}^{-1} \text{ s}^{-2} = 1 \text{ N m}^{-2}$ (the SI unit)
Standard atmosphere (atm)	101,325 Pa exactly
Bar (bar)	100,000 Pa exactly or 0.986923 atm
Torr (torr)	(101,325/760) Pa or (1/760) atm
Millimeter of mercury at 0°C (mm Hg)	(101,325/760) Pa or (1/760) atm
Pound of force per square inch (lbf in^{-2}, or psi)	6894.757 Pa or (1/14.69595) atm

EXAMPLE 5-2

Water starts to boil at 21°C (room temperature) if the air pressure above it is reduced to 18.65 torr. Express this pressure in atmospheres, pascals, and pounds of force per square inch.

Strategy

The question requires conversion among some of the different units for pressure. Construct appropriate unit-factors from the facts in Table 5–2.

Solution

To convert from torr to atmospheres:

$$18.65 \text{ torr} \times \left(\frac{1 \text{ atm}}{760 \text{ torr}} \right) = 0.02454 \text{ atm}$$

The conversion from torr to pascals proceeds similarly, but with a different unit-factor:

$$18.65 \text{ torr} \times \left(\frac{101{,}325 \text{ Pa}}{760 \text{ torr}} \right) = 2.486 \times 10^3 \text{ Pa}$$

A pair of unit-factors converts the pressure to pounds per square inch:

$$18.65 \text{ torr} \times \left(\frac{1 \text{ atm}}{760 \text{ torr}} \right) \times \left(\frac{14.69595 \text{ psi}}{1 \text{ atm}} \right) = 0.3606 \text{ psi}$$

Check

The three units of pressure increase in size in the order Pa, psi, atm. The numerical answers should therefore increase in the *reverse* order. They do. Also, one of the conversions can be completed using another choice of unit-factors from Table 5–2:

$$18.65 \text{ torr} \times \left(\frac{101{,}325 \text{ Pa}}{760 \text{ torr}} \right) \times \left(\frac{1 \text{ psi}}{6894.757 \text{ Pa}} \right) = 0.3606 \text{ psi}$$

EXERCISE

Barium is a solid at 510°C. If the pressure on a sample of barium at this temperature is increased to 61 kbar, the barium melts. Further compression (still at 510°C) to 68 kbar causes the barium to solidify once again. Express these two transition pressures in megapascals (MPa) and in torr.

Answer: 6.1×10^3 MPa and 6.8×10^3 MPa; 4.6×10^7 torr and 5.1×10^7 torr.

Boyle's Law: The Effect of Pressure on Gas Volume

Robert Boyle, an English natural philosopher and theologian, studied the properties of confined gases. He noted the familiar fact that when a gas is compressed or expanded, it tends to spring back to its original volume. These properties are much like those of metal springs, which were being investigated by his collaborator Robert Hooke. Boyle's experiments on the compression and expansion of air were reported in 1662 in a monograph, *Touching the Spring of the Air and Its Effects.*

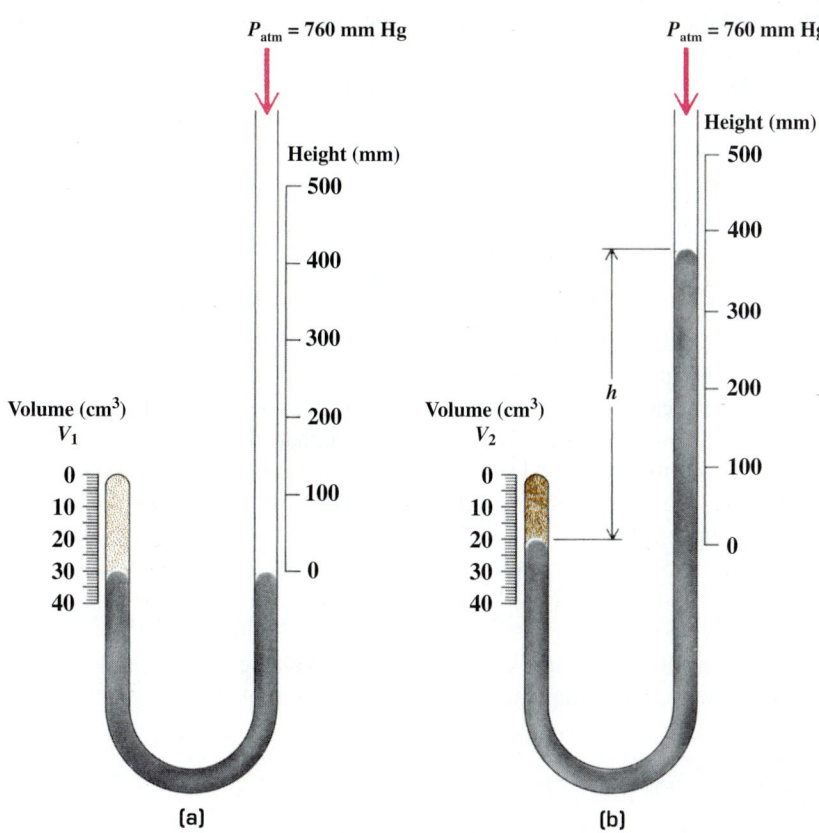

(a)

(b)

Figure 5–5 • Boyle's J-tube. When the height of mercury on the two sides of the tube is the same (*a*), the pressure of the confined gas must be equal to that of the atmosphere, 1 atm or 760 mm Hg. After mercury has been added (*b*), the pressure of the gas is increased by the number of millimeters of mercury in the height difference *h*. The higher pressure causes the gas to occupy a smaller volume.

Boyle worked with a simple piece of apparatus—a J-tube, in which air was trapped at the closed end by a column of mercury (Fig. 5–5). If the difference in height *h* between the two mercury levels in such a tube is zero, then the pressure of the air in the closed part exactly balances that of the atmosphere. The pressure *P* on the trapped air is therefore 1 atm, or 760 mm Hg. If mercury is poured into the open end of the tube, the pressure of the confined air is increased by 1 mm Hg for every millimeter by which the level of mercury on the open side of the tube exceeds the level on the closed side. That is,

$$P = (760 + h) \text{ mm Hg}$$

$$P = (760 + h) \text{ mm Hg} \times \left(\frac{1 \text{ atm}}{760 \text{ mm Hg}} \right) = \frac{(760 + h)}{760} \text{ atm}$$

The volume of the confined air is read off a scale on the previously calibrated tube. Squeezing a gas (increasing the pressure) certainly should force it into a smaller volume. Boyle confirmed exactly just such an inverse relationship. His data allowed a generalization:

> The product of the pressure and volume, *PV*, of a sample of gas is a constant at a constant temperature:
>
> $$PV = C \qquad \text{(at a fixed temperature and for a fixed amount of gas)}$$

This result is known as **Boyle's law** (Fig. 5–6).

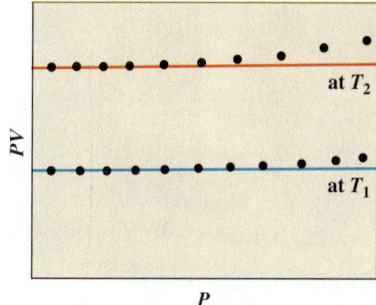

Figure 5–6 • If the product of the pressure (*P*) and volume (*V*) of a sample of gas is plotted against *P* at a fixed temperature T_1, a horizontal straight line results; that is, $PV = C$. At a higher temperature T_2, the value of *PV* is larger, but the line is again straight. At both temperatures, deviations from the straight line set in at higher pressures.

The conditions on Boyle's law are crucial. If either the temperature or the amount of gas changes during the experiment, then Boyle's law does not apply. The constant C depends strongly on the temperature and amount of gas, but, importantly, C is essentially independent of the *identity* of the gas trapped in the J-tube. Experiment shows that at 0°C and for 1.00 mol of any gas (for example, 32.0 g of O_2, 28.0 g of N_2, or 2.02 g of H_2)

$$PV = C = 22.4 \text{ L atm} \qquad \text{(at 0°C for 1.00 mol of any gas)}$$

A pressure of 1.00 atm and temperature of 0°C are chosen for this illustration because 1.00 atm and 0°C are readily achieved and reproduced in the laboratory. These conditions are called **standard temperature and pressure (STP).** The volume of 1.00 mol of a gas at STP is 22.4 L.

• Visualize 22.4 L as the contents of a cube that is about 28 cm (11 in) on an edge.

In many practical cases, a numerical value of C is not needed to compute useful results, as in the manometer of Figure 5–7. Suppose that a confined gas has a pressure P_1 and a volume V_1. We adjust the pressure to a new value, P_2, keeping the temperature and the amount of gas unchanged. Boyle's law applies both before and after the adjustment; that is, $P_1V_1 = C$ and $P_2V_2 = C$. The value of C is the same in the two cases, so it follows that

$$P_1V_1 = P_2V_2 \qquad \text{(at a fixed temperature and for a fixed amount of gas)}$$

This useful form of Boyle's law also applies if the volume of the gas is adjusted from V_1 to V_2 at constant temperature for a fixed amount of gas.

EXAMPLE 5–3

When you push down as hard as you can on the plunger of a sealed plastic syringe, you are able to compress the air inside into a volume of 11 cm³ (see Fig. 5–8). When you let go, the plunger pops up as the pressure inside equalizes with the pressure outside, which is 1.0 atm. The final volume of the trapped air is 42 cm³. The temperature of the air inside the syringe does not change. Calculate the pressure inside the syringe just before you release the plunger.

Strategy

Use Boyle's law in the form $P_1V_1 = P_2V_2$. Solve for the initial pressure, and substitute the values of the final pressure and of the initial and final volumes.

Solution

$$P_1 = P_2\left(\frac{V_2}{V_1}\right) = 1.0 \text{ atm} \left(\frac{42 \text{ cm}^3}{11 \text{ cm}^3}\right) = 3.8 \text{ atm}$$

Check

The answer of 3.8 atm exceeds 1.0 atm. This fits with the expansion of the contents of the syringe when the plunger is released.

EXERCISE

The long cylinder of a bicycle pump has a volume of 1131 cm³ and is filled with air at a pressure of 1.02 atm. The outlet valve is sealed shut, and the pump handle is pushed down until the volume of the air is 517 cm³. The temperature of the air trapped inside does not change. Compute the pressure inside the pump.

Answer: 2.23 atm.

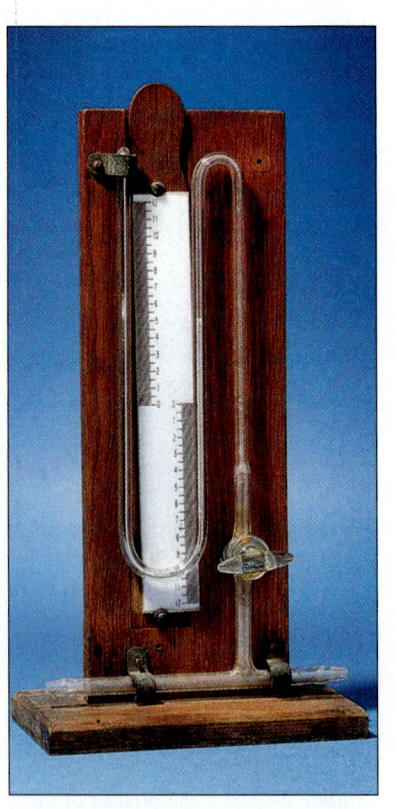

Figure 5–7 • A manometer is used to measure the pressure of a gas. The glass tube at the bottom is connected to the gas sample, and the difference in height between the levels of the mercury in the two arms of the tube is read off on a calibrated scale. This modern device has essentially the same design as Boyle's J-tube. *(Leon Lewandowski)*

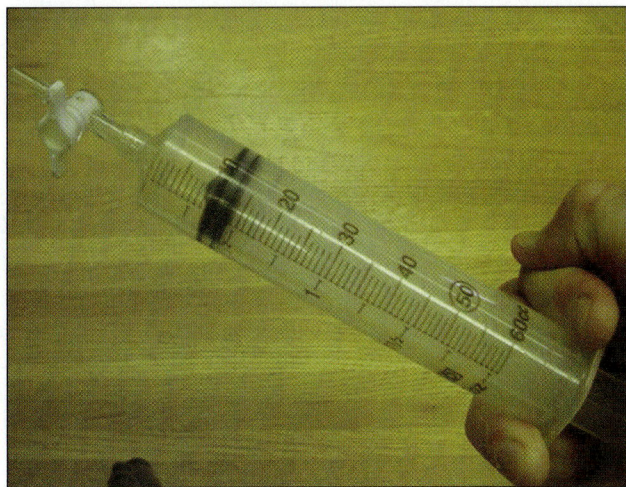

Figure 5-8 • Air compressed by finger pressure in a plastic syringe.

Careful measurements on real gases at pressures near 1 atm reveal small deviations from the predictions of Boyle's law. For gases at high pressure (> 50–100 atm), substantial deviations occur (see Fig. 5–6). Boyle's law is an idealization that real gases follow closely at low pressure but increasingly poorly as the pressure is raised.

5-3 Temperature and Charles's Law

Temperature is one of those elusive properties that is readily understood in a general way but is very difficult to pin down in a quantitative fashion. We have an instinctive feeling (through the sense of touch) of *hot* and *cold;* for example, water at its freezing point is colder than at its boiling point and thus should have a lower temperature. When the Celsius scale of temperature was devised, the freezing point of water was assigned a temperature of 0°C, and the boiling point (at 1 atm pressure), a temperature of 100°C. On the Fahrenheit scale, the same two temperatures are 32°F and 212°F, respectively.

Assigning two fixed points in this way does not tell how to assign other temperatures. For example, simple observation of a sample of diethyl ether at 1 atm quickly establishes that it boils at a temperature warmer than 0°C but cooler than 100°C. But what number should be assigned to this boiling temperature? Further arbitrary choices are certainly not the answer. The problem is that temperature is not a mechanical quantity like pressure; thus, it is more difficult to define.

Fortunately, some mechanical properties *depend* on temperature. For example, liquid mercury expands by an easily measured amount as it is heated from 0°C to 100°C. This change in volume could be used as the basis for the definition of a temperature scale. Temperature could be *assumed* to be linearly related to the volume of mercury. Then, temperatures could be read simply from the volume of mercury in a tube (a mercury thermometer). The problem with this definition, of course, is that it is tied to the properties of a single substance, mercury. If the volume of another substance is measured on the mercury scale, it is *not* linearly related to the

• Recall that to convert a temperature from degrees Celsius to degrees Fahrenheit, one first multiplies by 9/5 and then adds 32. To convert a temperature from degrees Fahrenheit to degrees Celsius, one first subtracts 32 and then multiplies by 5/9. That is: $t_F = 9/5 \, t_C + 32$ and $t_C = (5/9)(t_F - 32)$. See Appendix B.

• The molecular structure of diethyl ether appears on page 144.

Figure 5-9 • When a balloon is cooled with liquid nitrogen, its volume shrinks drastically as the air inside cools. *(Charles D. Winters)*

temperature. Why should mercury be chosen? And how is temperature defined below the freezing point or above the boiling point of mercury? We shall soon see how these questions are resolved, but, for the time being, let us simply adopt a *provisional* temperature scale based on a mercury thermometer and turn to a study of how the properties of gases change with temperature.

Charles's Law: The Effect of Temperature on Gas Volume

Boyle observed that the product of the pressure and volume of a confined gas changes upon heating and thus depends on temperature (Fig. 5–9), but the first quantitative experiments were performed by the French scientist Jacques Charles more than a century later. Charles confined gases in such a way that they could be heated or cooled at constant pressure (Fig. 5–10). Under these conditions, the gases expanded in volume when heated and contracted when cooled. A plot of the volume of the gas V versus its temperature t gave a straight line (Fig. 5–11).

> At constant pressure, the volume of a sample of gas is a linear function of its temperature **(Charles's law).**

Heating a sample of N_2 from the temperature at which water freezes to the temperature at which it boils causes it to expand to 1.366 times its original volume; the same 36.6% increase in volume is found for O_2, CO_2, and other gases. In fact, at sufficiently low pressures, *all* gases expand by the same relative amount if they start at the same initial temperature and end at the same final temperature. Writing Charles's law in the following mathematical form

$$V = V_0 + \alpha V_0 t \qquad \text{(at a fixed pressure and for a fixed amount of gas)}$$

allows further investigation of this striking result. In this equation, V represents the observed volume of a gas, V_0 represents its volume at 0°C (the "ice point"), t stands for the temperature measured in degrees Celsius, and α (the Greek letter "alpha") represents a new quantity, the "coefficient of thermal expansion." All

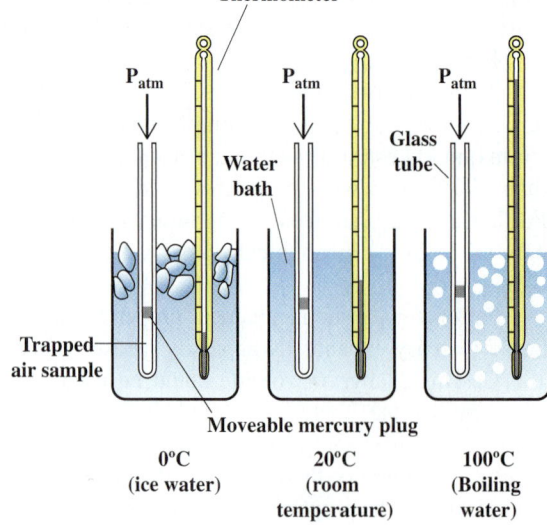

Figure 5-10 • To investigate Charles's law, a glass tube is sealed at one end and mounted vertically with the open end up. A small amount of mercury is placed down the opening to trap a bubble of gas. From then on, the pressure on the gas bubble remains equal to 760 mm Hg (atmospheric pressure) plus the height of the mercury "plug" above it. The temperature of the gas bubble is varied by immersing the tube in baths of various temperatures, and its volume is read from previously made calibration marks on the tube.

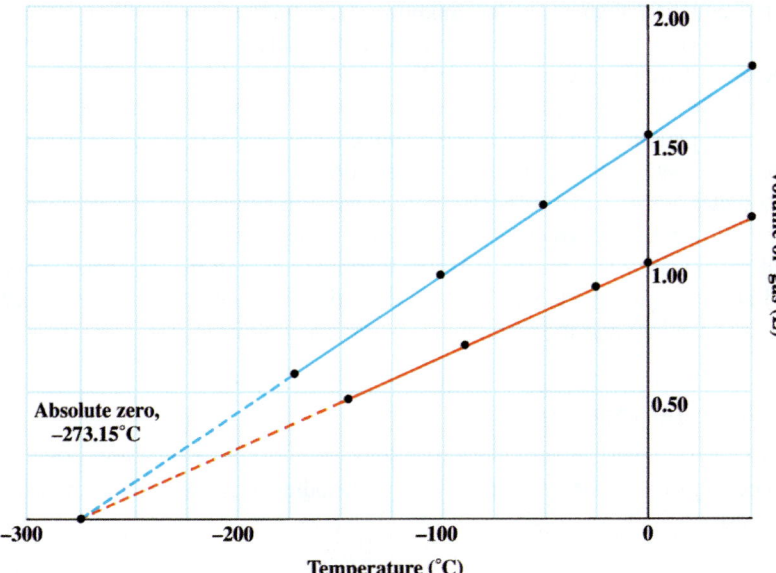

Figure 5–11 • The results of a Charles's law experiment for two samples of gas. The first (*red line*) has a volume of 1.0 L at a temperature of 0°C. It shrinks as it is cooled at constant pressure. The second (*blue line*) held at the same pressure takes up more volume at 0°C but shrinks faster as it is cooled. Its *percentage* change in volume is the same as that of the first sample for every degree of temperature change. Extrapolating the trends, as shown by the dashed lines, predicts that the volumes of both samples go to zero at a temperature of −273.15°C.

straight-line relationships have the mathematical form "$y = mx + b$," where the constant m is the slope of the line and the constant b is its y intercept (see Appendix C–2). Comparison shows that the equation for Charles's law is a straight-line relationship with V_0 equal to the y intercept and αV_0 equal to the slope. The observation that all gases expand by the same relative amount when heated between a set pair of temperatures at sufficiently low pressures now translates to saying that the constant α has the same value for all gases. In 1802, Gay-Lussac reported a value of $1/267$ $(°C)^{-1}$ for α. Subsequent experiments have refined this result to give $\alpha = 1/273.15$ $(°C)^{-1}$. Among liquids and solids, by contrast, the coefficient of thermal expansion varies widely from substance to substance.

Basing an Improved Temperature Scale on Charles's Law

The fact that all gases have the same coefficient of thermal expansion at low pressures suggests an improved way to measure temperature. Instead of relying on the thermal expansion of the single substance mercury, we can turn Charles's law around and use the thermal expansion of an entire *class* of substances, the gases, to define temperature (in degrees Celsius). We solve Charles's law for t:

$$t = \frac{V - V_0}{\alpha V_0} = \frac{\dfrac{V}{V_0} - 1}{\alpha} = \frac{1}{\alpha}\left(\frac{V}{V_0} - 1\right)$$

and substitute the observed value of α

$$t = 273.15°C\left(\frac{V}{V_0} - 1\right)$$

This equation underlies the operation of the gas thermometer. A measurement is made by noting the volume (V) of a low-pressure sample of a gas at the unknown temperature and also at the ice point (V_0) under the same low pressure. Substituting

• Gas thermometers are more than mere theoretical toys. They can give excellent practical results.

into the equation then gives the unknown temperature. Atmospheric pressure is sufficiently low for most measurements, but highly accurate determinations of temperature may require resorting to lower pressures.

Once we have defined temperature in this way, we can go back to our provisional scale of temperature and evaluate the *actual* changes in the volume of mercury with temperature. Suppose that a mercury thermometer is calibrated to match the gas thermometer at 0°C and 100°C and that the scale in between is divided evenly into 100 parts to mark off degrees. A temperature of exactly 40°C on the gas thermometer would read as 40.11°C on the mercury thermometer, and discrepancies would be found at other temperatures as well. The relation between the volume of mercury (and other liquids) and the temperature is *not* exactly linear.

Absolute Temperature

Charles's law suggests a very interesting lower limit to the temperatures that can be obtained physically. We cast the law in yet another form by solving the previous equation for V:

$$V = V_0\left(1 + \frac{t}{273.15°C}\right)$$

Negative temperatures on the Celsius scale correspond to temperatures below the freezing point of water and, of course, are physically meaningful. But what happens as t in this equation approaches $-273.15°C$? The term in parentheses on the right-hand side of the equation approaches zero, and the volume (V) approaches zero as well (Fig. 5–11). If t could go below $-273.15°C$, the volume would become negative, clearly an impossible result. We therefore surmise that $t = -273.15°C$ is a fundamental limit, a floor below which temperatures cannot be lowered. All real gases condense to a liquid or solid before they reach this **absolute zero** of temperature, but more rigorous arguments show that no substance (gas, liquid, or solid) can be cooled below $-273.15°C$. In fact, it becomes increasingly difficult to cool a substance as absolute zero is approached. The coldest temperature reached to date is within nanodegrees (1 nanodegree = $10^{-9}°C$) of absolute zero.

The absolute zero of temperature is a compellingly logical choice as the zero point of a temperature scale. The easiest way to create such a temperature scale is to add 273.15°C to the Celsius temperature. This gives the **Kelvin temperature scale:**

$$T\,(\text{Kelvin}) = 273.15 + t\,(\text{Celsius})$$

• The unit is not the "degree Kelvin," but simply the kelvin. It is named after William Thomson, Lord Kelvin, who first pointed out the existence of absolute zero.

The capital T signifies that this is an absolute scale. The unit on the Kelvin scale is named the "kelvin" (K). A kelvin is equal in size to a Celsius degree, but the Kelvin and Celsius scales have different zero points (Fig. 5–12). The formal conversion of a temperature of 25.00°C to the Kelvin scale therefore runs this way:

$$(25.00°C + 273.15°C) \times \left(\frac{1\,K}{1°C}\right) = 298.15\,K$$

Expressing the temperature on an absolute temperature scale simplifies the mathematical form of Charles's law considerably. It becomes

$$V \propto T \qquad \text{or} \qquad V = aT$$

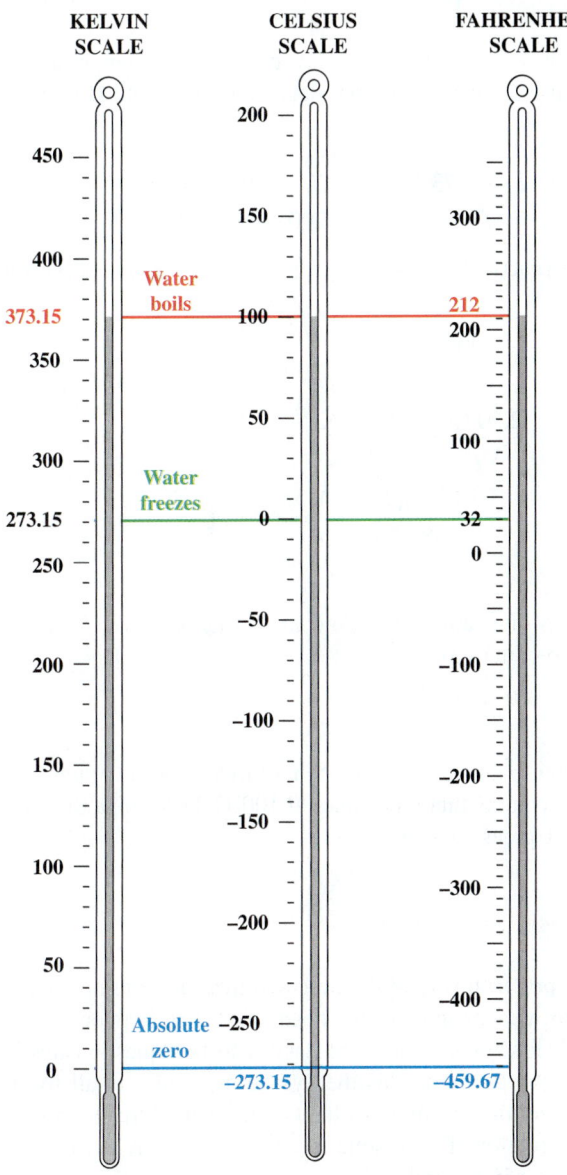

KELVIN SCALE **CELSIUS SCALE** **FAHRENHEIT SCALE**

Figure 5-12 • The Kelvin, the Celsius, and the Fahrenheit scales all find common use. The size of the degree is the same on the Kelvin and Celsius scales, but the zero points are offset by 273.15°C. Fahrenheit degrees are smaller by a factor of 5/9 than kelvins or Celsius degrees, so a 9°F change is counted for every 5°C or 5 K change in temperature. Also, the Fahrenheit zero is offset from the Celsius zero by 32°F.

where the proportionality constant, a, is determined by the pressure and the amount of gas present. Ratios of the volumes occupied at two different absolute temperatures by a fixed amount of gas at constant pressure have the simple and useful form

$$\frac{V_1}{V_2} = \frac{T_1}{T_2} \quad \text{(at a fixed pressure and for a fixed amount of gas)}$$

EXAMPLE 5-4

A scientist sets about to study the properties of gaseous hydrogen at low temperatures. She starts with a volume of 2.50 L of hydrogen at atmospheric pressure and a temperature of 25.00°C and cools the gas at constant pressure to −200.00°C. Predict the volume of the hydrogen at −200.00°C.

Strategy

Use Charles's law. The simplest form of Charles's law uses absolute temperatures. Therefore, start by converting temperatures from degrees Celsius into kelvins.

Solution

$$t_1 = 25.00°C \Rightarrow T_1 = 273.15 + 25.00 = 298.15 \text{ K}$$
$$t_2 = -200.00°C \Rightarrow T_2 = 273.15 - 200.00 = 73.15 \text{ K}$$

Write Charles's law in the practical "two-point" form, substitute, and solve for the unknown volume.

$$\frac{V_1}{V_2} = \frac{T_1}{T_2}$$
$$\frac{2.50 \text{ L}}{V_2} = \frac{298.15 \text{ K}}{73.15 \text{ K}}$$
$$V_2 = \frac{(73.15 \text{ K})(2.50 \text{ L})}{298.15 \text{ K}} = 0.613 \text{ L}$$

Check

The ratio V_1/V_2 should equal the ratio T_1/T_2. And in fact (2.50 L/0.613 L) and (298.15 K/73.15 K) both do equal 4.08.

EXERCISE

The gas in a gas thermometer that has been placed in a furnace has a volume that is 2.56 times larger than the volume that it occupies at 100°C. Determine the temperature in the furnace (in degrees Celsius).

Answer: 682°C.

Although we now have a practical way of defining and measuring temperature, we have come no closer to understanding its meaning. What happens on a microscopic scale when a substance is heated? The answer to this question lies in the **kinetic theory of matter.** According to this theory, the molecules in all forms of matter (gases, liquids, and solids) are in a constant state of motion; the higher the temperature, the faster the motion. This kinetic theory makes many predictions concerning the behavior of matter in general and of gases in particular. We consider some of them in Section 5–7.

5–4 The Ideal Gas Law

Boyle's law tells how the volume of a gas depends on the pressure if the temperature and amount of the gas do not change:

$$V \propto \frac{1}{P} \quad \text{(at constant temperature, fixed amount of gas)}$$

Charles's law tells how the volume of a gas depends on the temperature if the pressure and amount of the gas do not change:

$$V \propto T \quad \text{(at constant pressure, fixed amount of gas)}$$

where T is the absolute temperature (*not* the Celsius temperature). A third equation should relate the volume of a gas to its amount, at constant temperature and pressure. Under these conditions, the volume of a sample of *any* substance (whether solid, liquid, or gas) depends only on its amount. If this were not so, then substances would not have the constant densities that are observed. Therefore, we can certainly conclude that for a gas

$$V \propto n \qquad \text{(at constant temperature and pressure)}$$

where n is the chemical amount of the gas.

These three statements may be combined in the form

$$V \propto \frac{nT}{P}$$

We now introduce a constant of proportionality, which we label R:

$$V = R\frac{nT}{P} \qquad \text{or} \qquad PV = nRT$$

The highlighted equations are two forms of the **ideal gas law,** an empirical law that holds approximately for all gases near atmospheric pressure and that becomes increasingly accurate as the pressure is decreased.

Suppose a gas undergoes a change from some initial condition (described by P_1, V_1, T_1, and n_1) to a final condition (described by P_2, V_2, T_2, and n_2). We solve the ideal gas equation under the initial conditions for R:

$$R = \frac{P_1V_1}{n_1T_1}$$

and do the same for the equation under the final conditions

$$R = \frac{P_2V_2}{n_2T_2}$$

Because R is a constant, we conclude

$$\frac{P_1V_1}{n_1T_1} = \frac{P_2V_2}{n_2T_2}$$

a useful alternative form of the ideal gas law.

EXAMPLE 5-5

A weather balloon filled with helium has a volume of 1.0×10^4 L at 1.00 atm and 30°C. It ascends to an altitude at which the pressure is only 0.60 atm and the temperature is -20°C. What is the volume of the balloon after the ascent? Assume that the balloon stretches in such a way that the pressure inside stays close to the pressure outside.

Strategy

Start with the ideal gas law in the form

$$\frac{P_1V_1}{n_1T_1} = \frac{P_2V_2}{n_2T_2}$$

The initial state (case 1) is on the ground; the final state (case 2) is high in the air. Organize the information: P_1, P_2 and T_1, T_2 are known or readily computed, and V_1 is known. The final volume V_2 and the initial and final chemical amounts n_1 and n_2 are *not* known.

Solution

The chemical amount of helium inside the balloon is not known, but it stays constant during the ascent (no leaks). Therefore, n_1 equals n_2. Replace n_1 with n_2 in the preceding equation and solve for V_2. The n_2's cancel out, and the result is

$$V_2 = V_1\left(\frac{P_1}{P_2}\right)\left(\frac{T_2}{T_1}\right)$$

Temperatures must be expressed on the absolute scale. Here, $T_1 = 273.15 + 30 = 303$ K and $T_2 = 273.15 + (-20) = 253$ K. Inserting these T's, the two P's, and V_1 into the equation for V_2 gives

$$V_2 = 1.0 \times 10^4 \,\text{L}\left(\frac{1.00 \,\text{atm}}{0.60 \,\text{atm}}\right)\left(\frac{253 \,\text{K}}{303 \,\text{K}}\right) = 1.4 \times 10^4 \,\text{L}$$

Check

This result has units of volume, as it must. Lower pressure favors expansion; lower temperature favors contraction. In this case, expansion wins.

EXERCISE

At one point during its ascent, the same weather balloon has the same volume as when it was launched, although its temperature is $-10°C$. Compare the pressure at that point.

Answer: 0.868 atm.

• Memorizing the many useful forms of the ideal gas law is unwise. It is far better to understand how to derive what is needed for a given problem from the basic equation $PV = nRT$.

The approach outlined in Example 5–5 can be extended to cases in which other combinations of the four variables (P, V, T, and n) remain constant (Fig. 5–13). For example, when n and T remain constant, they can be canceled from the equation, and Boyle's law, which was used in solving Example 5–3, results. When n and P remain constant, this relationship reduces to Charles's law, used in Example 5–4.

We are not quite finished. We have shown that for a given gas, PV/nT is a constant, called R, but we have not yet established whether R is a *universal* constant or whether it changes from one gas to another (for example, from oxygen to nitrogen to hydrogen). To deal with this issue, we invoke Avogadro's hypothesis, now better called Avogadro's law because its validity was established by the success of the laws of chemical combination considered in Chapter 1. According to this law, fixed volumes of different gases at the same temperature and pressure contain the same number of molecules (and therefore of moles). In other words, for fixed V, P, and T, the value of n is fixed as well and is independent of the particular gas studied. Therefore, R is a **universal gas constant** that is the same for all gaseous substances.

The numerical value of R depends on the units chosen for P, V, n, and T. We use the Kelvin scale for T and moles for n. Experiments show that at the ice point ($T = 273.15$ K), the product PV for 1.0000 mol of gas approaches 22.414 L atm at low pressures. Hence,

Figure 5-13 • In a hot-air balloon, the volume remains nearly constant (the balloon is rigid) and the pressure is nearly constant as well (unless the balloon rises very high). Thus, n is inversely proportional to T: As the air inside the balloon is heated, its amount decreases and the density of the air in the balloon falls. This lower density gives the balloon its lift. *(Comstock)*

$$R = \frac{PV}{nT} = \frac{(22.414 \text{ L atm})}{(1.0000 \text{ mol})(273.15 \text{ K})} = 0.082057 \text{ L atm mol}^{-1} \text{ K}^{-1}$$

If P is expressed in the SI unit of the pascal (equivalent to 1 newton per square meter, N m^{-2}) and V is measured in cubic meters (m^3), then R has the value

$$R = \frac{(1.01325 \times 10^5 \text{ N m}^{-2})(22.414 \times 10^{-3} \text{ m}^3)}{(1.0000 \text{ mol})(273.15 \text{ K})} = 8.3145 \text{ N m mol}^{-1} \text{ K}^{-1}$$

One newton-meter (1 N m) is the work done by a force of one newton acting through a distance of one meter and is defined as one *joule* (the SI unit of energy, or work, abbreviated J) (see Appendix B–1). The gas constant is then

$$R = 8.3145 \text{ J mol}^{-1} \text{ K}^{-1}$$

EXAMPLE 5-6

What mass of helium is needed to fill the weather balloon from Example 5–5?

Strategy

Solve the ideal gas equation for n. Insert R along with the given values of temperature (on the Kelvin scale), volume and pressure to obtain the number of moles of helium. Convert this answer from moles to grams.

Solution

Solving $PV = nRT$ for n gives

$$n = \frac{PV}{RT}$$

Because P is given in atmospheres and V in liters, use $R = 0.08206$ L atm $\text{mol}^{-1}\,\text{K}^{-1}$ to cause the units to cancel:

$$n_{\text{He}} = \frac{(1.00\;\cancel{\text{atm}})(1.00 \times 10^4\;\cancel{\text{L}})}{(0.08206\;\cancel{\text{L}}\;\cancel{\text{atm}}\;\text{mol}^{-1}\;\cancel{\text{K}^{-1}})(303.15\;\cancel{\text{K}})} = 402\;\text{mol}$$

The molar mass of helium is 4.00 g mol^{-1}; hence, the mass of helium required is

$$m_{\text{He}} = 402\;\cancel{\text{mol He}} \times \left(\frac{4.00\;\text{g He}}{1\;\cancel{\text{mol He}}}\right) = \boxed{1610\;\text{g He}}$$

Check

The conditions are fairly close to standard temperature and pressure (STP) at which one mole of gas occupies 22.4 L. The number of moles of helium contained in 10^4 L at STP is

$$n_{\text{He}} = 10^4\;\cancel{\text{L}} \times \frac{1\;\text{mol}}{22.4\;\cancel{\text{L}}} = 450\;\text{mol}$$

which is fairly close to 402 mol.

EXERCISE

The weather balloon in Example 5–5 is filled with H_2 instead of He. Compute the mass of H_2 needed to fill it to the same volume under the same conditions.

Answer: 810 g.

Gas Density and Molar Mass

Example 5–6 shows how the mass m of a sample of a gas can be computed using the ideal gas law and given the temperature, pressure, volume, and the molar mass M of the gas. Using the fact that $n = m/M$, the ideal gas law can be rewritten as

$$PV = \frac{m}{M}RT$$

Rearranging this equation gives

$$\frac{m}{V} = \frac{P}{RT}M \qquad \text{or} \qquad d = \frac{P}{RT}M$$

because the density d of any matter equals its mass m divided by its volume V. The following example illustrates the calculation of gas densities.

EXAMPLE 5–7

Calculate the density of gaseous carbon tetrafluoride, CF_4, at room pressure (1.00 atm) and 50.0°C.

Strategy

Compute the molar mass of CF_4 and insert it, together with the pressure and absolute temperature, into the equation for the density of the gas. Choose R in units that ensure that pressure and temperature cancel out.

Carbon tetrafluoride, CF_4, is an exceptionally stable gas that condenses to a liquid at −128.5°C.

Solution

The molar mass of CF_4 is 88.00 g mol^{-1}, based on 12.01 g mol^{-1} for 1 C plus 4(18.998) g mol^{-1} for 4 F's. The absolute temperature is $T = 273.15 + 50.0 = 323.15$ K. Because the gas pressure is given in atmospheres, the easiest value of R to use is in L atm mol^{-1} K^{-1}. Then

$$d = \frac{P}{RT}M$$

$$= \left(\frac{1.00 \text{ atm}}{0.08206 \text{ L atm mol}^{-1}\text{ K}^{-1} \times 323.15 \text{ K}}\right)\frac{88.00 \text{ g CF}_4}{1 \text{ mol CF}_4} = 3.32 \text{ g L}^{-1}$$

Check

One mole of $CF_4(g)$ would occupy 22.4 L at STP. In this problem, P is 1.00 atm, the standard pressure, but T, at 50.0°C, exceeds the standard temperature. The volume of one mole of $CF_4(g)$ accordingly exceeds 22.4 L in proportion:

$$V_{CF_4} = 22.4 \text{ L} \times \frac{(273.15 + 50.0) \text{ K}}{273.15 \text{ K}} = 26.5 \text{ L}$$

The mass of one mole of CF_4 is 88.00 g. Therefore, $d = (88.00 \text{ g}/26.5 \text{ L}) = 3.32$ g L^{-1}, which is the same answer.

EXERCISE

Calculate the density of gaseous hydrogen at a pressure of 1.32 atm and a temperature of −45.0°C.

Answer: 0.142 g L^{-1}.

A rearrangement of the equation that was used in the preceding example gives the molar mass M of a gas in terms of its density d:

$$M = d\frac{RT}{P}$$

Recall from Section 1-8 that the molar mass M is just the fact needed to figure out a substance's molecular formula from its empirical formula. For years, measurements of gas density were exceedingly important in chemistry because they provided a route to the molar masses of newly synthesized substances. The method works whenever a substance is already gaseous at room conditions or can be vaporized (without decomposition) by heating or by reduction of the pressure or both. Gas densities are easily measured across a range of pressures and temperatures. A correct molecular formula does not require an exactly correct density, as illustrated in the following example.

EXAMPLE 5-8

The welding fuel in Example 1-11 had the empirical formula CH. A volume of 10.0 L of this gas at a pressure of 0.986 atm and a temperature of 65.0°C is found to have a mass of 8.9 g. Determine the approximate molar mass of the gas and assign its molecular formula.

Figure 5-14 • An oxy-acetylene torch is used to weld metals because of the high temperature of the flame produced by burning acetylene in oxygen. *(Charles D. Winters)*

Solution

Convert the temperature to kelvins and substitute it and the other data into the equation

$$M = d\frac{RT}{P}$$

$$M = \left(\frac{8.9 \text{ g}}{10.0 \text{ L}}\right)\left(\frac{0.08206 \text{ L atm mol}^{-1} \text{ K}^{-1} \times 338.15 \text{ K}}{0.986 \text{ atm}}\right) = 25 \text{ g mol}^{-1}$$

The formula unit CH has a molar mass of 13.02 g mol^{-1}. The answer is fairly close to two times 13.02 g mol^{-1}. The molecule consequently contains two CH units for a molecular formula of C_2H_2.

Check

The discrepancy between 25 g mol^{-1} and 26.04 g mol^{-1}, which is the correct molar mass of C_2H_2, does not hinder the conclusion. Other possible molecular formulas, such as CH or C_3H_3, have molar masses that differ grossly from 25 g mol^{-1}. The compound is in fact acetylene (Fig. 5–14).

EXERCISE

Fluorocarbons are compounds of fluorine and carbon. A 45.60 g sample of a gaseous fluorocarbon contains 7.94 g of carbon and 37.66 g of fluorine and occupies 7.40 L at STP ($P = 1.00$ atm and $T = 273.15$ K). Determine the approximate molar mass of the fluorocarbon and give its molecular formula.

Answer: $M = 138$ g mol^{-1}; C_2F_6.

5-5 Chemical Calculations for Gases

Judicious use of the gas laws allows chemists to measure the volumes of gases in chemical reactions instead of their masses. Volumes of gases are easier to measure than masses, so this is a useful technique. The required concurrent measurement of temperature and pressure adds no trouble. This application of the gas laws is illustrated in the following example.

EXAMPLE 5-9

A key step in the production of sulfuric acid is the oxidation of sulfur dioxide:

$$2 \text{ SO}_2(g) + \text{O}_2(g) \longrightarrow 2 \text{ SO}_3(g)$$

Suppose that 6.37 L of O_2 (measured at 420°C and a pressure of 1.22 atm) reacts completely with an excess of SO_2. The SO_3 produced is cooled to 80.0°C at $P = 0.970$ atm. Calculate the volume of the SO_3.

Strategy

Use the ideal gas law to get from V_{O_2}, the volume of O_2, to n_{O_2}, the chemical amount of O_2. Determine n_{SO_3} using the 2 : 1 ratio of coefficients in the balanced equation. Use the ideal gas law a second time to get from n_{SO_3} to V_{SO_3}. The volumes of $O_2(g)$ and $SO_3(g)$ are *not* in the ratio of the substances' coefficients in the balanced equation (as are gas volumes in Example 2–5) because the conditions of temperature and pressure change.

Solution

$$n_{O_2} = \frac{PV_{O_2}}{RT} = \frac{1.22 \text{ atm} \times 6.37 \text{ L}}{0.08206 \text{ L atm mol}^{-1} \text{ K}^{-1} \times (420 + 273.15) \text{ K}} = 0.1366 \text{ mol O}_2$$

The balanced equation shows 2 mol of SO_3 out per 1 mol O_2 in. Accordingly,

$$n_{SO_3} = 0.1366 \text{ mol O}_2 \times \left(\frac{2 \text{ mol SO}_3}{1 \text{ mol O}_2}\right) = 0.2732 \text{ mol SO}_3$$

Solve the ideal gas law for V and use it again

$$V_{SO_3} = \frac{n_{SO_3}RT}{P} =$$

$$\frac{0.2732 \text{ mol} \times 0.08206 \text{ L atm mol}^{-1} \text{ K}^{-1} \times (80.0 + 273.15) \text{ K}}{0.970 \text{ atm}} = 8.16 \text{ L}$$

Check

Write one ideal gas equation for the SO_3 and a second for the O_2, using subscripts to keep the variables straight. Divide the first by the second to obtain

$$\frac{P_{SO_3}V_{SO_3}}{P_{O_2}V_{O_2}} = \frac{n_{SO_3}RT_{SO_3}}{n_{O_2}RT_{O_2}}$$

But $n_{SO_3} = 2n_{O_2}$. Insert this fact and solve for V_{SO_3}. Then insert the values of the different variables:

$$V_{SO_3} = \frac{2n_{O_2}}{n_{O_2}}\left(\frac{P_{O_2}}{P_{SO_3}}\right)\left(\frac{T_{SO_3}}{T_{O_2}}\right)V_{O_2}$$

$$= 2\left(\frac{1.22 \text{ atm}}{0.97 \text{ atm}}\right)\left(\frac{353.15 \text{ K}}{693.15 \text{ K}}\right)6.37 \text{ L} = 8.16 \text{ L}$$

Note that R canceled out. It is *not* necessary to have a numerical value for R to solve the problem.

EXERCISE

Ethylene burns in oxygen:

$$C_2H_4(g) + 3 O_2(g) \longrightarrow 2 CO_2(g) + 2 H_2O(g)$$

A volume of 3.51 L of $C_2H_4(g)$ at a temperature of 25°C and a pressure of 4.63 atm reacts completely with $O_2(g)$. The water vapor is collected at a temperature of 130°C and a pressure of 0.955 atm. Calculate the volume of the water vapor.

Answer: $V_{H_2O} = 46.0$ L.

The convenience of measuring volumes of gases instead of masses is available even when some of the reactants and products in a reaction are solids or liquids. This is shown in the next example.

Figure 5–15 • When concentrated nitric acid is added to a copper penny, the copper is oxidized and an aqueous solution of copper(II) nitrate is formed. In addition, some of the nitrate ion is reduced to brown gaseous nitrogen dioxide, which bubbles off. (*Charles D. Winters*)

EXAMPLE 5–10

Copper reacts with concentrated nitric acid to give gaseous nitrogen dioxide and dissolved copper ions (Fig. 5–15). The balanced net ionic equation is

$$Cu(s) + 4\,H^+(aq) + 2\,NO_3^-(aq) \longrightarrow 2\,NO_2(g) + Cu^{2+}(aq) + 2\,H_2O(\ell)$$

A total of 6.80 g of copper is consumed according to this equation. All of the NO_2 is collected at a pressure of 0.970 atm and a temperature of 45°C. Calculate the volume that the NO_2 occupies.

Strategy

Start as in a "mass-to-mass" stoichiometry calculation (flow-charted in Fig. 2–3), but at the last stage use the ideal gas law to obtain the volume of the gaseous product instead of its mass.

Solution

Convert from the mass of copper to its chemical amount by dividing by the molar mass:

$$n_{Cu} = 6.80\text{ g Cu} \times \left(\frac{1\text{ mol Cu}}{63.55\text{ g Cu}}\right) = 0.107\text{ mol Cu}$$

Form the unit factor "2 mol of NO_2 per 1 mol of Cu" from the balanced equation and multiply n_{Cu} by this factor:

$$n_{NO_2} = 0.107\text{ mol Cu} \times \left(\frac{2\text{ mol }NO_2}{1\text{ mol Cu}}\right) = 0.214\text{ mol }NO_2$$

The ideal gas law gives the volume of $NO_2(g)$ under the stated conditions of temperature and pressure:

$$V_{NO_2} = \frac{n_{NO_2}RT}{P}$$

$$= \frac{(0.214\text{ mol})(0.08206\text{ L atm mol}^{-1}\text{ K}^{-1})(273.15 + 45)\text{ K}}{0.970\text{ atm}} = 5.76\text{ L}$$

Check

At STP, 0.214 mol of gas would occupy 4.79 L (equal to 0.214 mol × 22.4 L mol^{-1}). The volume of the NO_2 in the problem was measured under slightly different conditions: 0.970 atm and 318 K. Correct for the change from STP as follows:

$$V_{NO_2} = 4.79\text{ L} \times \frac{1.00\text{ atm}}{0.970\text{ atm}} \times \frac{(273.15 + 45)\text{ K}}{273.15\text{ K}} = 5.76\text{ L}$$

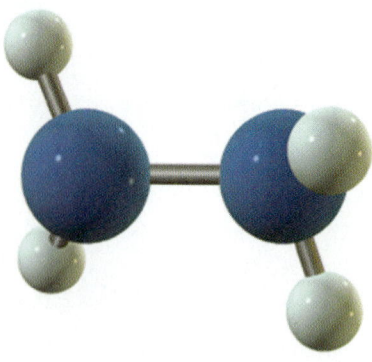

Hydrazine (N_2H_4) is a colorless liquid that boils at 113.5°C and freezes at 2°C. In addition to its use as a rocket fuel, it is a versatile reducing agent. It is used extensively to remove oxygen from boiler water, giving water and nitrogen as products:

$$N_2H_4(aq) + O_2(aq) \longrightarrow$$
$$2\,H_2O(\ell) + N_2(g)$$

EXERCISE

Hydrazine (N_2H_4), a rocket fuel, is prepared by the reaction of ammonia with a solution of sodium hypochlorite:

$$2\,NH_3(g) + NaOCl(aq) \longrightarrow N_2H_4(aq) + NaCl(aq) + H_2O(\ell)$$

What volume of gaseous ammonia at a temperature of 10°C and a pressure of 3.63 atm is required to produce 15.0 kg of hydrazine according to this equation?

Answer: 5.99 × 10³ L.

5-6   Mixtures of Gases

Mixtures of two or more non-reacting gaseous substances come up frequently in chemistry; for example, when a reaction generates more than one gaseous product. The **partial pressure** of a gas in a mixture is defined as the pressure that it would exert if it were the only gas present in the same container at the same temperature. Partial pressures are indicated by subscripting the symbol P with a suitable formula. Thus, P_{O_2} indicates the partial pressure of oxygen. **Dalton's law of partial pressures** relates partial pressures and the total pressure:

• This is the same Dalton who proposed the atomic theory.

> The total pressure of a mixture of gases equals the sum of the partial pressures of the individual gases.

Dalton's law holds under the same conditions as the ideal gas law: It is approximate at moderate pressures and becomes increasingly accurate as the pressure is lowered. A mixture of gases that follows Dalton's law is called an **ideal mixture.**

At low pressures, each gas independently obeys the ideal gas law, so that for a mixture of two gases A and B in a container of volume V

$$P_A V = n_A RT$$
$$P_B V = n_B RT$$

The total pressure is

$$P_{tot} = P_A + P_B = \frac{n_A RT}{V} + \frac{n_B RT}{V} = (n_A + n_B)\frac{RT}{V}$$

The total number of moles (n_{tot}) equals $n_A + n_B$. Recognizing this gives

$$P_{tot} = n_{tot}\frac{RT}{V}$$

This equation means that the ideal gas law applies to ideal mixtures (and not just pure gases), provided that the *total* chemical amount (n_{tot}) is used.

• The best example of a mixture of gases is the air. See Table 5–1.

Mole Fractions and Partial Pressures

The **mole fraction X** of a component in a mixture is defined as the number of moles of the component that are in the mixture divided by the total number of moles present. For a mixture of two components A and B

$$\text{mole fraction of A} = X_A = \frac{n_A}{n_{tot}} = \frac{n_A}{n_A + n_B}$$

and

$$\text{mole fraction of B} = X_B = \frac{n_B}{n_{tot}} = \frac{n_B}{n_A + n_B}$$

Mole fractions have no units (the units "mol" in the numerator and denominator cancel out), and their values always lie between 0 and 1. A **mole percentage** of a component simply equals a mole fraction multiplied by 100% (put on a basis of 100). The total of the mole fractions of all the components of a mixture equals 1, and the total of the mole percentages equals 100%. The mole fraction of component A in a mixture equals the fraction of all of the molecules in the mixture that are molecules of A. This is easily proved by multiplying both numerator and denominator of the first fraction in the preceding equations by Avogadro's number, which converts them from moles to molecules.

For a mixture of gases that follows the ideal gas law and includes gas A as a component

$$P_A = n_A \frac{RT}{V} \qquad \text{and} \qquad P_{tot} = n_{tot} \frac{RT}{V}$$

Dividing the first equation by the second gives

$$\frac{P_A}{P_{tot}} = \frac{n_A}{n_{tot}} \qquad \text{or} \qquad P_A = \frac{n_A}{n_{tot}} P_{tot}$$

Recognizing $\dfrac{n_A}{n_{tot}}$ as the mole fraction of A gives

$$\boxed{P_A = X_A P_{tot}}$$

The partial pressure of a single component in a mixture of ideal gases equals the total pressure multiplied by the mole fraction of that component.

EXAMPLE 5–11

Some sulfur is burned in excess oxygen (this reaction is used in the manufacture of sulfuric acid). The gaseous mixture that results contains 23.2 g of $O_2(g)$, 53.1 g of $SO_2(g)$, and no other gases. Its total pressure equals 2.13 atm. Calculate the partial pressure of $SO_2(g)$.

Strategy

Obtain the chemical amounts of $O_2(g)$ and $SO_2(g)$ in the mixture (using their molar masses) and use them to compute the mole fraction of $SO_2(g)$. The partial pressure of $SO_2(g)$ equals the total pressure multiplied by its mole fraction.

Solution

$$n_{O_2} = 23.2 \text{ g } O_2 \times \left(\frac{1 \text{ mol } O_2}{32.00 \text{ g } O_2} \right) = 0.725 \text{ mol } O_2$$

$$n_{SO_2} = 53.1 \text{ g } SO_2 \times \left(\frac{1 \text{ mol } SO_2}{64.06 \text{ g } SO_2} \right) = 0.829 \text{ mol } SO_2$$

$$X_{SO_2} = \frac{0.829 \text{ mol}}{(0.725 + 0.829) \text{ mol}} = 0.533$$

$$P_{SO_2} = X_{SO_2} P_{tot} = 0.533 \times 2.13 \text{ atm} = 1.14 \text{ atm}$$

Check

Slightly more than half of the particles in the mixture are molecules of sulfur dioxide. Therefore, the partial pressure of sulfur dioxide is slightly more than half the total pressure.

EXERCISE

A solid hydrocarbon is burned in air in a closed container, producing a mixture of gases having a total pressure of 3.34 atm. Analysis of the mixture shows it to contain 0.340 g of water vapor, 0.792 g of carbon dioxide, 0.288 g of oxygen,

3.790 g of nitrogen, and no other gases. Calculate the mole fraction and partial pressure of carbon dioxide in this mixture.

Answer: $X_{CO_2} = 0.0993$; $P_{CO_2} = 0.332$ atm.

5-7   The Kinetic Theory of Gases

The ideal gas law summarizes certain physical properties of gases at low pressures. It is an empirical law, the consequence of experimental observations, but its simplicity and generality make us ask whether it has some theoretical explanation that involves the properties of the atoms and molecules that make up gases. The best theory would predict other properties of gases in a natural way and explain why real gases deviate from the ideal gas law to a small but measurable extent. A very successful theory along these lines was developed in the 19th century, most notably by the physicists Rudolf Clausius, James Clerk Maxwell, and Ludwig Boltzmann, who worked in Germany, Scotland, and Austria, respectively. The kinetic theory of gases is one of the great milestones of science, and its success provides strong evidence for the atomic theory of matter discussed in Chapter 1. The postulates of the kinetic theory of gases are very simple:

1. A pure gas consists of a large number of identical molecules separated by distances that are large compared with their size.
2. The molecules of a gas are constantly moving in random directions (Fig. 5–16) with a distribution of speeds.
3. The molecules of a gas exert no forces on one another except during collisions, so that between collisions they move in straight lines with constant velocities.
4. The collisions of the molecules with each other and with the walls of the container are elastic; no energy is lost during a collision.

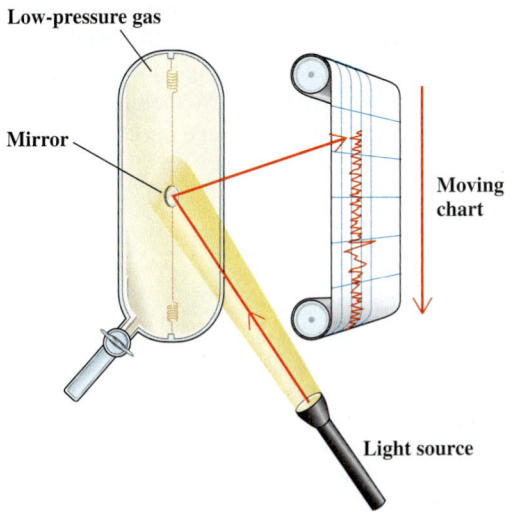

Low-pressure gas

Mirror

Moving chart

Light source

Figure 5–16 • A visible demonstration of the existence and random motion of molecules in a gas. A small mirror is suspended from a very thin flexible fiber inside a glass container of a gas at low pressure. A spot of light shines on the mirror and is reflected at an angle. Even very small motions of the mirror are detectable as movements of the reflected spot on a distant screen. The graph shows how the regular twisting movement of the mirror is perturbed by the random impact of gas molecules. If the gas is completely removed from the container, these perturbations stop.

Temperature and Molecular Motion

We now use the kinetic theory of gases to establish a relationship between the pressure and volume of an ideal gas and the mass and speed of its molecules as they fly about in their postulated random motion. Comparing this relationship with the ideal gas law ($PV = nRT$) then provides insight into the meaning of temperature. Because the goal is not mathematical rigor but physical understanding, we use arguments based on proportionality, without worrying about the specific numerical constants that appear in the equations.

We begin by recalling that the pressure of a gas is the total force exerted by its molecules on its containing walls divided by the area of the walls. In the kinetic theory, this force results from a ceaseless series of abrupt *impulses* as molecules bombard the container walls. The push (force) that a molecule exerts on a wall during its collision starts from zero, rises to a peak, and then falls back to zero a short time later as the molecule bounces away. The impulse per collision is proportional to the *momentum* ($m \times u$) carried by the molecule, where m is the mass of the molecule and u is its speed. The heavier the molecule and the faster it is moving, the greater its impulse when it strikes the wall.

The collision rate (the number of collisions per unit time) under most conditions is so large that the effect becomes that of a smooth push on the walls. We reason, then, that the pressure is proportional to the impulse per collision multiplied by the rate of collisions of molecules with the walls:

pressure $\propto$ (impulse per collision) $\times$ (rate of collisions with the walls)

We must now relate the rate of collision of molecules with the walls to properties of the molecules. First, the rate of collision must be proportional to the number of molecules per unit volume (N/V) because increasing the number of molecules without changing the volume of the container would certainly increase the number of wall collisions in direct proportion. Next, the rate of collision must be proportional to the speed of the molecules (u) because the faster the molecules move, the more often they collide with the wall. We can put all of this together (Fig. 5–17) and conclude that

$$P \propto (m \times u) \times \left[\left(\frac{N}{V} \right) \times u \right] \qquad \text{which is equivalent to} \qquad PV \propto Nmu^2$$

We have referred to the speed of the molecules as u, but actually there is no reason to believe that all molecules have the *same* speed. In fact, the second postulate of the kinetic theory refers to a distribution of speeds. The preceding equation applies then to the whole collection of molecules in an *average* sense. We therefore replace u^2 by the average of the squares of the speeds, the **mean-square speed** of all the molecules, denoted with an overbar: $\overline{u^2}$. The equation becomes

$$PV \propto Nm\overline{u^2}$$

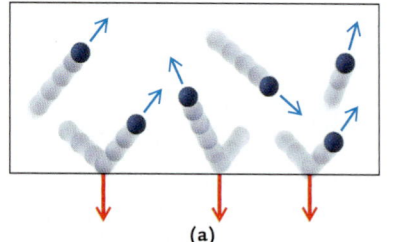

(a)

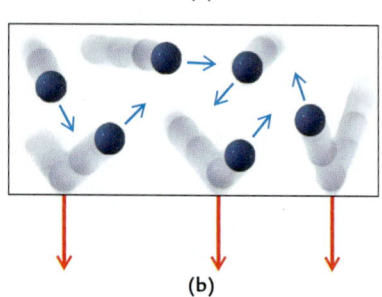

(b)

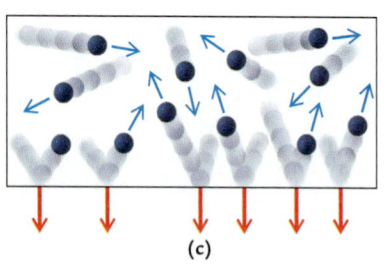

(c)

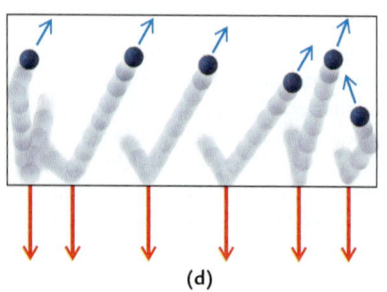

(d)

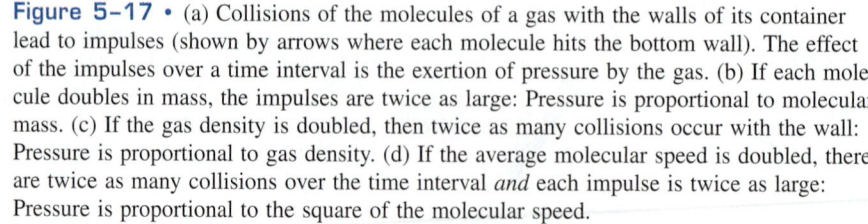

Figure 5–17 • (a) Collisions of the molecules of a gas with the walls of its container lead to impulses (shown by arrows where each molecule hits the bottom wall). The effect of the impulses over a time interval is the exertion of pressure by the gas. (b) If each molecule doubles in mass, the impulses are twice as large: Pressure is proportional to molecular mass. (c) If the gas density is doubled, then twice as many collisions occur with the wall: Pressure is proportional to gas density. (d) If the average molecular speed is doubled, there are twice as many collisions over the time interval *and* each impulse is twice as large: Pressure is proportional to the square of the molecular speed.

A more detailed argument shows that the proportionality constant in this relationship is $\frac{1}{3}$, so that

$$PV = \tfrac{1}{3}Nm\overline{u^2}$$

This equation is the desired relationship between pressure and volume of a gas and the motion of its molecules. Comparison to the ideal gas law

$$PV = nRT$$

allows us to write

$$\tfrac{1}{3}Nm\overline{u^2} = nRT$$

which has considerable interest—it links the speeds of the molecules of a gas and the temperature of the gas. We can tidy things a little, however. The equation has the number of molecules (N) on the left and the number of moles (n) on the right, but N is just n multiplied by Avogadro's number (N_0). Substituting nN_0 for N and simplifying gives

$$\tfrac{1}{3}N_0m\overline{u^2} = RT$$

Let us look at this result in two different ways. First, the **kinetic energy** of a particle of mass m moving at velocity u is defined in physics as $\frac{1}{2}mu^2$. Consequently, the *average* kinetic energy of N_0 molecules (one mole) is $\frac{1}{2}N_0m\overline{u^2}$. This is exactly the left side of our equation, but with the factor $\frac{1}{2}$ instead of $\frac{1}{3}$:

$$\text{average kinetic energy per mole} = \tfrac{1}{2}N_0m\overline{u^2} = \tfrac{3}{2} \times (\tfrac{1}{3}N_0m\overline{u^2}) = \tfrac{3}{2}RT$$

The outcome is that the average kinetic energy of the molecules of a gas depends only on the temperature and is independent of the mass of the molecules or the gas density.

A second way to look at the equation is to recall that if m is the mass of a single molecule, then N_0m is the mass of one mole of molecules: the molar mass M. We can then solve the equation for the mean-square speed:

$$\overline{u^2} = \frac{3RT}{M}$$

The higher the temperature and the less massive the molecules, the greater is the mean-square speed.

Distribution of Molecular Speeds

We define the **root-mean-square speed** (u_{rms}) as the square root of the mean-square speed $3RT/M$:

$$u_{\text{rms}} = \sqrt{\overline{u^2}} = \sqrt{\frac{3RT}{M}}$$

This equation can be used successfully only if all the terms in it are expressed in a coherent set of units such as the SI units. The value for R in SI units is

$$R = 8.3145 \text{ J mol}^{-1}\text{ K}^{-1}$$

Recall that in terms of the base units of the SI (see Appendix B),

$$1 \text{ J} = 1 \text{ kg (m/s)}^2 = 1 \text{ kg m}^2 \text{ s}^{-2}$$

where s is the abbreviation for second. R then has units of $kg\ m^2\ s^{-2}\ mol^{-1}\ K^{-1}$ when expressed in SI base units. We usually prefer to express molar masses in grams per mole; here, we must convert to kilograms per mole before using the equation:

$$\frac{kg}{mol} = \frac{g}{mol} \times \left(\frac{1\ kg}{1000\ g}\right)$$

By putting all of the units into the equation for u_{rms}, we can confirm that the final result comes out in the expected SI unit for a speed, which is meters per second. This is illustrated in the following example.

EXAMPLE 5–12

Calculate u_{rms} for (a) a helium atom and (b) an oxygen molecule at 298 K.

Strategy

Use the formula for the root-mean-square speed that was just derived. The factor $3RT$ appears in both expressions for u_{rms}. Calculate it first. Then combine it with the molar masses of the two gases, taking care with the units.

Solution

$$3RT = (3)\left(8.3145\frac{kg\ m^2}{s^2\ mol\ K}\right)(298\ K) = 7.43 \times 10^3\frac{kg\ m^2}{s^2\ mol}$$

The molar masses of helium and oxygen are 4.00 g mol^{-1} and 32.00 g mol^{-1}, respectively. Convert these to 4.00×10^{-3} kg mol^{-1} and 32.00×10^{-3} kg mol^{-1}, respectively, to ensure proper cancellation of units in the formula for u_{rms}.

$$u_{rms}(He) = \sqrt{\frac{7.43 \times 10^3\frac{kg\ m^2}{s^2\ mol}}{4.00 \times 10^{-3}\frac{kg}{mol}}} = 1363\ m\ s^{-1} = 1.36 \times 10^3\ m\ s^{-1}$$

$$u_{rms}(O_2) = \sqrt{\frac{7.43 \times 10^3\frac{kg\ m^2}{s^2 mol}}{32.00 \times 10^{-3}\frac{kg}{mol}}} = 482\ m\ s^{-1}$$

Check

The ratio of the root-mean-square speeds is $(1363/482) = 2.83$, which equals $\sqrt{32/4}$, as it should. The lighter He molecules move faster, which compensates for their lesser mass and gives them the same average kinetic energy as the heavier O_2 molecules. The root-mean-square speeds are both large—3050 and 1080 miles per hour, respectively.

EXERCISE

At a certain temperature, the root-mean-square speed of the molecules of H_2 in a sample of gas is 1055 m s^{-1}. Compute the root-mean-square speed of molecules of O_2 at the same temperature.

Answer: $u_{rms}(O_2) = 264.8$ m s^{-1}.

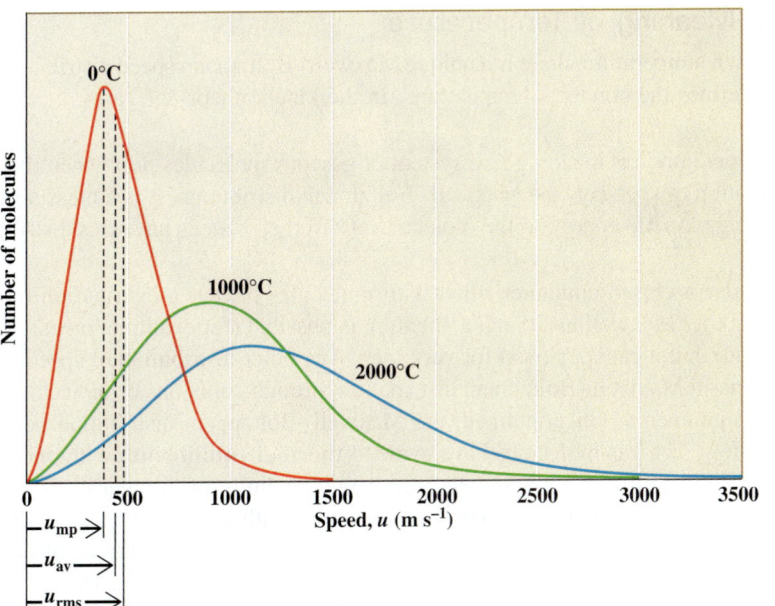

Figure 5–18 • The Maxwell–Boltzmann distribution of molecular speeds in nitrogen at three temperatures. The peak in each curve gives the most probable speed (u_{mp}), which is slightly smaller than the root-mean-square speed (u_{rms}). The average speed (u_{av}) (obtained simply by adding the speeds and dividing by the total number of molecules in the sample) lies in between. All three measures give comparable estimates of typical molecular speeds. All three increase with increasing temperature.

The root-mean-square speed gives some idea of the typical speed of molecules in a gas, but averages of any type can be deceptive. We need to know how many molecules in the gas have each speed. Figure 5–18 shows this for nitrogen at three temperatures. Such plots are called **speed distribution curves.** They indicate that the number of very fast and very slow molecules at all three temperatures is small and that the most probable speed is an intermediate value. They also indicate that higher speeds are more likely at higher temperatures.

The shape of the speed distribution curve is similar for all gases. The mathematical function that describes the curve was independently deduced from kinetic theory by Maxwell (1860) and Boltzmann (1872) and is called the **Maxwell–Boltzmann speed distribution.** It was confirmed by direct experiment in 1955 using an apparatus similar to the one shown in Figure 5–19.

• Remember the story of the man who drowned in a lake having an average depth of one foot.

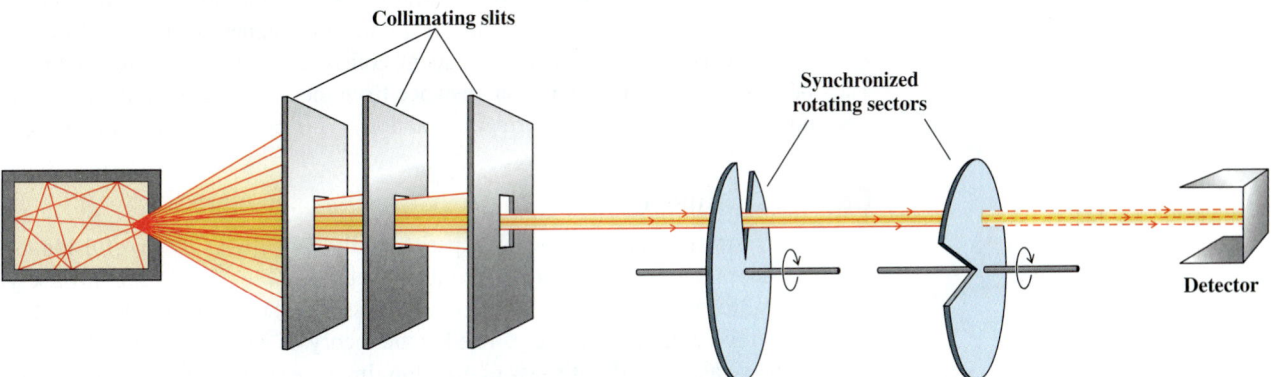

Figure 5–19 • This device can be used to measure the distribution of molecular speeds experimentally. Only those molecules with the correct speed to pass through *both* rotating sectors reach the detector, where their numbers can be counted. By changing the rate of rotation of the sectors, one can determine speed distribution.

The Meaning of Temperature

For each temperature, there is a unique Maxwell–Boltzmann speed distribution curve that defines the concept "temperature" in the kinetic theory of gases.

> Temperature has meaning in a system of gaseous molecules only when their distribution of speeds is the Maxwell–Boltzmann distribution. It is a measure of the average kinetic energy of the molecules when their speeds have this distribution.

Consider a closed container filled with molecules having a distribution of speeds that is not "Maxwellian." Such a situation is possible (for example, just after an explosion), but it cannot persist for very long. Any other distribution of speeds quickly becomes a Maxwell–Boltzmann distribution through collisions of molecules, which exchange energy. Once attained, the Maxwell–Boltzmann distribution persists indefinitely. The gas molecules have come to **thermal equilibrium** with one another, and a temperature can be defined only if the condition of thermal equilibrium exists. Finally, we cannot associate a temperature with a single molecule, or even a group of 10 or 20 molecules. Temperature is a property of collections of molecules having enough members to give meaning to a distribution of speeds.

Mean Free Path and Diffusion

The root-mean-square speeds of molecules in gases are high at ordinary temperatures, on the order of hundreds or thousands of meters per second (see Example 5–12). Suppose an odorous gas is released in one corner of a room. If the molecules followed straight-line trajectories, the odor would reach other points in the room almost instantly. Instead, it takes some time before the smell arrives. Why does it take so long for the molecules of odorant to spread out through the large volume?

The explanation is that molecules undergo many collisions with one another, even under conditions at which the gas follows the ideal gas law quite closely. A molecule in a typical gas at atmospheric pressure experiences about 10^{10} collisions per second; in other words, molecules move only for a time of about 10^{-10} second between collisions. We define the **mean free path** as the average distance traveled between collisions. It is the average speed multiplied by the average time between collisions, and it is on the order of 10^{-7} m (10^{-5} cm). Although this is large compared with the size of the molecules (several times 10^{-8} cm, as shown in Section 1–9), it is small compared with the size of a typical container. Molecules follow a zig-zag course (Fig. 5–20) as their frequent collisions incessantly redirect them. They take far longer to travel a given distance from their starting point than if there were no collisions. This type of irregular motion of molecules is called **diffusion.**

- The average time between repeating events (such as collisions) is the reciprocal of their average frequency. If a bell sounds 20 times per hour, the average duration between rings is 1/20 hour (3 minutes).

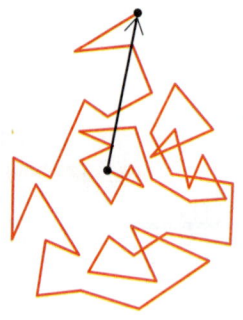

Figure 5–20 • A molecule in a gas at ordinary conditions follows a straight-line path only for a very short time before undergoing a collision. Its overall path is a zig-zag one.

Gaseous Effusion

In 1846, Thomas Graham showed experimentally that the rates of effusion of different gases through a small hole into a vacuum (Fig. 5–21) are inversely proportional to the square roots of the molar masses if the gases are all studied at the same temperature and pressure. The kinetic theory of gases explains this result, which is now called **Graham's law of effusion.** Imagine a container that has a small leak that allows its contents to effuse into a vacuum. The container can be filled at will with any gas, and different gases effuse from it at different rates. By Avogadro's law, at a fixed temperature and pressure the container holds the same number of molecules no matter which gas is selected. This leaves only one factor to explain the differing rates of effusion of the different gases—the average speeds of their molecules. It is reasonable to expect the rate of effusion to be proportional to the average speed (and to the root-mean-square speed) of the molecules.

For gases A and B then

$$\text{rate of effusion of A} \propto u_{\text{rms}}(\text{A}) \quad \text{and} \quad \text{rate of effusion of B} \propto u_{\text{rms}}(\text{B})$$

Dividing the first proportion by the second gives the ratio of the rates of effusion

$$\frac{\text{rate of effusion of A}}{\text{rate of effusion of B}} = \frac{u_{\text{rms}}(\text{A})}{u_{\text{rms}}(\text{B})}$$

Inserting the formulas for u_{rms} and canceling out the factors of $3RT$ then gives

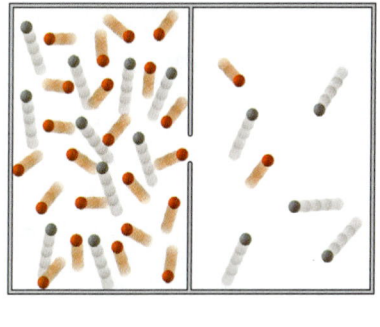

$$\frac{\text{rate of effusion of A}}{\text{rate of effusion of B}} = \frac{\sqrt{\dfrac{3RT}{M_A}}}{\sqrt{\dfrac{3RT}{M_B}}} = \sqrt{\frac{M_B}{M_A}}$$

Figure 5–21 • A small hole in the box permits molecules to effuse out into a vacuum. The lighter particles (here, helium atoms) effuse at a greater rate than the heavier oxygen molecules because their speeds are greater on the average.

The kinetic theory of gases thus predicts that the rate of effusion varies as to the inverse square root of the molar mass—exactly what was observed by Graham. The agreement provides additional support for the kinetic theory.

Graham's law can be applied to the effusion of a mixture of two gases through a small hole. The gas that emerges is enriched in the lighter component because lighter molecules effuse more rapidly than heavier ones. If B is heavier than A, the **enrichment factor** is $\sqrt{\dfrac{M_B}{M_A}}$.

EXAMPLE 5-13

Calculate the enrichment factor from effusion for a mixture of $^{235}\text{UF}_6$ and $^{238}\text{UF}_6$, uranium hexafluoride with two different uranium isotopes (^{235}U has a molar mass of 235.04 g mol^{-1}, and ^{238}U has a molar mass of 238.05) g mol^{-1}.

Strategy

Look up the molar mass of F in the table on the inside back cover. Add up the molar masses of $^{235}\text{UF}_6$ and $^{238}\text{UF}_6$. The enrichment factor is the square root of the ratio of the larger molar mass to the smaller.

Solution

$$M(^{238}\text{UF}_6) = 238.05 \text{ g mol}^{-1} + 6(18.998 \text{ g mol}^{-1}) = 352.04 \text{ g mol}^{-1}$$
$$M(^{235}\text{UF}_6) = 235.04 \text{ g mol}^{-1} + 6(18.998 \text{ g mol}^{-1}) = 349.03 \text{ g mol}^{-1}$$

$$\text{enrichment factor} = \sqrt{\frac{M(^{238}\text{UF}_6)}{M(^{235}\text{UF}_6)}} = \sqrt{\frac{352.05 \text{ g mol}^{-1}}{349.04 \text{ g mol}^{-1}}} = 1.0043$$

EXERCISE

A gas mixture contains equal numbers of molecules of N_2 and SF_6. A small portion of it is passed through a gaseous effusion apparatus. Calculate how many molecules of N_2 are present in the product gas for every 100 molecules of SF_6.

Answer: 228.

Graham's law of effusion holds true only if the opening in the vessel is small enough and the pressure low enough that most molecules follow straight-line trajectories through the opening without undergoing collisions with one another. A related but more complex phenomenon is gaseous diffusion through a porous barrier. This differs from effusion in that molecules undergo many collisions with one another and with the barrier during their passage through it. Interestingly, the rate of **gaseous diffusion** is also inversely proportional to the square root of the molar mass of the gas. If a mixture of a heavier gas B and a lighter gas A is placed in contact with a porous barrier, the gas passing through is enriched in the lighter component by a factor of $\sqrt{\dfrac{M_B}{M_A}}$, and the remaining gas is enriched in the heavier component.

During World War II, it was necessary to separate the isotope ^{235}U from ^{238}U in order to build the atomic bomb. Because ^{235}U has a natural abundance of only 0.7%, its isolation in nearly pure form posed a daunting task. The procedure that succeeded was to synthesize the relatively volatile compound UF_6 (boiling point, 56°C). Quantities of this compound were enriched in ^{235}U by passage through a porous barrier. As established in Example 5–13, a single passage gives an enrichment factor of only 1.0043. Consequently, thousands of passages in succession were necessary to provide sufficient enrichment in ^{235}U. A large gaseous diffusion plant was constructed in a very short time at the Oak Ridge National Laboratory, and the ^{235}U-enriched uranium was used in the Hiroshima atomic bomb. Similar methods are still used today to separate the isotopes of uranium (Fig. 5–22).

• The uranium that remains after the removal of the desired ^{235}U is called "depleted uranium."

Figure 5–22 • A gaseous diffusion plant for separation of uranium isotopes in uranium hexafluoride at Oak Ridge National Laboratory. If the uranium is to be used as fuel in nuclear power plants, a total enrichment over many stages, from 0.7% to 3.0% ^{235}U, is necessary. *(Oak Ridge National Laboratories)*

5–8   Real Gases

The ideal gas law

$$PV = nRT$$

is a particularly simple example of an **equation of state:** an equation showing the relationship among the pressure, temperature, chemical amount, and volume of a sample. Equations of state can be obtained either from theory or from experiments. They are useful not only for ideal gases but for real gases, liquids, and solids.

We have stressed that real gases obey the ideal gas law only at sufficiently low pressures. Deviations appear in a variety of forms. Boyle's law ($P \propto 1/V$ at constant T) is no longer satisfied at high pressures, and Charles's law ($V \propto T$ at constant P) begins to break down when the temperature becomes low. Avogadro's law also does not strictly hold for real gases at moderate pressures. Consider some concrete data. The volume of one mole of ideal gas at exactly 1 atm and 273.15 K equals 22.414 L. The *observed* volumes of one mole samples of several real gases under these conditions are:

H_2	22.433 L	O_2	22.397 L
He	22.434 L	CO_2	22.260 L
N_2	22.404 L	NH_3	22.079 L

None equals 22.414 L. The volumes of hydrogen and helium deviate to the high side, and those of carbon dioxide and ammonia are significantly low (by 0.7% for CO_2 and 1.5% for NH_3).

Figure 5–23 displays deviations from ideality in a different way. It shows a plot of the behavior of PV/nRT, which equals 1 for an ideal gas, for some real gases as the pressure increases at 25°C. The curve for hydrogen (red) never goes below the horizontal black line that represents ideal behavior. Hydrogen experiences a **posi-**

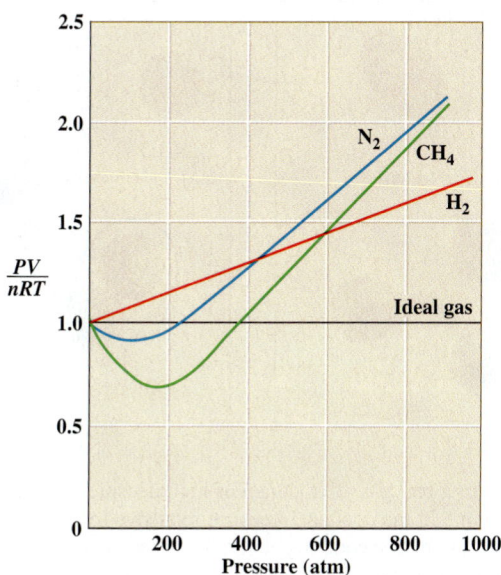

Figure 5–23 • A plot of PV/nRT against pressure shows deviations from the ideal gas law quite clearly. For a gas that follows the ideal gas law exactly, the plot is a horizontal straight line. The behavior of several real gases at 25°C deviates as shown.

tive deviation from ideality at this temperature. Methane and nitrogen experience **negative deviations** from ideality at moderate pressure, but positive deviations grow in and predominate at higher pressure. These observations show that the ideal gas law is inadequate at moderate and high pressures. Indeed, it is exactly correct for real gases only in the limit of zero pressure. Any improved equation of state should account for the two competing kinds of deviations.

Van der Waals Equation of State

In 1873, the Dutch physicist Johannes van der Waals proposed one of the earliest and most important improvements on the ideal gas equation of state. The **van der Waals equation of state** has the form

$$\left(P + a\frac{n^2}{V^2}\right)(V - nb) = nRT$$

The P, V, and T in the van der Waals equation are experimentally measured quantities. Comparison with the ideal gas law shows two modifications: an addition to P and a subtraction from V. These changes can be viewed as "fix-up factors" to the ideal gas law. They arise because of the existence of forces between molecules (Fig. 5–24), which are repulsive at short distances and attractive at large distances.

Repulsive forces keep molecules from overlapping. Each molecule denies other molecules entry into the volume that it occupies, so the effective volume available to any test molecule is not V, the true volume of the container, but $V - nb$, where b is a constant that equals the volume excluded per mole of molecules. Repulsive forces cause positive deviations from ideality.

Attractive forces confer a degree of "stickiness" on molecules. Sticky molecules linger together a while after collisions. Any tendency to group together reduces the effective number of independent molecules in the gas and therefore reduces the number of collisions made with the walls of the container. Fewer wall collisions means

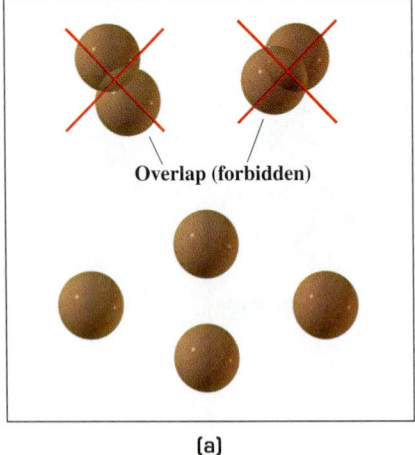
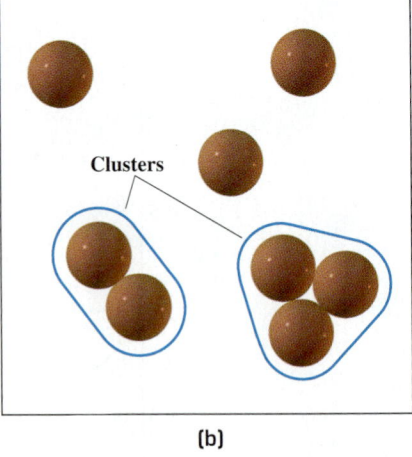

(a)   (b)

Figure 5–24 • (a) In a real gas at moderate or high densities, a test particle is excluded from some of the volume in the container because repulsive forces between molecules prevent it from sitting on top of a second molecule. (b) A second effect arises from attractive forces. When temporary associations of molecules form through these attractions, there are in effect fewer particles present in the gas and the pressure is lower.

TABLE 5-3	Van der Waals Constants of Several Gases		
Name	**Formula**	**a (atm L^2 mol^{-2})**	**b (L mol^{-1})**
Ammonia	NH_3	4.170	0.03707
Argon	Ar	1.345	0.03219
Carbon dioxide	CO_2	3.592	0.04267
Hydrogen	H_2	0.2444	0.02661
Hydrogen chloride	HCl	3.667	0.04081
Methane	CH_4	2.253	0.04278
Nitrogen	N_2	1.390	0.03913
Nitrogen dioxide	NO_2	5.284	0.04424
Oxygen	O_2	1.360	0.03183
Sulfur dioxide	SO_2	6.714	0.05636
Water	H_2O	5.464	0.03049

a lower pressure than the gas would exert if there were no attractions. Van der Waals argued that this effect, because it depends on attractions between pairs of molecules, should be proportional to the *square* of the number of molecules per unit volume $(N/V)^2$ or, equivalently, to the square of the number of moles per unit volume $(n/V)^2$. The effect therefore reduces the pressure by an amount $a \times (n^2/V^2)$, where a is a positive constant that depends on the strength of the attractive forces. By lowering P from its hypothetical ideal value, attractive forces cause negative deviations from ideality. Putting these two corrections into the ideal gas law gives the van der Waals equation.

The only way to get values for the constants a and b is to fit experimental PVT data for each and every real gas—the two constants differ from gas to gas (Table 5-3). They have the units

$$a: \quad \text{atm L}^2 \text{ mol}^{-2} \qquad b: \quad \text{L mol}^{-1}$$

when R has units of L atm mol^{-1} K^{-1}. For nitrogen, it is found that $b = 0.03913$ L mol^{-1} and $a = 1.390$ atm L^2 mol^{-2}. The constant b is the volume excluded by one mole of molecules; it should therefore be close to the volume per mole in the liquid state, where molecules are essentially in contact with each other. The density of liquid nitrogen is 0.808 g cm^{-3}. One mole of N_2 weighs 28.0 g, so

$$\left(\frac{V}{n}\right)_{N_2(\ell)} = \frac{1 \text{ cm}^3 \text{ N}_2(\ell)}{0.808 \text{ g N}_2(\ell)} \times \left(\frac{28.0 \text{ g N}_2}{1 \text{ mol N}_2}\right) \times \left(\frac{1 \text{ L}}{1000 \text{ cm}^3}\right) = 0.0347 \text{ L mol}^{-1}$$

This is reasonably close to the van der Waals b parameter of 0.03913 L mol^{-1}.

EXAMPLE 5-14

A sample consists of 8.00 kg of gaseous nitrogen and fills a 100 L flask at 300°C. What is the pressure of the gas, using the van der Waals equation of state?

Strategy

Express the amount of nitrogen in moles. Solve the van der Waals equation for P and substitute n_{N_2}, the given T and P, and a and b from Table 5-3.

Solution

$$n_{N_2} = 8.00 \times 10^3 \text{ g N}_2 \times \left(\frac{1 \text{ mol N}_2}{28.0 \text{ g N}_2}\right) = 286 \text{ mol N}_2$$

Divide both sides of the van der Waals equation by $V - nb$ and then subtract $a\dfrac{n^2}{V^2}$ from both sides:

$$P + a\frac{n^2}{V^2} = \frac{nRT}{V - nb}$$

$$P = \frac{nRT}{V - nb} - a\frac{n^2}{V^2}$$

All of the quantities on the right side of this equation are known:

$$P = \frac{(286 \text{ mol})(0.08206 \text{ L atm mol}^{-1}\text{K}^{-1})(273.15 + 300\ )\text{ K}}{100 \text{ L} - (286 \text{ mol})(0.03913 \text{ L mol}^{-1})} -$$

$$\frac{1.390 \text{ atm L}^2}{\text{mol}^2}\left(\frac{(286 \text{ mol})^2}{(100 \text{ L})^2}\right)$$

$$P = 151 \text{ atm} - 11.4 \text{ atm} = 140 \text{ atm}$$

Check

Compute P using the ideal gas law $P = nRT/V$. The answer is 134 atm, which is fairly close. The deviations of real gases from the predictions of the ideal gas law become even more sizeable at higher pressures.

EXERCISE

Find the difference between the pressure exerted by 1.50 mol of an ideal gas in a 2.00 L container at 600 K and the pressure exerted by 1.50 mol of sulfur dioxide in the same container at the same temperature. Obtain the pressure of the sulfur dioxide using the van der Waals equation.

Answer: The pressure exerted by the ideal gas is larger by 2.15 atm.

SUMMARY

5–1   The sources and chemical reactions of gases are quite varied. Air consists of a mixture of several gaseous substances, principally oxygen and nitrogen.

5–2   The physical behavior of gases at low density follows simple laws. The **pressure** of any gas is the force that it exerts on a unit area of the walls of its container; the air around us exerts a pressure that can be measured with a **barometer.** **Boyle's law** states that the pressure exerted by a sample of a gas is inversely proportional to its volume at fixed temperature.

5–3   Charles's law states that when a sample of gas is heated or cooled at constant pressure, its volume depends linearly on its temperature. Charles's law allows the definition of a universal temperature scale based on ratios of gas volumes. An **absolute zero** of temperature exists below which it is impossible to cool any substance, and it is logical to measure temperatures upward from that point. The **kelvin,** which is equal in size to a Celsius degree, is the unit of temperature for measurements referred to absolute zero.

5–4 Boyle's law, Charles's law, and Avogadro's law (which states that a given volume contains the same number of molecules of any gas at fixed pressure and temperature) can all be combined into a single mathematical statement, the **ideal gas law** ($PV = nRT$). The proportionality constant in this law is the **universal gas constant, R**.

5–5 The ideal gas law is the basis for many calculations of practical importance in chemistry because it relates the volumes of gases in reactions or other processes to chemical amounts or masses.

5–6 **Dalton's law** extends the ideal gas law to mixtures. Each component in an ideal gas mixture has a **partial pressure** equal to the product of its **mole fraction** and the total pressure. The partial pressures add up to the total pressure because the mole fractions must add up to unity.

5–7 The **kinetic theory of gases** provides a microscopic picture that accounts for properties of gases at low densities. It assumes that the molecules in a gas move in random directions with a distribution of speeds and travel in straight lines between collisions with other molecules or with the walls of their container. From this kinetic theory, in combination with the ideal gas law, it can be shown that the **average kinetic energy** of the molecules of a gas is proportional to the absolute temperature; therefore, the **mean-square speed** is as well. The full distribution of molecular speeds in a gas at **thermal equilibrium** is given by the Maxwell–Boltzmann equation. The kinetic theory can be used to estimate the diffusive motion of molecules in a gas, including their **mean free path,** and to account for the relative rates of effusion of gases through small holes. It is found under the same conditions of temperature and pressure that lighter molecules effuse more rapidly than heavier ones by a factor equal to the inverse of the ratio of the square root of their molar masses (**Graham's law).**

5–8 **Real gases** obey the ideal gas law only at sufficiently low pressures. At higher pressures, the **van der Waals equation of state** accounts for deviations from ideality in an approximate fashion by considering the repulsions and attractions among molecules. Calculations using the van der Waals equation give better agreement with experiment for gases at high pressures.

PROBLEMS

Note: Answers to blue-numbered problems are given in Appendix F. Problems that are more challenging are indicated with asterisks.

The Chemistry of Gases

1. Predict the gaseous product of the reaction of ammonium hydrosulfide ($NH_4HS(s)$) with the acid $HCl(aq)$.

2. Predict the gaseous product of the reaction of $NH_4HS(s)$ with the base $NaOH(aq)$.

3. Solid ammonium hydrosulfide (NH_4HS) decomposes to give a mixture of gases when it is heated. Write a balanced chemical equation representing this change.

4. Solid ammonium carbamate ($NH_4CO_2NH_2$) decomposes to give a mixture of gases when it is heated. Write a balanced chemical equation representing this change.

5. Ammonia (NH_3) is an important and useful gas. Suggest a way to generate it from ammonium bromide (NH_4Br). Include a balanced chemical equation.

6. Hydrogen cyanide (HCN) is a poisonous gas. Explain why solutions of potassium cyanide (KCN) should never be acidified. Include a balanced chemical equation.

7. Solid sodium peroxide (Na_2O_2) reacts when heated with $CO_2(g)$ to give $Na_2CO_3(s)$ and a single gas. Identify the gas and write a balanced equation to represent the change.

8. Solid ammonium dichromate (($NH_4)_2Cr_2O_7$) reacts when heated to give $Cr_2O_3(s)$, water vapor, and a second gas. Write a balanced chemical equation to represent this change.

Pressure and Boyle's Law

9. (See Example 5–1.) Estimate (in atmospheres) the pressure of a gas in a container if the pressure supports a column of water 44.5 cm high.

10. (See Example 5–1.) The metallic element gallium is a liquid with a density of 6.114 g cm^{-3} at 30°C. A barometer is constructed using liquid gallium in a vertical tube. On a

particular day when the temperature is 30°C, the height of the column of gallium is 168 cm. Compute that day's atmospheric pressure in atmospheres.

11. Determine the height of a column of heavy oil (density = 1.60 g cm^{-3}) in a tube if its mass is the same as that of a 305 mm column of mercury (density = 13.6 g cm^{-3}) in an identical tube.

12. Marine organisms use $Ca^{2+}(aq)$ ion from the ocean to synthesize $CaCO_3(s)$ for skeletons and shells. When the organisms die, their remains fall to the bottom. The amount of calcium carbonate that can be dissolved in seawater depends on the pressure. At great depths, where the pressure exceeds approximately 414 atm, the shells slowly redissolve. This prevents all the world's calcium from being tied up as insoluble $CaCO_3(s)$ at the bottom of the sea. Estimate the depth (in meters) of water that exerts a pressure great enough to dissolve seashells.

13. Suppose that an ocean of mercury replaced all the air of the Earth. How deep would this ocean have to be to exert the same pressure as the air?

14. Suppose that the atmosphere were perfectly uniform, with a density throughout equal to that of air at STP, 1.3 g L^{-1}. Calculate the depth of such an atmosphere that would cause a pressure of exactly 1 atm at the Earth's surface.

15. (See Example 5–2.) The "critical pressure" (see Section 6–5) of mercury is 172.00 MPa. Express this pressure in atmospheres and in bars.

16. (See Example 5–2.) Experimental studies of solid surfaces and the chemical reactions that occur on them require very low gas pressures to avoid surface contamination. A high-vacuum apparatus for such experiments can routinely reach pressures of 5×10^{-10} torr. Express this pressure in atmospheres and in pascals.

17. A sample of gaseous N_2 is trapped in a J-tube (as in Fig. 5–5). The level of the mercury on the side open to the atmosphere is 67.5 mm higher than the level on the other side of the tube. The atmospheric pressure in the room is 1.000 atm and the temperature is 0.00°C. Determine the pressure of the trapped N_2 (in atm).

18. A sample of 12.07 cm^3 of gaseous CO_2 is trapped in a J-tube like the one in the previous problem. The pressure on the CO_2 is 1.106 atm. More mercury is poured into the open side of the tube, and the mercury level rises on both sides. The volume of the gas is reduced to 11.51 cm^3. Compute the pressure (in atm) of the CO_2 after the new mercury has been added.

19. (See Example 5–3.) Some nitrogen is held in a 3.00 L tank at a pressure of 4.00 atm. The tank is connected to a 6.00 L tank that is completely empty (evacuated), and a valve is opened to connect the two. No temperature change occurs in the process. Determine the total pressure in this two-tank system after the nitrogen stops flowing.

20. (See Example 5–3.) The Stirling engine, a heat engine invented by a Scottish minister, has been considered for use in automobile engines because of its efficiency. In such an engine, a gas goes through a four-step cycle of (1) expansion at constant T, (2) cooling at constant V, (3) compression at constant T back to its original volume, and (4) heating at constant V back to its original temperature. Suppose that the gas starts at a pressure of 1.23 atm, and its volume changes from 0.350 L to 1.31 L during the expansion at constant T. Calculate the pressure of the gas at the end of this step in the cycle.

21. Each of the following sets of pressure/volume data applies to a fixed amount of a gas held at constant temperature. In each case, calculate the missing quantity.
    (a) $V = 0.587$ L at 1.00 atm; $V =$ __ at 2.37 atm
    (b) $V = 785$ mL at 1.75 atm: $V =$ __ at 110.0 kPa
    (c) $V = 604$ mL at 995 torr; $V = 312$ mL at __ torr
    (d) $V = x$ mL at 3.25 atm; $V = x/2$ mL at __ atm

22. Each of the following sets of pressure/volume data applies to a fixed amount of a gas held at constant temperature. In each case, calculate the missing quantity.
    (a) $V = 485$ L at 512 torr; $V =$ __ at 75.0 kPa
    (b) $V = 652$ mL at 1.50 atm; $V =$ __ at 110.0 kPa
    (c) $V = 60.0$ cm^3 at 742 torr; $V = 90.0$ mL at __ atm
    (d) $V = y$ mL at 3.25 atm; $V = 3.25y$ mL at __ kPa

Temperature and Charles's Law

23. Ordinary internal body temperature in healthy humans averages 98.6°F. Express this temperature on the Celsius and Kelvin temperature scales.

24. Temperature readings on the Celsius and Fahrenheit scales are numerically equal at only one temperature. Determine that temperature.

25. The absolute temperature of a 3.50 L sample of gas is tripled at constant pressure. Determine the volume of the gas after this change.

26. The Celsius temperature of a 3.50 L sample of gas is tripled from 20.0°C to 60.0°C at constant pressure. Determine the volume of the gas after this change.

27. (See Example 5–4.) The gill is an obscure unit of volume. If a fixed amount of $H_2(g)$ has a volume of 17.4 gills at 100°F, what volume would it have if the temperature were reduced to 0°F, assuming that its pressure stays constant?

28. (See Example 5–4.) A gas originally at a temperature of 26.5°C is cooled at constant pressure. Its volume decreases from 5.52 L to 5.26 L. Determine its new temperature in degrees Celsius.

29. (See Example 5–4.) Calcium carbide (CaC_2) reacts with water to produce acetylene (C_2H_2) according to the equation

$$CaC_2(s) + 2\,H_2O(\ell) \longrightarrow Ca(OH)_2(s) + C_2H_2(g)$$

A certain mass of CaC_2 reacts completely with water to give 64.5 L of C_2H_2 at 50°C and $P = 1$ atm. If the same mass of CaC_2 reacts completely at 400°C and $P = 1$ atm, what volume of C_2H_2 can be collected at the higher temperature?

30. (See Example 5–4.) A convenient laboratory source for oxygen of very high purity is the decomposition of potassium permanganate at 230°C:

$$2\ KMnO_4(s) \longrightarrow K_2MnO_4(s) + MnO_2(s) + O_2(g)$$

Suppose 3.41 L of oxygen is needed at atmospheric pressure and a temperature of 20°C. What volume of oxygen should be collected at 230°C and the same pressure in order to give this volume when cooled?

31. Each of the following sets of volume/temperature data applies to a sample of a gas held at constant pressure. In each case, use the ideal gas law to calculate the missing quantity.
(a) $V = 164$ L at 0.0°C; $V =$ __ at 100.0°C
(b) $V = 26$ quarts at 400 K; $V =$ __ at 200 K
(c) $V = 1.000 \times 10^3$ mL at 1000 K; $V =$ __ mL at 153°C
(d) $V = y$ mL at 200°C; $V = 6.4y$ mL at __ °C

32. Each of the following sets of volume/temperature data applies to a sample of a gas held at constant pressure. In each case, use the ideal gas law to calculate the missing quantity.
(a) $V = 26.8$ L at 25.0°C; $V =$ __ at 0.0°C
(b) $V = 619$ mL at 150 K; $V =$ __ at 450 K
(c) $V = 619$ cm^3 at 150 K; $V =$ __ mL at 150°C
(d) $V = y$ mL at 200 K; $V = 6.50y$ mL at __ K

The Ideal Gas Law

33. (See Example 5–5.) A gas is in a container with movable walls under a pressure of 5.50 atm. The volume of the container is increased by a factor of 3.25 by moving its walls. The absolute temperature is simultaneously doubled. Calculate the pressure of the gas after these changes.

34. (See Example 5–5.) Some gaseous helium is kept under a pressure of 20.0 atm in a steel tank at a temperature of 20.0°C. The tank is moved outside on a cold day when the temperature is 0.0°C. Determine the pressure inside the tank after it cools down to 0.0°C.

35. (See Example 5–5.) A 20.6 L sample of "pure" air is collected in Greenland at a temperature of −20.0°C and a pressure of 1.01 atm and is pumped into a 1.05 L bottle for shipment to Europe for analysis.
(a) Compute the pressure inside the bottle just after it is filled.
(b) Compute the pressure inside the bottle as it is opened in the 21.0°C comfort of the European laboratory.

36. (See Example 5–5.) Iodine heptafluoride can be made at elevated temperatures by the reaction

$$I_2(g) + 7\ F_2(g) \longrightarrow 2\ IF_7(g)$$

Suppose 63.6 L of IF$_7$(g) is synthesized at 300°C and a pressure of 0.459 atm. Calculate the volume that the gas occupies when heated to 400°C at a pressure of 0.980 atm, assuming that no decomposition or other reactions take place.

37. Certain thin-walled glass bottles generally withstand a pressure difference across their walls of 1.25 atm before they burst. A typical bottle of this type is sealed on a day when the temperature is 68°F and the pressure is 0.98 atm, and is then heated up. Estimate the temperature in degrees Fahrenheit at which it explodes.

38. The pressure of a poisonous gas inside a sealed container is 1.47 atm at 20°C. The barometric pressure is 0.96 atm. To what temperature (in degrees Celsius) must the container and its contents be cooled so that the container can be opened without any risk of gas spurting out?

39. (See Example 5–6.) What temperature (in kelvins) does 29.8 g of O$_2$ gas have at a pressure of 2.00 atm in a 10.0 L tank?

40. (See Example 5–6.) What mass (in grams) of gaseous neon exerts a pressure of 2.00 atm in a 35.0 L tank at 25.0°C?

41. According to a reference handbook, "The weight of one liter [of H$_2$Te(g)] is 6.234 g." Tell why this information is nearly valueless. Assume that H$_2$Te(g) follows the ideal gas law, and calculate the temperature (in degrees Celsius) at which this statement is true if the pressure is 1.00 atm.

42. A scuba diver's tank contains 0.30 kg of oxygen (O$_2$) compressed into a volume of 2.32 L.
(a) Use the ideal gas law to estimate the gas pressure inside the tank at 5°C, and express it in atmospheres and in pounds of force per square inch.
(b) What volume would this oxygen occupy at 30°C and a pressure of 0.98 atm?

43. Use the data in Table 5–1 and the ideal gas law to compute the number of atoms of krypton present in 15 cm^3 (a gasp) of clean, dry air at 21.0°C and a pressure of 0.95 atm. Take the molar mass of air to be 29 g mol^{-1}.

44. Like carbon dioxide, methane in the atmosphere exerts a greenhouse effect. Although the increase in atmospheric carbon dioxide concentration from the burning of fossil fuels attracts more attention, methane levels are also on the rise. In a recent survey, the average worldwide concentration of methane in dry air in the lower 7 to 10 miles of the atmosphere was 1.657 parts per million by volume. Estimate (to within a power of ten) the average number of molecules of methane in 1 L of this portion of the atmosphere.

45. (See Example 5–7.) Calculate the density of gaseous SF$_6$ at a temperature of 15°C and a pressure of 0.942 atm.

46. (See Example 5–7.) Calculate the density of gaseous N$_2$H$_4$ at a temperature of 225°C and a pressure of 1.38 atm.

47. (See Example 5–8.) The empirical formula of a gaseous fluorocarbon is CF$_2$. A mass of 1.55 g of the compound occupies 0.174 L at STP. Determine the molecular formula of this compound.

48. (See Example 5–8.) A sample of a gaseous binary compound of boron and chlorine weighing 2.842 g occupies 0.153 L at STP. This sample is decomposed to give solid boron and gaseous chlorine (Cl$_2$). At STP, the chlorine occupies 0.688 L. Determine the molecular formula of the compound.

Chemical Calculations for Gases

49. (See Example 5–10.) The classic method for manufacturing hydrogen chloride—still in use today to a small extent—involves the reaction of sodium chloride with excess sulfuric acid at elevated temperatures. The overall equation for this process is

$$NaCl(s) + H_2SO_4(aq) \longrightarrow NaHSO_4(s) + HCl(g)$$

What is the maximum volume of hydrogen chloride that can be produced at 550°C and a pressure of 742.3 torr from the reaction of 2500 kg of sodium chloride?

50. (See Example 5–10.) The laboratory-scale preparation of pure oxygen can be carried out by the Brin process, which consists of decomposing barium peroxide (BaO_2) by heating it:

$$2 BaO_2(s) \longrightarrow 2 BaO(s) + O_2(g)$$

Suppose that 5.00 g of BaO_2 is decomposed according to this equation.
(a) What volume of oxygen is produced (at 25°C and 1.00 atm)?
(b) What mass of BaO is formed?

51. (See Example 5–10.) Hydrogen is produced by the complete reaction of 6.24 g of sodium with an excess of gaseous hydrogen chloride.
(a) Write a balanced chemical equation for the reaction that occurs.
(b) How many liters of hydrogen are produced at a temperature of 20.0°C and a pressure of 655 torr?

52. (See Example 5–10.) In 1783, the French physicist Jacques Charles supervised and took part in the first human flight in a hydrogen balloon. Such balloons rely on the lower density of hydrogen relative to air to give them their buoyancy. In Charles's balloon ascent, the hydrogen was produced (together with iron(II) sulfate) from the action of aqueous sulfuric acid on iron filings.
(a) Write a balanced chemical equation for this reaction.
(b) What volume of hydrogen is produced at 300 K and a pressure of 1.0 atm when 300 kg of sulfuric acid is consumed in this reaction?

53. Elemental chlorine was first produced by Scheele in 1774 using the reaction of the mineral pyrolusite (MnO_2) with sulfuric acid and sodium chloride:

$$MnO_2(s) + 2 H_2SO_4(aq) + 4 NaCl(s) \longrightarrow$$
$$MnCl_2(s) + 2 Na_2SO_4(s) + 2 H_2O(\ell) + Cl_2(g)$$

Calculate the minimum mass of MnO_2 required to generate 4.48 L of gaseous chlorine, measured at a pressure of 735.5 torr and a temperature of 35.3°C.

54. Aluminum reacts with excess aqueous hydrochloric acid to produce hydrogen.
(a) Write a balanced chemical equation for the reaction. (*Hint*: Water-soluble $AlCl_3$ is the stable chloride of aluminum.)

(b) Calculate the mass of pure aluminum that furnishes 10.0 L of hydrogen at a pressure of 0.750 atm and a temperature of 30.0°C.

55. Elemental sulfur can be recovered from gaseous hydrogen sulfide (H_2S) through the reaction

$$2 H_2S(g) + SO_2(g) \longrightarrow 3 S(s) + 2 H_2O(\ell)$$

(a) What volume of H_2S (in liters at STP) is required to produce 2.00 kg (2000 g) of sulfur according to this equation?
(b) What minimum mass and volume (at STP) of SO_2 are required to produce 2.00 kg of sulfur according to this equation?

56. When ozone (O_3) is placed in contact with dry, powdered KOH at −15°C, the red-brown solid potassium ozonide (KO_3) forms, according to the balanced equation

$$5 O_3(g) + 2 KOH(s) \longrightarrow 2 KO_3(s) + 5 O_2(g) + H_2O(s)$$

Calculate the minimum volume of ozone needed (at a pressure of 0.134 atm and a temperature of −18.0°C) to produce 555 mg of KO_3.

57. A gaseous binary compound of carbon and hydrogen, in a volume of 300.0 cm^3 at a temperature of 300.0 K and a pressure of 1.200 atm, furnishes 1.054 g of H_2O and 1.930 g of CO_2 when burned completely in an excess of oxygen. Determine the molecular formula of this hydrocarbon.

58. A gaseous hydrocarbon, in a volume of 25.4 L at 400 K and a pressure of 3.40 atm, reacts in an excess of oxygen to give 47.4 g of H_2O and 231.6 g of CO_2. Determine the molecular formula of the hydrocarbon.

Mixtures of Gases

59. The partial pressure of $O_2(g)$ in a mixture of $O_2(g)$ and $N_2(g)$ is 1.02 atm. The total pressure of the mixture is 2.43 atm. Determine the partial pressure of $N_2(g)$ and the mole fraction of each gas in the mixture.

60. In the gas mixture described in the previous problem, exactly half of the $O_2(g)$ is removed. No change in temperature or volume occurs. Determine the partial pressure of the remaining $O_2(g)$ and its mole fraction.

61. (See Example 5–11.) Sulfur dioxide reacts with oxygen in the presence of platinum to give sulfur trioxide:

$$2 SO_2(g) + O_2(g) \longrightarrow 2 SO_3(g)$$

Suppose at one stage in the reaction 263 g SO_2, 837 g O_2, and 179 g SO_3 are present in the reaction vessel at a total pressure of 0.950 atm. Calculate the mole fraction of SO_3, the volume percent of SO_3, and the partial pressure of SO_3 at that stage.

62. (See Example 5–11.) The synthesis of ammonia from the elements is carried out at high pressures and temperatures.

$$N_2(g) + 3 H_2(g) \longrightarrow 2 NH_3(g)$$

Suppose at one stage in the reaction 149 g NH_3, 765 g N_2, and 164 g H_2 are present in the reaction vessel at a total pressure of 210 atm. Calculate the mole percentage of NH_3 and the partial pressure of NH_3 at that stage.

63. The atmospheric pressure at the surface of Mars is 5.92×10^{-3} atm. The Martian atmosphere is 95.3% CO_2 and 2.7% N_2 by volume, with small amounts of other gases also present. Compute the mole fraction and partial pressure of N_2 in the atmosphere of Mars.

64. The atmospheric pressure at the surface of Venus is 90.8 atm. The Venusian atmosphere is 96.5% CO_2 and 3.5% N_2 by volume, with small amounts of other gases also present. Compute the mole fraction and partial pressure of N_2 in the atmosphere of Venus.

65. A gas mixture at room temperature contains 10.0 mol of CO and 12.5 mol of O_2.
   (a) Compute the mole fraction of CO in the mixture.
   (b) When the mixture is heated, the CO starts to react with the O_2 to give CO_2:

   $$2\,CO(g) + O_2(g) \longrightarrow 2\,CO_2(g)$$

   The reaction is stopped when 3.0 mol of CO_2 is present. Determine the mole fraction of CO in the new mixture.

66. A gas mixture contains 4.5 mol of Br_2 and 33.1 mol F_2.
   (a) Compute the mole fraction of Br_2 in the mixture.
   (b) The mixture is heated above 150°C and starts to react to give BrF_5:

   $$Br_2(g) + 5\,F_2(g) \longrightarrow 2\,BrF_5(g)$$

   The reaction is stopped when 2.2 mol of BrF_5 is present. Determine the mole fraction of Br_2 in the mixture at that point.

The Kinetic Theory of Gases

67. (See Example 5–12.)
   (a) Compute the root-mean-square speed of H_2 molecules in gaseous hydrogen at a temperature of 300 K.
   (b) Repeat for SF_6 molecules in gaseous sulfur hexafluoride at 300 K.

68. (See Example 5–12.) Researchers report that the temperature of a gas consisting of 500 sodium atoms confined in a tiny atomic trap is 0.00024 K. Compute the root-mean-square speed of the atoms in this confinement.

69. Compare the root-mean-square speed of helium atoms near the surface of the sun, where the temperature is approximately 6000 K, with that of helium atoms in an interstellar cloud, where the temperature is 100 K.

70. The "escape velocity" for objects to leave the gravitational field of the Earth is 11.2 km s^{-1}. Calculate the ratio of the escape velocity to the root-mean-square speed of helium, argon, and xenon atoms at 2000 K. Does the result help to explain the low abundance of the light gas helium in the atmosphere? Explain.

71. Chlorine dioxide (ClO_2) is used for bleaching wood pulp. In a gaseous sample held at thermal equilibrium at a particular temperature, 35.0% of the molecules have speeds exceeding 400 m s^{-1}. The sample is heated slightly. Is the percentage of molecules with speeds in excess of 400 m s^{-1} then greater than or less than 35%? Explain.

72. The ClO_2 from Problem 71 is heated further until it explodes, yielding Cl_2, O_2, and other gaseous products. The mixture is then cooled until the original temperature is reached. Is the percentage of chlorine molecules with speeds in excess of 400 m s^{-1} greater than or less than 35%? Explain.

73. Methane (CH_4) effuses through a small opening in the side of a container at the rate of 1.30×10^{-8} mol s^{-1}. An unknown gas effuses through the same opening at the rate of 5.42×10^{-9} mol s^{-1} when maintained at the same temperature and pressure as the methane. Determine the molar mass of the unknown gas.

74. (See Example 5–13.) Equal chemical amounts of two gases, fluorine and bromine pentafluoride, are mixed. Determine the initial ratio of the rates of effusion of the two gases through a very small opening in their container.

Real Gases

75. (See Example 5–14.) Oxygen is supplied to hospitals and chemical laboratories under pressure in large steel cylinders. Typically, such a cylinder has an internal volume of 28.0 L and contains 6.80 kg of oxygen. Use the van der Waals equation to estimate the pressure inside such a cylinder at 20°C. Express it in atmospheres and in pounds of force per square inch.

76. (See Example 5–14.) Steam at high pressures and temperatures is used to generate electrical power in utility plants. A large utility boiler has a volume of 2500 m^3 and contains 140 metric tons (1 metric ton = 10^3 kg) of steam at a temperature of 540°C. Use the van der Waals equation to estimate the pressure of the steam under these conditions, in atmospheres and in pounds per square inch.

77. Careful measurements show that 0.7500 mol of CO_2 gas confined in a volume of 1.000 L at a temperature of 240.0 K exerts a pressure of 12.5655 atm.
   (a) Compute the ideal gas pressure (the pressure that a gas following the ideal gas law would exert under these conditions).
   (b) Compute the van der Waals pressure (see Table 5–3 for the van der Waals constants for CO_2).
   (c) Explain why neither of these pressures equals the observed pressure.

78. A careful experiment establishes that a 0.5000 mol sample of CO_2 occupies a volume of 0.5866 L at a temperature of 250.0 K and a pressure of 14.804 atm. These values do not exactly fit in the van der Waals equation. Suppose the van der Waals constant b remains unchanged at 0.04267 L mol^{-1}. Compute the value of a that makes the van der Waals equation exactly correct for these particular conditions.

ADDITIONAL PROBLEMS

79. The density of mercury is 13.5955 g cm^{-3} at 0.0°C but only 13.5094 g cm^{-3} at 35°C. Suppose that a mercury barometer is read on a hot summer day when the temperature is 35°C. The column of mercury is 760.0 mm long. Correct for the expansion of the mercury, and compute the true pressure in atmospheres.

80. The barometric pressure in the center of one of the worst hurricanes of the 20th century was reported to be 886 mbar. Express this pressure in atmospheres.

81. The following arrangement of flasks is set up.

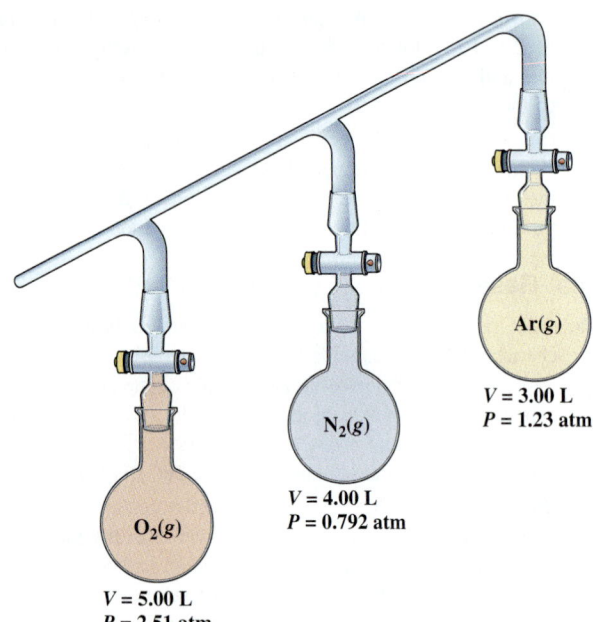

Ar(g)

V = 3.00 L
P = 1.23 atm

N₂(g)

V = 4.00 L
P = 0.792 atm

O₂(g)

V = 5.00 L
P = 2.51 atm

Assuming no temperature change, determine the final pressure inside the system after all stopcocks are opened. Ignore the volume of the connecting tube.

82. When a certain sample of a gas is cooled at constant pressure, it is found that the volume decreases linearly:

$$V = 209.4 \text{ L} + \left(0.456\frac{\text{L}}{\text{°F}}\right) \times t_F$$

where t_F is the temperature in degrees Fahrenheit. Estimate the absolute zero of temperature in degrees Fahrenheit from this relationship.

83. The Tire Industry Safety Council annually reminds motorists that tires can become dangerously underinflated when temperatures begin to dip below freezing. They say that every time the outside temperature drops by 10°F, the air pressure in a car's tires goes down approximately 1 pound of force per square inch (psi). Recall that 14.7 psi equals 1.00 atm.
(a) Explain why the Council's rule about tire pressure is at best only a rule of thumb.

(b) Assume that a typical tire with a constant volume is filled with a gas at 70°F to a gauge pressure of 28 psi. The temperature then falls to 20°F. What is the new pressure in the tire?

84. Amontons's law relates pressure to absolute temperature. Consider the ideal gas law and then write a statement of Amontons's law in a form that is analogous to the statements of Charles's law and Boyle's law in the text.

85. When 1 volume of a gaseous compound that contains only carbon, hydrogen, and nitrogen is burned in oxygen, 2 volumes of $CO_2(g)$, 4 volumes of water vapor, and 2 volumes of $NO_2(g)$ are produced.
(a) What volume of O_2 is required for the combustion?
(b) What is the molecular formula of the compound?
Assume that all volumes are measured at the same temperature and pressure.

86. Complete combustion (burning) of 2.00 volumes of a gaseous hydrocarbon to $CO_2(g)$ and $H_2O(g)$ required 9.00 volumes of pure $O_2(g)$. Both volumes were measured at the same temperature and pressure. The reaction yielded 0.135 g of H_2O and 0.330 g of CO_2. Determine the molecular formula of the hydrocarbon.

87. Compute the minimum masses (in grams) of potassium nitrate and silicon dioxide that must be mixed with 125.0 g of NaN_3 to ensure that no Na, Na_2O, or K_2O remains unreacted when the mixture is detonated in an automobile air bag. (*Note*: See *Chemistry in Your Life: How an Air Bag Works*.)

88. Determine the mass (in grams) of NaN_3 that would have to be decomposed to Na and N_2 to generate 70.0 L of nitrogen in an automobile air bag at 25°C and 1 atm pressure. (*Note*: See *Chemistry in Your Life: How an Air Bag Works*.)

89. A sample of limestone (calcium carbonate, $CaCO_3$) is heated at 950 K until it is completely converted into calcium oxide (CaO) and gaseous CO_2. The CaO is then all converted into calcium hydroxide by addition of water, yielding 8.47 kg of solid $Ca(OH)_2$. Use the ideal gas law to calculate the volume of CO_2 produced in the first step at 950 K and a pressure of 0.976 atm.

90. A gas exerts a pressure of 0.740 atm in a certain container. Suddenly a chemical change occurs that consumes half of the molecules originally present and forms two new molecules for every three consumed. Determine the new pressure in the container if the volume of the container and the temperature are unchanged.

91. Exactly 1.0 lb of Hydrone, an alloy of sodium with lead, yields 2.6 ft^3 (at 0.0°C and 1.00 atm) of hydrogen when it is treated with water. All of the sodium reacts according to the equation

$$2 \text{ Na}(alloyed) + 2 \text{ H}_2\text{O}(\ell) \longrightarrow 2 \text{ NaOH }(aq) + \text{H}_2(g)$$

and the lead does not react with water. Compute the percentage by mass of sodium in the alloy.

92. Baseball reporters say that long fly balls that would have carried for home runs in July "die" in the cool air of October and are caught. The idea behind this reasoning is that a baseball carries better when the air is less dense. Dry air is a mixture of gases with an effective molar mass of 29.0 g mol^{-1}.
(a) Compute the density of dry air on a July day when the temperature is 95.0°F and the pressure is 1.00 atm.
(b) Compute the density of dry air on an October evening when the temperature is 50.0°F and the pressure is 1.00 atm.
(c) Suppose that the humidity on the July day is 100%, so that the air is saturated with water vapor. Would the density of the hot, moist air be less than, equal to, or more than the density of the hot, dry air computed in part (a)? In other terms, does high humidity favor the home run?

93. A mixture of $CS_2(g)$ and excess $O_2(g)$ in a 10.0 L reaction vessel at 100.0°C is under a pressure of 3.00 atm. A spark ignites the mixture, and it explodes. The vessel successfully contains the explosion, in which all of the $CS_2(g)$ reacts to give $CO_2(g)$ and $SO_2(g)$. The vessel is cooled back to its original temperature of 100.0°C, and the pressure of the mixture of the two product gases and the unreacted $O_2(g)$ is found to be 2.40 atm. Calculate the mass (in grams) of $CS_2(g)$ originally present.

94. In an optical bottle, beams of laser light replace the physical walls of conventional containers to confine gases. The laser beams are tightly focused. Suppose that they briefly (for 0.5 s) exert enough pressure to confine 500 sodium atoms in a volume of $1.0 \times 10^{-15} m^3$ at a temperature of 0.00024 K.
(a) Use the ideal gas law to compute the pressure exerted on the "walls" of the optical bottle.

(b) In this gas, the mean free path (the average distance that the sodium atoms travel between collisions) is 3.9 m. Compare this with the mean free path of the atoms in gaseous sodium at room conditions.

95. Molecules of UF_6 are approximately 175 times more massive than H_2 molecules; however, Avogadro's number of H_2 molecules confined at a set temperature exert the same pressure on the walls of a certain container as the same number of UF_6 molecules. Explain how this is possible.

96. The number density of atoms (chiefly hydrogen) in interstellar space is about ten per cubic centimeter, and the temperature is about 100 K.
(a) Calculate the pressure of the gas in interstellar space, and express it in atmospheres.
(b) Under these conditions, an atom of hydrogen collides with another atom once every 1×10^9 seconds (that is, once every 30 years). By using the root-mean-square speed, estimate the distance traveled by a hydrogen atom between collisions. Compare this distance with the distance from the Earth to the sun (150 million km).

97. Naturally occurring fluoride is entirely ^{19}F, but suppose that it were 50% ^{19}F and 50% ^{20}F. Discuss whether gaseous diffusion of UF_6 would then work to separate ^{235}U from ^{238}U.

98. A 0.165 g sample of impure sodium chlorate ($NaClO_3$) is heated until all of the sodium chlorate decomposes completely into sodium chloride and gaseous oxygen. The impurities generate no gases during this operation. A total of 34.6 mL of dry oxygen gas is collected at a pressure of 751.3 torr and a temperature of 18.6°C. Determine the percentage of sodium chlorate in the impure sample.

CUMULATIVE PROBLEM

Ammonium Perchlorate

Ammonium perchlorate (NH_4ClO_4) is a solid rocket fuel used to propel the space shuttle. When heated above 200°C, it decomposes to several gaseous products, of which the most important are nitrogen, chlorine, oxygen, and water vapor.

(a) Write a balanced chemical equation for the decomposition of NH_4ClO_4, assuming that the products listed are the only ones generated.

(b) The sudden appearance of hot gaseous products in a small initial volume leads to a rapid increase in pressure and temperature that gives the rocket its thrust. What total pressure of gas would be produced at 800°C by igniting 7.00×10^5 kg of NH_4ClO_4 (a typical charge of the booster rockets in the space shuttle) and allowing it to expand to fill a volume of 6400 m^3 (6.40×10^6 L)? Use the ideal gas law.

(c) Calculate the mole fraction of chlorine and its partial pressure in the mixture of gases produced.

(d) The van der Waals equation applies only to pure real gases, not to mixtures. For a mixture like that resulting from the reaction of part (a), it may still be

A space shuttle taking off. *(NASA)*

possible to define effective a- and b-parameters to relate total pressure, volume, temperature, and total chemical amount. Suppose the gas mixture has $a = 4.00$ atm L^2 mol^{-2} and $b = 0.0330$ L mol^{-1}. Recalculate the pressure of the gas mixture in part (b) using the van der Waals equation.

(e) Calculate and compare the root-mean-square speeds of water and chlorine molecules under the conditions of part (b).

(f) The gas mixture from part (b) cools and expands until it reaches a temperature of 200°C and a pressure of 3.20 atm. Calculate the volume occupied by the gas mixture after this expansion has occurred. Use the ideal gas law.

6

Condensed Phases and Phase Transitions

CHAPTER OUTLINE

Solid iodine converts directly to a vapor (sublimes) when warmed at room pressure. Here, the purple iodine vapor is redeposited as solid on the cooled surface above.
(Charles D. Winters)

In the condensed phases of matter, attractive forces hold the constituent particles much closer together than in a gas. In this chapter, we consider some of the properties of the condensed phases in the light of the kinetic–molecular theory.

Transitions between phases (melting, boiling, freezing, and so forth) occur constantly in daily life, in industrial processes, and in research laboratories. Under the right conditions, two, or even three, phases of a substance can coexist. We discuss the dynamic processes that prevail when this happens and introduce a standard graphical display of how temperature and pressure affect phase transitions.

Dissolved species affect phase transitions in a solvent by altering the play of the forces among the particles of the solvent. We discuss some important practical applications of these effects, such as the measurement of the molar masses of unknown substances and purification through distillation.

6-1 Intermolecular Forces: Why Condensed Phases Exist

Matter consists of vast assemblies of molecules, atoms, and other particles that exert all kinds of forces on each other. The discussion of the kinetic theory of gases (in Section 5–7) does not consider these **intermolecular forces,** stating " . . . molecules of a gas exert no forces on one another except during collisions." This approach succeeds for gases at low pressure and low density. It starts to fail for gases at higher pressures, but fairly simple modifications such as those used in the van der Waals equation (see Section 5–8) can compensate. It collapses entirely for liquids and solids. Comparisons of **molar volume,** the space occupied per mole of substance, illustrate the problem. A gas at room conditions occupies approximately $24{,}000 \text{ cm}^3 \text{ mol}^{-1}$, but a solid or a liquid under the same conditions occupies only 10 to $100 \text{ cm}^3 \text{ mol}^{-1}$. The same number of particles occupies much less room because their mutual attractions hold them close together. This is why liquids and solids are called **condensed phases.**

We distinguish *inter*molecular forces from *intra*molecular forces; the latter are the chemical bonds discussed in Chapter 3. Covalent bonds among atoms establish and maintain the structure of discrete molecules and are *strong, directional,* and comparatively *short range* in their effect. Ionic bonds are as strong as (or stronger than) covalent bonds but non-directional and take hold at longer range than covalent bonds.

Intermolecular forces

- are always weaker than chemical bonds, usually much weaker;
- are less directional than covalent chemical bonds but often more directional than ionic bonds;
- operate at longer range than covalent chemical bonds but at shorter range than ionic bonds.

Potential Energy Curves

Graphs of the potential energy of pairs of particles illustrate the distinction between chemical bonds and intermolecular interactions. The existence of attractions among the particles in an assembly means that they have **potential energy** (energy of position) in addition to their kinetic energy (energy of motion). Consider the Moon and the Earth, which attract each other. To move the moon farther from the Earth would require energy to overcome the gravitational attraction between the two. After the move, the pair of bodies would have a higher potential energy. Moving the Moon closer to the Earth would lower the Earth–Moon potential energy. The lowering of potential energy with closer approach would ultimately be interrupted, however. If the surface of the Moon were to contact the surface of the Earth, a very strong repulsion would set in to oppose the closer approach of the bodies.

Figure 6–1 shows how the potential energy depends on the center-to-center distance for several pairs of ions, atoms, and molecules. The curves all have a central dip, but the dips differ in width and depth. At infinite separation, the potential energy of any pair of attracting particles equals zero. The dips show that the potential energy becomes negative as the two particles approach each other. It requires energy from outside to separate the two against their mutual attraction, and so their potential energy as a pair is less. Eventually, the particles "touch," and the potential energy shoots up as repulsive forces become dominant. The repulsive forces, which are very strong and short range, come from the extreme reluctance of core (non-valence) electrons of different particles to occupy the same region of space.

The bottom of the dip in each curve represents a state in which attractive and repulsive forces balance one another. If left alone, a pair of particles settles down

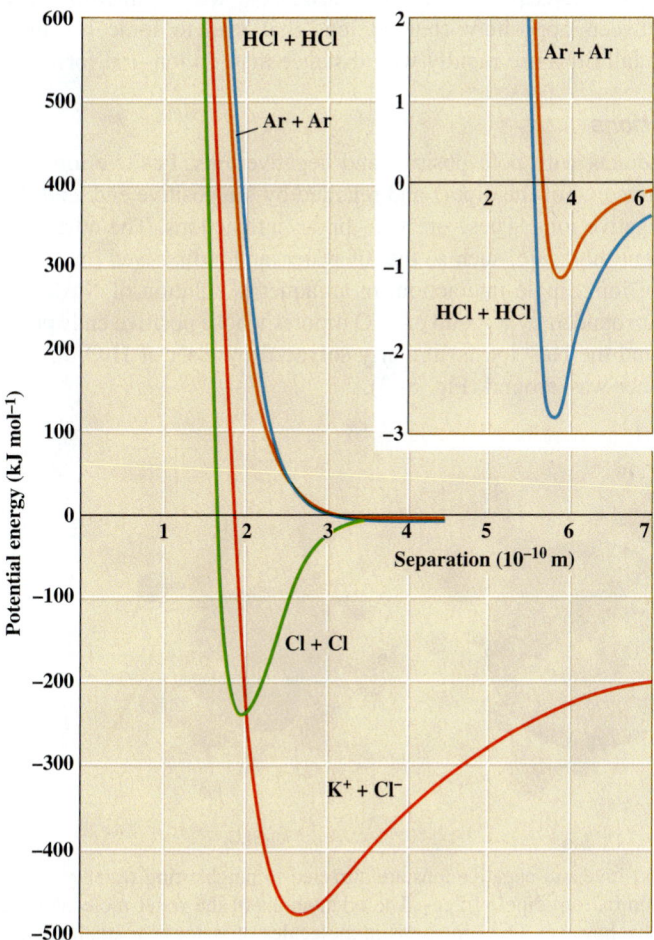

Figure 6–1 • The potential energy of a pair of atoms, ions, or molecules depends on the distance between the members of a pair. Here, the potential energy at large separations (to the right side of the graph) is set to zero. As two particles approach each other, the potential energy becomes negative because attractive forces come into effect. The lowest point in each curve occurs at the distance at which attractive and repulsive forces exactly balance. Strong attractive forces make for a deep dip in the potential energy curve. Note the very shallow potential energy minimum for HCl and Ar. The inset shows these same two curves with the vertical scale expanded by a factor of 100. (The HCl–HCl curve was computed for a pair of molecules lying side by side with the H end of one opposite the Cl end of the other.)

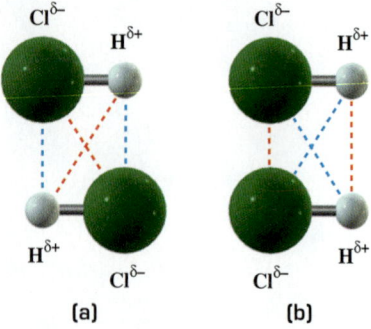

Figure 6–2 • A molecule of HCl is a dipole; it has a small net negative charge on the Cl end (symbolized δ^-), and a small net positive charge on the H end (symbolized δ^+). The interaction between two HCl molecules depends on their orientations. In (a), the oppositely charged ends (*dotted blue lines*) are closer than the ends with the same charge (*dotted red lines*). This gives a net attraction. In (b), the opposite is true, and the molecules repel each other.

• Ion–dipole forces require a mixture of two (or more) substances: one to provide the ions, another to provide the polar molecules.

"in contact" at this separation. The deeper the potential energy well (as the dip is known), the stronger the interaction between the members of the pair. The ion–ion interaction of K^+ with Cl^- (the red line in Fig. 6–1) starts at longest range and is the strongest on the graph, stronger even than the covalent interaction in Cl_2 (the green line). Both rate as true chemical bonds. The HCl–HCl and Ar–Ar interactions are hundreds of times weaker. They are **non-bonded attractions.** Most non-bonded attractions have potential wells no deeper than approximately 5 kJ mol^{-1}. Routine collisions with other particles at room temperature easily supply enough energy to lift interacting pairs out of such shallow wells (and so end the interaction). Chemical bonds have potential wells ranging from 100 to more than 1000 kJ mol^{-1} deep. Collisions at room temperature do not break bonds this strong.

Types of Non-Bonded Attractions

Chemists distinguish types of non-bonded interactions just as they distinguish types of chemical bonds.

Dipole–Dipole Interactions

Polar molecules interact through dipole–dipole forces. As shown in Figure 6–2, two polar molecules may repel each other as well as attract, depending on their orientation. Dipole–dipole forces are moderately strong but weaker than the very strong attractions between oppositely charged ions that lead to ionic bonding. Dipole–dipole forces fall off more rapidly with distance than do ion–ion forces.

Ion–Dipole Interactions

A polar molecule interacts with both positive and negative ions: Positive ions are attracted by the negative end of the dipole and repelled by the positive end, and the reverse is true for negative ions. These are ion–dipole interactions. The attraction between a polar solvent molecule, such as that of water, and a dissolved ion is the most common case of ion–dipole interaction. In an aqueous solution of NaCl, for example, the Cl^- is surrounded by a group of H_2O dipoles whose positive ends point toward the Cl^- ion, and the Na^+ ion is similarly **solvated** by a set of H_2O dipoles that are turned the other way around (Fig. 6–3).

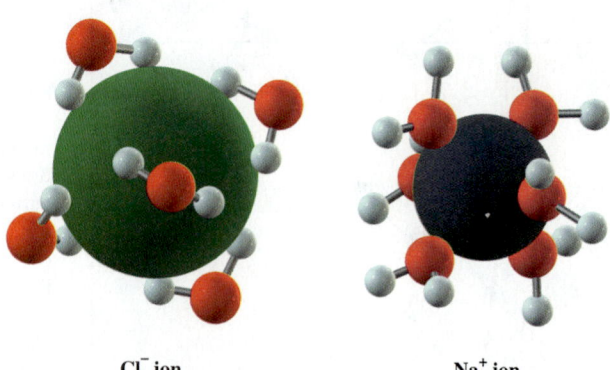

Cl^- ion Na^+ ion

Figure 6–3 • Both positive and negative ions are attracted to neighboring water molecules in aqueous solution by ion–dipole forces. The orientations of the water molecules are reversed in the two cases, however. Oxygen atoms in molecules of water bear small negative charges; hydrogen atoms bear small positive charges.

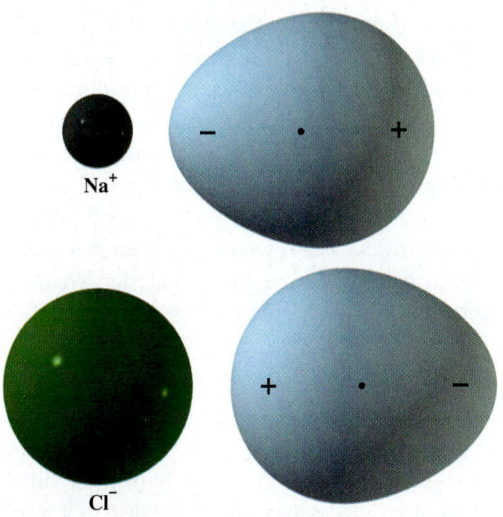

Figure 6–4 • As an ion approaches an atom or molecule, its electrostatic field distorts the distribution of the outer electrons. The effect of this distortion is to create a dipole moment that exerts an attractive force back on the ion. Here, the average electron densities are shown by shading.

Induced Dipole Attractions

The electrons in a non-polar molecule or atom are distributed symmetrically, but an approaching electric charge can distort the symmetry and so induce a dipole. An argon atom, for example, has no dipole moment, but an approaching Na^+ ion (with its positive charge) attracts the electrons on its near side more strongly than the ones on its far side. By tugging the nearer electrons harder, the Na^+ ion induces polarity in the argon atom where none existed before (Fig. 6–4). Once the induced dipole is present, the situation resembles the ion–dipole case just described. Negative ions and other dipoles also induce dipoles in non-polar molecules. Induced dipole attractions are weak and are effective only at short range.

Dispersion Forces

Helium atoms are uncharged and non-polar, so nothing mentioned so far explains the observation that helium atoms attract each other weakly but measurably. Such attractions derive from dispersion forces, which exist between all atoms and molecules. They arise from fluctuations over time in the distribution of the electrons on two neighboring atoms or molecules. A fluctuation in one molecule (making a temporary dipole) induces a second temporary dipole in the other. The interaction of these two dipoles then causes an attractive force between the two molecules. Figure 6–5

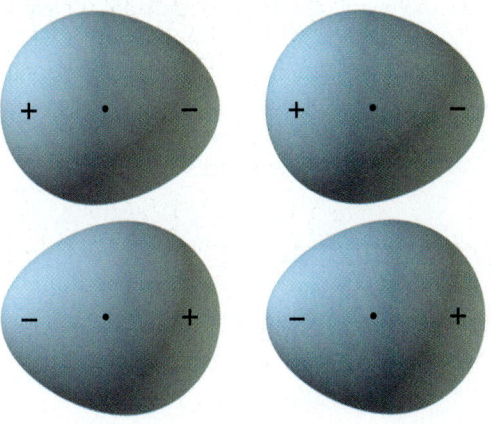

Figure 6–5 • A temporary fluctuation of the electron distribution on one atom induces a temporary dipole moment on a neighboring atom. The two dipole moments interact to give a net attractive force, called a "dispersion force."

provides an impression of the source of this interaction. The strength of dispersion forces increases with the number of electrons in the atom or molecule, so larger atoms or molecules interact more strongly than smaller ones. Among dispersion-"bonded" diatomic systems, such as He_2, Ne_2, and Ar_2, the interaction grows stronger going down a group in the periodic table. This contrasts with true covalently bonded diatomic molecules, in which the bonding generally becomes weaker going down a group.

The whole set of non-bonded interactions is often referred to as **van der Waals forces** in recognition of the early contributions of Johannes van der Waals to understanding how intermolecular forces influence the physical behavior of solids, liquids, and gases.

Hydrogen Bonding

One particularly strong type of dipole–dipole attraction is the **hydrogen bond.** Hydrogen bonds form when a hydrogen atom that is covalently bonded to a nitrogen, oxygen, or fluorine atom interacts with the lone electron pair of a second such atom nearby. Hydrogen bonds are the strongest kind of non-bonded attractions but are still much weaker than are regular covalent and ionic bonds.

For example, water molecules in the gas phase form a weakly associated $(H_2O)_2$ dimer (double molecule) by means of hydrogen bonding. Experimental studies have established the structure of this entity. As Figure 6–6 shows, a hydrogen atom from one water molecule slightly penetrates the "hard edge" of the oxygen atom of the other. This hydrogen in the middle keeps a regular covalent bond with its own oxygen atom but forms a longer, weaker bond (the hydrogen bond) with the second oxygen atom. It lies almost exactly on a straight line joining the two oxygen atoms and significantly closer to the second oxygen than would be expected from the usual steepness of repulsive forces. In the H_2O–H_2O hydrogen bond, the hydrogen atom has a fractional positive charge and is attracted to the fractional negative charge on the neighboring oxygen atom, as would be expected in any dipole–dipole attraction. The interaction is stronger than other dipole–dipole attractions for two reasons. First, the nucleus of a bonded H atom in water is only poorly shielded by electrons because, unlike all other elements, hydrogen has no core electrons. As a result, the bonded H atom can approach very close to the lone-pair electrons on the oxygen atom of a second molecule, leading to a large electrostatic interaction between the two. Second, a small amount of covalent bonding arises from the sharing of electrons between the two oxygen atoms and the intervening hydrogen atom. Covalent bonding explains why the hydrogen atom penetrates the oxygen atom. These details are typical of most hydrogen bonds.

Hydrogen bonds may be intermolecular (between molecules) or intramolecular (within a molecule). The latter shape the structures of many important and complex biological substances, as discussed in Chapter 25. The former are important in the liquid-state and solid-state structures of many simple substances.

• Hydrogen bonding also occurs when the hydrogen is bonded to other atoms. "Weak hydrogen bonds" involving hydrogen that is bonded to carbon are important in some biomolecules.

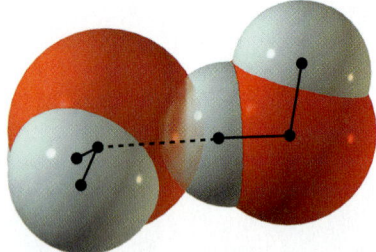

Figure 6–6 • A single hydrogen bond linking two otherwise isolated molecules of water. Hydrogen bonds are weaker than ordinary covalent bonds but stronger than other non-bonded attractions: it takes approximately 23 kJ to break a mole of the bonds shown here. Upwards of 150 kJ is required to break a mole of covalent bonds, but only 2 to 3 kJ is required to overcome a mole of typical dipole–dipole non-bonded attractions.

EXAMPLE 6–1

State which attractive forces are likely to predominate in the associations among molecules in the following substances:
(a) F_2 (b) HBr   (c) KCl   (d) HF

Solution

(a) Molecules of F_2 are non-polar, so the predominant attractive forces between molecules in F_2 are dispersion forces.

(b) The HBr molecule has a permanent dipole moment. The predominant forces between molecules are dipole–dipole. Dispersion forces also contribute to associations, especially because Br has a rather large number of electrons.

(c) The potassium ions are attracted to the chloride ions primarily by ion–ion forces.

(d) Liquid hydrogen fluoride has hydrogen bonds (dipole–dipole attractions) between HF molecules. Dispersion forces contribute as well.

EXERCISE

List the types of attractive intermolecular forces that exist among the particles in each of these substances: (a) NaH (b) ClBr (c) Rn (d) NH_3

Answer: (a) Ion–ion forces and dispersion forces. (b) Dipole–dipole forces and dispersion forces. (c) Dispersion forces. (d) Hydrogen bonding (dipole–dipole) and dispersion forces.

6-2 The Kinetic Theory of Liquids and Solids

The term "intermolecular attractions" refers to all attractions among the constituent particles of a material, whether they are molecules, atoms, or ions. Such attractions are responsible for the existence of solids and liquids. At very high temperatures, the high kinetic energy of the molecules overpowers all attractions: all materials are gaseous at a high enough temperature. Lowering the temperature lowers the average kinetic energy, and intermolecular attractions, which do not change with temperature, gain significance. They cause particles to congregate in small, loose clusters that grow slowly in size until, as the temperature drops further, droplets of liquid suddenly condense. At low enough temperatures, liquids congeal into solids. The temperatures at which liquefaction and freezing occur depend on the strengths of the intermolecular attractions, which naturally vary from substance to substance.

Because each molecule in a liquid or solid interacts strongly with many neighboring molecules, simple equations of state, such as the ideal gas equation or the van der Waals equation, are not available. A vivid picture of motions at the molecular level in solids and liquids has, however, emerged from experiment and calculation.

The Structure of Solids and Liquids

The similar molar volumes of solid and liquid forms of the same substance suggest that the distances between neighboring molecules in the two states must be approximately the same. Density measurements (see Section 1–9) show that *intermolecular contacts*, the shortest distances between the nuclei of atoms belonging to adjacent molecules, usually range between 3×10^{-10} m and 5×10^{-10} m in solids and liquids. At these distances, longer-range attractive forces and shorter-range repulsive forces just balance one another, giving a minimum in the potential energy. Although these intermolecular separations are significantly longer than most chemical bond distances (which range from approximately 0.5 to 2.5×10^{-10} m), they are much shorter than the intermolecular separations in gases, which average around 30×10^{-10} m under room conditions. The distinction is shown schematically in Figure 6–7.

Figure 6–7 • Intermolecular forces create structure in liquids and solids. If a single atom is removed from a snapshot of the atomic arrangement in a solid (a), it is easy to figure out exactly where to put it back in. For a liquid (b), the choices are limited. If a single atom is removed from a gas (c), no clue remains to tell where it came from.

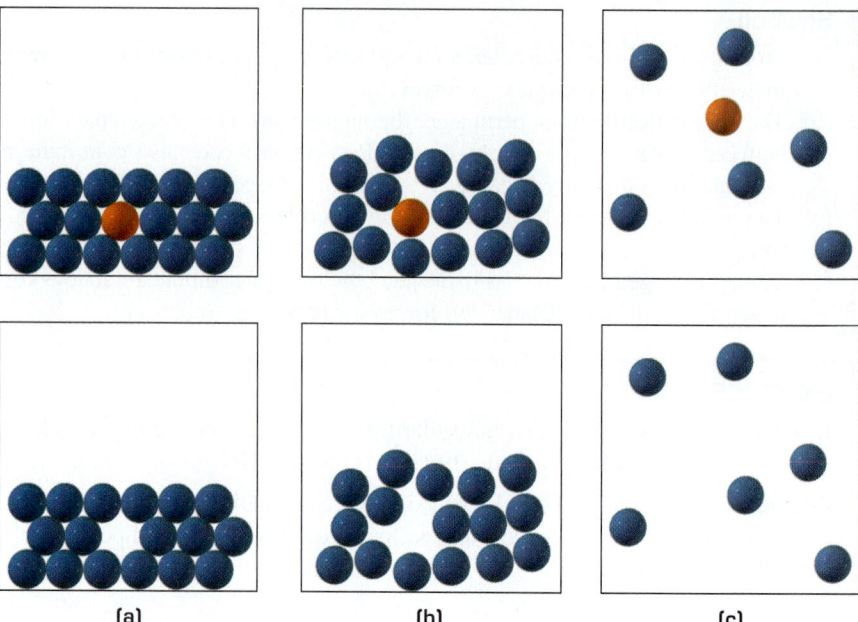

(a)   (b)   (c)

Gases change their volumes dramatically with changes in temperature, but liquids and solids do not. This observation is consistent with the presence of strong intermolecular attractions in the condensed states and their absence in the gaseous state. An increase in volume in a solid or liquid requires that attractive forces between molecules and their neighbors be partially overcome. A solid or liquid has stronger intermolecular attractions and does not expand as much for a given temperature increase as a gas, in which attractions play a much smaller role.

The long stretches of empty space between gas molecules explain why gases are so easily compressed compared with liquids and solids. It costs relatively little in energy to bring widely separated molecules closer together. Attempts to compress liquids and solids, however, must operate against the strong repulsive forces that come into effect once the molecules are in contact.

Motion in Solids and Liquids

Figure 6–7 shows a "snapshot" of a liquid and a solid, fixing the positions of the particles that make it up at a particular instant in time. The paths followed by the particles in these two states can also be examined over a short time interval (Fig. 6–8). In liquids, molecules have the freedom to travel through the sample, changing neighbors constantly in the course of their wanderings. In a solid, the molecules vibrate around their home positions, but mostly they stay close to that position. This explains why liquids flow more or less readily in response to an external stress: Their molecules can quickly change neighbors and tumble past each other. The rigidity of solids against external stress suggests, in contrast, a durable arrangement of neighbors around any given particle. This is the crucial difference between the solid and liquid states.

In a liquid, individual molecules experience interactions with neighbors that lead, at any instant, to a local environment that may closely resemble that in a solid, but then they quickly move on. Their trajectories consist of "rattling" motions in a

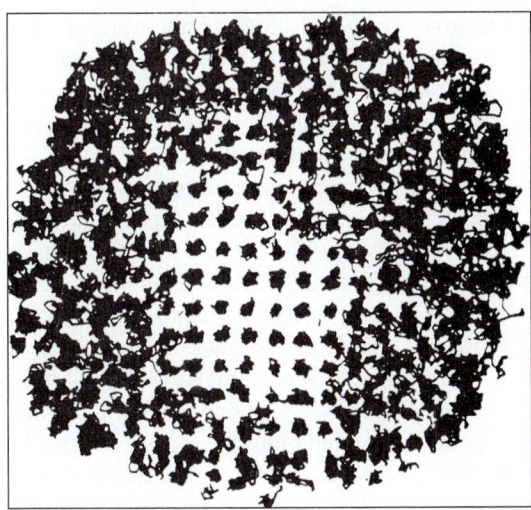

Figure 6-8 • In this computer-simulated picture of the motions of atoms in a tiny melting crystal, the atoms at the center (in the solid) move erratically around particular sites. The atoms at the surface (in the liquid) move over much greater distances.

temporary cage formed by neighbors and superimposed on erratic displacements over larger distances. In this respect, a liquid is intermediate between a gas and a solid. A gas (see Fig. 5–20) provides no temporary cages, so each molecule of a gas travels a longer distance before colliding with a second molecule. Consequently, diffusion is faster in a gas than in a liquid. In a solid, the cages are nearly permanent, so diffusion is very slow. When a solid is heated sufficiently, the thermal energy increases the amplitude of vibration of the molecules in the solid to the point that they are set free to make major excursions; the solid becomes a liquid. This process is called **melting** or **fusion.**

These ideas on motion in liquids and solids fit nicely with the observed rapidity with which liquids can dissolve substances compared with solids. Individual molecules of a liquid quickly wander into contact with those of an added second substance, and new attractions between the unlike particles have an early chance to replace those existing originally in the pure liquid.

6-3 Phase Equilibrium

Recall that a phase is a sample of matter that is uniform throughout, both in its chemical constitution and in its physical state (see Section 1–2). Our classification and discussion of the states of matter of a substance have so far focused on one phase at a time (solid, liquid, or gas), but two or more phases can coexist as well (for example, an ice cube floating in liquid water or mercury vapor confined above a column of liquid mercury in a thermometer). In this section, we examine the properties of coexisting phases of substances and the dynamic processes by which phase equilibrium is reached.

Vapor Pressure

Suppose we put a quantity of liquid water into an evacuated flask. We hold its temperature at 25.0°C by placing the flask in a constant-temperature bath, and we use a pressure gauge to monitor how the pressure inside the flask changes with time. Immediately after the water enters the flask, the pressure in the space above it begins to increase from zero. It increases rapidly at first but soon levels off at a value of

0.03126 atm, which is the **vapor pressure** of water at 25°C. Liquid water remains visible inside the flask, which now contains water vapor as well. Nothing new happens as long as the flask is left to itself at 25°C. The **system** (the contents of the flask) has reached **equilibrium,** a condition in which no further changes in macroscopic properties tend to occur. The passage toward equilibrium that was just described is a spontaneous process, occurring in the closed system without any external prompting. If we remove some of the water vapor that has formed, the vapor pressure in the flask is temporarily less than its equilibrium value, but it increases with time to re-establish itself at $P_{vap}(H_2O) = 0.03126$ atm. If we admit air to the flask and then reseal it, $P_{vap}(H_2O)$ stays very close to 0.03126 atm.

What happens on a microscopic scale to cause the spontaneous movement of the system toward equilibrium? According to kinetic theory, the molecules of water in the liquid are in constant motion, a state of thermal agitation. Some of those near the surface move fast enough to escape from the liquid into the space above it; this process of **evaporation** leads to the observed increase in pressure. Evaporation is the familiar process by which water (and other liquids) left out in open containers eventually disappears completely. Of course, the water evaporated from a puddle in the sunshine does not cease to exist; rather, it becomes widely dispersed in the air. In our experiment, the closed container keeps the evaporated water in the space above the liquid. As the number of water molecules in the space increases, some are recaptured by the liquid when they chance to strike its surface; this is **condensation.** As the pressure of the vapor increases, the rate of condensation increases until it balances the rate of evaporation from the surface (Fig. 6–9). Once this occurs, there is no further net flow of matter from one phase to the other. The system has reached an equilibrium that is characterized by a particular value of the vapor pressure. Water molecules continue to evaporate from the surface of the liquid, but other water molecules return to the liquid from the vapor at an equal rate.

The vapor pressure of the water is independent of the size and shape of the container. If the experiment is duplicated in a larger (or smaller) flask, then a greater (or lesser) *amount* of water evaporates on the way to equilibrium, but the final pressure in the flask at 25°C still equals 0.03126 atm as long as some liquid water is

• It is essential to distinguish between two similar terms: the "pressure of a vapor" and the "vapor pressure of a liquid." At a given temperature, the (equilibrium) vapor pressure above a liquid has a unique value (see Chapter 7), but the pressure of a vapor may equal any value.

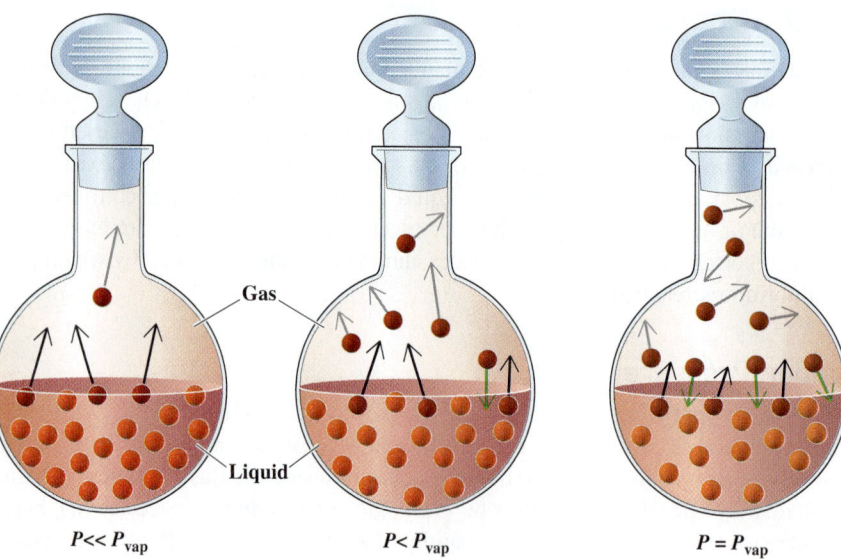

Figure 6–9 • The approach to equilibrium in evaporation and condensation. Initially, the pressure above the liquid is very low, and many more molecules leave the surface of the liquid than return to it. As time passes, more molecules fill the gas phase until the equilibrium vapor pressure (P_{vap}) is approached; the rates of evaporation and condensation then become equal.

Gas

Liquid

$P \ll P_{vap}$ $P < P_{vap}$ $P = P_{vap}$

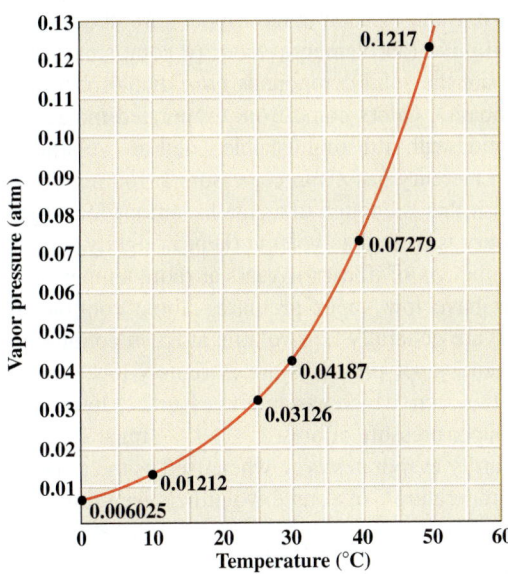

Figure 6-10 • The vapor pressure of water, like that of all other substances, increases exponentially with increasing temperature.

present. If the experiment is repeated at a temperature of 30.0°C, everything happens as just described, except that the pressure in the space above the water settles at a final value of 0.04187 atm. A higher temperature corresponds in the kinetic theory to a larger average kinetic energy among the water molecules. A new balance between the rates of evaporation and condensation is struck, but at a higher vapor pressure. The vapor pressure of water, and of all other substances, increases with increasing temperature (Fig. 6-10).

The random jostling of thermal motion eventually gives a few molecules in a liquid *or* solid whatever speed and direction they need to evaporate. Molecules in all liquids and solids thus have escaping tendencies: All liquids and solids possess a vapor pressure. *Volatile* liquids and solids have vapor pressures large enough that

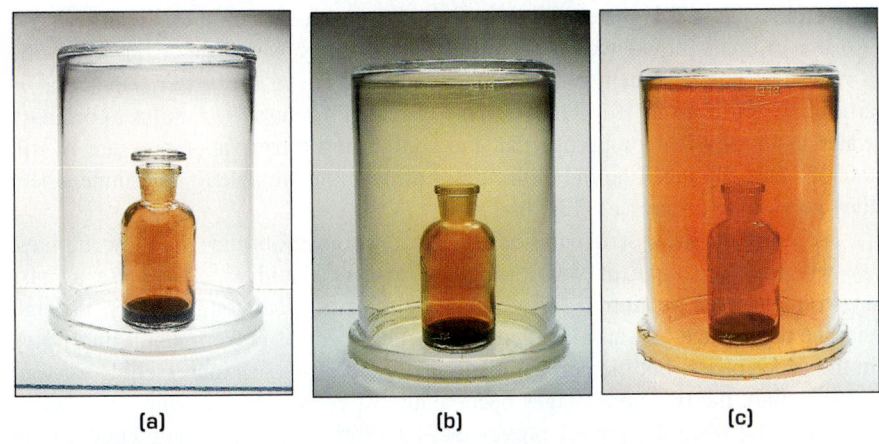

(a)   (b)   (c)

The volatile liquid bromine (Br_2) as it comes to equilibrium with its vapors at room temperature. (a) The bottle on the inside contains the element bromine (Br_2), in both liquid and vapor form. (b) After the bottle top is removed, reddish brown bromine vapors begin to fill the outer jar. (c) The reddish brown color of the vapor has stopped deepening. Liquid bromine (still visible inside the bottle) and bromine vapor are in equilibrium within the larger jar. *(Charles D. Winters)*

they evaporate from open containers reasonably soon at the temperature in question. At room temperature, the liquids bromine (an element), water (a compound), and gasoline (a mixture) are volatile, as are the solid compounds *para*-dichlorobenzene and naphthalene (both used in mothballs). Odors come from evaporated molecules that reach the nose, so liquids or solids with a distinguishable smell are volatile or contain a volatile component. Liquid mercury has a vapor pressure at room temperature of only 1.6×10^{-6} atm. Because this is enough to create hazardous local concentrations of mercury vapors, mercury is treated as volatile (kept in tightly capped bottles) despite the fact that open containers of it last for years at room temperature.

Non-volatile liquids and solids have low vapor pressures. Ionic compounds ($NaCl$, Na_2SO_4, $CaBr_2$, and so forth) are generally non-volatile at room conditions, as are most metals and minerals. Some vapor pressures are extremely low. One of the least volatile of all substances is tungsten, which has an immeasurably low vapor pressure at room temperature and a vapor pressure of only 2×10^{-25} atm at 1000°C. This property means that tungsten hardly evaporates even when used as the glowing filament in a lightbulb. Very high temperatures, of course, volatilize even tungsten.

- This vapor pressure corresponds to 1 atom of tungsten in a volume of 868 L.

The Dynamic Nature of Phase Equilibrium

The equilibrium coexistence between two phases, such as liquid and vapor, is the time-independent state achieved after the initial effects of the method of preparation of the system have died away. Once a system is at equilibrium, no further macroscopic changes can be seen, but activity continues at the molecular level. Phase equilibrium is a *dynamic* process that is quite different from the static equilibrium reached as a marble rolls to a stop after being spun into a bowl. In the equilibrium between liquid water and water vapor, the partial pressure levels off, not because evaporation and condensation stop, but rather because their rates become equal. In the equilibrium between ice and water, molecules likewise continue to pass between the two phases, although no change is observable macroscopically.

An elegant experiment demonstrates the dynamic nature of phase equilibrium. We inject some liquid water at 25°C into an evacuated flask. At the same time, we inject enough water vapor to give a pressure of 0.03126 atm. (This particular value is chosen because it is the equilibrium vapor pressure of water at this temperature.) We use ordinary liquid water, but instead of ordinary water vapor, in which the isotope ^{18}O has a percent abundance of only 0.2%, we use water vapor that is artificially enriched to 100% in ^{18}O. This "labelled" water is so similar to ordinary water in its physical properties that the equilibrium water-vapor pressure is still 0.03126 atm. By labelling the molecules, however, we are able to determine where they go as time progresses (Fig. 6–11).

After letting the system equilibrate (start toward equilibrium) for a few minutes, we withdraw a sample of water vapor and one of liquid water and use a mass spectrometer (see Section 1–6) to determine their isotopic compositions. We find that the percentage of water labelled with ^{18}O has decreased from its original value in the vapor phase, but the percentage in the liquid phase has increased. If we wait long enough, the two percentages become almost equal. The exchange of labelled molecules between the phases proves that equilibrium is not a static condition in which evaporation and condensation cease. The two processes continue unabated but balance each other at equilibrium.

- Not *exactly* equal because there is a very slight difference between the vapor pressures of $H_2{}^{16}O$ and $H_2{}^{18}O$. A larger effect on vapor pressure is seen if hydrogen is replaced by its heavy isotope deuterium.

Another important feature of equilibrium that can be observed from the liquid–vapor system is that its properties are independent of the direction from which it is approached. If we inject enough water vapor into an empty flask to make the initial pressure higher than the vapor pressure of liquid water, $P_{vap}(H_2O)$, then liquid water

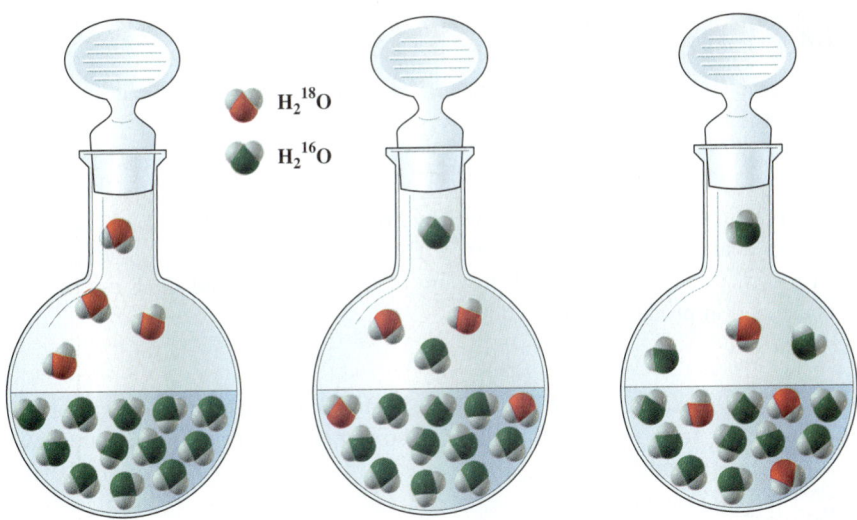

H$_2$^{18}O

H$_2$^{16}O

Figure 6-11 • In this experiment, the liquid water has the ordinary isotopic composition with only 0.2% ^{18}O atoms, whereas the vapor consists entirely of water with ^{18}O (*left*). As time goes on, the fraction of ^{18}O in the liquid phase increases, whereas that in the gas phase decreases, until eventually the two become almost equal (*right*).

condenses until $P_{vap}(H_2O)$ is reached (0.03126 atm at 25°C). The equilibrium is then indistinguishable from one set up starting with liquid water. Of course, if we do not use enough water vapor to exceed a pressure of $P_{vap}(H_2O)$, all the water remains in the vapor phase, and the two-phase equilibrium does not occur.

Correcting for the Vapor Pressure of Water in "Wet" Gases

Chemical reactions carried out in aqueous solution often generate gaseous products. It is sometimes necessary, for analytical purposes, to collect a product gas in such situations and measure its volume, with a view to then computing its chemical amount. A gas collected in this way, *over water,* is humid, or "wet." It contains a partial pressure of water vapor. According to Dalton's law of partial pressures (see Section 5–6), subtracting this partial pressure from the total pressure of the wet gas gives the pressure that the product gas *would* exert if it were dry. This is the pressure that must be used in calculations. The vapor pressure of water, the quantity that is subtracted, depends only on the temperature, and accurate values are widely available in tables (Table 6–1). Correcting for the vapor pressure of water is essential in quantitative work involving the collection of gases over water.

TABLE 6-1	Vapor Pressure of Water at Various Temperatures	
Temperature (°C)	**Vapor Pressure (atm)**	
15.0	0.01683	
17.0	0.01912	
19.0	0.02168	
21.0	0.02454	
23.0	0.02772	
25.0	0.03126	
30.0	0.04187	
50.0	0.1217	
100.0	1.0000	

EXAMPLE 6-2

Zinc reacts with dilute aqueous sulfuric acid to produce hydrogen according to the net ionic equation:

$$Zn(s) + 2\,H^+(aq) \longrightarrow Zn^{2+}(aq) + H_2(g)$$

The reaction is run at 30°C, and wet hydrogen having a volume of 17.2 mL is collected at a pressure of 757.1 torr. The vapor pressure of water at 30°C is 31.8 torr. Calculate the partial pressure of the hydrogen and the mass of zinc consumed in producing this hydrogen.

Strategy

Obtain P_{H_2} by subtracting P_{H_2O} from the total pressure. Use the ideal gas law to compute n_{H_2} from P_{H_2}. Use the stoichiometry of the equation to obtain the mass of Zn that reacted.

Solution

$$P_{H_2} = P_{total} - P_{H_2O} = 757.1\ \text{torr} - 31.8\ \text{torr} = 725.3\ \text{torr}$$

$$P_{H_2} = 725.3\ \text{torr} \times \left(\frac{1\ \text{atm}}{760\ \text{torr}}\right) = 0.9543\ \text{atm}$$

Convert the 17.2 mL to liters (0.0172 L) and the temperature to kelvins (303 K). This allows use of the gas constant in units of L atm mol^{-1} K^{-1}:

$$n_{H_2} = \frac{P_{H_2}V}{RT} = \frac{(0.9543\ \text{atm})(0.0172\ \text{L})}{\left(\dfrac{0.08206\ \text{L atm}}{\text{mol K}}\right)(303\ \text{K})} = 6.60 \times 10^{-4}\ \text{mol}\ H_2$$

$$m_{Zn} = 6.60 \times 10^{-4}\ \text{mol}\ H_2 \times \left(\frac{1\ \text{mol Zn}}{1\ \text{mol}\ H_2}\right) \times \left(\frac{65.39\ \text{g Zn}}{1\ \text{mol Zn}}\right)$$

$$= 4.32 \times 10^{-2}\ \text{g Zn} = 43.2\ \text{mg Zn}$$

If the vapor pressure of water is not subtracted, the (incorrect) answer is 45.1 mg, which is approximately 4% high, a serious error.

EXERCISE

Passage of an electric current through a dilute aqueous solution of sodium sulfate produces a mixture of gaseous hydrogen and oxygen according to the equation:

$$2\,H_2O(\ell) \longrightarrow 2\,H_2(g) + O_2(g)$$

One liter of the mixed gases is collected over water at 25°C and under a total pressure of 755.3 torr. Calculate the mass of oxygen that is present. The vapor pressure of water is 23.8 torr at 25°C.

Answer: $m_{O_2} = 0.421$ g.

6-4   Phase Transitions

Suppose that one mole of $SO_2(g)$ is compressed in a device that maintains the temperature of the sample at exactly 30°C (near room temperature) and records the volume of the sample continuously (red line in Fig. 6–12). At low pressures,

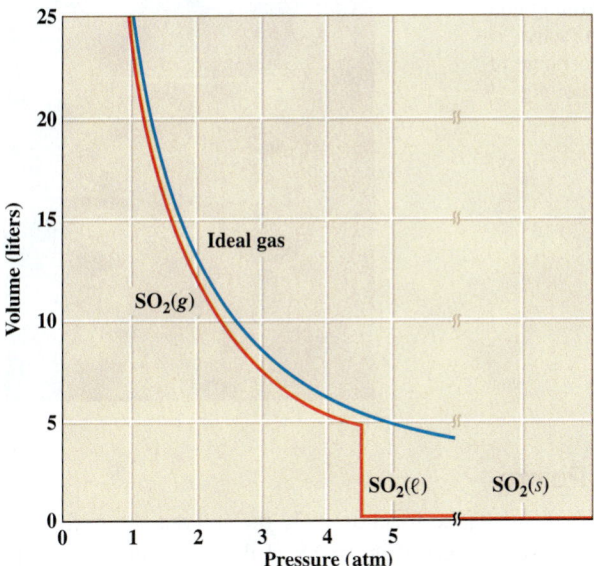

Figure 6-12 • As one mole of gaseous SO_2 is compressed at a constant temperature of 30°C, the volume at first falls somewhat below its ideal gas value. Then at 4.52 atm, the volume falls abruptly as the gas condenses to a liquid. At a much higher pressure, a further transition to the solid takes place.

the observations closely follow the inverse dependence that is predicted by the ideal gas law:

$$V_{SO_2} = \frac{nRT}{P} = \frac{(1 \text{ mol})\left(0.08206 \dfrac{\text{L atm}}{\text{mol K}}\right)(303.15 \text{ K})}{P} = 24.88 \text{ L atm}\left(\frac{1}{P}\right)$$

At higher pressures, deviations from ideality start to appear. The observed volume falls below the prediction of the ideal gas equation, but is still fairly well predicted by the van der Waals equation described in Section 5–8. Suddenly, at 4.52 atm, the volume of the sample drops abruptly by a factor of 100! What has happened? The SO_2 has been liquefied solely by the application of pressure. Further compression causes another abrupt (but smaller) change in volume as the liquid SO_2 freezes to a solid (still at a temperature of 30°C). Both events are called **phase transitions.** Phase transitions also occur with changes in temperature at constant pressure. If steam is cooled at 1 atm pressure, it condenses to liquid water at 100°C (Fig. 6–13) and freezes to solid ice at 0°C. Six phase transitions are conceivable among the three usual states of matter, and all occur (Fig. 6–14).

Figure 6-13 • Steam (water vapor at a temperature exceeding 100°C) in the hot gases from a flame condenses to liquid water on the cool interior surface of a beaker. *(Charles D. Winters)*

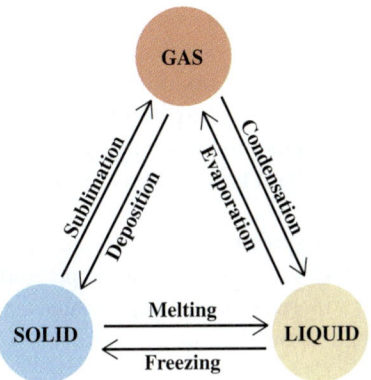

Figure 6-14 • Direct transitions between all of the three common states of matter are possible and are observed in everyday life.

Figure 6–15 • Water at a rolling boil. Only in this state of *ebullition* (active bubbling agitation) is the temperature of the whole of the liquid at the boiling point. *(Charles D. Winters)*

Boiling and Melting Points

In **boiling,** bubbles of gaseous substance form actively throughout the body of a liquid and rise up to escape at the surface (Fig. 6–15). Like evaporation, boiling originates in the escaping tendency of the molecules in a liquid. However, the two phenomena do differ. Boiling occurs throughout the body of a liquid; evaporation takes place only at the surface. Boiling can occur only when the vapor pressure of a liquid exceeds the external pressure under which it is held; evaporation takes place regardless of the external pressure. The **boiling point** is the temperature at which the vapor pressure of a liquid equals the external pressure. The external pressure influences boiling points quite strongly; for example, water boils merrily at 25°C if the external pressure is reduced below 0.03126 atm (recall that the vapor pressure of water at 25°C is just this number), but water must be heated to 121°C to boil under an external pressure of 2.0 atm. The **normal boiling point** of a liquid is defined as the temperature at which the vapor pressure of the liquid equals 1 atm.

In melting, a solid loses its fixed shape, slumping into a liquid. A **normal melting point** is the temperature at which a solid and its liquid are in equilibrium under a pressure of 1 atm of air. The normal melting point of ice is 0.00°C. Liquid water and ice coexist indefinitely (are in equilibrium) at this temperature at a pressure of 1 atm. If the temperature is lowered by even a small amount, then all of the water eventually freezes; if the temperature is raised slightly, then all of the ice eventually melts.

Normal boiling points and melting points are easily determined *characterizing* properties of pure substances. Observing the boiling and melting points of an unknown can confirm an identification made by other means. Caution is required because impurities in substances often markedly depress melting points and elevate boiling points (see Section 6–6). Fortunately, impurities nearly always betray themselves by blurring the sharp melting or boiling points exhibited by pure substances into melting or boiling ranges. For this reason, an examination of melting or boiling behavior frequently provides a helpful check of the purity of a sample.

Phase transitions are not always prompt but can require some time to take place. An example is the **superheating** of a liquid. Liquid water can reach a temperature somewhat above 100°C if heated rapidly. When vaporization of a superheated liquid *does* occur, it can be quite violent, with liquid thrown out of the container. To

• Cold tap water contains dissolved air. Heating it expels the air as very small bubbles. This is not boiling. In true boiling, the escaping gas and the liquid are the same substance.

• The qualifying term "normal" is often omitted in talking about melting points because they depend only weakly on the pressure.

avoid such unpleasant events in the laboratory, small porous particles (called "boiling chips") may be added to the liquid before it is heated. The chips help to initiate boiling as soon as the normal boiling point is reached by providing sites where bubbles of gas can form. **Undercooling** of liquids below their freezing points is also possible. In very careful experiments, undercooled liquid water has been studied at temperatures below $-30°C$ (at room pressure).

> • Undercooling is also called "supercooling."

Many materials react chemically when heated, before they have a chance to melt or boil. Such materials include household goods such as wood, cloth, paper, and some plastics (Fig. 6–16). Substances that change their chemical identity before changing state lack normal melting points or boiling points (or both). Cooking oil, for example, smokes and darkens (signs of chemical decomposition) when heated strongly at room pressure, but it does not boil. Sucrose (table sugar) melts, but it quickly begins to darken and eventually chars at its melting point (Fig. 6–17). A temperature hot enough to overcome the intermolecular attractions in sugar is also sufficient to break apart the molecules of sugar themselves.

Intermolecular Forces and Phase Transitions

The recorded normal melting and boiling points of substances have a prodigious range. Among the elements, the lowest melting point is near absolute zero (for helium) and the highest boiling point is over 5600°C (for tungsten).

> The stronger the intermolecular attractions in a substance, the lower the vapor pressure at any temperature.

Figure 6-16 • These materials have no melting or boiling points. They undergo chemical changes (decomposition) when heated, before undergoing physical changes (melting or boiling). *(Leon Lewandowski)*

Substances with strong intermolecular attractions require high temperatures to make their vapor pressures equal 1 atm; therefore, they have high boiling points. For this reason, ionic compounds generally boil at higher temperatures than substances with polar molecules, which in turn generally boil at higher temperatures than substances with non-polar molecules. In a group of similar substances in which the molar mass increases, such as the noble gases, the normal boiling point increases with the increasing strength of the dispersion forces. Thus, radon has the highest boiling point of the noble gases and helium the lowest.

Melting points depend on molecular shapes and on details of the molecular interactions more than do boiling points, so they vary less systematically with the strength of the attractive forces.

Figure 6-17 • When sugar is heated at room pressure, it melts and simultaneously partly decomposes to give a dark-colored "caramelized" mixture. *(Leon Lewandowski)*

The Special Case of Water

The physical and chemical properties of water shape every detail of the terrestrial environment. Many of these properties, especially those involving phase transitions, differ sharply from what would be expected by comparison with the properties of other substances of low molar mass. In particular:

1. Water expands when it freezes. Consequently, ice has a lower density than liquid water and floats in it. Most substances contract when they freeze.
2. Water has a boiling point much higher than would be expected from the trend set by the boiling points of other Group VI hydrides (H_2S, H_2Se, and H_2Te). It also has an abnormally high melting point.
3. The boiling of water (converting liquid water to steam) requires input of an exceptionally large amount of heat per mole (approximately 45 kJ mol^{-1}). The reverse transition, the condensation of steam (or water vapor) to liquid water, releases the same large amount of heat.

These properties derive from the many hydrogen bonds that exist among molecules in ice and liquid water. A water molecule can accept two hydrogen bonds at its oxygen end and also donate two hydrogen bonds. If all possible hydrogen bonds form in a mole (N_0 molecules) of H_2O, then every oxygen atom is surrounded tetrahedrally by four hydrogen atoms—its own two and two from neighboring molecules. The hydrogen bonds number $2 N_0$ and form a three-dimensional network. Such a network exists in ice. The small number of water molecules surrounding a given water molecule in such a network gives ice an unusually open structure (Fig. 6–18) and a lower density than would be expected in the absence of hydrogen bonds. When ice melts, some hydrogen bonds break, and the open structure of Figure 6–18 partially collapses. The result is a liquid with a smaller volume (higher density), as diagrammed in Figure 6–19.

• Not 4 N_0. Consider that every hydrogen bond has a water molecule at each of its ends.

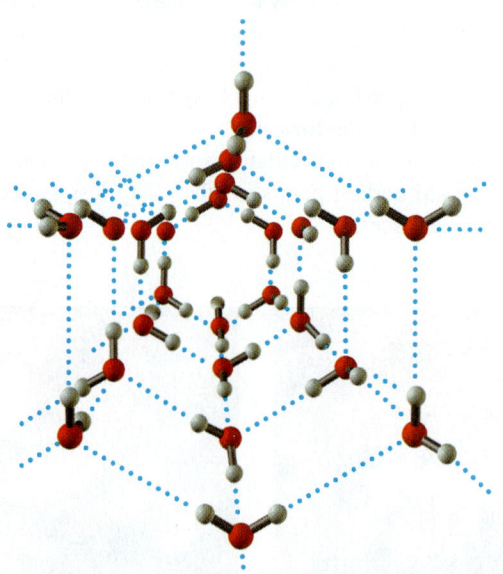

Figure 6–18 • The structure of ice. Each water molecule has four nearest neighbors. A given O atom "owns" two H atoms and is hydrogen-bonded to another two. The structure is quite open because of the hydrogen bonding. The array of linked hexagonal rings in this structure is echoed by the characteristic six-fold symmetry of snowflakes.

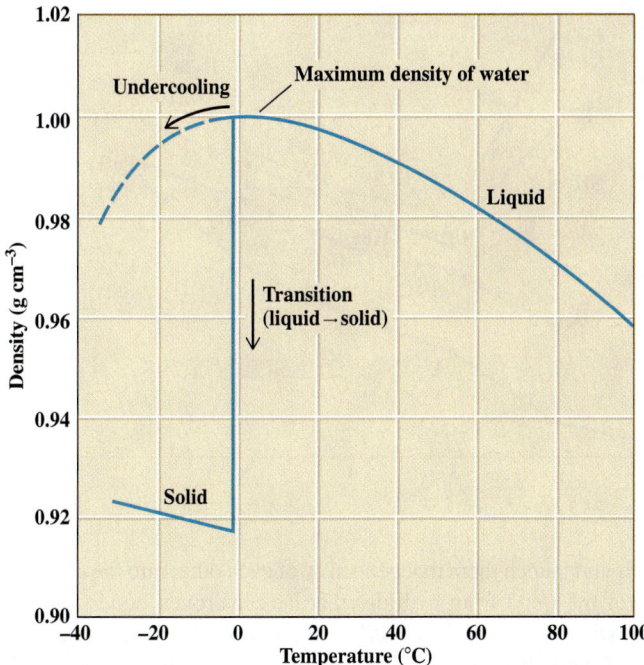

Figure 6–19 • The density of water rises to a maximum as it is cooled to 3.98°C and then slowly starts to decrease. Undercooled water (water chilled below its freezing point but not yet converted to ice) continues the smooth decrease in density. When liquid water freezes, the density drops suddenly (and the volume increases). For most substances, the density of the liquid increases steadily as temperature is lowered, and the density of the solid is greater than that of the liquid.

The reverse transformation, the freezing of water, is accompanied by an increase in volume. The expansion of water (Fig. 6–20) upon freezing can cause bursting of water pipes and freeze–thaw cracking of rocks and concrete. It has beneficial aspects as well, however. If ice were more dense than water, the winter ice that forms at the surface of a lake would sink to the bottom, and the lake would freeze solid from the bottom up. Instead, the ice remains at the surface, and the liquid water near the bottom achieves a stable wintertime temperature near 4°C, which permits the survival of aquatic life.

Hydrogen bonding also accounts for the anomalously high boiling point of water: A high temperature is required to overcome the strong attractions among the

(a)   (b)

Figure 6–20 • Unlike most substances, liquid water expands as it freezes. (a) A tightly sealed, brimful jar of water. (b) The expansion caused by freezing the water has shattered the jar. Although this volume change has profound significance in daily life, it still amounts to only 8.3% of the total volume. *(Charles D. Winters)*

Figure 6-21 • Trends in the boiling points of some hydrides of main-group elements and of the noble gases. The boiling points of HF, H_2O, and NH_3 are all higher than would be expected from the trend in boiling point among related compounds.

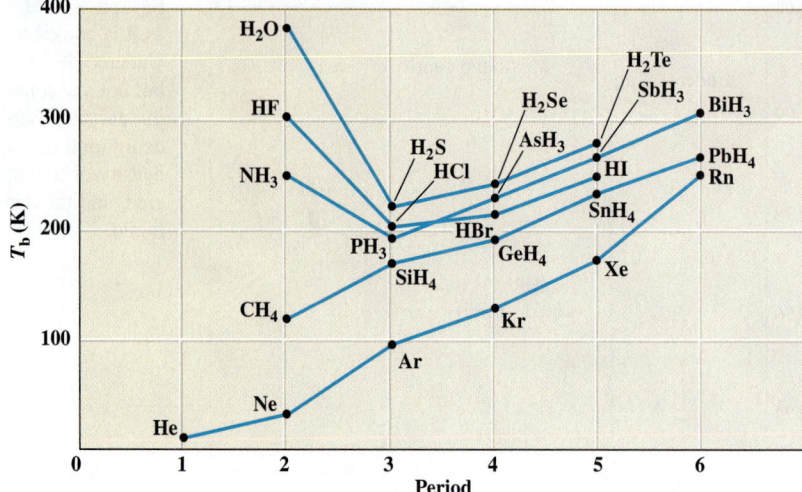

molecules in liquid water and launch them independently of each other into the gaseous state. An extrapolation of the trend from hydrides that lack hydrogen bonds gives a boiling point for "water without hydrogen bonds" near 150 K ($-123°C$). Life as we know it would not exist if water boiled at this low temperature. The boiling points of the compounds NH_3 and HF are also high compared with those of the hydrides of other elements of Groups V, VI, and VII (Fig. 6–21), and for the same reason.

Finally, hydrogen bonding is responsible for the large amount of heat required to boil or evaporate a quantity of water compared with the same quantity of other substances. The extra energy goes to overcome the hydrogen bonds in liquid water, many of which persist right up to the boiling point. Evaporation of water from oceans and lakes consumes a large fraction of the sun's energy and protects the environment against major swings in temperature. When large amounts of water condense all at once (as in a rainstorm), the energy previously absorbed in evaporation is released, often into a restricted region of the atmosphere. This energy fuels the violent winds of tornadoes and hurricanes.

EXAMPLE 6-3

Predict the order of increase in the normal boiling points of the following substances from Example 6–1: F_2, HBr, KCl, and HF.

Solution

As an ionic substance, KCl should have the highest boiling point of the four. HF should have a higher boiling point than HBr because its molecules form hydrogen bonds (Fig. 6–21) that are stronger than the dipole–dipole interactions in HBr. Fluorine, F_2, is non-polar and should have only weak dispersion forces. It should have the lowest boiling point of the four substances.

EXERCISE

Rank the following from lowest to highest in terms of expected normal boiling point: NaBr, Ar, HCl.

Answer: Ar < HCl < NaBr.

6-5 Phase Diagrams

By keeping the temperature of a substance constant, changing its pressure, and watching for phase transitions, we can determine the range of pressures over which each state of matter (gas, liquid, or solid) of the substance is stable. Repeating the process at many different temperatures gives the data necessary to create the **phase diagram** of that substance: a graph of pressure against temperature that shows the state of matter that is stable for every pressure–temperature combination. The phase diagrams for argon, carbon dioxide, and water are shown in Figure 6–22.

A great deal of information can be read from phase diagrams. Each substance has a unique combination of pressure and temperature at which the gas, liquid, and solid phases coexist in equilibrium. This combination is called the **triple point** (marked "T" in Fig. 6–22). Extending from the triple point are three lines, each of which denotes the conditions for the equilibrium coexistence of two phases: solid and gas, solid and liquid, and liquid and gas. The areas between the lines denote regions of pressure and temperature in which only one phase exists at equilibrium. The gas–liquid coexistence curve extends upward in temperature and pressure from the triple point. This line, stretching from T to C in the phase diagrams, is the vapor-pressure curve of the liquid substance. It displays the increasing pressure observed in an enclosed space above the liquid as it is warmed up very slowly. (Slow warming ensures that the liquid and its vapor stay in equilibrium.) Eventually, the vapor pressure reaches 1 atm. This occurs at a temperature equal to the normal

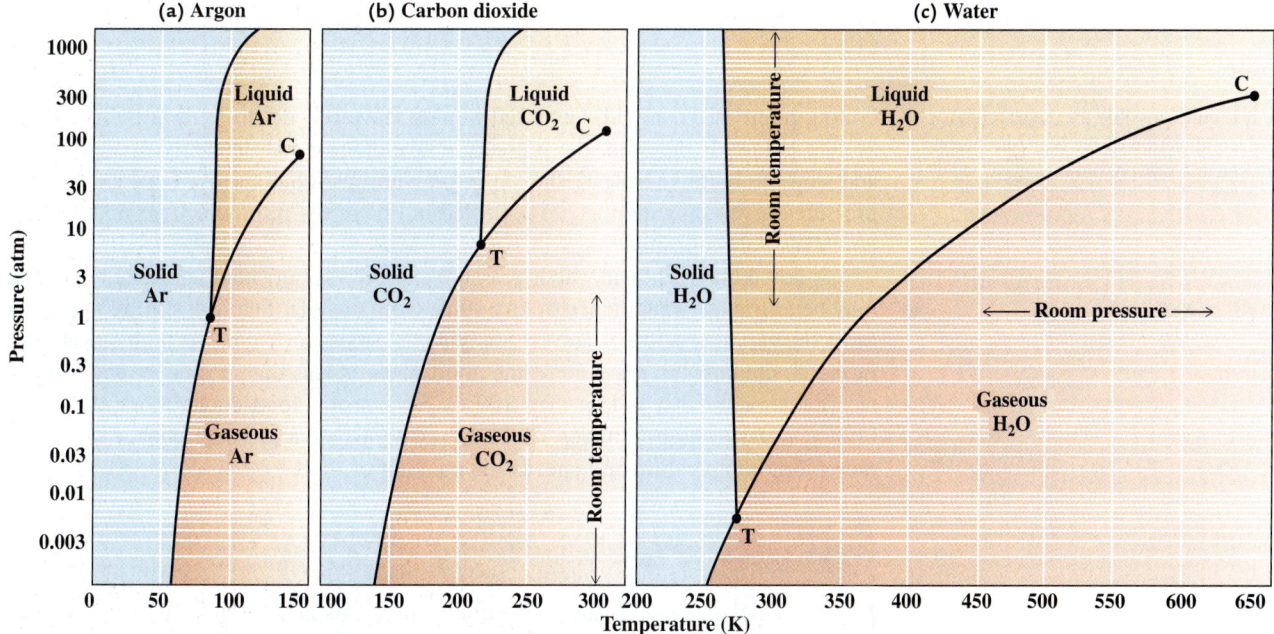

Figure 6–22 • In these phase diagrams, the pressure increases by a factor of 10 at regular intervals along the vertical axis. This method of graphing allows large ranges of pressure to be plotted. The marked horizontal and vertical lines are at a pressure of 1 atm and a temperature of 298.15 K, or 25°C (room conditions). Argon and carbon dioxide are gases at room conditions, but water is a liquid. The letter "T" marks the triple points of the substances, and the letter "C" marks their critical points. The region of stability of the liquid is larger for water than for either carbon dioxide or argon.

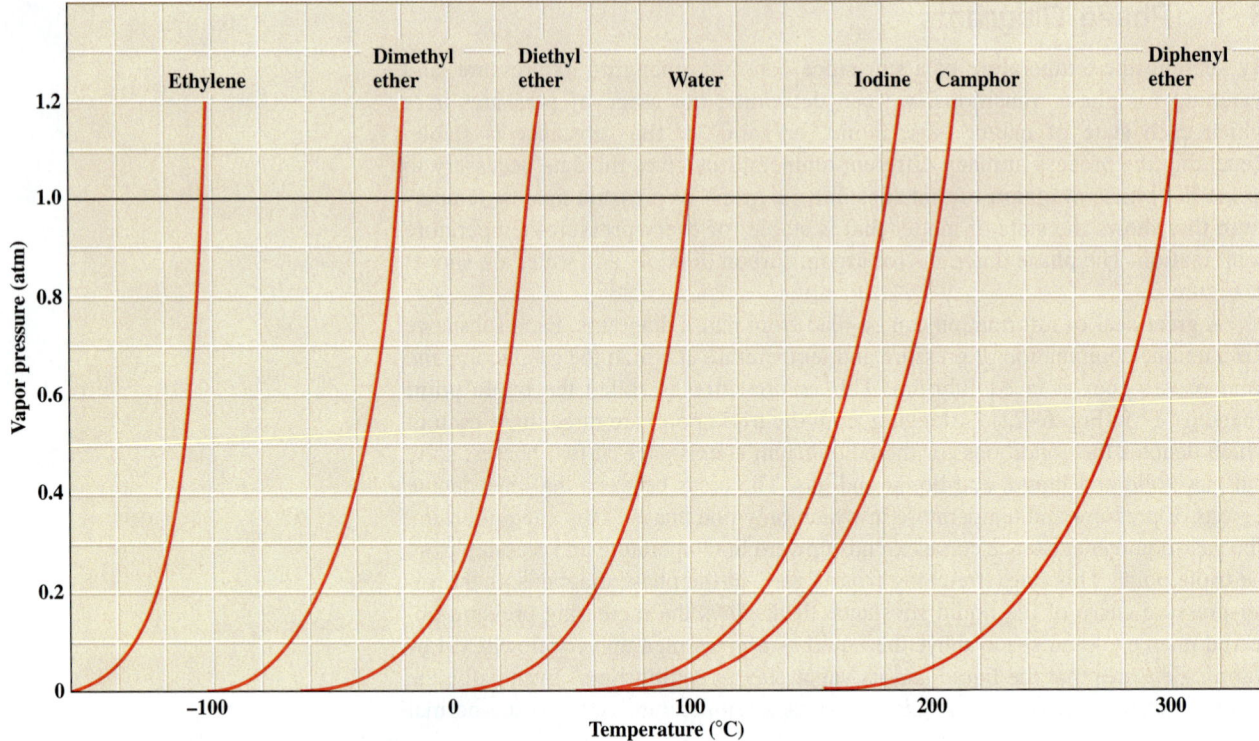

Figure 6–23 • The vapor pressure of a solid or liquid depends strongly on temperature. The temperature at which the vapor pressure becomes equal to 1 atm defines the normal boiling point of the liquid and the normal sublimation point of a solid.

• It is quite possible for the chemical decomposition of the liquid to intervene and prevent boiling.

boiling point of the liquid. Confined liquids, however, do not boil at their normal boiling points. If they did, the pressure above them would instantly exceed 1 atm and stop the boiling; instead, the vapor pressure just continues to increase. Vapor-pressure curves showing this are collected in Figure 6–23, in which normal boiling points also are indicated.

A liquid boils when its vapor pressure exceeds the surrounding pressure, which may be artificially maintained at any value (for example, below 1 atm by pumping away the vapor, or above 1 atm by applying an excess pressure). At high elevations, in the mountains, the pressure of the atmosphere is less than 1 atm, so water boils at a temperature that is less than 100°C. As a result, food cooks more slowly in boiling water at high altitudes. In contrast, the use of a pressure cooker increases the boiling temperature and speeds the cooking of food.

EXAMPLE 6–4

Atop a tall mountain, the atmospheric pressure is 0.50 atm. Use the vapor-pressure curve of water to estimate the boiling point of water under these conditions.

Solution

Find the position on the pressure axis corresponding to 0.5 atm in Figure 6–23 and move horizontally across to the curve for water. Reading directly down from this point locates a temperature of 82°C.

EXERCISE

Estimate the boiling temperature of diethyl ether ($C_2H_5OC_2H_5$) atop the same mountain (under a pressure of 0.50 atm), using the vapor pressure curve of diethyl ether in Figure 6–23.

Answer: 18°C.

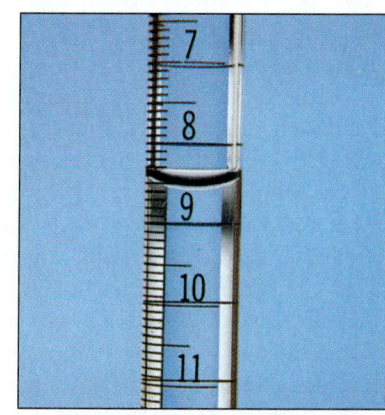

Figure 6–24 • When a gas and liquid coexist, the interface between them is clearly visible as a lens-like meniscus. The meniscus between coexisting liquid and gaseous forms of a pure substance disappears at the critical point. *(Leon Lewandowski)*

The gas–liquid coexistence curve does not continue forever but instead terminates at a point called the **critical point** (point C in Fig. 6–22). The pressure at the critical point of a substance is its **critical pressure,** and the temperature is its **critical temperature.** All along the liquid–gas coexistence curve, there is an abrupt, discontinuous change in the density and other properties between the two phases. However, the differences between the properties of the liquid and the gas become smaller as the critical point is approached, until they disappear altogether when the point is reached. If a liquid is put in a closed container and gradually heated, a **meniscus** initially appears at the boundary between liquid and gas (Fig. 6–24); at the critical point, this meniscus disappears! In other terms, at temperatures above the critical temperature (373.9°C for water), no amount of compression can cause a liquid to condense from a gaseous sample. The substance just gets denser and denser. Similarly, at pressures above the critical pressure, no amount of heating can cause a liquid to boil. When both T and P exceed their critical values, a substance becomes a **supercritical fluid,** so called because "fluid" includes both gases and liquids and "supercritical" means beyond the critical point.

The liquid–solid coexistence curve does not terminate, as the gas–liquid curve does at a critical point, but continues to indefinitely high pressures. In practice, such curves shoot up almost vertically because very large changes in pressure are necessary to change the freezing temperature of a liquid. For most substances, this curve inclines slightly to the right (see Fig. 6–22a, b)—an increase in pressure increases the freezing point of the liquid. Another way of saying this is that at constant temperature, an increase in pressure leads to the formation of a phase with higher density (smaller volume for a given mass), and for most substances the solid is denser than the liquid. Water and a few other substances are anomalous (see Fig. 6–22c). For them, the liquid–solid coexistence curve slopes up slightly to the *left,* meaning that an increase in pressure causes the solid to melt. This follows from the anomalous densities of the liquid and solid phases; ice is less dense than water (see Fig. 6–19), so when ice is compressed at 0°C, it melts.

• Some substances other than water that expand on freezing are the elements silicon, gallium, bismuth, and plutonium, and the compound gallium arsenide (GaAs).

For most substances, including water, atmospheric pressure occurs between the triple-point pressure and the critical pressure. Consequently, in our ordinary experience, we see three phases—gas, liquid, and solid—if we raise and lower the temperature enough. For a few substances, however, the triple-point pressure lies *above* $P = 1$ atm. In these cases, the phase transition called **sublimation** (see Fig. 6–14) occurs as the substance is warmed at atmospheric pressure. Sublimation takes the solid directly to a gas without passing through the liquid state. Carbon dioxide is such a substance (see Fig. 6–22b): its triple-point pressure is 5.117 atm (at a triple-point temperature of −56.57°C). Solid carbon dioxide (dry ice) sublimes to gaseous carbon dioxide at atmospheric pressure. In our daily experience, ordinary ice melts before it evaporates. However, ice does sublime at pressures below its triple-point pressure of 0.0060 atm (4.6 torr). This fact is used in freeze-drying, in which foods are frozen and then put in a vacuum chamber at a pressure of less than 0.0060 atm. The ice crystals that formed on freezing then sub-

An Exotic (But Useful) State of Matter

No known substance has a critical point at or even near atmospheric pressure; all occur at elevated pressures (Fig. 6–A). Consequently, supercritical fluids strike us as peculiar at first and exotic. However, the required temperature and pressure are not hard to attain for many substances; supercritical fluids are routinely prepared and have important uses. Like liquids, supercritical fluids can act as solvents to draw out and dissolve particular substances from mixtures. Moreover, their properties as solvents can be manipulated over a wide range by changing the temperature and pressure. This offers a considerable advantage over ordinary solvents: Supercritical fluids easily "de-solvate" from a product when the temperature

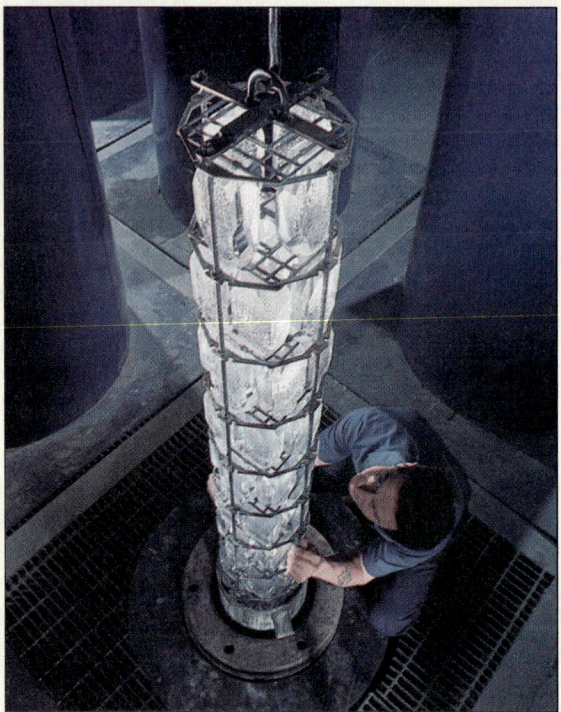

Figure 6–B • A rack of synthetic quartz crystals grown from silicon dioxide dissolved in supercritical water. *(Yoav Levy/Phototake)*

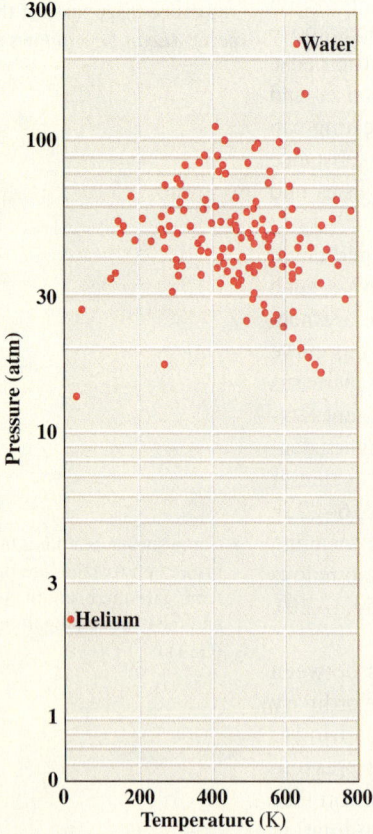

Figure 6–A • Each point on this pressure-versus-temperature graph is the critical point of a different substance. Note the critical point of water, which has strong intermolecular attractions, and the critical point of helium, which has the weakest intermolecular attractions of any substance known. All of the critical points are at pressures exceeding 1 atm, ordinary room pressure. This fact makes supercritical fluids alien to our experience.

and pressure are returned to ordinary values, but traditional solvents are commonly difficult to separate.

In recent years, supercritical fluid extraction has been widely developed in the food, drug, and chemical industries. Caffeine is extracted from green coffee beans by immersing them in supercritical fluid carbon dioxide. (Carbon dioxide has a critical temperature of 31°C and a critical pressure of 73 atm.) Careful control of the temperature and pressure allows for nearly complete removal of caffeine into the carbon dioxide while the flavorsome components remain in the coffee. A variation of this technique can be used to extract cholesterol from eggs and fats from potato chips and chicken.

Oxygen dissolves readily in supercritical water. Water, which becomes a *non*-polar solvent when it is supercritical, also dissolves many organic pollutants. This makes supercritical water an excellent medium for the destruction of such pollutants by oxygenation. Silicon dioxide, the major component of ordinary sand, dissolves to an extent in supercritical water. Large specimens of crystalline silicon dioxide (quartz) are prepared from these solutions for use in the electronics industry (Fig. 6–B).

lime, leaving a dried food (with a much lower mass) that can be reconstituted by adding water.

Many substances exhibit more than one solid phase as the temperature and pressure are varied (Fig. 6–25). At ordinary pressures, the stable state of carbon is graphite, a rather soft dark solid, but at high enough pressures, the hard, transparent form of carbon called diamond is stable. Below 13.2°C (and at room pressure), elemental tin undergoes a slow transformation from the metallic white form to a powdery gray form, a process referred to as "tin disease." No fewer than 13 solid phases of water (with names such as "ice II") are known, some of which exist only over a very limited range of temperatures and pressures.

Figure 6-25 • Red phosphorus and white phosphorus (stored under water in the tube at the top) are two different solid phases of the single substance phosphorus. The white and red forms coexist at room conditions because their phase transitions to black phosphorus, the most stable form of phosphorus under room conditions, are very slow. Black phosphorus is not shown. *(Charles D. Winters)*

EXAMPLE 6-5

Consider a sample of carbon dioxide at 250 K and 500 atm (see Fig. 6–22b).
(a) What phase(s) is (are) present at equilibrium?
(b) Suppose this sample of carbon dioxide is cooled at constant pressure. What happens?
(c) Describe a procedure to convert the sample into gaseous CO_2 without changing the temperature.

Solution

(a) This combination of T and P is well within the region of the phase diagram labelled liquid, so the sample is a liquid at equilibrium.
(b) A decrease in temperature at constant pressure corresponds to moving to the left in the figure. The liquid carbon dioxide freezes to a solid at approximately 225 K.
(c) If the pressure is lowered sufficiently at constant temperature (below the triple-point pressure of 5.1 atm, for example), the sample is converted completely to $CO_2(g)$.

EXERCISE

A sample of solid argon is heated at a constant pressure of 20 atm from 50 K to 150 K. Describe any phase transitions, and give the approximate temperatures at which they occur.

Answer: The solid argon melts completely to a liquid near 85 K, and the liquid vaporizes completely to a gas near 125 K.

6-6   Colligative Properties of Solutions

The physical properties of solutions differ from those of either the pure solute or pure solvent. For example, the vapor pressure, melting point, and boiling point of saltwater all differ from those of either salt or water. For dilute solutions, the numerical values of physical properties naturally lie close to those of the pure solvent. Remarkably, for certain properties, the amount of difference between pure solvent and dilute solution depends entirely on the *number* of solute particles that are present and not at all on their chemical identity. Such properties are called **colligative properties.** We discuss four colligative properties of solutions, the three just mentioned and a less-familiar fourth, the osmotic pressure. First, however, we introduce some additional measures of solution composition to aid the discussion. These units differ from the molarity (introduced in Section 2–4) but are related to it.

• "Colligative" comes from the Latin word meaning "to bind together, fasten, or attach."

Mass Fraction, Mole Fraction, and Molality

Giving a **mass fraction** (also called "weight fraction") is perhaps the simplest way to state the composition of a solution or other mixture. For component 1

$$w_1 = \text{mass fraction of component 1} = \frac{\text{mass of component 1}}{\text{total mass of solution}} = \frac{m_1}{m_{\text{total}}}$$

The **mass percentage** of a component equals the mass fraction multiplied by 100%:

$$\text{mass percentage of component 1} = \frac{m_1}{m_{\text{total}}} \times 100\%$$

For example, a solution of KCl in water that is 4.3% KCl by mass has w_{KCl} equal to 0.043. The solution contains 4.3 g KCl for every 100 g of solution. This composition is expressed in the unit-factors

$$\frac{4.3 \text{ g KCl}}{100 \text{ g solution}} \quad \text{and} \quad \frac{100 \text{ g solution}}{4.3 \text{ g KCl}}$$

and also in the unit-factors

$$\frac{4.3 \text{ g KCl}}{95.7 \text{ g water}} \quad \text{and} \quad \frac{95.7 \text{ g water}}{4.3 \text{ g KCl}}$$

The **mole fraction** of a substance in a solution or other mixture is the chemical amount (the number of moles) of that substance divided by the total chemical amount of all the substances present in the mixture:

$$X_1 = \text{mole fraction of component 1}$$

$$= \frac{\text{chemical amount of component 1}}{\text{total chemical amount of all components}} = \frac{n_1}{n_{\text{total}}}$$

Mole fractions were used in the discussion of gas mixtures and Dalton's law in Section 5–6. In a binary mixture containing n_1 mol of component 1 and n_2 mol of component 2, the mole fractions X_1 and X_2 are

$$X_1 = \frac{n_1}{n_1 + n_2} \quad \text{and} \quad X_2 = \frac{n_2}{n_1 + n_2} = 1 - X_1$$

The **mole percentage** of a component equals the mole fraction multiplied by 100%.

The mole fractions (or mass fractions) of all the components of a solution or any other mixture add up to 1. Similarly, the mole percentages (or mass percentages) of all the components add up to 100%.

Antifreeze being added to the radiator of a car. *(Courtesy of Union Carbide)*

EXAMPLE 6–6

An antifreeze solution is prepared by mixing 457 g of ethylene glycol ($C_2H_6O_2$) with 521 g of water. Calculate the mass fraction and the mole fraction of ethylene glycol in the mixture. The molar mass of ethylene glycol is 62.07 g mol^{-1}, and the molar mass of water is 18.02 g mol^{-1}.

Strategy

Compute the chemical amounts of the components in the solution and the total mass of the solution. Then apply the definitions of mass fraction and mole fraction.

Solution

$$m_{\text{solution}} = 457 \text{ g} + 521 \text{ g} = 978 \text{ g}$$

$$n_{C_2H_6O_2} = 457 \text{ g } C_2H_6O_2 \times \left(\frac{1 \text{ mol}}{62.07 \text{ g } C_2H_6O_2}\right) = 7.363 \text{ mol}$$

$$n_{H_2O} = 521 \text{ g } H_2O \times \left(\frac{1 \text{ mol}}{18.02 \text{ g } H_2O}\right) = 28.91 \text{ mol}$$

$$w_{C_2H_6O_2} = \frac{m_{C_2H_6O_2}}{m_{\text{solution}}} = \frac{457 \text{ g}}{978 \text{ g}} = 0.467$$

$$X_{C_2H_6O_2} = \frac{n_{C_2H_6O_2}}{n_{\text{solution}}} = \frac{7.363 \text{ mol}}{7.363 \text{ mol} + 28.91 \text{ mol}} = 0.203$$

Check

Ethylene glycol ($C_2H_6O_2$) contributes nearly half of the mixture based on mass, but only approximately a fifth of the mixture based on chemical amount. This makes sense. Ethylene glycol molecules have larger masses than water molecules so *fewer* of them are present in a given mass.

EXERCISE

A solution is prepared by mixing 37.9 g of methanol (CH_3OH) with 103.4 g of water. Calculate the mass percentage and mole percentage of methanol.

Answer: 26.8%, 17.1%.

Although the mole fraction can be used to discuss colligative properties, another unit, the **molality,** is generally preferred. The molality is defined as the number of moles of solute per kilogram of *solvent:*

$$\text{molality}_{\text{solute}} = m_{\text{solute}} = \frac{\text{moles}_{\text{solute}}}{\text{kg}_{\text{solvent}}}$$

• Using m for "molality" unfortunately invites confusion with the use of m for "mass." To overcome this, the use of b for "molality" has been officially suggested. It is rarely used.

The unit of molality is the mol kg^{-1}. A 1.0 *molal* solution of HCl contains 1.0 mol of HCl per *kilogram* of *solvent*. Such a solution differs from a 1.0 *molar* (1.0 M) solution, which contains 1.0 mol of HCl per *liter* of *solution*. Molarities work admirably for solution stoichiometry (Section 2–4), in which the emphasis is on the amount of a solute delivered by different quantities of solution. Molarities, however, change with temperature because solutions change their volume when heated or cooled. Molalities do *not* change with temperature and are preferable when the focus is on the effect of the solute on the properties of the solvent.

• *Alert*: The similarity between the terms molality and molarity (and molar and molal) often causes confusion.

Water has a density of 1.00 g cm^{-3} at 20°C. This means that 1.00 L of water (1000 cm^3) has a mass of 1.00×10^3 g, or 1.00 kg. As a consequence, in dilute aqueous solutions the number of moles of solute per liter of solution is *approximately* the same as the number of moles per kilogram of water. In other words, the molarity and molality of a solute in a dilute aqueous solution are nearly equal. This approximate equality fails for non-aqueous solutions, for concentrated aqueous solutions, and at temperatures significantly higher than room temperature.

EXAMPLE 6–7

A solution for washing glass is prepared by dissolving 84.6 g of ammonia (NH_3) in 822 g of water. The total volume of the solution is 0.945 L. Calculate the molality and the molarity of ammonia ($M = 17.03$ g mol^{-1}) in this solution.

Strategy

Rely on the definitions of the terms. Figure the chemical amount of the ammonia in the solution. Divide it by the mass of the *solvent* (in kg) for the first answer and then by the volume of the *solution* (in liters) for the second answer.

Solution

$$n_{NH_3} = 84.6 \text{ g NH}_3 \times \left(\frac{1 \text{ mol NH}_3}{17.03 \text{ g NH}_3} \right) = 4.968 \text{ mol NH}_3$$

$$\text{molality}_{NH_3} = m_{NH_3} = \frac{4.968 \text{ mol NH}_3}{822 \text{ g solvent}} \times \left(\frac{1000 \text{ g solvent}}{1 \text{ kg solvent}} \right)$$

$$= \frac{6.04 \text{ mol NH}_3}{\text{kg solvent}} = 6.04 \text{ mol kg}^{-1}$$

$$\text{molarity}_{NH_3} = c_{NH_3} = \frac{4.968 \text{ mol NH}_3}{0.945 \text{ L solution}} = \frac{5.26 \text{ mol NH}_3}{\text{L solution}} = 5.26 \text{ mol L}^{-1}$$

Check

The molality and molarity of the solution differ substantially. This is to be expected because the solution is far from dilute.

EXERCISE

Suppose that 32.6 g of acetic acid (CH_3COOH) is dissolved in 83.8 g of water, giving a total solution volume of 112.1 mL. Calculate the molality and molarity of acetic acid ($M = 60.05$ g mol^{-1}) in this solution.

Answer: $m_{CH_3COOH} = 6.48$ mol kg^{-1}; $c_{CH_3COOH} = 4.84$ mol L^{-1}.

Converting among Mass Fraction, Mole Fraction, Molality, and Molarity

We now have four measures of solution composition: mass fraction, mole fraction, molarity (see Section 2–4), and molality. Figure 6–26 shows how to convert from one to any of the others. For example, the masses that define the mass fraction can be converted into chemical amounts by using the molar masses (at the center of figure) and the chemical amounts then used to compute the mole fractions as in Example 6–6. The molality, on the other hand, is obtained by setting up a ratio of a chemical amount (the number of moles of solute) to a mass (the mass of the solvent in kilograms).

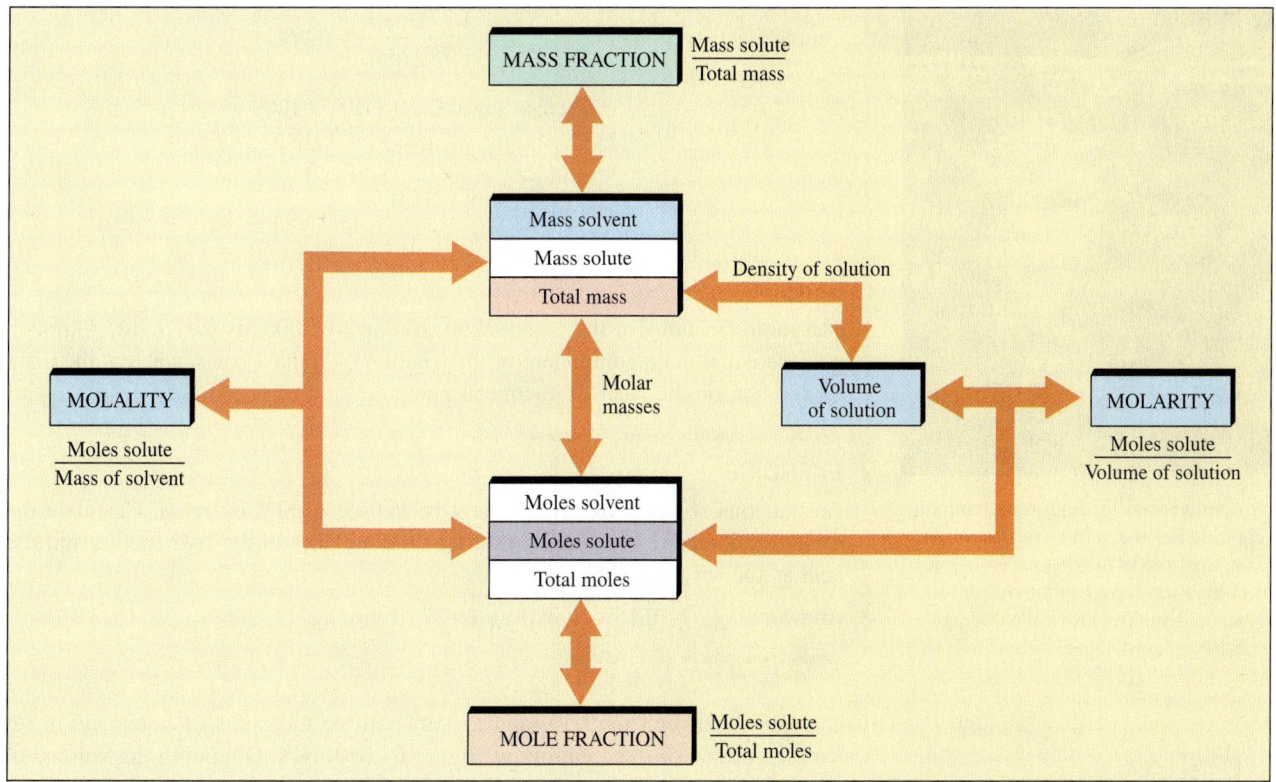

of solution

Figure 6–26 • Conversions between different measures of solution composition. Three of these (mass fraction, mole fraction, and molality) are completely interconvertible: once one is known, the other two can be calculated by following the arrows. Calculations relating molarity to the other concentration units require a knowledge of the density of the solution (not the solvent) as well. In a dilute aqueous solution, this density can be taken to be 1.00 g cm^{-3}.

EXAMPLE 6–8

A concentrated solution of glucose ($C_6H_{12}O_6$) in water is 75.3% glucose by mass. Calculate the mole fraction and the molality of glucose in this solution.

Strategy

Measures of solution composition do not depend on the size of the sample because they are ratios. For convenience, then, consider 100.0 g of the solution. This amount of solution contains 75.3 g of glucose and 24.7 g of water. Compute the chemical amounts of the glucose and water and apply the definitions.

Solution

$$n_{C_6H_{12}O_6} = 75.3 \text{ g } C_6H_{12}O_6 \times \left(\frac{1 \text{ mol } C_6H_{12}O_6}{180.16 \text{ g } C_6H_{12}O_6}\right)$$

$$= 0.418 \text{ mol } C_6H_{12}O_6$$

$$n_{H_2O} = 24.7 \text{ g } H_2O \times \left(\frac{1 \text{ mol } H_2O}{18.02 \text{ g } H_2O}\right) = 1.371 \text{ mol } H_2O$$

$$n_{tot} = 1.371 \text{ mol } + 0.418 \text{ mol } = 1.789 \text{ mol}$$

mole fraction of glucose $= X_{C_6H_{12}O_6} = \dfrac{0.418 \text{ mol}}{1.789 \text{ mol}} = 0.234$

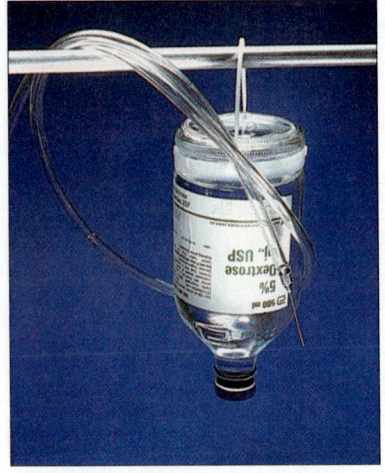

An intravenous feeding liquid containing glucose in a saline (salt) solution. The label reads "dextrose," an older, no longer preferred synonym for glucose, which is also called "grape sugar" or "corn sugar."

$$\text{mole fraction of water} = X_{H_2O} = \frac{1.371 \text{ mol}}{1.789 \text{ mol}} = 0.766$$

$$\text{molality}_{C_6H_{12}O_6} = \frac{0.418 \text{ mol } C_6H_{12}O_6}{24.7 \text{ g solvent}} \times \left(\frac{1000 \text{ g}}{1 \text{ kg}}\right)$$

$$= \frac{16.9 \text{ mol } C_6H_{12}O_6}{\text{kg solvent}} = 16.9 \text{ mol kg}^{-1}$$

Check

The mole fractions of the components add up to 1.000, as they must. Although glucose is the main component of the solution on a mass basis, water is the main component on the basis of chemical amount.

EXERCISE

An aqueous solution of sulfuric acid is 96.00% H_2SO_4 by mass. Calculate the mole fraction of H_2O in the solution and the molality of the H_2O (taking sulfuric acid as the solvent).

Answer: $X_{H_2O} = 0.185$; $\text{molality}_{H_2O} = 2.31 \text{ mol kg}^{-1}$.

Conversions between molarity and the other three measures of composition require the volume of the solution, as Figure 6–26 shows. Obtaining the volume of a solution from its mass is straightforward if the density of the solution (*not* the density of the solvent) is known. As a practical matter, such densities must be measured experimentally. The calculation then proceeds as shown in the following examples.

EXAMPLE 6-9

Creatinine ($C_4H_7N_3O$) is a component of muscle tissue and blood and a metabolic waste product. An 8.00% (by mass) solution of creatinine in water is prepared. An experiment gives the density of the solution as 1.0194 g cm^{-3} at 25°C. Calculate the concentration of creatinine (which has a molar mass of 113.12 g mol^{-1}) in mol L^{-1} at 25°C.

Strategy

Pick any specific mass of solution. A good choice is 100.0 g. This amount of solution contains 8.00 g of creatinine and 92.00 g of H_2O. Set up conversions to obtain the chemical amount of the creatinine in moles and the volume of the solution in liters. Then use the definition of molarity (that is, divide the first result by the second).

Solution

$$n_{C_4H_7N_3O} = 8.00 \text{ g } C_4H_7N_3O \times \left(\frac{1 \text{ mol } C_4H_7N_3O}{113.12 \text{ g } C_4H_7N_3O}\right) = 0.0707 \text{ mol } C_4H_7N_3O$$

$$V_{solution} = 100.0 \text{ g solution} \times \left(\frac{1 \text{ cm}^3}{1.0194 \text{ g solution}}\right) \times \left(\frac{1 \text{ mL}}{1 \text{ cm}^3}\right) \times \left(\frac{1 \text{ L}}{1000 \text{ mL}}\right)$$

$$= 0.09810 \text{ L}$$

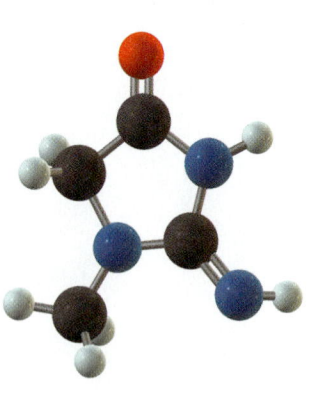

The structure of creatinine. The five-membered ring is nearly flat. The molecule is polar enough that creatinine is substantially soluble in water.

$$c_{C_4H_7N_3O} = \frac{n_{C_4H_7N_3O}}{V_{solution}} = \frac{0.0707 \text{ mol}}{0.09810 \text{ L}} = 0.721 \text{ mol L}^{-1}$$

EXERCISE

A solution prepared by mixing 20.00 g of $CdCl_2$ with 80.00 g of water has a density at 20°C of 1.1988 g cm^{-3}. Compute the molarity and molality of $CdCl_2$ in this solution.

Answer: $c_{CdCl_2} = 1.308$ mol L^{-1}; molality$_{CdCl_2} = 1.364$ mol kg^{-1}.

EXAMPLE 6–10

A 4.98 mol L^{-1} aqueous solution of sulfuric acid has a density of 1.2855 g cm^{-3}. Calculate the molality of the sulfuric acid in this solution.

Strategy

Recognize that the composition of the solution does not depend on its amount. Consider any convenient amount of solution, say 1 L. Compute the mass of this amount of solution and then the mass of the sulfuric acid that it contains. Subtract the mass of the sulfuric acid from the mass of the solution to obtain the mass of the solvent (water). Then use the definition of molality.

Solution

$$\text{mass}_{solution} = 1 \text{ L solution} \times \left(\frac{1000 \text{ cm}^3}{1 \text{ L}}\right) \times \left(\frac{1.2855 \text{ g}}{1 \text{ cm}^3 \text{ solution}}\right) = 1285.5 \text{ g}$$

$$\text{mass}_{H_2SO_4} = 1 \text{ L solution} \times \left(\frac{4.98 \text{ mol } H_2SO_4}{1 \text{ L solution}}\right) \times \left(\frac{98.08 \text{ g } H_2SO_4}{1 \text{ mol } H_2SO_4}\right)$$

$$= 488.4 \text{ g } H_2SO_4$$

$$\text{mass}_{solvent} = 1285.5 \text{ g} - 488.4 \text{ g} = 797.1 \text{ g} = 0.7971 \text{ kg}$$

$$\text{molality}_{H_2SO_4} = \frac{4.98 \text{ mol } H_2SO_4}{0.7971 \text{ kg solvent}} = \frac{6.25 \text{ mol } H_2SO_4}{\text{kg solvent}} = 6.25 \text{ mol kg}^{-1}$$

EXERCISE

An aqueous solution of HNO_3 has a concentration of 3.34 mol L^{-1} and a density of 1.1087 g cm^{-3}. Compute the molality of the HNO_3.

Answer: molality$_{HNO_3} = 3.72$ mol kg^{-1}.

Ideal Solutions and Raoult's Law

Consider a solution made by dissolving a single *non-volatile* solute in a solvent. A nonvolatile solute is one that has a negligible vapor pressure at the temperature under consideration. An example is a solution of sucrose (cane sugar) in water. The vapor pressure of sucrose above such a solution is effectively zero throughout the range of temperature in which water is a liquid.

• We consider the case of a volatile solute in Section 6–7.

The *solvent* vapor pressure is not zero and can be studied as the composition of the solution is changed at a fixed temperature. If the mole fraction of solvent (X_1)

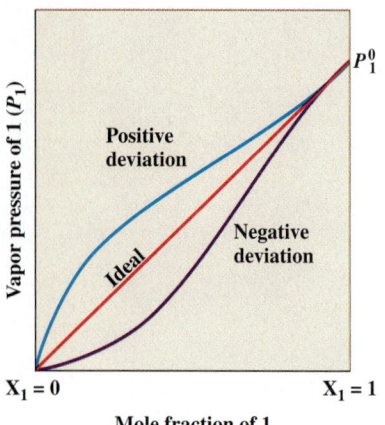

Figure 6–27 • A graph of the vapor pressure of the solvent (P_1) versus the mole fraction of solvent (X_1) is a straight line for an ideal solution. Non-ideal solutions behave differently; examples of positive and negative deviations from the behavior of an ideal solution are shown. The vapor pressure of pure solvent is P_1^0.

equals 1, then the vapor pressure is P_1^0, the vapor pressure of pure solvent at the temperature of the experiment. When X_1 approaches zero (which corresponds to pure solute), the vapor pressure P_1 of the solvent must go to zero also, because solvent is no longer present. As the mole fraction X_1 changes from 1 to zero, P_1 drops from P_1^0 to zero. What is the shape of the curve if P_1 is plotted versus X_1?

The French chemist François Marie Raoult found that, for some solutions, a plot of the vapor pressure of the solvent against its mole fraction comes very close to being a straight line (Fig. 6–27). Solutions that conform to this straight-line relationship obey the simple equation:

$$P_1 = X_1 P_1^0$$

which is known as **Raoult's law.** Such solutions are called **ideal solutions.** Only ideal solutions obey Raoult's law over the whole range of X_1, but *all* solutions with non-dissociating solutes approach the behavior predicted by Raoult's law as X_1 approaches 1 (that is, as X_2 approaches zero), just as all real gases obey the ideal gas law at sufficiently low densities. Solutions that deviate from straight-line behavior are **non-ideal** solutions. They may show negative deviations (with vapor pressures lower than those predicted by Raoult's law) or positive deviations (with higher vapor pressures).

On a molecular level, an ideal solution is one in which the solute–solvent attractions are equal in strength to the solvent–solvent attractions that existed before any solute was added. Negative deviations from ideality arise when the solute attracts solvent molecules especially strongly, reducing their tendency to escape into the vapor phase. Positive deviations result from the opposite case, in which solvent and solute molecules are not strongly attracted to each other, and the attractive forces between solvent molecules are disrupted by the presence of solute.

Raoult's law helps in explaining the four colligative properties.

Lowering of Vapor Pressure

The *change* in the vapor pressure of the solvent when a non-volatile solute is added is

$$\Delta P_1 = P_1 - P_1^0$$

Raoult's law consists of an expression for P_1. Substituting it into the expression for ΔP_1 gives

$$\Delta P_1 = X_1 P_1^0 - P_1^0$$

Because $X_1 = 1 - X_2$, we can rewrite this as

$$\Delta P_1 = X_1 P_1^0 - P_1^0 = (X_1 - 1)P_1^0 = -X_2 P_1^0$$

The negative sign means that the vapor pressure of solvent above a dilute solution is always *less* than the vapor pressure above the pure solvent; that is, ΔP_1 is always negative.

EXAMPLE 6–11

At 25°C, the vapor pressure of pure benzene is 0.1252 atm. Suppose 6.40 g of the solid hydrocarbon naphthalene, $C_{10}H_8$ (molar mass 128.17 g mol^{-1}), is dissolved in 78.0 g of pure benzene (molar mass 78.0 g mol^{-1}). Calculate the vapor pressure of benzene above the solution.

Strategy

Calculate the chemical amounts of the solvent benzene (n_1) and the solute naphthalene (n_2) and use them to obtain the mole fraction of benzene (X_1). Substitute this result into Raoult's law.

Solution

$$n_1 = 78.0 \text{ g } C_6H_6 \times \left(\frac{1.00 \text{ mol } C_6H_6}{78.0 \text{ g } C_6H_6} \right) = 1.00 \text{ mol } C_6H_6$$

$$n_2 = 6.40 \text{ g } C_{10}H_8 \times \left(\frac{1.00 \text{ mol } C_{10}H_8}{128.17 \text{ g } C_{10}H_8} \right) = 0.0499 \text{ mol } C_{10}H_8$$

$$X_1 = \frac{n_1}{n_1 + n_2} = \frac{1.00 \text{ mol}}{(1.00 + 0.0499) \text{ mol}} = 0.9525$$

$$P_1 = X_1 P_1^0 = 0.9525 \times 0.1252 \text{ atm} = 0.119 \text{ atm}$$

Check

The vapor pressure of benzene above the solution is *less* than the vapor pressure above pure benzene, as it must be.

EXERCISE

The vapor pressure of water at 80°C is 0.4672 atm. Determine the mass (in grams) of sucrose ($C_{12}H_{22}O_{11}$) that must be dissolved in 180.2 g of water at 80°C to reduce the vapor pressure above the solution to 0.4640 atm.

Answer: 24 g of sucrose.

Elevation of the Boiling Point

The normal boiling point of a pure liquid or a solution is defined as the temperature at which its vapor pressure reaches 1 atm. Because a dissolved solute reduces a solvent's vapor pressure, the temperature of the solution must be increased to bring it to a boil (Fig. 6–28). The boiling point of a solution of a non-volatile solute in a volatile solvent always exceeds the boiling point of the pure solvent.

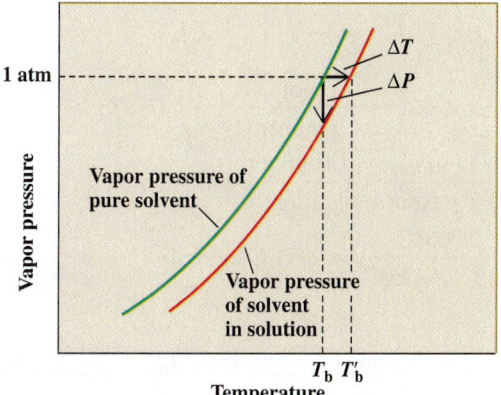

Figure 6-28 • The vapor pressure of the solvent above a dilute solution is lower than that of the pure solvent at all temperatures. As a result, in order for the solution to boil (that is, for the vapor pressure to reach 1 atm), a higher temperature is required than for the pure solvent. This amounts to elevation of the boiling point.

TABLE 6-2	Boiling-Point Elevation and Freezing-Point Depression Constants[a]				
Solvent	Chemical Formula	T_b (°C)	K_b (K kg mol^{-1})	T_f (°C)	K_f (K kg mol^{-1})
Acetic acid	CH_3COOH	118.1	3.07	17	3.9
Benzene	C_6H_6	80.1	2.53	5.5	4.9
Carbon tetrachloride	CCl_4	76.7	5.03	−22.9	32
Diethyl ether	$C_4H_{10}O$	34.7	2.02	−116.2	1.8
Ethanol	C_2H_5OH	78.4	1.22	−114.7	—
Naphthalene	$C_{10}H_8$	—	—	80.5	6.8
Water	H_2O	100.0	0.512	0.0	1.86

[a]At 1 atm pressure.

For sufficiently dilute solutions, the elevation of the boiling point ΔT_b is directly proportional to the concentration of solute in the solution. Because temperature changes are involved, molality m (moles of solute per kilogram of solvent), which is independent of temperature, is used to express the composition of the solution. In equation form

$$\Delta T_b = K_b m$$

in which K_b is a constant of proportionality. This equation is quite useful because K_b is a property of the solvent only; it does not depend on the identity of the solute. The K_b's of numerous solvents have been obtained by measuring the elevation of the boiling point for dilute solutions of known molality (see Table 6–2). The following example shows how a K_b is determined.

EXAMPLE 6-12

When 5.50 g of biphenyl ($C_{12}H_{10}$), a non-volatile compound, is dissolved in 100.0 g of benzene, the boiling point of the solution exceeds the boiling point of pure benzene by 0.903°C. Determine K_b for benzene.

Strategy

Compute the chemical amount of the solute biphenyl. Use it and the mass of the solvent (benzene) to figure the molality of the solution. Solve the boiling-point-elevation equation for K_b. Obtain the value of K_b by inserting the molality and the observed boiling-point elevation.

Solution

$$n_{biphenyl} = 5.50 \text{ g biphenyl} \times \left(\frac{1 \text{ mol}}{154 \text{ g biphenyl}}\right) = 0.0357 \text{ mol}$$

$$\text{molality}_{biphenyl} = m = \frac{0.0357 \text{ mol}}{100.0 \text{ g solvent}} \times \left(\frac{1000 \text{ g}}{1 \text{ kg}}\right) = 0.357 \text{ mol kg}^{-1}$$

$$K_b = \frac{\Delta T_b}{m} = \frac{0.903°C}{0.357 \text{ mol kg}^{-1}} = 2.53°C \text{ kg mol}^{-1}$$

Check

The answer equals the value in Table 6–2. Note that 2.53°C kg mol^{-1} equals 2.53 K kg mol^{-1} because a degree Celsius and a kelvin are equal in size.

EXERCISE

A sample consisting of 4.58 g of non-volatile picric acid ($C_6H_3N_3O_7$) is dissolved in 240.0 g of chloroform. The boiling point of the solution exceeds the boiling point of pure chloroform by 0.302 K. Determine K_b for chloroform.

Answer: 3.63 K kg mol^{-1}.

Knowing the K_b of a solvent allows determination of the molar masses of unknown substances. A solution of known composition is prepared and its boiling point measured. The following example shows a typical calculation.

EXAMPLE 6-13

When 6.30 g of a non-volatile hydrocarbon of unknown molar mass is dissolved in 150.0 g of benzene, the boiling point of the solution is 80.696°C. The boiling point of pure benzene is 80.099°C. Compute the molar mass of the hydrocarbon.

Strategy

Figure the change in the boiling point. Dividing it by the K_b of benzene gives the molality of the unknown in the solution. Use the molality to compute the chemical amount of the unknown that is present. Dividing the mass of the unknown by its chemical amount gives the molar mass of the unknown.

Solution

$$\Delta T_b = 80.696°C - 80.099°C = 0.597°C$$

$$\text{molality}_{unknown} = \frac{\Delta T_b}{K_b} = \frac{0.597°\cancel{C}}{2.53°\cancel{C} \text{ kg mol}^{-1}} = 0.236 \text{ mol kg}^{-1}$$

$$n_{unknown} = \left(\frac{0.236 \text{ mol unknown}}{1 \text{ kg } \cancel{\text{solvent}}}\right) \times 0.150 \text{ kg } \cancel{\text{solvent}} = 0.0354 \text{ mol}$$

$$M_{unknown} = \frac{6.30 \text{ g } \cancel{\text{unknown}}}{0.0354 \text{ mol } \cancel{\text{unknown}}} = 178 \text{ g mol}^{-1}$$

Check

The molar mass is reasonable: neither less than about 15 g mol^{-1}, which is impossible, nor exceedingly large. The unknown hydrocarbon might be anthracene ($C_{14}H_{10}$), which has a molar mass of 178.24 g mol^{-1}.

EXERCISE

When 1.744 g of a non-volatile solute of unknown molar mass is dissolved in 122.2 g of pure acetic acid, the boiling point of the acetic acid is elevated by 0.246°C. In another experiment, 0.0100 mol of a different non-volatile solute elevates the boiling point of 1.00 kg of acetic acid by 0.0307°C. Calculate the molar mass of the first solute.

Answer: $M_{unknown} = 178$ g mol^{-1}.

The Effect of Dissociation

So far we have considered only non-dissociating solutes. If a solute dissociates (as does NaCl, when it dissolves in water to furnish Na^+ and Cl^- ions), then the effective number of solute particles increases. Because colligative properties depend on the number of dissolved particles, the equations describing them must be adjusted to take dissociation into account. The required change is the insertion of a multiplying factor, the **van't Hoff _i_.** The equation for boiling point elevation becomes

$$\Delta T_b = imK_b$$

- The van't Hoff _i_ is named after Jacobus van't Hoff, the Dutch physical chemist who introduced it.

The van't Hoff _i_ equals the number of particles released into solution per formula unit of solute. Its minimum value equals 1 (no dissociation). Its maximum (symbolized by ν, the Greek letter nu) equals the number of particles generated by complete dissociation of a formula unit of solute. Values of ν for common electrolytes are usually obvious from the formula and name; thus ν clearly equals 2 for NaCl and 3 for K_2SO_4. Because aquated ions of opposite charge associate into ion-pairs, often to a significant extent, the van't Hoff _i_ for an electrolyte in water is usually less than ν. The amount of association depends on the concentration of the solution, so _i_ must be determined experimentally.

EXAMPLE 6–14

Lanthanum(III) chloride ($LaCl_3$) dissolves by dissociation into ions:

$$LaCl_3(s) \longrightarrow La^{3+}(aq) + 3\,Cl^-(aq)$$

0.2453 g of $LaCl_3$ is dissolved in 100.0 g of H_2O. What is the boiling point of the solution at atmospheric pressure, assuming no association among ions and that the solution behaves ideally?

Strategy

Obtain the molality of the solution and the van't Hoff _i_. Substitute them in the equation for boiling-point elevation along with the K_b of water.

Solution

$$n_{LaCl_3} = 0.2453\ \text{g LaCl}_3 \times \left(\frac{1\ \text{mol LaCl}_3}{245.3\ \text{g LaCl}_3}\right) = 0.001000\ \text{mol LaCl}_3$$

$$\text{molality}_{LaCl_3} = m_{LaCl_3} = \frac{0.001000\ \text{mol LaCl}_3}{0.1000\ \text{kg solvent}} = 0.01000\ \text{mol kg}^{-1}$$

Each mole of $LaCl_3$ gives four moles of particles in solution: three of Cl^- ions and one of La^{3+} ions. Hence $\nu = 4$. "No association among ions" means $i = \nu$. Substitution in the equation for boiling-point elevation gives

$$\Delta T_b = imK_b = 4\,(0.01000\ \text{mol kg}^{-1})(0.512\ \text{K kg mol}^{-1}) = 0.0205\ \text{K}$$

The predicted boiling point of the solution is ΔT_b (converted from K to °C) added to exactly 100°C, or 100.0205°C.

Check

The actual boiling point has been measured. It is slightly lower than this because the solution is non-ideal and ion-pairs in fact do form.

EXERCISE

Lead(II) nitrate dissociates to Pb^{2+} and NO_3^- ions in aqueous solution. Determine ν. Assuming no association among ions and ideal-solution behavior, calculate the boiling point of a solution prepared by mixing 0.194 mol of $Pb(NO_3)_2$ and 1.171 kg of water.

Answer: $\nu = 3$; boiling point $= 100.254°C$.

Depression of the Freezing Point

The addition of a solute *increases* the boiling point of a solvent but *decreases* its freezing point. We consider here only the simplest case of freezing-point depression, that in which the solid that freezes out of solution is the pure solvent. Cases in which solute crystallizes out in combination with the solvent are more complicated.

For sufficiently small concentrations of a non-dissociating solute, the freezing-point depression ΔT_f is related to the molality by the equation

$$\Delta T_f = -mK_f$$

where K_f is another positive constant that depends only on the properties of the solvent (Table 6–2). This formula is quite similar to the one describing boiling-point elevation. The minus sign appears because the change in freezing point ΔT_f (defined as the freezing point of the solution minus that of the pure solvent) is always negative; a solution always has a *lower* freezing temperature than the pure solvent. For dissociating solutes, the preceding equation becomes

$$\Delta T_f = -imK_f \quad \text{(for dissociating solutes)}$$

where the i again accounts for the extra particles created in solution by the dissociation.

The phenomenon of freezing-point depression explains the fact that seawater, which contains dissolved salts, has a lower freezing point than fresh water. Concentrated salt solutions have still lower freezing points. Spreading salt on an icy road creates a solution with a lower freezing point than pure water and causes the ice to melt. Antifreeze added to a car radiator lowers the freezing point of the coolant and keeps the cooling system from freezing (and perhaps cracking the engine block) in winter. Solutes depress the freezing point of their solvent by interfering slightly with the ability of solvent molecules to organize into a solid. Lowering the temperature compensates by reducing the average kinetic energy of the solvent molecules.

Measurements of the depression of the freezing point, like those of the elevation of the boiling point, can be used to determine molar masses of unknown substances.

EXAMPLE 6-15

The chemical amounts of the major dissolved species in a 1.000 L sample of seawater are given below. Estimate the freezing point of the seawater, given that $K_f = 1.86$ K kg mol^{-1} for water and assuming that no association occurs among the dissolved ions.

Na^+	0.458 mol	Cl^-	0.533 mol
Mg^{2+}	0.052 mol	SO_4^{2-}	0.028 mol
Ca^{2+}	0.010 mol	HCO_3^-	0.002 mol
K^+	0.010 mol	Br^-	0.001 mol
Neutral species	0.001 mol		

Strategy

Compute a total molarity (mol L^{-1}), based on the number of moles of all the particles listed in the table divided by the volume of the solution. Assume that the solution is sufficiently dilute that this molarity approximates the total molality (mol kg^{-1}) of the solution, which cannot be computed from the given facts. Then use the freezing-point depression equation.

Solution

The total molarity equals the sum of the chemical amounts of the nine species divided by 1.000 L of solution. This is 1.095 mol L^{-1}. Take this as the approximate molality of the seawater. Then

$$\Delta T_f = -mK_f \approx -(1.095 \text{ mol kg}^{-1})(1.86 \text{ K kg mol}^{-1}) = -2.04 \text{ K} = -2.04°C$$

The freezing point of pure water is 0.00°C, so the seawater should freeze at approximately −2°C.

Check

Oceanographers report that open-ocean seawater begins to freeze at −1.9 to −2.0°C (depending on the exact salinity).

EXERCISE

An aqueous solution is 0.040 molal in $CaCl_2$, 0.030 molal in NaCl, and 0.100 molal in sucrose. Assuming no association among the dissolved particles, predict the freezing point of this solution.

Answer: −0.52°C.

Both freezing-point depression and boiling-point elevation can be used to ascertain whether a species of known molar mass dissociates in solution (see Fig. 6–29), as the following example shows.

EXAMPLE 6–16

When 0.494 g of $K_3Fe(CN)_6$ is dissolved in 100.0 g of water, the freezing point is reduced to −0.093°C. How many ions are apparently present for each formula unit of $K_3Fe(CN)_6$ dissolved?

Strategy

Realize that the problem in effect asks for computation of the van't Hoff i for this solution. Compute all the other quantities in the freezing-point depression equation. Solve the equation for i; substitute to obtain a numerical answer.

Solution

The molar mass of $K_3Fe(CN)_6$ is 329.25 g mol^{-1}.

$$n_{K_3Fe(CN)_6} = 0.494 \text{ g } \cancel{K_3Fe(CN)_6} \times \left(\frac{1 \text{ mol } K_3Fe(CN)_6}{329.25 \text{ g } \cancel{K_3Fe(CN)_6}} \right)$$

$$= 0.00150 \text{ mol } K_3Fe(CN)_6$$

$$\text{molality}_{K_3Fe(CN)_6} = m = \frac{0.00150 \text{ mol}}{0.1000 \text{ kg solvent}} = 0.0150 \text{ mol kg}^{-1}$$

$$i = \frac{-\Delta T_f}{m K_f} = \frac{-(-0.093 - 0.000)°C}{1.86 \text{ K kg mol}^{-1}(0.0150 \text{ mol kg}^{-1})} \times \left(\frac{1 \text{ K}}{1°C} \right) = 3.3$$

Thus, 1 formula unit of $K_3Fe(CN)_6$ generates an apparent 3.3 ions in this solution.

Check

Section 3–8 identifies the $[Fe(CN)_6]^{3-}$ ion as a coordination complex, capable of independent existence in solution. It does not dissociate, and so the dissolution equation is

$$K_3[Fe(CN)_6](s) \longrightarrow 3 K^+(aq) + [Fe(CN)_6]^{3-}(aq)$$

for which v equals 4. The fact that i is less than v means that the aquated ions associate to an extent.

EXERCISE

Mannitol has a molar mass of 182.17 g mol^{-1}. When 1.00 g of mannitol is dissolved in 99.00 g of water, the freezing point of the resulting solution is $-0.103°C$. Determine how many moles of particles are formed in solution by mannitol for each mole dissolved.

Answer: One.

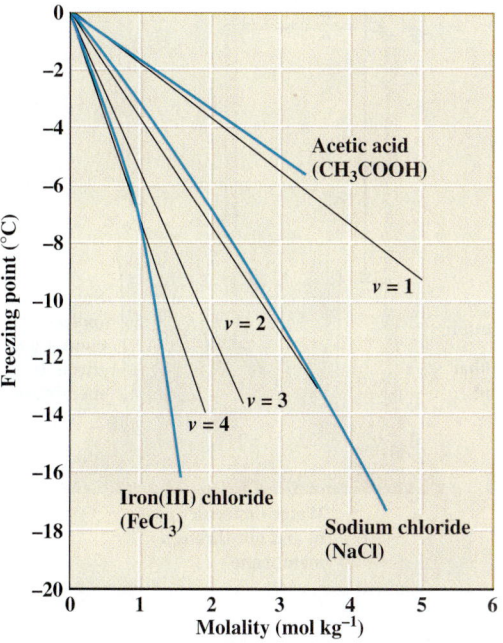

Figure 6–29 • The heavy, blue lines give the observed depression of the freezing point of water by CH_3COOH (acetic acid), NaCl (sodium chloride), and $FeCl_3$ (iron(III) chloride) as the molalities of the solutions increase. The black lines indicate the predicted ideal behavior for $v = 1$, 2, 3, and 4. The experimental ΔT's deviate in all three cases from any prediction, particularly at higher molalities. The curves indicate nevertheless that NaCl gives two moles of particles per mole dissolved, that $FeCl_3$ gives four moles of particles per mole dissolved, and that CH_3COOH gives one mole of particles per mole dissolved.

Osmotic Pressure

The fourth colligative property, osmotic pressure, plays a vital role in the transport of molecules across cell membranes in living organisms. Such membranes are **semipermeable.** They allow small molecules, such as those of water, to pass through, but they block the passage of large molecules, such as those of proteins or carbohydrates. Semipermeable membranes are fairly common natural and synthetic materials; a sheet of cellophane is an example. An ideal semipermeable membrane passes solvent molecules only and prevents the passage of every kind of solute molecules. Real membranes are less than ideally efficient but still can be used to separate solutes from solvents.

Suppose that one end of an open-ended glass tube is covered with a semipermeable membrane, clamped vertically with the membrane end down, and filled with a solution. When the membrane end of the tube is inserted in a beaker of pure solvent, solvent flows from the beaker into the tube (Fig. 6–30). The driving force for the flow is the tendency of the pure solvent to mix with the solution and dilute it, just as any solutions of differing composition tend to mix to a uniform composition when put in contact with each other. Because molecules of solute cannot pass down through the semipermeable membrane, the tendency can be expressed only by solvent molecules passing up. The solution rises within the tube to some final height h above the level of the pure solvent in the beaker. The pressure on the solution side of the membrane is greater than the atmospheric pressure on the surface of the pure solvent by an amount that is the **osmotic pressure Π** of the solution contained in the vertical tube. This osmotic pressure is computed from h by the equation

- The osmotic pressure of a solution measures its ability to pull additional solvent molecules into itself by osmosis.

$$\Pi = gdh$$

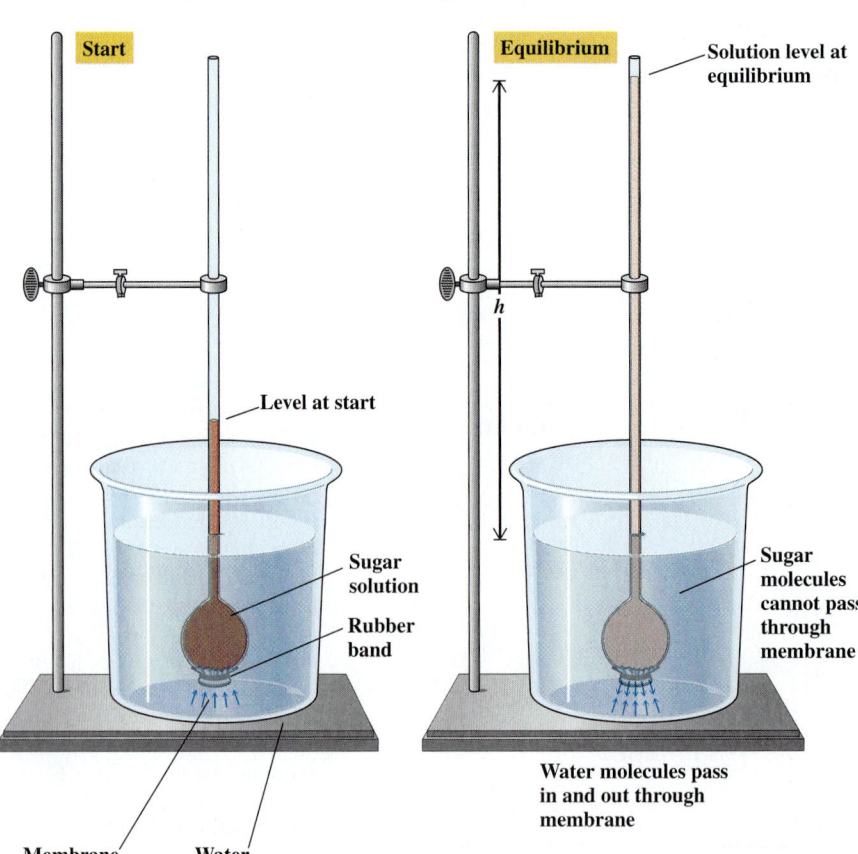

Figure 6–30 • In this device to measure osmotic pressure, the semipermeable membrane allows solvent, but not solute, molecules to pass through. This results in a net flow of solvent into the tube until an equilibrium is achieved, with the level of the solution at a height h above that of the solvent in the beaker. Once this happens, the solvent molecules pass through the membrane at the same rate in both directions.

where d is the density of the solution in the tube and g is the acceleration due to the gravitational field of the Earth. To use this equation in consistent SI units, d must be in kg m^{-3}, h in meters, and g in m s^{-2} (the value of g is 9.80665 m s^{-2}).

Typical dilute aqueous solutions are able to raise columns 10 to 30 cm high in this experiment. The densities of such solutions are close to the density of pure water, which is 1.00 g cm^{-3}, or 1000 kg m^{-3}. Suppose, then, that the final height of the column of solution in the tube is 0.22 m. The osmotic pressure of this solution is

$$\Pi = \left(\frac{9.81 \text{ m}}{s^2}\right) \times \left(\frac{1000 \text{ kg}}{m^3}\right) \times (0.22 \text{ m}) = 2.16 \times 10^3 \frac{\text{kg}}{\text{m s}^2} = 2.2 \times 10^3 \text{ Pa}$$

Converting from pascals to atmospheres gives

$$\Pi = 2.16 \times 10^3 \text{ Pa} \times \left(\frac{1 \text{ atm}}{1.01325 \times 10^5 \text{ Pa}}\right) = 0.021 \text{ atm}$$

This computation shows that easily measured heights of aqueous solutions correspond to quite small osmotic pressures.

In 1887 van't Hoff discovered an important approximate relationship between a solution's osmotic pressure (Π), its concentration (c), and its absolute temperature (T):

$$\Pi = cRT$$

where R is the gas constant. Taking c in mol L^{-1}, R in L atm mol^{-1} K^{-1}, and T in K in this equation gives Π in atmospheres. The concentration c of a solute equals n/V, where n is the chemical amount of the solute and V is the volume of the solution. Using this definition, van't Hoff's equation can be rewritten as

$$\Pi V = nRT$$

which has exactly the form of the ideal gas law and so suggests an analogy between the osmotic pressure of a solution and the ordinary pressure of a gas.

In the case of a dissociating solute, the van't Hoff i is inserted to compensate, just as with the other colligative properties.

$$\Pi = icRT \quad \text{(for dissociating solutes)}$$

Measurements of osmotic pressure provide a method of determining the molar mass of a dissolved substance, as the next example shows.

EXAMPLE 6-17

A biochemist dissolves 2.00 g of a (non-dissociating) protein in 0.100 L of water and finds the osmotic pressure to be 0.021 atm at 25°C. Estimate the molar mass of the protein.

Strategy

Obtain the concentration of the solute protein (in mol L^{-1}) by substitution in the equation for osmotic pressure. The solution contains 2.00 g solute/0.100 L. Divide this ratio by the molarity to obtain a result that has units of grams per mole, the units of molar mass.

Solution

$$c_{protein} = \frac{\Pi}{RT} = \frac{0.021 \text{ atm}}{(0.08206 \text{ L atm mol}^{-1} \text{ K}^{-1})(298 \text{ K})} = 8.6 \times 10^{-4} \text{ mol L}^{-1}$$

$$M_{protein} = \frac{\dfrac{2.00 \text{ g protein}}{0.100 \text{ L solution}}}{\dfrac{8.6 \times 10^{-4} \text{ mol protein}}{1 \text{ L solution}}} = 2.3 \times 10^4 \text{ g mol}^{-1}$$

EXERCISE

A dilute aqueous solution of a non-dissociating compound contains 1.19 g of the compound per liter of solution and has an osmotic pressure of 0.0288 atm at a temperature of 37.0°C. Compute the molar mass of the compound.

Answer: $M = 1.05 \times 10^3 \text{ g mol}^{-1}$.

Fresh water is scarce in some parts of the world. The method of **reverse osmosis** is often used in such environments to remove dissolved salts from brackish water for drinking or other domestic use. In reverse osmosis, pressure is applied to the surface of a saline solution, forcing pure water to pass from the solution through a membrane that is impermeable to ions. Reverse osmosis is being used increasingly in both domestic and industrial applications, especially for control of water pollution. The method is usually cheaper than distillation (Section 6–7).

Colligative Properties of Solutions Containing Several Solutes

Important categories of solutions, such as biological fluids, contain numerous solutes, some fully dissociated, others partly dissociated, yet others non-dissociated. Measurements of the colligative properties of such solutions can do no more than give *effective* concentrations of the solute particles taken as a whole. Such a concentration is called either an **osmolarity** or an **osmolality,** in parallel to molarity and molality:

$$\text{osmolarity} = \frac{\text{effective moles}_{\text{solute particles}}}{\text{liters}_{\text{solution}}} \qquad \text{osmolality} = \frac{\text{effective moles}_{\text{solute particles}}}{\text{kg}_{\text{solvent}}}$$

The osmolality of a solution equals the effective molality of all solutes in a solution. It equals im in the boiling-point elevation and freezing-point depression equations $\Delta T_f = -imK_f$ and $\Delta T_b = imK_b$. The osmolarity equals the effective molarity of all solutes and equals ic in the van't Hoff equation $\Pi = icRT$.

Osmometers measure osmolality or osmolarity. Currently available instruments rely on three of the four colligative properties; boiling-point elevation is not favored, partly because heating solutions often causes unwanted chemical reactions. Membrane osmometers use a high-precision gauge to measure the pressure change as solvent molecules osmose from a sealed solvent compartment into an open solution compartment. Vapor-pressure osmometers measure vapor-pressure lowering; freezing point osmometers measure freezing-point depression. Modern instruments operate rapidly and automatically and only require small samples. Their usefulness in chemistry lies in the determination of the molar masses of solutes, as discussed throughout this section. Osmometers also find wide use in diagnostic tests on phys-

CHEMISTRY IN YOUR LIFE

Osmosis, Dialysis, and Your Kidneys

The walls of living cells are semipermeable membranes through which the solvent (water) can pass. As the concentration of dissolved species inside and outside the cell changes, the osmotic pressure difference across the wall of the cell changes. This causes water to flow into or out of the cell, and the cell swells or shrinks (Fig. 6–C).

Red blood cells can be regarded as little bags that hold a solution of approximately 0.15 molal NaCl. They circulate in a blood fluid having the same osmotic pressure. Consequently, the proper medium for intravenous injections is 0.15 molal aqueous NaCl, *not* water. This ensures that the osmotic pressure of the blood fluid stays equal to the osmotic pressure inside the red cells. Injecting pure water into the bloodstream causes the red cells to swell and perhaps burst as water osmoses into them to dilute their contents to the same concentration as their surroundings. The consequences are catastrophic.

It is sometimes said that the elimination of wastes from the bloodstream proceeds by osmosis. In fact, kidney function is more akin to **dialysis,** in which some solute molecules as well as solvent molecules pass through a membrane. A dialyzing membrane has pores large enough to allow water, small ions, and small molecules to pass through but too small for large molecules. The kidneys contain millions of structures called "nephrons," each consisting of a cluster of capillaries along a long, narrow tube. The walls of the capillaries serve as dialyzing membranes. Blood pressure, which is twice as high in the capillaries of the nephrons as in the rest of the body, forces ions and small molecules (including the waste products of metabolism) through the capillary walls and into the tubes of the nephrons. Large molecules, such as proteins, stay in the blood. Molecules of waste compounds are then selectively re-

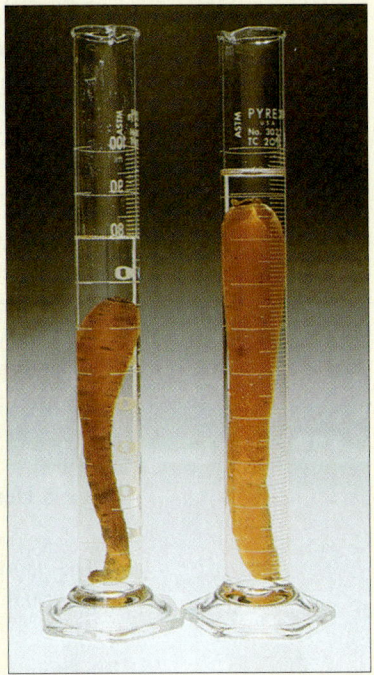

Figure 6–C • When a carrot is immersed in saltwater (*left*), the osmotic pressure outside the cells of the vegetable is higher than that inside. Water flows out into the solution, causing the carrot to shrink. A carrot left to stand in pure water (*right*) does not shrivel. (*Charles D. Winters*)

moved by dialytic transport into the cells of the tubular wall and ultimately secreted in the urine. Most of the water and solutes that have use in the body, such as glucose, amino acids, and ions of various salts, are reabsorbed from the nephron through the capillaries and returned to the bloodstream.

iological fluids, such as blood serum or urine. Blood serum in a healthy person has an osmolality between 0.28 and 0.30 mol kg^{-1}. An elevated serum osmolality may indicate dehydration, diabetes, stroke, shock, or certain other conditions. A depressed serum osmolality may indicate excess fluid intake or possibly lung cancer.

• The unit "osmol kg^{-1}" is often used to emphasize that this is an effective mole, as measured in an osmometer.

6-7 Mixtures and Distillation

The concept of an ideal solution is introduced in the preceding section in terms of one volatile component (the solvent) and one non-volatile component (the solute).

- The ideas discussed here can be extended to solutions of three or more volatile components as well.

We now consider solutions containing two volatile components. Such solutions are ideal only if the vapor pressure *of each component* follows Raoult's law over the whole range of mole fraction. In mathematical terms,

$$P_1 = X_1 P_1^0 \quad \text{and} \quad P_2 = X_2 P_2^0 \qquad \text{(for an ideal solution of components 1 and 2)}$$

The total vapor pressure above an ideal solution of two components is, from Dalton's law, the sum of the vapor pressures of the components.

$$P_{\text{tot}} = P_1 + P_2 = X_1 P_1^0 + X_2 P_2^0$$

The vapor pressures P_1 and P_2 and their sum P_{tot} are graphed as a function of mole fraction in Figure 6–31.

The simplicity of the preceding equations does not carry over to real solutions. Deviations from ideality cause the straight lines in Figure 6–31 to curve in various ways, depending on the relative strengths of the solute–solvent and solvent–solvent forces. Nevertheless, for sufficient dilute solutions (sufficiently small values of X_2), the vapor pressure of component 2 (even in a non-ideal solution) becomes directly proportional to X_2:

$$P_2 = k_H X_2$$

This is **Henry's law:**

> The vapor pressure of a volatile solute in a sufficiently dilute solution is proportional to the mole fraction of the solute in the solution

The constant of proportionality k_H is the **Henry's law constant.** It differs for every solute–solvent pair and depends on the temperature as well. The use of Henry's law is straightforward. The k_H for CO_2 in water equals 1650 atm at 25°C. Maintaining a pressure of, say, 3.0 atm of $CO_2(g)$ at 25°C over a quantity of water (in a soft-drink bottle, perhaps) forces the CO_2 into solution until its mole fraction equals

$$X_{CO_2} = \frac{P_{CO_2}}{1650 \text{ atm}} = \frac{3.0 \text{ atm}}{1650 \text{ atm}} = 1.8 \times 10^{-3}$$

A higher pressure of CO_2 squeezes even more CO_2 into solution. Figure 6–32 illustrates.

Raoult's law and Henry's law are identical for ideal solutions. In those cases, the Henry's law constant k_H is equal to P_2^0. For non-ideal solutions, Henry's law,

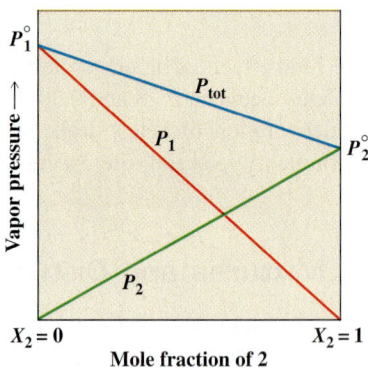

Figure 6–31 • In an ideal solution of two volatile liquids, the vapor pressures P_1 and P_2 of each component, as well as their sum, vary linearly with mole fraction. P_1^0 and P_2^0 are the vapor pressures of the pure component liquids.

like Raoult's law, is a *limiting law* that is valid only for dilute solutions. Whenever Raoult's law is valid for the solvent, Henry's law is valid for the solute.

EXAMPLE 6-18

The Henry's law constant for oxygen dissolved in water is 4.34×10^4 atm at 25°C. If the partial pressure of oxygen in air is 0.20 atm under ordinary atmospheric conditions, calculate the concentration (in moles per liter) of dissolved oxygen in water in equilibrium with air at 25°C.

Strategy

Assume exactly 1 L of solution. Solve Henry's law for X_{O_2} and use the result to calculate the mole fraction of oxygen in the solution. Then convert from mole fraction to molarity (Fig. 6–26).

Solution

$$X_{O_2} = \frac{P_{O_2}}{k_H} = \frac{0.20 \text{ atm}}{4.34 \times 10^4 \text{ atm}} = 4.6 \times 10^{-6}$$

This solution is nearly pure water. One liter of it therefore has a mass of 1000 g and contains

$$1000 \text{ g water} \times \left(\frac{1 \text{ mol water}}{18.02 \text{ g water}} \right) = 55.5 \text{ mol water}$$

Because X_{O_2} is so small, $n_{H_2O} + n_{O_2}$ is very close to n_{H_2O}, and

$$X_{O_2} = \frac{n_{O_2}}{n_{H_2O} + n_{O_2}} \approx \frac{n_{O_2}}{n_{H_2O}}$$

$$4.6 \times 10^{-6} = \frac{n_{O_2}}{55.5 \text{ mol}}$$

$$n_{O_2} = (4.6 \times 10^{-6})(55.5 \text{ mol}) = 2.6 \times 10^{-4} \text{ mol}$$

$$c_{O_2} = \frac{n_{O_2}}{V_{solution}} = \frac{2.6 \times 10^{-4} \text{ mol}}{1 \text{ L}} = 2.6 \times 10^{-4} \text{ mol L}^{-1}$$

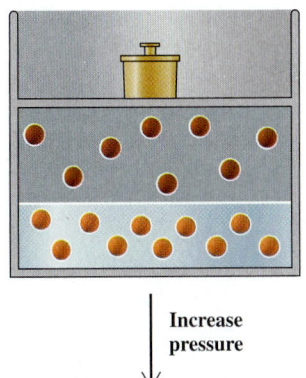

Increase pressure

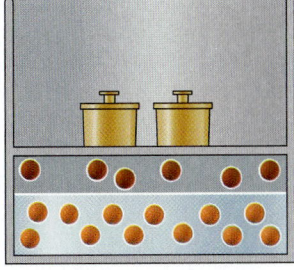

Figure 6-32 • According to Henry's law, as the pressure of a gas above a solution is increased, the mole fraction of the gas in solution increases in direct proportion.

EXERCISE

When the partial pressure of nitrogen over a sample of water at 19.4°C is 9.20 atm, then the concentration of nitrogen in the water is 5.76×10^{-3} mol L^{-1}. Compute the Henry's law constant for nitrogen in water at this temperature.

Answer: 8.86×10^4 atm.

Distillation

The vapor pressures of the pure components of a solution generally differ from each other. Consequently, the composition of the vapor above a solution generally differs from the composition of the solution itself. The case of a mixture of hexane (C_6H_{14}) and heptane (C_7H_{16}) illustrates the point well. These two hydrocarbons form a nearly ideal solution over the whole range of mole fractions. At 25°C, the vapor pressure of pure hexane is $P_1^0 = 0.198$ atm and the vapor pressure of pure heptane is $P_2^0 = 0.0600$ atm. Suppose that a solution contains 4.00 mol of hexane and 6.00

mol of heptane, so that its mole fractions are $X_1 = 0.400$ and $X_2 = 0.600$. Raoult's law gives the two partial pressures:

$$P_{\text{hexane}} = P_1 = X_1 P_1^0 = (0.400)(0.198 \text{ atm}) = 0.0792 \text{ atm}$$
$$P_{\text{heptane}} = P_2 = X_2 P_2^0 = (0.600)(0.0600 \text{ atm}) = 0.0360 \text{ atm}$$

We now use Dalton's law, adding the two pressures to find the total pressure:

$$P_{\text{tot}} = P_1 + P_2 = 0.1152 \text{ atm}$$

Let X_1' and X_2' represent the mole fractions of hexane and heptane in the vapor. Then

$$X_1' = \frac{0.0792 \text{ atm}}{0.1152 \text{ atm}} = 0.688$$

$$X_2' = 1 - X_1' = \frac{0.0360 \text{ atm}}{0.1152 \text{ atm}} = 0.312$$

The liquid and the vapor with which it is in equilibrium have different compositions (Fig. 6–33). The vapor is enriched in the more volatile component.

Such behavior is general. It underlies an important technique for the separation of mixtures called **fractional distillation.** The details of fractional distillation are best illustrated by replacing the plot of vapor pressure versus mole fraction in Figure 6–33 with the plot of boiling temperature versus mole fraction in Figure 6–34. Note that pure component 2, which has the *lower* vapor pressure (P_2^0 in Fig. 6–33) has the *higher* boiling point ($T_{\text{b,2}}$ in Fig. 6–34). Component 2 is less volatile and so requires a higher temperature to overcome the attractions among its molecules. In Figure 6–34, an "original" solution of 1 and 2 starts to boil when it is heated to the temperature given by the blue curved line. The vapor in equilibrium with the boiling solution is enriched in the more volatile component 1. The composition of the vapor is read from the plot by extending a line horizontally across the page from the blue curve until it intersects the red curve, and then tracing vertically down to the

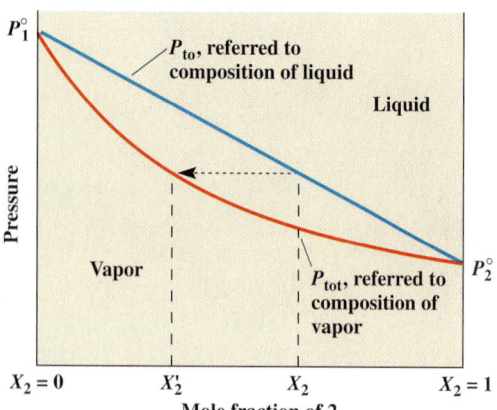

Figure 6–33 • The composition of the vapor above a solution differs from the composition of the liquid with which it is in equilibrium. Here, the straight blue line is the total pressure of the vapor in equilibrium with an ideal solution having mole fraction X_2 of component 2. By moving horizontally from that line to a point of intersection with the red curve, we can locate the mole fraction X_2' of component 2 in the vapor.

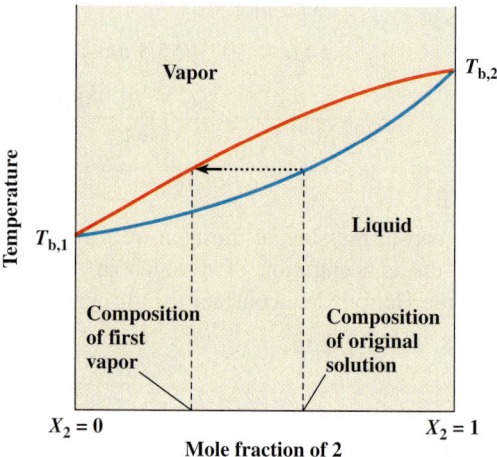

Figure 6–34 • The boiling temperature of an ideal solution varies with the composition of the solution. The blue curve is the boiling temperature referred to the liquid composition, and the red curve is the boiling temperature referred to the vapor composition. The vapors boiling off a solution are enriched in the more volatile component to an extent given by moving horizontally from the blue line to the red line.

composition axis. As vapors that are enriched in component 2 leave the solution, the remainder is enriched in component 1. The boiling temperature of the remaining solution increases, sliding up the blue curved line toward the boiling temperature of pure component 2.

We can manage the vaporization of the solution in different ways. If we simply boil it until it all boils away, the final composition of the collected vapor is obviously the same as the composition of the original solution. If we stop the boiling before all the solution has turned to vapor, collect the vapor that has boiled off to that point, and recondense it, the result differs. The condensed liquid (*distillate*) is now richer in component 1 than the original liquid. By starting over with this distillate and repeating the process again and again, we obtain mixtures that are successively richer in component 1. This is the principle behind a **fractional distillation column** (Fig. 6–35). Actually collecting vapors, condensing them, and reboiling would be incredibly laborious. Instead, the fractionating column contains a packing of glass beads or other inert materials that provides numerous condensation sites. All along the column, repeated vapor-to-liquid-to-vapor cycles take place essentially automatically. The contents of the column become stratified according to composition, with the more volatile component very prevalent at the top. Taking off the vapors from the top and cooling them furnishes a nearly pure portion of the more volatile component.

Fractional distillation is used in oil refineries to separate gasoline from other petroleum products. The technique is also used to separate nitrogen and oxygen from air: The air is liquefied and then fractionally distilled. The lower boiling nitrogen ($T_b = -196°C$) vaporizes before the oxygen ($T_b = -183°C$) and is greatly enriched in the "top cut" from the column.

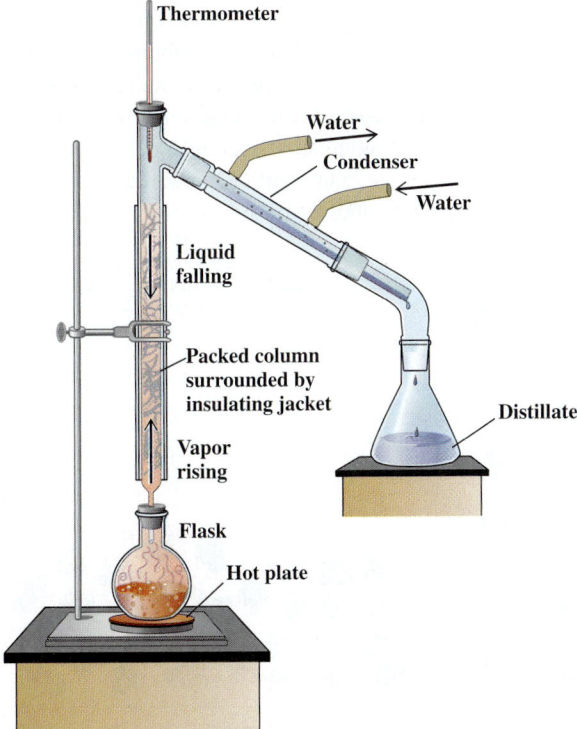

Thermometer

Water

Condenser

Water

Liquid falling

Packed column surrounded by insulating jacket

Distillate

Vapor rising

Flask

Hot plate

Figure 6–35 • In a distillation column, the temperature decreases moving up the column. The less volatile components condense and fall back to the flask, but the more volatile ones continue up the column to pass into the water-cooled condenser, where they condense and are recovered in the receiver.

6-8 Colloidal Dispersions

• A colloidal dispersion is also called a "colloid." The word is derived from a Greek word meaning "glue-like."

Colloidal dispersions resemble solutions in some ways, but the dispersed particles in a colloid are larger than the solvated molecules or ions found in true solutions. They range upward from 10^{-9} m to about 1000×10^{-9} m in diameter and consist of aggregates of many smaller molecules and ions or sometimes of single very large molecules (for example, some globular protein molecules have diameters on the order of 500×10^{-9} m).

The dispersed substance and the background medium in a colloid may be any combination of gas, liquid, or solid. Examples of colloids are extremely varied: aerosol sprays (liquid dispersed in gas), smoke (solid particles in air), milk (fat droplets and solids in water), mayonnaise (water droplets in oil), paint (solid pigment particles in oil for oil-based paints, or pigment and oil dispersed in water for latex paints), and biological fluids (proteins and fats in water).

Colloids with larger particles look translucent, cloudy, or milky, but those with smaller particles look clear, even under a microscope. The best evidence for the presence of colloidal particles is the **Tyndall effect:** Rays of visible light passing into a colloid are scattered, making the path of a beam of light through the dispersion visible from the side (Fig. 6–36). A familiar example is the visible cone marking the passage of light from a movie projector though a dispersion of small dust particles in air. The gemstone opal has remarkable optical properties that arise from colloidal water suspended in the solid silicon dioxide that makes up most of the mass of the stone.

Figure 6–36 • The Tyndall effect. The colloidal particles suspended in the liquid scatter light to the side. Neither a pure liquid nor a true solution exhibits the effect. (*Charles D. Winters*)

The dispersed particles in colloids usually acquire electric charges—some positive, some negative. Because all the particles of a certain kind of colloid acquire the same kind of charge, they tend to repel each other. This helps to keep them dispersed in the suspension and prevents them from settling. Some colloids settle out into two separate phases if left standing long enough; others can persist indefinitely. A suspension of gold particles in water prepared by Michael Faraday in 1857 shows no settling to date. The settling out of colloids is speeded by dissolving salts in the background medium. The added ions reduce the repulsive electrostatic forces between the dispersed particles, leading to their **coagulation** (aggregation into larger particles) and sedimentation (Fig. 6–37). An interesting example occurs in river deltas. Where river water that contains suspended clay particles meets the salty water of the ocean, the colloidal clay coagulates and settles out as flocculent (open,

Figure 6–37 • When a salt is added to a colloidal dispersion (a), the repulsive forces between the colloidal particles are reduced and coagulation (sticking together) occurs (b). Eventually, the aggregated particles fall to the bottom of the container as a sediment or precipitate (c).

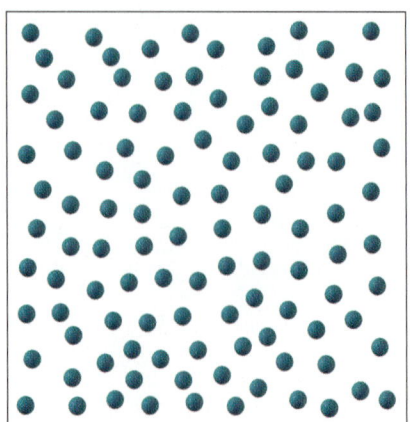

(a) Dispersion

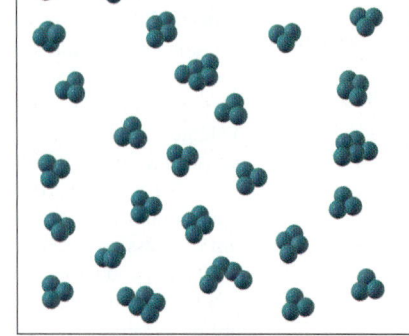

(b) Coagulation

(c) Sedimentation

(a)   (b)

Figure 6-38 • The colloidal dispersion of $PbCrO_4$ in (a) appears cloudy. After coagulation, the particles of $PbCrO_4$ soon settle to the bottom (b). (*Leon Lewandowski*)

low-density) sediments. Flocculating agents are deliberately added to paints so that the pigment settles in a loosely packed sediment. When the paint is stirred, the pigment is then easily redispersed through the medium. In the absence of such agents, the suspended particles tend to form compact sediments that are difficult to resuspend. Reversal of coagulation, a process called **peptization,** occurs frequently upon stirring or other treatment of coagulated colloids.

In some cases, such as in the precipitation of a solid from solution (see Section 4–2) the formation of a colloid is not desirable. If the particles of a newly formed precipitate are large, they fall quickly to the bottom of the vessel, where they can easily be separated from the liquid by filtration. Sometimes (especially with metal sulfides) the precipitate may appear as a colloidal dispersion, with particles small enough to pass through ordinary filter paper (Fig. 6–38). If this happens, separation can be achieved only by coagulation, by centrifugation, or by forcing the dispersion through a membrane, such as cellophane, that permits passage of only the small solvent molecules.

Many biological fluids are colloids. Blood plasma consists of quantities of colloidal lipids (fatty, water-insoluble compounds) and colloidal albumin, a protein, as well as conventionally dissolved salts, glucose, amino acids, vitamins, and hormones. The cytosol, which is the viscous, semifluid, semitransparent content of a cell, is a conventional solution of sugars, soluble proteins, and potassium, sodium, and other salts of small-molecule organic acids. This medium then contains water-insoluble proteins and other large biomolecules in colloidal dispersion.

Colloidal particles are in a state of constant motion called **Brownian motion** (after Robert Brown, a Scottish botanist who used a microscope to observe the motion of pollen particles in water). This motion results from the constant random buffeting of the particles by molecules of solvent. In 1905, Albert Einstein showed how the motion of Brownian particles could be described on a microscopic level; his work provided one of the most convincing verifications of the molecular hypothesis and of the kinetic theory of matter and led to a fairly accurate determination of Avogadro's number.

SUMMARY

6-1 Intermolecular forces give the molecules in an assembly **potential energy** in addition to their kinetic energy; such forces are responsible for the existence of condensed phases. **Ion–ion forces** dominate the interactions of ionic solids and liquids, whereas weaker **dipole–dipole forces** are important for **polar** molecules, in which the two ends of a molecule exhibit opposite charges. **Dispersion forces,** which arise from fluctuating dipole moments on neighboring atoms or molecules, cause attractions among atoms and non-polar molecules. **Hydrogen bonds** are particularly strong, dipole–dipole interactions between certain hydrogen-containing molecules, such as those of water. Hydrogen bonding causes many of the unexpected properties of water. Any two ions, atoms, or molecules exhibit **repulsive forces** at sufficiently short distances.

6-2   A kinetic model for liquids and solids that incorporates intermolecular forces can describe the qualitatively different structures and dynamics of the three states of matter.

6-3   Two phases of a pure substance can coexist in **phase equilibrium.** At a given temperature, a liquid or solid coexists with a vapor inside a closed container, and the **vapor pressure** is the final pressure that persists indefinitely above the solid or liquid when the rates of **evaporation** from and **condensation** onto a liquid or a solid become equal. The vapor pressure of the water above an aqueous solution must be taken into account in accurate measurements of the amounts of gas produced in a chemical reaction when the gas is collected over water. The **normal boiling point** is the temperature at which the vapor pressure of a liquid equals 1 atm. The **normal melting point** is the temperature at which solid and liquid coexist indefinitely at a pressure of 1 atm of air.

6-4   **Melting** and **freezing** are the solid-to-liquid and liquid-to-solid phase transitions, respectively. **Evaporation** and **condensation** describe the liquid-to-gas transition and its reverse, and **sublimation** and **deposition** describe the direct solid-to-gas transition and its reverse. The melting and boiling points of water are abnormally high compared with those of other substances of low molar mass. Also, the density of ice is less than that of liquid water, and the heat required to vaporize water is abnormally large. These exceptional characteristics derive from the strong intermolecular forces (hydrogen bonds) acting between molecules of water.

6-5   A **phase diagram** of a substance shows which phase is stable at any given pressure and temperature. The lines in such a diagram define conditions of coexistence of two phases in equilibrium, and a **triple point** represents conditions that allow the coexistence of three phases in equilibrium. The fluid–solid coexistence curve continues to arbitrarily high pressures, but the gas–liquid curve terminates at the **critical point** as the distinction between liquid and gas ceases to exist. At temperatures and pressures above its **critical temperature** and **critical pressure,** a substance is a **supercritical fluid.** For many substances, more than one solid phase appears in the phase diagram, especially at elevated pressures.

6-6   The presence of a solute changes the properties of a solvent. It **lowers the vapor pressure** of the solvent below that of the pure solvent in proportion to the mole fraction of solute **(Raoult's law).** Also, a solute **elevates the boiling point** and **depresses the freezing point** by amounts proportional to its **molality,** a fact that permits the measurement of solute molar masses. The **osmotic pressure** of a solution is the pressure difference that exists across a semipermeable membrane between the solution and a supply of pure solvent. It, too, depends on the concentration of solute and can be measured to determine the molar masses of solutes. These **colligative properties** depend only on the ratio of the number of solute particles to the number of solvent particles, and not on the identity of the solute. When solutes dissociate, additional particles form; the effect of solute dissociation on the colligative properties is taken into account by insertion of the **van't Hoff** *i* into the appropriate equations. Measurements of colligative properties on complex solutions give **osmolalities** or **osmolarities,** which are the effective molality or molarity of all the solutes.

6–7   The laws governing the properties of ideal solutions can be extended to mixtures of two volatile substances. **Henry's law** states that the partial pressure of a volatile solute is proportional to its mole fraction in solution at sufficiently low mole fractions. When two components in a liquid mixture have different vapor pressures, they can be separated by a process of repeated vaporization and condensation called **fractional distillation.**

6–8   **Colloids** are mixtures of substances in which one phase is dispersed as small particles (between approximately 10^{-6} and 10^{-9} m in diameter) in a second phase. Colloids exhibit the **Tyndall effect** and **Brownian motion.** Colloidal particles tend not to settle out, but colloids can be **coagulated** (often by the addition of solutions of salts). The reverse of coagulation is **peptization.** Colloidal particles also can be separated from the dispersing medium by centrifugation.

PROBLEMS

Note: Answers to blue-numbered problems are given in Appendix F. Problems that are more challenging are indicated with asterisks.

Intermolecular Forces: Why Condensed Phases Exist

1. (a) Use Figure 6–1 to estimate the length of the covalent bond in Cl_2 and the length of the ionic bond in K^+Cl^-. (*Note*: The latter corresponds to the distance between the atoms in an isolated single molecule of K^+Cl^-, not in solid potassium chloride.)
   (b) A book states: "the shorter the bond, the stronger the bond." What features of Figure 6–1 show that this is not always true?

2. True or false: Any two atoms held together by non-bonded attractions must be farther apart than any two atoms held together by a chemical bond. Explain.

3. (See Example 6–1.) Name the types of attractive forces that contribute to the interactions between atoms, molecules, or ions in the following substances. Indicate the one(s) you expect to predominate. (a) KF (b) HI (c) CH_3OH (d) Rn (e) N_2

4. (See Example 6–1.) Name the types of attractive forces that contribute to the interactions between atoms, molecules, or ions in the following substances. Indicate the one(s) you expect to predominate. (a) Ne (b) ClF (c) F_2 (d) $BaCl_2$ (e) NH_2OH

5. Predict whether a sodium ion is most strongly attracted to a bromide ion, a molecule of hydrogen bromide, or an atom of krypton.

6. Predict whether an atom of argon is most strongly attracted to another atom of argon, to an atom of neon, or to an atom of krypton.

7. As a vapor, methanol exists to an extent as a tetramer,

$(CH_3OH)_4$. In the tetramer, four

$$H-\overset{\overset{\displaystyle H}{|}}{\underset{\underset{\displaystyle H}{|}}{C}}-O-H$$

molecules are held together by hydrogen bonds. Propose a reasonable structure for this tetramer.

8. Hypofluorous acid (HOF) is the simplest possible compound that allows comparison between fluorine and oxygen in their ability to form hydrogen bonds. Solid HOF unexpectedly contains no H⋯F hydrogen bonds, but many H⋯O hydrogen bonds. Propose and draw a structure for chains of HOF molecules linked by O⋯H hydrogen bonds. The bond angle in HOF is 101°.

The Kinetic Theory of Liquids and Solids

9. The *diffusion constant* is a measure of the rapidity of mixing between substances caused by the chaotic motions of their molecules. Do you expect the diffusion constant to increase or decrease as the density of a liquid is increased (by compressing it) at constant temperature? Explain. What happens to the diffusion constant of a gas and a solid as the density increases?

10. Do you anticipate that the diffusion constant (see previous problem) would increase as the temperature of a liquid increases at constant pressure? Why or why not? Will the diffusion constant increase with temperature for a gas and a solid? Explain.

11. The mean free path of the molecules in a gas equals the average distance they travel between collisions (see Section 5–7). Compare the mean free path of molecules in a liquid to the mean free path of molecules in a gas at the same temperature.

12. Surface tension, which makes the surfaces of liquids behave like weak elastic skins, is caused by intermolecular interactions. Is the surface tension of molten sodium chlo-

ride higher than or lower than that of liquid carbon tetrachloride? Explain.

Phase Equilibrium

13. Hydrogen at a pressure of 1 atm condenses to a liquid at 20.3 K and solidifies at 14.0 K. The vapor pressure of liquid hydrogen is 0.213 atm at 16.0 K. Calculate the volume of 1.00 mol of H_2 vapor under these conditions, and compare it with the volume of 1.00 mol of H_2 at STP.

14. Helium condenses to a liquid at 4.224 K under atmospheric pressure and remains a liquid down to the absolute zero of temperature. The vapor pressure of liquid helium at 2.20 K is 0.05256 atm. Calculate the volume occupied by 1.000 mol of helium vapor in equilibrium with liquid helium under these conditions, and compare it with the volume of the same amount of helium at STP.

15. (See Example 6–2.) Calcium carbide reacts with water to produce acetylene (C_2H_2) and calcium hydroxide. The acetylene is collected over water at 40.0°C under a total pressure of 756.2 torr. The vapor pressure of water at this temperature is 55.3 torr. Calculate the mass of acetylene per liter of "wet" acetylene collected in this way, assuming ideal gas behavior.

16. (See Example 6–2.) When an excess of sodium hydroxide is added to an aqueous solution of ammonium chloride, gaseous ammonia is produced:

$$NaOH(aq) + NH_4Cl(aq) \longrightarrow$$
$$NaCl(aq) + NH_3(g) + H_2O(\ell)$$

   (a) Suppose 3.68 g of ammonium chloride reacts in this way at 30°C and a total pressure of 751.2 torr. At this temperature, the vapor pressure of water is 31.8 torr. Predict the volume of ammonia saturated with water vapor that would be produced under these conditions, assuming no leaks or other losses of gas.
   (b) The actual volume of gas collected is *much* smaller than the volume calculated above. Suggest the reason why.

Phase Transitions

17. (See Example 6–3.) In Chapter 3, it was stated that, other things being equal, ionic character in compounds of metals decreases with increasing oxidation number.
   (a) Rank the following compounds from lowest to highest in normal melting point: MnO_2, Mn_3O_4, Mn_2O_7.
   (b) Although potassium permanganate, $KMnO_4$, contains manganese in the +7 oxidation state, it behaves as an ionic compound. Explain why.

18. (See Example 6–3.) Rank the following from lowest to highest in normal boiling point: AsF_5, SbF_3, SbF_5, F_2, and Cl_2.

19. High in the Andes, an explorer notices that the water for tea is boiling vigorously at a temperature of 90°C. Use Figure 6–23 to estimate the atmospheric pressure at the altitude of the camp. What fraction of the Earth's atmosphere lies below the level of the explorer's camp?

20. The total pressure in a pressure cooker filled with water rises to 4.0 atm when it is heated, and this pressure is maintained by the periodic operation of a relief valve. Use Figure 6–22c to estimate the temperature of the water in the pressure cooker.

21. It is essential to boil a sample of a certain organic liquid. Unfortunately, the liquid starts to smoke and char before it boils. Suggest a way to cause the liquid to boil at a lower temperature and so avoid this decomposition.

22. Aluminum melts at a temperature of 660°C and boils at 2470°C, whereas thallium melts at a temperature of 304°C and boils at 1460°C. Which metal is more volatile at room temperature?

Phase Diagrams

23. At its melting point (624°C), solid plutonium has a density of 16.24 g cm^{-3}. The density of liquid plutonium is 16.66 g cm^{-3}. A small sample of liquid plutonium at 625°C is strongly compressed. Predict what phase changes, if any, occur.

24. Phase changes occur between different solid forms, as well as from solid to liquid, liquid to gas, and solid to gas. When white tin at 1.00 atm is cooled below 13.2°C, it spontaneously changes (over a period of weeks) to gray tin. The density of gray tin is *less* than the density of white tin (5.75 g cm^{-3} versus 7.31 g cm^{-3}). Some white tin is compressed to a pressure of 2.00 atm. At this pressure, should the temperature be higher or lower than 13.2°C for the conversion to gray tin to take place? Explain your reasoning.

25. (See Example 6–5.) Determine whether argon is a solid, a liquid, or a gas at each of the following combinations of temperature and pressure (use Fig. 6–22).
   (a) 50 atm and 100 K (b) 8 atm and 150 K (c) 1.5 atm and 25 K (d) 0.25 atm and 120 K.

26. (See Example 6–5.) Some water starts out at a temperature of 298 K and a pressure of 1 atm. It is compressed to 500 atm at constant temperature and then heated to 750 K at constant pressure. Next it is decompressed at 750 K back to 1 atm and finally cooled to 400 K at constant pressure.
   (a) What is the state (solid, liquid, or gas) of the water at the start of the experiment?
   (b) What is the state (solid, liquid, or gas) of the water at the end of the experiment?
   (c) Do any phase transitions occur during the four steps described? If so, at what temperature and pressure do they occur? (*Hint*: Trace out the various changes on the phase diagram of water [Fig. 6–22].)

27. The vapor pressure of solid acetylene at −84.0°C is 760 torr.
   (a) Does the triple-point temperature lie above or below −84.0°C? Explain.
   (b) Suppose a sample of solid acetylene is held under an external pressure of 0.80 atm and heated from 10 K to 300 K. What phase change, if any, occurs?

28. The triple point of hydrogen occurs at a temperature of 13.8 K and a pressure of 0.069 atm.
(a) What is the vapor pressure of solid hydrogen at 13.8 K?
(b) Suppose a sample of solid hydrogen is held under an external pressure of 0.030 atm and heated from 5 K to 300 K. What phase change, if any, occurs?

29. The density of nitrogen at its critical point is 0.3131 g cm^{-3}. At a very low temperature, 0.3131 g of solid nitrogen is sealed into a thick-walled glass tube with a volume of 1.000 cm^3. Describe what happens inside the tube as the tube is warmed past the critical temperature, 126.19 K.

30. At its critical point, ammonia has a density of 0.235 g cm^{-3}. You have a special thick-walled glass tube that has a 10.0 mm outside diameter, a wall thickness of 4.20 mm, and a length of 155 mm. How much ammonia must you seal into the tube if you wish to observe the disappearance of the meniscus as you heat the tube and its contents past 132.23°C, the critical temperature?

Colligative Properties of Solutions

31. (See Examples 6–6 and 6–8.) A disinfectant solution is made by dissolving 5.10 g of hydrogen peroxide (H_2O_2) in 164.6 g of water. Calculate the mass fraction, mole fraction, and molality of hydrogen peroxide in the solution.

32. (See Examples 6–6 and 6–8.) A solution of purified vinegar is determined to contain 32.1 g of acetic acid (CH_3COOH), 602 g of water, and no other components. Calculate the mass percentage, mole fraction, and molality of acetic acid in the solution.

33. (See Examples 6–8 and 6–9.) An aqueous solution of hydrochloric acid is 38.00% HCl by mass. The solution has a density of 1.1886 g cm^{-3} at 20°C. Compute the molality and the molarity of the HCl.

34. (See Examples 6–8 and 6–9.) An aqueous solution of acetic acid is 20.00% CH_3COOH by mass. The solution has a density of 1.0250 g cm^{-3} at a certain temperature. Compute the molality and molarity of the acetic acid.

35. (See Example 6–10.) Concentrated aqueous phosphoric acid sold for use in the laboratory contains 12.2 mol of H_3PO_4 per liter of solution at 25°C and has a density of 1.33 g cm^{-3}. Calculate the molality of the phosphoric acid.

36. (See Example 6–10.) A 9.20 mol L^{-1} perchloric acid ($HClO_4$) solution has a density of 1.54 g cm^{-3} at 25°C. Calculate the molality of the perchloric acid.

37. (See Example 6–11.) The vapor pressure of pure acetone (CH_3COCH_3) at 30°C is 0.3720 atm. Suppose 15.0 g of benzophenone ($C_{13}H_{10}O$), which does not dissociate in acetone, is dissolved in 50.0 g of acetone. Calculate the vapor pressure of acetone above the resulting solution.

38. (See Example 6–11.) The vapor pressure of diethyl ether ($M = 74.12 \text{ g mol}^{-1}$) at 30°C is 0.8517 atm. Suppose 1.800 g of maleic acid ($C_4H_4O_4$), which does not dissociate in diethyl ether, is dissolved in 100.0 g of diethyl ether at 30°C. Calculate the vapor pressure of diethyl ether above the resulting solution.

39. The vapor pressure of carbon tetrachloride (CCl_4) at 50°C is 0.437 atm. When 7.42 g of a pure, non-volatile, and non-dissociating substance is dissolved in 100.0 g of carbon tetrachloride, the vapor pressure of the solution at 50°C is 0.411 atm. Calculate the mole fraction of solute in the solution.

40. At 25°C, the vapor pressure of benzene (C_6H_6) is 0.1252 atm. When 10.00 g of an unknown non-volatile substance is dissolved in 100.0 g of benzene, the vapor pressure of the solution at 25°C is 0.1199 atm. Calculate the mole fraction of solute in the solution, assuming no dissociation by the solute.

41. (See Example 6–13.) When 39.8 g of a non-dissociating, non-volatile sugar is dissolved in 200.0 g of water, the boiling point of the water is raised by 0.30°C. Estimate the molar mass of the sugar.

42. (See Example 6–13.) When 2.60 g of a non-dissociating, non-volatile compound of indium and chlorine is dissolved in 50.0 g of tin(IV) chloride, the normal boiling point of the tin(IV) chloride is raised from 114.1°C to 116.3°C. If $K_b = 9.43 \text{ K kg mol}^{-1}$ for $SnCl_4$, what are the approximate molar mass and the molecular formula of the solute?

43. (See Example 6–15.) The following four substances are all soluble in benzene and do not dissociate in solution: C_6H_5Cl, N_2H_4, $(C_2H_5)_2O$, CCl_4. In separate experiments, 1.00 g of each is added to 100.0 g of benzene. Which substance gives the solution with the highest freezing point?

44. (See Example 6–15.) Sodium chloride (NaCl), calcium chloride ($CaCl_2$), and urea (NH_2CONH_2) all have been used for melting ice on streets. Estimate the lowering of the freezing point caused by dissolving 50.0 g of each in 1.00 kg of solvent (water). Assume that urea does not dissociate significantly in aqueous solution and that the two ionic substances dissociate completely into ions.

45. Ice cream is made by freezing a liquid mixture that, to a first approximation, can be considered a solution of sucrose ($C_{12}H_{22}O_{11}$) in water. Estimate the temperature at which the first ice crystals begin to appear in a mix that consists of 34% (by mass) sucrose in water. As ice crystallizes out, the remaining solution becomes more concentrated. What happens to its freezing point?

46. The solution to Problem 45 shows that to make homemade ice cream, temperatures ranging downward from −3°C are needed. Ice cubes from a freezer have a temperature of approximately −12°C (+10°F), which is cold enough, but contact with the warmer ice-cream mixture causes them to melt to liquid at 0°C, which is too warm. To obtain a liquid that is cold enough, salt (NaCl) is dissolved in water and ice is added to the salt water. The salt lowers the freezing point of the water enough so that it can freeze liquid ice-cream mixture. The instructions for an ice-cream maker say to add one part of salt to eight parts water (by mass). What is the freezing point of this solution (in degrees Celsius and degrees Fahrenheit)? Assume that the NaCl dissociates fully into ions and that the solution behaves ideally.

47. (See Example 6–16.) A solution is prepared by dissolving 1.55 g of acetone (CH_3COCH_3) in 50.00 g of water. Its freezing point equals $-0.970°C$. Does acetone dissociate into ions when dissolved in water? Explain your answer.

48. (See Example 6–16.) When 2.02 g of baking soda (sodium hydrogen carbonate, $NaHCO_3$) is dissolved in 200 g of water, the freezing point is $-0.396°C$. Determine the van't Hoff i for this solution. Write a single chemical equation to represent the dissolution of sodium hydrogen carbonate in water.

49. (See Example 6–16.) An aqueous solution is 0.8402 molal in Na_2SO_4. It has a freezing point of $-4.218°C$. Determine the effective number of particles arising from each Na_2SO_4 formula unit in this solution.

50. (See Example 6–16.) The freezing-point depression constant of pure H_2SO_4 is $6.12°C\ kg\ mol^{-1}$. When 2.3 g of ethanol (C_2H_5OH) is dissolved in 1.00 kg of pure sulfuric acid, the freezing point of the solution is $0.92°C$ lower than the freezing point of pure sulfuric acid. Determine how many particles are formed as one molecule of ethanol goes into solution in sulfuric acid.

51. (See Example 6–17.) A polymer of large molar mass is dissolved in water at $15°C$, and the resulting solution rises to a final height of 16.3 cm above the level of the pure water, as water molecules pass through a semipermeable membrane into the solution. If the solution contains 5.73 g of polymer per liter, estimate the molar mass of the polymer.

52. (See Example 6–17.) Suppose 0.117 g of a protein is dissolved in 10.0 cm^3 of ethyl alcohol (C_2H_5OH), which has a density of 0.789 g cm^{-3} at $20°C$. The solution rises to a height of 27.4 cm in an osmometer in the setup shown in Figure 6–30. What is the approximate molar mass of the protein? State all assumptions.

53. Careful measurements using a vapor-pressure osmometer fix the osmolality of a sample of urine at 1.40 mol kg^{-1}. Predict the freezing point and boiling point of this sample (assuming that no chemical change occurs when it is heated). Estimate the osmotic pressure of this sample at $25°C$.

54. When 0.494 mg of $K_3Fe(CN)_6$ is dissolved in 100.0 mg of water, the freezing point is found to be $-0.093°C$. Compute the osmolality of this solution and the molality of this solution. Explain why the two differ.

Mixtures and Distillation

55. (See Example 6–18.) The Henry's law constant at $25°C$ for carbon dioxide dissolved in water is 1.65×10^3 atm. A carbonated beverage is bottled under a carbon dioxide pressure of 5.0 atm.
    (a) Calculate the chemical amount of carbon dioxide dissolved per liter of water under these conditions, using 1.00 g cm^{-3} as the density of water.
    (b) Explain what happens on a microscopic level after the bottle cap is removed.

56. (See Example 6–18.) The Henry's law constant at $25°C$ for nitrogen dissolved in water is 8.57×10^4 atm; the constant for oxygen is 4.34×10^4 atm; and the constant for helium is 1.7×10^5 atm.
    (a) Calculate the number of moles of nitrogen and oxygen dissolved per liter of water at $25°C$ in equilibrium with air at $25°C$. Use Table 5–1.
    (b) Air is dissolved in blood and other bodily fluids. As a deep-sea diver descends, the pressure increases and the concentration of dissolved air in the diver's blood increases. If the diver returns to the surface too quickly, gas bubbles out of solution within the body so rapidly that it can cause a dangerous condition called "the bends." Show, using Henry's law, why divers sometimes use helium in combination with oxygen in place of compressed air.

57. At $25°C$, some water is added to a sample of gaseous methane (CH_4) at 1.00 atm pressure in a closed vessel and shaken until as much methane as possible dissolves. Then 1.00 kg of the solution is removed and boiled to expel the methane, yielding a volume of 3.01 L of $CH_4(g)$ at STP. Determine the Henry's law constant for methane in water. (*Note*: The solution is dilute enough that 1.00 kg of solution contains 1.00 kg of water.)

58. When exactly the same procedure as in Problem 57 is carried out using benzene (C_6H_6) in place of water, the volume of methane that results is 0.510 L at STP. Determine the Henry's law constant for methane in benzene.

59. At $20°C$, the vapor pressure of toluene is 0.0289 atm, and the vapor pressure of benzene is 0.0987 atm. Equal chemical amounts (equal numbers of moles) of toluene and benzene are mixed and form an ideal solution. Compute the mole fraction of benzene in the vapor in equilibrium with this solution at $20°C$.

60. At $90°C$, the vapor pressure of toluene is 0.534 atm, and the vapor pressure of benzene is 1.34 atm. Benzene (0.300 mol) is mixed with toluene (0.800 mol). The two form an ideal solution. Compute the mole fraction of toluene in the vapor in equilibrium with this solution at $90°C$.

61. At $40°C$, the vapor pressure of pure carbon tetrachloride (CCl_4) is 0.293 atm, and the vapor pressure of pure dichloroethane ($C_2H_4Cl_2$) is 0.209 atm. A nearly ideal solution is prepared by mixing 41.5 g of carbon tetrachloride with 22.6 g of dichloroethane.
    (a) Calculate the mole fraction of CCl_4 in the solution.
    (b) Calculate the total vapor pressure of the solution at $40°C$.
    (c) Calculate the mole fraction of CCl_4 in the vapor in equilibrium with the solution at $40°C$.

62. At 300 K, the vapor pressure of pure benzene (C_6H_6) is 0.1355 atm, and the vapor pressure of pure hexane (C_6H_{14}) is 0.2128 atm. Mixing 50.0 g of benzene with 50.0 g of hexane gives a solution that is nearly ideal.
    (a) Calculate the mole fraction of benzene in the solution.
    (b) Calculate the total vapor pressure of the solution at 300 K.

(c) Calculate the mole fraction of benzene in the vapor in equilibrium with the solution.

Colloidal Dispersions

63. What is the difference between a solution and a colloidal suspension? Give examples of each, and show how, in some cases, it may be difficult to classify a mixture as one or the other.

64. When hot water is mixed with a concentrated aqueous solution of iron(III) chloride, a colloid of hydrated iron(III) oxide forms. Attached to the surface of the colloidal particles are Fe^{3+} ions. As the number of Fe^{3+} ions on the surface of the colloidal particles increases, does the rate of coagulation increase or decrease?

65. The water in your beaker is boiling furiously. As you examine it, you exclaim, "Wow, these white clouds of steam are really hot." Your chemistry teacher hears you and comments that the white clouds, although hot, are not steam. You think it over and say, "All right, the white clouds of water vapor are really hot." "Wrong again," says your teacher. Explain what the white clouds above the beaker really are.

66. In a modern sewage-treatment plant, wastewater is allowed to stand in settling pools so that sand and other fine particles can settle out. "Floc" is then added to the pool. What purpose does the "floc" serve?

ADDITIONAL PROBLEMS

67. Heating sodium chlorate ($NaClO_3$) causes it to break down to sodium chloride (NaCl), gaseous oxygen (O_2), and no other products. A 265 mg sample of *impure* sodium chlorate is heated until all of the sodium chlorate is decomposed. The impurities do not react during this procedure. The gas that is evolved from this reaction has a volume of 35.12 mL when collected over water (collected wet) at 25.0°C and 742 torr. Determine the percentage by mass of sodium chlorate in the original impure sample. Take data from Table 6–1.

68. Moderate heating of sodium nitrate decomposes it to sodium nitrite ($NaNO_2$), gaseous oxygen (O_2), and no other products. A 327 mg sample of impure sodium nitrate is decomposed in this way, and the gas is collected over water at 23.0°C. It has a volume of 47.65 mL at 745 torr. Determine the percentage by mass of sodium nitrate in the original impure sample. Assume that heating the impurities does not generate any gases.

69. At 25°C, the equilibrium vapor pressure of water is 0.03126 atm. A humidifier is placed in a room of volume 112 m^3 and is operated until the air becomes saturated with water vapor. Assuming that no water vapor leaks out of the room and that initially there was no water vapor in the air, calculate the number of grams of water that have passed into the air.

70. The air over an unknown liquid is saturated with the vapor of that liquid at 25°C and a total pressure of 0.980 atm. Suppose that a sample of 6.00 L of the saturated air is collected, and the vapor of the unknown liquid is removed from that sample by cooling and condensation. The pure air remaining occupies a volume of 3.75 L at −50°C and 1.000 atm. Calculate the vapor pressure of the unknown liquid at 25°C.

71. If it is true that all solids and liquids have vapor pressures, then at low enough external pressures, every substance should start to boil. In space, there is effectively zero external pressure. Explain why spacecraft do not just boil away as vapors when placed into orbit.

72. Oxygen melts at 54.8 K and boils at 90.2 K at atmospheric pressure. At the normal boiling point, the density of the liquid is 1.14 g cm^{-3}, and the vapor can be approximated as an ideal gas. The critical point is defined by $T_c = 154.6$ K, $P_c = 49.8$ atm, and (density) $d_c = 0.436$ g cm^{-3}. The triple point is defined by $T_t = 54.4$ K, $P_t = 0.0015$ atm, a liquid density equal to 1.31 g cm^{-3}, and a solid density of 1.36 g cm^{-3}. At 130 K, the vapor pressure of the liquid is 17.25 atm. Use this information to construct a phase diagram showing P versus T for oxygen. You need not draw the diagram to scale, but you should give numerical labels to as many points as possible on both axes.

73. It can be shown that if a gas obeys the van der Waals equation, its critical temperature, its critical pressure, and its molar volume at the critical point are given by the equations

$$T_c = \frac{8a}{27\,Rb}, \quad P_c = \frac{a}{27b^2}, \quad \text{and} \quad \left(\frac{V}{n}\right)_c = 3b$$

where a and b are the van der Waals constants of the gas. Use the van der Waals constants for oxygen, carbon dioxide, and water (from Table 5–3) to estimate the critical point properties of these substances. Compare with the observed values given in Figure 6–22 and in Problem 72.

74. Section 5–8 establishes that the van der Waals constant b (with units of L mol^{-1}) is related to the volume per molecule in the liquid and thus to the sizes of the molecules. The combination of van der Waals constants a/b has units of L atm mol^{-1}. Because the L atm is a unit of energy (1 L atm = 101.325 J), a/b is proportional to the energy per mole for interacting molecules and thus to the strength of the attractive forces between molecules, as shown in Figure 6–1. Use the van der Waals constants in Table 5–3 to rank the attractive forces in the following from strongest to weakest: N_2, H_2, SO_2, and HCl.

75. The item on the chemistry examination was, "Define the critical temperature." The following answers all received zero credit. In each case explain what is wrong with the answer and what the author of the answer might have had in mind.
(a) "The temperature at which an object changes state."
(b) "The point defined by T_c, P_c, V_c."
(c) "The temperature where the substance can be either a liquid or a vapor."

(d) "The temperature at which one phase exists no matter what the pressure is."

76. The melting point of water decreases by approximately 1.0°C for each 100 atm increase in pressure.
   (a) Estimate the temperature at which liquid water freezes under a pressure of 400 atm.
   (b) One possible explanation of why a skate moves smoothly over ice is that the pressure exerted by the skater on the ice lowers its freezing point and causes it to melt. The pressure exerted by an object is the force (its mass times the acceleration of gravity, 9.8 m s^{-2}) divided by the area of contact. Calculate the change in the freezing point of ice when a skater with a mass of 75 kg stands on a blade of area 8.0×10^{-5} m^2 in contact with the ice. Is this sufficient to explain the ease of skating at a temperature of, for example, $-5°C$ (23°F)?

77. Sulfur is soluble in carbon disulfide, but iron is not. A mixture of powdered sulfur and powdered iron weighs 195.4 g. The components are separated by treatment with 1171 g of carbon disulfide. This gives a solution that is 0.0304 molal in sulfur.
   (a) Compute the mass of the iron in the original mixture.
   (b) Carbon disulfide is smelly and hard to handle. Suggest an alternate means of separating these components.

78. A recipe for a sodium carbonate solution calls for the use of either 1 part by mass of Na_2CO_3 or 2.7 parts by mass of $Na_2CO_3 \cdot 10H_2O$ dissolved in 5 parts by mass of water. Compute the molality of the solution in each case.

79. Silver nitrate is prepared by the oxidation of silver with hot nitric acid:

$$Ag(s) + 2\,HNO_3(aq) \longrightarrow$$
$$AgNO_3(aq) + NO_2(g) + H_2O(\ell)$$

   A drop of a 1% (by mass) $AgNO_3$ solution is placed in the eyes of newborn babies to prevent infection. Calculate the mass of silver required to prepare 1.00 L of a 1.00% $AgNO_3$ solution. The density of this solution is 1.007 g cm^{-3}.

*80. Veterinarians use Donovan's solution to treat skin diseases in animals. The solution is prepared by mixing 1.00 g of $AsI_3(s)$, 1.00 g of $HgI_2(s)$, and 0.900 g of $NaHCO_3(s)$ in enough water to make a total volume of 100.0 mL.
   (a) Compute the total mass of iodine per liter of Donovan's solution in units of g L^{-1}.
   (b) You need a lot of Donovan's solution to treat an outbreak of rash in an elephant herd. You have plenty of mercury(II) iodide and sodium hydrogen carbonate, but the only arsenic(III) iodide you can find is 1.50 L of a 0.100 mol L^{-1} aqueous solution. Explain how to prepare 3.50 L of Donovan's solution starting with these materials.

81. A saturated aqueous solution of NaCl in contact with excess NaCl is placed in an enclosed space at a temperature of 25°C. The presence of this solution maintains the rela-

tive humidity inside the space at 75.3%. If pure water replaces the NaCl solution, the relative humidity inside the enclosed space is maintained at 100%. Explain why there is a difference.

82. The vapor pressure of pure liquid CS_2 is 0.3914 atm at 20°C. When 40.0 g of rhombic sulfur is dissolved in 1.00 kg of CS_2 at 20°C, the vapor pressure of CS_2 falls to 0.3868 atm. Determine the molecular formula for the sulfur molecules dissolved in CS_2.

*83. A certain solvent has a freezing point of $-22.465°C$. Dilute (0.050 molal) solutions of four common acids are prepared in this solvent, and their freezing points are measured, with these results:

Acid	Freezing Point (°C)
HCl	-22.795
H_2SO_4	-22.788
$HClO_4$	-22.791
HNO_3	-22.632

   Determine K_f for this solvent and advance a reason why one of the acids differs so much from the others in its power to depress the freezing point.

84. The expressions for boiling-point elevation and freezing-point depression apply accurately to *dilute* solutions only (see Fig. 6–29). A saturated aqueous solution of NaI (sodium iodide) in water has a boiling point of 144°C. The mole fraction of NaI in the solution is 0.390. Compute the molality of this solution. Compare the boiling point elevation predicted by the expressions in this chapter with the elevation actually observed.

85. You take a bottle of a carbonated soft drink out of your refrigerator. The contents are liquid and stay liquid, even when you shake them. Presently, you remove the cap, and the liquid freezes solid. Offer a possible explanation for this observation.

*86. Mercury(II) chloride ($HgCl_2$) freezes at 276.1°C and has a freezing-point depression constant (K_f) of 34.3 K kg mol^{-1}. When 1.36 g of solid mercury(I) chloride (empirical formula HgCl) is dissolved in 100 g of $HgCl_2$, the freezing point is lowered by 0.99°C. Calculate the molar mass of the dissolved solute species, and give its molecular formula.

87. A new compound has the empirical formula $GaCl_2$. This surprises some chemists, who expect a chloride of gallium, based on the position of gallium in the periodic table, to have the formula $GaCl_3$ or possibly GaCl. They suggest that the molecular formula of the new compound is $Ga[GaCl_4]$, in which the bracketed group behaves as a unit with a -1 charge. Suggest experiments to test this hypothesis.

88. Ethylene glycol (CH_2OHCH_2OH) is used in antifreeze because, when mixed with water, it lowers the freezing

point below 0°C. What mass percentage of ethylene glycol in water must be used to reduce the freezing point of the mixture to −5.0°C, assuming ideal solution behavior?

89. The diagram represents a U-tube osmometer. The membrane is permeable to water only. Determine whether the level of the solution rises on left side, on the right side, or on neither side of the U-tube when the osmometer is left to itself. Explain. (*Hint*: See Example 6–16 for information about mannitol and Section 4–1 for information about fructose.)

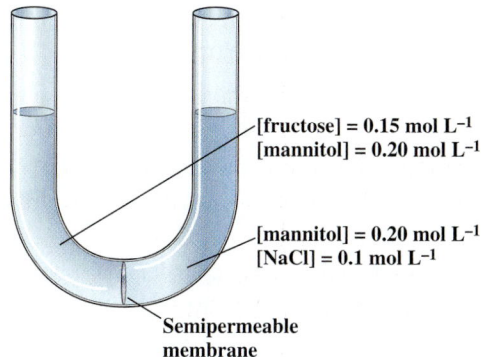

[fructose] = 0.15 mol L^{-1}
[mannitol] = 0.20 mol L^{-1}

[mannitol] = 0.20 mol L^{-1}
[NaCl] = 0.1 mol L^{-1}

Semipermeable membrane

90. A thermometer put into a bucket of water mixed with ice gives a temperature of 0°C. When some salt is added to the mixture, the temperature in the bucket drops several degrees. Your 14-year-old brother is astonished to see this. He says, "They put salt on icy streets in the winter to melt ice! Why does adding salt lower the temperature in the bucket? It should raise it." Explain.

91. The formula for osmotic pressure, $\Pi = cRT$ is only approximate, so the formula $\Pi = m'RT$ is derived and offered as an improvement. The quantity m' in this formula is the "volume molality," the number of moles of solute per liter of solvent. The measured osmotic pressure of a 0.50 molal solution of sucrose in water is 12.75 atm at 20°C. Compute the osmotic pressure of this solution according to both formulas to see which is better. The density of the sucrose solution is 1.061 g cm^{-3}.

*92. Two solutions are prepared. One contains 5 g of NaCl and 95 g of water, and the other contains 10 g of NaCl and 90 g of water. The two are placed in open beakers next to each other and are enclosed under a bell jar that is just big enough to fit over the beakers. After several weeks, which beaker has more liquid in it? Explain. (*Hint*: Think about the analogy to osmotic pressure.)

*93. The walls of erythrocytes (red blood cells) are permeable to water but not to sodium or chloride ions. In an experiment at 25°C, an aqueous solution of NaCl that has a freezing point of −0.406°C causes erythrocytes neither to swell nor to shrink, indicating that the osmotic pressure of their contents is equal to that of the NaCl solu-

tion. Calculate the osmotic pressure of the solution inside the erythrocytes under these conditions, assuming that its molarity and molality are equal.

94. In the days before refrigeration, people relied on pickling in brine (concentrated salt solution) to preserve meat and vegetables; they also preserved fruit in syrup that contained lots of sugar.
    (a) Is a pickled cucumber larger or smaller than a fresh cucumber. Why?
    (b) Why does pickling work to preserve food?

95. The Henry's law constant for the dissolution of $N_2(g)$ in $H_2O(\ell)$ is 8.57×10^4 atm, and for the dissolution of $N_2(g)$ in $CCl_4(\ell)$, it is 3.52×10^9 atm. Determine the mole fraction of saturated $N_2(aq)$ and saturated $N_2(CCl_4)$ if the pressure of $N_2(g)$ is 5.0 atm.

96. Henry's law is important in environmental chemistry, where it predicts the distribution of pollutants between water and the atmosphere. Benzene (C_6H_6) emitted in wastewater streams, for example, can pass into the air, where it is degraded by processes induced by light from the sun. The Henry's law constant for benzene in water at 25°C is 301 atm. Calculate the partial pressure of benzene vapor in equilibrium with a solution of 2.0 g of benzene per 1000 L of water. How many benzene molecules are present in each cubic centimeter of air?

*97. Methyl butyrate and ethyl acetate are completely miscible (soluble in each other in all proportions) and form ideal solutions. At 50°C, methyl butyrate (molar mass, 102.13 g mol^{-1}) has a vapor pressure of 0.1443 atm. At the same temperature, ethyl acetate (molar mass, 88.10 g mol^{-1}) has a vapor pressure of 0.3713 atm. The total vapor pressure at 50°C of a certain solution of the two substances is 0.2400 atm.
    (a) Compute the mole fraction of ethyl acetate in the solution.
    (b) Compute the mole fraction of ethyl acetate in the vapor phase.

*98. Refer to the data of Problem 60. Calculate the mole fraction of toluene in a mixture of benzene and toluene that boils at 90°C under atmospheric pressure.

99. During your research, you isolate an important new protein having a molar mass of approximately 100,000 g mol^{-1}. You have an osmometer that measures osmotic pressure with particular accuracy in the range of 50 to 150 Pa, and you tune it up in the hope of getting a highly accurate value of the protein's molar mass. For the best results, what mass of your protein should you dissolve in 100 mL of water? The experiment will be run at 25°C.

100. Gold–ruby glass has a deep red color produced by colloidal gold particles suspended in the glass. To make it, a small amount of gold is dissolved in molten glass, and the glass is then cooled to a temperature of 600 to 700°C, at which point the gold is not soluble and forms a suspension of small spherical gold crystals. Suppose a sample of gold–ruby glass contains 0.0080% by mass of

gold. The glass has a density of 2.8 g cm^{-3}, and the density of the gold itself is 19.3 g cm^{-3}. If the gold spheres have a radius of 3.0 nm (3.0 × 10^{-7} cm), calculate the number of colloidal particles per cubic centimeter of glass.

Yellow-green gaseous chlorine bubbles into water as it is produced from the oxidation of NaCl with K$_2$Cr$_2$O$_7$ in acidic solution. *(Charles D. Winters)*

CUMULATIVE PROBLEM

Chlorine and Chlorides

At room conditions, chlorine (Cl$_2$) is a yellow-green, poisonous gas with a suffocating odor. Chlorides, obtained from the reduction of chlorine, are more innocuous. Indeed, the aqueous chloride ion is essential in living tissues. Chlorine melts at −101°C and boils at −34°C at a pressure of 1 atm. Its critical point is at 143.8°C and 78.9 atm, and its triple point is at −101.5°C and 0.012 atm.

(a) Classify the forces that operate among neighboring Cl$_2$ molecules in liquid chlorine.

(b) Use Figure 6–1 to estimate the depth of the potential energy well (in kJ mol^{-1}) as two Cl$_2$ molecules approach each other. Estimate the approximate distance of closest approach.

(c) Chlorine is transported under high pressure in steel cylinders. You buy a cylinder and attach a take-off valve. Surprisingly, the pressure of chlorine at the valve does not drop as you remove gas! Only when nearly all of the chlorine has been removed does the pressure finally drop. Explain.

(d) A sample of chlorine at 25°C is heated to 150°C at a constant pressure of 1 atm, compressed to 100 atm, and then cooled back to 25°C at a constant pressure of 100 atm. What phase changes, if any, are observed?

(e) At 25°C and under a pressure of 1 atm, a maximum of 0.641 g of Cl$_2$ dissolves in 100.00 g of water. The density of this saturated "chlorine water" is 1.022 g cm^{-3}. Compute the mass fraction, mole fraction, molality, and molarity of the Cl$_2(aq)$, assuming no dissociation or other reaction in water.

(f) Use the data in part (e) to estimate the Henry's law constant for Cl$_2$ in water at 25°C, again assuming no dissociation or other reaction in water.

(g) Aqueous chlorine in fact does react with the solvent:

$$Cl_2(aq) + H_2O(\ell) \longrightarrow HOCl(aq) + H^+(aq) + Cl^-(aq)$$

The reaction stops after generating 0.326 mol of each product per 1.000 mol of Cl$_2(aq)$ originally dissolved. Determine ν and i for this solution of Cl$_2$. How does this situation differ from the case of the dissociation of a salt or a strong acid or base in water?

(h) Estimate the osmotic pressure of saturated Cl$_2(aq)$ at 25°C.

(i) Adding NaOH(s) to the chlorine water drives the reaction given in part (g) to completion and neutralizes the acidic HOCl(aq) and HCl(aq). The result is a solution of sodium chloride and sodium hypochlorite with neither unreacted Cl$_2$ nor NaOH left over. Determine the freezing point of this solution, assuming that the NaCl and NaOCl are completely dissociated in Na$^+$, Cl$^-$, and OCl$^-$ ions.

7

Chemical Equilibrium

CHAPTER OUTLINE

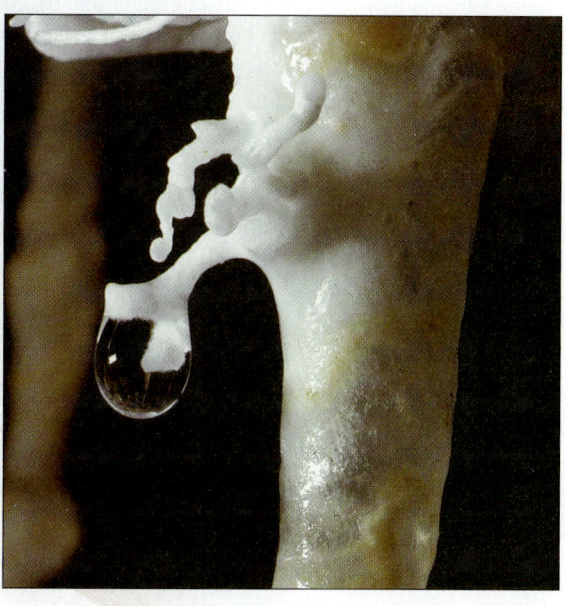

"Soda-straw" stalactites. Stalactites such as these form in caves as the loss of dissolved carbon dioxide shifts a heterogeneous equilibrium to favor precipitation of calcium carbonate from groundwater. *(W. Palmer/Visuals Unlimited)*

Chemists use balanced chemical equations to calculate the quantities of products expected when known quantities of reactants undergo chemical change. The assumption behind such stoichiometric calculations is that the reactions go to completion. The truth is that chemical reactions often do not actually reach completion, but instead stop short in an intermediate state of **chemical equilibrium,** in which unconsumed reactants are mixed with the products but no further net change occurs (Fig. 7–1).

Important questions immediately come to mind. What is the nature of this equilibrium? What governs the extent of a reaction? Can the extent of a reaction be altered? How can a reaction that has truly come to equilibrium be distinguished from one that is still proceeding, but very slowly? In this chapter, we answer these questions and also show how to calculate the amounts of reactants and products present at equilibrium. In Chapters 8 and 9, we focus on two particularly important types of chemical equilibria in aqueous solution—those involving acids and bases and those between ionic solids and their dissolved ions.

7–1 Chemical Reactions and Equilibrium

Suppose that some liquid water is sealed into an evacuated container and held at constant temperature. The water starts to evaporate as fast-moving molecules break free of the attractions of their neighbors in the bulk of the liquid and enter the space above. The pressure of this water vapor grows rapidly at first but then more slowly as molecules start to make the reverse transfer from vapor into liquid. When the rate of condensation equals the rate of evaporation, the vapor pressure settles to an easily measured final value that depends solely on the temperature. This pressure holds unchanged as long as the container and its contents are left alone. The liquid–vapor mixture has come to **equilibrium,** a condition in which all net tendency for change has been exhausted.

- This equilibrium is a phase equilibrium. See Section 6–3.

We represent this condition by the equilibrium chemical equation:

$$H_2O(\ell) \rightleftharpoons H_2O(g)$$

The double arrows emphasize the dynamic nature of the equilibrium: Liquid water continues to evaporate to form water vapor while the vapor simultaneously condenses to give liquid, even though the pressure of the water vapor does not change.

Comparison between the equilibrium vapor pressure recorded in this experiment and the vapor pressure above *any* enclosed sample of water held at the same temperature tells all that there is to know about the progress of the sample toward equilibrium. The shape and size of the container[1] do not matter, nor do the amounts of liquid and vapor, as long as some of each is present. Rather, water evaporates as long as the pressure inside the container is less than the experimental equilibrium vapor pressure; water vapor condenses whenever its pressure exceeds the equilibrium value.

---

[1] If the container is very large, then all of the water might evaporate without raising the pressure of the vapor up to the equilibrium value. The result is not equilibrium between the liquid and vapor because one of the essential components (the liquid) is absent.

Remarkably, a similar point applies to all chemical reactions:

> The equilibrium condition for *every* reaction can be summed up in a single equation in which a number, the **equilibrium constant** (K) of the reaction, equals an **equilibrium expression,** a function of properties of the reactants and products.

The equilibrium condition for the evaporation of water may be written as

$$\text{vapor pressure of } H_2O = P_{H_2O} = K$$

The value of K is numerically equal to the vapor pressure (in atmospheres) of pure water at the temperature in question. Equilibrium constants depend on temperature. At 25°C, K for this reaction is 0.03126, whereas at 30°C, it is 0.04187:

$$H_2O(\ell) \rightleftharpoons H_2O(g) \qquad K = 0.03126 \text{ at } 25°C$$
$$H_2O(\ell) \rightleftharpoons H_2O(g) \qquad K = 0.04187 \text{ at } 30°C$$

In a full, correct treatment of chemical equilibrium, equilibrium constants must be dimensionless (unitless) numbers. How can the dimensionless quantity K equal a pressure, which has units of atmospheres? Strictly speaking, it cannot. To make the equation $P_{H_2O} = K$ entirely correct, the vapor pressure of the water must be replaced by a ratio of its value to a reference pressure $P_{ref} = 1$ atm, so that the equation reads

$$\frac{P_{H_2O}}{P_{ref}} = K$$

in which both sides are dimensionless. Because P_{ref} equals 1 atm, its presence serves only to make the equation dimensionally correct provided that we express all pressures in atmospheres.

> The convention in this book is to express all pressures in atmospheres and to omit factors of P_{ref} because their value is unity. An equilibrium constant K is a pure number.

The equilibrium between a liquid and its vapor is so simple that introducing the equilibrium constant seems unnecessary. Chemical equilibria are usually more complicated than simple evaporation, however. Equilibrium equations provide a unified understanding of how even complex **reaction systems** (mixtures of many species taking part in concurrent chemical reactions) move toward equilibrium. Even in the most complex reaction systems, the physical principles parallel those at work in the evaporation and condensation of water.

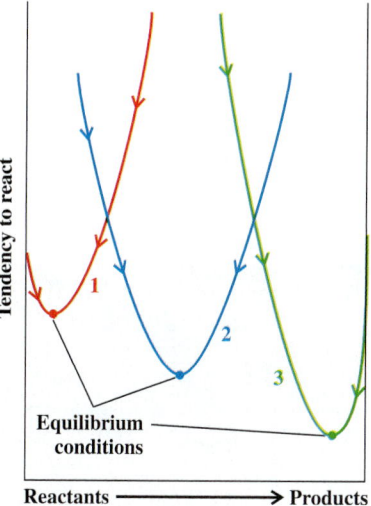

Figure 7-1 • Chemical reactions slow down and stop at intermediate points in their progress, as their products build up. The low points on the lines represent conditions in which no tendency remains to react in either direction. This is the equilibrium state. In case 1, the reaction barely begins before reaching equilibrium and stopping. In case 2, the reaction gets about halfway to completion. In case 3, the reaction very nearly reaches completion.

• This includes chemical equilibria and phase equilibria of all types.

A Simple Example of Chemical Equilibrium

When gaseous nitrogen dioxide ($NO_2(g)$) is cooled, its red-brown color fades. The reason for the change is the chemical conversion of the NO_2 to colorless dinitrogen tetraoxide (N_2O_4)

$$2\,NO_2(g) \longrightarrow N_2O_4(g)$$
$$\text{red-brown} \qquad\qquad \text{colorless}$$

• The NO_2 molecules *dimerize* (form a molecule with a doubled formula) to N_2O_4.

The reverse reaction

$$N_2O_4(g) \longrightarrow 2\,NO_2(g)$$

colorless   red-brown

restores the color when the temperature is raised (Fig. 7–2). At high temperature, the red-brown color reaches a maximum in intensity, which means that essentially no N_2O_4 is present. If a flask is filled with pure NO_2 at a high temperature and then cooled, the total pressure drops below the pressure predicted by the ideal gas law for NO_2 alone. Producing one molecule of N_2O_4 uses up two molecules of NO_2. This decreases the total number of molecules in the flask, which reduces the rate of molecular collisions with the interior walls and so reduces the pressure.

What proportion of the NO_2 molecules is converted to N_2O_4 at some intermediate temperature, such as 25°C? One way to find out is to measure the partial pressures of the two gases. An interesting result emerges when the partial pressures of NO_2 and N_2O_4 are studied in a series of experiments that start with different quantities of pure NO_2 sealed into a flask at 25°C: $P_{N_2O_4}$ *varies as the square of* P_{NO_2} (Fig. 7–3). In equation form the result is

$$P_{N_2O_4} = K(P_{NO_2})^2 \qquad \text{or} \qquad \frac{P_{N_2O_4}}{(P_{NO_2})^2} = K$$

where K is a constant. The value of K is 8.8 at 25°C, provided that the partial pressures of N_2O_4 and NO_2 are measured in atmospheres. Putting more NO_2 into the flask raises the final partial pressures of both gases, but the preceding equation is always satisfied when the pressures stop changing. Experiments at 30°C confirm the relationship, but with K equal to 6.0. The mixtures of NO_2 and N_2O_4 in these experiments are in chemical equilibrium. This is indicated by the use of double arrows in the chemical equation to indicate that both the forward and reverse reactions take place

$$2\,NO_2(g) \rightleftharpoons N_2O_4(g)$$

The K is the **equilibrium constant** of the reaction.

Chemical equilibrium, like the equilibrium between phases, is a *dynamic* state. Mixing a set of pure reactants gives products because the partial pressures of the reactants are high and productive collisions between their molecules are frequent. As the products build up, their partial pressures increase, and the reverse reaction, which regenerates the reactants from the products, begins to occur.

> As an equilibrium state is approached, the rates of the forward and backward reactions approach equality. At equilibrium, neither reaction stops, but their rates are equal, and no further change occurs in the partial pressures of reactants or products.

In the case of $NO_2(g)$ and $N_2O_4(g)$, chemical equilibrium occurs when the rate of formation of molecules of N_2O_4 from NO_2 molecules is exactly balanced by the rate of dissociation of N_2O_4 molecules back into NO_2 molecules. If isotopically labeled nitrogen dioxide ($^{15}NO_2$) and unlabelled N_2O_4 are mixed in their exact equilibrium proportions, the labeled nitrogen atoms become distributed between the two compounds, even though the amounts of the two in the equilibrium mixture do not change. This proves that the forward and reverse reactions continue after equilibrium is reached. Similar experiments give similar results for all types of chemical

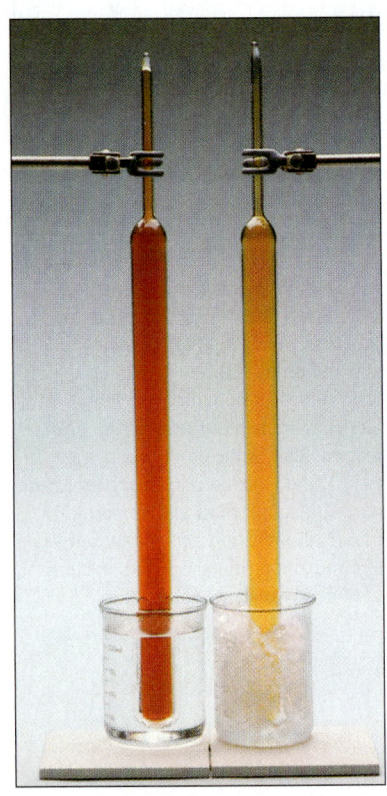

Figure 7–2 • Cooling a mixture of NO_2 and N_2O_4 (*right*) causes its color to fade as red-brown NO_2 is converted to colorless N_2O_4. Heating (*left*) reverses the conversion and intensifies the color. The two tubes contain the same amount of matter, which is distributed in different ways between NO_2 and N_2O_4. (*Charles D. Winters*)

• Similar experiments show the dynamic nature of equilibrium in the coexistence of the liquid and vapor phases, as discussed in Section 6–3.

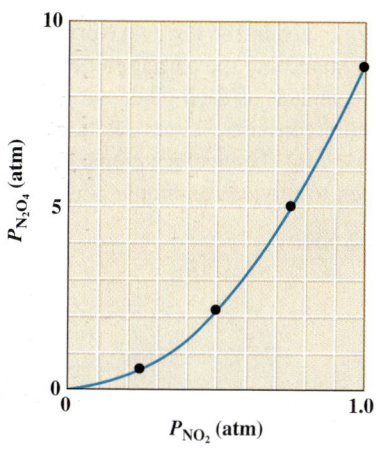

Figure 7-3 • A graph of the partial pressure of N_2O_4 at equilibrium versus that of NO_2 traces an arc of a parabola if the temperature is constant. The points on this curve are found by placing different amounts of pure NO_2 in a flask at 25°C, waiting a while, and then measuring the partial pressure of NO_2 and of N_2O_4.

Pure NO₂

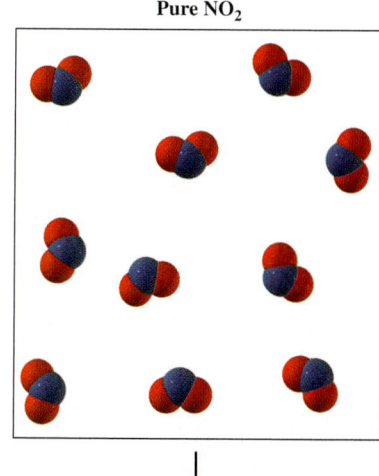

reactions. Chemical equilibrium is not a static condition, although macroscopic properties, such as partial pressures, do stop changing when equilibrium is attained. Equilibrium is the consequence of a dynamic balance between forward and backward reactions.

An equilibrium state can be reached starting with the products as well as with the reactants. When the study of NO_2 and N_2O_4 at 25°C is begun not with pure NO_2 but with an equal mass of pure N_2O_4, the ultimate equilibrium state is indistinguishable from the one established starting with pure NO_2 (Fig. 7–4). This characteristic of the equilibrium state is used to test whether a chemically reacting mixture is truly at equilibrium. Such a test is important because many reactions proceed exceedingly slowly. A very slow reaction may require months or even years to display signs of change. In the meantime, it is far from equilibrium, and its temporary changelessness is deceptive. If the same state can be reached beginning with either reactants or products, it is a true equilibrium state. In summary:

Equilibrium mixture

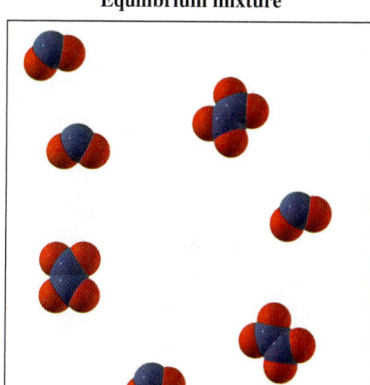

Equilibrium states in systems that are isolated from outside interference

1. Display no macroscopic evidence of change;
2. Are reached through spontaneous processes;
3. Show a dynamic balance of forward and reverse processes;
4. Can be reached starting with products as well as with reactants.

Equilibrium mixture

The Form of Equilibrium-Constant Expressions

The relationship between the partial pressures of NO_2 and N_2O_4 at equilibrium

$$\frac{P_{N_2O_4}}{(P_{NO_2})^2} = K$$

Figure 7–4 • The same equilibrium state is reached at a given temperature, regardless of whether the initial state consists entirely of NO_2 molecules, which dimerize, or of the same mass of N_2O_4 molecules, which dissociate.

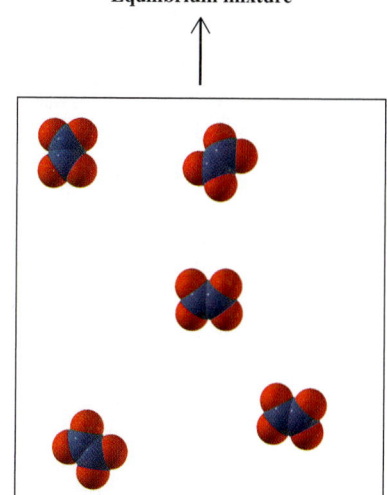

Pure N₂O₄

is a particular example of a general law that was formulated in 1864 by two Norwegian scientists, C. M. Guldberg and P. Waage. This law states that the partial pressures of reacting gases present at equilibrium are not independent of each other but are related through the temperature-dependent equilibrium constant. Similar relationships have been determined experimentally for other gas-phase chemical reactions. For example, the reaction between hydrogen and nitrogen to produce ammonia

$$N_2(g) + 3\,H_2(g) \rightleftharpoons 2\,NH_3(g)$$

is characterized by the relationship

$$\frac{(P_{NH_3})^2}{(P_{N_2})(P_{H_2})^3} = K$$

where K is equal to 6.8×10^5 at 25°C and 1.9×10^{-4} at 400°C, with pressures expressed in atmospheres. Notice that the form of this expression differs significantly from the one discovered for the equilibrium between NO_2 and N_2O_4. Is there any way other than experimentation to know the mathematical form of the relationship between the partial pressures and the equilibrium constant? The answer, fortunately, is "yes." Its details are given in this generalization:

> In a chemical reaction in which a moles of species A and b moles of species B react to form c moles of species C and d moles of species D,
>
> $$a\,A(g) + b\,B(g) \rightleftharpoons c\,C(g) + d\,D(g)$$
>
> the partial pressures at equilibrium are related through
>
> $$\frac{(P_C)^c(P_D)^d}{(P_A)^a(P_B)^b} = K$$
>
> provided that all species are present as low-pressure gases.

If more than two products or reactants appear in the chemical equation, additional partial pressures are inserted into the equilibrium-constant expression so that there is one for every product and reactant. The partial pressures of products appear in the numerator, and those of reactants appear in the denominator. Each partial pressure is raised to a power that is equal to its coefficient in the balanced chemical equation. In Section 7–6, we extend this generalization, called the **law of mass action,** to include solutions and heterogeneous equilibria.

• *Heterogeneous* equilibria involve chemical reactants and products in two or more phases.

The law of mass action holds even if multiple equilibria are going on simultaneously. For example, the relationship between the partial pressures of NO_2 and N_2O_4 at equilibrium

$$\frac{P_{N_2O_4}}{(P_{NO_2})^2} = K$$

is correct not only when $NO_2(g)$ and $N_2O_4(g)$ are the only gases present but also for complex mixtures containing any number of other species, even ones that react with the $NO_2(g)$ or $N_2O_4(g)$. Thus, although the nitrogen oxides $N_2O(g)$, $N_2O_3(g)$, and $NO(g)$ all may be present, with additional equilibrium equations relating their partial pressures to those of $NO_2(g)$ and $N_2O_4(g)$, the relationship between $P_{N_2O_4}$ and P_{NO_2} remains in force, with K unchanged. At high enough densities, when the gas mixture is not ideal and Dalton's law begins to fail, deviations do appear.

EXAMPLE 7-1

Write equilibrium-constant expressions for each of the following gas-phase chemical equilibria:

(a) $PCl_5(g) \rightleftharpoons PCl_3(g) + Cl_2(g)$
(b) $2\,NOCl(g) \rightleftharpoons 2\,NO(g) + Cl_2(g)$
(c) $CO(g) + \frac{1}{2}\,O_2(g) \rightleftharpoons CO_2(g)$

Solution

(a) $\dfrac{(P_{PCl_3})(P_{Cl_2})}{P_{PCl_5}} = K$.

Recall that products appear in the numerator and reactants in the denominator.

(b) $\dfrac{(P_{NO})^2(P_{Cl_2})}{(P_{NOCl})^2} = K$.

The exponents come from the coefficients in the balanced equation.

(c) $\dfrac{P_{CO_2}}{(P_{CO})(P_{O_2})^{1/2}} = K$.

Fractional exponents appear in the equilibrium-constant expression whenever fractional coefficients are present in the balanced equation.

EXERCISE

Write equilibrium-constant expressions for the reactions defined by the following equations:

(a) $3\,H_2(g) + SO_2(g) \rightleftharpoons H_2S(g) + 2\,H_2O(g)$
(b) $2\,C_2F_5Cl(g) + 4\,O_2(g) \rightleftharpoons Cl_2(g) + 4\,CO_2(g) + 5\,F_2(g)$

Answer: (a) $\dfrac{(P_{H_2S})(P_{H_2O})^2}{(P_{H_2})^3(P_{SO_2})} = K$. (b) $\dfrac{(P_{Cl_2})(P_{CO_2})^4(P_{F_2})^5}{(P_{C_2F_5Cl})^2(P_{O_2})^4} = K$.

7-2 Calculating Equilibrium Constants

Calculating a numerical equilibrium constant is straightforward when the equilibrium partial pressures of the reactants and products in a gas-phase chemical reaction can all be measured: Write down the equilibrium-constant expression and substitute in the experimental values (expressed in atmospheres but with the unit removed). For example, at a certain temperature, a mixture containing hydrogen, nitrogen, and ammonia reaches equilibrium with these partial pressures of the three gases:

$$P_{H_2} = 0.12 \text{ atm}, \ P_{N_2} = 0.70 \text{ atm}, \text{ and } P_{NH_3} = 0.84 \text{ atm}$$

The equilibrium constant for the reaction

$$N_2(g) + 3\,H_2(g) \rightleftharpoons 2\,NH_3(g)$$

equals

$$K = \frac{(P_{NH_3})^2}{(P_{N_2})(P_{H_2})^3} = \frac{(0.84)^2}{(0.70)(0.12)^3} = 5.8 \times 10^2$$

Phosgene ($COCl_2$) is an important chemical intermediate. Although it is highly toxic, observing standard safety precautions reduces the danger in handling it to a minimum. Phosgene is used in making polyurethane, a kind of polymer employed both in foams (for furniture cushions and bedding) and as a coating for wood.

• These tables are sometimes called I-C-E tables (initial, change, equilibrium).

In most cases, it is impractical to measure the equilibrium partial pressure of every reactant and product directly. Fortunately, the equilibrium constant usually can be derived from other available data. Consider the formation of phosgene, $COCl_2$, from CO and Cl_2 at 600°C:

$$CO(g) + Cl_2(g) \rightleftharpoons COCl_2(g)$$

This reaction, which reaches equilibrium quickly, is the only reaction among these substances at 600°C. Suppose that a mixture of CO and Cl_2 that has *initial* partial pressures (before reaction) of 0.60 atm for CO and 1.10 atm for Cl_2 is prepared. After the mixture reaches equilibrium, measurements reveal that the partial pressure of $COCl_2$ equals 0.10 atm. The equilibrium partial pressures of the CO and Cl_2 (and from them the K of the reaction) can be determined from these facts. To do so, set up a three-line table of changes similar to the ones used in Section 2–3; the difference is that now the final state is *equilibrium* (when the reaction has stopped by itself) rather than *completion* (when a limiting reactant has been used up). As before, the first two lines add to give the third in each column. The result is:

	CO(g) + P_{CO} (atm)	Cl_2(g) $\rightleftharpoons$ P_{Cl_2} (atm)	$COCl_2$(g) P_{COCl_2} (atm)
Initial	0.60	1.10	0
Change	?	?	+0.10
Equilibrium	?	?	0.10

The coefficients in the balanced chemical equation relate the entries on the "change" line. Production of 1 mol of $COCl_2$ consumes 1 mol of CO and 1 mol of Cl_2 (because all of the coefficients in the balanced equation equal 1). In accord with the ideal gas law and Dalton's law of partial pressures (Section 5–6), the change in the partial pressure of a reacting gas is proportional to the change in its chemical amount as long as the temperature and volume do not change. If the partial pressure of $COCl_2$ increases by 0.10 atm, then the partial pressures of CO and Cl_2 must each decrease by 0.10 atm. Inserting these values into the table and completing the addition in each column gives

	P_{CO} (atm)	P_{Cl_2} (atm)	P_{COCl_2} (atm)
Initial	0.60	1.10	0
Change	−0.10	−0.10	+0.10
Equilibrium	0.50	1.00	0.10

The values from the bottom line go into the equilibrium-constant expression, as follows:

$$K = \frac{(P_{COCl_2})}{(P_{CO})(P_{Cl_2})} = \frac{(0.10)}{(0.50)(1.00)} = 0.20$$

Completing a table of changes when the coefficients in the balanced equation do not all equal 1 is only a bit harder. Difficulties are minimized by avoiding fractions in the "change" line, as the following example shows.

EXAMPLE 7-2

Consider the equilibrium

$$4 NO_2(g) \rightleftharpoons 2 N_2O(g) + 3 O_2(g)$$

The three gases are introduced into a container at partial pressures of 3.6 atm (for NO_2), 5.1 atm (for N_2O), and 8.0 atm (for O_2). They soon react to reach equilibrium at a fixed temperature. The equilibrium partial pressure of the NO_2 is measured and found to equal 2.4 atm. Calculate the equilibrium constant of the reaction at this temperature, assuming that no competing reactions occur.

Strategy

Set up a table of changes to display the progress of the reaction toward equilibrium. Define the unknown changes in partial pressures in terms of an x such that the change in P_{NO_2} is $-4x$ atm, the change in P_{N_2O} is $+2x$ atm, and the change in P_{O_2} is $+3x$ atm. This avoids fractions in the table. Obtain K by substituting the equilibrium partial pressures into the equilibrium-constant equation.

Solution

	4 NO$_2$(g) $\rightleftharpoons$ P_{NO_2} (atm)	2 N$_2$O(g) P_{N_2O} (atm)	+	3 O$_2$(g) P_{O_2} (atm)
Initial	3.6	5.1		8.0
Change	$-4x$	$+2x$		$+3x$
Equilibrium	2.4	$5.1 + 2x$		$8.0 + 3x$

Doing the addition in the first column gives $3.6 - 4x = 2.4$. The solution of this equation is $x = 0.3$, and the table becomes

	P_{NO_2} (atm)	P_{N_2O} (atm)	P_{O_2} (atm)
Initial	3.6	5.1	8.0
Change	$-4(0.3)$	$+2(0.3)$	$+3(0.3)$
Equilibrium	2.4	5.7	8.9

Then

$$K = \frac{(P_{N_2O})^2(P_{O_2})^3}{(P_{NO_2})^4} = \frac{(5.7)^2(8.9)^3}{(2.4)^4} = 6.9 \times 10^2$$

EXERCISE

The compound $GeWO_4(g)$ forms at high temperature in the reaction

$$2 GeO(g) + W_2O_6(g) \rightleftharpoons 2 GeWO_4(g)$$

Some $GeO(g)$ and $W_2O_6(g)$ are mixed. Before they start to react, their partial pressures both equal 1.000 atm. After their reaction at constant temperature and volume, the equilibrium partial pressure of $GeWO_4(g)$ is 0.980 atm. Assuming that this is the only reaction that takes place, (a) determine the equilibrium partial pressures of GeO(g) and of W_2O_6(g), and (b) determine the equilibrium constant for the reaction.

Answer: (a) $P_{GeO} = 0.020$ atm; $P_{W_2O_6} = 0.510$ atm. (b) $K = 4.7 \times 10^3$.

Relationships Among the K's of Related Reactions

Consider these two chemical equations, which are accompanied by equilibrium-constant expressions:

$$2\,H_2(g) + O_2(g) \rightleftharpoons 2\,H_2O(g) \qquad \frac{(P_{H_2O})^2}{(P_{H_2})^2(P_{O_2})} = K_1$$

$$2\,H_2O(g) \rightleftharpoons 2\,H_2(g) + O_2(g) \qquad \frac{(P_{H_2})^2(P_{O_2})}{(P_{H_2O})^2} = K_2$$

According to the first equation, gaseous hydrogen and oxygen combine to form water vapor. According to the second equation, water vapor decomposes into gaseous hydrogen and oxygen. The second is clearly the reverse of the first. Inspection shows that the expressions for K_1 and K_2 are reciprocals of each other—the numerator of the first is the denominator of the second and vice versa. This means in algebraic terms

$$K_2 = \frac{1}{K_1} = (K_1)^{-1} \quad \text{or} \quad K_1 K_2 = 1$$

This result is general and can be stated in these words:

> **Rule 1.** The equilibrium constant associated with a reversed chemical equation equals the reciprocal of the equilibrium constant for the forward equation.

Suppose that we multiply all of the coefficients in the first of the preceding chemical equations by a factor, for example $\frac{1}{2}$. The result is

$$H_2(g) + \tfrac{1}{2}O_2(g) \rightleftharpoons H_2O(g)$$

The new equation states that one mole of hydrogen reacts with one-half mole of oxygen to yield one mole of water vapor. It is balanced and just as valid as the first equation (see Section 2–1). Its equilibrium-constant expression is

$$\frac{P_{H_2O}}{(P_{H_2})(P_{O_2})^{1/2}} = K_3$$

Comparison with the expression for K_1 shows that

$$K_3 = (K_1)^{1/2} = \sqrt{K_1}$$

This result generalizes to the following rule:

> **Rule 2.** When the coefficients in a balanced chemical equation are multiplied by a numerical factor, the equilibrium constant is raised to a power equal to that factor.

Instances in which the multiplying factor equals -1 merit special mention. Rule 2 provides that K is then raised to the -1 power, which is the same as taking its reciprocal. Rule 1 states that the reciprocal of K goes with the reversed equation. It follows that multiplying all of the coefficients in a balanced equation by -1 corresponds to writing the equation in reverse.

What happens to the K's if we add equations together? Consider these two reactions:

$$2\,BrCl(g) \rightleftharpoons Cl_2(g) + Br_2(g) \qquad \frac{(P_{Cl_2})(P_{Br_2})}{(P_{BrCl})^2} = K_4$$

$$Br_2(g) + I_2(g) \rightleftharpoons 2\,IBr(g) \qquad \frac{(P_{IBr})^2}{(P_{Br_2})(P_{I_2})} = K_5$$

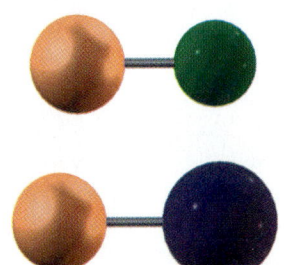

Adding the equations gives

$$2\,BrCl(g) + Br_2(g) + I_2(g) \rightleftharpoons 2\,IBr(g) + Cl_2(g) + Br_2(g)$$

$Br_2(g)$ appears on both sides and can be cancelled out. The result (and the corresponding K) is

$$2\,BrCl(g) + I_2(g) \rightleftharpoons 2\,IBr(g) + Cl_2(g) \qquad \frac{(P_{IBr})^2(P_{Cl_2})}{(P_{BrCl})^2(P_{I_2})} = K_6$$

Bromine chloride (BrCl) and iodine bromide (IBr) are two of a rather large group of compounds called the *interhalogens*, which are combinations of two or more halogen elements. Bromine chloride dissociates so readily into Br_2 and Cl_2 that it cannot be prepared in pure form. Iodine bromide is more stable, forming a black crystalline solid that melts at 41°C.

But, multiplying K_4 by K_5 gives

$$K_4 K_5 = \left(\frac{(P_{Cl_2})(P_{Br_2})}{(P_{BrCl})^2}\right)\left(\frac{(P_{IBr})^2}{(P_{Br_2})(P_{I_2})}\right) = \frac{(P_{IBr})^2(P_{Cl_2})}{(P_{BrCl})^2(P_{I_2})}$$

The expression on the right is identical with the expression for K_6. Therefore

$$K_6 = K_4 K_5$$

This general rule is:

> **Rule 3.** When chemical equations are added to give a new equation, their equilibrium constants are multiplied to give the equilibrium constant associated with the new equation.

Finally comes subtraction. Subtracting one chemical equation from another amounts to multiplying all of the coefficients in the equation that is being subtracted by -1 and then adding. Multiplying all of the coefficients by -1 means taking the reciprocal of its K. It follows that:

> **Rule 4.** When equation 2 is subtracted from equation 1 to give a new equation, the equilibrium constant for equation 1 is divided by (multiplied by the reciprocal of) the equilibrium constant for equation 2 to give the equilibrium constant associated with the new equation.

EXAMPLE 7-3

Nitrogen dioxide (NO_2) is a serious air pollutant. The question arises: To what extent does NO_2 tend to form from N_2 and O_2, which are normal components of the air? To find an answer, scientists study the following reactions involving oxides of nitrogen and establish their equilibrium constants (all measurements at 25°C):

$$NO(g) + \tfrac{1}{2}O_2(g) \rightleftharpoons NO_2(g) \qquad K_1 = 1.3 \times 10^6$$

$$\tfrac{1}{2}N_2(g) + \tfrac{1}{2}O_2(g) \rightleftharpoons NO(g) \qquad K_2 = 6.5 \times 10^{-16}$$

Find the equilibrium constant K_3 for the following NO_2-forming reaction at 25°C:

$$N_2(g) + 2\,O_2(g) \rightleftharpoons 2\,NO_2(g)$$

Strategy

Construct the target equation as a combination of the given equations. Get the reactants in the target equation onto the *left* of the given equations (by reversing equations if necessary). Get the products in the target equation onto the *right* of the given equations. Arrange for the cancellation of substances that appear in the given equations but not in the target equation by multiplying by numerical constants as necessary. Obtain the K of the target equation from the K's of the given equations by following Rules 1 to 4 for the reversals, multiplications, additions, or subtractions just used.

Solution

Addition of the two given equations puts N_2 and O_2 on the left of a new equation, where they are needed, and allows for the cancellation of $NO(g)$:

$$\tfrac{1}{2}N_2(g) + O_2(g) \rightleftharpoons NO_2(g)$$

By Rule 3, the K for this equation equals K_1K_2. The target equation is the same as this but with all of the coefficients doubled. By Rule 2, K_3, the equilibrium constant for the target equation, equals K_1K_2 squared (raised to the power 2):

$$K_3 = (K_1K_2)^2 = ((1.3 \times 10^6)(6.5 \times 10^{-16}))^2 = 7.1 \times 10^{-19}$$

Check

Write the equilibrium-constant expressions corresponding to the three equations

$$\frac{P_{NO_2}}{(P_{NO})(P_{O_2})^{1/2}} = K_1 \qquad \frac{P_{NO}}{(P_{N_2})^{1/2}(P_{O_2})^{1/2}} = K_2 \qquad \frac{(P_{NO_2})^2}{(P_{N_2})(P_{O_2})^2} = K_3$$

Compute $(K_1K_2)^2$ in terms of the first two expressions.

$$(K_1K_2)^2 = \left(\frac{P_{NO_2}}{(P_{NO})(P_{O_2})^{1/2}} \times \frac{P_{NO}}{(P_{N_2})^{1/2}(P_{O_2})^{1/2}} \right)^2$$

Cancellation of P_{NO} and simplification confirm that the right side of this equation equals the expression for K_3:

$$(K_1K_2)^2 = \left(\frac{P_{NO_2}}{(P_{O_2})^{1/2}} \times \frac{1}{(P_{N_2})^{1/2}(P_{O_2})^{1/2}} \right)^2 = \frac{(P_{NO_2})^2}{P_{O_2}} \times \frac{1}{P_{N_2}P_{O_2}} = \frac{(P_{NO_2})^2}{P_{N_2}(P_{O_2})^2} = K_3$$

EXERCISE

At 100°C, K for the reaction $CF_4(g) + 2\,H_2O(g) \rightleftharpoons CO_2(g) + 4HF(g)$ is 5.9×10^{23}, and K for the reaction $CO(g) + \tfrac{1}{2}O_2(g) \rightleftharpoons CO_2(g)$ is 1.3×10^{35}. Compute the equilibrium constant at 100°C for the reaction $2\,CF_4(g) + 4\,H_2O(g) \rightleftharpoons 2\,CO(g) + 8\,HF(g) + O_2(g)$.

Answer: 2.1×10^{-23}.

7-3 The Reaction Quotient

A quantity called the **reaction quotient** (Q) aids in the use of the law of mass action in many situations. For the general gas-phase reaction

$$a\,A(g) + b\,B(g) \rightleftharpoons c\,C(g) + d\,D(g)$$

the reaction quotient is defined as

$$Q = \frac{(P_C)^c (P_D)^d}{(P_A)^a (P_B)^b}$$

The form of the expression that defines Q is the same as the form of the equilibrium-constant expression. The (crucial!) difference is that for Q we use whatever partial pressures happen to prevail in the reaction system, but for K we *must* use *equilibrium* partial pressures. As long as the temperature does not change, K stays constant; Q changes as the amounts of the reactants and products change. Reactions always tend toward a state of equilibrium. This means that Q always tends to become equal to K. Clearly, Q increases if products form and reactants are consumed, but decreases if reactants form and products are consumed (if the reaction runs in reverse). A simple comparison between Q and K, as shown in Figure 7–5, therefore tells the direction in which a reaction runs to reach equilibrium.

• Or "Q chases K."

> At constant temperature, a reaction runs from left to right if its Q is less than its K, but runs from right to left if its Q is greater than its K.

A reaction stops when its Q becomes equal to its K because it has reached equilibrium.

EXAMPLE 7-4

The reaction between nitrogen and hydrogen to produce ammonia

$$N_2(g) + 3\,H_2(g) \rightleftharpoons 2\,NH_3(g)$$

is essential in making nitrogen-containing fertilizers. This reaction has an equilibrium constant equal to 1.9×10^{-4} at 400°C. Suppose that 1.0 mol of N_2, 0.20 mol of H_2, and 0.40 mol of NH_3 are sealed in a 1.00 L vessel at 400°C. In which direction does the reaction proceed?

Strategy

Assume the three gases form an ideal mixture (until they begin to react). Compute the initial partial pressures of each, using the ideal gas law. Use these numbers in the expression for Q to obtain an initial numerical Q. Compare Q to K.

Solution

$$P_{N_2(\text{init})} = n_{N_2(\text{init})} \frac{RT}{V}$$

$$= 1.0 \text{ mol} \frac{(0.08206 \text{ L atm mol}^{-1} \text{ K}^{-1})(400 + 273.15) \text{ K}}{1.00 \text{ L}} = 55 \text{ atm}$$

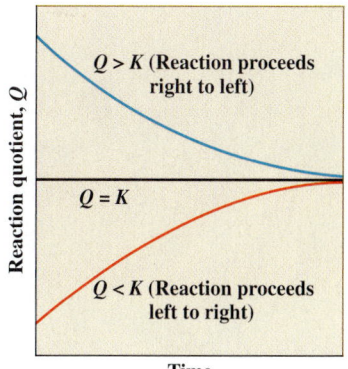

Figure 7–5 • The reaction quotient Q approaches the equilibrium constant K as reactions proceed over time. If Q starts out less than K, it increases (*red line*), and the reaction proceeds from left to right. If Q starts out larger than K, it decreases (*blue line*), and the reaction proceeds from right to left. When Q equals K (*black line*), no further change occurs.

By similar calculations,

$$P_{H_2(\text{init})} = 11 \text{ atm} \quad \text{and} \quad P_{NH_3(\text{init})} = 22 \text{ atm}$$

Therefore,

$$Q_{(\text{init})} = \frac{(P_{NH_3(\text{init})})^2}{(P_{N_2(\text{init})})(P_{H_2(\text{init})})^3} = \frac{(22)^2}{(55)(11)^3} = 6.6 \times 10^{-3}$$

Because $Q_{(\text{init})} > K$, the reaction proceeds from right to left. As it does, Q diminishes toward K. Ammonia decomposes until equilibrium is reached, at which point Q equals K.

EXERCISE

The equilibrium constant for the reaction

$$P_4(g) \rightleftharpoons 2\,P_2(g)$$

is 1.39 at 400°C. Suppose that 2.75 mol of $P_4(g)$ and 1.08 mol of $P_2(g)$ are mixed in a closed 25.0 L container at 400°C. Compute $Q_{(\text{init})}$ (the Q at the moment of mixing) and state the direction in which the reaction proceeds.
Answer: $Q_{(\text{init})} = 0.937$. The reaction proceeds from left to right, consuming $P_4(g)$.

If the equilibrium constant K of a reaction is very large, then, at equilibrium, the partial pressure of at least one of the reactant species must be so small that it is essentially zero. *In this case only,* the reaction reaches completion, as defined in Section 2–3. The reactant with the very small equilibrium partial pressure is the limiting reactant, and the methods of Chapter 2 may be used to calculate the chemical amounts or the masses of the products formed. If K is very small, the reactant partial pressures are large compared with those of the products at equilibrium, and the extent of reaction is small. There is essentially no reaction at equilibrium. For intermediate values of K, both reactants and products are present in significant proportions at equilibrium. The next section shows how to deal with such situations.

7–4 Calculation of Gas–Phase Equilibria

The law of mass action furnishes an equilibrium-constant expression for any gas-phase reaction. These expressions provide a means to compute the amounts of reactants and products that are present when a reaction reaches equilibrium. The details of such computations depend on the experimental data available. Suppose, for example, that we know K and all of the equilibrium partial pressures but one. Finding the missing partial pressure is straightforward, as is illustrated in the following example.

EXAMPLE 7–5

Hydrogen is made from natural gas (methane) for immediate consumption in industrial processes, such as the production of ammonia for fertilizers. The first step in this process is the "steam reforming of methane."

$$CH_4(g) + H_2O(g) \rightleftharpoons CO(g) + 3\,H_2(g)$$

The equilibrium constant for this reaction equals 0.172 at 900 K. Determine the partial pressure of $H_2(g)$ in an equilibrium mixture of these gases at 900 K if the partial pressures of $CH_4(g)$ and $H_2O(g)$ are both 0.400 atm and the partial pressure of $CO(g)$ is 0.872 atm.

Strategy

Set up the equilibrium-constant expression, and put in the different partial pressures and the value of K. Do the algebra to solve for the missing partial pressure.

Solution

$$\frac{(P_{CO})(P_{H_2})^3}{(P_{CH_4})(P_{H_2O})} = K$$

$$\frac{(0.872)(P_{H_2})^3}{(0.400)(0.400)} = 0.172$$

$$(P_{H_2})^3 = \left(\frac{(0.400)(0.400)}{0.872}\right)0.172 = 0.0316$$

$$P_{H_2} = \sqrt[3]{0.0316} = 0.316 \text{ atm}$$

Check

Put the four partial pressures back into the equilibrium-constant expression. Evaluating the expression gives the correct numerical K, as it must. Note that it is correct to insert the unit "atm" in the answer, as explained in Section 7–1.

EXERCISE

Carbon monoxide reacts with water to give hydrogen:

$$CO(g) + H_2O(g) \rightleftharpoons CO_2(g) + H_2(g)$$

At 900 K, the equilibrium constant for this reaction, the so-called shift reaction, equals 0.64. Suppose the partial pressures of three of the gases at equilibrium at 900 K are

$$P_{CO} = 2.00 \text{ atm}, P_{CO_2} = 0.80 \text{ atm, and } P_{H_2} = 0.48 \text{ atm}$$

Calculate the partial pressure of H_2O under these conditions.

Answer: $P_{H_2O} = 0.30$ atm.

Very commonly, the problem is to determine the equilibrium partial pressures of reactants or products from measurements of the partial pressures before any reaction occurs, as in the following example.

EXAMPLE 7–6

Some $I_2(g)$ and $H_2(g)$ are sealed in a flask at 400.0 K with partial pressures $P_{I_2} = 1.320$ atm and $P_{H_2} = 1.140$ atm. The sealed flask is then heated to 600.0 K, a temperature at which the gases react speedily to reach equilibrium.

$$I_2(g) + H_2(g) \rightleftharpoons 2 HI(g) \qquad K = 92.6 \text{ at } 600.0 \text{ K}$$

No other reactions take place. Determine the equilibrium partial pressures of the I_2, H_2, and HI at 600.0 K.

Strategy

Compute the initial partial pressures of the reactants and product. Set up a table of changes to obtain the equilibrium partial pressures of the reactants and product in terms of a single unknown. Evaluate the unknown by substituting these expressions in the equilibrium-constant equation for this reaction.

Solution

The initial partial pressures are the pressures just after the I_2 and H_2 are mixed and heated but before they react. From the ideal gas law at constant volume, the partial pressures of the reactants are

$$P_{I_2(init)} = 1.320 \text{ atm} \times \left(\frac{600.0 \text{ K}}{400.0 \text{ K}}\right) = 1.980 \text{ atm}$$

$$P_{H_2(init)} = 1.140 \text{ atm} \times \left(\frac{600.0 \text{ K}}{400.0 \text{ K}}\right) = 1.710 \text{ atm}$$

• Because $P_{gas} = n_{gas}(RT/V)$, the partial pressures of the gases are directly proportional to their chemical amounts in the mixture as long as T and V do not change.

The initial partial pressure of the HI is zero. The table of changes for the reaction is

	$I_2(g)$ P_{I_2} (atm)	+	$H_2(g)$ P_{H_2} (atm)	⇌	2 HI(g) P_{HI} (atm)
Initial	1.980		1.710		0
Change	$-x$		$-x$		$+2x$
Equilibrium	$1.980 - x$		$1.710 - x$		$2x$

Write the equilibrium-constant expression and substitute the numerical K and the information on the last line of the table into it:

$$\frac{(P_{HI})^2}{(P_{I_2})(P_{H_2})} = K$$

$$\frac{(2x)^2}{(1.980 - x)(1.710 - x)} = 92.6$$

Rearrangement results in the quadratic equation

$$88.6x^2 - 341.694x + 313.525 = 0$$

Solving for x using the quadratic formula (see Appendix C–3) gives

$$x = \frac{-(-341.694) \pm \sqrt{(341.694)^2 - 4(88.6)(313.525)}}{2(88.6)}$$

$$= 1.5044 \text{ atm} \quad \text{or} \quad 2.3522 \text{ atm}$$

The second root is not physically meaningful because it leads to negative partial pressures. The first root gives these partial pressures:

$$P_{I_2} = 1.980 \text{ atm} - 1.5044 \text{ atm} = 0.4756 \text{ atm}$$
$$P_{H_2} = 1.710 \text{ atm} - 1.5044 \text{ atm} = 0.2056 \text{ atm}$$
$$P_{HI} = 2 \times 1.5044 \text{ atm} = 3.0088 \text{ atm}$$

Round off to the correct number of significant digits: $P_{I_2} = 0.476 \text{ atm}$, $P_{H_2} = 0.206 \text{ atm}$, $P_{HI} = 3.01 \text{ atm}$.

Check

Substitute the answers into the original equilibrium-constant expression:

$$\frac{(3.01)^2}{(0.476)(0.206)} = 92.4$$

This differs slightly from 92.6 because of rounding off. Substitution of un-rounded partial pressures gives 92.6.

EXERCISE

At 100°C, the equilibrium constant for the reaction

$$H_2(g) + Si_2H_6(g) \rightleftharpoons 2\ SiH_4(g)$$

equals 89.6. The two reactant gases are mixed in a sealed flask at 100°C. At the moment of mixing (before any reaction takes place) the partial pressure of each is 0.560 atm. Compute the partial pressure of $SiH_4(g)$ when this reaction comes to equilibrium (no other reactions take place in the flask).

Answer: $P_{SiH_4} = 0.925$ atm.

The calculation of gas-phase equilibria occasionally requires the solution of cubic, quartic, and higher-order equations. Powerful programs running on both calculators and computers are available to solve such equations. Highly automated methods are not unconditionally the best choice, however. Simpler approaches often provide chemical insight that is absent in "black-box" methods and protect against errors, as illustrated in the following example.

EXAMPLE 7–7

At 330°C, the reaction

$$2\ NOCl(g) \rightleftharpoons 2\ NO(g) + Cl_2(g)$$

has an equilibrium constant of 4.0×10^{-4}. Find the equilibrium partial pressures of the gases if a reaction vessel is initially charged with 2.00 atm of $NO(g)$ and 1.50 atm of $Cl_2(g)$ at 330°C. Assume that no other reactions take place.

Strategy

Realize that this reaction runs in reverse to reach equilibrium (because the original mixture consists wholly of products). Recognize this in a table of changes by making the changes in the partial pressures of the products negative. Insert the equilibrium partial pressures from the table into the equilibrium-constant equation.

Solution

	2 NOCl(g) $\rightleftharpoons$ P_{NOCl} (atm)	2 NO(g) + P_{NO} (atm)	Cl_2(g) P_{Cl_2} (atm)
Initial	0	2.00	1.50
Change	$+2x$	$-2x$	$-x$
Equilibrium	$2x$	$2.00 - 2x$	$1.50 - x$

Nitrosyl chloride (NOCl) is a corrosive, reddish yellow gas. The N is bonded to the O and Cl and the molecule is bent, in accord with the valence shell electron-pair repulsion concepts developed in Chapter 3.

$$K = \frac{(P_{NO})^2 (P_{Cl_2})}{(P_{NOCl})^2}$$

$$4.0 \times 10^{-4} = \frac{(2.00 - 2x)^2(1.50 - x)}{(2x)^2}$$

This equation is a cubic equation in x. It can be rearranged to the standard form,

$$-4x^3 + 13.9984x^2 - 16x + 6 = 0$$

entered into a calculator, and solved. Solving the equation another way is more instructive. The K is small, indicating that the equilibrium mixture contains mostly NOCl. Hence, $2x$ must be relatively large and $(2.00 - 2x)$ or $(1.50 - x)$ must be relatively small. Also, x cannot exceed 1.00 (if it did, the partial pressure of NO would be negative). With these facts in mind and a calculator in hand, trying out plausible values of x in the equilibrium-constant equation and checking whether they give the correct K soon provides the answer. If $x = 0.90$, then

$$\frac{(2.00 - 2x)^2(1.50 - x)}{(2x)^2} = \frac{(2.00 - 1.80)^2(1.50 - 0.90)}{(1.80)^2} = 0.0074 = Q$$

The answer is labeled Q because it is in fact a reaction quotient. It exceeds K (0.0074 > 0.00040), which means that the equilibrium lies farther to the left than guessed. Trying $x = 0.99$ gives a new Q of 5.2×10^{-5}, which is smaller than K. Further trials lead to $x = 0.9732$, for which Q equals 4.0×10^{-4}. The equilibrium partial pressures are then

$$P_{NO} = 2.00 - 2x = 2.00 - 2(0.9732) = 0.05 \text{ atm}$$
$$P_{Cl_2} = 1.50 - x = 1.50 - 0.9732 = 0.53 \text{ atm}$$
$$P_{NOCl} = 2x = 2(0.9732) = 1.95 \text{ atm}$$

Check

The method of computation of x is self-checking, a big advantage. Putting the (un-rounded) equilibrium partial pressures into the equilibrium-constant expression gives 4.0×10^{-4}, as it should. Using a calculator to solve the cubic equation gives three roots:

$$x = 1.4964 \quad \text{and} \quad 1.0301 \quad \text{and} \quad 0.9732$$

Substituting 1.4964 or 1.0301 for x in computing the partial pressures gives a negative P_{NO}, confirming that 0.9732 is the only physically meaningful root.

EXERCISE

Find the equilibrium partial pressures of NO(g), NOCl(g), and $Cl_2(g)$ if the flask in the preceding example is initially charged with 1.000 atm of NOCl(g) at 330°C.

Answer: $P_{NO} = 0.088$ atm; $P_{Cl_2} = 0.044$ atm; $P_{NOCl} = 0.912$ atm.

Using the Concentrations of Gases in Equilibrium Calculations

Sometimes the data for a calculation of gas-phase equilibrium may be given as concentrations rather than partial pressures. In such a case, there are two options: (1) convert all concentrations to partial pressures before carrying out the calculations or (2) rewrite the equilibrium-constant expression in terms of concentration variables. Both options require the use of the ideal gas equation, which relates the concentration [A] of an ideal gas A to its partial pressure P_A, as follows:

• Recall from Section 2–4 that square brackets denote concentration in moles per liter.

$$[A] = \frac{n_A}{V} = \frac{P_A}{RT}$$

Multiplying both sides of the equation by RT gives

$$P_A = RT[A]$$

Let us substitute one such relation for every species appearing in the equilibrium-constant expression for a specific reaction. We choose the familiar $2\,NO_2(g) \rightleftharpoons N_2O_4(g)$ equilibrium. It is best to put the factors of $P_{ref} = 1$ atm back into the equilibrium-constant expression in order to examine the units of the resulting equations. With these factors included, the equilibrium-constant equation is

$$\frac{(P_{N_2O_4}/P_{ref})}{(P_{NO_2}/P_{ref})^2} = K$$

Substituting $P_{NO_2} = RT[NO_2]$ and $P_{N_2O_4} = RT[N_2O_4]$ gives

$$\frac{[N_2O_4](RT/P_{ref})}{[NO_2]^2(RT/P_{ref})^2} = \frac{[N_2O_4]}{[NO_2]^2}\left(\frac{RT}{P_{ref}}\right)^{-1} = K$$

Multiplying both sides of this equation by RT/P_{ref} gives

$$\frac{[N_2O_4]}{[NO_2]^2} = K\left(\frac{RT}{P_{ref}}\right)$$

For the general gas-phase reaction

$$a\,A(g) + b\,B(g) \rightleftharpoons c\,C(g) + d\,D(g)$$

a similar algebraic manipulation gives

$$\frac{[C]^c[D]^d}{[A]^a[B]^b} = K\left(\frac{RT}{P_{ref}}\right)^{a+b-c-d}$$

• In some books, the expression on the right side of this equation is called K_c (c for "concentration"), and what we call K is called K_p (p for "pressure"). This is misleading, however, because there is only one equilibrium constant, K.

In the general case, RT/P_{ref} is raised to a power that equals the number of moles of gas-phase reactants minus the number of moles of gas-phase products in the balanced chemical equation. This relationship allows us to work with concentrations of gas-phase species instead of partial pressures. The most useful unit for R in this context is the L atm mol^{-1} K^{-1} because partial pressures are commonly given in atmospheres and concentrations in moles per liter: $R = 0.08206$ L atm mol^{-1} K^{-1}.

If $a + b - c - d = 0$, the number of moles of gases does not change as reactants convert to products, and the quantity $(RT/P_{ref})^{a+b-c-d} = 1$. In this case only, the same numerical equilibrium constant may be used with expressions employing concentrations and with expressions employing partial pressures.

Phosphorus pentachloride, PCl_5, has trigonal bipyramidal geometry. It is made on an industrial scale by the reaction of Cl_2 with PCl_3. It reacts violently with an excess of water to form HCl and H_3PO_4; when PCl_5 and water are mixed in a $1:1$ molar ratio, $POCl_3$ is formed.

EXAMPLE 7-8

At elevated temperatures, PCl_5 dissociates extensively according to the equation

$$PCl_5(g) \rightleftharpoons PCl_3(g) + Cl_2(g)$$

At 300°C, the equilibrium constant for this reaction is $K = 11.5$. The concentrations of PCl_3 and Cl_2 at equilibrium in a container at 300°C both equal $0.0100 \text{ mol L}^{-1}$. Calculate the concentration of PCl_5 at equilibrium. The preceding is the only reaction taking place in the container.

Strategy

Plan to use the data in the form of concentrations. To allow this, multiply K by (RT/P_{ref}) raised to the proper power. Insert the given concentrations in the equilibrium-constant expression, set the expression equal to the number just arrived at, and solve for the missing concentration.

Solution

In the equation $PCl_5(g) \rightleftharpoons PCl_3(g) + Cl_2(g)$, two moles of gaseous products form for each mole of gas that is consumed, so (RT/P_{ref}) must be raised to the power $1 - 2 = -1$. Therefore,

$$\frac{[PCl_3][Cl_2]}{[PCl_5]} = K\left(\frac{RT}{P_{ref}}\right)^{-1}$$

Evaluate the right side of this equation:

$$K\left(\frac{RT}{P_{ref}}\right)^{-1} = K\left(\frac{P_{ref}}{RT}\right)$$

$$= 11.5\left(\frac{1 \text{ atm}}{(0.08206 \text{ L atm mol}^{-1} \text{ K}^{-1})(573 \text{ K})}\right) = 0.245 \text{ mol L}^{-1}$$

from which

$$\frac{[PCl_3][Cl_2]}{[PCl_5]} = 0.245 \text{ mol L}^{-1}$$

Solve for $[PCl_5]$ and insert the two known concentrations:

$$[PCl_5] = \frac{[PCl_3][Cl_2]}{0.245 \text{ mol L}^{-1}}$$

$$= \frac{(0.0100 \text{ mol L}^{-1})(0.0100 \text{ mol L}^{-1})}{0.245 \text{ mol L}^{-1}} = 4.08 \times 10^{-4} \text{ mol L}^{-1}$$

EXERCISE

The equilibrium constant for the reaction

$$CH_4(g) + H_2O(g) \rightleftharpoons CO(g) + 3 H_2(g)$$

equals 0.172 at 900 K. The concentrations of $H_2(g)$, $CO(g)$, and $H_2O(g)$ in an equilibrium mixture of gases at 900 K all equal $0.00642 \text{ mol L}^{-1}$. Calculate the concentration of $CH_4(g)$ in the mixture, assuming that this is the only reaction taking place.

Answer: $0.00839 \text{ mol L}^{-1}$.

7-5 The Effect of External Stresses on Equilibria: Le Châtelier's Principle

If a reaction system that is at equilibrium is left to itself, it simply stays there, unchanging. What happens if such a system is subjected to some stress from outside, such as a change in volume or temperature or a change in the concentration or partial pressure of one of its reactants or products? How does the system respond? The qualitative answer is embodied in a principle stated by Henri Le Châtelier in 1884:

> A system in equilibrium that is subjected to a stress reacts in a way that tends to counteract the stress.

Le Châtelier's principle provides a way to predict the response of an equilibrium system to an external perturbation.

Effects of Adding or Removing Reactants or Products

Consider what happens when a small quantity of a reactant is added to a reaction mixture that is at equilibrium while all other conditions are kept unchanged. The added reactant lowers the reaction quotient Q to a value that is less than K. Suddenly the mixture is no longer at equilibrium. It tends to return to equilibrium. To do so, it must generate products and consume reactants. Only in this way can Q be raised to again equal K. This response is described by saying that the equilibrium *shifts* to the right or that it shifts in the forward direction. Shifting to the right partially counteracts the stress imposed by adding the reactant.

• The system can never return to the way it was before the addition of the extra reactant. The perturbing stress is partially counteracted, not canceled.

Adding a *product* to an equilibrium reaction mixture causes Q to exceed K. The reaction must generate reactants and consume products to get back to equilibrium because that is the only way to lower Q. The equilibrium shifts to the left (or backward, or toward the reactants) when perturbed by the addition of a product.

Removing reactants or products from an equilibrium reaction mixture also imposes a stress. The kind of reasoning just illustrated leads to these predictions: Removal of a reactant from an equilibrium mixture causes the equilibrium to shift backward (to the left); removal of a product causes it to shift forward (to the right). Removals cause shifts in the reverse direction of those caused by additions.

EXAMPLE 7-9

An equilibrium mixture of $I_2(g)$, $H_2(g)$, and $HI(g)$ at 600.0 K has

$$P_{I_2} = 0.4756 \text{ atm}, P_{H_2} = 0.2056 \text{ atm, and } P_{HI} = 3.009 \text{ atm}$$

This is essentially the final equilibrium state of Example 7–6. The reaction system is stressed by addition of enough $I_2(g)$ momentarily to increase P_{I_2} to 2.000 atm. Use Le Châtelier's principle to predict the direction of the shift required to reestablish equilibrium. Determine the partial pressures of the three gases after equilibrium is re-established.

Strategy

Set up a table of changes in which "initial" refers to the moment after the addition of the new I_2 but before it starts to react. Then follow the approach used in Example 7–6.

Solution

Adding I_2, a reactant, shifts the equilibrium toward the right. The partial pressure of I_2 must drop from its momentary high of 2.000 atm. Therefore, assign a negative sign to the change in the partial pressure of the I_2 in the table of changes:

	$I_2(g)$ P_{I_2} (atm)	+	$H_2(s)$ P_{H_2} (atm)	$\rightleftharpoons$	2 HI(g) P_{HI} (atm)
Initial	2.000		0.2056		3.009
Change	$-x$		$-x$		$+2x$
Equilibrium	$2.000 - x$		$0.2056 - x$		$3.009 + 2x$

$$\frac{(P_{HI})^2}{(P_{H_2})(P_{I_2})} = \frac{(3.009 + 2x)^2}{(2.000 - x)(0.2056 - x)} = 92.6$$

Expanding the last equation gives the quadratic equation

$$88.6x^2 - 216.27x + 29.02 = 0$$

which has the roots

$$x = 0.1425 \quad \text{and} \quad 2.298$$

The second root leads to negative partial pressures of H_2 and I_2 and so is rejected as not physically meaningful. Substituting the first root into the expressions from the table of changes gives

$$P_{I_2} = 2.000 - x = 2.000 - 0.1425 = 1.86 \text{ atm}$$
$$P_{H_2} = 0.2056 - x = 0.2056 - 0.1425 = 0.063 \text{ atm}$$
$$P_{HI} = 3.009 + 2x = 3.009 + 2(0.1425) = 3.29 \text{ atm}$$

Check

The partial pressure of $I_2(g)$ is in the proper range. It exceeds 0.4756 atm (the value just before the stress), but is less than 2.000 atm (just after the stress). Also,

$$\frac{(3.29)^2}{(1.86)(0.063)} = 92.4, \text{ essentially equal to the given } K.$$

EXERCISE

A sealed flask held at 100°C contains the equilibrium

$$H_2(g) + Si_2H_6(g) \rightleftharpoons 2 SiH_4(g) \qquad K = 89.6$$

with $P_{H_2} = 0.0977$ atm, $P_{Si_2H_6} = 0.0977$ atm, and $P_{SiH_4} = 0.925$ atm (as in Exercise 7–6). Enough $SiH_4(g)$ is suddenly added to raise the partial pressure of $SiH_4(g)$ to 2.000 atm. Predict the direction in which the equilibrium shifts in response to this stress. Compute the partial pressure of $H_2(g)$ after equilibrium is re-established.

Answer: The equilibrium shifts to the left; $P_{H_2} = 0.1915$ atm.

Figure 7–6 diagrams the evolution over time of the partial pressures of the H_2, I_2, and HI in the reaction system treated in Examples 7–6 and 7–9. The figure illustrates a crucial point. When the equilibrium responds to the stress imposed by adding more I_2, the partial pressures do *not* return to their original values but settle at new values that also satisfy the equilibrium-constant equation. Adding even more I_2 would shift the equilibrium to produce more even more HI. Removing some HI would do the same thing. Chemical engineers take advantage of this kind of behavior in designing industrial syntheses. Whenever possible, they set things up to take off products and add reactants continuously. In this way, they force high overall yields even from reactions with small equilibrium constants.

Effects of Changing the Volume of the System

Suppose that we stress a gaseous reaction system that is at equilibrium by decreasing its volume. Pushing in on a movable container wall would do this. According to Le Châtelier's principle, the equilibrium shifts to favor whichever side of the reaction has the smaller volume. By retreating to a smaller volume, the equilibrium partially counteracts the stress imposed by compression. For example, the equilibrium

$$2\,P_2(g) \rightleftharpoons P_4(g)$$

shifts toward the product side if the volume of the container is decreased. It goes in that direction because one mole of gas occupies less volume in a container than two moles. The direction of the shift also can be understood in terms of pressures. Decreasing the volume increases the pressure of the system. The equilibrium shifts to the right because for every two P_2 molecules that react, only one P_4 molecule

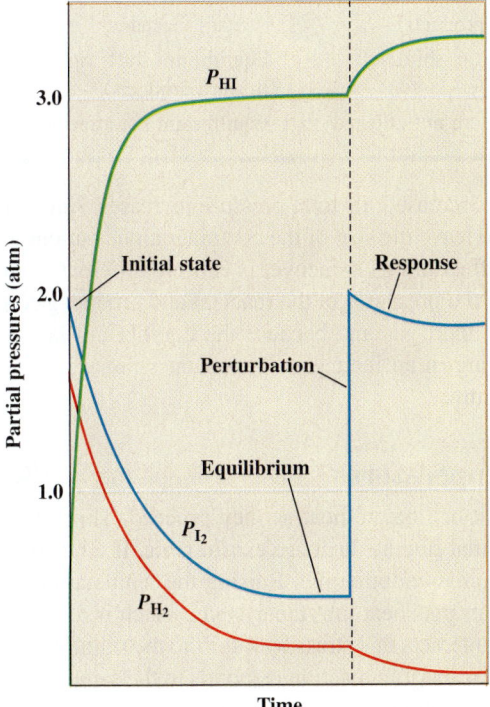

Figure 7–6 • Partial pressures versus time for the equilibrium

$$I_2(g) + H_2(g) \rightleftharpoons 2\,HI(g)$$

The left part of the figure shows the attainment of equilibrium starting from the initial conditions of Example 7–6. Then the equilibrium state is abruptly perturbed by an increase in the partial pressure of I_2 to 2.000 atm. In accordance with Le Châtelier's principle, the system responds (Example 7–9) in such a way as to decrease the partial pressure of I_2—that is, partially to counteract the perturbation that moved it away from equilibrium in the first place.

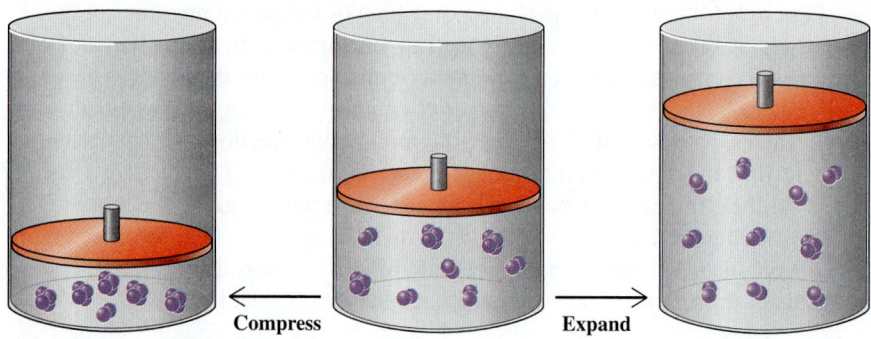

Figure 7–7 • An equilibrium mixture of P_2 and P_4 (*middle panel*) is forced into a smaller volume (*left*). Some P_2 molecules combine to give P_4 molecules in order to reduce the total number of molecules and thus the total pressure. If the volume is increased (*right*), some P_4 molecules dissociate to pairs of P_2 molecules to increase the total number of molecules and thereby the pressure.

forms, and fewer molecules exert a lower pressure. An *increase* in the volume of this system, on the other hand, shifts the equilibrium in favor of the reactant. Some $P_4(g)$ dissociates to form $P_2(g)$ if the volume is increased (Fig. 7–7) because more volume is available and two moles of gas occupies more volume than one mole does. If the total volume of the reactants equals the total volume of the products, then changing the volume has no effect upon the equilibrium. In summary:

	Volume Decreased (Pressure Increased)	Volume Increased (Pressure Decreased)
$V_{reactants} > V_{products}$	Equilibrium shifts right (toward products)	Equilibrium shifts left (toward reactants)
$V_{reactants} < V_{products}$	Equilibrium shifts left (toward reactants)	Equilibrium shifts right (toward products)
$V_{reactants} = V_{products}$	Equilibrium not affected	Equilibrium not affected

When the volume of a system is decreased, its total pressure increases. Another way to increase the total pressure is to leave the size of the container alone but pump in an inert gas, such as helium. The effect of this maneuver is entirely different from changing the volume. Because the partial pressures of the reactant and product gases are unchanged by the addition of an inert gas and because the equilibrium law is independent of total pressure, pumping in an inert gas at constant volume has no effect on the position of the equilibrium.

Effects of Changing the Temperature

Chemical reactions either absorb heat or liberate heat as they proceed. Those that absorb heat are **endothermic;** those that liberate heat are **exothermic.** If a reaction is exothermic, then its reverse reaction is endothermic. Raising the temperature of a reaction system that is at equilibrium puts heat into the system, which is a stress. An endothermic reaction partially counteracts this stress by a shift to the right, which takes up some of the added heat. An exothermic reaction responds to the same stress by a shift to the left. An exothermic reaction is endothermic when running backward, so this shift also takes up some of the added heat. Lowering the temperature has the reverse effects, favoring the products of exothermic reactions and the reactants of endothermic reactions.

	Temperature Raised	**Temperature Lowered**
Endothermic reaction	Equilibrium shifts right (toward products)	Equilibrium shifts left (toward reactants)
Exothermic reaction	Equilibrium shifts left (toward reactants)	Equilibrium shifts right (toward products)

Dissolving cobalt(II) chloride ($CoCl_2$) in warm aqueous HCl gives a blue solution as the cobalt(II) ion reacts with chloride ions to form the blue complex ion $[CoCl_4]^{2-}$. An identically prepared solution becomes pink when cooled to 0°C (Fig. 7–8). The reaction

$$[CoCl_4]^{2-}(aq) + 6\,H_2O(\ell) \rightleftharpoons [Co(H_2O)_6]^{2+}(aq) + 4\,Cl^-(aq)$$
$$\text{blue} \qquad\qquad\qquad\qquad \text{pink}$$

is responsible. This reaction is exothermic because lowering the temperature shifts it to favor the products.

One way to predict the effect of a temperature change on the position of an equilibrium is to write an explicit "heat" term into the equation, putting it on the left for endothermic reactions and on the right for exothermic reactions. Then treat this "heat" as if it were a material species that is added by raising the temperature and removed by lowering the temperature. Writing

$$[CoCl_4]^{2-}(aq) + 6\,H_2O(\ell) \rightleftharpoons [Co(H_2O)_6]^{2+} + 4\,Cl^-(aq) + \text{heat}$$
$$\text{blue} \qquad\qquad\qquad\qquad \text{pink}$$

Figure 7–8 • Shifting an equilibrium by changing the temperature. The equilibrium

$$[CoCl_4]^{2-}(aq) + 6\,H_2O(\ell) \rightleftharpoons$$
$$[Co(H_2O)_6]^{2+}(aq) + 4\,Cl^-(aq)$$

is operating in both test tubes. In a warm-water bath, the solution is blue, the color of the $[CoCl_4]^{2-}$ ion. In an ice bath, the solution is pink, the color of the $[Co(H_2O)_6]^{2+}$ ion. Lower temperature shifts this exothermic reaction to favor the products. *(Charles D. Winters)*

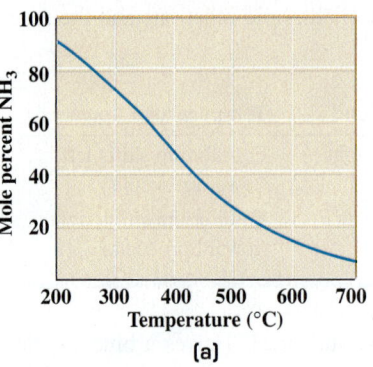
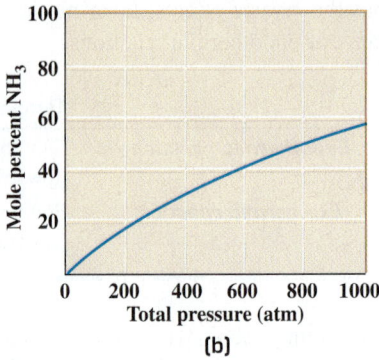

Figure 7–9 • (a) The equilibrium mole percentage of ammonia in a $1:3$ mixture of N_2 and H_2 varies with temperature; low temperatures favor a high yield of NH_3. The data shown correspond to a fixed total pressure of 300 atm. (b) At a fixed temperature (here, 500°C), the equilibrium amount of NH_3 increases with increasing total pressure.

makes it clear why the cold solution in Figure 7–8 is pink. The cold surroundings absorb some heat from the system so that the equilibrium shifts toward the right to compensate.

Another way to predict the effects of temperature changes is to learn how they change the values of equilibrium constants. As Section 11–8 establishes, the K's of exothermic reactions always decrease if the temperature is raised. Consequently, the equilibrium yield of the products of exothermic reactions decreases if the temperature is raised. The K's of endothermic reactions, on the other hand, always increase if the temperature is raised, causing an increase in the equilibrium yield of the products at higher temperature.

Driving Reactions to Completion

A high-pressure converter used in the synthesis of ammonia from nitrogen and hydrogen. *(E. R. Degginger/FPSA)*

In the chemical industry, high yields translate into less waste and pollution. Le Châtelier's principle guides efforts to increase yields of desired reactions by tuning the reaction conditions. As an example, consider the reaction

$$N_2(g) + 3\,H_2(g) \rightleftharpoons 2\,NH_3(g)$$

which is the basis of the industrial synthesis of ammonia. The reaction is exothermic, so working at the lowest practicable temperature increases the equilibrium yield of ammonia (Fig. 7–9a). At too low a temperature, the reaction is too slow, so a compromise temperature near 500°C usually is employed. The volume of gas decreases from reactants to product. This means that equilibrium yields of ammonia are increased by decreasing the volume of the reaction vessel (increasing the partial pressures of the reactants and product). Typically, total pressures of 150 to 300 atm are used (Fig. 7–9b), although some plants work at up to 900 atm. Even at high pressures, the equilibrium yield is usually only 15% to 20% because of the unfavorably small equilibrium constant. To overcome this, ammonia plants use a cyclic process in which the gas mixture is cooled so that the ammonia liquefies (its boiling point is much higher than those of nitrogen and hydrogen) and is removed. Continuous removal of the product helps to drive the reaction to completion. The industrial production of ammonia is discussed further in Chapter 22.

CHEMISTRY IN PROGRESS

Synthesis Gas

The production of ammonia described in Section 7–5 requires large supplies of hydrogen and nitrogen. Nitrogen is readily obtained from the distillation of air, as explained in Section 6–7. Hydrogen is another story. Currently, its principal source is the methane in natural gas, and the method of production involves two major steps. The first is the **reforming reaction:**

$$CH_4(g) + H_2O(g) \rightleftharpoons CO(g) + 3 H_2(g)$$

This reaction is endothermic and proceeds with an increase in the chemical amount of gas from two moles to four. It is accordingly carried out at high temperature (750 to 1000°C) and low total pressure to increase yield. Also, an excess of steam is added to drive the reaction toward completion. The gas mixture that it produces is called **synthesis gas** and is used directly for the production of methanol (CH_3OH, Problem 7–44) and other chemicals. If a feedstock other than methane is used, synthesis gas is obtained with different ratios of carbon monoxide to hydrogen. With propane, for example, the reforming reaction is

$$C_3H_8(g) + 3 H_2O(g) \rightleftharpoons 3 CO(g) + 7 H_2(g)$$

The second step in the production of hydrogen is the **shift reaction.** In it, additional steam reacts with the carbon monoxide from the first step:

$$CO(g) + H_2O(g) \rightleftharpoons CO_2(g) + H_2(g)$$

This reaction is exothermic so that low temperature favors the products. As is commonly the case with exothermic equilibria, however, the temperature cannot be lowered too far because the reaction then becomes too slow. The best compromise comes at a temperature of approximately 350°C. The shift reaction furnishes hydrogen beyond that from the reforming reaction, and also carbon dioxide as a by-product. It is in fact the major commercial source of carbon dioxide, which is frozen for refrigeration (dry ice), liquefied to fill fire extinguishers, or dissolved in water to make sparkling beverages. Gaseous CO_2 also is used in chemical synthesis, as in the production of salicylic acid for aspirin (see Fig. 22–2).

Figure 7–A • Coal reserves are much more extensive than those of petroleum or natural gas. The production of hydrocarbon gases from coal for use as fuel or in chemical synthesis is being studied in this demonstration plant in Texas. *(Courtesy of Shell Oil Co.)*

At present, natural gas is the primary raw material for making hydrogen and carbon dioxide. As reserves of natural gas are consumed, other sources must be developed. One older process that is again attracting attention is the reaction of coal with steam. Coal is a complex mixture that contains a large proportion of carbon by mass; for our purposes, it is sufficient to represent it as entirely carbon. The reaction with steam is

$$C(s) + H_2O(g) \rightleftharpoons CO(g) + H_2(g)$$

This mixture of product gases is called water gas and burns with a blue flame because of its carbon monoxide content. It has less hydrogen proportionately than does the synthesis gas produced from methane or higher hydrocarbons. Water gas can react further, in the shift reaction, to give additional hydrogen and carbon dioxide. Once a mixture of CO and H_2 is prepared with the appropriate proportions, the reforming reaction given at the beginning of this section can be reversed to make methane for use as a fuel; the overall process is referred to as the gasification of coal (Fig. 7–A).

EXAMPLE 7-10

For each of the following reactions, state whether a higher equilibrium yield of products is favored by higher or lower total volume, and by higher or lower temperature.

(a) $PCl_3(g) + Cl_2(g) \rightleftharpoons PCl_5(g)$ (exothermic)

(b) $CH_3OH(g) \rightleftharpoons CO(g) + 2 H_2(g)$ (exothermic)

(c) $N_2(g) + O_2(g) \rightleftharpoons 2 NO(g)$ (endothermic)

Solution

Higher yield in (a) is favored by lower total volume because the product has smaller volume than the reactants (1 mol of gaseous product versus 2 mol of gaseous reactants). Higher yield in (b) is favored by higher total volume (3 mol of gases produced from 1 mol of gas). Yields of (c) are unaffected by changes in the total volume because the volume of the reactants and products are equal (2 mol of gas gives 2 mol of gas).

Higher yield in (a) and (b) is favored by lower temperature because the reactions are exothermic. Higher yield in (c) is favored by higher temperature.

EXERCISE

State the effect of an increase in temperature and also of a decrease in volume on the equilibrium yield of the products of each of the following reactions:

(a) $CH_3OCH_3(g) + H_2O(g) \rightleftharpoons 2 CH_4(g) + O_2(g)$ (endothermic)

(b) $H_2O(g) + CO(g) \rightleftharpoons HCOOH(g)$ (exothermic)

Answer: (a) Higher temperature increases yield, and lower volume decreases yield. (b) Higher temperature decreases yield, and lower volume increases yield.

Le Châtelier's principle predicts certain properties of equilibrium states in a qualitative way. It is a precursor of and a valuable adjunct to the more complete thermodynamic analysis in Chapters 10 and 11.

7-6 Heterogeneous Equilibrium

Heterogeneous equilibria are equilibria that occur across phase boundaries. They involve solids, liquids, gases, and dissolved species in any combination. The equilibrium between liquid water and water vapor in a closed container (discussed in Section 7-1) is a heterogeneous equilibrium. It is represented

$$H_2O(\ell) \rightleftharpoons H_2O(g) \qquad P_{H_2O(g)} = K$$

As long as *some* liquid water is in the container, the pressure of water vapor adjusts to a constant, equilibrium value (which equals 0.03126 atm at 25°C). The position of this equilibrium is not affected by the amount of liquid water present, and no mention of the liquid appears in the equilibrium-constant equation. A similar situation is observed in the equilibrium between solid iodine and iodine in aqueous solution:

$$I_2(s) \rightleftharpoons I_2(aq) \qquad [I_2] = K$$

The equilibrium concentration of $I_2(aq)$ at 25°C is always 1.34×10^{-3} mol L^{-1}, regardless of the amount of solid iodine present, as long as there is some; the pure solid does not appear in the equilibrium-constant equation. Dissolved species, such as $I_2(aq)$, enter equilibrium-constant expressions through their concentrations. These "concentrations" are really dimensionless ratios equal to the concentration in mol L^{-1} divided by a reference concentration of 1.00 mol L^{-1}. They are similar to the dimensionless ratios obtained when partial pressures of gases in atmospheres are divided by a reference pressure of 1 atm.

Consider now a third heterogeneous equilibrium, one involving a chemical reaction. Solid calcium carbonate decomposes to solid calcium oxide and gaseous carbon dioxide:

$$CaCO_3(s) \rightleftharpoons CaO(s) + CO_2(g) \qquad P_{CO_2} = K$$

High temperature favors the reaction of calcium carbonate according to this equation (Fig. 7–10); low temperature favors the reverse reaction. When the equilibrium is studied experimentally, it is found that, at a fixed temperature, the equilibrium pressure of $CO_2(g)$ is a constant, independent of the amounts of $CaCO_3(s)$ and $CaO(s)$, as long as some of each is present (Fig. 7–11). The two pure solids are represented by unity in the equilibrium-constant expression, and the equilibrium constant is numerically equal to the partial pressure of carbon dioxide (expressed in units of atmospheres).

We are now in a position to state the **law of mass action** in a general form:

Figure 7–10 • Burning lime. The conversion of limestone to lime is important in the chemical industry because calcium oxide (lime) is a cheap base. The reaction is carried out in giant rotary lime kilns, ranging up to 150 m in length and 5 m in diameter. *(Courtesy of Chemical Lime Company)*

1. Gases enter equilibrium-constant expressions as partial pressures in atmospheres.

2. Dissolved species enter as concentrations, in units of moles per liter.

3. Pure solids and pure liquids are represented in equilibrium-constant expressions by the number 1 (unity); a solvent taking part in a chemical reaction also is represented by 1, provided that the solution is dilute.

4. Partial pressures and concentrations of products appear in the numerator, and those of reactants, in the denominator. Each is raised to a power equal to its coefficient in the balanced chemical equation.

(Continued on page 328.)

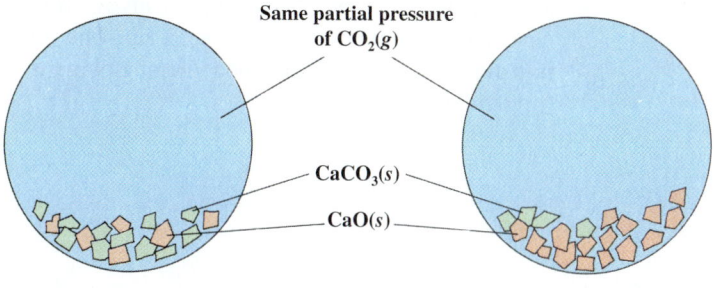

Same partial pressure of $CO_2(g)$

$CaCO_3(s)$

$CaO(s)$

Much $CaCO_3(s)$, little $CaO(s)$ **Little $CaCO_3(s)$, much $CaO(s)$**

Figure 7–11 • As long as both $CaO(s)$ and $CaCO_3(s)$ are present at equilibrium in a closed container at a given temperature, the partial pressure of $CO_2(g)$ is fixed. It does not depend on the relative amounts of the two solids present.

CHEMISTRY IN YOUR LIFE

Hemoglobin and Oxygen Transport

The cells of your body need oxygen to live, and your bloodstream satisfies this need with oxygen transported from the lungs. Whole blood can carry as much as 0.01 mol of O_2 per liter because the compound hemoglobin (Hb) in the red blood cells binds O_2 chemically. By contrast, blood plasma, which contains no hemoglobin, dissolves only about 0.0001 mol of O_2 per liter, a value that is close to the solubility of O_2 in ordinary water at room conditions (see Example 6–18). As highly oxygenated blood from the lungs passes through the capillaries, the binding of oxygen to hemoglobin loosens, and the freed O_2 is taken up by a different oxygen-binding compound, called myoglobin (Mb), that is found in nearby cells. Myoglobin then carries the oxygen on toward its ultimate fate, which is to oxidize a variety of compounds in the cell and be transported away in the form of carbon dioxide.

How can hemoglobin bind oxygen so aggressively in the lungs and yet release it so meekly to myoglobin in the capillaries? The answer starts with a comparison of the heterogeneous equilibria by which hemoglobin and myoglobin bind oxygen. Hemoglobin consists of a large protein (the globin part), in which four iron-containing heme groups are embedded (see Fig. 25–11). Oxygen can bind at each **heme** group, so one molecule of hemoglobin binds up to four molecules of O_2 in a series of *consecutive* equilibria:

$$Hb(aq) + O_2(g) \rightleftharpoons Hb(O_2)(aq)$$

$$K_1 = \frac{[Hb(O_2)]}{[Hb]\,P_{O_2}} = 4.88$$

$$Hb(O_2)(aq) + O_2(g) \rightleftharpoons Hb(O_2)_2(aq)$$

$$K_2 = \frac{[Hb(O_2)_2]}{[Hb(O_2)]\,P_{O_2}} = 15.4$$

$$Hb(O_2)_2(aq) + O_2(g) \rightleftharpoons Hb(O_2)_3(aq)$$

$$K_3 = \frac{[Hb(O_2)_3]}{[Hb(O_2)_2]\,P_{O_2}} = 6.49$$

$$Hb(O_2)_3(aq) + O_2(g) \rightleftharpoons Hb(O_2)_4(aq)$$

$$K_4 = \frac{[Hb(O_2)_4]}{[Hb(O_2)_3]\,P_{O_2}} = 1750$$

The four K's are *not* equal, and K_4 is far larger than the other three.

Myoglobin is different. The globin portion of myoglobin embeds just one heme group so that each myoglobin molecule can bind just one O_2:

$$Mb(aq) + O_2(g) \rightleftharpoons Mb(O_2)(aq)$$

$$K = \frac{[Mb(O_2)]}{[Mb]\,P_{O_2}} = 271$$

According to Le Châtelier's principle, changes in the partial pressure of O_2 (P_{O_2}) should affect the myoglobin equilibrium and all four hemoglobin equilibria similarly. Increasing P_{O_2} should shift all of the equilibria to the right; decreasing P_{O_2} should shift them all to the left. One way to check this is to plot the fraction of the binding sites occupied by O_2 as a function of P_{O_2}. Figure 7–B shows such **fractional saturation** plots for hemoglobin and myoglobin. Clearly, a reduction in P_{O_2} in fact causes less O_2 to be bound. But hemoglobin and myoglobin respond to the same lowering of P_{O_2} to very different degrees. At $P_{O_2} = 0.13$ atm (100 torr), both hemoglobin and myoglobin are more than 95% saturated with oxygen at equilibrium. In contrast, at $P_{O_2} = 0.040$ atm (30 torr), the saturation of hemoglobin is sharply less (55%), whereas that of myoglobin stays high (> 90%). The first of these partial pressures prevails in arterial blood from the lungs and the second in venous blood. Thus, as P_{O_2} diminishes in the capillaries, hemoglobin releases a substantial amount of oxygen, but myoglobin in the nearby cells remains able to take it up. The outcome is a transfer of O_2 from blood (hemoglobin) to cells (myoglobin).

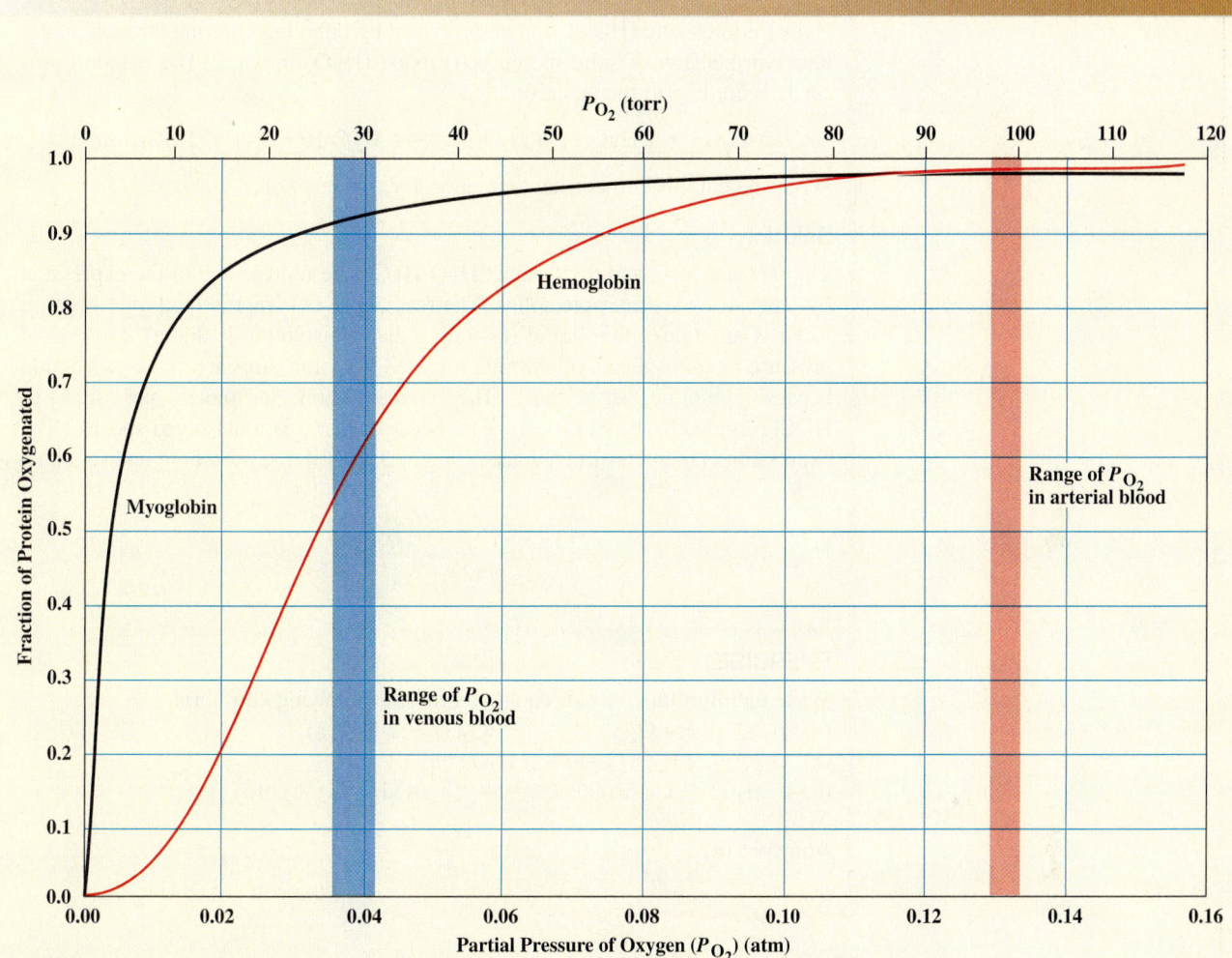

Figure 7–B • A plot of the fraction of the binding sites of hemoglobin (*red*) and myoglobin (*black*) that are occupied by O_2 as a function of the partial pressure of O_2.

The S-shaped curve for hemoglobin in Figure 7–B can be derived mathematically from the different K's for the four equilibria by which hemoglobin binds O_2. The large K_4 means that the affinity of hemoglobin for oxygen rises with the uptake of oxygen. Reversing the perspective, we see that the *loss* of the first O_2 from $Hb(O_2)_4$ sets up the easier loss of more O_2, and the fractional saturation curve for Hb dips below the curve for Mb. This phenomenon, called **positive cooperativity,** derives from structural changes within the hemoglobin molecule as it binds successive molecules of oxygen.

EXAMPLE 7–11

Hypochlorous acid (HOCl) can be produced by bubbling chlorine through an agitated suspension of solid mercury(II) oxide (HgO) in water. The reaction proceeds according to the equation

$$2\,Cl_2(g) + 2\,HgO(s) + H_2O(\ell) \rightleftharpoons HgO \cdot HgCl_2(s)' + 2\,HOCl(aq)$$

Write the equilibrium-constant equation for this reaction.

Solution

The reactant HgO and the product $HgO \cdot HgCl_2$ are represented in the expression by 1 because they are pure solids. The reactant H_2O is represented by 1 because water is an almost pure liquid (assuming that the solution is dilute). The partial pressure in atmospheres of chlorine (divided by 1 atm) appears in the expression because chlorine is a gas. The concentration in moles per liter of HOCl (divided by $1\ mol\ L^{-1}$) appears because HOCl is a dissolved species. The expression is then constructed according to Rule 4 in the preceding and set equal to K:

$$\frac{(1)[HOCl]^2}{(P_{Cl_2})^2(1)^2(1)} = \frac{[HOCl]^2}{(P_{Cl_2})^2} = K$$

EXERCISE

Write equilibrium-constant equations for the following equilibria:

(a) $Si_3N_4(s) + 4\,O_2(g) \rightleftharpoons 3\,SiO_2(s) + 2\,N_2O(g)$
(b) $O_2(g) + 2\,H_2O(\ell) \rightleftharpoons 2\,H_2O_2(aq)$
(c) $CaH_2(s) + 2\,C_2H_5OH(\ell) \rightleftharpoons Ca(OC_2H_5)_2(s) + 2\,H_2(g)$

Answer: (a) $\dfrac{(P_{N_2O})^2}{(P_{O_2})^4} = K.$ (b) $\dfrac{[H_2O_2]^2}{P_{O_2}} = K.$ (c) $(P_{H_2})^2 = K.$

These rules give equilibrium-constant equations that are reasonably accurate when the pressures of gases do not exceed several atmospheres and the concentrations of solutes do not exceed $0.1\ mol\ L^{-1}$. At higher pressures or concentrations, where gases and solutions depart seriously from ideality, more advanced methods of treatment are required.

EXAMPLE 7–12

Graphite (a form of solid carbon) is added to a vessel that contains pure $CO_2(g)$ at a pressure of 0.824 atm at a certain high temperature. The pressure in the vessel rises to an equilibrium value of 1.366 atm as a single reaction occurs to produce $CO(g)$ and no other products. Write a balanced chemical equation for the process, and determine the value of its corresponding equilibrium constant.

Strategy

Write an equation for the single reaction. Set up a table of changes and use the fact that the total pressure equals the sum of the partial pressures of the CO and CO_2 to figure out the partial pressures of the two gases at equilibrium. Substitute these partial pressures in the equilibrium-constant expression to obtain K.

Solution

The reaction involves three specified substances and no others. It can only be the combination of graphite with CO_2 to give CO. The equation

$$C(s, \text{graphite}) + CO_2(g) \rightleftharpoons 2\,CO(g)$$

represents this reaction. The table of changes is

	P_{CO_2} (atm)	P_{CO} (atm)
Initial	0.824	0
Change	$-x$	$+2x$
Equilibrium	$0.824 - x$	$2x$

Graphite does not appear in the table. As a pure solid, it is represented in the equilibrium-constant expression by a 1, both initially and at equilibrium.

At equilibrium the total pressure equals the sum of the partial pressures of the two gases:

$$P_{tot} = P_{CO_2} + P_{CO} = (0.824 \text{ atm} - x) + 2x$$
$$1.366 \text{ atm} = 0.824 + x$$
$$x = 1.366 - 0.824 = 0.542 \text{ atm}$$

The equilibrium partial pressures of the two gases are

$$P_{CO} = 2x = 1.084 \text{ atm} \quad \text{and} \quad P_{CO_2} = 0.824 - 0.542 = 0.282 \text{ atm}$$

Substituting these numbers in the equilibrium-constant expression gives K:

$$K = \frac{(P_{CO})^2}{1\,(P_{CO_2})} = \frac{(1.084)^2}{0.282} = 4.17$$

The numerical value of K depends on the way the equation is written, as explained in Section 7–2. If the balanced equation is written "$2\,C(s) + 2\,CO_2(g) \rightleftharpoons 4\,CO(g)$" (with doubled coefficients), the correct K is 17.4 (the square of 4.17). If the balanced equation is written "$\frac{1}{2}\,C(s) + \frac{1}{2}\,CO_2(g) \rightleftharpoons CO(g)$" (halved coefficients), the correct K is 2.04 (the square root of 4.17).

EXERCISE

A vessel holds pure $CO(g)$ at a pressure of 1.282 atm and a temperature of 354 K. A quantity of nickel is added, and the partial pressure of $CO(g)$ drops to an equilibrium value of 0.709 atm because of the reaction

$$Ni(s) + 4\,CO(g) \rightleftharpoons Ni(CO)_4(g)$$

Compute the equilibrium constant for this reaction at 354 K.

Answer: $K = 0.567$.

7-7 Extraction and Separation Processes

A very important type of heterogeneous equilibrium involves the partitioning of a solute between two mutually insoluble solvent phases. Such equilibria are used in many separation processes in chemical research and in industry.

• A familiar example is the separation of the oil from the water phase in salad dressing. In salad dressing, the water phase is denser, and the oil floats on top.

Suppose that the two immiscible liquids water and carbon tetrachloride are placed together in a container. **Immiscible** means "mutually insoluble." Immiscible liquids, if shaken together, mix temporarily but eventually separate into distinct phases, with the more dense on the bottom and the less dense on the top. In this case, water floats as a layer on top of the denser carbon tetrachloride. A visible boundary, the meniscus, separates the two phases. If a small quantity of iodine is added and the vessel is shaken in order to distribute the iodine thoroughly, some of the iodine dissolves in the carbon tetrachloride, and the rest dissolves in the water (Fig. 7–12). An equilibrium-constant equation governs the distribution of the iodine between the two immiscible solvents. Thus, if a little more iodine is added to the container (which is again shaken thoroughly), the iodine concentration in each solvent increases, but the *ratio* of the two concentrations remains the same as long as the temperature remains constant. The ratio of the concentrations of iodine in the two phases is the partition coefficient, the equilibrium constant K for the process

$$I_2(aq) \rightleftharpoons I_2(CCl_4)$$

It can be written as

$$K = \frac{[I_2]_{CCl_4}}{[I_2]_{aq}}$$

in which $[I_2]_{CCl_4}$ and $[I_2]_{aq}$ are the concentrations (in moles per liter) of I_2 in the CCl_4 (carbon tetrachloride) and aqueous phases, respectively. At 25°C, K has the value 85 for this equilibrium. The fact that K is larger than unity shows that iodine is more soluble in CCl_4 than it is in water.

Extraction takes advantage of the partitioning of a solute between two immiscible solvents to remove that solute from one solvent into another. Suppose that iodine is present as a contaminant in water or an aqueous solution. Most of the iodine could be removed by shaking the aqueous solution with CCl_4, allowing the two phases to separate, and then pouring off (decanting) the water layer from the denser layer of carbon tetrachloride. The larger the equilibrium constant for the partition of a solute from the original solvent into the extracting (that is, added) solvent, the more complete is the separation.

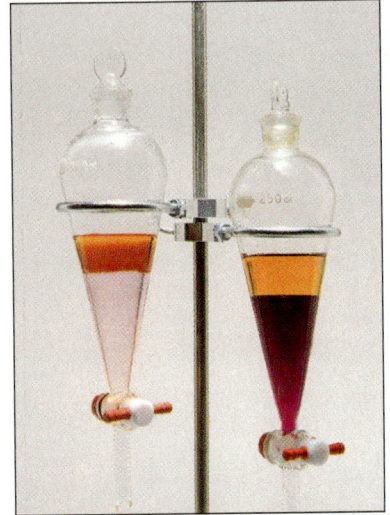

Figure 7–12 • Aqueous iodine has been poured on top of carbon tetrachloride in a separatory funnel (*left*). After the funnel is shaken (*right*), the partition of iodine between the upper (aqueous) phase and the lower (CCl_4) phase is complete. The iodine dissolves preferentially in the denser CCl_4. (*Leon Lewandowski*)

EXAMPLE 7–13

An aqueous solution has an iodine concentration of 2.00×10^{-3} mol L^{-1}. Calculate the percentage of iodine removed by extraction of 0.100 L of this aqueous solution with 0.050 L of CCl_4 at 25°C.

Strategy

Figure out the amount of iodine present in the system. Rely on the fact that this *amount* does not change, even though iodine is extracted from water into CCl_4, changing the *concentration* of iodine in both.

Solution

The chemical amount of I_2 originally present is

$$n_{I_2} = \left(\frac{2.00 \times 10^{-3} \text{ mol } I_2}{\text{L water}} \right) \times 0.100 \text{ L water} = 2.00 \times 10^{-4} \text{ mol } I_2$$

Shaking the aqueous solution with CCl_4 extracts some of this I_2 into the CCl_4 phase. If y mol of I_2 transfers to the CCl_4, then $2.00 \times 10^{-4} - y$ mol of I_2 remains in the water. At equilibrium

$$K = 85 = \frac{[I_2]_{CCl_4}}{[I_2]_{aq}} = \frac{\dfrac{y \text{ mol } I_2}{0.050 \text{ L}}}{\dfrac{(2.00 \times 10^{-4} - y) \text{ mol } I_2}{0.100 \text{ L}}}$$

$$85 = \frac{2.0y}{(2.00 \times 10^{-4} - y)}$$

• It is important to note that the changes in *concentration* of I_2 in the two phases do not have the same magnitude because the volumes are different. It is only the changes in numbers of moles that can be written as $-y$ and $+y$.

The volumes of the two solvents in the calculation do not change because they do not mix. Solving gives $y = 1.954 \times 10^{-4}$ mol I_2. Then

$$\text{percentage } I_2 \text{ removed} = \frac{1.954 \times 10^{-4} \text{ mol}}{2.000 \times 10^{-4} \text{ mol}} \times 100\% = 98\%$$

Most of the iodine is transferred, which is consistent with the fairly large partition coefficient. A second extraction with a fresh 0.050 L of CCl_4 would remove 98% of the 2% left after the first extraction.

EXERCISE

The Parkes process takes advantage of the immiscibility of molten zinc and molten lead. In this process, silver is extracted from lead into zinc. A 105 L sample of molten lead that contains 0.0306 mol L^{-1} of dissolved silver is shaken with 21.0 L of molten zinc. Although the actual mode of extraction is quite complicated, assume that the equilibrium concentration of silver in the molten zinc is 3.0×10^2 times that in the lead. What mass of silver (in grams) is originally present and how much of it remains dissolved in the lead at equilibrium?

Answer: 347 g of silver is present; 5.7 g of silver remains in the lead.

The most important use of extraction is in purification. In favorable cases, a desired solute has such a high partition coefficient compared with its impurities that an excellent separation is effected in a single pass. Impurities also may be extracted from a desired product. This approach is used industrially on a large scale to purify sodium hydroxide for use in the manufacture of rayon. The sodium hydroxide produced by electrolysis (see Section 23–2) typically contains 1% sodium chloride and 0.1% sodium chlorate as impurities. If a concentrated aqueous solution of this impure sodium hydroxide is extracted with liquid ammonia, the NaCl and $NaClO_3$ are partitioned into the ammonia phase in preference to the aqueous phase. The heavier aqueous phase is added to the top of an extraction vessel that is filled with liquid ammonia, and equilibrium is reached as droplets of it settle through the ammonia to the bottom. The concentrations of the impurities in the sodium hydroxide solution are reduced to approximately 0.08% NaCl and 0.0002% $NaClO_3$ by this procedure.

Chromatography

Partition equilibria form the basis for an important class of separation techniques called **chromatography.** This word comes from the Greek roots *chromato* and *graphia*, meaning "color" and "writing." It was chosen because the first

Figure 7–13 • In this demonstration of chromatography, a solvent travels upward along a strip of paper, separating the colored components of a sample of ink. *(Charles D. Winters)*

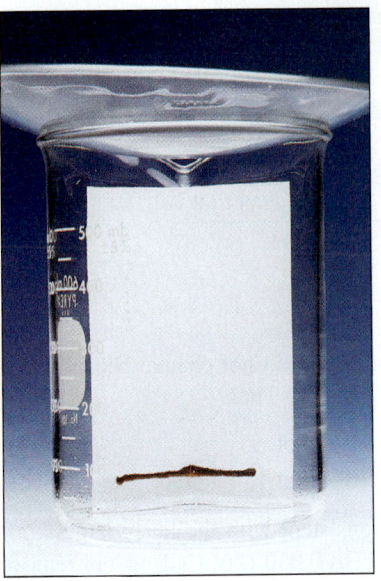 

chromatographic separations involved colored substances (Fig. 7–13). Chromatography is a continuous extraction process in which solute species are exchanged between two phases. One phase, the *mobile phase,* moves with respect to a second *stationary* phase. Numerous types of chromatography are distinguished, according to the different mobile and stationary phases used. A few of the more important ones are listed in Table 7–1. Some of the techniques listed are used principally to analyze complex mixtures in order to learn the identities and amounts of their components; others are well suited for the isolation of a substance from a mixture.

Column chromatography (Fig. 7–14) uses a tube packed with a porous material, frequently a silica gel onto which water has been adsorbed. Water is the stationary phase in this case. The mixture to be separated is introduced at the top of the column. Another solvent is used as the mobile phase and is then added from the top. As the mobile phase passes down the column, the components of the solute mixture are removed from the stationary phase and move with the mobile phase,

TABLE 7–1		Chromatographic Separation Techniques
Name	**Mobile Phase**	**Stationary Phase**
Gas–liquid	Gas	Liquid adsorbed on a porous solid in a tube
Gas–solid	Gas	Porous solid in a tube
Column	Liquid	Liquid adsorbed on a porous solid in a tubular column
Paper	Liquid	Liquid held in the pores of a thick paper
Thin layer	Liquid	Liquid or solid; solid is held on a glass plate and liquid may be adsorbed on it
Ion exchange	Liquid	Solid (finely divided ion-exchange resin) in a tubular column

Adapted from D. A. Skoog and D. M. West, *Analytical Chemistry,* Philadelphia, Saunders College Publishing, 1980, Table 18–1.

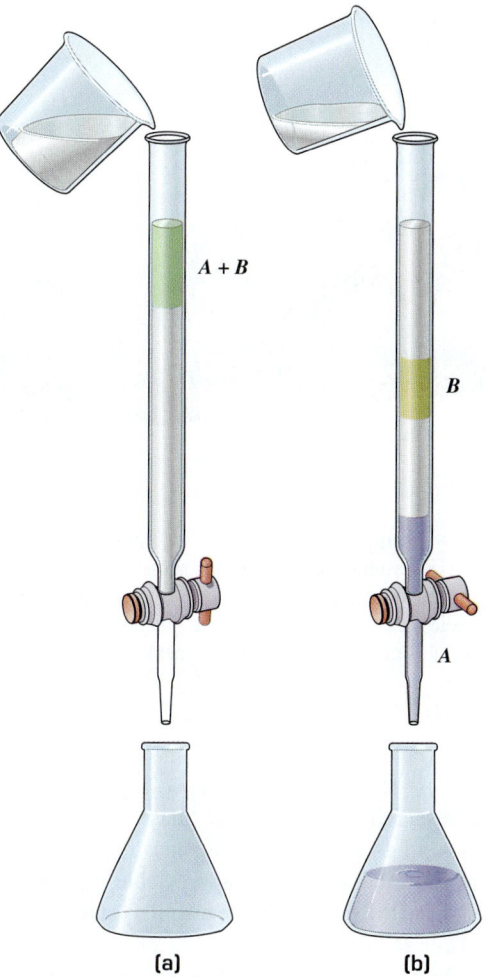

A + B

B

A

(a)   (b)

Figure 7–14 • (a) In a column chromatograph, the top of the column is loaded with a mixture of solutes to be separated (*green*). (b) With addition of solvent, the different solutes travel at different rates, giving rise to bands. In this case, both solutes are colored. The separated fractions can be collected in different flasks for use or analysis.

but they are retarded to differing extents by their interactions with the water adsorbed on the packing. The components separate as bands along the column and reach the bottom of the column at different times, where they are collected separately for analysis or use. In some procedures, it is most efficient to use different solvents in succession as mobile phases. The partition coefficient K of a solute, such as acetic acid (CH_3COOH), equals the ratio of its concentration in the stationary phase (for example, water) to that in the mobile phase (for example, benzene):

$$\frac{[CH_3COOH]_{aq}}{[CH_3COOH]_{benzene}} = K$$

As the mobile benzene passes over the stationary water, the acetic acid molecules move between the two phases. True equilibrium is never fully established in column chromatography because the motion of the mobile phase continually brings fresh solvent into contact with the stationary phase. Nevertheless, the partition coefficient K provides a good guide to the behavior of a particular solute on the column. The larger K is, the greater is the affinity of the solute for the stationary phase and therefore the slower is its progress down the chromatographic column.

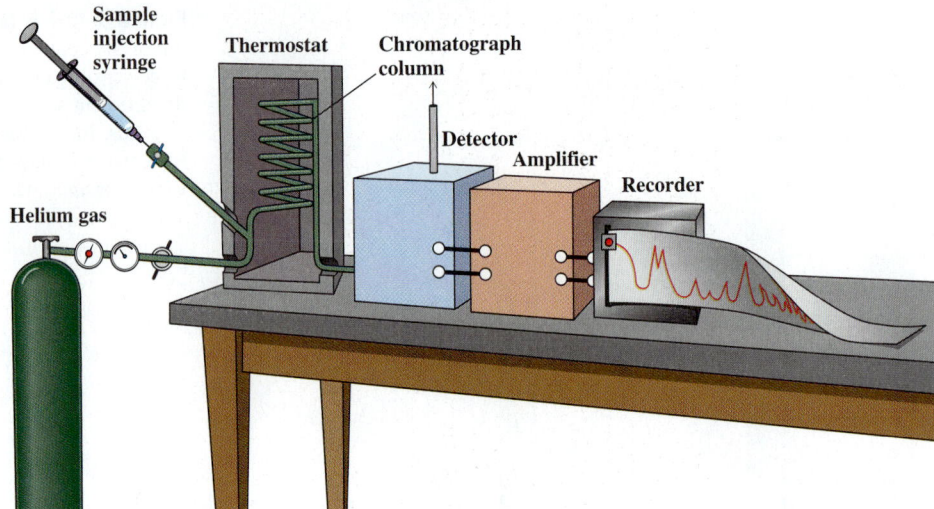

Figure 7–15 • In a gas–liquid chromatograph, the sample (which may be a complex mixture) is vaporized and passes through a column, carried in a stream of an inert gas such as helium or nitrogen. The residence time of any substance on the column depends on its partition coefficient from the vapor to the liquid in the column. A species leaving the column at a given time can be detected by a variety of techniques. The result is a gas chromatogram, with a peak corresponding to each substance in the mixture.

Gas–liquid chromatography (Fig. 7–15) is one of the most important separation techniques in modern chemical research. The stationary phase is once again a liquid adsorbed on a porous solid or held as a film on the inner wall of a quartz or metal capillary tube. The solutes are substances that have appreciable vapor pressures, and the mobile phase is a gas that flows through the capillary tube. Gas–liquid chromatography is widely used for separating the products of organic reactions. It also finds much use in determining the purity of substances because even very small amounts of impurities appear clearly as separate peaks in chromatograms. The technique is important in separating and identifying possibly toxic substances in trace amounts in a variety of environmental and biological samples. Amounts on the order of parts per trillion by mass (10^{-12} g in a 1 g sample) can be detected and identified.

SUMMARY

7–1 **Equilibrium** is a dynamic condition in which forward and reverse reactions occur at equal rates so that partial pressures and concentrations remain constant. The equilibrium condition is summed up in an equation in which **K,** the temperature-dependent **equilibrium constant,** equals an **equilibrium-constant expression.** For gas-phase reactions, the equilibrium-constant expression is the product of the partial pressures of the gases produced, each raised to a power equal to its coefficient in the balanced chemical equation, divided by the product of the partial pressures of the gases that react, each raised to a power equal to its coefficient in the balanced chemical equation.

7–2 Equilibrium constants (*K*'s) can be calculated from experimental partial pressure data by substitution into the appropriate equilibrium-constant equation. A table of changes is a standard device to assist in the calculation of equilibrium partial pressures of reactants and products in a reaction system when initial partial pressures and the K are known. If a chemical equation is obtained from other chemical equations by adding them, subtracting them, or multiplying them by a constant, the corresponding equilibrium constant is found by the multiplication, the division, or the raising to a power of their equilibrium constants.

7–3 A value of the **reaction quotient** Q results from inserting any arbitrary partial pressures into the form of the equilibrium-constant expression. As a reaction mixture approaches equilibrium, its reaction quotient approaches the equilibrium constant. If Q is less than K, the reaction shifts to the right to reach equilibrium; if Q is greater than K, the reaction shifts to the left.

7–4 Equilibrium laws can be used to compute the partial pressure of one reactant or product from those of the others, or to calculate the final partial pressures reached from a given initial state.

7–5 **Le Châtelier's principle** describes how a reaction mixture at equilibrium responds to external perturbation or stress. If a reactant or product gas is added, the reaction shifts in the direction that reduces the partial pressure of that gas. If the total pressure is increased by a decrease in the volume of the container, then the reaction shifts to reduce the total number of gas molecules, if possible, and thus reduces the total pressure. If the temperature of an equilibrium mixture is raised, K increases if the reaction is **endothermic** (absorbs heat as it proceeds), and the reaction shifts to the right. However, K decreases and the reaction shifts to the left if the reaction is **exothermic** (liberates heat as it proceeds).

7–6 **Heterogeneous equilibria** involve reactants and products in two or more phases. The **law of mass action** governs the form of equilibrium-constant expressions for all equilibria, including heterogeneous equilibria: Gases enter equilibrium-constant expressions as partial pressures; dissolved species enter as concentrations; pure solids, pure liquids, and solvents taking part in a reaction enter as the number 1. Partial pressures and concentrations of products appear in the numerator, and those of reactants in the denominator, with each raised to a power equal to its coefficient in the balanced chemical equation.

7–7 One particularly important type of heterogeneous equilibrium involves the **partition** of a solute species between two immiscible solvents. The equilibrium constant for such a process is called a **partition coefficient.** Procedures for the extraction of a dissolved species from one solvent into another depend on such equilibria, which also form the basis of **chromatography.** Numerous chromatographic techniques (including **column chromatography** and **gas–liquid chromatography**) allow chemical compounds to be separated and identified.

PROBLEMS

Note: Answers to blue-numbered problems are given in Appendix F. Problems that are more challenging are indicated with asterisks.

Chemical Reactions and Equilibrium

1. (See Example 7–1.) Write equilibrium-constant expressions for each of the following gas-phase reactions:
   (a) $H_2(g) + \frac{1}{2} O_2(g) \rightleftharpoons H_2O(g)$
   (b) $Xe(g) + 2 F_2(g) \rightleftharpoons XeF_4(g)$
   (c) $4 C_5H_5(g) + 25 O_2(g) \rightleftharpoons 20 CO_2(g) + 10 H_2O(g)$

2. (See Example 7–1.) Write equilibrium-constant expressions for each of the following gas-phase reactions:
   (a) $2 Cl_2(g) + 7 O_2(g) \rightleftharpoons 2 Cl_2O_7(g)$
   (b) $\frac{1}{2} N_2(g) + \frac{1}{2} O_2(g) + \frac{1}{2} Br_2(g) \rightleftharpoons NOBr(g)$
   (c) $C_5H_{12}(g) + 8 O_2(g) \rightleftharpoons 5 CO_2(g) + 6 H_2O(g)$

3. At a moderately elevated temperature, phosphoryl chloride ($POCl_3$) can be produced in the vapor phase from the gaseous elements. Write a balanced chemical equation and an equilibrium-constant expression for this system. Note that

gaseous phosphorus consists of P_4 molecules, at least at moderate temperatures.

4. If confined at high temperature, gaseous ammonia and oxygen quickly react and come to equilibrium with their products, water vapor and gaseous nitrogen monoxide. Write a balanced chemical equation and an equilibrium-constant expression for this system.

Calculating Equilibrium Constants

5. At 454 K, $Al_2Cl_6(g)$ reacts to form $Al_3Cl_9(g)$ according to the equation

$$3 Al_2Cl_6(g) \rightleftharpoons 2 Al_3Cl_9(g)$$

In an experiment at this temperature, the equilibrium partial pressure of $Al_2Cl_6(g)$ is 1.00 atm, and the equilibrium partial pressure of $Al_3Cl_9(g)$ is 1.02×10^{-2} atm. Compute the equilibrium constant corresponding to the above equation at 454 K.

6. At 298 K, $F_3SSF(g)$ decomposes partially to $SF_2(g)$. At equilibrium, the partial pressures of $SF_2(g)$ and $F_3SSF(g)$ are 1.1×10^{-4} atm and 0.0484 atm, respectively.
   (a) Write a balanced chemical equation to represent this reaction.
   (b) Compute the equilibrium constant corresponding to this equation.

7. The compound 1,3-di-t-butylcyclohexane exists in two forms that are known as the "chair" and "boat" conformations because the molecular structures resemble these objects. Equilibrium exists between these forms, represented by the equation

$$chair \rightleftharpoons boat$$

At 580 K, 6.42% of the molecules are in the chair form. Calculate the equilibrium constant for the reaction as written.

8. At 248°C and a total pressure of 1.000 atm, $SbCl_5$ dissociates according to the equation

$$SbCl_5(g) \rightleftharpoons SbCl_3(g) + Cl_2(g)$$

The fractional dissociation at equilibrium is 0.718. That is, 718 of every 1000 molecules of $SbCl_5$ originally present have dissociated at equilibrium. Calculate the equilibrium constant.

9. (See Example 7–2.) An experiment is run at 425°C to determine the equilibrium constant for the reaction

$$H_2(g) + I_2(g) \rightleftharpoons 2 HI(g)$$

at that temperature. Some $H_2(g)$ at a partial pressure of 5.73 atm is mixed with $I_2(g)$ at the same initial partial pressure. When the reaction comes to equilibrium, the partial pressure of HI is found to equal 9.00 atm. Compute the equilibrium constant, assuming that no side reactions occur.

10. (See Example 7–2.) A second experiment is run at 425°C to check the equilibrium constant for the reaction

$$H_2(g) + I_2(g) \rightleftharpoons 2 HI(g)$$

obtained in the previous problem. Some $H_2(g)$ at a partial pressure of 2.75 atm is mixed with $I_2(g)$ at a partial pressure of 1.50 atm. The reaction comes to equilibrium, and the partial pressure of HI is measured to be 2.79 atm. Compute the equilibrium constant, assuming that no other reactions occur.

11. Sulfuryl chloride (SO_2Cl_2) is a colorless liquid that boils at 69°C. Above this temperature, the vapors dissociate into sulfur dioxide and chlorine:

$$SO_2Cl_2(g) \rightleftharpoons SO_2(g) + Cl_2(g)$$

This reaction is slow at 100°C, but it is accelerated by the presence of some $FeCl_3$ (which does not affect the final position of the equilibrium). In an experiment, 3.174 g of $SO_2Cl_2(g)$ and a small amount of solid $FeCl_3$ are put into an evacuated 1.000 L flask, which is then sealed and heated to 100°C. At equilibrium, the total pressure in the flask equals 1.30 atm.
   (a) Calculate the partial pressure of each of the three gases present.
   (b) Calculate the equilibrium constant at this temperature.

12. A certain amount of $NOBr(g)$ is sealed in a flask, and the temperature is raised to 350 K. The following equilibrium is soon established:

$$NOBr(g) \rightleftharpoons NO(g) + \tfrac{1}{2} Br_2(g)$$

The total pressure in the flask when equilibrium is reached at this temperature is 0.675 atm, and the vapor density is 2.219 g L^{-1}.
   (a) Calculate the partial pressure of each species. (*Hint:* The vapor density cannot change in a gas-phase reaction if the volume is fixed.)
   (b) Calculate the equilibrium constant at this temperature.

13. At a certain temperature, the value of the equilibrium constant for the reaction

$$4 NH_3(g) + 7 O_2(g) \rightleftharpoons 6 H_2O(g) + 4 NO_2(g)$$

is K_1. How is K_1 related to the equilibrium constant K_2 for the related reaction

$$2 NH_3(g) + \tfrac{7}{2} O_2(g) \rightleftharpoons 3 H_2O(g) + 2 NO_2(g)$$

at the same temperature?

14. At 25°C, the equilibrium constant for the reaction

$$6 ClO_3F(g) \rightleftharpoons$$
$$2 ClF(g) + 4 ClO(g) + 7 O_2(g) + 2 F_2(g)$$

is 32.6. Calculate the equilibrium constant at 25°C for the reaction

$$\tfrac{1}{3} ClF(g) + \tfrac{2}{3} ClO(g) + \tfrac{7}{6} O_2(g) + \tfrac{1}{3} F_2(g) \rightleftharpoons ClO_3F(g)$$

15. (See Example 7–3.) Suppose that K_1 and K_2 are the respective equilibrium constants for the two reactions

$$XeF_6(g) + H_2O(g) \rightleftharpoons XeOF_4(g) + 2 HF(g)$$
$$XeO_4(g) + XeF_6(g) \rightleftharpoons XeOF_4(g) + XeO_3F_2(g)$$

Give the equilibrium constant for the reaction

$$XeO_4(g) + 2\,HF(g) \rightleftharpoons XeO_3F_2(g) + H_2O(g)$$

in terms of K_1 and K_2.

16. (See Example 7–3). At 1330 K, germanium(II) oxide (GeO) and tungsten(VI) oxide (W_2O_6) are both gases. The following two chemical equilibria are established simultaneously:

$$2\,GeO(g) + W_2O_6(g) \rightleftharpoons 2\,GeWO_4(g)$$
$$GeO(g) + W_2O_6(g) \rightleftharpoons GeW_2O_7(g)$$

The equilibrium constants for the two are, respectively, 7.0×10^3 and 38×10^3.
Compute K for the reaction

$$GeO(g) + GeW_2O_7(g) \rightleftharpoons 2\,GeWO_4(g)$$

The Reaction Quotient

17. (See Example 7–4.) The reaction

$$2\,NO(g) + Br_2(g) \rightleftharpoons 2\,NOBr(g)$$

has an equilibrium constant of 116.6 at 25°C. In each of the following mixtures, compute the reaction quotient Q and use it to decide the direction the reaction shifts (right or left) to come to equilibrium. The initial partial pressures (all in atm) in the gas mixture are
(a) $P_{NO} = 0.300$; $P_{Br_2} = 0.225$; $P_{NOBr} = 0.855$
(b) $P_{NO} = 0.225$; $P_{Br_2} = 0.300$; $P_{NOBr} = 0.855$
(c) $P_{NO} = 0.600$; $P_{Br_2} = 0.450$; $P_{NOBr} = 1.70$

18. (See Example 7–4.) Suppose the reaction

$$SbF_5(g) + 4\,Cl_2(g) \rightleftharpoons SbCl_3(g) + 5\,ClF(g)$$

has an equilibrium constant of 0.0200 at 300°C. Portions of $SbF_5(g)$, $Cl_2(g)$, $SbCl_3(g)$, and $ClF(g)$ are mixed at 300°C and have the following partial pressures (all in atm) immediately after mixing but before reaction. In each case, compute the reaction quotient Q, and use it to decide the direction the reaction proceeds (right or left) to come to equilibrium. Assume that this is the only reaction that takes place.
(a) $P_{SbCl_3} = 0.455$; $P_{ClF} = 0.225$; $P_{SbF_5} = 0.945$; $P_{Cl_2} = 0.330$
(b) $P_{SbF_5} = 0.300$; $P_{Cl_2} = 6.00$; $P_{SbCl_3} = 0.175$; $P_{ClF} = 0.675$
(c) $P_{SbF_5} = 0.600$; $P_{Cl_2} = 0.600$; $P_{SbCl_3} = 0.600$; $P_{ClF} = 0.600$

19. (See Example 7–4.) Some Al_2Cl_6 (at a partial pressure of 0.473 atm) is placed in a closed container at 454 K with some Al_3Cl_9 (at a partial pressure of 1.02×10^{-2} atm).
(a) Calculate the initial reaction quotient for the reaction

$$3\,Al_2Cl_6(g) \rightleftharpoons 2\,Al_3Cl_9(g)$$

(b) As the gas mixture reaches equilibrium, is there net production or consumption of Al_3Cl_9? (Use the data given in Problem 5.)

20. Some SF_2 (at a partial pressure of 2.3×10^{-4} atm) is placed in a closed container at 298.15 K with some F_3SSF (at a partial pressure of 0.0484 atm).

(a) Calculate the initial reaction quotient for the decomposition of F_3SSF to SF_2.
(b) As the gas mixture reaches equilibrium, is there net formation or dissociation of F_3SSF? (Use the data given in Problem 6.)

21. The progress of the reaction

$$H_2(g) + Br_2(g) \rightleftharpoons 2\,HBr(g)$$

can be monitored visually by following changes in the color of the reaction mixture (Br_2 is red-brown, and H_2 and HBr are colorless). A gas mixture is prepared at 700 K, in which 0.40 atm is the initial partial pressure of both H_2 and Br_2 and 0.90 atm is the initial partial pressure of HBr. The color of this mixture then fades as the reaction progresses toward equilibrium. Give a condition that must be satisfied by the equilibrium constant K (for example, it must be greater than or smaller than a given number).

22. Gaseous NO_2 is brown. At elevated temperatures, it reacts with CO according to

$$NO_2(g) + CO(g) \rightleftharpoons NO(g) + CO_2(g)$$

The other three gases taking part in this reaction are colorless. A gas mixture is prepared in which 3.4 atm is the initial partial pressure of both NO_2 and CO and 1.4 atm is the partial pressure of both NO and CO_2. The brown color of the mixture is observed to intensify as the reaction progresses toward equilibrium. Give a condition that must be satisfied by the equilibrium constant K (for example, it must be greater than or smaller than a given number).

Calculation of Gas-Phase Equilibria

23. (See Example 7–5.) Phosgene ($COCl_2(g)$) is produced by the reaction

$$CO(g) + Cl_2(g) \rightleftharpoons COCl_2(g)$$

for which the equilibrium constant equals 0.20 at 600°C. Calculate the partial pressure of phosgene that is in equilibrium with a mixture of 0.050 atm of carbon monoxide and 0.0045 atm of chlorine at 600°C.

24. (See Example 7–6). At 1107 K, the equilibrium constant of the reaction

$$H_2(g) + I_2(g) \rightleftharpoons 2\,HI(g)$$

equals 38.6. Calculate the partial pressure of hydrogen that is present at equilibrium in this system if the equilibrium partial pressure of iodine is 0.0484 atm and the equilibrium partial pressure of hydrogen iodide is 0.378 atm.

25. (See Example 7–6). The reaction

$$SO_2Cl_2(g) \rightleftharpoons SO_2(g) + Cl_2(g)$$

has an equilibrium constant at 100°C of 2.40.
(a) Suppose that the initial partial pressure of SO_2Cl_2 in a reaction tank with rigid walls is 2.500 atm and that no other species are present. Calculate the reaction quotient

Q, and state whether the total pressure increases or decreases as the reaction tends toward equilibrium.

(b) Calculate the partial pressures of SO_2Cl_2, SO_2, and Cl_2 when equilibrium is reached.

26. (See Example 7–6.) Bromine and chlorine react to give bromine monochloride:

$$Br_2(g) + Cl_2(g) \rightleftharpoons 2\ BrCl(g) \qquad K = 58.0 \text{ at } 25°C$$

(a) Suppose that some bromine and chlorine are mixed at 25°C and that the partial pressure of each is 0.125 atm before any reaction occurs. Calculate Q.

(b) Calculate the partial pressure of BrCl after the reaction mixture from part (a) reaches equilibrium at 25°C.

27. The dehydrogenation of benzyl alcohol gives the flavoring agent benzaldehyde:

$$C_6H_5CH_2OH(g) \rightleftharpoons C_6H_5CHO(g) + H_2(g)$$
$$K = 0.558 \text{ at } 523 \text{ K}$$

(a) Suppose 1.20 g of benzyl alcohol is placed in a 2.00 L vessel and heated to 523 K. What is the partial pressure of benzaldehyde when equilibrium is attained?

(b) What fraction of benzyl alcohol is dissociated into products at equilibrium at 523 K?

28. Heating isopropyl alcohol causes it to break down into acetone and hydrogen:

$$(CH_3)_2CHOH(g) \rightleftharpoons (CH_3)_2CO(g) + H_2(g)$$
$$K = 0.444 \text{ at } 179°C$$

(a) If 10.00 g of isopropyl alcohol is placed in a 10.00 L vessel and heated to 179°C, what is the partial pressure of acetone when equilibrium is attained?

(b) What fraction of isopropyl alcohol is broken down at equilibrium?

29. A weighed quantity of $PCl_5(s)$ is sealed in a 100.0 cm^3 glass bulb to which a pressure gauge is attached. The bulb is heated to 250°C, and the gauge shows that the pressure in the bulb rises from 0 atm to 0.776 atm. At this temperature, the solid PCl_5 is all vaporized and also partially dissociated into $Cl_2(g)$ and $PCl_3(g)$ according to the equation

$$PCl_5(g) \rightleftharpoons PCl_3(g) + Cl_2(g) \qquad K = 2.15 \text{ at } 250°C$$

Calculate the partial pressures of the three different chemical species in the vessel. Assume that this is the only reaction that takes place and that the contents of the bulb are at equilibrium.

30. Suppose 93.0 g of $HI(g)$ is placed in a glass vessel and heated to 1107 K. At this temperature, equilibrium is quickly established between $HI(g)$ and its decomposition products, $H_2(g)$ and $I_2(g)$:

$$2\ HI(g) \rightleftharpoons H_2(g) + I_2(g) \qquad K = 0.0259 \text{ at } 1107 \text{ K}.$$

The total pressure at equilibrium is observed to equal 6.45 atm. Calculate the equilibrium partial pressures of $HI(g)$, $H_2(g)$, and $I_2(g)$. No other reactions take place.

31. The combination reaction

$$Br_2(g) + I_2(g) \rightleftharpoons 2\ IBr(g) \qquad K = 322 \text{ at } 350 \text{ K}$$

is used to make iodine monobromide. Bromine at an initial partial pressure of 0.0500 atm is mixed with iodine at an initial partial pressure of 0.0400 atm and held at 350 K until equilibrium is reached. Calculate the equilibrium partial pressure of each of the gases.

32. Fluorine oxidizes oxygen to form oxygen difluoride (OF_2) according to the equation

$$F_2(g) + \tfrac{1}{2}O_2(g) \rightleftharpoons OF_2(g) \qquad K = 40.1 \text{ at } 298 \text{ K}$$

Some OF_2 is introduced into an evacuated container at 298 K and dissociates according to the above equation until its partial pressure reaches an equilibrium value of 1.10 atm. Calculate the equilibrium partial pressures of F_2 and O_2 in the container. No other reactions occur in the container.

33. (See Example 7–7.) Find the equilibrium partial pressures of $Al_2Cl_6(g)$ and $Al_3Cl_9(g)$ in the reaction

$$3\ Al_2Cl_6(g) \rightleftharpoons 2\ Al_3Cl_9(g)$$

if $K = 50$ and a flask is charged with enough Al_2Cl_6 to exert an initial partial pressure of 0.050 atm. Assume that no other reactions take place.

34. (See Example 7–7.) The reaction

$$3\ S_8(g) \rightleftharpoons 4\ S_6(g)$$

has $K = 1.6 \times 10^{-4}$ at 500 K. Suppose that a container initially holds 0.225 atm of $S_8(g)$ at 500 K and that this is the only reaction that takes place. Compute the partial pressure of $S_8(g)$ when equilibrium is reached.

35. For the mixtures of gases in Problem 17, compute the initial concentration of each gas in moles per liter.

36. For the mixtures of gases in Problem 18, compute the initial concentration of each gas in moles per liter.

37. (See Example 7–8.) Calculate the concentration (in mol L^{-1}) of phosgene ($COCl_2$) that is present at 600°C in equilibrium with carbon monoxide (at a concentration of 3.2×10^{-4} mol L^{-1}) and chlorine (at a concentration of 1.3×10^{-2} mol L^{-1}). (Use the data of Problem 23.)

38. (See Example 7–8.) Calculate the concentration (in mol L^{-1}) of SO_2 that is present at 100°C in equilibrium with SO_2Cl_2 (at a concentration of 2.5×10^{-4} mol L^{-1}) and chlorine (at a concentration of 5.7×10^{-3} mol L^{-1}). (Use the data of Problem 25.)

The Effect of External Stresses on Equilibria: Le Châtelier's Principle

39. (See Example 7–9.) Some $H_2(g)$ at a partial pressure of 5.73 atm is mixed with $I_2(g)$ at that same initial partial pressure, and the two react:

$$H_2(g) + I_2(g) \rightleftharpoons 2\ HI(g)$$

The equilibrium constant of this reaction equals 53.7 at the temperature of the experiment. After the reaction comes to equilibrium, the partial pressure of $HI(g)$ is suddenly reduced by 1.50 atm. Subsequently, the equilibrium re-establishes itself. Compute the partial pressure of the $HI(g)$ when the equilibrium is re-established.

40. (See Example 7–9.) Some $H_2(g)$ at a partial pressure of 5.73 atm is mixed with $I_2(g)$ at that same initial partial pressure, and the two react:

$$H_2(g) + I_2(g) \rightleftharpoons 2\,HI(g)$$

The equilibrium constant of this reaction equals 53.7 at the temperature of the experiment. After the reaction comes to equilibrium, the experimenter increases the partial pressure of $H_2(g)$ by 1.00 atm. Subsequently, the equilibrium re-establishes itself. Compute the partial pressure of the $HI(g)$ when the equilibrium is re-established.

41. (See Example 7–10.) The following reaction is exothermic:

$$3\,NO(g) \rightleftharpoons N_2O(g) + NO_2(g)$$

Explain the effect of each of the following stresses on the position of the equilibrium.
(a) $NO(g)$ is added to the equilibrium mixture without change of volume or temperature.
(b) The volume of the equilibrium mixture is increased at constant temperature.
(c) The temperature of the equilibrium mixture is increased.
(d) Gaseous argon (which does not react) is added to the equilibrium mixture while both the total gas pressure and the temperature are kept constant.
(e) Gaseous argon is added to the equilibrium mixture without changing the volume.

42. (See Example 7–10.) The following reaction is endothermic:

$$SO_3(g) \rightleftharpoons SO_2(g) + \tfrac{1}{2}\,O_2(g)$$

Explain the effect of each of the following stresses on the position of equilibrium.
(a) $SO_3(g)$ is added to the equilibrium mixture without a change of the volume or temperature.
(b) The mixture is expanded at constant temperature.
(c) The equilibrium mixture is heated.
(d) An inert gas is pumped into the equilibrium mixture while the total gas pressure and the temperature are kept constant.
(e) An inert gas is added to the equilibrium mixture without changing the volume.

43. (See Example 7–10.) The most extensively used organic compound in the chemical industry is ethylene (C_2H_4). The two equations

$$C_2H_4(g) + Cl_2(g) \rightleftharpoons C_2H_4Cl_2(g)$$
$$C_2H_4Cl_2(g) \rightleftharpoons C_2H_3Cl(g) + HCl(g)$$

represent the process by which vinyl chloride (C_2H_3Cl) is synthesized for eventual use in polymeric plastics (polyvinyl chloride, PVC). The by-product of the second reaction, HCl,

is now most inexpensively made by this and similar reactions, rather than by the direct combination of H_2 and Cl_2. Heat is given off in the first reaction and taken up in the second. Describe how you would design an industrial process to maximize the yield of vinyl chloride.

44. (See Example 7–10.) Methanol is made from synthesis gas via the exothermic reaction

$$CO(g) + 2\,H_2(g) \rightleftharpoons CH_3OH(g)$$

Describe how you would control the temperature and pressure to maximize the yield of methanol.

45. In a gas-phase reaction, it is observed that the equilibrium yield of products is increased by lowering the temperature and by reducing the volume.
(a) Is the reaction exothermic or endothermic?
(b) Is there a net increase or a net decrease in the number of gas molecules in the reaction?

46. The equilibrium constant of a gas-phase reaction is observed to increase as the temperature is increased. When the non-reacting gas neon is admitted to the reaction mixture (holding the temperature and the total pressure fixed and increasing the volume of the reaction vessel), the product yield is observed to decrease.
(a) Is the reaction exothermic or endothermic?
(b) Is there a net increase or a net decrease in the number of gas molecules in the reaction?

Heterogeneous Equilibrium

47. (See Example 7–11.) Using the law of mass action, write the equilibrium-constant expression for each of the following reactions:
(a) $8\,H_2(g) + S_8(s) \rightleftharpoons 8\,H_2S(g)$
(b) $C(s) + H_2O(\ell) + Cl_2(g) \rightleftharpoons COCl_2(g) + H_2(g)$
(c) $CaCO_3(s) \rightleftharpoons CaO(s) + CO_2(g)$
(d) $3\,C_2H_2(g) \rightleftharpoons C_6H_6(\ell)$

48. (See Example 7–11.) Using the law of mass action, write the equilibrium-constant expression for each of the following reactions:
(a) $3\,C_2H_2(g) + 3\,H_2(g) \rightleftharpoons C_6H_{12}(\ell)$
(b) $CO_2(g) + C(s) \rightleftharpoons 2\,CO(g)$
(c) $CF_4(g) + 2\,H_2O(\ell) \rightleftharpoons CO_2(g) + 4\,HF(g)$
(d) $K_2NiF_6(s) + TiF_4(s) \rightleftharpoons K_2TiF_6(s) + NiF_2(s) + F_2(g)$

49. (See Example 7–11.) Using the law of mass action, write the equilibrium-constant expression for each of the following reactions:
(a) $Zn(s) + 2\,Ag^+(aq) \rightleftharpoons Zn^{2+}(aq) + 2\,Ag(s)$
(b) $VO_4^{3-}(aq) + H_2O(\ell) \rightleftharpoons VO_3(OH)^{2-}(aq) + OH^-(aq)$
(c) $2\,As(OH)_6^{3-}(aq) + 6\,CO_2(g) \rightleftharpoons$
$$As_2O_3(s) + 6\,HCO_3^-(aq) + 3\,H_2O(\ell)$$

50. (See Example 7–11.) Using the law of mass action, write the equilibrium-constant expression for each of the following reactions:
(a) $6\,I^-(aq) + 2\,MnO_4^-(aq) + 4\,H_2O(\ell) \rightleftharpoons$
$$3\,I_2(aq) + 2\,MnO_2(s) + 8\,OH^-(aq)$$

(b) $2\,Cu^{2+}(aq) + 4\,I^-(aq) \rightleftharpoons 2\,CuI(s) + I_2(aq)$

(c) $\frac{1}{2}\,O_2(g) + Sn^{2+}(aq) + 3\,H_2O(\ell) \rightleftharpoons$
$$SnO_2(s) + 2\,H_3O^+(aq)$$

51. N_2O_4 is soluble in the solvent cyclohexane; however, dissolution does not prevent N_2O_4 from breaking down to give NO_2 according to the equation

$$N_2O_4(cyclohexane) \rightleftharpoons 2\,NO_2(cyclohexane)$$

An effort to compare this solution equilibrium to the similar equilibrium in the gas phase gave the following experimental data at 20°C:

$[N_2O_4]$ (mol L^{-1})	$[NO_2]$ (mol L^{-1})
0.190×10^{-3}	2.80×10^{-3}
0.686×10^{-3}	5.20×10^{-3}
1.54×10^{-3}	7.26×10^{-3}
2.55×10^{-3}	10.4×10^{-3}
3.75×10^{-3}	11.7×10^{-3}
7.86×10^{-3}	17.3×10^{-3}
11.9×10^{-3}	21.0×10^{-3}

(a) Graph the *square* of the concentration of NO_2 versus the concentration of N_2O_4.
(b) Compute the average equilibrium constant of this reaction.

52. NO_2 is soluble in carbon tetrachloride (CCl_4). As it dissolves, it dimerizes to give N_2O_4 according to the equation

$$2\,NO_2(CCl_4) \rightleftharpoons N_2O_4(CCl_4)$$

A study of this equilibrium gave the following experimental data at 20°C:

$[N_2O_4]$ (mol L^{-1})	$[NO_2]$ (mol L^{-1})
0.192×10^{-3}	2.68×10^{-3}
0.721×10^{-3}	4.96×10^{-3}
1.61×10^{-3}	7.39×10^{-3}
2.67×10^{-3}	10.2×10^{-3}
3.95×10^{-3}	11.0×10^{-3}
7.90×10^{-3}	16.6×10^{-3}
11.9×10^{-3}	21.4×10^{-3}

(a) Graph the concentration of N_2O_4 versus the *square* of the concentration of NO_2.
(b) Compute the average equilibrium constant of this reaction.

53. The "water gas" reaction

$$C(s) + H_2O(g) \rightleftharpoons CO(g) + H_2(g)$$
$$K = 2.6 \text{ at } 1000 \text{ K}$$

is industrially important. Calculate the reaction quotient Q for each of the following conditions (all partial pressures are given in atm), and state which direction the reaction shifts in coming to equilibrium at 1000 K:

(a) $P_{H_2O} = 0.600$; $P_{CO} = 1.525$; $P_{H_2} = 0.805$
(b) $P_{H_2O} = 0.724$; $P_{CO} = 1.714$; $P_{H_2} = 1.383$

54. The reaction

$$H_2S(g) + I_2(g) \rightleftharpoons 2\,HI(g) + S(s)$$
$$K = 0.0023 \text{ at } 110°C$$

is being studied as a means of removing foul-smelling $H_2S(g)$ from a waste stream. Calculate the reaction quotient Q for each of the following conditions, and determine whether solid sulfur is consumed or produced as the reaction comes to equilibrium at 110°C:

(a) $P_{I_2} = 0.461$ atm; $P_{H_2S} = 0.050$ atm; $P_{HI} = 0.0$ atm
(b) $P_{I_2} = 0.461$ atm; $P_{H_2S} = 0.050$ atm; $P_{HI} = 9.0$ atm

55. (See Example 7–12.) Pure solid NH_4HSe is placed in an evacuated container at 24.8°C. It reacts according to

$$NH_4HSe(s) \rightleftharpoons NH_3(g) + H_2Se(g)$$

and this is the only reaction that takes place. The equilibrium pressure in the container is 0.0184 atm.
(a) Calculate the equilibrium constant of this reaction at 24.8°C.
(b) In a similar container, the partial pressure of $NH_3(g)$ in equilibrium with $NH_4HSe(s)$ at 24.8°C is 0.0252 atm. What is the partial pressure of $H_2Se(g)$?

56. (See Example 7–12.) $NaHCO_3$ is used in dry chemical fire extinguishers because the products of its decomposition smother the fires.

$$2\,NaHCO_3(s) \rightleftharpoons Na_2CO_3(s) + H_2O(g) + CO_2(g)$$

(a) Calculate the equilibrium constant at 110°C if the total pressure of the gases in equilibrium with solid sodium hydrogen carbonate at 110°C is 1.648 atm. Assume that all of the gases come from the decomposition.
(b) What is the partial pressure of water vapor in equilibrium with $NaHCO_3(s)$ at 110°C if the partial pressure of $CO_2(g)$ is 0.800 atm?

57. The reaction

$$CO_2(g) + H_2(g) \rightleftharpoons H_2O(\ell) + CO(g)$$
$$K = 3.22 \times 10^{-4} \text{ at } 25°C$$

is under study. Compute the initial partial pressures of CO_2 and H_2 that must be mixed to produce CO at a pressure of 0.100 atm if the pressures of CO_2 and H_2 are required to be equal. Assume that all of the gases obey the ideal gas law and that no side-reactions occur.

58. The valuable chemical urea is formed from ammonia and carbon dioxide according to the equation

$$2\,NH_3(g) + CO_2(g) \rightleftharpoons CO(NH_2)_2(s) + H_2O(g)$$
$$K = 0.615 \text{ at } 25°C$$

What are the equilibrium partial pressures of ammonia and carbon dioxide at 25°C if crystalline urea is placed in a closed container in which water vapor is maintained at a constant pressure of 0.03126 atm?

Extraction and Separation Processes

59. An aqueous solution that contains 1.00×10^{-2} mol L^{-1} of I_2 is shaken with an equal volume of the immiscible solvent CCl_4. The iodine distributes itself between the aqueous and CCl_4 layers. When equilibrium is reached at 27°C, the concentration of I_2 in the aqueous layer is 1.30×10^{-4} mol L^{-1}. Calculate the partition coefficient K at 27°C for the reaction

$$I_2(aq) \rightleftharpoons I_2(CCl_4)$$

60. An aqueous solution that contains 2.50×10^{-2} mol L^{-1} of I_2 is shaken with an equal volume of the immiscible solvent CS_2. The iodine distributes itself between the aqueous and CS_2 layers. When equilibrium is reached at 25°C, the concentration of I_2 in the aqueous layer is 4.16×10^{-5} mol L^{-1}. Calculate the partition coefficient K at 25°C for the reaction

$$I_2(aq) \rightleftharpoons I_2(CS_2)$$

61. Benzoic acid (C_6H_5COOH) dissolves in water to the extent of 2.00 g L^{-1} at 15°C and in diethyl ether to the extent of 6.6×10^2 g L^{-1} at the same temperature.
    (a) Calculate the equilibrium constants at 15°C for the two reactions

$$C_6H_5COOH(s) \rightleftharpoons C_6H_5COOH(aq)$$

and

$$C_6H_5COOH(s) \rightleftharpoons C_6H_5COOH(ether)$$

    (b) From your answers to part (a), calculate the partition coefficient K for the reaction

$$C_6H_5COOH(aq) \rightleftharpoons C_6H_5COOH(ether)$$

62. Citric acid ($C_6H_8O_7$) dissolves in water to the extent of 720 g L^{-1} at 15°C and in diethyl ether to the extent of 22 g L^{-1} at the same temperature.
    (a) Calculate the equilibrium constants at 15°C for the two reactions

$$C_6H_8O_7(s) \rightleftharpoons C_6H_8O_7(aq)$$

and

$$C_6H_8O_7(s) \rightleftharpoons C_6H_8O_7(ether)$$

    (b) From your answers to part (a), calculate the partition coefficient K for the reaction

$$C_6H_8O_7(aq) \rightleftharpoons C_6H_8O_7(ether)$$

ADDITIONAL PROBLEMS

63. At 298 K, unequal amounts of $BCl_3(g)$ and $BF_3(g)$ were mixed in a container. The gases reacted to form $BFCl_2(g)$ and $BClF_2(g)$. When equilibrium was finally reached, the four gases were present in these relative chemical amounts:

    $BCl_3(90)$, $BF_3(470)$, $BClF_2(200)$, $BFCl_2(45)$.

    (a) Determine the equilibrium constants at 298 K for the two reactions

$$2\,BCl_3(g) + BF_3(g) \rightleftharpoons 3\,BFCl_2(g)$$
$$BCl_3(g) + 2\,BF_3(g) \rightleftharpoons 3\,BClF_2(g)$$

    (b) Determine the equilibrium constant of the reaction at 298 K

$$BCl_3(g) + BF_3(g) \rightleftharpoons BFCl_2(g) + BClF_2(g)$$

    and explain why knowing this equilibrium constant really adds nothing to what you knew in part (a).

64. Methanol can be synthesized by means of the reaction

$$CO(g) + 2\,H_2(g) \rightleftharpoons CH_3OH(g)$$

    for which the equilibrium constant at 225°C is 6.08×10^{-3}. Assume that the ratio of the pressures of $CO(g)$ and $H_2(g)$ is $1:2$. What values must they have for the partial pressure of methanol to equal 0.500 atm?

65. At equilibrium at 425.6°C, a sample of gaseous *cis*-1-methyl-2-ethylcyclopropane is 73.6% converted into the *trans* form, which is also gaseous, according to

$$cis \rightleftharpoons trans$$

    (a) Compute the equilibrium constant K for this reaction.
    (b) Suppose that 0.525 mol of the *cis* compound is placed in a 15.00 L vessel and heated to 425.6°C. Compute the equilibrium partial pressure of the *trans* compound.

66. The equilibrium constant for the reaction

$$(CH_3)_3COH(g) \rightleftharpoons (CH_3)_2CCH_2(g) + H_2O(g)$$

    is equal to 2.42 at 450 K.
    (a) A pure sample of the reactant, which is named "*tert*-butanol," is confined in a container of fixed volume at a temperature of 450 K and at an original pressure of 0.100 atm. Calculate the fraction of this starting material that is converted to products at equilibrium.
    (b) A second sample of the reactant is confined, this time at an original pressure of 5.00 atm. Again, calculate the fraction of the starting material that is converted to products at equilibrium.

67. Consider the reaction

$$2\,SO_2(g) + O_2(g) \rightleftharpoons 2\,SO_3(g)$$

    The equilibrium constant of this reaction rises from 0.25 at 1100 K to 0.70 at 523 K. A container holds a mixture of SO_2 and O_2, initially at partial pressures of 2.0 and 3.0 atm, respectively.
    (a) Find the equilibrium partial pressure of SO_3 in the flask if it is heated to 1100 K and its contents come to equilibrium at that temperature.
    (b) Repeat the calculation for a temperature of 523 K.
    (c) Show that the percentage of SO_2 that has reacted is greater at 523 K than it is at 1100 K.

68. At elevated temperatures, phosphorus pentachloride decomposes to give phosphorus trichloride and chlorine:

$$PCl_5(g) \rightleftharpoons PCl_3(g) + Cl_2(g)$$

    An equilibrium mixture of the three gases, obtained from the decomposition of initially pure phosphorus pentachloride in a vessel of fixed volume, has a density of 1.46 g L^{-1} at 523 K and 0.569 atm. Find the value of K for the reaction at 523 K.

69. Quantities of $I_2(g)$, $Br_2(g)$, and $IBr(g)$ are mixed at 400 K in such a way that their initial partial pressures are 2.00 atm, 4.00 atm, and 3.00 atm, respectively. They then react

$$I_2(g) + Br_2(g) \rightleftharpoons 2\,IBr(g) \qquad K = 131 \text{ at } 400 \text{ K}$$

No other reactions occur. Calculate the partial pressures of the three gases when equilibrium is reached.

70. Acetic acid in the vapor phase at 110°C consists of both monomeric and dimeric forms in equilibrium:

$$2\,CH_3COOH(g) \rightleftharpoons (CH_3COOH)_2(g)$$
$$K = 3.72 \text{ at } 100°C$$

This is the only reaction that takes place in the vapor under the conditions.
(a) Calculate the partial pressure of the dimer if the total pressure is 0.725 atm at equilibrium.
(b) What percentage of the acetic acid is dimerized?

71. Much of the content of Le Châtelier's principle can be understood by using the reaction quotient Q. Consider, for example, the reaction used to make cyclohexane (C_6H_{12}), a starting material in the production of nylon. In the presence of a small amount of nickel or platinum, benzene and hydrogen react rapidly to produce cyclohexane.

$$C_6H_6(g) + 3\,H_2(g) \rightleftharpoons C_6H_{12}(g)$$

(a) Write the expression for the reaction quotient Q for this reaction, and then rewrite it (using the ideal gas law) in terms of numbers of moles, volume, and temperature.
(b) The reaction comes to equilibrium. Then, a small chemical amount of H_2 is added while the volume and temperature are held fixed. Does Q become larger or smaller than K? In which direction does the reaction shift to restore equilibrium?
(c) The reaction comes to a fresh equilibrium. Then, the temperature is held fixed and the volume is decreased. Does Q become larger or smaller than K? In which direction does the reaction shift to restore equilibrium?

72. Reconsider the equilibrium

$$SO_2Cl_2(g) \rightleftharpoons SO_2(g) + Cl_2(g) \qquad K = 2.40 \text{ at } 100°C$$

which appears in Problems 11 and 25. Find the equilibrium partial pressure of $SO_2Cl_2(g)$ at 100°C if this is the sole reaction in a tank that is charged with an initial partial pressure of $SO_2Cl_2(g)$ equal, in three successive experiments, to:
(a) 0.500 atm, 1.00 atm, and 1.50 atm.
(b) How does the fraction of SO_2Cl_2 decomposed vary as the initial amount of SO_2Cl_2 increases?
(c) Explain the trend in terms of Le Châtelier's principle.

*73. Chlorine is only slightly soluble in water. Under a pressure of 1.00 atm of $Cl_2(g)$ at 298 K, 1.00 L of water at equilibrium dissolves just 0.091 mol of Cl_2.

$$Cl_2(g) \rightleftharpoons Cl_2(aq)$$

In such solutions, the concentration of $Cl_2(aq)$ concentration is 0.061 mol L^{-1}, and the concentrations of $Cl^-(aq)$ and $HOCl(aq)$ are both 0.030 mol L^{-1}. These two additional species are formed by the reaction

$$Cl_2(aq) + H_2O(\ell) \rightleftharpoons H^+(aq) + Cl^-(aq) + HOCl(aq)$$

There are no other Cl^--containing species. Compute the equilibrium constants for the two reactions at 298 K.

74. At 400°C, the reaction

$$BaO_2(s) + 4\,HCl(g) \rightleftharpoons BaCl_2(s) + 2\,H_2O(g) + Cl_2(g)$$

has an equilibrium constant equal to K_1. How is K_1 related to the equilibrium constant K_2 of the reaction

$$2\,Cl_2(g) + 4\,H_2O(g) + 2\,BaCl_2(s) \rightleftharpoons$$
$$8\,HCl(g) + 2\,BaO_2(s)$$

at 400°C?

75. Ammonium hydrogen sulfide, a solid, decomposes to give $NH_3(g)$ and $H_2S(g)$. At 25°C, some $NH_4HS(s)$ is placed in an evacuated container. A portion of it decomposes, and the total pressure at equilibrium is 0.659 atm. Extra $NH_3(g)$ is then injected into the container and, when equilibrium is reestablished, the partial pressure of $NH_3(g)$ is 0.750 atm.
(a) Compute the equilibrium constant for the decomposition of ammonium hydrogen sulfide.
(b) Determine the final partial pressure of $H_2S(g)$ in the container.
(c) Additional $NH_4HS(s)$ is added to the container with the intention of improving the yield of $H_2S(g)$ by shifting the equilibrium to the right. The partial pressure of $H_2S(g)$, however, does not change from the value calculated in part (b). Explain why.

76. The following reaction lies far to the right at equilibrium

$$KOH(s) + CO_2(g) \rightleftharpoons KHCO_3(s)$$
$$K = 6 \times 10^{15} \text{ at } 25°C$$

(a) Suppose that 7.32 g of $KOH(s)$ and 9.41 g of $KHCO_3(s)$ are placed in a closed 2.00 L evacuated container and allowed to reach equilibrium. Calculate the pressure of $CO_2(g)$ at equilibrium.
(b) An additional 0.50 g of $KOH(s)$ is added to the container after the equilibrium in part (a) is reached. Describe the effect of this addition on the pressure of $CO_2(g)$ in the container.

77. Carbon monoxide reduces nickel(II) oxide to nickel according to

$$NiO(s) + CO(g) \rightleftharpoons Ni(s) + CO_2(g)$$
$$K = 255.4 \text{ at } 754°C$$

If the total pressure of the system at 754°C is 2.50 atm, calculate the partial pressures of $CO(g)$ and $CO_2(g)$. No other reactions occur.

78. Both glucose (corn sugar) and fructose (fruit sugar) taste sweet, but fructose tastes sweeter. Each year in the United

States, tons of corn syrup destined to sweeten food are treated to convert glucose as fully as possible to the sweeter fructose. The reaction reaches this equilibrium:

$$glucose(aq) \rightleftharpoons fructose(aq)$$

(a) A solution containing glucose at a concentration of 0.2564 mol L^{-1} is treated at 25°C with an enzyme (catalyst) that causes equilibrium to be reached quickly. The final concentration of fructose is 0.1175 mol L^{-1}. In another experiment at the same temperature, a 0.2666 mol L^{-1} solution of pure fructose is treated with the same enzyme, and the final concentration of glucose is 0.1415 mol L^{-1}. Compute an average equilibrium constant for the above reaction.

(b) What percentage of glucose is converted to fructose at equilibrium at 25°C?

79. Stearic acid tends to dimerize when dissolved in hexane:

$$2 \, C_{17}H_{35}COOH(hexane) \rightleftharpoons (C_{17}H_{35}COOH)_2(hexane)$$

The equilibrium constant for this reaction is 40 at 48°C. Stearic acid (15.0 g) is dissolved in 1.250 L of hexane, and the dimerization comes to equilibrium at 48°C. Calculate the concentrations of both the monomer and the dimer.

80. Superheated steam ($H_2O(g)$) at a pressure of 22.2 atm is pumped into an iron vessel at 800 K (527°C). The iron and steam react

$$3 \, Fe(s) + 4 \, H_2O(g) \rightleftharpoons Fe_3O_4(s) + 4 \, H_2(g)$$
$$K = 12.3 \text{ at } 800 \text{ K}$$

If the steam is given enough time to come to equilibrium with the walls of the vessel, what partial pressure of hydrogen is generated?

81. Polychlorinated biphenyls (PCBs) are a major environmental problem. These oily substances have many uses, but they resist breakdown by bacterial action when spilled in the environment and, being fat-soluble, can accumulate to dangerous concentrations in the fatty tissues of fish and other animals. One little-appreciated complication in controlling the problem is that there are 209 different PCBs, all now in the environment. The various PCBs are generally similar, but their solubilities in fats differ considerably. The best measure of this is K_{OW}, the equilibrium constant for the partition of a PCB between the fat-like solvent octanol and water:

$$PCB(aq) \rightleftharpoons PCB(octanol)$$

An equimolar mixture of PCB-2 and PCB-11 in water is treated with an equal volume of octanol. Determine the ratio between the amounts of PCB-2 and PCB-11 in the water at equilibrium. At room temperature, K_{OW} is 3.98×10^4 for PCB-2 and 1.26×10^5 for PCB-11.

82. Refer to the data in Problems 61 and 62. Suppose that 2.00 g of a solid consisting of 50.0% benzoic acid and 50.0% citric acid by mass is added to 100.0 mL of water and 100.0 mL of diethyl ether and that the whole assembly is shaken. When the immiscible layers are separated and the solvents are removed by evaporation, two solids result. Calculate the percentages (by mass) of the major component in each solid.

83. At 25°C, the partition coefficient for the equilibrium

$$I_2(aq) \rightleftharpoons I_2(CCl_4)$$

equals 85. To 0.100 L of an aqueous solution that contains 2×10^{-3} mol L^{-1} of I_2, we add 0.025 L of CCl_4. The mixture is shaken in a separatory funnel, allowed to separate into two phases, and the CCl_4 phase is withdrawn.

(a) Calculate the fraction of the I_2 remaining in the aqueous phase.

(b) The 0.100 L aqueous phase is equilibrated with a fresh 0.025 L of CCl_4 and again separated. What fraction of the I_2 from the original aqueous solution is now in the aqueous phase?

(c) Compare your answer with that of Example 7–13, in which the same total amount of CCl_4 (0.050 L) was used in a single extraction. For a given total amount of extracting solvent, which is the more efficient way to remove iodine from water?

CUMULATIVE PROBLEM

Sulfuric Acid

Sulfuric acid is produced in larger volume than any other chemical. Its uses range from fertilizer manufacture to metal treatment and the synthesis of organic and medicinal chemicals. Its production and properties are discussed in detail in Chapter 22.

The modern industrial production of sulfuric acid takes place in three steps, for which the balanced chemical equations are

1. $S(s) + O_2(g) \rightleftharpoons SO_2(g)$

2. $SO_2(g) + \frac{1}{2} O_2(g) \rightleftharpoons SO_3(g)$

3. $SO_3(g) + H_2O(\ell) \rightleftharpoons H_2SO_4(\ell)$

Sulfur mined for conversion to sulfuric acid. *(David Nunuk/Science Photo Library/Photo Researchers, Inc.)*

(a) Write equilibrium-constant expressions for each of these three steps, with equilibrium constants K_1, K_2, and K_3.

(b) If these reactions could be carried out at 25°C, the equilibrium constants would be 3.9×10^{52}, 2.6×10^{12}, and 2.2×10^{14}. Write a balanced equation for the overall reaction, and calculate its equilibrium constant at 25°C.

(c) Although the products of all three reactions are strongly favored at 25°C (part (b)), reactions 1 and 2 occur too slowly to be practical; they must be carried out at elevated temperatures. At 700°C, the partial pressures of SO_2, O_2, and SO_3 in an equilibrium mixture are measured to be 2.23 atm, 1.14 atm, and 6.26 atm, respectively. Calculate K_2 at 700°C.

(d) At 600°C, K_2 has the value 9.5. Is reaction 2 exothermic or endothermic?

(e) Some SO_2 is placed in a flask and heated with oxygen to 600°C. At equilibrium, 62% of it has reacted to give SO_3. Calculate the partial pressure of oxygen at equilibrium in this reaction mixture.

(f) Equal chemical amounts of SO_2, O_2, and SO_3 are mixed and heated quickly to 600°C. Their total pressure before reaction is 0.090 atm. Does reaction 2 occur from right to left or from left to right? Does the total pressure increase or decrease during the course of the reaction?

(g) Reactions 1 and 3 are both exothermic. State the effects on equilibria 1 and 3 of an increase in temperature and of a decrease in volume. (*Note*: A change in volume has little effect on liquids and solids taking part in a reaction.)

8

Acid–Base Equilibria

CHAPTER OUTLINE

Fabrication of a specialized electrode for use in pH measurements. *(Courtesy of Orion Research, Inc.)*

Acid–base reactions comprise one of the major categories of reactivity established in Chapter 4. This chapter shows how the ideas about chemical equilibrium that are developed in Chapter 7 are applied to these reactions.

The focus in this chapter is on acid–base reactions in aqueous solutions, despite the fact that numerous acid–base reactions take place in the gaseous and solid states and in non-aqueous solutions. There are two reasons. First, aqueous acids and bases are overwhelmingly important in the daily events of life, from the properties of vinegar or ammonia-containing cleansers to the biochemical roles of the amino acids in proteins and the nucleic acids in genes. Second, acid–base reactions in aqueous solutions reach equilibrium rapidly (within moments). One need never wait for such a system to settle down to equilibrium.

8–1 Brønsted-Lowry Acids and Bases

Pure water contains equal, small concentrations of hydrogen ions ($H^+(aq)$) and hydroxide ions ($OH^-(aq)$) generated by autoionization

$$H_2O(\ell) \rightleftharpoons H^+(aq) + OH^-(aq)$$

Acidic solutions have higher concentrations of hydrogen ions than does pure water; basic solutions have higher concentrations of hydroxide ions. Late in the 19th century, Svante Arrhenius for this reason defined acids as $H^+(aq)$ donors and bases as $OH^-(aq)$ donors (see Section 4–3). However, many substances that contain neither hydrogen nor oxygen (and that cannot donate H^+ or OH^-) are found to give acidic or basic solutions. For example, aqueous $CuCl_2$ is acidic and aqueous NaCN is basic. Such observations prompted a broadening of the original definition. Acids were redefined as substances that increase the concentration of hydrogen ions in water and bases were redefined as substances that increase the concentration of hydroxide ions in water. The increases might come directly, by release of H^+ or OH^-, or indirectly, through reaction with water, as in

$$Cu^{2+}(aq) + H_2O(\ell) \rightleftharpoons CuOH^+(aq) + H^+(aq)$$

and

$$CN^-(aq) + H_2O(\ell) \rightleftharpoons HCN(aq) + OH^-(aq)$$

The enlarged definition was still not entirely satisfactory. It did not apply in non-aqueous solvents or in the absence of a solvent. It offered little insight as to *why*, in terms of bonding and structure, some substances are acidic and others basic. It provided no good basis for comparing the relative strengths of acids and bases. Two new definitions, both proposed in the early 1920s, helped deal with these issues.

The first, the Brønsted-Lowry definition (named for its originators, Johannes Brønsted and Thomas Lowry, who worked independently) identifies the transfer of H^+ ions from acid to base as the essence of the acid–base reaction. The second, the Lewis definition (named for *its* originator, Gilbert Lewis), identifies bases as electron-pair donors and acids as electron-pair acceptors and focuses on the making and breaking of bonds by the donation and acceptance of electron pairs. We take up the Brønsted-Lowry concept now and the Lewis concept in Section 8–8.

The Brønsted-Lowry definition of acid and base has proved extraordinarily helpful in organizing information about the relative strengths of acids and bases. It states:

A **Brønsted-Lowry acid** is a substance that can donate a hydrogen ion; a **Brønsted-Lowry base** is a substance that can accept a hydrogen ion.

Acids and bases occur as **conjugate acid–base pairs** in the Brønsted-Lowry picture. The formula of a conjugate base is obtained by subtracting a hydrogen ion (H^+) from the formula of the acid, and the formula of a conjugate acid is obtained by adding H^+ to the formula of the base. For example

$$OH^- \qquad\qquad H_2O \qquad\qquad H_3O^+$$

conjugate base of H_2O conjugate acid of OH^- conjugate acid of H_2O

and

conjugate base of H_3O^+

- Brønsted-Lowry acids and bases are very frequently called **proton donors** and **proton acceptors.** This is somewhat inexact. A proton is an $^1H^+$ ion, but ordinary hydrogen contains ^2H, as well as ^1H.

The conjugate acid of water, H_3O^+, is called the **oxonium** or **hydronium** ion. It consists of three H atoms bonded to a central O atom with a net charge of $+1$.

Acid–base reactions in the Brønsted-Lowry scheme consist of the transfer of an H^+ ion from donor to acceptor. The products are the conjugate base of the donor (the acid) and the conjugate acid of the acceptor (the base). For example, as acetic acid (CH_3COOH) dissolves in water, H_2O acts as a Brønsted-Lowry base to accept a hydrogen ion donated by the acetic acid:

$$\underset{\text{acid}_1}{CH_3COOH(aq)} + \underset{\text{base}_2}{H_2O(\ell)} \rightleftharpoons \underset{\text{acid}_2}{H_3O^+(aq)} + \underset{\text{base}_1}{CH_3COO^-(aq)}$$

The subscripts 1 and 2 designate the two conjugate acid–base pairs. The reaction creates a new acid, hydronium ion, and a new base, acetate ion (CH_3COO^-). Depending on the desired emphasis, one can say that CH_3COO^- is the conjugate base of CH_3COOH or that CH_3COOH is the conjugate acid of CH_3COO^-.

The autoionization of water fits the Brønsted-Lowry definition. To see this, consider the reaction

$$\underset{\text{acid}_1}{H_2O(\ell)} + \underset{\text{base}_2}{H_2O(\ell)} \rightleftharpoons \underset{\text{acid}_2}{H_3O^+(aq)} + \underset{\text{base}_1}{OH^-(aq)}$$

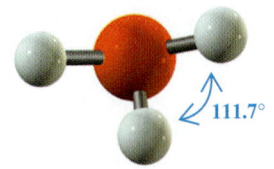

The tripod-like structure of the hydronium ion (H_3O^+).

in which one H_2O molecule functions as an acid and the other as a base. Now, compare the formula $H_3O^+(aq)$ on the right side of this equation with the formula $H^+(aq)$. The first singles out one molecule of water among the several that aquate the H^+ and shows it explicitly. The second lumps all the aquating molecules into the "(aq)." Both represent the same thing, a dissolved hydrogen ion. Hence, the preceding equation is equivalent to

$$H_2O(\ell) \rightleftharpoons H^+(aq) + OH^-(aq)$$

The H_3O^+ notation is usually used when discussing the Brønsted-Lowry theory because it highlights transfers of H^+ to H_2O, a common event. At other times, the briefer notation $H^+(aq)$ is preferred.

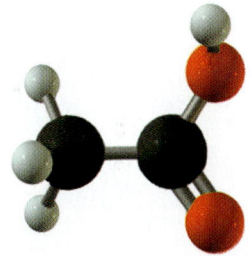

Acetic acid (CH_3COOH) is a weak organic acid. It is the active ingredient in vinegar, which is a 3% to 5% solution of acetic acid in water. Pure acetic acid is called "glacial acetic acid" because it freezes easily (at 16.7°C) into layered crystals.

EXAMPLE 8–1

When carbon dioxide is dissolved in water to make a carbonated beverage, it reacts with the water to form carbonic acid (H_2CO_3). Give the name and formula of the conjugate base of H_2CO_3.

Solution

Conjugate bases are formed by the removal of a hydrogen ion. Subtracting H^+ from H_2CO_3 leaves HCO_3^-. This is the hydrogen carbonate ion.

EXERCISE

Trimethylamine (C_3H_9N) is a soluble weak base with a foul odor (it contributes to the smell of rotten fish). Write the formula of its conjugate acid.

Answer: The conjugate acid is $C_3H_9NH^+$.

In a Brønsted-Lowry acid–base reaction, a hydrogen ion is transferred from an acid to a base. A fruitful alternative view of this interaction is that it is a competition between two bases for a hydrogen ion. In the reaction of ammonia with water,

$$\underset{\text{acid}_1}{H_2O(\ell)} + \underset{\text{base}_2}{NH_3(aq)} \rightleftharpoons \underset{\text{acid}_2}{NH_4^+(aq)} + \underset{\text{base}_1}{OH^-}$$

the two bases $NH_3(aq)$ and $OH^-(aq)$ compete for hydrogen ions. The outcome of the competition depends on the relative strengths of the two. $NH_3(aq)$ is a weaker base than $OH^-(aq)$, as shown by the fact that the forward reaction occurs only to a limited extent before equilibrium is reached. One mole of NH_3 produces much less than 1 mol of OH^- ion in solution.

EXAMPLE 8–2

Sodium cyanide is an ionic compound that dissolves in water to give a solution that turns litmus blue. Other indicators confirm that aqueous NaCN is basic. Write equations explaining these observations.

Solution

Sodium cyanide dissociates into $Na^+(aq)$ and $CN^-(aq)$ ions as it dissolves:

$$NaCN(s) \longrightarrow Na^+(aq) + CN^-(aq)$$

The solution is basic because the $CN^-(aq)$ ion takes hydrogen ion away from the solvent according to the equation and so generates $OH^-(aq)$ ion:

$$CN^-(aq) + H_2O(\ell) \rightleftharpoons OH^-(aq) + HCN(aq)$$

Cyanide ion acts as a Brønsted-Lowry base in this reaction.

EXERCISE

A solution of NH_4Cl in water gives colors with indicators that show that it is acidic. Write a net ionic equation that explains these observations.

Answer: $NH_4^+(aq) + H_2O(\ell) \rightleftharpoons NH_3(aq) + H_3O^+(aq)$

The Brønsted-Lowry approach is not limited to aqueous solutions. With liquid ammonia as the solvent, for example, we can write

$$\underset{\text{acid}_1}{HCl(am)} + \underset{\text{base}_2}{NH_3(\ell)} \rightleftharpoons \underset{\text{acid}_2}{NH_4^+(am)} + \underset{\text{base}_1}{Cl^-(am)}$$

NH_3 acts here as a base, even though the hydroxide ion OH^- is not present.

• The "(*am*)" means solvated by ammonia molecules.

Another advantage of the Brønsted-Lowry approach is the insight it gives into the phenomenon of **amphoterism.** An amphoteric molecule or ion can function as either an acid or a base, depending on reaction conditions (see Section 4–3). A good example is water itself. In the presence of ammonia, water serves as an acid, donating a hydrogen ion to NH_3; in the presence of acetic acid, it serves as a base, accepting a hydrogen ion from CH_3COOH. In the same way, the hydrogen carbonate ion can act as an acid

$$HCO_3^-(aq) + H_2O(\ell) \rightleftharpoons H_3O^+(aq) + CO_3^{2-}(aq)$$

or as a base:

$$HCO_3^-(aq) + H_2O(\ell) \rightleftharpoons H_2CO_3(aq) + OH^-(aq)$$

The Brønsted-Lowry definition of acid implies that any molecule or ion that contains hydrogen can serve as an acid if a strong enough base is available to accept (or extract) a hydrogen ion from it. Furthermore, any molecule or ion can act as a base if a strong enough acid is in the vicinity to force a hydrogen ion upon it. Thus, acetic acid acts as a base when dissolved in anhydrous sulfuric acid:

$$H_2SO_4(\ell) + CH_3COOH(sulfuric\ acid) \rightleftharpoons$$
$$CH_3COOH_2^+(sulfuric\ acid) + HSO_4^-(sulfuric\ acid)$$

$$\text{acid}_1 \qquad\qquad \text{base}_2 \qquad\qquad \text{acid}_2 \qquad\qquad \text{base}_1$$

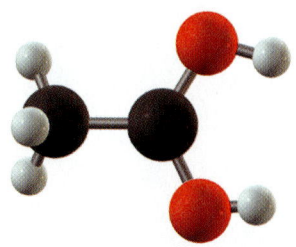

The structure of the $CH_3COOH_2^+$ ion, the product when acetic acid is forced to act as a base and accept a hydrogen ion. The H^+ ion is accepted at an oxygen atom.

8-2 Water and the pH Scale

The autoionization of water proceeds to only a slight extent under ordinary conditions, but it always happens and is responsible for a small but measurable presence of $H_3O^+(aq)$ and $OH^-(aq)$ ions in even the purest water. The equilibrium-constant expression for this reaction is

$$\frac{[H_3O^+][OH^-]}{(1)^2} = [H_3O^+][OH^-] = K_w$$

in accordance with the rules given in Section 7–6. The constant K_w is a small number because the autoionization reaction does not proceed far to the right before reaching equilibrium. Like all equilibrium constants, K_w depends on the temperature (Table 8–1).

- From this point on in this chapter, 1's representing pure solids, pure liquids, and the solvent in dilute solutions in equilibrium expressions will simply be omitted.

TABLE 8–1	Temperature Dependence of K_w
Temperature (°C)	**K_w**
0	0.114×10^{-14}
10	0.292×10^{-14}
20	0.681×10^{-14}
25	1.01×10^{-14}
30	1.47×10^{-14}
40	2.92×10^{-14}
50	5.47×10^{-14}
60	9.61×10^{-14}

Figure 8–1 • The H_3O^+–OH^- equilibrium in water operates like a see-saw: if [OH^-] is up, then [H_3O^+] is down, and vice versa.

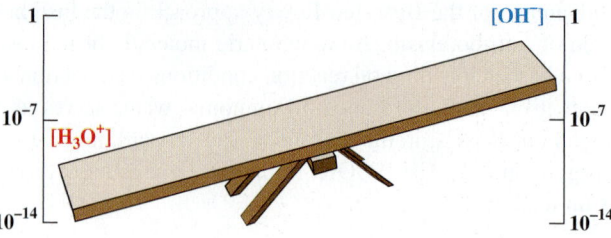

Figure 8–1 • The H_3O^+–OH^- equilibrium in water operates like a see-saw: if [OH^-] is up, then [H_3O^+] is down, and vice versa.

Its value at 25°C comes up frequently in computations:

$$K_w = 1.0 \times 10^{-14} \qquad \text{(at 25°C)}$$

Pure water contains no ions other than $H_3O^+(aq)$ and $OH^-(aq)$. Because bulk samples of water are electrically neutral, the number of positive charges they contain must equal the number of negative charges. Consequently,

$$[H_3O^+] = [OH^-] = y$$

Substituting into the equilibrium expression gives

$$y^2 = K_w = 1.0 \times 10^{-14}$$
$$y = 1.0 \times 10^{-7}$$

In pure water at 25°C, the concentrations of $H_3O^+(aq)$ and $OH^-(aq)$ ions are 1.0×10^{-7} mol L^{-1}. If something is dissolved in water that raises the concentration of $H_3O^+(aq)$ above 1.0×10^{-7} mol L^{-1}, then the concentration of $OH^-(aq)$ immediately drops below 1.0×10^{-7} mol L^{-1} so that the product [H_3O^+] × [OH^-] stays equal to 1.0×10^{-14} (Fig. 8–1).

Strong Acids and Bases

An aqueous solution is acidic if it contains an excess of H_3O^+ over OH^- ions. Acidic species generate excesses of $H_3O^+(aq)$ by reactions with water. A **strong acid** is an acid that reacts essentially completely with water to produce $H_3O^+(aq)$. Hydrochloric acid (HCl) is a strong acid. As it dissolves in water, essentially every molecule transfers a hydrogen ion to a water molecule to form $H_3O^+(aq)$:

$$\underset{\text{acid}_1}{HCl(aq)} + \underset{\text{base}_2}{H_2O(\ell)} \longrightarrow \underset{\text{acid}_2}{H_3O^+(aq)} + \underset{\text{base}_1}{Cl^-(aq)} \qquad \text{(reaction essentially complete)}$$

Perchloric acid ($HClO_4$) is another strong acid, one that is even stronger that hydrochloric acid:

$$\underset{\text{acid}_1}{HClO_4(aq)} + \underset{\text{base}_2}{H_2O(\ell)} \longrightarrow \underset{\text{acid}_2}{H_3O^+(aq)} + \underset{\text{base}_1}{ClO_4^-(aq)} \qquad \text{(reaction essentially complete)}$$

The difference in strength between HCl and $HClO_4$ cannot be observed in water solution. The reactions of these two acids with water both lie so far to the right at equilibrium that the difference between them is undetectably small. Water is said to have a **leveling effect** on strong acids. The concentration of H_3O^+ in a 0.10 mol L^{-1} solution of *any* strong acid that donates one hydrogen ion per molecule is simply

0.10 mol L^{-1}, because in every case H_2O molecules accept essentially all of the H^+ ions from the acid. The OH^- concentration is

$$[OH^-] = \frac{K_w}{[H_3O^+]} = \frac{1.0 \times 10^{-14}}{0.10} = 1.0 \times 10^{-13} \text{ mol L}^{-1}$$

A **strong base** is a base that reacts essentially completely to give $OH^-(aq)$ ion when put in water. The amide ion, NH_2^-, and the hydride ion, H^-, are both strong bases. For every mole of either of these species that is added to water, 1 mol of $OH^-(aq)$ is formed. The other products are NH_3 and H_2, respectively:

• Because H^- and NH_2^- ions are electrically charged, they are available only in combination with positive ions, as in the compounds NaH and KNH_2.

$$\underset{\text{acid}_1}{H_2O(\ell)} + \underset{\text{base}_2}{NH_2^-(aq)} \longrightarrow \underset{\text{acid}_2}{NH_3(aq)} + \underset{\text{base}_1}{OH^-(aq)} \qquad \text{(reaction essentially complete)}$$

$$\underset{\text{acid}_1}{H_2O(\ell)} + \underset{\text{base}_2}{H^-(aq)} \longrightarrow \underset{\text{acid}_2}{H_2(g)} + \underset{\text{base}_1}{OH^-(aq)} \qquad \text{(reaction essentially complete)}$$

The important base NaOH, a solid, increases the OH^- concentration in water by dissociating completely, as discussed in Section 4–3:

$$NaOH(s) \longrightarrow Na^+(aq) + OH^-(aq) \qquad \text{(reaction essentially complete)}$$

For every mole of NaOH that dissolves in water, essentially 1 mol of $OH^-(aq)$ forms. This makes NaOH a strong base, too. Strong bases are leveled in aqueous solution in the same way that strong acids are leveled, and for the same reason—the interaction between strong bases and water is so extensive that the intrinsic differences between strong bases are drowned out.

Calculating the concentration of $OH^-(aq)$ or $H_3O^+(aq)$ in a solution of a strong base or a strong acid is straightforward in any case of practical interest. If 0.10 mol of NaOH or NH_2^- ion or H^- ion is dissolved in enough water to make 1.0 L of solution, then in every case, $[OH^-] = 0.10$ mol L^{-1}. If 0.10 mol of HCl or $HClO_4$ or HBr is dissolved in enough water to make 1.0 L of solution, then in every case $[H_3O^+] = 0.10$ mol L^{-1}. The $OH^-(aq)$ or $H_3O^+(aq)$ from the autoionization of water is negligible in these practical cases. If an extremely small amount of strong base or acid were added to pure water (for example, 10^{-7} mol of base or acid per liter), then the autoionization of water would have to be taken into account as a source of $OH^-(aq)$ or $H_3O^+(aq)$.

The pH

The concentration of hydronium ion can range from approximately 10 mol L^{-1} to 10^{-15} mol L^{-1} in aqueous solutions. It is convenient to compress this enormous range by introducing a logarithmic scale for the intensity of acidity in aqueous solutions: the **pH scale**. The pH of an aqueous solution is given by the equation

$$pH = -\log_{10}[H_3O^+]$$

Pure water at 25°C has $[H_3O^+] = 1.0 \times 10^{-7}$ mol L^{-1}, so that

$$pH = -\log_{10}(1.0 \times 10^{-7}) = -(-7.00) = 7.00$$

A 0.10 M solution of HCl has $[H_3O^+] = 0.10$ mol L^{-1}, so

$$pH = -\log_{10}(0.10) = -\log_{10}(1.0 \times 10^{-1}) = -(-1.00) = 1.00$$

• The "concentration" of the H_3O^+ in this calculation is really a dimensionless ratio equal to 1.0×10^{-7} mol L^{-1} divided by a reference concentration of 1 mol L^{-1}. See Section 7–6.

whereas a 0.10 mol L^{-1} solution of NaOH has

$$pH = -\log_{10}\frac{1.0 \times 10^{-14}}{0.10} = -\log_{10}(1.0 \times 10^{-13}) = -(-13.00) = 13.00$$

- See Appendix C–3 for a discussion of significant figures when logarithms are involved.

As the examples show, calculating the pH is especially easy when the concentration of H_3O^+ is a power of 10, because the logarithm is then just the power to which 10 is raised. In other cases, a calculator is needed.

EXAMPLE 8-3

Calculate the pH (at 25°C) of an aqueous solution that has an $OH^-(aq)$ concentration of 1.2×10^{-6} mol L^{-1}.

Solution
The concentration of $H_3O^+(aq)$ in this sample is

$$[H_3O^+] = \frac{K_w}{[OH^-]} = \frac{1.0 \times 10^{-14}}{1.2 \times 10^{-6}} = 8.3 \times 10^{-9} \text{ mol L}^{-1}$$

Using a calculator, we find that

$$\log_{10}(8.3 \times 10^{-9}) = -8.08$$

The pH is the negative of this, or 8.08 .

EXERCISE
Compute the pH (at 25°C) of an aqueous solution in which the hydrogen ion concentration is twice the hydroxide ion concentration: $[H_3O^+] = 2.0 \times [OH^-]$.

Answer: pH = 6.85.

Solutions with $H_3O^+(aq)$ concentrations that exceed 1 mol L^{-1} have negative pHs; for example, if $[H_3O^+] = 2.0$ mol L^{-1} then pH $= -0.30$. However, $H_3O^+(aq)$ concentrations are most often lower than 1 mol L^{-1}, so the pH of most solutions is positive. A change in pH by one unit implies a change in the concentrations of H_3O^+ and OH^- by a factor of 10 (that is, by one order of magnitude). The direction of the change is sometimes confusing. Remember that

- *Increasing* the pH means *lowering* the concentration of H_3O^+;
- *Decreasing* the pH means *raising* the concentration of H_3O^+.

Also,

- Strictly speaking, these ranges apply only at 25°C. Variations of a few degrees in the temperature are often unimportant, however.

pH < 7	acidic solution	$[H_3O^+] > [OH^-]$
pH $= 7$	neutral solution	$[H_3O^+] = [OH^-]$
pH > 7	basic solution	$[H_3O^+] < [OH^-]$

The pH of a solution is easily measured with a **pH meter** (Fig. 8–2). In this instrument, the solution being tested is made a part of an electrochemical cell having an output voltage that is sensitive to the hydronium-ion concentration. (The principles behind such electrochemical measurements are considered in Chapter 13.) The values of the pH for several common fluids are shown in Figure 8–3.

The prefix "p" in pH signals us to take the negative logarithm of the hydronium-ion concentration. This usage has been generalized to give definitions such as

$$pOH = -\log_{10}[OH^-] \qquad \text{and} \qquad pK_w = -\log_{10}K_w$$

The notation has advantages. For example, the product of $[H_3O^+]$ and $[OH^-]$ in water always equals K_w; that is,

$$[H_3O^+][OH^-] = K_w$$

Taking the logarithm of both sides and multiplying through by -1 gives

$$-\log_{10}[H_3O^+] - \log_{10}[OH^-] = -\log_{10}K_w$$

These three negative logarithms are "p" quantities, so the equation becomes

$$pH + pOH = pK_w$$

At 25°C, $pK_w = -\log_{10}(1.0 \times 10^{-14}) = 14.00$. Hence,

$$pH + pOH = 14.00 \qquad \text{(at 25°C)}$$

Shifting from "p" notation back to regular notation requires raising 10 to the negative of the "p" quantity. If pH is given, the concentration of $H_3O^+(aq)$ is 10 raised to the power $(-pH)$; if pOH is given, the concentration of $OH^-(aq)$ is 10 to the power $(-pOH)$. Raising 10 to a power is accomplished on a calculator by using the 10^x key or the INV LOG key.

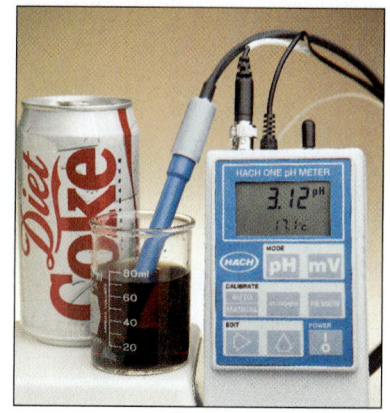

Figure 8-2 • Checking the pH of a soft drink with a pH meter equipped with a digital readout. (Charles D. Winters)

EXAMPLE 8-4

The pH of some grape juice at 25°C is 2.85. Calculate $[H_3O^+]$, $[OH^-]$, and pOH.

Solution

$$pH = 2.85 = -\log_{10}[H_3O^+]$$
$$[H_3O^+] = 10^{-pH} = 10^{-2.85}$$

The exponential number can be evaluated by using a calculator. The result is

$$[H_3O^+] = 1.4 \times 10^{-3} \text{ mol L}^{-1}$$

The concentration of the hydroxide ion is linked to that of the hydronium ion through the water autoionization equilibrium

$$[OH^-] = \frac{K_w}{[H_3O^+]} = \frac{1.0 \times 10^{-14}}{1.4 \times 10^{-3}} = 7.1 \times 10^{-12} \text{ mol L}^{-1}$$
$$pOH = -\log_{10}[OH^-] = -\log_{10}(7.1 \times 10^{-12}) = 11.15$$

Check

The sum of pH and pOH equals 14.00 (the value of pK_w at 25°C), as it must.

EXERCISE

Calculate $[H_3O^+]$, $[OH^-]$ and pOH in saliva that has a pH of 6.60 at 25°C.

Answer:

$[H_3O^+] = 2.5 \times 10^{-7} \text{ mol L}^{-1}$; $[OH^-] = 4.0 \times 10^{-8} \text{ mol L}^{-1}$; pOH = 7.40.

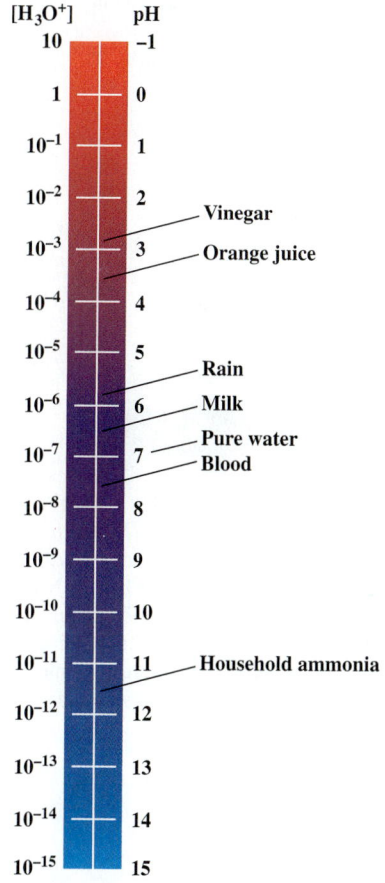

Figure 8-3 • Many everyday materials are acidic or basic aqueous solutions with a wide range of pH values.

8-3   The Strengths of Acids and Bases

Acids are classified as strong or weak depending on whether their reactions with water to give $H_3O^+(aq)$ go essentially to completion or reach an equilibrium somewhere short of completion. These reactions are often called **acid ionization** reactions, because many acids create ions in solution as they transfer H^+ ions to water, or **acid dissociation** reactions, because the acid breaks apart as it donates an H^+ ion. The proportion of an acid that has reacted with water in a particular solution is called the **fraction ionized** (or **fraction dissociated**) of that acid. The fraction ionized of a strong acid is essentially equal to 1; the fraction ionized of a weak acid is substantially less than 1. There are seven common strong acids. They are listed in Section 4–3 and at the top of Table 8–2.

Nearly all other acids are weak acids. The weaker the acid, the less it dissociates when dissolved in water at a given concentration and the higher the pH of its solution. A compound that is a weak acid is also a weak electrolyte (see Section 4–1). Solutions of weak electrolytes conduct electricity poorly compared with solutions of strong electrolytes at the same concentration because fewer ions are present as charge carriers.

- Weak electrolytes also depress the freezing point of solutions less than do strong electrolytes at the same concentration, as illustrated in Figure 6–29.

The Brønsted-Lowry model provides a basis for a quantitative scale of acid strength. The ionization of a generic acid (symbolized as "HA") in aqueous solution can be written as

$$HA(aq) + H_2O(\ell) \rightleftharpoons H_3O^+(aq) + A^-(aq)$$

where A^- is the conjugate base of HA. The corresponding equilibrium-constant equation is

$$\frac{[H_3O^+][A^-]}{[HA]} = K_a$$

where the subscript "a" in K_a stands for "acidity" and K_a is the **acidity constant** (also called the *acid ionization constant* or *acid dissociation constant*) of the acid. For example, if HA refers to hydrogen cyanide (HCN), then

$$HCN(aq) + H_2O(\ell) \rightleftharpoons H_3O^+(aq) + CN^-(aq) \qquad \frac{[H_3O^+][CN^-]}{[HCN]} = K_a$$

Hydrogen cyanide (HCN) is a highly toxic gas that dissolves in water to form equally toxic solutions of hydrocyanic acid. Its conjugate base, CN^-, forms complexes with gold and silver and is used in the treatment of mine tailings to extract these precious metals by bringing them into solution.

Table 8–2 gives values of K_a for a number of acids in aqueous solution at 25°C. The useful quantity $pK_a = -\log_{10}K_a$ is also given.

An acidity constant is a quantitative measure of the strength of the acid in a given solvent (in this case, water). A weak acid reacts incompletely with the solvent. The products of the reaction have low concentrations at equilibrium; thus, K_a is small. A strong acid reacts essentially completely with the solvent, so [HA] in the denominator is close to zero and K_a is large. Determining K_a's for strong acids is difficult because the amount of unreacted acid at equilibrium is too small to be measured accurately. Relative acid strengths are estimated in these cases by replacing water with a solvent that is a weaker base. If the chosen solvent is a poor enough base (that is, has substantially less tendency to accept H^+ ions than H_2O has), then even a relatively strong acid may donate hydrogen ions incompletely. For example, HCl donates H^+ ions less extensively than $HClO_4$ in the solvent diethyl ether ($C_2H_5OC_2H_5$); therefore, HCl is a weaker acid than $HClO_4$.

TABLE 8-2	Acidity Constants in Water at 25°C			
Acid	**Formula**	**Conjugate Base**	**K_a**	**pK_a**
Hydriodic	HI	I^-	$\approx 10^{11}$	≈ -11
Hydrobromic	HBr	Br^-	$\approx 10^9$	≈ -9
Perchloric	$HClO_4$	ClO_4^-	$\approx 10^7$	≈ -7
Hydrochloric	HCl	Cl^-	$\approx 10^7$	≈ -7
Chloric	$HClO_3$	ClO_3^-	$\approx 10^3$	≈ -3
Sulfuric (1)	H_2SO_4	HSO_4^-	$\approx 10^2$	≈ -2
Nitric	HNO_3	NO_3^-	≈ 20	≈ -1.3
Hydronium ion	H_3O^+	H_2O	1	0.0
Urea acidium ion	$(NH_2)CONH_3^+$	$(NH_2)_2CO$ (urea)	6.6×10^{-1}	0.18
Iodic	HIO_3	IO_3^-	1.6×10^{-1}	0.80
Oxalic (1)	$H_2C_2O_4$	$HC_2O_4^-$	5.9×10^{-2}	1.23
Sulfurous (1)	H_2SO_3	HSO_3^-	1.5×10^{-2}	1.82
Sulfuric (2)	HSO_4^-	SO_4^{2-}	1.2×10^{-2}	1.92
Chlorous	$HClO_2$	ClO_2^-	1.1×10^{-2}	1.96
Phosphoric (1)	H_3PO_4	$H_2PO_4^-$	7.5×10^{-3}	2.12
Arsenic (1)	H_3AsO_4	$H_2AsO_4^-$	5.0×10^{-3}	2.30
Chloroacetic	$ClCH_2COOH$	$ClCH_2COO^-$	1.4×10^{-3}	2.85
Hydrofluoric	HF	F^-	6.6×10^{-4}	3.18
Nitrous	HNO_2	NO_2^-	4.6×10^{-4}	3.34
Formic	HCOOH	$HCOO^-$	1.8×10^{-4}	3.74
Benzoic	C_6H_5COOH	$C_6H_5COO^-$	6.5×10^{-5}	4.19
Oxalic (2)	$HC_2O_4^-$	$C_2O_4^{2-}$	6.4×10^{-5}	4.19
Hydrazoic	HN_3	N_3^-	1.9×10^{-5}	4.72
Acetic	CH_3COOH	CH_3COO^-	1.8×10^{-5}	4.74
Propionic	CH_3CH_2COOH	$CH_3CH_2COO^-$	1.3×10^{-5}	4.89
Pyridinium ion	$HC_5H_5N^+$	C_5H_5N (pyridine)	5.6×10^{-6}	5.25
Carbonic (1)	H_2CO_3	HCO_3^-	4.3×10^{-7}	6.37
Sulfurous (2)	HSO_3^-	SO_3^{2-}	1.0×10^{-7}	7.00
Arsenic (2)	$H_2AsO_4^-$	$HAsO_4^{2-}$	9.3×10^{-8}	7.03
Hydrosulfuric	H_2S	HS^-	9.1×10^{-8}	7.04
Phosphoric (2)	$H_2PO_4^-$	HPO_4^{2-}	6.2×10^{-8}	7.21
Hypochlorous	HClO	ClO^-	3.0×10^{-8}	7.52
Hydrocyanic	HCN	CN^-	6.2×10^{-10}	9.21
Ammonium ion	NH_4^+	NH_3	5.6×10^{-10}	9.25
Carbonic (2)	HCO_3^-	CO_3^{2-}	4.8×10^{-11}	10.32
Methylammonium ion	$CH_3NH_3^+$	CH_3NH_2	2.3×10^{-11}	10.64
Arsenic (3)	$HAsO_4^{2-}$	AsO_4^{3-}	3.0×10^{-12}	11.52
Hydrogen peroxide	H_2O_2	HO_2^-	2.4×10^{-12}	11.62
Phosphoric (3)	HPO_4^{2-}	PO_4^{3-}	2.2×10^{-13}	12.66
Water	H_2O	OH^-	1.0×10^{-14}	14.00

Base Strength

Consider the reaction between the base ammonia and water:

$$NH_3(aq) + H_2O(\ell) \rightleftharpoons OH^-(aq) + NH_4^+(aq) \qquad \frac{[NH_4^+][OH^-]}{[NH_3]} = K_b$$

In this reaction, ammonia accepts H^+ ion from water. The equilibrium constant K_b is a **basicity constant.** Basicity constants in aqueous solution measure the ability of bases to accept (remove) hydrogen ions from water and so generate OH^-. They measure the strength of bases, just as acidity constants, K_a's, measure the strength of acids. The larger K_b is, the stronger the base. Now consider this reaction:

$$NH_4^+(aq) + H_2O(\ell) \rightleftharpoons H_3O^+(aq) + NH_3(aq) \qquad \frac{[H_3O^+][NH_3]}{[NH_4^+]} = K_a$$

Here, water accepts H^+ ion from ammonium ion, the conjugate acid of ammonia. This reaction is *not* the reverse of the one preceding it. Adding the two and canceling out identical terms exposes the true relationship:

$$NH_3(aq) + H_2O(\ell) \rightleftharpoons OH^-(aq) + NH_4^+(aq)$$
$$NH_4^+(aq) + H_2O(\ell) \rightleftharpoons H_3O^+(aq) + NH_3(aq)$$
$$\overline{H_2O(\ell) + H_2O(\ell) \rightleftharpoons H_3O^+(aq) + OH^-(aq)}$$

The sum is the autoionization of water. According to the rules in Section 7–2, equilibrium expressions are multiplied when their chemical equations are added. Doing this gives

$$K_aK_b = \frac{[H_3O^+][NH_3]}{[NH_4^+]} \times \frac{[NH_4^+][OH^-]}{[NH_3]} = [H_3O^+][OH^-] = K_w$$

The derivation is easily generalized. For any acid and its conjugate base in aqueous solution:

$$K_aK_b = K_w$$

> The strength of a base is inversely related to the strength of its conjugate acid; the weaker the acid, the stronger its conjugate base, and vice versa.

Figure 8–4 illustrates this inverse relationship.

Taking the logarithm of both sides of the equation $K_aK_b = K_w$ and multiplying through by -1 gives

$$(-\log_{10}K_a) + (-\log_{10}K_b) = -\log_{10}K_w$$

Using the p-notation (a prefix "p" on any quantity stands for the negative logarithm of that quantity), this equation becomes

$$pK_a + pK_b = pK_w \qquad \text{(for a conjugate acid–base pair)}$$

This relationship also shows that the weaker the acid (K_a small and pK_a large), the stronger the conjugate base (K_b large and pK_b small).

A helpful consequence of these relationships is that separate tables of K_b and pK_b values are unnecessary. When K_b's or pK_b's are required, they are easily calculated from K_a's or pK_a's using one of the preceding expressions. For ex-

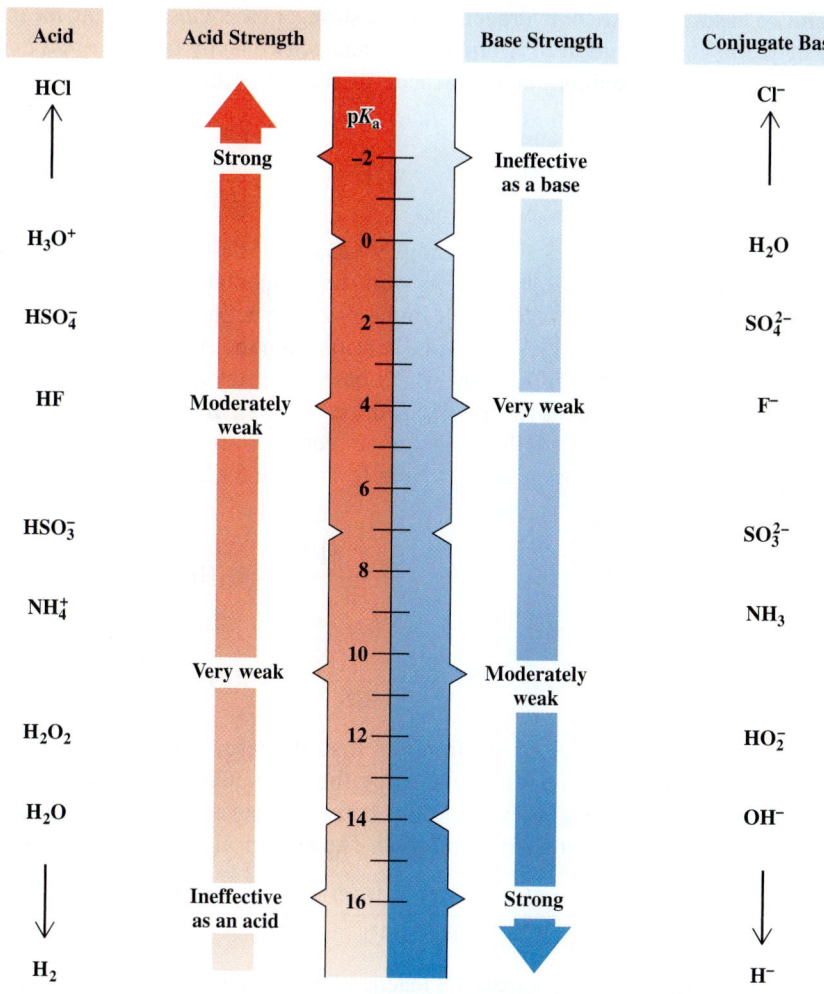

Figure 8-4 • The relative strengths of some acids and their conjugate bases. Strong acids have weak conjugate bases; strong bases have weak conjugate acids. The "one side high/other side low" pattern recalls the seesaw of Figure 8–1.

ample, K_a for the ammonium ion NH_4^+ equals 5.6×10^{-10} at 25°C (Table 8–2). Therefore

$$K_b(NH_3) = \frac{K_w}{K_a(NH_4^+)} = \frac{1.0 \times 10^{-14}}{5.6 \times 10^{-10}} = 1.8 \times 10^{-5}$$

The ammonium ion is a very weak acid. Ammonia, its conjugate base, is a moderately weak base.

Stronger bases generate more $OH^-(aq)$ ion in solution than equal concentrations of weaker bases. The base PO_4^{3-} is stronger than CN^-, which is stronger than Cl^-. At equilibrium at 25°C

- 1.0 mol of PO_4^{3-} ions per liter of solution produces 0.19 mol of $OH^-(aq)$;
- 1.0 mol of CN^- ions per liter of solution produces 0.0040 mol of $OH^-(aq)$;
- 1.0 mol of Cl^- ions per liter of solution produces less than 10^{-7} mol of $OH^-(aq)$.

The strong base NaOH produces the maximum possible amount of $OH^-(aq)$ ion: 1.0 mol of OH^- per mole of NaOH that dissolves.

If two bases compete for hydrogen ions, the stronger base wins. Winning means holding the larger proportion of hydrogen ions and consequently being present in a lesser amount when equilibrium is reached. To see this, consider the acid–base equilibrium

$$HF(aq) + CN^-(aq) \rightleftharpoons HCN(aq) + F^-(aq) \qquad \frac{[HCN][F^-]}{[HF][CN^-]} = K$$

In this reaction, the two bases $F^-(aq)$ and $CN^-(aq)$ compete for hydrogen ions. It is only an experimental detail whether we start the competition by mixing $HF(aq)$ with $CN^-(aq)$ or $F^-(aq)$ with $HCN(aq)$. In the first case action starts from the left side of the equation, and in the second case it starts from the right, but the equilibrium outcome is the same. To find out which base competes more effectively, we need the equilibrium constant for the competition reaction. To obtain it, we write the chemical equations and equilibrium-constant equations for the acid ionization of $HF(aq)$ and $HCN(aq)$, taking the K_a values from Table 8–2:

- Recall that this is a fundamental characteristic of the equilibrium state (see Section 7–1).

$$HF(aq) + H_2O(\ell) \rightleftharpoons H_3O^+(aq) + F^-(aq)$$
$$\frac{[H_3O^+][F^-]}{[HF]} = K_a(HF) = 6.6 \times 10^{-4}$$
$$HCN(aq) + H_2O(\ell) \rightleftharpoons H_3O^+(aq) + CN^-(aq)$$
$$\frac{[H_3O^+][CN^-]}{[HCN]} = K_a(HCN) = 6.2 \times 10^{-10}$$

If the second equation is subtracted from the first, the result equals the chemical equation for the reaction of interest. Accordingly, the first equilibrium-constant equation is divided by the second (as explained in Section 7–2) to obtain the equilibrium-constant equation for the competition reaction. The numerical value of K at 25°C is

$$K = \frac{K_a(HF)}{K_a(HCN)} = \frac{6.6 \times 10^{-4}}{6.2 \times 10^{-10}} = 1.1 \times 10^6$$

Hydrogen fluoride (HF) is a colorless liquid that boils at 19.5°C. It reacts with water as a weak acid. This contrasts with the other hydrogen halides, which are strong acids in water. Hydrogen fluoride and its aqueous solution, hydrofluoric acid, are highly corrosive materials that can etch glass.

This large K means that the competition reaction lies far to the right at equilibrium. Most of the hydrogen ions end up associated with the $CN^-(aq)$ in preference to the $F^-(aq)$. The $CN^-(aq)$ wins the competition against the $F^-(aq)$. It is the stronger base. The predominant dissolved species at equilibrium are $HCN(aq)$, which is a weaker acid than $HF(aq)$, and $F^-(aq)$, which is a weaker base than $CN^-(aq)$.

Similar manipulation of chemical equations and their associated equilibrium expressions gives valid predictions of the direction and extent of the transfer of hydrogen ions in all Brønsted-Lowry acid–base reactions.

Indicators

An **indicator** is a soluble compound that changes its color noticeably over a fairly short range of pH. The typical indicator is a weak organic acid that has a different color than its conjugate base (Fig. 8–5). Methyl red, for example, is red when a single H^+ has been accepted (to form the acid) but yellow when that H^+ has been donated (leaving the base). Litmus changes color from red to blue as its acidic form is converted to its basic form. Good indicators have such intense colors that just two or three drops of a dilute solution of indicator suffice to give a strong color to the solution under investigation. This makes changes in color easy to see, and because the concentration of indicator molecules is then very small, the pH of the solution under study is practically unaffected by the presence of the indicator. The color changes of the indicator follow the dictates of the other acids and bases present in the solution.

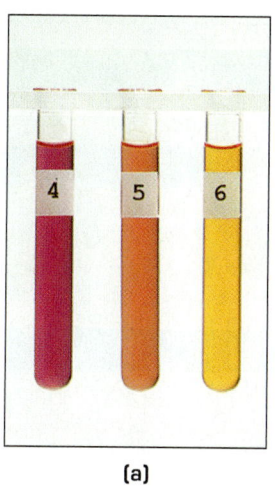
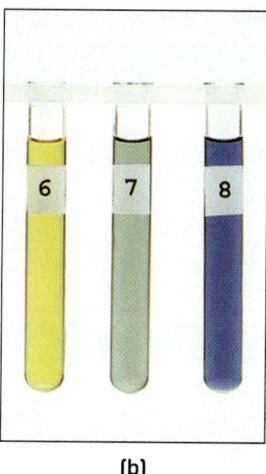
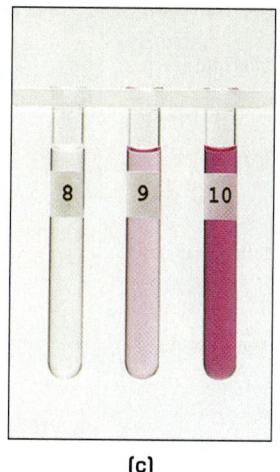

(a)   (b)   (c)

Figure 8–5 • Color changes in three indicators. Methyl red (a) goes from red at low pH to orange at about pH 5 to yellow at high pH. Bromothymol blue (b) is yellow at low pH, blue at high pH, and green at about pH 7. Phenolphthalein (c) goes from colorless to pink at about pH 9. (*Marna G. Clarke*)

Let us represent the acid form of a given indicator as HIn, and the conjugate base form as In⁻. The acid–base reaction involving the two is

$$HIn(aq) + H_2O(\ell) \rightleftharpoons H_3O^+(aq) + In^-(aq) \qquad \frac{[H_3O^+][In^-]}{[HIn]} = K_a$$

where K_a is the acidity constant for the indicator. This equilibrium-constant equation can be rearranged to give

$$\frac{[H_3O^+]}{K_a} = \frac{[HIn]}{[In^-]}$$

If the concentration of hydronium ion, $[H_3O^+]$, is large relative to K_a, this ratio is large and [HIn] is large compared with [In⁻]. The solution is the color of the acid form of the indicator because most molecules of the indicator are in the acid form. The indicator litmus, for example, has a K_a near 10^{-7}. If the pH is 5, then

$$\frac{[H_3O^+]}{K_a} = \frac{10^{-5}}{10^{-7}} = 100 \qquad \text{(for litmus at pH 5)}$$

so 100 times more indicator molecules are in the acid form than in the base form, and the solution is red, the color of the acid form of litmus.

As the concentration of hydronium ion is reduced, more molecules of HIn(*aq*) lose an H⁺ ion and go over to In⁻(*aq*). When $[H_3O^+]$ is near K_a, almost equal amounts of the two forms are present, and the color is an almost equal mixture of the colors of HIn(*aq*) and In⁻(*aq*). A further decrease in $[H_3O^+]$ to a value much smaller than K_a then leads to a predominance of the base form, with the corresponding color being observed.

• For example, the intermediate color of litmus is purple, the sum of red (the color of the acid form) and blue (the color of the base).

Different indicators have different values for K_a and change color at different pH values (Fig. 8–6). The weaker an indicator is as an acid, the higher is the pH at which the color change takes place.

The color changes of indicators occur over a range of 1 to 2 pH units. Methyl red, for example, is red when the pH is below 4.8 and yellow when the pH is above 6.0; shades of orange are seen for intermediate pH values. This limits the precision to which the pH can be determined through the use of indicators; for high-precision work, a pH meter must be used.

Figure 8-6 • Indicators change their colors at very different values of the pH, so the best choice of indicator in a titration depends on what is being titrated, on what the titrant is, and on the particular experimental conditions.

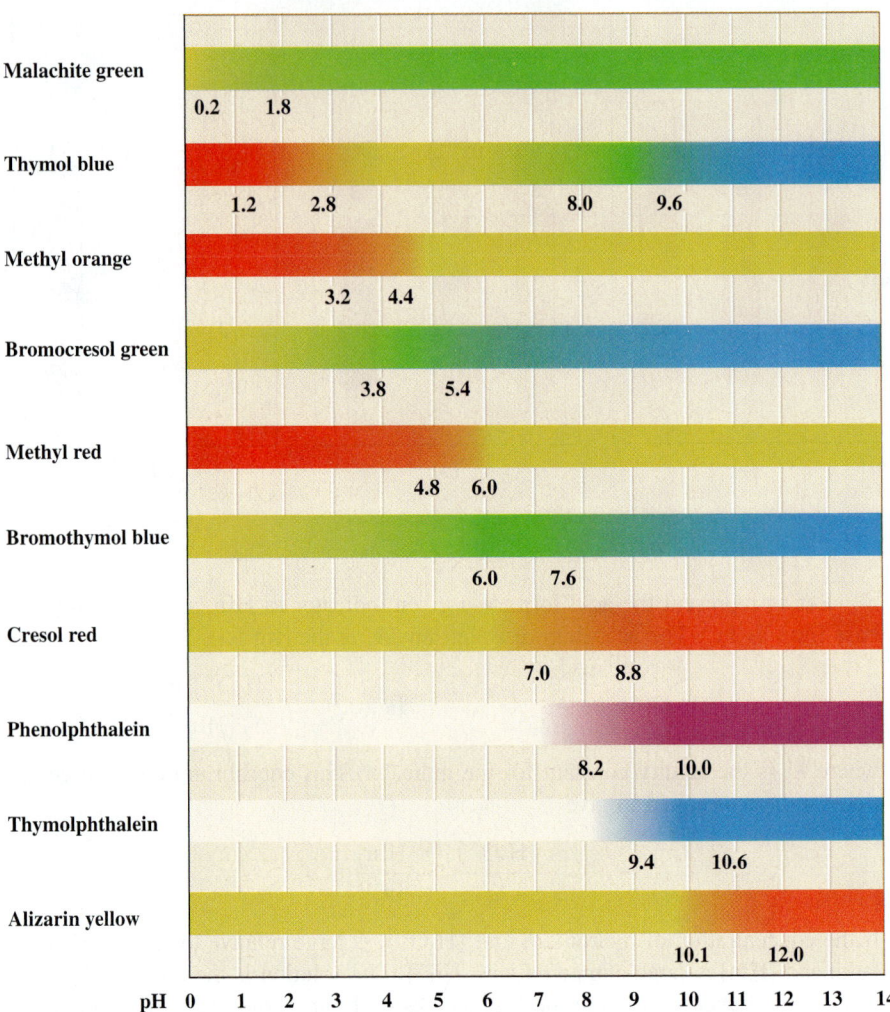

Indicator	pH range
Malachite green	0.2 1.8
Thymol blue	1.2 2.8 8.0 9.6
Methyl orange	3.2 4.4
Bromocresol green	3.8 5.4
Methyl red	4.8 6.0
Bromothymol blue	6.0 7.6
Cresol red	7.0 8.8
Phenolphthalein	8.2 10.0
Thymolphthalein	9.4 10.6
Alizarin yellow	10.1 12.0

pH 0 1 2 3 4 5 6 7 8 9 10 11 12 13 14

EXAMPLE 8-5

A sample of vinegar is tested with two different indicators. Thymol blue turns yellow, and methyl orange turns red. Estimate the pH of the vinegar.

Solution

If thymol blue is yellow, then the pH must be above about 2.8 (according to Fig. 8–6), but if methyl orange is red, the pH must be below about 3.2. The pH is therefore 3.0, to within about 0.2 pH units.

EXERCISE

Rainwater that has leached through a pile of building materials and puddled underneath turns bromothymol blue from yellow to blue, but it leaves phenolphthalein colorless. Estimate the pH of this water.

Answer: 7.6 to 8.2.

(a)   **(b)**   **(c)**

Figure 8–7 • (a) Red cabbage extract is a natural pH indicator. In acidic solution it is red. (b) Upon addition of base, several colors, including green and blue, appear as incomplete mixing causes different regions of the sample to have different pH. (c) Once a uniformly basic pH is established, the color of the cabbage extract is blue. *(Leon Lewandowski)*

Indicators abound in foods and other natural products (see Figs. 4–7 and 8–7). When a small amount of lemon juice is added to a cup of tea, the color of the tea changes from brown to light yellow. This change has nothing to do with the rather faint color of the lemon juice itself but arises from the decrease in pH of the tea caused by the citric acid that the juice contains. The acid acts on the natural indicators in the tea to change their color. The single naturally occurring indicator cyanidin causes both the red color of the poppy and the blue color of the cornflower. The sap of the poppy is acidic enough to turn the cyanidin red, but the sap of the cornflower is basic enough to make the cyanidin blue. Anthocyanins (Fig. 8–8) are natural dyes related to cyanidin. They contribute to the colors of strawberries, blueberries, cherries, oranges, and red cabbage.

Figure 8–8 • An anthocyanin dye gives these strawberries their characteristic red color. *(Charles D. Winters)*

8-4 Equilibria Involving Weak Acids and Bases

Weak acids and bases react only partially with water before coming to equilibrium. Consequently, calculations of the pH of their solutions involve the use of K_a or K_b and the laws of chemical equilibrium. Calculating $[H_3O^+]$ or $[OH^-]$ follows the pattern set in solving Example 7–6 for gas equilibria. In that case, we knew the initial partial pressures of the reactants, and we calculated the equilibrium pressures of the products of the incomplete reaction. Now, we know the initial concentration of an acid or base, and we calculate the equilibrium concentrations of the products of its incomplete reaction with water.

Weak Acids

A weak acid in water has a K_a that is smaller than 1, which is the K_a of the hydronium ion (see Table 8–2). The weaker the acid, the smaller its K_a, and the larger its pK_a. Values of pK_a start at zero for the strongest weak acid and range upward. If the pK_a of a compound or ion is greater than 14, then that species is ineffective as an acid in aqueous solution.

• Again, "HA" stands for a general weak acid.

When a weak acid HA is mixed in water, the original concentration of acid is almost always known, but reaction with water consumes some HA molecules and generates A^- and H_3O^+ ions.

$$HA(aq) + H_2O(\ell) \rightleftharpoons H_3O^+(aq) + A^-(aq)$$

We wish to calculate the concentrations of H_3O^+, A^-, and HA at equilibrium. The only complication relative to examples from Chapter 7 is that one of the products (H_3O^+) has a second source, the autoionization of the water. In cases of practical interest, the contribution from this source is almost always small because K_w is small and can be neglected. It is however good practice to confirm this at the end of each calculation by verifying that the $[H_3O^+]$ from the acid ionization alone exceeds 10^{-7} mol L^{-1} by at least one order of magnitude.

EXAMPLE 8–6

Acetic acid (CH_3COOH) donates a single hydrogen ion in water with a K_a of 1.8×10^{-5} at 25°C. A 1.00 mol quantity of acetic acid is dissolved in enough water at 25°C to give 1.00 L of solution. Calculate the pH of the solution and the fraction of the acetic acid that is ionized at equilibrium.

Solution

Acetic acid donates hydrogen ions in water according to the equation

$$CH_3COOH(aq) + H_2O(\ell) \rightleftharpoons H_3O^+(aq) + CH_3COO^-(aq)$$

Let y equal the number of moles per liter of acetic acid that reacts. The table of changes (or I-C-E table) for this reaction system is

	$[CH_3COOH]$ (mol L^{-1})	$[H_3O^+]$ (mol L^{-1})	$[CH_3COO^-]$ (mol L^{-1})
Initial	1.00	≈ 0	0
Change	$-y$	$+y$	$+y$
Equilibrium	$1.00 - y$	y	y

The table omits $H_2O(\ell)$ because it is the solvent. Some H_3O^+ comes from the autoionization of water, but the concentration from that source is very small compared with y for all but the weakest of acids or the most dilute of solutions. Therefore it is neglected (subject to a later check). The equilibrium-constant equation is

$$\frac{[H_3O^+][CH_3COO^-]}{[CH_3COOH]} = K_a$$

which becomes

$$\frac{y^2}{1.00 - y} = 1.8 \times 10^{-5}$$

It is possible to rearrange this equation and solve it for y using the quadratic formula or a pre-programmed calculator (see Appendix C–3). A quicker way uses chemical insight. The fraction of acetic acid that ionizes is likely to be small because it is a weak acid. That is, the equilibrium concentration of CH_3COOH is likely to be close to its initial concentration. If so, y can be neglected relative to 1.00. Neglecting it gives

$$1.8 \times 10^{-5} = \frac{y^2}{1.00 - y} \approx \frac{y^2}{1.00}$$

$$y \approx \sqrt{1.00 \times (1.8 \times 10^{-5})} = 4.24 \times 10^{-3}$$

From the table of changes, $[H_3O^+]$ equals y. Therefore

$$[H_3O^+] \approx 4.24 \times 10^{-3} \text{ mol L}^{-1}$$

$$pH \approx -\log(4.24 \times 10^{-3}) = 2.37$$

The fraction of the acetic acid that is ionized at equilibrium equals the concentration of CH_3COO^- ions present at equilibrium divided by the initial concentration of CH_3COOH:

$$\text{fraction ionized} = \frac{4.24 \times 10^{-3} \text{ mol L}^{-1}}{1.00 \text{ mol L}^{-1}} = 0.00424$$

This rounds off to 0.0042, which is only 0.42%. Fewer than 1 in 100 of the acetic acid molecules in this solution dissociate into ions.

Check

Substitution of the final concentrations into the equilibrium-constant expression gives back the correct value for K_a:

$$\frac{[H_3O^+][CH_3COO^-]}{[CH_3COOH]} = \frac{(4.24 \times 10^{-3})(4.24 \times 10^{-3})}{(1.00 - 0.00424)} = 1.8 \times 10^{-5}$$

Failing to subtract 0.00424 from 1.00 in the denominator barely changes the value of the fraction. This fact confirms the validity of the approximation used here. Also, the concentration of H_3O^+ from the reaction of the acetic acid (4.2×10^{-3} mol L^{-1}) exceeds 1×10^{-7} mol L^{-1} by several orders of magnitude so the neglect of the autoionization of water is justified.

EXERCISE

Propionic acid (CH_3CH_2COOH) has $K_a = 1.3 \times 10^{-5}$ at 25°C. (a) Compute the pH at equilibrium in a 0.65 mol L^{-1} solution of propionic acid. (b) Compute the fraction of the propionic acid ionized at equilibrium.

Answer: (a) pH = 2.54. (b) Fraction ionized = 0.0045.

Adding water to a solution of a weak acid causes the concentration of H_3O^+ to decrease toward 10^{-7} mol L^{-1}. The diluted solution more nearly resembles pure water, in which $[H_3O^+]$ equals 10^{-7} mol L^{-1}. Dilution also causes the fraction of the weak acid that is ionized to *increase*. Dilution is just like an expansion in the volume of a gas-phase reaction system (see Section 7–5). Le Châtelier's principle predicts a shift toward the side of the equilibrium having more particles. The following example traces in detail the changes caused by dilution.

EXAMPLE 8–7

Calculate the pH of $1.00 \times 10^{-4} \, \text{mol L}^{-1}$ solution of acetic acid and the fraction of the acetic acid that is ionized at 25°C. Note that this solution of acetic is 10,000 times more dilute than the solution in Example 8–6.

Solution

Set up the same table of changes as in Example 8–6, but with a different initial concentration of CH_3COOH. Once again, assume (subject to a later check) that the contribution of the autoionization of water to $[H_3O^+]$ is negligible.

$$CH_3COOH(aq) + H_2O(\ell) \rightleftharpoons H_3O^+(aq) + CH_3COO^-(aq)$$

	$[CH_3COOH]$ (mol L^{-1})	$[H_3O^+]$ (mol L^{-1})	$[CH_3COO^-]$ (mol L^{-1})
Initial	0.000100	≈ 0	0
Change	$-y$	$+y$	$+y$
Equilibrium	$0.000100 - y$	y	y

$$\frac{[H_3O^+][CH_3COO^-]}{[CH_3COOH]} = K_a$$

$$\frac{y^2}{0.000100 - y} = 1.8 \times 10^{-5}$$

Assuming that y is much smaller than 0.000100 and neglecting it in the denominator gives

$$\frac{y^2}{0.000100} \approx 1.8 \times 10^{-5}$$

from which $y \approx 4.2 \times 10^{-5}$. This corresponds to ionization of 42% of the CH_3COOH molecules, a sizeable proportion: y is *not* negligible compared with 0.000100, so 4.2×10^{-5} is *not* a correct answer for y.

Rearranging the equation and using either the quadratic formula or a pre-programmed calculator gives a correct y. It is more instructive, however, to proceed in a "chemical" way. Simply adjust y up or down to fit the equation better, just as reactions shift to left or right to come to equilibrium. The reaction quotient Q is the perfect vehicle for this procedure. If $y = 4.2 \times 10^{-5}$, then

$$Q = \frac{(4.2 \times 10^{-5}) \times (4.2 \times 10^{-5})}{(0.000100 - 4.2 \times 10^{-5})} = 3.0 \times 10^{-5}$$

This Q exceeds K_a. Adjusting y to equal 2.1×10^{-5} (one-half the previous y) gives

$$Q = \frac{(2.1 \times 10^{-5}) \times (2.1 \times 10^{-5})}{(0.000100 - 2.1 \times 10^{-5})} = 0.56 \times 10^{-5}$$

This new Q is less than K_a. The true y must lie between 4.2×10^{-5} and 2.1×10^{-5}. Adjusting y to 3.2×10^{-5} gives a Q that is slightly smaller than K_a; adjusting y to 3.6×10^{-5} gives a Q that is slightly larger. A final slight shift to

3.44×10^{-5} gives $Q = 1.80 \times 10^{-5}$, which is the same as K_a. From the table of changes, $[H_3O^+]$ and $[CH_3COO^-]$ equal y. Therefore

$$[H_3O^+] = [CH_3COO^-] = 3.44 \times 10^{-5} \text{ mol L}^{-1}$$
$$pH = \log_{10}(3.44 \times 10^{-5}) = 4.46$$
$$\text{fraction ionized} = \frac{[CH_3COO^-]}{0.000100 \text{ mol L}^{-1}} = \frac{3.44 \times 10^{-5} \text{ mol L}^{-1}}{0.00010 \text{ mol L}^{-1}} = 0.34$$

Check

This computational approach is self-checking. If y is obtained using the quadratic formula or a pre-programmed calculator, the final concentrations of H_3O^+, CH_3COO^-, and CH_3COOH should be substituted into the original equilibrium equation to catch data-entry or algebraic errors. The neglect of the autoionization of water as a source of $[H_3O^+]$ is justified because $[H_3O^+]$ is hundreds of times larger than $10^{-7} \text{ mol L}^{-1}$.

EXERCISE

For propionic acid (CH_3CH_2COOH), K_a is 1.3×10^{-5} at 25°C. (a) Calculate the pH of a 0.00010 mol L^{-1} solution of propionic acid in water at 25°C. (b) Calculate the fraction of the propionic acid ionized.

Answer: (a) pH = 4.52. (b) Fraction ionized = 0.30.

What is the cutoff on the approximation that was acceptable in Example 8–6, but not acceptable in Example 8–7? A good rule of thumb in this type of calculation is that a quantity that is neglected must turn out to be less than 4% to 5% of the quantity from which it is subtracted (or to which it is added). This rule recognizes that the accuracy of the results in equilibrium calculations is limited not only by the accuracy of the input data (the values of K_a and the initial concentrations) but also by the non-ideality of the solutions, which causes deviations from the mass action law itself. It is pointless and, in fact, wrong to calculate equilibrium concentrations to any higher degree of accuracy than 1% to 3% without evaluating and correcting for the effects of non-ideality.

• Such corrections fall outside of the scope of this text.

Weak Bases

The definition and description of weak acids, their K_a's, and the production of $H_3O^+(aq)$ ion when they are dissolved in water can be translated to apply to weak bases, their K_b's, and the production of $OH^-(aq)$ ion. For example, a weak base such as ammonia reacts incompletely with water to produce $OH^-(aq)$:

• Recall that a strong base reacts essentially completely to give $OH^-(aq)$.

$$H_2O(\ell) + NH_3(aq) \rightleftharpoons NH_4^+(aq) + OH^-(aq) \qquad \frac{[NH_4^+][OH^-]}{[NH_3]} = K_b$$

The K_b of a weak base is smaller than 1, and the weaker the base, the smaller the K_b. If the K_b of a compound or ion is smaller than 1×10^{-14}, then that species is ineffective as a base in aqueous solution.

The analogy to weak acids also holds in the calculation of the aqueous equilibria of weak bases, as the following example shows.

EXAMPLE 8–8

Calculate the pH of a solution made by dissolving 0.010 mol of ammonia (NH_3) in enough water to give 1.00 L of solution at 25°C. Also calculate the fraction of the NH_3 that is ionized. The K_b of NH_3 in water is 1.8×10^{-5} at this temperature.

Strategy

Plan to compute pOH and obtain pH from it. Write an equation for the reaction and set up a table for the changes that occur as the reaction goes to equilibrium. Neglect (subject to a check) the small contribution to [OH^-] that comes from the autoionization of water.

Solution

$$H_2O(\ell) + NH_3(aq) \rightleftharpoons NH_4^+(aq) + OH^-(aq)$$

	[NH_3] (mol L^{-1})	[NH_4^+] (mol L^{-1})	[OH^-] (mol L^{-1})
Initial	0.010	0	≈0
Change	$-y$	$+y$	$+y$
Equilibrium	$0.010 - y$	y	$+y$

$$\frac{[NH_4^+][OH^-]}{[NH_3]} = K_b$$

$$\frac{y^2}{0.010 - y} = 1.8 \times 10^{-5}$$

This equation can be solved rapidly by neglecting y in comparison to 0.01:

$$y \approx \sqrt{0.01(1.8 \times 10^{-5})} = 4.24 \times 10^{-4}$$
$$[OH^-] = [NH_4^+] \approx 4.24 \times 10^{-4} \text{ mol L}^{-1}$$
$$pOH = -\log_{10}(4.24 \times 10^{-4}) = 3.37$$
$$pH = 14.00 - pOH = \boxed{10.63}$$

Use the definition of fraction ionized:

$$\text{fraction ionized} = \frac{[NH_4^+]_{\text{equilibrium}}}{[NH_3]_{\text{initial}}} = \frac{4.24 \times 10^{-4} \text{ mol L}^{-1}}{0.010 \text{ mol L}^{-1}} = \boxed{0.042}$$

Check

The pH is greater than 7, as expected for a solution of a base, and [OH^-] greatly exceeds 10^{-7} mol L^{-1}, which justifies neglect of the autoionization of water as a source of [OH^-]. The y obtained by the approximate calculation is 4% of the quantity from which it is subtracted, justifying the approximation under the 4% to 5% rule of thumb. Further justification comes from running an exact solution for y using a pre-programmed calculator. This answer is $y = 4.15 \times 10^{-4}$, which leads to a pOH of 10.62 and a fraction ionized of 0.042.

EXERCISE

Cocaine is a weak base with a K_b of 2.6×10^{-7} at 25°C. It is not very soluble in water, and a nearly saturated solution at 25°C contains 0.0055 mol L^{-1}. Compute the pH of this solution.

Answer: 9.58.

Hydrolysis

Most of the acids considered up to this point have been uncharged species, with the general formula HA. Nothing in the Brønsted-Lowry picture, however, requires that acids consist of electrically neutral particles. For example, the salt ammonium chloride (NH_4Cl) dissolves readily in water:

$$NH_4Cl(s) \longrightarrow NH_4^+(aq) + Cl^-(aq)$$

Some of the $NH_4^+(aq)$ ions donate hydrogen ions to molecules of water, a straight-forward Brønsted-Lowry acid–base reaction:

$$\underset{\text{acid}_1}{NH_4^+(aq)} + \underset{\text{base}_2}{H_2O(\ell)} \rightleftharpoons \underset{\text{acid}_2}{H_3O^+(aq)} + \underset{\text{base}_1}{NH_3(aq)} \qquad \frac{[H_3O^+][NH_3]}{[NH_4^+]} = K_a$$

The $NH_4^+(aq)$ ion is weaker than many other weak acids (Table 8–2 gives its K_a as 5.6×10^{-10}), but solutions of ammonium chloride neutralize bases (Fig. 8–9) and have pH's well below 7.

The special term **hydrolysis** is applied to any reaction between water and an ion (not a molecule) that produces either hydrogen ion or hydroxide ion and so changes the pH from 7. Hydrolysis reactions are special cases of Brønsted-Lowry acid–base reactions. As just shown, the hydrolysis of the $NH_4^+(aq)$ ion is adequately described as a Brønsted-Lowry reaction in which water acts as a base and $NH_4^+(aq)$ acts as an acid. Similarly, the hydrolysis of an anion such as the fluoride ion (F^-) is simply another Brønsted-Lowry acid–base reaction, with water now acting as an acid:

$$H_2O(\ell) + F^-(aq) \rightleftharpoons HF(aq) + OH^-(aq)$$

This reaction raises the pH of a solution of the salt sodium fluoride (NaF).

Most ions hydrolyze to a detectable extent. It follows that most ions are acids or bases. The hydrolysis of anions typically raises the pH; the hydrolysis of cations typically lowers the pH. When both the cation and anion of a dissolved salt hydrolyze, the outcome for the pH depends on the relative strengths of the cation as an acid (given by its K_a) and the anion as a base (given by its K_b).

| (a) | (b) | (c) |

Figure 8–9 • Proof that a solution of ammonium chloride is acidic. (a) A solution of sodium hydroxide, with the pink color of the indicator phenolphthalein indicating its basic character. (b, c) As aqueous ammonium chloride is added, a neutralization reaction takes place and the indicator turns colorless. The change proceeds from the bottom up because the mixture is not well stirred and the solution of the ammonium chloride is denser than that of the sodium hydroxide. *(Leon Lewandowski)*

Just seven common anions have K_b's effectively equal to 0 and so do not raise the pH by hydrolysis. They are the chloride (Cl^-), bromide (Br^-), iodide (I^-), hydrogen sulfate (HSO_4^-), nitrate (NO_3^-), chlorate (ClO_3^-), and perchlorate (ClO_4^-) ions. These non-hydrolyzing anions are the conjugate bases of the seven common strong acids identified in Section 4–3 and Table 8–2.

The list of common non-hydrolyzing cations is also short, consisting of only ten: Li^+, Na^+, K^+, Rb^+, Cs^+, Mg^{2+}, Ca^{2+}, Sr^{2+}, Ba^{2+}, and Ag^+ ion. Most other cations, including such metal ions as Ni^{2+} and Fe^{2+}, hydrolyze detectably.

EXAMPLE 8–9

A student dissolves 0.100 mol of sodium acetate ($NaCH_3COO$) in enough water at 25°C to make 1.00 L of solution. Compute the pH of the solution.

Strategy

Determine whether the anion or the cation (or both) hydrolyzes by checking the list of anions and cations that do not hydrolyze. Obtain a K_a for the cation or a K_b for the anion from the data in Table 8–2. Formulate equilibrium-constant equations for concentrations of the ions involved in the hydrolysis reaction and find the pH as usual.

Solution

The salt dissolves completely according to the equation

$$NaCH_3COO(s) \longrightarrow Na^+(aq) + CH_3COO^-(aq)$$

The sodium ion does not hydrolyze

$$Na^+(aq) + H_2O(\ell) \longrightarrow NR$$

but the acetate ion does:

$$H_2O(\ell) + CH_3COO^-(aq) \rightleftharpoons CH_3COOH(aq) + OH^-(aq)$$
$$\frac{[CH_3COOH][OH^-]}{[CH_3COO^-]} = K_b$$

Assume that this is the only important source of $OH^-(aq)$ ion (that the autoionization of water can be neglected as a source of $OH^-(aq)$ ion). The table of changes for the preceding reaction is

	[CH$_3$COO$^-$] (mol L^{-1})	**[CH$_3$COOH]** (mol L^{-1})	**[OH$^-$]** (mol L^{-1})
Initial	0.100	0	≈ 0
Change	$-y$	$+y$	$+y$
Equilibrium	$0.100 - y$	y	y

The K_b for the acetate ion is

$$K_b = \frac{K_w}{K_a} = \frac{1.0 \times 10^{-14}}{1.8 \times 10^{-5}} = 5.6 \times 10^{-10}$$

in which the K_a is for CH_3COOH, the conjugate acid of CH_3COO^-. Substituting K_b and the values on the bottom line of the table of changes into the equilibrium-constant equation gives

$$\frac{y^2}{0.100 - y} = 5.6 \times 10^{-10}$$

Neglecting y relative to 0.100 and solving the resulting equation gives $y = 7.48 \times 10^{-6}$. This means $[OH^-] = 7.48 \times 10^{-6}$ mol L^{-1}. Consequently

$$pOH = 5.13 \quad \text{and} \quad pH = 14.00 - pOH = 8.87$$

Check

Substitution of the final concentrations into the equilibrium-constant equation gives the correct value of K_a:

$$\frac{[CH_3COOH][OH^-]}{[CH_3COO^-]} = \frac{(7.48 \times 10^{-6})(7.48 \times 10^{-6})}{(0.100 - 7.48 \times 10^{-6})} = 5.6 \times 10^{-5}$$

Note how small the value of y is relative to 0.100 (it is much less than 1% of it), while still large relative to 10^{-7}.

EXERCISE

Compute the pH of a 0.095 mol L^{-1} solution of NaF at 25°C. The K_a of HF is 6.6×10^{-4}.

Answer: pH = 8.08.

How does a species that is devoid of hydrogen, such as a metal ion, act as a Brønsted-Lowry acid when dissolved? Recall that all ions become aquated in water solution, as acknowledged by adding (*aq*) to their formulas. A positive ion, particularly one with +2 or +3 charge, attracts electrons to itself from its surrounding water molecules, as shown in Figure 6–3. A strong attraction weakens the O—H bonds in those water molecules so that their hydrogen ions are more easily donated than ordinarily. In this way it makes their hydrogen atoms acidic. For example, a Ni^{2+} ion in aqueous solution is closely associated with six molecules of water in a complex ion formulated $Ni(H_2O)_6^{2+}$. This ion acts as a Brønsted-Lowry acid according to the equation

$$Ni(H_2O)_6^{2+}(aq) + H_2O(\ell) \rightleftharpoons H_3O^+(aq) + Ni(H_2O)_5(OH)^+(aq)$$

The equilibrium can be treated in the same way as the acid ionization of HCN(*aq*) or $NH_4^+(aq)$ ion.

Knowing which ions hydrolyze and which do not hydrolyze allows one to predict whether a salt gives an acidic, basic, or neutral solution. Aqueous NaCl gives neutral solutions because neither Na^+ nor Cl^- ion hydrolyzes. Aqueous Na_3PO_4 is basic (Na^+ does not hydrolyze, but PO_4^{3-} ion does), and aqueous $NiCl_2$ is acidic (Ni^{2+} hydrolyzes, but Cl^- does not). When both the cation and anion in a salt hydrolyze (as in solutions of NH_4CN), the pH is raised if K_b for the anion exceeds K_a for the cation and lowered if K_b is less than K_a.

- Additional aspects of the interaction of metal ions with water are discussed in Section 9–5.

8–5 Buffer Solutions

A **buffer solution** (or **buffer**) is any solution that maintains a nearly constant pH despite small additions of acid or base. Typically, a buffer solution contains a mixture of a weak acid and its conjugate base (or a weak base and its conjugate acid). A buffer resists changes in pH because the acid reacts away added base and the conjugate base reacts away added acid. Buffer action is crucial in numerous chemical, biochemical, and physiological processes. Human blood, for example, has a pH near 7.4 that is maintained by a combination of carbonate, phosphate, and protein buffers. A blood pH below 7.0 or above 7.8 leads quickly to death.

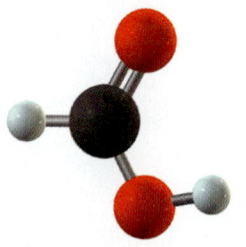

Formic acid (HCOOH) is the simplest carboxylic acid, with only a hydrogen atom attached to the —COOH group. It is used in curing leather, in dyeing, and in finishing paper and textiles. It is also one of the irritants in bee stings.

The pH of Buffer Solutions

Consider a buffer solution prepared by mixing some formic acid (HCOOH), a typical weak acid, with some sodium formate (NaHCOO) in water. The sodium formate dissociates as it dissolves to give sodium ions (Na^+) and formate ions ($HCOO^-$). An acid–base equilibrium is then quickly established involving formate ions and the formic acid

$$HCOOH(aq) + H_2O(\ell) \rightleftharpoons H_3O^+(aq) + HCOO^-(aq)$$

$$\frac{[H_3O^+][HCOO^-]}{[HCOOH]} = K_a$$

for which the acidity constant equals 1.8×10^{-4} at 25°C. What is the pH of this solution? Section 8–4 shows how to compute the pH of a solution containing a single weak acid such as HCOOH or a single weak base such as $HCOO^-$ ion. Now, a weak acid and its conjugate base are *both* present initially. The table-of-changes approach handles the situation easily. The calculations resemble those in Example 7–9, in which an equilibrium is established starting from a mixture in which both reactants and products are present, as the following example shows.

EXAMPLE 8–10

Suppose that 1.00 mol of HCOOH and 0.50 mol of NaHCOO are added to water and diluted to 1.00 L at 25°C. Calculate the pH of the solution.

Solution

The NaHCOO dissolves to give 0.50 M $Na^+(aq)$, which (because it does not hydrolyze) plays no role in determining the pH, and 0.50 M $HCOO^-(aq)$. The HCOOH dissolves to give 1.00 M HCOOH, which donates some H^+ ions according to the equilibrium

$$HCOOH(aq) + H_2O(\ell) \rightleftharpoons H_3O^+(aq) + HCOO^-(aq)$$

Set up a table of changes for this equilibrium:

	[HCOOH] (mol L^{-1})	[H$_3$O$^+$] (mol L^{-1})	[HCOO$^-$] (mol L^{-1})
Initial	1.00	≈0	0.50
Change	$-y$	$+y$	$+y$
Equilibrium	$1.00 - y$	y	$0.50 + y$

$$\frac{[H_3O^+][HCOO^-]}{[HCOOH]} = K_a$$

$$\frac{y(0.50 + y)}{1.00 - y} = 1.8 \times 10^{-4}$$

Because y is likely to be small relative to 0.50 (and to 1.00), this equation can be approximated as

$$1.8 \times 10^{-4} = \frac{y(0.50)}{1.00}$$

$$y = 3.6 \times 10^{-4}$$

$$[H_3O^+] = 3.6 \times 10^{-4} \quad \text{and} \quad pH = -\log_{10}(3.6 \times 10^{-4}) = 3.44$$

Check

The pH of 1.00 M HCOOH(aq) without added sodium formate equals 1.87 (calculated as illustrated in Example 8–6). Adding formate ion, a base, raises the pH, which makes sense.

Another check is to substitute the final concentrations into the equilibrium-constant expression to see whether the correct value of K_a results

$$\frac{[H_3O^+][HCOO^-]}{[HCOOH]} = \frac{(3.6 \times 10^{-4})(0.50 + 3.6 \times 10^{-4})}{(1.00 - 3.6 \times 10^{-4})} = 1.8 \times 10^{-4}$$

The addition and subtraction (but not the multiplication!) involving 3.6×10^{-4} can be neglected. Doing the computation twice, once with neglect and once without, shows this. Finally, the concentration of $H_3O^+(aq)$ is much larger than 10^{-7} mol L^{-1}, confirming that neglect of the autoionization of water was justified.

EXERCISE

Household bleach contains hypochlorous acid mixed with sodium hypochlorite. Suppose that 0.88 mol of hypochlorous acid (HClO) and 1.20 mol of sodium hypochlorite (NaClO) are mixed in enough water to give 1.50 L of solution. Compute the pH of the solution at 25°C.

Answer: pH = 7.66.

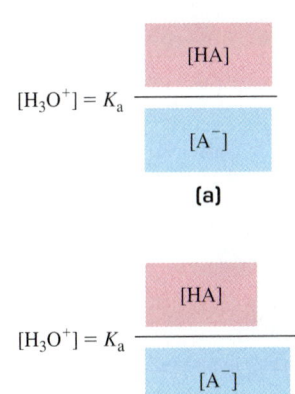

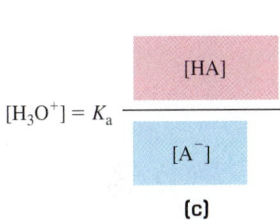

How Buffer Solutions Work

Writing the equilibrium-constant equation that governs the ionization of a weak acid HA as

$$[H_3O^+] = K_a \frac{[HA]}{[A^-]}$$

highlights the fact that the concentration of hydronium ion depends on the ratio of the concentration of the weak acid to the concentration of its conjugate base. A solution acts as a buffer if both of these concentrations are fairly large. Adding a small amount of base to an effective buffer takes away only a few percent of the HA molecules by converting them into A$^-$ ions and adds only a few percent to the amount of A$^-$ that was originally present. The ratio [HA]/[A$^-$] decreases, but only slightly. Added acid consumes a small fraction of the base A$^-$ to generate a bit more HA. The ratio [HA]/[A$^-$] now increases but, again, the change is only slight. Because the concentration of H_3O^+ is tied directly to this ratio, it also changes only slightly. The effect is diagrammed in Figure 8–10. The following example shows the quantitative workings of buffer action.

Figure 8-10 • The $[H_3O^+]$ in a buffer depends on the ratio [HA]/[A$^-$]. If both the numerator and denominator of this fraction are large, then small additions of base or acid change the ratio only very slightly. (a) Initial ratio. (b) The ratio after addition of a small amount of base. (c) The ratio after the addition of a small amount of acid.

EXAMPLE 8–11

Suppose 0.10 mol of a strong acid such as HCl is included as the solution in Example 8–10 is mixed. Calculate the pH of the resulting solution.

Strategy

Imagine that the 0.10 mol of H_3O^+ that comes from the dissolution of the HCl reacts completely with HCOO$^-$ ions, giving HCOOH (and H$_2$O), and that some of this HCOOH then reacts with H$_2$O to give HCOO$^-$ ion and H_3O^+. This route to equilibrium is one of many that might be imagined. The position of the final equilibrium does not depend on the route by which it is attained.

Solution

The 0.10 mol of H_3O^+ ions from HCl consumes 0.10 mol of $HCOO^-$. The concentrations of $HCOO^-$ and HCOOH after complete reaction of the HCl and $HCOO^-$ but before any subsequent reaction are

$$[HCOO^-]_{(init)} = 0.50 - 0.10 = 0.40 \text{ mol L}^{-1}$$
$$[HCOOH]_{(init)} = 1.00 + 0.10 = 1.10 \text{ mol L}^{-1}$$

Now, HCOOH reacts according to the equation

$$HCOOH(aq) + H_2O(\ell) \rightleftharpoons H_3O^+(aq) + HCOO^-(aq)$$

Set up a table of changes for this reaction:

	[HCOOH] (mol L^{-1})	[H$_3$O$^+$] (mol L^{-1})	[HCOO$^-$] (mol L^{-1})
Initial	1.10	≈ 0	0.40
Change	$-y$	$+y$	$+y$
Equilibrium	$1.10 - y$	y	$0.40 + y$

$$\frac{[H_3O^+][HCOO^-]}{[HCOOH]} = K_a$$
$$\frac{y(0.40 + y)}{1.10 - y} = 1.8 \times 10^{-4}$$

The unknown y is likely to be small relative to both 0.40 and 1.10. Neglect it in both the numerator and denominator of the fraction, and solve for y:

$$y \approx \left(\frac{1.10}{0.40}\right)(1.8 \times 10^{-4}) = 4.9 \times 10^{-4}$$
$$pH = -\log_{10}(4.9 \times 10^{-4}) = 3.31$$

- The small change is naturally in the direction of lower pH because acid was added.

Even though 0.10 mol of a strong acid was added, the pH decreased by only 0.13 pH units, from 3.44 to 3.31. The same amount of hydrochloric acid in a liter of pure water lowers the pH from 7.0 to 1.0.

Check

The value of y is about 0.1% of 0.40, which justifies neglecting y relative to 0.40 and 1.10. Also, the equilibrium concentration of $H_3O^+(aq)$ is much larger than 10^{-7} mol L^{-1}, confirming that the neglect of the autoionization of water was justified.

EXERCISE

A total of 0.20 mol of HCl is added to the bleach solution from Exercise 8–10. The volume of the solution does not change. Compute the final pH.

Answer: pH = 7.49.

The calculation is similar if a strong base such as $OH^-(aq)$ is mixed in instead of a strong acid. The strong base converts some formic acid to formate ions. Adding 0.10 mol of $OH^-(aq)$ to the $HCOOH/HCOO^-$ buffer of Example 8–10 *increases* the initial concentration of $HCOO^-$ ion from 0.50 mol L^{-1} to 0.60 mol L^{-1} and *decreases* the initial concentration of $HCOOH$ from 1.00 mol L^{-1} to 0.90 mol L^{-1}. After restoration of the acid–base equilibrium, the pH equals 3.57, a rise of 0.13 pH units. The same amount of $OH^-(aq)$ raises the pH of a liter of pure water from 7.0 to 13.0.

Designing Buffers

The pH of the solution strongly influences the success of numerous reactions and purifications. Fortunately, it is possible to design buffers to maintain any desired pH. Design considerations start with the acidity-constant equation in the form

$$[H_3O^+] = K_a \frac{[HA]}{[A^-]}$$

In the buffer systems in Examples 8–10 and 8–11, the quantities on the "change" line in the table of changes are negligibly small compared with the initial concentrations of the acid HA and its conjugate base A^-. In other words, the equilibrium concentrations of HA and A^- stay quite close to their initial values. When this is the case, which it is in most practical buffers, the preceding relationship can be approximated as

$$[H_3O^+] \approx K_a \frac{[HA]_{(init)}}{[A^-]_{(init)}}$$

Taking the negative logarithm of both sides and then using the definitions of pH and pK_a gives

$$pH \approx pK_a - \log_{10} \frac{[HA]_{(init)}}{[A^-]_{(init)}}$$

This result is known as the Henderson–Hasselbalch equation. It gives the correct pH of the solutions in Examples 8–10 and 8–11 as follows:

• Use care when employing this equation because it is approximate. It works only if both $[H_3O^+]$ and $[OH^-]$ are small relative to the original concentrations $[HA]_{(init)}$ and $[A^-]_{(init)}$.

Example 8–10: $\quad pH \approx -\log_{10}(1.8 \times 10^{-4}) - \log_{10} \frac{[HCOOH]_{(init)}}{[HCOO^-]_{(init)}}$

$$= 3.74 - \log_{10} \frac{1.00}{0.500} = 3.44$$

Example 8–11: $\quad pH \approx -\log_{10}(1.8 \times 10^{-4}) - \log_{10} \frac{[HCOOH]_{(init)}}{[HCOO^-]_{(init)}}$

$$= 3.74 - \log_{10} \frac{1.10}{0.40} = 3.31$$

The Henderson–Hasselbalch equation is helpful in designing buffers. If $[HA]_{(init)}$ and $[A^-]_{(init)}$ in a buffer solution differ a great deal, then their ratio $\frac{[HA]_{(init)}}{[A^-]_{(init)}}$ in the Henderson–Hasselbalch equation is either very large or very small. Also, their ratio changes rapidly when acid or base is added. The pH of the solution consequently changes rapidly, and the buffer is ineffective. The pH changes *least* rapidly when the ratio of $[HA]_{(init)}$ to $[A^-]_{(init)}$ equals 1. *An optimal buffer is a buffer in which the weak acid and its conjugate base have equal concentrations.* To construct a useful buffer:

1. Select a weak acid that has its pK_a as close as possible to the desired pH.
2. Substitute pH and pK_a in the Henderson–Hasselbalch equation to obtain the ratio of $[HA]_{(init)}$ and $[A^-]_{(init)}$ that gives exactly the desired pH.
3. Use *amounts* of acid and conjugate base that exceed the anticipated amount of acid or base that the buffer might be called upon to resist.

The last step recognizes that very dilute buffers have no practical use. Moreover, the Henderson–Hasselbalch equation does not give the correct pH for very dilute buffers.

EXAMPLE 8–12

Design a buffer system with pH 4.60 at 25°C.

Solution

From a list of pK_a values (Table 8–2), we find that the pK_a for acetic acid is 4.74, so the CH_3COOH/CH_3COO^- buffer is a suitable one. The ratio of concentrations required to give the desired pH can be computed as follows:

$$pH \approx pK_a - \log_{10} \frac{[CH_3COOH]_{(init)}}{[CH_3COO^-]_{(init)}}$$

$$4.60 \approx 4.74 - \log_{10} \frac{[CH_3COOH]_{(init)}}{[CH_3COO^-]_{(init)}}$$

$$\log_{10} \frac{[CH_3COOH]_{(init)}}{[CH_3COO^-]_{(init)}} \approx 4.74 - 4.60 = 0.14$$

$$\frac{[CH_3COOH]_{(init)}}{[CH_3COO^-]_{(init)}} \approx 10^{0.14} = 1.4$$

Such a ratio could be established by dissolving 0.10 mol of sodium acetate and 0.14 mol of acetic acid in 1.0 L of water, or 0.20 mol $NaCH_3COO$ and 0.28 mol CH_3COOH in 1.0 L of water, and so on. As long as the ratio of the concentrations is 1.4 (and the concentrations are not too small), the solution is buffered at pH 4.60.

EXERCISE

Design a buffer to maintain a pH of 9.20 at 25°C, and suggest how to prepare it.

Answer: A NH_3/NH_4^+ buffer with a $[NH_4^+]$ to $[NH_3]$ ratio of 1.12 would work. The buffer solution could be prepared by dissolving 0.100 mol of NH_3 and 0.112 mol of NH_4Cl in a reasonable quantity of water.

As Example 8–12 shows, the absolute concentrations of acid and conjugate base in a buffer are less important than their ratio in determining the pH. However, the absolute concentrations *do* affect the capacity of the solution to resist changes in pH induced by added acid or base. The higher the concentrations of the buffer-

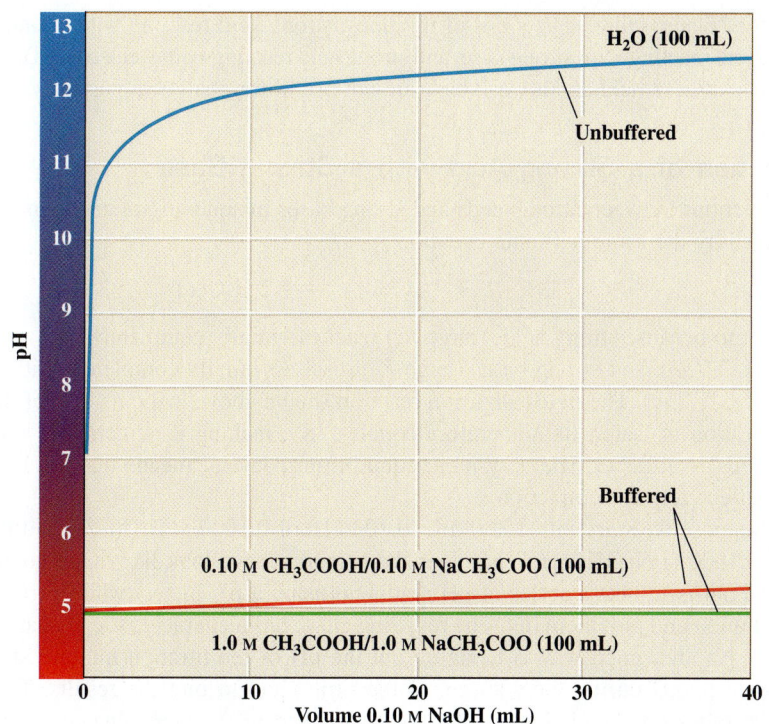

ing species, the smaller the change in pH when a fixed amount of a strong acid or base is added. In Example 8–10, for example, changing the concentrations of the buffering species from 1.00 mol L^{-1} and 0.500 mol L^{-1} to 0.500 mol L^{-1} and 0.250 mol L^{-1}, respectively, does not alter the original pH of 3.44 because the ra- tio of acid-to-base concentrations is unchanged. The pH after 0.100 mol of HCl is added (see Example 8–11) does change. It is 3.14 rather than 3.31. The buffer at lower concentration resists pH change less well (Fig. 8–11). The buffering capac- ity of any buffer is exhausted if sufficient strong acid (or strong base) is added to consume all of the weak base (or the weak acid) originally present.

8-6 Acid–Base Titration Curves

An acid–base titration consists of the controlled addition of a dissolved base to a dissolved acid (or the reverse). The acid and base react rapidly to neutralize each other. At the **equivalence point,** enough **titrant,** the solution being added, has gone in to make the chemical amounts of the acid and base exactly equal (assuming that the two react in a 1 : 1 molar ratio, which is the most common case). The pH of a titration mixture changes every time a drop of titrant is added, but the *rate* of this change varies enormously. A **titration curve,** a graph of pH as a function of the volume of titrant, displays in detail how the pH changes over the course of an acid– base titration. Significantly, the pH changes most rapidly near the equivalence point.

The exact shape of a titration curve depends on the K_a and K_b of the acid and base and on their concentrations. Experimental titration curves furnish K_a's or K_b's and the concentration of one of the reactants if the concentration of the other is known. Analytical chemists perform acid–base titrations and plot titration curves to obtain such information.

We consider three categories of titrations: strong acid reacting with strong base, weak acid reacting with strong base, and strong acid reacting with weak base. Titrations between a weak acid and a weak base are not useful for analytical purposes.

Titration of a Strong Acid with a Strong Base

All reactions between strong acids and strong bases in aqueous solution can be represented by the same net ionic equation:

$$H_3O^+(aq) + OH^-(aq) \rightleftharpoons 2\,H_2O(\ell)$$

This is so because strong acids (large K_a) react essentially completely with water to give $H_3O^+(aq)$, and strong bases (large K_b) react essentially completely with water to give $OH^-(aq)$. The **neutralization** reaction shown above is the reverse of the autoionization of water. Its K therefore equals $1/K_w$, making it numerically equal to $1.0 \times 10^{+14}$ at 25°C. The very large equilibrium constant means that the reaction goes effectively to completion.

Suppose we start with 100.0 mL (0.1000 L) of 0.1000 M HCl(*aq*) and titrate it with 0.1000 M NaOH(*aq*). What does the titration curve look like? One good way to find out is to take a series of pH readings (using a pH meter) while performing the titration and plot them out. The resulting curve has a characteristic S-shape (Fig. 8–12). Another good way is to figure out the pH of the titration mixture at many different points during the addition of the titrant and to plot the results. This approach relies on calculations of the type explained in the preceding two sections. Let us examine several of these calculations.

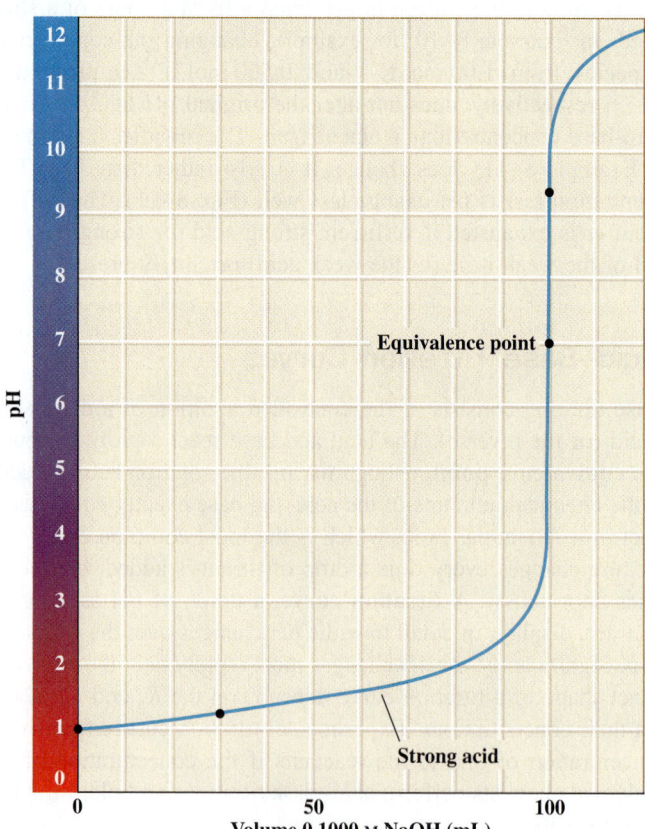

Figure 8–12 • An experimental titration curve for the titration of 100.0 mL of 0.1000 M HCl, a strong acid, with 0.1000 M NaOH, a strong base. The pH increases continuously as the base is added, but the rate of the increase changes dramatically, giving rise to a characteristic S-shaped curve. The black dots mark the four points on the curve the locations of which are specifically calculated in this section.

1. $V = 0$ mL of NaOH(aq) (titration not yet started)

The concentration of $H_3O^+(aq)$ is 0.1000 mol L^{-1} because HCl, a strong acid (large K_a), reacts essentially to completion with water to give $H_3O^+(aq)$. The pH is easily calculated:

$$pH = -\log_{10}[H_3O^+] = -\log_{10}(0.1000) = 1.000$$

It is helpful to know the chemical amount of H_3O^+ that must be neutralized during the course of the titration:

$$n_{H_3O^+} = (0.1000 \text{ L}) \times \left(\frac{0.1000 \text{ mol } H_3O^+}{1 \text{ L}}\right) = 10.00 \times 10^{-3} \text{ mol } H_3O^+$$

2. $V = 30.00$ mL of NaOH(aq) (mid-titration)

Dissolved NaOH dissociates essentially completely to give Na^+ ions and OH^- ions. The chemical amount of OH^- ion furnished by 30.00 mL of NaOH(aq) is

$$n_{OH^-} = 0.03000 \text{ L} \times \left(\frac{0.1000 \text{ mol } OH^-}{1 \text{ L}}\right) = 3.000 \times 10^{-3} \text{ mol } OH^-$$

According to the balanced equation, this chemical amount of OH^- ion neutralizes an equal amount of H_3O^+ ion. The amount of hydronium ion that remains unreacted is

$$n_{H_3O^+} = 10.00 \times 10^{-3} \text{ mol} - 3.000 \times 10^{-3} \text{ mol}$$
$$= 7.00 \times 10^{-3} \text{ mol}$$

Neutralization is not the only factor changing the concentration of H_3O^+ ion. Adding titrant dilutes the whole solution as well. The addition of 30.00 mL of titrant raises the volume of the mixture from 100.0 mL to 130.0 mL (that is, from 0.1000 L to 0.1300 L). The chemical amount of H_3O^+ ion must be divided by this *total* volume to obtain its concentration:

$$[H_3O^+] = \frac{n_{H_3O^+}}{V_{total}} = \frac{7.00 \times 10^{-3} \text{ mol}}{0.1300 \text{ L}} = 0.0538 \text{ mol L}^{-1}$$

The preceding calculations also can be performed quite conveniently using millimoles and milliliters:

$$n_{H_3O^+} = 10.00 \text{ mmol} - 3.000 \text{ mmol} = 7.00 \text{ mmol}$$

$$[H_3O^+] = \frac{n_{H_3O^+}}{V_{total}} = \frac{7.00 \text{ mmol}}{130.0 \text{ mL}} = 0.0538 \text{ mmol mL}^{-1}$$

$$= 0.0538 \frac{\text{mmol}}{\text{mL}} \times \left(\frac{1000 \text{ mL}}{1 \text{ L}}\right) \times \left(\frac{1 \text{ mol}}{1000 \text{ mmol}}\right) = 0.0538 \text{ mol L}^{-1}$$

The pH at this point in the titration is $-\log(0.0538)$, which equals 1.269.

3. $V = 100.00$ mL NaOH(aq) (the equivalence point)

The chemical amount of OH^- that has been added is

$$n_{OH^-} = 100.00 \text{ mL} \times \left(\frac{0.1000 \text{ mmol } OH^-}{1 \text{ mL}}\right) = 10.00 \text{ mmol } OH^-$$

This is just enough exactly to neutralize the 10.00 mmol (10.00×10^{-3} mol) of H_3O^+ that was originally present. The volume of titrant that just brings a titration to its equivalence point is the **equivalent volume** (V_e).

Midway in the titration of HCl(aq) with NaOH(aq). The base is being added to the acid from a buret. The swirls of pink come from the indicator phenolphthalein turning color in regions in which the concentration of OH^- temporarily exceeds the concentration of H_3O^+. *(Charles D. Winters)*

- The pH equals 7.0 at the equivalence point *only* in strong acid/strong base titrations and *not* in titrations involving weak acids or weak bases.

At the equivalence point of a strong acid/strong base titration, the mixture consists of a dissolved salt (in this case, NaCl) in which neither the cation nor anion hydrolyzes. The concentrations of OH^- and H_3O^+ ion in the solution therefore equal each other, and the pH equals 7.0 (at 25°C).

4. **$V = 100.05$ mL NaOH(aq) (beyond the equivalence point)**

The strong acid was all neutralized by the first 100.0 mL of titrant. The OH^- ion in the next 0.05 mL of titrant, which is approximately the volume of a single small drop of solution let in from a buret, has nothing to react with. It accumulates in the mixture. Its chemical amount equals the volume of titrant added beyond V_e multiplied by its concentration:

$$n_{OH^-} = 0.05 \text{ mL} \times \left(\frac{0.1000 \text{ mmol } OH^-}{1 \text{ mL}} \right) = 0.005 \text{ mmol } OH^-$$

The volume of the titration mixture is

$$V_{total} = 100.00 \text{ mL} + 100.05 \text{ mL} = 200.05 \text{ mL}$$

so that the concentration of OH^- is

$$[OH^-] = \frac{0.005 \text{ mmol}}{200.05 \text{ mL}} = 2.5 \times 10^{-5} \text{ mmol mL}^{-1} = 2.5 \times 10^{-5} \text{ mol L}^{-1}$$

The easiest way to obtain the pH is to get the pOH first:

$$pOH = -\log_{10}(2.5 \times 10^{-5}) = 4.60$$
$$pH = pK_w - pOH = 14.00 - 4.6 = 9.40$$

Systematically using different volumes of titrant in the "mid-titration" and the "beyond the equivalence point" calculations illustrated here generates a theoretical titration curve that matches the experimental curve (Fig. 8–12) very closely. Such calculations show that the concentration of H_3O^+ in this titration decreases by four orders of magnitude between 99.98 mL and 100.02 mL NaOH (see Problem 8–53). Because the pH changes so dramatically near the equivalence point, any indicator whose color changes between pH = 5.0 and pH = 9.0 signals the equivalence point of this titration to an accuracy of 0.02 mL in 100.0 mL, or 0.02%.

Calculating points on a titration curve when a strong base is titrated with a strong acid requires only slight adaptation of the methods just illustrated. The roles of acid and base, and of H_3O^+ and OH^-, are simply reversed. A plot of pOH versus the volume of titrant looks just like the pH plot that was just computed. The pH, which always equals $pK_w - pOH$, starts high and drops rapidly in the vicinity of the equivalence point.

Titration of a Weak Acid with a Strong Base

The classic example is the titration of a solution of acetic acid with a solution of sodium hydroxide. The two react according to the net ionic equation

$$CH_3COOH(aq) + OH^-(aq) \rightleftharpoons H_2O(\ell) + CH_3COO^-(aq)$$

This reaction is the reverse of the K_b reaction of the acetate ion (CH_3COO^-). Its equilibrium constant is therefore the reciprocal of K_b of CH_3COO^- ion:

- The K_b of the acetate ion is computed from the K_a of acetic acid in Example 8–9.

$$K = \frac{1}{K_b} = \frac{1}{5.6 \times 10^{-10}} = 2 \times 10^9$$

The large K means that the reaction goes essentially to completion.

Consider now the titration of 0.1000 M CH₃COOH with 0.1000 M NaOH. These numbers are deliberately chosen to equal the numbers in the strong acid/strong base titration that was just discussed. As in that titration, 10.00 mmol of acid is present at the start. As in that titration, the acid and base react in a 1:1 molar ratio so that addition of 10.00 mmol of OH^- ion brings the titration to the equivalence point. As in that titration, the equivalent volume is 100.0 mL. Yet despite these similarities, the experimental titration curve for this titration differs substantially from the curve for that titration, as shown by the blue and red lines in Figure 8–13. The following calculations of several points on the weak acid/strong base titration curve show why the differences arise.

1. **V = 0 mL of NaOH(aq) (titration not yet started)**

Before the addition of any titrant, the titration mixture is simply a 0.1000 M solution of acetic acid. Its pH is readily found by the method used in Example 8–6. It equals 2.88. This answer exceeds 1.000, the pH of 0.1000 M HCl, because the weak CH₃COOH is a less effective donor of hydrogen ions than the strong HCl.

2. **V = 30.00 mL of NaOH(aq) (mid-titration)**

Dissolved NaOH dissociates essentially completely to give Na^+ ions and OH^- ions. The chemical amount of OH^- ion furnished by 30.00 mL of NaOH(aq) is therefore

$$n_{OH^-} = 30.00 \text{ mL} \times \left(\frac{0.1000 \text{ mmol OH}^-}{1 \text{ mL}} \right) = 3.000 \text{ mmol OH}^-$$

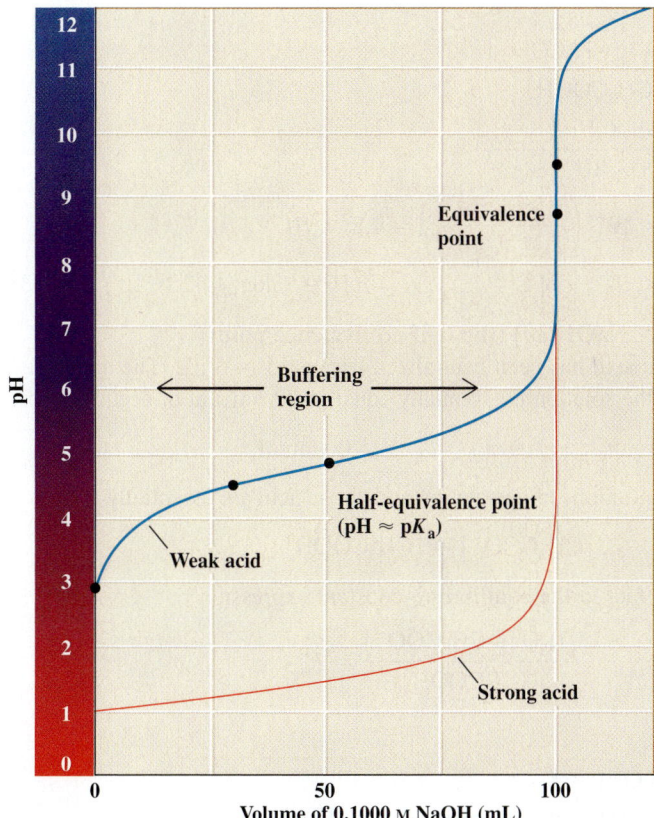

Figure 8-13 • A titration curve for the titration of a weak acid by a strong base (*upper curve*). The curve shown is for 100.0 mL of 0.1000 M CH₃COOH titrated with NaOH. The five marked points are points at which the pH is calculated in this section. For comparison, the thin red line shows the titration curve for a strong acid of the same amount and concentration.

The acetic acid solution starts out containing 10.00 mmol of CH_3COOH. The acid–base reaction converts 1 mmol of CH_3COOH into CH_3COO^- ion for every 1 mmol of OH^- ion that is added. Hence, 3.000 mmol of CH_3COO^- ion is now present

$$n_{CH_3COO^-} = 3.000 \text{ mmol } CH_3COO^-$$

and 7.00 mmol of CH_3COOH remains unreacted:

$$n_{CH_3COOH} = 10.00 \text{ mmol} - 3.000 \text{ mmol} = 7.00 \text{ mmol } CH_3COOH$$

The total volume of the solution is 130.0 mL. The concentrations of the CH_3COOH and CH_3COO^- are

$$[CH_3COOH] = \frac{7.00 \text{ mmol}}{130.0 \text{ mL}} = 5.38 \times 10^{-2} \text{ mol L}^{-1}$$

$$[CH_3COO^-] = \frac{3.00 \text{ mmol}}{130.0 \text{ mL}} = 2.31 \times 10^{-2} \text{ mol L}^{-1}$$

- Again, it is perfectly correct (and often very convenient) to obtain the molarities by dividing millimoles by milliliters.

The titration mixture at this point is nothing other than a buffer solution that might have been prepared by mixing a solution to contain acetic acid at a concentration of 5.38×10^{-2} mol L^{-1} and sodium acetate at a concentration of 2.31×10^{-2} mol L^{-1}. Its pH can be found by substitution in the Henderson–Hasselbalch equation, but the method used in Example 8–10 is more instructive. As in that example, y is the equilibrium concentration of H_3O^+ ion, and the dissociation of the acetic acid adds only negligibly to the concentration of acetate ion while subtracting only negligibly from the concentration of acetic acid:

$$K_a = \frac{[H_3O^+][CH_3COO^-]}{[CH_3COOH]}$$

$$1.8 \times 10^{-5} = \frac{y(2.31 \times 10^{-2} + y)}{(5.38 \times 10^{-2} - y)}$$

$$y = 1.8 \times 10^{-5}\left(\frac{5.38 \times 10^{-2}}{2.31 \times 10^{-2}}\right) = 4.2 \times 10^{-5}$$

$$[H_3O^+] \approx 4.2 \times 10^{-5} \text{ mol L}^{-1} \quad \text{and} \quad pH = -\log_{10}(4.2 \times 10^{-5}) = 4.38$$

3. **$V = 50.00$ mL of NaOH(aq) (the half-equivalence point)**

Half of the acetic acid has been neutralized because $V = V_e/2$. The amount of acetate ion in the solution is essentially equal to the amount of acetic acid

$$n_{CH_3COO^-} \approx n_{CH_3COOH} = 5.00 \text{ mmol}$$

so that the concentrations of acetate ion and acetic acid are essentially equal:

$$[CH_3COO^-] \approx [CH_3COOH]$$

Substituting this fact in the equilibrium-constant expression

$$\frac{[H_3O^+][CH_3COO^-]}{[CH_3COOH]} = K_a$$

gives

$$[H_3O^+] = K_a$$
$$pH = pK_a = -\log_{10}(1.8 \times 10^{-5}) = 4.74$$

The useful equation $pH = pK_a$ holds at half-equivalence in all titrations of a weak acid with a strong base.

In the vicinity of the half-equivalence point, the pH rises only slowly as the NaOH solution is added. The buffer solution created by the progress of the titration is at its most effective at half-equivalence. The titration curve (blue line in Fig. 8–13) shows how buffering action develops and is then overcome.

4. $V = 100.00$ mL NaOH(*aq*) added (the equivalence point)

10.00 mmol of OH^- has been added and has reacted with the original 10.00 mmol of CH_3COOH. The total volume is 200.00 mL. An identical solution could be prepared by mixing 10.00 mmol of $NaCH_3COO$ in enough water to give 200.00 mL of solution. The calculation of the pH is identical to the calculation of pH in Example 8–9 except that the original concentration of the CH_3COO^- ion is 0.05000 mol L^{-1}. The pH equals 8.72. Note well that when a weak acid is titrated with a strong base, the pH does *not* equal 7.0 at the equivalence point but is higher.

5. $V = 100.05$ mL NaOH(*aq*) (beyond the equivalence point)

The original 10.00 mmol of CH_3COOH was all reacted away by the 10.00 mmol of OH^- ion contained in the first 100.0 mL of titrant. The OH^- ion in the next 0.05 mL of titrant amounts to 0.005 mmol. It simply joins the solution of $NaCH_3COO$ produced by the reaction. Its concentration is

$$[OH^-] = \frac{0.005 \text{ mmol}}{200.05 \text{ mL}} = 2.5 \times 10^{-5} \text{ mmol mL}^{-1} = 2.5 \times 10^{-5} \text{ mol L}^{-1}$$

The reaction

$$CH_3COO^-(aq) + H_2O(\ell) \rightleftharpoons CH_3COOH(aq) + OH^-(aq)$$
$$K_b = 5.6 \times 10^{-10}$$

does very little to change this concentration. Its K_b (which is taken from Example 8–9) is small, and the added OH^- shifts it to the left. In other terms, the hydrolysis of the CH_3COO^- ion is nearly completely suppressed by the added OH^- ion. The pH beyond the equivalence point in this titration is consequently very close to the pH beyond the equivalence point in a strong acid/strong base titration. It is computed in the same way and equals 9.4.

If the equivalent volume V_e were not known in this titration, it could be determined experimentally by putting an indicator that changes color at a pH near 8.72 into the titration mixture. Phenolphthalein, which changes from colorless to red over a pH range from 8.2 to 10.0, would be suitable.

Titration of a Weak Base with a Strong Acid

The titration of a solution of ammonia with a solution of hydrochloric acid is a good example. The two react according to the net ionic equation

$$NH_3(aq) + H_3O^+(aq) \rightleftharpoons H_2O(\ell) + NH_4^+(aq) \qquad K = 1.8 \times 10^9$$

This reaction is the reverse of the K_a reaction of the ion NH_4^+. Its large K, the reciprocal of $K_a(NH_4^+)$ in Table 8–2, means that the reaction goes essentially to completion. Points on the titration curve are calculated by the methods used for the titration of a weak acid with a strong base, but modified by replacing K_a with K_b

and reversing the roles of OH^- and H_3O^+. Thus, the equation $pOH = pK_b$ holds at the half-equivalence point instead of $pH = pK_a$, and pOH, not pH, exceeds 7.0 at the equivalence point. A plot of pOH versus the volume of titrant closely resembles the pH plot just computed. When pH is needed, it is easiest to compute pOH first and then subtract it from pK_w.

Using Titration Curves

The pH changes a bit less sharply near the equivalence point when a weak base or weak acid is titrated compared with the case of a strong acid or base. This lowers the precision of determinations of the equivalent volume slightly, but not enough to prevent the very widespread use of titrations and experimental titration curves to determine the concentrations of weak acids or weak bases in solution and the K_a's or K_b's of weak acids or weak bases. The following example illustrates.

EXAMPLE 8–13

Exactly 50.00 mL of a solution of the weak acid propionic acid (CH_3CH_2COOH) of unknown concentration is titrated with a 0.1000 M solution of NaOH. The reaction is

$$CH_3CH_2COOH(aq) + OH^-(aq) \longrightarrow H_2O(\ell) + CH_3CH_2COO^-(aq)$$

The equivalence point is reached after 39.30 mL of NaOH solution has been added. At the half-equivalence point (19.65 mL), the pH is 4.85. Calculate the original concentration and the acidity constant, K_a, of propionic acid.

Solution

Abbreviate propionic acid and propionate ion as HOPr and OPr$^-$. It takes 1 mol of OH^- to react with 1 mol of HOPr, so the chemical amount of HOPr originally present equals the chemical amount of $OH^-(aq)$ added to reach the equivalence point:

$$n_{HOPr} = (0.1000 \text{ mol L}^{-1})(0.03930 \text{ L})$$

The original concentration of HOPr equals this amount divided by the original volume of the solution of HOPr:

$$c_{HOPr} = \frac{n_{HOPr}}{V} = \frac{(0.1000 \text{ mol L}^{-1})0.03930 \text{ L}}{0.05000 \text{ L}} = 0.07860 \text{ mol L}^{-1}$$

The approximate equation $pH = pK_a$ holds at half-equivalence, and the pH at half-equivalence equals 4.85. Therefore

$$pK_a = 4.85 \quad \text{(at half-equivalence)}$$
$$K_a = 10^{-4.85} = 1.4 \times 10^{-5}$$

Check

The acidity constant K_a is quite close to the value tabulated for CH_3CH_2COOH in Table 8–2.

EXERCISE

Exactly 25.00 mL of a solution of the weak base methylamine (CH_3NH_2) is titrated with 0.1000 M hydrochloric acid. They react according to the net ionic equation
$$CH_3NH_2(aq) + H_3O^+ \rightleftharpoons CH_3NH_3^+(aq) + H_2O(\ell).$$

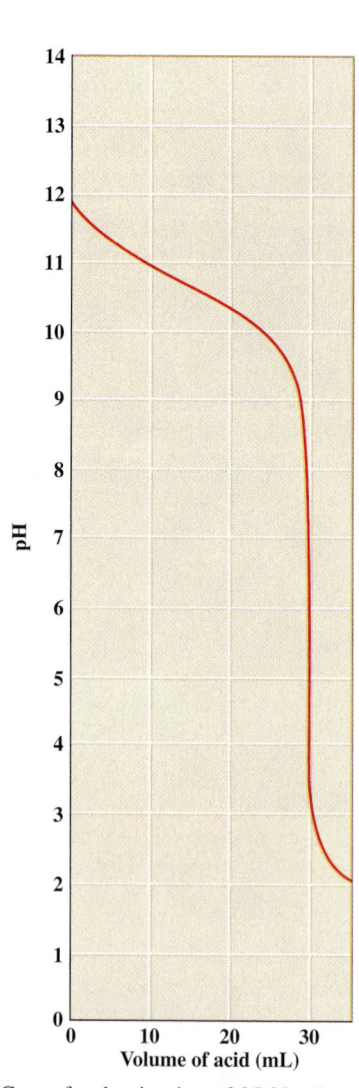

Curve for the titration of 25.00 mL of methylamine solution with 0.1000 M HCl.

The equivalence point is reached after 29.64 mL of HCl solution has been added. At the half-equivalence point (14.82 mL), the pH is 10.64. Calculate the original concentration of the methylamine and its basicity constant, K_b.

Answer: $[CH_3NH_2] = 0.1186$ mol L^{-1}; $K_b = 4.4 \times 10^{-4}$.

8-7 Polyprotic Acids

So far we have considered only **monoprotic acids,** acids capable of donating only a single hydrogen ion to acceptors. **Polyprotic acids** also exist. These acids can donate two or more hydrogen ions. One mole of a *diprotic* acid neutralizes double the amount of a strong base that a monoprotic acid neutralizes under the same circumstance. One mole of a *triprotic* acid neutralizes three times the amount. Sulfuric acid is a diprotic acid. One mole of it neutralizes 2 mol of OH^- ion, a strong base:

$$H_2SO_4(aq) + 2\ OH^-(aq) \longrightarrow 2\ H_2O(\ell) + SO_4^{2-}(aq)$$

Sulfuric acid reacts less completely with water than it does with $OH^-(aq)$ ion because water is a far weaker base than $OH^-(aq)$. The reaction is best described by two equations and two K_a's:

$$H_2SO_4(aq) + H_2O(\ell) \rightleftharpoons H_3O^+(aq) + HSO_4^-(aq) \qquad K_a = K_{a1} \approx 100$$
$$HSO_4^-(aq) + H_2O(\ell) \rightleftharpoons H_3O^+(aq) + SO_4^{2-}(aq) \qquad K_a = K_{a2} = 1.2 \times 10^{-2}$$

- The successive acidity constants for polyprotic acids are designated $K_{a1}, K_{a2},$ and so forth.

These two equilibria are established simultaneously. The hydrogen sulfate ion, which appears in blue on the right side of the first equation and in red on the left side of the second, is amphoteric. It is a base in the first reaction (its conjugate acid is H_2SO_4) and an acid in the second (its conjugate base is SO_4^{2-}). The magnitude of the K's show that H_2SO_4 is a strong acid but that HSO_4^- is a weak acid. The H_3O^+ ion in an aqueous solution of H_2SO_4, therefore, comes mostly from the first equilibrium.

Weak Polyprotic Acids

Weak polyprotic acids have small K_a's for the donation of even their first hydrogen ion. Examples are carbonic acid (H_2CO_3), formed from the reaction of CO_2 and water (Fig. 8–14), and phosphoric acid (H_3PO_4). Carbonic acid is diprotic, forming first HCO_3^- (hydrogen carbonate ion) and then CO_3^{2-} (carbonate ion). Phosphoric acid is triprotic, forming successively $H_2PO_4^-$, HPO_4^{2-}, and PO_4^{3-} as it donates its hydrogen ions. A weak acid with several hydrogen atoms in its formula does not necessarily donate them all. Citric acid ($C_6H_8O_7$) is a polyprotic acid, but it donates a maximum of three hydrogen ions per molecule, not eight. The formula of citric acid is often written $H_3C_6H_5O_7$ to segregate the acidic and non-acidic hydrogen atoms. Similarly, acetic acid ($C_2H_4O_2$) is a monoprotic (not tetraprotic) weak acid. Writing the formula $HC_2H_3O_2$ or CH_3COOH is meant to convey this.

Two simultaneous equilibria are involved in the interaction of the weak diprotic acid H_2CO_3 with water:

$$H_2CO_3(aq) + H_2O(\ell) \rightleftharpoons H_3O^+(aq) + HCO_3^-(aq)$$
$$\frac{[H_3O^+][HCO_3^-]}{[H_2CO_3]} = K_{a1} = 4.3 \times 10^{-7}$$

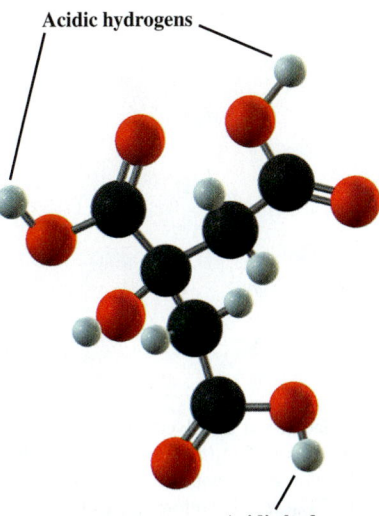

Acidic hydrogens

Acidic hydrogen

Citric acid ($C_6H_8O_7$) gives citrus fruits their tart taste. Only three of the eight hydrogens are acidic. Acidic hydrogens, the ones that can be donated, generally are bonded to oxygen or another electronegative element. Hydrogens bonded to carbon are rarely acidic.

and

$$HCO_3^-(aq) + H_2O(\ell) \rightleftharpoons H_3O^+(aq) + CO_3^{2-}(aq)$$

$$\frac{[H_3O^+][CO_3^{2-}]}{[HCO_3^-]} = K_{a2} = 4.8 \times 10^{-11}$$

Similar equilibria and equilibrium-constant equations can be written for other polyprotic acids. Two crucial points arise:

1. The H_3O^+ produced in the first equilibrium cannot be distinguished from the H_3O^+ produced in subsequent equilibria. The $[H_3O^+]$ in every equilibrium-constant equation is the same number.
2. A K_{a2} is invariably smaller than a K_{a1} because the negative charge left behind by the loss of a hydrogen ion in the first stage causes the second hydrogen ion to be more tightly bound. A K_{a3} is smaller than a K_{a2} for the same reason.

Exact calculations of simultaneous equilibria can be mathematically complex. Considerable simplification results when the original acid concentration is not too small and the acidity constants K_{a1}, K_{a2}, K_{a3} ... differ substantially in magnitude (by a factor of 100 or more). The latter condition is satisfied for very many polyprotic acids. Under such conditions, the equilibria can be treated one at a time, as in the following example.

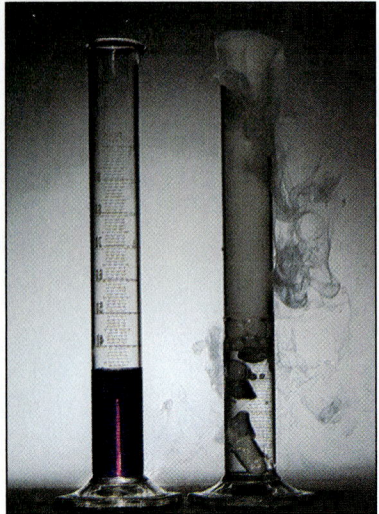

Figure 8–14 • The indicator phenolphthalein is pink in basic solution (*left*). When solid carbon dioxide is placed in the bottom of the cylinder (*right*), it dissolves to form carbonic acid. This causes the indicator to change to its colorless acidic form. (*Leon Lewandowski*)

EXAMPLE 8–14

Calculate the concentrations at equilibrium of H_2CO_3, HCO_3^-, CO_3^{2-}, and H_3O^+ in a saturated aqueous solution of H_2CO_3 in which the original concentration of H_2CO_3 is 0.034 mol L^{-1}.

Strategy

H_3O^+ ion comes from the ionization of H_2CO_3, from the subsequent dissociation of HCO_3^-, and, as always, from the autoionization of water. Because $K_{a1} >> K_{a2}$, and also $K_{a1} >> K_w$, plan to neglect (subject to a check) the H_3O^+ contributed by the acid dissociation of HCO_3^- and by the autoionization of water. Also, plan to neglect the reduction in the concentration of HCO_3^- caused by its donation of an H^+ ion. Use two I-C-E tables, one for each of the K_a's.

Solution

If y moles per liter of H_2CO_3 ionizes,

$$H_2CO_3(aq) + H_2O(\ell) \rightleftharpoons H_3O^+(aq) + HCO_3^-(aq)$$

	$[H_2CO_3]$ (mol L^{-1})	$[H_3O^+]$ (mol L^{-1})	$[HCO_3^-]$ (mol L^{-1})
Initial	0.034	≈ 0	0
Change	$-y$	$+y$	$+y$
Equilibrium	$0.034 - y$	y	y

where setting y equal to the change in both $[HCO_3^-]$ and $[H_3O^+]$ rests on neglecting the effect of the K_{a2} reaction on the concentration of the HCO_3^-. Then

$$K_{a1} = \frac{[H_3O^+][HCO_3^-]}{[H_2CO_3]}$$

• In fact, most dissolved CO_2 actually remains as $CO_2(aq)$, and only a small fraction reacts with water to give $H_2CO_3(aq)$. Therefore, $[H_2CO_3]$ indicates the total concentration of both of these species. Approximately 0.034 mol of CO_2 dissolves per liter of water at 25°C under a $CO_2(g)$ pressure of 1 atm.

$$4.3 \times 10^{-7} = \frac{y^2}{0.034 - y}$$

$$y = 1.2 \times 10^{-4} \text{ mol L}^{-1} = [H_3O^+] = [HCO_3^-]$$

$$[H_2CO_3] = 0.034 - y = 0.034 \text{ mol L}^{-1}$$

We now turn to the second-stage donation of a hydrogen ion. The equation and I-C-E table are

$$HCO_3^-(aq) + H_2O(\ell) \rightleftharpoons H_3O^+(aq) + CO_3^{2-}(aq)$$

	$[HCO_3^-]$ (mol L^{-1})	$[H_3O^+]$ (mol L^{-1})	$[CO_3^{2-}]$ (mol L^{-1})
Initial	1.2×10^{-4}	1.2×10^{-4}	0
Change	$-z$	$+z$	$+z$
Equilibrium	$1.2 \times 10^{-4} - z$	$1.2 \times 10^{-4} + z$	z

$$K_{a2} = \frac{[H_3O^+][CO_3^{2-}]}{[HCO_3^-]}$$

$$4.8 \times 10^{-11} = \frac{(1.2 \times 10^{-4} + z)z}{(1.2 \times 10^{-4} - z)}$$

$$z = 4.8 \times 10^{-11} \text{ mol L}^{-1} = [CO_3^{2-}]$$

The concentration of the base produced by the second equilibrium, $[CO_3^{2-}]$, is numerically equal to K_{a2}.

Check

Substituting the concentrations of H_2CO_3, HCO_3^-, CO_3^{2-}, and H_3O^+ back into the two equilibrium-constant expressions gives the correct values of K_{a1} and K_{a2}, as it must. The second equilibrium reduces $[HCO_3^-]$ by no more than approximately 10^{-11} mol L^{-1}, far less than the 1.2×10^{-4} mol L^{-1} that is present. The contribution to $[H_3O^+]$ from dissociation of HCO_3^- is also negligible (10^{-11} mol L^{-1}). Finally, $[H_3O^+]$ is much larger than 10^{-7} mol L^{-1}, so neglect of the autoionization of water is justified.

EXERCISE

Salicylic acid ($H_2C_7H_4O_3$) is a diprotic acid with $K_{a1} = 1.1 \times 10^{-3}$ and $K_{a2} = 3.6 \times 10^{-14}$ (at 25°C). It is sometimes taken as an analgesic drug instead of aspirin (acetylsalicylic acid), but its greater acidity can cause bleeding in the stomach. Calculate the concentrations at equilibrium of $H_2C_7H_4O_3(aq)$, $HC_7H_4O_3^-(aq)$, $C_7H_4O_3^{2-}(aq)$, $H_3O^+(aq)$, and $OH^-(aq)$ in 0.065 M salicylic acid.

Answer: $[H_2C_7H_4O_3] = 0.057$ mol L^{-1}; $[HC_7H_4O_3^-] = [H_3O^+] = 7.9 \times 10^{-3}$ mol L^{-1}; $[C_7H_4O_3^{2-}] = 3.6 \times 10^{-14}$ mol L^{-1}; $[OH^-] = 1.3 \times 10^{-12}$ mol L^{-1}.

Phosphoric acid (H_3PO_4) is a triprotic acid that is used in the manufacture of phosphate fertilizers. Because it is odorless and nontoxic, phosphoric acid has many uses in the food industry. Deoxyribonucleic acid (DNA) is a derivative of phosphoric acid (see Fig. 25–21).

If we had been working with a triprotic acid such as H_3PO_4, we could have gone on to calculate the concentration of the base (PO_4^{3-}) resulting from the third acid-dissociation step.

A similar procedure works in studying the reactions of a base that is capable of accepting two or more hydrogen ions. In a solution of sodium carbonate (Na_2CO_3), for example, the carbonate ion reacts as a base with water to form first HCO_3^- and then, in a second stage, H_2CO_3:

$$H_2O(\ell) + CO_3^{2-}(aq) \rightleftharpoons HCO_3^-(aq) + OH^-(aq)$$

$$\frac{[OH^-][HCO_3^-]}{[CO_3^{2-}]} = K_{b1} = \frac{K_w}{K_{a2}} = 2.1 \times 10^{-4}$$

$$H_2O(\ell) + HCO_3^-(aq) \rightleftharpoons H_2CO_3(aq) + OH^-(aq)$$

$$\frac{[OH^-][H_2CO_3]}{[HCO_3^-]} = K_{b2} = \frac{K_w}{K_{a1}} = 2.3 \times 10^{-8}$$

- Note how the *first* basicity constant equals K_w divided by the *second* acidity constant, and the *second* basicity constant equals K_w divided by the *first* acidity constant.

In this case $K_{b1} >> K_{b2}$, so that essentially all the OH^- originates from the first reaction. Of course, there is only one $OH^-(aq)$ concentration in the solution, and it is common to both reactions. The ensuing calculation for the concentrations of species present at equilibrium follows the approach used in Example 8–14.

Effect of pH on Solution Composition

Changing the pH shifts the positions of all acid–base equilibria in a solution, including those involving polyprotic acids. Le Chatelier's principle allows only qualitative predictions of the effects of a change in pH. To calculate the actual amount of the change, we use the appropriate acid–base equilibrium-constant equations. For example, we can rewrite the two equilibrium-constant equations that govern $H_2CO_3/HCO_3^-/CO_3^{2-}$ solutions as

$$\frac{[HCO_3^-]}{[H_2CO_3]} = \frac{K_{a1}}{[H_3O^+]} \qquad \frac{[CO_3^{2-}]}{[HCO_3^-]} = \frac{K_{a2}}{[H_3O^+]}$$

At a given pH, the right-hand sides are known, and the relative amounts of the three carbonate species can be calculated. This is illustrated in the following example.

EXAMPLE 8–15

Calculate the fractions of the total carbonate present as H_2CO_3, HCO_3^-, and CO_3^{2-} at pH 10.00 (at 25°C).

Solution

At this pH, $[H_3O^+] = 1.0 \times 10^{-10}$ mol L^{-1}. Insert that value and the values of K_{a1} and K_{a2} at 25°C in the equations above

$$\frac{[HCO_3^-]}{[H_2CO_3]} = \frac{4.3 \times 10^{-7}}{1.0 \times 10^{-10}} = 4.3 \times 10^3 \quad \text{and} \quad \frac{[CO_3^{2-}]}{[HCO_3^-]} = \frac{4.8 \times 10^{-11}}{1.0 \times 10^{-10}} = 0.48$$

Rewrite these ratios with the same species (say, HCO_3^-) in the denominator. The second requires no change. The first becomes

$$\frac{[H_2CO_3]}{[HCO_3^-]} = \frac{1}{4.3 \times 10^3} = 2.3 \times 10^{-4}$$

The fraction of the total carbonate present in each of the three forms equals the concentration of that form divided by the sum of the three concentrations. For example,

$$\text{fraction present as } H_2CO_3 = \frac{[H_2CO_3]}{[H_2CO_3] + [HCO_3^-] + [CO_3^{2-}]}$$

The right side of the equation is evaluated by dividing numerator and denominator by $[HCO_3^-]$ and inserting the ratios just calculated:

$$\text{fraction present as } H_2CO_3 = \frac{\dfrac{[H_2CO_3]}{[HCO_3^-]}}{\dfrac{[H_2CO_3]}{[HCO_3^-]} + 1 + \dfrac{[CO_3^{2-}]}{[HCO_3^-]}}$$

$$= \frac{2.3 \times 10^{-4}}{2.3 \times 10^{-4} + 1 + 0.48} = 1.6 \times 10^{-4}$$

The fractions present as HCO_3^- and CO_3^{2-} equal 0.68 and 0.32, respectively.

EXERCISE

Calculate the fractions of salicylic acid ($H_2C_7H_4O_3(aq)$), hydrogen salicylate ion ($HC_7H_4O_3^-(aq)$), and salicylate ion ($C_7H_4O_3^{2-}(aq)$) present when the drug is at equilibrium in a stomach that has a pH of 1.50. Use the K_{a1} and K_{a2} values from Exercise 8–14.

Answer: Fraction of $H_2C_7H_4O_3(aq) = 0.966$; fraction of $HC_7H_4O_3^-(aq) = 0.034$; fraction of $C_7H_4O_3^{2-}(aq) = 3.8 \times 10^{-14}$.

If we repeat the calculation of Example 8–15 at a series of different pH values, we can plot the graph shown in Figure 8–15. At high pH, CO_3^{2-} predominates; at low pH, H_2CO_3 predominates. At intermediate pH (near 8, the approximate pH of seawater), the hydrogen carbonate ion HCO_3^- is most prevalent.

Variations in the composition of sedimentary rocks in different locations trace back to variations in the pH of ancient carbonate-containing waters. Sediments deposited from alkaline (high pH) waters contain mostly carbonates because CO_3^{2-} ion predominated in solution. Sediments deposited from lakes and oceans with intermediate pH contain hydrogen carbonates or mixtures of carbonates and hydrogen carbonates. An example of the latter is trona ($(Na_2CO_3)_2 \cdot NaHCO_3 \cdot H_2O$), an ore

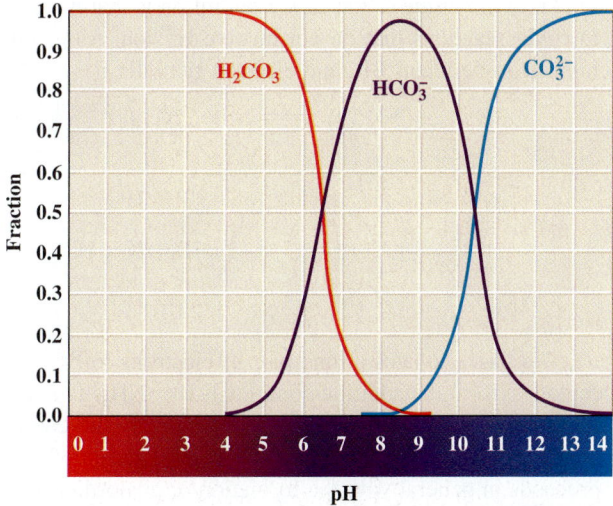

Figure 8-15 • The equilibrium fractions of H_2CO_3, HCO_3^-, and CO_3^{2-} present in aqueous solution at different values of the pH. Low pH favors H_2CO_3, and high pH favors CO_3^{2-}.

found in the western United States that is an important source of both carbonates of sodium. Acidic waters did not deposit carbonates but instead released $CO_2(g)$ to the atmosphere.

8-8   Lewis Acids and Bases

Suppose that we draw Lewis structures for all the participants in a typical Brønsted-Lowry acid–base reaction:

$$HF(aq) + H_2O(\ell) \rightleftharpoons F^-(aq) + H_3O^+(aq)$$

Two bases, the F^- ion and the H_2O molecule, compete for H^+ ions by offering lone pairs of electrons that can be accepted by those ions to form bonds. This kind of picture suggested to Gilbert Lewis a more general definition of acid and base than the Brønsted-Lowry definition. We now say:

> A **Lewis base** is a species that donates lone pairs of electrons; a **Lewis acid** is a species that accepts such electron pairs.

As the Lewis structures show, the fluoride ion (F^-) and the H_2O molecule in the preceding reaction both serve as Lewis bases; the hydrogen ion (H^+) can serve only as a Lewis acid. The sharing of a lone pair of electrons from the F^- ion to the H^+ ion gives HF, just as the sharing of lone pairs of electrons from H_2O to H^+ ion gives H_3O^+ ion. This process is often called **coordination** of an electron-pair donor to an electron-pair acceptor. The product of coordination is an **addition complex** because its formula equals the sum of the formulas of the Lewis acid and Lewis base that produced it.

• The term "coordination" is used to describe the sharing of lone pairs of electrons from ligand molecules (electron-pair donors) onto metal ions (electron-pair acceptors) in Section 3–8.

By their definition, Brønsted-Lowry acid–base reactions all involve just one Lewis acid, the H^+ ion. Every Brønsted-Lowry acid–base reaction is a competition between two Lewis bases to form an addition complex by donating an electron pair to the Lewis acid H^+. One virtue of the Lewis definition is that it expands the acid–base concept to cover reactions that do not involve H^+ ion. An example is the reaction between electron-deficient BF_3 and electron-rich NH_3:

$$BF_3(g) + NH_3(g) \rightleftharpoons F_3B\text{—}NH_3(s)$$

Ammonia, the Lewis base, donates a lone pair of electrons to BF_3, the Lewis acid or electron acceptor, to form the addition complex $F_3B\text{—}NH_3$. The B—N bond can be called a **coordinate covalent bond** because it forms by the coordination of an electron-pair donor to an electron-pair acceptor. Neutralization of a Lewis acid by a Lewis base proceeds in general with the formation of a coordinate covalent bond.

Octet-deficient compounds of Group III elements such as boron and aluminum are often strong Lewis acids because the Group III atoms can achieve valence octets by accepting an electron pair. For example

$$AlBr_3(g) + Br^-(g) \rightleftharpoons AlBr_4^-(g)$$

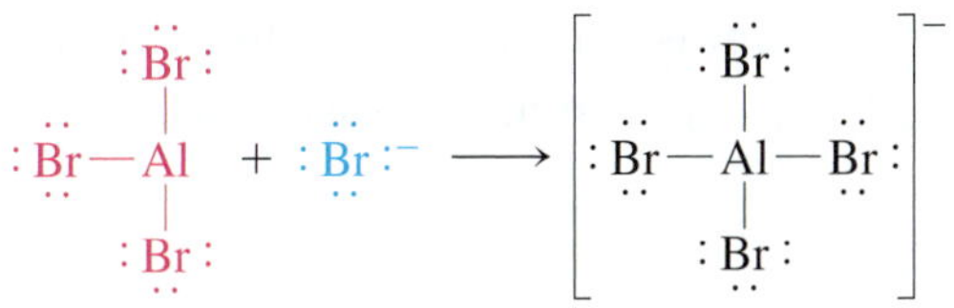

Atoms and ions from Groups V through VII have the necessary lone pairs to act as Lewis bases.

Compounds of main-group elements from the later periods can act as Lewis acids through valence expansion. In such reactions, the central atom accepts a share in additional lone pairs beyond the eight electrons needed to satisfy the octet rule. For example

$$SnCl_4(\ell) + 2\,Cl^-(aq) \rightleftharpoons SnCl_6^{2-}(aq)$$

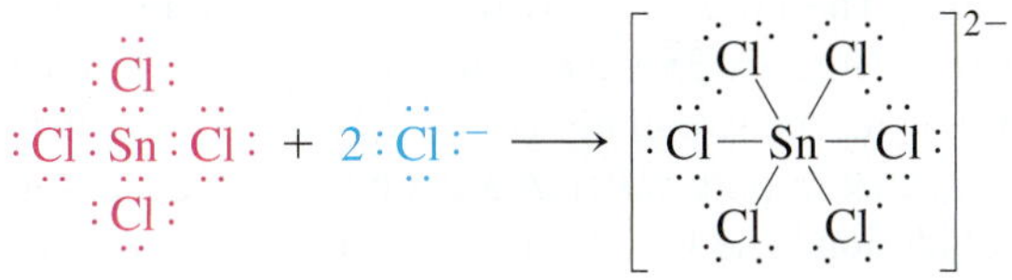

After the reaction, each atom of tin is surrounded by 12 rather than 8 valence electrons. The formation of a coordination complex by an ion of a transition metal is also a Lewis acid–base reaction. For example

$$Zn^{2+}(aq) + 4\,NH_3(aq) \rightleftharpoons Zn(NH_3)_4^{2+}(aq)$$

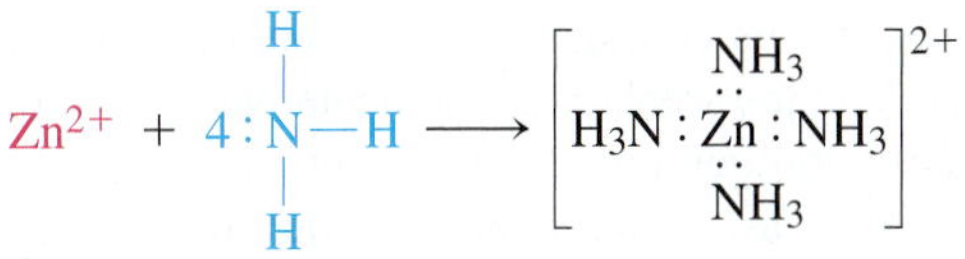

EXAMPLE 8–16

In the following reactions, identify the Lewis acid and the Lewis base.
(a) $AlCl_3(g) + Cl^-(g) \longrightarrow AlCl_4^-(g)$
(b) $CH_3COOH(aq) + NH_3(aq) \longrightarrow CH_3COO^-(aq) + NH_4^+(aq)$
(c) $Co^{3+}(aq) + 6\,F^-(aq) \longrightarrow [CoF_6]^{3-}(aq)$

Solution

(a) In $AlCl_3$ the Al atom has only six valence electrons. Thus, $AlCl_3$ acts as a Lewis acid, accepting electrons at the Al atom from lone pairs on the Lewis base, Cl^-.

(b) The Lewis base is NH_3. An NH_3 lone pair of electrons coordinates to the Lewis acid, H^+, displacing a Lewis base (the CH_3COO^- group). Acetic acid (CH_3COOH) is not a Lewis acid here because the molecule as a whole does not accept a lone pair.

(c) The transition metal ion Co^{3+} is the Lewis acid, forming coordinate covalent bonds by sharing electron pairs donated by the Lewis base, F^-.

EXERCISE

Gaseous PCl_5 molecules condense to form an ionic solid containing PCl_4^+ and PCl_6^- ions. Identify the Lewis acid and base in this process.

Answer: Lewis acid: one of the PCl_5 molecules. Lewis base: Cl^- ion.

SUMMARY

8-1 In the Brønsted-Lowry definition, an acid is a donor of a hydrogen ion and a base is an acceptor of a hydrogen ion. When a species acts as an acid, it is transformed to its **conjugate base;** when a species acts as a base, it gives rise to its **conjugate acid.** An acid–base reaction can be viewed as a competition between two bases for a hydrogen ion.

8-2 In its **autoionization,** water acts simultaneously as both acid and base, which leads to small concentrations of **hydronium** (H_3O^+) and hydroxide (OH^-) ions in water at equilibrium. The product of these two concentrations is K_w, the autoionization constant for water. At 25°C, K_w equals 1.0×10^{-14}. When a **strong acid** dissolves in water, it reacts essentially completely, increasing the concentration of hydronium ion over that in pure water. A **strong base** reacts essentially completely to give the hydroxide ion. The hydronium-ion concentration is conveniently represented on the logarithmic **pH scale** (pH $= -\log_{10}[H_3O^+]$). Pure water has a pH of 7.0 at 25°C. Lower values of pH indicate acidic solutions; higher values of pH indicate basic solutions.

8-3 The strength of an acid in water is shown by its **acidity constant,** K_a, the equilibrium constant for the reaction in which that acid donates a hydrogen ion to water. The strength of a base is shown by its **basicity constant,** K_b. The product of K_a for an acid and K_b for its conjugate base equals K_w. The larger the value of K_a, the stronger the acid and the weaker its conjugate base. In acid–base equilibria, the stronger acid donates hydrogen ions to the stronger base, giving more of the weaker acid and the weaker base as the reaction goes toward equilibrium. An **indicator** is a weak acid (or weak base) that has a noticeably different color from its conjugate weak base (or weak acid). Indicators can be used to estimate the pH of solutions.

8-4 The pH and concentration of dissolved species can be calculated at equilibrium from the initial concentration of a weak acid and its acidity constant K_a. It is usually a good approximation to neglect the contribution to $[H_3O^+]$ from the autoionization of water. Calculations of pH for solutions of weak bases proceed analogously to those of weak acids. **Hydrolysis** refers to the acid or base reaction of a dissolved cation or anion with water.

8-5 A **buffer solution** results from mixing a weak acid with its conjugate weak base. Such a solution resists changes in pH; adding a small amount of either strong acid or base has only a small effect on the buffer's pH. In the design of a buffer solution, the best choice of acid is one having a pK_a that is close to the desired pH of the buffer.

8-6 An **acid–base titration curve** is a plot of pH versus the volume of titrant added. An **equivalence point** on a titration curve is a point at which the chemical amount of base is equal to the chemical amount of acid. The volume of titrant to reach an equivalence point is the **equivalent volume.** If the acid and base are both strong, the pH at the equivalence point is 7.0 at 25°C. When a weak acid is titrated with a strong base, a buffering region exists in the vicinity of the half-equivalence point over which the pH changes slowly with added base, and the pH at the equivalence point is greater than 7.0. The computation of pH in the four distinct ranges of a titration curve corresponds to standard pH calculations. Acid–base titration is useful for determining concentrations and values of K_a for solutions of acids (or bases) of unknown concentration or unknown identity.

8-7 **Polyprotic acids** donate H^+ ion in several steps. Here it is usually a good approximation to assume that the pH is determined entirely by the first step; the extent of subsequent steps then can be calculated sequentially. From the successive acidity constants, the equilibrium concentrations of all the species present can be calculated at any value of the pH.

8-8 A **Lewis base** is a species that donates lone pairs of electrons; a **Lewis acid** is a species that accepts such electron pairs. The bond that forms between a Lewis acid and base is a **coordinate covalent bond,** in which both electrons are supplied by a lone pair on the Lewis base. Neutralization between a Lewis acid and base proceeds with the formation of such a bond and the creation of an **addition complex.**

PROBLEMS

Note: Answers to blue-numbered problems are given in Appendix F. Problems that are more challenging are indicated with asterisks.

Brønsted-Lowry Acids and Bases

1. (See Example 8–1.) Determine which of the following can act as Brønsted-Lowry acids. Give the formula of the conjugate base of those that can.
(a) Cl^- (d) NH_3
(b) HSO_4^- (e) H_2O
(c) NH_4^+

2. (See Example 8–1.) Determine which of the following can act as Brønsted-Lowry bases. Give the formula of the conjugate acid of those that can.
(a) F^- (d) OH^-
(b) SO_4^{2-} (e) H_2O
(c) O^{2-}

3. (See Example 8–2.) The following solutions are found to be acidic when tested with indicators. Write net ionic equations to explain their acidity.
(a) $NH_4Br(aq)$
(b) $H_2S(aq)$
(c) $(NH_4)_2SO_4(aq)$

4. (See Example 8–2.) The following solutions are found to be basic when tested with indicators. Write net ionic equations to explain their basicity.
(a) $Na_2CO_3(aq)$
(b) $RbOH(aq)$
(c) $Na_2HPO_4(aq)$

Water and the pH Scale

5. (See Example 8–3.) The concentration of H_3O^+ ion in a sample of wine is 2.0×10^{-4} mol L^{-1}. Calculate the pH and pOH of the wine at 25°C.

6. (See Example 8–3.) The concentration of OH^- ion in a solution of household bleach is 3.6×10^{-2} mol L^{-1}. Calculate the pOH and pH of the bleach at 25°C.

7. (See Example 8–4.) The pH of normal human urine is in the range of 5.5 to 6.5. Compute the range of the concentration of H_3O^+ and of OH^- at 25°C.

8. (See Example 8–4.) The pH of normal human blood is in the range of 7.35 to 7.45. Compute the range of the concentration of H_3O^+ and of OH^- in normal blood at 40°C. (*Hint*: Use the data in Table 8–1.)

9. A solution of nitric acid with pH 2.32 at 25°C is diluted with water to eight times its original volume. Compute the pH and the concentration of OH^- ion in the resulting solution.

10. A solution of potassium hydroxide with pH 11.65 at 25°C is diluted with water to six times its original volume. Compute the pH and the concentration of OH^- ion after the dilution.

11. Suppose that 11.74 mL of 0.071 M NaOH is added to 15.78 mL of 0.094 M HCl. Determine the pH of the resulting solution at room temperature, assuming that the volumes are additive.

12. Suppose that 264.9 mL of 0.065 M KOH is added to 127.3 mL of 0.073 M HCl. Determine the pOH and pH of the resulting solution at room temperature, assuming that the volumes are additive.

The Strengths of Acids and Bases

13. List the following acids in order of increasing strength: formic acid ($K_a = 1.8 \times 10^{-4}$); benzoic acid ($K_a = 6.5 \times 10^{-5}$); ammonium ion ($pK_a = 9.25$).

14. List the following acids in order of increasing strength: pyruvic acid ($pK_a = 2.49$); lactic acid ($K_a = 1.41 \times 10^{-4}$); hypochlorous acid ($pK_a = 7.53$).

15. Ephedrine ($C_{10}H_{15}ON$) is a weak base that is used in nasal sprays as a decongestant.
   (a) Write an equation for the acid–base reaction between ephedrine and water.
   (b) The basicity constant of ephedrine equals 1.4×10^{-4} at room temperature. Calculate the acidity constant of its conjugate acid.
   (c) Is ephedrine a weaker or a stronger base than ammonia?

16. Niacin (C_5H_4NCOOH), one of the B vitamins, is a weak acid.
   (a) Write an equation for its reaction with water.
   (b) The acidity constant of niacin equals 1.5×10^{-5} at room temperature. Calculate the basicity constant of the conjugate base of niacin.
   (c) Is the conjugate base of niacin a stronger or a weaker base than pyridine (the conjugate base of the pyridinium ion in Table 8–2)?

17. (a) Write a balanced equation for the acid–base equilibrium that is established when sodium hypochlorite (NaClO) is dissolved in water.
   (b) Calculate the equilibrium constant for the reaction in part (a) using Table 8–2.

18. (a) Write a balanced equation for the acid–base equilibrium that results when potassium nitrite (KNO_2) is dissolved in water.
   (b) Calculate the equilibrium constant for the reaction in part (a) using Table 8–2.

19. Phenol (C_6H_5OH) has a K_a of 1.1×10^{-10}.
   (a) Write the formula of the phenolate ion, the conjugate base of phenol.
   (b) Is the phenolate ion a weak or a strong base?
   (c) Write a series of balanced equations to represent the chemical events when solid sodium phenolate is mixed with water.

20. The pK_a of ethane (C_2H_6), a gas at room conditions, reportedly equals 50.6.
   (a) Is ethane a weak or a strong acid? Is that what you would expect, based on the formula of ethane?
   (b) Write the formula of the conjugate base of ethane.
   (c) Write a series of balanced equations to represent the chemical events if solid NaC_2H_5 were mixed with water.

21. Use the data in Table 8–2 to determine the equilibrium constant for the reaction

$$HClO_2(aq) + NO_2^-(aq) \rightleftharpoons HNO_2(aq) + ClO_2^-(aq)$$

22. Use the data in Table 8–2 to determine the equilibrium constant for the reaction

$$HPO_4^{2-}(aq) + HCO_3^- \rightleftharpoons PO_4^{3-}(aq) + H_2CO_3(aq)$$

23. (See Example 8–5.) (a) Which is the stronger acid: the acidic form of the indicator bromocresol green or the acidic form of methyl orange?
   (b) A solution gives a green color with bromocresol green and an orange color with methyl orange. Estimate the pH of the solution.

24. (See Example 8–5.) (a) Which is the stronger base: the basic form of the indicator cresol red or the basic form of thymolphthalein?
   (b) A solution gives a red color with cresol red and is colorless with thymolphthalein. Estimate the pH of the solution.

Equilibria Involving Weak Acids and Bases

25. (See Example 8–6.)
   (a) Find the pH and fraction of the acid that is dissociated in 0.20 mol L^{-1} solutions of formic acid, benzoic acid, and ammonium ion.
   (b) How does the pH vary in going from the solution containing the strongest acid to the one containing weakest?
   (c) How does the fraction of acid dissociated in these solutions vary in going from the strongest acid to the weakest?

26. (See Example 8–6.)
   (a) Find the pH and fraction dissociated of the acid in 0.20 mol L^{-1} solutions of pyruvic acid ($pK_a = 2.49$), lactic acid ($pK_a = 3.85$), and hypochlorous acid ($pK_a = 7.53$).
   (b) How does the pH vary in going from the solution containing the strongest acid to the one containing the weakest?
   (c) How does fractional dissociation vary in going from the strongest acid to the weakest?

27. (See Example 8–7.) The active component of aspirin is acetylsalicylic acid ($HC_9H_7O_4$), which has a K_a of 3.0×10^{-4} at room temperature.
   (a) Calculate the pH of a solution made by dissolving 0.65 g of acetylsalicylic acid in water and diluting to 50.0 mL.
   (b) Repeat the calculation, assuming that the same mass of acetylsalicylic acid is dissolved and diluted to 100.0 mL of solution.
   (c) Is the pH lower or higher in the more dilute solution? Which solution has the higher fraction of its aspirin ionized?

28. (See Example 8–7.) Vitamin C is ascorbic acid ($HC_6H_7O_6$), for which $K_a = 8.0 \times 10^{-5}$ at room temperature.
   (a) Calculate the pH of a solution made by dissolving a 500 mg tablet of pure vitamin C in water and diluting to 100.0 mL.
   (b) Calculate the pH of the same tablet dissolved in enough water to make 200.0 mL of solution.

29. (a) Calculate the pH of a 0.35 mol L^{-1} solution of propionic acid at 25°C.
   (b) How many moles of formic acid must be dissolved per liter of solution to obtain the same pH as the solution from part (a)?

30. (a) Calculate the pH of a 0.027 mol L^{-1} solution of benzoic acid at 25°C.
   (b) How many moles of acetic acid must be dissolved per liter of solution to obtain the same pH as the benzoic acid solution in part (a)?

31. (See Example 8–7.) Iodic acid (HIO_3) is strong for a weak acid, having a K_a equal to 0.16 at 25°C. Compute the pH of a 0.100 mol L^{-1} solution of HIO_3.

32. (See Example 8–7.) At 25°C, the K_a of pentafluorobenzoic acid (C_6F_5COOH) is 0.033. Suppose that 0.100 mol of pentafluorobenzoic acid is dissolved in enough water to make 1.00 L of solution. Compute the pH of the solution.

33. Papaverine hydrochloride (papHCl) is a drug used as a muscle relaxant. It is a weak acid. At 25°C, a 0.205 mol L^{-1} solution of papHCl has a pH of 3.31. Compute the K_a of papHCl.

34. 2-Germaacetic acid (GeH_3COOH) is a weak acid that is related structurally to acetic acid (CH_3COOH). At 25°C, a 0.050 mol L^{-1} solution of 2-germaacetic acid has a pH of 2.42. Compute the K_a of 2-germaacetic acid. State whether it is stronger or weaker than acetic acid.

35. (See Example 8–8.) Morphine is a weak base having a K_b of 8×10^{-7} at room temperature. Calculate the pH of a solution made by dissolving 0.040 mol of morphine in water and diluting to 600 mL.

36. (See Example 8–8.) Methylamine (CH_3NH_2) has a basicity constant of 4.4×10^{-4} at 25°C. Calculate the pH of a solution made by dissolving 2.17 g of methylamine in water and diluting to 800 mL at 25°C.

37. Determine the concentration of $H_3O^+(aq)$ and $NO_2^-(aq)$ in a solution in which $HNO_2(aq)$ is 4.50% ionized. Calculate the concentration of $HNO_2(aq)$ at equilibrium in this solution.

38. If 0.0145 mol of a certain weak base is placed in 1.000 L of water, then 0.012% of the molecules react, at equilibrium, to give OH^- ion. Compute K_b of this weak base.

39. Predict whether an aqueous solution of each of the following is acidic, basic, or neutral.
   (a) KCN
   (b) $CuBr_2$
   (c) $RbClO_4$
   (d) $Mg(NO_3)_2$

40. Predict whether an aqueous solution of each of the following is acidic, basic, or neutral.
   (a) $AlCl_3$
   (b) NH_4ClO_4
   (c) AgF
   (d) $SrCl_2$

41. In aqueous solutions of NH_4CN, the NH_4^+ ion is weakly acidic and the CN^- ion is weakly basic. Use data from Table 8–2 to determine the equilibrium constant for the reaction

$$NH_4^+(aq) + CN^-(aq) \rightleftharpoons NH_3(aq) + HCN(aq)$$

42. In aqueous solutions of NH_4F, the NH_4^+ ion is weakly acidic and the F^- ion is weakly basic. Use data from Table 8–2 to determine the equilibrium constant for the reaction

$$NH_4^+(aq) + F^-(aq) \rightleftharpoons NH_3(aq) + HF(aq)$$

43. You have 50.00 mL of a solution that is 0.100 M in acetic acid, and you neutralize it by adding 50.00 mL of a solution that is 0.100 M in sodium hydroxide. The pH of the resulting solution is not 7.0. Explain why. Determine whether the pH of the solution is greater than or less than 7.0.

44. A 75.00 mL portion of a solution that is 0.0460 M in $HClO_4$ is mixed with 150.00 mL of 0.0230 M $KOH(aq)$. State whether the pH of the resulting mixture is greater than, less than, or equal to 7.0. Explain.

Buffer Solutions

45. (See Examples 8–10 and 8–11.)
   (a) Calculate the equilibrium concentration of H_3O^+ and the pH in a solution prepared by dissolving 0.060 mol of formic acid and 0.045 mol of sodium formate in water and adjusting the volume to 500 mL at 25°C.
   (b) Suppose 0.010 mol of NaOH is added to the buffer from part (a). Calculate the pH of the solution that results (at 25°C).

46. (See Examples 8–10 and 8–11.)
   (a) Calculate the equilibrium concentration of H_3O^+ in a solution prepared by dissolving 3.62 g of NH_4Cl in 400 mL of a solution that is 0.100 M in ammonia (NH_3) and diluting to a total volume of 600 mL at 25°C.
   (b) Suppose 0.040 mol of HCl is added to the buffer from part (a). Calculate the pH of the solution that results (at 25°C).

47. The weak base "tris" (tris(hydroxymethyl)aminomethane) is widely used in biochemical research for the preparation of buffers. It offers low toxicity and a pK_b (equal to 5.92 at 25°C) that is convenient for the control of pH in clinical applications. A buffer solution is prepared by mixing 0.050 mol of tris with 0.025 mol of HCl in a volume of 2.00 L. Compute the pH of the solution at 25°C.

48. "Bis" is short for bis(hydroxymethyl)aminomethane. It is a weak base that is closely related to tris (see Problem 47) and has similar properties and uses. Its pK_b is 8.8 at 25°C.

A buffer solution is prepared by mixing 0.050 mol of bis with 0.025 mol of HCl in a volume of 2.00 L (the same proportions as in the previous problem). Compute the pH of the solution at 25°C.

49. (See Example 8–12.) A physician wishes to prepare a buffer solution at pH = 3.82 that efficiently resists changes in pH yet contains minimal concentrations of the buffering agents. Which one of the following weak acids, together with its sodium salt, would probably be best to use: *m*-chlorobenzoic acid, $K_a = 1.0 \times 10^{-4}$; *p*-chlorocinnamic acid, $K_a = 3.9 \times 10^{-5}$; 2,5-dihydroxybenzoic acid, $K_a = 1.1 \times 10^{-3}$; acetoacetic acid, $K_a = 2.6 \times 10^{-4}$? Explain.

50. (See Example 8–12.) Suppose you are designing a buffer system for imitation blood and you want the buffer to maintain the blood at the realistic pH of 7.40. All other things being equal, which buffer system would be preferable: H_2CO_3/HCO_3^- or $H_2PO_4^-/HPO_4^{2-}$? Explain.

Acid–Base Titration Curves

51. A total of 26.38 mL of 0.1439 M HBr is titrated at 25°C with 0.1219 M NaOH. Compute the pH before any base is added, when the titration is 1.00 mL short of the equivalence point, when the titration is at the equivalence point, and when the titration is 1.00 mL past the equivalence point.

52. A total of 100.0 mL of a 0.3750 M solution of the strong base $Ba(OH)_2$ is titrated at 25°C with a 0.4540 M solution of the strong acid $HClO_4$. Compute the pH of the titration solution before any acid is added, when the tritration is 1.00 mL short of the equivalence point, when the titration is at the equivalence point, and when the titration is 1.00 mL past the equivalence point. (*Caution*: Remember that each mole of $Ba(OH)_2$ gives 2 mol of OH^- in solution.)

53. A total of 100.00 mL of 0.1000 M $HClO_4$ is titrated with 0.1000 M KOH at 25°C. Compute the *change* in the pH when 0.04 mL of titrant is added to a titration mixture that has already received 99.98 mL of titrant.

54. A total of 50.00 mL of 0.2000 M NaOH is titrated with 0.1000 M HCl at 25°C. Compute the *change* in the pH when 0.04 mL of titrant is added to a titration mixture that has already received 99.98 mL of titrant.

55. A total of 50.00 mL of 0.1000 M hydrazoic acid (HN_3) is titrated with 0.1000 M NaOH at 25°C. Compute the pH before any NaOH is added, after the addition of 25.00 mL of NaOH, after the addition of 50.00 mL of NaOH, and after the addition of 51.00 mL of NaOH.

56. A total of 140.0 mL of 0.175 M aqueous ammonia is titrated with 0.106 M HCl at 25°C. Compute the pH before any HCl is added, when the titration is at the half-equivalence point, when the titration is at the equivalence point, and when the titration is 1.00 mL past the equivalence point.

57. A total of 100.00 mL of 0.1000 M HF (a weak acid) is titrated with 0.1000 M KOH at 25°C. Compute the *change*

in the pH when 0.04 mL of titrant is added to a titration mixture that has already received 99.98 mL of titrant.

58. A total of 50.00 mL of 0.2000 M aqueous ammonia is titrated with 0.1000 M HCl at 25°C. Compute the *change* in the pH when 0.04 mL of titrant is added to a titration mixture that has already received 99.98 mL of titrant.

59. Determine the mass of codeine ($C_{18}H_{21}O_3N$) in 100.0 mL of an aqueous solution if the solution requires 15.90 mL of 0.0750 M HCl to titrate it to the equivalence point. What is the pH at the equivalence point in this titration if the K_b of codeine equals 9.0×10^{-7} at 25°C? Recommend a suitable indicator for this titration.

60. A chemist who works in the process laboratory of the Athabasca Alkali Company makes frequent analyses for ammonia recovered from the Solvay process for making sodium carbonate. What is the pH at the equivalence point if she titrates the aqueous ammonia solution (approximately 0.10 mol L^{-1}) with a strong acid of comparable concentration? Select an indicator that would be suitable for the titration.

61. (See Example 8–13.) A total of 40.00 mL of a solution of cacodylic acid of unknown concentration is titrated with 0.2000 M KOH at 25°C. The equivalence point is reached after 23.28 mL of the KOH solution has been added. At the half-equivalence point (11.64 mL), the pH was 6.19. Calculate the original concentration of the acid and its acidity constant K_a at this temperature.

62. (See Example 8–13.) A total of 50.00 mL of a solution of aniline (a base) of unknown concentration is titrated with 0.1800 M HCl at 25°C. The equivalence point is reached after 37.32 mL of the HCl solution has been added. At the half-equivalence point (18.66 mL), the pH was 10.67. Calculate the original concentration of the aniline and its base ionization constant K_b at this temperature.

63. Examine the titration curve below.

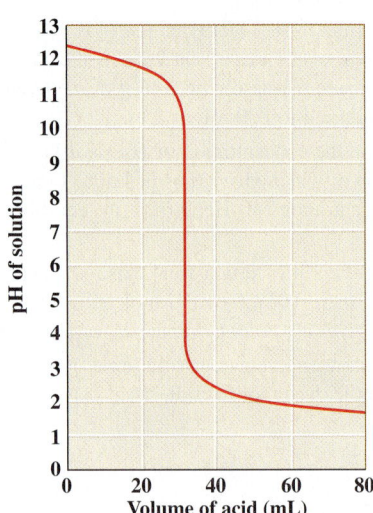

(a) Which of the following titrations could it represent: HCl by KOH, RbOH by HBr, NH_3 by HNO_3?

(b) Choose a suitable indicator for signaling the end-point of the titration. Justify your answer.

(c) Suppose that the figure represents the titration of 100.0 mL of NaOH(aq) by a solution of 0.065 M HNO_3. Calculate the concentration of the NaOH in the original solution.

64. Examine the titration curve below.

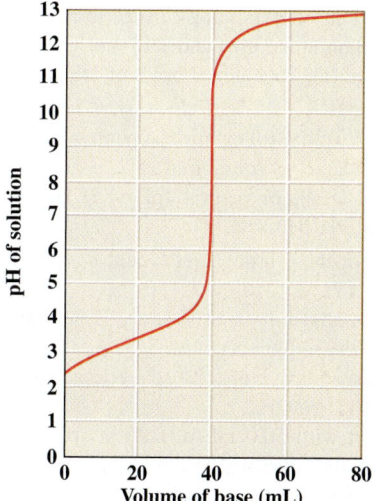

(a) Which of the following titrations could it represent: HCl by NaOH, NH_3 by HBr, HNO_2 by KOH?

(b) Choose a suitable indicator for signaling the end-point of the titration. Justify your answer.

(c) Estimate the K_a of the species in the original solution.

Polyprotic Acids

65. (See Example 8–14.) Arsenic acid (H_3AsO_4) is a weak triprotic acid. Given the three acidity constants from Table 8–2, calculate the equilibrium concentrations of H_3AsO_4, $H_2AsO_4^-$, $HAsO_4^{2-}$, AsO_4^{3-}, and H_3O^+ in a solution prepared by dissolving 0.100 mol of H_3AsO_4 in 1.000 L of solution.

66. (See Example 8–14.) Phthalic acid ($C_8H_4O_4$, abbreviated H_2Ph) is a diprotic acid. Its ionization in water takes place in two steps:

$$H_2Ph(aq) + H_2O(\ell) \rightleftharpoons H_3O^+(aq) + HPh^-(aq)$$
$$K_{a1} = 1.3 \times 10^{-3} \text{ at } 25°C$$
$$HPh^-(aq) + H_2O(\ell) \rightleftharpoons H_3O^+(aq) + Ph^{2-}(aq)$$
$$K_{a2} = 3.1 \times 10^{-6} \text{ at } 25°C$$

If 0.0100 mol of phthalic acid is dissolved per liter of solution, calculate the equilibrium concentrations of H_2Ph, HPh^-, Ph^{2-}, and H_3O^+.

67. A solution as initially prepared contains 0.050 mol L^{-1} of phosphate ion (PO_4^{3-}) at 25°C. Given the three acidity con-

stants from Table 8–2, calculate the equilibrium concentrations of PO_4^{3-}, HPO_4^{2-}, $H_2PO_4^-$, H_3PO_4, and OH^-.

68. Oxalic acid ($H_2C_2O_4$) ionizes in two stages in aqueous solution (Table 8–2). Calculate the equilibrium concentrations of $C_2O_4^{2-}$, $HC_2O_4^-$, $H_2C_2O_4$, and OH^- in a 0.10 M solution of sodium oxalate ($Na_2C_2O_4$) at 25°C.

69. (See Example 8–15.) The pH of a normal raindrop at 25°C is 5.6. Compute the concentrations of $H_2CO_3(aq)$, $HCO_3^-(aq)$, and $CO_3^{2-}(aq)$ in this raindrop if the total concentration of dissolved carbonates is 1.0×10^{-5} mol L^{-1}.

70. (See Example 8–15.) The pH of a drop of acid rain at 25°C is 4.0. Compute the concentration of $H_2CO_3(aq)$, $HCO_3^-(aq)$, and $CO_3^{2-}(aq)$ in the acid raindrop if the total concentration of dissolved carbonates is 3.6×10^{-5} mol L^{-1}.

Lewis Acids and Bases

71. (See Example 8–16.) Indicate which species is the Lewis acid and which the Lewis base in each of the following reactions:

(a) $Ag^+(aq) + 2\,NH_3(aq) \longrightarrow Ag(NH_3)_2^+(aq)$

(b) $N(CH_3)_3(g) + BF_3(g) \longrightarrow F_3BN(CH_3)_3(g)$

(c) $B(OH)_3(aq) + OH^-(aq) \longrightarrow B(OH)_4^-(aq)$

72. (See Example 8–16.) Indicate which species is the Lewis acid and which the Lewis base in each of the following reactions:

(a) $HgI_2(s) + 2\,I^-(aq) \longrightarrow HgI_4^{2-}(aq)$

(b) $Co^{2+} + 6\,H_2O \longrightarrow [Co(H_2O)_6]^{2+}$

(c) $Fe^{3+}(aq) + 6\,CN^- \longrightarrow [Fe(CN)_6]^{3-}(aq)$

73. An important step in many industrial processes (including the Solvay process discussed in the Cumulative Problem in Chapter 4) is the slaking of lime, in which water is added to calcium oxide to make calcium hydroxide.

(a) Write the balanced equation for this process.

(b) Can this be considered a Lewis acid–base reaction? If so, what is the Lewis acid and what is the Lewis base?

74. The impurity silica (SiO_2) must be removed from metal oxide or metal sulfide ores as the ore is being reduced to elemental metal. To do this, lime (CaO) is added. It reacts with the silica to form a slag of calcium silicate ($CaSiO_3$), which can be separated and removed from the ore.

(a) Write the balanced equation for this process.

(b) Can this be considered a Lewis acid–base reaction? If so, what is the Lewis acid and what is the Lewis base?

75. Chemists working with fluorine and its compounds sometimes find it helpful to think in terms of acid–base reactions in which the fluoride ion (F^-) is donated and accepted.

(a) Is the acid in this system the fluoride donor or the fluoride acceptor?

(b) Identify the acid and base in these reactions:

$$ClF_3O_2 + BF_3 \longrightarrow ClF_2O_2 \cdot BF_4$$
$$TiF_4 + 2\,KF \longrightarrow K_2[TiF_6]$$

76. Researchers working with ceramics often think of acid–base reactions in terms of oxide-ion donors and oxide-ion acceptors. The oxide ion is O^{2-}.
(a) In this system, is the base the oxide donor or the oxide acceptor?
(b) Identify the acid and base in each of these reactions:

$$2\ CaO + SiO_2 \longrightarrow Ca_2SiO_4$$
$$Ca_2SiO_4 + SiO_2 \longrightarrow 2\ CaSiO_3$$
$$Ca_2SiO_4 + CaO \longrightarrow Ca_3SiO_5$$

ADDITIONAL PROBLEMS

77. Calculate the pH of a solution obtained by mixing 1.00 L of 1.00×10^{-5} M NaOH, which has a pH of 9.00, and 1.00 L of 1.00×10^{-6} M NaOH, which has a pH of 8.00 at 25°C. (*Hint:* The answer 8.50 is incorrect.)

78. When placed in water, potassium starts to react instantly and continues to react with great vigor. On the basis of this information, select the better of the following two equations to represent the reaction:

$$2\ K(s) + 2\ H_2O(\ell) \longrightarrow 2\ KOH(aq) + H_2(g)$$
$$2\ K(s) + 2\ H_3O^+(aq) \longrightarrow 2\ K^+(aq) + H_2(g) + 2\ H_2O(\ell)$$

State the reason for your preference.

79. $Cl_2(aq)$ reacts with $H_2O(\ell)$ as follows:

$$Cl_2(aq) + 2\ H_2O(\ell) \rightleftharpoons$$
$$H_3O^+(aq) + Cl^-(aq) + HOCl(aq)$$

For an experiment to succeed, $Cl_2(aq)$ must be present, but the amount of Cl^- in the solution must be minimized. For this purpose, should the pH of the solution be high, low, or neutral? Explain.

80. In pure nitric acid, this autoionization reaction occurs:

$$2\ HNO_3(\ell) \rightleftharpoons NO_2^+(sln) + NO_3^-(sln) + H_2O(sln)$$

At equilibrium at 25°C, the molality of the NO_3^- ion is 0.25 mol kg^{-1}.
(a) Calculate the molarity of the NO_3^- ion. The density of pure nitric acid at 25°C is 1.503 g cm^{-3}.
(b) Compute the autoionization constant of nitric acid at 25°C. (This is "K_{HNO_3}" and is the constant for nitric acid that is analogous to K_w for water.)

81. Compute the pH of pure water at 60°C. See the K_w values in Table 8–1.

82. Calculate the pH of neutral water at three different temperatures (taking data from Table 8–1) and use the results, in conjunction with Le Chatelier's principle, to decide whether the autoionization of water is exothermic or endothermic.

83. Use the acidity constants in Table 8–2 to calculate equilibrium constants for the following reactions. In each case, identify the stronger Brønsted-Lowry acid and the stronger Brønsted-Lowry base.

(a) $H_2SO_3(aq) + CH_3COO^-(aq) \rightleftharpoons$
$$CH_3COOH(aq) + HSO_3^-(aq)$$
(b) $HF(aq) + CH_2ClCOO^-(aq) \rightleftharpoons$
$$CH_2ClCOOH(aq) + F^-(aq)$$
(c) $HCN(aq) + NO_2^-(aq) \rightleftharpoons HNO_2(aq) + CN^-(aq)$

84. Exactly 1.0 L of a solution of acetic acid gives the same color with methyl red indicator as 1.0 L of a solution of hydrochloric acid. Which solution neutralizes the greater amount of 0.10 M NaOH(aq)? Explain.

85. Urea (NH_2CONH_2) is a component of urine. Compute the equilibrium concentration of urea in a solution that starts out containing no urea and 0.15 mol L^{-1} of the urea acidium ion, the conjugate acid of urea. Use data from Table 8–1.

86. For each of the following compounds, indicate whether a 0.100 M aqueous solution at 25°C is acidic (pH < 7), basic (pH > 7), or neutral (pH = 7): HCl, NH_4Cl, KNO_3, Na_3PO_4, $NaCH_3COO$.

87. Which of these procedures would *not* make a pH = 4.74 buffer at 25°C?
(a) Mix 50.0 mL of 0.10 M acetic acid and 50.0 mL of 0.10 M sodium acetate.
(b) Mix 50.0 mL of 0.20 M acetic acid and 50.0 mL of 0.10 M NaOH.
(c) Start with 50.0 mL of 0.20 M acetic acid and add a solution of strong base until the pH equals 4.74.
(d) Start with 50.0 mL of 0.20 M HCl and add a solution of strong base until the pH equals 4.74.
(e) Start with 100.0 mL of 0.20 M sodium acetate and add 50.0 mL of 0.20 M HCl.

88. A buffer solution is 0.683 M in HN_3 (hydrazoic acid) and 0.593 M in NaN_3 (sodium azide).
(a) Calculate the pH of the solution at 25°C.
(b) Write the chemical equation that describes how the buffer responds when a small amount of NaOH(aq) is added.
(c) What happens to the pH of the solution when a small amount of NaOH(aq) is added?

89. A buffer solution is prepared by mixing 1.00 L of 0.050 M pentafluorobenzoic acid (C_6F_5COOH) and 1.00 L of 0.060 M sodium pentafluorobenzoate (NaC_6F_5COO). The K_a of this weak acid is 0.033 at 25°C. Determine the pH of the buffer solution at this temperature.

90. Compute the $[NH_4^+]/[NH_3]$ ratio in an aqueous mixture of ammonium chloride with ammonia having a pH = 4.00 at 25°C. Why is this solution a poor acid–base buffer?

91. Estimate the ratio of the concentration of the acid form of the indicator methyl red (HIn) to the base form (In$^-$) at pH 6.50 and 25°C. The K_a of methyl red equals 9.3×10^{-6} at 25°C.

92. It takes 4.71 mL of 0.0410 M NaOH to titrate a 50.00 mL sample of flat (no dissolved CO_2) GG's Cola to a pH of 4.9. At this point, the addition of one more drop (0.02 mL) of NaOH raises the pH to 6.1. The only acid in GG's Cola is phosphoric acid. Compute the concentration of phosphoric acid in this cola. Assume that the 4.71 mL of base removes only the first hydrogen from the H_3PO_4; that is, assume that the reaction is

$$H_3PO_4(aq) + OH^-(aq) \longrightarrow H_2O(\ell) + H_2PO_4^-(aq)$$

93. A total of 159.4 mL of 0.251 M H_2SO_4 is added to 202.3 mL of 0.451 M NaOH. Determine the pH of the resulting solution at 25°C, assuming that the volumes are additive.

94. The chief chemist of Victory Vinegar Works, Ltd., interviews two chemists for employment. He states, "Our high-grade vinegar must contain between 4.99 and 5.01% acetic acid by mass. How would you analyze our product to ensure that it meets this specification?"

Anne Dalton says, "I would titrate a 50.00 mL sample of the vinegar with 1.000 M NaOH, using phenolphthalein as an indicator to signal the equivalence point within 0.02 mL of base."

Charlie Cannizzarro says, "I would use a pH meter to determine the pH to 0.01 pH units and interface it with a computer to print out the mass percentage of acetic acid digitally." Which candidate did the chief chemist hire? Why?

95. Sodium benzoate, the sodium salt of benzoic acid, is used as a food preservative. A sample containing solid sodium benzoate mixed with sodium chloride is dissolved in 50.0 mL of 0.500 M HCl, giving an acidic mixture (benzoic acid mixed with HCl). This mixture is then titrated with 0.393 M NaOH. After the addition of 46.50 mL of the NaOH solution, an end-point (sudden rise in pH) is observed. Further addition of the NaOH solution to a total of 63.61 mL causes a *second* end-point. Calculate the mass of sodium benzoate (NaC_6H_5COO) in the original sample. (*Hint*: Decide which acid is neutralized first when NaOH is added.)

*96. An antacid tablet (like Tums or Rolaids) weighs 1.3259 g. The only acid-neutralizing ingredient in this brand of antacid is $CaCO_3$. When placed in 12.07 mL of 1.070 M HCl, the tablet fizzes merrily as $CO_2(g)$ is given off. After all of the CO_2 has left the solution, an indicator is added, followed by 11.74 mL of 0.5310 M NaOH. The indicator shows that, at this point, the solution is definitely basic. Addition of 5.12 mL of 1.070 M HCl makes the solution acidic again. Then, 3.17 mL of the 0.5310 M NaOH brings the titration exactly to an end-point, as signaled by the indicator. Compute the percentage by mass of $CaCO_3$ in the tablet.

97. Baking soda (sodium hydrogen carbonate, $NaHCO_3$) reacts with acids in foods to form carbonic acid (H_2CO_3), which in turn decomposes to water and carbon dioxide. In a batter, the carbon dioxide appears as gas bubbles that cause the bread or cake to rise.

(a) A rule of thumb in cooking is that 12 teaspoons of baking soda is neutralized by 1 cup of sour milk. The acid component of sour milk is lactic acid ($HC_3H_5O_3$). Write an equation for the neutralization reaction.

(b) If the density of baking soda is 2.16 g cm^{-3}, calculate the concentration of lactic acid in the sour milk, in moles per liter. Take 1 cup = 236.6 mL = 48 teaspoons.

(c) Calculate the volume of carbon dioxide that is produced at 1 atm pressure and 350°F (177°C) from the reaction of 12 teaspoons of baking soda.

98. Egg whites contain dissolved carbon dioxide and water, which react to give carbonic acid (H_2CO_3). In the first days after an egg is laid, it loses carbon dioxide through its shell. Does the pH of the egg white increase or decrease during this period?

99. Phosphonocarboxylic acid

$$\begin{array}{ccccc} & :\!\overset{..}{O} & H & \overset{..}{O}\!: & \\ & \| & | & \| & \\ H\!-\!\overset{..}{O}\!-\!P\!-\!C\!-\!C\!-\!\overset{..}{O}\!-\!H \\ & | & | & & \\ & H\!-\!\overset{..}{O}\!: & H & & \end{array}$$

inhibits the replication of the herpes virus. Structurally, it is a combination of phosphoric acid and acetic acid. Each molecule can donate three hydrogen ions. The stepwise acidity constants are $K_{a1} = 1.0 \times 10^{-2}$, $K_{a2} = 7.8 \times 10^{-6}$, and $K_{a3} = 2.0 \times 10^{-9}$ at body temperature. Enough phosphonocarboxylic acid is injected into the bloodstream of a patient to make its total concentration 1.0×10^{-5} mol L^{-1}. The pH of the blood does not change from 7.40. Determine the concentrations of all four forms of the acid in the patient's blood.

100. A reference book states that a saturated aqueous solution of potassium hydrogen tartrate is a buffer with a pH of 3.56. Write two chemical equations that show the buffer action of this solution. (Tartaric acid is a diprotic acid with the formula $H_2C_4H_4O_6$. Potassium hydrogen tartrate is $KHC_4H_4O_6$.)

*101. A solution of 1.0×10^{-8} M HNO_3 does *not* have a pH of 8.00 at 25°C.

(a) Explain why 1.0×10^{-8} M HNO_3 must have a pH below 7.

(b) Find the pH of the solution using these facts: The nitric acid is essentially completely dissociated; the total concentration of positive ions equals the total concentration of negative ions, that is, $[H_3O^+] = [OH^-] + [NO_3^-]$; the K_w equilibrium equation is fulfilled.

102. Solid aluminum hydroxide, $Al(OH)_3$, dissolves in both acidic and basic aqueous solutions. Write balanced equations for these reactions and show that they are acid–base reactions according to the Lewis definition. The aluminum species formed when aluminum hydroxide dissolves in basic solution is $Al(OH)_4^-$.

CUMULATIVE PROBLEM

Acid Rain

Power plant emissions cause acid rain. *(Simon Fraser/Science Photo Library/ Photo Researchers, Inc.)*

Acid rain is a major environmental problem throughout the industrialized world. It results largely from the burning of fossil fuels (coal, oil, and natural gas) that contain sulfur. The combustion of sulfur gives sulfur dioxide, the acid anhydride of the weak acid sulfurous acid, H_2SO_3. More serious is that sulfur dioxide may be oxidized to sulfur trioxide, the acid anhydride of the strong acid sulfuric acid, H_2SO_4. Both anhydrides react with water droplets in the air to increase the acidity of the rain, which damages trees, kills fish in lakes, and eats away stone and metal surfaces.

(a) A sample of rainwater is tested for pH by the use of two indicators. Addition of methyl orange to half of the sample gives a yellow color, while addition of methyl red to the other half gives a red color. Estimate the pH of the sample.

(b) The pH in acid rain can drop to 3 or even a bit lower in heavily polluted areas. Calculate the concentrations of H_3O^+ and OH^- in a raindrop at pH 3.30 at 25°C.

(c) When sulfur dioxide (SO_2) dissolves in water to form sulfurous acid ($H_2SO_3(aq)$), that acid can donate a hydrogen ion to water. Write a balanced chemical equation for this reaction, and identify the stronger Brønsted-Lowry acid and base in the equation.

(d) Ignore the further ionization of HSO_3^-, and calculate the pH of a solution in which the initial concentration of H_2SO_3 is 4.0×10^{-4} mol L^{-1}. (*Hint:* Use the quadratic formula or a series of approximations to solve the equation obtained from the equilibrium expression.)

(e) Now suppose that all of the dissolved SO_2 from part (d) is oxidized to SO_3, so that 4.0×10^{-4} mol of H_2SO_4 is dissolved per liter. Calculate the pH in this case. (*Hint:* Because the first ionization of H_2SO_4 is that of a strong acid, the concentration of H_3O^+ equals 4.0×10^{-4} plus the amount of hydrogen ion from the ionization of $HSO_4^-(aq)$.)

(f) Lakes have a natural buffering capacity, especially in regions where limestone in contact with the lake water gives rise to dissolved calcium carbonate. Write an equation for the reaction that occurs when a small amount of acid rain containing sulfuric acid falls into a lake containing carbonate (CO_3^{2-}) ions. Discuss how the lake resists further pH changes. What happens if a large excess of acid rain falls in the lake?

(g) A sample of 1.00 L of rainwater known to contain only sulfurous (and not sulfuric) acid is titrated with 0.0100 M NaOH. The equivalence point of the H_2SO_3/HSO_3^- titration is reached after 31.6 mL has been added. Calculate the original concentration of sulfurous acid in the sample, again ignoring any effect of SO_3^{2-} on the equilibria.

(h) Calculate the pH at the half-equivalence point. (*Hint:* Sulfurous acid is a strong enough acid that the Henderson–Hasselbalch equation cannot be used.)

9

Dissolution and Precipitation Equilibria

CHAPTER OUTLINE

Sugar dissolving in water. *(Richard Megna/Fundamental Photographs)*

Dissolution and its reverse, precipitation, are reactions in which solids pass into and out of solution. Such reactions match acid–base reactions in practical importance. On the global scale, they lead to the formation of rocks, mountains, and caves and profoundly affect the ecologies of natural waters. On a smaller scale, knowledge of such reactions helps engineers to prevent mineral build-up in boilers and physicians to reduce the incidence of painful kidney stones. Chemists use dissolution and precipitation to isolate single products from reaction mixtures and to purify impure solid samples.

In this chapter, we consider some quantitative aspects of the equilibria that govern dissolution and precipitation. The central theme is the manipulation of equilibria to control the solubilities of ionic solids.

9–1 The Nature of Solubility Equilibria

Often, the most difficult task in the chemical synthesis of a drug or other useful substance is not to make the compound but to purify it. Side-reactions can generate serious amounts of impurities. Impurities also enter with the starting materials, fall in accidentally, or are put in deliberately to make a preparative reaction go faster. Synthesizing a compound may take only hours, but the *work-up* (separating crude product) and subsequent purification may require weeks. **Recrystallization** is one of the most powerful methods for purifying solids. It relies on differences between the solubilities of the desired substance and its contaminants. An impure product is dissolved and reprecipitated, repeatedly if necessary. A successful purification by recrystallization depends on close control of the factors that influence solubility. Foremost among these are the chemical equilibria that exist between an undissolved substance and its solution.

In recrystallization, a solution begins to deposit a solid when it is brought to the point of saturation with respect to that substance. A **saturated solution** is one in which a dissolution–precipitation (solubility) equilibrium exists between the solid substance and its dissolved form. For example, the equilibrium between solid iodine and dissolved iodine can be written as

$$I_2(s) \rightleftharpoons I_2(aq)$$

In dissolution, the solvent attacks the solid and **solvates** it at the level of individual particles. To continue the example, this means that I_2–H_2O attractions replace the I_2–I_2 interactions formerly present in the solid, and the iodine molecules go off into the bulk of the H_2O. In precipitation, the reverse occurs; solute-to-solute attractions are re-established as the solute leaves the solution. Often, solute-to-solvent attractions persist right through the process of precipitation, and solvent incorporates itself into the solid. When lithium sulfate (Li_2SO_4) precipitates from water, the crystal that forms contains one molecule of water per formula unit:

$$2\,Li^+(aq) + SO_4^{2-}(aq) + H_2O(\ell) \longrightarrow Li_2SO_4 \cdot H_2O(s)$$

Such loosely bound solvent is **solvent of crystallization:** Lithium sulfate precipitates from water with one water of crystallization. Dissolving and then reprecipitating a compound may thus furnish material with a changed chemical formula.

Solubility equilibria resemble the equilibria that volatile liquids establish with their vapors in a closed container. In both cases, particles from a condensed phase tend to escape and to spread throughout a larger but limited volume. In both cases, equilibrium is a dynamic compromise in which the rate of escape of particles from

• The first separation of the element ytterbium (Yb) from the chemically very similar element lutetium (Lu) was achieved by a laborious series of 15,000 cycles of dissolution and reprecipitation of a mixture of $Yb(NO_3)_3$ and $Lu(NO_3)_3$, using aqueous nitric acid as the solvent.

• A saturated solution was defined previously (see Section 4–1) in slightly different, but equivalent, terms.

the condensed phase is equal to their rate of return. In a vaporization–condensation equilibrium, the vapor above the condensed phase is assumed to be an ideal gas. The analogous starting assumption for a dissolution–precipitation equilibrium is that the solution above the undissolved solid is an ideal solution.

• An ideal solution is a solution that obeys Raoult's law (see Section 6–6).

The equilibrium-constant equation for the dissolution of solid iodine has a very simple form

$$[I_2] = K$$

This equation indicates that the concentration of $I_2(aq)$ in a saturated solution of I_2 is constant (at constant temperature). This equilibrium law closely resembles the equilibrium law for vaporization–condensation, which states that the pressure of the vapor above a volatile substance is a constant (again, at constant temperature). No mention of the undissolved solute appears in the equation because the amount of undissolved solid present at the bottom of a flask has no effect on the position of the equilibrium, as long as some of it is present. The equilibrium expressions for many solubility equilibria are more complex than this, as we shall see.

Le Châtelier's principle applies to solubility equilibria. One way to exert a stress on a solubility equilibrium is to change the amount of solvent. Increases are particularly easy; just pour in some more. Adding solvent reduces the concentration of dissolved substance. More solid then tends to dissolve to restore the concentration of the dissolved substance to its equilibrium value. If so much solvent is added that *all* of the solid dissolves, then solid cannot simultaneously be dissolving and precipitating. The equilibrium ceases to exist, and the solution is **unsaturated.** This corresponds in a vaporization–condensation equilibrium to the complete evaporation of the condensed phase. Removing solvent from an already saturated solution forces additional solid to precipitate in order to maintain a constant concentration. A volatile solvent is easily removed by letting it evaporate. When conditions are right, the solid forms as crystals on the bottom and sides of the container (Fig. 9–1).

Figure 9–1 • The evaporating dish originally contained an aqueous solution of $(NH_4)_2Cr_2O_7$. It was loosely covered to keep out dust and allowed to stand. As the water evaporated, the solution became saturated and then deposited orange $(NH_4)_2Cr_2O_7$ in the form of long needle-like crystals. *(Charles D. Winters)*

The equilibrium constants for dissolution reactions depend on the temperature. If a dissolution reaction is exothermic, then raising the temperature of a saturated solution shifts the equilibrium to favor the solid, reducing the concentration of solute in solution. If a dissolution reaction is endothermic, then raising the temperature causes more solid to dissolve at saturation—that is, it shifts the solubility equilibrium toward the side of the products. In a survey of more than 500 solid inorganic compounds, 86% were found to increase in solubility in water with increasing temperature, and only 7% were found to decrease in solubility. (The remaining 7% had solubilities that increased and then decreased again, or vice versa.) In practice, when the behavior of a substance is not known, the odds favor heating the solution to dissolve more solid and cooling it to increase the amount of precipitate. Figure 9–2 displays the dependence of solubility on temperature for several substances.

• Thus, changes in temperature affect these equilibria in the same way that they affect other kinds of equilibria (see Section 7–5).

Dissolution–precipitation reactions frequently come to equilibrium slowly because it takes time to transfer material across the phase boundary between solid and solution. It can require days or even weeks of shaking a solid in contact with a solvent before the solution becomes saturated. Moreover, solutions sometimes become **supersaturated,** a condition in which the concentration of dissolved solid exceeds its equilibrium value. The delay then is in forming, rather than dissolving, the solid. Supersaturated solutions may persist for months or years and require extraordinary measures to induce formation of a precipitate, although the tendency toward equilibrium is always there (Fig. 9–3). The often sluggish approach to equilibrium in dissolution–precipitation reactions contrasts sharply with the rapid rates at which acid–base reactions come to equilibrium.

Figure 9-2 • The aqueous solubilities of a majority of solids increase with increasing temperature, but decreases and mixed behavior are common. Solubilities do not necessarily increase or decrease smoothly as the temperature rises. Sharp changes in slope occur if water of crystallization is lost or gained by the solid that is in contact with the solution. AgF (*green line*) has two such changes. The formulas of the solids are $AgF \cdot 4H_2O$ (below 19°C), $AgF \cdot 2H_2O$ (19–40°C), and AgF (above 40°C).

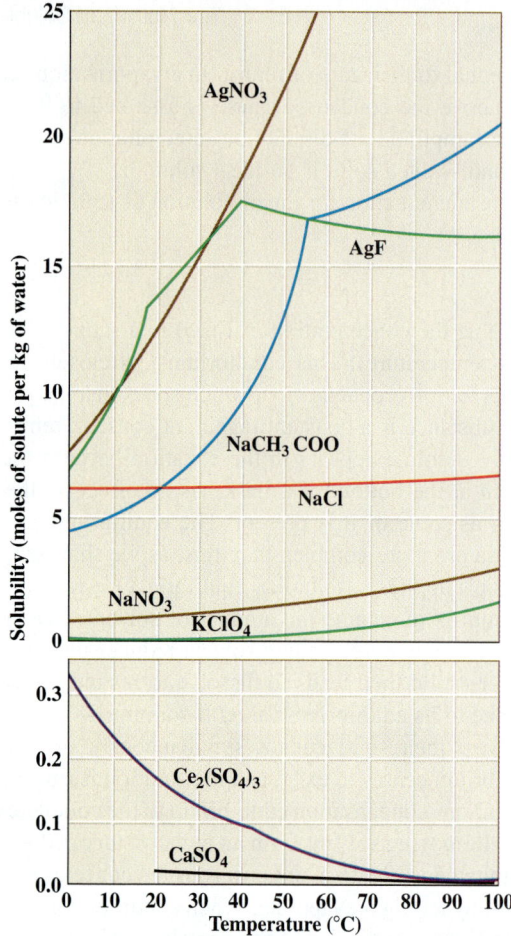

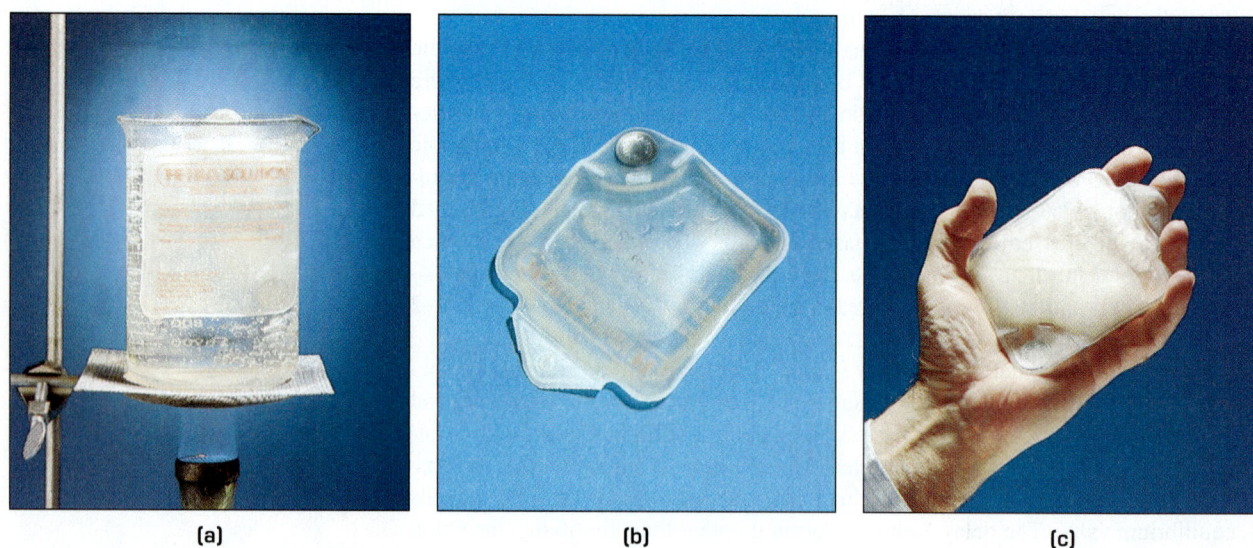

(a)	(b)	(c)

Figure 9-3 • (a) A concentrated solution of aqueous sodium acetate ($NaCH_3COO$) is contained in a plastic pouch. At the temperature of boiling water, all of the sodium acetate dissolves. (b) After cooling to room temperature, the sodium acetate tends to precipitate, but the process is quite slow. The solution becomes supersaturated, and the pouch can be stored almost indefinitely with no solid forming. (c) Flexing a metal disk that is in contact with the solution initiates precipitation. Much solid then forms, until the remaining solution is saturated. The change evolves a comfortable heat for soothing minor aches. *(Leon Lewandowski)*

9-2 The Solubility of Ionic Solids

When ionic solids (also known as *salts*) dissolve in water, they dissociate more or less completely into aquated ions (see Section 4–1). Chemists recognize the fact of dissociation by showing ions as products in the equations representing these dissolutions. For example, the dissolution of solid cesium chloride (CsCl) is written

$$CsCl(s) \rightleftharpoons Cs^+(aq) + Cl^-(aq)$$

If dissolution gave aquated *molecules* of CsCl, which it does not, then a better equation would be

$$CsCl(s) \rightleftharpoons CsCl(aq)$$

Ionic solids have a wide range of solubilities (see Fig. 9–2). For a highly soluble salt (such as CsCl), the concentrations of the ions in saturated aqueous solution are so large that the ions inevitably associate. The solution is non-ideal. Ion pairs (temporary pairs of oppositely charged ions, such as $Cs^+ \cdots Cl^-$) and larger ion clusters as well exist in such solutions. A full description of dissolution requires additional chemical equations to describe these interactions. We therefore restrict our attention to slightly soluble and insoluble salts. The concentrations of saturated solutions of these substances are less than about 0.1 mol L^{-1}. They are dilute enough that associations among the aquated ions are negligible. A single equation, one that shows complete dissociation into ions, describes them adequately.

• Table 4–1 defines "slightly soluble" and "insoluble" in terms of the mass that dissolves in a given mass of water.

The Solubility Product

The sparingly soluble ionic solid silver chloride establishes the following equilibrium when placed in water:

$$AgCl(s) \rightleftharpoons Ag^+(aq) + Cl^-(aq)$$

The equilibrium-constant equation for this reaction is written by following the general rules for heterogeneous equilibria from Section 7–6:

$$\frac{[Ag^+][Cl^-]}{1} = [Ag^+][Cl^-] = K_{sp}$$

where the subscript "sp," which stands for **solubility product,** distinguishes the K as referring to the dissolution of a slightly soluble ionic solid in water. The numerical value of $K_{sp}(AgCl)$ is 1.6×10^{-10} at 25°C. Like other equilibrium constants, a K_{sp} is a dimensionless number. All concentrations used in solubility-product expressions are understood to be divided by a reference concentration equal to 1.0 mol L^{-1}. As in other equilibrium expressions, we omit the reference concentrations.

In the AgCl solubility-product expression, the concentrations of the two ions produced are raised to the first power because their coefficients are 1 in the chemical equation. Silver chloride, a pure solid, is accounted for by the 1 in the denominator. Obviously, dividing by 1 does not change an algebraic expression. In chemical terms, this means that the amount of AgCl(s) does not affect the solubility equilibrium as long as *some* is present. If *no* solid is present, then the product of the two ion concentrations is no longer constrained to equal the constant.

• From this point on in this chapter, the 1's representing pure solids, pure liquids, and dilute solvents in equilibrium expressions will simply be omitted.

EXAMPLE 9–1

Write the equilibrium-constant equation for the dissolution of lead(II) iodide $(PbI_2(s))$ in water.

Strategy

Recognize lead(II) iodide as a compound that dissolves by dissociation into ions. Write a chemical equation for the dissolution, and then apply the rules in Section 7–6 to obtain the answer.

Solution

One mole of $PbI_2(s)$ produces 1 mol of Pb^{2+} ions and 2 mol of I^- ions in solution:

$$PbI_2(s) \rightleftharpoons Pb^{2+}(aq) + 2\,I^-(aq)$$
$$[Pb^{2+}][I^-]^2 = K_{sp}$$

The pure solid $PbI_2(s)$ does not appear explicitly.

EXERCISE

Write a K_{sp} equation for the dissolution of aluminum hydroxide $(Al(OH)_3)$ in water.

Answer: $[Al^{3+}][OH^-]^3 = K_{sp}$.

Solubility and K_{sp}

The solubility of an ionic solid in water is not the same as its solubility-product constant, but a relationship, often a simple one, exists between the two. For example, if s mol of $AgCl(s)$ dissolves in 1 L of solution, then the concentration of each ion at equilibrium equals s mol L^{-1}. We then write

$$[Ag^+][Cl^-] = s^2 = K_{sp}$$

At 25°C, K_{sp} is known to equal 1.6×10^{-10}. Therefore

$$s = \sqrt{1.6 \times 10^{-10}} = 1.26 \times 10^{-5} \text{ mol } L^{-1}$$

which rounds off to 1.3×10^{-5} mol L^{-1}. This is the **molar solubility** of AgCl at this temperature. Molar solubilities are molarities—the maximum possible molarities that substances have under defined conditions. They can be converted to mass fractions, mole fractions, or molalities, as explained in Section 6–6.

Solubilities are also expressed as **mass solubilities.** In the case of AgCl,

• Solubility information comes in many different units. Using the information effectively requires frequent conversions of units.

$$\text{mass solubility} = \frac{1.26 \times 10^{-5}\,\text{mol AgCl}}{\text{L solution}} \times \frac{143.3 \text{ g AgCl}}{1 \text{ mol AgCl}} = \frac{1.8 \times 10^{-3}\,\text{g AgCl}}{\text{L solution}}$$

A liter of saturated aqueous silver chloride at 25°C contains 1.8×10^{-3} g of AgCl. When the unit of mass is the gram, a mass solubility is often termed a *gram solubility*.

Solubility-product constants (and solubilities) generally depend strongly on temperature. At 100°C, the K_{sp} of silver chloride is about 140 times larger than at 25°C. This means that a liter of boiling water dissolves about 12 times more silver chloride than a liter of water at 25°C. For this reason, temperatures are always quoted in listings of K_{sp} values. Table 9–1 lists the solubility-product constants at 25°C of a number of important sparingly soluble salts.

TABLE 9-1	Solubility-Product Constants (K_{sp}'s) at 25°C		
Compound	K_{sp}	Compound	K_{sp}
Fluorides		Mn(OH)$_2$	2×10^{-13}
BaF$_2$	1.7×10^{-6}	Zn(OH)$_2$	4.5×10^{-17}
CaF$_2$	3.9×10^{-11}	**Iodates**	
MgF$_2$	6.6×10^{-9}	AgIO$_3$	3.1×10^{-8}
PbF$_2$	3.6×10^{-8}	Cu(IO$_3$)$_2$	1.4×10^{-7}
SrF$_2$	2.8×10^{-9}	Pb(IO$_3$)$_2$	2.6×10^{-13}
Chlorides		**Chromates**	
AgCl	1.6×10^{-10}	Ag$_2$CrO$_4$	1.9×10^{-12}
CuCl	1.0×10^{-6}	BaCrO$_4$	2.1×10^{-10}
PbCl$_2$	1.6×10^{-5}	PbCrO$_4$	1.8×10^{-14}
Hg$_2$Cl$_2$	2×10^{-18}	**Carbonates**	
Bromides		Ag$_2$CO$_3$	6.2×10^{-12}
AgBr	7.7×10^{-13}	BaCO$_3$	8.1×10^{-9}
CuBr	4.2×10^{-8}	CaCO$_3$	8.7×10^{-9}
PbBr$_2$	4.6×10^{-6}	PbCO$_3$	3.3×10^{-14}
Hg$_2$Br$_2$	1.3×10^{-21}	MgCO$_3$	4.0×10^{-5}
Iodides		SrCO$_3$	1.6×10^{-9}
AgI	1.5×10^{-16}	**Oxalates**	
CuI	5.1×10^{-12}	CuC$_2$O$_4$	2.9×10^{-8}
PbI$_2$	1.4×10^{-8}	FeC$_2$O$_4$	2.1×10^{-7}
Hg$_2$I$_2$	1.2×10^{-28}	MgC$_2$O$_4$	8.6×10^{-5}
Hydroxides		PbC$_2$O$_4$	2.7×10^{-11}
AgOH	1.5×10^{-8}	SrC$_2$O$_4$	5.6×10^{-8}
Al(OH)$_3$	3.7×10^{-15}	**Sulfates**	
Fe(OH)$_3$	1.1×10^{-36}	BaSO$_4$	1.1×10^{-10}
Fe(OH)$_2$	1.6×10^{-14}	CaSO$_4$	2.4×10^{-5}
Mg(OH)$_2$	1.2×10^{-11}	PbSO$_4$	1.1×10^{-8}

EXAMPLE 9-2

The K_{sp} of calcium fluoride equals 3.9×10^{-11} at 25°C. Calculate the concentrations of calcium and fluoride ions in a saturated solution of CaF$_2$ and the mass solubility of CaF$_2$ (in g L^{-1}) at 25°C.

Strategy

Plan to obtain the molar solubility and then convert to the mass solubility. Realize that calcium fluoride dissolves by dissociation into ions. Write the chemical equation and corresponding K_{sp} expression for the dissolution. Recognize that the molar solubility equals the equilibrium concentration of the Ca^{2+} ion.

Solution

$$\text{CaF}_2(s) \rightleftharpoons \text{Ca}^{2+}(aq) + 2\,\text{F}^-(aq) \qquad [\text{Ca}^{2+}][\text{F}^-]^2 = K_{sp}$$

Let s equal the molar solubility of CaF_2. Then, according to the balanced equation, the equilibrium concentration of Ca^{2+} ion equals s and the equilibrium concentration of F^- ion equals $2s$. Therefore,

$$[Ca^{2+}][F^-]^2 = s(2s)^2 = 4s^3 = K_{sp}$$

$$s^3 = \frac{K_{sp}}{4}$$

$$s = \sqrt[3]{\frac{3.9 \times 10^{-11}}{4}} = 2.14 \times 10^{-4} \text{ mol L}^{-1}$$

$$[Ca^{2+}] = 2.1 \times 10^{-4} \text{ mol L}^{-1} \quad \text{and} \quad [F^-] = 4.3 \times 10^{-4} \text{ mol L}^{-1}$$

Obtaining the mass solubility requires use of the molar mass of CaF_2:

$$\text{mass solubility} = \left(\frac{2.14 \times 10^{-4} \text{ mol Ca}^{2+}}{L}\right) \times \left(\frac{1 \text{ mol CaF}_2}{1 \text{ mol Ca}^{2+}}\right) \times \left(\frac{78.1 \text{ g CaF}_2}{1 \text{ mol CaF}_2}\right)$$

$$= 0.017 \text{ g L}^{-1}$$

Check

Substituting the concentrations of the Ca^{2+} and F^- into the K_{sp} expression gives 3.88×10^{-11}, which rounds off to the correct K_{sp}. Technical references report 0.018 g L^{-1} as the experimental mass solubility of CaF_2 at 25°C.

EXERCISE

Determine the mass of lead(II) iodate dissolved in 2.50 L of a saturated aqueous solution of $Pb(IO_3)_2$ at 25°C. The K_{sp} of $Pb(IO_3)_2$ is 2.6×10^{-13}.

Answer: $m_{Pb(IO_3)_2} = 0.056$ g.

• Values of K_{sp} also can be obtained from thermodynamic and electrochemical studies, as shown in Chapters 11 and 13.

The calculation of K_{sp}'s from measured solubilities requires the reversal of the procedure just outlined, as the following example illustrates.

EXAMPLE 9-3

At 25°C the mass solubility of silver chromate (Ag_2CrO_4), a red solid, equals 0.029 g L^{-1}. Calculate its K_{sp} and compare the answer with the value in Table 9–1.

Strategy

Realize that dissolution of Ag_2CrO_4 gives Ag^+ and CrO_4^{2-} ions in a 2:1 molar ratio. Determine the concentrations of the ions in the saturated solutions assuming that neither gets involved in further reactions. Substitute these concentrations into the equilibrium-constant equation.

Solution

$$[CrO_4^{2-}] = \left(\frac{0.029 \text{ g Ag}_2CrO_4}{\text{L solution}}\right)\left(\frac{1 \text{ mol Ag}_2CrO_4}{331.73 \text{ g Ag}_2CrO_4}\right)\left(\frac{1 \text{ mol CrO}_4^{2-}}{1 \text{ mol Ag}_2CrO_4}\right)$$

$$= 8.7 \times 10^{-5} \text{ mol L}^{-1}$$

$$[Ag^+] = 2 \times [CrO_4^{2-}] = 2 \times (8.7 \times 10^{-5}) \text{ mol L}^{-1}$$

$$K_{sp}(Ag_2CrO_4) = [Ag^+]^2[CrO_4^{2-}] = (2 \times 8.7 \times 10^{-5})^2(8.7 \times 10^{-5})$$

$$= 2.6 \times 10^{-12}$$

This exceeds 1.9×10^{-12}, the value in Table 9–1, by roughly 40%.

The discrepancy between $K_{sp}(Ag_2CrO_4)$ in Table 9–1 and the value obtained in Example 9–3 is a matter of obvious concern. Computing a solubility from a K_{sp} (or a K_{sp} from a solubility) is valid if the solution is ideal and if further reactions do not reduce the concentrations of the ions after they enter solution. When such reactions occur, as they do with Ag_2CrO_4, they cause higher solubilities than the K_{sp} predicts (or apparent K_{sp}'s that exceed the true values). To take another example, the solubility of $PbSO_4(s)$ computed from its K_{sp} at 25°C by the method just illustrated is 0.032 g L^{-1}, but the experimentally measured solubility is 0.0425 g L^{-1}, which is 33% higher. The difference is caused mainly by hydrolysis of the $Pb^{2+}(aq)$ and $SO_4^{2-}(aq)$ ions to form $PbOH^+(aq)$ and $HSO_4^-(aq)$ (see Section 8–4). These reactions shift the dissolution equilibrium to favor the products. A solubility or K_{sp} obtained without considering these complications, which are discussed further in Sections 9–4 and 9–5, must be regarded as only an estimate.

9-3 Precipitation and the Solubility Product

Up to now, the focus has been on a single, slightly soluble ionic solid that attains equilibrium with its constituent ions in water. In such solutions, the relative concentrations of the cations and anions echo the relative number of moles of each in the original salt. Thus, when AgCl dissolves, equal numbers of $Ag^+(aq)$ and $Cl^-(aq)$ ions result, and when Ag_2CrO_4 dissolves, twice as many $Ag^+(aq)$ ions as $CrO_4^{2-}(aq)$ ions result. An equilibrium-constant equation such as

$$[Ag^+][Cl^-] = K_{sp}$$

is more general than this, however, and continues in force even if the relative chemical amounts of the two ions in solution differ from those in the pure solid compound. Such situations often result when two solutions are mixed to give a precipitate or when a second ionic substance has been dissolved in the same solution.

Precipitation from Solution

Suppose that a solution of one soluble salt, such as $AgNO_3$, is mixed with a solution of a second, such as NaCl. Does a precipitate of silver chloride form? To answer this, we use the concept of the reaction quotient Q that was introduced in connection with gas-phase reactions (Section 7–3). The reaction quotient is evaluated at the moment that the mixing of the solutions is complete but before any reaction occurs. This is the initial reaction quotient

$$Q_{(init)} = [Ag^+]_{(init)}[Cl^-]_{(init)}$$

If $Q_{(init)} < K_{sp}$, no solid silver chloride can appear. On the other hand, if $Q_{(init)} > K_{sp}$, solid silver chloride precipitates until Q becomes equal to K_{sp} (Fig. 9–4).

• The reaction quotient in this case has a denominator equal to 1, which is omitted.

Figure 9–4 • Some solid silver chloride is in contact with a solution containing $Ag^+(aq)$ and $Cl^-(aq)$ ions. If the system is at equilibrium, then the product Q of the concentrations of the ions equals K_{sp} (*curved line*). If Q exceeds K_{sp}, solid silver chloride tends to precipitate until equilibrium is attained. If Q is less than K_{sp}, then additional solid tends to dissolve.

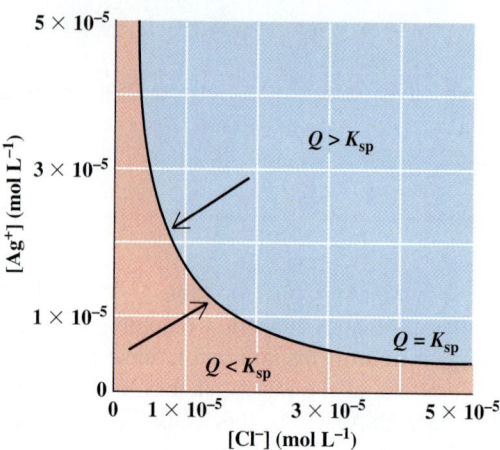

EXAMPLE 9–4

Suppose that 500 mL of a 0.0080 M solution of NaCl is mixed with 300 mL of a 0.040 M solution of $AgNO_3$. Does solid AgCl form at 25°C?

Strategy

Two effects change the concentration of the two reactants: dilution, as the solutions mix, and (possible) reaction to precipitate $AgCl(s)$. Separate the effects by choosing the initial state of the reaction as the state that exists immediately after the two solutions have been thoroughly mixed but before the reaction starts.

Solution

After dilution by mixing, but before reaction

$$[Ag^+]_{(init)} = 0.040 \text{ mol L}^{-1} \times \left(\frac{300 \text{ mL}}{800 \text{ mL}}\right) = 0.015 \text{ mol L}^{-1}$$

$$[Cl^-]_{(init)} = 0.0080 \text{ mol L}^{-1} \times \left(\frac{500 \text{ mL}}{800 \text{ mL}}\right) = 0.0050 \text{ mol L}^{-1}$$

The initial reaction quotient is

$$Q_{(init)} = [Ag^+]_{(init)}[Cl^-]_{(init)} = (0.015)(0.0050) = 7.5 \times 10^{-5}$$

Because $Q_{(init)}$ is much larger than $K_{sp}(AgCl)$, which equals 1.6×10^{-10}, according to Table 9–1, solid silver chloride is present at equilibrium.

EXERCISE

The K_{sp} of thallium(I) iodate ($TlIO_3$) is 3.1×10^{-6} at 25°C. Suppose that 555 mL of a 0.0022 M solution of $TlNO_3$ is mixed with 445 mL of a 0.0022 M solution of $NaIO_3$ at 25°C. Does $TlIO_3$ precipitate at equilibrium?

Answer: $TlIO_3$ does not precipitate.

If a precipitate forms, the next step is to calculate the equilibrium concentrations of the ions left behind in solution. Knowing such concentrations has considerable practical importance (for example, in water treatment and purification). The next example illustrates such a calculation.

EXAMPLE 9–5

Suppose that 500 mL of a 0.0080 M solution of NaCl is mixed with 300 mL of a 0.040 M solution of $AgNO_3$ at 25°C (as in Example 9–4). Calculate the equilibrium concentrations of silver and chloride ions.

Strategy

Get the concentrations of Ag^+ and Cl^- after mixing but before reaction. Then imagine that precipitation goes to completion (using up one of the ions) and subsequent dissolution restores some of that ion to solution. This works because the equilibrium state is the same regardless of the direction from which it is attained. A similar strategy was used in Example 8–11 (the case of the addition of a strong acid to a buffer solution).

Solution

After mixing but before reaction, the Ag^+ concentration is 0.015 mol L^{-1}, and the Cl^- concentration is 0.0050 mol L^{-1} (see Example 9–4). The limiting reactant is clearly the Cl^- ion. If the reaction went to completion, all of the chloride ion would be precipitated, and the concentration of the excess silver ion would be

$$[Ag^+] = 0.015 - 0.0050 = 0.010 \text{ mol } L^{-1} \quad \text{(at completion)}$$

Take this (hypothetical) state of completion as the *initial* state as the dissolution reaction proceeds:

$$AgCl(s) \rightleftharpoons Ag^+(aq) + Cl^-(aq)$$

	$[Aq^+]$ $(\text{mol } L^{-1})$	$[Cl^-]$ $(\text{mol } L^{-1})$
Initial	0.010	0
Change	$+y$	$+y$
Equilibrium	$0.010 + y$	y

Inserting the equilibrium values into the solubility-product expression gives

$$[Ag^+][Cl^-] = K_{sp}$$
$$(0.010 + y)y = 1.6 \times 10^{-10}$$

The quadratic equation can be rearranged to the standard form

$$y^2 + 0.010y - 1.6 \times 10^{-10} = 0$$

and solved using a calculator. A faster, more "chemical" approach is to observe that y should be much smaller than 0.010 because AgCl is only slightly soluble. Why not then simply neglect y when it is to be added to 0.010? Then the equation simplifies to

$$(0.010)y = 1.6 \times 10^{-10}$$

and is easily solved:

$$y = 1.6 \times 10^{-8} \text{ mol } L^{-1} = [Cl^-]$$

The thinking about the relative size of y is clearly correct. The equilibrium concentration of silver ion is

$$[Ag^+] = (0.010 + 1.6 \times 10^{-8}) \text{ mol } L^{-1} = 0.010 \text{ mol } L^{-1}$$

Check

The product of the answers for $[Ag^+]$ and $[Cl^-]$ equals K_{sp}, as it must. A calculator gives $y = 1.59999744 \times 10^{-8}$ and -0.010000016 as the roots of the quadratic equation. The positive root confirms the answer. It does not differ from 1.6×10^{-8} until its eighth digit, far beyond the precision of the original data. The negative root is not meaningful as a concentration.

EXERCISE

The K_{sp} of thallium(I) iodate ($TlIO_3$) is 3.1×10^{-6} at 25°C. Suppose that 125 mL of a 0.0400 mol L^{-1} solution of $TlNO_3$ is mixed with 375 mL of a 0.0400 mol L^{-1} solution of $NaIO_3$. Determine the concentration of $Tl^+(aq)$ in the solution at equilibrium.

Answer: $[Tl^+] = 1.5 \times 10^{-4}$ mol L^{-1}.

The Common-Ion Effect

Suppose that a small amount of $NaCl(s)$ is added to a saturated solution of AgCl. What happens? Sodium chloride is quite soluble in water. It dissolves to give $Na^+(aq)$ and $Cl^-(aq)$ ions, raising the concentration of chloride ion. The reaction quotient $Q = [Ag^+][Cl^-]$ then exceeds the K_{sp} of silver chloride, and silver chloride precipitates until the K_{sp} equation once again is satisfied.

Adding $AgNO_3(s)$ to a saturated solution of AgCl causes a similar precipitation of AgCl. Silver nitrate is also quite soluble in water. The added $Ag^+(aq)$ reduces the concentration of $Cl^-(aq)$ permitted at equilibrium and so reduces the solubility of AgCl. Figure 9–5 shows graphically how the presence in solution of either

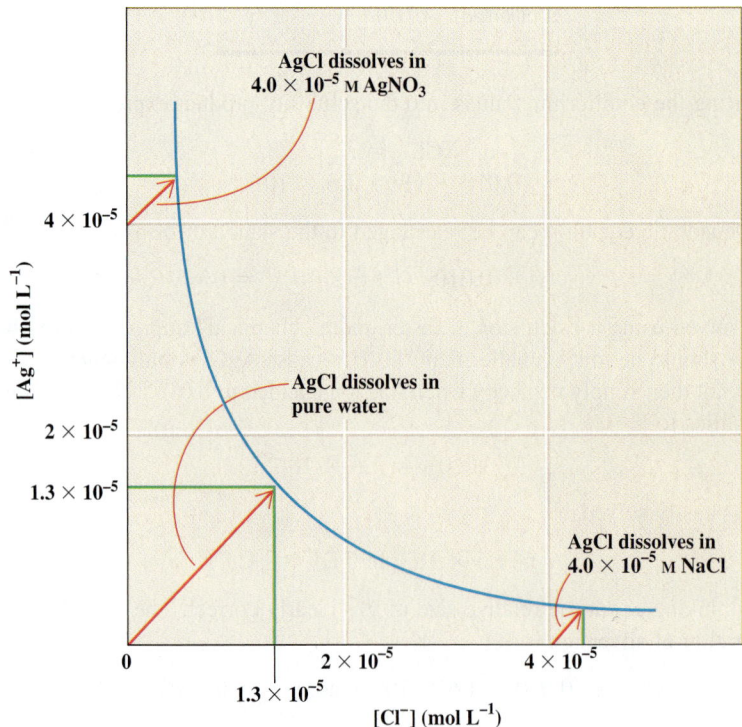

Figure 9–5 • The presence of a previously dissolved common ion reduces the solubility of a salt. In all three cases shown, as the AgCl dissolves, the concentrations of the ions follow the paths shown by the red arrows until they reach the blue equilibrium line. The molar solubilities are proportional to the lengths of the green lines shown: 1.3×10^{-5} mol L^{-1} for AgCl in pure water, but only 0.37×10^{-5} mol L^{-1} in either 4.0×10^{-5} mol L^{-1} $AgNO_3$ or 4.0×10^{-5} mol L^{-1} NaCl.

$Ag^+(aq)$ or $Cl^-(aq)$ from another source depresses the solubility of $AgCl(s)$ compared with its value in pure water. A similar effect occurs with all ionic solids and is referred to as the **common-ion effect:**

> If a solution and an ionic solid to be dissolved in it have an ion in common, then the solubility of the ionic solid is depressed compared with its solubility in pure water.

The common ion effect works according to Le Châtelier's principle. Added product (the common ion) shifts the solubility equilibrium to the left. Let us examine the effect quantitatively. Suppose some $AgCl(s)$ is added to a 0.100 mol L^{-1} NaCl solution. What is its solubility? If s equals the molar solubility of the AgCl, then the concentration of $Ag^+(aq)$ is $s \text{ mol L}^{-1}$, and the concentration of $Cl^-(aq)$ is

$$[Cl^-] = (0.100 + s) \text{ mol L}^{-1}$$

because the chloride ion has two sources: the 0.100 mol L^{-1} NaCl and the dissolution of AgCl. The expression for the solubility product is

$$[Ag^+][Cl^-] = K_{sp}$$
$$s(0.100 + s) = 1.6 \times 10^{-10}$$

The solubility of $AgCl(s)$ in this solution must be smaller than its solubility in pure water, which is known to be much smaller than 0.100 mol L^{-1}. That is,

$$s < 1.3 \times 10^{-5} \text{ mol L}^{-1} \ll 0.100 \text{ mol L}^{-1}$$

Because s is so small, we can approximate $(0.100 + s)$ as 0.100:

$$(0.100)s = 1.6 \times 10^{-10}$$
$$s = 1.6 \times 10^{-9} \text{ mol L}^{-1}$$

Clearly, s is in fact much smaller than 0.100 mol L^{-1}, so the approximation is a good one. At equilibrium

$$[Ag^+] = s = 1.6 \times 10^{-9} \text{ mol L}^{-1}$$
$$[Cl^-] = (0.100 + s) = (0.100 + 1.6 \times 10^{-9}) = 0.100 \text{ mol L}^{-1}$$

The molar solubility of AgCl equals $1.6 \times 10^{-9} \text{ mol L}^{-1}$. The mass solubility is

$$\text{mass solubility} = \frac{1.6 \times 10^{-9} \text{ mol AgCl}}{1 \text{ L solution}} \times \left(\frac{143.3 \text{ g AgCl}}{1 \text{ mol AgCl}} \right) = 2.3 \times 10^{-7} \text{ g L}^{-1}$$

The solubility of AgCl in a 0.100 mol L^{-1} solution of NaCl is about 8000 times smaller than its solubility in pure water.

EXAMPLE 9-6

Estimate the solubility (in g L^{-1}) of $CaF_2(s)$ in a 0.100 mol L^{-1} solution of NaF at 25°C.

Strategy

Realize that the only source of $Ca^{2+}(aq)$ in the solution is the dissolution of CaF_2 but that the $F^-(aq)$ has two sources, CaF_2 and NaF. The concentration of $Ca^{2+}(aq)$ at equilibrium therefore equals the molar solubility of the CaF_2.

Solution

Write the chemical equation and set up a table of changes. Let the change in the concentration of $Ca^{2+}(aq)$ as the reaction comes to equilibrium equal s:

$$CaF_2(s) \rightleftharpoons Ca^{2+}(aq) + 2\,F^-(aq)$$

	$[Ca^{2+}]$ (mol L^{-1})	$[F^-]$ (mol L^{-1})
Initial	0 mol L^{-1}	0.100 mol L^{-1}
Change	$+s$	$+2s$
Equilibrium	s	$0.100 + 2s$

$$[Ca^{2+}][F^-]^2 = K_{sp}$$
$$s(0.100 + 2s)^2 = K_{sp}$$

The molar solubility s is certain to be small, so try approximating $(0.100 + 2s)$ as 0.100:

$$s(0.100)^2 = 3.9 \times 10^{-11}$$
$$s = 3.9 \times 10^{-9} \text{ mol L}^{-1}$$

The resulting s is negligible compared with 0.100 mol L^{-1}, so the approximation is acceptable. The mass solubility is

$$\text{mass solubility} = \frac{3.9 \times 10^{-9} \text{ mol CaF}_2}{1 \text{ L solution}} \times \left(\frac{78.1 \text{ g CaF}_2}{1 \text{ mol CaF}_2}\right) = 3.0 \times 10^{-7} \text{ g L}^{-1}$$

Check

The solubility of CaF_2 in a 0.100 mol L^{-1} solution of NaF is about 54,000 times smaller than its solubility in pure water (computed in Example 9–2). This checks with the fact that the concentration of fluoride ion is about 232 times larger in this problem—the common-ion effect in this dissolution goes as to the square of the concentration of fluoride ion, and 232 squared equals about 54,000.

EXERCISE

The K_{sp} of thallium(I) iodate ($TlIO_3$) is 3.1×10^{-6} at 25°C. Determine the molar solubility of $TlIO_3$ in 0.050 mol L^{-1} KIO$_3$ at 25°C.

Answer: $s_{TlIO_3} = 6.2 \times 10^{-5}$ mol L^{-1}.

9–4   The Effects of pH on Solubility

Some solids are only poorly soluble in plain water but dissolve readily in more acidic solutions. The sulfides of copper and nickel in ores, for example, can be brought into solution with strong acids, a fact that aids greatly in the separation and recovery of these valuable metals in elemental form. A less desirable effect of pH on solubility can be seen dramatically in the damage caused by acid rain to buildings and monuments (Fig. 9–6). Both marble and limestone consist of small crystals of calcite ($CaCO_3$). This mineral has a slow rate of dissolution in

Figure 9–6 • The calcium carbonate in marble and limestone is very slightly soluble in neutral water. Its solubility is much larger in acidic water. Objects carved of these materials dissolve relatively rapidly in areas where rain, snow, and fog are acidified from air pollution. Shown here is the damage to a gargoyle at the Lincoln Cathedral in England between 1910 (*left*) and 1984 (*right*). (*Courtesy of the Dean of Lincoln*)

"natural" rainwater (with a pH of about 5.6) but dissolves much more rapidly as the rainwater becomes more acidic. The reaction

$$CaCO_3(s) + H_3O^+(aq) \longrightarrow Ca^{2+}(aq) + HCO_3^-(aq) + H_2O(\ell)$$

causes this increase. In this section, we examine the role of pH in solubility.

Solubility of Hydroxides

A direct effect of pH on solubility occurs with the numerous metal hydroxides. The concentration of OH^- appears explicitly in the K_{sp} expressions for such compounds. For example, zinc hydroxide is only very slightly soluble in pure water.

$$Zn(OH)_2(s) \rightleftharpoons Zn^{2+}(aq) + 2\,OH^-(aq)$$
$$[Zn^{2+}][OH^-]^2 = K_{sp} = 4.5 \times 10^{-17}$$

If the solution is acidified, the concentration of hydroxide ion decreases, which obliges an increase in the equilibrium concentration of $Zn^{2+}(aq)$. Zinc hydroxide is thus more soluble in acidic solution.

EXAMPLE 9–7

Estimate the solubility (in g L^{-1}) of $Zn(OH)_2$ in pure water at 25°C. Compare it with the solubility of the same compound in water buffered at a pH of 6.0.

Strategy

Solve the first part using the method of Example 9–2. Assume in the second part that the buffer consumes all of the OH^- ion that comes from the dissolution of $Zn(OH)_2$ and keeps the pH at 6.0.

Solution

At equilibrium

$$[Zn^{2+}][OH^-]^2 = K_{sp}$$

Let $[Zn^{2+}] = s$. Then $[OH^-] \approx 2s$ when $Zn(OH)_2$ dissolves in pure water. Substitute in the equilibrium expression

$$s(2s)^2 = 4s^3 = K_{sp} = 4.5 \times 10^{-17}$$
$$s = 2.2 \times 10^{-6} \text{ mol L}^{-1}$$

so the molar solubility is 2.2×10^{-6} mol L^{-1}. Multiplying by the molar mass, which is 99.4 g mol^{-1}, gives a mass solubility of 2.2×10^{-4} g L^{-1}.

If the pH remains 6.0 after dissolution of the zinc hydroxide, then pOH = 8.0 and

$$[OH^-] = 1.0 \times 10^{-8} \text{ mol L}^{-1}$$

Again, the molar solubility equals the equilibrium concentration of the Zn^{2+} ion

$$s = [Zn^{2+}] = \frac{K_{sp}}{[OH^-]^2} = \frac{4.5 \times 10^{-17}}{(1.0 \times 10^{-8})^2} = 0.45 \text{ mol L}^{-1}$$

so that 0.45 mol $Zn(OH)_2$, which is 45 g of $Zn(OH)_2$, should dissolve per liter of pH 6.0 buffer.

Check

The saturated solution of $Zn(OH)_2$ in pure water has $[OH^-] = 2s = 4.5 \times 10^{-6}$ mol L^{-1}. Therefore it has a pH of 8.65. The buffer lowers the pH to 6.0 and keeps it there, which means $[OH^-]$ is hundreds of times smaller. This should increase the solubility, and it does. *Note:* When ionic concentrations become as high as 0.45 mol L^{-1}, the simple form of the K_{sp} equation is likely to break down, but the qualitative result remains valid: $Zn(OH)_2$ is far more soluble at pH 6.0 than in pure water.

EXERCISE

Estimate the molar solubility of $Fe(OH)_3$ in a solution that is buffered to a pH of 2.9.

Answer: 2.2×10^{-3} mol L^{-1}.

Solubility of Salts of Weak Bases

Metal hydroxides can be described as salts of a strong base, the hydroxide ion. The solubility of salts in which the anion is a weak base also is affected by pH. For example, in saturated aqueous calcium fluoride, the following equilibrium exists:

$$CaF_2(s) \rightleftharpoons Ca^{2+}(aq) + 2 F^-(aq) \qquad K_{sp} = 3.9 \times 10^{-11}$$

If the solution is acidified, then some of the fluoride ion reacts with hydronium ion:

$$H_3O^+(aq) + F^-(aq) \rightleftharpoons HF(aq) + H_2O(\ell)$$

This reaction is the reverse of the acid ionization of HF so that its equilibrium constant is the reciprocal of K_a (HF), or $1/(3.5 \times 10^{-4}) = 2.9 \times 10^3$. Adding acid reduces the concentration of fluoride ion. The calcium ion concentration must increase in order to maintain the solubility-product equilibrium for CaF_2. As a result, the solubility of calcium fluoride increases. All ionic substances in which the anion is a weak base become more soluble when an acid is added. By contrast, the solubility of a salt like AgCl is only very slightly affected by a decrease in pH. The reason is that HCl is a strong acid, so that Cl^-, its conjugate base, is ineffective as a base; the reaction

• Why the reciprocal of K_a? See Section 7–2.

$$H_3O^+(aq) + Cl^-(aq) \longrightarrow HCl(aq) + H_2O(\ell)$$

occurs to a negligible extent.

Selective Precipitation of Ions

The analysis of a mixture of elements usually requires separation of the mixture into its components. One way to do this is to exploit the differences in the solubilities of compounds of the elements. To separate barium ion from calcium ion, for example, a search is made for ionic compounds of the two elements that (1) have a common anion and (2) have widely different solubilities. Scanning Table 9–1 reveals that the fluorides BaF_2 and CaF_2 are two such compounds: the solubility equilibria and K_{sp}'s for the two are

$$BaF_2(s) \rightleftharpoons Ba^{2+}(aq) + 2\,F^-(aq) \qquad K_{sp} = 1.7 \times 10^{-6}$$
$$CaF_2(s) \rightleftharpoons Ca^{2+}(aq) + 2\,F^-(aq) \qquad K_{sp} = 3.9 \times 10^{-11}$$

Consider a solution that is 0.10 mol L^{-1} in both Ba^{2+} and Ca^{2+} ion. Is it possible to add enough F^- to precipitate almost all of the Ca^{2+} ions while leaving the Ba^{2+} ions in solution? If so, an essentially complete separation of the two species can be achieved. In order for Ba^{2+} to remain in solution, the reaction quotient for the first reaction must remain smaller than K_{sp}:

$$Q = [Ba^{2+}][F^-]^2 < K_{sp}$$

Inserting K_{sp} and the concentration of Ba^{2+} into this expression and rearranging gives

$$[F^-]^2 < \frac{K_{sp}}{[Ba^{2+}]} = \frac{1.7 \times 10^{-6}}{0.10} = 1.7 \times 10^{-5}$$

Taking the square root on both sides of the inequality gives

$$[F^-] < 0.0041 \text{ mol } L^{-1}$$

As long as the concentration of fluoride ion remains smaller than 0.0041 mol L^{-1}, no BaF_2 should precipitate. To separate the barium and calcium ions as cleanly as possible (that is, to precipitate out as much calcium fluoride as possible but no barium fluoride), the fluoride ion concentration should be made as high as possible while not exceeding 0.0041 mol L^{-1}. If exactly this concentration of $F^-(aq)$ is chosen, then, at equilibrium,

$$[Ca^{2+}] = \frac{K_{sp}}{[F^-]^2} = \frac{3.9 \times 10^{-11}}{(0.0041)^2} = 2.3 \times 10^{-6} \text{ mol } L^{-1}$$

TABLE 9-2

Equilibrium Constants for the Dissolution of Some Metal Sulfides at 25°C

Metal Sulfide	K_{ss}^a
CuS	5×10^{-37}
PbS	3×10^{-28}
CdS	7×10^{-28}
SnS	9×10^{-27}
ZnS	2×10^{-25}
FeS	5×10^{-19}
MnS	3×10^{-14}
HgS (red)	5×10^{-54}
HgS (black)	2×10^{-53}

$^a K_{ss}$ is the equilibrium constant for the "sulfide solubility" reaction $MS(s) + H_2O(\ell) \rightleftharpoons M^{2+}(aq) + OH^-(aq) + HS^-(aq)$.

At an equilibrium concentration of F^- equal to 0.0041 mol L^{-1}, the concentration of Ca^{2+} is reduced from 0.10 to 2.3×10^{-6} mol L^{-1}. All of the Ba^{2+} ions remain in solution, but almost all of the Ca^{2+} ions are tied up in the precipitate of calcium fluoride at the bottom of the container. A good separation has been achieved.

Metal Sulfides

Controlling the solubility of metal sulfides has important applications. According to Table 4–1, most metal sulfides are "insoluble" in water. This means only a very small amount of a compound such as $ZnS(s)$ dissolves in water. Although it is tempting to write the dissolution reaction as

$$ZnS(s) \rightleftharpoons Zn^{2+}(aq) + S^{2-}(aq) \qquad K_{sp} = ?$$

in analogy to the dissolution of other weakly soluble salts, this is not correct. The S^{2-} ion, like the O^{2-} ion, is a very strong base (stronger than OH^-) and reacts almost quantitatively with water:

$$S^{2-}(aq) + H_2O(\ell) \rightleftharpoons HS^-(aq) + OH^-(aq) \qquad \text{(reaction essentially complete)}$$

The K_b for this reaction is on the order of 10^5, so virtually no S^{2-} is present in aqueous solution. A more realistic equation for the dissolution of zinc sulfide is found by adding the two preceding equations:

$$ZnS(s) + H_2O(\ell) \rightleftharpoons Zn^{2+}(aq) + OH^-(aq) + HS^-(aq)$$

for which the equilibrium-constant equation is

$$[Zn^{2+}][OH^-][HS^-] = K_{ss}$$

The subscript "ss" stands for "sulfide solubility." Measurements establish that K_{ss} for $ZnS(s)$ equals 2×10^{-25} at 25°C. Table 9–2 gives K_{ss} for a few other metal sulfides. In all of these sulfide solubility reactions, as the pH decreases (as the solution becomes more acidic), the concentration of OH^- decreases. Simultaneously, the concentration of HS^- decreases as the equilibrium

$$HS^-(aq) + H_3O^+(aq) \rightleftharpoons H_2S(aq) + H_2O(\ell)$$

is shifted to the right by the additional H_3O^+. When both $[OH^-]$ and $[HS^-]$ decrease, the concentration of the metal ion must increase in order to maintain a constant value for the product $[M^{2+}][OH^-][HS^-]$. Adding acid to the water *increases* the solubility of metal sulfides. The quantitative calculation of this effect requires treating several simultaneous equilibria, as the following example illustrates.

Hydrogen sulfide (H_2S) is a poisonous, foul-smelling gas. It arises in nature in the gases vented from volcanoes and also from the action of bacteria; it is most familiar as the odor of rotten eggs. When dissolved in water, it gives a weak acid—hydrosulfuric acid.

EXAMPLE 9-8

In saturated solutions of H_2S at 25°C, the concentration of $H_2S(aq)$ equals 0.1 mol L^{-1}. Calculate the molar solubility of $FeS(s)$ in a solution that is kept saturated in H_2S and is buffered at pH 2.0 at 25°C.

Strategy

Assume that the equilibrium concentration of $Fe^{2+}(aq)$ equals the solubility of the $FeS(s)$. Recognize that more than one reaction is taking place at once, and write equations and equilibrium expressions for the different reactions. Look up numerical K's in Table 8–2, 9–2, or elsewhere. The statement of the problem fixes some concentrations. Insert these knowns into the equilibrium equations and obtain the $Fe^{2+}(aq)$ concentration by working through the algebra.

Solution

Dissolution occurs by the reaction

$$FeS(s) + H_2O(\ell) \rightleftharpoons Fe^{2+}(aq) + HS^-(aq) + OH^-(aq)$$
$$[Fe^{2+}][HS^-][OH^-] = K_{ss} = 5 \times 10^{-19}$$

Concurrently, H_2S donates H^+ ion to water

$$H_2S(aq) + H_2O(\ell) \rightleftharpoons H_3O^+(aq) + HS^-(aq)$$
$$\frac{[H_3O^+][HS^-]}{[H_2S]} = K_a = 9.1 \times 10^{-8}$$

and water undergoes autoionization:

$$2\,H_2O(\ell) \rightleftharpoons H_3O^+(aq) + OH^-(aq) \qquad [H_3O^+][OH^-] = K_w = 1.0 \times 10^{-14}$$

An unspecified acid–base reaction buffers the solution at pH 2.0, which means that $[H_3O^+]$ stays equal to 1×10^{-2} mol L^{-1}. The concentration of the hydroxide ion at pH 2.0 is

$$[OH^-] = \frac{K_w}{[H_3O^+]} = \frac{1.0 \times 10^{-14}}{1 \times 10^{-2}} = 1 \times 10^{-12} \text{ mol L}^{-1}$$

The concentration of H_2S stays at 0.1 mol L^{-1} because the solution is kept saturated in H_2S. Substituting $[H_2S] = 0.1$ mol L^{-1} and $[H_3O^+] = 1 \times 10^{-2}$ mol L^{-1} into the K_a equation for H_2S gives

$$\frac{[H_3O^+][HS^-]}{[H_2S]} = \frac{(1 \times 10^{-2})[HS^-]}{0.1} = 9.1 \times 10^{-8}$$

Solving for $[HS^-]$ gives

$$[HS^-] = 9.1 \times 10^{-7} \text{ mol L}^{-1}$$

Now, all but one of the concentrations are known in the K_{ss} equation

$$[Fe^{2+}][HS^-][OH^-] = 5 \times 10^{-19}$$

Insert the known values and solve for the unknown:

$$[Fe^{2+}](9.1 \times 10^{-7})(1 \times 10^{-12}) = 5 \times 10^{-19}$$
$$[Fe^{2+}] = 0.5 \text{ mol L}^{-1} = \text{molar solubility}$$

EXERCISE

Calculate the solubility of $CdS(s)$ at 25°C in a solution that is saturated with H_2S and buffered at a pH of 2.0.

Answer: 8×10^{-10} mol L^{-1}.

The preceding shows that adjusting the pH properly keeps the ions of one metal (Fe^{2+} in the example) entirely in solution and causes those of a second (Cd^{2+} in the exercise) to precipitate almost entirely as metal sulfide. Setting the right pH is crucial for separating metal ions in qualitative analysis (described in Section 9–6).

9-5   Complex Ions and Solubility

In a **complex ion,** a central metal ion is bound to one or more ligand molecules or ions that are usually capable of independent existence in solution. In $Ag(NH_3)_2^+$, for example, a central Ag^+ ion is **coordinated** to the lone-pair electrons of two different ammonia molecules. In $Cu(NH_3)_4^{2+}$, each Cu^{2+} ion is surrounded by *four* ammonia molecules (Fig. 9–7). We now examine the effects of the formation of complex ions on equilibria in aqueous solutions. The structure and bonding of complex ions are considered in Chapter 19.

• Coordination complexes are introduced in Section 3–8; coordination is a Lewis acid–base reaction.

Complex-Ion Equilibria

Complex ions form by the stepwise addition of ligands to metal ions. When a soluble silver salt is dissolved in an aqueous solution containing ammonia, the following equilibria are established:

$$Ag^+(aq) + NH_3(aq) \rightleftharpoons Ag(NH_3)^+(aq)$$
$$\frac{[Ag(NH_3)^+]}{[Ag^+][NH_3]} = K_1 = 2.1 \times 10^3$$

$$Ag(NH_3)^+(aq) + NH_3(aq) \rightleftharpoons Ag(NH_3)_2^+(aq)$$
$$\frac{[Ag(NH_3)_2^+]}{[Ag(NH_3)^+][NH_3]} = K_2 = 8.2 \times 10^3$$

where the K's were determined at 25°C. Because the two K's are both fairly large, we anticipate that if a silver salt is dissolved in water that contains an excess of ammonia, most of the silver ends up in the complex ion $[Ag(NH_3)_2]^+$ at equilibrium. Let us work out a quantitative example.

EXAMPLE 9-9

Suppose that 0.100 mol of $AgNO_3$ is mixed with 1.00 L of a 1.00 mol L^{-1} solution of NH_3. Calculate the concentrations of the Ag^+, $Ag(NH_3)^+$, and $Ag(NH_3)_2^+$ ions at equilibrium at 25°C.

Strategy

Assume that most of the Ag^+ is complexed as $Ag(NH_3)_2^+$ (but plan to check this). Think of the two-stage complexation reaction as going to completion (forming $Ag(NH_3)_2^+$ until all of the Ag^+ is used up), and then reversing to a small extent to form first $Ag(NH_3)^+$ and then Ag^+. Take the imagined state after complete reaction but before dissociation of the complex as the initial state.

Solution

The initial (after complexation, before back-reaction) concentrations of the three silver-containing species and the ligand are

$$[Ag(NH_3)_2^+]_{(init)} = 0.100 \text{ mol L}^{-1}$$
$$[Ag^+]_{(init)} = 0 \text{ mol L}^{-1}$$
$$[Ag(NH_3)^+]_{(init)} = 0 \text{ mol L}^{-1}$$
$$[NH_3]_{(init)} = 1.00 - (2 \times 0.100) = 0.80 \text{ mol L}^{-1}$$

The two stages of the dissociation of the $Ag(NH_3)_2^+$ ion are the reverses of the complexation reactions. Consequently, their equilibrium constants are the reciprocals of K_2 and K_1, respectively:

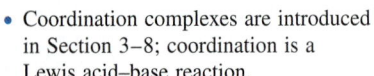

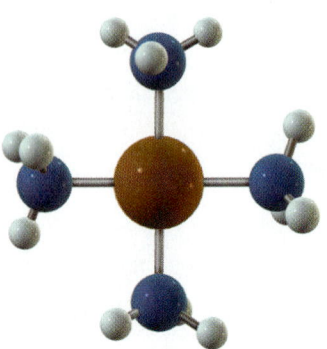

Figure 9–7 • The structures of the complex ions $Ag(NH_3)_2^+$ (*top*) and $Cu(NH_3)_4^{2+}$.

$$Ag(NH_3)_2^+(aq) \rightleftharpoons Ag(NH_3)^+(aq) + NH_3(aq)$$

$$\frac{[Ag(NH_3)^+][NH_3]}{[Ag(NH_3)_2^+]} = \frac{1}{K_2} = \frac{1}{8.2 \times 10^3}$$

$$Ag(NH_3)^+(aq) \rightleftharpoons Ag^+(aq) + NH_3(aq)$$

$$\frac{[Ag^+][NH_3]}{[Ag(NH_3)^+]} = \frac{1}{K_1} = \frac{1}{2.1 \times 10^3}$$

Suppose that y mol L^{-1} of the $Ag(NH_3)_2^+$ ion dissociates at equilibrium according to the first equation. The table of changes to reach equilibrium is then

$$Ag(NH_3)_2^+(aq) \rightleftharpoons Ag(NH_3)^+(aq) + NH_3(aq)$$

	$[Ag(NH_3)_2^+]$ (mol L^{-1})	$[Ag(NH_3)^+]$ (mol L^{-1})	$[NH_3]$ (mol L^{-1})
Initial	0.100	0	0.80
Change	$-y$	$+y$	$+y$
Equilibrium	$0.100 - y$	y	$0.80 + y$

$$\frac{1}{K_2} = \frac{[Ag(NH_3)^+][NH_3]}{[Ag(NH_3)_2^+]}$$

$$\frac{1}{8.2 \times 10^3} = \frac{y(0.80 + y)}{0.100 - y}$$

$$y = 1.5 \times 10^{-5} \text{ mol } L^{-1} = [Ag(NH_3)^+]$$

We next calculate the concentration of uncomplexed Ag^+ ion by substitution in the equilibrium-constant equation for the second step of the dissociation of the complex ion:

$$\frac{[Ag^+][NH_3]}{[Ag(NH_3)^+]} = \frac{1}{K_1} = \frac{1}{2.1 \times 10^3}$$

$$\frac{[Ag^+](0.80)}{1.5 \times 10^{-5}} = \frac{1}{2.1 \times 10^3}$$

$$[Ag^+] = 9 \times 10^{-9} \text{ mol } L^{-1}$$

The answer is small. Essentially all of the silver ion is tied up in the $Ag(NH_3)_2^+$ complex, the concentration of which remains 0.100 mol L^{-1}.

EXERCISE

Suppose that 0.25 mol of $CuSO_4$ is added to 1.0 L of 2.4 M NH_3 solution. At 25°C, the equilibrium constants for the addition of four successive NH_3 ligands to a Cu^{2+} ion are 1×10^4, 2×10^3, 5×10^2, and 9×10^1. Determine the concentrations of $Cu(NH_3)_4^{2+}$, $Cu(NH_3)_3^{2+}$, $Cu(NH_3)_2^{2+}$, $Cu(NH_3)^{2+}$, and Cu^{2+} at equilibrium at this temperature.

Answer: 0.25, 0.002, 3×10^{-6}, 1×10^{-9}, and 7×10^{-14} mol L^{-1}.

The calculation in Example 9–9 is simplified by two circumstances: An excess of ligand (ammonia) was present in solution, and the successive ligand association constants K_1 and K_2 are large compared with 1. The latter means that $1/K_2$ and $1/K_1$ are small compared with 1 and permits us to treat the equilibria one at a time, just

The thiosulfate ion ($S_2O_3^{2-}$) is related to the sulfate ion (SO_4^{2-}) by the replacement of one oxygen atom by one sulfur atom. It is prepared, however, by the reaction of elemental sulfur with sulfite ion (SO_3^{2-}).

as we treat the steps in the ionization of a polyprotic acid separately. When the association constants are not large (as often happens), a calculation that recognizes the concurrent action of the simultaneous equilibria is necessary. This calculation is beyond the scope of this text.

The formation of coordination complexes can have a large effect on the solubility of a compound in water. Silver bromide is very poorly soluble in water

$$AgBr(s) \rightleftharpoons Ag^+(aq) + Br^-(aq) \qquad K_{sp} = 7.7 \times 10^{-13}$$

but when a source of thiosulfate ion ($S_2O_3^{2-}$) is added, the complex ion $Ag(S_2O_3)_2^{3-}$ forms:

$$AgBr(s) + 2\,S_2O_3^{2-}(aq) \rightleftharpoons Ag(S_2O_3)_2^{3-}(aq) + Br^-(aq) \qquad K = 22$$

This greatly increases the solubility of the silver bromide, which goes into solution (Fig. 9–8). The formation of this complex ion is an important step in the development of photographic images; thiosulfate ion is a component of the "fixer" that brings silver bromide into solution from the unexposed portion of the film.

The same anion that precipitates a cation can re-dissolve the precipitate through complexation. The effect is illustrated by the addition of iodide ion to a solution containing mercury(II) ion. After a moderate amount of iodide ion has been added, an orange precipitate forms (Fig. 9–9), according to the equation

$$Hg^{2+}(aq) + 2\,I^-(aq) \rightleftharpoons HgI_2(s)$$

Further addition of iodide ion causes the orange solid to redissolve as complex ions form:

$$HgI_2(s) + I^-(aq) \rightleftharpoons HgI_3^-(aq)$$
$$HgI_3^-(aq) + I^-(aq) \rightleftharpoons HgI_4^{2-}(aq)$$

Silver chloride dissolves in a concentrated solution of sodium chloride in a similar way because of the formation of the $AgCl_2^-$ complex ion.

Hydrolysis and Amphoterism of Complex Ions

Many metal ions act as weak acids when dissolved in water, decreasing the pH of the solution (see Section 8–4). The Fe^{3+} ion is an example. A dissolved Fe^{3+} ion accepts electron-pairs from six water molecules, leading to the complex ion $Fe(H_2O)_6^{3+}$. This complex ion can act as a Brønsted-Lowry acid, donating hydrogen ions to the solvent

$$\underset{\text{acid}}{Fe(H_2O)_6^{3+}(aq)} + \underset{\text{base}}{H_2O(\ell)} \rightleftharpoons \underset{\text{acid}}{H_3O^+(aq)} + \underset{\text{base}}{Fe(H_2O)_5OH^{2+}(aq)}$$

$$\frac{[H_3O^+][Fe(H_2O)_5OH^{2+}]}{[Fe(H_2O)_6^{3+}]} = K_a = 7.7 \times 10^{-3} \qquad \text{(at 25°C)}$$

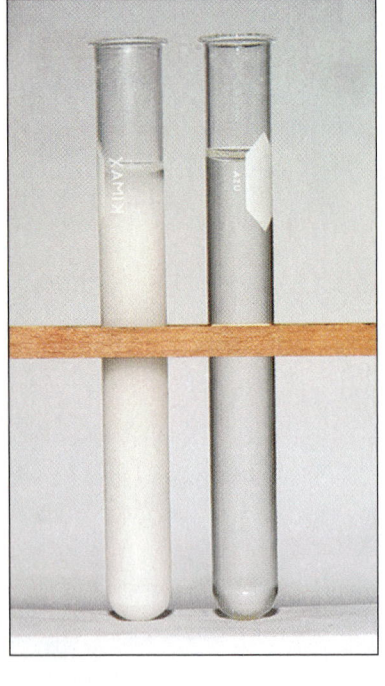

Figure 9–8 • An illustration of the effect of complex-ion formation on solubility. Both test tubes contain 2.0 g of AgBr, but the one on the right also contains dissolved thiosulfate ($S_2O_3^{2-}$) ion, which reacts with AgBr ions to give a soluble complex. Hardly any white solid AgBr dissolves in the pure water, but all of it dissolves in the solution containing thiosulfate. *(Leon Lewandowski)*

Figure 9-9 • The "orange tornado" is a striking demonstration of the effects of complexation on solubility. A solution is prepared with an excess of $I^-(aq)$ over $Hg^{2+}(aq)$, so that the Hg^{2+} is complexed as $HgI_3^-(aq)$ and $HgI_4^{2-}(aq)$. A magnetic stirrer is used to create a vortex in the solution. Addition of a solution containing Hg^{2+} down the center of the vortex then causes the orange solid HgI_2 to form (by the reaction $HgI_4^{2-}(aq) + Hg^{2+}(aq) \longrightarrow 2\ HgI_2(s)$, for example) in the layer at the edges of the vortex, giving the appearance of a tornado. *(Leon Lewandowski)*

Interestingly, the Fe^{3+} ion is an acid–base indicator. The $Fe^{3+}(aq)$ ion is almost colorless, but the $Fe(H_2O)_5OH^{2+}(aq)$ complex ion is brown. Hydrolysis colors aqueous $Fe(NO_3)_3$ pale brown. When strong acid is added, the equilibrium is driven back to the left, and the color fades.

Table 9–3 gives values of the pH of $0.1\ \text{mol L}^{-1}$ solutions of several metal ions. Those that form strong complexes with hydroxide ion have a low pH, whereas those that do not form such complexes give neutral solutions (pH 7).

EXAMPLE 9-10

Calculate the pH of a solution that is 0.100 M in $Fe(NO_3)_3$ at 25°C.

Strategy

Recognize that the Fe^{3+} ion in aqueous solution is complexed by water molecules and that the complex can donate hydrogen ions. Use the K_a (given in the preceding) of this reaction to compute the hydrogen ion concentration by the approach used in Example 8–6.

Solution

The complexed Fe^{3+} ion reacts as a weak acid:

$$Fe(H_2O)_6^{3+}(aq) + H_2O(\ell) \rightleftharpoons H_3O^+(aq) + Fe(H_2O)_5OH^{2+}(aq)$$

The table of changes for this reaction is

	$Fe(H_2O)_6^{3+}(aq)$ $(mol\ L^{-1})$	$H_3O^+(aq)$ $(mol\ L^{-1})$	$Fe(H_2O)_5OH^{2+}(aq)$ $(mol\ L^{-1})$
Initial	0.100	≈ 0	0
Change	$-y$	$+y$	$+y$
Equilibrium	$0.100 - y$	y	y

If y moles per liter of $Fe(H_2O)_6^{3+}$ reacts, then (neglecting the autoionization of water)

$$\frac{[H_3O^+][Fe(H_2O)_5(OH)^{2+}]}{[Fe(H_2O)_6^{3+}]} = K_a$$

$$\frac{y^2}{0.100 - y} = 7.7 \times 10^{-3}$$

$$y^2 + (7.7 \times 10^{-3})y - (7.7 \times 10^{-4}) = 0$$

$$y = 0.024\ \text{mol L}^{-1} = [H_3O^+]$$

The pH is $-\log_{10}[H_3O^+]$, which equals 1.62.

TABLE 9-3

pH of 0.1 Molar Aqueous Metal Nitrate Solutions at 25°C

Metal Nitrate	pH
$Fe(NO_3)_3$	1.6
$Pb(NO_3)_2$	3.6
$Cu(NO_3)_2$	4.0
$Zn(NO_3)$	5.3
$Ca(NO_3)_2$	6.7
$NaNO_3$	7.0

Check

Putting $y = 0.0242$ (the unrounded value) into the equilibrium-constant equation gives the correct K_a, and the pH equals the value tabulated in Table 9–3. Note that y cannot be neglected in comparison to 0.100. Doing so leads to a pH of 1.56, a substantial error. Finally, the concentration of H_3O^+ from the acid dissociation is much larger than from the autoionization of water, which was therefore properly neglected.

EXERCISE

The acidity constant at 25°C for the $Cu(H_2O)_4^{2+}$ ion is $K_a = 1 \times 10^{-7}$. Calculate the pH of a 0.100 mol L^{-1} solution of $Cu(NO_3)_2$ at 25°C.

Answer: 4.0.

Equations representing hydrolysis reactions of aquated metal ions are often simplified by omitting the specific number of water molecules surrounding the metal ion. The equation for the reaction of $Fe(H_2O)_6^{3+}(aq)$ ion as a weak acid then becomes

$$Fe^{3+}(aq) + 2 H_2O(\ell) \rightleftharpoons H_3O^+(aq) + FeOH^{2+}(aq)$$

This version of the equation is noncommittal about how many water molecules are coordinated, a number that is in fact sometimes uncertain.

Different cations behave differently as water ligands are replaced by hydroxide ions in an increasingly basic solution. The behavior of Zn^{2+} ion is particularly interesting. It forms a series of four hydroxo complexes:

$$Zn^{2+}(aq) + 2 H_2O(\ell) \rightleftharpoons H_3O^+(aq) + ZnOH^+(aq)$$
$$ZnOH^+(aq) + 2 H_2O(\ell) \rightleftharpoons H_3O^+(aq) + Zn(OH)_2(s)$$
$$Zn(OH)_2(s) + 2 H_2O(\ell) \rightleftharpoons H_3O^+(aq) + Zn(OH)_3^-(aq)$$
$$Zn(OH)_3^-(aq) + 2 H_2O(\ell) \rightleftharpoons H_3O^+(aq) + Zn(OH)_4^{2-}(aq)$$

Raising the pH shifts all of these equilibria to the right. In Brønsted-Lowry language, the polyprotic Brønsted-Lowry acid $Zn^{2+}(aq)$ donates four hydrogen ions. All of the products except the last are amphoteric (capable of reacting as both an acid and a base). The second, $Zn(OH)_2$, is also only slightly soluble in pure water (its K_{sp} is 1.9×10^{-17}). If enough acid is added to a quantity of undissolved $Zn(OH)_2$, OH^- ions are removed, forming the $Zn^{2+}(aq)$ ion in solution. If enough base is added to the same quantity of undissolved $Zn(OH)_2$, OH^- ions are added to form the $Zn(OH)_4^{2-}(aq)$ (zincate) ion in solution. Thus, $Zn(OH)_2$ dissolves at low pH and at high pH but remains almost completely undissolved at intermediate pHs (Fig. 9–10). This amphoterism can be used to separate Zn^{2+} from other cations that do not share the property. For example, Mg^{2+} ion adds a maximum of only two OH^- ions to form $Mg(OH)_2$, a sparingly soluble hydroxide. If a mixture of Mg^{2+} and Zn^{2+} ions is made sufficiently basic, the Mg^{2+} precipitates as $Mg(OH)_2$, but the zinc remains in solution as $Zn(OH)_4^{2-}$, allowing the two to be separated.

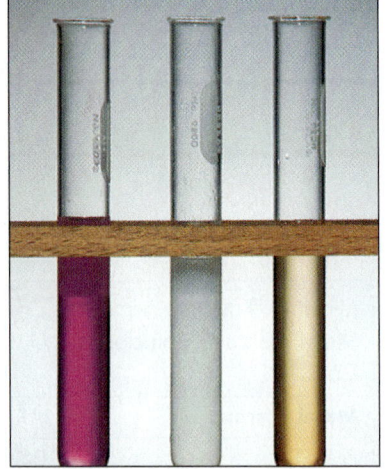

Figure 9–10 • Zinc hydroxide is insoluble in water as is evident from its milky appearance (*center*), but it dissolves readily in acid (*left*) and base (*right*). The indicator used is bromocresol red, which is red in acid and yellow in base. (*Leon Lewandowski*)

9-6   Controlling Solubility in Qualitative Analysis

The goal of the **qualitative analysis** of inorganic substances in aqueous solution is to determine the identities of ions that are present, without concern for the exact amounts of each. In this section, we show how the principles of solubility control underlie a scheme for the qualitative analysis of the common metal cations in aqueous solution. The same principles can be applied to identify anions. Qualitative analysis impresses some as a musty 19th-century technique that has no place in modern chemistry. No one today (or even 100 years ago) outside of introductory chemistry laboratories carries out a full qualitative analysis of the type we present. So why do it? The answer is simple: Learning qualitative analysis is the best way to study the chemistry of ions in solution, which is an immensely practical subject in modern research and technology. In this section, we can only begin to relate the basic ideas behind qualitative analysis to the principles of solubility presented earlier in the chapter. An entire book would be required to do justice to the subject. It would be even more instructive to spend a few weeks in a laboratory exploring the qualitative behavior of inorganic compounds.

We begin by listing the three basic tools that are used to separate cations from one another.

1. **Different solubilities of cations with the same anion.** For example, if HCl is added to a solution containing both $Ag^+(aq)$ and $Ni^{2+}(aq)$ ions, AgCl precipitates but $NiCl_2$ does not.

2. **pH control of relative solubilities.** Strong acids can bring insoluble hydroxides, carbonates, and many sulfides back into solution; pH control by means of buffers at intermediate pH can separate metal ions from one another.

3. **Complex-ion formation that brings a metal ion back into solution.** The dissolution of AgCl in an aqueous solution of ammonia, for example, takes place through the formation of the $Ag(NH_3)_2^+$ complex ion.

Once a cation or a group of cations has precipitated from a solution, it must be physically separated from the residual solution and the cations that remain in that solution. This is accomplished either by **filtration,** in which a filter catches the solid but lets the solution pass through, or by **centrifugation,** in which rapid spinning of the solution causes the solid to collect in a compact form at the bottom of a tube, allowing the **supernatant solution** above it simply to be poured off or withdrawn with a pipet.

In any qualitative-analysis scheme, the first step is to separate the ions into groups that contain a smaller number of ions (typically from three to seven). A different attack can then be applied to each group to separate and identify its ions, after which a confirmatory test, specific to the ion in question, can be carried out to verify the tentative conclusion from the separation process. One important point to remember is that group separations work only when done in the proper sequence. If a group of ions is only partially removed by precipitation in an early step, it can wreak havoc in subsequent separations. Reactions must therefore be brought as close to completion as possible. A second point is that reactants added to a solution remain there unless they are chemically removed. The procedure must be viewed as a whole, therefore, and if sulfide, for example, will interfere with a subsequent chloride reaction, the sulfide must be quantitatively removed before proceeding. Despite its name, qualitative analysis relies strongly on quantitative separations and on the precise control of reaction conditions.

An Overview of Group Separations

The first step in the scheme of qualitative analysis that we present here is to separate the cations that may be present in solution into five groups, each of which contains a smaller number of ions (Fig. 9–11). Each group can then be subjected to further analysis.

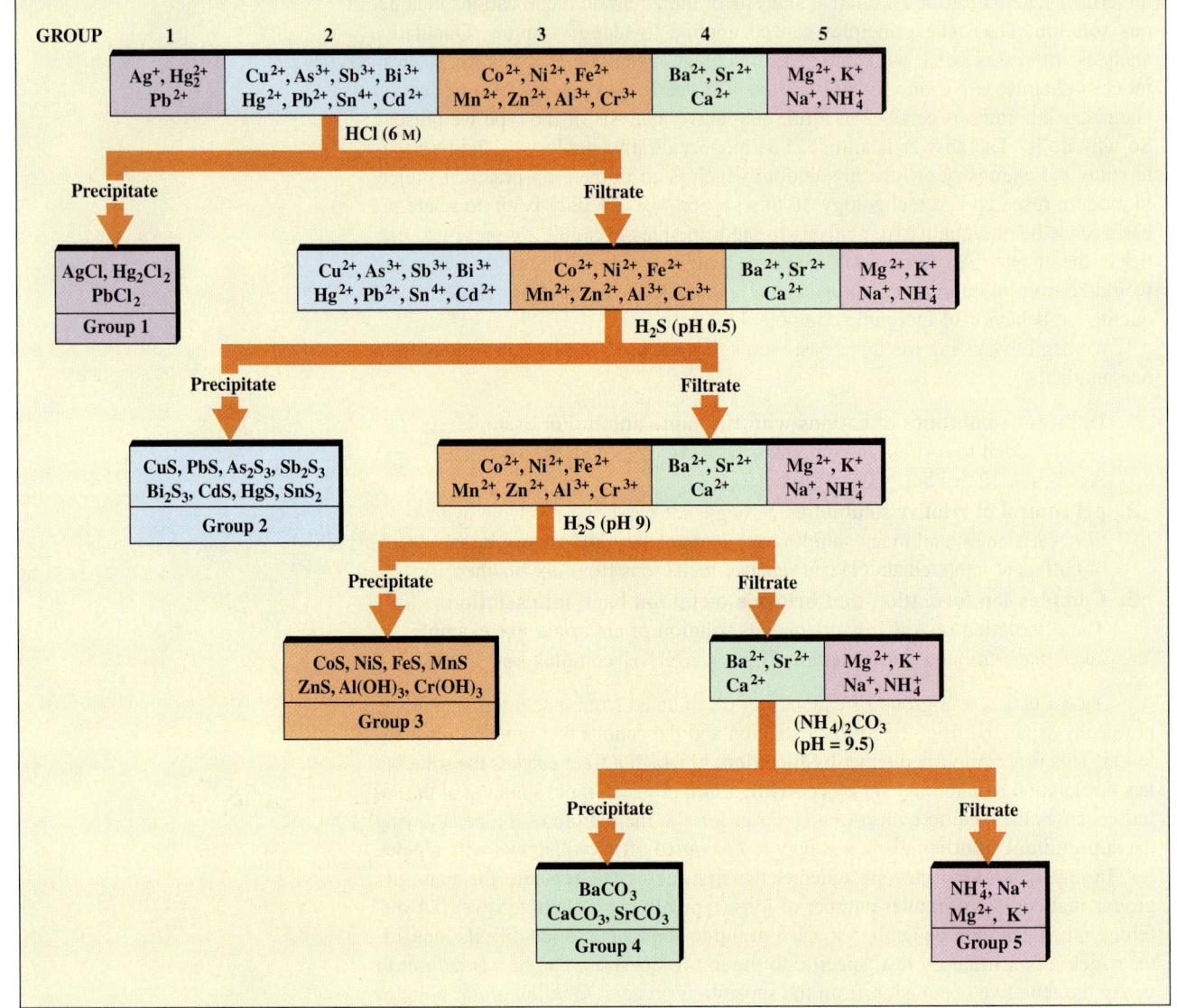

Figure 9–11 • A general outline of a qualitative analysis scheme. It is important to avoid confusing the groups in the periodic table with the five groups in this scheme of qualitative analysis. The two are unrelated.

Group 1: Ag^+, Hg_2^{2+}, Pb^{2+}

Addition of 6 M HCl to the mixture of cations listed in Figure 9–11 causes the precipitation of three insoluble white chlorides: $AgCl$, $PbCl_2$, and Hg_2Cl_2. All other cations remain in solution. A slight excess of chloride ion is used to force out as many of these three ions from solution as possible, but a large excess is avoided because it would cause the ions to redissolve as complex ions such as $AgCl_2^-$ and $PbCl_3^-$. The mercury(I) ion, Hg_2^{2+}, is unusual in that it contains two mercury atoms with a covalent bond between them. It differs from the mercury(II) ion, Hg^{2+}, which has a soluble chloride and remains in solution at this stage. Lead(II) chloride ($PbCl_2$) is sufficiently soluble in water that detectable concentrations of Pb^{2+} stay in solution after the addition of the HCl in this step. The Pb^{2+} that remains precipitates in the next step with Group 2, so it must also be considered in the scheme developed for the analysis of that group.

Group 2: As(III), Bi³⁺, Cd²⁺, Cu²⁺, Pb²⁺, Hg²⁺, Sb³⁺, Sn⁴⁺

The solution that remains after filtration of the insoluble Group 1 chlorides is neutralized with base and carefully reacidified to a pH of 0.5 ($[H_3O^+] = 0.3 \text{ mol L}^{-1}$). It is then saturated with H_2S. Under these highly acidic conditions, only the least soluble of the sulfides precipitate; these comprise the Group 2 cations. Most Group 2 ions do not exist in the solution from step 1 as simple aquated ions. Arsenic(III) is present in the covalent compound arsenious acid (H_3AsO_3), and several of the other ions form partially covalent complex ions with the Cl^- ion added in the first step; $BiCl_4^-$, $SbCl_4^-$, and $SnCl_6^{2-}$ are examples. Their sulfides are sufficiently insoluble, however, that these complex ions cannot prevent the sulfides from precipitating in this step. The colors of the sulfides (Fig. 9–12) can be a clue to the composition of an unknown sample.

Group 3: Al³⁺, Co²⁺, Cr³⁺, Fe³⁺ or Fe²⁺, Mn²⁺, Ni²⁺, Zn²⁺

After the Group 2 sulfides have precipitated, the pH of the solution is adjusted to 9 with an NH_4^+/NH_3 buffer, but the solution remains saturated in H_2S. Under these conditions, the acid-soluble Group 3 sulfides precipitate. These are the black CoS, NiS, and FeS, the salmon-colored MnS, and the white ZnS (Fig. 9–13). Aluminum and chromium precipitate as gelatinous hydroxides: $Al(OH)_3$ is white and $Cr(OH)_3$ is green. The two oxidation states of iron are not differentiated in this procedure. Solutions of Fe^{2+} oxidize to Fe^{3+} upon standing in air, but the Fe^{3+} is then reduced to Fe^{2+} by H_2S, and all of the iron precipitates in the +2 oxidation state as FeS.

Group 4: Ba²⁺, Ca²⁺, Sr²⁺

The alkali- and alkaline-earth sulfides are quite soluble and remain in solution after the precipitation of the Group 3 sulfides. The H_2S and NH_4^+ with which the solution has been treated must now be removed to prevent them from interfering with subsequent steps. The H_2S is removed by adding a few drops of 12 M HCl and evaporating almost to dryness. Then some 16 M HNO_3 is added, and strong heating removes the ammonium ion by decomposing the NH_4Cl and NH_4NO_3 to gaseous products. The resulting solid is then redissolved; it contains Group 4 and Group 5 cations. When the pH is adjusted to 9.5 and $(NH_4)_2CO_3$ is added, the three Group 4 cations precipitate as carbonates. Because $MgCO_3$ is more soluble than the heavier alkaline-earth carbonates, it remains in solution under the usual conditions of qualitative analysis.

Figure 9–12 • Some of the Group 2 sulfides. From left: CuS, Bi_2S_3, HgS, CdS, SnS_2, and Sb_2S_3. As_2S_3 (*not shown*) is yellow, orange, or red; PbS is black. (*Marna G. Clarke*)

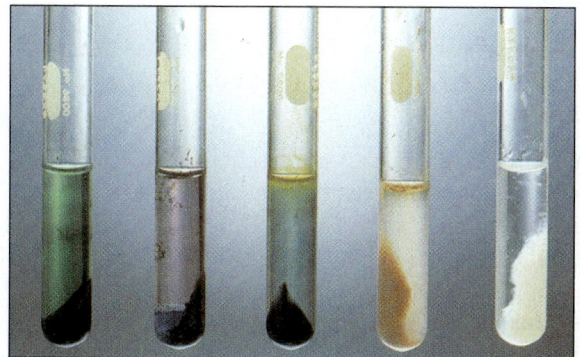

Figure 9–13 • Five Group 3 cations precipitate as sulfides when a solution saturated in H_2S is made basic. From left: NiS, CoS, FeS, MnS, and ZnS. Two precipitate as hydroxides: $Al(OH)_3$ and $Cr(OH)_3$. (*Marna G. Clarke*)

Group 5: K⁺, Mg²⁺, Na⁺, NH₄⁺

The alkali-metal cations remain in solution after this last step, together with Mg^{2+} and some ammonium ion, which often is also included in a qualitative-analysis unknown. This solution is not used, however, because it is almost impossible to avoid contamination with sodium ions in the reagents added in the earlier separation steps; of course, ammonium ion has been deliberately added as well. To identify Group 5 cations, several portions of the original unknown are subjected to tests that are specific to the particular ion in question, even in the presence of all the other cations in the qualitative-analysis procedure. For example, a positive identification of sodium can be made by adding a solution of magnesium uranyl acetate and observing the precipitation of $NaMg(UO_2)_3(CH_3COO) \cdot 9H_2O(s)$, one of the few salts of sodium that is only slightly soluble.

EXAMPLE 9–11

An unknown solution may contain ions from Groups 1 through 5. Addition of 6 M HCl gives a white precipitate. Saturation of the solution separated from this precipitate with H_2S at pH 0.5 gives no further precipitation, but when the pH is increased to 9, a precipitate forms. After this precipitate is filtered off, the H_2S and NH_4^+ ion are removed from the solution and the pH is adjusted to 9.5. No precipitate then forms upon addition of $(NH_4)_2CO_3$. What ions may be present in the original solution from each qualitative-analysis group?

Solution

Group 1: Either Ag^+ or Hg_2^{2+} (or both) could be present, giving a white chloride precipitate, but Pb^{2+} cannot be present because it would show up in Group 2 as well.
Group 2: None present (would have precipitated as sulfides at pH 0.5).
Group 3: Co^{2+}, Ni^{2+}, Fe^{2+}, Fe^{3+}, Mn^{2+}, Zn^{2+}, Al^{3+}, and Cr^{3+} all may be present because a precipitate forms at pH 9.
Group 4: None present (would have a carbonate precipitate with $(NH_4)_2CO_3$).
Group 5: Any Group 5 cation could be present.

EXERCISE

A solution may contain ions from Groups 1 through 5. Neither the addition of 6 M HCl nor the addition of H_2S at a pH of 0.5 causes a precipitate. What ions may be present from each group?

Answer: No Group 1 or Group 2 ions are present. Any ions from Groups 3 through 5 may be present.

Analysis for Group 1

The systematic analysis of each of the five groups is not described here. Instead, we discuss only Group 1, the members of which precipitate as chloride salts in acidic solution (Fig. 9–14). The members of this group, Ag^+, Hg_2^{2+}, and Pb^{2+}, differ sufficiently in their chemistry that their separation is straightforward.

Lead(II) chloride ($PbCl_2$) is the most soluble of the three chlorides, and its solubility increases strongly with temperature. When hot water (near 100°C) is added to the mixed chloride precipitate, the lead(II) chloride dissolves, accompanied by

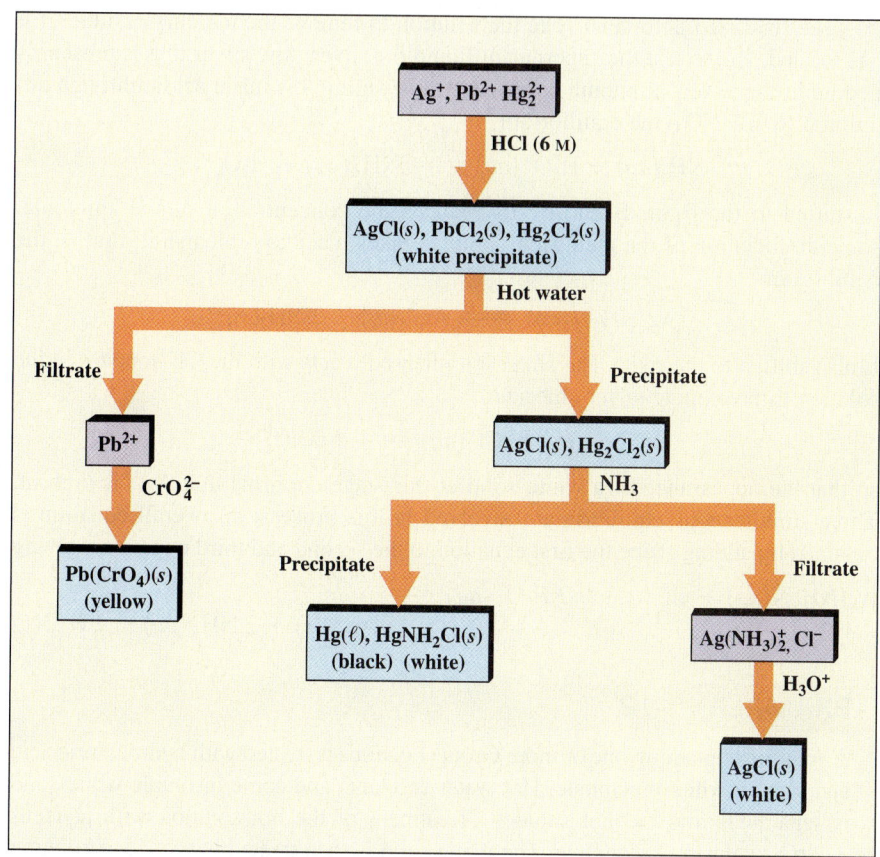

Figure 9-14 • A scheme for the qualitative analysis of the Group 1 cations. Precipitates at each stage are shown in blue boxes.

only small amounts of the other two chlorides. If the solution is centrifuged while still hot, the solid contains the silver chloride and mercury(I) chloride, and the lead(II) chloride is dissolved in the supernatant liquid. The presence of lead(II) ion can be confirmed by adding a soluble chromate salt to the supernatant liquid after pouring it off from the solids. If lead(II) ion is present, it precipitates as bright yellow lead(II) chromate (Fig. 9–15):

$$Pb^{2+}(aq) + CrO_4^{2-}(aq) \rightleftharpoons PbCrO_4(s)$$

To separate and identify the silver and mercury(I) that may be present, one treats the precipitate that remains after extraction with hot water with a concentrated solution of ammonia. Silver chloride dissolves in aqueous ammonia by complex ion formation:

$$AgCl(s) + 2\,NH_3(aq) \rightleftharpoons Ag(NH_3)_2^+(aq) + Cl^-(aq)$$

The equilibrium constant for this reaction is not very large ($K = 3 \times 10^{-3}$), so the dissolution is helped by using the highest possible concentration of ammonia (15 mol L^{-1}) to drive the equilibrium to the right.

By contrast, any mercury(I) chloride that may be present in the chloride precipitate does not dissolve when treated with a solution of ammonia, but reacts according to the equation

$$Hg_2Cl_2(s) + 2\,NH_3(aq) \longrightarrow HgNH_2Cl(s) + Hg(\ell) + NH_4^+(aq) + Cl^-(aq)$$

to leave a dark, solid residue. A gray to black residue after the addition of the ammonia is a positive indication for Hg_2^{2+}.

Figure 9-15 • Brilliant yellow lead(II) chromate PbCrO$_4$ (the artist's pigment chrome yellow) precipitates when a solution of a soluble chromate salt is added to one of a soluble lead(II) salt. *(Charles Steele)*

The final step is to centrifuge the solution to remove the mercury residue, if it has formed, and to test the supernatant liquid for silver. Any silver that is present is tied up in the silver–ammonia complex. As the solution is made acidic through addition of 6 M HNO_3, the equilibrium

$$NH_3(aq) + H_3O^+(aq) \rightleftharpoons NH_4^+(aq) + H_2O(\ell)$$

is shifted to the right. Because this reduces the concentration of the ammonia, some dissociation of the metal-ion complex takes place to replenish it; that is, the equilibrium

$$Ag(NH_3)_2^+(aq) \rightleftharpoons Ag^+(aq) + 2\,NH_3(aq)$$

is also shifted to the right. The silver ion released reacts with the Cl^- ion still in the solution to precipitate silver chloride:

$$Ag^+(aq) + Cl^-(aq) \rightleftharpoons AgCl(s)$$

so that the appearance of a white solid at this stage confirms that Ag^+ is present. Three simultaneous equilibria are involved in this process; an overall equation is obtained by adding twice the first equation to the second and third equations, giving

$$Ag(NH_3)_2^+(aq) + Cl^-(aq) + 2\,H_3O^+(aq) \rightleftharpoons$$
$$AgCl(s) + 2\,NH_4^+(aq) + 2\,H_2O(\ell)$$

EXAMPLE 9-12

A solution containing one or more Group 1 cations is treated with hydrochloric acid and gives a white precipitate. Hot water is added, and some insoluble white solid is separated from the hot solution. Treatment of the hot solution with aqueous K_2CrO_4 yields no precipitate. The white solid is then treated with a 15 M solution of $NH_3(aq)$ and dissolves completely. What metal ions are present in the unknown?

Solution

No Pb^{2+} is present because no yellow lead(II) chromate precipitates. The white solid contains no Hg_2Cl_2 because it dissolves completely when the $NH_3(aq)$ is added (any Hg_2Cl_2 would give an insoluble gray to black residue). Because some Group I ion must be present in the white solid in the first place, Ag^+ is present. Treating the white solid with concentrated aqueous ammonia would confirm this conclusion by dissolving it entirely.

EXERCISE

A solution containing one or more Group 1 cations is treated with 6 M HCl that is heated to 95°C. It gives a white precipitate that, after cooling to room temperature, dissolves completely in 15 M NH_3. What ions are definitely present and what ions are definitely absent from the solution?

Answer: Ag^+ is definitely present; Hg_2^{2+} and Pb^{2+} are definitely absent.

Other Methods of Analysis

The procedure we have outlined for qualitative analysis is a "classic" one, based on solubility differences under particular conditions of pH with particular reactants present. Although portions of this procedure are still in use in modern qualitative analysis, other methods are more widely used today.

CHEMISTRY IN YOUR LIFE

Lead All Around You

Like several other heavy metals, lead is quite toxic. Part of the reason appears in Table 9–2. The small K_{ss} of $PbS(s)$ means that lead(II) has a strong affinity for sulfide ions. This affinity extends to sulfur in the -2 oxidation state in other compounds as well, including many important proteins in the body that contain —SH groups. Once mobilized into a living system, lead bonds at —SH groups in proteins, disrupting their normal functions. Sufferers of lead poisoning experience neurological symptoms, anemia, liver and kidney damage, and hearing loss. Death can ensue. Lead poisoning in children slows mental and physical development and can cause learning and behavior problems.

Lead contamination is encountered throughout the home—in paint, house dust, and drinking water. Paints manufactured before 1960 often used white lead $(Pb_3(OH)_2(CO_3)_2)$ as a pigment. Remnants of such paint, which contained as much as 50% lead by mass, are still flaking from the walls of older buildings. Lead continues to be used in paint, but in much reduced amounts, because it makes the paint last longer. The U.S. Environmental Protection Agency allows up to 600 parts per million (ppm) by mass of lead in household paint. The compound tetraethyllead was used for decades in the United States as an anti-knock gasoline additive (see Section 24–1) before being phased out in the mid-1980s. This use released many millions of kilograms of lead and lead compounds into the environment through the exhaust pipes of cars.

Window sills often gather dust that contains lead from crumbling paint and from the emissions of smelters, incinerators, and foundries. Such dust can get on children's hands and toys and then into their mouths through normal behavior, such as thumb-sucking. This makes children particularly susceptible to lead poisoning.

Lead leaches into the water supply from old lead pipes in city systems and home plumbing and from the lead-containing solder used to connect pipes. Because the solubility of many lead compounds increases with temperature (see Fig. 9–2), health agencies urge against drinking hot water straight from the tap. They advise that

The claim on this bean can lid appears because joints in cans are sometimes closed with lead-containing solder. *(George Semple)*

hot water for cooking, baking, or brewing be drawn cold from the tap and heated on a stove.

Metallic lead (sometimes called "black lead") is a malleable gray-black substance that is coated with an oxide that rubs off easily. People who work with metallic lead dirty their hands with the oxide and may absorb lead through their skin. When lead-containing objects (for example, fishing sinkers, sash weights, and certain plumbing supplies) are handled, skin contact should be minimized and the hands washed thoroughly before food is prepared or eaten.

Recent findings suggest that lead has ill effects at blood levels once thought safe. Lower IQ scores, slower development, and more attention deficits than normal have been observed in children with concentrations in their blood as low as 10^{-4} g L^{-1}. Lead concentrations in the blood exceeding about 7×10^{-4} g L^{-1} constitute a medical emergency. The treatment is "chelation therapy." A solution of a ligand having a very strong affinity for Pb^{2+} ion is administered. The lead is tied up in the form of a complex ion, and, by Le Châtelier's principle, the concentration of lead in the blood and tissues is reduced. Successful therapy naturally requires dealing with the adverse effects of the complexing agent as well.

One of the most powerful methods of qualitative analysis is **emission spectroscopy,** in which a sample is heated to a temperature high enough to vaporize it, and the wavelengths of the light emitted are recorded. As we shall see in Chapter 16, each element emits light of a characteristic set of wavelengths. The light emitted

Figure 9-16 • Flame tests. *(Larry Cameron)*

TABLE 9-4	
Colors in the Flame Test	
Element	**Color**
Li	Red
Na	Yellow
K	Violet
Rb	Red
Cs	Blue
Ca	Brick red
Cu	Green
Sr	Crimson
Ba	Green

can be analyzed (automatically, using modern spectrometers) to find which elements are present. The method is useful for all of the elements in the periodic table. Emission spectroscopy is a 20th-century elaboration of a much earlier technique for qualitative analysis—the flame test. When certain elements are placed in a flame, they give off characteristic colors of visible light (Table 9–4 and Fig. 9–16). Visual observation then provides a confirmatory test for certain elements; crimson for strontium, green for barium, and pale violet for potassium are examples. Sodium gives a very intense yellow flame in this test.

A second method for qualitative analysis uses **ion-exchange chromatography.** In this type of column chromatography (see Section 7–7), a column is packed with an ion-exchange resin, an insoluble substance of high molar mass that possesses negatively or positively charged sites that can interact with ions in solution. **Cation-exchange resins** have negatively charged sites that can attract cations. **Anion-exchange resins** have positively charged sites that attract negative ions such as Cl^- or negatively charged complex ions such as $AsCl_4^-$. As a solution under analysis percolates down such a column, the ions in it displace whatever species occupy the sites on the resin, with varying degrees of effectiveness. They spend different amounts of time on the column and can be separated and identified by confirmatory tests.

SUMMARY

9–1   When a solution is in equilibrium with undissolved solute, then it is **saturated** with respect to that solute. In an **unsaturated** solution, the concentration of the solute is less than its equilibrium value; in a **supersaturated** solution, the concentration of the solute exceeds its equilibrium value.

9–2   The **solubility-product** expression is used for equilibrium calculations on the saturated aqueous solutions of slightly soluble ionic solids. It relates the concentrations of the cations and anions that make up a dissolved salt to an equilibrium constant, K_{sp}, and can be used to calculate the **molar solubility** or gram solubility of the salt, provided that no side-reactions occur involving the dissolved ions.

9–3   A solubility-product expression can be used to predict whether precipitation occurs when two solutions are mixed: If the initial reaction quotient $Q_{(init)}$ exceeds K_{sp}, precipitation is favored, but if $Q_{(init)}$ is less than K_{sp}, none can occur. The K_{sp}

equation can be used in quantitative calculations of the **common-ion effect:** the reduction in solubility of a salt in a solution in which one of its component ions is already present.

9–4 The pH of the solution often affects the solubility of salts. Salts in which the anion is a base, such as metal hydroxides and sulfides, increase in solubility as acid is added. The anion is removed by reaction with hydronium ion, and more of the salt dissolves to maintain the solubility-product equilibrium. The pH dependence of the solubility of metal sulfides is particularly useful in permitting the quantitative separation of different metal ions in aqueous solution through pH control.

9–5 The formation of soluble complex ions increases the solubility of salts. Solubility equilibria are shifted to favor dissolution as products are tied up (removed) in interactions with ligands. Certain metal ions that are strongly complexed with water act as Brønsted-Lowry acids by donating hydrogen ions to surrounding water molecules and thereby reducing the pH of the solution.

9–6 The guiding principle behind the **qualitative analysis** for ions in aqueous solution is the separation of different ionic species from one another on the basis of their different solubilities, followed by specific tests to determine which are present in a sample. The separation relies on the different solubilities of different metal ions with the same anion, on pH control of solubility, and on the use of ligands to bring metal ions selectively into solution in the form of complex ions. Many different schemes are possible, depending on the ions that may be present in a sample. Modern qualitative analysis emphasizes two techniques in addition to the classic methods of solution chemistry: **emission spectroscopy** and **ion-exchange chromatography.**

PROBLEMS

Note: Answers to blue-numbered problems are given in Appendix F. Problems that are more challenging are indicated with asterisks.

The Nature of Solubility Equilibria

1. Copper selenate precipitates from solution as $CuSeO_4 \cdot 5H_2O$. When heated above 150°C, all of the water of crystallization is lost. What mass of solid results from heating 64.8 g of copper selenate pentahydrate above 150°C?

2. Gypsum has the formula $CaSO_4 \cdot 2H_2O$. Plaster of Paris has the chemical formula $CaSO_4 \cdot \frac{1}{2}H_2O$. In making wall plaster, water is added to plaster of Paris, and the mixture then hardens into solid gypsum. How much water (in liters, at a density of 1.00 kg L^{-1}) should be added to a 25.0 kg sack of plaster of Paris to turn it into gypsum, assuming no loss due to evaporation?

3. When heated above 420°C, the mineral bieberite loses its water of crystallization to become cobalt(II) sulfate. When 38.4 g of bieberite is heated in this way, 21.2 g of $CoSO_4$ results. Give the chemical formula of bieberite.

4. A 1.00 g sample of magnesium sulfate is dissolved in water, and the water is then evaporated away until the residue is bone dry. If the temperature of the water is kept between 48°C and 69°C, the solid that remains weighs 1.898 g. If the experiment is repeated with the temperature held between 69°C and 100°C, however, the solid has a mass of 1.150 g. Determine how many waters of crystallization are present per $MgSO_4$ in each of these two solids.

5. The following graph shows the aqueous solubility of KBr as a function of temperature. If 80 g of KBr is added to 100 g of water at 10°C and the mixture is heated slowly, at what temperature does the last KBr dissolve?

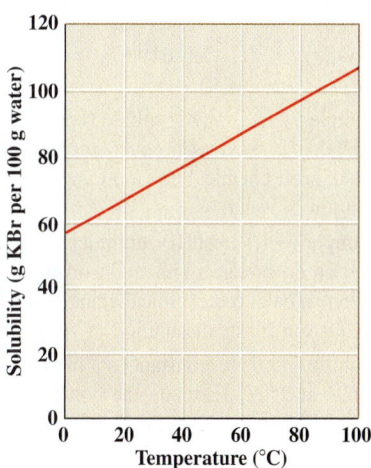

6. Figure 9–2 shows the aqueous solubility of $AgNO_3$ in mol kg^{-1}. If 255 g of $AgNO_3$ is mixed with 100 g of water at 95°C and cooled slowly, at what temperature does the solution become saturated?

7. The dissolution of ammonium chloride in water is quite endothermic. Predict the effect of increased temperature on the solubility of ammonium chloride.

8. Codeine has the molecular formula $C_{18}H_{21}NO_3$. It is used in cough medicines but has sometimes been abused, leading to addiction. It is soluble in water to the following extents: 1.00 g per 120 mL of water at room temperature, 1.00 g per 60 mL of water at 80°C. Compute the molal solubility (the solubility in moles of codeine per kilogram of water) at both temperatures, taking the density of water to be fixed at 1.00 g cm^{-3}. Is the dissolution of codeine endothermic or exothermic?

The Solubility of Ionic Solids

9. A solution consists of 0.156 mol of $SrBr_2$ dissolved in 4.15 L of solution. Assuming that dissociation is complete:
   (a) Determine the concentration of $SrBr_2$ in this solution in mol L^{-1}.
   (b) Determine the concentration of Br^- in this solution in mol L^{-1}.

10. A total of 5.0 g of $La_2(SO_4)_3$ is dissolved in 40.0 L of water. Determine the total concentration (in mol L^{-1}) of charged particles (negative ions and positive ions) in the solution, assuming complete dissociation.

11. (See Example 9–1.) Write balanced chemical equations to describe the dissolution of the following substances in water. Also, write an expression describing the equilibrium established between undissolved solid and the resulting solution.
   (a) $SrI_2(s)$ (c) $Ca_3(PO_4)_2(s)$
   (b) $I_2(s)$ (d) $BaCrO_4(s)$

12. (See Example 9–1.) Write balanced chemical equations to describe the dissolution of the following substances in water, and write the equilibrium-constant equation corresponding to each.
   (a) $Ag_2SO_4(s)$ (c) $Cu(OH)_2(s)$
   (b) $Br_2(\ell)$ (d) $CaF_2(s)$

13. (See Example 9–1.) Iron(III) sulfate $(Fe_2(SO_4)_3)$ is a yellow compound that is used as a coagulant in water treatment. Write a balanced chemical equation and a K_{sp} equation for its dissolution in water.

14. (See Example 9–1.) Lead(II) antimonate $(Pb_3(SbO_4)_2)$ has been used as an orange pigment in oil-base paints and in glazes. Write a balanced chemical equation and a solubility-product equation for its dissolution in water.

15. The solubility-product constant of mercury(I) iodide is 1.2×10^{-28} at 25°C. Estimate the concentrations of Hg_2^{2+} and I^- in equilibrium with solid Hg_2I_2 at this temperature.

16. The solubility-product constant of $BaSO_4$ is 1.1×10^{-10} at 25°C. Estimate the concentrations of Ba^{2+} ions and SO_4^{2-} ions in equilibrium with solid $BaSO_4$ at this temperature.

17. (See Example 9–2.) Thallium(I) iodate $(TlIO_3)$ is only slightly soluble in water. Its K_{sp} at 25°C is 3.1×10^{-6}. Estimate the mass solubility of thallium(I) iodate (in g L^{-1}) at 25°C.

18. (See Example 9–2.) Thallium thiocyanate $(TlSCN)$ is only slightly soluble in water. Its K_{sp} at 25°C is 1.8×10^{-4}. Estimate the mass solubility of thallium thiocyanate (in g L^{-1}) at 25°C.

19. The solubility-product constant for $PbCl_2$ at 25°C is given in Table 9–1.
   (a) Estimate the concentration of Pb^{2+} and Cl^- in equilibrium with solid $PbCl_2$ at 25°C.
   (b) The Environmental Protection Agency requires that water released into sewers contain less than 50 ppm Pb by mass. Is the concentration of Pb^{2+} in saturated $PbCl_2(aq)$ low enough that the solution might legally be flushed down the drain? (*Note*: "ppm" stands for "parts per million"; the water must contain less than 50 g of Pb per 10^6 g of solution.)

20. The solubility-product constant for Hg_2Cl_2 is 2×10^{-18} at 25°C.
   (a) Estimate the concentrations of Hg_2^{2+} and Cl^- ions in equilibrium with solid Hg_2Cl_2 at 25°C. List the assumptions made in this estimate.
   (b) Express the concentration of Hg_2^{2+} in saturated $Hg_2Cl_2(aq)$ in units of ppb. (*Note*: The unit "ppb" corresponds to 1 nanogram of solute per gram of solution.)

21. (See Example 9–3.) The solubility of silver chromate (Ag_2CrO_4) in 500 mL of water at 50°C is 0.027 g. Estimate $K_{sp}(Ag_2CrO_4)$ at this temperature.

22. (See Example 9–3.) At 25°C, 400 mL of water dissolves 0.00896 g of lead(II) iodate $(Pb(IO_3)_2)$ and no more. Estimate K_{sp} for lead(II) iodate at this temperature.

23. (See Example 9–3.) At 100°C, the mass solubility of AgCl is 1.8×10^{-2} g L^{-1}. Estimate $K_{sp}(AgCl)$ at this temperature.

24. (See Example 9–3.) A maximum of 0.017 g of silver dichromate $(Ag_2Cr_2O_7)$ dissolves in 300 mL of water at 25°C. Estimate the K_{sp} of silver dichromate at this temperature.

25. The following table gives the K_{sp}'s and experimental molar solubilities of three lead(II) halides.

Compound	K_{sp} (at 25°C)	Molar Solubility (at 25°C) (mol L^{-1})
PbF_2	3.3×10^{-8}	0.00657
$PbBr_2$	6.6×10^{-6}	0.026
PbI_2	9.8×10^{-9}	0.0016

In no case does the experimental molar solubility equal the molar solubility computed from the K_{sp} by the method of

Example 9–3. Explain why. Include chemical equations in the explanation. (*Hint*: See the discussion of hydrolysis in Section 8–4.)

26. You calculate the mass solubilities of potassium perchlorate ($KClO_4$) and nickel(II) carbonate ($NiCO_3$), two poorly soluble ionic compounds, from their K_{sp}'s. For which salt is the answer likely to be closer to the experimental mass solubility? Explain.

Precipitation and the Solubility Product

27. A solution of barium chromate ($BaCrO_4$) is prepared by dissolving 6.3×10^{-3} g of this yellow solid in 1.00 L of hot water. Will solid barium chromate precipitate upon cooling to 25°C? Explain.

28. A solution is prepared by dissolving 0.090 g of PbI_2 in 1.00 L of hot water. Will solid lead(II) iodide precipitate upon cooling to 25°C?

29. (See Example 9–4.) A solution is prepared by mixing 250.0 mL of 2.0×10^{-3} M $Ce(NO_3)_3$ and 150.0 mL of 1.0×10^{-2} M HIO_3 at 25°C. Determine whether $Ce(IO_3)_3(s)$ ($K_{sp} = 1.9 \times 10^{-10}$) tends to precipitate from this mixture.

30. (See Example 9–4.) A mixture is prepared by stirring 100.0 mL of a 0.0010 M $CaCl_2$ solution with 50.0 mL of a 6.0×10^{-5} M NaF solution at 25°C. Determine whether $CaF_2(s)$ tends to precipitate from this mixture.

31. (See Example 9–4.) Suppose that 140 mL of 0.0010 M $Sr(NO_3)_2$ is mixed with enough 0.0050 M NaF to make 1000 mL of solution. Does $SrF_2(s)$ precipitate at equilibrium? Explain.

32. (See Example 9–4.) Suppose that 140 mL of 0.0010 M $Sr(NO_3)_2$ is mixed with enough 0.0050 M Na_2CrO_4 to make 1000 mL of solution. Does $SrCrO_4(s)$ ($K_{sp} = 3.6 \times 10^{-5}$) precipitate at equilibrium? Explain.

33. (See Example 9–5.) A total of 50.0 mL of a 0.0500 M solution of $Pb(NO_3)_2$ is mixed with 40.0 mL of a 0.200 M solution of $NaIO_3$ and gives 90.0 mL of solution at 25°C. Calculate the $[Pb^{2+}]$ and $[IO_3^-]$ when the mixture comes to equilibrium. At this temperature, K_{sp} for $Pb(IO_3)_2$ is 2.6×10^{-13}.

34. (See Example 9–5.) Silver iodide (AgI) is used in place of silver chloride for the fastest photographic film because it is more sensitive to light and so can form an image in a very short exposure time. A silver-iodide emulsion is prepared by mixing 6.60 L of 0.10 M NaI solution with 1.50 L of 0.080 M $AgNO_3$ solution at 25°C. Calculate the concentration of silver ion in the 8.10 L of solution when the mixture comes to equilibrium. Calculate the fraction of silver ion removed from solution by the precipitation reaction.

35. (See Example 9–5.) Silver chromate (Ag_2CrO_4) precipitates when 50.0 mL of 0.100 M $AgNO_3$ and 30.0 mL of 0.0600 M Na_2CrO_4 are mixed. Estimate the $[Ag^+]$ and $[CrO_4^{2-}]$ remaining in solution at equilibrium at 25°C.

36. (See Example 9–5.) When 40.0 mL of 0.0800 M $Sr(NO_3)_2$ and 80.0 mL of 0.0500 M KF are mixed, a precipitate of strontium fluoride (SrF_2) is formed. The K_{sp} of strontium fluoride at 25°C is 2.8×10^{-9}. Calculate the $[Sr^{2+}]$ and $[F^-]$ remaining in solution at equilibrium, neglecting all other interactions between ions and assuming that the volumes are additive.

37. (See Example 9–6.) Calculate the molar solubility of $CaF_2(s)$ at 25°C in a 0.040 M aqueous solution of NaF.

38. (See Example 9–6.) Calculate the maximum mass of AgCl that can dissolve in 100 mL of 0.150 M NaCl at 25°C. Neglect complex-ion formation.

39. (See Example 9–6.) The solubility product of nickel(II) hydroxide ($Ni(OH)_2$) at 25°C equals 1.6×10^{-16}.
(a) Estimate the molar solubility of $Ni(OH)_2$ in pure water at 25°C.
(b) Estimate the molar solubility of $Ni(OH)_2$ in 0.100 M NaOH.

40. (See Example 9–6.) Silver arsenate (Ag_3AsO_4) is a slightly soluble salt:

$$Ag_3AsO_4(s) \rightleftharpoons 3\,Ag^+(aq) + AsO_4^{3-}(aq)$$
$$K_{sp} = 1.0 \times 10^{-22} \text{ at } 25°C$$

(a) Estimate the molar solubility of silver arsenate in pure water at 25°C.
(b) Estimate the molar solubility of silver arsenate in 0.100 M $AgNO_3$ at the same temperature.

The Effects of pH on Solubility

41. (See Example 9–7.) Compare the molar solubility of AgOH in pure water with that in a solution buffered at pH 7. Note the difference between the two: When AgOH is dissolved in pure water, the pH does not remain at 7.

42. (See Example 9–7.) Compare the molar solubility of $Mg(OH)_2$ in pure water with that in a solution buffered at pH 9.

43. To what pH must a solution be adjusted to maintain the molar solubility of $Fe(OH)_2$ at 0.001 mol L^{-1} at 25°C?

44. To what pH must a solution be adjusted to maintain the molar solubility of AgOH at 0.001 mol L^{-1} at 25°C?

45. An aqueous solution at 25°C is 0.10 M in both Mg^{2+} and Pb^{2+} ions. You wish to separate the two metal ions by taking advantage of the different solubilities of the oxalate salts, MgC_2O_4 and PbC_2O_4.
(a) What is the highest possible oxalate-ion concentration such that only one solid oxalate salt is present at equilibrium? Which ion is present in the solid—Mg^{2+} or Pb^{2+}?
(b) What fraction of the less-soluble ion still remains in solution under the conditions of part (a)?

46. An aqueous solution at 25°C contains Pb^{2+} ions at a concentration of 0.10 mol L^{-1} and Ag^+ ions at a concentration of 0.50 mol L^{-1}. You need to separate the two by taking advantage of the different solubilities of their chlorides, $PbCl_2$ and $AgCl$.
   (a) What is the highest possible chloride-ion concentration such that only one solid chloride salt is present at equilibrium? Which ion is present in the solid—Pb^{2+} or Ag^+?
   (b) What fraction of the less-soluble ion still remains in solution under the conditions of part (a)?

47. (See Example 9–8.) Calculate the equilibrium concentration of Zn^{2+} ions in a solution that is in contact with solid ZnS and in which $[H_3O^+] = 1.0 \times 10^{-5}$ mol L^{-1} and $[H_2S] = 0.10$ mol L^{-1} at 25°C.

48. (See Example 9–8.) Calculate the $[Cu^{2+}]$ in a solution that is in equilibrium with $CuS(s)$ at 25°C and in which $[H_3O^+] = 1.0 \times 10^{-3}$ mol L^{-1} and $[H_2S] = 0.10$ mol L^{-1}.

49. Find the highest pH at which $FeS(s)$ will *not* tend to precipitate from a solution that contains Fe^{2+} at a concentration of 0.10 mol L^{-1} and is saturated with H_2S at 25°C. Determine the concentration of $Pb^{2+}(aq)$ ion that is present in equilibrium with solid PbS at this pH in a solution that is saturated with H_2S.

50. Compute the highest pH at which Mn^{2+} ions at a concentration of 0.050 mol L^{-1} will remain unprecipitated in an aqueous solution containing 0.10 M H_2S. Determine the concentration of $Cd^{2+}(aq)$ present at equilibrium with solid CdS at this pH in a similar solution.

Complex Ions and Solubility

51. (See Example 9–9.) A total of 0.10 mol of $Cu(NO_3)_2$ and 1.5 mol of NH_3 are dissolved in water and diluted to a final volume of 1.00 L. Calculate the concentrations of the $Cu(NH_3)_4^{2+}$ ion and the Cu^{2+} ion at equilibrium, using the equilibrium constants given in Exercise 9–9.

52. (See Example 9–9.) The equilibrium constant for the formation of the $TlCl_4^-$ complex ion from Tl^{3+} and Cl^- is 1×10^{18}. Suppose that 0.080 mol of $Tl(NO_3)_3$ is dissolved in 1.00 L of 0.50 M NaCl. Calculate the concentration at equilibrium of $TlCl_4^-$ and of Tl^{3+}. Assume that almost all of the Tl^{3+} is bound up as $TlCl_4^-$ at equilibrium.

53. The pH of a 0.2 M solution of $CuSO_4$ at room temperature is 4.0. Write chemical equations to explain why a solution of this salt is neither basic (from the reaction of $SO_4^{2-}(aq)$ with water) nor neutral, but acidic.

54. Is a 0.05 M solution of $FeCl_3$ acidic, basic, or neutral? Explain your answer by writing chemical equations to describe any reactions taking place.

55. (See Example 9–10.) The acidity constant for $Co(H_2O)_6^{2+}(aq)$ is 3×10^{-10}. Calculate the pH of a 0.10 M solution of $Co(NO_3)_2$.

56. (See Example 9–10.) The acidity constant for $Fe(H_2O)_6^{2+}(aq)$ is 3×10^{-6}. Calculate the pH of a 0.10 mol L^{-1} solution of $Fe(NO_3)_2$ and compare it with the pH of a 0.10 mol L^{-1} solution of iron(III) nitrate solution (see Table 9–3).

57. A 0.15 mol L^{-1} aqueous solution of the chloride salt of the complex ion $Pt(NH_3)_4^{2+}$ is found to be weakly acidic, with a pH of 4.92. This is initially puzzling because the Cl^- ion in water is not acidic and NH_3 in water is basic, not acidic. Finally, it is suggested that the $Pt(NH_3)_4^{2+}$ ion as a group donates hydrogen ions. Compute the K_a of this acid, assuming that just one hydrogen ion is donated.

58. The pH of 0.10 M $Ni(NO_3)_2$ is 5.0. Calculate the acidity constant of $Ni(H_2O)_6^{2+}(aq)$.

Controlling Solubility in Qualitative Analysis

59. The cations in an aqueous solution that contains 0.100 M $Hg_2(NO_3)_2$ and 0.0500 M $Pb(NO_3)_2$ are to be separated by taking advantage of the difference in the solubilities of their iodides. What should be the concentration of iodide for the best separation (the case in which one of the cations remains entirely in solution, and the other precipitates as fully as possible)?

60. The cations in an aqueous solution that contains 0.150 M $Ba(NO_3)_2$ and 0.0800 M $Ca(NO_3)_2$ are to be separated by taking advantage of the difference in the solubilities of their sulfates. What should be the concentration of sulfate ion for the best separation?

61. Cesium nitrate and barium nitrate both form colorless crystals with similar densities. Suggest two ways of deciding the correct chemical identity of a crystal known to be one or the other.

62. A colorless crystal is known to be either potassium nitrate or calcium nitrate. Suggest two ways of deciding which is the correct chemical composition.

63. An aqueous solution contains one or more of the following types of metal ions (and no others): Hg^{2+}, Ni^{2+}, Sr^{2+}. Outline a procedure to determine which are present and which are not.

64. An aqueous solution contains one or more of the following types of metal ions (and no others): Ag^+, Cr^{3+}, Hg_2^{2+}. Outline a procedure to determine which are present and which are not.

65. (See Example 9–11.) A solution containing ions from the scheme outlined in Section 9–6 is treated with hydrochloric acid, and no precipitate forms. The pH is adjusted to 9, and the solution is then saturated with H_2S, causing a precipitate to appear. State and justify any conclusions that can be drawn about which qualitative-analysis groups are present and which are not.

66. (See Example 9–11.) A solution containing ions from the scheme outlined in Section 9–6 is treated with hydrochloric acid, and a white precipitate forms. This precipitate is removed, and the remaining solution is saturated with H_2S at pH 9. No precipitate forms. After removing ammonium chloride and H_2S, the pH of the solution is adjusted to 9.5, and ammonium carbonate is added, at which point another white precipitate forms. State and justify any conclusions that can be drawn about which groups are present and which are not present.

ADDITIONAL PROBLEMS

67. Write a chemical equation for the dissolution of mercury(I) chloride in water, and give its solubility-product expression.

68. Magnesium ammonium phosphate has the formula $MgNH_4PO_4 \cdot 6H_2O$. It is only slightly soluble in water (its K_{sp} is 2.3×10^{-13} at 25°C). Write a chemical equation and the corresponding equilibrium-constant equation for the dissolution of this compound in water.

69. The relative strengths of acids can be judged by direct comparison of their K_a's. Explain why it is not possible to judge the relative solubilities of salts by direct comparison of their K_{sp}'s.

70. Explain each of the following phenomena:
(a) Silver chloride is less soluble in aqueous sodium chloride than in water.
(b) Silver chloride is more soluble in aqueous ammonia than in water.
(c) Silver chloride is more soluble in hot water than in cold water.

71. Soluble barium compounds are poisonous, but barium sulfate is routinely ingested as a suspended solid in a "barium cocktail" to improve the contrast in X-ray images. Calculate the concentration of dissolved barium per liter of water in equilibrium with solid barium sulfate.

72. At 25°C, the K_{sp} of SrF_2 is 2.8×10^{-9}, and the K_{sp} of $SrSO_4$ is 2.8×10^{-7}. Although the K_{sp} of SrF_2 is 100 times smaller than that of $SrSO_4$, its molar solubility and mass solubility are both larger than those of $SrSO_4$! Compute the solubilities of both compounds, and explain the apparent paradox.

73. The mass solubilities of the carbonates of the Group II elements in water at 20°C are $CaCO_3$, 14×10^{-3} g L^{-1}; $SrCO_3$, 10×10^{-3} g L^{-1}; $BaCO_3$, 17×10^{-3} g L^{-1}. Note how the solubility goes down and then up again moving down the column of Group II elements in the periodic table.
(a) Compute the molar solubilities of the three compounds, and comment on the presence or absence of a periodic trend in the results.
(b) Compute K_{sp} for each of the compounds from these data. Why do the values for K_{sp} computed here differ somewhat from those in Table 9–1? Is there a periodic trend in the K_{sp}'s?

74. A saturated aqueous solution of silver perchlorate ($AgClO_4$) at room temperature contains 84.8% by mass $AgClO_4$, but a saturated solution of $AgClO_4$ in 60% aqueous perchloric acid contains only 5.63% by mass $AgClO_4$. Explain this large difference, using chemical equations.

75. The K_{sp} of $RbClO_4$ at 25°C equals 3.0×10^{-3}. Estimate the molar solubility of $RbClO_4$ at that temperature in
(a) pure water.

(b) 0.075 M aqueous Rb_2SO_4.
(c) 0.075 M aqueous $AgClO_4$.
(d) 0.035 M aqueous NaCl.

76. The concentration of calcium ion in a town's supply of drinking water is 0.0020 mol L^{-1}. This water is hard water because it contains such a large concentration of Ca^{2+}. Suppose that the water is to be fluoridated by the addition of NaF for the purpose of reducing tooth decay. What is the maximum concentration of fluoride ion that can be attained before precipitation of CaF_2 begins? Can the drinking water supply the concentration of fluoride ion recommended by the U.S. Public Health Service, of about 5×10^{-5} mol L^{-1} (1 mg fluorine per liter)?

77. The two solids CuBr(s) and AgBr(s) are only very slightly soluble in water. (K_{sp}(CuBr) = 4.2×10^{-8} and K_{sp}(AgBr) = 7.7×10^{-13} at 25°C.) Some CuBr(s) and AgBr(s) are mixed into a quantity of water that is then stirred until it is saturated with respect to both solutes. Next, a small amount of KBr is added and dissolves completely. Compute the ratio of $[Cu^+]$ to $[Ag^+]$ after the system reestablishes equilibrium.

78. The two salts $BaCl_2$ and Ag_2SO_4 are both far more soluble in water than either $BaSO_4$ ($K_{sp} = 1.1 \times 10^{-10}$) or AgCl ($K_{sp} = 1.6 \times 10^{-10}$) at 25°C. Suppose that 50.0 mL of 0.040 M $BaCl_2(aq)$ is added to 50.0 mL of 0.020 M $Ag_2SO_4(aq)$. Calculate the concentrations of $SO_4^{2-}(aq)$, $Cl^-(aq)$, $Ba^{2+}(aq)$, and $Ag^+(aq)$ that remain in solution at equilibrium.

79. The Mohr method of analysis for chloride ion exploits the difference in solubility between silver chloride and silver chromate. The analyst adds a small amount of chromate ion to a solution with unknown chloride concentration and then titrates with $AgNO_3(aq)$. The end-point is the appearance of a red precipitate of silver chromate. Consider a solution that is 0.100 M in Cl^- ion and 0.00250 M in CrO_4^{2-} ion to which 0.100 M $AgNO_3$ solution is being added. Compute the fraction of the Cl^- ion that remains in solution when $Ag_2CrO_4(s)$ first appears.

80. Oxide ion, like sulfide ion, is a strong base. Write an equation for the dissolution of CaO(s) in water, and give the corresponding equilibrium-constant equation. Write an equation for the dissolution of CaO(s) in an aqueous solution of a strong acid, and relate its equilibrium-constant equation to the previous one.

81. A liter of pure water dissolves about 0.16 mg of AgBr(s), a volume of solid smaller than the head of a pin. A liter of 0.10 M $NH_3(aq)$, which contains one molecule of NH_3 for every 555 molecules of H_2O, manages to dissolve 68 mg of AgBr(s).
(a) Explain how such a tiny change in the composition of the solvent can have such a large effect on the solubility of AgBr.
(b) Explain why AgCl is soluble in 6 M $NH_3(aq)$ while AgI is not.

82. Potassium ion reacts with "cryptands," which are large organic molecules that enfold potassium ions in a pocket. Let the abbreviation "crypt" represent a certain cryptand. The cryptand and the potassium ion (K^+) interact according to the equilibrium

$$\text{crypt} + K^+ \rightleftharpoons K\text{-crypt}^+ \qquad K = 3.1 \times 10^{11} \text{ at } 25°C$$

Compute the equilibrium concentration of K^+ if 0.10 mol of K^+ and 0.10 mol of crypt are mixed in 1.0 L of solvent at 25°C.

83. The organic compound "18-crown-6" binds with the alkali metals in aqueous solution by wrapping around and enfolding the ion. It presents a niche that nicely accommodates the K^+ ion but is too small for the Rb^+ ion and too large for the Na^+ ion. The values of the equilibrium constants at 25°C tell the story:

$$Na^+(aq) + 18\text{-crown-}6(aq) \rightleftharpoons Na\text{-crown}^+(aq) \quad K = 6.6$$
$$K^+(aq) + 18\text{-crown-}6(aq) \rightleftharpoons K\text{-crown}^+(aq) \quad K = 111.6$$
$$Rb^+(aq) + 18\text{-crown-}6(aq) \rightleftharpoons Rb\text{-crown}^+(aq) \quad K = 36$$

An aqueous solution contains $0.0080 \text{ mol L}^{-1}$ of 18-crown-6 and also $0.0080 \text{ mol L}^{-1}$ of K^+. Compute the equilibrium concentration of free K^+. (*Note:* "Free" means not tied up with the 18-crown-6.) Compute the concentration of free Na^+ if the solution contains 0.0080 M $Na^+(aq)$ instead of $K^+(aq)$.

84. (a) Use the data in Table 9–3 to obtain the K_a of $Pb^{2+}(aq)$ ion at 25°C.
(b) Estimate the pH of 0.20 M $Pb(NO_3)_2$ at 25°C.

85. An aqueous solution of $K_2[Pt(OH)_6]$ has a pH greater than 7. Explain this fact by writing an equation showing the $Pt(OH)_6^{2-}$ ion acting as a Brønsted-Lowry base and accepting a hydrogen ion from water.

*86. A reference book defines K_{spa} as the equilibrium constant of the reaction

$$MS(s) + 2 H_3O^+(aq) \rightleftharpoons$$
$$M^{2+}(aq) + H_2S(aq) + 2 H_2O(\ell) \qquad \frac{[M^{2+}][H_2S]}{[H_3O^+]^2} = K_{spa}$$

where $MS(s)$ stands for a metal sulfide. The book goes on to give a K_{spa} of 8×10^{-7} for $CdS(s)$ at 25°C.
(a) Show that this information is consistent with the K_{ss} for $CdS(s)$ that is given in Table 9–2. (*Hint:* Use data from Table 8–2.)
(b) Use the K_{spa} from the reference book to compute the solubility of $CdS(s)$ at 25°C in a solution that is saturated with H_2S and buffered at a pH of 2.0. (*Note:* This problem is identical to Exercise 9–8.)

CUMULATIVE PROBLEM

Carbonate Minerals

The carbonates are among the most abundant and important minerals in the Earth's crust. When these minerals come into contact with fresh water or seawater, solubility equilibria are established that greatly affect the chemistry of these natural waters. Calcium carbonate ($CaCO_3$), the most important natural carbonate, makes up limestone and other rocks such as marble. Other carbonate minerals include dolomite ($CaMg(CO_3)_2$) and magnesite ($MgCO_3$). These compounds are sufficiently soluble that their solutions are non-ideal, so calculations based on solubility-product expressions have only approximate validity.

(a) The rare mineral nesquehonite contains $MgCO_3$ together with water of crystallization. A sample containing 21.7 g of nesquehonite is acidified and heated, and the volume of $CO_2(g)$ produced is found to equal 3.51 L at STP. Assuming that all of the carbonate has reacted to form CO_2, give the chemical formula for nesquehonite.

(b) Write a chemical equation and a solubility-product expression for the dissolution of dolomite in water.

(c) In a sufficiently basic solution, the carbonate ion does not react significantly with water to form hydrogen carbonate ion. Estimate the mass solubility (in g L^{-1}) of limestone (calcium carbonate) in 0.10 M sodium hydroxide. Use the K_{sp} from Table 9–1.

(d) A solution of Na_2CO_3 is also strongly basic and has a CO_3^{2-} concentration of 0.10 mol L^{-1}. What is the mass solubility of limestone in this solution? Compare your answer with that to part (c).

(e) In a mountain lake having a pH of 8.1, the total concentration of carbonate species, $[CO_3^{2-}] + [HCO_3^-]$, is measured to be 9.6×10^{-4} mol L^{-1}, and the concentration of Ca^{2+} is 3.8×10^{-4} mol L^{-1}. Calculate the concentration of CO_3^{2-} in this lake, using $K_a = 4.8 \times 10^{-11}$ for the acidity constant of HCO_3^-. Is the lake unsaturated, saturated, or supersaturated with respect to $CaCO_3$?

(f) Does acid rainfall into the lake increase or decrease the solubility of limestone rocks in the lake's bed?

(g) Seawater contains a high concentration of Cl^- ions, which can form weak complexes with calcium ions such as $CaCl^+$. Does the presence of such complexes increase or decrease the equilibrium solubility of $CaCO_3$ in seawater?

(h) Among the major cationic impurities found in limestone rock are Fe^{2+} and Mn^{2+}. Outline a scheme of analysis to determine whether either or both of these ions are present in a given rock sample.

Carbonate minerals. Calcite (*top*) and aragonite (*middle*) are both $CaCO_3$; smithsonite (*bottom*) is $ZnCO_3$. (*Julius Weber*)

10

Thermochemistry

CHAPTER OUTLINE

The production of steel is made possible by the coupling of exothermic reactions to drive endothermic chemical reactions. *(Courtesy of Bethlehem Steel)*

Heat plays a central role in chemistry. Some chemical reactions, such as the combustion of oil or natural gas in a furnace or the dramatic reaction of iron(III) oxide with aluminum in the thermite process (Fig. 10–1), release large amounts of heat. Other reactions absorb heat. **Thermochemistry** is the measurement and prediction of such heat effects.

Heat is only one means by which energy is transferred in chemical and physical processes. The other is work. This chapter develops the relationship between heat and work as manifestations of energy and uses it to formulate the first law of thermodynamics, which states that energy is conserved in all processes.

10-1 Heat

Routine use of fire and ice gives everyone an intuitive notion of temperature and the nature of heat. Children discover that objects of different temperature reach a common temperature when placed in contact with each other. A piece of metal held over a flame gets hot. When the metal is plunged into water, it cools down and the water warms up until both come to the same intermediate temperature. Early scientists tried to explain these commonplace observations by describing heat as a fluid. This fluid, referred to from the Middle Ages through the 19th century as *caloric*, was assumed to flow from a hot to a cold body until the amounts in the two bodies became equal. The problem with this simple picture was that attempts to isolate caloric or establish its exact nature failed. All that was known was that caloric was massless; objects had the same mass whether hot or cold. Benjamin Thompson, Count Rumford carried out a significant experiment in 1798. In the course of his work as military advisor to the king of Bavaria, Thompson observed that boring out cannons produced heat. Heat could be produced indefinitely, as long as the cannon

(a)	(b)	(c)

Figure 10–1 • The thermite reaction $2\,Al(s) + Fe_2O_3(s) \longrightarrow 2\,Fe(\ell) + Al_2O_3(s)$ is among the most exothermic of all reactions, liberating approximately 16 kJ of heat for every gram of aluminum that reacts. The iron(III) oxide and aluminum are mixed in finely divided form. Like the combustion of paper, the reaction requires a source of ignition but then continues on its own. *(Charles D. Winters)*

A bolt glows red hot after being heated in the flame from a torch. *(Charles D. Winters)*

was rotated against a cutting tool. He concluded (correctly) that heat was not a substance contained in the brass of the cannon but was a form of motion generated by friction. Nonetheless, throughout the first half of the 19th century, the prevailing view was that heat was a fluid.

In the second half of the 19th century, the kinetic theory established temperature as a measure of the average kinetic energy of the atoms or molecules of which an object is composed. When two bodies are placed in contact, the collisions of their atoms or molecules cause such kinetic energy to be exchanged, and eventually the temperatures are equalized. In this theory, which is the one accepted today,

> **Heat** is a means by which energy is transferred from a hot body to a colder body when the two are placed in thermal contact.

"Heat flow" is merely a convenient way of summarizing the effect of the many collisions that take place when different collections of molecules come in contact.

If heat is a means by which energy is transferred, the proper unit for it is the same as the unit of energy itself, the joule. In many chemical processes, large amounts of heat are involved, and the kilojoule ($1 \text{ kJ} = 10^3 \text{ J}$) and the megajoule ($1 \text{ MJ} = 10^6 \text{ J}$) become convenient units.

• The joule is quite a small unit for practical chemistry. The combustion of only 2×10^{-5} g of natural gas, which occupies a volume of 0.03 cm^3 at STP, produces a joule of heat.

Sources of Heat

The Sun has always been the most important source of energy for the Earth, maintaining the temperatures at which life can exist. Thermonuclear reactions (Chapter 15) taking place in the Sun generate radiant energy that strikes the Earth and transfers heat to it. The average solar radiative input to the surface of the Earth is about 1368 J s^{-1} m^{-2}. The amount reaching a given locality depends on the latitude and obviously varies with the weather as well.

Fuels, materials with energy stored in the bonds of molecules, allow for the local production of heat. In most cases, the ultimate source of this heat is also the Sun (Fig. 10–2). The combustion of wood releases as heat the chemical potential energy that was accumulated from light by the tree as it grew. Burning coal releases energy from deposits of carbonaceous material derived from plants buried for millennia and modified by geological processes. The wires in an electric toaster emit heat when an electric current passes through them, but the source of the electrical energy is most likely a coal-fired generating plant. Even hydroelectric power comes from the Sun, which evaporates seawater that later condenses as rain. Rainfall in the mountains gathers in streams whose flow can be harnessed to operate electrical generators. The Sun provides, directly or indirectly, a very large proportion of the heat used today. One important exception is heat from nuclear fission, the splitting of nuclei of certain heavy elements. These reactions liberate much more energy per mole than do chemical reactions. Nuclear reactors generated about 23% of the electrical energy used in the United States in the year 2000. This energy was deposited in the nuclei as they were synthesized in the explosions of stars that existed before the Sun and the solar system formed. The heavy elements were subsequently incorporated into the rocks of the Earth.

• Nuclear fission is discussed in Chapter 15.

Heat and Work

A flow of heat is one of two means by which energy is transferred. The other is a flow of *work* to run machines, lift weights, or raise buildings. Heat is a transfer of energy by means of random chaotic motions on a microscopic (atomic or molecu-

Figure 10-2 • One of the oldest sources of heat energy (a wood fire) and one of the newest (wind turbines). The energy for both can be traced ultimately to the Sun. Trees extract energy from sunlight to grow, and air currents spring into motion because the solar heating varies at different localities on the surface of the Earth. *(Left, Tom Slack/Tom Slack & Associates; right, John Meade/Science Photo Library/Photo Researchers, Inc.)*

lar) scale. Work is a transfer of energy by means of organized, directed motions on a macroscopic (laboratory) scale. Sources of heat are also sources of work because the two are easily interconverted (a steam engine converts heat to work). Both heat and work can be used to synthesize high-energy chemical compounds for energy storage. The chemical potential energy in such compounds is tapped by running chemical reactions to reconvert it to heat and work. An automobile engine converts energy stored in its fuel simultaneously to work (to drive up a hill) and to heat (in the exhaust). All forms of energy are interconvertible. The term "energy crisis" refers to the depletion of sources of energy that can easily be harnessed, not to a shortage of energy itself.

10-2 Calorimetry

A **system** is a portion of the universe that we designate for study. A quantity of water or air, a single biological cell, and a mixture of substances poised to react chemically are all systems. Systems have boundaries, either real or conceptual, that separate them from their **surroundings,** which consist of everything that is not the system. **Calorimetry** concerns itself with measuring flows of heat across such boundaries. In thermochemistry, the symbol q stands for such a quantity of heat. The heat absorbed by system 1 during an experiment is q_1, the heat absorbed by system 2 is q_2, and so forth.

If a system absorbs or emits heat without undergoing an internal chemical or physical change, its temperature rises or falls. The change is easily measured with a thermometer, but the result is *not* the amount of heat gained or lost. Heat is *not* the same as temperature—after all, increasing the temperature of a swimming pool of water by 5°C requires a lot more heat than doing the same to a cup of water. The observed temperature change of the system must be multiplied by a factor called its **heat capacity** to obtain the amount of heat that it absorbs. A system with a large heat capacity (the swimming pool of water) absorbs more heat per degree of temperature change than a system with a small heat capacity (the cup of water).

Heat capacities are determined experimentally. Two types of experiments are possible: at constant pressure (in containers open to the pressure of the atmosphere) and at constant volume (in sealed containers with strong, inflexible walls). In experiments at constant pressure, systems can expand or contract when heated, but at constant volume they cannot. This affects the way they absorb heat and obliges the definition of two kinds of heat capacity. The **heat capacity at constant pressure** C_P equals the amount of heat required to raise the temperature of a system by 1°C (or by 1 K) at constant pressure, and the **heat capacity at constant volume** C_V equals the amount of heat required to raise the temperature of a system by 1°C (or by 1 K) at constant volume. Calorimetry experiments are most often performed at constant (atmospheric) pressure. Consequently, the plain term "heat capacity" almost always refers to heat capacity at constant pressure. The mathematical definition of this heat capacity is

- The difference between C_P and C_V is generally small for systems containing only liquids and solids, but large for those containing gases.

$$\text{heat capacity} = C_P = \frac{q}{\Delta T}$$

The ΔT ("delta-T") in this equation represents the *change* in the temperature, computed by subtracting the initial from the final temperature (never the reverse!). The unit of heat capacity must be a unit of energy (heat is a form of energy) divided by a unit of temperature change. The recommended unit of heat capacity is the joule per kelvin ($J\ K^{-1}$).

The **molar heat capacity** is the heat capacity of a sample of a substance divided by the chemical amount of substance comprising the sample:

$$c_P = \frac{C_P}{n}$$

- The convention is that molar heat capacities are indicated with a lowercase c, reserving C_P for the total heat capacity of a system at constant pressure.

Note the lowercase c. Combining the two preceding definitions gives the useful equations

$$c_P = \frac{q}{n\Delta T} \quad \text{and} \quad q = nc_P\Delta T$$

The unit of molar heat capacity is the joule per kelvin per mole ($J\ K^{-1}\ mol^{-1}$). Putting the heat capacity on a per-mole basis removes the effect of sample size. The molar heat capacity consequently is a useful characterizing property of substances. Table 10–1 gives molar heat capacities of several substances. As a rule, the molar heat capacities of solid compounds increase with the number of atoms contained in the molecule.

Another useful quantity is the **specific heat capacity,** or, more briefly, the *specific heat.* The word "specific" before the name of a physical quantity often (but not

TABLE 10-1	Some Molar Heat Capacities and Specific Heat Capacities at 25°C	
Substance or Material	**Molar Heat Capacity** c_P (J K^{-1} mol^{-1})	**Specific Heat Capacity** c_S (J K^{-1} g^{-1})
$O_2(g)$	28.91	0.903
$CO_2(g)$	37.12	0.843
$Hg(\ell)$	27.94	0.1393
$Cu(s)$	24.46	0.3849
$Fe(s)$	25.09	0.4493
$Zn(s)$	25.39	0.3883
$SiO_2(s)$	44.57	0.742
$H_2O(\ell)$	75.38	4.184
$CCl_2F_2(g)$	72.36	0.598
$CCl_2F_2(\ell)$	126.8 (at 17°C)	1.047 (at 17°C)
$KCl(s)$	51.29	0.6880
$CsCl(s)$	52.43	0.3114
$C_5H_{12}(\ell)$ (pentane)	167.19	2.317
Granite	—	0.82
Marble	—	0.88
Olive oil	—	1.97
Wood	—	1.7 to 1.8

Note: More molar heat capacities are available from the U.S. National Institute of Standards and Technology on the World Web Web at http://webbook.nist.gov/chemistry.

always) means "divided by the mass." Accordingly, the specific heat capacity of a system is its heat capacity at constant pressure divided by its mass:

$$c_s = \frac{C_P}{m}$$

This division also takes sample size into account, although in a different way. The natural SI unit for the specific heat capacity c_s is the J K^{-1} kg^{-1}. However, the J K^{-1} g^{-1} is more widely used because the specific heat capacities of many common materials are on the order of 1 J K^{-1} g^{-1}. A specific heat capacity of 1 J K^{-1} g^{-1} means that it takes 1 J to raise the temperature of 1 g of the material by 1 K (at constant pressure). For a mass m of material, the relation between the temperature change and the amount of heat absorbed is

$$q = mc_s\Delta T$$

As can be proved from the definitions and confirmed by comparing the columns in Table 10–1, the molar heat capacity of a substance equals its specific heat capacity multiplied by its molar mass:

$$c_P = Mc_s$$

Strictly speaking, molar heat capacities cannot be defined for materials that are not substances (for example, olive oil or wood) because molar masses do not exist for mixtures. The specific heat capacity remains useful for such materials.

- A reasonable *effective* molar mass can often be estimated for a mixture. An effective molar heat capacity then makes sense.

• The large heat capacity of the water in a lake or ocean *buffers* its surroundings against changes in temperature.

Campers fill this bag with water and leave it in sunlight all day. It absorbs enough solar energy to provide a comfortably warm evening shower. *(REI Photography)*

The specific heat capacity of water is considerably larger than the specific heat capacity of common rocks, such as granite, at the Earth's surface (see Table 10–1). Large bodies of water consequently should experience a smaller rise in temperature than the neighboring land during the summer and a smaller drop in temperature during the winter. The moderating effect of the oceans on the climate in coastal areas is indeed thoroughly documented. Water is also an excellent medium in which to store large amounts of energy collected with solar panels during the day for use at night.

EXAMPLE 10–1

An architect is evaluating the use of large beds of granite to store thermal energy in a solar-heated home.
(a) Calculate the minimum amount of heat required to raise the temperature of 45 kg of granite by 15.0 K. Use data from Table 10–1.
(b) The use of water in place of granite is a possibility. Calculate the temperature change of 45 kg of water if it stores the same amount of thermal energy as the 45 kg of granite.

Strategy

Use the relationship $q = mc_s \Delta T$. In part (a), obtain q using the given mass, temperature change, and the specific heat capacity of granite. In part (b), use the q from part (a) together with the specific heat capacity of water and the given mass of water to arrive at ΔT.

Solution

(a) The heat absorbed by the granite is

$$q = mc_s \Delta T = (45 \times 10^3 \text{ g}) \times \left(\frac{0.82 \text{ J}}{\text{K g}} \right) \times (15.0 \text{ K}) = 5.5 \times 10^5 \text{ J} = \boxed{550 \text{ kJ}}$$

This is a minimum because some heat is surely lost to the surroundings as the rocks heat up.
(b) The temperature change of the water is

$$\Delta T = \frac{q}{c_s m} = \frac{5.5 \times 10^5 \text{ J}}{(4.184 \text{ J K}^{-1} \text{ g}^{-1})(45 \times 10^3 \text{ g})} = \boxed{2.9 \text{ K}}$$

Check

The water and the granite have the same mass and absorb the same amount of heat. The water should heat up substantially less than the granite because the specific heat capacity of water is substantially larger than the specific heat capacity of granite. The answer fits this expectation.

EXERCISE

Exactly 500.0 kJ of heat is absorbed at a constant pressure by a sample of gaseous helium. The temperature increases by 15.0 K. (a) Compute the heat capacity of the sample. (b) The mass of the helium sample is 6.42 kg. Compute the specific heat capacity and the molar heat capacity of helium.
Answer: (a) $3.3 \times 10^4 \text{ J K}^{-1}$. (b) $5.19 \text{ J K}^{-1} \text{ g}^{-1}$ and $20.8 \text{ J K}^{-1} \text{ mol}^{-1}$.

Another unit for heat is in common use: the **calorie.** During the period when the connection between heat and energy was not fully understood, the calorie was defined as the amount of heat needed to raise the temperature of 1 g of water by 1°C. In other words, the specific heat capacity of water was taken to equal 1.0 cal °C^{-1} g^{-1} (which equals 1.0 cal K^{-1} g^{-1} because a *change* of 1°C equals a *change* of 1 K). A separate unit for heat became unnecessary when the nature of heat as a form of energy was understood, and the thermochemical calorie is now defined in terms of the joule:

> • The calorie used in measuring food energy is the "great calorie," or Calorie (with a capital "C"). It equals 1000 small calories, or 4184 J. (See Chemistry in Your Life: Food Calorimetry, later in this chapter.)

$$1 \text{ calorie} = 4.184 \text{ J exactly}$$

The Temperature at Thermal Equilibrium

When two non-reacting bodies at different temperatures touch, energy passes between them in the form of a flow of heat until they reach a common final temperature. If we know the heat capacities of the bodies and their original temperatures, we can compute their temperature at this point of thermal equilibrium, assuming that no heat is lost to or gained from the surroundings. The assumption means that

amount of heat gained by the cooler body =
amount of heat lost by the warmer body

We now explicitly adopt an important convention, one that is used throughout thermochemistry:

> A positive heat flow (positive q) corresponds to heat *gained* by a body or a system.

Under this convention, a heat loss is regarded as a negative gain, just as a book-keeper might regard a money loss as equivalent to the acquisition of a debt. The q's, the amounts of heat gained by body 1 and body 2, are thus equal in magnitude but opposite in sign:

$$q_1 = -q_2$$

Both q's can be related to the specific heat capacity of the substance composing the body and the temperature change of that body, giving

$$m_1 c_{s1} \Delta T_1 = -m_2 c_{s2} \Delta T_2$$
$$m_1 c_{s1} (T_f - T_i)_1 = -m_2 c_{s2} (T_f - T_i)_2$$

The second version of the equation emphasizes the rule that a temperature change (ΔT) is always the final temperature minus the initial and never the reverse. Equivalent equations can be written in terms of the chemical amounts and the molar heat capacities of the two bodies if they are substances and not mixtures:

$$n_1 c_{P1} \Delta T_1 = -n_2 c_{P2} \Delta T_2$$
$$n_1 c_{P1} (T_f - T_i)_1 = -n_2 c_{P2} (T_f - T_i)_2$$

The temperatures of body 1 and body 2 are equal at thermal equilibrium. The final temperature T_f is the only unknown quantity if the amounts of substance, the molar or specific heat capacities, and the initial temperatures are known.

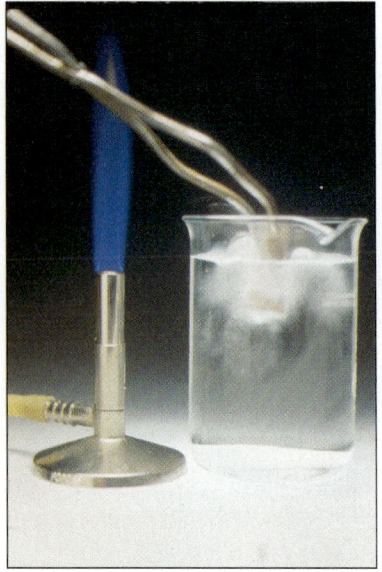

Plunging hot metal into cool water. In this case some heat is clearly being lost to the surroundings as a small amount of water boils off. *(Charles D. Winters)*

- Lowercase t's are now used for the temperatures because capital T's are reserved to represent absolute temperatures.

EXAMPLE 10-2

A piece of iron with a mass of 72.4 g is heated to 363 K (90°C) and then plunged into 100.0 g of water that is initially at 283 K (10°C). Calculate the temperature of thermal equilibrium, assuming that no heat is lost to the surroundings. Use specific heat capacities from Table 10–1.

Strategy

Write the expression for the heat gained by the water (body 1) in terms of its mass, specific heat capacity, and temperature change. Set it equal to the negative of a similar expression for the heat gained by the iron (body 2) and solve for the final or equilibrium temperature.

Solution

$$m_{water}(c_s)_{water}\Delta T_{water} = -m_{iron}(c_s)_{iron}\Delta T_{iron}$$

$$(100.0 \text{ g } H_2O)\left(\frac{4.184 \text{ J}}{\text{K g } H_2O}\right)(T_f - 283) \text{ K} = -(72.4 \text{ g Fe})\left(\frac{0.449 \text{ J}}{\text{K g Fe}}\right)(T_f - 363) \text{ K}$$

This is a linear equation in one unknown. Solve for T_f, which has units of K:

$$418.4T_f - 118{,}407 = -32.51T_f + 11{,}800$$
$$450.9T_f = 130{,}207$$
$$T_f = \boxed{289 \text{ K}}$$

Check

Redo the calculation using temperatures in degrees Celsius. Specific heat capacities in $J \, K^{-1} g^{-1}$ equal specific heat capacities in $J \, °C^{-1}g^{-1}$, because a temperature difference is the same number in K and °C. The equations become

$$m_{water}(c_s)_{water}\Delta t_{water} = -m_{iron}(c_s)_{iron}\Delta t_{iron}$$

$$(100.0 \text{ g } H_2O)\left(\frac{4.184 \text{ J}}{°C \text{ g } H_2O}\right)(t_f - 10)°C = -(72.4 \text{ g Fe})\left(\frac{0.449 \text{ J}}{°C \text{ g Fe}}\right)(t_f - 90)°C$$

$$418.4t_f - 4184 = -32.51t_f + 2926$$
$$t_f = 16°C$$

Converting to the Kelvin scale gives 289 K, which is correct. Note that exchanging the labels (so that the water is body 2 and the iron is body 1) does not change the answer.

EXERCISE

The specific heat capacities of hafnium and ethanol are $0.146 \text{ J } K^{-1} g^{-1}$ and $2.45 \text{ J } K^{-1} g^{-1}$, respectively. A piece of hot hafnium weighing 15.6 g and at a temperature of 160.0°C is dropped into 125 g of ethanol that has an initial temperature of 20.0°C. Calculate the temperature at thermal equilibrium, assuming no heat loss to the surroundings.

Answer: 21.0°C.

Calorimeters

A **calorimeter** is a container having an interior space that is thermally insulated from the surroundings and a means to measure the temperature in the interior. The idea of the device is to measure the temperature at thermal equilibrium among well-defined systems on the inside while preventing flows of heat to or from the outside. A perfect calorimeter would have walls completely impervious to the flow of heat. Real calorimeters are designed to minimize the inevitable leaks of heat and make them reproducible.

Figure 10–3 shows a simple calorimeter. It consists of a lidded Styrofoam cup (a coffee cup) that minimizes heat exchange with the outside (heat transfer through Styrofoam is slow), a thermometer to measure the interior temperature, and a stirrer to speed up attainment of thermal equilibrium on the inside. This type of calorimeter is frequently used to measure the absorption or evolution of heat accompanying a chemical reaction in aqueous solution, such as dissolution. Therefore, water fills the cup, and a solid is pictured on the bottom of the cup, ready to start dissolving. The water is the main source of heat to the reaction (or acceptor of heat from the reaction), but the thermometer, the stirrer, and the interior walls of the container also absorb or evolve heat. This complication is dealt with by defining the effective heat capacity of the interior parts of the calorimeter as the **calorimeter constant,** $C_{\text{calorimeter}}$. Use of the equation

$$q = C_{\text{calorimeter}} \Delta T$$

then tells how much heat the interior parts absorb. Calorimeter constants are determined by adding a known amount of heat to a calorimeter (for example, by passing a known electric current through a calibrated heating coil for a measured period of time) and observing the temperature rise. The calibrated calorimeter can then be used for further measurements, as the following example illustrates.

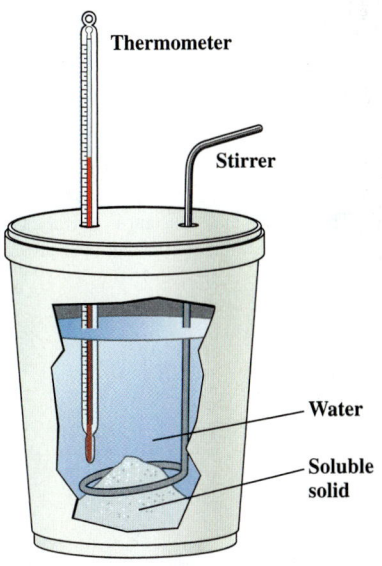

Thermometer

Stirrer

Water

Soluble solid

Figure 10-3 • A Styrofoam-cup calorimeter. Dissolution of the solid, which is just beginning, will cause the temperature inside the calorimeter to change. The change depends on how much heat the dissolution reaction absorbs, the heat capacity of the water, and the calorimeter constant (see Example 10–3).

EXAMPLE 10-3

(a) A newly constructed Styrofoam cup calorimeter is filled with 150.0 g of room-temperature water. Adding 1312 J of heat to these contents raises the interior temperature by 1.93 K (1.93°C). Calculate the calorimeter constant.

(b) Dissolving some lithium chloride in 150.0 g of room-temperature water contained in the same calorimeter causes a temperature rise of 3.46 K (3.46°C). Calculate the amount of heat evolved in the dissolution of the lithium chloride if the specific heat capacity and mass of the LiCl solution are the same as those of the water.

Strategy

Assume that the calorimeter does not leak heat. Think of the water, the interior parts of the calorimeter, and, in part (b), the dissolution reaction as different sub-systems that, taken together, comprise a whole system. Write expressions for the heat absorbed by each sub-system. The sum of these q's must equal the heat absorbed by the whole system.

Solution

(a) 1312 J of heat from outside is absorbed by the whole system, which consists of 150.0 g of water and the interior parts of the calorimeter:

$$q_{\text{calorimeter}} + q_{\text{water}} = 1312 \text{ J}$$

$$C_{\text{calorimeter}} \Delta T_{\text{calorimeter}} + m_{\text{water}} (c_s)_{\text{water}} \Delta T_{\text{water}} = 1312 \text{ J}$$

Both ΔT's equal 1.93 K because the interior parts of the calorimeter and the water are in thermal contact. Inserting the ΔT's, the mass of the water, and the specific heat capacity of water (Table 10–1) gives the equation

$$C_{\text{calorimeter}}(1.93 \text{ K}) + (150.0 \text{ g H}_2\text{O})\left(\frac{4.184 \text{ J}}{\text{K g H}_2\text{O}}\right)(1.93 \text{ K}) = 1312 \text{ J}$$

$$C_{\text{calorimeter}} = \frac{(1312 - 1211) \text{ J}}{1.93 \text{ K}}$$

$$= 52.2 \text{ J K}^{-1}$$

• This calorimeter constant is a heat capacity measured at constant pressure, a C_P.

(b) The heat absorbed by the interior parts of the calorimeter, by the water, and by the dissolution reaction must add up to zero because no heat passes into or out of the whole system. Mathematically,

$$C_{\text{calorimeter}}\Delta T + m_{\text{solution}}(c_s)_{\text{solution}}\Delta T_{\text{solution}} + q_{\text{reaction}} = 0$$

Inserting the calorimeter constant from part (a) and the other data gives

$$(52.2 \text{ J K}^{-1})(3.46 \text{ K}) + (150.0 \text{ g})\left(\frac{4.184 \text{ J}}{\text{K g}}\right)(3.46 \text{ K}) + q_{\text{reaction}} = 0$$

$$q_{\text{reaction}} = -2352 \text{ J}$$

The dissolution reaction *absorbs* -2.35×10^3 J. The heat that it *evolves* is the negative of this because evolving heat is the reverse of absorbing it. Hence, the dissolution of the lithium chloride evolves $+2.35 \times 10^3$ J.

EXERCISE

The dissolution of a small mass of $NH_4NO_3(s)$ in 150.0 g of water absorbs 321 J of heat. (a) Calculate the temperature change from dissolving the same mass of $NH_4NO_3(s)$ in 150.0 g of water inside the Styrofoam-cup calorimeter described in the example. Take the specific heat capacity and mass of the NH_4NO_3 solution to be the same as those of the water. (b) Calculate the temperature change from dissolving the same mass of $NH_4NO_3(s)$ in 150.0 g of water in a hypothetical calorimeter having interior parts that absorb zero heat.

Answer: (a) $\Delta T = -0.472$ K. (b) $\Delta T = -0.511$ K.

Calorimeters designed to study the heat evolved or absorbed in other types of chemical reactions, such as combustion reactions, are more elaborate but use the same fundamental principles. A combustion calorimeter is shown in Figure 10–4.

10-3   Enthalpy

Heat is a means by which energy is transferred among systems, not an inherent property of the systems themselves. To refer to the "heat content"of a system therefore has no meaning. For one important class of processes, however, the heat transferred is a measure of a change in a fundamental physical property of the system itself. For a process carried out at constant pressure, the heat absorbed by a system equals the change in its **enthalpy** (H):

• The enthalpy was formerly called the "heat content." This term, which is still encountered from time to time, is the source of the symbol H.

$$q_P = \Delta H$$

where the subscript "P"on the q means "at constant pressure" and ΔH symbolizes the final enthalpy minus the initial enthalpy, $H_f - H_i$.

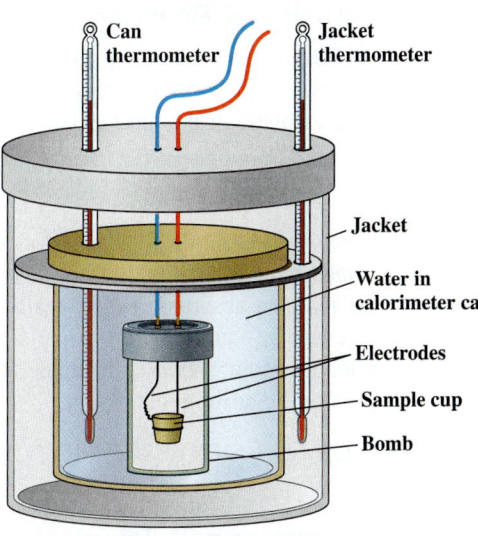

Figure 10-4 • The combustion calorimeter is also called a "bomb calorimeter"; it confines a combustion reaction to a fixed volume. (*Left, Leon Lewandowski*)

The measurement of enthalpy changes is a principal experimental activity in the field of thermochemistry. Workers run reactions in calorimeters at constant (atmospheric) pressure obtaining q_P (and thus ΔH) by the methods illustrated in Example 10–3. These enthalpy changes are sometimes called "heats of reaction." Yet, enthalpy and heat are actually quite different. Enthalpy is what is called a **state property** of a system. It depends only on the current state of the system (for example, on its temperature, pressure, volume, and composition as they are) and not on its history (the path followed in reaching the present state). When a system is brought from some initial state to some final state (by heating, cooling, expansion, compression, or reaction, for example), the enthalpy change depends only on the initial and final states, and not on the particular path followed between them (Fig. 10–5). Heat flow on the other hand is not a state property. The amount of heat that is transferred during a change depends on how the change was accomplished. It equals the enthalpy change *only* in the special case of a constant-pressure path. This is the reason for the subscript "*P*" in the preceding equation.

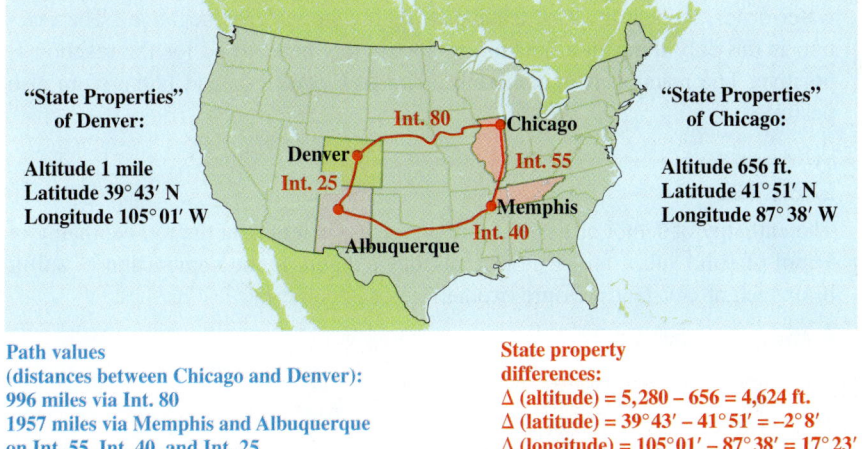

Figure 10-5 • Differences in state properties (such as the difference in altitude between two points) are independent of the path traced going from one to the other. Other properties (such as the total distance traveled) depend on the particular path that is followed.

Enthalpy of Reaction

• The term "enthalpy change of reaction" is also used. Leaving out the word "change" is justified by the fact that chemical reactions are by definition changes.

An **enthalpy of reaction** is the change in the enthalpy when a chemical reaction occurs. It equals the difference between the total enthalpy of the products and the total enthalpy of the reactants

$$\Delta H_r = H_{products} - H_{reactants}$$

where the subscript "r" stands for reaction. Chemical change is generally accompanied by some thermal effect; that is, ΔH_r is rarely zero. When ΔH_r is negative, the reaction absorbs a negative amount of heat if carried out at constant pressure (because q_P equals ΔH_r). In simpler words, the reaction evolves heat. It is **exothermic.** The combustion of methane (a major component of natural gas) is strongly exothermic at room conditions:

$$CH_4(g) + 2\,O_2(g) \longrightarrow CO_2(g) + 2\,H_2O(\ell) \qquad \Delta H_r = q_P < 0, \text{exothermic}$$

When ΔH_r is positive, the reaction absorbs heat from the surroundings if carried out at constant pressure. This means q_P is positive, and the reaction is **endothermic.** The formation of the pollutant nitrogen dioxide from the nitrogen and oxygen of the air, a reaction that occurs in automobile engines, is endothermic:

$$N_2(g) + 2\,O_2(g) \longrightarrow 2\,NO_2(g) \qquad \Delta H_r = q_P > 0, \text{endothermic}$$

EXAMPLE 10-4

When ammonium nitrate (NH_4NO_3) dissolves in water at room conditions, the temperature of the water drops. See Figure 10–6. Is the dissolution reaction

$$NH_4NO_3(s) \longrightarrow NH_4^+(aq) + NO_3^-(aq)$$

endothermic or exothermic?

Solution

Regard the water and the dissolution reaction as two systems in intimate thermal contact with each other but insulated from the outside. The temperature of the water drops, so heat must be leaving it (q_P for the water is negative). The reaction is the only thing in a position to absorb this heat, so q_P for the reaction is positive. The reaction is endothermic. The ΔH_r equals q_P and is therefore also positive.

EXERCISE

The enthalpy of 1 mol of gaseous sulfur dioxide is less than the total enthalpy of 1 mol of solid sulfur and 1 mol of gaseous oxygen. Is the combustion of sulfur in oxygen at constant pressure exothermic or endothermic?

Answer: Exothermic.

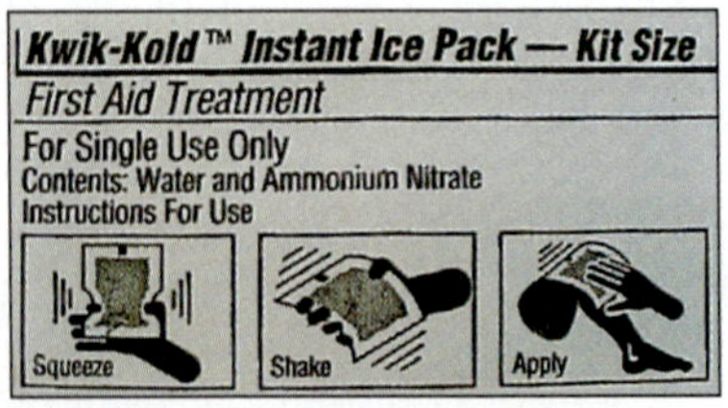

Kwik-Kold™ Instant Ice Pack — Kit Size
First Aid Treatment
For Single Use Only
Contents: Water and Ammonium Nitrate
Instructions For Use
Squeeze   Shake   Apply

Figure 10–6 • A dissolution reaction cools down the "Kwik-Cold Instant Ice Pack" very quickly once the solid and the water inside the pack come in contact. See Example 10–4.

Calorimetry experiments give numerical enthalpies of reaction. Burning carbon monoxide in oxygen generates carbon dioxide and evolves heat. The amount of the heat is determined by performing the reaction in a calorimeter. Repeated measurements establish that if 1.000 mol of $CO(g)$ is combusted in this way, starting with the reactants at 25°C and 1 atm pressure and ending with the product at 25°C and 1 atm pressure, then the enthalpy change caused by the reaction is

$$\Delta H = q_P = -2.830 \times 10^5 \text{ J} = -283.0 \text{ kJ}$$

Enthalpies of reaction are written in association with a balanced equation:

$$CO(g) + \tfrac{1}{2} O_2(g) \longrightarrow CO_2(g) \qquad \Delta H_r = -283.0 \text{ kJ mol}^{-1}$$

• In such a context the subscript "r" is not really needed and is frequently omitted.

The mol^{-1} ("per mole") in the units reflects the practice of reckoning enthalpies of reaction *per mole of the reaction as it is written*. Enthalpies of reaction are sometimes called *molar* enthalpies of reaction to emphasize this point. If all of the coefficients in the preceding equation are doubled, then ΔH_r doubles as well:

$$2 CO(g) + O_2(g) \longrightarrow 2 CO_2(g) \qquad \Delta H_r = -566.0 \text{ kJ mol}^{-1}$$

The doubling occurs because 1 mol of *this* reaction involves 2 mol of $CO(g)$ and 1 mol of $O_2(g)$ combining to give 2 mol of $CO_2(g)$. Clearly, it is essential to specify a balanced equation when giving a numerical enthalpy of reaction.

Enthalpies of reaction can be used to construct stoichiometric unit-factors like the ones used in Chapter 2. Suppose that 0.832 mol of $CO(g)$ burns at constant pressure, and the problem is to find the amount of heat that is absorbed. A good way to proceed is

$$q_P = \Delta H = 0.832 \text{ mol CO} \times \left(\frac{1 \text{ mol reaction}}{1 \text{ mol CO}}\right) \times \left(\frac{-283.0 \text{ kJ}}{\text{mol reaction}}\right) = -235 \text{ kJ}$$

The first unit-factor derives from the balanced equation, and the second from the experimental enthalpy of reaction quoted in the preceding. A computation based on the doubled equation and its ΔH_r goes as follows:

$$q_P = \Delta H = 0.832 \text{ mol CO} \times \left(\frac{1 \text{ mol reaction}}{2 \text{ mol CO}}\right) \times \left(\frac{-566.0 \text{ kJ}}{\text{mol reaction}}\right) = -235 \text{ kJ}$$

As shown, it gives the same answer.

EXAMPLE 10-5

Red phosphorus, a solid, reacts exothermically with liquid bromine (Fig. 10–7):

$$2 P(s) + 3 Br_2(\ell) \longrightarrow 2 PBr_3(g) \qquad \Delta H = -243 \text{ kJ mol}^{-1}$$

Calculate the enthalpy change when 2.63 g of phosphorus reacts completely with bromine.

Strategy

Realize that the units kJ mol^{-1} means "kilojoules per mole of reaction" and that the answer should be in kilojoules. Construct unit-factors using the molar mass of P, the amount of P in the reaction as it is written, and the enthalpy change associated with the reaction as it is written.

Solution

$$\Delta H = 2.63 \text{ g P} \times \left(\frac{1 \text{ mol P}}{30.97 \text{ g P}}\right) \times \left(\frac{1 \text{ mol reaction}}{2 \text{ mol P}}\right) \times \left(\frac{-243 \text{ kJ}}{\text{mol reaction}}\right)$$

$$= -10.3 \text{ kJ}$$

Figure 10–7 • Red phosphorus reacts vigorously and exothermically with liquid bromine. The fumes are a mixture of the product PBr_3 and unreacted bromine that has boiled off. *(Charles D. Winters)*

The structure of phosphorus tribromide (PBr_3).

EXERCISE

Hydrazine reacts with chlorine according to the equation

$$N_2H_4(\ell) + 2\,Cl_2(g) \longrightarrow 4\,HCl(g) + N_2(g) \qquad \Delta H = -420\ kJ\ mol^{-1}$$

Calculate the enthalpy change (a) when 25.4 g of hydrazine reacts completely with chlorine; (b) when 1.45 mol of HCl(g) is generated.

Answer: (a) $\Delta H = -333$ kJ. (b) $\Delta H = -152$ kJ.

If a chemical reaction (or any process) is reversed, the associated ΔH has exactly the same magnitude but the opposite sign. If this were not true, it would be possible to obtain an unlimited supply of heat by running a reaction cyclically at constant pressure:

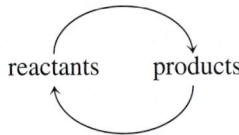

reactants   products

Each pass around the cycle would give back the reactants unchanged plus some free heat, an impossible outcome. Accordingly, ΔH_r for "un-combustion" of $CO_2(g)$

$$CO_2(g) \longrightarrow CO(g) + \tfrac{1}{2}\,O_2(g)$$

which is the reverse of the combustion of CO(g), certainly equals $+283.0\ kJ\ mol^{-1}$, even if un-combustion is never actually carried out in a calorimeter.

Phase Changes

Phase changes are not chemical reactions, but the enthalpy changes that accompany them are measured the same way as enthalpies of reaction, and they are routinely treated as if they were. The melting of ice is a familiar phase change. Heat must be transferred to ice in order to melt it to liquid water, so the process is endothermic, with ΔH positive. The melting of ice and the associated enthalpy change are represented

$$H_2O(s) \longrightarrow H_2O(\ell) \qquad \Delta H_{fus} = +6.007\ kJ\ mol^{-1}\ at\ 0°C$$

Here ΔH_{fus} is the **molar enthalpy of fusion,** the heat that is absorbed when 1 mol of substance melts at constant pressure. If the water freezes, the process is reversed, and an equal amount of heat is *given off* to the surroundings; that is, $\Delta H_{freez} = -\Delta H_{fus}$. Molar enthalpies of fusion have been measured for many thousands of substances and are listed in reference books and databases, in $kJ\ mol^{-1}$ or $kcal\ mol^{-1}$.

• Enthalpies of fusion are also often reported in kilocalories per gram or kilojoules per gram.

The vaporization of a mole of a liquid at constant pressure absorbs an amount of heat called the **molar enthalpy of vaporization** (ΔH_{vap}). For water, the "reaction" and the associated ΔH are

$$H_2O(\ell) \longrightarrow H_2O(g) \qquad \Delta H_{vap} = +40.66\ kJ\ mol^{-1}\ at\ 100°C$$

The reverse of this process is condensation. It is exothermic because $\Delta H_{cond} = -\Delta H_{vap}$. Like molar enthalpies of fusion, molar enthalpies of vaporization are important, characteristic properties of substances that are routinely measured.

TABLE 10-2	Enthalpy Changes of Fusion and Vaporization[a]	
Substance	ΔH_{fus} (kJ mol^{-1})	ΔH_{vap} (kJ mol^{-1})
NH_3	5.65	23.35
HCl	1.992	16.15
CO	0.836	6.04
CCl_4	2.5	30.0
H_2O	6.007	40.66
NaCl	28.81	170

[a]Measured at the normal melting point and the normal boiling point, respectively.

Table 10-2 lists the molar enthalpies of fusion and vaporization of several compounds. Note that vaporization generally requires more enthalpy per mole of substance than fusion does.

EXAMPLE 10-6

The vaporization of 100.0 g of carbon tetrachloride (CCl_4) absorbs 19.5 kJ of heat at its normal boiling point of 76.7°C. Calculate the molar enthalpy of vaporization of CCl_4 at 76.7°C.

Solution

The molar mass of CCl_4 is 153.8 g mol^{-1}, so the chemical amount in 100.0 g is

$$n_{CCl_4} = 100.0 \text{ g } CCl_4 \times \left(\frac{1 \text{ mol } CCl_4}{153.8 \text{ g } CCl_4} \right) = 0.6502 \text{ mol } CCl_4$$

The molar enthalpy of vaporization equals the enthalpy change per mole of CCl_4:

$$\Delta H_{vap} = \frac{19.5 \text{ kJ}}{0.6502 \text{ mol}} = 30.0 \text{ kJ mol}^{-1}$$

The result also can be reported by writing a balanced equation (with implied coefficients of 1) and its associated ΔH:

$$CCl_4(\ell) \longrightarrow CCl_4(g) \qquad \Delta H_{vap} = +30.0 \text{ kJ mol}^{-1}$$

EXERCISE

The melting of 0.140 g of Br_2 (which occurs at −7.2°C at 1 atm pressure) absorbs 9.43 J of heat. Compute the enthalpy change in 2.00 mol of Br_2 when it freezes.

Answer: $\Delta H = -2.15 \times 10^4$ J.

Hess's Law

Many reactions cannot be run cleanly or completely in a calorimeter. For example, attempts to measure the enthalpy of the reaction

$$C(s, \textit{graphite}) + \tfrac{1}{2} O_2(g) \longrightarrow CO(g)$$

in a calorimeter fail. When the reactants are mixed and ignited, about half of the graphite burns to $CO_2(g)$, and about half remains unreacted. How are the enthalpies of reactions such as this obtained? The answer relies on the fact that an enthalpy of a reaction is independent of the path by which reactants are converted to products. Measuring ΔH's of several steps along an indirect route and adding them up is just as good as making a single measurement along the direct route—as long as the routes start and end at the same places. A possible step 1 in an indirect route from graphite and oxygen to CO is

$$C(s, graphite) + O_2(g) \longrightarrow CO_2(g) \qquad \Delta H_1 = -393.5 \text{ kJ mol}^{-1}$$

This reaction runs well in a calorimeter, and the indicated enthalpy of reaction is readily measured. A possible step 2 is

$$CO_2(g) \longrightarrow CO(g) + \tfrac{1}{2}O_2(g) \qquad \Delta H_2 = +283.0 \text{ kJ mol}^{-1}$$

• The actual calorimetric experiment is performed by running the reverse of this reaction. See the discussion following Example 10–5.

for which the indicated enthalpy of reaction is also experimentally accessible. Because the sum of the reactions in the two steps equals the target reaction, the sum of ΔH_1 and ΔH_2 equals the ΔH of the target reaction. This is diagrammed in Figure 10–8. Writing out the equations and adding them confirms it:

$$\begin{aligned}
C(s, graphite) + O_2(g) &\longrightarrow CO_2(g) & \Delta H_1 &= -393.5 \text{ kJ mol}^{-1} \\
CO_2(g) &\longrightarrow CO(g) + \tfrac{1}{2}O_2(g) & \Delta H_2 &= +283.0 \text{ kJ mol}^{-1} \\
\hline
C(s, graphite) + \tfrac{1}{2}O_2(g) &\longrightarrow CO(g) & \Delta H &= \Delta H_1 + \Delta H_2 \\
& & &= -110.5 \text{ kJ mol}^{-1}
\end{aligned}$$

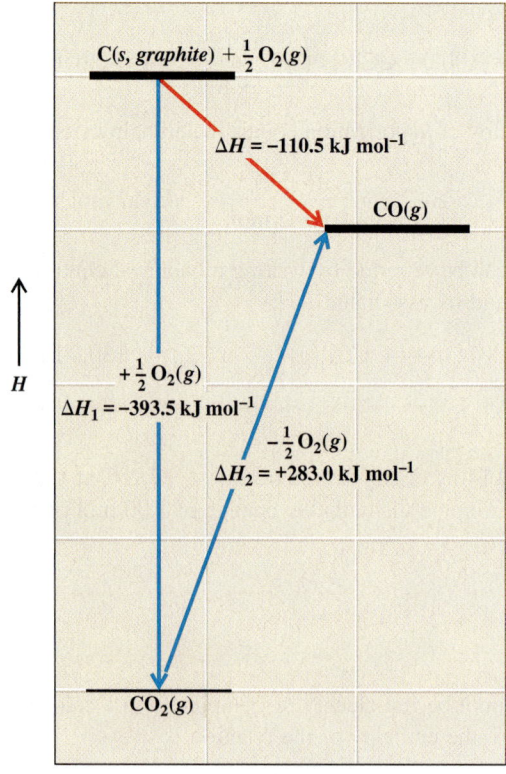

Figure 10–8 • The enthalpy change ΔH for the reaction of graphite with oxygen to give carbon monoxide (*red arrow*) can be determined by calorimetry measurements along an indirect (but easier to study) path (*blue arrows*) because the enthalpy is a state function. This indirect path consists of the reaction of an additional half mole of O_2 and then its removal.

A similar procedure always succeeds, no matter how complicated the path may be. The point is summarized in **Hess's law:**

> If two or more chemical equations are added to give a new equation, then adding the enthalpies of the reactions that they represent gives the enthalpy of the new reaction.

EXAMPLE 10-7

Hydrazine (N_2H_4) has been used as a rocket fuel because its reaction with oxygen is extremely exothermic:

$$N_2H_4(\ell) + O_2(g) \longrightarrow N_2(g) + 2\,H_2O(g) \qquad \Delta H = -534 \text{ kJ mol}^{-1}$$

Calculate the enthalpy of reaction when the water is produced in liquid rather than gaseous form:

$$N_2H_4(\ell) + O_2(g) \longrightarrow N_2(g) + 2\,H_2O(\ell)$$

Take 40.7 kJ mol^{-1} as the molar enthalpy of vaporization of water.

Strategy

Imagine that the reaction takes place in these steps: (1) the hydrazine and oxygen react to give gaseous nitrogen and water vapor; (2) the water vapor condenses to liquid water. Determine the enthalpy changes in each step and add them to find the enthalpy change of the overall process.

Solution

The individual steps and their associated enthalpies are:

$$N_2H_4(\ell) + O_2(g) \longrightarrow N_2(g) + 2\,H_2O(g) \qquad \Delta H_1 = -534 \text{ kJ mol}^{-1}$$
$$2\,H_2O(g) \longrightarrow 2\,H_2O(\ell) \qquad \Delta H_2 = -81.4 \text{ kJ mol}^{-1}$$

ΔH_2 equals two times the negative of the molar enthalpy of vaporization of water because each mole of reaction involves *two* moles of water and the process is condensation, which is the *reverse* of vaporization. Adding the two equations and their associated enthalpy changes gives

$$N_2H_4(\ell) + O_2(g) \longrightarrow N_2(g) + 2\,H_2O(\ell)$$
$$\Delta H = \Delta H_1 + \Delta H_2 = -615 \text{ kJ mol}^{-1}$$

EXERCISE

Calculate the enthalpy of the reaction

$$2\,N_2H_4(\ell) + 7\,O_2(g) \longrightarrow 2\,N_2O_5(s) + 4\,H_2O(g)$$

using the information in the example and the following:

$$N_2(g) + \tfrac{5}{2}O_2(g) \longrightarrow N_2O_5(s) \qquad \Delta H = -43 \text{ kJ mol}^{-1}$$

Answer: $\Delta H_r = -1154$ kJ mol^{-1}.

10-4 Standard Enthalpies of Reaction

The enthalpies of substances depend, often strongly, on their temperature, pressure, and physical state. These variables must be specified for every species that takes part in a change before observed enthalpy changes can be compiled and compared in any meaningful way. The issue is one of standardization. The agreed-upon means of dealing with it is to define thermodynamic **standard states** for every species that might take part in a reaction:

- For solids and liquids, the standard state is the state of the pure solid or liquid under a pressure of 1 atm and at a specified temperature.
- For gases, the standard state is the gaseous phase under a pressure of 1 atm, at a specified temperature, and exhibiting ideal-gas behavior.
- For dissolved species, the standard state is the state at a concentration of 1 mol L^{-1} under a pressure of 1 atm, at a specified temperature, and exhibiting ideal-solution behavior.

- Example 10–7 highlights the effect of physical state. The enthalpy of reaction changes by 15% when the product water forms as a liquid rather than a gas.

The definitions do not anoint any specific temperature as *the* standard-state temperature, a feature that actually simplifies the practical use of standard states. By far the most commonly specified temperature for thermodynamic standard states is 298.15 K (25°C exactly), but other choices are completely legitimate. If the temperature of a standard state is not explicitly indicated, 298.15 K should be assumed.

- In gas-law problems (Chapter 5), 1 atm and 273.15 K (0°C) are defined as "standard temperature and pressure" (STP). This definition has nothing to do with thermodynamic standard states.

Standard-state enthalpies (and other quantities) are designated by attaching a superscript ° (pronounced "naught") to the symbol for the quantity and writing the absolute temperature as a subscript. The formula of the chemical species and its physical state are then added in parentheses, if not clear from the context. Thus, $H_{300}^{\circ}(N_2(g))$ stands for the molar enthalpy of gaseous nitrogen in its standard state at 300 K, and $H_{500}^{\circ}(N_2(g))$ stands for the molar enthalpy of gaseous nitrogen in its standard state at 500 K. These two quantities differ by the amount of heat that is required at constant pressure to raise the temperature of a mole of $N_2(g)$ from 300 K to 500 K.

Gases cannot actually be prepared in standard states because real gases do not behave ideally. Gaseous water ($H_2O(g)$) at a temperature of 25°C and a pressure of 1 atm immediately condenses to $H_2O(\ell)$ because of the attractions among its molecules. The standard state of $H_2O(g)$ at 25°C is the hypothetical state that the water would have if the intermolecular attractions did not exist. The idealized nature of standard states of gases (and of solutions as well) presents no problems because it is always possible to relate the enthalpy of the real substance to the enthalpy of the substance in the idealized standard state (the exact methods are beyond the scope of this book).

A **standard enthalpy (change) of reaction** equals the enthalpy change when a reaction converts reactants in standard states into products in standard states at a specified temperature:

$$\Delta H_r^{\circ} = H_{products}^{\circ} - H_{reactants}^{\circ}$$

Any temperature can be chosen, but the most common choice is 298.15 K. By eliminating the obscuring effects of variations in temperature, pressure, and physical state, standard enthalpies of reaction provide a clear view of the thermal effects of bond rearrangements, a matter of prime concern to chemists, biochemists, and engineers. Standard enthalpies of reaction are the essential currency of thermochemistry.

Standard Molar Enthalpy of Formation

Despite their great importance, standard enthalpies of reaction (ΔH_r°'s) are not tabulated in any general way because the list would be too long. When they are needed, which is often, they are calculated from tables of a different quantity, one associated with individual chemical species rather than with reactions.

> A **standard molar enthalpy of formation** equals the enthalpy change when 1 mol of a species is formed in a standard state at a specified temperature from the most stable forms of the elements that constitute it, in standard states at the same temperature.[1]

The restriction "most stable" recognizes that elements change physical state (solid to liquid to gas) as the temperature changes and also exist in allotropic forms (see Section 3–1). The most stable form of an element is the form that is favored at equilibrium at 1 atm and the temperature that is specified. For example, oxygen exists as O_2 (dioxygen) and O_3 (ozone, or trioxygen), both of which can be solid, liquid, or gaseous. The most stable form at 298.15 K is $O_2(g)$, but the most stable form at 70 K (very cold) is liquid oxygen ($O_2(\ell)$). Solid carbon can be prepared in numerous allotropic forms (graphite, diamond, and the fullerenes). All differ in structure and physical properties, including enthalpy. The identity of the most stable at 1 atm pressure and a specified temperature must be determined experimentally.

• Figures 20–25 and 20–30 show the structures of diamond and graphite; Figure 18–A shows the structure of the fullerene C_{60}.

Experiments establish that $O_2(g)$ and graphite are the most stable forms of oxygen and carbon at 298.15 K and 1 atm. The standard molar enthalpy of formation of gaseous carbon dioxide at 298.15 K therefore equals the standard enthalpy of the reaction

$$C(s, \textit{graphite}) + O_2(g) \longrightarrow CO_2(g)$$

It is symbolized $\Delta H_f^\circ(CO_2(g))$, where the superscript $^\circ$ specifies standard states, the subscript "f" stands for formation, and the temperature of 298.15 K is understood. A table of standard molar enthalpies of formation at 298.15 K appears in Appendix D.

The ΔH_f° of the most stable form of an element in a standard state equals zero, because forming something from itself involves no change at all.

$$C(s, \textit{graphite}) \longrightarrow C(s, \textit{graphite}) \qquad \Delta H^\circ = 0 \text{ kJ mol}^{-1}$$

The ΔH_f°'s of other forms of an element are non-zero. At 298.15 K, the standard enthalpy of a mole of diamond exceeds the standard enthalpy of a mole of graphite by 1.895 kJ:

$$C(s, \textit{graphite}) \longrightarrow C(s, \textit{diamond}) \qquad \Delta H^\circ = 1.895 \text{ kJ mol}^{-1}$$

The difference exists because the relative positions of the C atoms differ in the graphite and diamond structures and the bonding is different (compare Figs. 20–25 and 20–30). Formation of diamond from graphite requires a small amount of heat as indicated by writing

$$\Delta H_f^\circ(C(s, \textit{diamond})) = 1.895 \text{ kJ mol}^{-1}$$

• Input of heat is necessary to make diamond from graphite at room conditions. It is *not* sufficient.

---

[1] There is one exception. White phosphorus (P_4 (s, white)) is less stable than both red and black phosphorus. It is nevertheless used in obtaining standard enthalpies of formation of phosphorus-containing species because it is more reproducible in its properties.

The standard enthalpy of formation of the free atoms of an element is frequently a much larger positive quantity. The following is found experimentally:

$$C(s, \text{ graphite}) \longrightarrow C(g) \qquad \Delta H° = 716.682 \text{ kJ mol}^{-1}$$

The $\Delta H°$ is large because production of gaseous carbon requires rupturing the bonds in graphite, not just rearranging them. The preceding reaction produces 1 mol of $C(g)$. Hence

$$\Delta H_f°(C(g)) = 716.682 \text{ kJ mol}^{-1}$$

Breaking up $N_2(g)$ molecules (the most stable form of nitrogen at 298.15 K and 1 atm) into $N(g)$ atoms is also strongly endothermic:

$$N_2(g) \longrightarrow 2 N(g) \qquad \Delta H° = 945.40 \text{ kJ mol}^{-1}$$

This reaction produces 2 mol of $N(g)$. The standard molar enthalpy of formation of $N(g)$ is the enthalpy change to produce 1 mol of $N(g)$, not two. It equals one half of 945.40 kJ mol^{-1}. This can be confirmed by using unit-factors, as follows:

$$\Delta H_f°(N(g)) = \frac{945.40 \text{ kJ}}{1 \text{ mol reaction}} \times \left(\frac{1 \text{ mol reaction}}{2 \text{ mol N}} \right) = 472.70 \text{ kJ mol}^{-1}$$

Calculating $\Delta H_r°$'s from $\Delta H_f°$'s

The calculation of standard enthalpies of reaction ($\Delta H_r°$'s) from standard molar enthalpies of formation ($\Delta H_f°$'s) relies on Hess's law. Here is how it works. A standard-state reaction, no matter how complex, can always be imagined to take place by way of these two steps:

1. Decomposition ("un-formation") of the reactants in standard states to give their constituent elements in standard states
2. Formation of the products in standard states from their constituent elements in standard states

The $\Delta H°$ of the second step equals the sum of the $\Delta H_f°$'s of the products of the reaction. This sum is obtained by multiplying the $\Delta H_f°$ of each substance by its coefficient in the balanced equation and adding up the results. The $\Delta H°$ of the first step equals a similar sum involving the $\Delta H_f°$'s of the reactants but multiplied by -1 because un-formation is the reverse of formation. By Hess's law, adding together the $\Delta H°$'s of these two steps gives $\Delta H_r°$. In summary, for the general reaction

$$aA + bB \longrightarrow cC + dD$$

$$\Delta H_r° = c\Delta H_f°(C) + d\Delta H_f°(D) - a\Delta H_f°(A) - b\Delta H_f°(B)$$

The following example illustrates the derivation of this equation in a specific case.

EXAMPLE 10–8

Use the $\Delta H_f°$ values in Appendix D to calculate the standard enthalpy of the following reaction at 25°C (298.15 K).

$$2 NO(g) + O_2(g) \longrightarrow 2 NO_2(g)$$

Solution

Imagine the reaction to proceed along the path shown in blue in Figure 10–9. This path consists of two steps, and $\Delta H°$'s are available for both. Step 1 is the standard-state decomposition of 2 mol of $NO(g)$ into its constituent elements:

$$2\,NO(g) \longrightarrow N_2(g) + O_2(g) \qquad (\Delta H°)_1 = -2\Delta H_f°(NO(g))$$

The factor of 2 appears because 2 mol of NO is decomposed; the minus sign reflects the fact that step 1 is the reverse of the formation of NO. Step 2 is the standard-state formation of the product from its constituent elements:

$$N_2(g) + 2\,O_2(g) \longrightarrow 2\,NO_2(g) \qquad (\Delta H°)_2 = 2\Delta H_f°(NO_2(g))$$

Adding these two steps gives the correct overall equation. Note how 1 mol of $O_2(g)$ cancels out:

$$
\begin{aligned}
2\,NO(g) &\longrightarrow N_2(g) + O_2(g) &\quad \text{Step 1}\\
\underline{N_2(g) + 2\,O_2(g)} &\underline{\longrightarrow 2\,NO_2(g)} &\quad \text{Step 2}\\
2\,NO(g) + O_2(g) &\longrightarrow 2\,NO_2(g) &
\end{aligned}
$$

By Hess's law, the standard enthalpy of the overall reaction equals the sum of the standard enthalpies of the two steps:

$$
\begin{aligned}
\Delta H_r° &= (\Delta H°)_1 + (\Delta H°)_2\\
&= -2\Delta H_f°(NO(g)) + 2\Delta H_f°(NO_2(g))
\end{aligned}
$$

Inserting values from Appendix D

$$
\begin{aligned}
\Delta H_r° &= -2(90.25\ \text{kJ mol}^{-1}) + 2(33.18\ \text{kJ mol}^{-1})\\
&= -180.50\ \text{kJ mol}^{-1} + 66.36\ \text{kJ mol}^{-1} = \boxed{-114.14\ \text{kJ mol}^{-1}}
\end{aligned}
$$

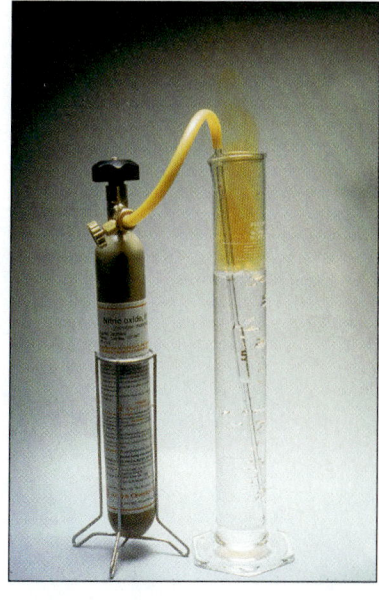

Nitrogen monoxide (NO), a colorless gas, bubbles up through water to react instantly with oxygen (O_2) in the air. The product of the reaction is brown nitrogen dioxide (NO_2). The reaction is substantially exothermic, as Example 10–8 establishes. *(Charles D. Winters)*

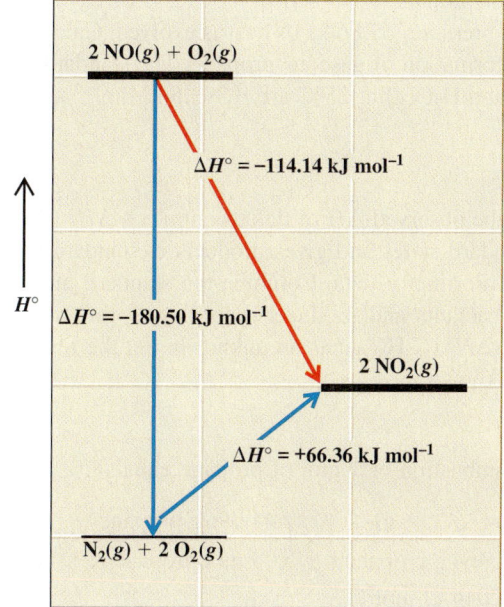

Figure 10–9 • The oxidation of nitrogen monoxide to nitrogen dioxide can be carried out directly *(red arrow)*. Alternatively, one can imagine a path in which the reactants are decomposed to the elements, which then combine to give products *(blue arrows)*. The enthalpy change is the same regardless of which path is used.

Check

Immediate substitution of the values from Appendix D into the general formula (without tracing through the steps) gives the same answer:

$$\Delta H_r^\circ = c\Delta H_f^\circ (C) + d\Delta H_f^\circ (D) - a\Delta H_f^\circ (A) - b\Delta H_f^\circ (B)$$
$$\Delta H_r^\circ = 2\Delta H_f^\circ (NO_2(g)) - 2\Delta H_f^\circ (NO(g)) - 1\Delta H_f^\circ (O_2(g))$$
$$= 2(33.18 \text{ kJ mol}^{-1}) - 2(90.25 \text{ kJ mol}^{-1}) - 1(0 \text{ kJ mol}^{-1})$$
$$= -114.14 \text{ kJ mol}^{-1}$$

EXERCISE

Use the ΔH_f° values in Appendix D to calculate ΔH_r° of the following reaction at 298.15 K:

$$2 H_2S(g) + 3 O_2(g) \longrightarrow 2 SO_2(g) + 2 H_2O(g)$$

Answer: $\Delta H_r^\circ = -1036.04 \text{ kJ mol}^{-1}$.

Direct synthesis of a compound from its elements is usually not practical in a calorimetry experiment. Lists of standard enthalpies of formation are instead compiled by Hess's law calculations employing data gathered on reactions that *are* practical. Such a calculation is illustrated by the following example.

EXAMPLE 10-9

Propane ($C_3H_8(g)$) burns in oxygen according to the equation

$$C_3H_8(g) + 5 O_2(g) \longrightarrow 3 CO_2(g) + 4 H_2O(g)$$

When 1.464 g of propane is burned in an excess of oxygen in a calorimeter at 25°C and 1 atm pressure, 67.86 kJ of heat is evolved. Calculate the standard molar enthalpy of formation of gaseous propane. The standard enthalpies of formation of $CO_2(g)$ and $H_2O(g)$ at 25°C are $-393.5 \text{ kJ mol}^{-1}$ and $-241.82 \text{ kJ mol}^{-1}$, respectively.

Strategy

Recognize that the observed ΔH of this reaction is a ΔH° (the reaction consumes reactants in standard states and gives products in standard states). Compute this ΔH_r° from the calorimetry data. Combine the standard molar enthalpies of formation of the reactants and products according to the highlighted equation on p. 458, keeping $\Delta H_f^\circ (C_3H_8(g))$ as an unknown. Set the result equal to ΔH_r°, and solve for the unknown.

Solution

The ΔH_r° for combustion of 1 mol of propane equals

$$\Delta H_r^\circ = \left(\frac{-67.86 \text{ kJ}}{1.464 \text{ g propane}}\right) \times \left(\frac{44.096 \text{ g propane}}{1 \text{ mol propane}}\right) \times \left(\frac{1 \text{ mol propane}}{1 \text{ mol reaction}}\right)$$
$$= -2044 \text{ kJ mol}^{-1}$$

The negative signs appear because the reaction is exothermic. Write the general equation

$$\Delta H^\circ_r = c\Delta H^\circ_f (C) + d\Delta H^\circ_f (D) - a\Delta H^\circ_f (A) - b\Delta H^\circ_f (B)$$

and specialize it to fit this case:

$$\Delta H^\circ_r = 3\Delta H^\circ_f (CO_2(g)) + 4\Delta H^\circ_f (H_2O(g)) - 5\Delta H^\circ_f (O_2(g)) - 1\Delta H^\circ_f (C_3H_8(g))$$

The ΔH°_r and all of the ΔH°_f 's, with the exception of the last, are known. Substitute them and solve for the unknown:

$$-2044 \text{ kJ mol}^{-1} = 3(-393.5 \text{ kJ mol}^{-1}) + 4(-241.82 \text{ kJ mol}^{-1})$$
$$- 5(0 \text{ kJ mol}^{-1}) - 1\Delta H^\circ_f (C_3H_8(g))$$
$$\Delta H^\circ_f (C_3H_8(g)) = -104 \text{ kJ mol}^{-1}$$

Check

Appendix D gives $-103.85 \text{ kJ mol}^{-1}$, which rounds off to -104 kJ mol^{-1}.

EXERCISE

Calcium sulfate reacts with carbon (in the form of graphite) as follows:

$$2 \text{ CaSO}_4(s) + C(s, \textit{graphite}) \longrightarrow 2 \text{ CaO}(s) + 2 \text{ SO}_2(g) + CO_2(g)$$

The reaction of 204.2 g of $CaSO_4(s)$ with excess graphite at 25°C and a constant pressure of 1 atm absorbs 456 kJ of heat. Calculate the standard molar enthalpy of formation of $CaSO_4(s)$. Take standard molar enthalpies of formation of the other compounds from Appendix D.

Answer: $\Delta H^\circ_f (CaSO_4(s)) = -1433 \text{ kJ mol}^{-1}$.

10–5 Bond Enthalpies

The breaking of chemical bonds in stable substances often generates highly reactive products. Consider the products when a C—H bond in methane is broken:

$$CH_4(g) \longrightarrow CH_3(g) + H(g)$$

Both $CH_3(g)$ and $H(g)$ are odd-electron species. They lack favored Lewis structures and react avidly with unconsumed CH_4 or with other CH_3 or H radicals. The bond-breaking reaction in CH_4 cannot be studied by calorimetry because these rapid subsequent reactions interfere. Fortunately, it and reactions like it can be studied by spectrometric methods (see Chapter 18). A particularly important quantity available from such experiments is the **bond enthalpy,** the enthalpy change when a specific bond is broken in a gas-phase reaction. This is invariably positive, because heat must be added to a collection of stable molecules to break their bonds. The bond enthalpy of a C—H bond in methane is $+439 \text{ kJ mol}^{-1}$. It is the measured standard enthalpy change for the reaction

$$CH_4(g) \longrightarrow CH_3(g) + H(g) \qquad \Delta H^\circ = +439 \text{ kJ mol}^{-1}$$

in which 6.022×10^{23} C—H bonds (Avogadro's number) are broken, one in each molecule in 1 mol of methane. The following gas-phase reactions all involve the breaking of 1 mol of C—H bonds:

$$C_2H_6(g) \longrightarrow C_2H_5(g) + H(g) \qquad \Delta H° = +410 \text{ kJ mol}^{-1}$$
$$CHF_3(g) \longrightarrow CF_3(g) + H(g) \qquad \Delta H° = +429 \text{ kJ mol}^{-1}$$
$$CHCl_3(g) \longrightarrow CCl_3(g) + H(g) \qquad \Delta H° = +380 \text{ kJ mol}^{-1}$$
$$CHBr_3(g) \longrightarrow CBr_3(g) + H(g) \qquad \Delta H° = +377 \text{ kJ mol}^{-1}$$

The approximate constancy of the standard enthalpies (all lie within 8% of their average value) persists in hundreds of analogous bond-breaking reactions in compounds containing C—H bonds. A similar near constancy develops in experiments on the disruption of O—H, N—H, and other types of bonds. It is therefore useful to tabulate *average* bond enthalpies (Table 10–3). Any given bond enthalpy probably deviates from the average value, but the deviations are mostly small.

Certain atoms (including C, N, O, and S) sometimes form double or triple bonds in order to achieve octets as their valence-electron configurations. In their Lewis structures, two such atoms may share two or three pairs of electrons rather than the single pair shared in a single bond. Experiment shows that the average bond enthalpy of a double bond significantly exceeds that of a single bond between atoms of the same two elements, and that the average bond enthalpy of a triple bond is higher still. For example, the average bond enthalpy of a C—C single bond is 348 kJ mol^{-1}, but that of a C=C double bond is 612 kJ mol^{-1} and that of a C≡C triple bond is 812 kJ mol^{-1}. The larger number of shared electron pairs leads to stronger bonds.

TABLE 10–3		Average Bond Enthalpies[a] (kJ mol^{-1})								
	H	**C**	**N**	**O**	**F**	**Cl**	**Br**	**I**	**S**	**Si**
H	436									
C	412	348 (single) 612 (double) 812 (triple)								
N	388	305 (single) 613 (double) 890 (triple)	163 (single) 409 (double) 946 (triple)							
O	463	360 (single) 743 (double)	157	146 (single) 497 (double)						
F	565	484	270	185	158					
Cl	431	338	200	203	254	243				
Br	366	276				219	193			
I	299	238				210	178	151		
S	338	259			496	250	212		264	
Si	318			466						226

[a]Most values from D. F. Shriver and P. W. Atkins, *Inorganic Chemistry*, 3ed., New York, W. H. Freeman & Co., 1999, p. 72.

Applications of Bond Enthalpies

Average bond enthalpies can be used to estimate the standard enthalpies of gas-phase reactions (ΔH_r°'s). Any such reaction can be imagined as a two-step process:

1. Breaking all the bonds in the reactant molecules to generate free atoms in the gas phase. This is the *atomization* of the reactants; the enthalpy change equals the sum of the enthalpies of the bonds that are broken.

2. Making the product molecules from the free atoms; the enthalpy change equals the sum of the bond enthalpies of all of the bonds but now with minus signs because bonds are being *formed,* not broken.

By Hess's law, a ΔH_r° equals the sum of the standard enthalpy changes of the two steps. Using average bond enthalpies (as in Table 10–3) gives only an estimate of ΔH_r° because the bond enthalpies of each type of bond vary a bit from compound to compound. The following example shows the details of a typical calculation.

EXAMPLE 10-10

Use a table of average bond enthalpies to estimate the standard enthalpy of the reaction

$$CCl_2F_2(g) + 2\,H_2(g) \longrightarrow CH_2Cl_2(g) + 2\,HF(g)$$

Dichlorodifluoromethane ($CCl_2F_2(g)$) has the Lewis structure

$$\begin{array}{c} :\!\ddot{C}l\!: \\ | \\ :\!\ddot{F}\!-\!C\!-\!\ddot{C}l\!: \\ | \\ :\!\ddot{F}\!: \end{array}$$

and CH_2Cl_2 has the same structure, with H atoms replacing the F atoms.

Strategy

Write two equations: one to show the break-up of the reactants into free atoms and the second to show the combination of the free atoms into products. Identify the number of bonds of every type, and use average bond enthalpies to estimate the ΔH°'s of the two steps. Add the two ΔH°'s to give the overall ΔH_r° (on the basis of Hess's law).

Solution

The atomization (Step 1) is

$$CCl_2F_2(g) + 2\,H_2(g) \longrightarrow C(g) + 2\,Cl(g) + 2\,F(g) + 4\,H(g)$$

It requires breaking 2 mol of C—Cl bonds, 2 mol of C—F bonds, and 2 mol of H—H bonds. Taking average bond enthalpies from Table 10–3 gives:

$$(\Delta H^\circ)_1 \approx \underset{\text{C—Cl}}{2(338\text{ kJ mol}^{-1})} + \underset{\text{C—F}}{2(484\text{ kJ mol}^{-1})} + \underset{\text{H—H}}{2(436\text{ kJ mol}^{-1})} = 2516\text{ kJ mol}^{-1}$$

Recombining the free atoms to create the products goes according to the equation

$$C(g) + 2\,Cl(g) + 2\,F(g) + 4\,H(g) \longrightarrow CH_2Cl_2(g) + 2\,HF(g)$$

Dichlorodifluoromethane (CCl_2F_2) is also known as Freon-12. Its low reactivity and high volatility make it useful as a refrigerant. Production of Freon-12 and related chlorofluorocarbons (CFCs) has been largely phased out because they accelerate depletion of the ozone layer in the outer atmosphere when lost to the air.

Figure 10–10 • The enthalpy ΔH for the overall reaction (*red arrow*) is the sum of the enthalpy changes for bond breaking (*black, green, and yellow arrows*) and the enthalpy changes for bond formation (*blue, orange, and purple arrows*). The final result is only approximate because average bond enthalpies have been used.

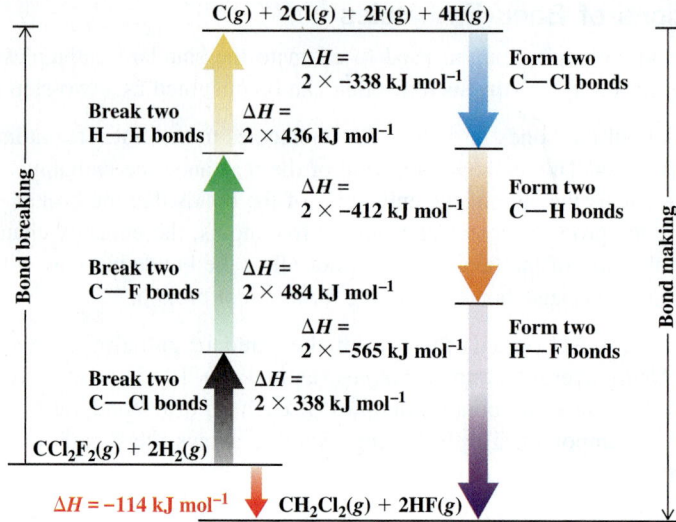

Recombination (step 2) involves the formation (with release of heat and, therefore, negative enthalpy change) of 2 mol of C—Cl bonds, 2 mol of C—H bonds, and 2 mol of H—F bonds.

$$(\Delta H^{\circ})_2 \approx \underset{\text{C—Cl}}{2(-338 \text{ kJ mol}^{-1})} + \underset{\text{C—H}}{2(-412 \text{ kJ mol}^{-1})} + \underset{\text{H—F}}{2(-565 \text{ kJ mol}^{-1})}$$

$$\approx -2630 \text{ kJ mol}^{-1}$$

$$\Delta H_{\mathrm{r}}^{\circ} = (\Delta H^{\circ})_1 + (\Delta H^{\circ})_2 \approx (2516 \text{ kJ mol}^{-1}) + (-2630 \text{ kJ mol}^{-1})$$

$$\approx -114 \text{ kJ mol}^{-1}$$

Figure 10–10 diagrams the solution.

Check

The experimental $\Delta H_{\mathrm{r}}^{\circ}$ is -158 kJ mol^{-1}. Expecting better agreement is unrealistic as long as average bond enthalpies are used. Note how the enthalpy of the reaction is the difference of two large numbers. Small relative errors in either can cause a large error in the final result.

EXERCISE

Estimate the standard enthalpy of reaction for the gas-phase reaction that forms methanol from methane and water:

$$CH_4(g) + H_2O(g) \longrightarrow CH_3OH(g) + H_2(g)$$

and compare it with the $\Delta H_{\mathrm{r}}^{\circ}$ obtained from the data in Appendix D.

Answer: Estimated: $\Delta H_{\mathrm{r}}^{\circ} = +79$ kJ mol^{-1}; using Appendix D: $\Delta H_{\mathrm{r}}^{\circ} = +116$ kJ mol^{-1}.

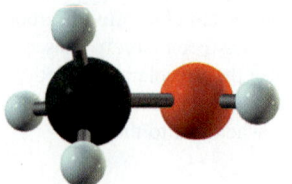

The structure of methanol.

Bond enthalpies are useful in situations in which hard calorimetric data are scarce or unavailable. An example is determination of the enthalpy change in breaking a specific bond or group of bonds in a complex biological molecule, such as the hydrogen bonds that hold the two strands of a DNA molecule together (see Fig. 25–20).

10-6 The First Law of Thermodynamics

The absorption of heat is not the only means by which a system can gain energy (Fig. 10–11). When a pitcher hurls a baseball, the ball gains kinetic energy (energy of motion) that has nothing to do with heat or temperature. When a person or a machine lifts a heavy object, the object gains potential energy (energy of position) that is readily convertible into other forms of energy (if dropped, the object falls and gains kinetic energy) but has nothing to do with heat or temperature. Energy transfers often involve *work* as well as heat. A fundamental connection exists between heat transfers, work transfers, and energy changes. This connection is the first law of thermodynamics, an empirical law that is essential to **chemical thermodynamics**. Thermodynamics is the study of heat, work, and energy; chemical thermodynamics is that part of the field in which calorimetric data are used to predict the positions of equilibria in chemical reactions.

• Potential energy is also discussed in Section 6–1.

• The remarkable way in which the results of calorimetry experiments are connected to the values of equilibrium constants is discussed in Chapter 11.

Work

Work is defined in mechanics as the product of the external force F on a body multiplied by the displacement d that the force causes the body to undergo. A force is best visualized as a push. Pushing on a massive object (such as a building) that does not move is work in a physiological sense because the effort tires the muscles. Mechanical work does not occur, however, because no motion (displacement) occurs. If a body moves over a distance d impelled by a constant force F applied along the direction of the path, then the work done on the body is

$$w = F \cdot d \qquad \text{(force along direction of path)}$$

Pushing a body in this way causes it to go faster. If the body is already in motion and if the push (force) opposes the direction of the motion, then the relationship includes a negative sign

$$w = -F \cdot d \qquad \text{(force opposite to direction of path)}$$

and the push slows the body down.

What is the relationship between work and energy in mechanical systems? Expending mechanical work on a body in the manner just described changes its kinetic energy (symbolized E_k). The kinetic energy of a body is its energy of motion, defined as $E_k = \frac{1}{2}mv^2$, where m is the mass of the body and v is its velocity. A push that does work on such a body either speeds it up or slows it down, which means that the push changes the velocity of the body and therefore changes its kinetic energy. In the absence of opposing forces (such as frictional forces), mechanical work expended on a body is entirely manifested in a change in the kinetic energy of the body:

$$w = \Delta E_k = \Delta\left(\tfrac{1}{2}mv^2\right) \qquad \text{(no frictional forces)}$$

Energy, including kinetic energy, is measured in joules, which are units of work as well. Indeed, the newton (the SI unit of force) multiplied by the meter (the SI unit of distance) equals the joule.

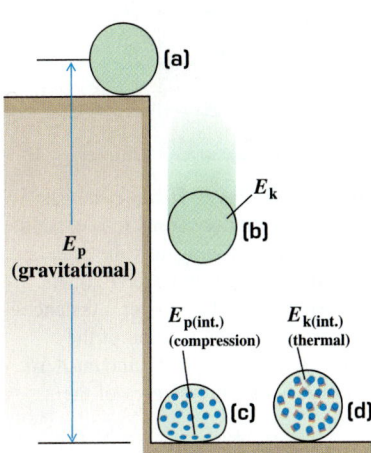

Figure 10–11 • A ball dropped from a height begins with a high potential energy (a) due to the gravitational attraction of the Earth. As it falls (b), this potential energy is converted into kinetic energy of motion of the ball. After impact with the ground (c), the molecules near the surface of the ball are pushed against each other, increasing the internal potential energy of the ball. As the ball stops bouncing, the molecules readjust their positions, but they move a little faster (d). The ball has a higher internal kinetic energy and a higher temperature.

Next, consider the work done in lifting objects in a gravitational field. Here, an opposing force operates. To raise a body of mass m from height h_1 to height h_2, an upward force sufficient to overcome the downward force of gravity (mg) must be exerted. Thus, the displacement is $h_2 - h_1$, and the work performed on the object is

$$w = mg(h_2 - h_1) = mg\Delta h$$

This work leads to a change in the potential energy E_p of the object ($\Delta E_p = mg\Delta h$), and we see once again a connection between the performance of mechanical work and a change in energy.

The most important kind of mechanical work in chemistry is **pressure–volume work.** This work occurs when systems expand against an outside pressure or are compressed by the operation of an outside pressure. Large changes in volume occur when gases are generated or consumed in chemical reactions, and large amounts of pressure–volume work may be transferred to or from the surroundings. The expansion of newly generated hot gases drives the pistons in an automobile engine to do work, for example. Let us analyze the pressure–volume work accompanying the volume changes of a gaseous system in a highly idealized, frictionless-cylinder piston engine. We imagine that the gas has pressure P_1 and is confined in the cylinder by a piston of cross-sectional area A and of negligible mass (Fig. 10–12). The force exerted on the piston face by the gas is $F_1 = P_1A$ (obtained by rearranging $P_1 = F_1/A$, the defining equation for pressure). Suppose that there is a gas on the other side of the piston with pressure P_{ext} ("ext" for "external"). If $P_{ext} = P_1$, the piston experiences equal forces on its two sides, and it does not move; if P_{ext} exceeds P_1, the gas in the cylinder is compressed and the piston moves inward; if P_{ext} is less than P_1, the gas in the cylinder expands and the piston moves outward. Suppose that expansion of the confined gas lifts the piston from h_1 to h_2. The work in this case is

$$w = -F_{ext}(h_2 - h_1) = -F_{ext}\Delta h$$

The negative sign is inserted because the force from the gas outside opposes the expansion of the gas inside the cylinder. Substitution for F_{ext} gives

$$w = -P_{ext}A\Delta h$$

But the product $A\Delta h$ is equal to the volume change of the system, ΔV (Fig. 10–12). Hence, the work is

$$w = -P_{ext}\Delta V$$

For an expansion, ΔV is positive, so that w is negative. A negative w means that the system performs work on the surroundings. It does this by pushing the surroundings back. If the gas is compressed (because P_{ext} exceeds P_1), the surroundings perform work on the system. Again, the applicable equation is $w = -P_{ext}\Delta V$, but now ΔV is negative so that w is positive. If no volume change occurs, then $\Delta V = 0$ and no pressure–volume work is done. Finally, if there is no mechanical link to the surroundings (that is, if $P_{ext} = 0$), then pressure–volume work cannot be performed.

If the pressure P_{ext} is expressed in newtons per square meter (pascals) and the volume change in cubic meters, then $P_{ext}\Delta V$ comes out in joules. These choices correspond to the recommended SI units. For many purposes, however, it is more convenient to express pressures in atmospheres and volumes in liters. The product of the two, the liter-atmosphere (L atm), then appears as a unit of work. It is related to the joule as follows:

$$1\ \text{L atm} = (10^{-3}\ \text{m}^3)(1.01325 \times 10^5\ \text{newton m}^{-2}) = 101.325\ \text{J}$$

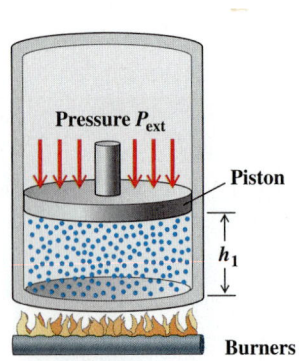

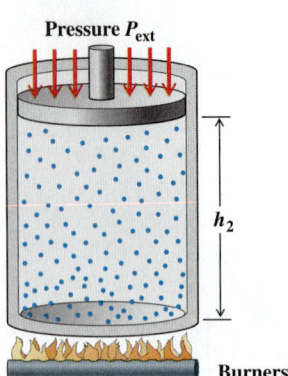

Figure 10–12 • Heating the gas inside the cylinder causes it to expand, pushing the piston against the pressure exerted by the gas outside (P_{ext}). As the piston is displaced over a distance $h_2 - h_1 = \Delta h$, the volume of the cylinder increases by an amount $A\Delta h$, where A is the surface area of the piston.

EXAMPLE 10-11

Suppose that 2.00 L of gas is confined in a cylinder with frictionless walls at a pressure of 1.00 atm. The external pressure is also 1.00 atm. The gas is then heated slowly. Because the piston slides perfectly freely, the pressure of the gas remains close to 1.00 atm as the volume of the gas increases (according to the ideal gas law, $PV = nRT$). The heating continues until a final volume of 3.50 L is reached. Calculate the work done on the gas, and express it in joules.

Solution

The system expands from 2.00 L to 3.50 L against a constant external pressure of 1.00 atm. Substitute into the equation for pressure-volume work:

$$w = -P_{ext}\Delta V$$
$$= -(1.00 \text{ atm})(3.50 \text{ L} - 2.00 \text{ L})$$
$$= -1.50 \text{ L atm}$$

Converting to joules gives

$$w = -1.50 \text{ L atm} \times \frac{101.325 \text{ J}}{\text{L atm}} = -152 \text{ J}$$

Work flows from the system to its surroundings. This can be expressed by saying that -152 J of work is done *on* the gas, or by saying that $+152$ J of work is done *by* the gas.

EXERCISE

Suppose that the gas in the cylinder described in the example is cooled slowly at a constant pressure of 1.00 atm from an initial volume of 2.00 L to a final volume of 1.20 L. Calculate the work done on the gas in joules.

Answer: $w = +81.1$ J.

The First Law

As Figure 10–11 emphasizes, energy comes in several forms. In chemistry (unlike physics), we usually are not interested in the *overall* kinetic and potential energies of systems. If we are studying the chemical reactions inside an automobile engine, for example, we do not care whether the automobile is moving or is stopped (that is, has high or low total kinetic energy), nor do we care whether it is at the top or the bottom of a hill (that is, has high or low total potential energy). The forms of energy we care about involve the *internal* kinetic and potential energies of systems: how fast their molecules are moving *relative* to the container that encloses them and how they interact with each other and the walls. We therefore focus our discussion exclusively on changes in the **internal energy** of systems, which we symbolize ΔE.

Both heat and work are means in which energy is transferred into and out of systems. If a change in internal energy is caused by the operation of a *mechanical* contact between a system and its surroundings, then *work* is done. If the change in internal energy is caused by the establishment of a *thermal* contact (leading to an equalization of temperature differences between system and surroundings), then *heat* is transferred. In many processes, both heat and work cross the boundary of the

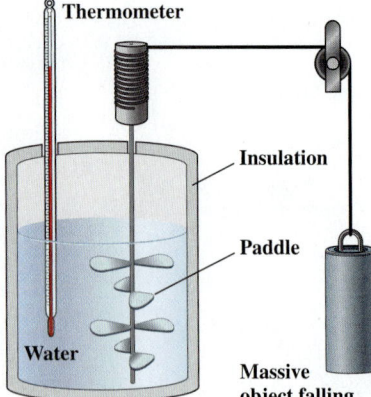

Thermometer

Insulation

Paddle

Water

Massive object falling

Figure 10–13 • As the mass falls, it turns a paddle that does work on the system (the water), causing an increase in its temperature.

system, and the change in the internal energy is the sum of the two contributions. This statement, called the first law of thermodynamics, takes the mathematical form[2]

$$\Delta E = q + w$$

We do not speak of systems as containing work or containing heat because work and heat become manifest only during *processes* that take systems from one state into another.

In the 1840s, the English physicist James Joule showed that the temperature of water could be increased by doing work on it. Figure 10–13 shows an apparatus in which a paddle, driven by a falling weight, churns the water in an insulated tank. Work is performed on the water, and its temperature increases. The work done is $-mg\Delta h$, where Δh is the (negative) change in the height of the weight and m is its mass. If all of this work goes to increase the temperature of the water (that is, if none is lost to friction in the mechanism), then the heat capacity of the water can be found. Two easily visualized paths connect the same initial state (cold water) and final state (warm water). On one of these paths (simple heating), heat is transferred to the system, but not work. On the second path (Joule's experiment), work is transferred to the system, but not heat. Additional imaginable paths involve various combinations of heat and work transfer. Because q and w depend on the particular process (or path) connecting the states, they are not state properties. Their sum ($\Delta E = q + w$), however, is independent of path, so that internal energy is a state property.

> The fundamental physical content of the first law of thermodynamics is the observation that q and w depend individually on the path followed between a given pair of states, but their sum does not.

In any process, the heat *added to* the system is *removed from* the surroundings, so that

$$q_{sys} = -q_{surr}$$

In the same way, the work done on the system is done by the surroundings, so that

$$w_{sys} = -w_{surr}$$

If we add these two, we conclude that

$$\Delta E_{sys} = -\Delta E_{surr}$$

so that the energy changes of system and surroundings have the same magnitude but opposite signs. The energy change of the universe (system plus surroundings) is then

$$\Delta E_{univ} = \Delta E_{sys} + \Delta E_{surr} = 0$$

---

[2]In some books, especially older ones, work is defined as positive when it is done by the system. This convention, the reverse of current recommended practice, arose because in many engineering applications the focus is the work done by a particular heat engine. Following this convention, the work is given by

$$w = P_{ext}\Delta V \quad \text{(older convention)}$$

and the first law reads

$$\Delta E = q - w \quad \text{(older convention)}$$

This convention is not used in this text. When using other books, take care to check which convention is employed.

In any process, the total energy of the universe remains unchanged: Energy is conserved.

This is another statement of the first law of thermodynamics.

Enthalpy and Energy

As long as the volume of a system is held constant, no pressure–volume work can be performed either *on* or *by* the system, as we have seen. We can write the first law of thermodynamics for the constant-volume case as

$$\Delta E = q + w = q + 0 \qquad \text{(at constant volume)}$$
$$\Delta E = q_V \qquad \text{(constant volume)}$$

We have already stated that

$$\Delta H = q_P \qquad \text{(constant pressure)}$$

• The subscript "V" emphasizes that the volume of the system is constant during the change. The pressure may change. The subscript "P" means that the pressure of the system is constant, but its volume may change.

The similarity of these two equations provokes immediate questions about the connection between enthalpy and internal energy. In fact, the enthalpy is defined as

$$H = E + PV$$

To see that this definition is consistent with the previous relations, consider an enthalpy change ΔH. According to the definition, that change is given by

$$\Delta H = \Delta E + \Delta(PV)$$

If the pressure is a constant external pressure (such as that exerted by the atmosphere), this can be written as

$$\Delta H = \Delta E + P_{ext}\Delta V \qquad \text{(constant pressure)}$$

If we substitute the first-law relation for ΔE, we find

$$\Delta H = (q_P + w) + P_{ext}\Delta V \qquad \text{(constant pressure)}$$

• Delta stands for a difference. Therefore $\Delta(PV) = P_2V_2 - P_1V_1$. Replacing P_2 and P_1 in this equation with P_{ext} and then factoring the P_{ext} out gives $\Delta(PV) = P_{ext}(V_2 - V_1) = P_{ext}\Delta V$.

The discussion of pressure–volume work develops the equation $w = -P_{ext}\Delta V$. Substitution gives

$$\Delta H = q_P - P_{ext}\Delta V + P_{ext}\Delta V = q_P \qquad \text{(constant pressure)}$$

The definition of enthalpy ensures that it is a state property and that heat transfers that are measured under conditions of constant pressure give enthalpy changes.

Thermochemistry experiments carried out at constant volume are best described using the internal energy because changes in internal energy equal the amount of heat gained at constant volume. Experiments at constant pressure are easier to describe using enthalpy because changes in enthalpy equal the amount of heat gained at constant pressure. Because constant-pressure conditions predominate in chemistry, the enthalpy is the more useful property for chemists.

A good estimate of the internal energy of a reaction (ΔE_r) can be obtained from the enthalpy of the reaction (ΔH_r) by making a few very reasonable assumptions. Suppose that the reaction is

$$C(s, \text{graphite}) + \tfrac{1}{2}O_2(g) \longrightarrow CO(g) \qquad \Delta H_r = -110.5 \text{ kJ mol}^{-1}$$

occurring at 1 atm and 25°C. We rearrange the general equation

$$\Delta H = \Delta E + \Delta(PV)$$

CHEMISTRY IN YOUR LIFE

Food Calorimetry

Your body is fueled by the metabolic oxidation of the carbohydrates, fats, and proteins that make up your food. "Calorie count" data on different foods derive directly or indirectly from calorimetric measurements. A portion of food (perhaps even an entire hamburger) or an edible chemical compound is sealed in a bomb calorimeter (see Fig. 10–4), and oxygen is pumped in to a pressure of perhaps 30 atm. Combustion is initiated with a spark, and the food burns. The quantity of heat that is generated is computed from the change in the temperature and the known heat capacities of the water and other materials surrounding the reaction. For example, the complete combustion of 1.00 mol of the carbohydrate sucrose (table sugar) liberates 5640 kJ in a bomb calorimeter:

$$C_{12}H_{22}O_{11}(s) + 12\ O_2(g) \longrightarrow$$
$$12\ CO_2(g) + 11\ H_2O(\ell) \qquad \Delta E_r^\circ = -5640\ \text{kJ mol}^{-1}$$

The ΔH_r° of this reaction also equals -5640 kJ mol^{-1} because $\Delta n_g = 0$.

The Calorie (1 Calorie = 1000 calories = 4.184 kJ) is the traditional unit of food energy in the United States, and food is commonly sold by mass. Conversion from kJ mol^{-1} to Calories per gram of sucrose is straightforward:

$$\Delta H^\circ = \left(\frac{-5640\ \text{kJ}}{\text{mol reaction}}\right) \times \left(\frac{1\ \text{mol reaction}}{1\ \text{mol sucrose}}\right)$$
$$\times \left(\frac{1\ \text{mol sucrose}}{342.3\ \text{g sucrose}}\right) \times \left(\frac{1\ \text{Calorie}}{4.184\ \text{kJ}}\right)$$
$$= \frac{-3.94\ \text{Calorie}}{\text{g sucrose}}$$

This is a specific enthalpy because it is on a per-unit-mass basis.

Calorimetry experiments have established that the specific enthalpies of combustion of carbohydrates vary only about 5% to 10% from one compound to the next. The carbohydrate starch, a very common polymeric food (see Fig. 25–19a) that has properties quite different from those of sucrose, liberates 4.19 Calories per gram when burned in a bomb calorimeter. This is only a 6% difference from sucrose. Food carbohydrates as a group release an average of 4.15 Calories per gram when burned to carbon dioxide and liquid water at a constant pressure of 1 atm at or near room temperature.

Calorimetry has established the specific enthalpies of combustion of numerous fats and proteins as well. They are also approximately constant:

- Fats of all kinds liberate approximately 9.45 Calories per gram
- Proteins liberate approximately 5.65 Calories per gram

Your body derives less energy than this from the carbohydrates, fats, and proteins that you eat. There are two reasons. First, the gastrointestinal tract is slightly inefficient in absorbing food. Carbohydrate absorption in a healthy adult is estimated at 97%, fat absorption at 95%, and protein absorption at 92%, on the basis of calorimetric measurements on fecal matter. Second, although the human metabolism converts absorbed carbohydrates and fats essentially completely to carbon dioxide and wa-

to obtain

$$\Delta E = \Delta H - \Delta(PV)$$

This equation is a concise rendering of

$$(E_{\text{products}} - E_{\text{reactants}}) = (H_{\text{products}} - H_{\text{reactants}}) - (P_{\text{products}}V_{\text{products}} - P_{\text{reactants}}V_{\text{reactants}})$$

The verbose version emphasizes that the difference is between the products and the reactants. Graphite is a solid. Its volume on the reactant side is small compared with the volume of the $O_2(g)$ and may reasonably be ignored in computing $\Delta(PV)$. If we then assume that the $O_2(g)$ and $CO_2(g)$ obey the ideal-gas law, we can write

$$\Delta(PV) \approx \Delta(nRT) = RT\Delta n_g$$

because the temperature is constant. Here, Δn_g is the change in the chemical amount of gases per mole of reaction

$$\Delta n_g = n_{\text{product gases}} - n_{\text{reactant gases}} = 1 - \tfrac{1}{2} = \tfrac{1}{2}$$

Nutrition Facts

Serving Size 1 Container (227g)

Amount Per Serving

Calories 240   Calories from Fat 25

% Daily Value*

Total Fat 3g	5%
Saturated Fat 1.5g	8%
Cholesterol 15mg	5%
Sodium 150mg	6%
Potassium 450mg	13%
Total Carbohydrate 45g	15%
Dietary fiber 1g	4%
Sugars 42g	
Protein 9g	

Vitamin A 2%	•	Vitamin C 6%
Calcium 35%	•	Iron 0%

*Percent Daily Values are based on a 2,000 calorie diet.

KEEP REFRIGERATED

Figure 10–A • Facts about yogurt.

ter, it does not convert protein to the same products (namely, $CO_2(g)$, $H_2O(\ell)$, $N_2(g)$, and $H_2SO_4(aq)$) that form when protein is combusted in a bomb calorimeter. The body discards compounds that have caloric value, such as urea and creatinine, in the urine. Calorimetric measurements on these wastes show that the enthalpy change in protein metabolism in healthy human beings equals only 77% of the enthalpy change in protein combustion. The average effective calorie values of the three categories of food thus come out to be

Carbohydrates $0.97 \times 4.15 = 4.0$ Calories g^{-1}

Fats $0.95 \times 9.45 = 9.0$ Calories g^{-1}

Proteins $0.92 \times 0.77 \times 5.65$
      $= 4.0$ Calories g^{-1}

Figure 10–A shows a "Nutritional Facts" panel from the side of container of yogurt. The data were obtained by analyzing the yogurt for carbohydrates, fats, and proteins. The fats, for example, were extracted from the aqueous yogurt mixture into diethyl ether or a similar solvent, as discussed in Section 7–7, and then weighed separately. One 227 g (0.5 lb) serving was found to contain 45 g of carbohydrates, 3 g of fat, and 9 g of protein (the rest was water and other non-digestibles). The effective Calorie content in human consumption of the serving of yogurt is then

$$45 \text{ g carbohydrate}\left(\frac{4.0 \text{ Calorie}}{\text{g carbohydrate}}\right)$$

$$+ 3 \text{ g fat}\left(\frac{9.0 \text{ Calorie}}{\text{g fat}}\right)$$

$$+ 9 \text{ g protein}\left(\frac{4.0 \text{ Calorie}}{\text{g protein}}\right) = 243 \text{ Calories}$$

This rounds off to 240 Calories, the value stated on the information panel.

Inserting the temperature, which is 298.15 K (25°C), gives

$$\Delta(PV) \approx RT\Delta n_{\text{g}} = \left(\frac{8.314 \text{ J}}{\text{K mol}}\right) \times 298.15 \text{ K} \times \tfrac{1}{2} = 1.24 \times 10^3 \text{ J mol}^{-1}$$

$$= 1.24 \text{ kJ mol}^{-1}$$

Now,

$$\Delta E_{\text{r}} = \Delta H_{\text{r}} - \Delta(PV)$$

$$\Delta E_{\text{r}} \approx -110.5 \text{ kJ mol}^{-1} - 1.24 \text{ kJ mol}^{-1} = -111.7 \text{ kJ mol}^{-1}$$

Note that R must be expressed in $J K^{-1} \text{mol}^{-1}$ if a result in joules mol^{-1} is desired. For reactions involving only liquids and solids, or reactions in which the chemical amount of gas does not change (those for which $\Delta n_{\text{g}} = 0$), the enthalpy of reaction ΔH_{r} and the internal energy of reaction ΔE_{r} are so nearly equal that the difference can be neglected. Even when the chemical amount of gas changes (as in this example), the $\Delta(PV)$ term is usually small compared with ΔH_{r}.

EXAMPLE 10-12

Calculate the internal energy of vaporization of CCl_4 at its normal boiling point, 349.9 K, and compare it with the enthalpy of vaporization at the same temperature (determined in Example 10–6).

Solution

This phase change can be described by the equation

$$CCl_4(\ell) \longrightarrow CCl_4(g)$$

Write the relationship

$$\Delta E_{vap} = \Delta H_{vap} - \Delta(PV) \approx \Delta H_{vap} - RT\Delta n_g$$

Here the change in the number of moles of gas per mole of reaction is $\Delta n_g = 1$ because 1 mol of gaseous products forms per mole of reaction and the reactant is a liquid. Inserting $T = 349.9$ K, $\Delta H_{vap} = 30.0$ kJ mol^{-1}, and $\Delta n_g = 1$ gives

$$\Delta E_{vap} \approx 30.0 \text{ kJ mol}^{-1} - \left(\frac{8.314 \text{ J}}{\text{K mol}}\right) \times \left(\frac{10^{-3} \text{ kJ}}{1 \text{ J}}\right) \times 349.9 \text{ K} \times 1$$

$$\Delta E_{vap} \approx (30.0 - 2.9) \text{ kJ mol}^{-1} = 27.1 \text{ kJ mol}^{-1}$$

Of 30.0 kJ mol^{-1} of thermal energy transferred from the surroundings to vaporize $CCl_4(\ell)$, 27.1 kJ mol^{-1} goes to increase the internal energy of the system (ΔE_{vap}), and 2.9 kJ mol^{-1} goes to expand the vapor against the pressure exerted by the surroundings ($\Delta(PV)$).

EXERCISE

Compute the internal energy of vaporization of water, and compare it with the enthalpy of vaporization from Table 10–2. Water boils at 373.15 K.

Answer: $\Delta E_{vap} \approx 40.66$ kJ $mol^{-1} - 3.10$ kJ $mol^{-1} = 37.56$ kJ mol^{-1}; ΔE is about 8% less than ΔH.

SUMMARY

10-1 **Thermochemistry** is the study of the heat effects associated with chemical reactions. **Heat** is a form of motion at the molecular level by which energy is transferred between materials. The units for heat flow are the same as the units of energy; important units are the joule, kilojoule, and calorie.

10-2 **Calorimetry** is the quantitative study of heat: the measurement of how much heat is evolved or absorbed in chemical reactions or phase changes, or changes in temperature. The **heat capacity** of any body is the amount of heat needed to raise its temperature by one unit of temperature. The **molar heat capacity** equals the heat capacity per mole of substance; the **specific heat capacity** equals the heat capacity per unit mass of material. These quantities are used in the calculation of the temperature change caused by flow of a quantity of heat to or from a body and the determination of the final temperature reached when two bodies at different temperatures are placed in contact with one another and insulated from their surroundings.

10-3 When heat is transferred to an object at constant pressure, the amount absorbed is equal to the change in the **enthalpy** ΔH of that object. Enthalpy is a **state property.**

It depends only on the state of the system and not on the path followed to reach that state. Reactions and other processes in which heat is evolved at constant pressure are **exothermic,** and ΔH is negative. **Endothermic** reactions absorb heat from the surroundings (at constant pressure) and have a positive ΔH. Enthalpy changes can be treated using the principles of stoichiometry and are proportional to the amount of substance reacting. **Hess's law** states that when several chemical equations are added or subtracted to give an overall equation, addition or subtraction of the corresponding enthalpy changes gives the enthalpy change associated with the overall equation.

10-4   Use of thermodynamic **standard states** allows meaningful comparison of observed enthalpy changes in different reactions. A **standard molar enthalpy of formation** (ΔH_f°) of a species equals the enthalpy change when 1 mol of a species is formed in a standard state at a specified temperature from the most stable forms of the elements that constitute it, in standard states at the same temperature. Tables of standard molar enthalpies of formation allow for the calculation of standard enthalpy of reactions (ΔH_r°'s) with the use of Hess's law.

10-5   Average **bond enthalpies** can be used to estimate the standard enthalpies (ΔH_r°'s) of gas-phase reactions.

10-6   In classical mechanics, **work** done on a body increases its energy. This energy can take many forms, such as kinetic energy (as when an object is accelerated to a greater velocity by a force acting on it) or potential energy (as when an object is lifted in a gravitational field). One important type of work in chemistry is **pressure–volume work,** the expansion of a system against, or its compression by, an external pressure. The **first law of thermodynamics** states that the change in **internal energy** of a system is the sum of the work done on it and the heat transferred to it. Both heat and work depend on the way the change occurs and thus are not functions of state, but their sum depends only on the initial and final states of the system. The first law is equivalent to the statement that the total energy of the universe remains constant because energy changes in a system are compensated for by equal and opposite changes in the surroundings: Energy is conserved. Enthalpy is defined by the equation $H = E + PV$, but only changes in enthalpy can be measured or calculated. This makes $\Delta H = \Delta E + \Delta(PV)$ a more useful form of the definition. For reactions involving only solids and liquids, ΔH and ΔE are almost equal. For reactions involving gases, ΔH exceeds ΔE by $RT\Delta n_g$ (at constant temperature). The quantity Δn_g is the change in the number of moles of gas per mole of reaction.

PROBLEMS

Note: Answers to blue-numbered problems are given in Appendix F. Problems that are more challenging are indicated by asterisks.

Heat

1. While using an electric drill to bore a hole in a board, you smell burning wood and see a plume of smoke coming from the hole. Where does the heat to start this fire come from?

2. Petroleum deposits in porous rocks under the Earth's surface currently supply the fuel for the heating of many homes and other buildings. The petroleum is thought to have come from decaying organic matter. Can the heat from oil be traced back to the Sun? Explain.

3. A report published on the World Wide Web about a proposed new engine states that at 298.15 K, the heat content of the engine block (consisting of 93 kg of steel and aluminum) is 11,000 MJ. Criticize this statement.

4. When hot metal is plunged into cold water, the metal cools down and the water heats up. Explain this fact in terms of the kinetic molecular theory.

Calorimetry

5. (See Example 10–1.) Rank the following from lowest heat capacity to highest: 1 kg water, 2 kg water, a 1 kg iron ball, a 2 kg iron plate.

6. (See Example 10–1.) Before the days of electricity, flatirons literally were made of iron. They were heated on a stove and had to be quite heavy, despite the fact that anyone could show that it was the heat and not the weight of a flatiron that took the wrinkles out of clothes. Explain why flatirons had to be heavy.

7. The specific heat capacities of Li(s), Na(s), K(s), Rb(s), and Cs(s) at 25°C are 3.57, 1.23, 0.756, 0.363, and 0.242 J K^{-1} g^{-1}, respectively. Compute the molar heat capacities of these elements, and identify any periodic trend. If there is a trend, use it to predict the molar heat capacity of Fr(s).

8. The specific heat capacities of $F_2(g)$, $Cl_2(g)$, $Br_2(g)$, and $I_2(g)$ are 0.824, 0.478, 0.225, and 0.145 J K^{-1} g^{-1}, respectively. Compute the molar heat capacities of these elements and identify any periodic trend. If there is a trend, use it to predict the molar heat capacity of $At_2(g)$.

9. The specific heat capacities of the elements nickel, zinc, rhodium, tungsten, and gold at 25°C are 0.444, 0.388, 0.243, 0.132, and 0.129 J K^{-1} g^{-1}, respectively. Calculate the molar heat capacities of these six metals. (*Note:* The rule of Dulong and Petit states that the molar heat capacities of the metallic elements all equal approximately 25 J K^{-1} mol^{-1}.)

10. Use the rule of Dulong and Petit stated in Problem 9 to estimate the specific heat capacities of vanadium, gallium, and silver.

11. The specific heat capacity of white phosphorus, $P_4(s)$, at 25°C is 0.757 J K^{-1} g^{-1}. Calculate the molar heat capacity of white phosphorus at this temperature.

12. The molar heat capacity of uranium at 25°C is 27.49 J K^{-1} mol^{-1}. Calculate the specific heat capacity of uranium at 25°C.

13. (See Example 10–2.) A piece of zinc at 20.0°C with a mass of 60.0 g is dropped into 200.0 g of water at 100.0°C. Calculate the temperature of thermal equilibrium of this mixture, assuming no heat losses to the container or other surroundings. The specific heat capacity of zinc is 0.389 J K^{-1} g^{-1}, and that of water near 100°C is 4.22 J K^{-1} g^{-1}.

14. (See Example 10–2.) Iron pellets weighing 17.0 g at a temperature of 92.0°C are mixed in an insulated container with 17.0 g of water at a temperature of 20.0°C. The specific heat capacity of water is 9.31 times greater than the specific heat capacity of iron. Compute the final temperature of this mixture. Assume that heat losses to and gains from the container and other surroundings are negligible.

15. (See Example 10–3.) A new Styrofoam-cup calorimeter has been constructed.

   (a) The temperature inside the calorimeter increases by 1.67 K when 1770 J of heat is added to 200.0 g of water contained in this calorimeter. Calculate the calorimeter constant.

   (b) When some NaOH(s) is dissolved in 225.0 g water inside this calorimeter, the temperature rises by 2.64 K. Calculate the amount of heat evolved in the dissolution of the sodium hydroxide.

16. (See Example 10–3.) A student puts 100.0 g of water at 20.0°C into a calorimeter (also at 20.0°C) that has a calorimeter constant of 117 J K^{-1}. Determine the mass of 60.0°C water that must be added to bring the temperature inside the calorimeter to 40.0°C. Assume that no heat leaks. Take data from Table 10–1 as necessary.

17. (See Example 10–3.) (a) An electric current passed through the resistance coil inside the bomb calorimeter diagrammed in Figure 10–4 generates 5682 J of heat. The temperature indicated by the can thermometer rises by 2.31°C, and the temperature indicated by the jacket remains unchanged. Calculate the calorimeter constant of this calorimeter. (b) When some benzene ($C_6H_6(\ell)$) is burned in oxygen in this calorimeter, the temperature indicated by the can thermometer rises by 4.40°C, again with no change in the temperature indicated by the jacket thermometer. Calculate the heat evolved in this combustion of benzene. Assume that the heat absorbed by the products of the reaction is negligible.

18. Water can be heated in tightly woven baskets by dropping in hot rocks. The technique is useful for cooking in the absence of metal or pottery vessels. Estimate the size (in liters) of basket needed to hold the rocks and water if 5 L of boiling-hot (100°C) water is required. The water comes from a well at 10°C, and the rocks (granite with a density of 2.7 g cm^{-3}) come from a hot fire at 900°C. Take data from Table 10–1. List the assumptions required to make the estimate.

Enthalpy

19. (See Example 10–4.) When slaked lime is added to water in the production of cement, the reaction

$$Ca(OH)_2(s) \longrightarrow Ca^{2+}(aq) + 2\,OH^-(aq)$$

takes place, causing the temperature to increase. Determine the sign of the enthalpy change ΔH of this reaction.

20. (See Example 10–4.) A commercial hand-warmer (see Fig. 9–3) consists of a plastic pouch containing a chemical dissolved in water. A small piece of flexible metal is also in the pouch. When the metal piece is flexed, crystals start to form near it and soon spread throughout the solution. At the same time, the pouch and its contents become pleasantly warm. Determine the sign of the enthalpy change ΔH for the process taking place in the pouch at room temperature.

21. (See Example 10–5.) Liquid bromine dissolves readily in aqueous NaOH:

$$Br_2(\ell) + 2\,NaOH(aq) \longrightarrow$$
$$NaBr(aq) + NaOBr(aq) + H_2O(\ell)$$

Suppose that 2.88×10^{-3} mol of $Br_2(\ell)$ is sealed in a glass capsule that is then immersed in a solution containing excess NaOH(aq). The capsule is broken, the mixture is stirred, and a measured 121.3 J of heat evolves. In a separate experiment, simply breaking an empty capsule and stirring the solution in the same way evolves 2.34 J of heat. Compute the enthalpy change when 1.00 mol of $Br_2(\ell)$ dissolves in excess NaOH(aq).

22. (See Example 10–5.) A chemist mixes 1.00 g of $CuCl_2$ with an excess of $(NH_4)_2HPO_4$ in dilute aqueous solution. He measures the evolution of 670 J of heat as the two substances react to give $Cu_3(PO_4)_2(s)$. Compute ΔH for the reaction of 1.00 mol of $CuCl_2$ with an excess of $(NH_4)_2HPO_4$.

23. (See Example 10–5.) Calculate the enthalpy change when 1.00 g of the underlined substance is consumed or produced in the following reactions.
    (a) $4\,Na(s) + O_2(g) \longrightarrow \underline{2\,Na_2O(s)}$
    $$\Delta H = -828 \text{ kJ mol}^{-1}$$
    (b) $CaMg(CO_3)_2(s) \longrightarrow CaO(s) + \underline{MgO(s)} + 2\,CO_2(g)$
    $$\Delta H = +302 \text{ kJ mol}^{-1}$$
    (c) $H_2(g) + \underline{2\,CO(g)} \longrightarrow H_2O_2(\ell) + 2\,C(s)$
    $$\Delta H = +33.3 \text{ kJ mol}^{-1}$$

24. (See Example 10–5.) Calculate the enthalpy change when 1.00 g of the underlined substance is consumed or produced in the following reactions.
    (a) $Ca(s) + \underline{Br_2(\ell)} \longrightarrow CaBr_2(s)$
    $$\Delta H = -683 \text{ kJ mol}^{-1}$$
    (b) $6\,Fe_2O_3(s) \longrightarrow \underline{4\,Fe_3O_4(s)} + O_2(g)$
    $$\Delta H = +472 \text{ kJ mol}^{-1}$$
    (c) $2\,\underline{NaHSO_4(s)} \longrightarrow 2\,NaOH(s) + 2\,SO_2(g) + O_2(g)$
    $$\Delta H = +806 \text{ kJ mol}^{-1}$$

25. Predict the sign of the enthalpy change for the following processes at constant pressure:
    (a) $H_2O(\ell) \longrightarrow H_2O(g)$
    (b) $Cl_2(g) \longrightarrow Cl_2(s)$
    (c) $Hg(\ell) \longrightarrow Hg(s)$

26. Predict the sign of the enthalpy change for the following processes at constant pressure:
    (a) $CO_2(s) \longrightarrow CO_2(g)$
    (b) $OsO_4(\ell) \longrightarrow OsO_4(g)$
    (c) $BaCl_2(s) \longrightarrow BaCl_2(\ell)$

27. (See Example 10–6.) The vaporization of 10.00 g of $CBr_4(\ell)$ (carbon tetrabromide) requires 1.359 kJ at its normal boiling point and a pressure of 1 atm. Calculate ΔH for the vaporization of 2.500 mol of $CBr_4(\ell)$.

28. (See Example 10–6.) The condensation of 1.00 g of methane (CH_4) to liquid methane at its normal boiling point and a pressure of 1 atm generates 555 J.
    (a) Calculate ΔH for the condensation of 2.00 mol of $CH_4(g)$ to $CH_4(\ell)$.
    (b) Calculate ΔH for the vaporization of 2.00 mol of $CH_4(\ell)$.

29. (See Example 10–6.) Calculate the enthalpy change when 2.38 g of carbon monoxide (CO) vaporizes at its normal boiling point. Use data from Table 10–2.

30. (See Example 10–6.) Molten sodium chloride is used for making elemental sodium and chlorine. Suppose that the electrical power to heat a vat containing 56.2 kg of molten sodium chloride is cut off and the salt crystallizes (without changing its temperature). Calculate the enthalpy change, using data from Table 10–2.

31. (See Example 10–7.) When solid cesium ($Cs(s)$) reacts with $O_2(g)$ to give $CsO_2(s)$, the ΔH equals -266.1 kJ mol^{-1}.

When *liquid* cesium ($Cs(\ell)$) reacts with $O_2(g)$ to give the same product, ΔH equals -268.8 kJ mol^{-1}. Determine ΔH for the process $Cs(\ell) \longrightarrow Cs(s)$.

32. (See Example 10–7.) The following data are found in a reference book:

    $$P_4(g) + 5\,O_2(g) \longrightarrow P_4O_{10}(s) \quad \Delta H = -3035.7 \text{ kJ mol}^{-1}$$
    $$4\,P(\ell) + 5\,O_2(g) \longrightarrow P_4O_{10}(s) \quad \Delta H = -2977.0 \text{ kJ mol}^{-1}$$

    Determine ΔH for the process $4\,P(\ell) \longrightarrow P_4(g)$.

33. (See Example 10–7.) The enthalpies of the following reactions were measured calorimetrically at 25°C and 1 atm with the indicated results.

    $$CH_2CO(g) + 2\,O_2(g) \longrightarrow 2\,CO_2(g) + H_2O(g)$$
    $$\Delta H = -981.1 \text{ kJ mol}^{-1}$$
    $$CH_4(g) + 2\,O_2(g) \longrightarrow CO_2(g) + 2\,H_2O(g)$$
    $$\Delta H = -802.3 \text{ kJ mol}^{-1}$$

    Calculate the enthalpy change at 25°C and 1 atm for the reaction

    $$2\,CH_4(g) + 2\,O_2(g) \longrightarrow CH_2CO(g) + 3\,H_2O(g)$$

34. (See Example 10–7.) Given the following two reactions and corresponding enthalpy changes,

    $$CO(g) + SiO_2(s) \longrightarrow SiO(g) + CO_2(g)$$
    $$\Delta H = +520.9 \text{ kJ mol}^{-1}$$
    $$8\,CO_2(g) + Si_3N_4(s) \longrightarrow$$
    $$3\,SiO_2(s) + 2\,N_2O(g) + 8\,CO(g)$$
    $$\Delta H = +461.05 \text{ kJ mol}^{-1}$$

    compute the ΔH of the reaction

    $$5\,CO_2(g) + Si_3N_4(s) \longrightarrow$$
    $$3\,SiO(g) + 2\,N_2O(g) + 5\,CO(g)$$

35. The enthalpy of a mole of diamonds exceeds the enthalpy of a mole of graphite. Which gives off more heat when burned— a pound of diamonds or a pound of graphite? Explain.

36. Under certain conditions, the enthalpy change during the combustion of monoclinic sulfur to $SO_2(g)$ is -9.376 kJ g^{-1}. Under the same conditions, the enthalpy change of combustion of the rhombic form of sulfur to $SO_2(g)$ is -9.293 kJ g^{-1}. Compute the ΔH for the reaction

    $$S(s,\ monoclinic) \longrightarrow S(s,\ rhombic)$$

Standard Enthalpies of Reaction

37. (See Example 10–8.) Calculate the standard enthalpy (ΔH_r°) at 25°C of the reaction

    $$N_2H_4(\ell) + 3\,O_2(g) \longrightarrow 2\,NO_2(g) + 2\,H_2O(\ell)$$

    using the standard molar enthalpies of formation (ΔH_f°) of reactants and products listed in Appendix D.

38. (See Example 10–8.) Calculate ΔH_r°'s for each of the following at 25°C. Use data from Appendix D.
    (a) $4\,NO(g) + 2\,O_2(g) \longrightarrow 4\,NO_2(g)$
    (b) $C(s) + CO_2(g) \longrightarrow 2\,CO(g)$

(c) $2 NH_3(g) + \frac{7}{2} O_2(g) \longrightarrow 2 NO_2(g) + 3 H_2O(g)$
(d) $C(s) + H_2O(g) \longrightarrow CO(g) + H_2(g)$

39. (See Example 10–8.) Zinc is found in nature in the form of the mineral sphalerite (ZnS). A step in the smelting of zinc is the roasting of sphalerite with oxygen to produce zinc oxide:

$$2 ZnS(s) + 3 O_2(g) \longrightarrow 2 ZnO(s) + 2 SO_2(g)$$

(a) Calculate the standard enthalpy of this reaction (ΔH_r°), using data from Appendix D.
(b) Estimate the heat absorbed when 3.00 metric tons (1 metric ton = 10^3 kg) of sphalerite is roasted at high temperature and a pressure of 1 atm.
(c) The correct answer to part (b) differs from the estimate. Explain why.

40. (See Example 10–8.) The thermite process (see Fig. 10–1) is used for welding steel rails. In this reaction, aluminum reduces iron(III) oxide to metallic iron:

$$2 Al(s) + Fe_2O_3(s) \longrightarrow 2 Fe(s) + Al_2O_3(s)$$

Burning a small charge of barium peroxide mixed with aluminum ignites the reaction of a mixture of aluminum powder and iron(III) oxide. Molten iron is produced, flows into the space between the rail ends, and solidifies.
(a) Calculate the standard enthalpy of this reaction, using data from Appendix D.
(b) Calculate the heat given off to the surroundings when 3.21 g of iron(III) oxide is reduced by aluminum at a constant pressure of 1 atm and the products cool down to room temperature (25°C).

41. The dissolution of calcium chloride in water

$$CaCl_2(s) \longrightarrow Ca^{2+}(aq) + 2 Cl^-(aq)$$

is used in first-aid hot packs. In these packs, an inner pouch containing the salt is broken, allowing the salt to dissolve in the surrounding water.
(a) Calculate the standard enthalpy of this reaction, using data from Appendix D.
(b) Suppose that 11.1 g of $CaCl_2$ is dissolved in 0.100 L of water at 20.0°C. Estimate the maximum temperature that can be reached inside the pack. Assume that the solution is an ideal solution with a heat capacity close to that of 100 g of pure water (418 J K^{-1}).

42. Ammonium nitrate dissolves in water according to the reaction

$$NH_4NO_3(s) \longrightarrow NH_4^+(aq) + NO_3^-(aq)$$

(a) Calculate the standard enthalpy of this dissolution using data from Appendix D.
(b) Suppose that 10.0 g of NH_4NO_3 is dissolved in 0.100 L of water at 20.0°C. Estimate the minimum temperature that can be reached by the solution. Assume that the solution is ideal and has a heat capacity of 418 J K^{-1}.
(c) From a comparison with the results of Problem 41, can you suggest a practical application of this dissolution reaction?

43. (See Example 10–9.) At 25°C the standard enthalpy of combustion (to $CO_2(g)$ and $H_2O(\ell)$) of the organic liquid cyclohexane ($C_6H_{12}(\ell)$) is -3923.7 kJ mol^{-1}. Determine the ΔH_f° of $C_6H_{12}(\ell)$. Use data from Appendix D.

44. (See Example 10–9.) At 25°C the standard enthalpy of combustion (to $CO_2(g)$ and $H_2O(\ell)$) of the organic liquid cyclohexene ($C_6H_{10}(\ell)$) is -3731.7 kJ mol^{-1}. Determine the ΔH_f° of $C_6H_{10}(\ell)$.

45. The standard molar enthalpies of formation of liquid and gaseous chloroform ($CHCl_3$) at 25°C are -134.5 kJ mol^{-1} and -103.1 kJ mol^{-1}, respectively. Determine the standard enthalpy change in condensing 2.5 mol of chloroform at 25°C.

46. The standard molar enthalpies of formation of $UF_6(s)$ and $UF_6(g)$ at 25°C are -2197.0 kJ mol^{-1} and -2063.7 kJ mol^{-1}, respectively. Determine the standard enthalpy of sublimation (solid-to-gas transition) of 0.5610 mol of UF_6 at 25°C.

47. The standard molar enthalpy of formation of octane $C_8H_{18}(\ell)$ is -269.1 kJ mol^{-1}.
(a) Use this fact and data in Appendix D to confirm that the combustion of propane $C_3H_8(g)$ to water and carbon dioxide produces more energy per gram of fuel than the combustion of octane to the same products.
(b) Taking octane as a model for commercial gasoline, discuss some advantages and disadvantages of replacing gasoline with propane as automobile fuel.

48. The standard molar enthalpy of formation of octane ($C_8H_{18}(\ell)$) is -269.1 kJ mol^{-1}. Compare the standard specific enthalpy of the combustion of octane with the standard specific enthalpy of the combustion of $C_2H_5OH(\ell)$. A specific enthalpy of combustion is the enthalpy change per unit mass of the fuel. Taking octane as a model for commercial gasoline, decide whether gasohol (a mixture of ethanol and gasoline) is a higher-energy fuel on a mass basis than gasoline.

Bond Enthalpies

49. A second chlorofluorocarbon used as a refrigerant and in aerosols (besides that discussed in Example 10–10) is CCl_3F. Use the average bond enthalpies from Table 10–3 to estimate the enthalpy change when 1 mol of this compound is broken into separate atoms in the gas phase.

50. Because it decomposes more quickly in the atmosphere and is much less liable to reduce the concentration of ozone in the stratosphere, the compound CF_3CHCl_2 (with a C—C bond) is being used as a substitute for CCl_3F and CCl_2F_2. Use average bond enthalpies from Table 10–3 to estimate the enthalpy change when 1 mol of CF_3CHCl_2 is broken into atoms in the gas phase.

51. (See Example 10–10.) Use average bond enthalpies (Table 10–3) to estimate ΔH_r° of the following reactions:
(a) $H—C≡N(g) \longrightarrow H(g) + C(g) + N(g)$
(b) $H—C≡N(g) \longrightarrow \frac{1}{2} H_2(g) + C(g) + \frac{1}{2} N_2(g)$
(c) $H—C≡N(g) \longrightarrow H—N≡C(g)$

52. (See Example 10–10.) Use average bond enthalpies (Table 10–3) to estimate ΔH_r° of these reactions:
(a) $3 H_2(g) + N_2(g) \longrightarrow 2 NH_3(g)$
(b) $2 H_2(g) + N_2(g) \longrightarrow H_2N{-}NH_2(g)$
(c) $H_2(g) + N_2(g) \longrightarrow HN{=}NH(g)$

53. (See Example 10–10.) Propane (C_3H_8) has the structure $H_3C{-}CH_2{-}CH_3$.
(a) Write the Lewis structure for propane.
(b) Use average bond enthalpies from Table 10–3 to estimate the standard change in enthalpy (ΔH°) for the following reaction:

$$C_3H_8(g) + 5 O_2(g) \longrightarrow 3 CO_2(g) + 4 H_2O(g)$$

54. (See Example 10–10.) Use average bond enthalpies from Table 10–3 to estimate ΔH_r° in each case.
(a) $CH_4(g) + CH_3Cl(g) \longrightarrow C_2H_5Cl(g) + H_2(g)$
(b) $C_2H_4(g) + H_2(g) \longrightarrow C_2H_6(g)$
(*Hint*: Refer to Section 3–4 to draw Lewis structures for ethylene (C_2H_4) and ethane (C_2H_6). The structure of chloroethane (C_2H_5Cl) has a Cl atom in place of one of the H atoms in ethane (C_2H_6).)

55. The reaction

$$BBr_3(g) + BCl_3(g) \longrightarrow BBr_2Cl(g) + BCl_2Br(g)$$

has $\Delta H_r^\circ \approx 0$. Explain why. (*Hint*: Sketch the Lewis structures of the four compounds.)

56. At 381 K, the following reaction takes place:

$$Hg_2Cl_4(g) + Al_2Cl_6(g) \longrightarrow 2 HgAlCl_5(g)$$
$$\Delta H_r^\circ = +10 \text{ kJ mol}^{-1}$$

(a) Offer an explanation for the very small ΔH_r° in terms of the known structures of the compounds

(b) Explain why the small ΔH_r° is evidence against

as the structure of $Hg_2Cl_4(g)$.

57. Methane ($CH_4(g)$) burns exothermically in oxygen to give gaseous carbon dioxide and water vapor. By analogy, carbon tetrafluoride ($CF_4(g)$) might be expected to burn exothermically in oxygen to give gaseous carbon dioxide and oxygen difluoride ($OF_2(g)$). It does not. Use bond enthalpies from Table 10–3 to explain why.

58. The average enthalpy of B—O single bonds equals 523 kJ mol^{-1}, but the enthalpy of the bond in BO(g) equals 788 kJ mol^{-1}. Explain the high enthalpy of this bond. (*Hint*: Draw Lewis structures of this odd-electron molecule.)

The First Law of Thermodynamics

59. (See Example 10–11.) Some nitrogen is heated slowly, maintaining the external pressure close to the internal pressure of 50.0 atm, until its volume has increased from 542 L to 974 L. Calculate the work done on the nitrogen as it is heated (in joules).

60. (See Example 10–11.) The gas mixture inside one of the cylinders of an automobile engine expands against a constant external pressure of 0.98 atm, from an initial volume of 150 mL (at the end of the compression stroke) to a final volume of 800 mL. Calculate the work done on the gas mixture during this process (in joules).

61. (See Example 10–11.) A constant pressure of 1.14 atm is exerted on a gas. The gas is cooled and contracts from an initial volume of 4.00 L to a final volume of 0.40 L. Compute the work done on the gas during the contraction (in liter-atmospheres and in joules).

62. (See Example 10–11.) A 1.25 L sample of a gas is heated and expands against a constant pressure of 0.86 atm to a final volume of 3.75 L. Compute the work done on the gas during the expansion (in liter-atmospheres and in joules).

63. (See Example 10–11.) Suppose that 2.00 L of a gas is confined at a pressure of 1.00 atm. The external pressure is zero atmospheres. The gas is heated and expands to a final volume of 3.50 L. Determine the work done on the gas during the expansion.

64. (See Example 10–11.) A 7.00 L sample of a gas is confined in a strong, rigid vessel. It is heated so that its pressure increases from 1.50 atm to 407 atm. Determine the work done on the gas during this process.

*65. A chemical system that is not an ideal gas is sealed in a strong, rigid container at room temperature and then heated vigorously.
(a) State whether ΔE, q, and w of the system are positive, negative, or zero during the heating process.
(b) Next, the container is cooled to its original temperature. Determine the signs of ΔE, q, and w for the cooling process.
(c) Designate heating as step 1 and cooling as step 2. Determine the signs of $(\Delta E_1 + \Delta E_2)$, $(q_1 + q_2)$, and $(w_1 + w_2)$, if possible.

66. A battery harnesses a chemical reaction to extract energy in the form of useful electrical work.
(a) A certain battery runs a toy truck and becomes partially discharged. In the process, it performs a total of 117.0 J of work on its immediate surroundings. It also gives off 3.0 J of heat, which the surroundings absorb. No other work or heat is exchanged with the surroundings. Compute q, w, and ΔE of the battery, making sure each quantity has the proper sign.
(b) The same battery is now recharged exactly to its original condition. This requires 210.0 J of electrical work from an outside generator. Determine q for the battery in this process. Explain why q has the sign that it does.

67. Calculate the difference between the enthalpy and internal energy of 2.00 mol of $N_2(g)$ at 400 K. Assume that $N_2(g)$ is an ideal gas.

68. Trinitrotoluene (TNT) is a solid. When it explodes, it forms several gases:

$$2\ C_7H_5N_3O_6(s) \longrightarrow$$
$$3\ N_2(g) + 7\ CO(g) + 5\ H_2O(g) + 7\ C(s)$$

Compute the *difference* between ΔH_r° and ΔE_r° at 298 K.

69. A sample of pure solid naphthalene ($C_{10}H_8$) weighing 0.6410 g is burned completely with oxygen to $CO_2(g)$ and $H_2O(\ell)$ in a constant-volume calorimeter at 298.15 K. The amount of heat evolved equals 25.79 kJ.
  (a) Write and balance a chemical equation for the combustion reaction.
  (b) Calculate the standard change in internal energy (ΔE°) associated with this reaction at 298.15 K.
  (c) Calculate the standard enthalpy change (ΔH°) for the same reaction as in part (b).
  (d) Calculate the standard molar enthalpy of formation of naphthalene, using data for the standard enthalpies of formation of $CO_2(g)$ and $H_2O(\ell)$ from Appendix D.

70. A sample of solid benzoic acid (C_6H_5COOH) that has a mass of 0.800 g is burned in an excess of oxygen to $CO_2(g)$ and $H_2O(\ell)$ in a constant-volume calorimeter at 25°C. The temperature rise equals 2.15°C. The heat capacity of the calorimeter and its contents is known to be 9382 J K^{-1}.
  (a) Write and balance an equation for the combustion of benzoic acid.
  (b) Calculate the standard change in internal energy (ΔE°) associated with this reaction at 25°C.
  (c) Calculate the standard enthalpy change (ΔH°) for the same reaction as in part (b).
  (d) Calculate the standard molar enthalpy of formation of benzoic acid, using data for the standard enthalpies of formation of $CO_2(g)$ and $H_2O(\ell)$ from Appendix D.

Additional Problems

71. A reference book quotes the specific heat capacity (at constant pressure) of aluminum as 0.215 cal $K^{-1}\ g^{-1}$.
  (a) Determine the molar heat capacity of aluminum in J $K^{-1}\ mol^{-1}$.
  (b) Compute the heat capacity of 1000 kg (a metric ton) of aluminum in J K^{-1}.

72. At one time, it was thought that the molar mass of indium was approximately 76 g mol^{-1}. Use the rule of Dulong and Petit (Problem 9) to show how the specific heat capacity of metallic indium, 0.233 J $K^{-1}\ g^{-1}$, makes this molar mass unlikely.

73. Suppose 61.0 g of hot metal that is initially at 120.0°C is plunged into 100.0 g of water that is initially at 20.00°C. The metal cools down and the water heats up until they reach a common temperature of 26.39°C. No heat is lost to the surroundings. Calculate the specific heat capacity of the metal, using 4.18 J $K^{-1}\ g^{-1}$ as the specific heat capacity of the water.

74. In their *Memoir on Heat,* published in 1783, Lavoisier and LaPlace reported (in translation): "The heat necessary to melt ice is equal to three quarters of the heat that can raise the same mass of water from the temperature of the melting ice to that of boiling water." Use this 18th-century observation to compute the amount of heat (in joules) needed to melt 1.00 g of ice. Assume that heating 1.00 g of water requires 4.18 J of heat for each 1.00°C throughout the range from 0°C to 100°C.

75. When glucose, a sugar, reacts fully with oxygen, carbon dioxide and water are produced:

$$C_6H_{12}O_6(s) + 6\ O_2(g) \longrightarrow 6\ CO_2(g) + 6\ H_2O(\ell)$$
$$\Delta H^\circ = -2820\ kJ\ mol^{-1}$$

Suppose that a person weighing 50 kg (mostly water, with specific heat capacity 4.18 J $K^{-1}\ g^{-1}$) eats a candy bar containing 14.3 g of glucose. If all the glucose reacts with oxygen and the heat produced is used entirely to increase the person's body temperature, what temperature increase would be produced? (In fact, most of the heat produced is lost to the surroundings before such a temperature increase results.)

*76. In walking a kilometer, you expend 100 kJ of energy. Compare this to the energy expended in driving a car that gets 8.0 km L^{-1} of gasoline (19 miles per gallon) the same distance. The density of gasoline is 0.68 g cm^{-3}, and its standard enthalpy of combustion is -48 kJ g^{-1}.

*77. An ice cube weighing 36.0 g and having a temperature of 0.0°C is dropped into 360 g of water that has a temperature of 20.0°C. Calculate the final temperature that is reached by the mixture, assuming no heat loss to the surroundings. The enthalpy of fusion of ice is $\Delta H_{fus} = 333$ J g^{-1}, and the specific heat capacity of water is 4.18 J $K^{-1}\ g^{-1}$. (*Hint:* Consider first the heat absorbed as the ice melts and its effect on the temperature of the surrounding water. Then calculate the final temperature attained from the mixture of liquid water at two different temperatures.)

*78. You have a supply of ice at 0.0°C and a beaker containing 150 g of water at 25°C. The enthalpy of fusion of ice is $\Delta H_{fus} = 333$ J g^{-1}, and the specific heat capacity of water is 4.18 J $K^{-1}\ g^{-1}$. Assuming that no heat is lost to or gaining from the surroundings, compute the mass of ice (in grams) that must be added to the beaker (and melted) to reduce the temperature of the water to 0°C.

79. A given substance can have more than one standard state at 1 atm and 298.15 K. True or false? Explain.

80. Which is larger: H_{300}° of $H_2O(\ell)$ or H_{400}° of $H_2O(g)$? Explain.

81. The following table summarizes some properties of liquid helium and liquid nitrogen, both of which are used as coolants:

	He(ℓ)	N$_2$(ℓ)
Boiling point	4.21 K	77.35 K
Specific heat capacity	4.25 J $K^{-1}\ g^{-1}$	1.95 J $K^{-1}\ g^{-1}$
Enthalpy of vaporization	25.1 J g^{-1}	200.3 J g^{-1}

Which liquid is the better coolant (on a per-gram basis) near its boiling point, and which is better at its boiling point? Explain.

82. The ΔH_f° of $Hg_2Br_2(s)$ at 25°C is -206.77 kJ mol^{-1}, and the ΔH_f° of $HgBr(g)$ is 96.23 kJ mol^{-1}. Compute the standard enthalpy of the following reaction at 25°C:

$$Hg_2Br_2(s) \longrightarrow 2\,HgBr(g)$$

83. Consider the precipitation reaction

$$Ag^+(aq) + Cl^-(aq) \longrightarrow AgCl(s)$$

is being studied.
(a) A coffee cup calorimeter has a calorimeter constant of 192 J K^{-1} and contains 100.0 mL of a 0.200 M aqueous solution of $AgNO_3$. The temperature of the calorimeter and its contents is 22.30°C. Then, 100.0 mL of a 0.200 M solution of HCl, also at a temperature of 22.30°C, is poured in. The temperature inside the calorimeter increases to 23.61°C as $AgCl(s)$ precipitates. Compute ΔH_r (in kJ mol^{-1}). List the approximations and assumptions that must be made.
(b) Compute ΔH_r° using the standard molar enthalpies of formation tabulated in Appendix D.
(c) The answers to parts (a) and (b) differ slightly and would do so even if the experiment were error-free and no approximations were made in the calculations. Explain why.

84. The cooling power of the dissolution reaction

$$NH_4NO_3(s) \longrightarrow NH_4^+(aq) + NO_3^-(aq)$$

is being evaluated.
(a) A calorimeter is constructed and found to have a calorimeter constant of 84.5 J K^{-1}. A 100.0 g quantity of water is placed inside. The device and its contents come to room temperature, which is 25.00°C. Then 3.20 g of $NH_4NO_3(s)$ is added to the water. It dissolves, causing the temperature inside the calorimeter to fall to 22.93°C. Compute ΔH_r (in kJ mol^{-1}) for the dissolution. List the approximations and assumptions that must be made.
(b) Compute ΔH_r° using the standard molar enthalpies of formation tabulated in Appendix D.
(c) List and discuss the reasons for the slight difference between the answers to parts (a) and (b).

*85. The gas most commonly used in welding is acetylene ($C_2H_2(g)$). When acetylene is burned in oxygen, the reaction that takes place is

$$C_2H_2(g) + \tfrac{5}{2}O_2(g) \longrightarrow 2\,CO_2(g) + H_2O(g)$$

(a) Using data from Appendix D, calculate ΔH_r°.
(b) Calculate the total heat capacity of 2.00 mol of $CO_2(g)$ and 1.00 mol of $H_2O(g)$, using $c_P(CO_2(g)) = 37$ J K^{-1} mol^{-1} and $c_P(H_2O(g)) = 36$ J K^{-1} mol^{-1}.
(c) When this reaction is carried out in an open flame, almost all the heat produced in part (a) goes to raise the temperature of the products. Calculate the maximum flame temperature attainable in an open flame burning acetylene in oxygen. Actual flame temperatures are lower than this because of heat losses to the surroundings.

86. The standard molar enthalpy of formation of $(NH_4)_2PtF_6(s)$ at 25°C is given in a table. Write the chemical equation for the reaction that would have to occur in the calorimeter in a direct measurement of this value.

87. Silicon nitride ($Si_3N_4(s)$) has physical, chemical, and mechanical properties that make it a useful industrial material. It is crucial to know its standard molar enthalpy of formation. A clever experiment allows the direct determination of ΔH_r° for

$$3\,CO_2(g) + Si_3N_4(s) \longrightarrow 3\,SiO_2(s) + 2\,N_2(g) + 3\,C(s)$$

State what additional data must either be looked up or measured before the ΔH_f° of $Si_3N_4(s)$ can be computed.

88. Use average bond enthalpies from Table 10–3 to estimate ΔH_r° for the following reactions. Draw Lewis structures as necessary. In C_8H_{18}, the carbon atoms form a continuous chain with single bonds; in C_2H_5OH the two carbon atoms are bonded to one another.
(a) $CH_4(g) + 2\,O_2(g) \longrightarrow CO_2(g) + 2\,H_2O(g)$
      (burning methane, or natural gas)
(b) $C_8H_{18}(g) + \tfrac{25}{2}O_2(g) \longrightarrow 8\,CO_2(g) + 9\,H_2O(g)$
      (burning octane, in gasoline)
(c) $C_2H_5OH(g) + 3\,O_2(g) \longrightarrow 2\,CO_2(g) + 3\,H_2O(g)$
      (burning ethanol, in gasohol)

89. When a ball of mass m is dropped through a height difference Δh, its potential energy changes by an amount $mg\Delta h$, where g is the acceleration of gravity, equal to 9.81 m s^{-2}. Suppose that all of that energy is converted into heat, increasing the temperature of the ball, when the ball hits the ground. If the specific heat capacity of the fabric of the ball is 0.850 J K^{-1} g^{-1}, calculate the height from which the ball must be dropped to increase the temperature of the ball by 1.00°C. (*Hint:* The mass of the ball does not matter because it cancels out once the problem is set up. Make sure to convert properly between grams and kilograms.)

90. During his honeymoon in Switzerland, James Joule is said to have used a thermometer to measure the temperature difference between the water at the top and at the bottom of a waterfall. Take the height of the waterfall to be Δh and the acceleration of gravity g to be 9.81 m s^{-2}. Assuming that all the potential energy change $mg\Delta h$ of a mass m of water is used to heat that water by the time it reaches the bottom, calculate the temperature difference between the top and the bottom of a waterfall 100 m high. Take the specific heat capacity of water to be 4.18 J K^{-1} g^{-1}.

91. The gas inside a cylinder expands against a constant external pressure of 1.00 atm from a volume of 5.00 L to a volume of 13.00 L. In doing so, it turns a paddle immersed in 1.00 L of water. Calculate the temperature rise of the water, assuming no loss of heat to the surroundings or frictional losses in the mechanism. Take the density of water to be 1.00 g cm^{-3} and its specific heat capacity to be 4.18 J K^{-1} g^{-1}.

*92. One mole of argon (assumed to be an ideal gas) is confined in a strong, rigid container of volume 22.41 L at 273.15 K.

The system is heated until 3.000 kJ (3000 J) of heat has been added. The molar heat capacity of the gas does not change during the heating and equals 12.47 J K^{-1} mol^{-1}.

(a) Calculate the original pressure inside the vessel (in atmospheres).

(b) Determine q for the system during the heating process.

(c) Determine w for the system during the heating process.

(d) Compute the temperature in degrees Celsius of the gas after the heating. Assume that the container has zero heat capacity.

(e) Compute the pressure (in atmospheres) inside the vessel after the heating.

(f) Compute ΔE of the gas during the heating process.

(g) Compute ΔH of the gas during the heating process.

(h) The correct answer to part (g) exceeds 3.000 kJ. The increase in enthalpy (which at one time was misleadingly called the "heat content") in this system exceeds the amount of heat actually added. Why is this not a violation of the law of conservation of energy?

*93. A runner generates 100 Calories (418 kJ) of energy per mile from the oxidation of food. The runner's body must dispose of essentially all this energy to avoid overheating.

(a) Suppose that the evaporation of sweat is the only way that energy is lost. Estimate the volume of sweat (in liters) that must be evaporated by the runner in the course of a marathon (just over 26 miles). Take the enthalpy of vaporization of water at the runner's body temperature as 44 kJ mol^{-1}.

(b) Does the answer seem high or low (based on experience viewing or running footraces)? List three other mechanisms that the body uses to give off energy.

CUMULATIVE PROBLEM

Methanol as a Gasoline Substitute

Methanol (CH_3OH) is used as a substitute for gasoline in certain high-performance vehicles. Its use as a fuel in ordinary cars has been seriously considered. Obviously, the thermochemistry of methanol must be thoroughly understood in order to design engines to burn it efficiently.

A methanol-powered bus. (*Vanessa Vick/Photo Researchers, Inc.*)

(a) Methanol in an automobile engine must be in the gas phase before it can react. Calculate the heat (in kilojoules) that must be added to 1.00 kg of liquid methanol to raise its temperature from 25.0°C to its normal boiling point, 65.0°C. The molar heat capacity of liquid methanol stays close to 81.6 J K^{-1} mol^{-1} over this range of temperature.

(b) Once methanol has reached its boiling point, it must be vaporized. The molar enthalpy of vaporization of methanol is 38 kJ mol^{-1}. How much heat must be added to vaporize 1.00 kg of methanol?

(c) Once it is in the vapor phase, the methanol can react with oxygen in the air according to

$$CH_3OH(g) + \tfrac{3}{2}O_2(g) \longrightarrow CO_2(g) + 2\,H_2O(g)$$

Use average bond enthalpies to estimate the enthalpy of this reaction. (*Note:* The structure of methanol is given on page 464.)

(d) Use data from Appendix D to calculate the standard enthalpy of this reaction, assuming it to be the same at 65°C as at 25°C.

(e) Calculate the amount of heat released when 1.00 kg of gaseous methanol is burned in air at constant pressure. Use the more accurate result of part (d) rather than that of part (c).

(f) Calculate the difference between the change in enthalpy and the change in internal energy when 1.00 kg of gaseous methanol is oxidized to gaseous CO_2 and gaseous H_2O at 65°C.

(g) Suppose now that the methanol is burned inside the cylinder of an automobile engine. Taking the radius of the cylinder to be 4.0 cm and the distance the piston moves during one stroke to be 12 cm, calculate the work done on the gas per stroke as it expands against an external pressure of 1.00 atm. Express the answer in liter-atmospheres and in joules.

11

Spontaneous Change and Equilibrium

CHAPTER OUTLINE

Highly organized symmetrical objects—snow crystals— form spontaneously from water vapor in clouds. *(Richard C. Walters/Visuals Unlimited)*

A spontaneous change is one that occurs by itself, given enough time, without outside intervention. One of the most striking features of spontaneous change is that it has a direction: Gases expand into empty containers but never spontaneously contract out of a container to leave it empty. Heat flows from a hot body to a cold one when the two come into thermal contact but never flows spontaneously from cold to hot. When a teakettle is started on a gas flame, the water always gets hotter; it never cools or freezes while the flame gets hotter. When a spark touches a mixture of hydrogen and oxygen, the two gases spontaneously (and explosively) react to produce water, but water never decomposes spontaneously to hydrogen and oxygen. One of the goals of thermodynamics must be to account for this *directionality* of spontaneous change Fig. 11–1.

This chapter introduces two important new state properties. The first, the entropy, is connected with the extent of disorder present in a system and in its surroundings. An examination of the way in which this property changes during real and hypothetical processes leads to valid predictions about the direction of spontaneous change and the equilibrium state that is ultimately achieved. The second property, the Gibbs energy, facilitates making such predictions in many chemically important situations.

11–1 Enthalpy and Spontaneous Change

Mechanical systems reach equilibrium in the state of lowest energy: A ball released from the rim of a bowl rolls around until it dissipates its kinetic energy and comes to rest in a state of lowest potential energy at the bottom of the bowl. During the 19th century, chemists sought a similar principle that would predict the equilibrium states of chemical reactions.

Two chemists, Marcellin Berthelot in France and Julius Thomsen in Denmark, suggested that the enthalpy could play this role in processes occurring at constant

Figure 11–1 • A bullet hitting a steel plate at a speed of 500 m s^{-1} melts as its kinetic energy is converted to heat. Molten lead sprays in all directions. These three photographs make sense only in the order shown. The reverse process is unmistakably impossible. (*©The Harold E. Edgerton 1992 Trust/Palm Press, Inc.*)

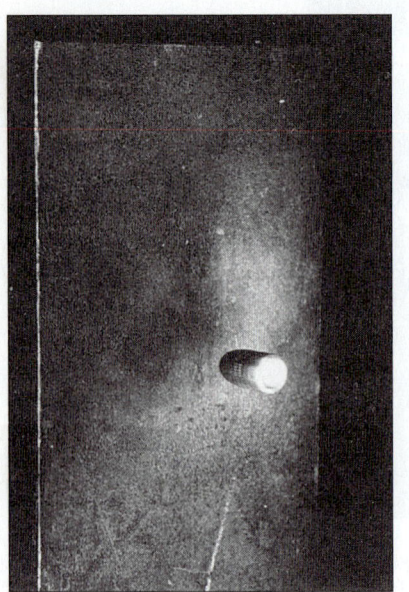

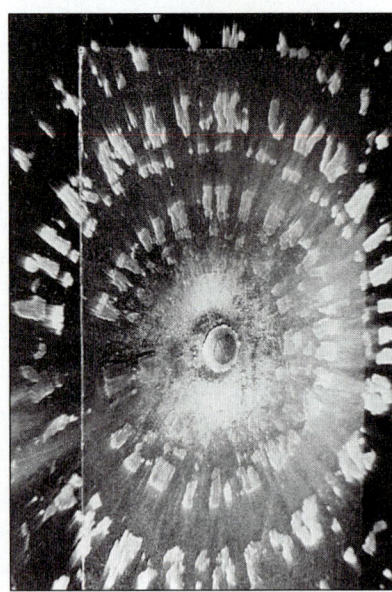

pressure. By 1878, they had become convinced that spontaneous chemical reactions "tend toward the production of the body or the system of bodies that sets free the most heat." They were wrong, however. If their principle were valid, one could predict the spontaneity of a reaction at constant pressure from the sign of ΔH. If the reaction were exothermic (ΔH negative), then it would be spontaneous. Spontaneous endothermic reactions would not be permitted.

Most spontaneous chemical reactions *are* exothermic, and, to this extent, the principle of Berthelot and Thomsen is a valid summary of much experimental data. There are many exceptions, however. The dissolution of ammonium chloride is endothermic

$$NH_4Cl(s) \longrightarrow NH_4^+(aq) + Cl^-(aq) \qquad \Delta H_r^\circ = +14.8 \text{ kJ mol}^{-1}$$

but the reaction is nevertheless spontaneous. The acid–base reaction

$$Ba(OH)_2 \cdot 8H_2O(s) + 2\,NH_4^+(aq) \longrightarrow Ba^{2+}(aq) + 2\,NH_3(g) + 10\,H_2O(\ell)$$

is also endothermic ($\Delta H_r^\circ = +119 \text{ kJ mol}^{-1}$), yet it is spontaneous.

Some reactions that are non-spontaneous at ordinary temperature become spontaneous at elevated temperatures, even though their enthalpy change remains positive. An example is the decomposition of chalk (a form of limestone) to lime and carbon dioxide at high temperature:

$$CaCO_3(s) \longrightarrow CaO(s) + CO_2(g) \qquad \Delta H_r^\circ = +178.3 \text{ kJ mol}^{-1}$$

This reaction becomes spontaneous above about 800°C even though the standard enthalpy change remains positive. The chemist Joseph Black observed this fact as early as 1755.

Many phase changes also disobey the principle of Berthelot and Thomsen. It is true that, below 100°C, water vapor spontaneously condenses to liquid in an exothermic process:

$$H_2O(g) \longrightarrow H_2O(\ell) \qquad \Delta H^\circ = -40.66 \text{ kJ mol}^{-1} \text{ (at 373.15 K)}$$

However, the reverse process, the boiling of liquid water

$$H_2O(\ell) \longrightarrow H_2O(g) \qquad \Delta H^\circ = +40.66 \text{ kJ mol}^{-1} \text{ (at 373.15 K)}$$

becomes spontaneous *above* 100°C, even though it is endothermic. The melting of a solid, such as ice, is another endothermic process that occurs spontaneously.

Clearly, the sign of the enthalpy change fails as a general criterion for the spontaneity of a chemical or a physical change. Exothermicity is one factor that favors spontaneous change, but there is another. As the temperature becomes higher, this second factor grows in importance, and the predictive power of the principle of Berthelot and Thomsen diminishes. A new state property is needed to establish a criterion for spontaneity that is valid at all temperatures.

11–2 Entropy

To search for this new criterion, consider a spontaneous process in which neither internal energy nor enthalpy plays a role: the free expansion of an ideal gas into a vacuum, which is shown in Figure 11–2. Suppose that 1.0 mol of an ideal gas is initially held in the left bulb in a volume $V/2$ while the right bulb is evacuated. After the stopcock is opened, the gas expands spontaneously to fill the entire volume V.

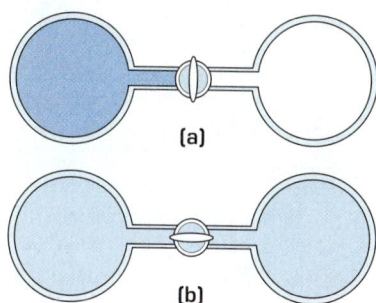

(a)

(b)

Figure 11–2 • The free expansion of a gas into a vacuum. (a) Stopcock closed; all gas is in the left bulb. (b) Stopcock opened; half of the gas is in each bulb.

Neither the internal energy nor the enthalpy of the gas changes during the expansion. So why does it occur? According to the kinetic-molecular theory, the N_0 molecules of the gas move around chaotically, colliding at random with one other and the walls of their container. Once the stopcock is opened, a particular molecule may stay for a time in the left bulb, but also may cross to the right, stay there for a while, then go back to the left, and so forth. No reason exists for it to prefer one bulb to the other. Over the long term, the molecule should spend equal time in the two bulbs. Put in other terms, the probability that a given molecule is in the left bulb at any moment drops from 1 to 1/2 when the stopcock is opened, and the probability that it is in the right bulb rises from 0 to 1/2.

• A probability is a likelihood or chance. The probability of an event is a fraction ranging from 0 (utterly impossible) to 1 (absolutely certain).

This concept of probability provides the key to understanding spontaneous change. Let us calculate the probability that the gas does *not* expand when the stopcock is opened. After the opening, the probability that molecule 1 is in the left bulb in Figure 11–2 equals 1/2. The same is true of molecule 2. The probability that molecules 1 and 2 both are on the left equals $1/2 \times 1/2 = 1/4$ (the probabilities are multiplied because the two events are independent; molecule 1 does not influence molecule 2). The probability of three specific molecules being on the left after the opening is $1/2 \times 1/2 \times 1/2 = 1/8$, and the probability of four specific molecules being there is $1/2 \times 1/2 \times 1/2 \times 1/2 = 1/16$. Figure 11–3 shows all 16 of the possible distributions of four molecules between the two bulbs. Continuing this argument, the probability that N_0 (6.0×10^{23}) molecules remain on the left after the stopcock is opened is

$$\text{probability} = \underbrace{\frac{1}{2} \times \frac{1}{2} \times \frac{1}{2} \times \cdots \times \frac{1}{2}}_{6.0 \times 10^{23} \text{ terms}} = \left(\frac{1}{2}\right)^{6.0 \times 10^{23}} = \frac{1}{2^{6.0 \times 10^{23}}}$$

Converting to a power of ten gives

$$\text{probability} = \frac{1}{10^{1.8 \times 10^{23}}}$$

This is a vanishingly small probability because $10^{1.8 \times 10^{23}}$ is an unimaginably large number. It is vastly larger than 1.8×10^{23}, which is itself a large number. To realize this, think about writing out these numbers. The number 1.8×10^{23} is fairly easy. It contains 22 zeros:

$$180,000,000,000,000,000,000,000$$

But $10^{1.8 \times 10^{23}}$ is 1 followed by 1.8×10^{23} zeros. To write out these zeros would require the whole human race—man, woman, and child—to write a zero every second for over a million years!

Nothing in the laws of mechanics prevents the gas from not expanding when the stopcock is opened, but such events do not occur because they are overwhelmingly improbable. Similar principles apply to all spontaneous change:

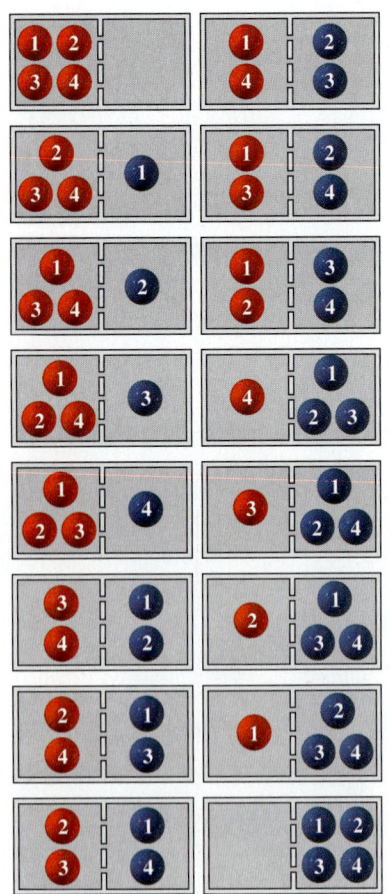

Figure 11–3 • The 16 possible ways four molecules may occupy a two-sided container. In only one of these are all four molecules on the left side.

> The directionality of spontaneous change is a consequence of the random behavior of the large numbers of molecules in macroscopic systems.

These are the systems in which we see inexorable, directional change every day.

EXAMPLE 11-1

Suppose that the gas in the preceding discussion has only 6 molecules instead of 6.0×10^{23}. What is the probability that at a given instant all 6 are located on the left side of the stopcock in Figure 11-2?

Solution

The probability that any given molecule is on the left is 1/2. The probability that all 6 are simultaneously on the left is

$$\frac{1}{2} \times \frac{1}{2} \times \frac{1}{2} \times \frac{1}{2} \times \frac{1}{2} \times \frac{1}{2} = \left(\frac{1}{2}\right)^6 = \frac{1}{64}$$

which is one in sixty-four. Finding 6 out of 6 molecules on the left would be uncommon, but not startling. Finding 6.0×10^{23} out of 6.0×10^{23} molecules on the same side is another story!

EXERCISE

Calculate the probability that at a given moment all the molecules of a 20-molecule gas in the right bulb in Figure 11-2b.

Answer: 1 in 2^{20}, or 1 in 1,048,576.

Microstates and Entropy

Spontaneous changes occur when constraints are removed from systems. In the free expansion of an ideal gas, the molecules were initially kept in one part of the container (the left bulb) by a closed stopcock. After this constraint was removed, each molecule had new scope for its motions. The number of states available to the system increased. A microscopic state, or **microstate,** is a particular way of arranging molecules among the positions accessible to them while keeping the total energy fixed. Figure 11-3 shows 16 equally probable microstates for a four-molecule gas in a two-sided container. By counting the pictures, we see that the "2-2" gas is six times more probable than the "4-0" gas because it can happen in six times more ways:

4 left 0 right	3 left 1 right	2 left 2 right	1 left 3 right	0 left 4 right
4-0	**3-1**	**2-2**	**1-3**	**0-4**
1 way	4 ways	6 ways	4 ways	1 way

At the start (a), the right-hand test-tube contains bromine vapor, which is yellow-brown, and the left-hand test-tube contains air. (b) Mixing is spontaneous after the constraint (the barrier separating the mouths of the two test-tubes) is removed. *(Charles D. Winters)*

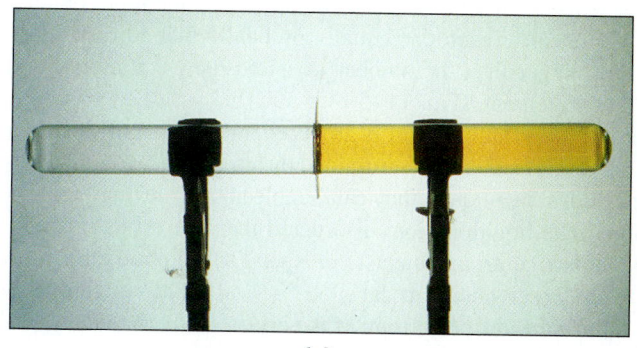

(a)

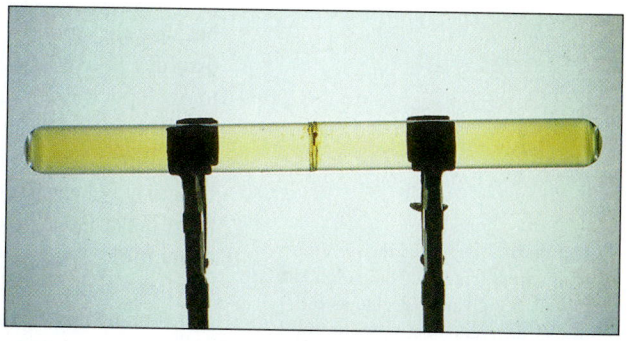

(b)

Figure 11–4 • The tomb of Ludwig Boltzmann displays his famous contribution to statistical thermodynamics. Nowadays, the logarithm to the base e is usually written as "ln," rather than "log." *(David W. Oxtoby)*

In other terms, the "2-2" gas has six times more microstates than the "4-0" gas. Even at this very small scale, chance favors the "even-split" gas more than any other single arrangement.

A new function helps in dealing with the statistical nature of spontaneous change. The **entropy** (S) of a system depends on the number of microstates available to its molecules according to the equation

$$S = k \ln W$$

where k is a constant and W is the number of available microstates. Ludwig Boltzmann (Figure 11–4) derived this equation in 1868, and k is called the *Boltzmann constant*. It equals the ratio of the universal gas constant R to Avogadro's number N_0:

$$k = \frac{R}{N_0} = \frac{8.314472 \text{ J K}^{-1} \text{ mol}^{-1}}{6.022142 \times 10^{23} \text{ mol}^{-1}} = 1.308630 \times 10^{-23} \text{ J K}^{-1}$$

In principle, one obtains the entropy of a system by counting the available microstates, taking the natural logarithm, and multiplying by k. The answer comes out in joules divided by kelvins, which is the SI unit of entropy. In practice, chemists usually compute the relative number of microstates before and after a constraint is removed. This gives the sign and magnitude of the *change* in the entropy.

The discussion of free expansion of a gas in this section makes it clear that spontaneous change in an *isolated* system (a system that is not interacting with the outside either mechanically or thermally) always goes to increase the number of microstates available to the molecules. It therefore follows that:

> Spontaneous change in an isolated system proceeds with an increase in the entropy of the system.

The restriction "isolated" is very important. Spontaneous change often decreases the entropy of systems that are *not* isolated, as we shall see.

Entropy and Disorder

Increases in entropy are commonly discussed as increases in disorder or randomness, as in the statement "Entropy is a measure of disorder." Making this connection requires some care because "disorder" and "random" are hard terms to define. Systems in which the molecules are constrained to occupy only certain positions in space can reasonably be thought of as having a low degree of disorder. Removing constraints frees the molecules to occupy more locations, and disorder among the molecules increases. An isolated system does not spontaneously become more ordered because the number of microstates (each one equally probable) that correspond to disorder is overwhelmingly larger than the number of microstates that correspond to any specified ordered arrangement. (Fig. 11–5).

The melting (fusion) of a crystalline solid is an excellent example of this. In a crystalline solid, the atoms or molecules are constrained by attractions to their neighbors to stay near fixed positions. In a liquid, they can jumble about, moving far away from these fixed positions. The liquid is more disordered than the crystal because many more microscopic molecular arrangements correspond to the observable state labelled "liquid" than to the observable state labelled "crystal." The conclusion is that the entropy of the liquid exceeds the entropy of the crystal. Experimentally, entropy changes of melting (symbolized ΔS_{fus}) indeed turn out to be positive.

• Section 6–2 identifies the *durability* of the arrangement of the neighbors around any given molecule as the crucial difference between the solid and liquid states of matter.

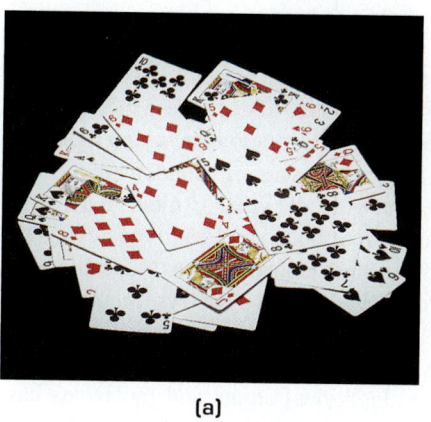

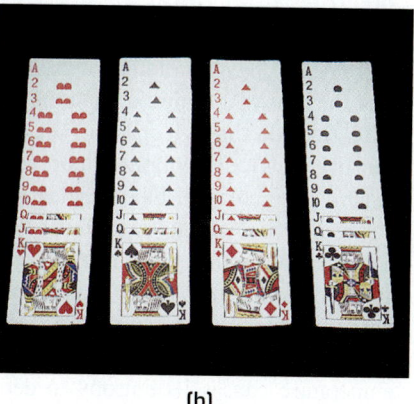

(a)   (b)

Figure 11-5 • (a) If a deck of cards is thrown down, they land in a disorderly heap. The number of ordered arrangements (as in (b)) of a deck of cards is far smaller than the number of disordered ones. *(Charles D. Winters)*

The evaporation of a liquid to a gas or vapor is another example. The molecules in a liquid mostly remain at the bottom of their container, but those in a gas move freely throughout the whole enclosed volume. A gas is more disordered than a liquid, so entropy changes of vaporization (ΔS_{vap}'s) are expected to be positive. Experiment confirms that they are.

Although associating entropy with disorder is helpful, the fundamental factor governing the entropy in any collection of atoms or molecules is the number of available microstates. In some cases (such as certain transitions between two solid phases), it is very difficult to say which phase is more disordered. A calculation of the relative number of microstates available, however, correctly predicts the sign of the entropy change (ΔS) in such cases.

• The same iron-clad convention holds for entropy as for temperature, volume, and other quantities. A change in entropy equals $S_f - S_i$ (the final value minus the initial value) and never the reverse.

EXAMPLE 11-2

Predict whether the entropy change of the system (ΔS) is positive or negative for each of the following processes:
(a) Steam condenses to liquid water.
(b) Oxygen and nitrogen are contained in separate volumes at the same pressure, with a membrane between the two. The membrane is pierced, and the gases mix in the combined volume.

Solution

(a) The ΔS is negative because the number of microstates available in the liquid is less than in the steam (a gas).
(b) The ΔS is positive because each molecule can be found in either volume after the constraint is removed; the number of available microstates is greater.

EXERCISE

Predict whether the entropy change is positive or negative for the system in each of the following processes: (a) compression of $O_2(g)$ at 1 atm to $O_2(g)$ at 5 atm; (b) shuffling a deck of cards that had been arranged by suit.

Answer: (a) negative. (b) positive.

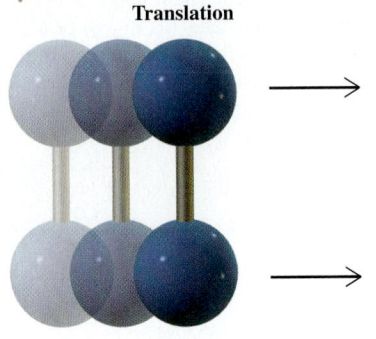

Figure 11–6 • The three types of molecular motion of a diatomic molecule.

Translation

Rotation

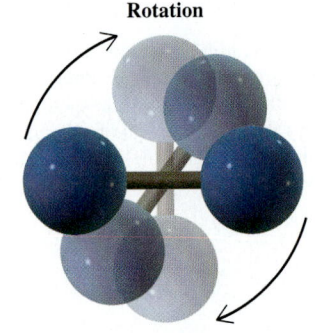

Vibration

11–3 Absolute Entropies and Chemical Reactions

How does a change in temperature affect the entropy of a system? To answer, consider the types of motion undergone by molecules (Fig. 11–6). These include translation (motion through space of the whole molecule), rotation (in which molecules spin like tops), and vibration (stretching and compression of bonds). As the temperature is increased, the energies associated with all types of molecular motion increase. Moreover, the number of ways in which this energy can be parcelled out among the molecules also increases. This means that more microstates become available to the molecules as the temperature rises: *The entropy of a body increases with increasing temperature.*

By the same logic, cooling a pure crystalline solid toward the absolute zero of temperature causes its entropy to decrease. Individual atoms and molecules have fewer and fewer alternatives in terms of their positions and energies as the temperature falls. This reduces the number of available microstates. In a perfectly ordered crystal, one in which the atoms hold motionless at specific sites, there would be only one microstate, and the entropy would equal zero:

$$S = k \ln W = k \ln 1 = 0 \, \text{J K}^{-1} \qquad \text{(in a perfectly ordered crystal)}$$

Cessation of motion on the atomic scale is associated with the absolute zero of temperature. From such considerations, we state the following:

> The entropy of a crystalline substance at equilibrium approaches zero as the absolute zero of temperature is approached.

This is one form of the **third law of thermodynamics,** which was discovered in the late 19th century by the German chemist Walther Nernst. It gives the baseline from which entropies at other temperatures can be measured.[1]

Measuring Entropies

Entropies can be measured, just like masses, volumes, or densities. One method uses calorimetry (Section 10–2). First, the molar heat capacity (c_P) of a substance is measured at a number of different temperatures, starting near absolute zero and going up as high as desired or possible. Some results for the element platinum are

T (K)	0	100	150	200	250	298.15	350
c_P (J K^{-1} mol^{-1})	0	19.6	23.0	24.6	25.1	25.8	26.1

These data are typical. The heat capacities of many substances equal zero at absolute zero, rise rapidly with the first increase in temperature, and then level off. Next, the quantity c_P/T, the molar heat capacity divided by the absolute temperature at which it was determined, is graphed as a function of T. Figure 11–7 shows the curve for platinum. It can be proved that the area under such a curve is proportional to the increase in the molar entropy of the substance from 0 K up to T. The increase equals

---

[1]This form of the third law is actually due to the German physicist Max Planck. Nernst stated only that entropy changes become zero at the absolute zero of temperature. Although historically the third law was discovered after the second law, we introduce it earlier in our treatment. The second law follows soon.

the **molar absolute entropy** of the substance at temperature T, because the entropy of substances near 0 K approaches zero, according to the third law. The final step is therefore simply to measure this area. The unit of the shaded area under the curve in Figure 11–7 equals the unit on the horizontal axis multiplied by the unit on the vertical axis (area equals length times width). Multiplying the kelvin (K) by the $J K^{-2} mol^{-1}$ gives the $J K^{-1} mol^{-1}$. Molar entropies are indeed measured in joules per kelvin per mole.

If a substance melts, boils, or undergoes any other phase transition before reaching the temperature T, the entropy change associated with the phase transition must be added to the result from the measurement of the area. This is not difficult because the ΔS of a phase transition equals the ΔH of the transition (readily measured calorimetrically) divided by the absolute temperature at which the transition takes place. For the fusion of a solid at the normal melting point T_{fus} and for vaporization of a liquid at its normal boiling point T_b, for example,

$$\Delta S_{fus} = \frac{\Delta H_{fus}}{T_{fus}} \qquad \Delta S_{vap} = \frac{\Delta H_{vap}}{T_b}$$

Note that the unit of both ΔS_{fus} and ΔS_{vap} is the joule per kelvin per mole ($J K^{-1} mol^{-1}$).

The absolute entropy of 1 mol of a substance in a standard state equals its **standard molar entropy** (S°). The superscript "naught" means "in a standard state," as explained in Section 10–4. The definitions of standard states are unchanged from that section: for solids and liquids, the state of the pure solid or liquid under a pressure of 1 atm and at a specified temperature; for gases, the gaseous phase under a pressure of 1 atm, at a specified temperature, and exhibiting ideal-gas behavior; for dissolved species, the state at a concentration of 1 mol L^{-1} under a pressure of 1 atm, at a specified temperature, and exhibiting ideal-solution behavior. Values of S° at 298.15 K (25°C) appear in Appendix D for a number of elements and compounds. Standard molar entropies at 298.15 K often are symbolized S°_{298} or $S^\circ_{298.15}$. They of course differ from standard molar entropies at other temperatures. When a temperature is not stated, 298.15 K should be assumed. Figure 11–8 displays values of S°_{298} for several elements both as solids and as gases.

Entropies of Reaction

An **entropy of reaction** is really the entropy *change* of reaction, the difference between the entropy of the products and the entropy of the reactants

$$\Delta S_r = S_{products} - S_{reactants}$$

The entropies of gases exceed those of liquids and solids under the same conditions of temperature and pressure, as anticipated from the greater number of microstates associated with the gaseous state. The ΔS_r when gases are produced from liquids or solids, as in the reaction

$$CaCO_3(s) \longrightarrow CaO(s) + CO_2(g)$$

is therefore positive unless the temperature and pressure change drastically. Figure 11–9 gives another example. More generally, the entropy of the reaction system increases when the total chemical amount of gas increases and the temperature and pressure do not change too much. Thus, ΔS_r for

$$2 H_2O(g) \longrightarrow 2 H_2(g) + O_2(g)$$

is positive because there is a net increase from 2 mol of gas to 3 mol.

• Notice that molar entropy has the same unit as molar heat capacity.

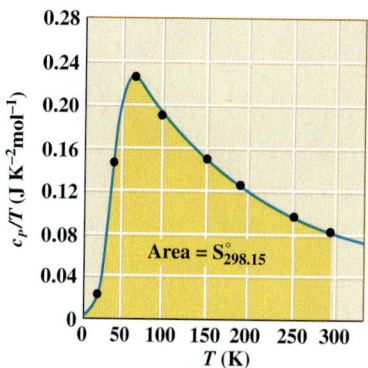

Figure 11–7 • A graph of c_P/T for platinum. The black dots represent experimental measurements at $P = 1$ atm. The area under the curve up to any temperature is proportional to the molar entropy at that temperature. Here the area has been shaded up to 298.15 K corresponding to S°, the standard molar entropy at 298.15 K.

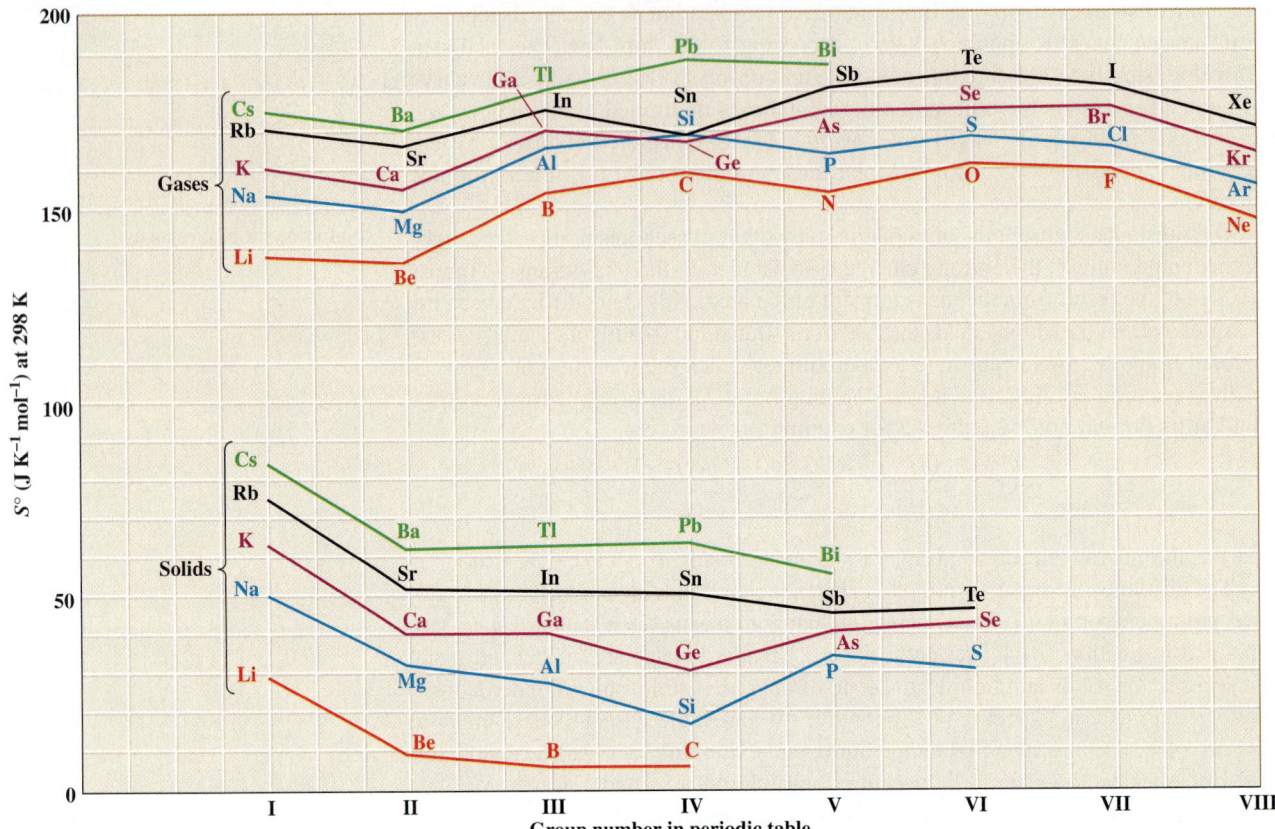

Figure 11-8 • The standard molar entropies of some elements at 298.15 K. Note the large differences in molar entropy between the solid and gaseous states.

Dissolution and precipitation reactions are major types of chemical reactions. It might seem that dissolution should always lead to an increase in entropy ($\Delta S_r > 0$) and precipitation to a decrease in entropy ($\Delta S_r < 0$) in the reaction system. The reasoning is that collections of molecules or ions certainly have fewer microstates organized in a crystal than dispersed through a solution. For many compounds, this is true. Dissolved sodium chloride has higher entropy than solid sodium chloride plus pure water for exactly this reason. In other cases, however, dissolution leads to a *decrease* in the entropy of a system. An example is the dissolution of magnesium chloride at room conditions:

$$MgCl_2(s) \longrightarrow Mg^{2+}(aq) + 2\,Cl^-(aq) \qquad \Delta S_r < 0$$

The reason for the negative ΔS_r is that the electric field exerted by the ions imposes order on the water molecules around them in solution (recall Fig. 6–3). This effect is large enough with $MgCl_2$ to outweigh the entropy increase accompanying the breakdown of the solid and cause a net decrease in the entropy of the reaction system.

A **standard entropy of reaction** equals the difference between the entropy of the products of a reaction in standard states at a specified temperature and the entropy of the reactants in standard states at the same temperature.

$$\Delta S_r^\circ = S_{products}^\circ - S_{reactants}^\circ$$

(a)

(b)

Figure 11-9 • (a) The decomposition of the orange solid $(NH_4)_2Cr_2O_7$ provides a spectacular example of a spontaneous reaction in which gases are produced from a solid starting material

$$(NH_4)_2Cr_2O_7(s) \longrightarrow$$
$$N_2(g) + 4\ H_2O(g) + Cr_2O_3(s)$$

The green dust that forms in the course of the reaction, together with the sparks and smoke, make this reacting system resemble a volcano. (b) The solid reactant and product. *(Charles D. Winters)*

For the general reaction

$$aA + bB \longrightarrow cC + dD$$

the standard entropy of reaction is

$$\Delta S_r^\circ = cS^\circ(C) + dS^\circ(D) - aS^\circ(A) - bS^\circ(B)$$

This equation can be used to calculate standard entropies of reaction (ΔS_r°'s) from standard-state entropies (S°'s) such as those tabulated in Appendix D. Values of ΔS_r° change with temperature, but only slightly. A ΔS_r° at 298.15 K is therefore a good approximation for one at another temperature, especially if the other temperature is not too different, a fact that is exploited heavily in practical chemistry (see Section 11–8). A calculation of ΔS_r° resembles a calculation of a standard enthalpy of reaction (ΔH_r°) from standard enthalpies of formation (ΔH_f°'s) (see Section 10–4). The (very important!) difference is that the standard molar entropy of an element in its most stable form at a specified temperature (its S° at that temperature) does *not* equal zero. This point is illustrated in the following example.

• Such approximations are used extensively later in this chapter.

• A substance in a standard state at 298.15 K (room temperature), whether a compound or an element, has plenty of disorder. The entropy of a substance can approach zero only at temperatures near absolute zero.

EXAMPLE 11-3

(a) Use standard molar entropy data from Appendix D to calculate ΔS_r° at 25°C for

$$2\ NO(g) + O_2(g) \longrightarrow 2\ NO_2(g)$$

(b) Calculate the standard change in entropy under the same conditions when 40.00 g of NO(g) reacts with excess $O_2(g)$ to given $NO_2(g)$.

Solution

(a) Specialize the general equation for computing ΔS_r° to fit this case:

$$\Delta S_r^\circ = 2\ S^\circ(NO_2(g)) - 2\ S^\circ(NO(g)) - 1\ S^\circ(O_2(g))$$

Locate the standard molar entropies in Appendix D, substitute them into this equation, and complete the arithmetic:

$$\Delta S_r^{\circ} = 2\left(\frac{239.95 \text{ J}}{\text{K mol}}\right) - 2\left(\frac{210.65 \text{ J}}{\text{K mol}}\right) - 1\left(\frac{205.03 \text{ J}}{\text{K mol}}\right) = -146.43 \text{ J K}^{-1} \text{ mol}^{-1}$$

(b) The problem concerns the entropy change in a specific system rather than "per mole of reaction as written." Set up unit-factors to make the required conversion:

$$\Delta S^{\circ} = 40.00 \text{ g NO}(g) \times \left(\frac{1 \text{ mol NO}(g)}{30.00 \text{ g NO}(g)}\right) \times \left(\frac{1 \text{ mol reaction}}{2 \text{ mol NO}(g)}\right)$$
$$\times \left(\frac{-146.43 \text{ J K}^{-1}}{\text{mol reaction}}\right) = -97.62 \text{ J K}^{-1}$$

Check

The decrease in entropy is consistent with the 3-to-2 decrease in the chemical amount of gases going from reactants to products.

EXERCISE

(a) Use data from Appendix D to calculate ΔS_r° at 298.15 K for the reaction

$$2 \text{ H}_2\text{S}(g) + 3 \text{ O}_2(g) \longrightarrow 2 \text{ SO}_2(g) + 2 \text{ H}_2\text{O}(g)$$

(b) Calculate ΔS° of the system when 26.71 g of $\text{H}_2\text{S}(g)$ reacts with excess $\text{O}_2(g)$ to give $\text{SO}_2(g)$ and $\text{H}_2\text{O}(g)$ and no other products at 298.15 K.

Answer: (a) $\Delta S_r^{\circ} = -152.79 \text{ J K}^{-1} \text{ mol}^{-1}$. (b) $\Delta S^{\circ} = -59.88 \text{ J K}^{-1}$.

• A negative absolute entropy implies more order in the system than perfect order—an obvious impossibility.

Some S° values in Appendix D, such as that of 1 M $\text{Mg}^{2+}(aq)$, are negative. These values are not absolute entropies but are referenced to another standard, because ions in solution do not become pure crystals at absolute zero. This fact has no effect on the calculation of the standard enthalpies (ΔS_r°'s) of changes that involve such ions. This is shown in the following example.

EXAMPLE 11–4

Compute ΔS_r° for the dissolution of $\text{MgCl}_2(s)$ in water at 298.15 K. Use data from Appendix D.

Solution

If the chemical reaction is

$$\text{MgCl}_2(s) \longrightarrow \text{Mg}^{2+}(aq) + 2 \text{ Cl}^-(aq)$$

then

$$\Delta S_r^{\circ} = 1 \; S^{\circ}(\text{Mg}^{2+}(aq)) + 2 \; S^{\circ}(\text{Cl}^-(aq)) - 1 \; S^{\circ}(\text{MgCl}_2(s))$$
$$= 1\left(\frac{-138.1 \text{ J}}{\text{K mol}}\right) + 2\left(\frac{56.5 \text{ J}}{\text{K mol}}\right) - 1\left(\frac{89.62 \text{ J}}{\text{K mol}}\right) = -114.7 \text{ J K}^{-1} \text{ mol}^{-1}$$

EXERCISE

Compute ΔS_r° for the reaction $\text{HCl}(g) \longrightarrow \text{H}^+(aq) + \text{Cl}^-(aq)$ at 298.15 K.

Answer: $\Delta S_r^{\circ} = -130.3 \text{ J K}^{-1} \text{ mol}^{-1}$.

11-4 The Second Law of Thermodynamics

The sign of ΔH, the change in the enthalpy of a system during a process, fails as a criterion for spontaneity. Many endothermic processes take place spontaneously, as discussed in Section 11–1, although some link between exothermicity and spontaneous change surely must exist. Does the entropy provide the desired criterion? At first, it appears that it does not. Although some spontaneous processes, such as the free expansion of a gas, clearly proceed with an increase in the entropy of the system, other spontaneous processes do not. Below 0°C, water undergoes a spontaneous phase transition, freezing, and *decreases* its entropy. Indeed, all freezing and condensation transitions have negative ΔS's. The entropy change from reactants to products is also negative for many spontaneous chemical reactions. When solutions of silver nitrate and sodium chloride are mixed, solid silver chloride precipitates spontaneously according to the equation

$$Ag^+(aq) + Cl^-(aq) \longrightarrow AgCl(s)$$

even though the entropy of $AgCl(s)$ is lower than that of the original ions (ΔS_r° for the preceding equals -33 J K^{-1} mol^{-1}).

Arriving at a general criterion for spontaneous change requires broadening the view to consider not only the system (the water that freezes in a beaker, or the solid silver chloride and its aqueous solution) but the surroundings as well. When water freezes or silver chloride precipitates, heat is liberated from the system into the surroundings (the beaker, the table, and the surrounding air, for example). The entropy of the surroundings, as well as that of the system, thereby changes. The **second law of thermodynamics** states:

> In any spontaneous process, the entropy of the universe (that is, of the system plus its surroundings) must increase.

Put in the form of an equation, the second law reads

$$\Delta S_{univ} = (\Delta S_{sys} + \Delta S_{surr}) > 0 \qquad \text{(spontaneous process)}$$

Any process leading to a *decrease* in the entropy of the universe cannot occur. Its *reverse* occurs spontaneously, however, because if the forward process has a negative ΔS_{univ}, then the reverse automatically has a positive ΔS_{univ}. If $\Delta S_{univ} = 0$, the forward and reverse processes occur equally well, and the process is at equilibrium.

With the aid of the second law of thermodynamics, we can begin to understand the different types of spontaneous processes discussed in this chapter. In the free expansion of an ideal gas, no heat is exchanged with the surroundings. Because the surroundings are unaffected, their entropy change is zero, and the direction of spontaneous change is determined entirely by the entropy of the system—it is the direction in which the entropy of the system increases. In an exothermic chemical reaction or phase change, heat flows to the surroundings, increasing their entropy. When a positive ΔS_{surr} is large enough, it compensates for a decrease in entropy of the system, as in the freezing of water. The second law explains why the principle of Berthelot and Thomsen linking spontaneous change to exothermicity sometimes worked but sometimes failed. Spontaneous endothermic changes may occur, despite an entropy decrease in the surroundings, if there is sufficient compensation from an entropy increase in the system.

The second law of thermodynamics profoundly affects the way in which we look at nature and at physical processes. It provides an arrow to time: a direction for the evolution of physical systems. Energy is conserved in both forward and reverse processes, so a process and its reverse are equally possible as far as the first law of thermodynamics is concerned. The entropy of the universe, however, increases in only one of the two directions. Real processes that decrease ΔS_{univ} are impossible or, rather, improbable beyond all conception.

The second law, like the first, has enormous practical importance. If a proposed process leads to a decrease in the entropy of the universe, it violates the second law. Trying to get it to run spontaneously is wasted effort, just as it is wasted effort trying to get processes to run that violate the first law by creating energy. Attempts by inventors to build perpetual-motion machines always fail because the machines violate one or another of the laws of thermodynamics.

11-5   The Gibbs Energy

In the previous section, the sign of the change in entropy of a system plus its surroundings (that is, the change of total entropy ΔS_{univ}) emerged as the ultimate criterion for deciding whether a process can occur:

$$\Delta S_{univ} > 0 \quad \text{(spontaneous)}$$
$$\Delta S_{univ} = 0 \quad \text{(equilibrium)}$$
$$\Delta S_{univ} < 0 \quad \text{(impossible)}$$

Using this criterion requires calculating the entropy change of the surroundings as well as the entropy change of the system, something that is often impracticable. It would certainly be desirable to have a state property that gives the feasibility of a process without reference to the surroundings. Such a state property exists *if consideration is restricted to processes at constant temperature and pressure.* Such changes are very important in chemistry and biochemistry, and this state property has corresponding importance. It is the **Gibbs energy** (G) and is defined in terms of the enthalpy, temperature, and entropy:

- "Constant temperature and pressure" allows both variables to change *during* a process, as long as they come back to their original values at the end.

$$G = H - TS$$

The Gibbs energy is measured in joules, just like the enthalpy (H). It is also called the "Gibbs free energy" and "Gibbs function."

- The Gibbs energy is named after J. Willard Gibbs, a pioneering mathematical physicist at Yale University who developed the field of statistical thermodynamics in the late 19th century.

To see how the Gibbs energy provides a criterion for spontaneity at constant pressure and temperature we focus on the *change* in the Gibbs energy of a system during a process. From the preceding definition,

$$\Delta G_{sys} = \Delta H_{sys} - \Delta(T_{sys}S_{sys})$$

If the temperature of the system is constant and equal to the temperature of the surroundings, then T_{sys} can be taken outside the parentheses and abbreviated to T:

$$\Delta G_{sys} = \Delta H_{sys} - T\Delta S_{sys}$$

The connection between ΔG_{sys} and ΔS_{univ}, the total entropy change, is made by noting (without derivation) that at constant pressure and temperature

$$\Delta S_{surr} = -\frac{\Delta H_{sys}}{T}$$

The negative sign is important in this equation. If the process is exothermic, ΔH_{sys} is negative, and the entropy change of the surroundings is positive. The total entropy change in a spontaneous process is then

$$\Delta S_{\text{univ}} = (\Delta S_{\text{sys}} + \Delta S_{\text{surr}}) = \left(\Delta S_{\text{sys}} - \frac{\Delta H_{\text{sys}}}{T}\right) > (0)$$

Multiplying both sides of the inequality by the absolute temperature T, which is always positive, gives

$$T\Delta S_{\text{sys}} - \Delta H_{\text{sys}} > 0 \qquad \text{(spontaneous process at constant } T \text{ and } P\text{)}$$

or, equivalently,

$$\Delta H_{\text{sys}} - T\Delta S_{\text{sys}} < 0 \qquad \text{(spontaneous process at constant } T \text{ and } P\text{)}$$

The right-hand side of this inequality equals ΔG_{sys}. Hence,

$$\Delta G_{\text{sys}} < 0 \qquad \text{(spontaneous process at constant } T \text{ and } P\text{)}$$

• The direction of the inequality changes when we multiply both sides of it by -1 in this step.

This result means that if the Gibbs energy of a system is forecast (by a computation, perhaps) to decrease during a process at constant temperature and pressure, then the process is spontaneous. If the Gibbs energy is forecast to increase, the process cannot occur under the conditions specified (because then ΔS_{univ} would be negative), but the reverse of the process is spontaneous under those conditions. If the Gibbs energy is forecast to stay constant, then ΔS_{univ} equals zero for the process and no change in either direction can occur under the conditions specified. This amounts to a state of equilibrium. In summary,

For a change at constant temperature and pressure
$$\begin{cases} \Delta G_{\text{sys}} < 0 & \text{(spontaneous)} \\ \Delta G_{\text{sys}} = 0 & \text{(equilibrium)} \\ \Delta G_{\text{sys}} > 0 & \text{(non-spontaneous, but reverse spontaneous)} \end{cases}$$

The Gibbs Energy and Phase Transitions

The freezing of liquids nicely illustrates the use of the Gibbs energy. The most familiar example is the water-to-ice transition:

$$H_2O(\ell) \longrightarrow H_2O(s)$$

Examine first what happens when this process occurs at the freezing point of water under atmospheric pressure, 273.15 K. The measured enthalpy change (the heat absorbed at constant pressure) associated with freezing water is

$$\Delta H_{273.15} = -6007 \text{ J mol}^{-1}$$

The entropy change of the water when it is frozen is

$$\Delta S_{273.15} = \frac{\Delta H_{273.15}}{T_{\text{freez}}} = \frac{-6007 \text{ J mol}^{-1}}{273.15 \text{ K}} = \frac{-21.99 \text{ J}}{\text{K mol}}$$

• Why is $\Delta H_{273.15}$ negative? Heat leaves the water when it freezes; the heat that the water absorbs is therefore negative.

The change in the Gibbs energy of the system per mole of H_2O that it contains is then

$$\Delta G_{273.15} = \Delta H_{273.15} - T\Delta S_{273.15} = -6007 \text{ J mol}^{-1} - (273.15 \text{ K})\left(\frac{-21.99 \text{ J}}{\text{K mol}}\right)$$

$$= 0 \text{ J mol}^{-1}$$

Finding that $\Delta G_{273.15}$ equals 0 J mol^{-1} is no great surprise. At the normal freezing point, the difference in the Gibbs energy between water and ice equals zero because the two are in equilibrium at this temperature.

Now consider what happens as water is cooled below 273.15 K to 263.15 K ($-10.00°C$). Liquid water can exist temporarily under these conditions; it is under-cooled (see Section 6–4). To calculate the change in the Gibbs energy of water as it freezes at this lower temperature, we assume that neither ΔH nor ΔS for the freezing process changes significantly with the drop in temperature. Then

$$\Delta G_{263.15} = -6007 \text{ J mol}^{-1} - (263.15 \text{ K})\left(\frac{-21.99 \text{ J}}{\text{K mol}}\right) = -220 \text{ J mol}^{-1}$$

Because $\Delta G < 0$, water at 263.15 K tends to freeze spontaneously (without outside intervention). A similar calculation using any temperature *greater* than 273.15 K gives a ΔG that exceeds zero, showing that freezing of water is non-spontaneous in that range of temperature. This accords with daily experience. Liquid water at atmospheric pressure never freezes when the temperature exceeds 273.15 K (equivalent to 0°C or 32°F). Rather, the reverse process occurs, and solid water (ice) melts.

Inspection of the equation

$$\Delta G = \Delta H - T\Delta S \qquad \text{(\textit{T} and \textit{P} constant)}$$

reveals that negative ΔG's are favored by negative ΔH's and by positive ΔS's. For the freezing of a liquid, ΔH is always negative (heat must be extracted to freeze things at constant pressure), but ΔS is also always negative (a substance in the solid state has less entropy than in the liquid state). Whether a liquid freezes at a given temperature and pressure depends on the outcome of a competition between two factors: an enthalpy effect that favors freezing and an entropy effect that opposes it (Fig. 11–10). At low temperatures (below T_{freez}), the enthalpy effect prevails and the liquid freezes spontaneously. At higher temperatures (above T_{freez}), the entropy effect prevails and freezing does not occur. At T_{freez}, the Gibbs energies of the solid and liquid are equal and the two phases coexist in a state of equilibrium. Similar analysis can be carried out for other phase transitions, such as boiling a liquid.

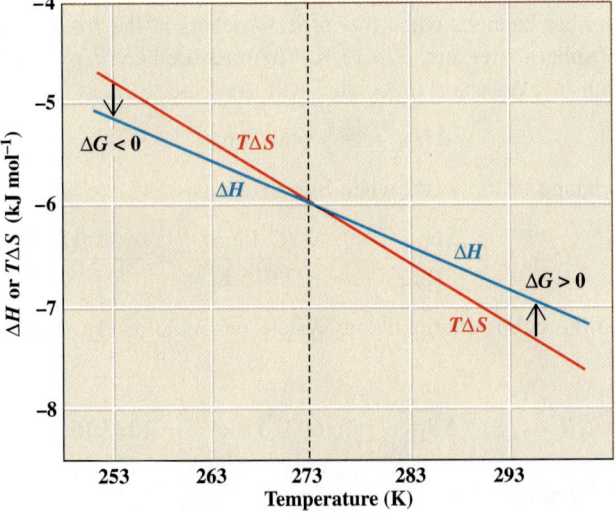

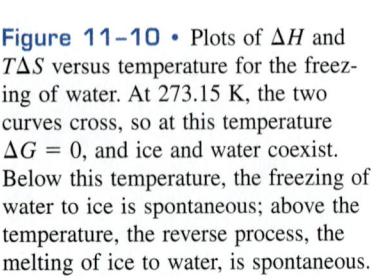

Figure 11–10 • Plots of ΔH and $T\Delta S$ versus temperature for the freezing of water. At 273.15 K, the two curves cross, so at this temperature $\Delta G = 0$, and ice and water coexist. Below this temperature, the freezing of water to ice is spontaneous; above the temperature, the reverse process, the melting of ice to water, is spontaneous.

EXAMPLE 11–5

Carbon disulfide (CS_2) is a liquid at room temperature.
(a) Calculate the ΔG_{vap}, the Gibbs energy for the vaporization of CS_2, at 25°C and 1 atm given that ΔS_{vap} equals 86.39 J K^{-1} mol^{-1} and ΔH_{vap} equals 27.66 kJ mol^{-1}.
(b) Compute the normal boiling point of CS_2 assuming that ΔS_{vap} and ΔH_{vap} at the normal boiling point equal their values at 25°C and 1 atm.

Strategy

Adapt the considerations developed for freezing. For part (a), write the equation $\Delta G_{vap} = \Delta H_{vap} - T\Delta S_{vap}$ and insert the given values. For part (b), use the same equation, setting ΔG_{vap} equal to zero and computing the boiling temperature.

Solution

(a) $\Delta G_{vap} = \Delta H_{vap} - T\Delta S_{vap}$

$$\Delta G_{vap} = 27.66 \text{ kJ mol}^{-1} - (298.15 \text{ K})(86.39 \text{ J K}^{-1}\text{ mol}^{-1})\left(\frac{1 \text{ kJ}}{1000 \text{ J}}\right)$$

$$= 1.90 \text{ kJ mol}^{-1}$$

(b) $\Delta G_{vap} = \Delta H_{vap} - T_b\Delta S_{vap} = 0$

$$T_b = \frac{\Delta H_{vap}}{\Delta S_{vap}} = \frac{27.66 \times 10^3 \text{ J mol}^{-1}}{86.39 \text{ J K}^{-1}\text{ mol}^{-1}} = 320.2 \text{ K or } 47.0°C$$

• Avoid the common error of failing to have $T\Delta S$ and ΔH in the same unit (kJ mol^{-1} or J mol^{-1}, it does not matter) before subtracting.

Check

(a) The answer is positive. This fits with the stated fact that CS_2 is a liquid at room temperature (298.15 K). (b) Reference books give the experimental normal boiling point of CS_2 as 319.5 K or 46.3°C. The discrepancy arises because ΔH_{vap} and ΔS_{vap} at 298.15 K in fact do differ from ΔH_{vap} and ΔS_{vap} at 319.5 K.

EXERCISE

The ΔS_{vap} and ΔH_{vap} of $PCl_3(\ell)$ equal 94.6 J K^{-1} mol^{-1} and 32.7 kJ mol^{-1}, respectively, at 298.15 K and 1 atm.
(a) Calculate ΔG for the vaporization of $PCl_3(\ell)$ at 298.15 K and 1 atm. Does PCl_3 boil under these conditions? (b) Estimate the normal boiling point of $PCl_3(\ell)$.

Answer: (a) $\Delta G_{vap} = +4.5$ kJ mol^{-1}; PCl_3 does not boil at 298.15 K and 1 atm. (b) $T_b \approx 346$ K (73°C).

Trouton's Rule

Experiments show that most liquids have approximately the same molar entropy of vaporization at their normal boiling points. That is, the increase in disorder in changing a mole of a liquid to a gas is nearly the same for every substance. Specifically,

$$\Delta S_{vap} = 88 \pm 5 \text{ J K}^{-1}\text{ mol}^{-1} \qquad \text{(for most liquids)}$$

This remarkable result is called **Trouton's rule.** Note that the ΔS_{vap} of carbon disulfide, in Example 11–5, lies well within this range. Trouton's rule comes about because vaporizations occur with an increase in molar volume that is very large and also nearly the same from substance to substance. The huge entropy originating from the increased volume drowns out other contributions to ΔS. The relationship between the entropy and enthalpy of vaporization,

$$\Delta S_{vap} = \frac{\Delta H_{vap}}{T_b}$$

and the observed (near) constancy of ΔS_{vap} mean that, although ΔH_{vap} and T_b vary widely from substance to substance, they do so in about the same proportion. Trouton's rule finds use in estimating enthalpies of vaporization from measurements of boiling points. Exceptions exist among substances that have especially large amounts of order as liquids. The ΔS_{vap} of water, for instance, is 109 J K^{-1} mol^{-1} because extensive hydrogen bonding organizes the liquid (see Section 6–1).

11–6   The Gibbs Energy and Chemical Reactions

• This definition parallels the definitions of enthalpy of reaction (Section 10–3) and entropy of reaction (Section 11–3).

The **Gibbs energy of reaction** equals the difference between the Gibbs energy of the products and the Gibbs energy of the reactants:

$$\Delta G_r = G_{products} - G_{reactants}$$

The results of the preceding section are applied to a chemical reaction by defining the reaction (subscript "r") as the system of interest (subscript "sys"). A chemical reaction can occur at constant temperature and pressure only if it loses Gibbs energy to its surroundings (ΔG_r negative) (Fig. 11–11). A reaction that would gain Gibbs energy from its surroundings at constant temperature and pressure (ΔG_r positive) cannot occur. The idea can be expressed somewhat more dramatically:

> The negative of the Gibbs energy of reaction ($-\Delta G_r$) is a measure of the *driving force* of a reaction under conditions of constant temperature and pressure.

"Constant temperature and pressure" means that the final temperature equals the initial temperature and the final pressure equals the initial pressure. Both variables may go up and down during the course of the reaction. Reactions under these conditions are easy to arrange. Consequently, ΔG_r's have great importance in discussions of chemical reactivity.

Gibbs energies of reaction (ΔG_r's) are measured in kJ mol^{-1} (or J mol^{-1}). The mol^{-1} ("per mole") in the unit means *per mole of the reaction as it is written*. They are thus really *molar* Gibbs energies of reaction. A numerical ΔG_r is ambiguous unless accompanied by a balanced chemical equation or other indication of the amounts of substances that react. The situation is identical to the one that exists with numerical molar enthalpies of reaction (ΔH_r's, see Section 10–3).

Figure 11–11 • A chemical reaction in the act of losing Gibbs energy. The reaction is the dissolution of HCl(g) in water. The upper flask was filled with HCl(g) and a small amount of water was injected into it. Now HCl(g) is dissolving spontaneously, lowering the pressure in the upper flask. The resulting partial vacuum is drawing more water up the tube from the lower flask. The change is so fast that a vigorous fountain of water plays into the upper flask. Some of the Gibbs energy of the reaction is appearing as work to lift the water. *(Leon Lewandowski)*

Standard Gibbs Energy of Reaction

Imagine running a chemical reaction at constant temperature and pressure. The change in Gibbs energy of the reaction equals

$$\Delta G_r = \Delta H_r - T\Delta S_r$$

where ΔH_r and ΔS_r are the enthalpy and entropy changes of the reaction. All three variables of reaction will differ in a subsequent run if the constant pressure or constant temperature differs or if the state of even one reactant or product differs. If the reactants and products are in standard states (Section 10–4), the situation is much less loose. We are now dealing with the *standard reaction*. The preceding equation becomes

$$\Delta G_r^\circ = \Delta H_r^\circ - T\Delta S_r^\circ$$

where the ° signs signify the standard states. The quantity on the left is the **standard Gibbs energy of reaction.** It equals the Gibbs energy change when reactants in standard states give products in standard states at a specified temperature:

$$\Delta G_r^\circ = G_{products}^\circ - G_{reactants}^\circ$$

The standard Gibbs energy of reaction depends strongly on temperature, so the temperature must be stated or implied in some fashion when discussing it. Often a subscript is used. If the temperature is 400 K, then

$$\Delta G_{r, 400}^\circ = G_{400, products}^\circ - G_{400, reactants}^\circ$$

If the temperature is not indicated explicitly, a temperature of 298.15 K (25°C) can be assumed. Computing the standard Gibbs energy of a reaction at a specified temperature is straightforward if the standard enthalpy and entropy of reaction at that temperature are known.

EXAMPLE 11-6

Compute the standard Gibbs energy of the reaction

$$2\ NO(g) + O_2(g) \longrightarrow 2\ NO_2(g)$$

at 298.15 K. The standard enthalpy of this reaction equals -114.14 kJ mol^{-1} at 298.15 K (as determined in Example 10–8); the standard entropy equals -146.43 J K^{-1} mol^{-1} (as determined in Example 11–3).

Strategy

Use the highlighted equation. Take care that the units of ΔH_r° and $T\Delta S_r^\circ$ are the same before subtracting.

Solution

$$\Delta G_r^\circ = \Delta H_r^\circ - T\Delta S_r^\circ$$

$$\Delta G_{r, 298.15}^\circ = -114.14 \text{ kJ mol}^{-1} - (298.15 \text{ K})(-146.43 \text{ J K}^{-1} \text{ mol}^{-1})\left(\frac{1 \text{ kJ}}{1000 \text{ J}}\right)$$

$$\Delta G_{r, 298.15}^\circ = -70.48 \text{ kJ mol}^{-1}$$

This standard reaction has a moderate driving force because the quantity $-\Delta G_r^\circ$ is a fairly large positive number.

EXERCISE

Compute ΔG_r° of the reaction

$$2\,H_2S(g) + 3\,O_2(g) \longrightarrow 2\,SO_2(g) + 2\,H_2O(g)$$

at 298.15 K. The ΔH_r° and ΔS_r° of this reaction at 298.15 K were obtained in Exercises 10–8 and 11–3, respectively.

Answer: $\Delta G_{r, 298.15}^\circ = -990.44$ kJ mol^{-1}.

Standard Molar Gibbs Energy of Formation

It is not possible to determine the absolute value of the Gibbs energy of a mole of substance in a practical experiment. For this reason, chemists define a **standard molar Gibbs energy of formation, ΔG_f°**. The definition of ΔG_f° parallels the definition of the standard molar enthalpy of formation ΔH_f° in Section 10–4:

> The standard molar Gibbs energy of formation is the change in Gibbs energy when 1 mol of a substance forms in a standard state at a specified temperature from the most stable forms of its constituent elements in standard states at the same temperature.

The equation

$$\Delta G_f^\circ = \Delta H_f^\circ - T\Delta S_f^\circ$$

can be used to compute ΔG_f° of any substance at any temperature. This relationship is a special case of the equation used in Example 11–6 because formation (subscript "f") is a particular kind of reaction (subscript "r"). For example, the standard molar Gibbs energy of formation of gaseous carbon dioxide, which is represented $\Delta G_f^\circ\,(CO_2(g))$, equals the ΔG_r° for

$$C(s,\ graphite) + O_2(g) \longrightarrow CO_2(g)$$

just as $\Delta H_f^\circ\,(CO_2(g))$ and $\Delta S_f^\circ\,(CO_2(g))$ equal the ΔH_r° and ΔS_r° for this reaction. It is instructive to complete the calculation of $\Delta G_f^\circ\,(CO_2(g))$ at 25°C. Appendix D gives

$$\Delta H_f^\circ\,(CO_2(g)) = -393.51\ \text{kJ mol}^{-1}$$

This value was determined calorimetrically. The ΔS_r° for the formation of $CO_2(g)$ is obtained by combining the standard molar entropies of graphite, gaseous oxygen, and gaseous carbon dioxide at 25°C:

$$\Delta S_r^\circ = S^\circ(CO_2(g)) - S^\circ(C(s,\ graphite)) - S^\circ(O_2(g))$$

• Recall the standard entropy of an element does *not* equal zero at 25°C.

The required numbers also come from calorimetry experiments and are tabulated in Appendix D.

$$\Delta S_r^\circ = 1(213.63\ \text{J K}^{-1}\ \text{mol}^{-1}) - 1(5.74\ \text{J K}^{-1}\ \text{mol}^{-1}) - 1(205.03\ \text{J K}^{-1}\ \text{mol}^{-1})$$
$$= 2.86\ \text{J K}^{-1}\ \text{mol}^{-1}$$

This answer equals ΔS_f° for $CO_2(g)$ because the reaction is one of formation. Then

$$\Delta G_f^\circ\,(CO_2(g)) = \Delta H_f^\circ\,(CO_2(g)) - T\Delta S_f^\circ\,(CO_2(g))$$

$$= -393.51\ \text{kJ mol}^{-1} - (298.15\ \text{K})\left(\frac{2.86\ \text{J}}{\text{K mol}}\right)\left(\frac{1\ \text{kJ}}{1000\ \text{J}}\right)$$

$$= -394.36\ \text{kJ mol}^{-1}$$

The ΔG_f° for an element in its most stable form at a specified temperature is zero because preparing an element from itself is no change at all. The formation of $Cl_2(g)$ at 298.15 K illustrates this fact. The balanced equation is

$$Cl_2(g) \longrightarrow Cl_2(g)$$

for which

$$\Delta G_f^\circ(Cl_2(g)) = \Delta H_f^\circ(Cl_2(g)) - T\Delta S_f^\circ(Cl_2(g))$$
$$= 0 - T(0) = 0 \text{ kJ mol}^{-1}$$

Standard molar Gibbs energies of formation of substances (ΔG_f°'s) are easily obtainable from ΔH_f° and S° data, as the foregoing illustrates. They are used so frequently, however, that they are generally tabulated alongside ΔH_f° and S° data, as in Appendix D. A look in Appendix D confirms that $\Delta G_f^\circ(CO_2(g))$ and $\Delta G_f^\circ(Cl_2(g))$ at 298.15 K equal -394.36 kJ mol^{-1} and 0 kJ mol^{-1}, respectively.

For the general reaction

$$aA + bB \longrightarrow cC + dD$$

$$\Delta G_r^\circ = c\Delta G_f^\circ(C) + d\Delta G_f^\circ(D) - a\Delta G_f^\circ(A) - b\Delta G_f^\circ(B)$$

This equation is used to compute a standard Gibbs energy of reaction in exactly the same way that the similar equation given in Section 10–4 is used to compute a standard enthalpy of reaction. The equation is valid because G, like H, is a state property. A relatively short table of values of ΔG_f°, such as that in Appendix D, can be used to calculate ΔG_r°'s for a host of chemical reactions at 298.15 K.

EXAMPLE 11–7

Calculate ΔG_r° for the following reaction at 298.15 K, using values of ΔG_f° from Appendix D.

$$3 \text{ NO}(g) \longrightarrow N_2O(g) + NO_2(g)$$

Solution

$$\Delta G_{r, 298.15}^\circ = \Delta G_f^\circ(N_2O) + \Delta G_f^\circ(NO_2) - 3 \Delta G_f^\circ(NO)$$

$$= 1\left(\frac{104.18 \text{ kJ}}{\text{mol}}\right) + 1\left(\frac{51.29 \text{ kJ}}{\text{mol}}\right) - 3\left(\frac{86.55 \text{ kJ}}{\text{mol}}\right)$$

$$= -104.18 \text{ kJ mol}^{-1}$$

EXERCISE

Calculate ΔG_r° for the following reaction at 298.15 K using ΔG_f° values from Appendix D:

$$Fe_2O_3(s, \text{ hematite}) + 3 \text{ Co}(s) \longrightarrow 3 \text{ CoO}(s) + 2 \text{ Fe}(s)$$

Answer: $\Delta G_{r, 298.15}^\circ = +99.5$ kJ mol^{-1}.

Effects of Temperature on $\Delta G°$

The standard Gibbs energy of a reaction depends strongly on the temperature. For this reason, a $\Delta G_r°$ calculated using $\Delta G_f°$'s from Appendix D is correct only at 298.15 K (25°C) and is quite wrong at other temperatures. Estimation of $\Delta G_r°$'s at temperatures other than 298.15 K is possible using the relationship

$$\Delta G_{r,T}° \approx \Delta H_{r,298.15}° - T\Delta S_{r,298.15}°$$

Such estimates are approximately correct because $\Delta H_r°$ and $\Delta S_r°$ do not depend strongly on temperature in most reactions (Fig. 11–12). Interestingly, both the standard absolute entropy $S°$ and standard absolute enthalpy $H°$ of substances *do* depend strongly on temperature.

Standard Gibbs energies of reaction depend strongly on temperature because T appears explicitly in the defining equation. Four different types of behavior are possible (Fig. 11–13), according to the relative signs of $\Delta H_r°$ and $\Delta S_r°$. If $\Delta H_r°$ is negative and $\Delta S_r°$ is positive, then $\Delta G_r°$ is always negative, and the standard reaction is spontaneous at all temperatures. If the reverse is true ($\Delta H_r°$ is positive and $\Delta S_r°$ is negative), then the standard reaction is never spontaneous. The other two cases are more interesting. In both, a temperature exists at which $\Delta G_r°$ equals zero. This temperature equals $\Delta H_r°$ divided by $\Delta S_r°$:

$$T = \frac{\Delta H_r°}{\Delta S_r°}$$

- If $\Delta H_r°$ and $\Delta S_r°$ have opposite signs in this formula, then T is negative. But negative absolute temperatures are meaningless, confirming that a crossover temperature exists only if $\Delta H_r°$ and $\Delta S_r°$ have the same sign.

At this temperature, the reactants and products at partial pressures of 1 atm or concentrations of 1 mol L^{-1} are in equilibrium. If $\Delta H_r°$ and $\Delta S_r°$ are both negative, the reaction is spontaneous at temperatures below this crossover point and not spontaneous above. If both are positive, the reaction is spontaneous above this temperature and not spontaneous below. An example of the latter is the decomposition of $CaCO_3$,

$$CaCO_3(s) \longrightarrow CaO(s) + CO_2(g)$$

in which both $\Delta H_r°$ and $\Delta S_r°$ are positive. The reaction occurs spontaneously, for a CO_2 gas pressure of 1 atm, only at elevated temperatures.

Figure 11–12 • The $\Delta S_r°$ (*green line*) of the reaction $3 NO(g) \longrightarrow N_2O(g) + NO_2(g)$ varies less than 5% between 0 and 300°C; the $\Delta H_r°$ (*blue line*) is even closer to being constant. $\Delta G_r°$ (*red line*), however, depends strongly on the temperature. The distinction is general: $\Delta S_r°$ and $\Delta H_r°$ change only slightly as the temperature changes, but $\Delta G_r°$ changes considerably.

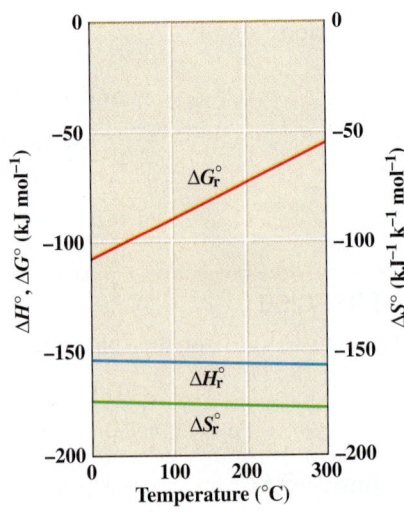

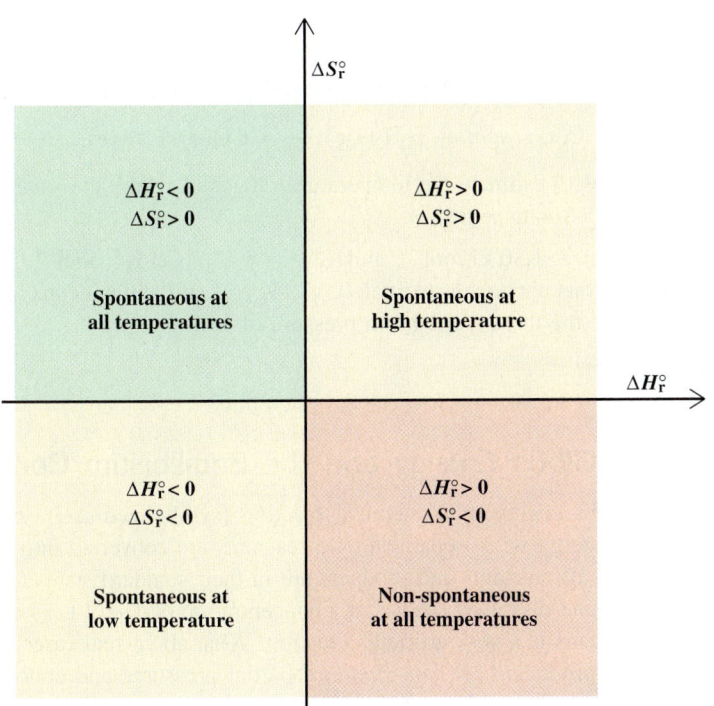

ΔS_r°

$\Delta H_r^\circ < 0$
$\Delta S_r^\circ > 0$

**Spontaneous at
all temperatures**

$\Delta H_r^\circ > 0$
$\Delta S_r^\circ > 0$

**Spontaneous at
high temperature**

ΔH_r°

$\Delta H_r^\circ < 0$
$\Delta S_r^\circ < 0$

**Spontaneous at
low temperature**

$\Delta H_r^\circ > 0$
$\Delta S_r^\circ < 0$

**Non-spontaneous
at all temperatures**

Figure 11–13 • The spontaneity of a reaction at different temperatures depends on the signs of ΔH_r° and ΔS_r°. When both signs are the same, the temperature determines the spontaneity of reaction.

EXAMPLE 11–8

A slow reaction in the Earth's interior combines dolomite with quartz to produce diopside and carbon dioxide:

$$CaMg(CO_3)_2(s) + 2\ SiO_2(s) \longrightarrow CaMgSi_2O_6(s) + 2\ CO_2(g)$$

dolomite   quartz   diopside   carbon dioxide

The standard enthalpy and standard entropy of this reaction at 298.15 K are

$$\Delta H_r^\circ = +155\ \text{kJ mol}^{-1} \qquad \Delta S_r^\circ = +331\ \text{J K}^{-1}\ \text{mol}^{-1}$$

Assume that the CO_2 produced is at atmospheric pressure (which is not very accurate for the interior of the Earth), and estimate the temperature ranges in which this reaction is and is not spontaneous.

Strategy

Assume that ΔH_r° and ΔS_r° do not change with temperature. Find the temperature at which $\Delta H_r^\circ = T\Delta S_r^\circ$.

Solution

$$T = \frac{\Delta H_r^\circ}{\Delta S_r^\circ} = \frac{155{,}000\ \text{J mol}^{-1}}{331\ \text{J K}^{-1}\ \text{mol}^{-1}} = 468\ \text{K} = 195°\text{C}$$

Because ΔH_r° and ΔS_r° are both positive, the reaction becomes spontaneous above about 195°C and remains non-spontaneous below about 195°C.

EXERCISE

Obtain data from Appendix D and compute ΔH_r° and ΔS_r° at 298.15 K for

$$C(s, graphite) + H_2O(g) \longrightarrow CO(g) + H_2(g)$$

Use these values to estimate the temperature ranges in which the standard reaction is and is not spontaneous.

Answer: $\Delta H_r^\circ = 131.30$ kJ mol^{-1}, and $\Delta S_r^\circ = +133.67$ J K^{-1} mol^{-1}. The reaction is spontaneous above approximately 980 K and non-spontaneous below 980 K when each of the three gases has a pressure of 1 atm.

11-7   The Gibbs Energy and the Equilibrium Constant

The standard Gibbs energy of a chemical reaction (symbolized ΔG_r°) equals the change in Gibbs energy when separated pure reactants are converted into separated pure products and all reactants and products are in their standard states (gases at a pressure of 1 atm and dissolved species at a concentration of 1 mol L^{-1}) at a specified temperature. This is a very artificial situation. What about real cases, in which the reactants and products have *non-standard* partial pressures and concentrations and are mixed together? To answer this question, recall the reaction quotient Q (defined in Section 7–3). The reaction quotient gives the progress of a reaction. It varies from zero (pure reactants) to infinity (pure products) and becomes equal to the equilibrium constant (K) at equilibrium. The relative values of Q and K establish the direction that a reaction takes in coming to equilibrium. If $Q < K$, then the reaction shifts toward the product side to reach equilibrium; if $Q > K$, it shifts toward the reactant side. It is therefore not surprising to discover a connection between ΔG_r and Q (Table 11–1). The exact relationship between the two is given by the equation

$$\Delta G_r = \Delta G_r^\circ + RT \ln Q$$

- The natural (or base e) logarithm of a number is written "ln." Scientific calculators have an "ln x" key for natural logarithms and an "e^x" key (or the equivalent) for the inverse operation, calculating the exponential of x (Appendix C–1).

This important equation merits close attention. If all the reactants and products are in standard states, then the reaction quotient Q equals 1. Because the logarithm of 1 is zero, the last term vanishes, and $\Delta G_r = \Delta G_r^\circ$, as is to be expected from the definitions of standard states. Whenever at least one partial pressure is not 1 atm or at least one concentration is not 1 mol L^{-1}, we simply insert the actual values into the expression for Q and proceed to evaluate the sign and mag-

TABLE 11–1	Criteria for Spontaneity in a Chemical Reaction		
Spontaneous Processes	**Equilibrium Processes**	**Non-spontaneous Processes**	**Conditions**
$\Delta S_{univ} > 0$	$\Delta S_{univ} = 0$	$\Delta S_{univ} < 0$	All conditions
$\Delta G_r < 0$	$\Delta G_r = 0$	$\Delta G_r > 0$	Constant P and T[a]
$Q < K$	$Q = K$	$Q > K$	Constant P and T[a]

[a]Indicates that $\Delta P = 0$ and $\Delta T = 0$. Pressure and temperature may vary during the course of the process as long as they attain their original values at the end.

nitude of ΔG_r. At equilibrium, ΔG_r becomes zero because no further tendency exists for the reaction to occur in either direction. At the same time, Q becomes equal to the equilibrium constant K. Substituting these two values into the preceding equation gives

$$0 = \Delta G_r^\circ + RT \ln K$$

or

$$-\Delta G_r^\circ = RT \ln K$$

This result provides a way to the calculate equilibrium constants from standard Gibbs energies of reaction (ΔG_r°'s), which are themselves readily obtained from tabulated standard Gibbs energies of formation (ΔG_f°'s) as illustrated in Example 11–7. On a deeper level, it shows that the extent to which any chemical reaction proceeds before reaching equilibrium can be deduced from calorimetric data alone.

- Note that ΔG_r° (unlike ΔG_r) depends only on the intrinsic properties of the products and reactants and does not change during the course of the reaction.

- This linkage of the results of thermochemical experiments to the nature of the equilibrium state is at the heart of chemical thermodynamics.

EXAMPLE 11–9

The ΔG_r° at 25°C for

$$3 \text{ NO}(g) \longrightarrow \text{N}_2\text{O}(g) + \text{NO}_2(g)$$

was obtained in Example 11–7. Now, calculate the equilibrium constant of this reaction at 25°C.

Solution

The ΔG_r° equals $-104.18 \text{ kJ mol}^{-1}$. Rearrange the equation $-\Delta G_r^\circ = RT \ln K$ and substitute to compute K:

$$\ln K = \frac{-\Delta G_r^\circ}{RT} = -\frac{-104{,}180 \text{ J mol}^{-1}}{(8.3145 \text{ J K}^{-1} \text{ mol}^{-1})(298.15 \text{ K})} = 42.03$$

$$K = \text{antiln } 42.03 = e^{42.03} = 1.8 \times 10^{18}$$

If we had written the reaction with doubled coefficients (6 NO $\longrightarrow$ 2 N$_2$O + 2 NO$_2$), then ΔG_r° would have been doubled, to $-208{,}360 \text{ kJ mol}^{-1}$ (because each mole of reaction converts 6 mol of NO into 2 mol of N$_2$O and 2 mol of NO$_2$). In the calculation, $\ln K$ would have been doubled and K would have been squared, exactly suiting the equilibrium-constant expression that is written for an equation with doubled coefficients (see Section 7–2).

EXERCISE

Calculate the equilibrium constants at 25°C for the reaction that is the reverse of the above reaction and also for the reaction that is the reverse multiplied through by 1/2:

$$\text{N}_2\text{O}(g) + \text{NO}_2(g) \longrightarrow 3 \text{ NO}(g) \text{ and } \tfrac{1}{2} \text{N}_2\text{O}(g) + \tfrac{1}{2} \text{NO}_2(g) \longrightarrow \tfrac{3}{2} \text{NO}(g)$$

State the relationship between these answers and the answer to Example 11–9.

Answer: 5.6×10^{-19} and 7.5×10^{-10}, respectively. The first is the reciprocal of the answer in Example 11–9; the second is the square root of the reciprocal of that answer.

The conversion of NO(g) to $N_2O(g)$ plus $NO_2(g)$ is spontaneous when all three gases are present at 1 atm and at 25°C. Its reverse is therefore non-spontaneous. The forward reaction is scarcely observed because it is so slow at room temperature, but its equilibrium constant nonetheless can be calculated. Such calculations often have enormous practical impact. All three of the nitrogen oxides in Example 11–9, for example, play roles in air pollution, but N_2O is perhaps the least noxious. This calculation ensures that a plan to exploit the reaction to cut down the amount of NO in cooled automobile exhaust would not be thermodynamically doomed at the outset. The fundamental driving force is there. Success of the plan would hinge on making the reaction go faster.

• Although much less poisonous than NO or NO_2, N_2O is still a serious atmospheric pollutant because it acts as a greenhouse gas (see Section 18–5).

The relationship between ΔG_r and Q can be used to calculate the Gibbs energy for a reaction under non-standard conditions. Start with the equation

$$\Delta G_r = \Delta G_r^\circ + RT \ln Q$$

and replace ΔG_r° with $-RT \ln K$ (the two are equal):

$$\Delta G_r = -RT \ln K + RT \ln Q$$
$$= RT\,(\ln Q - \ln K)$$

Using the property of logarithms that

$$\ln a - \ln b = \ln \frac{a}{b}$$

the equation becomes

$$\Delta G_r = RT \ln \frac{Q}{K}$$

Here the connection between the two different criteria for spontaneous reaction is quite clear. It is summarized in Figure 11–14. If the reaction quotient Q is less than K, then the quantity Q/K is less than 1. The logarithm of any positive number that is less than 1 is negative, so ΔG_r is less than 0. The reaction proceeds as written, from left to right. If Q exceeds K, then ΔG_r is greater than 0, and the reverse reaction (right to left) occurs until equilibrium is reached. The reaction-quotient criterion for the direction of reaction is a consequence of the second law of thermodynamics.

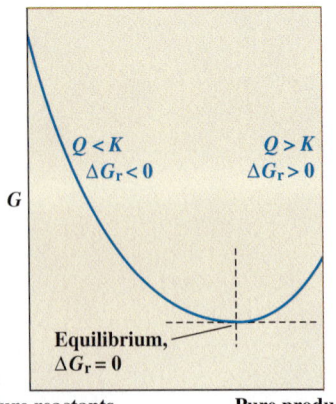

$Q < K$ $Q > K$
$\Delta G_r < 0$ $\Delta G_r > 0$

G

Equilibrium,
$\Delta G_r = 0$

Pure reactants   Pure products

Figure 11–14 • The Gibbs energy of a reaction is plotted against its progress from pure reactants (*left*) to pure products (*right*). The quantity ΔG_r is best understood as the slope of this line. Equilibrium comes at the lowest point of the line, where $\Delta G_r = 0$. To the reactant side of equilibrium, the slope is negative, $\Delta G_r < 0$, and $Q < K$. To the product side, the slope is positive, $\Delta G_r > 0$, and $Q > K$.

EXAMPLE 11–10

Calculate the Gibbs energy of reaction (ΔG_r) at 25°C for

$$3\,NO(g) \longrightarrow N_2O(g) + NO_2(g)$$

when the partial pressures of the three gases are all 0.0010 atm rather than 1 atm. This is the same reaction considered in Example 11–9.

Solution

Under conditions in which all partial pressures are 0.0010 atm, the reaction quotient is

$$Q = \frac{(P_{N_2O})(P_{NO_2})}{(P_{NO})^3} = \frac{(0.0010)(0.0010)}{(0.0010)^3} = 1.0 \times 10^3$$

Substituting this Q and the K from Example 11–9 into the equation that was just derived gives

$$\Delta G_{r,\,298.15} = RT \ln \frac{Q}{K}$$

$$= \left(\frac{8.3145\ \text{J}}{\text{K mol}} \right)(298.15\ \text{K}) \ln \left(\frac{1.0 \times 10^3}{1.8 \times 10^{18}} \right)$$

$$= -8.71 \times 10^4 \frac{\text{J}}{\text{mol}} = \boxed{-87.1\ \text{kJ mol}^{-1}}$$

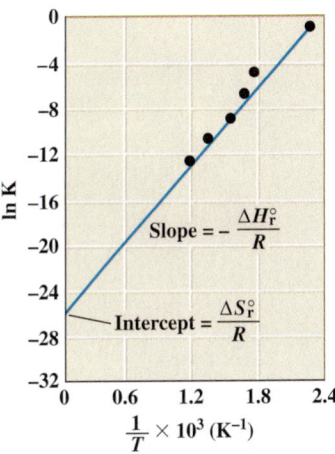

Check

The quantity $-\Delta G_r$ measures the driving force of the reaction at constant temperature and pressure. In this reaction, lower pressures reduce the magnitude of the driving force. This agrees with the prediction from Le Chatelier's principle (Chapter 7) that in a reaction in which the chemical amount of gas decreases, an increase in volume shifts the equilibrium toward the reactants.

EXERCISE

Calculate the Gibbs energy of reaction (ΔG_r) at 25°C for

$$N_2O(g) + NO_2(g) \longrightarrow 3\ NO(g)$$

when the partial pressures of the three gases are all 2.00 atm.

Answer: $\Delta G_{r,\,298.15} = +105.9\ \text{kJ mol}^{-1}$.

Figure 11–15 • The temperature dependence of the natural logarithm of the equilibrium constant for the reaction

$$N_2(g) + 3\ H_2(g) \rightleftharpoons 2\ NH_3(g)$$

Experimental data are shown by points.

11-8 The Temperature Dependence of Equilibrium Constants

Chemists must often estimate the value of an equilibrium constant for a reaction that is to take place at a constant temperature other than 25°C. Equilibrium constants depend strongly on temperature, so using $K_{298.15}$ is unacceptable. Fortunately, ΔH_r° and ΔS_r° depend only weakly on temperature, at least over a limited temperature range (recall Fig. 11–12). To the extent that the temperature dependence of ΔH_r° and ΔS_r° can be neglected, $\ln K$ is a linear function of $1/T$:

$$\ln K = -\frac{\Delta G_r^\circ}{RT} = -\frac{\Delta H_r^\circ}{RT} + \frac{\Delta S_r^\circ}{R}$$

A graph of $\ln K$ against $1/T$ is approximately a straight line with a slope of $-\Delta H_r^\circ /R$ and an intercept of $\Delta S_r^\circ /R$ (Fig. 11–15).

Besides giving equilibrium constants at various temperatures, this equation opens a route for calculating enthalpies and entropies of reaction from experimentally measured equilibrium constants. This is particularly useful for solution-phase reactions because entropies cannot be measured calorimetrically for such reactions. If K_1 and K_2 are the equilibrium constants of a reaction at temperatures T_1 and T_2, respectively, then

$$\ln K_2 = -\frac{\Delta H_r^\circ}{RT_2} + \frac{\Delta S_r^\circ}{R}$$

$$\ln K_1 = -\frac{\Delta H_r^\circ}{RT_1} + \frac{\Delta S_r^\circ}{R}$$

• Appendix C–2 reviews the slope-intercept form of the equation for a straight line.

Subtracting the second equation from the first gives

$$\ln K_2 - \ln K_1 = \frac{\Delta H_r^\circ}{R}\left(\frac{1}{T_1} - \frac{1}{T_2}\right)$$

$$\ln\frac{K_2}{K_1} = \frac{\Delta H_r^\circ}{R}\left(\frac{1}{T_1} - \frac{1}{T_2}\right)$$

This is known as the **van't Hoff equation,** after the Dutch chemist Jacobus van't Hoff. If ΔH_r° and K are known at one temperature, the van't Hoff equation allows for the calculation of K at a second temperature, in the approximation that ΔH_r° and ΔS_r° are constant. Alternatively, it can be used to estimate ΔH_r° without using a calorimeter by measuring the value of K at two different temperatures.

The van't Hoff equation shows that if ΔH_r° is negative (that is, if the standard reaction is exothermic, giving off heat), then an increase in temperature reduces K. If ΔH_r° is positive (if the standard reaction is endothermic, taking up heat), then an increase in temperature *increases* K. These conclusions about the effect of temperature on equilibrium constants are the same as those stated in Section 7–5 in connection with Le Châtelier's principle.

EXAMPLE 11–11

The manufacture of sulfuric acid requires the oxidation of sulfur dioxide to sulfur trioxide:

$$SO_2(g) + \tfrac{1}{2}O_2(g) \longrightarrow SO_3(g)$$

At 25°C, the equilibrium constant for this reaction equals 2.6×10^{12}, but the reaction occurs very slowly. Calculate K for this reaction at 550°C, assuming ΔH_r° and ΔS_r° to be independent of temperature over the range from 25 to 550°C.

Strategy

Find ΔH_r° at 298.15 K using ΔH_f° data from Appendix D. Substitute ΔH_r° in the van't Hoff equation along with the given T's and $K_{298.15}$.

Solution

$$\Delta H_r^\circ = 1(-395.72 \text{ kJ mol}^{-1}) - 1(-296.83 \text{ kJ mol}^{-1}) - \tfrac{1}{2}(0 \text{ kJ mol}^{-1})$$
$$= -98.89 \text{ kJ mol}^{-1}$$

$$\ln\left(\frac{K_{823}}{K_{298.15}}\right) = \frac{\Delta H_r^\circ}{R}\left(\frac{1}{298.15 \text{ K}} - \frac{1}{823 \text{ K}}\right)$$

$$= \left(\frac{-98,890 \text{ J mol}^{-1}}{8.315 \text{ J K}^{-1} \text{ mol}^{-1}}\right)\left(\frac{1}{298.15 \text{ K}} - \frac{1}{823 \text{ K}}\right) = -25.44$$

$$\frac{K_{823}}{K_{298.15}} = e^{-25.44} = 8.9 \times 10^{-12}$$

$$K_{823} = (8.9 \times 10^{-12})(2.6 \times 10^{12}) = 23$$

Check

The increase from room temperature to 823 K reduces K, as it should for this exothermic reaction. Experimentally, K_{823} equals 21.5. The ΔH_r° and ΔS_r° in fact do change somewhat over this large range of temperature.

EXERCISE

The reaction

$$2 \text{ Al}_3\text{Cl}_9(g) \longrightarrow 3 \text{ Al}_2\text{Cl}_6(g)$$

has an equilibrium constant of 8.8×10^3 at 443 K and a ΔH_r° at 443 K of 39.8 kJ mol^{-1}. Estimate the equilibrium constant at a temperature of 600 K.

Answer: $K_{600} = 1.5 \times 10^5$.

The Variation of Vapor Pressures with Temperature

Equilibria between pure liquids and their vapors are simple. They all have the form

$$\text{liquid}(\ell) \rightleftharpoons \text{vapor}(g) \qquad P_{\text{vapor}} = K$$

Like other equilibrium constants, K's for vaporizations depend strongly on the temperature. When the temperature dependence follows the van't Hoff equation, easy predictions of vapor pressures, which are technically important quantities, become possible. To see this, write the van't Hoff equation for a vaporization at two different temperatures:

$$\ln\left(\frac{K_2}{K_1}\right) = \ln\left(\frac{P_{\text{vapor}(2)}}{P_{\text{vapor}(1)}}\right) = \left(\frac{\Delta H_{\text{vap}}^\circ}{R}\right)\left(\frac{1}{T_1} - \frac{1}{T_2}\right)$$

Again, the assumption is that ΔH° and ΔS° of the process (this time a vaporization) are nearly independent of temperature. The vapor pressure at any given temperature can be estimated if it is known at some other temperature and if the enthalpy of vaporization also is known. The above equation is the **Clausius–Clapeyron equation.**

At the normal boiling point of a substance T_b, the vapor pressure is 1 atm. Taking T_1 to correspond to T_b and T_2 to some other temperature T, we find

$$\ln P_{\text{vapor}, T} = \frac{\Delta H_{\text{vap}}^\circ}{R}\left(\frac{1}{T_b} - \frac{1}{T}\right)$$

where $P_{\text{vapor}, T}$ is the vapor pressure expressed in atmospheres at the temperature T.

EXAMPLE 11–12

The molar enthalpy of vaporization of water is 40.66 kJ mol^{-1} at the normal boiling point $T_b = 373.15$ K. Assuming $\Delta H_{\text{vap}}^\circ$ and $\Delta S_{\text{vap}}^\circ$ are independent of temperature from 50 to 100°C, calculate the vapor pressure of water at 50°C (323 K).

Solution

$$\ln P_{\text{vapor}, 323} = \left(\frac{40,660 \text{ J mol}^{-1}}{8.315 \text{ J K}^{-1}\text{ mol}^{-1}}\right)\left(\frac{1}{373 \text{ K}} - \frac{1}{323 \text{ K}}\right) = -2.03$$

$$P_{\text{vapor}, 323} = e^{-2.03} = 0.131 \text{ atm}$$

Check

Table 6–1 (page 251) gives 0.1217 atm as the experimental vapor pressure of water at 50°C. This estimate is off by about 8% because $\Delta H_{\text{vap}}^\circ$ changes with temperature, an effect that is neglected in the Clausius–Clapeyron equation.

EXERCISE

The molar enthalpy of vaporization of butane at its normal boiling point of -0.50°C is 21.93 kJ mol^{-1}. Estimate the vapor pressure of butane at 25.0°C.

Answer: 2.29 atm.

CHEMISTRY IN YOUR LIFE

Life Does Not Violate the Second Law of Thermodynamics

According to the second law, the entropy (and, therefore, in a loose sense, the disorder) of the universe is constantly increasing. How is this reconciled with the evolution and existence of highly sophisticated life forms that are clearly in states of considerable order?

The answer is implied early in this chapter. A decrease in entropy is possible in a system if it is coupled to another system in which the entropy increases by a larger amount. Life on Earth, and the localized decrease in entropy that goes with it, is coupled to nuclear reactions in the Sun that are strongly exothermic and cause increases in entropy that more than compensate. The medium of coupling is sunlight.

The most direct use of sunlight on Earth is photosynthesis by green plants to convert carbon dioxide and water to more complex chemical structures. Cellulose and starch are important plant constituents that are made up of chains of sugar molecules. One such sugar is glucose, which forms, after many steps, according to the overall equation

$$6\,CO_2(g) + 6\,H_2O(\ell) \longrightarrow C_6H_{12}O_6(aq) + 6\,O_2(g)$$

At 298.15 K, the standard entropy of this reaction is negative ($\Delta S_r^\circ = -207$ J K^{-1} mol^{-1}) and the standard Gibbs energy is positive ($\Delta G_r^\circ = +2872$ kJ mol^{-1}). Glucose forms in plants against these barriers only by the intervention of light energy. Once substances of high Gibbs energy such as glucose are available, they can be decomposed (metabolized) to drive other, coupled chemical reactions. The overall scheme is outlined in Figure 11–A. Terrestrial organisms metabolize sugars and other substances to drive the production of a specific compound of high Gibbs energy, adenosine triphosphate (ATP), from

its precursor, adenosine diphosphate (ADP). Energy from subsequent reconversion of ATP to ADP is then harnessed to carry out processes that otherwise would not occur: the synthesis of proteins, the building of cell walls, and the replication of genetic material. The flow of energy from the Sun drives the local build-up of order represented by living organisms on Earth.

A similar analysis applies to the industrial production of materials of high Gibbs energy. The metals are an important example. The change in the Gibbs energy to produce a metal, let us say iron, from an oxide ore is positive:

$$Fe_2O_3(s) \longrightarrow 2\,Fe(s) + \tfrac{3}{2}O_2(g)$$
$$\Delta G^\circ = +742 \text{ kJ mol}^{-1}$$

If it were possible, we would couple this reaction directly to the nuclear reactions in the sun, producing iron through the use of sunlight. Because we cannot yet do this, we drive it by coupling to another chemical reaction with a negative ΔG:

$$\tfrac{3}{2}C(s, \textit{graphite}) + \tfrac{3}{2}O_2(g) \longrightarrow \tfrac{3}{2}CO_2(g)$$
$$\Delta G^\circ = -592 \text{ kJ mol}^{-1}$$

As the numbers show, even the coupling of this strongly spontaneous oxidation does not suffice to favor the overall process at 25°C. The reactants must be heated to a high temperature to produce metallic iron. The carbon used is in the form of coke, which is obtained from coal and ultimately from decayed plant matter, a resource of high G stored up from sunshine over many millions of years. A continuing major concern for human civilization is the rate at which petroleum, coal, and other resources of high Gibbs energy are being depleted.

SUMMARY

11-1 The first law of thermodynamics is a principle of constancy (the energy of the universe is always conserved). It says nothing about why **spontaneous** change (change taking place without outside intervention) proceeds in one direction only. The sign of ΔH, the change in enthalpy, does not predict the direction of spontaneous change in a process.

11-2 **Entropy**, a state property of chemical and physical systems, is related to the number of microscopic ways in which a particular state of a system can be attained. Higher entropy corresponds to a larger number of available microscopic states that have identical macroscopic characteristics. When a constraint is removed from

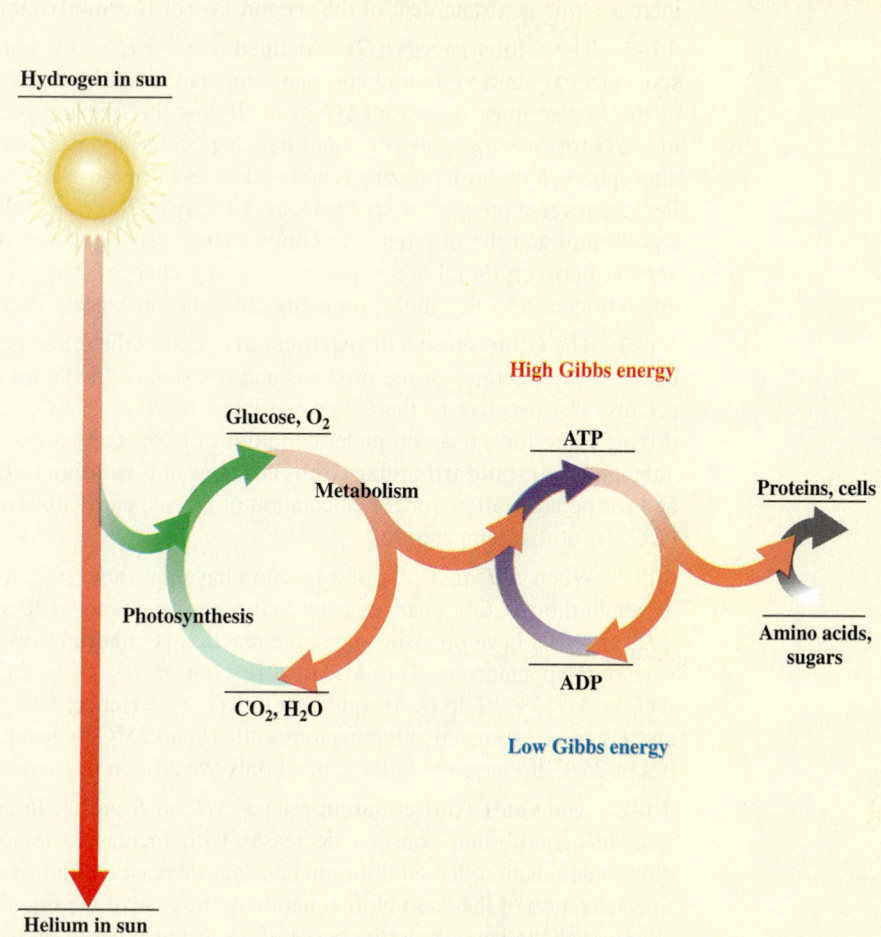

Figure 11-A • The synthesis of glucose and other sugars in plants, the production of ATP from ADP, and the elaboration of proteins and other biological molecules are all processes in which the Gibbs energy of the system must increase. They occur only through coupling to other processes in which the Gibbs energy decreases by an even larger amount.

a system, the entropy of the system increases because the number of accessible microscopic states (**microstates**) has increased. The direction of change is rooted in probability. The reverse of a spontaneous change is overwhelmingly unlikely to occur in the macroscopic systems dealt with in chemistry and in daily life.

11-3   According to the **third law of thermodynamics,** the entropy of any crystalline substance approaches zero as the absolute zero of temperature is approached. The absolute entropy of 1 mol of a substance in a standard state is a **standard molar entropy.** From tabulated standard molar entropies of the reactants and products, the standard entropy of a chemical reaction can be calculated. Reactions in which the total chemical amount of gas increases almost invariably have positive entropy changes.

11-4 The entropy of the system may increase or decrease in a spontaneous process, but when the entropy of the surroundings is added, the total must always increase; this is a statement of the **second law of thermodynamics.**

11-5 The **Gibbs energy** (G) is defined as $G = H - TS$. For changes to occur spontaneously in a system at constant temperature and pressure, the Gibbs energy of the system must decrease ($\Delta G < 0$). Below the freezing point (T_{freez}) of a liquid, ΔG for freezing is negative and freezing occurs spontaneously. Above T_{freez} (at atmospheric pressure), freezing is non-spontaneous because ΔG is positive, but melting (the reverse process) is spontaneous. Exactly at T_{freez}, the liquid and solid are in equilibrium and the difference in Gibbs energy between them is zero. In the conversion between liquid and vapor, the entropy change (ΔS_{vap}) at the boiling point approximates 88 J K^{-1} mol^{-1} for many different substances. This is **Trouton's rule.**

11-6 The **Gibbs energy of reaction** (ΔG_r) equals the difference between the sum of the Gibbs energies of the products and the sum of the Gibbs energies of the reactants. The negative of the Gibbs energy of reaction ($-\Delta G_r$) is a measure of the driving force for a reaction under conditions of constant temperature and pressure. Tabulations of **standard molar Gibbs energies of formation** (ΔG_f°'s) of substances at a temperature allow for the calculation of the standard Gibbs energies of reaction (ΔG_r°'s) at that temperature.

11-7 When ΔH_r° and ΔS_r° of a reaction have the same sign, a temperature exists at which the reaction changes from being non-spontaneous to spontaneous. When ΔH_r° and ΔS_r° have opposite signs, the reaction is either always spontaneous or always non-spontaneous. The ΔG_r of a reaction is related to ΔG_r° by the equation $\Delta G_r = \Delta G_r^{\circ} + RT \ln Q$. At equilibrium, $\Delta G_r = 0$. Hence, $\Delta G_r^{\circ} = -RT \ln K$. ΔG_r° and K can be estimated at temperatures other than 25°C by using ΔH_r° and ΔS_r° values at 25°C because the latter depend only weakly on temperature.

11-8 The **van't Hoff equation** relates ΔH_r° to K and T. In an exothermic reaction, the equilibrium constant decreases with increasing temperature; in an endothermic reaction, the equilibrium constant increases with increasing temperature. Specialization of the van't Hoff equation to the case of the liquid–vapor phase transition gives the vapor pressure curve of a substance based on its molar enthalpy of vaporization and its normal boiling point. This relationship is the **Clausius–Clapeyron equation.**

PROBLEMS

Note: Answers to blue-numbered problems are given in Appendix F. Problems that are more challenging are indicated with asterisks.

Enthalpy and Spontaneous Change

1. Write a balanced equation for (a) a process that is exothermic and spontaneous and (b) one that is endothermic and spontaneous.

2. Write a balanced equation for (a) a process that is exothermic and *not* spontaneous and (b) one that is endothermic and *not* spontaneous.

3. According to the erroneous principle of Berthelot and Thomsen, spontaneous change occurs in systems in such a way that the maximum heat is released. Give three examples of spontaneous changes that absorb heat and therefore violate this principle (state the conditions of each change).

4. As the temperature is increased, does the principle of Berthelot and Thomsen become more nearly or less nearly correct?

Entropy

5. (See Example 11–1).
   (a) How many "microstates" are there for the numbers that come up on a pair of dice?
   (b) What is the probability that a roll of a pair of dice shows two sixes?

6. (See Example 11–1.)
   (a) Suppose a volume is divided into *three* equal parts. How many microstates exist for distributing four molecules among the three parts?
   (b) What is the probability that all four molecules are located in the left-most third of the volume at the same time?

7. (See Example 11–2.) Predict the sign of the entropy change of the system in each of the following processes.
   (a) Sodium chloride melts.
   (b) A building is demolished.
   (c) A volume of air is divided into three separate volumes of nitrogen, oxygen, and argon, each at the same pressure and temperature as the original air.

8. (See Example 11–2.) Predict the sign of the entropy change of the system in each of the following processes:
   (a) A computer is constructed from iron, copper, carbon, hydrogen, silicon, gallium, lead, and arsenic.
   (b) A container holding a compressed gas develops a leak and the gas enters the atmosphere.
   (c) Solid carbon dioxide (dry ice) sublimes to gaseous carbon dioxide.

Absolute Entropies and Chemical Reactions

9. Which member of each pair has the higher standard entropy $(S°)$? Consult Appendix D if necessary.
   (a) 1 mol $F_2(g, 300 \text{ K})$ or 1 mol $F(g, 300 \text{ K})$
   (b) 1 mol $H_2O(g, 300 \text{ K})$ or 1 mol $H_2O(\ell, 300 \text{ K})$
   (c) 2 mol $O_2(g, 300 \text{ K})$ or 2 mol $O_2(g, 400 \text{ K})$
   (d) 10 g $O_2(g, 298.15 \text{ K})$ or 10 g $O_3(g, 298.15 \text{ K})$

10. Which member of each pair has the higher standard entropy $(S°)$? Consult Appendix D if necessary.
   (a) 10 g $P_2(g, 298.15 \text{ K})$ or 5 g $P_2(g, 298.15 \text{ K})$
   (b) 1 mol $F_2(g, 298.15 \text{ K})$ or 1 mol $I_2(g, 298.15 \text{ K})$
   (c) 1 mol $O_2(g, 298.15 \text{ K})$ or 1 mol $O_3(g, 298.15 \text{ K})$
   (d) 34 g $H_2O_2(g, 298.15 \text{ K})$ or 32 g $O_2(g, 298.15 \text{ K})$

11. Without consulting Appendix D, predict the sign of the entropy change (ΔS) for the following processes:
   (a) $SF_4(g) + F_2(g) \longrightarrow SF_6(g)$ at 298.15 K and 1 atm
   (b) $H_2S(g) + NH_3(g) \longrightarrow NH_4HS(s)$ at 298.15 K
   (c) $O_2(g) \longrightarrow 2 O(g)$ at 900 K and 2 atm
   (d) $CH_3OH(\ell) \longrightarrow CH_3OH(s)$ at 298.15 K and 1 atm

12. Without consulting Appendix D, predict the sign of the entropy change (ΔS) for the following processes:
   (a) $Ar(g, 600 \text{ K}) \longrightarrow Ar(g, 200 \text{ K})$ at 1 atm
   (b) $2 NH_4ClO_4(s) \longrightarrow$
       $N_2(g) + Cl_2(g) + 4 H_2O(g) + 2 O_2(g)$ at 400 K and 4 atm
   (c) $P_4(g) \longrightarrow 2 P_2(g)$ at 298.15 K and 1 atm
   (d) $C_6H_5CH_3(g) \longrightarrow C_6H_5CH_3(\ell)$ at 298.15 K and 1 atm

13. (See Example 11–3.) (a) Use data from Appendix D to calculate the standard entropy change $(\Delta S_r°)$ at 25°C for

$$N_2H_4(\ell) + 3 O_2(g) \longrightarrow 2 NO_2(g) + 2 H_2O(\ell)$$

   (b) Calculate the standard entropy change when 120.00 g of $N_2H_4(\ell)$ reacts with excess $O_2(g)$ to give the same products.

14. (See Example 11–3.) (a) Use data from Appendix D to calculate $\Delta S_r°$, the standard entropy change accompanying the reaction

$$CH_3COOH(g) + NH_3(g) \longrightarrow$$
$$CH_3NH_2(g) + CO_2(g) + H_2(g)$$

   at 25°C.
   (b) Calculate the standard entropy change when 120.00 g of $CH_3NH_2(g)$ is formed by this reaction.

15. (See Example 11–3.) The solid alkali metals (Li(s), Na(s), K(s), Rb(s), Cs(s)) react with $Cl_2(g)$ to give solid salts. Write and balance the equations for these reactions. Using the data in Appendix D, compute $\Delta S_r°$ at 25°C for each reaction and identify a periodic trend, if any.

16. (See Example 11–3.) The gaseous halogens ($F_2(g)$, $Cl_2(g)$, $Br_2(g)$, $I_2(g)$) all react with $H_2(g)$ to give binary compounds. Write and balance equations for these reactions. Using the data in Appendix D, compute $\Delta S_r°$ at 25°C for each reaction and identify a periodic trend, if any.

17. (See Example 11–4.) Compute $\Delta S_r°$ at 25°C for

$$Ba^{2+}(aq) + 2 Cl^-(aq) \longrightarrow BaCl_2(s)$$

18. (See Example 11–4.) Compute $\Delta S_r°$ for

$$I^-(aq) + I_2(aq) \longrightarrow I_3^-(aq)$$

The Second Law of Thermodynamics

19. When $H_2O(\ell)$ and $D_2O(\ell)$ are mixed, the following reaction occurs spontaneously:

$$H_2O(\ell) + D_2O(\ell) \longrightarrow 2 HOD(\ell)$$

   There is little difference between the enthalpy of an O—H and an O—D bond. What is the main driving force of this reaction?

20. The two gases $BF_3(g)$ and $BCl_3(g)$ are mixed in equal molar amounts. All B—F bonds have about the same bond enthalpy, as do all B—Cl bonds. Explain why the mixture tends to react to form $BF_2Cl(g)$ and $BCl_2F(g)$.

21. In a recent book about entropy, the author writes, "The second law, the Entropy Law, states that matter and energy can only be changed in one direction, that is, from usable to unusable, or from available to unavailable, or from ordered to disordered." This statement is glaringly wrong. Identify the errors.

22. The following advertisement is taken from a recent edition of a major metropolitan newspaper.

FREE ELECTRICITY EXTRACTED
FROM THE AIR
We have discovered how to produce torque with a closed loop Rankine cycle turbine powered by free energy taken from the air with an advanced heat pump, even in sub-zero weather.

EXPLOSIVE INCOME

You can manufacture or sell this technology including an air conditioner that operates for free, a water heater that can do 3000 gallons of free hot water per day and torque for a variety of purposes.

Do the claims in this advertisement violate the first law of thermodynamics? Do the claims violate the second law of thermodynamics?

23. The dissolution of calcium chloride,

$$CaCl_2(s) \longrightarrow Ca^{2+}(aq) + 2\,Cl^-(aq)$$
$$\Delta S^\circ = -44.7 \text{ J K}^{-1} \text{ mol}^{-1}$$

is spontaneous at 25°C. What conclusion can you draw about the change in entropy of the surroundings in this process?

24. Quartz ($SiO_2(s)$) does *not* spontaneously decompose to silicon and oxygen at 25°C in the reaction

$$SiO_2(s) \longrightarrow Si(s) + O_2(g)$$
$$\Delta S^\circ = +182.02 \text{ J K}^{-1} \text{ mol}^{-1}$$

Why?

25. Graphite is the most stable form of elemental carbon at a temperature of 298.15 K and a pressure of 1 atm. Explain why Appendix D gives zero as the ΔH_f° of C(s, graphite) at 298.15 K but lists a non-zero value for S°.

26. A classmate points to the equation $\Delta G_f^\circ = \Delta H_f^\circ - T\Delta S_f^\circ$ in the text and then to the following data taken from Appendix D:

	ΔH_f° (kJ mol^{-1})	S° (J K^{-1} mol^{-1})	ΔG_f° (kJ mol^{-1})
NaCl(s)	-411.15	72.13	-384.15

The classmate asks why the entry in the third column does not equal the entry in the first column minus 298.15 times the entry in the second column. Answer fully.

The Gibbs Energy

27. (See Example 11–5.) Tungsten melts at 3410°C and has an enthalpy of fusion of 35.4 kJ mol^{-1} at that temperature. Calculate the entropy of fusion of tungsten.

28. (See Example 11–5.) Tetraphenylgermane (($C_6H_5)_4$Ge) has a melting point of 232.5°C, and its enthalpy increases during fusion by 106.7 J g^{-1}. Calculate the molar enthalpy of fusion and the molar entropy of fusion of tetraphenylgermane.

29. (See Example 11–5.) The enthalpy of fusion of solid ammonia at 170 K is 5.65 kJ mol^{-1}, and the entropy of fusion is 28.9 J K^{-1} mol^{-1}.
   (a) Calculate the change in the Gibbs energy for the melting of 1.00 mol of ammonia at 170 K.
   (b) Calculate the change in the Gibbs energy for the conversion of 3.60 mol of solid ammonia to liquid ammonia at 170 K.

(c) Does ammonia melt spontaneously at 170 K?
   (d) Estimate the temperature at which solid and liquid ammonia are in equilibrium at a pressure of 1 atm.

30. (See Example 11–5.) Solid tin exists in two forms: white and gray. For the phase change

$$Sn(white) \longrightarrow Sn(gray)$$

the ΔH° is -2.1 kJ mol^{-1} and the ΔS° is -7.4 J K^{-1} mol^{-1} at -30°C.
   (a) Calculate the ΔG° of the conversion of white tin to gray tin at -30°C.
   (b) Calculate the ΔG° accompanying the conversion of 2.50 mol of white tin to gray tin at -30°C.
   (c) Does white tin convert spontaneously to gray tin at -30°C and 1 atm?
   (d) Estimate the temperature at which white tin and gray tin are in equilibrium at a pressure of 1 atm.

31. The normal boiling point of the solvent acetone is 56.2°C. Use Trouton's rule to estimate its molar enthalpy of vaporization.

32. At its boiling point, methanol has a molar enthalpy of vaporization of 35.28 kJ mol^{-1} and a molar entropy of vaporization of 104.5 J K^{-1} mol^{-1}.
   (a) Determine the boiling point of methanol in degrees Celsius.
   (b) Suggest a reason for the abnormally high molar entropy of vaporization of methanol.

The Gibbs Energy and Chemical Reactions

33. (See Example 11–6.) Use standard molar enthalpies of formation (ΔH_f°'s) and standard molar entropies (S°'s) from Appendix D to compute ΔG_f° at 298.15 K for the following species:
   (a) LiCl(s)
   (b) Na(g)
   (c) $CH_3OH(\ell)$

   (*Note*: The fourth column of the table in Appendix D contains the answers, which should however be obtained from data in the first two columns.)

34. (See Example 11–6.) Use ΔH_f° and S° data from Appendix D to compute ΔG_f° at 298.15 K for the following species:
   (a) FeS(s)
   (b) Na(g)
   (c) $CH_3OH(\ell)$

   (*Note*: The fourth column of the table in Appendix D contains the answers, which should however be obtained from data in the first two columns.)

35. (See Example 11–7.) Use data for Appendix D to calculate ΔG_r° at 25°C for the following:
   (a) $SrCl_2(s) + Ca(s) \longrightarrow CaCl_2(s) + Sr(s)$
   (b) $Fe_2O_3(s, hematite) + 2\,Al \longrightarrow 2\,Fe(s) + Al_2O_3(s)$

36. (See Example 11–7.) Use data in Appendix D to calculate ΔG_r° at 25°C for the following:
   (a) $CaCO_3(s) + 2\,H_3O^+(aq) \longrightarrow$
$$Ca^{2+}(aq) + CO_2(g) + 3\,H_2O(\ell)$$

(b) $CaCO_3(s) + 2 H^+(aq) \longrightarrow$
$$Ca^{2+}(aq) + CO_2(g) + H_2O(\ell)$$

37. For $CoCl_2 \cdot 6H_2O(s)$ at 25°C, ΔH_f° is -2115.4 kJ mol^{-1}, ΔG_f° is -1725.2 kJ mol^{-1}, and S° is 343 J K^{-1} mol^{-1}. Compute ΔH_r°, ΔG_r°, and ΔS_r° for

$$CoCl_2(s) + 6 H_2O(\ell) \longrightarrow CoCl_2 \cdot 6H_2O(s)$$

at 25°C. Take data from Appendix D as necessary.

38. Use data from the previous problem and Appendix D to compute ΔH_r°, ΔG_r°, and ΔS_r° at 25°C for

$$CoCl_2(s) + 6 H_2O(g) \longrightarrow CoCl_2 \cdot 6H_2O(s)$$

39. Predict the sign of ΔG_r° for reactions at low temperature for which
(a) ΔH_r° is negative and ΔS_r° is negative.
(b) ΔH_r° is negative and ΔS_r° is positive.
(c) ΔH_r° is positive and ΔS_r° is negative.
(d) ΔH_r° is positive and ΔS_r° is positive.

40. Predict the sign of ΔG_r° for reactions at high temperature for which
(a) ΔH_r° is negative and ΔS_r° is negative.
(b) ΔH_r° is negative and ΔS_r° is positive.
(c) ΔH_r° is positive and ΔS_r° is negative.
(d) ΔH_r° is positive and ΔS_r° is positive.

41. (See Example 11–8.) Over approximately what range of temperatures is each of the following processes spontaneous? Assume that all gases are at a pressure of 1 atm.
(a) The rusting of iron, a complex process that can be approximated as

$$4 Fe(s) + 3 O_2(g) \longrightarrow 2 Fe_2O_3(s)$$

(b) The preparation of $SO_3(g)$ from $SO_2(g)$, a step in the manufacture of sulfuric acid:

$$SO_2(g) + \tfrac{1}{2}O_2(g) \longrightarrow SO_3(g)$$

(c) The production of the anesthetic dinitrogen monoxide (nitrous oxide) through the decomposition of ammonium nitrate:

$$NH_4NO_3(s) \longrightarrow N_2O(g) + 2 H_2O(g)$$

(*Hint*: Use data from Appendix D to calculate ΔH_r° and ΔS_r° (assumed independent of temperature) and then use the definition of ΔG_r°.)

42. (See Example 11–8.) Use data from Appendix D to estimate the range of temperatures over which each of the following processes is spontaneous.
(a) The preparation of the gas phosgene ($COCl_2(g)$):

$$CO(g) + Cl_2(g) \longrightarrow COCl_2(g)$$

(b) The laboratory-scale production of oxygen from the decomposition of potassium chlorate:

$$2 KClO_3(s) \longrightarrow 2 KCl(s) + 3 O_2(g)$$

(c) The reduction of iron(II) oxide (wüstite) by coke (carbon), a step in the production of iron in a blast furnace:

$$FeO(s) + C(s) \longrightarrow Fe(s) + CO(g)$$

43. (See Example 11–8.) Explain why it is possible to reduce tungsten(VI) oxide (WO_3) to metal with hydrogen at an elevated temperature. Estimate the temperature at which ΔG_r° for the conversion becomes equal to zero. Use the data in Appendix D.

44. (See Example 11–8.) Tungsten(VI) oxide can also be reduced to tungsten by heating it with carbon in an electric furnace:

$$2 WO_3(s) + 3 C(s) \longrightarrow 2 W(s) + 3 CO_2(g)$$

(a) Calculate the standard Gibbs energy of this reaction at 25°C, and comment on the feasibility of the process at room conditions.
(b) Estimate the temperature required to cause this reaction to proceed spontaneously. List the assumptions that you make.

45. The following data are gathered on the reaction of hydrogen with carbon dioxide to give formic acid at 298.15 K:

	ΔH_r° (kJ mol^{-1})	ΔG_r° (kJ mol^{-1})
$H_2(g) + CO_2(g) \longrightarrow$ HCOOH (g)	$+30.9$	$+58.6$
$H_2(g) + CO_2(g) \longrightarrow$ HCOOH (ℓ)	-31.2	$+32.9$

Compute ΔH°, ΔG°, and ΔS° for the vaporization of liquid formic acid at 298.15 K.

46. According to the data in Problem 45, the reaction of $H_2(g)$ with $CO_2(g)$ to give gaseous formic acid is *endo*thermic and non-spontaneous, and the reaction to give liquid formic acid is exothermic but also non-spontaneous. Explain these facts from the point of view of the making and breaking of bonds during the reaction and the relative entropies of the reactants and products.

The Gibbs Energy and the Equilibrium Constant

47. (See Examples 11–7 and 11–9.) Calculate ΔG_r° and the equilibrium constant K at 25°C for the reaction

$$2 NH_3(g) + \tfrac{7}{2}O_2(g) \rightleftharpoons 2 NO_2(g) + 3 H_2O(g)$$

using data from Appendix D.

48. (See Examples 11–7 and 11–9.) Write an equation representing the dehydrogenation of gaseous ethane (C_2H_6) to acetylene (C_2H_2). Calculate ΔG_r° and the equilibrium constant for this reaction at 25°C, using data from Appendix D.

49. (See Examples 11–7 and 11–9.) Use the thermodynamic data in Appendix D to calculate equilibrium constants at

25°C for the following reactions:

(a) $SO_2(g) + \frac{1}{2}O_2(g) \rightleftharpoons SO_3(g)$

(b) $3\,Fe_2O_3(s) \rightleftharpoons 2\,Fe_3O_4(s) + \frac{1}{2}O_2(g)$

(c) $CuCl_2(s) \rightleftharpoons Cu^{2+}(aq) + 2\,Cl^-(aq)$

Write an equilibrium expression in each case.

50. (See Examples 11–7 and 11–9.) Use the thermodynamic data from Appendix D to calculate equilibrium constants at 25°C for the following reactions:

(a) $H_2(g) + N_2(g) + 2\,O_2(g) \rightleftharpoons 2\,HNO_2(g)$

(b) $Ca(OH)_2(s) \rightleftharpoons CaO(s) + H_2O(g)$

(c) $Zn^{2+}(aq) + 4\,NH_3(aq) \rightleftharpoons Zn(NH_3)_4^{2+}(aq)$

Write an equilibrium expression in each case.

51. (See Examples 11–7 and 11–9.) Use the ΔG_f° values in Appendix D to compute the equilibrium constant K at 25°C for the reaction

$$H_2PO_4^-(aq) + H_2O(\ell) \longrightarrow HPO_4^{2-}(aq) + H_3O^+(aq)$$

at 25°C. Compare the answer to the value for K_{a2} for $H_3PO_4(aq)$ in Table 8–2.

52. (See Examples 11–7 and 11–9.) Use the ΔG_f° values in Appendix D to compute the equilibrium constant K at 25°C for the reaction

$$H_3PO_4(aq) + H_2O(\ell) \longrightarrow H_2PO_4^-(aq) + H_3O^+(aq)$$

at 25°C. Compare the answer to the value for K_{a1} for $H_3PO_4(aq)$ in Table 8–2.

53. Although the dissolution of helium gas in water is energetically favored, helium is only very slightly soluble in water. What keeps this gas from dissolving in great quantities in water?

54. The molar solubility of rhombic sulfur ($S(s,\,rhombic)$) in water at 298.15°C is very slight, only $1.9 \times 10^{-8}\ mol\,L^{-1}$.

(a) Compute ΔG_r° for the dissolution of rhombic sulfur in water.

(b) Despite the fact that the ΔG_r° computed in part (a) is positive, rhombic sulfur placed in water at 25°C still spontaneously dissolves to the extent given above. Explain.

The Temperature Dependence of Equilibrium Constants

55. Calculate the equilibrium constant K for the following reaction at 590 K:

$$3\,CuCl(s) + \frac{1}{2}Al_2Cl_6(g) \rightleftharpoons Cu_3AlCl_6(g)$$
$$\Delta G_r^\circ = +29.4\ kJ\,mol^{-1} \qquad \text{at 590 K}$$

56. The compounds $MgFe_2Cl_8$ and $CaFe_2Cl_8$ can be prepared by the reactions

$$MgCl_2(s) + 2\,FeCl_3(g) \rightleftharpoons MgFe_2Cl_8(g)$$
$$K = 1.32 \qquad \text{at 873 K}$$

$$CaCl_2(s) + 2\,FeCl_3(g) \rightleftharpoons CaFe_2Cl_8(g)$$
$$K = 0.0250 \qquad \text{at 873 K}$$

(a) Calculate ΔG_r° for each of these reactions at 873 K.

(b) Which compound is more stable at 873 K with respect to decomposition to $FeCl_3(g)$ and magnesium (or calcium) chloride?

57. Oxygen dissolves in water to an extent, and fish extract the dissolved O_2 in their gills to support their metabolism. Without dissolved oxygen in their water, fish die. Heating the water in rivers and lakes has led on occasion to massive fish kills by suffocation. From these facts, determine the sign of ΔH_r for the equilibrium

$$O_2(g) \rightleftharpoons O_2(aq)$$

58. (a) The equation

$$Cl_2(g) \rightleftharpoons Cl_2(aq)$$

represents the dissolution of gaseous chlorine in water. The equilibrium constant for this dissolution doubles when the temperature is increased from 10 to 30°C. Is the production of $Cl_2(aq)$ by this reaction an endothermic or an exothermic process?

(b) Estimate ΔH_r° for this reaction.

59. One way to manufacture ethanol is by the reaction

$$C_2H_4(g) + H_2O(g) \rightleftharpoons C_2H_5OH(g)$$

The ΔH_f° of $C_2H_4(g)$ is $+52.3\ kJ\,mol^{-1}$; of $H_2O(g)$, $-241.8\ kJ\,mol^{-1}$; and of $C_2H_5OH(g)$, $-235.3\ kJ\,mol^{-1}$. Without doing detailed calculations, suggest the conditions of pressure and temperature that maximize the yield, at equilibrium, of ethanol.

60. Dimethyl ether (CH_3OCH_3) is a good substitute for environmentally harmful propellants in aerosol spray cans. It is produced by the dehydration of methanol:

$$2\,CH_3OH(g) \rightleftharpoons CH_3OCH_3(g) + H_2O(g)$$

Describe reaction conditions that favor the equilibrium production of this valuable chemical. As a basis for your answer, compute ΔH_r° and ΔS_r° for the preceding reaction at 298.15 K from the data in Appendix D.

61. (See Example 11–11.) For the synthesis of ammonia from its elements,

$$3\,H_2(g) + N_2(g) \rightleftharpoons 2\,NH_3(g)$$

calculate the equilibrium constant at 600 K, assuming no change in ΔH_r° and ΔS_r° between 298.15 K and 600 K. Use data from Appendix D.

62. (See Example 11–11.) The reaction of $Cu_2O(s)$ with $H_2(g)$ to yield $Cu(s)$ is represented by the chemical equation

$$Cu_2O(s) + H_2(g) \rightleftharpoons 2\,Cu(s) + H_2O(\ell)$$

Calculate the equilibrium constant for this reaction at 373 K, assuming ΔH_r° and ΔS_r° are approximately independent of temperature. Use data from Appendix D.

63. The reaction

$$2\,NO_2(g) \rightleftharpoons N_2O_4(g)$$

has an equilibrium constant at 25°C of 6.8. At 200°C, the equilibrium constant is 1.21×10^{-3}. Calculate the enthalpy change (ΔH_r°) assuming that ΔH_r° and ΔS_r° are constant over the temperature range from 25 to 200°C.

64. The acidity constant of chloroacetic acid ($ClCH_2COOH$) in water is 1.528×10^{-3} at 0°C and 1.230×10^{-3} at 40°C. Calculate the enthalpy of dissociation of the acid in water, assuming that ΔH_{dissoc}° and ΔS_{dissoc}° are constant over this range of temperature.

65. Stearic acid dimerizes when dissolved in hexane:

$$2\, C_{17}H_{35}COOH(hexane) \rightleftharpoons (C_{17}H_{35}COOH)_2(hexane)$$

The equilibrium constant for this reaction is 2900 at 28°C, but it drops to 40 at 48°C. Estimate ΔH_r° and ΔS_r°.

66. Stearic acid also dimerizes when dissolved in carbon tetrachloride:

$$2\, C_{17}H_{35}COOH(CCl_4) \rightleftharpoons (C_{17}H_{35}COOH)_2(CCl_4)$$

The equilibrium constant for this reaction is 2780 at 22°C, but it drops to 93.1 at 42°C. Estimate ΔH_r° and ΔS_r°.

67. The equilibrium constant for the reaction

$$\tfrac{1}{2} Cl_2(g) + \tfrac{1}{2} F_2(g) \rightleftharpoons ClF(g)$$

equals 9.3×10^9 at 298.15 K and 3.3×10^7 at 398.15 K.
(a) Calculate ΔG_r° at 298.15 K.
(b) Calculate ΔH_r° and ΔS_r°, assuming the enthalpy and entropy changes to be independent of temperature between 298.15 and 398.15 K.

68. The autoionization constant of water (K_w) is 1.139×10^{-15} at 0°C, but 9.614×10^{-14} at 60°C.
(a) Calculate the enthalpy of autoionization of water:

$$2\, H_2O(\ell) \rightleftharpoons H_3O^+(aq) + OH^-(aq)$$

(b) Calculate the entropy of autoionization of water.
(c) From these data, at what temperature is the pH of pure water 7.00?

69. The vapor pressure of ammonia at −50°C is 0.4034 atm, and at 0°C it is 4.2380 atm.
(a) Estimate the molar enthalpy of vaporization (ΔH_{vap}) of ammonia.
(b) Estimate the normal boiling temperature of $NH_3(\ell)$.

70. The vapor pressure of butyl alcohol (C_4H_9OH) at 70.1°C is 0.1316 atm; at 100.8°C, it is 0.5263 atm.
(a) Estimate the molar enthalpy of vaporization (ΔH_{vap}) of butyl alcohol.
(b) Estimate the normal boiling point of butyl alcohol.

71. Estimate ΔH_{vap} of a liquid if reducing the pressure on the liquid from 1.00 atm to 0.63 atm lowers its boiling point from 124 to 107°C.

72. Estimate the boiling point of water at a pressure of 0.50 atm, assuming that ΔH_{vap} is 40.66 kJ mol^{-1}.

73. Calculate the ΔH_r° for the formation of 1.00 mol of aqueous glucose from gaseous carbon dioxide and liquid water, using data from the Chemistry in Your Life box in Section 11–8.

74. Calculate the ΔH_r° for the formation of aqueous glucose from its elements in their standard states, using the data available in Appendix D and the results of the previous problem.

75. Solubilities change with temperature. Use the data in Appendix D to
(a) Compute K_{sp} for the dissolution of $PbI_2(s)$ in water at 25°C.
(b) Estimate K_{sp} for the dissolution of $PbI_2(s)$ in water at 100°C (boiling water).
(c) Estimate the temperature at which the reaction $PbI_2(s) \longrightarrow Pb^{2+}(aq) + 2\, I^-(aq)$ becomes spontaneous (has $\Delta G_r^\circ = 0$).

76. Many lead salts are more soluble in hot water than in cold water, but $PbSO_4(s)$ is an exception. Its solubility barely changes with temperature. Use the data in Appendix D to explain this fact.

Additional Problems

*77. A computer is programmed to select at random any whole number in the range 1 to 100 (this includes 1 and 100 themselves) and print it every time a key is pressed.
(a) What is the probability that ten keystrokes generate six even numbers, then one odd number, and then three even numbers?
(b) What is the probability that ten keystrokes generate nine even numbers and one odd number irrespective of the order?

*78. The computer described in Problem 77 is used again.
(a) Compute the probability that ten keystrokes generate ten numbers less than or equal to 50.
(b) Compute the probability that pushing a key 1 billion (10^9) times generates 1 billion numbers less than or equal to 50.
(c) Compute the probability that pushing a key ten times generates ten numbers less than or equal to 99.
(d) Compute the probability that pushing a key 10^9 times generates 1 billion numbers less than or equal to 99.

79. The N_2O molecule has the structure $N\!\!=\!\!N\!\!-\!\!O$. In an ordered crystal of N_2O, the molecules are lined up in a regular fashion, with the orientation of each determined by its position in the crystal. In a random crystal (formed on rapid freezing), each molecule has two equally likely orientations ("up" and "down").
(a) In which form is the entropy higher?
(b) What is the number of possible microstates available to a random crystal containing N_0 (Avogadro's number) of molecules?

*80. Problem 94 in Chapter 5 describes an optical atomic trap. In one experiment, a gas of 500 sodium atoms is confined in a volume of 1000 μm^3. The temperature of the system is 0.00024 K. Compute the probability that, by chance, these 500 slowly moving sodium atoms would all congregate in the left half of the available volume. Express your answer in scientific notation.

81. The following graphs are similar to Figure 11–7.

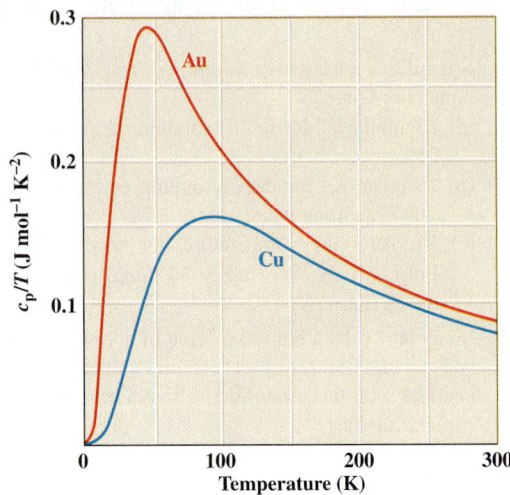

(a) Which element—copper or gold—has the higher standard molar entropy at 200 K?

(b) Estimate the standard molar entropy of copper and gold at 298.15 K by figuring the area under the blue curve and under the red curve. Compare the estimates to the values tabulated in Appendix D. (*Hint:* Find the area of one grid square in units that equal the unit on the horizontal axis multiplied by the units on the vertical axis. Then count the number of squares under each curve, including the contributions of partial squares.)

82. Without consulting Appendix D, predict the sign of ΔS_r° for each of the following.
(a) $4\ NH_3(g) + 5\ O_2(g) \longrightarrow 4\ NO(g) + 6\ H_2O(g)$
(b) $4\ NH_3(g) + 5\ O_2(g) \longrightarrow 4\ NO(g) + 6\ H_2O(\ell)$

83. Comment on the truth or falsity of each of these statements. Revise all of the false statements so that they are true.
(a) Energy is created in power plants.
(b) Scientists have found only a few exceptions to the second law of thermodynamics.
(c) The laws of thermodynamics indicate that we will absolutely never be able to make perpetual-motion machines.
(d) The second law of thermodynamics allows chemists to predict the rates of reactions.
(e) Classical thermodynamics is derived from Dalton's atomic theory of matter.

84. A quantity of ice is mixed with a quantity of hot water in a sealed, rigid, insulated container. The insulation prevents any heat exchange between the ice-water mixture and the surroundings. The contents of the container soon reach equilibrium. Tell whether the total *internal energy E* of the contents decreases, remains the same, or increases in this process. Make a similar statement about the total *entropy S* of the contents. Explain your answers.

85. Sulfur vapors include both cyclic S_8 and cyclic S_6 molecules. The ΔH_f°'s of $S_8(g)$ and $S_6(g)$ are 101.55 kJ mol^{-1} and 103.72 kJ mol^{-1}, respectively.
(a) Determine which of the two forms is favored on the basis of energetics by calculating ΔH_r° for

$$4\ S_6(g) \rightleftharpoons 3\ S_8(g)$$

(b) Which of the two forms do you expect to be favored on the basis of the entropy change in the reaction?

86. Sulfur is easily prepared in two different crystalline forms, the rhombic and the monoclinic. At atmospheric pressure, rhombic sulfur undergoes a transition to monoclinic when it is heated above 368.5 K.

$$S(s,\ rhombic) \longrightarrow S(s,\ monoclinic)$$

(a) What is the sign of the entropy change (ΔS) for this transition?
(b) $|\Delta H^\circ|$ for this transition is 400 J mol^{-1}. Calculate ΔS° for the transition.

87. Ethanol (CH_3CH_2OH) has a normal boiling point of 78.4°C and a measured molar enthalpy of vaporization of 38.74 kJ mol^{-1}. Calculate the molar entropy of vaporization of ethanol, and compare it with the prediction of Trouton's rule.

88. Use data from Appendix D to predict the normal boiling point of water (the temperature at which liquid water is in equilibrium with 1 atm of steam). Why does the answer differ from 100.00°C (373.15 K)?

89. (a) Estimate the normal boiling point of benzene (C_6H_6) using the data in Appendix D.
(b) The normal boiling point of benzene equals 80.1°C, and the ΔH_{vap} of benzene equals 30.8 kJ mol^{-1} at this temperature. These are experimental facts. Find ΔS_{vap} of benzene at 80.1°C.
(c) Is it correct to say that ΔH_{vap}(benzene) varies with temperature comparably to ΔS_{vap}(benzene)?

*90. Use data from Appendix D to estimate the temperature at which $I_2(g)$ and $I_2(s)$ are in equilibrium at a pressure of 1 atm. Can this equilibrium actually be achieved? Iodine melts at 113.5°C and boils at 184.4°C.

91. Given that ΔG_f° of $Si_3N_4(s)$ is -642.6 kJ mol^{-1}, use the data in Appendix D to compute ΔG_r° for

$$3\ CO_2(g) + Si_3N_4(s) \longrightarrow$$
$$3\ SiO_2(s,\ quartz) + 2\ N_2(g) + 3\ C(s,\ graphite)$$

92. The compound $Pt(NH_3)_2I_2$ comes in two forms, the *cis* and the *trans*, which differ in their molecular structures (see Chapter 19 for details). The following data are obtained at 298.15 K:

	ΔH_f° (kJ mol^{-1})	ΔG_f° (kJ mol^{-1})
cis	−286.56	−130.25
trans	−316.94	−161.50

Combine these data with data from Appendix D to compute the standard entropy ($S°$) of both of these compounds at 25°C.

93. Consider the reaction

$$2\ CuCl_2(s) \rightleftharpoons 2\ CuCl(s) + Cl_2(g)$$

(a) Use data from Appendix D to calculate $\Delta H_r°$ and $\Delta S_r°$ at 25°C.
(b) Calculate $\Delta G_{r,590}°$ (the $\Delta G°$ of the reaction at 590 K) assuming that $\Delta H_r°$ and $\Delta S_r°$ are independent of temperature.
(c) Careful high-temperature measurements establish that $\Delta H_{r,590}°$ is 158.36 kJ mol^{-1} and $\Delta S_{r,590}°$ is 177.74 J K^{-1} mol^{-1}. Use these facts to compute an improved value of $\Delta G_{r,590}°$.
(d) Determine the percentage error in $\Delta G_{r,590}°$ caused by using 298.15 K values in place of 590 K values.

94. Calculate the equilibrium pressure (in atmospheres) of $O_2(g)$ over a sample of pure $NiO(s)$ in contact with pure $Ni(s)$ at 25°C. The $NiO(s)$ decomposes according to the equation

$$NiO(s) \longrightarrow Ni(s) + \tfrac{1}{2}O_2(g)$$

Use data from Appendix D.

95. Although iodine is not very soluble in pure water, it dissolves readily in water that contains $I^-(aq)$ ion, thanks to the reaction

$$I_2(aq) + I^-(aq) \rightleftharpoons I_3^-(aq)$$

The equilibrium constant of this reaction was measured as a function of temperature, with these results:

T:	3.8°C	15.3°C	25.0°C	35.0°C	50.2°C
K:	1160	841	689	533	409

(a) Plot $\ln K$ against $1/T$, the reciprocal of the absolute temperature.
(b) Estimate $\Delta H_r°$ for this reaction.

96. Barium nitride vaporizes slightly at high temperature by means of the dissociation

$$Ba_3N_2(s) \rightleftharpoons 3\ Ba(g) + N_2(g)$$

At 1000 K, the equilibrium constant is 4.5×10^{-19}. At 1200 K, the equilibrium constant is 6.2×10^{-12}.
(a) Estimate $\Delta H_r°$ for this reaction.
(b) The equation is rewritten as

$$2\ Ba_3N_2(s) \rightleftharpoons 6\ Ba(g) + 2\ N_2(g)$$

Now the equilibrium constant is 2.0×10^{-37} at 1000 K and 3.8×10^{-23} at 1200 K. Estimate $\Delta H_r°$ for *this* reaction.

*97. The triple bond in the N_2 molecule is very strong, but at high enough temperatures, even it breaks down. At 5000 K, when the total pressure exerted by a sample of nitrogen is 1.00 atm, $N_2(g)$ is 0.65% dissociated at equilibrium.

$$N_2(g) \rightleftharpoons 2\ N(g)$$

At 6000 K with the same total pressure, the proportion of $N_2(g)$ dissociated at equilibrium rises to 11.6%. Use the van't Hoff equation to estimate the $\Delta H_r°$ for this reaction.

*98. The sublimation pressure of solid NbI_5 is the pressure of gaseous NbI_5 present in equilibrium with the solid. It is given by the empirical equation

$$\log_{10}P = -6762/T + 8.566$$

The vapor pressure of liquid NbI_5, on the other hand, is given by

$$\log_{10}P = -4653/T + 5.43$$

In these two equations, T is the absolute temperature in kelvins and P is the pressure in atmospheres.
(a) Determine the enthalpy and entropy of sublimation of $NbI_5(s)$.
(b) Determine the enthalpy and entropy of vaporization of $NbI_5(\ell)$.
(c) Calculate the normal boiling point of $NbI_5(\ell)$.
(d) Calculate the normal melting point of $NbI_5(s)$.
(*Hint*: At the normal melting point of a substance, the liquid and solid are in equilibrium and must have the same vapor pressure. If they did not, then vapor would continually escape from the phase with the higher vapor pressure and collect in the phase with the lower vapor pressure.)

99. "Equilibrium is like death." Explain and criticize this statement.

100. Consider the total entropy of the upper kilometer of the Earth's crust, together with the oceans and atmosphere. Discuss ways in which human activity has served to increase or decrease the entropy of that portion of the Earth. Consider such aspects as burning fossil fuels, reclaiming metals from their ores, air and water pollution, and building cities.

101. The standard change in the Gibbs energy at 25°C and pH 7 for the conversion of 1 mol of aqueous ATP to aqueous ADP is -34.5 kJ. The standard enthalpy change of the conversion under these conditions is -19.7 kJ.
(a) Calculate the standard entropy change for this process.
(b) The metabolism of 1 mol of glucose is coupled to the formation of 38 mol of ATP from ADP. What fraction of the high Gibbs energy of glucose is saved in ATP? Use data from the Chemistry in Your Life box in Section 11–8.

102. The reaction of nitrogen oxides (NO_x) with NH_3 to give N_2 has gained commercial acceptance as a means to reduce NO_x emissions from power plants.
(a) The reaction of NO_2 in this way is represented

$$6\ NO_2(g) + 8\ NH_3(g) \longrightarrow 7\ N_2(g) + 12\ H_2O(g)$$

Calculate $\Delta G_r°$ and K at 298.15 K.
(b) Write and balance similar equations for the conversion of $NO(g)$ and $N_2O(g)$ to $N_2(g)$ and $H_2O(g)$ by means of reaction with ammonia. Determine the equilibrium constants for these reactions at 298.15 K.

(c) Discuss the problem of finding the best temperature and pressure for running this reaction when dealing with a stream of gases that contains a mixture of NO, N_2O, and NO_2.

103. A report on the first synthesis of difluoromethanimine appeared recently. This compound exists in two isomeric forms:

and

When a sample of the isomer diagrammed on the left was dissolved in chloroform at 22°C, it slowly reacted to give the isomer in the diagram on the right. After seven days, 95% of the original amount was found to be converted. No further conversion occurred thereafter.

(a) Compute the equilibrium constant and ΔG_r° (at 22°C) for this isomerization reaction.

(b) Suppose that some of the isomer diagrammed on the right is prepared, purified, and then dissolved in chloroform at 22°C. Predict the relative amounts of the two isomers in this solution after seven days.

CUMULATIVE PROBLEM

Purifying Nickel

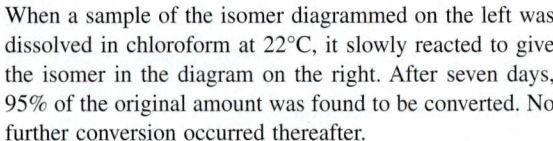

A huge reducer kiln at the heart of a nickel carbonyl refinery in Wales. *(Courtesy of INCO)*

Impure nickel, obtained from the smelting of nickel sulfide ores in a blast furnace, can be converted into metal of 99.90 to 99.99% purity by the Mond process, which relies on the equilibrium

$$Ni(s) + 4\,CO(g) \rightleftharpoons Ni(CO)_4(g)$$

The standard molar enthalpy of formation ΔH_f° of nickel tetracarbonyl ($Ni(CO)_4(g)$) is -602.9 kJ mol^{-1}, and its standard molar entropy S° is 410.6 J K^{-1} mol^{-1} at 25°C.

(a) Predict (without referring to a table) whether the entropy change of the system (the reacting atoms and molecules) is positive or negative in this process.

(b) At a temperature for which this reaction is spontaneous, predict whether the entropy change of the surroundings during the reaction is positive or negative.

(c) Use data from Appendix D to calculate ΔH_r° and ΔS_r° for this reaction.

(d) At what temperature is $\Delta G_r^\circ = 0$ (and $K = 1$) for this reaction?

(e) The first step in the Mond process is the equilibration of impure nickel with CO and $Ni(CO)_4$ at about 50°C. In this step, the equilibrium constant should be as large as possible, to draw as much nickel as possible into the vapor-phase compound. Calculate the equilibrium constant for the above reaction at 50°C.

(f) In the second step of the Mond process, the mixture of gases is removed from the reaction chamber and heated to about 230°C. At high enough temperatures, the sign of ΔG_r° is reversed and the reaction goes in reverse, depositing pure nickel. In this step, the equilibrium constant for the forward reaction should be as small as possible. Calculate the equilibrium constant for the above reaction at 230°C.

(g) The Mond process relies on the volatility of $Ni(CO)_4$ for its success. Under room conditions, this compound is a liquid, but it boils at 42.2°C with an enthalpy of vaporization of 29.0 kJ mol^{-1}. Calculate the entropy of vaporization of $Ni(CO)_4$, and compare it with that predicted by Trouton's rule.

(h) A recently developed variation of the Mond process carries out the first step at higher pressures and at a temperature of 150°C. Estimate the maximum pressure of $Ni(CO)_4(g)$ that can be attained before the gas liquefies at this temperature (that is, estimate the vapor pressure of $Ni(CO)_4(\ell)$ at 150°C).

12

Redox Reactions and Electrochemistry

CHAPTER OUTLINE

Copper ingots. *(James Foote/Photo Researchers, Inc.)*

The burning of fuel for heat involves oxidation–reduction (redox) reactions, as do the winning of iron and other metals from ores, the growth of trees and crops, the contraction and extension of muscles, and, indeed, all metabolic changes. Redox reactions were introduced in Chapter 4 as one of three broad categories of chemical change. They are reactions that occur with the *transfer* of one or more electrons. When a chemical entity loses electrons, it is *oxidized*; when it gains electrons, it is *reduced*. In this chapter, we consider redox reactions in greater depth. We show how any redox reaction can be thought of as the sum of simultaneous reduction and oxidation half-reactions. The concept of half-reactions proves helpful in balancing equations for redox reactions and in describing chemical events in electrochemical cells, which are defined and discussed. We next show how redox chemistry is exploited in practical processes for extracting and refining metals, taking copper and iron as examples. Finally, we outline ways in which electrochemical methods are used to isolate and purify metals such as aluminum, magnesium, and chromium.

12-1 Balancing Redox Equations

All valid chemical equations can be balanced either by inspection or, in complex cases, algebraically. The details are explained in Section 2–1. Both methods start with the full formulas of all substances on both sides of the reaction. Both then apply the condition of *material balance* (that the two sides of a balanced equation must represent equal numbers of moles of each element) to generate a set of balancing coefficients. Balancing chemical equations is a routine mathematical task that requires little chemical knowledge beyond the ability to interpret chemical formulas.

This section presents another method for balancing redox equations, one that highlights the transfer of electrons. This *half-equation method* makes it easy to determine where electrons come from in redox reactions, where they go, and in what numbers. Such details are vital in electrochemistry. The half-equation method also assists in the *completion* of equations. Completion consists of putting in reactants and products not mentioned in an unbalanced equation, but "obviously" required. Completing an equation (as opposed to merely balancing one) requires some knowledge of chemical reactivity. For example, the species $H_2O(\ell)$, $H^+(aq)$, and $OH^-(aq)$ are always potential participants in reactions that take place in aqueous solutions. The facts of acid–base reactivity make it clear that in acidic solutions, $H^+(aq)$ is much more likely as a reactant than $OH^-(aq)$, and $H_2O(\ell)$ is more likely as a product than $OH^-(aq)$. Similarly, in basic solutions, $OH^-(aq)$ is more likely as a reactant than $H^+(aq)$, and $H_2O(\ell)$ is more likely as a product than $H^+(aq)$.

- For simplicity, we now use the symbol $H^+(aq)$ rather than $H_3O^+(aq)$. The latter can always be obtained by replacing H^+ by H_3O^+ wherever it appears and inserting a compensating number of water molecules on the other side of the equation.

Balancing Redox Equations Using Half-Equations

Redox reactions consist of a reduction and a simultaneous oxidation. Each of these **half-reactions** requires the other in order to occur. A **half-equation** represents a half-reaction by showing electrons (e^-) explicitly. In a reduction half-equation, electrons appear on the left side (are gained by the reactants). In an oxidation half-equation, electrons appear on the right side (are lost by the reactants). The freedom to insert electrons on either side of the arrow makes half-equations easier to balance than whole equations.

The half-equation method consists of writing and balancing half-equations separately and then combining them to make a whole. A correct balanced chemical equation may never include "naked" electrons on either side; electrons must cancel out when the half-equations are combined.

A step-by-step procedure for balancing and completion of redox equations in aqueous solution follows. The steps lead to satisfaction of two criteria:

- *Material balance,* which requires that the same number of each type of atom appear on the two sides of the equation
- *Charge balance,* which requires that equal net electric charges appear on the two sides of the equation

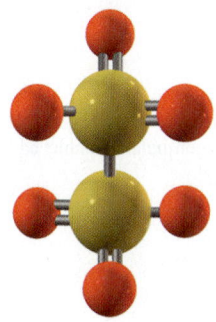

The structure of the dithionate ion ($S_2O_6^{2-}$). Note the sulfur–sulfur bond. Knowledge of oxidation numbers ($+5$ for the two sulfurs and -2 for the oxygens in this ion) is not needed to balance redox equations by the method presented here.

The steps are illustrated for the case of the dithionate ($S_2O_6^{2-}$) ion reacting with chlorous acid ($HClO_2$) in aqueous acidic solution:

$$S_2O_6^{2-}(aq) + HClO_2(aq) \longrightarrow SO_4^{2-}(aq) + Cl_2(g)$$

Step 1. *Write two unbalanced half-equations, one for the species that is oxidized and its product and one for the species that is reduced and its product.*

The half-reaction involving the dithionate ion is

$$S_2O_6^{2-} \longrightarrow SO_4^{2-}$$

and the half-reaction involving the chlorous acid is

$$HClO_2 \longrightarrow Cl_2$$

It is not necessary at this point to decide which species is oxidized and which is reduced.

The structure of chlorous acid ($HClO_2$).

Step 2. *Insert coefficients to make the numbers of atoms of all elements except oxygen and hydrogen equal on the two sides of each half-equation.*

Balancing sulfur in the first half-equation and chlorine in the second gives

$$S_2O_6^{2-} \longrightarrow 2\,SO_4^{2-} \quad \text{and} \quad 2\,HClO_2 \longrightarrow Cl_2$$

These are the only elements requiring attention in this step.

Step 3. *Balance oxygen by inserting H_2O on the side deficient in O in each half-equation.*

There are 6 O's on the left of the first half-equation but 8 O's on the right. Hence 2 H_2O's are inserted on the left:

$$2\,H_2O + S_2O_6^{2-} \longrightarrow 2\,SO_4^{2-}$$

The second half-equation has the shortage of O on the right, so H_2O's go in on the right:

$$2\,HClO_2 \longrightarrow Cl_2 + 4\,H_2O$$

It is legitimate to put H_2O on either side of the arrow because the reaction takes place in water. It is not correct to introduce O_2 or any other oxygen-containing substance.

Step 4. *Balance hydrogen. For a half-reaction in acidic solution, insert H^+ ion on the side deficient in hydrogen. For a half-reaction in basic solution, insert one H_2O for every missing H^+ on the side that is deficient in hydrogen and also insert an equal number of OH^-'s on the other side.*

- Note that doing this does not disrupt the oxygen balance achieved in Step 3.

The first equation lacks a total of 4 H's on the right and the solution is acidic; 4 H^+ ions are inserted on the right:

$$2\,H_2O + S_2O_6^{2-} \longrightarrow 2\,SO_4^{2-} + 4\,H^+$$

Hydrogen is deficient on the left in the second half-equation. Hence 6 H^+ ions are inserted on the left:

$$6\,H^+ + 2\,HClO_2 \longrightarrow Cl_2 + 4\,H_2O$$

Step 5. *Balance charge by inserting electrons as reactants or products in each half-equation.*

In the first half-equation, the net charge on the left is -2, and the net charge on the right is $2(-2) + 4(+1) = 0$. Correction of the imbalance requires two electrons on the right:

$$2\,H_2O + S_2O_6^{2-} \longrightarrow 2\,SO_4^{2-} + 4\,H^+ + 2\,e^- \qquad \text{(oxidation)}$$

This half-equation is labelled an oxidation because the electrons appear among the products. In the other half-equation, the net charge equals $+6$ on the left and 0 on the right. The imbalance is corrected by inserting six electrons on the left, making it a reduction:

$$6\,H^+ + 2\,HClO_2 + 6\,e^- \longrightarrow Cl_2 + 4\,H_2O \qquad \text{(reduction)}$$

- The number of electrons appearing in a balanced half-equation equals in magnitude the total change in oxidation number experienced by all the atoms in the half-equation.

Step 6. *Multiply the two half-equations by numbers that cause the electron loss in the oxidation to equal the electron gain in the reduction. Then add the two half-equations and cancel out the electrons. If H^+ ion, OH^- ion, or H_2O appears on both sides of the final equation, cancel out the duplications.*

Here, multiplying the oxidation half-equation by 3 and the reduction half-equation by 1 gives six electrons on the left in the reduction and six on the right in the oxidation:

$$6\,H_2O + 3\,S_2O_6^{2-} \longrightarrow 6\,SO_4^{2-} + 12\,H^+ + 6\,e^- \qquad \text{(oxidation)}$$
$$6\,H^+ + 2\,HClO_2 + 6\,e^- \longrightarrow Cl_2 + 4\,H_2O \qquad \text{(reduction)}$$

Addition of the balanced half-equations and cancellation of species that are duplicated on the two sides gives

$$2\,H_2O(\ell) + 3\,S_2O_6^{2-}(aq) + 2\,HClO_2(aq) \longrightarrow$$
$$6\,SO_4^{2-}(aq) + 6\,H^+(aq) + Cl_2(g)$$

Electrons must not appear in the final equation.

Step 7. *Check for material balance; check for charge balance.*

The final equation has 2 Cl, 6 S, 6 H, and 24 O atoms on each side. Also, the net charge on the left side is $3(-2) = -6$, which is equal to the net charge on the right side $6(-2) + 6 = -6$.

If the check fails, look first for omitted superscripts and subscripts. Particularly common errors are writing H or H_2 when H^+ is intended or OH instead of OH^-.

Figure 12-1 • Black particles of copper(II) sulfide react with concentrated nitric acid to liberate gaseous nitrogen monoxide and $SO_4^{2-}(aq)$ ion. The green color here comes from a complex of NO with Cu^{2+} ion. It fades to blue as NO is lost, and the final product $Cu^{2+}(aq)$ forms. *(Charles D. Winters)*

EXAMPLE 12-1

Complete and balance the following equation, which represents the dissolution of copper(II) sulfide in aqueous nitric acid (Fig. 12–1):

$$CuS(s) + NO_3^-(aq) \longrightarrow Cu^{2+}(aq) + SO_4^{2-}(aq) + NO(g)$$

Solution

Step 1. Write two half-equations:

$$CuS \longrightarrow Cu^{2+} + SO_4^{2-}$$
$$NO_3^- \longrightarrow NO$$

Step 2. Balance elements other than H and O. This balance already exists in both of the half-equations.

Step 3. Balance oxygen by inserting H_2O:

$$CuS + 4 H_2O \longrightarrow Cu^{2+} + SO_4^{2-}$$
$$NO_3^- \longrightarrow NO + 2 H_2O$$

Step 4. Balance hydrogen by inserting H^+ (acidic solution):

$$CuS + 4 H_2O \longrightarrow Cu^{2+} + SO_4^{2-} + 8 H^+$$
$$NO_3^- + 4 H^+ \longrightarrow NO + 2 H_2O$$

Step 5. Balance charge by inserting electrons:

$$CuS + 4 H_2O \longrightarrow Cu^{2+} + SO_4^{2-} + 8 H^+ + 8 e^- \quad \text{(oxidation)}$$
$$NO_3^- + 4 H^+ + 3 e^- \longrightarrow NO + 2 H_2O \quad \text{(reduction)}$$

Step 6. Make the numbers of electrons represented in the two half-equations equal; add the half-equations; cancel out duplicated terms. Here, if the first half-equation is multiplied through by 3 and the second by 8, then each ends up with 24 electrons:

$$3 CuS + 12 H_2O \longrightarrow 3 Cu^{2+} + 3 SO_4^{2-} + 24 H^+ + 24 e^-$$
$$8 NO_3^- + 32 H^+ + 24 e^- \longrightarrow 8 NO + 16 H_2O$$

Addition followed by cancellation of 24 e^-, 24 H^+, and 12 H_2O's gives

$$3\,CuS(s) + 8\,NO_3^-(aq) + 8\,H^+(aq) \longrightarrow$$
$$3\,Cu^{2+}(aq) + 3\,SO_4^{2-}(aq) + 8\,NO(g) + 4\,H_2O(\ell)$$

Step 7. Check. There are 3 Cu, 3 S, 8 N, 24 O, and 8 H atoms on each side; the net charge on each side is zero: $8(-1) + 8(+1) = 0 = 3(+2) + 3(-2)$.

EXERCISE

Complete and balance the following equation representing a reaction that takes place in acidic aqueous solution:

$$SO_2(aq) + Cr_2O_7^{2-}(aq) \longrightarrow Cr^{3+}(aq) + SO_4^{2-}(aq)$$

Answer: $3\,SO_2(aq) + Cr_2O_7^{2-}(aq) + 2\,H^+(aq) \longrightarrow$
$$2\,Cr^{3+}(aq) + 3\,SO_4^{2-}(aq) + H_2O(\ell).$$

When the reaction takes place in basic solution, remember to add H_2O and OH^-, rather than H^+, at Step 4.

EXAMPLE 12–2

Complete and balance the following equation. It represents a reaction that takes place in basic aqueous solution.

$$Ag(s) + HS^-(aq) + CrO_4^{2-}(aq) \longrightarrow Ag_2S(s) + Cr(OH)_3(s)$$

Solution

Step 1. The two unbalanced half-equations are

$$Ag + HS^- \longrightarrow Ag_2S$$
$$CrO_4^{2-} \longrightarrow Cr(OH)_3$$

Step 2. Balance the elements other than hydrogen and oxygen. Here, only the first half-equation is affected. It becomes

$$2\,Ag + HS^- \longrightarrow Ag_2S$$

Step 3. Balance oxygen by inserting H_2O. Here, all that is required is a single H_2O on the right side of the second half-equation:

$$CrO_4^{2-} \longrightarrow Cr(OH)_3 + H_2O$$

Step 4. Balance hydrogen by inserting H_2O on the side that is deficient in hydrogen and an equal amount of OH^- on the other side (basic solution). The first half-equation becomes

$$2\,Ag + HS^- + OH^- \longrightarrow Ag_2S + H_2O$$

and the second becomes

$$CrO_4^{2-} + 5\,H_2O \longrightarrow Cr(OH)_3 + H_2O + 5\,OH^-$$

Step 5. Balance charge by inserting electrons. The first half-reaction is revealed as the oxidation:

$$2\,Ag + HS^- + OH^- \longrightarrow Ag_2S + H_2O + 2\,e^- \qquad \text{(oxidation)}$$

and the second as the reduction:

$$CrO_4^{2-} + 5\,H_2O + 3\,e^- \longrightarrow Cr(OH)_3 + H_2O + 5\,OH^- \qquad \text{(reduction)}$$

Step 6. Make the number of e^-'s in the two half-equations equal; add the half-equations; cancel out duplicated terms. The first half-equation is multiplied by 3 so that it has six electrons on the right and the second is multiplied by 2 so that it has six electrons on the left:

$$6\,Ag + 3\,HS^- + 3\,OH^- \longrightarrow 3\,Ag_2S + 3\,H_2O + 6\,e^- \qquad \text{(oxidation)}$$
$$2\,CrO_4^{2-} + 10\,H_2O + 6\,e^- \longrightarrow 2\,Cr(OH)_3 + 2\,H_2O + 10\,OH^- \qquad \text{(reduction)}$$

Addition followed by cancellation of 6 e^-'s, 3 OH^-'s, and 5 H_2O's gives

$$6\,Ag(s) + 3\,HS^-(aq) + 2\,CrO_4^{2-}(aq) + 5\,H_2O(\ell) \longrightarrow$$
$$3\,Ag_2S(s) + 2\,Cr(OH)_3(s) + 7\,OH^-(aq)$$

Step 7. Check. There are 6 Ag, 13 H, 3 S, 2 Cr, and 13 O atoms on each side. The net charge on the left is $3(-1) + 2(-2) = -7$; the net charge on the right is $7(-1) = -7$.

EXERCISE

Complete and balance the following equation for a redox reaction that takes place in basic solution.

$$AsO_3^{3-}(aq) + Br_2(aq) \longrightarrow AsO_4^{3-}(aq) + Br^-(aq)$$

Answer: $AsO_3^{3-}(aq) + Br_2(aq) + 2\,OH^-(aq) \longrightarrow$
$$AsO_4^{3-}(aq) + 2\,Br^-(aq) + H_2O(aq)$$

Balancing Equations Representing Disproportionation

In a **disproportionation** reaction, the same chemical species is both oxidized and reduced; it reacts with itself (see Section 4–4). An example is the fate of chlorine dissolved in acidic solution:

$$Cl_2(aq) \longrightarrow ClO_3^-(aq) + Cl^-(aq) \qquad \text{(unbalanced)}$$

Balancing disproportionation equations is straightforward by the half-equation method once it is realized that the same species may appear on the left in both half-equations. In this case Step 1 gives:

$$Cl_2 \longrightarrow ClO_3^-$$
$$Cl_2 \longrightarrow Cl^-$$

Proceeding with Steps 2 to 5 yields

$$Cl_2 + 6\,H_2O \longrightarrow 2\,ClO_3^- + 12\,H^+ + 10\,e^- \qquad \text{(oxidation)}$$
$$Cl_2 + 2\,e^- \longrightarrow 2\,Cl^- \qquad \text{(reduction)}$$

Multiplying the reduction half-equation by 5 causes the electrons to cancel out when it is added to the oxidation half-equation:

$$6\,Cl_2 + 6\,H_2O \longrightarrow 2\,ClO_3^- + 10\,Cl^- + 12\,H^+$$

When the coefficients are all divided by 2, this becomes:

$$3\,Cl_2(aq) + 3\,H_2O(\ell) \longrightarrow ClO_3^-(aq) + 5\,Cl^-(aq) + 6\,H^+(aq)$$

CHEMISTRY IN YOUR LIFE

Redox Reactions Color Gems and Pigments

The pure, bright blue of sapphire comes from a simple redox reaction. If a small amount of titanium (mostly in the +4 oxidation state) is accidentally or deliberately doped into a crystal of corundum (Al_2O_3), no change in color results. Similarly, introduction of iron(II) or iron(III) impurities in corundum gives, at most, a very pale yellow color. When impurities of *both* titanium(IV) and iron(II) are present (even at the level of a few hundredths of 1%), however, the beautiful transparent blue of sapphire arises as electrons are transferred from iron-impurity sites to neighboring titanium-impurity sites:

$$Ti^{4+} + Fe^{2+} \longrightarrow Ti^{3+} + Fe^{3+}$$

This reaction requires energy to make it occur (ΔE is positive). This energy can be supplied by red, orange, or yellow light. When ordinary white light (made up of all colors) passes through a crystal of sapphire, these colors are strongly absorbed as the redox reaction takes place, and only the blue light passes through, giving the color that we see (Fig. 12–A). The reverse reaction occurs quickly without emitting colored light, and so the blue color persists.

A redox reaction also causes the intense color of the artist's pigment called Prussian blue (see Fig. 12–B).

Figure 12–A • A synthetic sapphire. *(Richard J. Green, Photo Researchers, Inc.)*

Figure 12–B • The dye "Prussian blue" forms upon mixing dilute solutions of $FeCl_3(aq)$ and $K_4[Fe(CN)_6](aq)$. *(Leon Lewandowski)*

Prussian blue is a hydrate of the compound iron(III) hexacyanoferrate(II), with formula $Fe_4[Fe(CN)_6]_3 \cdot xH_2O$. Absorption of light by this compound causes transfer of an electron from an Fe(II) located inside an $[Fe(CN)_6]^{4-}$ complex ion to an Fe(III) that is outside. The half-equations are

$$[Fe(CN)_6]^{4-} \longrightarrow [Fe(CN)_6]^{3-} + e^- \quad \text{(oxidation)}$$
$$Fe^{3+} + e^- \longrightarrow Fe^{2+} \quad \text{(reduction)}$$

This reaction is induced by red, orange, and yellow light and makes the dye appear blue. The reverse reaction occurs quickly without emitting or absorbing light, so the blue color persists. Both oxidation states of iron must be present to give this reaction; neither iron(II) hexacyanoferrate(II) nor iron(III) hexacyanoferrate(III) is blue because no redox reaction can occur in these cases. The intense colors of other mixed-valence oxides, such as black Fe_3O_4 and red Mn_3O_4, which can be formulated as $Fe^{II}(Fe^{III})_2O_4$ and $Mn^{II}(Mn^{III})_2O_4$, respectively, result from similar redox reactions.

12-2 Electrochemical Cells

The separation of a redox reaction into two half-reactions is achieved experimentally in a device called an electrochemical cell. As a first example, consider the oxidation–reduction reaction between metallic copper and an aqueous solution of silver nitrate:

$$Cu(s) + 2\,Ag^+(aq) \longrightarrow Cu^{2+}(aq) + 2\,Ag(s)$$

The nitrate ions are spectator ions. They are omitted in the preceding net ionic equation because they take no part in the reaction. Two electrons are transferred from each reacting copper atom to a pair of silver ions. The copper is thus oxidized while the silver ions are reduced. This redox reaction can be carried out by simply putting a piece of copper in an aqueous solution of silver nitrate (Fig. 12–2a). Metallic silver immediately begins to plate out on the copper, the concentration of silver ion decreases, and blue $Cu^{2+}(aq)$ ion appears in solution and increases in concentration as time passes. The displacement reaction is clearly spontaneous at constant temperature and pressure. Hence, the free energy of the reaction (ΔG_r) is less than zero. The reverse reaction does not take place spontaneously (Fig. 12–2b).

This same spontaneous reaction can be carried out without ever putting the two reactants in contact. The set-up starts with dipping a copper strip in a solution of $Cu(NO_3)_2$ and a silver strip in a solution of $AgNO_3$, as Figure 12–3 illustrates. This creates two **half-cells** in which the metallic strips are **electrodes.** Next, the two half-cells are connected by a **salt bridge,** which is an inverted U-shaped tube containing a solution of a salt such as $NaNO_3$. The ends of the bridge are stuffed with porous plugs that prevent extensive mixing of the solutions but allow ions to pass. Linking the half-cells in this way creates a whole cell. Finally, the two electrodes are connected by wires to an **ammeter,** an instrument that measures both the direction and the magnitude of electric current through it. A current passes through the ammeter during the operation of the cell.

As copper is oxidized on the left side, Cu^{2+} ions enter the solution:

$$Cu(s) \longrightarrow Cu^{2+}(aq) + 2\,e^-$$

The electrons that are released pass through the external circuit from left to right, as shown by the deflection of the ammeter needle. The electrons flow into the silver strip and are picked up by Ag^+ ions at the metal–solution interface:

$$Ag^+(aq) + e^- \longrightarrow Ag(s)$$

(a)

(b)

Figure 12–2 • (a) When copper is placed in a solution of silver nitrate, loose deposits of silver form on the surface of the copper and the solution turns blue as Cu^{2+} ion forms. (*Charles D. Winters*) (b) No reaction occurs between a silver spoon and a copper nitrate solution. (*Leon Lewandowski*)

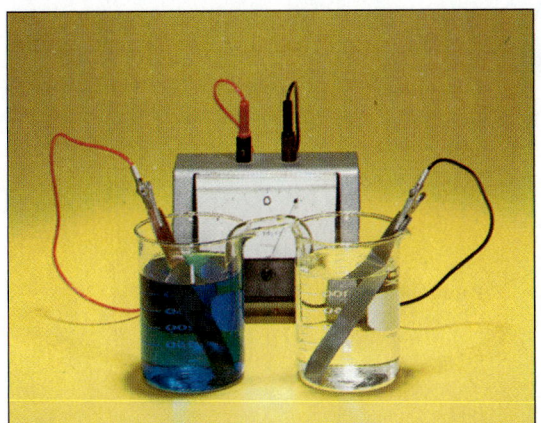

Figure 12–3 • An electrochemical cell running the reaction between copper ($Cu(s)$) and silver ions ($Ag^+(aq)$). The half-cells occupy the separate beakers. The solutions in the half-cells are connected by a salt bridge (the bent glass tube). The electrodes in the half-cells are connected electrically to an ammeter. The deflection of the needle on the ammeter dial shows the passage of the electric current. (*Leon Lewandowski*)

The resulting silver atoms plate out as a coating on the submerged surface of the strip. By itself, this process would lead to an extreme build-up of positive charge in the left-hand beaker and negative charge in the right-hand beaker. The salt bridge permits a compensating flow of positive ions into the right-hand beaker and of negative ions into the left-hand beaker, preserving electrical neutrality in each. Figure 12–4 diagrams the cell in full operation. Electrochemical action continues until the copper electrode is corroded away at the waterline or essentially all of the Ag^+ ion is plated out of solution.

The net chemical reaction in this simple electrochemical cell is the same as the reaction when a copper rod is immersed in an aqueous solution of silver nitrate. The difference is that the half-reactions occur at sites remote from each other. Direct transfer of electrons from copper atoms to silver ions is impossible because the reactants are isolated from each other. If the reaction is to occur, electrons must pass through the wire. The current can light up a lightbulb if one is connected into the external circuit, a conversion of chemical potential energy into radiant energy. It can turn a small electric motor, a conversion of chemical potential energy into mechanical work, or actuate any number of electronic devices.

The **electric current** just described is a direct current (DC) because it always flows in one direction (alternating current (AC) changes direction). The strength of a direct current is the amount of charge that flows past a point in a circuit per unit time. If Q is the magnitude of the charge in coulombs (abbreviation: C) and t is the time in seconds that it takes to pass by the test point in the circuit, then the current I is

$$I = \frac{Q}{t}$$

The SI unit for I is the ampere (abbreviation: A). An ampere equals a coulomb per second ($C\ s^{-1}$).

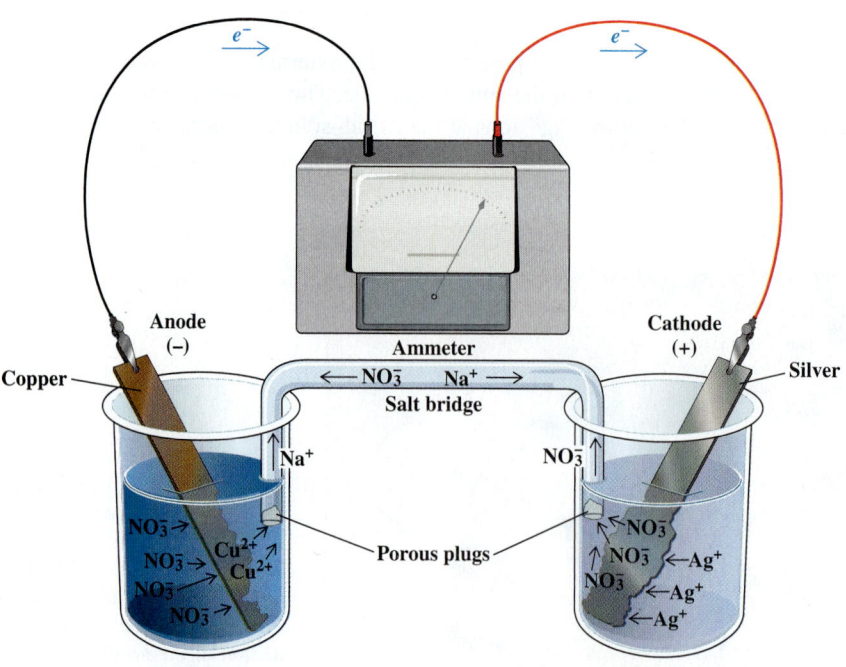

Figure 12–4 • Diagram of the operation of the cell shown in Figure 12–3. Electrons flow from the copper to the silver electrode through the wire. In solution, anions migrate toward the copper electrode (the anode) and cations toward the silver electrode (the cathode).

EXAMPLE 12–3

A battery delivers a steady current of 1.25 A for a period of 1.50 h. Calculate the total charge (Q) in coulombs that passes through the circuit.

Solution

$$Q = I \times t = 1.25 \, \frac{\text{C}}{\text{s}} \times 1.50 \, \text{h} \times \left(\frac{60 \, \text{min}}{1 \, \text{h}} \right) \times \left(\frac{60 \, \text{s}}{1 \, \text{min}} \right) = 6.75 \times 10^3 \, \text{C}$$

EXERCISE

Compute the time (in minutes) required for 12,600 C of electrical charge to pass through a lightbulb in a DC circuit if the current in the circuit is a steady 0.850 A.

Answer: $t = 247$ min.

Galvanic (Voltaic) Cells and Electrolytic Cells

The copper–silver cell diagrammed in Figure 12–4 operates because the two metals have different intrinsic affinities for electrons. This difference sets up an **electric potential difference** ($\Delta \mathcal{E}$) between the electrodes. An electric potential difference pushes electrons to flow, just as a height difference pushes water to flow downhill. Indeed, an alternative name is "electromotive force." Differences in electric potential are measured in volts (V) and are frequently called **voltages.** Voltages developed by cells are measured with a potentiometer. This instrument contains a calibrated, variable source of voltage. It is connected to the cell in such a way that its voltage opposes the voltage from the cell. It pushes electrons in the opposite direction. The external voltage is adjusted to make the current flowing from the cell fall to zero. At this point of balance, the external voltage equals the cell voltage and is read from a dial on the potentiometer.

An electrochemical cell that uses a spontaneous chemical reaction to develop a measurable voltage is called a **galvanic cell** or a **voltaic cell.**

• Luigi Galvani and Allessandro Volta were pioneers in the study of electricity and electrochemistry at the start of the 19th century.

> A galvanic cell (voltaic cell) effects a conversion of chemical potential energy into electrical energy that can be used to perform work.

If the opposing external voltage from a potentiometer or other electrical source is made to *exceed* the potential difference developed by a galvanic cell, the flow of electrons is forced into reverse. In the copper–silver cell, copper ions in solution then accept electrons and deposit as metallic copper while silver dissolves to furnish additional Ag^+ ions. The net chemical reaction that occurs is the reverse of the spontaneous reaction, namely,

$$2 \, \text{Ag}(s) + \text{Cu}^{2+}(aq) \longrightarrow 2 \, \text{Ag}^+(aq) + \text{Cu}(s)$$

A cell in which an external electromotive force drives a reaction to occur in the reverse of its spontaneous direction is called an **electrolytic cell:**

> An electrolytic cell uses electrical energy to carry out a chemical reaction that would otherwise not occur.

Anode and Cathode

An oxidation–reduction reaction is the combination of two half-reactions. In the copper–silver galvanic cell, the spontaneous change in the left-hand beaker of Figure 12–3 is

$$Cu(s) \longrightarrow Cu^{2+}(aq) + 2\,e^- \qquad \text{(oxidation)}$$

and the spontaneous change in the right-hand beaker is

$$Ag^+(aq) + e^- \longrightarrow Ag(s) \qquad \text{(reduction)}$$

The essence of an electrochemical cell is that the sites at which oxidation and reduction occur are remote from each other. In both galvanic and electrolytic cells:

• Remember this by noting that "anode" and "oxidation" both start with vowels and "cathode" and "reduction" both start with consonants.

> The site at which oxidation occurs in an electrochemical cell is called the **anode.**
> The site at which reduction occurs is called the **cathode.**

In the copper–silver galvanic cell, oxidation takes place at the copper electrode, which is therefore the anode, and reduction takes place at the silver electrode, which is therefore the cathode. Electrons flow through the external circuit from the anode to the cathode (see Fig. 12–4). The anode is electrically negative ($-$) relative to the cathode, which is (obviously) electrically positive ($+$) relative to the anode. The electron flow is from negative to positive. Inside the half-cell compartments, $NO_3^-\,(aq)$ anions migrate toward the anode, and $Ag^+(aq)$ and $Na^+(aq)$ cations migrate toward the cathode. Anions and cations move toward the anode and cathode, respectively, within the salt bridge as well. In Figure 12–4, negatively charged particles circulate in a clockwise loop from anode (copper) to cathode (silver) and through the salt bridge back toward the anode.

A sufficient external voltage applied to the electrodes forces this cell to run in reverse. The outside electromotive force pushes electrons into the copper electrode and draws them out of the silver electrode. The cell becomes an electrolytic cell. Oxidation takes place at the silver electrode, which, by definition, is now the anode, and reduction takes place at the copper electrode, which is now the cathode. Electrons still move through the external circuit from anode to cathode, but they now go (or are forced) from silver toward copper. Inside the half-cells, negative ions (anions) still move toward the anode, and positive ions (cations) still move toward the cathode, but the label switch means a reversal of their directions. The circulation of negatively charged particles is now counter-clockwise, from new anode (silver) to new cathode (copper) and through the salt bridge back toward the new anode.

Rather than diagramming cells in full (as in Figure 12–4), chemists frequently resort to a compact representation in which boundaries between phases are shown by a single vertical bar and a salt bridge by a double dashed bar. In these abbreviated diagrams, the anode always appears on the left, and the cathode on the right. The consequence of this convention is that electrons always flow through the (unshown) external circuit from left to right while anions move from right to left within the cell. The diagram of the copper–silver galvanic cell (Fig. 12–4) is abbreviated to

$$Cu(s)|Cu^{2+}(aq)||Ag^+(aq)|Ag(s)$$

• Indicators of state are often omitted in these diagrams. Ions should then be assumed to be aqueous (aq) and other substances solid (s).

The same cell, if forced by an external potential difference to run in reverse, as an electrolytic cell, is represented

$$Ag(s)|Ag^+(aq)||Cu^{2+}(aq)|Cu(s)$$

in which the order of all the symbols is reversed.

In many cells, the electrode does not react but serves solely as a conduit to deliver electrons to or remove electrons from the solution, where a reaction involving other species takes place. Platinum and graphite are inert in many (but not all) electrochemical reactions. A platinum electrode might be used to remove electrons as one dissolved species is oxidized to give a second dissolved species, as in the half-reaction

$$Fe^{2+}(aq) \longrightarrow Fe^{3+}(aq) + e^-$$

or to deliver electrons if this half-reaction is reversed. Its role in the above half-reaction would be diagrammed

$$Pt|Fe^{2+}(aq), Fe^{3+}(aq)$$

If a gas is bubbled over an immersed platinum electrode (or other inert electrode), half-reactions between the gas and dissolved ions may occur at the surface of the electrode. For example, a platinum electrode could deliver electrons to reduce gaseous chlorine to chloride ion:

$$Cl_2(g) + 2\,e^- \longrightarrow 2\,Cl^-(aq)$$

Such a cathode half-cell would be diagrammed as $Cl_2(g)|Cl^-(aq)|Pt$. If it is combined with the previous half-cell through a salt bridge, the following cell results:

$$Pt|Fe^{2+}(aq), Fe^{3+}(aq)||Cl_2(g)|Cl^-(aq)|Pt$$

• It was the unexpected dissolution of an "inert" platinum electrode in an electrochemical experiment that led to the discovery of the platinum-containing anticancer drug cisplatin.

EXAMPLE 12-4

The final step in the production of metallic magnesium from seawater (see Section 12–5) is the electrolysis of molten magnesium chloride in a large cell using a steel cathode and a graphite anode. The overall reaction is

$$Mg^{2+}(melt) + 2\,Cl^-(melt) \longrightarrow Mg(\ell) + Cl_2(g)$$

Write equations for the half-reactions occurring at the anode and at the cathode and indicate the direction in which electrons flow through the external circuit. Diagram the cell.

Solution

The anode is the site at which oxidation takes place—that is, where electrons are given up. The anode half-reaction must be

$$2\,Cl^-(melt) \longrightarrow Cl_2(g) + 2\,e^-$$

Reduction takes place at the cathode. The half-reaction must include electrons on its left:

$$Mg^{2+}(melt) + 2\,e^- \longrightarrow Mg(\ell)$$

Electrons move from the anode, where chlorine is liberated, through the external circuit to the cathode, where molten magnesium is produced. In a conventional cell diagram, the anode is always at the left.

$$graphite|Cl_2(g)|Cl^-(melt), Mg^{2+}(melt)|Mg(\ell)|steel$$

This cell does not use a salt bridge so none is indicated.

EXERCISE

The commercial production of fluorine relies on the electrolysis of a solution of potassium fluoride in liquid hydrogen fluoride using a steel cathode and a carbon anode. The overall reaction is

$$2\,HF(\ell) \longrightarrow F_2(g) + H_2(g)$$

Write equations for the half-reactions occurring at the anode and the cathode, and describe the direction of motion of the electrons in the external circuit. (*Hint:* Fluoride ion (F^-) is involved in the reaction.)

Answer: Anode: $2\,F^- \longrightarrow F_2(g) + 2\,e^-$. Cathode: $2\,HF + 2\,e^- \longrightarrow H_2(g) + 2\,F^-$. The electrons are pumped by an outside electromotive force from the anode, where fluorine is generated, to the cathode, where HF is reduced, and H_2 is generated.

12-3 Stoichiometry in Electrochemical Cells

In both galvanic and electrolytic cells, electrons depart from the anode, pass through an external circuit, and are recaptured at the cathode. The number of electrons lost invariably equals the number of electrons gained. This creates a quantitative relationship between the amounts of substance that react at the two electrodes. Michael Faraday reported this "doctrine of definite electrochemical action" on the basis of his experiments in 1834. Using the doctrine is straightforward. Electrons are treated as participants in chemical reactions, and ordinary stoichiometry calculations are carried out, but based on balanced half-equations. The only difficulty is in figuring out the quantity of electrons. Electrons cannot be stored or weighed in the same way as other reactants or products.

To overcome this difficulty, the **Faraday constant** ($\mathcal{F}$) is defined as the electric charge carried by 1 mol of electrons. The magnitude of the charge on a single electron, symbolized e, has been very accurately determined:

$$e = 1.60217646 \times 10^{-19}\ C$$

The Faraday constant equals this charge multiplied by Avogadro's number:

$$\mathcal{F} = \left(1.60217646 \times 10^{-19}\ \frac{C}{electron}\right) \times \left(6.0221420 \times 10^{23}\ \frac{electron}{mol\ e^-}\right)$$

$$\mathcal{F} = 9.6485342 \times 10^4\ \frac{C}{mol\ e^-}$$

Now, suppose that a current of I amperes ($C\ s^{-1}$) flows steadily for t seconds through an electrochemical cell. A total of It coulombs of charge passes through the cell. Dividing It by the Faraday constant gives the chemical amount, in moles, of electrons passing through the cell:

$$n_{e^-} = (It)\ C \times \left(\frac{1\ mol\ e^-}{96{,}485.342\ C}\right) = \frac{It}{96{,}485.342}\ mol\ e^-$$

All that is needed to measure the chemical amount of electrons transferred in an electrochemical cell is an ammeter (to measure I) and a stopwatch (to measure t).

Once n_{e^-} is known, standard stoichiometric methods give the chemical amounts and the masses of substances reacting at the electrodes in a cell. Suppose that we connect up our copper–silver galvanic cell one more time. The anode half-reaction is

$$Cu(s) \longrightarrow Cu^{2+}(aq) + 2\, e^-$$

and the cathode half-reaction is

$$Ag^+(aq) + e^- \longrightarrow Ag(s)$$

According to these balanced half-equations, 1 mol of electrons passing through the external circuit causes the oxidation of 1/2 mole of $Cu(s)$ (because each copper atom gives up two electrons) and the reduction of 1 mol of silver ions. From the molar masses of copper we calculate that $63.55/2 = 31.77$ g of copper is dissolved at the anode and $107.87/1 = 107.87$ g of silver is deposited at the cathode for each mole of electrons passing through the circuit. The same relationships hold if the cell is operated as an electrolytic cell, but in that case, silver is dissolved and copper is deposited (both half-reactions are reversed).

Figure 12–5 charts the flow of stoichiometric calculations that include electrons. An example follows.

EXAMPLE 12–5

A galvanic cell is constructed in which $Ag^+(aq)$ ions are reduced to silver at the cathode and zinc is oxidized to $Zn^{2+}(aq)$ ions at the anode. A steady current of 0.500 A passes through the cell for 101 min. Calculate the mass of zinc dissolved and the mass of silver deposited.

Strategy

Use the definition of an ampere $(1\ C\ s^{-1})$ to determine first the charge that passes through the cell and then (by use of the Faraday constant) the chemical amount of electrons. Use this amount with the balanced half-equations to obtain the chemical amounts of the two metals. Convert these amounts to masses.

Solution

$$Q = It = \left(\frac{0.500\ C}{s}\right) \times 101\ \text{min} \times \left(\frac{60\ s}{1\ \text{min}}\right) = 3030\ C$$

$$n_{e^-} = 3030\ C \times \left(\frac{1\ \text{mol}\ e^-}{96{,}485\ C}\right) = 3.14 \times 10^{-2}\ \text{mol}\ e^-$$

The balanced half-reactions are

$$Ag^+(aq) + e^- \longrightarrow Ag(s) \qquad \text{(reduction, cathode)}$$
$$Zn(s) \longrightarrow Zn^{2+}(aq) + 2\, e^- \qquad \text{(oxidation, anode)}$$

This leads to

$$m_{Ag} = 3.14 \times 10^{-2}\ \text{mol}\ e^- \times \left(\frac{1\ \text{mol Ag}}{1\ \text{mol}\ e^-}\right) \times \left(\frac{107.87\ \text{g Ag}}{1\ \text{mol Ag}}\right) = 3.39\ \text{g Ag}$$

$$m_{Zn} = 3.14 \times 10^{-2}\ \text{mol}\ e^- \times \left(\frac{1\ \text{mol Zn}}{2\ \text{mol}\ e^-}\right) \times \left(\frac{65.39\ \text{g Zn}}{1\ \text{mol Zn}}\right) = 1.03\ \text{g Zn}$$

Check

Both masses are reasonable "laboratory-size" quantities of metal.

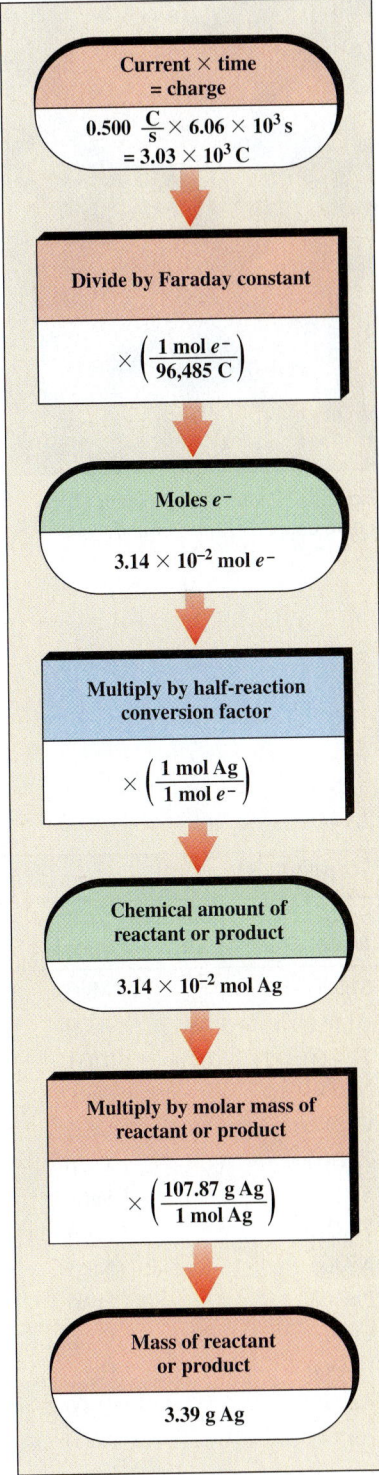

Figure 12–5 • A flow chart for a stoichiometric calculation involving electrons. Note the similarities to the flow chart in Figure 2–3.

Figure 12-6 • A specimen of native copper. *(Leon Lewandowski)*

EXERCISE

A galvanic cell generates an average current of 0.121 A for 15.6 min. The half-reaction at the cathode is $Pb^{2+}(aq) + 2\,e^- \longrightarrow Pb(s)$. What mass of lead is deposited at the cathode?

Answer: $m_{Pb} = 0.122$ g.

12-4 Metals and Metallurgy

The recovery of metals from their sources in the Earth relies heavily on oxidation–reduction reactions. **Extractive metallurgy,** a blend of chemistry, physics, and engineering, is devoted to learning the most efficient and environmentally safe ways of winning essential metals from their ores. As a science, it is comparatively recent, but its beginnings, thought to have occurred in the Near East about 6000 years ago, marked the emergence of humanity from the Stone Age. The earliest known metals were undoubtedly gold, silver, and copper because they could be found in their native (elemental) states (Fig. 12–6). Gold and silver were valued for their ornamental uses, but they were too soft to be made into tools. Iron also was found in elemental form, but only rarely, in meteorites.

Most metals are combined with other elements such as oxygen and sulfur in ores, and chemical processes are required to free them. As Table 12–1 shows, the Gibbs energies of formation of most metal oxides are negative. The reverse reactions, which yield the free metal and oxygen, therefore mostly have positive ΔG_f°'s; they are non-spontaneous. Non-spontaneous chemical reactions can be accomplished, but

---

TABLE 12-1			Metal Oxides Arranged According to Ease of Reduction	
Metal Oxide	**Metal**	n^a	$\Delta G_f^\circ / n^a$ **(kJ mol^{-1})**	**A Method of Production of the Metal**
$MgO(s)$	$Mg(s)$	2	-285	Electrolysis of $MgCl_2$
$Al_2O_3(s)$	$Al(s)$	6	-264	Electrolysis
$TiO_2(s)$	$Ti(s)$	4	-222	Reaction with Mg
$Na_2O(s)$	$Na(s)$	2	-188	Electrolysis of NaCl
$Cr_2O_3(s)$	$Cr(s)$	6	-176	Electrolysis, reduction by Al
$ZnO(s)$	$Zn(s)$	2	-159	Smelting of ZnS
$SnO_2(s)$	$Sn(s)$	4	-130	Smelting
$Fe_2O_3(s)$	$Fe(s)$	6	-124	Smelting
$NiO(s)$	$Ni(s)$	2	-106	Smelting of nickel sulfides
$PbO(s)$	$Pb(s)$	2	-94	Smelting of PbS
$CuO(s)$	$Cu(s)$	2	-65	Smelting of $CuFeS_2$
$HgO(s)$	$Hg(\ell)$	2	-29	Moderate heating of HgS
$Ag_2O(s)$	$Ag(s)$	2	-6	Found in elemental form
$Au_2O_3(s)$	$Au(s)$	6	$+13$	Found in elemental form

^{a}The standard Gibbs energies of formation of the metal oxides in kJ mol^{-1} are adjusted for fair comparison by dividing each by n, the number of moles of electrons required to reduce 1 mol of the metal oxide to the uncombined metal. Example: reducing 1 mol of Cr_2O_3 to Cr requires a total of 6 mol of electrons, so that $n = 6$.

only by coupling them with a second, spontaneous reaction. The greater the rise in the Gibbs energy, the more thermodynamically difficult it is to force a non-spontaneous reaction to take place. Thus, metals like silver or gold (at the bottom of Table 12–1) exist in nature as elements, and mercury can be released from its oxide or sulfide ore (cinnabar) simply by moderate heating (see Fig. 1–10). Winning copper, zinc, and iron requires more stringent conditions; the ores of these metals are reduced in chemical reactions at high temperatures, the collective term for which is **pyrometallurgy.** These reactions are carried out in huge furnaces, in which coke (coal from which volatile components have been expelled, leaving mostly carbon) serves both as the ultimate reducing agent and as fuel to heat the system to the required high temperature

$$C(s) + O_2(g) \longrightarrow CO_2(g) \qquad \Delta G° = -394 \text{ kJ mol}^{-1}$$

The process, called **smelting,** involves both chemical change and melting. Even smelting is not sufficient to win the metals at the top of Table 12–1, which have oxides (and also sulfides) with very negative $\Delta G_f°$'s. Electrometallurgical methods, which we consider in Section 12–5, are needed.

The free energies of formation in Table 12–1 are divided by the total change in the oxidation number when the oxide of a metal is reduced to elemental form. This allows for a fair comparison of the relative ease of reduction of two compounds that have the metals in different oxidation states. Reducing 1 mol of Fe_2O_3 to elemental iron, for example, requires oxidizing three times as much carbon (to CO_2) as the similar reduction of 1 mol of ZnO. Thus, the molar Gibbs energy of formation of Fe_2O_3 is divided by 6 in the table, whereas the molar Gibbs energy of formation of ZnO is divided by only 2. The change in the Gibbs energy at 1 atm and 298.15 K for the reaction

$$Cr_2O_3(s) + 3 \text{ Mg}(s) \longrightarrow 2 \text{ Cr}(s) + 3 \text{ MgO}(s)$$

is

$$\Delta G_r° = 3 \Delta G_f°(\text{MgO}) - \Delta G_f°(Cr_2O_3)$$

which can be rewritten as

$$\Delta G_r° = 6 \times \left(\tfrac{1}{2}\Delta G_f°(\text{MgO}) - \tfrac{1}{6}\Delta G_f°(Cr_2O_3)\right)$$

• Recall that the $\Delta G_f°$ of the most stable form of an element at a given temperature is zero. This is why only two terms appear on the right side of this equation.

Because MgO lies above Cr_2O_3 in the table, the difference enclosed in large parentheses is negative, and this reaction is spontaneous. A similar computation linking any pair of lines in the table confirms that a free metal spontaneously reduces any oxide below it in the table when the reactants and products are in standard states at 298.15 K. Table 12–1 amounts to an activity series because the order of entries ranks the metals by their ability to displace each other from their compounds with oxygen. A comparison to Table 4–2 (page 178), which gives an activity series for the displacement of metal ions from aqueous solution, reveals a strong general similarity and also some reversals of order. The reversals occur because this series uses $\Delta G_f°$'s of metal oxides while the other uses $\Delta G_f°$'s of metal ions in aqueous solution. The two sets of $\Delta G_f°$'s differ because the solid-state and solution environments differ. Thus, generalizations such as "sodium is a better reducing agent than magnesium," although often helpful, can be wrong. Relative reactivity depends on the environment in which substances are used and on the products that form.

• For example, Na(s) displaces $Mg^{2+}(aq)$ from aqueous solution, but Mg(s) displaces Na from $Na_2O(s)$.

According to Table 12–1, magnesium should react with titanium(IV) oxide (TiO_2) to give elemental titanium and MgO, which it does, and aluminum should reduce iron(III) oxide to give elemental iron and Al_2O_3, a process that proceeds

Figure 12–7 • Several gems from copper minerals. Clockwise from top: malachite with azurite, malachite, azurite with chalcopyrite, and turquoise. *(Leon Lewandowski)*

spectacularly once it is initiated (see Fig. 10–1). The metals at the top of the table are, in thermodynamic terms, the most difficult to produce, and they are the most reactive once made. It is no accident that, historically, the most reactive metallic elements were the last to be isolated in elemental form.

In the remainder of this section, we examine the production and uses of two of the most important metals produced by chemical reduction, copper and iron. In Section 12–5, we consider two other important metals, aluminum and magnesium, that require electrochemical processing.

Copper and Its Alloys

Copper occurs in a variety of minerals in the form of veins that are frequently exposed as outcrops at the Earth's surface. Chalcopyrite ($CuFeS_2$), chalcocite (Cu_2S), and bornite (Cu_5FeS_4) are chief among the copper-bearing ores. On exposure to the weathering action of the atmosphere, they are oxidized to beautiful green or blue basic copper carbonates such as malachite ($Cu_2CO_3(OH)_2$) and azurite ($Cu_3(CO_3)_2(OH)_2$), both of which are semiprecious stones (Fig. 12–7). Metallurgy may well have begun when a potter introduced an ore such as malachite into a kiln, perhaps as a colored glaze, and found a globule of a red metal after heating the ore in the reducing atmosphere generated by burning charcoal (carbon):

$$Cu_2CO_3(OH)_2(s) + C(s) \longrightarrow 2\,Cu(s) + 2\,CO_2(g) + H_2O(g)$$

Pure copper is not hard enough to make serviceable cutting tools. Some copper daggers and axes that date to the third millennium B.C. contain a few percent of arsenic, which confers hardness upon copper. It is not certain whether it is present by design or by the accidental use of ores that contained arsenic as an impurity. But the presence of tin with copper in ancient tools, at first in trace amounts and increasing to about 10% as the centuries passed, surely resulted from intent and marked the start of Bronze Age civilization. Bronze (Fig. 12–8), an alloy that contains about 10% tin and 90% copper, is hard without being brittle, can be cast in a mold, and melts at a temperature (950°C) substantially below the melting point of pure copper (1084°C).

Brass, which combines copper with up to 42% of zinc, is a second important alloy of copper. This alloy was first discovered by the Romans and is prized to this day in the manufacture of precision instruments because it is hard, easily machined, and corrosion resistant.

Modern demand for copper stems largely from its use in wires for conducting electricity. This has led to worldwide exploitation of copper ores, of which the most abundant is chalcopyrite ($CuFeS_2$). Open-pit mining operations on a mammoth scale (Fig. 12–9) remove and crush tons of rock that may contain less than 1% copper. **Froth flotation** is used to concentrate the copper ores (Fig. 12–10). In this process, crushed low-grade ore is mixed with water, a special oil, and a detergent. A stream of compressed air produces a foamy, frothy mixture in which oil-wetted particles of ore are buoyed up by clinging to air bubbles while water-wetted earth and rock sink to the bottom. The copper-rich froth is skimmed off, the oil is separated for reuse, and (after drying) a solid mass enriched in $CuFeS_2$ (typically to 25% to 30% copper) results.

The next step is the **roasting** of this enriched ore, in which it reacts with air at high temperatures in a redox reaction that converts the iron to an oxide, while leaving the copper as a sulfide:

$$6\,CuFeS_2(s) + 13\,O_2(g) \longrightarrow 3\,Cu_2S(s) + 2\,Fe_3O_4(s) + 9\,SO_2(g)$$

Figure 12–8 • This bronze of Greek origin is about 2500 years old. *(S. Tselentis/Superstock)*

 **Figure 12-9** • An open-pit copper mine in Utah. The ore contains less than 1% copper, but the metal can nevertheless be extracted economically, depending on fluctuations in the world market price. *(Don Green/ Phototake)*

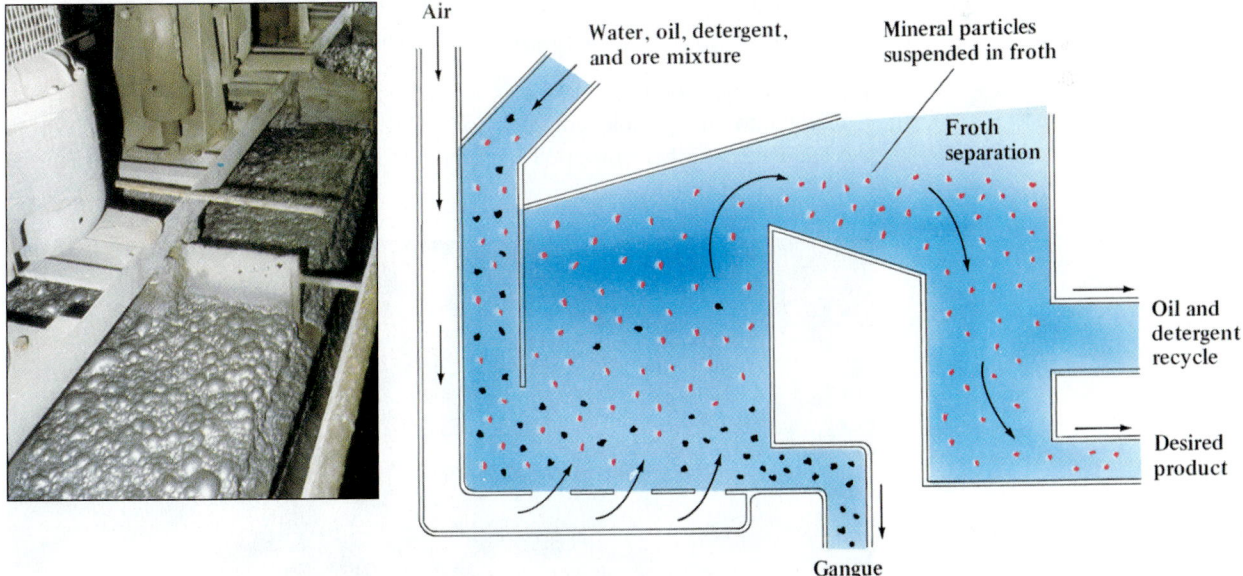

Figure 12-10 • Froth flotation for the enrichment of ore. Frothing agents are added to a water suspension of the pulverized ore. As compressed air is bubbled in, the desired mineral ($CuFeS_2$, in the case of copper production) is concentrated at the surface of a foam and borne upward while the unwanted rocky material settles to the bottom.

In the next stage, the mixture of Cu_2S and Fe_3O_4 is heated to 1100°C in a special furnace, and limestone ($CaCO_3$) and silica (SiO_2) are added as a **flux** to remove the iron in the reaction

$$CaCO_3(s) + Fe_3O_4(s) + 2\,SiO_2(s) \longrightarrow CO_2(g) + CaO \cdot FeO \cdot Fe_2O_3 \cdot (SiO_2)_2(\ell)$$
slag, composition variable

- This slag could also be formulated as a mixture of $CaSiO_3(\ell)$ (equivalent to $CaO \cdot SiO_2$) and $Fe_2SiO_5(\ell)$ (equivalent to $Fe_2O_3 \cdot SiO_2$) with FeO dissolved in it; additional amounts of all four oxides can be dissolved by such a mixture.

Because $Cu_2S(\ell)$ is not soluble in the calcium–iron slag and exceeds it in density, it settles to the bottom, and the molten slag layer can be poured off. The $Cu_2S(\ell)$ is run into another furnace (the converter), through which additional air is blown to cause the redox reaction

$$Cu_2S(\ell) + O_2(g) \longrightarrow 2\,Cu(\ell) + SO_2(g)$$

The liquid copper is then cooled and cast into large slabs of "blister copper" for further purification. The final refinement routinely uses an electrochemical method (see Section 12–5) that allows for the recovery of the small amounts of gold and silver that accompany the copper to this stage.

Three major problems afflict the smelting of copper: the pollution that can be caused by the SO_2 by-product (Fig. 12–11), the disposal of the calcium–iron–silicate slag, which can contain toxic heavy-metal impurities such as cadmium, and the energy costs associated with operating high-temperature furnaces. Emission-control regulations have required operators to devise new processes in which nearly all of the SO_2 is captured and converted to commercially important sulfuric acid (see Chapter 22). Careful attention to the recycling of water in newer smelters has greatly reduced the amount of heavy metals leaching out from the slag. Finally, energy costs are reduced by new **hydrometallurgical** methods for separating the copper and iron in the ore. When an aqueous mixture of copper(II) chloride and iron(III) chloride is added to the ore, the copper-bearing minerals are reduced in the reactions

$$CuFeS_2(s) + 3\,CuCl_2(aq) \longrightarrow 4\,CuCl(s) + FeCl_2(aq) + 2\,S(s)$$
$$CuFeS_2(s) + 3\,FeCl_3(aq) \longrightarrow CuCl(s) + 4\,FeCl_2(aq) + 2\,S(s)$$

Excess aqueous sodium chloride is added to the solids that form to convert the $CuCl(s)$ into the soluble complex ion $[CuCl_2]^-(aq)$, which is separated from the sulfur and further converted to metallic copper and copper(II) chloride in the

Figure 12–11 • The Ducktown region of Tennessee was laid waste in the 19th century by SO_2 from copper smelting. The copper is all mined out, but denuded hillsides remain. Modern methods of copper smelting have virtually eliminated pollution from SO_2. *(William Allard)*

spontaneous disproportionation reaction

$$2\,[CuCl_2]^-(aq) \longrightarrow Cu(s) + CuCl_2(aq) + 2\,Cl^-(aq)$$

The copper(II) chloride is recycled to reduce more ore.

Iron and Steel

The chief sources of iron are the minerals hematite (Fe_2O_3), magnetite (Fe_3O_4), goethite (FeO(OH)), limonite (FeO(OH)·nH_2O), and siderite ($FeCO_3$). The metallurgy of iron requires considerably higher temperatures than that of copper, and even though forced drafts were used to fan the fires, the molten iron produced in early times merely formed globules interspersed in a semiliquid slag. Repeated hammering squeezed the more fluid slag out of the red-hot mass to yield a product known as "wrought iron." Such iron was fairly soft and usually inferior to bronze until artisans learned that prolonged heating of iron in contact with charcoal caused it to absorb carbon. When iron containing several percent of dissolved carbon was quenched in water, it produced an extremely hard metal, a form of steel. This technology, developed by the Hittites around 1200 B.C., was further refined in Greco-Roman times. It was little changed until the **blast furnace** was devised in the late 14th and early 15th centuries.

Figure 12–12 illustrates the construction of a blast furnace for smelting iron ore. It consists of a cylindrical steel shell some 10 m in diameter at its widest part

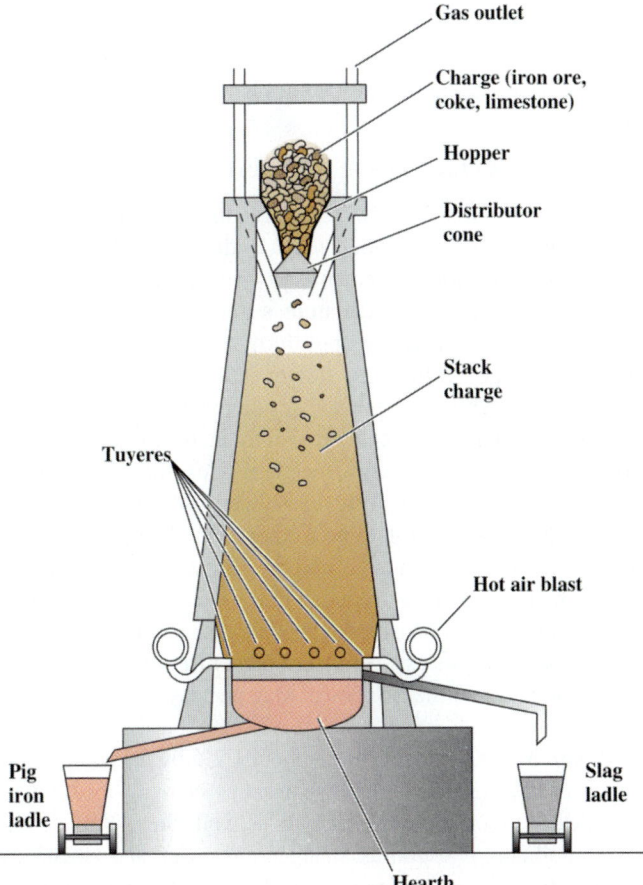

Figure 12-12 • Pig iron flows from a blast furnace. The sparks mark the oxidation of splashed droplets of the iron by the air. *(Science VU/AISI/Visuals Unlimited)*

and about 30 m high, lined with a refractory "fire brick." Designed for continuous operation, it is typically run for a period of about five years, after which its lining must be torn out and replaced. A charge of iron ore, coke, and limestone ($CaCO_3$) or dolomite ($CaMg(CO_3)_2$) enters the top of the stack at intervals through a hopper and works its way downward as the reaction proceeds. Preheated air enters the stack through a number of nozzles called **tuyeres** above the **hearth,** where the molten iron collects.

High-grade iron ore is not pure Fe_2O_3 (approximately 70% Fe) but contains up to about 27% silica, alumina, clay, and minor impurities that must be converted into a fluid slag for the process to be continuous. This is the reason that limestone or dolomite is added with the charge. Either converts silica and alumina (melting points approximately 1700 and 2050°C, respectively) into lower-melting calcium aluminum silicates (melting point approximately 1550°C).

$$CaCO_3(s) \longrightarrow CaO(s) + CO_2(g)$$
$$CaO(s) + Al_2O_3(s) + 2\,SiO_2(s) \longrightarrow CaO \cdot Al_2O_3 \cdot (SiO_2)_2(\ell) \qquad \text{composition variable}$$

Hematite is not reduced by direct reaction with carbon itself but rather by carbon monoxide in the hot gases as they pass up through the charge under forced draft:

$$Fe_2O_3(s) + 3\,CO(g) \longrightarrow 2\,Fe(s) + 3\,CO_2(g)$$

The carbon monoxide is produced in reactions between hot coke and the preheated air that enters through the tuyeres. At this level in the blast furnace, the temperature is at a maximum (about 1875°C) because of the highly exothermic reaction

$$C(s) + O_2(g) \longrightarrow CO_2(g)$$

At this high temperature, carbon dioxide rapidly comes to equilibrium with carbon monoxide in the reaction

$$C(s) + CO_2(g) \rightleftharpoons 2\,CO(g)$$

The equilibrium constant of this reaction at 1875°C is readily calculated from data in Appendix D, if we assume temperature-independent enthalpies and entropies. It is large, about 1×10^5. The reduction of hematite to iron occurs as a gas–solid reaction fairly high in the stack, where the temperature is still well below the melting point of pure iron (1535°C). The result is a spongy form of iron interspersed with the other components of the charge. As the mass slowly descends through the stack, it encounters increasingly higher temperatures until, just above the tuyeres, both the iron and the slag components liquefy and trickle down into the hearth of the furnace. The density of the molten calcium aluminum silicate (slag) is less than that of molten iron, so the slag floats as a layer on the iron. Periodically, the two liquids are separately drawn off into ladle cars, the iron to be further refined and the slag to be solidified into a clinker (fused ash), which is ground up for use in Portland cement or crushed for use as road ballast. The iron produced in the blast furnace, called **pig iron,** contains a total of 5% to 10% carbon, silicon, phosphorus, and other impurities, which are removed in a subsequent process. A modern blast furnace produces about 1500 metric tons of pig iron and 750 metric tons of slag per day while consuming about 1200 metric tons of coke.

Steel is iron mixed with a small amount (up to about 1.7% by mass) of carbon. The metallurgical properties of a steel depend on the exact percentage of carbon and on the levels and identities of other alloying elements as well. Steel is made from pig iron by the removal of silicon, phosphorus, manganese, sulfur, and other impurities, and the reduction of the carbon content to a target value that depends on the

• The fate of slag is important. If it had no good use, it would present troublesome disposal problems. Its use in Portland cement is treated in Section 21–4.

intended end use. Modern methods of conversion rely on the oxidation of the unwanted elements in the molten pig iron. The oxides escape as gases or react with a flux, which is added for this purpose, to form a slag phase that separates from the hot metal. At present, the largest quantities of steel are produced in **basic oxygen converters.** These are typically barrel-shaped furnaces open at one end and lined with refractory bricks. They stand perhaps 10 m high, have a diameter of 5 to 6 m, and are tiltable to allow the tipping in of starting materials and pouring out of product. Such a converter is charged with some 150 to 250 tons of molten pig iron transferred in insulated ladles from a blast furnace (Fig. 12–13). A quantity of cold scrap steel is also often added. Then the converter is turned upright under a hood (to collect gases), and a high-velocity jet of pure oxygen is blown down onto the charge through nozzles in a water-cooled retractable lance positioned 1 or 2 m above the surface of the molten charge (Fig. 12–14). More oxygen is blown in through bottom-mounted tuyeres, which are protected from burning by injection of a small amount of propane gas through an annular space at their ends. The propane forms a sheath around the tuyeres. Its endothermic decomposition

$$C_3H_8(g) \longrightarrow 3\,C(s) + 4\,H_2(g)$$

in this region cools and protects the tuyeres.

Figure 12–13 • Molten iron being charged into a basic oxygen converter. The basic oxygen process for steel-making is an *oxygen* process because it uses a blast of oxygen, rather than air, to oxidize impurities. It is a *basic* process because the flux contains a base, lime, to capture the acidic oxides (such as SiO_2 and P_4O_{10}) of those impurities. (*Courtesy of Bethlehem Steel*)

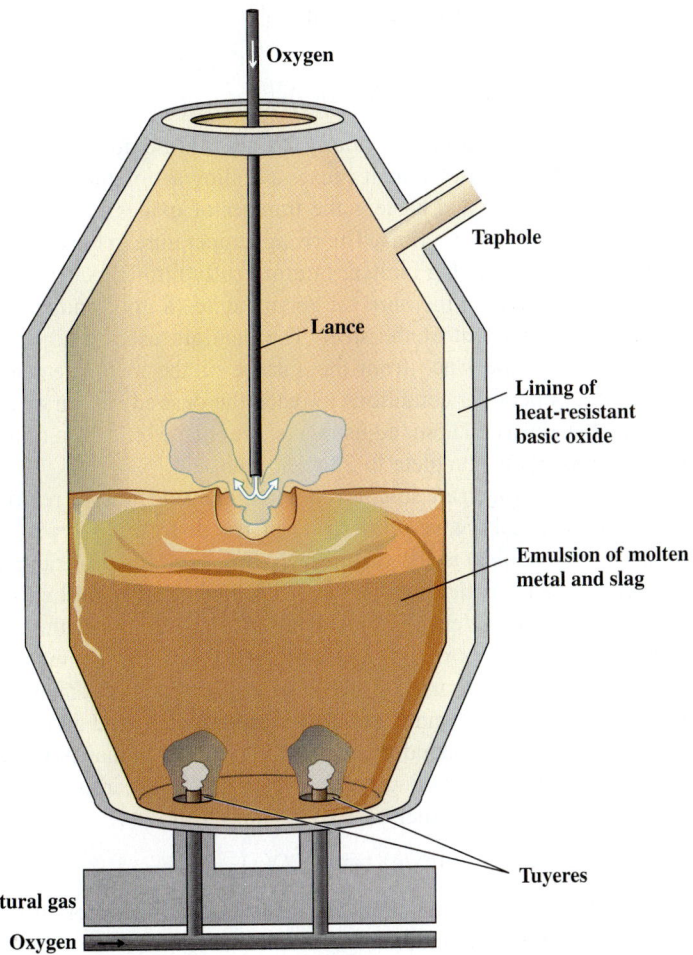

Figure 12–14 • The operation of a basic oxygen converter. Blasts of pure oxygen from above (the lance) and below (the tuyeres) refine pig iron into steel. After processing, the converter is tipped on its side, and the molten steel is tapped off through the tap-hole. The impurity-containing slag floats atop the steel and is easily separated.

Within seconds after blowing starts, strong oxidation commences in the pig iron. The flux, which is powdered lime ($CaO(s)$) or calcined (burnt) dolomite ($CaO \cdot MgO(s)$), is blown into the mixture along with the oxygen. Some iron is oxidized to $FeO(s)$ and is rapidly distributed throughout the charge:

$$2\,Fe(\ell) + O_2(g) \longrightarrow 2\,FeO(s)$$

Silicon burns directly to silicon dioxide, SiO_2, or reacts with FeO to give the same product:

$$Si(sln) + O_2(g) \longrightarrow SiO_2(s)$$
$$2\,FeO(s) + Si(sln) \longrightarrow 2\,Fe(\ell) + SiO_2(s)$$

The acidic SiO_2 then reacts with the basic flux and forms a product that separates from the hot metal as the slag:

$$SiO_2(s) + CaO(s) \longrightarrow CaO \cdot SiO_2(\ell)$$

Oxidation of the phosphorus and its removal into the slag proceed similarly, the overall equation being

• *"sln"* means "in solution." See Table 2–1 (page 52).

$$5\,FeO(s) + 2\,P(sln) + 3\,CaO(s) \longrightarrow 5\,Fe(\ell) + (CaO)_3 \cdot P_2O_5(\ell)$$

The slag also takes up MnO as it forms by oxidation of manganese in the pig iron, and some FeO; both of these metal oxides are acidic. Carbon, the major impurity in the pig iron, is oxidized to gaseous carbon monoxide:

$$C(sln) + FeO(s) \longrightarrow Fe(\ell) + CO(g)$$
$$C(sln) + \tfrac{1}{2} O_2(g) \longrightarrow CO(g)$$

Under the conditions in the converter, essentially no carbon dioxide ($CO_2(g)$) is produced. The escape of $CO(g)$ from the melt causes a boiling action that mixes the slag and metal phases thoroughly and hastens the transfer of oxides into the slag. All of the oxidation reactions are exothermic. The rising temperature in the converter melts the scrap steel and keeps the slag molten. Intermittently throughout the blow, a probe is plunged into the melt to measure the temperature (a good indicator of carbon content) and oxygen content of the steel. The data are used to adjust the blowing rate, the distance of the lance from the surface of the melt, the rate of addition of flux, and other operating conditions to reach the desired carbon content and temperature simultaneously. These adjustments are made under automatic computer control. The process is complete in 25 to 30 min.

Oxygen, hydrogen, and nitrogen are soluble in molten steel and must be removed before the steel can be cast. Otherwise, the escape of the dissolved gases would cause large void spaces (blowholes) in the casting. Dissolved oxygen also can react with alloying elements to form particles of oxides that remain in the finished steel as sources of internal and surface defects. The next stage of steel refining is therefore **vacuum degassing.** The pressure above the steel is reduced by large vacuum pumps as argon gas, which is not soluble in molten steel, is blown through the melt from the bottom. This circulates the liquid and speeds the escape of dissolved gases. Any desired alloying elements also are added during degassing. Additional oxygen is blown in briefly at the start of degassing to oxidize remaining carbon. During degassing, the low pressure above the melt shifts the equilibrium

$$C(sln) + \tfrac{1}{2} O_2(sln) \rightleftharpoons CO(g)$$

to the right and favors low levels of both carbon and oxygen in the steel. Toward the end of the degassing, aluminum is added. This further reduces the level of oxygen in the steel by the reactions

$$2\,Al(\ell) + \tfrac{3}{2}\,O_2(sln) \longrightarrow Al_2O_3(s)$$
$$2\,Al(\ell) + 3\,FeO(s) \longrightarrow Al_2O_3(s) + 3\,Fe(\ell)$$

The insoluble aluminum oxide forms an easily separated layer.

The carbon content of fully refined steel is far less than that of the original pig iron. For example, in steel intended for automobile body sheets, carbon is reduced to ultra-low levels (0.002% C by mass) on a routine basis. Ultra-low carbon steels have better press-forming characteristics than ordinary low-carbon or mild steel (0.1% to 0.4% C by mass). Steel-making by the basic oxygen process allows the production of large batches of steel of controlled chemical make-up at short intervals. These are major advantages over other ways of making steel, particularly when the steel is taken directly for the continuous casting of sheets. The basic oxygen process has entirely supplanted the formerly dominant open-hearth process, in which pig iron was cooked in oxygen-enriched air with a basic flux for periods of several hours.

12-5 Electrometallurgy

The methods of pyrometallurgy and hydrometallurgy (Section 12–4) work poorly to produce aluminum and the alkali and alkaline-earth elements as free metals. These metals have such high Gibbs energies relative to their ores that only **electrometallurgy,** or electrolytic production, can produce them in large quantity. Electrochemical cells also are used to purify these metals as well as others.

Aluminum

Aluminum is the third most abundant element in the Earth's crust (after oxygen and silicon), accounting for 8.2% of the total mass. The most important ore for the production of aluminum is bauxite, a hydrated aluminum oxide that contains 50% to 60% Al_2O_3; 1% to 20% Fe_2O_3; 1% to 10% silica; minor concentrations of transition metal oxides; and 20% to 30% water. Bauxite is not pure hydrated alumina (Al_2O_3) and must be purified before reduction to elemental aluminum. This is accomplished by the **Bayer process,** which takes advantage of the fact that the amphoteric oxide, alumina, is soluble in strong bases, whereas iron(III) oxide is not. Bauxite is treated with aqueous sodium hydroxide, which dissolves the alumina

$$Al_2O_3(s) + 2\,OH^-(aq) + 3\,H_2O(\ell) \longrightarrow 2\,Al(OH)_4^-(aq)$$

and allows for its separation from hydrated iron oxide and other insoluble impurities by filtration. Pure hydrated aluminum oxide precipitates when the solution is cooled and seeded with crystals of the product to induce rapid precipitation:

$$2\,Al(OH)_4^-(aq) \longrightarrow Al_2O_3 \cdot 3H_2O(s) + 2\,OH^-(aq)$$

The solid is separated, and the water of hydration is driven off at high temperature (1200°C).

Aluminum in its elemental form is a relative novelty. Sir Humphrey Davy obtained it as an alloy of iron and proved its metallic nature in 1809. It was first prepared in relatively pure form by H. C. Oersted in 1825 by reduction of aluminum chloride with potassium dissolved in mercury

$$AlCl_3(s) + 3\,K(sln) \longrightarrow 3\,KCl(s) + Al(sln)$$

after which the mercury was removed by distillation. Aluminum remained largely a laboratory curiosity until 1886, when Charles Hall in the United States (then a 21-year-old recent graduate of Oberlin College) and Paul Héroult (a Frenchman of the same age) independently invented an efficient process for its production. In 1999, the worldwide annual production of aluminum by the **Hall–Héroult process** was approximately 2.1×10^7 metric tons. The process consists of the electrolytic deposition of aluminum from aluminum oxide (Al_2O_3) dissolved in molten cryolite (Na_3AlF_6). Industrial plants use large, rectangular steel boxes (perhaps 6 m long, 2 m wide, and 1 m high) as cells. The negative pole of an external voltage source is connected to the walls of the cell, making them the cathode, and the positive pole of the same source is connected to massive graphite rods that dip into the molten cryolite, making them the anode (Fig. 12–15). Enormous currents (50,000 to 100,000 A) are forced through the cell. Molten cryolite is completely dissociated into Na^+ and AlF_6^{3-} ions. It is an excellent solvent for Al_2O_3, which is present in an equilibrium distribution of ions such as Al^{3+}, AlF^{2+}, AlF_2^+, ..., AlF_6^{3-}, and O^{2-}. The aluminum-containing species are reduced to elemental aluminum when the voltage is applied. Simultaneously O^{2-} is oxidized to $O_2(g)$ that rapidly reacts with the graphite electrode to produce $CO_2(g)$. Cryolite melts at 1000°C, but its melting point is lowered by the dissolved Al_2O_3, so that the operating temperature of the cell is about 950°C. Compared with the melting point of pure Al_2O_3 (2050°C), this is a low temperature, and it is the reason the Hall–Héroult process is economically feasible. The aluminum appears as a liquid that is somewhat denser than the molten cryolite solution and insoluble in it. It collects at the bottom of the cell, from which it is tapped periodically. The overall cell reaction is therefore

$$2\ Al_2O_3(sln) + 3\ C(s) \longrightarrow 4\ Al(\ell) + 3\ CO_2(g)$$

- This is second only to the production of steel (about 7.6×10^8 metric tons).

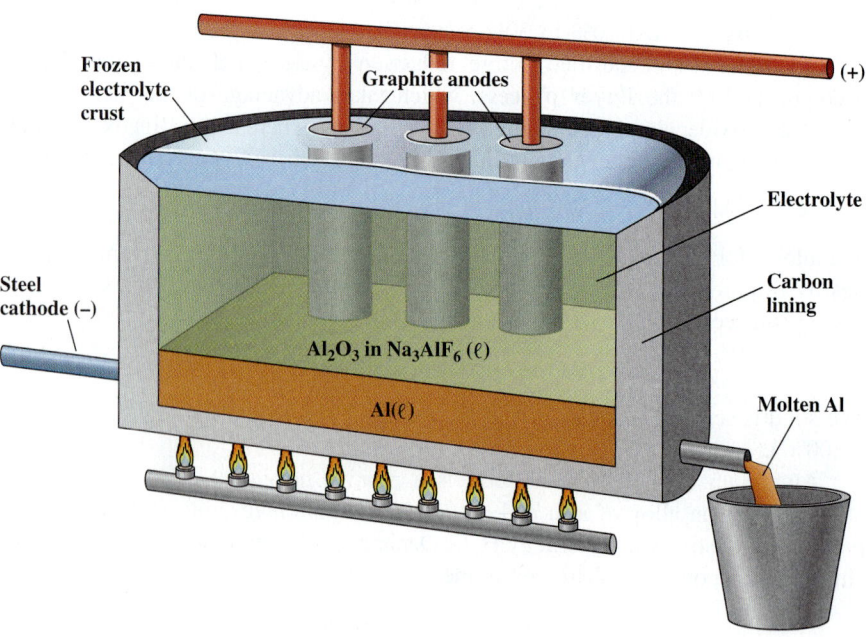

Figure 12–15 • An electrolytic cell used in the Hall–Héroult process for making aluminum.

Molten aluminum is being tapped from an electrolytic cell. *(Courtesy of Alcoa)*

Aluminum and its alloys have a tremendous variety of applications. Many of these make use of its low density (Table 12–2), an advantage over iron or steel in situations in which weight savings are desirable. This includes the transportation industry, in which aluminum is used in vehicles from automobiles to Moon rockets. Its high electrical conductivity combines with its low density to make aluminum useful for electrical transmission lines. For structural and building applications, its resistance to corrosion is a positive feature, as is the fact that it becomes stronger at subzero temperatures that can make steel and iron brittle. Household applications such as aluminum foil, aluminum soft-drink cans, and cooking utensils are well known.

Magnesium

Like aluminum, magnesium is a very abundant element on the surface of the Earth, but one that is difficult to prepare in elemental form. Although ores, such as dolomite ($CaMg(CO_3)_2$) and carnallite ($KCl \cdot MgCl_2 \cdot 6H_2O$) exist, the major commercial source of magnesium and its compounds is seawater. Magnesium forms the second most abundant positive ion in the sea, and Mg^{2+} is separated from the other cations in seawater (Na^+, Ca^{2+}, and K^+, in particular) by taking advantage of the fact that magnesium hydroxide is the least soluble hydroxide of the group. To recover magnesium economically requires an inexpensive base to treat large volumes of seawater and efficient methods for separating the $Mg(OH)_2(s)$ that precipitates from the solution. One base that is used in this way is calcined dolomite, which is prepared by heating dolomite to high temperatures to drive off carbon dioxide:

$$CaMg(CO_3)_2(s) \longrightarrow CaO \cdot MgO(s) + 2\,CO_2(g)$$

When $CaO \cdot MgO(s)$ is added to seawater, the greater solubility of calcium hydroxide ($K_{sp} = 5.5 \times 10^{-6}$) relative to magnesium hydroxide ($K_{sp} = 1.2 \times 10^{-11}$) leads to the reaction

$$CaO \cdot MgO(s) + Mg^{2+}(aq) + 2\,H_2O(\ell) \longrightarrow 2\,Mg(OH)_2(s) + Ca^{2+}(aq)$$

TABLE 12–2

Densities of Selected Metals

Metal	Density (g cm^{-3}) at Room Conditions
Li(s)	0.534
Na(s)	0.971
Mg(s)	1.738
Al(s)	2.702
Ti(s)	4.54
Zn(s)	7.133
Fe(s)	7.874
Ni(s)	8.902
Cu(s)	8.96
Ag(s)	10.50
Pb(s)	11.35
Hg(ℓ)	13.53
U(s)	18.95
Au(s)	19.32
Pt(s)	21.45

An advantage of this process is that the magnesium hydroxide that is produced includes not only the magnesium from the seawater but also that from the dolomite.

An interesting alternative to dolomite as a base for magnesium production is employed in a process used on the coast of Texas (Fig. 12–16). In this process, oyster shells (composed largely of $CaCO_3$) are calcined to give lime (CaO), which is added to the seawater to yield magnesium hydroxide. The $Mg(OH)_2$ slurry (a suspension in water) is washed and filtered in huge nylon filters. The addition of hydrochloric acid neutralizes the magnesium hydroxide and yields aqueous magnesium chloride:

$$Mg(OH)_2(s) + 2\ HCl(aq) \longrightarrow MgCl_2(aq) + 2\ H_2O(\ell)$$

After the water is evaporated, the solid magnesium chloride can be melted (melting point 708°C) in a steel electrolysis cell that might hold as much as 10 tons of the molten salt. The steel in the cell acts as the cathode during electrolysis, with graphite anodes suspended from the top. The cell reaction is

$$MgCl_2(\ell) \longrightarrow Mg(\ell) + Cl_2(g)$$

The molten magnesium liberated at the cathode floats to the surface and is dipped out periodically, while the chlorine generated at the anodes is collected and reacted with steam at high temperatures to produce hydrochloric acid. This is recycled for further reaction with magnesium hydroxide.

Until 1918, the major use of elemental magnesium was in fireworks and flash-bulbs, which took advantage of its great reactivity with the oxygen in air and the bright light that reaction gives off. Since that time, many further uses for the metal and its alloys have been developed. Magnesium is even less dense than aluminum and is used in alloys with that metal to lower its density and improve its resistance to corrosion under basic conditions. As we discuss in Section 13–7, magnesium can

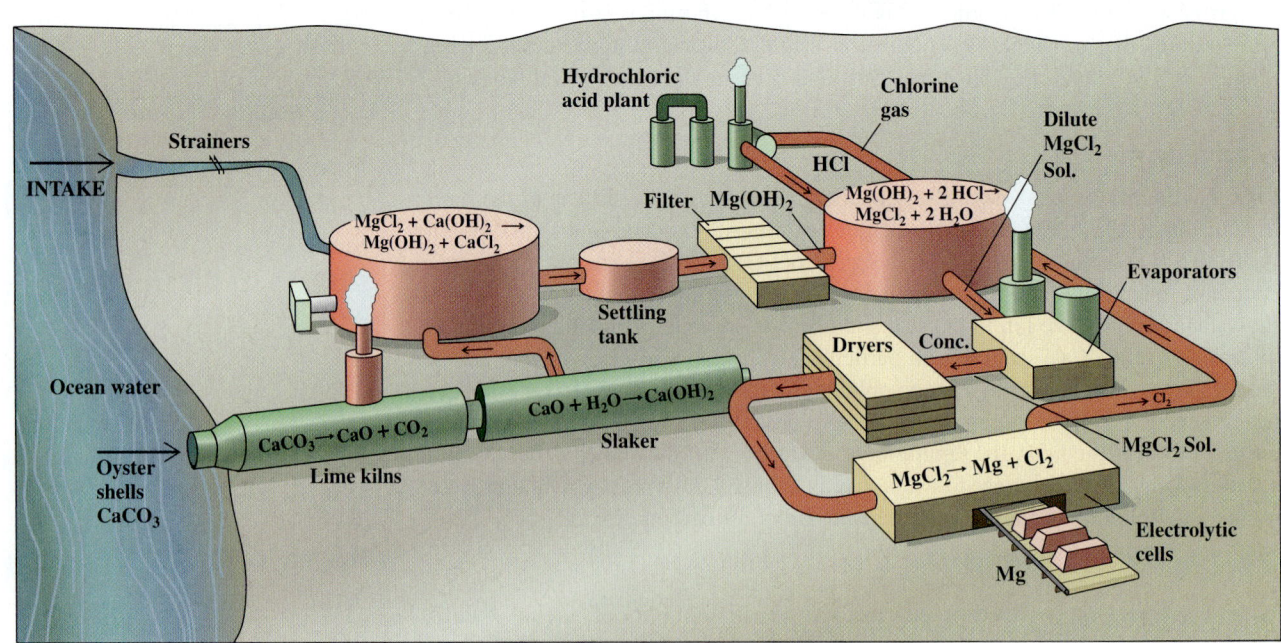

Figure 12–16 • Magnesium hydroxide is produced starting with the addition of lime (CaO) to seawater. Reaction of the magnesium hydroxide with hydrochloric acid produces magnesium chloride, which, after drying, is electrolyzed to give metallic magnesium.

CHEMISTRY IN YOUR LIFE

Redox Reactions in the Mouth

Dentists install temporary crowns so that patients can chew during the time it takes to fabricate a permanent crown. Aluminum often is used for this purpose because it is inexpensive and easily manipulated. It can, however, present problems. Aluminum is an active metal; it combines spontaneously with oxygen and displaces hydrogen from water at room conditions. Ordinarily, neither reaction proceeds because aluminum is protected by a coating of aluminum oxide that is so effective that aluminum can be melted in the air (at 660°C) without severe oxidation. The reactive tendency of aluminum is thwarted, but *not* removed.

The tendency can manifest itself if a tooth with an aluminum crown opposes one with a gold crown. The mouth then becomes the site of a galvanic cell:

$$Al(s) \longrightarrow Al^{3+}(aq) + 3\,e^-$$
(anode)

$$O_2(aq) + 4\,H^+(aq) + 4\,e^- \longrightarrow 2\,H_2O(\ell)$$
(cathode)

When the dissimilar metals touch, electrons pass from the aluminum to the gold. Simultaneously, cations in the oral fluid move toward the gold and anions move toward the aluminum (Fig. 12–C). The electrical activity generally stimulates a nerve and causes pain. Moreover, the $Al^{3+}(aq)$ ion has an unpleasant metallic taste. The same effect occurs if a piece of aluminum foil from a candy wrapper lodges between two teeth and touches a gold crown or inlay.

Another dental scenario for unleashing aluminum's true reactivity involves mercury. Mercury forms alloys called "amalgams" with many metals, including aluminum, gold, copper, and silver. Because aluminum oxide does not adhere to an amalgamated surface, amalgamated aluminum reacts well with water or air. "Silver"

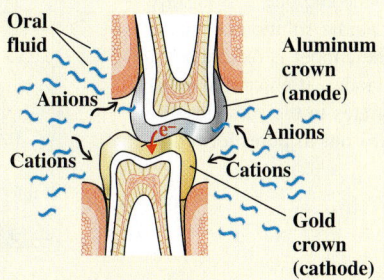

Figure 12–C • A galvanic cell operates when aluminum touches gold in the mouth. Electrons (*shown in red*) pass directly from the aluminum to the gold. Aluminum is oxidized and oxygen is reduced.

fillings are prepared by grinding together an alloy of silver, tin, copper, and sometimes zinc, with liquid mercury. The fresh mixture, which is called "dental amalgam," is packed into the cavity in the tooth while still in a pasty semi-liquid state. It then hardens. If a dentist allows a temporary aluminum crown to come in contact with a new amalgam filling, mercury from the filling may amalgamate the aluminum. The protective coating of aluminum oxide then fails, and the aluminum crown starts to react with oxygen and water. The reactions are exothermic, and the heat can cause severe pain.

Dentists have to be cautious about another redox reaction when using a zinc-containing amalgam. If water contaminates the fresh amalgam during manipulation, it soon reacts with zinc:

$$Zn(s) + H_2O(\ell) \longrightarrow Zn^{2+}(aq) + 2\,OH^-(aq) + H_2(g)$$

The trapped gaseous hydrogen causes the filling to expand unduly as it hardens. The consequence is again pain and perhaps a cracked tooth.

---

be used as a "sacrificial anode" to prevent the oxidation of another metal with which it is in contact. It is also used as a reducing agent to produce other metals such as titanium, uranium, and beryllium from their compounds. Both of these uses reflect the ease with which magnesium can be oxidized.

Electrorefining and Electroplating

Metals such as copper, silver, nickel, and tin that have been produced by pyrometallurgical methods are too impure for many purposes. The methods of **electrorefining** are adopted to purify them further. Crude metallic copper, for example, is

Figure 12–17 • In the electrolytic refining of copper, a large number of slabs of impure copper, which serve as anodes, alternate with thin sheets of pure copper (the cathodes). Both dip into a dilute acidic solution of copper sulfate. As the copper is oxidized from the impure anodes, it enters the solution and migrates to the cathodes, where it plates out in purer form.

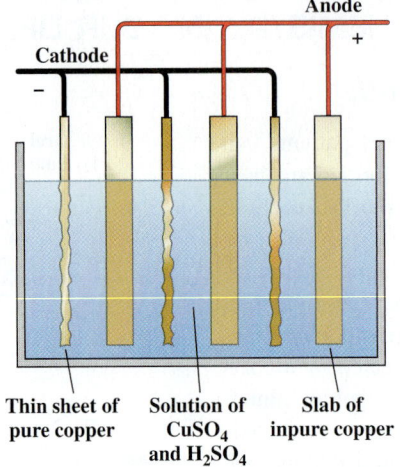

Cathode

Anode +

−

Thin sheet of pure copper   Solution of CuSO₄ and H₂SO₄   Slab of inpure copper

• About one quarter of the silver produced in the United States and one eighth of the gold are by-products of the electrolytic refining of copper.

cast into slabs, which are used as anodes in electrolysis cells that contain a solution of $CuSO_4$ in aqueous H_2SO_4. Thin sheets of pure copper serve as cathodes. Copper that dissolves at the anodes passes through the solution to be deposited in purer form on the cathodes (Fig. 12–17). Impurities that are more easily oxidized than copper, such as nickel, go into solution along with the copper but then remain there; elements such as silver and gold that are less easily oxidized do not dissolve but fall away from the anode as a metallic slime. Periodically, the anode slime and the solution are removed and processed further to recover the valuable elements they contain.

A related process is **electroplating,** in which electrolysis is used to plate out a thin layer of a metal on top of another material, frequently a second metal (Fig. 12–18). In chrome plating, for example, the piece of metal to be plated is placed in

Figure 12–18 • Silver being plated onto tableware at the cathode of an electrolytic cell. *(Courtesy Reed & Barton Silversmiths)*

a bath of hot sulfuric acid (H_2SO_4) mixed with chromic acid (H_2CrO_4) and is made the cathode in an electrolytic cell. As current passes through the cell, chromium is reduced from the $+6$ oxidation state in chromic acid to metallic chromium and plates out on the cathode. When such plating is used for decorative purposes, the chromium layer can be as thin as 2.5×10^{-5} cm (corresponding to 2 g of chromium to cover each square meter of surface). Thicker layers ranging up to 10^{-2} cm are found in hard chromium plate, which is prized for its resistance to wear. Electroplating is used with many other metals as well. Steel can be plated with cadmium to improve its resistance to corrosion in marine environments. Gold and silver plating are used both for decorative purposes and in electronic devices because of the low electrical resistance of these metals.

EXAMPLE 12-6

A layer of metallic chromium of thickness 3.0×10^{-3} cm is to be plated onto an automobile bumper from the chromic acid bath that was just described. The bumper has a surface area of 2.0×10^3 cm^2, and a steady current of 250 A is to be used. Determine how long the current must run through to achieve the desired thickness. The density of chromium is 7.2 g cm^{-3}.

Strategy

Calculate the mass of chromium that must be deposited, and then reverse the flow shown in Figure 12–5.

Solution

The volume of the chromium is the product of the thickness of the layer and the area of the surface:

$$V_{Cr} = (3.0 \times 10^{-3} \text{ cm})(2.0 \times 10^3 \text{ cm}^2) = 6.0 \text{ cm}^3$$

The mass of chromium is the product of this volume and the density of the element:

$$m_{Cr} = (6.0 \text{ cm}^3) \times \left(\frac{7.2 \text{ g}}{\text{cm}^3}\right) = 43.2 \text{ g}$$

$$n_{Cr} = 43.2 \text{ g Cr} \times \left(\frac{1 \text{ mol Cr}}{52.00 \text{ g Cr}}\right) = 0.831 \text{ mol Cr}$$

The balanced half-equation for the production of elemental chromium from H_2CrO_4 is

$$H_2CrO_4(aq) + 6 \text{ H}^+(aq) + 6 \text{ } e^- \longrightarrow Cr(s) + 4 \text{ H}_2O(\ell)$$

Hence 6 mol of electrons is required for each mole of chromium deposited. The number of moles of electrons is then

$$n_{e^-} = 0.831 \text{ mol Cr} \times \left(\frac{6 \text{ mol } e^-}{1 \text{ mol Cr}}\right) = 4.98 \text{ mol } e^-$$

The charge (in coulombs) is the product of this and the Faraday constant:

$$Q = (4.98 \text{ mol } e^-) \times \left(\frac{9.6485 \times 10^4 \text{ C}}{\text{mol } e^-}\right) = 4.81 \times 10^5 \text{ C}$$

The time required for this quantity of electric charge to pass through the cell at the given current is

$$t = \frac{Q}{I} = \frac{4.81 \times 10^5 \ C}{250 \ C \ s^{-1}} = 1.9 \times 10^3 \ s = 32 \ min$$

EXERCISE

A layer of gold 1.00 mm thick is to be plated from a solution that contains gold in the $+3$ oxidation state onto a specialized electronic part with a surface area of 4.50 cm^2. A steady current of 0.880 A is used. How long does the plating run last? The density of gold is 19.32 g cm^{-3}.

Answer: $t = 242$ min.

SUMMARY

12-1 In **oxidation–reduction (redox) reactions,** one chemical species gains electrons (it is reduced) while another species loses electrons (it is oxidized). Chemical equations to represent redox reactions in aqueous solution can be both completed (with H^+, OH^-, and H_2O) and balanced by the **half-equation method.** The oxidation and reduction steps are conceived as **half-reactions** and represented by **half-equations** in which electrons appear explicitly. When the two half-equations are combined to give the overall balanced equation, the electrons must cancel out. **Disproportionation** reactions, in which different atoms of the same chemical element are oxidized and reduced in the course of a single reaction, can be balanced by the same method.

12-2 In an **electrochemical cell,** oxidation and reduction half-reactions are carried out at physically separated **electrodes.** The electrons released by oxidation at the **anode** pass through an external circuit and are taken up by reduction at the **cathode.** At the same time, ions migrate within the cell to maintain charge neutrality throughout. This includes passage through a **salt-bridge** if necessary. When a spontaneous reaction takes place in an electrochemical cell, the cell is a **galvanic cell (voltaic cell);** the electric current produced by such a cell can be used to do work. In an **electrolytic cell,** an external source forces a current through the cell and nonspontaneous reactions are caused to occur, allowing the generation of products possessing a higher Gibbs energy than the reactants.

12-3 Michael Faraday's "doctrine of definite electrochemical action" implies a quantitative relationship between the mass of a chemical species consumed or produced at an electrode in an electrochemical cell and the amount of charge passing through the circuit. The **Faraday constant** gives the charge per mole of electrons and serves as an essential conversion in stoichiometric calculations based on half-equations.

12-4 The extraction of metals from their ores provides a most important example of redox chemistry. Different methods are used for different metals, ranging from gentle heating to very energy-intensive electrochemical methods. Between the two limits are the techniques of **pyrometallurgy,** in which the desired non-spontaneous reaction to obtain the metal is coupled to a spontaneous oxidation reaction, typically the oxidation of carbon to carbon dioxide, in a **smelter** at high temperature. Frequently, a **flux** of limestone or other substances is added to the ores during smelting to remove impurities from the reaction mixture.

12-5   Electrolytic cells are used in producing and purifying metals, especially those that have high Gibbs energies relative to their common ore sources. The **electrometallurgy** of aluminum and magnesium, which uses such cells, leads to metals and alloys that are particularly useful because of their low density and good electrical conductivity. Electrolytic cells also are used to plate a thin layer of one metal onto a second metal.

PROBLEMS

Note: Answers to blue-numbered problems are given in Appendix F. Problems that are more challenging are indicated with asterisks.

Balancing Redox Equations

1. The following balanced equations represent reactions that occur in aqueous acid. Break them down into balanced oxidation and reduction half-equations.
   (a) $2\,H^+(aq) + H_2O_2(aq) + 2\,Fe^{2+}(aq) \longrightarrow$
      $2\,Fe^{3+}(aq) + 2\,H_2O(\ell)$
   (b) $H^+(aq) + 2\,H_2O(\ell) + 2\,MnO_4^-(aq) +$
      $5\,SO_2(aq) \longrightarrow 2\,Mn^{2+}(aq) + 5\,HSO_4^-(aq)$
   (c) $5\,ClO_2^-(aq) + 4\,H^+(aq) \longrightarrow$
      $4\,ClO_2(g) + Cl^-(aq) + 2\,H_2O(\ell)$

2. The following balanced equations represent reactions that occur in aqueous base. Break them down into balanced oxidation and reduction half-equations.
   (a) $4\,PH_3(g) + 4\,H_2O(\ell) + 4\,CrO_4^{2-}(aq) \longrightarrow$
      $P_4(s) + 4\,Cr(OH)_4^-(aq) + 4\,OH^-(aq)$
   (b) $NiO_2(s) + 2\,H_2O(\ell) + Fe(s) \longrightarrow$
      $Ni(OH)_2(s) + Fe(OH)_2(s)$
   (c) $2\,OH^-(aq) + 2\,NO_2(g) \longrightarrow$
      $NO_3^-(aq) + NO_2^-(aq) + H_2O(\ell)$

3. (See Example 12–1.) Complete and balance the following equations for redox reactions taking place in aqueous acid:
   (a) $VO_2^+(aq) + SO_2(g) \longrightarrow VO^{2+}(aq) + SO_4^{2-}(aq)$
   (b) $Br_2(\ell) + SO_2(g) \longrightarrow Br^-(aq) + SO_4^{2-}(aq)$
   (c) $HCOOH(aq) + MnO_4^-(aq) \longrightarrow CO_2(g) + Mn^{2+}(aq)$
   (d) $Pb_3O_4(s) \longrightarrow Pb^{2+}(aq) + PbO_2(s)$
   (e) $Cr_2O_7^{2-}(aq) + Np^{4+}(aq) \longrightarrow Cr^{3+}(aq) + NpO_2^{2+}(aq)$
   (f) $Hg_2HPO_4(s) + Au(s) + Cl^-(aq) \longrightarrow$
      $Hg(\ell) + H_2PO_4^-(aq) + AuCl_4^-(aq)$

4. Balance the following redox equations either by inspection or algebraically. (Review Section 2–1 if necessary.)
   (a) $(VO_2)_2SO_4(s) + SO_2(g) \longrightarrow VOSO_4(s)$
   (b) $Br_2(\ell) + SO_2(g) + H_2O(\ell) \longrightarrow$
      $HBr(aq) + H_2SO_4(aq)$
   (c) $HCOOH(aq) + HMnO_4(aq) + HCl(aq) \longrightarrow$
      $CO_2(g) + MnCl_2(aq) + H_2O(\ell)$
   (d) $Pb_3O_4(s) + HCl(aq) \longrightarrow$
      $PbCl_2(aq) + PbO_2(s) + H_2O(\ell)$
   (*Note*: The answers here should resemble the answers to parts (a) through (d) in Problem 3.)

5. Balance the following redox equations either by inspection or algebraically. (Review Section 2–1 if necessary.)
   (a) $HMnO_4(aq) + H_2S(aq) + H_2SO_4(aq) \longrightarrow$
      $MnSO_4(aq) + H_2O(\ell)$
   (b) $Zn(s) + HNO_3(aq) + HCl(aq) \longrightarrow$
      $ZnCl_2(aq) + NH_4Cl(aq) + H_2O(\ell)$
   (c) $Sn(s) + HNO_3(aq) + HCl(aq) \longrightarrow$
      $SnCl_4(aq) + N_2O(g) + H_2O(\ell)$
   (d) $H_2O(\ell) + KMnO_4(aq) + H_2SO_4(aq) \longrightarrow$
      $MnSO_4(aq) + K_2SO_4(aq) + H_2O_2(aq)$

6. (See Example 12–1.) Complete and balance the following equations for reactions taking place in acidic solution:
   (a) $MnO_4^-(aq) + H_2S(aq) \longrightarrow Mn^{2+}(aq) + SO_4^{2-}(aq)$
   (b) $Zn(s) + NO_3^-(aq) \longrightarrow Zn^{2+}(aq) + NH_4^+(aq)$
   (c) $Sn(s) + NO_3^-(aq) \longrightarrow Sn^{4+}(aq) + N_2O(g)$
   (d) $H_2O_2(aq) + MnO_4^-(aq) \longrightarrow O_2(g) + Mn^{2+}(aq)$
   (e) $UO_2^{2+}(aq) + Te(s) \longrightarrow U^{4+}(aq) + TeO_4^{2-}(aq)$
   (f) $PbSO_4(s) \longrightarrow Pb(s) + PbO_2(s) + SO_4^{2-}(aq)$

7. (See Example 12–2.) Complete and balance the following equations for redox reactions taking place in basic solution:
   (a) $Cr(OH)_3(s) + Br_2(aq) \longrightarrow CrO_4^{2-}(aq) + Br^-(aq)$
   (b) $ZrO(OH)_2(s) + SO_3^{2-}(aq) \longrightarrow Zr(s) + SO_4^{2-}(aq)$
   (c) $HPbO_2^-(aq) + Re(s) \longrightarrow Pb(s) + ReO_4^-(aq)$
   (d) $HXeO_4^-(aq) \longrightarrow XeO_6^{4-}(aq) + Xe(g)$
   (e) $Ag_2S(s) + Cr(OH)_3(s) \longrightarrow$
      $Ag(s) + HS^-(aq) + CrO_4^{2-}(aq)$
   (f) $N_2H_4(aq) + CO_3^{2-}(aq) \longrightarrow N_2(g) + CO(g)$

8. (See Example 12–2.) Complete and balance the following equations for redox reactions taking place in basic solution:
   (a) $OCl^-(aq) + I^-(aq) \longrightarrow IO_3^-(aq) + Cl^-(aq)$
   (b) $SO_3^{2-}(aq) + Be(s) \longrightarrow S_2O_3^{2-}(aq) + Be_2O_3^{2-}(aq)$
   (c) $P_4(s) \longrightarrow HPO_3^{2-}(aq) + PH_3(g)$
   (d) $H_2BO_3^-(aq) + Al(s) \longrightarrow BH_4^-(aq) + H_2AlO_3^-(aq)$
   (e) $O_2(g) + Sb(s) \longrightarrow H_2O_2(aq) + SbO_2^-(aq)$
   (f) $Sn(OH)_6^{2-}(aq) + Si(s) \longrightarrow HSnO_2^-(aq) + SiO_3^{2-}(aq)$

9. "Doctor solution" for sweetening gasoline is prepared by the reaction

   $$PbS(s) + O_2(g) \longrightarrow PbO_2^{2-}(aq) + SO_4^{2-}(aq)$$

   in aqueous base. Complete and balance this equation.

10. Elemental copper is attacked by solutions of nitric acid and oxidized.

(a) Complete and balance each of the following equations, which have been written to represent this reaction:

$$Cu(s) + HNO_3(aq) \longrightarrow Cu^{2+}(aq) + NO_2(g)$$
$$Cu(s) + HNO_3(aq) \longrightarrow Cu^{2+}(aq) + NO(g)$$

(b) Recently, it was suggested that this reaction is actually

$$Cu(s) + HNO_3(aq) \longrightarrow Cu^{2+}(aq) + HNO_2(aq)$$

(The $HNO_2(aq)$ then reacts further.) Complete and balance this equation.

11. Write balanced chemical equations for the following reactions:
    (a) Solid calcium plus gaseous chlorine gives solid calcium chloride.
    (b) Iron(III) ion mixed with tin(II) ion in aqueous solution gives iron(II) ion and tin(IV) ion.
    (c) Solid iron(III) oxide reacts with gaseous hydrogen to give water vapor and solid iron.
    (d) Solid potassium reacts with aqueous hydrogen peroxide to give aqueous potassium hydroxide.

12. Write balanced equations for the following reactions:
    (a) Solid barium reacts with gaseous oxygen to give solid barium peroxide.
    (b) Cerium(IV) ion plus iodide ion in aqueous solution gives cerium(III) ion and aqueous iodine.
    (c) Solid manganese(III) hydroxide reacts with solid chromium to give solid manganese(II) hydroxide and solid chromium(III) hydroxide.
    (d) Solid mercury(I) chloride reacts with solid cobalt to give liquid mercury and solid cobalt(II) chloride.

13. Nitrous acid (HNO_2) disproportionates in acidic solution to nitrate ion (NO_3^-) and nitrogen monoxide (NO). Write a balanced equation for this reaction.

14. Thiosulfate ion ($S_2O_3^{2-}$) disproportionates in acidic solution to give solid sulfur and aqueous hydrogen sulfite ion (HSO_3^-). Write a balanced equation for this reaction.

Electrochemical Cells

15. (See Example 12–3.) An automobile battery delivers a current of 45 A for a period of 1.2 s to start the engine. Calculate the quantity of charge (in coulombs) passing through the battery.

16. (See Example 12–3.) A lightbulb in a flashlight draws a steady current of 3.00×10^{-3} A. Calculate the quantity of charge (in coulombs) passing through the bulb in 1.00 h.

17. A total of 6.5×10^3 C of charge passes through an electric iron in 18 min. Calculate the average current drawn by the iron in amperes.

18. Calculate the time required for 4.87×10^5 C to pass through an electric circuit if the current is 41.2 A.

19. (See Example 12–4.) Acidified water can be electrolyzed to produce hydrogen and oxygen (see Fig. 1–9). Write equa-

tions for the half-reactions that occur at the anode and at the cathode, and verify that they combine correctly to give the overall decomposition of water to its elements.

20. (See Example 12–4.) A galvanic cell operates to produce $Br^-(aq)$ and $Ni(OH)_2(s)$ from metallic nickel and bromate ion (BrO_3^-) in a basic aqueous solution. Write equations for the half-reactions that occur at the anode and at the cathode and for the overall reaction.

21. Diagram an electrochemical cell using the reaction

$$2\,Mn^{3+}(aq) + H_2(aq) \longrightarrow 2\,Mn^{2+}(aq) + 2\,H^+(aq)$$

22. Diagram an electrochemical cell in which an external voltage source is used to produce nickel and iodine from a solution of nickel(II) iodide, according to the equation

$$Ni^{2+}(aq) + 2\,I^-(aq) \longrightarrow Ni(s) + I_2(aq)$$

23. Sketch the following galvanic cell, indicating the direction of the flow of electrons in the external circuit and the direction of motion of the ions in the salt bridge.

$$Pt(s)|Cr^{2+}(aq), Cr^{3+}(aq)||Cu^{2+}(aq)|Cu(s)$$

Write a balanced equation for the overall reaction in this cell.

24. Sketch the following galvanic cell, indicating the direction of the flow of electrons in the external circuit and the direction of motion of ions in the salt bridge.

$$Ni(s)|Ni^{2+}(aq)||HCl(aq)|H_2(g)|Pt(s)$$

Write a balanced equation for the overall reaction in this cell.

25. Sketch the following electrochemical cells. Indicate in the sketch the sites of oxidation and reduction and the direction of flow of electrons. Write a balanced equation to represent the reaction taking place in each cell.
    (a) $Pt(s)|Fe^{2+}(aq), Fe^{3+}(aq)||MnO_4^-(aq), Mn^{2+}(aq)|Pt(s)$
    (b) $C(s, graphite)|I^-(aq), I_2(aq)||Ag^+(aq)|Ag(s)$

26. Sketch the following electrochemical cells. Indicate in the sketch the sites of oxidation and reduction and the direction of flow of electrons.
    (a) $Pb(s)|Pb^{2+}(aq)||Cu^{2+}(aq)|Cu(s)$
    (b) $Ag(s)|Ag^+(aq)||I_2(aq), I^-(aq)|C(graphite)$

Stoichiometry in Electrochemical Cells

27. An electric circuit from a battery carries a steady current of 2.5 A. How many electrons pass a given point in the wire per minute?

28. In a television set, the picture is created by an electron beam that sweeps over the screen, causing scintillations (flashes of light) where it strikes. A typical current for the electron beam is 60 microamperes (μA). How many electrons strike the screen per second?

29. Compute the quantity of electricity (in coulombs) that is required for each of the following reductions:

(a) 1.00 mol of $Ag^+(aq)$ to $Ag(s)$

(b) 2.00 mol of $H_2O(\ell)$ to $H_2(g)$

(c) 1.50 mol of $Mn^{3+}(aq)$ to $Mn^{2+}(aq)$

(d) 3.00 mol of $ClO_4^-(aq)$ to $Cl^-(aq)$

30. What number of coulombs of electricity is produced in each of the following oxidations?

(a) 2.00 mol of $Zn(s)$ to $Zn^{2+}(aq)$

(b) 1.00 mol of $H_2O_2(aq)$ to $O_2(g)$

(c) 2.50 mol of $MnO_2(s)$ to $MnO_4^-(aq)$

(d) 1.70 mol of $H_2S(aq)$ to 0.850 mol of $H_2S_2O_3(aq)$

31. (See Example 12–5.) A quantity of electricity equal to 6.95×10^4 C passes through an electrolytic cell that contains a solution of $Sn^{4+}(aq)$ ions. Compute the maximum chemical amount, in moles, of $Sn(s)$ that can be deposited at the cathode. What mass of Sn is this?

32. (See Example 12–5.) A quantity of electricity equal to 9.263×10^4 C passes through a galvanic cell that has an $Ni(s)$ anode. Compute the maximum chemical amount, in moles, of $Ni^{2+}(aq)$ that can be formed in solution. What maximum mass of $Ni(s)$ is lost from the anode?

33. A galvanic cell uses a zinc anode immersed in a $Zn(NO_3)_2$ solution and a platinum cathode immersed in a NaCl solution and in contact with $Cl_2(g)$ at 1 atm and 25°C.

(a) Write a balanced equation for the cell reaction.

(b) A steady current of 0.800 A is observed to flow for a period of 25.0 min. How much charge passes through the circuit during this time? How many moles of electrons is this charge equivalent to?

(c) Calculate the change in mass of the zinc electrode.

(d) Calculate the volume of gaseous chlorine consumed as a result of the reaction.

34. A galvanic cell uses a cadmium cathode immersed in a $CdSO_4$ solution and a zinc anode immersed in a $ZnSO_4$ solution.

(a) Write a balanced equation for the cell reaction.

(b) A steady current of 1.45 A is observed to flow for a period of 2.60 h. How much charge passes through the circuit during this time? How many moles of electrons is this charge equivalent to?

(c) Calculate the change in mass of the zinc electrode.

(c) Calculate the change in mass of the cadmium electrode.

35. Two electrolytic cells are connected in series. (In this arrangement, equal numbers of electrons pass through the two cells.) In the first cell, $Ag^+(aq)$ is reduced to $Ag(s)$. In the second, $Cd^{2+}(aq)$ is reduced to $Cd(s)$. These are the only reduction reactions taking place in either cell. After a time, electrolysis is stopped, and it is found that 0.475 g of solid silver has been deposited in the first cell. Determine the mass of cadmium deposited in the second cell.

36. Two electrolytic cells are connected in series. In the first cell, H_2O is oxidized to $O_2(g)$, and in the second $Cl^-(aq)$ is oxidized to $Cl_2(g)$. These are the only oxidation reactions taking place in the cells. The gases are collected as they

evolve. After a time, electrolysis is stopped, and it is found that 2.65 L of oxygen has been evolved from the first cell. Determine the volume of chlorine evolved from the second cell. The volume measurements are performed at the same temperature and pressure.

37. An acidic solution containing copper ions is electrolyzed, producing gaseous oxygen (from water) at the anode and metallic copper at the cathode. For every 16.0 g of oxygen that is generated, 63.5 g of copper plates out. What is the oxidation state of the copper in the solution?

38. Michael Faraday reported that passing electricity through one solution liberated 1 mass of hydrogen at the cathode and 8 masses of oxygen at the anode. The same quantity of electricity liberated 36 masses of chlorine at the anode and 58 masses of tin at the cathode from a second solution. What were the oxidation states of hydrogen, oxygen, chlorine, and tin in these solutions?

Metals and Metallurgy

39. The principal ore of manganese is pyrolusite (MnO_2), which can be reduced to elemental manganese with aluminum. Write a balanced equation for the reaction, and calculate the standard enthalpy and standard Gibbs energy of reaction, using data in Appendix D. The aluminum is oxidized to aluminum(III) oxide.

40. Uranium can be prepared in elemental form by reducing U_3O_8 with calcium. Write a balanced equation for the reaction, and calculate ΔH_r° and ΔG_r° using the following thermodynamic data:

	$\Delta H_f^\circ(\text{kJ mol}^{-1})$	$S^\circ(\text{J K}^{-1}\,\text{mol}^{-1})$
$U_3O_8(s)$	-3575	282.6
$U(s)$	0	50.2
$CaO(s)$	-635	39.8
$Ca(s)$	0	41.4

41. As mentioned in the text, mercury(II) oxide (HgO) decomposes to mercury and oxygen when heated. Use data from Appendix D to estimate the temperature at which this change becomes spontaneous—that is, the temperature at which ΔG_r° becomes negative.

42. Use data from Appendix D to predict the temperature above which CuO decomposes spontaneously to $Cu(s)$ and $O_2(g)$ at atmospheric pressure.

43. Analysis of a copper-bearing seam of rock shows it to contain three copper-containing substances: 1.1% chalcopyrite ($CuFeS_2$), 0.42% covellite (CuS), and 0.51% bornite (Cu_5FeS_4) by mass. Calculate the total mass of copper present in 1.0 metric ton (1.0×10^3 kg) of this ore.

44. Calculate (a) the mass of copper and (b) the volume of sulfur dioxide at STP produced in the smelting of 1.0 metric ton (1.0×10^3 kg) of chalcopyrite ($CuFeS_2$).

45. What properties of bronze make it better for use in tools than copper?

46. How does brass differ from bronze? Why is brass used in precision instruments?

47. What is the purpose of froth flotation in the preparation of copper? Describe the process.

48. What are the three major problems affecting the smelting of copper?

49. (a) Write balanced equations for the reduction of hematite (Fe_2O_3), magnetite (Fe_3O_4), and siderite ($FeCO_3$) to iron with carbon monoxide. In each case, let the coefficient of Fe equal 1 (to allow easier comparison).
(b) Calculate ΔG_r° at 298.15 K for each of the three reactions in part (a).
(c) From your results, predict which iron compound should be easiest to reduce with $CO(g)$ from the standpoint of thermodynamics.

50. Calculate ΔG_r° at 298.15 K for these reactions:

$$Fe_2O_3(s) + \tfrac{3}{2}C(s) \longrightarrow 2\,Fe(s) + \tfrac{3}{2}CO_2(g)$$
$$Fe_2O_3(s) + 3\,C(s) \longrightarrow 2\,Fe(s) + 3\,CO(g)$$
$$Fe_2O_3(s) + 3\,CO(g) \longrightarrow 2\,Fe(s) + 3\,CO_2(g)$$

Which is the better reducing agent for $Fe_2O_3(s)$: C (being oxidized to CO or CO_2) or CO?

51. Write a balanced equation for the reaction of the mineral bornite (Cu_5FeS_4) with aqueous $FeCl_3$ if the products of the reaction are the same as in the reaction of chalcopyrite with $CuCl_2$.

52. Write a balanced equation to represent the redox reaction of the mineral bornite (Cu_5FeS_4) with aqueous copper(II) chloride ($CuCl_2$) if the products of the reaction are the same as in the reaction of chalcopyrite with $CuCl_2$.

53. Set up an entry for $CoO(s)/Co(s)$ for inclusion in Table 12–1; indicate where in the table it should be placed. Use data from Appendix D.

54. Calculate ΔG_r°'s at 298.15 K for the reactions

$$2\,Na(s) + Mg^{2+}(aq) \longrightarrow Mg(s) + 2\,Na^+(aq)$$
$$Na_2O(s) + Mg(s) \longrightarrow MgO(s) + 2\,Na(s)$$

to confirm that, at this temperature, Na(s) spontaneously reduces $Mg^{2+}(aq)$ but that Mg(s) spontaneously reduces $Na_2O(s)$.

55. Why is limestone or dolomite used in freeing iron from its ores?

56. What is pig iron? What must be removed from pig iron to change it into steel?

57. When pig iron is refined into steel, silicon and phosphorus impurities in the iron are oxidized to acidic oxides that are then trapped in the slag. Write balanced chemical equations for the oxidation of each of these two elements by oxygen. Write balanced equations for the reactions by which these products are trapped in the slag. Classify each reaction as a redox or an acid–base reaction.

58. What is the advantage of the basic oxygen process compared with other methods of making steel?

Electrometallurgy

59. In the Downs process, molten sodium chloride is electrolyzed to produce sodium and chlorine. Write equations representing the processes taking place at the anode and at the cathode in the Downs process.

60. The first element to be prepared by electrolysis was potassium. In 1807, Humphrey Davy passed an electric current through molten potassium hydroxide (KOH), obtaining liquid potassium at one electrode and water and oxygen at the other. Write equations representing the processes taking place at the anode and at the cathode.

61. A current of 55,000 A is passed through a series of 100 Hall–Héroult cells for a period of 24 h. Calculate the maximum theoretical mass of aluminum that can be recovered.

62. A current of 75,000 A is passed through an electrolysis cell containing molten $MgCl_2$ for a period of 7.0 d. Calculate the maximum theoretical mass of magnesium that can be recovered.

63. What is the advantage of using molten cryolite, rather than molten alumina, in the Hall–Héroult process?

64. What property of alumina does the Bayer process exploit?

65. An important use for magnesium is to make titanium. In the Kroll process, magnesium reduces titanium(IV) chloride to elemental titanium in a sealed vessel at 800°C. Write a balanced chemical equation for this reaction. What mass of magnesium is needed in theory to produce 100 kg of titanium from titanium(IV) chloride?

66. Calcium is used to reduce vanadium(V) oxide to elemental vanadium in a sealed steel vessel. Vanadium is used in vanadium steel alloys for jet engines, high-quality knives, and tools. Write a balanced chemical equation for this process. What mass of calcium is needed in theory to produce 20.0 kg of vanadium from vanadium(V) oxide?

67. What are two advantages of using dolomite in separating magnesium ions from seawater?

68. Describe the use of oyster shells in obtaining magnesium from seawater.

69. (See Example 12–6.) Galvanized steel consists of steel with a thin coating of zinc to slow corrosion. The zinc can be deposited by making the steel object the cathode and a block of zinc the anode in an electrolytic cell containing a dissolved zinc salt. Suppose a steel garbage can is to be galvanized and requires a total mass of 7.32 g of zinc to be coated to the required thickness. For how long should a current of 8.50 A be passed through the cell to achieve this?

70. (See Example 12–6.) In the electroplating of a silver spoon, the spoon acts as the cathode, and a piece of pure silver as the anode. Both dip into a solution of silver cyanide (AgCN). Suppose that a current of 1.5 A is passed through such a cell for a period of 22 min and that the spoon has a surface area of 16 cm². Calculate the average thickness of the silver layer deposited on the spoon, taking the density of silver to be 10.5 g cm⁻³.

ADDITIONAL PROBLEMS

71. Balance the following redox equations. Determine in each case which species is oxidized and which is reduced.
(a) $N_2O_5(\ell) + Au(s) \longrightarrow (NO_2)Au(NO_3)_4(s) + NO_2(g)$
(b) $TiCl_4(\ell) + NO_2Cl(g) \longrightarrow$
$$Cl_2(g) + TiOCl_2(s) + (NO)_2TiCl_6(s)$$
(c) $NHBr_2(s) \longrightarrow NH_4Br(s) + N_2(g) + Br_2(g)$
(d) $NaN_3(s) + NO_2Cl(g) \longrightarrow$
$$N_2O(g) + NaNO_3(s) + Cl_2(g)$$

72. The following "balanced" equation appeared recently in the pages of the *Journal of the American Chemical Society*:

$$[Co(H_2O)_6]^{2+} + HSO_5^- \longrightarrow$$
$$[Co(H_2O)_5(OH)]^{2+} + 2 H^+ + SO_4^-$$

In what respect is this equation not balanced? Balance the equation. (*Note*: The formulas HSO_5^- and SO_4^- are correct.)

73. Balance the following pair of related redox equations, one of which takes place between gases and the other in solution. Identify which species is oxidized and which species is reduced. Explain the similarity between the two.

$$Cl_2O(g) + N_2O_5(g) \longrightarrow NOCl(g) + O_2(g)$$
$$Cl_2O(aq) + HNO_3(aq) \longrightarrow$$
$$NO^+(aq) + Cl^-(aq) + O_2(g) + H_2O(\ell)$$
(in acidic solution)

74. Balance the following pair of related redox equations, one of which takes place between gases and the other in solution. Identify which species is oxidized and which species is reduced. Explain the similarity between the two.

$$Cl_2O_3(g) + N_2O_3(g) \longrightarrow NO_2Cl(g) + O_2(g)$$
$$Cl_2O_3(aq) + HNO_2(aq) \longrightarrow$$
$$NO_2^+(aq) + Cl^-(aq) + O_2(g) \quad \text{(in acidic solution)}$$

75. The drain cleaner Drano consists of aluminum turnings mixed with sodium hydroxide. When it is added to water, the sodium hydroxide dissolves and releases heat. The aluminum reacts with the basic water to generate bubbles of hydrogen and aqueous $Al(OH)_4^-$ ions. Write a balanced net ionic equation for this reaction.

76. Sodium hypochlorite (NaOCl) is the major ingredient of household bleach. Suppose its aqueous solution reacts with an aqueous solution of sodium thiosulfate ($Na_2S_2O_3$) under basic conditions, generating aqueous sodium sulfate and sodium chloride. Write a balanced net ionic equation for this reaction.

77. The following equation is written to describe a reaction event in aqueous base:

$$NO(g) + NH_2OH(aq) \longrightarrow N_2O(g) + N_2(g) + H_2O(\ell)$$

This is an invalid equation (see Section 2–1). Prove this by writing balanced half-equations for two different oxidations that could proceed concurrently in the solution.

78. Tarnish forms on silver spoons when airborne sulfur compounds react with the silver to produce a dark coating of silver sulfide (Ag_2S). Tarnish can be removed by putting the corroded spoons in an aluminum container (a pie plate works) containing a warm solution of baking soda (sodium hydrogen carbonate, $NaHCO_3$). The $NaHCO_3(aq)$ does not react but serves only as an electrolyte in an electrochemical cell in which the aluminum container is the anode. Write a balanced chemical equation for the tarnish-removing reaction.

79. A current passed through inert electrodes immersed in an aqueous solution of sodium chloride produces chlorate ion ($ClO_3^-(aq)$) at the anode and gaseous hydrogen at the cathode. Given this fact, write a balanced equation for the most likely chemical reaction if gaseous hydrogen and aqueous sodium chlorate are mixed and allowed to react spontaneously until they reach equilibrium.

80. A newly discovered bacterium is able to reduce selenate ion ($SeO_4^{2-}(aq)$) to elemental selenium ($Se(s)$) in reservoirs. This is significant because the soluble selenate ion is potentially toxic; elemental selenium, however, is insoluble and harmless. Assume that water is oxidized to oxygen as the selenate ion is reduced. Compute the mass of oxygen produced if all the selenate in a 10^{12} L reservoir contaminated with 100 mg L⁻¹ of selenate ion is reduced to selenium.

81. Thomas Edison invented an electric meter that was nothing more than a simple coulometer, a device to measure the amount of electricity passing through a circuit. In this meter, a small, fixed fraction of the total current supplied to a household was passed through an electrolytic cell, plating out zinc at the cathode. Each month, the cathode was removed and weighed to determine the amount of electricity used. If 0.25% of a household's electricity passed through such a coulometer, and the cathode increased in mass by 1.83 g in a month, how many coulombs of electricity were used during that month?

*82. The chief chemist of the Brite-Metal Electroplating Co. is required to certify that the rinse solutions that are discharged from its tin-plating process into the municipal sewer system contain no more than 10 parts per million (ppm) by mass of Sn^{2+}. The chemist devises the following analytical procedure to determine the concentration. At regular intervals, a 100 mL (= 100 g) sample is withdrawn from the waste stream and acidified to pH 1.0. A starch solution and 10 mL of 0.1 M potassium iodide are added, and a 25.0 mA current is passed through the solution between platinum electrodes. Iodine appears as a product of electrolysis at the anode when the oxidation of Sn^{2+} to Sn^{4+} is practically complete and signals its appearance by the deep blue color of a complex formed with starch. What is the maximum time of electrolysis to the appearance of the blue color that ensures that the concentration of Sn^{2+} does not exceed 10 ppm?

83. The galvanic cell $Zn(s)|Zn^{2+}(aq)||Ni^{2+}(aq)|Ni(s)$ is constructed using a completely immersed zinc electrode that weighs 32.68 g and a nickel electrode immersed in 575 mL of 1.00 M $Ni^{2+}(aq)$ solution. The cell furnishes an average current of 0.0715 A from the moment it is connected until one of its reactants is used up.
    (a) Which reactant is the limiting reactant in this cell?
    (b) How long (in seconds) does the cell furnish a current?
    (c) How much mass has the nickel electrode gained when the cell stops delivering a current?
    (d) What is the concentration of the $Ni^{2+}(aq)$ when the cell stops delivering a current?

84. Compute the equilibrium constant of the reaction

$$CO_2(g) + C(s) \rightleftharpoons 2\,CO(g)$$

at room temperature (25°C) and at 1875°C.

85. The principal ore of chromium is the mineral chromite ($FeCr_2O_4$), for which ΔG_f° equals -1344 kJ mol^{-1} at 25°C. Because the main use of chromium is in chromium-containing steels, the ore is generally reduced directly to the alloy ferrochrome by the reaction

$$FeCr_2O_4(s) + 4\,C(s) \longrightarrow Fe(s) + 2\,Cr(s) + 4\,CO(g)$$

Calculate ΔG_r° for the production of ferrochrome according to this equation, using Appendix D for other information you may need. (*Note*: The ΔG° in mixing iron and chromium in the alloy is small and can be neglected.)

86. Compare and contrast the smelting of copper with that of iron. Include in your discussion the differences in starting materials, in temperatures and reducing agents employed, and in the by-products and their disposal.

87. Two students are discussing the equation that represents the roasting of the ore chalcopyrite:

$$6\,CuFeS_2(s) + 13\,O_2(g) \longrightarrow$$
$$3\,Cu_2S(s) + 2\,Fe_3O_4(s) + 9\,SO_2(g)$$

One says that Cu starts in a +2 oxidation state, Fe in a +2 oxidation state, and S in a -2 oxidation state. He concludes that roasting oxidizes S and Fe but, strangely, reduces Cu. Another student disputes this, saying that Cu starts in a 0 oxidation state, Fe in a +2 oxidation state, and S in a -1 oxidation state. He concludes that roasting oxidizes Cu and Fe but causes S to disproportionate. Resolve this dispute.

88. Use data from Appendix D to predict whether elemental magnesium should react with water on purely thermodynamic grounds. What is observed in practice? Explain.

89. Compare and contrast the production of aluminum with that of magnesium. Include in your discussion the differences in starting materials, in reactions preparatory to the electrolysis, and in reaction by-products.

90. A 55.5 kg slab of crude copper from a smelter has a copper content of 98.3%. Estimate the length of time it takes to purify it electrochemically if it is used as the anode in a cell that has acidic copper(II) sulfate as its electrolyte and a current of 2.00×10^3 A is passed through the cell.

*91. Chromium is separated from iron in the ore chromite ($FeCr_2O_4(s)$) by fusing the ore with sodium carbonate:

$$FeCr_2O_4(s) + Na_2CO_3(\ell) + O_2(g) \longrightarrow$$
$$Fe_2O_3(s) + Na_2CrO_4(\ell) + CO_2(g)$$

    (a) Balance the preceding chemical equation.
    (b) Calculate ΔH_r°, assuming that the enthalpies of fusion of Na_2CO_3 and Na_2CrO_4 are approximately equal. The standard molar enthalpy of formation of $FeCr_2O_4(s)$ is -1445 kJ mol^{-1} and that of $Na_2CrO_4(s)$ is -1342 kJ mol^{-1}. Use Appendix D for the additional data you require.
    (c) The sodium chromate produced in the reaction is soluble in water and can be leached from the solidified melt. Suggest a procedure for converting it to elemental chromium.

*92. A steady current of 2.0 A is passed through a molten salt for 12.1 min, resulting in the deposition of 260 mg of a metallic element at the cathode. Name the two elements that could produce this behavior.

93. Seawater furnishes the "ore" for the preparation of metallic magnesium, and the cost of the product is accordingly almost completely determined by the cost of the energy used in the electrolytic reduction process. An engineer suggests reducing MgO directly with carbon (from coal).
    (a) Estimate the temperature at which K equals 1 for the reaction

$$2\,MgO(s) + C(s) \rightleftharpoons CO_2(g) + 2\,Mg(\ell)$$

    (b) Tests of this reaction establish that a practical process would have to be run at around 2500°C, a temperature that is difficult to reach and maintain. Moreover, it is difficult to separate the two products, and the back-reaction between them proves hazardously rapid. To reduce the possibility of back-reaction, the process is redesigned in such a way that that the product Mg is produced in solution in liquid antimony.

$$2\,MgO(s) + C(s) \rightleftharpoons CO_2(g) + 2\,Mg(sln)$$

    Predict whether this change raises or lowers the temperature required for a practical process. Explain.

CUMULATIVE PROBLEM

Lead

The metallurgy of lead parallels that of iron and copper in many respects. Lead ores generally contain lead sulfide (PbS) in some form, sometimes as the mineral galena.

(a) Balance the following equations, which represent reactions occurring in the smelting of lead ores:

   1. $PbS(s) + O_2(g) \longrightarrow PbSO_4(s)$
   2. $PbS(s) + PbO(s) \longrightarrow Pb(s) + SO_2(g)$
   3. $PbS(s) + PbSO_4(s) \longrightarrow Pb(s) + SO_2(g)$

(b) Metallic lead sometimes is prepared by heating ores containing lead sulfide in a furnace with scrap iron. Write a balanced equation for this reaction. Determine whether this is a spontaneous reaction at 298.15 K.

(c) Lead can be purified by passing an electric current through an aqueous solution of $PbSiF_6$ (which contains $Pb^{2+}(aq)$ and $SiF_6^{2-}(aq)$ ions) using the crude lead as the anode and a piece of pure lead as the cathode. Write equations for the half-reactions that take place at the anode and at the cathode.

(d) Suppose that, by mistake, the source of electric current in the purification cell described in the previous part is connected so that the crude lead is the cathode and the pure lead is the anode. Does an electric current flow? Does the electrochemical purification work? Explain.

(e) Some lead ores are high in bismuth(III) oxide (Bi_2O_3), which has a standard Gibbs energy of formation of -497 kJ mol^{-1}. Between which two entries would bismuth(III) oxide be inserted in Table 12–1?

(f) A major use of lead is in the manufacture of lead–sulfuric acid storage batteries for automobiles. The reaction occurring when such a battery is discharged is

$$Pb(s) + PbO_2(s) + SO_4^{2-}(aq) \longrightarrow PbSO_4(s)$$

Complete and balance this equation for acidic conditions.

(g) A lead–acid storage battery delivers an average current of 0.705 A for a period of 17.4 min. Determine the mass of $PbSO_4$ that is formed inside the battery.

Molten lead being poured to make lead soldiers. *(Dan McCoy/Rainbow)*

13

Electrochemistry and Cell Voltage

CHAPTER OUTLINE

Lead–acid batteries connected in series in an electric vehicle. (*Tom Myers/Photo Researchers, Inc.*)

Electrochemistry combines the chemistry of oxidation–reduction reactions with the physics of the flow of electric currents. Galvanic cells use the decrease in Gibbs energy of a spontaneous chemical reaction to push electrons through a wire from anode to cathode. The potential difference (voltage) developed between the electrodes in such cells drives this transfer. In electrolytic cells, the chemical reactions are non-spontaneous, but controlled intervention with an outside voltage forces them along to give valuable compounds with high Gibbs energies.

Electrochemistry is a supremely practical field, dealing with the storage of energy and its efficient conversion from readily available sources (such as solar or chemical energy) into forms that are useful for technological applications. Although the thermodynamic considerations governing electrochemical transformations have been thoroughly worked out, the design of efficient electrochemical cells must contend with such a multitude of other, conflicting factors that it remains very much an art as well as a science. In this chapter, we present the fundamental theory and show its application in cases in which electrochemical action is desired (batteries and electrolysis cells) as well as in cases in which it is not desired (corrosion).

13–1 The Gibbs Energy and Cell Voltage

Differences in electric potential ($\Delta\mathcal{E}$'s) drive the flow of current in electrical devices. If two regions of different potential are physically separated, no current flows unless a path (a copper wire, for example) is in place to conduct the flow of electrons. Without a path, the potential difference simply abides; hence, the word "potential" is well chosen. Passage of an electric current can perform a type of work not previously discussed, **electrical work.** Recall that work is one means (the other is heat) by which energy is transferred between a system and its surroundings. The transfer of pressure–volume work, considered in Chapters 10 and 11, requires a *mechanical* connection between a system and its surroundings. Similarly, the transfer of electrical work requires an *electrical* connection between system and surroundings. The apparatus of electrodes, salt bridge, and wires discussed in Section 12–2 is such a connection. If an amount of charge (Q) moves through an electrical connection, driven by a potential difference ($\Delta\mathcal{E}$), the electrical work is

- An electric potential difference is measured with a potentiometer. Zero electric current flows during such measurements.

$$w_{elec} = -Q\Delta\mathcal{E}$$

The minus sign appears in this equation because of the convention (adopted in Chapters 10 and 11) that positive work is work that is performed *on* a system. If the system is a galvanic cell, then the potential difference, $\Delta\mathcal{E}$, is positive, and w_{elec} is negative. The cell performs electrical work in the surroundings (acts as a battery). If the system is an electrolytic cell, $\Delta\mathcal{E}$ is negative and w_{elec} is positive; an energy source in the surroundings, such as an electrical generator, performs work on the cell.

If the preceding equation is used with SI units, then w_{elec} is measured in joules and Q in coulombs. It follows that $\Delta\mathcal{E}$ is measured in joules per coulomb. One joule per coulomb is defined as one **volt** (V). The volt (or derivatives such as the millivolt and kilovolt) is the only unit used to measure potential difference, which is often called "voltage," as pointed out in Section 12–2. The Q in the preceding equation is the amount of charge passing through the electrical connection. It equals the average current (I) multiplied by the time (t) during which the current is flowing. The equation for the electrical work therefore also can be written

$$w_{elec} = -It\,\Delta\mathcal{E}$$

One further unit for electrical work or energy that is in common use is the kilowatt-hour (kW-h). The **watt** (W) is a unit of power, or the rate of transfer of energy. It is defined as one joule per second ($J\,s^{-1}$). The kilowatt is 1000 watts, or $10^3\,J\,s^{-1}$. The kilowatt-hour, then, is the total energy transferred (or work done) in 1 h at a rate of $10^3\,J\,s^{-1}$. It is equal to

$$1\;kW\text{-}h = \left(\frac{10^3\;J}{s}\right)(3600\;s) = 3.6 \times 10^6\;J\;\text{(exactly)}$$

because 1 h equals exactly 3600 s.

EXAMPLE 13-1

The battery in Example 12–3 transfers 6750 C of charge. Assume that it maintains a constant voltage of 6.00 V during this process. Calculate the total work done *by* the battery in joules and in kilowatt-hours.

Strategy

The w_{elec} in the formula "$w_{elec} = -Q\Delta\mathcal{E}$" represents work done *on* the battery. Work done *by* the battery is the negative of this quantity. Compute w_{elec} and reverse its sign.

Solution

$$w_{elec} = -Q\Delta\mathcal{E} = -(6750\;C)(6.00\;V) = -40{,}500\;C\;V$$
$$= -4.05 \times 10^4\;J$$

$$\text{work done by battery} = -w_{elec} = -(-4.05 \times 10^4\;J) = +4.05 \times 10^4\;J$$

$$= 4.05 \times 10^4\;J \times \left(\frac{1\;kW\text{-}h}{3.60 \times 10^6\;J}\right) = 0.0112\;kW\text{-}h$$

EXERCISE

Calculate the total electrical work done *on* an electrolytic cell by the surroundings if an average current of 0.741 A is run through it at a voltage of 2.53 V for 4150 s.

Answer: work done on cell $= w_{elec} = +7.78 \times 10^3\;J$.

Consider a galvanic cell at constant temperature and pressure that is ready to operate but is disconnected so that no electric current is actually flowing. Thermodynamics provides a fundamental relationship between the change in Gibbs energy accompanying whatever process is poised to take place within this cell (ΔG) and the maximum electrical work that can be transferred:

$$\Delta G = w_{elec,\,max} \qquad \text{(at constant } T \text{ and } P\text{)}$$

According to this equation, if the cell is connected for a while, the change in the Gibbs energy of its contents equals the maximum amount of energy that can appear in the surroundings as electrical work.

How does the transfer occur? The spontaneous decrease in Gibbs energy inside the cell first manifests itself as an increase in the Gibbs energy of electrons as they are caused to move through the wire. This Gibbs energy then can be harvested by connecting the wire to a motor that is set up to do work, or it can be wasted by simply allowing electrical resistance to dissipate it in the form of heat.

The subscript "max" appears in the preceding equation because real galvanic cells inevitably waste some of their decrease in Gibbs energy as heat when the circuit is closed and a measurable current flows. If the current flows very slowly, essentially one electron at a time, then this wastage is negligible and

$$w_{\text{elec, max}} = -Q\Delta\mathcal{E}$$

although the transfer of the charge then requires a prohibitively long time. Combining this equation with the previous one gives

$$\Delta G = -Q\Delta\mathcal{E} \qquad \text{(at constant } T \text{ and } P)$$

This equation states that the potential difference of a cell, an easily measured property, is proportional to the change in the Gibbs energy of the cell. If $\Delta\mathcal{E}$ is positive, then ΔG is negative: The change within the cell is spontaneous and can produce electrical work in the surroundings. This is the case of a galvanic cell, which we have been discussing. On the other hand, if $\Delta\mathcal{E}$ is negative, then ΔG is positive: The change within the cell is non-spontaneous and requires the input of electrical work from an external source before it occurs. This is the case of an electrolytic cell.

Suppose now that the process inside the cell is a redox reaction that involves the passage though the external circuit of n moles of electrons per mole of reaction. This corresponds to $n\mathcal{F}$ coulombs of charge per mole of reaction (where $\mathcal{F}$ is the Faraday constant, 96,485 C mol^{-1}). The last equation becomes

$$\Delta G_r = -Q\Delta\mathcal{E} = -n\mathcal{F}\Delta\mathcal{E} \qquad \text{(at constant } T \text{ and } P)$$

The n here has units of mol mol^{-1} (moles of electrons per mole of reaction). This equals 1, meaning that n has no unit. Consequently, ΔG_r has the units J mol^{-1} when $\Delta\mathcal{E}$ is measured in volts. It is indeed a molar free energy of reaction.

Because differences in electric potential are very easily measured, the preceding relationship provides an attractive way to determine changes in the molar Gibbs energy of reaction. Dividing both sides by $-n\mathcal{F}$ gives

$$\Delta\mathcal{E} = \frac{-\Delta G_r}{n\mathcal{F}} \qquad \text{(at constant } T \text{ and } P)$$

This form allows an important conclusion: The potential difference of an electrochemical cell does *not* depend on the size of the cell. If the amount of reactants that is converted to products is, for example, doubled (signified by doubling the coefficients in the equation that represents the reaction going on in the cell), then ΔG_r is doubled, but n, the number of moles of electrons transferred per mole of reaction, is also doubled, and $\Delta\mathcal{E}$ remains unchanged.

Standard States and Cell Voltages

Once again, recall the definitions of standard states of substances: for solids and liquids, the state of the pure solid or liquid under a pressure of 1 atm and at a specified temperature; for gases, the gaseous phase under a pressure of 1 atm, at a specified temperature, and exhibiting ideal-gas behavior; for dissolved species, the state at a

• The Gibbs energy of a reaction is often called the "*free* energy" or "Gibbs free energy" because it is the energy that is available (free) to do useful work (at constant temperature and pressure).

concentration of 1 mol L^{-1} under a pressure of 1 atm, at a specified temperature, and exhibiting ideal-solution behavior. Any temperature can be specified. The most common choice is 298.15 K (25°C exactly). The standard Gibbs energy (ΔG_r°) for a reaction is the change in Gibbs energy when reactants in standard states give products in standard states.

For redox reactions, all of which can in principle be carried out in electrochemical cells, a new quantity, the **standard cell voltage** ($\Delta \mathcal{E}^\circ$) is useful. It is related to the standard Gibbs energy of reaction by the previous equation, with added superscript zeros ("naughts") to indicate that standard states are being used:

$$\Delta \mathcal{E}^\circ = \frac{-\Delta G_r^\circ}{n\mathscr{F}}$$

The standard cell voltage is the potential difference developed by a galvanic cell in which all reactants and products in the operative reaction are in standard states. Such a cell is called a **standard cell,** and the half-cells that make it up are called **standard half-cells.** The standard cell voltage is readily calculated from the standard Gibbs energy of the reaction by means of the preceding equation. The reverse statement is equally true and perhaps more important. The ΔG_r° (and from it the equilibrium constant) of any redox reaction can be determined by a single measurement of voltage whenever an electrochemical cell can be constructed that uses the reaction and has the reactants and products in standard states:

$$\Delta G_r^\circ = -n\mathscr{F}\Delta \mathcal{E}^\circ$$

EXAMPLE 13–2

A standard Ni^{2+}|Ni half-cell is connected to a standard Cu^{2+}|Cu half-cell to make a galvanic cell. The voltage of the cell equals 0.602 V at 25°C, and elemental copper plates out in the Cu^{2+}|Cu half-cell compartment if the reaction is allowed to proceed. Write an equation to represent the reaction taking place in the cell and calculate ΔG_r° for this reaction.

Strategy

For simplicity, write a balanced equation having the smallest possible whole-number coefficients. Break the equation into half-equations to determine n, the number of moles of electrons transferred per mole of reaction as written. Recognize that the two standard half-cells combine to make a standard cell and that the equation $\Delta G_r^\circ = -n\mathscr{F}\Delta \mathcal{E}^\circ$ applies.

Solution

The cell reaction is

$$\text{Ni}(s) + \text{Cu}^{2+}(aq) \longrightarrow \text{Ni}^{2+}(aq) + \text{Cu}(s)$$

which combines the half-reactions

$$\text{Ni}(s) \longrightarrow \text{Ni}^{2+}(aq) + 2\,e^- \qquad \text{(oxidation)}$$
$$\text{Cu}^{2+}(aq) + 2\,e^- \longrightarrow \text{Cu}(s) \qquad \text{(reduction)}$$

The half-equations show that 2 mol of e^- passes through the external circuit per mole of the reaction that is written. Therefore $n = 2$ for this reaction.

$$\Delta G_r^\circ = -n\mathscr{F}\Delta\mathcal{E}^\circ$$

$$= -2 \times \left(\frac{96,485 \text{ C}}{\text{mol}}\right) \times (0.602 \text{ V}) = -1.16 \times 10^5 \text{ C V mol}^{-1}$$

$$= -1.16 \times 10^5 \text{ C V mol}^{-1} \times \left(\frac{1 \text{ J}}{1 \text{ C V}}\right) \times \left(\frac{1 \text{ kJ}}{1000 \text{ J}}\right) = -116 \text{ kJ mol}^{-1}.$$

This reaction is spontaneous running in a beaker (Fig. 13–1) as well as in a cell.

EXERCISE

A standard $Cr^{3+}|Cr$ half-cell and a standard $Co^{2+}|Co$ half-cell are connected to make a galvanic cell. The voltage of the cell equals 0.464 V at 25°C, and Co^{2+} ion is reduced to elemental Co if the reaction is allowed to proceed. Write an equation to represent the reaction taking place in the cell, and calculate its ΔG_r°.

Answer: $2 \text{ Cr}(s) + 3 \text{ Co}^{2+}(aq) \longrightarrow 2 \text{ Cr}^{3+}(aq) + 3 \text{ Co}(s)$

$$\Delta G_r^\circ = -269 \text{ kJ mol}^{-1}.$$

Figure 13–1 • The reduction of $Cu^{2+}(aq)$ ion by nickel (see Example 13–2) occurs spontaneously when a bar of metallic nickel is put into a solution containing $Cu^{2+}(aq)$ ions, which are blue. Copper plates out, and the solution turns green as $Ni^{2+}(aq)$ forms. In this experiment, Gibbs energy of reaction is wasted as heat evolves into the surroundings. In an electrochemical cell, it can be harnessed to do useful electrical work. *(Marna G. Clarke)*

13-2   Half-Cell Potentials

We could tabulate all the conceivable galvanic cells and their standard potential differences, but the list would be very long. To avoid this, we introduce the **standard half-cell reduction potential,** a quantity that expresses the intrinsic tendency of a reduction half-reaction to occur when both the reactant and product are in standard states.

The galvanic cell described in Example 13–2 is made up of $Ni^{2+}|Ni$ and $Cu^{2+}|Cu$ standard half-cells (the concentrations of Ni^{2+} and Cu^{2+} ions are both 1 M, and the two metals are in standard states as well). We write the half-reactions taking place in the two half-cells as reductions:

$$Ni^{2+}(aq) + 2 e^- \longrightarrow Ni(s)$$
$$Cu^{2+}(aq) + 2 e^- \longrightarrow Cu(s)$$

The standard reduction potential for the first half-reaction is symbolized $\mathcal{E}^\circ(Ni^{2+}|Ni)$. It measures the electromotive force, or push, favoring the half-reaction when the $Ni^{2+}(aq)$ and $Ni(s)$ are in standard states. The standard reduction potential $\mathcal{E}^\circ(Cu^{2+}|Cu)$ does the same for the second half-reaction. A more positive standard reduction potential implies a greater tendency for the reduction to take place.

Reduction does not take place in isolation. The electrons must come from somewhere, and the source can only be another half-cell. When two half-cells are connected to make a galvanic cell, the one with the more positive reduction potential "wins" and reduction occurs there; the one with the less positive reduction potential runs in reverse, as an oxidation, to supply electrons. The potential *difference* that is measured between the two electrodes of the cell (the cell voltage) then equals the *difference* between the two standard reduction potentials. The standard cell of Example 13–2 has $\Delta\mathcal{E}^\circ = 0.602$ V with $\mathcal{E}^\circ(Cu^{2+}|Cu)$ more positive than $\mathcal{E}^\circ(Ni^{2+}|Ni)$. Therefore,

$$\Delta\mathcal{E}^\circ = \mathcal{E}^\circ(Cu^{2+}|Cu) - \mathcal{E}^\circ(Ni^{2+}|Ni) = 0.602 \text{ V}$$

• In algebraic comparisons, the signs of the numbers are considered. The number 2 is algebraically greater than -3 although smaller than -3 in absolute value.

When two standard half-cells are combined to make a galvanic cell, reduction occurs in the half-cell having the algebraically greater standard reduction potential (making it the cathode). Oxidation occurs in the other half-cell (making it the anode). Therefore,

$$\Delta \mathcal{E}° = \mathcal{E}°(\text{cathode}) - \mathcal{E}°(\text{anode})$$

Standard half-cell reduction potentials ($\mathcal{E}°$'s) cannot be measured experimentally. Only differences in standard reduction potential ($\Delta\mathcal{E}°$'s) are measurable. Chemists have consequently selected a particular half-reaction and arbitrarily assigned it a standard reduction potential $\mathcal{E}°$ of zero exactly. This tactic makes possible the efficient tabulation of other half-cell reduction potentials.

By international convention, the $\mathcal{E}°$ for the reduction of $H^+(aq)$ to give $H_2(g)$ equals zero at all temperatures:

$$2\,H^+(aq, 1\ \text{M}) + 2\,e^- \longrightarrow H_2(g, P = 1\ \text{atm}) \qquad \mathcal{E}° = 0\ \text{V} \qquad \text{(by convention)}$$

Standard reduction potentials are determined relative to this reference by building galvanic cells that combine their standard half-cells with the standard $H^+|H_2$ half-cell and measuring the potential difference. Although it seems strange to use a gas as a cathode or anode, it is fairly easy to arrange (Fig. 13–2). The experimental voltages (the $\Delta\mathcal{E}°$'s) that come from such devices are then substituted in the equation

$$\Delta\mathcal{E}° = \mathcal{E}°(\text{cathode}) - \mathcal{E}°(\text{anode})$$

to compute the standard reduction potential of the half-cell in question.

The $\mathcal{E}°$ of a standard $H^+|H_2$ half-cell is zero whether it serves as the anode or cathode. If another standard half-cell functions as the cathode when hooked up in a galvanic cell with a standard $H^+|H_2$ half-cell, then the $\mathcal{E}°$ of the second half-cell is positive. If it functions as the anode, then its $\mathcal{E}°$ is negative. A standard

Gaseous hydrogen evolves from an illuminated photo-electrode. *(Nathan S. Lewis/California Institute of Technology)*

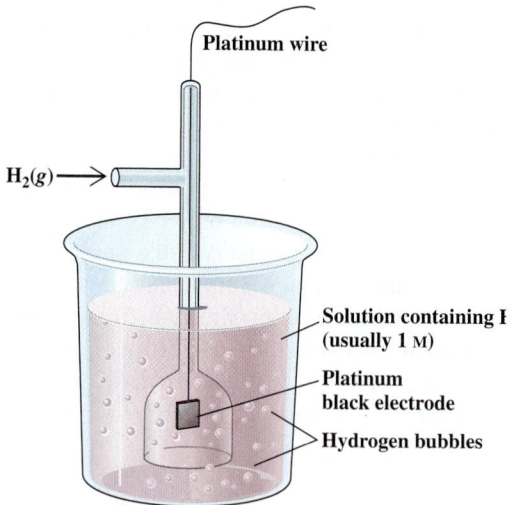

Platinum wire

$H_2(g) \longrightarrow$

Solution containing H
(usually 1 M)

Platinum
black electrode

Hydrogen bubbles

Figure 13–2 • Hydrogen is a gas at room conditions, and electrodes cannot be constructed from it directly. In this hydrogen half-cell, a foil of platinum covered with a coating of finely divided platinum is dipped into the solution, and a stream of hydrogen is passed over the surface. The platinum does not react but provides a support surface for the reaction of H_2 to give H^+, or the reverse. It also provides the necessary electrical connection to carry away or bring in electrons.

reduction half-reaction has a positive $\mathcal{E}°$ if it has a greater intrinsic tendency to occur than the reduction of $H^+(aq, 1 \text{ M})$ to $H_2(g, 1 \text{ atm})$. It has a negative $\mathcal{E}°$ if it has a lesser tendency. For example, when a standard $Cu^{2+}|Cu$ half-cell is connected to a standard $H^+|H_2$ half-cell, elemental copper plates out. This shows that the copper half-cell is the cathode and the hydrogen half-cell is the anode. The cell is represented

$$Pt(s)|H_2(g, 1 \text{ atm})|H^+(aq, 1 \text{ M}) \vdots Cu^{2+}(aq, 1 \text{ M})|Cu(s)$$

with the anode half-cell to the left of the salt bridge and the cathode half-cell to the right, according to convention (see Section 12–2). The observed cell voltage at 298.15 K is 0.345 V, so

$$\Delta\mathcal{E}° = \mathcal{E}°(\text{cathode}) - \mathcal{E}°(\text{anode})$$
$$= \mathcal{E}°(Cu^{2+}|Cu) - \mathcal{E}°(H^+|H_2)$$
$$0.345 \text{ V} = \mathcal{E}°(Cu^{2+}|Cu) - 0$$
$$\mathcal{E}°(Cu^{2+}|Cu) = 0.345 \text{ V}$$

The standard potential of the $Cu^{2+}|Cu$ half-cell equals 0.345 V on a scale on which the standard $H^+|H_2$ half-cell potential is 0 V:

$$Cu^{2+} + 2\,e^- \longrightarrow Cu(s) \qquad \mathcal{E}° = 0.345 \text{ V}$$

When a standard $Ni^{2+}|Ni$ half-cell is connected to a standard $H^+|H_2$ half-cell and the resulting cell operates spontaneously, the concentration of $Ni^{2+}(aq)$ rises as $Ni(s)$ dissolves. This observation identifies the nickel half-cell as the anode half-cell. Oxidation occurs there. The conventional cell diagram

$$Ni(s)|Ni^{2+}(aq, 1 \text{ M}) \vdots H^+(aq, 1 \text{ M})|H_2(g, 1 \text{ atm})|Pt(s)$$

represents this. The measured cell voltage at 298.15 K is 0.257 V, so that

$$\Delta\mathcal{E}° = \mathcal{E}°(\text{cathode}) - \mathcal{E}°(\text{anode})$$
$$0.257 \text{ V} = 0 - \mathcal{E}°(Ni^{2+}|Ni)$$
$$\mathcal{E}°(Ni^{2+}|Ni) = -0.257 \text{ V}$$

When a standard nickel and a standard copper half-cell are hooked together, the copper half-cell is the cathode half-cell because it has the more positive reduction potential. The cell is represented

$$Ni(s)|Ni^{2+}(aq, 1 \text{ M}) \vdots Cu^{2+}(aq, 1 \text{ M})|Cu(s)$$

Its voltage at 298.15 K should be

$$\Delta\mathcal{E}° = \mathcal{E}°(\text{cathode}) - \mathcal{E}°(\text{anode})$$
$$= \mathcal{E}°(Cu^{2+}|Cu) - \mathcal{E}°(Ni^{2+}|Ni)$$
$$= 0.345 \text{ V} - (-0.257 \text{ V}) = 0.602 \text{ V}$$

as Figure 13–3 shows diagrammatically. Experiment confirms this prediction.

The numbers of electrons appearing in half-equations do not figure in the computation of the potential difference of a cell. Suppose a standard $Al^{3+}|Al$ half-cell is connected to a standard $Pb^{2+}|Pb$ half-cell. Lead plates out at the cathode. The cell is diagrammed

$$Al(s)|Al^{3+}(aq, 1 \text{ M}) \vdots Pb^{2+}(aq, 1 \text{ M})|Pb(s)$$

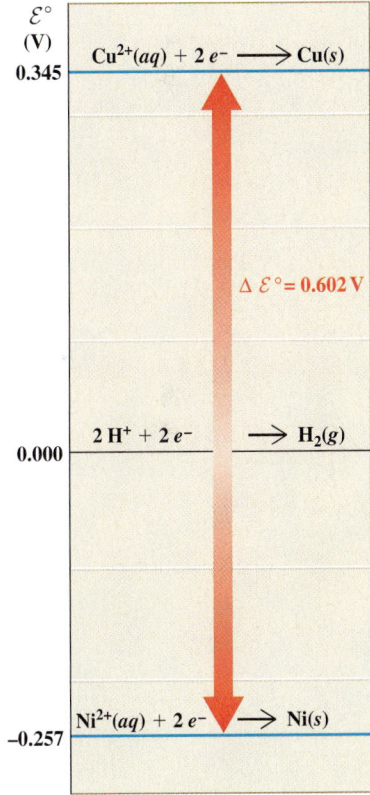

Figure 13–3 • Determining $\Delta\mathcal{E}°$'s from $\mathcal{E}°$'s. Here, the vertical distance between a pair of standard reduction potentials is proportional to the $\Delta\mathcal{E}°$ generated by an electrochemical cell that combines the two. The reduction half-reaction with the higher reduction potential forces the other half-reaction to run in reverse.

The reduction half-reactions associated with the two half-cells are

$$Pb^{2+}(aq) + 2\,e^- \longrightarrow Pb(s)$$
$$Al^{3+}(aq) + 3\,e^- \longrightarrow Al(s)$$

Multiplying the first half-equation by 3 and the second by 2 and then subtracting the second from the first causes the electrons to cancel out and gives the balanced equation

$$2\,Al(s) + 3\,Pb^{2+}(aq) \longrightarrow 2\,Al^{3+}(aq) + 3\,Pb(s)$$

This equation represents the cell reaction. Parallel multiplications are *not* performed in computing the potential difference. In this case

$$\Delta\mathcal{E}° = \mathcal{E}°(Pb^{2+}|Pb) - \mathcal{E}°(Al^{3+}|Al)$$
$$= -0.126\ \text{V} - (-1.662\ \text{V}) = 1.536\ \text{V}$$

The standard reduction potentials used in this calculation come from Table 13–1. A more extensive list of standard reduction potentials appears in Appendix E.

TABLE 13–1	Some Standard Reduction Potentials in Aqueous Solution at 25°C
Reduction Half-Reaction	**Standard Reduction Potential $\mathcal{E}°$ (V)**
$F_2(g) + 2\,e^- \longrightarrow 2\,F^-(aq)$	2.866
$Au^{3+}(aq) + 3\,e^- \longrightarrow Au(s)$	1.498
$Cl_2(g) + 2\,e^- \longrightarrow 2\,Cl^-(aq)$	1.3583
$Pt^{2+}(aq) + 2\,e^- \longrightarrow Pt(s)$	1.18
$Br_2(\ell) + 2\,e^- \longrightarrow 2\,Br^-(aq)$	1.066
$Hg^{2+}(aq) + 2\,e^- \longrightarrow Hg(\ell)$	0.851
$Ag^+(aq) + e^- \longrightarrow Ag(s)$	0.800
$I_2(s) + 2\,e^- \longrightarrow 2\,I^-(aq)$	0.536
$Cu^{2+}(aq) + 2\,e^- \longrightarrow Cu(s)$	0.345
$2\,H^+(aq) + 2\,e^- \longrightarrow H_2(g)$	0 exactly
$Pb^{2+}(aq) + 2\,e^- \longrightarrow Pb(s)$	−0.126
$Sn^{2+}(aq) + 2\,e^- \longrightarrow Sn(s)$	−0.138
$Ni^{2+}(aq) + 2\,e^- \longrightarrow Ni(s)$	−0.257
$Co^{2+}(aq) + 2\,e^- \longrightarrow Co(s)$	−0.28
$Cd^{2+}(aq) + 2\,e^- \longrightarrow Cd(s)$	−0.403
$Fe^{2+}(aq) + 2\,e^- \longrightarrow Fe(s)$	−0.447
$Cr^{3+}(aq) + 3\,e^- \longrightarrow Cr(s)$	−0.744
$Zn^{2+}(aq) + 2\,e^- \longrightarrow Zn(s)$	−0.762
$Mn^{2+}(aq) + 2\,e^- \longrightarrow Mn(s)$	−1.185
$Al^{3+}(aq) + 3\,e^- \longrightarrow Al(s)$	−1.662
$Mg^{2+}(aq) + 2\,e^- \longrightarrow Mg(s)$	−2.372
$Na^+(aq) + e^- \longrightarrow Na(s)$	−2.71
$Ca^{2+}(aq) + 2\,e^- \longrightarrow Ca(s)$	−2.868
$K^+(aq) + e^- \longrightarrow K(s)$	−2.931
$Li^+(aq) + e^- \longrightarrow Li(s)$	−3.040

The half-reactions in Table 13–1 and Appendix E are arranged in order of decreasing reduction potential. The most easily reduced species has the most positive reduction potential and is at the top. In any galvanic cell that is constructed by combining standard half-cells at 25°C, the half-reaction that is listed higher in the table proceeds as written at the cathode, and the half-reaction that is listed lower proceeds in reverse at the anode.

EXAMPLE 13-3

A standard $(Pt|MnO_4^-, H^+, Mn^{2+})$ half-cell consists of an inert platinum electrode in contact with a solution containing $MnO_4^-(aq)$, $H^+(aq)$, and $Mn^{2+}(aq)$ ions in standard states. Such a half-cell is assembled and connected to a standard $Zn^{2+}|Zn$ half-cell. Use data from Appendix E to
(a) calculate the standard potential difference $(\Delta \mathcal{E}^\circ)$ of the cell at 25°C.
(b) write a balanced equation for the overall cell reaction.

Strategy

Scan the half-equations in the table in Appendix E, which is not in alphabetical order, for the ones that represent the two half-reactions going on in this cell. Realize that the correct choices must contain all of the reactants and products appearing in the half-cell diagrams, not just some of them. Assign the half-equation that lies higher in the table as the cathode (reduction) half-equation, and compute the standard potential difference. Obtain a balanced overall equation by combining the balanced half-equations in such a way that electrons do not appear as either reactants or products, as explained in Step 6 of the method for balancing redox equations in Section 12–1.

Solution

(a) Appendix E lists

$$MnO_4^-(aq) + 8\,H^+(aq) + 5\,e^- \longrightarrow Mn^{2+}(aq) + 4\,H_2O(\ell) \quad \mathcal{E}^\circ = 1.507\ V$$
$$Zn^{2+}(aq) + 2\,e^- \longrightarrow Zn(s) \qquad\qquad\qquad \mathcal{E}^\circ = -0.762\ V$$

Several half-equations in Appendix E involve $Mn^{2+}(aq)$ or $MnO_4^-(aq)$ or $H^+(aq)$, but only one involves all three. $\mathcal{E}^\circ(Pt|MnO_4^-, H^+, Mn^{2+})$ lies *above* $\mathcal{E}^\circ(Zn^{2+}|Zn)$ in the table, so the $(Pt|MnO_4^-, H^+, Mn^{2+})$ half-cell is the cathode half-cell. Therefore

$$\Delta\mathcal{E}^\circ = \mathcal{E}^\circ(Pt|MnO_4^-, H^+, Mn^{2+}) - \mathcal{E}^\circ(Zn^{2+}|Zn)$$
$$= 1.507\ V - (-0.762\ V) = +2.269\ V$$

(b) Multiplying the cathode half-equation by 2 and the anode half-equation by 5 and subtracting the second from the first causes the electrons to cancel out and gives a balanced equation for the overall reaction

$$2\,MnO_4^-(aq) + 16\,H^+(aq) + 10\,e^- \longrightarrow 2\,Mn^{2+}(aq) + 8\,H_2O(\ell)$$
$$\underline{\quad - (5\,Zn^{2+}(aq) + 10\,e^- \longrightarrow 5\,Zn(s))\quad}$$

$$2\,MnO_4^-(aq) + 16\,H^+(aq) + 5\,Zn(s) \longrightarrow$$
$$2\,Mn^{2+}(aq) + 8\,H_2O(\ell) + 5\,Zn^{2+}(aq)$$

The same answer comes by reversing the second half-equation (because the $Zn^{2+}|Zn$ half-cell runs as an oxidation) and adding it to the first.

Check

Calculate ΔG_r° for the overall reaction from thermochemical data. The calculation uses the method explained in Section 11–6 and data from Appendix D:

$$\Delta G_r^\circ = 2\ \Delta G_f^\circ(\text{Mn}^{2+}(aq)) + 8\ \Delta G_f^\circ(\text{H}_2\text{O}(\ell)) + 5\ \Delta G_f^\circ(\text{Zn}^{2+}(aq))$$
$$- 2\ \Delta G_f^\circ(\text{MnO}_4^-(aq)) - 16\ \Delta G_f^\circ(\text{H}^+(aq)) - 5\ \Delta G_f^\circ(\text{Zn}(s))$$
$$= 2(-228.1\ \text{kJ mol}^{-1}) + 8(-237.18\ \text{kJ mol}^{-1}) + 5(-147.06\ \text{kJ mol}^{-1})$$
$$- 2(-447.2\ \text{kJ mol}^{-1}) - 16(0\ \text{kJ mol}^{-1}) - 5(0\ \text{kJ mol}^{-1})$$
$$= -2194.5\ \text{kJ mol}^{-1}$$

The equation $\Delta G_r^\circ = -n\mathscr{F}\Delta\mathcal{E}^\circ$ then provides an independent value of $\Delta\mathcal{E}^\circ$:

$$\Delta\mathcal{E}^\circ = \frac{\Delta G_r^\circ}{-n\mathscr{F}} = \frac{-2194.5 \times 10^3\ \text{J mol}^{-1}}{-10(96{,}485\ \text{C mol}^{-1})} = 2.27\ \text{J C}^{-1} = 2.27\ \text{V}$$

The check confirms the answer to three significant figures.

EXERCISE

A galvanic cell is constructed using a standard $\text{Mg}^{2+}|\text{Mg}$ half-cell and a standard $\text{Sc}^{3+}|\text{Sc}$ half-cell. Use information in Appendix E to

(a) calculate the standard potential difference ($\Delta\mathcal{E}^\circ$) developed by the cell at 25°C.

(b) write a balanced equation for the overall cell reaction.

Answer: (a) $\Delta\mathcal{E}^\circ = 0.295$ V; (b) $2\ \text{Sc}^{3+}(aq) + 3\ \text{Mg}(s) \longrightarrow$
$$2\ \text{Sc}(s) + 3\ \text{Mg}^{2+}(aq).$$

Disproportionation

Reduction potentials allow for a straightforward assessment of whether a species is stable with respect to *disproportionation*. Recall from Sections 4–4 and 12–1 that in disproportionation a single chemical species is both oxidized and reduced. For a species to be susceptible to disproportionation, it must be able both to give up electrons *and* to accept electrons. In a listing of standard reduction potentials, a species that disproportionates must appear on the right side of one half-reaction and on the left side of another.

This alone is not enough, however. For a species to disproportionate spontaneously, the half-reaction in which it is reduced must lie higher in the table (have a higher reduction potential) than the half-reaction in which it is oxidized. If this is the case, the first half-reaction drives the second to go in reverse, and spontaneous disproportionation occurs. As an example, consider two half-reactions involving copper in oxidation states 0, +1, and +2:

$$\text{Cu}^+(aq) + e^- \longrightarrow \text{Cu}(s) \qquad \mathcal{E}^\circ = 0.521\ \text{V}$$
$$\text{Cu}^{2+}(aq) + e^- \longrightarrow \text{Cu}^+(aq) \qquad \mathcal{E}^\circ = 0.153\ \text{V}$$

The first reaction occurs as a reduction and drives the second as an oxidation, giving the net reaction

$$2\ \text{Cu}^+(aq) \longrightarrow \text{Cu}^{2+}(aq) + \text{Cu}(s) \qquad \Delta\mathcal{E}^\circ = 0.521\ \text{V} - 0.153\ \text{V} = 0.368\ \text{V}$$

$Cu^+(aq)$ ion in its standard state disproportionates spontaneously to $Cu^{2+}(aq)$ and $Cu(s)$. Whenever $\Delta\mathcal{E}° > 0$, we have $\Delta G° < 0$, and the standard reaction (the one in which reactants and products are in standard states) occurs spontaneously, although in some cases the rate may be very slow.

EXAMPLE 13-4

Use data in Appendix E to decide whether $Fe^{2+}(aq)$ ion in its standard state at 25°C is unstable with respect to disproportionation to products in standard states at 25°C.

Strategy

Scan Appendix E for half-equations involving $Fe^{2+}(aq)$ and $Fe^{3+}(aq)$. Combine them pair-wise to make overall equations in which Fe^{2+} is the reactant and other iron-containing species are the products. Compute $\Delta\mathcal{E}°$ to determine in each case whether the equation represents a spontaneous reaction.

Solution

The only relevant half-equations, together with their standard reduction potentials, are

$$Fe^{3+}(aq) + e^- \longrightarrow Fe^{2+}(aq) \qquad \mathcal{E}° = 0.771 \text{ V}$$
$$Fe^{2+}(aq) + 2\,e^- \longrightarrow Fe(s) \qquad \mathcal{E}° = -0.447 \text{ V}$$

The $Fe^{2+}(aq)$ appears on the rightt in the first and on the left in the second half-reaction. For Fe^{2+} to disproportionate, the first half-reaction would have to be driven backward as an oxidation. The overall reaction would be

$$3\,Fe^{2+}(aq) \longrightarrow Fe(s) + 2\,Fe^{3+}(aq)$$

The standard potential difference for this disproportionation reaction at 25°C is

$$\Delta\mathcal{E}° = \mathcal{E}°(Fe^{2+}|Fe) - \mathcal{E}°(Fe^{3+}, Fe^{2+})$$
$$\Delta\mathcal{E}° = -0.447 \text{ V} - 0.771 \text{ V} = -1.218 \text{ V}$$

A negative voltage means that a reaction is thermodynamically not favored. The $Fe^{2+}(aq)$ ion in its standard state at 25°C is stable against disproportionation to these products.

• Aqueous solutions of Fe^{2+} ion are still notoriously contaminated by Fe^{3+} ion. The problem is not that the Fe^{2+} ion disproportionates but that it is easily oxidized by oxygen (from the air).

EXERCISE

Use data from Appendix E to decide whether $Au^+(aq)$ ion in its standard state at 25°C is unstable with respect to disproportionation to products in standard states at 25°C. If it is, then write an equation for the disproportionation reaction and compute the associated $\Delta\mathcal{E}°$.

Answer: Standard-state $Au^+(aq)$ ion is unstable at 25°C:

$$3\,Au^+(aq) \longrightarrow 2\,Au(s) + Au^{3+}(aq) \qquad \Delta\mathcal{E}° = +0.40 \text{ V}.$$

13-3   Oxidizing and Reducing Agents

• The chemistry of fluorine is discussed in some detail in Chapter 23.

A strong **oxidizing agent** is a chemical species that is itself easily reduced. Strong oxidizing agents have large positive standard reduction potentials. They appear as reactants in half-equations toward the top of Table 13–1 (and Appendix E). Fluorine has the largest reduction potential listed, and fluorine molecules indeed avidly accept electrons to produce fluoride ions. Other strong oxidizing agents are hydrogen peroxide (H_2O_2) and the permanganate ion (MnO_4^-) in solution.

A strong **reducing agent,** on the other hand, is itself very easily oxidized. Strong reducing agents are *products* in half-reactions having large *negative* standard reduction potentials. The best reducing agents appear on the right-hand side of half-equations toward the bottom of Table 13–1 or Appendix E. Metallic lithium and potassium (not Li^+ ion or K^+ ion) are potent reducing agents in aqueous systems. Arranging the products of the half-reactions in Table 13–1 from the last to the first generates an activity series of reducing agents ranked from strongest to weakest. The order of the metals in this activity series is the same as in Table 4–2, which also deals with relative reducing power in aqueous systems. Table 13–1 and Appendix E, however, include both metals and non-metals. The elements to the left in the periodic table (the metals) are good reducing agents; the elements to the right are good oxidizing agents.

Oxygen as an Oxidizing Agent

The most important oxidizing agent for terrestrial life is elemental oxygen itself. Gaseous dioxygen (O_2) has a fairly high reduction potential in standard acidic solution (in which $[H^+] = 1$ M and the pH is 0):

$$O_2(g) + 4\,H^+(aq) + 4\,e^- \longrightarrow 2\,H_2O(\ell) \qquad \mathcal{E}° = 1.229\text{ V}$$

and is accordingly a good oxidizing agent. Ozone (trioxygen, O_3) is even better, as is shown by the larger standard reduction potential in aqueous acidic solution for the half-reaction

$$O_3(g) + 2\,H^+(aq) + 2\,e^- \longrightarrow O_2(g) + H_2O(\ell) \qquad \mathcal{E}° = 2.076\text{ V}$$

Because of its great oxidizing power, ozone is used commercially to bleach wood pulp and to disinfect and sterilize water. It oxidizes algae and organic impurities in water but leaves no undesirable residue.

The oxygen in hydrogen peroxide has the intermediate oxidation state -1, so that hydrogen peroxide can either be reduced to water (in which the O has oxidation state -2) or oxidized to molecular oxygen (in which the O has oxidation state 0). These processes are represented by the following reduction half-equations (if the reactions take place in acidic solution):

$$H_2O_2(aq) + 2\,H^+(aq) + 2\,e^- \longrightarrow 2\,H_2O(\ell) \qquad \mathcal{E}° = 1.776\text{ V}$$
$$O_2(g) + 2\,H^+(aq) + 2\,e^- \longrightarrow H_2O_2(aq) \qquad \mathcal{E}° = 0.695\text{ V}$$

Hydrogen peroxide can act as either an oxidizing agent or a reducing agent in acidic solution, although it is far stronger as an oxidizing agent. This oxidizing power makes hydrogen peroxide useful as a bleach and germicide. As a reducing agent, it reduces only chemical species with reduction potentials greater than 0.695 V. The preceding pair of reduction potentials also shows that acidic solutions of H_2O_2 disproportionate spontaneously into O_2 and water. This reaction is also spontaneous in neutral solution, but it is slow enough in that case that aqueous solutions of hydrogen peroxide can be stored for long periods of time if they are kept out of the light.

In basic aqueous solution, the half-reactions of oxygen-containing species involve hydroxide ions instead of hydrogen ions, and the standard reduction potentials change considerably. At pH 14, the pH at which $OH^-(aq)$ is in the standard state ($[OH^-] = 1$ M),

$$O_2(g) + 2\,H_2O(\ell) + 4\,e^- \longrightarrow 4\,OH^-(aq) \qquad \mathcal{E}° = 0.401\ \text{V}$$
$$O_3(g) + H_2O(\ell) + 2\,e^- \longrightarrow O_2(g) + 2\,OH^-(aq) \qquad \mathcal{E}° = 1.24\ \text{V}$$

Both oxygen and ozone are less effective oxidizing agents in basic solution than in acidic solution. The same is true of hydrogen peroxide (Fig. 13–4).

Nitrogen, Sulfur, and Phosphorus

Figure 13–5 shows standard reduction potentials for several half-reactions involving the oxoacids of nitrogen, sulfur, and phosphorus and their conjugate bases. The species listed are those that predominate at the pH given (0 or 14). For example, in acidic solution, nitrous acid exists as HNO_2 molecules, but in basic solution, almost all these molecules have reacted to form NO_2^- ions. Nitric acid, on the other hand, is a strong acid and is almost fully ionized to nitrate ion and hydrogen ion, even at pH 0. Because of the participation of hydrogen ion or hydroxide ion in the reduction half-reactions, the half-cell potentials depend strongly on pH, as can be seen by comparing the left side of Figure 13–5 with the right side.

Figure 13–4 • Hydrogen peroxide is a reasonably strong oxidizing agent even in basic solution. Here a solution containing $Cr^{3+}(aq)$ ion (*left*) is made basic (*center*). When H_2O_2 is added, it still oxidizes the $Cr^{3+}(aq)$ to the orange-yellow chromate ($CrO_4^{2-}(aq)$) ion (*right*).

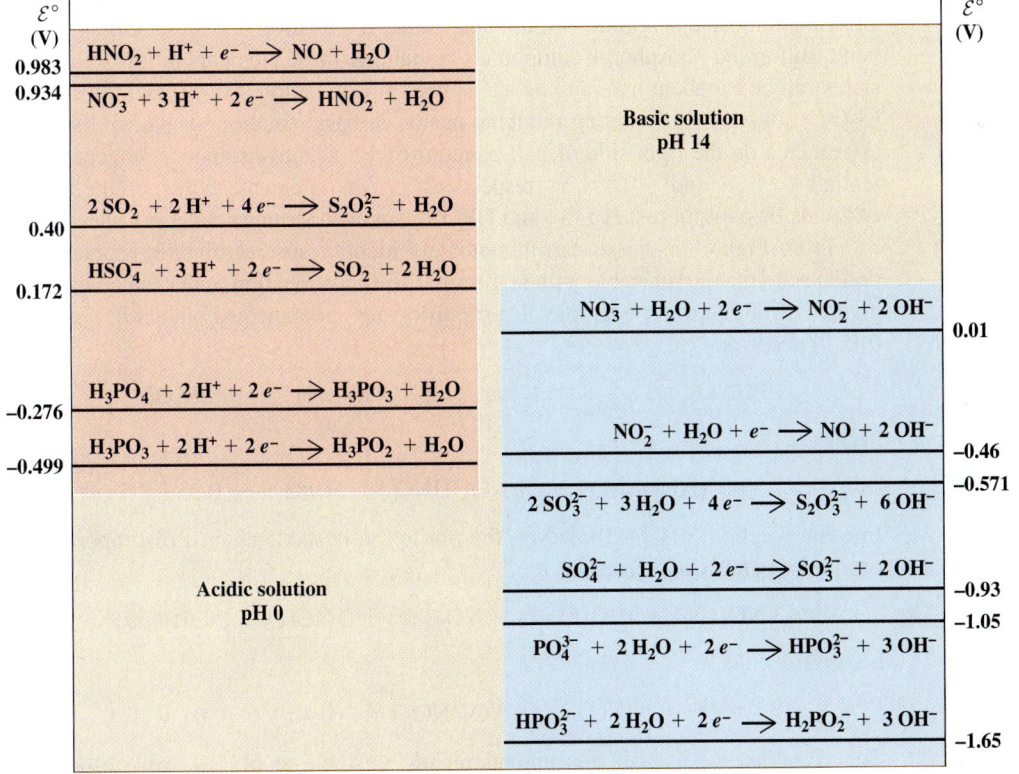

Figure 13–5 • The standard reduction potentials for oxides of nitrogen, sulfur, and phosphorus are higher in acidic solution (*left*, pH 0) than in basic solution (*right*, pH 14).

Figure 13-6 • When a copper wire is inserted into concentrated nitric acid, copper ions are produced. The yellow fumes are $NO_2(g)$, one of the products from the reduction of HNO_3. *(Marna G. Clarke)*

Comparison of the reduction potentials of the nitrogen, sulfur, and phosphorus oxoacids on the left side of the figure shows immediately that the phosphorus oxoacids are the most difficult to reduce and therefore are the least effective oxidizing agents. The oxoacids of sulfur are intermediate in this regard, and nitrate ion in acidic solution (nitric acid) and nitrous acid are the strongest oxidizing acids listed. In concentrated form, nitric acid attacks and dissolves many metals, such as copper (Fig. 13–6), which phosphoric acid does not attack. In applications for which an acid that is *not* a strong oxidizing agent is needed, phosphoric acid is preferable to nitric or sulfuric acid (Fig. 13–7).

The right side of Figure 13–5 shows that the oxidizing strengths of all of these oxoacids are drastically curtailed when their basic forms at pH 14 replace them. The highest standard reduction potential (that for nitrate ion) is only +0.01 V at pH 14, showing that basic conditions should be avoided if oxidation is desired. On the other hand, sulfur and phosphorus compounds containing those elements in low oxidation states can be excellent *reducing* agents at pH 14. Both sulfite (SO_3^{2-}) and thiosulfate ($S_2O_3^{2-}$) ions are fairly strong reducing agents in basic solution, as shown by their appearance on the right side of half-equations with negative standard reduction potentials (-0.93 and -0.571 V, respectively). Two conjugate bases of the lower oxoacids of phosphorus, $H_2PO_2^-$ and HPO_3^{2-}, are even stronger reducing agents.

• Quick identification relies on finding a pattern: the same species appearing "above and on the left" and "below and on the right" in the figure.

From Figure 13–5, we can immediately identify species of intermediate oxidation state that are unstable with respect to disproportionation under the conditions given. Nitrous acid, for example, disproportionates spontaneously in acidic solution (pH 0) according to

$$3 \, HNO_2(aq) \longrightarrow NO_3^-(aq) + 2 \, NO(g) + H^+(aq) + H_2O(\ell)$$

because

$$\Delta\mathcal{E}° = \mathcal{E}°(HNO_2|NO) - \mathcal{E}°(NO_3^-|HNO_2) = 0.983 \, V - 0.934 \, V > 0$$

In basic solution (pH 14), however, the nitrite ion is stable against disproportionation to NO and NO_3^- according to

$$3 \, NO_2^-(aq) + H_2O(\ell) \longrightarrow NO_3^-(aq) + 2 \, NO(g) + 2 \, OH^-(aq)$$

because

$$\Delta\mathcal{E}° = \mathcal{E}°(NO_2^-|NO) - \mathcal{E}°(NO_3^-|NO_2^-) = -0.46 \, V - 0.01 \, V < 0$$

The $NO_2^-(aq)$ ion is still thermodynamically unstable at pH 14 with respect to $NO_3^-(aq)$ and $N_2O(g)$ or $N_2(g)$ (not shown in Fig. 13–5), but these reactions occur so slowly that $NO_2^-(aq)$ can be kept indefinitely in basic solution.

It is frequently supposed that a species that contains an element in a high oxidation state is automatically a strong oxidizing agent. Some highly oxidized species (such as aqueous permanganate ion, which contains Mn in the $+7$ oxidation state) are indeed potent oxidizing agents, but Figure 13–5 shows that a general correlation does not exist. Nitrous acid (which has N in the $+3$ oxidation state) would be a stronger oxidizing agent than nitric acid (N in the $+5$ oxidation state) at pH 0 if it were stable against disproportionation. Aqueous sulfur dioxide (S in the $+4$ oxidation state) is a stronger oxidizing agent than hydrogen sulfate ion (S in the $+6$ oxidation state). The decisive factor is not the magnitude of the oxidation number but rather the relative stabilities of the redox pair of chemical species. Nitrous acid is a strong oxidizing agent precisely because it is much less stable than its reduced form, nitrogen monoxide. Thermodynamic stability changes with oxidation state in an irregular fashion.

Figure 13–7 • Concentrated nitric acid reacts with sugar after some time (*left*), and sulfuric acid (*center*) reacts immediately. By contrast, no reaction is seen with phosphoric acid (*right*). Soft drinks contain both phosphoric acid and dissolved sugar. (*Leon Lewandowski*)

EXAMPLE 13-5

(a) Determine whether 1.00 M $H_3PO_3(aq)$ (phosphorous acid) and 1.00 M $SO_2(aq)$ (sulfur dioxide) are stable with respect to disproportionation in acidic solution (pH 0) at 25°C. (b) Determine which of the two is the stronger reducing agent under these conditions. Refer to Figure 13–5 for necessary data.

Solution

(a) The only half-reactions involving $H_3PO_3(aq)$ in Figure 13–5 are

$$H_3PO_3(aq) + 2\,H^+(aq) + 2\,e^- \longrightarrow$$
$$H_3PO_2(aq) + H_2O(\ell) \qquad \mathcal{E}° = -0.499\ \text{V}$$

$$H_3PO_4(aq) + 2\,H^+(aq) + 2\,e^- \longrightarrow$$
$$H_3PO_3(aq) + H_2O(\ell) \qquad \mathcal{E}° = -0.276\ \text{V}$$

Subtracting the second from the first gives

$$2\,H_3PO_3(aq) \longrightarrow H_3PO_4(aq) + H_3PO_2(aq)$$

$H_3PO_3(aq)$ is stable against this disproportionation because

$$\Delta\mathcal{E}° = -0.499\ \text{V} - (-0.276\ \text{V}) < 0$$

For sulfur dioxide, the pertinent half-reactions are:

$$2\,SO_2(aq) + 2\,H^+(aq) + 4\,e^- \longrightarrow S_2O_3^{2-}(aq) + H_2O(\ell) \qquad \mathcal{E}° = 0.40\ \text{V}$$
$$HSO_4^-(aq) + 3\,H^+(aq) + 2\,e^- \longrightarrow SO_2(aq) + 2\,H_2O(\ell) \qquad \mathcal{E}° = 0.172\ \text{V}$$

Doubling the second and subtracting it from the first gives

$$4\,SO_2(aq) + 3\,H_2O(\ell) \longrightarrow 2\,HSO_4^-(aq) + S_2O_3^{2-}(aq) + 4\,H^+(aq)$$

$SO_2(aq)$ is unstable with respect to this disproportionation because

$$\Delta\mathcal{E}° = 0.40\ \text{V} - 0.172\ \text{V} > 0$$

(b) $H_3PO_3(aq)$ is the stronger reducing agent because the half-reaction in which it appears on the right side,

$$H_3PO_4(aq) + 2\,H^+(aq) + 2\,e^- \longrightarrow$$
$$H_3PO_3(aq) + H_2O(\ell) \qquad \mathcal{E}° = -0.276\ \text{V}$$

has a more negative reduction potential than (lies below) the half-reaction involving SO_2:

$$HSO_4^-(aq) + 3\,H^+(aq) + 2\,e^- \longrightarrow SO_2(aq) + 2\,H_2O(\ell) \qquad \mathcal{E}° = 0.172\ V$$

EXERCISE

By referring to Figure 13–5, determine (a) whether $HPO_3^{2-}(aq)$ and $SO_3^{2-}(aq)$ ions are stable with respect to disproportionation in basic solution and (b) which is the stronger oxidizing agent at pH 14.

Answer: (a) $HPO_3^{2-}(aq)$ is stable, but $SO_3^{2-}(aq)$ is not; (b) $SO_3^{2-}(aq)$ is the stronger oxidizing agent (although neither one is very strong).

13-4   Concentration Effects and the Nernst Equation

The voltages developed by galvanic cells depend on the concentrations of solutes and the partial pressures of gases that are reactants or products in the cell reaction. If any of these deviate from standard-state values (1 M for solutes, 1 atm for gases), then the cell is no longer a standard cell, and its voltage deviates from the standard voltage. Fortunately, it is not difficult to compute the effect of non-standard states on cell voltages. From Chapter 11, we know that the Gibbs energy of a chemical reaction is related to its reaction quotient (Q) by the equation

$$\Delta G_r = \Delta G_r° + RT \ln Q$$

Combining this equation with the equations

$$\Delta G_r = -n\mathscr{F}\Delta\mathcal{E} \qquad \text{and} \qquad \Delta G_r° = -n\mathscr{F}\Delta\mathcal{E}°$$

from Section 13–1 gives

$$-n\mathscr{F}\Delta\mathcal{E} = -n\mathscr{F}\Delta\mathcal{E}° + RT \ln Q$$

Division by $-n\mathscr{F}$ then gives

$$\Delta\mathcal{E} = \Delta\mathcal{E}° - \left(\frac{RT}{n\mathscr{F}}\right)\ln Q$$

This relationship, known as the **Nernst equation,** gives the voltage of a cell as a function of the concentrations and pressures of species entering into the reaction quotient (Q). In the Nernst equation, the standard and non-standard voltages $\Delta\mathcal{E}°$ and $\Delta\mathcal{E}$ are measured in volts, so the unit of the subtracted term $(RT/n\mathscr{F})\ln Q$ is the volt. The reaction quotient Q and its natural logarithm $\ln Q$ have no units. The unit of $(RT/n\mathscr{F})$ must therefore be the volt. The following units-only calculation confirms that it is

$$\left(\frac{RT}{n\mathscr{F}}\right) \sim \frac{(\text{J mol}^{-1}\,\text{K}^{-1})\text{K}}{1(\text{C mol}^{-1})} = \frac{\text{J}}{\text{C}} = \text{V}$$

Note that n, the number of moles of electrons transferred per mole of cell reaction as written, is represented by a 1 in the preceding because its unit, "mol mol^{-1}," equals 1.

The Nernst equation must be specialized for every different cell. Suppose that the galvanic cell

$$Al(s)\,|\,Al^{3+}(aq)\,\|\,Zn^{2+}(aq)\,|\,Zn(s)$$

• The Nernst equation is valid only when the cell has the same temperature and pressure at the conclusion of the cell reaction as at the start.

is set up. The overall reaction is

$$3 \, Zn^{2+}(aq) + 2 \, Al(s) \longrightarrow 2 \, Al^{3+}(aq) + 3 \, Zn(s)$$

for which

$$Q = \frac{[Al^{3+}]^2}{[Zn^{2+}]^3}$$

The cell transfers 6 mol of electrons through the external circuit per mole of reaction as written, so n equals 6. The Nernst equation for this electrochemical cell is

$$\Delta\mathcal{E} = \Delta\mathcal{E}^\circ - \frac{RT}{6\,\mathscr{F}} \ln \frac{[Al^{3+}]^2}{[Zn^{2+}]^3}$$

Suppose that someone chooses to balance the chemical equation for this cell reaction using a different set of coefficients. What happens to its Nernst equation? If all of the coefficients in the preceding chemical equation are doubled

$$6 \, Zn^{2+}(aq) + 4 \, Al(s) \longrightarrow 4 \, Al^{3+}(aq) + 6 \, Zn(s)$$

then Q is different:

$$Q = \frac{[Al^{3+}]^4}{[Zn^{2+}]^6}$$

But the cell now transfers 12 mol of electrons through the external circuit per mole of reaction as written, so n equals 12. The Nernst equation for the electrochemical cell looks different

$$\Delta\mathcal{E} = \Delta\mathcal{E}^\circ - \frac{RT}{12\,\mathscr{F}} \ln \frac{[Al^{3+}]^4}{[Zn^{2+}]^6}$$

but is in fact mathematically equivalent because

$$\ln \frac{[Al^{3+}]^4}{[Zn^{2+}]^6} = 2 \ln \frac{[Al^{3+}]^2}{[Zn^{2+}]^3}$$

The standard voltages ($\Delta\mathcal{E}^\circ$'s) of electrochemical cells often depend sensitively on the temperature. If a cell operates at 25°C (298.15 K), then the correct $\Delta\mathcal{E}^\circ$ can be obtained for use in the Nernst equation by combining the standard half-cell reduction potentials in Table 13–1 or Appendix E. If it operates at a temperature other than 25°C, then a $\Delta\mathcal{E}^\circ$ that has been measured at that temperature (or computed for that temperature) should be used.

The Nernst equation applies to half-cells as well as to complete electrochemical cells. The reduction potential of a non-standard half-cell is given by

$$\mathcal{E} = \mathcal{E}^\circ - \left(\frac{RT}{n_{hc}\,\mathscr{F}} \right) \ln Q_{hc}$$

where n_{hc} equals the number of electrons appearing in the half-equation and Q_{hc} is the reaction quotient for the half-cell reaction written as a reduction. Electrons do not appear in Q_{hc}. The equation for reduction in a $Zn^{2+}|Zn$ half-cell is

$$Zn^{2+}(aq) + 2 \, e^- \longrightarrow Zn(s)$$

So the Nernst equation for this half-cell is

$$\mathcal{E} = \mathcal{E}^\circ - \left(\frac{RT}{2\,\mathscr{F}} \right) \ln \left(\frac{1}{[Zn^{2+}]} \right)$$

The value of $\mathcal{E}°$ depends on the temperature. If the temperature is 298.15 K, then $\mathcal{E}°$ equals -0.762 V, as listed in Table 13–1 and Appendix E.

When two half-cells in which reactants or products are in non-standard states are connected to make a galvanic cell, the one with the more positive value of $\mathcal{E}$ (not $\mathcal{E}°$) is the cathode, where reduction takes place, and the cell voltage is

$$\Delta\mathcal{E} = \mathcal{E}(\text{cathode}) - \mathcal{E}(\text{anode})$$

EXAMPLE 13–6

The cell

$$\text{Zn}(s)\,|\,\text{Zn}^{2+}(aq)\;||\;\text{MnO}_4^-(aq),\,\text{H}^+(aq),\,\text{Mn}^{2+}(aq)\,|\,\text{Pt}(s)$$

(the same cell used in Example 13–3) is set up at 298.15 K with the following non-standard concentrations: $[\text{H}^+] = 0.010$ M, $[\text{MnO}_4^-] = 0.12$ M, $[\text{Mn}^{2+}] = 0.0010$ M, and $[\text{Zn}^{2+}] = 0.015$ M. Calculate the cell voltage.

Strategy

Realize that the computation of the cell voltage requires the Nernst equation because at least one reactant or product is not in a standard state. Write the overall cell reaction and the Nernst equation for this cell. Insert the given data in the Nernst equation.

Solution

From Example 13–3, the balanced equation for cell reaction is

$$2\,\text{MnO}_4^-(aq) + 5\,\text{Zn}(s) + 16\,\text{H}^+(aq) \longrightarrow$$
$$2\,\text{Mn}^{2+}(aq) + 5\,\text{Zn}^{2+}(aq) + 8\,\text{H}_2\text{O}(\ell)$$

The general Nernst equation

$$\Delta\mathcal{E} = \Delta\mathcal{E}° - \left(\frac{RT}{n\mathscr{F}}\right)\ln Q$$

is specialized to this cell as follows:

$$\Delta\mathcal{E} = 2.269 \text{ V} - \left(\frac{RT}{10\,\mathscr{F}}\right)\ln\frac{[\text{Mn}^{2+}]^2[\text{Zn}^{2+}]^5}{[\text{MnO}_4^-]^2[\text{H}^+]^{16}}$$

in which the values of n and $\Delta\mathcal{E}°$ come from Example 13–3. Inserting the temperature, the constants R and $\mathscr{F}$, and the four concentrations and using the fact that 1 V equals 1 J C^{-1} as a unit-factor gives

$$\Delta\mathcal{E} = 2.269 \text{ V} - \left(\frac{(8.315 \text{ J mol}^{-1}\text{ K}^{-1})298.15 \text{ K}}{10\,(96{,}485 \text{ C mol}^{-1})}\right)\left(\frac{1 \text{ V}}{1 \text{ J C}^{-1}}\right)\ln\frac{(0.0010)^2(0.015)^5}{(0.12)^2(0.010)^{16}}$$

$$= 2.269 \text{ V} - \left(\frac{0.02569 \text{ V}}{10}\right)\ln(5.3 \times 10^{18}) = \boxed{2.16 \text{ V}}$$

EXERCISE

A hobby shop sells the "Alupower Aluminum-Air Power Cell" (Figure 13–8). This cell can be represented

$$\text{Al}(s)\,|\,\text{Al}^{3+}(aq),\,\text{OH}^-(aq)\,|\,\text{O}_2(g)\,|\,\text{inert metal}(s)$$

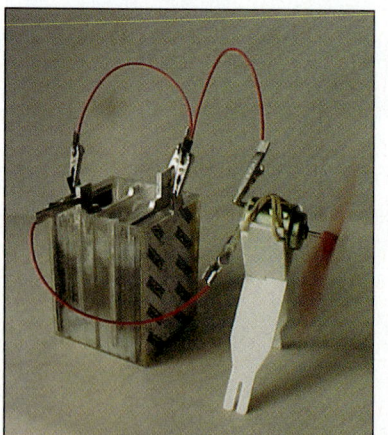

Figure 13–8 • An aluminum-air battery provides Gibbs energy to drive a propellor. The reaction is
$$4\,\text{Al}(s) + 3\,\text{O}_2(g) + 6\,\text{H}_2\text{O}(\ell) \longrightarrow$$
$$4\,\text{Al}^{3+}(aq) + 12\,\text{OH}^-(aq).$$ The anode is a plate of aluminum. Oxygen enters the cell through the side wall which is porous to gas. When it reaches the metallic mesh coated with a catalyst, it is reduced to hydroxide ion. The catalyst speeds up the reduction of oxygen but does not affect the theoretical voltage of the cell.

It has a standard voltage of 2.063 V at 25°C, but the precipitation of $Al(OH)_3(s)$ inside the cell lowers the concentrations of the products of the cell reaction substantially. Suppose that $[Al^{3+}] = 1.1 \times 10^{-4}$ mol L^{-1}, $[OH^-] = 3.3 \times 10^{-4}$ mol L^{-1}, and that $P_{O_2} = 0.20$ atm. Compute the cell voltage at 25°C.

Answer: $\Delta\mathcal{E} = 2.34$ V.

The Nernst equation is quite frequently employed with $T = 298.15$ K. The solution to the preceding example provides a useful computational short-cut for such cases:

$$\left(\frac{RT}{n\mathscr{F}}\right) = \left(\frac{0.02569 \text{ V}}{n}\right) \qquad \text{when } T = 298.15 \text{ K}$$

Non-standard concentrations and partial pressures can affect cell voltages enough to reverse the direction of spontaneous change. Consider the following cell, which is a standard cell, ready to operate at 298.15 K:

$$Zn(s)|Zn^{2+}(aq, 1.0 \text{ M})||Cr^{3+}(aq, 1.0 \text{ M})|Cr(s)$$

The cell reaction is

$$3 \text{ Zn}(s) + 2 \text{ Cr}^{3+}(aq) \longrightarrow 2 \text{ Cr}(s) + 3 \text{ Zn}^{2+}(aq)$$

in which 6 mol of electrons is transferred per mole of reaction. The standard potential difference is

$$\Delta\mathcal{E}° = \mathcal{E}°(Cr^{3+}|Cr) - \mathcal{E}°(Zn^{2+}|Zn) = -0.744 - (-0.762) = 0.018 \text{ V}$$

• The standard reduction potentials come from Table 13-1.

Now, suppose that the solution in the $Cr^{3+}|Cr$ half-cell compartment is diluted from 1 M in Cr^{3+} ion to 0.00010 M, but everything else is kept the same. The cell voltage changes, and the Nernst equation for this cell tells the new cell voltage:

$$\Delta\mathcal{E} = 0.018 \text{ V} - \left(\frac{RT}{n\mathscr{F}}\right)\ln Q$$

$$= 0.018 \text{ V} - \left(\frac{0.02569 \text{ V}}{n}\right)\ln \frac{[Zn^{2+}]^3}{[Cr^{3+}]^2}$$

$$= 0.018 \text{ V} - \left(\frac{0.02569 \text{ V}}{6}\right)\ln \frac{(1.0)^3}{(0.00010)^2}$$

$$= 0.018 \text{ V} - 0.0789 \text{ V} = -0.061 \text{ V}$$

This negative voltage means the cell reaction, which was spontaneous when the reactants and products were in standard states, is now *non*-spontaneous. The reverse reaction is now spontaneous. The conventional diagram for the galvanic cell after the dilution is

$$Cr(s)|Cr^{3+}(aq, 0.00010 \text{ M})||Zn^{2+}(aq, 1.0 \text{ M})|Zn(s)$$

and the potential difference for this galvanic cell is the negative of -0.061 V, which is $+0.061$ V.

Let us check whether $+0.061$ V is correct by computing the (non-standard) reduction potential for the half-reaction

$$Cr^{3+}(aq, 0.00010 \text{ M}) + 3 e^- \longrightarrow Cr(s)$$

and subtracting it from the reduction potential for the $Zn^{2+}(aq, 1.0 \text{ M})|Zn(s)$ half-cell to obtain $\Delta\mathcal{E}$. Clearly, n_{hc} is 3 in the reduction of $Cr^{3+}(aq)$. Using the Nernst equation for a half-cell:

$$\mathcal{E} = \mathcal{E}° - \left(\frac{RT}{n_{hc}\mathscr{F}}\right) \ln Q_{hc}$$

$$\mathcal{E}(Cr^{3+}|Cr) = \mathcal{E}°(Cr^{3+}|Cr) - \left(\frac{RT}{3\mathscr{F}}\right)\ln \frac{1}{[Cr^{3+}]}$$

$$= -0.744 \text{ V} - \left(\frac{0.02569 \text{ V}}{3}\right)\ln \frac{1}{0.00010}$$

$$= -0.823 \text{ V}$$

The potential difference developed by the cell is

$$\Delta\mathcal{E} = \mathcal{E}(\text{cathode}) - \mathcal{E}(\text{anode})$$
$$= \mathcal{E}(Zn^{2+}(1 \text{ M})|Zn) - \mathcal{E}(Cr^{3+}(0.00010 \text{ M})|Cr)$$
$$= -0.762 \text{ V} - (-0.823 \text{ V}) = 0.061 \text{ V}$$

The two ways of computing the voltage of the non-standard cell give the same answer. Figure 13–9 shows the downward shift of the $Cr^{3+}|Cr$ reduction potential when the $Cr^{3+}(aq)$ is diluted.

Measuring Concentrations with Electrochemical Cells

As the preceding discussion shows, the Nernst equation allows us to calculate the potential differences developed by non-standard galvanic cells if we know the partial pressures (for gases) and concentrations (for solutes) of the different reactants and products in the cell reaction. This is certainly important, but the reverse calculation, obtaining concentrations by measuring cell voltages, is even more important. If we know the reaction taking place in a cell, then we know the standard cell voltage $\Delta\mathcal{E}°$ (from a table such as that in Appendix E). If we then measure the actual $\Delta\mathcal{E}$, the Nernst equation gives the reaction quotient Q. From Q we can calculate an unknown concentration if we know the concentrations (or partial pressures) of the other species taking part in the reaction. The following example illustrates this.

Figure 13–9 • The standard reduction potential for the $Cr^{3+}|Cr$ half-cell is -0.744 V, slightly above that of the standard $Zn^{2+}|Zn$ half-cell, which is -0.762 V. Dilution of the Cr^{3+} ion in the $Cr^{3+}|Cr$ half-cell from the standard 1 M to 0.00010 M lowers the reduction potential to -0.823 V, which is below the reduction potential of the standard $Zn^{2+}|Zn$ half-cell.

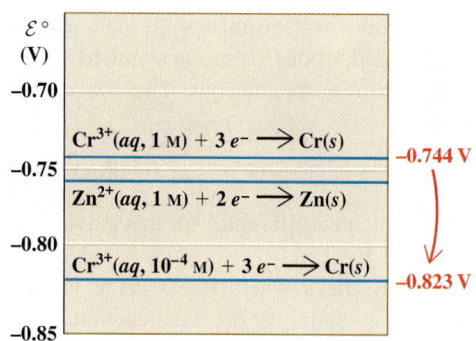

EXAMPLE 13-7

One half of a galvanic cell consists of a zinc anode immersed in a 1.00 M solution of $Zn(NO_3)_2$. The other half consists of a platinum cathode that has gaseous hydrogen bubbling over it at a pressure of 1.00 atm. This cathode is immersed in a solution of unknown hydrogen-ion concentration. The voltage of the cell is 0.473 V at 298.15 K.

(a) Diagram the cell.

(b) Write an equation for the reaction taking place in the cell, and calculate the reaction quotient Q at 25°C.

(c) Calculate the unknown concentration of hydrogen ions.

Solution

(a) The conventional cell diagram puts the anode to the left:

$$Zn(s)|Zn^{2+}(aq, 1.00 \text{ M})||H^+(aq, ? \text{ M})|H_2(g, 1 \text{ atm})|Pt(s)$$

(b) The half-reactions are

$$2 H^+(aq) + 2 e^- \longrightarrow H_2(g) \qquad \text{(reduction, cathode)}$$
$$Zn(s) \longrightarrow Zn^{2+}(aq) + 2 e^- \qquad \text{(oxidation, anode)}$$

The simplest equation for the overall reaction is therefore

$$Zn(s) + 2 H^+(aq) \longrightarrow Zn^{2+}(aq) + H_2(g)$$

with $n = 2$. The voltage of the standard cell is

$$\Delta\mathcal{E}° = \mathcal{E}°(\text{cathode}) - \mathcal{E}°(\text{anode}) = 0.0000 \text{ V} - (-0.7618 \text{ V}) = +0.7618 \text{ V}$$

The measured cell voltage differs from 0.7618 V because $H^+(aq)$ is not in a standard state. The Nernst equation relates Q to $\Delta\mathcal{E}$ and $\Delta\mathcal{E}°$:

$$\Delta\mathcal{E} = \Delta\mathcal{E}° - \left(\frac{RT}{n\mathcal{F}}\right)\ln Q$$

Solve for $\ln Q$ and substitute the two known voltages:

$$\ln Q = \left(\frac{n\mathcal{F}}{RT}\right)(\Delta\mathcal{E}° - \Delta\mathcal{E}) = \left(\frac{n\mathcal{F}}{RT}\right)(0.7618 \text{ V} - 0.473 \text{ V})$$

It is known that $(RT/n\mathcal{F})$ equals $(0.02569 \text{ V}/n)$ at 298.15 K. Hence, $(n\mathcal{F}/RT)$ equals $(n/0.02569 \text{ V})$ at 298.15 K, and

$$\ln Q = \left(\frac{2}{0.02569 \text{ V}}\right)(0.7618 - 0.473) \text{ V} = 22.48$$

Take the anti-logarithm of both sides of the equation

$$Q = e^{22.48} = 5.8 \times 10^9$$

(c) Compute $[H^+]$ by substituting into the reaction-quotient expression:

$$Q = \frac{[Zn^{2+}]P_{H_2}}{[H^+]^2}$$

$$5.8 \times 10^9 = \frac{(1.00)(1.00)}{[H^+]^2}$$

$$[H^+] = 1.3 \times 10^{-5} \text{ M}$$

EXERCISE

Suppose that the concentration of H^+ in the cell in the preceding example equals 1.0×10^{-3} M, that the partial pressure of $H_2(g)$ equals 1.00 atm, and that the cell voltage equals 0.851 V. Determine the concentration of Zn^{2+} ion.

Answer: $[Zn^{2+}] = 9.6 \times 10^{-10}$ M.

The preceding example and exercise show the extraordinary sensitivity that the voltage developed by an electrochemical cell can have to small concentrations of dissolved species. No other analytical technique could so easily measure a concentration of $Zn^{2+}(aq)$ on the order of 10^{-9} mol L^{-1}.

pH Meters

• Recall that $H^+(aq)$ and $H_3O^+(aq)$ refer to the same thing. See Section 8–1.

The electrochemical cell described in Example 13–7 is a **pH meter,** a device to measure the H^+ concentration of an unknown sample and so obtain its pH by the use of equation pH $= -\log_{10}[H_3O^+] = -\log_{10}[H^+]$. Commercial pH meters (see Fig. 8–2) are constructed somewhat differently, however. A $Zn^{2+}|Zn$ half-cell is not used as the reference, and it is a nuisance to have to bubble hydrogen through the half-cell containing H^+ at the unknown concentration. Instead, a more portable and miniaturized pair of electrodes is used. In a typical pH meter, two electrodes are dipped into the solution of unknown pH. One of these, the **glass electrode,** usually consists of an AgCl-coated silver wire in contact with a solution of HCl of known concentration (for example, 1.000 mol L^{-1}) contained in a thin-walled glass bulb. The concentration of H^+ ion in this bulb is constant and known. A pH-dependent potential develops across the thin glass membrane when the glass electrode is immersed in a solution of different, unknown $[H^+]$. The second half-cell is frequently a **saturated calomel electrode** that consists of a platinum wire in electrical contact with a paste of liquid mercury, calomel ($Hg_2Cl_2(s)$), and a saturated solution of potassium chloride. Figure 13–10 diagrams a glass electrode and a calomel electrode in-

• Commercial pH meters very often combine the two electrodes into a single probe for greater convenience.

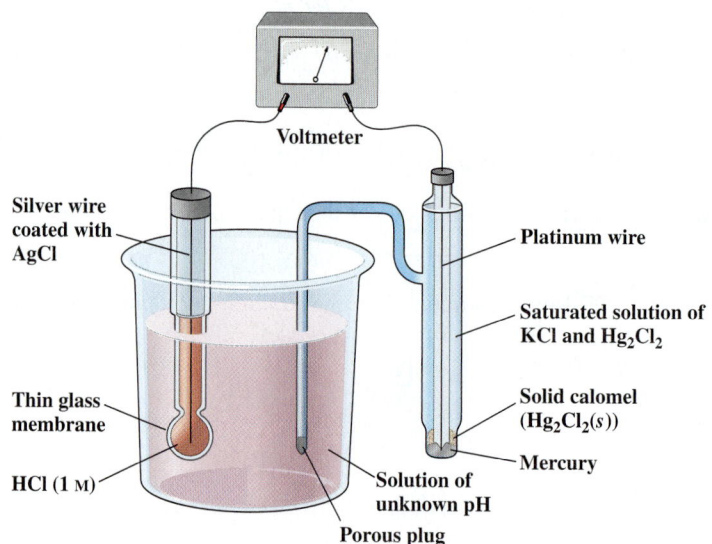

Voltmeter

Silver wire coated with AgCl

Platinum wire

Saturated solution of KCl and Hg_2Cl_2

Solid calomel ($Hg_2Cl_2(s)$)

Mercury

Thin glass membrane

HCl (1 M)

Solution of unknown pH

Porous plug

Figure 13–10 • A pH meter consists of a glass electrode (*left*) and a calomel electrode (*right*), here shown dipping into a solution of unknown hydrogen ion concentration.

serted in a solution of unknown pH. A pH meter is calibrated by measuring voltages for solutions buffered at known pHs.

The glass electrode has a number of advantages. It responds only to differences in $[H^+]$ and does so over a wide range of pH. It is unaffected by substances that may be present (such as strong oxidizing agents) that make a hydrogen electrode unreliable. Strongly colored solutions that obscure the color changes of acid–base indicators do not interfere with the glass electrode. Finally, the glass electrode can be miniaturized to an astonishing extent, enough to permit insertion into individual living cells, for example.

Other types of electrodes are sensitive to the concentrations of ions other than H^+ ion. The simplest example of such an **ion-selective electrode** is a metal wire that can be used to detect the concentration of the corresponding metal ion in solution. Silver and copper wires can be used in this way to determine the concentrations of Ag^+ and Cu^{2+} in solutions that do not contain interfering ions. Many other electrodes have been developed for detecting and determining the concentrations of specific ions. Glass of chemically modified composition, for example, can be used to construct membrane electrodes that are sensitive to the concentrations of potassium, sodium, or any of the halide ions.

13-5   Equilibrium Constants from Electrochemistry

The standard potential difference in a cell gives the standard free energy of the cell reaction. The relationship, which is established in Section 13–1, is

$$\Delta G_r^\circ = -n\mathscr{F}\Delta\mathscr{E}^\circ$$

But ΔG_r° is related to the equilibrium constant (K) by the following equation, which is developed in Section 11–7:

$$\Delta G_r^\circ = -RT \ln K$$

Combining the two equations gives

$$RT \ln K = n\mathscr{F}\Delta\mathscr{E}^\circ$$

$$\boxed{\ln K = \left(\frac{n\mathscr{F}}{RT}\right)\Delta\mathscr{E}^\circ}$$

The same result can be obtained in a more provocative way. Imagine a standard galvanic cell set up to run a redox reaction. The Nernst equation for the cell is

$$\Delta\mathscr{E} = \Delta\mathscr{E}^\circ - \left(\frac{RT}{n\mathscr{F}}\right)\ln Q$$

All reactants and products are present in standard states, so Q equals 1, $\ln Q$ equals 0, and the actual cell voltage $(\Delta\mathscr{E})$ equals the standard cell voltage $(\Delta\mathscr{E}^\circ)$. Now, a wire is connected between the anode and cathode, allowing a current to flow. The cell reaction generates products and uses up reactants. This means that Q increases, which means that $\Delta\mathscr{E}$ decreases. Eventually, $\Delta\mathscr{E}$ falls to 0. At this point the reaction stops because no more current can flow. The cell is dead, or, more precisely, at equilibrium. At equilibrium, Q equals K. Replacing Q with K and replacing $\Delta\mathscr{E}$ with 0 in the Nernst equation gives

$$\Delta\mathscr{E}^\circ = \left(\frac{RT}{n\mathscr{F}}\right)\ln K \qquad \text{which rearranges to} \qquad \ln K = \left(\frac{n\mathscr{F}}{RT}\right)\Delta\mathscr{E}^\circ$$

EXAMPLE 13-8

Calculate the equilibrium constant of the redox reaction

$$2\,MnO_4^-(aq) + 5\,Zn(s) + 16\,H^+(aq) \longrightarrow$$
$$2\,Mn^{2+}(aq) + 5\,Zn^{2+}(aq) + 8\,H_2O(\ell)$$

at 298.15 K using the standard potential difference established in Example 13–3.

Solution

From Example 13–3, $\Delta\mathcal{E}° = 2.269$ V and $n = 10$. Therefore,

> • When the value of n is not obvious upon inspection of a balanced redox equation, try rewriting the equation as the sum of two half-equations.

$$\ln K = \left(\frac{n\mathcal{F}}{RT}\right)\Delta\mathcal{E}° = \left(\frac{10\,(96{,}485\;C\;mol^{-1})}{(8.315\;J\;mol^{-1}\;K^{-1})\;298.15\;K}\right)\!\left(\frac{1\;J\;C^{-1}}{1\;V}\right)\!(2.269\;V)$$
$$= 883.1$$
$$K = e^{883.1} = 3.4 \times 10^{383}$$

This overwhelmingly large equilibrium constant reflects the great strength of permanganate ion as an oxidizing agent and the great strength of elemental zinc as a reducing agent. It means that, for all practical purposes, no MnO_4^- ions are present at equilibrium. Such a large equilibrium constant is impossible to determine from direct measurements of concentration but is easy to obtain using an electrochemical cell.

Check

Calculate the $\Delta G_r°$ for the reaction at 298.15 K using the $\Delta G_f°$ data in Appendix D. The details appear in Example 13–3: The result is $\Delta G_r° = -2194.5\;kJ\;mol^{-1}$. The K can then be obtained independently using the method explained in Section 11–8:

$$\ln K = \frac{-\Delta G_r°}{RT} = \frac{-(-2194.5 \times 10^3\;J\;mol^{-1})}{(8.315\;J\;mol^{-1}\;K^{-1})(298.15\;K)} = 885.2$$
$$K = e^{885.2} = 2.7 \times 10^{384}$$

The difference between the thermochemical $\ln K$ and the electrochemical $\ln K$, which are based on completely different kinds of experiments, is small.

EXERCISE

Calculate the equilibrium constant for the redox reaction

$$2\,Sc^{3+}(aq) + 3\,Mg(s) \longrightarrow 2\,Sc(s) + 3\,Mg^{2+}(aq)$$

at 25°C using the standard potential difference that was calculated in Exercise 13–3.

Answer: $K = 8 \times 10^{29}$.

Acid–Base and Solubility Equilibria

The example just presented illustrates how equilibrium constants for overall cell reactions can be determined electrochemically. It happened to have been a redox equilibrium, but related procedures can be used to measure acidity and basicity constants and solubility-product constants. Electrochemical measurements provide many equilibrium constants that would be difficult to determine by other means.

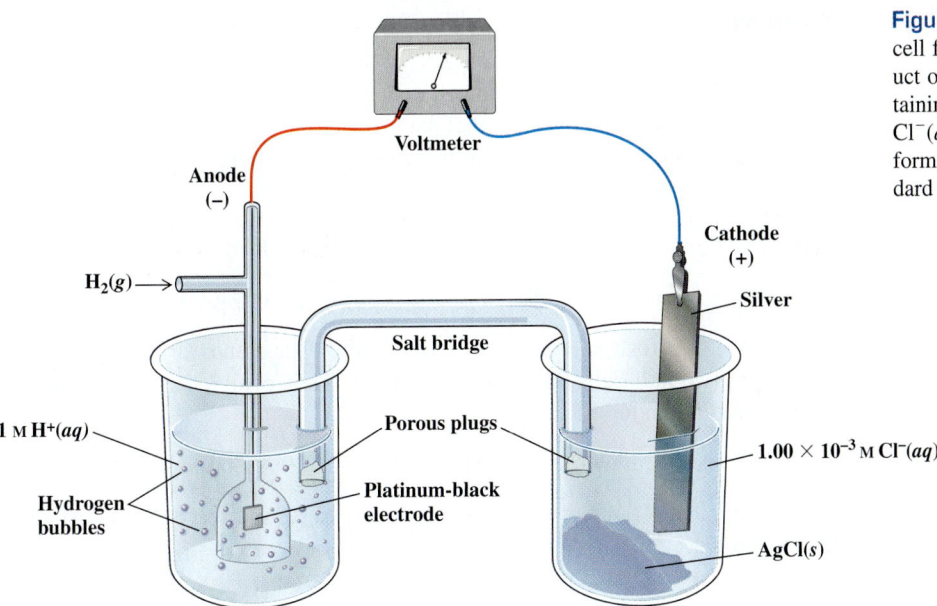

Figure 13–11 • An electrochemical cell for measuring the solubility product of AgCl. An $Ag^+|Ag$ half-cell containing AgCl(s) in contact with $Cl^-(aq)$ at a known concentration forms the cathode; the anode is a standard $H^+(aq)|H_2(g)$ half-cell.

It is straightforward to determine the K_a of a weak acid electrochemically with a pH meter. We prepare a buffer solution (see Section 8–5) in which the concentrations of both the acid [HA] and its conjugate base [A$^-$] are known. A measurement with a pH meter determines [H$^+$], and we then calculate the acidity constant by substituting the three concentrations into the expression

$$K_a = \frac{[H^+][A^-]}{[HA]}$$

The electrochemical determination of solubility-product constants is also straightforward. Suppose that we need the K_{sp} of AgCl. We construct a half-cell that contains AgCl(s) in equilibrium with a known concentration of $Cl^-(aq)$ (established with 0.00100 M NaCl, for example) and an unknown concentration of $Ag^+(aq)$. A silver electrode is used so that the half-reaction involved is either the reduction of $Ag^+(aq)$ or the oxidation of Ag(s). This is, in effect, a non-standard $Ag^+|Ag$ half-cell with a reduction potential to be determined. We combine the $Ag^+|Ag$ half-cell with any other half-cell that is convenient, as long as its reduction potential is accurately known. In the following example, the second half-cell is a standard $H^+|H_2$ half-cell (Fig. 13–11).

EXAMPLE 13–9

A galvanic cell consists of a standard hydrogen half-cell (with platinum electrode) as the anode, and a silver half-cell as the cathode:

$$Pt(s)|H_2(g, 1\ atm)|H^+(aq), 1\ M)\|Ag^+(aq, ?\ M)|Ag(s)$$

The $Ag^+(aq)$ ion in the cathode compartment is in equilibrium with AgCl(s) and $Cl^-(aq)$ ion that has a concentration of 1.00×10^{-3} mol L^{-1}. The measured cell voltage is $\Delta\mathcal{E} = 0.398$ V. Calculate the concentration of $Ag^+(aq)$ in the cell and the K_{sp} of silver chloride at 25°C.

Strategy

Write the cell reaction and compute $\Delta\mathcal{E}°$ of the cell, using data from Table 13–1. Set up the Nernst equation for the cell and solve it for Q at 298.15 K. Use Q to obtain $[Ag^+]$ in the cell. Realize that the product of $[Ag^+]$ and $[Cl^-]$ in the cell equals K_{sp} because the two ions are in equilibrium with $AgCl(s)$.

Solution

The half-cell reactions combine to give this overall cell reaction:

$$H_2(g) + 2\,Ag^+(aq) \longrightarrow 2\,H^+(aq) + 2\,Ag(s)$$

Note that $n = 2$ for the reaction as written here. The standard reduction potentials of the two half-cells appear in Table 13–1. They combine as follows to give the standard potential difference:

$$\Delta\mathcal{E}° = \mathcal{E}°(\text{cathode}) - \mathcal{E}°(\text{anode}) = 0.800\ \text{V} - 0.000\ \text{V} = 0.800\ \text{V}$$

Use the fact that $RT/n\mathcal{F} = 0.02569\ \text{V}/n$ at 298.15 K, and solve the Nernst equation for $\ln Q$:

$$\Delta\mathcal{E} = \Delta\mathcal{E}° - \left(\frac{RT}{n\mathcal{F}}\right)\ln Q$$

$$\ln Q = \frac{n}{0.02569\ \text{V}}(\Delta\mathcal{E}° - \Delta\mathcal{E})$$

Insert the information that is specific to this cell:

$$\ln\left(\frac{[H^+]^2}{[Ag^+]^2\,P_{H_2}}\right) = \frac{2}{0.02569\ \text{V}}(0.800 - 0.398)\ \text{V} = 31.3$$

$$\left(\frac{[H^+]^2}{[Ag^+]^2\,P_{H_2}}\right) = e^{31.3} = 3.9 \times 10^{13}$$

Substitute 1 for $[H^+]$ and 1 for P_{H_2}, and solve for $[Ag^+]$:

$$\frac{1^2}{[Ag^+]^2\,1} = 3.9 \times 10^{13}$$

$$[Ag^+] = 1.6 \times 10^{-7}\ \text{mol L}^{-1}$$

Compute K_{sp} of AgCl:

$$K_{sp} = [Ag^+][Cl^-]$$
$$= (1.6 \times 10^{-7})(1.00 \times 10^{-3}) = 1.6 \times 10^{-10}$$

Check

The K_{sp} equals the value given in Table 9–1.

EXERCISE

The following cell is constructed:

$$Zn(s)|Zn^{2+}(aq,\ 1\ \text{м})||Hg_2^{2+}(aq,\ ?\ \text{м})|Hg(\ell)$$

The solution in the cathode compartment is in contact with solid mercury(I) bromide, so that the solubility equilibrium

$$Hg_2Br_2(s) \rightleftharpoons Hg_2^{2+}(aq) + 2\,Br^-(aq)$$

is established there. The concentration of $Br^-(aq)$ in the cathode compartment equals 1.0×10^{-4} mol L^{-1}, and the cell voltage is 1.178 V at 25°C. (a) Write an equation for the cell reaction. (b) Calculate the concentration of $Hg_2^{2+}(aq)$ in the cell. (c) Calculate K_{sp} for Hg_2Br_2 at 25°C.

Answer: (a) $Zn(s) + Hg_2^{2+}(aq) \longrightarrow Zn^{2+}(aq) + 2\ Hg(\ell)$.
(b) $[Hg_2^{2+}] = 1.3 \times 10^{-13}$ mol L^{-1}. (c) $K_{sp} = 1.3 \times 10^{-21}$.

13-6 Batteries and Fuel Cells

Electrochemical cells have numerous applications in industry and in everyday life. In this section, we focus on two of the most important: batteries to store energy and fuel cells to convert chemical energy into electrical energy.

Batteries

The origin of the **battery** is lost in history, but it is believed that Persian artisans must have used some form of battery to gold-plate jewelry as long ago as the second century B.C. The modern development of the battery began with Alessandro Volta in 1800. Volta constructed a "pile" that consisted of a stack of alternating zinc and silver disks, between pairs of which were inserted disks of paper moistened with an acidic solution. This device was found to generate a potential difference across its ends that became larger as more pairs of disks were added. It could deliver a severe shock if the stack contained a great many disks.

In Volta's invention, a large number of galvanic cells were arranged in **series** (cathode to anode) so that their individual voltages added together. Such an arrangement is, strictly speaking, a **battery of cells,** but modern usage makes no distinction between a single cell and such a combination, and the word "battery" has come to mean the cell itself. A distinction is usually made between cells that are discarded when their electrical energy has been spent and those that can be recharged. The former are called **primary** cells, and the latter are called **secondary** cells.

Batteries vary in size and chemistry. *(Charles D. Winters)*

Figure 13-12 • Much of the zinc used in the ordinary flashlight battery is produced by electrolysis of aqueous zinc sulfate solutions. In this cell house, there are hundreds of cells, each containing a series of cathodes where zinc plates out. *(Charles D. Winters)*

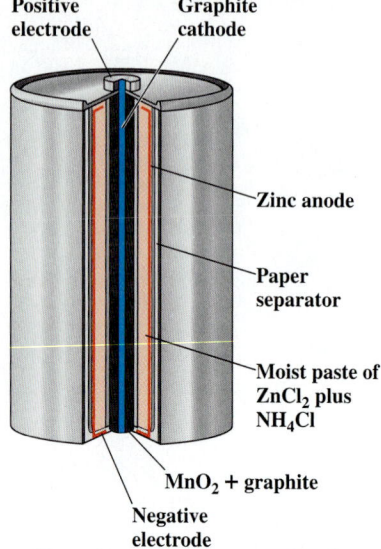

Positive electrode

Graphite cathode

Zinc anode

Paper separator

Moist paste of $ZnCl_2$ plus NH_4Cl

MnO_2 + graphite

Negative electrode

Figure 13-13 • In a Leclanché dry cell, electrons are released to an external circuit at the anode and enter the cell again at the cathode, where the reduction of MnO_2 occurs.

The most familiar primary cell is the **Leclanché cell** (also called a "zinc–carbon dry cell"), which is used for flashlights, portable radios, and a host of other purposes. Each year, more than 5 billion such dry cells are used worldwide, and estimates place the quantity of zinc reacted in dry cells at over 30 metric tons per day (Fig. 13–12). The "dry cell" is not really dry. Its electrolyte is a moist powder that contains ammonium chloride and zinc chloride. Figure 13–13 is a cutaway illustration of a dry cell, which consists of a zinc shell for an anode (negative electrode) and an axial graphite rod for a cathode (positive electrode); the rod is surrounded by a densely packed layer of graphite and manganese dioxide. Each of these components performs an interesting and essential function. At the zinc anode, oxidation takes place:

$$Zn(s) \longrightarrow Zn^{2+}(aq) + 2\,e^- \qquad \text{(oxidation, anode)}$$

The moist salt mixture permits ions to transport charge within the cell equal in quantity to that carried by electrons in the external circuit, just as a salt bridge does in the galvanic cells considered up to this point. Manganese dioxide is the ultimate electron acceptor and is reduced to Mn_2O_3 by electrons conducted to it by the graphite rod:

$$2\,MnO_2(s) + 2\,NH_4^+(aq) + 2\,e^- \longrightarrow$$
$$Mn_2O_3(s) + 2\,NH_3(aq) + H_2O(\ell) \qquad \text{(reduction, cathode)}$$

In addition, some ammonium ion is reduced to gaseous ammonia and hydrogen:

$$2\,NH_4^+(aq) + 2\,e^- \longrightarrow 2\,NH_3(g) + H_2(g) \qquad \text{(reduction, cathode)}$$

A build-up of gases is, however, prevented by formation of a complex ion

$$Zn^{2+}(aq) + 2\,NH_3(g) \longrightarrow [Zn(NH_3)_2]^{2+}(aq)$$

and the re-oxidation of the hydrogen:

$$2\,MnO_2(s) + H_2(g) \longrightarrow Mn_2O_3(s) + H_2O(\ell)$$

The ingenious device of mixing powdered graphite with powdered MnO_2 greatly increases the effective surface area of the cathode and enables currents of several amperes to flow. The overall cell reaction is

$$Zn(s) + 2\,MnO_2(s) + 2\,NH_4^+(aq) \longrightarrow$$
$$[Zn(NH_3)_2]^{2+}(aq) + Mn_2O_3(s) + H_2O(\ell) \qquad \text{(overall)}$$

The cell components are hermetically sealed in a steel shell that, because it is in contact with the zinc, is the negative terminal of the battery. A fresh Leclanché dry cell has a potential difference of 1.5 V.

The Leclanché cell has several disadvantages. Concentrations change as the cell reaction progresses, so the cell voltage falls with use. Rapid withdrawal of energy may cause the production of ammonia and hydrogen to outpace their removal, leading to a sharp drop in voltage (a cell in this condition recovers on standing). Unused cells deteriorate because of a slow reaction between $Zn(s)$ and $NH_4^+(aq)$; thus, they can go dead while sitting on the shelf. Finally, the cells are not rechargeable because the $Zn^{2+}(aq)$ ions migrate away from the anode and are trapped in the complex ion.

In an **alkaline dry cell,** the ammonium chloride is replaced by potassium hydroxide. Here the half-cell reactions become

$$Zn(s) + 2\,OH^-(aq) \longrightarrow Zn(OH)_2(s) + 2\,e^- \qquad \text{(oxidation, anode)}$$
$$2\,MnO_2(s) + H_2O(\ell) + 2\,e^- \longrightarrow Mn_2O_3(s) + 2\,OH^-(aq) \qquad \text{(reduction, cathode)}$$

• Lowering the temperature slows most reactions. The shelf life of a dry cell can be increased by 100% to 200% by storage in a freezer or refrigerator.

• Recall that NH_4Cl is acidic in aqueous solution, and KOH is basic (alkaline).

The overall reaction is then

$$Zn(s) + 2\,MnO_2(s) + H_2O(\ell) \longrightarrow Zn(OH)_2(s) + Mn_2O_3(s) \qquad \text{(overall)}$$

Because no dissolved species take part in the overall reaction, the reaction quotient Q does not change as the reaction progresses, and a steadier voltage results. The standard voltage of "alkaline batteries" is 1.54 V; they replace Leclanché cells without problems.

A third primary dry cell is the **zinc–mercuric oxide cell,** or **mercury battery,** depicted in Figure 13–14. It frequently is given the shape of a small button and is used in automatic cameras, hearing aids, digital calculators, and quartz-electric watches. The anode in this cell is a mixture of mercury and zinc; the cathode is steel in contact with solid mercury(II) oxide (HgO). The electrolyte is a 45% potassium hydroxide solution that saturates an absorbent material. The anode half-reaction is the same as that in an alkaline dry cell:

$$Zn(s) + 2\,OH^-(aq) \longrightarrow Zn(OH)_2(s) + 2\,e^- \qquad \text{(oxidation, anode)}$$

The cathode half-reaction is now

$$HgO(s) + H_2O(\ell) + 2\,e^- \longrightarrow Hg(\ell) + 2\,OH^-(aq) \qquad \text{(reduction, cathode)}$$

The overall reaction is

$$Zn(s) + HgO(s) + H_2O(\ell) \longrightarrow Zn(OH)_2(s) + Hg(\ell) \qquad \text{(overall)}$$

In the mercury cell, mercury serves also to alloy, or amalgamate, the zinc (alloys of mercury with other metals are called "amalgams"). This cell has a very stable output of 1.34 V, a fact that makes it especially valuable for use in communication equipment and scientific instruments. A disadvantage is that mercury and mercury compounds are toxic. Careless disposal of mercury batteries leads to environmental contamination.

Rechargeable Batteries

Chemical change at the electrodes always accompanies the extraction of energy from a battery. When the redox reaction in a battery reaches equilibrium, this change has gone to its fullest extent: The battery is completely discharged. In some battery designs, the electrodes can be regenerated to their original form (or near it) by imposing an external potential difference to force current through the battery in a direction opposite to the way it flowed during discharge. Such batteries are **secondary batteries,** and the process of reconstituting them to their original state is called "recharging." To recharge a run-down secondary battery, one uses an external source with voltage larger than that of the battery in its original state and, of course, opposite in polarity.

The **nickel–cadmium cell** (or nicad battery, Fig. 13–15) is used in hand-held electronic calculators, cordless electric shavers, camcorders, and portable radios. Its half-reactions during discharge are

$$Cd(s) + 2\,OH^-(aq) \longrightarrow Cd(OH)_2(s) + 2\,e^- \qquad \text{(oxidation, anode)}$$

$$2\,NiO(OH)(s) + 2\,H_2O(\ell) + 2\,e^- \longrightarrow 2\,Ni(OH)_2(s) + 2\,OH^-(aq)$$
$$\text{(reduction, cathode)}$$

This battery gives a fairly constant voltage of 1.4 V. When the battery is connected to an external voltage source, the preceding reactions are reversed as the battery is recharged. Unfortunately, the nicad battery suffers from "discharge memory." If the battery is frequently recharged after only partial discharge, it starts to require recharging on that basis. In addition, cadmium, like mercury, is a toxic substance.

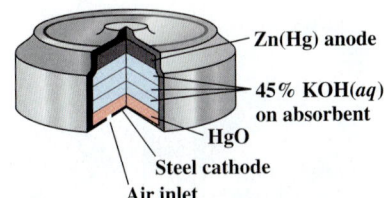

Figure 13–14 • A zinc–mercuric oxide dry cell of the type used in electric watches and cameras.

Figure 13–15 • A rechargeable nickel–cadmium battery.

• A secondary battery is converted from a galvanic cell into an electrolytic cell during recharging. The terms "discharge" and "recharge" are misleading because the actual quantity of electrical charge in the battery does not change in either process. Perhaps "de-energize" and "re-energize" would be better.

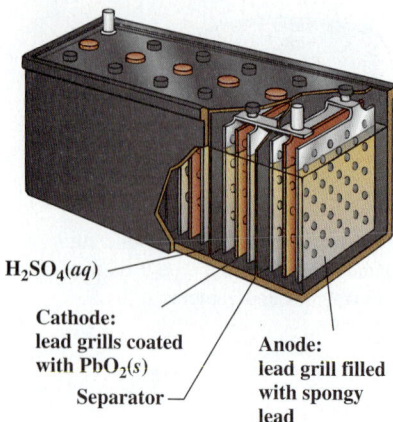

$H_2SO_4(aq)$

Cathode:
lead grills coated
with PbO$_2$(s)

Separator

Anode:
lead grill filled
with spongy
lead

Figure 13–16 • In a lead–acid storage battery, anodes made of Pb alternate with cathodes of Pb coated with PbO$_2$; the electrolyte is sulfuric acid (sometimes therefore called "battery acid").

• When a dead battery is to be recharged, the wire from the charger that is marked "+" must always connect to the battery terminal marked "+," and "−" to "−." In this way, the external source opposes the normal flow in the battery during discharge.

Perhaps the best-known secondary battery is the **lead-acid battery,** which is universally used in automobiles. A 12 V lead storage battery consists of six cells (Fig. 13–16) that are connected in series (cathode to anode) by an internal lead linkage and housed together in a hard rubber or plastic case. When cells are connected in series, their voltages add, so the six cells in this battery generate 2 V each. In each cell, the anode consists of porous metallic lead. The sponge-like texture of this electrode maximizes its area of contact with the electrolyte. The cathode is of similar design, but its lead has been converted to lead dioxide. A sulfuric acid solution (37% by mass) serves as the electrolyte.

When the external circuit is completed, electrons are released from the anode. The resulting Pb^{2+} ions immediately react with SO_4^{2-} ions to precipitate a layer of insoluble lead sulfate on the surface of the electrode. At the cathode, the electrons from the external circuit reduce PbO_2 to water and Pb^{2+} ions, which also react immediately with the sulfate ions in the electrolyte to precipitate $PbSO_4$ on the electrode. The half-cell reactions during discharge are

$$Pb(s) + SO_4^{2-}(aq) \longrightarrow PbSO_4(s) + 2\ e^-$$

(oxidation, anode)

$$PbO_2(s) + SO_4^{2-}(aq) + 4\ H^+(aq) + 2\ e^- \longrightarrow PbSO_4(s) + 2\ H_2O(\ell)$$

(reduction, cathode)

The anode and cathode are both largely converted to $PbSO_4(s)$ when the storage battery is fully discharged. Because sulfuric acid is a reactant, its concentration falls as the battery is discharged. Measuring the density of the electrolyte provides a quick way of estimating the state of charge of the battery.

When a voltage exceeding 2 V is applied across the terminals of a single cell (or 12 V for the entire battery) in such a direction as to make the electrode that was originally the cathode into the anode and the electrode that was originally the anode into the cathode, the half-cell reactions are reversed. Continued charging of the cell restores it to its initial state, ready for another work-producing discharge half-cycle. Lead storage batteries endure many thousands of cycles of discharge and recharge before they finally fail because $PbSO_4$ flakes off from the electrodes or internal short circuits develop. These short circuits occur when elongated crystals of lead and lead oxide, formed during recharging, finally connect the electrodes. In automobiles, lead-acid batteries usually are not designed to undergo complete discharge–recharge cycles. Instead, a DC generator converts some of the kinetic energy of the vehicle into electrical energy for continuous or intermittent charging. About 1.8×10^7 J of energy can be obtained in the discharge of an average automobile battery, enough to light a 100 W lightbulb for 50 h. Currents as large as 100 A are drawn from the battery for the short time needed to start the engine. Of course, when such huge currents flow, the operation of a battery is very far from equilibrium, and the potential difference across its electrodes is much smaller than the theoretical potential difference. The Nernst equation is of no value in calculating the voltage of a battery during this non-equilibrium process.

Fuel Cells

A battery is a closed system that delivers electrical energy by electrochemical reactions. Once the chemicals originally present are consumed, the battery must be either recharged or discarded. In contrast, a **fuel cell** is designed for continuous operation, with reactants (fuel) being supplied and products removed continuously.

Fuel cells based on the reaction

$$2\,H_2(g) + O_2(g) \longrightarrow 2\,H_2O(\ell)$$

are used on space vehicles (Fig. 13–17). No spare batteries burden the vehicle, and the astronauts drink the water that is produced in the operation of the fuel cell. The hydrogen–oxygen fuel cell is represented schematically in Figure 13–18. The electrodes can be any non-reactive conductor (graphite, for example). Their function is merely to conduct electrons into and out of the cell and to facilitate the exchange of electrons between the gases and the ions in solution. The electrolyte transports charge through the cell, and the ions dissolved in it participate in the half-reactions at each electrode. Acidic solutions present corrosion problems, so an alkaline solution (1 M KOH, for instance) is preferable. The following half-reactions take place:

$$H_2(g) + 2\,OH^-(aq) \longrightarrow 2\,H_2O(\ell) + 2\,e^- \qquad \text{(oxidation, anode)}$$
$$\tfrac{1}{2}O_2(g) + H_2O(\ell) + 2\,e^- \longrightarrow 2\,OH^-(aq) \qquad \text{(reduction, cathode)}$$

The standard reduction potentials at 25°C are

$$\mathcal{E}°(H_2|H_2O) = -0.828\ \text{V} \qquad \text{and} \qquad \mathcal{E}°(O_2|OH^-) = 0.401\ \text{V}$$

and the overall cell reaction is the production of water:

$$H_2(g) + \tfrac{1}{2}O_2(g) \longrightarrow H_2O(\ell)$$

The standard cell voltage is therefore

$$\Delta\mathcal{E}° = \mathcal{E}°(\text{cathode}) - \mathcal{E}°(\text{anode}) = 0.401 - (-0.828) = 1.229\ \text{V}$$

Another practical fuel cell employs the overall reaction

$$CO(g) + \tfrac{1}{2}O_2(g) \longrightarrow CO_2(g)$$

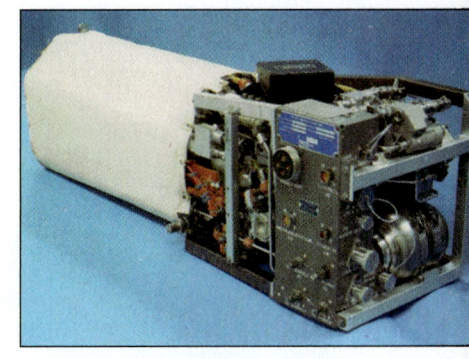

Figure 13–17 • This hydrogen–oxygen fuel cell was used on U.S. space missions. *(Courtesy of United Technologies)*

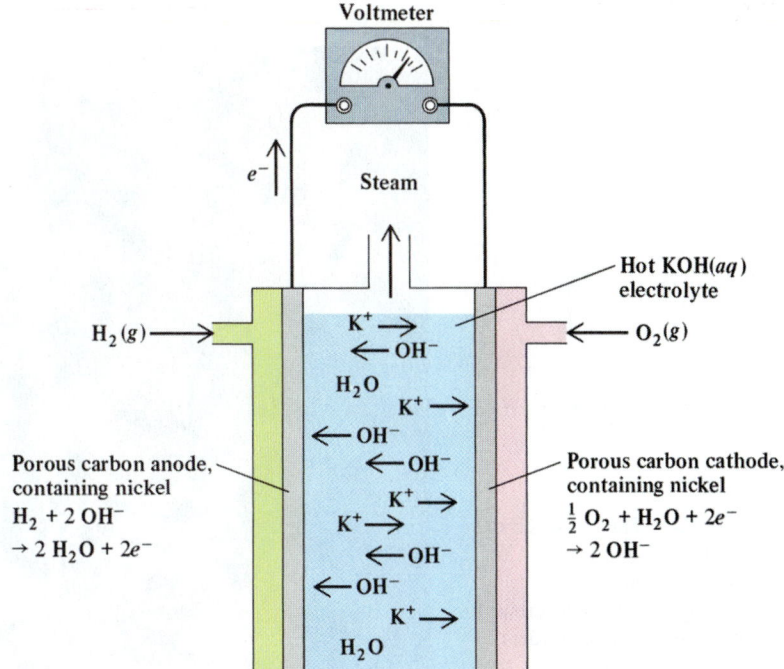

Figure 13–18 • In the hydrogen–oxygen fuel cell, the two gases are fed in separately and are oxidized or reduced at the electrodes. A hot solution of potassium hydroxide between the electrodes completes the circuit, and the steam produced in the reaction evaporates from the cell continuously.

for which the half-reactions are

$$CO(g) + H_2O(\ell) \longrightarrow CO_2(g) + 2\,H^+(aq) + 2\,e^- \qquad \text{(oxidation, anode)}$$
$$\tfrac{1}{2}O_2(g) + 2\,H^+(aq) + 2\,e^- \longrightarrow H_2O(\ell) \qquad \text{(reduction, cathode)}$$

The electrolyte usually is concentrated phosphoric acid, and the operating temperature is between 100°C and 200°C. Platinum is the electrode material of choice.

Natural gas (largely CH_4) and even fuel oil can be "burned" electrochemically to provide electrical energy by either of two approaches. They can be converted into CO and H_2 or to CO_2 and H_2 prior to use in a fuel cell by reforming reactions with steam

$$CH_4(g) + H_2O(g) \longrightarrow CO(g) + 3\,H_2(g)$$
$$CO(g) + H_2O(g) \longrightarrow CO_2(g) + H_2(g)$$

at temperatures of about 500°C. Alternatively, they can be used directly in a fuel cell at higher temperatures (up to 750°C) with a molten alkali-metal carbonate as the electrolyte. Either type of fuel cell is attractive as an electrochemical energy converter in areas where hydrocarbon fuels are readily available, but large-scale power plants (conventional or nuclear) are remote.

13-7   Corrosion and Its Prevention

The **corrosion** of metals (Fig. 13–19) is a significant problem. It has been estimated that, in the United States alone, the annual cost of corrosion is on the order of tens of *billions* of dollars. The effects of corrosion are both visible (the formation of rust on exposed iron and steel surfaces) and invisible (the cracking and resulting loss of strength of metal beneath the surface). The corrosion of the various structural metals under ordinary terrestrial conditions is thermodynamically spontaneous and

Figure 13–19 • The corrosion of iron and steel is a major problem for industrial society, ranging in scale from the rusting of nails and screws to the structural weakening of girders and bridges. *(M. D. Ippolito)*

therefore cannot, in the long run, be stopped. Iron, steel, and aluminum eventually corrode. What really matters, in human terms, is how long such processes take and how they can be prevented from happening prematurely and catastrophically. Before ways to achieve this kind of prevention can be developed, the mechanisms of corrosion must be studied and understood.

Although corrosion is a serious problem for many metals, we focus on the spontaneous electrochemical reactions of iron. The corrosion of iron can be pictured as resulting from the operation of a galvanic cell, in which some regions of the metal surface act as cathodes and others as anodes, with the electrical circuit being completed by electron flow through the iron itself. The electrochemical half-cells appear in parts of the metal containing impurities or in regions that are subject to stress. The anode reaction is

$$\text{Fe}(s) \longrightarrow \text{Fe}^{2+}(aq) + 2\,e^- \qquad \text{(oxidation, anode)}$$

Various cathode reactions are possible. In the absence of oxygen (for example, at the bottom of a lake), the corrosion reactions are

$$\text{Fe}(s) \longrightarrow \text{Fe}^{2+}(aq) + 2\,e^- \qquad \text{(oxidation, anode)}$$
$$\underline{2\,\text{H}_2\text{O}(\ell) + 2\,e^- \longrightarrow 2\,\text{OH}^-(aq) + \text{H}_2(g)} \qquad \text{(reduction, cathode)}$$
$$\text{Fe}(s) + 2\,\text{H}_2\text{O}(\ell) \longrightarrow \text{Fe}^{2+}(aq) + 2\,\text{OH}^-(aq) + \text{H}_2(g) \qquad \text{(overall)}$$

However, these (and similar reactions) are generally slow and do not cause serious amounts of corrosion unless accelerated by the intervention of bacteria (see following discussion). In Figure 13–20, indicators make it possible to distinguish cathodic and anodic regions in nails corroding in the absence of oxygen. More extensive corrosion takes place when the iron is in contact with both oxygen and water. In this case, the cathode reaction is

$$\tfrac{1}{2}\text{O}_2(g) + 2\,\text{H}^+(aq) + 2\,e^- \longrightarrow \text{H}_2\text{O}(\ell) \qquad \text{(reduction, cathode)}$$

The Fe^{2+} ions formed concurrently at the anode can migrate to the cathode, where they are further oxidized by O_2 to the $+3$ oxidation state to form rust ($\text{Fe}_2\text{O}_3 \cdot x\text{H}_2\text{O}$), a hydrated form of iron(III) oxide:

$$2\,\text{Fe}^{2+}(aq) + \tfrac{1}{2}\text{O}_2(g) + (2 + x)\,\text{H}_2\text{O}(\ell) \longrightarrow \text{Fe}_2\text{O}_3 \cdot x\text{H}_2\text{O} + 4\,\text{H}^+(aq)$$

The hydrogen ions produced in this reaction allow the corrosion cycle to continue.

The nature of the cathodic and anodic regions can be illustrated by the way in which a piece of iron or steel corrodes when a portion of the paint that protects it is chipped off (Fig. 13–21). The exposed area acts as the cathode because it is open to the atmosphere (air and water) and is therefore rich in oxygen, whereas oxygen-poor areas *underneath* the paint act as anodes. The result is that rust forms on the cathode (the visible, exposed region) and pitting occurs at the anode (loss of metal through oxidation of iron and flow of metal ions to the cathode). This pitting can lead to loss of structural strength in girders and other supports. The most serious harm done by corrosion is not the visible rusting but the damage done underneath the painted surface.

A number of factors speed up corrosion. Dissolved salt improves the flow of charge through solution; a well-known example is the more rapid rusting of cars in areas where salt is spread to combat icy roads. Higher acidity also favors corrosion (note that H^+ is a reactant in the half-reaction at the cathode). Acidity is increased by dissolved CO_2 (which produces H^+ and HCO_3^- ions) and, more seriously, by dissolved oxides of sulfur from air pollution brought down in acid rain.

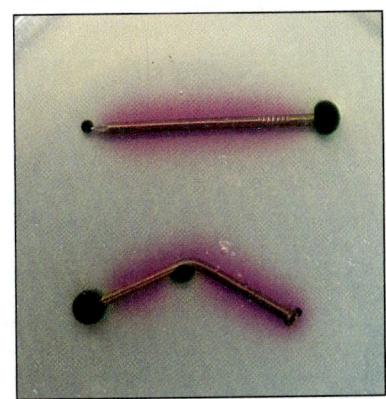

Figure 13–20 • Nails corroding in water in the absence of oxygen. Dark blue-green indicates anodic regions where $\text{Fe}(s)$ has been oxidized to $\text{Fe}^{2+}(aq)$ ion. Red indicates cathodic regions where water has been reduced to $\text{OH}^-(aq)$ ion. The colors appear because the water contains $[\text{Fe}(\text{CN})_6]^{3+}$ ion, which reacts with Fe^{2+} to give a blue-green complex, and phenolphthalein, which turns red in the presence of OH^- ion; the solution also contains a thickening gel to slow diffusion. Note how oxidation occurs at the bend and at the ends of the nails; these are regions of stress. *(Charles D. Winters)*

Figure 13–21 • The corrosion of iron. Note that pitting occurs at the anodic region and rust appears at the cathodic region.

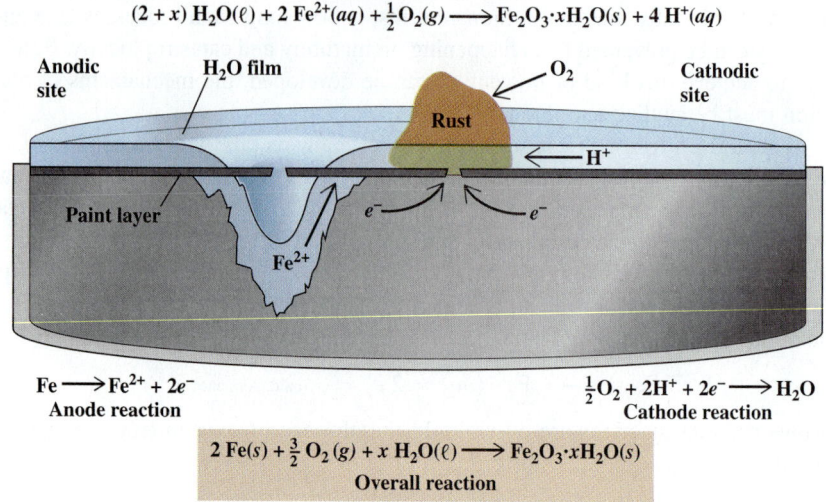

Secondary reaction

$$(2 + x)\,H_2O(\ell) + 2\,Fe^{2+}(aq) + \tfrac{1}{2}O_2(g) \longrightarrow Fe_2O_3{\cdot}xH_2O(s) + 4\,H^+(aq)$$

$$Fe \longrightarrow Fe^{2+} + 2e^-$$
Anode reaction

$$\tfrac{1}{2}O_2 + 2H^+ + 2e^- \longrightarrow H_2O$$
Cathode reaction

$$2\,Fe(s) + \tfrac{3}{2}O_2(g) + x\,H_2O(\ell) \longrightarrow Fe_2O_3{\cdot}xH_2O(s)$$
Overall reaction

Biocorrosion

Although corrosion is usually thought of as occurring when a metal is in contact with air and water (*aerobic* conditions), significant corrosion also occurs under conditions that completely exclude oxygen from the metal (*anaerobic* conditions). Quite often, microorganisms accelerate these processes in a mechanism referred to as **biocorrosion.**

One major group of agents for biocorrosion is the sulfate-reducing bacteria. These organisms greatly speed up the corrosion of iron in pipes buried in clay soils or covered by polluted water. Such environments contain sulfate ion ($SO_4^{2-}(aq)$) but no $O_2(g)$ or $O_2(aq)$, so that corrosion by the reactions just described cannot occur; however, the iron is oxidized, and the ultimate oxidizing agent is the sulfate ion. Again, the process is best viewed as a short-circuited galvanic cell with the half-reactions taking place at separate locations on the metal surface. The half-reaction at the anodic sites is

$$4\,Fe(s) + HS^-(aq) + 7\,OH^-(aq) \longrightarrow$$
$$3\,Fe(OH)_2(s) + FeS(s) + H_2O(\ell) + 8\,e^- \qquad \text{(oxidation, anode)}$$

At the cathodic sites, the sulfate ion is reduced:

$$SO_4^{2-}(aq) + 5\,H_2O(\ell) + 8\,e^- \longrightarrow HS^-(aq) + 9\,OH^-(aq) \qquad \text{(reduction, cathode)}$$

The overall reaction

$$4\,Fe(s) + SO_4^{2-}(aq) + 4\,H_2O(\ell) \longrightarrow$$
$$3\,Fe(OH)_2(s) + FeS(s) + 2\,OH^-(aq) \qquad \text{(overall)}$$

is thermodynamically spontaneous. In the absence of the bacteria, however, this reaction does not take place at a significant rate. The bacteria provide the necessary pathway for the reaction to occur.

Events in the cathodic regions, where the bacteria live, take place in two steps:

$$8\,H_2O(\ell) + 8\,e^- \longrightarrow 4\,H_2(g) + 8\,OH^-(aq)$$
$$SO_4^{2-}(aq) + 4\,H_2(g) \longrightarrow HS^-(aq) + OH^-(aq) + 3\,H_2O(\ell)$$

In the absence of the bacteria, there is no rapid way for the second step to occur. Corrosion quickly stalls with the build-up of $H_2(g)$ generated in the first step. The

CHEMISTRY IN PROGRESS

Turning Corrosion to Good Use

Chlorinated hydrocarbons are the products, by-products, or wastes of numerous industrial processes. Some are quite noxious and have unfortunately escaped into the air (see Section 18–5) or the ground, where they resist decomposition and present long-term hazards. These chemicals have made the term "toxic waste" part of everyday language.

Dozens of schemes for the remediation of spills of liquid and solid chlorinated hydrocarbons have been suggested. The standard procedure is "pump and treat." Wells are drilled. Contaminated groundwater is brought up, cleansed chemically in a special plant, and pumped back into the ground. Other concepts range from *bioremediation* (encouraging existing microorganisms to consume specific pollutants or even breeding new ones) to a Star Wars approach in which electron-beam accelerators originally designed for weapons research blast the pollutants as the contaminated water flows by.

One new idea is stunning in its simplicity and apparent success. Tons of iron filings are buried in underground walls positioned to intercept the movement of contaminated groundwater. Chlorinated hydrocarbons in the groundwater percolate through the barrier and, over a period of several days, corrode the iron(0), oxidizing it to iron(II) or iron(III). They themselves are reduced to inorganic chloride (the aqueous chloride ion, which is innocuous in the environment) and a nontoxic or less toxic hydrocarbon.

Clean-up by corrosion is working right now to destroy spilled trichloroethylene (C_2HCl_3) at a site in Sunnyvale, California, where a slow, arduous course of remediation by the standard method had been anticipated. The water seeping through the barrier meets the standards of the U.S. Environmental Protection Agency, its operators say.

In clean-up by corrosion, the high Gibbs energy of elemental iron relative to its oxides and chlorides drives the reactions that consume the pollutants. The operation

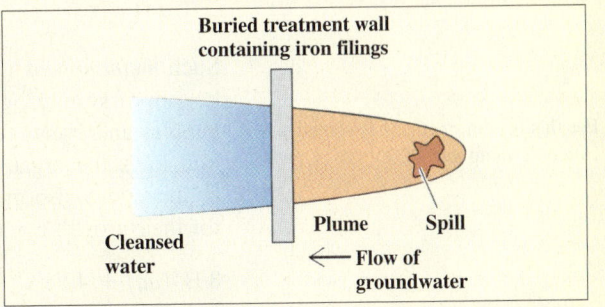

An underground wall of iron filings intercepts and destroys pollutants in groundwater.

consumes no energy once the iron is smelted. Although the land and its patterns of water flow must be surveyed very carefully to ensure that the flow of contaminated groundwater does not somehow miss the barrier containing the iron, researching a site and building an underground wall cost no more than constructing a water-treatment plant. The underground walls require little maintenance, cleanse the water more rapidly than a treatment plant, waste less water, and, it is thought, will remain effective for many years. Finally, the land above can be used for purposes other than a waste-treatment facility.

The planned corrosion of iron may help eliminate pollutants other than chlorinated hydrocarbons. Iron seems capable of degrading dye wastes from the textile industry. It also reduces chromium in the +6 oxidation state to chromium in the +3 oxidation state. This is particularly useful because many highly toxic compounds in which chromium is in the +6 state are water soluble and disperse rapidly in groundwater when spilled. Compounds in which chromium is in the +3 state are generally insoluble and stick on the iron barrier. Iron also may produce immobile compounds of technetium, a radioactive element that contaminates several sites operated by the Department of Energy.

bacteria accelerate the second step by combining the $H_2(g)$ with $SO_4^{2-}(aq)$ as it forms. The overall corrosive reaction then runs smoothly. Although the bacteria do not incorporate the iron into their own organisms, they still feed on the pipes in the sense that they use them as an energetically accessible source of electrons to reduce sulfate. Iron pipes corroded in this way become covered with black blotches of iron(II) sulfide. If this soft FeS(s) is wiped away, an anodic pit is revealed. Within the pit shines a bright surface of metallic iron.

A second example of biocorrosion involves a group of anaerobic microorganisms that live by oxidizing hydrogen with carbon dioxide to generate methane and water:

$$4\,H_2(g) + CO_2(g) \longrightarrow CH_4(g) + 2\,H_2O(\ell) \qquad \Delta G_r^\circ = -131\ \text{kJ mol}^{-1}$$

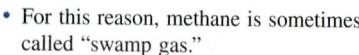

• For this reason, methane is sometimes called "swamp gas."

Such *methanogens* (methane-generating bacteria) thrive in the airless muck at the bottom of swamps, generating considerable amounts of methane, which collects in bubbles and eventually rises to the surface. The free energy of the preceding reaction fuels the organisms' growth. What is significant for corrosion is that some methanogens also grow in the complete absence of H_2 if $Fe(s)$ is present. Energy for their growth comes from the reaction

$$8\,H^+(aq) + 4\,Fe(s) + CO_2(g) \longrightarrow$$
$$CH_4(g) + 4\,Fe^{2+}(aq) + 2\,H_2O(\ell) \qquad \Delta G_r^\circ = -136\ \text{kJ mol}^{-1}$$

Observe that ΔG_r° is nearly the same as in the reaction in which H_2 is oxidized. In a sense, the bacteria can eat iron as a substitute for H_2. The iron corrodes as the microorganism uses it as a source of electrons to reduce carbon dioxide.

Inhibition of Corrosion

Corrosion of metals can be inhibited in a number of ways. Coatings of paint or plastics obviously protect the metal, but they can crack or sustain other damage, thereby localizing and accentuating the process. A very important method of protecting metals arises from the phenomenon of **passivation,** in which a thin metal-oxide layer forms on the surface and prevents further electrochemical reactions. Some metals become passivated spontaneously upon exposure to air; aluminum, for example, reacts with oxygen to form a thin, protective layer of Al_2O_3. A tightly adhering oxide layer less than 20 atoms deep protects magnesium, commonly used for transmission cases in automobiles and in other structural applications, from corroding in air. Special paints designed for rust prevention contain potassium dichromate ($K_2Cr_2O_7$) and lead oxide (Pb_3O_4), which cause the surface of iron to be oxidized and passivated. The presence of small quantities of alloying materials can greatly influence the amount of protection that a film provides. Stainless steel is an alloy of iron with chromium in which the chromium leads to passivation and prevents rusting. The integrity of a surface film is constantly assaulted by routine day-to-day events; the metal may be bent, shocked by blows, and scoured by abrasive particles. Protection against corrosion requires a surface film that can heal itself as fast as it is broken down.

A different way of preventing the corrosion of iron is to use a **sacrificial anode.** According to the standard reduction potentials of iron and magnesium,

$$Fe^{2+}(aq) + 2\,e^- \longrightarrow Fe(s) \qquad \mathcal{E}^\circ = -0.447\ \text{V}$$
$$Mg^{2+}(aq) + 2\,e^- \longrightarrow Mg(s) \qquad \mathcal{E}^\circ = -2.372\ \text{V}$$

Figure 13–22 • The iron of this ship's hull is protected against corrosion by the blocks of zinc bolted to it. In salt water, the zinc acts as a sacrificial anode, suffering oxidation in place of the ship's hull. (*Science VU/©NSRDC/Visuals Unlimited*)

$Mg^{2+}(aq)$ is much harder to reduce than $Fe^{2+}(aq)$. It follows that $Mg(s)$ is more easily oxidized than $Fe(s)$. A piece of magnesium in electrical contact with iron is oxidized in preference to the iron, and the iron is thereby protected. The magnesium is the sacrificial anode; once it is consumed by oxidation, it must be replaced. Sacrificial anodes are used to protect iron ship hulls, bridges, and water pipes from corrosion (Fig. 13–22). Magnesium plates are attached at regular intervals along a piece of buried pipe. It is far easier to replace them periodically than to replace the entire pipe.

SUMMARY

13-1 Thermodynamics demonstrates a direct connection between work, the Gibbs energy, and potential difference (voltage) in an electrochemical cell. In a cell operated at constant temperature and pressure, the maximum **electrical work** done is equal to the change in the Gibbs energy of the redox reaction and to the product (with a negative sign) of the total charge passing through the external circuit and the cell voltage. A galvanic cell can produce useful work from a spontaneous reaction; an electrolytic cell requires the input of work, from an outside source of electrical energy, to force a non-spontaneous chemical reaction to occur. The **standard cell voltage** is the voltage developed by an electrochemical cell in which all reactants and products are in thermodynamic standard states.

13-2 **Half-cell reduction potentials** are tabulated using the convention that the reduction potential of a standard $H^+(aq)|H_2(g)$ half-cell is exactly zero. The reduction potential of any other half-cell can be measured by setting it up in combination with a half-cell that has a reduction potential that is already known relative to this standard and measuring the potential difference (the cell voltage).

13-3 The relative strengths of oxidizing and reducing agents depend on the products they form; in aqueous systems, the pH of the solution is a particularly important influence. Tables of standard reduction potentials show which species are unstable with respect to disproportionation and which are strong oxidizing or reducing agents.

13-4 The **Nernst equation** describes the effect of changes of concentrations or pressures on cell voltages. It can be used to compute the concentration of a dissolved species taking part in a redox reaction from the cell voltage, provided that the concentrations of other species being consumed and produced are all known. The **pH meter** is an application of the equation.

13-5 The equilibrium constant (K) of a redox reaction can be calculated from the reaction's standard cell voltage. Electrochemical cells also can be used to measure K_a's, K_b's, and K_{sp}'s.

13-6 Electrochemical cells are used to generate and store electricity. In **batteries,** spontaneous reactions generate an electric current. In a **primary battery,** the cells are discarded and the materials recycled after the reactants are consumed and the voltage drops, but in a **secondary battery,** the original reactants can be regenerated by recharging. A **fuel cell** allows the continuous conversion of chemical to electrical energy as reactants are fed in and caused to react in an electrochemical cell.

13-7 **Corrosion** in metals often can be understood as the operation of short-circuited electrochemical cells, in which stress or exposure to water and air creates cathodic and anodic regions in a single material and permits the oxidation and dissolution of the metal. Biological organisms play a major role in accelerating corrosion in some settings. The effects of corrosion are reduced through **passivation** (the use of oxidized surface layers) and the provision of **sacrificial anodes** that are corroded away in preference to the metal that is being protected.

PROBLEMS

Note: Answers to blue-numbered problems are given in Appendix F. Problems that are more challenging are indicated with asterisks.

The Gibbs Energy and Cell Voltage

1. Electricity usually is sold by the kilowatt-hour (kW-h). If 1 kW-h of electrical energy costs 10 cents, compute the cost of leaving a 150 W porch light burning overnight (a period of 10 h).

2. (See Example 13–1.) A flashlight in which two batteries generate an average current of 1.1 A across a voltage of 2.9 V is left on for 10 h. How many kW-h of energy are expended? Use this answer and the answer to Problem 1 to comment on

the relative cost of producing electricity from a battery and from a power station, if the batteries cost $1.00 each and are essentially dead after the 10 h of use.

3. How long does it take a 1.34 V battery, from which a steady current of 0.800 A is drawn, to perform 1.00 J of electrical work in the surroundings?

4. A galvanic cell is operated extremely slowly in order to extract nearly the theoretical maximum electrical work. A current of 1.60×10^{-18} A (equal to 10.0 electrons per second) flows across a potential difference of 2.00 V. How long does it take to perform 1.00 J of electrical work in the surroundings?

5. (See Example 13–2.) Write a balanced chemical equation to represent the reaction in the following electrochemical cells. Compute the standard Gibbs energy of reaction ΔG_r° at 25°C for each case.
   (a) $Pb(s)|Pb^{2+}(aq)\|Cu^{2+}(aq)|Cu(s)$ $\Delta \mathcal{E}^\circ = 0.471$ V
   (b) $Pt(s)|Fe^{2+}(aq), Fe^{3+}(aq)\|$
       $MnO_4^-(aq), H^+(aq), Mn^{2+}(aq)|Pt(s)$ $\Delta \mathcal{E}^\circ = 0.736$ V
   (c) $Ag(s)|Ag^+(aq)\|Co^{2+}(aq)|Co(s)$ $\Delta \mathcal{E}^\circ = -1.08$ V

6. (See Example 13–2.) Write a balanced chemical equation to represent the reaction in the following electrochemical cells. Compute the standard Gibbs energy of reaction ΔG_r° at 25°C for each case.
   (a) $Cd(s)|Cd^{2+}(aq)\|Zn^{2+}(aq)|Zn(s)$ $\Delta \mathcal{E}^\circ = -0.359$ V
   (b) $Sn(s)|Sn^{2+}(aq)\|Fe^{3+}(aq), Fe^{2+}(aq)|Pt(s)$
       $\Delta \mathcal{E}^\circ = 0.909$ V
   (c) $Ag(s)|Ag^+(aq)\|I_2(s)|I^-(aq)|C(s, graphite)$
       $\Delta \mathcal{E}^\circ = -0.264$ V

   Check the answers by computing ΔG_r° independently using ΔG_f° data from Appendix D.

7. A $Ni(s)|Ni^{2+}(aq)\|Ag^+(aq)|Ag(s)$ galvanic cell is constructed in which the standard cell voltage is 1.057 V. Calculate the Gibbs energy change at 25°C per *gram* of metallic silver produced if all concentrations remain at their standard-state values of 1 M throughout the process. What is the maximum electrical work done *by* the cell on its surroundings per gram of silver produced in this experiment?

8. A $Zn(s)|Zn^{2+}(aq)\|Co^{2+}(aq)|Co(s)$ galvanic cell is constructed in which the standard cell voltage is 0.48 V. Calculate the Gibbs energy change at 25°C per gram of zinc dissolved at the anode, if all concentrations remain at their standard-state values of 1 M throughout the process. What is the maximum electrical work done by the cell on its surroundings per gram of zinc reacting in this experiment?

Half-Cell Potentials

9. (See Example 13–3.) A galvanic cell is constructed by connecting a standard $Br_2(\ell)|Br^-$ half-cell to a standard $Co^{2+}|Co$ half-cell.
   (a) Write balanced chemical equations for the half-reactions at the anode and the cathode and for the overall cell reaction.

(b) Calculate the cell voltage at 25°C (use information from Appendix E).

10. (See Example 13–3.) A galvanic cell is constructed by connecting a standard $Pt|Fe^{2+}, Fe^{3+}$ half-cell to a standard $Cd^{2+}|Cd$ cathode half-cell.
    (a) Write balanced chemical equations for the half-reactions at the anode and the cathode and for the overall cell reaction.
    (b) Calculate the cell voltage at 25°C (use information from Appendix E).

11. In a galvanic cell, one half-cell consists of a zinc strip dipped into a 1.00 M solution of $Zn(NO_3)_2$. In the second half-cell, solid indium adsorbed on graphite is in contact with a 1.00 M solution of $In(NO_3)_3$. Indium is observed to plate out as the galvanic cell operates, and the initial cell voltage equals 0.425 V at 25°C.
    (a) Write balanced equations for the half-reaction at the anode and the half-reaction at the cathode.
    (b) Calculate the standard reduction potential of the $In^{3+}|In$ half-cell. Consult Table 13–1 for the standard reduction potential of the $Zn^{2+}|Zn$ half-cell.

12. In a galvanic cell, one half-cell consists of gaseous chlorine bubbled over a platinum electrode at a pressure of 1.00 atm into a 1.00 M solution of NaCl. The second half-cell has a strip of solid gallium immersed in a 1.00 M $Ga(NO_3)_3$ solution. The initial cell voltage equals 1.918 V at 25°C, and the concentration of chloride ion increases as the cell operates.
    (a) Write balanced equations for the half-reaction at the anode and the half-reaction at the cathode.
    (b) Calculate the standard reduction potential of a $Ga^{3+}|Ga$ half-cell at 25°C. Consult Appendix E for the standard reduction potential of the $Cl_2|Cl^-$ half-cell.

13. (See Example 13–4.) By referring to Appendix E, predict whether $Mn^{2+}(aq)$ in its standard state at room temperature disproportionates spontaneously according to

   $$3\ Mn^{2+}(aq) \longrightarrow Mn(s) + 2\ Mn^{3+}(aq)$$

   Show calculations to support the prediction.

14. (See Example 13–4.) The following reduction potentials have been measured for the oxidation states of thallium:

   $$Tl^{3+}(aq) + e^- \longrightarrow Tl^{2+}(aq) \qquad \mathcal{E}^\circ = -0.37\ V$$
   $$Tl^{2+}(aq) + e^- \longrightarrow Tl^+(aq) \qquad \mathcal{E}^\circ = +2.87\ V$$

   Does $Tl^{2+}(aq)$ disproportionate spontaneously to $Tl^{3+}(aq)$ and $Tl^+(aq)$? Explain.

15. One method to reduce the concentration of unwanted $Fe^{3+}(aq)$ ion in solutions of $Fe^{2+}(aq)$ is to drop a piece of metallic iron into the storage container. Write the reaction that removes the $Fe^{3+}(aq)$, and compute its standard potential difference.

16. By referring to Problems 13 and 15, suggest a way to remove unwanted $Mn^{3+}(aq)$ ions from solutions of $Mn^{2+}(aq)$. Compute the standard potential difference for the reaction proposed for this use.

Oxidizing and Reducing Agents

17. Would you expect powdered solid aluminum to act as an oxidizing agent or a reducing agent?

18. Would you expect sodium perchlorate ($NaClO_4(aq)$) in a concentrated acidic solution to act as an oxidizing agent or a reducing agent?

19. Bromine sometimes is used in place of chlorine as a disinfectant in swimming pools. If the effectiveness of a chemical as a disinfectant depends solely on its strength as an oxidizing agent, do you expect bromine to be better or worse than chlorine as a disinfectant, at a given concentration?

20. Many bleaches, including chlorine and its oxides, oxidize dyes in cloth, destroying their color. Predict which of the following is the strongest bleach at a given concentration and pH 0: $NaClO_3(aq)$, $NaClO(aq)$, $Cl_2(aq)$. How does the strongest chlorine-containing bleach compare in strength with ozone ($O_3(g)$)?

21. Find in Appendix E the standard potential $\mathcal{E}°$ for the reduction in acidic solution of the permanganate ion (MnO_4^-) to manganese(II) ion (Mn^{2+}). Use this potential and other standard reduction potentials from Appendix E to decide which of the following tend to be oxidized by $MnO_4^-(aq)$ in acidic solution with the reactants and products in standard states at 298.15 K.
(a) $F^-(aq)$ (b) $Cl^-(aq)$ (c) $Cr^{3+}(aq)$ (d) $ClO_4^-(aq)$

22. The standard potential $\mathcal{E}°$ for the reduction in basic solution of platinum(II) hydroxide ($Pt(OH)_2(s)$) to elemental platinum ($Pt(s)$) is 0.16 V. Use this potential and standard reduction potentials from Appendix E to decide which of the following tend to be oxidized by $Pt(OH)_2(s)$ in basic solution when all reactants and products are in standard states at 298.15 K.
(a) $Ni(s)$ (b) $OH^-(aq)$ (c) $Pb(s)$ (d) $F^-(aq)$

23. (See Example 13–5.) Figure 13–5 can be extended to include species in which the oxidation states of nitrogen, sulfur, and phosphorus are zero or negative. In basic solution, for example, the following standard reduction potentials have been measured at pH 14:

$$P_4(s) + 12\ H_2O(\ell) + 12\ e^- \longrightarrow 4\ PH_3(g) + 12\ OH^-(aq)$$
$$\mathcal{E}° = -0.89\ V$$
$$4\ H_2PO_2^-(aq) + 4\ e^- \longrightarrow P_4(s) + 8\ OH^-(aq)$$
$$\mathcal{E}° = -2.05\ V$$

(a) Does $P_4(s)$ disproportionate spontaneously in basic solution (at pH 14) at 25°C?
(b) Which is the stronger reducing agent at pH 14: $P_4(s)$ or $PH_3(g)$?

24. (See Example 13–5.) Extending Figure 13–5, the following additional reduction potentials are measured at pH 14:
$$S(s) + H_2O(\ell) + 2\ e^- \longrightarrow HS^-(aq) + OH^-(aq)$$
$$\mathcal{E}° = -0.51\ V$$
$$S_2O_3^{2-}(aq) + 3\ H_2O(\ell) + 4\ e^- \longrightarrow 2\ S(s) + 6\ OH^-(aq)$$
$$\mathcal{E}° = -0.74\ V$$

(a) Does sulfur disproportionate spontaneously under standard basic conditions at 25°C?

(b) Which is the stronger reducing agent: $S(s)$ or $HS^-(aq)$?

25. Use data from Appendix E to add the elements Eu (europium) Ga (gallium), La (lanthanum), and Sm (samarium) in their correct places in the "Activity Series for Some Electropositive Elements" that appears as Table 4–2 (page 178).

26. Use data from Table 13–1 to arrange the Group VII elements (the halogens) in increasing order of the ability to displace other elements from chemical compounds.

Concentration Effects and the Nernst Equation

27. (See Example 13–6.) A galvanic cell employs the reaction
$$Pb^{2+}(aq) + 2\ Cr^{2+}(aq) \longrightarrow Pb(s) + 2\ Cr^{3+}(aq)$$

The initial concentrations of $Pb^{2+}(aq)$, $Cr^{2+}(aq)$, and Cr^{3+} (aq) are 0.15 M, 0.20 M, and 0.0030 M, respectively. Calculate the initial voltage generated by the cell at 25°C.

28. (See Example 13–6.) A galvanic cell employs the reaction
$$2\ Ag(s) + Cl_2(g) \longrightarrow 2\ Ag^+(aq) + 2\ Cl^-(aq)$$

The initial partial pressure of $Cl_2(g)$ is 1.00 atm, the initial concentration of $Ag^+(aq)$ is 0.25 M, and the initial concentration of $Cl^-(aq)$ is 0.016 M. Calculate the initial voltage generated by the cell at 25°C.

29. (See Example 13–6.) Use the Nernst equation to determine the potential difference at 25°C of the following cells when the dissolved ions have the concentrations indicated.
(a) $Pb(s)|Pb^{2+}(aq, 0.002\ M)||Cu^{2+}(aq, 0.075\ M)|Cu(s)$
(b) $Pt(s)|Fe^{2+}(aq, 0.020\ M), Fe^{3+}(aq, 0.010\ M)||MnO_4^-$ $(aq, 0.050\ M), Mn^{2+}(aq, 0.95\ M), H^+(aq, 0.10\ M)|Pt(s)$
(c) $Co(s)|Co^{2+}(aq, 0.14\ M)||(Ag^+(aq, 0.10\ M)|Ag(s)$

30. (See Example 13–6.) Use the Nernst equation to determine the potential difference at 25°C in the following cells when the dissolved ions have the concentrations indicated.
(a) $Cd(s)|Cd^{2+}(aq, 0.0020\ M)||Zn^{2+}(aq, 0.075\ M)|Zn(s)$
(b) $Sn(s)|Sn^{2+}(aq, 0.010\ M)||Fe^{3+}(aq, 0.050\ M), Fe^{2+}(aq, 0.950\ M)|Pt(s)$
(c) $Ag(s)|Ag^+(aq, 0.0020\ M)||I_2(s)|I^-(aq, 0.021\ M)|C$ (s, graphite)

31. Calculate the reduction potential at 25°C for a $Pt|Cr^{3+}, Cr^{2+}$ half-cell in which $[Cr^{3+}]$ is 0.15 M and $[Cr^{2+}]$ is 0.0019 M.

32. Calculate the reduction potential at 25°C for an $I_2(s)|I^-$ half-cell in which $[I^-]$ is 1.5×10^{-6} mol L^{-1}.

33. If the half-cell from Problem 31 is connected to a standard $Cd^{2+}|Cd$ half-cell to make a galvanic cell, which half-cell acts as the anode?

34. If the half-cell from Problem 32 is connected to a standard $Pt|Fe^{3+}, Fe^{2+}$ half-cell to make a galvanic cell, which half-cell acts as the cathode?

35. (See Example 13–7.) Working at 25°C, a student connects a standard $I_2(s)|I^-$ half-cell to a $Pt|H^+|H_2(g, 1\ atm)$ half-cell in which the concentration of H^+ is unknown. The measured cell voltage is 0.841 V, and the $I_2(s)|I^-$ half-cell is the cathode. What is the pH in the $Pt|H^+|H_2(g, 1\ atm)$ half-cell?

36. (See Example 13–7.) A standard $Cu^{2+}(aq)|Cu(s)$ half-cell is connected to a $Br_2(\ell)|Br^-$ half-cell in which the concentration of bromide ion is unknown. The measured cell voltage is 0.963 V at 25°C, and the $Cu(s)$ is the anode. What is the bromide ion concentration in the $Br_2(\ell)|Br^-$ half-cell?

37. (See Example 13–7.) The following reaction occurs in an electrochemical cell:

$$6\,HClO(aq) + 2\,Cr^{3+}(aq) + H_2O(\ell) \longrightarrow$$
$$3\,Cl_2(g) + Cr_2O_7^{2-}(aq) + 8\,H^+(aq)$$

(a) Calculate $\Delta\mathcal{E}°$ for this cell at 298.15 K.
(b) At pH 0, with $[Cr_2O_7^{2-}] = 0.80$ M, $P_{Cl_2} = 0.10$ atm, and $[HClO] = 0.20$ M, the cell voltage is 0.351 V (at 298.15 K). Calculate the concentration of $Cr^{3+}(aq)$ in the cell.

38. (See Example 13–7.) A galvanic cell is constructed in which the overall reaction is

$$Cr_2O_7^{2-}(aq) + 14\,H^+(aq) + 6\,I^-(aq) \longrightarrow$$
$$2\,Cr^{3+}(aq) + 3\,I_2(s) + 7\,H_2O(\ell)$$

(a) Calculate $\Delta\mathcal{E}°$ for this cell.
(b) The cell voltage at 25°C equals 0.772 V, with $[Cr_2O_7^{2-}] = 1.5$ M, $[I^-] = 0.40$ M, and $[H^+] = 1$ M. Calculate the concentration of $Cr^{3+}(aq)$ in the cell.

39. The following silver-cadmium cell is constructed:

$$Cd(s)|Cd^{2+}(aq, 0.75\text{ M})||Ag^+(aq, 0.0010\text{ M})|Ag(s)$$

(a) Compute the potential difference $(\Delta\mathcal{E})$ at 298.15 K. Take necessary data from Appendix E or Table 13–1.
(b) The standard potential difference for the cell reaction (its $\Delta\mathcal{E}°$) equals 1.177 V at 323.15 K. This differs from $\Delta\mathcal{E}°$ at 298.15 K. Compute the potential difference $(\Delta\mathcal{E})$ of the above cell at 323.15 K.

40. The following lanthanum-cerium cell is constructed:

$$La(s)|La^{3+}(aq, 1.000\text{ M})||Ce^{3+}(aq, 0.0150\text{ M})|Ce(s)$$

(a) Compute the potential difference of this cell at 298.15 K. Take necessary data from Appendix E.
(b) The standard potential difference $\Delta\mathcal{E}°$ of this particular cell is quite insensitive to changes in temperature, but the observed potential difference $\Delta\mathcal{E}$ is not. Compute the temperature at which $\Delta\mathcal{E}$ equals 0.00 V.

Equilibrium Constants from Electrochemistry

41. (See Example 13–8.) A galvanic cell consists of a silver electrode dipping into a 0.10 M $AgNO_3$ solution connected through a salt bridge and an external circuit to a nickel electrode dipping into a 0.10 M $Ni(NO_3)_2$ solution.
(a) Write balanced chemical equations for the half-reaction occurring at the anode, the half-reaction occurring at the cathode, and the overall cell reaction.
(b) Calculate the initial cell voltage at 298.15 K.
(c) Calculate the equilibrium constant at 298.15 K for the overall reaction as written in part (a).

42. (See Example 13–8.) A galvanic cell is composed of two half-cells. The first consists of a Pt electrode immersed in a solution of MnO_4^- and Mn^{2+} ions at pH = 0, and the second of a Pt electrode immersed in a solution of Fe^{3+} and Fe^{2+} ions at pH = 0. The concentrations of all these ions are 0.010 mol L^{-1}.
(a) Write balanced chemical equations for the half-reaction occurring at the anode, the half-reaction occurring at the cathode, and the overall cell reaction.
(b) Calculate the initial cell voltage at 25°C.
(c) Calculate the equilibrium constant at 25°C for the overall reaction as written in part (a).

43. (See Example 13–8.) Consider the reaction

$$3\,HClO_2(aq) + 2\,Cr^{3+}(aq) + 4\,H_2O(\ell) \longrightarrow$$
$$3\,HClO(aq) + Cr_2O_7^{2-}(aq) + 8\,H^+(aq)$$

(a) Use standard reduction potentials in Appendix E to calculate the equilibrium constant of this reaction at 25°C.
(b) Dichromate ion $(Cr_2O_7^{2-})$ is orange, and $Cr^{3+}(aq)$ ion is pink. If 2.00 L of 1.00 M $HClO_2$ solution is added to 2.00 L of 0.50 M $Cr^{3+}(aq)$ and the reaction goes to equilibrium, what color is the solution?

44. (See Example 13–8.) Consider the reaction

$$6\,Hg^{2+}(aq) + 2\,Au(s) \longrightarrow 3\,Hg_2^{2+}(aq) + 2\,Au^{3+}(aq)$$

(a) Use standard reduction potentials in Appendix E to calculate the equilibrium constant of this reaction at 25°C.
(b) If 1.00 L of a 1.00 M $Au(NO_3)_3$ solution is added to 1.00 L of a 1.00 M $Hg_2(NO_3)_2$ solution, calculate the concentrations of Hg^{2+}, Hg_2^{2+}, and Au^{3+} at equilibrium at 25°C.

45. The following standard reduction potentials at 25°C have been determined for the aqueous chemistry of indium:

$$In^{3+}(aq) + 2\,e^- \longrightarrow In^+(aq) \qquad \mathcal{E}° = -0.40\text{ V}$$
$$In^+(aq) + e^- \longrightarrow In(s) \qquad \mathcal{E}° = -0.21\text{ V}$$

Calculate the equilibrium constant (K) for the disproportionation of $In^+(aq)$ at 25°C. The equation is

$$3\,In^+(aq) \longrightarrow 2\,In(s) + In^{3+}(aq)$$

46. (a) Use data from Appendix E to compute the equilibrium constant at 25°C for the reaction

$$Hg^{2+}(aq) + Hg(\ell) \rightleftharpoons Hg_2^{2+}(aq)$$

(b) Ammonia reacts with an aqueous solution of Hg_2Cl_2 to form the white, very insoluble compound $HgNH_2Cl$ (see Section 9–6). Apply Le Châtelier's principle to the above reaction to explain why the addition of ammonia to such a solution also always produces black elemental mercury that mixes with the white $HgNH_2Cl$ to give a gray precipitate.

47. A galvanic cell is made up of a standard $Pt(s)|H^+(aq)|H_2(g)$ cathode connected to a non-standard $Pt(s)|H_2(g)|H^+(aq)$ anode in which the concentration of H^+ is unknown but is kept constant by the action of a buffer consisting of a weak acid

HA (0.10 M) mixed with its conjugate base A$^-$(aq, 0.10 M). The measured cell voltage is $\Delta\mathcal{E} = 0.150$ V at 25°C, with a hydrogen pressure of 1.00 atm at both electrodes. Calculate the pH in the buffer solution, and from it, determine the K_a of the weak acid. (*Hint:* Because the same species are taking part in the reactions at anode and cathode, the standard potential difference of the cell $\Delta\mathcal{E}°$ is 0.)

48. The overall cell reaction in a galvanic cell at 25°C is

$$Ag^+(aq) + \tfrac{1}{2} H_2(g) \longrightarrow Ag(s) + H^+(aq)$$

The reduction occurs in a standard Ag$^+$|Ag half-cell, and the oxidation occurs at a platinum wire that has hydrogen bubbling over it at 1.00 atm pressure and is immersed in a buffer solution containing benzoic acid and sodium benzoate. The concentration of benzoic acid (C$_6$H$_5$COOH) is 0.10 M, and the concentration of benzoate ion (C$_6$H$_5$COO$^-$) is 0.050 M. The measured cell voltage is 1.030 V. Calculate the pH in the buffer solution and determine the K_a of benzoic acid.

49. (See Example 13–9.) A galvanic cell runs the overall reaction

$$Br_2(\ell) + H_2(g) \longrightarrow 2\,Br^-(aq) + 2\,H^+(aq)$$

(a) Calculate $\Delta\mathcal{E}°$ for this cell at 25°C.
(b) Silver ions are added to the cathode compartment until AgBr precipitates and [Ag$^+$] reaches 0.060 M. The cell voltage is 1.710 V, the pH in the anode compartment is 0, and P_{H_2} in the anode compartment is 1.0 atm. Calculate the solubility-product constant (K_{sp}) for AgBr at 25°C.

50. (See Example 13–9.) A galvanic cell is constructed in which the overall reaction is

$$Pb(s) + 2\,H^+(aq) \longrightarrow Pb^{2+}(aq) + H_2(g)$$

(a) Calculate $\Delta\mathcal{E}°$ for this cell at 25°C.
(b) Chloride ions are added until PbCl$_2$(s) precipitates in the anode compartment and [Cl$^-$] reaches 0.15 M. At that point, the cell voltage is 0.22 V, the pH is 0, and $P_{H_2} = 1.0$ atm. Calculate [Pb^{2+}] under these conditions.
(c) Estimate the solubility-product constant K_{sp} of PbCl$_2$.

Batteries and Fuel Cells

51. Calculate the standard voltage $\Delta\mathcal{E}°$ of a lead–acid cell at 25°C. What is the voltage if six such cells are connected in series?

52. Calculate the standard voltage at 25°C of the zinc–mercuric oxide cell shown in Figure 13–14. Take $\Delta G_f°(\text{Zn(OH)}_2(s)) = -553.5$ kJ mol^{-1}. (*Hint:* Try calculating $\Delta G_r°$ for the overall reaction and then finding $\Delta\mathcal{E}°$ from it.)

53. What quantity of charge (in coulombs) is a fully charged 2.04 V lead–acid cell theoretically capable of furnishing if the spongy lead available for reaction at the anodes weighs 10 kg and there is excess PbO$_2$?

54. What quantity of charge (in coulombs) is a fully charged 1.34 V zinc–mercuric oxide watch battery theoretically capable of furnishing through an external circuit if the mass of HgO in the battery is 0.50 g?

55. What is the theoretical maximum amount of work (in joules) that can be obtained from the lead–acid storage cell in Problem 53?

56. What is the theoretical maximum amount of work (in joules) that can be obtained from the watch battery in Problem 54?

57. The concentration of the electrolyte, sulfuric acid, in a lead–acid storage battery diminishes as it is discharged. Is a discharged battery recharged by replacing the dilute H$_2$SO$_4$ with fresh, concentrated H$_2$SO$_4$? Explain.

58. One cold winter morning, the temperature is well below 0°F. In a futile effort to start your car, you run the battery down completely. Several hours later, you return to replace your fouled spark plugs and discover that the liquid in the battery has now frozen even though the temperature is actually a bit higher than it was in the morning. Explain how this can happen.

59. Explain why alkaline batteries produce a steadier voltage than dry cells do.

60. Dry cells and alkaline batteries cannot be recharged. Why not?

61. List two problems with the use of nicad batteries.

62. Why happens to prevent a lead–acid storage battery from undergoing an unlimited number of discharge–recharge cycles?

63. Consider a fuel cell that accomplishes the overall reaction

$$H_2(g) + \tfrac{1}{2}O_2(g) \longrightarrow H_2O(\ell)$$

If the cell operates with 60% efficiency, calculate the amount of electrical work generated per gram of water produced. The gas pressures are constant at 1 atm, and the temperature is 25°C.

64. Consider a fuel cell that accomplishes the overall reaction

$$CO(g) + \tfrac{1}{2}O_2(g) \longrightarrow CO_2(g)$$

Calculate the amount of electrical work that could be obtained from the conversion of 1.00 mol of CO(g) to CO$_2$(g) in such a fuel cell if it operates with 100% efficiency at 25°C and if the pressure of each gas equals 1 atm. (*Hint:* In the absence of half-cell voltages from Appendix E, use Appendix D to calculate $\Delta G_r°$ and relate that to $w_{elec,\,max}$.)

Corrosion and Its Prevention

65. Two half-reactions proposed for the corrosion of iron in the absence of oxygen are

$$\begin{aligned}
Fe(s) &\longrightarrow Fe^{2+}(aq) + 2\,e^- & \text{(oxidation)} \\
2\,H_2O(\ell) + 2\,e^- &\longrightarrow 2\,OH^-(aq) + H_2(g) & \text{(reduction)}
\end{aligned}$$

Calculate the standard voltage generated by a galvanic cell running this pair of half-reactions at 25°C. Is the overall reaction spontaneous? As the pH is decreased from 14, does the reaction become spontaneous?

66. In the presence of oxygen, the reduction half-reaction in the preceding problem is replaced by

$$\tfrac{1}{2}O_2(g) + 2\,H^+(aq) + 2\,e^- \longrightarrow H_2O(\ell)$$

but the anode half-reaction is unchanged. Calculate the standard cell voltage for *this* pair of half-reactions running in a galvanic cell. Is the overall reaction spontaneous when the reactants and products are present in standard states at 25°C? As the water becomes more acidic, does the driving force for the rusting of iron increase or decrease?

67. Could sodium be used as a sacrificial anode to protect the iron hull of a ship? Explain.

68. If it is shown that titanium can be used as a sacrificial anode to protect iron, what conclusion can be drawn about the standard reduction potential of its half-reaction?

$$Ti^{3+}(aq) + 3\,e^- \longrightarrow Ti(s)$$

69. You invent a galvanic cell that employs a catalyst modeled on biomolecules isolated from methanogenic bacteria to run the following reaction smoothly and cleanly:

$$4\,H_2(g) + CO_2(g) \longrightarrow CH_4(g) + 2\,H_2O(\ell)$$

(a) Compute the standard voltage of your cell at 298.15 K.
(b) Determine the expected potential difference of your cell if design considerations require you to supply $H_2(g)$ and $CO_2(g)$ at partial pressures of 0.50 atm and to take off $CH_4(g)$ at a pressure of 1 atm.

*70. Use data from Appendix E and the discussion of biocorrosion in Section 13–7 to obtain the standard reduction potential for the half-reaction

$$8\,H^+(aq) + CO_2(g) + 8\,e^- \longrightarrow CH_4(g) + 2\,H_2O(\ell)$$

71. (See Chemistry in Progress.) How is corrosion being used to clean up groundwater contaminated with chlorinated hydrocarbons?

72. (See Chemistry in Progress.) How might corrosion of iron help in cleaning up wastes that contain compounds of chromium in the +6 oxidation state?

ADDITIONAL PROBLEMS

73. Estimate the cost of the electrical energy needed to produce 2.1×10^{10} kg (a year's supply for the world) of aluminum from $Al_2O_3(s)$ if electrical energy costs 10 cents per kilowatt-hour (1 kW-h = 3.6 MJ = 3.6×10^6 J) and if the cell voltage is 5 V.

74. Sheet iron can be galvanized by passing a direct current through a cell containing a solution of zinc sulfate between a graphite anode and the iron sheet. Zinc plates out on the iron. The process can be made continuous if the iron sheet is a coil that unwinds as it passes through the electrolysis cell and coils up again after it emerges from a rinse bath. Calculate the cost of the electricity needed to deposit a 0.250 mm thickness of zinc on both sides of an iron sheet that is 1.00 m wide and 100 m long if a current of 25 A at a voltage of 3.5 V is used and the energy efficiency of the process is 90%. The cost of electricity is 10 cents per kilo-

watt-hour (1 kW-h = 3.6 MJ), and the density of zinc is 7.133 g cm^{-3}.

75. A half-cell has a graphite electrode immersed in an acidic solution (pH 0) of Mn^{2+} (concentration 1.00 M) in contact with solid MnO_2. A second half-cell has an acidic solution (pH 0) of H_2O_2 (concentration 1.00 M) in contact with a platinum electrode past which gaseous oxygen at a pressure of 1.00 atm is bubbled. The two half-cells are connected to form a galvanic cell.

(a) By referring to Appendix E, write balanced equations for the half-reactions at the anode and the cathode and for the overall cell reaction.
(b) Calculate the cell voltage at 298.15 K.

76. (a) Based only on the standard reduction potentials for the Cu^{2+}, Cu^+ and the $I_2(s)|I^-$ half-reactions, would you expect $Cu^{2+}(aq)$ to be reduced to $Cu^+(aq)$ by $I^-(aq)$?
(b) The formation of $CuI(s)$ plays a strong role in the interaction between $Cu^{2+}(aq)$ and $I^-(aq)$.

$$Cu^{2+}(aq) + I^-(aq) + e^- \longrightarrow CuI(s) \qquad \mathcal{E}° = 0.86 \text{ V}$$

Taking into account this added information, do you expect $Cu^{2+}(aq)$ to be reduced by $I^-(aq)$?

77. Calculate the minimum external voltage that must be applied to produce $Ni(s)$ and $I_2(s)$ in an electrolytic cell that is 1.00 M in $NiSO_4$ and 1.00 M in NaI. Write balanced chemical equations for the half-reactions that take place at the anode and at the cathode.

*78. In an old European church, the stained-glass windows have experienced such extreme darkening from corrosion that hardly any light comes through. A microprobe analysis shows that tiny cracks and defects on the glass surface are enriched in insoluble Mn(III) and Mn(IV) compounds. From Appendix E, suggest a reducing agent and conditions that might successfully convert these compounds to soluble Mn(II) without simultaneously reducing Fe(III), which colors the glass, to Fe(II). Take MnO_2 as representative of the insoluble Mn(III) and Mn(IV) compounds.

79. By referring to the half-equations in Problems 23 and 24, decide whether $PH_3(g)$ or $HS^-(aq)$ is the stronger reducing agent in basic aqueous solution.

80. (a) Calculate the half-cell potential for the reaction

$$O_2(g) + 4\,H^+(aq) + 4\,e^- \longrightarrow 2\,H_2O(\ell)$$

at pH 7 with the oxygen pressure at 1 atm.
(b) Explain why aeration of solutions of $I^-(aq)$ leads to oxidation of the $I^-(aq)$ ion. Write a balanced equation for the redox reaction that occurs.
(c) Does the same problem arise with solutions containing Br^- or Cl^-? Explain.
(d) Is oxidation of the halide ions favored or opposed by increasing acidity?

81. An engineer needs to prepare a galvanic cell that uses the reaction

$$2\,Ag^+(aq) + Zn(s) \longrightarrow Zn^{2+}(aq) + 2\,Ag(s)$$

and generates an initial voltage of 1.50 V. She has 0.010 M $AgNO_3(aq)$ and 0.100 M $Zn(NO_3)_2(aq)$ solutions as well as electrodes of zinc and silver, wires, containers, water, and a KNO_3 salt bridge. Sketch the cell. Clearly indicate the concentrations of all solutions.

82. Consider a galvanic cell for which the anode reaction is

$$Pb(s) \longrightarrow Pb^{2+}(aq,\,0.010\;M) + 2\,e^-$$

and the cathode reaction is

$$VO^{2+}(aq,\,0.10\;M) + 2\,H^+(aq,\,0.10\;M) + e^- \longrightarrow$$
$$V^{3+}(aq,\,1.0 \times 10^{-5}\;M) + H_2O(\ell)$$

The measured cell voltage is 0.640 V.
(a) Calculate $\mathcal{E}°$ for the (VO^{2+}, V^{3+}) half-cell, taking $\mathcal{E}°$ for the $Pb^{2+}|Pb$ half-cell from Appendix E.
(b) Calculate the equilibrium constant (K) at 25°C for the reaction

$$Pb(s) + 2\,VO^{2+}(aq) + 4\,H^+(aq) \longrightarrow$$
$$Pb^{2+}(aq) + 2\,V^{3+}(aq) + 2\,H_2O(\ell)$$

83. A galvanic cell operating at 25°C is made up of two half-cells. The first half-cell contains a Pb rod immersed in a 0.10 M $Pb(NO_3)_2$ solution. The second half-cell contains a Pb rod immersed in a 0.010 M $Pb(NO_3)_2$ solution. A salt bridge connects the two half-cells.
(a) Calculate the cell voltage at 25°C.
(b) Concentrated H_2SO_4 is added to the first half-cell, causing insoluble lead sulfate ($K_{sp} = 1.1 \times 10^{-8}$ at 25°C) to precipitate. The equilibrium concentration of SO_4^{2-} ion is 0.10 M. Calculate the cell voltage after the addition of the H_2SO_4.

84. Write a balanced equation for the reduction of $Cu^{2+}(aq)$ by $I^-(aq)$ ion, based on the discussion in Problem 76. Calculate the equilibrium constant for this reaction at 25°C.

85. A wire is fastened across the terminals of the Leclanché cell in Figure 13–13. Indicate in which direction the electrons flow in the wire.

86. The zinc–silver cell is used in miniature "button" batteries for a variety of specialized uses. The cell is represented

$$Zn(s)|ZnO(s)|KOH(aq)|Ag_2O(s)|Ag(s)$$

The KOH electrolyte is held on an absorbent disc between the two electrodes (as shown in Figure 13–14); hence, no salt bridge is needed in this cell.
(a) Write half-equations for the oxidation and reduction taking place in this cell.
(b) What is the overall reaction taking place in this cell?
(c) The $\Delta G_f°$ of $ZnO(s)$ is -318.32 kJ mol^{-1}, and the $\Delta G_f°$ of $Ag_2O(s)$ is -10.84 kJ mol^{-1} at 25°C. Calculate $\Delta\mathcal{E}°$ for the zinc–silver cell at this temperature.

87. As a junior executive trainee employed by the Bunny Battery Company, you are assigned to lead a group of important vis-

itors on a tour of the factory. After seeing every stage of the fabrication process, one visitor asks, "Why is there no provision for charging up the batteries before they are shipped? When do you put the electricity into them?" Write a brief statement explaining to the visitor exactly why such a procedure is unnecessary and telling where the energy in Bunny batteries comes from.

88. Overcharging a lead–acid storage battery can generate hydrogen. Write a balanced equation to represent the reaction that takes place.

89. The rights to manufacture a new fuel cell in which hydrazine is oxidized to nitrogen and water according to the equation

$$N_2H_4(aq) + O_2(g) \longrightarrow N_2(g) + 2\,H_2O(\ell)$$

are for sale. The sellers point out that the products of the reaction are completely innocuous in the environment. Your firm asks you to evaluate this new invention. As a start, check whether the cell is thermodynamically feasible. Do this by using data from Appendix D to compute its $\Delta\mathcal{E}°$.

*90. The following reaction is carried out in a fuel cell at 1200 K:

$$CO(g) + \tfrac{1}{2}\,O_2(g) \longrightarrow CO_2(g)$$

The electrolyte is molten Na_2CO_3, and the reactant and product gases are all at 1.0 atm pressure. Compute $\Delta G_r°$ at this temperature (use Appendix D, and assume $\Delta H_r°$ and $\Delta S_r°$ are independent of temperature). What voltage should the fuel cell produce?

91. Iron or steel is often covered by a thin layer of a second metal to prevent rusting: Tin cans consist of steel covered with tin, and galvanized iron is made by coating iron with a layer of zinc. If the protective layer is broken, however, iron rusts more readily in a tin can than in galvanized iron. Explain this observation by comparing the reduction potentials of iron, tin, and zinc.

92. The Bobay process is used to produce $Cl_2(g)$ by the direct electrolysis of HCl dissolved in water. This reaction requires a potential difference of 1.4 V when the concentration of the HCl is about 1 M. A new process replaces water as the solvent with a molten mixture of KCl (59 mol %) and LiCl (41 mol %) at 400°C. Hydrogen chloride dissolves without dissociation in this mixture, and applying a potential difference of about 1 V generates gaseous Cl_2 and H_2 at the electrodes.
(a) Write half-equations for the chemical events at the anode and at the cathode in both the original and the new process. Take care to show the state of dissolution of each reactant and product.
(b) Estimate the energy savings per mole of Cl_2.
(c) A savings in energy is one reason the new process is deemed an improvement, despite requiring an elevated temperature. Suggest at least one other reason.
(d) Hydrogen chloride is soluble in both molten lithium chloride ($LiCl(\ell)$) and potassium chloride ($KCl(\ell)$). Explain why a mixture is used.

93. A reference book lists the following standard reduction potentials:

$$AgCl(s) + e^- \longrightarrow Ag(s) + Cl^-(aq) \qquad \mathcal{E}° = 0.222 \text{ V}$$
$$AgBr(s) + e^- \longrightarrow Ag(s) + Br^-(aq) \qquad \mathcal{E}° = 0.071 \text{ V}$$
$$AgI(s) + e^- \longrightarrow Ag(s) + I^-(aq) \qquad \mathcal{E}° = -0.152 \text{ V}$$
$$AgAt(s) + e^- \longrightarrow Ag(s) + At^-(aq) \qquad \mathcal{E}° = 0.64 \text{ V}$$

Discuss the trend in K_{sp} of the silver salts of the Group VII elements.

94. Arrange the compounds Hg_2Cl_2, Hg_2Br_2, and Hg_2I_2 in order of increasing solubility in water using the following data:

$$Hg_2Cl_2(s) + 2 e^- \longrightarrow 2 Hg(\ell) + 2 Cl^-(aq)$$
$$\mathcal{E}° = 0.268 \text{ V}$$
$$Hg_2Br_2(s) + 2 e^- \longrightarrow 2 Hg(\ell) + 2 Br^-(aq)$$
$$\mathcal{E}° = 0.134 \text{ V}$$

$$Hg_2I_2(s) + 2 e^- \longrightarrow 2 Hg(\ell) + 2 I^-(aq)$$
$$\mathcal{E}° = -0.040 \text{ V}$$

95. Use the information in the first two entries in the table in Appendix E to compute K_a for hydrofluoric acid (HF) at 25°C.

*96. A galvanic cell combines these half-cells: a silver wire in contact with an equilibrium mixture of equal volumes of 0.10 M $AgNO_3$ and 3.0 M NH_3; a silver wire in contact with 0.10 M aqueous $AgNO_3$. The cell develops a voltage of 0.56 V at 25°C. Determine K for the reaction

$$Ag^+(aq) + 2 NH_3(aq) \rightleftharpoons Ag(NH_3)_2^+(aq)$$

CUMULATIVE PROBLEM

Manganese

Manganese is the 12th most abundant element on the Earth's surface. Its most important ore is pyrolusite (MnO_2). The preparation and uses of manganese and its compounds are intimately bound up with electrochemistry.

Black crystals of manganite, MnOOH, an ore of manganese. (*Biophoto Associates/Photo Researchers, Inc.*)

(a) Elemental manganese in a state of high purity can be prepared by electrolyzing aqueous solutions of Mn^{2+}. At which electrode (anode or cathode) does the manganese appear? Electrolysis also is used to make MnO_2 in high purity from Mn^{2+} solutions. At which electrode does the MnO_2 appear?

(b) The Winkler method is an analytical procedure for determining the amount of oxygen dissolved in water. In the first step, $Mn(OH)_2(s)$ is oxidized by dissolved oxygen to $Mn(OH)_3(s)$ in basic aqueous solution. Write the oxidation and reduction half-equations for this step, and write the balanced overall equation. Then use Appendix E to calculate the standard voltage that would be measured if this reaction were carried out in a galvanic cell at 25°C.

(c) Calculate the equilibrium constant at 25°C for the reaction in part (b).

(d) In the second step of the Winkler method, the $Mn(OH)_3(s)$ is acidified to give $Mn^{3+}(aq)$, and iodide ion is added. Does $Mn^{3+}(aq)$ spontaneously oxidize $I^-(aq)$? Write a balanced equation for its reaction with $I^-(aq)$, and use data from Appendix E to calculate the equilibrium constant of the reaction at 25°C. Titration of the I_2 produced completes the use of the Winkler method.

(e) Manganese(IV) oxide is an even stronger oxidizing agent than $Mn(OH)_3$. It oxidizes zinc in the "dry cell," or flashlight battery, found in every home. Such a battery has a cell voltage of 1.5 V. Calculate the electrical work done by a dry cell in 1.00 h if it produces a steady current of 0.70 A at 1.5 V. Express your answer in joules and in kilowatt-hours.

(f) The reduction of permanganate ion (in which Mn has the +7 oxidation state) in acidic aqueous solution is represented

$$MnO_4^-(aq) + 8 H^+(aq) + 5 e^- \longrightarrow$$
$$Mn^{2+}(aq) + 4 H_2O(\ell) \qquad \mathcal{E}° = 1.507 \text{ V at } 298.15 \text{ K}$$

The reduction of the analogous fifth-period species, pertechnetate ion, follows a similar course:

$$TcO_4^-(aq) + 8\,H^+(aq) + 5\,e^- \longrightarrow$$
$$Tc^{2+}(aq) + 4\,H_2O(\ell) \qquad \mathcal{E}° = 0.500\text{ V at }298.15\text{ K}$$

Which is the stronger oxidizing agent: permanganate ion or pertechnetate ion?

(g) A galvanic cell is made from two half-cells. In the first, a platinum electrode is immersed in a solution at pH 2.00 that is 0.100 M in both $MnO_4^-(aq)$ and $Mn^{2+}(aq)$. In the second, a zinc electrode is immersed in a 0.0100 M solution of $Zn(NO_3)_2$. Calculate the theoretical cell voltage at 25°C.

14 Chemical Kinetics

Magnesium ribbon burning in air. *(Charles D. Winters)*

Why do some chemical reactions proceed with lightning speed, while others take days, months, or years? How do changes in temperature affect the rate of chemical change? How can a chemical reaction speed up in the presence of one substance but slow down or stop in the presence of another? Such questions are the province of chemical kinetics.

Chemical kinetics is the study of the rates at which reactions proceed and of the detailed events along the way as reactants are transformed to products. Kinetics is of enormous practical importance. Thus, the rates at which stratospheric ozone-depleting species form, react, and are converted into innocuous species are items of urgent study in atmospheric chemistry.

Questions of rate typically hinge on the **mechanisms** of reactions. These are the detailed pathways by which reactants combine. For example, the redox reaction

$$5\,Fe^{2+}(aq) + MnO_4^-(aq) + 8\,H^+(aq) \longrightarrow 5\,Fe^{3+}(aq) + Mn^{2+}(aq) + 4\,H_2O(\ell)$$

certainly does *not* involve the simultaneous collision of five $Fe^{2+}(aq)$ ions and one $MnO_4^-(aq)$ ion with eight $H^+(aq)$ ions. Such a collision would be exceedingly rare. Instead, it proceeds through a series of elementary steps, each of which involves a collision of two or, at most, three species. Early steps in the series give intermediates that later react further to generate the final products. A major goal of chemical kinetics is to deduce the mechanisms of reactions from experimental knowledge of their rates under different conditions.

14-1 Rates of Chemical Reactions

Numerous factors influence the rates of chemical reactions, and competing influences play out in complex ways. For example, the majority of reactions slow down as they progress, but such behavior is by no means universal. Numerous reactions start slowly but then pick up speed as they go along. Temperature almost always exerts a strong effect on rates, a fact that makes careful monitoring and control of temperature critical for quantitative measurements in chemical kinetics. Increasing the concentration of a reactant often speeds up a reaction (Fig. 14–1) but sometimes has no effect and can even slow the reaction. The concentrations of non-reactants (including reaction products) also influence rates, sometimes profoundly. Rates often depend on the physical form of the reactants. A heated steel nail oxidizes only very slowly in dry air to iron oxide, but heated steel wool burns spectacularly in oxygen (Fig. 14–2).

Measuring Reaction Rates

A kinetics experiment measures how fast the concentration of a substance that is taking part in a chemical reaction changes. Many clever ways to monitor changing concentrations exist, but much scope for ingenuity remains in inventing new ones. If a reaction is slow enough, it can be allowed to proceed for a measured time and then abruptly *quenched* (stopped) by rapid cooling to a low temperature. Subsequent leisurely chemical analysis gives the concentration of a particular reactant or product at the moment of cooling. Quenching is not useful for rapid reactions, especially

Figure 14–1 • The rate of reaction
of zinc with aqueous sulfuric acid depends on the concentration of the acid.
The dilute solution reacts slowly (*left*),
and the more concentrated solution reacts rapidly (*right*). (*Charles Steele*)

Figure 14–1 • The rate of reaction
of zinc with aqueous sulfuric acid depends on the concentration of the acid.
The dilute solution reacts slowly (*left*),
and the more concentrated solution reacts rapidly (*right*). (*Charles Steele*)

those involving gas mixtures, which are difficult to cool quickly. An alternative is
to use the absorption of light as a probe. Substances differ in the wavelengths of
light they absorb. If a reactant (or product) absorbs a particular wavelength without
overlap by other species, then measurements at that wavelength show the changes
in concentration of the reactant over time.

Determining the average rate of a reaction closely resembles determining the
average speed of a traveling car. On a road trip

$$\text{average speed} = \frac{\text{distance traveled}}{\text{elapsed time}} = \frac{\text{change in location}}{\text{change in time}}$$

The average speed has units of miles per hour, meters per second, or any other convenient combination of distance unit divided by time unit. The **average reaction
rate** is obtained by dividing the change in concentration of a reactant or product by
the time interval over which the change occurs:

$$\text{average reaction rate} = \frac{\text{change in concentration}}{\text{change in time}}$$

If concentration is measured in moles per liter (mol L^{-1}) and time in seconds (s),
then the rate of a reaction has the unit $\text{mol L}^{-1}\,\text{s}^{-1}$.

Consider a specific example. In the gas-phase reaction

$$NO_2(g) + CO(g) \longrightarrow NO(g) + CO_2(g)$$

$NO_2(g)$ and $CO(g)$ are consumed as $NO(g)$ and $CO_2(g)$ are produced. Suppose that
a probe monitors the concentration of $NO(g)$ as it is produced. The average rate of
the reaction over any interval of time equals the change in the concentration of
$NO(g)$ (symbolized $\Delta[NO]$) divided by the duration of the interval (Δt):

$$\text{average rate} = \frac{\Delta[NO]}{\Delta t} = \frac{[NO]_f - [NO]_i}{t_f - t_i}$$

This average rate depends on exactly which interval of time is selected because the
rate at which NO is produced changes from one minute to the next, just as the av-

Figure 14–2 • Steel wool burning
in oxygen. (*Leon Lewandowski*)

erage speed of a car on a cross-country trip changes with time. Results like these are obtained when NO_2 and CO are mixed at an elevated temperature and the concentration of NO is monitored over time:

Time (s)	[NO] (mol L^{-1})
0	0
50	0.0155
100	0.0223
150	0.0262
200	0.0287
250	0.0304

The average rate of reaction during the first 50 s is

$$\text{average rate} = \frac{\Delta[NO]}{\Delta t} = \frac{(0.0155 - 0) \text{ mol L}^{-1}}{(50 - 0) \text{ s}} = 31 \times 10^{-5} \text{ mol L}^{-1} \text{ s}^{-1}$$

The average rate during the second 50 s (the interval from 50 s to 100 s) is $16 \times 10^{-5} \text{ mol L}^{-1} \text{ s}^{-1}$, and the average rate during the third 50 s is $5.8 \times 10^{-5} \text{ mol}$ $L^{-1} \text{ s}^{-1}$. Clearly, this reaction slows down as it progresses. The blue curve in Figure 14–3 plots a full profile of the concentration of NO as this reaction proceeds, including all of the information in the preceding table. Thus, the slope of the green line segment in Figure 14–3 equals the average rate of reaction during the first 50 s. Note how this line segment connects the point (50 s, 0.0155 mol L^{-1}) and the point (0 s, 0 mol L^{-1}) on the blue curve.

The **instantaneous rate** of a reaction is its rate at a particular moment in its course. An instantaneous rate is obtained by taking the average rate over shorter and shorter intervals (smaller Δt's) that run from just before until just after the particular instant in question. The trend in these averages is followed to an infinitely brief Δt. Graphically, this is done by drawing a straight line that is tangent to the curve at time t (see the red line segment in Fig. 14–3). The slope of the tangent is the in-

• A traveler may drive 480 mi in 8 h, for an average speed of 60 miles per hour (mph). An 8 h stop then lowers the average speed over the 16 h trip to 30 mph. Another 480 mi in 8 h raises the average speed on the trip to 40 mph (960 mi in 24 h).

• A focus on the relative magnitude of these rates (by thinking of them as "31," "16," and "6") helps to show what is going on.

• A tangent line is a straight line that touches a curve at one point without crossing it.

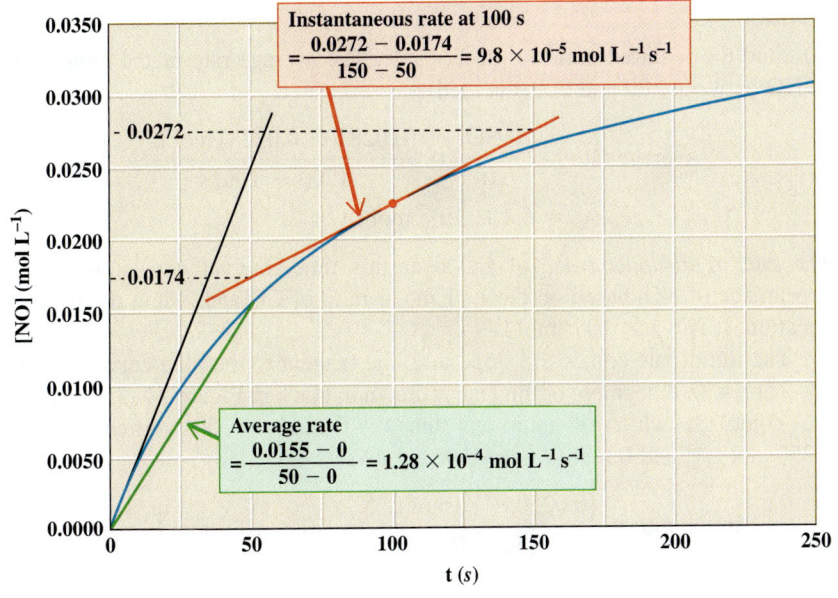

Figure 14–3 • The blue curve plots the concentration of NO as a function of time in the reaction $NO_2 + CO \rightarrow NO + CO_2$. The slope of the green line gives the *average rate* of the reaction over the time interval from 0 s to 50 s. It equals the change in NO concentration divided by the duration of the interval. The red line is tangent to the blue curve at $t = 100$ s. Its slope, which is calculated in the red box, is the *instantaneous rate* 100 s after the start of the reaction. An instantaneous rate exists at any moment in the progress of the reaction. The slope of the black line gives the *initial rate*, which is the rate at the instant the reaction starts (see Example 14–1).

stantaneous rate at that moment. At time $t = 100$ s, for example, the slope of the tangent line in the figure is

$$\text{slope} = \frac{(0.0272 - 0.0174) \text{ mol L}^{-1}}{(150 - 50) \text{ s}} = 9.8 \times 10^{-5} \text{ mol L}^{-1} \text{ s}^{-1}$$
$$= \text{instantaneous rate}$$

The concentrations used here were read from the graph using the points where the red line intersects the $t = 150$ and $t = 50$ vertical lines; any pair of points on the red line could have been used (see Appendix C–2).

Chemical kinetics uses mainly instantaneous rates. Therefore, the instantaneous rate is referred to from now on simply as the *rate*. The rate of a reaction at the moment that it begins (at $t = 0$) is often of special interest. It is the (instantaneous) **initial rate** of the reaction.

The rate of this reaction could just as well have been measured by monitoring changes in the concentrations of $CO_2(g)$, $NO_2(g)$, or $CO(g)$. The compounds NO and CO_2 are produced in a $1:1$ molar ratio. Consequently, the rate of increase of the concentration of CO_2 must equal that of NO. The concentrations of the two reactants, NO_2 and CO, *decrease* at the same rate that the concentrations of the products increase because their coefficients in the balanced equation also both equal 1. This is summarized as

$$\text{rate} = -\frac{\Delta[NO_2]}{\Delta t} = -\frac{\Delta[CO]}{\Delta t} = \frac{\Delta[NO]}{\Delta t} = \frac{\Delta[CO_2]}{\Delta t}$$

EXAMPLE 14–1

Use data from the table on page 609 and Figure 14–3 to calculate the following:
(a) The average rate of disappearance of CO between $t = 100$ and $t = 200$ s.
(b) The instantaneous rate of appearance of NO at $t = 0$ (the initial rate of the reaction).

Solution

(a) Find the required numbers in the table. The average rate of the *appearance* of NO between 100 and 200 s is then

$$\text{average rate} = \frac{\Delta[NO]}{\Delta t} = \frac{(0.0287 - 0.0223) \text{ mol L}^{-1}}{(200 - 100) \text{ s}}$$
$$= 6.4 \times 10^{-5} \text{ mol L}^{-1} \text{ s}^{-1}$$

The rate of *disappearance* of CO over this time interval equals the rate of appearance of NO, based on the $1:1$ molar ratio of CO and NO in the balanced equation. It is $6.4 \times 10^{-5} \text{ mol L}^{-1} \text{ s}^{-1}$.

(b) The initial rate equals the slope of a line tangent to the blue curve in Figure 14–3 at $t = 0$. A segment of this line is drawn in black in Figure 14–3. Estimating the points at which it intersects the $t = 0$ and $t = 50$ vertical lines as 0.250 mol L^{-1} and 0 mol L^{-1} respectively gives

$$\text{initial rate} = \frac{(0.0250 - 0) \text{ mol L}^{-1}}{(50 - 0) \text{ s}} = 50 \times 10^{-5} \text{ mol L}^{-1} \text{ s}^{-1}$$

Check

The blue curve in Figure 14–3 flattens out moving from left to right. This means that the reaction slows down as it proceeds. The initial rate therefore should *exceed* all other average and instantaneous rates calculated in the discussion, but it should not differ from them grossly. The answer is satisfactory on both points.

EXERCISE

(a) Use the same data to calculate the average rate of appearance of CO_2 between $t = 50$ s and $t = 150$ s. (b) Use the graph (Figure 14–3) to estimate the instantaneous rate of the reaction at $t = 50$ s.

Answer: (a) 11×10^{-5} mol L^{-1} s^{-1}; (b) 19×10^{-5} mol L^{-1} s^{-1}.

The chemical equation

$$2 \, NO_2(g) + F_2(g) \longrightarrow 2 \, NO_2F(g)$$

contains coefficients that differ from 1. According to this equation, 2 mol of NO_2 disappear and 2 mol of NO_2F appear for each mole of F_2 that reacts. The concentration of NO_2 changes twice as fast as the concentration of F_2. The concentration of NO_2F also changes twice as fast as that of F_2, and the change has the opposite sign. Such complications are dealt with by defining a single rate for this reaction, as follows:

$$\text{rate} = -\left(\frac{1}{2}\right)\frac{\Delta[NO_2]}{\Delta t} = -\frac{\Delta[F_2]}{\Delta t} = \left(\frac{1}{2}\right)\frac{\Delta[NO_2F]}{\Delta t}$$

Here, the rate of change of the concentration of each species is divided by its coefficient in the balanced chemical equation. Rates of change of reactants appear with negative signs, and rates of change of products, with positive signs. When this definition is used, the rate of the reaction comes out the same regardless of which reactant or product is being monitored. For the general reaction

$$aA + bB \longrightarrow cC + dD$$

the rate is

$$\text{rate} = -\frac{1}{a}\frac{\Delta[A]}{\Delta t} = -\frac{1}{b}\frac{\Delta[B]}{\Delta t} = \frac{1}{c}\frac{\Delta[C]}{\Delta t} = \frac{1}{d}\frac{\Delta[D]}{\Delta t}$$

These relations hold true provided that there are no transient intermediate species or, if there are intermediates, their concentrations are independent of time for most of the reaction period.

14–2 Reaction Rates and Concentrations

In general, both forward and reverse reactions can occur in a chemical system: Once products are formed, they can react to give back the original reactants. The net rate of a chemical reaction equals the difference between the forward and reverse rates:

$$\text{net rate} = \text{forward rate} - \text{reverse rate}$$

Strictly speaking, measurements of concentration versus time give the net rate rather than simply the forward rate. Near the beginning of a reaction that starts from pure reactants, however, the concentrations of reactants are far higher than the concentrations of products, and the reverse rate can be neglected. Furthermore, many reactions go to completion (have very large equilibrium constants) and so have a measurable rate only in the forward direction, or else the experiment can be arranged in such a way that the products are removed as fast as they are formed. In this section, the focus is on forward rates exclusively.

Order of a Reaction

- The forward rate may depend on the concentrations of non-reactants as well.

The forward rate of a homogeneous chemical reaction nearly always depends on the concentrations of one (or more) of the reactants. As an example, consider the decomposition of dinitrogen pentaoxide (N_2O_5). This compound is a white solid that is stable below 0°C but decomposes when vaporized:

$$N_2O_5(g) \longrightarrow 2\,NO_2(g) + \tfrac{1}{2}\,O_2(g)$$

The rate of the decomposition depends on the concentration of $N_2O_5(g)$. In fact, as Figure 14–4 shows, a graph of the rate versus the concentration of $N_2O_5(g)$ is a straight line that passes through the origin when extrapolated to zero concentration. The equation of the straight line is

$$\text{rate} = k[N_2O_5]$$

- The distinction between the *rate* of a reaction and the *rate constant* for that reaction is crucial.

Such a relation between the rate of a reaction and concentration is called a **rate law,** and the proportionality constant (k) is called the **rate constant** (or rate coefficient) for the reaction. Rate constants, like equilibrium constants, are independent of concentration but depend strongly on temperature.

For many (but not all) reactions with a single reactant, the rate is proportional to the concentration of that reactant raised to a power; that is, the rate law for

$$aA \longrightarrow \text{products}$$

frequently has the form

$$\text{rate} = k[A]^n$$

The power n in such a rate law is *not,* in general, equal to the coefficient a in the balanced chemical equation. For example, for the decomposition of ethane at high temperatures and low pressures

$$C_2H_6(g) \longrightarrow 2\,CH_3(g)$$

the rate law has the form

$$\text{rate} = k[C_2H_6]^2$$

so that $n = 2$ even though the coefficient of C_2H_6 in the chemical equation is 1.

The power to which the concentration is raised is called the **order** of the reaction with respect to that reactant. Thus, the decomposition of N_2O_5 is **first order** with respect to N_2O_5, whereas the decomposition of C_2H_6 is **second order** with respect to C_2H_6. Some processes are zeroth order over a range of concentrations. Because $[A]^0 = 1$, such reactions have rates that are independent of concentration.

$$\text{rate} = k \qquad \text{(zeroth-order kinetics)}$$

A reaction order does not have to be an integer; fractional powers are sometimes found. At 450 K, the decomposition of acetaldehyde (CH_3CHO) is described by the rate law

$$\text{rate} = k[CH_3CHO]^{3/2}$$

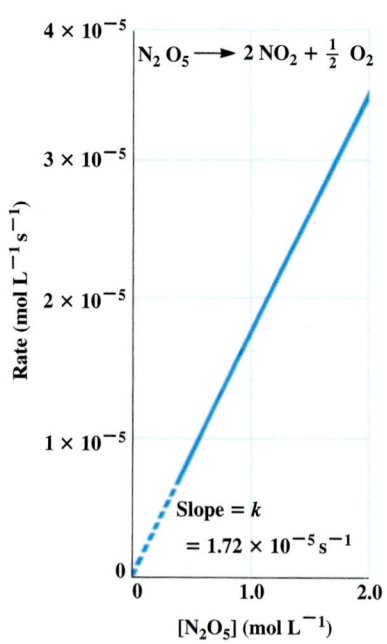

Figure 14–4 • The rate of decomposition of $N_2O_5(g)$ at 25°C is proportional to its concentration. The slope of this line is equal to the rate constant (k) for the reaction.

The following example illustrates the way that the order of a reaction with respect to a reactant is deduced from experimental data.

EXAMPLE 14-2

At high temperatures, HI reacts according to the equation

$$2\,HI(g) \longrightarrow H_2(g) + I_2(g)$$

At 443°C, the initial rate of the reaction increases with the concentration of HI, as shown in the following table:

Experiment No.	[HI] (mol L^{-1})	Initial Rate $(\text{mol L}^{-1}\,\text{s}^{-1})$
1	0.0050	7.5×10^{-4}
2	0.010	3.0×10^{-3}
3	0.020	1.2×10^{-2}

(a) Determine the order of the reaction with respect to HI and write the rate law.
(b) Calculate the rate constant and give its units.
(c) Calculate the instantaneous rate of the reaction when the concentration of HI equals 0.0020 M.

Strategy

Write the rate law for each of two different concentrations, $[HI]_1$ and $[HI]_2$:

$$\text{rate}_1 = k([HI]_1)^n \quad \text{and} \quad \text{rate}_2 = k([HI]_2)^n$$

Divide the second by the first, and obtain the value of n by substituting values from the data table. After n has been evaluated, obtain the value of k by substituting the data from one of the experiments into the rate law.

Solution

When the second equation in the preceding is divided by the first, the rate constant k cancels out, leaving the order n as the only unknown quantity:

$$\frac{\text{rate}_2}{\text{rate}_1} = \left(\frac{[HI]_2}{[HI]_1}\right)^n$$

Putting the numbers from Experiments 1 and 2 into this equation allows solution for n

$$\frac{3.0 \times 10^{-3}}{7.5 \times 10^{-4}} = \left(\frac{0.010}{0.0050}\right)^n$$

which simplifies to

$$4.0 = (2.0)^n$$

Clearly, $n = 2$. The reaction is second order in HI. In a case in which the solution of the equation is less obvious, the proper procedure is to take the logarithm of both sides, giving (in this case)

$$\ln 4.0 = n \ln 2.0$$

$$n = \frac{\ln 4.0}{\ln 2.0} = \frac{1.386}{0.693} = 2$$

The rate law has the form

$$\text{rate} = k[\text{HI}]^2$$

(b) The rate constant (k) can be calculated by inserting any of the sets of data into the preceding equation. Using the data from Experiment 1 gives

$$7.5 \times 10^{-4} \text{ mol L}^{-1} \text{ s}^{-1} = k(0.0050 \text{ mol L}^{-1})^2$$

$$k = \frac{7.5 \times 10^{-4} \text{ mol L}^{-1} \text{ s}^{-1}}{2.5 \times 10^{-5} \text{ mol}^2 \text{ L}^{-2}} = 30 \text{ L mol}^{-1} \text{ s}^{-1}$$

(c) Finally, the instantaneous rate when $[\text{HI}] = 0.0020$ M is computed by substitution in the rate law

$$\text{rate} = k[\text{HI}]^2 = (30 \text{ L mol}^{-1} \text{ s}^{-1})(0.0020 \text{ mol L}^{-1})^2 = 1.2 \times 10^{-4} \text{ mol L}^{-1} \text{ s}^{-1}$$

Check

(a) Duplicate the calculation using the data from Experiments 1 and 3:

$$\frac{1.2 \times 10^{-2}}{7.5 \times 10^{-4}} = \left(\frac{0.020}{0.0050}\right)^n$$

$$16 = (4.0)^n \qquad \text{Again, clearly, } n = 2.$$

(b) Use data from Experiment 2:

$$3.0 \times 10^{-3} \text{ mol L}^{-1} \text{ s}^{-1} = k(0.010 \text{ mol L}^{-1})^2$$

$$k = \frac{3.0 \times 10^{-3} \text{ mol L}^{-1} \text{ s}^{-1}}{1.0 \times 10^{-4} \text{ mol}^2 \text{ L}^{-2}} = 30 \text{ L mol}^{-1} \text{ s}^{-1}$$

The data from Experiment 3 also give $k = 30 \text{ L mol}^{-1} \text{ s}^{-1}$. Such consistency is remarkable. Usually, rate constants obtained from independent experiments differ at least slightly.

EXERCISE

Ammonium cyanate reacts to form urea according to the equation $\text{NH}_4\text{CNO}(aq) \rightarrow \text{NH}_2\text{CONH}_2(aq)$. The initial rate of the reaction is measured at 65°C for three different initial concentrations of ammonium cyanate:

[NH₄CNO] (mol L⁻¹)	Initial Rate (mol L⁻¹ s⁻¹)
0.100	3.60×10^{-2}
0.201	1.44×10^{-1}
0.400	5.75×10^{-1}

(a) Determine the order of the reaction with respect to ammonium cyanate and write the rate law.
(b) Calculate the rate constant, and give its units.
(c) Calculate the initial reaction rate if the concentration of NH_4CNO is 0.0500 M.

Answer: (a) Second order, rate = $k[\text{NH}_4\text{CNO}]^2$; (b) $k = 3.59 \text{ L mol}^{-1} \text{ s}^{-1}$; (c) $9.00 \times 10^{-3} \text{ mol L}^{-1} \text{ s}^{-1}$.

• This is a very famous reaction. In 1828, Wöhler showed that urea prepared in this way is identical to urea from urine. This helped to overthrow "vitalism," the doctrine holding that compounds produced by living organisms differ from synthetic compounds by possessing a "vital force."

Rates That Depend on Two or More Concentrations

In the cases considered so far, each reaction rate depended on only a single concentration. In some reactions, however, the rate depends on the concentrations of two or more different chemical species. When this is so, the rate law has the form

$$\text{rate} = k[A]^m[B]^n$$

with an additional factor put in for every additional substance found to affect the rate. The exponents m and n do not derive from the coefficients that A and B have in a balanced equation representing the reaction. Indeed, A and B are not necessarily even reactants or products in the reaction. Any substance might influence the rate of a particular reaction. Discovering the species that affect the rate requires experimentation; determining the order of the reaction with respect to each such species requires more experimentation.

The exponents m, n..., which are usually integers or half-integers, give the order of the reaction. The preceding reaction is "mth order in A." This means that a change in the concentration of A by a certain factor leads to a change in the rate equal to that factor raised to the mth power. It is nth order in B, and its overall order is $m + n$. Experiment establishes that at room temperature the redox reaction

$$H_2PO_2^-(aq) + OH^-(aq) \longrightarrow HPO_3^{2-}(aq) + H_2(g)$$

follows the rate law

$$\text{rate} = k[H_2PO_2^-][OH^-]^2$$

Therefore, this reaction is first order in $H_2PO_2^-(aq)$, second order in $OH^-(aq)$, and third order overall. The units of the rate constant k depend on the overall reaction order. If all concentrations are expressed in mol L^{-1} and if $p = m + n + \cdots$ is the overall reaction order, then k has units of mol$^{-(p-1)}$ L$^{(p-1)}$ s^{-1}.

EXAMPLE 14–3

Use the rate equation given in the preceding discussion to determine the effect of the following changes on the rate of oxidation of $H_2PO_2^-(aq)$:
(a) Tripling the concentration of $H_2PO_2^-(aq)$ at constant pH.
(b) Changing the pH from 13 to 14 at a constant concentration of $H_2PO_2^-(aq)$.

Solution

(a) Because the reaction is first order in $H_2PO_2^-(aq)$, tripling this concentration causes a tripling of the reaction rate .
(b) Raising the pH from 13 to 14 means increasing the $OH^-(aq)$ concentration by a factor of 10. Because the reaction is second order in $OH^-(aq)$ (that is, this term is squared in the rate law), this causes an increase in reaction rate by a factor of 10^2, or 100 .

EXERCISE

Sucrose breaks down to a mixture of fructose and glucose

$$\text{sucrose}(aq) + H_2O(\ell) \longrightarrow \text{fructose}(aq) + \text{glucose}(aq)$$

in acidic solution. The rate of this reaction is

$$\text{rate} = k[H^+][\text{sucrose}]$$

• This reaction is important in the beverage industry because the products are sweeter than sucrose. Note how the concentration of a non-reactant exerts a strong influence on its rate.

What is the effect on the rate of (a) cutting the concentration of sucrose in half at constant pH; (b) keeping the concentration of sucrose the same but lowering the pH from 1.5 to 0.5?

Answer: (a) Rate is cut in half. (b) Rate increases by a factor of 10.

Rate laws that contain two or more concentrations are more difficult to obtain experimentally than those that contain only one. A good way to proceed is to observe the initial rate of the reaction in a series of experiments in which a single concentration is systematically varied while all the others are held constant. Then a new series is run, changing a different concentration. The following example illustrates this procedure.

EXAMPLE 14–4

Nitrogen monoxide reacts rapidly with oxygen to give nitrogen dioxide:

$$2\,NO(g) + O_2(g) \longrightarrow 2\,NO_2(g)$$

The initial rate of the reaction is measured three times at the same temperature but with different initial concentrations of NO and O_2:

Experiment No.	[NO] (mol L^{-1})	[O$_2$] (mol L^{-1})	Initial Rate (mol L^{-1} s^{-1})
1	1.0×10^{-4}	1.0×10^{-4}	2.8×10^{-6}
2	1.0×10^{-4}	3.0×10^{-4}	8.4×10^{-6}
3	2.0×10^{-4}	3.0×10^{-4}	3.4×10^{-5}

Determine the rate law and the value of the rate constant.

Strategy

The initial concentrations of the NO and O_2 clearly affect the rate. Non-reactants often affect rates as well. Assume that any non-reactants either do not affect the rate or affect it equally in all three experiments. The rate law then has the form

$$\text{rate} = k[O_2]^m[NO]^n$$

where the exponents m and n must be determined. Write such an equation for each experiment. Divide one equation by another to obtain m and n. Compute k by substitution of the concentration and rate data from one of the experiments into the final rate equation.

Solution

$$\text{rate}_1 = k(1.0 \times 10^{-4}\,\text{mol L}^{-1})^m(1.0 \times 10^{-4}\,\text{mol L}^{-1})^n$$
$$\text{rate}_2 = k(3.0 \times 10^{-4}\,\text{mol L}^{-1})^m(1.0 \times 10^{-4}\,\text{mol L}^{-1})^n$$

The concentration of NO is the same in the first and second experiments. It cancels out when the second equation is divided by the first:

$$\frac{\text{rate}_2}{\text{rate}_1} = \frac{8.4 \times 10^{-6}\,\text{mol L}^{-1}\text{s}^{-1}}{2.8 \times 10^{-6}\,\text{mol L}^{-1}\text{s}^{-1}} = 3.0$$

$$= \frac{k(3.0 \times 10^{-4}\,\text{mol L}^{-1})^m(1.0 \times 10^{-4}\,\text{mol L}^{-1})^n}{k(1.0 \times 10^{-4}\,\text{mol L}^{-1})^m(1.0 \times 10^{-4}\,\text{mol L}^{-1})^n}$$

$$3.0 = \left(\frac{3.0 \times 10^{-4}}{1.0 \times 10^{-4}}\right)^m$$

$$3.0 = (3.0)^m$$

$$m = 1.0$$

A similar ratio of the equations for second and third experiments gives the order with respect to NO:

$$\frac{\text{rate}_3}{\text{rate}_2} = \frac{3.4 \times 10^{-5} \text{ mol L}^{-1}\text{s}^{-1}}{8.4 \times 10^{-6} \text{ mol L}^{-1}\text{s}^{-1}} = 4.05$$

$$= \frac{k(3.0 \times 10^{-4} \text{mol L}^{-1})^m (2.0 \times 10^{-4} \text{ mol L}^{-1})^n}{k(3.0 \times 10^{-4} \text{ mol L}^{-1})^m (1.0 \times 10^{-4} \text{ mol L}^{-1})^n}$$

$$4.05 = \left(\frac{2.0 \times 10^{-4}}{1.0 \times 10^{-4}}\right)^n$$

$$4.05 = 2^n$$

$$n = 2.0$$

The rate law is

$$\text{rate} = k[O_2][NO]^2$$

Insert data from the third experiment into the rate law:

$$3.4 \times 10^{-5} \text{ mol L}^{-1} \text{ s}^{-1} = k(3.0 \times 10^{-4} \text{ mol L}^{-1})(2.0 \times 10^{-4} \text{ mol L}^{-1})^2$$

$$k = \frac{3.4 \times 10^{-5} \text{ mol L}^{-1} \text{ s}^{-1}}{(3.0 \times 10^{-4} \text{ mol L}^{-1})(2.0 \times 10^{-4} \text{ mol L}^{-1})^2}$$

$$k = 2.8 \times 10^6 \text{ L}^2 \text{ mol}^{-2} \text{ s}^{-1}$$

Check

The data from the first experiment and second experiment give the same value of k. Note that experimental errors in the determination of rates often prevent such excellent confirmation.

EXERCISE

Ozone (O_3) also reacts with $NO(g)$:

$$NO(g) + O_3(g) \longrightarrow NO_2(g) + O_2(g)$$

The table gives the initial rate observed at different initial concentrations of NO and O_3:

Experiment No.	[NO] (mol L^{-1})	[O$_3$] (mol L^{-1})	Initial Rate (mol L^{-1} s^{-1})
1	1.10×10^{-5}	1.10×10^{-5}	1.32×10^{-3}
2	1.10×10^{-5}	3.40×10^{-5}	4.08×10^{-3}
3	6.00×10^{-5}	3.40×10^{-5}	2.22×10^{-2}

Determine the rate law and the value of the rate constant.

Answer: Rate $= k[NO][O_3]$ and $k = 1.09 \times 10^7$ L mol^{-1} s^{-1}.

14-3   The Dependence of Concentrations on Time

Measuring the initial rate or other instantaneous rate of a chemical reaction involves determining small changes in concentration ($\Delta[A]$) occurring during a short time interval (Δt). Obtaining sufficiently precise experimental data for these small changes often poses a real challenge. An alternative is to fit all the data over a longer time interval with an equation that expresses the concentration of a species directly in terms of the elapsed time. For any given simple rate law, a corresponding equation for the dependence of concentration on time can be obtained.

First-Order Reactions

Consider again the reaction

$$N_2O_5(g) \longrightarrow 2\,NO_2(g) + \tfrac{1}{2}\,O_2(g)$$

The experimental rate law is

$$\text{rate} = -\frac{\Delta[N_2O_5]}{\Delta t} = k[N_2O_5]$$

This is a first-order reaction. The concentration of N_2O_5 dwindles from its original value, symbolized $[N_2O_5]_0$, as time passes. We want an explicit equation that gives $[N_2O_5]$, the "running concentration" of N_2O_5, in terms of the elapsed time (t) and $[N_2O_5]_0$, the original concentration of N_2O_5. Such an equation can be derived using the methods of calculus, and we give only the result here. It is

$$[N_2O_5] = [N_2O_5]_0 e^{-kt}$$

This equation is called the **integrated rate law** for this first-order reaction. Note that the term "rate" does not appear in it. The term "e^{-kt}" is called the "exponential of $-kt$" and is often written "$\exp(-kt)$." Taking the natural logarithm (the "ln") of both sides of the equation and following the rules for the manipulation of logarithms reviewed in Appendix C–1 gives the useful form

$$\ln[N_2O_5] = \ln[N_2O_5]_0 - kt$$

- Recall that "ln" stands for "natural logarithm," 2.303 times larger than the common, or base-10, logarithm. See Appendix C–1.

With either form of the integrated rate law (and knowing the value of k), we can determine the remaining concentration of N_2O_5 at any time after pure N_2O_5 starts to react, or we can compute how long it takes the concentration of N_2O_5 to fall from its original value to any specified value. Similar equations can be written for any first-order reaction:

$$[A] = [A]_0 e^{-kt} \qquad \text{and} \qquad \ln[A] = \ln[A]_0 - kt$$

- The general form of a linear equation is $y = mx + b$ (see Appendix C–2). Here, $\ln[A]$ plays the part of y; $-k$, the part of m; t, the part of x; and $\ln[A]_0$, the part of b.

A plot of the natural logarithm of the concentration $[A]$ against time (t) is a straight line with slope $-k$ and intercept $\ln[A]_0$ (Fig. 14–5).

A useful concept in discussions of first-order reactions is the **half-life** ($t_{1/2}$)—the time it takes for the concentration of a reactant A to fall to one half of its original value. This is, of course, $[A]_0/2$. The preceding equation can be rewritten as

$$\ln[A] - \ln[A]_0 = -kt$$

$$\ln\!\left(\frac{[A]}{[A]_0}\right) = -kt$$

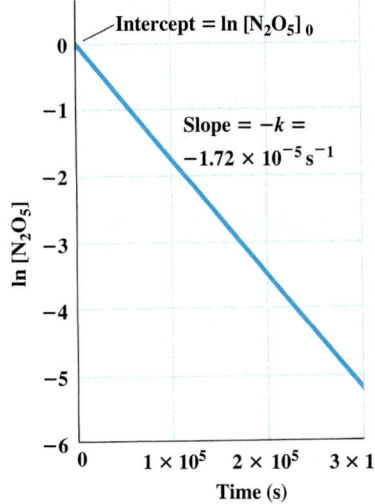

Figure 14-5 • In a first-order reaction such as the decomposition of N_2O_5, a graph of the natural logarithm of the concentration against time is a straight line, the slope of which gives the rate constant for the reaction.

At the moment that t becomes equal to $t_{1/2}$, the concentration [A] has fallen to $[A]_0/2$. Substituting these values:

$$\ln\frac{\left(\dfrac{[A]_0}{2}\right)}{[A]_0} = \ln \tfrac{1}{2} = -\ln 2 = -kt_{1/2}$$

$$t_{1/2} = \frac{\ln 2}{k}$$

If the rate constant of a reaction is given in s^{-1}, then the half-life comes out in seconds. During the first half-life of the reaction, the concentration of A falls to one half of its initial value; in two half-lives, it falls to one quarter (which is $\tfrac{1}{2} \times \tfrac{1}{2}$) of its initial value; in three half-lives, it falls to one eighth (which is $\tfrac{1}{2} \times \tfrac{1}{2} \times \tfrac{1}{2}$), and so forth (Fig. 14–6).

• The concept of a half-life proves useful in the discussion of nuclear decay in Chapter 15.

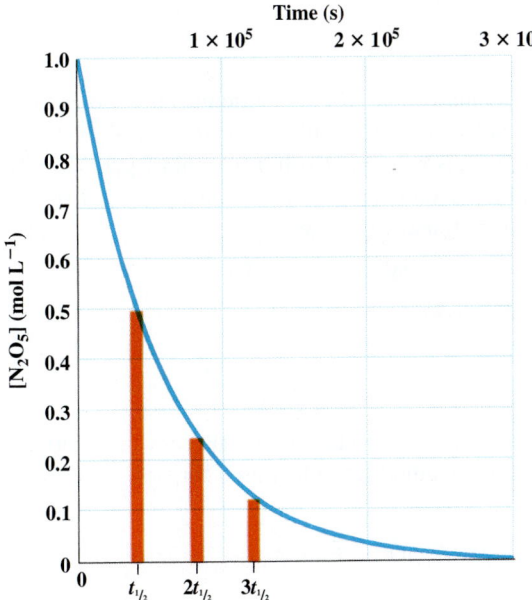

Figure 14-6 • The same data as in Figure 14–5 are graphed in a concentration-versus-time picture. The half-life ($t_{1/2}$) is the time it takes for the initial concentration to be reduced to one half of its original value. In two half-lives, the concentration falls to one quarter of its initial value.

EXAMPLE 14-5

The half-life of $N_2O_5(g)$ at 25°C is 4.03×10^4 s.
(a) Find the rate constant (k) for the first-order decomposition of $N_2O_5(g)$ at 25°C.
(b) Determine the fraction of N_2O_5 molecules that remain unreacted after 24 h.

Solution

(a) Solve the equation that defines the first-order half-life for k and substitute the given half-life:

$$k = \frac{\ln 2}{t_{1/2}} = \frac{0.6931}{4.03 \times 10^4 \text{ s}} = 1.72 \times 10^{-5} \text{ s}^{-1}$$

(b) The explicit equation for the concentration of the reactant as a function of time is

$$[N_2O_5] = [N_2O_5]_0 e^{-kt}$$

Put in the value for k and substitute $t = 8.64 \times 10^4$ s (24 h equals 86,400 s):

$$[N_2O_5] = [N_2O_5]_0 \exp[-(1.72 \times 10^{-5} \text{ s}^{-1})(8.64 \times 10^4 \text{ s})] = [N_2O_5]_0 e^{-1.486}$$

All scientific calculators provide for raising e to a power: Simply enter the power in the display and push the "e^x" button.

$$\frac{[N_2O_5]}{[N_2O_5]_0} = e^{-1.486} = 0.226$$

This means that 22.6% of the N_2O_5 molecules remain unreacted after 24 h at 25°C.

Check

The half-life is about 11 h, so 24 h is somewhat longer than two half-lives. After 11 h, about half of the N_2O_5 remains unreacted. After 22 h, about one quarter (half of half) remains. After 24 h, somewhat less than a quarter should remain, and does.

EXERCISE

Sulfuryl chloride ($SO_2Cl_2(g)$) decomposes at an elevated temperature to $SO_2(g)$ and $Cl_2(g)$. The reaction is first order in SO_2Cl_2 with a half-life of 1.48×10^5 s. (a) Determine the rate constant of this reaction. (b) If 1.00 mol of pure SO_2Cl_2 is sealed in a container at this temperature, how much remains after 1.00 h? Assume that no recombination of SO_2 and Cl_2 occurs.

Answer: (a) $k = 4.68 \times 10^{-6} \text{ s}^{-1}$. (b) 0.983 mol of SO_2Cl_2.

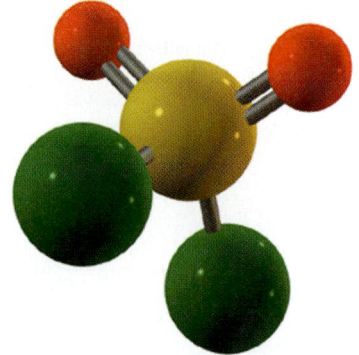

The structure of sulfuryl chloride (SO_2Cl_2). This compound is used as a chlorinating agent.

Second-Order Reactions

Explicit equations giving the concentration of a reactant as time passes can be derived for second-order reactions by the same method (calculus) used for first-order reactions. An example is the reaction

$$2\,NO_2(g) \longrightarrow 2\,NO(g) + O_2(g)$$

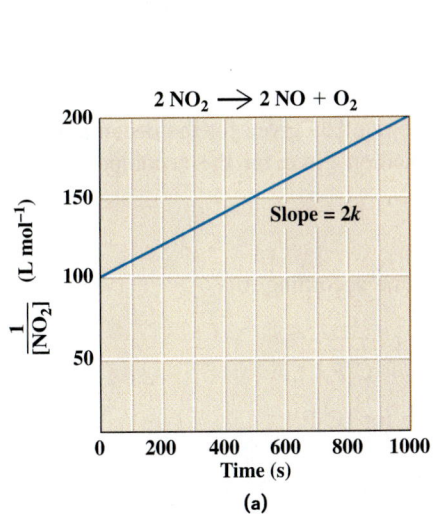

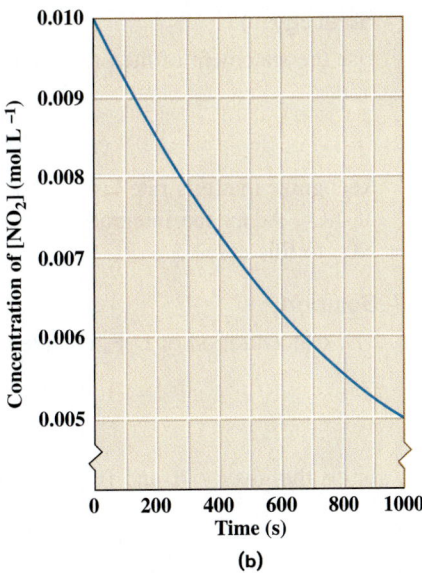

Figure 14–7 • (a) For a second-order reaction such as $2\,NO_2 \rightarrow 2\,NO + O_2$, a graph of the reciprocal of the concentration against time is a straight line with slope $2k$; k in this case equals $0.050\ L\ mol^{-1}s^{-1}$. (b) The same data graphed in a concentration-versus-time picture.

for which the observed rate law is

$$\text{rate} = -\frac{1}{2}\frac{\Delta[NO_2]}{\Delta t} = k[NO_2]^2$$

The desired explicit equation (an integrated rate law) is

$$\frac{1}{[NO_2]} = 2kt + \frac{1}{[NO_2]_0}$$

In this case, plotting the reciprocal of the concentration of NO_2 against t gives a straight line. The straight line has a slope of $2k$ and an intercept of 1 divided by the initial concentration (Fig. 14–7). The factor of 2 in this equation comes from the coefficient 2 in the balanced equation.

The concept of half-life has little use for second-order reactions. Setting $[NO_2]$ equal to $[NO_2]_0/2$ in the preceding equation and solving for t, which is now $t_{1/2}$, is certainly possible:

$$\frac{2}{[NO_2]_0} = 2kt_{1/2} + \frac{1}{[NO_2]_0}$$

$$t_{1/2} = \frac{1}{2k[NO_2]_0}$$

However, the second-order half-life depends on the initial concentration of the NO_2, a property of the experimental set-up, and not just on k, a property of the reaction.

EXAMPLE 14–6

The dimerization of tetrafluoroethylene (C_2F_4) to octafluorocyclobutane (C_4F_8) is second order in the reactant C_2F_4 and second order overall. At 450 K, the rate constant (k) equals $0.0448\ L\ mol^{-1}\ s^{-1}$. If the initial concentration of C_2F_4 is $0.100\ mol\ L^{-1}$, what is its concentration after 203 s?

Strategy

Use the statement of the problem to write the rate law for the dimerization reaction:

$$\text{rate} = -k[C_2F_4]^2$$

Recognize that this rate law is just like the rate law given for the decomposition of NO_2. Adapt the integrated rate law equation given for the decomposition of NO_2 to this case.

Solution

The concentration of C_2F_4 changes with time according to

$$\frac{1}{[C_2F_4]} = 2kt + \frac{1}{[C_2F_4]_0}$$

Insert the given t, k, and $[C_2F_4]_0$ and solve for $[C_2F_4]$:

$$\frac{1}{[C_2F_4]} = 2\left(\frac{0.0448\ \text{L}}{\text{mol s}}\right)(203\ \text{s}) + \frac{1}{0.100\dfrac{\text{mol}}{\text{L}}}$$

$$= 18.2\ \text{L mol}^{-1} + 10.0\ \text{L mol}^{-1}$$

$$= 28.2\ \text{L mol}^{-1}$$

$$[C_2F_4] = \frac{1}{28.2\ \text{L mol}^{-1}} = 3.55 \times 10^{-2}\ \text{mol L}^{-1}$$

EXERCISE

In another dimerization experiment with C_2F_4, all conditions are the same except that the initial concentration of C_2F_4 is 0.200 mol L^{-1}. Compute the concentration of C_2F_4 203 s after the reaction starts.

Answer: 4.31×10^{-2} mol L^{-1}.

In a real experimental study, the reaction order may be unknown. A standard procedure is to plot first the logarithm of the concentration against time, and then the reciprocal of the concentration against time (as in Fig. 14–8). A straight line in the first plot indicates first-order kinetics; a straight line in the second indicates second-order kinetics. If neither plot gives a straight line, the kinetics is more complex.

14–4   Reaction Mechanisms

Most reactions proceed through a series of two or more steps, each of which is an **elementary step.** An elementary step results from a direct collision of atoms, ions, or molecules. The number of reacting species in an elementary step is called the **molecularity** of the step. The molecularity is simply the number of particles that must collide for the step to happen. The rate of an elementary step is proportional to the product of the concentrations of the reacting species, each raised to a power equal to its coefficient in the balanced equation that represents the step. Only when a reaction is known by some means to occur as an elementary step

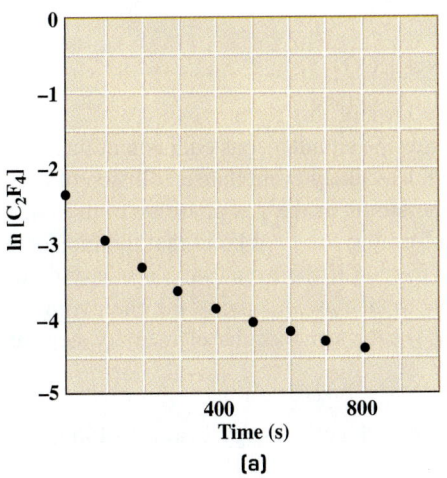

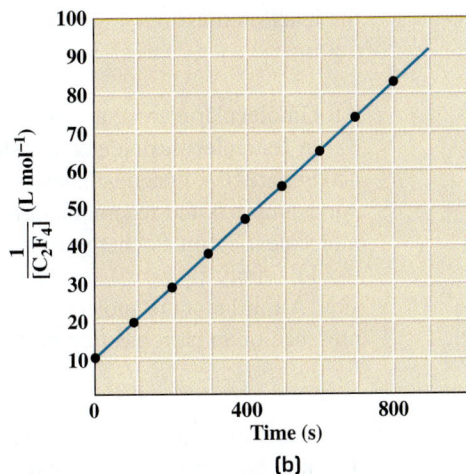

Figure 14–8 • Suppose that the rate of the reaction $2\,C_2F_4 \rightarrow C_4F_8$ is being studied for the first time, under the conditions given in Example 14–6. Plotting the logarithm of the concentration of C_2F_4 against time (graph a) tests whether the reaction is first order in C_2F_4; the curving line shows that it is not. Plotting the reciprocal of the concentration of C_2F_4 against time (graph b) tests whether the reaction is second order in C_2F_4. The nearly perfect straight-line relationship strongly suggests that it is.

can its rate law be determined by inspection of a chemical equation. The rate law for multistep reactions, which are the vast majority of reactions, must be determined experimentally.

Types of Elementary Reactions

A **unimolecular** elementary step involves only a single reactant molecule (or ion) that breaks apart or shakes its atoms into a new relationship. An example is the dissociation of excited-state molecules of N_2O_5 in the gas phase:

$$N_2O_5^*(g) \longrightarrow NO_2(g) + NO_3(g)$$

The asterisk indicates that the reacting molecules are in high-energy states, usually as the result of prior collisions with other molecules. The rate law for the step is

$$\text{rate} = k[N_2O_5^*]$$

Another example is the redox decomposition of mercury(II) sulfite in aqueous solution:

$$HgSO_3(aq) \longrightarrow Hg(aq) + SO_3(aq) \qquad \text{rate} = k[HgSO_3]$$

in which a pair of electrons moves from the sulfite ion to the mercury(II) ion.

The most common type of elementary step involves the collision of two atoms, ions, or molecules and is called **bimolecular.** An example is the reaction that is discussed in Section 14–1:

$$NO_2(g) + CO(g) \longrightarrow NO(g) + CO_2(g)$$

The rate at which a *single* NO_2 molecule collides with CO molecules is clearly proportional to the number of CO molecules per unit volume, that is, to [CO]. The rate of collision of *all* the NO_2 molecules is proportional to $[NO_2]$, so the rate law for

• Chemists call the process by which a molecule gains energy an "excitation," and the result, an "excited state"; the loss of energy is a "relaxation."

• The species Hg(*aq*) does exist. This reaction is the subject of Problem 85.

the reaction must be

$$\text{rate} = k[NO_2][CO]$$

All bimolecular elementary steps have rate laws of this form.

A **termolecular** elementary step requires the simultaneous collision of three reacting particles, which is inherently a much less likely event than a collision of two. An example is the recombination of iodine atoms in the gas phase to form iodine molecules. So much energy is released in forming the I—I bond that the molecule would simply fly apart as soon as it was formed if the event were a binary collision. A third atom or molecule is necessary to take away some of the energy. In the presence of an inert gas (argon, for instance) such termolecular elementary steps as

- The ΔE_r° for the combination $2 I(g) \rightarrow I_2(g)$ is -149 kJ mol^{-1}.

$$I(g) + I(g) + Ar(g) \longrightarrow I_2(g) + Ar(g)$$

occur in which the inert atom leaves with more kinetic energy than it had initially. The rate law for this termolecular reaction is

$$\text{rate} = k[I]^2[Ar]$$

Elementary steps involving collisions of four or more molecules are not observed, and even termolecular collisions are rare, particularly if other pathways are possible.

Elementary steps in liquid solvents involve encounters among dissolved species. If the solution is ideal, the rates of these processes are proportional to the product of the concentrations of the colliding species. Solvent molecules are always present and may influence the outcome of an encounter among solutes. Even so, they do not appear in the rate law because the concentration of the solvent does not vary appreciably. A reaction such as the recombination of iodine atoms occurs readily in solution. It appears to be second order with the rate law

- This invariance of solvent concentration is the reason that the concentrations of solvents can be replaced by a 1 in equilibrium expressions (see Section 7–6).

$$\text{rate} = k[I]^2$$

only because the third body involved is a solvent molecule. In the same way, a reaction between a solvent and a solute appears to be unimolecular, and only the concentration of solute molecules enters the rate laws.

Putting Elementary Steps Together

A **reaction mechanism** consists of a series of elementary steps, together with their rate constants or an indication of their relative rates, that can be combined to yield the overall reaction. A mechanism tells exactly how the reactants are transformed to products in a chemical reaction, including the existence of transitory intermediate species and the participation of catalysts.

Consider a typical mechanism. The gas-phase reaction of nitrogen dioxide with carbon monoxide to give nitrogen monoxide and carbon dioxide at low temperatures is thought to proceed by these elementary steps:

$$NO_2(g) + NO_2(g) \longrightarrow NO_3(g) + NO(g) \qquad \text{(slow)}$$
$$NO_3(g) + CO(g) \longrightarrow NO_2(g) + CO_2(g) \qquad \text{(fast)}$$

Both steps are bimolecular. Numerical rate constants do not appear; only the relative terms "fast" and "slow" do. This is common in stating mechanisms when one step is substantially slower than all the others. Adding the two steps gives

$$2 NO_2(g) + NO_3(g) + CO(g) \longrightarrow NO_3(g) + NO(g) + NO_2(g) + CO_2(g)$$

The $NO_3(g)$ and one molecule of $NO_2(g)$ from each side cancel out, giving the overall equation

$$NO_2(g) + CO(g) \longrightarrow NO(g) + CO_2(g)$$

The steps of a reaction mechanism must add up to give the correct overall reaction, just as they do in this example. If each elementary step occurs the same number of times in the course of the reaction, the addition is straightforward. If one step occurs, say, twice as often as the others, it must be multiplied by 2 before the steps are added.

This mechanism contains a **reaction intermediate.** An intermediate is a species that is formed and then completely consumed in the reaction and does not appear among the products in the overall balanced chemical equation. Clearly, the intermediate here is NO_3. A major challenge in chemical kinetics is to identify intermediates, which are frequently so short-lived that they are difficult to detect experimentally.

It is usually possible to suggest several mechanisms by which a given reaction occurs. Indeed, some reactions proceed by different mechanisms under different conditions, or even by two different mechanisms concurrently. One of the goals of chemical kinetics is to use the observed rate law for a reaction and suitable additional experiments to eliminate as many as possible of the various conceivable reaction mechanisms.

EXAMPLE 14-7

Consider the following reaction mechanism:

$$Cl_2(g) \longrightarrow 2\,Cl(g)$$
$$Cl(g) + CHCl_3(g) \longrightarrow HCl(g) + CCl_3(g)$$
$$CCl_3(g) + Cl(g) \longrightarrow CCl_4(g)$$

(a) Determine the molecularity of each elementary step.
(b) Write the equation for the overall reaction.
(c) Identify the reaction intermediate(s).

Solution

(a) The first step is unimolecular (one reacting particle); the other two are bimolecular (two reacting particles).
(b) Adding the three steps gives

$$Cl_2(g) + 2\,Cl(g) + CHCl_3(g) + CCl_3(g) \longrightarrow$$
$$2\,Cl(g) + HCl(g) + CCl_3(g) + CCl_4(g)$$

The two species that appear in equal amounts on both sides cancel out to leave

$$Cl_2(g) + CHCl_3(g) \longrightarrow HCl(g) + CCl_4(g)$$

(c) The two reaction intermediates are $Cl(g)$ and $CCl_3(g)$.

EXERCISE

Consider the mechanism

$$H_2(g) + 2\,NO(g) \longrightarrow N_2O(g) + H_2O(g)$$
$$N_2O(g) + H_2(g) \longrightarrow N_2(g) + H_2O(g)$$

(a) Determine the molecularity of each elementary step. (b) Write the equation for the overall reaction. (c) Identify the reaction intermediate(s).

Answer: (a) The first step is termolecular and the second is bimolecular. (b) The overall reaction is $2\,H_2(g) + 2\,NO(g) \longrightarrow N_2(g) + 2\,H_2O(g)$. (c) The only intermediate is $N_2O(g)$.

Kinetics and Chemical Equilibrium

A connection exists between the equilibrium constant of a reaction that takes place in a sequence of steps and the rate constants in each step. To appreciate this important link between thermodynamics and kinetics, consider a general reaction such as

$$2\,A(g) + B(g) \rightleftharpoons C(g) + D(g)$$

and suppose that it proceeds by some mechanism such as

$$A(g) + A(g) \longrightarrow A_2(g) \qquad\qquad \text{rate} = k_1[A]^2$$
$$A_2(g) + B(g) \longrightarrow C(g) + D(g) \qquad \text{rate} = k_2[A_2][B]$$

The constants k_1 and k_2 are the rate constants for the two forward elementary steps. As this (or any) reaction nears equilibrium, its reverse becomes important. The reverse reaction proceeds by the reverse mechanism, which is

$$C(g) + D(g) \longrightarrow A_2(g) + B(g) \qquad \text{rate} = k_{-2}[C][D]$$
$$A_2(g) \longrightarrow A(g) + A(g) \qquad\qquad \text{rate} = k_{-1}[A_2]$$

Here k_{-1} and k_{-2} are the rate constants for the two reverse steps. The **principle of detailed balance** states that, at equilibrium, the rate of every elementary step is balanced by (equal to) the rate of its reverse. For the first step in the above mechanism, this means that

$$\text{forward rate at equilibrium} = \text{reverse rate at equilibrium}$$
$$k_1[A]^2_{eq} = k_{-1}[A_2]_{eq}$$

and for the second step

$$\text{forward rate at equilibrium} = \text{reverse rate at equilibrium}$$
$$k_2[A_2]_{eq}[B]_{eq} = k_{-2}[C]_{eq}[D]_{eq}$$

where the subscript "eq" emphasizes that each concentration is an equilibrium concentration. It follows by algebraic rearrangement that

$$\frac{k_1}{k_{-1}} = \frac{[A_2]_{eq}}{[A]^2_{eq}} \qquad \text{and} \qquad \frac{k_2}{k_{-2}} = \frac{[C]_{eq}[D]_{eq}}{[A_2]_{eq}[B]_{eq}}$$

Note that the equilibrium constants[1] K_1 and K_2 for the two elementary steps equal the same quantities:

$$K_1 = \frac{[A_2]_{eq}}{[A]^2_{eq}} \qquad \text{and} \qquad K_2 = \frac{[C]_{eq}[D]_{eq}}{[A_2]_{eq}[B]_{eq}}$$

---

[1]Gas-phase equilibrium constants involve partial pressures expressed in atmospheres, whereas the convention in chemical kinetics is to use concentrations even for gaseous species. Therefore, the constants such as K_1, which we introduce here, are related to true dimensionless equilibrium constants by multiplicative factors of the form (RT/P_{ref}) raised to a power, as discussed in Section 7–4. Nevertheless, we refer to constants such as K_1 in this section as equilibrium constants.

Adding the steps in this mechanism gives the overall reaction. When reactions are added, their equilibrium constants are multiplied (see Section 7–2). Therefore, the equilibrium constant for the overall reaction is

$$K = K_1 K_2 = \frac{k_1}{k_{-1}} \frac{k_2}{k_{-2}} = \frac{[A_2]_{eq}[C]_{eq}[D]_{eq}}{[A]^2_{eq}[A_2]_{eq}[B]_{eq}} = \frac{[C]_{eq}[D]_{eq}}{[A]^2_{eq}[B]_{eq}}$$

The concentrations of the intermediate A_2 cancel out, giving the usual mass-action expression. This result can be generalized to any reaction:

> The product of the forward rate constants for the elementary steps in any chemical reaction divided by the product of the reverse rate constants equals the equilibrium constant of the reaction.

If a reaction has several possible mechanisms, the forward and reverse rate constants of the various steps are still consistent in this way with the equilibrium constant.

14–5 Reaction Mechanisms and Rate Laws

In many reaction mechanisms, one step is significantly slower than all the others. Such a step is called the **rate-determining step** because an overall reaction occurs only as fast as its slowest step, just as diners go through a cafeteria line only as fast as they pass the "bottleneck" of a slow cashier. Figure 14–9 shows passage through a literal bottleneck as the rate-determining step in the three-step "reaction" in which water runs from one soft-drink bottle to another through a connector.

When the Rate-Determining Step Occurs First

If the rate-determining step is the first one, the analysis is particularly simple. An example is the reaction

$$2\,NO_2(g) + F_2(g) \longrightarrow 2\,NO_2F(g)$$

for which the experimental rate law is

$$\text{rate} = k_{obs}\,[NO_2][F_2]$$

A possible mechanism for the reaction is

$$NO_2(g) + F_2(g) \xrightarrow{k_1} NO_2F(g) + F(g) \qquad \text{(slow)}$$
$$NO_2(g) + F(g) \xrightarrow{k_2} NO_2F(g) \qquad \text{(fast)}$$

The first step is slow and determines the rate:

$$\text{rate} = k_1[NO_2][F_2]$$

The subsequent fast step does not affect the reaction rate because F atoms join with NO_2 molecules essentially as soon as they are produced.

When the Rate-Determining Step Comes after One or More Fast Steps

Mechanisms in which the rate-determining step occurs after one or more fast steps often are signaled by a reaction order greater than two, by a non-integral reaction order, or by an inverse dependence of the rate on the concentration of a species

(a)   **(b)**

Figure 14–9 • Metaphor becomes reality as a true bottleneck controls the rate of transfer of water. (*a*) The first step is too quick to photograph: Water runs down to the neck of the top bottle following inversion. The second step, which is stalled at this moment, is passage through the bottleneck. The third step, water flowing to the bottom of the lower bottle, will certainly go fast. (*b*) Swirling creates a vortex, speeding up the bottleneck step and thereby accelerating the entire process.

• A k_{obs} is an observed (experimental) rate constant.

taking part in the reaction. An example is the reaction

$$2\,NO(g) + O_2(g) \longrightarrow 2\,NO_2(g)$$

for which the experimental rate law is

$$\text{rate} = k_{obs}[NO]^2[O_2]$$

One possible mechanism would be a single-step termolecular reaction of two NO molecules with one O_2 molecule. This mechanism is consistent with the experimental rate law, but three-way collisions are quite rare. If an alternative pathway exists, it is usually followed. One such alternative is the two-step mechanism:

$$NO(g) + NO(g) \underset{k_{-1}}{\overset{k_1}{\rightleftharpoons}} N_2O_2(g) \qquad \text{(fast equilibrium)}$$

$$N_2O_2(g) + O_2(g) \overset{k_2}{\longrightarrow} 2\,NO_2(g) \qquad \text{(slow)}$$

The second, slow step determines the overall rate of the reaction. The rate of this step is

$$\text{rate} = k_2[N_2O_2][O_2]$$

The concentration of an intermediate such as N_2O_2 cannot be varied at will, however. The N_2O_2 has to wait to react with O_2 in the slow step. This gives the *reverse* of the first step plenty of time to take place. At equilibrium in the first step, the rate of formation of N_2O_2 equals the rate of its decomposition back to NO:

$$k_1[NO]^2 = k_{-1}[N_2O_2]$$

$$\frac{[N_2O_2]}{[NO]^2} = \frac{k_1}{k_{-1}} = K_1$$

$$[N_2O_2] = K_1[NO]^2$$

Substituting this relationship into the rate law gives

$$\text{rate} = k_2K_1\,[NO]^2[O_2]$$

This result is consistent with the observed reaction order, with $k_2K_1 = k_{obs}$.

> In general, a fast elementary step occurs before a slow (rate-determining) step reaches equilibrium, with its forward and reverse occurring at the same rate.

EXAMPLE 14–8

In basic aqueous solution, the redox reaction

$$I^-(aq) + OCl^-(aq) \longrightarrow Cl^-(aq) + OI^-(aq)$$

generally is agreed to occur according to the following mechanism:

$$OCl^-(aq) + H_2O(\ell) \underset{k_{-1}}{\overset{k_1}{\rightleftharpoons}} HOCl(aq) + OH^-(aq) \qquad \text{(fast equilibrium)}$$

$$I^-(aq) + HOCl(aq) \overset{k_2}{\longrightarrow} HOI(aq) + Cl^-(aq) \qquad \text{(slow)}$$

$$OH^-(aq) + HOI(aq) \overset{k_3}{\longrightarrow} H_2O(\ell) + OI^-(aq) \qquad \text{(fast)}$$

What rate law is predicted by this mechanism?

• The hypochlorite ion OCl^- in basic solution is the active ingredient in household bleach. Here it oxidizes the iodide ion.

Strategy

Write the rate law for the rate-determining step, which is the step labelled "slow." Use the equilibrium-constant equation for the fast equilibrium to eliminate the concentrations of intermediates from the rate law for the rate-determining step.

Solution

For the second step, the slow step,

$$\text{rate} = k_2[\text{I}^-][\text{HOCl}]$$

The concentration of HOCl is controlled by its equilibrium with OCl^- and OH^- (the first step):

$$\frac{[\text{HOCl}][\text{OH}^-]}{[\text{OCl}^-]} = K_1 = \frac{k_1}{k_{-1}}$$

• Recall that H_2O does not appear in the equilibrium-constant expression because it is the solvent.

Solving this equation for [HOCl] and substituting in the previous equation gives

$$\text{rate} = k_2 K_1 \frac{[\text{I}^-][\text{OCl}^-]}{[\text{OH}^-]}$$

This is, in fact, the experimentally observed rate law.

The OH^- ion is produced in the first step of the mechanism and consumed in the last. It is an intermediate, yet its concentration has not been eliminated from the rate law. The OH^- ion acts as a *negative* catalyst; the higher its concentration, the slower the reaction. The rate law could be rewritten

$$\text{rate} = \frac{k_2 K_1}{K_w}[\text{I}^-][\text{HOCl}][\text{H}_3\text{O}^+]$$

using the fact that $K_w = [\text{H}_3\text{O}^+][\text{OH}^-]$. The version of the rate equation shows that H_3O^+ ion catalyzes the reaction (makes it go faster).

EXERCISE

Suppose the mechanism for the reaction given in the preceding example were

$$\text{OCl}^-(aq) + \text{H}_2\text{O}(\ell) \underset{k_{-1}}{\overset{k_1}{\rightleftharpoons}} \text{HOCl}(aq) + \text{OH}^-(aq) \qquad \text{(fast equilibrium)}$$

$$\text{I}^-(aq) + \text{HOCl}(aq) \underset{k_{-2}}{\overset{k_2}{\rightleftharpoons}} \text{HOI}(aq) + \text{Cl}^-(aq) \qquad \text{(fast equilibrium)}$$

$$\text{OH}^-(aq) + \text{HOI}(aq) \overset{k_3}{\longrightarrow} \text{H}_2\text{O}(\ell) + \text{OI}^-(aq) \qquad \text{(slow)}$$

What rate law is now predicted?

Answer: rate $= k_3 K_2 K_1 \dfrac{[\text{I}^-][\text{OCl}^-]}{[\text{Cl}^-]}$.

The rate of the reaction between $\text{I}^-(aq)$ and $\text{OCl}^-(aq)$ (in Example 14–8) depends inversely on the concentration of $\text{OH}^-(aq)$ ion. Such a form is often a clue that a rapid equilibrium occurs in the first steps of a reaction, preceding the rate-determining step. Fractional orders of reaction provide a similar clue, as in the gas-phase reaction of H_2 with Br_2 to form HBr:

$$\text{H}_2(g) + \text{Br}_2(g) \longrightarrow 2\,\text{HBr}(g)$$

The rate law before very much HBr builds up is

$$\text{rate} = k_{\text{obs}}[\text{H}_2][\text{Br}_2]^{1/2}$$

How can such a fractional power appear? One reaction mechanism that predicts this rate law is

$$\text{Br}_2(g) + \text{M}(g) \underset{k_{-1}}{\overset{k_1}{\rightleftharpoons}} \text{Br}(g) + \text{Br}(g) + \text{M}(g) \qquad \text{(fast equilibrium)}$$

$$\text{Br}(g) + \text{H}_2(g) \xrightarrow{k_2} \text{HBr}(g) + \text{H}(g) \qquad \text{(slow)}$$

$$\text{H}(g) + \text{Br}_2(g) \xrightarrow{k_3} \text{HBr}(g) + \text{Br}(g) \qquad \text{(fast)}$$

Here M stands for a second molecule that does not combine with the bromine molecules but supplies the energy to break them up. The reaction rate is determined by the slow step:

$$\text{rate} = k_2[\text{Br}][\text{H}_2]$$

However, [Br] is fixed by the establishment of equilibrium in the first step:

$$\frac{[\text{Br}]^2}{[\text{Br}_2]} = K_1 = \frac{k_1}{k_{-1}}$$

so that

$$[\text{Br}] = K_1^{1/2}[\text{Br}_2]^{1/2}$$

The rate law predicted by this mechanism is thus

$$\text{rate} = k_2 K_1^{1/2}[\text{H}_2][\text{Br}_2]^{1/2}$$

This is in accord with the observed fractional power in the rate law. By contrast, the mechanism

$$\text{H}_2(g) + \text{Br}_2(g) \xrightarrow{k_1} 2\,\text{HBr}(g) \qquad \text{(slow)}$$

in which HBr forms in a single bimolecular step, predicts the rate law

$$\text{rate} = k_1[\text{H}_2][\text{Br}_2]$$

This disagrees with the observed rate law, so the simple bimolecular mechanism must be ruled out as the major contributor to the measured rate.

Difficulty of Obtaining Mechanisms from Rate Laws Alone

Obtaining a rate law from a proposed mechanism is relatively straightforward; doing the reverse is much harder. In fact, rival mechanisms often give rise to the same rate law, and only some independent type of measurement enables a choice between them.

A reaction mechanism can never be proved from an experimental rate law; it can only be disproved if it is inconsistent with the experimental behavior. A classic example is the combination reaction

$$\text{H}_2(g) + \text{I}_2(g) \longrightarrow 2\,\text{HI}(g)$$

for which the observed rate law is

$$\text{rate} = k_{\text{obs}}[\text{H}_2][\text{I}_2]$$

(contrast this with the rate law given above for the combination of H_2 with Br_2). This is one of the most extensively studied reactions in chemical kinetics. Until 1967 it was believed to occur entirely by a simple bimolecular collision. At that time,

J. H. Sullivan investigated the effect of illuminating the reacting sample with light, which splits some of the I_2 molecules into iodine atoms. If the mechanism proposed above is correct, the effect of the light on the reaction should be small because it leads only to a small decrease in the concentration of I_2.

Instead, Sullivan observed a dramatic *increase* in the rate of reaction under illumination. The increase could be explained only by the participation of iodine atoms in the reaction mechanism. One such mechanism is

$$I_2(g) \underset{k_{-1}}{\overset{k_1}{\rightleftharpoons}} I(g) + I(g) \qquad \text{(fast equilibrium induced by illumination)}$$

$$I(g) + H_2(g) \underset{k_{-2}}{\overset{k_2}{\rightleftharpoons}} IH_2(g) \qquad \text{(fast equilibrium)}$$

$$IH_2(g) + I(g) \overset{k_3}{\longrightarrow} 2\,HI(g) \qquad \text{(slow)}$$

for which the rate law is

$$\begin{aligned}
\text{rate} &= k_3[H_2I][I] \\
&= k_3 K_2[H_2][I]^2 \\
&= k_3 K_2 K_1[H_2][I_2]
\end{aligned}$$

This mechanism not only gives the same rate law that is observed experimentally but also explains the effect of light on the reaction. The first reaction mechanism also appears to contribute significantly to the overall rate.

This example illustrates the hazards of trying to determine reaction mechanisms from rate laws: Several mechanisms can fit any given empirical rate law, and it is always possible that a new piece of information will come along to suggest a different mechanism. The problem is that, under ordinary conditions, the reaction intermediates can only be postulated and not isolated and studied like the reactants and products. This state of affairs is changing, however, with the development of experimental techniques that allow for direct study of the transient species that form in small concentrations during the course of a chemical reaction.

Chain Reactions

A **chain reaction** is one that proceeds through a series of elementary steps, some of which are repeated many times. Chain reactions consist of three stages:

1. **Initiation,** in which two or more reactive intermediates are generated
2. **Propagation,** in which the reactive intermediates react to give products and, simultaneously, furnish new supplies of reactive intermediates
3. **Termination,** in which two reactive intermediates combine to give a stable product

An example of a chain reaction is the reaction of methane with fluorine to give CH_3F and HF:

$$CH_4(g) + F_2(g) \longrightarrow CH_3F(g) + HF(g)$$

Although in principle this reaction could occur in a one-step bimolecular process, that route turns out to be too slow to contribute significantly under normal conditions. Instead, the main mechanism is this chain reaction:

$$\begin{aligned}
CH_4(g) + F_2(g) &\longrightarrow \cdot CH_3(g) + HF(g) + F\cdot(g) \qquad &\text{(initiation)} \\
\cdot CH_3(g) + F_2(g) &\longrightarrow CH_3F(g) + F\cdot(g) \qquad &\text{(propagation)} \\
CH_4(g) + F\cdot(g) &\longrightarrow \cdot CH_3(g) + HF(g) \qquad &\text{(propagation)} \\
\cdot CH_3(g) + F\cdot(g) + M(g) &\longrightarrow CH_3F(g) + M(g) \qquad &\text{(termination)}
\end{aligned}$$

In the initiation step, two reactive intermediates ($\cdot CH_3$ and $F\cdot$) are produced. These species are **radicals,** species that have one or more unpaired valence electrons. Dots in the formulas of radicals represent unpaired electrons. Radicals tend to transfer or share electrons to attain valence octets. Consequently, they are usually highly reactive. During the propagation steps, $\cdot CH_3$ and $F\cdot$ react to form products but are immediately replaced. The two propagation steps occur again and again, churning out quantities of products (CH_3F and HF) as reactants (CH_4 and F_2) are consumed. Eventually, two reactive intermediates come together in a termination step. Chain reactions are important in building up the long-chain molecules called "polymers," as discussed in Chapter 25.

• In a chain reaction, the equation for the overall reaction equals the sum of the propagation steps only. These steps occur far more times than the initiation or termination steps, so they dominate the overall stoichiometry of the reaction.

The chain reaction between CH_4 and F_2 proceeds at a constant rate because each propagation step both consumes and produces a reactive intermediate. The concentrations of the reactive intermediates remain approximately constant and are determined by the rates of chain initiation and termination. Another type of chain reaction is possible in which the number of reactive intermediates increases during one or more propagation steps. This is called a **branching chain reaction.** An example is the reaction of oxygen with hydrogen. The mechanism is complex and can be initiated in various ways, causing several reactive intermediates such as $\cdot O\cdot$, $H\cdot$, and $\cdot OH$ to form. Some propagation steps are of the type we have seen already for CH_4 and F_2, such as

$$\cdot OH(g) + H_2(g) \longrightarrow H_2O(g) + H\cdot(g)$$

in which one reactive intermediate ($\cdot OH$) is used up and one ($H\cdot$) is produced. Other propagation steps are branching, however:

$$H\cdot(g) + O_2(g) \longrightarrow \cdot OH(g) + \cdot O\cdot(g)$$
$$\cdot O\cdot(g) + H_2(g) \longrightarrow \cdot OH(g) + H\cdot(g)$$

In these steps, each reactive intermediate that is used up leads to the generation of two others. This leads to a rapid growth in the number of reactive species, speeding the rate further and possibly causing an explosion.

14-6 The Effect of Temperature on Reaction Rates

The rates of most chemical reactions depend strongly on the temperature. A knowledge of numerical rate constants and the details of their temperature dependence is crucial in enterprises ranging from the timely production of useful chemicals to tracing the fate of pollutants in the environment.

Rate Constants for Gas-Phase Reactions

Molecules, atoms, or ions clearly must collide before they can react. The kinetic theory of gases (Section 5–7) provides an estimate of the frequency of collisions between molecules in gases. A typical small molecule (one of oxygen, for example) in a gas at room conditions strikes another molecule on the order of 10^9 times per second. If every collision led to a reaction, then the reacting molecules would be nearly consumed in about 10^{-9} second. Some reactions do proceed almost this fast. An example is the bimolecular reaction between two molecules of ClO:

• The role played by chlorine monoxide, ClO, in the depletion of ozone in the outer atmosphere is discussed in Section 18–5.

$$2\,ClO(g) \longrightarrow Cl_2(g) + O_2(g)$$

for which the observed rate constant at 273 K is 6×10^{10} L mol^{-1} s^{-1}. If the initial pressure of ClO(g) could be raised as high as 1 atm, then at 273 K the initial concentration of ClO(g) would equal about 0.04 M. Using the second-order rate law (see Section 14–3), after 1 nanosecond (1 × 10^{-9} s), the concentration of ClO(g) would equal 0.007 mol L^{-1}, and after 10 nanoseconds, it would equal 0.0008 mol L^{-1}. Most reactions proceed far more slowly than this. Indeed, rates can be slower by factors of 10^{12} or more. The naïve idea that "to collide is to react" clearly requires modification if such rates are to be understood.

The relative orientation of the colliding molecules certainly must play a role in determining whether a particular collision results in a reaction. In a bimolecular reaction such as

$$2\,NOCl(g) \longrightarrow 2\,NO(g) + Cl_2(g)$$

it seems obvious that the two NOCl molecules must approach each other in such a way that the chlorine atoms are close together in order for a Cl$_2$ molecule to split off (Fig. 14–10). The collision frequency must therefore be multiplied by a **steric factor** (less than unity) to account for the fact that only a fraction of the collisions occurs with the proper orientation to lead to reaction. For small molecules, however, such a steric factor could hardly reduce the reaction rate by much more than one order of magnitude, so some further explanation must be sought for the slow rates that are frequently observed.

A clue is found in the observed temperature dependence of rate constants. The rates of many reactions increase extremely rapidly with increases in temperature. In fact, a crude rule of thumb sometimes quoted by chemists is that reaction rates double for every 10°C rise in temperature. In 1887, Svante Arrhenius suggested that rate constants vary exponentially with the reciprocal of the absolute temperature,

$$k = A\,e^{-E_a/RT} = A\exp\left(-E_a/RT\right)$$

where the constant E_a has units of energy per mole, and the pre-exponential factor A is a constant having the same units as k. This relationship is now known as the **Arrhenius equation,** and the two constants are known as the Arrhenius parameters (or factors). Taking the natural logarithm of both sides of the equation gives

$$\ln k = \ln A - \frac{E_a}{RT}$$

which is equivalent to

$$\ln\left(\frac{k}{A}\right) = -\frac{E_a}{RT} \qquad \text{and} \qquad \ln\left(\frac{A}{k}\right) = \frac{E_a}{RT}$$

According to the Arrhenius equation, a plot of ln k against $1/T$ should be a straight line with slope $-E_a/R$ and intercept ln A. Many rate constants show just this kind of temperature dependence (Fig. 14–11).

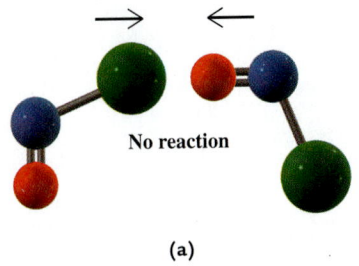

No reaction

(a)

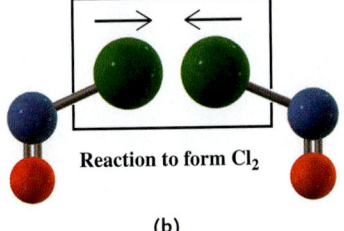

Reaction to form Cl$_2$

(b)

Figure 14–10 • The steric effect. (a) This collision of NOCl molecules, although energetic, produces no Cl$_2$; the Cl atoms (green) are too far away from each other. (b) When two fast-moving NOCl molecules collide with two chlorine atoms close together, Cl$_2$ forms (along with NO(g)).

EXAMPLE 14–9

The decomposition of hydroxylamine (NH$_2$OH) in the presence of oxygen has the rate law

$$\text{rate} = k_{obs}[NH_2OH][O_2]$$

where k_{obs} is 0.237 × 10^{-4} L mol^{-1} s^{-1} at 0°C and rises to 2.64 × 10^{-4} L mol^{-1} s^{-1} at 25°C. Calculate E_a and A for this reaction.

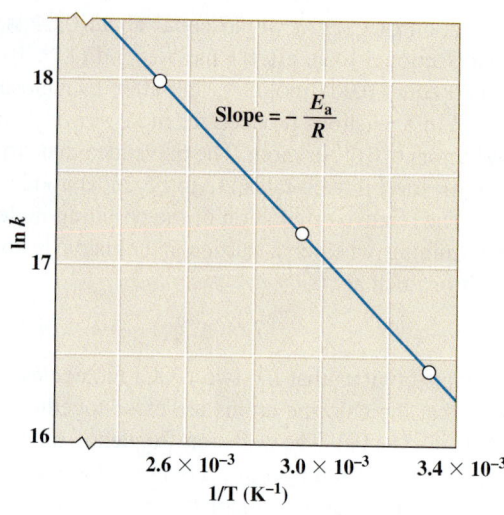

Figure 14–11 • An Arrhenius plot of ln k against $1/T$ for the reaction of benzene vapor with oxygen atoms. A considerable extrapolation to $1/T = 0$ would be necessary to determine the constant ln A from the intercept of this line.

• Notice the similarity in form between this equation and the van't Hoff equation (see Section 11–8), which describes the temperature dependence of *equilibrium* constants. Also, compare Figure 14–11 with Figure 11–15.

Solution

Write the Arrhenius equation at two different temperatures T_1 and T_2:

$$\ln k_1 = \ln A - \frac{E_a}{RT_1} \qquad \text{and} \qquad \ln k_2 = \ln A - \frac{E_a}{RT_2}$$

If the first equation is subtracted from the second, the term ln A cancels out, leaving

$$\ln k_2 - \ln k_1 = \ln\frac{k_2}{k_1} = -\frac{E_a}{R}\left(\frac{1}{T_2} - \frac{1}{T_1}\right)$$

In this case,

$$T_1 = 273.15 \text{ K} \qquad k_1 = 0.237 \times 10^{-4} \text{ L mol}^{-1}\text{ s}^{-1}$$
$$T_2 = 298.15 \text{ K} \qquad k_2 = 2.64 \times 10^{-4} \text{ L mol}^{-1}\text{ s}^{-1}$$

Substitution of these values gives

$$\ln\left(\frac{2.64 \times 10^{-4} \text{ L mol}^{-1}\text{ s}^{-1}}{0.237 \times 10^{-4} \text{ L mol}^{-1}\text{ s}^{-1}}\right) = \frac{-E_a}{R}\left(\frac{1}{298.15 \text{ K}} - \frac{1}{273.15 \text{ K}}\right)$$

$$2.410 = -\frac{E_a}{8.3140 \text{ J K}^{-1}\text{ mol}^{-1}}(-3.07R \times 10^{-4} \text{ K}^{-1})$$

$$E_a = 6.528 \times 10^4 \text{ J mol}^{-1} = \boxed{65.3 \text{ kJ mol}^{-1}}$$

Once E_a is known, the parameter A can be calculated by using data at either temperature. At 273.15 K

$$\ln\left(\frac{A}{k_1}\right) = \frac{E_a}{RT_1}$$

$$\ln\left(\frac{A}{0.237 \times 10^{-4} \text{ L mol}^{-1}\text{ s}^{-1}}\right) = \frac{6.528 \times 10^4 \text{ J mol}^{-1}}{8.3145 \text{ J K}^{-1}\text{ mol}^{-1}(273.15 \text{ K})} = 28.74$$

$$\frac{A}{0.237 \times 10^{-4} \text{ L mol}^{-1}\text{ s}^{-1}} = e^{28.74}$$

$$A = (0.237 \times 10^{-4} \text{ L mol}^{-1}\text{ s}^{-1})(3.03 \times 10^{12})$$
$$= \boxed{7.2 \times 10^7 \text{ L mol}^{-1}\text{ s}^{-1}}$$

A more accurate determination of E_a and A would use measurements at a series of temperatures rather than at only two, with a fit to a plot such as that in Figure 14–11.

EXERCISE

The reaction

$$N(SO_3)_3^{3-}(aq) + H_2O(\ell) \longrightarrow HN(SO_3)_2^{2-}(aq) + HSO_4^-(aq)$$

follows the rate law

$$\text{rate} = k_{obs}\left[N(SO_3)_3^{3-}\right]\left[H^+\right]$$

The rate increases by a factor of 23.1 when the temperature is raised from 283 K to 313 K. Compute E_a for the reaction.

Answer: 77.1 kJ mol^{-1}.

Arrhenius believed that to react upon collision, molecules must become "activated," and the parameter E_a became known as the **activation energy.** His ideas were refined by scientists who followed. In 1915, A. Marcelin pointed out that, although molecules make many collisions, not all collisions are reactive. Only those collisions in which the collision energy (that is, the relative translational kinetic energy of the colliding molecules) exceeds some critical minimum can result in reaction.

> The activation energy (E_a) is the minimum collision energy that reactants must have in order to form products.

The exponential dependence of rate constants on temperature that appears in the Arrhenius equation derives from the Maxwell–Boltzmann distribution of molecular kinetic energies. At a low temperature, only a small fraction of pairs of colliding molecules possesses the minimum energy to react. This fraction corresponds to the blue shaded portion of the whole area under the distribution curve T_1 in Figure 14–12. As the temperature is increased, the Maxwell–Boltzmann distribution curve flattens and spreads toward higher energies. At a higher temperature T_2, a much larger fraction of pairs has energy equal to or exceeding E_a. This fraction corresponds to the red shaded portion of the area under curve T_2 in Figure 14–12. The fraction of pairs having energies above E_a is proportional to $\exp(-E_a/RT)$. The reaction rate is therefore proportional to $\exp(-E_a/RT)$, and both the strong temperature dependence and the order of magnitude of the experimental rate constants can be understood.

The Reaction Coordinate and the Activated Complex

Why should there be a threshold collision energy E_a for a reaction between two molecules? To understand this, consider the analogy in Figure 14–13. In a landslide, a boulder shaken loose on one slope may roll up a neighboring hill. As it does, its potential energy increases and its kinetic energy decreases (it gets higher and slows down). If the boulder reaches the top of the hill, it starts down the other side, with its kinetic energy then increasing and its potential energy decreasing. Not every boulder makes it over the top, however. If its initial speed (and therefore kinetic energy)

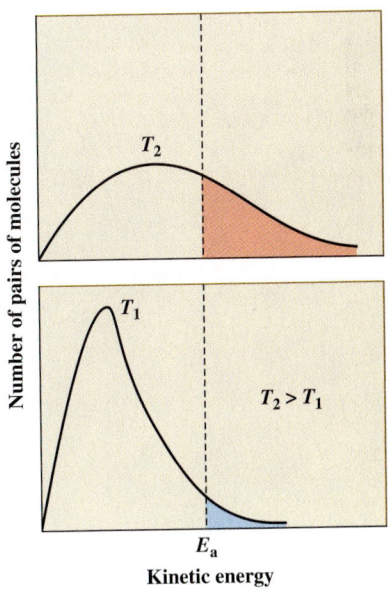

Figure 14–12 • The Maxwell–Boltzmann distribution of molecular kinetic energies at two different temperatures. A relatively small temperature change has a large effect on the fraction of molecules having enough kinetic energy to react.

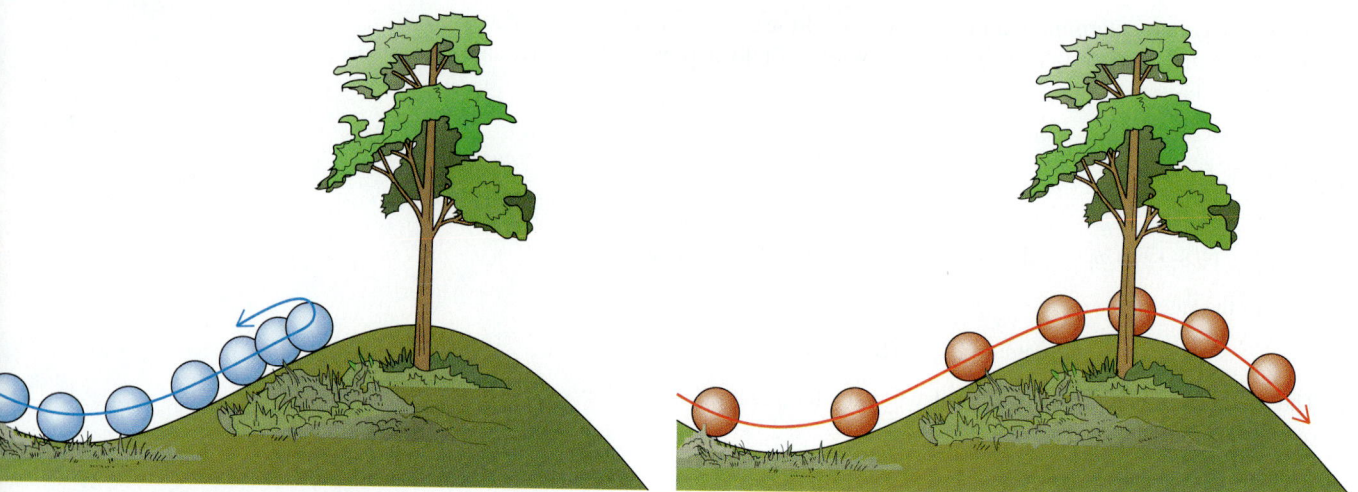

Figure 14–13 • An earthquake has set some large boulders rolling. Their fates are analogous to those of molecules in a chemical reaction. The blue boulder rolls up a hill and then back. It does not have enough energy to make it over. The red boulder has more kinetic energy. It makes it over the hill to reach the valley beyond.

is too small, a boulder rolls partway up, stops, and then rolls back. Only a boulder with initial kinetic energy higher than a critical threshold passes over the barrier.

This model applies to molecular collisions and reactions. As two reactant molecules, atoms, or ions approach each other along a **reaction path,** their potential energy increases as the bonds within them distort. At some maximum potential energy (the top of the hill), they become associated in an unstable entity called an **activated complex.** The activated complex is a **transition state.** In the transition state, the smooth ascent in potential energy as the reactants come together crosses over to a smooth descent as the products separate. Just as a rolling boulder can stop short of the top of a hill, some pairs of reactants can fail to react. Only those pairs with sufficient kinetic energy can stretch bonds and rearrange atoms sufficiently to reach the transition state that separates reactants from products. If the barrier is too high, almost all colliding pairs of reactant molecules bounce apart without reacting. The height of the barrier is close to the measured activation energy for the reaction.

Figure 14–14 is a graph of potential energy versus position along the reaction path for the reaction first considered in Section 14–1.

$$NO_2(g) + CO(g) \longrightarrow NO(g) + CO_2(g)$$

The horizontal axis represents the course of an individual reaction event in this gas-phase bimolecular reaction and is called the **reaction coordinate.** Two activation energies are shown: $E_{a, \text{for}}$ for the forward reaction, and $E_{a, \text{rev}}$ for the reverse reaction, in which NO takes an oxygen atom from CO_2 to form NO_2 and CO. The difference between the two turns out to equal ΔE_r, the change in internal energy of the chemical reaction:

$$E_{a, \text{for}} - E_{a, \text{rev}} = \Delta E_r$$

The ΔE_r in this equation is a thermodynamic quantity, obtainable from calorimetric measurements. The $E_{a, \text{for}}$ and $E_{a, \text{rev}}$, however, must be found from the temperature dependence of the rate constants for the forward and reverse reactions. In this reaction, the forward and reverse activation energies are 132 and 358 kJ mol^{-1}, respectively, and ΔE_r from thermodynamics is -226 kJ mol^{-1}.

Some elementary reactions are barrierless and so have an activation energy of zero and no well-defined transition state. In a reaction between an ion and a mole-

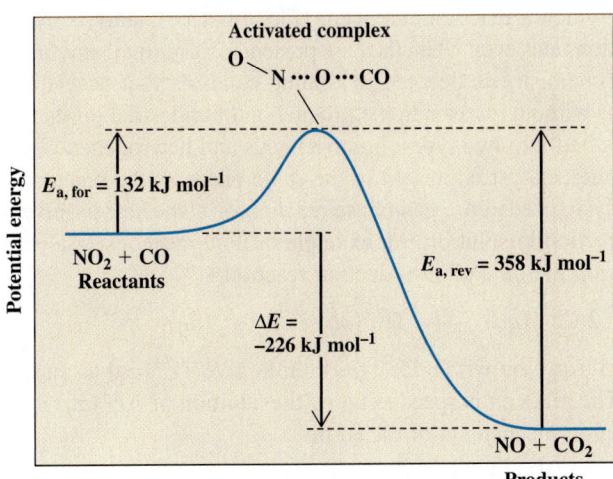

Figure 14–14 • The energy profile along the reaction coordinate for the reaction $NO_2 + CO \rightarrow NO + CO_2$, which proceeds by the bimolecular collision of reactant molecules at higher temperatures (above about 500 K).

cule, the attractive ion-dipole or ion-induced dipole force (see Section 6–1) may more than counteract any increase in potential energy due to bond reorganization. Some reactions between neutral particles are also barrierless. For example, the rate constant for the bimolecular elementary step

$$H(g) + NO_2(g) \longrightarrow OH(g) + NO(g)$$

does not change over the temperature range from 195 to 2000 K. This implies that E_a equals zero.

Remarkably, multistep reactions sometimes even have negative activation energies, reflecting the fact that the overall rate of reaction slows at higher temperature. How can this be? Let us examine a specific example, the reaction of NO with oxygen:

$$2\,NO(g) + O_2(g) \longrightarrow 2\,NO_2(g)$$

The observed rate law is

$$\text{rate} = k_{\text{obs}}\,[NO]^2[O_2]$$

and k_{obs} decreases with increasing temperature. In Section 14–5, it was shown that this rate law could be accounted for with a two-step mechanism. The first step is a rapid equilibrium (with equilibrium constant K_1) of two NO molecules with their dimer, N_2O_2. The second step is the slow reaction (with rate constant k_2) between N_2O_2 and O_2 to form products. The observed rate constant is therefore the product of k_2 and K_1. Although k_2 increases with increasing temperature, K_1 is an equilibrium constant. Provided that the corresponding reaction is sufficiently exothermic (as it is in this case), K_1 decreases so rapidly with increasing temperature that the product $k_2 K_1$ decreases as well.

14–7 Kinetics of Catalysis

A **catalyst** is a substance that takes part in a chemical reaction and speeds it up but itself undergoes no permanent chemical change. Catalysts therefore do not appear in overall balanced chemical equations, but their presence very much affects the rate law, modifying and speeding existing pathways or, more commonly, providing

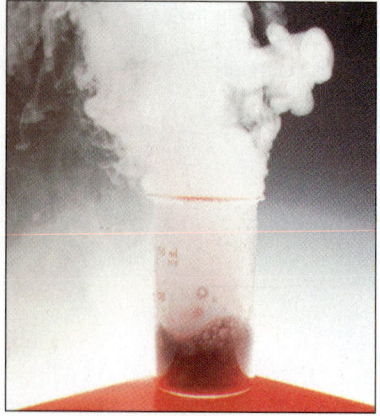

Figure 14–15 • The decomposition of hydrogen peroxide (H_2O_2) to water and oxygen is catalyzed by solid MnO_2. Here the water evolves as steam because of the heat generated by the reaction.

• Sulfuric acid production is examined in Section 22–2.

completely new paths by which a reaction can occur (Fig. 14–15). Catalysts exert significant effects on reaction rates even when they are present in very small amounts. Great effort in industrial chemistry is devoted to finding catalysts that accelerate particular desired reactions without increasing the production of undesired products.

Catalysis can be classified into two types: homogeneous and heterogeneous. In **homogeneous catalysis,** the catalyst is present in the same phase as the reactants, as when a gas-phase catalyst speeds up a gas-phase reaction or a species dissolved in solution speeds up a reaction in solution. An example of homogeneous catalysis is the effect of silver ions on the oxidation–reduction reaction

$$Tl^+(aq) + 2\,Ce^{4+}(aq) \longrightarrow Tl^{3+}(aq) + 2\,Ce^{3+}(aq)$$

The direct reaction of a $Tl^+(aq)$ ion with a $Ce^{4+}(aq)$ ion to give $Tl^{2+}(aq)$ as an intermediate is quite slow. The reaction is speeded up by the addition of $Ag^+(aq)$ ion, which takes part in a reaction mechanism of the form

$$Ag^+(aq) + Ce^{4+}(aq) \underset{k_{-1}}{\overset{k_1}{\rightleftharpoons}} Ag^{2+}(aq) + Ce^{3+}(aq) \qquad \text{(fast equilibrium)}$$

$$Tl^+(aq) + Ag^{2+}(aq) \overset{k_2}{\longrightarrow} Tl^{2+}(aq) + Ag^+(aq) \qquad \text{(slow)}$$

$$Tl^{2+}(aq) + Ce^{4+}(aq) \overset{k_3}{\longrightarrow} Tl^{3+}(aq) + Ce^{3+}(aq) \qquad \text{(fast)}$$

The $Ag^+(aq)$ ions are not permanently transformed in this sequence because those used up in the first step are regenerated in the second. They play the role of catalyst in significantly speeding the rate of the overall reaction. The $H^+(aq)$ and $OH^-(aq)$ ions often act as homogeneous catalysts, especially in biochemistry. Recall from Exercise 14–3 that $H^+(aq)$ speeds the conversion of sucrose to fructose and glucose; it can play a similar role in breaking proteins into smaller molecules.

In **heterogeneous catalysis,** the catalyst is present as a distinct phase. The most important case is the catalytic action of certain solid surfaces (Fig. 14–16) on gas-phase and solution-phase reactions. A crucial step in the production of sulfuric acid, for example, involves the use of a solid oxide of vanadium ($V_2O_5(s)$) as catalyst. Many other solid catalysts are used in industrial processes. One of the best-studied of such reactions is the addition of hydrogen to ethylene (C_2H_4) to form ethane (C_2H_6):

$$C_2H_4(g) + H_2(g) \longrightarrow C_2H_6(g)$$

The process occurs extremely slowly in the gas phase but is catalyzed by the presence of a platinum surface. Platinum speeds the reaction by causing H_2 molecules to dissociate to hydrogen atoms attached to its surface (Fig. 14–17). The atoms can then react sequentially with ethylene molecules to form ethane. Heterogeneous catalysis is also important in the atmosphere, where small solid particles (aerosols) can speed up reactions that would occur slowly in the gas phase.

A catalyst speeds up a reaction by increasing the pre-exponential factor A in the Arrhenius equation or, more often, by providing an alternative mechanism with a lower activation energy (E_a). The catalyst lowers E_a by providing a new activated complex of lower potential energy (Fig. 14–18). A corollary follows: The same catalyst speeds both the forward and reverse reactions because it lowers both the forward and reverse activation energies equally. However:

A catalyst has no effect on the thermodynamics of the overall reaction.

Figure 14–16 • Several solid catalysts used by the petrochemical industry. *(Harshaw/Filtrol Partnership)*

The change in the Gibbs energy (ΔG_r) is independent of the path followed; therefore, the equilibrium constant is never changed by catalytic action. Reaction products that

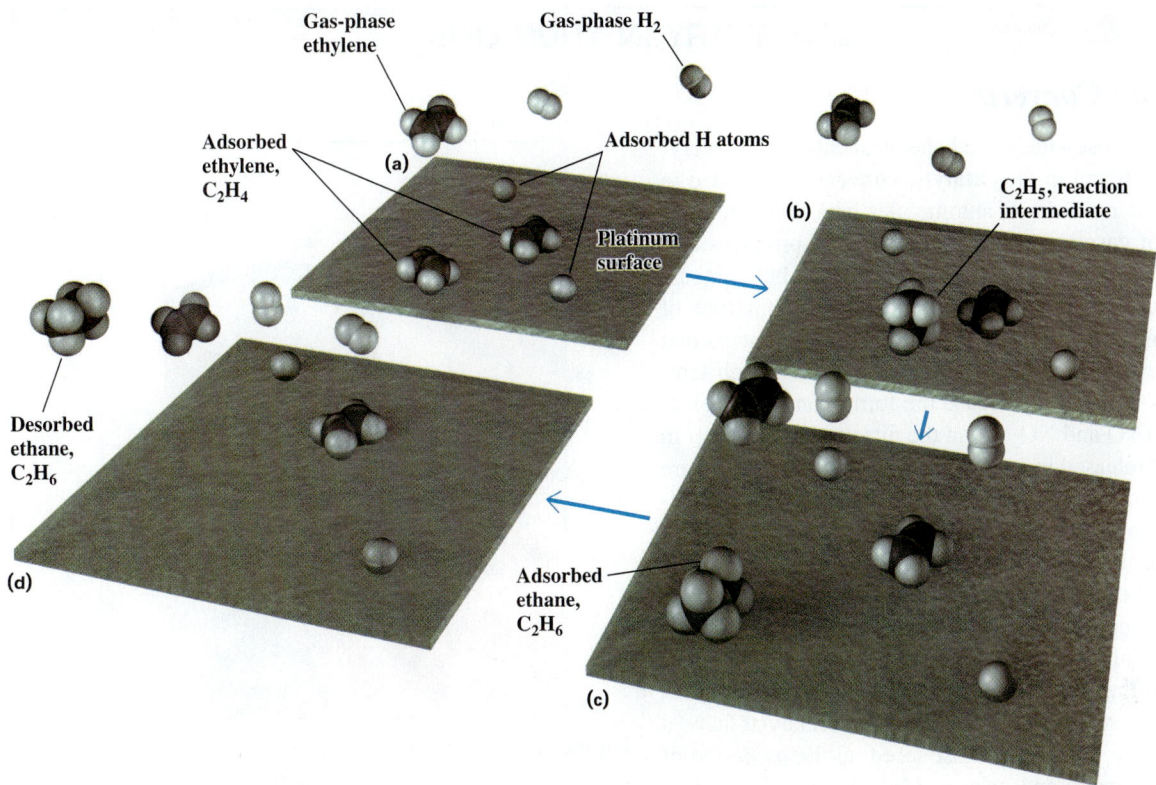

Figure 14–17 • (a–d) Platinum catalyzes the reaction $H_2 + C_2H_4$ by providing a surface that promotes the dissociation of H_2 to H atoms, which can then add stepwise to the C_2H_4 to give ethane (C_2H_6).

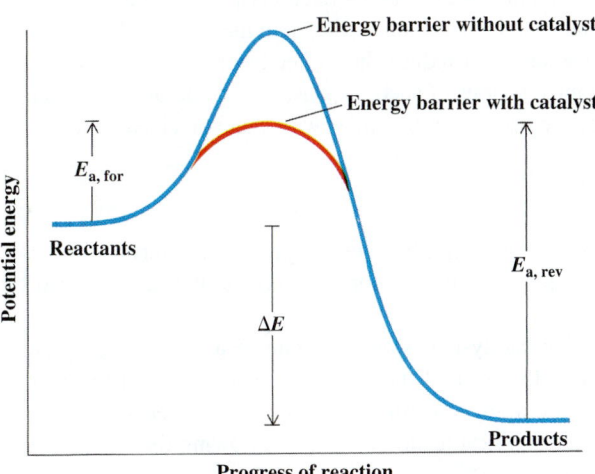

Figure 14–18 • The most important way in which catalysts speed up reactions is by reducing the activation energy.

CHEMISTRY IN YOUR LIFE

Catalytic Converters

One of the most widely used chemical catalysts in everyday life is found in the **catalytic converters** installed in the exhaust manifolds of automobiles. The complete combustion of gasoline in the engine of a car produces carbon dioxide and water as primary products, but combustion is always incomplete. The result is a mixture of carbon monoxide (CO) and hydrocarbon fragments that can be represented as C_mH_n. In addition, the high temperatures in the engine cause the formation of oxides of nitrogen (NO and NO_2) from the nitrogen and oxygen in the air. If released in quantity into the air, these gases are further converted into photochemical smog, the notorious and unpleasant haze that obscures many cities and can impair respiratory health.

A number of changes in automobile design have been used to reduce this pollution. One major step was the introduction of catalysts into the exhaust stream of the car, to convert the pollutant gases into less harmful forms before releasing them into the air. These catalysts have a dual purpose. First, they must speed up the *oxidation* of carbon monoxide and unburned hydrocarbons to carbon dioxide and water:

$$CO(g), C_mH_n(g), O_2(g) \xrightarrow{\text{catalyst}} CO_2(g), H_2O(g)$$

Second, they must accelerate the *reduction* of nitrogen oxides to nitrogen and oxygen:

$$NO(g), NO_2(g) \xrightarrow{\text{catalyst}} N_2(g), O_2(g)$$

Both of these processes are favored by thermodynamics at low temperatures but are far too slow to take place before the gases leave the tailpipe.

Catalysts that are good at speeding up oxidation reactions are usually poor at speeding the reduction of nitrogen oxides. A combination of two catalysts is therefore used. The noble metals, although expensive, are particularly useful catalysts. Typically, platinum, palladium, and

Figure 14–A • A catalytic converter used to reduce automobile exhaust pollution. The converter has been cut open to reveal the platinum, palladium, and rhodium catalysts. In this design, a heating element raises the interior temperature to 400°C in seconds. This activates the catalysts and so reduces emissions of harmful gases in the first minutes after the car is started. (*United Technologies*)

rhodium are deposited on a fine honeycomb mesh of alumina (Al_2O_3), giving a large surface area that increases the contact time of the exhaust gas with the catalysts (Fig. 14–A). The platinum serves primarily as an oxidation catalyst and the rhodium as a reduction catalyst. Catalytic converters can be poisoned with certain metals that block their active sites and reduce their effectiveness. Lead is one of the most serious of such poisons; as a result, unleaded fuel must be used in automobiles with catalytic converters.

are not favored thermodynamically cannot be caused to form by catalysis. The role of catalysts is to speed up the rate of production of products that are allowed by thermodynamics.

An **inhibitor** is a negative catalyst. It slows the rate of a reaction, frequently by barring access to a path of low E_a and thereby forcing the reaction to proceed by a path of higher E_a. Inhibitors are also important industrially because they can be used to reduce the rates of undesirable side-reactions, allowing desired products to be removed in greater yield.

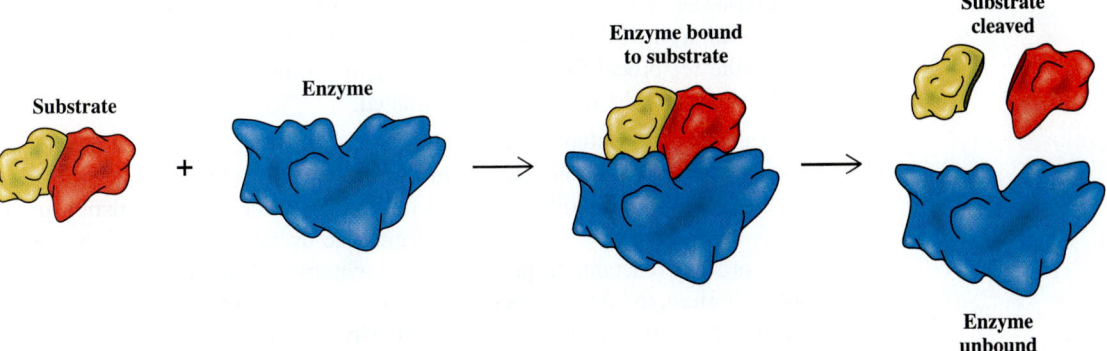

Figure 14–19 • The binding of a substrate to an enzyme, and the subsequent reaction of the substrate.

Catalysts called enzymes accelerate many chemical reactions in living systems. An **enzyme** is a large protein molecule (typically of molar mass 20,000 g mol^{-1} or more) that is both very effective and selective in its catalytic action. One or more reactant molecules (called **substrates**) bind to an enzyme at its active sites. These are regions of the enzyme molecule where local structure favors quick chemical transformation to a specific product, followed by release of that product (Fig. 14–19). The molecules of most other substances are not accommodated at the active sites. The enzyme urease, for example, catalyzes only the hydrolysis of urea, $(NH_2)_2CO$:

$$H^+(aq) + (NH_2)_2CO(aq) + 2\,H_2O(\ell) \xrightarrow{\text{urease}} 2\,NH_4^+(aq) + HCO_3^-(aq)$$

• The ending *-ase* in the name of a substance signals that the substance is an enzyme.

It does not bind most other kinds of molecules, even those of very similar structure. In some cases, a second species does bind to the enzyme and can act as an inhibitor, preventing the enzyme from carrying out its usual role as catalyst. Much modern chemical research is aimed at designing catalysts that are as powerful as enzymes in speeding up selected chemical reactions.

SUMMARY

14–1 The **average rate** of a chemical reaction over a period equals the change in the concentration of a reactant or product divided by the elapsed time. In general, the rates of reactions change from moment to moment. An **instantaneous rate** (or simply a rate) is a rate at a particular moment. An instantaneous rate is obtained from a plot of the concentration of a reactant or product against time. It equals the slope of the line tangent to the plot at the specified moment divided by the coefficient of the chemical species in the balanced equation (and with a minus sign for reactants). An **initial rate** is the rate at the instant a reaction begins.

14–2 An **experimental rate law** gives the dependence of the observed rate of a reaction on the concentrations of various species. In a rate law, the rate equals a **rate constant** (k) multiplied by the concentrations of reactant, product, or other species raised to integral or fractional powers. The powers are the **orders** of the reaction with respect to each chemical species, and their sum is the **overall reaction order.**

14–3 The **half-life** of a reaction is the time it takes for the concentration of a reacting species to fall to half its initial value. This concept is most useful for first-order reactions. From a rate law, explicit equations giving the dependence of reactant or product concentrations on time can be devised for zero-order and (using calculus) for first-order and second-order reactions. By graphing appropriate functions of the concentration (the concentration itself for zero-order reactions, the logarithm for first-order, the reciprocal for second-order) against time, the order of the reaction can be deduced, and the rate constant evaluated.

14–4 An **elementary step** involves a direct collision of atoms, ions, or molecules. Such steps may be **unimolecular, bimolecular,** or **termolecular,** according to whether one, two, or three colliding bodies take part. A reaction **mechanism** consists of a series of elementary steps that add to give the overall reaction and that detail the progress from reactants to products. The **principle of detailed balance** states that, at equilibrium, the rate of every elementary step in a reaction mechanism equals the rate of its reverse. Therefore, the product of the forward rate constants in a reaction mechanism divided by the product of the reverse rate constants gives the equilibrium constant for the reaction; the reaction **intermediates** cancel out of the final expression.

14–5 A major goal of chemical kinetics is to determine reaction mechanisms from the experimental rate laws for the reactions. If one step is much slower than the others, it is possible to write the form of the rate law for any given mechanism and compare it with an experiment. A particularly important type of reaction is the **chain reaction,** the mechanism of which involves three types of steps: **initiation** (in which highly reactive intermediates are produced), **propagation** (in which reactants are converted to products as reactive intermediates are regenerated), and **termination** (in which the reactive intermediates combine to form stable species and the reaction ends).

14–6 Reaction rates usually depend strongly on temperature. In an elementary step, the reactants come together to form a high-energy **activated complex.** In this **transition state,** bonds are distorted to make an entity that is intermediate between reactants and products. Formation of an activated complex requires both sufficient energy in the collision of the reacting particles and proper orientation (the **steric factor**). The difference between the energy of this activated complex and the energy of the reactants is the **activation energy** for the forward reaction. The activation energy and the temperature are related to the rate constant by the **Arrhenius equation.**

14–7 A catalyst speeds up a reaction by increasing the pre-exponential factor (A) in the Arrhenius equation or, more often, by providing an alternative mechanism with a lower activation energy (E_a). **Inhibitors** have the opposite effect. Catalysis can be **homogeneous** (occurring within a single phase) or **heterogeneous** (occurring at the boundary between two phases, as on the surface of a solid). An especially important group of catalysts are **enzymes,** which selectively speed up certain reactions in living systems. Enzymes are large, complex molecules that are capable of highly specific interactions (at their **active sites**) with the reacting **substrate** molecules. The interactions promote specific reaction pathways.

PROBLEMS

Note: Answers to blue-numbered problems are given in Appendix F. Problems that are more challenging are indicated with asterisks.

Rates of Chemical Reactions

1. A car going north on an interstate highway passes mile marker 11.5 at 11:46 A.M. and mile marker 30.0 at 12:02:24 P.M. Determine its average rate of speed in miles per minute and in miles per hour.

2. (See Example 14–1.) A reaction generates carbon dioxide in a vessel of constant volume at a temperature of 25°C. The partial pressure of the $CO_2(g)$ is 0.520 atm at 11:00:00 A.M. and 0.645 atm at 11:30:00 A.M. the same day. Compute the

average rate of appearance of $CO_2(g)$ during this period of time (in mol L^{-1} s^{-1}).

3. A cross-country hitchhiker gets a ride in New York that takes him 550 mi west in 11 h and lets him out at a truck stop. It takes 2 h before he gets a new ride, which takes him another 300 mi in 5.5 h. Determine his average rate of travel after 11 h, after 12 h, after 13 h, and after 18.5 h.

4. (See Example 14–1.) A chemical reaction generates 1.0 mol L^{-1} of product in the first minute after the reactants are mixed. In the second minute, it generates 0.5 mol L^{-1} of the product, and in the third minute, it generates 0.25 mol L^{-1} of product. Determine the average rate of the reaction in mol L^{-1} s^{-1} during the first 60 s, during the first 120 s, and during the last 120 s.

5. (See Example 14–1.) Use Figure 14–3 to estimate graphically the instantaneous rate of production of NO at $t = 200$ s.

6. (See Example 14–1.) Use Figure 14–3 to estimate graphically the instantaneous rate of production of NO at $t = 50$ s.

7. Give three related expressions for the rate of the reaction

$$N_2(g) + 3 H_2(g) \longrightarrow 2 NH_3(g)$$

assuming that the concentrations of any intermediates are constant and that the volume of the reaction vessel does not change.

8. Give four related expressions for the rate of the reaction

$$2 H_2CO(g) + O_2(g) \longrightarrow 2 CO(g) + 2 H_2O(g)$$

assuming that the concentrations of any intermediates are constant and that the volume of the reaction vessel does not change.

9. At one instant in the course of the reaction $N_2(g) + 3 H_2(g) \rightarrow 2 NH_3(g)$, the rate of appearance of $NH_3(g)$ is 0.150 mol L^{-1} s^{-1}.
(a) What is the rate of disappearance of $H_2(g)$ at that instant?
(b) What is the rate of disappearance of $N_2(g)$ at that instant?

10. In the reaction $SO_2(g) + \frac{1}{3} O_3(g) \rightarrow SO_3(g)$, the rate of disappearance of the reactant $SO_2(g)$ at a certain instant is 0.100 mol L^{-1} s^{-1}.
(a) What is the rate of disappearance of $O_3(g)$ at that instant?
(b) What is the rate of appearance of $SO_3(g)$ at that instant?

Reaction Rates and Concentrations

11. Determine the overall order of the reaction, given the rate law
(a) rate $= k[NO_2]^2$.
(b) rate $= k[I_2]$.

12. Determine the overall order of the reaction, given the rate law
(a) rate $= k[CH_3CHO]^{3/2}$.
(b) rate $= k$.

13. Give the units for the rate constant k of a third-order reaction if concentrations are measured in mol L^{-1} and time is measured in seconds.

14. Give the units for the rate constant k of a one-half–order reaction if concentrations are measured in mol L^{-1} and time is measured in seconds.

15. (See Example 14–2.) The following data were obtained for the reaction $2 NOBr(g) \rightarrow 2 NO(g) + Br_2(g)$ at a certain temperature:

[NOBr] (mol L^{-1})	Rate (mol L^{-1} s^{-1})
0.0130	1.35×10^{-4}
0.0088	6.2×10^{-5}
0.0049	1.92×10^{-5}

(a) Write the rate law for this reaction.
(b) Calculate the rate constant.

16. (See Example 14–2.) The following data were obtained for the reaction $C_4H_8(g) \rightarrow$ products at 750 K:

[C$_4$H$_8$] (mol L^{-1})	Rate (mol L^{-1} s^{-1})
0.0095	3.42×10^{-5}
0.0070	2.52×10^{-5}
0.0025	9.0×10^{-6}

(a) Write the rate law for this reaction.
(b) Calculate the rate constant.

17. (See Example 14–3.) At a fixed pH of 2.0, the rate law for the reaction $2 H^+(aq) + 2 I^-(aq) + H_2O_2(aq) \rightarrow I_2(aq) + 2 H_2O(\ell)$ is

$$\text{rate} = k[H_2O_2][I^-]$$

State the effect on the rate of
(a) increasing the concentration of I^- by a factor of 10.
(b) decreasing the concentration of H_2O_2 by a factor of 4.

18. (See Example 14–3.) The reaction

$$2 ClO_2(aq) + 2 OH^-(aq) \longrightarrow$$
$$ClO_3^-(aq) + ClO_2^-(aq) + H_2O(\ell)$$

has the rate law

$$\text{rate} = k[ClO_2]^2[OH^-]$$

State the effect on the rate of
(a) increasing the concentration of ClO_2 by a factor of 3.
(b) changing the pH from 12.0 to 11.0.

19. Nitrogen monoxide reacts with hydrogen at elevated temperatures according to the following chemical equation:

$$2 NO(g) + 2 H_2(g) \longrightarrow N_2(g) + 2 H_2O(g)$$

When the concentration of H_2 is cut in half, the rate of the reaction also is cut in half; when the concentration of NO is multiplied by 10, the rate of the reaction increases by a factor of 100.
(a) Write the rate law for this reaction, and give the units of the rate constant (k).

(b) If [NO] is multiplied by 3, and $[H_2]$ by 2, what change in the rate will be observed?

20. In the presence of vanadium(V) oxide, $SO_2(g)$ reacts with an excess of oxygen to give $SO_3(g)$:

$$SO_2(g) + \tfrac{1}{2} O_2(g) \xrightarrow{V_2O_5} SO_3(g)$$

This reaction is an important step in the manufacture of sulfuric acid. It is observed that tripling the SO_2 concentration increases the rate by a factor of 3 but that tripling the SO_3 concentration *decreases* the rate by a factor of $1.7 \approx \sqrt{3}$. The rate is insensitive to the concentration of O_2 as long as an excess of O_2 is present.
(a) Write the rate law for this reaction, and give the units of the rate constant (k).
(b) If $[SO_2]$ is multiplied by 2, and $[SO_3]$, by 4, but all other conditions are unchanged, what change in the rate is observed?

21. (See Example 14–4.) In a study of the reaction of pyridine (C_5H_5N) with methyl iodide (CH_3I) in a benzene solution, the following initial reaction rates were measured at 25°C for different initial concentrations of the two reactants:

$[C_5H_5N]$ (mol L^{-1})	$[CH_3I]$ (mol L^{-1})	Initial Rate (mol L^{-1} s^{-1})
1.00×10^{-4}	1.00×10^{-4}	7.5×10^{-7}
2.00×10^{-4}	2.00×10^{-4}	3.0×10^{-6}
2.00×10^{-4}	4.00×10^{-4}	6.0×10^{-6}

(a) Write the rate law for this reaction.
(b) Calculate the rate constant (k) and give its units.
(c) Predict the initial reaction rate that would be seen in a solution in which $[C_5H_5N]$ is 5.0×10^{-5} M and $[CH_3I]$ is 2.0×10^{-5} M.

22. (See Example 14–4.) The initial rate for the oxidation of iron(II) by cerium(IV)

$$Ce^{4+}(aq) + Fe^{2+}(aq) \longrightarrow Ce^{3+}(aq) + Fe^{3+}(aq)$$

is measured at several different initial concentrations of the two reactants:

$[Ce^{4+}]$ (mol L^{-1})	$[Fe^{2+}]$ (mol L^{-1})	Initial Rate (mol L^{-1} s^{-1})
1.1×10^{-5}	1.8×10^{-5}	2.0×10^{-7}
1.1×10^{-5}	2.8×10^{-5}	3.1×10^{-7}
3.4×10^{-5}	2.8×10^{-5}	9.5×10^{-7}

(a) Write the rate law for this reaction.
(b) Calculate the rate constant (k) and give its units.
(c) Predict the initial reaction rate for a solution in which $[Ce^{4+}]$ is 2.6×10^{-5} M and $[Fe^{2+}]$ is 1.3×10^{-5} M.

23. (See Example 14–4.) For the reaction $Cr(H_2O)_6^{3+}(aq) + SCN^-(aq) \rightarrow Cr(H_2O)_5SCN^{2+}(aq) + H_2O(\ell)$, the following data were obtained:

$[Cr(H_2O)_6^{3+}]$ (mol L^{-1})	$[SCN^-]$ (mol L^{-1})	Initial Rate (mol L^{-1} s^{-1})
0.021	0.045	1.9×10^{-9}
0.021	0.084	3.5×10^{-9}
0.053	0.084	8.9×10^{-9}

(a) Determine the rate law for this reaction.
(b) Calculate the rate constant (k) and give its units.

24. (See Example 14–4.) For the reaction $2\,HgCl_2(aq) + C_2O_4^{2-}(aq) \rightarrow Hg_2Cl_2(s) + 2\,Cl^-(aq) + 2\,CO_2(g)$, the following data were obtained:

$[HgCl_2]$ (mol L^{-1})	$[C_2O_4^{2-}]$ (mol L^{-1})	Initial Rate (mol L^{-1} s^{-1})
0.096	0.13	2.1×10^{-7}
0.096	0.21	5.5×10^{-7}
0.171	0.21	9.8×10^{-7}

(a) Determine the rate law for this reaction.
(b) Calculate the rate constant (k) and give its units.

The Dependence of Concentrations on Time

25. (See Example 14–5.) The reaction $SO_2Cl_2(g) \rightarrow SO_2(g) + Cl_2(g)$ is first order, with a half-life of 4.5×10^4 s at 320°C.
(a) Calculate the rate constant (k).
(b) What fraction of the $SO_2Cl_2(g)$ initially present remains unreacted after 16.2 h at 320°C?

26. (See Example 14–5.) At 322°C, the reaction $FClO_2(g) \rightarrow FClO(g) + O(g)$ is first order, with a half-life of 1.48×10^3 s.
(a) Calculate the rate constant (k).
(b) What fraction of the $FClO_2(g)$ initially present has reacted after 2.30 h at 322°C?

27. Chloroethane decomposes at elevated temperatures according to the equation

$$C_2H_5Cl(g) \longrightarrow C_2H_4(g) + HCl(g)$$

This reaction obeys first-order kinetics. After 340 s at 800 K, a measurement shows that the concentration of $C_2H_5Cl(g)$ has decreased from 0.0098 mol L^{-1} to 0.0016 mol L^{-1}. Calculate the rate constant and the half-life ($t_{1/2}$) at 800 K.

28. The isomerization reaction

$$CH_3NC(g) \longrightarrow CH_3CN(g)$$

obeys the first-order rate law

$$\text{rate} = k[CH_3NC]$$

in the presence of an excess of argon. Measurements at 500 K reveal that in 520 s the concentration of $CH_3NC(g)$ decreases to 71% of its original value. Calculate the rate constant (k) and the half-life ($t_{1/2}$) at 500 K.

29. The first-order reaction A → products has the following rate law: rate = $k[A]$, with $k = 2.0$ s^{-1}. Indicate which of the following statements about this reaction are true and which are false. Explain your reasoning.
    (a) The reaction slows down as time goes on.
    (b) After 2.0 s the reaction is over.
    (c) The rate of the reaction doubles if the concentration of A is doubled while all other conditions remain unchanged.
    (d) The half-life of the reaction is 2.0 s.
    (e) The half-life of the reaction is 0.50 s.

30. The second-order reaction 2 B → products has the following rate law: rate = $k[B]^2$, with $k = 2.0$ L mol^{-1} s^{-1}. Indicate which of the following statements about this reaction are true and which are false. Explain your reasoning.
    (a) The concentration of B increases as time goes on.
    (b) After 1.0 s the reaction is half over.
    (c) The rate of the reaction quadruples if the concentration of B is doubled while all other conditions remain unchanged.
    (d) A plot of [B] versus time gives a straight line.
    (e) A plot of 1/[B] versus time gives a straight line.

31. (See Example 14–6.) At 25°C in CCl$_4$ solution, the reaction

    $$I + I \longrightarrow I_2$$

    is second order in the concentration of the iodine atoms and the rate constant (k) equals 8.2×10^9 L mol^{-1} s^{-1}. Suppose that the initial concentration of I atoms is 1.00×10^{-4} mol L^{-1}. Calculate their concentration after 2.0×10^{-6} s.

32. (See Example 14–6.) HO$_2$ is a highly reactive chemical species that plays a role in atmospheric chemistry. The reaction

    $$HO_2(g) + HO_2(g) \longrightarrow H_2O_2(g) + O_2(g)$$

    is second order in [HO$_2$], with a rate constant at 25°C of 1.4×10^9 L mol^{-1} s^{-1}. Suppose some HO$_2$ could be confined at an initial concentration of 2.0×10^{-8} mol L^{-1} at 25°C. Calculate the concentration that would remain after 1.0 s, assuming that no other reactions take place.

33. The decomposition of N$_2$O$_5$ dissolved in CCl$_4$ has a rate constant of 5.5×10^{-4} s^{-1} at a certain temperature. A 0.200 M solution of N$_2$O$_5$ is prepared at this temperature. Plot the concentration of N$_2$O$_5$ as a function of time for the first hour (3600 s) after mixing, assuming that no other reactions involving N$_2$O$_5$ occur (including back-reaction of the products).

34. At a certain temperature, the dimerization of C$_2$F$_4(g)$ to C$_4$F$_8(g)$ has a rate constant of 0.0600 L mol^{-1} s^{-1}. A container is prepared in which the initial concentration of C$_2$F$_4(g)$ is 0.200 mol L^{-1}.
    (a) Plot the concentration of C$_2$F$_4$ as a function of time for the first 180 s after the reaction starts, assuming that no other reactions involving C$_2$F$_4$ occur (including back-reaction of the product).

    (b) On the same graph paper, plot the concentration of C$_4$F$_8(g)$ as a function of time over the same period, making the same assumption.

Reaction Mechanisms

35. Identify each of the following elementary reactions as unimolecular, bimolecular, or termolecular, and write the rate law:
    (a) $HCO(g) + O_2(g) \longrightarrow HO_2(g) + CO(g)$
    (b) $CH_3(g) + O_2(g) + N_2(g) \longrightarrow CH_3O_2(g) + N_2(g)$
    (c) $HO_2NO_2(g) \longrightarrow HO_2(g) + NO_2(g)$

36. Identify each of the following elementary reactions as unimolecular, bimolecular, or termolecular, and write the rate law.
    (a) $BrONO_2(g) \longrightarrow BrO(g) + NO_2(g)$
    (b) $HO(g) + NO_2(g) + Ar(g) \longrightarrow HNO_3(g) + Ar(g)$
    (c) $O(g) + H_2S(g) \longrightarrow HO(g) + HS(g)$

37. (See Example 14–7.) Consider the following reaction mechanism:

    $$H_2O_2(g) \longrightarrow H_2O(g) + O(g)$$
    $$O(g) + CF_2Cl_2(g) \longrightarrow ClO(g) + CF_2Cl(g)$$
    $$ClO(g) + O_3(g) \longrightarrow Cl(g) + 2\,O_2(g)$$
    $$Cl(g) + CF_2Cl(g) \longrightarrow CF_2Cl_2(g)$$

    (a) What is the molecularity of each elementary step?
    (b) Write the overall equation for the reaction.
    (c) Identify the reaction intermediates.

38. (See Example 14–7.) Consider the following reaction mechanism:

    $$NO_2Cl(g) \longrightarrow NO_2(g) + Cl(g)$$
    $$Cl(g) + H_2O(g) \longrightarrow HCl(g) + OH(g)$$
    $$OH(g) + NO_2(g) + N_2(g) \longrightarrow HNO_3(g) + N_2(g)$$

    (a) What is the molecularity of each elementary step?
    (b) Write the overall equation for the reaction.
    (c) Identify the reaction intermediate(s).

39. The rate constant of the elementary reaction

    $$BrO(g) + NO(g) \longrightarrow Br(g) + NO_2(g)$$

    is 1.3×10^{10} L mol^{-1} s^{-1} at 25°C, and its equilibrium constant is 5.0×10^{10} at this temperature. Calculate the rate constant at 25°C of the elementary reaction

    $$Br(g) + NO_2(g) \longrightarrow BrO(g) + NO(g)$$

40. The compound IrH$_3$(CO)(P(C$_6$H$_5$)$_3$)$_2$ exists in two forms: the meridional ("mer") and the facial ("fac"). At 25°C in a nonaqueous solvent, the reaction mer → fac has a rate constant of 2.33 s^{-1}, and the reaction fac → mer has a rate constant of 2.10 s^{-1}. What is the equilibrium constant of the mer-to-fac reaction at 25°C?

Reaction Mechanisms and Rate Laws

41. (See Example 14–8.) Write the overall reaction and the rate laws that correspond to the following reaction mechanisms.

Be sure to eliminate intermediates from the answers:

(a) $A + B \underset{k_{-1}}{\overset{k_1}{\rightleftharpoons}} C + D$ (fast equilibrium)

$C + E \xrightarrow{k_2} F$ (slow)

(b) $A \underset{k_{-1}}{\overset{k_1}{\rightleftharpoons}} B + C$ (fast equilibrium)

$C + D \underset{k_{-2}}{\overset{k_2}{\rightleftharpoons}} E$ (fast equilibrium)

$E \xrightarrow{k_3} F$ (slow)

42. (See Example 14–8.) Write the overall reaction and the rate laws that correspond to the following mechanisms. Be sure to eliminate intermediates from the answers:

(a) $2 A + B \underset{k_{-1}}{\overset{k_1}{\rightleftharpoons}} D$ (fast equilibrium)

$D + B \xrightarrow{k_2} E + F$ (slow)

$F \xrightarrow{k_3} G$ (slow)

(b) $A + B \underset{k_{-1}}{\overset{k_1}{\rightleftharpoons}} C$ (fast equilibrium)

$C + D \underset{k_{-2}}{\overset{k_2}{\rightleftharpoons}} F$ (fast equilibrium)

$F \xrightarrow{k_3} G$ (slow)

43. HCl reacts with propene (CH_3CHCH_2) according to

$$HCl(g) + CH_3CHCH_2(g) \longrightarrow CH_3CHClCH_3(g)$$

The experimental rate law is

$$\text{rate} = k[HCl]^3[CH_3CHCH_2]$$

Which, if any, of the following mechanisms are consistent with the observed rate law?

(a) $HCl + HCl \rightleftharpoons H + HCl_2$ (fast equilibrium)
$H + CH_3CHCH_2 \longrightarrow CH_3CHCH_3$ (slow)
$HCl_2 + CH_3CHCH_3 \longrightarrow CH_3CHClCH_3 + HCl$ (fast)

(b) $HCl + HCl \rightleftharpoons H_2 + Cl_2$ (fast equilibrium)
$HCl + CH_3CHCH_2 \rightleftharpoons CH_3CHClCH_3^*$ (fast equilibrium)
$CH_3CHClCH_3^* + H_2Cl_2 \longrightarrow CH_3CHClCH_3 + 2 HCl$ (slow)

(c) $HCl + CH_3CHCH_2 \rightleftharpoons H + CH_3CHClCH_2$
    (fast equilibrium)
$H + HCl \rightleftharpoons H_2Cl$ (fast equilibrium)
$H_2Cl + CH_3CHClCH_2 \longrightarrow HCl + CH_3CHClCH_3$ (slow)

44. Chlorine reacts with hydrogen sulfide in aqueous solution:

$$Cl_2(aq) + H_2S(aq) \longrightarrow S(s) + 2 H^+(aq) + 2 Cl^-(aq)$$
$$\text{rate} = k[Cl_2][H_2S]$$

Which, if any, of the following mechanisms are consistent with the observed rate law?

(a) $Cl_2 + H_2S \longrightarrow H^+ + Cl^- + Cl^+ + HS^-$ (slow)
$Cl^+ + HS^- \longrightarrow H^+ + Cl^- + S$ (fast)

(b) $H_2S \rightleftharpoons HS^- + H^+$ (fast equilibrium)
$HS^- + Cl_2 \longrightarrow 2 Cl^- + S + H^+$ (slow)

(c) $H_2S \rightleftharpoons HS^- + H^+$ (fast equilibrium)
$H^+ + Cl_2 \rightleftharpoons H^+ + Cl^- + Cl^+$ (fast equilibrium)
$Cl^+ + HS^- \longrightarrow H^+ + Cl^- + S$ (slow)

45. Nitryl chloride is a reactive gas. Experiments establish that the following equation and rate law describe its decomposition over a wide range of temperatures:

$$2 NO_2Cl(g) \longrightarrow 2 NO_2(g) + Cl_2(g) \qquad \text{rate} = k[NO_2Cl]$$

Which, if any, of the following mechanisms are consistent with the observed rate law?

(a) $NO_2Cl \longrightarrow NO_2 + Cl$ (slow)
$Cl + NO_2Cl \longrightarrow NO_2 + Cl_2$ (fast)

(b) $2 NO_2Cl \rightleftharpoons N_2O_4 + Cl_2$ (fast equilibrium)
$N_2O_4 \longrightarrow 2 NO_2$ (slow)

(c) $2 NO_2Cl \rightleftharpoons ClO_2 + N_2O + ClO$ (fast equilibrium)
$N_2O + ClO_2 \rightleftharpoons NO_2 + NOCl$ (fast equilibrium)
$NOCl + ClO \longrightarrow NO_2 + Cl_2$ (slow)

46. Ozone in the upper atmosphere is decomposed by nitrogen monoxide through the reaction

$$O_3(g) + NO(g) \longrightarrow O_2(g) + NO_2(g)$$

The experimental rate law for this reaction is

$$\text{rate} = k[O_3][NO]$$

Which, if any, of the following mechanisms are consistent with the observed rate law?

(a) $O_3 + NO \longrightarrow O + NO_3$ (slow)
$O + O_3 \longrightarrow 2 O_2$ (fast)
$NO_3 + NO \longrightarrow 2 NO_2$ (fast)

(b) $O_3 + NO \longrightarrow O_2 + NO_2$ (slow)

(c) $NO + NO \rightleftharpoons N_2O_2$ (fast equilibrium)
$N_2O_2 + O_3 \longrightarrow NO + NO_2 + O_2$ (slow)

The Effect of Temperature on Reaction Rates

47. (See Example 14–9.) The rate of the elementary reaction

$$Ar(g) + O_2(g) \longrightarrow Ar(g) + O(g) + O(g)$$

has been studied as a function of temperature between 5000 and 18,000 K. The following data were obtained for the rate constant:

T (K)	k (L mol^{-1} s^{-1})
5,000	5.49×10^6
10,000	9.86×10^8
15,000	5.09×10^9
18,000	8.60×10^9

(a) Calculate the activation energy of this reaction.
(b) Calculate the factor A in the Arrhenius equation for the temperature dependence of the rate constant.

48. (See Example 14–9.) The reaction

$$H(g) + D_2(g) \longrightarrow HD(g) + D(g)$$

is the exchange of isotopes of hydrogen of mass number 1 (H) and 2 (D, deuterium). The following data were obtained for the rate constant of this reaction:

T (K)	k (L mol^{-1} s^{-1})
299	1.56×10^4
327	3.77×10^4
346	7.6×10^4
440	1.07×10^6
549	8.7×10^6
745	8.7×10^7

(a) Calculate the activation energy of this reaction.
(b) Calculate the Arrhenius parameter (A) for this reaction.

49. The rate constant of the elementary reaction

$$BH_4^-(aq) + NH_4^+(aq) \longrightarrow BH_3NH_3(aq) + H_2(g)$$

is $k = 1.94 \times 10^{-4}$ L mol^{-1} s^{-1} at 30.0°C, and the reaction has an activation energy of 161 kJ mol^{-1}.
(a) Compute the rate constant (k) of the reaction at 40.0°C.
(b) Equal concentrations of $BH_4^-(aq)$ and $NH_4^+(aq)$ are mixed at 30.0°C. After 1.00×10^4 s half of each has reacted. How long will it take to consume half of the reactants if an identical experiment is performed at 40.0°C?

50. Dinitrogen tetraoxide (N_2O_4) decomposes spontaneously at room temperature in the gas phase:

$$N_2O_4(g) \longrightarrow 2\, NO_2(g)$$

At 30°C, $k = 5.1 \times 10^6$ s^{-1}, and the activation energy of the reaction is 54.0 kJ mol^{-1}.
(a) Calculate the time it takes, in seconds, for the partial pressure of $N_2O_4(g)$ to decrease from 0.100 atm to 0.099 atm at 30°C.
(b) Repeat the calculation of part (a) for the reaction at 60°C.

51. The activation energy of the isomerization reaction of CH_3NC in Problem 28 is 161 kJ mol^{-1} and the reaction rate constant at 600 K is 0.41 s^{-1}.
(a) Calculate the Arrhenius parameter (A) for this reaction.
(b) Calculate the rate constant for this reaction at 1000 K.

52. Cyclopropane isomerizes to propylene in a first-order process:

$$\text{cyclopropane} \longrightarrow \text{propylene}$$

The activation energy is 272 kJ mol^{-1}. At 500°C, the rate constant is 6.1×10^{-4} s^{-1}.
(a) Calculate the Arrhenius parameter (A) for this reaction.
(b) Calculate the rate constant for this reaction at 25°C.

53. The activation energy of the gas-phase reaction

$$OH(g) + HCl(g) \longrightarrow H_2O(g) + Cl(g)$$

is 3.5 kJ mol^{-1}, and the change in the internal energy of reaction ΔE_r is -66.8 kJ mol^{-1}. Calculate the activation energy of the *reverse* of the reaction.

54. The compound HOCl is known, but the related compound HClO, in which the atoms are connected in a different order

in the molecule, is not known. Calculations indicate that the activation energy of the conversion $HOCl(g) \rightarrow HClO(g)$ is 311 kJ mol^{-1} and that the activation energy of the conversion $HClO(g) \rightarrow HOCl(g)$ is 31 kJ mol^{-1}. Estimate ΔE_r for the reaction $HOCl(g) \rightarrow HClO(g)$.

Kinetics of Catalysis

55. How would you describe the role of the CF_2Cl_2 in the reaction mechanism of Problem 37?

56. Compare homogeneous catalysis with heterogeneous catalysis. Give an example of each.

57. One of the mottos of "green chemistry" (environmentally benign chemistry) is that "catalytic reactants are superior to stoichiometric reactants." Explain this statement.

58. (See "Chemistry in Your Life: Catalytic Converters.") Why are the platinum and rhodium in an automobile's catalytic converter deposited in thin layers on a honeycomb of a supporting substance? Why must unleaded gasoline be used in automobiles equipped with catalytic converters?

ADDITIONAL PROBLEMS

59. A freight train traveling west starts up a long mountain slope at a speed of 60.0 mi per hour (mph). By the end of the first minute, its speed has slowed to 59.0 mph, and it slows down by 1.0 mph every minute thereafter because of the steepness of the grade. Compute its average speed during the first minute, during the first 3 min, during the second 3 min, and during the first 20 min.

60. Jones writes a reaction as $C_2H_6(g) \rightarrow C_2H_4(g) + H_2(g)$ and concludes that it is a first-order reaction. Smith writes the same reaction as $2\, C_2H_6(g) \rightarrow 2\, C_2H_4(g) + 2\, H_2(g)$ and concludes that it is a second-order reaction. Both are amazed to learn that the reaction is 3/2 order. Explain the misunderstanding.

61. Hemoglobin (Hb) molecules in blood bind oxygen and carry it to cells, where it takes part in metabolism. The first step in the binding of oxygen is

$$Hb(aq) + O_2(aq) \longrightarrow (HbO_2)(aq)$$
$$\text{rate} = 4 \times 10^7 \text{ L mol}^{-1} \text{ s}^{-1} [Hb][O_2]$$

Calculate the initial rate at which oxygen is bound to hemoglobin if the concentration of hemoglobin is 2×10^{-9} mol L^{-1} and that of oxygen is 5×10^{-5} mol L^{-1}.

62. A compound called di-t-butyl peroxide (abbreviation: DTBP) decomposes to give acetone and ethane:

$$(CH_3)_3COOC(CH_3)_3(g) \longrightarrow 2\,(CH_3)_2CO(g) + C_2H_6(g)$$

The total pressure of the reaction mixture changes with time, as shown by the following data at 147.2°C:

Time (min)	P_{tot} (atm)	Time (min)	P_{tot} (atm)
0	0.2362	26	0.3322
2	0.2466	30	0.3449
6	0.2613	34	0.3570
10	0.2770	38	0.3687
14	0.2911	40	0.3749
18	0.3051	42	0.3801
20	0.3122	46	0.3909
22	0.3188	—	—

(a) Calculate the partial pressure of DTBP at each time from these data. Assume that at time 0, DTBP is the only gas present. Recall that each decrease by x atm in the partial pressure of DTBP is accompanied by an increase of $2x$ in the total pressure because every gas molecule that dissociates gives three new gas molecules.

(b) Are the data better described by a first-order or a second-order rate law with respect to DTBP concentration?

63. A total of 8.23×10^{-3} mol of InCl(s) is placed in 1.00 L of 0.010 M HCl(aq) at 75°C. The InCl(s) dissolves quite quickly, and then the following reaction occurs:

$$3 \, In^+(aq) \longrightarrow 2 \, In(s) + In^{3+}(aq)$$

As this disproportionation proceeds, the solution is analyzed at intervals to determine the concentration of In$^+$(aq) that remains:

Time (s)	[In$^+$] (mol L^{-1})
0	8.23×10^{-3}
240	6.41×10^{-3}
480	5.00×10^{-3}
720	3.89×10^{-3}
1,000	3.03×10^{-3}
1,200	3.03×10^{-3}
10,000	3.03×10^{-3}

(a) Plot ln [In$^+$] versus time, and determine the apparent rate constant for this first-order reaction.

(b) Determine the half-life of this reaction.

(c) Determine the equilibrium constant (K) for the reaction under the experimental conditions.

64. Carbon dioxide reacts with ammonia to give ammonium carbamate, a solid. The reverse reaction also occurs:

$$CO_2(g) + 2 \, NH_3(g) \rightleftharpoons NH_4OCONH_2(s)$$

The forward reaction is first order in $CO_2(g)$ and second order in $NH_3(g)$. Its rate constant is 0.238 atm^{-2} s^{-1} at 0.0°C (expressed in terms of partial pressures rather than concentrations). The reaction in the reverse direction is zero order,

and its rate constant, at the same temperature, is 1.60×10^{-7} atm s^{-1}. Experimental studies show that, at all stages in the progress of this reaction, the net rate is equal to the forward rate minus the reverse rate. Compute the equilibrium constant of this reaction at 0.0°C.

65. A reaction has order of negative one (-1) with respect to H$^+$ ion. How is its rate affected if the pH of the reaction mixture is raised from 3.00 to 4.00?

66. Would the reaction considered in Problem 65 go faster in a buffer containing equal concentrations of ammonia and ammonium ions or in a buffer containing equal concentrations of acetic acid and sodium acetate? (See Table 8–2 for values of K_a).

67. The reaction of OH$^-$ with HCN in aqueous solution at 25°C has a forward rate constant (k_{for}) of 3.7×10^9 L mol^{-1} s^{-1}. Using this information and the measured acidity constant of HCN (see Table 8–2), calculate the rate constant (k_{rev}) in the first-order rate law

$$rate = k_{rev}[CN^-]$$

for the transfer of hydrogen ions to CN$^-$ from surrounding water molecules:

$$H_2O(\ell) + CN^-(aq) \longrightarrow OH^-(aq) + HCN(aq)$$

*68. When pure solid magnesium perchlorate (Mg(ClO$_4$)$_2$) is heated to 440°C, it first breaks down to give magnesium chloride (MgCl$_2$) and oxygen. Then, as the product magnesium chloride builds up, it starts to react with the magnesium perchlorate to generate magnesium oxide, chlorine, and oxygen. Pure magnesium chloride undergoes no changes at 440°C.

(a) Write balanced chemical equations to represent the two processes.

(b) There is always some magnesium chloride mixed with the magnesium oxide residue from the decomposition of magnesium perchlorate at 440°C. On the basis of this fact, determine which of the two reactions in part (a) goes faster.

69. The kinetics of the reaction

$$I_2(aq) + CH_3COCH_3(aq) \longrightarrow CH_3COCH_2I(aq) + HI(aq)$$

are being studied. Iodine (I$_2$) forms a dark blue complex with starch. Putting a small amount of starch in the reaction mixture colors it blue until all the iodine is used up. Use the data reported below to find the rate law for this reaction, including an approximate value of k, the rate constant. All experiments take place at 25°C. Assume that the average rate of the reaction in each experiment equals its initial rate.

Experiment No.	Initial [I$_2$] (mol L^{-1})	Initial [CH$_3$COCH$_3$] (mol L^{-1})	Time Required for Blue Color to Fade (s)
1	2.0×10^{-4}	0.80	100
2	1.6×10^{-4}	0.80	98
3	1.2×10^{-4}	0.40	201

70. Millions of kilograms of acetone (CH_3COCH_3), an important industrial solvent, evaporate each year. Acetone in the air is attacked by the very reactive hydroxyl radical:

$$CH_3COCH_3(g) + OH(g) \longrightarrow H_2O(g) + \text{other products}$$
$$\text{rate} = k[CH_3COCH_3][OH]$$

Recent experiments have established that the rate constant for this reaction equals 1.10×10^8 L mol^{-1} s^{-1} at 300 K. Other studies establish that the OH radical has a steady average concentration of 8.4×10^8 molecules per liter in the lower atmosphere. Estimate the half-life (in days) of acetone in the lower atmosphere.

71. The gas-phase decomposition of acetaldehyde can be represented by the overall chemical equation

$$CH_3CHO(g) \longrightarrow CH_4(g) + CO(g)$$

It is thought to occur through this sequence of elementary steps:

$$CH_3CHO \longrightarrow CH_3 + CHO$$
$$CH_3 + CH_3CHO \longrightarrow CH_4 + CH_2CHO$$
$$CH_2CHO \longrightarrow CO + CH_3$$
$$CH_3 + CH_3 \longrightarrow CH_3CH_3$$

Show that this reaction mechanism corresponds to a chain reaction, and identify the initiation, propagation, and termination steps.

72. Lanthanum(III) phosphate crystallizes as a hemihydrate, $LaPO_4 \cdot \frac{1}{2} H_2O$. When it is heated, it loses water to give anhydrous lanthanum(III) phosphate:

$$2\,LaPO_4 \cdot \tfrac{1}{2} H_2O(s) \longrightarrow 2\,LaPO_4(s) + H_2O(g)$$

This reaction is first order in the chemical amount of the reactant. The rate constant varies with temperature as follows:

Temperature (°C)	k (s^{-1})
205	2.3×10^{-4}
219	3.69×10^{-4}
246	7.75×10^{-4}
260	12.3×10^{-4}

Compute the activation energy of this reaction.

73. The water in a pressure cooker boils at a temperature greater than 100°C because it is under pressure. At this higher temperature, the chemical reactions associated with the cooking of food take place at a greater rate.

(a) Some food cooks fully in 5 min in a pressure cooker at 112°C and in 10 min in an open pot at 100°C. Estimate the average activation energy for the reactions associated with the cooking of this food.

(b) How long does the same food take to cook in an open pot of boiling water in Denver, where the average atmospheric pressure is 0.818 atm and the boiling point of water is 94.4°C?

*74. The figure below shows the decrease in the concentration of A as it reacts from an initial concentration of 1.00 mol L^{-1} according to three rate laws: one zeroth-order, one first-order in A, and one second-order in A. The numerical values of the rate constants are the same in each case.

(a) From the graph, estimate the first half-life of the reaction according to each rate law. This equals the time it takes for the concentration of A to reach half of its original value. Estimate the second half-life of each reaction (the time it takes for the concentration of A to reach to half the value it had after the first half-life). Use these

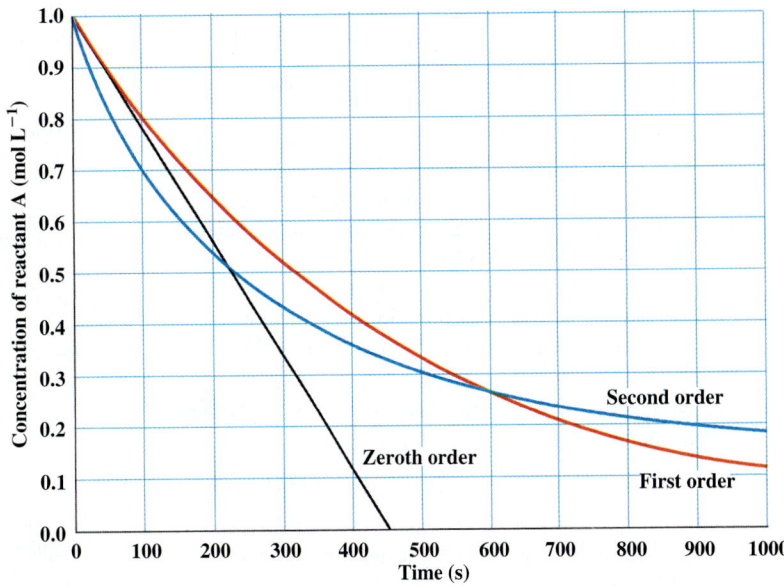

estimates to predict the third half-life of the reaction according to each rate law.

(b) Determine the rate constant for each rate law. Take care to include the proper units.

(c) Determine the rate of the reaction according to each rate law after 300 s has elapsed.

(d) Explain, on the basis of molecular collisions, why the zeroth-order reaction does not start fastest but finishes first and why the second-order reaction starts out fastest but slows down the most.

75. A stream of gaseous H_2 is directed onto finely divided platinum powder in the open air. The metal immediately glows white-hot and continues to do so as long as the stream continues. Explain.

76. (a) A certain first-order reaction has an activation energy of 53 kJ mol^{-1}. It is run twice, first at 298 K and then at 308 K (10°C higher). All other conditions are identical. Show that, in the second run, the reaction occurs at double its rate in the first run.

(b) The same reaction is run twice more, but these runs take place at 398 K and 408 K. Show that the reaction goes 1.5 times faster at 408 K than it does at 398 K.

*77. The gas-phase reaction between hydrogen and iodine

$$H_2(g) + I_2(g) \underset{k_{-1}}{\overset{k_1}{\rightleftharpoons}} 2\,HI(g)$$

proceeds with a forward rate constant at 1000 K of $240 \text{ L mol}^{-1} \text{ s}^{-1}$ and an activation energy of 165 kJ mol^{-1}. By using this information and data from Appendix D, calculate the activation energy for the reverse reaction and the value of k_{-1} at 1000 K. Assume that ΔH_r° and ΔS_r° for this reaction are independent of temperature between 298.15 and 1000 K.

78. The following reaction mechanism has been proposed for a chemical reaction:

$$A_2 \underset{k_{-1}}{\overset{k_1}{\rightleftharpoons}} A + A \qquad \text{(fast equilibrium)}$$

$$A + B \underset{k_{-2}}{\overset{k_2}{\rightleftharpoons}} AB \qquad \text{(fast equilibrium)}$$

$$AB + CD \overset{k_3}{\longrightarrow} AC + BD \qquad \text{(slow)}$$

(a) Write a balanced equation for the overall reaction.

(b) Write the rate law that corresponds to the preceding mechanism. Express the rate in terms of concentrations of reactants only (A_2, B, CD).

(c) Suppose that the first two steps in the preceding mechanism are endothermic and the third one is exothermic. Does an increase in temperature *increase* the reaction rate constant, *decrease* it, or cause *no change*? Explain.

79. (See "Chemistry in Your Life: Catalytic Converters.") The combustion of octane

$$2\,C_8H_{18}(g) + 25\,O_2(g) \longrightarrow 16\,CO_2(g) + 18\,H_2O(g)$$

often is used to represent the burning of gasoline. Give three arguments against a one-step elementary reaction as the mechanism for this reaction.

80. Chlorine molecules, Cl_2, tend to be fairly unreactive, but chlorine atoms, $Cl\cdot$, like most free radicals, are very reactive. The reaction $Cl_2(g) \rightleftharpoons 2\,Cl\cdot(g)$ reaches equilibrium very quickly. Propose a mechanism that explains why.

81. Glycine is oxidized by permanganate ion in acidic solution

$$NH_2CH_2COOH(aq) + MnO_4^-(aq) \longrightarrow$$
$$MnO_2(s) + CH_2O(aq) + CO_2(g) + NH_4^+(aq)$$

(a) Complete and balance the preceding equation.

(b) The rate of the reaction increases during the first 2500 s, reaches a maximum, and then falls off toward zero. Usually reactions slow down as they proceed. Suggest a reason why this reaction speeds up for a while before it slows down.

82. The nerve gas Soman is a cholinesterase inhibitor. Antidotes to poisoning by Soman are cholinesterase reactivators. What kind of substance is cholinesterase? Suggest what happens to it when it is inhibited by Soman.

83. You and your lab partner measure the rate constant for the reaction

$$S_2O_8^{2-}(aq) + 2\,I^-(aq) \longrightarrow 2\,SO_4^{2-}(aq) + I_2(aq)$$

at 12 different temperatures ranging from 15 to 40°C. You prepare an Arrhenius plot (a plot of $\ln k$ versus $1/T$ as in Fig. 14–11) and calculate the activation energy of the reaction, which comes out to 54 kJ mol^{-1}. You then repeat the 12 runs under identical conditions except that you add a small amount of $Fe^{2+}(aq)$ ion as a catalyst. The reaction runs faster in every case. A plot of $\ln k$ versus $1/T$ gives an activation energy of 63 kJ mol^{-1}. Your lab partner says, "The catalyzed reaction has to have a lower activation energy, not a higher one, because catalysts work by lowering the activation energy of reactions." He accuses you of doing the calculations wrong. You accuse him of blunders in setting up the experiments! In fact, neither the experiments nor the calculations contain any errors. Explain.

*84. The burning of fossil fuels increases the concentration of CO_2 in the atmosphere and may lead to global warming (see Chapter 18). Alarm over this prospect has prompted research into the possible disposal of CO_2 on the deep-ocean floor. In this concept, the pressure of the overlying water maintains large lakes of liquid CO_2 safely on the bottom. CO_2 might, however, leak out of the lakes by the reaction

$$CO_2(\ell) \longrightarrow CO_2(aq)$$

with unknown consequences for the deep-sea environment. Accordingly, the rate of the preceding reaction was studied. Droplets of liquid CO_2 were found to shrink slowly when held under water at high pressure:

Time (h)	Diameter (mm)	
	Droplet 1 (at 28 MPa)	Droplet 2 (at 35 MPa)
0	14.4	13.0
1	12.2	10.5
2	10.1	9.0
3	9.0	7.3
4	7.5	5.9
5	4.6	3.0
6	3.0	—

(a) Plot the diameters of the two droplets as a function of time. Does the rate of dissolution of the liquid CO_2 depend on the pressure in these experiments?

(b) Determine the order of the dissolution reaction in these experiments.

(c) Compute the rate constant for this dissolution. Explain the choice of units in your answer.

85. At 25°C, aqueous hydrogen sulfite ion ($HSO_3^-(aq)$) reacts very rapidly with aqueous mercury(II) ion ($Hg^{2+}(aq)$) to form $HgSO_3(aq)$, which then decomposes as follows:

$$HgSO_3(aq) \longrightarrow Hg(aq) + SO_3(aq) \quad k_1 = 0.0106 \text{ s}^{-1}$$

As it is formed, the $Hg(aq)$ reacts rapidly with additional $Hg^{2+}(aq)$ to form mercury(I) ion:

$$Hg(aq) + Hg^{2+}(aq) \longrightarrow Hg_2^{2+}(aq)$$
$$k_2 = 5.9 \times 10^8 \text{ L mol}^{-1}\text{ s}^{-1}$$

A solution containing 0.0400 mmol L^{-1} of $HSO_3^-(aq)$ and 0.400 mmol L^{-1} of $Hg^{2+}(aq)$ is prepared at 25°C.

(a) Graph the concentration of $HgSO_3(aq)$ as a function of time for the first 360 s after the solution is mixed.

(b) Graph the concentration of $Hg_2^{2+}(aq)$ as a function of time for the same 360 s.

86. According to a recent report, the concentration of atmospheric methane (CH_4) is increasing at a rate of about 0.9% per year. This is a matter of concern because methane in the atmosphere traps solar heat, causing global warming. Make a reasonable assumption about the kinetics of methane pro-

duction and estimate how long it may take for the concentration of methane in the atmosphere to double. Be sure to state the assumption.

*87. Oxygen oxidizes nitrogen monoxide in aqueous solution at pH 1.5 in a reaction that has third-order kinetics:

$$4 \text{ NO}(aq) + O_2(aq) + 2 \text{ H}_2O(\ell) \longrightarrow 4 \text{ HNO}_2(aq)$$
$$\text{rate} = k[\text{NO}]^2[O_2]$$

The third-order rate constant equals $6.4 \times 10^6 \text{ L}^2\text{ mol}^{-2}\text{ s}^{-1}$ at 23°C. A solution containing 6.0×10^{-4} M $NO(aq)$ and 6.0×10^{-4} M $O_2(aq)$ at pH 1.5 is allowed to react at 23°C.

(a) Compute the initial rate of the reaction.

(b) Assume that the reaction continues at its initial rate for 0.01 s. Compute the concentrations of the two reactants after 0.01 s.

(c) Substitute the concentrations computed in part (b) into the rate law to compute the rate at $t = 0.01$ s.

(d) Assume that the reaction continues at the rate computed in part (c) for the next 0.01 s. Compute the rate at $t = 0.02$ s.

(e) By continuing in this way (using 0.01 s intervals), estimate the time required for the concentration of $NO(aq)$ to fall to 3.0×10^{-4} M, which is half of its original value.

88. A fish swims in a lake that contains pollutant Z at a concentration of 2.0×10^{-6} mmol L^{-1}. The fish takes up Z and stores it in its body in a process that has a rate constant equal to $1.0 \times 10^{-4} \text{ s}^{-1}$. The fish eliminates Z in a process that has a rate constant equal to $1.0 \times 10^{-5} \text{ s}^{-1}$. Assuming that the fish survives, compute the concentration of Z in its body when the rates of uptake and elimination balance.

89. The compound azulene ($C_{10}H_8$) is stable at room temperature, but at high temperature it rapidly rearranges to naphthalene (also $C_{10}H_8$). Researchers report that the rearrangement "is a clean and normal unimolecular reaction" between 1300 K and 1900 K. They further report the following experimental Arrhenius equation:

$$k = 8.51 \times 10^{12} \text{ s}^{-1} \exp(-263 \text{ kJ mol}^{-1}/RT)$$

Calculate the temperature at which the half-life of this reaction is half of its value at 1300 K.

CUMULATIVE PROBLEM

Sulfite and Sulfate Kinetics

The pollutant sulfur dioxide dissolves in water droplets (fog, clouds, and rain) in the atmosphere and reacts according to the equation

$$SO_2(aq) + H_2O(\ell) \longrightarrow HSO_3^-(aq) + H^+(aq)$$

The $HSO_3^-(aq)$ is then, much more slowly, oxidized by dissolved oxygen that is also in the droplets:

$$2 \text{ HSO}_3^-(aq) + O_2(aq) \longrightarrow 2 \text{ SO}_4^{2-}(aq) + 2 \text{ H}^+(aq)$$

Sulfur burns readily in oxygen to generate SO_2, which is eventually converted in the atmosphere to sulfuric acid. *(General Motors Corporation)*

Although the second reaction has been studied for many years, only recently was it discovered that it could proceed by the steps

$$2\,HSO_3^-(aq) + O_2(aq) \longrightarrow S_2O_7^{2-}(aq) + H_2O(\ell) \qquad \text{(fast)}$$
$$S_2O_7^{2-}(aq) + H_2O(\ell) \longrightarrow 2\,SO_4^{2-}(aq) + 2\,H^+(aq) \qquad \text{(slow)}$$

The previously undetected intermediate, $S_2O_7^{2-}(aq)$, is well known in other reactions. It is the *disulfate* ion.

In an experiment at 25°C, a solution was mixed with the realistic initial concentrations of 0.270 M $HSO_3^-(aq)$ and 0.0135 M $O_2(aq)$. The initial pH was 3.90. The following table tells what happened in the solution, beginning from the moment of mixing:

Time (s)	$[HSO_3^-]$ (mol L^{-1})	$[O_2]$ (mol L^{-1})	$[S_2O_7^{2-}]$ (mol L^{-1})	$[HSO_4^-] + [SO_4^{2-}]$ (mol L^{-1})
0.000	0.270	0.0135	0.000	0.000
0.010	0.243	0.000	13.5×10^{-3}	0.000
10.0	0.243	0.000	11.8×10^{-3}	3.40×10^{-3}
45.0	0.243	0.000	7.42×10^{-3}	12.2×10^{-3}
90.0	0.243	0.000	4.08×10^{-3}	18.8×10^{-3}
150.0	0.243	0.000	1.84×10^{-3}	23.3×10^{-3}
450.0	0.243	0.000	0.034×10^{-3}	26.9×10^{-3}
600.0	0.243	0.000	0.005×10^{-3}	27.0×10^{-3}

The structure of the disulfate ion, $S_2O_7^{2-}$.

(a) Determine the average rate of increase of the total of the concentrations of the sulfate plus hydrogen sulfate ions during the first 10 s of the experiment.

(b) Determine the average rate of disappearance of hydrogen sulfite ion during the first 0.010 s of the experiment.

(c) Explain why the hydrogen sulfite ion (HSO_3^-) stops disappearing after 0.010 s.

(d) Plot the concentration of disulfate ion versus time on a piece of graph paper. Use the graph to estimate the instantaneous rate of disappearance of disulfate ion 90.0 s after the reaction starts.

(e) Plot the total of the concentrations of sulfate plus hydrogen sulfate ions versus time and estimate the rate of increase of this total 90.0 s after the reaction starts.

(f) Use the graphs of the previous two parts to estimate the rate of the reaction 150 s after the reactants are mixed.

(g) Determine the order with respect to the disulfate ion of the second step of the conversion and also the rate constant of the second step.

(h) Determine the half-life of the second step of the conversion process.

(i) Explain why nothing definite can be concluded from this experiment about the order and rate constant of the first step of the conversion.

(j) At 15°C, the rate constant of the second step of the conversion is only 62% of its value at 25°C. Compute the activation energy of the second step.

(k) The first step of the conversion occurs much faster when 1.0×10^{-6} M $Fe^{2+}(aq)$ ion is added. What role does Fe^{2+} play?

(l) Write a balanced equation for the overall reaction that gives sulfuric acid from SO_2 dissolved in water droplets in the air.

15

Nuclear Chemistry

CHAPTER OUTLINE

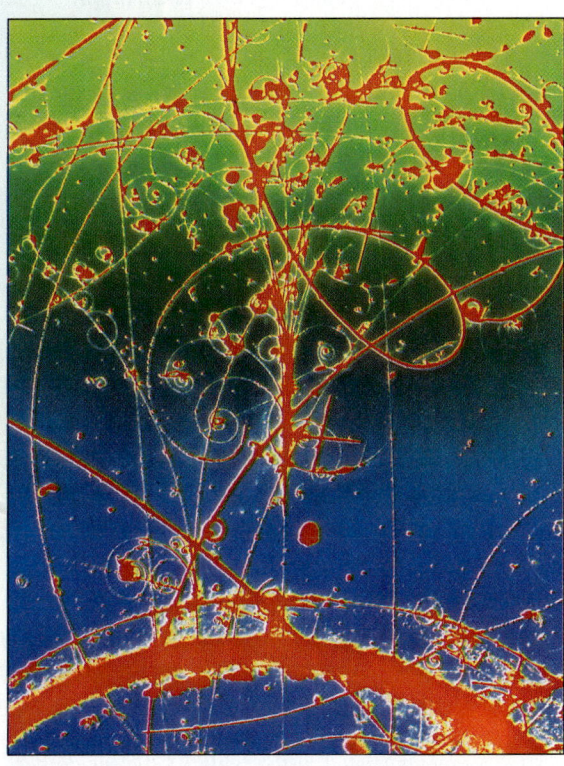

The tracks of subatomic particles in a bubble chamber.
(Science Photo Library/Photo Researchers Inc.)

This book has so far focused on the bulk properties of matter: the masses of reactants and yields of products; melting, boiling, and sublimation; and the drift of reaction systems toward equilibrium. Until the end of the 19th century, chemistry was principally concerned with such *macroscopic* properties, and it was appropriate to begin with such a viewpoint. Concepts about the *microscopic* nature of matter do of course appear in the first 14 chapters. The law of definite proportions and the law of combining volumes, macroscopic observations that they are, demanded interpretation in terms of atoms and molecules; the relationship between the energy and temperature of ideal gases required explanation in terms of kinetic-molecular theory. Such excursions into the microscopic world were aimed at clarifying macroscopic chemical phenomena, however, rather than at discovering their ultimate causes.

The point of view now shifts. In this and the following six chapters, we examine matter and chemical phenomena from the inside out, considering the ways in which the building blocks introduced in Section 1–5 (the proton, neutron, and electron) fit together to give matter its properties. We start here with the atomic nucleus and move on in the subsequent chapters to atoms, molecules, and the aggregates of molecules that form liquids and solids.

15–1 Mass–Energy Relationships in Atomic Nuclei

Recall from Section 1–5 that almost all the mass of an atom derives from protons and neutrons contained in a very small volume in its central kernel, or nucleus. The nucleus occupies less than one *trillionth* of the space in the atom and is held together by strong forces that isolate it from the influence of nearby electrons and other nuclei. The term **nuclide** is applied to an atom when the exact constitution of its nucleus is known or is important. A nuclide is characterized by the number of protons, Z, and the number of neutrons, N, that its nucleus contains. The atomic number, Z, determines the charge $+Ze$ on the nucleus and controls the chemical identity of the nuclide; the sum $Z + N = A$ is the mass number of the nuclide and equals the integer closest to the relative atomic mass of the nuclide. Symbols of the form $^A_Z El$, where El is the chemical symbol of the atom, represent nuclides fully. Isotopes are nuclides that have the same atomic number Z but different mass numbers A.

The kilogram and gram are inconveniently large units for measuring the masses of single atoms (single nuclides) and the elementary particles that make them up. The atomic mass unit (u) is better suited for measurements on this scale. The definition of this tiny unit of mass has been arranged deliberately so that Avogadro's number of atomic mass units equals 1 g of mass:

$$6.0221420 \times 10^{23}\ u = 1\ g$$

As a consequence, the mass of a single atom expressed in atomic mass units is numerically equal to the mass of 1 mol of such atoms expressed in grams. Table 15–1 gives the masses of elementary particles and of selected nuclides in atomic mass units. Each atomic mass ("nuclidic mass") includes the contribution from the surrounding electrons in the atom as well as the nucleus. It is instructive to compare these atomic masses in Table 15–1 to the chemical relative atomic masses of the elements (listed in the table on the inside back cover of this book). The numbers are identical when the nuclide listed is the only one that occurs in nature (the case with

- The term "isotope" often is used when "nuclide" is intended. All atoms are nuclides. Isotopes are nuclides that have a relationship. The distinction is the same as the one between "people" and "sisters."

- It follows that $1\ u = 1.6605387 \times 10^{-24}\ g = 1.6605387 \times 10^{-27}\ kg$.

- The data, which are quite precise, come from mass spectrometry experiments (see Fig. 1–20).

TABLE 15–1	Masses of Selected Elementary Particles and Nuclides		
Elementary Particle	**Symbol**	**Mass (u)**	**Mass (kg)**
Electron (beta particle)	$_{-1}^{0}e^-$, $_{-1}^{0}\beta^-$	$5.48579911 \times 10^{-4}$	$9.1093819 \times 10^{-31}$
Positron	$_{1}^{0}e^+$, $_{1}^{0}\beta^+$	$5.48579911 \times 10^{-4}$	$9.1093819 \times 10^{-31}$
Proton	$_{1}^{1}p^+$	1.0072764669	$1.6726216 \times 10^{-27}$
Neutron	$_{0}^{1}n$	1.0086649158	$1.6749272 \times 10^{-27}$

Nuclide	Mass (u)	Nuclide	Mass (u)
$_{1}^{1}H$	1.007825032	$_{14}^{30}Si$	29.97377022
$_{1}^{2}H$	2.014101778	$_{15}^{30}P$	29.9783138
$_{1}^{3}H$	3.016049268	$_{16}^{32}S$	31.9720707
$_{2}^{3}He$	3.016029310	$_{17}^{35}Cl$	34.96885271
$_{2}^{4}He$	4.002603250	$_{17}^{37}Cl$	36.96590260
$_{3}^{7}Li$	7.0160040	$_{19}^{40}K$	39.9639987
$_{4}^{8}Be$	8.00530509	$_{20}^{40}Ca$	39.9625912
$_{4}^{9}Be$	9.0121821	$_{22}^{49}Ti$	48.947871
$_{4}^{10}Be$	10.0135337	$_{35}^{81}Br$	80.916291
$_{5}^{8}B$	8.024607	$_{37}^{87}Rb$	86.909183
$_{5}^{10}B$	10.0129370	$_{38}^{87}Sr$	86.908879
$_{5}^{11}B$	11.0093055	$_{38}^{90}Sr$	89.907738
$_{6}^{11}C$	11.011433	$_{53}^{127}I$	126.904468
$_{6}^{12}C$	12 exactly	$_{86}^{222}Rn$	222.017572
$_{6}^{13}C$	13.003354838	$_{88}^{226}Ra$	226.025403
$_{6}^{14}C$	14.003241988	$_{88}^{228}Ra$	228.031064
$_{7}^{14}N$	14.003074005	$_{89}^{228}Ac$	228.031015
$_{8}^{16}O$	15.994914622	$_{90}^{232}Th$	232.038050
$_{8}^{17}O$	16.9991315	$_{90}^{234}Th$	234.043595
$_{8}^{18}O$	17.999160	$_{91}^{231}Pa$	231.035879
$_{9}^{19}F$	18.9984032	$_{92}^{231}U$	231.036289
$_{11}^{23}Na$	22.9897697	$_{92}^{234}U$	234.040945
$_{12}^{24}Mg$	23.9850419	$_{92}^{235}U$	235.043923
		$_{92}^{238}U$	238.050783

sodium, fluorine, and iodine), but the numbers differ when two or more isotopes contribute to a weighted-average atomic mass (the case with hydrogen, carbon, chlorine, and uranium).

The Einstein Mass–Energy Relationship

Most nuclides exist indefinitely, but some are **radioactive** and decay at detectable rates by processes considered in Section 15–2. All nuclides (except $_{1}^{1}H$) have one or more neutrons in their nucleus. The neutrons provide short-range attractive interactions that help to contain the enormous electrostatic repulsions among the closely packed protons in a nucleus. How many neutrons are required? Figure 15–1 shows a plot of neutron number $N \, (= A - Z)$ against atomic number Z (protons in the

Figure 15–1 • A plot of N versus Z for all stable (*black*) and many unstable (*red*) nuclides. The colored arrows show the transforming effects of decay of unstable nuclides via alpha emission, beta emission, and positron emission or electron capture.

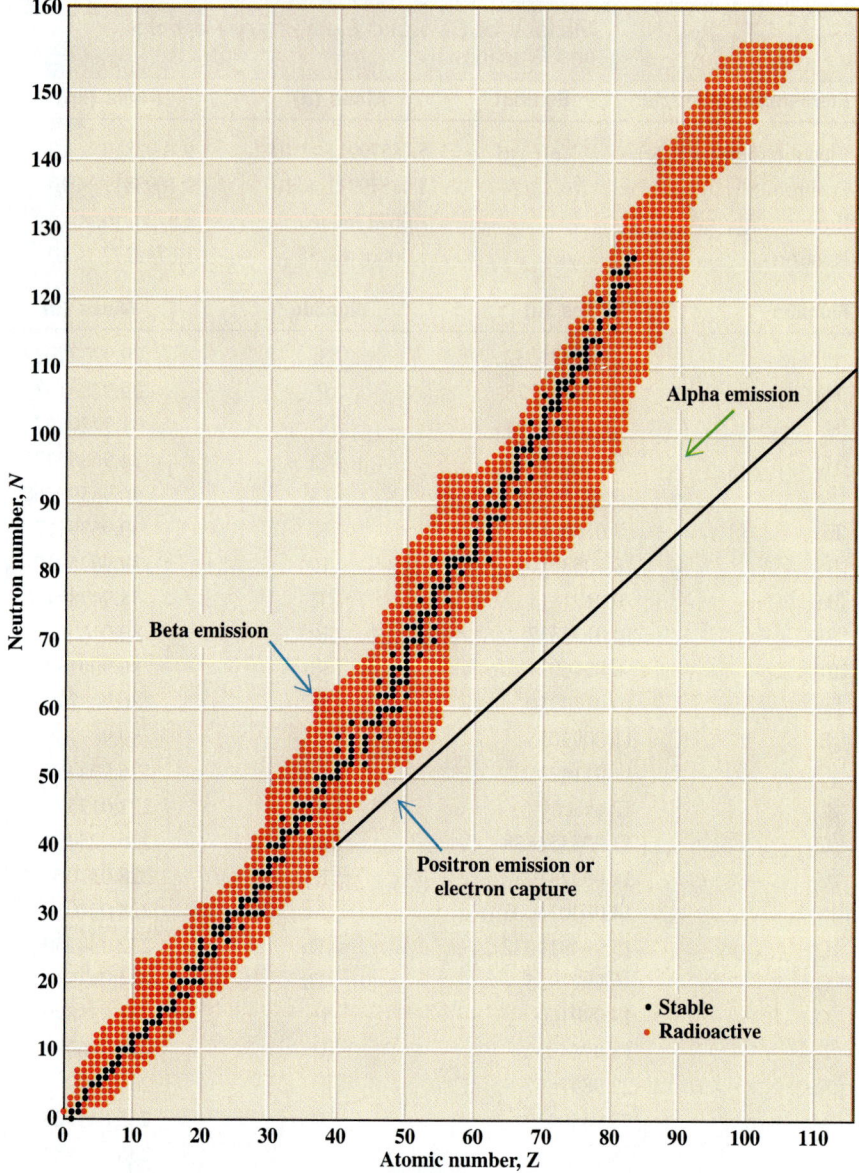

• Some of the heavier nuclides are *nearly* stable: The half-life of ^{238}U is 4.5 billion years.

nucleus) for the known nuclides of the elements. Stable nuclides are shown in black so that a *belt of stability* is visible in black. Up to about $Z = 20$, the ratio N/Z is close to 1 for stable nuclides; nearly equal numbers of protons and neutrons are required. As Z increases, this ratio grows; progressively more neutrons than protons are required for stability in the nucleus. The belt of stability ends at $^{209}_{83}Bi$, the heaviest stable nuclide. It appears that holding more than 83 protons in the same nucleus indefinitely is just not possible, no matter how many neutrons are present.

Although neutrons lend stability to the nuclei of atoms, free neutrons are not stable. They decay, with a half-life of about 14.6 minutes, into a proton and an electron. The change is represented by the balanced **nuclear equation**

$$^{1}_{0}n \longrightarrow {}^{1}_{1}p^{+} + {}^{0}_{-1}e^{-}$$

Like chemical equations, nuclear equations must be balanced. The criteria for balance in nuclear reactions differ because nuclear reactions change the chemical identities of nuclides. In balanced nuclear equations:

- The total mass number (the sum of the A's, the left superscripts) of the particles on the two sides must be equal.
- The total atomic number (the sum of the Z's, the left subscripts) of the particles on the two sides must be equal.

Mass is *not* conserved in nuclear reactions. The decay of a free neutron is a good example. The masses of the neutron, proton, and electron all appear in Table 15–1. Subtracting the mass of the reactant (a neutron) from the masses of the products (a proton and an electron) gives

$$\Delta m = \underbrace{1.0072765}_{\text{proton mass}} + \underbrace{0.0005486}_{\text{electron mass}} - \underbrace{1.0086649}_{\text{neutron mass}} = -0.0008398 \text{ u}$$

$$\Delta m = (-8.398 \times 10^{-4} \text{ u}) \times \left(\frac{1 \text{ g}}{6.022142 \times 10^{23} \text{ u}} \right) = -1.395 \times 10^{-27} \text{ g}$$

$$= -1.395 \times 10^{-30} \text{ kg}$$

- The high-precision mass data in Table 15–1 are rounded to seven decimal places here simply for ease in computation.

If mass were conserved, Δm would equal zero. The negative sign means that the mass of the system decreases. The apparent violation of the law of conservation of mass is explained by appealing to the special theory of relativity, which was formulated by Albert Einstein in 1905. In this theory, energy and mass are merely different manifestations of the same quantity and are related by the famous equation

$$E = mc_0^2$$

in which c_0 stands for the speed of light in a vacuum. This equivalency implies that the transfer of energy by any means (heat, work, emission of radiation, and so on) into or out of a system requires the transfer of a proportional amount of mass.

- The subscript in c_0 emphasizes that this is the speed of light *in a vacuum*, exactly $2.99792458 \times 10^8 \text{ ms}^{-1}$. Light slows down in a medium (such as water or glass). The symbol c stands for the speed of light in a medium.

$$\Delta E = c_0^2 \, \Delta m$$

Any process (including a chemical reaction) that releases energy from a system (ΔE negative) thus is accompanied by a decrease in the mass of the system (Δm negative). The observed loss of mass when a neutron decays simply means that the decay releases energy to the surroundings. The energy appears as kinetic energy of the particles that are produced. If its amount is to be computed in joules, c_0 must be in meters per second, and Δm, in kilograms:

$$\Delta E = c_0^2 \, \Delta m = (2.99792 \times 10^8 \text{ m s}^{-1})^2 (-1.395 \times 10^{-30} \text{ kg})$$

$$= -1.254 \times 10^{-13} \text{ J}$$

- It helps in checking the units to recall that $1 \text{ J} = 1 \text{ kg m}^2 \text{ s}^{-2}$. See Appendix B.

The energy change of the system when 1 mol of neutrons decays is obtained by multiplying by Avogadro's number:

$$\Delta E = \left(\frac{-1.254 \times 10^{-13} \text{ J}}{\text{neutron}} \right) \times \left(\frac{6.022142 \times 10^{23} \text{ neutron}}{1 \text{ mol}} \right)$$

$$= -7.552 \times 10^{10} \text{ J mol}^{-1} = -7.552 \times 10^7 \text{ kJ mol}^{-1}$$

This ΔE is hundreds of thousands of times larger than the molar energies of chemical reactions. For example, the ΔE in the combustion of carbon to CO_2 is only about -400 kJ mol^{-1}. Nuclear reactions without exception involve energy changes orders of magnitude larger than those of chemical reactions (Fig. 15–2).

Figure 15-2 • Self-heating in a radioactive metal. Relatively few atoms of plutonium (about 2.5 per second out of every 10 billion) are taking part in nuclear reactions in this pellet of nearly pure plutonium-238, but the energy released is enough to make it glow red-hot. The pellet was manufactured as a compact power source for a space probe. *(Savannah River Site, photo used with permission.)*

The changes in energy that accompany nuclear reactions nearly always are expressed in more convenient energy units than the joule—namely, the **electron volt** (eV) and the **megaelectron volt** (MeV). The electron volt is the energy given to an electron when it is accelerated through a potential difference of exactly 1 V:

$$\Delta E \text{ (joules)} = Q \text{ (coulombs)} \times \Delta\mathcal{E} \text{ (volts)}$$
$$1 \text{ eV} = 1.602176 \times 10^{-19} \text{ C} \times 1 \text{ V} = 1.602176 \times 10^{-19} \text{ J}$$

The megaelectron volt is 1 million times larger than an electron volt:

$$1 \text{ MeV} = 10^6 \text{ eV} = 1.602176 \times 10^{-13} \text{ J}$$

Let us use Einstein's equation to compute (in MeV) the energy change associated with a mass change equal to 1 u. This value provides a useful stepping-stone in later computations of mass–energy equivalencies. We substitute $\Delta m = 1.660539 \times 10^{-27}$ kg (1 u) into Einstein's equation:

$$\Delta E = c_0^2 \, \Delta m = (2.997925 \times 10^8 \text{ m s}^{-1})^2 (1.660539 \times 10^{-27} \text{ kg})$$
$$= 1.492418 \times 10^{-10} \text{ J}$$

and then convert to MeV:

$$\Delta E = (1.492418 \times 10^{-10} \text{ J}) \times \left(\frac{1 \text{ MeV}}{1.602176 \times 10^{-13} \text{ J}} \right) = 931.494 \text{ MeV}$$

That is, the **energy equivalent** of 1 u of mass is 931.494 MeV. In the decay of a neutron

$$\Delta E = (-8.398 \times 10^{-4} \text{ u}) \times \left(\frac{931.494 \text{ MeV}}{\text{u}} \right) = -0.7823 \text{ MeV}$$

Binding Energies of Nuclei

The **binding energy** E_B of a nucleus is the negative of the energy change ΔE that occurs when the nucleus is formed from its component protons and neutrons. For the ^4_2He nucleus, for example,

$$2\,^1_1p^+ + 2\,^1_0n \longrightarrow {}^4_2\text{He}^{2+} \qquad E_B = -\Delta E = ?$$

Binding energies of nuclei are calculated using Einstein's equation and precise mass data from a mass spectrometer. Most mass-spectrometric experiments measure the mass of the atom, including most or all of the extranuclear electrons, and *not* the mass of the bare nucleus. For this reason, we take the binding energy E_B of a nucleus to equal the negative of the energy change ΔE for the formation of the neutral atom from hydrogen atoms and neutrons. For the ^4_2He nucleus, this corresponds to the nuclear reaction

$$2\,^1_1\text{H} + 2\,^1_0n \longrightarrow {}^4_2\text{He}$$

The correction for the differences in binding energies of the electrons that are present in the hydrogen atoms and the helium atom is small and can be neglected.

EXAMPLE 15-1

Calculate the binding energy of $_2^4$He from the data in Table 15–1, and express it in joules, in kilojoules per mole, and in megaelectron volts.

Solution

The mass change in forming one atom of $_2^4$He from its constituent particles, first in atomic mass units and then in kilograms, is

$$\Delta m = m(_2^4\text{He}) - 2m(_1^1\text{H}) - 2m(_0^1n)$$

$$\Delta m = (4.002603250 \text{ u}) - 2(1.007825032 \text{ u}) - 2(1.0086649158 \text{ u})$$

$$= -0.030376646 \text{ u}$$

$$= (-0.030376646 \text{ u}) \times \left(\frac{1 \text{ g}}{6.0221420 \times 10^{23} \text{ u}}\right) \times \left(\frac{1 \text{ kg}}{1000 \text{ g}}\right)$$

$$= -5.0441596 \times 10^{-29} \text{ kg}$$

Substitution of the mass change into Einstein's equation gives

$$\Delta E = c_0^2 \Delta m = (2.9979246 \times 10^8 \text{ m s}^{-1})^2(-5.0441596 \times 10^{-29} \text{ kg})$$

$$= -4.5334647 \times 10^{-12} \text{ J}$$

$$E_B = -\Delta E = +4.5334647 \times 10^{-12} \text{ J}$$

The binding energy of 1 mol of $_2^4$He atoms exceeds the binding energy of a single atom by a factor of Avogadro's number:

$$E_B = \left(\frac{+4.5334647 \times 10^{-12} \text{ J}}{_2^4\text{He atom}}\right) \times \left(\frac{6.0221420 \times 10^{23} \, _2^4\text{He atom}}{\text{mol}}\right) \times \left(\frac{1 \text{ kJ}}{1000 \text{ J}}\right)$$

$$= 2.7301168 \times 10^9 \text{ kJ mol}^{-1}$$

The easiest way to compute the binding energy in megaelectron volts is to use the energy equivalent of 1 u that was just established:

$$E_B = -\Delta E = -(-0.030376646 \text{ u}) \times \left(\frac{931.494 \text{ MeV}}{1 \text{ u}}\right) = 28.2957 \text{ MeV}$$

EXERCISE

Calculate the binding energy of $_2^3$He in joules, in kilojoules per mole, and in megaelectron volts. Round off the answers to five significant digits.

Answer: $E_B = 1.2366 \times 10^{-12} \text{ J} = 7.4468 \times 10^8 \text{ kJ mol}^{-1} = 7.7181 \text{ MeV}$.

$_2^4$He has a mass number A of 4, meaning that there are four **nucleons** (protons plus neutrons) in the $_2^4$He nucleus. Its **binding energy per nucleon** is

$$\frac{E_B}{A} = \frac{28.2957 \text{ MeV}}{4} = 7.07392 \text{ MeV}$$

The binding energy per nucleon is a direct measure of the stability of the nucleus. It varies with the mass number among the nuclides, increasing to a maximum of about 8.8 MeV per nucleon for iron and nickel, as Figure 15–3 indicates.

Figure 15–3 • The binding energy per nucleon of most known nuclides as a function of the mass number. A total of 2212 nuclides is shown. Of these, only 274 are stable; the rest undergo some type of radioactive decay. Maximum stability occurs in the vicinity of ^{56}Fe. Note that ^{4}He (*red circle*) is unusually stable for its mass number.

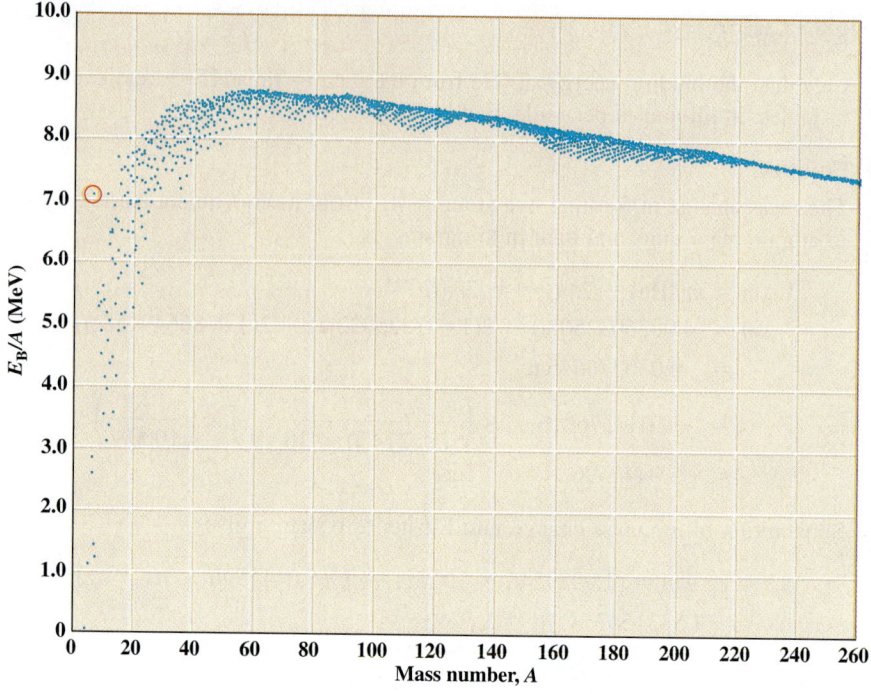

15–2 Nuclear Decay Processes

Why are some nuclides radioactive but others stable? The thermodynamic criterion for spontaneity in a process at constant T and P is that $\Delta G < 0$. For nuclear reactions, the change in the Gibbs energy is essentially equal to the energy change (ΔE). Consequently, the thermodynamic criterion for a spontaneous nuclear reaction simplifies to

$$\Delta E < 0 \qquad \text{or equivalently} \qquad \Delta m < 0$$

The equivalence of these criteria follows from Einstein's equation.

Nuclear reactions with $\Delta m < 0$ always increase the binding energy per nucleon among the products. In terms of Figure 15–3, this means that nuclear reactions that produce nuclides positioned higher up the page are thermodynamically spontaneous. These include reactions that combine the nuclei of light elements to produce heavier elements (fusion reactions) as well as those that split apart the nuclei of the heaviest elements (fission reactions). Thus, nearly all nuclides are thermodynamically unstable with respect to either fusion or fission. Fortunately, most nuclear reactions have exceedingly large activation energies and proceed at zero rates except at extremely high temperatures (see Section 15–6). Otherwise, all matter would long since have reacted to give the most stable nuclides in Figure 15–3. The nuclear reactions that *do* occur with detectable rates under ordinary conditions have been the subject of intense experimental study since the discovery of radioactivity in 1896. They proceed along relatively few pathways.

Before discussing these pathways, it is necessary to mention **antiparticles.** Each subatomic particle is thought to have an antiparticle of the same mass but opposite charge. Thus, the antiparticle of an electron is a **positron** (e^{+}), and the antiparticle of a proton is a negatively charged particle called an **antiproton.** Antiparticles have only transient existences because matter and antimatter quickly annihilate one another when they come close together, emitting radiation carrying an equivalent quantity of energy. In the case of an electron and a positron, two oppositely directed **gamma rays** (high-energy electromagnetic waves), called **annihilation radiation,** are emitted. Another important particle–antiparticle pair is the **neutrino** (symbol ν) and **antineutrino** (symbol $\bar{\nu}$), both of which are uncharged and almost massless. They carry angular momentum (see Section 16–3) and interact only weakly with ordinary matter. Elaborate equipment is required to detect the rare occasions when neutrinos interact with certain nuclides.

• Recent experimental results indicate that neutrinos have masses on the order of one billionth of the mass of the proton.

Pathways for Decay

If an unstable nuclide contains fewer protons than do stable nuclides of the same mass number, it is termed "proton deficient." Such a nuclide may decay by transforming one of its neutrons into a proton and emitting a high-energy electron, also called a **beta particle.** The transformation simultaneously emits an antineutrino, $\bar{\nu}$. The daughter nuclide has the same mass number A as the parent, but its atomic number Z has been increased by one, corresponding to the conversion of a neutron to a proton. This is **beta decay,** or **beta emission.** An electron (beta particle) is symbolized $_{-1}^{0}e^-$ (or $_{-1}^{0}\beta^-$) in equations for nuclear reactions. The superscript 0 indicates a mass number of 0, which is in fact the integer closest to the atomic mass of the electron. The subscript -1 is not a true atomic number but accounts for the change in the charge of a nucleus when an electron leaves or joins it. Finally, as in chemical equations, the superscript "$-$" gives the charge of the species. Two equations for the beta decay of unstable nuclides are

$$_{6}^{14}\text{C} \longrightarrow {}_{7}^{14}\text{N} + {}_{-1}^{0}e^- + \bar{\nu} \qquad \text{and} \qquad {}_{19}^{40}\text{K} \longrightarrow {}_{20}^{40}\text{Ca} + {}_{-1}^{0}e^- + \bar{\nu}$$

These equations are balanced as to mass number (left superscript) and atomic number (left subscript).

• Antineutrinos and neutrinos do not affect balance.

When a nuclide has too *many* protons for stability relative to the number of neutrons it contains, it may decay by emitting a positron, symbolized $_{1}^{0}e^+$ (or $_{1}^{0}\beta^+$). Like its antiparticle, the electron, a positron has a mass number of zero, but its charge is $+1$ instead of -1. In **positron emission,** a proton is converted into a neutron, which is retained in the nucleus, and a high-energy positron is emitted together with a neutrino (ν). The mass number of the nuclide remains the same, but the atomic number Z *decreases* by one. Examples of positron emission are

$$_{10}^{19}\text{Ne} \longrightarrow {}_{9}^{19}\text{F} + {}_{1}^{0}e^+ + \nu \qquad \text{and} \qquad {}_{29}^{64}\text{Cu} \longrightarrow {}_{28}^{64}\text{Ni} + {}_{1}^{0}e^+ + \nu$$

EXAMPLE 15–2

Write balanced nuclear equations for (a) beta decay by $_{11}^{24}\text{Na}$ and (b) positron emission by $_{11}^{21}\text{Na}$.

Strategy

Write full symbols for the beta particle (electron) and positron. Subtract the mass number (A) and atomic number (Z) given in these symbols from the A and Z of the decaying nuclide to obtain the A and Z of the product nuclide. Consult a periodic table to learn the chemical symbol associated with the new Z. Include an antineutrino (for beta decay) or a neutrino (for positron emission).

Solution

(a) A beta particle is symbolized $_{-1}^{0}e^{-}$. Beta emission from $_{11}^{24}$Na increases Z by 1 and leaves A unchanged. Because Z goes to 12, the product nuclide is Mg:

$$_{11}^{24}\text{Na} \longrightarrow _{12}^{24}\text{Mg} + _{-1}^{0}e^{-} + \bar{\nu}$$

(b) A positron is symbolized $_{1}^{0}e^{+}$; emission of a positron by $_{11}^{21}$Na decreases Z by 1 and leaves A unchanged. Because Z goes to 10, the product nuclide is Ne. The nuclear equation is

$$_{11}^{21}\text{Na} \longrightarrow _{10}^{21}\text{Ne} + _{1}^{0}e^{+} + \nu$$

Check

In both nuclear equations, the sums of the mass numbers (left superscripts) are the same on the two sides, as are the sums of the nuclear charges (left subscripts).

EXERCISE

Write balanced nuclear equations for (a) beta decay by $_{4}^{10}$Be and (b) positron emission by $_{5}^{8}$B.

Answer: (a) $_{4}^{10}\text{Be} \longrightarrow _{5}^{10}\text{B} + _{-1}^{0}e^{-} + \bar{\nu}$.   (b) $_{5}^{8}\text{B} \longrightarrow _{4}^{8}\text{Be} + _{1}^{0}e^{+} + \nu$.

Another pathway exists by which a nuclide with an excess of protons can attain stability: Its nucleus may capture one of its own surrounding electrons and use it to convert a proton to a neutron. As in positron emission, the mass numbers of the parent and daughter atoms are the same and the atomic number decreases by one. In this case, however, only a neutrino is emitted from the nucleus. If neutrino emission were the only sign of electron capture, it might go unnoticed because neutrinos are so hard to detect. Capture of an electron into the nucleus, however, is followed by drastic rearrangement of the remaining extranuclear electrons. These changes lead to emission of high-energy electromagnetic radiation (**X-rays** or **gamma rays**) that is detected easily. Examples of **electron capture** (E.C.) are

$$_{13}^{26}\text{Al} \xrightarrow{\text{E.C.}} _{12}^{26}\text{Mg} + \nu \qquad _{17}^{36}\text{Cl} \xrightarrow{\text{E.C.}} _{16}^{36}\text{S} + \nu$$

These nuclear equations are not balanced. Balancing them requires showing the captured electrons explicitly:

$$\left(_{13}^{26}\text{Al}^{+} + _{-1}^{0}e^{-}\right) \xrightarrow{\text{E.C.}} _{12}^{26}\text{Mg} + \nu \qquad \left(_{17}^{36}\text{Cl}^{+} + _{-1}^{0}e^{-}\right) \xrightarrow{\text{E.C.}} _{16}^{36}\text{S} + \nu$$

where the parentheses emphasize that the electrons that are captured start out as part of the atom whose nucleus absorbs them. Electron capture is quite common for heavier, neutron-deficient nuclei.

The three paths for nuclear decay discussed so far lead to changes in the atomic number (Z) and neutron number (N) but not in the mass number (A). They are indicated by two blue arrows running "northwest–southeast" in Figure 15–1. By contrast, the process of **alpha decay** consists of the emission of an alpha particle (a $_{2}^{4}\text{He}^{2+}$ ion, also written $_{2}^{4}\alpha^{2+}$). This causes a decrease in mass number A by 4 (the atomic number Z and the neutron number N each decrease by 2). Alpha decay is indicated in Figure 15–1 by the green arrow running "northeast–southwest." Alpha decay occurs chiefly among the heavier elements, with atomic numbers in the region of unstable nuclei beyond bismuth ($Z = 83$) in the periodic table. Examples are

$$_{88}^{226}\text{Ra} \longrightarrow _{86}^{222}\text{Rn} + _{2}^{4}\text{He} \qquad \text{and} \qquad _{92}^{238}\text{U} \longrightarrow _{90}^{234}\text{Th} + _{2}^{4}\text{He}$$

When nuclei are very proton deficient (have a large excess of neutrons) or very neutron deficient (have a large excess of protons), an excess particle may "boil off," that is, may be ejected directly from the nucleus. These decay modes are called **neutron emission** and **proton emission,** respectively. Emission of a neutron reduces N but not Z and produces daughter nuclides lying one step lower than the parent on the N axis in Figure 15–1. Proton emission reduces Z but not N and produces daughter nuclides lying to the left of the parent in the figure. Finally, certain nuclides of high mass number are known spontaneously to fission (split into two chunks of roughly equal size).

• Fission reactions are discussed more extensively in Section 15–5.

Mass Changes in Nuclear Decay

In nuclear reactions, a negative Δm corresponds to a negative ΔE, which corresponds to the liberation of energy. In beta decay, this liberated energy appears as kinetic energy of the beta particle (emitted electron), the antineutrino, and the daughter nuclide. The daughter nuclide is massive compared with the beta particle and neutrino, and the recoil kinetic energy it carries away from the event is small and can be neglected. The kinetic energy of the beta particle plus that of the antineutrino therefore equals $-\Delta E$. The antineutrino is not easily observed, but the kinetic energy of the emitted beta particle varies in a continuous range from essentially zero up to a maximum, as shown in Figure 15–4, reflecting all the ways that the liberated energy can be parceled out between it and the antineutrino.

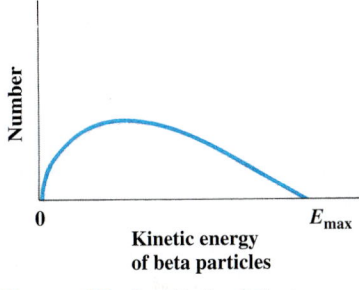

Figure 15–4 • Emitted electrons (beta particles) in beta decay have a distribution of kinetic energies up to a cutoff value of E_{max}.

EXAMPLE 15–3

Calculate the maximum kinetic energy (in MeV) of the beta particle in the decay

$$^{24}_{11}\text{Na} \longrightarrow\ ^{24}_{12}\text{Mg} +\ ^{0}_{-1}e^- + \overline{\nu}$$

considered in Example 15–2. The masses of *neutral atoms* of ^{24}Na and ^{24}Mg are 23.99096 and 23.98504 u, respectively.

Strategy

Use the fact that the maximum kinetic energy of the beta particle equals $-\Delta E$ for the beta emission reaction. Plan to compute ΔE by substitution in the Einstein equation, which requires Δm, the mass of the products minus the mass of the reactants. Realize that the given masses include the mass of the electrons surrounding the nucleus. Plan an adjustment to account for the change in the number of extranuclear electrons between parent and daughter nuclide.

Solution

It is tempting to compute the change in mass by adding the mass of the $^{24}_{12}\text{Mg}$ and the beta particle and subtracting the mass of the $^{24}_{11}\text{Na}$. This leads to an error. A neutral $^{24}_{11}\text{Na}$ atom has only 11 electrons outside of its nucleus, whereas a neutral $^{24}_{12}\text{Mg}$ atom has 12 electrons. The quoted mass for $^{24}_{12}\text{Mg}$ includes this extra electron, which has no source on the reactant side. Therefore, the mass of one electron must be subtracted from the mass of the $^{24}_{12}\text{Mg}$ atom as Δm is computed:

$$\Delta m = m_{\text{products}} - m_{\text{reactants}}$$

$$= [m(^{24}_{12}\text{Mg}) - m(^{0}_{-1}e^-)] + m(^{0}_{-1}e^-) - m(^{24}_{11}\text{Na})$$

$$= (23.98504 - 0.0005485799)\ \text{u} + 0.0005485799\ \text{u} - 23.99096\ \text{u}$$

$$= 23.98504\ \text{u} - 23.99096\ \text{u} = -0.00592\ \text{u}$$

Note in the arithmetic that 0.0005485799 u is subtracted and then added so that a numerical value is not actually required. The energy change is

$$\Delta E = (-0.00592 \text{ u}) \times \left(\frac{931.494 \text{ MeV}}{1 \text{ u}} \right) = -5.51 \text{ MeV}$$

The maximum kinetic energy of the beta particle is $+5.51 \text{ MeV}$.

Check

The beta decay energy of $^{24}_{11}\text{Na}$ has been determined experimentally. The Center for Nuclear Spectrometry and Mass Spectrometry lists it as 5.5158 MeV, which is essentially the same as the answer. This listing is on the World Wide Web at csnwww.in2p3.fr/AMDC/masstables/Ame1995/mass_rmd.mas95.

EXERCISE

One atom of the nuclide $^{10}_{4}\text{Be}$ undergoes beta decay (Exercise 15–2). Use data from Table 15–1 to calculate the maximum kinetic energy (in MeV) of the emitted electron.

Answer: 0.5558 MeV.

• Nuclear physicists sometimes emphasize this symmetry by using the name "negatron" for the beta particle (electron).

Positron emission is the "anti-process" of beta decay, a symmetry that suggests that emitted positrons should, like emitted electrons in beta decay, have kinetic energies that range from zero to a maximum of ΔE. This is exactly what is found. However, positron emission *lowers* Z, whereas beta emission raises it. This causes the computation of the maximum kinetic energy of an emitted positron to proceed slightly differently.

EXAMPLE 15–4

Calculate the maximum kinetic energy of the positron emitted in the decay

$$^{11}_{6}\text{C} \longrightarrow {}^{11}_{5}\text{B} + {}^{0}_{1}e^{+} + \nu$$

The masses of neutral atoms of $^{11}_{6}\text{C}$ and $^{11}_{5}\text{B}$ are 11.011433 and 11.009306 u, respectively.

Strategy

Use the approach of Example 15–3. Again, the given nuclidic masses include the masses of the electrons that surround the nucleus. Take care to account for the change in the number of electrons between parent and daughter nuclide.

Solution

A neutral $^{11}_{6}\text{C}$ parent atom has six electrons surrounding its nucleus, whereas a neutral $^{11}_{5}\text{B}$ daughter has only five. The stated masses of the two atoms include these electrons. To avoid counting an extra electron on the parent side of the nuclear equation, it is necessary to subtract one electron mass from the mass of the $^{11}_{6}\text{C}$. Then compute the change in mass:

$$\Delta m = m_{\text{products}} - m_{\text{reactants}}$$

$$\Delta m = m(^{11}_{5}\text{B}) + m(^{0}_{1}e^{+}) - [m(^{11}_{6}\text{C}) - m(^{0}_{1}e^{-})]$$

$$\Delta m = 11.009306 \text{ u} + 0.0005485799 \text{ u} - 11.011433 \text{ u} + 0.0005485799 \text{ u}$$

$$= -0.001030 \text{ u}$$

The energy change is

$$\Delta E = (-0.001030\ u) \times \left(\frac{931.494\ \text{MeV}}{1\ u}\right) = -0.9594\ \text{MeV}$$

The maximum kinetic energy of the positron is $+0.9594\ \text{MeV}$.

EXERCISE

The nuclide $^{8}_{5}B$ decays by positron emission (see Exercise 15–2). Use data from Table 15–1 to calculate the maximum kinetic energy (in MeV) of the emitted positron.

Answer: 16.958 MeV.

The energy that is liberated in alpha emission also is computed from the difference in mass between the products and the reactant. For the alpha decay event

$$^{226}_{88}\text{Ra} \longrightarrow {}^{222}_{86}\text{Rn} + {}^{4}_{2}\text{He}$$

the mass difference is

$$\Delta m = m(^{222}_{86}\text{Rn}) + m(^{4}_{2}\text{He}) - m(^{226}_{88}\text{Ra})$$

Taking mass data from Table 15–1 gives

$$\Delta m = 222.017572\ u + 4.002603250\ u - 226.025403\ u = -0.005228\ u$$

The masses in the preceding equation are for neutral atoms. They include 88 electrons on the left and 88 electrons on the right of the nuclear equation; no correction of the type used for beta decay or positron emission is required. The change in energy is

$$\Delta E = -0.005229\ u \times \left(\frac{931.494\ \text{MeV}}{1\ u}\right) = -4.870\ \text{MeV}$$

The energy from this alpha decay reaction causes radium to glow in the dark.

Detecting and Measuring Radioactivity

Many methods have been developed to detect, identify, and quantitatively measure the products formed in nuclear reactions. Some are quite simple, but others require complex electronic instrumentation. Perhaps the simplest radiation detector is the photographic emulsion, which was first used by Henri Becquerel, the discoverer of radioactivity. In 1896, Becquerel reported his observation that potassium uranyl sulfate ($K_2UO_2(SO_4)_2 \cdot 2H_2O$) could expose a photographic plate even in the dark. Such detectors are still in use in the form of film badges, worn to monitor personal exposure to radiation. The degree of darkening of the film is proportional to the quantity of radiation received.

Rutherford and his students used a screen coated with zinc sulfide to detect the location of alpha particles by the pinpoint flashes or scintillations of light they produce (see Section 1–5). The modern **scintillation counter** is an outgrowth of that technique. The zinc sulfide scintillation screen is replaced by a sodium bromide crystal in which Tl^+ ions substitute for a small fraction of the Na^+ ions. Such a crystal emits a pulse of light when it absorbs a beta particle or a gamma ray. A photomultiplier tube detects and counts the light pulses.

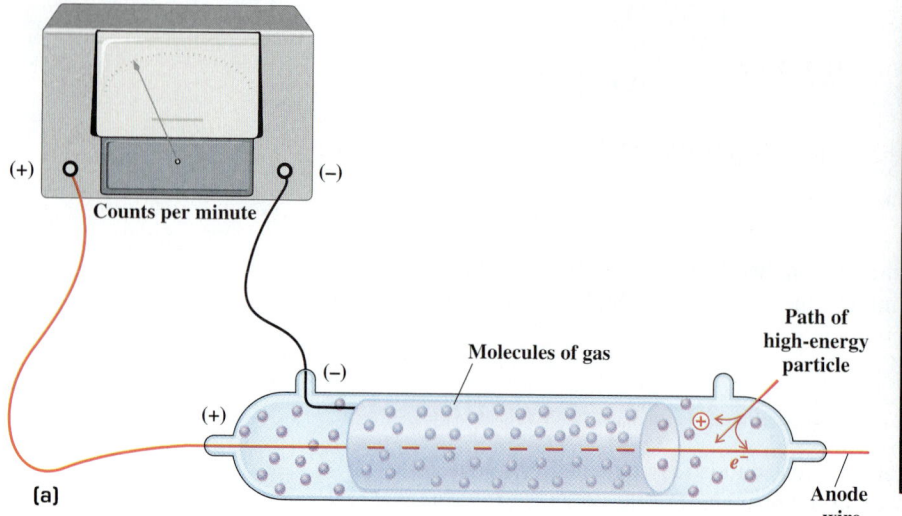

Figure 15-5 • (a) In a Geiger tube, radiation ionizes gas in the tube, freeing electrons that are accelerated to the anode wire in a cascade. Their arrival creates an electrical pulse. (b) A Geiger counter measuring the emissions from a sample of carnotite, a uranium-containing mineral. *(Dave Davidson/Tom Stack & Associates)*

The Geiger counter (Fig. 15–5) measures radiation by a different method. A cylindrical tube, usually of glass, is coated internally with metal to provide a negative electrode, and a wire is run down the center of the tube to serve as a positive electrode. The tube is filled to a pressure of about 0.1 atm with a mixture of 90% argon and 10% ethyl alcohol vapor. Then a potential difference of about 1000 V is applied across the electrodes. When a high-energy beta particle enters the tube, it produces ion pairs (electrons and positive ions). The electrons, being much lighter than the positive ions, are quickly accelerated toward the wire, which has a positive charge. As they advance, they encounter and ionize other neutral atoms. An avalanche of electrons builds up, and a large electron current flows into the central wire. The sudden increase in the electrical conductance of the tube causes a drop in the potential difference, which is recorded, and the multiplicative electron discharge is quenched by the alcohol molecules. In this way, single beta particles produce electrical pulses that can be amplified and counted.

15-3 Kinetics of Radioactive Decay

Nuclear decay events occur at random, making it impossible to predict exactly when a particular nucleus among a group of identical unstable nuclei will disintegrate. Nevertheless, the laws of probability ensure that, if the group is large, a definite fraction will decay within a given period of time. Processes of nuclear decay follow first-order kinetics. They are described by the same equations developed in Chapter 14 for the rates of first-order chemical reactions. The integrated first-order rate law has the form

• An analogy: Life insurance companies cannot predict which policyholders in a large group will die by a given age but can establish, from experience, the fraction who will die.

$$N = N_i e^{-kt}$$

where N_i is the number of nuclei initially present at $t = 0$, and N is the number that remain after a time, t. The first-order rate constant (k), which is now called a **decay constant,** is related to the half-life ($t_{1/2}$) through

$$t_{1/2} = \frac{\ln 2}{k} = \frac{0.6931}{k}$$

just as in first-order chemical kinetics. The half-life is the time required for half of the nuclei in a sample to decay. Known half-lives of unstable nuclei range from 10^{-21} s to more than 10^{24} years. Table 15–2 lists the half-lives and decay modes for a number of unstable nuclides.

There is one important practical difference between chemical kinetics and nuclear kinetics. In chemical kinetics, the concentration of a reactant or product is monitored over time, and the rate of the reaction then is found from the rate of change of that concentration. In nuclear kinetics, the rate of occurrence of decay events, $-\Delta N/\Delta t$, is measured directly with a Geiger counter or other radiation detector. This decay rate is the **activity,** the average disintegration rate, in numbers of nuclei per unit of time.

$$\text{activity} = A = \frac{-\Delta N}{\Delta t} = kN$$

• ΔN is the change in the number of radioactive atoms, and Δt is the interval of time over which the change occurs. As Δt gets shorter and shorter, the ratio approaches the instantaneous rate discussed in Section 14–1.

TABLE 15–2	Decay Characteristics of Some Radioactive Nuclides		
Nuclide	$t_{1/2}$	**Decay Mode**	**Daughter**
$^3_1$H (tritium)	12.26 yr	Beta	$^3_2$He
$^8_4$Be	$\approx 1 \times 10^{-16}$ s	Alpha	$^4_2$He
$^{14}_6$C	5730 yr	Beta	$^{14}_7$N
$^{22}_{11}$Na	2.601 yr	Positron emission	$^{22}_{10}$Ne
$^{24}_{11}$Na	15.02 h	Beta	$^{24}_{12}$Mg
$^{32}_{15}$P	14.28 d	Beta	$^{32}_{16}$S
$^{35}_{16}$S	87.2 d	Beta	$^{35}_{17}$Cl
$^{36}_{17}$Cl	3.01×10^5 yr	Beta	$^{36}_{18}$Ar
$^{40}_{19}$K	1.28×10^9 yr	Beta (89.3%)	$^{40}_{20}$Ca
		Electron capture (10.7%)	$^{40}_{18}$Ar
$^{59}_{26}$Fe	44.6 d	Beta	$^{59}_{27}$Co
$^{60}_{27}$Co	5.27 yr	Beta	$^{60}_{28}$Ni
$^{90}_{38}$Sr	29 yr	Beta	$^{90}_{39}$Y
$^{109}_{48}$Cd	453 d	Electron capture	$^{109}_{47}$Ag
$^{125}_{53}$I	59.7 d	Electron capture	$^{125}_{52}$Te
$^{131}_{53}$I	8.021 d	Beta	$^{131}_{54}$Xe
$^{127}_{54}$Xe	36.41 d	Electron capture	$^{127}_{53}$I
$^{137}_{57}$La	$\approx 6 \times 10^4$ yr	Electron capture	$^{137}_{56}$Ba
$^{222}_{86}$Rn	3.824 d	Alpha	$^{218}_{84}$Po
$^{226}_{88}$Ra	1600 yr	Alpha	$^{222}_{86}$Rn
$^{232}_{90}$Th	1.40×10^{10} yr	Alpha	$^{228}_{88}$Ra
$^{235}_{92}$U	7.04×10^8 yr	Alpha	$^{231}_{90}$Th
$^{238}_{92}$U	4.468×10^9 yr	Alpha	$^{234}_{90}$Th
$^{239}_{93}$Np	2.350 d	Beta	$^{239}_{94}$Pu
$^{239}_{94}$Pu	2.411×10^4 yr	Alpha	$^{235}_{92}$U

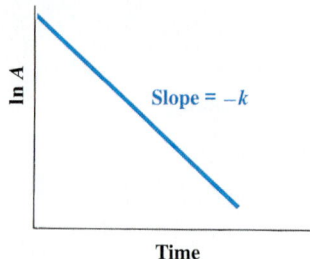

Figure 15–6 • A graph of the logarithm of the activity of a radioactive nuclide against time is a straight line with a slope equal to the negative of the decay constant.

Because the activity is proportional to the number of nuclei (N), the activity also falls off exponentially with time:

$$A = A_i\, e^{-kt}$$

A plot of $\ln A$ against t is linear, with slope $-k = -(\ln 2)/t_{1/2}$, as Figure 15–6 shows. The activity A is reduced to half its initial value in a time $t_{1/2}$. Once k and A of a particular sample are known, the number of nuclei (N) in the sample can be calculated:

$$N = \frac{A}{k}$$

The SI unit of activity is the **becquerel** (Bq), which equals one radioactive disintegration per second ($1\ \text{Bq} = 1\ \text{s}^{-1}$). The becquerel supersedes an older and much larger unit of activity called the **curie** (Ci), which is defined as 3.7×10^{10} disintegrations per second.

EXAMPLE 15–5

Tritium (^{3}H) decays by beta emission to ^{3}He with a half-life of 12.26 years. A sample of a tritium-containing compound has an initial activity of 0.833 Bq. Calculate the decay constant k, the number of tritium nuclei present in the sample initially (N_i), and the activity of the sample after 2.50 years.

Strategy

The decay constant k equals $\ln 2$ divided by the half-life. Compute k in both reciprocal years (yr^{-1}) and in reciprocal seconds (s^{-1}) because the latter is easier to use with the initial activity, which is given in becquerels ($1\ \text{Bq} = 1\ \text{s}^{-1}$). Use the initial activity and k to compute the initial number of nuclei. Use the elapsed time and k to compute the activity after 2.50 years.

Solution

$$k = \frac{\ln 2}{t_{1/2}} = \frac{0.69315}{12.26\ \text{yr}} = 0.05654\ \text{yr}^{-1}$$

$$t_{1/2} = 12.26\ \text{yr} \times \left(\frac{365.2\ \text{d}}{1\ \text{yr}}\right) \times \left(\frac{24\ \text{h}}{1\ \text{d}}\right) \times \left(\frac{3600\ \text{s}}{1\ \text{h}}\right) = 3.868 \times 10^8\ \text{s}$$

$$k = \frac{\ln 2}{t_{1/2}} = \frac{0.69315}{3.868 \times 10^8\ \text{s}} = 1.792 \times 10^{-9}\ \text{s}^{-1}$$

The number of nuclei of tritium initially present is

$$N_i = \frac{A_i}{k} = \frac{0.833\ \text{s}^{-1}}{1.792 \times 10^{-9}\ \text{s}^{-1}} = 4.65 \times 10^8\ ^3\text{H nuclei}$$

The radioactivity of these nuclei weakens with time according to

$$A = A_i\, e^{-kt}$$

After 2.50 yr

$$A = (0.833\ \text{Bq})\, e^{-(0.05654\ \text{yr}^{-1})(2.50\ \text{yr})} = (0.833\ \text{Bq})\, e^{-0.1413} = 0.723\ \text{Bq}$$

EXERCISE

The ^{60}Co used in cancer therapy emits high-energy radiation (gamma rays) that kills cells. Cobalt-60 decays with a half-life of 5.27 years. A 0.19 g pellet of cobalt contains ^{60}Co mixed with stable ^{59}Co and has an activity of $4.0 \times$

10^{11} Bq. (a) Determine the percentage by mass of ^{60}Co in the pellet. Take the molar mass of ^{60}Co to be 60 g mol^{-1}. (b) Calculate the activity of the pellet after 10.0 years.

Answer: (a) 5.0%. (b) 1.1×10^{11} Bq.

Radioactive Dating

The decay of radioactive nuclides enables geochemists to measure the ages of rocks from their isotopic composition. Suppose that a uranium-bearing mineral was formed 2 billion (2.00×10^9) years ago and that it has remained geologically unaltered. The ^{238}U atoms in the mineral decay, with a half-life of 4.468×10^9 years, to form a series of short-lived intermediates, ending in the stable isotope of lead ^{206}Pb (Fig. 15–7). The fraction of ^{238}U remaining after 2.00×10^9 years should be

$$\frac{N}{N_i} = e^{-kt} = \exp\left(\frac{-\ln 2}{4.468 \times 10^9 \text{ yr}} \times 2.00 \times 10^9 \text{ yr}\right) = 0.733$$

One ^{206}Pb atom forms for each ^{238}U that decays, so the number of ^{206}Pb atoms should be $1.00 - 0.733$, or 0.267 times the original number of ^{238}U atoms. The present-day ratio of the abundances of the lead and uranium isotopes is then 0.267/0.733,

Figure 15–7 • The radioactive nuclide ^{238}U decays by a series of alpha (α) and beta (β) emissions to the stable nuclide ^{206}Pb.

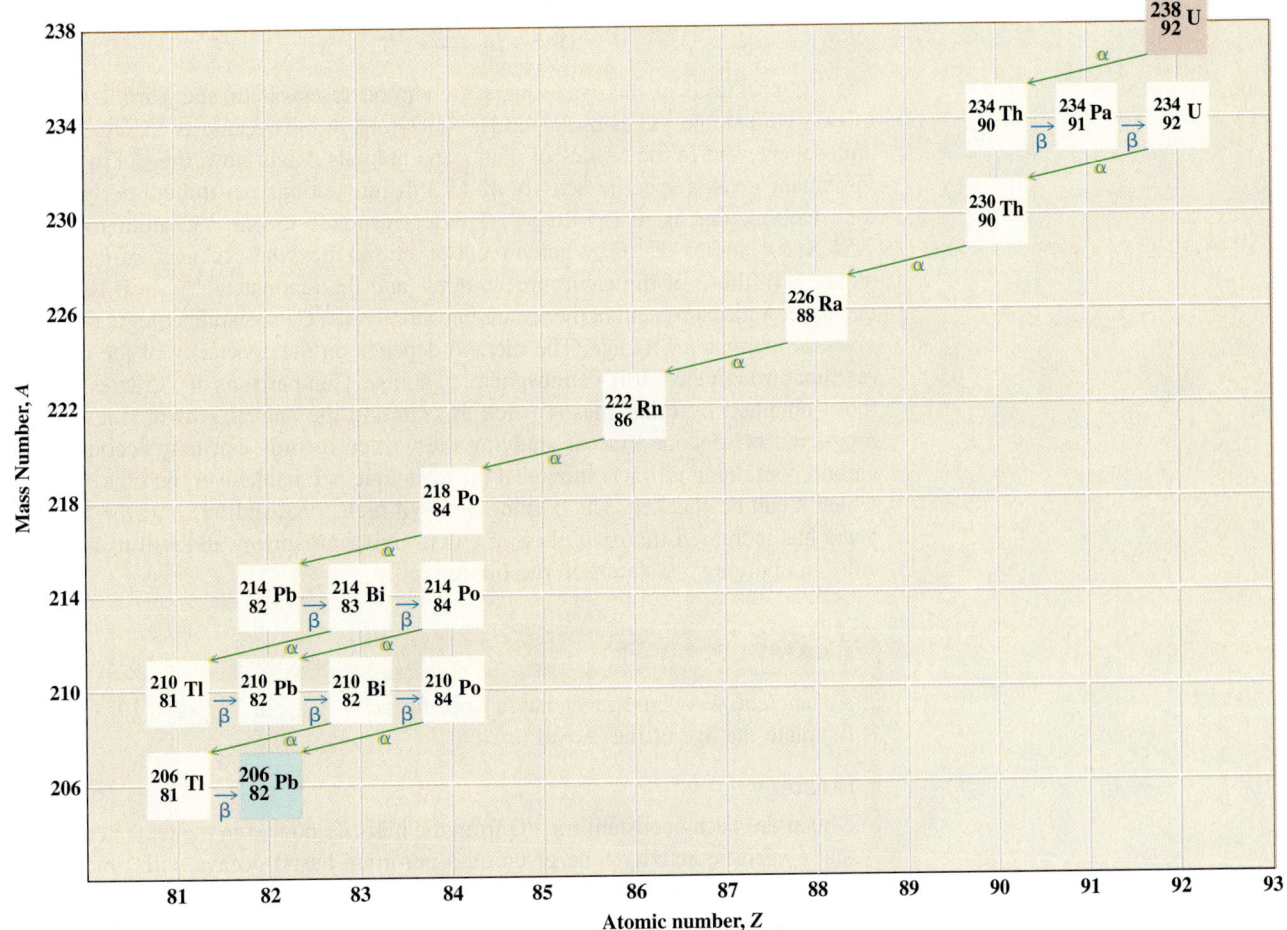

which equals 0.364. This ratio is fixed by the time elapsed since the mineral was formed. Of course, to calculate the age of a mineral, we work backward from the measured $N(^{206}\text{Pb}):N(^{238}\text{U})$ ratio.

This method of dating assumes that the stable nuclide generated at the end of the decay series (^{206}Pb in this example) derives exclusively from the parent species (^{238}U in this case) subsequent to the formation of the rock and that neither it nor the parent entered or left the rock over the course of geological time. Whenever possible, geochemists determine the ratios of several different parent–daughter pairs of nuclides from the same rock sample. For example, ^{40}K decays to ^{40}Ar (a gas, but trapped in the rock matrix) with a half-life of 1.28×10^9 years, and ^{87}Rb decays to ^{87}Sr with a half-life of 4.9×10^{11} years. Study of each pair ideally should yield the same age. Analysis of a large number of rock samples suggests that the oldest surface rocks on Earth are about 3.8 billion years old. An estimate of 4.55 billion years for the age of the Earth and the solar system comes from isotopic analysis of meteorites, which are believed to have formed at the same time.

Another type of radioactive dating, one that has been of particular value to archaeologists and anthropologists, is **carbon-14 dating,** which relies on the decay of ^{14}C. This nuclide decays by beta emission:

$$^{14}_{6}\text{C} \longrightarrow {}^{14}_{7}\text{N} + {}^{0}_{-1}e^- + \bar{\nu} \qquad t_{1/2} = 5730 \text{ yr}$$

All ^{14}C on Earth would have decayed away long ago were it not for the fact that the supply is replenished continuously in the Earth's atmosphere by the reaction of ^{14}N in the air with high-energy neutrons generated by cosmic rays:

$$^{14}_{7}\text{N} + {}^{1}_{0}n \longrightarrow {}^{14}_{6}\text{C} + {}^{1}_{1}\text{H}$$

The ^{14}C produced in this way enters the carbon reservoir on the Earth's surface, mixing with stable ^{12}C as dissolved $\text{H}(^{14}\text{C})\text{O}_3^-(aq)$ in the oceans, as $^{14}\text{CO}_2(g)$ in the atmosphere, and in the tissues of plants and animals. Right now, the ^{14}C in the environment gives a specific activity of 15.3 disintegrations per minute per gram of total carbon, that is, 0.255 Bq g^{-1}. This corresponds to one ^{14}C atom for every 7.54×10^{11} atoms of ^{12}C. When a plant or animal dies, the exchange of its carbon atoms with those of the environment stops, and the amount of ^{14}C in it begins to decrease. A measurement of the remaining activity of ^{14}C in an archaeological sample gives an estimate of its age. The method depends on the constancy of the cosmic-ray flux through the Earth's atmosphere, of course. Comparisons of ^{14}C dates against those obtained by other means (such as counting the annual growth rings of the long-lived bristlecone pine or studying the written records that may accompany a carbon-containing artifact) show that the technique is reliable over the time span for which it can be checked. The burning of fossil fuels (coal and oil) over the last 100 years has increased the proportion of ^{12}C in living organisms and will cause difficulty in applying ^{14}C dating in the future.

EXAMPLE 15–6

An ancient wooden shovel has a specific activity from ^{14}C of 0.195 Bq g^{-1}. Estimate the age of the shovel.

Strategy

Obtain the decay constant for ^{14}C from the half-life quoted in the text. Recognize that a specific activity (one given on a per-gram basis) decays with time in the same fashion as a total activity.

Solution

The decay constant for ^{14}C is

$$k = \frac{0.6931}{5730 \text{ yr}} = 1.21 \times 10^{-4} \text{ yr}^{-1}$$

The specific activity of the shovel has fallen from 0.255 Bq g^{-1} to 0.195 Bq g^{-1}. These values are on a per-gram basis, but the mass of the shovel does not change, so they can be used for A_i and A. Substitute them into the equation for the decay of the activity in the form

$$\frac{A}{A_i} = e^{-kt}$$

$$\frac{0.195 \text{ Bq g}^{-1}}{0.255 \text{ Bq g}^{-1}} = e^{-kt} = \exp\left(-(1.21 \times 10^{-4} \text{ yr}^{-1})t\right)$$

Taking the natural logarithm (ln) of both sides gives

$$\ln\left(\frac{0.195}{0.255}\right) = -(1.21 \times 10^{-4} \text{ yr}^{-1})t$$

$$t = \boxed{2200 \text{ yr}}$$

The shovel was made with wood from a tree that died approximately 2200 years ago.

Check

The age of the shovel is substantially less than 5730 years, the half-life of ^{14}C. This is consistent with the specific activity not yet having fallen to $\frac{1}{2}$ of its original value.

EXERCISE

Linen cloth from a religious relic contains one ^{14}C atom for every 8.25×10^{11} atoms of ^{12}C. Estimate the age of the cloth.

Answer: 740 years.

15-4 Radiation in Biology and Medicine

Radiation has both harmful and beneficial effects for living organisms. All forms of radiation (alpha particles, beta particles, positrons, gamma rays, X-rays) cause damage in proportion to the amount of energy that they deposit in cells and tissues. The damage takes the form of chemical changes in cellular molecules that alter their functions and may lead either to uncontrolled multiplication and growth of cells or to their death. Alpha particles lose their kinetic energy over very short distances in matter (typically 10 cm in air and 0.05 cm in water and tissues), producing intense ionization in their wakes until they accept electrons and become harmless helium atoms. Radium, for example, is an alpha emitter that substitutes for calcium in bone tissue. Its alpha emissions then disrupt production of blood cells in the nearby bone marrow. Beta particles and gamma rays penetrate better than do alpha particles and so present a radiation hazard even when their source is well outside an organism.

A person who is accidentally or deliberately exposed to radiation receives a *dose* of that radiation. Radiation of whatever type that deposits 1 J of energy per 1 kg of tissue delivers an absorbed dose equal to 1 **gray** (Gy):

$$1 \text{ Gy} = 1 \text{ J kg}^{-1}$$

Two 1 Gy doses of radiations of different type may differ in their toxicity, depending on the actual volume of affected tissue, the type of tissue (some organs are more susceptible than others), and the dosage rate (a slow dose does not do the same harm as does a sudden one). To take all of these factors into account, the **sievert** (Sv) has been defined. This unit measures the *effective* dosages of radiation received by humans. An absorbed dose (physical dose) of 1 Gy of beta radiation or gamma radiation delivers a human dose of approximately 1 Sv. Alpha radiation is more toxic; a physical dose of 1 Gy of alpha radiation equals about 10 Sv.

Exposure to radiation is unavoidable. The average person in the United States receives an effective radiation dose of about 3 millisieverts (mSv) annually from natural sources, which include cosmic radiation and natural radioactive nuclides such as ^{14}C, ^{40}K, and ^{222}Rn in the air and soil. Another 0.5 to 1 mSv comes from human activities (including dental and medical X-rays and airplane flights, which cause greater exposure to cosmic rays higher in the atmosphere). The safe level of exposure to radiation is a controversial issue among biologists because the consequences of exposure may become apparent only after many years. Also, it is necessary to distinguish between tissue damage experienced by the exposed individual and genetic damage, which may not become apparent for several generations. It is much easier to define the radiation level that will, with high probability, cause death to an exposed person. The LD_{50} level in humans (50% probability that death will result in 30 days after a single exposure) is 5 Gy.

- The gray is an SI unit. Radiation dosage also is measured in a unit called the "rad" (radiation absorbed dose), which is widely used in the United States. 1 rad = 0.01 Gy.

- Like the gray, the sievert is an SI unit. Effective human dosage also is measured in a unit called the "rem" (radiation equivalent in man). 1 rem = 0.01 Sv.

EXAMPLE 15–7

The beta decay of ^{40}K, which is a natural part of the body, makes all humans slightly radioactive. The relative abundance of ^{40}K in ordinary potassium is 0.0118%, its half-life is 1.28×10^9 years, and its beta particles have an average kinetic energy of 0.55 MeV. A 70 kg man contains about 170 g of potassium. (a) Estimate how radioactive this man is by calculating the total activity (in becquerels) of his ^{40}K. (b) Use this activity to estimate (in Gy) the annual absorbed dose of radiation from beta decay of the man's internal ^{40}K.

Strategy

Use the definition of activity (A). Here, it equals the number of ^{40}K atoms (N) multiplied by their decay constant k. Calculate N and k by constructing unit-factors from the given data and multiplying them together. Do not bother to use the highly precise nuclidic mass of ^{40}K listed in Table 15–1 because other data are given much less precisely. Use 40.0 u (which implies 40.0 g mol^{-1}) instead. Obtain the annual absorbed dose by figuring the rate at which decay of the ^{40}K atoms generates energy and assuming that all of this energy is deposited within the man's body.

Solution

(a)

$$N_{^{40}\text{K}} = 170 \text{ g K} \times \left(\frac{1 \text{ mol K}}{40.0 \text{ g K}}\right) \times \left(\frac{6.022 \times 10^{23} \text{ atoms K}}{1 \text{ mol K}}\right)$$
$$\times \left(\frac{0.0118 \text{ atoms } ^{40}\text{K}}{100 \text{ atoms K}}\right) = 3.02 \times 10^{20} \text{ atoms } ^{40}\text{K}$$

$$k = \frac{\ln 2}{t_{1/2}} = \frac{0.6931}{1.28 \times 10^9 \, \text{yr} \times \left(\frac{365 \, \text{d}}{1 \, \text{yr}}\right) \times \left(\frac{24 \, \text{h}}{1 \, \text{d}}\right) \times \left(\frac{3600 \, \text{s}}{1 \, \text{h}}\right)}$$

$$= 1.72 \times 10^{-17} \, \text{s}^{-1}$$

$$A = kN = (1.72 \times 10^{-17} \, \text{s}^{-1})(3.02 \times 10^{20}) = 5.19 \times 10^3 \, \text{s}^{-1}$$

$$= 5.19 \times 10^3 \, \text{Bq}$$

About 5190 atoms of ^{40}K disintegrate every second. At this low rate of decay, the total number of ^{40}K atoms stays essentially the same from year to year, even if new ^{40}K is not ingested from food, drink, or the air.

(b) The rate at which energy is generated from the decay of the man's ^{40}K is

$$\text{rate} = \left(\frac{5.19 \times 10^3 \, \text{disint.}}{\text{s}}\right) \times \left(\frac{0.55 \, \text{MeV}}{\text{disint.}}\right) \times \left(\frac{3600 \, \text{s}}{1 \, \text{h}}\right) \times \left(\frac{24 \, \text{h}}{1 \, \text{d}}\right) \times \left(\frac{365 \, \text{d}}{1 \, \text{yr}}\right)$$

$$= 9.00 \times 10^{10} \, \text{MeV yr}^{-1}$$

Convert this to joules per year:

$$9.00 \times 10^{10} \, \frac{\text{MeV}}{\text{yr}} \times \left(\frac{1.602 \times 10^{-13} \, \text{J}}{1 \, \text{MeV}}\right) = 0.0144 \, \text{J yr}^{-1}$$

Each kilogram of the man's body tissue receives $\frac{1}{70}$ of this energy every year. The annual physical dosage is thus $0.21 \times 10^{-3} \, \text{J kg}^{-1}$, which is $0.21 \times 10^{-3} \, \text{Gy}$, or 0.21 mGy.

Check

A physical dose of 0.21 mGy converts to a human dose of 0.21 mSv (because for beta radiation, 1 Gy equals approximately 1 Sv). The average human dose from natural sources is about 3 mSv annually. By the calculation, ^{40}K accounts for roughly 7% of this, which is plausible.

EXERCISE

Strontium-90 decays by beta emission with a half-life of 28.1 years. Assume that the average kinetic energy of the beta particles is 0.22 MeV. How much ^{90}Sr (in grams) administers a 1.25 mGy dose of radiation to a 65 kg woman in 60 s if the woman absorbs all of the decay energy internally?

Answer: 0.0012 g.

Although radiation can do great harm, it confers great benefits in medical applications. The diagnostic importance of X-ray imaging hardly needs to be mentioned. Both X-rays and gamma rays are used selectively to destroy malignant cells in cancer therapy. The beta-emitting ^{131}I nuclide finds use in the treatment of cancer of the thyroid because iodine is taken up preferentially by the thyroid gland. Heart pacemakers use the energy generated by the decay of tiny amounts of radioactive ^{238}Pu (see Fig. 15–2), converted into electrical energy. The element technetium has an important radioactive isotope, ^{99m}Tc, an excited state of ^{99}Tc, that decays to the nuclear ground state by emission of gamma rays with a half-life of 6 h. It is taken up by healthy heart tissue, allowing a gamma-ray detector to provide an image of the heart that is useful diagnostically.

An advanced generator for ^{99m}Tc, an isotope used in nuclear medicine to study heart tissue. The *m* stands for "metastable" and refers to a short-lived, excited nuclear state. *(Courtesy of DuPont Pharmaceuticals)*

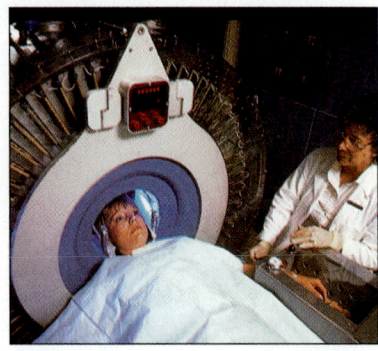

Figure 15-8 • With her head centered in a ring of radiation detectors, a patient is about to undergo a positron emission tomography (PET) scan. *(Hank Morgan/Science Source/Photo Researchers Inc.)*

Positron emission tomography (PET) is another important diagnostic technique that relies on radiation (Fig. 15–8). It uses radioactive nuclides such as ^{11}C (half-life, 20.4 min) and ^{15}O (half-life, 2.0 min) that emit positrons when they decay. These nuclides are incorporated (quickly, because of their short half-lives) into substances such as glucose, which are injected into the patient. By following the pattern of annihilation radiation issuing from the body (generated as positrons encounter normal matter), researchers can study blood flow and glucose metabolism in healthy and diseased individuals. Computer-constructed pictures of positron emissions are useful in understanding how the brain works; for example, the location and level of glucose metabolism in the brain change as Alzheimer's disease progresses (Fig. 15–9).

Less direct benefits are also numerous. An example is the study of the mechanism of photosynthesis, in which light drives the combination of carbon dioxide and water to form glucose in the green leaves of plants:

$$6 \, CO_2(g) + 6 \, H_2O(\ell) \longrightarrow C_6H_{12}O_6(s) + 6 \, O_2(g)$$

Exposure of plants to carbon dioxide containing an artificially increased proportion of ^{14}C as a tracer permits the mechanism of the reaction to be followed. At intervals, the plants are analyzed to find what compounds have incorporated ^{14}C in their

---

CHEMISTRY IN PROGRESS

Radioimmunoassay

One of the most dangerous radioactive nuclides in the fallout from the 1986 meltdown and fire at Chernobyl was iodine-131, which undergoes beta decay with a half-life of 8.021 days. Microgram quantities of ^{131}I deliver a lethal whole-body dose of radiation (see Problem 45 in this chapter). The much smaller amounts in the fallout hundreds of miles downwind from Chernobyl posed a hazard because the thyroid gland concentrates iodine in the course of synthesizing the hormone thyroxin, $C_{15}H_{11}I_4NO_4$. People in the path of the fallout were told to take potassium iodide pills to protect their thyroids against radiation damage. The basis for the intervention was chemical: Binding sites in the thyroid gland would pick up very little outside iodine, radioactive or not, against the flood of competition offered by the iodine in the pills.

The concept of competition for binding sites underlies the important technique called **radioimmunoassay** (RIA). This method permits the measurement of the concentration of substances of biological interest, such as hormones, at concentrations as low as 10^{-12} mol L^{-1}. In RIA, the binding sites are on *antibodies,* which are proteins produced by the immune system in response to the presence of a foreign protein. Antibodies bind quite specifically to foreign proteins as part of the body's defense against disease. Suppose that we wish to assay for protein H, a hormone, in a sample. We add H*, a radioisotopically labelled form of H, together with an H-binding antibody. This antibody naturally also binds H*, which is chemically equivalent to H. If any H is in the sample, it prevents some H* from binding to the antibody because it competes for binding sites. The radioactive emissions from the H*–antibody complex are then lower than if no H were in the sample. By comparing the rate of decay events in H*–antibody complex obtained from the sample with the rate in H*–antibody complex obtained from a series of standards that were prepared with known amounts of H, the amount of H in the sample can be inferred. The method is sensitive, highly specific, and easily automated.

RIA can be adapted to analyze for non-hormonal substances. Suppose that a sample of urine is being tested for cocaine. A known amount of radioisotopically labelled cocaine is complexed to a carrier protein. This complex is added to the sample, and the mixture is allowed to equilibrate. Cocaine, if present in the sample, replaces some of the radioactive cocaine in the complex. The complex then is separated, and its radioactivity is measured. The observed count rate is inversely related to the concentration of cocaine in the sample.

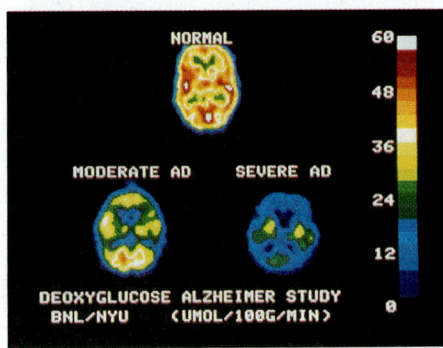

Figure 15–9 • Positron emission tomographs of the brains of a healthy older person and of patients suffering from Alzheimer's disease. Lighter colors indicate regions having higher concentrations of radioactive glucose and therefore more active metabolism. *(Courtesy of Dr. Mony de Leon/NYU Medical Center and the National Institute on Aging)*

molecules and thereby to identify the intermediates in photosynthesis. Radioactive tracers also are used widely in medical diagnosis. The *radioimmunoassay* technique, invented by Nobel laureate Rosalyn Yalow, allows for determination of very low levels of drugs and hormones in body fluids (see Chemistry in Progress: Radio-immunoassay).

15–5 Nuclear Fission

By the 1920s, physicists and chemists began using particle accelerators to bombard samples with high-energy particles in order to *induce* nuclear reactions. One of the first results of this program was the 1932 discovery by James Chadwick of the neutron as a product of the reaction of alpha particles with light nuclides such as $^9_4$Be:

$$^4_2He + {^9_4}Be \longrightarrow {^1_0}n + {^{12}_6}C$$

Shortly after Chadwick's discovery, a group of physicists in Rome, led by Enrico Fermi, began to study the interaction of neutrons with nuclei of various elements. The experiments produced a number of radioactive species, and it was evident that the absorption of a neutron increased the N-to-Z ratio in target nuclei above the stability line (see Fig. 15–1). One of the targets used was uranium, the heaviest naturally occurring element. Several radioactive products resulted, but none had chemical properties resembling those of the elements lying between $Z = 86$ (radon) and $Z = 92$ (uranium). It appeared to the Italian scientists in 1934 that several new transuranic elements ($Z > 92$) had been synthesized, and an active period of investigation followed.

In Berlin in 1938, Otto Hahn and Fritz Strassmann sought to characterize the supposed transuranic elements. To their bewilderment, they found that barium ($Z = 56$) appeared among the products of the bombardment of uranium by neutrons. Hahn informed his former colleague Lise Meitner, and she suggested that the products of the bombardment of uranium by neutrons were not transuranic elements but fragments of uranium atoms resulting from a process she termed **fission.** The implication of this idea—the possible release of enormous amounts of energy from the splitting of the nucleus—was instantly apparent. In August 1939, with the outbreak of World War II imminent, Albert Einstein wrote to U.S. president Franklin Roosevelt to inform him of the possible military uses of nuclear fission and his concern that Germany might develop a nuclear explosive. As a result, in 1942 President Roosevelt authorized the Manhattan District Project, an intense, coordinated effort by a large number of physicists, chemists, and engineers to make a fission bomb of unprecedented destructive power.

The operation of the first atomic bomb hinged on the fission of uranium in a chain reaction induced by absorption of neutrons. The two most abundant isotopes of uranium are ^{235}U and ^{238}U, which have natural abundances of 0.720% and 99.275%, respectively. Both isotopes undergo fission on absorbing a neutron, ^{238}U only with "fast" neutrons but ^{235}U with both "fast" and "slow" neutrons. In the early days of neutron research, it was not known that the absorption of neutrons by nuclei depends strongly on neutron velocity. Fermi and his co-workers observed that neutron absorption experiments conducted on a wooden table led to a much higher yield of radioactive products than did those performed on a marble-topped table. Fermi then repeated the irradiation experiments with a block of paraffin wax interposed between the radium–beryllium neutron source and the target, with the startling result that the level of induced radioactivity was greatly enhanced. Within hours, Fermi had found the explanation: The high-energy (fast) neutrons emitted from the radium–beryllium source were slowed down by collision with the nuclei of the paraffin molecules. Because of their lower kinetic energies, their probability of reaction with ^{235}U was greater and a higher yield was achieved. Hydrogen nuclei and the nuclei of other light elements such as ^{12}C (in graphite) are very effective in slowing down high-velocity neutrons and are called **moderators.**

The fission of ^{235}U follows many different patterns, and some 34 elements have been detected among the products. In any single fission event, two particular nuclides are produced, together with two or three neutrons; collectively, they carry away about 200 MeV of kinetic energy. Three of the many pathways are

$$
{}^{1}_{0}n + {}^{235}_{92}U \rightleftharpoons
\begin{cases}
{}^{92}_{36}Kr + {}^{142}_{56}Ba + 2\,{}^{1}_{0}n \\
{}^{101}_{41}Nb + {}^{133}_{51}Sb + 2\,{}^{1}_{0}n \\
{}^{94}_{38}Sr + {}^{140}_{54}Xe + 2\,{}^{1}_{0}n
\end{cases}
$$

In the electrorefining of uranium for nuclear-reactor fuel, uranium(III) chloride is deposited on the anode of an electrolytic cell. Here, it comprises 3% of the mass in a matrix of LiCl and KCl and causes the amethyst color of the deposit. *(Courtesy of Argonne National Laboratories)*

The fission process is a branching chain reaction because two or three new neutrons are produced for each neutron absorbed. Figure 15–10 illustrates the self-propagating nuclear chain reaction and its accompanying exponential growth of numbers of neutrons.

Such growth results in an explosion if the rate of production of neutrons is not soon matched by the rate of their loss. Fermi and his associates demonstrated, on December 2, 1942, that a controlled, self-sustaining neutron chain reaction occurred in a uranium "pile" that used cadmium control rods to absorb neutrons, thereby balancing the rates of neutron generation and loss. This proved the feasibility of having nuclear reactors that operate at a controlled and steady neutron concentration and, hence, a constant rate of energy production.

Although Fermi and his associates were the first scientists to demonstrate a self-sustaining nuclear chain reaction, a natural uranium fission reactor "went critical" about 1.8 billion years ago in a place now called Oklo in the Gabon Republic of equatorial Africa. The clue that this had happened came when French scientists discovered that the ^{235}U content of ore from a site in the open-pit mine at Oklo was only 0.7171%; the normal ^{235}U content in other areas of the mine was 0.7207%. Although this was not a large deviation, it was significant, and investigation revealed that other elements were also present in the ore in the exact proportions expected if nuclear fission had occurred. This established beyond all doubt that a self-sustaining nuclear reaction had occurred there. From the size of the active ore mass and the depletion of ^{235}U, it is estimated that the reactor generated about 15,000 megawatt-years (5×10^{17} J) of energy over a period of about 100,000 years.

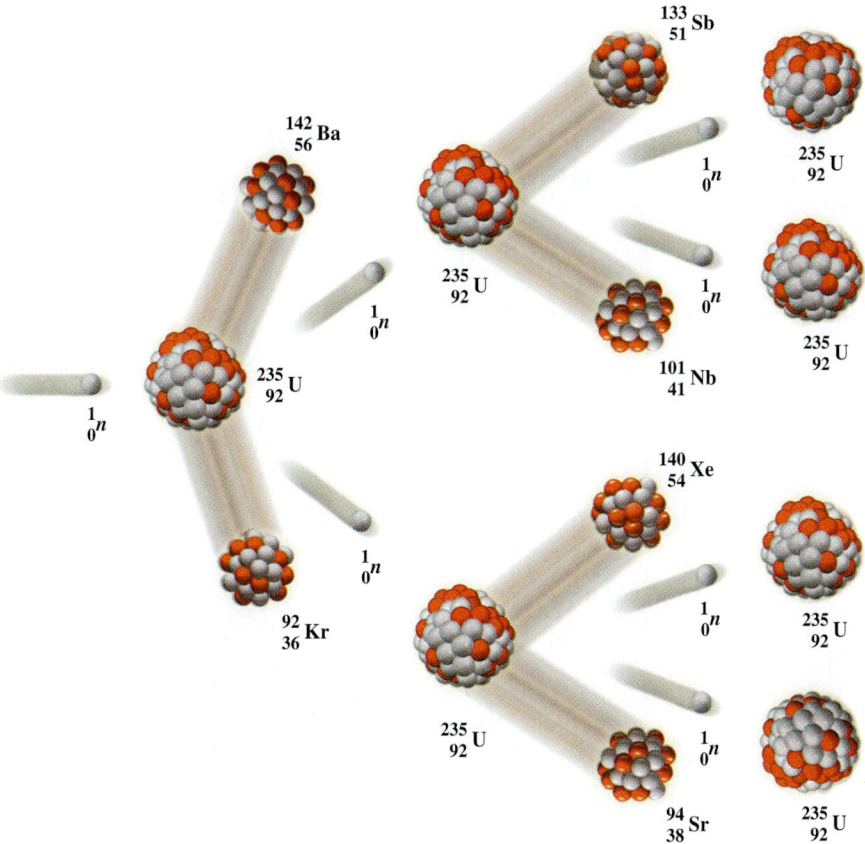

Figure 15–10 • In a self-propagating nuclear chain reaction, the number of neutrons grows exponentially during fission.

Nuclear Power Reactors

Most nuclear power reactors in the United States use rods containing oxides of uranium as fuel. The uranium is primarily ^{238}U, but the amount of ^{235}U is enriched above natural abundance to a level of about 3%. The moderator used to slow the neutrons (to increase the efficiency of the fission) is ordinary water in most cases, so these reactors are called "light water" reactors. The controlled release of energy by nuclear fission in power reactors demands a delicate balance between neutron generation and neutron loss. As mentioned earlier, this is accomplished by means of steel control rods containing ^{112}Cd or ^{10}B, nuclides that have a very large neutron-capture probability. These rods are automatically inserted into or withdrawn from the fissioning system in response to changes in the neutron flux. As the nuclear reaction proceeds, the moderator (water) transfers the heat that it releases to a steam generator. The steam then drives turbines that generate electricity (Fig. 15–11).

The risks associated with the operation of fission reactors are small but not negligible, as the accident at the Three Mile Island reactor in March 1979 and the disaster at Chernobyl in April 1986 attest. If a reactor has to be shut down quickly, there is danger of a meltdown, in which the heat from the continuing fission processes melts the uranium fuel. Coolant must be circulated until heat from the decay of short-lived unstable nuclides has been dissipated. The Three Mile Island accident

• A different reactor design employs "heavy water," in which the protons of ordinary water are replaced by deuterons, giving $(^2H)_2O$ (or D_2O). Heavy water is an efficient but expensive moderator.

Figure 15–11 • A schematic diagram of a pressurized-water nuclear power reactor.

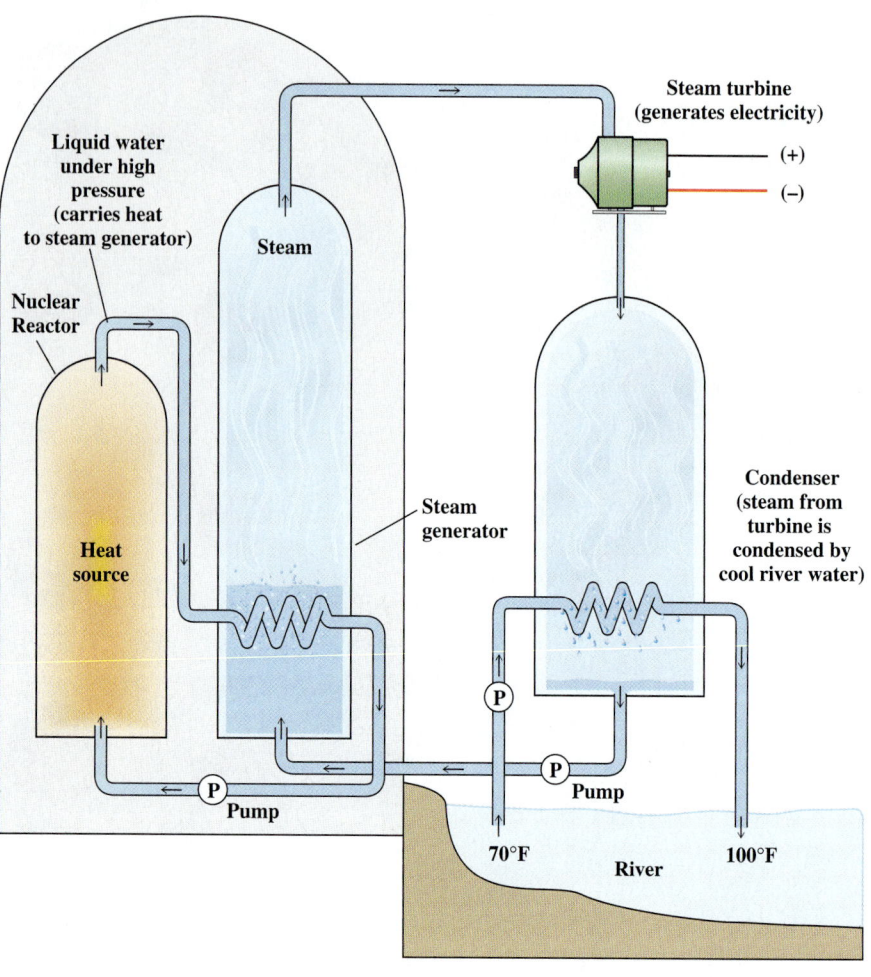

Steam turbine (generates electricity)

(+)

(−)

Liquid water under high pressure (carries heat to steam generator)

Steam

Nuclear Reactor

Heat source

Steam generator

Condenser (steam from turbine is condensed by cool river water)

P

P

P

Pump

Pump

70°F

100°F

River

resulted in a partial meltdown because some water-coolant pumps were inoperative and others were shut down too soon, causing damage to the core and slight release of radioactivity in the environment. The Chernobyl disaster was caused by a failure of the water-cooling system and a meltdown. The rapid and uncontrolled nuclear reaction that took place set the graphite moderator on fire and burst through the reactor building, spreading radioactive nuclides with an activity estimated at 2×10^{20} Bq into the environment. A major part of the problem was that the reactor at Chernobyl, unlike those in the United States, was not in a massive containment building (Fig. 15–12).

The safe disposal of the radioactive wastes from nuclear reactors is also an important but controversial matter that has not yet been resolved. A variety of proposals have been advanced, including burial of radioactive waste in deep mines on either a recoverable or permanent basis, burial at sea, and launching the waste into outer space. The first alternative is the only one that appears credible. The essential requirement is that the disposal site(s) be stable with respect to possible earthquakes and invasion by underground water movements. For a nuclide such as plutonium-

Figure 15–12 • An aerial view of the Three Mile Island nuclear power plant in Pennsylvania. The containment buildings for the two power reactors are the smaller, cylindrical buildings to the right. The four large structures are cooling towers. *(Comstock)*

Inside a containment building. The reactor core sits under a pool of water. A blue glow emanates as high-energy particles from the core are slowed by the water.

239, which has a half-life of 24,000 years, a storage site that is stable over 240,000 years is needed before the activity drops to 0.1% of its original value. Some shorter-lived nuclides are more hazardous over short periods of time, but the threat diminishes more quickly.

15–6 Nuclear Fusion and Nucleosynthesis

Nuclear **fusion** is the union of two light nuclei to form a heavier nucleus with the release of energy. Fusion reactions often are called thermonuclear reactions because the colliding particles must possess very high kinetic energies, corresponding to temperatures of millions of degrees, before the reactions are initiated. Thermonuclear reactions are the processes that occur in the Sun and other stars. In 1939, Hans Bethe (and, independently, Carl von Weizsäcker) proposed that in normal stars the following reactions take place:

$$\begin{aligned}
{}_1^1\text{H}^+ + {}_1^1\text{H}^+ &\longrightarrow {}_1^2\text{H}^+ + {}_1^0e^+ + \nu \\
{}_1^2\text{H}^+ + {}_1^1\text{H}^+ &\longrightarrow {}_2^3\text{He}^{2+} + \gamma \\
{}_2^3\text{He}^{2+} + {}_2^3\text{He}^{2+} &\longrightarrow {}_2^4\text{He}^{2+} + 2\,{}_1^1\text{H}^+
\end{aligned}$$

In the first reaction, two protons fuse to form a deuteron ($^2\text{H}^+$), with the emission of a positron and a neutrino that have a total of 1.026 MeV of kinetic energy. In the second reaction, a deuteron combines with a proton to form a helium nucleus of mass 3, together with a gamma ray. The third reaction forms a normal helium nucleus and regenerates two protons. The net result is conversion of four protons in a helium nucleus. Each of the reactions is exothermic, but hard to initiate: Up to 1.25 MeV is required to overcome the electrostatic repulsion between the positively charged nuclei. Because the overall result of the sequence is to convert hydrogen into helium, the process is called **hydrogen burning.**

Fusion reactions in the interior of the Sun are the ultimate source of most of our energy resources on Earth. Our Sun is still in its hydrogen-burning period, generating helium from hydrogen. *(NASA)*

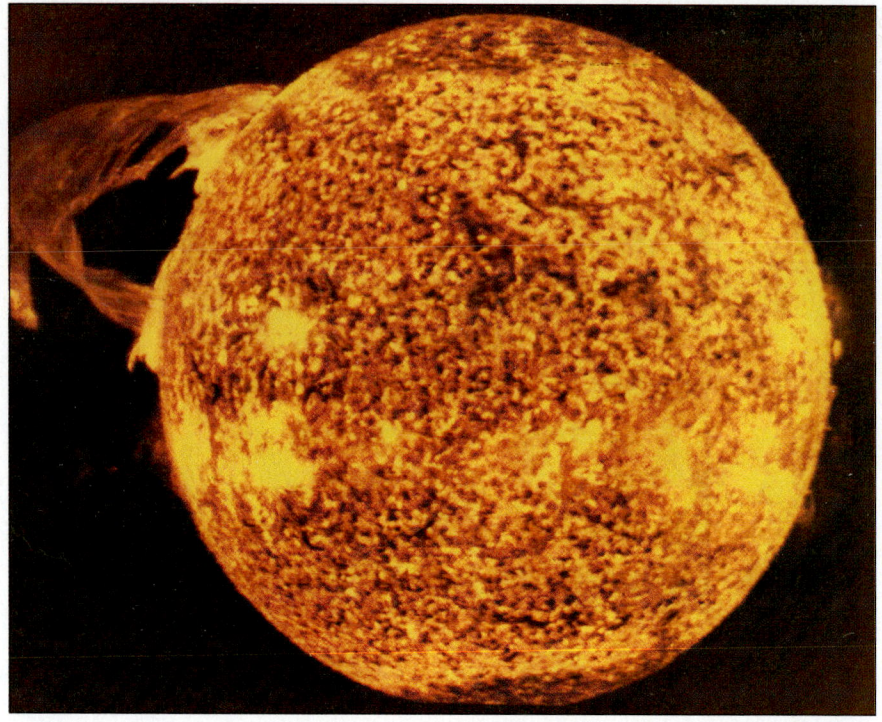

As a star ages and accumulates helium, it begins to contract under the influence of its immense gravity. The star's helium core begins to heat up. When the core reaches a temperature of about 10^8 K, another kind of **nucleosynthesis** begins, the first reaction of which is

$$2\,{}_2^4\text{He}^{2+} \rightleftharpoons {}_4^8\text{Be}^{4+}$$

This reaction is written as an equilibrium because the ^{8}Be nucleus reverts to helium nuclei with a half-life of only 2×10^{-16} s. Even with this short half-life, the ^{8}Be nuclei are believed to react with helium nuclei to form ^{12}C nuclei in the exothermic reaction

$$\text{{}_4^8Be}^{4+} + {}_2^4\text{He}^{2+} \longrightarrow {}_6^{12}\text{C}^{6+}$$

The density of the core of a star that is burning helium in this way is on the order of 10^5 g cm^{-3}.

This process of nucleosynthesis continues beyond the formation of ^{12}C nuclei to produce additional, heavier nuclei: those of ^{13}N, ^{13}C, ^{14}N, ^{15}O, ^{15}N, and ^{16}O. Stars in this stage have moved away from the main sequence that characterizes the younger stars and are classified as red giants. Even here, the process does not end. Similar cycles occur until the temperature of the core is about 4×10^9 K, its density is about 3×10^6 g cm^{-3}, and the nuclei are those of ^{56}Fe, ^{59}Co, and ^{60}Ni. These nuclei have the maximum binding energy per nucleon (see Fig. 15–3). It is thought that the synthesis of still heavier nuclei occurs in the immense explosions of supernovas.

Heavy elements also can be produced in **particle accelerators,** which accelerate electrons, protons, and ions to high speeds. The resulting collisions can cause new elements to form. Technetium, for example, is not found in nature but was first produced in 1937 by directing high-energy deuterons at a molybdenum target:

$$\text{{}_{42}^{96}Mo} + {}_1^2\text{H} \longrightarrow {}_{43}^{97}\text{Tc} + {}_0^1 n$$

and the binding energy per nucleon of the following nuclides, using data from Table 15–1.
(a) $^{40}_{20}Ca$　(b) $^{87}_{37}Rb$　(c) $^{238}_{92}U$

4. (See Example 15–1.) Calculate the total binding energy, in both kilojoules per mole and megaelectron volts per atom, and the binding energy per nucleon of the following nuclides, using data from Table 15–1.
(a) $^{10}_{4}Be$　(b) $^{35}_{17}Cl$　(c) $^{49}_{22}Ti$

5. The nuclide hassium-269, first synthesized in 1994, decays rapidly according to the equation

$$^{269}_{108}Hs \longrightarrow ^{265}_{106}Sg + ^{4}_{2}He$$

An experiment establishes that $\Delta E = -9.23$ MeV for this reaction. Compute the difference in mass (in u) between an atom of hassium-269 and one of seaborgium-265.

6. An atom of nobelium-257 has a mass of 257.096853 u. It emits an alpha particle to give fermium-253:

$$^{257}_{102}No \longrightarrow ^{253}_{100}Fm + ^{4}_{2}He$$

The energy change of the system is -8.45 MeV. Compute the mass (in u) of an atom of fermium-253.

7. Use data from Table 15–1 to predict which is more stable: four protons, four neutrons, and four electrons organized as two $^{4}_{2}He$ atoms or as one $^{8}_{4}Be$ atom. What is the mass difference?

8. (a) Use data from Table 15–1 to predict which is more stable: 16 protons, 16 neutrons, and 16 electrons organized as two $^{16}_{8}O$ atoms or as one $^{32}_{16}S$ atom. What is the mass difference?
(b) If the answer is that $^{32}_{16}S$ is more stable, what keeps O_2 molecules in the atmosphere, most of which contain two $^{16}_{8}O$ atoms, from combining into sulfur atoms? If the answer is that two $^{16}_{8}O$ atoms are more stable, what prevents $^{32}_{16}S$ atoms in rocks and minerals from breaking down into $^{16}_{8}O$ atoms?

Nuclear Decay Processes

9. (See Example 15–2.) Write balanced equations that represent the following nuclear reactions:
(a) Beta emission by $^{37}_{17}Cl$
(b) Positron emission by $^{22}_{11}Na$
(c) Alpha emission by $^{224}_{88}Ra$
(d) Electron capture by $^{82}_{38}Sr$

10. (See Example 15–2.) Write balanced equations that represent the following nuclear reactions:
(a) Alpha emission by $^{155}_{70}Yb$
(b) Positron emission by $^{26}_{14}Si$
(c) Electron capture by $^{65}_{30}Zn$
(d) Beta emission by $^{100}_{41}Nb$

11. Of ^{11}C, ^{237}Np, and ^{15}C, which is most likely to decay by:
(a) beta emission?
(b) alpha emission?
(c) positron emission?

12. Of ^{31}Si, ^{252}Fm, and ^{24}Al, which is most likely to decay by:
(a) beta emission?
(b) alpha emission?
(c) positron emission?

13. The natural abundance of ^{30}Si is 3.1%. On irradiation with neutrons, this nuclide is converted into ^{31}Si, which decays to form the stable nuclide ^{31}P. This provides a way of doping silicon with phosphorus in a much more uniform fashion than is possible by ordinary mixing of silicon and phosphorus and gives semiconductor devices capable of handling much higher levels of power. Write balanced nuclear equations for the two steps in the preparation of ^{31}P from ^{30}Si.

14. The most convenient way to prepare the element polonium is to expose ordinary bismuth (which is 100% ^{209}Bi) to a beam of neutrons. Write balanced nuclear equations for the two steps in the preparation of polonium.

15. (See Example 15–3.) The nuclide $^{40}_{19}K$ undergoes spontaneous decay to $^{40}_{20}Ca$ with emission of a beta particle and an antineutrino.
(a) Calculate the maximum kinetic energy of the beta particle (in MeV) using masses from Table 15–1.
(b) Explain why the answer to part (a) differs from the average kinetic energy of the beta particle quoted in Example 15–6 for this decay.

16. (See Example 15–3.) The nuclide $^{14}_{6}C$ decays to $^{14}_{7}N$ with emission of a beta particle and an antineutrino. Calculate the maximum kinetic energy of the beta particle (in MeV) using masses from Table 15–1.

17. (See Example 15–4.) Find the maximum kinetic energy (in MeV) of the positron that is emitted in the decay of neon-19 (mass of neutral atom 19.0018798 u) to fluorine-19 (mass of neutral atom 18.9984032 u):

$$^{19}_{10}Ne \longrightarrow ^{19}_{9}F + ^{0}_{1}e^+ + \nu$$

18. (See Example 15–4.) Find the mass (in u) of a neutral atom of oxygen-15 if the maximum kinetic energy of the positrons emitted in the reaction

$$^{15}_{8}O \longrightarrow ^{15}_{7}N + ^{0}_{1}e^+ + \nu$$

is 1.732 MeV. The mass of a neutral atom of nitrogen-15 is 15.0001089 u.

19. Compute ΔE (in MeV) for the following electron capture:

$$^{231}_{92}U \xrightarrow{\text{E.C.}} ^{231}_{91}Pa + \nu$$

The masses of the neutral atoms are given in Table 15–1.

20. Most $^{231}_{92}U$ atoms decay by electron capture (see the previous problem). However, some decay by alpha emission:

$$^{231}_{92}U \longrightarrow ^{227}_{92}Ac + ^{4}_{2}He$$

Calculate ΔE (in MeV) if the mass of a neutral atom of actinium-227 equals 227.0277470 u. Consult Table 15–1 for other masses.

21. What is the purpose of wearing a film badge when working with radioactive substances?

22. How did Rutherford and his students detect alpha particles in the famous experiments that led to the postulation of the nuclear atom?

Kinetics of Radioactive Decay

23. How many radioactive disintegrations occur per minute in a 0.0010 g sample of ^{209}Po that has been freshly separated from its decay products? The half-life of ^{209}Po is 103 years.

24. How many alpha particles are emitted per minute by a 0.0010 g sample of ^{238}U that has been freshly separated from its decay products? Assume that each decay emits one alpha particle. The half-life of ^{238}U is 4.47×10^9 yr.

25. (See Example 15–5.) The nuclide ^{19}O, prepared by neutron irradiation of ^{19}F, has a half-life of 29 s.
   (a) How many ^{19}O atoms are in a freshly prepared sample if its activity is 2.5×10^4 Bq?
   (b) After 2.00 min, how many ^{19}O atoms remain?

26. (See Example 15–5.) The nuclide ^{35}S decays by beta emission with a half-life of 87.1 days.
   (a) Calculate the mass (in grams) of a sample of pure ^{35}S that has a beta emission activity of nuclide of 3.70×10^2 Bq.
   (b) Calculate the mass (in grams) of ^{35}S that remains after 365 days.

27. The activity of a 1.00 mg sample of pure ^{137}Cs is 3.19×10^9 Bq. Determine the half-life of ^{137}Cs in years. Take the molar mass of this nuclide to equal 137 g mol^{-1}.

28. A sample of mixed metals weighing 125.3 mg is found to be radioactive, with an activity of 3.0×10^7 Bq. A study of the energy of the emitted beta particles establishes that the radioactivity is due entirely to ^{59}Fe. What percentage by mass of the sample is ^{59}Fe, assuming that its molar mass is 59.0 g mol^{-1}? The half-life of ^{59}Fe is 44.6 days.

29. The half-life of a free neutron was thought to be about 1100 s until a new experiment established it as 876 ± 21 s. Suppose that a procedure starts with 1.00 mol of free neutrons. Calculate how much time is required for the quantity of neutrons to be reduced to 0.90 mol according to the old half-life and the new half-life.

30. Gallium citrate that contains the radioactive nuclide ^{67}Ga is used in medicine as a tumor-seeking agent. Gallium-67 decays with a half-life of 77.9 h. How much time (in hours) is required for the activity to decay to 5.0% of its initial activity?

31. Astatine is the rarest naturally occurring element, with ^{219}At appearing as the product of a very minor side-branch in the decay of ^{235}U (itself not a very abundant nuclide). It is estimated that the mass of all the naturally occurring ^{219}At in the upper kilometer of the Earth's surface has a steady value of only 44 mg. Calculate the total activity (in becquerels) of all the naturally occurring astatine in this part of the Earth. The half-life of ^{219}At is 54 s, and its mass is 219.01 u.

32. Technetium has not been found in nature. It can be obtained readily as a product of uranium fission in nuclear power

plants, however, and is now produced in quantities of many kilograms per year. Calculate the total activity (in becquerels) of 1.0 mg of ^{99m}Tc (an excited nuclear state of ^{99}Tc), which has a half-life of 6.0 h.

33. Uranium-238 decays over several steps to lead-206, as diagrammed in Figure 15–7. An experiment on a uranium-bearing rock establishes that the ratio of the number of atoms of ^{206}Pb to the number of atoms of ^{238}U equals 0.333. All the ^{206}Pb in the rock comes from the decay of the ^{238}U, and ^{238}U is lost from the rock only by its radioactive decay.
   (a) Determine the fraction of the original amount of ^{238}U that has decayed.
   (b) The half-life of ^{238}U is 4.468×10^9 years. How old is the rock?

34. The nuclide ^{87}Rb decays to ^{87}Sr with a half-life of 4.9×10^{10} years. A mineral that contained ^{87}Rb but no ^{87}Sr when it was originally formed now is found to contain one ^{87}Sr atom for each 120 ^{87}Rb atoms. Estimate the age of the mineral, in years.

35. (See Example 15–6.) The specific activity of ^{14}C in the biosphere is 0.255 Bq g^{-1}. What is the age of a piece of papyrus from an Egyptian tomb if its beta counting rate is 0.153 Bq g^{-1}? The half-life of ^{14}C is 5730 years.

36. (See Example 15–6.) The charcoal residues at an ancient hearth have only 0.125 of the ^{14}C activity of growing wood. Estimate the age of the residues.

37. (See Example 15–6.) Detecting ^{14}C activity at levels lower than 5×10^{-4} Bq g^{-1} is very difficult experimentally. If 5×10^{-4} Bq g^{-1} is the lowest activity that can be measured, determine the maximum age that can be assigned to an object by ^{14}C methods.

38. (See Example 15–6.) The ^{14}C activity of carbon extracted from the wood of an ancient bow is counted with great difficulty and found to equal $5 \pm 2 \times 10^{-4}$ Bq g^{-1}. Determine the maximum age of the bow (corresponding to the lower end of the range of activity) and the minimum age of the bow (corresponding to the higher end of the range of activity).

39. How is the level of ^{14}C in a living organism kept constant?

40. What assumptions are made in using the ratio $N(^{206}\text{Pb})/N(^{238}\text{U})$ to estimate the age of a uranium-containing mineral specimen?

Radiation in Biology and Medicine

41. Write balanced equations for the radioactive decays of ^{11}C and ^{15}O, both of which are used in positron emission tomography to scan the uptake of glucose in the body.

42. Write balanced equations for the radioactive decays of ^{13}N and ^{18}F, two other radioactive nuclides that are also used in positron emission tomography. What is the ultimate fate of the positrons?

43. The positrons emitted by ^{11}C have a maximum kinetic energy of 0.99 MeV, and those emitted by ^{15}O have a maximum kinetic energy of 1.72 MeV. Calculate the relative severity (a ratio) of the radiation doses caused by ingesting a given fixed chemical amount (equal numbers of atoms) of each of these radioactive nuclides.

44. Compare the relative health risks of contact with a given amount of ^{226}Ra, which decays with a half-life of 1622 years and emits 4.78 MeV alpha particles, with contact with the same chemical amount of ^{14}C, which decays with a half-life of 5730 years and emits beta particles with energies of up to 0.155 MeV.

45. (See Example 15–7.) The radioactive nuclide ^{131}I undergoes beta decay with a half-life of 8.021 days. Quantities of this nuclide, which is a product of nuclear fission, were released into the environment in the Chernobyl accident. A victim of radiation poisoning has absorbed 5.0×10^{-6} g ($5.0 \ \mu$g) of ^{131}I.
    (a) Calculate the activity in becquerels of the ^{131}I in this person, taking the molar mass of the nuclide to equal 131 g mol^{-1}.
    (b) Calculate the physical absorbed dose of radiation (in Gy) caused by this nuclide during the first 1.0 s after its ingestion. Assume that beta particles emitted by ^{131}I have an average kinetic energy of 0.40 MeV, that all of this energy is deposited within the victim's body, and that the victim weighs 60 kg.
    (c) Is this poisoning incident likely to be fatal? Remember that the ^{131}I diminishes its activity as it decays.

46. (See Example 15–7.) The nuclide ^{239}Pu undergoes alpha decay with a half-life of 2.411×10^4 years. An atomic energy worker breathes in 5.0×10^{-6} g ($5.0 \ \mu$g) of ^{239}Pu, which lodges permanently in a lung.
    (a) Calculate the activity in becquerels of the inhaled ^{239}Pu, taking the molar mass of the nuclide to equal 239 g mol^{-1}.
    (b) Calculate the physical absorbed dose of radiation (in Gy) administered by this poisoning during the first year. Assume that alpha particles emitted by ^{239}Pu have an average kinetic energy of 5.24 MeV, that all of this energy is deposited within the worker's body, and that the worker weighs 60 kg.
    (c) Is this dose likely to be lethal?

Nuclear Fission and Fusion

47. Strontium-90 is one of the most hazardous products of atomic-weapons testing because of its long half-life ($t_{1/2} = $ 28.1 years) and its chemical tendency to accumulate in the bones.
    (a) Write nuclear equations for the decay of ^{90}Sr via the successive emission of two beta particles.
    (b) The nuclidic mass of ^{90}Sr is 89.9073 u, and the nuclidic mass of ^{90}Zr is 89.9043 u. Calculate the energy released per ^{90}Sr atom (in MeV) in decaying to ^{90}Zr.
    (c) What is the initial activity of 1.00 g of ^{90}Sr released into the environment, in becquerels?
    (d) What activity would the material from part (c) show after 100 years?

48. Plutonium-239 is the fissionable nuclide produced in breeder reactors; it also is produced in ordinary nuclear plants and in weapons tests. It is an extremely poisonous substance with a half-life of 24,100 years.
    (a) Write a nuclear equation for the alpha decay of ^{239}Pu.
    (b) The nuclidic mass of ^{239}Pu is 239.05216 u, and the nuclidic mass of ^{235}U is 235.043923 u. Calculate the energy released per ^{239}Pu atom in alpha decay (in MeV).
    (c) What is the initial activity, in becquerels, of 1.00 g of ^{239}Pu buried in a disposal site for radioactive wastes?
    (d) What activity would the material from part (c) show after 100,000 years?

49. The three naturally occurring isotopes of uranium are ^{234}U (half-life 2.5×10^5 years), ^{235}U (half-life 7.0×10^8 years), and ^{238}U (half-life 4.5×10^9 years). As time passes, does the average molar mass of the uranium in a sample taken from nature increase, decrease, or remain constant? Explain.

50. Natural lithium consists of 7.42% ^{6}Li and 92.58% ^{7}Li. Much of the tritium ($^3_1$H) used in experiments with fusion reactions is made by the capture of neutrons by ^{6}Li atoms.
    (a) Write a balanced nuclear equation for the process. What is the other particle produced?
    (b) After ^{6}Li is removed from natural lithium, the remainder is sold for other uses. Is its molar mass greater or smaller than that of natural lithium?

51. Calculate the amount of energy released, in kilojoules per gram of uranium, in the fission reaction

$$^{235}_{92}U + ^1_0n \longrightarrow ^{94}_{36}Kr + ^{139}_{56}Ba + 3 \ ^1_0n$$

Take $m(^{94}Kr)$ and $m(^{139}Ba)$ as 93.919 u and 138.909 u, respectively, and find other required masses in Table 15–1.

52. Calculate the amount of energy released, in kilojoules per gram of deuterium (^{2}H), in the fusion reaction

$$^2_1H + ^2_1H \longrightarrow ^4_2He$$

Use the masses given in Table 15–1. Compare your answer with that from the preceding problem.

53. What is the purpose of a moderator in a nuclear reactor? What kinds of substances have been used as moderators?

54. How does ^{238}U differ from ^{235}U in undergoing fission?

55. Describe the fusion processes that are associated with the various stages of a star's life.

56. Describe the plasma in which fusion takes place. What methods are under consideration for containing this plasma?

ADDITIONAL PROBLEMS

57. Tritium is $^3_1$H, an isotope of hydrogen. It is radioactive, emitting beta particles with a half-life of 12.26 years. Iodine-131 is one of many radioactive isotopes of iodine. It also decays by beta emission and has a half-life of 8.021 days. Ingestion of iodine-131 is far more dangerous than ingestion of tritium. Explain why.

58. A recent report states that the tau neutrino has (with 95% confidence) a maximum mass corresponding to 76 eV. Express this upper limit of mass in kilograms and in atomic mass units.

59. Hydrazine ($N_2H_4(\ell)$) reacts with oxygen in a rocket engine to form nitrogen and water vapor:

$$N_2H_4(\ell) + O_2(g) \longrightarrow N_2(g) + 2\,H_2O(g)$$

(a) Calculate ΔH_r^0 and ΔE_r^0 of this reaction at 25°C. Use data from Appendix D.

(b) Calculate the loss in mass (in grams) during the reaction of 1.000 mol of hydrazine.

60. Working in Rutherford's laboratory in 1932, Cockcroft and Walton bombarded a lithium target with 0.70 MeV protons and observed the following reaction:

$$^{7}_{3}\text{Li} + {}^{1}_{1}\text{H} \longrightarrow 2\,{}^{4}_{2}\text{He}$$

The alpha particles were each found to have a kinetic energy of 8.5 MeV. This research provided the first experimental confirmation of Einstein's $\Delta E = c_0^2 \Delta m$ relationship. Use this result and masses from Table 15–1 to calculate c_0.

61. The radioactive nuclide ${}^{64}_{29}\text{Cu}$ decays by emission of positrons to ${}^{64}_{28}\text{Ni}$. The mass of a ${}^{64}_{29}\text{Cu}$ atom is 63.92976 u, and the mass of a ${}^{64}_{28}\text{Ni}$ atom is 63.92796 u. Calculate the maximum kinetic energy of the positrons emitted (in MeV).

62. The nuclide ${}^{64}\text{Cu}$ (see previous problem) sometimes decays by emission of beta particles having a maximum kinetic energy of 0.573 MeV. Identify the daughter nuclide in this decay, and compute its mass in atomic mass units.

63. The nuclide ${}^{231}_{92}\text{U}$ converts spontaneously to ${}^{231}_{91}\text{Pa}$.

(a) Write two balanced nuclear equations, one for this conversion proceeding by electron capture and the other for it proceeding by positron emission.

(b) Using the masses given in Table 15–1, calculate the change in mass for each process. Explain why in this case electron capture can occur spontaneously, but positron emission cannot.

64. Selenium-82 undergoes double beta decay:

$$^{82}_{34}\text{Se} \longrightarrow {}^{82}_{36}\text{Kr} + 2\,{}^{0}_{-1}e^- + 2\,\bar{\nu}$$

This low-probability process occurs with a half-life of 3.5×10^{27} s, one of the longest half-lives ever measured. Estimate the activity in an 82.0 g (1.00 mol) sample of this nuclide. How many ${}^{82}\text{Se}$ nuclei decay in a day?

65. The half-life of ${}^{14}\text{C}$ is $t_{1/2} = 5730$ years, and carbon separated from modern wood charcoal has a specific activity of 0.255 Bq g^{-1}.

(a) Calculate the number of ${}^{14}\text{C}$ atoms per gram of carbon in modern wood charcoal.

(b) Calculate the fraction of carbon atoms in the biosphere that are ${}^{14}\text{C}$.

66. Carbon-14 is produced in the upper atmosphere by the reaction

$$^{14}_{7}\text{N} + {}^{1}_{0}n \longrightarrow {}^{14}_{6}\text{C} + {}^{1}_{1}\text{H}$$

where the neutrons come from nuclear processes induced by cosmic rays. It is estimated that the steady-state ${}^{14}\text{C}$ activity in the biosphere is 1.1×10^{19} Bq.

(a) Estimate the total mass of carbon in the biosphere, using the data in Problem 65.

(b) The Earth's crust has an average carbon content of 250 parts per million by mass, and the total crustal mass is 2.9×10^{25} g. Estimate the fraction of the carbon in the Earth's crust that is part of the biosphere. Speculate on the whereabouts of the rest of the carbon in the Earth's crust.

67. Over geologic time scales, an atom of ${}^{238}\text{U}$ decays to a stable ${}^{206}\text{Pb}$ atom in a series of eight alpha emissions, each of which leads to the formation of one helium atom. A geochemist analyzes a rock and finds that it contains 9.0×10^{-5} cm^3 of helium (at STP) per gram and 2.0×10^{-7} g ${}^{238}\text{U}$ per gram. Estimate the age of the mineral, given that $t_{1/2}$ of ${}^{238}\text{U}$ is 4.468×10^9 years.

68. The half-lives of ${}^{235}\text{U}$ and ${}^{238}\text{U}$ are 7.04×10^8 years and 4.468×10^9 years, respectively. The current terrestrial abundance ratio of the two isotopes is ${}^{238}\text{U}:{}^{235}\text{U} = 137.7$. Estimate how long it has been since a hypothetical supernova explosion in the early universe produced ${}^{238}\text{U}$ and ${}^{235}\text{U}$ in equal abundance. (*Note:* The age of the Earth is about 4.5×10^9 years.)

69. Cobalt-60 and iodine-131 are used in treatments of different types of cancer. Cobalt-60 decays with a half-life of 5.27 years, emitting beta particles with a maximum energy of 0.32 MeV. Iodine-131 decays with a half-life of 8.021 days, emitting beta particles with a maximum energy of 0.60 MeV.

(a) Equal small chemical amounts (not enough to be lethal) of these nuclides are ingested and remain in the body indefinitely. Find the ratio of the total lifetime radiation doses from the two radioactive nuclides.

(b) The two nuclides are ingested but then purged out after a short time (1 h). Find the ratio of doses from the two radioactive nuclides in this case.

70. Figure 15–2 shows how plutonium-238 self-heats (to red heat!) from the disintegration of about 2.5 atoms per second out of every 10 billion. Estimate the half-life of ${}^{238}\text{Pu}$ (in years).

71. Boron is used in control rods in nuclear power reactors because it is a good neutron absorber. When the nuclide ${}^{10}\text{B}$ captures a neutron, an alpha particle (helium nucleus) is emitted. What other nuclide is formed? Write a balanced equation.

72. A puzzling observation that led to the discovery of isotopes was the fact that lead obtained from uranium-containing ores had a molar mass lower by two full units than did lead obtained from thorium-containing ores. Explain this result, using the fact that decay of radioactive uranium and thorium to stable lead occurs via alpha and beta emission.

73. By 1913, the elements radium, actinium, thorium, and uranium all had been discovered, but element 91, between thorium and uranium in the periodic table, was not yet known.

The approach used by Meitner and Hahn was to look for the parent that decays to form actinium. Alpha and beta emission are the most important decay pathways among the heavy radioactive elements. What elements would decay to actinium by each of these two pathways? If radium salts show no sign of actinium, what does this suggest about the parentage of actinium? What is the origin of the name of element 91, discovered by Meitner and Hahn in 1918?

74. The average energy released in the fission of a ^{235}U nucleus is about 200 MeV. Suppose that the conversion of this energy into electrical energy is 40% efficient. What mass of ^{235}U is converted into its fission products in a year's operation of a 1000-megawatt nuclear power station? Recall that $1\ \text{W} = 1\ \text{J s}^{-1}$.

75. The radiant energy received by the Earth from the Sun is approximately $3.4 \times 10^{17}\ \text{J s}^{-1}$. This energy represents approximately 4.5×10^{-10} times the total energy put out by the Sun.
  (a) Calculate the change in mass of the Sun per second.
  (b) Assuming that the overall hydrogen-burning reaction in the Sun is the conversion of four $^1_1$H atoms to one $^4_2$He atom and that the entire energy change appears as radiant energy, calculate the mass of hydrogen reacting per second in the Sun. Use the data of Table 15–1.

76. Compare fission and fusion nuclear reactions with regard to the nature of the starting materials, their use to generate electrical energy, the technical problems that need to be overcome in their use, and their by-products.

CUMULATIVE PROBLEM

Radon

Radioactive ^{222}Rn and ^{220}Rn constantly form from the decay of uranium and thorium in rocks and soil and, being gaseous, seep out of the ground. The radon isotopes decay fairly quickly, but their products, which also are radioactive, are then in the air and attach themselves to dust particles. Thus, airborne radioactivity can accumulate to worrisome levels in poorly ventilated basements dug in ground that is rich in uranium and thorium.

(a) Describe the composition of an atom of ^{222}Rn, and compare it with that of an atom of ^{220}Rn.

(b) Although ^{222}Rn is a decay product of ^{238}U, ^{220}Rn comes from ^{232}Th. How many alpha particles are emitted in forming these radon isotopes from their uranium or thorium starting points? (*Hint*: Alpha decay changes the mass number A, but other decay processes do not.)

(c) Can alpha decay alone explain the formation of these radon isotopes from ^{238}U and ^{232}Th? If not, state what other types of decay must occur.

(d) Can $^{222}_{86}$Rn and $^{220}_{86}$Rn decay by emission of an alpha particle? Write balanced nuclear equations for these two decay processes, and calculate the change in mass that would result. The masses of ^{222}Rn and ^{220}Rn are 222.01757 and 220.01138 u, respectively; those of ^{218}Po and ^{216}Po are 218.00897 and 216.001905 u, respectively.

(e) Calculate the change in energy in the alpha decay of one ^{220}Rn nucleus, in megaelectron volts and in joules.

(f) The half-life of ^{222}Rn is 3.82 days. Calculate the initial activity of 2.00×10^{-8} g of ^{222}Rn, in becquerels.

(g) What is the activity of the ^{222}Rn from part (f) after 14 days?

(h) The half-life of ^{220}Rn is 54 s. Are the health risks for exposure to a given amount of radon for a given short period of time greater or smaller for ^{220}Rn than for ^{222}Rn? Explain.

Radon most commonly enters houses by seeping in from the ground through foundations or basement walls.

16

Quantum Mechanics and the Hydrogen Atom

CHAPTER OUTLINE

Traveling water waves exhibit interference and diffraction, just as do waves associated with light and matter. (*Comstock*)

Mechanics is the branch of physics that deals with the interactions and motions of particles and the collections of particles that make up working mechanisms. It uses concepts such as mass, velocity, and force. The 200 years that followed the seminal work of Isaac Newton were the classical period in the study of mechanics (and of physics in general). By the end of this period (about 1900), physicists had achieved a deep and fruitful understanding that successfully dealt with problems ranging from the motions of the planets in their orbits to the design of a bicycle. These attainments now form the subject called *classical mechanics* and are eminently useful. In classical mechanics, a fundamental role is played by energy, of which there are two types: (1) kinetic energy, arising from the motion of particles, and (2) potential energy, arising from the interactions of particles with each other or with an external field (such as gravity) and depending on the *positions* of the particles. In classical mechanics, one may, in principle, predict the future positions of a group of particles from a knowledge of their present positions and momenta and of the forces between them. A major triumph of classical mechanics was the kinetic theory of gases, which is outlined in Section 5–7.

At the end of the 19th century, physicists and chemists naturally thought that the motion of elementary particles (such as the recently discovered electron) could be described by classical mechanics. They reasoned that, once the correct laws of force were discovered, the properties of atoms and molecules could be predicted to any desired degree of accuracy by solving Newton's equations of motion. The perception grew that all of the fundamental laws of physics had been discovered. At the dedication of the Ryerson Physics Laboratory at the University of Chicago in 1894, A.A. Michelson said, "Our future discoveries must be looked for in the sixth decimal place." Little did he realize the far-reaching changes that would shake physics and chemistry during the subsequent 30 years. Central to those changes was the discovery that all particles have wave-like properties, the effects of which are most pronounced for low-mass particles such as electrons. The incorporation of wave and particle aspects of matter into a single comprehensive theory is the achievement of *quantum mechanics,* the great 20th-century extension of classical mechanics. In this chapter, we discuss the origins of the quantum theory and the implications of that theory for the structure of the hydrogen atom, the simplest of all atoms. In subsequent chapters, we consider the consequences of the quantum nature of matter for atoms containing two or more electrons, for chemical bonds, and for the liquid and solid states.

16–1 Waves and Light

Many kinds of waves appear in physics and chemistry. The waves in water are a familiar example, whether they are stirred up by the winds over the oceans, set off by a stone dropped into a quiet pool, or created by a laboratory water-wave machine. Sound waves, periodic compressions of the air that move from a source to a detector such as the human ear, are another example. Light waves, as we shall see, are the consequence of oscillating electric and magnetic fields moving through space. Even some chemical reactions occur in such a way that waves of color move through the sample as the reaction proceeds. Common to all these wave phenomena is the oscillatory variation with time of some property (Table 16–1) at a fixed location in

TABLE 16–1	
Some Kinds of Waves	
Wave	**Oscillating Quantity**
Water	Height of water surface
Acoustic (sound)	Density of medium
Light	Electric and magnetic fields

Figure 16–1 • A snapshot of a traveling wave moving out from a central point.

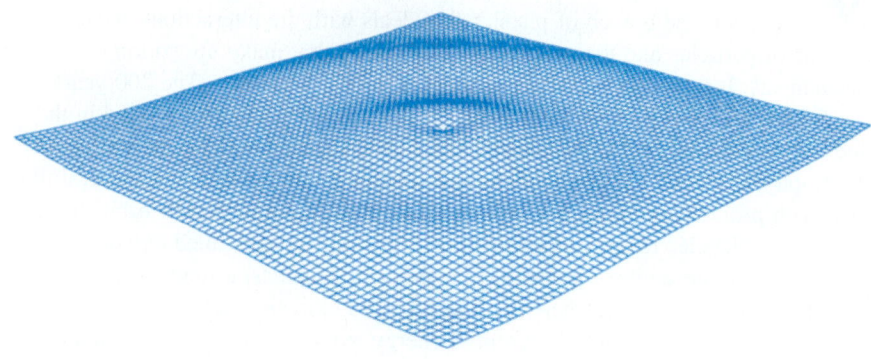

space. This is visualized most easily for a wave in water. A snapshot of such a wave (Fig. 16–1) records the crests and troughs present at some instant in time. The **amplitude** of the wave at any point is the height of the water surface over the level of the water when it is undisturbed. The distance between two successive crests (or troughs) is called the **wavelength** of the wave, provided that this distance is reproducible from crest to crest (Fig. 16–2). Wavelengths usually are given the symbol λ (Greek lambda). The **frequency** of a water wave is measured by counting the number of crests passing a fixed point in space every second. The frequency has units of waves (or cycles) per second, or simply s^{-1}. If 12 water-wave crests pass a certain point in 30 s, for example, the frequency is

• The "per second" (s^{-1}) in the context of the frequency of waves is often called a *hertz* (Hz). In Chapter 15, the same unit (s^{-1}) was used to measure the frequency of radioactive disintegrations and was, in that context, called a *becquerel* (Bq).

$$\text{frequency} = \nu = \frac{12}{30 \text{ s}} = 0.40 \text{ s}^{-1}$$

where the Greek letter ν (nu) represents the frequency. If the wavelength and frequency of a traveling wave are both known, its speed, which is the rate at which a particular wave crest moves along through the medium, is readily calculated. As Figure 16–2 shows, in a time interval $\Delta t = \nu^{-1}$, the wave moves through one wavelength, so the speed (the distance traveled divided by the time elapsed) is

$$\text{speed} = \frac{\text{distance traveled}}{\text{time elapsed}} = \frac{\lambda}{\nu^{-1}} = \lambda\nu$$

The speed of a wave is the product of its wavelength and its frequency.

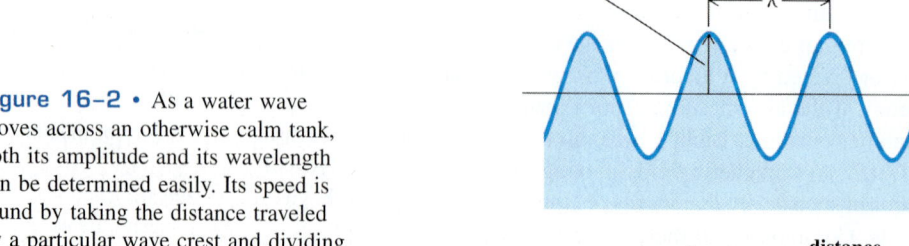

Figure 16–2 • As a water wave moves across an otherwise calm tank, both its amplitude and its wavelength can be determined easily. Its speed is found by taking the distance traveled by a particular wave crest and dividing by the time elapsed.

James Clerk Maxwell had demonstrated by 1865 that light is **electromagnetic radiation.** A beam of light emitted by a laser (Fig. 16–3) consists of electric and magnetic fields oscillating perpendicular to the direction in which the light is propagating (Fig. 16–4). As in the case of a water wave, the wavelength is called λ, and the frequency, ν. The speed of light passing through a vacuum (c_0) is equal to the product $\lambda\nu$.

$$c_0 = \lambda\nu = 2.99792458 \times 10^8 \text{ m s}^{-1}$$

- The constancy of c_0 allows definition of the meter in terms of the second. Officially, 1 m equals "the length of path traveled by light in a vacuum during 1/299,792,458 of a second."

The speed of light in a vacuum is a universal constant, the same for all types of light, although the wavelength and frequency depend on the color of the light. Green light has a range of frequencies near $5.7 \times 10^{14} \text{ s}^{-1}$ and wavelengths near 5.3×10^{-7} m (530 nm). Red light has a lower frequency and a longer wavelength than does green light, and violet light, a higher frequency and shorter wavelength than green light (Fig. 16–5). A laser such as that shown in Figure 16–3 emits nearly monochromatic light (light with a single frequency and wavelength), but white light or simple daylight contains the full range of visible wavelengths. White light can be resolved into its component wavelengths with a prism or a diffraction grating.

- The nanometer (abbreviation: nm) is frequently used to measure the wavelength of visible light. Recall that 1 nm is 10^{-9} m.

The light visible to the eye is only a very small part of the entire electromagnetic spectrum (see Fig. 16–5). Light can have wavelengths longer than red light or shorter than violet. Such radiation is not visible to human eyes but can be detected by other means. The warmth felt radiating from a stone pulled out of a fire consists largely of **infrared** radiation, which has wavelengths that are longer than those of visible light. Microwave ovens use radiation of longer wavelength than infrared, and radio communication uses still longer waves. The position of a radio station on the dial is given by its frequency in hertz (Hz); 1 Hz is one cycle per second. Thus, FM stations typically broadcast at frequencies in the region of tens to hundreds of megahertz (1 MHz = 10^6 s^{-1}); AM stations have much lower broadcast frequencies, from hundreds to thousands of kilohertz (1 kHz = 10^3 s^{-1}). Types of light with shorter wavelengths than those of visible light include **ultraviolet** light, X-rays, and gamma rays. The latter two kinds of radiation, with their high frequencies and short wavelengths, are more penetrating than is visible or infrared radiation and so can be used to image bones or organs beneath the skin. The high energy and penetrating power of X-rays and gamma rays also make them hazardous, so exposure to these forms of radiation must be monitored carefully.

Figure 16–3 • A laser emits a sharply focused beam of light with a very narrow range of wavelengths. The direction of motion of a laser beam is easily manipulated with mirrors. *(Hank Morgan/Photo Researchers, Inc.)*

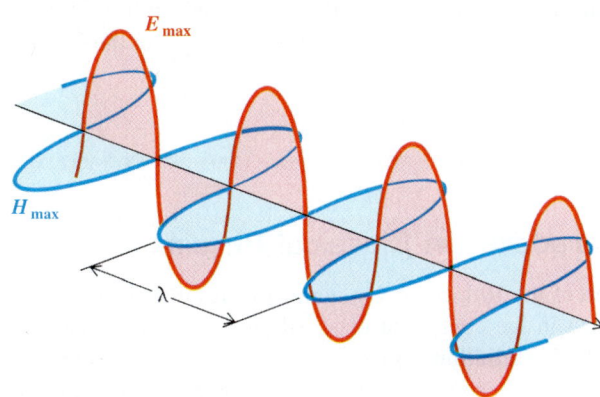

Figure 16–4 • Light consists of waves of oscillating electric (E) and magnetic (H) fields that are perpendicular to one another and to the direction of propagation of the light.

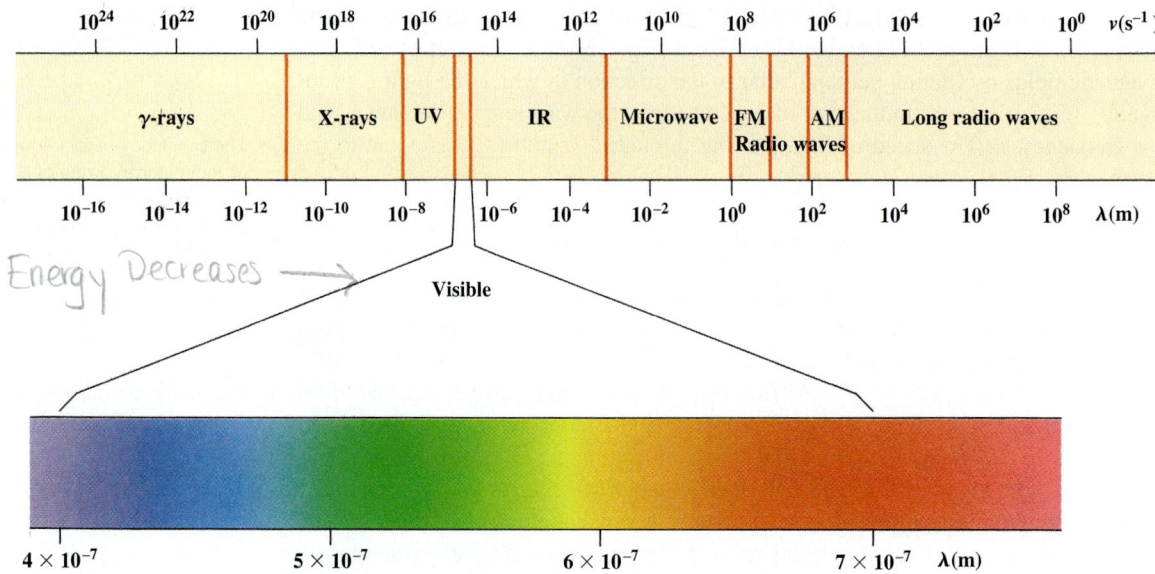

Energy Decreases ⟶

Figure 16-5 • The electromagnetic spectrum. The frequency doubles 80 times from right to left in the diagram as the wavelength doubles 80 times from left to right. Visible light is confined within a single doubling near the middle.

EXAMPLE 16-1

Many cellular telephones in the United States operate in the frequency range of 869 to 894 MHz. Compute the wavelength of the radiation from a cellular telephone emitting at a frequency of 879 MHz.

Solution

The wavelength is related to the frequency through

$$\lambda = \frac{c_0}{\nu} = \frac{3.00 \times 10^8 \text{ m s}^{-1}}{879 \times 10^6 \text{ s}^{-1}} = 0.341 \text{ m}$$

so the wavelength is 34.1 cm (about 13 inches).

EXERCISE

An AM radio station calls itself "Radio-89" because it broadcasts at a frequency of 890 kHz ($8.90 \times 10^5 \text{ s}^{-1}$). Compute the wavelength of the wave broadcast from this station.

Answer: 337 m (about 0.2 mi).

16-2 Paradoxes in Classical Physics

Toward the end of the 19th century, a number of results began to appear that could not be explained in the context of classical physics. One of these arose from the study of **blackbody radiation.** As a solid body (for example, a bar of iron) is heated, it emits radiation and becomes first red, then orange, then white as its temperature increases. The distribution of frequencies of the radiated light changes with the tem-

• Red-hot is surely too hot to touch, but white-hot is hotter still.

CHEMISTRY IN YOUR LIFE

Incandescent Light

The production of light by a hot object is called *incandescence*. Many hot objects behave much like ideal black-bodies in their emission of light at different wavelengths. Light from the sun, for example, closely follows the blackbody radiation curve for its surface temperature of 5500°C. The peak of this curve lies near the center of the visible region of the spectrum. Human vision in fact evolved toward efficient detection of this, the most abundant light present on Earth. Stars other than the Sun, with different surface temperatures, vary in color from cooler red giants to hotter blue-white stars.

Wood fires represent, after the Sun, our most ancient source of light. The red glow of the coals comes from a blackbody temperature of about 1200°C, and the yellow-orange of the flames results from small particles of soot (carbon) inside the flame emitting at a temperature of about 1500°C. Most modern incandescent lights use tungsten filaments heated to about 2500°C; tungsten is the metal of choice because of its high melting point and small rate of evaporation. In the quartz–halogen light bulb, a quartz (SiO_2) bulb permits still higher temperatures. The filament is again made of tungsten, and bromine or iodine is put inside the bulb as well. These halogens react with evaporated tungsten atoms and redeposit them on the filament. A whiter and more natural light results from the higher temperature.

An *optical pyrometer* is a device that measures the intensity of radiated light. Measurements at two or more wavelengths permit the temperature of the radiating object to be estimated. Such instruments are needed to measure high temperatures at which ordinary thermometers cannot be used. An experienced observer can estimate the temperature of a glowing object to within 50°C to 100°C just from the color of the light it emits.

Figure 16–A • A candle flame. *(George Semple)*

perature of the body. A blackbody is an idealized version of the iron bar. It consists of an enclosure lined with black and having a small hole from which radiation can escape. A stream of radiation exiting from the hole must consist entirely of radiation that is emitted by the interior walls because perfectly black walls reflect no radiation. What is the intensity of the light emitted through the hole at each wavelength λ for a blackbody at temperature T? Two typical experimental curves are shown in Figure 16–6. Also shown is the prediction of the classical theory of radiation:

$$\text{intensity} = \left(\frac{8\pi R}{N_0}\right)\left(\frac{T}{\lambda^4}\right)$$

which severely disagrees with experiment. Some theorists called this the **ultraviolet catastrophe:** A heated body does *not* emit radiation approaching infinite intensity at short wavelengths (the violet end of the spectrum), but classical theory predicts that it does. The paradox could not be resolved within classical physics.

Figure 16–6 • Experimental results on the dependence of the intensity of blackbody radiation on wavelength (frequency) for two different temperatures: 5000 K (red) and 7000 K (blue). The classical theory (*dashed curves*) disagrees badly with observations at shorter wavelengths (higher frequencies). The Sun has a blackbody temperature near 5780 K. Its intensity curve lies between the red and blue curves.

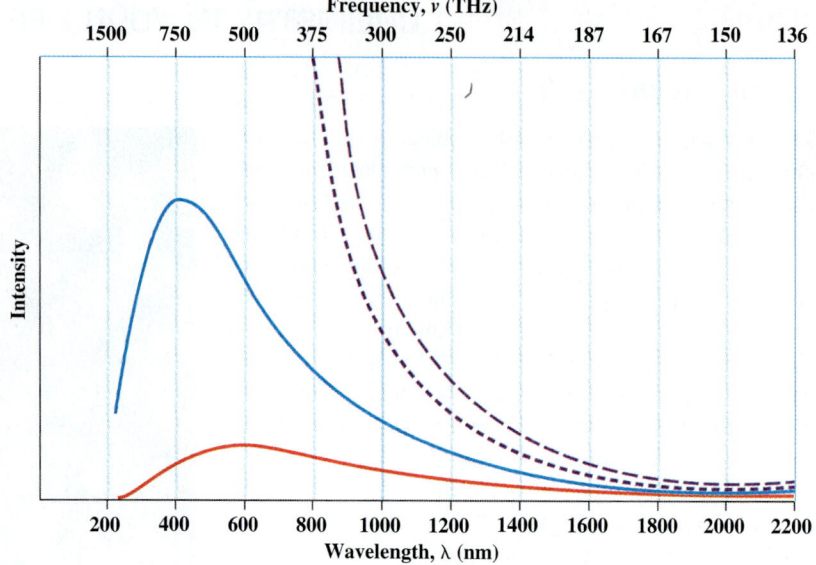

Figure 16–7 • (a) Photoelectric effect. Blue light is effective in ejecting electrons from the surface of this metal, but red light is not. (b) The maximum kinetic energy of the ejected electrons varies linearly with the frequency of light used.

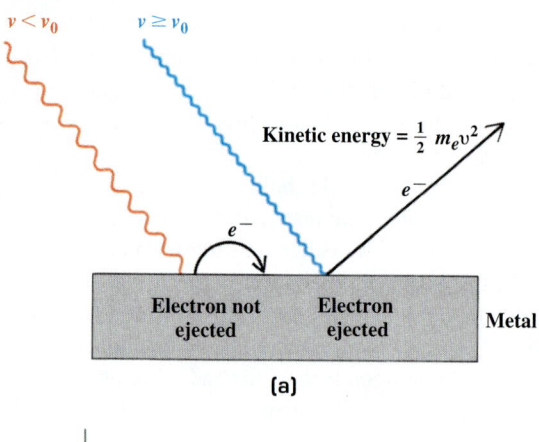

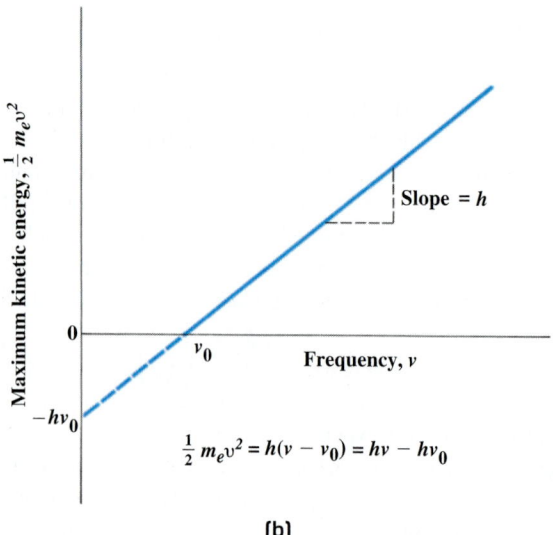

$$\tfrac{1}{2} m_e v^2 = h(v - v_0) = hv - hv_0$$

A second paradox arose from the discovery of the **photoelectric effect.** In this effect, a beam of light falling on a metal surface in an evacuated space ejects electrons from the surface, causing an electric current (called a *photocurrent*) to flow. This fact was in itself not difficult to understand because electromagnetic radiation carries energy. The problem came in explaining the observed dependence of the effect on the frequency of the light. It was found that, for frequencies less than a certain threshold frequency ν_0, no electrons were ejected. Once the frequency exceeded the threshold, the photocurrent increased rapidly. According to classical theory, the energy associated with electromagnetic radiation depends on the *intensity* of the radiation (proportional to the square of the magnitude of the electric field) and not on the frequency. How, then, did a very weak beam of blue light eject electrons from sodium, while a very intense red beam had no effect (Fig. 16–7a)? Further experiments measured the kinetic energy of the ejected electrons and revealed a linear dependence of the maximum kinetic energy on frequency, as shown in Figure 16–7b.

$$\text{maximum kinetic energy} = \tfrac{1}{2}m_e v^2 = h(\nu - \nu_0)$$

where m_e is the mass of the electron and h is a constant. This behavior also was inexplicable in classical physics.

The third, and in some ways most fundamental, paradox was the stability of the atom. Rutherford's experiments (see Section 1–5) had shown that an atom consists of a very small, massive nucleus surrounded by electrons. If the negatively charged electrons are stationary, what keeps them from falling into the nucleus in response to the electrostatic attraction of the positively charged nucleus? If the electrons move in orbits around the nucleus, like planets in a solar system, what keeps them from emitting radiation, as classical theory predicts they must, and so spiraling into the nucleus (Fig. 16–8)?

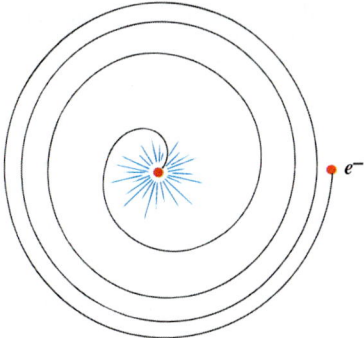

Figure 16–8 • In classical theory, atoms constructed according to the Rutherford model are not stable. The motion of the electrons in their orbits would cause them to radiate energy and so quickly spiral into the nucleus.

Atomic Spectra

Closely related to the question of atomic stability were the results of early experiments on how atoms actually do gain and lose energy by interacting with light. Light consisting of more than one frequency can be resolved according to frequency by passing it through a glass prism. Thus, a prism breaks down a beam of white light, which contains a mixture of all the visible frequencies, into a multicolored band, a **spectrum.** Components of higher frequency (toward the blue end in Fig. 16–5) interact differently with the glass than do those of lower frequency (toward the red end) and are bent to greater angles by the prism. In the instrument called a **spectrograph,** light enters through a narrow, vertical slit, passes through a prism, and goes on to a photographic plate or other detector (Fig. 16–9). If the light entering the spectrograph contains all frequencies, then the detector records the full spectrum, made up of overlapping lines that are images of the slit. If the incoming light consists of a mixture of discrete frequencies, the detector records an array of single lines, each having a different color; these are **spectroscopic lines.**

Early spectrographic experiments measured **emission spectra,** as diagrammed in Figure 16–9a. The sample was heated strongly in a flame or by passage of an electric current (Fig. 16–10). This heating broke up any chemical bonds and provided energy to the atoms in the sample. The energy *excited* the atoms, which quickly relaxed by emitting energy in the form of light (and by other means as well). Investigators found that atoms do not emit a full range of frequencies; instead, they

• The detector in this arrangement may be a human eye. An observer sees a series of differently colored lines if the light consists of only certain frequencies. If the light is white light, the observer sees a rainbow.

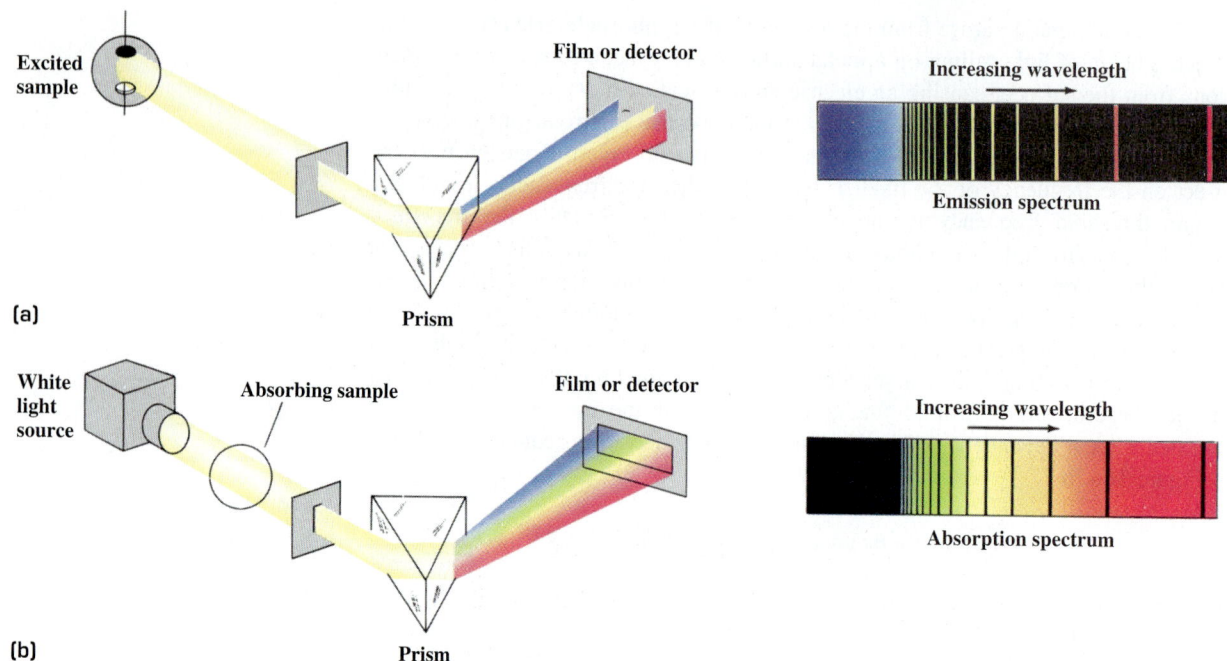

(a)

Increasing wavelength

Emission spectrum

(b)

Increasing wavelength

Absorption spectrum

Figure 16–9 • (a) Emission spectroscopy. The light given off by an excited sample passes through a prism that separates it according to wavelength. The dispersed light then is recorded photographically or by another detector. The frequencies of the emission lines are obtained by comparison to standards. (b) Absorption spectroscopy. White light from a source passes through the unexcited sample, which absorbs certain discrete wavelengths of light from it. The result is the appearance of dark lines superimposed on a continuous bright background.

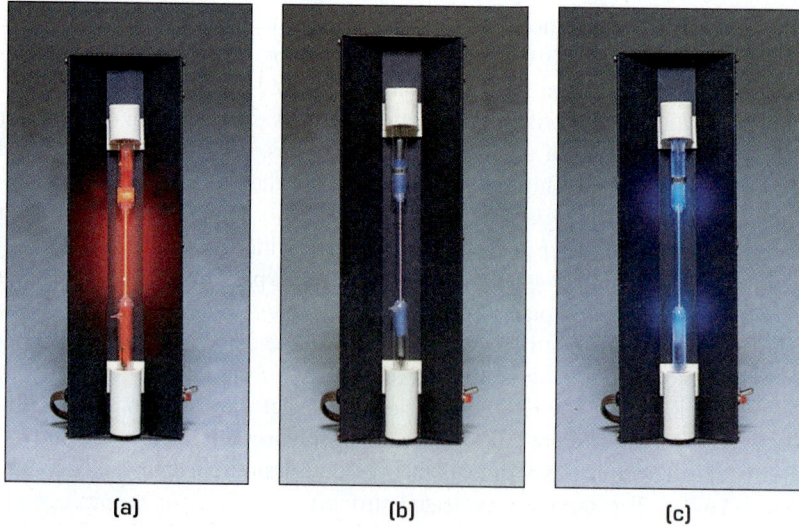

(a)   **(b)**   **(c)**

Figure 16–10 • When gaseous atoms are excited in an electrical discharge, they emit light. Here the colors of the light emitted by three elements are shown: (a) Ne(g), (b) Ar(g), and (c) Hg(g). Each emission consists of several frequencies (wavelengths) of light, and the perceived color depends on which predominate. *(Leon Lewandowski)*

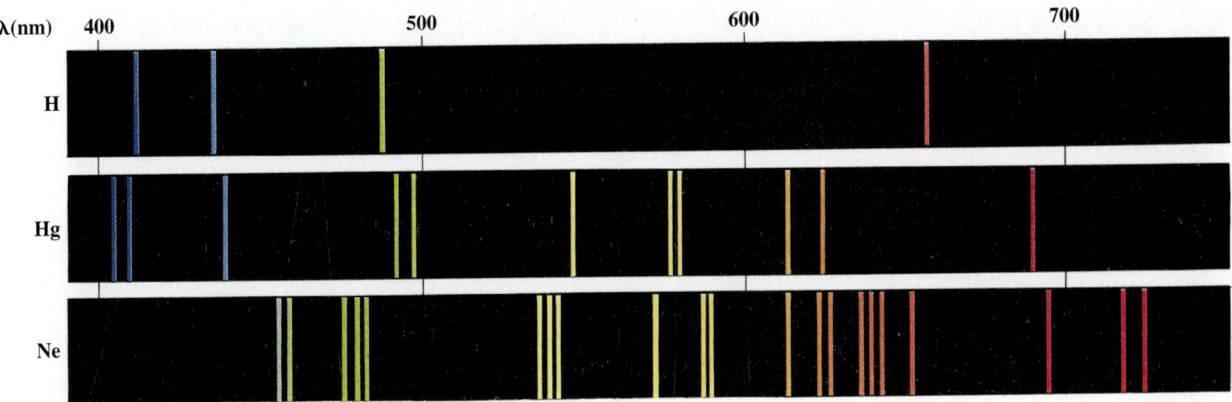

Figure 16-11 • Atoms of hydrogen, mercury, and neon emit light at specific wavelengths. The pattern seen is characteristic of the element under study.

emit light at numerous *discrete* frequencies that differ depending on their chemical identity (Fig. 16–11): Atomic spectra are line spectra. Further experiments established that atoms also absorb discrete frequencies when irradiated with white light (Fig. 16–9b) and that the frequencies in **absorption spectra** equaled those in emission spectra for a given kind of atoms.

The patterns of emission or absorption sometimes could be fitted to simple formulas. In 1885, Johann Balmer published a paper pointing out that excited H atoms emit light in the visible region of the spectrum with frequencies given by the formula

$$\nu = \left(\frac{1}{4} - \frac{1}{n^2}\right)(3.29 \times 10^{15} \text{ s}^{-1}) \qquad n = 3, 4, 5, \ldots$$

Note how requiring n to be a whole number restricts the possible ν's to discrete values. These lines are now called the Balmer series. Other spectroscopists discovered additional series in the H-atom spectrum as improvements in spectrographic techniques made it possible to explore the ultraviolet and infrared regions of the spectrum. The Lyman series occurs in the ultraviolet; the Paschen series, Brackett series, and Pfund series occur in the infrared. The following formula (originally advanced by Rydberg and Ritz in a somewhat different form) accounts for the lines in all of these series:

$$\nu = \left(\frac{1}{n_2^2} - \frac{1}{n_1^2}\right)(3.29 \times 10^{15} \text{ s}^{-1}) \qquad n_1 > n_2; \qquad n_1 \text{ and } n_2 \text{ both integers}$$

where each of the five series had a different n_2: 1 (Lyman), 2 (Balmer), 3 (Paschen), 4 (Brackett), and 5 (Pfund). Explaining these remarkable series of discrete frequencies was a major theoretical challenge in the early part of the 20th century.

Atoms emit at discrete frequencies in regions of the electromagnetic spectrum quite remote from the visible, if they are properly excited. In 1913, Henry Moseley reported that metallic elements emit X-rays of characteristic frequency when excited by bombardment with high-energy electrons. He found that the frequency ν of each element's characteristic X-ray emission fit the formula

$$\sqrt{\nu} = (Z - 1)(4.98 \times 10^7 \text{ s}^{-1/2})$$

where Z is an integer having a different value for each element (Fig. 16–12). At the time, this result offered strong support for Rutherford's nuclear model of the atom

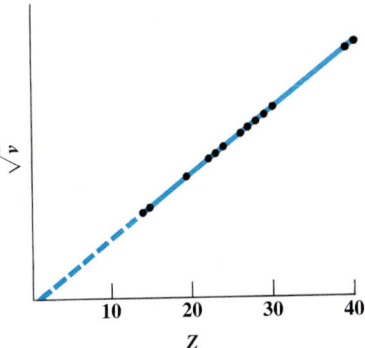

Figure 16-12 • The results of Henry Moseley's experiments on the emission of X-radiation by metallic elements. The square root of the frequency of the characteristic X-rays emitted in response to bombardment by high-energy electrons is a straight-line function of atomic number. The elements range from aluminum ($Z = 13$) through zirconium ($Z = 40$).

• The Moseley experiment is still routinely performed under the name "X-ray fluorescence spectroscopy." The characteristic frequencies of the X-rays emitted by a sample of a solid unknown bombarded by an electron beam reveal the identities of the elements present.

(Section 1–5) because the Z in Moseley's law was clearly the same as Rutherford's Z, the atomic number. Squaring both sides of Moseley's empirical equation gives

$$\nu = (Z - 1)^2 (2.48 \times 10^{15} \text{ s}^{-1})$$

This equation is provocative because $2.48 \times 10^{15} \text{ s}^{-1}$ is nearly equal to $\frac{3}{4}$ of the constant in the Rydberg-Ritz formula ($3.29 \times 10^{15} \text{ s}^{-1}$), and $\frac{3}{4}$ equals $\left(\dfrac{1}{1^2} - \dfrac{1}{2^2}\right)$.

16–3 Planck, Einstein, and Bohr

Early philosophers and scientists assumed that nature is continuous, that nature does not make jumps. On a macroscopic scale, this seems true enough. One can measure out 9 kg, or 8.23 kg, or 6.4257 kg of a substance such as carbon-12, and it appears that the mass can have any value, to any number of decimal places, provided that the balance is sufficiently sensitive.

On an atomic scale, this apparent seamless continuity breaks down, just as individual grains of sand become distinguishable when a smooth beach is closely inspected. "Graininess" pervades matter on the atomic scale. The fact is that carbon-12 comes in "packets," each of which has a mass close to 1.9926×10^{-26} kg. One might, in principle, be able to "weigh out" two, three, or any whole number of such packets, but one cannot obtain $1\frac{1}{2}$ or any other non-integral number of packets. Carbon-12 is not a continuous material but comes in chunks, each of which contains the minimum measurable mass of the substance. The chunks are in fact ^{12}C atoms. Similarly, electric charge comes in packets of size e, called electrons or protons. We cannot obtain non-integral numbers of such packets in experiments involving chemical energies.

A central idea of quantum theory is that energy also is not continuous but comes in discrete packets. A packet of energy is called a **quantum** of energy. The quantization of the energy of electromagnetic radiation and of matter leads to a resolution of the paradoxes presented in the last section.

• At very high energies (high enough to probe the structure of the proton itself), there is evidence for the existence of *quarks*, packets having a minimum charge of $\pm\frac{1}{3}e$ rather than e.

Planck's Constant and the Photon

Max Planck published a resolution of the paradox of blackbody radiation in 1901. Prior to Planck, radiation from a blackbody or other solid object was thought to arise from the oscillations of groups of atoms on the surface of the body. If the atoms oscillated with a frequency ν, then radiation of frequency ν would be emitted. In this theory, oscillators with high frequencies greatly outnumber those with low frequencies. Even a small amount of energy placed in these oscillators (by a rise in temperature) shifts the radiation from the blackbody drastically to high frequencies, leading to the ultraviolet catastrophe. Planck made the daring hypothesis that it was *not* possible to put an arbitrarily small amount of energy into an oscillator of frequency ν. He suggested instead that energy is gained or lost by an oscillator in packets, or quanta, of magnitude $h\nu$, where h is a small constant. The energy in an oscillator must then be an integral multiple of $h\nu$—that is, it must go up or down in steps: $1 \times (h\nu)$, $2 \times (h\nu)$, and so forth. In this scheme, most of the high-frequency oscillators cannot gain energy because packets of energy that are large enough to be accepted by them are scarce. Furthermore, an oscillator with its energy at the allowed minimum cannot lose energy by emitting radiation. The mathematical development of Planck's hypothesis predicts a fall-off in intensity at high frequencies in the blackbody emission spectrum. The detailed development describes the experimental observations quite accurately. The constant h in the relation

$$E = h\nu$$

is now referred to as **Planck's constant** and has been measured to high precision:

$$h = 6.62606876 \times 10^{-34} \text{ J s} = 6.62606876 \times 10^{-34} \text{ kg m}^2 \text{ s}^{-1}$$

- Planck's constant is a fundamental constant of nature, occurring again and again in quantum theory.

In 1905, Einstein used Planck's quantum hypothesis to explain the photoelectric effect. He suggested that light consists of a stream of packets of energy called **photons,** each of which carries energy $E = h\nu$. A photon of low-frequency light that strikes an electron on a metal surface does not deliver enough energy to eject the electron, but a photon of high-frequency light does. If a threshold energy of $h\nu$ is required to overcome the binding of the electron to the atoms of the surface, then, from the law of conservation of energy, any excess energy $h\nu - h\nu_0$ should appear as kinetic energy of the outgoing electron:

- The symbol Φ is an uppercase Greek "phi." Also note the difference between v ("vee") for velocity and ν ("nu") for frequency.

$$\text{maximum kinetic energy of electron} = (\tfrac{1}{2}m_e v^2)_{max} = h(\nu - \nu_0) = h\nu - \Phi$$

That is what is observed experimentally (see Fig. 16–7b). The constant h in the photoelectron effect experiment equals Planck's constant, the constant in the blackbody radiation experiment. The quantity $h\nu_0$ equals the energy binding the ejected electron to the surface and is called the *work function* (symbolized Φ) of the material.

EXAMPLE 16–2

Light with a wavelength of 400 nm strikes a surface of metallic cesium in a photoelectric cell and ejects electrons with a maximum kinetic energy of 1.54×10^{-19} J. Calculate the work function of cesium and the longest wavelength of light that is capable of ejecting electrons from metallic cesium.

Solution

The frequency of the light is

$$\nu = \frac{c_0}{\lambda} = \frac{2.9979 \times 10^8 \text{ m s}^{-1}}{4.00 \times 10^{-7} \text{ m}} = 7.495 \times 10^{14} \text{ s}^{-1}$$

The work function Φ can be calculated from Einstein's formula:

$$(\tfrac{1}{2}m_e v^2)_{max} = h\nu - \Phi$$
$$1.54 \times 10^{-19} \text{ J} = (6.6261 \times 10^{-34} \text{ J s})(7.495 \times 10^{14} \text{ s}^{-1}) - \Phi$$
$$= 4.966 \times 10^{-19} \text{ J} - \Phi$$
$$\Phi = (4.966 - 1.54) \times 10^{-19} \text{ J} = 3.43 \times 10^{-19} \text{ J}$$

The minimum frequency ν_0 for the light to eject electrons is the frequency that supplies just enough energy to shake electrons loose from the surface:

$$h\nu_{min} = \Phi$$
$$\nu_{max} = \frac{3.426 \times 10^{-19} \text{ J}}{6.6261 \times 10^{-34} \text{ J s}} = 5.17 \times 10^{14} \text{ s}^{-1}$$

From this, the maximum wavelength λ_0 can be calculated:

$$\lambda_{max} = \frac{c_0}{\nu_{min}} = \frac{2.9979 \times 10^8 \text{ m s}^{-1}}{5.17 \times 10^{14} \text{ s}^{-1}} = 5.80 \times 10^{-7} \text{ m} = 580 \text{ nm}$$

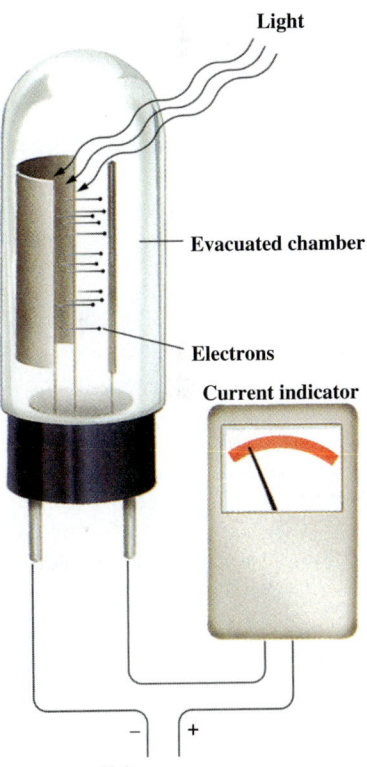

Light

Evacuated chamber

Electrons

Current indicator

Voltage source

In a photoelectric cell (photocell), light strikes a metal surface and ejects electrons. The electrons are attracted to a positively charged collector, and a current flows through the cell. If the light stops or the frequency falls too low, this photocurrent stops. Photocells are used widely for detecting and measuring light.

EXERCISE

A photon with 6.94×10^{-19} J of energy barely ejects an electron from a polished zinc surface. (a) Does a photon with a wavelength of 210 nm suffice to do this? (b) If so, what is the maximum kinetic energy of the ejected electron?

Answer: (a) Yes. (b) 2.53×10^{-19} J.

Planck's and Einstein's results suggest that light behaves like a wave in some circumstances but like a stream of particles (photons) in others. This **wave–particle duality** is not a contradiction in terms but rather a part of the fundamental nature of light.

The Bohr Atom

In 1913, Niels Bohr proposed a model to account for the stability of the atom and the experimental spectra of the H atom and one-electron ions such as He^+, Li^{2+}, and Be^{3+}. His model builds on the Rutherford nuclear model (Section 1–5). In it, an electron of mass m_e is assumed to move in a circular orbit of radius r around a fixed nucleus, like a planet in orbit around a star. Classical physics predicts that an electron moving in this way emits light of continuously increasing frequency. Bohr avoided this difficulty by in effect taking it as a postulate that a set of n stable, discrete orbits (characterized by radius r_n and energy E_n) occurs and that light is emitted or absorbed only when the electron "jumps" from one stable orbit to another.

The *linear* momentum of a particle is the product of its mass and its velocity; this equals $m_e v$ for an electron. The **angular momentum** is a different quantity that describes a turning motion around an axis. For the circular paths of the Bohr model, the angular momentum of the electron is the product of its mass, its velocity, and the radius of the orbit ($m_e vr$). Bohr's actual postulate was that the angular momentum of the electron was quantized and equal to some whole-number multiple of $h/2\pi$, where h is Planck's constant:

$$\text{angular momentum} = m_e vr = n\frac{h}{2\pi} \qquad n = 1, 2, 3, \ldots$$

• This is frequently the way in which new theories are developed.

This hypothesis cannot be derived, and in fact it is likely that Bohr adopted it after working backward from the known spectrum of atomic hydrogen. Having assumed this condition of quantization, Bohr used the classical equations of circular motion to calculate the allowed radius r_n for each integral value of n. He found

$$r_n = \frac{n^2}{Z} a_0$$

where Z is the positive charge on the nucleus (1 for a hydrogen atom, 2 for a He^+ ion, and so forth), and a_0 is the **Bohr radius,** a constant. Bohr was able to calculate a_0 as a combination of Planck's constant, the charge of the electron, and the mass of the electron:

$$a_0 = 5.29177 \times 10^{-11} \text{ m}$$

• Chemists also use the Bohr radius itself for discussions of atomic size.

Chemists often use the **ångström** (abbreviation: Å), a unit of length equal to 10^{-10} m, in discussions of atomic size. The Bohr radius, 0.529 Å, is the predicted distance of the electron from the nucleus in the state $n = 1$ of the H atom. In H atoms having higher values of n, the electron is farther away from the nucleus. When $n = 2$, it is (4×0.529) Å away; when $n = 3$, it is (9×0.529) Å away, and so forth. In the He^+ ion, which like the H atom has just one electron, the nuclear charge is larger $(Z = 2)$. Its stronger attraction contracts the entire pattern of distances by a factor of 2.

Bohr also calculated the total energy of the electron (the sum of its kinetic and potential energies) in the stable orbits. It equals

$$E_n = -\frac{Z^2}{n^2}\left(\frac{h^2}{8\pi^2 m_e a_0^2}\right) \qquad n = 1, 2, 3, \ldots$$

The combination in parentheses of the Bohr radius, Planck's constant, the mass of the electron, and other constants is an amount of energy. It occurs so frequently that it is defined as a (non-SI) unit of energy called the **rydberg** (Ry). Substitution of the various constants inside the parentheses in the preceding equation shows that 1 Ry = 2.179872×10^{-18} J. Hence

$$E_n = -\frac{Z^2}{n^2}\,\text{Ry} = -\frac{Z^2}{n^2}(2.18 \times 10^{-18}\,\text{J})$$

The allowed values of the energy are shown in Figure 16–13 for hydrogen, for which $Z = 1$. The state with $n = 1$ is called the **ground state** because it is the state of lowest energy for the system of nucleus plus electron. An H atom in this state has an energy of -1 Ry, or -2.18×10^{-18} J. States with higher values of n are **excited states** and have higher energy. As n becomes large, the radius of the atom becomes

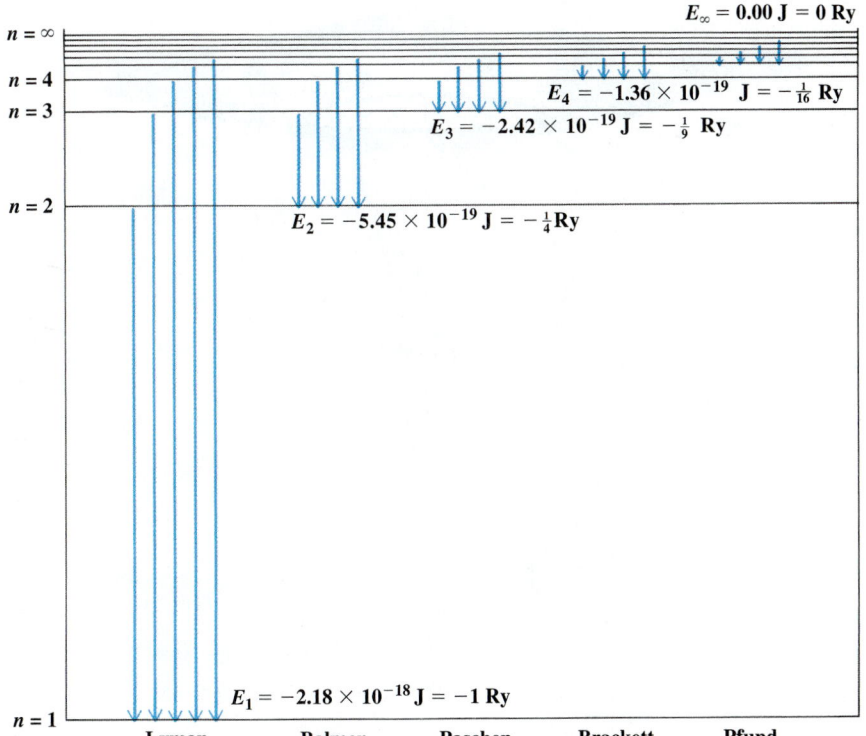

Figure 16–13 • In the energy levels of the hydrogen atom, the separated electron and proton are assigned zero energy, and all other energies are more negative than that. Atoms absorb light by jumping from lower to higher energy levels and emit light by falling from higher to lower energy levels (*blue arrows*). Each series of related emissions is named after its discoverer.

very large, and the energy of the atom rises from -1 Ry toward 0 Ry. The energy E_n reaches 0 Ry only when n becomes infinite, at which the electron and proton are at rest and separated by an infinite distance. The energy of the atom is negative for all smaller distances (Fig. 16–14). It may seem strange to refer to a "negative energy," but the negative sign is solely a result of choosing the zero of energy to correspond to the infinitely separated electron and nucleus.

• This convention was previously used in Figure 6–1, which displays potential energy diagrams for chemical bonding.

The **ionization energy** of an atom is the minimum energy needed to remove an electron from the atom when it is in its ground state. Removing an electron from an H atom in its ground state involves a change from an initial state in which the quantum number n equals 1 to a final state in which the quantum number n equals infinity. The energy change equals the final energy minus the initial energy:

$$\Delta E = E_2 - E_1 = \left(-\frac{1^2}{\infty^2}\right) \text{Ry} - \left(-\frac{1^2}{1^2}\right) \text{Ry}$$

$$= 0 - (-1)\,\text{Ry} = 1\,\text{Ry} = 2.18 \times 10^{-18}\,\text{J}$$

This is the ionization energy of a single H atom. Multiplying it by Avogadro's number gives the ionization energy (IE) per mole of H atoms:

$$IE = (6.022 \times 10^{23}\,\text{mol}^{-1})(2.18 \times 10^{-18}\,\text{J})$$

$$= 1.31 \times 10^6\,\text{J mol}^{-1} = 1310\,\text{kJ mol}^{-1}$$

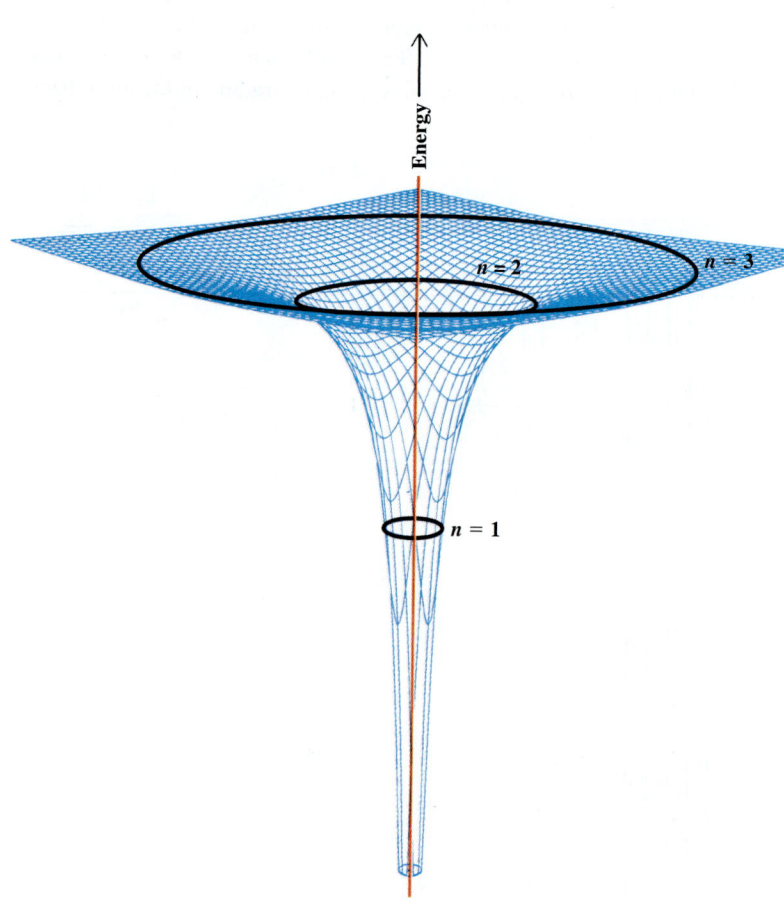

Figure 16–14 • The interaction between the electron and nucleus has its lowest energy when the electron is closest to the nucleus. Moving the electron away from the nucleus can be seen as moving it up the sides of a steep energy well. In the Bohr theory, it can "catch" and stick to the sides only at certain allowed values of r, the radius, and of E, the energy. The first three of these are shown by rings.

This value agrees with the experimentally observed ionization energy of hydrogen atoms.

EXAMPLE 16–3

Consider the $n = 2$ state of the Li^+ ion. Use the Bohr theory to calculate the distance of the electron from the nucleus and the energy of the ion relative to the energy of the separated nucleus and electron.

Solution

Because $Z = 3$ for a Li^+ ion (the nuclear charge is $+3e$) and because $n = 2$ in this case, the radius is

$$r_n = \frac{n^2}{Z} a_0 = \frac{2^2}{3} a_0 = \frac{4}{3}(0.529 \text{ Å}) = 0.705 \text{ Å}$$

The energy is

$$E_n = -\frac{Z^2}{n^2}\text{Ry} = -\frac{(3)^2}{(2)^2}\text{Ry} = -\frac{9}{4}(2.18 \times 10^{-18} \text{ J}) = -4.90 \times 10^{-18} \text{ J}$$

EXERCISE

A certain one-electron ion has a radius of 2.12 Å and an energy of -2.18×10^{-18} J. Name the ion and tell what quantum state it is in.

Answer: The ion is beryllium(III), Be^{3+}. It is in the $n = 4$ quantum state.

Atomic Spectra

The Bohr theory brings the concept of energy quantization to the problem of explaining the absorption and emission spectra of the H atom and one-electron ions. Suppose that a collection of H atoms are in an initial energy state characterized by the quantum number n_1. A beam of white light striking these atoms contains photons of many frequencies. Some deliver just the correct energy to excite some atoms to a final, higher-energy state n_2. These photons are removed from the incoming beam, giving rise to an absorption line in the spectrum. Photons not delivering the correct energy for an excitation are not absorbed. The change in energy of an absorbing atom equals its energy in the final state minus its energy in the initial state:

$$\Delta E = E_2 - E_1 = \left(-\frac{Z^2}{n_2^2}\right)\text{Ry} - \left(-\frac{Z^2}{n_1^2}\right)\text{Ry} = -Z^2\left(\frac{1}{n_2^2} - \frac{1}{n_1^2}\right)\text{Ry}$$

This change corresponds to an *upward* transition in Figure 16–13. The energy transferred by a photon equals Planck's constant times its frequency, or $h\nu$. Therefore

$$h\nu = -Z^2\left(\frac{1}{n_2^2} - \frac{1}{n_1^2}\right)\text{Ry}$$

Substituting the numerical value of h and replacing 1 Ry with its equivalent in joules gives

$$(6.626 \times 10^{-34} \text{ J s})\nu = -Z^2\left(\frac{1}{n_2^2} - \frac{1}{n_1^2}\right)(2.18 \times 10^{-18} \text{ J})$$

$$\nu = +Z^2\left(\frac{1}{n_1^2} - \frac{1}{n_2^2}\right)\left(\frac{2.18 \times 10^{-18} \text{ J}}{6.626 \times 10^{-34} \text{ J s}}\right)$$

$$\nu = Z^2\left(\frac{1}{n_1^2} - \frac{1}{n_2^2}\right)(3.29 \times 10^{15} \text{ s}^{-1}) \qquad n_2 > n_1 \qquad \text{(absorption)}$$

This formula gives the frequencies of all the absorption lines in the spectrum.

In emission spectroscopy, the atoms *lose* energy, undergoing transitions *downward* on the energy level diagram in Figure 16–13. The frequencies of the emission lines are given by the formula

$$\nu = Z^2\left(\frac{1}{n_2^2} - \frac{1}{n_1^2}\right)(3.29 \times 10^{15} \text{ s}^{-1}) \qquad n_1 > n_2 \qquad \text{(emission)}$$

which differs from the formula for absorption only in the exchange of subscripts between the n's. Note that the term involving the n's is positive in both formulas.

For hydrogen atoms, which have $Z = 1$, the formula for emission is identical to the empirical Rydberg-Ritz formula in Section 16–2. This means that the Bohr theory correctly predicts the observed emission spectrum of hydrogen from the ultraviolet, through the visible, and into the infrared region, a considerable triumph. Similar agreement exists between the predictions of Bohr theory and spectroscopic results on various one-electron ions. As a bonus, the formula for spectroscopic emission goes over to Moseley's empirical formula for the frequencies of X-rays emitted by metallic elements if Z is replaced by $Z - 1$ and the transition in every case is from $n_1 = 2$ down to $n_2 = 1$.

The Bohr theory is part of what is now called the old quantum mechanics. Despite its successes, Bohr theory could not be extended to predict the energy levels and spectra of most atoms and ions having more than one electron. A new theory was necessary. Modern quantum mechanics retains the concept of discrete energy states and electrons making transitions between them. It also retains the quantization of angular momentum, although in a form slightly different from that advanced by Bohr. On the other hand, it dispenses entirely with the circular orbits of the Bohr theory.

16–4 Waves, Particles, and the Schrödinger Equation

In his theory of the photoelectric effect, Einstein treated light as having particle-like as well as wave-like properties. In 1924, the young French physicist Louis de Broglie posed this question: If light, which we think of as a wave, can have particle-like properties, then why cannot particles of matter have wave-like properties? He demonstrated that Bohr's assumption about the quantization of angular momentum in the hydrogen atom could be rationalized by ascribing such wave-like properties to the electron.

Up to this point we have considered only one type of wave: a **traveling wave.** Electromagnetic radiation (light, X-rays, and gamma rays) is described by such traveling waves, which move through space at speed c_0. Another type of wave that arises in physical situations is a **standing wave.** The vibrations set up by plucking a guitar string stretched taut between two fixed pegs are standing waves. A plucked string starts to vibrate, but not all oscillations are possible. Because the ends are fixed, the only oscillations that can persist are those in which an integral number of half-wavelengths fits into the length (L) of the string (Fig. 16–15). The condition on the allowed wavelengths is

$$\frac{n\lambda}{2} = L \qquad n = 1, 2, 3, \ldots$$

It is impossible to create a wave with any other value of λ if the ends of the string are fixed. The oscillation with $n = 1$ is called the **fundamental** or **first harmonic,** and the higher values of n correspond to higher harmonics; for example, the oscillation with $n = 2$ is the second harmonic. In the higher ($n > 1$) harmonics, the points on the string at which half-wavelength sections meet are special. The wave moves up and down on the two sides of these points, but no vibration occurs at the points themselves. Regions of no vibration in a standing wave are called **nodes.** (The fixed ends do not count as nodes.) In a set of standing waves:

> The higher the number of the harmonic then the greater the number of nodes, the shorter the wavelength, the larger the frequency, and the higher the energy.

A simple experiment with a guitar confirms the last point. Plucking a string midway along its length excites mainly the first harmonic and requires minimal effort. Plucking near one end excites higher harmonics and requires distinctly more effort. The presence of higher frequencies in the second case is evident from the harsh timbre of the note.

De Broglie realized that standing waves are examples of quantization: Only certain discrete states of vibration are allowed for the vibrating string, each one characterized by a different integer n in the preceding equation. He suggested that the quantization of a one-electron atom might have the same origin and that the electron might be associated with a standing wave—in this case, a *circular* standing wave about the nucleus of the atom (Fig. 16–16). Standing waves in guitar strings encounter friction and die away with time. If a standing wave is set up and there is no damping to make it fade away, then it persists indefinitely. Such situations are called **stationary states.** If electrons are wave-like, then they must exist in stable atoms as standing "matter-waves," stationary states that form a pattern about the nucleus. The admittedly uncomfortable notion of an electron in an atom as a motionless, standing matter-wave succeeds in explaining why electrons in atoms do not continuously radiate energy. The electrons do *not* move in curving paths, as in the Bohr planetary model.

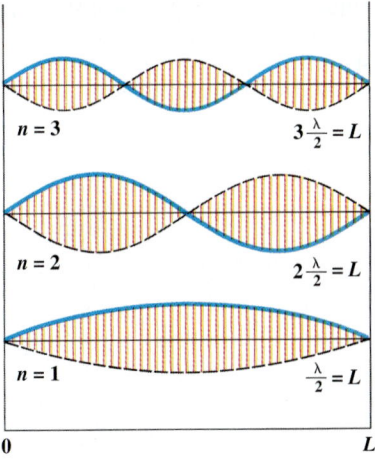

Figure 16–15 • A string of length L with fixed ends can vibrate in only a restricted set of ways. The positions of largest amplitude of vibration are shown here for the largest three of these harmonics. In a standing wave like this, the whole string is in motion except at its ends and at the nodes.

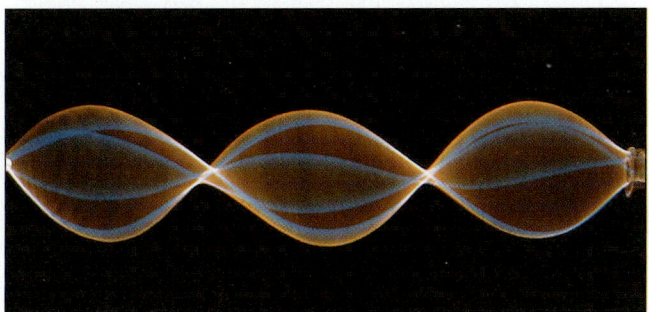

The third harmonic ($n = 3$) of vibration in a stretched rubber tube fastened at its ends. This standing wave has two nodes and a wavelength equal to two thirds of the tube's length. *(©1991 Richard Megna/Fundamental Photographs)*

Figure 16–16 • A circular standing wave on a closed loop. The state shown has $n = 7$, with seven full wavelengths around the circle.

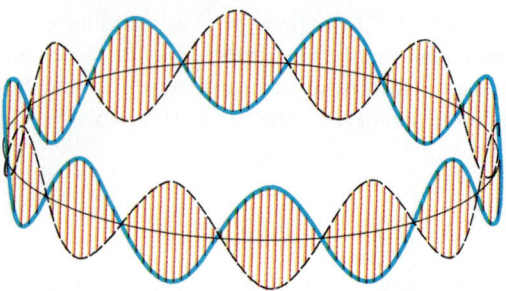

• In this case, there is an integral number of wavelengths rather than of half-wavelengths, as for the guitar string. The reason is that the circular loop comes back on itself, rather than reaching a fixed end.

For a circular standing wave to persist, a whole number of wavelengths must fit into the circumference of the circle $2\pi r$. The condition on the allowed wavelengths is

$$2\pi r = n\lambda \qquad n = 1, 2, 3, \ldots$$

Bohr's assumption about quantization of the angular momentum of the electron is

$$m_e v r = n\frac{h}{2\pi}$$

which can be rewritten as

$$2\pi r = n\left(\frac{h}{m_e v}\right)$$

Setting the two quantities that are equal to $2\pi r$ equal to each other (and canceling out the n) shows that the wavelength of the standing wave is related to the linear momentum $p = m_e v$ of the electron by the formula

$$\lambda = \frac{h}{m_e v} = \frac{h}{p}$$

De Broglie showed from the theory of relativity that just this relationship exists between the wavelength and the momentum of a photon. He therefore proposed as a generalization that any particle moving with linear momentum p has associated with it wave-like properties and a wavelength λ given by h/p.

Under what circumstances does the wave-like nature of particles become apparent? When waves emanating from two sources pass through the same region of space, they *interfere* with each other. Visualize two water waves encountering each other. When two crests meet, **constructive interference** occurs and a higher crest (larger amplitude) appears; where a crest of one wave meets a trough of the other, **destructive interference** (smaller amplitude) occurs (Fig. 16–17). When X-rays, which are waves, are sent into a crystal, they are scattered by the atoms of the crystal (picture a water wave encountering a vertical post embedded in the bottom of a lake). The X-rays scattered from each atom in the crystal interfere with each other to give a characteristic X-ray **diffraction pattern.** According to de Broglie's generalization, such diffraction also should be observed when a beam of *electrons* is sent into a crystal. In 1927, C. Davisson and L.H. Germer showed that a crystal in fact diffracts electrons and that de Broglie's relationship correctly predicts their wavelengths. The experiment provided a brilliant confirmation of de Broglie's hypothesis

• We discuss X-ray diffraction further in Chapter 20.

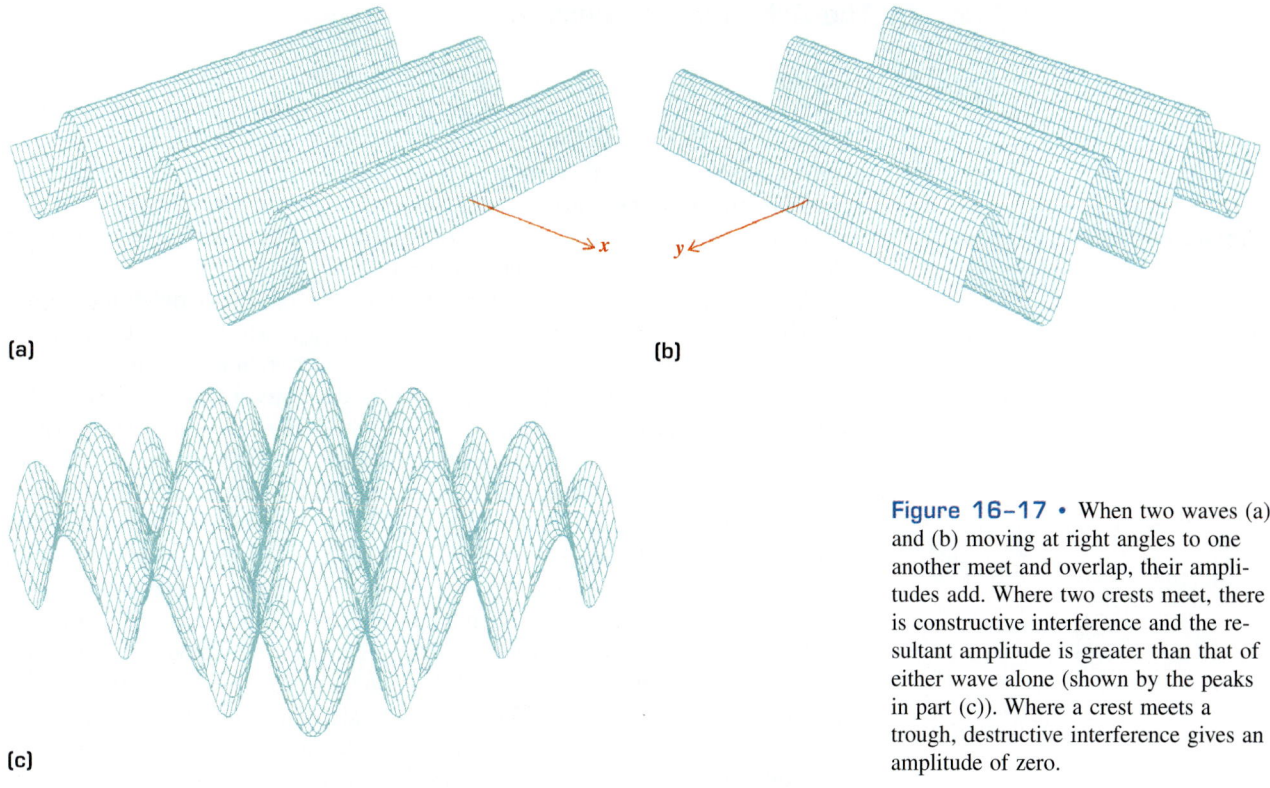

(a)

(b)

(c)

Figure 16–17 • When two waves (a) and (b) moving at right angles to one another meet and overlap, their amplitudes add. Where two crests meet, there is constructive interference and the resultant amplitude is greater than that of either wave alone (shown by the peaks in part (c)). Where a crest meets a trough, destructive interference gives an amplitude of zero.

that electrons have wave-like properties. "Waves" and "particles" are idealized limits; protons, electrons, and photons all possess both wave and particle aspects.

EXAMPLE 16–4

Calculate the wavelength of an electron moving with velocity 1.0×10^6 m s^{-1}.

Solution

$$\lambda = \frac{h}{p} = \frac{h}{m_e v} = \frac{6.626 \times 10^{-34} \text{ kg m}^2 \text{ s}^{-1}}{(9.11 \times 10^{-31} \text{ kg})(1.0 \times 10^6 \text{ m s}^{-1})} = 7.3 \times 10^{-10} \text{ m}$$
$$= 7.3 \text{ Å}$$

This wavelength is about a dozen times larger than the radius of a ground-state hydrogen atom. Clearly, the wave-like properties of electrons are essential to an understanding of atomic structure.

EXERCISE

Calculate the de Broglie wavelength of a baseball of mass 0.17 kg that is thrown with a velocity of 30 m s^{-1}.

Answer: 1.3×10^{-34} m $= 1.3 \times 10^{-24}$ Å. This wavelength is far too small to be observed and need not be considered in studying the motion of a thrown baseball.

The Schrödinger Equation

De Broglie's work attributes wave properties to electrons. By their nature, waves do not have exact positions and trajectories. Classical mechanics, which is based on just such concepts, is therefore inadequate to describe electrons. Modern **wave mechanics** starts with the particle-as-wave notion and develops a way to determine the *probabilities* of electrons (and other particles) having certain positions and momenta. Wave mechanics is one of two equivalent mathematical formalisms developed in the 1920s to express the ideas of quantum mechanics.

- The other formalism is called *matrix mechanics*.

Wave mechanics starts with a fundamental equation, the **Schrödinger equation,** which was formulated by the Austrian physicist Erwin Schrödinger in 1925. Schrödinger, a recognized authority on the theory of vibrations and the associated quantization of standing waves, knew how **wave functions** are used to describe the behavior of ordinary standing waves (for example, water waves in a tank). These mathematical expressions give the heights (amplitudes) of the waves at different positions along a line or on a surface. He reasoned that an electron (or other particle with wave-like properties) should be described by a similar wave function that has different values at different positions in space. A wave function for an electron or other material particle gives the "height" (amplitude) of the matter-wave at each point in three-dimensional space. Such amplitudes are virtually impossible to visualize but can be dealt with mathematically. Schrödinger symbolized quantum-mechanical wave functions by the Greek letter ψ (psi). The wave function $\psi(x, y, z)$ evaluated at point 1 (x_1, y_1, z_1) equals the amplitude of a matter-wave at that point; at point 2 (x_2, y_2, z_2), the wave function has a different amplitude. Wave functions take positive values in some regions of space (where they are said to have *positive phase*) and negative values in other regions (where they have *negative phase*). The points at which a wave function passes through zero and changes sign are called *nodes,* just as with a standing wave on a guitar string.

- The phase of wave functions proves important in chemical bonding theory. Wave functions must have the same phase (that is, constructively reinforce one another) if they are to combine to form a bond between atoms.

According to quantum mechanics, only functions that are solutions to the Schrödinger equation when it is set up for a particular atom provide meaningful descriptions of the electrons in that atom. This means that most mathematical functions are unsuitable for describing matter-waves. Moreover, good wave functions exist at all only for certain values of the energy of an atom. Most energies are excluded. Quantization of the energy of atoms thus arises as a natural consequence of requiring a wave function that fits the Schrödinger equation. The ψ that corresponds to the lowest value of the energy of an atom is the ground-state wave function; ψ's that correspond to higher energies are excited-state wave functions. Finally, an allowed stationary-state wave function tells all that can be known about *all* observable properties of an atom, not just the energy of the atom.

- In the Born interpretation, the SI unit for ψ is the $m^{-3/2}$. Then ψ^2 has the unit m^{-3}, and ψ^2 multiplied by the small volume (measured in m^3) is a pure number, as a probability must be.

Wave functions describing material particles have no easy physical interpretation. By 1927, the German physicist Max Born and others had developed the idea that the *square* of the wave function for a particle (ψ^2) could meaningfully be interpreted as a **probability density.** In this approach, every point in space is regarded as the center of a tiny box with edges Δx, Δy, and Δz. If the value of ψ^2 at a point is multiplied by the volume of its surrounding tiny box, the result is the probability of finding the particle inside the box. In this way, a positive number equal to $\psi^2(x, y, z) \cdot (\Delta x \, \Delta y \, \Delta z)$ becomes associated with every point in space. The pattern of these numbers through space is a probability density. The particle exists mostly in regions of space in which the probability density is large. This probabilistic interpretation of the wave function is now generally accepted because it provides a consistent picture of particle motion on a microscopic scale.

16–5 The Hydrogen Atom

A hydrogen atom consists of a single electron interacting with a nucleus of charge $+1e$. The solutions of the Schrödinger equation for this arrangement of particles have great importance in chemistry. Recall that the hydrogen atom is the simplest example of a one-electron atom or ion; others are He^+, Li^{2+}, and all other ions in which all but one electron have been stripped off. They differ in the charge (Ze) on the nucleus and therefore in the attractive force felt by the electron. Mathematical analysis shows that wave functions that satisfy the Schrödinger equation for hydrogen atoms and one-electron ions exist only for these values of the energy:

$$E_n = -\frac{Z^2}{n^2}\, Ry \qquad n = 1, 2, 3, \ldots$$

which are exactly the energies found in the Bohr theory. Here, however, quantization of the energy comes about naturally from the Schrödinger description of the electron as a standing wave rather than through an arbitrary assumption about the angular momentum.

The energy of one-electron atoms or ions depends only on the quantum number n, which is the **principal quantum number.** Two more quantum numbers appear from the mathematics when the wave functions that satisfy the H-atom Schrödinger equation are investigated systematically. Each can be associated with observable properties of the atom. The first is the **angular momentum quantum number** (l), which takes on any integral value from 0 up to and including $n - 1$. Substituting l into a simple formula gives the angular momentum of the atom arising from the motion of the electron. The second is the **magnetic quantum number** (m_l), which may take on any integral value (including 0) from $-l$ to $+l$. It is called the magnetic quantum number because its value governs the behavior of the atom in magnetic fields. For $n = 1$ (the ground state), the only allowed values of l and m_l are ($l = 0$, $m_l = 0$). For $n = 2$, there are $n^2 = 4$ sets of allowed values:

$$(l = 0, m_l = 0) \quad (l = 1, m_l = 1) \quad (l = 1, m_l = 0) \quad (l = 1, m_l = -1)$$

The restrictions on l and m_l cause exactly n^2 solutions of the Schrödinger equation to exist for every value of n.

A wave function ψ that satisfies the one-electron Schrödinger equation with quantum numbers n, l, and m_l is called an **orbital.** Although the word "orbital" recalls the orbits of the Bohr atom, there is no real resemblance. An orbital is *not* a trajectory traced by an individual electron. The word refers to a function ψ that gives the amplitude of a standing matter-wave:

> An orbital is a wave function.

A wave function ψ has different values at different points in space, and ψ^2 tells the probability of finding an electron in the vicinity of each point. For this reason, orbitals often are visualized as regions of space:

> An orbital is a region of space in which an electron is most likely to be found.

It is not necessary to write down actual mathematical functions when referring to orbitals. They can be identified instead by their three quantum numbers: n, l, and m_l. For example, "$(n = 2, l = 1, m_l = -1)$" designates a specific orbital in the H atom. The convention in chemistry is to replace the angular momentum quantum number with a lower-case letter:

- The first four letters come from early spectroscopy in which certain spectroscopic lines were referred to as "sharp," "principal," "diffuse," and "fundamental." Subsequent letters are assigned alphabetically.

Angular Momentum Quantum Number *l*	Corresponding Letter
$l = 0$	s
$l = 1$	p
$l = 2$	d
$l = 3$	f
$l = 4$	g
$\vdots$	$\vdots$

A wave function with $n = 1$ and $l = 0$ is a $1s$ orbital under this labelling convention. One with $n = 3$ and $l = 1$ is a $3p$ orbital, one with $n = 4$ and $l = 3$ is a $4f$ orbital, and so forth. The quantum number m_l usually is omitted from such labels. Orbitals that have the same n and l are distinguished from each other when it is necessary to do so by adding subscripts, as discussed later. Figure 16–18 gives some allowed combinations of quantum numbers for H-atom orbitals and their conventional designations.

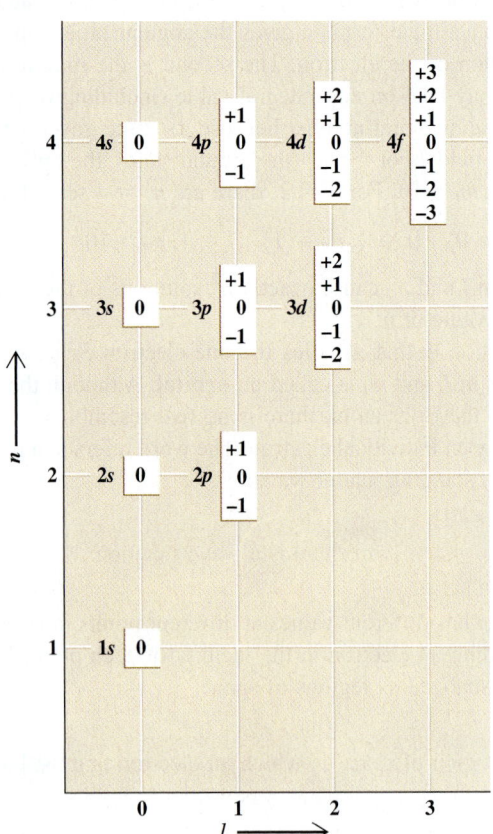

Figure 16–18 • Some sets of allowed quantum numbers for orbitals in one-electron atoms or ions. The allowed values of the m_l quantum number are shown to the right of each (n, l) symbol.

EXAMPLE 16–5

Give the labels of all of the groups of orbitals with $n = 4$, and state how many m_l values are present in each group.

Solution

The quantum number l may range from 0 to $n - 1$, so its allowed values in this case ($n = 4$) are 0, 1, 2, and 3. The labels for the groups of orbitals are then

$$\begin{array}{ll} l = 0 & 4s \\ l = 1 & 4p \\ l = 2 & 4d \\ l = 3 & 4f \end{array}$$

The quantum number m_l ranges from -1 to $+1$, so the number of m_l values is $2l + 1$. This gives

one 4s orbital
three 4p orbitals
five 4d orbitals
seven 4f orbitals

for a total of $16 = 4^2 = n^2$ orbitals with $n = 4$. All have the same energy but differ in their shapes or orientations.

EXERCISE

(a) How many orbitals are there with $n = 5$? (b) How many of these are 5g orbitals? (c) How many are 5h orbitals?

Answer: (a) 25 orbitals. (b) 9 orbitals. (c) Zero, because 5h orbitals are not possible.

Sizes and Shapes of Orbitals

The shapes and sizes of the hydrogen-atom orbitals are important in chemistry. All **s orbitals** are spherically symmetrical about the nucleus. This means that for s orbitals the values of $\psi (x, y, z)$ and $\psi^2(x, y, z)$ depend solely on the distance r of point (x, y, z) from the nucleus. The direction does not matter. There are several ways to visualize s orbitals. One (shown in Fig. 16–19a for the 1s, 2s, and 3s orbitals) is to represent the probability density ψ^2 with shading, making the shading heavier where ψ^2 is larger in order to convey the way the probability density changes. A second way is to plot the wave functions themselves versus r, the distance to the nucleus. Figure 16–19b does this for ψ_{1s}, ψ_{2s}, and ψ_{3s}. Note that the 2s and 3s wave functions have regions of negative phase, although the 1s does not. A third way, perhaps the most useful in chemistry, is to plot a **radial probability distribution** ($r^2\psi^2$) (Fig. 16–19c). This function gives the probability of finding the electron at the distance r from the nucleus. Recall again that $\psi^2(x, y, z)$ by itself equals the probability of finding the electron within a tiny volume surrounding point (x, y, z) in space. Multiplying it by r^2 accounts for the greater number of points that becomes available as the distance to the nucleus increases. These are the points located on the

Figure 16–19 • Three representations of hydrogen s orbitals. (a) An electron-density representation of a hydrogen atom in its $1s$, $2s$, and $3s$ states. The spheres are cut off at a radius that encloses 90% of the total electron density. (b) The wave functions graphed against distance from the nucleus r. (c) The radial probability distribution, equal to $r^2\psi^2$. The distance a_0 is the Bohr radius (0.529 Å).

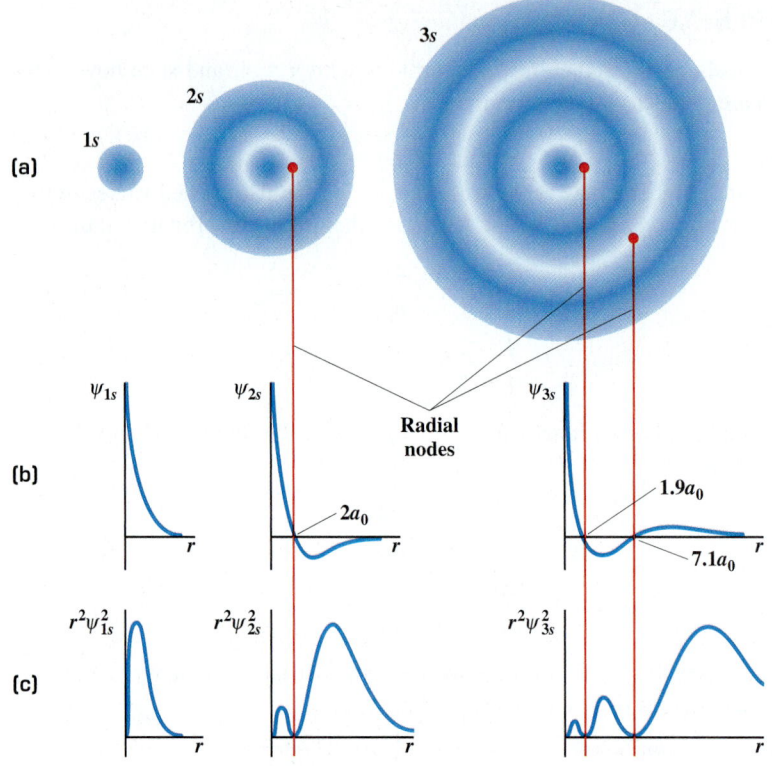

• The blue lines in the graphs in Figure 16–19 approach the horizontal axis asymptotically as r increases. They never reach it.

surface of a sphere of radius r. The value of ψ^2 for s orbitals is at its largest within the nucleus (at $r = 0$), but the interior of the nucleus offers only a few tiny volumes for an electron to occupy. At larger r, many more points meet the distance specification, which makes many more tiny volumes available. As shown in Figure 16–19c, the radial probability distribution functions $r^2(\psi_{1s})^2$, $r^2(\psi_{2s})^2$, and $r^2(\psi_{3s})^2$ reach their highest values away from the nucleus, unlike the functions $(\psi_{1s})^2$, $(\psi_{2s})^2$, and $(\psi_{3s})^2$, which do not include the factor r^2.

What is meant by the *size* of an orbital? Unlike bowling balls or dumbbells, orbitals do not have sharp boundaries. The allowed wave functions in an H atom (and their squares) have non-zero values very far away from the nucleus. This obliges setting an arbitrary cut-off. A common choice is the radius of a boundary sphere (like the skin of a balloon) inside of which the electron has a large probability, say 90%, of being found. Figure 16–19a shows cross-sections of spheres that enclose 90% of the electron probability for the H-atom $1s$, $2s$, and $3s$ orbitals. The radius of a 90% boundary sphere is 1.41 Å for a $1s$ orbital, 3.29 Å for a $2s$ orbital, and 10.28 Å for a $3s$ orbital. The increase means that the electron is more likely to be found farther from the nucleus for larger n, which is the quantum-mechanical version of the increase in radius of the Bohr orbits with increasing n.

Figure 16–19 also shows (in red) the location of nodes in $2s$ and $3s$ orbitals. Nodes in orbitals are surfaces, such as planes or spheres. Studying the number, location, and shape of their nodes is an excellent way to understand orbitals. For example:

> The more numerous the nodes in an orbital, the higher the energy of the orbital.

This statement recalls the case of standing waves on a guitar string, in which higher frequency (and, thus, higher energy) vibrations have more nodes. An ns orbital (an orbital with principal quantum number n and angular momentum quantum number $l = 0$) has $n - 1$ **radial nodes** that are spherical shells on which ψ (and ψ^2) equal zero. An electron in a $2s$ orbital can exist both inside and outside of a spherical shell of radius 1.06 Å ($2a_0$; see Fig. 16–19b) without ever existing *at* the surface of the shell, an interesting consequence of its wave aspect. An electron in a $3s$ orbital can exist within a first spherical node, between this node and second spherical node, and outside of the second node, but never at either node. As we shall see, the chemical behavior of atoms is dominated by electron density outside the outermost nodes in their wave functions.

Orbitals in which electrons have angular momentum (that is, ones in which the quantum number l exceeds zero) have more complex shapes than the spheres associated with $l = 0$. Three orbitals exist for $l = 1$, corresponding to the three possible values of m_l. These **p orbitals** resemble flattened dumbbells, as shown in Figure 16–20. The flattened balls, or lobes, in these diagrams enclose regions in which the probability of finding the electron exceeds a certain cut-off value. A p_x orbital is oriented along the x axis, with one lobe of high probability density on one side of the yz plane and an identical lobe on the other side. A p_y orbital has the same shape but is oriented along the y axis so that the lobes are on either side of the xz plane. A p_z orbital completes the triad with the same shape oriented along the z axis. An electron in a p_x orbital has zero probability of being found in the yz plane because this plane is a node, an **angular node**. All p orbitals have one angular node. Recall that wave functions change phase (go from positive to negative or the reverse) across a node. The positive and negative signs positioned on the lobes in the drawings in Figure 16–20 indicate this change. They have nothing to do with the electrical charge of the particle that is being represented.

The angular node in p orbitals passes through the nucleus so that the probability density at the nucleus equals zero in p orbitals. This is true, in fact, of all types of orbitals except s orbitals. The electron can never exist at the nucleus in orbitals with $l > 0$ ($p, d, f, \ldots$).

A p orbital may have radial nodes, like those in $2s$ or $3s$ orbitals. The rule is that an np wave function has $n - 2$ radial nodes. Thus, a $2p$ wave function has zero radial nodes, a $3p$ wave function has one, a $4p$ wave function has two, and so forth. Because all p orbitals possess one angular node, their total count of nodes works out this way:

$$(n - 2) \text{ radial nodes} + 1 \text{ angular node} = (n - 1) \text{ total nodes} \quad \text{(for } p \text{ orbitals)}$$

The total number of nodes in a p orbital is the same as in an s orbital that has the same principal quantum number n. The energies of the H-atom orbitals depend on the total number of nodes $n - 1$ but not on the type of node (radial or angular). The ns and np orbitals in H (and in single-electron ions) have the same energy.

As in the case of s orbitals, the radial nodes in p orbitals occur so close to the nucleus that they have little effect on the chemical bonding of atoms. For the purposes of chemistry, we may therefore represent all p orbitals (not just the $2p$ orbitals) by pictures like those in Figure 16–20.

The shapes of **d orbitals** are more complicated. Figure 16–21 depicts the five $3d$ orbitals. A d_{xy} orbital has four lobes of high electron probability density directed between the x and y axes and with highest density in the xy plane. These lobes are separated by angular nodes, which are the xz and yz planes. The sign of a d_{xy} wave function changes four times as these planes are crossed in a full-circle tour around

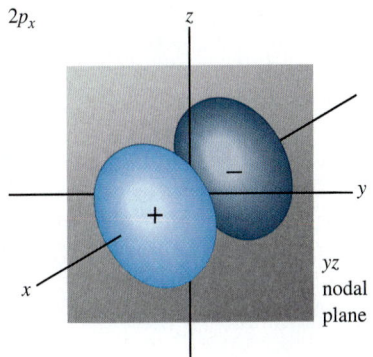

$2p_x$

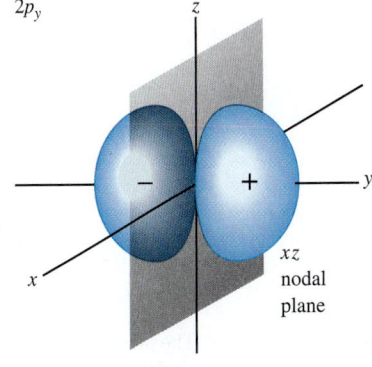

$2p_y$

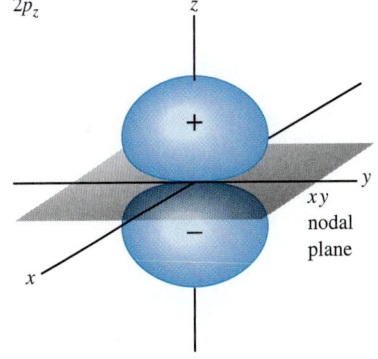

$2p_z$

Figure 16–20 • The distribution of electron density in the three $2p$ orbitals, with the relative phase of the wave function and the nodal planes indicated.

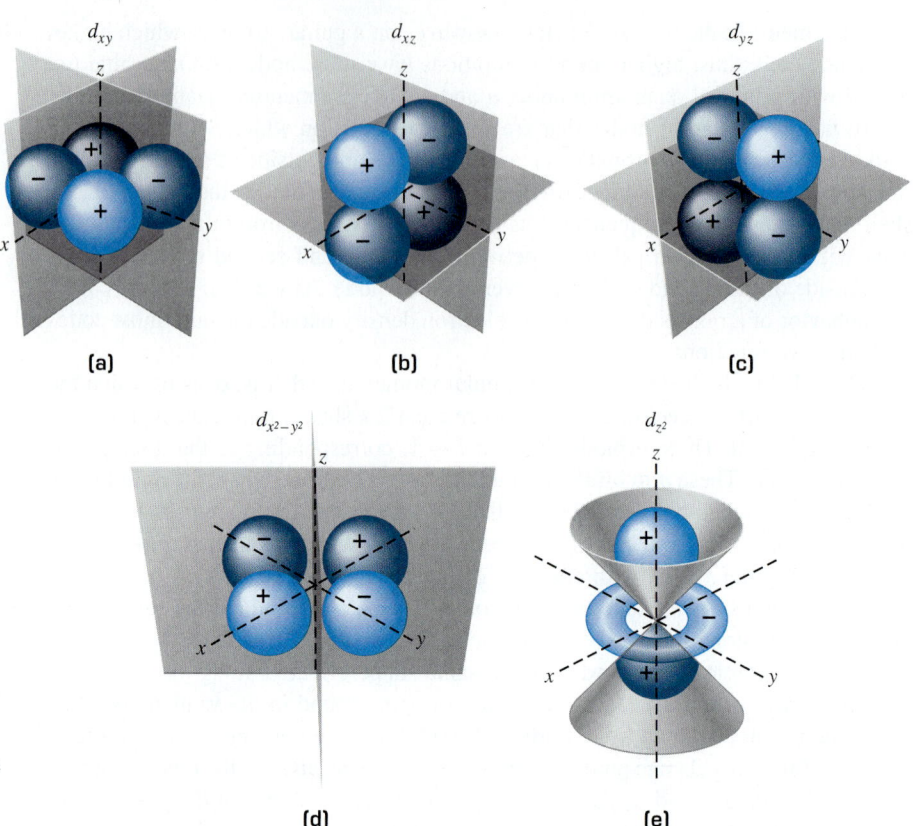

Figure 16–21 • The shapes of the five $3d$ orbitals, with relative phases and nodal surfaces indicated.

the z axis. The changes are indicated by the alternating positive and negative signs on the four lobes in Figure 16–21. The d_{xz} and d_{yz} orbitals are identical to the d_{xy} in shape but differ in orientation. A d_{xz} has its four lobes directed between the x and z axes, and a d_{yz} orbital has its four lobes directed between the y and z axes. Note how the subscripts give the axial orientation. A $d_{x^2-y^2}$ orbital is also identical in shape to a d_{xy}, but its four lobes of high electron probability density are directed along the x and y axes, not between them. The angular nodes in a $d_{x^2-y^2}$ orbital are planes bisecting the angles between the x and y axes. Finally, a d_{z^2} orbital has a different shape, with high probability density along the z axis and in a "doughnut" of high probability density in the xy plane. The phase of a d_{z^2} orbital changes four times in a circuit around either the x axis or the y axis. This shows that a d_{z^2} orbital, like all other d orbitals, has two angular nodes. The difference is that the angular nodes are cones in a d_{z^2} orbital, not planes.

The total count of nodes in d orbitals works out this way:

$$(n-3) \text{ radial nodes} + 2 \text{ angular nodes} = (n-1) \text{ total nodes} \qquad \text{(for } d \text{ orbitals)}$$

No radial nodes appear in Figure 16–21 because the diagrams are for $3d$ ($n=3$) orbitals. The radial nodes in $4d$ and $5d$ orbitals and higher d orbitals have little effect on the chemical bonding of atoms. For the purposes of chemistry, we may represent all d orbitals (not just the $3d$ orbitals) by pictures like those in Figure 16–21.

The wave functions for f orbitals ($l = 3$) and orbitals of higher l can be calculated, but they are hard to draw or to visualize. They play a smaller role in chemistry than the s, p, and d orbitals. We therefore pause at this point and summarize the important features of orbital shapes and sizes:

1. For a given value of l, an increase in n leads to an increase in the average distance of the electron from the nucleus and therefore in the size of the orbital. A 2s orbital, for example, is larger than a 1s orbital, and a 3s orbital is larger than a 2s (see Fig. 16–19).
2. An orbital with quantum numbers n and l has l angular nodes and $n - l - 1$ radial nodes, giving a total of $n - 1$ nodes. For a one-electron atom or ion, the energy depends only on the number of nodes; that is, on n but not on l or m_l.
3. As r approaches zero, the wave function goes to zero for all orbitals except the s orbitals; thus, only an electron in an s orbital can "penetrate to the nucleus"—that is, have a finite probability of being found at the nucleus.

These general statements prove helpful in studying the electronic structure of many-electron atoms (in Chapter 17), even though they are obtained for the one-electron case.

EXAMPLE 16-6

Compare the 3p and 4d orbitals of a hydrogen atom with respect to (a) number of radial and angular nodes and (b) energy of the corresponding atom.

Solution

(a) The 3p orbital has a total of $n - 1 = 3 - 1 = 2$ nodes. Of these, one is angular ($l = 1$) and one is radial. The 4d orbital has $4 - 1 = 3$ nodes. Of these, two are angular ($l = 2$) and one is radial.
(b) The energy of a one-electron atom depends only on n. The energy of an atom with an electron in a 4d orbital is then higher than is that of an atom with an electron in a 3p orbital because $4 > 3$.

EXERCISE

Compare the 3d and 5d orbitals of a He^+ ion with respect to (a) number of radial and angular nodes and (b) size of the orbital.

Answer: (a) Each has two angular nodes, but the 5d orbital has two radial nodes as well. (b) The 5d orbital is larger.

Electron Spin

If a beam of hydrogen atoms in their ground state (with $n = 1, l = 0, m_l = 0$) is directed through a non-uniform magnetic field, the beam splits into two beams, each containing half of the atoms (Fig. 16–22). The atoms behave as if they were tiny magnets. This result is explained by introducing a *fourth* quantum number to complete the description of the wave function of the electron. This quantum number, m_s, may take on only two values, $\frac{1}{2}$ and $-\frac{1}{2}$. The behavior of the H-atoms in the

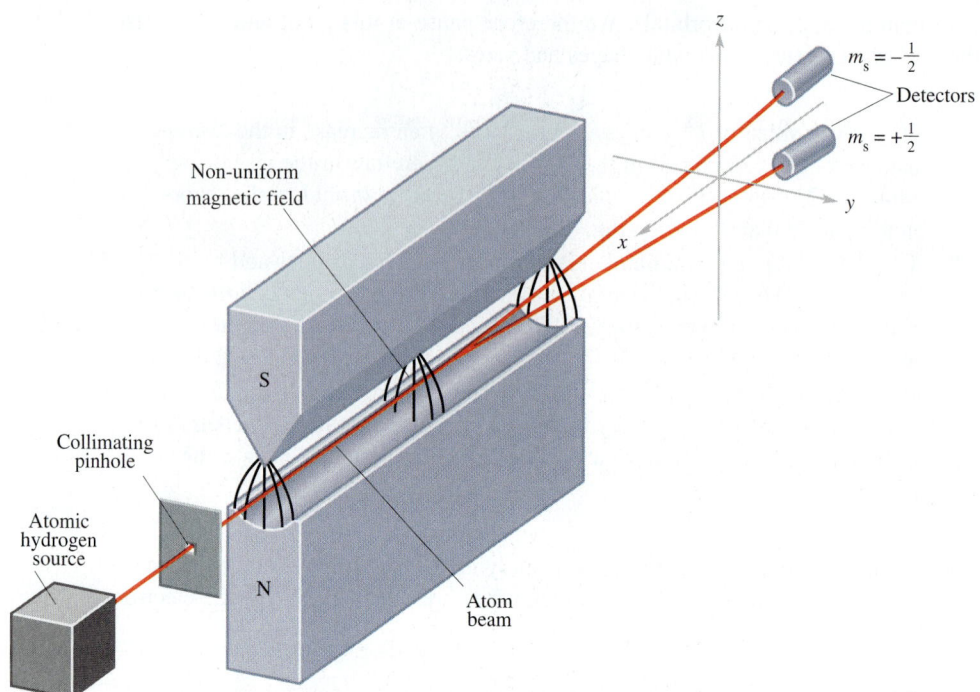

Figure 16–22 • A beam of hydrogen atoms is split into two beams when traversing a non-uniform magnetic field. Atoms with spin quantum number $m_s = +\frac{1}{2}$ comprise one beam, and those with $m_s = -\frac{1}{2}$, the other.

magnetic field can be explained by classical theory if their electrons are regarded as balls of charge spinning around their own axes. For this reason, the fourth quantum number is referred to as the **spin quantum number.** When $m_s = +\frac{1}{2}$, the electron spin is in one direction; when $m_s = -\frac{1}{2}$, the spin is in the other. The picture of a spinning ball of charge is not literally correct, however. The spin quantum number comes from relativistic effects that are not included in the Schrödinger equation. For most purposes in chemistry, it is sufficient simply to solve the Schrödinger equation and then associate with each electron a spin quantum number $m_s = +\frac{1}{2}$ or $m_s = -\frac{1}{2}$, which does not affect the spatial probability density of the electron. The spin does affect the number of different quantum states with energy E, however, doubling it from n^2 to $2n^2$. This fact assumes considerable importance when we consider many-electron atoms in Chapter 17.

The single quantum number n of the Bohr theory is thus superseded by a set of four quantum numbers (n, l, m_l, and m_s) in modern quantum mechanics. For a one-electron atom or ion in free space, only n affects the energy, and only l affects the orbital shape. All four quantum numbers must be specified to give a complete description of a one-electron system.

SUMMARY

16–1   A traveling wave has **amplitude** (the extent of variation of the quantity that oscillates), **wavelength** (λ; the distance that the oscillating quantity covers before repeating itself), **frequency** (ν; the number of cycles passing a fixed point per second), and speed (the product of frequency and wavelength). **Electromagnetic waves** range in wavelength from the meter or kilometer scale of radio waves down through the micrometer scale of visible light to the nanometer scale of X-rays and beyond. All electromagnetic radiation propagates at the same speed c_0 in a vacuum (approximately 3.00×10^8 m s^{-1}).

16-2   Classical physics fails to explain how the intensity of electromagnetic radiation emitted by a heated object (idealized as a **blackbody**) varies with wavelength; instead, it predicts the **ultraviolet catastrophe.** Classical physics fails to explain the existence of a threshold frequency (in which light of a certain minimum frequency is required to eject electrons from a metallic surface) in the **photoelectric effect.** Classical physics also does not account for the stability of the nuclear atom but instead predicts that atoms with revolving electrons radiate continuously. Finally, classical physics does not explain the observation of discrete frequencies in the emission spectra of single atoms.

16-3   Light has particle properties as well as wave properties. This **wave–particle duality** means that the energy of electromagnetic waves comes in quantized packets called **photons.** Quantum theory allowed explanations of blackbody radiation (by Planck), the photoelectric effect (by Einstein), and the stability and characteristic spectra of atoms (by Bohr). In Bohr's planetary model of the hydrogen atom, only certain orbital radii and energies are allowed, and transitions between orbits are accompanied by the emission or absorption of photons of light having the frequency $\nu = \Delta E/h$, where ΔE equals the difference in the energies of the allowed orbits and h is **Planck's constant.** The Bohr model predicts the observed spectrum of hydrogen atoms and one-electron ions but cannot successfully be extended to atoms or ions with more than one electron.

16-4   Louis de Broglie proposed in 1924 that the wavelength of a moving particle of matter is inversely proportional to its momentum, with the proportionality constant (Planck's constant) so small that macroscopic objects have negligible wavelengths. For electrons and other particles on the atomic scale, wave properties play an essential role. The **Schrödinger wave equation** associates a **wave function** (symbolized by ψ) with the electrons in an atom. The square of the wave function gives the electron density in the close vicinity of a point in space, or, equivalently, the probability of finding an electron in that small region. The Schrödinger equation provides the starting point for modern quantitative prediction of the properties of atoms and molecules.

16-5   Solutions to the Schrödinger equation for one-electron atoms and ions correspond to the same energies as the Bohr model but replace circular orbits by **orbitals,** which are wave functions that give the distribution of electron probability density throughout the atom. The energies, shapes, and sizes of orbitals are characterized by three **quantum numbers.** A fourth quantum number (the spin quantum number m_s) describes the inherent magnetic properties of the electron. The energy of an electron in a one-electron atom or ion depends only on the first quantum number (the principal quantum number) n; the second and third quantum numbers, l and m_l, determine the shapes and orientations of the orbitals and, therefore, the electron-density distribution. The letters s, p, d, and f traditionally are associated with values of l from 0 to 3, respectively. Only s orbitals are spherically symmetrical; all others describe electron-density distributions that depend on direction as well as distance from the nucleus. The overall number of **nodes** in a wave function is $n - 1$ and the number of **angular nodes** is l. As a result, the number of lobes in orbitals increases with increasing l. Sizes and shapes can be associated with orbitals most easily by identifying boundary surface "balloon skins" inside which the electron has some large probability of being found.

PROBLEMS

Note: Answers to blue-numbered problems are given in Appendix F. Problems that are more challenging are indicated with asterisks.

Waves and Light

1. Some water waves reach the beach at a rate of one every 3.2 s, and the distance between their crests is 2.1 m. Calculate the speed of these waves.

2. The spacing between bands of color in a chemical wave from an oscillating reaction is measured to be 1.2 cm, and a new wave appears every 42 s. Calculate the speed of propagation of the chemical waves.

3. (See Example 16–1.) An FM radio station broadcasts at a frequency of $9.86 \times 10^7 \, s^{-1}$ (98.6 MHz). Calculate the wavelength of the radio waves.

4. (See Example 16–1.) The gamma rays emitted by ^{60}Co are used in radiation treatment of cancer. These gamma rays have a frequency of $2.83 \times 10^{20} \, s^{-1}$. Calculate their wavelength, expressing your answer in meters and in ångströms.

5. Radio waves of wavelength 6.00×10^2 m can be used to communicate with spacecraft over large distances.
   (a) Calculate the frequency of these radio waves.
   (b) Suppose that a radio message is sent home by the astronauts in a spaceship approaching the planet Mars at a distance of 8.0×10^{10} m from Earth. How long (in minutes) does it take for the message to travel from the spaceship to Earth?

6. An argon-ion laser emits light of wavelength 488 nm.
   (a) Calculate the frequency of the light.
   (b) Suppose that a pulse of light from this laser is sent from Earth, is reflected from a mirror on the Moon, and returns to its starting point. Calculate the time elapsed for the round trip, taking the distance from Earth to Moon to be 3.8×10^5 km.

7. The speed of sound in dry air at 20°C is $343.5 \, m \, s^{-1}$, and the frequency of the sound from the middle C note on a piano is $261.6 \, s^{-1}$, according to the American standard-pitch scale. Calculate the wavelength of this sound and the time it takes to travel 30.0 m across a concert hall.

8. Ultrasonic waves have frequencies too high to be detected by the human ear but can be produced and detected by vibrating crystals. Calculate the wavelength of an ultrasonic wave of frequency $5.0 \times 10^4 \, s^{-1}$ that is propagating through a sample of water at a speed of $1.5 \times 10^3 \, m \, s^{-1}$. Explain why ultrasound can be used to probe the size and position of the fetus inside the abdomen of the mother. Could audible sound with a frequency of $8000 \, s^{-1}$ be used for this purpose?

Paradoxes in Classical Physics

9. Both blue and green light eject electrons from the surface of potassium. In which case do the ejected electrons have the higher average kinetic energy?

10. When an intense beam of green light is directed on a copper surface, no electrons are ejected. What happens if the green light is replaced with red light?

11. Excited lithium atoms emit light strongly at a wavelength of 671 nm. This emission predominates when Li atoms are excited in a flame. Predict the color of the flame.

12. Excited mercury atoms emit light strongly at a wavelength of 454 nm. This emission predominates when Hg atoms are excited in a flame. Predict the color of the flame.

Planck, Einstein, and Bohr

13. Barium atoms in a flame emit light as they undergo transitions from one energy level to another one that lies 3.6×10^{-19} J lower in energy. Calculate the wavelength of light emitted and, by referring to Figure 16–5, predict the color visible in the flame.

14. Potassium atoms in a flame emit light as they undergo transitions from one energy level to another one that lies 4.9×10^{-19} J lower in energy. Calculate the wavelength of light emitted and, by referring to Figure 16–5, predict the color visible in the flame.

15. The sodium D-line is actually a pair of closely spaced spectroscopic lines seen in the emission spectrum of sodium atoms. The wavelengths are centered at 589.3 nm. The intensity of this emission makes it the major source of light in the sodium arc light and causes the yellow color of that light.
    (a) Calculate the energy change per sodium atom emitting a photon at the D-line wavelength.
    (b) Calculate the energy change per mole of sodium atoms emitting photons at the D-line wavelength.
    (c) If a sodium arc light produces 1.000 kW ($1000 \, J \, s^{-1}$) of radiant energy at this wavelength, how many moles of sodium atoms emit photons per second?

16. The power output of a laser is measured by its wattage, the number of joules of energy it radiates per second (1 watt = $1 \, J \, s^{-1}$). A 10 W laser produces a beam of green light with a wavelength of 520 nm (5.2×10^{-7} m).
    (a) Calculate the energy carried by each photon.
    (b) Calculate the number of photons emitted by the laser per second.

17. (See Example 16–2.) Cesium frequently is used in photocells because its work function (3.43×10^{-19} J) is the lowest of all the elements. Such photocells are efficient because the broadest range of wavelengths of light can eject electrons. What colors of light eject electrons from cesium? What colors of light eject electrons from selenium, which has a work function of 9.5×10^{-19} J?

18. (See Example 16–2.) Some alarm systems use the photoelectric effect. A beam of light strikes a piece of metal in a photocell, ejecting electrons continuously and causing a small electric current to flow. When someone steps into the light beam, the current is interrupted and the alarm is trig-

gered. What is the maximum wavelength of light that can be used in such an alarm system if the photocell metal is sodium, with a work function of 4.41×10^{-19} J?

19. Light having a wavelength of 2.50×10^{-7} m falls on the surface of a piece of chromium in an evacuated glass tube. If the work function of the chromium is 7.21×10^{-19} J, determine the following:
    (a) The maximum kinetic energy of the photoelectrons emitted from the chromium.
    (b) The speed of photoelectrons having this maximum kinetic energy.

20. Calculate the maximum wavelength of electromagnetic radiation if it is to cause detachment of electrons from the surface of metallic tungsten, which has a work function of 7.29×10^{-19} J. If the maximum speed of the emitted photoelectrons is to be 2.00×10^6 m s^{-1}, what should be the wavelength of the radiation?

21. (See Example 16–3.) Consider three hydrogen atoms in the quantum states $n = 3, n = 6$, and $n = 8$.
    (a) Compute the energies of these atoms. Express the answers both in joules (J) and in rydbergs (Ry). State which atom is in the highest energy state.
    (b) Compute the radii of these atoms. Express the answers both as multiples of a_0 and in meters. State which of the atoms is largest.

22. (See Example 16–3.) Consider three He$^+$ ions in the quantum states $n = 2, n = 4$, and $n = 7$.
    (a) Compute the energies of these ions in joules (J) and in rydbergs (Ry).
    (b) Compute the radii of these atoms as multiples of a_0 and in meters.

23. A hydrogen atom undergoes the following transitions. Determine the amount of energy lost by the atom in each case, and express it both in rydbergs and in joules.
    (a) $n = 7$ to $n = 4$
    (b) $n = 4$ to $n = 3$
    (c) $n = 3$ to $n = 4$
    (*Hint:* Recall that the loss of a negative is a gain.)

24. A Li^{2+} ion undergoes the following transitions. Determine the amount of energy lost by the ion in each case and express it both in rydbergs and in joules.
    (a) $n = 3$ to $n = 1$
    (b) $n = 6$ to $n = 2$
    (c) $n = 3$ to $n = 4$

25. A single photon is absorbed or emitted in each transition in Problem 23. Calculate its wavelength.

26. A single photon is absorbed or emitted in each transition in Problem 24. Calculate its wavelength.

27. Use the Bohr model to calculate the radius and the energy of the B^{4+} ion in the $n = 3$ state. How much energy is required to remove the electrons from 1 mol of B^{4+} ions in this state? What frequency and wavelength of light are emitted in a transition from the $n = 3$ to the $n = 2$ state of this ion? Express all results in SI units.

28. He$^+$ ions are observed in stellar atmospheres. Use the Bohr model to calculate the radius and the energy of the He$^+$ ion in the $n = 5$ state. How much energy is required to remove the electrons from 1 mol of He$^+$ in this state? What frequency and wavelength of light are emitted in a transition from the $n = 5$ to the $n = 3$ state of this ion? Express all results in SI units.

29. The radiation emitted in the transition from $n = 3$ to $n = 2$ in a neutral hydrogen atom has a wavelength of 656.1 nm. What is the wavelength of radiation emitted from a doubly ionized lithium atom (Li^{2+}) if a transition occurs from $n = 3$ to $n = 2$? In what region of the spectrum does this radiation lie?

30. The Be^{3+} ion has a single electron. Calculate the frequencies and wavelengths of light seen in its emission spectrum for the first three lines of each of the series analogous to the Lyman and Balmer series of neutral hydrogen. In what region of the spectrum does this radiation lie?

Waves, Particles, and the Schrödinger Equation

31. A guitar string is stretched between two pegs separated by 50 cm.
    (a) Calculate the wavelength of its fundamental mode of vibration (i.e., its first harmonic) and of its third harmonic.
    (b) How many nodes does the third harmonic have?

32. Suppose we picture an electron in a chemical bond as being a wave with fixed ends. Take the length of the bond to be 1.0 Å.
    (a) Calculate the wavelength of the electron wave in its ground state and in its first excited state.
    (b) How many nodes does the first excited state have?

33. (See Example 16–4.) Calculate the de Broglie wavelength of the following:
    (a) An electron moving at a speed of 1.00×10^3 m s^{-1}.
    (b) A proton moving at a speed of 1.00×10^3 m s^{-1}.
    (c) A baseball having a mass of 170.0 g and moving at a speed of 95 km h^{-1}.

34. (See Example 16–4.) Calculate the de Broglie wavelength of the following:
    (a) An electron that has been accelerated to a kinetic energy equivalent to 1.20×10^7 J mol^{-1}.
    (b) A helium atom moving at a speed of 353 m s^{-1}. This is the root-mean-square speed of helium atoms in gaseous helium at 20 K.
    (c) A krypton atom moving at a speed of 299 m s^{-1}. This is the root-mean-square speed of krypton atoms in gaseous krypton at 300 K.

The Hydrogen Atom

35. Give labels for the orbitals described by each of the following sets of quantum numbers:
    (a) $n = 4, l = 1$ (b) $n = 2, l = 0$ (c) $n = 6, l = 3$

36. Give labels for the orbitals described by each of the following sets of quantum numbers:
(a) $n = 3, l = 2$ (b) $n = 7, l = 4$ (c) $n = 5, l = 1$

37. (See Example 16–6.) How many radial nodes and how many angular nodes does each of the orbitals in Problem 35 have?

38. (See Example 16–6.) How many radial nodes and how many angular nodes does each of the orbitals in Problem 36 have?

39. Where is an electron in a $3p_y$ orbital most likely to be found? Where does the probability of finding it vanish?

40. Where is an electron in a $3d_{x^2 - y^2}$ orbital most likely to be found? Where does the probability of finding it vanish?

41. What is a quantum number? How many quantum numbers are needed to specify the state of the electron in a Cl^{16+} ion?

42. What is an atomic orbital? How does it resemble and how does it differ from the Bohr planetary orbits?

43. Complete the following table:

n	l	Label	No. of Orbitals
2	1	—	—
—	—	3d	—
4	—	—	7

44. Complete the following table:

		No. of Nodes	
n	l	Radial	Angular
1	0	—	—
2	—	1	—
—	—	3	1
—	—	1	3

45. Which of the following combinations of quantum numbers are allowed for the electron in a one-electron atom or ion, and which are not?
(a) $n = 2, l = 2, m_l = 1, m_s = \frac{1}{2}$
(b) $n = 3, l = 1, m_l = 0, m_s = -\frac{1}{2}$
(c) $n = 5, l = 1, m_l = 2, m_s = \frac{1}{2}$
(d) $n = 4, l = -1, m_l = 0, m_s = \frac{1}{2}$

46. Which of the following combinations of quantum numbers are allowed for the electron in a one-electron atom or ion, and which are not?
(a) $n = 3, l = 2, m_l = 1, m_s = 0$
(b) $n = 2, l = 0, m_l = 0, m_s = -\frac{1}{2}$
(c) $n = 7, l = 2, m_l = -2, m_s = \frac{1}{2}$
(d) $n = 3, l = -3, m_l = -2, m_s = -\frac{1}{2}$

ADDITIONAL PROBLEMS

47. A piano tuner uses a tuning fork that emits sound with a frequency of 440 s^{-1}. Calculate the wavelength of the sound from this tuning fork and the time the sound takes to travel 10.0 m across a large room. Take the speed of sound in air to be 343 m s^{-1}.

48. The distant galaxy called Cygnus A is one of the strongest sources of radio waves reaching the Earth. The distance of this galaxy from the Earth is 3×10^{24} m. How long (in years) does it take a radio wave from Cygnus A to reach the Earth? If the wavelength of the radio wave is 10 m, what is its frequency?

49. Hot objects can emit blackbody radiation that appears red, orange, white, or bluish white, but never green. Explain.

50. Describe briefly the ways in which classical physics at the end of the 19th century was unable to account for several important observations about the properties of light, electrons, and atoms.

51. Compare the energy (in joules) carried by an X-ray photon (wavelength ($\lambda = 0.20$ nm) with that carried by an AM radio wave photon ($\lambda = 200$ m). Calculate the energy of 1 mol of each type of photon. What effect do you expect each type of radiation to have in inducing chemical reactions in substances through which it passes?

52. When ultraviolet light of wavelength 131 nm strikes a polished nickel surface, the maximum kinetic energy of ejected electrons is measured to be 7.04×10^{-19} J. Calculate the work function of nickel.

53. Photons with frequencies in the Lyman series are emitted as hydrogen atoms undergo transitions from various excited states to the ground state. If ground-state He^+ ions are present in the same gas (near stars, for example), can they absorb these photons? Explain.

*54. Name a transition in the C^{5+} ion that leads to the absorption of green light.

55. In two-photon ionization spectroscopy, two photons are absorbed by an atom, ion, or molecule to remove an electron from it. Suppose two photons of the same wavelength are used to remove the single electron from a He^+ ion in its $2s$ state. Calculate the maximum wavelength that can be used.

56. Frequency-doubling crystals are used in optics to double the frequency of light that passes through them. Suppose that the output from such a crystal is to be used to excite a hydrogen atom from the $n = 2$ to the $n = 5$ state. What wavelength of input light should be used?

*57. The energies of macroscopic objects as well as those of microscopic ones are quantized, but the effects of the quantization are not seen because the difference in energy between adjacent states is so small. Apply Bohr's quantization of angular momentum to the revolution of the Earth (mass 6.0×10^{24} kg), which moves with a velocity of 3.0×10^4 m s^{-1} in a circular orbit (radius 1.5×10^{11} m) around the Sun, which can be treated as fixed. Calculate the value of the quantum number n for the present state of the Earth–Sun system. What would be the effect of an increase in n by 1?

*58. Sound waves, like light waves, can interfere with each other, giving maximum and minimum levels of sound. Suppose that a listener standing directly between two loudspeakers hears the same tone being emitted from both. This listener observes that, on moving one of the speakers a distance of 0.16 m farther away, the perceived intensity of the tone decreases from a maximum to a minimum.

(a) Calculate the wavelength of the sound.

(b) Calculate its frequency, using 343 m s^{-1} as the speed of sound.

59. A microwave oven uses radiation with a frequency of 2.45×10^9 s^{-1}. Calculate the number of such photons that must be absorbed to raise the temperature of 100.0 g of water from 15°C to its boiling point, 100°C. Take the specific heat capacity of water to be 4.2 J K^{-1} g^{-1}.

60. In a universe in which Planck's constant has the value $h = 1$ J s, what would be the de Broglie wavelength of a baseball of mass 170 g moving at a speed of 30 m s^{-1}?

61. Classical physics predicts that an electron orbiting a nucleus will emit light of continuously increasing frequency. How did the Bohr theory address this objection to the Rutherford model of the atom (the nuclear atom)? How does quantum mechanics deal with this issue?

62. Which properties of the hydrogen atom did the Bohr model predict correctly?

63. How does the $3d_{xy}$ orbital of an electron in an O^{7+} ion resemble, and how does it differ from, the $3d_{xy}$ orbital of an electron in a hydrogen atom?

64. What is the difference between the amplitude of its wave function and the average electron density in a hydrogen atom?

CUMULATIVE PROBLEM

Interstellar Space

The vast stretches of space between the stars are by no means empty. They contain both gases and dust particles at very low concentrations. These low-density contents can affect significantly the electromagnetic radiation that telescopes detect arriving from distant stars and other sources. The gas in interstellar space consists primarily of hydrogen (either neutral or ionized) at a concentration on the order of one atom per cubic centimeter. The dust (thought to be mostly solid water, methane, or ammonia) is even less concentrated, with typically only a few dust particles (10^{-4} to 10^{-5} cm in radius) per cubic *kilometer*.

(a) The hydrogen in interstellar space near a star is ionized largely by the high-energy photons from the star. Such regions are called "H II regions." Calculate the longest wavelength of light that ionizes a ground-state hydrogen atom.

(b) Suppose that a ground-state hydrogen atom absorbs a photon with a wavelength of 65 nm. Calculate the kinetic energy of the electron ejected. (*Note*: This is the gas-phase analogue of the photoelectric effect for solids.)

(c) What is the de Broglie wavelength of the electron from part (b)?

(d) Ionized electrons in H II regions can be recaptured by hydrogen nuclei. In such events, the atom emits a series of photons of increasing energy as the electrons cascade down through the quantum states of the hydrogen atom. Extremely high quantum states can be detected. In particular, the transition from the state $n = 110$ to $n = 109$ for the hydrogen atom has been detected. What is the Bohr radius of an electron in the state $n = 110$ for hydrogen?

(e) Calculate the wavelength of light emitted as an electron in a hydrogen atom undergoes a transition from level $n = 110$ to $n = 109$. In what region of the electromagnetic spectrum does this lie?

(f) The regions farther from stars are called "H I regions." In these regions, almost all of the hydrogen is neutral rather than ionized and is in its ground state. Do such hydrogen atoms absorb light in the Balmer series emitted from the atoms in the H II regions?

A cloud of luminous ionized hydrogen atoms surrounds a compact cluster of hot stars. *(Courtesy of Anglo-Australian Telescope Board)*

(g) Determine the number of angular nodes and the number of radial nodes in the $109g$ wave function.

(h) According to Section 16–5, the energy of the hydrogen atom depends only on the quantum number n. In fact, this is not quite true. The electron spin (m_s quantum number) couples very weakly with the spin of the nucleus, making the ground state split into two states with slightly different energies. The radiation emitted in a transition from the upper to the lower of these levels has a wavelength of 21.2 cm and is of great importance in astronomy because it allows the H I regions to be studied. What is the energy difference between these two levels (in kJ mol^{-1})?

(i) The gas and dust particles between a star and the Earth absorb the star's light more strongly in the blue region of the spectrum than in the red. As a result, stars appear slightly redder than they actually are. Does an estimate of the temperature of a star based on its apparent color give too high or too low a number?

17

Many-Electron Atoms and Chemical Bonding

CHAPTER OUTLINE

Types of compounds: covalent nitrogen dioxide, a brown gas, and ionic calcium fluoride, a solid. (*Charles D. Winters*)

Modern quantum mechanics makes quantitative, verifiable predictions about the properties of atoms and their interactions with one another, subjects that are obviously of the greatest importance to chemistry. Such predictions often require imposing calculations. Fortunately, quantum mechanics also provides a qualitative structure for describing the sizes, energies, oxidation states, and bonding properties of atoms throughout the periodic table. In this way, it gives a deep underpinning to the simple ideas of bonding used in the Lewis electron-dot model.

A question was raised at the beginning of Chapter 3: Why do some combinations of atoms form molecules of great stability, while others join together only for fleeting moments or not at all? We now return to this question, using the insights provided by quantum mechanics. The first step is to understand the properties of electrons in isolated atoms. As we shall show, the behavior of single atoms provides important clues to the ways in which the atoms combine to form molecules.

17–1 Many-Electron Atoms and the Periodic Table

In principle, setting up the Schrödinger equation for a particular atom and solving it mathematically provides a complete picture of the atom's distribution of electrons, irrespective of their number. In practice, solutions to the Schrödinger equation do not exist in explicit form for any many-electron atom, not even the helium atom (two electrons). Fortunately, much of what is known about the properties of hydrogen-atom wave functions carries over to many-electron atoms by use of the **orbital approximation:**

> The electron density of an isolated many-electron atom is approximately the sum of the electron densities of each of the individual electrons taken separately.

The electron densities of individual electrons in many-electron atoms are given by orbitals that differ from the hydrogen-atom orbitals in some ways and resemble them in others:

1. Their angular dependence is identical, so it is straightforward to associate quantum numbers l and m_l with each one.
2. Their radial dependence differs from the radial dependence of the hydrogen-atom orbitals, but a principal quantum number n still can be defined.
3. They are generated by the same pattern of allowed values of n, l, and m_l. That is, n may equal any positive integer, l may equal any integer from 0 up to and including $n - 1$, and m_l may equal any integer (including 0) from $-l$ to $+l$.

In the orbital approximation, every electron in an atom has a set of four quantum numbers (n, l, m_l, m_s) that describe its spatial distribution and spin state; every electron dwells in an atomic orbital with a characteristic size, shape, and energy and has a spin (up or down).

• Remember from Chapter 16: "An orbital is a wave function."

In many-electron atoms, the energies of the orbitals do *not* depend solely on n, as in one-electron atoms. They depend on both n and l (but not on m_l). For example, a $2p$ orbital has a somewhat higher energy than a $2s$, and a $3d$ orbital has a somewhat higher energy than a $3p$. Orbitals still, however, fall into groups according to their energies. A group of orbitals with exactly equal energies (such as the $2p_x$, $2p_y$, and $2p_z$ orbitals) comprises a **subshell**. The s orbitals are "groups of one" and are referred to as subshells as well. Subshells having similar energies (such as the $2s$ and the $2p$ subshell) make up a **shell** of orbitals. Electrons in a given shell have about the same average distance from the nucleus, as "shell" suggests.

The orbital approximation neglects direct interactions between electrons. These include electrical repulsion (particles with electrical charges of the same sign repel each other) and a more subtle quantum mechanical interaction: Electrons with **parallel spins** (both having the value of m_s) tend to stay apart better than those with **paired spins** (having different values of m_s). When electron–electron interactions become large enough, the orbital approximation starts to break down.

Building Up Electron Configurations

Electrons in many-electron atoms occupy orbitals according to certain rules, of which the most important is the **Pauli exclusion principle:**

> No two electrons in an atom may have the same set of four quantum numbers (n, l, m_l, m_s).

An equivalent statement is that every atomic orbital (characterized by a set of three quantum numbers: n, l, and m_l) in a many-electron atom can hold a maximum of two electrons, one with m_s equal to $+\frac{1}{2}$ and the other with m_s equal to $-\frac{1}{2}$. The Pauli principle imposes structure in many-electron atoms. Without it, all the electrons might crowd close up to the nucleus in some low-energy orbital, perhaps the $1s$. With it, electrons are organized. This organization is described by listing the occupied orbitals and the number of electrons in each. Such a list is called an **electron configuration.** Numerous electron configurations exist for every many-electron atom and ion, depending on their energy. The configurations of lowest energy, the **ground-state electron configurations,** are the most important because they govern chemical reactivity under ordinary conditions. Ground-state configurations are determined by imagining a bare nucleus having atomic number Z (and so having a charge of $+Z$), mentally surrounding it with empty atomic orbitals, and then feeding in Z electrons, starting with the lowest-energy orbital and obeying the Pauli principle at all times. The order of energy of the orbitals is found by moving from left to right in the following table, starting at the bottom row and working up:

- Two such electrons have paired spins. Unpaired spins are the same as parallel spins.

$5f$	$6d$	$7p$	$8s$
$4f$	$5d$	$6p$	$7s$
	$4d$	$5p$	$6s$
	$3d$	$4p$	$5s$
		$3p$	$4s$
		$2p$	$3s$
			$2s$
			$1s$

- As will be established, the order of energies in this table is implicit in the structure of the periodic table. It need not be memorized.

Starting at the energetic bottom and working up embodies what is called the **auf-bau principle** (*Aufbau* is German for "building-up"). In the following, we use the aufbau principle, the Pauli principle, and some additional rules to build up the ground-state electron configurations of the different elements.

Building Up from Helium to Argon

Helium ($Z = 2$) has two electrons. The lowest-energy orbital, the $1s$ orbital, accommodates both of these electrons (with opposite spins) when they are fed in during the building-up process. The resulting electron configuration is shown in an orbital diagram such as in Figure 17–1, or by the notation "$1s^2$," where the superscript 2 means that two electrons occupy the $1s$ orbital. The $1s$ orbital in the helium atom is significantly larger than the $1s$ orbital in the helium ion, He^+. In the ion, the single electron feels the full nuclear charge, which is $+2e$, but in the atom, each electron partially shields the other electron from the nuclear charge. The $1s$ orbital in the helium atom can be described approximately in terms of an "effective" nuclear charge Z_{eff} of 1.7, which lies between $+1$ (the value for complete shielding by the other electron) and $+2$ (no shielding).

- The superscript in "$1s^2$" is not an exponent. The notation is read "one s two," not "one s squared" and not "one s to the second power."

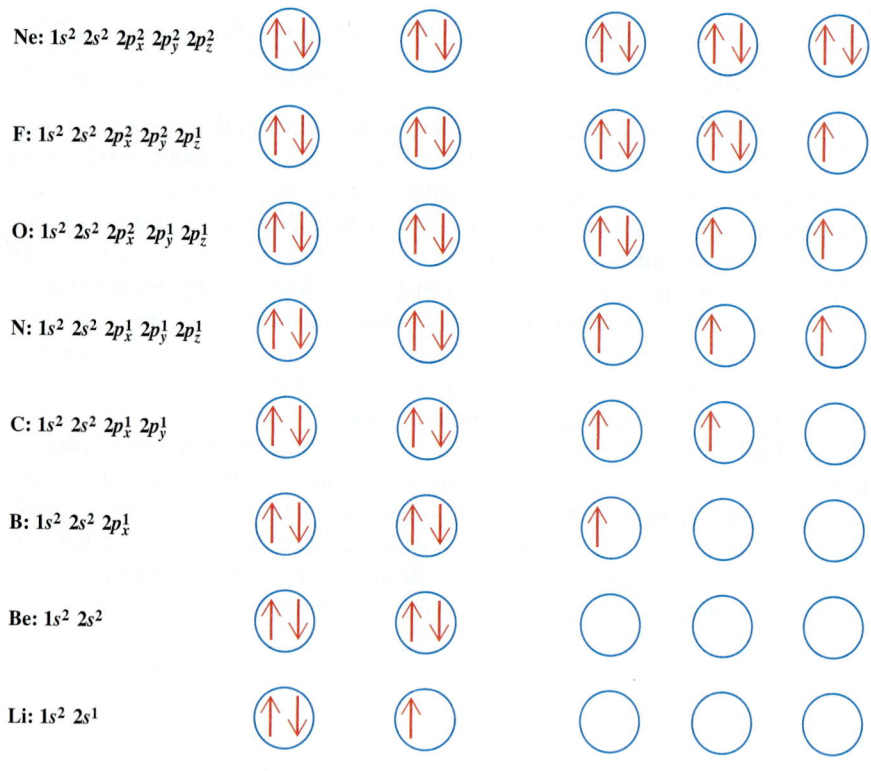

Figure 17–1 • The ground-state electron configurations of first- and second-period atoms. The red arrows represent electrons with spin quantum numbers $m_s = +\frac{1}{2}$ or $m_s = -\frac{1}{2}$; each blue circle represents an orbital. The $2p$ orbitals have the same energy. The force of their subscripts is merely to designate a *different* orbital, not a *specific* orbital.

Lithium has three electrons. The third electron cannot join the first two in the $1s$ orbital, because that would violate the Pauli principle. It goes into the $2s$ orbital, which is the next orbital up in order of energy. Why is the $2s$ orbital at lower energy than a $2p$? A $2s$ orbital puts some electron probability right at the nucleus, as discussed in Section 16–5, but a $2p$ orbital does not. In other words, a $2s$ orbital penetrates toward the nucleus better than does a $2p$ orbital. The $2s$ orbital experiences (part of the time) an almost unshielded nuclear attraction. This lowers its energy relative to a $2p$ orbital, which experiences the smaller attraction of a more highly shielded nucleus. The ground state of the lithium atom is therefore $1s^2 2s^1$. The configuration $1s^2 2p^1$ is not impossible but is an excited state of Li.

The next two elements present no difficulties. Beryllium has the ground-state configuration $1s^2 2s^2$, with the $2s$ orbital now filled. In boron, with five electrons, the latest addition occupies a $2p$ orbital; boron has the ground-state configuration $1s^2 2s^2 2p^1$. Because the three $2p$ orbitals of boron (designated the $2p_x$, $2p_y$, and $2p_z$) are part of the same subshell (have the same energy), each has an equal chance for occupancy. It is not necessary (or possible) to specify which of these $2p$ orbitals is occupied when writing the electron configuration.

With carbon, the sixth element in the periodic table, comes a new question. Does the sixth electron go into the same $2p$ orbital as the fifth (for example, both in a $2p_x$ orbital) with opposite spin, or does it go into a fresh orbital of equal energy with the same spin? The answer is found in the general observation that two electrons having the same spin and occupying different orbitals in a subshell repel each other less strongly than two having opposite spins and occupying the same orbital. This allows the separately occupied orbitals to be somewhat smaller, increasing the attractive interaction of the electrons with the nucleus, and so leading to lower energy. The situation is summarized in **Hund's rule:**

> When electrons occupy orbitals of the same energy, the lowest energy state corresponds to the configuration with the greatest number of unpaired electrons.

Hund's rule means that the ground-state configuration of carbon is $1s^2 2s^2 2p_x^1 2p_y^1$ as shown in Figure 17–1, or $1s^2 2s^2 2p_x^1 2p_z^1$, or $1s^2 2s^2 2p_y^1 2p_z^1$, all of which are equivalent, and is *not* $1s^2 2s^2 2p_x^2$, $1s^2 2s^2 2p_y^2$, or $1s^2 2s^2 2p_z^2$, which are equivalent excited-state configurations. Usually the ground-state configuration is written $1s^2 2s^2 2p^2$, with the understanding that two different $2p$ orbitals are intended.

The presence of unpaired electrons in atoms of carbon has implications for the physical behavior of the element. A substance is **paramagnetic** if it is attracted into a magnetic field. All substances that have one or more unpaired electrons in the atoms, molecules, or ions that compose them are paramagnetic. Thus, the ground-state carbon atom is paramagnetic. Substances in which all the electrons are paired are **diamagnetic.** They are pushed *out* of a magnetic field, although the force they experience is orders of magnitude smaller than the force that pulls a typical paramagnetic substance *into* a magnetic field. Of the atoms examined so far, H, Li, B, and C, are paramagnetic, but He and Be have all of their electron spins paired and so are diamagnetic.

The ground-state electron configurations of nitrogen, oxygen, fluorine, and neon come from the step-wise filling of the $2p$ subshell, consistent with the Pauli principle and Hund's rule. They are given in Figure 17–1. The six elements from boron to neon are called *p*-**block** elements because building up their configurations involves the filling of a p subshell. The two elements that precede them (lithium and

• Effects of magnetic fields on substances are particularly important among compounds of the transition elements and are discussed more extensively in Section 19–3.

beryllium) are called **s-block** elements, as are hydrogen and helium, for a similar reason.

The building-up of configurations for the elements in the third row of the periodic table, from sodium to argon, echoes the building-up in the second. The single orbital in the $3s$ subshell is filled, and then the three orbitals in the $3p$ subshell. Writing full-length electron configurations with more than 10 or 12 electrons becomes tiresome. A common shortcut is to list explicitly only electrons that are added in building up beyond the last preceding noble-gas element, indicating the others by writing the chemical symbol of the noble gas in brackets. The ground-state electron configuration of silicon is written $[\text{Ne}]3s^23p^2$ under this system.

EXAMPLE 17–1

Write the ground-state electron configurations of magnesium and sulfur. Are the gaseous atoms of these elements paramagnetic or diamagnetic?

Solution

The noble-gas element preceding magnesium and sulfur is element 10, neon. Magnesium has two electrons beyond neon's 10. These electrons go into a $3s$ orbital, the next highest in energy after the $2p$ orbitals, to give the ground-state configuration $[\text{Ne}]3s^2$. A gaseous atom of magnesium is diamagnetic because all of its electrons are paired.

Sulfur has six electrons beyond the neon core; the first two are in the $3s$ orbital, the next four in the $3p$ orbitals. Its ground-state configuration is $[\text{Ne}]3s^23p^4$.

When four electrons occupy three p orbitals, two electrons must pair in one of the orbitals. By Hund's rule, the lowest energy configuration for sulfur has the remaining two electrons in different orbitals with parallel spins. Because it has unpaired electrons, the gaseous sulfur atom is paramagnetic.

EXERCISE

The ground-state electron configuration of an atom is $1s^22s^22p^63s^23p^3$. Name the atom. Is it paramagnetic? If so, state how many unpaired electrons it has.

Answer: The atom is phosphorus. It is paramagnetic with three unpaired electrons.

Transition Elements and Beyond

For neutral atoms at the beginning of the fourth period, it is an experimental fact that the $4s$ orbital has a lower energy than the $3d$ orbitals, although the two energies are very close. This is explained by noting again that s orbitals penetrate to the nucleus better than any others. Here, a $4s$ orbital experiences an effective nuclear charge large enough to lower its energy below that of a $3d$ orbital. Filling the $4s$ orbital next gives $[\text{Ar}]4s^1$ and $[\text{Ar}]4s^2$ as the ground-state configurations of potassium ($Z = 19$) and calcium ($Z = 20$), respectively. After the $4s$ orbital is filled, building-up starts into the $3d$ orbitals. The ground-state configuration of scandium ($Z = 21$) is $[\text{Ar}]3d^14s^2$. Scandium is the first transition element. Interestingly, its $3d$ orbitals are actually at lower energy than its $4s$ orbital because the energy of the $3d$ orbitals declines rather faster than the energy of the $4s$ as the nuclear charge increases from 19 to 21. As a consequence, a scandium atom loses its $4s$ electrons more easily than its $3d$ electron. The ground state of the Sc^+ ion is $[\text{Ar}]3d^14s^1$ (*not* $[\text{Ar}]4s^2$), and the ground state of Sc^{2+} ion is $[Ar]3d^1$ (*not* $[Ar]4s^1$). The drop in the energy of the $3d$ subshell relative to the $4s$ continues across the transition series from scandium to zinc: $4s$ electrons are more easily removed than $3d$ electrons throughout. This energy drop has important chemical consequences. The elements copper and zinc rarely have oxidation states higher than $+2$, and in the elements beyond zinc, the $3d$ electrons are held so tightly that they play no chemical role. The ten elements from scandium to zinc are **d-block** elements because a d subshell is filled during the building-up of their configurations.

Spectroscopic evidence shows that two transition elements in the fourth period have ground-state electron configurations that differ from expectation based on the smooth filling of the $3d$ subshell. From its position between vanadium $[Ar]3d^34s^2$ and manganese $[Ar]3d^54s^2$, chromium might be expected to have the ground-state configuration $[Ar]3d^44s^2$. In fact, ground-state chromium has the configuration $[Ar]3d^54s^1$, and $[Ar]3d^44s^2$ is an excited state. Similarly, ground-state copper is $[Ar]3d^{10}4s^1$ rather than $[Ar]3d^94s^2$. These two "surprises" result from electron–electron interactions. Recall that the orbital approximation neglects these interactions. In chromium and in copper, the interaction energy becomes comparable to the energy difference between the $4s$ and $3d$ subshells, making expectations based on the approximation invalid.

Building-up continues in the fourth period of the periodic table with six more p-block elements (gallium through krypton). Then the fifth period echoes the build-up of the fourth period in its entirety. The effect of electron–electron interactions is observed repeatedly in the d-block elements of the fifth period. Six of the ten have ground-state configurations that are unexpected under the orbital approximation.

Electron configurations in the sixth period start with the filling of the $6s$ subshell and proceed through the $4f$ subshell (giving 14 **f-block** elements), the $5d$ subshell, and finally the $6p$, ending with the radioactive noble gas radon ($Z = 86$). The building-up of electron configurations for elements in the seventh period starts with the $7s$ orbital and continues with the $5f$ and $6d$ orbitals, following the same pattern as in the sixth period. The pattern of build-up suggests that the seventh period should conclude with the filling of the $7p$ orbitals as element 118 is reached. Synthesis of some of the elements with Z equal to 110 and above has been reported, but the ground-state electron configurations remain unknown.

Figure 17–2 summarizes the building-up of ground-state electron configurations. The configurations of most elements can be written by locating them in Figure 17–2, or indeed in any periodic table, and tracing the filling of the subshells to come up to that location. The elements for which this procedure fails appear in Figure 17–2 with the final portion of their experimentally determined ground-state configuration just under their symbol.

Shells and the Periodic Table

A shell is a set of orbitals that have similar energies. Shell structure in atoms is real, as confirmed by both experimental and computational studies of the distribution of electron probability density. The electron density in argon has three distinct shells, as Figure 17–3 shows, although the edges of the shells overlap. In argon, electrons occupy the $n = 1$ shell (consisting entirely of the $1s$ subshell), the $n = 2$ shell (the $2s$ and $2p$ subshells), and the $n = 3$ shell (the $3s$ and $3p$ subshells). During the building-up of electron configurations, a new shell is opened whenever an electron first goes into an ns orbital: the $n = 4$ shell is opened with the first filling of the $4s$ orbital, the $n = 5$ shell, with the first filling of the $5s$ orbital, and so forth.

Number of Electrons

Subshells being filled

	1	2	3	4	5	6	7	8	9	10	11	12	13	14
7p	113	114	115	116	117	118 (End of period 7)								
6d	103 Lr	104 Rf	105 Db	106 Sg	107 Bh	108 Hs	109 Mt	110	111	112				
5f	89 Ac $6d^17s^2$	90 Th $6d^27s^2$	91 Pa $5f^26d^17s^2$	92 U $5f^36d^17s^2$	93 Np $5f^46d^17s^2$	94 Pu	95 Am	96 Cm $5f^76d^17s^2$	97 Bk	98 Cf	99 Es	100 Fm	101 Md	102 No
7s	87 Fr	88 Ra												
6p	81 Tl	82 Pb	83 Bi	84 Po	85 At	86 Rn (End of period 6)								
5d	71 Lu	72 Hf	73 Ta	74 W	75 Re	76 Os	77 Ir	78 Pt $5d^96s^1$	79 Au $5d^{10}6s^1$	80 Hg				
4f	57 La $5d^16s^2$	58 Ce $4f^15d^16s^2$	59 Pr	60 Nd	61 Pm	62 Sm	63 Eu	64 Gd $4f^75d^16s^2$	65 Tb	66 Dy	67 Ho	68 Er	69 Tm	70 Yb
6s	55 Cs	56 Ba												
5p	49 In	50 Sn	51 Sb	52 Te	53 I	54 Xe (End of period 5)								
4d	39 Y	40 Zr	41 Nb $4d^45s^1$	42 Mo $4d^55s^1$	43 Tc	44 Ru $4d^75s^1$	45 Rh $4d^85s^1$	46 Pd $4d^{10}$	47 Ag $4d^{10}5s^1$	48 Cd				
5s	37 Rb	38 Sr												
4p	31 Ga	32 Ge	33 As	34 Se	35 Br	36 Kr (End of period 4)								
3d	21 Sc	22 Ti	23 V	24 Cr $3d^54s^1$	25 Mn	26 Fe	27 Co	28 Ni	29 Cu $3d^{10}4s^1$	30 Zn				
4s	19 K	20 Ca												
3p	13 Al	14 Si	15 P	16 S	17 Cl	18 Ar (End of period 3)								
3s	11 Na	12 Mg												
2p	5 B	6 C	7 N	8 O	9 F	10 Ne (End of period 2)								
2s	3 Li	4 Be												
1s	1 H	2 He (End of period 1)												

Figure 17–2 • In this "building-up" version of the periodic table, the lightest elements are at the bottom. Electrons fill subshells from bottom to top in order of energy as the atomic number of the atom increases. The numbers across the top give the number of electrons in each subshell. The ground-state electron configurations of most elements are apparent from their positions in the table. Those that are known to differ from expectation are indicated explicitly.

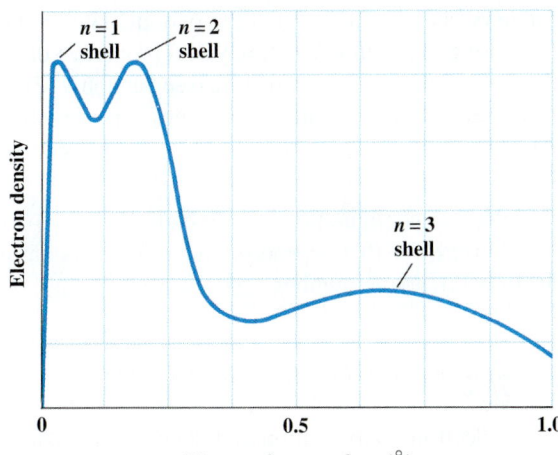

Figure 17-3 • The shell structure of the distribution of electrons is clearly visible in the three peaks in the radial electron density of argon.

Shells become filled (become **closed shells**) when an electron configuration belonging to one of the noble-gas elements (He, Ne, Ar, Kr, Xe, Rn) is attained. The filling of a shell in the building-up of electron configurations corresponds to the ending of a period (a row) in the periodic table. In closed-shell atoms and ions, the next available unoccupied orbital is separated from occupied orbitals by a significant gap in energy. Electrons in closed shells rarely become directly involved in chemical bonding. From the point of view of neighboring atoms with which they might interact chemically, they are "buried" close to the nucleus. The noble-gas elements have electrons in closed shells only, which is consistent with their poor chemical reactivity.

Valence electrons are electrons that can become directly involved in chemical bonding. They occupy the outermost (highest energy) shell of an atom, beyond the immediately preceding noble-gas configuration. Among the s-block and p-block elements (the representative elements), valence electrons include electrons in s and p subshells only. This means that the representative elements have a maximum of eight valence electrons. The octet rule that figures so heavily in writing Lewis electron-dot structures of second- and third-period atoms and ions arises from the eight available sites for valence electrons in one s and three p orbitals. Among the d-block and f-block elements (the transition elements), valence electrons usually consist of electrons in s orbitals plus electrons in unfilled d and f subshells (sometimes electrons in filled d and f subshells also serve as valence electrons). The special properties of the transition elements can be ascribed to the partial filling of valence d orbitals or f orbitals, a feature to which we return in Chapter 19. The term "valence" is used whenever the emphasis is on bonding. Thus, valence electrons occupy valence orbitals in valence subshells, which are in valence shells.

An atom's chemical behavior depends strongly on how many valence electrons it has and on their configuration. Elements in the same group (column) of the periodic table have the same number of valence electrons in related valence-electron configurations. They consequently resemble each other chemically. Sodium (configuration [Ne]$3s^1$) and potassium (configuration [Ar]$4s^1$), for example, both have a single valence electron in an s orbital outside a closed shell and are chemically similar. The metals titanium (Ti, configuration [Ar]$3d^24s^2$), zirconium (Zr, configuration [Kr]$4d^25s^2$), and hafnium (Hf, configuration [Xe]$5d^26s^2$) likewise have similar properties.

• Some of these terms were defined in Section 3–3.

Nineteenth-century chemists discovered the periodic law on the basis of similarities and trends among the chemical and physical properties of the elements before anything was known about electron structure (see Section 3–2). Now it develops that the periodic law derives from quantum mechanics, a remarkable result. The periodic law, first stated in Chapter 3, can be restated to emphasize this:

> The ground-state electron configurations of the elements vary periodically with atomic number; all properties that depend on these electron configurations tend to vary periodically with atomic number.

EXAMPLE 17–2

Write the valence-electron configuration and state the number of valence electrons in each of the following atoms and ions: (a) Os, (b) Pt, (c) Cr^{2+}.

Strategy

Refer to any copy of the periodic table to obtain the order of filling of orbitals during building-up. Use this order to write the ground-state electron configuration. Check for exceptions using Figure 17–2. Decide which electrons are core electrons and which are valence electrons by using the definitions given in the preceding.

Solution

(a) Osmium ($Z = 76$) has 76 electrons. The first 54 can be represented [Xe]. The remaining 22 occupy the $4f$, $5d$, and $6s$ subshells. By the time Os is reached in the periodic table, the $4f$ and $6s$ subshells have been filled and the $5d$ subshell is being filled. Its electron configuration is therefore $[Xe]4f^{14}5d^{6}6s^{2}$. The 14 f electrons are not valence electrons because the $4f$ subshell is filled, but the d and s electrons are valence electrons. The valence-electron configuration is therefore $5d^{6}6s^{2}$, and Os has **eight** valence electrons.

(b) Platinum ($Z = 78$) has 78 electrons. Its ground-state electron configuration is $[Xe]4f^{14}5d^{9}6s^{1}$, according to Figure 17–2. The electrons in the filled $4f$ subshell are not counted as valence electrons, so its valence-electron configuration is $5d^{9}6s^{1}$, and Pt has **ten** valence electrons.

(c) Chromium ($Z = 24$) has 24 electrons in the ground-state configuration $[Ar]3d^{5}4s^{1}$ (see Fig. 17–2). The Cr^{2+} ion has 22 electrons. Its ground-state configuration derives from that of Cr by the loss of the least tightly held electrons, which are the $4s$ electron and one $3d$ electron. Its valence-electron configuration is $[Ar]3d^{4}$, for a total of **four** valence electrons.

EXERCISE

Write the valence-electron configuration and state the number of valence electrons in each of the following atoms and ions: (a) Nd, (b) Lu, (c) La^{3+}.

Answer: (a) Nd (neodymium) has the valence-electron configuration $4f^{4}6s^{2}$ for six valence electrons. (b) Lu (lutetium) has the valence-electron configuration $5d^{1}6s^{2}$ for three valence electrons beyond a filled $4f$ subshell. (c) La^{3+} (lanthanum(III) ion) has the closed-shell configuration [Xe] for zero valence electrons.

17-2 Experimental Measures of Orbital Energies

An important technique called **photoelectron spectroscopy** gives accurate values for the energies of different orbitals in atoms across the periodic table. The data eliminate any possible questions about the order of orbitals to use in building up electron configurations, provide overwhelming evidence for the existence of shells and subshells in many-electron atoms, and furnish an experimental basis for the distinction between core and valence electrons. Photoelectron spectroscopy is simply the photoelectric effect discussed in Section 16–2 applied not to metals but to free atoms or molecules. A gaseous sample is subjected to a beam of high-energy radiation of fixed frequency. The photons in the beam have sufficiently high energy (given by $h\nu$, the product of Planck's constant and the photon's frequency) to eject electrons from atoms that they strike. At impact, part of the photon's energy goes to overcome the binding attraction between an electron and its atom. The rest appears as kinetic energy of the ejected electron. In mathematical terms, if the binding energy of an electron equals Φ, and its kinetic energy when it is ejected equals $\frac{1}{2}m_e v^2$, then

$$h\nu = \Phi + \tfrac{1}{2}m_e v^2$$

In photoelectron spectroscopy, the ejected electrons are detected, and their kinetic energies are measured. Because $h\nu$ is known, the binding energies of the different electrons in the atom are computed easily:

$$\Phi = h\nu - \tfrac{1}{2}m_e v^2$$

Figure 17–4 is a photoelectron spectrum of neon. The three major peaks in the spectrum correspond to ejection of an electron from the $2p$ subshell, the $2s$ subshell, and the $1s$ subshell, which are the three occupied subshells in a ground-state atom of neon. Figure 17–5, the product of hundreds of photoelectron spectroscopy experiments, graphs orbital energies in nearly all of the elements. The data abundantly confirm the existence of subshells. They show that subshells group themselves into shells of similar energies, but with some significant exceptions. The $3d$ subshell for elements 21 through 29 (scandium through copper) lies substantially above the $3s$ and $3p$ and only slightly below the $4s$. This is why the $3d$ orbitals are valence orbitals

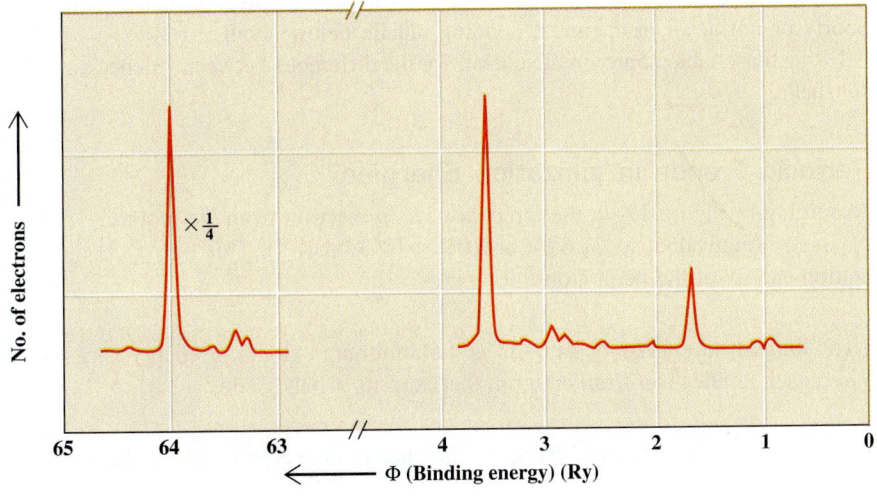

No. of electrons

$\times\frac{1}{4}$

65   64   63   4   3   2   1   0

$\longleftarrow$ Φ **(Binding energy) (Ry)**

Figure 17–4 • The photoelectron spectrum of neon shows three major peaks, confirming that the electrons in neon are organized in three subshells of different energies. It requires 1.59 Ry to eject a $2p$ electron from neon, 3.56 Ry to eject a $2s$ electron and 64.0 Ry to eject a $1s$ electron. (The height of the $1s$ peak has been reduced to fit in the picture.) The customary energy unit in photoelectron spectroscopy is the electron volt (1 Ry = 13.59 eV). Displaying increasing binding energy from right to left is customary in these spectra.

Figure 17–5 • The energies of different subshells in the first 97 elements, as determined by photoelectron spectroscopy. Negative values on the vertical axis correspond to bound states. For example, ejection of the $1s$ electron from hydrogen ($Z = 1$) requires 1 Ry of energy. Subshells having the same principal quantum number, such as the $2s$ and $2p$, have similar energies and are separated from orbitals of different n. Significant exceptions do exist, however.

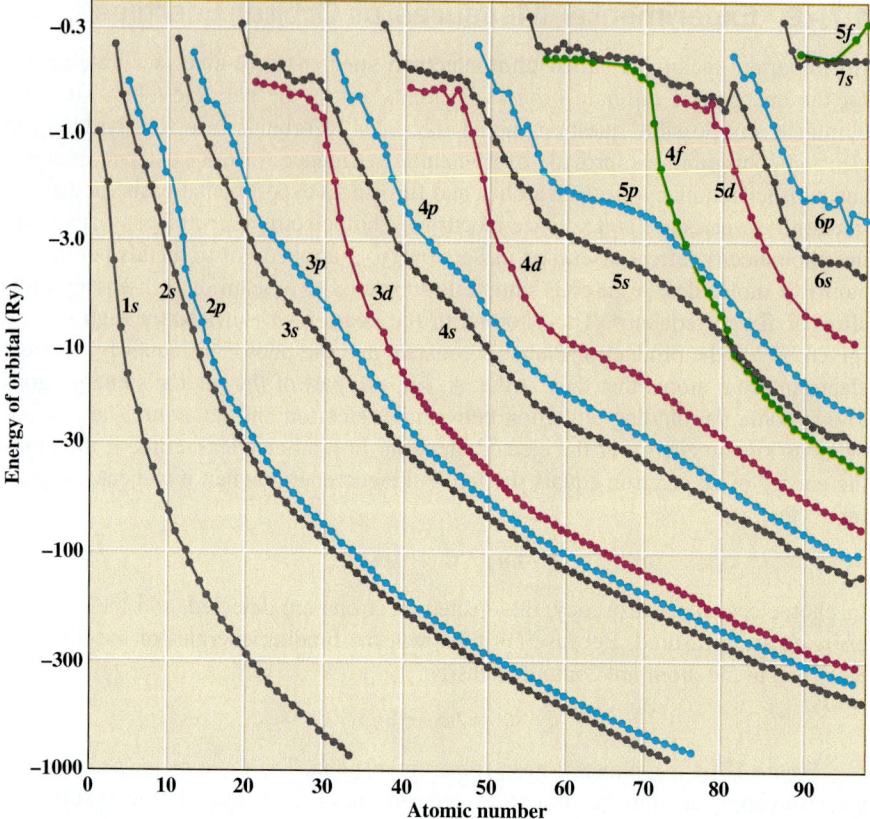

in these transition elements. The energy of the $3d$ subshell then rapidly diminishes as Z reaches 30, taking the $3d$ electrons out of the picture as valence electrons for elements 30 (zinc) and higher. Similar observations can be made about the $4d$ and $5d$ subshells and the $4f$ and $5f$ subshells, explaining why electrons in filled d and f subshells are not counted as valence electrons. The energies of the highest-energy subshell in the noble gases (elements 2, 10, 18, 36, 54, and 86), which participate poorly or not at all in chemical bonding, all lie below about -1 Ry. This makes -1 Ry a reasonable approximate cut-off for the difference between valence and core subshells.

Periodic Trends in Ionization Energies

According to Figure 17–4, the removal of a $2p$ electron from Ne requires 1.59 Ry of energy (equivalent to 21.6 eV or 2.08×10^3 kJ mol^{-1}). This energy is the ionization energy of the neon atom:

> The **ionization energy** of an atom is the minimum amount of energy necessary to detach an electron from an atom that is in its ground state.

In other terms, an ionization energy is the change in energy (ΔE) for the process

$$X(g) \longrightarrow X^+(g) + e^- \qquad \Delta E = IE_1$$

where X stands for any atom. Ionization energies are always positive because ground-state atoms are stable systems that require an input of energy to detach an electron. The abbreviation "IE" in the equation is subscripted with a 1 because this is the first ionization energy of the atom. The second ionization energy (IE_2) is defined as the minimum energy to remove a second electron, or ΔE, for the process

$$X^+(g) \longrightarrow X^{2+}(g) + e^- \qquad \Delta E = IE_2$$

The second ionization energy of an atom always exceeds the first because the second electron must be extracted against the attraction of a larger positive charge. The third, fourth, and higher ionization energies are defined similarly to the second and are larger still. A series of ionization energies differs from a series of photoelectron binding energies. The photoelectron spectroscopy experiment measures the energy required to punch electrons out of all of the different subshells in a neutral atom. Ionization energies (except for IE_1) are for the removal of electrons from ions, not neutral atoms, and the electron always comes from the highest-energy occupied orbital.

Ionization energies strongly influence how atoms interact with other atoms to form chemical bonds. Figure 17–6 gives the first and second ionization energies of the first 18 elements. Clearly, these properties are periodic functions of atomic number. The periodic trend in the first ionization energy (Figs. 17–6 and 17–7) gives valuable insights into the stability of various electron configurations. For example, IE_1 drops sharply from He to Li. It does so for two reasons: first, a $2s$ electron is on the average much farther from the nucleus than a $1s$ electron; second, the $1s$ electrons screen the nucleus in Li so effectively that the $2s$ electron "sees" a net positive charge close to $+1$ rather than the larger charge seen by the electrons in helium. Beryllium has a larger IE_1 than does lithium because its nuclear charge is $+4$ rather than $+3$ and the electron still is being removed from a $2s$ orbital. The IE_1 of boron is somewhat less than that of beryllium because the fifth electron resides in a higher-

• There are as many ionization energies as there are electrons. The element krypton ($Z = 36$) has IE_1 through IE_{36}, all of which have been measured.

• The ability of the positively charged nucleus to bind electrons is reduced by the repulsions among the electrons themselves. This is called screening.

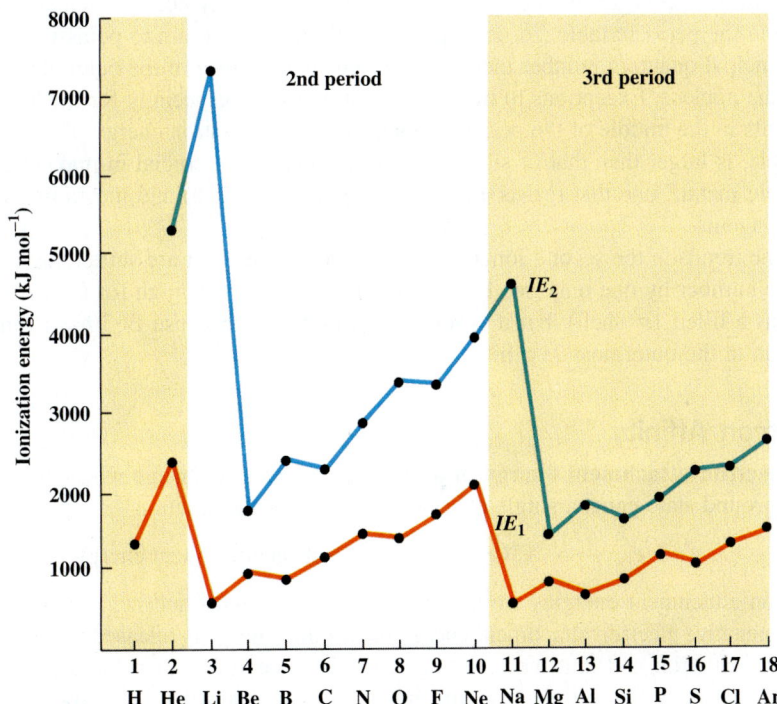

Figure 17-6 • First and second ionization energies of atoms of the first three periods.

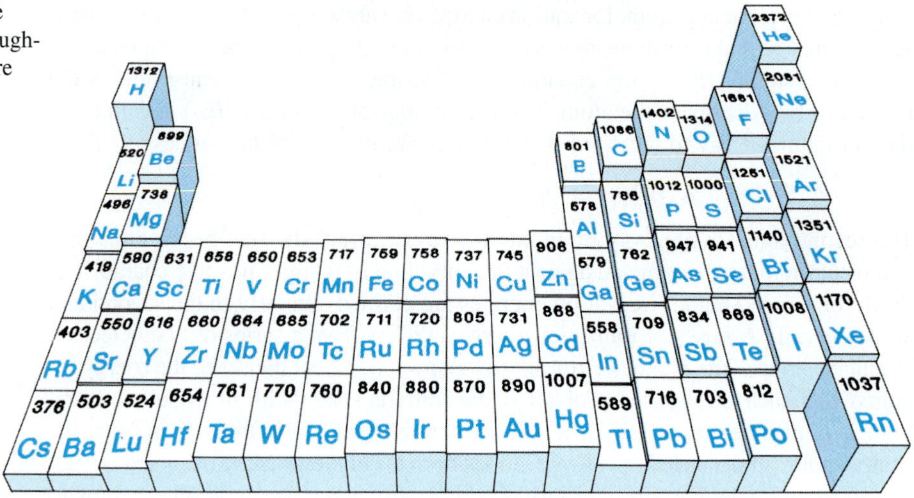

Figure 17–7 • The variation of the first ionization energies of atoms throughout the periodic table. The energies are given in kJ mol^{-1}.

Figure 17–8 • Copper (the penny) is oxidized easily, and silver (the spoon) tarnishes, but gold (the cufflink) resists oxidation. *(Leon Lewandowski)*

energy (and, therefore, less tightly bound) $2p$ orbital. In carbon and nitrogen, the additional electrons go into $2p$ orbitals as the nuclear charge increases to hold the outer electrons more tightly. Hence, IE_1 increases. This trend is reversed from nitrogen to oxygen. The rise in nuclear charge from N to O (which would by itself give O a higher ionization energy) is counteracted by another effect. The ground-state oxygen atom must accommodate two electrons in a single $2p$ orbital, leading to expansion of that $2p$ orbital and weaker electron–nucleus attractions. Fluorine and neon then show successively higher first ionization energies as the nuclear charge increases. Both the general periodic trends in IE_1 and the exceptions are explained in gratifying detail through the orbital description of many-electron atoms.

Ionization energies tend to decrease for the successively heavier elements in a group in the periodic table (for example, from lithium to sodium to potassium). As the principal quantum number increases, so does the distance of the outer electrons from the nucleus. Exceptions to this trend exist, however, especially for the heavier elements in the middle of the periodic table. The first ionization energy of gold, for example, is larger than that of silver or copper. This fact is crucial in making gold a "noble metal," one that resists attack by oxygen (Fig. 17–8) and most other oxidizing agents.

The trends in the second ionization energy are similar but are shifted higher in atomic number by one unit (see Fig. 17–6). Thus, IE_2 is very high for Li (because Li$^+$ has a filled $1s^2$ shell), but it is relatively low for Be because Be$^+$ has a single electron in the outermost $2s$ orbital.

Electron Affinity

The **electron attachment energy** of an atom is the energy change when the atom in its ground state gains a single electron. It is ΔE for the reaction

• This process is *not* the same as either the process that defines IE_1 or the reverse of that process.

$$X(g) + e^- \longrightarrow X^-(g) \qquad \Delta E = \text{electron attachment energy}$$

Electron attachment energies are positive for some elements and negative for others. A negative electron attachment energy means that energy is released as the atom gains an electron and that the resulting anion is more stable than the atom from which it came. A positive electron attachment energy means that the extra electron

CHEMISTRY IN YOUR LIFE

Fireworks and Neon Signs

The shower of colors in a fireworks display (Fig. 17–A) and the cold glare of a neon sign have a common origin: the emission of light from excited states of atoms and ions. In fireworks, strongly exothermic reactions such as the oxidation of magnesium

$$Mg(s) + \tfrac{1}{2}O_2(g) \longrightarrow MgO(s) \; \Delta H° = -602 \text{ kJ mol}^{-1}$$

not only provide brilliant flashes of light (see Fig. 4–13) but also provide enough energy to raise metal atoms from their ground states to excited states. These excited atoms then emit light as they return to their ground states. The same process occurs when metal atoms are placed in a flame (see Fig. 9–16). Sodium is excited from its ground state (configuration [Ne]$3s^1$) to a state with configuration [Ne]$3p^1$, in which the valence electron is in a higher-energy $3p$ orbital. The transition back to the ground state is accompanied by the emission of yellow light at a wavelength of 589 nm. Red colors in fireworks are produced by excited strontium atoms, violet by potassium, and green by barium, just as in the corresponding flame tests. Blue colors are more difficult to achieve.

Gas discharges are used in many types of lights. The neon sign is the direct descendant of the gas-discharge tubes that figured in the discovery of the electron (see Section 1–5). When an electric current is passed through a tube that has been evacuated to a low gas pressure, ionization occurs, and the atoms or molecules of the gas are excited to higher-energy states. The light emitted by neon is red, and the light emitted by argon is violet (see Fig. 16–10). Other colors are achieved by using colored glass because other gases are either too expensive or react too readily with the electrodes. In a sodium-vapor light, a small amount of neon is used to start the electrical discharge because sodium, the main emitter, is solid at room

Figure 17–A • Fireworks. *(Joseph Nettis/Photo Researchers, Inc.)*

temperature. As the temperature rises, the sodium is vaporized and excited to give the characteristic yellow light of the [Ne]$3p \longrightarrow$ [Ne]$3s$ transition as its atoms return to the ground state. Similar principles apply to the bluish white light produced by mercury-vapor lamps. Vapor-discharge lamps are more efficient sources of illumination than incandescent filaments (like those in common lightbulbs) because more of their output lies in the visible region of the spectrum. Their light is further from true white, however.

Excited states of atoms also figure in the production of light in lasers. In a helium–neon laser, light is emitted as neon atoms relax from one excited state ($1s^22s^22p^55s^1$) to a second, lower state ($1s^22s^22p^53p^1$). In a properly engineered laser, many excited atoms are made to emit light at the same moment, giving an intense coherent beam of light—a laser beam.

---

spontaneously flies away from the negative ion if it is in free space. Seventy-five percent of the elements have negative electron attachment energies. The gaseous atoms of these elements "prefer" to gain an electron to become anions. They have an affinity for an additional electron.

> The **electron affinity (*EA*)** of an atom equals the negative of the energy change of the system when a neutral atom in its ground state gains an electron:
>
> $$\text{electron affinity} = -\Delta E \text{ (electron attachment)}$$

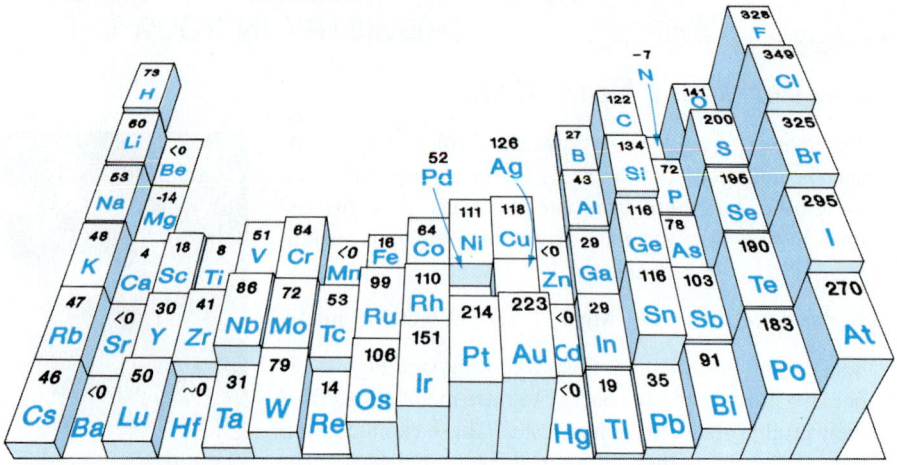

Figure 17-9 • The electron affinities (in kJ mol^{-1}) of gaseous atoms of the elements. Some of the elements have negative electron affinities; these elements include the noble gases, which are not visible at the extreme right.

The periodic trends in the electron affinity (Fig. 17–9) parallel those in the ionization energy for the most part, except that they are shifted one unit *lower* in atomic number. The reason is clear. Attaching an electron to F (fluorine) gives F$^-$ (fluoride ion) with the electron configuration $1s^2 2s^2 2p^6$, which is the same as the configuration of Ne. Fluorine has a large affinity for electrons because the resulting closed-shell configuration is very stable. This stability is, of course, the same reason that Ne has a large ionization energy. The noble gases have low (actually negative) electron affinities for the same reason that the alkali metals have small ionization energies: The last electron resides in a new shell far from the nucleus and is almost totally screened from the nuclear charge.

No atom has a positive affinity for a *second* electron. A gaseous ion with a net charge of -2 is always unstable with respect to the loss of an electron. Attaching a second electron means bringing it into a species that is already negatively charged. The two repel each other, and the energy rises. Doubly negative ions such as O^{2-}, however, can often be stabilized in crystalline environments by electrostatic interaction with neighboring positive ions, as in CaO(s).

EXAMPLE 17–3

Consider the elements selenium (Se) and bromine (Br). Which should have the higher first ionization energy and which the higher electron affinity?

Solution

These two elements adjoin one another in the periodic table. Bromine has one more electron in the $4p$ subshell, and this electron should be more tightly bound than the $4p$ electrons in Se because of the extra unit of positive charge on the nucleus. Thus, *IE*$_1$ should be greater for Br.

The additional electron goes into the $4p$ subshell when Se and Br gain an electron, but the nuclear charge is higher for Br, stabilizing the Br$^-$ negative ion more. Also the Br$^-$ ion has a closed-shell electron configuration and Se$^-$ does not. Hence bromine has a larger electron affinity than selenium.

17–3 Sizes of Atoms and Ions

The sizes of atoms and ions exert considerable influence on their chemistry. Sizes can be estimated in several ways. One is to measure or compute the electron density of the atom or ion, obtain a boundary sphere that encloses some percentage (such as 90% or 99%) of the electron density, and let the radius of the boundary sphere stand for the radius of the atom. This approach is illustrated in Section 16–5 using a 90% boundary sphere for an atom of hydrogen. Another method is to calculate the radial distance at which the electron density of the largest shell of the atom reaches a maximum and define *that* distance as the radius of the atom. This method gives a radius of approximately 0.65 Å for an atom of argon when applied to Figure 17–3, which is a diagram of the shell structure of argon.

Experimental data on size come from X-ray diffraction experiments that probe the packing of atoms and ions in crystalline solids. The X-ray method accurately measures *internuclear distances,* which are distances from the center of one atom to the center of another in a crystal (see Section 20–1). The **metallic radius** of an element equals one-half of the internuclear distance of the nearest-neighbor atoms in a metallic crystal (see Section 20–3). For non-metallic elements, which do not form metallic crystals, a **covalent radius** is used. It equals one-half of the internuclear distance between atoms of the element bound to each other by a single bond in a molecule. The required internuclear distances also are determined by X-ray diffraction experiments.

The term **atomic radius** refers to metallic radii for metals and covalent radii for non-metals. Atomic radii are always estimates. Exact values cannot be given because atoms of the same element have different sizes in different environments; atoms truly are *not* hard spheres of fixed size. Despite this, the use of atomic radii is well established and has benefits within its limitations. Figure 17–10 gives atomic radii that have been adjusted to account for environmental effects. These *intrinsic* atomic radii put the elements on the same footing for the study of periodic trends in atomic size. Interestingly, they closely approximate atomic radii estimated by calculating the radial distance at which the electron density in the outermost shell of the atom reaches a maximum.

Ionic radii are estimated from internuclear distances obtained by X-ray diffraction experiments on crystalline ionic compounds. This method gives the sum of the radii of adjoining cations and anions in an ionic compound with high precision. It is however hard to say just where an anion ends and an adjoining cation begins. Different choices in dealing with this issue lead to different answers, so caution is required when comparing ionic radii from different sources. Also, ionic radii vary depending on the exact surroundings of the ions in the crystal.

The sizes of ions differ, usually markedly, from the sizes of their parent atoms. Cations are always smaller than their parent atoms because fewer electrons screen

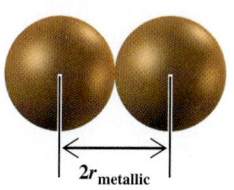

An element's metallic radius is half the internuclear distance between nearest-neighbor atoms of the element in a metallic crystal.

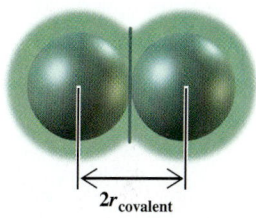

An element's covalent radius is half the internuclear distance between two atoms of the element bound by a single bond in a molecule.

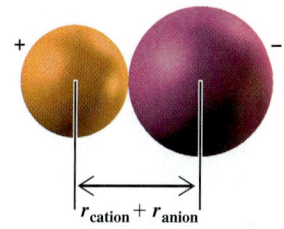

The ionic radii of an adjoining cation and anion add up to the internuclear distance. The exact boundary between the two ions is harder to define than this hard-sphere picture suggests.

Li 1.57	Be 1.12											B 0.88	C 0.77	N 0.74	O 0.66	F 0.64
Na 1.91	Mg 1.60											Al 1.43	Si 1.18	P 1.10	S 1.04	Cl 0.99
K 2.35	Ca 1.97	Sc 1.64	Ti 1.47	V 1.35	Cr 1.29	Mn 1.37	Fe 1.26	Co 1.25	Ni 1.25	Cu 1.28	Zn 1.37	Ga 1.53	Ge 1.22	As 1.21	Se 1.17	Br 1.14
Rb 2.50	Sr 2.15	Y 1.82	Zr 1.60	Nb 1.47	Mo 1.40	Tc 1.35	Ru 1.34	Rh 1.34	Pd 1.37	Ag 1.44	Cd 1.52	In 1.67	Sn 1.58	Sb 1.41	Te 1.37	I 1.33
Cs 2.72	Ba 2.24	Lu 1.72	Hf 1.59	Ta 1.47	W 1.41	Re 1.37	Os 1.35	Ir 1.36	Pt 1.39	Au 1.44	Hg 1.55	Tl 1.71	Pb 1.75	Bi 1.82		

La 1.88	Ce 1.82	Pr 1.82	Nd 1.81	Pm 1.81	Sm 1.80	Eu 2.06	Gd 1.79	Tb 1.76	Dy 1.75	Ho 1.74	Er 1.73	Tm 1.72	Yb 1.94

Figure 17–10 • Intrinsic atomic radii in ångströms ($1 \text{ Å} = 1 \times 10^{-10}$ m) of some elements. Intrinsic means data from crystal structure studies have been adjusted to account for different packing effects in different structures, as recommended by Goldschmidt. (Source: A. F. Wells, *Structural Inorganic Chemistry,* 5th ed., Clarendon Press, Oxford (1984).)

each other less from the central attraction of the nucleus, allowing the system to shrink. Anions are always larger than their parent atoms. As a rough guideline, the radius of an ion with a noble-gas electron configuration is approximately 0.85 Å less than its parent atom if it is a cation and 0.85 Å greater than its parent atom if it is an anion.

Figure 17–10 and Figure 17–11 show that

- Atomic size generally increases moving down a group in the periodic table.
- Among the *s*- and *p*-block elements, atomic size generally decreases moving from left to right across a period.

Size increases going down a group because the added electrons cannot go into the close-in orbitals, which are filled by the core electrons, but must occupy more distant electron shells. Size generally decreases from left to right across a period because the nuclear charge steadily increases while electrons successively join the same shell at about the same distance from the nucleus. Electrons added in the same shell are relatively ineffective in shielding each other from the growing electrostatic attraction of the nucleus, which draws the electrons in closer.

Superimposed on these broad trends are some more subtle effects that can have significant consequences for chemistry. For example, the atomic radii of the elements in periods 4, 5, and 6 go to a relative minimum about half-way across the *d*-block (at or near Fe, Ru, and Os) and then *increase* into the following *p* block. Another such effect is shown dramatically in Figure 17–12, in which several sets of ionic radii are plotted. The radii increase with atomic number in all sets because the added electrons must occupy a more distant shell in each case. This is the same reason that atoms increase in size going down a group in the periodic table. The crucial point here is that the *rate* of this increase changes considerably when ions containing the same number of electrons as atoms of argon are reached. The change in size per step from Li^+ to Na^+ to K^+ is quite large, for example, but the subsequent changes, to Rb^+ and Cs^+, are smaller. The intervening filling of a *d* subshell, which first takes place in the fourth period, causes this. Electrons in a *d* subshell shield the charge on the nucleus less effectively than *s* or *p* electrons. This causes an increased effective nuclear charge that draws in all of the electrons. The radius

- These species are S^{2-}, Cl^-, Ar, K^+, Ca^{2+}, Sc^{3+}, and Ti^{4+}.

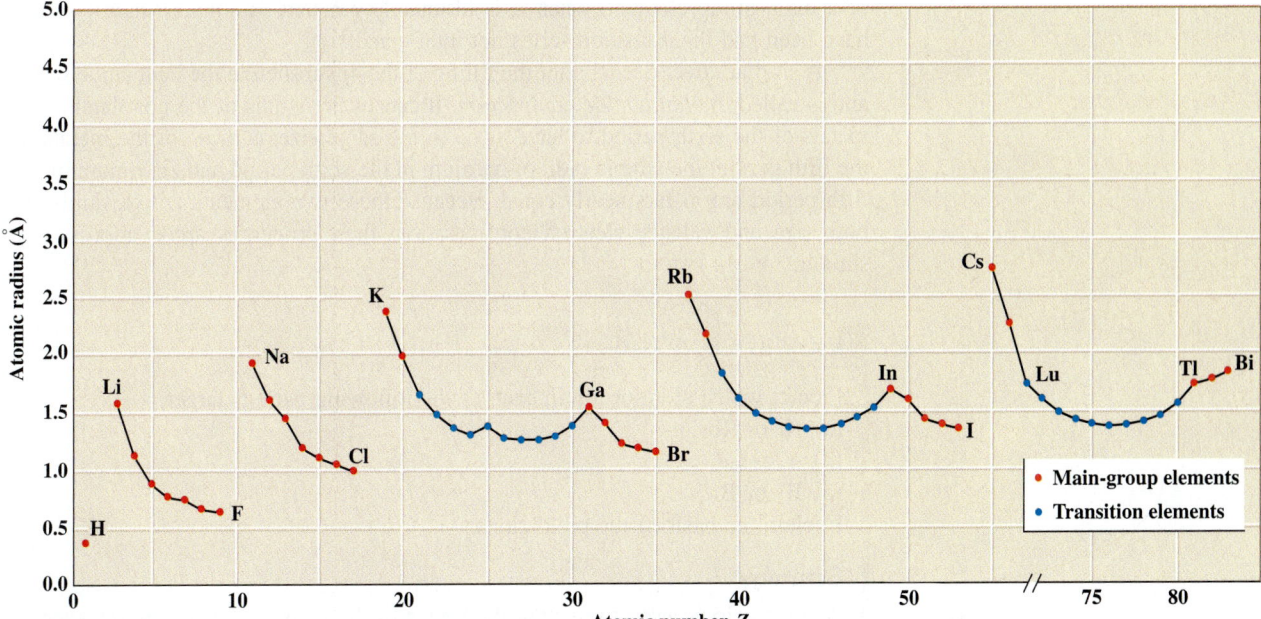

Figure 17-11 • Periodic trends in atomic radius. Each series of connected points corresponds to a row in the periodic table. Points for *s*-block and *p*-block elements are red; points for *d*-block elements are blue. Noble gases and *f*-block elements are omitted.

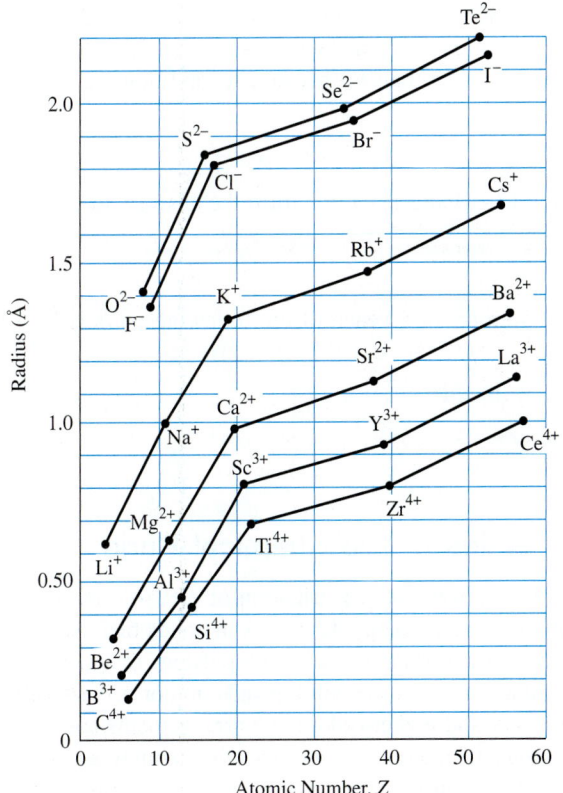

Figure 17-12 • Ionic radii plotted versus atomic number. Each line connects a set of atoms or ions having the same charge; all species have noble gas electron configurations.

of a main-group element, when it is ultimately reached, is smaller than it would have been had the transition series not intervened.

A similar effect occurs after the filling of the 4*f* subshell in the lanthanide series and is called the *lanthanide contraction*. It causes the atoms of the post-lanthanide metals of the sixth period to have close to the same size as those of the metals of the fifth period: the atomic radii of hafnium in the sixth period and zirconium in the fifth period are in fact nearly equal. Because these two elements are so similar in both size and valence electron configuration, their properties are extraordinarily similar.

EXAMPLE 17–4

Predict which atom or ion in each of the following pairs is larger:
(a) Kr or Rb
(b) Y or Cd
(c) F^- or Br^-
Explain the basis of the predictions.

Strategy

Adapt the logic used to explain the trends in the atomic sizes of the representative elements to these cases.

Solution

(a) Rb should be larger than Kr. It contains one more electron than Kr, and this electron occupies a 5*s* orbital that is outside the Kr closed shell.
(b) Y should be larger than Cd. The nuclear charge increases through the transition series from Y ($Z = 39$) to Cd ($Z = 48$), but the added electrons go into the same subshell and do not screen the nuclear charge very effectively.
(c) Br^- should be larger than F^- because its additional electrons occupy more distant electron shells.

Check

Data in Figures 17–10 and 17–12 confirm the answers to parts (b) and (c).

EXERCISE

Predict which atom or ion in each of the following pairs is larger: (a) Lu or Ir. (b) Zn^{2+} or Cd^{2+}. (c) Cs or Kr.

Answer: (a) Lu. (b) Cd^{2+}. (c) Cs.

17–4 Properties of the Chemical Bond

Chemical bonds result from the redistribution of electron density among atoms to arrive at a situation of lower energy. Electrons in bonds feel the attraction of two or more nuclei. These attractions outweigh the repulsions that also originate as atoms come together and so maintain the bond. Bonds do not occur solely between pairs of atoms; three-center and higher-order bond arrangements are well known. In fact every atom in a molecule interacts with every other atom to an extent. Nevertheless, bonding is most often discussed with a focus on pairs of atoms. Two experimentally measurable properties characterize bonded pairs: bond length and bond enthalpy. These properties are related to a third, theoretical, property: the bond order.

Bond Lengths

The first question to ask about a molecule is: What is its structure? For a diatomic molecule, the question is answered by giving the **bond length,** which is the distance between the nuclei of the two bonded atoms. Describing the structure of a polyatomic molecule requires additional bond lengths and bond angles as well. Millions of bond lengths and angles in hundreds of thousands of substances have been measured to varying degrees of precision by spectroscopic techniques (discussed in Section 18–4) and by X-ray diffraction (Section 20–1). Proper understanding of this information requires an appreciation of the fact that molecules are not rigid, unchanging bodies. Bond lengths and angles change from moment to moment as molecules rotate, vibrate, and interact with their surroundings.

Table 17–1 lists the bond lengths in a number of diatomic molecules. Some trends are immediately obvious. Within a group in the periodic table, bond lengths usually increase with increasing atomic number. Thus, the bond length increases in the series F_2 to Cl_2 to Br_2 to I_2. Similarly, Cl—F has a shorter bond length than does Cl—Br. These trends are directly related to the corresponding trends in atomic radius, which generally increases with increasing atomic number in a given group of the periodic table. This qualitative connection cannot be made quantitative, however, because bond lengths are not simply sums of atomic radii. Too much rearrangement of electron density occurs during bond formation to allow any such approximation.

TABLE 17–1	Properties of Some Diatomic Molecules	
Molecule	**Bond Length (Å)**	**Bond Enthalpy[a] (kJ mol^{-1})**
H_2	0.751	436
N_2	1.100	945
O_2	1.211	498
F_2	1.417	158
Cl_2	1.991	243
Br_2	2.286	193
I_2	2.669	151
HF	0.926	568
HCl	1.284	432
HBr	1.424	366
HI	1.620	298
ClF	1.632	255
BrF	1.759	285
BrCl	2.139	219
ICl	2.324	211
NO	1.154	632
CO	1.131	1076

[a]ΔH^0 for the reaction $XY(g) \longrightarrow X(g) + Y(g)$ at 298.15 K.

- This approximate constancy of bond lengths allows for great simplification in thinking about chemical bonding. One must be alert for exceptions, however, as in multiple bonding (see later discussion).

A striking result from the extensive experimental studies of bond lengths is that the length of the bond between two specific elements usually changes very little from one substance to the next. A C—H bond has almost the same length in acetylsalicylic acid (aspirin, $C_9H_8O_4$) as in methane (CH_4). Table 17–2 shows the lengths for O—H, C—C, and C—H bonds in a number of molecules. They are constant to within a few percent. The shortest known chemical bond is the one between the two hydrogen atoms in the hydrogen molecule. Its length is 0.751 Å. The lengths of the longest chemical bonds range upward of 3.0 Å or more.

Bond Enthalpies

Molecules form because the bonded atoms have a lower energy when they are close to each other than when they are far apart. The observed molecular geometry is the one that gives the molecule the lowest energy. The energy change ΔE_d when a bond is broken ("d" here stands for dissociation) to give two fragments of a molecule indicates the strength of that bond. It is more common to report the enthalpy change ΔH_d for breaking the bond (that is, the heat absorbed at constant pressure) than ΔE_d because ΔH_d is measured directly in calorimetric experiments at constant pressure. This is the **bond enthalpy.** The ΔE_d and ΔH_d per mole of bonds broken are related through

$$\Delta E_d = \Delta H_d - RT$$

- At room temperature, RT is only 2 to 3 kJ mol^{-1}.

where R is the gas constant, 8.3145 J K^{-1} mol^{-1}.

Table 17–1 lists bond enthalpies for selected diatomic molecules. Again, certain systematic trends with changes in atomic number are evident. Bonds generally grow weaker with increasing atomic number, as shown by the decrease in the bond enthalpies of the hydrogen halides in the order HF > HCl > HBr > HI. Note, however, the unusual weakness of the bond in the fluorine molecule F_2 (its bond enthalpy is significantly *smaller* than that of Cl_2 and is comparable to that of I_2). Bond strength decreases dramatically in the diatomic molecules moving from N_2 (945 kJ mol^{-1}) to O_2 (498 kJ mol^{-1}) to F_2 (158 kJ mol^{-1}). What accounts for this

TABLE 17–2	Constancy of Bond Lengths from Substance to Substance	
Bond	**Molecule**	**Bond Length (Å)**
O—H	H_2O	0.958
	H_2O_2	0.960
	HCOOH	0.95
	CH_2OH	0.956
C—C	diamond	1.5445
	C_2H_6	1.536
	CH_3CHF_2	1.540
	CH_3CHO	1.50
C—H	CH_4	1.091
	C_2H_6	1.107
	C_2H_4	1.087
	C_6H_6	1.084
	CH_3Cl	1.11
	CH_2O	1.06

behavior? A successful theory of bonding must explain both general trends and the reasons for particular exceptions. We examine these questions further in Chapter 18.

The ΔH_d of a bond X—Y between an atom of element X and an atom of element Y is approximately constant (within about 10%) from one compound to another. It is therefore possible to tabulate *average* bond enthalpies from measurements on a series of compounds (see Table 10–3).

• Average bond enthalpies can be used to estimate the standard enthalpies of formation of compounds, as explained in Section 10–5.

Chemical bonds come in many different kinds and exhibit a considerable range of strength. A rough categorization of bonds by their strength (their ΔH_d) is

Weak chemical bonds	Up to approximately 200 kJ mol^{-1}
Average chemical bonds	Centered at 500 kJ mol^{-1}
Strong chemical bonds	Greater than 800 kJ mol^{-1}

One of the very strongest chemical bonds is that between carbon and oxygen in carbon monoxide ($\Delta H_d = 1076$ kJ mol^{-1}). No exact lower limit to the strength of chemical bonds exists. It is a matter of judgment and circumstances when an interaction between two atoms becomes weak enough to demote it from a chemical bond to a "non-bonded attraction."

Bond Order

Sometimes the length and enthalpy of a bond X—Y are *not* reproduced from one chemical compound to another. Table 17–3 shows the bond lengths and bond enthalpies of carbon–carbon bonds in ethane (H_3CCH_3), ethylene (H_2CCH_2), and acetylene (HCCH). The differences are great. Carbon–carbon bonds from many other molecules fall into one of the three classes given in the table (that is, some carbon–carbon bond lengths are close to 1.54 Å, others are close to 1.34 Å, and still others are close to 1.20 Å). This confirms the existence of not one, but three types of carbon–carbon bonds. The weakest and longest (as in ethane) is a single bond and is indicated by C–C; that of intermediate strength (as in ethylene) is a double bond, C=C; and the strongest and shortest (as in acetylene) is a triple bond, C≡C. The **bond order** is defined as the number of shared electron pairs, so the bond orders of these three types of bonds are 1, 2, and 3.

• As shown in Section 3–4, the different carbon–carbon bonds have Lewis structures with different numbers of electron pairs shared between the atoms.

Even these three types do not cover all the carbon–carbon bonds found in nature, however. In benzene (C_6H_6), the experimental carbon–carbon bond lengths all equal to 1.397 Å, and the carbon–carbon bond enthalpies all equal 505 kJ mol^{-1}. These carbon–carbon bonds are intermediate between a single and a double bond (bond order $1\frac{1}{2}$). Such facts about benzene, which is discussed further in Section 18–3, support the following generalizations:

The bond enthalpy of a given bond X—Y increases as its bond order increases. The bond length of a given bond decreases as the bond order increases.

TABLE 17–3	Three Types of Carbon–Carbon Bonds		
Bond	**Molecule**	**Bond Length (Å)**	**Bond Enthalpy (kJ mol^{-1})**
C—C	C_2H_6 (H_3C—CH_3)	1.536	348
C=C	C_2H_4 (H_2C=CH_2)	1.337	615
C≡C	C_2H_2 (HC≡CH)	1.204	812

17-5 Ionic and Covalent Bonds

Chemical bonds arise from the sharing or transfer of electrons among two or more atoms. When electron density is mainly shared between two atoms, then the bond between them is **covalent.** When electron density is mainly transferred from one atom to another, the resulting chemical bond is **ionic.** Ionic and covalent bonds are idealized limits. Real bonds between unlike atoms are neither fully ionic nor fully covalent but possess a mixture of ionic and covalent character. Bonds in which the mixture of ionic and covalent character is intermediate between the extremes are **polar covalent.** In this section, we present a classical picture of chemical bonding, deferring a quantum-mechanical description to Chapter 18. To start, we re-introduce a most useful concept—the electronegativity. Considerations of electronegativity served in Chapter 3 as a rough guide to distinguish between ionic and covalent bonds. Now we establish the electronegativity as an atomic property on a more exact basis.

Electronegativity

The electronegativity of an atom is a measure of its ability in a molecule to attract electrons to itself. Elements toward the lower left corner of the periodic table have low ionization energies and small electron affinities. They give up electrons readily and accept electrons poorly. They tend to act as electron *donors* in interactions with other elements. In contrast, elements in the upper right corner of the periodic table have high ionization energies and also (except for the noble gases) large electron affinities. As a result, these elements accept electrons easily but give them up only reluctantly. Such atoms act as electron *acceptors.*

These facts suggested to chemist Robert Mulliken in 1934 a simple way to assign numerical electronegativities: Make them proportional to the average of the ionization energy and the electron affinity of the atom:

$$\text{electronegativity (Mulliken)} \propto \tfrac{1}{2}(IE_1 + EA)$$

• The electronegativity of an atom sometimes is confused with its electron affinity. This equation makes it clear that the two properties are not the same.

Electron acceptors (like the halogens) have high ionization energies and electron affinities and are thus highly electronegative. Electron donors (like the alkali metals) have low ionization energies and electron affinities and therefore low electronegativities; they are electropositive.

By the Mulliken formula, the electronegativity is an invariant atomic property, as accurate as the measured ionization energies and electron affinities from which it is calculated. The "power of an atom in a molecule to attract electrons to itself" in fact depends on the number and kind of the atom's neighbors, its oxidation state, and other factors. The Mulliken formula can be fine-tuned to deal with some of these factors, but detailed information about atomic environments is often lacking. Tables of *average* electronegativities, in which some kind of typical environment is assumed for each type of atom, suffice for most purposes. The most widely used set of average electronegativities (Fig. 17–13) was derived using a method devised by Linus Pauling. On the Pauling scale, fluorine, the most electronegative element, is assigned a value of 3.98, and cesium, the least electronegative element (and accordingly the most electro*positive* element) has an electronegativity of 0.79. Average electronegativities are useful in exploring periodic trends and making semi-quantitative comparisons. They should not be treated as high-precision experimental results.

The periodic trends in the electronegativity, as shown in Figure 17–13, are quite interesting. High electronegativity is favored by a small atomic radius and a large

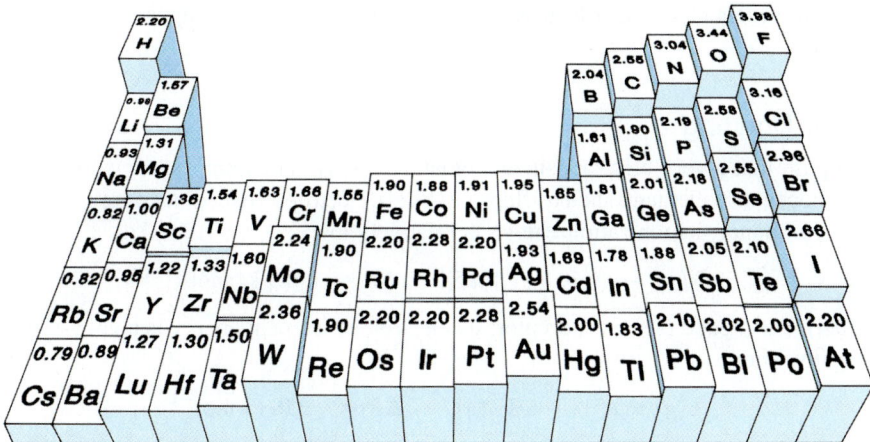

Figure 17–13 • Average electronegativities of atoms, computed according to the method developed by Linus Pauling. Electronegativity values have no units.

effective nuclear charge felt by the outer electrons. The electronegativity thus decreases going down the periodic table in the groups at the left and right sides of the table, such as the alkali metals and the halogens. In these groups, atomic radius increases sufficiently rapidly with atomic number to make it the predominant factor and to cause the decrease in electronegativity. For a group of late transition elements such as copper, silver, and gold, however, the atomic radius changes only slowly. The rise in effective nuclear charge then predominates so that the electronegativity *increases* going down the group's column in the table.

> • The late transition metals are found toward the end of each transition series; the early transition metals are found toward the beginning.

The absolute value of the difference in the electronegativity of two bonded atoms tells the degree of polarity to be found in their bond. A large difference means that the bond is ionic and that electrons are transferred nearly completely to the more electronegative atom. A small difference means that the bond is covalent and that the electrons in the bond are nearly evenly shared. Intermediate values of the difference signify intermediate character in the bond, that is, a polar covalent bond.

EXAMPLE 17–5

Using Figure 17–13, arrange the following bonds in order of decreasing polarity: H—C, O—O, H—F, I—Cl, Cs—Au.

Solution

The differences in electronegativity between the atoms taking part in the five bonds are (without regard to sign) 0.35, 0.00, 1.78, 0.50, and 1.75, respectively. The order of decreasing polarity is the order of decrease in this difference. H—F ≈ Cs—Au > I—Cl > H—C > O—O. The O—O bond is non-polar.

EXERCISE

Which bond has the greatest ionic character: H—Li, N—P, or N—S?

Answer: The H—Li bond.

The noble gases participate only rarely in chemical bonding. Their high ionization energies and negative electron affinities mean that they are reluctant either to give up or to accept electrons. Electronegativities generally are not assigned to the noble gases.

Ionic Bonding

Suppose that a very electropositive element, such as potassium, reacts with a very electronegative element, such as fluorine. Being electropositive, potassium is a good electron donor; its ionization energy is relatively low:

$$K(g) \longrightarrow K^+(g) + e^- \qquad \Delta E = +419 \text{ kJ mol}^{-1}$$

Fluorine is a good electron acceptor; its electron attachment energy is a good-sized, negative number:

$$F(g) + e^- \longrightarrow F^-(g) \qquad \Delta E = -328 \text{ kJ mol}^{-1}$$

Adding these two equations gives an equation representing the transfer of an electron from potassium to fluorine with the two atoms held far apart. The energy change for the transfer is the sum of the ΔE's:

$$K(g) + F(g) \longrightarrow K^+(g) + F^-(g) \qquad \Delta E_\infty = +91 \text{ kJ mol}^{-1}$$

The subscript "∞" on ΔE emphasizes that the reactant atoms are infinitely distant from each other. They are free atoms, just as the product ions are free ions. Even in this favorable case (with very different electronegativities), it *costs* energy to transfer an electron. Checking the ionization energies and electron attachment energies of the elements shows that this is always the case. For atoms separated by large distances in the gas phase, electron transfer to form ions is never favored energetically.

- The smallest ionization energy (for cesium, 376 kJ mol^{-1}) and the most negative electron attachment energy (for chlorine, -349 kJ mol^{-1}) still add up to a positive ΔE_∞.

How, then, does an ionic bond form? As the two ions approach each other, the electrostatic or *Coulomb* force between them gives rise to an interaction energy proportional to the product of the charges divided by the separation R between their centers:

$$\Delta E_{Coulomb} = k\frac{Q_1 Q_2}{R}$$

where k is a constant of proportionality. Because Q_1 and Q_2 have opposite signs in this case, this Coulomb energy is negative, and the total energy is lowered as the ions move toward each other. At short enough distances, the electrostatic attraction more than compensates for the energy required to transfer the electron. An ionic bond has formed. Figure 17–14 shows the variation of ionic and neutral-state energies with separation R for the case of potassium fluoride, which is typical. For large separations, the pair of neutral atoms is more stable, but at shorter distances, the ionic species becomes favored because of the electrostatic attraction. At very short distances, the electrons of the ions begin to repel each other, and the energy rises steeply. The length of an ionic bond is determined by a balance of attractive and repulsive forces. In KF(g), this is at $R = 2.13$ Å. It requires 489 kJ mol^{-1} to break the ionic bond in KF(g).

- Potential energy curves of this type also appear in Section 6–1.

Gaseous molecules of potassium fluoride exist separately only at high temperatures. Under most conditions, a collection of gaseous molecules of KF interacts further to form an ionic solid. The exothermic change

$$KF(g) \longrightarrow KF(s) \qquad \Delta E^\circ = -241 \text{ kJ mol}^{-1}$$

occurs as each positive ion is surrounded by several negative ions and vice versa. The solid product has a three-dimensional structure in which K^+ and F^- ions oc-

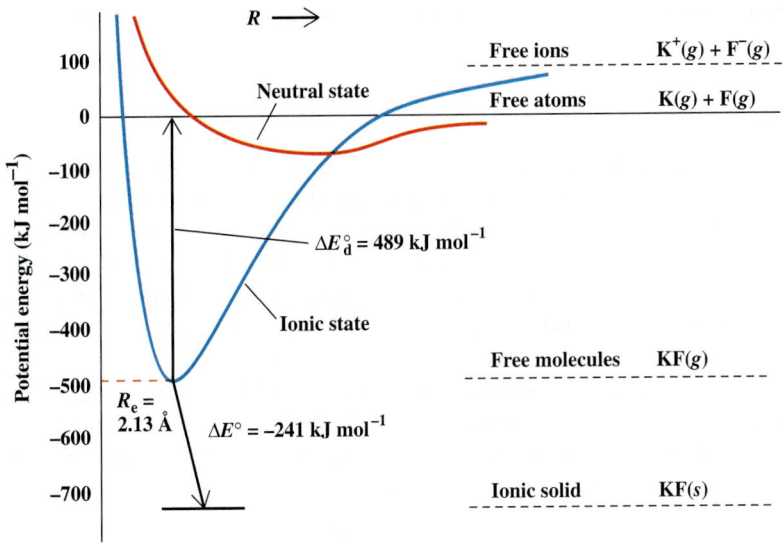

Figure 17–14 • The energetics of ionic bonding in potassium fluoride. The curves show the potential energies of the ionic and neutral states of K + F as a function of the distance R between the nuclei. A pair of neutral atoms has lower energy at large separations, but a pair of ions has lower energy at small separations. The lowest energy comes when *many* pairs of ions interact in an ionic solid.

cupy alternate adjacent sites. The ionic solid KF(*s*) has the lowest energy in Figure 17–14.

Covalent Bonding

Charge transfer followed by electrostatic attraction accounts for the formation of ionic bonds. What happens when two atoms with similar or equal electronegativities interact? The H_2 molecule is quite stable, for example, with a bond dissociation energy of 432 kJ mol^{-1}, yet it consists of two identical atoms. Charge transfer from one H atom to the other to form an ionic bond is out of the question. The stability of H_2 comes instead from a different type of bonding, covalent bonding, which arises from the *sharing of electrons* between atoms.

Consider the bonding of the H_2^+ molecular ion. This, the simplest of all possible polyatomic species, is stable but rather reactive. Its properties have been studied both experimentally and theoretically. The H—H bond length in this ion is 1.06 Å, and the dissociation energy ΔE_d is 255.5 kJ mol^{-1}. A simple classical model of covalent bonding explains the existence of H_2^+. In this model, the electron is a point negative charge that exerts an electrostatic attraction on the two hydrogen nuclei, which meanwhile repel each other. If the electron lies between the two (Fig. 17–15a), then its attractions pull the two nuclei together, leading to a bond. The electron screens the positive-to-positive electrostatic repulsion of the nuclei by interposing its negative charge, to which both are attracted. If the electron in H_2^+ lies *outside* the region between the nuclei (Fig. 17–15b), then the attractions between it and the nuclei push the two nuclei apart.

This picture of covalent bonding can be extended to other molecules. Covalent bonds arise when electrons spend most of their time located in the regions between nuclei, that is, when electrons are shared between nuclei. Electrons that spend most of their time outside of such regions actively oppose bonding because they add to the repulsion between the nuclei. The shared valence-electron pairs of the Lewis model (see Chapter 3) are located between two nuclei, where they experience net attractive interactions with each nucleus and strengthen the bond through the mechanism illustrated in Figure 17–15. This model of covalent bonding is developed along quantum-mechanical lines in Chapter 18. The fundamental conclusion reached

• The structures of ionic solids are examined in Section 20–3.

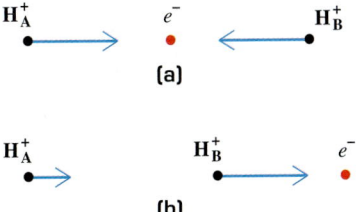

Figure 17–15 • Electrons attract hydrogen nuclei because the two have opposite charges. (a) An electron between two hydrogen nuclei causes a net force pulling the nuclei together. (b) An electron outside the internuclear region attracts the nearer nucleus more strongly than the farther, causing a net force that pulls the two nuclei apart.

there is the same: Covalent bonding arises from electrons being shared in the region between two nuclei.

Percent Ionic Character

In a bond that is almost purely ionic, such as that of KF, nearly complete transfer of an electron from the electropositive to the electronegative species takes place. Hence, KF can be represented fairly accurately as K^+F^-, with a charge $+e$ on the positive ion and $-e$ on the negative ion. The charge distribution for a molecule such as HF, with significant covalent character, is more complex. If charges are assigned to its two atoms, then it is best described as $H^{\delta+}F^{\delta-}$, a notation that means that some fraction δ of the full charge $\pm e$ is located at each atom.

- Here, δ is a lowercase Greek delta.

Dipole moments (see Section 3–7) provide a useful measure of ionic character and of electronegativity differences, especially for bonds in diatomic molecules. If a distance R separates two charges of equal magnitude and opposite sign, $+e\delta$ and $-e\delta$, the dipole moment μ (Greek "mu") is

$$\mu = (e\delta)R$$

Dipole moments can be measured experimentally by electrical and spectroscopic methods. In H—F, for example, the value of δ calculated from the measured dipole moment is 0.41, substantially less than the value of 1 for a purely ionic bond. We convert to a "percent ionic character" by multiplying by 100% and say that the bond in H—F is 41% ionic. Less than 100% ionic bonding occurs for two reasons: (1) incomplete transfer of electrons between atoms, which is to say polar covalent bonding, and (2) distortion of the electron charge distribution around one ion by the electric field of the other ion. This distortion, or **polarization,** of the electron charge alters the dipole moment of the molecule. When polarization is extreme, it is no longer a good approximation to regard the ions as point charges. A more accurate description of the distribution of electron charge is necessary.

Table 17–4 gives a scale of ionic character for diatomic molecules, based on the definition of δ. The proportion of ionic character computed from observed dipole moments and bond lengths generally parallels the difference in Pauling electronegativity (Fig. 17–16): High ionic character usually corresponds to large differences in electronegativity, with the more electropositive atom carrying the charge $+\delta$.

TABLE 17–4	Ionic Character of Diatomic Molecules		
Molecule	**Percent Ionic Character (100 δ)**	**Molecule**	**Percent Ionic Character (100 δ)**
H_2	0	CsF	70
CO	2	LiCl	73
NO	3	LiH	76
HI	6	KBr	78
ClF	11	NaCl	79
HBr	12	KCl	82
HCl	18	KF	82
HF	41	LiF	84
		NaF	88

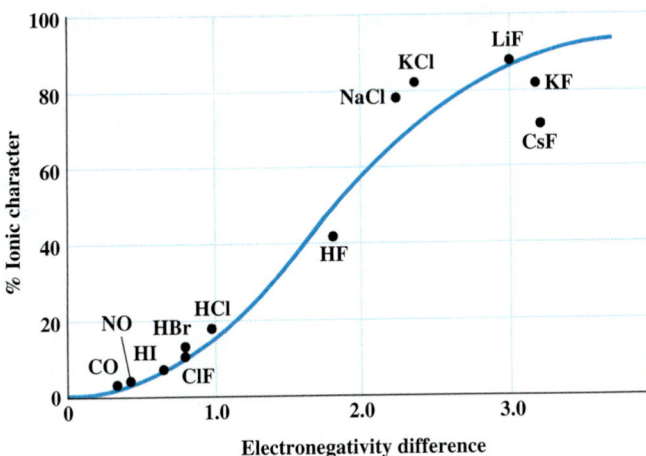

Figure 17–16 • Two measures of
ionic character for diatomic molecules
are the electronegativity difference
(from Fig. 17–13) and the percent
ionic character 100δ (calculated from
the observed dipole moments and bond
lengths). The black points represent
the actual data, and the blue curve is
drawn to guide the eye. The two measures correlate approximately.

Exceptions to the general trend exist, however. Carbon is less electronegative than
is oxygen, so a charge distribution of $C^{\delta+}O^{\delta-}$ would be predicted for the CO molecule. In fact, the measured dipole moment is quite small in magnitude and in the
other direction: $C^{\delta-}O^{\delta+}$, with $\delta = 0.02$. This discrepancy arises because of the high
electron density of the lone pair on the carbon atom. Clearly, tabulated average electronegativities are not appropriate for every different environment that an atom may
have.

EXAMPLE 17–6

The OH radical is a highly reactive species found in flames. Predict whether the
oxygen or the hydrogen atom should carry a fractional positive charge.

Strategy

Use the definition of electronegativity to conclude that the more electronegative
atom gets extra negative charge. Check the tabulated electronegativities.

Solution

The tabulated electronegativities reveal that the hydrogen should carry the positive charge $+\delta$. Its electronegativity (2.20) is smaller than that of oxygen (3.44).

Check

Oxygen lies to the right of hydrogen in the periodic table and so should have a
larger electronegativity.

EXERCISE

Another radical found in flames is CH. (a) Predict which atom in CH should carry
a fractional positive charge. (b) Does this radical have more ionic character than
the OH radical (in the example) or less?

Answer: (a) Hydrogen should carry the positive charge. (b) Less ionic character
than OH.

17-6   Oxidation States and Chemical Bonding

• A refresher definition of oxidation number: the electrical charge that an atom in a molecule would have if all shared electrons were transferred completely to the more electronegative element.

Oxidation states, also called oxidation numbers (see Section 3–8), are useful in the discussion of the chemical bonds formed by the elements. Figure 17–17 shows the most common oxidation numbers of the elements of the main groups and transition metals. The oxidation numbers range up to $+8$ and down to -3, although oxidation numbers of -1, -2, and -3 are seen only for the more electronegative elements on the right side of the periodic table. The diagonal red lines in Figure 17–17a show the tendency of the main-group elements to achieve oxidation states that correspond to closed-shell electron configurations. For example, the most important oxidation states of sulfur are -2 and $+6$, and the ions S^{2-} and S^{6+} have the electron configurations $[Ne]3s^23p^6$ and $[Ne]$, respectively, both of which are closed-shell configurations. Although oxidation state is *not* in general the true charge on an atom in a compound, the correlation is worth noting.

Further study of Figure 17–17a reveals a second interesting trend. The common positive oxidation state for compounds of the heavier elements in Groups III, IV, V, VI, and VII is often two steps lower than the maximum. Thus, the chemistry of carbon, silicon, and germanium is dominated by the $+4$ oxidation state, but the $+2$ oxidation state is more important for tin and lead, the later members of the group. In the $+2$ state of these elements, the valence p electrons participate in bonding, but the valence s electrons do not. The halides of the Group IV elements also show the point. The tetrachlorides CCl_4, $SiCl_4$, and $GeCl_4$ are all covalent compounds. The next heavier element, tin, forms both $SnCl_2$ and $SnCl_4$. The former, with its $+2$ oxidation state, has mainly ionic bonding and is a solid at room temperature. The latter ($+4$ oxidation state) has largely covalent bonding and is a volatile liquid. Lead, the last element of the group, forms lead(II) chloride, a fairly soluble ($1 \text{ g}/100 \text{ g } H_2O$) ionic compound, but lead(IV) chloride does not exist. A related trend occurs among the Group III elements. Boron, aluminum, and gallium show few signs of a $+1$ oxidation state, but this state is known in indium and is important in the chemistry of thallium, as in thallium(I) chloride (TlCl), a slightly soluble ionic compound.

• One familiar tin(II) compound is stannous fluoride (systematic name tin(II) fluoride, SnF_2), which was formerly added to toothpaste to prevent tooth decay. In most brands, it has been replaced by other fluorine-containing compounds.

The transition metals have a great variety of oxidation states in their compounds (Fig. 17–17b). The early members of each period show maximum oxidation numbers that correspond to the participation of all the outer s orbital and d orbital electrons in ionic or covalent bonds. Thus, elements in the scandium group have the valence electron configuration $(n-1)d^1ns^2$ and show only the $+3$ oxidation state. Manganese has the valence electron configuration $3d^54s^2$ and has a maximum oxidation state of $+7$, as do the other elements in its group; the group forms compounds such as perrhenic acid ($HReO_4$) and manganese(VII) oxide (Mn_2O_7), a dark-red liquid. Lower oxidation states occur as well, of course, including the green solid MnO ($+2$ oxidation state) and the black solids Mn_2O_3 ($+3$) and MnO_2 ($+4$) (Fig. 17–18).

Among the transition elements, the *heavier* elements tend to form stable compounds in *higher* oxidation states. This is the opposite of the trend among the representative elements. Thus, the $+2$ oxidation state is common in the compounds of all the elements of the first transition series (except Sc) but is less prevalent in the second and third transition series. The chemistry of iron is dominated by the $+2$ and $+3$ oxidation states, as in the common oxides FeO and Fe_2O_3, but the $+8$ state, which is nonexistent for iron, is possible for the later members of the iron group, ruthenium and osmium. The oxide OsO_4, for example, is a volatile yellow solid that melts at $41°C$ and boils at $131°C$. The chemistry of nickel is almost entirely that of

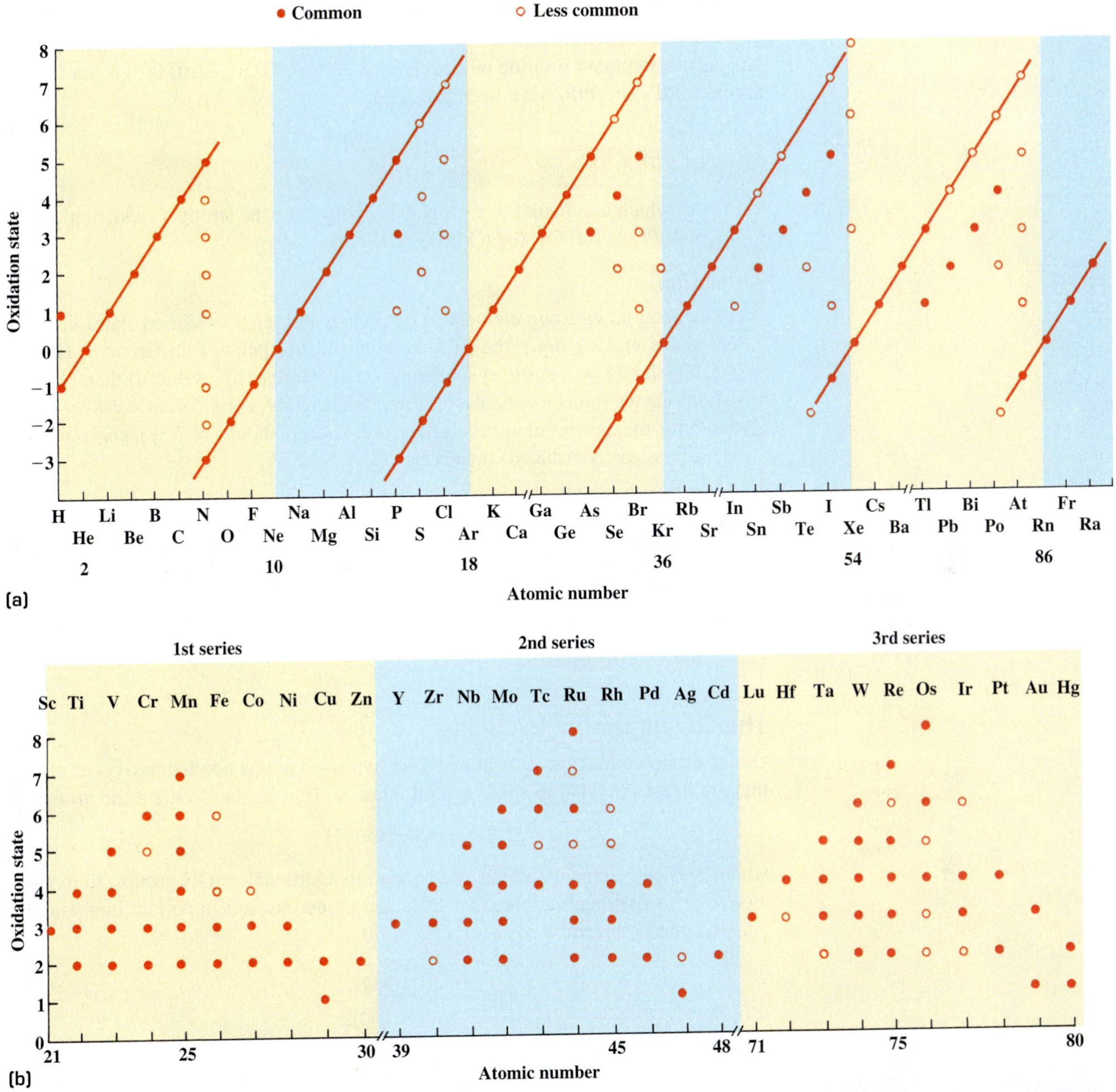

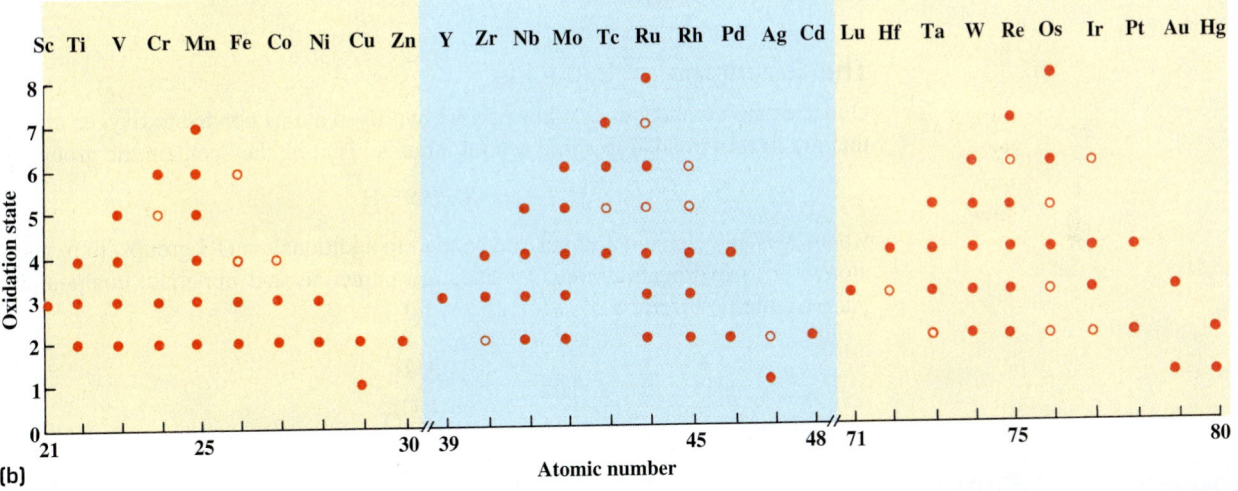

(a)

(b)

Figure 17–17 • Some of the oxidation states found in compounds of (a) the representative elements and (b) the transition elements. All elements have oxidation states of zero in their elemental form. These states are omitted except for the noble gases.

Figure 17–18 • Several oxides of manganese. They are arranged in order of increasing oxidation number of the Mn, with MnO at the bottom; Mn_2O_3 and Mn_3O_4 at the left and right, respectively; and MnO_2 at the top. A compound of still higher oxidation state, Mn_2O_7, is a dark red liquid that explodes easily. *(Charles D. Winters)*

the +2 oxidation state, but the chemistry of palladium and platinum, which lie below nickel in the periodic table, is increasingly dominated by the +4 state. Thus, NiF_2 is the dominant fluoride of nickel; both PdF_2 and PdF_4 exist; PtF_2 is *not* found, but both PtF_4 and PtF_6 have been prepared.

EXAMPLE 17-7

Predict which compound in each pair is likely to be the stronger oxidizing agent: (a) $NaBiO_3$ or $NaPO_3$; (b) $KMnO_4$ or $KReO_4$.

Solution

(a) Among main-group elements, compounds in higher oxidation states become less stable moving *down* the periodic table. Bi lies below P in Group V. Hence, $NaBiO_3$ should be a stronger oxidizing agent (more easily reduced) than $NaPO_3$. (b) Among the transition elements, compounds in higher oxidation states are more stable for the heavier elements. Therefore, $KReO_4$ should be a weaker oxidizing agent (less easily reduced) than $KMnO_4$.

EXERCISE

Which compound in the following pairs is more likely to be preparable in stable form: (a) $Co_2(SO_4)_3$ or $Rh_2(SO_4)_3$; (b) $P(NO_3)_3$ or $Bi(NO_3)_3$?

Answer: (a) $Rh_2(SO_4)_3$. (b) $Bi(NO_3)_3$.

The Strengths of Oxoacids

Oxoacids are compounds that have acidic hydrogen atoms bonded to oxygen atoms that are in turn bonded to some central atom X. That is, they contain the grouping

$$-X-O-H$$

where X is any element and may be bonded to additional —OH groups, to oxygen atoms, or to hydrogen atoms. Oxoacids are numerous and important compounds. An example is sulfuric acid:

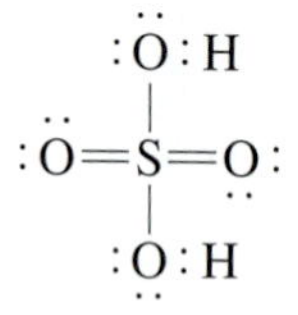

It acts as an acid as its $-\overset{|}{\underset{|}{S}}-O-H$ groups donate H^+ ions to water molecules, yielding first HSO_4^- and then SO_4^{2-} ion. Sulfuric acid is a strong acid because it donates the first H^+ ion quite effectively, as shown by its large acidity constant (see Table 8-2).

The strength of an oxoacid depends on the electronegativity of the central atom X. This fact can be understood by considering the distribution of charge across the $-X-O-H$ grouping for different X's. If X has a low electronegativity, then the X—O bond must be almost completely ionic because the electronegativity difference between X and O is large. The oxoacid grouping is then best represented —X$^+$ (O—H)$^-$. Putting a net negative charge on the O—H portion of the grouping holds the H$^+$ tightly to the oxygen and prevents donation of H$^+$ ions. In fact, Na—O—H, K—O—H, and other compounds in which X is highly electropositive tend to donate an OH$^-$ ion and not an H$^+$ ion. They are basic, not acidic. As the electronegativity of the central atom X increases, the difference in electronegativity between X and O becomes less, making the X—O bond in the —X—O—H grouping more nearly covalent. This puts less negative charge on the oxygen atom, and the oxoacid releases H$^+$ more readily (Fig. 17–19). The conclusion is:

Oxoacids of the same structure show increasing acid strength as the electronegativity of the central atom increases.

The qualification about the same structure is required because the effect of *other* O's that might be bonded to the central atom X has yet to be considered. In other words, the statement predicts (correctly) that the oxoacid Cl—O—H has a larger K_a than Br—O—H (Cl is more electronegative than Br), but says nothing about the relative strengths of Cl—O—H and OCl—O—H.

Suppose that several oxoacids have the same central atom X but different numbers of O's bonded to the atom X. The series of oxoacids Cl—O—H, OCl—OH, O$_2$Cl—OH, O$_3$Cl—OH is an example. Each additional O increases the oxidation state of X by 2. An atom in a higher oxidation state has an increased electronegativity because having lost electrons it has a greater power to attract electrons to itself in the molecule. This means that:

The acid strength of oxoacids with a given central element increases with the oxidation state of the central atom, or, equivalently, with the number of lone oxygen atoms attached to the central atom.

If the formulas of oxoacids are rewritten as XO$_n$(OH)$_m$, then the acid strengths fall into distinct classes, according to the value of n, which is the number of lone oxygen atoms (see Table 17–5). Each increase by 1 in n leads to an increase in the acidity constant K_a by a factor of about 10^5. Another way to understand this effect is to focus on the stability of the conjugate base, XO$_{n+1}$(OH)$^-_{m-1}$, of the oxoacid. The larger the number of lone O atoms attached to the central atom, the more easily the net negative charge can be spread out over the ion (remember that the electronegative O atom accepts negative charge well), and, therefore, the more stable the base is. This leads to a larger K_a.

• Note that Table 17–5 includes many familiar acids written in unfamiliar forms. Thus, H$_2$SO$_4$ is listed as SO$_2$(OH)$_2$, HNO$_3$ is NO$_2$(OH), and so forth.

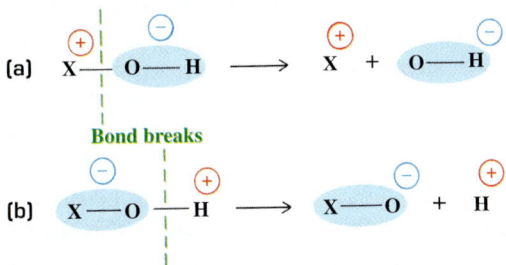

(a) X—O—H $\longrightarrow$ X + O—H

Bond breaks

(b) X—O—H $\longrightarrow$ X—O + H

Figure 17–19 • In (a), the atom X is electropositive, so extra electron density (*blue*) accumulates on the OH group. The X—O bond then can break easily, making the compound a base. In (b), the atom X is electronegative, so electron density is withdrawn from the H atom to the X—O bond. In this case, it is the O—H bond that breaks easily, and the compound is an acid.

TABLE 17-5		Acid Ionization Constants for Oxoacids of the Non-metals					
$X(OH)_m$ Very Weak	K_a	$XO(OH)_m$ Weak	K_a	$XO_2(OH)_m$ Strong	K_a	$XO_3(OH)_m$ Very Strong	K_a
Cl(OH)	3×10^{-8}	$H_2PO(OH)$	8×10^{-2}	$SeO_2(OH)_2$	1×10^3	$ClO_3(OH)$	2×10^7
$Te(OH)_6$	2×10^{-8}	$IO(OH)_5$	2×10^{-2}	$ClO_2(OH)$	5×10^2		
Br(OH)	2×10^{-9}	$SO(OH)_2$	2×10^{-2}	$SO_2(OH)_2$	1×10^2		
$As(OH)_3$	6×10^{-10}	ClO(OH)	1×10^{-2}	$NO_2(OH)$	20		
$B(OH)_3$	6×10^{-10}	$HPO(OH)_2$	1×10^{-2}	$IO_2(OH)$	1.6×10^{-1}		
$Ge(OH)_4$	4×10^{-10}	$PO(OH)_3$	8×10^{-3}				
$Si(OH)_4$	2×10^{-10}	$AsO(OH)_3$	5×10^{-3}				
I(OH)	4×10^{-11}	$SeO(OH)_2$	3×10^{-3}				
		$TeO(OH)_2$	3×10^{-3}				
		NO(OH)	5×10^{-4}				

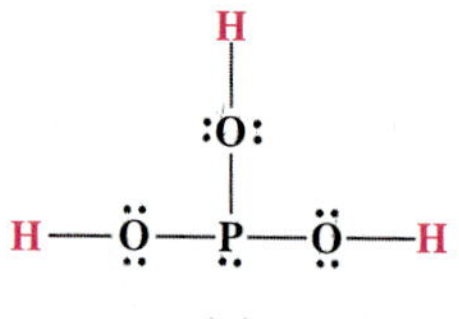

(a)

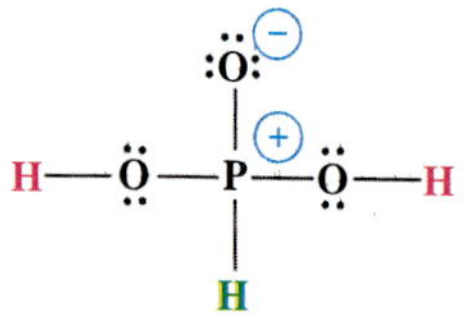

(b)

Figure 17-20 • (a) The simplest Lewis diagram that can be drawn for H_3PO_3 is plausible but not consistent with the facts of the substance's chemical behavior. This acid would be triprotic, like H_3PO_4. (b) The observed structure of H_3PO_3. The hydrogen atom attached to the phosphorus atom is not released into acidic solution, so the acid is diprotic.

EXAMPLE 17-8

Predict the stronger acid in each of the following pairs:
(a) H_2SO_3, H_2SeO_3
(b) HIO_3, HIO

Solution

(a) The two oxoacids have the same structure. Because sulfur is more electronegative than selenium, H_2SO_3 should be stronger than H_2SeO_3.
(b) The oxoacid $IO_2(OH)$, with two lone oxygen atoms on the iodine atom, should be stronger than I(OH), with zero lone oxygen atoms.

Check

The K_a data in Table 17-5 confirm these relative strengths. Note however that the electronegtivity of S is only slightly more than that of Se and that H_2SeO_4 is *stronger* than the comparable H_2SO_4.

EXERCISE

Identify the stronger acid in each pair: (a) H_2MoO_4, H_2CrO_4; (b) $HMnO_4$, H_2MnO_4.

Answer: (a) H_2MoO_4. (b) $HMnO_4$.

An unusual and interesting structural result can be obtained from Table 17-5. The simplest Lewis structure for the oxoacid with formula H_3PO_3 is shown in Figure 17-20a. Note that each atom in it achieves an octet configuration. Such a structure could be formulated $P(OH)_3$ and is analogous to $As(OH)_3$, which has no lone oxygen atoms bonded to the central atom ($n = 0$). The expected value of K_a for $P(OH)_3$ based on this analogy is on the order of 10^{-9} (a very weak acid). In fact, however, H_3PO_3 is a much stronger acid ($K_a = 1 \times 10^{-2}$). It fits better into the class of acids with one lone oxygen atom bonded to the central atom. This analysis based on chemical properties is shown to be correct by X-ray diffraction measurements. The true structure of H_3PO_3 is best represented by Figure 17-20b! This corresponds either to a Lewis structure with more than eight electrons around the central phosphorus atom or to one with formal charges on the central phosphorus and lone oxygen atoms. The formula of this acid is written as $HPO(OH)_2$ in Table 17–5. Unlike phosphoric acid (H_3PO_4), which is a triprotic acid, H_3PO_3 is a diprotic acid. The third hydrogen atom, the one bonded directly to the phosphorus atom, is not lost even in strongly basic aqueous solution.

Further Comparisons of Acid Strength

Consider the trends in the acidity constants of the binary hydrides in Groups V through VII of the periodic table (these are not oxoacids, but compounds such as NH_3, H_2O, and HF). The K_a's increase from left to right in the periodic table and also increase from top to bottom. The horizontal trend is understood by noting that a central atom of higher electronegativity withdraws more electron density from the vicinity of the hydrogen atom, making it easier to donate an H^+ ion. However, the vertical trend in the K_a's does *not* match the decrease in electronegativity going down the columns in this part of the table: H_2S is *more* acidic than H_2O, although S is less electronegative than O. Similarly, in Group VII, the K_a's increase in the order $HF < HCl < HBr < HI$, but the electronegativity of the halogen decreases. The unconsidered factor is the greater **polarizability** (susceptibility to polarization) of larger atoms. In large atoms, the distribution of the electrons is more easily distorted by approaching electric charges or dipoles than it is in small atoms. Thus, the bonding electrons in HI are more readily pushed back by negative charge on an approaching base than are the bonding electrons in HF. The result is that H^+ ion is more easily extracted, and K_a is larger.

SUMMARY

17–1 According to the **orbital approximation,** electrons in many-electron atoms occupy orbitals resembling those of the hydrogen atom in their shapes and the number of their nodes. The **Pauli exclusion principle** states that no more than two electrons (with opposite spin) can occupy any such atomic orbital. The ground-state electron configurations of successive atoms in the periodic table are built up (according to the **aufbau principle**) by the systematic filling of **subshells** and **shells** of orbitals, starting with those of lowest energy. **Closed-shell** ions or atoms (such as the noble gases) are those for which the next atomic orbitals available for occupancy by electrons are separated by a large energy gap from the highest occupied orbitals. Such species are comparatively unreactive. Atoms tend to interact chemically to attain closed-shell configurations.

17–2 Orbital energies can be measured by **photoelectron spectroscopy,** in which the kinetic energies of electrons ejected by high-energy photons are determined. The minimum energy to eject an electron is the first ionization energy, which generally increases left to right across a period and decreases down a group in the periodic table. The **electron attachment energy** of an atom is the energy change that occurs when the atom gains an electron. The **electron affinity** is the negative of the electron attachment energy. Periodic trends in the electron affinity generally parallel those in the **first ionization energy.**

17–3 The **atomic radius** of an element is either a **metallic radius** (for the metals) or a **covalent radius** (for the non-metals). Atomic radii equal half the distance between the nuclei of adjoining identical atoms. They are obtained experimentally from X-ray diffraction studies and vary somewhat for the same element, depending on the structure. **Ionic radii** are estimated from internuclear distances in ionic compounds. The interpolation of *d*-block and *f*-block elements leads to a relative contraction in size of the atoms or ions that follow them in a row of the periodic table.

17–4   A **bond length** is an internuclear distance between bonded atoms; a **bond enthalpy** is the enthalpy change in breaking a bond. Bond lengths and enthalpies involving a given pair of elements stay constant to within a few percent. Multiple bonds cause exceptions, but even then, a similar constancy prevails within the set of doubly or triply bonded atoms.

17–5   The electronegativity of an atom is a measure of its power in a molecule to attract electrons to itself. Atoms of very different electronegativities form ionic bonds. In ionic bonding, electrons are transferred and stability is achieved from electrostatic (Coulomb) attraction. Atoms of similar electronegativities form polar covalent bonds in which the sharing of electrons between the two nuclei is responsible for stability. The **percent ionic character** in a polar covalent bond is related to difference in electronegativity of the bonded atoms.

17–6   The most important oxidation states of the elements vary systematically through the periodic table. For main-group elements, lower oxidation states predominate among the heavier elements, but the reverse is true for transition elements. Trends in electronegativity and oxidation states can account for many relationships in chemical behavior. An example is oxoacid strength, which increases with increasing electronegativity of the central atom and with an increasing number of lone oxygen atoms attached to the central atom. High **polarizability** in the atom to which a hydrogen atom is bonded increases the acidity of that hydrogen in a binary acid.

PROBLEMS

Note: Answers to blue-numbered problems are given in Appendix F. Problems that are more challenging are indicated with asterisks.

Many-Electron Atoms and the Periodic Table

1. (See Example 17–1.) Give the ground-state electron configurations of the following elements: (a) Si (b) S (c) Co

2. (See Example 17–1.) Give the ground-state electron configurations of the following elements: (a) N (b) Mn (c) Eu

3. Write ground-state electron configurations for the following ions: Be^+, C^-, Ne^{2+}, Mg^+, P^{2+}, Cl^-, As^+, I^-. Which are paramagnetic due to the presence of unpaired electrons?

4. Write ground-state electron configurations for the following ions: Li^-, B^+, F^-, Al^{3+}, S^-, Ar^+, Br^+, Te^-. Which are paramagnetic due to the presence of unpaired electrons?

5. Identify the atom or ion corresponding to each of the following descriptions:
   (a) An atom with ground-state electron configuration $[Kr]4d^{10}5s^25p^1$
   (b) An ion with charge -3 and ground-state electron configuration $[Ne]3s^23p^6$
   (c) An ion with charge $+2$ and ground-state electron configuration $[Ar]3d^3$

6. Identify the atom or ion corresponding to each of the following descriptions:
   (a) An atom with ground-state electron configuration $[Xe]4f^{14}5d^66s^2$
   (b) An ion with charge -2 and ground-state electron configuration $[He]2s^22p^6$
   (c) An ion with charge $+3$ and ground-state electron configuration $[Kr]4d^6$

7. Predict the atomic number of the (as yet unknown) element in the seventh period that is a halogen.

8. Predict the atomic number of the (as yet unknown) alkaline earth element in the eighth period. Suppose that the eighth-period alkaline earth element is discovered and turns out to have atomic number 138. Explain. (*Hint*: Recall that the atomic number of radium is only 88.)

9. Suppose that the spin quantum number did not exist and that only one electron could occupy each orbital of a many-electron atom. Give the atomic numbers of the first three noble-gas atoms in this case.

10. Suppose that the spin quantum number had three allowed values ($m_s = 0, +\frac{1}{2}, -\frac{1}{2}$). Give the atomic numbers of the first three noble-gas atoms in this case.

11. State the maximum number of electrons that can be placed in the $n = 6$ shell of an atom.

12. State the maximum number of electrons that can be placed in the $5f$ subshell of an atom.

13. Determine the atomic number of the first element that would be expected to have an electron in a $6f$ orbital when in its ground state.

14. Determine the atomic number of the first element that would be expected to have an electron in a $5g$ orbital in its ground state.

Experimental Measures of Orbital Energies

15. (See Example 17–3.) Decide which atom in each of the following pairs of atoms should have the higher first ionization energy:
    (a) Rb or Sr   (c) Xe or Cs
    (b) Po or Rn   (d) Ba or Sr

16. (See Example 17–3.) Decide which atom in each of the following pairs of atoms should have the higher first ionization energy:
    (a) Bi or Xe   (c) Rb or Y
    (b) Se or Te   (d) K or Ne

17. (See Example 17–3.) Decide which atom in each of the following pairs of atoms should have the greater electron affinity:
    (a) Xe or Cs   (c) Ca or K
    (b) Pm or F   (d) Po or At

18. (See Example 17–3.) Decide which atom in each of the following pairs of atoms should have the greater electron affinity:
    (a) Rb or Sr   (c) Ba or Te
    (b) I or Rn   (d) Bi or Cl

19. Sketch how the first ionization energy would vary with the atomic number Z from $Z = 1$ through 8 if the spin quantum number did not exist and only one electron could occupy each orbital of a many-electron atom.

20. Suppose that the spin quantum number had three allowed values ($m_s = 0, +\frac{1}{2}, -\frac{1}{2}$). Sketch how the first ionization energy would vary as Z varied from 1 through 12.

21. The cesium atom has the lowest ionization energy of all the neutral atoms in the periodic table, $375.5 \text{ kJ mol}^{-1}$. What is the longest wavelength of light that can ionize a cesium atom? In which region of the electromagnetic spectrum does this fall?

22. Until recently, it was thought that the Ca^- ion was unstable, so that the calcium atom had a negative electron affinity. Some new experiments have established an electron affinity of $+4.1 \text{ kJ mol}^{-1}$ for Ca. What is the longest wavelength of light that can remove an electron from a Ca^- ion? In which region of the electromagnetic spectrum does this wavelength fall?

23. First ionization energies (IE_1's) generally increase from left to right across a period; nevertheless, IE_1 is less for boron (801 kJ mol^{-1}) than for beryllium (899 kJ mol^{-1}). Suggest a reason.

24. The IE_1 of N is 1402 kJ mol^{-1}, and the IE_1 of O is only 1314 kJ mol^{-1}. Explain this exception to the periodic trend in IE_1, which generally increases from left to right across a period.

25. The IE_2 (second ionization energy) of B is bigger than the IE_2 of Be, but the IE_1 of B is smaller than the IE_1 of Be. Explain.

26. The first three ionization energies of magnesium are $IE_1 = 735 \text{ kJ mol}^{-1}$, $IE_2 = 1445 \text{ kJ mol}^{-1}$, and $IE_3 = 7730 \text{ kJ mol}^{-1}$. Explain why the increase from IE_2 and IE_3 is so much larger than the increase from IE_1 to IE_2.

Sizes of Atoms and Ions

27. Explain why atomic radii increase from top to bottom within a group in the periodic table.

28. Explain why atomic radii generally decrease from left to right within a period of s- and p-block elements.

29. (See Example 17–4.) State which species in each of the following pairs is expected to have the larger radius:
    (a) Li or Rb   (d) K or Ca
    (b) K or K^+ (e) Ne or O^{2-}
    (c) Rb^+ or Kr

30. (See Example 17–4.) State which species in each of the following pairs is expected to have the larger radius:
    (a) Mn or Mn^{2+} (d) Ge or As
    (b) Mg or Ca   (e) Ba^+ or Cs
    (c) I^- or Xe

31. (See Example 17–4.) Predict the larger ion in each of the following pairs:
    (a) O^-, S^{2-} (c) Mn^{2+}, Mn^{4+}
    (b) Co^{2+}, Ti^{2+} (d) Ca^{2+}, Sr^{2+}
    Give reasons for your answers.

32. (See Example 17–4.) Predict the larger ion in each of the following pairs:
    (a) S^{2-}, Cl^- (c) Ce^{3+}, Dy^{3+}
    (b) Tl^+, Tl^{3+} (d) S^-, I^-
    Give reasons for your answers.

33. The elements manganese ($Z = 25$) and europium ($Z = 63$) have abnormally large atomic radii based on the general trend among their neighbors in the periodic table (see Figs. 17–10 and 17–11). Write the electron configurations of these atoms and identify similarities that might explain the exceptions.

34. Atoms of the elements gallium, indium, and thallium are larger than atoms of zinc, cadmium, and mercury, respectively, not smaller. Write the electron configurations of all these atoms and identify similarities that might explain why.

Properties of the Chemical Bond

35. Predict how the bond length and bond enthalpy vary in a series of diatomic compounds involving one element in combination with successive elements that are members of a group in the periodic table. Are there exceptions to these general trends?

36. Describe how the bond length and bond enthalpy vary through a series of compounds in which the bonded atoms remain the same but the bond order increases.

37. The bond lengths of the X—H bond in NH_3, PH_3, and SbH_3 are 1.02, 1.42, and 1.71 Å, respectively. Estimate the length of the As—H bond in AsH_3, the gaseous compound that decomposes on a heated glass surface in Marsh's test for arsenic. Which of these four hydrides has the weakest X—H bond?

38. Arrange the following covalent diatomic molecules in order of the length of the bond: BrCl, ClF, IBr. Which of the three has the weakest bond (the smallest bond enthalpy)?

39. The bond length in H—I (1.62 Å) is close to the sum of the atomic radii of H (0.37 Å) and I (1.33 Å). What does this indicate about the polarity of the HI bond?

40. The bond length in the F_2 molecule is 1.417 Å, instead of 1.28 Å, which is twice the atomic radius of F. What might account for the unexpected length of the F—F bond?

Ionic and Covalent Bonds

41. Ionic compounds tend to have higher melting and boiling points and to be less volatile (that is, have lower vapor pressures) than covalent compounds. Use electronegativity differences to predict which of the following pairs of compounds has the higher vapor pressure at room temperature:
    (a) CI_4 and KI
    (b) BaF_2 and OF_2
    (c) SiH_4 and NaH

42. Use electronegativity differences to predict which compound in each of the following pairs has the higher boiling point:
    (a) $MgBr_2$ and PBr_3
    (b) OsO_4 and SrO
    (c) Cl_2O and Al_2O_3

43. Use the data in Figures 17–7 and 17–9 to compute the energy changes (ΔE) of the following pairs of reactions:
    (a) $K(g) + Cl(g) \longrightarrow K^+(g) + Cl^-(g)$ and
    $K(g) + Cl(g) \longrightarrow K^-(g) + Cl^+(g)$
    (b) $Na(g) + Cl(g) \longrightarrow Na^+(g) + Cl^-(g)$ and
    $Na(g) + Cl(g) \longrightarrow Na^-(g) + Cl^+(g)$

44. Use the data in Figures 17–7 and 17–9 to compute the energy changes (ΔE) of the following pairs of reactions:
    (a) $Na(g) + I(g) \longrightarrow Na^+(g) + I^-(g)$ and
    $Na(g) + I(g) \longrightarrow Na^-(g) + I^+(g)$
    (b) $Rb(g) + Br(g) \longrightarrow Rb^+(g) + Br^-(g)$ and
    $Rb(g) + Br(g) \longrightarrow Rb^-(g) + Br^+(g)$
    Explain why Na^+I^- and Rb^+Br^- form in preference to Na^-I^+ and Rb^-Br^+.

45. The percent ionic character of a bond can be approximated by the formula $16\Delta + 3.5\Delta^2$, where Δ is the magnitude of the difference in the electronegativities of the atoms (see Fig. 17–13). Calculate the percent ionic character of HF, HCl, HBr, HI, and CsF, and compare the results with those of Table 17–4.

46. The percent ionic characters of the bonds in several interhalogen molecules (as estimated from their measured dipole moments and bond lengths) are ClF (11%), BrF (15%), BrCl (5.6%), ICl (5.8%), and IBr (10%). Estimate the percent ionic characters for each of these molecules using the equation in the preceding problem, and compare them with the given values.

47. (See Examples 17–5 and 17–6.)
    (a) Use electronegativities to arrange the following bonds in order of decreasing polarity: N—O, N—N, N—P, and C—N.
    (b) Predict the atom that carries the fractional positive charge in each case.

48. (See Examples 17–5 and 17–6.) Among the diatomic molecules formed by the halogen atoms are IF, ICl, ClF, BrCl, and Cl_2.
    (a) Use electronegativities from Figure 17–13 to rank the bonds in these compounds in order from least ionic to most ionic in character.
    (b) Predict the atom that carries the fractional positive charge in each case.

Oxidation States and Chemical Bonding

49. Predict the highest oxidation state for the element in question in the compounds of each of the following elements: V, P, I, Sr.

50. Predict the highest oxidation state for the element in question in the compounds of each of the following elements: W, At, Xe, Fr.

51. Suggest why the $6s$ electrons in lead tend not to participate in bonding (making the $+2$ oxidation state more prevalent than the $+4$ state) but the $2s$ electrons in carbon, which is in the same column of the periodic table, generally do participate in bonding.

52. Iron almost invariably occurs in its compounds in the $+2$ or $+3$ oxidation state, but osmium does exhibit the $+8$ oxidation state in its compounds. Why is it easier for osmium to share its $6s$ and $5d$ electrons than it is for iron to share its $4s$ and $3d$ electrons?

53. Explain why Mn_2O_7, RuO_4, and OsO_4 have quite low melting points (6°C, 25°C, and 41°C, respectively) compared with the melting points of most transition-element oxides.

54. Tin(IV) chloride is a liquid at room temperature and pressure, whereas tin(II) chloride is a solid under the same conditions. Explain why.

55. In formic acid (HCOOH), one hydrogen atom and both oxygen atoms are bonded directly to the central carbon atom. Predict a range of K_{a1} for this acid, based on the correlations shown in Table 17–5. Compare your prediction with the measured value from Table 8–2. At pH 14, do you expect HCOOH(aq), HCOO$^-$(aq), or COO^{2-}(aq) to be the predominant species present? Explain.

56. In carbonic acid (H_2CO_3), both hydrogen atoms are bonded to oxygen atoms, and all three oxygen atoms are bonded directly to the central carbon atom. In what range do you predict K_{a1} for this acid to occur, based on the correlations shown in Table 17–5? The measured value of 4.3×10^{-7} quoted in Table 8–2 actually applies to the equilibrium

$$CO_2(aq) + 2\,H_2O(\ell) \rightleftharpoons HCO_3^-(aq) + H_3O^+(aq)$$

Explain how this information provides evidence that only a small fraction of the dissolved CO_2 is present as $H_2CO_3(aq)$.

57. Some oxoacids contain several central atoms of the same chemical element. An example is $H_2B_4O_7$, which can be written as $B_4O_5(OH)_2$. In such a case, we expect acid strength to correlate approximately with the ratio of the number of lone oxygen atoms to the number of central atoms (this ratio is 5:4 for $H_2B_4O_7$, for example). Rank the following in order of increasing acid strength: $H_2B_4O_7$, H_3BO_3, $H_5B_3O_7$, and $H_6B_4O_9$.

58. Use the approach suggested in the previous problem to rank the following in order of increasing acid strength: H_3PO_4, $H_3P_3O_9$, $H_4P_2O_6$, $H_4P_2O_7$, $H_5P_3O_{10}$. Assume that no hydrogen atoms are bonded directly to phosphorus in these compounds. Sodium salts of these polyphosphoric acids are used as "builders" in detergents to improve cleaning power.

59. Draw Lewis structures that satisfy the octet rule for H_2SO_3 and H_2SO_4. Decide which acid is stronger. Explain.

60. Oxoacids of the transition elements fall into classes similar to those shown in Table 17–5 for the non-metal oxoacids, although the values of K_a are smaller by about a factor of 10^4 for the compounds of the transition metals. Use this information to estimate the first acidity constant K_a for H_2MnO_4, H_4ZrO_4, $HMnO_4$, and H_3MnO_4. In none of these oxoacids is a hydrogen atom directly bonded to the central metal atom.

ADDITIONAL PROBLEMS

61. An atom of sodium has the electron configuration $[Ne]6s^1$. Explain how this is possible.

62. An atom or ion in which each electron shell is filled or half-filled has a total electron density that is spherically symmetrical (that is, the electron density varies with distance from the nucleus but not with direction). Identify the atoms and ions in the following list that are spherically symmetrical in their ground states: F^-, Na, Si, S^{2-}, Ar^+, Ni, Cu, Mo, Rh, Sb, W, Au.

63. Chromium(IV) oxide is used in making magnetic recording tapes because of its paramagnetic properties. It can be described as a solid made up of Cr^{4+} and O^{2-} ions. Give the electron configuration of Cr^{4+} in CrO_2, and determine the number of unpaired electrons on each chromium ion.

*64. Which is higher, the third ionization energy of lithium or the energy required to eject a $1s$ electron from a lithium atom in a photoelectron spectroscopy experiment? Explain.

65. Consider the elements Sr, Li, P, and Si.
   (a) Which has the greatest difference between its first and second ionization energy? Explain.
   (b) Which has the greatest difference between its second and third ionization energy?

66. The radii of the ions N^{3-}, O^{2-}, and F^- equal 1.71, 1.40, and 1.36 Å, respectively. The contraction is explained as the effect of increased nuclear charge from nitrogen ($Z = 7$) to fluorine ($Z = 9$) on the same electron configuration. Suggest why the contraction is much larger from N^{3-} to O^{2-} than from O^{2-} to F^-.

67. Arrange the following six atoms or ions in order of size, from the smallest to the largest: K, F^+, Rb, Co^{25+}, Br, F, Rb^-.

68. Use the guideline given in the text to estimate the radii of the Na^+ and Cl^- ions and the internuclear distance between neighboring Na^+ and Cl^- ions in sodium chloride. One reference source gives the Na—Cl bond distance as 2.36 Å, and another gives it as 2.81 Å. Explain how both sources might be correct.

69. Consult Figure 17–13, and compute the difference in electronegativity between the atoms in LiCl and in HF. Based on their physical properties (see following table), are the two similar or different in their bonding?

	LiCl	HF
Melting point	605°C	−83.1°C
Boiling point	1350°C	19.5°C

70. Ordinarily, two metals, when mixed, form alloys that maintain their metallic character. If the two metals differ sufficiently in electronegativity, they can form compounds with significant ionic character. Consider the solid produced by mixing equal chemical amounts of Cs and Rb compared with that produced by mixing Cs and Au. Compute the electronegativity difference in each case, and determine whether either has significant ionic character. If either compound is ionic or partially ionic, which atom carries the net negative charge? Are there alkali-metal halides with similar or smaller electronegativity differences?

*71. The electronegativities of the elements in Group IV of the periodic table are as follows: C, 2.55; Si, 1.90; Ge, 2.01; Sn, 1.88; Pb, 2.10. Explain the lack of a clear-cut trend. Is any trend seen in Group III? (*Hint*: Consider the relative influence of the size and the nuclear charge of the elements on the electronegativity.)

*72. A stable triatomic molecule containing one atom each of nitrogen, sulfur, and fluorine can be formed. Three bonding structures are possible, depending on which is the central atom: NSF, SNF, and SFN.
   (a) Write a Lewis structure for each of these molecules, indicating the formal charges on each atom.
   (b) Often the structure with the least separation of formal charge is the most stable. Is this statement consistent with

the observed structure for this molecule, namely NSF, with a central sulfur atom?

(c) Does consideration of the electronegativities of N, S, and F from Figure 17–13 help to rationalize this observed structure? Explain.

73. Compare the mechanism of covalent bond formation with that of ionic bond formation. Can both mechanisms contribute simultaneously to the strength of a bond?

74. A reference book tabulates five different values of the electronegativity of molybdenum, a different one for each oxidation state from $+2$ to $+6$. Predict which electronegativity is highest and which is lowest.

75. Nothing is known of the chemistry of element 114. Predict the maximum oxidation state of this element. Based on the trends in the oxidation states of other members of its group, is it likely that this oxidation state will be the dominant one?

*76. H_3PO_2 is an oxoacid that has a first acidity constant of 8×10^{-2}. Predict the molecular structure H_3PO_2. Is this acid monoprotic, diprotic, or triprotic in aqueous solution?

77. Fluorine chemists sometimes say, "Fluorine gives wings to the metals," meaning that metal fluorides, particularly those of transition elements in high oxidation states, are volatile. Explain why this is so.

CUMULATIVE PROBLEM

Iodine

The shiny purple-black crystals of elemental iodine were first prepared in 1811 from the ashes of seaweed. Several species of seaweed concentrate the iodine that is present in very low proportions in seawater. For many years, seaweed continued as the major practical source of this element. Today, iodine is produced from natural brines via oxidation of iodide ion with chlorine.

(a) Write the ground-state electron configuration for the iodine atom.

(b) Is the first ionization energy of an iodine atom larger or smaller than the first ionization energy of its immediate neighbors in the periodic table, tellurium and xenon? Make the same comparison for the electron affinity.

(c) Iodine is an essential trace element in the human diet. Iodine deficiency causes goiter, the enlargement of the thyroid gland. To prevent goiter, much salt intended for human consumption is "iodized" by the addition of small quantities of sodium iodide. Calculate the electronegativity difference between sodium and iodine. Is sodium iodide an ionic or a covalent compound? What is its chemical formula?

(d) Iodine is an important reactant in synthetic organic chemistry because bonds form readily between carbon and iodine. Use electronegativities to determine whether the C—I bond is ionic, purely covalent, or polar covalent in character.

(e) The highest oxidation state observed for iodine is $+7$. A synthetic route to this state begins with the disproportionation, by heating, of barium iodate (in which iodine is in the $+5$ oxidation state):

$$5\ Ba(IO_3)_2 \longrightarrow Ba_5(IO_6)_2 + 4\ I_2 + 9\ O_2$$

Treatment of $Ba_5(IO_6)_2$ with concentrated nitric acid yields white crystals of orthoperiodic acid (H_5IO_6). Dehydration of this acid gives periodic acid (HIO_4) but not the acid anhydride (I_2O_7). Why is the highest oxidation state of iodine $+7$ and not a higher or lower number?

(f) Rank the three oxoacids HIO_3, H_5IO_6, and HIO_4 according to their expected acid strength.

Iodine sublimes from the bottom and recondenses at the top of this flask. *(Larry Cameron)*

18

Molecular Orbitals, Spectroscopy, and Atmospheric Chemistry

CHAPTER OUTLINE

Water waves in a wave tank. Interference between standing waves in water can be used to model the combination of atomic orbitals to give molecular orbitals. *(Courtesy of Central Scientific Company)*

Gaseous nitrogen and oxygen make up 99% of the Earth's atmosphere. The two substances resemble each other in some ways (both are diatomic molecular substances of low molar mass that liquefy only below 100 K) but differ in fundamental respects. The chemical bond in nitrogen is significantly stronger than the bond in oxygen; oxygen is paramagnetic, but nitrogen is not; oxygen reacts far more readily than nitrogen. What causes these differences, and how do they shape the roles the two gases play in the structure and dynamics of the atmosphere?

In this chapter, we answer these questions by broadening the ideas presented in Chapters 16 and 17. We show how *molecular* orbitals, a natural extension of atomic orbitals, help to explain the stability of molecules and the nature of their bonds. We explore the electron configurations of molecules and present an aufbau principle for molecules that is similar to the one for atoms. We examine the rearrangement of electrons in polyatomic molecules that leads to the valence shell electron-pair repulsion (VSEPR) model for molecular geometry. Finally, we consider the interactions between light (radiant energy) and molecules. Light provides our most important probe for measuring the properties of molecules: their geometries, bond lengths, and bond energies. Light also can play a crucial role in changing molecules, causing bonds to break or rearrange, and inducing chemical reactions both in the laboratory and in the environment.

18-1   Diatomic Molecules

The role of valence electrons in covalent chemical bonds is touched on at several points in this book. In the Lewis model (see Section 3–4), covalent bonding is introduced as originating quite simply from the tendency of atoms to achieve noble-gas electron configurations by sharing electrons. Section 17–5 points out that electrons located between nuclei exert electrostatic forces to pull the nuclei together. Chapter 16 establishes the part-wave, part-particle nature of electrons: An electron is not simply "at" some point in space; instead, an electron is best described by a characteristic wave function, the square of which gives the *probability* of finding the electron in a very small element of volume surrounding the point. This applies to electrons in molecules as well as in atoms. In molecules, the wave functions, or orbitals, may be spread out, or *delocalized,* over several atoms. Orbitals that span two or more atoms are called **molecular orbitals,** in distinction to atomic orbitals, which are localized on single atoms. In this section, we discuss the molecular orbitals of diatomic molecules. Polyatomic molecules are considered in the next section.

• Our molecular orbital model is a simplified quantum-mechanical description that contains most, although not all, of the physical factors that lead to the formation of chemical bonds.

Molecular Orbitals and Covalent Bonding

The simplest possible bonded species is the one-electron molecular ion H_2^+. The wave functions of the electron in H_2^+ are the solutions of the Schrödinger equation for an electron bound in the electrostatic field of *two* nuclei of charge $+e$ separated by a distance R. A full calculation of the ground-state and higher-energy molecular orbitals in H_2^+ has been carried out numerically, but the answer is complex. For many purposes, a pictorial description is sufficient and provides useful insight. In this description, the molecular orbitals in H_2^+ are regarded as constructed by *superposition* of the atomic orbitals centered on each of the two nuclei. As mentioned previously (and shown in Fig. 16–17), the **overlap** (superposition) of waves leads to two kinds of interference: If the waves are *in phase,* interference is constructive, and the amplitudes add to give a larger total amplitude; if the waves are *out of phase,* interference is destructive, and the amplitudes cancel to give smaller amplitude or zero amplitude (a node is formed). Thus, when two $1s$ atomic orbitals (wave functions) on neighboring centers are superposed, two different molecular orbitals form. In the first, constructive interference gives a higher amplitude (and therefore a greater average electron density) between the two nuclei. In the second, destructive interference gives a lower amplitude, and a node appears between the nuclei.

The two molecular orbitals arising from overlap of two $1s$ atomic orbitals are designated as σ_{1s} (constructive interference) and σ_{1s}^* (destructive interference).

> The Greek letter σ (sigma) in a label indicates that the electron density in the molecular orbital in question is distributed symmetrically about a bond axis.

The subscripts in the designations tell which atomic orbitals are combined, that is, the parentage of the molecular orbitals. The two modes of combination and the electron probability densities in the resulting molecular orbitals are shown in Figure 18–1. The figure makes it clear that an electron in the σ_{1s} orbital has an enhanced probability of being found between the nuclei, so that the σ_{1s} orbital is a **bonding orbital.** In contrast, an electron in the σ^* orbital has a *reduced* probability of being found between the nuclei, so that the σ_{1s}^* orbital is an **antibonding orbital.** The σ_{1s}^* antibonding molecular orbital has a higher energy because it has a node. Therefore, the single electron occupies the σ_{1s} molecular orbital in the ground state of H_2^+. If the electron is excited into the σ_{1s}^* orbital, then the H_2^+ ion dissociates.

A molecular orbital, just like an atomic orbital, can hold two electrons if they are spin-paired (one with spin quantum number $+\frac{1}{2}$ and the other with spin quantum number $-\frac{1}{2}$). The H_2 molecule has the same molecular orbitals as does the H_2^+

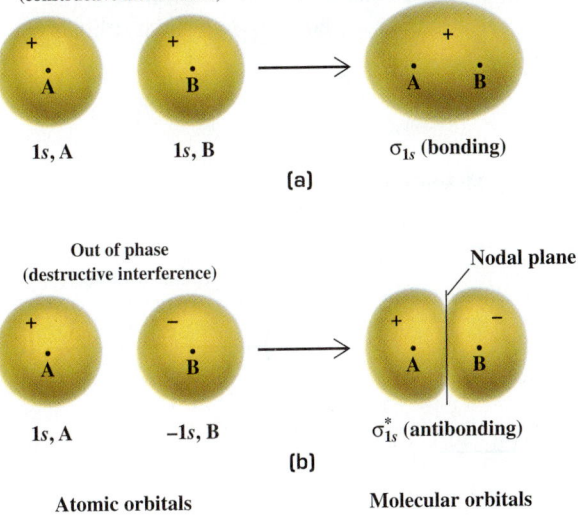

In phase
(constructive interference)

$1s, A$ $1s, B$ σ_{1s} (bonding)

(a)

Out of phase
(destructive interference)

Nodal plane

$1s, A$ $-1s, B$ σ_{1s}^* (antibonding)

(b)

Atomic orbitals   **Molecular orbitals**

Figure 18–1 • The $1s$ atomic orbitals on atoms A and B overlap in two ways. (a) If the two have the same phase (either $+, +$ or $-, -$), they are *in phase* and interfere constructively to give a σ_{1s} bonding molecular orbital that has *increased* electron density between the nuclei. (b) If the two have different phases (either $+, -$ or $-, +$) they are *out of phase* and interfere destructively to give a σ_{1s}^* antibonding molecular orbital that has *decreased* electron density (a nodal plane) between the nuclei.

ion but has two electrons. In its ground state, H_2 accommodates the two with opposing spins in a σ_{1s} bonding molecular orbital. Quite generally, covalent bonding arises from the sharing of electrons (most often electron pairs with opposite spins) in bonding molecular orbitals. In these orbitals, the electron probability density is largest between the nuclei and tends to pull the nuclei together. Electron sharing *alone* is not sufficient for chemical-bond formation. Electrons that are shared in an *antibonding* molecular orbital tend to force the nuclei apart, reducing the bond strength.

> • The name "antibonding" was well chosen. An electron in an antibonding orbital in a molecule diminishes the stability of the molecule.

Correlation Diagrams

The relative energies and the parentage of molecular orbitals are often displayed in **correlation diagrams,** of which Figure 18–2 is an example. This diagram shows how two $1s$ atomic orbitals of equal energy on different centers (atom A and atom B) mix to give a lower-energy σ_{1s} molecular orbital and a higher-energy σ_{1s}^* molecular orbital. It lists the molecular orbitals in order of energy and correlates them with the atomic orbitals from which they derive.

Correlation diagrams are used to arrive at *molecular* electron configurations. Available electrons are fed into the molecular orbitals, starting with the one of lowest energy and obeying the Pauli principle (in effect, a maximum of two electrons per orbital). The process is completely analogous to the aufbau of the electron configurations of atoms and results in a ground-state electron configuration for the molecule as a whole. The correlation diagram in Figure 18–2 works for H_2^+, which has the electron configuration $(\sigma_{1s})^1$, for H_2, which has the electron configuration $(\sigma_{1s})^2$, and for other first-period diatomic species as well. More complex molecules require more complex correlation diagrams.

In the Lewis theory of chemical bonding, a shared pair of electrons corresponds to a single bond; two shared pairs, to a double bond; and so forth (see Section 17–4). In the molecular-orbital theory, electrons can be shared in antibonding orbitals as well as bonding orbitals. Antibonding electrons reduce the bond strength, however, and also the bond order. Therefore, the definition of bond order is expanded as follows:

$$\text{bond order} = \tfrac{1}{2}(\text{number of electrons in bonding molecular orbitals} - \text{number of electrons in antibonding molecular orbitals})$$

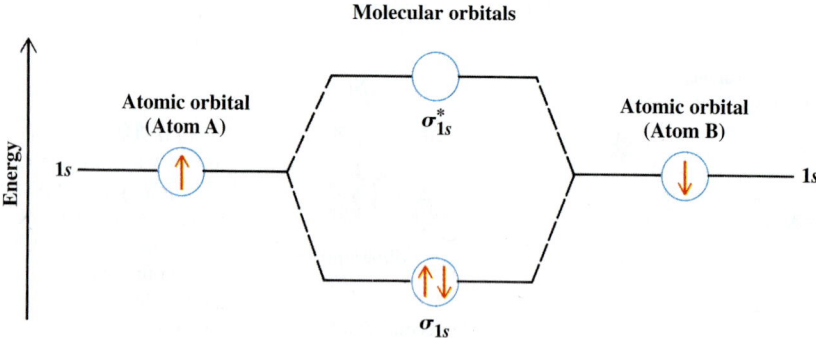

Figure 18–2 • A correlation diagram for first-period diatomic molecules and ions. In the ground-state H_2 molecule (shown here), two electrons (represented by red arrows) occupy the σ_{1s} molecular orbital.

EXAMPLE 18-1

Give the ground-state electron configuration and the bond order of the He_2^+ molecular ion.

Solution

The He_2^+ ion has three electrons. Feeding three electrons into the molecular orbitals represented in the center of the correlation diagram in Figure 18–2 gives the ground-state configuration $(\sigma_{1s})^2(\sigma_{1s}^*)^1$. This notation indicates that the ion has a doubly occupied σ_{1s} orbital (bonding) and a singly occupied σ_{1s}^* orbital (antibonding). The bond order is

$$\text{bond order} = \tfrac{1}{2}(2 \text{ electrons in } \sigma_{1s} - 1 \text{ electron in } \sigma_{1s}^*) = \tfrac{1}{2}$$

This should be a weaker bond than that in H_2.

EXERCISE

Give the ground-state electron configuration and the bond order of the H_2^{2-} molecular ion.

Answer: Electron configuration: $(\sigma_{1s})^2(\sigma_{1s}^*)^2$. The bond order is zero (and H_2^{2-} is not observed as a stable molecular ion).

Homonuclear Diatomic Molecules

The ground-state electron configurations of **homonuclear** (that is, having identical nuclei) diatomic molecules and molecular ions made from first-period elements are shown in Table 18–1. These configurations are simply a listing of the occupied molecular orbitals in order of increasing energy, together with the number of electrons in each orbital. The observed bond lengths and bond enthalpies also are given. Higher bond order corresponds to larger bond enthalpies and shorter bond lengths. The species He_2 has a bond order of zero and has never been detected.

A general prescription for obtaining a molecular-orbital description of the bonding in molecules now can be written:

1. Combine the atomic orbitals to form molecular orbitals. The total number of molecular orbitals formed in this way must equal the number of atomic orbitals used.

• Very weak attractive interactions do exist between helium atoms. These are due to dispersion forces, as mentioned in Section 6–1. They are not sufficient to sustain He_2 molecules as distinct entities.

TABLE 18-1	Configurations and Bond Orders for First-Row Homonuclear Diatomic Molecules			
Species	**Electron Configuration**	**Bond Order**	**Bond Enthalpy (kJ mol⁻¹)**	**Bond Length (Å)**
H_2^+	$(\sigma_{1s})^1$	$\tfrac{1}{2}$	255	1.06
H_2	$(\sigma_{1s})^2$	1	431	0.74
He_2^+	$(\sigma_{1s})^2(\sigma_{1s}^*)^1$	$\tfrac{1}{2}$	251	1.08
He_2	$(\sigma_{1s})^2(\sigma_{1s}^*)^2$	0	Not observed	

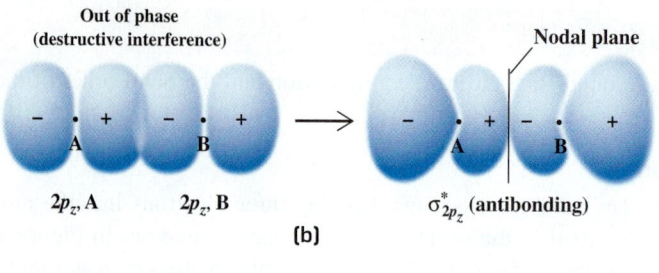

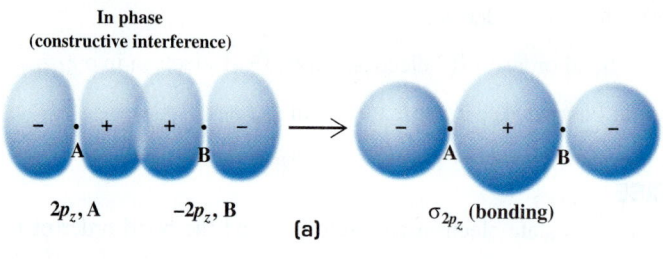

Figure 18-3 • The overlap of $2p_z$ orbitals on neighboring atoms A and B. (a) Constructive (in-phase) interference gives a σ_{2p_z} bonding molecular orbital. (b) Destructive (out-of-phase) interference gives a $\sigma_{2p_z}^*$ antibonding molecular orbital. The signs on the orbitals give relative phase, not electric charge.

2. Arrange the molecular orbitals in order from lowest to highest energy.
3. Put in electrons (at most two per molecular orbital), starting from the orbital of lowest energy. Apply Hund's rule where appropriate.

This procedure is readily applied to second-period homonuclear diatomic molecules. The $2s$ atomic orbitals of the two atoms are mixed in the same fashion as $1s$ orbitals, giving one σ_{2s} bonding orbital and one σ_{2s}^* antibonding orbital. The $2p$ orbitals are of two different types, depending on whether they are oriented parallel to or perpendicular to the internuclear (bond) axis. The z axis customarily is taken to lie along the bond. The lobes of the $2p_z$ orbitals of the two atoms then are oriented along this axis, pointing directly at each other. The orbitals combine "end to end" to give a bonding σ_{2p_z} molecular orbital and an antibonding $\sigma_{2p_z}^*$ molecular orbital (Fig. 18–3). These orbitals are σ molecular orbitals because, like the σ_{1s} orbital and the σ_{1s}^* orbital, their electron density is distributed symmetrically around the internuclear axis.

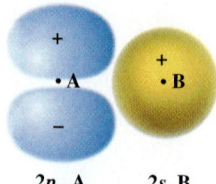

Figure 18-4 • Non-overlap of a $2p_z$ orbital on atom A with a $2s$ orbital on neighboring atom B. Constructive interference (+ phase with + phase) in the region above the axis connecting the two atoms is cancelled by destructive interference (− phase against + phase) in the region below.

The situation differs with the $2p_x$ and $2p_y$ orbitals. The lobes of these orbitals are oriented perpendicular to the bond axis. Neither can combine with the $2s$ orbital on the other atom because constructive interference on the side of the bond axis on which they match the $2s$ in phase always is exactly cancelled by destructive interference on the other side, where their phase opposes that of the $2s$ (Fig. 18–4). On the other hand, two $2p_x$ orbitals, which are oriented "side by side" perpendicular to the bond axis, *can* combine because their phases match, plus to plus and minus to minus. They form a bonding and an antibonding molecular orbital (Fig. 18–5). These orbitals position maximum electron density on either side of the internuclear axis, with that axis lying in a nodal plane (in this case, the yz plane). They are designated by the Greek letter π rather than σ.

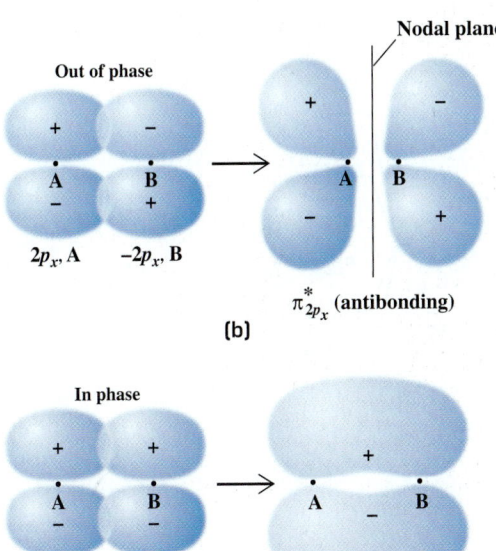

Figure 18–5 • The overlap of $2p_x$-orbitals on neighboring atoms A and B. These orbitals lie side by side but still interact. (a) In-phase interference gives a π_{2p_x} bonding molecular orbital. (b) Out-of-phase interference gives a $\pi^*_{2p_x}$ antibonding molecular orbital. The signs on the orbitals give relative phase, not electric charge.

> The Greek letter π (pi) in a label indicates that the electron density in the molecular orbital in question has a nodal plane that contains a bond axis.

The π_{2p_x} orbital is bonding, and the $\pi^*_{2p_x}$ is antibonding. The two $2p_y$ orbitals also can combine "side by side," giving a π_{2p_y} orbital and a $\pi^*_{2p_y}$ orbital. Their lobes project above and below the xz plane, which is the plane of the page in Figure 18–5.

It bears repeating that the $+$ and $-$ signs in Figures 18–1, 18–3, 18–4, and 18–5 do *not* refer to positive or negative electric charges but to the phase of the wave function in the various regions of space. The relative phase of the two atomic orbitals as they overlap determines whether the resulting molecular orbital is bonding or antibonding. Bonding orbitals come from the overlap of wave functions having the same phase in the same region; antibonding orbitals come from the superposition of wave functions having opposite phase in the same region.

The next step is to determine the energy ordering of the molecular orbitals. In general, this requires a calculation that is beyond the scope of this text, but these qualitative observations can be offered:

1. The *average* energy of a bonding–antibonding pair of molecular orbitals lies approximately at the energy of the original atomic orbitals.
2. The difference in energy between members of a bonding–antibonding pair increases as the degree of overlap of the atomic orbitals increases.

In second-period diatomic molecules, the σ_{1s} bonding orbitals and σ^*_{1s} antibonding orbitals are equally occupied (by two electrons each). Therefore, these core electrons have little net effect on bonding properties and need not be considered. Once again, we see that chemical bonding is dominated by valence-electron behavior. Correlation diagrams for the molecular orbitals formed by the overlap of $2s$ and $2p$ orbitals are shown in Figure 18–6. Two diagrams are required because two different orderings of energy are found for molecular orbitals in diatomic molecules

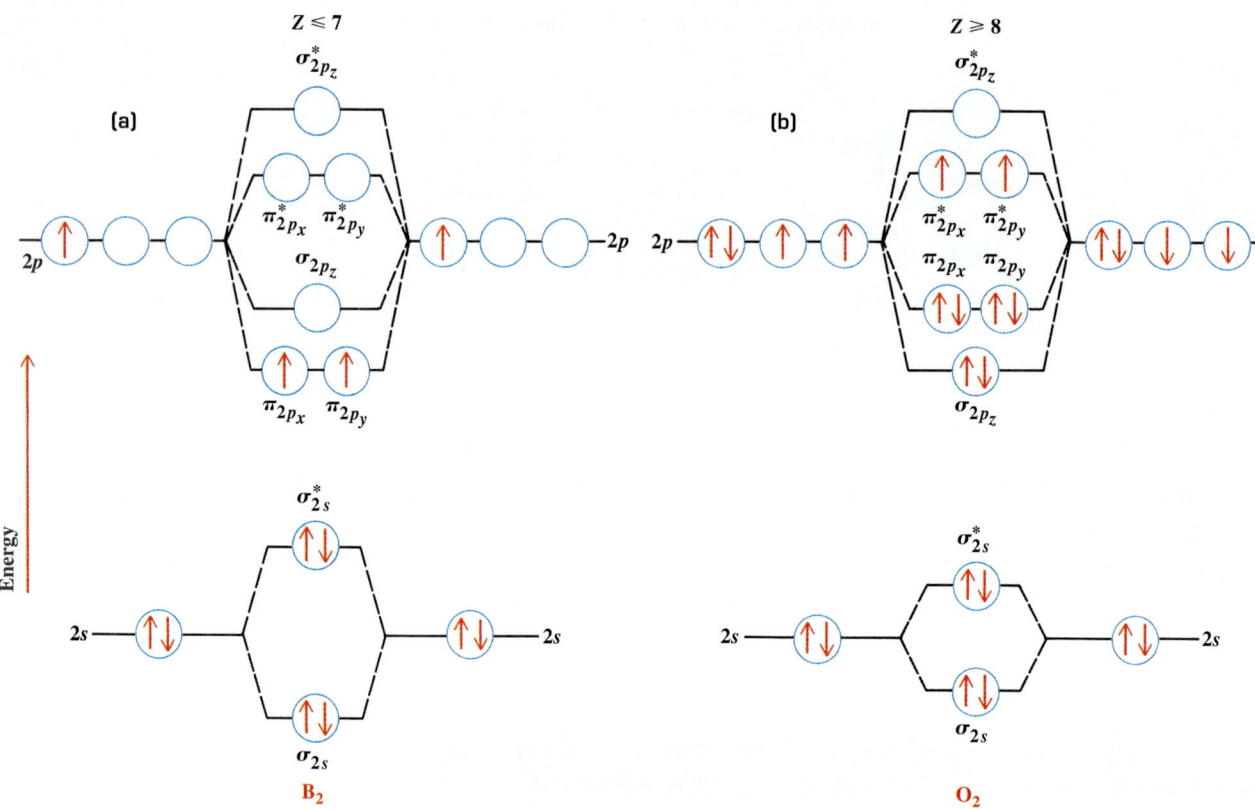

Figure 18-6 • (a and b) Correlation diagrams for second-period diatomic molecules. In each case, the atomic orbitals at the sides combine to give the molecular orbitals in the center. The red arrows represent the valence electrons in B_2 and O_2. Each of these two molecules is paramagnetic, having two unpaired electrons.

formed from second-period elements. The first (Fig. 18–6a) applies to the molecules with atoms lithium through nitrogen (that is, the first part of the period) and their positive and negative molecular ions. The second (Fig. 18–6b) applies to molecules of the later elements oxygen, fluorine, and neon and their positive and negative molecular ions. The distinction between the two is with the π_{2p} and σ_{2p} orbitals, which are quite close in energy and reverse their order once oxygen is reached.

Valence electrons are now put into the correlation diagram, starting with the orbital of lowest energy and working up. The maximum number of electrons per orbital remains two, in accord with the Pauli principle. Placement of all available valence electrons gives the ground-state molecular orbital configuration. The overall bond orders can be calculated as before or expressed in terms of **σ bonds** and **π bonds**:

• It is important in counting electrons at this stage to include *only* valence electrons.

$$\text{no. of } \sigma \text{ bonds} = \tfrac{1}{2}(\text{number of electrons in } \sigma \text{ molecular orbitals}$$
$$- \text{ number of electrons in } \sigma^* \text{ molecular orbitals})$$

$$\text{no. of } \pi \text{ bonds} = \tfrac{1}{2}(\text{number of electrons in } \pi \text{ molecular orbitals}$$
$$- \text{ number of electrons in } \pi^* \text{ molecular orbitals})$$

$$\text{overall bond order} = \text{number of } \sigma \text{ bonds} + \text{number of } \pi \text{ bonds}$$

EXAMPLE 18-2

Determine the ground-state electron configuration, bond order, and number of σ and π bonds in the F_2 molecule.

Strategy

Use the correlation diagram in Figure 18–6b. Recall that an atom of fluorine has 7 valence electrons and so take care to represent 14 electrons in the final electron configuration of F_2.

Solution

The electron configuration for the ground state of F_2 is:

$$(\sigma_{2s})^2(\sigma_{2s}^*)^2(\sigma_{2p_z})^2(\pi_{2p})^4(\pi_{2p}^*)^4$$

Eight valence electrons occupy bonding orbitals, and six occupy antibonding orbitals. The bond order is

$$\text{bond order} = \tfrac{1}{2}(8 - 6) = 1$$

and the F_2 molecule has a single bond. This bond arises from electrons in a σ molecular orbital and is therefore a σ bond. There are no π bonds because there are equal numbers of electrons in π bonding and π^* antibonding orbitals.

EXERCISE

Determine the ground-state valence electron configuration, bond order, and number of σ and π bonds in the B_2^- molecular ion.

Answer: Ground-state valence electron configuration: $(\sigma_{2s})^2(\sigma_{2s}^*)^2(\pi_{2p})^3$. Bond order $= \tfrac{1}{2}(5 - 2) = \tfrac{3}{2}$. Zero σ bonds and $\tfrac{3}{2}\pi$ bonds.

Table 18–2 summarizes the bonding formalism and the properties of first- and second-period homonuclear diatomic molecules. The relationship between bond order, bond length, and bond strength is clear. The bond orders obtained from molecular-orbital theory agree completely with the results of the Lewis electron-dot model.

TABLE 18-2	Molecular Orbitals of Homonuclear Diatomic Molecules				
Species	Number of Valence Electrons	Ground-State Valence-Electron Configuration	Bond Order	Bond Length (Å)	Bond Enthalpy (kJ mol^{-1})
H_2	2	$(\sigma_{2s})^2$	1	0.746	436
He_2	4	$(\sigma_{1s})^2(\sigma_{1s}^*)^2$	0	Not observed	
Li_2	2	$(\sigma_{2s})^2$	1	2.67	106
Be_2	4	$(\sigma_{2s})^2(\sigma_{2s}^*)^2$	0	2.45	9
B_2	6	$(\sigma_{2s})^2(\sigma_{2s}^*)^2(\pi_{2p})^2$	1	1.59	297
C_2	8	$(\sigma_{2s})^2(\sigma_{2s}^*)^2(\pi_{2p})^4$	2	1.24	607
N_2	10	$(\sigma_{2s})^2(\sigma_{2s}^*)^2(\pi_{2p})^4(\sigma_{2p_z})^2$	3	1.10	945
O_2	12	$(\sigma_{2s})^2(\sigma_{2s}^*)^2(\sigma_{2p_z})^2(\pi_{2p})^4(\pi_{2p}^*)^2$	2	1.21	498
F_2	14	$(\sigma_{2s})^2(\sigma_{2s}^*)^2(\sigma_{2p_z})^2(\pi_{2p})^4(\pi_{2p}^*)^4$	1	1.41	158
Ne_2	16	$(\sigma_{2s})^2(\sigma_{2s}^*)^2(\sigma_{2p_z})^2(\pi_{2p})^4(\pi_{2p}^*)^4(\sigma_{2p_z}^*)^2$	0	Not observed	

• Hund's rule states that when electrons occupy orbitals of the same energy, the lowest energy state corresponds to the configuration with the greatest number of unpaired electrons (see Section 17–1).

An important prediction, one that is not made by the Lewis model, comes out of the molecular-orbital approach. In the ground state of O_2, the two highest-energy electrons should follow Hund's rule and have parallel spins. Thus, the molecular-orbital approach predicts that O_2 is paramagnetic because it contains unpaired electrons. This is exactly what is found experimentally (Fig. 18–7). In contrast, the Lewis electron-dot structure for O_2,

$$:\overset{..}{O}::\underset{..}{O}:$$

shows all the electrons paired. Molecular-orbital theory also explains the high chemical reactivity of dioxygen (O_2) in a plausible way: The two π^* electrons, which are unpaired and located in different regions of space, readily interact with other molecules.

Heteronuclear Diatomic Molecules

Diatomic molecules, such as carbon monoxide (CO), nitrogen monoxide (NO), and hydrogen fluoride (HF), are **heteronuclear** because they contain two different kinds of atoms. Molecular orbitals can be constructed from their atomic orbitals, just as in the homonuclear case. In the correlation diagrams, the energies of the starting sets of atomic orbitals are no longer the same because the atoms are no longer the same. The energy levels of the more electronegative atom are displaced *downward* because the atom attracts valence electrons more strongly. For many heteronuclear diatomic molecules of second-period elements (those in which the electronegativity difference is not too large), the correlation diagram shown in Figure 18–8 results. It resembles Figure 18–6a, except for the difference just mentioned. This diagram guides assignment of the electron configurations of heteronuclear diatomic molecules and molecular ions of the second period. As an example, take CO, which has ten valence electrons (four from C, six from O). Its ground-state valence-electron configuration is

$$(\sigma_{2s})^2(\sigma_{2s}^*)^2(\pi_{2p})^4(\sigma_{2p_z})^2$$

The CO molecule has eight electrons in bonding orbitals and two in antibonding orbitals for a bond order of $\frac{1}{2}(8 - 2) = 3$ (one σ bond and two π bonds). Carbon

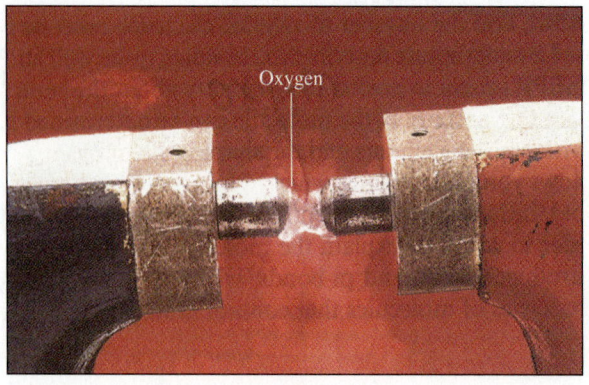

(a)

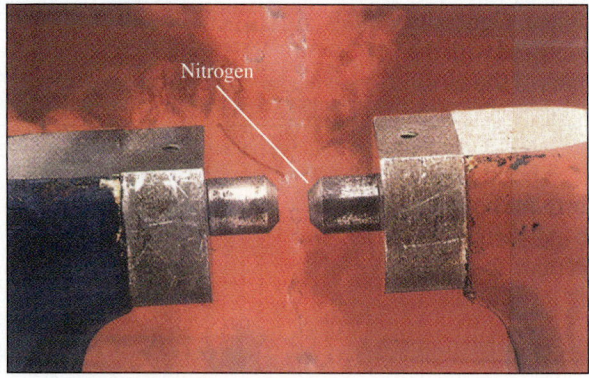

(b)

Figure 18–7 • (a) Oxygen is paramagnetic; liquid oxygen poured between the pole faces of a magnet is attracted and held there. (b) When the experiment is repeated with liquid nitrogen, which is diamagnetic, the liquid pours straight through. *(Larry Cameron)*

Molecular orbitals

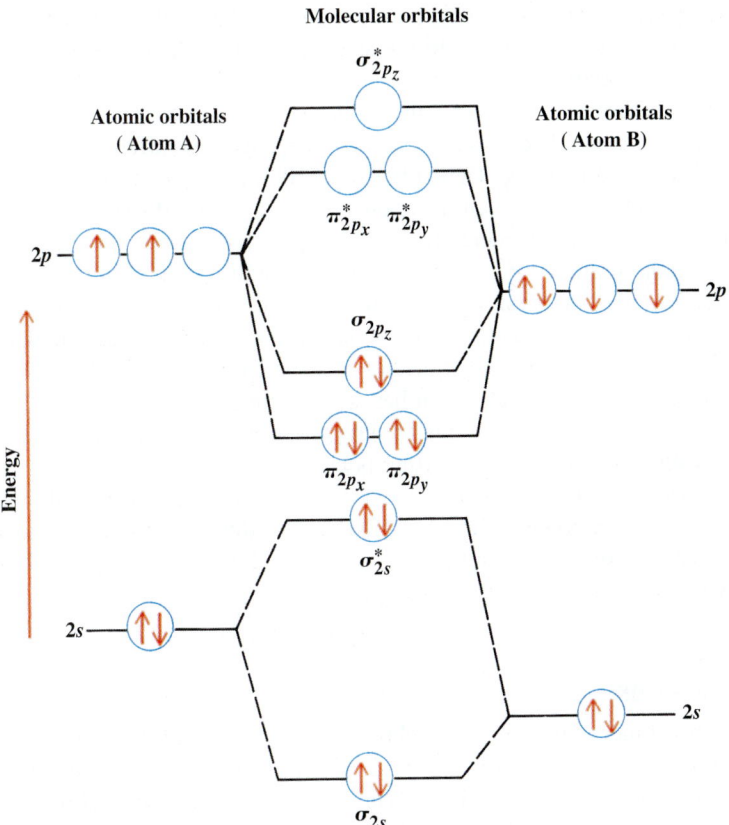

Figure 18-8 • Correlation diagram for second-period heteronuclear diatomic molecules of the general formula AB. The atomic orbitals for the more electronegative atom (B) are displaced downward because they have lower energies than do those of the less electronegative atom (A). The red arrows show the orbital filling for CO in its ground state.

monoxide is indeed tightly bound; it has the largest bond enthalpy (1076 kJ mol^{-1}) of all second-period diatomic molecules. Nitrogen monoxide resembles carbon monoxide somewhat but possesses an additional electron in an antibonding orbital (a π_{2p}^* orbital), giving the ground-state configuration:

$$(\sigma_{2s})^2(\sigma_{2s}^*)^2(\pi_{2p})^4(\sigma_{2p_z})^2(\pi_{2p}^*)^1$$

Hence, NO is paramagnetic and has a bond order of only $2\frac{1}{2}$. Its bond enthalpy is 631 kJ mol^{-1}.

18-2 Polyatomic Molecules

Bonding in polyatomic molecules can be described in two ways. The first uses *delocalized* molecular orbitals, with electron wave functions that are spread out over the entire molecule. These molecular orbitals are first constructed from valence atomic orbitals, then placed in order of increasing energy, and finally filled with the available valence electrons. Complete delocalization encounters a fundamental *chemical* objection. Bonds of a given type have approximately constant properties such as enthalpy and length (see Section 17–4). If electrons are described by molecular orbitals that spread over the entire molecule, then why do the properties of a given type of bond depend only weakly on the particular molecule in which it finds itself? Another theoretical approach to bonding develops from this consideration—one that uses orbitals that are *localized*, either as bonding orbitals between particular pairs of atoms or as lone-pair orbitals on individual atoms. This approach, **valence-bond theory,** is related to the qualitative Lewis theory, in which a similar localization of electrons is assumed.

Which description is better—the delocalized molecular-orbital picture or the localized valence-bond approach? Each has advantages and affords insights into the nature of chemical bonding. They differ primarily in their treatment of the *correlation* of the positions of different electrons. In the delocalized molecular-orbital picture, the electrons are considered to move independently of each other, so that there is a reasonable probability that several electrons will be found in the same region of space. The valence-bond approach minimizes this by localizing electron distributions in individual bonds or on individual atoms. In most cases, the delocalized picture underestimates the importance of electron correlation, but the valence-bond approach overestimates it. The truth lies somewhere in between, and accurate computational techniques for predicting the properties of small molecules have been developed using each model as a starting point.

In this section, localized (valence-bond) orbitals are used for σ bonds and, wherever possible, for π bonds. Delocalized molecular orbitals are brought in to describe π bonds as required in certain bonding situations. This approach gives physical insight without mathematical complication.

Hybridization

Valence-bond theory uses **hybrid orbitals** that are created by combining (mixing) the familiar s, p, and d atomic orbitals of Chapter 16 in various ways. Recall that atomic orbitals are wave functions. Hybridization is a mathematical operation performed on these functions to give new atomic orbitals. Chemists hybridize orbitals because the orientations and shapes of the hybrids explain the geometry and bonding in many molecules better than do the orientations and shapes of the standard set of atomic orbitals. Note that hybridization goes on in the mind of the chemist. It is *not* a physical change in an atom. We show the use of hybridization in valence-bond theory graphically (rather than mathematically) in a series of examples.

BeH₂.   The molecule of beryllium hydride has four valence electrons (two from the Be atom and one from each H atom). The valence shell electron-pair repulsion (VSEPR) theory (see Section 3–7) assigns a steric number of 2 to the Be atom and so predicts that BeH_2 is linear, with the H atoms on exactly opposite sides of the central Be atom. This geometry is observed experimentally.

The ground-state valence-electron configuration of Be is $2s^2$; that is, Be has two spin-paired electrons in a spherically symmetric orbital. How can such an atom form two bonds by sharing electrons with hydrogen atoms? The answer starts with the recognition that the $2p$ subshell lies close above the $2s$ subshell in energy. It costs only a small amount of energy to promote one valence electron from the $2s$ subshell to the $2p$ subshell, giving the Be atom the excited-state valence configuration $2s^1 2p^1$. The next step is to hybridize the half-filled $2s$ with the half-filled $2p$ to generate two new, half-filled atomic orbitals, the **sp hybrid atomic orbitals** (Fig. 18–9a). This mathematical operation does *not* alter the overall distribution of the electrons. An electron in an sp orbital is, however, more likely to be found on one side of the nucleus than on the other, as Figure 18–9 shows. The directional character of the sp hybrid orbitals suggests the next step in the construction of BeH_2. The H atoms approach from opposite sides. The half-filled $1s$ orbital of the first H atom overlaps with one half-filled sp hybrid orbital to give a σ bond while the $1s$

• An sp hybrid orbital is a "daughter" of an s orbital and a p orbital and is named after them.

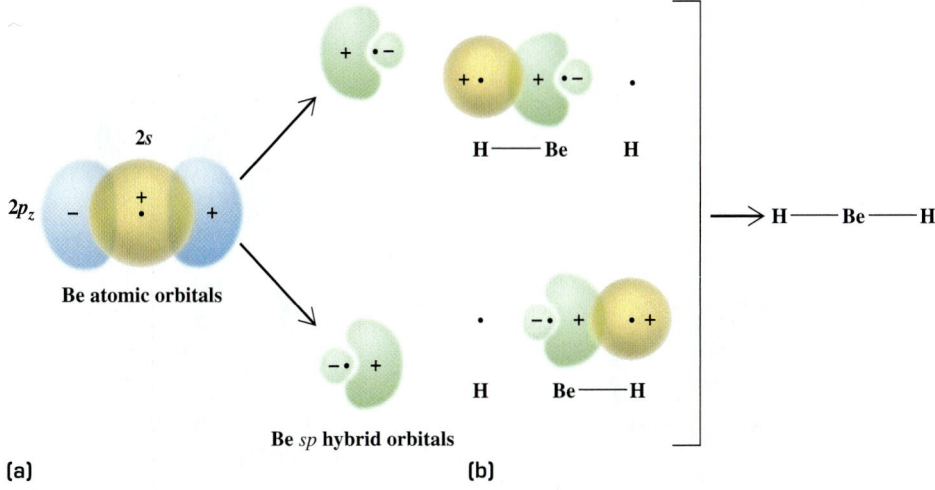

Figure 18-9 • (a) Formation of two *sp* hybrid orbitals (*green*) from the 2*s* orbital (*yellow*) and a 2*p* orbital (*blue*) on a Be atom. The use of the $2p_z$-orbital is strictly a convention. Using the $2p_x$ or $2p_y$ gives the same picture. (b) Two *σ* bonds form. The first comes from overlap of the first *sp* orbital on Be with a 1*s* orbital on an H atom. The second comes from a similar overlap involving the second *sp* orbital.

orbital of the second H does the same with the second *sp* hybrid orbital (Fig. 18–9b). The formation of the bonds more than makes up for the energy investment in promoting an electron in the Be atom from the 2*s* orbital to the 2*p* orbital. The directional character of the *sp* hybrid orbitals correctly accounts for the observed molecular geometry.

BF_3. Because the central B atom has steric number 3, the VSEPR theory predicts a trigonal planar structure. The ground-state valence-electron configuration of the B atom is $2s^2 2p^1$. Promotion of a 2*s* electron to the 2*p* subshell gives the B atom the excited-state valence configuration $2s^1 2p_x^1 2p_y^1$. Hybridization of these three atomic orbitals gives three *sp^2* **hybrid atomic orbitals** (Fig. 18–10). The angles between the regions of maximum electron density (the large lobes) of the sp^2 orbitals all equal 120°, a result that originates naturally in the mathematics when an *s* orbital and two *p* orbitals are hybridized. Each sp^2 orbital on the B atom contains one electron. Overlap of these half-filled orbitals with half-filled 2*p* atomic orbitals on three F atoms gives three *σ* bonds directed from the B atom toward F atoms at the corners of an equilateral triangle. The BF_3 molecule has exactly this geometry.

CH_4. The central C has a steric number of 4, and the VSEPR theory accordingly predicts a tetrahedral structure. The ground-state valence-electron configuration of C is $2s^2 2p^2$. Promotion of a 2*s* electron to the empty 2*p* orbital gives the excited-state configuration $2s^1 2p_x^1 2p_y^1 2p_z^1$. Hybridization of the 2*s* orbital with all three 2*p* orbitals gives four equivalent *sp^3* **hybrid atomic orbitals.** The large lobes of these hybrid orbitals point from the center of a tetrahedron toward the corners (Fig. 18–11). Again, the directional character of the hybrid set originates naturally in the mathematics when the starting set is hybridized. Each half-filled sp^3 hybrid orbital overlaps with a half-filled 1*s* orbital on an H atom to give a *σ* bond. This accounts for all eight valence electrons. Formation of the four *σ* bonds releases more energy than was required to promote the 2*s* electron. The result is a tetrahedral structure for CH_4.

• This notation breaks a previous pattern. The superscript "2" in "sp^2" gives the number of parent *p* orbitals, *not* the number of electrons that occupy it. Thus "sp^2" refers to a single orbital, and "$(sp^2)^2$" would signify the occupancy of that orbital by two electrons.

Figure 18–10 • Shapes and relative spatial dispositions of the three sp^2 hybrid orbitals, formed from $2s$, $2p_z$, and $2p_y$ atomic orbitals on a single atom.

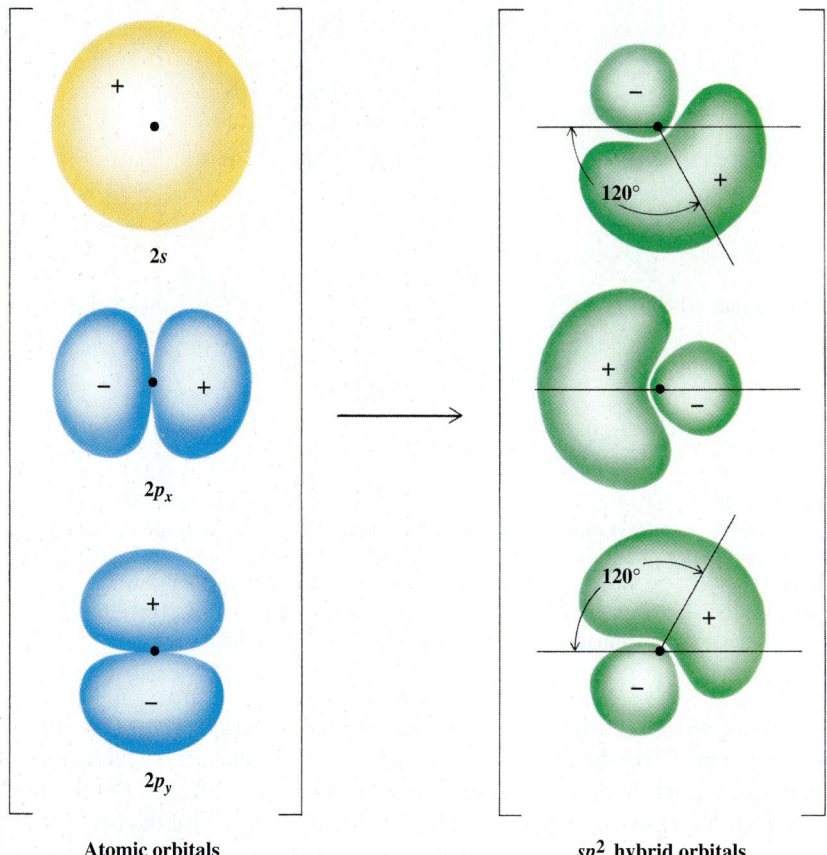

Atomic orbitals   **sp^2 hybrid orbitals**

NH$_3$.   The nitrogen atom in NH_3 has a steric number of 4 (like the C atom in methane). If the four valence orbitals on the N atom are sp^3 hybridized, then three sp^3 hybrid orbitals can overlap with $1s$ orbitals of the three H atoms to give three σ bonds (using six valence electrons) while the fourth sp^3 orbital accommodates the other two valence electrons, which are the lone pair on the N atom. The approach predicts that all three H-N-H angles in ammonia should equal 109.5° (the tetrahedral angle). Experimentally, they equal 107.3°.

H$_2$O.   The O atom has a steric number of 4. Using sp^3 hybridization on the O atom predicts an H-O-H bond angle of 109.5° as two σ bonds form by overlap of sp^3 orbitals with $1s$ orbitals on the H atoms. The experimental bond angle is 104.5°.

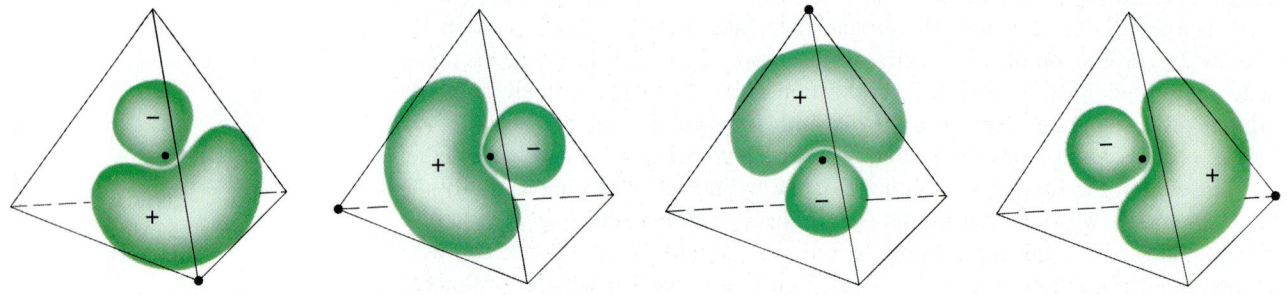

Figure 18–11 • Shapes and relative spatial dispositions of the four sp^3 hybrid orbitals, which point to the corners of a tetrahedron.

The discussion establishes these relationships between the steric numbers used in the VSEPR theory and mode of hybridization:

Steric Number	Hybridization	Geometry
2	sp	Linear
3	sp^2	Trigonal planar
4	sp^3	Tetrahedral

This table helps in developing valence-bond pictures of molecules and making predictions of molecular geometries. A step-by-step procedure for making such predictions follows:

1. Draw a Lewis structure of the molecule to determine the steric number of each of its atoms.

2. Hybridize the orbitals on the atoms as dictated by their steric numbers. (Note that the number of hybrid atomic orbitals formed always equals the number of atomic orbitals that is mixed.)

3. Generate σ bonds that are directed along the axes connecting atoms by allowing each atom's hybrid orbitals to overlap with the atomic orbitals, either hybrid or unhybridized, on its neighbors.

4. Combine knowledge of the three-dimensional geometry at each atom to arrive at a picture of the overall shape of the molecule.

EXAMPLE 18–3

Predict the three-dimensional structure of ethane (C_2H_6).

Strategy

Draw a Lewis structure for ethane. Determine the steric number, the hybrid orbitals, and the local geometry at each atom. Combine the geometries to arrive at the molecular structure.

Solution

The Lewis structure of ethane is

$$\begin{array}{ccc} & H & H \\ & \ddot{} & \ddot{} \\ H:\ddot{C}::\ddot{C}:H \\ & \ddot{} & \ddot{} \\ & H & H \end{array}$$

This diagram tells the connections of the atoms, but not their spatial arrangement. The two C atoms have four bonds and zero lone pairs. Their steric numbers are 4. (The steric numbers of the six H atoms are 1, but this is immaterial to the structure of the molecule.) Steric number 4 goes with sp^3 hybridization. Bonds point outward from the C atoms toward the corners of a tetrahedron. The six H-C-C bond angles and six H-C-H bond angles thus should approximate 109.5°. Figure 18–12 shows the structure, which can be described as two tetrahedra joined at a corner. Rotation about the σ bond that connects the C atoms takes place readily because the electron density in a σ bond is symmetrical about the bond axis; the rotation does not affect the bond strength. Related structures that have rotated relative positions of the two sets of H atoms therefore are possible.

Figure 18–12 • The structure of ethane. The four atoms surrounding each carbon lie at the corners of a tetrahedron.

EXERCISE

Predict the three-dimensional structure of hydrazine (H_2NNH_2).

Answer: Both N atoms have steric number 4 and are sp^3 hybridized. The four H-N-H bond angles and two H-N-H bond angles should approximate 109.5°. The structure of hydrazine appears on page 216.

The valence-bond theory and the VSEPR theory obviously are connected closely. In fact, the valence-bond theory can be thought of as an attempt to model the shapes from the VSEPR theory using the mathematical formalism of orbitals. Ultimately, neither is more than a device to rationalize observed bond angles and the observed equivalence of bonds in some molecules. They sometimes encounter difficulties. For example, the bond angles in the heavier hydrides of the Group V elements do not follow the example of ammonia. All are much closer to 90° than to 109.5°:

Compound	Bond Angles
NH_3	107.3°
PH_3	93.6°
AsH_3	91.8°
SbH_3	91.3°

These molecules are all pyramidal and the bonds within each are equivalent. The rationalization of the observed angles is that the lone pairs in these compounds must have more *s* orbital character than assigned by standard sp^3 hybridization and the σ bonds correspondingly more *p* orbital character.

It is possible to extend the valence-bond picture to deal with steric numbers greater than 4 by hybridizing selected *d* orbitals with *s* orbitals and *p* orbitals. This creates new sets of hybrid orbitals that can be used to explain trigonal bipyramidal and octahedral structures. The extension is not very useful, however. Like the simple Lewis electron-dot model, the hybridization approach has limits and is most useful in describing the bonding of the lighter elements with steric numbers up to 4.

Double and Triple Bonds

In molecules containing double or triple bonds, electrons take part in π bonds over and above the underlying framework of σ bonds. As an example, consider ethylene (C_2H_4), which has the Lewis structure

$$\begin{matrix} H & & H \\ \cdot\cdot & & \cdot\cdot \\ C & :: & C \\ \cdot\cdot & & \cdot\cdot \\ H & & H \end{matrix}$$

Both C atoms have steric numbers of 3 (obtained by counting three attachments to other atoms and zero lone pairs). The valence orbitals at each C atom are accordingly sp^2 hybridized. Four of the six half-filled sp^2 hybrid orbitals overlap head-to-head (σ overlap) with 1s orbitals on four H atoms. The other two, one on each C atom, overlap head-to-head with each other. The resulting framework of σ bonds accounts for 10 of the 12 valence electrons. Assume that all of these σ bonds lie in the same plane. If the orbitals used in sp^2 hybrid orbitals are the $2s$, $2p_x$, and $2p_y$ orbitals, then this is the xy plane, and the $2p_z$ orbitals project side by side perpendicular to this plane. They can combine to make a π_{2p_z} bonding molecular orbital and a $\pi^*_{2p_z}$ antibonding orbital, as illustrated previously (Fig. 18–5). The π bonding orbital accommodates the last two valence electrons and so raises the bond order of the link between the C atoms to 2. The π^* antibonding molecular orbital remains empty.

It was assumed that all of the atoms in ethylene lie in a plane. Can one CH_2 group rotate around the C=C double bond relative to the other, so that its two hydrogen atoms depart from the xy plane (Fig. 18–13)? Such a rotation reduces the mixing of the two $2p_z$ orbitals as their regions of high amplitude move out of alignment. When the rotation reaches 90°, the mixing vanishes because the region in which one p_z orbital has largest amplitude lies in exact conflict with the zero-amplitude nodal plane of the other. The π bond ceases to exist. The C—C σ bond remains intact during any rotation, but even partial loss of the π bond raises the energy of the molecule substantially. Therefore, a planar molecule is favored. Experimentally, the molecule of ethylene is indeed planar with both H-C-H bond angles equal to 115.5° and all four H-C-C bond angles equal to 122.2°.

A similar procedure can be applied to the molecule of acetylene (C_2H_2), which contains a triple bond. The Lewis structure is

$$H:C\!\equiv\!C:H$$

The molecule has ten valence electrons, and the steric number of each C atom is 2. Label the axis connecting the two C atoms as the z axis. Then, mixing the $2s$ orbital with the $2p_z$ orbital on each C atom gives two half-filled sp hybrid orbitals, as in BeH_2. These overlap with one another and with the $1s$ orbitals on the two H atoms to form three σ bonds that use six valence electrons. The $2p_x$ atomic orbitals on the two C atoms overlap side by side to form a π_{2p_x} and a $\pi^*_{2p_x}$ molecular orbital, while the $2p_y$ orbitals do the same to give a π_{2p_y} and a $\pi^*_{2p_y}$ molecular orbital. Filling the two bonding π molecular orbitals with the remaining four valence electrons raises the order of the C—C bond to 3. It is a triple bond. The two antibonding π^*

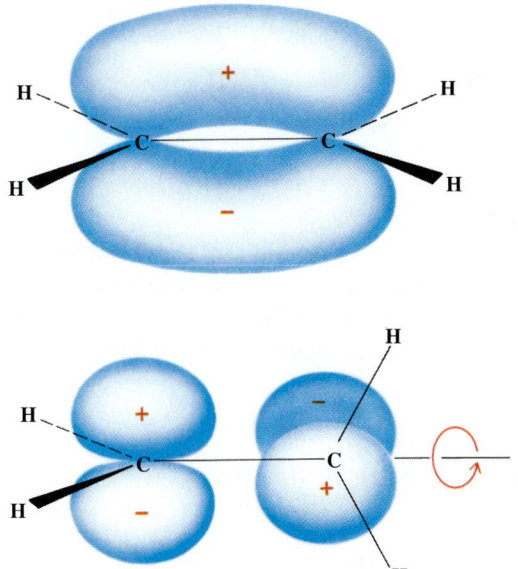

Figure 18–13 • If the ethylene molecule is twisted around the C=C axis, the overlap of the two p_z orbitals decreases, giving a higher energy. The most stable conformation of ethylene is planar.

molecular orbitals, which are at higher energy, remain empty. The directions of the σ bonds lead to a prediction that the molecule is linear. This is confirmed experimentally.

• The naming of organic compounds is discussed in Chapter 24.

Figure 18–14 • The molecular structure of propene. Valence orbitals are sp^2 hybridized on two carbons and sp^3 hybridized on the third.

EXAMPLE 18–4

Discuss the bonding and predict the geometry of the molecule of propene (CH_2CHCH_3). Its Lewis structure is

$$
\begin{array}{c}
\text{H}\quad\text{H}\quad\text{H} \\
\ddot{\text{C}}::\ddot{\text{C}}:\ddot{\text{C}}:\text{H} \\
\ddot{\text{H}}\qquad\ddot{\text{H}}
\end{array}
$$

Strategy

Determine the steric number of each C atom to obtain its hybridization scheme. The hybridization scheme indicates the geometry at each atom. Realize that the overall shape of the molecule is a consequence of these geometries.

Solution

Identify the C atoms for reference: C-1, C-2, and C-3 from left to right. Atoms C-1 and C-2 have steric number 3 and so have sp^2 hybridization. Atom C-3 has steric number 4 and has sp^3 hybridization. The bond geometry is accordingly trigonal planar at C-1 and C-2 and tetrahedral at C-3. The σ bonding framework (six C—H and two C—C bonds) uses 16 of the 18 valence electrons.

The p_z orbitals of C-1 and C-2 overlap to give one π bonding molecular orbital (as well as a π^* antibonding orbital at higher energy). Both remaining valence electrons go into the bonding π molecular orbital, giving a double bond between C-1 and C-2. As in the case of ethylene, the mixing of the two p_z orbitals is best (and the energy is lowest) when the lobes of the two orbitals are parallel to each other. Thus, the hydrogen atoms on C-1 and C-2 should lie in the same plane as the C-C-C skeleton. The predicted (and observed) three-dimensional structure is shown in Figure 18–14.

EXERCISE

The Lewis structure of a molecule of propyne (H_3C—$C\equiv CH$) is

$$
\begin{array}{c}
\text{H} \\
\ddot{\text{C}} \\
\text{H}:\ddot{\text{C}}\text{—}\text{C}\equiv\text{C}:\text{H} \\
\ddot{\text{H}}
\end{array}
$$

Discuss its bonding and predict its overall geometry.

Answer: One carbon atom in the structure is sp^3 hybridized, and the other two carbon atoms are sp hybridized. The atoms in the molecule lie on a single straight line, with the exception of the three hydrogen atoms on the sp^3 hybridized carbon atom, which point outward toward three of the corners of a tetrahedron (Fig. 18–15).

Figure 18–15 • The molecular structure of propyne. Valence orbitals are sp hybridized on two carbons and sp^3 hybridized on the third.

The compound 2-butene has the formula $CH_3CHCHCH_3$ and the Lewis structure

$$
\begin{array}{c}
\text{H}\quad\text{H}\qquad\text{H} \\
\text{H}:\ddot{\text{C}}:\ddot{\text{C}}::\ddot{\text{C}}:\ddot{\text{C}}:\text{H} \\
\ddot{\text{H}}\qquad\ddot{\text{H}}\ \ddot{\text{H}}
\end{array}
$$

The bonding at the outer two carbon atoms in this structure can be described by sp^3 hybridization, and at the inner two, by sp^2 hybridization. The bonding molecular orbital formed from the mixing of the p_z orbitals on the two central atoms has its lowest energy when the lobes of these atomic orbitals are parallel, giving a planar structure for the carbon skeleton of the molecule. However, the outer CH_3 (methyl) groups can be placed in two ways relative to the double bond (Fig. 18–16). In one, the *cis* form, the two CH_3 groups lie on the same side of the double bond. In the other, the *trans* form, they lie on opposite sides. Conversion of one form to the other requires breaking the central π bond (by rotating the two p_z orbitals 90° with respect to each other, as in Fig. 18–13) and then reforming it as the rotation continues from 90° to 180°. Both the *cis* form and *trans* form of 2-butene have been prepared. Because breaking a π bond costs a significant amount of energy, interconversion between the two is very slow at room temperature. Molecules that have the same chemical formula but different structures are isomers; this type of isomerism is called **cis–trans isomerism.**

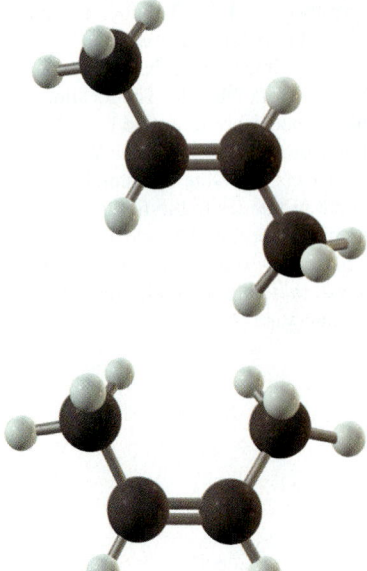

Figure **18–16** • The two *cis–trans* isomers of 2-butene.

18–3 The Conjugation of Bonds and Resonance Structures

When double or triple bonds alternate with single bonds in a molecule, the π electrons often interact, a phenomenon called **conjugation.** The molecular orbital picture, with its emphasis on delocalization, explains this effect better than does the valence-bond approach. Conjugation occurs in 1,3-butadiene ($H_2CCHCHCH_2$), which has the Lewis structure

$$\begin{array}{ccccc} H & H & & H \\ \ddot{} & \ddot{} & & \ddot{} \\ C & :: C & : C & :: C \\ \ddot{} & & \ddot{} & \ddot{} \\ H & & H & H \end{array}$$

All four carbon atoms have steric numbers of 3, so all are sp^2 hybridized. The remaining p_z orbitals have maximum overlap when the four carbon atoms lie in the same plane; accordingly, this molecule is predicted to be planar. From these four p_z atomic orbitals, four π molecular orbitals can be constructed by combining phases as shown in Figure 18–17. The four electrons that remain after the σ bonding is complete go into the two lowest-energy π orbitals. The first of these is bonding among all four carbon atoms, and the second is bonding between the outer carbon-atom pairs but antibonding between the central pair; 1,3-butadiene therefore has stronger and shorter bonds between the outer carbon pairs than between the two central carbon atoms. It is a **conjugated π electron system.** Conjugated systems have lower energy than do comparable non-conjugated systems.

Resonance Hybrids

Conjugation also occurs in molecules that, in the language of the Lewis theory of Chapter 3, are "resonance hybrids." Benzene is a ring molecule with the formula C_6H_6. Two Lewis structures can be written for it:

Figure 18–17 • The four π molecular orbitals formed from four $2p_z$ atomic orbitals in 1,3-butadiene, viewed from the side. The locations of planar nodes are indicated by red lines; orbitals with more nodes are of higher energy. Note the similarity in nodal properties to the first four harmonics of a vibrating string (*right*). Only the two lowest-energy orbitals are occupied in the molecule in the ground state (*left*).

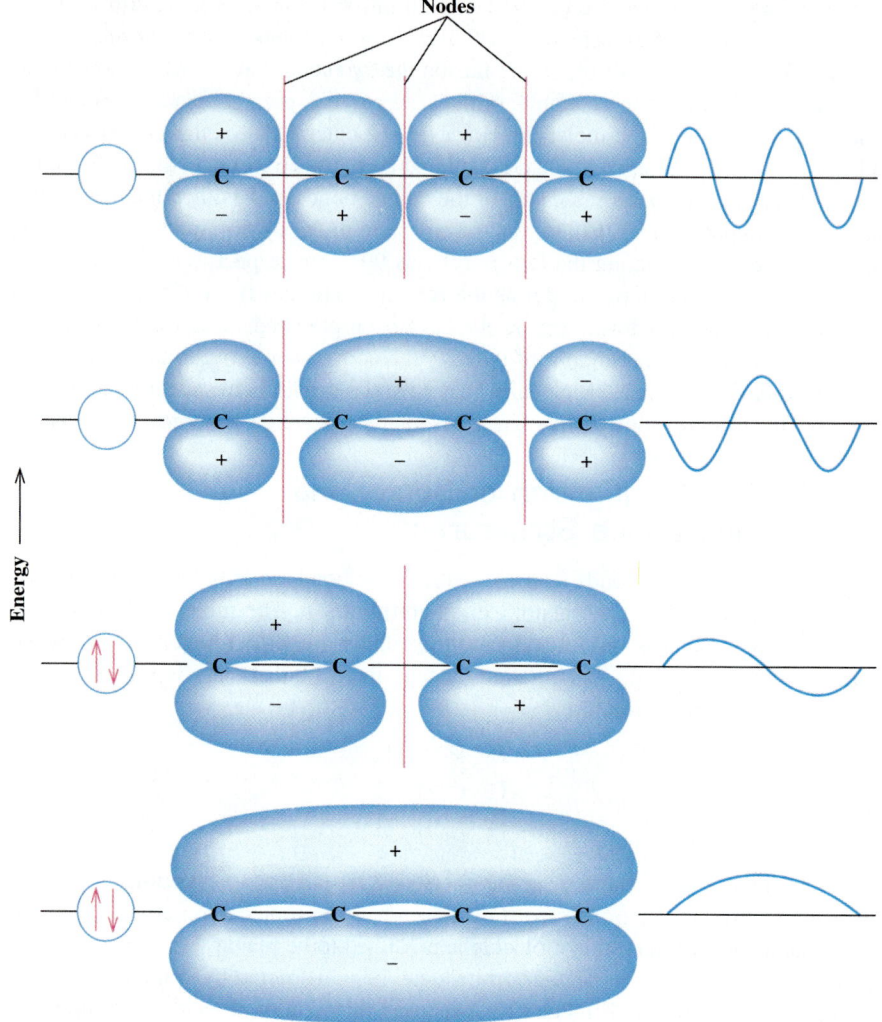

These structures are often called "Kekulé structures," after the 19th-century German chemist August Kekulé, who first formulated the ring structure of benzene. The modern view is that the molecule does not flip-flop from moment to moment between the two structures but rather has a single, time-independent electron distribution in which the π bonding takes place through delocalized molecular orbitals. Each carbon atom has sp^2 hybridization, which accounts for three of its four valence orbitals. The remaining six p orbitals (one from each carbon atom and conventionally labelled p_z) overlap to give six molecular orbitals delocalized over the entire molecule. These six π orbitals and their energy-level diagram are shown in Figure 18–18. The lowest-energy π orbital has one nodal plane (the plane of the molecule). The next two have two nodal planes each (the plane of molecule and a plane perpendicular to it). The next two have three nodal planes each, and the highest-energy π orbital (which is strongly antibonding) has four nodal planes. The C_6H_6 molecule has 30 valence electrons, of which 24 maintain the σ bond framework (six C—C single bonds and six C—H single bonds). The other six go into the three lowest-energy π orbitals. The shapes of these three orbitals are such that the six C—C bonds are completely equivalent (see Fig. 18-18). Benzene in its ground state has

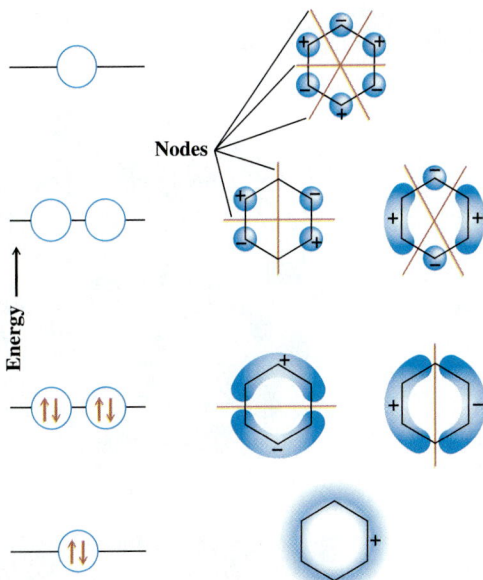

Figure 18-18 • The shapes of the six π molecular orbitals in benzene, viewed from above. The six molecular orbitals lie perpendicular to the plane of the molecule. Note the similarity in nodal properties to the standing waves on a loop shown in Figure 16–16. Only the three lowest-energy molecular orbitals are occupied by electrons in benzene in the ground state.

six C—C bonds of equal length that are intermediate between single and double bonds in their properties.

Other resonance structures can be described using delocalized molecular orbitals. For example, as is discussed in Section 3–5, the bent molecule SO_2 has the two resonance Lewis structures

$$\left[\ddot{O}{=}\ddot{S}{-}\ddot{\ddot{O}}{:} \longleftrightarrow :\ddot{O}{-}\ddot{S}{=}\ddot{\ddot{O}} \right]$$

The alternative to (or amplification of) the concept of resonance in this case is to say that sp^2 hybridization on the central sulfur atom leads to a σ bond between the S and each O at 120° angles and that p orbitals on the three atoms also overlap to give a set of π orbitals. A pair of electrons occupies the lowest energy π orbital, which is a bonding orbital, and is spread out over all three atoms. This delocalization adds a bond order of $\frac{1}{2}$ to each link. Consequently, the bond order of each of the S—O bonds is 1 (from the σ bond) $+ \frac{1}{2}$(from the π bond) $= \frac{3}{2}$. The molecular orbital description accounts for the experimentally observed equivalency of the two S—O bonds quite naturally.

18-4 The Interaction of Light with Molecules

Molecules, like atoms, emit and absorb electromagnetic radiation at discrete frequencies as they undergo transitions between one quantum state and another. **Molecular spectroscopy** concerns itself with the observation and interpretation of these interactions.

Excited electronic states in molecules arise from promotion of electrons into higher-energy molecular orbitals (just as excited electronic states in atoms arise from promotion of electrons into higher-energy atomic orbitals). Electronic transitions in molecules fit in the framework of the molecular-orbital theory developed so far. Molecules also rotate and vibrate, which single atoms cannot do, and the energies of these motions are quantized. Rotation and vibration add a great number of allowed quantum states to the energy-level diagrams of even simple molecules.

CHEMISTRY IN PROGRESS

The Fullerenes

Until 1985, just two crystalline forms of carbon, diamond and graphite, were known (see Section 3–1). In that year, a new form was discovered: buckminsterfullerene, which consists of ball-like molecules with the formula C_{60}. These molecules have a highly symmetrical structure: 60 carbon atoms arranged in a closed net with 20 hexagonal faces and 12 pentagonal faces. Each pentagonal face touches five hexagonal faces, and each hexagonal face touches alternating hexagonal and pentagonal faces (Fig. 18–A). This is exactly the design on the surface of a soccer ball.

C_{60} is named after the architect Buckminster Fuller, who invented the geodesic dome, which the molecular structure of C_{60} also resembles. In C_{60}, every C atom has two single bonds and one double bond for a steric number of 3; all 60 atoms are accordingly sp^2 hybridized. As in benzene, the π electrons of the double bonds are delocalized. This means that the 60 p orbitals (one each from the 60 carbon atoms) do not overlap merely in local double bonds but mix to give 60 molecular orbitals spread over both sides of the entire closed surface of the molecule. The lowest 30 of these molecular orbitals are occupied by the 60 π electrons. Delocalization makes up for the strain imposed in this structure by the distortion of one of the three bond angles at each carbon atom from the 120° associated with sp^2 hybridization down to 108°. The first conjecture that this structure might be chemically stable came in 1970 and was guided by just such bonding theory.

In 1985, Harold Kroto, Robert Curl, and Richard Smalley were interested in certain long-chain carbon molecules that had been discovered spectroscopically by

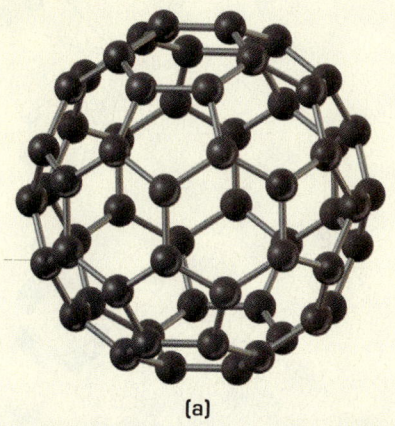

(a)

(b)

Figure 18–A • (a) The structure of C_{60}, buckminsterfullerene. Note the pattern of hexagons and pentagons. (b) The design on the surface of a soccer ball has the same pattern as the structure of C_{60}. (*George Semple*)

Consequently, the spectra of molecules are complex. Molecular-spectroscopy experiments in different regions of the spectrum provide various kinds of information about molecular structure (bond lengths, bond angles, and bond energies). Molecular spectroscopy is in fact a principal source of data about bonding and structure.

Different Kinds of Molecular Spectroscopy

The type of information furnished by a spectroscopy experiment depends on the frequency of the radiation that is used. Table 18–3 names the regions of the spectrum used in different kinds of molecular spectroscopy and their frequencies, wavelengths, and energies. The information obtained in the different regions includes the following:

X-ray Region.   Very high-frequency (high-energy) photons are called X-rays or gamma rays (see Figure 16–5). They convey tens of thousands of kilojoules per mole of energy. When absorbed by molecules, high-energy photons lift core elec-

astronomers in the vicinity of red giant stars. They sought to duplicate the conditions near these stars by vaporizing a graphite target with a laser beam. Analysis by mass spectrometry (see Section 1–6) of the debris blasted out of the target serendipitously revealed (along with what they hoped for) the presence of a large proportion of molecules of molar mass 721 g mol^{-1}, which corresponds to the molecular formula C_{60}. Although the amount of C_{60} present in their beam was far too small for isolation and direct determination of structure, they correctly suggested the cage structure shown in Figure 18–A and named it. In 1990, other workers succeeded in synthesizing C_{60} in gram quantities by striking an electric arc between two carbon rods held under an inert atmosphere. The carbon vapors condensed to a soot. Extraction of the soot with an organic solvent and chromatography (see Section 7–7) allowed separation of C_{60} from various impurities. Since 1990, it has been found that C_{60} forms in sooting flames when hydrocarbons are burned. Thus, the new form of carbon was (in the words of one of its discoverers) "under our noses since time immemorial," but it was not seen until a clue came from outer space. Kroto, Curl, and Smalley shared the 1996 Nobel Prize in chemistry for their contribution.

Buckminsterfullerene is not the only new form of carbon that emerges from the chaos of carbon vapor condensing at high temperature. A whole family of closed-cage carbon molecules, the *fullerenes,* has been synthesized. These molecules have formulas like C_{70} or C_{84} or even C_{400}. Proper conditions favor *nanotubes,* which consist of seamless, cylindrical shells of thousands of sp^2 hybridized carbon atoms arranged in hexagons (Fig. 18–B). The ends of the tubes are capped by the introduction of pentagons into the hexagonal network. These structures all have a delocalized π electron system that covers the inner and outer surfaces of the cage or cylinder. Nanotubes and the fullerenes offer exciting prospects for materials science and technical applications.

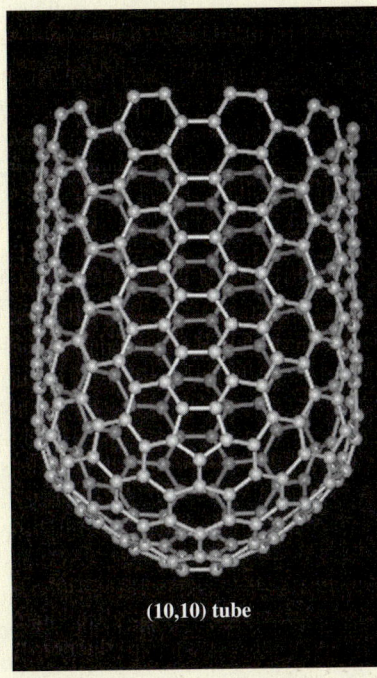

(10,10) tube

Figure 18–B • The structure of a single-walled nanotube. *(Center for Nanoscale Science and Technology at Rice University)*

trons out of stable orbitals close to the nuclei, putting them into high-energy outer orbitals or detaching them entirely from the molecule (as in photoelectron spectroscopy, Section 17–2). The exact frequencies of X-ray absorptions depend on the environment of the absorbing atoms as well as their identity. **X-ray absorption spectroscopy** (XAS) therefore provides information about bonding.

Visible and Ultraviolet Region. Photons in the ultraviolet (UV) and visible regions of the spectrum carry less energy (typically 150–1200 kJ mol^{-1}), which is usually not enough to unseat tightly bound core electrons. Such radiation can excite *valence* electrons to higher unoccupied electronic energy levels, detach valence electrons entirely, and break chemical bonds. Studies of suitable **visible-UV absorption spectra** reveal the energy ordering of σ and π bonding and antibonding molecular orbitals.

Infrared Region. Infrared (IR) radiation is less energetic yet. Its photons have energies in the approximate range of 2 to 150 kJ mol^{-1}, which is insufficient to

• Note that the energies conveyed by photons in the UV and visible region indeed match the typical energies of chemical bonds.

TABLE 18–3	Different Kinds of Molecular Spectroscopy				
Name	**Region of Spectrum**	**Approximate Frequency Range (s^{-1})**	**Approximate Energy ($kJ\ mol^{-1}$)**	**Energy Levels Involved**	**Information Obtained**
Nuclear magnetic resonance	Radio	10^7–10^9	5×10^{-6} to 5×10^{-4}	Nuclear spin states	Connectivity of atoms in molecules
Microwave	Microwave	10^9–10^{12}	0.001 to 1	Rotational	Bond lengths and bond angles
Infrared	Infrared	10^{13}–10^{14}	2 to 150	Vibrational	Stiffness of bonds, identification of compounds
Ultraviolet-visible	Visible, ultraviolet	10^{14}–10^{15}	150 to 1200	Valence electron	Electron configurations
X-ray absorption spectroscopy	X-ray	10^{16}–10^{19}	>10,000	Core electron	Core-electron energies

perturb the electrons of most molecules significantly; however, the absorption of IR radiation can change the *vibrational* states of molecules. The nuclei in molecules do not stand still relative to each other. They oscillate in various motions that depend on the length, strength, and arrangement in space of the bonds that connect them. Bonds stretch like springs; groups of atoms wag or rock relative to the other groups. Incoming radiation with a frequency equal to the natural frequency of an oscillation can cause a molecule to absorb energy and reach a higher vibrational state. For polyatomic molecules, numerous modes of vibrational motion are possible (Fig. 18–19). A typical IR spectrum has tens or even hundreds of absorption peaks because of the many modes of vibration that exist in even fairly simple molecules. If two samples have identical IR spectra, they almost surely contain the same compounds. One stretch of the IR spectrum is in fact called the fingerprint region. At the same time, many *functional groups* (common subgroups in molecules such as C=O double bonds or C—H single bonds) have vibrational frequencies that do not change much from compound to compound. Finding absorptions at the right frequencies strongly suggests the presence in the molecule of certain functional groups.

Microwave Region.   Photons in the microwave region and in the far (long-wavelength) infrared region of the spectrum have energies that range from 0.001 to 1 kJ mol^{-1}. Such radiation excites the low-energy rotational states of a molecule. We think of objects as being able to rotate at any speed, but in fact only certain rotational speeds and energies are possible. Rotation is quantized, and for very small

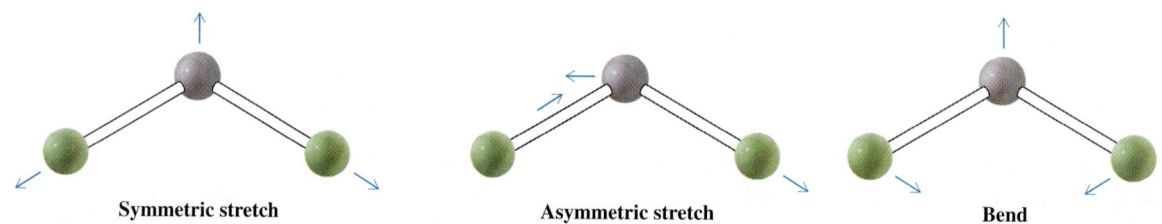

Symmetric stretch   Asymmetric stretch   Bend

Figure 18–19 • The three types of vibrational motion possible for a non-linear tri-atomic molecule. The displacements of each atom during each type of vibration are shown by arrows.

objects such as molecules, this quantization becomes important. The rotational energy depends on the masses of the atoms and on the bond lengths and angles, so **microwave spectroscopy** provides an excellent method of determining molecular structure to high precision.

Radio Region. Finally, **nuclear magnetic resonance** (NMR) spectroscopy employs low-energy (between 0.005 and 0.5 J mol^{-1}) radio waves to "tickle" the nuclei in a molecule. Nuclei have spin (just as electrons have spin), and the energies of different nuclear spin states are split apart by a magnetic field: the stronger the field, the greater the splitting. Nuclei in a magnetic field can change their spin state ("spin-flip") by absorbing or emitting photons of FM radio frequencies. Nuclear spins are sensitive both to the chemical environment of a nucleus (thus to the bonding of the atom) and to the other nuclear spins nearby. NMR spectrometry accordingly offers a way to identify what bonding groups are present in a molecule. For example, ^{13}C nuclei have two spin states. When acetic acid (CH_3COOH) is put into an NMR spectrometer, the magnetic field in the device raises the energy of one state and lowers the energy of the other in the ^{13}C atoms that are present. Irradiation with photons of the right frequency delivers energy to spin-flip ^{13}C nuclei from the lower-energy spin state to the higher. The exact split in energy between spin states depends on the local chemical environment of the nucleus. Therefore, the ^{13}C NMR spectrum of acetic acid has two peaks: (1) a methyl resonance, in which the ^{13}C nuclei in —CH_3 groups absorb energy, and (2) a carboxylic acid resonance, in which the ^{13}C nuclei in —$COOH$ groups absorb energy. The ^{13}C NMR experiment indicates the chemical environment of all the different carbon atoms in even rather complex molecules.

An adaptation of NMR spectrometry that is used in medical diagnosis is called **magnetic resonance imaging** (MRI). It relies on emission of radio-frequency radiation by the ^{1}H nuclei in the water contained by the body's organs. The patient is placed in the opening of a large magnet (Fig. 18–20), and a radio transmitter raises the ^{1}H nuclear spins in the relevant part of the body to their high-energy state. A radio receiver coil detects the radio-frequency photons subsequently emitted. The amplitude of the signal indicates the concentration of water present. Thus, MRI can identify tumors by the excess water in their cells. The time delay until emission occurs is related to the type of tissue. Figure 18–21 shows an MRI of the brain.

Although the modes of interaction of radiation with molecules have been described separately, they usually do not occur separately. Most experimental spectroscopic lines correspond to changes in more than one kind of energy. For example, a line in the visible region of the absorption spectrum may correspond to

• A microwave oven operates by exciting the water molecules in the food to higher energy states. The water molecules gain rotational energy. As they dissipate this energy in collisions, the food is heated.

• The ^{13}C NMR experiment succeeds in compounds containing naturally occurring carbon despite the low fractional abundance (0.0111) of ^{13}C.

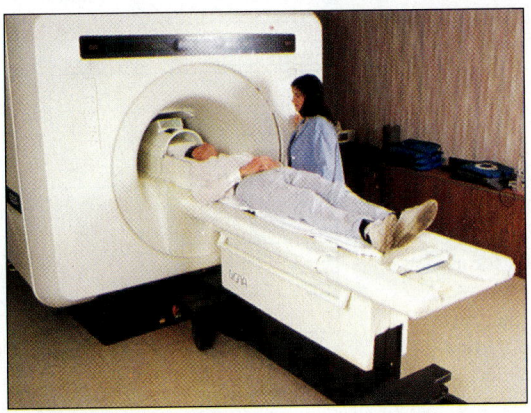

Figure 18–20 • A magnetic resonance imaging (MRI) machine. The patient is placed on the platform and rolled forward into the opening in the magnet. *(Peter Arnold, Inc.)*

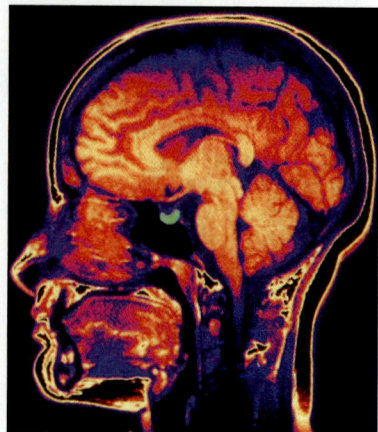

Figure 18–21 • A computer-enhanced MRI scan of a normal human brain with the pituitary gland highlighted. *(Scott Camazine/Photo Researchers, Inc.)*

a change in the molecules' vibrational and rotational energies as well as their electronic energies (Fig. 18–22).

Absorption Spectra and Color

The absorption of visible or UV light causes changes in the valence-electronic states of molecules. Consider the case of ethylene (C_2H_4). The bonding of the two carbon atoms in ethylene is described as the overlap of sp^2 hybridized atomic orbitals on the two carbon atoms, with the two remaining carbon $2p$ orbitals combined into a π bonding molecular orbital and a π^* antibonding molecular orbital (Fig. 18–23). The lowest-energy state of ethylene has two electrons in the π bonding orbital. Exciting an electron from the π orbital to the π^* orbital gives a higher-energy state of the molecule. An ethylene molecule in this excited state has quite different properties from those of one in the ground state. The electron in the π^* antibonding orbital cancels the effect of the electron in the π bonding orbital, and the net bond order is reduced to 1 (the σ bond). Consequently, the molecule in the excited state is more easily dissociated: It has a lower bond enthalpy and a longer C—C bond length. The wavelength of light required to excite this $\pi \rightarrow \pi^*$ transition in ethylene is 162 nm. Samples of ethylene strongly absorb light of this wavelength, which is well into the UV region of the spectrum. Because ethylene does not absorb strongly in the visible region of the spectrum, visible light passes almost unaffected through samples of the substance, which is therefore colorless. As a general rule, small molecules with isolated π bonds (or with σ bonds only) do not absorb light in the visible region and so are colorless.

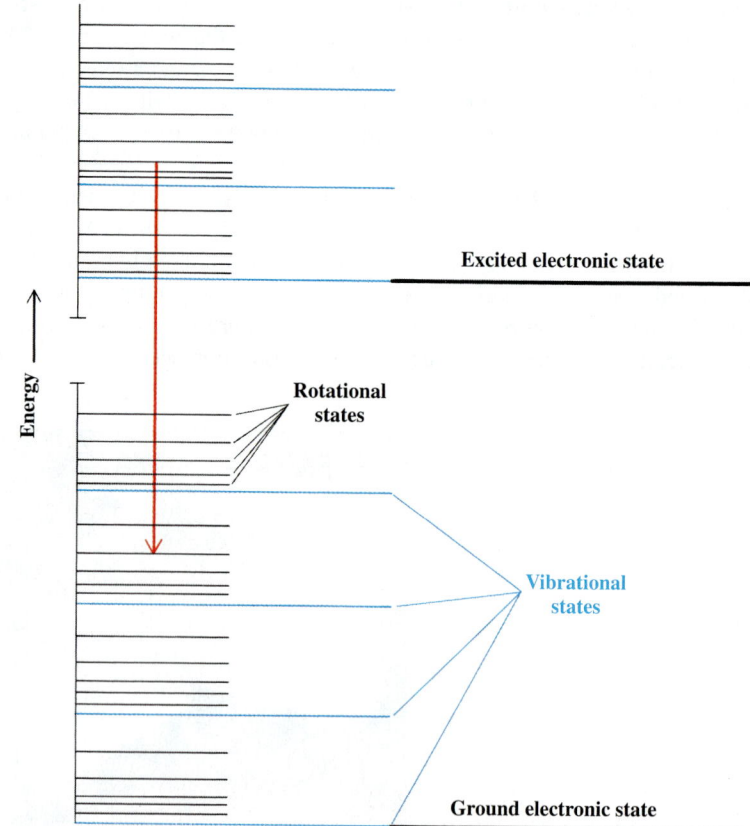

Figure 18–22 • An energy-level diagram for a diatomic molecule, showing the electronic, vibrational, and rotational levels. The red arrow indicates one of the many possible transitions; in this particular transition the rotational, vibrational, and electronic states of the molecules all change.

TABLE 18–4	Absorption of Light by Molecules with π Electron Systems	
Molecule	Number of C=C Bonds	Wavelength of Absorption Maximum (nm)
C_2H_4	1 (non-conjugated)	162
C_4H_6	2 (conjugated)	217
C_6H_8	3 (conjugated)	251
C_8H_{10}	4 (conjugated)	304

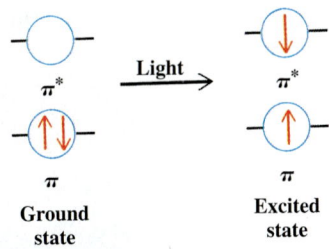

Figure 18–23 • Occupancy of the π molecular orbitals in the ground state and first excited state of ethylene (C_2H_4). Ethylene in the excited state is produced by irradiation of ground-state ethylene with UV light in the appropriate frequency range.

Now, suppose that a molecule has a conjugated π bonding system, that is, bonding in which double bonds alternate with single bonds, as in butadiene or benzene (see Section 18–3). Conjugation lowers the energy of π orbitals by allowing the orbitals to span several atoms, as in butadiene (see Fig. 18–17), rather than remain localized between a pair of atoms, as in ethylene (see Figure 18–13). Computations of the degree of lowering show (and experiments confirm) that conjugation in general lowers the energy of excited-state orbitals *more* than it does the energy of ground-state π orbitals. Consequently, the energy differences surmounted in $\pi \rightarrow \pi^*$ transitions in conjugated molecules are less than those in non-conjugated ones. Increasing the degree of conjugation (adding more alternating single and double bonds) shrinks the difference in energy further. Thus, increased conjugation shifts the energy and the frequency of the absorbed light lower and the wavelength of that light higher (Table 18–4).

For a long enough chain, the first absorption is shifted into the visible region of the spectrum, and the substance takes on color. The color of a material is related to its absorption spectrum in an indirect way. We see the light that is transmitted *through* the material or reflected by the material, *not* the light that is absorbed. In other words, the eye reports the color **complementary** to that which is most strongly absorbed by the sample (Fig. 18–24). An example is the dye beta-carotene (Fig. 18–25), which is responsible for the orange color of carrots, the yellow and orange in certain bird feathers, and the colors of some processed foods. The strength of its absorption at different wavelengths is graphed in Figure 18–26 in comparison with the absorption of the dye indigo. The beta-carotene absorbs strongly in the blue and violet region of the spectrum, and so it appears orange or yellow (the colors complementary to blue and violet) to an observer. Indigo absorbs at longer wavelengths in the yellow-orange region and appears bluish violet.

EXAMPLE 18–5

Suppose that you set out to design a new green dye. Over what range of wavelengths would you want your trial compound to absorb light?

Solution

A good green dye must transmit green light and absorb other colors of light. This can be achieved by a molecule that absorbs *both* in the violet-blue and in the orange-red regions of the spectrum. Thus, you would look for a dye with strong absorptions in both of these regions.

The naturally occurring substance chlorophyll, which is responsible for the green color of grass and leaves, absorbs light over just these wavelength ranges; the plant converts the absorbed solar energy into chemical energy for growth.

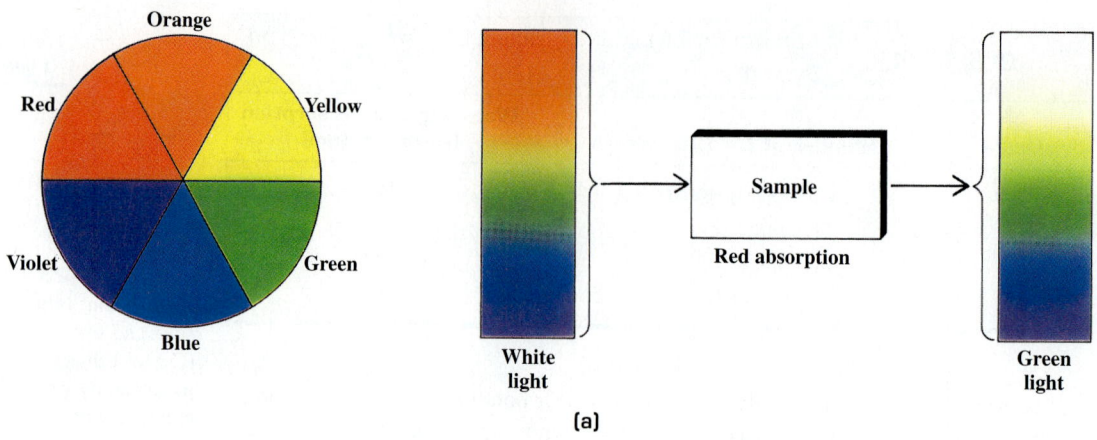

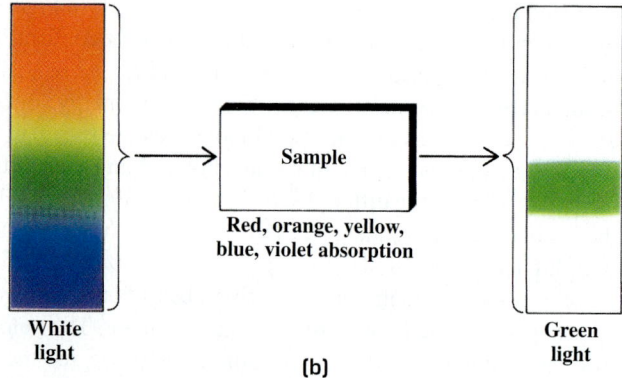

Figure 18–24 • Two ways in which the color green can be produced. (a) If the sample strongly absorbs red light but transmits yellow and blue light, the sample appears green. Green is the complementary color of red, shown opposite to it on the color wheel. (b) If all visible light *except* green is absorbed by the sample, the sample also appears green.

EXERCISE

A translucent layer of red plastic is held over a sheet of white paper so that sunlight shines through it, casting a red light onto the paper. The red plastic then is covered with a translucent layer of yellow plastic. (a) What color is seen on the paper? (b) What wavelengths of light does the yellow plastic absorb?

Answer: (a) Orange. (b) Violet (400–500 nm, approximately).

Photochemistry

What happens to molecules after they absorb radiation? Some emit one or more photons after a longer or shorter delay and so return to their original states. Others lose all of their added energy in collisions with neighboring molecules that raise the local temperature. Yet others "relax" to ground states by a combination of emissions and collisions. An additional outcome exists if a molecule absorbs enough energy to reach an excited electronic state—it may react chemically. Molecules in excited states often fragment or rearrange to give new molecules. **Photochemistry,** the study of the chemical reactions that follow the excitation of molecules to higher electronic states through absorption of photons, is an active area of modern research.

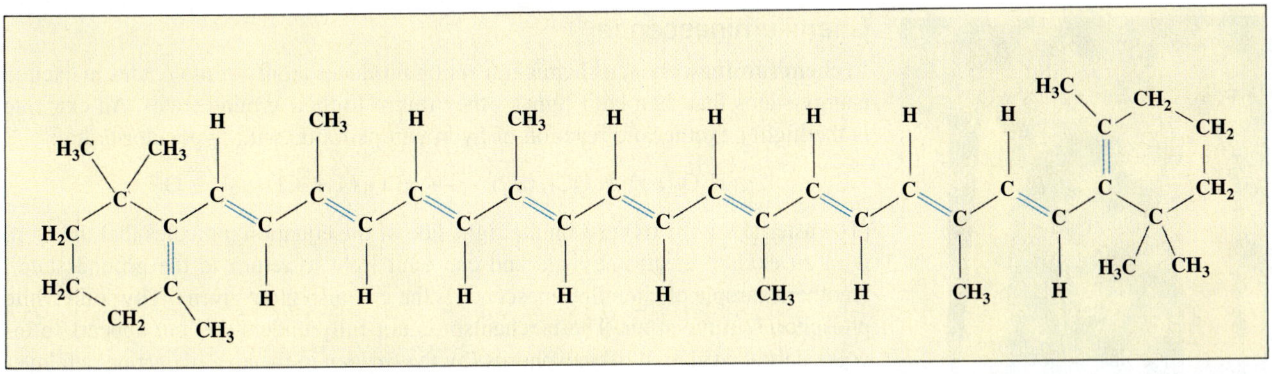

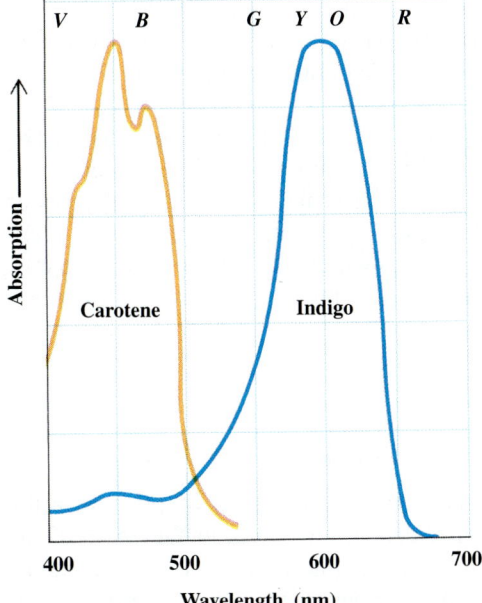

Figure 18–25 • The molecular structure of the dye beta-carotene. Eleven double bonds alternate with single bonds in this conjugated molecule.

Some compounds must be stored in the dark because of their great photolytic sensitivity: They break down rapidly when exposed to light. An example is anhydrous hydrogen peroxide, which reacts explosively when illuminated to give water and oxygen:

$$H_2O_2 \longrightarrow 2\,OH \longrightarrow H_2O + \tfrac{1}{2}O_2$$

The rather weak O—O bond in H_2O_2 breaks apart when the energy of a visible photon is added to the molecule in a process called **photodissociation.**

Another type of photochemical reaction involves changes in molecular structure without the complete breaking of bonds. Molecules like those of *trans*-2-butene (shown in Fig. 18–16) absorb UV light by the excitation of an electron from a π molecular orbital, just as ethylene does (see Fig. 18–23). In the excited electronic state of *trans*-2-butene, the carbon–carbon double bond is effectively reduced to a single bond, and one CH_3 group can rotate relative to the other to form *cis*-2-butene, which is locked in when the energy of excitation is lost subsequently. Absorption of UV light has led to the formation of an isomer of the original molecule.

• The common 3% aqueous solution of hydrogen peroxide is completely safe, but it is usually stored in dark brown bottles to slow down its gradual degradation by photochemical reaction.

Figure 18–26 • The absorption spectra of the two dyes, beta-carotene and indigo, differ in the visible region, giving them different colors. The letters stand for the color of the light at each wavelength (violet, blue, green, yellow, orange, and red).

Chemiluminescence

In **chemiluminescence,** a chemical reaction produces atoms or molecules in excited energy states that then emit light as they return to their ground states. An example is the highly exothermic reaction of hydrogen peroxide with hypochlorite ion:

$$H_2O_2(aq) + OCl^-(aq) \longrightarrow H_2O(\ell) + Cl^-(aq) + O_2^*$$

(The asterisk on the oxygen on the right side of the equation indicates that the atom is in an excited electronic state and can emit light to return to the ground state.) Another example of chemiluminescence is the greenish glow given off by solid white phosphorus in moist air. The mechanism is not fully understood but appears to involve partial oxidation of phosphorus (by the oxygen in the air) in a vapor-rich layer just over the solid to produce excited states of unstable molecules such as $(PO)_2$ and HPO that then emit light.

A third example of chemiluminescence is the reaction used in novelty lightsticks (Fig. 18–27). A molecule derived from oxalic acid (by replacement of its two hydrogen atoms with hydrocarbon groups indicated as R) reacts with hydrogen peroxide according to the equation

$$
\begin{array}{c}
\underset{\displaystyle \parallel}{O}\;\; \underset{\displaystyle \parallel}{O} \\[-0.2em]
RO-C-C-OR + H_2O_2 \longrightarrow \underset{\underset{O-O}{|}}{C} - \underset{|}{C} + 2\,ROH
\end{array}
$$

The product of this reaction, C_2O_4, is a high-energy species that has not been isolated. It decomposes rapidly and exothermically to form carbon dioxide. It can also pass some of its energy along to a molecule of some fluorescent substance (represented by the symbol "Fl") if such is present:

$$Fl + C_2O_4 \longrightarrow 2\,CO_2 + Fl^*$$

where the asterisk indicates that the fluorescer is in an excited electronic state. As molecules of the fluorescer return to their ground states, they emit visible light. In commercial lightsticks, the fluorescer is mixed with the oxalic-acid derivative, and the hydrogen peroxide and a catalyst are kept separate in an inner, sealed tube with thin glass walls. Bending the lightstick breaks the glass tube, the reactants mix, and the reaction begins.

Bioluminescence is the chemical generation of light by living organisms. The most familiar example is the light produced by the firefly, which uses chemical energy from ATP (see Fig. 11–A) in a complex mechanism to excite molecules of substances called luciferins. The flash of the firefly comes as luciferin in an excited state returns to its ground state by emitting a photon. The process is so efficient that luciferins can be used in biological analysis to measure small concentrations of ATP.

Figure 18–27 • These lightsticks and lighttubes emit light through chemiluminescence. The different colors arise from the use of different fluorescers in combination with the same energizing chemical reaction. *(Charles D. Winters)*

18–5   Atmospheric Chemistry and Air Pollution

Nowhere is the ability of light to cause chemical reactions so apparent as in the Earth's atmosphere. Although the chemical composition given in Table 5–1 is correct for the average make-up of the portion of the atmosphere closest to the Earth's surface, it does not do justice to the strong variations in atmospheric properties with

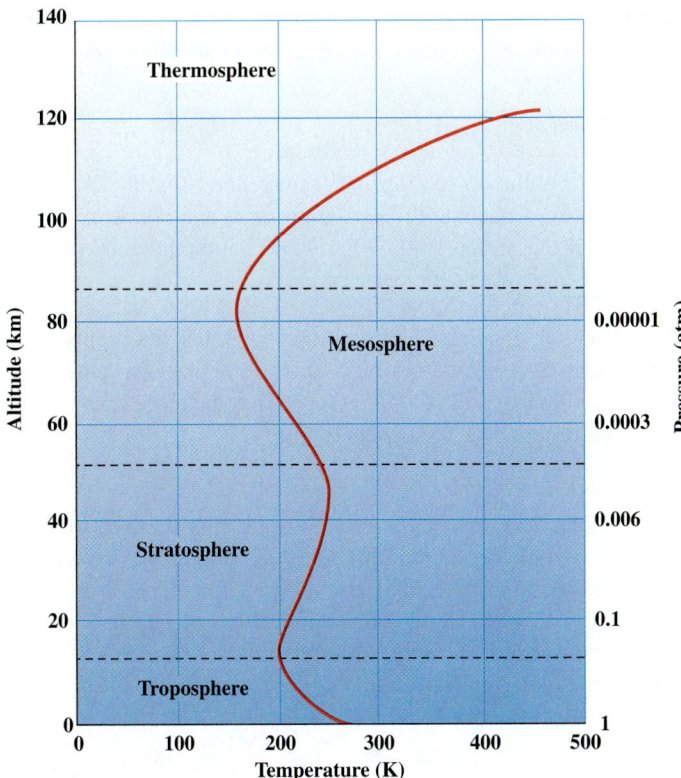

altitude, the dramatic role of local fluctuations in the concentration of trace gases, and the dynamics underlying the average concentrations seen. The atmosphere is a complex chemical system that is far from equilibrium. Its properties are determined by an intricate combination of thermodynamic and kinetic factors. It is a multi-layered structure (Fig. 18–28), bathed in solar radiation and interacting at the bottom with the oceans and land masses. At least four layers are identifiable, each with a characteristic variation of temperature. In the outer two layers (the **thermosphere** and the **mesosphere**), atmospheric density is low and high-energy solar radiation leads to extensive ionization of the particles that are present. The third layer, the **stratosphere,** is the region between 12 and 50 km (approximately) above the Earth's surface, and the **troposphere** is the lowest 12 km. In the troposphere (like the mesosphere), warmer air lies beneath cooler air. This is a dynamically unstable situation because warm air is less dense and tends to rise. Convection currents constantly mix the gases in the troposphere, contributing greatly to the weather. In the stratosphere (like the thermosphere), the temperature increases with height, and there is little vertical mixing from convection. Mixing across the borders between the four layers is also slow, so most of the chemical processes in the layers can be described separately.

Atmospheric chemistry dates back to the 18th century. Cavendish, Priestley, Lavoisier, and Ramsay were the first scientists to study the composition of the atmosphere. In recent years, the field has developed in two different, but related, directions. First, the sensitivity of chemical analysis has greatly improved, and analyses for substances at concentrations below the nmol L^{-1} level are now carried out routinely. Airplanes and satellites allow the full distribution of trace substances to be mapped on a global scale. Second, advances in gas-phase chemical kinetics have led to a better quantitative understanding of the ways in which substances in the atmosphere react. Much of the impetus for these studies of atmospheric chemistry has come from concern over the effect of air pollution on life.

EXAMPLE 18-6

A mobile pollution monitoring team reports finding 200 ppbv of SO_2 at an industrial site. Calculate the partial pressure and concentration of SO_2 in the air at the site if the pressure is 1.00 atm and the temperature is 27.0°C.

Strategy

Assume that the ideal gas law and Dalton's law apply. Recall or review (in Appendix C–4) the definition of ppbv (200 parts per billion by volume). Use the definition to construct a unit factor to find the mole fraction of SO_2. Use the ideal gas law to compute the concentration.

Solution

The phrase "200 ppb by volume" means that the SO_2 in 1 billion L of the polluted air would have a volume of 200 L (under the same conditions of temperature and pressure). By Avogadro's law (see Section 1–4), 200 molecules of SO_2 must be present for every billion molecules of polluted air. This is also the ratio of the number of moles of SO_2 to the number of moles of all gases in the air. The mole fraction of SO_2 is then

$$X_{SO_2} = \frac{n_{SO_2}}{n_{tot}} = \frac{200}{10^9} = 2.00 \times 10^{-7}$$

According to Dalton's law of partial pressures (Section 5–6), the partial pressure of a gas is the product of its mole fraction and the total pressure (1.00 atm here), so

$$P_{SO_2} = 2.00 \times 10^{-7} \text{ atm}$$

The concentration in moles per liter (n/V) is obtained using the ideal gas law

$$[SO_2] = \frac{n_{SO_2}}{V} = \frac{P_{SO_2}}{RT} = \frac{2.00 \times 10^{-7} \text{ atm}}{(0.08206 \text{ L atm mol}^{-1} \text{ K}^{-1})(273.15 + 27.0) \text{ K}}$$
$$= 8.12 \times 10^{-9} \text{ mol L}^{-1}$$

EXERCISE

Cigarette smoke contains 200 to 400 ppmv (parts per million by volume) of carbon monoxide (CO). Find the range of partial pressure (in atmospheres) and the range of concentration (in *molecules* per liter) of CO in a puff of cigarette smoke (at ordinary room conditions).

Answer: 2×10^{-4} to 4×10^{-4} atm; 5×10^{18} to 1×10^{19} molecules of CO per liter.

Stratospheric Chemistry

The Sun emits light over a broad range of wavelengths, with the highest intensity at approximately 500 nm, near the center of the visible region of the spectrum. Its emissions are intense at wavelengths far into the UV. Because the energy $h\nu$ carried by a photon is inversely proportional to the wavelength λ ($h\nu = hc/\lambda$), the UV photons carry much more energy than do photons of visible light. If substantial numbers of these photons penetrated to the Earth's surface, much harm would be done to living organisms. Fortunately, the outer portions of the atmosphere (especially the thermosphere) absorb UV radiation through the photodissociation of oxygen molecules

$$O_2 + h\nu \longrightarrow 2\,O$$

where "$h\nu$" symbolizes a photon. This reaction reduces the number of high-energy photons, especially those with wavelengths of less than 200 nm, reaching the lower parts of the atmosphere.

EXAMPLE 18 – 7

The bond-dissociation energy (ΔE_d) of O_2 is 496 kJ mol^{-1}. Calculate the maximum wavelength of light that can photodissociate an oxygen molecule.

Strategy

Convert the dissociation energy from a *molar* basis to a *molecular* basis. Use the relationships developed in Chapter 17 to compute the wavelength of a photon that delivers this amount of energy to a molecule. Realize that longer-wavelength photons deliver *less* energy and therefore must fail to dissociate the molecule.

Solution

Because 496 kJ dissociates 1 mol of O_2 molecules, the energy to dissociate one molecule equals 496 kJ divided by Avogadro's number:

$$\Delta E_d = \frac{496 \times 10^3 \text{ J mol}^{-1}}{6.022 \times 10^{23} \text{ mol}^{-1}} = 8.24 \times 10^{-19} \text{ J}$$

The energy delivered by a photon is related inversely to its wavelength (λ)

$$E = h\nu = \frac{hc_o}{\lambda}$$

so that

$$\lambda = \frac{hc_o}{8.24 \times 10^{-19} \text{ J}} = \frac{(6.626 \times 10^{-34} \text{ J s})(2.998 \times 10^8 \text{ m s}^{-1})}{8.24 \times 10^{-19} \text{ J}}$$

$$= 2.41 \times 10^{-7} \text{ m} = 241 \text{ nm}$$

A photon with a wavelength *shorter* than 241 nm is energetic enough to dissociate oxygen molecules. Photons with wavelengths shorter than 200 nm are the most efficient in causing photodissociation.

Check

The wavelength 241 nm is near the middle of the UV region of the spectrum (see Figure 16–5). Radiation in this region is known to excite valence electrons (Table 18–3) and so should be able to break chemical bonds.

EXERCISE

The bond-dissociation energy of nitrogen is 943 kJ mol^{-1}. Compute the minimum frequency of light that can photodissociate nitrogen. Calculate the maximum wavelength of such light.

Answer: $\nu = 2.36 \times 10^{15}$ s^{-1}; $\lambda = 127$ nm. The photodissociation of nitrogen requires higher-frequency (shorter-wavelength) light than oxygen does. Nitrogen lets through UV light that oxygen absorbs.

Rather few photons with wavelengths of less than 200 nm are able to penetrate to the stratosphere, but those that do establish a small concentration of oxygen atoms in that layer. These atoms can collide with the much more prevalent oxygen molecules to form excited-state molecules of ozone (O_3):

$$O(g) + O_2(g) \rightleftharpoons O_3^*(g)$$

The asterisk indicates that the ozone that is formed is in a highly excited state. It has 106 kJ mol^{-1} more energy than does ground-state ozone. The excited-state ozone O_3^* often dissociates back to O and O_2, as indicated by the reverse arrow. Sometimes, if another atom or molecule collides with it soon enough, it loses its excess energy to that atom or molecule and attains its ground state:

$$O_3^*(g) + M(g) \longrightarrow O_3(g) + M(g)$$

where M stands for the atom or molecule with which it collides (the most likely ones are oxygen and nitrogen molecules because they are the most abundant species in the stratosphere). The result of these two reactions is a small concentration of ground-state ozone in the stratosphere.

Under laboratory conditions, ozone is an unstable compound. Its conversion to oxygen

$$O_3(g) \longrightarrow \tfrac{3}{2}O_2(g)$$

is thermodynamically favored but takes place quite slowly in the absence of light. In the stratosphere, ozone photodissociates readily to O_2 and O in a photolytic reaction that requires much less energy than the dissociation of O_2:

$$O_3(g) + h\nu \longrightarrow O_2(g) + O(g) \qquad \Delta E_d = 104 \text{ kJ mol}^{-1}$$

This process occurs most efficiently for wavelengths between 200 and 350 nm. The energy of light of these wavelengths is too small to be absorbed by molecular oxygen but quite large enough to damage organisms at the Earth's surface. The ozone layer shields the Earth's surface from 200 to 350 nm UV radiation in sunlight. The balance between formation and photodissociation leads to a steady-state concentration of more than 10^{15} molecules of ozone per liter in the stratosphere.

Depletion of Stratospheric Ozone

Certain types of air pollution give rise to *radicals* that catalyze the depletion of ozone in the stratosphere. Radicals are chemical species containing unpaired electrons. They usually are formed by the breaking of a covalent bond to form a pair of neutral species. One pressing concern involves chlorofluorocarbons, compounds of chlorine, fluorine, and carbon that are used as refrigerants and in some aerosol sprays.

• Radicals were mentioned previously in connection with reaction mechanisms (see Section 14–5).

CHEMISTRY IN YOUR LIFE

Wear Sunscreen

Ultraviolet radiation of wavelengths between 290 and 320 nm (sometimes called UV-B) penetrates to the epidermal layer of the skin, where it causes sunburn. Repeated exposure to UV-B appears to cause certain cancers. Death rates from skin melanomas in the United States correlate positively with southerly latitude: A white male from Florida is about twice as likely to develop this form of cancer as is one from Montana. Ozone in the stratosphere absorbs much UV-B, but some penetrates even an undepleted ozone layer. Hence the growing popularity of sunscreen lotions and creams among those working or playing outdoors.

The active ingredients in these products are compounds that absorb UV photons by undergoing a transition to an excited electronic state. For example, the compound *para*-aminobenzoic acid (see Fig. 18–C) absorbs strongly at wavelengths from 283 to 289 nm. This molecule has numerous electronic energy levels and many transitions connecting them. In one, an electron is promoted from a π molecular orbital mainly associated with the C=O group to a π^* molecular orbital in the same region. The resulting excited-state molecule quickly relaxes to its ground state, either by emitting radiation or by collisions with other molecules that transfer the energy

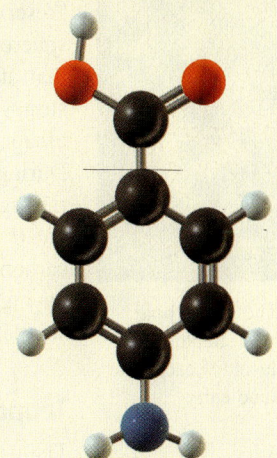

Figure 18–C • The structure of *para*-aminobenzoic acid. Other sunscreen compounds have related structures.

to the surroundings as heat. This step makes it ready to absorb another UV photon. The re-emitted radiation is at longer wavelengths than the UV and is not harmful to the skin. Certain sunscreens, however, give the skin a faint bluish cast as their active ingredients re-emit absorbed energy at a visible wavelength.

These compounds are non-reactive at sea level but are photodissociated in the stratosphere in reactions such as

$$CCl_2F_2(g) + h\nu \longrightarrow CClF_2(g) + Cl(g)$$

The atomic chlorine thus released soon reacts with O atoms or O_3 to give ClO:

$$Cl(g) + O(g) \longrightarrow ClO(g)$$
$$Cl(g) + O_3(g) \longrightarrow ClO(g) + O_2(g)$$

The ClO radical is the immediate culprit in the destruction of stratospheric ozone: Local decreases in ozone correlate directly with local increases in ClO. It catalyzes the conversion of O_3 to O_2, probably by the mechanism

$$2\ ClO(g) + M(g) \longrightarrow ClOOCl(g) + M(g)$$
$$ClOOCl(g) + h\nu \longrightarrow ClOO(g) + Cl(g)$$
$$ClOO(g) + M(g) \longrightarrow Cl(g) + O_2(g) + M(g)$$
$$\underline{2 \times (Cl(g) + O_3(g) \longrightarrow ClO(g) + O_2(g))}$$

net reaction: $\quad 2\ O_3(g) \longrightarrow 3\ O_2(g)$

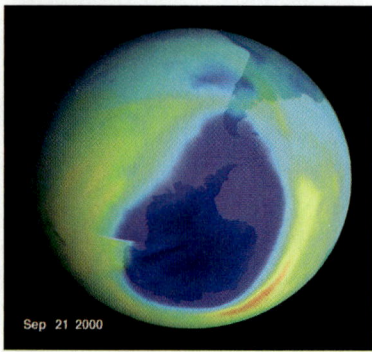

Figure 18–29 • The ozone hole (blue) over Antarctica on September 21, 2000. The hole develops each year between late August and early October. *(NASA)*

• Exposure to UV radiation appears to cause certain skin cancers. A predicted consequence of depletion of the stratospheric ozone layer is an increased worldwide incidence of skin cancer.

where M stands for N_2 and O_2 molecules. This catalytic cycle ordinarily is disrupted rather quickly as Cl reacts with other stratospheric species to form less reactive "reservoir molecules" such as HCl and $ClONO_2$. The lack of mixing in the stratosphere keeps the reservoir molecules around for long periods, however. It is believed that atomic Cl escapes its reservoirs through heterogeneous reactions on the icy stratospheric clouds that form during the intense cold of the Antarctic winter. The escaped Cl gives rise to the annual "ozone holes" above the Antarctic (Fig. 18–29). During these episodes, more than 70% of the ozone is depleted before the values rise again. After years of worsening, the degree of depletion seems to be leveling out (Fig. 18–30). This suggests that the phase-out of production of chlorofluorocarbons in the industrial countries begun in 1992 (the Montreal protocol) may be having a positive effect. Measurements over other parts of the globe have shown smaller, but still severe, depletion of the ozone layer.

Tropospheric Chemistry

The troposphere is the part of the atmosphere in contact with the Earth's surface. It is therefore most directly and immediately influenced by human activities, especially by the gases or small particles put into the air by automobiles, power plants, and factories. Some air pollutants have long lifetimes and are spread fairly evenly over the Earth's surface; others are more ephemeral and attain large concentrations only around particular cities or industrial areas.

The oxides of nitrogen are major air pollutants. Their persistence in the atmosphere shows the importance of kinetics, as opposed to thermodynamics, in the chemistry of the atmosphere. All of the oxides of nitrogen are thermodynamically unstable with respect to the elements at 25°C, as shown by their positive standard Gibbs energies of formation at that temperature. They form by direct reaction of nitrogen and oxygen whenever air is heated to high enough temperatures, either in an industrial process or in the engine of a car. They accumulate to much higher than equilibrium concentrations in the atmosphere because they decompose slowly. Interconversions among NO, NO_2, and N_2O_4, the most important nitrogen oxides in

Figure 18–30 • Trend in ozone depletion over Halley Bay, Antarctica. These measurements were all taken in the Antarctic spring (October), when depletion is at its worst. The unit on the vertical axis, the Dobson, is equivalent to a 0.01 mm thick layer of pure ozone gas at STP (273 K and 1 atm pressure).

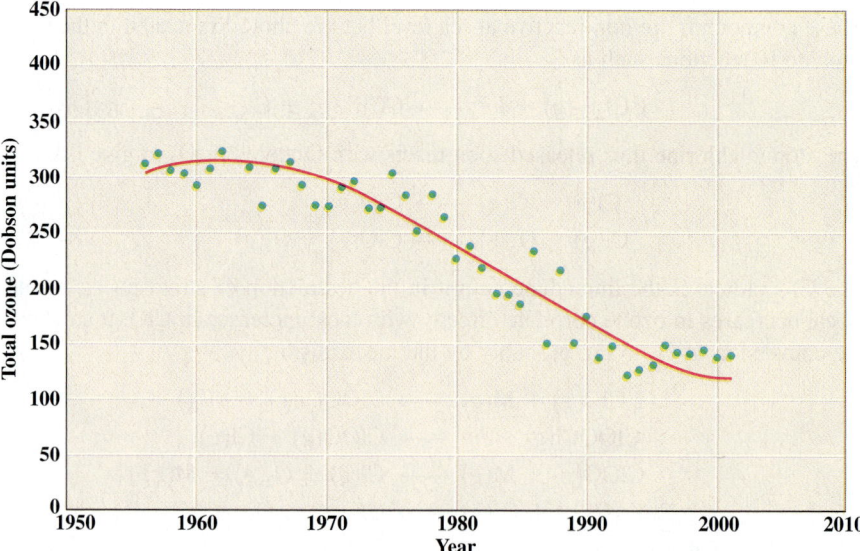

air pollution, are rapid and depend strongly on temperature, and so in pollution reports they are generally grouped together as "NO_x." Photochemical smog (Fig. 18–31) is formed by the action of light on nitrogen dioxide, followed by subsequent reaction to produce ozone:

$$NO_2(g) + h\nu \longrightarrow NO(g) + O(g)$$
$$O(g) + O_2(g) + M(g) \longrightarrow O_3(g) + M(g) \qquad (M = N_2 \text{ or } O_2)$$

It may seem paradoxical that the problems with ozone in the stratosphere involve its *depletion,* but in the troposphere, its *production.* In fact, the different effects arise from the different concentrations of species in the two layers of the atmosphere. Although ozone is beneficial in reducing the intensity of the UV radiation penetrating to the Earth's surface, it is quite harmful in direct contact with organisms because of its strong oxidizing power. Levels of 10 to 15 ppmv are sufficient to kill small mammals, and a level as low as 3 ppmv in urban air triggers an "ozone alert." In addition, ozone can react with incompletely oxidized organic compounds from gasoline and with nitrogen oxides in the air to produce harmful irritants such as methyl nitrate (CH_3NO_3).

Figure 18–31 • Photochemical smog casts a pall over Mexico City, which is said to have the highest level of air pollution in the world. *(Tom McHugh/Photo Researchers, Inc.)*

The oxides of sulfur create global pollution problems because they have longer lifetimes in the atmosphere than do the oxides of nitrogen. A certain fraction of the SO_2 and SO_3 in the air originates from biological processes and from volcanoes, but most comes from the oxidation of sulfur present in petroleum and in coal that is burned for fuel. If the sulfur is not removed from the fuel or the exhaust gas, SO_2 enters the atmosphere as a stable but reactive pollutant. Further oxidation by radicals leads to sulfur trioxide:

$$SO_2(g) + OH(g) \longrightarrow SO_2OH(g)$$
$$SO_2OH(g) + O_2(g) \longrightarrow SO_3(g) + OOH(g)$$
$$OOH(g) \longrightarrow O(g) + OH(g)$$

The atomic oxygen produced can then react with molecular oxygen to form ozone.

Acid rain results from the reaction of NO_2 and SO_3 with hydroxyl radicals and water vapor in the air to form nitric acid (HNO_3) and sulfuric acid (H_2SO_4), which are soluble in water and return to the Earth in the rain. This process occurs naturally to a certain extent and is beneficial in providing the crucial nutrients nitrogen and sulfur to soils that are poor in the two elements. Large-scale emission of sulfur oxides from burning coal and from metal smelting, however, has altered the atmospheric balance and increased the acidity of rainfall in much of the northern hemisphere to harmful levels. The effects are to reduce fish populations in lakes, to damage forests severely (Fig. 18–32), to speed the corrosion of metals, and to erode stone and marble structures. To combat acid rain, we must reduce emissions of SO_2. The methods currently used involve powdered limestone, which decomposes at high temperatures to lime and carbon dioxide.

$$CaCO_3(s) \longrightarrow CaO(s) + CO_2(g)$$

Figure 18–32 • The effects of acid rain on a stand of trees and the natural regeneration of young spruce trees in the Karkonosze National Park, southwestern Poland. Here, forests have experienced acid rain in which the pH was as low as 1.7. *(Simon Fraser/ Science Photo Library/ Photo Researchers, Inc.)*

The lime can combine with sulfur dioxide (in an acid–base reaction) to produce solid calcium sulfite:

$$CaO(s) + SO_2(g) \longrightarrow CaSO_3(s)$$

The enormous problem of the disposal of the $CaSO_3$ solid waste that is produced by this process has slowed its adoption as an antipollution measure. The best option would be to convert the SO_2 to economically useful sulfuric acid, and some progress has been made in this direction, as discussed in Chapter 22.

Figure 18–33 • The greenhouse effect. Sunlight conveys energy to the Earth's surface. Much of this energy is radiated back into space in the longer-wavelength infrared region of the spectrum, but infrared-absorbing molecules, both in clouds and in the clear atmosphere, block some of the loss, keeping the lower atmosphere warmer.

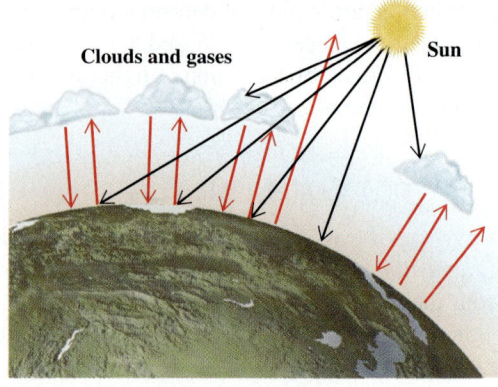

The Greenhouse Effect

Most short-wavelength photons from the Sun are absorbed by the outer atmosphere and do not reach the Earth's surface. The radiation that does reach the surface maintains a livable temperature, in balance with re-radiation from the Earth back into space. Certain gases in the troposphere play a crucial role in this balance because they absorb infrared radiation emitted by the warm surface of the Earth rather than letting it pass out to space (Fig. 18–33). Water vapor and water in the form of clouds are the most obvious of these gases. On cloudy winter nights, the temperature does not fall as low as it does on clear nights because clouds provide a thermal blanket that absorbs outgoing radiation with wavelengths near 20,000 nm.

Two other gases that absorb infrared radiation to a significant extent are carbon dioxide and methane. Both are distributed uniformly in fairly low concentrations throughout the troposphere. The concentrations of both of these gases have increased steadily over the 200 to 300 years since the beginning of the Industrial Revolution. The tremendous increase in the burning of fossil fuels as energy sources is without a doubt the reason for this trend (Fig. 18–34). It is estimated that by the year 2050, the concentration of CO_2 in the atmosphere will have doubled over premodern values.

Figure 18–34 • Worrisome trends: Growth in annual CO_2 emissions from human activities, 1950–1998 (*black*); average monthly atmospheric concentration of CO_2 at Mauna Loa, Hawaii, 1955–2001 (*red*). The latter varies seasonally.

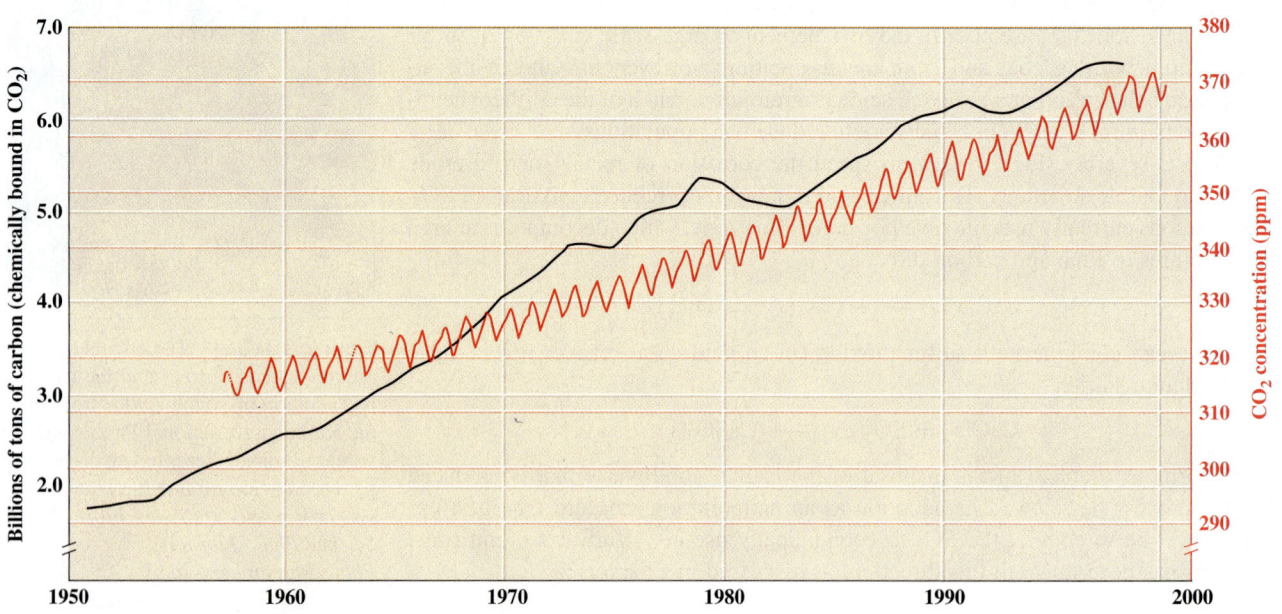

Such changes are viewed with alarm because of the **greenhouse effect.** This term refers to a global increase in average surface temperature that will occur if heat given off by the Earth's surface is prevented from escaping to space by higher concentrations of gases that absorb IR radiation. An increase in average surface temperature of 2°C to 5°C over the 21st century would have profound climatic repercussions, including the melting of some polar ice, a consequent increase in the level of the sea, and the conversion of arable land to desert. To prevent these undesirable changes, it will be necessary to develop new energy sources that are not based on the burning of fossil fuels.

SUMMARY

18-1   **Molecular orbitals** of diatomic molecules can be formed from superpositions (**overlap**) of atomic orbitals: Constructive (in-phase) interference gives **bonding orbitals** and destructive (out-of-phase) interference gives **antibonding orbitals.** In **correlation diagrams** (energy-level diagrams for molecules), the molecular orbitals are arranged in order of increasing energy and filled with available electrons to determine the ground-state electron structure of the molecule. The molecular orbital description of bonding agrees with the Lewis model in predicting **bond orders** and goes beyond the Lewis model to explain the paramagnetism of O_2 and other species.

18-2   **Valence-bond theory** uses the concept of **hybridization** of atomic orbitals. Atomic orbitals on central atoms in molecules are first mixed to form new hybrid atomic orbitals that contain both s and p character (sp, sp^2, sp^3) and have directional character. These hybrid orbitals project from the central atoms in such a way as to minimize electron–electron repulsion, as in the VSEPR model. As the other atoms of the molecule are brought up, σ **bonds** that are largely localized between pairs of atoms are formed. Multiple bonds result from the occupancy of π **molecular orbitals,** which are formed from p orbitals that are left over after hybridization is completed. Rotation about double bonds requires breaking the π bond, which requires energy. This allows the existence of *cis–trans* isomers in compounds with double bonds.

18-3   The conjugation of bonds occurs when single and double (or triple) bonds alternate. They are best described by delocalized molecular orbitals. The bonding in molecules represented by resonance Lewis structures can be understood in terms of a valence-bond framework of σ bonds overlain by π bonds that are spread over the entire molecule. The recently discovered fullerenes are examples.

18-4   X-rays are energetic enough to remove core electrons from molecules; UV and visible radiation can remove valence electrons from molecules or raise molecules to excited electron states. Lower-energy radiation excites vibrations of atoms relative to each other or excites rotations of the whole molecule. In **nuclear magnetic resonance** (NMR), a magnet splits the energies of nuclear spin states in molecules, and transitions between these states are observed at radio frequencies. The exact energy of the split depends on the chemical environment of the nucleus. Measurements of absorptions and emissions of radiation by molecules provide information on bond lengths, bond energies, and bond angles. Light can induce chemical reactions, exciting molecules to higher-energy electronic states from which they break apart or rearrange. The study of such processes is **photochemistry.** In **chemiluminescence,** chemical reactions generate light as they proceed.

18-5   In the outer layers of the atmosphere, **photodissociation** of oxygen molecules absorbs much of the light of highest frequency. In the **stratosphere,** the resulting oxygen atoms collide with oxygen molecules to produce ozone. Ozone in turn absorbs at somewhat longer wavelengths and plays a crucial role in shielding the surface of the Earth from harmful UV light. The destruction of stratospheric ozone by atmospheric pollutants, especially chlorofluorocarbons, is a very strong concern. Other pollution concerns focus on the **troposphere,** the lowest level of the atmosphere. Emissions of oxides of nitrogen and sulfur can lead to irritating and dangerous photochemical smog and to acid rain. Burning of coal, oil, and natural gas has increased the amounts of carbon dioxide in the atmosphere and may cause a severe increase in global temperatures over the next century through the **greenhouse effect.**

PROBLEMS

Note: Answers to blue-numbered problems are given in Appendix F. Problems that are more challenging are indicated with asterisks.

Diatomic Molecules

1. (See Example 18–2.) Write the ground-state valence-electron configurations for the nine diatomic ions X_2^{2+}, in which X stands for the elements from He to Ne in the periodic table. Give the bond order of each species.

2. (See Example 18–2.) Write the ground-state valence-electron configurations for the seven diatomic ions X_2^-, in which X stands for the elements from Li to F in the periodic table. Give the bond order of each species.

3. If an electron is removed from a fluorine molecule, an F_2^+ molecular ion is formed.
   (a) Give the ground-state valence-electron configurations for F_2 and F_2^+.
   (b) Give the bond order of each species.
   (c) Predict which species should be paramagnetic.
   (d) Predict which species has the larger bond energy.

4. When one electron is added to an oxygen molecule, a superoxide ion (O_2^-) is formed. The addition of two electrons gives a peroxide ion (O_2^{2-}). Removal of an electron from O_2 leads to O_2^+.
   (a) Construct the correlation diagram for O_2^{2-}.
   (b) Give the ground-state electron configuration for each of the following: O_2^+, O_2, O_2^-, O_2^{2-}.
   (c) Give the bond order of each species.
   (d) Predict which species should be paramagnetic.
   (e) Predict the order of increasing bond energy among the species.

5. Predict the ground-state valence-electron configuration and the total bond order for the molecule S_2, which forms in the gas phase when sulfur is heated to a high temperature. Is S_2 paramagnetic or diamagnetic? (*Hint*: Make use of the group relationship between sulfur and oxygen.)

6. Predict the ground-state valence-electron configuration and the total bond order for the molecule I_2. Is I_2 paramagnetic or diamagnetic? (*Hint*: Make use of the group relationship between iodine and fluorine.)

7. For each of the following valence-electron configurations of a homonuclear diatomic molecule or molecular ion, identify the element X, Q, or Z:
   (a) X_2: $(\sigma_{2s})^2(\sigma_{2s}^*)^2(\sigma_{2p_z})^2(\pi_{2p})^4(\pi_{2p}^*)^4$
   (b) Q_2^+: $(\sigma_{2s})^2(\sigma_{2s}^*)^2(\pi_{2p})^4(\sigma_{2p_z})^1$
   (c) Z_2^-: $(\sigma_{2s})^2(\sigma_{2s}^*)^2(\sigma_{2p_z})^2(\pi_{2p})^4(\pi_{2p}^*)^3$

8. For each of the following valence-electron configurations of a homonuclear diatomic molecule or molecular ion, identify the element X, Q, or Z:
   (a) X_2: $(\sigma_{2s})^2(\sigma_{2s}^*)^2(\sigma_{2p_z})^2(\pi_{2p})^4(\pi_{2p}^*)^2$
   (b) Q_2^-: $(\sigma_{2s})^2(\sigma_{2s}^*)^2(\pi_{2p})^3$
   (c) Z_2^{2+}: $(\sigma_{2s})^2(\sigma_{2s}^*)^2(\sigma_{2p_z})^2(\pi_{2p})^4(\pi_{2p}^*)^2$

9. For each of the electron configurations in Problem 7, determine the total bond order of the molecule or molecular ion.

10. For each of the electron configurations in Problem 8, determine the total bond order of the molecule or molecular ion.

11. For each of the electron configurations in Problem 7, determine whether the molecule or molecular ion is paramagnetic or diamagnetic.

12. For each of the electron configurations in Problem 8, determine whether the molecule or molecular ion is paramagnetic or diamagnetic.

13. Following the pattern of Figure 18–8, work out the correlation diagram for the CN molecule, showing the relative energy levels of the atoms and the bonding and antibonding orbitals of the molecule. Indicate the occupation of the molecular orbitals by arrows. State the order of the bond, and comment on the magnetic properties of CN.

14. Following the pattern of Figure 18–8, work out the correlation diagram for the BeN molecule, showing the relative energy levels of the atoms and the bonding and antibonding orbitals of the molecule. Indicate the occupation of the molecular orbitals by arrows. State the order of the bond, and comment on the magnetic properties of BeN.

15. The bond in the transient diatomic molecule CF is longer than the bond in the related ion CF^+. Explain why. Refer to the proper molecular-orbital correlation diagram.

16. The compound nitrogen monoxide (NO) forms when the nitrogen and oxygen in air are heated. Predict whether the

nitrosyl ion (NO^+) has a shorter or a longer bond than the NO molecule. Is NO^+ paramagnetic like NO or diamagnetic?

17. Propose an electron configuration for the HeH^- molecular ion and predict the bond order of this ion. Comment on the stability of this ion.

18. The molecular ion HeH^+ has an equilibrium bond length of 0.774 Å. Draw an electron-correlation diagram for this ion, indicating the occupied molecular orbitals. Is HeH^+ paramagnetic? When the HeH^+ ion dissociates, is a lower-energy state reached by forming $He + H^+$ or $He^+ + H$?

Polyatomic Molecules

19. (See Example 18–3.) Use valence-bond theory to predict the three-dimensional structure of the amide ion (NH_2^-).

20. (See Example 18–3.) Use valence-bond theory to predict the three-dimensional structure of the hydronium ion (H_3O^+).

21. (See Example 18–3.) Write a Lewis electron-dot structure for each of the following molecules and ions. Formulate the hybridization for the central atom in each case, and predict the molecular geometry of each molecule or ion.
(a) CCl_4 (b) CO_2 (c) OF_2 (d) CH_3^- (e) BeH_2

22. (See Example 18–3). Write a Lewis electron-dot structure for each of the following molecules and ions. Formulate the hybridization for the central atom in each case, and predict the molecular geometry of each molecule or ion.
(a) BF_3 (b) BH_4^- (c) PH_3 (d) CS_2 (e) CH_3^+

23. Describe the hybrid orbitals used by the chlorine atom in the ClO_3^+ and ClO_2^+ molecular ions. Sketch the expected geometries of these ions.

24. Describe the hybrid orbitals used by the chlorine atom in the ClO_4^- and ClO_3^- molecular ions. Sketch the expected geometries of these ions.

25. The sodium salt of the unfamiliar orthonitrate ion (NO_4^{3-}) has been prepared. What hybridization is expected on the N atom at the center of this ion? Predict the geometry of the NO_4^{3-} ion.

26. Describe the hybrid orbitals used by the carbon atom in $N\equiv C-Cl$. Predict the geometry of the molecule.

27. How many σ bonds and how many π bonds does each of the following molecules contain?
(a) ethylene (C_2H_4)
(b) fluorine (F_2)
(c) propane (C_3H_8)
(d) carbon tetrachloride (CCl_4)

28. How many σ bonds and how many π bonds does each of the following molecules contain?
(a) nitrogen (N_2)
(b) formaldehyde (CH_2O)
(c) boron trifluoride (BF_3)
(d) acetylene (C_2H_2)

29. (See Example 18–4.) The compound acetone has the chemical formula $CH_3(CO)CH_3$. Two of the carbon atoms and the oxygen atom are all bonded to the third carbon atom. Draw

the Lewis structure for this molecule, give the hybridization of each carbon atom, and describe the π orbitals and their occupation by electrons. Sketch the three-dimensional structure of the molecule, showing all angles.

30. (See Example 18–4.) The compound acetic acid has the chemical formula CH_3COOH. Both oxygen atoms are bonded to the second carbon atom. Draw the Lewis structure for this molecule, give the hybridization of each carbon atom, and describe the π orbitals and their occupation by electrons. Draw the three-dimensional structure of the molecule, showing all angles.

31. (See Example 18–4.) Molecules of acetaldehyde (CH_3CHO) have a CH_3 group, an oxygen atom, and a hydrogen atom attached to a central carbon atom. Draw the Lewis structure for this molecule, give the hybridization of each carbon atom, and describe the π orbitals and their occupation by electrons. Draw the three-dimensional structure of the molecule, showing all angles.

32. (See Example 18–4.) Acrylic fibers are polymers made from a starting material called acrylonitrile ($H_2C(CH)CN$). In acrylonitrile a $C\equiv N$ group replaces an H atom on ethylene. Draw the Lewis structure for this molecule, give the hybridization of each carbon atom, and describe the π orbitals and the number of electrons that occupy each one. Draw the three-dimensional structure of the molecule, showing all angles.

33. In Section 18–2, it was explained that the compound 2-butene exists in two isomeric forms that can be interconverted only by breaking the central double bond. A third isomer (called 2-methylpropene) also exists. In it, both CH_3 groups are attached to the *same* carbon atom, with two H atoms on the other carbon atom. Bearing this fact in mind, how many isomers are possible for each of the following molecular formulas, if each molecule contains a central $C=C$ double bond? Remember that rotation of the molecule as a whole does not give a different isomer.
(a) $C_2H_2Br_2$
(b) C_2H_2BrCl

34. Repeat the determination of the number of isomers as discussed in the preceding problem for the following molecular formulas.
(a) C_2Cl_2BrF
(b) $C_2HBrClF$

The Conjugation of Bonds and Resonance Structures

35. Predict the three-dimensional structure of the nitrite ion (NO_2^-). A good Lewis structure of this ion includes resonance structures. Show how resonance structures are avoided by introducing delocalized π molecular orbitals. Determine the bond order of the N—O bonds in the nitrite ion.

36. Predict the three-dimensional structure of the nitrate ion (NO_3^-). A good Lewis structure of this ion includes resonance structures. Show how resonance structures are avoided by introducing delocalized π molecular orbitals. Determine the bond order of the N—O bonds in the nitrate ion.

The Interaction of Light with Molecules

37. Discuss the effect of the absorption of microwave radiation on a molecule. What information is gained from a study of the absorption spectrum in this region?

38. Discuss the effect of the absorption of X-ray radiation on a molecule. What information is gained from a study of the absorption spectrum in this region?

39. Suppose that the ethylene molecule gains an additional electron to give the $C_2H_4^-$ ion. Does the bond order of the carbon–carbon bond increase or decrease? Explain.

40. Suppose that the ethylene molecule is ionized by a photon to give the $C_2H_4^+$ ion. Does the bond order of the carbon–carbon bond increase or decrease? Explain.

41. (See Example 18–5.) The color of the dye indanthrene brilliant orange is evident from its name. In what wavelength range would you expect the maximum in the absorption spectrum of this molecule to lie? Refer to the color spectrum in Figure 16–5.

42. (See Example 18–5.) In what wavelength range would you expect the maximum in the absorption spectrum of the dye crystal violet to lie?

43. The structure of the molecule of cyclohexene is:

Does the absorption of UV light by cyclohexene occur at longer or at shorter wavelengths than in benzene? Explain.

44. The naphthalene molecule has a structure that corresponds to two benzene molecules fused together:

The π electrons in this molecule are delocalized over the entire molecule. The wavelength of maximum absorption in the UV-visible part of the spectrum in benzene is 255 nm. Is the corresponding wavelength shorter or longer than 255 nm for naphthalene?

Atmospheric Chemistry and Air Pollution

45. (See Example 18–7.) The bond-dissociation energy of a typical C—F bond in a chlorofluorocarbon is approximately

440 kJ mol^{-1}. Calculate the maximum wavelength of light that can photodissociate a molecule of CCl_2F_2, breaking such a C—F bond.

46. (See Example 18–7.) The bond-dissociation energy of a typical C—Cl bond in a chlorofluorocarbon is approximately 330 kJ mol^{-1}. Calculate the maximum wavelength of light that can photodissociate a molecule of CCl_2F_2, breaking such a C—Cl bond.

47. Draw Lewis structures (including resonance structures) for the ozone molecule (O_3). Determine the steric number and hybridization of the central oxygen atom, and identify the molecular geometry. Describe the nature of the π bonds in ozone, and give the bond order of its O—O bonds.

48. Carbon dioxide (CO_2) and sulfur dioxide (SO_2) are formed in the burning of coal. Determine the shapes of these two molecules, identify the hybridization at the central atom, and compare the nature of their π bonds.

ADDITIONAL PROBLEMS

49. (a) Sketch the shapes of occupied molecular orbitals of the valence shell for the N_2 molecule. Label the orbitals as σ orbitals or π orbitals, and specify which are bonding and which are antibonding.

(b) One electron is removed from the highest occupied molecular orbital of N_2. Does the equilibrium N—N distance become longer or shorter? Explain briefly.

50. Calcium carbide (CaC_2) is an intermediate in the manufacture of acetylene (C_2H_2). It is the calcium salt of the carbide (also called acetylide) ion (C_2^{2-}). Give the electron configuration of this diatomic molecular ion and determine its bond order.

51. Show how the fact that the B_2 molecule is paramagnetic indicates that the energy ordering of the orbitals in this molecule is given by Figure 18–6a rather than 18–6b.

52. The Be_2 molecule has been detected experimentally. It has a bond length of 2.45 Å and a bond dissociation enthalpy of 9.46 kJ mol^{-1}. Write the ground-state electron configuration of Be_2 and predict its bond order, using the theory developed in the text. Compare the experimental bonding data on Be_2 with the data given for B_2, C_2, N_2, and O_2 in Table 18–2. Is the prediction of the simple theory seriously incorrect?

*53. (a) The ionization energy of molecular hydrogen (H_2) is *higher* than that of atomic hydrogen (H), whereas the ionization energy of molecular oxygen (O_2) is *lower* than that of atomic oxygen (O). Explain. (*Hint*: Think about the stability of the molecular ion that forms, in relation to bonding and antibonding electrons.)

(b) What prediction would you make for the relative ionization energies of atomic and molecular fluorine (F and F_2)?

54. The stable molecular ion $H_3^+(g)$ is triangular, with H—H distances of 0.87 Å. Sketch the structure of the ion, and locate the region of greatest electron density of the lowest-energy molecular orbital on the sketch.

55. Sketch correlation diagrams for the molecules F_2, HF, and NaF. How does molecular orbital theory account for the increasing polarity of the bonds in these molecules?

56. Discuss the bonding in O_3, O_3^-, and O_3^{2-} in terms of the various bonding theories (Lewis theory, valence-bond theory, molecular orbital theory). Include predictions of the bond angles and of the relative lengths of the O—O bonds in the three.

57. According to recent spectroscopic results, the nitramide molecule is non-planar.

Previously, it had been thought to be planar.
(a) Predict the bond order of the N—N bond in the non-planar structure.
(b) If the molecule really were planar after all, what would be the bond order of the N—N bond?

58. *Trans*-tetrazene (N_4H_4) consists of a chain of four nitrogen atoms with the two end atoms bonded to two hydrogen atoms each. Use the concepts of steric number and hybridization to predict the overall geometry of the molecule. Give the expected structure of *cis*-tetrazene.

59. The conjugate base of formic acid (HCOOH) is the formate ion (HCOO$^-$). Both species have a central carbon atom bonded to the two oxygen atoms and to a hydrogen atom.
(a) Determine the molecular geometries of formic acid and its conjugate base.
(b) Explain how the π orbitals differ between formic acid and the molecular ion.
(c) The bond lengths of the C—O bonds in HCOOH are 1.23 Å (for the bond to the lone oxygen) and 1.36 Å (for the bond to the oxygen with a hydrogen atom attached). Give a likely range of lengths for C—O bond length in the formate ion.

60. (a) Use data from Appendix D to calculate the enthalpy change when 1 mol of gaseous benzene is formed from carbon and hydrogen *atoms,* all in the gas phase at 298.15 K.
(b) Compare this result with the enthalpy change that you calculate for forming one of the resonance structures for benzene, using the bond enthalpies for C=C, C—C, and C—H given in Table 10–3.
(c) The additional lowering of the enthalpy found experimentally (the more negative value of $\Delta H°$) is due to *resonance stabilization*: the delocalization of electrons over the whole carbon ring. How large is resonance stabilization in benzene?

61. It has been suggested that a compound of formula $C_{12}B_{24}N_{24}$ might exist and have a structure like that of C_{60} (buckminsterfullerene).
(a) Explain the logic of this suggestion by comparing the number of valence electrons in C_{60} and $C_{12}B_{24}N_{24}$.
(b) Propose the most symmetrical pattern of C, B, and N atoms in $C_{12}B_{24}N_{24}$ to occupy the 60 atom sites in the buckminsterfullerene structure.

62. The standard molar enthalpy of combustion of $C_{60}(s)$ equals $-25{,}891$ kJ mol^{-1}. Compute the standard enthalpy of formation of $C_{60}(s)$.

63. Figure 18–24 shows two ways in which green light emerges from an absorbing sample through which white light is passed. How could you use a prism to determine which of these two mechanisms is the origin of the green light seen in a particular case?

64. An electron in the π orbital of ethylene (C_2H_4) is excited by a photon to the π^* orbital. Will the bond in the excited ethylene molecule be longer or shorter than in ground-state ethylene? Explain.

65. One isomer of retinal is converted into a second isomer by the absorption of a photon:

This process is a key step in the chemistry of vision. Free retinal (in the form shown to the left of the arrow) has an absorption maximum at 376 nm, in the UV region of the spectrum, but this absorption shifts into the visible range when the retinal is bound in a protein, as it is in the eye.
(a) How many of the C=C double bonds are *cis* and how many are *trans* in each of the structures above? (Consider the relative positions of the two largest groups attached at each double bond when assigning labels.) Describe the motion that takes place on absorption of a photon.
(b) If the ring and the —CHO group in retinal were replaced by —CH$_3$ groups, would the absorption maximum of the molecule in the UV-visible portion of the spectrum shift to longer or to shorter wavelengths?

66. The ground-state electron configuration of the H_2^+ molecular ion is $(\sigma_{1s})^1$.
(a) An ion of H_2^+ absorbs a photon and is excited to the σ_{1s}^* molecular orbital. Predict what happens to the ion.
(b) Another ion of H_2^+ absorbs even more energy in an interaction with a photon and is excited to the σ_{4s} molecular orbital. Predict what happens to this ion.

67. Compare and contrast the chemical behaviors of ozone (O_3) and nitrogen dioxide (NO_2) in the stratosphere and in the troposphere.

68. Write balanced chemical equations that describe the formation of nitric and sulfuric acid in rain, starting with sulfur in coal, and oxygen, nitrogen, and water vapor in the atmosphere.

69. Describe the greenhouse effect and its mechanism of operation. Give three examples of energy sources that contribute to increased CO_2 in the atmosphere, and three that do not.

CUMULATIVE PROBLEM

Bromine

A chemical plant uses bromine to manufacture brominated flame retardants for plastics and fibers. *(Courtesy of Ethyl Corporation)*

Elemental bromine is a brownish red liquid that was first isolated in 1826. The current method of production is to oxidize bromide ions in natural brines with elemental chlorine.

(a) Bromine compounds have been known and used for centuries. The deep purple color symbolic of imperial power in Roman times originated from the compound dibromoindigo, which was extracted in tiny quantities from purple snails (about 800 snails per gram of compound). What color and maximum wavelength of *absorbed* light would give this deep purple color?

(b) What is the ground-state electron configuration of the valence electrons of bromine molecules, Br_2? Is bromine paramagnetic or diamagnetic?

(c) What is the electron configuration of the Br_2^+ molecular ion? Is its bond stronger or weaker than that in Br_2? What is its bond order?

(d) What excited electronic state is responsible for the brownish red color of bromine? Refer to Figures 18–6 and 18–23.

(e) One of the most extensively used compounds of bromine is an additive in leaded gasoline called "ethylene dibromide" (CH_2BrCH_2Br). What is the hybridization at the two carbon atoms in this compound? (*Note*: Each C atom is bonded to the other C atom, to two H atoms, and to one Br atom.)

(f) The action of light on bromine compounds released into the air (such as from leaded gasoline) causes the formation of the BrO radical. Give the bond order of this species by comparing it to the related radical OF.

(g) Synthetic bromine-containing compounds in the atmosphere contribute to the destruction of ozone in the stratosphere. The BrO (see part (f)) can take part with ClO in the following catalytic cycle:

$$Cl + O_3 \longrightarrow ClO + O_2$$
$$Br + O_3 \longrightarrow BrO + O_2$$
$$ClO + BrO \longrightarrow Cl + Br + O_2$$

Write the overall equation for this cycle, and explain why it would be important even in the Antarctic winter, when there is not enough sunlight to split apart many O_2 molecules.

19 Coordination Complexes

CHAPTER OUTLINE

The color of these crystals of rhodochrosite ($MnCO_3$) arises from the interaction of Mn^{2+} ions with their environment. *(Charles D. Winters)*

Solid copper(II) sulfate is made by reacting copper and hot concentrated sulfuric acid ("oil of vitriol"); its traditional name, "blue vitriol," recalls this origin and reports the color that is its most obvious property. There is more to this compound than copper and sulfate, however; it contains water as well. The water is important in blue vitriol because when it is driven away by strong heat, the blue color vanishes, leaving greenish white anhydrous copper(II) sulfate (Fig. 19–1). The blue of blue vitriol comes from a **coordination complex** in which H_2O molecules bond to Cu^{2+} ions to form composite ions with the formula $[Cu(H_2O)_4]^{2+}$. As a Lewis acid, the Cu^{2+} ion coordinates the four water molecules into a group by accepting electron density from the lone pairs on each water molecule. By acting as electron-pair donors (Lewis bases) and sharing electron density with the Cu^{2+} ion, the four water molecules, which in this interaction are called **ligands,** come into the **coordination sphere** of the Cu^{2+} ion. Blue vitriol has the chemical formula $Cu(H_2O)_4SO_4 \cdot H_2O$; the fifth water molecule is not coordinated directly to copper.

The positive ions of every metal in the periodic table accept electron density to some degree and therefore can coordinate surrounding electron donors, even if only weakly. The solvation of the K^+ ion by water molecules in aqueous solution (see Fig. 4–1) is an example of weak coordination. The ability to make fairly strong, *directional* bonds by accepting electron pairs from neighboring molecules or ions is characteristic of the transition elements. Coordination occupies a middle place energetically between the weak intermolecular attractions in solids and liquids (see Section 6–1) and covalent and ionic bonds, which are stronger. Thus, heating blue vitriol disrupts the $Cu—H_2O$ bonds at temperatures well below those required to break the covalent bonds in the SO_4^{2-} group. The enthalpies of the bonds between $+2$ transition-metal ions and coordinated water molecules range between 170 and 210 kJ mol^{-1}. This is far less than the enthalpy required to disrupt the strongest chemical bonds, but it is by no means small. The strengths of metal–ligand interactions do vary, of course, depending on the identities of both the positive ion and the ligand. Metal ions with $+3$ charges, for example, always have stronger coordinate bonds with water than those having $+2$ charges.

• These are the bond enthalpies discussed in Sections 10–5 and 17–4.

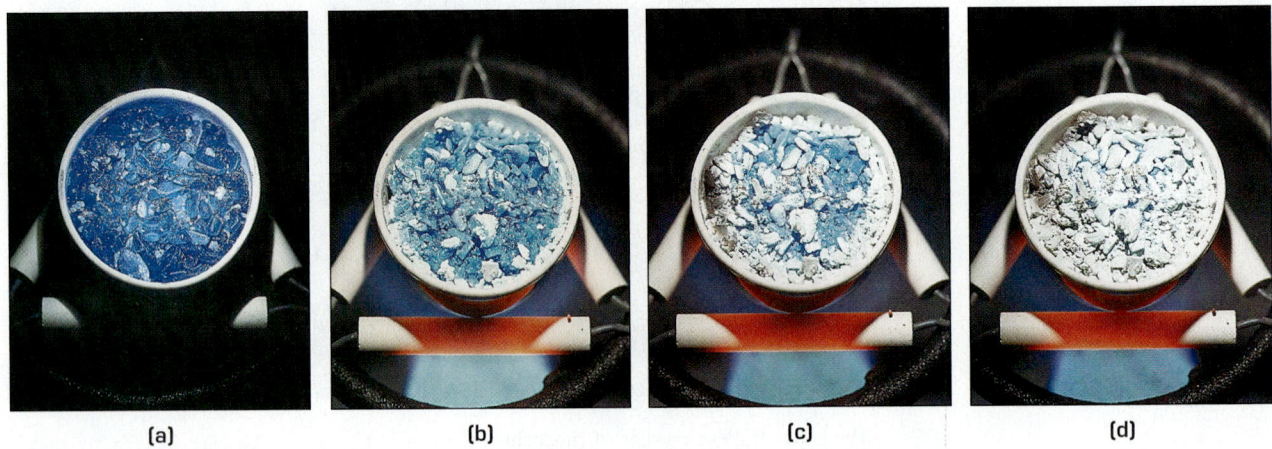

| (a) | (b) | (c) | (d) |

Figure 19–1 • Heating hydrated copper(II) sulfate drives off the water. The hydrated compound $(Cu(H_2O)_4SO_4 \cdot H_2O)$ is blue (*left*), but the anhydrous compound $(CuSO_4)$ is greenish white (*right*). (*Charles D. Winters*)

19-1   The Formation of Coordination Complexes

The formation of a coordination complex is a Lewis acid–base reaction (see Section 8–8):

$$\text{central atom} + \text{ligands} \longrightarrow \text{coordination complex}$$

$$\text{Lewis acid} + \text{Lewis bases} \longrightarrow \text{composite ion or molecule}$$

$$Fe^{2+} + 6\,CN^- \longrightarrow [Fe(CN)_6]^{4-}$$

$$Pt^{4+} + 6\,NH_3 \longrightarrow [Pt(NH_3)_6]^{4+}$$

$$Ni + 4\,CO \longrightarrow [Ni(CO)_4]$$

As the third example shows, the central atom in a complex need not have a positive charge to coordinate ligands. The total number of metal–ligand bonds in a complex (usually two to six) is the **coordination number** of the metal. The particular atom of a ligand molecule that actually provides the electron density to a central metal atom is a donor, or **ligating**, atom. Table 19–1 lists some common ligands and shows their donor atoms.

Brackets are used in the chemical formulas of coordination complexes to group together the symbols of the central atom and its coordinated ligands. In the formula $[Pt(NH_3)_6]Cl_4$, the portion in brackets represents a positively charged coordination complex in which Pt coordinates six NH_3 ligands. The brackets emphasize that complexes are distinct chemical entities with their own properties. The symbol of the central atom comes first within the brackets.

Coordination modifies the chemical and physical properties of both central atom and ligands. Consider the chemistry of aqueous cyanide (CN^-) and iron(II) (Fe^{2+}) ions. The former reacts immediately with acid to generate gaseous hydrogen cyanide (HeN), a deadly poison. The latter instantly precipitates a gelatinous hydroxide when

> • This use of brackets has nothing to do with signifying the concentrations of species in solution, as in $[Cu^{2+}]$. The intended meaning of brackets is usually clear from the context.

TABLE 19–1	Common Monodentate Ligands and Their Names	
Ligand	**Formula**	**Name**
Fluoride ion	:F$^-$	Fluoro
Chloride ion	:Cl$^-$	Chloro
Nitrite ion	:NO$_2^-$	Nitro
	:ONO$^-$	Nitrito
Carbonate ion	:OCO$_2^{2-}$	Carbonato
Cyanide ion	:CN$^-$	Cyano
Thiocyanate ion	:SCN$^-$	Thiocyanato
	:NCS$^-$	Isothiocyanato
Hydride ion	:H$^-$	Hydrido
Oxide ion	:O^{2-}	Oxido
Hydroxide ion	:OH$^-$	Hydroxo
Water	:OH$_2$	Aqua
Ammonia	:NH$_3$	Ammine
Carbon monoxide	:CO	Carbonyl
Nitrogen monoxide	:NO	Nitrosyl

The ligating atom is indicated by a pair of red dots representing a lone pair of electrons. In the CO_3^{2-} ligand, either one or two of the oxygen atoms can donate a lone pair to the metal.

mixed with aqueous base. Reaction between the two gives the complex ion $[Fe(CN)_6]^{4-}(aq)$, which undergoes neither of these two reactions nor any others considered diagnostic of simple CN^- ion and Fe^{2+} ion. Because coordination changes a ligand's chemical behavior, a given ligand may be present in multiple forms in the same compound. The two Cl^- ions in $[Pt(NH_3)_3Cl]Cl$ differ chemically because one is coordinated and one is not. Treatment of an aqueous solution of this substance with Ag^+ ion immediately precipitates the uncoordinated Cl^- as $AgCl(s)$ but not the coordinated Cl^-.

• This change, which involves the fragment —Pt—NH₂—H, recalls the increase in acidity of oxoacids as the electronegativity of X in the fragment —X—O—H increases. See Section 17–6.

Coordination also can change the acid–base behavior of a ligand. Ammonia is a base—a hydrogen ion acceptor (Brønsted-Lowry definition) or an electron-pair donor (Lewis definition). When NH_3 is coordinated in the $[Pt(NH_3)_6]^{4+}$ ion, the Pt(IV) accepts electrons from ammonia so strongly that the N—H bonds are weakened and the hydrogen atoms of the ammonia become acidic. The equilibrium constant of the reaction

$$[Pt(NH_3)_6]^{4+}(aq) + H_2O(\ell) \rightleftharpoons H_3O^+(aq) + [Pt(NH_3)_5(NH_2)]^{3+}(aq)$$

is 1.0×10^{-7}. This value of K_a means that the complex ion $[Pt(NH_3)_6]^{4+}$ is a weak acid, comparable to carbonic acid or sulfurous acid in strength. The acidity of aqueous solutions of transition-metal ions is a parallel phenomenon (see Section 9–5).

Types of Ligands

The ligands in Table 19–1 are capable of forming only a single bond to a central metal atom. They are called monodentate (from Latin *mono,* meaning "one," plus *dens,* meaning "tooth," indicating that they bind at only one point). Other ligands are capable of forming two or more such bonds and are referred to as bidentate, tridentate, and so forth.

• The Lewis structure of ethylenediamine is

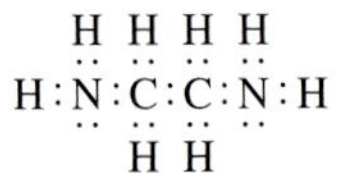

```
    H H H H
H:N:C:C:N:H
    H H
```

Ethylenediamine ($NH_2CH_2CH_2NH_2$), in which two NH_2 groups are held together by a carbon backbone, is a particularly important bidentate ligand. Both nitrogen atoms in ethylenediamine have lone electron pairs to share. If all the nitrogen donors of three ethylenediamine molecules bind to a single ion, say Co^{3+}, then that Co^{3+} ion has a coordination number of 6, and the formula of the resulting complex is $[Co(en)_3]^{3+}$ (where "en" is the accepted abbreviation for ethylenediamine). Complexes in which a ligand coordinates via two or more donors to the same central atom are called **chelates** (from Greek *chela,* meaning "claw," because the ligand bites onto the central atom like pincers). The structures of some important chelating ligands are given in Figure 19–2.

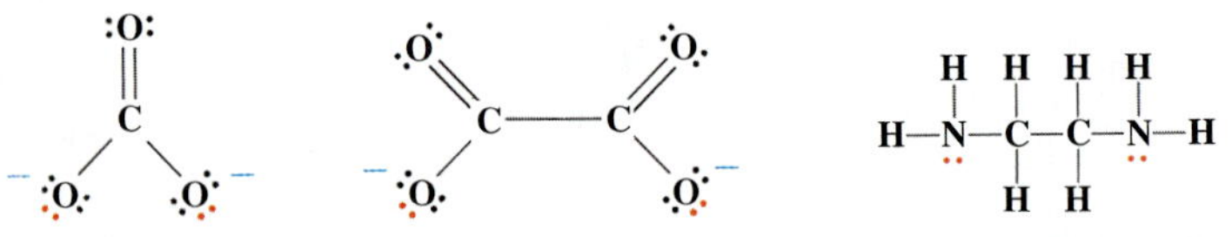

Figure 19–2 • (a–c) The Lewis structures of three bidentate ligands. Each is capable of ligating (donating a pair of electrons to an acceptor) at the sites marked in red.

Carbonate ion, CO_3^{2-} **(a)**

Oxalate ion, $C_2O_4^{2-}$ **(b)**

Ethylenediamine, $NH_2 CH_2 CH_2 NH_2$ **(c)**

Charge and Oxidation Number in Complexes

The overall electric charge on a coordination complex is the sum of the individual charges assigned to the metal ion and the ligands that surround it. Thus, the complex of Cu^{2+} (copper(II)) with four Br^- ions is an anion with a -2 charge, $[CuBr_4]^{2-}$.

EXAMPLE 19-1

Write formulas, including the overall electric charge, of coordination complexes containing the following metal ions and ligands:
(a) Co^{3+} with six F^- ligands.
(b) Pt^{4+} with two NH_3 ligands, two H_2O ligands, and two Cl^- ligands.
(c) Cr^{3+} with one NH_3 ligand, one ethylenediamine (en) ligand, and three NO_2^- ligands.

Solution

(a) The complex is an anion with the formula $[CoF_6]^{3-}$. The six -1 charges of the ligands combine with the $+3$ of the cobalt to give an overall charge of -3.
(b) The complex is a cation, $[Pt(NH_3)_2(H_2O)_2Cl_2]^{2+}$.
(c) The complex has the formula $[Cr(NH_3)(en)(NO_2)_3]$. There is no net charge because the charges of the three NO_2^- ligands and the Cr^{3+} metal ion add up to zero (and the other ligands are uncharged). Zero charges on complexes usually are not shown explicitly.

EXERCISE

Write formulas, including the overall electric charge, of coordination complexes containing the following metal ions and ligands: (a) Zn^{2+} with four Cl^- ligands; (b) Cr with six CO ligands; (c) Ni^{2+} with two H_2O ligands and two $C_2O_4^{2-}$ (oxalate) ligands.

Answer: (a) $[ZnCl_4]^{2-}$. (b) $[Cr(CO)_6]^0$. (c) $[Ni(H_2O)_2(C_2O_4)_2]^{2-}$.

The procedure used in Example 19–1 can be reversed to determine the oxidation number (oxidation state) of the central metal atom from the chemical formula of the coordination compound, a point that comes up frequently.

EXAMPLE 19-2

Determine the oxidation state of the coordinated metal atom in each of the following compounds:
(a) $K[Co(NH_3)_2(CN)_4]$
(b) $[Os(CO)_5]$
(c) $Na[Co(H_2O)_3(OH)_3]$

Strategy

Use the principle of charge neutrality (Section 3–3) together with knowledge of the charges of the ligands and non-ligating ions that are present. Plan to refer to Table 19–1 (and Tables 3–2 and 3–5 if necessary).

Solution

(a) The oxidation state of K is known to be $+1$, so the complex in brackets is an anion with a -1 charge, $[Co(NH_3)_2(CN)_4]^-$. The charge on the two NH_3 ligands is zero, and the charge on each of the four CN^- ligands is -1. The oxidation state of the Co must then be $+3$ because 4×-1 (for the CN^-) $+ 2 \times 0$ (for the NH_3) plus 3 (for Co) equals the required net charge on the complex, -1.

(b) The ligand CO has zero charge; the complex has zero charge as well. Therefore, the oxidation state of the osmium is zero.

(c) The complex contains three neutral ligands (the water molecules) and three ligands with -1 charges (the hydroxide ions). The Na^+ ion contributes only $+1$, so the oxidation state of the cobalt must be $+2$.

EXERCISE

Determine the oxidation state of the coordinated metal atom in each of the following compounds: (a) $K_3[Fe(CN)_6]$; (b) $[Co(en)_2(SCN)_2]Cl$; (c) $Na[Rh(NH_3)_3Cl_3]$.

Answer: (a) $+3$. (b) $+3$. (c) $+2$.

Naming Coordination Compounds

Up to now, only chemical formulas have been used to represent coordination compounds, but for many purposes, a name is needed. Some of these substances have names that were given to them before their structures were known. Thus, $K_3[Fe(CN)_6]$ was called potassium ferricyanide, and $K_4[Fe(CN)_6]$ was potassium ferrocyanide (these are complexes of Fe^{3+} (ferric) and Fe^{2+} (ferrous) ions, respectively). The older names still find some use but gradually are being replaced by systematic names based on the following set of rules:

1. The names of coordination complexes are written as single words, built from the names of the ligands, prefixes to indicate how many ligands are present, and a name for the central metal.

2. A coordination complex may be an ion or a neutral molecule. If it is ionic, the compound in which it is found is named according to the pattern in simple ionic compounds: the positive ion is named first, followed (after a space) by the name of the negative ion, regardless of which is the complex ion.

3. The names of anionic ligands are obtained by replacing the usual ending with the suffix *-o*. The names of neutral ligands are unchanged. Exceptions to the latter rule are *aqua* (for water), *ammine* (for NH_3), and *carbonyl* (for CO). See Table 19–1.

• Note that the NH_3 ligand is named an ammine (with a double *m*), but the NH_2 group in ethylenediamine is called an amine group (with a single *m*).

4. Greek prefixes (*di-, tri-, tetra-, penta-, hexa-*) are used to indicate the number of ligands of a given type attached to the central ion, if there is more than one. The prefix *mono-* is not used when there is only one ligand of a given type. If the name of the ligand itself contains the terms *mono-, di-,* and so forth (as in ethylenediamine), then the name of the ligand is placed in parentheses and the prefixes *bis-, tris-,* and *tetrakis-* are used instead of *di-, tri-,* and *tetra-*.

5. If more than one type of ligand is present, the ligands are listed in alphabetical order, ignoring the prefixes that tell how often each type of ligand occurs in the coordination sphere.
6. The oxidation state of the central metal atom is given by a Roman numeral enclosed in parentheses immediately following the name of the metal. If the complex ion has a net negative charge, the ending *-ate* is added to the stem of the name of the metal (if the metal's symbol is based on its Latin name, the Latin stem is used).

The following examples illustrate the systematic naming of several complexes:

$K_3[Fe(CN)_6]$	potassium hexacyanoferrate(III)
$K_4[Fe(CN)_6]$	potassium hexacyanoferrate(II)
$[Fe(CO)_5]$	pentacarbonyliron(0)
$[Co(NH_3)_5CO_3]Cl$	pentaamminecarbonatocobalt(III) chloride
$K_3[Co(NO_2)_6]$	potassium hexanitrocobaltate(III)
$[Cr(H_2O)_4Cl_2]Cl$	tetraaquadichlorochromium(III) chloride
$[Pt(NH_2CH_2CH_2NH_2)_3]Br_4$	tris(ethylenediamine)platinum(IV) bromide
$K_2[CuCl_4]$	potassium tetrachlorocuprate(II)

EXAMPLE 19-3

Interpret the names and write the formulas of these coordination compounds:
(a) Sodium tricarbonatocobaltate(III).
(b) Diamminediaquadichloroplatinum(IV) bromide.
(c) Sodium tetranitratoborate(III).

Solution

(a) In the anion, three carbonate ligands (with -2 charges) are coordinated to a cobalt atom in the $+3$ oxidation state. Because the complex ion thus has an overall charge of -3, three sodium cations are required, making the correct formula $Na_3[Co(CO_3)_3]$. Writing the formula of this compound requires a previous knowledge of the charge on the carbonate group.
(b) The ligands coordinated to one Pt(IV) are two ammonia molecules, two water molecules, and two chloride ions. Ammonia and water are electrically neutral, but the two chloride ions contribute a total charge of $2 \times (-1) = -2$ that adds with the $+4$ of the platinum and gives the complex ion a $+2$ charge. Two bromide anions are required to balance this, so the formula is $[Pt(NH_3)_2(H_2O)_2Cl_2]Br_2$.
(c) The complex anion has four nitrate ligands, each with a -1 charge, coordinated to a central boron(III). This gives a net charge of -1 on the complex ion and requires one sodium ion in the formula, $Na[B(NO_3)_4]$.

EXERCISE

(a) Name the compound with the chemical formula $[Ni(NH_3)Cl(en)_2]Cl$.
(b) Write the chemical formula for sodium diammineaquatrichlorocobaltate(II).

Answer: (a) amminechlorobis(ethylenediamine)nickel(II) chloride.
(b) $Na[Co(NH_3)_2(H_2O)Cl_3]$.

Ligand Substitution Reactions

An otherwise bewildering collection of information on chemical reactivity is easily rationalized in terms of one ligand substituting for another in coordination complexes. A good way to follow the progress of ligand substitution (as well as other reactions involving coordination complexes) is to watch for color changes. Anhydrous nickel(II) sulfate, for example, is a yellow crystalline solid. If exposed to moist air at room temperature, it takes up six water molecules per formula unit. The water molecules coordinate with the nickel ions to form a bright green complex:

$$\underset{\text{yellow}}{NiSO_4(s)} + \underset{\text{colorless}}{6\,H_2O(g)} \longrightarrow \underset{\text{green}}{[Ni(H_2O)_6]SO_4(s)}$$

Heating the green hexaaquanickel(II) sulfate sufficiently (a temperature well above the boiling point of water is required) drives off the water and regenerates the yellow $NiSO_4$ in the reverse of the reaction just written. A similar coordination reaction generates a product with a completely different color when yellow $NiSO_4(s)$ is exposed to gaseous ammonia. This time, the product is a blue-violet complex:

$$\underset{\text{yellow}}{NiSO_4(s)} + \underset{\text{colorless}}{6\,NH_3(g)} \longrightarrow \underset{\text{blue-violet}}{[Ni(NH_3)_6]SO_4(s)}$$

Heating the blue-violet product drives off ammonia, a process that can be followed by watching the color of the solid change back to yellow. Given these facts, it is not hard to explain the observation that a green $[Ni(H_2O)_6]^{2+}(aq)$ solution turns blue-violet when treated with $NH_3(aq)$ (Fig. 19–3). The NH_3 must be displacing the H_2O from the coordination sphere:

$$\underset{\text{green}}{[Ni(H_2O)_6]^{2+}(aq)} + \underset{\text{colorless}}{6\,NH_3(aq)} \longrightarrow \underset{\text{blue-violet}}{[Ni(NH_3)_6]^{2+}(aq)} + \underset{\text{colorless}}{6\,H_2O(\ell)}$$

This is a classic ligand substitution reaction.

Complexes that readily undergo substitution of one ligand for another are **labile,** whereas complexes in which substitution proceeds slowly or not at all are **inert.** In an inert complex, a large energy of activation (see Section 14–6) of ligand substitution prevents rapid reaction even though there is a thermodynamic tendency to proceed. In the substitution reaction

$$[Co(NH_3)_6]^{3+}(aq) + 6\,H_3O^+(aq) \longrightarrow [Co(H_2O)_6]^{3+}(aq) + 6\,NH_4^+(aq)$$

Figure 19–3 • When ammonia is added to the green solution of nickel(II) sulfate on the left (which contains $[Ni(H_2O)_6]^{2+}$ ions), ligand substitution occurs to give the blue-violet solution on the right (which contains $[Ni(NH_3)_6]^{2+}$ ions). *(Leon Lewandowski)*

the products are favored thermodynamically by an enormous amount (the equilibrium constant is about 10^{64}). Yet the inert $[Co(NH_3)_6]^{3+}$ complex ion lasts for weeks in acidic solution because no low-energy path for the reaction exists. The $[Co(NH_3)_6]^{3+}$ ion is thermodynamically unstable relative to $[Co(H_2O)_6]^{3+}$ but kinetically stable (that is, inert). The closely related cobalt(II) complex $[Co(NH_3)_6]^{2+}$ undergoes a similar substitution reaction:

$$[Co(NH_3)_6]^{2+}(aq) + 6\,H_3O^+(aq) \longrightarrow [Co(H_2O)_6]^{2+}(aq) + 6\,NH_4^+(aq)$$

in a few seconds. This cobalt(II) complex is thermodynamically unstable and also labile.

Substitution of one ligand for another can proceed in stages in the coordination sphere of a metal. By controlling the reaction conditions, substitution usually can be stopped at intermediate stages. For example, all of the possible four-coordinate compositions of Pt(II) with the two ligands NH_3 and Cl^-,

$$[Pt(NH_3)_4]^{2+} \quad [Pt(NH_3)_3Cl]^+ \quad [Pt(NH_3)_2Cl_2] \quad [Pt(NH_3)Cl_3]^- \quad \text{and} \quad [PtCl_4]^{2-}$$

can be prepared and isolated (the ions are isolated as salts with suitable negative or positive ions). Such mixed-ligand complexes add to the richness of coordination chemistry.

19-2 Structures of Coordination Complexes

The Alsatian-Swiss chemist Alfred Werner pioneered the field of coordination chemistry in the late 19th century. At the time, a number of complexes of cobalt(III) chloride with ammonia were known and had these chemical formulas and colors:

Compound 1	$CoCl_3 \cdot 6NH_3$	Orange-yellow
Compound 2	$CoCl_3 \cdot 5NH_3$	Purple
Compound 3	$CoCl_3 \cdot 4NH_3$	Green
Compound 4	$CoCl_3 \cdot 3NH_3$	Green

Treating these compounds with aqueous hydrochloric acid did not remove the ammonia, suggesting that it was somehow closely bound with the cobalt ions. Treatment with aqueous silver nitrate at 0°C, on the other hand, gave interesting results. With compound 1, all of the chloride present was precipitated as solid AgCl. With compound 2, however, only two thirds of the chloride was precipitated, and with compound 3, only one third was precipitated. Compound 4 did not react at all with silver nitrate. Werner accounted for these facts by postulating the existence of coordination complexes with six ligands (chloride ions or ammonia molecules or both kinds of ligands) attached to each Co^{3+} ion. Specifically, he wrote the formulas for compounds 1 to 4 as follows:

Compound 1	$[Co(NH_3)_6]^{3+}(Cl^-)_3$
Compound 2	$[Co(NH_3)_5Cl]^{2+}(Cl^-)_2$
Compound 3	$[Co(NH_3)_4Cl_2]^+(Cl^-)$
Compound 4	$[Co(NH_3)_3Cl_3]$

Only those chloride ions that were *not* ligands attached directly to cobalt were precipitated on the addition of cold aqueous silver nitrate.

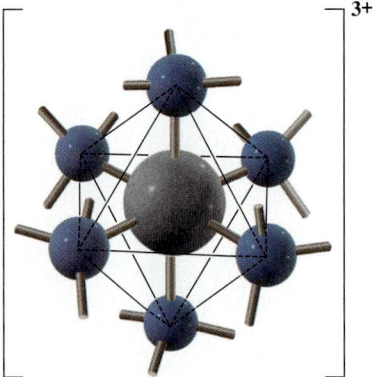

Figure 19–4 • The octahedral structure of the $[Co(NH_3)_6]^{3+}$ ion. All six corners of the octahedron are equivalent. The hydrogen atoms are suppressed for clarity.

Werner realized that his proposal gave predictions about the electrical conductivity of aqueous solutions of salts of these complex ions. Compound 1, for example, should have a molar electrical conductivity close to that of $Al(NO_3)_3$, which also yields one $+3$ ion and three -1 ions per formula unit dissolved in water. His experiments confirmed this resemblance and also showed that compound 2 resembled $Mg(NO_3)_2$ and that compound 3 resembled $NaNO_3$ in their electrical conductivities. Compound 4 behaved like a non-electrolyte, with a very low conductivity, as expected from the absence of ions in the formula.

Werner and other chemists studied other coordination complexes of different metals and ligands, using both physical and chemical techniques. This research has shown that the most common coordination number by far is 6, as in the cobalt complexes just discussed. Other coordination numbers ranging from 2 to 12 have been observed, however. Of these, the most important and interesting are 4 (as in $[PtCl_4]^{2-}$), 2 (as in $[Ag(NH_3)_2]^+$), and 5 (as in $[Ni(CN)_5]^{3-}$).

Three-Dimensional Structures

If the central cobalt ion in $[Co(NH_3)_6]^{3+}$ is bonded to six ammonia ligands, what is the geometrical structure of the complex? This question naturally occurred to Werner, who suggested that the arrangement should be the simplest and most symmetrical possible, with ligands located at the six vertices of an octahedron (Fig. 19–4). Modern methods of X-ray diffraction (see Sections 20–1 and 20–2) allow for very precise determinations of atomic positions in crystals and completely confirm Werner's hypothesis of octahedral coordination for this complex. Such techniques were unavailable a century ago, however, and so Werner turned to a study of the properties of substituted complexes to test his hypothesis.

If one ammonia ligand is replaced with a chloride ion, the resulting complex has the formula $[Co(NH_3)_5Cl]^{2+}$, with one vertex of the octahedron occupied by Cl^-, and the other five, by NH_3. Only one structure of this type is possible because all six vertices of the original octahedron are equivalent; all ways of drawing singly substituted octahedral structures $[MA_5B]$ (where $A = NH_3$, $B = Cl^-$, $M = Co^{3+}$, for example) become equivalent when the structure is imagined to rotate (tumble) around different axes in space. Now suppose that a second NH_3 ligand is replaced with Cl^-. The second Cl^- can lie in one of the positions closest to the first Cl^- in the plane perpendicular to the $Co—Cl^-$ line (Fig. 19–5a), or it can lie in the sixth position, on the opposite side of the central metal atom

Figure 19–5 • (a) The structure of the *cis*-$[Co(NH_3)_4Cl_2]^+$ ion; (b) the structure of the *trans*-$[Co(NH_3)_4Cl_2]^+$ ion. These isomeric complex ions have distinctly different properties. The *cis* complex is purple in solution, but the *trans* is green, for example. Hydrogen atoms are omitted to simplify the picture.

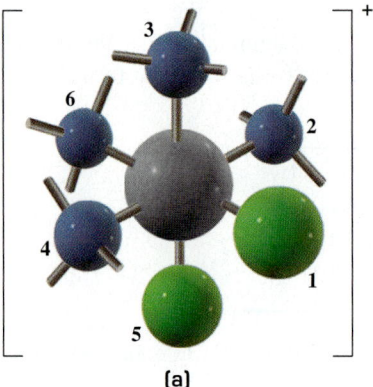

(a)

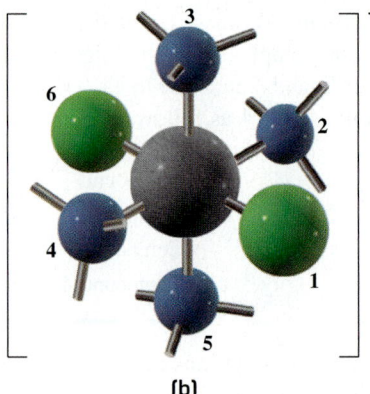

(b)

(Fig. 19–5b). The former structure, in which the two Cl^- ligands are closer to each other, is called a *cis* structure, *cis*-$[Co(NH_3)_4Cl_2]^+$, and the latter, with the two Cl^- ligands farther apart, is a *trans* structure, *trans*-$[Co(NH_3)_4Cl_2]^+$. The octahedral-structure model predicts that exactly two different ions with the chemical formula $[Co(NH_3)_4Cl_2]^+$ should exist. Such structurally different species are called ***cis–trans* isomers** (and are similar to the *cis–trans* isomers discussed in Section 18–2). When Werner began his work, only the green *trans* form was known, but by 1907, he had prepared the *cis* isomer and showed that it differed from the *trans* isomer in color (it was violet rather than green) and in other physical properties. The isolation of two, and only two, isomers of this complex was good (although not conclusive) evidence that the octahedral structure was correct. Almost all six-coordinated complexes of the transition elements have octahedral structures; hence, many compounds can display similar isomerism. The complex ion $[CoCl_2(en)_2]^+$, for example, also has a purple *cis* form and a green *trans* form (Fig. 19–6).

Figure 19–6 • The complex ion $[CoCl_2(en)_2]^+$ is an octahedral complex that has *cis* and *trans* isomers, according to the relative positions of the two Cl^- ligands. Salts of the *cis* isomer are purple, and those of the *trans* isomer are green. *(Charles Steele)*

EXAMPLE 19-4

How many structural isomers exist for the octahedral coordination compound $[Co(NH_3)_3Cl_3]$?

Strategy

Begin with the two isomers of the octahedral ion $[Co(NH_3)_4Cl_2]^+$ and see how many different structures can be made by replacing one more ammonia molecule with Cl^-. Focus on the geometrical relationships between the existing ligands and the replacing ion.

Solution

Replacing any of the four NH_3 ligands in the *trans* form (Fig. 19–5b) at sites 2, 3, 4, or 5 gives equivalent structures that can be superimposed by rotation about the axis connecting positions 1 and 6. The new Cl^- is unique. It is in the middle between the old Cl's, which are on the ends, as shown in Figure 19–7a. Replacing an NH_3 ligand in the *cis* form works differently. If the Cl^- goes in at either site 3 or 6 in Figure 19–5a, the result is simply a rotated version of Figure 19–7a. Replacement of the NH_3 at site 4, however, gives a different structure, which is shown in Figure 19–7b. The three Cl^-'s are all *cis* to each other; they are equivalent. Replacement at site 2 gives a rotated version of the same all-*cis* isomer. We conclude that two, and only two, isomers of $[Co(NH_3)_3Cl_3]$ can exist.

In fact, only one form of $[Co(NH_3)_3Cl_3]$ has been prepared to date, presumably because the two isomers interconvert rapidly, but two and only two isomers are known for the closely related octahedral coordination complex $[Cr(NH_3)_3(NO_2)_3]$.

EXERCISE

How many different isomers of the octahedral coordination complex $[Co(NH_3)_3(H_2O)Cl_2]^+$ are possible? *Hint:* Start with Figure 19–7 and replace one Cl^- ligand with an H_2O ligand.

Answer: 3.

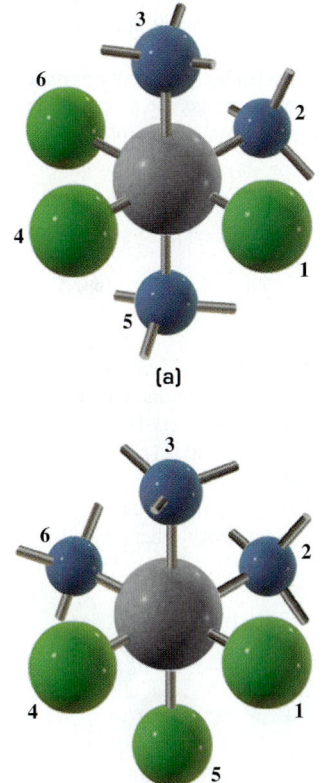

(a)

(b)

Figure 19–7 • The two structural isomers of $[Co(NH_3)_3Cl_3]$.

Square-Planar, Tetrahedral, and Linear Structures

Complexes with coordination numbers of 4 are typically either tetrahedral or square planar. The tetrahedral geometry (Fig. 19–8a) predominates for four-coordinate complexes of the early transition metals (those toward the left side of the *d* block of elements in the periodic table). There is no possibility of *cis–trans* isomerism for tetrahedral complexes of the general form MA_2B_2 because all such structures are superimposable.

The square-planar geometry (Fig. 19–8b, c) is common for four-coordinate complexes of Au^{3+}, Ir^+, and Rh^+ and, most especially, for ions with the d^8 valence-electron configuration: Ni^{2+}, Pd^{2+}, and Pt^{2+}. The Ni^{2+} ion forms a few tetrahedral structures, but four-coordinate Pd^{2+} and Pt^{2+} are nearly exclusively square planar. Square-planar complexes of the type MA_2B_2 can have isomers, as Figures 19–8b and c illustrate for *cis*- and *trans*-$[Pt(NH_3)_2Cl_2]$. The *cis* form of this compound is a potent and widely used anticancer drug called "cisplatin"; the *trans* form has no therapeutic properties.

Finally, linear complexes with coordination numbers of 2 exist, especially for ions with d^{10} configurations such as Cu^+, Ag^+, Au^+, and Hg^{2+}. In aqueous solution the central silver atom in a complex such as $[Ag(NH_3)_2]^+$ strongly attracts several water molecules as well. Thus its actual coordination number under these circumstances may exceed 2.

Chiral Structures

• Figure 16–4 shows a plane wave, which can represent plane-polarized light. When such a wave enters a sample containing one enantiomer and not its mirror image, the plane in which the wave oscillates is rotated.

The complex ion $[Co(en)_3]^{3+}$ displays a type of isomerism that differs from the isomerism discussed so far. It is called **chirality** and is illustrated by the two structures shown in Figure 19–9. These two structures are quite similar, yet not identical. Their ligands are connected in the same relative arrangement, but, taken as a unit, they bear the same relationship to each other that the left hand does to the right. The two are mirror images. No matter how these two structures are tumbled and turned about, they cannot be superimposed. Such mirror-image pairs are referred to as **enantiomers.** The difference between enantiomers is a subtle one, and many of their physical and chemical properties are the same. Members of enantiomeric pairs differ in the way they interact with other objects that have "handedness," just as a right hand, a chiral object, fits well into a right glove but poorly into a left glove. This leads to *stereoselective* reactions, in which one enantiomer reacts preferentially. Many reactions of biological importance are stereoselective. Enantiomers also can be distinguished by the way in which their solutions rotate the plane of polarization of plane-polarized light: If one does so in a clockwise sense, the other does so in a counterclockwise sense. For this reason, enantiomers are also called **optical isomers.**

Figure 19–8 • Four-coordinate complexes. (a) Tetrahedral $[FeCl_4]^-$. (b) Square-planar *cis*-$[Pt(NH_3)_2Cl_2]$ (in which the chlorides (green) are near to each other). (c) Square-planar *trans*-$[Pt(NH_3)_2Cl_2]$ (in which the chlorides are far from each other).

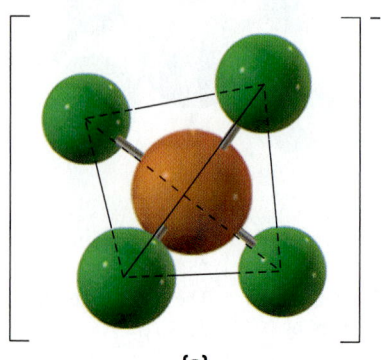

(a)

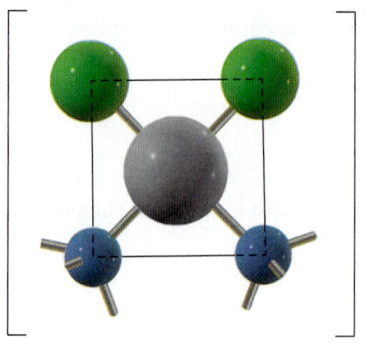

(b)

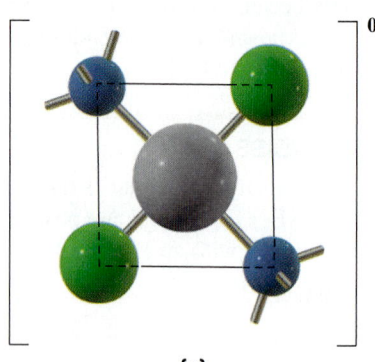

(c)

Mirror

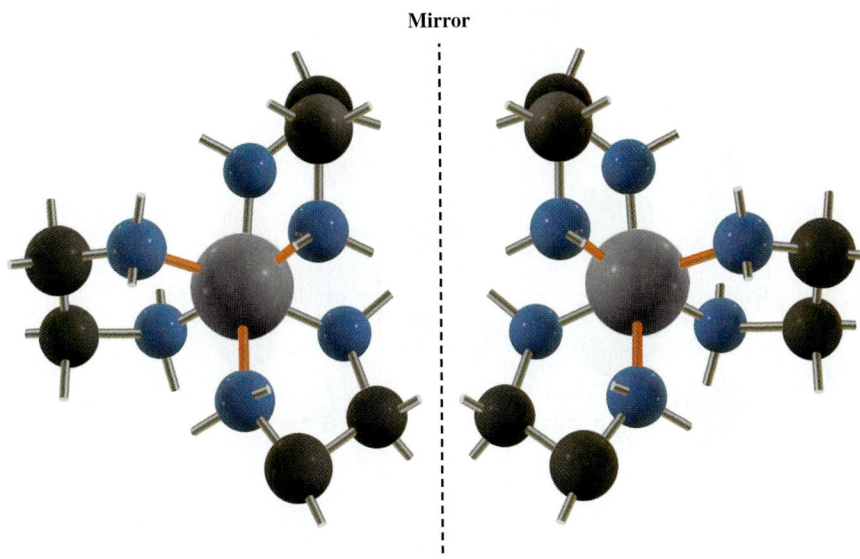

Figure 19-9 • The enantiomers of the $[Co(en)_3]^{3+}$ ion. Three coordinate bonds (highlighted in red) are in front of the central cobalt, and three are behind. The ligands wrap clockwise around the cobalt ion from front to back in the isomer on the right but wrap counterclockwise in the isomer on the left. Reflection through the mirror transforms one enantiomer into the other. The two cannot be superimposed by any rotation.

EXAMPLE 19-5

Suppose that the complex ion $[Co(NH_3)_2(H_2O)_2Cl_2]^+$ is synthesized with the two ammine ligands *cis* to each other, the two aqua ligands *cis* to each other, and the two chloro ligands *cis* to each other, as shown in Figure 19–10. Determine whether this complex is chiral—that is, whether the structure can be superimposed on its mirror image.

Strategy

When all else fails, a structure and its mirror image can be represented by drawing the structure on paper, viewing it in a small mirror, and copying the image. It is faster to indicate a mirror with a dotted line and create the mirror image by making each point in the structure on the far side of the dotted line lie the same distance from the dotted line as the generating point in the original structure (as in Fig. 19–10). Rotate the mirror image in space (either mentally or by making auxiliary drawings) until as many points as possible coincide in position with their generating points in the original structure. If *all* points can be made to coincide, the mirror image is superimposable.

Approximately enantiomeric seashells. The spiral body of the shell winds to the left in the shell on the left but to the right in the shell on the right. The mirror-image relationship is imperfect in other respects, however.

Solution

The complex ion *cis,cis,cis*-$[Co(NH_3)_2(H_2O)_2Cl_2]^+$ is chiral, because the two structures cannot be superimposed even after turning them.

EXERCISE

Is the square-planar complex $[Pt(NH_3)(H_2O)BrCl]$, in which Pt is surrounded by four different ligands, chiral? If the answer is yes, draw a representation of a pair of optical isomers of this geometry. If no, explain.

Answer: Forming the mirror image of this square-planar complex is equivalent to simply turning over the plane of the structure; hence, the answer is no.

Figure 19–10 • The structure of *cis,cis,cis*-$[Co(NH_3)_2(H_2O)_2Cl_2]^+$ ion, together with its mirror image.

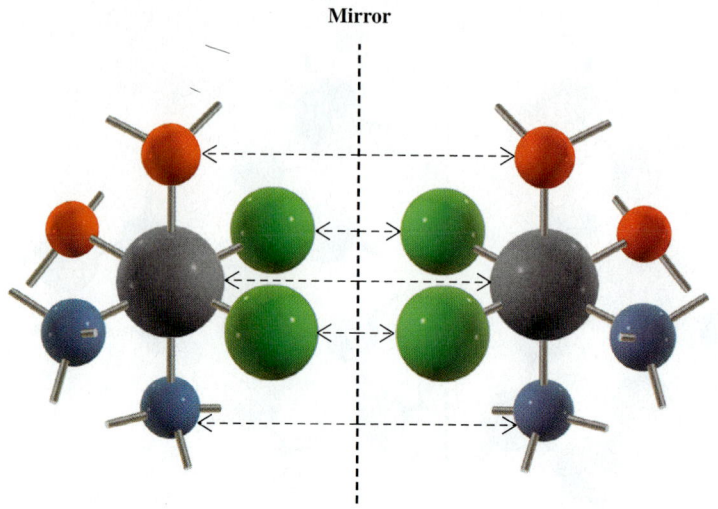

Mirror

CHEMISTRY IN PROGRESS

Sequestering Agents as Miracle Ingredients

The hexadentate ligand EDTA (ethylenediaminetetraacetate ion) has numerous practical applications that originate from its ability to chelate with metal ions. The structure of EDTA and its mode of chelation to a metal ion with octahedral coordination are shown in Figure 19–A. The ligand wraps around the metal ion as its two N donors and four O donors all attach simultaneously. This type of chelation gives very strongly bound complexes that lock up or **sequester** many kinds of metal ion from their ordinary reactions. For example, EDTA solubilizes the scummy precipitates that Ca^{2+} ions form with anionic constituents of soap by forming a stable complex with Ca^{2+}. Thus, it breaks up the main contributor to bathtub rings and is a "miracle ingredient" in some bathtub cleaners. EDTA has been used as an antidote to lead poisoning because its great affinity for Pb^{2+} ions prevents the Pb^{2+} ions from the destructive interactions they otherwise undergo in the body. This same high affinity allows for its use in the removal of metal ions that are trace contaminants of water. Complexes of EDTA with iron are used in plant foods to permit a slow release of iron to the plant. EDTA also sequesters copper and nickel ions in edible fats and oils. Because these metal ions catalyze the oxidation reactions that make oils rancid, the presence of EDTA retards spoilage.

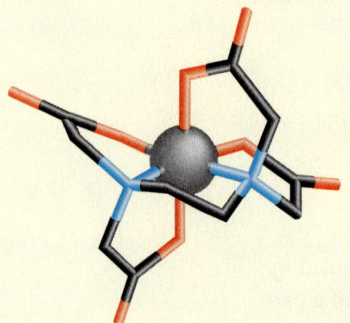

EDTA anion

Figure 19–A • The wire-frame structure on the left emphasizes the way in which EDTA enfolds a metal ion as it chelates; hydrogens are omitted for clarity. The diagram on the right gives the full structure of EDTA and shows (*in red*) the six pairs of electrons that it donates. (*Julius Weber*)

19-3 Crystal-Field Theory and Magnetic Properties

What is the nature of the bonding in coordination complexes of the transition elements? Why does Pt(IV) form only octahedral complexes, whereas Pt(II) forms almost exclusively square-planar ones, and under what circumstances does Ni(II) form octahedral, square-planar, or tetrahedral complexes? Can trends in the lengths and strengths of metal–ligand bonds be understood and predicted? Answering such questions requires a theoretical description of the bonding in coordination complexes.

A simple but useful model for bonding in coordination complexes is **crystal-field theory,** which starts from an ionic description of the metal–ligand bonds, omitting entirely the sharing of electrons from the ligands to the metal. Crystal-field theory considers the response of a central metal to the approach of negatively charged ligands. In an octahedral complex, six negative charges from the lone pairs of six ligands are brought up along the $\pm x$, $\pm y$, and $\pm z$ coordinate axes toward a metal atom or ion located at the origin. In an atom or ion of a transition element in free space, the energy of an electron is the same in each of the five d orbitals in a given shell. When the external charges are brought up, however, the energies of the d orbitals change by different amounts because of differing degrees of repulsion between the electrons in the d orbitals and the donor electron pairs on the ligands (Fig. 19–11). An electron in a $d_{x^2-y^2}$ or d_{z^2} orbital is most likely to be found along the coordinate axes, where it comes close to electrons from the ligand in an octahedral complex. This proximity raises the energies of these orbitals. By contrast, an electron in a d_{xy}, d_{yz}, or d_{xz} orbital has its greatest probability density between the coordinate axes, so it experiences less repulsion from an octahedral array of approaching charges and has lower energy than do the $d_{x^2-y^2}$ and d_{z^2} orbitals. In the octahedral field created by six ligand lone pairs, the d orbital energy levels split into

• In this model, all the d orbital energies increase, but the increase is smaller for some orbitals than for others.

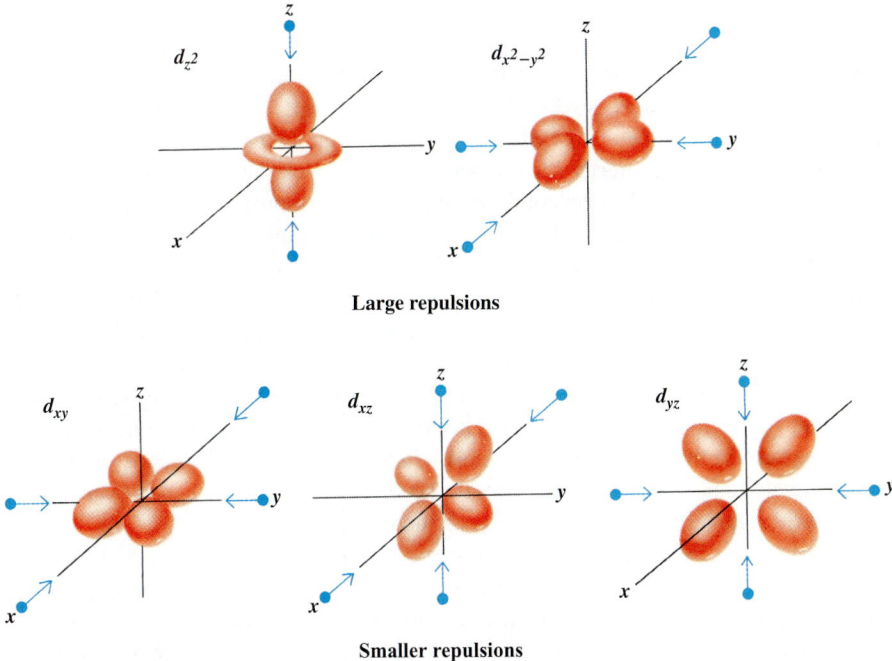

Figure 19–11 • Octahedral crystal-field splitting of 3d orbital energies by ligands. As the external charges approach the five 3d orbitals, the largest repulsions arise in the d_{z^2} and $d_{x^2-y^2}$ orbitals, which point directly at two or four of the approaching charges.

Figure 19–12 • Six point charges in an octahedral array about a central metal ion create an octahedral field about the metal ion. Such a field increases the energies of all five d orbitals on the metal ion, but the increase is larger for the d_{z^2} and $d_{x^2-y^2}$ orbitals. As a result, the d orbitals are split into two sets that differ by the energy Δ_o. The occupancy of these orbitals by the four d electrons of Mn(III) is shown for (a) the low-spin (large Δ_o) and (b) the high-spin (small Δ_o) case.

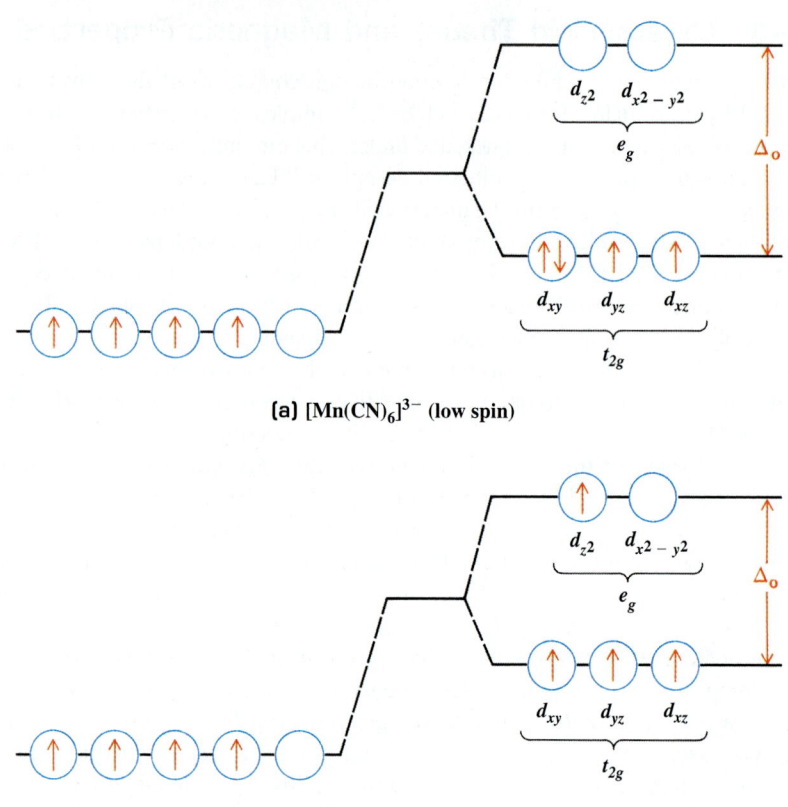

(a) $[Mn(CN)_6]^{3-}$ (low spin)

(b) $[Mn(H_2O)_6]^{3+}$ (high spin)

• The symbols t_{2g} and e_g come from a mathematical representation of the symmetries of the orbitals that does not concern us here. The subscript "o" on Δ_o stands for "octahedral."

• Recall from Section 17–1 that the Cr atom has the ground-state electronic configuration $[Ar]3d^5 4s^1$. When they form ions, the fourth-row transition elements lose their $4s$ electrons more easily than their $3d$ electrons, so the valence-electron configuration of Cr^{3+} is $3d^3$.

two groups (Fig. 19–12). The three lower-energy orbitals, called t_{2g} levels, correspond to the d_{xy}, d_{yz}, and d_{xz} orbitals of the transition-metal atom; the two higher-energy orbitals, called e_g levels, correspond to the $d_{x^2-y^2}$ and d_{z^2} orbitals of the transition-metal ion. The energy difference between the two sets of levels is Δ_o, the **crystal-field splitting energy** for the octahedral complex that is formed.

Now consider the electron configuration for a transition-metal ion experiencing such an octahedral crystal field. The Cr^{3+} ion, for example, has three d electrons. According to Hund's rules (see Section 17–1), one electron goes into each of the three t_{2g} orbitals with parallel spins. For an ion such as Mn^{3+}, which has a fourth d electron, there are two conceivable ground-state electron configurations. The fourth electron can occupy either a t_{2g} orbital, with spin opposite to that of the electron already in that orbital (Fig. 19–12a), or an e_g orbital, with spin parallel to those of the three t_{2g} electrons (Fig. 19–12b). The former happens when the splitting Δ_o is large because in that case the energy cost of placing an electron in an e_g orbital is prohibitively high. If the splitting is small, however, the e_g orbital is occupied in order to avoid having two electrons in the same orbital.

Two types of electron configurations are possible in octahedral complexes of central metal atoms or ions having four to seven d electrons. When Δ_o is large, **low-spin complexes** are formed in which electrons are paired in the lower-energy t_{2g} orbitals; the e_g orbitals are not occupied until the t_{2g} orbitals are filled. When Δ_o is small, **high-spin complexes** occur in which electrons are placed singly in both t_{2g} and e_g orbitals and remain unpaired to the fullest extent possible.

There is interesting direct evidence for the reality of the high-spin versus low-spin distributions of *d* electrons in coordination complexes. Precise experimental work in which crystals diffract X-rays (see Section 20–1) provides maps of the electron density within crystals. When this technique recently was applied to a compound having Co^{2+} in an octahedral environment of F^- ions, it was possible to detect the seven 3*d* electrons of the Co^{2+} and confirm that their distribution was that of the configuration $t_{2g}^5 e_g^2$ (high spin) and not $t_{2g}^6 e_g^1$ (low spin).

Magnetic Properties

The existence of high-spin and low-spin electron configurations accounts for the magnetic properties of many different coordination compounds. As discussed in Section 17–1, substances can be classified as paramagnetic or diamagnetic according to whether they are attracted into an inhomogeneous magnetic field. *All substances are either attracted or repelled by magnets*, although most such interactions are so weak that experiments with low-power magnets show nothing. Figure 19–13 depicts an experiment to demonstrate the universal susceptibility of substances to the influence of magnetic fields. A cylindrical sample is suspended so that its bottom is between the poles of a powerful magnet but its top extends out of the field. It is thus in a non-uniform magnetic field, the intensity of which varies from strong to weak. It is weighed very precisely and then reweighed in the absence of the magnet. The net force on the sample changes in the presence of the magnetic field. Substances that are repelled by a non-uniform magnetic field appear to weigh less when dipped into one and are *diamagnetic*. Substances that are attracted by a magnetic field appear to weigh more and are *paramagnetic*. The weighings just described, with due calibration and correction, give numerical values for the **magnetic susceptibility** of a substance, its tendency to be attracted by magnetic fields. The susceptibility of a diamagnet is negative and small (because the substance is repelled by the test magnet). The susceptibility of a paramagnet is positive and can be quite large.

• This is one place where the distinction between mass and weight is crucial. The mass of the sample does not change when the magnet is removed, but the weight does because the net force pulling the sample down is then different.

• Iron and steel are very strongly attracted into the magnetic field. They are ferromagnetic.

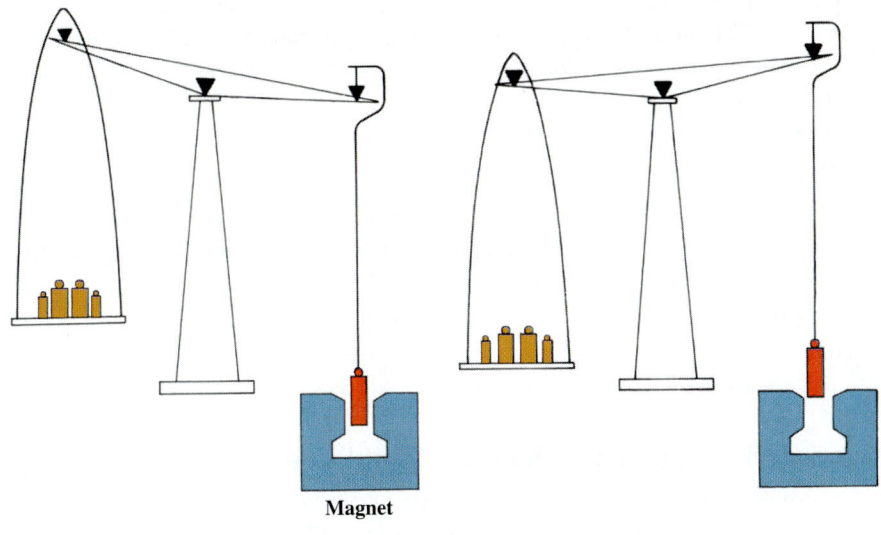

Magnet

(a) Paramagnetic

(b) Diamagnetic

Figure 19–13 • (a) If a sample of a paramagnetic substance is immersed partially into a magnetic field, it is attracted down into the field. (b) If a diamagnetic substance is put into the same field in the same way, it is buoyed up by the field.

Paramagnetism always is associated with atoms, ions, or molecules that contain one or more unpaired electrons. Diamagnetic substances have the spins of all of their electrons paired. Thus, measurements of magnetic susceptibility reveal which substances have one or more unpaired electrons and which have paired electrons only. In many cases, the number of unpaired electrons per molecule in a paramagnet can be counted, based on the magnitude of its magnetic susceptibility. A mole of a substance with two unpaired electrons per formula unit is pulled into a magnetic field more strongly than is a mole of a substance with only one unpaired electron per formula unit.

These facts emerge in connection with coordination complexes because paramagnetism is prevalent among complexes of the transition elements. The great majority of other chemical substances are diamagnetic. Among complexes of a given metal ion, the number of unpaired electrons, as observed by magnetic susceptibility, varies with the identity of the ligands. This can be understood through crystal-field theory. Both $[Mn(CN)_6]^{3-}$ and $[Mn(H_2O)_6]^{3+}$ ions, for example, have six ligands surrounding the central Mn^{3+} ion, yet the former has only two unpaired electrons (see Fig. 19–12a) and the latter has four (see Fig. 19–12b). The crystal-field splitting energy is larger for the CN^- ligands than for the H_2O ligands, giving a low-spin complex in the first case and a high-spin complex in the second. Similarly, $[Fe(CN)_6]^{4-}$ is diamagnetic, but $[Fe(H_2O)_6]^{2+}$ is paramagnetic to the extent of four unpaired electrons. In $[Fe(CN)_6]^{4-}$, the six d electrons on the Fe^{2+} ion all are paired in the lower t_{2g} orbitals (configuration $(t_{2g})^6$ shown in Fig. 19–14a); in $[Fe(H_2O)_6]^{2+}$, two electrons have left the t_{2g} orbital to occupy e_g orbitals (configuration $(t_{2g})^4(e_g)^2$ shown in Fig. 19–14b).

• Indeed, the Lewis-dot representation of bonding (see Chapter 3) calls for displaying all valence electrons in pairs.

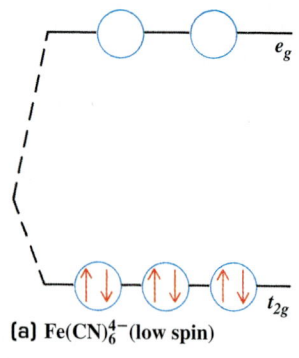

(a) $Fe(CN)_6^{4-}$ (low spin)

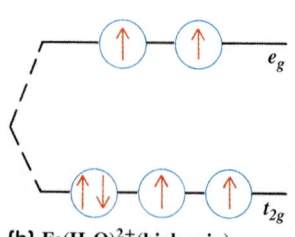

(b) $Fe(H_2O)_6^{2+}$ (high spin)

Figure 19–14 • The occupancy of the $3d$ orbitals in a low-spin and a high-spin octahedral complex of Fe(II).

EXAMPLE 19–6

The complex ion $[CoF_6]^{3-}$ is paramagnetic, but $[Co(NH_3)_6]^{3+}$ is not. Identify the d electron configurations in these two octahedral complex ions.

Solution

The Co^{3+} ion, like the Fe^{2+} ion, has six d electrons. Its high-spin complexes have four unpaired spins $(t_{2g})^4(e_g)^2$ and are paramagnetic (as in Fig. 19–14b); its low-spin complexes have no unpaired spins $(t_{2g})^6$ and are diamagnetic (as in Fig. 19–14a). Therefore, $[CoF_6]^{3-}$ must be a high-spin complex, and $[Co(NH_3)_6]^{3+}$, a low-spin complex, with the splitting Δ_o greater for ammonia than for fluoride-ion ligands.

EXERCISE

The octahedral complex ions $[FeCl_6]^{3-}$ and $[Fe(CN)_6]^{3-}$ are, respectively, high-spin and low-spin complexes. How many unpaired electrons are in each, and in which is the octahedral-field splitting greater?

Answer: Five in $[FeCl_6]^{3-}$ and one in $[Fe(CN)_6]^{3-}$. Δ_o is larger in the cyanide complex.

Square-Planar and Tetrahedral Complexes

Crystal-field theory can be applied to square-planar and tetrahedral complexes as well as to octahedral complexes. Consider a square-planar complex in which four ligands are brought up toward the metal ion along the $\pm x$ and $\pm y$ axes. A $d_{x^2-y^2}$ orbital on the metal has a higher energy than does a d_{z^2} orbital in such a field because an electron in the $d_{x^2-y^2}$ orbital has maximum probability density along the x and y axes, where it is most strongly repelled by ligand electrons. In the same way, the energy of the d_{xy} orbital lies above that of the d_{xz} and d_{yz} in a square-planar field because the former lies in the plane of the ligands, and the latter wave functions have a node in the xy plane. Figure 19–15a shows the resulting level structure.

Tetrahedral complexes result from bringing ligands to alternating corners of an imaginary cube (for a total of four out of eight) having the metal ion at its center. The $d_{x^2-y^2}$ and d_{z^2} orbitals point toward the centers of the cube faces, but the other three orbitals point toward the centers of the cube edges. The cube edges are closer to the ligands, and these three orbitals therefore are repelled more strongly than are the $d_{x^2-y^2}$ and d_{z^2} orbitals. The result is shown in Figure 19–15b and is the reverse of the energy-level ordering found for octahedral complexes. The magnitude of the splitting Δ_t is significantly smaller than Δ_o.

Octahedral complexes are more common than are square-planar or tetrahedral complexes because the formation of six bonds to ligands, rather than four, confers greater stability. Square-planar complexes are important primarily for low-spin d^8 electron configurations. Forming an octahedral complex in such cases forces two electrons into a high-energy e_g level; in the square-planar complex, they stay in a lower-energy, d_{xy} orbital. Square-planar complexes predominate for the d^8 ions Pt(II) and Pd(II) because the d orbitals of these ions are split strongly by most ligands. The Ni(II) ion is also a d^8 ion but experiences significantly smaller ligand-field splittings and consequently often forms octahedral complexes. Tetrahedral complexes are less common because of the lower coordination number, the smaller size of ligand-field splitting, and the necessity of placing electrons into a higher-energy level at an earlier stage in the filling process.

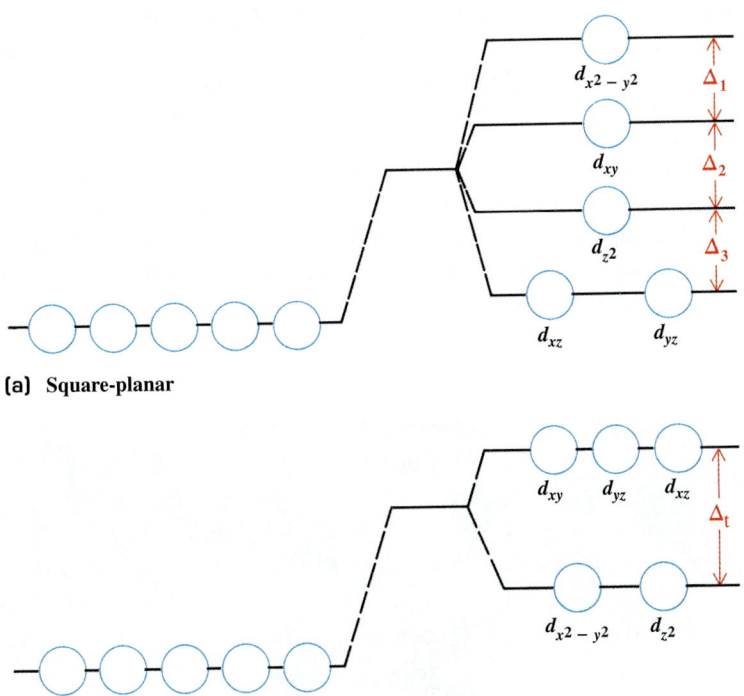

(a) **Square-planar**

(b) **Tetrahedral**

Figure 19–15 • Energy-level structures of the $3d$ orbitals in (a) square-planar and (b) tetrahedral crystal fields.

Figure 19–16 • Several coordination compounds. From the top left and moving clockwise, they are $[Cr(CO)_6]$ (*white*), $K_3[Fe(C_2O_4)_3]$ (*green*), $[Co(en)_3]I_3$ (*dark orange*), $[Co(NH_3)_5(H_2O)]Cl_3$ (*rose*), and $K_3[Fe(CN)_6]$ (*red-orange*). (*Charles D. Winters*)

• Aqueous solutions of the transition metals that we have indicated up to now as $Mn^{2+}(aq)$ or $Zn^{2+}(aq)$ also can be represented with the coordinated water molecules shown explicitly: $[Mn(H_2O)_6]^{2+}(aq)$ and $[Zn(H_2O)_4]^{2+}(aq)$. We follow this practice in this chapter to emphasize the nature of the coordination.

Figure 19–17 • The colors of the hexaaqua complexes of the metal ions (from left) Mn^{2+}, Fe^{3+}, Co^{2+}, Ni^{2+}, Cu^{2+}, and Zn^{2+} prepared from their nitrate salts. Note that the d^{10} Zn^{2+} complex is colorless. The green color of the Ni^{2+} is due to the absorption of both red and blue from the white light illuminating the solution. The yellow cast of the solution containing $Fe(H_2O)_6^{3+}$ ion is caused by a species produced by hydrolysis of that ion; if this reaction is suppressed, the solution of $[Fe(H_2O)_6]^{3+}$ is pale purple. (*Leon Lewandowski*)

19-4   The Colors of Coordination Complexes

The most striking properties of the complexes of the transition elements are their colors (Fig. 19–16). Complexes of Co(III), for example, have these colors:

$[Co(NH_3)_6]^{3+}$	Orange
$[Co(NH_3)_4Cl_2]^+$	A green form and a violet form
$[Co(NH_3)_5(H_2O)]^{3+}$	Red

Colors arise when complexes absorb light in some portion of the visible spectrum. As explained in Section 18–4, the color perceived in a sample is the color *complementary* to the color that is absorbed most strongly. Suppose that white light passes into an aqueous solution containing $[Co(NH_3)_5Cl]^{2+}(aq)$ ion. This ion absorbs strongly near a wavelength of 530 nm, in the yellow-green region of the spectrum. White light consists of photons with a broad distribution of wavelengths, but only the blue and red light is transmitted by the solution, which appears purple. A material that absorbs across all visible wavelengths appears gray or black, and one that absorbs weakly or not at all in the visible range is colorless.

The colors of many octahedral transition-metal compounds arise from the excitation of electrons from occupied t_{2g} levels to empty e_g levels. These are called *d-d transitions* because they occur as a rearrangement of electron density among the *d* orbitals of the transition element. The frequency (ν) of light that is capable of inducing a *d-d* transition is related to the energy difference between the two states, which is the crystal-field splitting energy:

$$h\nu = \Delta_o$$

The larger the crystal-field splitting, the higher the frequency of light absorbed most strongly and the shorter its wavelength. The $[Co(NH_3)_6]^{3+}$ ion, an orange species that absorbs most strongly in the violet region of the spectrum, has a larger crystal-field splitting Δ_o than does the $[Co(NH_3)_5Cl]^{2+}$ ion, a purple species that absorbs most strongly at lower frequencies (longer wavelengths) in the yellow-green region of the spectrum. A d^{10} complex (such as those of Zn^{2+} or Cd^{2+}) is colorless because all the *d* levels (both t_{2g} and e_g) are filled; consequently, a $t_{2g} \rightarrow e_g$ transition cannot take place, and the absorption in the visible region of the spectrum is very slight. High-spin d^5 complexes such as $[Mn(H_2O)_6]^{2+}$ also show only very weak absorption because processes in which a spin is reversed on excitation by light (required by the Pauli principle in these cases) occur only rarely. For this reason, the Mn^{2+} complex ion with six water ligands is very pale pink (Fig. 19–17).

CHEMISTRY IN YOUR LIFE

Ruby Red and Emerald Green

Crystal-field theory explains the dramatic colors of rubies and emeralds. Rubies result from the substitution of a small amount of chromium for aluminum in aluminum oxide (corundum) crystals. Pure corundum (Al_2O_3) is colorless, but when it is combined with about 1% of the green solid Cr_2O_3, the deep red hue of ruby results. The Cr^{3+} ions replace Al^{3+} ions in slightly distorted octahedral crystal sites, each defined by six oxygen ions. Free Cr^{3+} ions have a $3d^3$ valence electron configuration in their ground states, but at the octahedral sites in the crystal, this becomes $(t_{2g})^3$; three electrons with parallel spins occupy the lower-energy t_{2g} levels. Light of the proper wavelength excites electrons into the higher-energy e_g levels of the Cr^{3+} ions. Two strong absorption bands result, both in the visible region of the spectrum (Fig. 19–B, *top*). The lower-energy band corresponds to absorption of green and yellow light, and the other, to absorption of violet light. The two bands overlap, so that most blue light also is absorbed. The little that passes through combines with the strongly transmitted red light to give a characteristic purple tinge to the red of the ruby.

Emeralds also have Cr^{3+} impurities that substitute for Al^{3+} ions in a distorted octahedral site formed by six oxygen ions. What then causes the dramatic difference between ruby red and emerald green? The answer lies with a small change (about 10%) in the crystal-field splitting parameter Δ_o. The crystal into which chromium substitutes to give emerald is not corundum but beryl, with the formula $Be_3Al_2Si_6O_{18}$. Like pure corundum, pure beryl is colorless. The presence of beryllium and silicon makes the bonding in beryl weaker than in corundum. The weaker crystal field at the octahedral sites means that both of the Cr^{3+} absorption bands in emerald are shifted to lower energy (longer wavelength, Fig. 19–B, *bottom*). One absorption now overlaps the red part of the visible spectrum, blocking transmission of red, orange, and yellow light. The other absorption shifts downward from violet toward blue and also changes shape to open a transmission "window" between the two bands. Green light with a trace of blue is transmitted, giving a color that we call "emerald green." A subtle change in the crystal field completely changes the color of the gem.

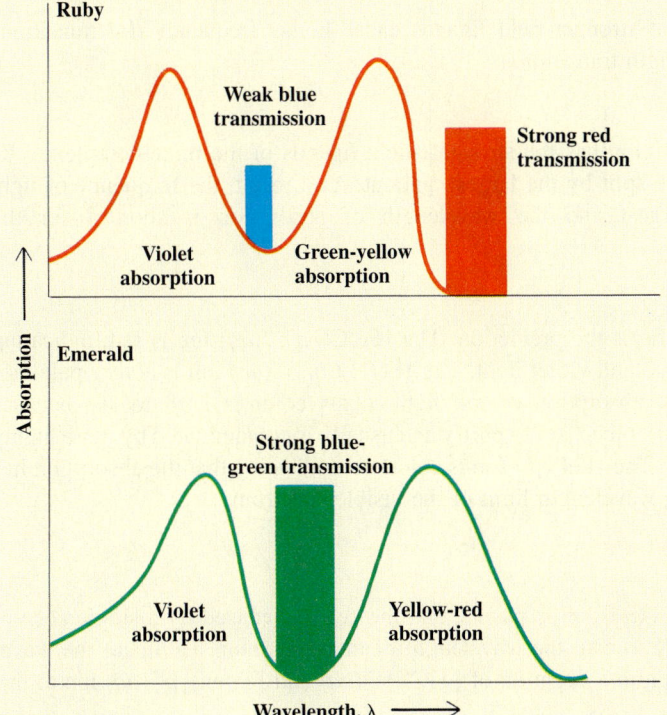

Figure 19–B • The absorption spectra of ruby and emerald. Small shifts in the shape of the spectrum have a dramatic effect on the perceived color. (*Julius Weber*)

The simple crystal-field model succeeds in accounting for the absorption spectra of transition-metal complexes, just as it accounts for the magnetic properties described in Section 19–3. From the spectra, it is possible to rank ligands from those that interact most weakly with the metal ion (and therefore give the smallest crystal-field splitting) to those that interact most strongly and give the largest splitting. Although such an ordering is not followed invariably for every ligand with every central metal atom, it is nonetheless useful. The ordering is called the **spectrochemical series.** For a selection of ligands from weakest to strongest, it is

$$I^- < Br^- < Cl^- < F^-, OH^- < H_2O < NCS^- < NH_3 < en < CO, CN^-$$

| weak-field ligands (high spin) | intermediate-field ligands | strong-field ligands (low spin) |

This ordering cannot be explained within the crystal-field model, and indeed its existence points up an unsatisfactory feature of crystal-field theory. The bonding in coordination compounds cannot be fully ionic, as assumed in the crystal-field description. If it were, it would be impossible for neutral ligands such as NH_3 and CO to give larger splittings than small, negatively charged ligands such as F^- ion. The crystal-field theory has been modified to include covalent as well as ionic aspects of coordination. The resulting extended theory, which uses molecular orbital methods of the type described in Chapter 18, successfully explains the spectrochemical series, but a discussion is beyond the scope of this book.

EXAMPLE 19–7

Predict which of the following octahedral coordination complexes has the shortest wavelength of absorption in the visible spectrum: $[FeF_6]^{3-}$, $[Fe(CN)_6]^{3-}$, $[Fe(H_2O)_6]^{3+}$.

Strategy

Use the fact that stronger-field ligands cause higher-frequency *d-d* transitions (shorter-wavelength transitions).

Solution

The $[Fe(CN)_6]^{3-}$ ion has the strongest-field ligands of the three complexes. Its energy levels are split by the largest amount. As a result, the frequency of light absorbed is largest and the wavelength of absorption is shortest for the $[Fe(CN)_6]^{3-}$ ion.

Check

Experiment confirms the prediction: The $[Fe(CN)_6]^{3-}(aq)$ ion is red, indicating absorption of blue and violet light. The $[Fe(H_2O)_6]^{3+}(aq)$ ion is a very pale violet due to weak absorption of red light. (This color is hard to see because $[Fe(H_2O)_6]^{3+}(aq)$ ion is hard to purify and usually is contaminated by more highly colored species.) The $[FeF_6]^{3-}$ ion is colorless, indicating that the absorption lies beyond the long wavelength limit of the visible spectrum.

EXERCISE

The $[Ti(H_2O)_6]^{3+}(aq)$ ion is purple, and the color is caused by a single absorption maximum. Estimate the wavelength of this maximum. Compare the wavelengths of maximum absorption of $[TiCl_6]^{3-}(aq)$ and $[Ti(en)_3]^{3+}(aq)$ ion to this estimate.

Answer: The absorption maximum should be in the yellow-green region (complementary to purple) in the vicinity of 530 nm (observed: 510 nm). The absorption maximum for $[TiCl_6]^{3-}$ should come at a longer wavelength than this; the absorption maximum for $[Ti(en)_3]^{3+}$ should come at a shorter wavelength.

19-5 Coordination Complexes in Biology

Coordination complexes, particularly chelates, play fundamental roles in the biochemistry of both plants and animals. Trace amounts of at least nine transition elements—vanadium, chromium, manganese, iron, cobalt, nickel, copper, zinc, and molybdenum—are essential to life.

Several of the most important complexes are based on the organic compound *porphine,* which has the approximately planar structure shown in Figure 19–18. The π bonds in these compounds are conjugated. Donation of the two acidic (N-bound) hydrogens from porphine to some base leaves (porphine)$^{2-}$, which has four nitrogen atoms ready to bind to a metal ion M^{2+} and form a chelate structure. This tetradentate ligand, modified by the addition of several side groups, gives a complex with Fe^{2+} ions called heme (Fig. 19–18b). The absorption of light by heme is responsible for the red color of blood. In *hemoglobin,* the compound that transports oxygen in the blood, the fifth coordination site of the iron(II) ion binds *globin* (a high-molar-mass protein), and the sixth is occupied by water or molecular oxygen. In cases of carbon monoxide poisoning, CO molecules occupy the sixth coordination site and block the binding and transport of oxygen. Hemoglobin has a complex structure that contains four heme groups (see Fig. 25–11).

• The equilibria by which hemoglobin in the blood binds and later releases O_2 are discussed on pages 326–327.

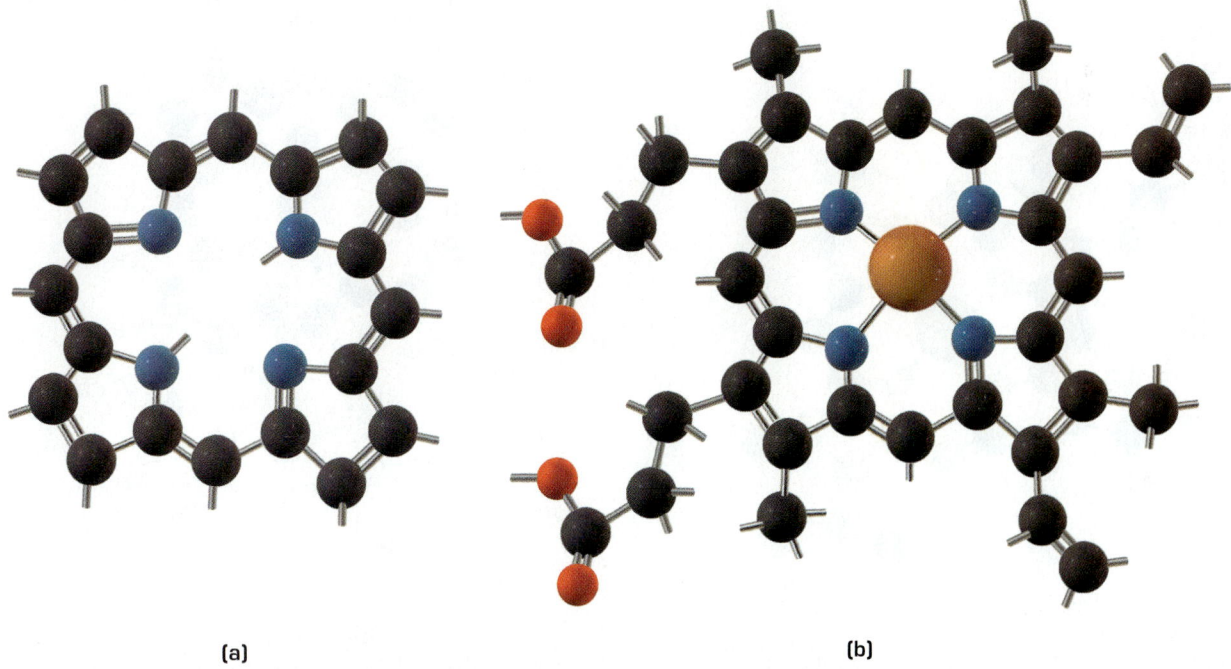

(a)   (b)

Figure 19–18 • Structures of (a) porphine ($C_{20}H_{14}N_4$) and (b) heme ($C_{34}H_{32}N_4O_4Fe$), a chelate based on porphine. Note the alternation of double and single bonds in the structures. Porphine is a dark red solid. Heme is isolated as brown needle-like crystals with a violet sheen.

Photosynthesis depends on the properties of chlorophyll, which contains a derivative of the porphine molecule with different side groups and with a magnesium ion at its center (Fig. 19–19). The aqueous Mg^{2+} ion does not absorb light in the visible region of the spectrum, but chlorophyll, in which Mg^{2+} is chelated, does. Absorption of light by this complex provides the energy to carry out photosynthesis, producing glucose from water and carbon dioxide:

$$6\,CO_2(g) + 6\,H_2O(\ell) \longrightarrow 6\,O_2(g) + C_6H_{12}O_6(aq) \qquad \Delta G° = 2872 \text{ kJ mol}^{-1}$$

The appearance of chlorophyll and related substances in the early biological history of the Earth provided the means for sunlight to drive the above reaction to the right. The accumulation of O_2 about 2 billion years ago profoundly changed the nature of the atmosphere from a reducing environment (dominated by CO_2, CH_4, and NH_3) to an oxidizing environment (dominated by oxygen). The harvesting of the Sun's energy using chlorophyll led to life as we know it on Earth.

Vitamin B_{12} is useful in the treatment of pernicious anemia and other diseases. Its structure has certain similarities to those of heme and chlorophyll. Again, a metal ion is coordinated by a planar tetradentate ligand, with two other donors completing the coordination octahedron by occupying *trans* positions. In vitamin B_{12}, the metal ion is cobalt and the planar ring system is not porphine but corrin (Fig. 19–20), in which two of the five-membered N-containing rings are joined directly rather than having a carbon atom between them. In the human body, enzymes derived from vitamin B_{12} accelerate a range of important reactions, including those producing red blood cells.

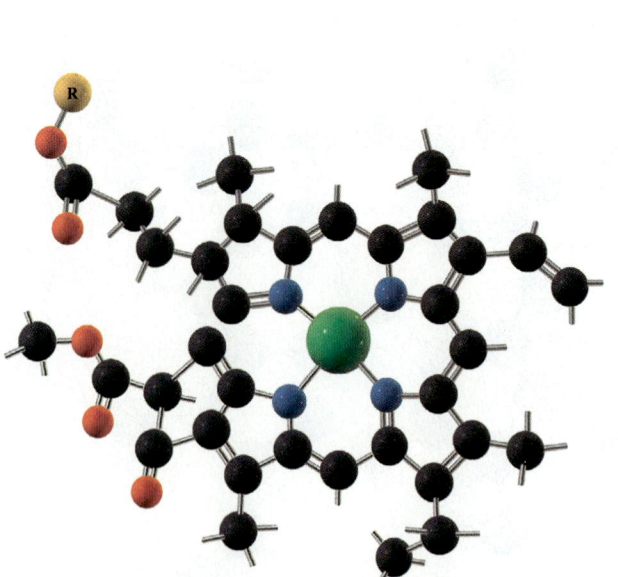

Figure 19–19 • The structure of chlorophyll *a*, $C_{55}H_{72}MgN_4O_5$. The R stands for the phytyl group ($C_{20}H_{39}$), a long hydrocarbon "tail" that increases the solubility of the compound in less polar environments. Solid chlorophyll *a* forms as waxy blue-black (not green) crystals; a solution in ethanol is blue-green.

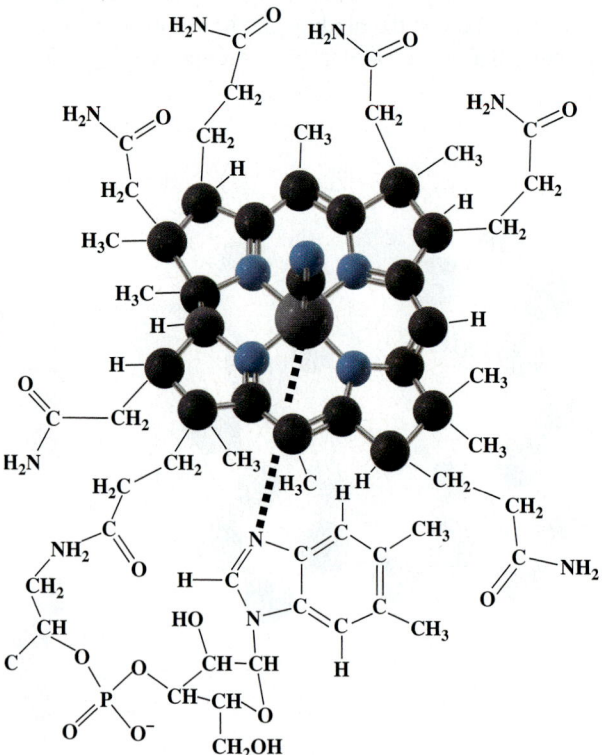

Figure 19–20 • The structure of the cobalt complex vitamin B_{12}, $Co(C_{62}H_{88}N_{13}O_{14}P)CN$. For clarity, the structure is "exploded"; atoms that are not part of the corrin ring are shown typographically.

CHEMISTRY IN PROGRESS

Odor Visualization

A novel technique called "smell-seeing" takes advantage of the color changes that metalloporphyrins undergo when they are exposed to vapors that contain molecules capable of serving as ligands. Metalloporphyrins are coordination complexes between metal ions and any of a number of derivatives of porphine (Fig. 19–18a). They are generally intensely colored. For example, zinc tetraphenylporphyrin ($ZnC_{44}H_{28}N_4$), which has the structure shown in Figure 19–C, is orange. It absorbs light very strongly at 423 nm and less strongly at 548 nm.

Other metalloporphyrins have other colors. These colors change, often radically, when additional ligands coordinate axially to the metal as it sits in the hole in the porphine doughnut. Many odoriferous and toxic materials are capable of just such coordination. Thus, butylamine, which reeks unpleasantly of ammonia, reacts with orange zinc tetraphenylporphyrin to turn it yellow. In smell-seeing, a sensing array is created by immobilizing numerous different metalloporphyrins as small spots on an inert plastic backing. The array is scanned with an ordinary flatbed scanner or an inexpensive digital camera to obtain the "before" colors of the spots. It is then exposed to a vapor, which reacts to give new colors at each spot (Fig. 19–D). The array is rescanned. Computerized subtraction of the "before" image from the "after" image then gives a color-change signature for that vapor. Comparison to a library of such signatures allows for the identification of the chemical compound. Potential applications include the detection of additives or spoilage in foods and the production of disposable general-purpose vapor dosimeters.

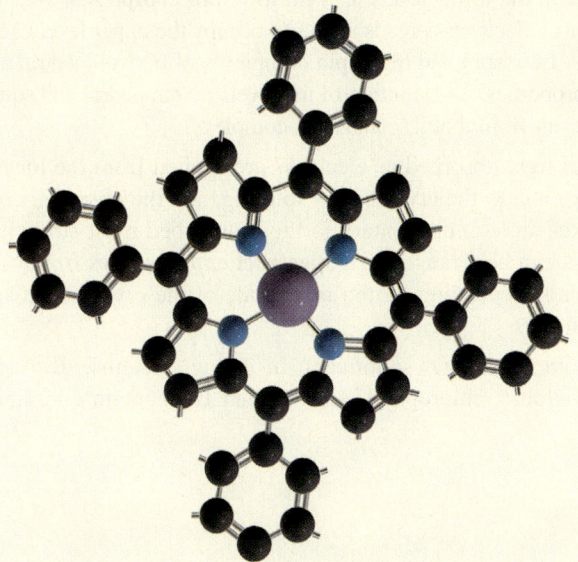

Figure 19–C • The structure of the coordination complex zinc tetraphenylporphyrin ($Zn(C_{44}H_{28}N_4)$). Tetraphenylporphyrin derives from porphine by the substitution of four phenyl groups (C_6H_5 groups) for H atoms in locations around the perimeter of the molecule.

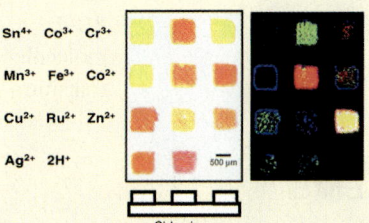

Figure 19–D • A miniaturized sensing array of metalloporphyrins. At the left is the array in its original colors. At the right is the color-change profile after exposure for 1 min to air containing vapors of butylamine. (*Ken Suslick, University of Illinois, Urbana–Champaign*)

SUMMARY

19–1 A **coordination complex** forms when a metal ion (or atom) coordinates with the lone-pair electrons of **ligand** ions or molecules in a Lewis acid–base reaction. The number of lone pairs accepted from ligands is the **coordination number** of the metal. One important class of ligands are those that **chelate** (bind to metal atoms or ions at two or more ligating atoms). Complexes differ in chemical and physical properties from their independent metal and ligand components. In aqueous solution, they exhibit either thermodynamic or kinetic stability, or both. Complexes that exchange ligands rapidly are said to be **labile,** but those that exchange slowly are **inert.**

19-2 The most common three-dimensional structures of coordination complexes are **octahedral** (for coordination number 6) and **square planar** or **tetrahedral** (for coordination number 4). These structures display two types of isomerism, in which the same ligands are attached to the central atom but in arrangements that cannot be superimposed no matter how the complex is rotated. In **optical isomerism,** the two isomers are mirror-image pairs and are called **enantiomers;** each is **chiral.** In *cis–trans* **isomerism,** the relative positions of the ligands as they coordinate to the metal are different.

19-3   Crystal-field theory explains many of the properties of coordination complexes. If charges (from ions or lone-pair electrons on molecules) are brought up toward a transition-metal atom or ion, the energies of the five *d* orbitals on the metal atom split apart. In the case of an octahedral complex, they form two groups: three t_{2g} orbitals (at lower energy) and two e_g orbitals (at higher energy). The energy difference between the two is the **octahedral crystal-field splitting** Δ_o for the complex. When Δ_o is large, electrons pair in the lower levels to form **low-spin complexes.** When Δ_o is small, the same number of electrons spreads across to occupy the upper levels, forming **high-spin complexes.** Low-spin and high-spin complexes of a given central atom have different magnetic properties. The splitting of the levels in tetrahedral and square-planar complexes is different from that in octahedral complexes.

19-4 The frequency of light absorbed as electrons are excited from the lower to the upper level is proportional to the crystal-field splitting, and the observed colors of coordination complexes are complementary to those absorbed most strongly by these transitions. Ligands can be arranged in a **spectrochemical series** from weak-field to strong-field ligands, depending on the magnitude of the crystal-field splitting they induce in metal ions.

19-5 Coordination complexes are significant in living systems. Biological molecules such as hemoglobin, chlorophyll, and vitamin B_{12} contain coordinated transition-metal atoms.

PROBLEMS

Note: Answers to blue-numbered problems are given in Appendix F. Problems that are more challenging are indicated with asterisks.

The Formation of Coordination Complexes

1. (See Example 19–1.) Write formulas, including overall electric charge, of coordination complexes containing the following metal ions and ligands:
   (a) Fe^{2+} with six CN^- ligands.
   (b) Mn^{3+} with one NH_3 ligand, two H_2O ligands, and three Cl^- ligands.
   (c) Pt^{2+} with one H_2O ligand, one Br^- ligand, and one ethylenediamine (en) ligand.

2. (See Example 19–1.) Write formulas, including overall electric charge, of coordination complexes containing the following metal ions and ligands:
   (a) Cr with four CO ligands.
   (b) Co^{3+} with two Cl^- ligands and two $C_2O_4^{2-}$ ligands.
   (c) Rh^{2+} with three NH_3 ligands and three Br^- ligands.

3. Is methylamine (CH_3NH_2) a monodentate or a bidentate ligand? With which of its atoms does it bind to a metal ion?

4. Show how the glycinate ion (H_2N—CH_2—COO^-) can act as a bidentate ligand. (Draw a Lewis structure if necessary.) Which atoms in the glycinate ion bind to a metal ion?

5. Calculate the pH of a 0.124 M solution of $[Pt(en)_3]Cl_4$ at 25°C. The cation is a weak acid with $K_{a1} = 4.5 \times 10^{-6}$ and $K_{a2} = 2.45 \times 10^{-10}$ at this temperature.

6. $[Au(NH_3)_4]PO_4$ has a K_{sp} of 1.0×10^{-11} at 25°C. A simple calculation predicts that the solubility of this substance is 3.2×10^{-6} mol L^{-1}. The observed solubility is greater than this. Use the fact that PO_4^{3-} ion is a weak base (see Chapter 8) and that $[Au(NH_3)_4]^{3+}$ is a weak acid (with a K_a of 3.3×10^{-8}) to write a reaction that would explain this enhanced solubility.

7. (See Example 19–2.) Determine the oxidation state of the metal in each of the following coordination complexes: $[V(NH_3)_4Cl_2]$, $[Mo_2Cl_8]^{4-}$, $[Co(H_2O)_2(NH_3)_3Cl_3]^-$, $[Ni(CO)_4]$.

8. (See Example 19–2.) Determine the oxidation state of the metal in the following complexes: $[Mn_2(CO)_{10}]$, $[Re_3Br_{12}]^{3-}$, $[Fe(H_2O)_4(OH)_2]^+$, $[Co(NH_3)_4Cl_2]^+$.

9. (See Example 19–3.) Give the chemical formula of each of the following compounds:
   (a) Sodium tetrahydroxozincate(II).
   (b) Dichlorobis(ethylenediamine)cobalt(III) nitrate.
   (c) Triaquabromoplatinum(II) chloride.
   (d) Tetraamminedinitroplatinum(IV) bromide.

10. (See Example 19–3.) Give the chemical formula of each of the following compounds:
    (a) Silver hexacyanoferrate(II).
    (b) Potassium tetraisothiocyanatocobaltate(II).
    (c) Sodium hexafluorovanadate(III).
    (d) Potassium trioxalatochromate(III).

11. Assign a systematic name to each of the following chemical compounds:
    (a) $NH_4[Cr(NH_3)_2(NCS)_4]$
    (b) $[Tc(CO)_5]I$
    (c) $K[Mn(CN)_5]$
    (d) $[Co(NH_3)_4(H_2O)Cl]Br_2$

12. Give the systematic name of each of the following chemical compounds:
    (a) $[Ni(H_2O)_4(OH)_2]$
    (b) $[HgClI]$
    (c) $K_4[Os(CN)_6]$
    (d) $[FeBrCl(en)_2]Cl$

13. Describe the behavior of a coordination complex that is both thermodynamically stable and labile.

14. Describe the behavior of a coordination complex that is thermodynamically unstable and kinetically inert.

15. Write an equation for the dissociation of tetraamminecopper(II) ion to $Cu^{2+}(aq)$ and $NH_3(aq)$ in basic aqueous solution and calculate ΔG_r° at 25°C. Use data from Appendix D. Is this complex thermodynamically stable with respect to this reaction under standard conditions? Why does it dissociate under acidic conditions? Write a balanced equation for the dissociation reaction at pH = 0, and calculate ΔG_r° at 25°C for that reaction.

16. The ΔG_f° of the tetracyanonickelate(II) anion is $+472.1$ kJ mol^{-1} in aqueous solution at 25°C. Use this information together with data from Appendix D to calculate the equilibrium constant at this temperature for the dissociation

$$[Ni(CN)_4]^{2-}(aq) \longrightarrow Ni^{2+}(aq) + 4\,CN^-(aq)$$

Isotopically labelled Na^{13}CN is added to an aqueous solution of $K_2[Ni(CN)_4]$. Subsequent examination of the $[Ni(CN)_4]^{2-}$ present reveals extensive incorporation of $^{13}CN^-$ into the complex ion. Comment on these observations, and discuss the difference between lability and thermodynamic stability as applied to complexes of nickel(II) with the cyanide ion.

Structures of Coordination Complexes

17. Suppose that 0.010 mol of each of the following compounds is dissolved in separate 1 L portions of water: KNO_3, $[Co(NH_3)_6]Cl_3$, $Na_2[PtCl_6]$, $[Cu(NH_3)_2Cl_2]$. Rank the resulting four solutions in order of their conductivity, from lowest to highest.

18. Suppose that 0.010 mol of each of the following compounds is dissolved (separately) in 1 L of water: $BaCl_2$, $K_4[Fe(CN)_6]$, $[Cr(NH_3)_4Cl_2]Cl$, $[Fe(NH_3)_3Cl_3]$. Rank the resulting four solutions in order of their conductivity, from lowest to highest.

19. (See Examples 19–4 and 19–5.) Draw the structures of all possible isomers for the following complexes. Indicate which isomers are mirror-image pairs (enantiomers).
    (a) Diamminebromochloroplatinum(II) (square planar).
    (b) Diaquachlorotricyanocobaltate(III) ion (octahedral).
    (c) Trioxalatovanadate(III) ion (octahedral).

20. (See Examples 19–4 and 19–5.) Draw the structures of all possible isomers for the following complexes. Indicate which isomers are enantiometric pairs.
    (a) Bromochloro(ethylenediamine)platinum(II) (square planar).
    (b) Tetraamminedichloroiron(III) ion (octahedral).
    (c) Amminechlorobis(ethylenediamine)iron(III) ion (octahedral).

21. (See Examples 19–4 and 19–5.) Iron(III) forms octahedral complexes. Sketch the structures of all the distinct isomers of $[Fe(en)_2Cl_2]^+$, indicating which pairs of structures are mirror images of each other.

22. (See Examples 19–4 and 19–5.) Platinum(IV) forms octahedral complexes. Sketch the structures of all the distinct isomers of $[Pt(NH_3)_2Cl_2F_2]$, indicating which pairs of structures are mirror images of each other.

Crystal-Field Theory and Magnetic Properties

23. For each of the following ions, draw diagrams like those in Figure 19–14 to show orbital occupancies in both weak and strong octahedral fields. Indicate the total number of unpaired electrons in each case.
    (a) Mn^{2+} (d) Mn^{3+}
    (b) Zn^{2+} (e) Fe^{2+}
    (c) Cr^{3+}

24. Repeat the work of the preceding problem for the following ions:
    (a) Cr^{2+} (d) Pt^{4+}
    (b) V^{3+} (e) Co^{2+}
    (c) Ni^{2+}

25. Show that transition metal ions having 1, 2, or 3 electrons in their d orbitals will have the same distribution of electrons in both weak and strong octahedral fields. Give examples of ions having these arrangements.

26. Show that transition metal ions having 8, 9, or 10 electrons in their d orbitals will have the same distribution of electrons in both weak and strong octahedral fields. Give examples of ions having these arrangements.

27. (See Example 19–6.) Experiments can measure not only whether a compound is paramagnetic but also the number of unpaired electrons. It is found that the octahedral complex ion $[Fe(CN)_6]^{3-}$ has fewer unpaired electrons than does the octahedral complex ion $[Fe(H_2O)_6]^{3+}$. How many unpaired

electrons are present in each species? Give the *d* electron configuration of each species.

28. (See Example 19–6.) The octahedral complex ion $[MnCl_6]^{3-}$ has more unpaired electrons than does the octahedral complex ion $[Mn(CN)_6]^{3-}$. How many unpaired electrons are present in each species? Give the *d* electron configuration of each species.

29. Explain why octahedral coordination complexes with 3 and 8 *d* electrons on the central metal atom are particularly stable. Under what circumstances would you expect complexes with 5 or 6 *d* electrons on the central metal atom to be particularly stable?

30. The following standard reduction potentials have been measured in acidic aqueous solution:

$$Mn^{3+}(aq) + e^- \longrightarrow Mn^{2+}(aq) \qquad \mathcal{E}° = 1.54 \text{ V}$$
$$Fe^{3+}(aq) + e^- \longrightarrow Fe^{2+}(aq) \qquad \mathcal{E}° = 0.77 \text{ V}$$
$$Co^{3+}(aq) + e^- \longrightarrow Co^{2+}(aq) \qquad \mathcal{E}° = 1.84 \text{ V}$$

Explain why the reduction potential for Fe^{3+} lies below those for the $+3$ oxidation states of the elements on either side of it in the periodic table.

31. The value of Δ_o is larger when the metal ion is in a higher oxidation state (e.g., complexes of Fe^{3+} are more likely to be strong field complexes than are complexes of Fe^{2+}). Explain why this is so.

32. The value of Δ_o is larger when the metal ion is part of the second or third transition series rather than the first transition series (e.g., complexes of W^{2+} are more likely to be strong field complexes than are complexes of Cr^{2+}). Explain why this is so.

The Colors of Coordination Complexes

33. An aqueous solution of zinc nitrate contains the $[Zn(H_2O)_6]^{2+}$ ion and is colorless. What conclusions can be drawn about the absorption spectrum of the $[Zn(H_2O)_6]^{2+}$ ion and the configuration of the *d* electrons of the central metal in this complex?

34. An aqueous solution of sodium hexaiodoplatinate(IV) is black. What conclusions can be drawn about the absorption spectrum of the $[PtI_6]^{2-}$ complex ion?

35. Estimate the wavelength of maximum absorption for the octahedral ion hexacyanoferrate(III) from the fact that light transmitted by a solution of it is red. Estimate the crystal-field splitting energy Δ_o (in kJ mol^{-1}).

36. Estimate the wavelength of maximum absorption for the octahedral ion hexaaquanickel(II) from the fact that its solutions transmit green light. Estimate the crystal-field splitting energy Δ_o (in kJ mol^{-1}).

37. The chromium(III) ion in aqueous solution has a blue-violet color.
(a) What is the color complementary to blue-violet?
(b) Estimate the wavelength of maximum absorption for a $Cr(NO_3)_3$ solution.

(c) Does the wavelength of maximum absorption increase or decrease if cyano ligands are substituted for the coordinated water? Explain.

38. An aqueous solution containing the hexaamminecobalt(III) ion has a yellow color.
(a) What is the color complementary to yellow?
(b) Estimate the wavelength of maximum absorption in the visible spectrum by this solution.

39. (a) A solution of $Fe(NO_3)_3$ has only a pale color, but one of $K_3[Fe(CN)_6]$ is bright red. Do you expect a solution of $K_3[FeF_6]$ to be brightly colored or pale? Explain your reasoning.
(b) Would you predict that a solution of $K_2[HgI_4]$ is colored or colorless? Explain.

40. (a) A solution of $Mn(NO_3)_2$ has a very pale pink color, but one of $K_4[Mn(CN)_6]$ is a deep blue. Explain why the two differ so much in the intensities of their colors.
(b) Which of the following does crystal-field theory predict to be colorless in aqueous solution: $K_2[Co(NCS)_4]$, $Zn(NO_3)_2$, $[Cu(NH_3)_4]Cl_2$, $CdSO_4$, $AgClO_3$, $Cr(NO_3)_2$?

41. Why are there fewer tetrahedral complexes than octahedral complexes?

*42. Predict the color of a d^5 tetrahedral complex. Explain the prediction.

43. The inertness of $[Co(NH_3)_6]^{3+}$ may be related to the stability of its six *d* electrons in a strong octahedral field. If this is the case, which iron(II) complex would be more likely to be inert, $[FeF_6]^{4-}$ or $[Fe(CN)_6]^{4-}$?

44. Complexes of Cr(III) tend to be inert whether they contain weak-field ligands, such as F^-, or strong-field ligands, such as CN^-. In contrast, $[Co(NH_3)_6]^{3+}$ is inert while $[CoF_6]^{3-}$ is labile. Explain these observations.

Coordination Complexes in Biology

45. Vitamin B_{12s} is 4.43% cobalt by mass. Determine the molar mass of vitamin B_{12s} if each molecule of it contains one atom of cobalt.

46. Hemoglobin, the compound that transports oxygen in the bloodstream, has a molar mass of 6.8×10^4 g mol^{-1}. It is also 0.33% iron by mass. Determine how many iron atoms are in a single hemoglobin molecule.

47. Oxygenated hemoglobin is red. Is oxygen a weak-field ligand or a strong-field ligand?

48. Deoxygenated hemoglobin, in which water has replaced oxygen in the sixth coordination site, is blue. Is water a weak-field or a strong-field ligand in this case?

ADDITIONAL PROBLEMS

49. Explain why ligands are usually negative or neutral in charge and only rarely positive.

50. If *trans*-$[Cr(en)_2(NCS)_2]SCN$ is heated, it reacts to form $[Cr(en)_2(NCS)_2]\ [Cr(en)(NCS)_4]$, a solid, and gaseous ethylenediamine. Write a balanced chemical equation for this reaction. What are the oxidation states of chromium in the reactant and in the two complex ions in the product?

51. A coordination complex has the molecular formula $[Ru_2(NH_3)_6Br_3](ClO_4)_2$. Determine the oxidation state of ruthenium in this complex.

52. Write balanced chemical equations for the production of $HCN(g)$ from $CN^-(aq)$ and $HCl(aq)$, for the production of iron(II) hydroxide from aqueous Fe^{2+} and aqueous NaOH, and for the production of hexacyanoferrate(II) ion from $CN^-(aq)$ and $Fe^{2+}(aq)$.

53. Heating 2.0 mol of a coordination compound gives 1.0 mol NH_3, 2.0 mol H_2O, 1.0 mol HCl, and 1.0 mol $(NH_4)_3[Ir_2Cl_9]$. Write the formula of the original (six-coordinated) coordination compound and name it.

54. Match each compound in the group on the left with the compound on the right most likely to have the same electrical conductivity per mole in aqueous solution.
    (a) $[Fe(H_2O)_5Cl]SO_4$ HCN
    (b) $[Mn(H_2O)_6]Cl_3$ $Fe_2(SO_4)_3$
    (c) $[Zn(H_2O)_3(OH)]ClO_4$ NaCl
    (d) $[Fe(NH_3)_6]_2(SO_4)_3$ $ZnSO_4$
    (e) $[Cr(NH_3)_3Br_3]$ $LaCl_3$

55. Three different compounds are known with the same empirical formula $CrCl_3 \cdot 6H_2O$. When exposed to a dehydrating agent, compound 1 (which is dark green) loses 2 mol of water per mole of compound, compound 2 (light green) loses 1 mol of water, and compound 3 (violet) loses no water. What are the probable structures of these compounds? If an excess of silver nitrate solution is added to 100.0 g of each of these compounds, what mass of silver chloride precipitates in each case?0

56. The octahedral structure is not the only possible six-coordinate structure. Other possibilities include a planar hexagonal structure and a triangular prism structure. In the latter, the ligands are arranged in two parallel triangles, one lying above the metal atom and the other below the metal atom with its corners directly in line with the corners of the first triangle. Show that the existence of two and only two isomers of $[Co(NH_3)_4Cl_2]^+$ provides evidence against both of these possible structures.

57. Cobalt(II) forms more tetrahedral complexes than any other ion except zinc(II). Draw the structure(s) of the tetrahedral complex $[CoCl_2(en)]$. Could this complex exhibit geometric or optical isomerism? If one of the Cl^- ligands is replaced by Br^-, what kinds of isomerism, if any, are possible in the resulting compound?

58. Crystal-field theory treats the energy-level splittings induced in the five d orbitals. The same procedure could be applied to p orbitals. Predict the level splittings (if any) induced in the three p orbitals by octahedral and square-planar crystal fields.

59. The complex ions $[Mn(CN)_6]^{5-}$, $[Mn(CN)_6]^{4-}$, and $[Mn(CN)_6]^{3-}$ all have been synthesized and all are low-spin octahedral complexes. For each complex, determine the oxidation number of Mn, the configuration of the d electrons (how many t_{2g} and how many e_g), and how many unpaired electrons are present.

60. Is the coordination compound $[Co(NH_3)_6]Cl_2$ diamagnetic or paramagnetic?

*61. The value of Δ_o is greater for complexes of iron(III) than for complexes of iron(II) with the same ligand. Explain how the observation that both $[Fe(H_2O)_6]^{3+}$ and $[Fe(H_2O)_6]^{2+}$ are high-spin complexes does not contradict this statement.

62. Rank the complex ions listed below in order of increasing strength of crystal-field splitting: $[CoF_6]^{3-}$, $[Co(NH_3)_6]^{3+}$, $[CoF_4]^-$, $[Ir(NH_3)_6]^{3+}$.

*63. A coordination compound has the empirical formula $PtBr(en)(SCN)_2$ and is diamagnetic.
    (a) Examine the d electron configurations on the metal atoms, and explain why the molecular formula $[Pt(en)_2(SCN)_2][PtBr_2(SCN)_2]$ is preferred for this substance.
    (b) Name this compound.

64. The coordination geometries of $[Mn(NCS)_4]^{2-}$ and $[Mn(NCS)_6]^{4-}$ are tetrahedral and octahedral, respectively. Explain why the two have the same room-temperature molar magnetic susceptibility.

65. The compound $Cs_2[CuF_6]$ is bright orange in color and paramagnetic. Determine the oxidation number of copper in this compound, the most likely geometry of the coordination around the copper, and the possible configurations of the d electrons of the copper.

66. In the coordination compound $(NH_4)_2[Fe(H_2O)F_5]$, the Fe is octahedrally coordinated.
    (a) Based on the fact that F^- is a weak-field ligand, predict whether this compound is diamagnetic or paramagnetic. If it is paramagnetic, tell how many unpaired electrons it has.
    (b) By comparison to other complexes discussed in the chapter, predict the likely color of this compound.
    (c) Determine the d electron configuration of the iron in this compound.
    (d) Name this compound.

67. Molecular nitrogen (N_2) can act as a ligand in certain coordination complexes. Predict the structure of $[V(N_2)_6]$, which is isolated by condensing V with N_2 at 25 K. Is this compound diamagnetic or paramagnetic? What is the formula of the carbonyl compound of vanadium that has the same number of electrons?

68. Discuss the role of complexes of the transition elements in biology. Consider such aspects as their absorption of light, the existence of many different structures, and the possibility of multiple oxidation states.

CUMULATIVE PROBLEM

Platinum

Crystals of potassium hexachloroplatinate(IV) $K_2[PtCl_6]$.

The precious metal platinum was first discovered by South American Indians who found impure deposits in the gold mines of what is now Ecuador and made it into small items of jewelry. Its high melting point (1772°C) makes platinum harder to work than gold (m.p. 1064°C) and silver (m.p. 962°C), but this same property, and its resistance to chemical attack, make platinum quite suitable as a material for high-temperature crucibles. Although platinum is a noble metal, it forms a wide variety of compounds in the +4 and +2 oxidation states, many of which are coordination complexes. Its coordinating abilities make it an important catalyst for organic and inorganic reactions.

(a) The anticancer drug cisplatin, *cis*-$[Pt(NH_3)_2Cl_2]$ (see Fig. 19–8b), can be prepared from K_2PtCl_6 via reduction with N_2H_4 (hydrazine), giving K_2PtCl_4, followed by replacement of two chloride-ion ligands with ammonia. Give systematic names to the three platinum complexes referred to in this statement.

(b) The coordination compound diamminetetracyanoplatinum(IV) has been prepared, but salts of the hexacyanoplatinate(IV) ion have not. Write the chemical formulas of these two species.

(c) In a certain compound, Pt(II) is coordinated to two chloride ions and to two molecules of ethylene (C_2H_4) to give an unstable yellow crystalline solid. Can this complex have more than one isomer? If so, describe the possible isomers.

(d) Platinum(IV) is readily complexed by ethylenediamine. Draw the structures of both enantiomers of the complex ion *cis*-$[Pt(Cl)_2(en)_2]^{2+}$. In this ion, the Cl^- ligands are *cis* to one another.

(e) In platinum(IV) complexes, the octahedral crystal-field splitting Δ_o is relatively large. Is K_2PtCl_6 diamagnetic or paramagnetic? What is its d electron configuration?

(f) Is cisplatin diamagnetic or paramagnetic?

(g) The salt $K_2[PtCl_4]$ is red, but $[Pt(NH_3)_4]Cl_2 \cdot H_2O$ is colorless. In what regions of the spectrum do the dominant electronic absorptions lie for these compounds?

(h) When the two salts from part (g) are dissolved in water and the solutions mixed, a green precipitate, called Magnus's green salt, forms. Propose a chemical formula for this salt, and assign the corresponding systematic name.

20

Structure and Bonding in Solids

CHAPTER OUTLINE

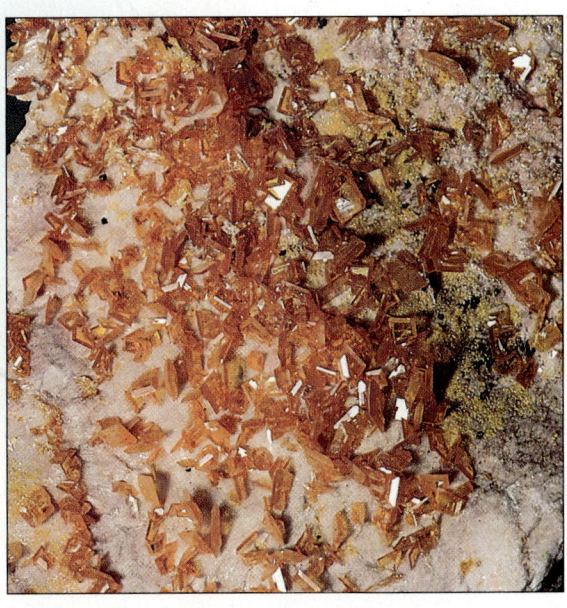

Microcrystals of realgar, As_4S_4. *(Brian Parker/ Tom Stack & Associates)*

Chemical bonds among atoms give rise to the structures called molecules. Intermolecular forces, which are weaker than bonds, act among molecules to organize them into the larger-scale structures of the liquid and solid states that were described in general terms in Chapter 6. In this chapter, the solid state is considered more closely, with emphasis on the diffraction experiments that have revealed so much about its nature.

Solids fall into two categories, depending on the *range* of microscopic order they possess. **Crystalline solids** possess long-range order in their structures. A distinct pattern of atoms repeats again and again over long distances in such solids. **Amorphous solids** lack long-range order. Although highly ordered, crystalline solids do not contain molecules (or atoms or ions) lined up in perfect infinite rows as if in a chemical cemetery. Defect and disorder play important roles in many technologically important crystalline materials. Other substances form "plastic crystals" and "liquid crystals" that are intermediate in their properties between the solid and liquid states.

20-1   Probing the Structure of Condensed Matter

In the condensed phases, the organization of the surroundings of the individual molecule, atom, or ion has enormous importance. How is such structure investigated?

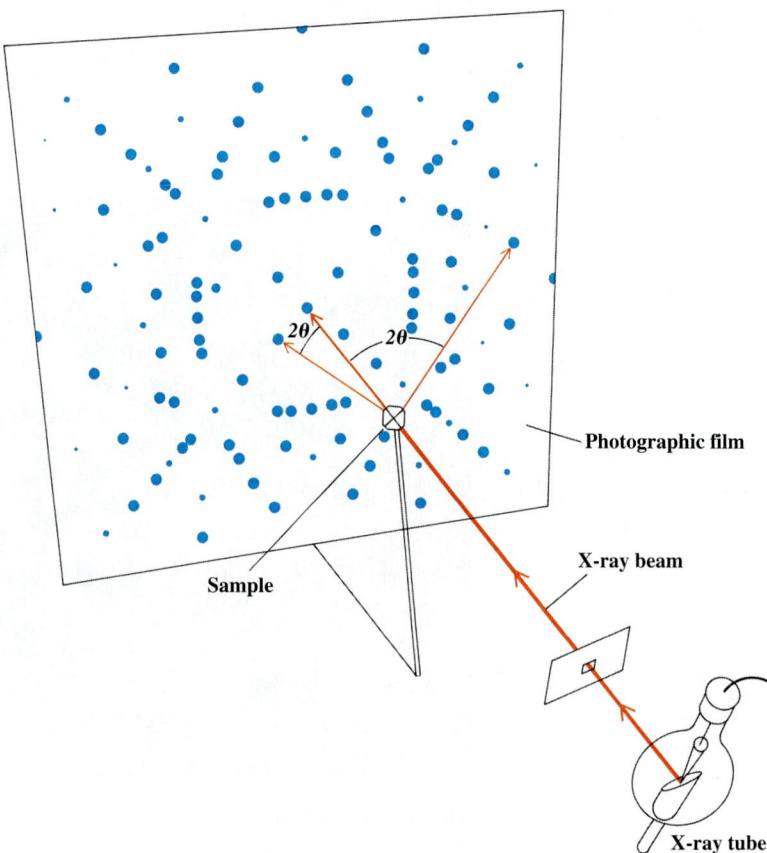

Figure 20–1 • A well-defined beam of "white" X-rays (X-rays having a range of wavelengths) strikes a thin sample of a crystalline solid. Some of the X-rays are absorbed, some pass through unchanged, and others are scattered at various angles 2θ to the straight-through beam.

Figure 20-2 • Scattering of visible light causes the halos that appear around streetlamps in the fog. Light rays exit the lamp, headed in all directions. In foggy conditions, they strike water droplets suspended in the air and are scattered toward the viewer. These redirected rays appear as a diffuse glow surrounding the "direct beam" that the viewer usually sees. *(Ray Nelson/Phototake)*

X-Ray and Neutron Diffraction

The most useful probe for the structure of the condensed states of matter is a beam of X-radiation. In **X-ray diffraction** experiments, X-rays strike samples of solids or liquids (or even gases), interact, and exit. Different arrangements of atoms in the sample affect the X-rays differently. X-rays are electromagnetic waves, similar to visible light but with wavelengths three orders of magnitude shorter. X-rays are not visible, but they register on photographic film and are detected easily by other means as well.

The set-up for a typical X-ray experiment on a crystalline solid is diagrammed in Figure 20–1. Much of the radiation passes straight through a thin sample; however, some is diverted in differing amounts at various angles from the main beam (Fig. 20–2). The intensity of a beam of scattered X-radiation is measured readily by how much it darkens a piece of photographic film; a record of an X-ray diffraction experiment is the **diffraction pattern** of the material under investigation. A complete diffraction record consists of the intensities of all possible diffracted beams and their directions relative to the main beam. The **scattering angle, 2θ,** is defined as the angle between the direction of the diffracted beam and the direction of the main beam (see Fig. 20–1).

X-rays are easy to generate. Electrons are accelerated by high voltage and directed against a metallic target inside an evacuated tube. As they hit, they slow suddenly, emitting *white X-radiation* (X-rays with a range of wavelengths). If energetic enough, the impacts excite atoms in the target to higher energy levels. These atoms subsequently emit nearly *monochromatic* (single-wavelength) X-rays as they return to their ground states. The wavelength of this *characteristic radiation,* which was first studied by Moseley (see Section 16–2), depends on the element used in the target. Tubes designed to produce characteristic radiation are used when monochromatic X-rays are required. X-ray diffraction patterns contain a maximum of structural information when the X-rays have wavelengths of about 0.5 to 3 Å because this is roughly the range of distances between neighboring atoms in liquids and solids. The size of the probe is well suited to the size of the objects being probed.

X-rays are not unique as a probe of the structure of condensed matter. **Neutron diffraction,** in which a beam of neutrons replaces the X-rays, is closely related. Like all particles, a neutron possesses wave properties, with an associated

• The X-ray portion of the electromagnetic spectrum includes wavelengths from approximately 0.1 to 100 Å (see Fig. 16–5).

wavelength given by the de Broglie relation, $\lambda = h/p$ (see Section 16–4). Applying this equation shows that neutrons moving at a speed of 2.8×10^3 m s^{-1} (a typical speed for neutrons at room temperature) have a wavelength of 1.4 Å, well matched to the distances between atoms in solids and liquids. The nuclei in condensed matter scatter such neutrons strongly, and a beam of neutrons makes an excellent structural probe. **Electron diffraction,** based on the same principle, is also possible and is used as a probe of structure.

How X-Rays Interact with Matter

When an X-ray beam strikes any material target, it causes the electrons in the target to move back and forth rapidly—that is, to oscillate as they are pushed and pulled by the changing electric field of the X-radiation. An oscillating electron emits electromagnetic radiation. Thus, each electron in the path of an X-ray beam becomes a minuscule, secondary X-ray source of its own, emitting an expanding sphere of X-radiation. This is the phenomenon called **scattering.**

Some scattered X-rays make fresh encounters with new electrons and are re-scattered. Far more important is the *interference* that the scattered waves experience among themselves. Interference occurs with all sorts of waves. It is apparent, for example, among the waves formed when a handful of pebbles is cast into a quiet pool of water. The expanding rings of ripples run into each other, interfering as they do so. The outcome of interference between two traveling waves moving in the same direction with the same wavelength depends on their relative *phase* (just as for standing waves, as discussed in Section 18–1). Constructive interference occurs if the two waves are exactly in phase, matching crest for crest and valley for valley as they go along; their amplitudes add to give a stronger resultant wave. Destructive interference occurs if two waves are exactly out of phase—that is, if crest matches trough and trough matches crest. Destructive interference leads to the cancellation of both waves. Interference between two waves that are neither exactly in nor exactly out of phase gives an intermediate result.

Suppose that X-radiation strikes two neighboring scattering centers. Each becomes a source of scattered radiation. The expanding spheres of scattered waves soon encounter each other and interfere. In some directions, the waves are in phase and reinforce each other (Fig. 20–3a); in other directions, they are out of phase and

> • The instability of the nuclear atom in classical physics (see Section 16–2) arises because orbiting charged particles are oscillating (from one side of the atom to the other) and must emit radiation.

Each pebble falling into a still pool starts a widening circle of waves. Where the circles overlap, the waves interfere. *(Yoav Levy/Phototake)*

Figure 20–3 • A beam of X-rays (not shown) is striking two scattering centers, which are emitting scattered radiation. The difference in the lengths of the paths followed by the scattered waves determines whether they interfere constructively (a) or destructively (b). This path difference depends on the distance between the centers and also on the direction in which the scattered waves are moving.

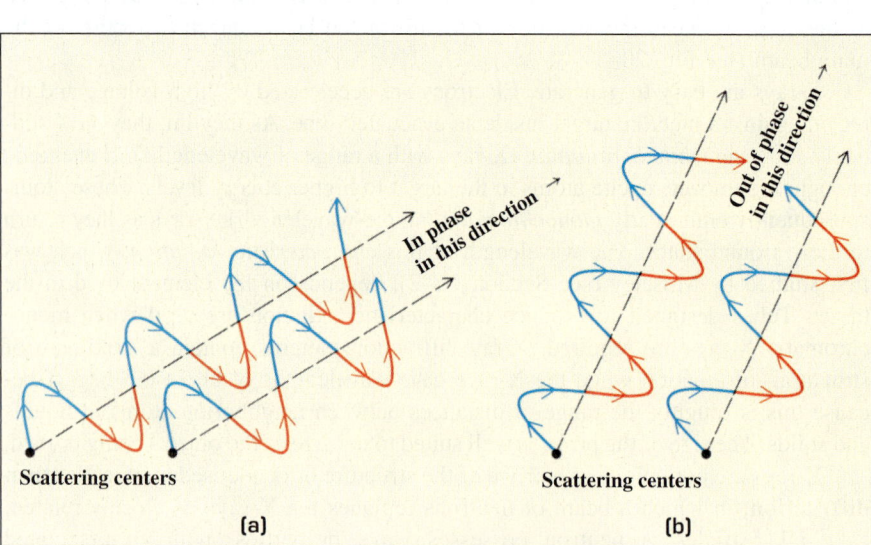

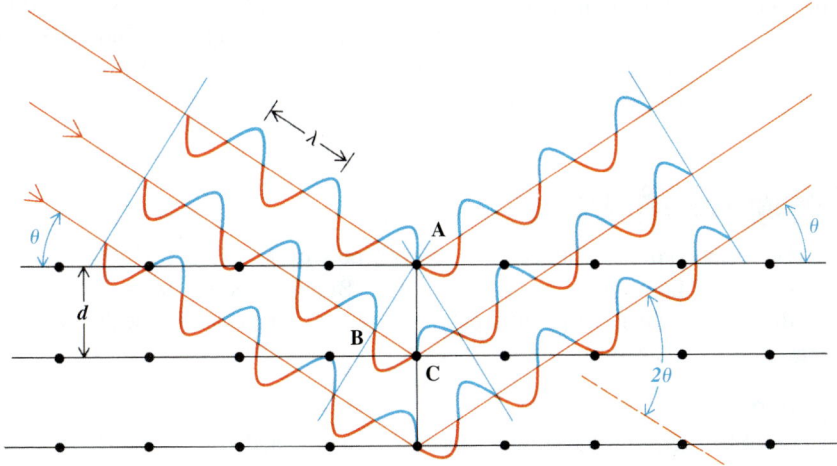

Figure 20–4 • The X-rays coming in from the upper left are all in phase, have the same wavelength λ, and make an angle θ with a set of planes of atoms in a crystalline solid. The X-rays scattered by atoms in the top plane must be reinforced fully by those scattered from all the lower planes (only two of which are shown). Otherwise, no scattered beam is observed. X-rays scattered by atoms in lower planes travel farther. One half of the difference in path length of the beams scattered by the two topmost planes is equal to the side BC of the triangle ABC. The side BC has length $d \sin \theta$, where d is the distance between the planes. The total difference in path length is then $2d \sin \theta$. For full reinforcement, the difference between the path lengths of the two scattered beams must equal an integral number n of wavelengths. Hence $n\lambda = 2d \sin \theta$ is the condition for Bragg scattering of X-rays by a crystal.

cancel each other out (Fig. 20–3b). Reinforcement (constructive interference) occurs when the paths traversed by two waves differ in length by a whole number of wavelengths. Thus, in Figure 20–3a, the scattered waves reinforce because one has traveled exactly one wavelength farther than the other. As Figure 20–3 shows, the difference in path length between the scattered waves depends on the distance between the centers and also on the direction considered. This is the physical basis of X-ray diffraction as a structural probe.

The Bragg Law

To illustrate the connection between X-ray diffraction patterns and the locations of atoms in a sample, consider the scattering by a large group of atoms that are arranged in a regular pattern of parallel planes, which is the way atoms are arranged in crystalline solids. The **Bragg law** sets an essential condition for the allowed angles of X-ray diffraction in such a case.

Crystalline solids often possess well-developed planar faces. To understand the Bragg law, suppose first that *visible* light strikes such a crystal face at an angle θ. The smooth face acts as a mirror, reflecting the light. The reflected beam goes off in a different direction but makes the same angle θ to the plane of the face as the incoming beam. The angle between the direction of the incoming beam and the direction of the reflected beam is 2θ (see Fig. 20–1).

The geometrical relationship between the incoming and outgoing beams is unchanged if single-wavelength (monochromatic) X-rays replace the visible light. Now, however, a flat exterior face is no longer needed because the X-rays penetrate the crystal, and "reflection" (actually scattering followed by interference) comes from evenly spaced stacks of atoms in parallel planes within the crystal. X-rays scattered by atoms in planes lower in the stack must travel longer paths than do those scattered by atoms in planes nearer the top. The traversal of different paths means different relative phase among "reflected" beams and gives rise to interference. Destructive interference among the beams restricts the number of observable X-ray "reflections" from a given stack of planes to a select few values of θ. The analysis shown in Figure 20–4 can be used to quantify this argument. The conclusion is

• This law was formulated by W. H. Bragg and W. L. Bragg (father and son) in 1914. The scattering of X-rays by crystals is now often called Bragg scattering.

$$n\lambda = 2d \sin \theta \qquad \text{(the Bragg law)}$$

In this equation, d is the perpendicular distance between the parallel planes in the stack, λ is the wavelength of the X-rays, θ is as just defined, and n is a whole number. The case $n = 1$ is called first-order Bragg diffraction (or the "first-order reflection"), $n = 2$ is second-order, and so forth.

EXAMPLE 20–1

A crystalline sample scatters a beam of X-rays of wavelength 0.7093 Å at an angle 2θ of 14.66°. If this is a second-order Bragg reflection ($n = 2$), compute the distance between the parallel planes of atoms from which the scattered beam appears to have been reflected.

Solution

The problem requires substituting in the Bragg law and solving for d, the interplanar spacing:

$$d = \frac{n\lambda}{2 \sin \theta} = \frac{2(0.7093 \text{ Å})}{2 \sin\left(\dfrac{14.66°}{2}\right)} = 5.559 \text{ Å}$$

EXERCISE

A crystal scatters a beam of X-rays of wavelength 1.936 Å at an angle 2θ of 88.30°. What is the order of this Bragg reflection if the distance between the parallel planes of atoms involved in the scattering is 5.559 Å?

Answer: Fourth-order.

EXAMPLE 20–2

In a crystal of silver, planes of silver atoms are separated by a distance of 1.4446 Å. Compute the θ's for all possible Bragg reflections for these planes if the wavelength of the incident X-radiation is 0.7093 Å.

Solution

Write down the Bragg law:

$$n\lambda = 2d \sin \theta$$

The problem gives the values of λ and d. Rearranging the equation and inserting these values gives

$$\sin \theta = n\frac{\lambda}{2d} = n\frac{0.7093 \text{ Å}}{2(1.4446) \text{ Å}} = 0.2455\, n$$

If $n = 1$, then $\sin \theta$ is 0.2445 and θ is $\sin^{-1}(0.2455) = 14.21°$. Substituting $n = 2$ and $n = 3$ and $n = 4$ gives θ's of 29.41°, 47.43°, and 79.11°. Larger values of n give values of $\sin \theta$ exceeding 1, which corresponds to no possible angle.

Experimental Results from X-Ray Diffraction

When a crystalline solid is put in an X-ray beam and rotated, one of two types of diffraction pattern is observed: (1) many sharply defined *spots* appear across the pattern at different angles from the primary beam (Fig. 20–5a) or (2) many sharply defined *rings* encircle the primary beam (Fig. 20–5b).

A pattern such as that shown in Figure 20–5a arises when the specimen is a **single crystal,** so that sets of planes of atoms run straight through the sample from one end to the other. Each spot in the diffraction pattern registers a different diffracted beam of X-rays. The angles that these diffracted beams make with the primary beam are the allowed angles (2θ's) for Bragg reflections from planes of atoms in the crystal. Rotation of the sample successively lines up different atomic planes so that the Bragg equation is satisfied, and different diffracted beams shoot out to create the diffraction pattern.

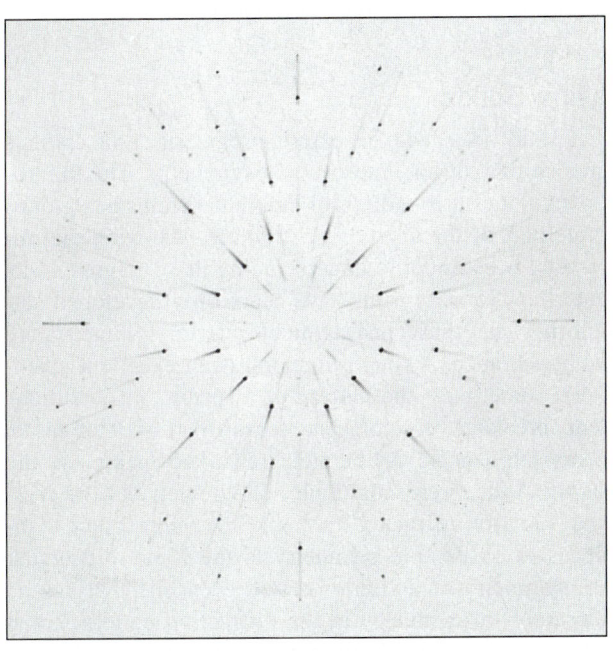

(a)

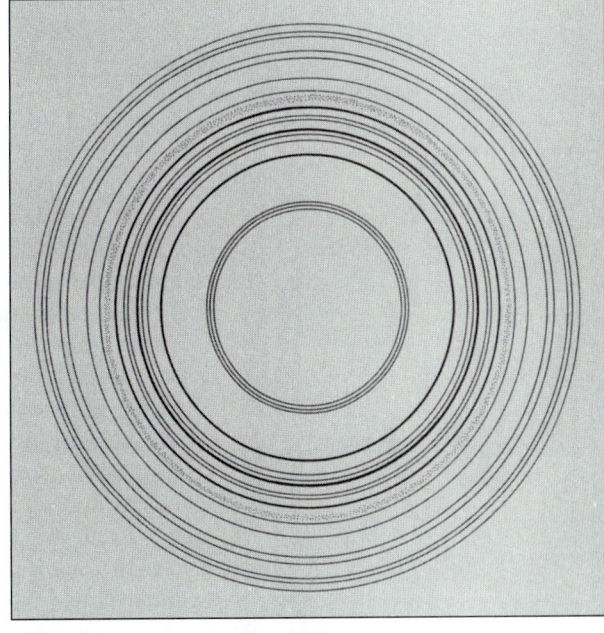

(b)

Figure 20–5 • Order in X-ray diffraction patterns derives from order in the structure of the material that is doing the scattering. (a) This pattern was obtained by irradiating a crystal of the mineral almandite ($Fe_3Al_2(SiO_4)_3$) with monochromatic X-rays (X-rays of a single wavelength) in a specialized camera. Note the symmetry of the pattern. (b) This pattern is observed by irradiating a polycrystalline powder of albite, $NaAlSi_3O_8$, with monochromatic X-rays and rotating the sample. *(Stephen J. Guggenheim/UIC Department of Geological Sciences)*

Figure 20-6 • A close-up of polycrystalline sphalerite (ZnS). (*Charles D. Winters*)

If the X-ray diffraction pattern consists of many sharply defined rings (see Fig. 20–5b), the solid is **polycrystalline.** It consists of randomly oriented, tiny crystalline grains either cemented together as a conglomerate or present as a loose powder (Fig. 20–6). Typical microcrystals have radii as small as 10^{-3} mm. Rings arise because the microcrystals are randomly oriented in the powder or conglomerate. Each tiny crystal gives its own set of sharply defined diffracted beams. Each ring in the diffraction pattern is caused by diffracted beams with the same 2θ coming from many different crystals that are lying tumbled in every possible orientation relative to the beam. Such X-ray diffraction patterns are called "powder patterns."

As the microcrystals in a polycrystalline solid become smaller in size, the rings in the diffraction pattern become more diffuse, and those at higher angles lose intensity. Extrapolation of this trend gives the diffraction pattern of an amorphous solid, which shows several diffuse, blurred rings encircling the primary X-ray beam. The rings mean that there is some structure, but their diffuseness means the structure is short-range only. The observation of *rings* instead of individual diffracted spots again arises from the random orientation of the individual scattering units. A liquid resembles an amorphous solid in its diffraction pattern and in its microscopic structure.

20-2 Symmetry and Structure

Symmetry and order are closely allied. In art and design, symmetry pleases the eye because it satisfies a human desire to detect or impose order (Fig. 20–7).

Symmetry in Crystalline Solids

Figure 20-7 • The symmetry of the pattern on this Navajo rug is an important aspect of its aesthetic appeal. (*Charles D. Winters*)

As suggested by Réné Just Haüy in 1784, the exterior regularities of crystals (Fig. 20–8) are a consequence of interior, microscopic symmetry. This microscopic symmetry is at the level of the surroundings of individual atoms and groups of atoms. It is an unalterable part of the crystalline structure. Knowing the microscopic symmetry of a crystal is essential to understanding its structure and its chemical and physical properties. Some crystals have beautifully developed natural faces; others, although they may have just as much interior symmetry, are formless lumps in external appearance. A rough diamond is an excellent example of the latter (Fig. 20–9). Sometimes, the history of a specimen (conditions of synthesis, heat treatment, breakage, erosion, and so forth) robs it of faces. Also, naturally occurring crystals can be deliberately reworked to almost any shape. Such actions do not affect the microscopic interior symmetry of the crystal. X-rays penetrate crystals and are diffracted by them. The microscopic symmetry inside the crystal leads to observable symmetry in the X-ray diffraction pattern. To learn about the symmetry of a single crystal, even one that has no well-developed faces, we search for symmetry in the diffraction pattern that it produces when bathed in a beam of X-rays. Figure 20–10 shows views of typical diffraction patterns, one from a crystal of the mineral hydroxyapatite ($Ca_5(PO_4)_3OH$) and the other from a crystal of an organic compound called fluorenone ($C_{13}H_8O$). Crystals give different *projections* of their total diffraction patterns, which are three-dimensional objects, when they are irradiated with X-rays from different angles. Hence, these striking patterns are only parts of the

Figure 20-8 • Crystals often, but not always, have well-developed planar faces. Some of these crystals of sodium chloride have well-developed faces; others have ill-formed exteriors but are still just as crystalline in their interior structure. *(Runk Shoenberger/Grant Heilman Photography, Inc.)*

total diffraction patterns of these crystals. The diffraction pattern from hydroxyapatite (Fig. 20–10a) has 6-fold **rotational symmetry.** This means that rotating the pattern by an angle of 60° (one sixth of a full circle) around its center gives back the original pattern. Similarly, a diffraction pattern has 2-fold rotational symmetry if rotation by 360°/2 = 180° around some axis takes it into itself. It has 3-fold rotational symmetry or 4-fold symmetry if a 360°/3 = 120° or a 360°/4 = 90° rotation achieves this self-coincidence. The pattern from fluorenone (Fig. 20–10b) has rotational symmetry (a 2-fold axis) and has a second kind of symmetry as well—**mirror symmetry.** Every spot on the left side of a vertical central line has a matching spot on the right side; every spot in the upper half of the pattern has a corresponding spot in the lower half: The fluorenone pattern has two lines of mirror symmetry. Figure 20–11 diagrams how rotational symmetry and mirror symmetry work.

• Symmetry disclosed in a diffraction pattern is a property of the crystal as a whole, not necessarily of the molecules that compose it.

Figure 20-9 • The facets that give gem-quality diamonds (*left*) their brilliance and appeal come from the carefully planned cutting and shaping of rough diamonds (*above*). *(Charles D. Winters)*

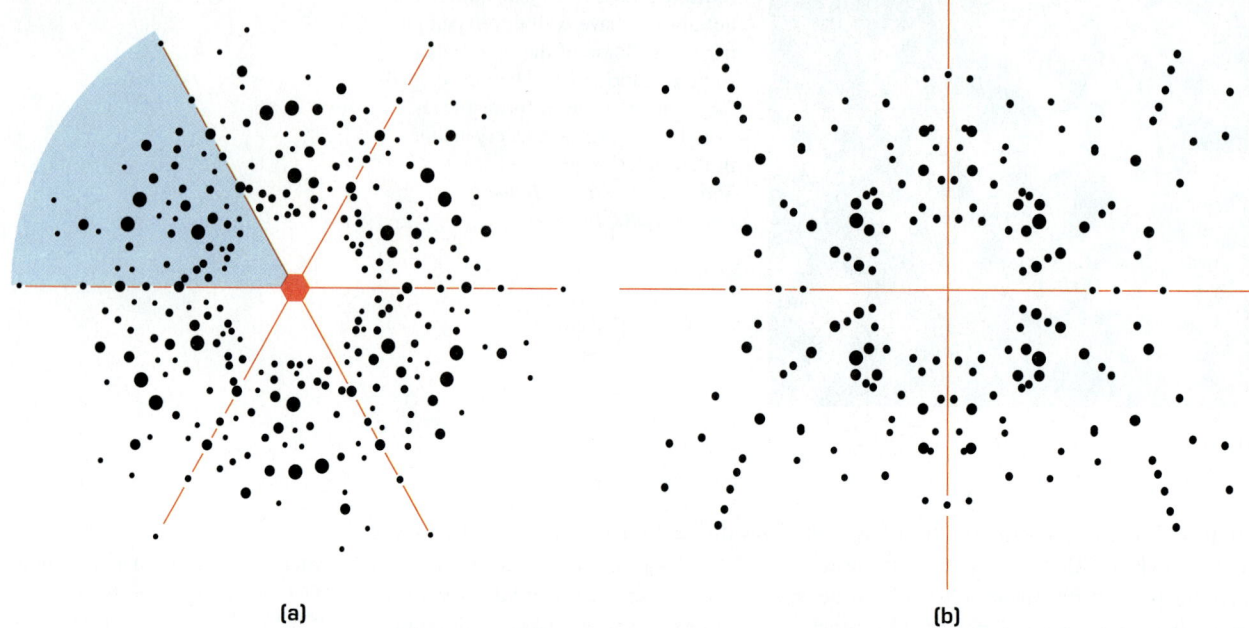

(a)   **(b)**

Figure 20-10 • Views of the diffraction patterns of crystals of (a) hydroxyapatite ($Ca_5(PO_4)_3(OH)$) and (b) fluorenone ($C_{13}H_8O$). The symmetry in these patterns implies the presence of symmetry in the crystals that cast them.

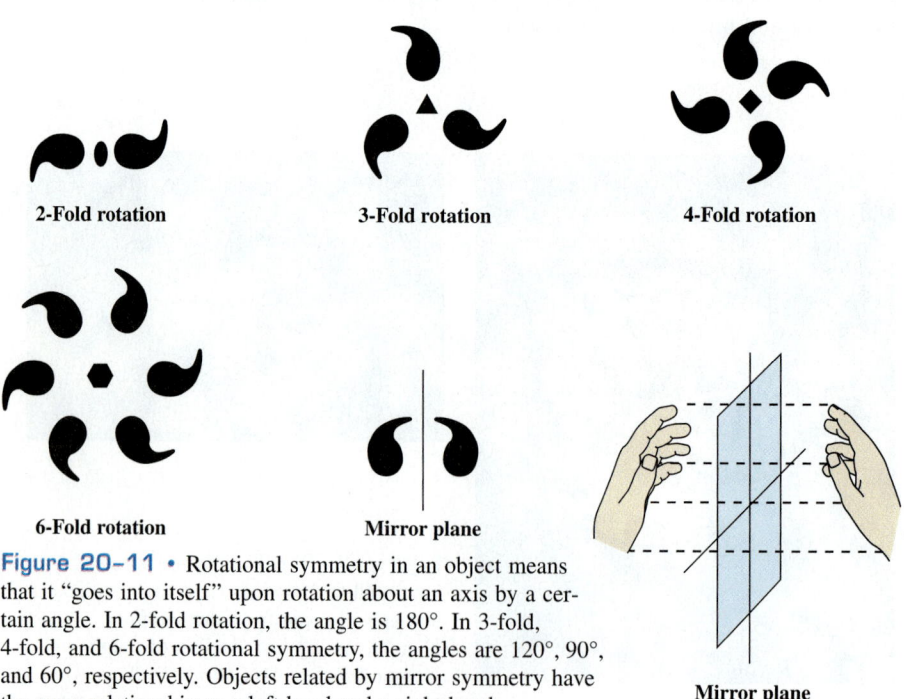

2-Fold rotation   **3-Fold rotation**   **4-Fold rotation**

6-Fold rotation   **Mirror plane**

Mirror plane

Figure 20-11 • Rotational symmetry in an object means that it "goes into itself" upon rotation about an axis by a certain angle. In 2-fold rotation, the angle is 180°. In 3-fold, 4-fold, and 6-fold rotational symmetry, the angles are 120°, 90°, and 60°, respectively. Objects related by mirror symmetry have the same relationship as a left hand and a right hand.

EXAMPLE 20-3

List the capital letters of the Roman alphabet that have a vertical mirror line of symmetry.

Solution

A vertical mirror line is present if the right side of the object "goes into" the left side when taken across the line. We carefully print each capital letter and examine it for symmetry. This reveals a vertical mirror line in the letters A, H, I, M, O, T, U, V, W, X, and Y.

EXERCISE

List the capital letters of the Roman alphabet that have a horizontal mirror line of symmetry.

Answer: B, C, D, E, H, I, K, O, X.

Only a few different kinds of symmetry have been found in the hundreds of thousands of diffraction patterns that have been recorded and studied. These consist of 2-fold, 3-fold, 4-fold, and 6-fold rotational symmetry in various combinations with each other and with mirror symmetry. Crystals are classified into one of seven crystal systems on the basis of the symmetry that is found in their diffraction patterns. The seven **crystal systems** and their minimum essential symmetry in diffraction are listed in Table 20–1. "Minimum essential symmetry" means that the symmetry *must* be present in the diffraction pattern in order to qualify the crystal for membership in the system. Crystals that belong to a system may and often do have additional symmetry beyond the listed minimum, but they cannot move up the table to join a higher system unless they have the minimum for that system. Also, symmetry is never ignored; crystals are classified as far toward the top of the list in Table 20–1 as possible. Several of the names of the crystal systems describe their minimum essential symmetry. Thus, hexagonal crystals (*hexa* means "six"), tetragonal crystals (*tetra* means "four"), and trigonal crystals (*tri* means "three") have at least 6-fold, 4-fold, and 3-fold interior symmetry, respectively. About 50% of all crystals studied

TABLE 20-1	The Seven Crystal Systems and the Minimum Essential Symmetry of Their Diffraction Patterns
Crystal System	**Minimum Essential Symmetry**
Hexagonal	One 6-fold rotation
Cubic	Four independent 3-fold rotations[a]
Tetragonal	One 4-fold rotation
Trigonal	One 3-fold rotation
Orthorhombic	Three mutually perpendicular 2-fold rotations
Monoclinic	One 2-fold rotation
Triclinic	No symmetry required

[a]Each of these axes passes through the diagonally opposite corners of a cube; each makes a 70.53° angle to the other three.

so far have been monoclinic; 25%, orthorhombic; and 15%, triclinic. The prevalence of the remaining, more symmetric, systems dwindles in the following order: cubic, tetragonal, trigonal, and hexagonal.

The dependence of the pattern of scattering by a crystalline solid on the direction of the X-ray beam is an example of *anisotropy,* the dependence of a property on the direction of its measurement. Many other properties of crystals are directional. If the value of a property changes with direction of measurement, then the crystal is anisotropic with respect to that property. A better-known anisotropy is the tendency of many crystals to cleave preferentially because they have different strengths in different directions. Crystals of mica resemble stacks of sheets of paper. It is easier to peel the sheets apart than to rip across them (Fig. 20–12). All crystalline solids are anisotropic with respect to the property of diffraction. Unlike crystals, amorphous solids are *isotropic* with respect to this and all other properties, as are liquids and gases.

Figure 20–12 • The mechanical properties of crystals of mica are quite anisotropic. Thin sheets can be peeled off of a crystal of mica easily with a razor blade, but the sheets resist stresses in other directions well. *(Charles D. Winters)*

The Crystal Lattice

The numerous sharply defined spots (for single crystals) or sharply defined rings (for polycrystalline solids) in the diffraction patterns of crystals are consequences of the long-range order in their organization at the atomic level. Identical sites within crystals recur regularly, at distances that are on the order of 10^{-10} m. The three-dimensional array made up of all the points within a crystal that have the same environment in the same orientation is a **crystal lattice.** A crystal lattice is an abstraction lifted away from a real crystal, embodying the scheme of repetition that is at work in that crystal. Figure 20–13 shows (in a two-dimensional example) how a lattice is separated from a repeating pattern. The lattice is imagined to extend infinitely in all directions.

Fundamental differences among lattices are found in their differing symmetries, not their sizes. If we construct a large number of different lattices, we find that the possible symmetries of the surroundings of each lattice point are restricted severely by the requirement that the surroundings of each and every lattice point be identical. It turns out that exactly seven three-dimensional lattices are distinguishable based

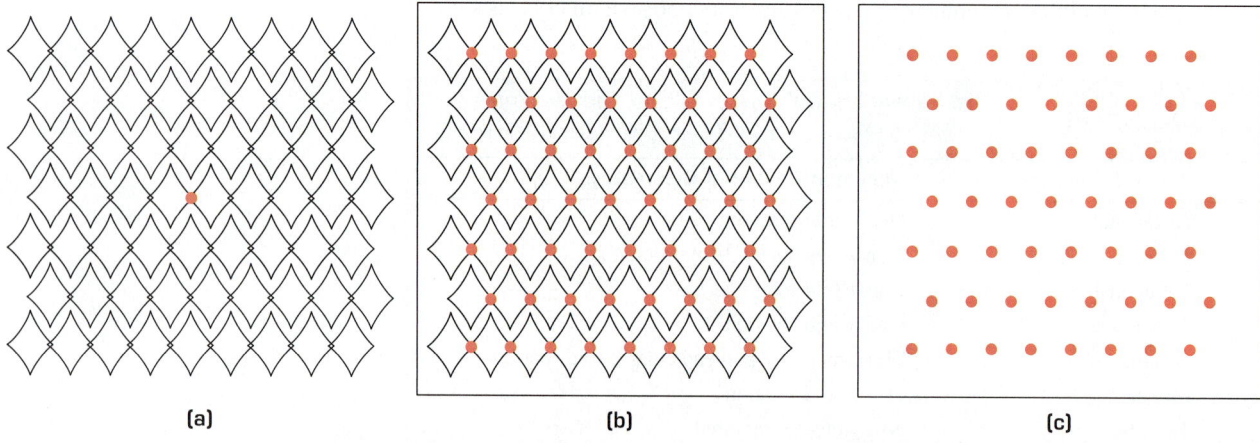

(a) (b) (c)

Figure 20–13 • Generating a lattice from a repeating pattern. (a) Select any point in the pattern and mark it. (b) Move out through the pattern, marking additional points that have the same surroundings in the same orientation. The targeted point recurs regularly, and the collection of marked points quickly grows to resemble a gridwork or lattice. (c) Discard the original pattern and keep the gridwork of points.

on the symmetry at their lattice points: the hexagonal, cubic, tetragonal, trigonal, orthorhombic, monoclinic, and triclinic lattices. It is not an accident that the names of these categories of lattices are the same as the names of the seven crystal systems. The correspondence follows from the fact that the symmetry of a particular crystal is always present in its lattice. Thus, a crystal in the tetragonal system always has a lattice of at least tetragonal symmetry.[1]

Unit Cells

Eight points of a crystal lattice define a **unit cell.** The eight points are the corners of the cell and must be chosen so that they define a box having three pairs of parallel faces. Solid figures of this type are **parallelepipeds** (Fig. 20–14). Every unit cell in a crystal is packed with atoms of exactly the same kinds, in the same numbers, and in the same relative positions. These contents of the unit cell are the repeating *motif* from which the crystal is constructed. Atoms need not occupy the corners of a unit cell, which are lattice points, although nothing prevents this. The sides and edges of unit cells may cut right through atoms. Only the portions of atoms actually within the walls of a unit cell count as belonging to that cell. Atoms on boundaries count as shared. Unit cells and crystal lattices, which are both products of the human imagination, allow a crystal's structure to be visualized as the result of a tidy stacking up (side by side, front to back, and bottom to top) of little building bricks, the unit cells. A single unit cell contains all structural information about its crystal.

As Figure 20–15 shows, a unit cell has 12 edges in addition to its 8 corners and 6 faces. The size and shape of a unit cell are fully described by three edge lengths (a, b, and c) and by the three angles between these edges (α, β, and γ). These are the *cell parameters* or *cell constants*.

> The symmetry that defines the seven different crystal systems imposes conditions on the shape of the unit cell.

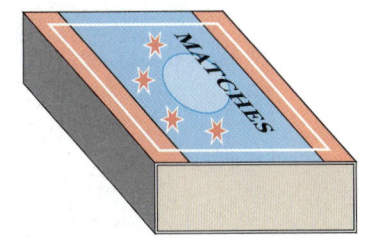

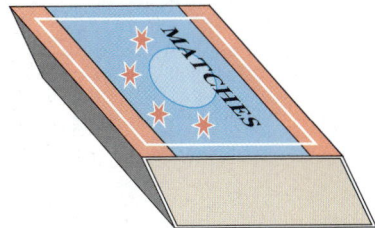

Figure 20–14 • The sliding outer cover of a matchbox is a parallelepiped and remains one even if it is squashed.

• The six faces of a unit cell (top, bottom, front, back, left, and right) are in general different. The structure comes from stacking up unit cells in exactly the same orientation.

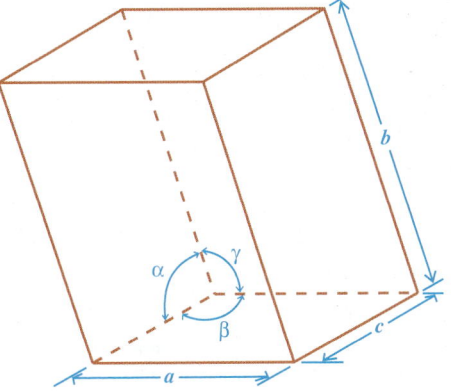

Figure 20–15 • Unit cells always have three pairs of mutually parallel faces. Only six pieces of information are required to construct a scale model of a unit cell—the three cell edges (a, b, and c) and the three angles between the edges (α, β, and γ). By convention, the angle between the edges a and b is labelled "γ," the angle between b and c is labelled "α," and the angle between a and c is labelled "β."

[1]The phrase "at least" appears because a lattice may have additional symmetry that the crystal does not have. A lattice contains only points, which are spherically symmetrical, but actual crystals contain atoms in groupings that may have shapes of low symmetry.

These conditions are given in Table 20–2, and the unit cells are shown in Figure 20–16.

TABLE 20-2	The Conditions Imposed by Symmetry on the Shapes of Unit Cells	
	Conditions on Cell Edges	**Conditions on Cell Angles**
Hexagonal	$a = b$; no conditions on c	$\alpha = \beta = 90°$, $\gamma = 120°$
Cubic	$a = b = c$	$\alpha = \beta = \gamma = 90°$
Tetragonal	$a = b$; no conditions on c	$\alpha = \beta = \gamma = 90°$
Trigonal	$a = b = c$	$\alpha = \beta = \gamma \neq 90° < 120°$
Orthorhombic	No conditions	$\alpha = \beta = \gamma = 90°$
Monoclinic	No conditions	$\alpha = \gamma = 90°$
Triclinic	No conditions	No conditions

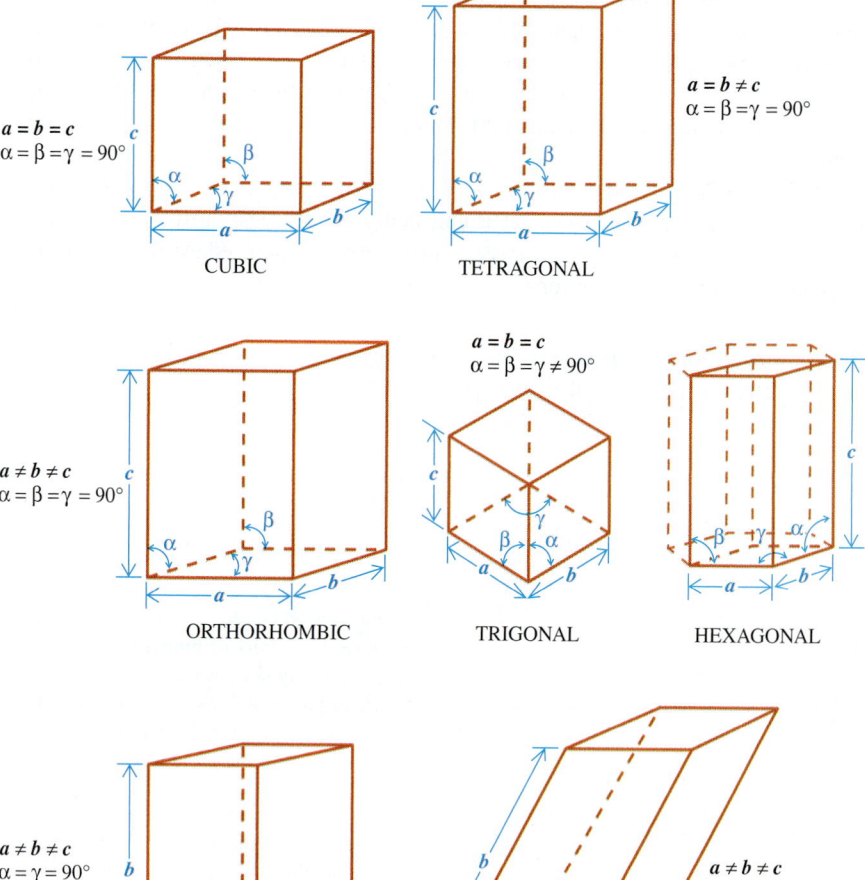

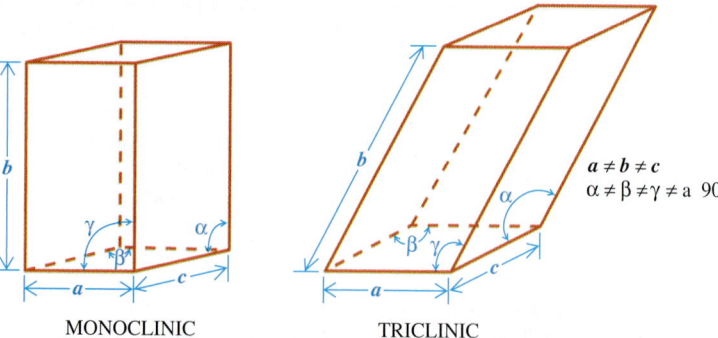

Figure 20–16 • Shapes of the unit cells in the seven crystal systems. Clearly, the more symmetrical crystal systems have more symmetrical cells, as shown by the conditions restricting their edges and angles.

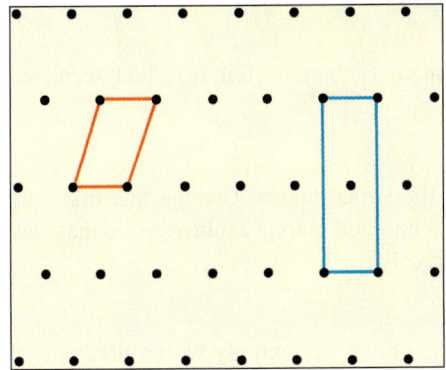

Figure 20–17 • In this two-dimensional lattice, every lattice point is at the intersection of a horizontal and vertical mirror line. It is possible to draw a primitive unit cell, such as the cell outlined in red, that contains $\frac{1}{4} \times 4 = 1$ lattice point. The larger centered unit cell outlined in blue, which contains two lattice points, is preferred, however. As a shape, the blue cell has two mirror lines—the full symmetry of the lattice—but the red cell does not.

Many choices of unit cell are possible in a given crystal lattice. The first rule in selecting a cell is to respect the symmetry of the crystal lattice. Table 20–2 shows how to do this. For example, an acceptable unit cell for a monoclinic crystal always has 90° angles between one edge and each of the other two (these are the angles α and γ in Figure 20–16). Other conceivable cells are not used because they ignore the presence of symmetry. The second rule in picking a unit cell is to find the *smallest* cell that still has the symmetry of the crystal lattice. Nothing is gained by using large cells when small ones will do. A unit cell that contains exactly one lattice point is called a **primitive** unit cell. Such a cell is the smallest possible. A primitive unit cell might seem to contain *eight* lattice points, one at each of its corners, but any corner lattice point actually is shared with seven neighboring cells that just meet at that corner. Only one eighth of each of the eight corner lattice points actually belongs to any unit cell. Because $\frac{1}{8} \times 8 = 1$, the eight shared lattice points at the corners always contribute exactly one lattice point to a unit cell.

Often, a primitive unit cell does not have the full symmetry of the crystal lattice. If so, a larger unit cell that does have the characteristic symmetry is chosen deliberately. Desirable larger cells usually have some combination of 90° or 120° angles between their edges because these angles are built into important symmetry operations (Fig. 20–17). The larger cells are *non-primitive* and are drawn on a given lattice by selecting unit cell corners in such a way that other lattice points are included either at the center of the cell or at the centers of some or all of its faces. Just three types of non-primitive cells suffice to describe most crystals. If an extra lattice point is at the center of the unit cell, then it is **body centered.** If extra lattice points are at the centers of all six faces of the unit cell (but not at the body center), then that cell is **face centered.** If extra lattice points are at the centers of a single pair of parallel faces of the unit cell (and nowhere else), then that cell is **side centered.** These non-primitive cells are shown in Figure 20–18.

No matter how a unit cell is selected, it contains a whole number of lattice points. A unit cell with two lattice points automatically has twice the volume of a primitive unit cell drawn on the same lattice. Similarly, if a unit cell contains four lattice points, then it has four times the volume of a primitive unit cell of the same lattice.

• Think of the armrests in a row of theater seats. Each theatergoer has armrests on both sides but must share them. Each (excluding those at the ends of rows) owns only $2 \times \frac{1}{2} = 1$ armrest.

• Unit cells on the very boundaries of crystals obviously are exceptions, but they are comparatively rare exceptions.

Primitive

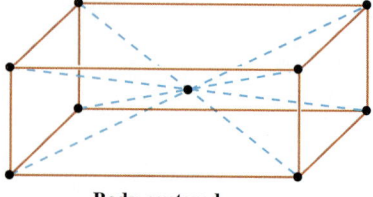

Body-centered

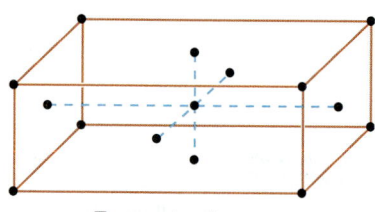

Side-centered

Face-centered

Figure 20–18 • Centered unit cells have lattice points in addition to the lattice points at their eight corners. A body-centered lattice has an additional lattice point at the center of the cell, compared with a primitive lattice. A face-centered lattice has additional lattice points at the centers of the six faces; a side-centered lattice has additional lattice points at the centers of two parallel sides of the unit cell. These lattices in this figure are orthorhombic, but centering can occur in other crystal systems as well. (*Note*: The dots in these diagrams are not atoms, but lattice points.)

EXAMPLE 20–4

Determine how many lattice points belong to each unit cell in (a) a body-centered lattice and (b) a face-centered lattice.

Strategy

Sketch the unit cells and locate all of the lattice points. Use the fact that only lattice points contained entirely within a unit cell belong exclusively to that cell. Others are shared with neighboring unit cells.

Solution

(a) In a body-centered lattice, one lattice point lies entirely within the body of each cell and belongs entirely to it. Each of the eight lattice points situated at the eight corners is shared with seven neighboring cells. Hence, each cell owns $1 + (8 \times \frac{1}{8}) =$ 2 lattice points.

(b) In a face-centered lattice, lattice points are located at the centers of the six faces of each cell, as well as its eight corners. Each of the points in the faces is shared with a single neighboring cell. Therefore, there are $(6 \times \frac{1}{2}) + (8 \times \frac{1}{8}) =$ 4 lattice points per cell.

EXERCISE

Determine how many lattice points belong to each unit cell in a side-centered lattice.

Answer: 2.

Unit-Cell Volume and Density

The volume of its unit cell tells a great deal about a crystal because it is the size of the repeating motif in a crystal. If the three angles α, β, and γ of a unit cell are 90°, the volume of the unit cell is

$$V = abc$$

If one or more angles is other than 90°, then a formula that takes into account the slanting sides of the nonrectangular box is required:

$$V = abc \sqrt{1 - \cos^2\alpha - \cos^2\beta - \cos^2\gamma + 2 \cos \alpha \cos \beta \cos \gamma}$$

This formula is easy to use, thanks to the availability of calculators. The quantity under the square root sign is always less than or equal to 1 but greater than 0.

EXAMPLE 20–5

The triclinic unit cell of crystalline malonic acid ($C_3H_4O_4$) has cell edges $a = 5.33$ Å, $b = 8.36$ Å, and $c = 5.14$ Å. The angles are $\alpha = 104°$, $\beta = 94.8°$, and $\gamma = 71.5°$. Compute the volume of this unit cell.

Solution

Use the COS key on a calculator to obtain the cosines of the three angles. They are -0.241922, -0.083678, and $+0.317305$, respectively. Substitute the three cosines into the expression in parentheses under the radical sign in the preceding formula. The result equals 0.846636. The positive square root of this is 0.920128. The product of the three lengths is 229.032 Å^3, so the actual volume of the unit cell is 0.920128 × 229.032 Å^3 = 211 Å^3, expressed to three significant digits.

EXERCISE

One form of the protein myoglobin, extracted from the sperm whale, crystallizes in the monoclinic crystal system. The unit cell has the dimensions $a = 61.6$ Å, $b = 26.9$ Å, and $c = 33.9$ Å. The angles α and γ exactly equal 90° and angle β is 105.5°. Compute the volume of this unit cell.

Answer: 5.41×10^4 Å^3.

The six cell constants—a, b, c, α, β, and γ—of the unit cell are determined from a study of the crystal's X-ray diffraction pattern. If the mass of the unit cell contents is known, the theoretical cell density can be computed. This density in principle equals the measured density of the crystal, a quantity that can be determined by completely independent experiments. Comparing the experimental density of a crystal with the density calculated from X-ray diffraction provides a valuable check on both methods. The following example illustrates this type of calculation.

EXAMPLE 20-6

Crystalline sodium chloride has a cubic unit cell with the cell edge $a = 5.6402$ Å at room temperature. Each unit cell contains four Na$^+$ and four Cl$^-$ ions. Compute the density of sodium chloride in grams per cubic centimeter (g cm^{-3}).

Strategy

Recall the definition of *density*, which is the mass of an object divided by its volume. Compute the mass and volume of a unit cell of NaCl. Divide the first by the second and then use suitable unit factors to convert to the specified unit.

Solution

According to Table 20–2, the cell angles in a cubic unit cell equal 90°, and the three edges are equal to each other ($a = b = c$). These facts, together with the fact that cos 90° = 0, simplify the formula for the volume of a cubic unit cell considerably:

$$V_{\text{cubic unit cell}} = a^3$$

For NaCl, this volume is

$$V_{\text{NaCl unit cell}} = (5.6402 \text{ Å})^3 = 179.43 \text{ Å}^3.$$

A unit cell holds four Na$^+$ ions and four Cl$^-$ ions, for a total mass 233.772 u. Compute the density of a unit cell:

$$d = \frac{m}{V} = \frac{233.772 \text{ u}}{179.43 \text{ Å}^3} = 1.3029 \text{ u Å}^{-3}$$

Converting this density to g cm^{-3} requires unit factors relating atomic mass units to grams and cubic ångstroms to cubic centimeters.

$$d = \frac{1.3029\ \text{u}}{\text{Å}^3} \times \left(\frac{(10^8)^3\ \text{Å}^3}{1\ \text{cm}^3} \right) \times \left(\frac{1\ \text{g}}{6.0221 \times 10^{23}\ \text{u}} \right) = 2.1635 \frac{\text{g}}{\text{cm}^3}$$

Check

Daily experience shows that sodium chloride (salt) sinks in water (as it dissolves). Its density therefore exceeds 1 g cm^{-3}, the density of water. Similarly, salt is less dense than are metals such as iron (density 7.87 g cm^{-3}). The answer is therefore reasonable. The experimental density of NaCl from conventional measurements at room temperature is 2.16 g cm^{-3}.

• Conventionally measured densities of crystals often fall slightly below densities observed from X-ray experiments due to defects in the crystals (see Section 20–4).

EXERCISE

Crystals of the compound NiAs$_2$ (nickel arsenide) have an orthorhombic unit cell with edges $a = 5.75$ Å, $b = 5.82$ Å, and $c = 11.43$ Å. Each unit cell contains 8 nickel atoms and 16 arsenic atoms. Compute the density of this compound in g cm^{-3}.

Answer: $d = 7.24$ g cm^{-3}.

The same ideas can be applied to determine the molar mass of a compound if it is not known by other means, as shown in the following example.

EXAMPLE 20–7

A hydrate of calcium chloride crystallizes in the hexagonal system and contains one formula unit per unit cell. The cell constants are $a = b = 7.8759$ Å and $c = 3.9545$ Å, and the observed density of the crystal is 1.71 g cm^{-3}. (a) Compute the molar mass of the hydrate. (b) Determine the formula of the hydrate.

Strategy

Plan to use the fact that the molar mass of a substance equals its molar volume multiplied by its density. Compute the volume of a single unit cell from the cell constants. The problem states that this volume is the volume occupied by a single formula unit. Therefore, the molar volume equals the cell volume multiplied by Avogadro's number.

Solution

A hexagonal unit cell has the angles $\alpha = 90°$, $\beta = 90°$, and $\gamma = 120°$ (see Table 20–2).

$$V_{\text{cell}} = (7.8759\ \text{Å})^2 (3.9545\ \text{Å})$$
$$\sqrt{1 - \cos^2 90 - \cos^2 90 - \cos^2 120 + 2\cos 90 \cos 90 \cos 120}$$
$$= 245.297\ \text{Å}^3\ (0.866025) = 212.43\ \text{Å}^3$$

The conversion

$$\frac{212.43\ \text{Å}^3}{\text{unit cell}} \times \left(\frac{1\ \text{unit cell}}{1\ \text{formula unit}} \right) = \frac{212.43\ \text{Å}^3}{\text{formula unit}}$$

gives the volume occupied by one formula unit of this hydrate. Compute the molar volume in cubic centimeters (cm^3) by another series of unit factors:

$$V_{\text{m}} = \frac{212.43\ \text{Å}^3}{\text{formula unit}} \times \left(\frac{1\ \text{cm}^3}{1 \times 10^{24}\ \text{Å}^3} \right) \times \left(\frac{6.0221 \times 10^{23}\ \text{formula units}}{1\ \text{mol}} \right)$$

$$= 127.93 \frac{cm^3}{mol}$$

$$M = V_m d = 127.93 \frac{cm^3}{mol} \times 1.71 \frac{g}{cm^3} = 219 \frac{g}{mol}$$

(b) The $CaCl_2$ part of the formula contributes $(2 \times 35.45) + 40.08 = 110.98$ g mol^{-1}. The difference of 108 g mol^{-1} must come from the water of hydration. But 108 g mol^{-1} is almost exactly six times the molar mass of H_2O. The formula of the hydrate is therefore $CaCl_2 \cdot 6H_2O$.

EXERCISE

Corundum is a crystalline material classified in the hexagonal crystal system. It contains six formula units per unit cell. The dimensions of the unit cell are $a = b = 4.7591$ Å and $c = 12.9894$ Å. The density of the crystal is 3.987 g cm^{-3}. Compute the molar mass of the compound that makes up corundum.

Answer: $M = 102.0$ g mol^{-1} (the compound is Al_2O_3).

The Intensities of the Diffracted Beams

An X-ray diffraction pattern reveals the interior symmetry of a crystal and allows the unit cell parameters—that is, the size and shape of the unit cell—to be determined. This information comes from the relative positions of the diffracted beams in the diffraction pattern. A record of a crystal's diffraction contains more information than this, however. Each diffracted beam has a characteristic *intensity;* some are bright, and others are dim. Variations in intensity are visible in the experimental diffraction patterns in Figure 20–5. The intensities of the diffracted beams depend on the distribution throughout the unit cell of the scattering power—that is, of the atoms with their electrons. These intensities are determined by the identities and locations of the atoms in the unit cell.

From an accurate record of the intensities and directions of the many X-ray beams diffracted by a crystal, it is nearly always possible to compute back to the locations and identities of the atoms in the unit cell. This is called "solving the crystal structure." Knowing exact atomic locations is the same as knowing the detailed molecular structure of the substance that makes up a given crystal, something that is obviously of tremendous value to chemists, biologists, physicists, and engineers. Solving a crystal structure requires highly accurate intensity data and much computation. The process is now automated, and the use of specially designed instruments and very efficient computer programs has made the determination of molecular structures by X-ray analysis quite routine. High-quality results identify the atoms and give the distances between them to the nearest 0.005 or 0.010 Å.

• Frequently, the most difficult part is growing good crystals of the substance of interest.

20–3 Bonding in Crystals

The nature of the cohesive forces acting among the particles of substances in condensed phases provides a good basis for classifying them and for understanding their properties. Four "bond types" of crystals usually are distinguished. The principal forces that maintain each structure are van der Waals forces, electrostatic (ionic) forces, metallic bonds, and covalent bonds. We consider each of these types in turn.

Molecular Crystals

Molecular crystals include the frozen noble gases, oxygen, nitrogen, the halogens, many covalent and partially covalent compounds such as CS_2, NO_2, Al_2Cl_6, $FeCl_3$, and $BiCl_3$, and the vast majority of organic compounds. In molecular crystals, it is always possible to single out groups of atoms in which the distance from each atom to at least one atom within the group is significantly smaller than the distance to any atom that is not within the group. The groups of atoms are, of course, the molecules, within which bonding occurs by means of strong, short covalent bonds (the sharing of electrons). In molecular crystals, the molecules are maintained in their positions in a lattice by van der Waals forces. This is a blanket term introduced in Section 6–1 for all attractive forces that involve complete molecules and neutral atoms. Van der Waals forces include dipole–dipole, ion–dipole, induced dipole, and dispersion forces but do not include attractions between oppositely charged ions or covalent bonds. Van der Waals forces pull neighboring molecules (or atoms) toward each other. Eventually, their effects are balanced by short-range repulsive forces as the electrons of closely neighboring particles start to interfere. These repulsions rise so rapidly as distance decreases that a fairly well-defined radius (called the **van der Waals radius**) can be assigned to every element (see Table 20–3). The van der Waals radius of an atom is effectively the radius it denies to other atoms to which it is not bonded; van der Waals attractions are not strong enough to pull the centers of non-bonded atoms closer together than the sum of their van der Waals radii.

The trade-off between attractive and repulsive forces among even small molecules in a molecular crystal is quite complex because so many atoms are involved. A useful simplifying approach is to picture a molecule as a superposition of a set of spheres, one centered at each nucleus. The radius of each sphere is the van der Waals radius of the element involved. Figure 20–19 shows such a "space-filling

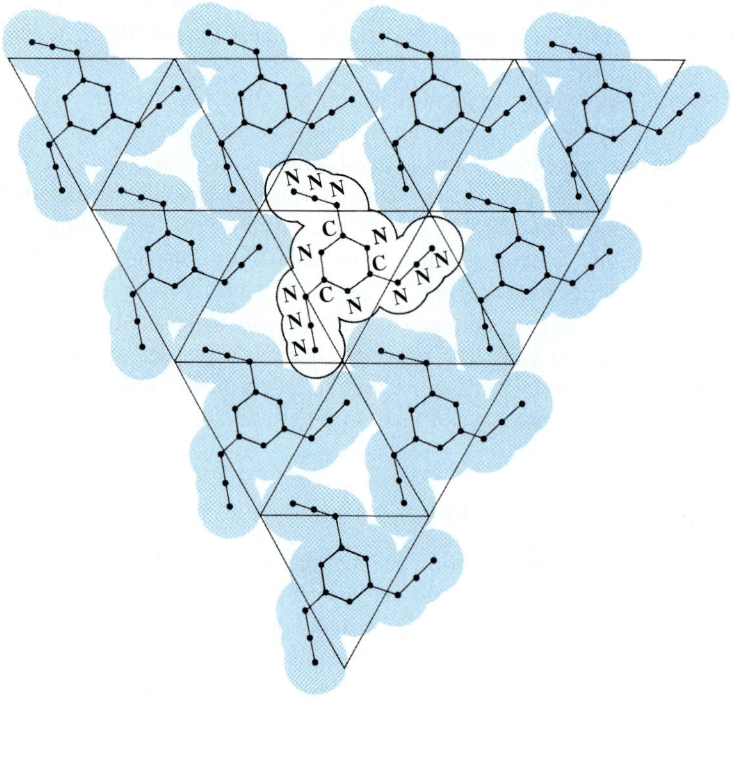

TABLE 20–3

Van der Waals Radii of Several Elements

Element	Radius (Å)
Fluorine	1.47
Chlorine	1.75
Carbon	1.70
Hydrogen	1.20
Nitrogen	1.55
Oxygen	1.52

Figure 20–19 • The van der Waals radii of the carbon and nitrogen atoms are superimposed (*in blue*) on an outline of the molecular structure of cyanuric triazide (C_3N_{12}). This shows the volume of space that is excluded by each molecule to others. Van der Waals forces in the molecular crystal hold the molecules in contact in a pattern that minimizes empty space. Note the 3-fold symmetry of the pattern.

model" of cyanuric triazide (C_3N_{12}). In molecular crystals, such shapes pack together in such a way that no molecules are suspended in empty space and none overlap. The figure shows how nature solves the problem of efficiently packing many copies of the rather complicated molecular shape of C_3N_{12} in a single layer. Many such layers stack up with a slight offset that minimizes unfilled space between layers to create the molecular crystal of C_3N_{12}.

Van der Waals forces are much weaker than are the cohesive forces in ionic, metallic, and covalent crystals. As a consequence, molecular crystals typically have low melting points and are soft and easily deformed. The frozen noble-gas elements, which have single-atom "molecules," crystallize in the highly symmetric face-centered cubic lattice at atmospheric pressure. More complex molecules usually form crystals of lower symmetry, in the monoclinic or triclinic systems. They usually do not display the striking physical properties of electrical conductivity and magnetism that are characteristic of metals. Their interesting properties are primarily those of their molecular units.

Molecular crystals are of great scientific value. Obtaining proteins and other macromolecules in the crystalline state allows their structures to be determined by X-ray diffraction. Knowing the three-dimensional structures of biological molecules is the starting point for understanding how they work.

- The 1988 Nobel Prize in chemistry was awarded for the isolation of molecular crystals of a key protein at the photosynthetic reaction center of *Rhodopseudomonas viridis*, a bacterium (see Problem 20–28). X-ray diffraction studies of the crystal structure of this substance, which contains about 630,000 atoms per molecule, have provided important insights into how photosynthesis works.

Ionic Crystals

Compounds formed by elements with significantly different electronegativities are largely ionic, and, to a first approximation, the ions may be treated as hard, charged spheres that occupy positions on the crystal lattice and interact electrostatically (see Section 17–5). The elements of Groups I and II of the periodic table react with those of Groups VI and VII to form ionic compounds, most of which crystallize in the cubic system. The high symmetry of these crystalline solids reflects their structural simplicity. Specifically, the alkali-metal halides (except for the cesium halides), the ammonium halides, and the oxides and sulfides of the alkaline-earth metals all crystallize in the simple and beautiful **rock salt,** or **sodium chloride,** structure, shown in Figure 20–20. It may be viewed as a face-centered cubic lattice with anions at the lattice points and cations occupying positions exactly between pairs of anions, or, equivalently, as the same lattice with cations at the lattice points and anions occupying positions exactly between. Either way, one cation and one anion are associated with each lattice point, and every ion is surrounded by six equidistant ions of the opposite charge.

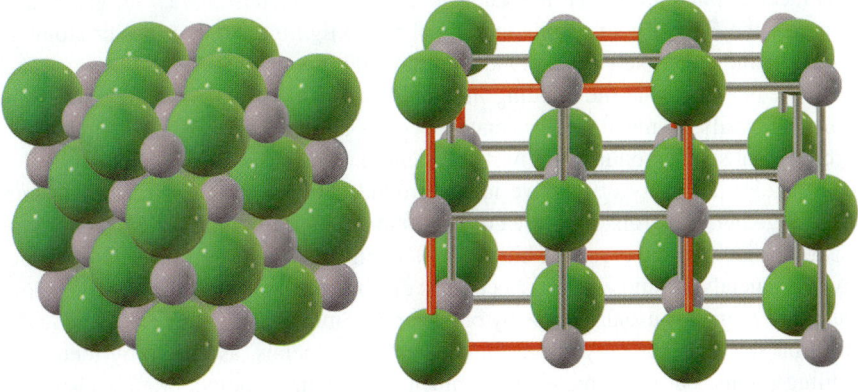

Figure 20–20 • The structure of sodium chloride. On the left, the sizes of the Na$^+$ (*purple-pink*) and Cl$^-$ (*green*) ions are shown to scale. On the right, the ions are reduced in size to allow a unit cell (*shown by red lines*) to be outlined clearly.

Figure 20–21 • The structure of
cesium chloride. On the left, the sizes
of the Cs⁺ (*purple-pink*) and Cl⁻
(*green*) ions are shown to scale. On
the right, they are reduced to allow a
unit cell (*shown by red lines*) to be
outlined clearly. Note that the crystal
lattice in this structure is simple
cubic—not body-centered cubic, as is
sometimes stated.

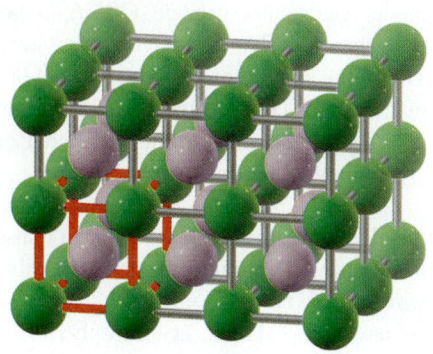

The favored crystal structure for an ionic substance is the one that brings each
ion as close as possible to the maximum number of ions of the opposite charge
(because opposite charges attract), while simultaneously keeping it as far away as
possible from ions of the same charge (because like charges repel). Different types
of ionic structures emerge, depending on the relative sizes of the cations and anions.
The cesium ion, for example, is large enough for eight chloride ions to come up
around it without contacting each other. Consequently, in cesium chloride, the chloride ions occupy the lattice points of a simple cubic lattice and the cesium ions sit
at the centers of the unit cells defined by these eight points (Fig. 20–21). This structure does not work for sodium chloride because the sodium ion is smaller than the
cesium ion; eight chloride ions would interfere seriously with each other if all were
brought in contact with a central sodium ion.

The strength and range of electrostatic forces make ionic crystals hard, high-
melting, and brittle solids. They are electrical insulators, but melting an ionic crystal disrupts the lattice and sets the ions free to move, so ionic liquids are good electrical conductors.

Metallic Crystals

The most characteristic property of metals is their good ability to conduct electricity and heat. Electrical conduction arises from the flow of electrons from regions of
high potential energy to those of low potential energy. Thermal conduction is as-
sisted by the flow of electrons from high-temperature regions (where their kinetic
energies are high) to low-temperature regions (where their kinetic energies are low).
The clear implication is that the valence electrons in metals move with great ease.
Why are the electrons so mobile in a metal but so tightly bound to ions or atoms in
an insulating solid such as diamond, sodium chloride, or sulfur?

The reason is that the bonding in metals is quite different from the bonding in
other materials. In metals, the valence electrons are delocalized in huge molecular
orbitals that span essentially unlimited distances. A metallic crystal can be regarded
as a regular assembly of positive ions surrounded by an "electron sea" that binds
them together. Electrons in the sea do not belong to specific metal atoms and move
easily throughout the crystal. This explains the high electrical and thermal conductivity of metals. Liquids as well as crystals can be metals; in fact, the electrical conductivity of metals usually drops by only a small amount when they are melted. The
electron sea provides very strong binding in some metals, as shown by their high
boiling points, but other metals have much lower boiling points. Metals also have

a very large range of melting points. Gallium melts at 29.78°C, below normal body temperature in humans (Fig. 20–22), and mercury stays liquid at temperatures that freeze water. On the other hand, many metals require temperatures in excess of 1000°C to melt, and tungsten, the highest-melting elemental metal, melts at 3410°C. This high melting point makes it particularly useful for the filaments in incandescent lightbulbs.

Most metallic elements have crystal structures of high symmetry and crystallize in body-centered cubic, face-centered cubic, or hexagonal lattices. Why is this? If the attractions between identical atoms in a crystal are non-directional, as is the case with many metals, lowest energy should favor structures that have the most efficient possible packing together of the atoms. A little experimentation with marbles quickly shows that the most efficient packing of equal-sized balls in a single layer occurs when each one contacts six others. Stacking such layers atop one another fills three-dimensional space. The most efficient packing of equal spheres in such a stack occurs when each sphere is surrounded by 12 neighbors—6 in its own layer, 3 from the layer above, and 3 from the layer below. Indeed, many metals do crystallize in exactly such **close-packed** arrangements in which each atom has 12 nearest neighbors. There are two types of close packing, depending on the way in which the layers are arranged relative to one another, as shown in Figure 20–23. The first generates a face-centered cubic unit cell (like that in Fig. 20–18 but with all cell edges equal) and is called **cubic close packing** because the cell has cubic symmetry. The second type of close packing, **hexagonal close packing,** generates a unit cell with hexagonal symmetry and also achieves a coordination number of 12 for every atom. The two close-packing schemes have the same efficiency but differ fundamentally in their symmetries.

Some crystalline metallic elements do not have close-packed structures, as Figure 20–24 shows. Sodium and the other alkali metals crystallize in body-centered cubic unit cells in which the atoms have a coordination number of only 8. Polonium crystallizes in a simple cubic cell that gives the atoms a coordination number of just 6. Likewise, each atom in metallic tin is surrounded by six close neighbors, although four neighbors are at one distance and two at a slightly greater distance. This arrangement, which generates a body-centered unit cell with tetragonal symmetry, stems from directional character in the Sn—Sn bonds.

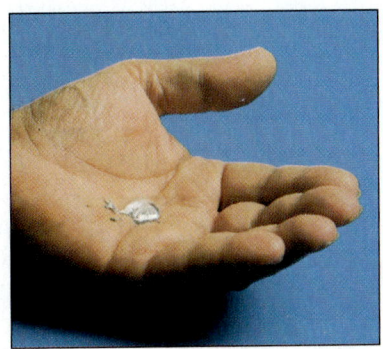

Figure 20–22 • Gallium has a low enough melting point that it melts from the heat of the body. *(Leon Lewandowski)*

• Both cubic close packing and hexagonal close packing fill 74.0% of the volume with spherical atoms and leave 26.0% empty.

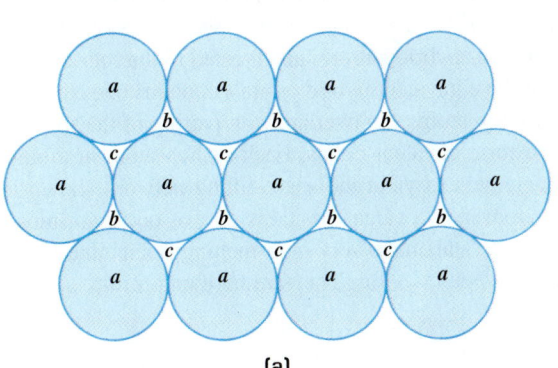

(a)

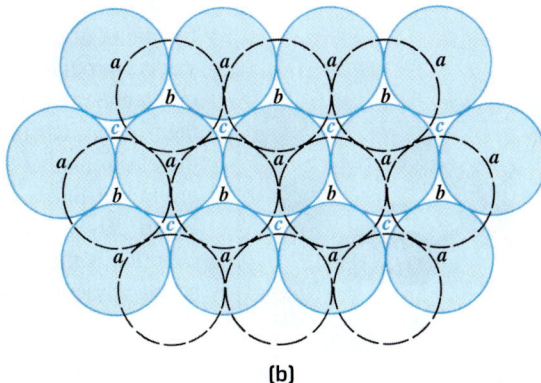

(b)

Figure 20–23 • Choices in the close packing of spheres. (a) A layer of spheres centered on sites labelled *a*. Gaps between spheres are labelled alternately *b* and *c*. (b) A second layer of spheres has been added, with centers directly over the sites labelled *b*. Positioning the spheres of a third layer over *c*-sites gives cubic close packing; positioning them over *a*-sites gives hexagonal close packing.

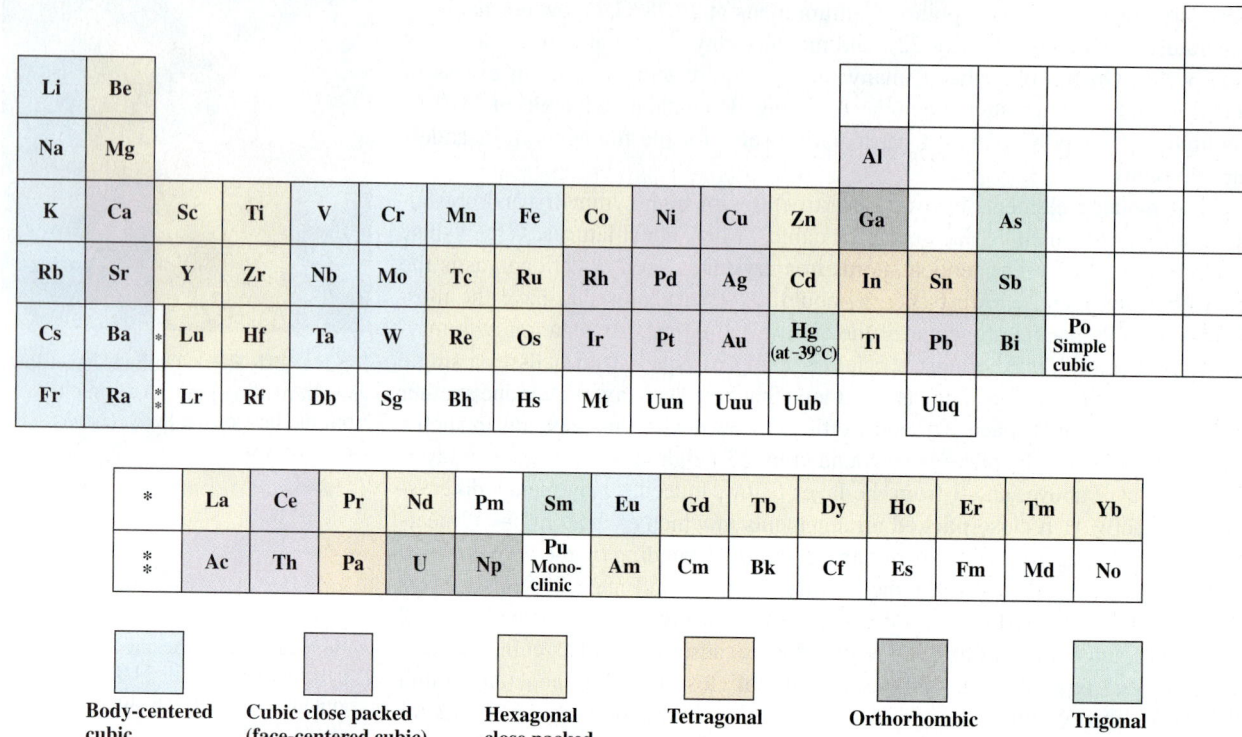

Li	Be																	
Na	Mg											Al						
K	Ca	Sc	Ti	V	Cr	Mn	Fe	Co	Ni	Cu	Zn	Ga		As				
Rb	Sr	Y	Zr	Nb	Mo	Tc	Ru	Rh	Pd	Ag	Cd	In	Sn	Sb				
Cs	Ba	* Lu	Hf	Ta	W	Re	Os	Ir	Pt	Au	Hg (at -39°C)	Tl	Pb	Bi	Po Simple cubic			
Fr	Ra	** Lr	Rf	Db	Sg	Bh	Hs	Mt	Uun	Uuu	Uub		Uuq					

*		La	Ce	Pr	Nd	Pm	Sm	Eu	Gd	Tb	Dy	Ho	Er	Tm	Yb
**		Ac	Th	Pa	U	Np	Pu Mono-clinic	Am	Cm	Bk	Cf	Es	Fm	Md	No

Body-centered cubic   Cubic close packed (face-centered cubic)   Hexagonal close packed   Tetragonal   Orthorhombic   Trigonal

Figure 20–24 • Lattice structures of the metallic elements at 25°C and a pressure of 1 atm. The majority of the metals adopt a body-centered cubic, a face-centered cubic, or a hexagonal close-packed structure.

Covalent Network Crystals

In the final type of crystalline solid, covalent network crystals, the atoms are linked by covalent bonds rather than by the electrostatic attractions of ionic crystals or the valence-electron interactions of metallic crystals. The archetype of the covalent network crystal is diamond, which belongs to the cubic system. The ground-state electron configuration of a carbon atom is $1s^22s^22p^2$, and, as established in Section 18–2, the four valence orbitals can be replaced by four hybrid sp^3 orbitals directed to the four corners of a regular tetrahedron. If each of the equivalent hybrid orbitals contains one electron, then sp^3 orbitals of neighboring carbon atoms can overlap, with pairing of the spins of the electrons that they contain, to form covalent bonds. Each carbon atom in a large group can link covalently to four others to yield the space-filling network shown in Figure 20–25. In a sense, every atom in a covalent crystal is part of one giant molecule that is the crystal itself. These crystals have very high melting points due to the strong attractions between covalently bound atoms. They are hard and brittle.

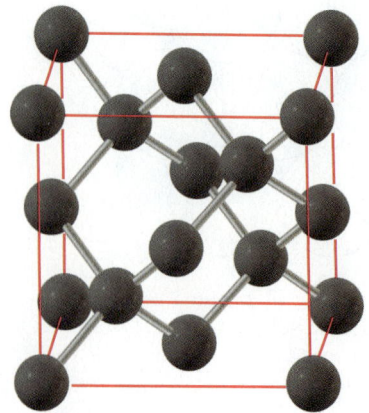

Figure 20–25 • The structure of diamond. The face-centered cubic unit cell (*solid red lines*) contains eight carbon atoms, four at lattice points, and four others completely inside the cell. Each atom has four covalently bonded nearest neighbors surrounding it at the corners of a tetrahedron. Silicon has the same organization of atoms, but the unit cell is larger.

EXAMPLE 20-8

Explain the hardness of diamond in terms of its crystal type.

Solution

Denting a diamond requires the simultaneous bending of many strong directional C—C bonds that support each other in sets of interlocking triangles.

EXERCISE

Explain the brittleness of diamond in terms of its crystal type.

Answer: The network is strong, but failure of one bond under extreme stress throws extra strain on neighboring bonds closely associated with it in the above-mentioned interlocking triangles. These bonds quickly fail, and the fracture propagates rapidly through the crystal—that is, the crystal shatters.

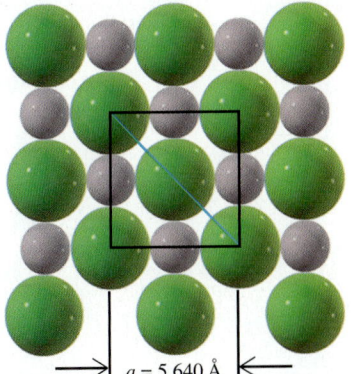

Figure 20-26 • Looking down on one face of the face-centered cubic unit cell of NaCl and its immediate surroundings. The blue line is a face diagonal of the unit cell. The distances between the centers of different ions can be computed using geometrical relationships.

Bond Lengths and Other Interatomic Distances in Crystals

Interatomic distances in all types of crystals can be computed once the size and shape of the unit cell and the locations of the atoms within it are known. The geometry is straightforward when the unit cell is cubic. We illustrate with three examples.

- In crystalline sodium chloride, Na^+ and Cl^- ions alternate along the edge of the face-centered cubic unit cell, as Figure 20–26 shows. The length of the cell edge is 5.640 Å. This makes the shortest $Na^+ - Cl^-$ distance equal to half the cell edge, or 2.820 Å. Chloride ions form rows along the *face diagonals* of the unit cells in sodium chloride. By the Pythagorean theorem, the distance between neighboring Cl^- ions along the face diagonals is $\sqrt{2}(5.640/2) = 3.988$ Å. This is the shortest $Cl^- - Cl^-$ distance in the crystal; others can be calculated by drawing appropriate triangles and doing the geometry. The shortest $Na^+ - Na^+$ distance also equals 3.988 Å, by a similar use of right triangles.

- Iron crystallizes in a body-centered cubic cell having an edge length of 2.866 Å and an Fe atom at every lattice point. Obviously, the Fe–Fe distance along the cell edge equals 2.866 Å, but this is *not* the shortest Fe–Fe distance. Rather, the closest contact is between an Fe at a corner and the Fe at the body center. This distance equals half of the *body diagonal* of the cube. The body diagonals of a cube (there are four) connect its opposite corners and pass through the body center. Their lengths equal $\sqrt{3}$ times the edge of the unit cube, as shown in Figure 20–27. The shortest Fe–Fe distance in crystalline iron is therefore $\sqrt{3}(2.866)/2 = 2.482$ Å.

- Diamond has a face-centered cubic unit cell with an edge of 3.567 Å. The distance from one carbon to the next along the cell edge is 3.567 Å (far), and the C–C distance along the face diagonal is $\sqrt{2}(3.567/2) = 2.522$ Å (still rather far). The shortest C–C distance is along a different direction. As Figure 20–25 shows, four of the eight carbon atoms in the unit cell of diamond are not positioned at lattice points, but one fourth of the way into the unit cell along the body diagonals. The distance from a carbon at a cell corner to one of these atoms equals the length of the body diagonal divided by four, or $\sqrt{3}(3.567)/4 = 1.544$ Å.

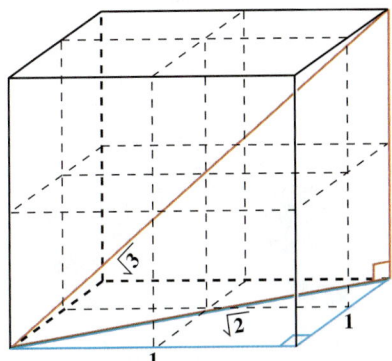

Figure 20-27 • Geometrical relationships in a cube. Applying the Pythagorean theorem to the right triangle in blue shows that the face diagonal of a cube equals $\sqrt{2}$ times the edge. Applying the theorem once more, this time to the triangle marked in red, shows that the body diagonal is $\sqrt{3}$ times longer than the edge.

Such calculations allow for the estimation of atomic sizes. The radius of an iron atom, for example, can be taken as half the distance between the nearest neighbor atoms in the crystalline element, or 2.48/2 = 1.24 Å. According to Figure 17–10, iron has an atomic radius of 1.26 Å. The slight difference arises because the environment of the Fe atoms differs: Here each Fe has 8 nearest neighbors, not 12, which is the case in Figure 17–10. Similarly, the radius of the carbon atom is nominally half the distance between nearest neighbor carbon atoms in diamond, or 0.77 Å. Before the determination of crystal structures by X-ray diffraction became routine, tables of atomic radii derived from a relatively few crystal structure determinations were used to estimate or predict bond lengths in new compounds. Nowadays, new compounds are studied by X-ray diffraction to obtain bond lengths.

The Structures of the Elements

The elements provide examples of three of the four classes of crystalline solids described in this section (atoms of a single element cannot form ionic bonds). Examples of the simple structures adopted by metallic elements already have been discussed. Non-metallic elements have more complex structures, reflecting a competition between intermolecular and intramolecular bonding in many cases and producing molecular or covalent solids with quite varied properties.

Atoms of the halogens have seven valence electrons. At room conditions, each atom combines with an identical atom to form a diatomic molecule. Once this single bond forms, there is no further bonding capacity: The halogen diatomic molecules interact with one another only through relatively weak van der Waals forces to form molecular solids with low melting and boiling points.

The Group VI elements oxygen, sulfur, and selenium display dissimilar structures in the solid state. Each oxygen atom (with six valence electrons) can form one double or two single bonds. Except in ozone (O_3), which is thermodynamically unstable relative to O_2, oxygen prefers to use up all of its bonding capacity with an intramolecular double bond, forming a liquid and a solid that are bound only weakly by van der Waals forces. Diatomic sulfur molecules, S=S, on the other hand, are relatively rare, being encountered experimentally only in high-temperature vapors. At room temperature the bonding goes from one S to another to give either rings or chains. The most stable form of sulfur at room temperature consists of S_8 molecules, with eight sulfur atoms arranged in a puckered ring (Fig. 20–28). The S_8 molecules in turn interact by van der Waals forces to make elemental sulfur the rather soft molecular solid that it is. Above 160°C, the rings in molten sulfur start to break open and relink to form long chains, producing a highly viscous liquid. An unstable ring form of selenium, Se_8, is known, but the thermodynamically stable form of this element is a gray crystal of metallic appearance that consists of very long, covalently bound, spiral chains, with weak interchain interaction. Crystalline tellurium has a similar structure. In the Group VI elements, then, a trend exists (moving down the periodic table) away from the formation of multiple bonds and toward the chains and rings characteristic of collections of atoms that each form two single bonds.

A similar trend emerges among the Group V elements. In this group, only nitrogen forms diatomic molecules with triple bonds, in which all the bonding capacity is used between pairs of atoms. Elemental phosphorus exists in three forms, in all of which the phosphorus atoms form three single bonds rather than one triple bond. White phosphorus (Fig. 20–29a) consists of tetrahedral P_4 molecules that interact with each other through weaker van der Waals forces. Black and red forms

Figure 20–28 • The structure of an S_8 molecule. The orthorhombic unit cell of rhombic sulfur, the most stable form of elemental sulfur at room temperature, is large. It contains 16 of these S_8 molecules, for a total of 128 atoms of sulfur.

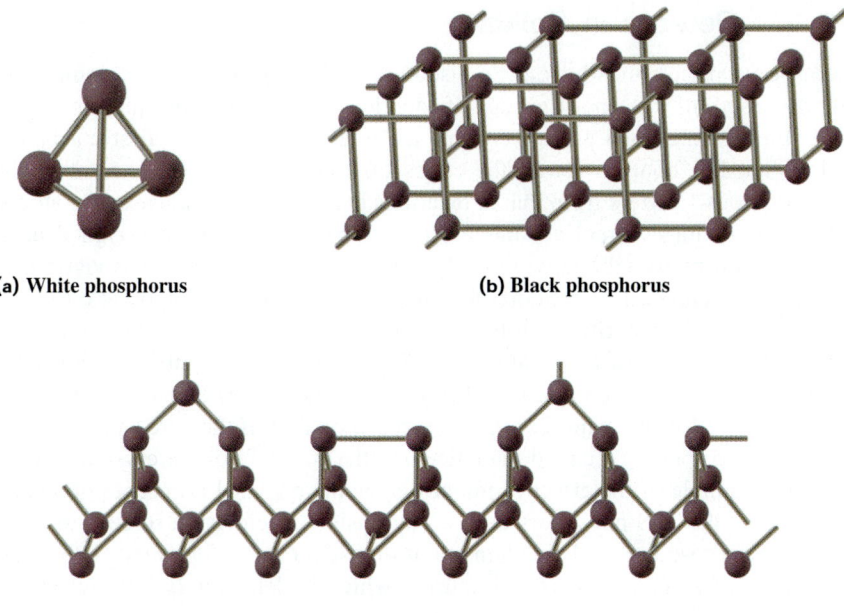

Figure 20-29 • Structures of elemental phosphorus.

(a) White phosphorus　　　　**(b) Black phosphorus**

(c) Red phosphorus

of phosphorus (Figs. 20–29b, c), on the other hand, are higher-melting network solids in which the three bonds formed by each atom connect it directly or indirectly with all the other atoms in the solid. Unstable, solid forms of arsenic and antimony that consist of As_4 or Sb_4 tetrahedra like those in white phosphorus can be prepared by rapid cooling of the vapor. The stable forms of these elements have structures related to that of black phosphorus.

　　The properties of the solids just described are on the borderline between covalent and molecular. Other elements, those of intermediate electronegativity, exist as solids with properties on the borderline between metallic and covalent and are called **semi-metals.** Antimony, for example, has a metallic luster but is a rather poor conductor of electricity and heat. Silicon and germanium are **semiconductors,** with electrical conductivities far lower than those of metals but still significantly higher than those of true insulators like diamond.

　　Some elements of intermediate electronegativity exist in two crystalline forms with very different properties. White tin has a tetragonal crystal structure and is a metallic conductor. Below 13°C, it crumbles slowly to form a powder of gray tin (with the diamond structure). The latter is a poor conductor and has few, if any, uses. Its formation at low temperature is known as "tin disease." The thermodynamically stable form of carbon at room conditions is not the insulator diamond, but graphite. Graphite consists of sheets of repeated hexagons, with only rather weak interactions between layers (Fig. 20–30). Each carbon atom shows sp^2 hybridization and bonds to three other carbon atoms, with its remaining p orbital (perpendicular to the graphite layers) taking part in extended π bonding interactions over the whole plane. Graphite can be pictured as a stack of parallel sheets of adjoining benzene rings, with significant π electron delocalization over each sheet. The delocalized electrons give graphite a conductivity approaching that of the metallic elements. The conductivity and chemical inertness of graphite make it useful for electrodes in electrochemistry.

• The classification of the elements as metals, non-metals, or semi-metals is discussed in Section 3–1.

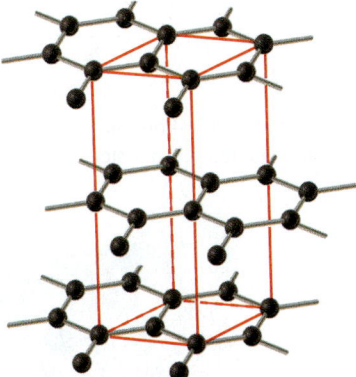

Figure 20-30 • The structure of graphite. The unit cell is outlined in red.

20-4 Defects in Solids

So far, crystalline solids have been discussed as if they consisted of completely orderly arrays of stationary groups of identical atoms extending indefinitely in three dimensions. Such perfect crystals are unattainable. In the first place, any real crystal has boundaries. At the edges of a crystal, the environment of the atoms obviously differs from that at locations buried in the bulk. Next, the atoms, molecules, or ions in real crystals are not pinned motionless at assigned positions, but constantly vibrate, vigorously at high temperatures and less vigorously at low temperatures. Furthermore, as a practical matter, it is impossible to rid a real crystal of all impurities. Atoms that do not fit the idealized pattern therefore crop up here and there in a real crystal. Even trace-level impurities can exert a large influence on the chemical and physical properties of a crystal. Impurities may substitute for the major constituents in a crystal or, if they are small enough, may fit in the openings (called interstices) in the structure of the crystal. One interesting example of the effect of impurities on the physical properties of a crystal concerns the ZnS phosphors used on the inside of television screens to make them glow. These materials contain about one silver atom for every 10,000 formula units of ZnS. Without this impurity, which is deliberately introduced, the ZnS does not fluoresce.

Some additional types of lattice imperfections would occur even in an utterly pure crystalline substance. These **point defects** include the simple **vacancy,** in which an atom is missing from a lattice site, and the **interstitial atom,** in which an atom is inserted in a site different from its normal site. In real crystals, a small fraction of the normal atom sites remain unoccupied. Such vacancies are called **Schottky defects.** Figure 20–31a illustrates Schottky defects in the crystal structure of an element (metal, noble gas, or nonmetal). Schottky defects also occur in ionic crystals, but with the restriction that the imperfect crystal remain electrically neutral. Thus, in sodium chloride, for every missing Na^+ ion, a Cl^- ion also must be missing (see Fig. 20–31b). Likewise, in $CaCl_2$, for each Ca^{2+} vacancy, there must be two Cl^- vacancies.

In certain kinds of crystals, atoms or ions are displaced from their regular lattice sites to interstitial sites, and the crystal defect consists of the lattice vacancy plus the interstitial atom or ion. Figure 20–31c illustrates this type of lattice imperfection, known as a **Frenkel defect.** The silver halides (AgCl, AgBr, AgI) are examples of crystals in which Frenkel defects are prevalent. The crystal structures of these compounds are established primarily by the anion lattice, and the silver ions occupy highly disordered, almost random sites, much as in a liquid. The rate of

• The notion of a perfect or ideal crystal is quite analogous to that of a perfect or ideal gas. Neither actually exists. Both provide starting points for understanding the states (crystalline solid and gas, respectively) that they represent.

Figure 20–31 • Point-lattice imperfections. The red X's denote vacancies.

(a) Schottky defects in a metal

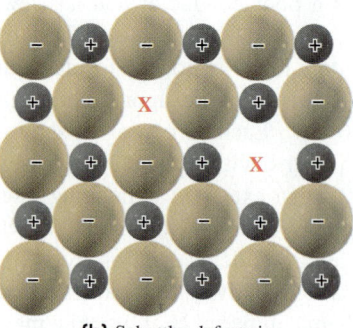

(b) Schottky defects in an ionic crystal

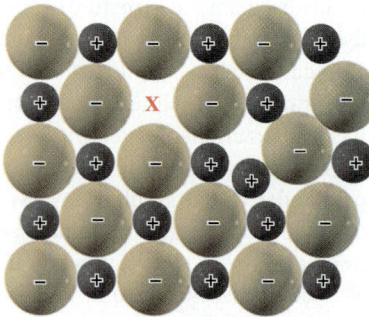

(c) Frenkel defect in an ionic crystal

diffusion of silver ions in these solids is exceptionally high, as studies using radioactive isotopes of silver have shown. Both Frenkel and Schottky defects in crystals are mobile, jumping from one place to another with frequencies that depend on the temperature and the strength of the atomic forces. Diffusion in crystalline solids largely arises from the presence and mobility of point defects.

Bombarding crystals with subatomic particles or high-energy radiation damages them. The simplest types of damage are the creation of an isolated interstitial atom and an isolated vacancy. The accumulation of defects and their eventual clustering or other interaction can cause marked deterioration in the electrical and mechanical properties of materials. Concerns of this type are fundamental in the engineering of materials for nuclear reactors, which must withstand large fluxes of neutrons. An unexpected loss of strength in a structural part in a nuclear reactor could be catastrophic.

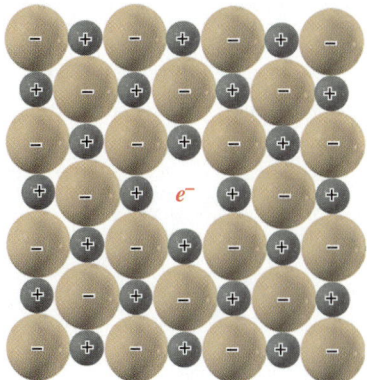

Figure 20–32 • A color center in a crystal.

If an alkali halide crystal such as NaCl(*s*) is irradiated with X-rays, ultraviolet radiation, or high-energy electrons, some Cl^- ions may lose an electron:

$$Cl^- + h\nu \longrightarrow Cl + e^-$$

The resulting Cl atom, being uncharged and much smaller than a Cl^- ion, is no longer strongly bound in the crystal and can diffuse to the surface and escape. The electron can migrate through the crystal quite freely until it encounters an anion vacancy and is trapped in the Coulomb field of the surrounding cations (Fig. 20–32). This creates a crystal defect called a **color center** (also called an F-center, from the German word for color, *Farbe*). It is the simplest of a family of electronic crystal defects and, as the name suggests, color centers impart colors to crystals of the alkali or alkaline earth halides (Fig. 20–33).

Non-Stoichiometric Compounds

In Chapter 1, the law of definite proportions was discussed as a principal piece of evidence that led to Dalton's atomic theory. Subsequent careful experiments have shown that in fact the law of definite proportions is violated for a great many solid-state compounds. These materials do *not*, after all, have fixed and unvarying compositions, but exist over a range of compositions in a single phase. The phenomenon is usually small, but dramatic instances do occur. FeO (wüstite) ranges in composition from $Fe_{0.85}O$ to $Fe_{0.95}O$, depending on the method of preparation. It is never found with its nominal 1:1 composition. NiO varies only from $Ni_{0.97}O$ to NiO in composition, but the variation is accompanied by major changes in properties. When the compound is prepared in the 1:1 composition, it is pale green and an electrical insulator. When it is prepared in an excess of oxygen, it is black and conducts electricity fairly well. The case of TiO is extreme: It ranges from $Ti_{0.75}O$ (a formula that is equivalent to Ti_3O_4) all the way up to $Ti_{1.45}O$ (a formula that is approximately equivalent to Ti_3O_2). Cu_2S also deviates considerably from its nominal stoichiometry, depending on the method of preparation.

The reason for the deviations is the availability of more than one oxidation state for the metal. In wüstite, the iron exists in both the $+2$ oxidation state and the $+3$ state. Imagine a sample having the composition $Fe_{1.00}O_{1.00}$, corresponding to equal numbers of Fe^{2+} cations and O^{2-} anions. Suppose that some Fe^{3+} ions are introduced. For every two Fe^{3+} ions that go in, three Fe^{2+} ions must leave in order to maintain overall charge balance. The amount of iron in the sample becomes less as vacancies occur at some of the iron sites in the crystal. Similarly, in the black form of nickel(II) oxide, some Ni^{2+} ions are replaced by Ni^{3+} ions, and compensating

Figure 20–33 • Pure calcium fluoride (CaF_2) is white, but the natural crystal of calcium fluoride (fluorite) shown here is clouded with purple because of the presence of color centers, lattice sites at which the F^- anion is replaced by an electron only.

• Non-stoichiometric compounds sometimes are called berthollides, in honor of Claude Berthollet, who believed that the proportions of the elements in compounds could vary over a range (see Section 1–3).

vacancies occur at nearby nickel atom sites in the crystal. In TiO, vacancies occur at both Ti sites and O sites: In the composition $Ti_{0.75}O$, nearly all of the O sites are occupied and a bit fewer than 25% of the Ti sites are vacant; in $Ti_{1.45}O$, nearly all of the Ti sites are filled, but 33% of the O sites are vacant.

• See Problem 69 for additional data on TiO.

EXAMPLE 20-9

The composition of a sample of wüstite is $Fe_{0.930}O$. What percentage of the iron is in the form of iron(III)?

Strategy

Apply the principle of charge balance: The total positive charge of the Fe^{2+} and Fe^{3+} cations must equal the total negative charge of the O^{2-} anions.

Solution

If the sample contains 1 mol of O, then it contains 0.930 mol of Fe. Suppose y mol of the iron is Fe^{3+}. Then, $0.930 - y$ mol of it is Fe^{2+}. The total charge on the two kinds of iron cations, measured in moles of electron charge, is

$$\text{total charge on iron cations} = 3y + 2(0.930 - y)$$

This positive charge is exactly balanced by the 2 mol of negative charge carried by the 1 mol of O^{2-} ions because the wüstite is electrically neutral:

$$3y + 2(0.930 - y) = 2$$

Solving for y gives

$$y = 0.140 \text{ mol}$$

The percentage of iron in the form of Fe^{3+} equals the number of moles of Fe^{3+} divided by the total number of moles of iron, multiplied by 100%:

$$\% \text{ iron in form of } Fe^{3+} = \frac{0.140}{0.930} \times 100\% = \boxed{15.1\%}$$

Check

Imagine that the sample of wüstite contains 1 mol of Fe. If the answer is correct, this sample contains 0.151 mol of Fe^{3+} and 0.849 mol of Fe^{2+}. The chemical amounts of oxygen required to balance these amounts of Fe^{3+} and Fe^{2+} are

$$n_O = 0.151 \text{ mol } Fe^{3+} \times \left(\frac{3 \text{ mol O}}{2 \text{ mol } Fe^{3+}}\right) = 0.2265 \text{ mol O}$$

$$n_O = 0.849 \text{ mol } Fe^{2+} \times \left(\frac{1 \text{ mol O}}{1 \text{ mol } Fe^{2+}}\right) = 0.849 \text{ mol O}$$

The total amount of O is 1.0755 mol. The formula of the sample is therefore $FeO_{1.0755}$. This formula is equivalent to $Fe_{0.930}O$, so the answer must be correct.

EXERCISE

A sample of wüstite, Fe_xO, contains one Fe^{3+} ion for every three Fe^{2+} ions. Calculate the value of x in its formula.

Answer: 0.889.

20–5   Liquid Crystals

Melting a molecular crystal actually involves two effects. The first is what is usually pictured in melting: The thermal energies of the molecules become so large that the lattice breaks down and, as its long-range order is lost, the crystal liquefies. The second effect is the reorientation of the molecules, which begin to tumble and spin in all directions. Usually the two processes occur at the same temperature, and the crystal melts to an ordinary liquid. Sometimes, however, one of the two occurs at a lower temperature than the other. If the molecules are naturally nearly spherical (an example is CCl_4), then they may start to tumble before the lattice is disrupted. The substance stays solid and is called a **plastic crystal.** If the molecules are long, stiff, and thin, there may not be enough room for them to tumble after the lattice breaks down. They instead retain some degree of orientational order, and the result is a **liquid crystal.** In both cases, a higher temperature brings on the effect that has been delayed, and the substance becomes an ordinary liquid.

Many organic materials do not show a single solid-to-liquid transition, but rather a cascade of transitions involving new intermediate phases of the types just described. Five percent of all organic compounds exhibit liquid-crystal behavior. Not unexpectedly, the X-ray diffraction patterns of liquid crystals combine characteristics of crystal and liquid diffraction. Thus, an X-ray diffraction pattern from a liquid crystal can have both diffuse rings (the signature of a liquid) and some fairly sharply defined spots (the signature of a crystal). In recent years, liquid crystals have found a wide variety of practical applications, which range from use as temperature sensors to displays on calculators and other electronic devices.

The Structure of Liquid Crystals

The compound terephthal-*bis*-(4-*n*-butylaniline) (TBBA) is a good example of a compound that forms liquid crystals. Its molecules are rod-like, with hydrocarbon groups at each end separated by a relatively rigid backbone of benzene rings and $N{=}C$ double bonds.

$$H_9C_4 - \bigcirc - N{=}C - \bigcirc - C{=}N - \bigcirc - C_4H_9$$

The molecules have a tendency to line up even in the liquid phase, as shown in Figure 20–34a. Ordering in the liquid phase persists only over small distances, however, and on average a given molecule is equally likely to take any orientation. The simplest type of liquid-crystal phase is the **nematic phase** (Fig. 20–34b). In a nematic liquid crystal, the molecules show a preferred orientation in a particular

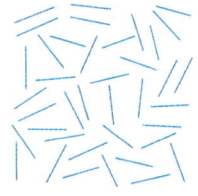
(a) Liquid

(b) Nematic phase

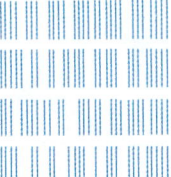
(c) Smectic A phase

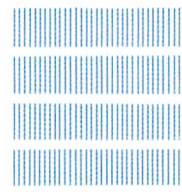
(d) Molecular crystal

Figure 20–34 • Different states of structural order for rod-shaped molecules. In a real sample, the lining up of the molecules would not be so nearly perfect.

direction, but their centers are distributed at random, as they would be in an ordinary liquid. Although liquid-crystal phases are characterized by a net orientation (or "lining up") of molecules over large distances, it is wrong to think that all the molecules point in exactly the same direction. Fluctuations in the orientation of each molecule exist. Only *on average* do the molecules have a greater probability of pointing in one direction.

Some liquid crystals can form one or more **smectic phases.** These show a variety of microscopic structures that are distinguished by the letters A, B, C, and so forth. One of them, the smectic A structure, is shown in Figure 20–34c: The molecules continue to show net orientational ordering, but now, unlike the nematic phase, the centers of the molecules also tend to lie in layers. Within each layer, however, these centers are distributed at random, as in an ordinary liquid.

CHEMISTRY IN YOUR LIFE

Liquid-Crystal Displays

Digital watches and many calculators use liquid-crystal displays. The operation of such displays relies on the fact that the particular orientation chosen by a liquid crystal is very sensitive to the position and nature of the surfaces with which it is in contact and to small electric or magnetic fields. In a liquid-crystal display, a nematic liquid crystal is contained in a small cell that has interior surfaces treated to specify the orientation of the liquid crystal it holds. Polarizing filters set up on each end of this cell allow only photons with particular polarizations to pass through the cell. In the absence of an electric field,

light passes through the liquid crystal and both filters and is reflected by an underlying mirror, so the display appears white (Fig. 20–A, *top*). If an electric field is applied across some portion of the display, the preferred orientation of the molecules in that area changes, causing a different polarization of light; the "rotated" light is blocked by the second filter, and that part of the display appears black (Fig. 20–A, *bottom*). When the electric field is turned off, the liquid-crystal molecules relax rapidly to their original orientations, and the display again turns white.

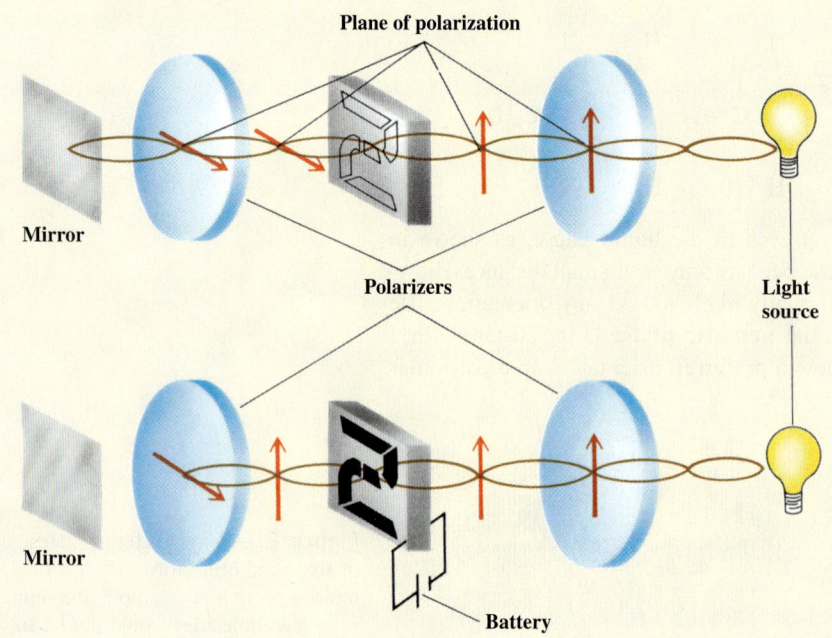

Figure 20–A • How a liquid-crystal display device works. The light has a polarization that permits it to pass through the second polarizer and strike the mirror, giving a display that appears bright (*top*). Imposition of a potential difference (*bottom*) causes the liquid-crystal molecules to reorient, so that the resulting polarized light does not reach the mirror and the display appears dark.

At low enough temperatures, a liquid crystal freezes to a crystalline solid (see Fig. 20–34d) in which the orientations of the molecules are ordered and their centers lie on a regular three-dimensional lattice. The meaning of the term "liquid crystal" can be seen from the progression of structures in Figure 20–34. Liquid crystals are solid-like in showing orientational ordering among their molecules but liquid-like in the random distribution of the centers of those molecules.

A third type of liquid crystal is called **cholesteric.** The name stems from the fact that many of these liquid crystals involve derivatives of the cholesterol molecule. In each layer, the molecules show a nematic type of ordering, but the direction of orientation changes from layer to layer.

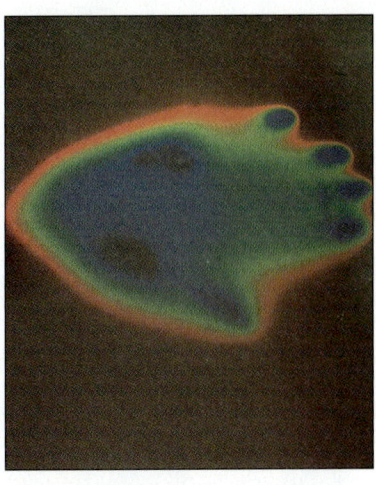

A hand pressed against a cholesteric liquid-crystal film on a black backing. The different colors reveal temperature variations. *(Kurt Nassau/Phototake)*

SUMMARY

20-1 In both **crystalline solids** and **amorphous solids,** structure is probed in diffraction experiments using X-rays, neutrons, or electrons. In **X-ray diffraction,** a portion of a beam of X-rays is scattered by the electrons of the material being studied. This interaction is followed by interference among the **scattered** waves. X-ray **diffraction patterns** convey information about the distribution of the scattering centers (atoms) because the distances between atoms control the phase differences of the scattered waves arising from the atoms. The **Bragg law** relates the wavelength of the incoming X-rays, the distance between parallel planes of scattering centers, and the scattering angles at which various "reflected" beams are observed.

20-2 All crystals have a characteristic symmetry that is revealed in X-ray diffraction experiments. Crystals are classified into seven different **crystal systems** that are based on the **symmetry** (including **rotational symmetry** and **mirror symmetry**) of their diffraction patterns. These are the **hexagonal, cubic, tetragonal, trigonal, orthorhombic, monoclinic,** and **triclinic** systems. The three-dimensional array made up of all of the points within a crystal that have the same environment in the same orientation is a **crystal lattice;** it represents the scheme of repetition at work in the construction of a crystal. There are seven types of crystal lattice, corresponding to the seven crystal systems. A proper selection of eight points of a crystal lattice defines a **unit cell,** a box that contains all the structural information about a crystal necessary to reproduce it. A crystal is created by stacking up unit cells to fill space. **Primitive** unit cells (the smallest possible unit cells) are the usual choice in discussing crystal structures. Non-primitive unit cells (including **body-centered, face-centered,** and **side-centered** cells) are used when no primitive cell has the full symmetry of the lattice. The volume of a unit cell can be computed from X-ray measurements of the cell edges (a, b, and c) and the cell angles (α, β, and γ). The density of a crystal is equal to the mass of the contents of a unit cell of that crystal divided by the volume of the cell. From accurate observations of the intensities as well as the directions of the many X-ray beams diffracted by a crystal, it is nearly always possible to compute back to the locations of the atoms in the unit cell.

20-3 Crystals can be classified, according to the nature of their bonding interactions, as molecular, ionic, metallic, or covalent network. **Molecular crystals** consist of discrete molecules held in positions on a lattice by **van der Waals forces.** These forces cause molecules to approach each other until atoms come into "contact" at their **van der Waals radii. Ionic crystals** are maintained by electrostatic attractions between cations and anions. They often have structures of high symmetry, such as the **rock salt** (sodium chloride) structure. **Metallic crystals** are held together by the delocalized valence electrons in huge molecular orbitals that extend across the entire crystal. Elemental metals often use either **cubic close packing** or **hexagonal close packing** to minimize the amount of empty space between the spherical atoms. **Covalent network crystals** contain atoms that are held in a three-dimensional network by covalent chemical bonds. In a sense, an entire covalent crystal is a single molecule. The structures of the chemical elements exemplify three of the four crystal bonding types. Knowledge of the locations of atoms in the unit cell and of the size and shape of the unit cell allows for the computation of bond distances and other interatomic distances in crystals.

20–4   Real crystals always contain **lattice imperfections** such as **point defects.** Vacant lattice sites are called **Schottky defects.** The displacement of an atom or ion from its regular lattice site to an interstitial site is a **Frenkel defect.** An **F-center** or **color center** occurs when an anion in an ionic solid is replaced at a lattice site by an electron. A **non-stoichiometric compound** violates the law of definite proportions by having a range of compositions but is readily understood in terms of lattice vacancies and multiple oxidation states.

20–5   In **liquid crystals,** the lattice structure of a crystal vanishes upon melting, but the molecules still are partially constrained by their neighbors and do not move as freely as they do in a true liquid. Liquid crystals form, especially with long, rod-like molecules. In **nematic phases,** the molecules show a preferred orientation in a particular direction, but their centers are distributed at random, as they would be in an ordinary liquid. In **smectic phases,** the molecules continue to show net orientational ordering, but now the centers of the molecules also lie in layers. In **cholesteric** liquid crystals, the molecules show a nematic type of ordering, but this orientation changes from plane to plane.

PROBLEMS

Note: Answers to blue-numbered problems are given in Appendix F. Problems that are more challenging are indicated with asterisks.

Probing the Structure of Condensed Matter

1. (See Example 20–1.) The second-order Bragg reflection of X-radiation with $\lambda = 1.660$ Å from a set of parallel planes in copper occurs at an angle $2\theta = 54.70°$. Calculate the distance between the scattering planes in the crystal.

2. (See Example 20–1.) The second-order Bragg reflection of X-radiation with $\lambda = 1.237$ Å from a set of parallel planes in aluminum occurs at an angle $2\theta = 35.58°$. Calculate the distance between the scattering planes in the crystal.

3. (See Example 20–2.) The distance between members of a set of equally spaced planes of atoms in crystalline lead is 4.950 Å. Some X-rays with $\lambda = 1.936$ Å are diffracted by this set of parallel planes. Calculate the angle 2θ at which the fourth-order Bragg reflection should be observed.

4. (See Example 20–2.) The distance between members of a set of equally spaced planes of atoms in crystalline sodium is 4.28 Å. Some X-rays with $\lambda = 1.539$ Å are diffracted by this set of parallel planes. Calculate the angle 2θ at which the second-order Bragg reflection should be observed.

5. (See Example 20–2.) The members of a series of equally spaced parallel planes of ions in crystalline LiCl are separated by 2.570 Å. Calculate all of the angles 2θ at which diffracted beams of various orders might be seen if the X-ray wavelength used is 2.167 Å.

6. (See Example 20–2.) A recent precise re-determination of the structure of vitamin B_{12} used X-rays with a wavelength of 0.6500 Å from a synchrotron source. In the crystal, the spacing between a certain set of planes containing cobalt atoms is 12.55 Å.
   (a) Calculate the smallest five angles 2θ at which reflected beams of various orders involving these planes might be seen.
   (b) Calculate n for the highest-order reflection possible from these planes.

7. Compare the X-ray diffraction pattern of a single crystal with that of a microcrystalline solid. Account for the differences.

8. Compare the X-ray diffraction pattern of an amorphous solid with that of a microcrystalline solid. Account for the differences.

Symmetry and Structure

9. (See Example 20–3.) Which of the following have 3-fold rotational symmetry? Explain.
   (a) An isosceles triangle
   (b) An equilateral triangle
   (c) A tetrahedron
   (d) A cube
   (e) A regular hexagon

10. (See Example 20–3.) Which of the following have 4-fold rotational symmetry? Explain.
    (a) A cereal box (exclusive of the writing on the sides)
    (b) A stop sign (not counting the writing)
    (c) A tetrahedron
    (d) A cube
    (e) A rectangular solid

11. (See Example 20–3.) Which of the following objects have at least one mirror line (or plane)? Explain.
    (a) The number 1881
    (b) A cereal box (exclusive of the writing on the sides)
    (c) A fishhook
    (d) A golf club

12. (See Example 20–3.) Which of the following objects have at least one mirror line (or plane)? Explain.
    (a) A baseball (counting the seam)
    (b) The letters "bood" taken as a group
    (c) The number 6009 printed in block digits
    (d) A closed textbook

13. Identify a symmetry element that must be present in the diffraction pattern of a crystal that belongs to the tetragonal crystal system.

14. Identify a symmetry element that must be present in the diffraction pattern of a crystal that belongs to the orthorhombic crystal system.

15. (See Example 20–5.) An orthorhombic crystal of the sugar D-fructose has a unit cell with edges a, b, and c equal to 8.06 Å, 9.12 Å, and 10.06 Å, respectively. Compute the volume of the unit cell in cubic ångströms ($Å^3$) and in cubic meters (m^3).

16. (See Example 20–5.) Compute the volume (in $Å^3$) of the unit cell of potassium hexacyanoferrate(III) $K_3[Fe(CN)_6]$, a substance that crystallizes in the monoclinic system with $a = 8.40$ Å, $b = 10.44$ Å, $c = 7.04$ Å, and $\beta = 107.5°$.

17. (See Example 20–6.) The unit cell of diamond contains eight carbon atoms and has a volume of 45.385 $Å^3$.
    (a) Compute the volume (in $Å^3$) per carbon atom in diamond.
    (b) Calculate the density of diamond (in g cm^{-3}). (*Hint*: Recall that 6.0221×10^{23} u = 1.000 g and that 1×10^8 Å = 1 cm.)

18. (See Example 20–6.) The unit cell of crystalline gold contains four gold atoms. The cell is cubical and has an edge that is 4.0786 Å long.
    (a) Compute the volume (in $Å^3$) of the unit cell of gold.
    (b) An atom of gold weighs 196.967 u. Compute the density of gold in units of u $Å^{-3}$.
    (c) Express the density of gold in units of g cm^{-3}. (*Hint*: Recall that 6.0221×10^{23} u = 1.000 g and that 1×10^8 Å = 1 cm.)

19. At 25°C, the unit cell of elemental silicon is cubic, with an edge of 5.431 Å, and contains eight Si atoms. The density of elemental silicon at this temperature is 2.328 g cm^{-3}.
    (a) Calculate the volume of one unit cell of silicon (in cm^3).
    (b) Calculate the mass (in grams) of silicon present in a unit cell.
    (c) Calculate the mass (in grams) of one atom of silicon.
    (d) The mass of an atom of silicon is 28.0855 u. Estimate Avogadro's number to four significant digits. (*Note*: The mass of any amount of a substance expressed in atomic mass units divided by its mass expressed in grams equals Avogadro's number.)

20. One form of crystalline iron has a body-centered cubic unit cell with an iron atom located at every lattice point. Its density at 25°C is 7.86 g cm^{-3}. The length of the edge of the cubic unit cell is 2.866 Å. Use these facts to estimate Avogadro's number. (*Hint*: Follow the procedure suggested in the preceding problem.)

21. The common and important mineral quartz is crystalline SiO_2. There are in fact two distinct forms of quartz: α-quartz (a trigonal crystal) and β-quartz (a hexagonal crystal). In each form, the unit cell contains three SiO_2 units, but the unit cell of α-quartz has a volume of 113.01 $Å^3$ and the unit cell of β-quartz is slightly larger, with a volume of 118.15 $Å^3$. The density of α-quartz is 2.648 g cm^{-3}. Compute the density of β-quartz.

22. Compute the density of nickel(II) oxide. This substance crystallizes in the cubic system. The unit cell has $a = 4.177$ Å, and there are four NiO units in each cell.

23. The compound $Pb_4In_3B_{17}S_{18}$ crystallizes in the monoclinic system with a unit cell having $a = 21.021$ Å, $b = 4.014$ Å, and $c = 18.898$ Å and the only non-90° angle equal to 97.07°. Every unit cell has two molecules. Compute the density of this substance.

24. Strontium chloride hexahydrate ($SrCl_2 \cdot 6H_2O$) crystallizes in the trigonal system in a unit cell with $a = 8.9649$ Å and $\alpha = 100.576°$. The unit cell contains three formula units. Compute the density of this substance.

25. (See Example 20–7.) Sodium sulfate (Na_2SO_4) crystallizes in the orthorhombic system in a unit cell with $a = 5.863$ Å, $b = 12.304$ Å, and $c = 9.821$ Å. The density of these crystals is 2.663 g cm^{-3}. Determine how many Na_2SO_4 formula units are present in the unit cell.

26. (See Example 20–7.) The density of turquoise $(CuAl_6(PO_4)_4(OH)_8(H_2O)_4)$ is 2.927 g cm^{-3}. This gemstone crystallizes in the triclinic system with cell constants $a = 7.424$ Å, $b = 7.629$ Å, and $c = 9.910$ Å, $\alpha = 68.61°$, $\beta = 69.71°$, and $\gamma = 65.08°$. Verify that the volume of the unit cell is 461.40 Å^3, and determine how many copper atoms are present in each unit cell of turquoise.

27. A certain crystal of sodium chloride is a perfect little cube measuring 1.00×10^{-3} m on an edge. The microscopic unit cell of crystalline sodium chloride is a cube that has an edge 5.6402 Å $= 5.6402 \times 10^{-10}$ m in length.
  (a) Compute the volume of a unit cell of sodium chloride.
  (b) Compute the number of unit cells in this particular crystal.
  (c) Compute the number of unit cells that are on the surface of this crystal and the percentage of all the unit cells that are surface unit cells.

28. A crucial protein at the photosynthetic reaction center of the purple bacterium *Rhodopseudomonas viridis* has been separated from the organism, crystallized, and studied by X-ray diffraction. This substance crystallizes with a primitive unit cell in the tetragonal system. The cell dimensions are $a = b = 223.5$ Å and $c = 113.6$ Å.
  (a) How many lattice points are in this unit cell?
  (b) Determine the volume (in Å^3) of this cell.
  (c) One of the crystals in this experiment was a rectangular prism (shaped like a box), with dimensions $1 \times 1 \times 3$ mm. Compute how many unit cells were in this crystal.

Bonding in Crystals

29. Classify each of the following crystalline solids as molecular, ionic, metallic, or covalent network:
  (a) $BaCl_2$
  (b) SiC
  (c) CO
  (d) Co

30. Classify each of the following crystalline solids as molecular, ionic, metallic, or covalent network:
  (a) Rb
  (b) C_5H_{12}
  (c) B
  (d) Na_2HPO_4

31. The melting point of cobalt is 1495°C, and the melting point of barium chloride is 963°C. Rank the four substances in Problem 29 from lowest to highest in melting point.

32. The boiling point of pentane (C_5H_{12}) is slightly less than the melting point of rubidium. Rank the four substances in Problem 30 from lowest to highest in melting point.

33. Rank the following elements from lowest to highest in electrical conductivity under room conditions: antimony, copper, and oxygen.

34. Rank the following elements from lowest to highest in thermal conductivity under room conditions: chromium, nitrogen, and germanium.

35. By examining Figure 20–21, determine the number of nearest neighbors and second-nearest neighbors of a Cs^+ ion in crystalline CsCl. The nearest neighbors of the Cs^+ ion are Cl^- ions, and the second-nearest neighbors are Cs^+ ions.

36. Repeat the determination of the preceding problem for the NaCl crystal, referring to Figure 20–20.

37. Draw a sketch of a body-centered cubic lattice and use it to determine the number of nearest neighbor and second-nearest neighbor atoms to a given sodium atom in crystalline sodium. Crystalline sodium has an atom at each point in this lattice.

38. Draw a sketch of a face-centered cubic lattice and use it to determine the number of nearest neighbor and second-nearest neighbor atoms to a given aluminum atom in the structure of crystalline aluminum, which has an atom at each point in this lattice.

39. White tin crystallizes in a body-centered *tetragonal* unit cell with edges $a = b = 5.8315$ Å and $c = 3.1813$ Å. Tin atoms sit at the lattice points (and at other locations in the unit cell as well). Compute the distance between an Sn atom at a cell corner and an Sn atom at the body center.

40. Thallium bromide crystallizes in the cesium chloride structure. The edge of the unit cell is 3.97 Å. Compute the shortest Tl-to-Br distance in this structure.

Defects in Solids and Liquid Crystals

41. Does the presence of Frenkel defects change the density of a crystal? Explain.

42. What effect does the (unavoidable) presence of Schottky defects have on the determination of Avogadro's number using the method described in Problems 19 and 20?

43. Iron(II) oxide is non-stoichiometric. A particular sample was found to contain 76.55% iron and 23.45% oxygen by mass.
  (a) Calculate the empirical formula of the compound (four significant digits).
  (b) What fraction of the iron in this sample is in the +3 oxidation state?

44. A sample of nickel oxide contains 78.23% Ni by mass.
  (a) What is the empirical formula of the nickel oxide to four significant digits?
  (b) What fraction of the nickel in this sample is in the +3 oxidation state?

45. Compare the nature and extent of order in liquid hydrogen chloride (HCl) with that in the plastic crystalline phase of HCl at the same temperature. Which has the higher entropy? Which has the higher enthalpy?

46. Compare the nature and extent of order in the smectic liquid crystal and isotropic liquid phases of a substance. Which has the higher entropy? Which has the higher enthalpy?

ADDITIONAL PROBLEMS

47. Define the terms *scattering* and *interference* with respect to the effects of a sample on an X-ray beam passing through it.

48. X-rays of wavelength $\lambda = 1.54$ Å are scattered by a crystal at an angle 2θ of 32.15°. Calculate the wavelength of the X-rays in another experiment if this same diffracted beam from the same crystal is observed at an angle of 34.46°.

49. Some water waves with a wavelength of 3.0 m are diffracted by an array of evenly spaced posts in the water. If the rows of posts are separated by a distance of 5.0 m, calculate the angle 2θ at which the first-order "Bragg diffraction" of these water waves is seen.

50. If the wavelength λ of the X-rays is too large relative to the spacing of planes in the crystal, no Bragg diffraction is seen because $\sin \theta$ is larger than 1 in the Bragg equation, even for $n = 1$. Calculate the longest wavelength of X-rays that can give Bragg diffraction from a set of planes separated by 4.20 Å.

51. Find all of the symmetry operations in the following:
    (a) the number 10801
    (b) the number 86198
    (c) a teacup

52. (a) You receive a valuable golden cube as a gift. To protect this treasure, you construct a tight-fitting wooden case with a plush lining. In how many different ways can you put your cube into its case?
    (b) Suppose that, in the preceding situation, the valuable bauble had a tetrahedral shape. In how many different ways could you put it into a tight-fitting case?
    (c) In how many different ways could you place an object with the diagrammed shape into a tight-fitting case?

53. Use Figure 20–25 to determine the number of carbon atoms per unit cell of diamond.

54. Use Figure 20–30 to determine the number of carbon atoms per unit cell of graphite.

55. Consider a collection of hard spherical atoms arranged at the lattice points of a primitive cubic lattice. A unit cell is con-structed so that one eighth of the corner atoms lie within the unit cell. Compute the fraction of the volume of the unit cell that is empty if the atoms are in contact along the edge of the cell.

56. Show that hard spherical atoms arranged at the lattice points of a body-centered cubic lattice and in contact along the body diagonal of the conventional unit cell occupy only 68% of the total volume available in the structure.

57. The number of beams diffracted by a single crystal depends on the wavelength λ of the X-rays used and on the volume associated with one lattice point in the crystal—that is, on the volume V_p of a primitive unit cell. An approximate formula is

$$\text{number of diffracted beams} = \frac{4}{3\pi}\left(\frac{2}{\lambda}\right)^3 V_p$$

   (a) Compute the volume of the conventional unit cell of crystalline sodium chloride. This cell is cubical and has an edge of 5.6402 Å.
   (b) The NaCl unit cell contains four lattice points. Compute the volume of a primitive unit cell for NaCl.
   (c) Use the formula given in this problem to estimate the number of diffracted rays if NaCl is irradiated with X-rays having a wavelength of 2.2896 Å.
   (d) Use the formula to estimate the number of diffracted rays that are observed if NaCl is irradiated with X-rays having the shorter wavelength of 0.7093 Å.

58. A compound contains three elements: sodium, oxygen, and chlorine. It crystallizes in a primitive cubic lattice. The oxygen atoms are at the corners of the unit cells, the chlorine atoms are at the centers of the unit cells, and the sodium atoms are at the centers of the faces of the unit cells. What is the formula of the compound?

59. At room temperature, monoclinic sulfur has the unit-cell dimensions $a = 11.04$ Å, $b = 10.98$ Å, $c = 10.92$ Å, and $\beta = 96.73°$. Each cell contains 48 atoms of sulfur. Compute the density of monoclinic sulfur in units of g cm^{-3}.

*60. White tin crystallizes in a body-centered tetragonal unit cell with edges $a = b = 5.8315$ Å and $c = 3.1813$ Å. Tin atoms occupy the following sites in the unit cell:

	x	y	z
Sn1	$0a$	$0b$	$0c$
Sn2	$\frac{1}{2}a$	$\frac{1}{2}b$	$\frac{1}{2}c$
Sn3	$\frac{1}{2}a$	$0b$	$\frac{1}{4}c$
Sn4	$0a$	$\frac{1}{2}b$	$\frac{3}{4}c$

Compute the distance between Sn2 and Sn1, the distance between Sn2 and Sn3, and the distance between Sn2 and Sn4. (*Hint*: Sn1 and Sn2 are the atoms referred to in Problem 39. Locate them and the other atoms on a sketch of the unit cell, and use the Pythagorean theorem.)

61. The following graphic appeared on a cardboard box to indicate that the cardboard was recyclable. The graphic does *not* have 3-fold symmetry. What symmetry element or elements does it possess? Redraw the graphic so that it has 3-fold symmetry.

62. Polonium is the only metallic element that crystallizes in the simple cubic lattice at room temperature.
(a) What is the distance between nearest-neighbor polonium atoms if the first-order reflection of X-radiation with $\lambda = 1.785$ Å from the parallel faces of its unit cells appears at an angle of $2\theta = 30.96°$ from these planes?
(b) Compute the density of polonium (in g cm^{-3}).

63. Within a recent six-month period, two completely independent reports on the structure of mercury(I) nitrite appeared in the scientific literature. The results were in reasonable agreement, with the following sets of unit-cell constants:

	Report 1	Report 2
a	4.4145 Å	4.435 Å
b	10.3334 Å	10.344 Å
c	6.2775 Å	6.301 Å
α	Exactly 90°	Exactly 90°
β	108.84°	108.74°
γ	Exactly 90°	Exactly 90°

The two reports agree that the unit cell contains four mercury ions and four nitrite ions.
(a) What is the crystal system of mercury(I) nitrite?
(b) Compute the density of mercury(I) nitrite according to both sets of data.
(c) Explain why there must be equal numbers of Hg atoms and NO_2^- ions in the unit cell.
(d) Comment on the use of significant digits in the two reports.

64. Name two elements that form molecular crystals, two that form metallic crystals, and two that form covalent network crystals. What generalizations can you make about the portions of the periodic table where each type is found?

65. Beryllium chloride is a covalent compound with a rather high melting point (405°C). Draw the Lewis structure of BeCl$_2$, and use it to suggest why the intermolecular attractions in solid BeCl$_2$, a molecular solid, are unusually strong.

66. Solid carbon dioxide (dry ice) and solid iodine both sublime at ordinary temperatures. What do these solids have in common structurally? Why do many other solids of the same type not sublime under the same conditions?

67. Sodium hydride (NaH) crystallizes in the rock salt structure, with four formula units of NaH per cubic unit cell. A beam of monoenergetic neutrons, selected to have a velocity of 2.639×10^3 m s^{-1}, is scattered in second order through an angle of $2\theta = 36.26°$ by the parallel faces of the unit cells of a sodium hydride crystal.
(a) Calculate the wavelength of the neutrons.
(b) Calculate the edge length of the cubic unit cell.
(c) Calculate the distance from the center of an Na$^+$ ion to the center of a neighboring H$^-$ ion.
(d) If the radius of an Na$^+$ ion is 0.98 Å, what is the radius of an H$^-$ ion, assuming that the two ions are in contact?

68. A crystal of sodium chloride has a density of 2.165 g cm^{-3} in the absence of defects. Suppose that a crystal of NaCl in which 0.15% of the sodium ions and 0.15% of the chloride ions are missing is grown. What is the density in this case?

69. A compound of titanium and oxygen contains 28.31% oxygen by mass.
(a) If its empirical formula is Ti$_x$O, calculate x to four significant digits.
(b) The non-stoichiometric composition Ti$_{0.75}$O has been determined by X-ray studies to have a Ti^{2+}–O^{2-} lattice in which 23.5% (not 25%) of the Ti sites are vacant. Calculate the fraction of O^{2-} sites that are also vacant in this material.

70. Classify the bonding in the following amorphous solids as molecular, ionic, metallic, or covalent network:
(a) amorphous silicon, used in photocells to collect light energy from the sun.
(b) polyvinyl chloride, a plastic of long-chain molecules composed of CH$_2$CHCl repeating units, and used in pipes and siding.
(c) soda-lime silica glass, used in windows.
(d) copper-zirconium glass, an alloy of the two elements with approximate formula Cu$_3$Zr$_2$, used for its high strength and good conductivity.

71. The chemical As$_2$S$_3$ is common in nature as the crystalline mineral orpiment. Melting a crystal of orpiment and recooling it always gives amorphous As$_2$S$_3$. Speculate on why crystals of orpiment cannot be made in the laboratory but are relatively common in nature.

72. Is methane (CH$_4$) more likely to form a plastic crystalline phase or a liquid crystalline phase? Should it be relatively easy or difficult to form an amorphous solid from methane?

CUMULATIVE PROBLEM

Phosphorus

Solid elemental phosphorus appears in a variety of forms, with crystals in all seven crystal systems reported under various conditions of temperature, pressure, and sample preparation.

(a) The thermodynamically most stable form of phosphorus under room conditions is black phosphorus. The unit cell of black phosphorus is orthorhombic, with edges of length 3.314, 4.376, and 10.48 Å. Calculate the volume of one unit cell and determine the number of phosphorus atoms per unit cell if the density of black phosphorus is 2.69 g cm^{-3}.

(b) The form of solid phosphorus that is easiest to prepare from the liquid or gaseous state is white phosphorus, which consists of P_4 molecules in a cubic lattice. When X-rays of wavelength 2.29 Å are scattered from the parallel faces of its unit cells, the second-order Bragg diffraction is observed at an angle 2θ of 14.23°. Calculate the length of the edge of the unit cell of white phosphorus. At what angle is fourth-order Bragg diffraction seen?

(c) When liquid white phosphorus is heated to 300°C in the absence of air, a red solid forms that shows diffuse diffraction rings. This material is used in the strip on a matchbox against which a safety match is struck. What is the nature of this red solid?

(d) Red phosphorus (part c) has been reported to convert into monoclinic, triclinic, tetragonal, and cubic red forms with different heat treatments. Identify the changes in the shape of the unit cell as a cubic lattice is converted first to tetragonal, then monoclinic, then triclinic.

(e) A monoclinic form of red phosphorus has been studied. The unit cell of this form has cell edges of 9.21, 9.15, and 22.60 Å and an angle β of 106.1° and contains 84 atoms of phosphorus. Estimate the density of this form of phosphorus.

(f) Phosphorus forms many compounds with other elements. Describe the nature of the bonding in the following solids: white elemental phosphorus (P_4), black elemental phosphorus, sodium phosphate (Na_3PO_4), and phosphorus trichloride (PCl_3).

Two forms of elemental phosphorus: white and red. *(Charles D. Winters)*

21

Silicon and Solid-State Materials

CHAPTER OUTLINE

Powders of silicon nitride, Si_3N_4, formed in the flame produced by a carbon dioxide laser. *(The MIT Report)*

The remainder of this book takes up some important applications of the principles treated in the first 20 chapters, beginning with aspects of materials science. This practical discipline focuses on the structures and properties of materials in useful applications. Numerous modern materials are designed at the molecular level to possess desired properties, to act for example as superconductors or semiconductors, or to retain strength, flexibility, and toughness despite heat or cold. Two Group IV elements are prominent in such engineered materials: silicon, which is essential in many ceramics (discussed in this chapter), and carbon, the polymer-builder (discussed in Chapter 25).

- The third important class of materials, besides ceramics and polymers, is the metals. Their preparation and uses are described in Chapter 12.

21-1 Semiconductors

The structure of solid silicon is the same as that of diamond and can be understood in the same way. The valence orbitals on each Si atom become sp^3 hybridized and undergo σ overlap with sp^3 hybrid orbitals on neighboring Si atoms. Each atom is thus covalently linked to four others that surround it at the corners of a regular tetrahedron. A unit cell of the resulting "giant molecule" network structure is shown in Figure 20–25. There are just enough valence electrons in this network to provide a single bond between each silicon atom and its four neighbors.

An alternative to this localized-orbital explanation of the bonding in silicon is to consider that the valence electrons occupy molecular orbitals (of the type discussed in Chapter 18) that are spread out, or delocalized, over the entire silicon crystal. The particular advantage of this approach is the insight that it gives into the way a silicon crystal absorbs energy. Recall how the $1s$ orbitals from two hydrogen atoms mix to form a lower-energy bonding molecular orbital and a higher-energy antibonding molecular orbital in the H_2 molecule (see Fig. 18–2). In a silicon crystal, the $4N_0$ valence atomic orbitals from 1 mol (N_0 atoms) of silicon all mix and split into two ranges or **bands** of orbital energies, each containing $2N_0$ very closely spaced energy levels (Fig. 21–1). The lower band is called the **valence band,** and the upper one, the **conduction band.** Electrons with energies in the valence band are essentially localized. Electrons with higher energies, in the conduction band, can migrate through the crystal and so conduct an electric current. Between the top of the valence band and the bottom of the conduction band is an energy region that is forbidden to electrons. The difference in energy between the top of the valence band and the bottom of the conduction band is called the **band-gap energy** (E_g). For pure silicon, the band-gap energy is 1.94×10^{-19} J. This is the amount of energy that an electron must gain to exit the valence band and enter the conduction band.

- Each Si atom contributes one $3s$ and three $3p$ valence orbitals, for a total of four orbitals. Recall that the number of molecular orbitals after mixing is equal to the number of atomic orbitals used.

The N_0 valence electrons in a mole of silicon are just enough to fill the valence band completely. These electrons can accept energy only if the amount offered exceeds what is required to jump the gap between the valence band and the conduction band (Fig. 21–1). But the band-gap energy in silicon is equivalent to 117 kJ mol^{-1}, a large amount of energy. If it were to be supplied by a thermal source, the temperature of the source would have to be on the order of

$$ T = \frac{E_g}{R} = \frac{117,000 \text{ J mol}^{-1}}{8.315 \text{ J mol}^{-1} \text{ K}^{-1}} = 14,000 \text{ K} $$

- Another unit used for band-gap energies is the electron volt, defined in Section 15–1 as 1.60218×10^{-19} J. The band-gap energy in pure Si is 1.21 eV.

At room temperature, only a few electrons per mole in the extreme tail of the Boltzmann distribution have enough energy to jump the gap, so the conduction band in pure silicon is very sparsely populated with electrons. The result is that

Figure 21–1 • The valence orbitals of the silicon atom combine in crystalline silicon to give two bands of very closely spaced levels that are separated energetically by a band gap. The higher energy conduction band is almost empty because the band gap is wide; nearly all of the valence electrons are in the valence band.

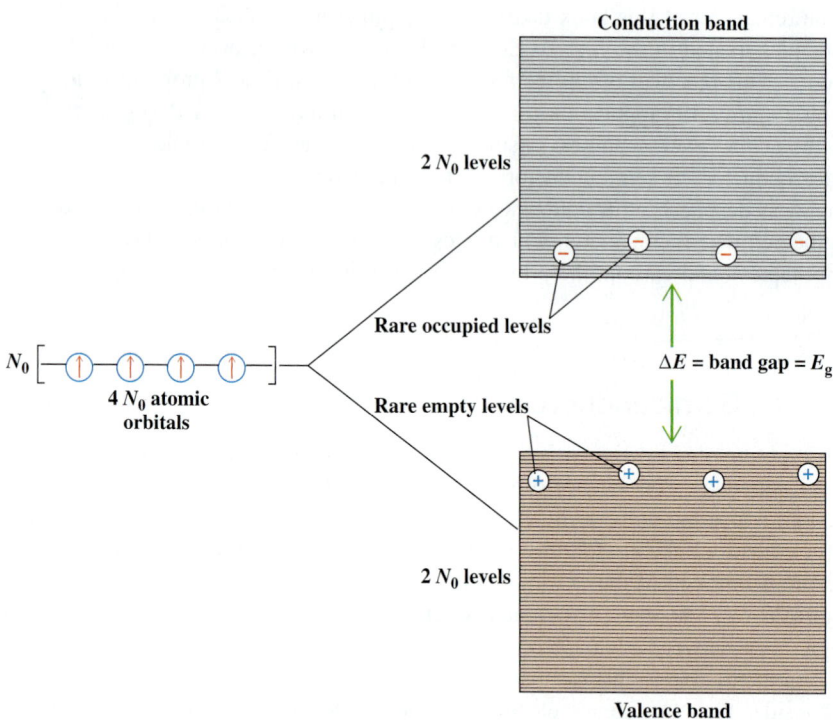

pure silicon is a poor conductor of electricity. Good electrical conductivity requires a net motion of many electrons under the impetus of a small electrical potential difference. The conductivity of silicon is roughly 10^{11} times smaller than that of a typical metal (copper) at room temperature. Silicon is called a **semiconductor** because its electrical conductivity, although less than that of a metal, is far larger than that of an **insulator** like diamond, which has an even larger band gap. Increasing the temperature increases the conductivity of a semiconductor because more electrons are excited into the conduction band. Another way to increase the conductivity of a semiconductor is to irradiate it with a beam of photons that are energetic enough (that is, of frequency high enough) to excite electrons from the valence band to the conduction band. This process resembles the photoelectric effect described in Section 16–2, with the difference that the electrons are not removed from the material, but only excited into the conduction band, ready to convey a current if a potential difference is imposed.

• Note that the conductivities of metals typically decrease slightly with increasing temperature.

EXAMPLE 21–1

Calculate the longest wavelength of light that can excite electrons from the valence band to the conduction band in silicon. In what region of the spectrum does this wavelength fall?

Strategy

Realize that if a photon just barely excites an electron across the band gap, its energy equals E_g, the band-gap energy. Then use the fact that the energy carried by a photon equals hc_0/λ, where h is Planck's constant; c_0, the speed of light in a vacuum; and λ, the wavelength of the photon.

Solution

Set hc_0/λ equal to E_g and solve for λ. Insert the values of the constants and the given value of E_g:

$$\frac{hc_0}{\lambda} = E_g$$

$$\lambda = \frac{hc_0}{E_g} = \frac{(6.626 \times 10^{-34} \text{ J s}) \times \left(2.998 \times 10^8 \frac{\text{m}}{\text{s}}\right)}{1.94 \times 10^{-19} \text{ J}}$$

$$= 1.02 \times 10^{-6} \text{ m} = 1020 \text{ nm}$$

This wavelength falls in the infrared region of the spectrum. Photons with shorter wavelengths (for example, visible light) carry more than enough energy to excite electrons to the conduction band in silicon.

EXERCISE

Calculate the longest wavelength of light that can excite electrons in diamond, in which the band-gap energy is 8.7×10^{-19} J.

Answer: 230 nm, in the ultraviolet (UV) region of the spectrum. Visible light does not carry enough energy to excite electrons from the valence band in diamond to the conduction band.

Doped Semiconductors

When certain other elements are added to pure silicon in a process called **doping,** the silicon acquires interesting electronic properties. If atoms of a Group V element such as arsenic or antimony are diffused into silicon, they substitute for silicon atoms in the network. Such atoms have five valence electrons so that each substitution introduces one more electron into the silicon crystal than is needed for bonding. The extra electrons occupy energy levels just below the lowest level of the conduction band and require little energy to jump from this **donor-impurity** level into the conduction band, so the electrical conductivity of the silicon crystal is increased. The change is roughly proportional to the amount of dopant that is added. Silicon doped with atoms of a Group V element exhibits increased conductivity and is called an **n-type semiconductor** to indicate that the charge carrier is negative.

On the other hand, if a Group III element such as gallium is used as a dopant, the valence band contains one fewer electron per dopant atom. One electron is missing in the network of chemical bonds for each Group III atom added. The doping creates **holes** (unoccupied energy levels) in the valence band, each with an effective charge of $+1$. If a voltage difference is impressed across a sample of silicon that is doped in this way, it causes the positively charged holes to move toward the negative source of potential. Equivalently, a valence-band electron next to the (positive) hole moves in the *opposite* direction (that is, toward the positive source of potential). Whether one thinks of holes or electrons as the mobile charge carrier, the result is the same; the difference is in the words, not the physical phenomenon. Silicon that has been doped with a Group III element is called a **p-type semiconductor** to indicate that the carrier has an effective positive charge.

Figure 21-2 • An integrated circuit consists of a series of connected electrical circuits printed onto a tiny silicon wafer. The Pentium chip shown here contains 3.3 million transistors and is about the size of a fingernail. *(Courtesy of Intel Corporation)*

• "Isoelectronic" means that it has the same number of electrons.

The manufacture of integrated circuits requires extraordinary care to avoid contamination. Here, metal connections are being deposited on a silicon wafer. The garments protect the work piece from the worker, not the reverse. *(Courtesy of Allied Signal Aerospace Company)*

A different class of semiconductors is based not on silicon but on equimolar compounds of Group III with Group V elements. Gallium arsenide, for example, is isoelectronic to the Group IV semiconductor germanium. Doping GaAs with the Group VI element tellurium produces an *n*-type semiconductor; doping with zinc, which has one valence electron *fewer* than gallium, gives a *p*-type semiconductor. Other III–V combinations have different band-gap energies and are useful in particular applications. Indium antimonide (InSb), for example, has a small enough band-gap energy that absorption of infrared radiation causes electrons to be excited from the valence to the conduction band; an electric current flows when a small potential difference is applied, and InSb is used as a detector of infrared radiation. Still other compounds formed between elements of the zinc group (zinc, cadmium, and mercury) and Group VI elements such as sulfur also have the same average number of valence electrons per atom as does silicon and make useful semiconductors.

Semiconductors perform a wide range of electronic functions, such as amplifying current and converting alternating current into direct current. These functions formerly required the use of vacuum tubes, which occupy much more space, generate larger amounts of heat, and require considerably more energy to operate than do **transistors,** their semiconductor counterparts. More importantly, semiconductors can be built into integrated circuits (Fig. 21–2) and made to store information and to process it at great speeds.

Solar cells based on doped silicon or gallium arsenide provide a way to convert radiant energy from the Sun directly into electrical work by a technology that is virtually non-polluting. The high capital costs of solar cells make them uncompetitive with conventional fossil-fuel sources of energy except in certain specialized applications (such as calculators, or remote sensing and safety devices). As reserves of fossil fuels dwindle, solar energy will become an important option.

Pigments and Phosphors

The band-gap energy of an insulator or semiconductor has a significant effect on its color. Pure diamond has a large band gap (see Exercise 21–1), so even blue light does not have enough energy to excite electrons from the valence band to the conduction band; visible light passes through diamond without being absorbed, so diamond is colorless. Cadmium sulfide, with a band-gap energy of 4.2×10^{-19} J, corresponding to a wavelength of 470 nm, absorbs violet and blue light but strongly transmits yellow, giving it a deep yellow color (Fig. 21–3). Cadmium sulfide is in fact the pigment called cadmium yellow. Cinnabar (HgS) has a smaller band-gap energy of 3.2×10^{-19} J and absorbs all light except red (Fig. 21–4). It has a deep red color and is the pigment vermilion. Semiconductors with band-gap energies of less than 2.8×10^{-19} J absorb all wavelengths of visible light and, when powdered to eliminate reflections, appear black. These include silicon (see Example 21–1), germanium, and gallium arsenide.

Doping silicon creates donor-impurity levels close enough to the conduction band or acceptor levels close enough to the valence band that thermal excitation can cause electrons to move into a conducting level. The corresponding doping of insulators or wide–band-gap semiconductors can bring donor or acceptor levels into positions where visible light can be absorbed or emitted. The doping thereby changes the colors and optical properties of the materials. Nitrogen doped into diamond creates a donor-impurity level in the band gap. Transitions to this level can absorb some blue light, giving the diamond a yellowish (undesirable) color. On the other hand, boron doped into diamond gives an acceptor level that absorbs red light most strongly and gives a rare and highly prized blue diamond.

Phosphors are wide–band-gap materials containing dopants that create new levels in the band gap. These levels are at energies tuned to suit the emission of a particular color of light. When electrons are excited to these levels, the phosphor emits light of the particular color as it returns to its ground state. A fluorescent lamp consists of a tube coated on its inside with different phosphors and containing a gas at low pressure together with a small amount of mercury. When a sufficiently large

• Fluorescent lamps are thus high-voltage, gas-discharge tubes (see Section 1–5).

Figure 21–3 • Mixed crystals of two semiconductors with different band-gap energies, CdS (*yellow*) and CdSe (*black*), show a range of colors, illustrating a decrease in the band-gap energy as the composition of the mixture becomes richer in Se. (*Kurt Nassau/Phototake*)

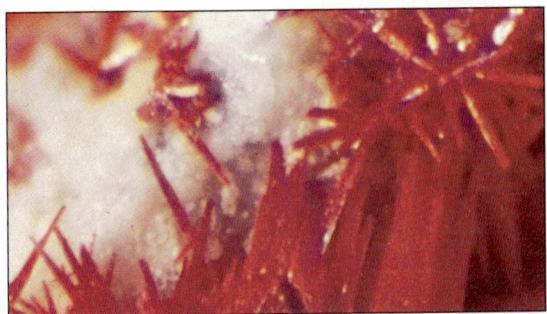

Figure 21–4 • Crystalline cinnabar, HgS. (*Julius Weber*)

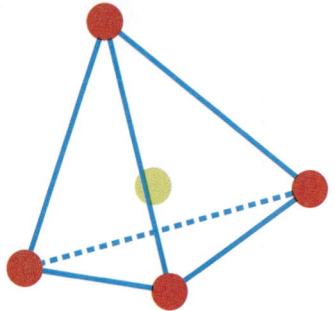

Figure 21-5 • The tetrahedral orthosilicate anion SiO_4^{2-}. It is often represented by drawing the tetrahedral outline defined by the oxygen atoms at the corners.

• Orthosilicates derive from orthosilicic acid H_4SiO_4 (equivalent to $SiO_2 \cdot 2H_2O$), and not from metasilicic acid H_2SiO_3 ($SiO_2 \cdot H_2O$). In the names of inorganic acids, the prefix *ortho-* designates the most highly hydrated acid. A *meta-* acid is less highly hydrated.

voltage is applied across the electrodes at the ends of the tube, a current flows. Mercury ions in the tube emit violet and UV light. The phosphors absorb these high-energy photons and emit new photons at lower energies and longer wavelengths, giving a nearly white light that is far more desirable than is the bluish light that comes from a mercury-vapor lamp without the phosphors.

Phosphors also are used in television screens. The picture is formed by scanning a beam of electrons (from an electron gun) over the inside of the screen. The electrons strike the phosphors coating the screen, exciting their electrons and causing them to emit light. In a black-and-white television tube, the phosphors are a mixture of silver doped into ZnS, which gives blue light, and silver doped into $Zn_xCd_{1-x}S$, which gives yellow light. The combination of the two provides a reasonable approximation of white. A color television tube or computer monitor uses three different electron guns, with three corresponding types of phosphor on the screen. Silver doped in ZnS gives blue light; manganese doped in Zn_2SiO_4 gives green light; and europium doped in YVO_4 gives red light. Masks are used to ensure that each electron beam encounters only the phosphors corresponding to the desired color. White on such a tube consists of a combination of small red, green, and blue lights, as can be verified by examining a white area on such a display with a magnifying glass.

21-2   Silicate Minerals

Silicon and oxygen make up most of the Earth's crust, with oxygen accounting for 47% and silicon 28% of its mass. The silicon–oxygen bond is partially ionic and strong. It forms the basis for a class of minerals called **silicates,** which make up the bulk of the rocks, clays, sands, and soils in the Earth's crust. From time immemorial, silicates have provided the ingredients for building materials such as bricks, cement, concrete, ceramics, and glass.

The structure-building properties of silicates originate in the tetrahedral orthosilicate anion, SiO_4^{4-}, in which both elements have their most common oxidation numbers: $+4$ for silicon and -2 for oxygen. The negative charge of the silicate ion is balanced by the compensating charge of one or more cations. The simplest silicates consist of individual SiO_4^{4-} anions (Fig. 21–5), with cations arranged around them on a regular crystalline lattice. Such silicates are properly called **orthosilicates.** Examples are forsterite (Mg_2SiO_4) and fayalite (Fe_2SiO_4), which are at the two extremes in a class of minerals called **olivines** ([Mg,Fe]$_2SiO_4$). The range of proportions of magnesium and iron in the olivines is continuous.

Other kinds of silicate structures form when two or more SiO_4^{4-} tetrahedra link together and share oxygen corners. The simplest are the **disilicates,** in which two SiO_4^{4-} tetrahedra are linked (Fig. 21–6), as in thortveitite ($Sc_2(Si_2O_7)$). Additional linkages of tetrahedra create the ring, chain, sheet, and network structures shown in Figure 21–7 and listed in Table 21–1.

Figure 21-6 • The disilicate anion $Si_2O_7^{6-}$. It consists of two silicate tetrahedra linked at a corner. In representations of silicate structures, linked tetrahedra often are projected onto a plane, as shown.

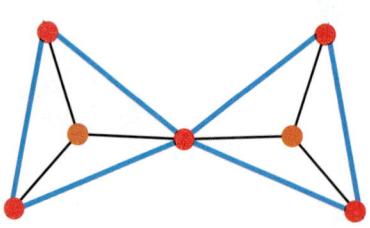

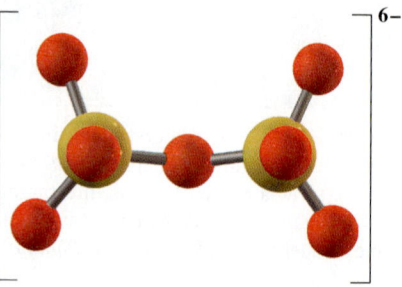

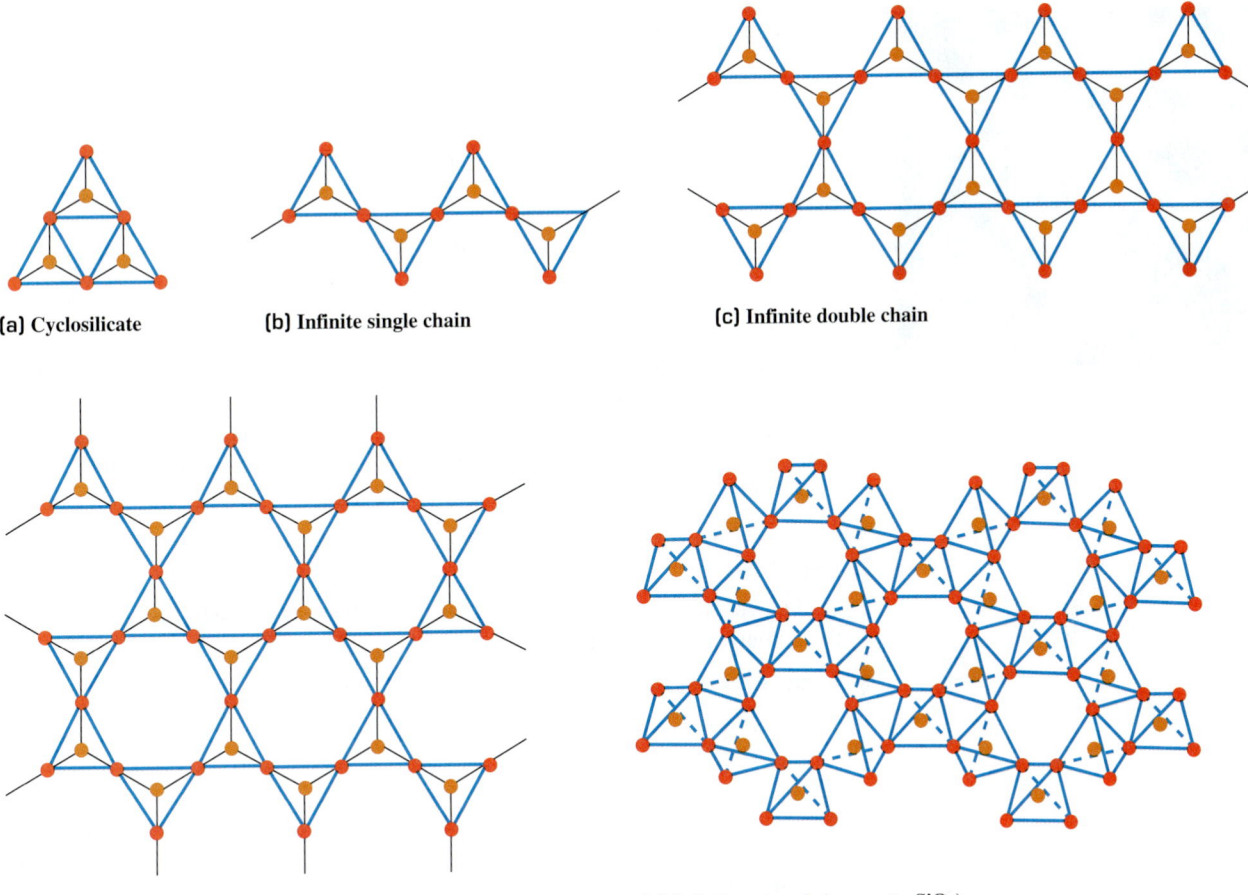

(a) Cyclosilicate

(b) Infinite single chain

(c) Infinite double chain

(d) Infinite sheet

(e) Infinite network (α-quartz, SiO₂)

Figure 21-7 • Five classes of silicate structures. All of these structures are three-dimensional. In (a) through (d), the SiO₄ tetrahedra are projected onto the plane, with the fourth oxygen positioned above the silicon, as shown in Figure 21-6. In (e), the tetrahedra are tilted at various angles and linked in 3-fold and 6-fold spirals along the line of viewing. In all, the fundamental structural unit is the same, but the overall O:Si ratio is less than 4 because oxygen atoms are shared between silicon atoms.

TABLE 21-1	Silicate Structures				
Structure	**Figure**	**Corners Shared at Each Si**	**Repeat Unit**	**O:Si Ratio**	**Example**
Tetrahedra	21-5	0	SiO_4^{4-}	4:1	Olivines
Pairs of tetrahedra	21-6	1	$Si_2O_7^{6-}$	$3\frac{1}{2}$:1	Thortveitite
Closed rings	21-7a	2	SiO_3^{2-}	3:1	Beryl
Infinite single chains	21-7b	2	SiO_3^{2-}	3:1	Pyroxenes
Infinite double chains	21-7c	$2\frac{1}{2}$	$Si_4O_{11}^{6-}$	$2\frac{3}{4}$:1	Amphiboles
Infinite sheets	21-7d	3	$Si_2O_5^{2-}$	$2\frac{1}{2}$:1	Talc
Infinite three-dimensional network	21-7e	4	SiO_2	2:1	Quartz

A crystal of talc. This mineral is peeled easily into broad, flat, lustrous plates that are translucent or transparent and have a greasy feel. Talc is white when pure; here, impurities give it an apple-green cast. *(Charles D. Winters)*

EXAMPLE 21-2

By referring to Table 21–1, predict the structural class in which the mineral Egyptian blue ($CaCu(Si_2O_5)_2$) belongs. Give the oxidation state of each of its atoms.

Strategy

Use the ratio of oxygen atoms to silicon atoms to locate the correct entry in the table.

Solution

Because the O:Si atom ratio is 5:2, this mineral should have an infinite sheet structure with the repeating unit $Si_2O_5^{2-}$. The oxidation state of Si is $+4$, the oxidation state of O is -2, and the oxidation state of Ca is $+2$. These oxidation states are consistent with the positions of the three elements in the periodic table. Therefore, the oxidation state of Cu must be $+2$.

EXERCISE

Predict the structural class of the mineral åkermanite ($Ca_2Mg(Si_2O_7)$), and give the oxidation states of its atoms.

Answer: Disilicate (pairs of tetrahedra). Ca and Mg, $+2$; Si, $+4$; O, -2.

Figure 21–8 • The mineral quartz is one form of silica (SiO_2). Pure quartz (rock-crystal) is colorless. Impurities give it many different colors: brown, green, red, yellow, purple (the last of these is the gemstone amethyst). Quartz is an important component in granite, gneiss, and other common rocks. *(Leon Lewandowski)*

The physical properties of the silicates derive from their structures. An example is talc, which has the composition $Mg_3(Si_4O_{10})(OH)_2$ and a structure consisting of layers and parallel infinite sheets (Fig. 21–7d). In talc, all of the strong bonding interactions among the atoms occur within the layers, which are attracted to neighboring layers only by van der Waals forces. These weaker forces permit one layer to slip easily across another, which accounts for the ease with which it can be pulverized to make talcum powder. When all four vertices of each tetrahedron are linked to other tetrahedra, three-dimensional network structures result (Fig. 21–7e). The simplest and most abundant example of a three-dimensional network structure is the hard-rock mineral quartz (Fig. 21–8). Note that the quartz network carries no charge; consequently, there are no cations in its structure. Three-dimensional network silicates such as quartz are much stiffer and harder than are the linear and layered silicates.

Asbestos is a generic term for a group of naturally occurring hydrated silicates that can be processed mechanically into long fibers (Fig. 21–9). Some of these, such as tremolite ($Ca_2Mg_5(Si_4O_{11})_2(OH)_2$), possess the infinite double-chain structure of Figure 21–7c. Others, such as a chrysotile ($Mg_3(Si_2O_5)(OH)_4$), have a sheet structure (see Fig. 21–7d), but the sheets are rolled up into long tubes. Asbestos minerals are fibrous because the bonds along the tubes are stronger than the bonds that hold different tubes together. Asbestos is an excellent thermal insulator that is non-combustible, acid-resistant, and strong. For many years, it was used extensively in cement for pipes and ducts, and woven into fabric to make fire-resistant roofing paper and floor tiles. Its use has decreased in recent years because inhalation of its small fibers during mining and manufacturing or during the removal of frayed and crumbling building materials can cause the lung disease asbestosis.

Aluminosilicates

An important class of minerals, called **aluminosilicates,** results from the replacement of some of the silicon atoms in silicates by aluminum atoms. Aluminum is the third most abundant element in the Earth's crust (8% by mass), occurring largely in the form of aluminosilicates. Aluminum in minerals can be a simple cation (Al^{3+}), or it can replace silicon in tetrahedral coordination. When it replaces silicon, it contributes only three electrons to the bonding framework in place of the four electrons of silicon atoms. The additional required electron is supplied by the ionization of a metal atom such as sodium or potassium; the resulting alkali-metal ions occupy nearby sites in the aluminosilicate structure.

The most abundant aluminosilicate minerals in the Earth's surface are the **feldspars,** which result from the substitution of aluminum for silicon in three-dimensional silicate networks like quartz. The aluminum ions must be accompanied by other cations such as sodium, potassium, or calcium to maintain overall charge neutrality. Albite is a feldspar with the chemical formula $NaAlSi_3O_8$. In the high-temperature form of this mineral, the aluminum and silicon atoms are distributed at random over the tetrahedral sites available to them. At lower temperatures, other crystal structures become thermodynamically stable, with partial ordering of Al and Si sites.

Replacing one of the four silicon atoms in the structural unit of talc ($Mg_3(Si_4O_{10})(OH)_2$) with an aluminum atom, and simultaneously inserting a potassium atom to supply the fourth electron needed for electrical neutrality, gives the composition $KMg_3(AlSi_3O_{10})(OH)_2$. This substance belongs to the family of **micas.** Micas are harder than talc, and their layers slide less readily over one another, although the crystals still cleave easily into sheets (see Fig. 20–12). The cations occupy sites between the infinite sheets, and the van der Waals attractions that hold adjacent sheets together in talc are augmented by an ionic contribution. The further replacement of the three Mg^{2+} ions in $KMg_3(AlSi_3O_{10})(OH)_2$ by two Al^{3+} ions gives the mineral **muscovite** ($KAl_2(AlSi_3O_{10})(OH)_2$). Writing its formula in this way indicates that two kinds of aluminum atoms exist in the structure: One Al atom per formula unit occupies a tetrahedral site, substituting for an Si atom; the other two Al atoms in the formula unit are located between two adjacent layers. The formulas that mineralogists and crystallographers use convey significantly more information than the usual empirical chemical formula of a compound.

Clay Minerals

Clays are minerals produced by the weathering of primary aluminosilicate minerals. They have similar structures, but their compositions can vary widely as one element is replaced by another. Invariably they are microcrystalline or powdered in form and usually are hydrated. They are used as supports for catalysts, as fillers in paint, and as ion-exchange vehicles. Some clays readily absorb water and swell as they do. This property makes them useful as lubricants and bore-hole sealers in the drilling of oil wells.

Clay structures derive formally from the structures of talcs and micas. Take the infinite sheet mica named pyrophyllite ($Al_2(Si_4O_{10})(OH)_2$) as an example. Replacing one of six Al^{3+} ions in the pyrophyllite structure by one Mg^{2+} ion and one Na^+ ion (which together carry the same charge) gives the composition $MgNaAl_5(Si_4O_{10})_3(OH)_6$. This clay, called montmorillonite, readily absorbs water, which infiltrates between the infinite sheets and associates by ion-dipole forces with the Mg^{2+} and Na^+ ions located there, causing the montmorillonite to swell (Fig. 21–10).

Figure 21–9 • The fibrous structure of asbestos is apparent in this sample. *(Leon Lewandowski)*

• The importance of X-ray diffraction for determining the structures of silicates cannot be overemphasized. From the empirical formula for muscovite ($KH_2Al_3Si_3O_{12}$), there is no indication of the structural roles of the aluminum atoms in their different sites or of the location of the hydrogen atoms in the structure.

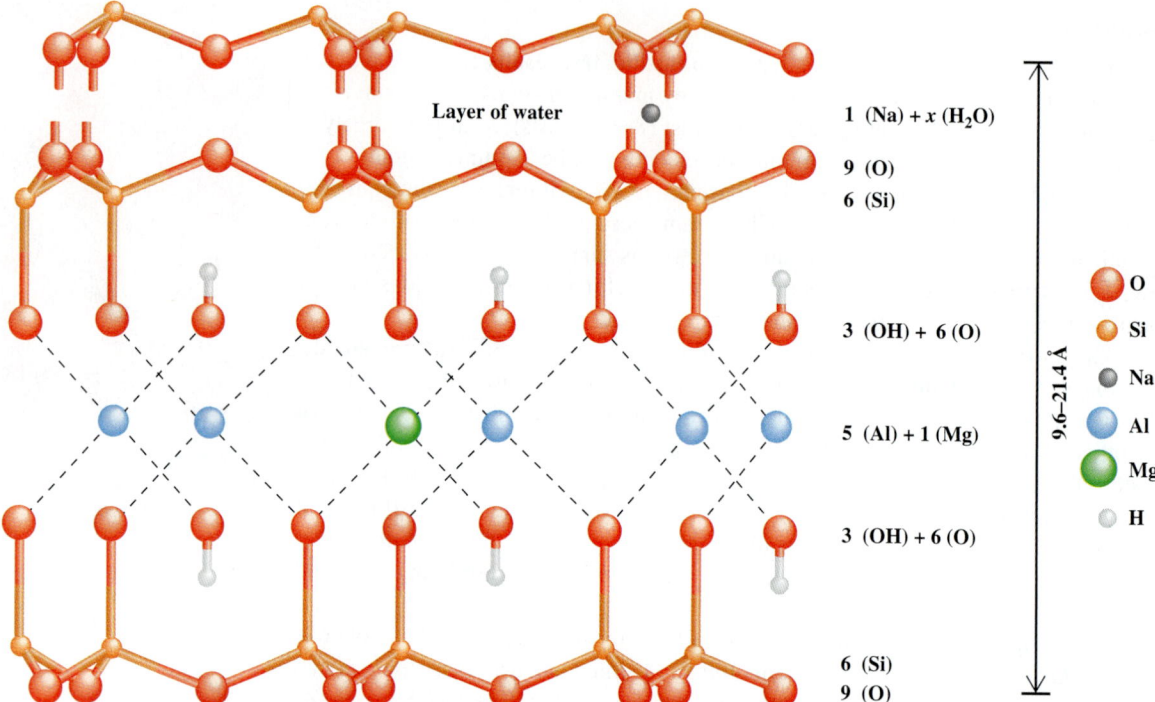

Layer of water

1 (Na) + x (H$_2$O)

9 (O)

6 (Si)

3 (OH) + 6 (O)

5 (Al) + 1 (Mg)

3 (OH) + 6 (O)

6 (Si)

9 (O)

9.6–21.4 Å

● O
● Si
● Na
● Al
● Mg
● H

Figure 21–10 • Structure of the clay mineral montmorillonite. Insertion of variable amounts of water causes the distance between layers to swell from 9.6 Å to more than 20 Å. When one Al^{3+} ion is replaced by an Mg^{2+} ion, an additional ion such as Na$^+$ is introduced into the water layers to maintain overall charge neutrality.

As noted previously, talc, Mg$_3$(Si$_4$O$_{10}$)$_3$(OH)$_2$, is a layered mineral. If water molecules are allowed to take up positions between the layers, the swelling clay vermiculite results. As is true of all clays, substitution is common, and vermiculite usually contains iron and aluminum, which replace magnesium and silicon in varying proportions. When heated, vermiculite pops like popcorn, as the steam generated by the vaporization of water between the layers puffs the flakes up into a light, fluffy material with air inclusions. Because of its open, porous structure, vermiculite is used for thermal insulation and as an additive to loosen compacted soils.

Zeolites

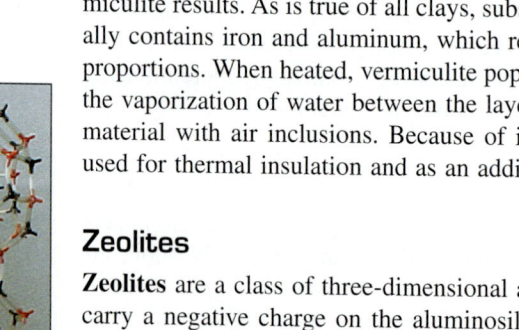

Figure 21–11 • A model of the structure of a zeolite showing the characteristic cavities and channels. Each aluminum and silicon atom (both colored black in the model) is at the center of a tetrahedron of oxygen atoms (red in the model).

Zeolites are a class of three-dimensional aluminosilicates. Like the feldspars, they carry a negative charge on the aluminosilicate framework that is compensated by neighboring cations such as the alkali-metal and alkaline-earth ions. Zeolites differ from the feldspars in having much more open structures consisting of polyhedral cavities connected by tunnels (Fig. 21–11). Many zeolites are found in nature, but they also can be synthesized under conditions controlled to favor the formation of cavities of very uniform size and shape. In most zeolites, water molecules are accommodated within these cavities, where they provide a mobile phase for the in-and-out migration of the charge-compensating cations. This enables zeolites to serve as ion-exchange materials and is the basis for their use in the softening of water. "Hardness" in water arises from dissolved calcium and magnesium salts such as Ca(HCO$_3$)$_2$ and Mg(HCO$_3$)$_2$. Such salts are converted to insoluble carbonates (boiler scale) when the water is heated and also form objectionable precipitates (bathtub

CHEMISTRY IN PROGRESS

Silicones

The chains of alternating silicon and oxygen atoms that characterize silicate minerals also form the basis for an important class of synthetic materials—the **silicones.** The simplest of these is dimethyl silicone, a chain molecule that can be derived (structurally) from the silicate chains shown in Figure 21–7b by replacing the two lone oxygen atoms on each silicon atom by methyl (—CH$_3$) groups. Dimethyl silicone is a *polymeric* molecule because it contains numerous repetitions of a single base unit (the —O—Si(CH$_3$)$_2$— grouping) that are connected by chemical bonds. We consider other polymers in Chapter 25.

The starting material for making dimethyl silicone is (CH$_3$)$_2$SiCl$_2$, which reacts with water to form a dihydroxo compound:

$$(CH_3)_2SiCl_2(aq) + 2\,H_2O(\ell) \longrightarrow$$
$$(CH_3)_2Si(OH)_2(aq) + 2\,H^+(aq) + 2\,Cl^-(aq)$$

Two molecules of (CH$_3$)$_2$Si(OH)$_2$ then can react, with loss of water, to form an Si—O—Si linkage.

Further condensation with additional (CH$_3$)$_2$Si(OH)$_2$ molecules leads to long-chain molecules with the overall chemical formula [(CH$_3$)$_2$SiO]$_n$. The Si—O bond lengths and Si—O—Si bond angles are much the same as in silicate minerals. Dimethyl silicone fluids are used widely as lubricating greases and in electrical transformers.

Silicones can be synthesized with cross-linkages between chains that lead to networks similar to the networks formed by silicates. Cross-linking requires the insertion of reactive side-groups on the silicon chain. For example, the molecules of the silicone sealant present in many bathtub caulks have acetate (—OCOCH$_3$) and hydroxide (—OH) side-groups in place of some of the methyl groups of dimethyl silicone. These groups react (Fig. 21–A), with the release of acetic acid, to give silicon atoms that have three, instead of two, oxygen linkages to other silicon atoms. The loss of the acetic acid accounts for the characteristic vinegary smell of acetic acid that develops after the caulk has been applied. The resulting cross-linked polymer materials are durable and water-resistant.

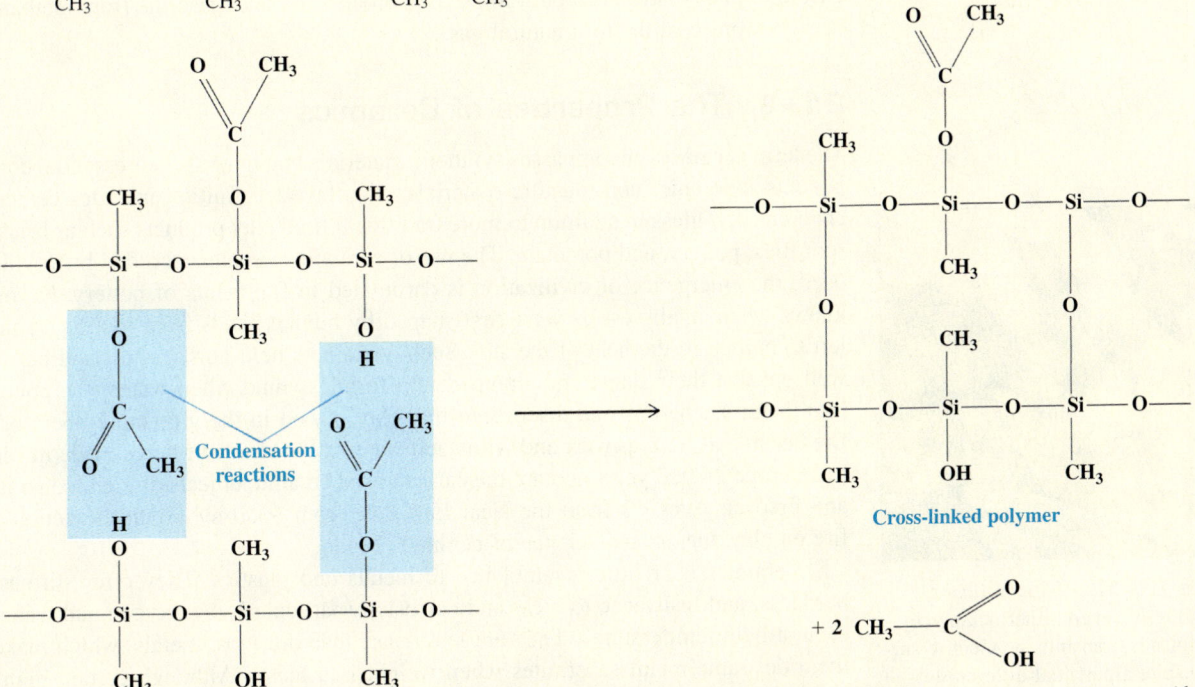

Figure 21–A • Two polymeric silicone chains with reactive side-groups (the —OH and —OCOCH$_3$ groups) cross-link, releasing acetic acid.

ring) with soaps. When hard water is passed through a column packed with a zeolite that has sodium ions in its structure, the calcium and magnesium ions exchange with the sodium ions and are removed from the water phase into the zeolite:

$$Na_2Z(s) + Ca^{2+}(aq) \rightleftharpoons CaZ(s) + 2\,Na^+(aq)$$

When the ion-exchange capacity of the zeolite is exhausted, this reaction can be reversed by passing a concentrated solution of sodium chloride through the zeolite to regenerate it in the sodium form. Ion exchange also works to separate chemically similar cations of the rare-earth elements or the actinides from one another on the basis of their slightly different tendencies to associate with zeolites used as the stationary phase in column chromatography (see Section 7–7).

A second use of zeolites derives from the ease with which they absorb small molecules. Their sponge-like affinity for water makes them useful as drying agents; they are put between the panes of double-pane glass windows to prevent stray moisture trapped there from condensing on the inner surfaces. The pore size of zeolites can be calibrated during their synthesis to allow molecules that are smaller than a certain size to pass through, but to hold back larger molecules. Such zeolites serve as "molecular sieves"; they have been used to capture nitrogen molecules from a gas stream while permitting oxygen molecules to pass through.

• This is the method used to separate oxygen from the air for sale at "oxygen bars."

Perhaps the most exciting use of zeolites is as catalysts. Molecules of various sizes and shapes have different rates of diffusion through a zeolite. These variations enable chemists to enhance the rates and yields of desired reactions and suppress unwanted reactions. The most extensive applications of zeolites at present are in the catalytic cracking of crude oil, a process that involves breaking down long-chain hydrocarbons and re-forming them into branched-chain molecules of lower molecular mass for use in high-octane unleaded gasoline. A relatively new process uses "shape-selective" zeolite catalysts to convert methanol (CH_3OH) to high-quality gasoline. Plants have been built to use this process to make gasoline from methanol derived from coal or from natural gas.

21–3   The Properties of Ceramics

The term **ceramics** encompasses synthetic materials that have as their essential components inorganic, non-metallic materials. This broad definition includes cement, concrete, and glass in addition to more traditional fired clay products such as bricks, roof tiles, pottery, and porcelain. The use of ceramics predates recorded history; indeed, the emergence of civilization is chronicled in fragments of pottery. No one knows when small vessels were first shaped by human hands from moist clay and left to harden in the heat of the sun. Such containers held nuts, grains, and berries well but lost their shape and slumped into formless mud when water was poured into them. Then someone discovered that clay placed in the glowing embers of a fire became as hard as rock and withstood water well. Molded figures made in central Europe 24,000 years ago are the earliest fired ceramic objects discovered so far, and fired clay vessels from the Near East date from 8000 B.C. With the action of fire on clay, the art and science of ceramics began.

Figure 21–12 • This insulator, consisting of several interlocking ceramic pieces, transmits no electric current despite a potential difference of hundreds of thousands of volts between one end and the other. *(Niagara Mohawk Power)*

Ceramics bear little resemblance to metals and plastics. They offer stiffness, hardness, and resistance to wear and corrosion (particularly by oxygen and water), even at high temperatures. They are less dense than are most metals, which makes them desirable metal substitutes when weight is a factor. Although certain highly specialized ceramics become electrical superconductors at low temperatures, most are good electrical insulators at ordinary temperatures, a property that is exploited in electronics and power transmission (Fig. 21–12). Ceramics retain their strength well at high temperatures. Several important structural metals soften or melt at temperatures a thousand degrees below the melting points of their chemical compounds in ceramics. Aluminum, for example, melts at 660°C, whereas aluminum oxide (Al_2O_3) does not melt until a temperature of 2051°C is reached.

Against these advantages must be listed some strong disadvantages. Ceramics are generally brittle and low in tensile strength. They tend to have high thermal expansion but low thermal conductivity, making them subject to **thermal shock,** in which a sudden local temperature change causes cracking or shattering. Metals and plastics dent or deform under stress, but ceramics cannot dissipate stress in this way. They break instead. A major drawback of ceramics as structural materials is their tendency to fail unpredictably and catastrophically in use. Moreover, some ceramics lose mechanical strength as they age.

Composition and Structure of Ceramics

Ceramics use a wide variety of chemical compounds, and useful ceramic bodies are nearly always mixtures of several compounds. **Silicate ceramics,** which include the traditional pots, dishes, and bricks, are made from aluminosilicate clay minerals. All contain the tetrahedral SiO_4 grouping discussed in Section 21–2. In **oxide ceramics,** silicon is a minor or non-existent component. Instead, a number of metals combine with oxygen to give compounds such as alumina (Al_2O_3), magnesia (MgO), or yttria (Y_2O_3). **Non-oxide ceramics** contain as principal components compounds that are free of oxygen. Some important examples are silicon nitride (Si_3N_4), silicon carbide (SiC), and boron carbide (approximate composition, B_4C).

A **ceramic phase** is any portion of the whole body that is physically homogeneous and bounded by a surface that separates it from other parts. Distinct phases are visible at a glance in coarse-grained ceramic pieces; in a fine-grained piece, phases can be seen under a microscope (Fig. 21–13). Many ceramics are porous; they have small openings into which air or water can infiltrate. Fully dense ceramics have no channels of this sort. Two ceramic pieces can have the same chemical composition but quite different densities if the first is porous and the second is not. When examined on a still finer scale, most ceramics, like metals, are microcrystalline. They consist of small crystalline grains cemented together and show this by interacting with X-rays to produce a diffraction pattern comprising numerous well-defined rings (see Section 20–1). The **microstructure** of such objects includes the sizes and shapes of the grains, the sizes and distribution of voids (openings between grains) and cracks, the identity and distribution of impurity grains, and the presence of stresses within the structure. Microstructural variations have enormous importance in ceramics because slight changes at this level strongly influence the properties of individual ceramic pieces. This is less true for plastic and metallic objects.

The microstructure of a ceramic body depends markedly on the details of its fabrication. The techniques of forming and firing a ceramic piece become as important as its chemical composition in determining ultimate behavior because they confer a unique microstructure. Thus inconsistent quality is the biggest problem with ceramics as structural materials. Ceramic engineers can produce parts that are stronger than steel, but not reliably because of the difficulties of monitoring and controlling microstructure. Gas-turbine engines fabricated of silicon nitride (Fig. 21–14), for example, run well at 1370°C, which is hot enough to soften or melt most metals. The higher operating temperature increases engine efficiency, and the ceramic turbines weigh less, further boosting fuel economy. Despite these advantages, ceramic gas turbines are not made commercially. Ceramic turbines

• The distinction between ceramics and plastics is not always clear-cut. Glass is a ceramic that can be shaped by plastic deformation at high temperatures, and some silicates behave like plastics upon the addition of water.

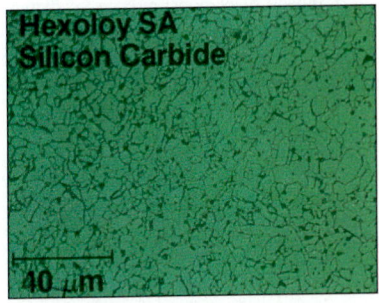

Figure 21–13 • A close-up of a silicon carbide ceramic, showing its granular texture. (*Carborundum Company*)

Figure 21–14 • A part freshly fabricated from silicon nitride for use in a gas-turbine engine. (*GTE Labs, Inc.*)

that work have to be built from selected, pretested components. The testing costs and rejection rates are so high that economical mass production has been impossible so far.

Making Ceramics

The manufacture of most ceramics involves four steps: (1) preparation of the raw material; (2) forming into the desired shape, often achieved by mixing a powder with water or other binder and molding the resulting plastic mass; (3) drying and firing the piece, also called its **densification,** because pores (voids) in the dried ceramic fill in; and (4) finishing the piece by sawing, grooving, grinding, or polishing.

The raw materials for traditional ceramics are natural clays that are mined as powders or thick pastes and become plastic enough, after adjustment of their water content, for forming freehand or on a potter's wheel. In contrast, modern specialized ceramics (both oxide and non-oxide) require chemically pure raw materials that are produced synthetically. Close control of the purity of the starting materials for these ceramics is essential to producing finished pieces with the desired properties. In addition to being formed by hand or in open molds, ceramic pieces can be shaped by the squeezing (compacting) of the dry or semi-dry powders in a strong, closed mold of the desired shape, at both ordinary and at elevated temperatures (hot pressing).

Firing a ceramic causes **sintering,** in which the fine particles of the ceramic start to merge together by diffusion at high temperatures. Atoms migrate out of their individual grains, and **necks** grow to connect adjoining particles. The density of the material increases because the voids between grains are partially filled. Sintering occurs below the melting point of the material and shrinks the ceramic body. In addition to the growing together of the grains of the ceramic, firing causes partial melting, chemical reactions among different phases, reactions with gases in the atmosphere of the firing chamber, and recrystallization of compounds with an accompanying increase in crystal size. All of these changes influence the microstructure of a piece and must be understood and controlled. Although firing accelerates physical and chemical changes, thermodynamic equilibrium in a fired ceramic piece rarely is reached. Kinetic factors, including the rate of heating, the length of time at which each temperature is held, and the rate of cooling, influence microstructure. The use of microwave radiation (as in microwave ovens) rather than a kiln to sinter ceramics is under development in ceramics factories. In conventional firing, the outside of a ceramic heats up before the inside; the uneven heating creates stresses that can weaken or even shatter the piece. Microwaves heat a ceramic quickly throughout its whole bulk.

- Some ceramic objects (such as those composed of cement) require no firing in their fabrication. In the hardening of cement, the heat required for chemical and structural transformations is supplied internally by chemical reactions.

21-4   Silicate Ceramics

The silicate ceramics include materials that vary widely in composition, structure, and uses. They range from simple earthenware bricks and pottery to fine porcelain, glass, and cement. Their structural strength is based on the same linking of silicate-ion tetrahedra that gives structure to silicate minerals in nature.

Pottery and Structural Clay Products

Section 21–2 notes that aluminosilicate clays are the products of the weathering of primary minerals. When water is added to such clays in moderate amount, the result is a thick paste that is easily molded into different shapes. Clays expand as water

A selection of useful ceramic pieces sintered in a microwave-assisted tunnel kiln in Great Britain. *(Courtesy of Copenhurst.tech Limited)*

invades the space between adjacent aluminosilicate sheets of the mineral, but release most of this water to a dry atmosphere and shrink back to their original size. A small fraction of the water or hydroxide ions remains rather tightly bound by ion-dipole forces to cations between the aluminosilicate sheets and is lost only when the clay is heated to a high temperature. The firing of aluminosilicate clays causes irreversible chemical changes to take place. The clay kaolinite ($Al_2Si_2O_5(OH)_4$) undergoes the following reaction:

$$\underset{\text{kaolinite}}{3\ Al_2Si_2O_5(OH)_4(s)} \longrightarrow \underset{\text{mullite}}{Al_6Si_2O_{13}(s)} + \underset{\text{silica}}{4\ SiO_2(s)} + \underset{\text{water}}{6\ H_2O(g)}$$

The fired ceramic body is a mixture of two phases, mullite and silica. Mullite, a rare mineral in nature, takes the form of needle-like crystals that interpenetrate and confer strength on the ceramic. When the temperature is above 1470°C, the silica phase forms as minute grains of cristobalite, one of the several crystalline forms of SiO_2.

If chemically pure $Al_2Si_2O_5(OH)_4$ is fired, the finished ceramic object is white. Purified clay minerals are the raw material for fine china. Naturally occurring clays contain impurities, such as transition-element oxides, that affect the color of both the unfired clay and the fired ceramic object if they are not removed. The colors of the oxides of the transition elements arise from their absorption of light at visible wavelengths, as explained by crystal-field theory (see Section 19–4). Common colors for natural ceramics are yellow or greenish yellow, brown, and red. Bricks are red when the clay used to make them has a high iron content (Fig. 21–15).

Before a clay is fired in a kiln, it must first be freed of casual moisture by slow heating to about 500°C. A clay body dried at room temperature and placed directly into a hot kiln crumbles or even explodes from the sudden expulsion of water. A fired ceramic shrinks somewhat as it cools, causing cracks to form. Cracks limit the strength of the fired object and are undesirable. Their occurrence can be reduced by coating the surface of a partially fired clay object with a **glaze,** a thin layer that holds the ceramic in a state of tension as it cools. Glazes have no sharp melting temperature; rather, they harden and develop resistance to shear stresses increasingly as the temperature of the high-fired clay object is reduced gradually. Glazes generally are aluminosilicates that have high aluminum content to raise their viscosity and thereby reduce the tendency to run off the surface during firing. Besides reducing cracking, glazes allow artisans to color and decorate fired objects and seal their surfaces against penetration by water or other liquids. Again, oxides of transition elements (particularly Ti, V, Cr, Mn, Fe, Co,

Figure 21–15 • Kaolinitic clays with substantial percentages of iron are common. These clays are used to make bricks, and the iron makes the bricks red, which explains the color scheme of many city streets. *(Herb Pressman/International Stock Photo)*

Ni, and Cu) are responsible for the colors. The oxidation state of the transition element in the glaze is crucial in determining the color. It is controlled by regulating the composition of the atmosphere in the kiln. Atmospheres rich in oxygen give high oxidation states, and those poor or lacking in oxygen give low oxidation states.

Glass

Glassmaking probably originated in the Near East about 3500 years ago. It is one of the oldest domestic arts, and its beginnings, like those of metallurgy, are obscure. Both required high-temperature, charcoal-fueled ovens and vessels made of materials that did not easily melt in order to initiate and contain the necessary chemical reactions. A great advance was the invention of glassblowing, which probably occurred in the first century B.C. The glassblower dipped a long iron tube into molten glass and rotated it to cause a ball of semisolid material to accumulate on the end. Blowing into the iron tube forced the soft glass to take the form of a hollow ball (Fig. 21–16) that could be further shaped into a vessel and severed from the blowing tube with a blade. Artisans in the Roman Empire developed glassblowing to a high degree, but with the decline of that civilization, glassmaking in Europe deteriorated until the Venetians redeveloped the lost techniques a thousand years later.

> • Glassblowing is possible only because glass has no sharp melting point. Instead, there is a reasonably wide temperature range over which glass is sufficiently plastic to be worked.

Glasses are amorphous solids of widely varying composition. Outwardly, glasses often resemble crystalline solids and have mechanical properties similar to those of crystals. On a molecular level, however, glasses resemble liquids in which the diffusional motions of the molecules have been brought to a halt. As in crystals, cohesion in glasses may depend on van der Waals forces, electrostatic (ionic) forces, covalent bonds, or metallic bonds; however, in this chapter, we use the term "glass" in the restricted and familiar sense to refer to materials formed from silica, usually in combination with metal oxides. The absence of long-range order in glasses makes them isotropic in their physical properties—that is, their properties are the same in all directions. This has certain advantages in technology, including, for example, the fact that glasses expand uniformly in all directions with an increase in temperature. The mechanical strength of glass is intrinsically very high, even exceeding the tensile strength of steel provided that the surface is free of scratches and other imperfections. Flaws in the surface provide sites where fractures can start when the glass is stressed. When a glass object of intricate shape and non-uniform thickness is

Figure 21–16 • Glassblowing. When softened by heat, masses of glass can be formed into "balloons" by blowing air into them. *(Leon Lewandowski)*

cooled suddenly, internal stresses are locked in. These stresses may cause the object to shatter when the object is later heated or struck (Fig. 21–17). Heating a strained object to a temperature somewhat below its softening point and holding it there for a while before allowing it to cool slowly, a technique called **annealing,** gives short-range diffusion of atoms a chance to occur and to eliminate internal stresses.

The softening and annealing temperatures of a glass and other properties, such as its density, depend on its chemical composition (Table 21–2). Silica itself (SiO_2) forms a glass if it is heated above its melting point and then cooled rapidly to avoid crystallization. The resulting **vitreous** (glassy) silica (Fig. 21–18), also called fused quartz, has many desirable properties but finds limited use because the high temperatures required to shape it make it quite expensive. Sodium silicate glasses form in the high-temperature reaction of silica sand with anhydrous sodium carbonate (soda ash, Na_2CO_3):

$$Na_2CO_3(s) + n\ SiO_2(s) \longrightarrow Na_2O \cdot nSiO_2(s) + CO_2(g)$$

The melting point of the non-volatile product is about 900°C, and the glassy state results if cooling through that temperature is rapid. The product, called "water glass," is water-soluble and thus unsuitable for making vessels. Its aqueous solutions are used in some detergents and as adhesives for sealing cardboard boxes, however.

An insoluble glass having useful structural properties results if sufficient lime (CaO) is added to the sodium carbonate–silica starting mixture. **Soda-lime glass** has the approximate composition $Na_2O \cdot CaO \cdot (SiO_2)_6$. This glass is easy to melt and shape and is used in many applications, ranging from bottles to window glass. It

Figure 21–17 • Internal stresses cause a delicate glass teardrop (called a "Prince Rupert's drop") to shatter when its tail is merely scratched. *(James L. Amos/Peter Arnold, Inc.)*

Newly formed automobile windows are inspected after careful annealing to eliminate internal stress. *(Courtesy of PPG Industries, Inc.)*

TABLE 21–2	Composition and Properties of Various Glasses			
Silica Glass	**Soda-Lime Glass**	**Borosilicate Glass**	**Aluminosilicate Glass**	**Leaded Glass**
Composition				
SiO_2, 99.9%	SiO_2, 73%	SiO_2, 81%	SiO_2, 63%	SiO_2, 56%
H_2O, 0.1%	Na_2O, 17%	B_2O_3, 13%	Al_2O_3, 17%	PbO, 29%
	CaO, 5%	Na_2O, 4%	CaO, 8%	K_2O, 9%
	MgO, 4%	Al_2O_3, 2%	MgO, 7%	Na_2O, 4%
	Al_2O_3, 1%		B_2O_3, 5%	Al_2O_3, 2%
Coefficient of Linear Thermal Expansion ($°C^{-1}/1 \times 10^{-7}$)[a]				
5.5	93	33	42	89
Softening Point (°C)				
1580	695	820	915	630
Annealing Point (°C)				
1050	510	565	715	435
Density (g cm^{-3})				
2.20	2.47	2.23	2.52	3.05

[a]The coefficient of linear thermal expansion is defined as the fractional increase in length of a body when its temperature is increased by 1°C.

Figure 21-18 • Glassy, non-crystalline vitreous silica results when molten SiO_2 is cooled rapidly. It has the lowest coefficient of linear thermal expansion of any known material (see Table 21–2). *(Charles D. Winters)*

accounts for over 90% of all of the glass manufactured today. Its structure, shown schematically in Figure 21–19, is a three-dimensional network of the type discussed in Section 21–2 but with random coordination of the silicate tetrahedra. The network is a giant polyanion, with Na^+ and Ca^{2+} ions distributed in the void spaces to compensate for the negative charge on the network.

Replacement of lime and some of the silica in a glass by other oxides (Al_2O_3, B_2O_3, K_2O, or PbO) modifies its properties noticeably. For example, the thermal conductivity of ordinary (soda-lime) glass is quite low, and its coefficient of thermal expansion is high. This means that strong internal stresses are set up when one part of a glass object is sharply heated or chilled. The object may then shatter. The coefficient of thermal expansion is much less in certain borosolicate glasses, in which many of the silicon sites are occupied by boron. Pyrex, the most familiar of these glasses, has a coefficient of linear expansion about one third that of soda-lime glass. It is the material of choice for laboratory glassware and household ovenware. A glass called Vycor, which has an even smaller coefficient of expansion (approximating that of fused silica), is made by chemical treatment of borosilicate glass to leach out its sodium. This process leaves a porous structure that is densified by raising the temperature and shrinking the glass to its final volume.

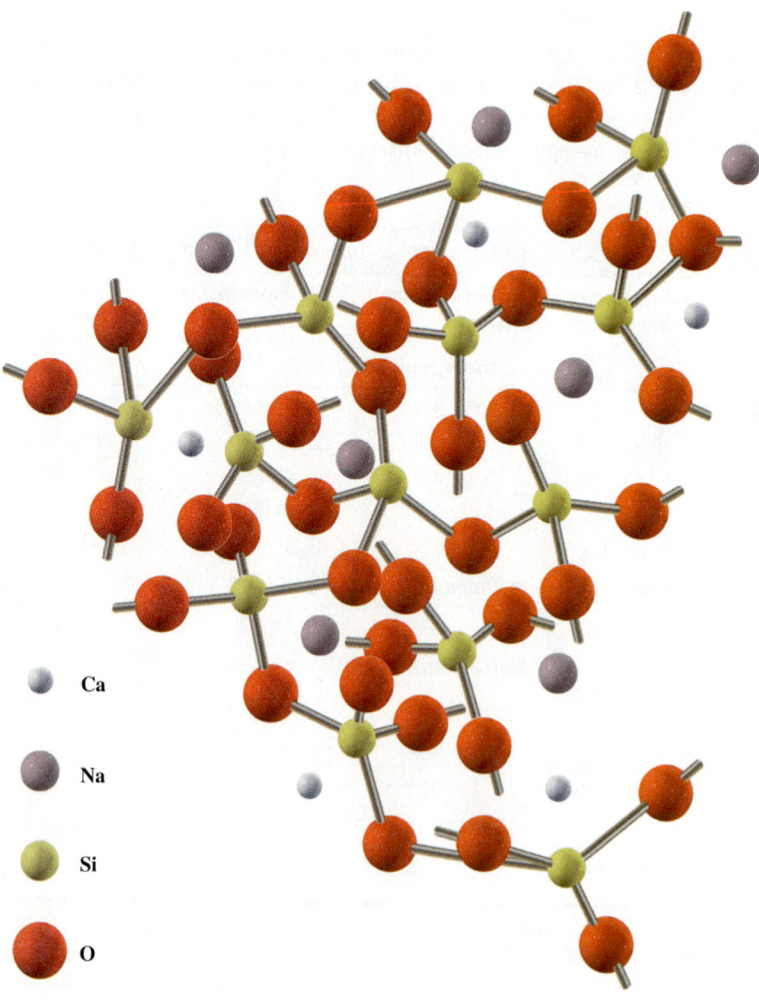

Ca

Na

Si

O

Figure 21–19 • The structure of a soda-lime glass. Note the tetrahedral coordination of oxygen atoms around each silicon atom.

A major use of glass is in fiber optics for the transmission of voice messages, television images, and data as light pulses. Tens of thousands of audio messages can be transmitted simultaneously through glass fibers no larger in diameter than a hair by encoding the audio signal into electronic impulses that modulate light from a laser source. The light passes down the glass fiber as though it were a tube (Fig. 21–20). Chemical control of the composition of the glass reduces light loss and permits messages to pass over distances of many kilometers without amplification.

Cements

Hydraulic cement was first developed by the Romans, who discovered that a mixture of lime (CaO) and dry volcanic ash reacts slowly with water, even at low temperatures, to form a durable solid. They used this knowledge to build the Roman Pantheon, a circular building in which the concrete dome, spanning 143 feet without internal support, still stands nearly 2000 years after its construction! The knowledge of cement making was lost for centuries after the fall of the Roman Empire. An English bricklayer, Joseph Aspdin, who patented a process for calcining a mixture of limestone and clay, rediscovered it in 1824. He called the product **Portland cement** because, when mixed with water, it hardened to a material that resembled a kind of limestone found on the Isle of Portland. Today, Portland cement is manufactured in every industrial country with global production exceeding that of any other material. Portland cement opened up a new age in the construction of highways and buildings. Rock could be crushed and then molded into cement rather than shaped with cutting tools.

Portland cement is a finely ground (powdered) mixture of compounds produced by the high-temperature reaction of lime, silica, alumina, and iron oxide. The lime (CaO) may come from limestone or chalk deposits. The silica (SiO_2) and alumina (Al_2O_3) often are obtained in clays or slags. The blast furnaces of steel mills are a common source of slag, which is a by-product of the smelting of iron ore (see Section 12–4). The composition of slag varies, but it can be represented as a calcium aluminum silicate of approximate formula $CaO \cdot Al_2O_3 \cdot (SiO_2)_2$. This material is crushed and ground to a fine powder, blended with lime in the correct proportion, and burned again in a horizontal rotary kiln at temperatures up to 1500°C to produce "cement clinker." A final stage of grinding and the addition of about 5% by mass of gypsum ($CaSO_4 \cdot 2H_2O$) to lengthen the setting time completes the process of manufacture. The composition of typical Portland cement is shown in Table 21–3, which gives percentages of the separate oxides. These simple compounds are

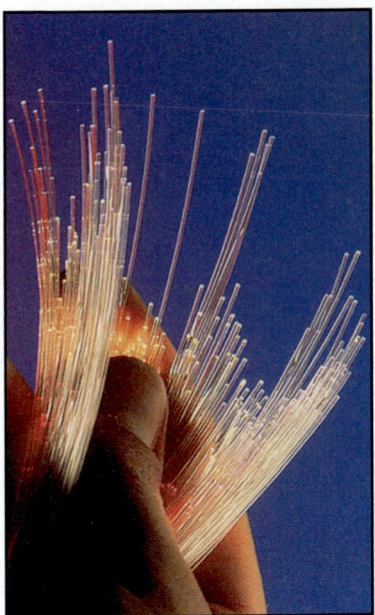

Figure 21–20 • Optical fibers conducting light. *(Phillip Hayson/Photo Researchers, Inc.)*

• Cement kilns in the United States also are used for the disposal by burning of wastes such as used oils and solvents.

TABLE 21–3	Typical Chemical Composition of Portland Cement
Oxide	**Percentage by Mass**
Lime (CaO)	63.8
Silica (SiO_2)	21.3
Alumina (Al_2O_3)	4.6
Sulfur trioxide (SO_3)	2.9
Magnesia (MgO)	2.8
Iron(III) oxide (Fe_2O_3)	2.6
Minor oxides (Na_2O, K_2O)	0.6
Water and other volatiles	1.3

the "elements" of cement making. They combine in the cement to make more complex compounds such as dicalcium silicate, $(CaO)_2 \cdot SiO_2$; tricalcium silicate, $(CaO)_3 \cdot SiO_2$; tricalcium aluminate, $(CaO)_3 \cdot Al_2O_3$; and tetracalcium aluminum ferrite, $(CaO)_4 \cdot Al_2O_3 \cdot Fe_2O_3$.

Cement *sets* when the semi-liquid slurry first formed by the addition of water to the powder becomes a solid of low strength. Subsequently, it gains strength in a slower *hardening* process. Setting and hardening involve a complex group of exothermic reactions in which several hydrated compounds form. Portland cement is a **hydraulic cement** because it hardens not by loss of water but by chemical reactions that incorporate water into the final body. The main reaction during setting is the hydration of the tricalcium aluminate, which can be approximated by the equation

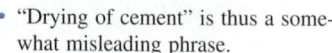

• "Drying of cement" is thus a somewhat misleading phrase.

$$(CaO)_3 \cdot Al_2O_3(s) + 3\,CaSO_4 \cdot 2H_2O(s) + 26\,H_2O(\ell) \longrightarrow$$
$$(CaO)_3 \cdot Al_2O_3 \cdot (CaSO_4)_3 \cdot 32H_2O(s)$$

The product forms after 5 or 6 hours as a microscopic forest of long crystalline needles that lock together to solidify the cement. Later, the calcium silicates react with water to harden the cement. One reaction is

$$6\,(CaO)_3 \cdot SiO_2(s) + 18\,H_2O(\ell) \longrightarrow (CaO)_5 \cdot (SiO_2)_6 \cdot 5H_2O(s) + 13\,Ca(OH)_2(s)$$

The hydrated calcium silicates develop as strong tendrils that coat and enclose unreacted grains of cement, each other, and other particles that may be present, binding them in a robust network. Most of the strength of cement comes from these entangled networks, which in turn ultimately depend on chains of $-O-Si-O-Si$ bonds for strength. In the final stages of hardening, the iron-containing compounds react with water. Hardening is slower than setting; it may take as long as a year for a cement to attain its ultimate strength.

Portland cement rarely is used alone. In general, it is combined with sand, water, and lime to make **mortar,** which is applied with a trowel to bond bricks or stones. When Portland cement is mixed with sand and aggregate (crushed stone or pebbles) in the proportions of $1 : 3.75 : 5$ by volume, the mixture is called **concrete.** Concrete is outstanding in its resistance to compressive forces. It therefore finds primary use for the foundations of buildings and the construction of dams, in which the compressive loads are enormous. The stiffness (resistance to bending) of concrete is high, but its fracture toughness (resistance to impact) is substantially lower, and its tensile strength is relatively poor. For these reasons, concrete usually is reinforced with steel rods when it is used in structural elements such as beams that are subject to transverse or tensile stresses.

The water required for chemical combination in the hardening of cement is only about 20% of its mass. It is customary to mix far more water than this with cement powders to make the slurry easier to work. As the excess water evaporates during hardening, it leaves voids or pores that typically comprise 25% to 30% of the volume of the solid. Pores weaken cement, the fracture strength of which is related inversely to the size of the largest pores. "Macro-defect-free" (MDF) cement, has pores only a few micrometers across (rather than about a millimeter, as in conventional cement). The smaller pores result from the addition of water-soluble polymers that make a dough-like "liquid" cement that is pliable with the use of far less water. Unset MDF cement is kneaded mechanically and extruded into the desired shape. The final result possesses substantially increased bending resistance and toughness. MDF cement even can be molded into springs and shaped on a conventional lathe. Reinforcement with fibers further increases its toughness.

Architectural forms executed in concrete in a California shopping mall. (*Coco McCoy/Rainbow*)

21–5 Non-Silicate Ceramics

Many useful ceramics that are not based on the Si—O bond and the SiO_4 tetrahedron exist. They have important uses in electronics, optics, and the chemical industry. Some of these materials are oxides, but others contain neither silicon nor oxygen.

Oxide Ceramics

Oxide ceramics are materials that contain oxygen in combination with any of a number of metals. These materials are named by adding an -*ia* ending to the stem of the name of whatever element is mainly present. Thus, if the main chemical component of an oxide ceramic is BeO, it is a *beryllia* ceramic; if the main component is Y_2O_3, it is an *yttria* ceramic; and if it is MgO, it is a *magnesia* ceramic. As Table 21–4 shows, the melting points of these and other oxides are substantially higher than the melting points of the elements themselves. Such high temperatures are hard to achieve and maintain, and the molten oxides corrode most container materials. Oxide ceramic bodies therefore are not shaped by melting the oxide and pouring it into a mold. Instead, these ceramics are fabricated by sintering, like the silicate ceramics. Oxide ceramic bodies are rarely single pure compounds. When ceramics are fabricated from alumina, for example, high-purity Al_2O_3 is prepared and then deliberately doped with about 1% MgO by mass. The MgO promotes a very fine-grained structure during firing, and the resulting piece is stronger.

Alumina (Al_2O_3) is the most important non-silicate ceramic material. It melts at a temperature of 2051°C, and it retains strength even at temperatures of 1500°C to 1700°C. Alumina has a large electrical resistivity and withstands both thermal shock and corrosion well. These properties make it a good material for spark-plug insulators, and most spark plugs now use a ceramic that is 94% alumina.

• By comparison, pure iron melts at 1535°C, and steel melts at a temperature well below 1500°C.

High-density alumina is fabricated in such a way that open pores between the grains are eliminated nearly completely, and the grains are small, having an average diameter as small as 1.5 μm. Unlike most other ceramics, high-density alumina has good mechanical strength, particularly against impact, which has led to its use in armor plating. The ceramic, suitably backed by a fracture-resistant framing, absorbs the energy of an impacting projectile by breaking, so penetration does not occur. High-density alumina also is used in high-speed cutting tools for machining metals. Such a use is perhaps surprising, but the temperature resistance of the ceramic allows for much faster cutting speeds, and a ceramic cutting edge has no tendency to weld to the workpiece, as metallic tools do. High-density alumina also is

TABLE 21–4		Melting Points of Some Metals and Their Oxides	
Metal	**Melting Point (°C)**	**Oxide**	**Melting Point (°C)**
Be	1287	BeO	2570
Mg	651	MgO	2800
Al	660	Al_2O_3	2051
Si	1410	SiO_2	1723
Ca	865	CaO	2572
Y	1852	Y_2O_3	2690

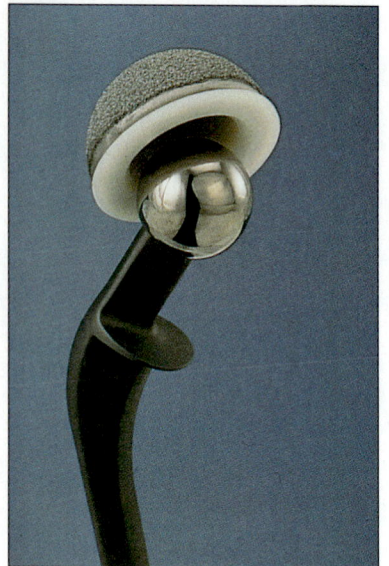

Figure 21–21 • High-density alumina forms the socket in this artificial hip. *(SIU/Photoresearchers, Inc.)*

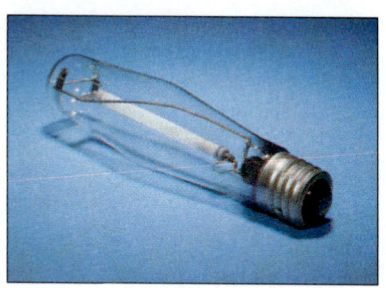

Figure 21–22 • The translucent, white inner envelope of high-density alumina in this high-intensity sodium-vapor lamp is visible through the outer envelope of glass. *(Leon Lewandowski)*

used in artificial joints (Fig. 21–21). If Al_2O_3 doped with a small percentage of MgO is fired in a vacuum or in a hydrogen atmosphere (instead of air) at a temperature of 1800°C to 1900°C, even very small pores, which scatter light and make the material white, are removed. The resulting ceramic is translucent. The invention of this material has changed the color of streetlights all over the world. It is employed to contain the sodium in high-intensity sodium-discharge lamps (Fig. 21–22). With envelopes of high-density alumina, such lamps can be operated at temperatures of 1500°C to give a whiter light and more of it. Old-style sodium-vapor lamps with glass envelopes were limited to a temperature of 600°C because the sodium vapor corroded the glass envelope at higher temperatures. The light from a low-temperature sodium-vapor lamp has an undesirable yellow color.

Magnesia (MgO) combines high thermal conductivity with excellent electrical resistance. It is used mainly as a **refractory,** which is a ceramic material that withstands a temperature of more than 1500°C without melting (MgO melts at 2800°C). Another use is as insulation in electrical heating devices. Magnesia is prepared from magnesite ores, which consist of $MgCO_3$ and a variety of impurities. When purified magnesite is heated to 800°C to 900°C, carbon dioxide is driven off to leave fine grains of $MgO(s)$:

$$MgCO_3(s) \longrightarrow MgO(s) + CO_2(g)$$

After cooling, fine-grained MgO reacts avidly with water to form magnesium hydroxide:

$$MgO(s) + H_2O(\ell) \longrightarrow Mg(OH)_2(s)$$

It therefore is called "caustic magnesia." Heating fine-grained $MgO(s)$ to 1700°C causes the MgO grains to sinter. Fully sintered MgO consists of large crystals and does not react with water. It is "dead-burned magnesia." The Gibbs energy $\Delta G°$ of the reaction between MgO and H_2O does not change when magnesia is dead burned. The altered microstructure (larger crystals versus small) slows the kinetics of the reaction with water exceedingly, however.

Beryllia (BeO) resembles alumina in many of its properties. It is less dense (3.00 g cm^{-3} versus 3.97 g cm^{-3}), however, which makes it attractive in aviation and space. Working with beryllia requires special care because the inhaled dust causes the lung disease berylliosis. Beryllia ceramics also tend to react with water vapor at high temperatures, forming poisonous vapors.

Non-Oxide Ceramics

In non-oxide ceramics, nitrogen or carbon takes the place of oxygen in combination with silicon or boron. Examples are boron nitride (BN), boron carbide (B_4C), the silicon borides (SiB_4 and SiB_6), silicon nitride (Si_3N_4), and silicon carbide (SiC). All of these compounds possess strong, short covalent bonds and find use as abrasives and cutting tools. They are hard and strong, but brittle. Forming these materials into useful, finished shapes requires high-purity starting materials and precise control of the temperature and atmosphere during firing.

Much research has aimed at making gas-turbine and other engines out of ceramics. Heat engines run more efficiently at higher temperatures, and above 1000°C, certain ceramics have better properties (strength, corrosion resistance, density) for engines than do metals in all respects except their intrinsic brittleness. Of the oxide ceramics, only alumina and zirconia (ZrO_2) are strong enough, but both resist thermal shock too poorly for this application. Attention therefore has turned to the non-oxide **silicon nitride** (Si_3N_4). Silicon nitride is a covalent network solid (Fig. 21–23).

In the structure, every silicon atom bonds to four nitrogen atoms that surround it at the corners of a tetrahedron, and each nitrogen atom is surrounded by three silicon atoms. The similarity to the way in which SiO_4 units join together in networks in the silicate minerals (see Section 21–2) is clear.

At first, silicon nitride seems unpromising as a high-temperature structural material. It is thermodynamically unstable in contact with water at room temperature, as shown by the negative standard Gibbs energy for the reaction

$$Si_3N_4(s) + 6\,H_2O(\ell) \longrightarrow 3\,SiO_2(s) + 4\,NH_3(g) \qquad \Delta G_r^{\circ} = -566\ \text{kJ mol}^{-1}$$

In fact, this reaction causes finely ground Si_3N_4 powder to give off a perceptible odor of ammonia in moist air. Silicon nitride is also thermodynamically unstable in air, reacting spontaneously with oxygen:

$$Si_3N_4(s) + 3\,O_2(g) \longrightarrow 3\,SiO_2(s) + 2\,N_2(g) \qquad \Delta G_r^{\circ} = -1923\ \text{kJ mol}^{-1}$$

Both of these reactions are thermodynamically favored at higher temperatures as well. In practice, however, neither reaction occurs at a perceptible rate when Si_3N_4 is in bulk form. Initial contact of oxygen or water with $Si_3N_4(s)$ forms a surface film of $SiO_2(s)$ that protects the bulk of the $Si_3N_4(s)$ from further attack. When *strongly* heated in air (to about 1900°C), Si_3N_4 does decompose, violently, but below that temperature, it resists attack.

- This is an excellent example of a kinetic barrier preventing a reaction. It is similar to the well-known corrosion protection that a layer of Al_2O_3 confers on metallic aluminum.

Fully dense silicon nitride parts are stronger than metallic alloys at high temperatures. Ball bearings made of fully dense silicon nitride work well without lubrication at temperatures up to 700°C and last longer than do steel ball bearings. Because the strength of silicon nitride shapes increases with the density attained in the production process, the trick is to form a dense piece of silicon nitride in the desired shape. **Reaction bonding** is one method for making useful shapes of silicon nitride. Powdered silicon is compacted in molds, removed, and then fired under an atmosphere of nitrogen at 1250°C to 1450°C. The reaction

$$3\,Si(s) + 2\,N_2(g) \longrightarrow Si_3N_4(s)$$

(text continues on page 904)

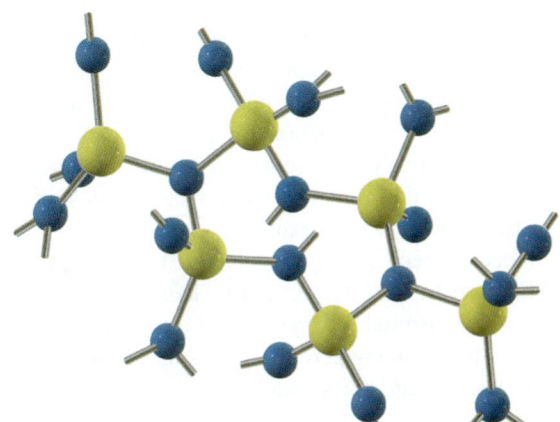

Figure 21–23 • The structure of silicon nitride (Si_3N_4). Each Si atom is bonded to four nitrogen atoms, and each N atom is bonded to three Si atoms. The result is a strong network.

CHEMISTRY IN PROGRESS

Superconducting Ceramics

The oxide ceramics discussed in this section are all oxides of single metals. A natural idea for new ceramics is to make materials containing two (or more) oxides in equal or nearly equal molar amounts. Thus, if $BaCO_3$ and TiO_2 are mixed and heated to high temperature, they react to give the ceramic barium titanate:

$$BaCO_3(s) + TiO_2(s) \longrightarrow BaTiO_3(s) + CO_2(g)$$

Barium titanate is a **mixed-oxide ceramic** (writing the formula as $BaO \cdot TiO_2$ shows this). It has the same structure as the mineral **perovskite** ($CaO \cdot TiO_2$) (Fig. 21–B), except, of course, that Ba replaces Ca. Perovskites typically have two metal atoms for every three oxygen atoms, giving them the general formula ABO_3, where A stands for a metal atom at the center of the unit cube, and B, for an atom of a different metal at the cube corners.

Figure 21–C • The levitation of a small magnet above a disk of superconducting material. A superconducting material cannot be penetrated by an external magnetic field. At room temperature, the magnet rests on the ceramic disk. When the disk is cooled with liquid nitrogen, it becomes superconducting and excludes the magnet's field, forcing the magnet into the air. Here the magnet floats above a disk of chilled 1-2-3 ceramic. *(Yoav Levy/Phototake)*

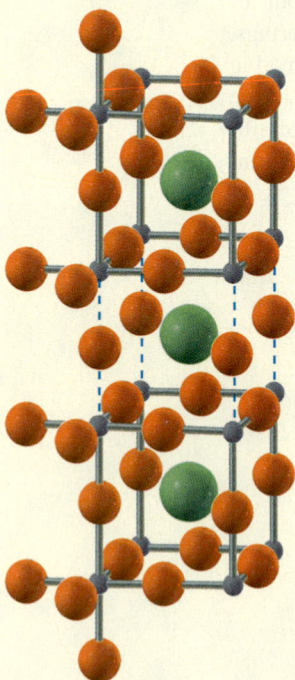

Figure 21–B • The structure of perovskite ($CaTiO_3$). A stack of three unit cells is shown, with some additional O atoms from neighboring cells. Each Ti (*gray*) is surrounded by 6 O's; each Ca (*green*) has 12 O's as nearest neighbors.

Interest in perovskite ceramic compositions surged in the late 1980s and early 1990s following the discovery that some of them become **superconducting** at relatively high temperatures. A superconducting material offers no resistance whatsoever to the flow of an electric current. The phenomenon was discovered by the Dutch physicist Heike Kammerlingh-Onnes in 1911 when he cooled mercury below its superconducting transition temperature of 4 K. Such very low temperatures are difficult to achieve and maintain, but if practical superconductors could be made to work at higher temperatures (or even at room temperature!), then power transmission, electronics, transportation, medicine, and many other aspects of human life would be transformed. Over 60 years of research with metallic systems culminated in 1973 in the discovery of a niobium–tin alloy with a record-high superconducting transition temperature of 23.3 K. Progress toward higher-temperature superconductors then stalled until 1986, when K. Alex Müller and J. Georg Bednorz, who had had the inspired notion to look for higher transition temperatures among perovskite ceramics, found a Ba-La-Cu-O perovskite phase that had a transition temperature of 35 K. This result motivated other scientists, who soon discovered another rare-earth–containing perovskite ceramic that became a superconductor at 90 K. This result was particularly exciting because 90 K exceeds the boiling point of liquid nitrogen (77 K), a relatively inexpensive refrigerant (Fig. 21–C). This 1-2-3 compound (so called because its formula $YBa_2Cu_3O_{(9-x)}$ has one yttrium, two barium, and three copper atoms per formula unit) is not an ideal perovskite. If it were, it would have nine oxygen atoms in combination with its

six metal atoms. Instead, a number of the oxygen sites in the solid are vacant. The deficiency makes x in the formula somewhat greater than 2, depending on the exact method of preparation. The 1-2-3 compound is a nonstoichiometric solid the structure of which is given in Figure 21–D.

Very substantial engineering effort has gone into developing ceramic superconductors for use in making cables, motors, generators, transformers, and electromagnets. The high-temperature superconducting ceramic $(Bi,Pb)_2Sr_2Ca_2Cu_3O_x$, which has a transition temperature of 110 K, has attracted particular attention as a practicable material. Recent improvements in the manufacture of superconducting tape made of "BSCCO-2223" have led to plans to test superconducting cables (Fig. 21–E) in electric utility networks and the construction of a 5000 horsepower electric motor that uses superconducting windings.

Figure 21–D • The structure of the 1-2-3 compound $YBa_2Cu_3O_{(9-x)}$ derives from that of perovskite, as can be seen by comparison to the structure in Figure 21–B. It is a layered perovskite in which one third of the layers contain yttrium (*gray*) and two thirds contain barium (*green*). Every third Ba in the hypothetical "$BaCuO_3$," is replaced by a Y, and the O's in the layer occupied by the Y are removed. Note the puckering in the Cu-O layers, as well.

Figure 21–E • Strands of high-temperature superconducting tape braided together to form a power transmission cable. The tapes consist of filaments of the high-temperature superconducting ceramic BSCCO-2223 imbedded in a matrix of silver. This composite material is much more flexible than the ceramic alone, which is brittle and fragile. *(Courtesy of American Superconductor)*

forms the ceramic. The parts neither swell nor shrink significantly during the chemical conversion from Si to Si_3N_4, making it possible to fabricate complex shapes reliably. Unfortunately, reaction-bonded Si_3N_4 is still somewhat porous and is too weak for many applications. In the "hot-pressed" forming process, Si_3N_4 powder is prepared in the form of exceedingly small particles by reaction of silicon tetrachloride with ammonia:

$$3\,SiCl_4(g) + 4\,NH_3(g) \longrightarrow Si_3N_4(s) + 12\,HCl(g)$$

The solid Si_3N_4 forms as a smoke. It is captured as a powder, mixed with a carefully controlled amount of MgO additive, placed in an enclosed mold, and sintered at 1850°C under a pressure of 230 atm. The resulting ceramic shrinks to nearly full density (no pores). Because the material does not flow well (to fill a complex mold completely), only simple shapes are possible. Hot-pressed silicon nitride is impressively tough and can be machined only with great difficulty, with diamond tools.

In the silicate minerals discussed in Section 21–2, AlO_4 units routinely substitute for SiO_4 tetrahedra as long as positive ions of some type are present to balance the electric charge. This fact suggests that in silicon nitride Al^{3+} could replace some Si^{4+} ions if a compensating replacement of O^{2-} for N^{3-} were made simultaneously. Experiments show that ceramic alloying of this type works well, giving many new ceramics with great potential called **sialons.** The discovery of these ceramics illustrates how a structural theme in a naturally occurring material can guide the search for new materials.

- The name "sialon" comes from the chemical symbols of the four elements involved: Si, Al, O, and N (Si-Al-O-N).

Boron has one fewer valence electron than carbon, and nitrogen has one more valence electron. **Boron nitride** (BN) is therefore isoelectronic with carbon, and it is not surprising that it has two structural modifications that resemble the structures of graphite and diamond. In hexagonal boron nitride, the boron and nitrogen atoms take alternate places in extended "chicken-wire" sheets. The sheets stack up in such a way that each boron atom has a nitrogen atom directly above it and directly below it, and vice versa. In graphite (see Fig. 20–30), the carbon atoms lie in similar sheets, but these sheets are offset so that half of the atoms lie on lines between the centers of the hexagons in the sheets above and below. The cubic form of boron nitride has the diamond structure, is comparable in hardness to diamond, and resists oxidation better. Boron nitride often is prepared by **chemical vapor deposition,** a procedure used in the fabrication of several other ceramics as well. In this method, a controlled chemical reaction of gases on a contoured, heated surface gives a solid product of the desired shape. If a cup made of BN is needed, a cup-shaped mold is heated to a temperature exceeding 1000°C, and a mixture of $BCl_3(g)$ and $NH_3(g)$ is passed over its surface. The reaction

$$BCl_3(g) + NH_3(g) \longrightarrow BN(s) + 3\,HCl(g)$$

deposits a cup-shaped layer of BN(s). Boron nitride cups and tubes are used to contain and evaporate molten metals.

Silicon carbide (SiC) is diamond in which half of the C atoms are replaced by Si atoms. Also known by its trade name of Carborundum, silicon carbide originally was developed as an abrasive, but it is now used primarily as a refractory and as an additive in steel manufacture. It is formed and densified by methods similar to those used with silicon nitride. It is a candidate material for high-temperature engines, and much research has been performed in this area. Silicon carbide often is produced in the form of small plates or whiskers. In these forms, it works well to reinforce other ceramics. Fired silicon carbide whiskers are quite small (0.5 mm in diameter and

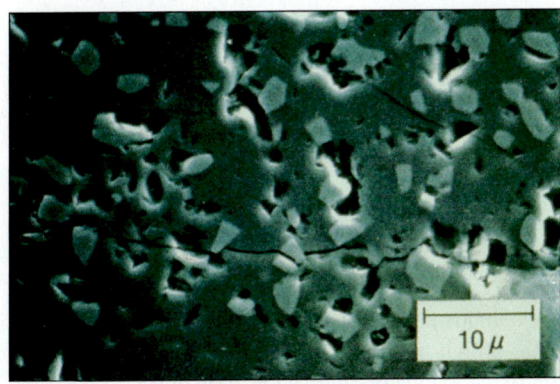

Figure 21-24 • Microstructure in a composite ceramic. The light-colored areas are TiB_2, the gray areas are SiC, and the dark areas are voids. The TiB_2 acts to toughen the SiC matrix. Note that the crack shown passing through from left to right is forced to deflect around the TiB_2 particles. (*Carborundum Company*)

50 mm long) but very strong. They are mixed with a second ceramic material before that material is formed. Firing then gives a **composite ceramic.** Such composites are stronger and tougher than are non-reinforced bodies of the same primary material. Whiskers serve to reinforce the main material by stopping cracks; they either deflect advancing cracks or soak up their energy, and a widening crack must dislodge them to proceed (Fig. 21-24).

SUMMARY

21-1 Elemental silicon forms a covalently bonded solid in which all of the electrons are localized between pairs of atoms. Their energies fill all the levels in the **valence band** and leave the **conduction band** virtually empty. The result is a low electrical conductivity, which can be increased by heating the solid or illuminating it, thereby exciting electrons across the **band gap.** Other **intrinsic semiconductors** are made by combining elements to give crystals with the same number of valence electrons per atom as silicon (four). An important example is gallium arsenide. An **insulator** such as diamond has a larger band-gap energy than does a semiconductor, so its conduction band is unoccupied by electrons at ordinary temperatures. By contrast, the conduction band in a metal is partly filled and electrons can change levels with minimal input of energy. The conductivities of silicon and other intrinsic semiconductors can be increased by **doping,** the deliberate addition of impurity atoms of selected types. If a Group V element is added to pure silicon, the excess electrons occupy levels just below the conduction band, giving an **n-type semiconductor.** If atoms of a Group III element substitute for silicon atoms, they make a **p-type semiconductor,** in which electrical conduction occurs by the hopping of positive **holes** in the valence band when a voltage is applied.

21-2 **Silicates** are minerals containing tetrahedral SiO_4 structural units. A variety of silicate structures occur in nature, ranging from single tetrahedra to pairs, rings, chains, sheets, and three-dimensional networks of linked tetrahedra, with cations to balance the total charge in the crystal. In **aluminosilicates,** aluminum atoms substitute for some of the tetrahedrally bonded silicon atoms, and additional cations are present as well to maintain a net charge of zero. In **clay** minerals, water can be absorbed between sheets of bonded atoms and cause swelling. **Zeolites** are network aluminosilicates having pores on a molecular scale. They are useful for exchanging ions, for separating mixtures of gases, and as catalysts.

21-3   **Ceramics** are synthetic, inorganic, non-metallic materials. They resist high temperatures and corrosion and are stiff and hard but also brittle, making them liable to sudden failure when stressed. The **microstructures** of these materials are complex, with grains of different phases, cracks, and pores. Ceramics are made by **firing** to a high temperature, which leads to partial melting, to crystal phase changes, and to **sintering,** in which grains grow together and the ceramic body shrinks or **densifies.**

21-4   Many **silicate ceramics** use natural clay minerals such as kaolin as their starting material. A **glaze** is a thin, glassy layer that coats the object, giving it strength and also often decorating it. Different compositions of silicate **glass** are used for different applications. Pure silica (SiO_2) has a high melting point that is lowered by the addition of sodium oxide and calcium oxide to give ordinary **soda-lime glass.** Another silicate ceramic is **Portland cement,** a complex mixture of compounds produced from oxides of calcium, aluminum, silicon, and iron, to which calcium sulfate is added.

21-5   **Oxide ceramics** that do not contain silicon have high melting points and resist corrosion, making them good **refractories. High-density alumina,** a material fabricated with minimal pore size, is useful in cutting tools and armor plating. **Magnesia** combines very high thermal conductivity and heat resistance with very low electrical conductivity. **Non-oxide ceramics** have strong, covalently bonded network structures. They include **silicon nitride,** a strong, tough material with potential uses in the structural components of gas turbines. **Sialons** are ceramic alloys in which some aluminum replaces silicon in silicon nitride with the simultaneous replacement of some nitrogen by oxygen. **Boron nitride** is used for high-temperature ceramic vessels. **Silicon carbide,** or Carborundum, is an abrasive that is now used to reinforce other ceramics. Some oxide ceramics have been found to be **superconducting,** losing all resistance to the flow of an electric current at a certain temperature.

PROBLEMS

Note: Answers to blue-numbered problems are given in Appendix F. Problems that are more challenging are indicated with asterisks.

Semiconductors

1. (See Example 21–1.) Electrons in a semiconductor can be excited from the valence band to the conduction band through the absorption of photons with energies exceeding the band-gap energy. At room temperature, indium phosphide (InP) is a semiconductor that absorbs light only at wavelengths of less than 920 nm. Calculate the band-gap energy in InP.

2. (See Example 21–1.) Both GaAs and CdS are semiconductors that are being studied for possible use in solar cells to generate electric current from sunlight. Their band-gap energies are 2.29×10^{-19} J and 3.88×10^{-19} J, respectively, at room temperature. Calculate the longest wavelength of light that is capable of exciting electrons across the band gap in each of these substances. In which region of the electromagnetic spectrum do these wavelengths fall? Use this result to explain why CdS-based sensors are used in some cameras to estimate the proper exposure conditions.

3. The number of electrons excited to the conduction band by random thermal agitation per cubic centimeter in a semiconductor can be estimated from the equation

$$n_e = (4.8 \times 10^{15} \text{ cm}^{-3} \text{ K}^{-3/2})\, T^{3/2}\, e^{-E_g/(2 RT)}$$

where T is the temperature in kelvins and E_g the band-gap energy. The band-gap energy of diamond at 300 K is 8.7×10^{-19} J. How many electrons are thermally excited to the conduction band at this temperature in a 1.00 cm^3 diamond crystal?

4. The band-gap energy of pure crystalline germanium is 1.1×10^{-19} J at 300 K. How many electrons are excited from the valence band to the conduction band in a 1.00 cm^3 crystal of germanium at 300 K? Use the equation given in the preceding problem.

5. Describe the nature of electrical conduction in (a) Si doped with P and (b) InSb doped with Zn.

6. Describe the nature of electrical conduction in (a) Ge doped with In and (b) CdS doped with As.

7. In a light-emitting diode (LED) of the type used in displays on electronic equipment, watches, and clocks, a voltage is imposed

across an *n-p* semiconductor junction. The electrons on the *n* side combine with the holes on the *p* side and emit light. This process also can be described as the emission of light as electrons fall from levels in the conduction band to empty levels in the valence band. It is the reverse of the production of electric current by illumination of a semiconductor. Many LEDs are made from semiconductors having the general composition $GaAs_{1-x}P_x$. When *x* is varied between 0 and 1, the band gap changes, and with it, the color of light emitted by the diode. When $x = 0.4$, the band-gap energy is 2.9×10^{-19} J. Predict the wavelength and color of the light emitted by this LED.

8. When the LED described in the previous problem has the composition $GaAs_{0.14}P_{0.86}$ (that is, $x = 0.86$), the band-gap energy has increased to 3.4×10^{-19} J. Predict the wavelength and color of the light emitted by this LED.

9. The pigment zinc white (ZnO) turns bright yellow when heated, but the white color returns when the sample is cooled. Does the band gap increase or decrease when the sample is heated?

10. Mercury(II) sulfide (HgS) exists in two different crystalline forms. In cinnabar, the band-gap energy is 3.2×10^{-19} J; in metacinnabar, it is 2.6×10^{-19} J. In some old paintings with improperly formulated paints, the pigment vermilion (cinnabar) has transformed to metacinnabar upon exposure to light. Describe the color change that results.

Silicate Minerals

11. Draw a Lewis electron-dot structure for the disilicate ion ($Si_2O_7^{6-}$). What changes in the structure you have drawn would be necessary to give the structure of the diphosphate ion ($P_2O_7^{4-}$) and the disulfate ion ($S_2O_7^{2-}$)? What is the analogous compound of chlorine?

12. Draw a Lewis electron-dot structure for the cyclosilicate ion $Si_6O_{18}^{12-}$. This ion forms part of the structure of beryl and emerald.

13. (See Example 21–2.) By referring to Table 21–1, predict the structure of each of the following silicate minerals (network, sheets, double chains, and so forth). Give the oxidation state of each atom.
    (a) Andradite ($Ca_3Fe_2(SiO_4)_3$)
    (b) Vlasovite ($Na_2ZrSi_4O_{10}$)
    (c) Hardystonite ($Ca_2ZnSi_2O_7$)
    (d) Chrysotile ($Mg_3Si_2O_5(OH)_4$)

14. (See Example 21–2.) By referring to Table 21–1, predict the structure of each of the following silicate minerals (network, sheets, double chains, and so forth). Give the oxidation state of each atom.
    (a) Tremolite ($Ca_2Mg_5(Si_4O_{11})_2(OH)_2$)
    (b) Gillespite ($BaFeSi_4O_{10}$)
    (c) Uvarovite ($Ca_3Cr_2(SiO_4)_3$)
    (d) Barysilite ($MnPb_8(Si_2O_7)_3$)

15. By referring to Table 21–1, predict the structure of each of the following aluminosilicate minerals (network, sheets, dou-

ble chains, and so forth). In each case, the aluminum atoms grouped with the silicon and oxygen in the formula substitute for Si in tetrahedral sites. Give the oxidation state of each atom.
    (a) Bikitaite ($Li(AlSi_2O_6)$)
    (b) Muscovite ($KAl_2(AlSi_3O_{10})(OH)_2$)
    (c) Cordierite ($Al_3Mg_2(AlSi_5O_{18})$)

16. By referring to Table 21–1, predict the structure of each of the following aluminosilicate minerals (network, sheets, double chains, and so forth). In each case, the aluminum atoms grouped with the silicon and oxygen in the formula substitute for Si in tetrahedral sites. Give the oxidation state of each atom.
    (a) Amesite ($Mg_2Al(AlSiO_5)(OH)_4$)
    (b) Phlogopite ($KMg_3(AlSi_3O_{10})(OH)_2$)
    (c) Thomsonite ($NaCa_2(Al_5Si_5O_{20}) \cdot 6H_2O$)

17. Calcite ($CaCO_3$) reacts with quartz to form wollastonite ($CaSiO_3$) and carbon dioxide.
    (a) Write a balanced chemical equation for this reaction.
    (b) Use the data in Appendix D to calculate ΔH_r° and ΔS_r° at 25°C.
    (c) Estimate the temperature at which $\Delta G_r^\circ = 0$ ($K = 1$).

18. Tremolite ($Ca_2Mg_5Si_8O_{22}(OH)_2$) reacts with calcite ($CaCO_3$) and quartz to give diopside ($CaMgSi_2O_6$), water vapor, and carbon dioxide.
    (a) Write a balanced chemical equation for this reaction.
    (b) At 25°C, the standard enthalpies of formation of tremolite and diopside are $-12,360$ and -3206.2 kJ mol^{-1}, respectively, and their standard entropies are 548.9 and 142.93 J K^{-1} mol^{-1}. Use this information, together with data from Appendix D, to calculate ΔH_r° and ΔS_r°.
    (c) Estimate the temperature at which $\Delta G_r^\circ = 0$ ($K = 1$).

The Properties of Ceramics

19. Ceramics and metals can each have great mechanical strength, but in different ways. Discuss.

20. Compare the effect of slow and rapid heating on a ceramic with the effects on a metal.

21. Compare the ways in which ceramic materials are shaped with the methods used for shaping a metal like steel.

22. Compare the microstructure of a typical ceramic material with that of a metal.

Silicate Ceramics

23. The soft mineral steatite (soapstone) has been much used for the carving of figurines. It is a hydrated magnesium silicate of composition $Mg_3Si_4O_{10}(OH)_2$. Firing steatite at about 1000°C transforms it into a hard two-phase composite of magnesium silicate ($MgSiO_3$) and quartz in much the same way that firing transforms clay minerals into a composite of mullite and cristobalite. Write a balanced chemical equation for the reaction attending the firing of steatite.

24. Pyrophyllite ($Al_2Si_4O_{10}(OH)_2$) is a clay mineral that frequently is used together with or in place of kaolinite. Write

a balanced chemical equation for the production of mullite and cristobalite on the firing of pyrophyllite.

25. Calculate the volume of carbon dioxide produced at 0°C and 1 atm pressure when a sheet of ordinary glass of mass 2.50 kg is made from its starting materials—sodium carbonate, calcium carbonate, and silica. Take the composition of the glass to be $Na_2O \cdot CaO \cdot (SiO_2)_6$.

26. Calculate the volume of steam produced at 600°C and a pressure of 1.00 atm when a 4.0 kg brick of pure kaolinite is dehydrated completely.

27. A sample of soda-lime glass for drinking glasses is analyzed and found to contain the following percentages by mass of oxides: SiO_2, 72.4%; Na_2O, 18.1%; CaO, 8.1%; Al_2O_3, 1.0%; MgO, 0.2%; BaO, 0.2%. (The elements are not actually present as binary oxides, but this is the way glass compositions usually are given.) Calculate the chemical amounts of silicon, sodium, calcium, aluminum, magnesium, and barium atoms per mole of oxygen atoms in this sample.

28. A sample of Portland cement is analyzed and found to contain the following percentages by mass of oxides: CaO, 64.3%; SiO_2, 21.2%; Al_2O_3, 5.9%; Fe_2O_3, 2.9%; MgO, 2.5%; SO_3, 1.8%; Na_2O, 1.4%. Calculate the chemical amounts of calcium, silicon, aluminum, iron, magnesium, sulfur, and sodium atoms per mole of oxygen atoms in this sample.

29. The most important contributor to the strength of hardened Portland cement is tricalcium silicate $((CaO)_3 \cdot SiO_2)$, for which the standard enthalpy of formation at 25°C is -2929.2 kJ mol^{-1}. Calculate the standard enthalpy change for the production of 1.00 mol of tricalcium silicate from quartz and lime.

30. One of the simplest of the heat-generating reactions that takes place when water is added to cement is the production of calcium hydroxide (slaked lime) from lime. Write a balanced chemical equation for this reaction, and use data from Appendix D to calculate the amount of heat generated by the reaction of 1.00 kg of lime with water at room conditions.

Non-Silicate Ceramics

31. Silicon carbide (SiC) is made by the high-temperature reaction of silica sand (quartz) with coke; the by-product is carbon monoxide.
    (a) Write a balanced chemical equation for this reaction.
    (b) Estimate the standard enthalpy change per mole of SiC(s) produced.
    (c) Predict (qualitatively) the following physical properties of silicon carbide: conductivity, melting point, and hardness.

32. Boron nitride (BN) is made by the reaction of gaseous boron trichloride with gaseous ammonia.
    (a) Write a balanced chemical equation for this reaction.
    (b) Calculate the standard enthalpy change per mole of BN produced, given that the standard molar enthalpy of formation of BN(s) equals -254.4 kJ mol^{-1} at 25°C.

(c) Predict (qualitatively) the following physical properties of boron nitride: conductivity, melting point, and hardness.

33. The standard Gibbs energy of formation of solid silicon carbide (SiC) at 25°C is -62.8 kJ mol^{-1}. Write an equation for the reaction of 1 mol of SiC(s) with $O_2(g)$ to form SiO_2 (*quartz*) and $CO_2(g)$ and compute ΔG_r°. Is silicon carbide thermodynamically stable in air at room conditions?

34. The standard Gibbs energy of formation of boron carbide (B_4C) is -71 kJ mol^{-1} at 25°C. Write an equation for the reaction of 1 mol of $B_4C(s)$ with $O_2(g)$ to form $B_2O_3(s)$ and $CO_2(g)$ and compute ΔG_r°. Is boron carbide thermodynamically stable in air at room conditions?

35. Calculate the average oxidation number of the copper in $YBa_2Cu_3O_{(9-x)}$ if $x = 2$. Assume that the rare-earth element yttrium is in its usual +3 oxidation state.

36. The mixed oxide ceramic $Tl_2Ca_2Ba_2Cu_3O_{(10+x)}$ has zero electrical resistance at 125 K. Calculate the average oxidation number of the copper in this compound if $x = 0.50$ and thallium is in the +3 oxidation state.

ADDITIONAL PROBLEMS

37. Compare the hybridization of silicon atoms in Si(s) with that of carbon atoms in graphite (see Fig. 20–30). If silicon were to adopt the graphite structure, would its electrical conductivity be high or low?

38. Describe how the band gap varies from a metal to a semiconductor to an insulator.

39. Suppose that some people are sitting in a row at a movie theater, with a single empty seat on the left end of the row. Every 5 minutes, a person moves into a seat on his or her left if it is empty. In what direction and with what speed does the empty seat "move" along the row? Comment on the connection with hole motion in *p*-type semiconductors.

40. A sample of silicon doped with antimony is an *n*-type semiconductor. Suppose that a small amount of gallium is added to such a semiconductor. Describe how the conduction properties of the solid vary with the amount of gallium added.

41. Silica (SiO_2) exists in several forms, including quartz (molar volume, 22.69 cm^3 mol^{-1}) and cristobalite (molar volume, 25.74 cm^3 mol^{-1}).
    (a) Use data from Appendix D to calculate ΔH°, ΔS°, and ΔG° for the conversion of quartz to cristobalite at 25°C.
    (b) Which form is thermodynamically stable at 25°C?
    (c) Which form is stable at very high temperatures, provided that melting does not take place first?

42. Predict the structure of each of the following silicate minerals (network, sheets, double chains, and so forth). Give the oxidation state of each atom.
    (a) Apophyllite $(KCa_4(Si_8O_{20})F \cdot 8H_2O)$
    (b) Rhodonite $(CaMn_4(Si_5O_{15}))$
    (c) Margarite $(CaAl_2(Al_2Si_2O_{10})(OH)_2)$

43. By referring to Table 21–1, predict the structure of manganpyrosmalite, a silicate mineral with the chemical formula $Mn_6Fe_2Si_6O_{15}(OH)_6Cl_4$. Give the oxidation state of each atom in this formula unit.

44. A reference book lists the chemical formula of one form of vermiculite as $[(Mg_{2.36}Fe_{0.48}Al_{0.16})(Si_{2.72}Al_{1.28})O_{10}(OH)_2]$ $[Mg_{0.32}(H_2O)_{4.32}]$. Determine the oxidation state of the iron in this mineral.

45. When talc ($Mg_3Si_4O_{10}(OH)_2$) reacts with forsterite (Mg_2SiO_4), enstatite ($MgSiO_3$) and water vapor form.
(a) Write a balanced chemical equation for this reaction.
(b) If these minerals are all present at equilibrium and the pressure is increased, is there net formation of products or net formation of reactants?

46. The most common feldspars are those containing potassium, sodium, and calcium cations. They are called, respectively, orthoclase ($KAlSi_3O_8$), albite ($NaAlSi_3O_8$), and anorthite ($CaAl_2Si_2O_8$). The solid solubility of orthoclase in albite is limited, and its solubility in anorthite is almost negligible. Albite and anorthite, however, are completely miscible at high temperatures and show complete solid solution. Give an explanation for these observations, based on the radii of the K^+, Na^+, and Ca^{2+} ions, which are 1.33 Å, 0.98 Å, and 0.99 Å, respectively.

47. The clay mineral kaolinite ($Al_2Si_2O_5(OH)_4$) is formed by the weathering action of water containing dissolved carbon dioxide on the feldspar mineral anorthite ($CaAl_2Si_2O_8$). Write a balanced chemical equation for the reaction that occurs. The CO_2 forms H_2CO_3 as it dissolves. As the pH is lowered, does the weathering occur to a greater or a lesser extent?

48. The formulas of certain zeolites can be written as $M_2O \cdot Al_2O_3 \cdot ySiO_2 \cdot wH_2O$, where M is an alkali metal such as Na or K, y is 2 or more, and w is any integer. Compute the mass percentage of Al in a zeolite that has M = K, $y = 4$, and $w = 6$.

49. Iron oxides are red when the average oxidation state of the iron is high and black when it is low. To impart each of these colors to a pot made from clay that contains iron oxides, would you use an air-rich or a smoky atmosphere in the kiln? Explain.

50. In what ways does soda-lime glass resemble, and in what ways does it differ from, a pot made from the firing of kaolinite? Include the following aspects in your discussion: composition, structure, physical properties, and method of preparation.

51. Leaded glass is used for fine crystal in place of soda-lime-silica glass. Table 21–2 shows that it is "heavier" (denser); it also has a higher refractive index. The latter property gives the leaded glass its attractive sparkle. What is the ratio of the number of lead atoms to silicon atoms for leaded glass of the composition listed in Table 21–2?

*52. The cement industry is one of the nation's largest consumers of energy because of the high temperatures needed to form Portland cement from its oxide starting materials. One way that has been proposed to reduce energy costs is to incorporate calcium fluoride as a flux in the oxide mixture, in much the same way that limestone is used in steel production (see Section 12–4). What effect would the CaF_2 have, and how would energy consumption be reduced as a result?

53. Burnt dolomite bricks are used in the linings of furnaces in the cement and steel industries. Pure dolomite contains 45.7% $MgCO_3$ and 54.3% $CaCO_3$ by mass. Determine the empirical formula of dolomite.

54. Refractories can be classified as acidic or basic, depending on the properties of the oxides in question. A basic refractory must not be used in contact with acid, and an acidic refractory must not be used in contact with base. Classify magnesia and silica as acidic or basic refractories.

*55. A book states correctly that beryllia (BeO) ceramics have some use but possess only poor resistance to strong acids and bases. It goes on to state that beryllia reacts with water vapor at high temperature to give off toxic vapors. Write likely chemical equations for the reaction of BeO with a strong acid, with a strong base, and with steam, $H_2O(g)$.

56. Some ceramics containing thoria (ThO_2) have been made. Identify a likely problem in the use of such materials.

57. Silicon nitride resists all acids except hydrofluoric, with which it reacts to give silicon tetrafluoride and ammonia. Write a balanced chemical equation for this reaction.

58. Compare oxide ceramics like alumina and magnesia, which have significant ionic character, with covalently bonded non-oxide ceramics like SiC and B_4C (Problems 33 and 34), with respect to thermodynamic stability at ordinary conditions.

59. A new inorganic polymer is developed. It is made from plentiful starting materials by simple mixing, followed by heating to high temperature. When hot, it is readily shaped and molded. When cooled, it solidifies to a hard, transparent, brittle solid that can be cut and polished easily. It is completely non-flammable, resists attack by most chemicals, can be sterilized easily for medical uses, and is far more durable than are organic polymers ("plastics"); however, its manufacture consumes much energy, and it is not biodegradable. Its brittleness leads to occasional catastrophic loss of structural integrity in use. These episodes create sharp-edged fragments. Sooner or later, everybody using this ceramic will probably suffer injuries from it. As a high government official, you must decide whether to license or forbid the production of this material. What do you do? Explain.

22 Chemical Processes

CHAPTER OUTLINE

A plant used in the production of ammonia from nitrogen and hydrogen via the Haber–Bosch process. *(Dresser Industries/Kellogg Company)*

$\mathbf{A}$t the heart of practical chemistry lie the processes that transform starting materials into desired products. This chapter introduces a **process-based** approach to understanding the operations of chemical industry, emphasizing the availability of raw materials and the effect of pollutants on the environment. The process-based approach then is applied in describing how certain important compounds of sulfur, phosphorus, and nitrogen are made; the modern uses of these compounds also are discussed.

22-1 The Chemical Industry

A firm grasp of chemical principles is essential to the design of clean and efficient manufacturing processes for chemical products. Industrial processes are under constant development to improve yields, reduce waste, and minimize expense. Chemical plants are costly, and a new process invariably passes through several **pilot-plant** stages after its development in the research laboratory. In these scaling-up operations, the process is optimized with respect to such variables as temperature, pressure, and time of reaction. Almost all laboratory-scale chemical processes are **batch processes;** the chemist mixes the reactants in a container under conditions that favor the desired reaction, then subsequently removes and separates the product. Many industrial-scale processes, in contrast, are **continuous processes,** designed so that reactants go in and products come out simultaneously. It is seldom easy to convert a process from batch to continuous mode, but continuous operation is generally more economical.

Processes in the chemical industry can be represented schematically (Fig. 22–1) as

$$\text{starting materials} \longrightarrow \text{desired products} + \text{by-products}$$

Energy, often a great deal of it, plays a part in these transformations. Sometimes energy must be put in to make the desired reaction happen. Sometimes energy is produced, usually as heat, and must be either recycled or safely discarded. Let us examine each of the terms in this schematic process in light of the basic principles of chemistry.

Starting Materials. In the ideal case, the starting materials for a chemical process are easily obtained and transported to the production site. Starting materials tend to have low Gibbs energies simply because substances of high Gibbs energy mostly have disappeared, over geological time, in spontaneous reactions to give substances of low Gibbs energy. Thus, aluminum is abundant in the terrestrial crust, not in elemental form but tied up in materials of much lower Gibbs energy such as the clay kaolinite, $Al_2Si_2O_5(OH)_4 (\Delta G_f^\circ = -3800 \text{ kJ mol}^{-1})$. Some compounds of high Gibbs energy do persist in nature because their rates of reaction are very slow or they are walled off in their geological deposits. Hydrocarbons such as octane occur in petroleum deposits and have high Gibbs energies relative to their combustion products:

$$C_8H_{18}(g) + \tfrac{25}{2} O_2(g) \longrightarrow 9 H_2O(\ell) + 8 CO_2(g) \qquad \Delta G_r^\circ = -5272 \text{ kJ mol}^{-1}$$

This makes them exceedingly valuable as chemical starting materials for polymers, fibers, detergents, and other goods, not just as fuels.

• The Earth, taken as a whole, is not at equilibrium. As long as energy in the form of solar radiation continues to arrive, new local build-ups of compounds of high Gibbs energy are possible.

• The processing and uses of hydrocarbons are considered in Chapter 24.

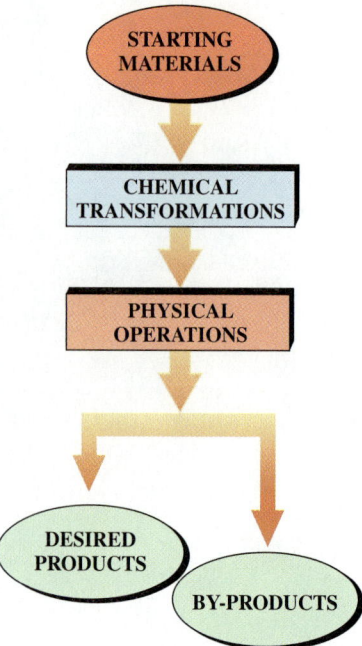

Figure 22-1 • A generalized chemical process. This figure employs the color conventions that are followed through the next chapters. Starting materials (*red*) are converted to products (*green*) via a series of chemical reactions (*blue*) and physical operations (*orange*).

Wallboard ("dry wall") has a core of calcium sulfate dihydrate (gypsum, $CaSO_4 \cdot 2H_2O$) held between layers of thick paper. (*Charles D. Winters*)

Energy. A desired but non-spontaneous reaction (one with $\Delta G_r > 0$) can in principle be caused to proceed by linking it to another reaction for which $\Delta G_r < 0$. In many chemical processes, the linkage is accomplished using electricity. The powerful oxidizing (bleaching) agent sodium chlorate ($NaClO_3$), for example, is made by the overall reaction

$$NaCl(aq) + 3\,H_2O(\ell) \longrightarrow NaClO_3(aq) + 3\,H_2(g) \qquad \Delta G_r^\circ = +835 \text{ kJ mol}^{-1}$$

The reaction is driven by electrical work that comes from burning fossil fuels in a conventional power plant or from a nuclear reactor. Many chemical transformations are practical only if inexpensive energy sources are available. It is very desirable to use solar energy directly to carry out chemical reactions, as nature does in photosynthesis, the process by which green plants convert water and carbon dioxide (which have low Gibbs energies) to glucose and other sugars (which have high Gibbs energies). Much current research is aimed at understanding photosynthesis and copying its principles to carry out other chemical transformations.

Desired Products. An ideal chemical process would give the desired product (1) in 100% yield at room temperature and pressure, (2) at a convenient and controllable rate, and (3) already pure or in an easily purified form. Obviously, compromises are necessary in real processes. The equilibrium yield of a chemical reaction is strongly influenced by the pressure and temperature. Thermodynamics provides a way to estimate the optimal conditions for a high yield, but expensive equipment may be required to achieve those conditions. The rate of a reaction also depends on pressure and temperature and is sensitive to the presence of catalysts as well. The control of reaction rates is so important in the chemical industry that researchers search constantly for catalysts that speed the formation of desired products without leading to unwanted by-products. At the same time, they must anticipate the consequences of an unexpected acceleration of a reaction or the sudden onset of runaway side-reactions (caused perhaps by catalytic impurities in the starting materials). Finally, the last point—separability of the desired product—cannot be overemphasized. A process that yields a product contaminated with large amounts of hard-to-separate (or dangerous) impurities may face abandonment, even if it gives good yields at a convenient rate.

By-products. Even when the by-products of chemical processes are relatively harmless, the question of what to do with them remains. Each year, a small mountain of $CaSO_4$ is produced in the course of synthesizing phosphate fertilizers. Although $CaSO_4$ is harmless in itself (it is used in making wallboard), large volumes of it create awkward and expensive disposal problems. Other by-products are toxic to plant and animal life, ranging from slightly irritating to deadly. The practice of burying toxic wastes in landfills risks the contamination of water sources and the direct exposure of nearby residents. It is now thoroughly regulated. Strong social, economic, and legal incentives now exist to reduce or eliminate worthless and hazardous by-products. The current concept is that pollution and the cost of its removal are product defects. The U.S. Pollution Prevention Act of 1990 was the first environmental law to focus on preventing pollution at the source rather than trying for so-called end-of-the pipe solutions. Since 1990, various agencies of the U.S. government have funded iniatives in "Green Chemistry," in "environmentally benign chemical synthesis and processing," and in "environmentally conscious manufacturing." These programs encourage avoidance of toxic feedstocks and solvents and the reduction of the number of process steps. Parts of the chemical industry have responded by modifying their processes to minimize the generation of wastes. Other measures that diminish adverse environmental impact are the recycling of materials such as glass and paper and the incineration of combustible wastes. When adequate precautions are taken to ensure complete combustion, incineration converts even very hazardous hydrocarbons into carbon dioxide and water vapor. One experimental method for the disposal of hazardous wastes injects them into an electrically generated plasma in which the temperature exceeds 10,000°C. Another uses oxygen dissolved in supercritical water to oxygenate pollutants and produce benign, smaller molecules (see Chemistry in Your Life: An Exotic (But Useful) State of Matter in Chapter 6).

The ideal way to dispose of by-products is to find a use for them. For example, the uranium fuel-processing industry uses nitric acid to oxidize scrap uranium that is contaminated with dirt and grease to uranyl nitrate ($UO_2(NO_3)_2$), a compound that is soluble in water and more easily purified. The oxidation reaction gives off the nitrogen oxides NO and NO_2, which are pollutants. Vented as waste, the gases would contribute to acid rain (see Section 18–5). Instead, they are caused to react first with oxygen and then with water to give nitric acid:

$$NO(g) + \tfrac{1}{2} O_2(g) \longrightarrow NO_2(g)$$
$$3\,NO_2(g) + H_2O(\ell) \longrightarrow 2\,HNO_3(aq) + NO(g)$$

These reactions reduce the amount of nitrogen oxides released to the atmosphere and regenerate nitric acid for use in the oxidation reaction.

Few chemical processes occur in one stage. Most require many steps, with numerous chemicals involved along the way that do not appear in the finished product. Figure 22–2 shows how aspirin is made from primary raw materials—air, water, salt, sulfur, petroleum, and natural gas—and how chlorine, acetone, hydrogen, and sodium sulfate are produced as by-products. The figure also shows the interdependency built into the chemical industry. An aspirin manufacturer does not start with the primary raw materials but instead buys intermediate chemicals that may have applications in a variety of other fields. He would buy sulfuric acid from a supplier who makes far larger quantities—for use by steel mills, for example.

Table 22–1 names the top 15 chemicals in the United States during 2000 and gives the mass of each that was produced. Few of these chemicals are used *directly* as consumer products. Household ammonia, for example, is a negligible fraction of the total ammonia produced, most of which undergoes further chemical transformations for use as fertilizer in forms that include ammonium nitrate, urea, and ammonium phosphate. This and the following chapters discuss the methods by which some of these chemicals are produced. Of course, chemicals do not have to be produced in mountainous quantities to be important. For example, pharmaceuticals are vital to health care even though they are produced in relatively small amounts.

Raw Materials

The raw materials of the heavy-chemicals industry (Fig. 22–3) are relatively few in number, chiefly air, water, limestone, salt, coal, natural gas and petroleum, sulfur, and phosphate rock, simply because these materials are abundantly available at low cost. Their sources are listed in Table 22–2. Both geochemical and biochemical transformations are responsible for the generation of economically useful quantities of these raw materials near the Earth's surface.

• Even complete combustion has its negative side because large-scale emissions of CO_2 contribute to the greenhouse effect, described in Section 18–5.

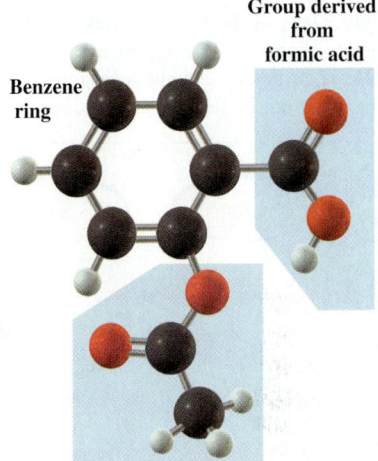

The structure of acetylsalicylic acid, the active component of aspirin. Note the substructures, which are those of three simpler compounds: benzene, formic acid, and acetic acid.

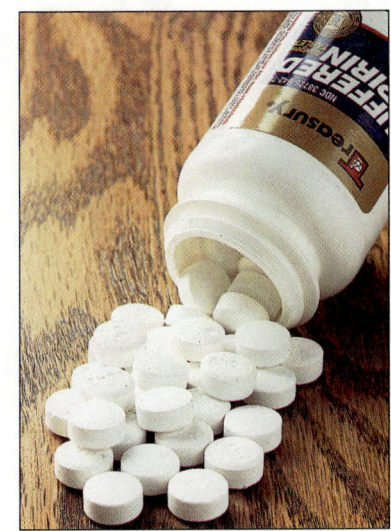

Aspirin tablets. *(George Semple)*

Geochemical transformations are slow. Time scales range from years (for the weathering of the softest rocks) to millions and even billions of years. One reason for this slowness is that geochemical processes are commonly solid-state reactions, and the diffusion of atoms through solids is many orders of magnitude slower than that through liquids and gases. Geochemical transformations begin with the crystallization of a **magma,** a molten silicate fluid forced out from beneath the Earth's surface, to form **igneous** rocks and to vent volatile components, such as water, sulfur dioxide, and the oxides of nitrogen, that might be trapped or dissolved in the magma. Although the composition of magmas varies, the elements oxygen, silicon, and aluminum predominate, with smaller amounts of iron, mag-

Figure 22–2 • A process flow diagram for the synthesis of aspirin, which derives ultimately from air, water, salt, sulfur, petroleum, and natural gas. Many of the intermediates are organic chemicals that appear in Chapter 24. Their names and chemical formulas need not be learned at this stage.

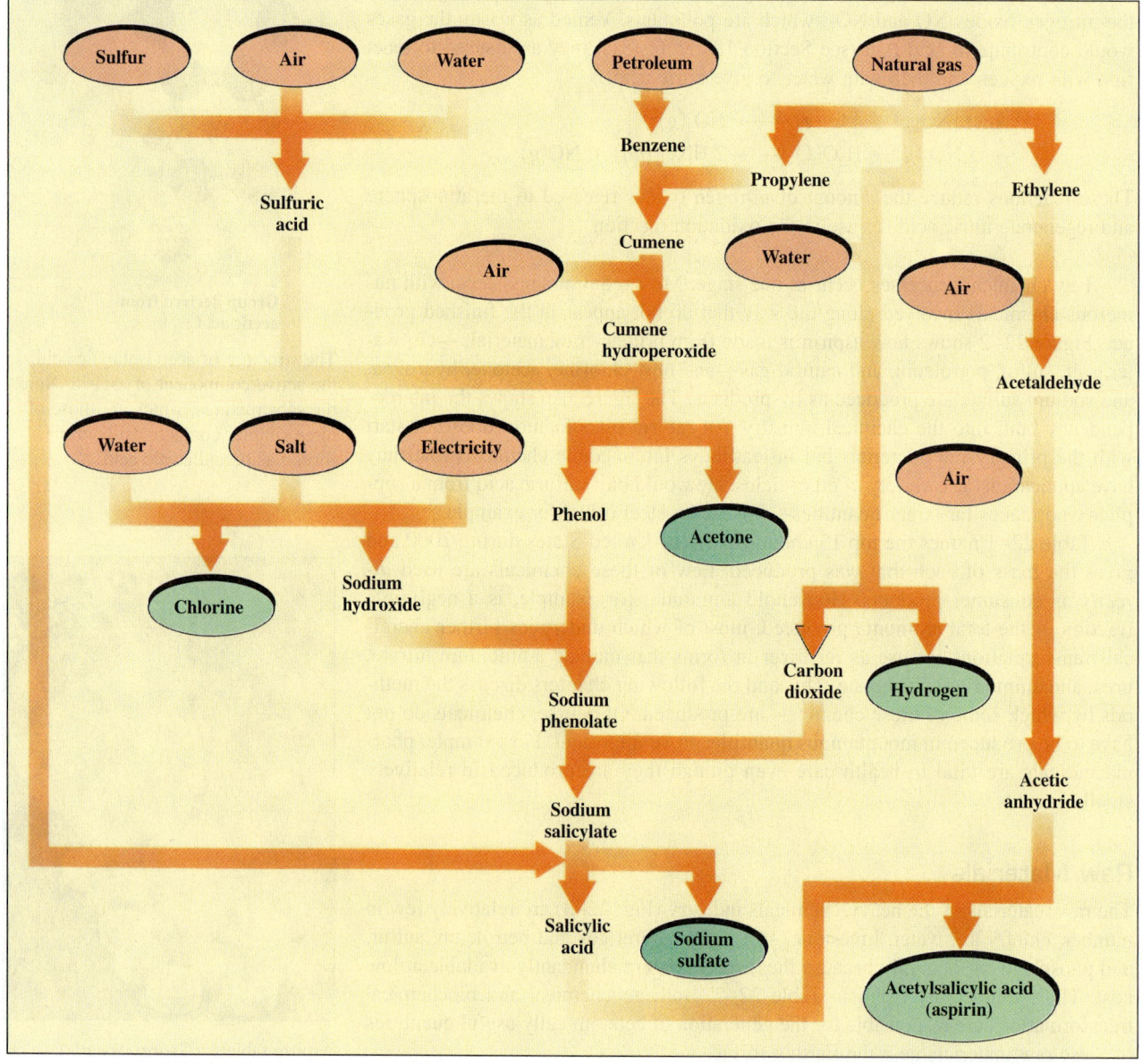

nesium, calcium, sodium, and potassium. As a magma cools, it crystallizes in stages, giving minerals with different compositions and structures. Although aluminum, iron, and magnesium are abundant in igneous rocks, they are so tightly bound in minerals that the rocks are not economic sources for these elements. Minerals in igneous rocks do provide some of the less common metals, however, such as molybdenum, tin, and uranium.

Sedimentary rocks form from the weathering of other rocks. Water (especially moving water in streams) breaks rocks into smaller pieces and dissolves them partially. Carbon dioxide, which forms carbonic acid when dissolved, causes further dissolution and chemical reaction. Quartz, a form of silica (SiO_2), is left behind in the weathering process and provides an important source of silicon. Aluminum precipitates in clays and in the mineral bauxite (hydrated Al_2O_3), and iron forms oxide ores. Calcium carbonate ($CaCO_3$) precipitates as limestone. More soluble salts remain in solution and can be found in the oceans, in salt brines in deep wells, and as solid deposits where the water has evaporated. Thus, sylvite (KCl) is primarily obtained as a mineral deposit, and $MgCl_2$ is extracted from brines or from the ocean. Both brines and minerals are important sources of sodium chloride.

TABLE 22-1	Top 15 Chemicals by Mass in the United States (2000)*		
Rank and Name	**Formula**	**U.S. Production (in 10^9 kg)**	**Principal End Uses**
1 Sulfuric acid	H_2SO_4	39.62	Fertilizers, chemicals, processing
2 Ethylene	C_2H_4	25.15	Plastics, antifreeze
3 Lime	CaO	20.12	Paper, chemicals, cement
4 Phosphoric acid	H_3PO_4	16.16	Fertilizers
5 Ammonia	NH_3	15.03	Fertilizers
6 Propylene	C_3H_6	14.45	Gasoline, plastics
7 Chlorine	Cl_2	12.01	Bleaches, plastics, water purification
8 Sodium hydroxide	NaOH	10.99	Chemical processing, aluminum production, soap
9 Sodium carbonate	Na_2CO_3	10.21	Glass
10 Ethylene dichloride	$C_2H_4Cl_2$	9.92	Plastics, drycleaning
11 Nitric acid	HNO_3	7.79	Fertilizers, explosives
12 Ammonium nitrate	NH_4NO_3	7.49	Fertilizers, mining
13 Urea	H_2NCONH_2	6.96	Fertilizers, animal feed, plastics
14 Ethyl benzene	$C_6H_5C_2H_5$	5.97	Intermediate in the production of styrene, solvent
15 Styrene	$C_6H_5CHCH_2$	5.41	Plastics

*The list includes homogeneous chemical substances only (and so excludes steel, concrete and other mixtures). It omits petrochemical feedstocks and substances such as salt (NaCl), gypsum ($CaSO_4$), sulfur (S), oxygen (O_2), and nitrogen (N_2) that do not require processing. Adapted from *Chemical and Engineering News,* June 25, 2001.

(a)   (b)   (c)

Figure 22–3 • Several of the important raw materials for industrial chemistry. (a) Coal, mined from both open pits and tunnels. *(Frank Grant/International Stock Photo)* (b) Limestone, being quarried here as a source of lime. *(Ireco, Inc.)* (c) Water, needed in good purity and large quantities. *(Arturo A. Wesley/Black Star)*

The minerals that make up the Earth's crust undergo continual structural and chemical transformations. **Metamorphic** rocks form in response to heat and pressure below the Earth's surface. An example is the conversion of coarsely crystalline limestone into the compact, coherent mass of small crystals known as marble. As metamorphism proceeds, the elemental composition of the metamorphosed rocks becomes more uniform. High-grade metamorphism merges eventually (at high temperatures) with magma formation.

TABLE 22–2	Sources of Major Raw Materials for Chemistry	
Atmosphere		**Hydrosphere**
Oxygen (O_2)		Water (H_2O)
Nitrogen (N_2)		Sodium bromide (NaBr)
Noble gases		Sodium iodide (NaI)
		Magnesium chloride ($MgCl_2$)
		Sodium chloride (NaCl)
Lithosphere		**Biosphere**
Silica (SiO_2)		Coal
Limestone ($CaCO_3$)		Petroleum
Sodium chloride (NaCl)		Natural gas
Potassium chloride (KCl)		Sulfur (S)
Sodium carbonate (Na_2CO_3)		Phosphate rock ($Ca_5(PO_4)_3F$)
Sodium sulfate (Na_2SO_4)		Organic natural products
Metal ores		

The atmosphere is the envelope of gases that surrounds the Earth. The lithosphere (from the Greek *lithos,* meaning "stone") includes the outer parts of the solid Earth. *Hydrosphere* refers to water in its many forms. *Biosphere* refers to living organisms, their immediate surroundings, and their products.

A lava flow at the Kilauea rift in Hawaii. Lava is a molten silicate fluid that forms igneous rock as it solidifies. *(Soames Summerhays/Science Source/PhotoResearchers, Inc.)*

If geochemical processes are characterized by their slowness, biochemical processes are notable for their speed and selectivity. Natural products obtained from plants and animals play a large role in chemistry, being used both directly and as precursors for chemical syntheses of other substances. The terpenes, for example, are a class of compounds found in almost all plants; they are a major component of turpentine, which is obtained from the resin of pine trees. Many terpenes are used directly as artificial scents in the perfume industry; others serve as precursors in vitamin A synthesis (Fig. 22–4).

The distinction between chemical and biological synthesis has become blurred. On the one hand, living agents such as specialized bacteria are finding increasing use in the synthesis of complex products that are difficult to produce by conventional chemical means. On the other hand, certain natural products are obtained more easily from non-biological materials than from their traditional sources. Methanol, for example, once was obtained by the destructive distillation

(a) β-ionone

(b) Vitamin A

Figure 22–4 • (a) The terpene β-ionone ($C_{13}H_{20}O$) is used extensively in perfumes. It is also a precursor in the synthesis of vitamin A. (b) The structure of vitamin A derives from that of β-ionone. It resembles the structures of retinal (crucial for vision, Problem 18–65) and the dye β-carotene (see Fig. 18–24). Carrots in the diet supply vitamin A for metabolic transformation to retinal. This explains why eating carrots can improve night vision.

• Hence the name "wood alcohol."

of wood. It now is produced mainly through the reaction of synthesis gas, outlined in Section 7–5.

The fossil fuels are among the major chemical raw materials produced by biological activity. The decay of plant matter mainly through the action of bacterial reducing agents creates first peat, then lignite, and finally bituminous coal. Crude oil (petroleum) is thought to have formed from the decomposition under reducing conditions of the remains of small marine organisms. Hydrocarbons of low molar mass formed in this way are gases rather than liquids. Pockets of natural gas (primarily methane, with smaller amounts of hydrogen, ethane, and propane) therefore are found in association with deposits of petroleum. Although all of these fossil fuels have a biological origin, it is obvious that geological transformations brought them to the form in which we find them today.

A chemical raw material with an unexpected biological origin is sulfur. Although sulfur is distributed widely in compounds, its most convenient commercial source is the large deposits of elemental sulfur often associated with underground geological structures called salt domes. It is thought that calcium sulfate precipitated in these settings and came in contact with natural gas and carbon dioxide. A reaction among these substances, catalyzed by bacterial enzymes, then gave elemental sulfur and calcium carbonate:

$$4\,CaSO_4(s) + 3\,CH_4(g) + CO_2(g) \longrightarrow 4\,CaCO_3(s) + 4\,S(s) + 6\,H_2O(\ell)$$

The origins of phosphate rock, which has the approximate chemical formula $Ca_5(PO_4)_3F$, are less clear. The rock is found in large deposits in certain parts of the world, notably Florida and northern Africa. Phosphate ions are present in rather low concentration in seawater, and the solubility of calcium phosphate is comparable to that of calcium carbonate. Therefore, it is surprising that the phosphate is found in such a concentrated form, rather than scattered through the world's limestone deposits. Deposits of phosphate rock may have resulted from extensive dissolution and reprecipitation of phosphate–carbonate mixtures, perhaps in the small, but critical, pH range in which the phosphate precipitates and the carbonate remains in solution. It is more likely, however, that the phosphate rock resulted from accumulation by marine organisms. Phosphate rock is the source of phosphorus and phosphate-based fertilizers. As such, it is the raw material used in greatest volume in the chemical industry (see Section 22–3).

The foxglove plant produces digitalis, an important drug in the treatment of heart disease. *(Pamela Harper)*

22-2 Production of Sulfuric Acid

The consumption of sulfuric acid, the chemical produced in the greatest amount in the world today, has come to indicate the health of a country's entire manufacturing base. The uses of this strongly acidic, extremely reactive compound range from fertilizer manufacture to metal treatment to the production of pharmaceuticals. Its central role in the modern chemical industry was established largely between 1750 and 1900. In 1750, sulfuric acid was produced and used on a very modest scale in metal assaying and treatment, and by some doctors for "cures" that had no scientific basis. New uses for the acid stimulated the search for new ways to produce it; as the price went down, additional uses were discovered and exploited. From 1750 to 1900, the price of sulfuric acid declined steadily, and the amount produced grew explosively. This mutual stimulation between new uses and new processes is the hallmark of the chemical industry, and the story of sulfuric acid is a good way to illustrate it.

In the Middle Ages alchemists heated crystalline green vitriol (iron(II) sulfate heptahydrate) to produce sulfuric acid for the first time:

$$FeSO_4 \cdot 7H_2O(s) \longrightarrow H_2SO_4(\ell) + FeO(s) + 6\,H_2O(g)$$

By the 17th century, sulfuric acid was being made on a limited commercial scale from iron pyrite ores. The FeS_2 in these ores was converted to $FeSO_4$ and then oxidized to iron(III) sulfate ($Fe_2(SO_4)_3$). This solid was broken up and strongly heated in small clay vessels, a treatment that decomposed it into iron(III) oxide and sulfur trioxide:

$$Fe_2(SO_4)_3(s) \longrightarrow Fe_2O_3(s) + 3\,SO_3(g)$$

The gaseous product was absorbed in water to make aqueous sulfuric acid:

$$SO_3(g) + H_2O(\ell) \longrightarrow H_2SO_4(aq)$$

Production of sulfuric acid was stimulated by the discovery in 1744 that the valued dyestuff indigo, used since antiquity to dye cotton, also could be used as a wool dye after treatment with concentrated sulfuric acid.

• The iron in FeS_2 is in the $+2$, not the $+4$, oxidation state because the sulfur occurs as S_2^{2-} ions, analogous to the peroxide ion O_2^{2-}.

• Blue jeans are dyed with indigo.

The Lead-Chamber Process for Sulfuric Acid

The lead-chamber process, developed in the mid-18th century, probably also had its origins in the laboratories of alchemists, who burned sulfur in earthenware vessels. The small amounts of SO_3 that formed (alongside the main product, SO_2) were condensed and absorbed into water to make sulfuric acid. An accidental discovery revealed that the addition of sodium nitrate ($NaNO_3$) or potassium nitrate (KNO_3) improved the yield of SO_3. These compounds decompose when heated to give nitrogen dioxide, which reacts with SO_2 to give SO_3:

$$SO_2(g) + NO_2(g) \longrightarrow SO_3(g) + NO(g)$$

In 1736, Joshua Ward took the next important step by replacing the earthenware vessels with large glass bottles arranged in series to speed up the process.

The scale of manufacture of sulfuric acid was increased further, and dramatically, by the development of the room-sized **lead chamber,** first used by John Roebuck in 1746. While earthenware vessels had outputs of several ounces, and Ward's glass bottles had outputs of several pounds, the lead-chamber method gave ton quantities of sulfuric acid at a reduced price. In the lead-chamber process, a mixture of sulfur and potassium nitrate was placed in a ladle and ignited inside a large lead-lined room, the floor of which was covered with water. The gases condensed on the lead walls and were absorbed in the water. After the process was repeated several times, the dilute sulfuric acid was removed and concentrated by boiling it. Later developments included blowing in steam to speed up the reaction with water and disperse the gases, and separating the burning chamber from the absorption chamber.

Joseph Gay-Lussac improved the process significantly in 1835, when he built a tower to recover the by-product NO and convert it back to NO_2 by reaction with oxygen. More precisely, in a Gay-Lussac tower, NO was converted to nitrous acid (HNO_2) dissolved in aqueous sulfuric acid

$$2\,NO(g) + \tfrac{1}{2}O_2(g) + H_2O(\ell) \longrightarrow 2\,HNO_2(aq)$$

which then reacted in a second tower, named after its developer, John Glover, to oxidize sulfur dioxide:

$$2\,HNO_2(aq) + SO_2(g) \longrightarrow H_2SO_4(aq) + 2\,NO(g)$$

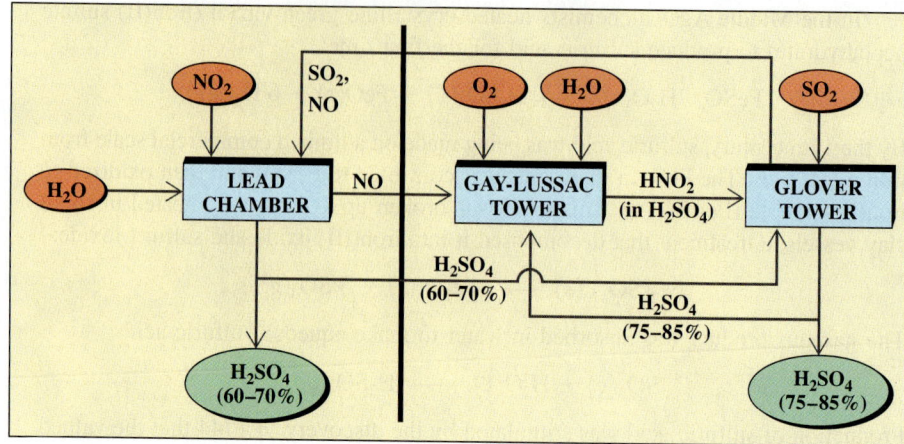

Figure 22–5 • The Glover–Gay-Lussac process for making sulfuric acid. The portion of the diagram to the left of the black line shows the original lead-chamber process, which produced a more dilute acid. The original process also consumed and emitted more nitrogen oxides than the full process as it later developed.

The overall reaction from the two steps was

$$SO_2(g) + \tfrac{1}{2} O_2(g) + H_2O(\ell) \longrightarrow H_2SO_4(aq)$$

Recycling the oxides of nitrogen greatly reduced the consumption of sodium or potassium nitrate, which now was needed only to make up for losses in the process. In addition, the Glover tower gave a more concentrated solution of sulfuric acid, one containing 75% to 85% H_2SO_4 by mass, compared with the 60% to 70% obtained by earlier methods. Figure 22–5 is a flow diagram for the full process after the development of the Glover tower in 1859.

The Contact Process

As early as 1831, the Englishman Peregrine Phillips observed that platinum could catalyze the conversion of SO_2 to SO_3 that lies at the heart of the production of sulfuric acid. Little attention was paid to this discovery until the 1870s, when the growth of the German dye industry stimulated a search for a method of producing sulfuric acid that was more concentrated than that from the Glover tower. Platinum catalysts were rediscovered and patented but initially had limited usefulness because they were poisoned by impurities in the sulfur dioxide feed. As a result, the early applications of this method used, somewhat paradoxically, sulfuric acid from the Glover tower as a raw material. It was decomposed by heating,

$$H_2SO_4(aq) \longrightarrow SO_2(g) + H_2O(\ell) + \tfrac{1}{2} O_2(g)$$

and the relatively pure SO_2 was converted into SO_3 over the catalyst and then back to H_2SO_4. In this way, a highly concentrated acid was obtained, but at great expense. Over the next 30 years, researchers recognized the role of arsenic and other impurities in poisoning the catalyst and found ways to remove the impurities from the SO_2 feed, allowing for the production of highly concentrated sulfuric acid without the intermediate lead-chamber step. Different catalysts were studied as well, leading eventually to the selection of an oxide of vanadium (V_2O_5), which is the principal catalyst in use today.

The catalytic production of sulfuric acid from SO_2 is called the **contact process** and is outlined in Figure 22–6. In the reaction of SO_2 with oxygen,

$$SO_2(g) + \tfrac{1}{2} O_2(g) \longrightarrow SO_3(g)$$

the chemical amount of gas decreases from 3/2 to 1 mol. This fact suggests running the reaction under high total pressures to increase the yield of product. However, the small advantage gained by using high pressures is more than offset by the greater cost of the equipment that would be needed, so atmospheric pressure is used. A thermodynamic analysis gives

$$\Delta H_r^\circ = -98.9 \text{ kJ mol}^{-1} \quad \text{and} \quad \Delta S_r^\circ = -94.0 \text{ J K}^{-1} \text{ mol}^{-1} \quad \text{(at 298.15 K)}$$

Because the reaction is exothermic, lower temperature favors a greater degree of conversion to products at equilibrium. The temperature at which ΔG_r° is zero (and the equilibrium constant K is 1) is readily estimated:

$$T \approx \frac{\Delta H_{r,\,298}^\circ}{\Delta S_{r,\,298}^\circ} = \frac{-98,900 \text{ J}}{-94.0 \text{ J K}^{-1}} = 1050 \text{ K}$$

• Review Section 11–6 for the derivation of this equation.

The temperature must be maintained substantially below this (approximately 780°C) point to achieve a significant equilibrium yield of SO_3. The problem is that the reaction slows down at lower temperature, although the catalyst does help to speed it up. A two- to four-stage process typically is used. The entering $SO_2(g)$ reaches the first catalyst at a temperature of 420°C. As the reaction begins, heat is evolved and the temperature of the reacting gas mixture rises. After a few seconds, the mixture has reached equilibrium at about 600°C, with conversion of 60% to 70% of the SO_2.

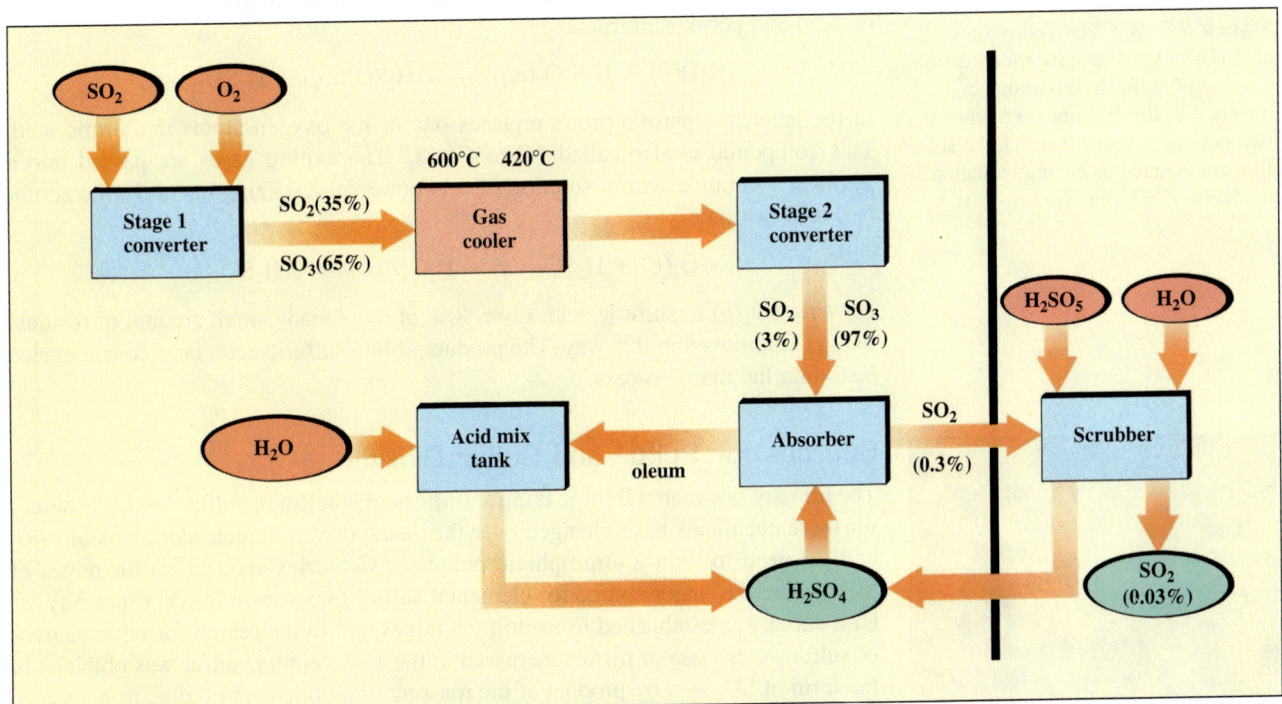

Figure 22-6 • A flow chart of the contact process for making sulfuric acid. The portion of the diagram to the right of the black line represents a process that is used in some modern plants to reduce air pollution from residual SO_2.

Figure 22–7 • The structure of disulfuric acid ($H_2S_2O_7$).

The gases then are cooled back to 420°C and allowed to react one or two more times over the catalyst, using lower temperatures and longer exposure periods. The result is conversion of about 97% of the SO_2 to SO_3. For even greater conversion, the gases then are passed into a tower where SO_3 dissolves in sulfuric acid. This step removes the reaction product, so that the equilibrium again shifts to the right when the un-reacted SO_2 is passed over the catalyst for a final time. The new SO_3 from this last stage is absorbed into sulfuric acid, giving an overall conversion of about 99.7% of the SO_2 introduced originally. Careful consideration of thermodynamics and kinetics has led to a highly efficient process. Almost all sulfuric acid manufactured today is made by the contact process.

If SO_3 were absorbed in water rather than in sulfuric acid, the product would be more dilute and less SO_3 would be absorbed. In addition, the direct reaction of SO_3 with water produces an acidic mist that is difficult to condense. Absorption of SO_3 into sulfuric acid gives fuming sulfuric acid, or **oleum,** which can be used directly or diluted with water to the desired strength. An equimolar amount of SO_3 dissolved in H_2SO_4 gives disulfuric acid ($H_2S_2O_7$), also called "pyrosulfuric acid" (Fig. 22–7).

The amount of SO_2 that escapes from the contact process to pollute the air is very small, but further removal of objectionable SO_2 from the tail gases can be achieved (at additional expense) in another step. Some H_2SO_4 is electrolytically oxidized to peroxodisulfuric acid ($H_2S_2O_8$):

$$2\,H_2SO_4(aq) \longrightarrow H_2S_2O_8(aq) + 2\,H^+(aq) + 2\,e^- \quad \text{(oxidation, anode)}$$
$$2\,H^+(aq) + 2\,e^- \longrightarrow H_2(g) \quad \text{(reduction, cathode)}$$

Figure 22–8 • Peroxodisulfuric acid ($H_2S_2O_8$). The apparent oxidation number of sulfur in this compound equals +7; this becomes +6 when the two peroxo oxygens that link the sulfurs are counted as having oxidation numbers of −1 (see Section 4–4).

In this compound, a peroxo group (O—O) replaces the bridging oxygen atom in disulfuric acid (Fig. 22–8). It quickly reacts with water to give a mixture of sulfuric acid and peroxosulfuric acid:

$$H_2O(\ell) + H_2S_2O_8(aq) \longrightarrow H_2SO_4(aq) + H_2SO_5(aq)$$

In the latter, the peroxo group replaces one of the oxygen atoms in sulfuric acid. This compound is also called "Caro's acid." The exiting gases are passed into a scrubber to mingle with a solution of this powerful oxidizing agent. In the scrubber, the reaction

$$SO_2(g) + H_2SO_5(aq) + H_2O(\ell) \longrightarrow 2\,H_2SO_4(aq)$$

converts $SO_2(g)$ to sulfuric acid. Over 90% of the already small amount of residual $SO_2(g)$ is removed in this way. The product, dilute sulfuric acid, is of course cycled back into the main process.

Sources for Sulfur and Sulfur Dioxide

The primary raw material for making sulfuric acid is sulfur or sulfur dioxide. Sources for these chemicals have changed over the years, driven by considerations of price and the need to reduce atmospheric pollution. Centuries ago, the sulfur mines of Sicily were the major source for elemental sulfur. Increases in price, especially after a cartel was established to exploit the mines, led to the search for other sources of sulfur. As the use of metals increased in the 19th century, sulfur was obtained in the form of SO_2 as a by-product of the roasting of sulfide ores of zinc, iron, or copper through reactions such as

A crystal of sulfur. (*C. B. Jones/Taurus Photo*)

$$ZnS(s) + \tfrac{3}{2}O_2(g) \longrightarrow ZnO(s) + SO_2(g)$$

Capturing the sulfur dioxide and converting it to sulfuric acid provided an inexpensive starting material and reduced air pollution from SO_2.

In the late 1890s, a new development shifted interest from sulfide ores back to elemental sulfur as a starting material. Herman Frasch discovered a new method to extract sulfur from underground deposits. The salt domes of Louisiana, Texas, and Mexico contained substantial reserves of sulfur in porous limestone, but extraction had proved difficult. The **Frasch process** (Fig. 22–9) is an ingenious mining method in which sulfur is liquefied by superheated water injected into the deposit and forced to the surface with compressed air. The sulfur usually is shipped as a liquid in heated vessels to sulfuric acid plants.

Since the 1970s, the Frasch process has declined in importance as a source of sulfur for sulfuric acid owing to a competing process that recovers sulfur impurities in natural gas and petroleum. The latter process was developed to reduce sulfur oxide pollution from the burning of oil and gas. Hydrogen sulfide and other sulfide impurities are acidic and so are preferentially extracted into solutions of basic potassium carbonate:

$$H_2S(g) + CO_3^{2-}(aq) \longrightarrow HS^-(aq) + HCO_3^-(aq)$$

After removal from this solution, a portion of the H_2S is burned in air at 1000°C over an aluminum oxide catalyst to yield sulfur dioxide:

$$H_2S(g) + \tfrac{3}{2}O_2(g) \longrightarrow SO_2(g) + H_2O(g)$$

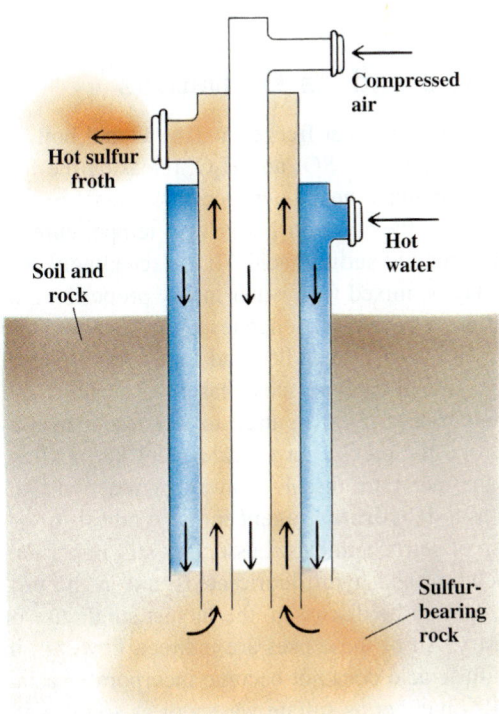

Compressed air

Hot sulfur froth

Soil and rock

Hot water

Sulfur-bearing rock

Figure 22–9 • The Frasch process uses three concentric pipes driven into the ground. Superheated water (at a temperature of 160°C) is pumped under pressure through the outermost pipe into the sulfur-bearing rock formation. This heats the rock above the melting point of sulfur, 119°C. The molten sulfur is heavier than water and collects in a pool, where heated compressed air pumped through the innermost pipe works it into a froth that rises to the surface through the third pipe.

Stairs cut into a gigantic block of solid sulfur. This block was formed by pumping molten sulfur into a mold and letting it cool. *(C. B. Jones/Taurus Photo)*

This can be used directly to make sulfuric acid, but an increasing percentage is now being converted back to elemental sulfur by reaction with additional H_2S in the **Claus process,**

$$SO_2(g) + 2\,H_2S(g) \longrightarrow 3\,S(\ell) + 2\,H_2O(g)$$

which uses an Fe_2O_3 catalyst. This method greatly reduces pollution from the burning of natural gas and petroleum. New processes to reduce sulfur pollution from the burning of coal are greatly needed now.

22-3   Uses of Sulfuric Acid and Its Derivatives

Pure sulfuric acid is a colorless, viscous liquid that freezes at 10.4°C and boils at 279.6°C. When heated, it partially decomposes to SO_3 and H_2O, releasing the former as fumes but retaining the latter. Continued heating reduces the mass percentage of H_2SO_4 to 98.33%, at which point the solution boils (at a temperature of 338°C). This material is sold as "concentrated sulfuric acid." It is exceedingly corrosive and reactive (Fig. 22–10). It may be mixed with water in any proportion, although the mixing must be done cautiously because it generates a good deal of heat. Despite its corrosive power, sulfuric acid is easily handled and transported in steel drums. This fact, together with its acid strength and low cost, has given sulfuric acid a tremendous range of uses. In metals processing, it is used to leach copper, uranium, and vanadium from their ores and to "pickle," or descale, steel. In pickling, layers of oxides on the surfaces of the metal are dissolved away by reaction with the acid. Much sulfuric acid is used as a **dehydrating agent** in the synthesis of organic chemicals and in the processing of petrochemicals. Lesser, but still important, amounts are used in making hydrochloric and hydrofluoric acids and in the production of TiO_2 pigments for paint. As will be discussed, the largest single use of sulfuric acid is in the fertilizer industry. All of these uses are indirect, however, in the sense that the sulfur from the sulfuric acid does not become incorporated as an ingredient in products. Rather, it ends up either as sulfate wastes or as **spent acid,** acid that has been diluted and contaminated by its use in processing. The *direct* uses of sulfuric acid are surprisingly few.

• The biggest direct use by consumers of sulfuric acid is in lead–acid batteries in their automobiles (see Section 13–6). This use consumes only a minute fraction of total sulfuric acid production.

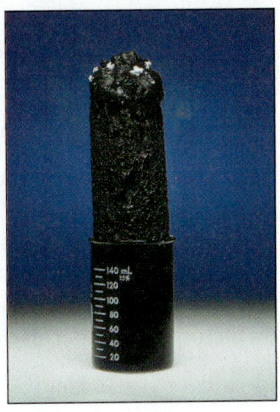

Figure 22-10 • Sulfuric acid reacts spectacularly with a quantity of sucrose ($C_{12}H_{22}O_{11}$), leaving a black char that consists mostly of carbon. The acid chars numerous other organic compounds similarly, and is therefore very dangerous to the unprotected skin. *(Charles D. Winters)*

An important chemical made from sulfuric acid is sodium sulfate (Na_2SO_4). In the past, it was produced in the Mannheim process, the reaction of sulfuric acid with sodium chloride:

$$2\, NaCl(s) + H_2SO_4(\ell) \longrightarrow Na_2SO_4(s) + 2\, HCl(g)$$

The Mannheim process is described in Section 23–1. It requires high temperatures (800–900°C) and is now little used because of its high energy cost. A method still in use is the **Hargreaves process,** in which the immediate reactant is SO_2 rather than H_2SO_4:

$$4\, NaCl(s) + 2\, SO_2(g) + 2\, H_2O(g) + O_2(g) \longrightarrow 2\, Na_2SO_4(s) + 4\, HCl(g)$$

The SO_2 for this reaction often comes from the decomposition of spent sulfuric acid. Sodium sulfate is also a by-product of the manufacture of rayon. Cellulose from cotton fiber or from wood reacts with carbon disulfide (CS_2) in the presence of NaOH to form a viscous solution, called *viscose*. The solution then reacts with sulfuric acid, regenerating carbon disulfide and cellulose in the form of long strands, suitable for being woven into cloth. The net chemical change that accompanies the production of the rayon strands is a neutralization reaction to produce sodium sulfate and water:

$$2\, NaOH + H_2SO_4 \longrightarrow Na_2SO_4 + 2\, H_2O$$

Sodium sulfate is used as a phosphate substitute in detergents and in the manufacture of paper products.

Wood Pulp and Paper Processing

One of the most important applications of sulfur chemistry is in processing wood to wood pulp for use in paper and cardboard. Wood consists of three major components: cellulose, lignin, and oils and resins. Cellulose is a fibrous polymeric compound of carbon, oxygen, and hydrogen that provides supporting structure. It is discussed in Section 25–2. Lignin, which is also a complex polymeric material, binds the fibers of cellulose together. To produce wood pulp, the cellulose must be separated from the lignin. The separation is achieved by digesting (cooking) wood chips in either an acidic or a basic aqueous medium that breaks down the lignin chemically but leaves the cellulose unchanged. The breakdown is a hydrolysis. Water in the digestion liquors attacks the lignin polymer and converts it to alcohols and organic acids (or salts of organic acids). The major methods of digestion are the sulfate (Kraft) process and the sulfite process.

The safe practice when mixing sulfuric acid and water (or an aqueous solution) is to add the acid to the water. Doing the reverse risks injury from spattering the acid. Spatters occur because less dense water collects temporarily atop the more dense acid. Heat developed by the dilution of the acid easily boils water in this restricted region. Adding acid to water mixes the two more effectively. Here, the indicator methyl orange changes color from yellow to red as the acid goes in. *(Leon Lewandowski)*

• The name "Kraft" comes from the German word meaning "strength."

The **sulfate process** (Fig. 22–11) is the more widely used of these two methods because it works for almost all kinds of wood and produces stronger paper. The digestion liquor is an aqueous solution of NaOH and Na_2S. After digestion for several hours at 175°C, the pulp (now primarily cellulose) is separated by filtration, and the liquor is treated with non-aqueous solvents to extract any desired organic materials (for example, turpentine, pine oil, rosin, soaps, and tannin). The next and crucial step is the removal and recycling of the inorganic components from the spent liquor. First, sodium sulfate is added to make up for the inevitable losses in the process. Then the liquor is concentrated and its organic component is burned, generating heat to run the paper mill. During the burning, the sodium sulfate is reduced by the organic matter in the liquor, a process indicated by the equation

• In this equation, the "[C]" stands for the many varieties of carbon-containing matter present.

$$Na_2SO_4(s) + 2[C] \longrightarrow Na_2S(s) + 2\,CO_2(g)$$

The sodium hydroxide in the liquor reacts with the carbon dioxide produced by the oxidation of the organic matter, giving sodium carbonate:

$$2\,NaOH(s) + CO_2(g) \longrightarrow Na_2CO_3(s) + H_2O(g)$$

To regenerate the sodium hydroxide for a new cycle of digestion, the solid residue from the burning is dissolved in water. Limestone ($CaCO_3$) is decomposed to lime (CaO) and carbon dioxide in a separate lime kiln, and the lime is hydrated with water to produce slaked lime, $Ca(OH)_2$. The addition of this base to the dissolved residue precipitates the carbonate ion that was produced in the previous step and generates hydroxide ion:

$$CO_3^{2-}(aq) + Ca(OH)_2(s) \longrightarrow 2\,OH^-(aq) + CaCO_3(s)$$

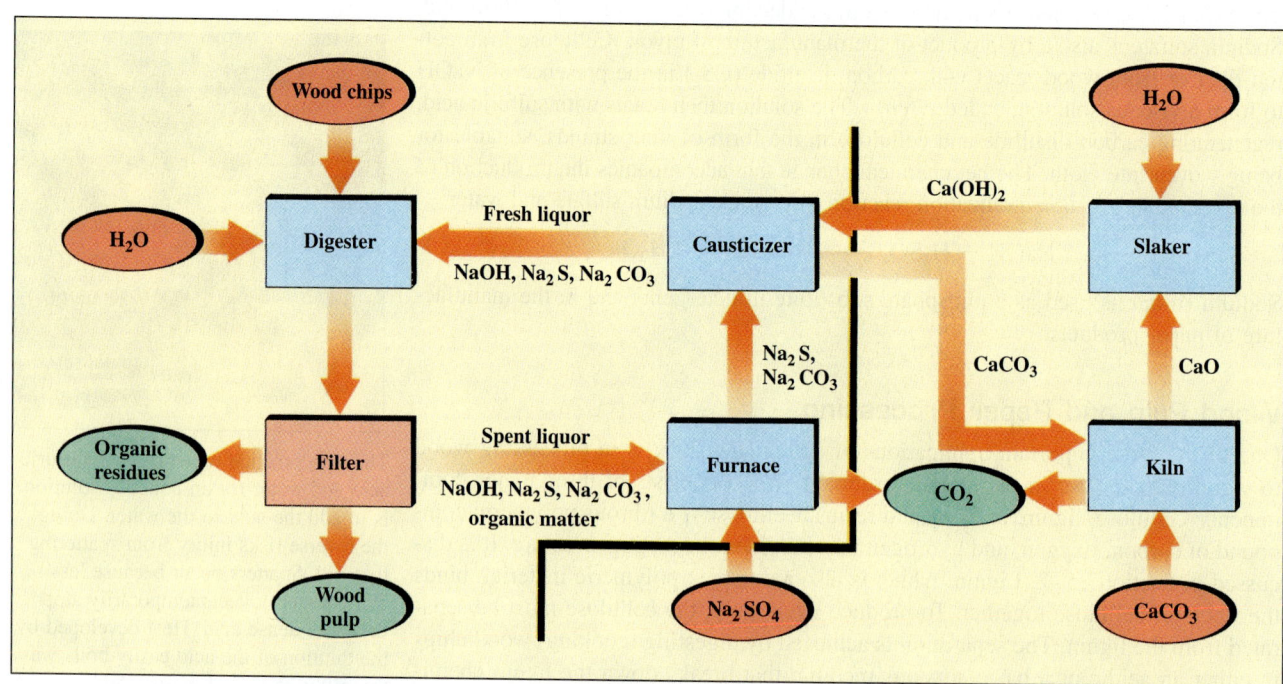

Figure 22–11 • The sulfate process for pulping wood. The overall reaction is wood chips + water $\longrightarrow$ wood pulp + organic residues + carbon dioxide. The entire part of the flow diagram to the right of the black line serves only to make up losses in the process. *(Manley Prim/FMC Corporation)*

The sodium hydroxide thus is restored to the digestion liquor. The calcium carbonate is itself separated and recycled to the kiln to produce additional lime. In this process, the only material that is consumed chemically is water; sodium sulfate and lime are added only as needed to make up for losses from what ideally would be a closed cycle. The major problem with the sulfate process is the occurrence of side-reactions that lead to the formation of H_2S and other sulfur compounds that smell vile and cause severe air and water pollution if they are not captured.

• The Kraft process is now banned in Germany for environmental reasons.

In contrast to the sulfate process, the **sulfite process** uses a mildly acidic medium, with a pH near 4.5. It is limited to certain soft woods such as spruce and hemlock, but it produces a lighter pulp that requires less bleaching and is suitable for high-quality writing paper. In the digestion liquor, gaseous sulfur dioxide is added to a solution containing hydroxide ion (generally prepared by reacting MgO with water) to produce hydrogen sulfite ions:

$$SO_2(g) + OH^-(aq) \longrightarrow HSO_3^-(aq)$$

The action of the hydrogen sulfite ion at 60°C over 6 to 12 hours hydrolyzes the cellulose–lignin complex and dissolves the lignin. After the process is complete, the pulp is recovered by filtration and the liquor is evaporated. The organic residues are again burned to recover heat; during this procedure, the $Mg(HSO_3)_2$ decomposes according to the equation

$$Mg(HSO_3)_2(s) \longrightarrow MgO(s) + H_2O(g) + 2\,SO_2(g)$$

Both the SO_2 and MgO are recovered and recycled through the process.

Phosphorus and Phosphate Fertilizers

At one time, no single use of sulfuric acid dominated the others. The situation changed with the growing production of phosphate fertilizers, which are processed with sulfuric acid and now account for well over half of the annual consumption of sulfuric acid in the United States. Phosphorus occurs in nature primarily as the PO_4^{3-} ion in phosphate rock, the main component of which is fluorapatite ($Ca_5(PO_4)_3F$). Originally, the rock was simply ground up and applied to fields to make them more fertile. A more active fertilizer is made by treating a slurry of ground-up fluorapatite with sulfuric acid to produce **superphosphate,** a mixture of calcium dihydrogen phosphate and gypsum:

$$2\,Ca_5(PO_4)_3F(s) + 7\,H_2SO_4(aq) + 17\,H_2O(\ell) \longrightarrow$$
$$3\,Ca(H_2PO_4)_2 \cdot H_2O(s) + 7\,CaSO_4 \cdot 2H_2O(s) + 2\,HF(g)$$

Superphosphate is more effective because the essential phosphorus nutrient is in a more soluble form. This mixture provides the important secondary nutrients calcium and sulfur as well as phosphorus to growing crops.

A variant of the superphosphate process uses an excess of sulfuric acid to produce phosphoric acid according to the **wet-acid process.** The chemical reaction in this case is

$$Ca_5(PO_4)_3F(s) + 5\,H_2SO_4(aq) + 10\,H_2O(\ell) \longrightarrow$$
$$3\,H_3PO_4(aq) + 5\,CaSO_4 \cdot 2H_2O(s) + HF(g)$$

The world's largest linear disc re-claimer, reclaiming phosphate shale at a phosphorus plant in Idaho. *(Manley Prim/FMC Corporation)*

The hydrogen fluoride that is liberated is carried to an absorption tower where it reacts with SiF_4 to give H_2SiF_6, a chemical that is used in aqueous solution to fluor-idate drinking water. The solid gypsum ($CaSO_4 \cdot 2H_2O$) is filtered out, and the dilute solution of phosphoric acid (H_3PO_4) is concentrated by evaporation.

The major use for wet-process phosphoric acid is again fertilizer manufacture, in which it takes the place of sulfuric acid to produce **triple superphosphate** fer-tilizer, which does not contain gypsum:

$$Ca_5(PO_4)_3F(s) + 7\ H_3PO_4(aq) + 5\ H_2O(\ell) \longrightarrow 5\ Ca(H_2PO_4)_2 \cdot H_2O(s) + HF(g)$$

This fertilizer contains a greater percentage of phosphorus, a significant advantage in reducing transportation costs. In recent years, the use of triple superphosphate fertilizer has declined. The major phosphorus-containing fertilizer now is ammo-nium phosphate, which supplies the necessary nutrient nitrogen as well as phos-phorus. Ammonium phosphate is obtained by the acid-base reaction of phosphoric acid with ammonia:

$$H_3PO_4(aq) + 3\ NH_3(aq) \longrightarrow (NH_4)_3PO_4(aq)$$

The manufacture of ammonia and other nitrogen-based fertilizers is discussed later in this chapter.

The major non-fertilizer use of wet-process phosphoric acid is in the manufac-ture of detergents and cleaning agents that are sodium salts of phosphoric acid and its derivatives. Phosphoric acid in excess reacts with sodium carbonate to give sodium dihydrogen phosphate, a weak acid:

$$2\ H_3PO_4(aq) + Na_2CO_3(aq) \longrightarrow 2\ NaH_2PO_4(aq) + CO_2(aq) + H_2O(\ell)$$

When the sodium carbonate is in excess, the product is sodium hydrogen phosphate, a weak base:

$$H_3PO_4(aq) + 2\,Na_2CO_3(aq) \longrightarrow 2\,NaHCO_3(aq) + Na_2HPO_4(aq)$$

Sodium carbonate is too weak a base to remove the last hydrogen ion from HPO_4^{2-}, so the addition of the stronger (but more expensive) base NaOH is required:

$$NaOH(aq) + Na_2HPO_4(aq) \longrightarrow Na_3PO_4(aq) + H_2O(\ell)$$

Sodium phosphate (Na_3PO_4), which is also called "trisodium phosphate," or TSP, gives strongly basic solutions that are excellent for washing and cleaning.

Heating solid sodium hydrogen phosphate (Na_2HPO_4) expels one molecule of water for every two formula units:

Figure 22-12 • The structure of the tripolyphosphate anion ($P_3O_{10}^{5-}$).

The product, sodium diphosphate (sodium pyrophosphate, $Na_4P_2O_7$), is the sodium salt of diphosphoric (pyrophosphoric) acid ($H_4P_2O_7$). The diphosphate anion has the same structure as the disilicate anion (see Fig. 21–6): Each phosphorus atom is coordinated tetrahedrally to four oxygen atoms (as in the original PO_4^{3-} ion), and the two tetrahedra share a single corner. Heating a mixture of sodium hydrogen phosphate and sodium dihydrogen phosphate causes the linkage of *three* phosphate tetrahedra:

$$2\,Na_2HPO_4(s) + NaH_2PO_4(s) \longrightarrow Na_5P_3O_{10}(s) + 2\,H_2O(g)$$

Again, one molecule of water is expelled for each link that forms. The product is called sodium tripolyphosphate or STPP (Fig. 22–12). It is used as a **builder** in detergents. It helps to sequester Ca^{2+} and Mg^{2+} ions, to maintain the pH in an appropriate range, to keep dirt in suspension, and to increase the efficiency of the detergent itself. The use of phosphates in detergents has been restricted heavily because of its adverse environmental impact. When released into the groundwater, phosphates fertilize the growth of algae in lakes. Excessive algal growth in a lake leads to its early destruction through **eutrophication.** The algae capture CO_2 from the air photosynthetically. When they die, their bodies fall to the bottom of the lake. The decay of this organic material recycles the phosphates (and other nutrients) to succeeding generations of algae. As the lake becomes shallower, bottom-rooted plants can take hold. Eventually the lake becomes a marsh and, finally, a meadow.

The phosphoric acid made by the wet-acid process contains residual gypsum and impurities from the phosphate rock. The **furnace process** is used for the production of purer phosphoric acid. Phosphate rock, SiO_2 (in the form of sand), and elemental carbon (in the form of coke) are fed continuously into an electric arc furnace that operates at 2000°C. The overall reaction is approximately

$$12\,Ca_5(PO_4)_3F(\ell) + 43\,SiO_2(\ell) + 90\,C(s) \longrightarrow$$
$$90\,CO(g) + 20\,Ca_3Si_2O_7(\ell) + 3\,SiF_4(g) + 9\,P_4(g)$$

• The $P_3O_{10}^{5-}$ ion chelates (see Section 19–1) the "hard-water ions" Ca^{2+} and Mg^{2+}. This keeps them in solution, improving the effectiveness of the detergent.

• For historical reasons, P_4O_{10} is often called "phosphorus pentaoxide" after its empirical formula, P_2O_5.

Figure 22–13 • The structure of P_4O_{10} is closely related to the tetrahedral P_4 structure in white phosphorus (see Fig. 20–29a).

• Many food labels call Na_2HPO_4 "disodium phosphate."

• This is the strongest bond in any diatomic molecule except carbon monoxide.

• Transport of nitrate to the soil is one of the few benefits of acid rain.

The elemental white phosphorus is burned in air to tetraphosphorus decaoxide (P_4O_{10}) (Fig. 22–13):

$$P_4(g) + 5 O_2(g) \longrightarrow P_4O_{10}(s)$$

This is a powerful drying agent because it reacts avidly with water. It dehydrates HNO_3 to N_2O_5, and even H_2SO_4 to SO_3. The ultimate product of its reaction with water is phosphoric acid:

$$P_4O_{10}(s) + 6 H_2O(\ell) \longrightarrow 4 H_3PO_4(aq)$$

The purer phosphoric acid from the furnace process is used widely in the food industry. Its acidity gives a pleasant tartness to soft drinks (called "phosphates" when they were first formulated). Calcium dihydrogen phosphate monohydrate ($Ca(H_2PO_4)_2 \cdot H_2O$) is mixed with $NaHCO_3$ in some baking powders. When water is added, these ionic compounds dissolve, and the acidic dihydrogen phosphate ions react with the basic hydrogen carbonate ions to generate carbon dioxide:

$$H_2PO_4^-(aq) + HCO_3^-(aq) \longrightarrow HPO_4^{2-}(aq) + CO_2(g) + H_2O(\ell)$$

The liberated gas makes the bread or other baked goods rise. Another phosphoric acid derivative used in the food industry is sodium hydrogen phosphate (Na_2HPO_4), which is used as an emulsifier to distribute the butterfat uniformly throughout pasteurized processed cheese.

22–4 Nitrogen Fixation

Nitrogen at the Earth's surface exists almost entirely (99.9%) as gaseous diatomic molecules (N_2), which make up 78% by volume of the atmosphere. The N—N bond is a triple bond with a ΔH_d° equal to 945 kJ mol^{-1}. The high energetic price for the dissociation of nitrogen means that the N_2 molecule is highly stable and quite unreactive. Consequently, most plants and animals cannot extract nitrogen, which is essential to life, from the abundant molecular nitrogen in the atmosphere, but take it instead from compounds of nitrogen with other elements. The formation of such compounds by both natural and artificial means is referred to as **nitrogen fixation.**

Natural Sources of Fixed Nitrogen

Figure 22–14 illustrates how nitrogen moves through a worldwide cycle involving biological, geological, and atmospheric processes. Herbivores and omnivores obtain fixed nitrogen from plants that they eat. They use it in a variety of essential biological functions, including the synthesis of proteins. Eventually, they excrete it in forms such as urea ($(NH_2)_2CO$) or return it to the soil upon their death. Plants then reuse a portion as they grow, but some also is converted to molecular nitrogen by bacteria and escapes to the atmosphere. To continue the cycle, a way is needed to fix the nitrogen again.

Nature has at least two ways of fixing atmospheric nitrogen. One is the action of lightning on air. The high temperature in a lightning stroke through the air causes the reaction of nitrogen with oxygen to form nitrogen monoxide (NO). The nitrogen monoxide subsequently is oxidized to nitrogen dioxide (NO_2), which reacts with hydroxyl radicals (OH) in the atmosphere to form dilute nitric acid. The HNO_3 falls to the ground in the rain and snow and provides a source of fixed nitrogen (in the form of the nitrate ion) for plant growth. Nitrogen fixed in this way is thought to amount to at least 10% of the annual total. A second way in which nitrogen is fixed is by the action of certain bacteria called diazatrophs that live on the roots of legumes such as soybeans, clover, and alfalfa. The bacteria reduce N_2 to NH_3, which the plants release to the soil. Farmers plant legumes every few years to fertilize the soil for other crops. Natural fixation of nitrogen generates the equivalent of 90 billion kg of ammonia annually, according to current estimates.

By the late 18th century, the growing population of Europe needed more food, but the limited supply of fixed nitrogen limited crop yields. The first step was to reuse animal wastes efficiently. Sailing ships went as far as the coastal regions of South America to haul back guano, the excrement dropped by seagulls, because it was rich in fixed nitrogen and easily mined. Mineral deposits of sodium nitrate in Chile were exploited for fertilizer. By the end of the 19th century, some fixed nitrogen was being recovered in the form of ammonia from the waste liquors of natural gas plants. It was generally converted to ammonium sulfate ($(NH_4)_2SO_4$) by reaction with sulfuric acid and applied in that form as a fertilizer. But the limitations on fertility imposed by a shortage of fixed nitrogen continued to loom.

• Sodium nitrate was called Chile saltpeter. Saltpeter, which was used in the making of gunpowder, is potassium nitrate.

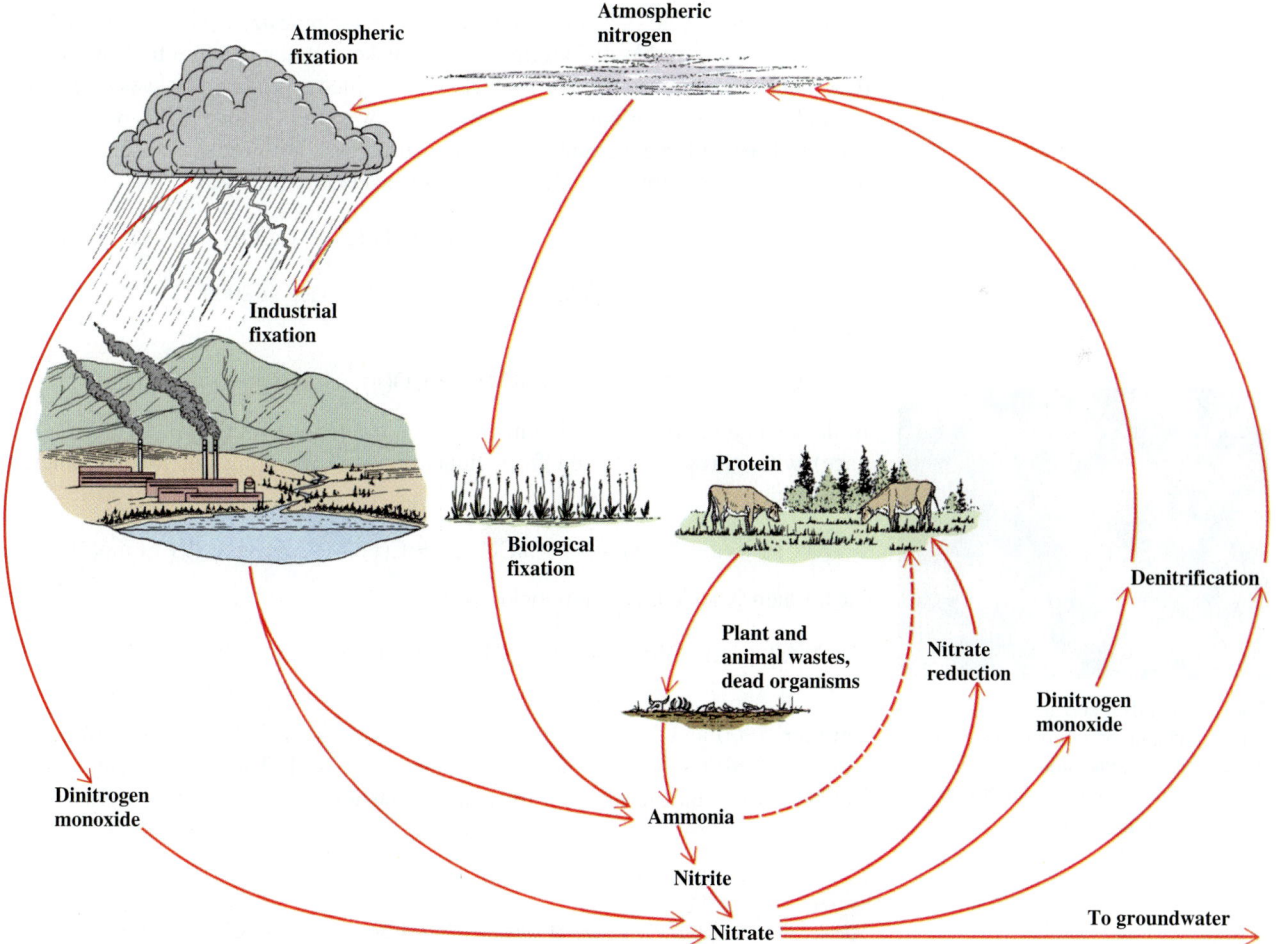

Figure 22-14 • The nitrogen cycle.

Early Industrial Processes

In 1898, Sir William Crookes addressed the British Association in these words:[1]

> England and all civilised nations stand in deadly peril of not having enough to eat. As mouths multiply, food resources dwindle. ... It is the chemist who must come to the rescue of the threatened communities. It is through the laboratory that starvation may ultimately be turned to plenty.

He added that "there is a gleam of light amid this darkness and despondency... the fixation of atmospheric nitrogen," which would be "one of the great discoveries awaiting the ingenuity of chemists." Within ten years of this prediction, three methods had been developed for fixing atmospheric nitrogen.

The first was the **electric-arc process,** which was based on the observation by Cavendish in the 1780s that a spark causes the combination of nitrogen and oxygen in air. Electrical technology was sufficiently advanced by 1900 to create electric arcs with temperatures of 2000°C to 3000°C. The NO that formed was cooled, oxidized to NO_2, and passed into an absorption tower, where it reacted with water to make dilute nitric acid (HNO_3). This process closely resembled the fixing of nitrogen by lightning discharges. However, the electric-arc process was never a significant factor for fixing nitrogen; energy costs were high, and the process did not produce large amounts of fixed nitrogen in a continuous manner.

The **cyanamide (Frank–Caro) process** was more promising. This method uses calcium carbide (CaC_2), a substance that was first synthesized accidentally in 1892, when an inventor in North Carolina tried to make metallic calcium by heating lime (CaO) and tar in an electric-arc furnace. He obtained a product that was clearly not calcium and threw it into a stream, where he saw to his surprise that it reacted with water to liberate large amounts of a combustible gas. The product was calcium carbide (CaC_2), which reacts with water to give acetylene (C_2H_2):

$$CaC_2(s) + 2\,H_2O(\ell) \longrightarrow C_2H_2(g) + Ca(OH)_2(aq)$$

Calcium carbide is now made in an electric-arc furnace at 2000°C to 2200°C from lime and coke:

$$CaO(s) + 3\,C(s) \longrightarrow CaC_2(s) + CO(g) \qquad \Delta H_r^\circ = +465 \text{ kJ mol}^{-1}$$

In the cyanamide process, calcium carbide is reacted (at 1100°C in an electric furnace) with nitrogen obtained from liquefying and distilling air, to give calcium cyanamide:

$$CaC_2(s) + N_2(g) \longrightarrow CaCN_2(s) + C(s) \qquad \Delta H_r^\circ = -291 \text{ kJ mol}^{-1}$$

Steam then is added, and ammonia forms:

$$CaCN_2(s) + 4\,H_2O(g) \longrightarrow Ca(OH)_2(s) + CO_2(g) + 2\,NH_3(g)$$
$$\Delta H_r^\circ = -154 \text{ kJ mol}^{-1}$$

Between 1900 and 1930, calcium cyanamide was used directly as a fertilizer because it furnishes both ammonia and lime to the soil. The major current use of $CaCN_2$ is as a starting material for making plastics.

This fire retardant contains ammonium sulfate and ammonium phosphate, both made from manufactured ammonia. After the fire, these compounds fertilize regrowth of vegetation. *(Phototake)*

---

[1]Quoted in L. F. Haber, *The Chemical Industry, 1900–1930.* Oxford, UK: Clarendon Press, 1971, p. 84.

The Haber–Bosch Process

The predominant source of industrial fixed nitrogen today is the **Haber–Bosch process** for the synthesis of ammonia. The German chemists Fritz Haber and Walther Nernst carried out studies from 1900 to 1910 on the effect of pressure and temperature on the equilibrium between ammonia and its component elements:

$$N_2(g) + 3\,H_2(g) \rightleftharpoons 2\,NH_3(g)$$

The practical ammonia synthesis subsequently developed by Haber with chemical engineer Kurt Bosch required both deep chemical insight and equipment capable of operating under conditions of high temperature and pressure. At 298 K, ΔG_r° and ΔH_r° for the preceding reaction are -33.0 kJ mol^{-1} and -92.2 kJ mol^{-1}, respectively. The value for the equilibrium constant at 298 K accordingly is

$$\ln K_{298} = \frac{-\Delta G_r^\circ}{RT} = \frac{-(-33{,}000\ \text{J mol}^{-1})}{(8.3145\ \text{J K}^{-1}\ \text{mol}^{-1})(298\ \text{K})} = 13.32$$

$$K_{298} = \text{antiln}\,(13.32) = e^{13.32} = 6 \times 10^5$$

Application of anhydrous ammonia to a soybean field. *(Thomas Hovland/Grant Heilman Photography, Inc.)*

The formation of the product (ammonia) is thus strongly favored at room temperature on thermodynamic grounds. Unfortunately, the reaction is very slow at 298 K, and the temperature must be raised to speed the reaction. Because the process is exothermic, increased temperature lowers the equilibrium constant, decreasing the yield of ammonia. Thus, a competition exists between thermodynamic factors, which favor a high yield at low temperature, and kinetic factors, which favor the use of a high temperature.

The compromise adopted in industrial production is a temperature between 700 K and 900 K. The equilibrium constant in this range of temperature can be estimated using the van't Hoff equation (see Section 11–8). If a temperature of 800 K is chosen:

$$\ln \frac{K_{800}}{K_{298}} = \frac{\Delta H_r^\circ}{R}\left(\frac{1}{T_1} - \frac{1}{T_2}\right)$$

$$= \frac{-92{,}200\ \text{J mol}^{-1}}{8.3145\ \text{J K}^{-1}\text{mol}^{-1}}\left(\frac{1}{298\ \text{K}} - \frac{1}{800\ \text{K}}\right) = -23.35$$

Solving for the ratio of equilibrium constants gives

$$\frac{K_{800}}{K_{298}} = e^{-23.35} = 7 \times 10^{-11}$$

and the equilibrium constant is

$$K_{800} = (7 \times 10^{-11})\,K_{298} = 4 \times 10^{-5}$$

This value is only approximate because ΔH_r° changes somewhat with temperature between 298 K and 800 K.

The qualitative result is that the equilibrium yield of ammonia is reduced greatly at the higher temperature. To increase this yield, high total pressures are used. Increased pressure favors the product because 2 mol of gaseous product are formed from 4 mol of gaseous reactants. At 800 K and a total pressure of 200 atm, the yield is 15% to 30%.

A catalyst increases the rate of this reaction. Haber used osmium, a rare and expensive metal, in his original work. Later, he developed a less expensive catalyst based on partially oxidized iron containing small amounts of aluminum, which has

• In light of the enormous economic importance of the Haber–Bosch process, the equilibrium constant of the reaction has been measured and remeasured at elevated temperatures. It is 0.95×10^{-5} at 800 K.

• Worldwide fixation of nitrogen by artificial means amounts to about 90 billion kg of ammonia annually, roughly equal to the amount fixed in natural processes.

been used ever since. The ammonia produced is liquefied to separate it from the unreacted nitrogen and hydrogen, which are recycled.

Although the principles of the Haber–Bosch process were developed before World War I, its implementation in Germany was speeded greatly by the need for fixed nitrogen to make military explosives when the supplies of nitrates from Chile were cut off. It is an irony of history that the commercialization of this process, which more than any other has expanded the food-producing capacity of the world, began as a military project that probably prolonged World War I by at least a year.

Despite the significant role that it plays in fertilizer production, the Haber–Bosch process for the fixing of nitrogen has real problems. Fertilizer factories consume large amounts of energy, require expensive structural materials to operate at high pressure and high temperature, and depend on hydrogen as a reactant. Because most hydrogen currently is obtained from petroleum and natural gas, both non-renewable resources, an alternative process using coal or, ultimately, water eventually must be found for the production of hydrogen. Scientists also are trying to design catalysts that imitate the action of nitrogen-fixing bacteria as an economical way of converting atmospheric nitrogen to ammonia or to some other fixed form.

22-5   Chemicals from Ammonia

Ammonia, a base, is highly reactive and forms many valuable compounds. It reacts with sulfuric, phosphoric, and nitric acids to give ammonium sulfate, ammonium phosphate, and ammonium nitrate, respectively. All of these have been used as chemical fertilizers, with the last two gaining particular popularity in recent years. Ammonia also is readily converted to urea ($(NH_2)_2CO$). The modern synthesis of urea uses the reaction of ammonia with the weak acid CO_2 in hot aqueous solution (180–200°C) under pressure to form the intermediate ammonium carbamate, NH_2COONH_4, which is then dehydrated by heat:

$$2\,NH_3(aq) + CO_2(aq) \longrightarrow NH_2COONH_4(aq)$$
$$NH_2COONH_4(aq) \longrightarrow (NH_2)_2CO(s) + H_2O(g)$$

Urea is a widely used, solid nitrogen fertilizer and an ingredient in other products ranging from glues to skin creams to disinfectants. Urea has historical significance as well, having been synthesized first by Friedrich Wöhler in 1828 from ammonia and cyanic acid (HCNO). Wöhler's work was important in demonstrating that an "organic" compound, formed in human and animal metabolism (and excreted in urine), could be synthesized from strictly inorganic starting materials.

Nitric Acid

One of the most important products made from ammonia is nitric acid, which is generated by the overall reaction

$$NH_3 + 2\,O_2 \longrightarrow HNO_3 + H_2O$$

This equation is deceptive because direct combination of ammonia with oxygen does not produce nitric acid. The most commonly used industrial process (the Ostwald process, summarized in Figure 22–15) involves three steps that, when added together, give the preceding overall reaction.

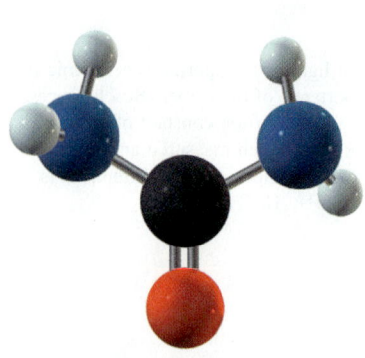

Urea ($NH_2)CO$ is an important solid component in some fertilizers.

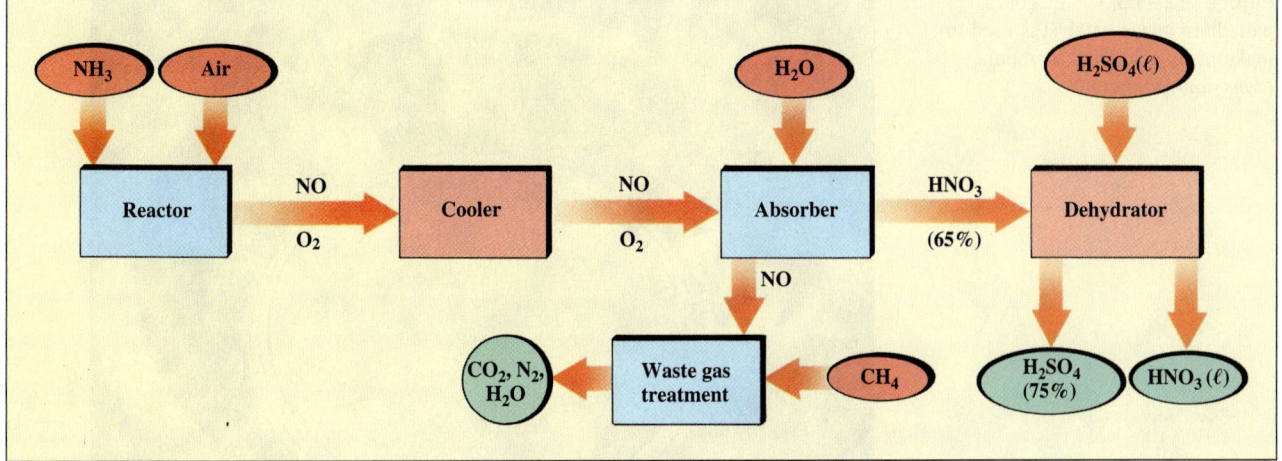

Figure 22-15 • Flow diagram for the synthesis of nitric acid from ammonia by the Ostwald process. Both the second step (oxidation of NO to NO_2) and the third step (absorption into water) described in the text take place in the absorber.

The first step in the Ostwald process is the partial oxidation of ammonia in air:

$$4\,NH_3(g) + 5\,O_2(g) \longrightarrow 4\,NO(g) + 6\,H_2O(g) \qquad \Delta H_r^\circ = -905.5\ \text{kJ mol}^{-1}$$

This strongly exothermic reaction occurs only very slowly at room temperature. In addition, once some NO is produced, a competing reaction occurs,

$$4\,NH_3(g) + 6\,NO(g) \longrightarrow 5\,N_2(g) + 6\,H_2O(g) \qquad \Delta H_r^\circ = -1808.0\ \text{kJ mol}^{-1}$$

which is undesirable because it returns nitrogen to its elemental state. To obtain practical yields of NO, the reaction is carried out at elevated temperatures (800–950°C), and a catalyst is used, consisting of a fine gauze made of noble metals such as platinum and rhodium, or gold and palladium (Fig. 22–16). This catalyst promotes the desired reaction in preference to the competing one, and the first step occurs with extraordinary speed. Excellent conversion is obtained with a mere 3×10^{-4} s of contact time between gases and the gauze.

The second, and rate-determining, step is the reaction of nitrogen monoxide with oxygen to produce nitrogen dioxide:

$$2\,NO(g) + O_2(g) \longrightarrow 2\,NO_2(g) \qquad \Delta H_r^\circ = -114.1\ \text{kJ mol}^{-1}$$

The kinetics of this reaction, which requires no catalyst, are discussed in Example 14–4 and Section 14–6. The rate law has the form

$$\text{rate} = k[NO]^2[O_2]$$

and the rate constant k *decreases* with increasing temperature. Because the reaction occurs faster at lower temperatures, this step is carried out at the lowest temperature conveniently reached by the cooling water available, 10°C to 40°C.

The third step is the absorption of the $NO_2(g)$ into water:

$$3\,NO_2(g) + H_2O(\ell) \longrightarrow 2\,H^+(aq) + 2\,NO_3^-(aq) + NO(g)$$
$$\Delta H_r^\circ = -113.5\ \text{kJ mol}^{-1}$$

The $NO(g)$ produced in this step is cycled back into the second step to give more $NO_2(g)$. The nitric acid has a concentration in the range of 50% to 65% by mass.

When a heated platinum wire is placed just above the surface of a solution of concentrated aqueous ammonia, it continues to glow as it catalyzes the continued exothermic oxidation of NH_3 to NO by oxygen from the air.

Figure 22–16 • This gold–palladium gauze catalyst is used to make nitric acid from ammonia. (*Johnston-Matthey*)

Concentrated nitric acid rapidly reacts with many proteins to give a yellow product. Here, the reaction turns a white feather yellow. The same reaction instantly stains the skin yellow when nitric acid is touched accidentally. (*Charles D. Winters*)

Distilling off the water does not increase the concentration beyond 69% HNO_3, which is the "concentrated nitric acid" in normal laboratory use. Adding a powerful dehydrating agent (concentrated sulfuric acid) and then distilling separates more water to give a solution that contains 95% to 98% nitric acid (fuming nitric acid).

The Ostwald process illustrates well the kind of problems involved in practical chemistry. The original overall reaction is quite simple and is strongly favored thermodynamically at room conditions:

$$NH_3(g) + 2\,O_2(g) \longrightarrow HNO_3(\ell) + H_2O(\ell) \quad \Delta G_r^\circ = -301.46 \text{ kJ mol}^{-1}$$

Nevertheless, kinetic effects and the presence of competing equilibria require the careful design of a multistep process to obtain practical yields.

Nitric acid in aqueous solution is an excellent donor of hydrogen ions; it is a strong acid. Its concentrated aqueous solution is a much stronger oxidizing agent than are solutions of either sulfuric or phosphoric acid. Thus, in dilute solution, nitric acid reacts more readily than do solutions of either sulfuric or phosphoric acid. In dilute solution, nitric acid reacts with copper and most other metals to give nitrates and NO:

$$8\,HNO_3(aq) + 3\,Cu(s) \longrightarrow 3\,Cu(NO_3)_2(aq) + 2\,NO(g) + 4\,H_2O(\ell)$$

In concentrated HNO_3, the reaction is

$$4\,HNO_3(aq) + Cu(s) \longrightarrow Cu(NO_3)_2(aq) + 2\,NO_2(g) + 2\,H_2O(\ell)$$

Only gold and the platinum metals can resist attack by concentrated nitric acid. The acid also reacts, often violently, with organic compounds. The major use for nitric acid is the manufacture of ammonium nitrate by its reaction with ammonia:

$$HNO_3(g) + NH_3(g) \longrightarrow NH_4NO_3(s)$$

Primarily used as a fertilizer, ammonium nitrate has an important secondary use as an industrial explosive, as explained later.

Hydrazine

Another important product made from ammonia is hydrazine (N_2H_4). In principle, direct oxygenation can produce hydrazine from ammonia:

$$2\,NH_3(g) + \tfrac{1}{2}\,O_2(g) \longrightarrow N_2H_4(\ell) + H_2O(\ell)$$

However, this fails as a practical method. Competing reactions such as

$$2\,NH_3(g) + \tfrac{3}{2}\,O_2(g) \longrightarrow N_2(g) + 3\,H_2O(\ell)$$

are thermodynamically favored because the N_2 molecule is so stable. The commercial production of hydrazine relies on the oxidation of ammonia with hypochlorite ion (OCl^-) in aqueous solution, a process known as the **Raschig synthesis:**

$$2\,NH_3(aq) + OCl^-(aq) \longrightarrow N_2H_4(aq) + H_2O(\ell) + Cl^-(aq)$$

Pure liquid hydrazine resembles hydrogen peroxide in many of its physical properties, boiling at 114°C and freezing at 2°C (H_2O_2 boils at 152°C and freezes at -1.7°C). Hydrazine has had some use as a rocket fuel because its reaction with oxygen is extremely exothermic:

$$N_2H_4(\ell) + O_2(g) \longrightarrow N_2(g) + 2\,H_2O(g) \qquad \Delta H_r^\circ = -534\ kJ\ mol^{-1}$$

A small mass of liquid hydrazine can produce a tremendous thrust as the very hot gaseous products rush out at high speed from the rocket engine.

Nitrogen in hydrazine has an oxidation state of -2, intermediate between those of molecular nitrogen (0) and nitrogen in ammonia (-3). As a result, hydrazine (like hydrogen peroxide) can act either as a reducing agent or as an oxidizing agent in aqueous solution. In basic solution (at pH 14), the standard reduction potential for the half-reaction

$$N_2(g) + 4\,H_2O(\ell) + 4\,e^- \longrightarrow N_2H_4(aq) + 4\,OH^-(aq)$$

is $\mathcal{E}^\circ = -1.16$ V. This means that hydrazine tends to be oxidized easily to nitrogen under these conditions and so should act as a good reducing agent. In a strongly acidic medium, hydrazine is converted almost completely to its conjugate acid, the hydrazinium ion $N_2H_5^+$. The standard reduction potential $\mathcal{E}^\circ = 1.275$ V at a pH of 0 for the half-reaction

$$N_2H_5^+(aq) + 3\,H^+(aq) + 2\,e^- \longrightarrow 2\,NH_4^+(aq)$$

suggests that $N_2H_5^+$ should be reduced easily to NH_4^+ and therefore should serve as an oxidizing agent under these conditions. This reaction is slow, however, and hydrazine is known mainly as a reducing agent.

Hydrazine is used to control corrosion in boilers through the reaction

$$6\,Fe_2O_3(s) + N_2H_4(aq) \longrightarrow 4\,Fe_3O_4(s) + N_2(g) + 2\,H_2O(\ell)$$

The red Fe_2O_3 (ordinary rust) is reduced to Fe_3O_4, which forms a protective black layer to hinder further rusting. Hydrazine also can be used to remove certain metal ions from the wastewaters of chemical plants. Chromate ion (CrO_4^{2-}), for example, is reduced to Cr^{3+} in the reaction

$$20\,H^+(aq) + 4\,CrO_4^{2-}(aq) + 3\,N_2H_4(aq) \longrightarrow$$
$$4\,Cr^{3+}(aq) + 3\,N_2(g) + 16\,H_2O(\ell)$$

Adding base then precipitates the Cr^{3+}:

$$Cr^{3+}(aq) + 3\,OH^-(aq) \longrightarrow Cr(OH)_3(s)$$

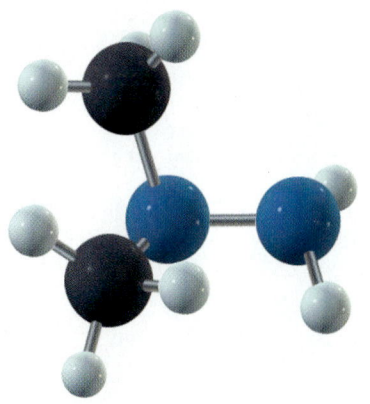

Dimethylhydrazine ($(CH_3)_2NNH_2$). This derivative of hydrazine fueled the lunar excursion modules in their departures from the surface of the moon. The oxidant was the nitrogen oxide N_2O_4. Ignition was a certainty because the two burst into flame on contact.

Hydrazine also removes halogens (X_2, X = F, Cl, Br, I) from industrial wastewater by the reaction

$$N_2H_4(aq) + 2\,X_2(aq) \longrightarrow N_2(g) + 4\,HX(aq)$$

Explosives

An **explosive** is a material that, when subjected to shock, decomposes rapidly and exothermically to produce a large volume of gas. Useful explosives need to be thermodynamically unstable but kinetically stable, so that they can be stored safely for long periods of time before they are detonated. Table 22–3 lists a number of explosives and their decomposition products. It is noteworthy that all of them contain nitrogen. The stability of N_2, a product of the detonation in all of these cases, contributes greatly to the exothermicity of the reactions. The evolution of heat alone, however, is not as important as the *detonation velocity*, a measure of the rate at which the heat is released. The combustion of gasoline, for example, releases about 48 kJ g^{-1}, far more than the heat released by the explosives listed in the table. The difference is that the detonation of an explosive is a branching chain reaction in which the release of heat takes place at a tremendous rate.

Explosives can be characterized as **primary** (or *initiating*) or **secondary.** Primary explosives such as mercury fulminate and lead azide are very sensitive and explode when shocked or heated (Fig. 22–17). They are dangerous to handle and are used in small amounts as detonators to start the explosion of larger amounts of secondary explosives. The latter are less sensitive to shock but explode when detonated by a primary explosive.

The earliest explosive was gunpowder, which was developed before 1000 A.D. in China. Early gunpowder (black powder) was a mixture of potassium nitrate, sulfur, and charcoal (carbon) that was moistened and ground up. The dried product was granulated and loaded into cartridges or bombs. It was used both for military and civilian applications until the middle of the 19th century. In 1845, nitrocellulose, or guncotton, was discovered as the product of the treatment of cotton with nitric acid. It is far more powerful than early gunpowder and is the major constituent of modern gunpowder. Trinitrotoluene (TNT) is used for large-scale military purposes. It is manufactured by reaction of toluene ($C_6H_5CH_3$) with a mixture

• One test for sensitivity in explosives is to drop a heavy weight on small samples and measure the average distance of fall required to cause an explosion.

TABLE 22–3	Explosives		
Name	**Formula**	**Products**	**ΔH of Explosion (kJ g^{-1})**
Gunpowder	$2\,KNO_3 + 3\,C + S$	$N_2 + 3\,CO_2 + K_2S$	−2.1
Nitrocellulose	$C_{24}H_{29}O_9(NO_3)_{11}$	$20.5\,CO + 3.5\,CO_2 + 14.5\,H_2O + 5.5\,N_2$	−4.5
Nitroglycerin	$C_3H_5(NO_3)_3$	$3\,CO_2 + 2.5\,H_2O + 1.5\,N_2 + 0.25\,O_2$	−6.4
Ammonium nitrate	NH_4NO_3	$H_2O + N_2 + 0.5\,O_2$	−1.6
TNT	$C_6H_2CH_3(NO_2)_3$	$3.5\,CO + 3.5\,C + 2.5\,H_2O + 1.5\,N_2$	−4.4
Picric acid	$C_6H_2(OH)(NO_2)_3$	$5.5\,CO + 1.5\,H_2O + 0.5\,C + 1.5\,N_2$	−4.4
Ammonium picrate	$(NH_4)(C_6H_2(NO_2)_3O)$	$4\,CO + 2\,C + 3\,H_2O + 2\,N_2$	−2.6
Tetryl	$C_7H_5N_5O_8$	$5.5\,CO + 2.5\,H_2O + 1.5\,C + 2.5\,N_2$	−4.7
Mercury fulminate	$Hg(ONC)_2$	$Hg + 2\,CO + N_2$	−1.5
Lead azide	PbN_6	$Pb + 3\,N_2$	−1.5

of nitric acid and sulfuric acid, leading to the substitution of three nitro groups (NO_2) on the benzene ring:

$$C_6H_5CH_3(\ell) + 3\ HNO_3\ (sln) \longrightarrow C_6H_2CH_3(NO_2)_3(\ell) + 3\ H_2O\ (sln)$$

Civilian uses of explosives include mining and road and railroad construction, fields that advanced rapidly with the development of dynamite in the second half of the 19th century. The major constituent of dynamite is nitroglycerin (glycerin trinitrate), made by the addition of high-purity glycerin to a mixture of concentrated nitric and sulfuric acids:

$$C_3H_5(OH)_3(\ell) + 3\ HNO_3(sln) \longrightarrow C_3H_5(ONO_2)_3(\ell) + 3\ H_2O\ (sln)$$

Molecular structure of TNT.

• The great dehydrating power of sulfuric acid drives this reaction (and the previous one) to the right.

The Swedish industrialist Alfred Nobel discovered that nitroglycerin could be exploded by shock, and he developed a detonator for it. Nitroglycerin exploded too easily for routine use, however, until Nobel found that absorbing it in diatomaceous earth (made up of the shells of tiny marine organisms) stabilized it greatly. This mixture was the first dynamite, and Nobel made a vast fortune selling it throughout the world, endowing the Nobel Prizes in his will. Modern dynamite still uses nitroglycerin, but mixed with other substances, such as ammonium nitrate or sodium nitrate.

In recent years, ammonium nitrate mixed with small amounts of fuel oil has increasingly supplanted dynamite as an industrial explosive. The advantage of this formulation is its inherent safety. If a misfire occurs and the charge is not set off, the volatile fuel oil evaporates, and the remaining ammonium nitrate is once again safe. Also, the fuel oil can be added after the ammonium nitrate has been placed in the borehole, avoiding the risks of transporting explosive materials. It is unfortunate that ammonium nitrate and fuel oil are readily obtained by terrorists; such a mixture was used in the 1995 bombing in Oklahoma City.

Ammonium nitrate decomposes non-explosively above 200°C,

$$NH_4NO_3(\ell) \longrightarrow N_2O(g) + 2\ H_2O(g) \qquad \Delta G_r^\circ = -169\ \text{kJ mol}^{-1}$$

providing a commercial source of dinitrogen monoxide (nitrous oxide), which is used as an anesthetic ("laughing gas") and an aerosol propellant. When the mass of

Sticks of dynamite. *(Charles D. Winters)*

CHEMISTRY IN PROGRESS

New High-Energy Materials

Some of the newest high-energy materials (the term includes both propellants and explosives) have extremely interesting molecular structures. In 2000, chemists at the University of Chicago announced the synthesis of octanitrocubane (Fig. 22–A). This compound derives from cubane C_8H_8, which was first synthesized in 1964. Cubane contains eight C atoms arranged at the corners of a cube. Each C is covalently bonded to three other C's and to a single H. Ordinarily, carbon forms four single bonds at angles close to $109.5°$ (sp^3 hybridization, see Figs. 18–10 and 18–11). In cubane, all of the C—C—C bond angles equal $90°$. Forcing the bonds into this arrangement adds a very considerable amount of *strain energy* to the structure, and cubane is a powerful explosive. Even more energy is packed in by replacing all eight hydrogen atoms with nitro (—NO_2) groups, which are structural components of most of the explosives in Table 22–3. Calculations indicate that octanitrocubane decomposes explosively, as follows:

$$C_8N_8O_{16}(s) \longrightarrow 4\,N_2(g) + 8\,CO_2(g)$$
$$\Delta H_r = -3180 \text{ kJ mol}^{-1}$$

This corresponds to a ΔH of explosion of -8.5 kJ g^{-1}, exceeding anything listed in Table 22–3. Other calculations suggest that the detonation velocity of octanitrocubane will greatly exceed that of TNT or HMX (high-melting explosive), the current preferred military explosive. Thus, octanitrocubane may be the world's most

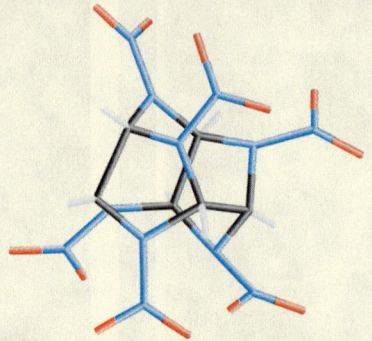

Figure 22–B • The structure of the high-energy compound HNIW ($C_6H_6N_{12}O_{12}$). This "wireframe" diagram emphasizes the cage formed by fused 7-, 6-, and 5-membered rings.

powerful non-nuclear explosive. Interestingly, it is not particularly shock sensitive, making it an ideal secondary explosive. According to its discoverers, it can be pounded with a hammer without effect.

Another high-energy compound with an intriguing structure was synthesized in 1986 by U.S. Navy researchers but not immediately announced. It is hexanitro-hexaazaisowurtzitane (HNIW, $C_6H_6N_{12}O_{12}$). Molecules of HNIW contain a strained cage of fused 7-, 6-, and 5-membered rings (Fig. 22–B) with six nitro groups attached. Its ΔH of explosion equals -5.3 kJ g^{-1}, assuming that its explosive decomposition is fairly closely represented by

$$C_6H_6N_{12}O_{12}(s) \longrightarrow$$
$$3\,CO(g) + 3\,CO_2(g) + 6\,N_2(g) + 3\,H_2O(g)$$

This equation cannot be entirely correct because some oxides of nitrogen do form in the experimental detonation of HNIW.

A third new high-energy material is a salt, ammonium dinitramide ($NH_4N(NO_2)_2$), which is remarkable in containing neither carbon nor chlorine. It is a possible replacement for ammonium perchlorate (NH_4ClO_4) in the solid-fuel boosters for the space shuttles. A replacement is needed because ammonium perchlorate gives chlorine (see page 237) and also some hydrochloric acid when burned during launches. These substances are expelled as part of the rocket exhaust and harm the environment.

The making of octanitrocubane required years of work and the development at some junctures of novel synthetic methods. Detailed testing of its high-energy properties awaits provision of larger supplies. HNIW is also not easy to make. At this time, it is too expensive for commercial or military use.

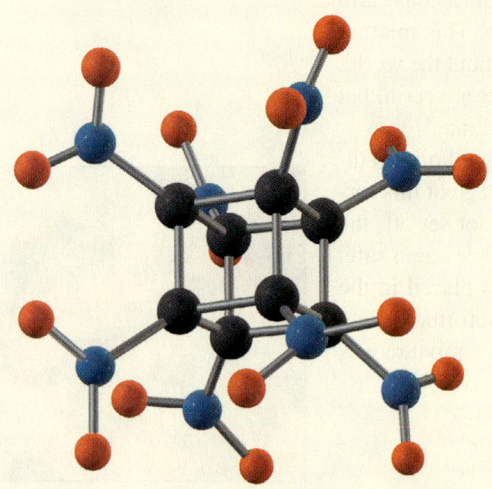

Figure 22–A • The structure of octanitrocubane ($C_8(NO_2)_8$).

ammonium nitrate exceeds a critical value (many tons) or when the ammonium nitrate contains chloride impurities, it may detonate (Fig. 22–18):

$$NH_4NO_3(s) \longrightarrow N_2(g) + 2\,H_2O(g) + \tfrac{1}{2}O_2(g) \qquad \Delta G_r^\circ = -273\ \text{kJ mol}^{-1}$$

The U.S. government set limits on the quantity of ammonium nitrate that may be stored in one place after a shipload of it blew up in the harbor of Texas City, Texas, in 1947, devastating the town.

SUMMARY

22–1 An ideal chemical process gives the desired product in 100% yield, at an acceptable and controllable rate, and in acceptable purity or in a form that is easily purified. It causes little pollution and provides for nonpolluting disposal of by-products as an integral part of its design. Chemical processes generally involve many steps and intermediate chemical compounds. Most large-volume chemicals are little used by consumers, but they are essential in making the products that consumers do use. Raw materials for industrial chemistry come from both geochemical and biochemical processes. Geochemical transformations start with the crystallization of a **magma,** a molten silicate fluid, to form **igneous** rocks. **Sedimentary** rocks come from the weathering, dissolution, and precipitation of other minerals. **Metamorphosis** of sedimentary rocks occurs under heat and pressure. The most important chemical raw materials that come from biological activity are coal, petroleum, natural gas, phosphate rock, and sulfur (in some deposits).

22–2 Sulfuric acid was formerly produced in the **lead-chamber** process, in which oxides of nitrogen serve as oxygen carriers for the oxidation of SO_2 to SO_3. The **contact process,** which uses the direct oxidation of SO_2 to SO_3 in the presence of a catalyst, followed by absorption of the SO_3 in H_2SO_4, is currently the main synthetic route to sulfuric acid. Sulfur for sulfuric acid comes from metal sulfides or from deposits of elemental sulfur that are mined by the **Frasch process.** Sulfur present as H_2S in natural gas and petroleum is another source. Most uses of sulfuric acid are indirect in that sulfur from the acid does not become a part of the product but ends up as a sulfate waste or as **spent acid.** Sodium sulfate is synthesized from SO_2, NaCl, and O_2 in the **Hargreaves process** and is a by-product of rayon manufacture.

22–3 The processing of wood to wood pulp for paper and cardboard uses the **sulfate process,** in which the digestion medium is basic, and the **sulfite process,** in which it is weakly acidic. The most important use of sulfuric acid is in the production of phosphate fertilizers. Phosphate rock (principally $Ca_5(PO_4)_3F$) is treated with H_2SO_4 to give **superphosphate** fertilizer. The same type of reaction using an excess of sulfuric acid produces **wet-process** phosphoric acid. A higher grade of phosphoric acid is afforded by the **furnace process,** in which phosphate rock first is reduced to elemental phosphorus, then burned in air to P_4O_{10} and reacted with water to give H_3PO_4. The reaction of phosphoric acid with ammonia gives ammonium phosphate, the major phosphorus-containing fertilizer in current use. Sodium phosphates from the neutralization of phosphoric acid are used in cleaning products and in detergents.

22–4 The formation of compounds between molecular nitrogen (N_2) and other elements is called **nitrogen fixation.** In the **cyanamide process,** calcium carbide (CaC_2) is generated by heating lime (CaO) and coke (C) in an electric furnace and then reacted with nitrogen to fix nitrogen in the compound calcium cyanamide ($CaCN_2$). The **Haber–Bosch process** uses the direct combination of N_2 with H_2 to yield NH_3. To be practical, the process requires a compromise between low temperature, which favors a high equilibrium yield of ammonia, and high temperature,

Figure 22–18 • Ammonium nitrate (NH_4NO_3) decomposes explosively in the presence of powdered zinc and a catalyst containing the chloride ion. *(Charles D. Winters)*

which increases the reaction's rate. A catalyst speeds up the reaction, and high pressure favors higher equilibrium yields of ammonia.

22-5   The oxidation of ammonia by air in the **Ostwald process** produces nitric acid (HNO_3). Its major use is in reaction with ammonia to give ammonium nitrate (NH_4NO_3), which is used for fertilizer and as an industrial explosive. The important reducing agent hydrazine (N_2H_4) is made from aqueous ammonia by oxidation with aqueous hypochlorite ion ($ClO^-(aq)$) in the **Raschig synthesis.** Many explosives are nitrogen-containing compounds. **Primary explosives** detonate when shocked or heated; **secondary explosives** are less sensitive and require a primary explosion to induce detonation. Gunpowder, nitrocellulose, TNT, and nitroglycerin are all nitrogen-containing explosives. Dynamite is a mixture of nitroglycerin with diatomaceous earth. Ammonium nitrate mixed with fuel oil has largely supplanted dynamite as an industrial explosive.

PROBLEMS

Note: Answers to blue-numbered problems are given in Appendix F. Problems that are more challenging are indicated with asterisks.

The Chemical Industry

1. Define the terms *batch process* and *continuous process*. How do the designs of the two types of processes differ from each other?

2. Define the terms *by-products* and *waste products*. Is there a distinction between the two?

3. In what ways are geochemical processes similar to and different from processes in the chemical industry? Draw a process diagram for the evolution of marble from calcium and carbonate ions separately present in an igneous rock.

4. In what ways are biochemical processes analogous to and different from processes in the chemical industry? Draw a process diagram for the production of sulfur, beginning with calcium sulfate, carbon dioxide, and marine bacteria.

5. Identify a useful source of each of the following chemical elements as specifically as you can, based on information in the chapter:
   (a) sulfur (b) carbon (c) molybdenum

6. Identify a useful source of each of the following chemical elements as specifically as you can, based on information in the chapter:
   (a) phosphorus (b) silicon (c) magnesium

Production of Sulfuric Acid

7. In the lead-chamber process, the sulfuric acid is contained in a lead container. Explain the great resistance of lead to corrosion by sulfuric acid using the information in Table 4–1.

8. Concentrated sulfuric acid can be stored for years in containers made of mild steel; however, mild steel pipes that are used to carry a stream of the same acid corrode quickly and burst. Suggest an explanation.

9. Propose several ways in which predictions from Le Châtelier's principle could be used to improve the yield of $SO_3(g)$ from $SO_2(g)$ and air.

10. Give a thermodynamic analysis of the Claus process; that is, determine the ΔH_r° and ΔS_r° of the main reaction and predict conditions that favor a high yield of sulfur. Take $S(s)$ as the physical form of sulfur.

11. Residual SO_2 can be removed from the effluent gases of sulfuric acid plants by oxidation with H_2SO_5, as explained in the text. Suppose that the electrolysis system that supplies the H_2SO_5 fails. Would it be thermodynamically feasible to substitute an aqueous solution of H_2O_2 as the oxidant for the SO_2? Consult Appendix D for necessary data.

12. Amounts of sulfur in a sample can be measured by burning the sample in oxygen and then collecting the oxides of sulfur in a dilute solution of hydrogen peroxide. What product forms from this reaction? Write a balanced equation. Suggest a good analytical method to determine the quantity of the product that is formed.

13. Write a balanced overall chemical equation for the production of sulfuric acid from zinc sulfide ore.

14. Write a balanced overall chemical equation for the production of sulfuric acid from hydrogen sulfide in natural gas.

Uses of Sulfuric Acid and Its Derivatives

15. Hot concentrated sulfuric acid reacts with phosphorus to give sulfur dioxide, phosphoric acid, and water. Write a balanced chemical equation for this reaction.

16. Hot concentrated sulfuric acid reacts with sulfur to give sulfur dioxide and water. Write a balanced chemical equation for this reaction.

17. Write chemical equations to represent the production of HF and of HCl, assuming that supplies of CaF_2 and of NaCl are available.

18. Write an equation to represent the oxidation of copper by hot concentrated sulfuric acid.

19. Describe (via chemical equations) how sodium sulfide for use in the sulfate (Kraft) process can be prepared starting from sulfuric acid and other easily accessible materials.

20. Describe (via chemical equations) how sodium tripolyphosphate for detergents can be prepared starting from phosphoric acid and other easily accessible materials.

21. Predict the product(s) of the reaction of disulfuric acid ($H_2S_2O_7$) with water.

22. Phosphoric acid (H_3PO_4) can be kept pure only in the crystalline state. Above its melting point of 42.4°C, pairs of H_3PO_4 units gradually lose water. What is the other product of this dehydration reaction? What is its structure?

Nitrogen Fixation

23. A minor source of fixed nitrogen is emission from the inside of the Earth via volcanoes. One estimate of the total amount of fixed nitrogen generated per year from this source is 2×10^8 kg. How many atoms of fixed nitrogen are emitted per second, on the average, by volcanoes?

24. Recent observations of lightning flashes using a new method suggest that the average lightning stroke produces 10^{27} molecules of NO. The rate of lightning flashes worldwide is about 100 per second. Use these data to estimate how much nitrogen (in metric tons) is fixed annually by lightning flashes.

25. Write an overall equation to represent the production of NH_3 by the cyanamide process, as described in the text. Assume that the starting materials are limestone ($CaCO_3$), coke (C), nitrogen (N_2), and water. Determine the ΔH_r° of the process represented by your equation.

26. Suppose that, in the previous problem, the calcium hydroxide and carbon dioxide that are produced are recycled to regenerate the calcium carbonate starting material. Write the overall equation in this case, and calculate the ΔH_r°.

27. Some chemical plants manufacture ammonia from the Haber–Bosch process using pressures as high as 700 atm, rather than the 200 atm described in the text. What are the advantages and disadvantages of working at these higher pressures? For a given yield of product, should a higher or a lower temperature be used if the pressure is raised? Is the reaction then faster or slower?

28. Suppose that a catalyst is developed that allows the rapid production of ammonia from the elements at 600 K. Would this be useful in the Haber-Bosch process? Estimate the equilibrium constant for the formation of 2 mol of NH_3 at this temperature, using the van't Hoff equation.

Chemicals from Ammonia

29. For each of the three steps in the Ostwald process for making nitric acid from ammonia, state whether a high yield of product is favored by low or by high temperature.

30. For each of the three steps in the Ostwald process for making nitric acid from ammonia, state whether a high yield of product is favored by low or by high total pressure.

31. Write balanced equations to represent the reaction of concentrated nitric acid with Cr(s) and with Fe(s). Assume that the nitric acid is reduced to $NO_2(g)$ and that Cr and Fe are oxidized to the +3 oxidation state.

32. Write balanced equations to represent the reaction of concentrated nitric acid with Ni(s) and Tl(s), assuming that the nitric acid is reduced to NO(g) and that Ni and Tl are oxidized to the +2 and +3 oxidation states, respectively.

33. *Trans*-tetrazene has the formula N_4H_4. It consists of a chain of four nitrogen atoms with the two terminal nitrogen atoms bonded to two hydrogen atoms each. Draw a Lewis structure for this compound.

34. Diimine (N_2H_2) cannot be isolated, but evidence points to its existence in solution.
   (a) Draw a Lewis structure for this compound.
   (b) Would the conversion of hydrazine to diimine in solution require an oxidizing agent or a reducing agent?

35. Two reduction half-reactions involving hydrazine (N_2H_4) are

$$N_2H_4(aq) + 2\,H_2O(\ell) + 2\,e^- \longrightarrow 2\,NH_3(aq) + 2\,OH^-(aq)$$
$$\mathcal{E}° = 0.1 \text{ V}$$
$$N_2(g) + 4\,H_2O(\ell) + 4\,e^- \longrightarrow N_2H_4(aq) + 4\,OH^-(aq)$$
$$\mathcal{E}° = -1.16 \text{ V}$$

Determine whether hydrazine in solution at pH 14 is thermodynamically stable against disproportionation to gaseous N_2 and aqueous NH_3.

36. In our discussion of conversions among N_2, N_2H_4, and NH_3, we did not mention a compound of nitrogen named hydroxylamine (NH_2OH).
   (a) What is the oxidation number of nitrogen in NH_2OH?
   (b) Two reduction half-reactions involving NH_2OH are

$$2\,NH_2OH(aq) + 2\,e^- \longrightarrow N_2H_4(aq) + 2\,OH^-(aq)$$
$$\mathcal{E}° = -0.73 \text{ V}$$
$$N_2(g) + 4\,H_2O(\ell) + 2\,e^- \longrightarrow 2\,NH_2OH(aq) + 2\,OH^-(aq)$$
$$\mathcal{E}° = -1.59 \text{ V}$$

Determine whether hydroxylamine in solution at pH 14 is thermodynamically stable against disproportionation to gaseous N_2 and aqueous N_2H_4.

37. The reaction between liquid hydrazine and liquid nitric acid is suggested to boost a specialized type of rocket because the only products, $N_2(g)$ and $H_2O(\ell)$, are completely innocuous in the atmosphere. Write a balanced equation and compute its ΔH_r°.

38. Write a balanced chemical equation for a reaction between hydrazine and dinitrogen tetraoxide that gives only nitrogen and water. Compute ΔH_r°.

39. Describe a practical reaction sequence by which $N_2O_4(s)$ could be obtained from the elements.

40. Describe a practical reaction sequence by which $N_2O(g)$ could be obtained from the elements.

41. The decomposition of TNT, according to Table 22–3, proceeds by the reaction

$$2\,C_6H_2(NO_2)_3CH_3 \longrightarrow 3\,N_2 + 7\,CO + 5\,H_2O + 7\,C$$

If the carbon produced could be converted to CO, additional heat would be given off, and the explosion would be still more powerful. Explain why ammonium nitrate (NH_4NO_3) sometimes is mixed with TNT.

42. During World War II, the Allies used to grind up downed Nazi war planes, which contained aluminum, and mix them in with ammonium nitrate to make bombs that then were dropped over Germany. Explain the reason for putting in the aluminum from the airplanes.

43. (a) Write a plausible equation for the explosive decomposition of the high-energy material ammonium dinitramide ($NH_4N(NO_2)_2$).
    (b) The ΔH_f° of ammonium dinitramide is -36 kJ mol^{-1}. Use your equation to estimate the ΔH of explosion of this compound (in kJ g^{-1}).

44. Using data given in the chapter and in Appendix D, estimate the enthalpy of formation of the explosive HNIW.

ADDITIONAL PROBLEMS

45. Identify the state of lower Gibbs energy in each of the following pairs:
    (a) Aspirin versus (coal, hydrogen, and air)
    (b) Seawater versus (pure water, sodium chloride, and magnesium carbonate)
    (c) (Vitamin A and oxygen) versus (carbon dioxide and water)

46. Define igneous, sedimentary, and metamorphic rocks, based on their different sources. Give an example of a mineral found in each kind of rock.

47. How do plants carry out chemical reactions such as the synthesis of sugars, which have a higher Gibbs energy than the starting materials of water and carbon dioxide? How do animals carry out similar processes, such as the building of proteins that have higher Gibbs energy than the starting materials?

48. In a recent year, the total consumption of carbonated beverages in the United States corresponded to 63 billion 12 oz cans or bottles. If a 12 oz can has a volume of 355 mL, and the concentration of dissolved CO_2 is 0.15 mol L^{-1}, calculate the total mass of CO_2 produced and used for making carbonated beverages during that year.

49. Identify four naturally occurring sources of sulfur or sulfur compounds that can be converted into sulfuric acid. How have these changed in importance throughout history?

50. Both SO_2 and oxides of nitrogen (NO and NO_2) are emitted in flue gases from coal-fired power plants. Write chemical equations representing a possible process in which the SO_2 pollutants are removed from smokestack emissions and sulfuric acid is produced.

51. Use thermodynamics to compare the oxidation of SO_2 in the absence of NO_2

$$SO_2(g) + O_2(g) \longrightarrow SO_3(g)$$

with the oxidation of SO_2 by NO_2:

$$SO_2(g) + NO_2(g) \longrightarrow SO_3(g) + NO(g)$$

Which has the larger thermodynamic driving force ($-\Delta G_r^\circ$) at 25°C? At low temperatures, does the NO_2 act by increasing the yield or by speeding up the reaction?

52. Identify the oxidizing agent in the Glover–Gay-Lussac reaction, which produces SO_3 from SO_2. What change in oxidation state does it undergo? Compare with the corresponding oxidizing agent in the earlier lead-chamber process.

53. The standard reduction potential for the half-reaction

$$S_2O_8^{2-}(aq) + 2\,e^- \longrightarrow 2\,SO_4^{2-}(aq)$$

is 2.0 V. Use the Nernst equation to compute the minimum voltage required to operate an electrolysis unit that converts 1 M SO_4^{2-} to 0.5 M $S_2O_8^{2-}$ and hydrogen at a pressure of 0.10 atm if the pH is 0 and T is 298.15 K.

54. The world production of H_2SO_4 in a recent year was 5.51×10^{10} kg. Suppose that all locations converted sulfur to H_2SO_4 at 99.97% efficiency but lost the rest of the sulfur as SO_2 to the atmosphere. What then was the total mass of emitted sulfur dioxide (in kilograms)?

55. Write a chemical equation for (a) a precipitation reaction, (b) an acid–base reaction, and (c) a redox reaction, in which aqueous sulfuric acid is a reactant.

56. When sulfuric acid is used to remove a layer of oxide from a piece of steel, where do the metallic elements that comprise part of the oxide scale go? Are they removed from the metal and carried away in the acid, or do they remain with the piece of metal?

57. Most sulfuric acid (other than that used in making fertilizer) does not form a part of the final product. It lingers inconveniently as spent sulfuric acid that is both diluted and contaminated. Spent acid must not be dumped in the environment. You propose to regenerate H_2SO_4 by heating spent acid to high temperatures (about 1000 K) in the presence of natural gas (CH_4) to make $SO_2(g)$ by the reaction

$$4\,SO_3(g) + CH_4(g) \longrightarrow CO_2(g) + 4\,SO_2(g) + 2\,H_2O(g)$$

and then recycling the $SO_2(g)$ to make more H_2SO_4.
    (a) Estimate ΔG_r° at 1000 K for this reaction.
    (b) A critic asserts that the reaction probably will proceed as follows:

$$SO_3(g) + CH_4(g) \longrightarrow CO_2(g) + H_2O(g) + H_2S(g)$$

Estimate ΔG_r° at 1000 K for the critic's proposed reaction.
(c) How can you tell which reaction will occur? Explain.

58. The Na_2S for the sulfate (Kraft) process can be made by heating Na_2SO_4 with coke (C). Write a balanced chemical equation for this reaction. Can this reaction be used to supply heat for the digestion of the wood, or does it consume heat?

59. A chain of P—O—P linkages can be constructed by adding additional phosphate groups to diphosphoric acid (through reaction with H_3PO_4 and elimination of water). When the chain is quite long, the product is called metaphosphoric acid. Determine the empirical formula of metaphosphoric acid.

60. Suppose that a fertilizer plant produces phosphoric acid through the wet-acid process and uses it to make triple superphosphate fertilizer ($Ca(H_2PO_4)_2 \cdot H_2O$). Write a balanced equation for the overall reaction. How many moles of gypsum are generated per mole of $Ca(H_2PO_4)_2 \cdot H_2O$?

61. The phosphorus content of fertilizers conventionally is reported by assuming all phosphorus to be present as P_2O_5 and then computing the percentage by mass of P_2O_5. For example, triple superphosphate has the formula $Ca(H_2PO_4)_2 \cdot H_2O$ and therefore contains 1 mol of P_2O_5 per mole of triple superphosphate. The fraction of P_2O_5 equals the molar mass of P_2O_5 divided by the molar mass of triple superphosphate. This fertilizer is "56.3% P_2O_5." Carry out the corresponding calculation for ordinary superphosphate fertilizer, which contains 7 mol of gypsum ($CaSO_4 \cdot 2H_2O$) for every 3 mol of $Ca(H_2PO_4)_2 \cdot H_2O$. Account for the origin of the word "triple" in "triple superphosphate."

62. The disposal of by-product gypsum ($CaSO_4 \cdot 2H_2O(s)$) is a significant problem for fertilizer manufacturers.
    (a) A typical fertilizer plant may produce 100 metric tons of phosphoric acid per day by the wet-acid process (1 metric ton = 10^3 kg). What mass of gypsum is produced per *year*?
    (b) The density of crystalline gypsum is 2.32 g cm^{-3}. What volume of gypsum is generated per year? (In fact, the volume is several times *larger* because of the high porosity of the gypsum produced.)

63. During the early 1970s, when the prices of petroleum and natural gas rose rapidly, the price of nitrogen-based fertilizers increased rapidly as well. Explain why the two commodities are so closely linked.

64. A piece of hot platinum foil is inserted into a container filled with a mixture of ammonia and oxygen. Brown fumes appear near the surface of the foil. Write balanced equations for the reactions taking place.

65. Nitrogen oxides (NO_x) from power stations, steam generators, and other fossil-fuel–burning sources contribute to air pollution. One promising technology (called NOxOUT) for removing NO_x from stack emissions injects an aqueous solution of urea (($NH_2)_2CO$) into various points of the combustion system. Urea reacts with the nitrogen oxides to give N_2, CO_2, and H_2O.

(a) Write a balanced chemical equation to represent the reaction of urea with NO_2.
(b) Write a second balanced chemical equation to represent the reaction of urea with NO.

66. The concentration of $NO_3^-(aq)$ in a solution can be determined by adding sulfuric acid and mercury and then collecting the gaseous NO that the reaction generates:

$$2\,NO_3^-(aq) + 8\,H^+(aq) + 3\,SO_4^{2-}(aq) + 6\,Hg(\ell) \longrightarrow$$
$$3\,Hg_2SO_4(s) + 4\,H_2O(\ell) + 2\,NO(g)$$

A 357 mL sample of a solution is treated in this way and generates 11.46 mL of NO(g), measured at a pressure of 0.965 atm and a temperature of 18.8°C. Compute the concentration of nitrate ion in the solution.

67. The formation of hydrazine in the Raschig synthesis is thought to be a two-step process that involves NH_2Cl as an intermediate. Write balanced equations for the two steps in the Raschig synthesis.

68. Molecules of hydrazine (N_2H_4) have a sizable dipole moment. Sketch a structure that is definitely *eliminated* as the structure of the hydrazine molecule by this information.

69. (a) Compute ΔG_r° (at 25°C) for

$$6\,Fe_2O_3(s) + N_2H_4(aq) \longrightarrow 4\,Fe_3O_4(s) + N_2(g) + 2\,H_2O(\ell)$$

(b) As explained in the text, this reaction is used to prevent further rusting of the walls of boilers. A boilermaker asks you about a proposal to improve this procedure by making the solution of hydrazine strongly acidic (to "burn off the worst of the rust as the hydrazine simultaneously seals off the rest"). Evaluate this idea.

70. Determine how much (in grams) of each ingredient to mix to prepare 100 g of gunpowder. Use the information in Table 22–3.

71. You have the following chemicals in the laboratory—water, iron(II) sulfide, concentrated sulfuric acid, barium chloride, and air. State what procedures you could use to prepare the following: H_2S, $BaSO_4$, SO_2. Include balanced chemical equations.

72. Nitric acid can be prepared in the laboratory by heating a mixture of sodium nitrate and sulfuric acid. A mixture of nitric acid with water distills out of the mixture and may be condensed in a beaker or flask cooled by ice water.
    (a) Write a balanced equation to represent the reaction that generates the nitric acid.
    (b) From the information in the chapter, what is the maximum concentration of HNO_3 in the liquid that distills out?

73. Minor impurities often react with nitric acid prepared by the method described in Problem 72 to turn it brown.
    (a) What substance is responsible for this color?
    (b) What kinds of reactions are taking place between the nitric acid and the impurities?

23

Chemistry of the Halogens

CHAPTER OUTLINE

Applying the perfluorinated polymer Teflon as electrical insulation on copper wire. (*Tom Carroll*)

Fluorine, chlorine, bromine, and iodine are all quite reactive; they form compounds with nearly all the elements in the periodic table, including, in the case of fluorine, the noble gases xenon, krypton, and argon. The 19th century saw the development of processes to obtain abundant supplies of elemental chlorine, along with important co-products such as sodium carbonate and sodium hydroxide, from sodium chloride. The ready availability of these materials has greatly influenced industrial development ever since. The 19th century also saw the first preparation of elemental fluorine, the most reactive of all the elements.

23-1 Chemicals from Salt

By the second half of the 18th century, several important industries in Europe and North America faced growing shortages of raw materials. The glass industry relied on soda ash (sodium carbonate, Na_2CO_3) or potash (potassium carbonate, K_2CO_3) as a flux to lower the viscosity of the molten glass (see Section 21–4). These compounds were derived from the leachings of ashes: soda ash from the ashes of sea plants, and potash from the ashes of trees and shrubs. By 1750, demand from the glass industry was outrunning the supplies from these sources, which varied in quality in any case.

Soapmakers faced a different problem of supply. They needed an inexpensive base to react with vegetable oils and animal fats to make soap. The base that was readily available was lime (CaO), produced by high-temperature decomposition of limestone:

$$CaCO_3(s) \xrightarrow{850°C} CaO(s) + CO_2(g)$$

In the 18th century, lime kilns were among the largest industrial structures in Europe. The lime could be **slaked** by the addition of water to give a sparingly soluble hydroxide:

$$CaO(s) + H_2O(\ell) \longrightarrow Ca(OH)_2(s)$$

In this form, it served well to neutralize acids but had no use in the manufacture of soap because the calcium salts that formed from its reaction with oils and fats were insoluble in water. A soap must be water-soluble.

Textile manufacturers meanwhile confronted an expanding demand for high-quality bleached cotton and linen to make good clothes. In early bleaching methods, cloth or fibers were repeatedly exposed to lactic acid from sour milk, bases from wood ashes, and direct sunlight. Cotton took up to three months to bleach, and linen, up to six months, and the process depended on the weather and the supply of milk. In the closing years of the 18th century, the lead-chamber process (see Section 22–2) began to supply sulfuric acid that could replace the sour milk and greatly hasten that part of the process. The lack of an inexpensive base and a bleaching agent more effective than sunlight created major bottlenecks.

This section describes how salt (sodium chloride, NaCl) provided answers to all of these problems. Processes were developed to convert salt into sodium carbonate and sodium hydroxide; the by-products of these processes included compounds of chlorine that could be made into effective bleaches. Several competing processes rose and fell over the course of the 19th century, with a central role played by the utilization of waste materials to achieve cost savings. These processes, together with the production of sulfuric acid (discussed in Chapter 22), dominated the 19th-century chemical industry and continue to play an important role in the 21st.

- The other starting materials for glass are lime (CaO) and silica (SiO_2).
- The soapmaking reaction is discussed in Section 24–2.

947

The Leblanc Process

In 1775, the French Academy of Sciences offered a prize for a satisfactory process to make sodium carbonate from sodium chloride. The problem was solved during the next 15 years by a French doctor and amateur chemist, Nicolas Leblanc. Figure 23–1 outlines the **Leblanc process.**

The first step is the production of salt cake, or sodium sulfate (Na_2SO_4). The process Leblanc used was the standard one at the time, adding sulfuric acid to sodium chloride at temperatures of 800°C to 900°C:

- This is the Mannheim process mentioned in Section 22–3.

$$2\,NaCl(s) + H_2SO_4(\ell) \longrightarrow Na_2SO_4(s) + 2\,HCl(g)$$

The sodium sulfate was ground up, mixed with limestone and charcoal or powdered coal, and put in a furnace. The reactions that then took place can be summarized in two equations:

$$Na_2SO_4(s) + 2\,C(s) \longrightarrow Na_2S(s) + 2\,CO_2(g)$$
$$Na_2S(s) + CaCO_3(s) \longrightarrow Na_2CO_3(s) + CaS(s)$$

The result was a porous gray-black mixture called **black ash,** which consisted primarily of calcium sulfide and sodium carbonate, with some unreacted calcium carbonate and other impurities. Filtering systems were developed to leach out the soluble sodium carbonate, leaving behind the solid calcium sulfide as a by-product. The liquid then was heated in a furnace to remove sulfur and carbon impurities as

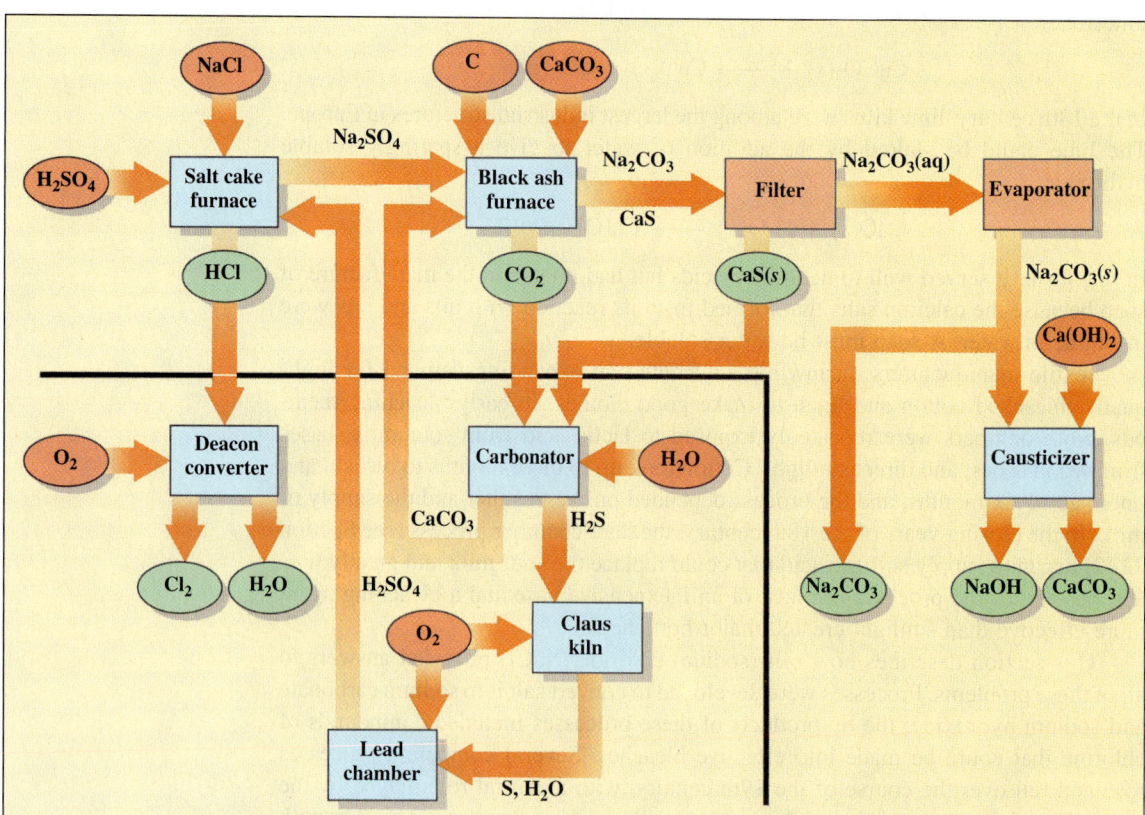

Figure 23–1 • In the original version of the Leblanc process, HCl, CO_2, and CaS were all waste products generated along with the desired Na_2CO_3. Later versions of the process (below the black line) converted HCl into useful Cl_2 and recovered sulfur from the CaS to make sulfuric acid.

$SO_2(g)$ and $CO_2(g)$ and leave behind the desired sodium carbonate. If sodium hydroxide was needed, the sodium carbonate was boiled with slaked lime:

$$Na_2CO_3(aq) + Ca(OH)_2(aq) \longrightarrow CaCO_3(s) + 2\,NaOH(aq)$$

The precipitation of the insoluble calcium carbonate left a solution of sodium hydroxide that could be concentrated further or evaporated to dryness.

The use of the Leblanc process spread rapidly, especially in Great Britain, and caused an increased demand for sulfuric acid as a starting material. A look at the overall equation for the process

$$2\,NaCl + H_2SO_4 + 2\,C + CaCO_3 \longrightarrow Na_2CO_3 + 2\,CO_2 + CaS + 2\,HCl$$

shows that, in addition to the primary product, sodium carbonate, and the carbon dioxide, two significant by-products result—calcium sulfide and hydrogen chloride. The former was dumped at sea or left in piles on land, where it weathered to produce noxious H_2S and SO_2 gases. Hydrogen chloride initially was vented as an acidic gas, which laid waste the vegetation for miles around the plants. Tall chimneys (hundreds of feet high) were built, but they only dispersed the HCl farther. In the next remedy attempted, the gaseous hydrogen chloride was passed into towers in which water trickled downward over a packing of coke. The hydrogen chloride dissolved to create a dilute solution of hydrochloric acid that was run off into streams and rivers, substituting water pollution for air pollution. In 1863, the Alkali Act prohibited such pollution in Great Britain. Although some condemned the law as excessive interference by the government in private business, it stimulated the search for ways of using the hydrochloric acid. Efficient processes soon were developed to convert hydrochloric acid to bleaching powder, making a noxious waste into a useful product. Before considering these processes, let us return to a somewhat earlier period to examine the role of chlorine and its compounds as bleaching agents.

• This law was perhaps the first major piece of environmental legislation.

Chlorine and Bleaches

Chlorine was discovered in 1774 by the Swedish chemist Carl Wilhelm Scheele, who prepared it in its elemental form by the reaction of hydrochloric acid with pyrolusite, a mineral composed of MnO_2:

$$4\,HCl(aq) + MnO_2(s) \longrightarrow Cl_2(g) + MnCl_2(aq) + 2\,H_2O(\ell)$$

Scheele did not realize that the greenish yellow gas he had produced was an element, and it remained for Humphrey Davy to identify it as such in 1811 and to name it (from the Greek word *chloros,* meaning "green"). By 1786 Berthollet and de Saussure had described chlorine's bleaching properties. These were unsatisfactory in several ways, however. Chlorine disintegrated the cloth unless monitored carefully; it was difficult to handle and, at the time, could not be transported. The Scottish chemist Charles Tennant made a significant advance in 1799 when he patented a material that he called **bleaching powder,** which was formed by saturating slaked lime with chlorine:

$$Ca(OH)_2(s) + Cl_2(g) \longrightarrow CaCl(OCl)(s) + H_2O(\ell)$$

Together with sulfuric acid, bleaching powder (also called "chlorinated lime") greatly reduced the time needed to bleach cotton and linen, so that by the 1830s, one week (rather than several months) was sufficient to bleach cotton goods.

Bleaching powder acts in aqueous solution by releasing hypochlorite ion (OCl^-):

$$CaCl(OCl)(s) \longrightarrow Ca^{2+}(aq) + Cl^-(aq) + OCl^-(aq)$$

Figure 23–2 • Household bleach contains about 5% sodium hypo-chlorite (NaClO) by mass. *(Charles D. Winters)*

which in turn decomposes, liberating oxygen to carry out the bleaching. One disadvantage of bleaching powder is that it decomposes on standing to calcium chloride and oxygen:

$$CaCl(OCl)(s) \longrightarrow CaCl_2(s) + \tfrac{1}{2} O_2(g)$$

This can be avoided by preparing calcium hypochlorite ($Ca(OCl)_2$), which also produces twice as much hypochlorite ion in solution, making it more effective on the basis of mass. Modern household bleach uses the sodium salt of hypochlorite ion (NaOCl) rather than the calcium salt (Fig. 23–2). Sodium chlorite ($NaClO_2$) is an even stronger bleach that is widely used in industrial applications, such as paper bleaching.

For almost a century after its discovery in 1774, the major method for making chlorine for bleach was the original reaction used by Scheele. This is an extremely wasteful method because all of the manganese and much of the chlorine is lost as $MnCl_2$. By the mid–19th century, hydrochloric acid (the noxious by-product of the Leblanc process) was in extensive use for bleach manufacture, and a less wasteful method of oxidizing it was needed. Between 1868 and 1874, the British chemist and industrialist Henry Deacon developed a process to convert gaseous hydrogen chloride into chlorine over a copper-chloride catalyst:

$$2\,HCl(g) + \tfrac{1}{2} O_2(g) \longrightarrow Cl_2(g) + H_2O(g)$$

The **Deacon process** was widely adopted; hydrochloric acid from the Leblanc process became the major source of chlorine for bleaches.

Decline of the Leblanc Process

As described later in this section, a competitive process for making sodium carbonate came into use in Belgium in the mid–19th century. As time went on, the sodium carbonate from the Leblanc process became a less important product economically than hydrogen chloride, and it was sold at a loss to meet this competition. Profits were still made from the Leblanc process only because of the co-production of bleaching powder from hydrogen chloride, the former waste product. This turnabout illustrates the way in which the chemical industry changes to meet changing circumstances.

With the successful capture and use of the hydrogen chloride from the Leblanc process, only one major waste product remained—calcium sulfide. This presented a disposal problem, but the sulfur in the CaS had value in making sulfuric acid if it could be freed from the compound. At the time, sulfur was obtained in elemental form from mines in Sicily and from ores containing pyrite (FeS_2). Supplies and prices were often uncertain. By 1888, a process had been developed in which a slurry of "tank waste" (primarily a water suspension of calcium sulfide) was passed through a series of cylinders, where it came into contact with carbon dioxide. The reaction produced hydrogen sulfide

$$CaS(s) + CO_2(aq) + H_2O(\ell) \longrightarrow H_2S(aq) + CaCO_3(s)$$

which could be oxidized to sulfur with air over an iron(III) oxide catalyst:

$$H_2S(g) + \tfrac{1}{2} O_2(g) \longrightarrow S(s) + H_2O(g)$$

Although ingenious, this discovery came too late to save the Leblanc process from the competition of the Solvay process and the electrolytic production of chlorine, the processes we consider next.

The Solvay Process

Figure 23–3 shows a flow diagram for the **Solvay process.** The first and most important reaction is

$$2\,NaCl(aq) + 2\,NH_3(aq) + 2\,CO_2(g) + 2\,H_2O(\ell) \longrightarrow$$
$$2\,NaHCO_3(s) + 2\,NH_4Cl(aq)$$

- The coefficients of 2 appear in this equation to avoid fractional coefficients in the equations for the subsequent steps.

This reaction is carried out in a "carbonating tower," a tall reactor in which brine (concentrated aqueous NaCl) is saturated with ammonia and carbon dioxide. The relatively insoluble sodium hydrogen carbonate (sodium bicarbonate, $NaHCO_3$) is removed by filtration on huge rotating drum filters. It then is decomposed thermally in a rotary dryer at over 270°C to give sodium carbonate as the product:

$$2\,NaHCO_3(s) \longrightarrow Na_2CO_3(s) + H_2O(g) + CO_2(g)$$

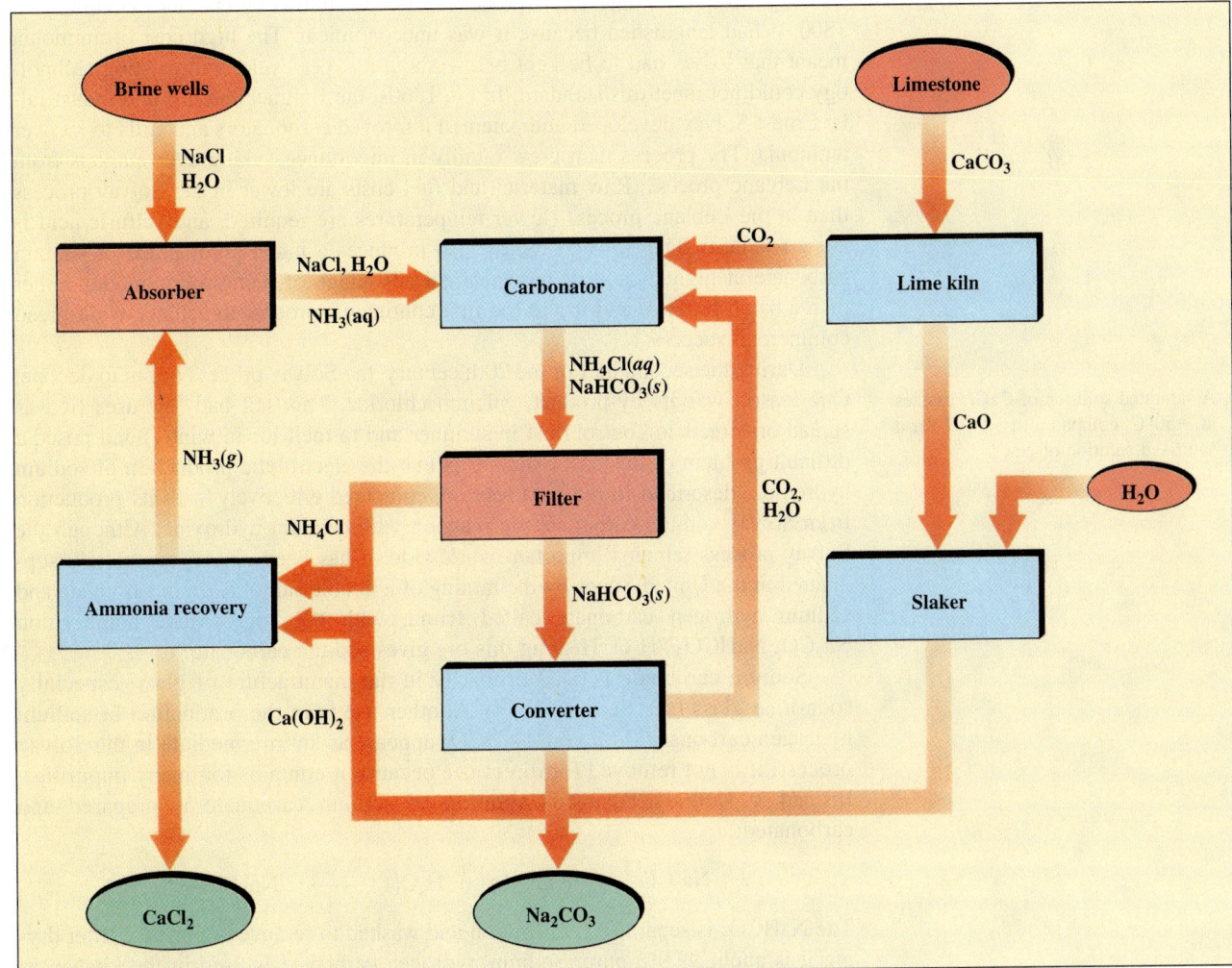

Figure 23–3 • The net reaction in the Solvay process is the production of sodium carbonate and calcium chloride from salt and calcium carbonate. Ammonia plays a crucial role in making the process work but is not consumed at all, except for inevitable small losses.

The carbon dioxide freed in this reaction is returned to the carbonating tower, where it is supplemented by CO_2 generated in a lime kiln:

$$CaCO_3(s) \longrightarrow CaO(s) + CO_2(g)$$

The lime (CaO) formed in this reaction is slaked:

$$CaO(s) + H_2O(\ell) \longrightarrow Ca(OH)_2(aq)$$

The calcium hydroxide is then used in a final step to recover ammonia for reuse in the carbonating tower:

$$2\,NH_4Cl(aq) + Ca(OH)_2(aq) \longrightarrow 2\,NH_3(aq) + CaCl_2(aq) + 2\,H_2O(\ell)$$

The sum of these five steps gives the remarkably simple overall reaction for the Solvay process:

$$2\,NaCl(aq) + CaCO_3(s) \longrightarrow Na_2CO_3(s) + CaCl_2(aq)$$

Although this "ammonia-soda" process had been studied by chemists beginning in 1800, it had languished because it was uneconomical. The high cost of ammonia meant that losses had to be kept below about 5% per cycle, and existing technology could not meet this standard. In the 1860s, the Belgian chemist and industrialist Ernest Solvay developed and patented improved carbonators and stills to recover ammonia. The process then grew rapidly in importance, taking over markets from the Leblanc process. Raw material and fuel costs are lower in the Solvay process than in the Leblanc process (lower temperatures are required, and sulfuric acid is not a reactant). The Solvay process also produces a much purer grade of sodium carbonate at higher yield. It has the final advantage of being a continuous rather than a batch process, and it was the first continuous process to achieve widespread commercial success.

• A saturated solution of $CaCl_2$ freezes at $-50°C$, compared to $-20°C$ for a saturated solution of NaCl.

During the second half of the 20th century, the Solvay process began to decline. One reason was its by-product, calcium chloride. This salt had few uses (it was spread on streets to control dust in summer and to melt ice in winter) and posed a difficult problem of disposal. Another is that the electrolytic production of sodium hydroxide, described in the next section, competed effectively with its production from Solvay sodium carbonate via reaction with calcium hydroxide. Although the Solvay process remains important worldwide, it has been almost completely supplanted in the United States by the mining of a double salt of sodium carbonate and sodium hydrogen carbonate called **trona,** with the approximate composition $Na_2CO_3 \cdot NaHCO_3 \cdot 2H_2O$. Heating this ore gives sodium carbonate.

Sodium carbonate is used primarily in the manufacture of glass, especially soda-lime glass (see Section 21–4). Another use is in the production of sodium hydrogen carbonate. Although $NaHCO_3$ appears as an intermediate in the Solvay process, it is not removed for direct use because it contains too many impurities. Instead, a saturated aqueous solution of sodium carbonate is prepared and carbonated:

$$Na_2CO_3(aq) + CO_2(g) + H_2O(\ell) \longrightarrow 2\,NaHCO_3(s)$$

The $NaHCO_3$ is separated by filtration and washed to remove impurities. After drying it is about 99.9% pure. Sodium hydrogen carbonate is used in the kitchen as baking soda because it reacts with acids, such as those present in milk, to generate carbon dioxide for the rising of dough. It is also used in dry chemical fire extinguishers (see Problem 56 in Chapter 7).

23-2 Electrolysis and the Chlor-Alkali Industry

In the 1890s, a new process for producing bases and chlorine was developed and grew rapidly in importance: the electrolytic generation of sodium hydroxide and chlorine from salt solutions. The dominant technique in the United States for this process is the **diaphragm cell** (Fig. 23–4). If an electric current is passed through a sufficiently concentrated solution of sodium chloride, the anode reaction is

$$2\,Cl^-(aq) \longrightarrow Cl_2(g) + 2\,e^-$$

and the cathode reaction is

$$2\,H_2O(\ell) + 2\,e^- \longrightarrow H_2(g) + 2\,OH^-(aq)$$

The overall reaction is the sum of the half-reactions:

$$2\,H_2O(\ell) + 2\,Cl^-(aq) \longrightarrow 2\,OH^-(aq) + H_2(g) + Cl_2(g) \qquad \Delta\mathcal{E}° = -2.186\ V$$

The gaseous hydrogen and chlorine bubble out and are collected and dried, while the hydroxide ion forms part of a mixed solution of sodium hydroxide and sodium chloride. The less soluble sodium chloride precipitates upon evaporation and can be removed by filtration, leaving a concentrated solution containing 50% by mass of sodium hydroxide in water. This solution may be used directly or evaporated further to make solid sodium hydroxide.

Early diaphragm cells used graphite anodes, but these had to be replaced frequently and have been superseded by anodes made of titanium coated with platinum, ruthenium, or iridium. The cathodes are steel boxes with perforated sides. The diaphragm prevents the H_2 and Cl_2 gases from mixing but allows ions to pass. During electrolysis, the contents of the anode compartment are kept under pressure to minimize the in-migration of $OH^-(aq)$ from the cathode compartment. This prevents the undesirable side-reaction

$$2\,OH^-(aq) + Cl_2(g) \longrightarrow OCl^-(aq) + Cl^-(aq) + H_2O(\ell)$$

• This standard potential difference corresponds to a standard Gibbs energy of reaction (ΔG_r°) of +422 kJ mol^{-1} (see Section 13–1). The reaction is *not* spontaneous.

• Sodium hydroxide often is referred to as caustic soda because its solutions are so corrosive. It is one of the few soluble inorganic bases (see Section 4–3) and is certainly the most important.

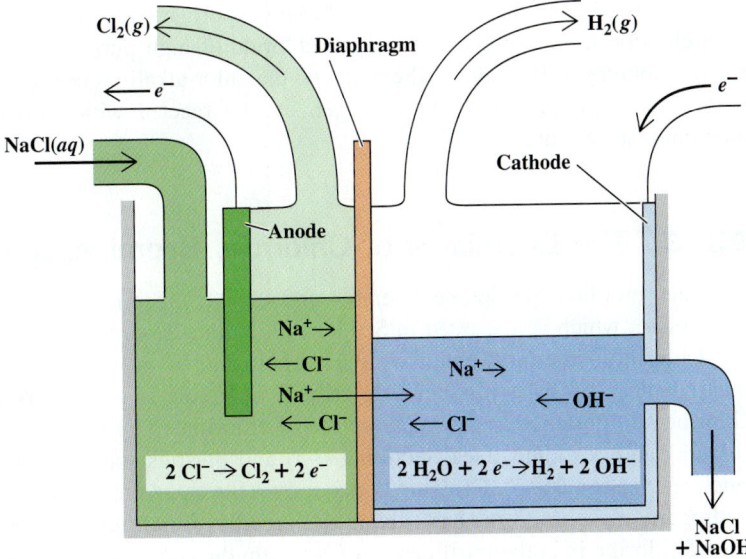

Figure 23-4 • In a diaphragm cell, the cathode is a perforated steel box. Note that the level of the solution deliberately is made higher in the anode compartment than in the cathode compartment in order to minimize migration of OH^- through the diaphragm to the anode, where it could react with the chlorine being generated.

Figure 23–5 • Chlor-alkali electrolysis cells. These use state-of-the-art membrane technology. (*Charles D. Winters*)

• Alternative pulp bleaches include hydrogen peroxide (H_2O_2), oxygen (O_2), and ozone (O_3).

The early chlor-alkali industry in the United States and Canada clustered around Niagara Falls to take best advantage of the inexpensive electricity generated from the waterfall. (*International Stock Photo*)

Asbestos was the first widely used material for these diaphragms, but new diaphragms have been developed based on fluorinated polymers. Another approach is to replace the diaphragm with an ion-exchange membrane that permits only sodium ions to pass to the cathode (Fig. 23–5).

The electrolytic production of sodium hydroxide and chlorine was proposed in the mid–19th century, but the only source for electricity at the time was the battery. This made the process so expensive that it was little more than a laboratory curiosity. By the 1890s, however, the development of the dynamo changed the situation dramatically. The electrolysis industry grew up rapidly around sources of inexpensive electricity from waterpower—in Norway and near Niagara Falls in the United States. The industry's first effect was to doom the Leblanc process, which had survived so long only on profits from the production of chlorine. With competing chlorine produced in greater purity electrolytically, the Leblanc process quickly disappeared. Over a longer period, the electrolytic chlor-alkali process has reduced the role of the Solvay process as well, by yielding a base (NaOH) that competes effectively with sodium carbonate in many applications.

Is the story now complete? Far from it. Consider that a major use of chlorine has long been to bleach wood pulp, giving a pure white cellulose fiber for paper making. Between 50 and 80 kg of chlorine whitens a metric ton of pulp from the sulfate process (see Section 22–3). About 10% of the bleaching chlorine combines with organic compounds present in the wood pulp to generate hundreds of organochlorine compounds. These compounds, some of which are carcinogens or mutagens, appear in the effluent from paper mills that bleach with chlorine and are difficult to remove by waste treatments. As a consequence, the chlorine bleaching of pulp is being phased out by the paper industry.

The loss of a major market for chlorine confronts the chlor-alkali industry with a severe problem of maintaining a balance between stable demand for sodium hydroxide and falling demand for chlorine, because the laws of stoichiometry require inexorably that every 2 mol of sodium hydroxide produced by the electrolysis of brine be accompanied by 1 mol of chlorine. If chlorine were to become an unwanted waste product instead of a valuable by-product, the electrolysis process would suffer heavily. Indeed, the demand for these two products has only rarely been in ideal balance, resulting in price fluctuations in both commodities.

A by-product of the electrolysis of brine is hydrogen. It can be reacted directly with chlorine to give gaseous hydrogen chloride of high purity, although less expensive sources of HCl exist. Alternatively, the chlor-alkali process can be coupled with a fertilizer plant and the hydrogen can be reacted with nitrogen to make ammonia (see Section 22–4).

23–3   The Chemistry of Chlorine, Bromine, and Iodine

Chlorine, bromine, and iodine resemble one another much more closely than they do fluorine, which is discussed in Sections 23–4 and 23–5.

Chlorine is the most extensively used halogen. Its production and its role in making bleaches have just been described. That role is now overshadowed by chlorine's importance in the manufacture of chlorinated hydrocarbons for use as solvents and pesticides (see Section 24–3) and plastics such as polyvinyl chloride (see Section 25–3). Growing concern over the toxic properties of chlorinated hydrocarbons (Fig. 23–6) has slowed the growth of the industry. Another major use of chlorine is in the purification of titanium dioxide. This substance is mined in impure form. It reacts above 900°C with chlorine and coke to give titanium tetrachloride:

$$TiO_2(s) + 2\,C(s) + 2\,Cl_2(g) \longrightarrow TiCl_4(g) + 2\,CO(g)$$

Because $TiCl_4$ is a volatile liquid (b.p. 136°C), it can easily be purified by distillation. It then is oxidized to pure titanium dioxide, regenerating chlorine:

$$TiCl_4(\ell) + O_2(g) \longrightarrow TiO_2(s) + 2\,Cl_2(g)$$

Titanium dioxide is the major white pigment used in paint. It is more opaque than other pigments, so paints containing it have high covering power.

Bromine is obtained largely from naturally occurring brines, where it occurs in the form of aqueous bromide ion (Br^-). Treatment of these brines with chlorine results in the reaction (Fig. 23–7)

$$2\,Br^-(aq) + Cl_2(g) \rightleftharpoons 2\,Cl^-(aq) + Br_2(aq) \qquad \Delta G^\circ_r = -52.4\,kJ\,mol^{-1}$$

This equilibrium lies far to the right: Chlorine, the stronger oxidizing agent, displaces bromide ion from solution. The bromine that is generated is swept out with steam, distilled, and purified. Until recently, the most important use for bromine was in the production of dibromoethane, $C_2H_4Br_2$, which is used in gasoline to scavenge the lead deposited from the breakdown of tetraethyl lead; the use of this "antiknock" additive has diminished sharply, however, with the phase-out of leaded gasoline. A major current use for bromine is in the production of flame-retardant organic compounds. Aqueous solutions of $CaBr_2$ and $ZnBr_2$ also are used as high-density fluids to recover oil from deep wells.

Iodine occurs in seawater to the extent of only $6 \times 10^{-7}\%$, but it is concentrated in certain species of kelp. Burning these plants provides ashes from which the recovery of iodine is commercially feasible. Iodine is nutritionally essential. It is present in the metabolism-stimulating hormone thyroxin, secreted by the thyroid gland. Most brands of table salt contain 0.01% NaI to prevent goiter, the enlargement of the thyroid gland, from iodine deficiency. Silver iodide is used in high-speed photographic film, whereas silver bromide and silver chloride are used in slower-speed film and in photographic printing paper.

The hydrogen halides HCl, HBr, and HI are all gases at room conditions, and all three dissolve in water to form strong acids. Hydrogen chloride usually is prepared by the direct combination of hydrogen and chlorine over a platinum catalyst or as a by-product of organic chemicals processing. Its aqueous solution, hydrochloric acid, is a major industrial acid used extensively for the cleaning of metal surfaces. Hydrogen bromide and hydrogen iodide have fewer uses, mostly in the area of organic chemical synthesis.

Solution Chemistry of the Halogens

When chlorine, bromine, or iodine is added to water, it disproportionates in part according to the equilibrium

$$X_2(aq) + H_2O(\ell) \rightleftharpoons HOX(aq) + X^-(aq) + H^+(aq)$$

The equilibrium constants for these reactions are small, but making the solution basic shifts the reactions to the right. The HOX is neutralized, and the overall reaction becomes

$$X_2(aq) + 2\,OH^-(aq) \rightleftharpoons OX^-(aq) + X^-(aq) + H_2O(\ell)$$

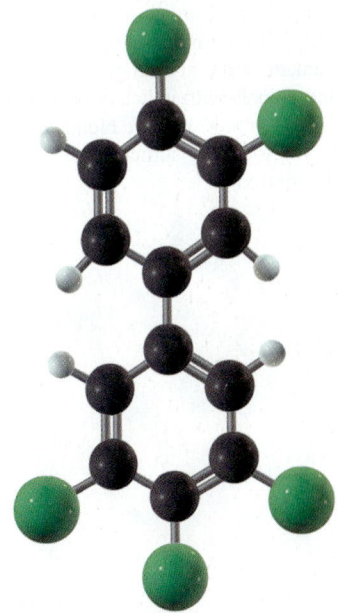

Figure 23–6 • One of the many polychlorinated biphenyls (PCBs), which once were used as heat exchange and insulating fluids. Manufacture of these compounds was discontinued in the United States in 1976 because of their toxic effects.

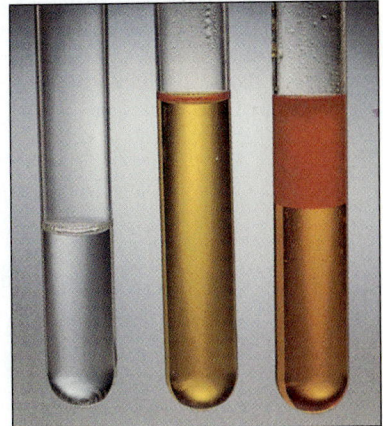

Figure 23–7 • Bromide ion in aqueous solution (*left*) is colorless. Oxidation by chlorine yields bromine (*center*), which is colored in aqueous solution. The color is more intense in an organic solvent (*upper layer on right*). (*Charles D. Winters*)

The structure of thyroxin. Goiters, swellings of the neck caused by enlargement of the thyroid gland, occur in individuals with iodine-poor diets as the gland processes more blood in attempting to obtain sufficient iodine to make thyroxin.

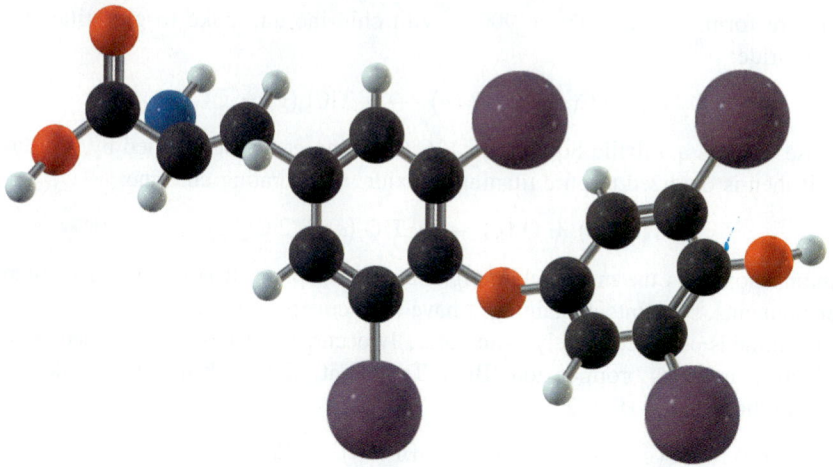

No complications arise with chlorine, and stable solutions of Cl^- (chloride) and OCl^- (hypochlorite) ions result when chlorine is mixed with aqueous base at around 25°C. Solutions of sodium hypochlorite (NaOCl) can be prepared in this way for use as a laundry bleach. The aqueous hypochlorite ion is, however, thermodynamically unstable with respect to a *further* disproportionation:

$$3\,ClO^-(aq) \longrightarrow 2\,Cl^-(aq) + ClO_3^-(aq) \qquad \Delta G_r^\circ = -90.7\,kJ\,mol^{-1}$$

At room temperature this reaction is very slow, but it becomes rapid at about 75°C. The situation is different with bromine. When bromine is mixed with aqueous base, the OBr^- ion forms but undergoes disproportionation (to Br^- and BrO_3^-) rapidly even at 25°C. Solutions containing OBr^- ion can be prepared only "in the cold" (near 0°C). In the case of iodine, the analogous disproportionation of OI^- ion (to I^- and IO_3^-) is so fast at all temperatures that OI^- ion is essentially unknown in solution.

• Astatine (At) is technically a halogen as well. Little is known of its chemistry because all of its isotopes are unstable and short-lived. It therefore is excluded from this discussion.

The solubilities of the halogens in neutral water at 25°C are 0.091 mol L^{-1} (Cl_2), 0.21 mol L^{-1} (Br_2), and 0.00134 mol L^{-1} (I_2). These values are much less than in base because the disproportionation reactions that were just described occur to a lesser extent than in base. The solubility of iodine is by far the least of the three, but it is increased considerably in the presence of iodide ion by a complexation reaction

$$I_2(s) + I^-(aq) \rightleftharpoons I_3^-(aq) \qquad K = 714 \quad \text{at 298.15 K}$$

This reaction can also be described as a Lewis acid–base reaction with I_2 as the acid and I^- as the base. The product, called the triiodide ion, has a linear structure that is symmetric in solution and in certain crystalline salts but unsymmetric (unequal I—I distances) in others. Similar reactions do not occur with chlorine and bromine.

• Potassium iodate has many uses in analytical chemistry that take advantage of its oxidizing power and other favorable properties (see Cumulative Problem, page 88).

Among the possible acids of formula HXO_3, only iodic acid exists in anhydrous form. It is a crystalline solid that is stable up to its melting point. The alkali-metal chlorates, bromates, and iodates have the generic formula MXO_3. They are well-characterized salts that can be crystallized from aqueous solution. Acidic solutions of the salts are powerful oxidizing agents that react quantitatively with their corresponding halide ion to yield the halogen. For example:

$$IO_3^-(aq) + 5\,I^-(aq) + 6\,H^+(aq) \longrightarrow 3\,I_2(aq) + 3\,H_2O(\ell)$$

When potassium chlorate is heated carefully in the absence of catalysts, it disproportionates to potassium perchlorate and potassium chloride:

$$4\ KClO_3(\ell) \longrightarrow 3\ KClO_4(s) + KCl(s)$$

Perchloric acid can be formed by heating potassium perchlorate with concentrated sulfuric acid, but this is a hazardous reaction that is prone to explosions. A safer procedure is to oxidize an aqueous solution of a chlorate salt in an electrolytic cell. The half-reaction at the anode is

$$ClO_3^-(aq) + H_2O(\ell) \longrightarrow ClO_4^-(aq) + 2\ H^+(aq) + 2\ e^-$$

Perchloric acid is a colorless liquid that freezes at $-112°C$. In its anhydrous form it is a powerful oxidizing agent, but in aqueous solution at moderate concentration, it displays little oxidizing ability. Aqueous perbromic acid, which is also synthesized electrolytically, is, by contrast, a powerful oxidizing agent. Aqueous periodic acid is a more useful oxidizing agent than either perbromic or perchloric acid because its behavior is more controllable.

Halogen Oxides

Chlorine, bromine, and iodine form several oxides, most of which are thermally unstable. Dichlorine monoxide is prepared by reacting chlorine with mercury(II) oxide:

$$2\ Cl_2(g) + 2\ HgO(s) \longrightarrow Cl_2O(g) + HgCl_2 \cdot HgO(s)$$

In basic solution it forms hypochlorites, as it is the anhydride of hypochlorous acid:

• Acid and base anhydrides are discussed in Section 4–3.

$$Cl_2O(g) + 2\ OH^-(aq) \longrightarrow 2\ OCl^-(aq) + H_2O(\ell)$$

Chlorine dioxide is an unstable gas that can be prepared by reacting chlorine with silver chlorate:

$$Cl_2(g) + 2\ AgClO_3(s) \longrightarrow 2\ ClO_2(g) + 2\ AgCl(s) + O_2(g)$$

It is a very powerful oxidizing agent that is used to bleach wheat flour and as a sterilizing agent. Like nitrogen dioxide, it consists of "odd-electron" molecules; unlike that molecule, it shows little tendency to dimerize. In basic solution, it disproportionates to form chlorite and chlorate ions and can be regarded as a mixed anhydride of chlorous and chloric acids:

• Gaseous chlorine dioxide was used to kill anthrax spores contaminating certain U.S. Senate offices after a terrorist attack in 2001.

$$2\ ClO_2(g) + 2\ OH^-(aq) \longrightarrow ClO_2^-(aq) + ClO_3^-(aq) + H_2O(\ell)$$

The most stable of the chlorine oxides is dichlorine heptaoxide (Cl_2O_7), a colorless liquid that boils without decomposing at $82°C$. It is prepared by dehydration of perchloric acid with P_4O_{10}:

$$12\ HClO_4(\ell) + P_4O_{10}(s) \longrightarrow 6\ Cl_2O_7(\ell) + 4\ H_3PO_4(\ell)$$

The oxides of bromine, Br_2O and BrO_2, can be prepared by methods that parallel those for the corresponding oxides of chlorine. The most important of the iodine oxides is diiodine pentaoxide, a white crystalline solid that can be prepared by thermal dehydration of iodic acid:

$$2\ HIO_3(s) \longrightarrow I_2O_5(s) + H_2O(g)$$

Figure 23-8 • Iodine reacts with starch to form a dark blue complex. (*Leon Lewandowski*)

Diiodine pentaoxide is useful in analytical chemistry for the detection and quantitative determination of carbon monoxide, which it readily oxidizes:

$$5\,CO(g) + I_2O_5(s) \longrightarrow I_2(s) + 5\,CO_2(g)$$

The iodine released in this reaction can be detected visually by means of the dark blue complex it forms with starch added as an indicator (Fig. 23–8).

23-4 Fluorine: Preparation and Uses

The chemistry of fluorine is full of surprises. Fluorine ranks 13th in abundance among the elements in the rocks of the Earth's crust, and its compounds were used as early as 1670 for the decorative etching of glass. Yet its isolation as a free element defied heroic efforts over many years before finally being accomplished in 1886. The element itself and some of its compounds are highly reactive and toxic, but other compounds are at the opposite extreme and are deliberately chosen for uses demanding inert, non-toxic materials.

The differences between fluorine and the other halogens are extreme enough to place it in a class by itself. Fluorine is far more reactive than chlorine, bromine, and iodine and is indeed the most reactive of all the elements. Elemental fluorine is among the strongest oxidizing agents. The other halogens are moderate to very good at accepting electrons in aqueous media, as shown by the standard reduction potentials of 0.536 V (for $I_2(s)$), 1.066 V (for $Br_2(\ell)$), and 1.358 V (for $Cl_2(g)$). Fluorine is avid for electrons:

$$F_2(g) + 2\,e^- \longrightarrow 2\,F^-(aq) \qquad \mathcal{E}° = 2.87\ V$$

Because the fluor*ine* atom in F_2 is so good at gaining an electron, fluor*ide* compounds (containing F^-) strongly resist oxidation, the loss of electrons.

The major source of fluorine is the ore fluorspar, which contains the mineral fluorite (CaF_2) (see Fig. 20–33). The cubic structure of CaF_2 is shown in Figure 23–9. Another source of fluorine is fluorapatite ($Ca_5(PO_4)_3F$), which is mined in great quantity to make phosphate fertilizer (see Section 22–3). At one time fluorine-containing by-products (HF, SiF_4, and H_2SiF_6) from fertilizer plants were valueless and created environmental pollution that was comparable to the hydrogen chloride pollution caused by the early Leblanc process (see Section 23–1). Now the fluorine from fluorapatite has economic value in making fluorine compounds.

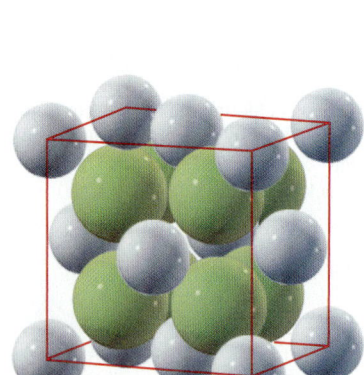

Figure 23-9 • Structure of CaF_2. The gray spheres are calcium, and the yellow spheres are fluorine. When heated (but not yet melted), crystals of fluorite become excellent conductors of electricity. The array of Ca^{2+} ions remains crystalline, but the F^- ions become disordered. They are in effect molten and serve to carry the current.

Hydrogen Fluoride

The starting point for the chemistry of fluorine is hydrogen fluoride, which is made by the reaction of fluorspar with concentrated sulfuric acid:

$$CaF_2(s) + H_2SO_4(aq) \longrightarrow 2\,HF(g) + CaSO_4(s)$$

Anhydrous hydrogen fluoride is a colorless liquid that freezes at $-83.37°C$ and boils at $19.54°C$, near room temperature. The boiling point far exceeds the value predicted by the trend among the other hydrogen halides (Fig. 6–21), which is strong evidence for association of the HF molecules (by hydrogen bonds) in the liquid. In this and other physical properties, HF is closer to water and ammonia than to HCl, HBr, or HI. As pure liquids, both water and hydrogen fluoride contain polar molecules but few ions; both are poor electrical conductors. Liquid hydrogen fluoride undergoes autoionization much as water does:

$$2\,HF(\ell) \rightleftharpoons H^+(sln) + FHF^-(sln)$$

The reaction has an equilibrium constant (at 19.5°C) of 2.0×10^{-14}, a value close to the room-temperature autoionization constant of water (1.0×10^{-14}). Hydrogen fluoride in aqueous solution is a weak acid ($K_a = 6.6 \times 10^{-4}$ at 25°C), quite unlike HCl, HBr, and HI, which are strong acids. Its weakness as an acid is due both to the strength of the H—F bond and to the extensive hydrogen bonding in its aqueous solutions. Fluoride ions interact so strongly with hydronium ions (H_3O^+) that they remain bound in $F^- \cdots H_3O^+$ complexes, rather than separating, as is the case with the chloride, bromide, and iodide ions.

Liquid hydrogen fluoride has excellent solvent properties; it dissolves polar inorganic salts such as NaF and many non-polar organic compounds as well. Using it as a solvent for chemical reactions requires more caution and preparation than using water because HF is quite poisonous, causes serious burns, and corrodes silicate glass by the reaction

$$4\,HF(\ell) + SiO_2(s) \longrightarrow SiF_4(g) + 2\,H_2O(\ell)$$

This reaction underlies one long-standing use of aqueous hydrogen fluoride: etching decorative patterns in glass (Fig. 23–10). Hydrogen fluoride does not attack Teflon and certain other fluorine-containing plastics. Their availability in recent years has made it easier to handle hydrogen fluoride safely.

A major use for hydrogen fluoride is in the aluminum industry. It is reacted with $NaAlO_2(aq)$, itself prepared from bauxite ore ($Al_2O_3(s)$), and NaOH(aq) to make synthetic cryolite according to the overall equation

$$NaAlO_2(aq) + 2\,NaOH(aq) + 6\,HF(g) \longrightarrow Na_3AlF_6(s) + 4\,H_2O(\ell)$$

Molten cryolite (m.p. 1012°C) dissolves $Al_2O_3(s)$. The very important Hall–Héroult process to produce aluminum (see Fig. 12–15) depends on the electrolysis of such solutions. An essential additive in the molten cryolite–Al_2O_3 solution is aluminum trifluoride (AlF_3), which is synthesized by the acid–base reaction of HF(g) with $Al(OH)_3(s)$.

Figure 23–10 • Glass etched with hydrofluoric acid. The artist covered areas with wax to create parts of the design and then applied the HF. The wax, which resists HF, was removed later. *(The Corning Museum of Glass)*

The Preparation of Fluorine

Elemental fluorine is produced from hydrogen fluoride. The first reproducible preparation was achieved by Henri Moissan in 1886 by the electrolysis of anhydrous hydrogen fluoride at −50°C. Today, the element is produced commercially by a similar method, the electrolysis of molten potassium hydrogen fluoride ($KF \cdot 2HF(\ell)$), which can be regarded as a solution of KF in liquid HF. In the solution, F^- ion associates strongly with HF to form FHF^- (the hydrogen difluoride ion), the species that actually loses the electrons:

$$
\begin{array}{ll}
FHF^-(sln) \longrightarrow F_2(g) + H^+(sln) + 2\,e^- & \text{(anode, oxidation)} \\
\underline{2\,H^+(sln) + 2\,e^- \longrightarrow H_2(g)} & \text{(cathode, reduction)} \\
H^+(sln) + FHF^-(sln) \longrightarrow F_2(g) + H_2(g) & \text{(overall)}
\end{array}
$$

This process resembles the electrolytic production of $Cl_2(g)$ (see Section 23–2). As fluorine and hydrogen evolve, HF(ℓ) is fed in continuously to keep the composition of the electrolyte constant at $KF \cdot 2HF$. Commercial fluorine cells use a steel cathode and carbon anode. The cells are engineered carefully to prevent contact between the product gases, which react explosively.

Fluorine is a very pale yellow gas. It condenses to a canary-yellow liquid at −188.14°C and solidifies at −219.62°C. It is an exceedingly toxic substance. Concentrations exceeding 25 ppm in air quickly damage the eyes, nose, lungs, and skin. Fluorine's pungent, irritating odor makes it first detectable in air at a concentration of about 3 ppm, well below the level of acute poisoning but above the recommended safe working level of 0.1 ppm. The ferocious reactivity of fluorine demands the utmost care in its use, and handling procedures that minimize the hazards have been developed. Certain metals (nickel, copper, steel, and the nickel–copper alloy Monel metal) resist attack by fluorine by forming a layer of a fluoride salt on their surfaces. Reactions can be carried out at room temperature in vessels made of these metals. The pure element is regularly packaged and shipped as a compressed gas in special cylinders but is also often generated at the point of use.

The role of elemental fluorine in the chemical industry has grown to major proportions since 1945. The motive in finding ways to prepare, purify, and handle it was the need to synthesize a volatile uranium compound to separate the isotopes ^{235}U and ^{238}U by gaseous diffusion, as described in Section 5–7. Uranium hexafluoride had the required properties, and gaseous diffusion still is used for preparing ^{235}U-enriched reactor fuel. At present, more than half of the world's production of $F_2(g)$ is used to make UF_6 from uranium(IV) oxide by these reactions:

$$UO_2(g) + 4\,HF(g) \longrightarrow UF_4(s) + 2\,H_2O(g)$$
$$UF_4(s) + F_2(g) \longrightarrow UF_6(g)$$

Most of the remaining fluorine production goes to make sulfur hexafluoride by the reaction

$$S(s) + 3\,F_2(g) \longrightarrow SF_6(g)$$

This unreactive gas is a good electrical insulator. Its presence within high-voltage electrical equipment minimizes arcing or sparking when the switches are thrown. Electrical discharges that do occur under an SF_6 atmosphere decompose the gas to an extent (into sulfur and fluorine), but the dissociation products rapidly recombine. Some triple-pane windows are manufactured with SF_6 sealed in the "dead-air" spaces between the panes because it cuts heat transmission and muffles sound better than air.

For many years, it was believed that no chemical agent could oxidize a fluoride to fluorine. The corollary was that electrochemical methods were essential in the preparation of elemental fluorine. Both ideas were shown to be wrong in 1986, when elemental fluorine was generated by completely chemical means in a reaction at 150°C involving metal fluorides:

$$2\,K_2MnF_6(s) + 4\,SbF_5(g) \longrightarrow 4\,KSbF_6(s) + 2\,MnF_3(s) + F_2(g)$$

• Neither starting compound in this remarkable reaction requires elemental fluorine for its synthesis, so the process is not an empty circular trick.

In this reaction, Mn(IV) is reduced to Mn(III) and F is oxidized from the −1 to the 0 oxidation state. Apparently, MnF_4 is formed as $SbF_5(s)$, acting as a Lewis acid, removes F^- from $K_2MnF_6(s)$. The MnF_4 is not stable and quickly decomposes to $F_2(g)$ and $MnF_3(s)$.

23–5   Compounds of Fluorine

In forming compounds, a fluorine atom either shares electrons in a single covalent bond or gains one electron to form the fluoride ion, F^-. In either case, it attains the noble-gas electron configuration $(1s^2 2s^2 2p^6)$. The other halogens tend to behave similarly but tolerate numerous exceptions; fluorine does not. The rules for assigning oxidation numbers (see Section 4–4) recognize the special character of fluorine by setting its oxidation number in compounds always equal to −1. Chlorine, bromine, and iodine may have positive oxidation numbers.

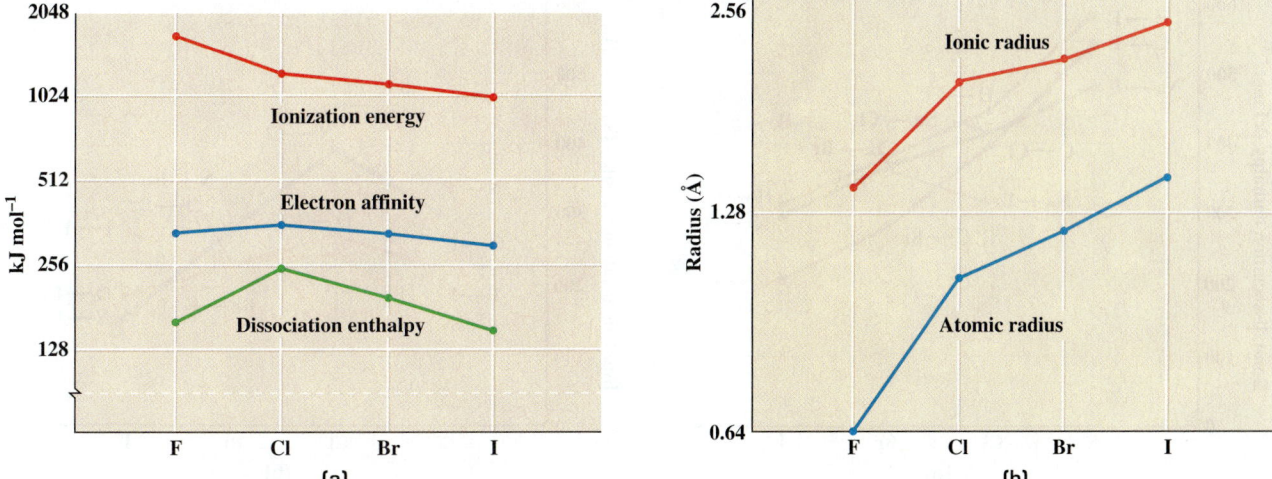

Figure 23-11 • The measured physical properties of fluorine often do not fall on the trend-lines defined by the behavior of the other halogens. (a) Here the ionization energy of fluorine is unexpectedly high, and its electron affinity and bond dissociation enthalpy are unexpectedly low. (b) The sizes of the fluorine atom and the fluoride ion are both smaller than predicted by comparison with the atoms or ions of the other halogens.

The chemical and physical properties of fluorine (and its compounds) deviate from the values predicted by extrapolating the trends among the other halogens. Figure 23–11 shows that the bond enthalpy of F_2 is smaller than would be expected from the trend among the other halogens, but fluorine's first ionization energy is *larger*. Despite its position as the most electronegative of the elements, electron-affinity data show that $F(g)$ accepts another electron (to form the gaseous anion) less readily than $Cl(g)$ and only slightly more readily than $Br(g)$. The unexpectedly low electron affinity of $F(g)$ is related to the contracted size of the atom. As the graphs of size in Figure 23–11b show, $F(g)$ and $F^-(g)$ are smaller than the extrapolation of the trends among the other halogens would predict, causing a relative crowding and greater mutual repulsion among the electrons in their $n = 2$ shell.

Strong bonds, both covalent and ionic, characterize the interaction of fluorine with most elements. Figure 23–12a shows that the ΔH_d°'s of H—F, C—F, and Na—F bonds exceed those of H, C, and Na combined with the other halogens. Fluorine's strong bonds with other elements and weak bond with itself (in F_2, the ΔH_d° is only 158 kJ mol^{-1}) together explain its terrific reactivity. Both factors are traceable to the small size of the fluorine atom. The two atoms in the fluorine molecule are only 1.417 Å apart. This puts their lone-pair electrons closer to each other than the lone pairs in the other halogens: Repulsion among the lone pairs weakens the bond. Small fluorine makes short bonds to non-fluorine atoms as well. Lone-pair repulsions of the type just described become important only with elements that, like fluorine itself, are small and electronegative. Hence, O—F and N—F bonds are, like the F—F bond, weaker than otherwise expected (Fig. 23–12b).

Because fluorine atoms are small, many of them fit around a given central atom before crowding each other. Therefore, atoms often attain a larger coordination number with fluorine than with other elements. The result is the existence of many fluorides, such as $K_2[NiF_6]$, $Cs[AuF_6]$, and PtF_6, in which the central element has an unexpectedly high oxidation number. The other halogens offer no comparable compounds.

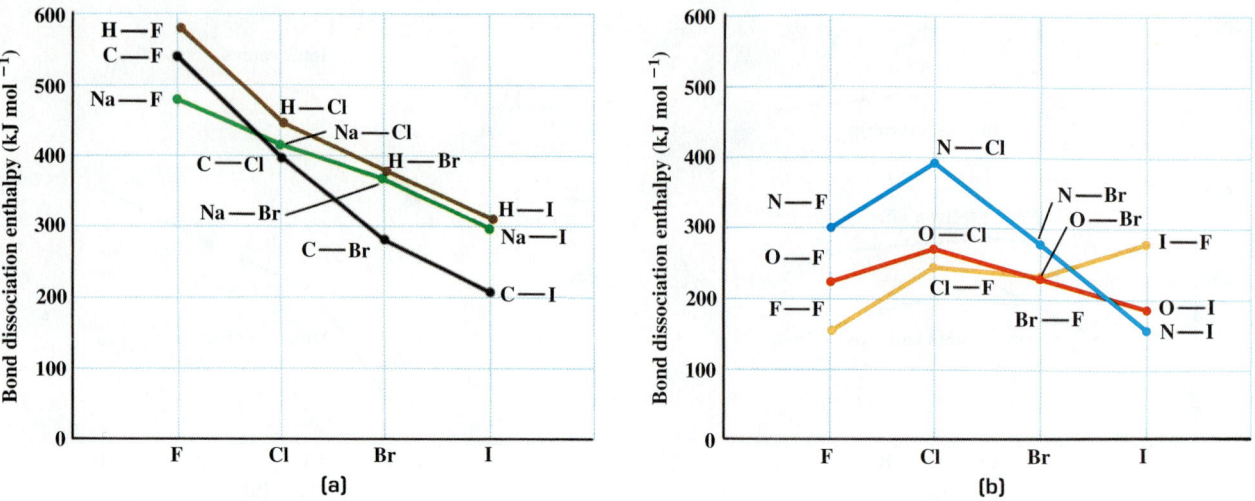

Figure 23–12 • (a) Fluorine forms strong bonds with more electropositive elements such as carbon, hydrogen, and sodium. (b) In contrast, fluorine forms unexpectedly weak bonds with the electronegative elements nitrogen, oxygen, and fluorine compared with the bonds formed by the other halogens.

Reactions of Fluorine

In general, fluorine reacts with other elements and with compounds by oxidizing them. It displaces the other, less reactive halogens from compounds, just as chlorine displaces bromine and iodine, the elements located below it in the periodic table, and bromine displaces iodine. An example is the reaction

$$2 \, NaCl(s) + F_2(g) \longrightarrow 2 \, NaF(s) + Cl_2(g)$$

Fluorine sometimes breaks this pattern by going much further. Thus, if $F_2(g)$ is heated with potassium chloride, the reaction is

$$KCl(s) + 2 \, F_2(g) \longrightarrow KClF_4(s)$$

Here, the chlorine is not displaced from the compound but is oxidized to the $+3$ state and surrounded by fluorine atoms in the "tetrafluorochlorate(III)" ion. Further heating of $KClF_4(s)$ with F_2 eventually does displace chlorine from potassium, but in an unexpected form, $ClF_5(g)$:

$$KClF_4(s) + F_2(g) \longrightarrow KF(s) + ClF_5(g)$$

In a variation, fluorine oxidizes iodine in potassium iodide and also displaces it:

$$KI(s) + F_2(s) \longrightarrow KF(s) + IF(g)$$

Under different conditions with the same reactants, the theme is carried to an extreme:

$$KI(s) + 4 \, F_2(g) \longrightarrow KF(s) + IF_7(g)$$

Fluorine also can displace oxygen from its compounds:

$$2 \, CaO(s) + 2 \, F_2(g) \longrightarrow 2 \, CaF_2(s) + O_2(g)$$

With other oxides, only a portion of the oxygen may be displaced, as in the reaction of $SO_2(g)$ with $F_2(g)$ to give thionyl fluoride (Fig. 23–13):

$$2 \, SO_2(g) + 2 \, F_2(g) \longrightarrow 2 \, SOF_2(g) + O_2(g)$$

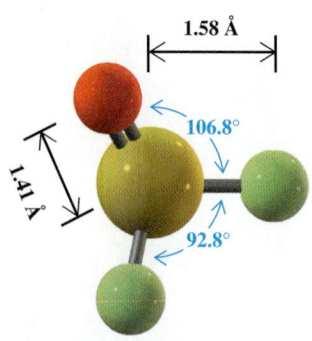

1.58 Å

106.8°

1.41 Å

92.8°

Figure 23–13 • In SOF₂, oxygen and fluorine both bond to the central sulfur atom.

The most important oxide of all is water. Water vapor at 100°C burns in fluorine with a weak, luminous flame. The reaction is

$$2 F_2(g) + 2 H_2O(g) \longrightarrow 4 HF(g) + O_2(g)$$

As is often the case, different conditions favor different products. An excess of fluorine in contact with ice can give hydrogen fluoride and highly unstable hydrogen oxyfluoride:

$$F_2(g) + H_2O(s) \longrightarrow HF(\ell) + HOF(\ell)$$

Oxygen difluoride also forms

$$2 F_2(g) + H_2O(s) \longrightarrow 2 HF(\ell) + OF_2(g)$$

and the reaction that gives O_2 and HF still takes place to a limited extent. In all three of these reactions, fluorine disrupts stable H_2O, taking oxygen from the -2 to the 0 or $+2$ state. Few materials other than fluorine can oxidize water in this way.

- Assigning an oxidation number of -1 to fluorine and $+1$ to hydrogen in hypofluorous acid (HOF) means that the oxidation number of oxygen in this compound is 0.

Metal Fluorides

The alkali-metal and alkaline-earth elements form ionic salts with fluorine with the expected stoichiometry. Except for lithium fluoride, the alkali-metal fluorides are all quite soluble in water. The differences between the alkaline-earth fluorides (BeF_2, MgF_2, CaF_2, SrF_2, and BaF_2) and the corresponding chlorides, bromides, and iodides are striking: BeF_2 is ionic and soluble in water, but $BeCl_2$ is covalent enough to sublime; CaF_2 is insoluble in water, but the other calcium halides are quite soluble in water. All of these fluorides are prepared most easily by reaction of the corresponding hydroxide or carbonate with aqueous hydrofluoric acid. In their lower oxidation states, the transition metals and Group II and IV metals also form fluoride salts with a range of solubilities; ZnF_2, SnF_2, and PbF_2 are examples.

Fluorine forms salts with most metals in their lower oxidation states but is capable of oxidizing many metals to exceptionally high oxidation states as well. The chemistry of cobalt centers on the $+2$ oxidation state, for example, except in complex ions, in which the $+3$ state is not uncommon. Cobalt(III) fluoride, which is obtained by direct fluorination of the metal, is one of the very few simple ionic salts of that element in which the cation has an oxidation number of $+3$. Platinum is oxidized directly to platinum(IV) by fluorine and (at higher temperatures) to platinum(VI) fluoride. The latter is a dark red solid with a low melting point (56.7°C) and is a covalent molecular compound. Other heavy metals that form volatile hexafluorides include osmium, iridium, molybdenum, tungsten, and uranium. Fluorides of lower oxidation state are known for all of these metals.

- These hexafluorides have significant covalent character and are volatile. Hence the adage among chemists: "Fluorine gives wings to the elements."

In the great majority of its compounds, silver has only a $+1$ oxidation state. Silver in the $+2$ oxidation state, however, can be formed by heating silver(I) fluoride in a stream of fluorine:

$$2 AgF(s) + F_2(g) \longrightarrow 2 AgF_2(s)$$

Even the $+3$ state of silver can be formed in a ternary compound when potassium fluoride is mixed with silver fluoride and heated in fluorine:

$$KF(s) + AgF(s) + F_2(g) \longrightarrow KAgF_4(s)$$

Low concentrations (about 1 mg L^{-1}) of fluoride ion in drinking water help to prevent tooth decay. Fluoride ion substitutes for hydroxide ion in tooth enamel, changing some of the hydroxyapatite ($Ca_5(PO_4)_3OH$) in the enamel to fluorapatite

• Similar F^- for OH^- replacements occur in many minerals. An example is topaz, a semiprecious mineral with formula $Al_2SiO_4(F,OH)_2$. The fluoride and hydroxide ions are grouped within parentheses because the degree of substitution of one ion for the other varies with the origin of the mineral.

$(Ca_5(PO_4)_3F)$. The replacement works because the F^- and OH^- ions have the same charge (-1) and similar radii (1.33 Å for F^- versus approximately 1.2 Å for OH^-). Fluorapatite is less soluble than is hydroxyapatite in the acidic oral environment fostered by consumption of sweets. The deliberate fluoridation of civic water supplies to cut down tooth decay is a common public-health measure in the United States, and fluoride salts often are added to toothpaste.

Covalent Fluorides

Fluorine forms compounds with all the elements in Groups III through VII. The Group III elements (boron, aluminum, gallium, indium, and thallium) all form trifluorides, and thallium forms TlF as well. Boron trifluoride and aluminum trifluoride are strong Lewis acids that readily form the complex ions $[BF_4]^-$ and $[AlF_6]^{3-}$ in solutions that contain F^- ion. Boron trifluoride is a gas at room temperature, but the trifluorides of aluminum and the other members of Group III are solids that melt near 1000°C. They are intermediate between ionic and covalent in their bonding, and in the crystalline state their structures consist of extended arrays of metal atoms in sixfold coordination with fluorine atoms. The trichlorides of these elements are much more covalent in character; $AlCl_3(s)$ sublimes as $Al_2Cl_6(g)$ when it is heated to only 178°C.

• BF_3 is widely used as a catalyst for polymerization reactions (see Chapter 25).

The transitional character of the bonding is even more evident in the fluorides of Group IV. At room temperature, carbon tetrafluoride and silicon tetrafluoride are gases in which the molecules are covalent; germanium tetrafluoride is a molecular liquid, but the tetrafluorides of tin and lead are solids with considerable ionic character, but not as ionic as the difluorides. This point illustrates a general rule: The halides of the heavier transition and post-transition elements are more covalent when the oxidation number of the element is high and more ionic when it is low. Carbon tetrafluoride is a kinetically stable compound. Although its reaction with water is a thermodynamically favored process at room conditions, the reaction does not occur at an appreciable rate below 500°C. Silicon and germanium tetrafluorides, however, are extensively hydrolyzed by water but can be distilled successfully from aqueous solutions containing hydrofluoric acid.

The fluorides of the Group V elements are just as varied in composition and behavior. Nitrogen forms fluorides having the composition N_2F_2 (dinitrogen difluoride), N_2F_4 (dinitrogen tetrafluoride), and NF_3 (nitrogen trifluoride). All are gases at room conditions. Nitrogen trifluoride has pyramidal molecules (Fig. 23–14), is rather inert, and has no electron-donor properties. The presence of the electronegative fluorine atoms in fact gives the nitrogen atom in NF_3 a charge of $+0.80$, according to one calculation. Like CF_4, it is not hydrolyzed by water. It fluorinates certain metals when heated:

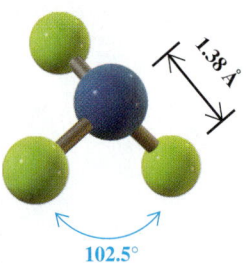

Figure 23–14 • The structure of NF_3. The three F—N—F angles are equal; the three N—F bond lengths are equal.

$$2\,NF_3(g) + Cu(s) \longrightarrow N_2F_4(g) + CuF_2(s)$$

and is used as an etchant in plasma technology for the fabrication of semiconductors. Dinitrogen tetrafluoride is much more reactive than is nitrogen trifluoride, undergoing hydrolysis in water and dissociating when it is heated. The PF_3 molecule has a trigonal pyramidal structure like NF_3, as the VSEPR model predicts. Unlike ammonia, however, it does not donate the lone pair of electrons on the central atom readily, presumably because the highly electronegative fluorine atoms withdraw electron density from the central atom. Like BF_3, PF_5 is a very powerful Lewis acid and is useful as a catalyst for polymerization.

• Recall that "VSEPR" stands for "valence shell electron-pair repulsion." This model for predicting molecular shapes is explained in Section 3–7.

The remaining Group V elements (arsenic, antimony, and bismuth) resemble phosphorus in forming both trifluorides and pentafluorides that readily hydrolyze in aqueous solution. They are powerful fluorinating agents, capable of introducing fluorine into other molecules by substitution reactions.

The highly toxic gas oxygen difluoride (OF_2) can be made by passing fluorine through aqueous alkaline solutions:

$$2 F_2(g) + 2 OH^-(aq) \longrightarrow OF_2(g) + 2 F^-(aq) + H_2O(\ell)$$

It is worthwhile to compare OF_2 with the analogous compound formed between chlorine and oxygen. Both Cl_2O and OF_2 are gases at room conditions, and both have oxygen atoms bonded to two halogens. In both compounds, the steric number of the oxygen atom is 4, with two bonds and two lone pairs on the central oxygen atom. Consequently, both have nonlinear molecules (Fig. 23–15). Despite these similarities, the two differ profoundly in their reactivity. Oxygen difluoride contains "positive oxygen" (note the +2 oxidation number for oxygen) because fluorine is highly electronegative. It reacts with water by oxidizing it to liberate O_2:

$$OF_2(g) + H_2O(\ell) \longrightarrow O_2(g) + 2 HF(aq)$$

Dichlorine oxide contains ordinary "negative oxygen" and behaves quite differently with water. It gives hypochlorous acid without oxidation or reduction:

$$Cl_2O(g) + H_2O(\ell) \longrightarrow 2 HOCl(aq)$$

The chemistries of the compounds diverge along the same lines in many other reactions.

The later Group VI elements also form fluorides. The two most important fluorides of sulfur are quite dissimilar in their properties. Both sulfur tetrafluoride and sulfur hexafluoride are gases at room temperature; the former is highly reactive, and the latter is extremely inert. Their seesaw and octahedral molecular structures are given in Figures 3–21b and 3–18, respectively. Both selenium and tellurium form a tetrafluoride and a hexafluoride with fluorine. These compounds, like the fluorides of their Group V neighbors arsenic and antimony, are excellent fluorinating agents. Their strikingly enhanced reactivity compared with sulfur hexafluoride is not easy to explain, except to note that the larger central atoms are less well shielded by their fluorine atoms.

Fluorine reacts with the other halogens to form several **interhalogen** compounds. In these compounds, fluorine atoms, as always in the −1 oxidation state, surround a central chlorine, bromine, or iodine atom as XF, XF_3, and XF_5, where X is Cl, Br, or I. The only XF_7 compound occurs with iodine. The VSEPR model provides accurate predictions of the structures of these compounds. In IF_7, iodine has a steric number of 7, and the structure is based on the pentagonal bipyramid (Fig. 23–16).

Fluorinated Hydrocarbons and Chlorofluorocarbons

Fluorine can substitute for hydrogen in hydrocarbons, giving rise to an important class of compounds, the fluorinated hydrocarbons. The C—F bond is short and strong, an average of 1.2 times stronger than the C—H bond and 1.4 to 2.2 times stronger than the bonds between carbon and other halogens; it is the strongest single bond. Fluorination of hydrocarbons in which no double or triple bonds exist

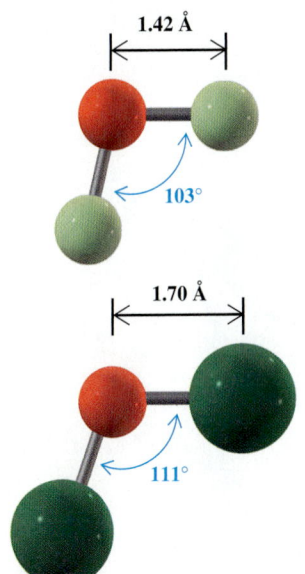

Figure 23–15 • The nonlinear structures of Cl_2O (*bottom*) and OF_2. The two are closely related, but the bond distances are substantially longer in Cl_2O and the bond angle is larger.

Figure 23–16 • The molecular structure of IF_7 is based on the pentagonal bipyramidal geometry. The two axial fluorines are 1.786 Å away from the central iodine; the five equatorial fluorines are farther (1.858 Å) away, presumably to relieve crowding in the equatorial plane.

Figure 23–17 • The structure of per-fluoroneopentane, C_5F_{12}, a perfluorocarbon. The fluorine atoms on the perimeter of the molecule hold their electrons tightly. This and other perfluorocarbons consequently tend to be unreactive.

Figure 23–18 • A Teflon-coated "no-stick" frying pan.

• The reverse is true in the reactions of the other halogens with hydrocarbons.

Figure 23–19 • The molecular structure of the polymer Teflon. The fluorine atoms shield the carbon chain against attack. In fact, the chain must twist to accommodate their bulk, completing a full spiral every 26 C atoms along the chain.

therefore generally increases thermal and chemical stability. When fluorine atoms replace all the hydrogen atoms in hydrocarbons, compounds called **fluorocarbons** (or **perfluorocarbons**) result (Fig. 23–17). The electron density on the fluorine atoms in fluorocarbons is held tightly and is not easily distorted by approaching atoms or ions. This reduces the reactivity of these compounds and weakens their intermolecular forces. The compounds tend to have low melting and boiling points and low enthalpies of vaporization.

One provocative use of fluorine compounds is in the preparation of blood substitutes. Many fluorocarbons are excellent solvents for oxygen. A milliliter of liquid C_7F_{16}, for example, dissolves 1900 times more oxygen at room conditions than does a milliliter of water. Small laboratory animals can survive total immersion in oxygenated fluorocarbon liquids because their lungs successfully extract the oxygen they need from the solution. Fluorocarbons are insoluble in water. For use in the bloodstream, they are prepared as fine-grained emulsions in water. Rats have survived for days without apparent harm after their blood was completely replaced with a concentrated emulsion of $C_8F_{17}Br$ in water. Over a period of time, the composition of the rats' blood reverted to normal as the blood substitute was excreted and new red blood cells grew. Practical fluorocarbon blood substitutes would need no matching of blood types, could be stored for emergency use, and would not transmit disease.

The best-known perfluorocarbon is the solid polytetrafluoroethylene (Teflon), formulated as —$(CF_2CF_2)_n$— where n is a large number. Chemically, this compound is nearly completely inert, resisting attack from boiling sulfuric acid, molten potassium hydroxide, fluorine gas, and other aggressive chemicals. Physically, it has excellent heat stability, is a very good electrical insulator, and has a low coefficient of friction ("no-stick" Teflon) that makes it useful for bearing surfaces in machines as well as coating frying pans (Fig. 23–18). In Teflon the carbon atoms lie in a long chain that is encased by tightly bound fluorine atoms (Fig. 23–19). Even reactants with a strong innate ability to disrupt C—C bonds (like fluorine itself) fail to attack Teflon at observable rates because no route of attack exists past the surrounding fluorine atoms and their tightly held electrons.

Chlorofluorocarbons (often called CFCs) contain one or more carbon atoms with both fluorine and chlorine atoms attached as side-groups. Chlorofluorocarbons are chemically inert, non-toxic gases or low-boiling liquids such as dichlorodifluoromethane, CF_2Cl_2 (Fig. 23–20). They are relatively inexpensive to make and have boiling points conveniently near room temperature. Their properties fit them well to serve as propellants in aerosol sprays and as refrigerants. Soon after their discovery, chlorofluorocarbons replaced hydrocarbons in the former application and SO_2 and NH_3 in the latter application. The inertness of chlorofluorocarbons has drawbacks as well as advantages because the molecules that escape into the air persist unchanged. In time, they mix into the stratosphere, where ultraviolet sunlight finally breaks them down. Chlorine atoms generated in this decomposition catalyze the destruction of ozone (O_3) in the high reaches of the atmosphere. The phase-out of CFCs under the Montreal Protocol (see Section 18–5) prompted an intense search for replacements. Substitutes must be reactive enough to break down in the troposphere, yet non-toxic and reasonably non-flammable. They also should have equal (or better) properties for their application, work in existing equipment, and be manufacturable at a reasonable price. Compounds containing at least one hydrogen atom per molecule are generally more reactive than are CFCs, and fluorine does not participate in the ozone destruction cycle. Therefore, attention has focused on HCFCs (hydrochlorofluorocarbons) and HFCs (hydrofluorocarbons) as CFC

• The term "halofluorocarbon" includes the possible presence of bromine and iodine.

Figure 23–20 • The structure of CF_2Cl_2. This compound, also called CFC-12, is being replaced in its major use in air-conditioning units. It harms the stratospheric ozone layer, and production has been ended in the developed nations.

CHEMISTRY IN YOUR LIFE

Fluorinated Drugs

The effective anticancer drug 5-fluorouracil ($C_4H_3N_2O_2F$, Fig. 23–A) is a planned drug, the result of rational drug design. Cancer cells use uracil ($C_4H_4N_2O_2$), which is essential in cellular reproduction, and other biochemicals more rapidly than normal cells because they reproduce faster. The idea to substitute fluorine for hydrogen in uracil came from the observation that fluorine-substituted organic compounds are often more toxic than unsubstituted compounds. Replacing a C—H bond in a molecule with a C—F bond is, from the point of view of altering molecular shape and size, only a minimal change. As a consequence, fluorine-substituted compounds often successfully mimic their unsubstituted versions and enter biochemical processes. However, C—F bonds are stronger than C—H bonds (see Table 10–3), strong enough to stop all metabolic steps that require them to break. Thus, 5-fluorouracil enters the cell and assumes the role of uracil. The fluorine atom blocks an essential change at the 5-position and prevents replication of the genetic material of the cell, which no longer reproduces. The drug acts preferentially against tumors because it concentrates in fast-growing cells. Still, 5-fluorouracil is harmful to all cells. This drug is rarely given alone. In

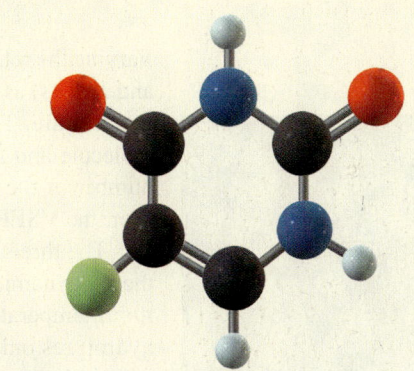

Figure 23–A • The structure of 5-fluorouracil, a drug that is much used in cancer chemotherapy.

combination with other drugs it usually achieves large synergistic effects; that is, its effects in combination with those of a second drug exceed the sum of the effects of either drug given separately.

Other fluorine-substituted compounds are effective against viruses such as herpes and against many bacteria. They are also useful in the treatment of the inflammation of rheumatoid arthritis and of malaria.

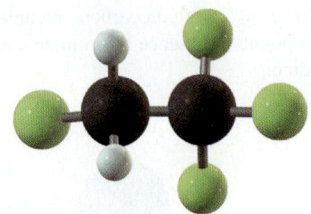

Figure 23–21 • The structure of the HFC 1,1,1,2-tetrafluoroethane, or "HFC-134a." This CFC substitute contains no chlorine and so does not harm the ozone layer.

substitutes. The compound "HFC-134a" (Fig. 23–21) works well in many applications, but acceptable substitutes have not been identified for every current use of CFCs.

23-6 Compounds of Fluorine and the Noble Gases

For many years, the noble gases were believed to be chemically inert. They failed to react with the strongest oxidizing agents available, and an early attempt (1895) by Moissan to make argon react with fluorine also failed. In 1933, Linus Pauling suggested on theoretical grounds that xenon and fluorine should form compounds. Soon thereafter, chemists subjected a mixture of the two gases to an electrical discharge, but no evidence of reaction was found. This result, other negative results, and the well-known dangers of working with elemental fluorine discouraged further experiments. It seemed clear that if fluorine could not force a noble gas into combination, then nothing could. Chemists' interest was further stifled because generally accepted simple bonding theory appeared to ratify the complete chemical inertness of the noble gases. Then, in 1962, Neil Bartlett treated gaseous xenon with the powerful fluorinating agent PtF_6, and a yellow-orange solid compound of platinum, fluorine, and xenon resulted. Beyond doubt, xenon had been forced to combine chemically. In 1965, an irony emerged: the same fluorine–xenon mixtures that did not react in the 1933 electrical discharge experiment *did* give $XeF_2(s)$ simply upon exposure to sunlight!

Bartlett's discovery spurred research on the reactions of the noble gases with fluorine, and soon afterward the synthesis of xenon tetrafluoride by direct combination of the elements at high pressure was reported (Fig. 23–22):

$$Xe(g) + 2\,F_2(g) \longrightarrow XeF_4(s)$$

Varying the relative amounts of xenon and fluorine permitted the synthesis of $XeF_2(s)$ and $XeF_6(s)$ as well. All three xenon fluorides are colorless crystalline solids at room temperature. The VSEPR theory correctly predicts the linear structure of the XeF_2 molecule and the square planar structure of the XeF_4 molecule. Because the steric number of the central Xe is 7 and not 6, XeF_6 molecules are not octahedral; however, the VSEPR theory does not correctly predict their complex structure.

The three xenon fluorides are moderate to powerful fluorinating agents but are thermodynamically stable with respect to decomposition into xenon and fluorine at room temperature. Their behavior toward water is quite diverse. Xenon difluoride hydrolyzes only very slowly in water, and solutions as concentrated as 0.1 M can be prepared at low temperatures. On the other hand, XeF_4 and XeF_6 react vigorously with water to form xenon trioxide (XeO_3):

$$3\,XeF_4(s) + 6\,H_2O(\ell) \longrightarrow XeO_3(aq) + 2\,Xe(g) + 12\,HF(aq) + \tfrac{3}{2}\,O_2(g)$$
$$XeF_6(s) + 3\,H_2O(\ell) \longrightarrow XeO_3(aq) + 6\,HF(aq)$$

Xenon trioxide is soluble in aqueous solution without ionization, but in basic solution it acts as a Lewis acid, accepting OH^-:

$$XeO_3(aq) + OH^-(aq) \longrightarrow HXeO_4^-(aq)$$

This hydrogen xenate ion then disproportionates to form xenon and the perxenate ion, in which the oxidation number of xenon is +8:

$$2\,HXeO_4^-(aq) + 2\,OH^-(aq) \longrightarrow XeO_6^{4-}(aq) + Xe(g) + O_2(g) + 2\,H_2O(\ell)$$

Figure 23–22 • Crystals of xenon tetrafluoride formed in making the first binary noble-gas compound. *(Argonne National Laboratories)*

Only one simple krypton halide has been prepared. Irradiating a mixture of krypton and fluorine at low temperature with electrons forms krypton difluoride, a thermodynamically unstable substance. It is a volatile, colorless, crystalline solid. Experiment confirms the linear structure predicted by the VSEPR theory. It forms salt-like compounds with strong fluoride-ion acceptors such as the Lewis acids AsF_5 and SbF_5:

$$KrF_2 + AsF_5 \longrightarrow [KrF]^+ + [AsF_6]^-$$

Krypton difluoride is a stronger oxidizing agent than is fluorine itself. It has a total bond enthalpy (for *both* Kr—F bonds) of only 96 kJ mol^{-1}; this is substantially less than the bond enthalpy of $F_2(g)$ (158 kJ mol^{-1}). It oxidizes Xe(g) to XeF$_6(s)$ at room temperature and oxidizes I_2 directly to IF$_7$. A solution of KrF$_2$ in HF(ℓ) even attacks gold:

$$7\,KrF_2(g) + 2\,Au(s) \longrightarrow 2\,KrF[AuF_6](s) + 5\,Kr(g)$$

Heating the KrF[AuF$_6$] to 60°C generates orange AuF$_5$, in which gold has a +5 oxidation number. By way of comparison, $F_2(g)$ itself requires a temperature of 250°C to act on powdered gold and then generates only AuF$_3$ (mixed with unreacted gold).

No simple fluorides of argon have been prepared; however, the first synthesis of a neutral compound containing argon was reported in 2000. This was HArF (argon fluorohydride), prepared by illuminating hydrogen fluoride that had been frozen into a solid argon matrix.

SUMMARY

23-1 Sodium chloride is the source for three important chemicals: sodium carbonate, sodium hydroxide, and chlorine. In the obsolete **Leblanc process,** sodium chloride was heated with sulfuric acid to form sodium sulfate. Heating sodium sulfate with a mixture of carbon (from coal) and limestone (CaCO$_3$) gave a mixture from which the water-soluble sodium carbonate could be extracted. The process had two noxious by-products—CaS and HCl. Oxidation of the HCl from the Leblanc process to Cl$_2$ either by reaction with the mineral pyrolusite (MnO$_2$) or by the **Deacon process** changed the HCl from a nuisance to an asset because chlorine, in the form of chlorinated lime, makes an excellent bleach for textiles. The **Solvay process** supplanted the Leblanc process. In it, the net reaction is the conversion of NaCl and CaCO$_3$ to Na$_2$CO$_3$ and the by-product CaCl$_2$. Ammonia and ammonium chloride are important intermediates that are recycled through this continuous process.

23-2 Sodium hydroxide and chlorine also can be produced from salt by the electrolysis of concentrated aqueous solutions, with hydrogen a by-product. In the **diaphragm cell,** the anode is made of titanium coated with a noble metal, and the cathode is steel. Sodium hydroxide, a strong soluble base, has many uses in chemical and materials processing. Chlorine is used to make bleach, chlorinated hydrocarbons, and plastics and in the purification of titanium for TiO$_2$ paint pigments.

23-3 Chlorine, bromine, and iodine all disproportionate in water, producing hydrohalic acid and hypohalous acids. Acidic solutions of chlorates, bromates, and iodates are powerful oxidizing agents. They react with solutions of chlorides, bromides, and iodides to produce chlorine, bromine, and iodine, respectively.

23-4 The chemistry of fluorine differs sharply from that of the other halogens. Fluorine atoms form weak bonds to other fluorine atoms but particularly strong bonds to most other kinds of atoms. These facts explain the high chemical reactivity of F_2. Modern methods for preparing fluorine are based on the original 1886 electrolytic process and use KF dissolved in HF. Most fluorine compounds are made directly or indirectly from hydrogen fluoride, a colorless hydrogen-bonded liquid. Elemental fluorine is used in uranium production and isotope separation and in the manufacture of sulfur hexafluoride.

23-5 Fluorine displaces other halogens from their compounds and oxidizes elements in compounds exposed to it. The compounds of fluorine with metals in their lower oxidation states are salts with strong ionic character. The higher oxidation states of many of the same elements yield volatile fluorides with significant covalent character. Compounds of fluorine with the nonmetals possess a wide range of reactivities and molecular geometries. Some fluorine compounds, such as SF_6, are quite inert toward chemical reactions; others, such as BF_3, are strong Lewis acids that readily accept a share in electron pairs from fluoride ions or other Lewis bases; still others, such as SbF_5, are strong fluorinating agents. **Fluorinated hydrocarbons** arise by the substitution of fluorine for hydrogen in hydrocarbons and often have greater thermal and chemical stability than the parent hydrocarbon. The release of chlorine upon eventual decomposition of **chlorofluorocarbons** in the upper atmosphere leads to depletion of the ozone in that region. The unique properties of fluorine shape the properties of the important fluorinated polymer polytetrafluoroethylene (Teflon).

23-6 Work with PtF_6, a strong fluorinating agent, opened up modern research on the compounds of the noble gases by its reaction with xenon; xenon and krypton fluorides are made by direct reaction with fluorine or with other fluorinating agents.

PROBLEMS

Note: Answers to blue-numbered problems are given in Appendix F. Problems that are more challenging are indicated with asterisks.

Chemicals from Salt

1. Calculate the equilibrium constant at 25°C for the production of sodium hypochlorite bleach from dichlorine oxide and sodium hydroxide:

$$Cl_2O(g) + 2\,OH^-(aq) \rightleftharpoons 2\,OCl^-(aq) + H_2O(\ell)$$

Relevant data appear in Appendix D.

2. Calculate the equilibrium constant at 25°C for the preparation of chlorine by the reaction of hydrochloric acid with pyrolusite, used by Scheele in 1774:

$$2\,Cl^-(aq) + 4\,H^+(aq) + MnO_2(s) \longrightarrow$$
$$Cl_2(g) + Mn^{2+}(aq) + 2\,H_2O(\ell)$$

Relevant data appear in Appendix D.

3. Compare the Leblanc process to the Solvay process with regard to (a) expense and availability of starting materials and (b) purity of the final product.

4. Compare the Leblanc process to the Solvay process with regard to the by-products of the two methods. What uses are made of these by-products?

5. Calculate the standard enthalpy of the Leblanc process per mole of Na_2CO_3 produced, using data from Appendix D. From this, what do you conclude about the overall heat requirements of that process?

6. Calculate the standard enthalpy of the Solvay process per mole of Na_2CO_3 produced, using data from Appendix D. From this, what do you conclude about the overall heat requirements of that process?

7. Which is the stronger acid, chlorous acid or hypochlorous acid? Which would be preferred to convert potassium fluoride to hydrofluoric acid? (*Hint*: Refer to Table 8–2.)

8. Which is the stronger base, potassium carbonate or potassium hydroxide? Which would be preferred to convert potassium hydrogen phosphate (K_2HPO_4) to potassium phosphate (K_3PO_4)? (*Hint*: Refer to Table 8–2.)

Electrolysis and the Chlor-Alkali Industry

9. Some pure molten sodium chloride is electrolyzed.
   (a) Write a balanced equation to represent the reaction that occurs.
   (b) A student uses data from Appendix E to conclude that a voltage source must supply 4.07 V to electrolyze molten sodium chloride. Explain why this conclusion is faulty.

10. A dilute aqueous solution of sodium chloride is electrolyzed. Gases form at both the anode and cathode, and neither gas is chlorine. Write a balanced equation to represent the reaction that occurs.

11. What impurity is likely to be present in NaOH that is made by mixing $Ca(OH)_2$ with Na_2CO_3 but absent in NaOH that is made electrolytically in a diaphragm cell?

12. What impurity is likely to be present in NaOH that is made electrolytically in a diaphragm cell but absent in NaOH that is made by mixing $Ca(OH)_2$ with Na_2CO_3?

The Chemistry of Chlorine, Bromine, and Iodine

13. Gaseous chlorine can be used to extract bromine from seawater via the reaction

$$2\,Br^-(aq) + Cl_2(g) \longrightarrow Br_2(\ell) + 2\,Cl^-(aq)$$

Calculate the volume of $Cl_2(g)$ at STP that would be needed to extract all the bromine from 1000 m^3 of seawater, assuming complete reaction and taking the $Br^-(aq)$ concentration in seawater to be 0.0036 M.

14. Bubbling chlorine through a basic solution containing iodide ions produces chloride ions and iodate (IO_3^-) ions.
    (a) Write a balanced chemical equation for this reaction.
    (b) Suppose that 0.216 L of gaseous chlorine (measured at STP) is needed to react completely with 50.0 mL of the iodide solution in part (a). Determine the initial concentration of iodide ion in that solution.

15. Dichlorine monoxide can be prepared by reaction of chlorine with mercury(II) oxide:

$$2\,Cl_2(g) + 2\,HgO(s) \longrightarrow Cl_2O(g) + HgCl_2\cdot HgO(s)$$

    (a) What is the oxidation state of chlorine in Cl_2O?
    (b) What oxoacid results when Cl_2O reacts with water?

16. (a) What is the oxidation state of chlorine in $HClO_4$?
    (b) The compound P_4O_{10} is a powerful enough dehydrating agent to remove water from $HClO_4$, forming phosphoric acid (H_3PO_4) and an oxide of chlorine. What is the chemical formula of this oxide of chlorine, the anhydride of perchloric acid?
    (c) Write a balanced chemical equation for the dehydration of perchloric acid by P_4O_{10}.

17. Given the following standard reduction potentials,

$$\begin{aligned}
ClO_2 + e^- &\longrightarrow ClO_2^- & \mathcal{E}° &= 0.954\ V \\
ClO_3^- + H_2O + e^- &\longrightarrow ClO_2 + 2\,OH^- & \mathcal{E}° &= -0.25\ V
\end{aligned}$$

    decide whether $ClO_2(g)$ is stable with respect to disproportionation in water at pH 14.

18. Given the following standard reduction potentials,

$$\begin{aligned}
ClO^- + H_2O + 2\,e^- &\longrightarrow Cl^- + 2\,OH^- & \mathcal{E}° &= 0.90\ V \\
ClO_3^- + 2\,H_2O + 4\,e^- &\longrightarrow ClO^- + 4\,OH^- & \mathcal{E}° &= 0.48V
\end{aligned}$$

    decide whether 1 M aqueous hypochlorite ion is stable with respect to disproportionation in aqueous solution at pH 14.

Fluorine Chemistry

19. Which bond is stronger: Si—F or Si—Cl? Give reasons for your answer.

20. Which reaction of phosphorus is more exothermic: that with fluorine to give phosphorus trifluoride, or that with chlorine to give phosphorus trichloride? Give reasons for your answer.

21. Write balanced equations for the reaction of elemental fluorine with the following:
    (a) $SrO(s)$ (c) $UF_4(s)$
    (b) $O_2(g)$ (d) $NaCl(s)$

22. Write balanced equations for the reaction of elemental fluorine with the following:
    (a) $SO_2(g)$ (c) $Na(s)$
    (b) $MgO(s)$ (d) $GeO(s)$

23. The alkali metals form one or more acid fluorides, with formulas $MF\cdot HF$, $MF\cdot 2HF$, and $MF\cdot 3HF$ (where M stands for the alkali metal). In these compounds, additional HF molecules link with F^- ions by means of hydrogen bonds, in a manner similar to the way water of hydration is incorporated in crystalline hydrates. Nothing similar is observed with the other alkali-metal halides.

    Suppose that a compound is found to contain 2.48 g of fluorine for every gram of sodium. What is the empirical formula of this compound? Assume that any hydrogen present is not detected in this analysis.

24. Carbon and fluorine can be combined in more ways than those mentioned in the text. When carbon in the form of graphite is treated with elemental fluorine, F atoms occupy interstitial spaces between the layers of C atoms, and compounds of empirical formula C_4F, C_2F, and CF result. The bonds between carbon and fluorine in all three of these materials are covalent. Graphite fluorides are used as electrodes in advanced primary batteries and as lubricants.

    Suppose that a graphite fluoride is analyzed and found to contain 2.53 g of carbon for every gram of fluorine. What is the empirical formula of this compound?

25. A gaseous binary compound of chlorine and fluorine has a density at STP of 4.13 g L^{-1} and is 38.35% chlorine by mass. Determine its molecular formula.

26. A gaseous compound has the molecular formula ClO_3F. Determine the volume occupied by 225 g of this compound at STP.

27. Predict the structures of the following fluorine-containing molecules:
    (a) OF_2 (d) BrF_5
    (b) BF_3 (e) IF_7
    (c) BrF_3 (f) SeF_6

28. Predict the structures of the following fluorine-containing molecular ions:
    (a) SiF_6^{2-} (d) AsF_6^-
    (b) ClF_2^+ (e) BF_4^-
    (c) IF_4^+ (f) BrF_6^+

29. Thionyl and selenyl difluoride have the compositions SOF_2 and $SeOF_2$, with the oxygen and fluorine atoms directly linked to the central sulfur or selenium atom. They can be prepared by reaction of a strong fluorinating agent, such as PF_5, with sulfur dioxide or selenium dioxide:

$$SO_2(g) + PF_5(g) \longrightarrow SOF_2(g) + POF_3(g)$$

(a) Use the VSEPR theory to predict the structures of SOF_2 and $SeOF_2$. Draw the structures.
(b) Thionyl difluoride can coordinate with BF_3 through its lone pair. Does it act as a Lewis acid or a Lewis base in this reaction?

30. When two oxygen atoms take the place of four fluorine atoms in SF_6, the compound has the composition SO_2F_2 and is called "sulfuryl difluoride." Like SF_6, SO_2F_2 is comparatively unreactive and hydrolyzes with difficulty.

(a) Use the VSEPR theory to predict the structure of SO_2F_2. Draw the structure.
(b) In the conversion of thionyl difluoride (SOF_2) to sulfuryl difluoride,

$$2\ SOF_2 + O_2 \longrightarrow 2\ SO_2F_2$$

does the thionyl difluoride act as a Lewis acid or a Lewis base?

31. Although BrF_3 is a covalent compound, $BrF_3(\ell)$ is slightly conductive due to the autoionization of BrF_3 molecules. This is analogous to the small conductivity of water arising from its autoionization. Write a balanced chemical equation for the autoionization of BrF_3. Can this reaction be described as a Lewis acid–base reaction?

32. The molecules in liquid BrF_3 are linked through "fluorine bonds" that are analogous to the hydrogen bonds in water. Do you expect the entropy of vaporization of $BrF_3(\ell)$ to be larger or smaller than the Trouton's rule value of $88\ J\ K^{-1}\ mol^{-1}$ (see Section 11–5)?

33. Rank the following compounds in order of normal boiling point, from lowest to highest: CaF_2, PtF_6, and PtF_4.

34. For each of the following pairs of fluorine compounds, predict which has the higher melting point:

(a) NaF or PF_3
(b) AsF_3 or AsF_5
(c) CF_4 or C_5F_{12}

35. One of the hazards in making Teflon is that the starting material, tetrafluoroethylene ($C_2F_4(g)$), can explode, giving C(*graphite*) and $CF_4(g)$. The standard enthalpy of formation of $C_2F_4(g)$ is $-651\ kJ\ mol^{-1}$. Use this fact, together with data from Appendix D, to estimate ΔH if a tank containing 1.00 kg of C_2F_4 were to explode.

36. Compute ΔG_r° at 25°C for

$$CF_4(g) + 2\ H_2O(\ell) \longrightarrow CO_2(aq) + 4\ HF(aq)$$

using data from Appendix D. Explain how it is that CF_4 spontaneously decomposes according to the above equation, yet in practice is stable up to 500°C.

Compounds of Fluorine and the Noble Gases

37. At 25°C, the reaction

$$XeF_6(g) + 3\ H_2(g) \longrightarrow Xe(g) + 6\ HF(g)$$

has $\Delta H_r^\circ = -1282\ kJ\ mol^{-1}$. Using this fact and information from Appendix D, calculate:

(a) the standard molar enthalpy of formation ΔH_f° of $XeF_6(g)$.
(b) the average enthalpy of an Xe—F bond in $XeF_6(g)$.

38. At 25°C, the reaction

$$XeF_4(g) + 2\ H_2(g) \longrightarrow Xe(g) + 4\ HF(g)$$

has $\Delta H_r^\circ = -887\ kJ\ mol^{-1}$. Using this fact and information from Appendix D, calculate:

(a) the ΔH_f° of $XeF_4(g)$.
(b) the average enthalpy of an Xe—F bond in $XeF_4(g)$.

39. Xenon oxotetrafluoride ($XeOF_4$) is a liquid at room temperature. It can be prepared by the controlled hydrolysis of xenon hexafluoride:

$$XeF_6(g) + H_2O(\ell) \longrightarrow XeOF_4(\ell) + 2\ HF(g)$$

At 25°C, the standard enthalpies of formation of the noble-gas compounds in this reaction are

$$\Delta H_f^\circ (XeF_6(g)) = -298\ kJ\ mol^{-1}$$
$$\Delta H_f^\circ (XeOF_4(\ell)) = +148\ kJ\ mol^{-1}$$

Calculate ΔH_r° for the hydrolysis. Consult Appendix D for additional data.

40. The energy of explosion of $XeO_3(s)$ into its gaseous elements was measured in a constant-volume calorimeter:

$$XeO_3(s) \longrightarrow Xe(g) + \tfrac{3}{2} O_2(g)$$

(a) A 2.763×10^{-4} mol sample released 112 J of heat when it was exploded. Calculate the standard *energy* of formation (ΔE_f°) of $XeO_3(s)$.
(b) Is the standard *enthalpy* of formation of $XeO_3(s)$ larger or smaller than your answer to part (a)?

41. The reduction of perxenate ion (XeO_6^{4-}) to xenon is so favored thermodynamically in acidic solution that it even oxidizes Mn^{2+} to MnO_4^-, one of the strongest oxidizing agents ordinarily encountered, with evolution of oxygen. Write balanced half-equations and an equation for the overall reaction in this process.

42. The following standard reduction potentials have been measured for the oxides of xenon in acidic aqueous solution:

$$H_4XeO_6(aq) + 2\ H^+(aq) + 2\ e^- \longrightarrow XeO_3(aq) + 3\ H_2O(\ell)$$
$$\mathcal{E}^\circ = +2.36\ V$$

$$XeO_3(aq) + 6\ H^+(aq) + 6\ e^- \longrightarrow Xe(g) + 3\ H_2O(\ell)$$
$$\mathcal{E}^\circ = +2.12\ V$$

(a) Would you classify perxenic acid (H_4XeO_6) as an oxidizing agent or as a reducing agent?
(b) In the preceding, XeO_3 acts as an oxidizing agent in one half-reaction and as a reducing agent in the other. Is XeO_3

stronger as an oxidizing agent or as a reducing agent at pH 0?

(c) Is XeO_3 stable with respect to disproportionation to $Xe(g)$ and $H_4XeO_6(aq)$ in acidic aqueous solution?

ADDITIONAL PROBLEMS

43. Chlorine-containing bleaches act by oxidizing, but other bleaches, like SO_2, are reducing bleaches. By referring to Appendix E, predict whether bromine is a stronger or a weaker bleach than chlorine.

44. You are stranded in a desert. You have a supply of water, but it is crawling with bacteria that will infect you fatally if you drink it as it is. The standard means of sterilizing water is to add $Cl_2(g)$. You have no chlorine. You do, however, have some $KMnO_4(s)$, $H_2S(aq)$, $NaCl(s)$, $SO_2(g)$, and $Na_2S_2O_3(aq)$. Which chemical (or combination of chemicals) should you test first as a substitute for Cl_2 in sterilizing the water?

45. In the Deacon process, an equilibrium is reached between chlorine, steam, hydrogen chloride, and oxygen at 450°C.

(a) Use the ΔH_f° and S° values from Appendix D to estimate an equilibrium constant at 450°C for the reaction

$$2\,HCl(g) + \tfrac{1}{2}\,O_2(g) \rightleftharpoons Cl_2(g) + H_2O(g)$$

(b) Suppose that bromine replaces chlorine in the preceding reaction. After referring to the appropriate bond enthalpies in Table 17–1, state whether the equilibrium constant at 450°C is larger or smaller. Explain.

46. It is estimated that a certain formation at Owens Lake, California, holds about 3.5×10^7 metric tons of trona. Estimate how long this formation could supply the United States with Na_2CO_3 at the current rate of consumption (see Table 22–1).

47. Calculate the theoretical minimum external cell voltage to drive a chlor-alkali cell in which all reactants and products are in their standard states at 25°C.

48. Suppose that a chlor-alkali plant has 250 cells, through each of which a current of 100,000 A passes continuously.

(a) Calculate the mass of chlorine that this plant can produce per 24-hour day.

(b) If the cell voltage is 3.5 V (higher than the theoretical minimum voltage from Problem 47), calculate the total electrical energy consumed by the plant in one day, and express it both in joules and in kilowatt-hours.

(c) If electricity costs $0.05 per kilowatt-hour, calculate the cost for the electricity to run the plant for one day.

49. A magazine article about the chlor-alkali process in the United States reports that "[annual] production of chlorine [recently] ... total[ed] about 10.9 million tons. ... Caustic soda production as usual [was] 4 to 5 percent higher than chlorine production." Explain the phrase "as usual" by computing the theoretical ratio of the yields of these two chemicals in the process.

50. Astatine is a little-studied element whose longest lived isotope, ^{210}At, has a half-life of only 8.3 hours.

(a) Predict whether HAt is a stronger or a weaker acid than is HI.

(b) The At^- ion is produced when elemental astatine reacts in aqueous solution with zinc, but no reaction is seen with $1\,M\,Fe^{2+}(aq)$. What range of reduction potentials for conversion of At_2 to At^- is consistent with these observations?

(c) If $Cl_2(g)$ is bubbled through a solution containing $At^-(aq)$, what reaction results?

(d) Write a balanced equation for the reaction that should occur when $At_2(s)$ is dissolved in basic aqueous solution.

51. Decide which halogen was in the water used by the camper who told this story: "Big signs told us to boil the water at the campsite before cooking with it or drinking it. We boiled some water and then used it to cook spaghetti, which tasted fine. When we cleaned up, we used some unboiled water. The dirty pot and our dishes turned blue-black as soon as the water touched them." Explain how boiling removed the halogen.

52. Propose a method to make the compound shown in Figure 21–6 from biphenyl and chlorine. Explain why *mixtures* of PCBs were always used in practical applications.

53. Write a series of reactions that leads to the production of elemental fluorine from calcium fluoride and other readily available chemicals.

54. A commercial fluorine cell generates 3.3 kg of fluorine per hour by electrolysis of $KF \cdot 2HF$. Compute the average current passing through this cell.

55. The reaction

$$K_2[NiF_6](s) + TiF_4(s) \longrightarrow$$
$$K_2[TiF_6](s) + NiF_2(s) + F_2(g)$$

generates elemental fluorine. Does $TiF_4(s)$ play the role of a Lewis acid or a Lewis base in this reaction? Explain.

56. (a) Compute the density of $HF(g)$ at 1.00 atm pressure and its normal boiling point, 19.54°C, assuming ideal gas behavior.

(b) The observed density of $HF(g)$ under the conditions from part (a) is $3.11\ g\,L^{-1}$. Explain the large discrepancy with the calculated result of part (a).

*57. The reaction

$$4\,HF(\ell) + SiO_2(s) \longrightarrow SiF_4(g) + 2\,H_2O(\ell)$$

can be used to release gold that is distributed in certain quartz veins of hydrothermal origin. If the quartz contains $1.0 \times 10^{-3}\%$ gold by mass and the gold has a market value of $350 per troy ounce, would the process be economically feasible if commercial (50% by mass) aqueous hydrogen fluoride (density $1.17\ g\,cm^{-3}$) costs $0.25 per liter? (1 troy ounce = 31.3 g)

58. Ingestion of 5 to 10 g of $NaF(s)$ is lethal to a 70 kg man. Poisoning from smaller doses can be treated with calcium therapy. Write chemical equations for the effect of $NaF(s)$ on the body.

59. Dioxygen difluoride, O_2F_2 (sometimes written "FOOF"), is a particularly potent fluorinating agent. It is made by irradiating a mixture of O_2 and F_2 at the temperature of liquid nitrogen. The O-to-O distance in O_2F_2 is nearly as short as the distance in O_2, and the O-to-F distances are quite long. Draw the Lewis structure for this molecule, determine the bond order of all bonds, and describe its geometry. What is the name of the analogous compound of oxygen and hydrogen?

60. The fluorinating agent dioxygen difluoride (O_2F_2) (see preceding problem) is important because it converts the plutonium in almost any plutonium-containing material to PuF_6 under mild conditions. The volatility of PuF_6 then allows the separation of radioactive Pu from various impurities. Write balanced chemical equations for the reaction of PuO_2 with FOOF and the reaction of Pu with FOOF.

61. When $SF_4(g)$ reacts with $CsF(s)$, the SF_5^- ion is formed. Use the VSEPR theory to predict its geometry.

62. The compound S_2F_{10} was not mentioned in the chapter but does exist.
   (a) Taking the known bond-forming tendencies of fluorine into account, suggest a probable Lewis structure for this compound.
   (b) Use the VSEPR model to predict the molecular geometry of S_2F_{10}.

63. The result of the accidental first preparation of Teflon immediately was subjected to elemental analysis for both chlorine and fluorine. The report on the new mysterious white powder was that it contained no chlorine and 48.4% fluorine by mass. Compare these first values to the actual percentages of fluorine and chlorine in Teflon.

64. The ionization energy of O_2 is 1180 kJ mol^{-1}. In 1962, Neil Bartlett reported that $O_2(g)$ reacts with $PtF_6(g)$ to form the solid ionic compound $O_2^+PtF_6^-$. The ionization energy of Xe is 1170.6 kJ mol^{-1}. Explain why it occurred to Bartlett that xenon also might form a compound with PtF_6. Give the formula of the compound.

65. Calculate the equilibrium constant for the reaction

$$Xe(g) + 2F_2(g) \rightleftharpoons XeF_4(s)$$

at 25°C if $\Delta G_f^\circ(XeF_4(s))$ is -134 kJ mol^{-1}.

66. Xenon difluoride reacts with arsenic pentafluoride to give an ionic compound:

$$2XeF_2(g) + AsF_5(g) \longrightarrow [Xe_2F_3][AsF_6]^-(s)$$

Identify which species are Lewis acids in this reaction and which are Lewis bases.

24

From Petroleum to Pharmaceuticals

CHAPTER OUTLINE

A single chemical plant can produce more than 1 billion lb of a product per year. This Texas plant produces ethylene (C_2H_4), a starting material for many plastics. *(Oxidental Petroleum; Robin Smith/Tony Stone Images)*

Carbon is unique among the elements in the large number of compounds it forms and in the variety of their structures. In combination with hydrogen, it forms molecules with single, double, and triple bonds; chains; rings; branched chains; branched and interlocked rings; and cages (Fig. 24–1). The thousands of stable hydrocarbons make a sharp contrast to the mere two stable compounds between oxygen and hydrogen (water and hydrogen peroxide). Even the rather versatile elements nitrogen and oxygen form only six nitrogen oxides.

The unique behavior of carbon relates to its position in the periodic table. As a second-period element, carbon has relatively small atoms, which allow it easily to form the double and triple bonds that are rare in the compounds of related elements, such as silicon. As a Group IV element, carbon can form four covalent bonds, more than the other second-period elements. This gives it wide scope for structural elaboration. Finally, as an element of intermediate electronegativity, carbon can form covalent compounds both with more electronegative elements such as oxygen, nitrogen, and the halogens and with more electropositive elements such as hydrogen and the heavy metals mercury and lead.

The study of the compounds of carbon is the discipline traditionally called **organic chemistry,** although the chemistry of carbon is intimately bound up with the chemistry of inorganic materials and with biochemistry as well.

24–1 Petroleum Refining and the Hydrocarbons

When the first oil well was drilled in 1859 near Titusville, Pennsylvania, the effects that the exploitation of petroleum would have on everyday life in the years to come could not have been anticipated. Today the petroleum and petrochemical industries span the world and touch the most minute details of our daily lives. In the early years of the 20th century, the development of the automobile, fueled by low-cost gasoline derived from petroleum, changed many people's lifestyles. The subsequent use of gasoline and fuel oil to power trains and planes, tractors and harvesters, pumps and coolers transformed travel, agriculture, and industry. Finally, the spectacular growth of the petrochemical industry since 1945 has led to the introduction of innumerable new products ranging from pharmaceuticals to plastics and synthetic fibers. Over half of the chemical compounds produced in largest volume stem directly or indirectly from petroleum.

The prospects for the continued enjoyment of inexpensive petroleum and petrochemicals in the 21st century are clouded. Many wells have been drained, and the remaining petroleum is more costly to extract and often of lower quality. Petroleum appears to have originated from the deposition and decay of organic matter (of animal or vegetable origin) in oxygen-poor marine sediments. Subsequently, this matter migrated to the porous sandstone rocks from which it is extracted today. In the past 100 years, humankind has consumed a significant fraction of the petroleum accumulated in the Earth over many millions of years. The imperative for the future is to save the reserves that remain for uses for which few substitutes are available (such as the manufacture of petrochemicals) while finding other sources for heat and energy.

Figure 24–1 • The structure of the hydrocarbon adamantane, $C_{10}H_{16}$. This wire-frame structure emphasizes the interlocking six-membered rings (carbon atoms lie at the black intersections; hydrogen atoms at the white ends). Note the similarity to the structure of diamond (Fig. 20–25).

• The sulfur content in petroleum is particularly significant because the burning of high-sulfur petroleum products releases quantities of sulfur oxides that pollute the air and cause acid rain.

Petroleum Distillation and the Straight-Chain Alkanes

Although crude petroleum oil contains small amounts of oxygen, nitrogen, and sulfur, its major constituents are **hydrocarbons,** compounds of carbon and hydrogen. The most prevalent hydrocarbons in petroleum are the **straight-chain alkanes** (also

Much of the oil pumped from wells today comes from beneath the ocean floor. It is extracted using offshore oil rigs, such as this one in the Gulf of Mexico. *(Tom Tracy/Black Star)*

called normal alkanes, or *n*-alkanes), which consist of chains of carbon atoms bonded to one another by single bonds, with enough hydrogen atoms on each carbon atom to bring it to the maximum bonding capacity of four. The simplest alkanes are methane (CH_4), with just one carbon atom, and ethane (C_2H_6). These alkanes have the generic formula C_nH_{2n+2}. Table 24-1 lists the names and formulas of the first few. The ends of the molecules are methyl (CH_3) groups, with methylene (CH_2) groups in between. We could write pentane (C_5H_{12}) as $CH_3CH_2CH_2CH_2CH_3$ to indicate the structure more explicitly, or in abbreviated fashion as $CH_3(CH_2)_3CH_3$ (Fig. 24-2).

Each carbon atom in a straight-chain alkane forms four single covalent bonds that point to the corners of a tetrahedron (exhibiting sp^3 hybridization). Rotation around these C—C single bonds occurs quite easily (Fig. 24-3), as explained in Section 18-2. Accordingly, a given hydrocarbon molecule in a gas or liquid constantly changes its conformation as the chain flexes and writhes. The term "straight

TABLE 24-1	
Straight-Chain Alkanes	
Methane	CH_4
Ethane	C_2H_6
Propane	C_3H_8
Butane	C_4H_{10}
Pentane	C_5H_{12}
Hexane	C_6H_{14}
Heptane	C_7H_{16}
Octane	C_8H_{18}
Nonane	C_9H_{20}
Decane	$C_{10}H_{22}$
Undecane	$C_{11}H_{24}$
Dodecane	$C_{12}H_{26}$
Tridecane	$C_{13}H_{28}$
Tetradecane	$C_{14}H_{30}$
Pentadecane	$C_{15}H_{32}$
Hexadecane	$C_{16}H_{34}$
Triacontane	$C_{30}H_{62}$

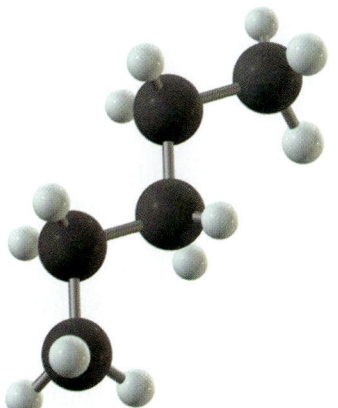

Figure 24-2 • The structure of the alkane pentane (C_5H_{12}). It consists of three methylene (—CH_2—) groups in a line capped at each end by methyl (—CH_3) groups.

Figure 24-3 • The two —CH_3 groups in ethane rotate easily about the bond that joins them.

Figure 24–4 • Two of the many possible conformations of the straight-chain alkane hexadecane, $C_{16}H_{34}$. Carbon atoms are at the black intersections; hydrogen atoms are at the white ends.

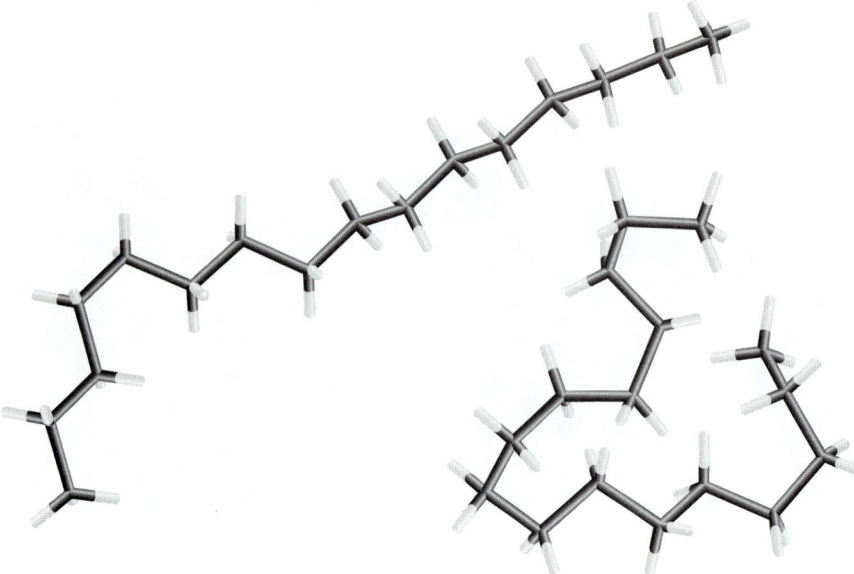

chain" refers only to the bonding pattern in which each carbon atom is bonded to the next one in a sequence; the carbon atoms are not located along a straight line. An alkane molecule with 15 or 16 carbon atoms looks quite different when it is extended to give a "stretched" molecule and when it turns back on itself (Fig. 24–4). These two conformations (and the many others) interconvert rapidly at room temperature.

Figure 24–5 shows the melting and boiling points of the straight-chain alkanes; both increase with the number of carbon atoms and thus with molecular mass. This is a consequence of the increasing strength of dispersion forces between heavier molecules, as discussed in Section 6–1. Methane, ethane, propane, and butane are all gases at room temperature, but the first several hydrocarbons that follow them in the alkane series are liquids. Mineral oil is a mixture of hydrocarbons that is liquid at room temperature. Alkanes beyond about $C_{17}H_{36}$ are waxy solids at room temperature, with melting points that increase with the number of carbon atoms present. Paraffin wax, a low-melting solid, is a mixture of alkanes that have 20 to 30 carbon atoms per molecule. Petrolatum (petroleum jelly, or Vaseline) is a different mixture that is semisolid at room temperature.

A mixture such as petroleum does not boil at a single, sharply defined temperature. Instead, as it is heated, the compounds of lower boiling point (the most volatile) boil off first, and as the temperature is raised, more and more of the material vaporizes. The existence of a boiling-point range permits components of a mixture to be separated by distillation, as Section 6–7 discusses. The earliest petroleum distillation was a simple batch process: The crude oil was heated in a still, the volatile fractions were removed at the top and condensed to gasoline, and the still was cleaned for another batch. Modern petroleum refineries use much more sophisticated and efficient distillation methods, in which crude oil is added continuously and fractions of different volatility are tapped off at various points up and down the distillation column (Figs. 24–6 and 24–7). To save on energy costs, heat exchangers capture and recycle the heat released as separated vapors condense to liquids.

Paraffin wax and mineral oil (also called mineral spirits, liquid paraffin, and paraffin oil) are rather unreactive mixtures of hydrocarbons. Their commercial uses are numerous: in polishes, packaging, cosmetics, candles, and laxative preparations. *(Charles D. Winters)*

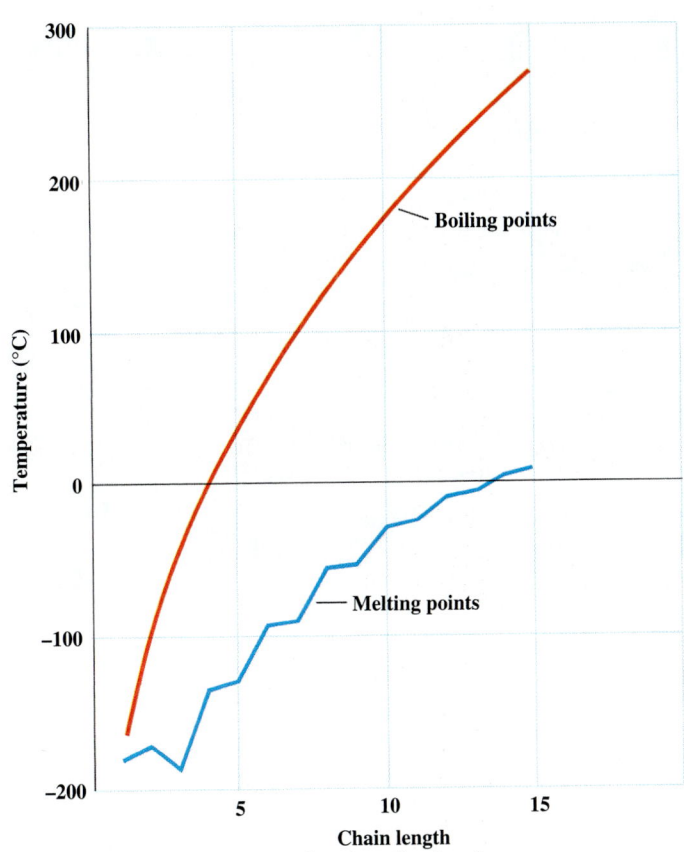

Figure 24-6 • In the distillation of petroleum, the lighter, more volatile hydrocarbon fractions are removed from higher up the column, the heavier fractions from lower down.

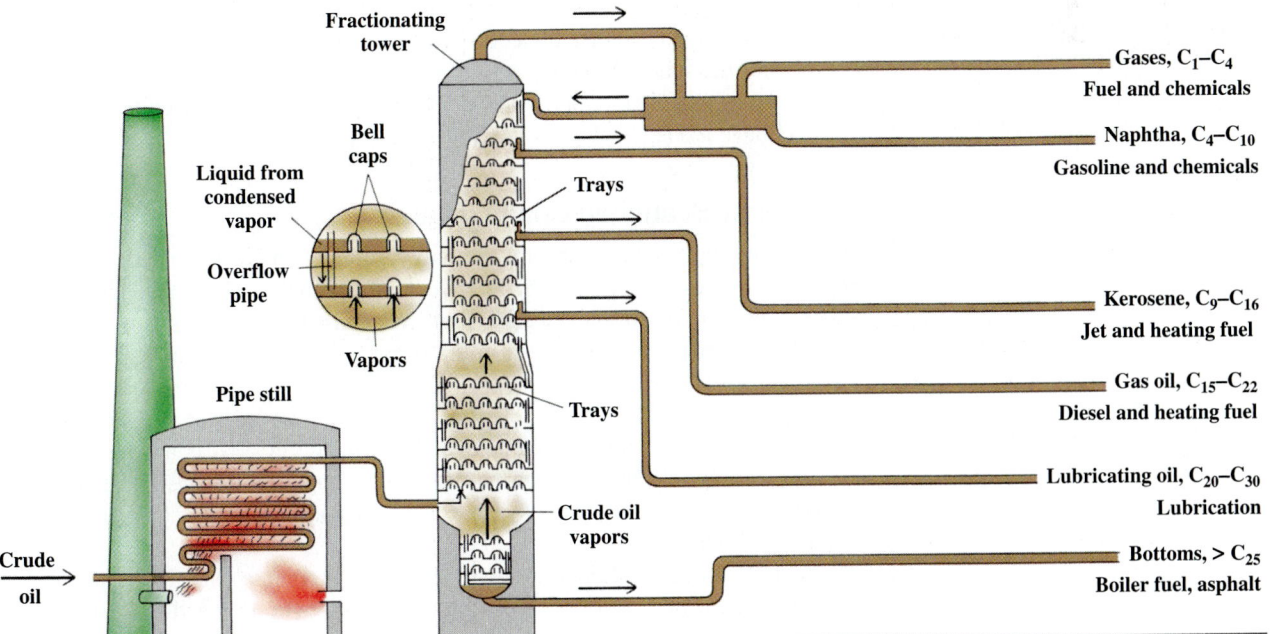

Figure 24–7 • Distillation towers in a modern oil refinery.

Figure 24–7 • Distillation towers in a modern oil refinery.

• The boiling range of naphtha is from about 40°C to 180°C. It is a mixture of dozens of hydrocarbons.

Distillation allows hydrocarbons to be separated by boiling point and thus by molecular mass. The gases that emerge from the top of the distillation column resemble the natural gas that collects in rock cavities above petroleum deposits. These gas mixtures can be separated further by redissolving the ethane, propane, and butane in a liquid solvent such as hexane. The methane-rich mixture of gases that remains is used for chemical synthesis or is shipped by pipeline to heat homes. Redistillation of the hexane and its dissolved gases allows for their separation and use as chemical starting materials. Propane and butane also are bottled under pressure as liquefied petroleum gas (LPG), which is used for fuel in rural areas. After the gases, the next fraction to emerge from the petroleum distillation column is **naphtha,** which is used primarily in the manufacture of gasoline. Subsequent fractions of successively higher molecular mass are used for jet and diesel fuel, heating oil, and machine lubricating oil. The thick, non-volatile sludge that remains at the bottom of the distillation unit is pitch or asphalt, which is used to roof buildings and pave roads.

Cyclic and Branched-Chain Alkanes

The straight-chain alkanes are not the only hydrocarbons in petroleum. Two other important classes of compounds, the cyclic and branched-chain alkanes, also are represented. A **cycloalkane** consists of at least one chain of carbon atoms attached at the ends to form a closed loop. In the formation of this additional C—C bond, two hydrogen atoms are eliminated, so the general formula for cycloalkanes with one ring is C_nH_{2n} (Fig. 24–8). The smallest cycloalkanes, cyclopropane and cyclobutane, are very strained compounds because the C—C—C bond angles in both are substantially less than the normal tetrahedral angle of 109.5°. Therefore, they are more reactive than are the heavier cycloalkanes or the straight-chain compounds propane and butane.

Branched-chain alkanes are hydrocarbons that contain only C—C and C—H single bonds but in which the carbon atom are no longer arranged in a single continuous chain. At least one carbon atom in each molecule is bonded to three or four

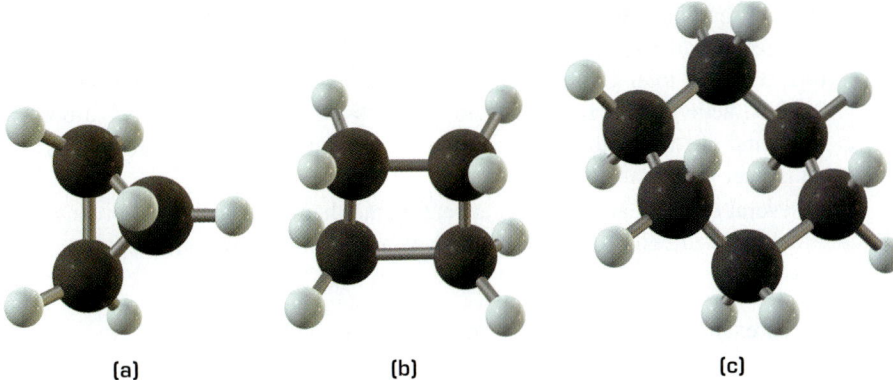

Figure 24-8 • Three cyclic hydrocarbons. (a) Cyclopropane, C_3H_6. (b) Cyclobutane, C_4H_8. (c) Cyclohexane, C_6H_{12}.

other carbon atoms, rather than to only one or two as in the straight-chain alkanes or the cycloalkanes. The simplest branched-chain molecule (Fig. 24–9b) is that of 2-methylpropane, sometimes referred to as isobutane. This substance has the same molecular formula as butane (C_4H_{10}), but its molecule has a different structure, in which a central carbon atom is bonded to three CH_3 groups and only one hydrogen atom. The two compounds butane and 2-methylpropane are isomers. The molecules can be interconverted only by breaking and reforming chemical bonds.

The number of possible isomers increases rapidly with increasing numbers of carbon atoms in the hydrocarbon molecule. Thus, butane and 2-methylpropane are the only two isomers of molecular formula C_4H_{10}, but C_5H_{12} has three isomers, C_6H_{14} has five, C_7H_{16} has nine, and $C_{30}H_{62}$ has millions of isomers. A systematic procedure has been developed to name these isomers and has been codified by the International Union of Pure and Applied Chemistry (IUPAC). The following set of rules is a part of that procedure:

1. Find the longest continuous chain of carbon atoms in the molecule. The molecule is named as a derivative of this alkane. Thus, in Figure 24–9b, the longest chain has three carbon atoms, so the molecule is named as a derivative of propane.

2. Determine the names of the hydrocarbon groups attached to the chain selected. These side groups are called **alkyl groups.** Their names are obtained by dropping the ending -*ane* from the corresponding alkane and replacing it with -*yl*. The methyl group (CH_3) is derived from methane (CH_4); the ethyl group (C_2H_5), from ethane (C_2H_6); the propyl group ($CH_2CH_2CH_3$), from propane (C_3H_8), and so on. Note also the *isopropyl* group ($CH(CH_3)_2$), which differs from the propyl group in that it attaches by its middle carbon atom rather than by one at an end.

3. Number the carbon atoms along the chain identified in Rule 1. Identify the alkyl groups by the number of the carbon atom at which they are attached to the chain. The carbon chain is numbered from the end that gives the lowest number for the position of the first attached group. The methyl group in the molecule in Figure 24–9b is attached to the second carbon atom of the three in the propane chain, so the molecule is called 2-methylpropane.

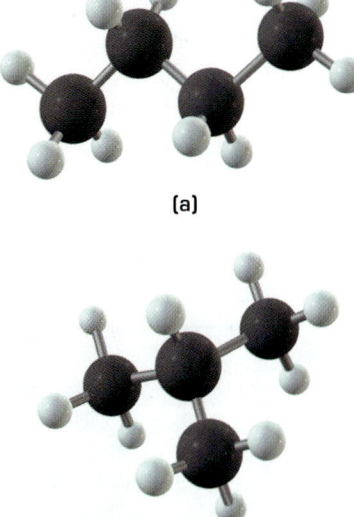

Figure 24-9 • Two isomeric hydrocarbons with the molecular formula C_4H_{10}. (a) Butane. (b) 2-Methyl propane.

4. If more than one alkyl group of the same type is attached to the chain, use the prefixes *di-* (two), *tri-* (three), *tetra-* (four), *penta-* (five), and so forth to specify the total number of such attached groups in the molecule. Thus, 2,2,3-trimethylbutane has two methyl groups attached to the second carbon atom and one to the third carbon atom of the four-atom butane chain. It is an isomer of heptane (C_7H_{16}).

5. If several types of alkyl groups appear, name them in alphabetical order. Ethyl is listed before methyl, which appears before propyl, and so forth.

As an example, let us name the following branched-chain alkane:

$$CH_3 \quad CH_2 \quad CH_2{-}CH_3$$
$$C \qquad C$$
$$CH_3 \quad H \quad CH_3 \quad CH_2{-}CH_3$$

The longest continuous chain links six carbon atoms, so this is a derivative of hexane. We number the carbon atoms starting from the left:

$$CH_3 \quad \overset{3}{CH_2} \quad \overset{5}{CH_2}{-}\overset{6}{CH_3}$$
$$\overset{2}{C} \qquad \overset{4}{C}$$
$$\overset{1}{CH_3} \quad H \quad CH_3 \quad CH_2{-}CH_3$$

Methyl groups are attached to carbon atoms 2 and 4, and an ethyl group is attached to atom 4. The name is thus 4-ethyl-2,4-dimethylhexane. Note that if we had started numbering from the right, the higher number 3 would appear for the position of the first methyl group; thus the numbering from the left is correct.

The fraction of branched-chain alkanes in a gasoline affects how it burns in an engine. Gasoline composed entirely of straight-chain alkanes burns very unevenly, causing a "knocking" that can damage the engine. Blends that are richer in branched-chain and cyclic alkanes burn with much less knocking. Smoothness of combustion is measured quantitatively through the **octane number** of the gasoline, which was defined in 1927 by selecting two compounds present in ordinary gasoline that lie at extremes in the knocking they caused. Pure 2,2,4-trimethylpentane (commonly known as "isooctane") burns very smoothly and was assigned an octane number of 100. Pure heptane gave the most knocking of the compounds examined at the time and was assigned octane number 0. Mixtures of heptane and isooctane cause intermediate amounts of knocking. Standard mixtures of these two compounds define a scale to evaluate the knocking caused by real gasolines, which are complex mixtures of branched-chain and straight-chain hydrocarbons. If a gasoline sample gives the same amount of knocking in a test engine as a mixture of 90% (by volume) 2,2,4-trimethylpentane and 10% heptane, it is assigned the octane number 90. Certain additives increase the octane rating of gasoline. The least expensive is tetraethyllead ($Pb(C_2H_5)_4$), a compound that has very weak bonds between the central lead atom and the ethyl carbon atoms. It readily releases ethyl radicals, $\cdot C_2H_5$, into the gasoline during combustion. These reactive species speed and smooth combustion, reduce knocking, and give better fuel performance. Unfortunately, the lead is released into the atmosphere, causing long-term health hazards. The use of lead compounds in gasoline has been phased out almost completely in the United States. The substitute additive "MTBE" (methyl *t*-butyl ether, see Section 24–2) has been used in some U.S.

Typical octane ratings of gasoline. Higher octane fuel costs more. *(Charles D. Winters)*

gasolines at low concentration since 1979 to increase octane ratings. The chemical processing of fuel to convert straight-chain to branched-chain compounds is another way to raise octane ratings.

Alkenes and Alkynes

The hydrocarbons discussed so far are referred to as **saturated** because all the carbon–carbon bonds are single bonds. Hydrocarbons that have carbon–carbon double and triple bonds are **unsaturated** (Fig. 24–10). Ethylene (C_2H_4) has a double bond between its carbon atoms and is the simplest **alkene**. The simplest **alkyne** is acetylene (C_2H_2), which has a triple bond between its carbon atoms. In naming these compounds, one replaces the *-ane* ending of the corresponding alkane with *-ene* when a double bond is present, and with *-yne* when a triple bond is present. Ethene is thus the systematic name for ethylene, and ethyne for acetylene. The nonsystematic names of these two compounds are so well established, however, that we shall use them. For compounds with a carbon backbone of four or more carbon atoms, it is necessary to specify the location of the double or triple bond. This is done by numbering the carbon–carbon bonds and putting the number of the lower-numbered carbon involved in the multiple bond before the name of the alkene or alkyne. Thus, two different, isomeric alkynes have the formula C_4H_6:

$$HC\equiv CCH_2CH_3 \qquad \text{(1-butyne)} \qquad CH_3C\equiv CCH_3 \qquad \text{(2-butyne)}$$

There is yet another subtlety for alkenes. Recall from Section 18–2 that rotation does not take place readily about a carbon–carbon double bond, as it does about a single bond. Many alkenes therefore exist in isomeric forms, depending on whether bonded side-groups are located on the same side or on opposite sides of the double bond. Only a single isomer of 1-butene exists, but two 2-butenes exist:

$$cis \qquad \underset{CH_3}{\overset{H}{}}C=C\underset{CH_3}{\overset{H}{}} \qquad trans \qquad \underset{CH_3}{\overset{H}{}}C=C\underset{CH_3}{\overset{CH_3}{}}$$

These compounds differ in melting point, boiling point, density, and other physical and chemical properties.

Compounds with two double bonds are called dienes, those with three are trienes, and so forth. The compound 1,3-pentadiene, for example, is a derivative of pentane with two double bonds:

$$CH_2=CHCH=CHCH_3$$

In such **polyenes**, each double bond may lead to *cis* and *trans* conformations, depending on its neighboring groups; thus several isomers with the same bonding patterns but different molecular geometries and physical properties may exist.

Alkenes are key compounds for the synthesis of valuable organic chemicals and polymers, but they are not present to a significant extent in crude petroleum. One way to obtain alkenes is through **cracking** petroleum by heat or with catalysts. In **catalytic cracking,** the heavier fractions from the distillation column (compounds of C_{12} or higher) are passed over a silica–alumina catalyst at temperatures of 450°C to 550°C. Reactions such as

$$CH_3(CH_2)_{12}CH_3 \longrightarrow CH_3(CH_2)_4CH=CH_2 + C_7H_{16}$$

occur to break the long chain into fragments. The shorter-chain hydrocarbons that result have lower boiling points and can be added to gasoline. In fact, the alkenes

- The generic formula for alkenes having one double bond is C_nH_{2n}, and the generic formula for alkynes having one triple bond is C_nH_{2n-2}.

- Alkenes are also referred to as "olefins."

- "Enynes," which contain both double and triple bonds within the same molecule, also exist.

Figure 24–10 • One way to distinguish alkanes from alkenes is by their differing behaviors with aqueous $KMnO_4$. This strong oxidizing agent undergoes no reaction with hexane, retaining its purple color (*left*). When $KMnO_4$ is placed in contact with 1-hexene, however, a redox reaction takes place in which the brown solid MnO_2 is formed (*right*) and —OH groups are added to either side of the double bond in the 1-hexene, giving a compound with the formula $CH_3(CH_2)_3CH(OH)CH_2OH$. (*Charles Steele*)

have higher octane numbers than do the corresponding alkanes and perform better in the engine. **Thermal cracking** uses higher temperatures of 850°C to 900°C and no catalyst. It produces shorter-chain alkenes such as ethylene and propylene through reactions such as

$$CH_3(CH_2)_{10}CH_3 \longrightarrow CH_3(CH_2)_8CH_3 + CH_2{=}CH_2$$

This is not so useful for gasoline production because the short-chain alkenes are too volatile. It is of great importance for the chemical industry, however, because ethylene and propylene are among the most important starting materials for synthesis.

Aromatic Hydrocarbons

One last group of hydrocarbons present in crude petroleum is the **aromatic hydrocarbons,** of which benzene is the simplest example. The benzene molecule consists of a six-carbon ring in which the delocalization of π electrons significantly increases the molecular stability (see Section 18–3). Benzene sometimes is represented simply by its chemical formula C_6H_6 and sometimes (to show structure) by a hexagon with a circle inside of it:

The six points of the hexagon represent the six carbon atoms, with the hydrogen atoms omitted for simplicity. The circle represents the delocalized π electrons, which are spread out evenly over the ring. The molecules of other aromatic compounds contain benzene rings with various side-groups, or two (or more) benzene rings linked by alkyl chains or fused side by side, as in naphthalene ($C_{10}H_8$):

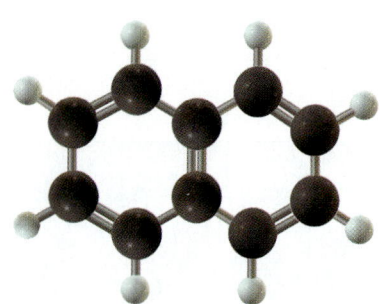

A less abstract representation of the structure and bonding in naphthalene.

Besides benzene, the most prevalent aromatic compounds in petroleum are toluene, in which a methyl group replaces one hydrogen atom on the benzene ring, and the xylenes, in which two such replacements are made:

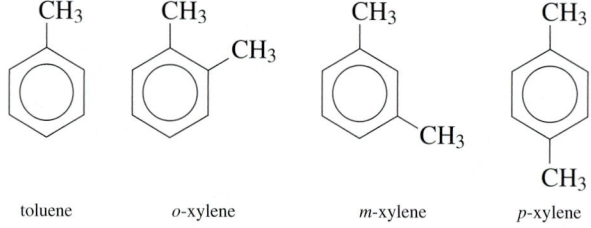

toluene   o-xylene   m-xylene   p-xylene

• The prefixes *o-*, *m-*, and *p-* stand for *ortho-*, *meta-*, and *para-* and are used to distinguish isomeric molecules with doubly substituted benzene rings.

This set of compounds is referred to as BTX (standing for *benzene-toluene-xylene*). The BTX in petroleum is of great importance to polymer synthesis (Chapter 25). These components also are very important in gasoline formulations because they significantly increase octane number. In fact, they can be used to make high-performance fuels with octane numbers above 100, as are required in modern aviation.

A major advance in petroleum refining has been the development of **reforming** reactions, which allow for the production of BTX aromatics from straight-chain alkanes that contain the same number of carbon atoms. A fairly narrow distillation fraction containing only C_6 to C_8 alkanes is taken as the starting material. The reforming reactions use high temperatures and transition-metal catalysts, such as platinum or rhenium, on alumina supports, and their detailed mechanisms are not fully understood. Apparently, a straight-chain alkane such as hexane is cyclized and dehydrogenated to give benzene as the primary product. Heptane yields mostly toluene, and octane yields a mixture of xylenes. Toluene is of increasing importance as a solvent because tests show it to be far less carcinogenic than benzene. Benzene is more important for chemical synthesis, however, so a large fraction of the toluene produced is converted to benzene by **hydrodealkylation:**

toluene   benzene

> • We can write the hydrodealkylation reaction from an interpretation of the name: "hydro" means the addition of hydrogen, and "de-alkyl" means the removal of the alkyl group.

This reaction is carried out at high temperatures (550°C to 650°C) and pressures of 40 to 80 atm.

24-2 Functional Groups and Organic Synthesis

The first section of this chapter described the diverse compounds formed from the two elements carbon and hydrogen and the extraction of these compounds from petroleum. The hydrocarbons are just the start of organic chemistry, however. Hydrocarbon chains can be modified extensively by attachment or insertion of non-carbon elements, such as oxygen, nitrogen, or a halogen, or of combinations of such elements. In considering the effects of these changes, attention is focused away from the structures of entire molecules to the structures of **functional groups,** which consist of the non-carbon atoms plus the portions of the molecule that adjoin them. Functional groups tend to be the reactive sites in organic molecules, and their chemical properties depend only rather weakly on the nature of the hydrocarbon chain to which they are linked. This fact permits us to regard an organic molecule as a hydrocarbon frame, which mainly governs size and shape, to which are attached functional groups that mainly determine the chemistry of the molecule. Table 24–2 shows some of the most important functional groups.

A small number of hydrocarbon building blocks from petroleum and natural gas (methane, ethylene, propylene, benzene, and xylene) are the starting points for the synthesis of most of the high-volume organic chemicals produced today. We discuss the ways in which the largest-volume organic chemicals are synthesized.

Halides

One of the simplest functional groups consists of a single halogen atom. **Alkyl halides** form when mixtures of alkanes and halogens (except iodine) are heated or exposed to light:

$$CH_4 + Cl_2 \xrightarrow{250°-400°C \text{ or light}} CH_3Cl \quad + \quad HCl$$

methane   chlorine   chloromethane   hydrogen chloride

> • For illustrative purposes, we take the halogen to be chlorine. Similar compounds of the other halogens exist as well.

TABLE 24–2	Some Important Functional Groups in Organic Compounds	
Functional Group	**Type of Compound**	**Examples**
—F, —Cl, —Br, —I	Alkyl or aryl halide	CH_3CH_2Br (bromoethane)
—OH	Alcohol	CH_3CH_2OH (ethanol)
	Phenol	(benzene ring)—OH HO (1,3,-dihydroxybenzene, or resorcinol)
—O—	Ether	CH_3—O—CH_3 (dimethyl ether)
(aldehyde group)	Aldehyde	$CH_3CH_2CH_2$—C(=O)—H (butyraldehyde, or butanal)
C=O	Ketone	CH_3—C(=O)—CH_3 (propanone, or acetone)
(carboxylic acid group)	Carboxylic acid	CH_3COOH (acetic acid, or ethanoic acid)
(ester group)	Ester	CH_3—C(=O)—O—CH_3 (methyl acetate or methyl ethanoate)
—N (amine group)	Amine	CH_3NH_2 (methylamine)
(amide group)	Amide	CH_3—C(=O)—NH_2 (acetamide)

This **substitution reaction** proceeds by a chain mechanism, as described in Section 14–5. Ultraviolet light initiates the reaction by dissociating a small number of chlorine molecules into highly reactive atoms. These take part in linked reactions of the form

$$Cl\cdot + CH_4 \longrightarrow HCl + \cdot CH_3$$
$$\cdot CH_3 + Cl_2 \longrightarrow CH_3Cl + Cl\cdot$$

(propagation)

The product of this reaction, chloromethane (also called methyl chloride) is used in synthesis to add methyl groups to organic molecules. The chlorination of methane does not produce CH_3Cl cleanly. Depending on conditions, variable quantities of more highly chlorinated compounds (and compounds that contain more than one carbon atom per molecule) appear. Chlorination with an excess of chlorine is the basis for the industrial synthesis of dichloromethane (CH_2Cl_2, also called "methylene chloride"), trichloromethane ($CHCl_3$, chloroform), and tetrachloromethane (CCl_4, carbon tetrachloride). All three are used as solvents and have vapors with an anesthetic or narcotic effect. Both chloroform and carbon tetrachloride are now known to be carcinogenic, which has eliminated the use of the latter as a household solvent for cleaning soiled clothing and removing grease.

A more important industrial route to alkyl chlorides than the free-radical reactions just described is the addition of chlorine to $C\!\!=\!\!C$ double bonds. In these **addition reactions,** the π bond between the C atoms is broken while the σ bond remains intact. Each C forms a σ bond to a new atom in the process. Thus, chlorine adds across the double bond in ethylene to produce 1,2-dichloroethane. Billions of kilograms of this product (commonly called "ethylene dichloride") are manufactured each year by mixing chlorine and ethylene over an iron(III) oxide catalyst at moderate temperatures (40°C to 50°C) either in the vapor phase or in a solution of 1,2-dibromoethane:

$$CH_2\!\!=\!\!CH_2 + Cl_2 \longrightarrow CH_2ClCH_2Cl$$

Almost all the 1,2-dichloroethane produced is used to make chloroethylene (vinyl chloride, $CH_2\!\!=\!\!CHCl$). This is accomplished by heating the 1,2-dichloroethane to 500°C over a charcoal catalyst to abstract HCl:

$$CH_2ClCH_2Cl \longrightarrow CH_2\!\!=\!\!CHCl + HCl$$

The HCl can be recovered and converted to Cl_2 for further production of 1,2-dichloroethane from ethylene. Vinyl chloride has a much lower boiling point than 1,2-dichloroethane (-13°C compared to 84°C), so the two are easily separated by fractional distillation. The only use of vinyl chloride, but a very important one, is in the production of polyvinyl chloride plastic, as discussed in Section 25–3.

Alcohols and Phenols

Alcohols have the $-OH$ functional group attached to an alkyl group. The simplest alcohol is methanol (CH_3OH), which is also known as wood alcohol because, until 1926, it was made entirely from the distillation of wood. Now it is made exclusively from synthesis gas, as described in Section 7–5. The next higher alcohol, ethanol (CH_3CH_2OH), can be produced from the fermentation of sugars. Although this is the major source of ethanol for alcoholic beverages and for "gasohol" (automobile fuel made up of 90% gasoline and 10% ethanol), it is not of significance for industrial production, which uses the direct hydration of ethylene. In this reaction water adds across the $C\!\!=\!\!C$ double bond:

$$CH_2\!\!=\!\!CH_2 + H_2O \longrightarrow CH_3CH_2OH$$

Temperatures of 300°C to 400°C and pressures of 60 to 70 atm are used, with phosphoric acid as a catalyst. Both methanol and ethanol are used widely as solvents and as intermediates for further chemical synthesis.

Two three-carbon alcohols exist, depending on whether the $-OH$ group is attached to a terminal carbon atom (1-propanol) or the central carbon atom (2-propanol):

$$CH_3CH_2CH_2OH \qquad CH_3\underset{\underset{\displaystyle OH}{|}}{C}HCH_3$$

　　　　　　1-propanol　　　　　　2-propanol

The two frequently are referred to as propyl alcohol and isopropyl alcohol, respectively. The systematic names of alcohols are obtained by replacing the *-ane* ending of the corresponding alkane by *-anol* and using a numerical prefix where necessary to identify the carbon atom to which the $-OH$ group is attached. Isopropyl alcohol is made from propylene by a hydration reaction that is catalyzed

• Ethylene dichloride is the largest-volume derived organic chemical in the petrochemical industry.

• This conversion of HCl to Cl_2 occurs through the Deacon process (see Section 23–1):
$2\,HCl + \frac{1}{2}O_2 \longrightarrow Cl_2 + H_2O$

• *Hydration* means addition of water.

• Contrast the names of alcohols with the names of alkyl halides. If the $-OH$ group is replaced by a chlorine atom, the names of the resulting compounds are 1-chloropropane and 2-chloropropane.

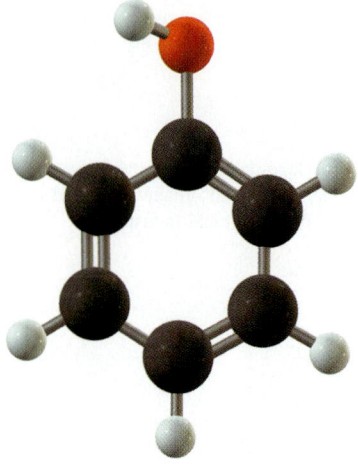

Figure 24–11 • The structure of phenol, C_6H_5OH.

by sulfuric acid. The reaction sequence is an interesting one. The first step is addition of H^+ to the double bond,

$$CH_3CH{=}CH_2 + H^+ \longrightarrow CH_3{-}CH^+{-}CH_3$$

producing a transient charged species in which the positive charge is centered on the central carbon atom. Attack by negative HSO_4^- ions occurs at this positive site and leads to a neutral intermediate that can be isolated.

$$\begin{array}{c} CH_3{-}CH{-}CH_3 \\ | \\ OSO_3H \end{array}$$

Further reaction with water then causes the replacement of the $-OSO_3H$ group by an $-OH$ group and the regeneration of the sulfuric acid:

$$H_2O + \begin{array}{c} CH_3{-}CH{-}CH_3 \\ | \\ OSO_3H \end{array} \longrightarrow \begin{array}{c} CH_3{-}CH{-}CH_3 \\ | \\ OH \end{array} + H_2SO_4$$

This mechanism explains why only 2-propanol, and no 1-propanol, is formed.

The compound 1-propanol is a **primary** alcohol: the carbon atom to which the $-OH$ group is bonded has exactly one other carbon atom attached to it. The isomeric compound 2-propanol is a **secondary** alcohol because the carbon atom to which the $-OH$ group is attached has two carbon atoms (in the two methyl groups) attached to it. The simplest **tertiary** alcohol (in which the carbon atom that is attached to the $-OH$ group is at the same time bonded to three other carbon atoms) is 2-methyl-2-propanol (also called "tertiary butyl alcohol"):

$$\begin{array}{c} OH \\ | \\ CH_3{-}C{-}CH_3 \\ | \\ CH_3 \end{array}$$

This structural classification of alcohols into three groups is made because it effectively divides the alcohols according to their chemical properties as well. For example, treatment of primary alcohols with oxidizing agents such as $KMnO_4$ or $K_2Cr_2O_7$ produces aldehydes and carboxylic acids. Secondary alcohols, treated in similar fashion, produce ketones, whereas tertiary alcohols are not oxidized by these reactants. Aldehydes, ketones, and carboxylic acids are discussed later in this section.

Phenols are organic compounds in which an $-OH$ group is attached directly to an aromatic ring. The simplest example is phenol itself, C_6H_5OH (Fig. 24–11). Phenols differ substantially from alcohols in both physical and chemical properties. One of the most important differences is in their acidity. Phenol (also called "carbolic acid") has an acidity constant of 1×10^{-10}, much larger than that of typical alcohols, which have K_a's ranging from 10^{-16} to 10^{-18}. The reason for this difference is the greater stability of the conjugate base (the phenolate ion, $C_6H_5O^-$) due to the spreading out of the negative charge over the aromatic ring. Phenol, although not a strong acid, does react readily with sodium hydroxide to form the salt sodium phenolate:

$$C_6H_5OH + NaOH \longrightarrow H_2O + C_6H_5O^- Na^+$$

The corresponding reaction between NaOH and alcohols does not occur to a significant extent.

• It might seem that alcohols and phenols would be bases rather than acids because of their $-OH$ groups. Compounds with $-X-O-H$ bonds act as bases however only if X is quite electropositive (see Section 17–6), which C is not.

Phenol synthesis uses reactions that are quite different from those employed to make alcohols. One method, introduced in 1924 and still used to a small extent today, involves the chlorination of the benzene ring followed by reaction with sodium hydroxide:

$$C_6H_6 + Cl_2 \longrightarrow C_6H_5Cl + HCl$$
$$C_6H_5Cl + 2\,NaOH \longrightarrow C_6H_5O^-\,Na^+ + NaCl + H_2O$$

and then the addition of a hydrogen-ion donor. The first step in this sequence illustrates a characteristic difference between the reactions of aromatics and alkenes. If chlorine reacts with an alkene, it *adds* across the double bond (as we have seen in the production of 1,2-dichloroethane). When an aromatic ring is involved, on the other hand, chlorine *substitutes* for hydrogen instead, preserving the aromatic π bonding structure.

Almost all phenol today is prepared by the acid-catalyzed reaction of benzene with propylene to give cumene, or isopropyl benzene:

$$C_6H_6 + \underset{\underset{\displaystyle CH_3\quad H}{}}{\overset{\overset{\displaystyle CH_2}{\|}}{C}} \xrightarrow{H^+} C_6H_5-\underset{\underset{\displaystyle CH_3}{|}}{\overset{\overset{\displaystyle CH_3}{|}}{C}}-H$$

As in the production of 2-propanol, the first step is the addition of H^+ to propylene to give $CH_3-CH^+-CH_3$. This ion then attaches to the benzene ring to give the cumene and regenerate the H^+ ion. Subsequent reaction of cumene with oxygen (Fig. 24–12) gives phenol and acetone, an important compound discussed later in this section. The ultimate products of the manufacture of phenol are mostly polymers and aspirin (see Fig. 22–2).

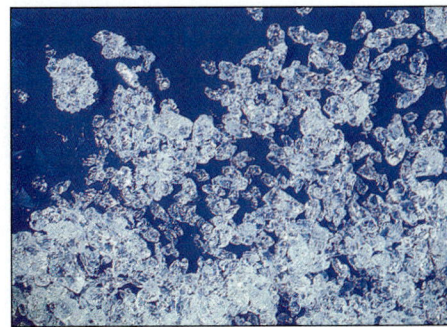

Crystals of phenol, which melts at 42.5°C. Pure phenol is white but takes on a pink or red color when exposed to the light. (*Charles D. Winters*)

Ethers

Ethers are characterized by the —O— functional group, in which an oxygen atom provides a link between two separate alkyl or aromatic groups. One important ether is diethyl ether, often simply called "ether," in which two ethyl groups are linked to the same oxygen atom:

$$C_2H_5-O-C_2H_5$$

Figure 24-12 • The production of phenol and acetone from cumene is a two-step process involving insertion of O_2 to make a peroxide, followed by acid-catalyzed migration of the —OH group to form the products.

This solvent is useful for organic reactions and formerly was used as an anesthetic. It can be produced by a **condensation reaction** (a reaction in which a small molecule such as water is split out) between two molecules of ethanol in the presence of concentrated sulfuric acid, which serves as a dehydrating agent:

$$CH_3CH_2\boxed{OH + H}OCH_2CH_3 \xrightarrow{H_2SO_4} CH_3CH_2-O-CH_2CH_3 + H_2O$$

If methyl bromide is reacted with tert-butyl alcohol, the product is methyl *t*-butyl ether (MTBE)

$$CH_3\boxed{Br + H}O-\underset{\underset{CH_3}{|}}{\overset{\overset{CH_3}{|}}{C}}-CH_3 \longrightarrow CH_3-O-\underset{\underset{CH_3}{|}}{\overset{\overset{CH_3}{|}}{C}}-CH_3 + HBr$$

This is an example of an *unsymmetrical* ether, so called because the two groups bonded to the oxygen atom differ. MTBE also can be made by heating a mixture of methanol and methylpropene (isobutene) in contact with a catalyst:

$$CH_3OH + \underset{\underset{CH_3}{|}}{\overset{\overset{CH_3}{|}}{C}}=CH_2 \xrightarrow{catalyst} CH_3-O-\underset{\underset{CH_3}{|}}{\overset{\overset{CH_3}{|}}{C}}-CH_3$$

The second reaction has been used in recent years to produce large quantities of MTBE as a fuel oxygenate, an ingredient to increase the oxygen content of gasoline and so to reduce the amounts of carbon monoxide and ozone in automobile exhaust. Unfortunately, it is more soluble in water than other components of gasoline, has a smaller molecular size, and is less biodegradable. When it gets into groundwater from leaking underground gasoline tanks or from outboard-motor emissions, it is more mobile than are other gasoline constituents and presents possible hazards to users of the water.

In a cyclic ether, oxygen forms part of a ring with carbon atoms, as in the molecule of tetrahydrofuran (Fig. 24–13), a common solvent. The smallest such ring has two carbon atoms bonded to each other and to the oxygen atom; it occurs in ethylene oxide,

$$\overset{\displaystyle O}{\overset{\diagup\diagdown}{CH_2-CH_2}}$$

which is made by the direct oxidation of ethylene over a silver catalyst:

$$CH_2{=}CH_2 + \tfrac{1}{2}O_2 \xrightarrow{Ag} \overset{\displaystyle O}{\overset{\diagup\diagdown}{CH_2-CH_2}}$$

Ethylene oxide and other cyclic ethers with three-atom rings are called **epoxides.** The major use of ethylene oxide is in making ethylene glycol:

$$\overset{\displaystyle O}{\overset{\diagup\diagdown}{CH_2-CH_2}} + H_2O \longrightarrow HO-CH_2-CH_2-OH$$

This reaction is carried out at 195°C under pressure, or at lower temperatures (50°C to 70°C), with sulfuric acid as a catalyst. Ethylene glycol is a dialcohol, or **diol,** in which two —OH groups are attached to adjacent carbon atoms. It is the main ingredient in automotive cooling system antifreeze mixtures.

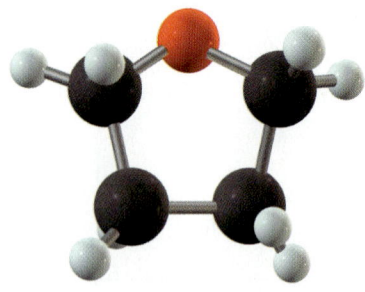

Figure 24–13 • The structure of tetrahydrofuran, C_4H_8O. The five atoms forming the ring are not coplanar.

Aldehydes and Ketones

An **aldehyde** contains the characteristic $-\overset{\displaystyle O}{\underset{\displaystyle H}{\overset{\parallel}{C}}}$ functional group in its molecules.

Aldehydes result from the dehydrogenation of (removal of H_2 from) primary alcohols. Formaldehyde, for example, results from the dehydrogenation of methanol at high temperatures over an iron oxide–molybdenum oxide catalyst:

- The name *aldehyde* is short for *al*cohol *dehyd*rogenated.

$$CH_3OH \longrightarrow \overset{H}{\underset{H}{\diagdown}}C{=}O + H_2$$

Another reaction giving the same product is the direct oxidation of the alcohol:

$$CH_3OH + \tfrac{1}{2}O_2 \longrightarrow \overset{H}{\underset{H}{\diagdown}}C{=}O + H_2O$$

Formaldehyde is readily soluble in water. A 40% aqueous solution, called formalin, is used to preserve biological specimens. Formaldehyde is a component of wood smoke and helps to preserve smoked meat and fish, probably by reacting with nitrogen-containing functional groups in the proteins of the cells of decay-producing bacteria. Its major use is in making polymeric adhesives and insulating foam.

The next aldehyde in the series is acetaldehyde, the dehydrogenation product of ethanol (Fig. 24–14). Acetaldehyde is not actually made industrially from ethanol but rather by the oxidation of ethylene, using a $PdCl_2$ catalyst.

Ketones have the $\diagup\!\!\diagdown C{=}O$ functional group in which a carbon atom forms a double bond to an oxygen atom and single bonds to two separate alkyl or aromatic groups. Such compounds can be described as the products of dehydrogenation or oxidation of secondary alcohols, just as aldehydes come from primary alcohols. The simplest ketone, acetone (Fig. 24–15), is in fact made commercially by the dehydrogenation of 2-propanol over a copper oxide or zinc oxide catalyst at 500°C:

$$CH_3-\underset{\underset{\displaystyle OH}{|}}{CH}-CH_3 \longrightarrow CH_3-\underset{\overset{\displaystyle \|}{O}}{C}-CH_3 + H_2$$

Acetone is also produced (in greater volume) as the co-product with phenol of the oxidation of cumene, as discussed earlier. It is a widely used solvent and the starting material for the synthesis of a number of polymers.

Acetone is an ingredient in many nail polish removers. It also dissolves glue residues left by bumper stickers and price tags, and removes ink, many paints, and temporary tattoos from the skin. (*Charles D. Winters*)

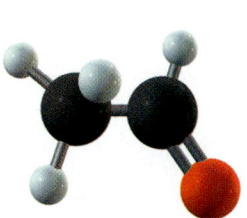

Figure 24–14 • The structure of acetaldehyde, CH_3CHO.

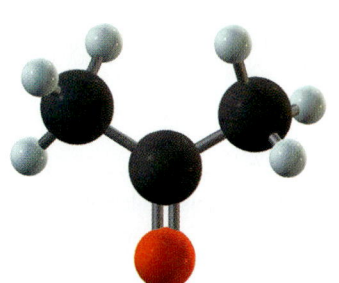

Figure 24–15 • The molecular structure of acetone, CH_3COCH_3.

Carboxylic Acids and Esters

Carboxylic acids contain the $-\overset{\displaystyle O}{\underset{\displaystyle OH}{C}}$ functional group (also written as —COOH).

They are the products of the oxidation of aldehydes, just as aldehydes are the products of the oxidation of primary alcohols. The most familiar carboxylic acid is acetic acid, which is responsible for the tart taste of vinegar. Industrially, acetic acid can be produced by the air oxidation of acetaldehyde over a manganese or cobalt acetate catalyst at 55°C to 80°C:

$$CH_3C\overset{O}{\underset{H}{}} + \tfrac{1}{2}O_2 \xrightarrow{Mn(CH_3COO)_2} CH_3C\overset{O}{\underset{OH}{}}$$

For reasons of economy, the reaction now preferred for acetic acid production is the combination of methanol with carbon monoxide (both derived from natural gas) over a catalyst containing rhodium and iodine:

$$CH_3OH + CO \xrightarrow{Rh, I_2} CH_3COOH$$

This is a **carbonylation,** the insertion of CO into a C—O bond.

Acetic acid is only one of a numerous group of monocarboxylic acids with formulas $H(CH_2)_nCOOH$. The simplest in this series, formic acid (HCOOH), has n equal to 0. This compound was first isolated from extracts of the crushed bodies of ants, and its name stems from the Latin word *formica,* meaning "ant." The longer-chain carboxylic acids are called "fatty acids." Their connections to fats and to soap are shown in Figure 24–16.

Carboxylic acids react with alcohols or phenols to give **esters,** forming water as by-product. An example is the reaction of acetic acid with methanol to give methyl acetate:

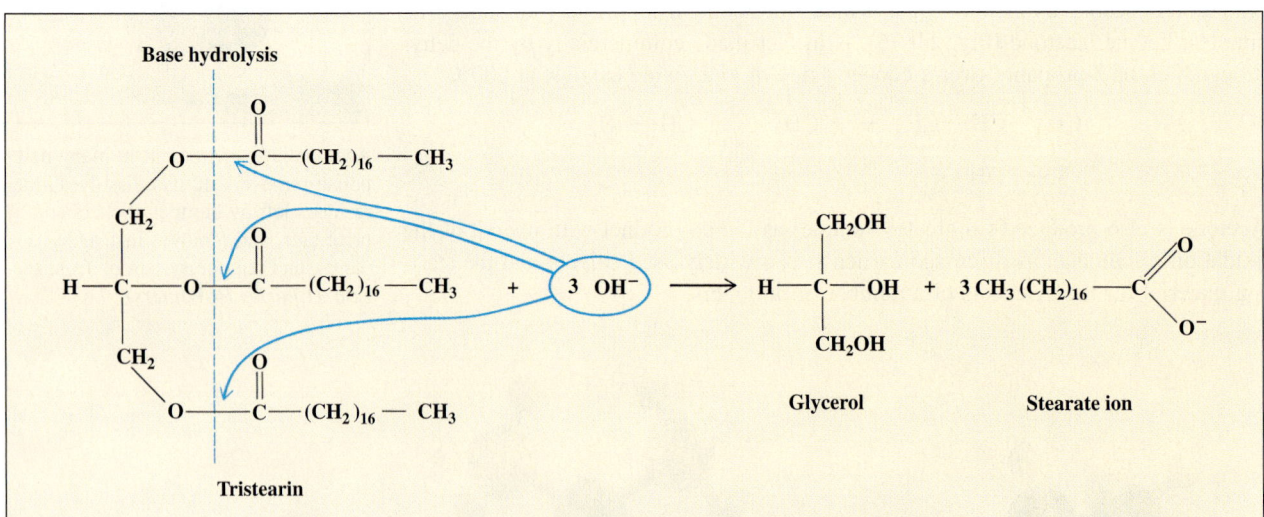

Figure 24–16 • Animal fats are largely mixtures of fatty acid triesters of glycerol. Tristearin (*left*) is an example. Treatment of tristearin with sodium hydroxide breaks the three ester bonds to form glycerol and sodium stearate (other fats react similarly). The resulting sodium salt of a fatty acid is a soap. Soaps cut grime by simultaneously interacting with grease particles at the hydrocarbon tail and with water at the carboxylate-ion end-group. This disperses the grease.

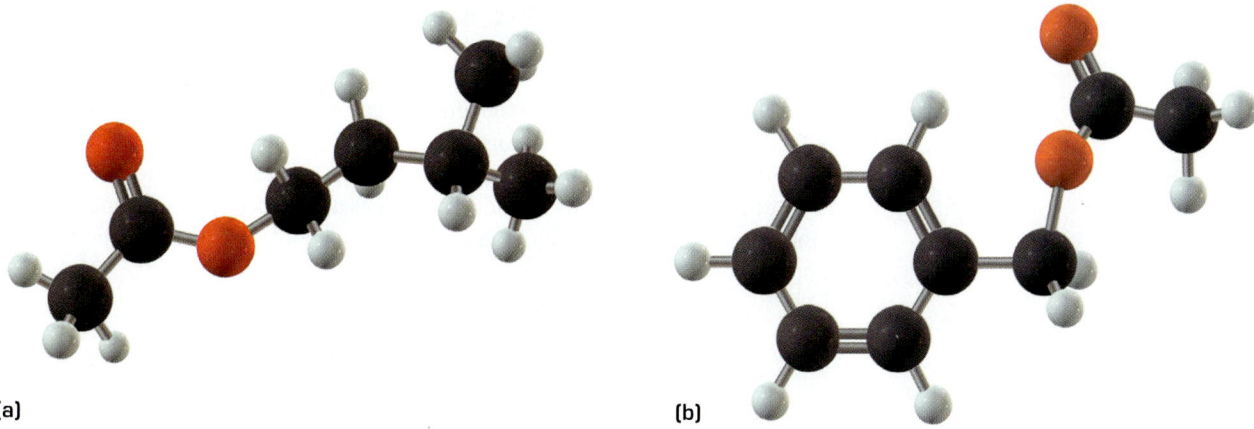

(a)　　　　　　　　　　　　　　　　　　　　**(b)**

Figure 24-17 • The structures of (a) isoamyl acetate, $CH_3COO(CH_2)_2CH(CH_3)_2$, and (b) benzyl acetate, $C_6H_5CH_2OOCCH_3$.

$$CH_3C\overset{O}{\overbrace{\qquad}}\boxed{OH + H}-OCH_3 \longrightarrow CH_3C\overset{O}{\overbrace{\qquad}}O-CH_3 + H_2O$$

Esters are named by first stating the name of the alkyl group of the alcohol (the methyl group in this case) followed by the name of the carboxylic acid with the ending *-ate* (acetate). One of the most important esters in commercial production is vinyl acetate, with the structure

$$CH_3-C\overset{O}{\underset{O-CH=CH_2}{\diagup}}$$

It is prepared by the reaction of acetic acid not with an alcohol but with ethylene and oxygen over a catalyst such as $CuCl_2$ or $PdCl_2$:

$$CH_3C\overset{O}{\overbrace{\qquad}}OH + CH_2{=}CH_2 + \tfrac{1}{2}O_2 \xrightarrow{CuCl_2} CH_3C\overset{O}{\overbrace{\qquad}}O-CH{=}CH_2 + H_2O$$

　　Esters are colorless, volatile liquids that frequently have pleasant odors. Many occur naturally in flowers and fruits. Isoamyl acetate (Fig. 24–17a), for example, is generated in apples as they ripen and contributes to the flavor and odor of the fruit. Benzyl acetate, the ester formed from acetic acid and benzyl alcohol (Fig. 24–17b), is a major component of oil of jasmine and is used in the preparation of perfumes.

- The reaction of a carboxylic acid with an alcohol to give an ester plus water is formally similar to the reaction of an inorganic acid with an inorganic base to give a salt plus water.

- The $-CH{=}CH_2$ side group is called a vinyl group.

- The scents of fine perfumes involve subtle blending of many different compounds, often in very small quantities.

Amines and Amides

The **amines** can be regarded as derivatives of ammonia (NH_3). They have the general formula R_3N, where R can represent a hydrocarbon group or hydrogen. If only one hydrogen atom in the ammonia molecule is replaced by a hydrocarbon group, the result is called a **primary** amine; examples are ethylamine and aniline:

ethylamine　　　aniline　　　dimethylamine　　　trimethylamine

If two hydrocarbon groups replace hydrogen atoms in the ammonia molecule, the compound is a **secondary** amine (such as dimethylamine). Replacement of all three hydrogen atoms gives a **tertiary** amine (such as trimethylamine). Amines are bases because the lone electron pair on the nitrogen can accept a hydrogen ion in the same way that the lone pair on the nitrogen in ammonia does.

A primary or secondary amine (or ammonia itself) can react with a carboxylic acid to form an **amide.** This is another condensation reaction and is analogous to the formation of an ester from the reaction of an alcohol with a carboxylic acid. An example of amide formation is

$$\text{CH}_3\text{C}\overset{\text{O}}{\underset{}{\|}}\boxed{\text{OH} + \text{H}}\text{—N(CH}_3)_2 \longrightarrow \text{CH}_3\text{C}\overset{\text{O}}{\underset{}{\|}}\text{—N(CH}_3)_2 + \text{H}_2\text{O}$$

If ammonia is the reactant, an —NH_2 group replaces the —OH group in the carboxylic acid as the amide is formed:

$$\text{CH}_3\text{C}\overset{\text{O}}{\underset{}{\|}}\text{—OH} + \text{NH}_3 \longrightarrow \text{CH}_3\text{C}\overset{\text{O}}{\underset{}{\|}}\text{—NH}_2 + \text{H}_2\text{O}$$
acetamide

Amide linkages are present in every protein molecule and therefore have great importance in biochemistry (see Section 25–2).

24-3   Pesticides and Pharmaceuticals

The organic compounds discussed so far all have relatively small molecules; most are produced in large volumes. Substances such as these provide starting materials for the synthesis of numerous, structurally more complex, organic compounds that find applications in agriculture, medicine, and consumer products. In this section, we discuss a selection of these compounds, all of which are used as pesticides or as pharmaceuticals. Some of the structures and syntheses of these compounds are intricate. In reading this section, it is not important to memorize complicated structures. A wiser aim is to note the hydrocarbon frames of the molecules, to recognize functional groups, and to begin to appreciate the extremely diverse structures and properties of organic compounds.

Chemists have developed a shorthand notation to represent the structures of complex organic molecules. This notation, which focuses the attention on the most important aspects of structure, is illustrated in Figure 24–18. In this notation, the symbol "C" for a carbon atom is omitted, and only the C—C bonds are shown. A carbon atom is assumed to lie at each end of the line segments that represent bonds. In addition, hydrogen atoms attached to carbon are omitted. Terminal carbon atoms (those at the ends of chains) and their associated hydrogen atoms are shown explicitly, however. To generate the full structure (and the molecular formula) from such a shorthand formula, carbon atoms must be inserted at the end of each bond, and enough hydrogen atoms must be attached to each carbon atom to give it a valence of four.

Insecticides

The chemical control of insect pests dates back thousands of years. The earliest insecticides were inorganic compounds of copper, lead, and arsenic, as well as some naturally occurring organic compounds such as nicotine (Fig. 24–19a). Very few of these "first-generation" insecticides are in use today because of their adverse side effects on plants, animals, and humans.

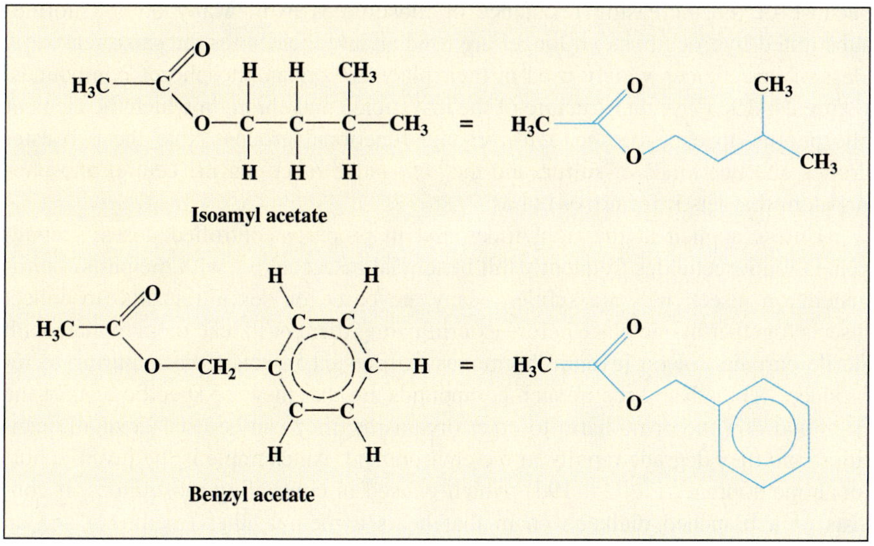

Figure 24-18 • In the shorthand notation for the structures of organic compounds illustrated on the right side of this figure, carbon atoms are assumed to lie where the lines indicating bonds intersect. Furthermore, enough hydrogen atoms are assumed to be bonded to each carbon atom to give it a total valence of four. Terminal carbon atoms (those at the ends of chains) and their associated hydrogen atoms are shown explicitly, however.

After World War II, controlled organic syntheses gave rise to a second generation of insecticides. The success of these agents fueled rapid growth in the use of chemicals for insect control. The leading insecticide of the 1950s and 1960s was DDT (an abbreviation for *di*chloro*di*phenyl*tri*chloroethane; see Fig. 24–19b). Although DDT was of worldwide importance in slowing the spread of typhus (transmitted by body lice) and malaria (transmitted by mosquitoes), its use was banned in the United States in 1972 because of its adverse effects on birds, fish, and other forms of life that can accumulate DDT to high concentrations and

Figure 24-19 • The structures of several insecticides.

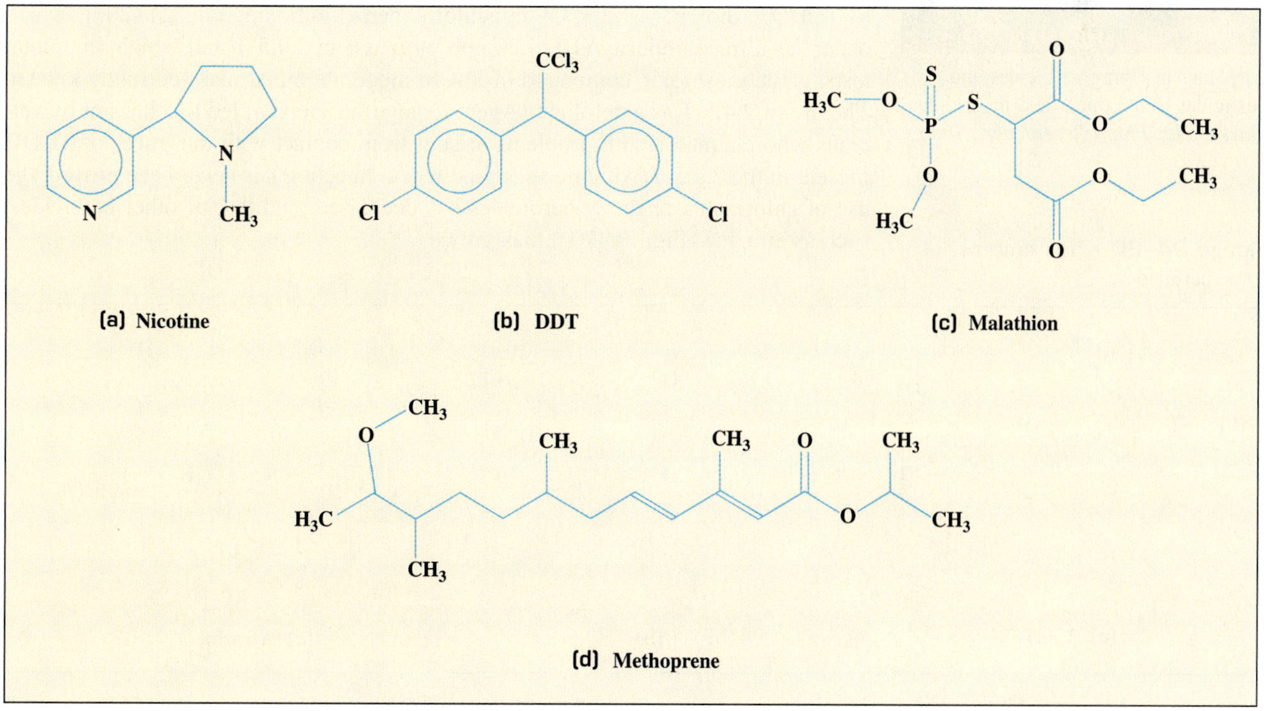

• The use of DDT against malaria-bearing mosquitoes caused the incidence of malaria in India to fall from about 75 million cases per year in the early 1950s to 50 *thousand* cases per year in 1961. After spraying stopped, the incidence of the disease increased to 6 million cases in 1976. It has since fallen with the development of other pesticides.

because of the increasing resistance of mosquitoes to it. Many other chlorine-substituted hydrocarbons no longer are used as insecticides for the same reason. A class of insecticides widely used in their place are organophosphorus compounds. Figure 24–19c gives the structure of the insecticide malathion, in which the element phosphorus appears together with organic functional groups. Note the two ester groups, the two kinds of sulfur, and the "expanded octet" on the central phosphorus atom that lets it form five bonds.

Unless applied at the right times and in properly controlled doses, second-generation insecticides frequently kill beneficial insects along with the pests. Third-generation insecticides are subtler. Many are based on sex attractants (to collect insects together in one place before exterminating them or to lead them to mate with sterile partners) or on juvenile hormones (to prevent insects from maturing to reproduce). The advantages of such compounds are that they are specific against the pests and do little or no harm to other organisms; they can be used in small quantities; and they degrade rapidly in the environment. An example is the juvenile hormone methoprene (Fig. 24–19d), which is used in controlling mosquitoes. It consists of a branched dialkene chain that bears a methyl ether (methoxy) and an isopropyl–ester functional group.

Herbicides

Chemical weed control has contributed (together with fertilizers) to the "green revolution" of the past 55 years, which have seen dramatic increases in agricultural productivity throughout the world. The first herbicide of major importance, introduced in 1945 and still in use today, was 2,4-D (2,4-dichlorophenoxyacetic acid; Fig. 24–20a), a derivative of phenol with chlorine and carboxylic acid functional groups. It kills broadleaf weeds in wheat, corn, and cotton without unduly persisting in the environment, as the chlorinated insecticides discussed earlier do. A related compound is 2,4,5-T (2,4,5-trichlorophenoxyacetic acid), in which a hydrogen atom in 2,4-D is replaced by a chlorine. In recent years, much attention has been given to TCDD ("dioxin," or 2,3,7,8-tetrachlorodibenzo-*p*-dioxin; Fig. 24–20b), which occurs as a trace impurity (10 to 20 ppb by mass) in 2,4,5-T and which, in animal tests, is the most toxic compound of low to moderate molar mass currently known. The use of 2,4,5-T as a defoliant (Agent Orange) in Vietnam led to a lawsuit by veterans who claimed health problems arising from contact with the traces of TCDD present in the 2,4,5-T. Although such a direct connection has never been proved, the use of chlorinated phenoxy herbicides has decreased, and that of other herbicides, such as atrazine (Fig. 24–20c), has grown.

Airplanes applying herbicides and pesticides to a lettuce field in Arizona. (*Jack Fields/Photo Researchers, Inc.*)

Figure 24–20 • The structures of some herbicides.

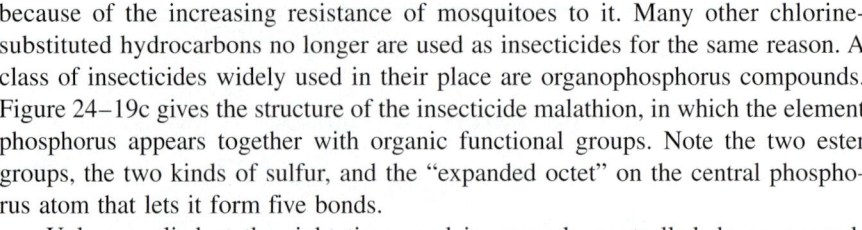

(a) 2,4-D   (b) TCDD   (c) Atrazine

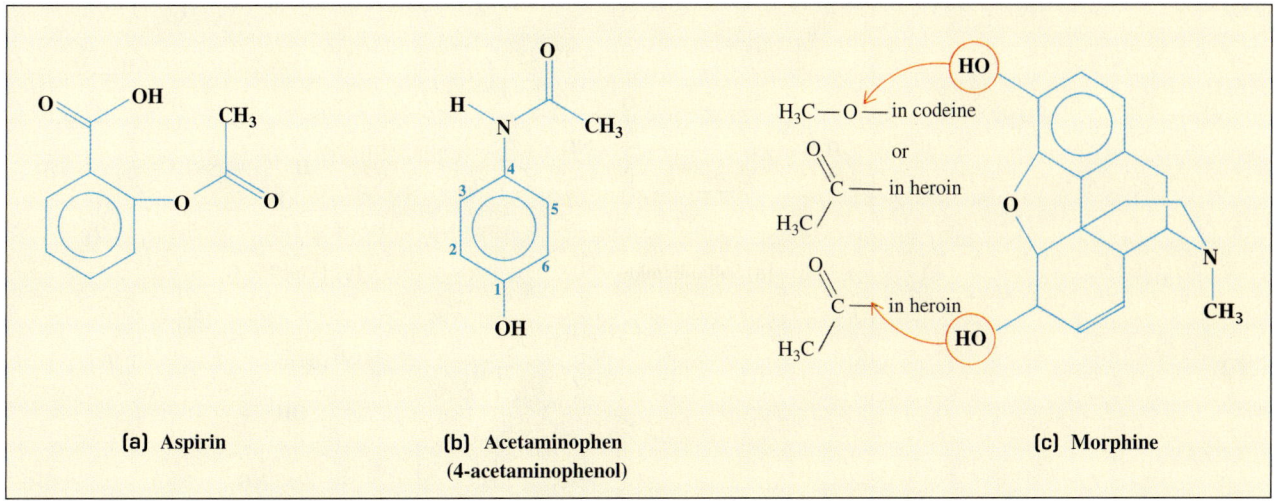

(a) Aspirin   **(b) Acetaminophen (4-acetaminophenol)**   **(c) Morphine**

Figure 24–21 • The molecular structures of some analgesics. Note how slight the differences are between morphine, codeine, and heroin.

Analgesics

Drugs that relieve pain are called **analgesics.** The oldest and most widely used analgesic is aspirin, which has the chemical name acetylsalicylic acid (Fig. 24–21a). Over 15 million kg of aspirin is synthesized each year, using the synthesis outlined in Figure 22–2. Aspirin acts to reduce fevers as well as to relieve pain. Some recent studies have suggested that regular consumption may reduce the chances of heart disease. As an acid, aspirin can irritate the stomach lining, a side-effect that can be reduced by combining it in a buffer with a weak base such as sodium hydrogen carbonate. Another important pain reliever is 4-acetaminophenol, or acetaminophen (Fig. 24–21b). This compound is sold under many trade names, most prominently Tylenol. Both acetylsalicylic acid and 4-acetaminophenol are derivatives of phenol, the former being converted to an acetic acid ester with an additional carboxylic acid functional group and the latter having an amide functional group.

A much more powerful pain reliever, which is available only by prescription because it is addictive, is morphine (Fig. 24–21c). Morphine acts on the central nervous system, apparently because its shape fits a receptor site on the nerve cell

Crystals of 4-acetaminophenol (Tylenol) viewed under polarized light. *(Phillip A. Harrington/Fran Heyl Associates)*

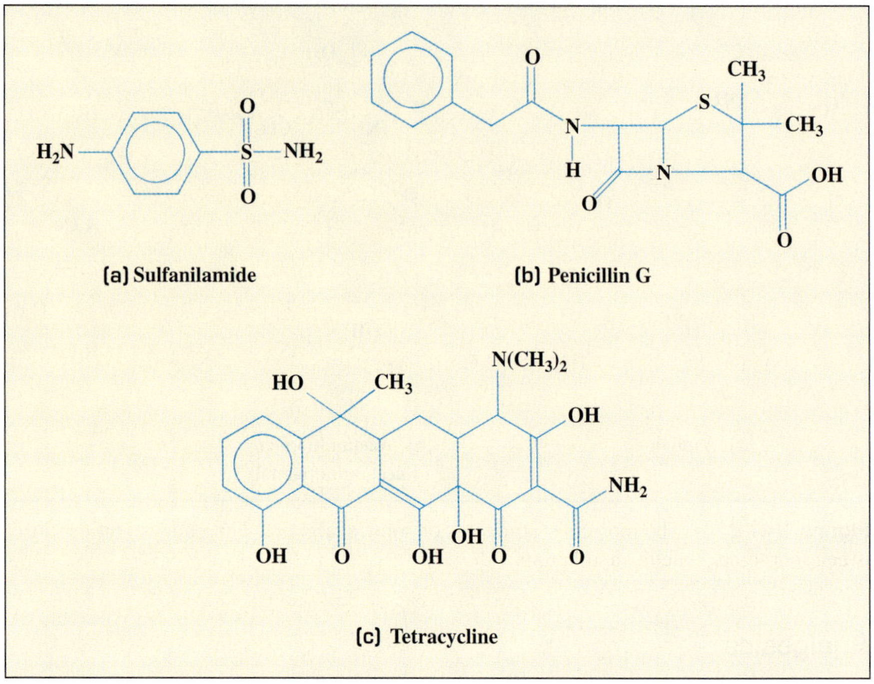

(a) Sulfanilamide

(b) Penicillin G

(c) Tetracycline

Figure 24–22 • The molecular structures of some antibiotics.

and blocks the transmission of pain signals to the brain. Its structure contains five interconnected rings. A very small change (the replacement of one —OH group by an —OCH₃ group, giving a methyl ether) converts morphine into codeine, a prescription drug used as a cough suppressant. Replacing *both* —OH groups by acetyl groups (—COCH₃) generates the notoriously addictive substance heroin.

Antibacterial Agents

The advent of antibacterial agents changed the treatment of bacterial diseases such as tuberculosis and pneumonia dramatically beginning in the 1930s. The first "wonder drug" was sulfanilamide (Fig. 24–22a), a derivative of aniline. Other "sulfa drugs" are obtained by replacing one of the hydrogen atoms on the sulfonamide group by other functional groups. Bacteria mistake sulfanilamide for *p*-aminobenzoic acid, a molecule with a closely similar shape but a —COOH carboxylic acid group in place of the —SO₂NH₂ group. The drug then interferes with the organism's synthesis of folic acid, an essential biochemical, and kills it. Mammals do not synthesize folic acid (they obtain it from their diet) and are not affected by sulfanilamide.

The molecule of penicillin (Fig. 24–22b) contains an amide linkage connecting a substituted double ring (including S and N atoms) to a benzyl (phenylmethyl) group. It is a natural product formed by certain molds. Although the total synthesis of penicillin was achieved in 1957, that chemical route is not competitive economically with biosynthesis via fermentation. The mold grows for several days in tanks that may hold up to 100,000 L of a fermentation broth (Fig. 24–23). The penicillin later is separated by solvent extraction. Penicillin works by deactivating enzymes in the bacteria that are responsible for building cell walls. Derivatives of natural penicillin have been developed and also are used.

Figure 24–23 • Fermentation tanks used in modern penicillin production.

A final group of antibiotics are the tetracyclines, which are derivatives of the four-ring aromatic compound represented in Figure 24–22c. These drugs have the broadest spectrum of antibacterial activity found to date.

• The name *tetracycline* reflects molecular structure: *tetra-* ("four") + *cycl-* ("ring").

Steroids

The **steroids** are naturally occurring compounds that derive formally from cholesterol. The structure of cholesterol, which is shown in Figure 24–24a, contains a group of four fused hydrocarbon rings (three six-atom rings and one five-atom ring). All steroids possess this "steroid nucleus." Cholesterol itself is present in all tissues of the human body. When present in excess in the bloodstream, it can accumulate in the arteries, restricting the flow of blood and leading to heart attacks. Its derivatives have widely different functions. The hormone cortisone (Fig. 24–24b), which is secreted by the adrenal glands, regulates the metabolism of sugars, fats, and proteins in all body cells. As a drug, cortisone acts to reduce inflammation and to moderate allergic responses. It is often prescribed against arthritic inflammation of the joints.

The human sex hormones also are derivatives of cholesterol. Here, the resourcefulness of nature in building compounds with quite different functions from the same starting material is particularly evident. The female sex hormone progesterone (Fig. 24–24c), for example, differs from the male sex hormone testosterone only by the exchange of an acetyl (—COCH₃) group for a hydroxyl (—OH) group. Oral contraceptives are synthetic compounds with structures that are closely related to progesterone.

Small crystals of the antibiotic tetracycline. *(Ulof Bjorg Christianson/Fran Heyl Associates)*

(a) Cholesterol

(b) Cortisone

R = —OH (testosterone)

R = —C(=O)CH₃ (progesterone)

(c) Progesterone and testosterone

Figure 24–24 • The molecular structures of some steroids.

SUMMARY

24-1 The **straight-chain alkanes** are hydrocarbons with the general formula C_nH_{2n+2} in which carbon atoms are linked one by one in extended chains by single covalent bonds. They are the major constituents of petroleum. The distillation of petroleum separates its constituents according to boiling point. In **branched-chain alkanes,** the chain of carbon atoms has branches so that at least one carbon atom bonds to three other carbon atoms. In **cycloalkanes,** a carbon chain closes a loop upon itself. The general formula for cycloalkanes with one closed ring is C_nH_{2n}. Alkanes are named as derivatives of the alkane that corresponds to the longest continuous chain of carbon atoms in the molecular formula. The naming system then uses the names of the various **alkyl groups** to specify side-chains and numbers to tell their locations. **Alkenes** have at least one carbon–carbon double bond; **alkynes** have at least one carbon–carbon triple bond. Both categories are unsaturated. Isomeric alkenes have double bonds at different locations along a chain of carbon atoms of the same length. *Cis–trans* isomerism in alkenes is a consequence of the lack of free rotation around the carbon–carbon double bond. In **polyenes,** two or more double bonds occur along a given chain of carbon atoms. **Catalytic cracking** and **thermal cracking** produce shorter-chain alkenes from the alkanes in the starting material. The **aromatic hydrocarbons** contain rings of carbon atoms in which delocalized π electrons significantly increase the molecular stability. Petroleum contains useful quantities of benzene, toluene, and the isomeric xylenes. In **reforming** reactions, six-, seven-, and eight-carbon alkanes are converted to the aromatic hydrocarbons, which are used in high-performance fuels and in synthesis.

24-2 **Functional groups** are sites of heightened reactivity in organic molecules. They arise from insertion of non-carbon atoms in a hydrocarbon chain or the attachment of non-carbon atoms to a chain. The replacement of a hydrogen atom by a halogen creates an **alkyl halide.** A similar replacement by an —OH group gives an **alcohol.** When an —OH group attaches to an aromatic hydrocarbon ring, the result is a **phenol. Ethers** are characterized by the —O— functional group, in which an oxygen atom links two alkyl or aromatic groups. Cyclic ethers with various ring sizes exist; those with three members are called **epoxides.** An **aldehyde** has the characteristic $-C{\overset{\displaystyle O}{\underset{\displaystyle H}{<}}}$ functional group and comes from the dehydrogenation of an alcohol.

A **ketone** also contains a carbon atom that is double bonded to an oxygen atom but has two other carbon atoms (and therefore no hydrogen atoms) bonded to that carbon atom. **Carboxylic acids** contain the $-C{\overset{\displaystyle O}{\underset{\displaystyle OH}{<}}}$ functional group and condense with alcohols to give **esters.** The **amines** are derivatives of NH_3 in which one, two, or all three hydrogen atoms are replaced by hydrocarbon groups. Such replacements give **primary, secondary,** and **tertiary** amines, respectively. The condensation of a primary or secondary amine with a carboxylic acid gives an **amide.**

24-3 The organic compounds produced from petroleum in large amounts have rather small molecules of simple structure. They are elaborated into materials with more complex molecular structures that serve a range of purposes. These include insecticides; herbicides; pharmaceuticals, including **analgesics** to relieve pain; and antibacterials and antibiotics to control infections. The **steroids** are biologically active compounds that are derivatives of cholesterol.

PROBLEMS

Note: Answers to blue-numbered problems are given in Appendix F. Problems that are more challenging are indicated with asterisks.

Petroleum Refining and the Hydrocarbons

1. Is it possible for a gasoline to have an octane number exceeding 100? Explain.

2. Is it possible for a motor fuel to have a negative octane rating? Explain.

3. A gaseous alkane is burned completely in oxygen. The volume of the carbon dioxide that forms equals twice the volume of the alkane burned (the volumes are measured at the same temperature and pressure). Name the alkane and write a balanced equation for its combustion.

4. A gaseous alkyne is burned completely in oxygen. The volume of the water vapor that forms equals the volume of the alkyne burned (the volumes are measured at the same temperature and pressure). Name the alkyne and write a balanced equation for its combustion.

5. (a) Use average bond enthalpies from Table 10–3 and the ΔH_f° of C(g) and H(g) from Appendix D to estimate the standard enthalpy of formation of cyclopropane ($C_3H_6(g)$).
   (b) Calorimetry experiments at 25°C establish the following:

   $$C_3H_6(g) + \tfrac{9}{2} O_2(g) \longrightarrow 3\, CO_2(g) + 3\, H_2O(g)$$
   $$\Delta H_r^\circ = -1959 \text{ kJ mol}^{-1}$$

   Combine this result with data from Appendix D to calculate the standard enthalpy of formation of cyclopropane.
   (c) By comparing the answers from (a) and (b), estimate the "strain enthalpy" associated with forming a ring of three carbon atoms (with C—C—C bond angles of only 60°) in cyclopropane.

6. (a) Use average bond enthalpies from Table 10–3 and the ΔH_f° of C(g) and H(g) from Appendix D to estimate the standard enthalpy of formation of cyclobutane ($C_4H_8(g)$).
   (b) The standard enthalpy at 25°C for the combustion of cyclobutane

   $$C_4H_8(g) + 6\, O_2(g) \longrightarrow 4\, CO_2(g) + 4\, H_2O(g)$$

   is $\Delta H_r^\circ = -2568$ kJ mol^{-1}. Use this, together with data from Appendix D, to calculate the standard enthalpy of formation of cyclobutane.
   (c) By comparing your answers from (a) and (b), estimate the "strain enthalpy" associated with forming a ring of four carbon atoms (with bond angles less than 109.5°) in cyclobutane.

7. (a) Write a chemical equation involving structural formulas for the catalytic cracking of decane into an alkane and an alkene that contain equal numbers of carbon atoms.

Assume that the reactant and both products have a straight chain of carbon atoms.
   (b) How many isomers of the alkene are possible?

8. (a) Write an equation involving structural formulas for the catalytic cracking of 2,2,3,4,5,5-hexamethylhexane. Assume that the cracking occurs between carbon atoms 3 and 4.
   (b) How many isomeric alkenes of formula C_6H_{12} are possible?

9. Write structural formulas for the following:
   (a) 2,3-dimethylpentane
   (b) 3-ethyl-2-pentene
   (c) methylcyclopropane
   (d) 2,2-dimethylbutane
   (e) 3-propyl-2-hexene
   (f) 3-methyl-1-hexene
   (g) 4-ethyl-2-methylheptane
   (h) 4-ethyl-2-heptyne

10. Write structural formulas for the following:
   (a) 2,3-dimethyl-1-cyclobutene
   (b) 2-methyl-2-butene
   (c) 2-methyl-1,3-butadiene
   (d) 2,3-dimethyl-3-ethylhexane
   (e) 4,5-diethyloctane
   (f) cyclooctene
   (g) propadiene
   (h) 2-pentyne

11. Write structural formulas for *trans*-3-heptene and *cis*-3-heptene.

12. Write structural formulas for *cis*-4-octene and *trans*-4-octene.

13. Name the following hydrocarbons:

   (a) $CH_2{=}C{=}\overset{\displaystyle H}{\underset{\displaystyle |}{C}}{-}CH_2{-}CH_2{-}CH_3$

   (b) $H_2C{=}\overset{H}{\underset{|}{C}}{-}\overset{H}{\underset{|}{C}}{=}C{-}C{=}CH_2$ with H, H below the third and fourth carbons

   (c) $H_2C{=}\overset{\displaystyle CH_3}{\underset{\displaystyle |}{C}}{-}CH_2{-}CH_2{-}CH_2{-}CH_3$

   (d) $CH_3{-}CH_2{-}C{\equiv}C{-}CH_2{-}CH_3$

14. Name the following hydrocarbons:

   (a) $CH_2{=}\overset{\displaystyle }{\underset{\displaystyle CH_3}{C}}{-}\overset{\displaystyle }{\underset{\displaystyle CH_3}{C}}{=}CH_2$

   (b) $CH_3{-}\overset{H}{\underset{|}{C}}{=}\overset{H}{\underset{|}{C}}{-}\overset{H}{\underset{|}{C}}{=}\overset{H}{\underset{|}{C}}{-}CH_3$

(c)
$$CH_3-\overset{\overset{\displaystyle CH_3}{|}}{\underset{\underset{\displaystyle CH_3}{|}}{C}}-CH_2-CH_3$$

(d)
$$CH_3-\overset{}{\underset{\underset{\displaystyle CH_2}{\|}}{C}}-CH_3$$

15. State the hybridization of each of the carbon atoms in the hydrocarbon structures in Problem 13.

16. State the hybridization of each of the carbon atoms in the hydrocarbon structures in Problem 14.

Functional Groups and Organic Synthesis

17. Write balanced equations for the following reactions. Use structural formulas to represent the organic compounds.
  (a) The production of butyl acetate from butanol and acetic acid
  (b) The conversion of ammonium acetate to acetamide and water
  (c) The dehydrogenation of 1-propanol
  (d) The complete combustion (to CO_2 and H_2O) of heptane

18. Write balanced equations for the following reactions. Use structural formulas to represent the organic compounds.
  (a) The complete combustion (to CO_2 and H_2O) of cyclopropanol
  (b) The reaction of isopropyl acetate with water to give acetic acid and isopropanol
  (c) The dehydration of ethanol to give ethylene
  (d) The reaction of 1-butyl iodide with water to give 1-butanol

19. Outline, using chemical equations, the synthesis of the following from easily available petrochemicals and inorganic starting materials:
  (a) Vinyl bromide (CH_2=CHBr)
  (b) 2-Butanol
  (c) Acetone (CH_3COCH_3)

20. Outline, using chemical equations, the synthesis of the following from easily available petrochemicals and inorganic starting materials:
  (a) Vinyl acetate (CH_3COOCH=CH_2)
  (b) Formamide ($HCONH_2$)
  (c) 1,2-Difluoroethane

21. Write a general equation (using the symbol R to represent a general alkyl group) for the formation of an ester by the condensation of a tertiary alcohol with a carboxylic acid.

22. Explain why it is impossible to form an amide by the condensation of a tertiary amine with a carboxylic acid.

23. In a recent year, the United States produced 6.26×10^9 kg of ethylene dichloride (1,2-dichloroethane) and 15.87×10^9 kg of ethylene. Assuming that all significant quantities of ethylene dichloride were produced from ethylene, what fraction of the ethylene production went to make ethylene dichloride? What mass of chlorine was required for this conversion?

24. In a recent year, the United States produced 6.26×10^9 kg of ethylene dichloride (1,2-dichloroethane) and 3.73×10^9 kg of vinyl chloride. Assuming that all significant quantities of vinyl chloride were produced from ethylene dichloride, what fraction of the ethylene dichloride production went to make vinyl chloride? What mass of hydrogen chloride was generated as a by-product?

25. Dimethyl ether and ethanol are isomers (both have the molecular formula C_3H_6O). Suggest a reason that dimethyl ether is less soluble in water than ethanol and has a lower boiling point.

26. The long-chain hydrocarbon portions of fats may be either saturated or unsaturated. Propose an explanation for the fact that saturated fats are mostly solids at room temperature while unsaturated fats are mostly liquids.

Pesticides and Pharmaceuticals

27. (a) The insecticide methoprene (see Fig. 24–19d) is an ester. Write the structural formulas for the alcohol and the carboxylic acid that react to form it. Name the alcohol.
  (b) Suppose that the carboxylic acid from part (a) is chemically changed so that the OCH_3 group is replaced by an H atom and the COOH group is replaced by a CH_3 group. Name the hydrocarbon that would result.

28. (a) The herbicide 2,4-D (see Fig. 24–20a) is an ether. Write the structural formulas of the two alcohol or phenol compounds that, upon condensation, would form this ether. (The usual method of synthesis is different, however.)
  (b) Suppose that hydrogen atoms replace by chlorine atoms and a —CH_3 group replaces the carboxylic acid group in the two compounds in part (a). Name the resulting compounds.

29. (a) Write the molecular formula of acetylsalicylic acid (see Fig. 24–21a).
  (b) An aspirin tablet contains 325 mg of acetylsalicylic acid. Calculate the chemical amount (in moles) of that compound in the tablet.

30. (a) Write the molecular formula of acetaminophen (see Fig. 24–21b).
  (b) A tablet of extra-strength Tylenol contains 500 mg of acetaminophen. Calculate the chemical amount (in moles) of that compound in the tablet.

31. Describe the changes in hydrocarbon structure and functional groups that are needed to make cortisone from cholesterol (see Fig. 24–24).

32. Describe the changes in hydrocarbon structure and functional groups that are needed to make testosterone from cortisone (see Fig. 24–24).

ADDITIONAL PROBLEMS

33. The boiling points of the straight-chain alkanes increase as the molar mass increases (Fig. 24–5). Suggest a reason.

34. Organic molecules that are symmetrical and rigid generally have higher melting points than do molecules of similar molar mass that are not symmetrical or rigid. Draw the structures of compound in the following pairs and decide which has the higher melting point:
  (a) *Tert*-butyl alcohol or 2-butanol (also called *sec*-butyl alcohol)
  (b) Propane or cyclopropane

35. At 25°C, the standard enthalpy of formation of heptane is -187.82 kJ mol^{-1}, and the standard enthalpy of formation of isooctane (2,2,4-trimethylpentane) is -1224.13 kJ mol^{-1}.
  (a) Compute the standard molar enthalpies of combustion of each of the two if they are burned to $CO_2(g)$ and $H_2O(g)$ in an internal-combustion engine.
  (b) Which compound delivers more heat per gallon? (Isooctane has a density of 5.77 lb per U.S. gallon, and heptane has a density of 5.71 lb per U.S. gallon.)

*36. Ethylene (C_2H_4) is an organic compound of great economic significance that often is transported by pipeline. It has a critical temperature of 282.65 K and a critical pressure of 50.096 atm.
  (a) Express the critical temperature and pressure of ethylene in degrees Fahrenheit and in pounds per square inch.
  (b) A pipeline contains ethylene at a pressure of 54 atm and a temperature of 15°C. Discuss the physical state of the ethylene within the pipe.
  (c) The pipeline is 100 km long and has an interior diameter of 250 mm. Estimate the mass of ethylene in the pipeline, assuming that it is an ideal gas.
  (d) The density of the ethylene within the pipe from part (c) is 0.20 g cm^{-3}. Compute the actual mass of ethylene in the pipe.
  (e) If the temperature of the entire pipeline and its contents falls to 0°C on a cold day, suggest what the operators could do to make sure that the ethylene in the pipe does not change its physical state.

37. The hydrocarbon cyclodecene has a *trans* isomer and a *cis* isomer. Write structural formulas for the two.

38. Compare catalytic cracking with thermal cracking of hydrocarbons. What is the purpose of each type of process?

39. Alkenes and aromatic compounds, as well as branched-chain alkanes, increase the octane number of gasoline. Describe the methods used to increase the proportions of these two types of compounds in gasoline.

*40. Consider the following proposed structures for benzene, each of which is consistent with the molecular formula C_6H_6:

(i)

(ii)

(iii)

(iv) $H_3CC \equiv C - C \equiv CCH_3$

(v) $H_2C = CH - C \equiv C - CH = CH_2$

  (a) When benzene reacts with chlorine to give C_6H_5Cl, only one isomer of that compound is formed. Which of the five proposed structures for benzene are consistent with this observation?
  (b) When C_6H_5Cl reacts further with chlorine to give $C_6H_4Cl_2$, exactly three isomers of the latter compound are formed. Which of the five proposed structures for benzene are consistent with this observation?

41. How does *dehydration* differ from *dehydrogenation*? Give an example of each.

42. When an ester forms from an alcohol and a carboxylic acid, an oxygen atom links the two parts of each ester molecule. This atom could have come originally from the alcohol, from the carboxylic acid, or randomly from either. Propose an experiment using isotopes to determine which is the case.

43. Glycerol ($C_3H_8O_3$) has a three-carbon chain with one $-OH$ group attached to each carbon atom.
  (a) Draw the structure of glycerol.
  (b) What kind of reaction does glycerol undergo with stearic acid to form tristearin (see Fig. 24–16)? What type of organic compound is tristearin?

44. (a) It is reported that ethylene is released when pure ethanol is passed over alumina (Al_2O_3) that is heated to 400°C, but that diethyl ether is obtained at a temperature of 230°C. Write balanced equations for both of these dehydration reactions.
  (b) If the temperature is raised well above 400°C, an aldehyde forms. Write a chemical equation for this reaction.

45. In what ways do the systematic development of pesticides and of pharmaceuticals resemble each other, and in what ways do they differ? In your answer, consider such aspects as "deceptor" molecules, which are mistaken by living organisms for other molecules, side effects, and the relative advantages of a broad versus a narrow spectrum of activity.

46. The steroid stanolone is an androgenic steroid (a steroid that develops or maintains certain male sexual characteristics). It is derived from testosterone by adding a molecule of hydrogen across the $C = C$ double bond in testosterone.
  (a) Using Figure 24–24c as a guide, draw the molecular structure of stanolone.
  (b) What is the molecular formula of stanolone?

25

Synthetic and Biological Polymers

CHAPTER OUTLINE

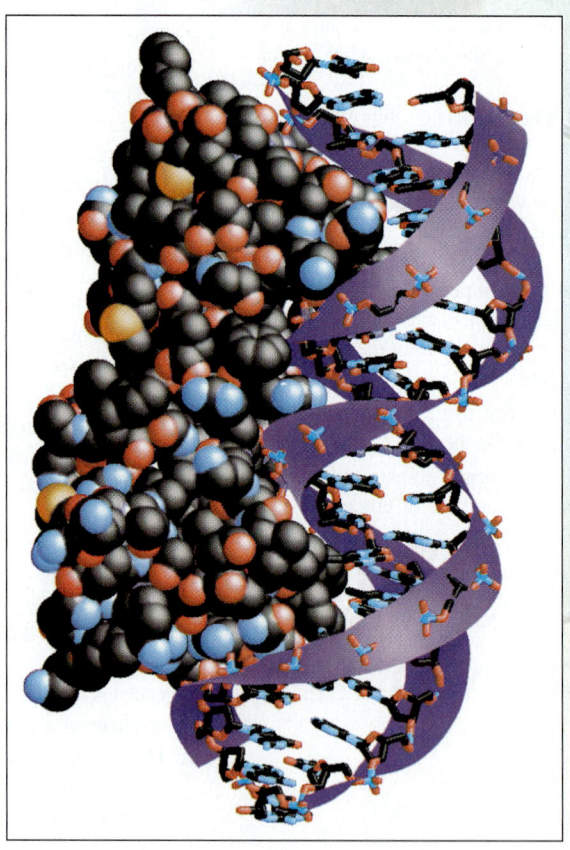

An interaction between biopolymers. Here, interaction with a protein distorts the famous double helix (*purple*) of DNA.

Hundreds of thousands of organic compounds possess molecules having 20 or fewer carbon atoms combined with hydrogen, oxygen, nitrogen, and the halogens in the functional groups discussed in Chapter 24. Hundreds of thousands more such compounds await isolation from natural sources or synthesis. All of this barely touches the full potential of carbon in compound formation, however, because carbon atoms can string together in stable chains of essentially unlimited length to make truly huge molecules, ones that contain hundreds of thousands or even millions of atoms. Such products, called **polymers,** form by the linking of numerous separate small **monomer units** in strands and webs.

Polymers are everywhere. Biological polymers (biopolymers) are crucial in life processes. They include energy-storage and structural materials (starch, cellulose, and protein), enzymes, and the genetic substances DNA and RNA, which store and transmit the information required for the replication of life. Most foods and fibers are polymers. The diverse, useful materials lumped under the name "plastic" are all synthetic (or semi-synthetic) carbon-based polymers, but polymers are not obliged to contain carbon. The chains of alternating Si and O atoms found in many silicate minerals (see Section 21–2) are true polymeric structures.

25-1 Making Polymers

To construct a polymer, a large number of monomer units must add serially to a starting molecule; the reaction must not falter after the first few have reacted. Continued growth is achieved by having the polymer molecule retain highly reactive functional groups at all times during its synthesis. The two major types of polymer growth are addition polymerization and condensation polymerization.

Addition Polymerization

In **addition polymerization,** monomers react to form a polymer chain without net loss of atoms. The most common type of addition polymerization involves free-radical chain reaction of monomers that have $C=C$ double bonds. As an example, consider the polymerization of vinyl chloride (chloroethylene, $CH_2=CHCl$) to polyvinyl chloride:

$$n\ CH_2=\underset{\underset{Cl}{|}}{\overset{\overset{H}{|}}{C}} \longrightarrow \left[CH_2-\underset{\underset{Cl}{|}}{\overset{\overset{H}{|}}{C}} \right]_n$$

As in the chain reactions considered in Section 14–5, the overall process consists of three steps: initiation, propagation (repeated many times to build up a long chain), and termination. The polymerization of vinyl chloride (Fig. 25–1) can be initiated by a small concentration of molecules that have bonds weak enough to be broken by the action of light or heat, giving radicals. An example of such an **initiator** is a peroxide, which can be represented as $R-O-O-R'$, where R and R' represent alkyl groups. The weak $O-O$ bonds can break

$$R-\overset{..}{\underset{..}{O}}-\overset{..}{\underset{..}{O}}-R' \longrightarrow R-\overset{..}{\underset{..}{O}}\cdot + \cdot\overset{..}{\underset{..}{O}}-R' \quad \text{(initiation)}$$

to give radicals. In this equation, the valence electrons of the oxygen atoms are represented explicitly to show that their valence shells are incomplete after the reaction. The radicals remedy this by reacting avidly with vinyl chloride, accepting electrons

(a)

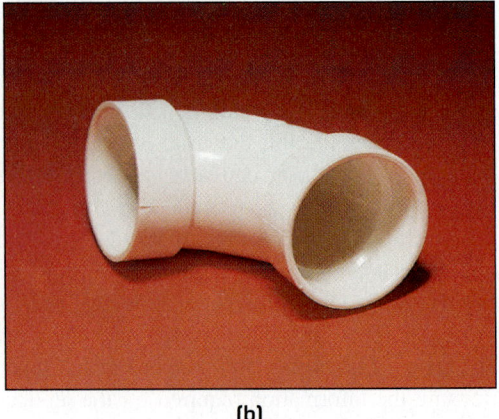

(b)

Figure 25-1 • (a) Several billion kilograms of polyvinyl chloride are produced each year in this chemical plant in Texas. (b) A pipe fitting of polyvinyl chloride. *(Occidental Petroleum)*

from the C=C double bonds to reestablish a closed-shell electron configuration on the oxygen atoms:

$$R—\overset{..}{\underset{..}{O}}\!\cdot\; +\; CH_2=CHCl \longrightarrow R—\overset{..}{\underset{..}{O}}—CH_2—\overset{\displaystyle H}{\underset{\displaystyle Cl}{\overset{|}{\underset{|}{C}}}}\!\cdot \quad \text{(propagation)}$$

One of the two π electrons in the vinyl chloride double bond has been used to form a single bond with the R—O· radical. The other remains on the second carbon atom, leaving it as a seven–valence-electron atom that reacts with another vinyl chloride molecule:

$$R—O—CH_2—\overset{\displaystyle H}{\underset{\displaystyle Cl}{\overset{|}{\underset{|}{C}}}}\!\cdot\; +\; CH_2=\overset{\displaystyle H}{\underset{\displaystyle Cl}{\overset{|}{\underset{|}{C}}}} \longrightarrow R—O—CH_2—\overset{\displaystyle H}{\underset{\displaystyle Cl}{\overset{|}{\underset{|}{C}}}}—CH_2—\overset{\displaystyle H}{\underset{\displaystyle Cl}{\overset{|}{\underset{|}{C}}}}\!\cdot \quad \text{(propagation)}$$

At each stage, the end-group of the lengthening chain is left one electron short of a valence octet and remains quite reactive. The reaction can continue, building up long-chain molecules of high molecular mass. The vinyl chloride monomers always attach to the growing chain with their CH_2 group, because the odd electron is more stable on a CHCl end-group. This gives the polymer a regular alternation of —CH_2— and —CHCl— groups.

Termination occurs when the radical end-groups on two different chains encounter each other and the two chains couple to give a longer chain:

$$R—O—(CH_2—CHCl)_m—CH_2—\overset{\displaystyle H}{\underset{\displaystyle Cl}{\overset{|}{\underset{|}{C}}}}\!\cdot\; +\; \cdot\overset{\displaystyle H}{\underset{\displaystyle Cl}{\overset{|}{\underset{|}{C}}}}—CH_2—(CHCl—CH_2)_n—O—R' \longrightarrow$$

$$R—O—(CH_2—CHCl)_m—CH_2—\overset{\displaystyle H}{\underset{\displaystyle Cl}{\overset{|}{\underset{|}{C}}}}—\overset{\displaystyle H}{\underset{\displaystyle Cl}{\overset{|}{\underset{|}{C}}}}—CH_2—(CHCl—CH_2)_n—O—R'$$

$$\text{(termination)}$$

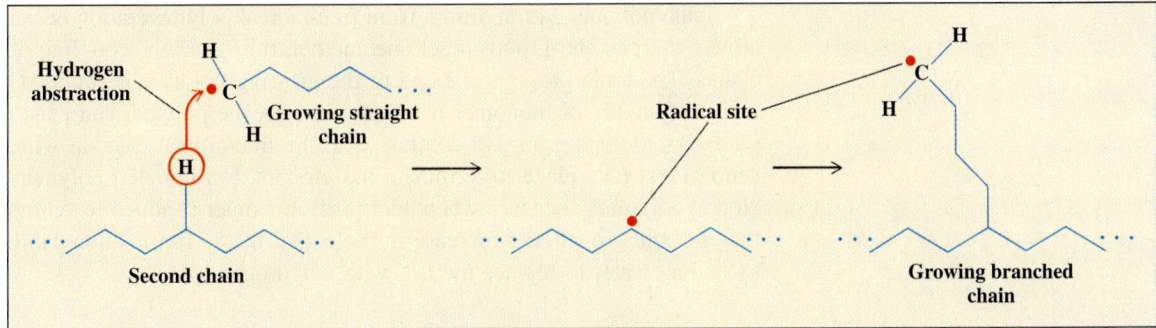

Figure 25–2 • A growing branched chain results during free-radical polymerization when a radical at one end of a growing straight chain by chance extracts a hydrogen atom from the middle of a second chain.

Alternatively, a hydrogen atom may transfer from one end-group to the other:

$$R—O—(CH_2—CHCl)_m—\overset{\displaystyle H}{\underset{\displaystyle H}{\overset{|}{\underset{|}{C}}}}—\overset{\displaystyle H}{\underset{\displaystyle Cl}{\overset{|}{\underset{|}{C}}}}\cdot + \cdot\overset{\displaystyle H}{\underset{\displaystyle Cl}{\overset{|}{\underset{|}{C}}}}—CH_2—(CHCl—CH_2)_n—O—R' \longrightarrow$$

$$R—O—(CH_2—CHCl)_m—CH{=}CHCl + CH_2Cl—CH_2—(CHCl—CH_2)_n—O—R'$$

(termination)

The latter termination step leaves a double bond on one chain end and a $—CH_2Cl$ group on the other. When polymer molecules are long, the exact nature of the end-groups has little effect on the physical and chemical properties of the material.

A different type of hydrogen-transfer step often has a much larger effect on the properties of the resulting polymer. Suppose that hydrogen-atom transfer occurs not from the monomer unit on the *end* of a second chain, but from a monomer unit in the *middle* of that chain (Fig. 25–2). Then the first chain stops growing but the radical site moves to the middle of the second chain, and growth resumes from that point, forming a *branched* polymeric chain with very different properties.

Addition polymerization can be initiated by ions as well as by free radicals. An example is the polymerization of acrylonitrile:

$$n\ CH_2{=}\underset{\underset{\displaystyle C{\equiv}N}{|}}{CH} \longrightarrow \left[CH_2\underset{\underset{\displaystyle C{\equiv}N}{|}}{CH}\right]_n$$

A suitable initiator for this process is butyl lithium $(CH_3CH_2CH_2CH_2)^-\ Li^+$. The butyl anion (abbreviated Bu^-) reacts with the end carbon atom in a molecule of acrylonitrile to give a new anion:

$$Bu^{\ominus}Li^{\oplus} + CH_2{=}\underset{\underset{\displaystyle C{\equiv}N}{|}}{CH} \longrightarrow Bu—CH_2—\underset{\underset{\displaystyle C{\equiv}N}{|}}{CH^{\ominus}Li^{\oplus}} \qquad \text{(inititation)}$$

The new anion then reacts with an additional molecule of acrylonitrile according to

$$Bu—CH_2—\underset{\underset{\displaystyle C{\equiv}N}{|}}{CH^{\ominus}Li^{\oplus}} + CH_2{=}\underset{\underset{\displaystyle C{\equiv}N}{|}}{CH} \longrightarrow Bu—CH_2—\underset{\underset{\displaystyle C{\equiv}N}{|}}{CH}—CH_2—\underset{\underset{\displaystyle C{\equiv}N}{|}}{CH^{\ominus}Li^{\oplus}}$$

(propagation)

The process continues, building up a long-chain polymer.

Ionic polymerization differs from free-radical polymerization because the negatively charged end-groups repel one another, ruling out the coupling of two active chains. The ionic group at the end of the growing polymer is stable at each stage. Once the supply of monomer has been used up, the polymer can exist indefinitely with its ionic end-group, in contrast with the free-radical case, in which some reaction must take place to terminate the process. Ion-initiated polymers are called "living polymers" because when additional monomer is added (even months later), they resume growth and increase in molecular mass. Termination can be achieved by adding water to replace the Li^+ with a hydrogen ion:

$$-(CH_2-CH)_n\!-CH_2-CH^{\ominus}Li^{\oplus} + H_2O \longrightarrow$$
$$\overset{|}{C\!\equiv\!N} \qquad \overset{|}{C\!\equiv\!N}$$

$$-(CH_2-CH)_n\!-CH_2-CH_2 + Li^{\oplus} + OH^{\ominus}$$
$$\overset{|}{C\!\equiv\!N} \qquad \overset{|}{C\!\equiv\!N} \qquad \text{(termination)}$$

Condensation Polymerization

• Condensation reactions appear in other contexts. For example, two molecules of sulfuric acid (H_2SO_4) condense to form disulfuric acid ($H_2S_2O_7$) (see Fig. 22–7), and a carboxylic acid condenses with an alcohol to form an ester (see Section 24–2).

A second important mechanism for polymerization is **condensation polymerization,** in which a small molecule (frequently water) is split off as each monomer unit is attached to the growing polymer. An example is the polymerization of 6-aminohexanoic acid. The first two molecules react upon heating according to

An amide linkage and water form in the reaction of an amine with a carboxylic acid. The new molecule still has an amine group on one end and a carboxylic acid group on the other end, so it can react with two more molecules of 6-aminohexanoic acid. The process repeats to build up a long-chain molecule. For each monomer unit added, one molecule of water is split off. The final polymer in this case is called "nylon-6." It is used in fiber-belted radial tires and in carpets.

Addition and condensation polymerization differ in the distribution of chain lengths over the course of the reaction (Fig. 25–3). In addition polymerization, reaction begins only at the initiator sites, so relatively few polymer chains are growing at any one time. If the reaction is interrupted short of completion, the result is a number of polymer molecules mixed with unreacted monomer. In condensation polymerization, every monomer is a potential initiator, and so dimers and trimers begin to form throughout the mixture almost immediately upon heating. If the reaction is stopped short of completion, the mixture contains a large number of "oligomers" (molecules with short polymer chains) and almost no monomer. If the reaction later resumes, the oligomers react with each other to give long-chain polymers.

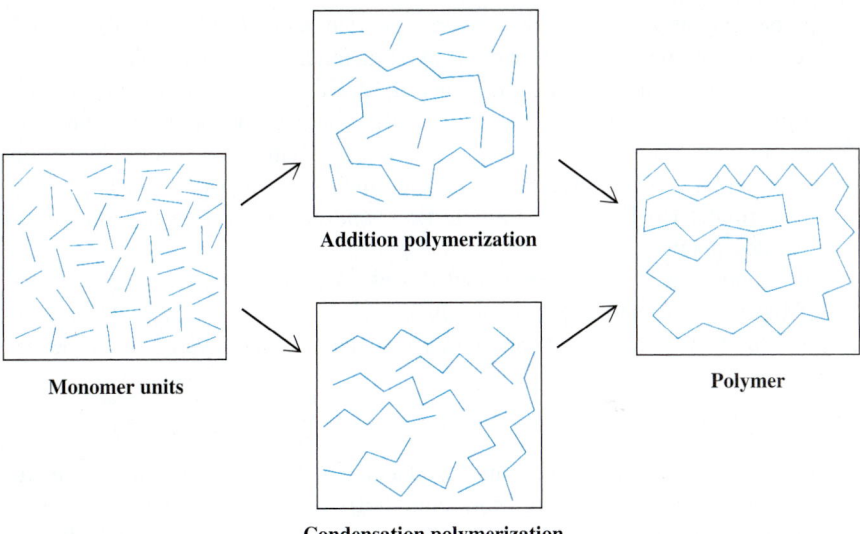

Addition polymerization

Monomer units

Condensation polymerization

Polymer

Figure 25–3 • The proportions of long- and short-chain molecules differ in addition and condensation polymerization throughout the course of the reaction, but similarly long chains are formed at completion.

Copolymers

A **copolymer** is a polymer formed from two or more different monomers (Fig. 25–4). Consider butadiene and styrene (vinyl benzene). Free-radical initiation causes each of these compounds to polymerize. The polymerization of styrene to polystyrene proceeds similarly to the polymerization of vinyl chloride outlined at the beginning of this section. Butadiene has two $C=C$ bonds per monomer, and its polymerization is different and interesting. The π electrons shift upon initiation and during propagation, causing the systematic relocation of double bonds:

$$R-O\cdot + CH_2=CH-CH=CH_2 \longrightarrow R-O-CH_2-CH=CH-CH_2\cdot$$

(initiation)

One π electron from each of the two double bonds has moved to the central $C-C$ bond, converting it into a double bond. A third π electron is used to make the $O-C$ bond to the initiator, and the fourth remains as an unpaired electron on the terminal $-CH_2$ group, ready to react further according to

$$R-O-CH_2-CH=CH-CH_2\cdot + CH_2=CH-CH=CH_2 \longrightarrow$$
$$R-O-CH_2-CH=CH-CH_2-CH_2-CH=CH-CH_2\cdot \quad \text{(propagation)}$$

In the polybutadiene that ultimately forms, $+CH_2-CH=CH-CH_2+_n$, all of the monomers' $C-C$ single bonds have been converted into double bonds, and all the double bonds have become single bonds.

Now suppose that polymerization is initiated in a *mixture* of styrene and butadiene. In this case, the growing polymer chain may encounter and react with either type of monomer. The result is a **random copolymer,** with butadiene (B) and styrene (S) units in an irregular sequence along the chain. A segment of this random copolymer might be symbolized as

$$-B-B-S-B-S-S-B-B-B-S-B-S-S-$$

Figure 25–4 • Lego and Duplo toy building bricks, which can themselves be clicked together, are made from a copolymer of acrylonitrile, butadiene, and styrene. *(Leon Lewandowski)*

The probability of attachment of a given type of monomer unit depends not only on its concentration in the mixture but also on its inherent reactivity. Thus, the fraction of B monomer units in a chain generally does not equal the fraction of B in the original mixture. Nonetheless, varying the composition of the mixture does change the ratio of the two monomer units in the polymer that results. For example, a 1:6 molar ratio of styrene to butadiene monomers is used to make styrene-butadiene rubber (SBR) for automobile tires, and a 2:1 ratio gives a copolymer used in latex paints.

A second type of copolymer is called a **block copolymer.** It consists of a series of blocks connected together, in which each block is a chain of one type of monomer unit. A typical block copolymer of butadiene and styrene might be represented as

$$-(\text{B}-\text{B}-\text{B}-\text{B}-\text{B}-\text{B}-\text{B})_m(\text{S}-\text{S}-\text{S}-\text{S}-\text{S}-\text{S}-\text{S})_n-$$

in which a large block of butadiene units and a large block of styrene units are linked to form the final polymer chain. More complicated block copolymers might have several alternating blocks. Such copolymers can be made by using the "living-polymer" process described earlier: polymerization in pure butadiene is initiated by butyl lithium and allowed to proceed until long chains are created and the butadiene is used up, after which some styrene is added and the chains grow further with styrene units. Block copolymers have very different properties from random ones. They also differ from simple mixtures of two polymers. A mixture of polystyrene and polybutadiene tends to separate into two phases, each rich in one of the polymers. This cannot occur in a copolymer because the polystyrene and polybutadiene portions of the chain are connected by a chemical bond.

A third type of copolymer is called a **graft copolymer.** It is made by polymerizing one component and then "grafting on" side-chains of the second to the chains of the first (Fig. 25–5). In high-impact polystyrenes, for example, a small amount of butadiene is polymerized first. Upon addition of styrene and a free-radical initiator, radical sites are formed along the polybutadiene chain and serve as starting points for polystyrene growth. The result is a polybutadiene chain onto which are grafted numerous polystyrene chains, just like branches on a tree limb in an orchard. The resulting plastic is significantly harder and more resistant to impact than ordinary polystyrene.

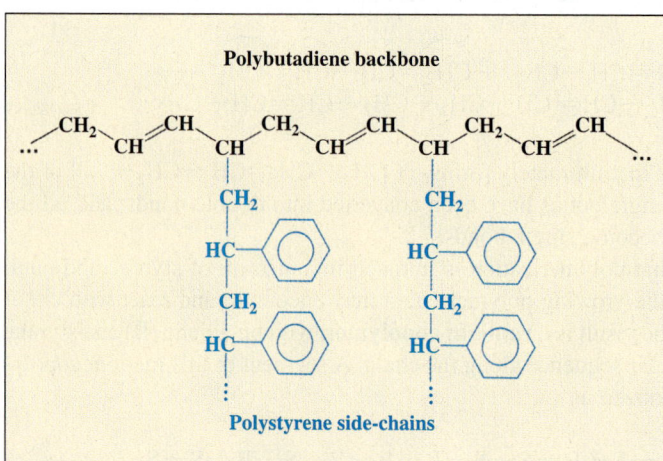

Figure 25–5 • The graft copolymer of butadiene with styrene shows excellent resistance to impact.

Cross-Linking

If every monomer forming a polymer has only two reactive sites, then only chains and rings can be made. The 6-aminohexanoic acid used in making nylon-6, for example, has one amine and one carboxylic acid group per molecule. When both functional groups react, one link is forged in the polymer chain, but that link cannot react further. If some or all of the monomers in a polymer have three or more reactive sites, however, then cross-linking is possible to form sheets or networks. This is often desirable in synthetic polymers because it leads to a stronger material.

• Note the analogy to the different solid-phase structures of the elements, among which the number of possible bonds per atom profoundly affects the nature of the solid (see Chapter 20).

One important example of cross-linking involves copolymers of phenol and formaldehyde (Fig. 25–6). When these two compounds are mixed with the phenol in excess (in the presence of an acid catalyst), straight chain polymers form. The first step is the addition of formaldehyde to phenol to give methylolphenol:

$$\text{(phenol)}\ \text{OH} + H_2C{=}O \longrightarrow \text{(ring)}\ CH_2OH,\ OH$$

Molecules of methylolphenol then condense (releasing water) to form a linear polymer called "novalac":

$$2n\ \left[\text{CH}_2\text{OH, OH ring}\right] \longrightarrow \left[\ \text{CH}_2 \text{ ring OH} \quad \text{CH}_2 \text{ ring OH}\ \right]_n + 2n\ H_2O$$

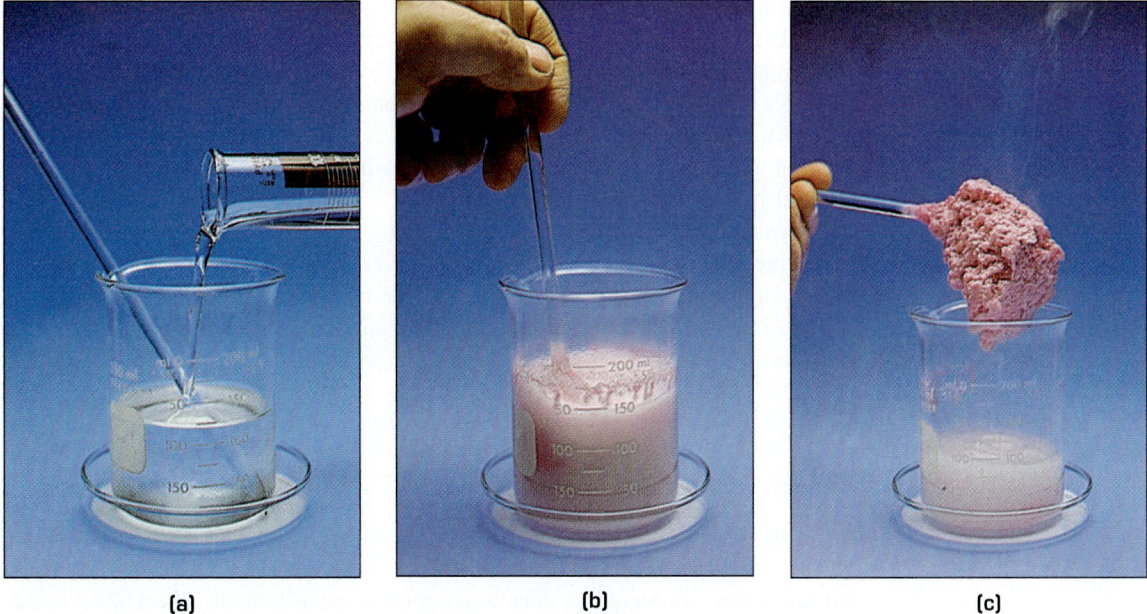

(a)   (b)   (c)

Figure 25–6 • When a mixture of phenol (C_6H_5OH) and formaldehyde (CH_2O) that is dissolved in acetic acid is treated with concentrated hydrochloric acid, a cross-linked phenol-formaldehyde polymer grows. (Charles D. Winters)

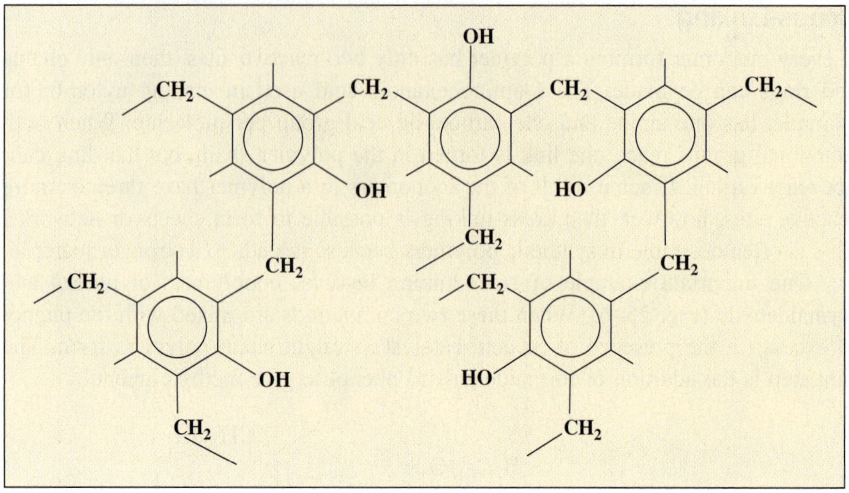

Figure 25–7 • Cross-linking in the
phenol-formaldehyde polymer.

Figure 25–7 • Cross-linking in the phenol-formaldehyde polymer.

If, on the other hand, the reaction is carried out with an excess of formaldehyde, then di- and trimethylolphenols form:

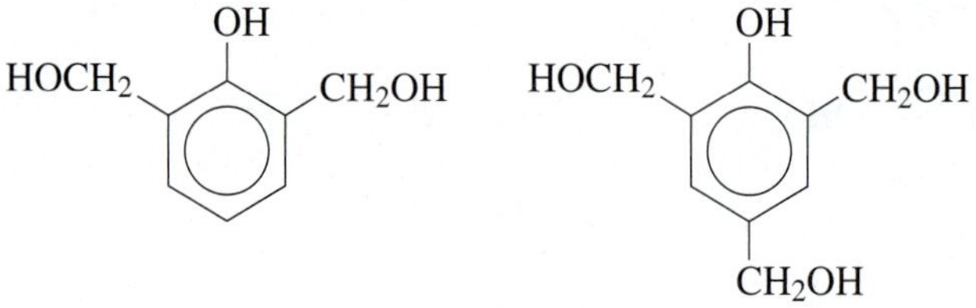

Each of these monomers has more than two reactive sites and can react with up to three others to form a cross-linked structure of the type shown in Figure 25–7. The cross-linked polymer is much stronger than the linear polymer. The very first synthetic plastic, Bakelite, was made in 1907 from cross-linked phenol and formaldehyde. Modern phenol–formaldehyde polymers are used extensively in the construction industry as adhesives for plywood, with more than 1 billion kg produced per year in the United States.

• One of the first uses of Bakelite was in the manufacture of billiard balls.

Cross-linking is often desirable because it leads to a stronger material. Sometimes cross-linking agents are added deliberately to form additional bonds between polymer chains. Polybutadiene, for example, contains double bonds that can be linked upon the addition of appropriate oxidizing agents. One especially important kind of cross-linking occurs through sulfur chains in rubber, as discussed in Section 25–3.

25-2   Biopolymers

All of the products of human ingenuity in the design of polymers pale beside the products of nature. Plants and animals employ a tremendous variety of long-chain molecules with different functions: some for structural strength, others to act as catalysts, and still others to provide instructions for the synthesis of vital components of the cell. In this section, we discuss three important classes of these biologically important polymers: proteins, carbohydrates, and nucleic acids.

Amino Acids

The monomeric building blocks of the important biopolymers called proteins are the α-amino acids. The simplest amino acid is glycine, which has the molecular structure shown in Figure 25–8. An amino acid, as indicated by the name, contains an amine group ($-NH_2$) and a carboxylic-acid group ($-COOH$). In α-amino acids, the two groups are bonded to the same carbon atom. Other types of amino acids exist, but we restrict our attention to α-amino acids. In acidic aqueous solution, the amine group of an amino acid gains a hydrogen ion to give an $-NH_3^+$ group; in basic solution, the carboxylic acid group loses a hydrogen ion to give a $-COO^-$ group. At intermediate pH, both reactions occur, that is, the hydrogen ion transfers from the carboxylic acid group to the amine group. The amino acid is then a *zwitterion* (a positive ion on one end and a negative ion on the other). The simple amino acid form shown in Figure 25–8 is rarely present.

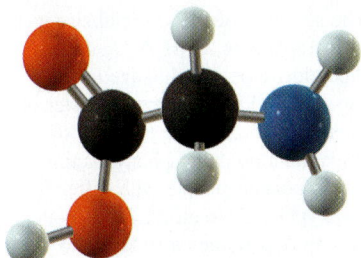

Figure 25–8 • The structure of glycine.

Two glycine molecules can condense with loss of water to form an amide:

$$\underset{\underset{OH}{|}}{\overset{\overset{O}{\|}}{C}}-CH_2-\underset{\overset{|}{H}}{N}\overset{H}{\diagup} \quad + \quad \underset{\underset{OH}{|}}{\overset{\overset{O}{\|}}{C}}-CH_2-NH_2 \quad \longrightarrow \quad \underset{\underset{OH}{|}}{\overset{\overset{O}{\|}}{C}}-CH_2-\underset{\overset{|}{H}}{N}-\overset{\overset{O}{\|}}{C}-CH_2-NH_2 + H_2O$$

The amide functional group connecting two amino acids is referred to as a **peptide linkage,** and the resulting molecule is called a *dipeptide*—in this case, diglycine. Because the two ends of the molecule still have carboxylic acid and amine groups, further condensation reactions to form a **polypeptide,** a polymer composed of many amino-acid groups, are possible. If glycine were the only amino acid available, the result would be polyglycine, a rather uninteresting protein. Of course, nature does not stop with glycine as a monomer unit. Any of 20 different α-amino acids are found in most natural polypeptides. In each of these, one of the two hydrogen atoms on the central carbon atom of glycine is replaced by another side-group.

- The $\Delta G°$ for the formation of the peptide linkage is positive; the biosynthesis of polypeptides requires Gibbs energy.

Alanine is the next simplest α-amino acid after glycine; it has a $-CH_3$ group in place of an $-H$ atom on the α-carbon. This substitution has a profound consequence. In alanine, four different groups are attached to the α-carbon: $-COOH$, $-NH_2$, $-CH_3$, and $-H$. Four different groups can be arranged in a tetrahedral structure around a central atom in two ways:

$$\underset{\underset{COOH}{|}}{\overset{\overset{CH_3}{|}}{H \blacktriangleright C \blacktriangleleft NH_2}} \qquad \underset{\underset{HOOC}{|}}{\overset{\overset{H_3C}{|}}{H_2N \blacktriangleright C \blacktriangleleft H}}$$

These two structures are mirror images (enantiomers) of each other. They cannot be interconverted without breaking and re-forming bonds, a very slow process. The above two forms of alanine are designated with the prefixes L and D for *levo* and *dextro* (from the Latin for "left" and "right," respectively). A carbon atom with four different groups attached to it is called a **chiral center.**

- Mirror-image isomers, or enantiomers, also occur in coordination complexes, as discussed in Section 19–2.

If a mixture of L- and D-alanine were caused to polymerize into chains of any great length, nearly every polymer molecule would have a different structure because the sequences of the D-alanine and L-alanine monomeric units would differ. To create polymers with definite, reproducible structures for particular roles, there is only one recourse: build all polypeptides from just one of the optical isomers. Nearly all naturally occurring α-amino acids are the L-form, and most earthly organisms have no use for D-α-amino acids in making polypeptides. Terrestrial life

The structure of the artificial sweetener Aspartame, $C_{14}H_{18}N_2O_5$. In this dipeptide, a peptide linkage unites aspartic acid with the methyl ester of phenylalanine. The metabolism of Aspartame produces phenylalanine, which poisons people who suffer from PKU (phenylketonuria). Foods containing Aspartame carry labels to warn phenylketonurics not to eat them.

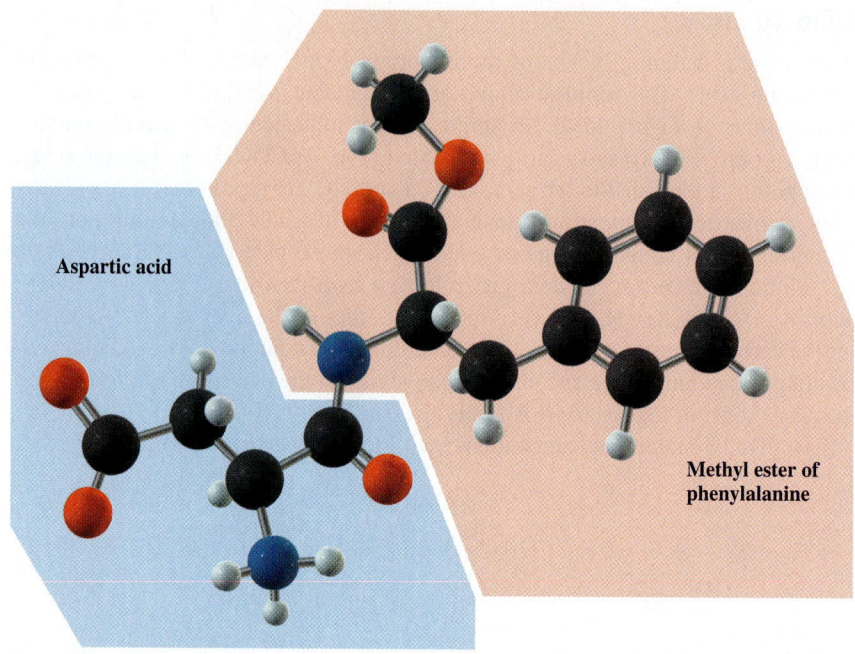

Aspartic acid

Methyl ester of phenylalanine

• The drug cyclosporin involves an exception. First isolated from an obscure fungus, it has saved thousands of transplanted organs from rejection. It contains a polypeptide ring with 11 amino-acid links, ten of which have the L-form, but one of which, an alanine, has the D-form.

presumably could go on equally well using mainly D-amino acids (all biomolecules would be mirror images of their present forms). The mechanism by which the established preference initially was selected is not known.

The —H group of glycine and the —CH₃ group of alanine give just two α-amino-acid building blocks. All 20 important α-amino acids are shown in Table 25–1 by listing their side-groups. Note the variety of their chemical and physical properties. Some side-groups are basic; others are acidic. Some are compact; others are bulky. Some can take part in hydrogen bonds; others can react readily with metal ions to form coordination complexes. These varied properties lead to even more variety in the polymers derived from the α-amino acids.

Proteins

A **protein** is a long polypeptide made by joining α-amino acids through peptide linkages. The term usually is applied to polymers having at least 50 amino-acid groups; large proteins may contain thousands of monomer units. Because any one of 20 α-amino acids may appear at each point in the chain, the number of possible sequences of amino acids in even small proteins is staggering. Moreover, the amino-acid sequence describes only the first aspect of the molecular structure of a protein. It contains no information about the three-dimensional conformation adopted by the protein. The $\diagup\!\!\!\!\diagdown C{=}O$ group and the $\diagup\!\!\!\!\diagdown N{-}H$ group in each amino acid along the protein chain are potential sites for hydrogen bonds. Hydrogen bonds also may involve functional groups on the amino-acid side-chains. Also, the cysteine side-groups (—CH₂—SH) can react with one another, with loss of hydrogen, to form —CH₂—S—S—CH₂— disulfide bridges between different cysteine groups in a single chain or between neighboring chains (the same kind of cross-linking by sulfur occurs in the vulcanization of rubber). As a result of these strong intrachain interactions, the molecules of a given protein have a rather well-defined conformation even in solution compared with the much more varied range of conformations

TABLE 25–1	α-Amino-Acid Side-Groups		
Name	**Symbol**	**Short Symbol**	**Structure of Side-Group**
Hydrogen side-group			
Glycine	Gly	G	—H
Alkyl side-groups			
Alanine	Ala	A	—CH$_3$
Valine	Val	V	—CH—CH$_3$ $\|$ CH$_3$
Leucine	Leu	L	—CH$_2$—CH—CH$_3$ $\|$ CH$_3$
Isoleucine	Ile	I	—CH—CH$_2$—CH$_3$ $\|$ CH$_3$
Proline	Pro	P	(structure of entire amino acid) CH$_2$ CH$_2$ CH$_2$ HN—CH COOH
Aromatic side-groups			
Phenylalanine	Phe	F	—CH$_2$—⬡
Tyrosine	Tyr	Y	—CH$_2$—⬡—OH
Tryptophan	Trp	W	—CH$_2$—C HC N H (indole ring)
Alcohol-containing side-groups			
Serine	Ser	S	—CH$_2$OH
Threonine	Thr	T	OH $\|$ —CH $\|$ CH$_3$
Basic side-groups			
Lysine	Lys	K	—CH$_2$CH$_2$CH$_2$CH$_2$NH$_2$
Arginine	Arg	R	—CH$_2$CH$_2$CH$_2$NH—C $\parallel$ NH NH$_2$
Histidine	His	H	—CH$_2$—C=C—H HN N CH

TABLE 25–1	Continued		
Name	**Symbol**	**Short Symbol**	**Structure of Side-Group**
Acidic side-groups			
Aspartic acid	Asp	O	$-CH_2COOH$
Glutamic acid	Glu	E	$-CH_2CH_2COOH$
Amide-containing side-groups			
Asparagine	Asn	N	$-CH_2\overset{\overset{\displaystyle O}{\|}}{C}-NH_2$
Glutamine	Gln	Q	$-CH_2CH_2\overset{\overset{\displaystyle O}{\|}}{C}-NH_2$
Sulfur-containing side-groups			
Cysteine	Cys	C	$-CH_2-SH$
Methionine	Met	M	$-CH_2CH_2-S-CH_3$

Figure 25-9 • The structure of silk. In silk, the amino-acid side-groups fit into a sheet-like structure. Here, all side-groups are shown as methyl ($-CH_3$) groups, but H and other small side-groups are possible.

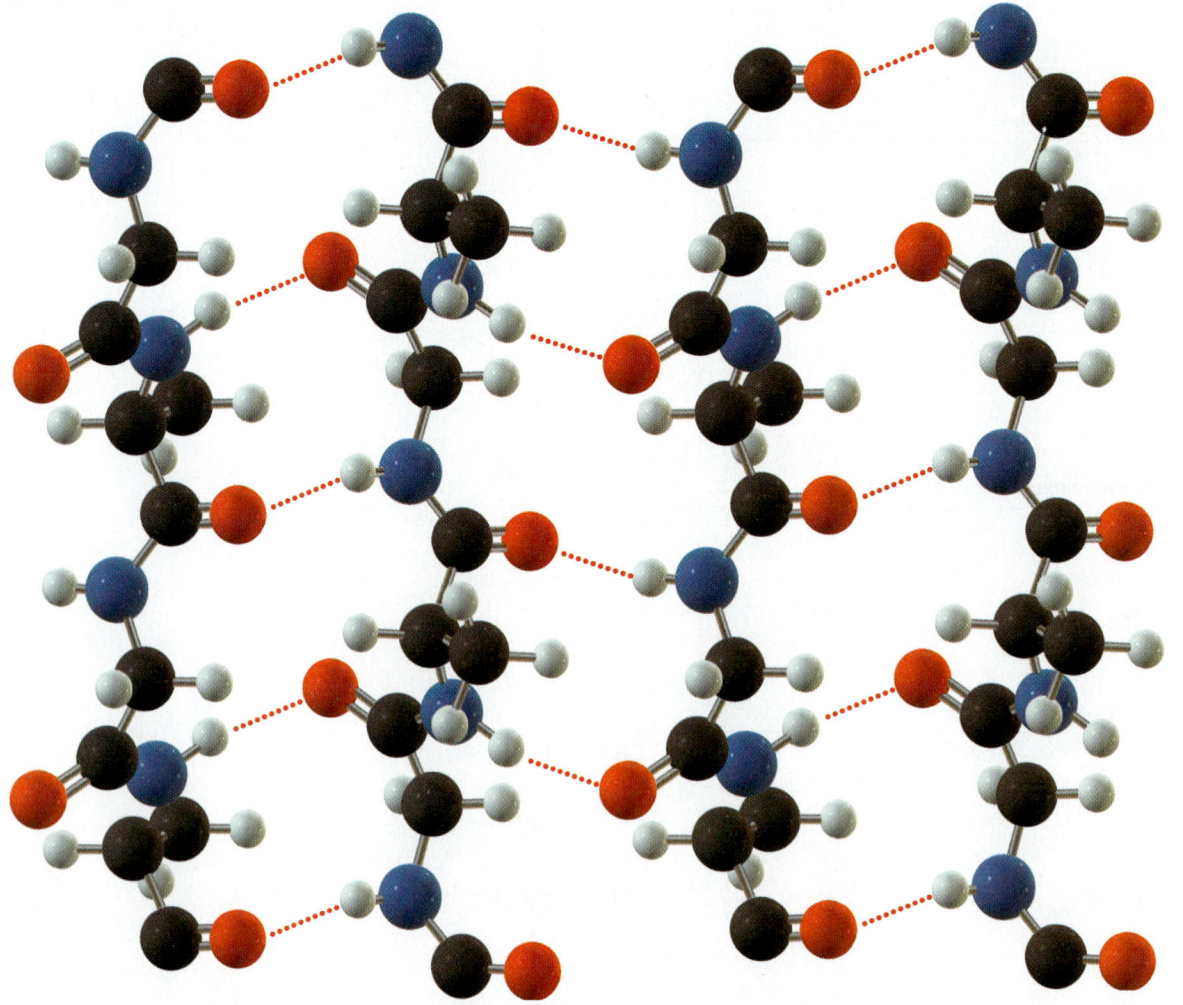

available to a simple alkane chain (see Fig. 24–4). The three-dimensional structures of many proteins have been determined by X-ray diffraction (see Section 20–1). This method relies on getting the protein to crystallize, something that is often hard to do. Nonetheless, these determinations are the main source of information about protein structure.

Proteins fall into two categories: fibrous and globular. **Fibrous proteins** usually are structural materials in nature. They consist of sheets of regularly linked polymer chains or else of long fibers comprising protein chains twisted into spirals. Silk is a fibrous protein in which the monomer units are primarily glycine and alanine, with smaller amounts of serine and tyrosine. The protein chains are cross-linked by hydrogen bonds to form sheet-like structures (Fig. 25–9) that are arranged so that the non-hydrogen side-groups all lie on one side of the sheet; the sheets then stack into layers. The relatively weak forces between sheets give silk its characteristic smooth feel. Wool and hair contain proteins that are composed predominantly of amino acids having side-chains that are larger, bulkier, and less regularly distributed than are those in silk. Instead of forming sheet structures, the protein molecules twist into a right-handed coil called an **α-helix** (Fig. 25–10). This is a spiral structure in which each $\diagdown$C$=$O group is hydrogen-bonded to the $\diagdown$N—H group of the fourth amino acid further along the chain; the bulky side-groups jut out from the helix so that they do not overlap and interfere with one another. In hair, three such α-helices are twisted in a *left*-handed coil to form a protofibril that is held together by sulfur bridges and hydrogen bonds. Many of these protofibrils are bound together in a hair cell.

The second type of protein is the **globular protein.** Globular proteins include the carriers of oxygen in the blood (hemoglobin) and in cells (myoglobin). They have irregular folded structures (Fig. 25–11) and typically consist of between 100 and 1000 amino-acid groups in one or more chains. Globular proteins commonly have parts of their structure in α-helices and sheets, with other portions in more disordered forms. Hydrocarbon side-groups tend to cluster in regions that exclude water, whereas charged and polar side-groups tend to remain in close contact with water. The sequence of amino-acid residues (structural units) has been worked out for many such proteins by cleaving them into smaller pieces and analyzing the structure of the fragments. Overlapping sequences are needed to obtain a complete picture, and the required analysis is quite a complex puzzle. It took Frederick Sanger ten years to complete the first such determination of sequence for the 51 amino acids in bovine (cattle-derived) insulin (Fig. 25–12). His persistence earned him the Nobel Prize in chemistry in 1958. Now, automated procedures allow for the rapid determination of amino-acid sequences in much longer protein molecules.

The ability of a protein to fulfill its biological function depends sensitively on amino-acid sequence. In sickle-cell anemia, the amino-acid sequence in hemoglobin suffers an apparently slight change: a valine (Val) replaces a glutamic acid (Glu) at one position in two of the chains represented by the colored ribbons in Figure 25–11. This change disrupts the packing of the chains and causes the red blood cells containing the hemoglobin to be brittle and bent (sickle-shaped) instead of supple and nearly spherical. Sufferers from the disease experience painful episodes in which their capillaries become clogged by broken red blood cells.

Enzymes constitute a very important class of globular proteins. They catalyze nearly every reaction in living cells, such as those synthesizing or breaking down other proteins, transporting substances across cell walls, or recognizing and resisting foreign bodies. Enzymes must have two characteristics: They must be effective in lowering the activation barrier for a reaction, and they must be selective so as to act only on a restricted group of substrates. As an example, consider the enzyme

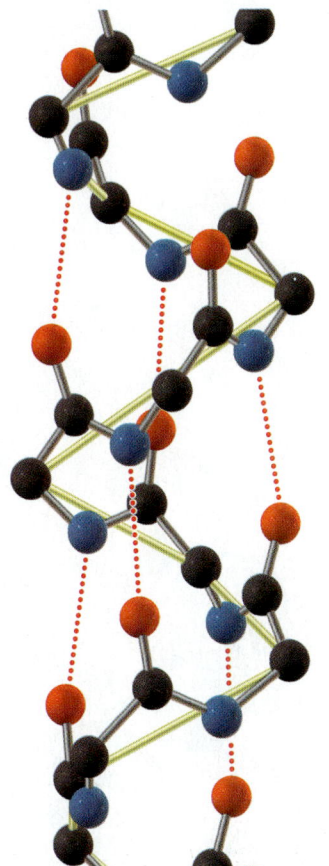

Figure 25–10 • The structure of the α-helix for a stretch of linked glycine monomer units. The superimposed yellow line highlights the helical structure, which is maintained by hydrogen bonds (*red dotted lines*). The hydrogen atoms themselves are omitted for clarity.

• Figure 14–19 gives a picture of enzyme-substrate binding and subsequent reaction.

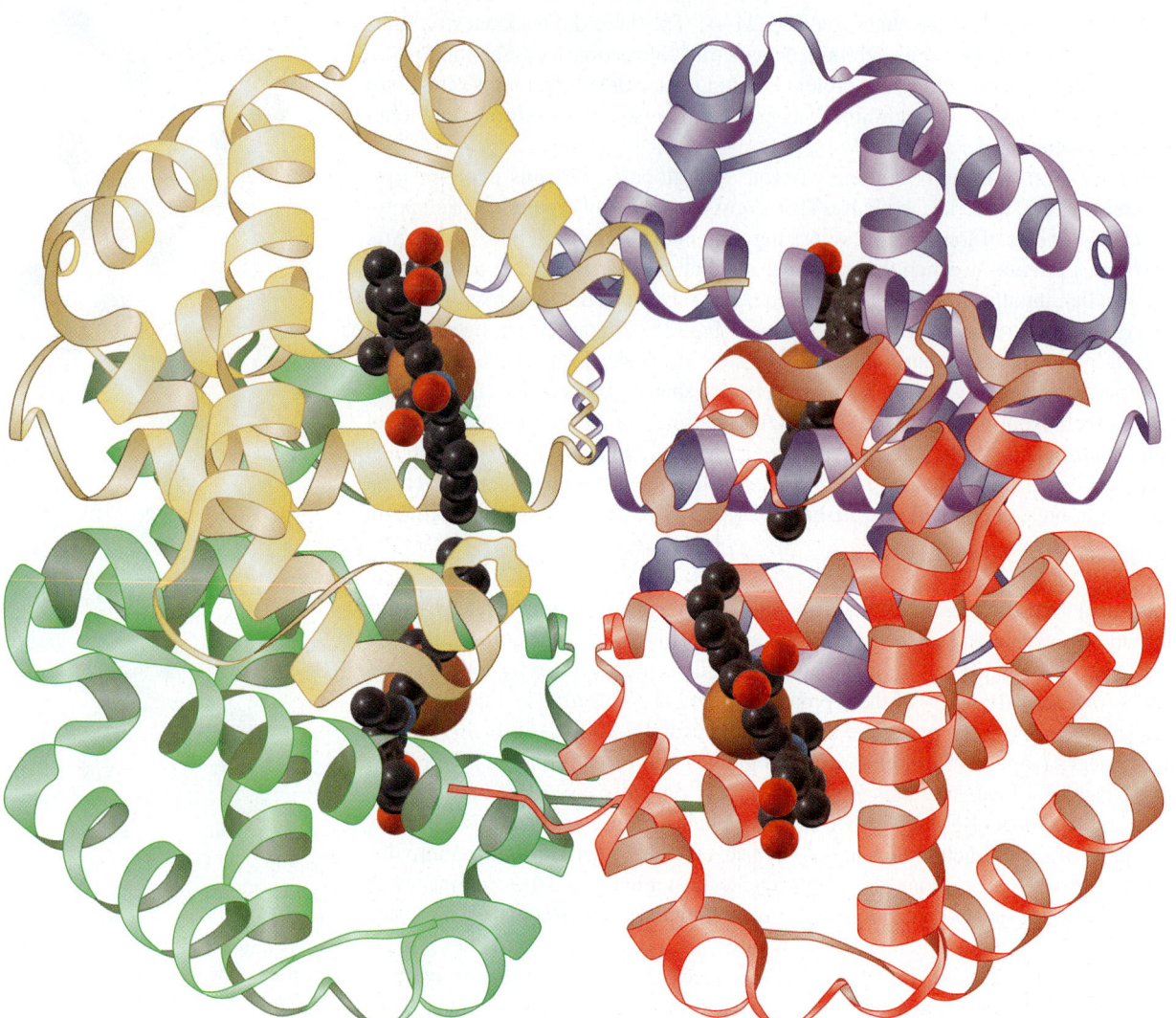

Figure 25–11 • A computer-generated model of the structure of hemoglobin. Much of this globular protein, represented by the colored ribbons, is coiled in stretches of an α-helix. The four heme groups have their central iron atoms enlarged for visibility. Oxygen binds at these atoms during oxygen transport by hemoglobin.

"carboxypeptidase A," the structure of which has been determined by X-ray diffraction. It acts to remove one amino acid at a time from the carboxylic-acid end of a polypeptide. Figure 25–13 shows the structure of the active site (with a peptide chain in place, ready to be cleaved). A special feature of this enzyme is the role played by the zinc ion, which is coordinated to two histidine residues in the enzyme and to a carboxylate group on a nearby glutamic-acid residue. The zinc ion, which has a positive charge, acts to remove electrons from the carbonyl group of the peptide linkage, making it more positive and thereby more susceptible to attack by water or by the carboxyl group of a second glutamic-acid residue. The side-chain on the outer amino acid of the peptide being cleaved is located in a hydrophobic cavity, which favors large aromatic or branched side-chains (such as that in tyrosine)

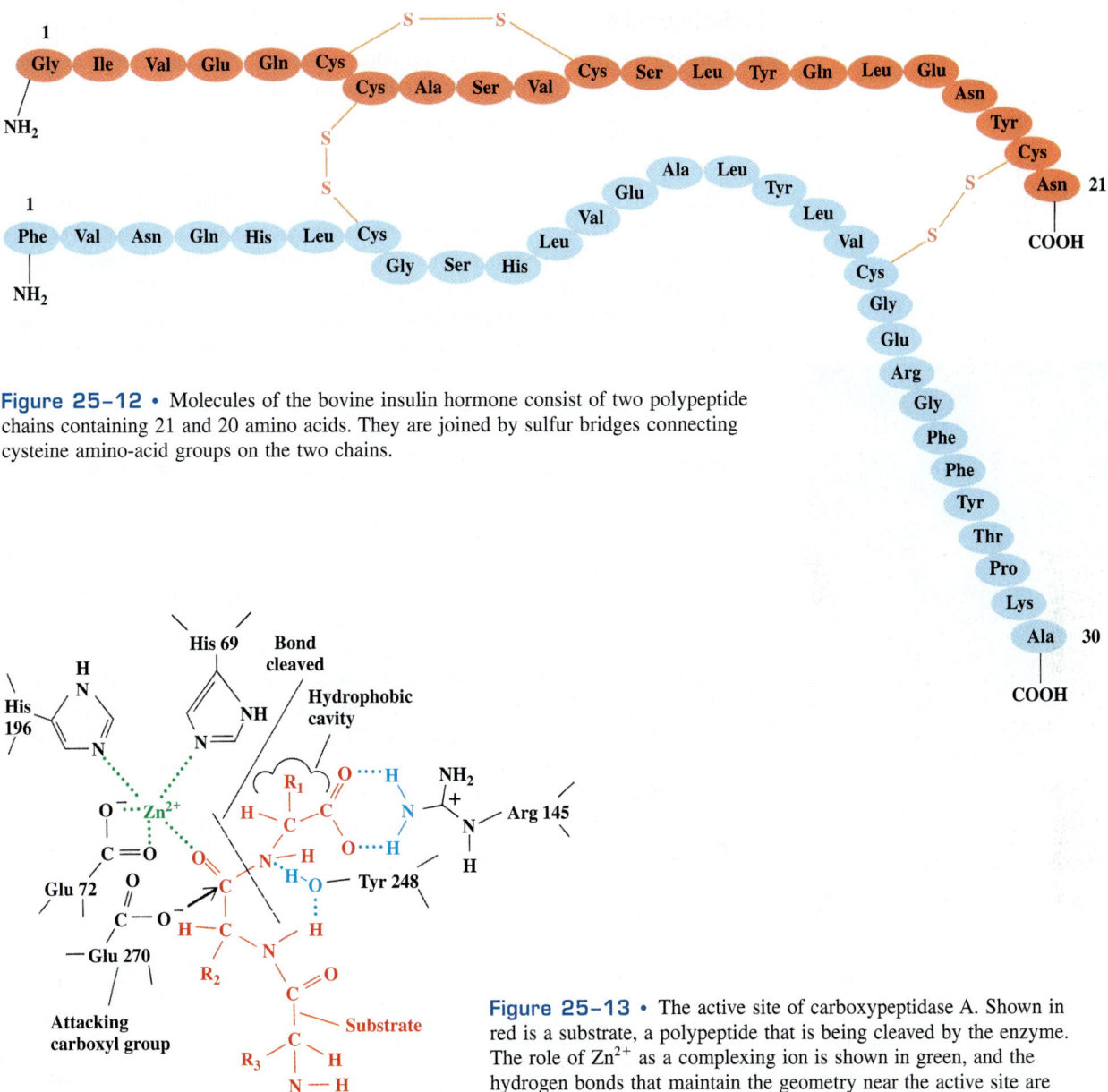

Figure 25-12 • Molecules of the bovine insulin hormone consist of two polypeptide chains containing 21 and 20 amino acids. They are joined by sulfur bridges connecting cysteine amino-acid groups on the two chains.

Figure 25-13 • The active site of carboxypeptidase A. Shown in red is a substrate, a polypeptide that is being cleaved by the enzyme. The role of Zn^{2+} as a complexing ion is shown in green, and the hydrogen bonds that maintain the geometry near the active site are shown in blue.

over smaller hydrophilic side-chains (such as that in aspartic acid). Carboxypeptidase A thus shows selectivity in the rates with which it cleaves peptide chains.

The "molecular engineering" that lies behind nature's design of carboxypeptidase A and other enzymes is truly remarkable. The amino-acid residues that form the active site and determine its catalytic properties are *not* adjacent to one another in the protein chain. As indicated by the numbers after the residues in Figure 25–13, the two glutamic-acid residues are the 72nd and 270th amino acids along the chain. The enzyme adopts a conformation in which the key residues, distant from one another in terms of chain position, are nonetheless quite close in their positions in three-dimensional space, allowing the enzyme to carry out its specialized function.

Carbohydrates

Carbohydrates are important natural products, most of which are produced photo-synthetically by green plants, according to the overall equation

$$n\ CO_2(g) + m\ H_2O(\ell) \longrightarrow C_n(H_2O)_m(s) + n\ O_2(g)$$

where n and m are whole numbers. The general formula $C_n(H_2O)_m$ suggests a hydrate of carbon and explains the name. Carbohydrates include both polymers, such as starch and cellulose, and the non-polymeric compounds that condense (with loss of water) to form these polymers. Non-polymeric carbohydrates are called sugars, or **saccharides.** Sugars are often edible, with a sweet taste (Fig. 25–14). Familiar examples include the food components cane sugar, fruit sugar, and milk sugar, which are named sucrose, fructose, and lactose, respectively (the ending for sugar names is -*ose*).

Monosaccharides (general formula: $C_nH_{2n}O_n$) are the simplest sugars. They contain either an aldehyde $-C\!\!\underset{H}{\overset{O}{\lozenge}}$ group attached to a chain of carbon atoms, each of which has an $-OH$ group attached to it, or a ketone $\diagdown\!C\!=\!O$ group somewhere in the middle of such a chain and are called **aldoses** or **ketoses,** depending on which group they contain. Monosaccharides with three, four, five, and six carbon atoms are called trioses, tetroses, pentoses, and hexoses, respectively.

D-Glucose ($C_6H_{12}O_6$), the most important monosaccharide, is an **aldohexose** (a six-carbon sugar containing an aldehyde group). Glucose exists in solution as an equilibrium mixture of the three structural forms shown in Figure 25–15. In aldohexoses, four of the carbon atoms (those numbered 2–5 in Figure 25–15b) each have four different groups attached. They are chiral centers. A switch in chirality at one of these carbon atoms corresponds to switching the $-H$ and $-OH$ attached to that carbon. Two different chiralities at four different carbon atoms means that 2^4, or 16, different isomeric aldohexoses exist. All 16 are known. D-Glucose is the

• Carbohydrates, however, do not contain water molecules as such.

Figure 25–14 • Sweets are carbohydrates. *(Charles D. Winters)*

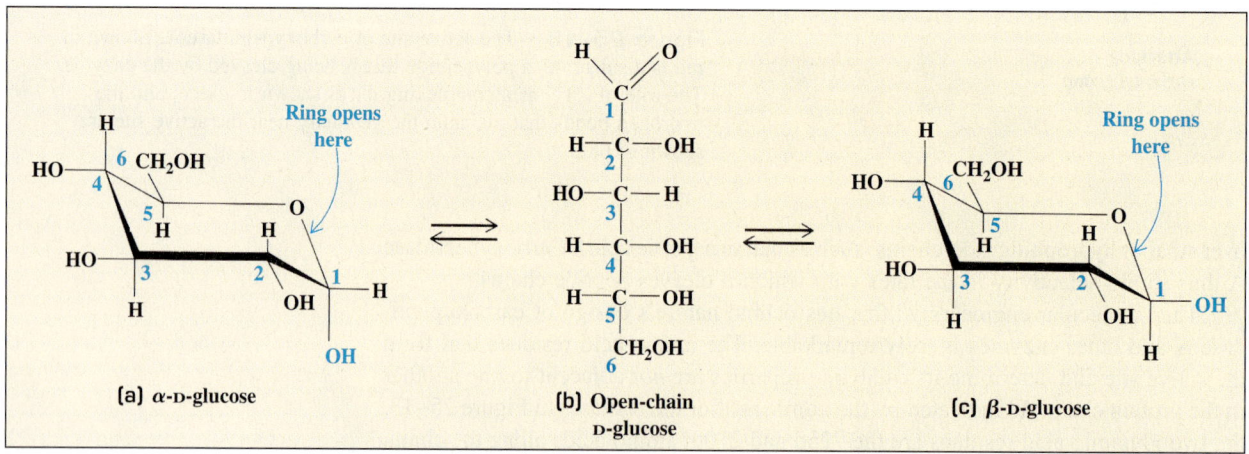

Figure 25–15 • In solution, D-glucose exists in two ring forms (a and c), which interconvert via an open-chain form (b). The two rings differ in the placement of the $-OH$ and $-H$ groups of carbon atom 1.

aldohexose that has exactly the chiralities at carbons 2 through 5 that are indicated in Figure 25–15. Its enantiomer (mirror-image isomer), L-glucose, which has the —H and —OH groups switched at carbons 2 through 5, is not found in nature. The other 14 aldohexoses arise with the different combinations of chirality at carbons 2 through 5. They have different names, such as D- and L-galactose, D- and L-mannose, and D- and L-sorbose.

Why does D-glucose have multiple forms in solution? The aldehyde group in straight-chain D-glucose (and other aldohexoses) reacts readily with the —OH group on carbon 5 to create a ring consisting of five carbon atoms and one oxygen atom and bearing one —CH₂OH and four —OH side-groups (Fig. 25–15a and c). The reaction creates two cyclic forms because the ring closure occurs in two ways. In one, the new —OH group (shown in blue in Fig. 25–15a and c) projects *down* from the ring. In the other, the new —OH group projects *out away* from the ring. Another way to understand this is to note that closing the ring creates a fifth chiral carbon atom in D-glucose. The two ring forms are called α-D-glucose and β-D-glucose. In aqueous solution, they interconvert rapidly via the open-chain form and cannot be separated. They can be isolated separately in crystalline form, however.

The monosaccharide D-ribose ($C_5H_{10}O_5$) is an aldopentose. The aldehyde group in D-ribose readily reacts with the —OH groups on carbon atom 4 to complete a ring, in this case, a five-atom ring, as shown in Figure 25–16a. Cyclic D-ribose is part of the genetically important biopolymer ribonucleic acid (RNA). The related compound D-deoxyribose ($C_5H_{10}O_4$) derives from D-ribose by the substitution of an —H for the —OH on carbon 2. It forms a five-atom ring by the same reaction as D-ribose (Fig. 25–16b). The resulting cyclic D-deoxyribose is an essential part of deoxyribonucleic acid (DNA), the famous biopolymer that stores genetic information.

D-Fructose, another important sugar, has the same molecular formula as D-glucose but is a ketohexose rather than an aldohexose. It has its C=O group at carbon atom 2 rather than carbon atom 1. In solution, the straight-chain form of fructose is in equilibrium with both a five-atom ring and a six-atom ring, both of which have α and β forms (Fig. 25–17). In all forms of this and other simple sugars, the exposed —OH groups interact strongly with water through hydrogen bonds, with the result that the sugars are quite soluble in water.

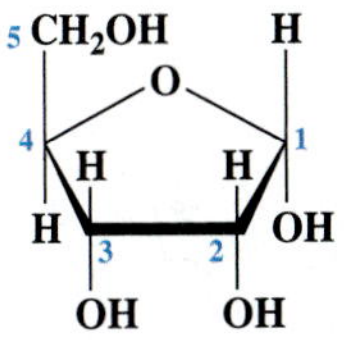

• L-sorbose is important commercially as the precursor of vitamin C (ascorbic acid).

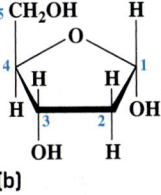

(a)

(b)

Figure 25–16 • The structures of the cyclic forms of (a) D-ribose and (b) D-deoxyribose.

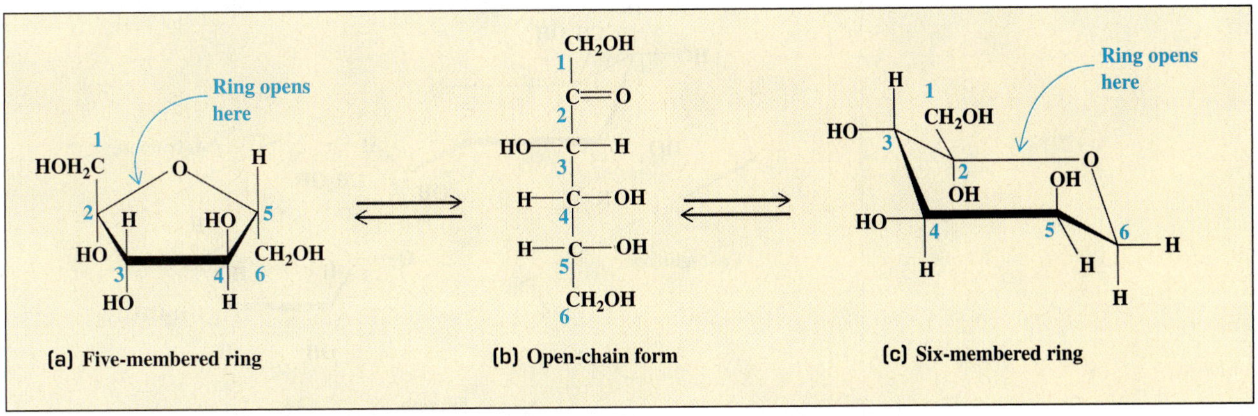

(a) Five-membered ring

(b) Open-chain form

(c) Six-membered ring

Figure 25–17 • The ketohexose sugar D-fructose exists in solution as an equilibrium mixture of five forms: a five-atom β ring, an open chain, a six-atom β ring, a five-atom α ring in which the —CH₂OH and —OH on carbon 2 are exchanged from what is shown here, and a six-atom α ring, in which the —CH₂OH and —OH on carbon 2 are similarly exchanged.

Green plants continue the process of photosynthesis past the stage of the simple sugars by linking monosaccharides together as monomer units to form more complex carbohydrates. **Disaccharides** are composed of two simple sugars linked together in a condensation reaction, with the elimination of water. Figure 25–18 shows the structures of lactose ($C_{12}H_{22}O_{11}$), formed by condensing β-D-galactose with α-D-glucose, and sucrose ($C_{12}H_{22}O_{11}$), formed by condensing α-D-glucose with β-D-fructose. These two disaccharides have the same molecular formula, but differ in digestibility and taste. Indeed sugars do differ in their sweetness. Fructose is sweeter than sucrose, which is sweeter than glucose. It follows that fructose is a lower-calorie natural sugar because less of it gives the same sweet taste. Sucrose solutions can be treated with an enzyme called *invertase,* which hydrolyzes the bond between the two rings and leaves a mixture of fructose and glucose, called *invert sugar,* which is sweeter than the original sucrose because of its fructose content.

Linking together a large number of saccharide units produces polymers called **polysaccharides.** The position of the oxygen atom linking the monomer units ex-

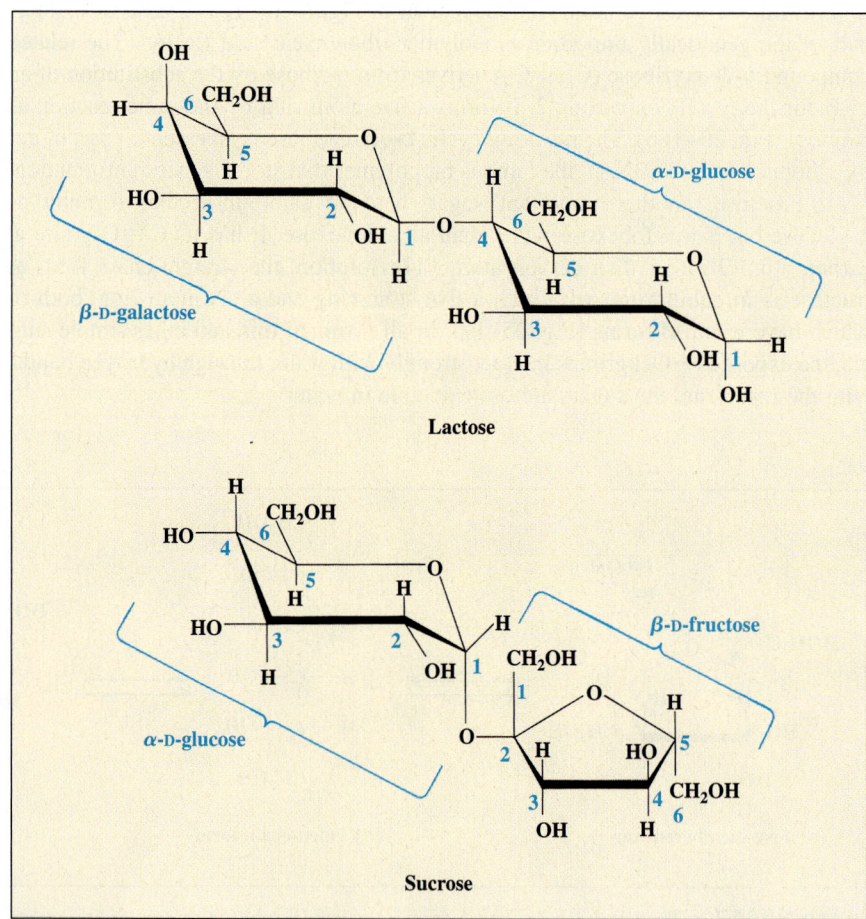

Figure 25–18 • Two disaccharides, showing how each derives from monosaccharide building blocks.

erts a profound effect on the properties and function of these biopolymers. Starch (Fig. 25–19a) is a polymer of α-D-glucose and is metabolized by humans and animals. Cellulose (Fig. 25–19b), a polymer of β-D-glucose, is digested only by certain bacteria that live in the digestive tracts of goats, cows, and other ruminants, and in some insects, such as termites. Cellulose forms the structural fiber of trees and plants and is present in linen, cotton, and paper. It is the world's most abundant organic compound. Despite the great solubility of D-glucose, the monomer unit in starch and cellulose, neither polymer is soluble in water.

• Thus, cows can digest grass, but humans cannot.

Nucleic Acids

As described earlier, proteins are copolymers made up typically of 20 types of monomer unit. Simply mixing the amino acids and dehydrating them to form polymer chains at random would never lead to the particular structures needed by living cells. How does the cell preserve information about the amino-acid sequences that make up its proteins, and how does it transmit this information to daughter cells when it reproduces? These questions lie in the field of molecular genetics, an area in which chemistry is of central importance.

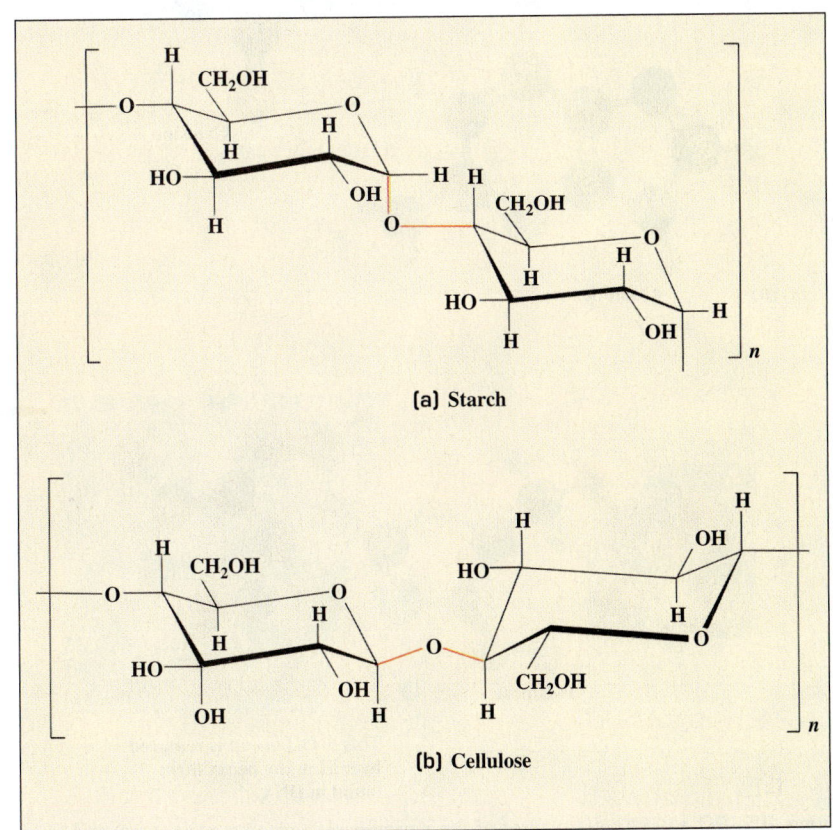

(a) Starch

(b) Cellulose

Figure 25–19 • Both (a) starch and (b) cellulose are polymers of glucose. In starch all the cyclic glucose units are α-D-glucose, but in cellulose, all the monomer units are β-D-glucose.

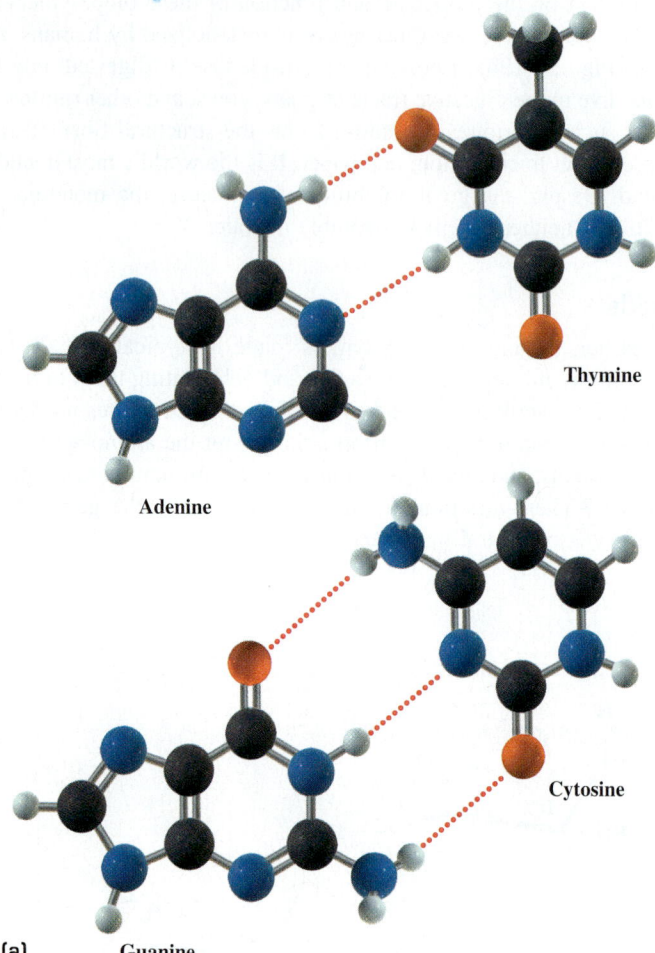

Thymine

Adenine

Cytosine

(a)   **Guanine**

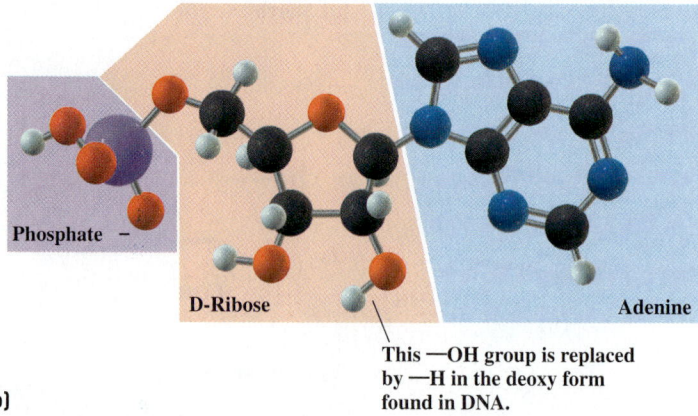

Phosphate   −

D-Ribose

Adenine

This —OH group is replaced
by —H in the deoxy form
found in DNA.

(b)

Figure 25-20 • (a) The structures of the purine bases (adenine or guanine) and pyrimidine bases (thymine and cytosine). Hydrogen bonding between pairs of bases is indicated by red dots. (b) The structure of the nucleotide adenosine monophosphate (AMP).

The primary genetic material is DNA. This biopolymer is made up of four types of monomer units called **nucleotides.** Each nucleotide is constructed of three parts:

1. One molecule of a pyrimidine or purine base. The four bases are thymine, cytosine, adenine, and guanine (Fig. 25–20a).
2. One molecule of the aldopentose D-deoxyribose ($C_5H_{10}O_4$).
3. One molecule of phosphoric acid (H_3PO_4).

The cyclic D-deoxyribose molecule links the base to the phosphate group, undergoing two condensation reactions with loss of water, one at the —OH on carbon 3 and the other at the —OH on carbon 5 (Fig. 25–16b), to form the nucleotide (Fig. 25–20b). The first clue to unlocking the structure of DNA was the observation that, although the proportions of the four bases in DNA from different organisms are quite variable, the chemical amount of cytosine (C) is always approximately equal to that of guanine (G), and the chemical amount of adenine (A) is always approximately equal to that of thymine (T). This suggested some type of "base-pairing" in DNA that could lead to association of C with G and A with T. The second crucial observation was an X-ray diffraction study by Rosalind Franklin and Maurice Wilkins that suggested the presence of helical structures of more than one chain in DNA.

These two pieces of information were put together by James Watson and Francis Crick in their famous 1953 proposal of a double-helix structure for DNA. They concluded that DNA consists of two interacting helical strands of nucleic-acid polymer (Fig. 25–21), with each cytosine on one strand linked through hydrogen bonds to a guanine on the other, and each adenine to a thymine. This accounted for the observed molar ratios of the bases, and it also provided a model for the replication of the molecule, which is crucial for passing on information during the reproductive process. One DNA strand serves as a template upon which a second DNA strand is synthesized. A DNA molecule replicates by starting to unwind at one end. As it does so, new nucleotides are guided into position opposite the proper bases on each of the two strands. If the nucleotide does not fit the template, it cannot link to the polymeric strand under construction. The result of the polymer synthesis is two double-helix molecules, each containing one strand from the original and one new strand, identical to the original in every respect.

Information is encoded in DNA in the order of the base pairs. Research since 1953 has succeeded in breaking this genetic code and establishing the connection between the base sequence in a segment of DNA and the amino-acid sequence of the protein synthesized according to the directions in that segment. The code in the nucleic acids consists of consecutive, non-overlapping triplets of bases, with each triplet standing for a particular amino acid. Thus, a polymeric nucleic-acid strand containing only the side-group cytosine was found to give a polypeptide of pure proline, meaning that the triplet CCC codes for proline. The nucleic-acid strand AGAGAGAGA . . . is read as the alternating triplets AGA and GAG and gives a polypeptide consisting of alternating arginine (coded by AGA) and glutamic acid (coded by GAG) monomer units. There are 64 (= 4^3) possible triplets, so more than one code exists for most of the 20 amino acids. Some triplets serve as signals to terminate a polypeptide chain. Remarkably, the genetic code appears to be universal, independent of the particular species of plant or animal, a finding that suggests a common origin of all terrestrial life.

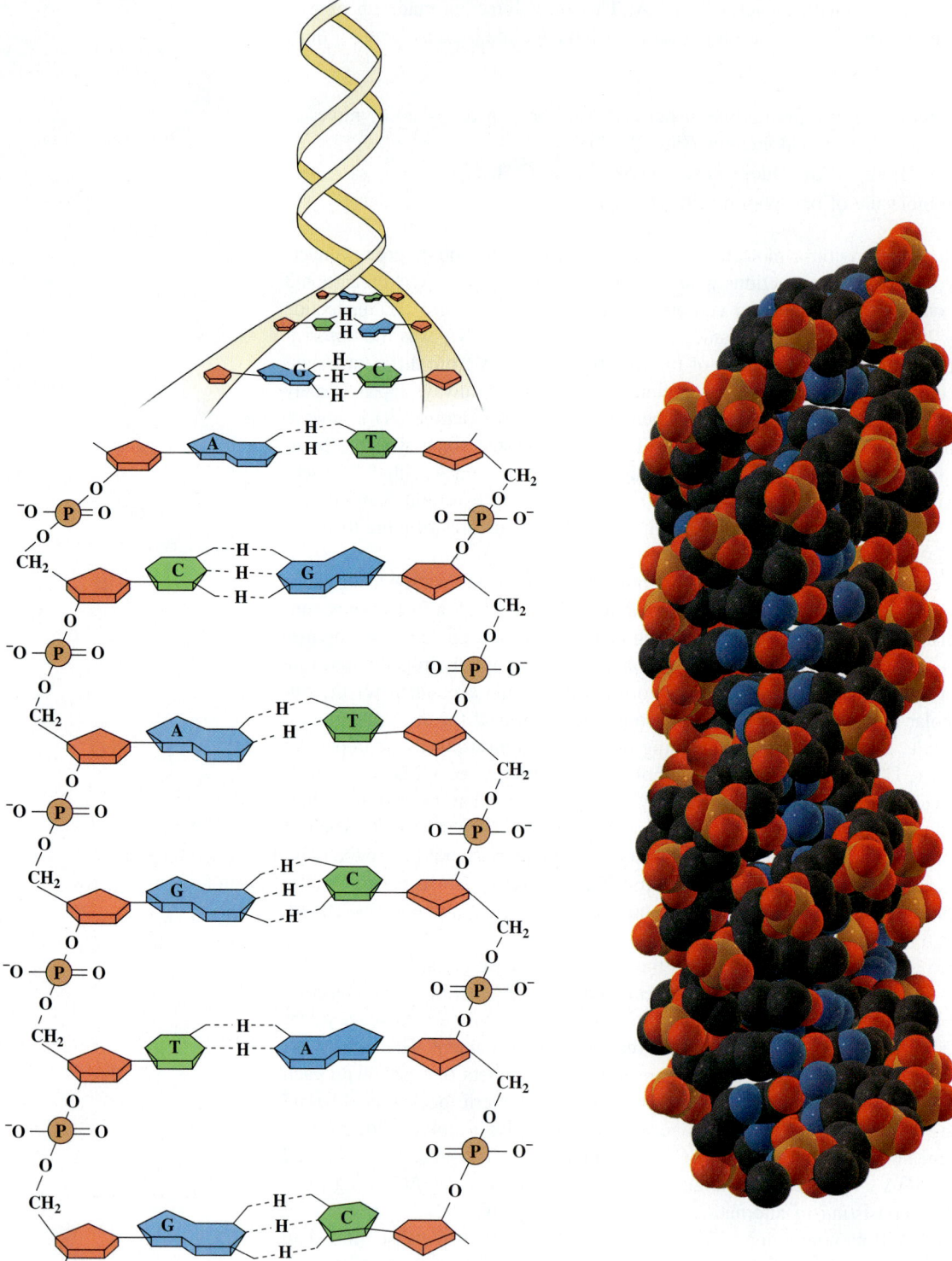

Figure 25–21 • The double helix structure of DNA.

25-3   Uses for Polymers

If food is excluded, the three largest uses for polymers are in fibers, plastics, and elastomers (rubbers). These three types of materials are distinguished on the basis of their physical properties, especially their resistance to stretching. A typical fiber strongly resists stretching and elongates by less than 10% before breaking. Plastics are intermediate in their resistance to stretching and elongate 20% to 100% before breaking. Elastomers stretch readily, with elongations of 100% to 1000% (that is, some rubbers can be stretched by a factor of ten without breaking). In this section, we examine the major kinds of polymers and their uses.

Fibers

Many important fibers, including cotton and wool, are biopolymers. The first commercially successful synthetic polymers were not made by polymerization reactions, but through the chemical transformation of the natural polymer cellulose (Fig. 25–19b). The origins of the modern synthetic fiber industry are said to lie in an accident in the laboratory of the German chemist Christian Schonbein, who in 1845 wiped up a spill of nitric and sulfuric acids with a cotton apron, hung up the apron to dry, and thereby produced the polymer cellulose trinitrate (guncotton). In guncotton, all three —OH groups in the glucose subunits of the cellulose (Fig. 25–19b) are replaced by —ONO$_2$ nitrate groups. The compound is an explosive (see Section 22–5); it subsequently was developed into cordite and continues in use today as a propellant in guns and rockets.

By varying the amounts of acid used, one can prepare a second nitrated derivative of cellulose. In nitrocellulose (cellulose dinitrate), only the two —OH groups directly attached to each ring are nitrated. When this compound is dissolved in a mixture of camphor and alcohol and the solvent is evaporated carefully, the product is celluloid, which was used for photographic film in the early motion-picture industry. The nitrate groups in celluloid make it quite flammable, as might be expected from a polymer closely related to guncotton. This fact, combined with the high price of camphor, led to the use of cellulose acetate instead of celluloid in photographic film. In cellulose acetate, all three —OH groups on the glucose rings are esterified by treatment with a mixture of acetic acid and sulfuric acid to

form —O—C side-groups.

The discovery of nitrocellulose also led to the production of the first synthetic fiber. In 1884, the French chemist Hilaire Bernigaud, Count of Chardonnet, showed how to remove the nitrate groups in nitrocellulose and to spin the resulting reconstituted cellulose into fibers. Modest commercial success for the resulting "Chardonnet silk" synthetic fiber ensued until the development of the alternative and cheaper **viscose rayon** some ten years later. In the viscose-rayon process (Fig. 25–22), which is still used today, cellulose is digested in a concentrated solution of NaOH to convert the —OH groups into —O$^-$Na$^+$ ionic groups. Reaction with CS$_2$ leads to the formation of about one "xanthate" group for every two glucose monomer units:

Filter paper (cellulose) dissolves in a concentrated ammonia solution containing [Cu(NH$_3$)$_4$]$^{2+}$ ions. When the solution is extruded into aqueous sulfuric acid, a dark-blue thread of rayon (regenerated cellulose) precipitates. (*Leon Lewandowski*)

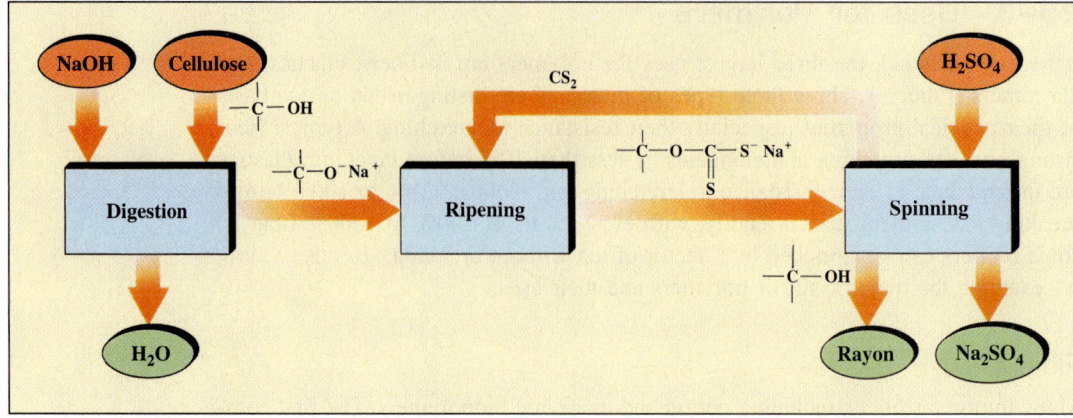

Figure 25–22 • The viscose-rayon process. The chemical state of the —OH group is shown at each stage. Sodium sulfate is a commercially significant by-product of the process.

Such substitutions weaken the hydrogen-bond forces holding polymer chains together. In the ripening step, some of these xanthate groups are removed with regeneration of CS_2, and others migrate to the —CH_2OH groups from the ring —OH groups. After this has taken place, sulfuric acid is added to neutralize the NaOH and to remove the remaining xanthate groups. At the same time, the viscose rayon is spun out to form fibers. The chemical composition of the completed rayon is very close to that of the cellulose that began the process, but the polymer molecules are stretched out and aligned in fibers.

• If the —OH groups are esterified with acetic acid to make "acetate" fiber, the reduced interchain hydrogen-bond interaction and the bulk of the acetate groups cause the fiber to be softer, but also weaker than rayon.

Rayon is a semi-synthetic fiber because it is prepared from a natural polymeric starting material. The first true synthetic polymeric fiber was nylon, which was developed in the 1930s by Wallace Carothers at the Du Pont company. He knew of the condensation of an amine with a carboxylic acid to form an amide linkage (see Section 24–2) and noted that if each molecule had *two* amine or carboxylic acid functional groups, then long-chain polymers could be formed. The specific starting materials upon which Carothers settled after a number of attempts were adipic acid and hexamethylenediamine:

$$\underset{\text{adipic acid}}{HO-\overset{\overset{\displaystyle O}{\|}}{C}-(CH_2)_4-\overset{\overset{\displaystyle O}{\|}}{C}-OH} \qquad \underset{\text{hexamethylenediamine}}{H_2N-(CH_2)_6-NH_2}$$

The two condense with loss of water according to the equation

$$HO-\overset{\overset{\displaystyle O}{\|}}{C}-(CH_2)_4-\overset{\overset{\displaystyle O}{\|}}{C}-\boxed{OH + H}-\underset{\overset{\displaystyle |}{H}}{N}-(CH_2)_6-NH_2 \longrightarrow$$

$$HO-\overset{\overset{\displaystyle O}{\|}}{C}-(CH_2)_4-\overset{\overset{\displaystyle O}{\|}}{C}-\underset{\overset{\displaystyle |}{H}}{N}-(CH_2)_6-NH_2 + H_2O$$

The resulting molecule has a carboxylic-acid group on one end (which can react with another molecule of hexamethylenediamine) and an amine group on the other end (which can react with another molecule of adipic acid). The process can continue indefinitely, leading to a polymer with the formula

$$\left[\begin{matrix} \overset{\displaystyle O}{\underset{\displaystyle \parallel}{C}}-(CH_2)_4-\overset{\displaystyle O}{\underset{\displaystyle \parallel}{C}}-\underset{\displaystyle H}{N}-(CH_2)_6-\underset{\displaystyle H}{N} \end{matrix}\right]_n$$

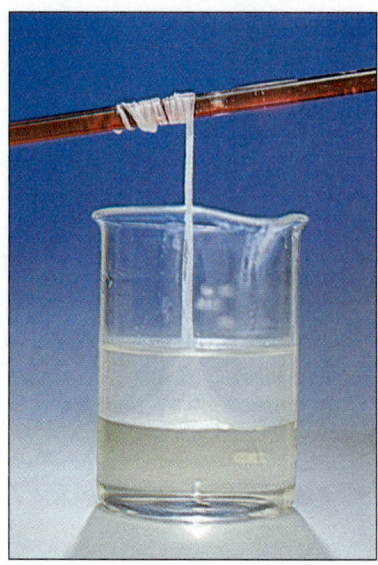

called nylon-66 (Fig. 25–23). The nylon is extruded as a thread or spun as a fiber from the melt. The combination of well-aligned polymer molecules and N—H ⋯ O hydrogen bonds between chains makes nylon one of the strongest materials known.

The designation "66" indicates that this nylon has six carbon atoms on the starting carboxylic acid and six on the diamine. Other nylons can be made with different numbers of carbon atoms. After nylon-66, the most important is called "nylon-6" and can be made from the polymerization of 6-aminohexanoic acid, as outlined in Section 25–1:

$$n\ HO-\overset{\displaystyle O}{\underset{\displaystyle \parallel}{C}}-(CH_2)_5-\underset{\displaystyle H}{\overset{\displaystyle H}{N}} \longrightarrow \left[-\overset{\displaystyle O}{\underset{\displaystyle \parallel}{C}}-(CH_2)_5-\underset{\displaystyle H}{N}-\right]_n + n\ H_2O$$

Figure 25–23 • Hexamethylenediamine is dissolved in water (*lower layer*) and adipoyl chloride, a derivative of adipic acid, is dissolved in hexane (*upper layer*). At the interface between the layers, nylon forms; the nylon can be drawn out and wound up on a stirring rod. (*Leon Lewandowski*)

This synthesis uses a monomer in which a single molecule has both a carboxylic acid and an amine group. Such a molecule can "bite its own tail" to form a cyclic molecule with a loss of water:

$$\underset{\displaystyle HO}{\overset{\displaystyle O}{\diagdown}}C-(CH_2)_5-NH_2 \longrightarrow \begin{matrix} H_2C & \overset{\displaystyle O}{\diagup}C-N\overset{\displaystyle H}{\diagdown} & CH_2 \\ H_2C & & CH_2 \\ & CH_2 \end{matrix} + H_2C$$

This compound, called "caprolactam," is the normal starting material in the production of nylon-6, which is a little less expensive but not quite as strong as nylon-66. The bonding in these nylons is just like the bonding in naturally occurring polypeptides and proteins. Polyglycine could in fact be referred to as "nylon-2," a simple polyamide in which each repeating unit contains two carbon atoms.

Just as a carboxylic acid reacts with an amine to give an amide, so it reacts with an alcohol to give an ester, which suggests the possible reaction of a dicarboxylic acid and a glycol (dialcohol) to form a polymer. The polymer produced most extensively in this way is polyethylene terephthalate, which is built up from terephthalic acid (a benzene ring with —COOH groups on either end) and ethylene glycol. The first two molecules react according to

A room full of nylon bobbins for use in making fire hoses. (*Tom Carroll*)

Strong, lightweight Mylar covered the wings of the *Gossamer Albatross,* which succeeded in a flight across the English Channel powered solely by a single man. *(Corbis–Bettmann)*

$$\underset{\text{terephthalic acid}}{\text{HO}-\overset{\displaystyle O}{\underset{}{C}}-\bigcirc-\overset{\displaystyle O}{\underset{}{C}}-OH} + \underset{\text{ethylene glycol}}{HO-CH_2-CH_2-OH} \longrightarrow$$

$$HO-\overset{\displaystyle O}{\underset{}{C}}-\bigcirc-\overset{\displaystyle O}{\underset{}{C}}-O-CH_2-CH_2-OH \quad + H_2O$$

Further reaction then builds up the polymer, which is referred to as "polyester" and sold under the trade name Dacron. The planar benzene rings in this polymer make it stiffer than nylon, which has no aromatic groups in its backbone, and help to make polyester fabrics crush-resistant. The same polymer, when formed in a thin sheet rather than a fiber, is known as Mylar, a very strong film used for audio- and video-tapes.

A final class of polymer fibers is the acrylics, which are built up from the free-radical polymerization of acrylonitrile:

$$n\;CH_2{=}\underset{\underset{\textstyle C{\equiv}N}{|}}{CH} \longrightarrow \left[CH_2-\underset{\underset{\textstyle C{\equiv}N}{|}}{CH}\right]_n$$

The resulting polymer then is dissolved and spun into fibers as the solvent evaporates. Pure polyacrylonitrile has an inconveniently high melting point and cannot be dyed, so most acrylics are copolymers with vinyl acetate, vinyl chloride, styrene, or other monomers. The presence of chlorine atoms in the copolymer reduces the flammability of the fabric.

Table 25–2 summarizes the structures, properties, and uses of some important fibers.

TABLE 25–2	Fibers		
Name	**Structural Units**	**Properties**	**Samples Uses**
Rayon	Regenerated cellulose	Absorbent, soft, easy to dye, poor wash and wear	Dresses, suits, coats, curtains, blankets
Acetate	Acetylated cellulose	Fast drying, supple, shrink-resistant	Dresses, shirts, draperies, upholstery
Nylon	Polyamide	Strong, lustrous, easy to wash, smooth, resilient	Carpeting, uphol-stery, tents, sails, hosiery, stretch fabrics, rope
Dacron	Polyester	Strong, easy to dye, shrink-resistant	Permanent-press fabrics, rope, sails, thread
Acrylic (Orlon)	$\left[CH_2-\underset{\underset{\textstyle C{\equiv}N}{|}}{CH}\right]_n$	Warm, lightweight, resilient, quick-drying	Carpeting, sweaters, baby clothes, socks

CHEMISTRY IN PROGRESS

Paint

Paint contains three major components: pigment, binder, and thinner. The pigment gives opaqueness and color to the paint and is frequently an inorganic solid such as cadmium yellow (CdS) or chrome green (Cr_2O_3) that is ground up into small particles before being dispersed in the paint. The thinner is a volatile solvent that is essential to spread the paint but then mostly evaporates. The binder serves to bring the pigment into the thinner; upon evaporation of the thinner, the binder cross-links to provide mechanical strength to the paint film.

Most binders in current use are polymeric. One broad class of binders are the "alkyd resins," which are polyesters with unsaturated hydrocarbon side-chains. As the thinner evaporates, these side-chains cross-link. Alkyd paints are quite versatile and can be formulated with flat, semigloss, and high-gloss finishes. Other binders are used in latex paints, in which water is the thinner. Copolymers of styrene and butadiene (in a 2:1 ratio) are low in cost and were the first binders to be used in water-thinned paints. Acrylic paints, which are more expensive but tougher and more durable, also use water as the thinner (Fig. 25–A). Binders in acrylic paints are copolymers that contain methylmethacrylate monomer units.

Figure 25–A • Acrylic paint. *(Comstock)*

Plastics

Plastics are loosely defined as polymeric materials that can be molded or extruded into desired shapes and that harden upon cooling or solvent evaporation. Rather than being spun into threads in which the molecules are aligned, as in fibers, plastics are cast into three-dimensional forms or spread into films for packaging applications. Although celluloid articles were fabricated by plastic processing by the late 1800s, the first important synthetic plastic was Bakelite, the cross-linked phenol–formaldehyde resin discussed in Section 25–1. Table 25–3 lists some of the most important organic plastics, along with some properties and uses. Also given are the symbols that appear on plastic food and beverage containers to aid the separation of the materials according to chemical constitution during the recycling process.

• Under this broad definition, ordinary glass, discussed in Section 21–4, is a plastic as well. This is quite correct. *Plastic* means "capable of being molded."

Ethylene ($CH_2{=}CH_2$) forms polyethylene through a free-radical–initiated addition polymerization mechanism at high pressures (1000 to 3000 atm) and temperatures (300°C to 500°C), conditions that require elaborate equipment (Fig. 25–24). The product is not the perfect linear chain implied by the simple equation

$$n\,CH_2{=}CH_2 \longrightarrow {+}CH_2CH_2{]}_n$$

Free radicals commonly abstract hydrogen from the middles of chains in this synthesis. Consequently the polyethylene is heavily branched with hydrocarbon side-chains of varying lengths. It is called **low-density polyethylene** (LDPE) because the difficulty of packing the irregular side-chains gives it a lower density than that of perfectly linear polyethylene. This irregularity also makes LDPE relatively soft, so its primary uses are in coatings, plastic packaging, trash bags, and squeeze bottles in which softness is an advantage, not a drawback.

TABLE 25–3	Plastics			
Name	**Structural Units**	**Properties**	**Sample Uses**	**Recycling Code**
Polyethylene terephthalate	$-(OCH_2CH_2O-\overset{O}{\overset{\|}{C}}-\underset{benzene}{\bigcirc}-\overset{O}{\overset{\|}{C}})_n$	High tensile strength, tear resistance	Clothing fiber, tire cord, plastic film, soft-drink bottles	♳ 1 PETE
Polyethylene	$-(CH_2-CH_2)_n$	High density; hard, strong, stiff	Molded containers, lids, toys, pipe, milk bottles	♴ 2 HDPE
		Low density; soft, flexible, clear	Packaging, trash bags, squeeze bottles	♶ 4 LDPE
Polyvinyl chloride	$-(CH_2-\underset{Cl}{CH})_n$	Nonflammable, resistant to chemicals	Water pipes, roofing, credit cards, phonograph records	♵ 3 PVC
Polypropylene	$-(CH_2-\underset{CH_3}{CH})_n$	Stiffer, harder than high-density polyethylene, higher melting point	Containers, lids, carpeting, luggage, rope	♷ 5 PP
Polystyrene	$-(CH_2-\underset{\bigcirc}{CH})_n$	Brittle, flammable, not resistant to chemicals, easy to process and dye	Furniture, toys, refrigerator linings, insulation	♸ 6 PS
Phenolics	Phenol-formaldehyde copolymer	Resistant to heat, water, chemicals	Plywood adhesive, fiberglass binder, circuit boards	♹ 7 OTHER

Figure 25–24 • A 720-million-lb-per-year polyethylene plant in Louisiana.

A major breakthrough occurred in 1954, when the German chemist Karl Ziegler showed that ethylene also could be polymerized with a catalyst consisting of $TiCl_4$ and an organoaluminum compound (for example, $Al(C_2H_5)_3$). The addition of ethylene takes place at each stage within the coordination sphere of the titanium atom, so that monomer units can add only at the end of the growing chain. The result is linear polyethylene, also called **high-density polyethylene** (HDPE). With such regular linear chains HDPE contains large crystalline regions at low temperatures, making it much harder than LDPE and suitable for molding into bowls, lids, and toys.

A third kind of polyethylene introduced in the late 1970s is called **linear low-density polyethylene** (LLDPE). It is made by the same metal-catalyzed reactions as HDPE, but it is a deliberate copolymer with other 1-alkenes such as 1-butene. It has some side-groups, which reduce the crystallinity and the density, but these are of a controlled short length, as opposed to the irregular, long side-branches in LDPE. LLDPE is stronger and more rigid than is LDPE; it is also less expensive because lower pressures and temperatures are used in its manufacture.

If one of the hydrogen atoms of the ethylene monomer unit is replaced with a different type of atom or functional group, the plastics that form upon polymerization have different properties. Substitution of a methyl group (that is, the use of propylene as monomer) leads to polypropylene:

$$\left[CH_2 - \underset{\underset{CH_3}{|}}{CH} \right]_n$$

The possibility of different relative placements of the methyl groups leads to three distinct forms of polypropylene (Fig. 25–25). In the **isotactic** form, all the methyl groups are arranged on the same side, whereas in the **syndiotactic** form, they alternate in a regular fashion. The **atactic** form shows a random positioning of methyl groups. Polymerization of propylene cannot be carried out successfully by a simple free-radical reaction. It was first performed in 1953–1954 by Ziegler and the Italian

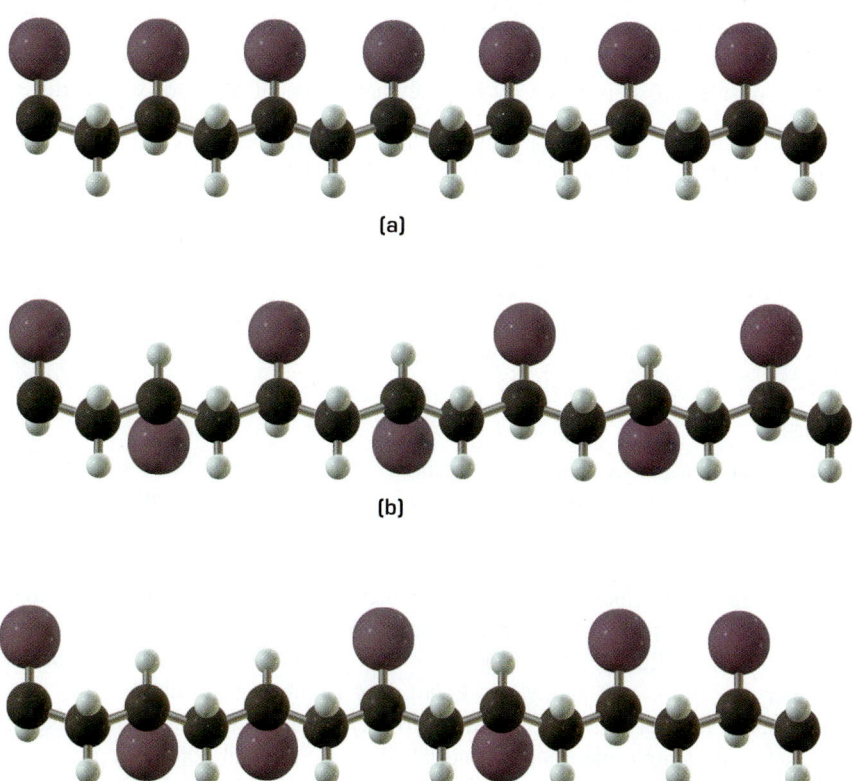

(a)

(b)

(c)

Figure 25–25 • The structures of (a) isotactic, (b) syndiotactic, and (c) atactic polypropylene. In these structures, the purple spheres represent —CH$_3$ (methyl) side groups on the long chain.

• Ziegler and Natta were awarded a Nobel Prize in 1963 for their development of polymerization catalysts.

chemist Giulio Natta, who used the Ziegler catalyst employed in making HDPE. Natta showed that the Ziegler catalyst led to isotactic polypropylene, and he developed another catalyst using VCl_4 that gave the syndiotactic form. Polypropylene plastic is stiffer and harder than HDPE and has a higher melting point, so it is particularly useful in applications in which somewhat higher temperatures are used (such as in the sterilization of medical instruments).

Substituting a chlorine atom for one of the hydrogen atoms in ethylene gives vinyl chloride, which polymerizes to polyvinyl chloride (PVC), according to the equation

$$n\ CH_2{=}CHCl \longrightarrow \left[CH_2{-}\underset{\underset{Cl}{|}}{CH}\right]_n$$

• Water plasticizes proteins, which are natural polymers, by masking some of the hydrogen bonds that would otherwise form between the long-chain molecules. A moist cake of soybean curd (a vegetable protein) has a jelly-like consistency. Thorough drying shrinks it and makes it hard and tough.

The largest uses of PVC are where a rigid and chemically resistant material is needed, for example, in pipes and architectural products such as gutters, downspouts, and siding. **Plasticizers** commonly are added to PVC to make it less rigid for other uses. A plasticizer is an additive that softens a plastic and makes it more flexible. In a rigid, amorphous plastic, non-bonded attractions between the tangled molecules of the polymer define a honeycomb-like structure. A plasticizer, which typically has small molecules, intervenes to mask some of these attractions. The result is fewer points of attachment between polymer chains and a reduction in the rigidity of the structure.

In polystyrene, a benzene ring replaces one hydrogen atom in the polyethylene molecule. Because of the great bulk of such a ring, atactic polystyrene does not crystallize to any significant extent. The most familiar use of this polymer is in polystyrene foam, which is used in disposable containers for food and drinks and as insulation. A volatile liquid or a compound that dissociates to gaseous products on heating is added to the molten polystyrene. It forms bubbles that remain as the polymer is cooled and molded. The large number of gas-filled pockets in the final product make it a good thermal insulator. Polyurethane is another polymer that is fabricated as a foam insulation. It is made by the polymerization of a glycol with a diisocyanate, as in the reaction

$$n\ HO{-}CH_2{-}CH_2{-}OH + n\ O{=}C{=}N{-}(CH_2)_m{-}N{=}C{=}O \longrightarrow$$

$$\left[O{-}CH_2{-}CH_2{-}O{-}\overset{\overset{O}{\|}}{C}{-}\underset{\underset{H}{|}}{N}{-}(CH_2)_m{-}\overset{\overset{H}{|}}{N}{-}\underset{\underset{O}{\|}}{C}\right]_n$$

Rubber

An **elastomer** is a plastic that can be deformed to a large extent and still recover its original shape when the deforming stress is removed. The term **rubber** was introduced by Joseph Priestley, who observed that such materials could be used to rub out pencil marks. Natural rubber, the specific elastomer Priestley used, occurs in over 200 plant species (including the common dandelion). The only important source is the tropical rubber tree (*Hevea braziliensis*). When its bark is cut, this tree exudes latex, a milky fluid that contains about 35% natural rubber. Natural rubber is a polymer of **isoprene** (2-methylbutadiene). The isoprene molecule contains two double bonds, and polymerization removes only one of these, so natural rubber is unsaturated, containing one double bond per isoprene unit. In polymeric isoprene, the geometry at each double bond can be either *cis* or *trans* (Fig. 25–26). Natural

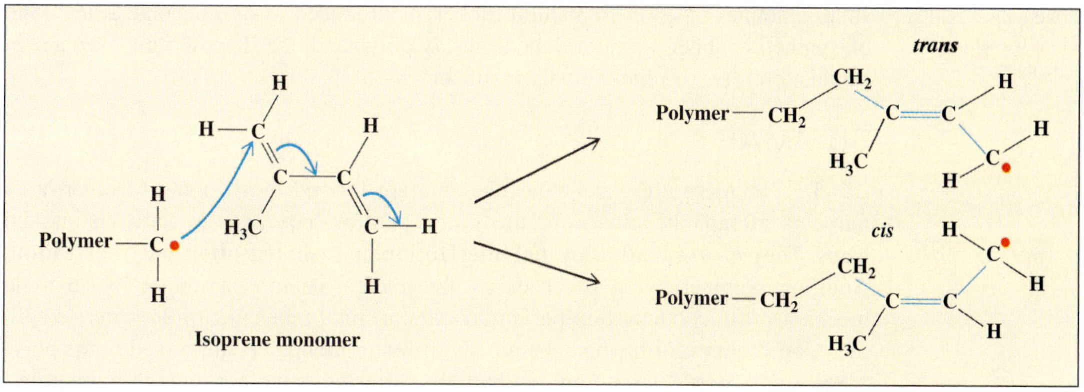

Figure 25-26 • In the polymerization of isoprene, *cis* or *trans* configurations can result at each double bond in the polymer. For the monomer conformation shown, a *trans* linkage appears most likely to result. Rotation about the central C—C bond in isoprene favors the *cis* linkage. The blue arrows show the redistribution of the electrons upon bond formation.

rubber is all-*cis* polyisoprene. The all-*trans* form also occurs in nature in the sap of certain trees; it is called *gutta-percha*. This material is used to cover golf balls because it is particularly tough. Isoprene can be polymerized by free-radical addition polymerization, but the resulting polymer contains a mixture of *cis* and *trans* double bonds and is useless as an elastomer.

Pure natural rubber has limited usefulness because it melts, is soft, and does not spring back fully to its original form after being stretched. In 1839, the American inventor Charles Goodyear discovered that adding sulfur to natural rubber and heating the mixture made the rubber harder, more resilient, and non-melting. Goodyear's process, **vulcanization,** involves the formation of sulfur bridges between the methyl side-groups on different chains. Small amounts of sulfur ($< 5\%$) yield an elastic material in which sulfur links between chains remain after stretching and enable the rubber to regain its original form when the stretching force is removed. Large amounts give the very hard, non-elastic material ebonite.

Research on synthetic substitutes for natural rubber began in the United States and Europe before World War II. An accidental discovery at the Du Pont company led to the production of neoprene, a rubbery polymer formed from monomeric chloroprene (2-chloro-1,3-butadiene), in which the methyl group on isoprene is replaced by chlorine. Further attention focused on copolymers of styrene with butadiene (now called SBR) and copolymers of acrylonitrile with butadiene (NBR). The Japanese occupation of the rubber-producing countries of Southeast Asia during World War II sharply curtailed the supply of natural rubber to the Allied nations, and rapid steps were taken to increase the production of synthetic rubbers. The initial production goal was 40,000 t yr^{-1} of SBR. By 1945, U.S. production had reached a nearly incredible 600,000 t yr^{-1}. During those few years, many advances were made in production techniques, quantitative analysis, and basic understanding of rubber elasticity. Styrene–butadiene rubber production continued after the war, and in 1950, SBR exceeded natural rubber in overall production volume for the first time. More recently, several factors have favored natural rubber: increasing cost of the hydrocarbon feed stock for synthetic rubber, gains in productivity of natural rubber, and the preference for radial-belted tires, which use more natural rubber.

The development of the Ziegler–Natta catalysts had an effect on rubber production as well. First, it facilitated the synthesis of all-*cis* polyisoprene and the demonstration that its properties were nearly identical to those of natural rubber. (A

• Chewing gum is made from chicle, which exudes from the cut bark of yet another tree. Chicle is a mixture of *cis*- and *trans*-polyisoprene and several other substances.

• Neoprene still is produced today as a specialty rubber because of its resistance to chemical attack and heat.

• The silicones, which were discussed at the end of Section 21–2, are another important class of elastomer.

small amount of "synthetic natural rubber" is produced today.) Second, a new kind of synthetic rubber was developed: all-*cis* polybutadiene. It now ranks second in production after styrene–butadiene rubber.

SUMMARY

25–1 **Polymers** are giant molecules that are formed by stringing together large numbers of separate **monomer units** in repetitive sequences to form chains and webs. They grow by **addition polymerization** and **condensation polymerization.** Addition polymerization proceeds by free-radical chain reaction or by an ionic mechanism. In condensation polymerization, a small molecule, often water, is split out. **Copolymers** form when chemically different monomers are mixed in the polymerization process. Depending on how the differing monomer units join the growing polymer chain, the result is a **random, block,** or **graft copolymer.** Polymer chains growing from monomers with more than two reactive sites can be **cross-linked** into sheets and networks.

25–2 In proteins, the polymer chain is formed of α-amino acid monomer units. The carboxylic-acid group of one amino acid joins with the amine group of a second in a **peptide linkage.** Each of the 20 commonly occurring α-amino acids (except glycine) has a carbon atom that is a **chiral center,** meaning that the molecule can be made in two forms that are mirror-image isomers. Natural proteins contain almost exclusively L-amino acids. The side-groups along a polymer chain determine how the protein is folded, coiled, looped back on itself by hydrogen bonds, or cross-linked to other protein chains. **Fibrous proteins** are structural materials that form sheets or fibers. In a fibrous protein, the polymer may be twisted into a right-handed coil called the α-**helix,** which is maintained by hydrogen bonds. **Globular proteins** such as hemoglobin have irregular folded structures. Enzymes are globular proteins in which folding of the polymer chain forms active sites at which reactions are catalyzed. Sugars are **carbohydrates,** compounds of the general formula $C_n(H_2O)_m$. **Monosaccharides,** or simple sugars, can be linked to form long chains, or **polysaccharides.** Cellulose and starch are both condensation polymers of glucose that differ only in the position of the oxygen atoms linking the monomer units. **Nucleic acids** maintain and transmit information for the synthesis of proteins in cells. Deoxyribonucleic acid (DNA) is made up of monomer units called **nucleotides.** Each nucleotide is formed by condensation of a pyrimidine (thymine or cytosine) or purine (adenine or guanine), a D-deoxyribose sugar, and a molecule of phosphoric acid. The sequence of the four types of nucleotide down the DNA polymer encodes the information for the sequence of the amino acids in proteins. The DNA molecule has two complementary strands in which base pairs are held together by hydrogen bonds.

25–3 Useful polymers occur in fibers, plastics, and elastomers (rubber). Rayon, a semi-synthetic fiber, is a regenerated form of cellulose. The first true synthetic polymeric fiber was nylon, a polyamide formed by the condensation of a dicarboxylic acid and a diamine. **Plastics** are polymeric materials that can be molded or extruded into appropriate shapes and that harden upon cooling or solvent evaporation. The polymerization of ethylene gives low-density polyethylene (LDPE), high-density polyethylene (HDPE), and linear low-density polyethylene (LLDPE). The difference concerns the number and length of side-chains projecting from the polymer molecules. If a methyl group replaces one of the hydrogen atoms in every monomer unit of polyethylene, the result is polypropylene. The relative positions of the methyl groups attached to the carbon backbone may be **isotactic** (all on the same side), **syndiotactic** (alternating in a regular pattern), or **atactic** (distributed at random). Substitution of a chlorine atom for one of the hydrogen atoms in ethylene gives vinyl chloride, which poly-merizes to polyvinyl chloride. The **elastomer** natural rubber is a polymer of **isoprene** (2-methylbutadiene), in which the geometry at every double bond along the polymer chain is *cis*. Natural rubber is cross-linked by S—S bonds in a process called **vul-canization,** giving a material that is used for automobile tires. Synthetic rubbers in-clude neoprene, in which a chlorine atom replaces the methyl group in natural rub-ber, and copolymers of butadiene with styrene or with acrylonitrile.

PROBLEMS

Note: Answers to blue-numbered problems are given in Appendix F. More challenging problems are indicated with asterisks.

Making Polymers

1. Write a balanced chemical equation to represent the addition polymerization of 1,1-dichloroethylene. The product of this reaction is Saran, used as a plastic wrap.

2. Write a balanced chemical equation to represent the addition polymerization of tetrafluoroethylene. The product of this re-action is Teflon.

3. A polymer produced by addition polymerization consists of —(CH_2—O)— groups joined in a long chain. What was the starting monomer?

4. The polymer polymethyl methacrylate is used to make Plexiglas. It has the formula

$$\left[CH_2 - \underset{\underset{\underset{O}{\overset{\|}{C}}\ OCH_3}{\overset{CH_3}{\overset{|}{C}}}{} \right]_n$$

   Draw the structural formula of the starting monomer.

5. The monomer glycine (NH_2—CH_2—COOH) can undergo condensation polymerization to form polyglycine, in which the structural units are joined by amide linkages.
   (a) What is the molecule split off in the formation of poly-glycine?
   (b) Draw the structure of the repeat unit in polyglycine.

6. The polymer $\left[NH - CH(CH_3) - \underset{\overset{\|}{O}}{C} \right]_n$ forms upon con-densation polymerization with loss of water. Draw the struc-ture of the starting monomer.

7. Sodium amide, $NaNH_2$, can act as an ionic initiator for ad-dition polymerization, with liquid ammonia as solvent. By making an analogy with the role of butyl lithium described in the text, write a chemical equation for the process by which $NaNH_2$ initiates the polymerization of acrylonitrile.

8. The polymerization initiated by butyl lithium is anionic be-cause the growing polymer retains an anionic functional group during its growth. Cationic polymerization is also pos-sible. Write a chemical equation for the cationic addition polymerization of propylene (CH_2CHCH_3) initiated by mo-lecular iodine, I_2. (*Hint*: I_2 dissociates into I^+ and I^-, and the I^+ attacks the monomer.)

9. The polymer nylon-66 forms immediately at room tempera-ture when adipoyl chloride and hexamethylenediamine come in contact (see Fig. 25–23), but the reaction between adipic acid and hexamethylenediamine proceeds differently. The first product is a "nylon salt." Strong heating of the nylon salt is required to split out the water and generate nylon-66. Give the structure of the nylon salt that results from the re-action between one molecule of adipic acid and one mole-cule of hexamethylenediamine.

10. Pure styrene polymerizes (often with dangerous rapidity) when a very small amount of a dibenzoyl peroxide is added. The product is a fibrous, tough solid having a molar mass of about 500,000 g mol^{-1}.
    (a) Write the (approximate) formula for this product.
    (b) Is the dibenzoyl peroxide in this reaction a catalyst?

Biopolymers

11. Draw the zwitterion structures of glycine, alanine, and glu-tamic acid.

12. Which one of the 20 amino acids whose side-groups are shown in Table 25–1 cannot exhibit chirality?

13. Study the structures of the side-groups of amino acids shown in Table 25–1 and decide which ones are likely to be hy-drophilic (water-loving, likely to interact strongly with water).

14. Decide which of the side-groups of amino acids shown in Table 25–1 are likely to be hydrophobic (water-fearing, not able to interact well with water).

15. How many tripeptides could be synthesized using just three different species of α-amino acid?

16. How many dipeptides could be formed using the 20 amino acids listed in Table 25–1?

17. Draw the structure of the pentapeptide alanine–leucine–phenylalanine–glycine–isoleucine. Assume that the free —NH_2 group is at the alanine end of the peptide chain.

Would this compound be more likely to dissolve in water or in octane? Explain.

18. Draw the structure of the pentapeptide aspartic acid–serine–lysine–glutamic acid–tyrosine. Assume that the free —NH$_2$ group is at the aspartic acid end of the peptide chain. Would this compound be more likely to dissolve in water or in octane? Explain.

19. By referring to Figure 25–18a, draw the structure of the ring form of β-D-galactose. How many chiral centers are in the molecule?

20. Refer to Figure 25–16. How many chiral centers are present in the ring form of D-ribose? How many chiral centers are present in the ring form of D-deoxyribose? Make a drawing that represents the three-dimensional structure of L-ribose.

21. Suppose that a long-chain polypeptide is constructed entirely from phenylalanine monomer units. What is its empirical formula? How many amino acids does it contain if its molar mass is 17,500 g mol^{-1}?

22. The aldohexose D-idose differs from D-glucose by having the opposite chirality at carbon atoms 2, 3, and 4. Write an open-chain structural formula for D-idose and also for L-idose, using Figure 25–15 as a starting point.

23. Write an open-chain structural formula for 6-deoxy-D-glucose and also for 2,6-dideoxy-D-glucose.

24. A typical bacterial DNA has a molar mass of 4×10^9 g mol^{-1}. Approximately how many nucleotides does it contain?

Uses for Polymers

25. (a) What is the empirical formula of starch?
    (b) What is the empirical formula of cellulose?

26. (a) What is the empirical formula of guncotton (cellulose trinitrate)?
    (b) What is the empirical formula of cellulose acetate?

27. Determine the mass of adipic acid and the mass of hexamethylenediamine needed to make 1.00×10^3 kg of nylon-66 fiber.

28. Determine the mass of terephthalic acid and the mass of ethylene glycol needed to make 10.0 kg of polyester fiber.

29. Diamond has been called "the ultimate branched-chain alkane." Explain why (refer to the structural diagram of diamond in Fig. 20–25).

30. High-density polyethylene has been called "the ultimate straight-chain alkane." Explain why.

31. In a recent year, 4.37 billion kg of low-density polyethylene was produced in the United States. What volume of gaseous ethylene at STP would give this amount?

32. In a recent year, 2.84 billion kg of polystyrene was produced in the United States. Polystyrene is the addition polymer formed from the styrene monomer, $C_6H_5CH=CH_2$. How many styrene monomer units were incorporated in that 2.84 billion kg of polymer?

ADDITIONAL PROBLEMS

33. In the addition polymerization of acrylonitrile, a very small amount of butyl lithium causes a reaction that can consume hundreds of pounds of the monomer; however, the butyl lithium is called an initiator, not a catalyst. Explain why.

34. Based on the fact that the free-radical polymerization of ethylene is spontaneous and the fact that polymer molecules are less disorganized than the starting monomers, decide whether the polymerization reaction is exothermic or endothermic. Explain.

35. Sketch the structures of random, block, and graft copolymers of vinyl chloride with butadiene.

36. Can vinyl chloride and styrene form a graft copolymer? Explain.

37. Polypeptides are synthesized from a 50:50 mixture of L-alanine and D-alanine. How many different isomeric molecules containing 22 monomer units are possible?

38. Based on the information in the text, explain the similarity in the odors of burning hair and burning fingernail clippings.

39. An osmotic-pressure measurement on a solution containing hemoglobin shows that the molar mass of that protein is approximately 65,000 g mol^{-1}. A chemical analysis shows it to contain 0.344% of iron by mass. How many iron atoms does each hemoglobin molecule contain?

40. At 25°C, the standard Gibbs energy of formation of aqueous glucose is −917.2 kJ mol^{-1}. Use this value, together with data from Appendix D, to calculate $\Delta G°$ at 25°C for the degradation of 1 mol of D-glucose to aqueous ethanol and aqueous carbon dioxide, a chemical reaction carried out by certain bacteria.

41. L-sucrose tastes sweet, but it is not metabolized. It has been suggested as a potential non-nutritive sweetener. Draw the molecular structure of L-sucrose, using Figure 25–18b as a starting point.

42. The sequence of bases found in one strand of DNA reads ACTTGACCG. Write the sequence of bases in the complementary strand.

43. Approximately 950 million lb of ethylene dichloride was exported from the United States in a recent year, according to a trade journal. The journal goes on to state that "between 500 million and 550 million pounds of PVC could have been made from that ethylene dichloride." Compute the range of percentage yields of PVC from ethylene dichloride that is implied by these figures.

44. The complete hydrogenation of natural rubber (the addition of H$_2$ to all double bonds) gives a product that is indistinguishable from the product of the complete hydrogenation of gutta-percha. Explain how this strengthens the conclusion that these two substances are isomers of each other.

45. What are the similarities and differences between the formation of isotactic, syndiotactic, and atactic polypropylene on the one hand, and the formation of all-*cis*-, all-*trans*-, and mixed *cis*- and *trans*-polyisoprene on the other hand?

Scientific Notation and Experimental Error

A-1 Scientific Notation

Very large and very small numbers are common in chemistry. Repeatedly writing down such numbers in the ordinary way (as in 602 214 199 000 000 000 000 000) is tedious and error-prone. **Scientific notation** offers a better way.

In scientific notation, a number is expressed as a coefficient between 1 and 10 multiplied by 10 raised to some power. The coefficient sometimes is called the *pre-exponential,* and the 10-to-a-power, the *exponential* part of the notation. The power may be negative or positive (or zero), but must be an integer. Any number can be represented in scientific notation, as the following examples show:

$$643.8 = 6.438 \times 10^2$$
$$19\ 000\ 000 = 1.9 \times 10^7$$
$$0.0236 = 2.36 \times 10^{-2}$$
$$-0.00297 = -2.97 \times 10^{-3}$$
$$602\ 214\ 199\ 000\ 000\ 000\ 000\ 000 = 6.02214199 \times 10^{23}$$

When a number is to be converted to scientific notation, the power to which 10 is raised is $+n$ if the decimal point is moved n places to the *left* in the process of repositioning it and $-n$ if the required movement is n places to the *right*. A minus sign in an exponent does *not* make a number negative. In scientific notation, negative numbers are signified by placing a minus sign in front of the coefficient, just as in ordinary notation. A negative exponent means that the number is a fraction (lies between 0 and 1 or between 0 and -1 if it has a negative sign in front of it).

- The power to which the 10 is raised when a number such as 6.42 is expressed in scientific notation equals 0, because no movement of the decimal point is required. Simply writing 6.42 is acceptable scientific notation for this number; few scientists would trouble to write 6.42×10^0.

Addition and Subtraction in Scientific Notation. When two or more numbers written in scientific notation are combined by addition or subtraction, they must first be re-expressed as multiples of the *same* power of 10:

$$
\begin{array}{rcl}
6.43 \times 10^4 & \longrightarrow & 6.43 \times 10^4 \\
+\,2.1 \times 10^3 & \longrightarrow & +\,0.21 \times 10^4 \\
+\,3 \times 10^2 & \longrightarrow & +\,0.03 \times 10^4 \\
\hline
? & & 6.67 \times 10^4
\end{array}
$$

Multiplication. The pre-exponential parts of the numbers are multiplied. Then the exponential parts are multiplied separately by adding the exponents. The answer is written as the combination of the new pre-exponential part and the new exponential part:

$$(1.38 \times 10^{-16}) \times (8.80 \times 10^3) = (1.38 \times 8.80) \times 10^{(-16+3)} = 12.1 \times 10^{-13}$$

If the new pre-exponential part equals or exceeds 10 or is less than 1, it is usually desirable to rewrite the answer to bring the pre-exponential part between 1 and 10. In this example, the pre-exponential part of the answer exceeds 10 and is adjusted as follows:

$$12.1 \times 10^{-13} = 1.21 \times 10^{1} \times 10^{-13} = 1.21 \times 10^{-12}$$

Division. The pre-exponential part of the first number is divided by the pre-exponential part of the second and the result is multiplied by 10 raised to a power equal to the first exponent minus the second:

$$\frac{6.63 \times 10^{-27}}{2.34 \times 10^{-16}} = \frac{6.63}{2.34} \times \frac{10^{-27}}{10^{-16}}$$
$$= 2.83 \times 10^{-27-(-16)}$$
$$= 2.83 \times 10^{-11}$$

Remember that subtracting a negative quantity is the same as adding a quantity of that magnitude ("minus a minus is a plus").

All calculators and computers equipped for scientific and engineering calculations can accept and display numbers in scientific notation; however, they cannot determine whether a key-punch error has occurred or whether the answer makes sense. These issues are the user's responsibility! Develop the habit of mentally estimating the order of magnitude of the answer as a rough check on the calculator's result.

A-2 Experimental Error

As in all other experimental sciences, quantitative measurements in chemistry are subject to some degree of error. Error can be reduced by carrying out additional measurements or by changing or improving the experimental apparatus, but it can never be eliminated altogether. The degree of error in an experiment obviously has a profound effect on the conclusions that can be drawn from it. It is therefore important to assess the error in every measurement. Error appears in two ways: (1) from a lack of precision (random errors), and (2) from a lack of accuracy (systematic errors).

Precision and Random Errors

Precision refers to the degree of agreement in a collection of experimental results obtained by repeating a measurement under conditions as close to identical as possible. If the conditions are truly identical, then the differences among the trials are due to random error. As a specific example, let us consider some actual results of an early, important experiment to measure the charge e on the electron. The American physicist Robert Millikan carried out the experiment in 1909. The experimental method, discussed in Section 1–5, involved a study of the motion of charged oil drops suspended in air in an electric field. Millikan made hundreds of measurements on many different oil drops; one set of results for one particular drop is quoted in Table A–1. Values that ranged from 4.894 to 4.941 $\times$ 10^{-10} esu were found. What should be reported as the best estimate for e? The proper procedure is first to examine the data to see whether any of the results are especially far from the rest (a value above 5 $\times$ 10^{-10} esu would fall into this category). Such values are likely to result from some mistake in carrying out or reporting that particular measurement

• Cases are on record, however, where just such exceptional results have been real, not errors, and have led to significant breakthroughs.

TABLE A-1				Thirteen Measurements of e $(10^{-10}$ esu*$)$								
1	2	3	4	5	6	7	8	9	10	11	12	13
4.915	4.920	4.937	4.923	4.931	4.936	4.941	4.902	4.927	4.900	4.904	4.897	4.894

From R.A. Millikan. *Phys. Rev.* 32:349, 1911.
*1 esu = 3.3356×10^{-10} C.

and therefore are excluded from further consideration. No such outliers appear in Millikan's data. The best estimate for e then is obtained by calculating the average or mean value of the data. The mean value of any property after a series of N measurements $x_1, x_2, \ldots, x_N$ is

$$\bar{x} = \frac{1}{N}(x_1 + x_2 + \cdots + x_N)$$

where the bar above the x means that this is the average value. In the present case, the average for e is 4.917×10^{-10} esu.

The average by itself does not convey any estimate of uncertainty. If all of the measurements gave results falling between 4.91 and 4.92×10^{-10} esu, the uncertainty surely would be less than if the results were scattered across the interval between 4×10^{-10} and 6×10^{-10} esu. Furthermore, an average of 100 measurements assuredly should have less uncertainty than an average of five measurements. How can these ideas be made quantitative? A statistical measure of the spread of data, called the **standard deviation σ,** is useful in this regard. It is given by the formula

$$\sigma = \sqrt{\frac{(x_1 - \bar{x})^2 + (x_2 - \bar{x})^2 + \cdots + (x_N - \bar{x})^2}{N - 1}}$$

The standard deviation is found by adding the squares of the deviations of the individual data points from the average value x, dividing by $N - 1$, and taking the square root. The reasons for this particular (and apparently rather complicated) formula are beyond the scope of this presentation. We merely state the conclusion— that the standard deviation σ is a useful measure of the uncertainty in any experimental result. For Millikan's data, $N = 13$ and $\sigma = 0.017 \times 10^{-10}$ esu, so the result of these 13 measurements of the charge on the electron is $e = (4.917 \pm 0.017) \times 10^{-10}$ esu.

Accuracy and Systematic Errors

The charge on the electron, e, has been measured by several different techniques since Millikan's day. The current best estimate for e is

$$e = (4.80320419 \pm 0.00000018) \times 10^{-10} \text{ esu}$$
$$= (1.60217646 \pm 0.00000006) \times 10^{-19} \text{ C}$$

This value lies well outside the range of uncertainty that was estimated from Millikan's original data. In fact, it is distinctly less than the smallest of his 13 measurements of e. Why? Every experiment has a second source of error: *systematic error* that causes a shift in the measured values from the true value and reduces the **accuracy** of the result. By making more measurements, a scientist can reduce the uncertainty due to *random* errors and improve the precision of a result, but if systematic errors are present, the average value continues to deviate from the true value. Systematic errors can result from a miscalibration of the experimental apparatus or from fundamental inadequacies in a specific technique for measuring a property. In the case of Millikan's experiment, the then-accepted value for the viscosity of air (used in calculating the charge *e*) was subsequently found to be in error. This caused Millikan's results to be systematically too high.

Error therefore arises from two sources. Lack of precision (from random errors) can be estimated by a statistical analysis of a series of measurements, but lack of accuracy (from systematic errors) is much more problematic. If a systematic error is known to be present, it should be corrected for before the result is reported (for example, if the apparatus has not been calibrated correctly, it should be recalibrated). The problem is that systematic errors of which we have no knowledge may be present. In this case, the experiment should be repeated with different apparatus to eliminate the systematic error caused by one particular piece of equipment; better still, a different and independent way of measuring the property might be devised. Only after independent confirmatory experimental data are available can a scientist be convinced of the accuracy of a result—that is, how closely it approximates the true result.

A–3 Significant Figures

The number of **significant figures** (significant digits) in a measured quantity, or in a number calculated from a measured quantity, is the number of digits used to express it, not counting zeros that are present for the sole purpose of positioning the decimal point.

• This estimate of uncertainty might come from our knowledge of the measuring instrument and our experience with similar measurements.

Suppose that an experimenter determines that the mass of a sample of sodium chloride equals 8.247 g with an estimated uncertainty of ±0.002 g. This measurement has four significant figures. The experimenter is confident of the first three digits (8, 2, and 4) but not so sure about the fourth digit (7), which nevertheless is meaningful. The experimenter is saying that the true mass of the sodium chloride lies between 8.245 g and 8.249 g. Under such circumstances, writing digits to the right of the 7 is meaningless.

The standard interpretation when an experimental quantity is reported is that the experimental uncertainty, which is unavoidable, resides in the last digit written. A volume recorded as 22.4 L implies that the uncertainty in the measurement is on the order of tenths of a liter ($V = 22.4 \pm 0.3$ L, for example). A volume recorded as 22.48 L, on the other hand, implies a far smaller uncertainty, on the order of hundredths of a liter ($V = 22.48 \pm 0.02$ L, for example). Similarly, reporting that a distance is 10.000 m is quite different from reporting 10.0 m: the implied uncertainty in "10.000 m" is on the order of 0.001 m, whereas the implied uncertainty in "10.0 m" is on the order of 0.1 m, which is 100 times greater. The second measurement could be completed easily with a common meter stick. The first would require a more complex method. Additional digits should not be written unless the precision of the determination justifies them.

• Experimental results appearing on digital display screens or printed by a computer are as subject to experimental uncertainty as any others.

The following rules are used to arrive at the number of significant figures in a measured number:

1. All non-zero digits are counted. Thus, 489 g has three significant figures, and 111.1 g has four significant figures.

2. Zeros that precede the first non-zero digit in the number are *never* counted. Such zeros serve only to position the decimal point. Thus, the zeros in "0.00463°C" are excluded in tallying up significant figures— "0.00463°C" has three significant figures.

3. Zeros surrounded by non-zero digits *always* are counted. The zeros in "1005 g" are significant because omitting them not only repositions the decimal point but also changes the meaning of the number in the same way that leaving out the 1 or the 5 would change it.

4. Zeros that follow the last non-zero digit *sometimes* are counted. The trailing zeros in the measurements "9.0 g" and "13.620 mL" do nothing to position the decimal point. The only possible reason for their presence is to communicate the amount of uncertainty in the measurement; they are definitely significant. In "700 m," however, the two trailing zeros may or may not be significant. They may be present solely to position the decimal point but also may be intended to convey the precision of the measurement. The uncertainty in the measurement is on the order of ± 1 m or ± 10 m or perhaps ± 100 m. It is impossible to tell which without further information. The scientific notation described in Section A–1 avoids this ambiguity. The measurement "700 m" translates in scientific notation to any of the following:

7.00×10^2 m   three significant figures

7.0×10^2 m   two significant figures

7×10^2 m   one significant figure

Significant Figures in Calculations

Often, several experimental quantities must be combined arithmetically to obtain a final result. For example, figuring a vehicle's speed on a track requires dividing the length of the track by the time required to traverse it, a combination of two independent measurements. Calculations may require any number of additions, subtractions, multiplications, and divisions in any combination. What number of significant figures should be retained in the results?

Suppose that the 8.247 g of sodium chloride mentioned in the preceding discussion is dissolved in 160.1 g of water. What is the mass of the resulting solution? It is tempting simply to write $160.1 + 8.247 = 168.347$ g, but this is not correct. Stating that the mass of a quantity of water is 160.1 g implies experimental uncertainty as to the number of tenths of a gram of water present. The true mass could be a few tenths of a gram larger, or few tenths smaller. This uncertainty also must apply to the sum of the masses, so that the last two digits in the sum are not significant and should be **rounded off,** leaving 168.3 g as the final result.

• Significant figures also must be considered in determining logarithms or inverse logarithms. The rules for these operations are presented best in the context of a review of logarithms and are given in Appendix C–1.

> Following addition or subtraction, round off so that the sum or difference has the same number of decimal places as there are in the measurement with the smallest number of decimal places.

Here are some more examples:

94.17 g	11.171 m
+ 0.023 g	− 2.1 m
94.193 g (round off to 94.19 g)	9.071 m (round off to 9.1 m)

Rounding off is a two-part operation: First discard the digits that are not significant; then adjust the last digit that remains.

1. If the first discarded digit (reading from left to right) is less than 5, the remaining digits are left as they are. For example, when 168.341 is rounded off to four significant figures, it is rounded down to 168.3, because the first discarded digit, 4, is less than 5.

2. If the first discarded digit is greater than 5, then the last digit is increased by 1. By this rule, 168.374 becomes 168.4 when rounded off to four digits; the first discarded digit, the 7, exceeds 5.

3. If the first digit discarded equals 5 and is followed by one or more non-zero digits, then the last digit is increased by 1. Thus, 168.3503 becomes 168.4 when rounded off to four digits. Similarly, 16.25001 becomes 16.3 when rounded off to three digits.

4. If the first digit discarded is 5 and *all* subsequent discarded digits are zeros, the last digit retained is rounded to the nearest even digit. By this rule, 168.350 becomes 168.4 when rounded off to four digits, and 168.450 also becomes 168.4. The change in rounding up numbers of this type exactly equals the change in rounding them down. No reason exists to favor either choice, and a policy of always rounding down (or up) might cause errors to accumulate. The nearest-even-digit rule ensures that, over many cases, as many numbers are rounded up as down. Other rules sometimes are used.

In multiplication or division, it is not the number of decimal places that matters (as in addition or subtraction) but the number of significant figures in the least precisely known quantity. Suppose, for example, that the volume of the sample of sodium chloride mentioned in the preceding is measured and found to equal 4.34 cm^3. The density is given by the mass divided by the volume:

$$d = \frac{8.247 \text{ g}}{4.34 \text{ cm}^3} = 1.90023 \ldots \text{g cm}^{-3}$$

How many significant figures should be reported? Because the volume is the less precisely known quantity (three significant figures rather than four for the mass), it controls the precision that may be reported properly in the answer. In this case, there are only three significant figures, so the result is rounded to 1.90 g cm^{-3}.

> In multiplication and division, the proper number of significant figures in the product or quotient equals the smallest number of significant figures in the measured quantities used as input.

The following examples show this rule in action:

$$2.3 \times 19.987 = 45.9701 \qquad \text{(round off to 46)}$$

$$\frac{0.0114}{0.0045} = 2.533333 \qquad \text{(round off to 2.5)}$$

$$\frac{11.187 \times 22.2}{17.76 \times 14.0} = 0.9988393 \qquad \text{(round off to 0.999)}$$

The rule also applies to computations in which multiplication and division are mixed, as the last example shows.

It is best to carry out arithmetical operations and then to round the final answer to the correct number of significant figures, rather than to round off the input data first. The difference is usually small, but following this recommendation can sometimes prevent wrong answers. For example, the correct way to add the three distances 15 m, 6.6 m, and 12.6 m is

$$
\begin{array}{c}
15 \ \ \text{m} \\
+6.6 \ \text{m} \\
+12.6 \ \text{m} \\
\hline
34.2 \ \longrightarrow \ 34 \ \text{m}
\end{array}
\qquad \text{rather than} \qquad
\begin{array}{c}
15 \ \ \text{m} \longrightarrow 15 \ \text{m} \\
6.6 \ \text{m} \longrightarrow \ \ 7 \ \text{m} \\
12.6 \ \text{m} \longrightarrow 13 \ \text{m} \\
\hline
35 \ \text{m}
\end{array}
$$

For the same reason, extra digits often are carried through the intermediate steps of a worked example and discarded only at the final stage. If a calculation is done entirely on a scientific calculator or a computer, several extra digits usually are carried along automatically. Before reporting the final answer, however, it is important to round off to the proper number of significant figures.

• Alert readers will notice that this practice is followed in this book.

Sometimes mathematical constants and **exact numbers,** numbers obtained by counting and not by measurements, appear in expressions. Exact numbers should be considered to have an infinite number of significant figures. The precision of the result then is controlled by the precision of the other input. If the mass of an object is 3.142 g, then the mass of three identical such objects is

$$
3 \times 3.142 = 9.426 \text{ g}
$$

This answer is *not* rounded off to one significant figure because the number 3 comes from counting the objects. It is known exactly and has an infinite number of significant digits, not just one. Similarly, the uncertainty in the volume of a sphere, computed from the formula $V = \frac{4}{3}\pi r^3$, depends only on the uncertainty in the radius r. The 4 and 3 are exact values (4.000 ... and 3.000 ..., respectively) and not the results of measurements, and π, which is known to many millions of digits ($\pi = 3.14159265358979...$), can be used as required. Note that mathematical constants and exact numbers are not necessarily whole numbers.

PROBLEMS

Note: Answers to blue-numbered problems are given in Appendix F.

Scientific Notation

1. Express the following in scientific notation:
   (a) 0.0000582
   (b) 1402
   (c) 7.93
   (d) −6593.00
   (e) 0.002530
   (f) 1.47

2. Express the following in scientific notation:
   (a) 4579
   (b) −0.05020
   (c) 2134.560
   (d) 3.825
   (e) 0.0000450
   (f) 9.814

3. Convert the following from scientific notation to ordinary decimal form:
   (a) 5.37×10^{-4}
   (b) 9.390×10^{6}
   (c) -2.47×10^{-3}
   (d) 6.020×10^{-3}
   (e) 2×10^{4}

4. Convert the following from scientific notation to ordinary decimal form:
   (a) 3.333×10^{-3}
   (b) -1.20×10^{7}
   (c) 2.79×10^{-5}
   (d) 3×10^{1}
   (e) 6.700×10^{-2}

5. Determine which of the following numbers is largest algebraically: $1.90 \times 10^{2}, 9.7 \times 10^{-2}, -4.90 \times 10^{2}, -4.10 \times 10^{-3}$.

6. Arrange the following from largest to smallest in the algebraic sense: 11.7×10^{-7}, -14.8×10^{-4}, -2.17×10^{3}, 3.19×10^{-2}.

7. A certain chemical plant produces 7.46×10^{8} kg of polyethylene. Express the production in decimal form.

8. A microorganism contains 0.0000046 g of vanadium. Express this amount in scientific notation.

Experimental Error

9. A group of students takes turns using a laboratory balance to weigh the water contained in a beaker. The results they report are 111.42 g, 111.67 g, 111.21 g, 135.64 g, 111.02 g, 111.29 g, and 111.42 g.
  (a) Should any of the data be excluded before calculating the average?
  (b) Calculate the average value of the mass of the water in the beaker from the valid measurements.
  (c) Calculate the standard deviation σ of the valid data.

10. By measuring the sides of a small box, a group of students make the following estimates for its volume: 544 cm³, 590 cm³, 523 cm³, 560 cm³, 519 cm³, 570 cm³, and 578 cm³.
  (a) Should any of the data be excluded before calculating the average?
  (b) Calculate the average value of the volume of the box from the valid measurements.
  (c) Calculate the standard deviation σ of the valid data.

11. Of the measurements in Problems 9 and 10, which is more precise?

12. A more accurate determination of the mass in Problem 9 (using a better balance) gives the value 104.67 g, and a more accurate determination of the volume in Problem 10 gives the value 553 cm³. Which of the two measurements in Problems 9 and 10 is more accurate, in the sense of having the smaller systematic error relative to the actual value?

Significant Figures

13. State the number of significant figures in each of the following measurements:
  (a) 13.604 L
  (b) −0.00345°C
  (c) 340 lb
  (d) 3.40×10^{2} mi
  (e) 6.248×10^{-27} J

14. State the number of significant figures in each of the following measurements:
  (a) −0.0025 in
  (b) 7000 g
  (c) 143.7902 s
  (d) 2.670×10^{7} Pa
  (e) 2.05×10^{-19} J

15. Round off each of the measurements in Problem 13 to two significant figures.

16. Round off each of the measurements in Problem 14 to two significant figures.

17. Round off the measured number 2,997,215.548 to nine significant figures.

18. Round off the measured number in Problem 17 to eight, seven, six, five, four, three, two, and one significant figures.

19. Express the results of the following additions and subtractions to the proper number of significant figures. All of the numbers are measured quantities and have the same units.
  (a) $67.314 + 8.63 - 243.198 =$
  (b) $4.31 + 64 + 7.19 =$
  (c) $3.2156 \times 10^{15} - 4.631 \times 10^{13} =$
  (d) $2.41 \times 10^{-26} - 7.83 \times 10^{-25} =$

20. Express the results of the following additions and subtractions to the proper number of significant figures. All of the numbers are measured quantities and have the same units.
  (a) $245.876 + 4.65 + 0.3678 =$
  (b) $798.36 - 1005.7 + 129.652 =$
  (c) $7.98 \times 10^{17} + 6.472 \times 10^{19} =$
  (d) $4.32 \times 10^{-15} + 6.257 \times 10^{-14} - 2.136 \times 10^{-13} =$

21. Express the results of the following multiplications and divisions to the proper number of significant figures. All of the numbers are measured quantities.
  (a) $\dfrac{-72.415}{8.62} =$
  (b) $52.814 \times 0.00279 =$
  (c) $(7.023 \times 10^{14}) \times (4.62 \times 10^{-27}) =$
  (d) $\dfrac{4.3 \times 10^{-12}}{9.632 \times 10^{-26}} =$

22. Express the results of the following multiplications and divisions to the proper number of significant figures. All of the numbers are measured quantities.
  (a) $129.578 \times 32.33 =$
  (b) $\dfrac{4.7791}{3.21 \times 5.793} =$
  (c) $\dfrac{10{,}566.9}{3.584 \times 10^{29}} =$
  (d) $(5.247 \times 10^{13}) \times (1.3 \times 10^{-17}) =$

23. Compute the area of a triangle if its base and altitude are measured to equal 42.07 cm and 16.0 cm, respectively. (The area of a triangle is $\frac{1}{2}$ its base multiplied by its altitude.) Explain your use of significant figures in the answer.

24. An inch is defined as exactly 2.54 cm. The length of a table is measured as 404.16 cm. Compute the length of the table in inches. Explain your use of significant figures in the answer.

Units and Symbols

B-1 The International System of Units

Most measurements in physics and chemistry require the use of units, which are reference standards to which experimentally observed quantities can be related. These measurements are really *ratios* between the thing being measured and the size of the "measuring stick" (the reference unit). If a zoologist tells us that a snake is 5 long, we still know nothing about the snake until we learn whether inches, feet, meters (heaven forbid!), or some other unit of length is intended. Whenever the magnitude of an experimental observation depends on the units selected to express it (as in the case of the snake), the observation is said to have **dimensions.** Over the course of history, different countries and regions evolved numerous sets of locally accepted units to express length, mass, and other physical dimensions of importance in commerce and industry. These diverse units gradually are being replaced by international standards that allow easy comparison of measurements made in different localities and avoid misunderstanding. The unified system of units that currently is recommended by international agreement is a metric system called the SI, which stands for Système International d'Unités, or International System of Units. This appendix outlines the use of SI units and discusses interconversions with other systems of units.

> • We say "a" metric system because several variant systems exist that use the metric ideas first adopted in France in 1790.

The SI uses seven **basic units,** which are listed in Table B–1. These units are the defined standards of reference for seven physical quantities that were chosen for their importance and the ease with which they are measured. Of these quantities, only luminous intensity is not used in this book. Length, mass, and time, the first three quantities in Table B–1, are familiar from everyday life. A **length** is a distance or a spatial separation; the **mass** of an object is the quantity of matter that it contains; and a period of **time** is a temporal separation. Measurements of **temperature,** the degree of hotness or coldness of an object

TABLE B–1	Base Units in the International System of Units	
Quantity	**Unit**	**Symbol**
Length	meter	m
Mass	kilogram	kg
Time	second	s
Temperature	kelvin	K
Chemical amount (of substance)	mole	mol
Electric current	ampere	A
Luminous intensity	candela	cd

on some defined scale, are of daily importance. The full definition of temperature requires some care (see Section 5–3). Many units of temperature exist in addition to the SI base unit, the kelvin. Of these, the Fahrenheit degree is the most familiar in the United States, and the Celsius degree is the most important in science. The relationships among the most important temperature scales and units of temperature are discussed later. The quantity called *amount of substance,* or **chemical amount,** is perhaps unfamiliar, but its name reveals its importance in chemistry. Chemical amount is less concerned with the total quantity of matter in a sample, which is given by its mass, than with the number of particles in that sample that exhibit the same behavior in a chemical reaction. This point is discussed in full in Section 1–7. Finally, the measurement of **electric current,** the rate of passage of electric charge through a circuit, is important in the study of electrochemistry (see Section 12–2).

The seven base units in the SI all have more or less complex technical definitions that are intended to allow for their precise reproduction in independent laboratories. We give none of these definitions here. Knowing the definitions is less important than is having a sense of the size of the units. Thus, a meter is about 10% longer than 1 yard (a length of 39.37 in.), and a kilogram is very close to 10% heavier than 2 lb (it is 2.205 lb). However, these comparisons are helpful only up to a point; mastery of the SI system requires learning to "think metric" instead of estimating and planning in terms of other units and then translating. A table of conversion factors between SI units, other metric units, and units in the U.S. customary system is given at the front of this book.

Units other than the seven SI base units also see a great deal of use. Depending on the scientific field, they are indeed often encountered more frequently than are the base units themselves. Such units are called **derived units,** and several are listed in Table B–2. All derived units are obtained as some combination (by means of multiplication or division) of the seven base units. The following are some important physical quantities that are measured in derived units.

TABLE B–2	Some Derived Units in the International System of Units		
Quantity	**Definition**	**Unit**	**Name/Abbreviation**
Area	Length × length	m^2	square meter
Volume	Length × length × length	m^3	cubic meter
Density	Mass/length^3	$kg \, m^{-3}$	kilogram per cubic meter
Velocity	Length/time	$m \, s^{-1}$	meter per second
Acceleration	Length/time^2	$m \, s^{-2}$	meter per second squared
Force	Mass × acceleration	$kg \, m \, s^{-2}$	newton (N)
Pressure	Force/area	$N \, m^{-2} = kg \, m^{-1} \, s^{-2}$	pascal (Pa)
Energy	Force × distance	$N \, m = kg \, m^2 \, s^{-2}$	joule (J)
Power	Energy/time	$W = kg \, m^2 \, s^{-3}$	watt (W)
Electric charge	Electric current × time	$A \, s$	coulomb (C)
Electric potential difference	Energy/electric charge	$J \, C^{-1} = kg \, m^2 \, s^{-3} \, A^{-1}$	volt (V)

Area and Volume.   Area is spatial extent in two directions, and volume is spatial extent in three directions. (Recall that length is extent in one direction.) It follows that area has dimensions of length × length, and volume has dimensions of length × length × length. The natural SI unit of area is formed simply by squaring the SI unit of length. It is the meter × meter, or square meter (m^2). The natural SI unit of volume is the meter × meter × meter, or cubic meter (m^3).

Density.   It is often important to know how tightly or loosely the matter that comprises a given sample is packed together. This information is given by the **density** of the sample, defined as its mass divided by its volume:

$$\text{density} = \frac{\text{mass}}{\text{volume}} \quad \text{or} \quad d = \frac{m}{V}$$

From this definition and the discussion of volume in the previous paragraph, it follows that density has dimensions of mass divided by length × length × length. The natural SI unit of density is accordingly the kilogram per cubic meter (abbreviated kg/m^3, or $kg\ m^{-3}$). Density is used widely to aid in the identification of materials because it is independent of sample size and is therefore an inherent physical characteristic of the sample. (Density does depend on sample temperature, especially for liquids and gases.) For example, the density of "fool's gold" (iron pyrites) is far less than is the density of true gold, although a big lump of fool's gold could easily have a larger mass than a small nugget of true gold.

• Note the use of a negative exponent to indicate the "m^3" in the denominator of a fraction. This useful notation and other aspects of exponents are reviewed in Appendix C.

Speed and Velocity.   The speed of a moving object is the rate of change of its position with time—that is,

$$\text{speed} = \frac{\text{change in position}}{\text{change in time}}$$

A change in position (the distance between position 1 and position 2) is a length and has the units of length. An elapsed time, of course, has units of time. Speed thus has the units of length divided by time, and the natural unit of speed in the International System is the meter per second, or $m\ s^{-1}$. Velocity has the same units as does speed. The difference between speed and velocity is not in their units but in the fact that speed has magnitude only but velocity has both magnitude and direction. Driving a car faster than the posted limit nets the driver a speeding ticket; driving slowly the wrong way on a one-way street is a different violation that gets what could be called a "velocity ticket," because it is the direction of travel that is forbidden, not the rate of travel.

Acceleration.   Acceleration is the rate of change of velocity with time:

$$\text{acceleration} = \frac{\text{change in velocity}}{\text{change in time}}$$

From this definition, acceleration must have the units of velocity, which are length divided by time, again divided by time. This is equivalent to length divided by time squared. The definition is used in just this way in evaluating automobiles. A powerful car is able to accelerate from a standstill up to a velocity of $60\ mi\ h^{-1}$ in 6 s. It gains speed at the rate of $10\ mi\ h^{-1}$ every second, for an average acceleration of $10\ mi\ h^{-1}\ s^{-1}$. This unit of acceleration uses a mixture of time units. The natural SI unit of acceleration uses seconds exclusively. It is the meter per second per second, equivalent to the meter per second squared, or $m\ s^{-2}$.

Force.   A physical force is easily understood as a push or pull. The force exerted by the Earth's gravity, which tugs all terrestrial objects downward, is the most familiar force in everyday experience. An unbalanced force acting on an object causes it to change its velocity—that is, to accelerate. This observation is part of Newton's second law of motion, which gives the size of the force as the product of the mass of the object and the observed acceleration:

$$\text{force} = \text{mass} \times \text{acceleration}$$

Newton's second law provides the definition of the SI unit of force. The SI unit of mass (kilogram) is multiplied by the SI unit of acceleration (the meter per second per second) to give the kg m s^{-2}. This derived unit has a special name; 1 kg m s^{-2} is called a *newton* (N), in honor of Sir Isaac Newton.

- One newton is approximately the gravitational force exerted by the Earth on an apple.

Pressure.   A pressure is the force exerted on an object (such as the wall of a container) divided by the area that receives that force:

$$\text{pressure} = \frac{\text{force}}{\text{area}}$$

Substituting the SI units of force and area into this definition establishes that the natural SI unit of pressure is the newton per square meter, or N m^{-2}. This unit is named a pascal (Pa), in honor of the French physicist Blaise Pascal. Combining the fact that a newton equals 1 kg m s^{-2} with the definition of the *pascal* gives the pascal in terms of SI base units:

$$\text{pascal} = \frac{N}{m^2} = \frac{\text{kg m s}^{-2}}{m^2} = \text{kg m}^{-1}\,\text{s}^{-2}$$

Energy.   Perhaps the most familiar manifestation of energy is mechanical work, and it is through the definition of mechanical work that one arrives at the units of energy most easily. Mechanical work is defined as the product of the external force on an object times the distance through which the force acts:

$$\text{work} = \text{force} \times \text{distance}$$

The natural SI unit of work is therefore the SI unit of force (the newton) multiplied by the SI unit of distance (the meter). It is the *newton-meter* (N m). Because mechanical work is a form of energy, the newton-meter is also the SI unit of energy. A newton-meter is called a *joule* (J), in honor of the English physicist James Joule. In terms of the SI base units, a joule is

$$\text{joule} = \text{newton} \times \text{meter} = \text{kg m s}^{-2} \times m = \text{kg m}^2\,\text{s}^{-2}$$

An object in motion has energy by virtue of that motion. This is called its kinetic energy. The kinetic energy of an object is defined by the equation

$$\text{kinetic energy} = \tfrac{1}{2} \times (\text{mass of the object}) \times (\text{velocity of the object})^2$$

To verify that kinetic energy is indeed an energy, simply multiply out the SI units of mass and velocity as required by the definition

$$\text{kg} \times (\text{m s}^{-1})^2 = \text{kg m}^2\,\text{s}^{-2} = J$$

The $\tfrac{1}{2}$ in the definition contributes no units and accordingly is omitted from the equation.

"Lo-cal cola" is common in the U.S.; this is the Australian equivalent. *(Charles D. Winters)*

TABLE B–3		Prefixes in SI				
Fraction	**Prefix**	**Symbol**	**Factor**	**Prefix**	**Symbol**	
10^{-1}	deci-	d	10	deca-	da	
10^{-2}	centi-	c	10^2	hecto-	h	
10^{-3}	milli-	m	10^3	kilo-	k	
10^{-6}	micro-	μ	10^6	mega-	M	
10^{-9}	nano-	n	10^9	giga-	G	
10^{-12}	pico-	p	10^{12}	tera-	T	
10^{-15}	femto-	f	10^{15}	peta-	P	
10^{-18}	atto-	a	10^{18}	exa-	E	

Prefixes in SI Units

Because scientists work on scales ranging from the microscopic to the astronomical, a tremendous range exists in the magnitudes of measured quantities. For this reason, a set of **prefixes** has been incorporated into the International System of Units to simplify the description of small and large quantities (Table B–3). The prefixes are used to specify various powers of 10 times the base and derived units. The natural SI unit of volume, the cubic meter (m^3), for example, is inconveniently large for ordinary laboratory work, in which measurements of volume are typically on the order of 10^{-4} to 10^{-6} m^3. Adding the prefix *centi-* to *meter* defines the centimeter (cm), a unit of length equal to exactly 10^{-2} m. The cubic centimeter (cm^3) is then $10^{-2} \times 10^{-2} \times 10^{-2} = 10^{-6}$ cubic meter. Typical laboratory-scale measurements of volume thus come out between 1 and 100 when expressed in cubic centimeters. Similarly, the natural SI unit of density is inconveniently small for many purposes. The room-temperature density of water, for example, is nearly 1000 kg m^{-3}. Removing the prefix on the unit of mass (switching to the gram instead of the kilogram) and inserting a prefix in the unit of volume (going to cubic centimeters from cubic meters) gives a new SI unit of density, the gram per cubic centimeter, that is much larger:

$$1 \text{ g cm}^{-3} = 1000 \text{ kg m}^{-3}$$

Most solids and liquids at ordinary conditions have densities between 0.2 and 20 g cm^{-3}.

Non-SI Units

In addition to base and derived SI units, certain additional units that are not part of the SI are used in this book. These units are summarized in Table B–4. The first is the liter, which, like the cm^3, has a convenient size for the measurement of the volumes used in laboratory-scale chemistry. One liter equals 10^{-3} m^3, or 1 dm^3:

$$1 \text{ L} = 1 \text{ dm}^3 = 10^{-3} \text{ m}^3 = 10^3 \text{ cm}^3$$

• It is particularly worth noting that 1 mL = 1 cm^3.

The SI prefixes can be combined with this unit (and other non-SI units as well). For example, the milliliter (1 mL = 10^{-3} L = 1 cm^3) is quite common in chemistry, and the deciliter (1 dL = 10^{-1} L = 100 cm^3) is much used in the health sciences.

TABLE B–4	Some Non-SI Units Used in This Book		
Physical Quantity	**Name of Unit**	**Symbol for Unit**	**Value in SI Units**
Volume	liter	L	$1 \text{ dm}^3 = 10^{-3} \text{ m}^3$
Length	ångström	Å	10^{-10} m
Mass	atomic mass unit	u	$1.66054 \times 10^{-27} \text{ kg}$
Mass	metric ton (tonne)	t	10^3 kg
Energy	electron volt	eV	$1.60218 \times 10^{-19} \text{ J}$
Pressure	standard atmosphere	atm	$1.01325 \times 10^5 \text{ Pa}$
Time	minute	min	60 s
Time	hour	h	3600 s
Time	day	d	$8.64 \times 10^4 \text{ s}$

Second, chemists frequently prefer the ångström (Å) as a unit of length in discussing sizes of atoms and molecules:

$$1 \text{ Å} = 10^{-10} \text{ m} = 100 \text{ pm} = 0.1 \text{ nm}$$

Most atomic sizes and chemical bond lengths fall in the range of one to several ångströms, and the use of either picometers or nanometers is slightly awkward. Similarly, the *atomic mass unit* is just right for expressing the masses of individual atoms (which range from 1 to 270 u), the *electron volt* is well-sized to express energy changes on the atomic scale, and the *metric ton* is convenient for talking about the industrial production of chemicals.

Finally, the *standard atmosphere* (atm) is used in this book as a unit of pressure. It is not a simple power of ten times the pascal, which is the natural SI unit of pressure, but is defined as 1.01325×10^5 Pa. This definition makes the standard atmosphere quite close to the observed sea-level atmospheric pressure. It is used because most chemistry is carried out at pressures near atmospheric pressure, for which the pascal is too small to be convenient. A slightly different unit of pressure, called the *bar* (defined as exactly 10^5 Pa), has been recommended to replace the standard atmosphere but is not used here.

Many non-SI units continue in common use in the United States. Most are now defined in terms of SI units. Table B–5 lists a few familiar units and their current official definitions in terms of SI units. Note that the number 3.78541 and the others like it in Table B–5 have an infinite number of significant figures because they are exact numbers (as explained in Appendix A–3) and not measurements.

The metric system has been widely adopted by makers of soft drinks in the U.S.

TABLE B–5	Some Familiar Units Defined in Terms of SI Units		
Physical Quantity	**Name of Unit**	**Symbol for Unit**	**Value in SI Units**
Volume	U.S. gallon	gal	3.78541 dm^3 exactly
Length	inch	in	$2.54 \times 10^{-2} \text{ m}$ exactly
Mass	pound avoirdupois	lb	0.45359237 kg exactly
Energy	calorie	cal	4.184 J exactly

Temperature Scales

The two most important units of temperature in the United States are the Celsius degree (°C) and the Fahrenheit degree (°F). The first is a derived SI unit that is exactly equal to the kelvin in magnitude; the second is a non-SI unit equal to $\frac{5}{9}$ of a kelvin:

$$1°C = 1\,K \qquad \text{and} \qquad 1°F = \tfrac{5}{9}\,K$$

- These equalities refer to the sizes of the different kinds of degrees and not to temperatures.

Measurements of temperature, unlike those of most other physical quantities, require a point of reference as well as a unit. This requirement leads to scales of temperature, of which three appear in this book: (1) the absolute scale, which uses the kelvin as its unit of temperature difference, (2) the Celsius scale, which uses the °C, and (3) the Fahrenheit scale, which uses the °F. On the Fahrenheit scale, the freezing point of water is 32°F and the boiling point of water under a pressure of 1 atm is 212°F. The interval between these two temperatures is 180°F. On the Celsius scale, these two points have values of 0°C and 100°C, respectively, making the interval 100°C, and on the absolute scale the two points have values of 273.15 K and 373.15 K, making the interval 100 K. Evidently, the kelvin and the Celsius degree are equal in size, as already stated. The two are larger than the Fahrenheit degree by the factor 180/100 because 100 Celsius degrees (or kelvins) are enough to span the same temperature interval as 180 Fahrenheit degrees (see Fig. 5–12). It follows that the Fahrenheit degree is $100/180 = \tfrac{5}{9}$ of the Celsius degree (or kelvin).

- This "Celsius poem" gives Fahrenheit users good advice: 30 is hot/20 is nice/10 put a coat on/0 is ice.

To convert a temperature given in degrees Fahrenheit to degrees Celsius, this fact is used in the formula

$$t_{°C} = \left(\frac{5°C}{9°F}\right)(t_{°F} - 32°F)$$

Solving the formula for $t_{°F}$ in terms of $t_{°C}$ gives

$$t_{°F} = \left(\frac{9°F}{5°C}\right)(t_{°C}) + 32°F$$

- In this formula, as well as the next two formulas, units are displayed to show the way that they cancel out. The virtue of this practice is discussed in the next section of this appendix.

The Kelvin scale uses absolute zero as its zero point. It is impossible for a temperature to reach or fall below absolute zero; temperatures on the Kelvin scale are always positive. Temperatures on the Kelvin scale are related to Celsius temperature by the formula

$$T_K = \left(\frac{1\,K}{1°C}\right)(t_{°C} + 273.15°C)$$

Note the practice of using an uppercase T for a temperature on an absolute scale and a lowercase t for temperatures that use other scales.

B-2 The Conversion of Units Using the Unit-Factor Method

If all the quantities in a calculation are inserted in units that are either base units of the International System or combinations of base units, the final result automatically comes out in SI units as well. This coherence is the great advantage of the SI. Nevertheless, it is essential to be able to convert at will among units because non-SI units are unavoidable and sometimes even desirable, both in scientific and non-scientific situations. The **unit-factor method** works to solve all problems of unit conversion. The basis of the method is in two facts: first, the measured value of a physical quantity is mathematically equivalent to a pure number multiplied by a unit (example: 64.3 g = 64.3 × grams); second, number × unit combinations may be manipulated under the ordinary rules of algebra. To see how these facts help in the conversion of units, suppose that a mass has been found to equal 64.3 g and must be expressed in kilograms (the SI base unit of mass) for use in a formula. To convert, start with the equation 1 kg = 1000 g and divide both sides by 1000 g:

$$\left(\frac{1\ \text{kg}}{1000\ \text{g}}\right) = 1$$

The 64.3 g is now multiplied by this newly created **unit-factor,** so called because it equals 1 (unity). This does not change the mass, but it does change the unit.

$$64.3\ \text{g} \times \left(\frac{1\ \text{kg}}{1000\ \text{g}}\right) = 0.0643\ \text{kg}$$

Canceling the unit of grams in the starting value against the grams in the denominator of the unit-factor gives the unit of the final result. The unit-factor was set up with kilograms in the numerator and grams in the denominator just so that this cancellation would work. If this unit-factor had been inserted upside down, the desired cancellation would have been impossible. On the other hand, it would have been equally correct to write 1 g = 10^{-3} kg and carry out the same conversion as

$$64.3\ \text{g} \times \left(\frac{10^{-3}\ \text{kg}}{1\ \text{g}}\right) = 0.0643\ \text{kg}$$

Attention to the cancellation of units in this method eliminates the common error of multiplying by a conversion factor instead of dividing, or vice versa.

Other conversions of units may involve more than just powers of 10, but they are equally easy to set up. For example, to express 16.4 in in meters, use the fact that 1 in = 0.0254 m to construct a unit-factor:

$$16.4\ \text{in} \times \left(\frac{0.0254\ \text{m}}{1\ \text{in}}\right) = 0.417\ \text{m}$$

More complicated combinations are possible as well. To convert from liter-atmospheres to joules (the SI unit of energy), multiply by two successive unit-factors to get units that are part of the SI:

$$1\ \text{L atm} \times \left(\frac{10^{-3}\ \text{m}^3}{1\ \text{L}}\right) \times \left(\frac{101{,}325\ \text{Pa}}{1\ \text{atm}}\right) = 101.325\ \text{Pa m}^3$$

Manipulation of the SI units on the right side of the equation confirms that a "Pa m^3" is truly equivalent to a joule:

$$101.325\ \text{Pa m}^3 = 101.325\ \frac{\text{N}}{\text{m}^2}\ \text{m}^3 = 101.325\ \text{N m} = 101.325\ \text{J}$$

The new unit-factor

$$\left(\frac{101.325\ \text{J}}{1\ \text{L atm}}\right)$$

or its reciprocal

$$\left(\frac{1\ \text{L atm}}{101.325\ \text{J}}\right)$$

can be used in any subsequent conversions between the liter-atmosphere and the joule.

It is very important to write out units explicitly when doing chemical calculations and to cancel out these units thoughtfully in intermediate steps. Cancellation provides an essential check that units have not been inadvertently changed without proper conversions and that an incorrect formula has not been used. If a result that is supposed to be a temperature comes out with units of $m^3 \, s^{-2}$, a mistake clearly has been made!

B-3 Symbols and Units for Certain Quantities in Chemistry

TABLE B-6	Symbols and Units for Certain Quantities in Chemistry[a]				
Quantity	Recommended Symbols	Other Symbols	SI Unit	Other Units	Section of Book
Fundamental Physical Quantities					
Mass	m	M	kg	g, u, Da, lb	1–3, 1–7, 15–1
Length, height, distance, radius	l, h, d, r		m	in, ft, yd, mi	5–2
Volume	V, v		m^3	L, gal	1–4, 1–9
Force	F		N	lbf	10–6
Energy	E		J	cal, kcal, kW-h, L atm, electron volt (eV), rydberg (Ry)	10–2, 10–6, 13–1, 15–1, 16–3
Potential energy	$E_p \, V, \Phi$	P. E.			10–6
Kinetic energy	E_k, T, K	K. E.			10–6
Area	A		m^2	in^2, ft^2, yd^2, acre	5–2
Pressure	P, p		$Pa = N \, m^{-2}$	torr, mm Hg, atm, psi	5–2
Speed	v, u, w, c		$m \, s^{-1}$	$mi \, h^{-1}, km \, h^{-1}$	5–7, 15–1
Speed of light in a vacuum	c_0	c	$m \, s^{-1}$	$mi \, s^{-1}$	15–1, 16–1
Velocity	$\boldsymbol{v, u, w, c}$	v, u, w, c	$m \, s^{-1}$	$ft \, s^{-1}$	
Acceleration	a		$m \, s^{-2}$		
Acceleration of gravity	g				5–2
Frequency	ν (Greek nu), f		$Hz = s^{-1}$	min^{-1}, yr^{-1}, cps	16–1
Atoms and Molecules					
Chemical amount	n		mol		1–7
of substance A	$n_A, n(A)$		mol		
Avogadro's number	N_A, L	N_0	mol^{-1}		1–7, 1–9
Molar mass	M	MW	$kg \, mol^{-1}$	$g \, mol^{-1}$	1–7
of substance A	$M_A, M(A)$				
Relative molecular mass	M_r	MW	1		1–7
Relative atomic mass	M_r	A	1		1–6
Number of atoms, molecules, ions or other entities	N		1		1–7
of entities A	$N_A, N(A)$		1		
Mass of entity A	$m_A, m(A)$		kg	g, u, Da, lb	1–8
Universal gas constant	R		$J \, K^{-1} \, mol^{-1}$	$L \, atm \, K^{-1} \, mol^{-1}$	5–4
Density	ρ	d	$kg \, m^{-3}$	$g \, cm^{-3}, lb \, ft^{-3}$	1–9, 5–4
Molar volume	V_m		$m^3 \, mol^{-1}$	$cm^3 \, mol^{-1}$	6–1

TABLE B–6	(continued)				
Quantity	**Recommended Symbols**	**Other Symbols**	**SI Unit**	**Other Units**	**Section of Book**
Mixtures and Solutions					
Concentration of component A	c c_A	C C_A	mol m^{-3}	mol L^{-1}	2–4
Mole fraction of component A	$x_A, x(A)$	X_A	1		6–6
Fractional abundance	x	p, X	1		1–6
Partial pressure of component A	P_A	p_A	Pa	atm, bar, torr	5–6
Mass fraction of component A	w_A		1		6–6
Volume fraction of component A	Φ_A		1		C–4
Molality of component A	m_A, b_A		mol kg^{-1}		6–6
Solubility of component A	s_A	S_A	mol m^{-3}	mol L^{-1}	9–2
Mass solubility		s, S	kg m^{-3}	g m^{-3}, g L^{-1}, g solute/100 g solvent	4–1, 9–2
Osmotic pressure	Π	π	Pa	atm bar, torr	6–6
pH	pH		1		8–2
Chemical Equilibrium					
Equilibrium constant	$K, K^{\ominus}$	K_{eq}			7–1
Acidity constant	K_a		1		8–3
Basicity constant	K_b		1		8–3
Solubility-product constant	K_{sol}, K_s	K_{sp}	1		9–2
Sulfide-solubility constant		K_{ss}	1		9–4
Thermodynamics					
Heat	q, Q		J	kJ, kcal	10–1
Work	w, W		J	kJ, kcal	10–6
Internal energy	U	E	J	kJ, kcal	10–6
Change in internal energy	ΔU	ΔE	J	kJ, kcal	10–6
Internal energy of reaction	$\Delta_r U$	ΔE_r	J mol^{-1}	kJ mol^{-1}, kcal mol^{-1}	10–6
Standard internal energy of reaction	$\Delta_r U^{\ominus}$	ΔE_r°	J mol^{-1}	kJ mol^{-1}, kcal mol^{-1}	
Gibbs energy	G		J	kJ, kcal	11–5
Change in Gibbs energy	ΔG		J	kJ, kcal	11–5
Gibbs energy of reaction	$\Delta_r G$	ΔG_r	J mol^{-1}	kJ mol^{-1}, kcal mol^{-1}	11–5
Standard Gibbs energy of reaction	$\Delta_r G^{\ominus}$	ΔG_r°	J mol^{-1}	kJ mol^{-1}, kcal mol^{-1}	11–6
Enthalpy	H		J	kJ, kcal	10–3
Change in enthalpy	ΔH		J	kJ, kcal	10–3
Enthalpy of reaction	$\Delta_r H$	ΔH_r	J mol^{-1}	kJ mol^{-1}, kcal mol^{-1}	10–3
Standard enthalpy of reaction	$\Delta_r H^{\ominus}$	ΔH_r°	J mol^{-1}	kJ mol^{-1}, kcal mol^{-1}	10–4
Absolute temperature	T		K	°R	5–3
Celsius temperature	t	T	°C		5–3
Boltzmann constant	k, k_B		J K^{-1}	cal K^{-1}	11–2
Entropy	S		J K^{-1}	cal K^{-1}	11–2

TABLE B–6	*(continued)*				
Quantity	**Recommended Symbols**	**Other Symbols**	**SI Unit**	**Other Units**	**Section of Book**
Change in entropy	ΔS		$J\,K^{-1}$	$cal\,K^{-1}$	11–3
Entropy of reaction	$\Delta_r S$	ΔS_r	$J\,K^{-1}\,mol^{-1}$	$cal\,K^{-1}\,mol^{-1}$	11–3
Standard entropy of reaction	$\Delta_r S^{\ominus}$	ΔS_r	$J\,K^{-1}\,mol^{-1}$	$cal\,K^{-1}\,mol^{-1}$	11–3
Molar entropy	S		$J\,K^{-1}\,mol^{-1}$	$cal\,K^{-1}\,mol^{-1}$	11–3
Standard molar entropy	S°		$J\,K^{-1}\,mol^{-1}$	$cal\,K^{-1}\,mol^{-1}$	11–3
Heat capacity					
At constant P	C_P		$J\,K^{-1}$	$cal\,K^{-1}, J\,(°C)^{-1},$ $cal\,(°C)^{-1}$	10–2
At constant V	C_V		$J\,K^{-1}$	$cal\,K^{-1}, J\,(°C)^{-1},$ $cal\,(°C)^{-1}$	10–2
Molar heat capacity					
At constant P		c_P	$J\,K^{-1}\,mol^{-1}$	$cal\,K^{-1}\,mol^{-1}$	10–2
At constant V		c_V	$J\,K^{-1}\,mol^{-1}$	$cal\,K^{-1}\,mol^{-1}$	10–2
Specific heat capacity		c_s	$J\,K^{-1}\,kg^{-1}$	$J\,K^{-1}\,g^{-1}$	10–2
Electrochemistry					
Electric potential difference	$E, \Delta V, U$	$\Delta\mathcal{E}$	$V = J\,C^{-1}$		13–1
Standard potential difference	$E^{\ominus}$	$\Delta\mathcal{E}°, \Delta E°$	V		
Half-cell potential	V_R, V_L	$\mathcal{E}, E_{hc}$	V		13–2
Standard half-cell potential	$E^{\ominus}, V_R^{\circ}, V_L^{\circ}$	$\mathcal{E}°, E_{hc}^{\circ}$	V		
Electric current	I		$A = C\,s^{-1}$		12–2
Electric charge	Q		C	e (electron charge)	12–2, 13–1
Faraday constant	F	$\mathscr{F}$	$C\,mol^{-1}$		12–3
Number of electrons transferred					
in a redox reaction	n		1		12–5
in a half-reaction	n	n_{hc}	1		12–5
Electrical power	P		$W = J\,s^{-1}$		13–1
Kinetics					
Rate (velocity)	v	Rate	$mol\,m^{-3}\,s^{-1}$	$mol\,L\,s^{-1}$	14–1
Overall order of reaction	n		1		14–3
Rate constant for order n	k		$(mol^{-1}\,m^3)^{n-1}\,s^{-1}$	$(mol^{-1}\,L)^{n-1}\,s^{-1}$	14–2
Pre-exponential factor	A		$(mol^{-1}\,m^3)^{n-1}\,s^{-1}$	$(mol^{-1}\,L)^{n-1}\,s^{-1}$	14–6
Activation energy	E_a		$J\,mol^{-1}$	$kJ\,mol^{-1},$ $kcal\,mol^{-1}$	14–6
Half-life	$t_{1/2}$	τ	s	min, h, d, yr	14–3, 15–3
Nuclear Chemistry					
Activity (radioactivity)	A		$Bq = s^{-1}$	Ci	15–3
Decay constant	k		s^{-1}	$min^{-1}, h^{-1}, d^{-1},$ yr^{-1}	15–3
Half-life	$t_{1/2}$	τ	s	min, h, d, yr	15–3
Absorbed dose of radiation			Gy	rad	15–4
Dose equivalent			Sv	rem	15–4

TABLE B-6	*(continued)*				
Quantity	Recommended Symbols	Other Symbols	SI Unit	Other Units	Section of Book
Electromagnetic Radiation, Spectrometry, and Atomic Structure					
Wavelength	λ		m		16–1
Work function	Φ	Φ	J	eV	16–3
Planck's constant	h		J s		16–3
Bohr radius	a_o		m	bohr, Å	16–3
Wave function (in 3-d)	ψ		$m^{-3/2}$		16–4
Ionization energy	E_i	IE	$J\ mol^{-1}$	$kJ\ mol^{-1}$, eV, Ry	17–2
Electron affinity	E_{ea}	EA	$J\ mol^{-1}$	$kJ\ mol^{-1}$, eV, Ry	17–2
Bonding and Bond Geometry					
Dipole moment (directed from − to + by definition)	$\boldsymbol{\mu}, \boldsymbol{p}$	μ	C m	Debye	3–7, 17–5
Bond length (bond distance)	r	l, d	m	Å	3–4, 17–4
Bond dissociation energy	$\Delta E_d, D$		$J\ mol^{-1}$	eV, $kJ\ mol^{-1}$	10–5, 17–2
Bond dissociation enthalpy	ΔH_d		$J\ mol^{-1}$	eV, $kJ\ mol^{-1}$	10–5, 17–2

[a]Recommendations come from the International Union of Pure and Applied Chemistry (Quantities, Units and Symbols in Physical Chemistry, 2nd ed, Oxford, Blackwell Scientific Publications, 1993).

PROBLEMS

Note: Answers to blue-numbered problems are given in Appendix F.

The International System of Units

1. Rewrite the following in scientific notation, using only the base units of Table B–1, without prefixes:
   (a) 65.2 nanograms
   (b) 88 picoseconds
   (c) 5.4 terawatts
   (d) 17 kilovolts

2. Rewrite the following in scientific notation, using only the base units of Table B–1, without prefixes:
   (a) 66 mK
   (b) 15.9 MJ
   (c) 0.13 mg
   (d) 62 GPa

3. A certain quantity is measured in an SI unit called the *weber* (Wb). A weber equals a volt-second (1 Wb = 1 V s). Use the information in Table B–2 to express the weber in terms of base SI units.

4. A certain quantity is measured in the SI system in joules per tesla ($J\ T^{-1}$). Use the information in Table B–2 and the fact that $1\ T = 1\ kg\ s^{-2}\ A^{-1}$ to show that $1\ J\ T^{-1}$ equals $1\ A\ m^2$.

5. Arrange the following volumes in order from smallest to largest: $100\ cm^3$, 500 mL, $100\ dm^3$, 150 L, 1.5 gal (US).

6. Arrange the following in order from smallest to largest: 10^6 mm, $10^6\ \mu m$, 10^{-4} Mm, 10^{-2} km, 1000 in.

7. Express the following temperatures in °C:
   (a) 9001°F
   (b) 98.6°F (the normal body temperature of humans)
   (c) 20°F above the boiling point of water at 1 atm pressure
   (d) −40°F

8. Express the following temperatures in °F:
   (a) 5000°C (b) 40.0°C (c) 212°C (d) −40°C

9. Express the temperatures given in Problem 7 on the Kelvin scale.

10. Express the temperatures given in Problem 8 on the Kelvin scale.

11. Construct an "inverted" version of Table B–5 in which the SI units are given in terms of the more familiar inch, pound, and so forth.

12. The following routines sometimes are recommended as easy-to-remember ways to convert a temperature on the Celsius scale to the Fahrenheit scale and vice versa:

    $t_{°C} \longrightarrow t_{°F}$: add 40, multiply the result by $\frac{9}{5}$, subtract 40
    $t_{°F} \longrightarrow t_{°C}$: add 40, multiply the result by $\frac{5}{9}$, subtract 40

    Show algebraically why these steps give the right answers.

The Conversion of Units Using the Unit-Factor Method

13. Express the following in SI units, either base or derived:
(a) 55.0 mi h^{-1} (1 mi $= 1609.344$ m)
(b) 1.51 g cm^{-3}
(c) 1.6×10^{-19} C Å
(d) 0.15 mol L^{-1}
(e) 5.7×10^3 L atm min^{-1}

14. Express the following in SI units, either base or derived:
(a) 67.3 atm
(b) 1.0×10^4 V cm^{-1}
(c) 7.4 Å h^{-1}
(d) 22.4 L mol^{-1}
(e) 14.7 lb in^{-2}

15. Which of the following is not a possible unit of density: lb in.$^{-3}$, mg L^{-1}, Pa m^2 s^{-2}, J m^{-1} s^{-2}.

16. Which of the following is not a possible unit of speed: mi h^{-1}, in. d^{-1}, N s kg^{-1}, kg J^{-1}.

17. The kilowatt-hour is a common unit in measurements of the consumption of electrical energy. What is the conversion factor between the kilowatt-hour and the joule? Express 15.3 kW-h in joules.

18. A car's rate of fuel consumption is often measured in miles per gallon.
(a) Determine the conversion factor between miles per gallon and the SI unit of meters per cubic decimeter.
(b) Express 30.0 mi gal^{-1} in SI units.
(c) Express the meter per cubic decimeter in base SI units (with no use of prefixes).

19. A certain V-8 engine has a displacement of 404 in^3. Express this volume in cubic centimeters (cm^3) and liters.

20. Light travels in a vacuum at a speed of 3.00×10^8 m s^{-1}.
(a) Convert this speed to miles per second.
(b) Express this speed in furlongs per fortnight, a little-used unit of speed. (A furlong, a distance used in horse racing, is 660 ft; a fortnight is exactly two weeks.)

C Mathematics for General Chemistry

Mathematics is essential in chemistry. This appendix reviews some important mathematical facts and techniques for general chemistry.

C-1 Powers and Logarithms

Taking a logarithm and its inverse, raising a number to a power, are important in many chemical problems. Although the ready availability of electronic calculators makes the mechanical execution of these operations quite routine, it remains important to understand what is involved in these "special functions."

Powers

The mathematical expression 10^4 is read as "ten raised to the fourth power." Such an expression implies multiplying 10 by itself four times to give 10,000:

$$10^4 = 10 \times 10 \times 10 \times 10 = 10,000$$

Ten, or any other number, raised to the power 0 equals 1:

$$10^0 = 1$$

Negative powers of 10 give numbers smaller than 1, and are equivalent to raising 10 to the corresponding *positive* power and then taking the reciprocal:

$$10^{-3} = \frac{1}{10^3} = 0.001$$

• It is obvious that when the power is not a positive integer, we can no longer talk sensibly about multiplying 10 by itself that number of times. The quantity 10^x still can be calculated, however.

The idea of raising to a power can be extended to include powers that are not whole numbers. Raising to the power 0.5, for example, is the same as taking the square root:

$$10^{0.5} = 10^{1/2} = \sqrt{10} = 3.1623\ldots$$

Scientific calculators have a 10^x (or INV LOG) key that can be used for calculating powers of 10 in cases in which the power is not a whole number.

Numbers other than 10 can be raised to powers as well; a number that is raised to a power is referred to as a **base.** Many calculators have a y^x key that gives the value of any base y raised to the power x. One of the most important bases in scientific problems is the transcendental number called e (2.7182818...). The e^x (or INV LN) key on a calculator is used to raise e to the power x. The quantity e^x is called the **exponential** of x.

When a base is raised to a power that is the sum of two numbers, the result equals the value of the base raised to the first number multiplied by the value of the base raised to the second number. For example

$$10^{21+6} = 10^{21} \times 10^6 = 10^{27}$$

The same type of relationship holds for any base, including the base e. This property of powers is very useful in chemistry.

Logarithms

Logarithms occur frequently in chemistry problems. The logarithm of a number is the exponent to which some base has to be raised to obtain the number. The base is almost always either 10 or e. Thus,

$$B^a = n \quad \text{and} \quad \log_B n = a$$

where a is the logarithm, B is the base, and n is the number.

Common logarithms are base-10 logarithms—that is, they are powers to which 10 has to be raised in order to give the numbers. For example, $10^3 = 1000$, so $\log_{10} 1000 = 3$. Explicit mention of the base 10 often is omitted when showing common logarithms so that this equation is written as $\log 1000 = 3$. On an electronic calculator, a common logarithm is found by putting the number in the display and pressing the "log" key. The answer appears in the display. Only the logarithms of 1, 10, 100, 1000, and so forth are whole numbers. The logarithms of other numbers are decimal fractions. The decimal point in a logarithm divides it into two parts. To the left of the decimal point is the *characteristic*; to the right is the *mantissa*. Thus, the logarithm in the equation

> • Or pressing "log," entering the number, and then pressing "equals." The exact key strokes depend on the make and model of calculator.

$$\log (7.310 \times 10^3) = 3.8639$$

has a characteristic of 3 and a mantissa of 0.8639. As may be verified with a calculator, the base-10 logarithm of the much larger number 7.310×10^{23} is 23.8639. This logarithm clearly is related to the previous one. It has the same mantissa, but a different characteristic. The fact is that the characteristic is determined solely by the location of the decimal point in the number. In other words, the characteristic derives from the power to which the 10 is raised when a number is expressed in scientific notation. The mantissa, on the other hand, derives from the pre-exponential part.

How is the issue of significant figures handled when the logarithm of an experimental number is taken? Information about precision is transferred entirely in the pre-exponential part of an experimental number when it is put into scientific notation. (It appears in the number of digits written in the pre-exponential part.) The power to which the 10 is raised is exact. Therefore, when taking the logarithm of a measured number, the mantissa is written with as many significant figures as are present in the pre-exponential part of the number.

A logarithm is truly an exponent; as such, it follows the same rules of multiplication and division as other exponents. In multiplication and division

> • "The characteristic comes free" in determining how many significant figures to write in the logarithm of a measured number. The four significant figures in 7.310×10^{23} give rise to the four digits after the decimal point (that is, in the mantissa) in 23.8639, which is the logarithm of 7.310×10^{23}.

$$\log (n \times m) = \log n + \log m$$
$$\log (n/m) = \log n - \log m$$

Thus, the logarithm of 15 is

$$\log 5 + \log 3 = 0.69897 + 0.47712 = 1.17609$$

because 5 times 3 is 15. Also, the logarithm of $\frac{5}{3}$ is

$$\log 5 - \log 3 = 0.69897 - 0.47712 = 0.22185$$

Furthermore,

$$\log n^m = m \log n$$

so that the logarithm of 3^5 is

$$\log 3^5 = 5 \log 3 = 5 \times 0.47712 = 2.3856$$

• Logarithms can be negative; negative numbers do not have logarithms. Trying to compute the logarithm of a negative number on a calculator gives an error message.

Logarithms do not exist for negative numbers because there is no power to which 10 (or any other base) can be raised to give a negative number. This point can be understood as follows: The numbers between 0 and 1 take the set of negative numbers, ranging from $-\infty$ (negative infinity) up to 0, as their logarithms; the logarithm of 1 equals 0; the numbers that exceed 1 have the set of positive numbers as their logarithms. Thus, no real numbers remain to serve as the logarithms of negative numbers.

Another frequently used base for logarithms is the number e ($e = 2.7182818\ldots$). The logarithm to the base e is called the **natural logarithm** and is indicated by $\log_e$ or ln. The problem "find the logarithm to the base e of 8.23" could be written as "find ln 8.23." On a calculator, use the "ln" key to get natural logarithms. For example,

$$\ln 8.23 = 2.108 \qquad \text{and} \qquad \ln 0.0147 = -4.220$$

Base-e logarithms are related to base-10 logarithms by the formula

$$\ln n = \ln 10 \,(\log n) = 2.3025851 \log n$$

As stated previously, calculating logarithms and powers are inverse operations. Thus, to find the number for which 3.8639 is the common logarithm, simply calculate

$$10^{3.8639} = 7.310 \times 10^3$$

To find the number that has a natural logarithm of 2.108, calculate

$$e^{2.108} = 8.23$$

C-2   Using Graphs

In many situations in science, it is important to know how one quantity (measured or predicted) depends on a second quantity. The position of a traveling car depends on the time, for example, or the pressure of a gas depends on its volume at a given temperature. One very useful way to show such a relation is through a **graph,** in which one quantity is plotted against another.

The usual convention in drawing graphs is to use positions on the horizontal axis to represent values of the variable that the experimenter controls, and to use positions on the vertical axis to represent values of the measured or calculated quantity. Suppose the volume of a sample of gas is adjusted to the value 2.60 L. On a pressure gauge the pressure reads 3.22 atm. The volume in this case is the independent variable, because it is controlled from outside the experiment, and the pressure is the dependent, or response, variable. To display the result of this experiment on a graph, locate a point by moving along the (horizontal) volume axis to 2.60 L and then, starting from that point, moving along the vertical (pressure) axis to 3.22 atm. Mark this experimental point. Further measurements at different volumes would generate further points on the graph. After a series of such measurements, the points on the graph will often lie along a recognizable curve. Graphs of experimental data very frequently show some degree of scatter from a smooth curve because every experimental measurement involves some degree of uncertainty. Nothing is gained by drawing in a sawtooth curve that passes precisely through every measured point. When a real relation exists between the quantities measured, a systematic trend develops to the points despite small errors. A curve should be drawn that represents

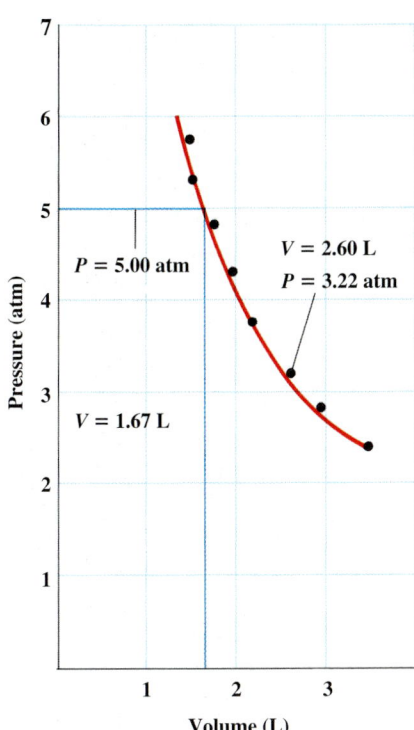

Figure C-1 • Connecting a set of experimentally determined points with a smooth curve on a graph gives a pictorial impression of the way one experimental quantity depends on another. Note the slight scatter of the points in this case.

this trend (Fig. C–1). Such a graph then can be used to predict the results of future measurements. Suppose, for example, that a gas pressure of 5.00 atm is needed for an experiment (see Fig. C–1). Read across from 5.00 atm to the curve that shows pressure against volume, and then move down to the horizontal axis to read off the volume (1.67 L). If the pressure gauge is to read 5.00 atm, the gas volume should be approximately 1.67 L. Alternatively, one could predict the pressure that will result from a given gas volume.

Straight-Line Graphs

The curves plotted on graphs can have many shapes. Of these, the simplest and most important is a straight line. A straight line is a graph of a relation such as

$$y = 4x + 7$$

or, more generally,

$$y = mx + b$$

where the dependent variable y is plotted along the vertical axis and the independent variable x along the horizontal axis (Fig. C–2). The m is the **slope** of the straight line, and the b is the **y-intercept.** The y-intercept is the point on the y-axis at which the straight line crosses that axis. This can be seen by setting x to zero and noting that y then equals b. The slope is a measure of the steepness of the straight line: the larger the value of m, the steeper the line. If the line goes *up* running from left to right, the slope is positive; if it goes *down* running from left to right, the slope is negative.

The slope of a straight line can be determined from the coordinates of two points on it. Suppose, for example, that when $x = 3$, then $y = 5$, and that when $x = 4$, then $y = 7$. These two points can be written in shorthand notation as (3, 5) and (4, 7).

Figure C–2 • A straight-line or linear relationship between two experiment variables is a very desirable result. It is easy to graph and easy to represent mathematically. The equation of this straight line ($y = 2x - 1$) fits the general form $y = mx + b$. Its y-intercept is -1, and its slope is 2.

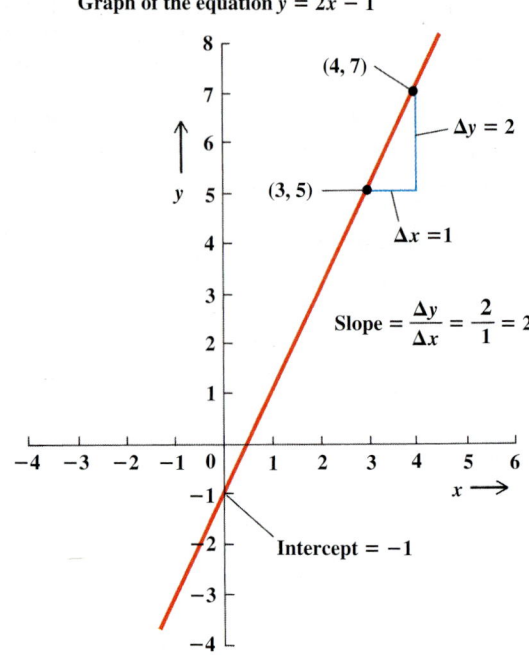

Graph of the equation $y = 2x - 1$

The slope of the line then can be defined as "the rise over the run," the change in the y-coordinate divided by the change in the x-coordinate:

> • This is true only if x increases from left to right on the horizontal axis and y increases from bottom to top on the vertical axis, which is the accepted practice.

$$\text{slope} = m = \frac{\Delta y}{\Delta x} = \frac{7 - 5}{4 - 3} = \frac{2}{1} = 2$$

The symbol Δ (capital Greek "delta") indicates the change in a quantity: the final value minus the initial value. If the two quantities being graphed have units, then the slope has units as well. If distance traveled (in meters) is graphed against time (in seconds) the units of the slope are meters per second (m s^{-1}).

C–3   Solutions of Algebraic Equations

Linear Equations

Many equations involving two variables give straight lines when one variable is graphed as a function of the other. Such equations are called linear equations. The idea has been generalized to equations with *more* than two variables. A **linear equation** is any equation that can be put into the form

$$a_1x_1 + a_2x_2 + a_3x_3 + \cdots = b$$

> • The unknowns could be symbolized x, y, z, and so forth, or with any other convenient symbols. Using subscripts guards against running out of symbols when many unknowns are involved.

where the a's are constants, and the subscripted x's are the variables, or unknowns. Clearly, the equation $y = mx + b$ can be put in this form. Linear equations never contain terms in which unknowns are multiplied together (such as x_1x_2) or raised to powers (such as x_1^2 or x_2^3). The preceding general form indeed has no provision for such terms.

A linear equation involving just one unknown has the form

$$ax = b$$

or can be rearranged to this form. Such equations are encountered frequently in chemistry. The constants a and b in such an equation fix the value of x. Determining that value is called solving the equation and is straightforward. For example,

$$5x + 9 = 0$$
$$5x = -9$$
$$x = -\tfrac{9}{5}$$

More generally, if $ax = b$, then $x = b/a$. Clearly however, the equation has no solution if $a = 0$.

Systems of Linear Equations

A **system** of linear equations is a group of two or more linear equations involving the same unknowns. The equations are simultaneous—all hold true at the same time. A system of linear equations may contain any number of equations (but a minimum of two) and any number of unknowns. The equations

$$2x_1 + x_2 - 3x_3 = 1$$
$$5x_1 + 2x_2 - 6x_3 = 5$$
$$3x_1 - x_2 - 4x_3 = 7$$

make up a system of three linear equations in the three unknowns x_1, x_2, and x_3. The three equations

$$2x_1^2 + 2x_2 - 3x_3 = 1$$
$$5x_1x_3 + 3x_2 - 6x_3 = 32$$
$$3x_1 - 2x_2 - 4x_3 = -22$$

form a system, but not a *linear* system because two of the three are non-linear. The first contains x_1^2, and the second contains the cross-term x_1x_3.

A **solution** of a system of linear equations consists of a list of numerical values for the unknowns. The values for the unknowns satisfy every equation in the system separately. A solution of the system of three linear equations

$$2x_1 + x_2 - 3x_3 = 1$$
$$5x_1 + 2x_2 - 6x_3 = 5$$
$$3x_1 - x_2 - 4x_3 = 7$$

is $x_1 = 3$, $x_2 = -2$, $x_3 = 1$. When these values are substituted into the equations

$$
\begin{array}{ccc}
x_1 & x_2 & x_3 \\
\downarrow & \downarrow & \downarrow
\end{array}
$$
$$(2 \times 3) + (1 \times -2) - (3 \times 1) = 1$$
$$(5 \times 3) + (2 \times -2) - (6 \times 1) = 5$$
$$(3 \times 3) - (1 \times -2) - (4 \times 1) = 7$$

simple arithmetic confirms that a true statement results in every case.

Not all systems of linear equations have solutions. Systems lack solutions when their equations are **inconsistent.** Inconsistent equations contradict each other. They cannot all hold true at the same time. The system

$$1x_1 + 3x_2 = 3$$
$$2x_1 + 6x_2 = -8$$

is inconsistent because the only way both equations hold true is for 6 to equal -8. (To see this, multiply all of the terms in the first equation by 2 and compare the result to the second equation.) The system

$$2x_1 + 2x_2 = 5$$
$$2x_1 - 2x_2 = 1$$
$$2x_1 + 3x_2 = 1$$

is also inconsistent. Satisfying the first and second equations (without reference to the third) requires that $x_1 = \frac{3}{2}$ and $x_2 = 1$. Satisfying the second and third equations (without reference to the first) requires that $x_1 = \frac{1}{2}$ and $x_2 = 0$, which is a contradiction. This inconsistent system has three equations but only two unknowns. Systems in which the number of equations exceeds the number of unknowns are often, but not always, inconsistent. Finally, a single *degenerate* equation causes inconsistency. Degenerate equations have zeros multiplying all of the unknowns, as in the example

$$0x_1 + 0x_2 + 0x_3 = 7$$

No possible values of x_1, x_2, and x_3 make this equation true. Hence, no system of equations in which it appears has a solution.

　　Inconsistent systems do not arise in chemically meaningful applications except as the result of errors in algebraic manipulation or transcription. They are mentioned here to help in troubleshooting such errors.

Solving Systems of Linear Equations.　Solving chemistry problems often comes down to solving a consistent system of linear equations. In a consistent system:

- A single solution exists if the number of unknowns *equals* the number of independent equations and
- An infinite number of solutions exists if the number of unknowns *exceeds* the number of independent equations

(Here, "independent" means that none of the equations may equal another simply multiplied through by a constant. An equation created in this way adds no new information to the system and cannot be counted as independent.) The first case is very important. If the consistent system is small (two to four or five equations), the unique solution quickly can be obtained by algebraic manipulations with pencil and paper. For example, the solution of the system

$$5x_1 + 3x_2 = 13$$
$$x_1 - x_2 = 1$$

is obtained by rearranging the second equation to isolate x_1 on the left: $x_1 = 1 + x_2$. This expression for x_2 is then substituted into the first equation to give

$$5(1 + x_2) + 3x_2 = 13$$

Carrying out the indicated operations gives

$$5 + 5x_2 + 3x_2 = 13$$
$$8x_2 = 13 - 5$$
$$x_2 = 1$$

Putting $x_2 = 1$ into the second equation then gives $x_1 = 2$. The solution is $x_1 = 2$; $x_2 = 1$.

Another example arises in Example 2–2 (page 57). The system of equations is

$$x + y = 1$$
$$x - w = 0$$
$$3x + y + z = 4$$
$$y + 2z - 3w = 4$$

From the second equation, $x = w$. Substituting x for w in the fourth equation gives the following system of three equations:

$$x + y = 1$$
$$3x + y + z = 4$$
$$-3x + y + 2z = 4$$

Adding the second and third of these equations eliminates x and gives

$$2y + 3z = 8$$

Multiplying the first equation by 3 and adding it to the third equation eliminates x in a different way, giving

$$4y + 2z = 7$$

The system is now down to these two equations in two unknowns:

$$2y + 3z = 8$$
$$4y + 2z = 7$$

Multiplying the first equation by 2 and leaving the second equation alone gives

$$4y + 6z = 16$$
$$4y + 2z = 7$$

Subtracting the second equation from the first gives

$$4z = 9$$

from which

$$z = \tfrac{9}{4}.$$

Inserting $z = \tfrac{9}{4}$ back into the set of two equations gives $y = \tfrac{5}{8}$. Inserting $z = \tfrac{9}{4}$ and $y = \tfrac{5}{8}$ back into the set of three equations gives $x = \tfrac{3}{8}$. And, because $w = x$, $w = \tfrac{3}{8}$ as well. The solution of this system of equations is

$$x = \tfrac{3}{8}, y = \tfrac{5}{8}, z = \tfrac{9}{4}, w = \tfrac{3}{8}$$

This is the only set of values that satisfies the four original equations simultaneously.

Solving Systems of Equations with a Calculator. Solving large systems of linear equations by hand becomes tiresome. Fortunately, powerful methods (*matrix* methods) exist to solve systems of practically unlimited size. These methods lend themselves well to automation, and many calculators have built-in programs that employ them. The first step in using such a program is to translate all the equations into the standard form

$$a_1x_1 + a_2x_2 + a_3x_3 + \cdots + a_nx_n = b$$

A system of three equations in three unknowns would then look like this:

$$a_{1,1}x_1 + a_{1,2}x_2 + a_{1,3}x_3 = b_1$$
$$a_{2,1}x_1 + a_{2,2}x_2 + a_{2,3}x_3 = b_2$$
$$a_{3,1}x_1 + a_{3,2}x_2 + a_{3,3}x_3 = b_3$$

Some of the a's and b's may, of course, equal zero or negative numbers. Next, launch the calculator program. The details depend on the model and make of the calculator. On Texas Instruments calculators such as the TI-85 or TI-86, keying $\boxed{\text{2nd}}$, then $\boxed{\text{SIMULT}}$, calls up the program for solving simultaneous equations. This program starts with a screen that asks for the number of simultaneous equations. Entering "2" brings up a list with room for three entries, $a_{1,1}$, $a_{1,2}$, and b_1; entering "3" brings a list with room for four entries, and so forth. After these numbers are entered from left to right for the first equation, a second screen appears for the second equation, a third screen for the third equation, and so on. All of the constants must be entered, and in the correct order. (This requires 3×4 or 12 entries for a system of three equations.) Pushing the $\boxed{\text{SOLVE}}$ key then causes the calculator to display the solution as a list in the order x_1, x_2, x_3.... The error message SINGULAR MAT (singular matrix) indicates that an inconsistent system of linear equations has been entered. Other error messages signal certain types of data entry error. An absence of error messages does *not* certify that the solution is correct. Only checking (by substitution into the original equations) ensures this.

When the number of unknowns in a consistent system *exceeds* the number of independent equations, an infinite number of solutions exist. Obtaining one of these solutions requires nothing more than arbitrarily assigning values to unknowns until the number of unknowns that remains equals the number of equations. This new system can then be solved either by hand or on a calculator.

Quadratic Equations

Non-linear equations are of many kinds. One of the most common in chemistry is the quadratic equation. Quadratic equations in the unknown x always can be rearranged into the form

$$ax^2 + bx + c = 0$$

in which the constant a is non-zero and the constants b and c may be positive, negative, or zero. Every quadratic equation has two **roots.** These are the values of x for which the equation is satisfied. One good way to obtain the roots of quadratic equations is simply to memorize and use the quadratic formula:

$$x = \frac{-b \pm \sqrt{b^2 - 4ac}}{2a}$$

• The symbol $\pm$ is a compact way to represent two solutions to this equation (called "roots" of the equation). One corresponds to using the $+$ sign, and the other to the $-$ sign.

The plus-or-minus sign in this formula causes it to generate the two roots. Suppose that the equation

$$x = 3 + \frac{7}{x}$$

arises in a chemistry problem. Multiplying by x and rearranging the terms gives

$$x^2 - 3x - 7 = 0$$

This is a quadratic equation having $a = 1$, $b = -3$, and $c = -7$. Inserting these values into the quadratic formula gives

$$x = \frac{3 \pm \sqrt{(-3)^2 - 4(1)(-7)}}{2} = \frac{3 \pm \sqrt{9 + 28}}{2} = \frac{3 \pm \sqrt{37}}{2}$$

The two roots of the equation are

$$x = 4.5414 \qquad \text{and} \qquad x = -1.5414$$

The choice of the proper root to use as the answer in a chemistry problem generally can be made on physical grounds. If x corresponds to a concentration, for example, a negative root is "unphysical" and can be discarded.

Electronic calculators routinely are pre-programmed to solve quadratic equations. For example, on many Texas Instruments calculators, the key sequence $\boxed{\text{2nd}}$, then $\boxed{\text{POLY}}$, calls up a screen asking for the "order" of the equation to be solved. Quadratic equations have an order of 2. Entering a 2 brings up a screen with space for the three constants a_2, a_1, and a_0 (which are this maker's labels for a, b, and c, respectively). Entering the constants and pushing the $\boxed{\text{SOLVE}}$ key gives the two roots of the equation.

- If the quantity $b^2 - 4ac$ under the radical sign is negative, *no* real solution exists because the square root of a negative number is imaginary. When this happens in the course of a chemical calculation, it is almost always an indication that an error has been made.

Other Non-Linear Equations

The solution of cubic or higher-order algebraic equations (or more complicated equations involving sines, cosines, logarithms, or exponentials) becomes more difficult. "Equation Solver" programs on many calculators easily solve most such equations. The availability of such programs is a mixed blessing, however. They require practice with a particular model of calculator. Their official-looking output can seduce users into uncritical acceptance of wrong answers that come from input errors. And, of course, they do not work when the battery fails. Approximate or numerical methods can give the user a better feel for the equation and often provide answers even when a calculator is unavailable. As an illustration, consider the equation

$$x^2\left(\frac{2.00 + x}{3.00 - x}\right) = 1.00 \times 10^{-6}$$

Equations similar to this arise frequently in equilibrium calculations in chemistry. With a bit of algebra, it can be rearranged to

$$x^3 + 2.00x^2 + (1.00 \times 10^{-6})x - (3.00 \times 10^{-6}) = 0$$

A calculator spits out three roots when this cubic (order 3) equation is entered: -1.9999987, -0.0012254, and $+0.001224$. For more insight (or if a working calculator is not available), try to reduce the equation to a simpler approximate equation. Thus, assume x to be small relative to both 2.00 and 3.00. The result is

$$x^2\left(\frac{2.00}{3.00}\right) \approx 1.00 \times 10^{-6}$$

- The symbol $\approx$ means "is approximately equal to."

Solving *this* equation, even by hand, is very quick, and leads to the roots $x = \pm 0.00122$. A glance confirms that the solutions obtained in this way are in fact small compared to 2.00 (and 3.00) and that the approximation was a good one. When a quantity (x, in this case) is added to or subtracted from a larger quantity in a complicated equation, it is usually worth a try to simplify the equation by simply neglecting the small quantity. Note that this tactic works only for addition and subtraction and never for multiplication or division.

- Whenever an equation is solved by approximation, a formal check should be run to confirm that the approximation is valid.

C-4 Ratios

One way of comparing two numbers is to form their **ratio.** A ratio is written by using a division sign (as in $7 \div 11$), by separating the two numbers with a colon (as in $7:11$), by constructing a fraction (as in $7/11$), or by computing a decimal equivalent ($7/11 = 0.636363...$). Two ratios come from every pair of numbers because it matters which number divides into which. The other ratio involving 7 and 11 is $11:7$, or $11/7$, which equals $1.57142857...$. This ratio is the **reciprocal** of the first ratio. The product of a number and its reciprocal equals 1.

In chemistry, the ratios among sets of numbers are quite often more important than are the numbers themselves. These ratios are routinely computed with electronic calculators, but calculators display their results as decimals, which creates a problem because ratios in the form of fractions are required for many purposes in chemistry. The easiest solution is to know and recognize the decimal equivalents of simple ratios such as the following:

$\frac{1}{8} = 0.125$	$\frac{1}{7} = 0.142857...$	$\frac{1}{6} = 0.166...$	$\frac{1}{5} = 0.2$
$\frac{1}{4} = 0.25$	$\frac{2}{7} = 0.285714...$	$\frac{1}{3} = 0.333...$	$\frac{3}{8} = 0.375$
$\frac{2}{5} = 0.4$	$\frac{3}{7} = 0.428571...$	$\frac{1}{2} = 0.5$	$\frac{4}{7} = 0.57142...$
$\frac{2}{3} = 0.666...$	$\frac{3}{5} = 0.6$	$\frac{5}{8} = 0.625$	$\frac{5}{7} = 0.714285...$
$\frac{3}{4} = 0.75$	$\frac{4}{5} = 0.8$	$\frac{5}{6} = 0.833...$	$\frac{6}{7} = 0.857142...$
$\frac{7}{8} = 0.875$			

Suppose a calculation gives 2.375 as a desired ratio. The preceding allows this to be recognized as equal to $2\frac{3}{8}$. It is then easy to express the ratio as a fraction: $2\frac{3}{8} = \frac{19}{8}$.

A common application of ratios in chemistry is to convey the relative amounts of different substances making up some mixture. To do this, the amount of each component of the mixture is in turn divided by the total amount of all of the components. When the various amounts are measured as masses, the result is a **mass fraction** w for each component.

$$\text{mass fraction of component 1} = w_1 = \frac{m_1}{\Sigma m_i}$$

The units in the numerator and denominator of this fraction are the same so that a mass fraction has no units. A **mass percentage** is a mass fraction put on a per hundred basis. Multiplying the ratio of the mass of the part to the mass of the whole by 100% does this:

$$\text{mass percentage} = (\text{mass fraction}) \times 100\%$$

When one component is present in a small or very small relative amount, its contribution to the whole is sometimes expressed on a **parts per million** (ppm) or **parts per billion** (ppb) basis:

$$\text{ppmm} = \text{mass fraction} \times 10^6$$
$$\text{ppbm} = \text{mass fraction} \times 10^9$$

where the final m's attached to the abbreviations mean that these are parts per million and parts per billion by mass. Use of parts per million and parts per billion avoids inconveniently small numbers.

If the amounts in a mixture are compared on the basis of their relative volumes, then the **volume fraction** ϕ of each component is computed

$$\text{volume fraction of component 1} = \phi_1 = \frac{V_1}{\Sigma V_i}$$

where the V's are the volumes of the components prior to mixing. The volume percentage, parts per millions by volume (ppmv), and parts per billion by volume (ppbv) of a component then are obtained by multiplying its volume fraction by 100%, 10^6, and 10^9, respectively.

C-5 Formulas for the Area and Volume of Simple Shapes

Geometry plays an important role in understanding the three-dimensional structures of molecules, as well as in other fields of chemistry. The following *mensuration formulas* are given for reference.

1. Area of a circle having radius r:

$$A = \pi r^2$$

2. Area of a triangle having base b and altitude a:

$$A = ba/2$$

3. Volume of a cube having edge e:

$$V = e^3$$

4. Volume of a rectangular prism having edges a, b, and c:

$$V = abc$$

5. Volume of a sphere having radius r:

$$V = \tfrac{4}{3}\pi r^3$$

PROBLEMS

Note: Answers to the blue-numbered problems are given in Appendix F.

Powers and Logarithms

1. Calculate each of the following expressions. Assume that all the numbers result from experiments and give the answers to the proper number of significant figures.
   (a) $\log (3.56 \times 10^4)$
   (b) $e^{-15.69}$
   (c) $10^{8.41}$
   (d) $\ln(6.893 \times 10^{-22})$

2. Calculate each of the following expressions. Assume that all the numbers result from experiments and give the answers to the proper number of significant figures.
   (a) $10^{-16.528}$
   (b) $\ln (4.30 \times 10^{13})$
   (c) $e^{14.21}$
   (d) $\log (4.983 \times 10^{-11})$

3. What number has a common logarithm of 0.4793?

4. What number has a natural logarithm of -15.824?

5. Determine the common logarithm of 3.00×10^{1211}. It is likely that your calculator will *not* give a correct answer. Explain why.

6. Compute the value of $10^{7107.8}$. It is quite likely that your calculator will *not* give the correct answer. Explain why.

7. The common logarithm of 5.64 is 0.751. Without using a calculator, determine the common logarithm of 5.64×10^7 and of 5.64×10^{-3}.

8. The common logarithm of 2.68 is 0.428. Without using a calculator, determine the common logarithm of 2.68×10^{192} and of 2.68×10^{-289}.

Using Graphs

9. A graph of distance traveled against time elapsed is found in one case to be a straight line. After an elapsed time of 1.5 h, the distance traveled was 75 mi. After an elapsed time of 3.0 h, the distance traveled was 150 mi. Calculate the slope of the graph of distance against time, and give its units.

10. The pressure of a sample of gas in a rigid container is measured at several different temperatures, and it is found that a graph of pressure against temperature is a straight line. At 20.0°C, the pressure is 4.30 atm, and at 100.0°C, the pressure is 5.47 atm. Calculate the slope of the graph of pressure against temperature, and give its units.

11. Rewrite each of the following linear equations in the form $y = mx + b$, and give the slope and intercept of the corresponding graph. Then draw the graph.
    (a) $y = 4x - 7$
    (b) $7x - 2y = 5$
    (c) $3y + 6x - 4 = 0$

12. Rewrite each of the following linear equations in the form $y = mx + b$, and give the slope and y-intercept of the corresponding graph. Then draw the graph.
    (a) $y = -2x - 8$
    (b) $-3x + 4y = 7$
    (c) $7y - 16x + 53 = 0$

13. Graph the equation

    $$y = 2x^3 - 3x^2 + 6x - 5$$

    from $x = -2$ to $x = +2$. Is the graph a straight line?

14. Graph the equation

    $$y = 8 - 10x - 3x^2/2 - 3x$$

    from $x = -3$ to $x = +3$. Is the graph a straight line?

Solutions of Algebraic Equations

15. Solve the following linear equations for x:
    (a) $7x + 5 = 0$
    (b) $-4x + 3 = 0$
    (c) $-3x = -2$

16. Solve the following linear equations for x:
    (a) $6 - 8x = 0$
    (b) $-2x - 5 = 0$
    (c) $-4x = -8$

17. Solve the following consistent systems of linear equations, using a calculator if necessary:
    (a) $2x - 5y = 1$
        $3x + 2y = 11$
    (b) $2x - y + z = 6$
        $2x + y + z = 3$
        $x + y + z = 2$
    (c) $2x - y + 2z = 6$
        $2x + y + z = 3$
        $x + y + z = 2$

18. Solve the following consistent systems of linear equations, using a calculator if necessary:

    (a) $3x - 2y = 7$
        $x + 3y = -16$
    (b) $1x + 1y + 0z + 0w = 1$
        $1x + 0y + 0z - 1w = 0$
        $3x + 1y + 1z + 0w = 4$
        $0x + 1y + 2z - 3w = 4$
    (c) $x - y + 3z = 6$
        $2x + 3y - z = 3$
        $5x + 7y + z = 7$

19. Solve the following quadratic equations for x:
    (a) $4x^2 + 7x - 5 = 0$
    (b) $2x^2 = -3 - 6x$
    (c) $2x + 3/x = 6$

20. Solve the following quadratic equations for x:
    (a) $6x^2 + 15x + 2 = 0$
    (b) $4x = 5x^2 - 3$
    (c) $1/2 - x + 3x = 4$

21. Solve each of the following equations for x using the approximation of small x. Check the answers using a calculator's equation-solving capability, if a suitable model is available.
    (a) $x(2.00 + x)^2 = 2.6 \times 10^{-6}$
    (b) $x(3.00 - 7x)(2.00 + 2x) = 0.230$
    (c) $2x^3 + 3x^2 + 12x = -16$

22. Solve each of the following equations for x using the approximation of small x. Check the answers using a calculator's equation-solving capability, if a suitable model is available.
    (a) $x(2.00 + x)(3.00 - x)(5.00 + 2x) = 1.58 \times 10^{-15}$
    (b) $x(3.00 + x)(1.00 - x)/2.00 - x = 0.122$
    (c) $12x^3 - 4x^2 + 35x = 10$

23. Use a calculator to solve the equation

    $$\log \ln x = -x$$

    for x. Give x to four significant figures.

24. Use a calculator to find a number that is equal to the reciprocal of its own natural logarithm. Report the answer to four digits.

Ratios

25. A mixture contains 24.60 g of water, 17.2 g of NaCl, and 0.0020 g of mercury. The volumes of these three components are 24.67 mL, 7.944 mL, and 1.47×10^{-4} mL, respectively.
    (a) Compute the mass fraction of all three components in the mixture.
    (b) Compute the volume fraction of all three components in the mixture.
    (c) Compute the parts per million by volume (ppmv) of mercury in the mixture.
    (d) Compute the parts per billion by mass (ppbm) of mercury in the mixture.

26. A gaseous mixture contains 13.6 ppm by volume of xenon. If the total of the volumes of all the components in the mixture equals 716 L, compute the volume of xenon in the mixture.

Standard Chemical Thermodynamic Properties

This table lists standard enthalpies of formation ΔH_f°, standard third-law entropies S°, and standard Gibbs energies of formation ΔG_f° for a variety of chemical species at 1 atm and 25°C (298.15 K). The table proceeds from the left to the right side of the periodic table. Binary compounds are listed under the element that occurs to the left in the periodic table, except that binary oxides and hydrides are listed with the other element. Thus, KCl is listed with potassium and its compounds, but ClO_2 is listed with chlorine and its compounds.

Note that the *solution-phase* entropies are not absolute entropies but are measured relative to the arbitrary standard $S^\circ(H^+(aq)) = 0$. It is for this reason that some of them are negative.

Most of the data are taken from the *NBS Tables of Chemical Thermodynamic Properties* (1982), or the *NIST-JANF Thermochemical Tables,* 4th Ed. (1998), changed where necessary from a standard pressure of 0.1 MPa to 1 atm.

Additional data are readily available. The U.S. National Institute of Standards and Technology (NIST) maintains a Web site (webbook.nist.gov/chemistry) with thermochemical data on more than 6000 substances. The Committee on Data for Science and Technology (CODATA) offers a shorter list of data (www.codata.org).

	Substance	ΔH_f° (25°C) (kJ mol^{-1})	S° (25°C) (J K^{-1} mol^{-1})	ΔG_f° (25°C) (kJ mol^{-1})
	H(g)	217.96	114.60	203.26
	H$_2$(g)	0	130.57	0
	H$^+$(aq)	0	0	0
	H$_3$O$^+$(aq)	−285.83	69.91	−237.18
I	Li(s)	0	29.12	0
	Li(g)	159.37	138.66	126.69
	Li$^+$(aq)	−278.49	13.4	−293.31
	LiH(s)	−90.54	20.01	−68.37
	Li$_2$O(s)	−597.94	37.57	−561.20
	LiF(s)	−615.97	35.65	−587.73
	LiCl(s)	−408.61	59.33	−384.39
	LiBr(s)	−351.21	74.27	−342.00
	LiI(s)	−270.41	86.78	−270.29
	Na(s)	0	51.21	0
	Na(g)	107.32	153.60	76.79
	Na$^+$(aq)	−240.12	59.0	−261.90
	Na$_2$O(s)	−414.22	75.06	−375.48

Substance	ΔH_f° (25°C) (kJ mol^{-1})	S° (25°C) (J K^{-1} mol^{-1})	ΔG_f° (25°C) (kJ mol^{-1})
NaOH(s)	−425.61	64.46	−379.53
NaF(s)	−573.65	51.46	−543.51
NaCl(s)	−411.15	72.13	−384.15
NaBr(s)	−361.06	86.82	−348.98
NaI(s)	−287.78	98.53	−286.06
NaNO$_3$(s)	−467.85	116.52	−367.07
Na$_2$S(s)	−364.8	83.7	−349.8
Na$_2$SO$_4$(s)	−1387.08	149.58	−1270.23
NaHSO$_4$(s)	−1125.5	113.0	−992.9
Na$_2$CO$_3$(s)	−1130.68	134.98	−1044.49
NaHCO$_3$(s)	−950.81	101.7	−851.1
K(s)	0	64.18	0
K(g)	89.24	160.23	60.62
K$^+$(aq)	−252.38	102.5	−283.27
KO$_2$(s)	−284.93	116.7	−239.4
K$_2$O$_2$(s)	−494.1	102.1	−425.1
KOH(s)	−424.76	78.9	−379.11
KF(s)	−567.27	66.57	−537.77
KCl(s)	−436.75	82.59	−409.16
KClO$_3$(s)	−397.73	143.1	−296.25
KBr(s)	−393.80	95.90	−380.66
KI(s)	−327.90	106.32	−324.89
KMnO$_4$(s)	−837.2	171.71	−737.7
K$_2$CrO$_4$(s)	−1403.7	200.12	−1295.8
K$_2$Cr$_2$O$_7$(s)	−2061.5	291.2	−1881.9
Rb(s)	0	76.78	0
Rb(g)	80.88	169.98	53.09
Rb$^+$(aq)	−251.17	121.50	−283.98
RbCl(s)	−435.35	95.90	−407.82
RbBr(s)	−394.59	109.96	−381.79
RbI(s)	−333.80	118.41	−328.86
Cs(s)	0	85.23	0
Cs(g)	76.06	175.49	49.15
Cs$^+$(aq)	−258.28	133.05	−292.02
CsF(s)	−553.5	92.80	−525.5
CsCl(s)	−443.04	101.17	−414.15
CsBr(s)	−405.81	113.05	−391.41
CsI(s)	−346.60	123.05	−340.58

	Substance	ΔH_f° (25°C) (kJ mol^{-1})	S° (25°C) (J K^{-1} mol^{-1})	ΔG_f° (25°C) (kJ mol^{-1})
II	Be(s)	0	9.44	0
	BeO(s)	−609.6	14.14	−580.3
	Mg(s)	0	32.68	0
	Mg(g)	147.70	148.54	113.13
	Mg^{2+}(aq)	−466.85	−138.1	−454.8
	MgO(s)	−601.70	26.94	−569.45

Substance	ΔH_f° (25°C) (kJ mol^{-1})	S° (25°C) (J K^{-1} mol^{-1})	ΔG_f° (25°C) (kJ mol^{-1})
MgCl$_2$(s)	−641.32	89.62	−591.82
MgSO$_4$(s)	−1284.9	91.6	−1170.7
Ca(s)	0	41.42	0
Ca(g)	178.2	154.77	144.33
Ca^{2+}(aq)	−542.83	−53.1	−553.58
CaH$_2$(s)	−186.2	42	−147.2
CaO(s)	−635.09	39.75	−604.05
CaS(s)	−482.4	56.5	−477.4
Ca(OH)$_2$(s)	−986.09	83.39	−898.56
CaF$_2$(s)	−1219.6	68.87	−1167.3
CaCl$_2$(s)	−795.8	104.6	−748.1
CaBr$_2$(s)	−682.8	130	−663.6
CaI$_2$(s)	−533.5	142	−528.9
Ca(NO$_3$)$_2$(s)	−938.39	193.3	−743.20
CaC$_2$(s)	−59.8	69.96	−64.9
CaCO$_3$(s, calcite)	−1206.92	92.9	−1128.84
CaCO$_3$(s, aragonite)	−1207.13	88.7	−1127.80
CaSO$_4$(s)	−1434.11	106.9	−1321.86
CaSiO$_3$(s)	−1634.94	81.92	−1549.66
CaMg(CO$_3$)$_2$(s, dolomite)	−2326.3	155.18	−2163.4
Sr(s)	0	52.3	0
Sr(g)	164.4	164.51	130.9
Sr^{2+}(aq)	−545.80	−32.6	−559.48
SrCl$_2$(s)	−828.9	114.85	−781.1
SrCO$_3$(s)	−1220.0	97.1	−1140.1
Ba(s)	0	62.8	0
Ba(g)	180	170.24	146
Ba^{2+}(aq)	−537.64	9.6	−560.77
BaCl$_2$(s)	−858.6	123.68	−810.4
BaCO$_3$(s)	−1216.3	112.1	−1137.6
BaSO$_4$(s)	−1473.2	132.2	−1362.3

III

Substance	ΔH_f° (25°C) (kJ mol^{-1})	S° (25°C) (J K^{-1} mol^{-1})	ΔG_f° (25°C) (kJ mol^{-1})
Sc(s)	0	34.64	0
Sc(g)	377.8	174.68	336.06
Sc^{3+}(aq)	−614.2	−255	−586.6
Ti(s)	0	30.63	0
Ti(g)	469.9	180.19	425.1
TiO$_2$(s, rutile)	−944.7	50.33	−889.5
TiCl$_4$(ℓ)	−804.2	252.3	−737.2
TiCl$_4$(g)	−763.2	354.8	−726.8
Cr(s)	0	23.77	0
Cr(g)	396.6	174.39	351.8
Cr$_2$O$_3$(s)	−1139.7	81.2	−1058.1
CrO$_4^{2-}$(aq)	−881.15	50.21	−727.75

Substance	ΔH_f° (25°C) (kJ mol^{-1})	S° (25°C) (J K^{-1} mol^{-1})	ΔG_f° (25°C) (kJ mol^{-1})
$Cr_2O_7^{2-}(aq)$	−1490.3	261.9	−1301.1
$W(s)$	0	32.64	0
$W(g)$	849.4	173.84	807.1
$WO_2(s)$	−589.69	50.54	−533.92
$WO_3(s)$	−842.87	75.90	−764.08
$Mn(s)$	0	32.01	0
$Mn(g)$	280.7	238.5	173.59
$Mn^{2+}(aq)$	−220.75	−73.6	−228.1
$MnO(s)$	−385.22	59.71	−362.92
$MnO_2(s)$	−520.03	53.05	−465.17
$MnO_4^-(aq)$	−541.4	191.2	−447.2
$Fe(s)$	0	27.28	0
$Fe(g)$	416.3	180.38	370.7
$Fe^{2+}(aq)$	−89.1	−137.7	−78.9
$Fe^{3+}(aq)$	−48.5	−315.9	−4.7
$Fe_{0.947}O(s, wüstite)$	−266.27	57.49	−245.12
$Fe_2O_3(s, hematite)$	−824.2	87.40	−742.2
$Fe_3O_4(s, magnetite)$	−1118.4	146.4	−1015.5
$Fe(OH)_3(s)$	−832.62	104.56	−705.47
$FeS(s)$	−100.0	60.29	−100.4
$FeCO_3(s)$	−740.57	93.1	−666.72
$Fe(CN)_6^{3-}(aq)$	561.9	270.3	729.4
$Fe(CN)_6^{4-}(aq)$	455.6	95.0	695.1
$Co(s)$	0	30.04	0
$Co(g)$	424.7	179.41	380.3
$Co^{2+}(aq)$	−58.2	−113	−54.4
$Co^{3+}(aq)$	92	−305	134
$CoO(s)$	−237.94	52.97	−214.22
$CoCl_2(s)$	−312.5	109.16	−269.8
$Ni(s)$	0	29.87	0
$Ni(g)$	429.7	182.08	384.5
$Ni^{2+}(aq)$	−54.0	−128.9	−45.6
$NiO(s)$	−239.7	37.99	−211.7
$Pt(s)$	0	41.63	0
$Pt(g)$	565.3	192.30	520.5
$PtCl_6^{2-}(aq)$	−668.2	219.7	−482.7
$Cu(s)$	0	33.15	0
$Cu(g)$	338.32	166.27	298.61
$Cu^+(aq)$	71.67	40.6	49.98
$Cu^{2+}(aq)$	64.77	−99.6	65.49
$CuO(s)$	−157.3	42.63	−129.7
$Cu_2O(s)$	−168.6	93.14	−146.0
$CuCl(s)$	−137.2	86.2	−119.88
$CuCl_2(s)$	−220.1	108.07	−175.7

Substance	ΔH_f° (25°C) (kJ mol^{-1})	S° (25°C) (J K^{-1} mol^{-1})	ΔG_f° (25°C) (kJ mol^{-1})
$CuSO_4(s)$	−771.36	109	−661.9
$Cu(NH_3)_4^{2+}(aq)$	−348.5	273.6	−111.07
$Ag(s)$	0	42.55	0
$Ag(g)$	284.55	172.89	245.68
$Ag^+(aq)$	105.58	72.68	77.11
$AgCl(s)$	−127.07	96.2	−109.81
$AgNO_3(s)$	−124.39	140.92	−33.48
$Ag(NH_3)_2^+(aq)$	−111.29	245.2	−17.12
$Au(s)$	0	47.40	0
$Au(g)$	366.1	180.39	326.3
$Zn(s)$	0	41.63	0
$Zn(g)$	130.73	160.87	95.18
$Zn^{2+}(aq)$	−153.89	−112.1	−147.06
$ZnO(s)$	−348.28	43.64	−318.32
$ZnS(s, sphalerite)$	−205.98	57.7	−201.29
$ZnCl_2(s)$	−415.05	111.46	−369.43
$ZnSO_4(s)$	−982.8	110.5	−871.5
$Zn(NH_3)_4^{2+}(aq)$	−533.5	301	−301.9
$Cd(s)$	0	51.76	0
$Cd^{2+}(aq)$	−75.90	−73.2	−77.612
$Hg(\ell)$	0	76.02	0
$Hg(g)$	61.32	174.85	31.85
$HgO(s)$	−90.83	70.29	−58.56
$HgCl_2(s)$	−224.3	146.0	−178.6
$Hg_2Cl_2(s)$	−265.22	192.5	−210.78
III $B(s)$	0	5.86	0
$B(g)$	562.7	153.34	518.8
$B_2H_6(g)$	35.6	232.00	86.6
$B_5H_9(g)$	73.2	275.81	174.9
$B_2O_3(s)$	−1272.77	53.97	−1193.70
$H_3BO_3(s)$	−1094.33	88.83	−969.02
$BF_3(g)$	−1137.00	254.01	−1120.35
$BF_4^-(aq)$	−1574.9	180	−1486.9
$BCl_3(g)$	−403.76	289.99	−388.74
$BBr_3(g)$	−205.64	324.13	−232.47
$Al(s)$	0	28.33	0
$Al(g)$	326.4	164.43	285.7
$Al^{3+}(aq)$	−531	−321.7	−485
$Al_2O_3(s)$	−1675.7	50.92	−1582.3
$AlCl_3(s)$	−704.2	110.67	−628.8
$Ga(s)$	0	40.88	0
$Ga(g)$	277.0	168.95	238.9
$Tl(s)$	0	64.18	0
$Tl(g)$	182.21	180.85	147.44

	Substance	ΔH_f° (25°C) (kJ mol^{-1})	S° (25°C) (J K^{-1} mol^{-1})	ΔG_f° (25°C) (kJ mol^{-1})
IV	C(s, graphite)	0	5.74	0
	C(s, diamond)	1.895	2.377	2.900
	C(g)	716.682	157.99	671.29
	CH$_4$(g)	−74.81	186.15	−50.75
	C$_2$H$_2$(g)	226.73	200.83	209.20
	C$_2$H$_4$(g)	52.26	219.45	68.12
	C$_2$H$_6$(g)	−84.68	229.49	−32.89
	C$_3$H$_8$(g)	−103.85	269.91	−23.49
	n-C$_4$H$_{10}$(g)	−124.73	310.03	−15.71
	C$_4$H$_{10}$(g) isobutane	−131.60	294.64	−17.97
	n-C$_5$H$_{12}$(g)	−146.44	348.40	−8.20
	C$_6$H$_6$(g) benzene	82.93	269.2	129.66
	C$_6$H$_6$(ℓ) benzene	49.03	172.8	124.50
	CO(g)	−110.52	197.56	−137.15
	CO$_2$(g)	−393.51	213.63	−394.36
	CO$_2$(aq)	−413.80	117.6	−385.08
	CS$_2$(ℓ)	89.70	151.34	65.27
	CS$_2$(g)	117.36	237.73	67.15
	H$_2$CO$_3$(aq)	−699.65	187.4	−623.08
	HCO$_3^-$(aq)	−691.99	91.2	−586.77
	CO$_3^{2-}$(aq)	−677.14	−56.9	−527.81
	HCOOH(ℓ)	−424.72	128.95	−361.42
	HCOOH(aq)	−425.43	163	−372.3
	HCOO$^-$(aq)	−425.55	92	−351.0
	CH$_2$O(g)	−108.57	218.66	−102.55
	CH$_3$OH(ℓ)	−238.66	126.8	−166.35
	CH$_3$OH(g)	−200.66	239.70	−162.01
	CH$_3$OH(aq)	−245.93	133.1	−175.31
	H$_2$C$_2$O$_4$(s)	−827.2	120	−697.9
	HC$_2$O$_4^-$(aq)	−818.4	149.4	−698.34
	C$_2$O$_4^{2-}$(aq)	−825.1	45.6	−673.9
	CH$_3$COOH(ℓ)	−484.5	159.8	−390.0
	CH$_3$COOH(g)	−432.25	282.4	−374.1
	CH$_3$COOH(aq)	−485.76	178.7	−396.46
	CH$_3$COO$^-$(aq)	−486.01	86.6	−369.31
	CH$_3$CHO(ℓ)	−192.30	117.3	−128.12
	C$_2$H$_5$OH(ℓ)	−277.69	160.7	−174.89
	C$_2$H$_5$OH(g)	−235.10	282.59	−168.57
	C$_2$H$_5$OH(aq)	−288.3	148.5	−181.64
	CH$_3$OCH$_3$(g)	−184.05	266.27	−112.67
	CF$_4$(g)	−925	261.50	−879
	CCl$_4$(ℓ)	−135.44	216.40	−65.28
	CCl$_4$(g)	−102.9	309.74	−60.62
	CHCl$_3$(g)	−103.14	295.60	−70.37
	COCl$_2$(g)	−218.8	283.53	−204.6

Substance	ΔH_f° (25°C) (kJ mol^{-1})	S° (25°C) (J K^{-1} mol^{-1})	ΔG_f° (25°C) (kJ mol^{-1})
$CH_2Cl_2(g)$	−92.47	270.12	−65.90
$CH_3Cl(g)$	−80.83	234.47	−57.40
$CBr_4(s)$	29.4	357.94	67
$CH_3I(\ell)$	−15.5	163.2	13.4
$HCN(g)$	135.1	201.67	124.7
$HCN(aq)$	107.1	124.7	119.7
$CN^-(aq)$	150.6	94.1	172.4
$CH_3NH_2(g)$	−22.97	243.30	32.09
$CO(NH_2)_2(s)$	−333.51	104.49	−197.44
$Si(s)$	0	18.83	0
$Si(g)$	455.6	167.86	411.3
$SiC(s)$	−65.3	16.61	−62.8
$SiO_2(s, quartz)$	−910.94	41.84	−856.67
$SiO_2(s, cristobalite)$	−909.48	42.68	−855.43
$Ge(s)$	0	31.09	0
$Ge(g)$	376.6	167.79	335.8
$Sn(s, white)$	0	51.55	0
$Sn(s, gray)$	−2.09	44.14	0.13
$Sn^{2+}(aq)$	−8.8	−17.0	−27.2
$Sn(g)$	302.1	168.38	267.3
$SnO(s)$	−285.8	56.5	−256.9
$SnO_2(s)$	−580.7	52.3	−519.6
$Sn(OH)_2(s)$	−561.1	155	−491.7
$Pb(s)$	0	64.81	0
$Pb(g)$	195.0	161.9	175.26
$Pb^{2+}(aq)$	−1.7	10.5	−24.43
$PbO(s, yellow)$	−217.32	68.70	−187.91
$PbO(s, red)$	−218.99	66.5	−188.95
$PbO_2(s)$	−277.4	68.6	−217.36
$PbS(s)$	−100.4	91.2	−98.7
$PbI_2(s)$	−175.48	174.85	−173.64
$PbSO_4(s)$	−919.94	148.57	−813.21
V $N_2(g)$	0	191.50	0
$N(g)$	472.70	153.19	455.58
$NH_3(g)$	−46.11	192.34	−16.48
$NH_3(aq)$	−80.29	111.3	−26.50
$NH_4^+(aq)$	−132.51	113.4	−79.31
$N_2H_4(\ell)$	50.63	121.21	149.24
$N_2H_4(aq)$	34.31	138	128.1
$NO(g)$	90.25	210.65	86.55
$NO_2(g)$	33.18	239.95	51.29
$NO_2^-(aq)$	−104.6	123.0	−32.2
$NO_3^-(aq)$	−205.0	146.4	−108.74
$N_2O(g)$	82.05	219.74	104.18

Substance	ΔH_f° (25°C) (kJ mol^{-1})	S° (25°C) (J K^{-1} mol^{-1})	ΔG_f° (25°C) (kJ mol^{-1})
$N_2O_4(g)$	9.16	304.18	97.82
$N_2O_5(s)$	−43.1	178.2	113.8
$HNO_2(g)$	−79.5	254.0	−46.0
$HNO_3(\ell)$	−174.10	155.49	−80.76
$NH_4NO_3(s)$	−365.56	151.08	−184.02
$NH_4Cl(s)$	−314.43	94.6	−202.97
$(NH_4)_2SO_4(s)$	−1180.85	220.1	−901.90
P(s, white)	0	41.09	0
P(s, red)	−17.5	22.85	−12.0
P(s, black)	−12.85	22.59	−7.34
$P(g)$	314.64	163.08	278.28
$P_2(g)$	144.3	218.02	103.7
$P_4(g)$	58.91	279.87	24.47
$PH_3(g)$	5.4	210.12	13.4
$H_3PO_4(s)$	−1279.0	110.50	−1119.2
$H_3PO_4(aq)$	−1288.34	158.2	−1142.54
$H_2PO_4^-(aq)$	−1296.29	90.4	−1130.28
$HPO_4^{2-}(aq)$	−1292.14	−33.5	−1089.15
$PO_4^{3-}(aq)$	−1277.4	−222	−1018.7
$PCl_3(g)$	−287.0	311.67	−267.8
$PCl_5(g)$	−374.9	364.47	−305.0
As(s, gray)	0	35.1	0
$As(g)$	302.5	174.10	261.0
$As_2(g)$	222.2	239.3	171.9
$As_4(g)$	143.9	314	92.4
$AsH_3(g)$	66.44	222.67	68.91
$As_4O_6(s)$	−1313.94	214.2	−1152.53
$Sb(s)$	0	45.69	0
$Sb(g)$	262.3	180.16	222.1
$Bi(s)$	0	56.74	0
$Bi(g)$	207.1	186.90	168.2

	Substance	ΔH_f° (25°C) (kJ mol^{-1})	S° (25°C) (J K^{-1} mol^{-1})	ΔG_f° (25°C) (kJ mol^{-1})
VI	$O_2(g)$	0	205.03	0
	$O(g)$	249.17	160.95	231.76
	$O_3(g)$	142.7	238.82	163.2
	$OH^-(aq)$	−229.99	−10.75	−157.24
	$H_2O(\ell)$	−285.83	69.91	−237.18
	$H_2O(g)$	−241.82	188.72	−228.59
	$H_2O_2(\ell)$	−187.78	109.6	−120.42
	$H_2O_2(aq)$	−191.17	143.9	−134.03
	S(s, rhombic)	0	31.80	0
	S(s, monoclinic)	0.30	32.6	0.096
	$S(g)$	278.80	167.71	238.28
	$S_8(g)$	102.30	430.87	49.66
	$S^{2-}(aq)$	33.1	−14.6	85.8

Substance	ΔH_f° (25°C) (kJ mol^{-1})	S° (25°C) (J K^{-1} mol^{-1})	ΔG_f° (25°C) (kJ mol^{-1})
H$_2$S(g)	−20.63	205.68	−33.56
H$_2$S(aq)	−39.7	121	−27.83
HS$^-$(aq)	−17.6	62.8	12.08
SO(g)	6.26	221.84	−19.87
SO$_2$(g)	−296.83	248.11	−300.19
SO$_3$(g)	−395.72	256.65	−371.08
H$_2$SO$_3$(aq)	−608.81	232.2	−537.81
HSO$_3^-$(aq)	−626.22	139.7	−527.73
SO$_3^{2-}$(aq)	−635.5	−29	−486.5
H$_2$SO$_4$(ℓ)	−813.99	156.90	−690.10
HSO$_4^-$(aq)	−887.34	131.8	−755.91
SO$_4^{2-}$(aq)	−909.27	20.1	−744.53
SF$_6$(g)	−1209	291.71	−1105.4
Se(s, black)	0	42.44	0
Se(g)	227.07	176.61	187.06

	Substance	ΔH_f° (25°C) (kJ mol^{-1})	S° (25°C) (J K^{-1} mol^{-1})	ΔG_f° (25°C) (kJ mol^{-1})
VII	F$_2$(g)	0	202.67	0
	F(g)	78.99	158.64	61.94
	F$^-$(aq)	−332.63	−13.8	−278.79
	HF(g)	−273.30	173.67	−275.40
	HF(aq)	−320.08	88.7	−296.82
	XeF$_4$(s)	−261.5	—	—
	Cl$_2$(g)	0	222.96	0
	Cl(g)	121.68	165.09	105.71
	Cl$^-$(aq)	−167.16	56.5	−131.23
	HCl(g)	−92.31	186.80	−95.30
	ClO$^-$(aq)	−107.1	42	−36.8
	ClO$_2$(g)	102.5	256.73	120.5
	ClO$_2^-$(aq)	−66.5	101.3	17.2
	ClO$_3^-$(aq)	−103.97	162.3	−7.95
	ClO$_4^-$(aq)	−129.33	182.0	−8.52
	Cl$_2$O(g)	80.3	266.10	97.9
	HClO(aq)	−120.9	142	−79.9
	ClF$_3$(g)	−163.2	281.50	−123.0
	Br$_2$(ℓ)	0	152.23	0
	Br$_2$(g)	30.91	245.35	3.14
	Br$_2$(aq)	−2.59	130.5	3.93
	Br(g)	111.88	174.91	82.41
	Br$^-$(aq)	−121.55	82.4	−103.96
	HBr(g)	−36.40	198.59	−53.43
	BrO$_3^-$(aq)	−67.07	161.71	18.60
	I$_2$(s)	0	116.14	0
	I$_2$(g)	62.44	260.58	19.36
	I$_2$(aq)	22.6	137.2	16.40
	I(g)	106.84	180.68	70.28

APPENDIX D   Standard Chemical Thermodynamic Properties

	Substance	ΔH_f° (25°C) (kJ mol^{-1})	S° (25°C) (J K^{-1} mol^{-1})	ΔG_f° (25°C) (kJ mol^{-1})
	I$^-$(aq)	−55.19	111.3	−51.57
	I$_3^-$(aq)	−51.5	239.3	−51.4
	HI(g)	26.48	206.48	1.72
	ICl(g)	17.78	247.44	−5.44
	IBr(g)	40.84	258.66	3.71
VIII	He(g)	0	126.04	0
	Ne(g)	0	146.22	0
	Ar(g)	0	154.73	0
	Kr(g)	0	163.97	0
	Xe(g)	0	169.57	0

Standard Half-Cell Reduction Potentials at 25°C

Half-Reaction	$\mathcal{E}°(V)$
$F_2(g) + 2\,H^+(aq) + 2\,e^- \longrightarrow 2\,HF(aq)$	3.053
$F_2(g) + 2\,e^- \longrightarrow 2\,F^-(aq)$	2.866
$O_3(g) + 2\,H^+(aq) + 2\,e^- \longrightarrow O_2(g) + H_2O(\ell)$	2.076
$H_2O_2(aq) + 2\,H^+(aq) + 2\,e^- \longrightarrow 2\,H_2O(\ell)$	1.776
$Au^+(aq) + e^- \longrightarrow Au(s)$	1.692
$PbO_2(s) + SO_4^{2-}(aq) + 4\,H^+(aq) + 2\,e^- \longrightarrow PbSO_4(s) + 2\,H_2O(\ell)$	1.691
$MnO_4^-(aq) + 4\,H^+(aq) + 3\,e^- \longrightarrow MnO_2(s) + 2\,H_2O(\ell)$	1.679
$HClO_2(aq) + 2\,H^+(aq) + 2\,e^- \longrightarrow HClO(aq) + H_2O(\ell)$	1.645
$HClO(aq) + H^+(aq) + e^- \longrightarrow \frac{1}{2}\,Cl_2(g) + H_2O(\ell)$	1.611
$2\,NO(g) + 2\,H^+(aq) + 2\,e^- \longrightarrow N_2O(g) + H_2O(\ell)$	1.591
$Mn^{3+}(aq) + e^- \longrightarrow Mn^{2+}(aq)$	1.542
$MnO_4^-(aq) + 8\,H^+(aq) + 5\,e^- \longrightarrow Mn^{2+}(aq) + 4\,H_2O(\ell)$	1.507
$Au^{3+}(aq) + 3\,e^- \longrightarrow Au(s)$	1.498
$BrO_3^-(aq) + 6\,H^+(aq) + 5\,e^- \longrightarrow \frac{1}{2}\,Br_2(\ell) + 3\,H_2O(\ell)$	1.482
$ClO_3^-(aq) + 6\,H^+(aq) + 5\,e^- \longrightarrow \frac{1}{2}\,Cl_2(g) + 3\,H_2O(\ell)$	1.47
$PbO_2(s) + 4\,H^+(aq) + 2\,e^- \longrightarrow Pb^{2+}(aq) + 2\,H_2O(\ell)$	1.455
$Au^{2+}(aq) + 2\,e^- \longrightarrow Au(s)$	1.401
$Cl_2(g) + 2\,e^- \longrightarrow 2\,Cl^-(aq)$	1.358
$Au^{3+}(aq) + 2\,e^- \longrightarrow Au^+(aq)$	1.29
$O_3(g) + H_2O(\ell) + 2\,e^- \longrightarrow O_2(g) + 2\,OH^-(aq)$	1.24
$Cr_2O_7^{2-}(aq) + 14\,H^+(aq) + 6\,e^- \longrightarrow 2\,Cr^{3+}(aq) + 7\,H_2O(\ell)$	1.232
$O_2(g) + 4\,H^+(aq) + 4\,e^- \longrightarrow 2\,H_2O(\ell)$	1.229
$MnO_2(s) + 4\,H^+(aq) + 2\,e^- \longrightarrow Mn^{2+}(aq) + 2\,H_2O(\ell)$	1.224
$ClO_4^-(aq) + 2\,H^+(aq) + 2\,e^- \longrightarrow ClO_3^-(aq) + H_2O(\ell)$	1.189
$Pt^{2+}(aq) + 2\,e^- \longrightarrow Pt(s)$	1.18
$Br_2(\ell) + 2\,e^- \longrightarrow 2\,Br^-(aq)$	1.066
$NO_3^-(aq) + 4\,H^+(aq) + 3\,e^- \longrightarrow NO(g) + 2\,H_2O(\ell)$	0.957
$Pd^{2+}(aq) + 2\,e^- \longrightarrow Pd(s)$	0.951
$2\,Hg^{2+}(aq) + 2\,e^- \longrightarrow Hg_2^{2+}(aq)$	0.920
$Hg^{2+}(aq) + 2\,e^- \longrightarrow Hg(\ell)$	0.851
$Ag^+(aq) + e^- \longrightarrow Ag(s)$	0.800
$Hg_2^{2+}(aq) + 2\,e^- \longrightarrow 2\,Hg(\ell)$	0.797
$Fe^{3+}(aq) + e^- \longrightarrow Fe^{2+}(aq)$	0.771
$O_2(g) + 2\,H^+(aq) + 2\,e^- \longrightarrow H_2O_2(aq)$	0.695
$BrO_3^-(aq) + 3\,H_2O(\ell) + 6\,e^- \longrightarrow Br^-(aq) + 6\,OH^-(aq)$	0.61

Half-Reaction	$\mathcal{E}°$(V)
$MnO_4^-(aq) + 2\,H_2O(\ell) + 3\,e^- \longrightarrow MnO_2(s) + 4\,OH^-(aq)$	0.595
$I_2(s) + 2\,e^- \longrightarrow 2\,I^-(aq)$	0.536
$Cu^+(aq) + e^- \longrightarrow Cu(s)$	0.521
$O_2(g) + 2\,H_2O(\ell) + 4\,e^- \longrightarrow 4\,OH^-(aq)$	0.401
$Cu^{2+}(aq) + 2\,e^- \longrightarrow Cu(s)$	0.345
$Hg_2Cl_2(s) + 2\,e^- \longrightarrow 2\,Hg(\ell) + 2\,Cl^-(aq)$	0.268
$PbO_2(s) + H_2O(\ell) + 2\,e^- \longrightarrow PbO(s) + 2\,OH^-(aq)$	0.247
$AgCl(s) + e^- \longrightarrow Ag(s) + Cl^-(aq)$	0.222
$SO_4^{2-}(aq) + 4\,H^+(aq) + 2\,e^- \longrightarrow H_2SO_3(aq) + H_2O(\ell)$	0.172
$Cu^{2+}(aq) + e^- \longrightarrow Cu^+(aq)$	0.153
$S_4O_6^{2-}(aq) + 2\,e^- \longrightarrow 2\,S_2O_3^{2-}(aq)$	0.08
$NO_3^-(aq) + H_2O(\ell) + 2\,e^- \longrightarrow NO_2^-(aq) + 2\,OH^-(aq)$	0.01
$2\,H^+(aq) + 2\,e^- \longrightarrow H_2(g)$	0.000 exactly
$Fe^{3+}(aq) + 3\,e^- \longrightarrow Fe(s)$	-0.037
$Pb^{2+}(aq) + 2\,e^- \longrightarrow Pb(s)$	-0.126
$Sn^{2+}(aq) + 2\,e^- \longrightarrow Sn(s)$	-0.138
$Ni^{2+}(aq) + 2\,e^- \longrightarrow Ni(s)$	-0.257
$Co^{2+}(aq) + 2\,e^- \longrightarrow Co(s)$	-0.28
$PbSO_4(s) + 2\,e^- \longrightarrow Pb(s) + SO_4^{2-}(aq)$	-0.359
$Mn(OH)_3(s) + e^- \longrightarrow Mn(OH)_2(s) + OH^-(aq)$	-0.40
$Cd^{2+}(aq) + 2\,e^- \longrightarrow Cd(s)$	-0.403
$Cr^{3+}(aq) + e^- \longrightarrow Cr^{2+}(aq)$	-0.407
$Fe^{2+}(aq) + 2\,e^- \longrightarrow Fe(s)$	-0.447
$Ga^{3+}(aq) + 3\,e^- \longrightarrow Ga(s)$	-0.549
$Fe(OH)_3(s) + e^- \longrightarrow Fe(OH)_2(s) + OH^-(aq)$	-0.56
$2\,SO_3^{2-}(aq) + 3\,H_2O(\ell) + 4\,e^- \longrightarrow S_2O_3^{2-}(aq) + 6\,OH^-(aq)$	-0.571
$PbO(s) + H_2O(\ell) + 2\,e^- \longrightarrow Pb(s) + 2\,OH^-(aq)$	-0.580
$Ni(OH)_2(s) + 2\,e^- \longrightarrow Ni(s) + 2\,OH^-(aq)$	-0.72
$Co(OH)_2(s) + 2\,e^- \longrightarrow Co(s) + 2\,OH^-(aq)$	-0.73
$Cr^{3+}(aq) + 3\,e^- \longrightarrow Cr(s)$	-0.744
$Zn^{2+}(aq) + 2\,e^- \longrightarrow Zn(s)$	-0.762
$2\,H_2O(\ell) + 2\,e^- \longrightarrow H_2(g) + 2\,OH^-(aq)$	-0.828
$Cr^{2+}(aq) + 2\,e^- \longrightarrow Cr(s)$	-0.913
$SO_4^{2-}(aq) + H_2O(\ell) + 2\,e^- \longrightarrow SO_3^{2-}(aq) + 2\,OH^-(aq)$	-0.93
$Mn^{2+}(aq) + 2\,e^- \longrightarrow Mn(s)$	-1.185
$Sm^{3+}(aq) + e^- \longrightarrow Sm^{2+}(aq)$	-1.55
$Mn(OH)_2(s) + 2\,e^- \longrightarrow Mn(s) + 2\,OH^-(aq)$	-1.56
$Al^{3+}(aq) + 3\,e^- \longrightarrow Al(s)$	-1.662
$Eu^{3+}(aq) + 3\,e^- \longrightarrow Eu(s)$	-1.991
$Sc^{3+}(aq) + 3\,e^- \longrightarrow Sc(s)$	-2.077
$H_2(g) + 2\,e^- \longrightarrow 2\,H^-(aq)$	-2.23
$Sm^{3+}(aq) + 3e^- \longrightarrow Sm(s)$	-2.304
$Ce^{3+}(aq) + 3\,e^- \longrightarrow Ce(s)$	-2.336
$Mg^{2+}(aq) + 2\,e^- \longrightarrow Mg(s)$	-2.372
$La^{3+}(aq) + 3\,e^- \longrightarrow La(s)$	-2.379

Half-Reaction	$\varepsilon°$(V)
$Mg(OH)_2(s) + 2\,e^- \longrightarrow Mg(s) + 2\,OH^-(aq)$	-2.690
$Na^+(aq) + e^- \longrightarrow Na(s)$	-2.71
$Ca^{2+}(aq) + 2\,e^- \longrightarrow Ca(s)$	-2.868
$Ba^{2+}(aq) + 2\,e^- \longrightarrow Ba(s)$	-2.912
$K^+(aq) + e^- \longrightarrow K(s)$	-2.931
$Rb^+(aq) + e^- \longrightarrow Rb(s)$	-2.98
$Cs^+(aq) + e^- \longrightarrow Cs(s)$	-3.026
$Li^+(aq) + e^- \longrightarrow Li(s)$	-3.040
$Sr^{2+}(aq) + 2\,e^- \longrightarrow Sr(s)$	-4.10

All voltages are standard reduction potentials (relative to the potential of the standard hydrogen half-cell) at 25°C and 1 atm pressure. Most data are taken from the *CRC Handbook of Chemistry and Physics*, 80th edition (1999).

Solutions to Selected Odd-Numbered Problems

Chapter 1

1. Table salt: homogeneous mixture; wood: heterogeneous mixture; mercury: homogeneous substance (element); lemonade: heterogeneous mixture; sodium chloride: homogeneous substance (compound); ketchup: heterogeneous mixture.

3. A mixture of ice cubes and water is a heterogeneous mixture that becomes homogeneous upon standing.

5. 0.011 g C

7. It has been lost to the surroundings in the form of gaseous carbon dioxide and water vapor.

9. 0.71 g SO_2 11. 28.4 g O

13. (a) 1:1   (b) 8:3   (c) 2:1   (d) 7:1   (e) 13:5

15. (a) 2.618:1   (b) 3.491:1

17. The ratio of the mass of a zinc atom to the mass of an iron atom is 1.171:1.

19. (a) The ratio of the mass of Si combined with 1.000 g of N in the first compound to the mass of Si combined with 1.000 g of N in the second compound is 4:3.
    (b) SiN (or some multiple)

21. 2.0 L N_2O and 3.0 L O_2

23.

Symbol	Z	N	A	No. electrons
$^{12}C^-$	6	6	12	7
$^{31}P^{2+}$	15	16	31	13
^{11}B	5	6	11	5
Bi^{3+}	83	126	209	80

25. Manhattan Island is 10,000 to 100,000 times larger than 10 cm.

27. (a) 145/94 = 1.543/1   (b) 95 electrons   29. 12.0005

31. 12.31 g. This answer is a weighted average that reflects the fact that the bag contains more green than red marbles.

33. 64.93

35. (a) 108.01   (b) 145.90   (c) 126.04   (d) 158.03   (e) 132.14

37. (a) 1.0314×10^{23} molecules SF_2 (b) 0.8439 g S
    (c) 0.51572×10^{23} molecules S_2F_4 and 0.8439 g S
    (d) The SF_2 has twice the number of molecules because each molecule has half the mass; the SF_2 and the S_2F_4 have the same proportion of S and F by mass because they contain the same relative number of atoms of S and F.

39. $2.1072979 \times 10^{-22}$ g; 126.90447 u

41. (a) 0.5375 mol   (b) 3.237×10^{23} molecules

43. (a) 12.3 g   (b) 12.3

45. 100 kegs of nails (from rounding up "99.2 kegs" to avoid fractional kegs)

47. 31 ng Hg   **49.** N_4H_6, H_2O, LiH, $C_{12}H_{26}$ **51.** 79.12%

53. 8.553% Cl; 36.67% F; 7.720% O; 47.06% Pt

55. $PbCrO_4$ is 64.11% Pb by mass; $PbCrO_4 \cdot PbO$ is 75.84% Pb by mass.

57. $SiCl_2$ **59.** Fe_3Si_7 **61.** BaN, Ba_3N_2 **63.** Na_2SO_4

65. C_7H_{10} **67.** C_2H_4S **69.** $C_{30}H_{50}$ **71.** 11.1 g; 0.0840 mol

73. 0.0920 g cm^{-3} **75.** 2540 cm^3 = 2.54 L   **77.** 2.95×10^{-23} cm^3

Chapter 2

1. (a) $N_2 + O_2 \longrightarrow 2\,NO$

Element	Left	Right
N	2	2
O	2	2

(b) $2\,N_2 + O_2 \longrightarrow 2\,N_2O$

Element	Left	Right
N	4	4
O	2	2

(c) $K_2SO_3 + 2\,HCl \longrightarrow 2\,KCl + H_2O + SO_2$

Element	Left	Right
K	2	2
S	1	1
O	3	3
H	2	2
Cl	2	2

(d) $2\,NH_3 + 3\,O_2 + 2\,CH_4 \longrightarrow 2\,HCN + 6\,H_2O$

Element	Left	Right
N	2	2
H	14	14
O	6	6
C	2	2

(e) $CaC_2 + 3\,CO \longrightarrow 4\,C + CaCO_3$

Element	Left	Right
Ca	1	1
C	5	5
O	3	3

3. (a) $6\,H_2 + P_4 \longrightarrow 4\,PH_3$ (b) $2\,K + O_2 \longrightarrow K_2O_2$
(c) $PbO_2 + Pb + 2\,H_2SO_4 \longrightarrow 2\,PbSO_4 + 2\,H_2O$
(d) $2\,BF_3 + 3\,H_2O \longrightarrow B_2O_3 + 6\,HF$ (e) $2\,KClO_3 \longrightarrow 2\,KCl + 3\,O_2$
(f) $2\,K_2O_2 + 2\,H_2O \longrightarrow 4\,KOH + O_2$
(g) $3\,PCl_5 + 5\,AsF_3 \longrightarrow 3\,PF_5 + 5\,AsCl_3$
(h) $2\,KOH + K_2Cr_2O_7 \longrightarrow 2\,K_2CrO_4 + H_2O$
(i) $P_4 + 4\,NaOH + 2\,H_2O \longrightarrow 2\,PH_3 + 2\,Na_2HPO_3$

5. (a) $2\,C_6H_6(\ell) + 15\,O_2(g) \longrightarrow 12\,CO_2(g) + 6\,H_2O(g)$
(b) $2\,F_2(g) + H_2O(\ell) \longrightarrow OF_2(g) + 2\,HF(g)$
(c) $CaC_2(s) + 2\,H_2O(\ell) \longrightarrow Ca(OH)_2(aq) + C_2H_2(g)$
(d) $7\,O_2(g) + 4\,NH_3(g) \longrightarrow 4\,NO_2(g) + 6\,H_2O(\ell)$
(e) $2\,Al(\ell) + 6\,NaOH(aq) \longrightarrow 3\,H_2(g) + 2\,Na_3AlO_3(aq)$

7.   (a) $Fe_2O_3(s) + 3\ H_2(g) \longrightarrow 2\ Fe(s) + 3\ H_2O(g)$
   (b) $3\ Fe_2O_3(s) + H_2(g) \longrightarrow 2\ Fe_3O_4(s) + H_2O(g)$

9.   (a) $3\ NO_2(g) + H_2O(\ell) \longrightarrow 2\ HNO_3(aq) + NO(g)$
   (b) $9\ KOH(aq) + 6\ V(s) + 5\ KClO_3(aq) \longrightarrow$
   $$3\ K_3HV_2O_7(aq) + 5\ KCl(aq) + 3\ H_2O(\ell)$$
   (c) $2\ Ag_2S(s) + 8\ KCN(aq) + O_2(g) + 2\ H_2O(\ell) \longrightarrow$
   $$4\ KAg(CN)_2(aq) + 2\ S(s) + 4\ KOH(aq)$$
   (d) $4\ Cl_2(g) + NaI(aq) + 8\ NaOH(aq) \longrightarrow NaIO_4(aq) + 8\ NaCl(aq) + 4\ H_2O(\ell)$
   (e) $2\ Cr(CO)_6(s) + 4\ KOH(aq) \longrightarrow$
   $$KHCr_2(CO)_{10}(aq) + K_2CO_3(aq) + KHCOO(aq) + H_2O(\ell)$$

11.   (a) $44.01\ g\ CO_2 + 12.01\ g\ C = 56.02\ g\ CO$
   (b) $172.35\ g\ C_6H_{14} + 607.98\ g\ O_2 = 528.12\ g\ CO_2 + 252.21\ g\ H_2O$
   (c) $221.87\ g\ Mn_2O_7 + 18.015\ g\ H_2O = 239.88\ g\ HMnO_4$

13. $9.88\ mol\ H_2O_2$

15.   (a) $0.2507\ g\ CH_4$ (b) $0.6262\ g\ TiCl_2$ (c) $20.41\ g\ Na_3VO_4$
   (d) $3.058\ g\ K_2O_2$

17. $418\ g\ KCl$; $199\ g\ Cl_2$

19. $7.10\ L\ C_2H_6$ **21.** $7.83\ g\ K_2Zn_3[Fe(CN)_6]_2$ **23.**   12

25. $0.134\ g\ SiO_2$ **27.** $85.77\%\ NaNO_3$

29.   250 cheeseburgers made; 100 hamburger patties left over

31.   75 tables   **33.** NH_3 limiting; $53.9\ g\ H_2O$ formed

35. $Ba(OH)_2 \cdot 8H_2O$ limiting; $23.18\ g\ BaCl_2$ formed

37. $1.126\ g\ H_2O$ **39.** $13.3\ L\ CO_2$ **41.** $303.0\ g\ Fe$; 83.93% yield

43.   (a) 2609 cookies   (b) 209 cookies

45. $120\ mmol = 1.20 \times 10^{-1}\ mol$ **47.** $0.083\ L = 83\ mL$

49. $84.90\%\ BaCl_2$ **51.** $1.07\ kg\ N_2H_4$ **53.** $0.0168\ L = 16.8\ mL$

55. $430\ mg\ (= 0.430\ g)\ AgCl$

57. $1.58\ mol$; 9.49×10^{23} molecules; $1580\ mmol\ H_2O$

59. $504\ mg\ Au$. Using the empirical formula gives the fact that the drug is 50.36% gold by mass; hence, every milligram of drug supplies 0.5036 mg of gold, regardless of the molecular formula.

61.   1180 metric tons or $1.18 \times 10^9\ g\ Na_3AlF_6$ **63.** $3 \times 10^{-9}\%$

65.   745 metric tons NaOH

67. $2.0\ kmol\ CO_2$; $4.0\ kmol\ O$ atoms; $2.4 \times 10^{27}\ O$ atoms

Chapter 3

1. PH_3, HCl, SiH_4, and H_2S **3.**   (a) Rb_3As (b) SrF_2 (c) CaS

5.   Melting points are highest in Group IV and fall off on both sides. Only sulfur is out of line.

7.   (a) M belongs in Group II (alkaline-earth elements).   (b) M is magnesium.

9.   (a) Hydrogen, nitrogen, oxygen, chlorine, fluorine, helium, neon, argon, krypton, xenon, and radon are gases at room temperature.   (b) Yes. Most are found in the upper-right-hand corner of the periodic table.

11.   Predicted melting point 1250°C (observed 1545°C); predicted boiling point 2400°C (observed 2831°C); predicted density $3\ g\ cm^{-3}$ (observed $2.99\ g\ cm^{-3}$)

13. AsH_3, HI, PbH_4, H_2Te

15. **(a)** Sb 46 core e^-, 5 valence e^-, · S̈b ·

(b) Br 28 core e^-, 7 valence e^-, : B̈r ·

(c) B 2 core e^-, 3 valence e^-, · Ḃ ·

(d) Ra 86 core e^-, 2 valence e^-, Ra ·

17. : Ẍe :   : S̈e :   · S̈e ·   Be ·

19. Xe 46 core e^-, 8 valence e^-; Se^{2-} 28 core e^-, 8 valence e^-;
Se^+ 28 core e^-, 5 valence e^-; Be^+ 2 core e^-, 1 valence e^-.

21. **(a)** Calcium sulfide (CaS) $Ca^{2+}[S]^{2-}$ Ċa ·   : S̈ : Ca^{2+} $\left(: \ddot{S} :\right)^{2-}$

(b) Cesium iodide (CsI) $Cs^+ [I]^-$ Cs ·   : Ï · Cs^+ $\left(: \ddot{I} :\right)^-$

(c) Francium sulfide (Fr_2S) $Fr^+ [S]^{2-}$ Fr ·   : S̈ : Fr^+ $\left(: \dot{\ddot{S}} :\right)^{2-}$

(d) Strontium nitride (Sr_2N_3) $Sr^{2+} [N]^{3-}$ · Sr ·   · N̈ · Sr^{2+} $\left(: \ddot{N} :\right)^{3-}$

(e) Lithium oxide (Li_2O) $Li^+ [O]^{2-}$ Li ·   : Ö : Li^+ $\left(: \ddot{O} :\right)^{2-}$

23. **(a)** aluminum sulfide   **(b)** cesium selenide   **(c)** lithium sulfide
(d) calcium nitride   **(e)** cesium oxide   **(f)** potassium bromide

25. **(a)** 1 e^- missing   **(b)** 1 e^- extra   **(c)** 2 e^- extra

27. **(a)** Formal charge on each O equals -1; formal charge on central S equals $+2$.
(b) Formal charge on each O equals -1; formal charge on central S equals $+1$; formal charge on other S equals zero.   **(c)** All formal charges equal zero.
(d) Formal charges on C and N equal zero; formal charge on S equals -1.

29.   H—N=O̤ and H—Ö=N̈

The first is preferred.

31. **(a)** Group IV CO_2 **(b)** Group VII Cl_2O_7 **(c)** Group V NO_3^-
(d) Group VI HSO_4^-

33. **(a)** H—S̈e—H **(b)** H—S̈b—H **(c)** H—Ö—Ï : **(d)** $\left[: C≡N—\dot{C}=\ddot{N} :\right]^{2-}$
       |
       H

35. : F̈—Ö—F̈ :

37.   : O :
             ‖
    H—N̈—C—N̈—H
       | |
       H   H

N—H 1.01×10^{-10} m; C—N 1.47×10^{-10} m; C=O 1.20×10^{-10} m

39. **(a)** $\left(: \ddot{O}—Cl—\ddot{O} : \right)^-$ **(b)** $\left(: \overset{:\ddot{Cl}:}{\underset{:\ddot{Cl}:}{\ddot{Cl}—Al—\ddot{Cl}}} : \right)^-$ **(c)** $\left(: Xe—\ddot{F} : \right)^+$
       : O :

41.

$$\overset{(0)\ \ (0)\ \ (-1)}{\ddot{O}=N-\ddot{O}:} \longleftrightarrow \overset{(-1)\ \ (0)\ \ (0)}{:\ddot{O}-N=\ddot{O}}$$

The N—O bond should be approximately 1.3×10^{-10} m (midway between 1.47×10^{-10} m for an N—O single bond and 1.18×10^{-10} m for an N=O double bond).

43.

$$\overset{(0)\ \ (0)\ \ (0)}{\ddot{O}=N-\ddot{F}:} \qquad \overset{(-1)\ \ (+1)\ \ (0)}{N=\ddot{O}-\ddot{F}:}$$

45. $:\ddot{S}-S-\ddot{S}:$ structure   **47.** (a) PF_5 structure   (b) SF_4 structure   (c) $:\ddot{O}=Xe=\ddot{O}$ structure

49. (a) selenium tetrafluoride   (b) iodine tribromide
(c) tetraphosphorus hexaoxide

51. Missing entries are $AgNO_3$, calcium hypochlorite, $KHSO_4$, gallium sulfite, $KClO_3$, sodium hydrogen carbonate

53. sodium hydrogen tartrate

55. $P_4O_{10} + 6\,PCl_5 \longrightarrow 10\,POCl_3$

57. (a) $SN = 4$ (b) $SN = 3$ (c) $SN = 6$ (d) $SN = 4$ (e) $SN = 5$

59. (a) tetrahedral   (b) bent (angle $< 120°$)   (c) octahedral
(d) pyramidal   (e) distorted T

61. (a) $SN = 6$, square planar   (b) $SN = 4$, bent, angle $< 109.5°$
(c) $SN = 4$, pyramidal, angle $< 109.5°$ (d) $SN = 2$, linear

63. (a) planar AB_3: SO_3 (b) pyramidal AB_3: NF_3
(c) bent AB_2^-: NO_2^- (d) planar AB_3^{2-}: CO_3^{2-}

65. CI_4 non-polar, SO_2 polar, SF_6 non-polar, $SOCl_2$ polar, IBr_3 polar

67. No, because the VSEPR theory predicts a bent molecule in both cases

69. (a) linear   (b) the N end

71. (a) iron(III) oxide   (b) titanium(IV) bromide   (c) tungsten(VI) oxide
(d) lead(IV) chloride   (e) manganese(III) fluoride   (f) silver(I) chlorate

73. (a) SnF_2 (b) Re_2O_7 (c) CoF_3 (d) WCl_5 (e) $Cu(NO_3)_2$
(f) $NiBrO_4 \cdot 6H_2O$

75. $Re_2S_7 + H_2 \longrightarrow 2\,ReS_3 + H_2S$

Chapter 4

1. Electrolytes give solutions that conduct electricity. Strong electrolytes dissociate essentially completely into ions. The solutions conduct electricity well. Weak electrolytes dissociate into ions only partially. Their solutions conduct electricity poorly. Non-electrolytes do not dissociate into ions. Their solutions do not conduct electricity.

3. $Mg(ClO_4)_2(s) \longrightarrow Mg^{2+}(aq) + 2\,ClO_4^-(aq)$

5. (a) Deoxyribose is a sugar and should dissolve in water.   (b) Table 4–1 reveals that $KClO_4$ is a sparingly soluble salt. It dissolves in water to only a limited extent. (c) Ethylene glycol, like most short-chain alcohols, should be soluble in water.

7. Sodium chloride, sodium bromide, potassium chloride, and potassium bromide

9. Add ethanol to the aqueous solution to force $RbCl(s)$ to precipitate.

11. (a) $Ag^+(aq) + Cl^-(aq) \longrightarrow AgCl(s)$
    (b) $K_2CO_3(s) + 2\,H^+(aq) \longrightarrow 2\,K^+(aq) + CO_2(g) + H_2O(\ell)$
    (c) $2\,Cs(s) + 2\,H_2O(\ell) \longrightarrow 2\,Cs^+(aq) + 2\,OH^-(aq) + H_2(g)$
    (d) $2\,MnO_4^-(aq) + 16\,H^+(aq) + 10\,Cl^-(aq) \longrightarrow$
    $$5\,Cl_2(g) + 2\,Mn^{2+}(aq) + 8\,H_2O(\ell)$$

13. overall equation: $NaBr(aq) + AgNO_3(aq) \longrightarrow AgBr(s) + NaNO_3(aq)$; net ionic equation: $Ag^+(aq) + Br^-(aq) \longrightarrow AgBr(s)$

15. A solution prepared by dissolving solid barium sulfate contains only Ba^{2+} and SO_4^{2-} ions. A solution prepared by mixing barium chloride and sodium sulfate contains Na^+ and Cl^- ions as well as Ba^{2+} and SO_4^{2-} ions.

17. (a) $Zn(NO_3)_2(aq) + K_2S(aq) \longrightarrow$
    $$ZnS(s) + 2\,KNO_3(aq);\ Zn^{2+}(aq) + S^{2-}(aq) \longrightarrow ZnS(s)$$
    (b) $2\,AgClO_4(aq) + CaCl_2(aq) \longrightarrow$
    $$2\,AgCl(s) + Ca(ClO_4)_2(aq);\ Ag^+(aq) + Cl^-(aq) \longrightarrow AgCl(s)$$
    (c) $3\,NaOH + Fe(NO_3)_3 \longrightarrow Fe(OH)_3 + 3\,NaNO_3;\ Fe^{3+} + 3\,OH^- \longrightarrow Fe(OH)_3$
    (d) $Ba(CH_3COO)_2 + Na_3PO_4 \longrightarrow NR$

19. (a) $Be(NO_3)_2(aq) + NaCH_3COO(aq) \longrightarrow NR$
    (b) $Ba(NO_3)_2(aq) + Ag_2SO_4(aq) \longrightarrow BaSO_4(s) + 2\,AgNO_3(aq)$
    $Ba^{2+}(aq) + SO_4^{2-}(aq) \longrightarrow BaSO_4(s)$
    (c) $2\,NaOH(aq) + CaCl_2(aq) \longrightarrow 2\,NaCl(aq) + Ca(OH)_2(s)$;
    $2\,OH^-(aq) + Ca^{2+}(aq) \longrightarrow Ca(OH)_2(aq)$
    (d) $NaCl(aq) + K_3PO_4(aq) \longrightarrow NR$

21. All ions remain in solution. The net ionic equation
    $Na^+(aq) + NO_3^-(aq) + K^+(aq) + Cl^-(aq) \longrightarrow$
    $$Na^+(aq) + Cl^-(aq) + K^+(aq) + NO_3^-(aq)$$
    is equivalent to NR.

23. The second solution should contain sulfate ions.

25. (a) hydrosulfuric acid   (b) periodic acid   (c) carbonic acid
    (d) hydrobromic acid

27. (a) I_2O_5; diiodine pentaoxide; acid anhydride   (b) BaO; barium oxide; base anhydride   (c) CrO_3; chromium(VI) oxide; acid anhydride   (d) dinitrogen monoxide; acid anhydride

29. (a) base anhydride of $Mg(OH)_2$, magnesium hydroxide   (b) acid anhydride of H_2SO_4, sulfuric acid   (c) acid anhydride of $HClO$, hypochlorous acid
    (d) base anhydride of $CsOH$, cesium hydroxide

31. $0.1160\ \text{mol L}^{-1}$ 33. $0.1105\ \text{mol L}^{-1}$

35. (a) $KNO_3(aq)$, $H_2O(\ell)$, and $CO_2(g)$ (b) $ZnBr_2(aq)$ and $H_2(g)$
    (c) $ZnSO_4(aq)$ and $H_2O(\ell)$

37. (a) $KClO_3(aq)$ and $H_2O(\ell)$ (b) $NaNO_3(aq)$ and $H_2O(\ell)$
    (c) $BaSO_4(s)$, $NH_3(g)$, and $H_2O(\ell)$

39. (a) $2\,HBr(aq) + Ca(OH)_2(s) \longrightarrow CaBr_2(aq) + 2\,H_2O(\ell)$
    (b) $2\,NH_3(aq) + H_2SO_4(aq) \longrightarrow (NH_4)_2SO_4(aq)$
    (c) $LiOH(aq) + HNO_3(aq) \longrightarrow LiNO_3(aq) + H_2O(\ell)$

41. (a) $Ca(OH)_2 + HF \longrightarrow CaF_2 + 2\,H_2O$. The acid is hydrofluoric acid; the base is calcium hydroxide; the salt is calcium fluoride.
    (b) $H_2SO_4 + 2\,RbOH \longrightarrow Rb_2SO_4 + 2\,H_2O$. The acid is sulfuric acid; the base is rubidium hydroxide; the salt is rubidium sulfate.

(c) $2\,HNO_3 + Zn(OH)_2 \longrightarrow Zn(NO_3)_2 + 2\,H_2O$. The acid is nitric acid; the base is zinc hydroxide; the salt is zinc nitrate.

(d) $CH_3COOH + KOH \longrightarrow KCH_3COO + H_2O$. The acid is acetic acid; the base is potassium hydroxide; the salt is potassium acetate.

43. (a) Phosphorus trifluoride; phosphoric acid; hydrofluoric acid

(b) $PF_3 + 3\,H_2O \longrightarrow H_3PO_3 + 3\,HF$

45. Sodium sulfide (Na_2S)

47. Tomato sauce is acidic and aluminum is an active metal:

$2\,Al + 6\,H^+ \longrightarrow 2\,Al^{3+} + 3\,H_2$

49. (a) $\overset{0\,+1\,-2}{CH_2O}$ (b) $\overset{+1\,+4\,-2}{H_2CO_3}$ (c) $\overset{+1\,-1}{RbH}$ (d) $\overset{+5\,-2}{N_2O_5}$

51. As(III) is oxidized to As(V) and I(0) is reduced to I(-1).
The increase of $2(+2)$ is balanced by a decrease of $4(-1)$.

53. (a) $2\,\overset{+3}{PF_2}I + 2\,\overset{0}{Hg} \longrightarrow \overset{+2}{P_2}F_4 + \overset{+1}{Hg_2}I_2$

(b) $2\,\overset{+5\,-2}{KClO_3} \longrightarrow 2\,\overset{-1}{KCl} + \overset{0}{O_2}$

(c) $4\,\overset{-3}{NH_3} + 5\,\overset{0}{O_2} \longrightarrow 4\,\overset{+2\,-2}{NO} + 6\,\overset{-2}{H_2O}$

(d) $2\,\overset{0}{As} + 6\,\overset{+1}{NaOH} \longrightarrow 2\,Na_3\overset{+3}{As}O_3 + 3\,\overset{0}{H_2}$

55. (a) P($+3$) is reduced to P($+2$); Hg(0) is oxidized to Hg($+1$): increase $2(+1)$; decrease $2(-1)$ (b) Cl($+5$) is reduced to Cl(-1); O(-2) is oxidized to O(0): increase $3(+2)$; decrease $1(-6)$ (c) O(0) is reduced to O(-2); N(-3) is oxidized to N($+2$): increase $4(+5)$; decrease $10(-2)$ (d) H($+1$) is reduced to H(0); As(0) is oxidized to As($+3$): increase $2(+3)$; decrease $6(-1)$

57. $Au \mid \overset{0\ +1\,+6\,-2}{H_2SeO_4} \mid \overset{+3}{Au_2}(\overset{+6\,-2}{SeO_4})_3 \mid \overset{+1\ +4\,-2}{H_2SeO_3} \mid \overset{+1\,-2}{H_2O}$

gold is oxidized; selenium (from H_2SeO_4) is reduced

59. $2\,Al_2O_3 + 3\,C \longrightarrow 3\,CO_2 + 4\,Al$. Carbon is oxidized; aluminum oxide is reduced.

61. (a) $2\,HCl(g) + 4\,O_2(g) \longrightarrow H_2O(g) + Cl_2O_7(g)$; oxygenation

(b) $H_2C{=}O(g) + H_2(g) \longrightarrow CH_3OH(\ell)$; hydrogenation (addition across double bond)   (c) $Mg(s) + 2\,HCl(aq) \longrightarrow MgCl_2(aq) + H_2(g)$; redox (displacement)

(d) $N_2(g) + 3\,H_2(g) \longrightarrow 2\,NH_3(g)$; redox (synthesis)

63. (a) $4\,\overset{+1\,+5\,-2}{HClO_3} \longrightarrow 4\,\overset{+4\,-2}{ClO_2} + \overset{0}{O_2} + 2\,\overset{+1\,-2}{H_2O}$

$2\,\overset{+1\,+5\,-2}{HClO_3} \longrightarrow 2\,\overset{+1\,-1}{HCl} + 3\,\overset{0}{O_2}$

$\overset{+1\,+5\,-2}{HClO_3} \longrightarrow 2\,\overset{+1\,+7\,-2}{HClO_4} + \overset{+1\,-1}{HCl}$

(b) The third reaction is a disproportionation (the first two are decompositions).

65. (a) $NH_4NO_2(\ell) \longrightarrow 2\,H_2O(g) + N_2(g)$ (b) $\overset{-3\,+1}{NH_4^+}\ \overset{+3\,-2}{NO_2^-}$; The overall oxidation number for nitrogen in ammonium nitrite is zero so the reaction may be viewed simply as a dehydration. However, it can also be considered a redox reaction in the sense that the nitrogen in the ammonium ion is oxidized and the nitrogen in the nitrite ion is reduced.

67. $Ba(s) + H_2(g) \longrightarrow BaH_2(s)$

$$BaH_2(s) + 2 H_2O(\ell) \longrightarrow Ba(OH)_2(aq) + 2 H_2(g)$$
$$BaH_2(s) + 2 HCl(aq) \longrightarrow BaCl_2(aq) + 2 H_2(g)$$
$$BaH_2(s) + ZnSO_4(aq) + 2 H_2O(\ell) \longrightarrow BaSO_4(s) + Zn(OH)_2 + 2 H_2(g)$$

69. (a) $H_2Te(g) \longrightarrow H_2Te(aq)$; dissolution

(b) $SrO(s) + CO_2(g) \longrightarrow SrCO_3(s)$ acid/base (acid anhydride and base anhydride)

(c) $2 HI(aq) + CaCO_3(s) \longrightarrow CaI_2(aq) + H_2O(\ell) + CO_2(g)$ acid/base

(d) $Na_2O(s) + 2 NH_4Br(aq) \longrightarrow 2 NaBr(aq) + 2 NH_3(g) + H_2O(\ell)$; acid/base

71. 151.6 mg Fe

Chapter 5

1. H_2S **3.** $NH_4HS(s) \longrightarrow NH_3(g) + H_2S(g)$

5. Treat ammonium bromide with strong base:
$$NH_4Br(s) + NaOH(aq) \longrightarrow NaBr(aq) + NH_3(g)$$

7. The gas is oxygen: $2 Na_2O_2(s) + 2 CO_2(g) \longrightarrow 2 Na_2CO_3(s) + O_2(g)$

9. 0.0431 atm   **11.** 2.59 m   **13.** 76 cm

15. 1697.5 atm; 1.7200×10^3 bar   **17.** 1.089 atm   **19.** 1.33 atm

21. (a) 0.248 L   (b) 1.27×10^3 mL   (c) 1.93×10^3 torr   (d) 6.50 atm

23. 37.0°C; 310.2 K   **25.** 10.5 L   **27.** 14.3 gill   **29.** 134 L

31. (a) 224 L   (b) 13 qt   (c) 426 mL   (d) 2.8×10^3°C

33. 3.38 atm   **35.** (a) 19.8 atm   (b) 23.0 atm   **37.** 741°F

39. 262 K

41. (a) The density of the gas depends on the temperature and pressure, which are not given.

(b) 253 K or −20°C

43. 4.1×10^{16} atoms Kr   **45.** 5.82 g L^{-1} **47.** C_4F_8

49. 3.0×10^6 L HCl

51. (a) $2 Na(s) + 2 HCl(g) \longrightarrow 2 NaCl(s) + H_2(g)$ (b) 3.79 L

53. 14.9 g MnO_2 **55.** (a) 932 L H_2S (b) 466 L SO_2; 1.33 kg SO_2

57. C_3H_8 **59.** (a) $P_{N_2} = 1.41$ atm   (b) $X_{O_2} = 0.420$; $X_{N_2} = 0.580$

61. $X_{SO_3} = 0.0688$; $V\% = 6.88\%$; $P_{SO_3} = 0.0654$ atm

63. $X_{N_2} = 0.027$; $P_{N_2} = 1.60 \times 10^{-4}$ atm

65. (a) $X_{CO} = 0.444$ (b) $X_{CO} = 0.33$

67. (a) 1.93×10^3 m s^{-1} (b) 226 m s^{-1}

69. (a) 6×10^3 m s^{-1} (6000 K)   (b) 8×10^2 m s^{-1} (100 K)

71. Greater. The higher temperature shifts the distribution to higher speeds.

73. 92.3 g mol^{-1} **75.** 162 atm; 2.39×10^3 psi

77. (a) 14.77 atm   (b) 13.24 atm   (c) Both equations of state are only approximate.

Chapter 6

1. (a) Cl—Cl 2.0×10^{-10} m; K^+—Cl^- 2.5×10^{-10} m

(b) The Cl-to-Cl bond length is shorter than the K^+-to-Cl^- bond but it is not as strong (240 kJ mol^{-1} versus 480 kJ mol^{-1}).

3. (a) *ion-ion;* induced dipole; dispersion   (b) *dipole-dipole;* dispersion

(c) *hydrogen bonding* (dipole-dipole); dispersion   (d) *dispersion*

(e) *dispersion*

5. Br^- (bromide ion)

7.
$$CH_3-O-H---O$$

(with CH$_3$ above the right O, H atoms below each O, and O---H—O—CH$_3$ with CH$_3$ below)

```
                    CH3
                     |
   CH3 — O — H --- O
         |         |
         H         H
         |         |
         O --- H — O — CH3
         |
        CH3
```

9. In all three phases, the diffusion constant should decrease as the density of the phase increases. At higher densities, molecules are closer together. In gases, they collide more often and travel shorter distances between collisions. In liquids and solids, less space is available for molecules to move around one another.

11. The mean free path in the gas phase should be longer than it is in the liquid phase.

13. 6.16 L mol^{-1}, several times smaller than the volume of 22.4 L mol^{-1} at STP

15. 0.9345 g L^{-1}

17. **(a)** $Mn_2O_7 < MnO_2 < Mn_3O_4$ (experimental: $5.9°C < 535°C < 1564°C$)
    (b) The ions are the potassium ion and the permanganate ion. The permanganate ion contains covalent bonds between Mn and O.

19. 0.69 atm; 31% lies below

21. Reduce the pressure on the sample so that it boils at a lower temperature.

23. No phase change occurs.

25. **(a)** Liquid   **(b)** Gas   **(c)** Solid   **(d)** Gas

27. **(a)** Above $-84°C$; the triple point is the highest temperature at which gas and solid can co-exist.   **(b)** The solid sublimes at some temperature below $-84°C$.

29. The meniscus between phases disappears at the critical temperature.

31. mass fraction = 0.0301; mol fraction = 0.0161; molality = $0.911 \text{ mol kg}^{-1}$

33. $16.81 \text{ mol kg}^{-1}$; 12.39 mol L^{-1} **35.** 91 mol kg^{-1}

37. 0.340 atm   **39.** 0.059   **41.** $3.4 \times 10^2 \text{ g mol}^{-1}$ **43.** CCl_4

45. $-2.8°C$; the freezing point of the solution decreases as its molality increases.

47. $i = 0.98$; this indicates essentially no dissociation.

49. 2.70 particles (complete dissociation would give 3 particles)

51. $8.59 \times 10^3 \text{ g mol}^{-1}$

53. Freezing point $-2.60°C$; boiling point $100.717°C$; osmotic pressure 34 atm (estimated assuming that molality and molarity are the same for the solution)

55. **(a)** $0.17 \text{ mol } CO_2$ **(b)** Because the partial pressure of carbon dioxide in the atmosphere is much less than 1 atm, the excess CO_2 bubbles out of the solution and escapes when the cap is removed.

57. 414 atm   **59.** 0.774   **61. (a)** 0.542   **(b)** 0.255 atm   **(c)** 0.623

63. A solution (such as ethanol in water) is homogeneous down to the molecular level, but a colloidal suspension (such as gold in water) contains larger particles of one phase dispersed in a second phase. Even in solutions, particles of one component can form clusters, blurring the distinction between the two types of mixtures.

65. Steam and water vapor are invisible. The white clouds are a colloidal suspension of droplets of liquid water that have condensed in the cool air above the beaker.

Chapter 7

1. **(a)** $\dfrac{P_{H_2O}}{P_{H_2}(P_{O_2})^{1/2}} = K$ **(b)** $\dfrac{P_{XeF_4}}{P_{Xe}(P_{F_2})^2} = K$ **(c)** $\dfrac{(P_{CO_2})^{20}(P_{H_2O})^{10}}{(P_{C_5H_5})^4(P_{O_2})^{25}} = K$

3. $P_4(g) + 2\,O_2(g) + 6\,Cl_2(g) \longrightarrow 4\,POCl_3(g);\ K = \dfrac{(P_{POCl_3})^4}{P_{P_4}(P_{O_2})^2(P_{Cl_2})^6}$

5. $K = 1.04 \times 10^{-4}$ **7.** $K = 14.6$ **9.** $K = 53.5$

11. **(a)** $P_{Cl_2} = P_{SO_2} = 0.58$ atm; $P_{SO_2Cl_2} = 0.14$ atm **(b)** $K = 2.4$

13. $K_1 = (K_2)^2$ **15.** $\dfrac{K_2}{K_1}$

17. **(a)** $Q = 36.1$, shifts right **(b)** $Q = 48.1$, shifts right **(c)** $Q = 17.8$, shifts right

19. **(a)** $Q = 9.83 \times 10^{-4}$, shifts left (consumes Al_3Cl_9)

21. $K < 5.1$ **23.** 4.5×10^{-5} atm

25. **(a)** $Q = 0$; pressure increases **(b)** $P_{Cl_2} = P_{SO_2} = 1.53$ atm; $P_{SO_2Cl_2} = 0.97$ atm

27. **(a)** 0.180 atm **(b)** 0.756 **29.** $P_{PCl_3} = P_{Cl_2} = 0.358$ atm; $P_{PCl_5} = 0.060$ atm

31. $P_{IBr} = 0.0768$ atm; $P_{Br_2} = 0.0116$ atm; $P_{I_2} = 0.0016$ atm

33. $P_{Al_2Cl_6} = 0.020$ atm; $P_{Al_3Cl_9} = 0.020$ atm

35.

	[NO] (mol L^{-1})	[Br$_2$] (mol L^{-1})	[NOBr] (mol L^{-1})
(a)	0.0123	0.00920	0.0350
(b)	0.00920	0.0123	0.0350
(c)	0.0245	0.0184	0.0695

37. 6.0×10^{-5} mol L^{-1} **39.** 7.82 atm

41. **(a)** shifts right (to counteract partially the addition of a reactant)
(b) shifts left (toward the side having a larger chemical amount of gas)
(c) shifts left (because it is harder for the reaction to evolve heat at increased T)
(d) shifts left (the pressure can remain constant only if the total volume increases)
(e) no effect

43. Use high P and low T for the first step; use low P and high T for the second step.

45. **(a)** exothermic **(b)** decrease

47. **(a)** $K = \dfrac{(P_{H_2S})^8}{(P_{H_2})^8}$ **(b)** $K = \dfrac{P_{COCl_2}P_{H_2}}{P_{Cl_2}}$ **(c)** $K = P_{CO}$ **(d)** $K = \dfrac{1}{(P_{C_2H_2})^3}$

49. **(a)** $K = \dfrac{[Zn^{2+}]}{[Ag^+]^2}$ **(b)** $K = \dfrac{[VO_3(OH)^{2-}][OH^-]}{[VO_4^{3-}]}$

(c) $K = \dfrac{[HCO_3^-]^6}{[As(OH)_6^{3-}](P_{CO_2})^6}$

51. **(a)** The graph is a straight line passing through the origin. **(b)** The experimental values of K range from 0.0342 to 0.0424, with an average value of 0.0384.

53. **(a)** $Q = 2.05$; reaction shifts right **(b)** $Q = 3.27$; reaction shifts left

55. **(a)** 8.46×10^{-5} atm **(b)** 0.00336 atm

57. 17.7 atm **59.** 76 **61.** **(a)** $K_1 = 0.0164; K_2 = 5.4$ **(b)** 330

Chapter 8

1. Only HSO_4^-, NH_4^+, NH_3, and H_2O can act as Brønsted–Lowry acids. Their conjugate bases are SO_4^{2-}, NH_3, NH_2^-, and OH^-, respectively.

3. (a) $NH_4^+(aq) + H_2O(\ell) \rightleftharpoons NH_3(aq) + H_3O^+(aq)$
   (b) $H_2S(aq) + H_2O(\ell) \rightleftharpoons H_3O^+(aq) + HS^-(aq)$
   (c) $NH_4^+(aq) + H_2O(\ell) \rightleftharpoons NH_3(aq) + H_3O^+(aq)$; this reaction evidently proceeds further than the reaction $SO_4^{2-}(aq) + H_2O(\ell) \rightleftharpoons HSO_4^-(aq) + OH^-(aq)$

5. pH = 3.70; pOH = 10.30

7. $3 \times 10^{-7}\,M < [H_3O^+] < 3 \times 10^{-6}\,M$ $3 \times 10^{-9}\,M < [OH^-] < 3 \times 10^{-8}\,M$

9. pH = 3.22; $[OH^-] = 1.7 \times 10^{-11}\,M$ 11. pH = 1.63

13. weakest   ammonium ion < benzoic acid < formic acid   strongest

15. (a) $C_{10}H_{15}ON(aq) + H_2O(\ell) \rightleftharpoons C_{10}H_{15}ONH^+(aq) + OH^-(aq)$
    (b) $K_a = 7.1 \times 10^{-11}$ (c) stronger base than ammonia

17. (a) $ClO^-(aq) + H_2O(\ell) \rightleftharpoons HOCl(aq) + OH^-(aq)$ (b) $K_b = 3.3 \times 10^{-7}$

19. (a) $C_6H_5O^-$ (b) weak base
    (c) $NaC_6H_5O(s) \longrightarrow Na^+(aq) + C_6H_5O^-(aq)$; $Na^+(aq) + H_2O(\ell) \longrightarrow NR$;
    $C_6H_5O^-(aq) + H_2O(\ell) \rightleftharpoons C_6H_5OH(aq) + OH^-(aq)$

21. $K = 24$

23. (a) The acid form of methyl orange is the stronger acid.   (b) between 3.8 and 4.4

25. (a)

formic acid	pH = 2.22	fraction dissociated = 0.030
benzoic acid	pH = 2.44	fraction dissociated = 0.018
ammonium ion	pH = 4.98	fraction dissociated = 5.3×10^{-5}

(b) The pH increases in going from the solution containing the strongest acid to the solution containing the weakest acid.   (c) The fraction dissociated decreases in going from the solution containing the strongest acid to the solution containing the weakest acid.

27. (a) pH = 2.35   (b) pH = 2.50   (c) Both the pH and the fraction dissociated are higher in the more dilute solution.

29. (a) pH = 2.67   (b) 0.027 mol formic acid   31. pH = 1.16

33. $K_a = 1.2 \times 10^{-6}$ 35. pH = 10.36

37. $[H_3O^+] = [NO_2^-] = 0.0099\,M$; $[HNO_2] = 0.21\,M$

39. (a) basic   (b) acidic   (c) neutral   (d) acidic   41. $K = 0.90$

43. At the equivalence point, the solution is essentially 0.05 M sodium acetate. Hydrolysis of the acetate ion raises the pH above 7.

45. (a) $[H_3O^+] = 2.4 \times 10^{-4}$; pH = 3.62   (b) $[H_3O^+] = 1.6 \times 10^{-4}$; pH = 3.79

47. pH = 8.08

49. The best acid to use is m-chlorobenzoic acid. Because its pK_a is close to the desired pH value, it will have a ratio of conjugate base-to-acid that is close to 1:1.

51. 0.842, 2.67, 7.00, 11.32   53. 4.00   55. 2.86, 4.72, 9.36, 11.00

57. 2.12   59. 0.357 g codeine; pH = 4.97 bromocresol green

61. $K_a = 6.5 \times 10^{-7}$; 0.1164 M

63. **(a)** RbOH by HBr   **(b)** pH = 7 at equivalence point; bromothymol blue and others   **(c)** 0.023 M

65. $[H_3AsO_4] = 0.080$ M; $[H_2AsO_4^-] = [H_3O^+]$
$$= 0.020 \text{ M}; [HAsO_4^{2-}] = 9.3 \times 10^{-8} \text{ M}; [AsO_4^{3-}] = 1.4 \times 10^{-17} \text{ M}$$

67. $[PO_4^{3-}] = 0.020$ M; $[HPO_4^{2-}] = [OH^-]$
$$= 0.030 \text{ M}; [H_2PO_4^-] = 1.6 \times 10^{-7} \text{ M}; [H_3PO_4] = 7.1 \times 10^{-18} \text{ M}$$

69. $[H_2CO_3] = 8.5 \times 10^{-6}$ M; $[HCO_3^-] = 1.5 \times 10^{-6}$ M; $[CO_3^{2-}] = 2.8 \times 10^{-11}$ M

71. **(a)** Lewis acid: Ag^+; Lewis base: NH_3 **(b)** Lewis acid: BF_3; Lewis base: $N(CH_3)_3$
(c) Lewis acid: $B(OH)_3$; Lewis base: OH^-

73. **(a)** $CaO(s) + H_2O(\ell) \longrightarrow Ca(OH)_2$ **(b)** Yes. The Lewis base is the oxide ion and the Lewis acid is an H^+ ion that is removed from the weaker Lewis base OH^- in the compound water.

75. **(a)** The acid is the fluoride ion acceptor.   **(b)** BF_3 is the acid and ClF_3O_2 is the base; TiF_4 is the acid and KF is the base.

Chapter 9

1. 45.1 g   **3.** $CoSO_4 \cdot 7H_2O$ **5.** About 45°C

7. Solubility should increase as temperature increases.

9. $[Sr^{2+}] = 0.0376$ mol L^{-1}; $[Br^-] = 0.0752$ mol L^{-1}

11. **(a)** $SrI_2(s) \rightleftharpoons Sr^{2+}(aq) + 2\,I^-(aq)$; $K_{sp} = [Sr^{2+}][I^-]^2$
(b) $I_2(s) \rightleftharpoons I_2(aq)$; $K = [I_2(aq)]$
(c) $Ca_3(PO_4)_2(s) \rightleftharpoons 3\,Ca^{2+}(aq) + 2\,PO_4^{3-}(aq)$; $K_{sp} = [Ca^{2+}]^3[PO_4^{3-}]^2$
(d) $BaCrO_4(s) \rightleftharpoons Ba^{2+}(aq) + CrO_4^{2-}(aq)$; $K_{sp} = [Ba^{2+}][CrO_4^{2-}]$

13. $Fe_2(SO_4)_3(s) \rightleftharpoons 2\,Fe^{3+}(aq) + 3\,SO_4^{2-}(aq)$; $K_{sp} = [Fe^{3+}]^2[SO_4^{2-}]^3$

15. $[Hg_2^{2+}] = 3.1 \times 10^{-10}$ mol L^{-1}; $[I^-] = 6.2 \times 10^{-10}$ mol L^{-1}

17. 0.67 g L^{-1}

19. **(a)** $[Pb^{2+}] = 0.016$ mol L^{-1}; $[Cl^-] = 0.032$ mol L^{-1} **(b)** The solution cannot lawfully be flushed down the drain.

21. $K_{sp} = 1.7 \times 10^{-11}$ **23.** $K_{sp} = 1.6 \times 10^{-8}$

25. The experimental solubilities exceed the computed solubilities by factors of 3.3 (for PbF_2), 2.2 (for $PbBr_2$), and 1.2 (for PbI_2) because the method of Example 9–3 neglects the hydrolysis of Pb^{2+} ion ($Pb^{2+}(aq) + 2\,H_2O(\ell) \rightleftharpoons PbOH^+(aq) + H_3O^+(aq)$) and F^- ion ($F^-(aq) + H_2O(\ell) \rightleftharpoons HF(aq) + OH^-(aq)$) and also assumes that the solutions are ideal.

27. Yes, but after the cooling, Q exceeds K_{sp} by only a factor of about 3. Hydrolysis may lower the concentrations of the ions enough to prevent precipitation (see discussion on page 407).

29. No; Q is less than K_{sp}.   **31.** No; Q is less than K_{sp}.

33. $[Pb^{2+}] = 2.3 \times 10^{-10}$ mol L^{-1}; $[IO_3^-] = 0.033$ mol L^{-1}

35. $[Ag^+] = 0.0175$ mol L^{-1}; $[CrO_4^{2-}] = 6.2 \times 10^{-9}$ mol L^{-1}

37. $s = 2.4 \times 10^{-8}$ mol L^{-1}

39. **(a)** $s = 3.4 \times 10^{-6}$ mol L^{-1} **(b)** $s = 1.6 \times 10^{-14}$ mol L^{-1}

41. AgOH is much more soluble in the solution buffered at pH 7. Saturating plain water with AgOH yields a solution with pH ~10.

43. pH = 8.6

45. **(a)** $[C_2O_4^{2-}] = 8.6 \times 10^{-4}$ M; Pb^{2+} ion is in the solid.   **(b)** 3.1×10^{-7}

47. $[Zn^{2+}] = 2 \times 10^{-13}$ mol L^{-1} **49.** pH = 2.4; $[Pb^{2+}] = 6 \times 10^{-11}$ M

51. $[Cu(NH_3)_4]^{2+} = 0.10$ M; $[Cu^{2+}] = 8 \times 10^{-14}$ M

53. $CuSO_4(s) \longrightarrow Cu^{2+}(aq) + SO_4^{2-}(aq)$ (dissolution)
$Cu^{2+}(aq) + H_2O(\ell) \rightleftharpoons CuOH^+(aq) + H^+(aq)$ (hydrolysis)
$SO_4^{2-}(aq) + H_2O(\ell) \rightleftharpoons HSO_4^-(aq) + OH^-(aq)$ (hydrolysis)
(The hydrolysis of $SO_4^{2-}(aq)$ occurs to a lesser extent than does the hydrolysis of the $Cu^{2+}(aq)$).

55. pH = 5.3   **57.** $K_a = 9.6 \times 10^{-10}$ **59.** $[I^-] = 5.3 \times 10^{-4}$

61. Dissolve the unknown nitrate salt in water. If a precipitate forms when a soluble sulfate, carbonate, or fluoride salt is added, the cation is Ba^{2+}. One could also distinguish between Cs^+ and Ba^{2+} by doing a flame test.

63. Acidify the solution to a pH 0.5 and saturate it with H_2S. A precipitate of HgS will form if Hg^{2+} ions are present. Separate any precipitate and adjust the pH of the remaining solution to pH 9 by adding ammonia to create a buffer system containing NH_4^+ ions and NH_3; a precipitate of NiS will form if Ni^{2+} ions are present. Separate any precipitate. Then, expel the hydrogen sulfide from the remaining solution by acidifying and heating it. Adjust the pH to 9.5 and add ammonium carbonate. A precipitate of $SrCO_3$ will form if Sr^{2+} ions are present.

65. The absence of precipitation upon treatment with HCl means that no Group 1 ions are present. At least one ion from Group 2 or Group 3 is present. There is no evidence concerning the presence or absence of ions from Groups 4 and 5.

Chapter 10

1. The heat is generated by friction between the drill bit and the wood.

3. The engine block has a *heat capacity* (measured in J K^{-1}). It is meaningless to refer to the heat content of an object at a fixed temperature; heat is a flow.

5. 1 kg iron ball < 2 kg iron plate < 1 kg water < 2 kg water

7. The molar heat capacities in J K^{-1} mol^{-1} are: 24.8 (Li); 28.3 (Na); 29.6 (K); 31.0 J (Rb); 32.2 J (Cs). Extrapolation gives the value 33.5 J K^{-1} mol^{-1} for Fr.

9. The molar heat capacities in J K^{-1} mol^{-1} are: 26.1 (Ni); 25.4 (Zn); 25.0 (Rh); 24.3 (W); 25.4 J (Au). The values follow the rule of Dulong and Petit within 5%.

11. 93.8 J K^{-1} mol^{-1} **13.** 97.8°C

15. **(a)** 2.23×10^2 J K^{-1} **(b)** 3.07 kJ

17. **(a)** 2.46×10^3 J K^{-1} **(b)** 10.8 kJ   **19.** ΔH_r is negative.

21. −41.3 kJ   **23.** **(a)** −6.68 kJ   **(b)** +7.49 kJ   **(c)** +0.594 kJ

25. **(a)** positive   **(b)** negative   **(c)** negative   **27.** +112.7 kJ

29. +0.513 kJ   **31.** −2.7 kJ mol^{-1} **33.** −623.5 kJ mol^{-1}

35. Diamond, which starts with more enthalpy than graphite but reacts to give the same products

37. −555.93 kJ mol^{-1}

39. **(a)** −878.26 kJ mol^{-1} **(b)** -1.35×10^7 kJ absorbed   **(c)** The smelting of sphalerite is not carried out under standard conditions.

41. **(a)** −81.4 kJ mol^{-1} **(b)** 39.5°C   **43.** −152.3 kJ mol^{-1} **45.** −78 kJ

47. **(a)** For combustion to $CO_2(g)$ and $H_2O(\ell)$: $\Delta H°(\text{propane}(g)) = -50.3 \text{ kJ g}^{-1}$; $\Delta H°$ $(\text{octane}(\ell)) = -47.7 \text{ kJ g}^{-1}$ **(b)** The combustion of propane is more exothermic per unit mass. However, propane is gaseous at ordinary conditions, and far more bulky. The propane could be liquefied under pressure to save transport space, but working with pressurized materials can be hazardous.

49. 1498 kJ   **51.** **(a)** 1302 kJ mol^{-1} **(b)** 611 kJ mol^{-1} **(c)** 24 kJ mol^{-1}

53. **(a)**

$$
\begin{array}{ccccccc}
 & H & & H & & H & \\
 & | & & | & & | & \\
H\!-\!\!&C&\!\!-\!\!&C&\!\!-\!\!&C&\!\!-\!H \\
 & | & & | & & | & \\
 & H & & H & & H &
\end{array}
$$

(b) $\Delta H_r° = -1685 \text{ kJ mol}^{-1}$

55. Each side of the equation has three B—Br and three B—Cl bonds.

57. More enthalpy (1936 kJ versus 1648 kJ) is required to break four C—F bonds than to break four C—H bonds; much more enthalpy is released (1852 kJ versus 740 kJ) by formation of four O—H bonds than by formation of four O—F bonds.

59. $-2.19 \times 10^6 \text{ J}$ **61.** $+4.10 \text{ L atm}$; $+416 \text{ J}$ **63.** w is zero.

65. **(a)** $\Delta E = q > 0$; $w = 0$ **(b)** $\Delta E = q < 0$; $w = 0$ **(c)** Although it is true that $(\Delta E_1 + \Delta E_2) = (q_1 + q_2)$, nothing is known about the value of the sums; they may be positive, negative, or zero. The sum $(w_1 + w_2)$ equals zero.

67. 6.65 kJ

69. **(a)** $C_{10}H_8(s) + 12\,O_2(g) \longrightarrow 10\,CO_2(g) + 4\,H_2O(\ell)$ **(b)** $-5157 \text{ kJ mol}^{-1}$ **(c)** $-5162 \text{ kJ mol}^{-1}$ **(d)** 84 kJ mol^{-1}

Chapter 11

1. **(a)** exothermic and spontaneous: $2\,C_8H_{18}(\ell) + 25\,O_2(g) \longrightarrow 8\,CO_2(g) + 9\,H_2O(g)$; many other answers are possible   **(b)** endothermic and spontaneous: $H_2O(s) \longrightarrow$ $H_2O(\ell)$ at room temperature; many other answers are possible

3. $H_2O(\ell) \longrightarrow H_2O(g)$ spontaneous at room pressure and temperatures above 100°C; $2\,NH_4Cl(s) + Ba(OH)_2 \cdot 8H_2O(s) \longrightarrow BaCl_2(aq) + 2\,NH_3(g) + 10\,H_2O(\ell)$ at room conditions
$NH_4Cl(s) \longrightarrow NH_4Cl(aq)$ at room conditions.

5. **(a)** 36   **(b)** 1/36   **7.** **(a)** positive   **(b)** positive   **(c)** negative

9. **(a)** 1 mol $F_2(g, 300 \text{ K})$ **(b)** 1 mol $H_2O(g, 300 \text{ K})$ **(c)** 2 mol $O_2(g, 400 \text{ K})$ **(d)** 10 g $O_3(g, 298.15 \text{ K})$

11. **(a)** negative   **(b)** negative   **(c)** positive   **(d)** negative

13. **(a)** $\Delta S_r° = -116.58 \text{ J K}^{-1} \text{ mol}^{-1}$ **(b)** $\Delta S° = -436.56 \text{ J K}^{-1}$

15. $2\,M(s) + Cl_2(g) \longrightarrow 2\,MCl(s)$;
$M = \text{Li}, \Delta S_r° = -162.54 \text{ J K}^{-1} \text{ mol}^{-1}$; $M = \text{Na}, \Delta S_r° = -181.12 \text{ J K}^{-1} \text{ mol}^{-1}$;
$M = \text{K}, \Delta S_r° = -186.14 \text{ J K}^{-1} \text{ mol}^{-1}$; $M = \text{Rb}, \Delta S_r° = -184.72 \text{ J K}^{-1} \text{ mol}^{-1}$;
$M = \text{Cs}, \Delta S_r° = -191.08 \text{ J K}^{-1} \text{ mol}^{-1}$
With the exception of Rb, $\Delta S_r°$ increases as the atomic number of the alkali metal increases.

17. $\Delta S_r° = 1.08 \text{ J K}^{-1} \text{ mol}^{-1}$

19. The reaction is driven by an increase in entropy; many more ways exist to arrange the atoms as $HOD(\ell)$ than as $H_2O(\ell)$ plus $H_2O(\ell)$.

21. The second law does not deal with matter. Both energy and entropy changes can be positive or negative in any process.

23. $\Delta S_{surr} > 44.7 \text{ J K}^{-1} \text{ mol}^{-1}$

25. ΔH_f° refers to a process $C(s, graphite) \longrightarrow C(s, graphite)$; S° is a property of the substance.

27. $9.61 \text{ J K}^{-1} \text{ mol}^{-1}$

29. **(a)** 737 J mol^{-1} **(b)** 2.65 kJ **(c)** no   **(d)** 196 K **31.** 29 kJ mol^{-1}

33. **(a)** $\Delta G_f^{\circ} = -384.38 \text{ kJ mol}^{-1}$ **(b)** $\Delta G_f^{\circ} = 76.79 \text{ kJ mol}^{-1}$
(c) $\Delta G_f^{\circ} = -166.3 \text{ kJ mol}^{-1}$

35. **(a)** 33.0 kJ mol^{-1} **(b)** $-840.1 \text{ kJ mol}^{-1}$

37. $\Delta H_r^{\circ} = -87.9 \text{ kJ mol}^{-1}$; $\Delta G_r^{\circ} = -32.3 \text{ kJ mol}^{-1}$; $\Delta S_r^{\circ} = -186 \text{ J K}^{-1} \text{ mol}^{-1}$

39. **(a)** negative   **(b)** negative   **(c)** positive   **(d)** positive

41. **(a)** spontaneous if $T < 3000 \text{ K}$ **(b)** spontaneous if $T < 1050 \text{ K}$
(c) spontaneous at all T

43. Raising the temperature lowers ΔG_r°, which becomes equal to zero at $T \approx 895 \text{ K}$.

45. $\Delta H_r^{\circ} = 62.1 \text{ kJ mol}^{-1}$; $\Delta G_r^{\circ} = 25.7 \text{ kJ mol}^{-1}$; $\Delta S_r^{\circ} = 122 \text{ J K}^{-1} \text{ mol}^{-1}$

47. $\Delta G_r^{\circ} = -550.23 \text{ kJ mol}^{-1}$; $K = 2.5 \times 10^{96}$

49. **(a)** 2.6×10^{12}; $K = \dfrac{P_{SO_3}}{P_{SO_2}(P_{O_2})^{1/2}}$ **(b)** 5.4×10^{-35}; $K = (P_{O_2})^{1/2}$
(c) 5.3×10^3; $K = [Cu^{2+}][Cl^-]^2$

51. $K = 6.2 \times 10^{-8}$; the same value appears in Table 8–2.

53. The entropy change associated with the dissolution of helium in water must be negative.

55. $K = 2.5 \times 10^{-3}$ **57.** negative   **59.** high pressure and low temperature

61. $K = 0.0046$ **63.** -58 kJ mol^{-1}

65. $\Delta H_r^{\circ} = -1.7 \times 10^5 \text{ J mol}^{-1}$; $\Delta S_r^{\circ} = -5 \times 10^2 \text{ J K}^{-1} \text{ mol}^{-1}$

67. **(a)** $\Delta G_r^{\circ} = -56.9 \text{ kJ mol}^{-1}$
(b) $\Delta H_r^{\circ} = -55.7 \text{ kJ mol}^{-1}$; $\Delta S_r^{\circ} = 4.1 \text{ J K}^{-1} \text{ mol}^{-1}$

69. **(a)** $\Delta H_{vap} = 23.8 \text{ kJ mol}^{-1}$ **(b)** 240 K

71. 34 kJ mol^{-1} **73.** $\Delta H_r^{\circ} = 2810 \text{ kJ mol}^{-1}$

75. **(a)** 8×10^{-9} (compare to 1.4×10^{-8} in Table 9–1)   **(b)** 1.5×10^{-6}
(c) 1100 K (unattainable because the solvent boils away)

Chapter 12

1. **(a)** $H_2O_2(aq) + 2 H^+(aq) + 2 e^- \longrightarrow 2 H_2O(\ell)$ and
$Fe^{2+}(aq) \longrightarrow Fe^{3+}(aq) + e^-$
(b) $SO_2(aq) + 2 H_2O(\ell) \longrightarrow HSO_4^-(aq) + 3 H^+(aq) + 2 e^-$ and
$MnO_4^-(aq) + 8 H^+(aq) + 5 e^- \longrightarrow Mn^{2+}(aq) + 4 H_2O(\ell)$
(c) $ClO_2^-(aq) \longrightarrow ClO_2(g) + e^-$ and
$ClO_2^-(aq) + 4 H^+(aq) + 4 e^- \longrightarrow Cl^-(aq) + 2 H_2O(\ell)$

3. **(a)** $2 VO_2^+(aq) + SO_2(g) \longrightarrow 2 VO^{2+}(aq) + SO_4^{2-}(aq)$
(b) $Br_2(\ell) + SO_2(g) + 2 H_2O(\ell) \longrightarrow 2 Br^-(aq) + SO_4^{2-}(aq) + 4 H^+(aq)$
(c) $5 HCOOH(aq) + 2 MnO_4^-(aq) + 6 H^+(aq) \longrightarrow$
$$5 CO_2(g) + 2 Mn^{2+}(aq) + 8 H_2O(\ell)$$
(d) $3 Pb_3O_4(s) + 12 H^+(aq) \longrightarrow 3 PbO_2(s) + 6 Pb^{2+}(aq) + 6 H_2O(\ell)$
(e) $Cr_2O_7^{2-}(aq) + 2 H^+(aq) + 3 Np^{4+}(aq) \longrightarrow$
$$2 Cr^{3+}(aq) + 3 NpO_2^{2+}(aq) + H_2O(\ell)$$
(f) $3 Hg_2HPO_4(s) + 2 Au(s) + 8 Cl^-(aq) + 3 H^+(aq) \longrightarrow$
$$6 Hg(\ell) + 3 H_2PO_4^-(aq) + 2 AuCl_4^-(aq)$$

5. **(a)** $8 \, HMnO_4(aq) + 5 \, H_2S(aq) + 3 \, H_2SO_4(aq) \longrightarrow 8 \, MnSO_4(aq) + 12 \, H_2O(\ell)$

(b) $4 \, Zn(s) + HNO_3(aq) + 9 \, HCl(aq) \longrightarrow 4 \, ZnCl_2(aq) + NH_4Cl(aq) + 3 \, H_2O(\ell)$

(c) $2 \, Sn(s) + 2 \, HNO_3(aq) + 8 \, HCl(aq) \longrightarrow 2 \, SnCl_4(aq) + N_2O(g) + 5 \, H_2O(\ell)$

(d) $2 \, H_2O(\ell) + 2 \, KMnO_4(aq) + 3 \, H_2SO_4(aq) \longrightarrow$
$$2 \, MnSO_4(aq) + K_2SO_4(aq) + 5 \, H_2O_2(aq)$$

7. **(a)** $2 \, Cr(OH)_3(s) + 3 \, Br_2(aq) + 10 \, OH^-(aq) \longrightarrow$
$$2 \, CrO_4^{2-}(aq) + 6 \, Br^-(aq) + 8 \, H_2O(\ell)$$

(b) $ZrO(OH)_2(s) + 2 \, SO_3^{2-}(aq) \longrightarrow Zr(s) + 2 \, SO_4^{2-}(aq) + H_2O(\ell)$

(c) $7 \, HPbO_2^-(aq) + 2 \, Re(s) \longrightarrow 7 \, Pb(s) + 5 \, OH^-(aq) + 2 \, ReO_4^-(aq) + H_2O(\ell)$

(d) $4 \, HXeO_4^-(aq) + 8 \, OH^-(aq) \longrightarrow 3 \, XeO_6^{4-}(aq) + 6 \, H_2O(\ell) + Xe(g)$

(e) $3 \, Ag_2S(s) + 2 \, Cr(OH)_3(s) + 7 \, OH^-(aq) \longrightarrow$
$$6 \, Ag(s) + 3 \, HS^-(aq) + 2 \, CrO_4^{2-}(aq) + 5 \, H_2O(\ell)$$

(f) $N_2H_4(aq) + 2 \, CO_3^{2-}(aq) \longrightarrow N_2(g) + 2 \, CO(g) + 4 \, OH^-(aq)$

9. $PbS(s) + 2 \, O_2(g) + 4 \, OH^-(aq) \longrightarrow PbO_2^{2-}(aq) + SO_4^{2-}(aq) + 2 \, H_2O(\ell)$

11. **(a)** $Ca(s) + Cl_2(g) \longrightarrow CaCl_2(s)$

(b) $2 \, Fe^{3+}(aq) + Sn^{2+}(aq) \longrightarrow 2 \, Fe^{2+}(aq) + Sn^{4+}(aq)$

(c) $Fe_2O_3(s) + 3 \, H_2(g) \longrightarrow 3 \, H_2O(s) + 2 \, Fe(s)$

(d) $2 \, K(s) + H_2O_2(aq) \longrightarrow 2 \, KOH(aq)$

13. $3 \, HNO_2 \longrightarrow NO_3^- + H^+ + 2 \, NO + H_2O$ **15.** 54 C   **17.** 6.0 C

19. $2 \, H_2O + 2 \, e^- \longrightarrow H_2 + 2 \, OH^-$; $2 \, H_2O \longrightarrow O_2 + 4 \, H^+ + 4e^-$;
$$2 \, H_2O \longrightarrow 2 \, H_2 + O_2$$

21. $Pt|H_2|H^+ \, \| \, Mn^{3+}, Mn^{2+}|Pt$

23. $2 \, Cr^{2+}(aq) + Cu^{2+}(aq) \longrightarrow 2 \, Cr^{3+}(aq) + Cu(s)$

25. **(a)** $5 \, Fe^{2+}(aq) + MnO_4^-(aq) + 8 \, H^+(aq) \longrightarrow 5 \, Fe^{3+}(aq) + 4 \, H_2O(\ell) + Mn^{2+}(aq)$

(b) $2 \, I^-(aq) + 2 \, Ag^+(aq) \longrightarrow I_2(s) + 2 \, Ag(s)$

27. $9.4 \times 10^{20} \, e^-$

29. **(a)** 9.65×10^4 C   **(b)** 3.86×10^5 C   **(c)** 1.45×10^5 C

(d) 2.32×10^6 C

31. 0.180 mol Sn $= 21.4$ g Sn

33. **(a)** $Cl_2(g) + Zn(s) \longrightarrow Zn^{2+}(aq) + 2 \, Cl^-(aq)$

(b) 1.20×10^3 C; 0.0124 mol e^- **(c)** anode loses 0.407 g Zn

(d) 0.152 L Cl_2

35. 0.248 g C   **37.** $+2$

39. $3 \, MnO_2(s) + 4 \, Al(\ell) \longrightarrow 3 \, Mn(s) + 2 \, Al_2O_3(s)$ $\Delta H_r^\circ = -1791.3$ kJ mol^{-1};
$\Delta G_r^\circ = -1769.1$ kJ mol^{-1} (Other equations with different ΔH_r° and ΔG_r° are correct as well; for example dividing all coefficients and ΔH_r° and ΔG_r° by 3.)

41. spontaneous at $T > 839$ K ($566°C$)   **43.** 9.8 kg Cu

45. Bronze is better for tools than copper because it is harder yet not brittle.

47. Froth flotation is used to concentrate copper ores. Crushed, low-grade ore is mixed with water, oil, and detergent. A stream of compressed air is passed through the mixture, producing a frothy mixture. The copper-rich froth is skimmed off and the oil is separated for reuse.

49. **(a)** and **(b)**

Equation	ΔG_r°
$\frac{1}{2} Fe_2O_3(s) + \frac{3}{2} CO(g) \longrightarrow Fe(s) + \frac{3}{2} CO_2(g)$	-14.7 kJ mol^{-1}
$\frac{1}{3} Fe_3O_4(s) + \frac{4}{3} CO(g) \longrightarrow Fe(s) + \frac{4}{3} CO_2(g)$	-4.4 kJ mol^{-1}
$FeCO_3(s) + CO(g) \longrightarrow Fe(s) + 2 \, CO_2(g)$	$+15.15$ kJ mol^{-1}

(c) Per mole of Fe obtained, Fe_2O_3 should be the easiest to reduce with $CO(g)$.

51. $Cu_5FeS_4(s) + 7\, FeCl_3(aq) \longrightarrow 5\, CuCl(s) + 8\, FeCl_2(aq) + 4\, S(s)$

53. $\Delta G_f^\circ/n = -107\ kJ\ mol^{-1}$; CoO should be placed between Fe_2O_3 and NiO.

55. Limestone and dolomite react with silica and alumina impurities in the ores to give calcium aluminum silicates (slag) for separation.

57. $Si(sln) + O_2(g) \longrightarrow SiO_2(s)$ (redox)
$4\, P(sln) + 5\, O_2(g) \longrightarrow P_4O_{10}(s)$ (redox)
$CaO(s) + 2\, SiO_2(s) \longrightarrow CaO{\cdot}(SiO_2)_2(\ell)$ (acid–base)
$6\, CaO(s) + P_4O_{10}(s) \longrightarrow 2\, (CaO)_3{\cdot}P_2O_5(\ell)$ (acid–base)

59. cathode $Na^+ + e^- \longrightarrow Na$; anode $Cl^- \longrightarrow \frac{1}{2}\, Cl_2 + e^-$

61. 4.4×10^4 kg Al

63. The melting point of cryolite is considerably lower than that of aluminum oxide. Its use saves energy costs.

65. $2\, Mg(\ell) + TiCl_4(\ell) \longrightarrow Ti(s) + MgCl_2(s)$; 102 kg Mg

67. Dolomite is an inexpensive base that also provides extra magnesium for refining.

69. 42.4 min

Chapter 13

1. 15 cents   **3.** 0.933 s

5. **(a)** $Pb(s) + Cu^{2+}(aq) \longrightarrow Pb^{2+}(aq) + Cu(s)$ $\Delta G_r^\circ = -90.9\ kJ\ mol^{-1}$
(b) $MnO_4^-(aq) + 8\, H^+(aq) + 5\, Fe^{2+}(aq) \longrightarrow Mn^{2+}(aq) + 5\, Fe^{3+}(aq) + 4\, H_2O(\ell)$
$$\Delta G_r^\circ = -355\ kJ\ mol^{-1}$$
(c) $2\, Ag(s) + Co^{2+}(aq) \longrightarrow 2\, Ag^+(aq) + Co(s)$ $\Delta G_r^\circ = +208\ kJ\ mol^{-1}$

7. $\Delta G_r^\circ = -945.4\ J\ g^{-1}$; $w_{elec,\,max} = +945.4\ J\ g^{-1}$

9. **(a)** anode: $Co(s) \longrightarrow Co^{2+}(aq) + 2\, e^-$; cathode: $Br_2(\ell) + 2\, e^- \longrightarrow 2\, Br^-(aq)$;
overall: $Co(s) + Br_2(\ell) \longrightarrow Co^{2+}(aq) + 2\, Br^-(aq)$ **(b)** 1.35 V

11. **(a)** anode: $Zn(s) \longrightarrow Zn^{2+}(aq) + 2\, e^-$; cathode: $In^{3+}(aq) + 3\, e^- \longrightarrow In(s)$
(b) -0.337 V

13. Disproportionation is not spontaneous; $\Delta\mathcal{E}^\circ = -2.727$ V

15. $Fe^{3+}(aq) + Fe(s) \longrightarrow 2\, Fe^{2+}(aq)$; $\Delta\mathcal{E}^\circ = 1.218$ V

17. reducing agent   **19.** Br_2 is a less effective oxidizing agent than is Cl_2.

21. **(a)** no   **(b)** yes   **(c)** yes   **(d)** no

23. **(a)** yes   **(b)** $P_4(s)$ is a stronger reducing agent (is more easily oxidized) than is $PH_3(g)$.

25. La/La^{3+} just slightly above Mg/Mg^{2+}; Eu/Eu^{3+} just above Al/Al^{3+}; Sm/Sm^{3+} just above Mn/Mn^{2+}; Ga/Ga^{3+} just above Fe/Fe^{2+}.

27. 0.365 V   **29.** **(a)** 0.52 V   **(b)** 0.644 V   **(c)** 1.05 V

31. -0.29 V

33. The $Cd(s)|Cd^{2+}(aq)$ half-cell acts as the anode.

35. pH = 5.15   **37.** **(a)** 0.379 V   **(b)** 0.13 M

39. **(a)** 1.029 V   **(b)** 0.989 V

41. **(a)** anode: $Ni(s) \longrightarrow Ni^{2+}(aq) + 2\, e^-$; cathode: $Ag^+(aq) + e^- \longrightarrow Ag(s)$;
overall: $Ni(s) + 2\, Ag^+(aq) \longrightarrow Ni^{2+}(aq) + 2\, Ag(s)$ **(b)** 1.057 V
(c) 5.4×10^{35}

43. 8×10^{41}; orange   **45.** 3×10^{6} **47.** pH $= 2.53$; $K_a = 0.0029$

49. (a) 1.066 V   (b) $K_{sp} = 7.8 \times 10^{-13}$

51. $\Delta\mathcal{E}° = 1.691$ V $- (-0.359$ V$) = 2.050$ V; for six cells, the voltage is 12.300 V

53. 9.3×10^{6} C   **55.** $w_{max} = 1.9 \times 10^{4}$ kJ

57. No. Recharging the battery requires more than replacing the sulfuric acid. It requires restoration of lead and lead(IV) oxide and removal of lead(II) sulfate.

59. The chemical reaction in an alkaline battery does not include any solutes. Thus, the value of Q does not change during discharge, and the potential remains constant.

61. discharge memory, disposal of toxic substance

63. 7.9×10^{3} J g^{-1}

65. $\Delta\mathcal{E}° = -0.381$ V. The reaction is not spontaneous at pH 14 (standard basic conditions). It becomes spontaneous if the pH is lowered to 7.56, assuming that $[Fe^{2+}]$ and P_{H_2} remains unchanged.

67. Sodium is active enough to serve as a sacrificial anode. However, sodium reacts violently with water and, therefore, would be more of a danger to a ship than a protector.

69. (a) $\Delta\mathcal{E}° = 0.169$ V (using $\Delta G_f°$ data from Appendix D)
(b) $\Delta\mathcal{E} = 0.158$ V (at 25°C)

71. Groundwater containing spilled chlorinated hydrocarbons is caused to seep through tons of buried iron filings. The iron is corroded to Fe^{2+} or Fe^{3+} while the pollutant is reduced to inorganic chloride and less toxic (or nontoxic) hydrocarbons.

Chapter 14

1. 1.18 mi min^{-1}; 70.9 mi h^{-1}

3. 50 mi h^{-1}; 46 mi h^{-1}; 42 mi h^{-1}; 46 mi h^{-1} **5.** 5.6×10^{-5} mol L^{-1} s^{-1}

7. $-\Delta[N_2]/\Delta t = -\frac{1}{3}\Delta[H_2]/\Delta t = -\frac{1}{2}\Delta[NH_3]/\Delta t$

9. (a) 0.225 mol L^{-1} s^{-1} (b) 0.0750 mol L^{-1} s^{-1}

11. (a) second order   (b) first order   **13.** L^2 mol^{-2} s^{-1}

15. (a) Rate $= k[NOBr]^2$ (b) $k = 0.800$ L mol^{-1} s^{-1}

17. (a) rate increases by a factor of ten   (b) rate decreases by a factor of four

19. (a) rate $= k[H_2][NO]^2$; units are L^2 mol^{-2} s^{-1} (b) rate would increase by a factor of 18

21. (a) rate $= k[C_5H_5N][CH_3I]$ (b) $k = 75$ L mol^{-1} s^{-1}
(c) rate $= 7.5 \times 10^{-8}$ mol L^{-1} s^{-1}

23. (a) rate $= k[Cr(H_2O)_6^{3+}][SCN^-]$ (b) $k = 2.0 \times 10^{-6}$ L mol^{-1} s^{-1}

25. (a) $k = 1.5 \times 10^{-5}$ s^{-1} (b) 0.41   **27.** $k = 0.0053$ s^{-1}; $t_{1/2} = 130$ s

29. (a) true   (b) false   (c) true because the reaction is first order in A
(d) false ($t_{1/2} = 0.35$ s^{-1})   (e) false

31. $[A] = 2.3 \times 10^{-5}$ mol L^{-1}

35. (a) bimolecular; rate $= k[HCO][O_2]$ (b) termolecular; rate $= k[CH_3][O_2][N_2]$
(c) unimolecular; rate $= k[HO_2NO_2]$

37. (a) The first step is unimolecular; the remaining three are bimolecular.
(b) $H_2O_2 + O_3 \longrightarrow H_2O + O_2$ (c) O, ClO, CF$_2$Cl, Cl

39. $k_{-1} = 0.26$ L mol^{-1} s^{-1}

41. (a) $A + B + E \longrightarrow D + F$; rate $= \dfrac{k_1 k_2}{k_{-1}} \dfrac{[A][B][E]}{[D]}$

   (b) $A + D \longrightarrow B + F$; rate $= \dfrac{k_1 k_2 k_3}{k_{-1} k_{-2}} \dfrac{[A][D]}{[B]}$

43. only the mechanism (b)   **45.** only the mechanism (a)

47. (a) $4.25 \times 10^5 \text{ J mol}^{-1}$ (b) $1.54 \times 10^{11} \text{ L mol}^{-1} \text{ s}^{-1}$

49. (a) $1.49 \times 10^{-3} \text{ L mol}^{-1} \text{ s}^{-1}$ (b) $1.30 \times 10^3 \text{ s}$

51. (a) $4.3 \times 10^{13} \text{ L mol}^{-1} \text{ s}^{-1}$ (b) $1.7 \times 10^5 \text{ s}^{-1}$ **53.** 70.3 kJ mol^{-1}

55. CF_2Cl_2 is a catalyst.

57. A catalyst speeds a reaction without being consumed. Hence, catalytic reactants are generally needed in only small quantities and can be used many times. Stoichiometric reactants are consumed in a reaction. Large amounts may be needed, and supplies must be replenished as the reaction proceeds.

Chapter 15

1. (a) $2\,{}^{12}_{6}\text{C} \longrightarrow {}^{23}_{12}\text{Mg} + {}^{1}_{0}n$ (b) ${}^{15}_{7}\text{N} + {}^{1}_{1}\text{H} \longrightarrow {}^{12}_{6}\text{C} + {}^{4}_{2}\text{He}$
   (c) $2\,{}^{3}_{2}\text{He} \longrightarrow 2\,{}^{1}_{1}\text{H} + {}^{4}_{2}\text{He}$

3. (a) $3.300 \times 10^{10} \text{ kJ mol}^{-1}$; 342.1 MeV/atom; 8.551 MeV/nucleon
   (b) $7.312 \times 10^{10} \text{ kJ mol}^{-1}$; 757.9 MeV/atom; 8.711 MeV/nucleon
   (c) $1.738 \times 10^{11} \text{ kJ mol}^{-1}$; 1801.7 MeV/atom; 7.570 MeV/nucleon

5. 4.01251 u

7. The two ${}^{4}_{2}\text{He}$ nuclides are more stable; their mass is less than that of one ${}^{8}_{4}\text{Be}$ by 0.0000987 u.

9. (a) ${}^{37}_{17}\text{Cl} \longrightarrow {}^{37}_{18}\text{Ar} + {}^{0}_{-1}e^- + \tilde{\nu}$ (b) ${}^{22}_{11}\text{Na} \longrightarrow {}^{22}_{10}\text{Ne} + {}^{0}_{+1}e^+ + \nu$
   (c) ${}^{224}_{88}\text{Ra} \longrightarrow {}^{220}_{86}\text{Rn} + {}^{4}_{2}\text{He}$ (d) $\left({}^{82}_{38}\text{Sr}^+ + {}^{0}_{-1}e^-\right) \longrightarrow {}^{82}_{37}\text{Rb} + \nu$

11. beta emission, ${}^{11}_{6}\text{C}$; alpha emission, ${}^{237}_{93}\text{Np}$; positron emission, ${}^{15}_{6}\text{C}$

13. ${}^{30}_{14}\text{Si} + {}^{1}_{0}n \longrightarrow {}^{31}_{14}\text{Si} \longrightarrow {}^{31}_{15}\text{P} + {}^{0}_{-1}e + \tilde{\nu}$

15. (a) 1.311 MeV   (b) A maximum kinetic energy is observed for the beta particle only when the decay event gives the anti-neutrino zero kinetic energy. Such decay events are rare.

17. 2.2164 MeV   **19.** -0.38 MeV

21. Film badges monitor personal exposure to radiation. The degree of darkening of the film is proportional to the quantity of radiation received.

23. $3.7 \times 10^{10} \text{ min}^{-1}$ **25.** (a) 1.0×10^6 atoms ${}^{19}_{8}\text{O}$ (b) 5.9×10^4 atoms

27. 30.3 yr   **29.** old: $t = 167$ s; new: $t = 133$ s   **31.** 1.6×10^{18} Bq

33. (a) 0.25 of the uranium has decayed   (b) $t = 1.85 \times 10^9$ yr

35. $t = 4220$ yr   **37.** 5.1×10^4 yr

39. ${}^{14}\text{C}$ is constantly generated by reaction of ${}^{14}\text{N}$ in the atmosphere with cosmic rays. Living organisms constantly exchange C atoms with the environment. This exchange stops when the organism dies but the decay of ${}^{14}\text{C}$ already in the body continues after death.

41. ${}^{11}_{6}\text{C} \longrightarrow {}^{11}_{5}\text{B} + {}^{0}_{+1}e^- + \nu$; ${}^{15}_{8}\text{O} \longrightarrow {}^{15}_{7}\text{N} + {}^{0}_{+1}e^- + \nu$

43. Exposure from ${}^{15}_{8}\text{O}$ is greater by a factor of $(1.72/0.99) = 1.74$.

45. (a) 2.3×10^{10} Bq   (b) 0.025 mGy s^{-1} (c) Yes. A dose of 5 Gy has a 50% chance of being lethal; this dose is greater than 8.6 Gy in the first 8 days.

47. (a) $^{90}_{38}\text{Sr} \longrightarrow \, ^{90}_{39}\text{Y} + \, _{-1}^{0}e^- + \tilde{\nu}$ and $^{90}_{37}\text{Rb} \longrightarrow \, ^{90}_{40}\text{Zr} + \, _{-1}^{0}e^- + \tilde{\nu}$
(b) 2.8 MeV released   (c) 5.24×10^{12} Bq   (d) 4.45×10^{11} Bq

49. The average molar mass of uranium increases with time because the lighter isotopes have shorter half-lives, assuming that even lighter isotopes of uranium are not among the products.

51. 7.59×10^7 kJ g^{-1}

53. Moderators slow neutrons, increasing the efficiency of the fission process. Certain light nuclides, such as ^1_1H (in water) and $^{12}_6\text{C}$ (in graphite) are effective moderators.

55. Normal stars fuse subatomic particles to make H and He. A star that has aged and accumulated helium may fuse He nuclei to make Be nuclei. Reaction of Be with He produces nuclei of C, N, and O. The oldest stars use fusion to produce still heavier nuclei, such as those of iron, cobalt, and nickel.

Chapter 16

1. 0.66 m s^{-1} **3.** 3.04 m   **5.** (a) 5.00×10^5 s^{-1} (b) 4.4 min

7. $\lambda = 1.313$ m; time $= 0.0873$ s

9. The electrons have higher kinetic energy when blue light is used.

11. orange-red   **13.** 550 nm; green

15. (a) 3.371×10^{-19} J   (b) 203.0 kJ mol^{-1} (c) 4.926×10^{-3} mol s^{-1}

17. cesium: $\lambda = 579$ nm, green, blue, or violet light; selenium: $\lambda = 209$ nm, no visible light works; ultraviolet light must be used.

19. (a) 7.4×10^{-20} J   (b) 4.0×10^5 m s^{-1}

21.

	(a) (energy)		(b) (radius)	
$n = 3$	-0.111 Ry	-2.42×10^{-19} J	$9a_0$	4.76×10^{-10} m
$n = 6$	-0.0278 Ry	-0.606×10^{-19} J	$36a_0$	19.0×10^{-10} m
$n = 8$	-0.0156 Ry	-0.341×10^{-19} J	$64a_0$	33.9×10^{-10} m
	(highest energy)		(largest atom)	

23. (a) 0.04209 Ry $= 9.175 \times 10^{-20}$ J   (b) 0.04861 Ry $= 10.60 \times 10^{-20}$ J
(c) -0.04861 Ry $= -10.60 \times 10^{-20}$ J

25. (a) 2164 nm   (b) 1874 nm (light is absorbed)
(c) 1874 nm (light is emitted)

27. $r_3 = 9.525 \times 10^{-11}$ m; $E_3 = -6.055 \times 10^{-18}$ J; $IE = +3.647 \times 10^6$ J;
$\nu = 1.142 \times 10^{16}$ s^{-1}; $\lambda = 2.624 \times 10^{-8}$ m

29. 72.90 nm (1/9 of 656.1 nm); ultraviolet light

31. (a) $n = 1, \lambda = 100$ cm; $n = 3, \lambda = 33.3$ cm   (b) 2 nodes

33. (a) 727 nm   (b) 0.396 nm   (c) 1.48×10^{-34} m

35. (a) $4p$ (b) $2s$ (c) $6f$

37. (a) 2 radial nodes; 1 angular node   (b) 1 radial; 0 angular
(c) 2 radial; 3 angular

39. It is most likely to be found along the y axis. The probability of finding it vanishes in the xz plane and on the sphere of its radial node.

41. A quantum number is a number arising in the quantum mechanical treatment of the atom that restricts the possible values of properties of the atom. Three quantum numbers in this one-electron Cl^{16+} ion.

43.

n	l	radial	angular
2	1	$2p$	3
3	2	$3d$	5
4	3	$4f$	7

45. **(a)** not allowed ($l = n$)   **(b)** allowed   **(c)** not allowed ($m_l > l$)   **(d)** not allowed ($l < 0$)

Chapter 17

1. **(a)** $1s^2 2s^2 2p^6 3s^2 3p^2$ or $[Ne]3s^2 3p^2$ **(b)** $1s^2 2s^2 2p^6 3s^2 3p^4$ or $[Ne]3s^2 3p^4$
(c) $[Ar]3d^7 4s^2$

3. Be^+: $[He]2s^1$ C^-: $[He]2s^2 2p^3$ Ne^{2+}: $[He]2s^2 2p^4$ Mg^+: $[Ne]3s^1$
P^{2+}: $[Ne]3s^2 3p^1$ Cl^-: $[Ne]3s^2 3p^6$ As^+: $[Ar]3d^{10} 4s^2 4p^2$
I^-: $[Kr]4d^{10} 5s^2 5p^6$ All but Cl^- and I^- are paramagnetic.

5. **(a)** In   **(b)** P^{3-} **(c)** V^{2+} **7.** 117   **9.** $Z = 1, 5$, and 9

11. 72   **13.** 121   **15.** **(a)** Sr   **(b)** Rn   **(c)** Xe   **(d)** Sr

17. **(a)** Cs   **(b)** F   **(c)** K   **(d)** At   **21.** 318.6 nm; near ultraviolet

23. The nuclear charge Z rises from 4 to 5 moving from beryllium to boron, but the outermost ($2p$) electron in boron is more effectively shielded from the nucleus than is the outermost ($2s$) electron in Be.

25. Boron's higher Z makes its IE_2 higher than for beryllium because the electron comes from the same ($2s$) subshell. IE_1 corresponds to the removal of a $2s$ electron in Be but a $2p$ electron for B (see Problem 23).

27. Atomic radius generally increases down a group because the valence electrons are located in shells with higher values of n. The core electrons shield the valence electrons from the higher nuclear charge.

29. **(a)** Rb   **(b)** K   **(c)** Kr   **(d)** K   **(e)** O^{2-}

31. **(a)** S^{2-} **(b)** Ti^{2+} **(c)** Mn^{2+} **(d)** Sr^{2+}

33. Mn: $[Ar]3d^5 4s^2$ Eu: $[Xe]4f^7 6s^2$ Both configurations include half-filled, inner subshells ($3d$ and $4f$).

35. Bond length should increase and bond enthalpy should decrease as size increases down a group. Exceptions do exist; the bond in BrF is longer than that in ClF, but BrF has the higher bond enthalpy.

37. As—H bond length should be ~1.56 Å. The Sb—H bond should be the weakest.

39. Bond lengths are often shorter than the sum of atomic radii in polar molecules because of the electrostatic attraction between the oppositely charged ends of the dipole. Such shortening is slight in HI, indicating that it is not very polar.

41. **(a)** CI_4 **(b)** OF_2 **(c)** SiH_4

43.

$$K(g) + Cl(g) \longrightarrow K^+(g) + Cl^-(g) \qquad \Delta E = 70 \text{ kJ mol}^{-1}$$
$$K(g) + Cl(g) \longrightarrow K^-(g) + Cl^+(g) \qquad \Delta E = 1203 \text{ kJ mol}^{-1}$$

(b)

$Na(g) + Cl(g) \longrightarrow Na^+(g) + Cl^-(g)$	$\Delta E = 147$ kJ mol^{-1}
$Na(g) + Cl(g) \longrightarrow Na^-(g) + Cl^+(g)$	$\Delta E = 1198$ kJ mol^{-1}

45. HF: 40%   HCl: 19%   HBr: 14%   HI: 8%   CsF: 87%
The formula overestimates the percent ionic character for CsF.

47. most polar: N—P > C—N > N—O > N—N :least polar

49. V: $+5$ P: $+5$ I: $+7$ Sr: $+2$

51. The nucleus in Pb attracts its $6s$ electrons more strongly than its $6p$ electrons because the $6s$ orbital penetrates the shielding provided by the core electrons better than the $6p$ orbital does. The $6s$ electrons become (almost) core electrons themselves. In C, shielding by core electrons is much less a factor, because far fewer core electrons exist.

53. The bonding in Mn_2O_7, RuO_4, and OsO_4 is covalent; forming ions with charges of $+7$ or $+8$ would require an extremely high input of energy.

55. $5 \times 10^{-4} < K_{a1} < 8 \times 10^{-2}$; Table 8–2 lists 1.8×10^{-4}. At pH $= 14$, the predominant form is $HCOO^-$.

57. weakest: $H_3BO_3 < H_5B_3O_7 < H_6B_4O_9 < H_2B_4O_7$: strongest

59. H_2SO_4 is stronger. It has more lone oxygens and the sulfur has a higher formal charge in H_2SO_4 than in H_2SO_3.

Chapter 18

1. He_2^{2+}: $(\sigma_{1s})^2$ bond order 1.
Li_2^{2+}: $(\sigma_{1s})^2(\sigma_{1s}^*)^2$; bond order 0.
Be_2^{2+}: $(\sigma_{1s})^2(\sigma_{1s}^*)^2(\sigma_{2s})^2$; bond order 1.
B_2^{2+}: $(\sigma_{1s})^2(\sigma_{1s}^*)^2(\sigma_{2s})^2(\sigma_{2s}^*)^2$; bond order 0.
C_2^{2+}: $(\sigma_{1s})^2(\sigma_{1s}^*)^2(\sigma_{2s})^2(\sigma_{2s}^*)^2(\pi_{2p})^2$; bond order 1.
N_2^{2+}: $(\sigma_{1s})^2(\sigma_{1s}^*)^2(\sigma_{2s})^2(\sigma_{2s}^*)^2(\pi_{2p})^4$; bond order 2.
O_2^{2+}: $(\sigma_{1s})^2(\sigma_{1s}^*)^2(\sigma_{2s})^2(\sigma_{2s}^*)^2(\sigma_{2p})^2(\pi_{2p})^4$; bond order 3.
F_2^{2+}: $(\sigma_{1s})^2(\sigma_{1s}^*)^2(\sigma_{2s})^2(\sigma_{2s}^*)^2(\sigma_{2p})^2(\pi_{2p})^4(\pi_{2p}^*)^2$; bond order 2.
Ne_2^{2+}: $(\sigma_{1s})^2(\sigma_{1s}^*)^2(\sigma_{2s})^2(\sigma_{2s}^*)^2(\sigma_{2p})^2(\pi_{2p})^4(\pi_{2p}^*)^4$; bond order 1.

3. **(a)** F_2: $(\sigma_{2s})^2(\sigma_{2s}^*)^2(\sigma_{2p})^2(\pi_{2p})^4(\pi_{2p}^*)^4$ F_2^+: $(\sigma_{2s})^2(\sigma_{2s}^*)^2(\sigma_{2p})^2(\pi_{2p})^4(\pi_{2p}^*)^3$
(b) F_2: 1, F_2^+: $\frac{3}{2}$ **(c)** F_2^+ should be paramagnetic   **(d)** $F_2 < F_2^+$

5. $(\sigma_{3s})^2(\sigma_{3s}^*)^2(\sigma_{3p_z})^2(\pi_{3p})^4(\pi_{3p}^*)^2$; bond order $= 2$; paramagnetic

7. **(a)** F   **(b)** N   **(c)** O   **9.** **(a)** 1   **(b)** $\frac{5}{2}$ **(c)** $\frac{3}{2}$

11. **(a)** diamagnetic   **(b)** paramagnetic   **(c)** paramagnetic

13. Bond order: $2\frac{1}{2}$; paramagnetic

15. The highest-energy electron in CF is in a π_{2p}^* molecular orbital. Removing it gives a stronger bond.

17. $(\sigma_{1s})^2(\sigma_{1s}^*)^2$, bond order 0, should be unstable

19. sp^3 hybridization of central N^-, bent molecular ion

21. **(a)** sp^3 on C, tetrahedral   **(b)** sp on C, linear   **(c)** sp^3 on O, bent
(d) sp^3 on C, pyramidal   **(e)** sp on Be, linear

23. ClO_3^+: sp^2 hybrid orbitals, trigonal planar; ClO_2^+, sp^2 hybrid orbitals, bent

25. sp^3 hybrid orbitals, tetrahedral

27. **(a)** 5 σ bonds, 1 π bond   **(b)** 1 σ bond   **(c)** 10 σ bonds   **(d)** 4 σ bonds

29.

$$
\begin{array}{ccc}
H & :\!O\!: & H \\
| & \| & | \\
H-C-&C-&C-H \\
| & & | \\
H & & H
\end{array}
$$

Central carbon atom is sp^2 hybridized; other carbon atoms are sp^3 hybridized. One π orbital involving $2p_z$ orbitals on the C and O is doubly occupied; one π^* orbital involving $2p_z$ orbitals on C and O is empty. Angles: H-C-C, H-C-H $\approx 109.5°$ and C-C-O, C-C-C $\approx 120°$.

31.

$$
\begin{array}{cc}
:\!O\!: & H \\
\| & | \\
H-C-&C-H \\
& | \\
& H
\end{array}
$$

The CH_3 carbon atom is sp^3 hybridized; the other carbon atom is sp^2 hybridized. One π orbital involving $2p_z$ orbitals on the C and O is doubly occupied; one π^* orbital involving $2p_z$ orbitals on C and O is empty. The three groups around the second carbon form an approximately trigonal planar structure, with bond angles near 120°. The geometry around the first carbon atom is approximately tetrahedral, with angles near 109.5°.

33. **(a)** 3   **(b)** 3

35.

$$
\left(:\!\overset{..}{O}\!=\!N\overset{..}{\underset{:\!O\!:}{}} \quad \longleftrightarrow \quad :\!\overset{..}{\overset{-1}{O}}\!-\!N\overset{..}{\underset{O\!:}{}}\right)^-
$$

$SN = 3$, sp^2 hybridization, bent molecule. The $2p_z$ orbitals perpendicular to the plane of the molecule can be combined into a π molecular orbital containing one pair of electrons. Complete occupancy of this orbital (and vacancy in the π^* orbital) adds bond order $\frac{1}{2}$ to each N—O bond, for a total bond order of $\frac{3}{2}$ per bond.

37. Absorption of microwave radiation causes molecules to rotate faster. Information is gained about bond lengths and angles and molecular geometry.

39. decrease, because the electron enters a π^* antibonding orbital

41. in the blue range, around 450 nm

43. shorter; the π bonding is localized (not conjugated) in cyclohexene

45. 270 nm

47.

$$
\left(\overset{..}{O}\!=\!\overset{..}{O}\overset{..}{\underset{O\!:}{}} \quad \longleftrightarrow \quad :\!\overset{..}{O}\!-\!\overset{..}{O}\overset{..}{\underset{O\!:}{}}\right)
$$

$SN = 3$, sp^2 hybridization, bent molecule. Two electrons in a π orbital formed from the three $2p_z$ orbitals perpendicular to the molecular plane, total bond order of $\frac{3}{2}$ for each O—O bond.

Chapter 19

1. **(a)** $[Fe(CN)_6]^{3-}$ **(b)** $[Mn(NH_3)(H_2O)_2(Cl)_3]$ **(c)** $[Pt(H_2O)Br(en)]^+$

3. monodentate, at the N-atom lone pair   **5.** 3.13   **7.** $+2, +2, +2, 0$

9. **(a)** $Na_2[Zn(OH)_4]$ **(b)** $[CoCl_2(en)_2]NO_3$ **(c)** $[Pt(H_2O)_3Br]Cl$
   (d) $[Pt(NH_3)_4(NO_2)_2]Br_2$

11. **(a)** ammonium diamminetetraisothiocyanatochromate(IV)
   (b) pentacarbonyliodotechnetium(I)

(c) potassium pentacyanomanganate(IV)

(d) tetraammineaquachlorocobalt(III) bromide

13. Such a complex persists in solution but exchanges ligands with other complexes, as can be verified by isotopic labeling experiments.

15. $[Cu(NH_3)_4]^{2+}(aq) \longrightarrow Cu^{2+}(aq) + 4 NH_3(aq)$ $\Delta G_r^{\circ} = +70.56$ kJ mol^{-1}, complex is stable against dissociation

   In acid,

   $[Cu(NH_3)_4]^{2+}(aq) + 4 H_3O^{+}(aq) \longrightarrow Cu^{2+}(aq) + 4 NH_4^{+}(aq) + 4 H_2O(\ell)$

   $\Delta G_r^{\circ} = -140.68$ kJ mol^{-1}, dissociates spontaneously

17. $[Cu(NH_3)_2Cl_2] < KNO_3 < Na_2[PtCl_6] < [Co(NH_3)_6]Cl_3$

23. (a) strong: 1, weak: 5   (b) strong or weak: 0   (c) strong or weak: 3

   (d) strong: 2, weak: 4   (e) strong: 0, weak: 4

25. In octahedral complexes having only 1, 2, or 3 electrons in d orbitals, only the t_{2g} orbitals will be occupied; there is no need to pair any electrons or to promote any to the e_g level. V^{4+} ions are d^1 ions; Ti^{2+} ions are d^2 ions; Cr^{3+} ions are d^3 ions.

27. $[Fe(CN)_6]^{3-}$ 1 unpaired t_{2g}^5 $[Fe(H_2O)_6]^{3+}$ 5 unpaired $t_{2g}^3 e_g^2$

29. In an octahedral environment, the d^3 and d^8 configurations always give a large excess of electrons (either 3 and 4) in the more stable t_{2g} level compared to the less stable e_g level. The d^5 and d^6 configurations give an excess of 5 and 6 electrons in the more stable t_{2g} level, but only in a strong crystal field.

31. The higher positive charge on the metal ion draws the ligands closer to the d orbitals, thus increasing the extent of interaction between d electrons and ligand electrons.

33. This ion does not absorb any significant amount of light in the visible range; it has a $t_{2g}^6 e_g^4$ (filled d subshell) electron configuration.

35. $\lambda \approx 480$ nm; $\Delta_o \approx 250$ kJ mol^{-1}

37. (a) orange-yellow   (b) approximately 600 nm (actual: 575 nm)

   (c) decrease because CN$^-$ is a strong-field ligand and increases Δ_o

39. (a) pale, because F$^-$ is an even weaker field ligand than H$_2$O, so it should be high-spin d^5. This is a half-filled subshell and the complex absorbs only weakly.

   (b) colorless, because Hg(II) is a d^{10} (filled-subshell) species

41. The formation of fewer bonds means that tetrahedral complexes are less stable than octahedral complexes.

43. $[Fe(CN)_6]^{4-}$ 45. 1330 g mol^{-1} 47. Oxygen is a strong-field ligand.

Chapter 20

1. 3.613 Å   3. 102.9°   5. 49.87° and 115.0°

7. A single crystal gives a diffraction pattern with spots, and a microcrystalline solid gives one with rings. The difference comes from the many random orientations of the small crystals in the second case.

9. (b), (c), (d), and (e)   11. (a), (b), (c), and (d)   13. one fourfold rotation

15. 739 Å^3 = 7.39 × 10^{-28} m^3

17. (a) 5.6731 Å^3 per C atom   (b) 3.5157 g cm^{-3}

19. (a) 1.602 × 10^{-22} cm^3 (b) 3.729 × 10^{-22} g   (c) 4.662 × 10^{-23} g

   (d) 6.025 × 10^{23} (in error by 0.05%)

21. 2.533 g cm^{-3} 23. 4.059 g cm^{-3} 25. 8

27. (a) $1.7943 \times 10^{-28} \text{ m}^3$ (b) 5.57×10^{18} (c) $1.89 \times 10^{13}, 3.38 \times 10^{-4}\%$

29. (a) Ionic   (b) Covalent network   (c) Molecular   (d) Metallic

31. $CO < BaCl_2 < Co < SiC$ **33.** $O_2 < Sb < Cu$ **35.** 8, 6

37. 8 nearest neighbors, 6 second-nearest neighbors   **39.** 4.4197 Å

41. Not by a significant amount, because each vacancy is accompanied by an interstitial. In large numbers, such defects could cause a small bulging of the crystal and lower its density.

43. (a) $Fe_{0.9352}O$ (b) 0.1386 of the iron

45. The liquid is less ordered than the plastic crystal because the molecular centers are free to move. It has a higher entropy and a higher enthalpy.

Chapter 21

1. $2.16 \times 10^{-19} \text{ J}$ **3.** 6.1×10^{-27} electrons; that is, no electrons

5. (a) electron movement (n-type)   (b) hole movement (p-type)

7. 680 nm, red   **9.** decrease

11.

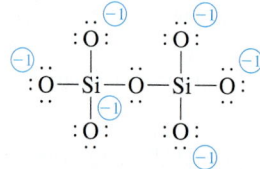

The structures of $P_2O_7^{4-}$ and $S_2O_7^{4-}$ have the same number of electrons, and are found by writing P or S in place of Si. The formal charges on the P's are $+1$ in the resulting structure, and those on the S's are $+2$. The chlorine compound is Cl_2O_7, dichlorine heptaoxide.

13. (a) tetrahedra. Ca: $+2$, Fe: $+3$, Si: $+4$, O: -2
     (b) infinite sheets: Na: $+1$, Zr: $+2$, Si: $+4$, O: -2
     (c) pairs of tetrahedra: Ca: $+2$, Zn: $+2$, Si: $+4$, O: -2
     (d) infinite sheets. Mg: $+2$, Si: $+4$, O: -2, H: $+1$

15. (a) infinite network. Li: $+1$, Al: $+3$, Si: $+4$, O: -2
     (b) infinite sheets. K: $+1$, Al: $+3$, Si: $+4$, O: -2, H: $+1$
     (e) closed rings or infinite single chains: Al: $+3$, Mg: $+2$, Si: $+4$, O: -2

17. (a) $CaCO_3(s) + SiO_2(s) \longrightarrow CaSiO_3(s) + CO_2(g)$
     (b) $\Delta H° = +89.41 \text{ kJ mol}^{-1}$, $\Delta S° = +160.8 \text{ J K}^{-1} \text{ mol}^{-1}$ (c) 556 K

19. The strength of typical ceramics is in their ability to resist compression (related to their hardness). Typical metals are stronger in their ability to resist tensile forces (stretching).

21. If the ceramic is a product of firing of clay, the starting clay material can be shaped by hand; other ceramic compounds are shaped in a mold by compression at high temperature. Metals like steel can be melted and poured into molds.

23. $Mg_3Si_4O_{10}(OH)_2(s) \longrightarrow 3\, MgSiO_3(s) + SiO_2(s) + H_2O(g)$ **25.** 234 L

27. 0.418 mol Si, 0.203 mol Na, 0.050 mol Ca, 0.0068 mol Al, 0.002 mol Mg, 0.0005 mol Ba

29. -113 kJ

31. (a) $SiO_2(s) + 3\, C(s) \longrightarrow SiC(s) + 2\, CO(g)$ (b) $+624.6 \text{ kJ mol}^{-1}$
     (c) low conductivity, high melting point, very hard

33. -1188.2 kJ mol^{-1}; unstable. The reaction is:

$SiC(s) + 2 O_2(g) \longrightarrow SiO_2(quartz) + CO_2(g)$

35. $\frac{7}{3}$

Chapter 22

1. A batch process is carried out by mixing reactants, letting them react, and then extracting products in a series of separate steps. In a continuous process, all the steps run simultaneously. A batch process can take place in a single container with one opening, but a continuous process requires either flow through several containers or at least different places into which reactants can be introduced and from which products are removed.

3. Both use high temperatures and pressures for certain processes, and both exploit solubility differences to separate products, but geochemical processes are much slower.

$Ca^{2+}, CO_3^{2-}(in\ igneous\ rocks) \xrightarrow{water} CaCO_3(aq) \xrightarrow{evaporation} CaCO_3\ (limestone)$

$\xrightarrow{heat\ and\ pressure} CaCO_3(marble)$

5. **(a)** in salt domes, produced by bacteria from $CaSO_4$, CH_4, and CO_2
(b) in coal, produced from plant matter by bacterial reduction, heat, and pressure
(c) in igneous rocks

7. The lead is protected (passivated) by a layer of lead(II) sulfate.

9. Operate at high total pressures and low temperatures; continuously remove SO_3 as it is produced.

11. Yes. The reaction that would occur, $H_2O_2(aq) + SO_2(g) \longrightarrow H_2SO_4(aq)$, has $\Delta G_r^\circ = -321.7$ kJ mol$^{-1} < 0$, assuming that $H_2SO_4(aq)$ is entirely $HSO_4^-(aq)$ and $H^+(aq)$. Thus, the process is thermodynamically feasible.

13. $ZnS(s) + 2 O_2(g) + H_2O(\ell) \longrightarrow ZnO(s) + H_2SO_4(\ell)$

15. $5 H_2SO_4(aq) + 2 P(s) \longrightarrow 5 SO_2(g) + 2 H_3PO_4(aq) + 2 H_2O(\ell)$

17. $CaF_2(s) + H_2SO_4(\ell) \longrightarrow 2 HF(g) + CaSO_4(s)$
$2 NaCl(s) + H_2SO_4(\ell) \longrightarrow 2 HCl(g) + Na2SO_4(s)$

19. $H_2SO_4(\ell) + 2 NaOH(s) \longrightarrow Na2SO_4(s) + 2 H_2O(\ell)$
$Na_2SO_4(s) + 2 C(s) \longrightarrow Na_2S(s) + 2 CO_2(g)$

21. $H_2S_2O_7(s) + H_2O(\ell) \longrightarrow 2 H_2SO_4(\ell)$ **23.** 3×10^{26} atoms per second

25. $CaCO_3(s) + 2 C(s) + N_2(g) + 4 H_2O(\ell) \longrightarrow 2 NH_3(g) + 2 CO_2(g) + CO(g) + Ca(OH)_2(s)$; $\Delta H_r^\circ = +374.4$ kJ mol^{-1}. Dividing by 2 gives $+187.2$ kJ per mole of $NH_3(g)$ produced.

27. Higher pressures mean higher yields but also more expensive equipment. At fixed yield, a higher temperature can be used at high pressure, giving a faster reaction.

29. High yield is favored by low temperature for all three steps.

31. $3 HNO_3(aq) + Cr(s) + 3 H^+(aq) \longrightarrow 3 NO_2(g) + Cr^{3+}(aq) + 3 H_2O(\ell)$
$3 HNO_3(aq) + Fe(s) + 3 H^+(aq) \longrightarrow 3 NO_2(g) + Fe^{3+}(aq) + 3 H_2O(\ell)$

33.   **35.** unstable

37. $5 N_2H_4(\ell) + 4 HNO_3(\ell) \longrightarrow 7 N_2(g) + 12 H_2O(\ell); \Delta H_r^\circ = -2986.71 \text{ kJ mol}^{-1}$

39. $N_2(g) + 3 H_2(g) \longrightarrow 2 NH_3(g)$ (Haber–Bosch)
$4 NH_3(g) + 5 O_2(g) \longrightarrow 4 NO(g) + 6 H_2O(g)$ (Pt catalyst, first step of Ostwald process)
$2 NO(g) + O_2(g) \longrightarrow 2 NO_2(g)$ (second step of Ostwald process)
$2 NO_2(g) \longrightarrow N_2O_4(s)$ (low temperature)

41. The decomposition of NH_4NO_3 liberates some O_2, which can react with the C from the TNT to give $CO_2(g)$.

43. $NH_4N(NO_2)_2(g) \longrightarrow 2 N_2(g) + 2 H_2O(g) + O_2(g)$

Chapter 23

1. $K = 3.2 \times 10^{16}$

3. The Leblanc process requires carbon (coke) and sulfuric acid, in addition to the salt and calcium carbonate required by the Solvay process. The sulfuric acid in particular adds to the cost. The product, Na_2CO_3, is purer from the Solvay process because it does not contain CaS impurities.

5. $2 NaCl(s) + H_2SO_4(\ell) + 2 C(s) + CaCO_3(s) \longrightarrow Na_2CO_3(s) + 2 CO_2(g) + CaS(s) + 2 HCl(g)$ $\Delta H_r^\circ = +259 \text{ kJ mol}^{-1}$. The LeBlanc process is endothermic, so heat must be supplied to make it run.

7. Chlorous acid is stronger and should be used.

9. The data in Appendix D are for the standard state at 298.15 K. The melting point of NaCl is much higher.

11. calcium hydroxide   **13.** $4.0 \times 10^4 \text{ L} = 40 \text{ m}^3$

15. (a) $+1$ (b) HClO (hypochlorous acid)   **17.**   not stable

19. Si—F, because the F atom is more electronegative and smaller

21. (a) $2 SrO(s) + 2 F_2(g) \longrightarrow 2 SrF_2(s) + O_2(g)$
(b) $O_2(g) + 2 F_2(g) \longrightarrow 2 OF_2(g)$
(c) $UF_4(s) + F_2(g) \longrightarrow UF_6(s)$
(d) $2 NaCl(s) + F_2(g) \longrightarrow 2 NaF(s) + Cl_2(g)$

23. $NaF \cdot 2HF$ **25.** ClF_3

27. (a) bent   (b) trigonal planar   (c) distorted T   (d) square pyramidal
(e) pentagonal bipyramidal   (f) octahedral

29. (a) pyramidal   (b) Lewis base

31. $2 BrF_3 \rightleftharpoons BrF_4^- + BrF_2^+$. Yes. BrF_2^+ ion and BrF_3 are Lewis acids (electron pair acceptors), competing for F^- ion, a Lewis base (electron pair donor).

33. $PtF_6 < PtF_4 < CaF_2$ **35.** $2.74 \times 10^3 \text{ kJ}$

37. (a) -345 kJ mol^{-1} (b) 136 kJ mol^{-1} **39.** $+190 \text{ kJ mol}^{-1}$

41. $XeO_6^{4-}(aq) + 12 H^+(aq) + 8 e^- \longrightarrow Xe(g) + 6 H_2O(\ell)$
$Mn^{2+}(aq) + 4 H_2O(\ell) \longrightarrow MnO_4^-(aq) + 8 H^+(aq) + 5e^-$
$5 XeO_6^{4-}(aq) + 8 Mn^{2+}(aq) + 2 H_2O(\ell) \longrightarrow 5 Xe(g) + 8 MnO_4^-(aq) + 4 H^+(aq)$

Chapter 24

1. Yes, if the amount of knocking is less than that of isooctane. Examples are the BTX compounds.

3. ethane; $2 C_2H_6(g) + 7 O_2(g) \longrightarrow 4 CO_2(g) + 6 H_2O(\ell)$

5. **(a)** -64 kJ mol^{-1} **(b)** $+53 \text{ kJ mol}^{-1}$ **(c)** $+117 \text{ kJ mol}^{-1}$

7. **(a)** $C_{10}H_{22} \longrightarrow C_5H_{10} + C_5H_{12}$ **(b)** two: 1-pentene and 2-pentene

9. **(a)**

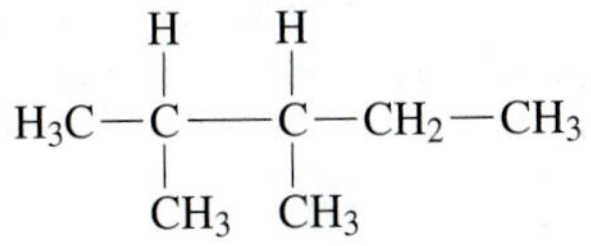

(b)

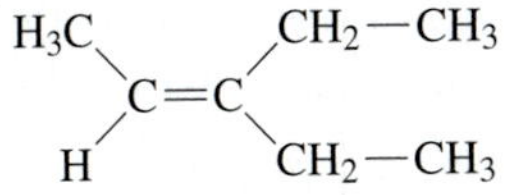

(c)

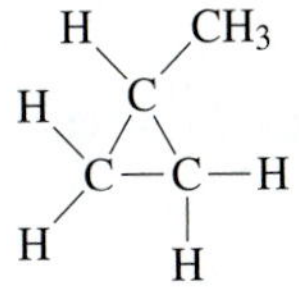

(d)

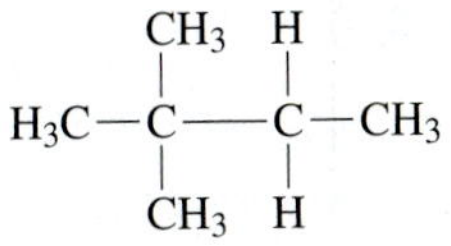

(e)

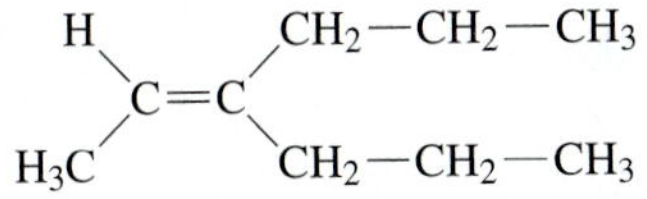

(f)

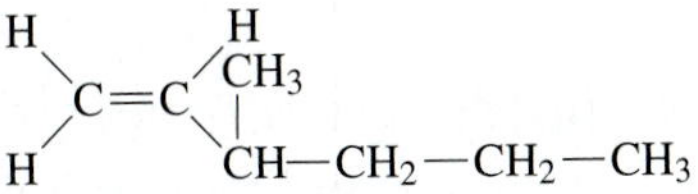

(g)

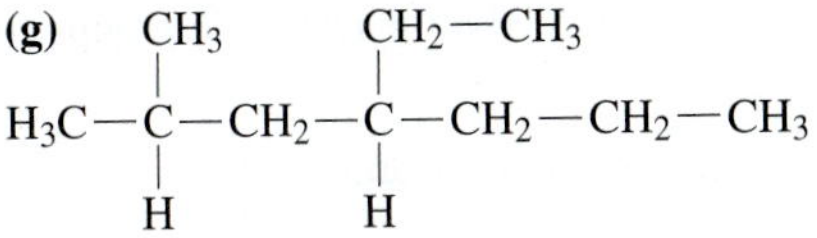

(h)

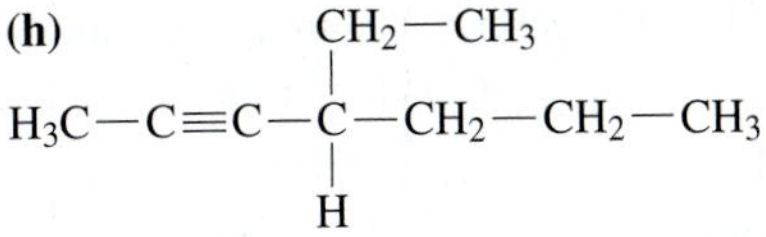

11.

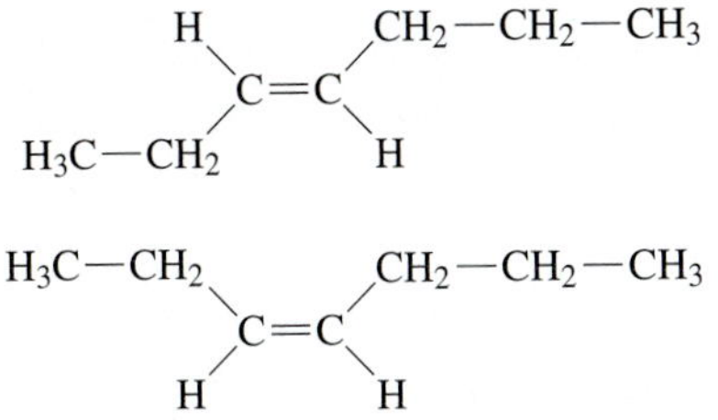

13. **(a)** 1,2-hexadiene   **(b)** 1,3,5-hexatriene   **(c)** 2-methyl-1-hexene
(d) 3-hexyne

15. **(a)** The first three are sp^2; the last three are sp^3.   **(b)** All sp^2
(c) The first two on the main chain are sp^2; all others are sp^3.
(d) The third and fourth are sp; all others are sp^3.

17. **(a)** $CH_3CH_2CH_2CH_2OH + CH_3COOH \longrightarrow CH_3COOCH_2CH_2CH_2CH_3 + H_2O$
(b) $NH_4CH_3COO \longrightarrow CH_3CONH_2 + H_2O$
(c) $CH_3CH_2CH_2OH \longrightarrow CH_3CH_2CHO$ (propionaldehyde)
(d) $CH_3(CH_2)_5CH_3 + 11 O_2 \longrightarrow 7 CO_2 + 8 H_2O$

19. **(a)** $CH_2{=}CH_2 + Br_2 \longrightarrow CH_2BrCH_2Br$; $CH_2BrCH_2Br \longrightarrow CH_2{=}CHBr + HBr$
(b) $CH_3CH_2CH{=}CH_2 + H_2O \longrightarrow CH_3CH_2CH(OH)CH_3$ (using H_2SO_4)
(c) $CH_3CH{=}CH_2 + H_2O \longrightarrow CH_3CH(OH)CH_3$ (using H_2SO_4);
$CH_3CH(OH)CH_3 \longrightarrow CH_3COCH_3 + H_2$ (copperoxide or zincoxide catalyst)

21.

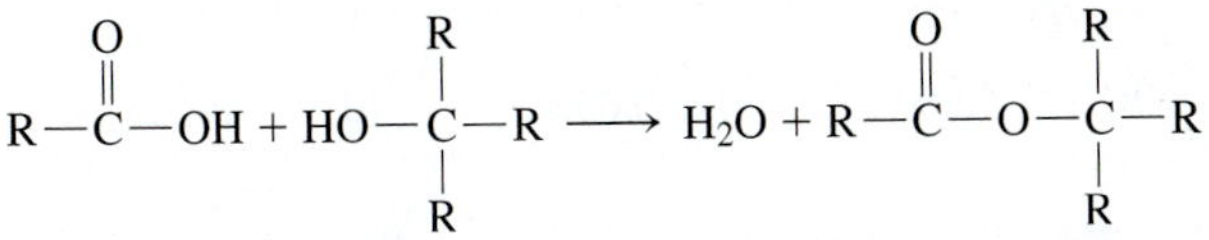

23. 11.2%; 4.49×10^9 kg

25. Molecules of dimethyl ether cannot form hydrogen bonds to each other or to water.

27. **(a)** Alcohol: $CH_3CH(OH)CH_3$, isopropyl alcohol; carboxylic acid:
$H_3CCH(OCH_3)CH(CH_3)(CH_2)_2CH(CH_3)CH_2CH{=}CHC(CH_3){=}CHCOOH$
(b) 3,7,10-trimethyl-2,4-dodecadiene

29. (a) $C_9H_8O_4$ (b) 1.80×10^{-3} mol

31. Dehydrogenate to make $C=O$ double bond from $C-OH$ bond on first ring; move $C=C$ bond from first to second ring; insert $C=O$ group on third ring; remove hydrocarbon side chain on fourth ring together with H atom and replace it with an $-OH$ group and a $-COCH_2OH$ group.

Chapter 25

1. $n\ CCl_2=CH_2 \longrightarrow +(CCl_2-CH_2)_n-$ 3. formaldehyde, $CH_2=O$

5. (a) H_2O (b) $+(NH-CH_2-CO)-$

7. $NH_2^{\ominus}Na^{\oplus} + CH_2=CH \longrightarrow NH_2-CH_2-CH^{\ominus}Na^{\oplus}$
$\qquad\qquad\qquad\quad |\qquad\qquad\qquad\qquad\qquad\qquad |$
$\qquad\qquad\qquad\ C\equiv N\qquad\qquad\qquad\qquad\qquad C\equiv N$

9. $HOOC(CH_2)_4COO^-$; $^+H_3N(CH_2)_6NH_2$

11. $^+H_3NCH_2COO^-$; $^+H_3NC(CH_3)HCOO^-$; $^+H_3NC(CH_2CH_2COOH)HCOO^-$

13. Hydrophilic side-groups contain atoms of O, N, or S. Hydrophobic side groups contain atoms of C and/or H only.

15. 27   17. in octane, because the side-groups are all hydrocarbon

19. 5 chiral centers   21. C_9H_9NO; 119 units

23.

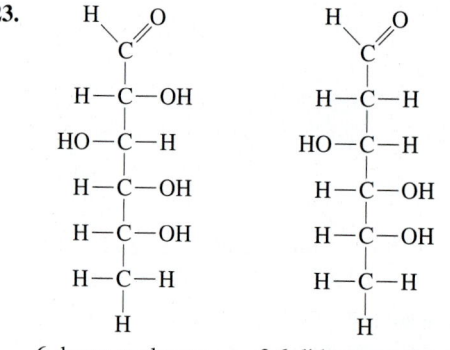

6-deoxy-D-glucose   2,6-dideoxy-D-glucose

25. (a) $C_6H_{10}O_5$ (b) $C_6H_{10}O_5$ (both the same)

27. 646 kg adipic acid and 513 kg of hexamethylenediamine

29. Every C atom has the maximum number of bonds to other C atoms (4).

31. 3.49×10^{12} L $= 3.49 \times 10^9$ m^3 = 3.49 cubic kilometers

Appendix A

1. (a) 5.82×10^{-5} (b) 1.402×10^3 (c) 7.93   (d) -6.59300×10^3
   (e) 2.530×10^{-3} (f) 1.47

3. (a) 0.000537   (b) 9,390,000   (c) -0.00247 (d) 0.006020
   (e) 20,000

5. 1.90×10^2 7. 746,000,000 kg

9. (a) the value 135.64 g   (b) 111.34 g   (c) 0.22 g

11. that of the mass in Problem 9

13. (a) 5   (b) 3   (c) either 2 or 3   (d) 3   (e) 4

15. (a) 14 L   (b) $-0.0034°C$ (c) 340 lb $= 3.4 \times 10^2$ lb
    (d) 3.4×10^2 mi   (e) 6.2×10^{-27} J

17. 2,997,215.55

19. (a) -167.25 (b) 76   (c) 3.1693×10^{15} (d) -7.59×10^{-25}

21. (a) -8.40 (b) 0.147   (c) 3.24×10^{-12} (d) 4.5×10^{13}

23. $A = 337$ cm^2

Appendix B

1. (a) 6.52×10^{-11} kg ("kilo" is a prefix, but kg is nevertheless a base unit)
    (b) 8.8×10^{-11} s   (c) 5.4×10^{12} kg m^2 s^{-3} (d) 1.7×10^4 kg m^2 s^{-3} A^{-1}

3. 1 Wb $= 1$ kg m^2 s^{-2} A^{-1}

5. 100 cm^3 < 500 mL < 1.5 gal (US) < 100 dm^3 < 150

7. (a) $4983°C$ (b) $37.0°C$ (c) $111°C$ (d) $-40°C$

9. (a) 5256 K   (b) 310.2 K   (c) 384 K   (d) 233 K

11.

1 m^3 $= 264.1722\ldots$ gal	1 m $= 39.37008\ldots$ in
1 kg $= 2.204623\ldots$ lb	1 J $= 0.2390057\ldots$ cal

13. (a) 24.6 m s^{-1} (b) 1.51×10^3 kg m^{-3} (c) 1.6×10^{-12} A s m
    (d) 1.5×10^2 mol m^{-3} (e) 9.6×10^3 kg m^2 s^{-3} $= 9.6 \times 10^3$ W

15. Pa m^2 s^{-2} and J m^{-1} s^{-1} are not units of density.

17. 1 kW-h $= 3.6 \times 10^6$ J; 15.3 kW-h $= 5.51 \times 10^7$ J

19. 6620 cm^3 or 6.62 L

Appendix C

1. (a) 4.551   (b) 1.53×10^{-7} (c) 2.6×10^8 (d) -48.7264

3. 3.015   5. 121.477   7. 7.751 and -2.249 9. 50 mi h^{-1}

11. (a) slope $= 4$, intercept $= -7$ (b) slope $= \frac{7}{2}$, intercept $= -\frac{5}{2}$
    (c) slope $= -2$, intercept $= \frac{4}{3}$

13. graph is not a straight line

15. (a) $x = -\frac{5}{7} = -0.7142857\ldots$ (b) $x = \frac{3}{4} = 0.75$ (c) $x = \frac{2}{3} = 0.6666\ldots$

17. (a) $x = 3, y = 1$ (b) $x = 1, y = -\frac{3}{2}, z = \frac{5}{2}$ (c) $x = 1, y = -\frac{2}{3}, z = \frac{5}{3}$

19. (a) $x = 0.5447$ and -2.295 (b) $x = -0.6340$ and -2.366
    (c) $x = 2.366$ and 0.6340

21. (a) $x = 6.5 \times 10^{-7}$ (also two complex roots)
    (b) $x = 4.07 \times 10^{-2}, 0.399$ and -1.011
    (c) $x = -1.3732$ (also two complex roots)

23. $x = 1.086$

25. (a) Use w to represent mass fraction: $w_{water} = 0.588$;
    $w_{NaCl} = 0.411$; $w_{mercury} = 4.8 \times 10^{-5}$
    (b) Use ϕ to represent volume fraction: $\phi_{water} = 0.7564$;
    $\phi_{NaCl} = 0.2436$; $\phi_{mercury} = 4.51 \times 10^{-6}$
    (c) ppmv of mercury $= 4.51$ (d) ppbm of mercury $= 4.8 \times 10^4$

ANSWERS TO CUMULATIVE PROBLEMS
Chapter 1

(a) The masses of gallium present per 1.000 g of sulfur in the five samples are: 1.450 g Ga in sample 1; 2.175 g Ga in sample 2; 4.349 g Ga in sample 3; 1.450 g Ga in sample 4; and 2.175 g Ga in sample 5. The small, whole-number ratios (2 to 3 to 6 to 2 to 3) among these masses are consistent with the law of definite proportions, which applies to compounds, not mixtures.

(b) There are at least three different compounds. Samples 1 and 4 have the same proportions of gallium and sulfur. They may be the same compound, but this is not certain because it is possible for different compounds to have the same elements in the same definite proportions. A similar statement applies to samples 2 and 5. Samples 1 and 2 must be different compounds, despite their similar appearance.

(c) If sample 1 is GaS, then sample 2 is Ga_3S_2 (or a multiple), sample 3 is Ga_3S (or a multiple), sample 4 is GaS (or a multiple), and sample 5 is Ga_3S_2 (or a multiple).

(d) If sample 1 is Ga_2S_3, then sample 2 is GaS (or a multiple), sample 3 is Ga_2S (or a multiple), sample 4 is Ga_2S_3 (or a multiple), and sample 5 is GaS (or a multiple).

(e) The relative atomic mass of gallium would be 46.49. (Note that this is two thirds of the accepted value.)

(f) Ga_2S_3, GaS, Ga_2S, Ga_2S_3, and GaS.

(g) ^{32}S and ^{69}Ga. The only way for the chemical relative atomic mass of S to come out so close to 32 is for ^{32}S, the lightest isotope, to dominate in abundance; the argument is the same in the case of gallium.

(h) 1.167×10^{22} gallium atoms and 1.750×10^{22} sulfur atoms. The ratio is 1.500:1 or 3:2.

(i) The densities of the five samples are 3.65, 3.86, 4.18, 3.65, and 3.86 g cm^{-3}, respectively. The equality of the densities of samples 1 and 4 supports but does not prove the conclusion that they are the same compound. The same is true for samples 2 and 5.

(j) Each atom occupies an average of 2.14×10^{-23} cm^3 (samples 1 and 4), 2.19×10^{-23} cm^3 (samples 2 and 5), and 2.26×10^{-23} cm^3 (sample 3).

Chapter 2

(a) 0.01000 mol L^{-1} **(b)** 0.0150 mol L^{-1}

(c) $KIO_3(aq) + 5\,KI(aq) + 6\,HCl(aq) \longrightarrow 3\,I_2(aq) + 3\,H_2O(\ell) + 6\,KCl(aq)$

(d) KIO_3 is the limiting reagent; 0.668 g KI and 1 g HCl will be left over.

(e) 90.36%   **(f)** 1.429 mL IF$_7(g)$ and 4.286 mL of OF$_2(g)$

(g) No, not with the available data. Avogadro's hypothesis applies only to gases.

Chapter 3

(a) The O^{-2} ion has a total of ten electrons, of which eight are valence electrons (an octet) and two are core electrons. TiO_2 is titanium(IV) oxide.

(b) The ion in BaO_2 must be the peroxide ion O_2^{-2}, and the compound is barium peroxide.

A Lewis structure, with formal charges, for the peroxide ion is $:\!\overset{\ominus}{\underset{\cdot\cdot}{O}}\!:\!\overset{\ominus}{\underset{\cdot\cdot}{O}}\!:$ The O—O bond is a single bond (length about 1.48×10^{-10} m in Table 3–3).

(c) The ion in KO_2 is the superoxide ion O_2^-; the compound is named potassium superoxide. The superoxide ion has 17 total electrons, of which 4 are core electrons. The best

Lewis structures one can draw are the resonance structures $\left(: \overset{..}{\underset{..}{O}} : \overset{..}{\underset{..}{O}} \overset{\ominus}{:} \quad \longleftrightarrow \quad \overset{\ominus}{:} \overset{..}{\underset{..}{O}} : \overset{..}{\underset{..}{O}} :\right)$

in which only one of the oxygen atoms attains an octet electron configuration.

(d) The Lewis structure of dihydrogen trioxide is $H : \overset{..}{\underset{..}{O}} : \overset{..}{\underset{..}{O}} : \overset{..}{\underset{..}{O}} : H$

(e) Table 3–3 gives the average length of O—O single bonds as 1.48×10^{-10} m; take this as the predicted length. (The experimental lengths both equal 1.445×10^{-10} m.) Table 3–3 gives the average length of O—H single bonds as 0.96×10^{-10} m; take this as the predicted length. (The experimental O—H lengths in H_2O_3 both equal 0.982×10^{-10} m.)

(f) The steric numbers of all three oxygen atoms equal 4, with two bonded pairs and two lone pairs. The H-O-O bond angles and the O-O-O bond angle should all be somewhat less than the tetrahedral angle of 109.5°. (Experiment gives values of 100.8° for both the H-O-O angles and 106.4° for the O-O-O angle.)

Chapter 4

(a) CO_2 is the anhydride of H_2CO_3.

(b) The second step is a precipitation reaction:
$$2\,Na^+(aq) + 2\,HCO_3^-(aq) \longrightarrow 2\,NaHCO_3(s)$$

(c) This decomposition reaction is not an acid–base reaction in the Arrhenius sense.

(d) CaO is the anhydride of $Ca(OH)_2$, calcium hydroxide, which is prepared in step 5.

(e) $2\,NH_4^+(aq) + 2\,OH^-(aq) \longrightarrow 2\,NH_3(aq) + 2\,H_2O(\ell)$

(f) None of the steps is a redox reaction because oxidation numbers do not change.

(g) $2\,NaCl(aq) + CaCO_3(s) \longrightarrow Na_2CO_3(s) + CaCl_2(aq)$. When calcium chloride and sodium carbonate are mixed in aqueous solution, solid calcium carbonate precipitates and sodium chloride remains in solution. These compounds are the net starting materials in the Solvay process.

Chapter 5

(a) $2\,NH_4ClO_4(s) \longrightarrow N_2(g) + Cl_2(g) + 2\,O_2(g) + 4\,H_2O(g)$

(b) 328 atm (c) $X_{Cl_2} = \frac{1}{8} = 0.125$ (exactly); $P_{Cl_2} = 41.0$ atm

(d) 318 atm (e) $u_{rms}(H_2O) = 1220$ m s^{-1}; $u_{rms}(Cl_2) = 614$ m s^{-1}

(f) 2.89×10^5 m^3

Chapter 6

(a) dispersion forces

(b) on the order of 1 to 2 kJ mol^{-1}; a distance of around 4×10^{-10} m (estimated by comparison to the Ar + Ar curve in Figure 6–1)

(c) The high pressure has partially liquefied the chlorine in the tank. Evaporation of the liquid keeps the interior pressure constant as gaseous chlorine is removed. The pressure drops only when the last of the liquid evaporates.

(d) The chlorine is heated and then compressed. During these operations the original gas becomes, by gradual degrees, a supercritical fluid. The supercritical fluid is then cooled at high pressure. It becomes, by gradual degrees, a liquid. The net change is from gas to liquid, but a sharp transition during which two phases are both present is never observed.

(e) mass percentage = 0.637%; mole fraction = 0.00163; molality = 0.0904 mol kg^{-1}; molarity = 0.0924 mol L^{-1}

(f) Henry's law constant k_H = 613 atm^{-1}

(g) ν = 3. The reaction of 1 particle gives $(1 - 0.326) + 3(0.326) = 1.652$ particles; i = 1.652. In this case, water serves as true reactant as well as a dispersing medium.

(h) P = 3.73 atm   **(i)** $-0.67°$

Chapter 7

(a) $\dfrac{P_{SO_2}}{P_{O_2}} = K_1 \quad \dfrac{P_{SO_3}}{P_{SO_2}(P_{O_2})^{1/2}} = K_2 \quad \dfrac{1}{P_{SO_2}} = K_3$

(b) $S(s) + \frac{3}{2}O_2(g) + H_2O(\ell) \rightleftharpoons H_2SO_4(\ell)$; $K = K_1K_2K_3 = 2.2 \times 10^{79}$

(c) K_2 = 2.63   **(d)** exothermic   **(e)** P_{O_2} = 0.029 atm

(f) Left to right; the pressure decreases.

(g) An increase in temperature shifts both equilibria to the left. A decrease in volume does not affect equilibrium 1 and shifts equilibrium 3 to the right.

Chapter 8

(a) 4.4 to 4.8   **(b)** $[H_3O^+] = 5.0 \times 10^{-4}$ M; $[OH^-] = 2.0 \times 10^{-11}$ M

(c) $H_2SO_3(aq) + H_2O(\ell) \longrightarrow HSO_3^-(aq) + H_3O^+(aq)$. The stronger acid is H_3O^+. The stronger base is HSO_3^-.

(d) pH = 3.41   **(e)** pH = 3.11

(f) $H_3O^+(aq) + CO_3^{2-} \longrightarrow HCO_3^-(aq) + H_2O(\ell)$. The HSO_4^- in the sulfuric acid reacts according to $HSO_4^-(aq) + CO_3^{2-}(aq) \longrightarrow SO_4^{2-}(aq) + HCO_3^{2-}(aq)$. This gives rise to an HCO_3^-/CO_3^{2-} buffer that resists further changes in pH. An excess of acid rain overwhelms the buffer and leads to the formation of H_2CO_3.

(g) 3.16×10^{-4} M   **(h)** The pH is 3.81.

Chapter 9

(a) $MgCO_3 \cdot 3H_2O$

(b) $CaMg(CO_3)_2(s) \longrightarrow Ca^{2+}(aq) + Mg^{2+}(aq) + 2\,CO_3^{2-}(aq)$

$$[Ca^{2+}][Mg^{2+}][CO_3^{2-}]^2 = K_{sp}$$

(c) 9.3×10^{-3} g L^{-1}

(d) 8.7×10^{-6} g L^{-1}, smaller than in part **(c)** because of the common-ion effect.

(e) $[CO_3^{2-}] = 5.8 \times 10^{-6}$ M; $Q = 2.2 \times 10^{-9} < K_{sp} = 8.7 \times 10^{-9}$, so the lake is slightly less than saturated.   **(f)** increase   **(g)** increase

(h) Dissolve the carbonate sample in acid and heat to remove carbonate ion in the form of carbon dioxide. Saturate with H_2S and adjust the pH to 9 with an NH_4^+/NH_3 buffer. A sulfide precipitate indicates the presence of Group 3 cations (here, Fe^{2+} or Mn^{2+}). Further analysis (not discussed in this chapter) would be required to decide whether both cations are present, or only one.

Chapter 10

(a) 102 kJ   **(b)** 1.2×10^3 kJ   **(c)** -507 kJ mol^{-1}

(d) -676.49 kJ mol^{-1} **(e)** 2.11×10^4 kJ   **(f)** $\Delta H - \Delta E = 43.9$ kJ

(g) -0.60 L atm = -61 J

Chapter 11

(a)   negative   (b)   positive

(c) $\Delta H_r^\circ = -160.8$ kJ mol^{-1}; $\Delta S_r^\circ = -409.5$ J K^{-1} mol^{-1} (d)   393 K = 120°C

(e) 4.0×10^4 (f) 2.0×10^{-5}

(g)   92.0 J K^{-1} mol^{-1}, close to the Trouton's rule value of 88 J K^{-1} mol^{-1}

(h)   16.7 atm

Chapter 12

(a) $PbS(s) + 2\,O_2(g) \longrightarrow PbSO_4(s)$
$PbS(s) + 2\,PbO(s) \longrightarrow 3\,Pb(s) + SO_2(g)$
$PbS(s) + PbSO_4(s) \longrightarrow 2\,Pb(s) + 2\,SO_2(g)$

(b) $PbS(s) + Fe(s) \longrightarrow Pb(s) + FeS(s)$; $\Delta G^\circ = -1.7$ kJ mol^{-1}, (barely) spontaneous

(c)   anode: $Pb(s,\ crude) \longrightarrow Pb^{2+}(aq) + 2\,e^-$
cathode: $Pb^{2+}(aq) + 2\,e^- \longrightarrow Pb(s,\ pure)$

(d)   An electric current still flows, but it deposits pure lead on the crude lead electrode, and lead from the pure electrode is oxidized and passes into solution.

(e)   between PbO and CuO

(f) $Pb(s) + PbO_2(s) + 2\,SO_4^{2-}(aq) + 4\,H^+(aq) \longrightarrow 2\,PbSO_4(s) + 2\,H_2O(aq)$

(g)   2.31 g

Chapter 13

(a)   Manganese appears at the cathode and MnO_2 at the anode.

(b)   oxidation: $Mn(OH)_2(s) + OH^-(aq) \longrightarrow Mn(OH)_3(s) + e^-$
reduction: $O_2(aq) + 2\,H_2O(aq) + 4\,e^- \longrightarrow 4\,OH^-(aq)$
total: $4\,Mn(OH)_2(s) + O_2(aq) + 2\,H_2O(aq) \longrightarrow 4\,Mn(OH)_3(s)$
$\Delta\mathcal{E}^\circ = 0.401 - (-0.40) = 0.80$ V

(c) $K = 1 \times 10^{54}$ (d) $Mn^{3+}(aq)$ spontaneously oxidizes $I^-(aq)$
$2\,Mn^{3+}(aq) + 2\,I^-(aq) \longrightarrow 2\,Mn^{2+}(aq) + I_2(s)$ $K = 9 \times 10^{32}$

(e) 3.8×10^3 J $= 1.0 \times 10^{-3}$ kW-h   (f)   permanganate ion   (g)   2.14 V

Chapter 14

(a) 3.40×10^{-4} mol L^{-1} s^{-1} (b)   2.7 mol L^{-1} s^{-1}

(c)   All of the oxygen is consumed, so the first step of the process is over.

(d) 5.43×10^{-5} mol L^{-1} s^{-1} (e) 1.09×10^{-4} mol L^{-1} s^{-1}

(f) 2.45×10^{-5} mol L^{-1} s^{-1} (g)   first order, $k = 0.0133$ s^{-1}

(h)   52.1 s   (i)   It happens too fast for study in this experiment.

(j)   34 kJ mol^{-1}

(k) Fe^{2+} acts as a catalyst.

(l) $2\,SO_2(aq) + O_2(aq) + 2\,H_2O(\ell) \longrightarrow 2\,SO_4^{2-}(aq) + 4\,H^+(aq)$

Chapter 15

(a)   An atom of ^{222}Rn has 86 electrons outside the nucleus. Inside the nucleus are 86 protons and $222 - 86 = 136$ neutrons. An atom of ^{220}Rn has the same number of electrons and protons, but only 134 neutrons in the nucleus.

(b) Four alpha particles are produced in making ^{222}Rn from ^{238}U; three are produced in making ^{220}Rn from ^{232}Th.

(c) If $^{238}_{92}$U were used to lose four alpha particles, $^{222}_{84}$Po would form instead of $^{222}_{86}$Rn. Somewhere along the way, two beta particles ($^{0}_{-1}e^{-}$) must be ejected from the nucleus to raise Z to 86. The same is true of the production of ^{220}Rn from ^{232}Th.

(d) $^{222}_{86}$Rn $\longrightarrow$ $^{218}_{84}$Po + $^{4}_{2}$He; $\Delta m = -0.00600\ u < 0$; allowed
$^{220}_{86}$Rn $\longrightarrow$ $^{216}_{84}$Po + $^{4}_{2}$He; $\Delta m = -0.00687\ u < 0$; allowed

(e) $\Delta E = -6.40\ \text{MeV} = -1.03 \times 10^{-12}\ \text{J}$

(f) $A = 1.14 \times 10^{8}\ \text{Bq}$ **(g)** $A = 9.0 \times 10^{6}\ \text{Bq}$

(h) greater

Chapter 16

(a) 91.1 nm   **(b)** $8.8 \times 10^{-19}\ \text{J}$ **(c)** 0.52 nm = 5.2 Å

(d) $6.40 \times 10^{-7}\ \text{m} = 6400$ Å

(e) $5.98 \times 10^{-2}\ \text{m} = 5.98\ \text{cm}$, in the microwave region

(f) No, because the lowest energy absorption for a ground-state hydrogen atom is in the ultraviolet region of the spectrum.

(g) 4 angular nodes and 104 radial nodes   **(h)** $9.37 \times 10^{-25}\ \text{J}$; 0.564 J mol^{-1}

(i) too low

Chapter 17

(a) $[\text{Kr}]4s^{2}5s^{2}5p^{5}$

(b) The first ionization energy is higher for I than for Te, but lower for I than for Xe. The electron affinity is higher for I than for Te or Xe.

(c) electronegativity difference = 1.73. The compound (NaI) is ionic.

(d) electronegativity difference = 0.11. The C—I bond is polar covalent.

(e) The +7 oxidation state corresponds to participation of all the $5s$ and $5p$ electrons in bonding. By the end of the fifth period, the $4d$ electrons are so low in energy and so close to the nucleus that they cannot participate in bonding (see Fig. 17–7).

(f) strongest $\text{HIO}_4 > \text{HIO}_3 > \text{H}_5\text{IO}_6$ weakest. Note that the strength of the acid with iodine in the +5 oxidation state falls between the strengths of the two acids in which iodine is in the +7 oxidation state.

Chapter 18

(a) yellow light, near 585 nm

(b) $(\sigma_{4s})^{2}(\sigma_{4s}^{*})^{2}(\sigma_{4p_z})^{2}(\pi_{4p})^{4}(\pi_{4p}^{*})^{4}$ diamagnetic

(c) $(\sigma_{4s})^{2}(\sigma_{4s}^{*})^{2}(\sigma_{4p_z})^{2}(\pi_{4p})^{4}(\pi_{4p}^{*})^{3}$ stronger. Bond order is $\frac{3}{2}$ versus 1.

(d) The lowest-energy excited state arises from the excitation of an electron from the filled π_{4p}^{*} orbital to the unfilled $\sigma_{4p_z}^{*}$ orbital.

(e) sp^{3} hybridization   **(f)** $\frac{3}{2}$ order

(g) overall: $2\ \text{O}_3 \longrightarrow 3\ \text{O}_2$. This mechanism does not require formation of atomic oxygen, as does the mechanism in Section 18–5 that is based only on chlorine.

Chapter 19

(a) *cis*-diamminedichloroplatinum(II), potassium hexachloroplatinate(IV), and potassium tetrachloroplatinate(II)

(b) $[Pt(NH_3)_2(CN)_4]$ and $[Pt(CN)_6]^{2-}$

(c) Pt(II) forms square-planar complexes. Consequently, two forms are possible, arising from *cis* or *trans* placement of the ethylene molecules.

(d)

The two structures derive from those shown in Figure 19–9 by replacement of an en ligand with a pair of Cl^- ligands.

(e) The six *d* electrons in Pt(IV) are all in the lower t_{2g} level in a low-spin, large Δ_o complex. All are paired, so the compound is diamagnetic.

(f) Diamagnetic, with the four lowest energy levels in the square-planar configuration (all but the $d_{x^2-y^2}$) fully occupied.

(g) Red transmission corresponds to absorption of green light by $K_2[PtCl_4]$. A colorless solution has absorptions at either higher or lower frequency than visible. Because Cl^- is a weaker field ligand than NH_3, the absorption frequency should be higher for the $[Pt(NH_3)_4]^{2+}$ complex, putting it in the ultraviolet region of the spectrum.

(h) $[Pt(NH_3)_4][PtCl_4]$, tetraammineplatinum(II), tetrachloroplatinate(II)

Chapter 20

(a) Volume is 152.0 $\mathring{A}^3$. Eight atoms per unit cell.

(b) 18.5 $\mathring{A}$. Angle $2\theta = 28.69°$ **(c)** It is amorphous.

(d) Cubic to tetragonal: one cell edge is stretched or shrunk. Tetragonal to monoclinic: the angles between two adjacent faces (and their opposite faces) are changed from 90°. Monoclinic to triclinic: the remaining two angles between faces are deformed from 90°.

(e) The density is 2.36 g cm^{-3}.

(f) P_4 (white) and PCl_3 are molecular solids; P (black) is a covalent network solid; Na_3PO_4 is an ionic solid.

Index/Glossary